# THE NEUROSCIENCES

## SECOND STUDY PROGRAM

# THE

## SECOND STUDY

PUBLISHED BY

The Rockefeller University Press

NEW YORK · 1970

# NEUROSCIENCES

PROGRAM  Francis O. Schmitt *Editor-in-Chief*

*Associate Editors:*  Gardner C. Quarton

Theodore Melnechuk

George Adelman

*Contributing Editors:*  Theodore Holmes Bullock

M. V. Edds, Jr.

Gerald M. Edelman

Robert Galambos

Francis O. Schmitt

José P. Segundo

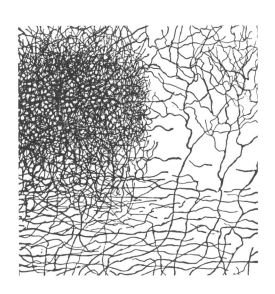

# INTRODUCTION

NEVER BEFORE in history has man been beset with such unrelenting and global problems as now confront him. The specters of war, poverty, hunger, overpopulation, pollution, and other social and ecological imbalances cast ominous shadows on all human progress. It is pointless to blame science for these pandemic ills or to call a moratorium on scientific research in a panic-stricken attempt to avert chaos, especially at a time when epoch-making scientific and technological accomplishments are pointing the way to the cure of many dread diseases and are opening new insights into nature. Rather, we should mobilize every resource of science and technology and develop vigorous civic, political, and spiritual leadership, which, acting in concert, would be capable of charting a course leading to even greater fulfillments of the human mind and spirit.

If the physical basis of brain function were better understood, substantial progress could be made in the alleviation of mental ills and in the search for an understanding of the nature of man as a cognitive individual. Concomitantly, new dimensions of mental capability would be available to solve the pressing survival problems facing man today and to open up unexpected opportunities of human accomplishment.

With a deep commitment to the achievement of such objectives and to an investigation of the nature of learning, memory, and thought processes, a group of eminent scientists from various disciplines joined forces in 1962 to establish a new idiom of productive interaction by the founding of an interuniversity organization, the Neurosciences Research Program (NRP) under the sponsorship of the Massachusetts Institute of Technology.

A major activity of NRP is to bridge intellectual and professional gaps between the disparate neural and behavioral sciences and to help establish a unified neuroscience. Accordingly, NRP holds triennial Intensive Study Programs (ISP) modeled after the program sponsored by the National Institutes of Health and held in 1958 at the University of Colorado in Boulder for the purpose of integrating and nucleating the disciplines of biophysics and biophysical chemistry. Publication of the resultant volume, *Biophysical Science—A Study Program,* and the subsequent

v

founding of the Biophysical Society aided greatly in merging several sciences that had been widely separated—in the traditional sense of university disciplines—into a unified field with new vigor.

The first ISP in neuroscience, convened in 1966, surveyed the field at each level from the molecular and cellular through the neural to the behavioral. The resultant volume, *The Neurosciences: A Study Program,* is now in its fourth printing. Attesting to the rapid growth of neuroscience and furthering the trend toward unification of the field in the United States, a national Society for Neuroscience has since been founded under the auspices of the National Academy of Sciences, and similar societies have recently been formed in other countries. The maturation curve of worldwide neuroscience has left the slow-rising induction period; the slope is increasing rapidly through what may well prove to be a transition to a curve comparable in its exponential rate of rise to that which characterized the recent golden decade of molecular genetics.

Whether neuroscience will indeed gel and lead to signal advances in an understanding of the human brain and behavior will depend on the imagination and industry of neuroscientists, but will also require appropriate professional societies, versatile publications, and strong support from public and private funding. The most important factor will be key discoveries comparable in importance for neuroscience with those made by molecular geneticists—namely, that microorganisms and viruses can be employed effectively in genetic research and that genetic information can be encoded, transcribed, and translated from one type of supermolecule, DNA and RNA, to another macromolecular type, protein. The latter theoretical line probably will not apply directly in neuroscience in the sense that psychological information is directly encoded in giant macromolecules, as is genetic information. However, macromolecular properties, such as recognition, specific conformations and their rapid alteration, and allosteric interaction will doubtless play a major role in neuroscience. In all likelihood, other macromolecular properties even more significant for brain function will be discovered, possibly involving giant macromolecular nets controlled cybernetically.

One of NRP's purposes is to scan the horizon for physical and biological theories that could lead to key discoveries that may resolve some of the mysteries and complexities of neuroscience, including those of the relationship of molecules to mind. However, NRP's primary concern during the present transition period will be to stimulate research on many fronts and to emphasize especially significant salients, some of which have already been perceived and made the primary themes of the 1969 ISP*:

1. Molecular neurobiology, including relevant recent advances in molecular genetics.
2. The ontogenetic development of the nervous system, including phylogenetic and behavioral changes.
3. The "language" (i.e., bioelectrical encoding and processing of information) of neurons, neuronal assemblies, nets, and other brain subsystems.
4. Determinants of neural and behavioral plasticity and of complex psychological functions.

The same high standards that characterized the 1966 ISP were applied to the selection of topics, speakers, and other participants for the 1969 ISP. Again, the long and intensive planning process included many discussions with panels of NRP Associates, Consultants, and Staff.

In 1969, as in 1966, some 50 Fellows from many parts of the United States and from 10 other countries participated. These young scientists were selected on the advice of eminent neuroscientists here and abroad and proved to be, as predicted, an extraordinarily able group. Their contributions during the lecture and symposium discussions enriched the program immensely and were included in many of the manuscripts.

For three weeks, from July 21 through August 8, 1969, 160 scientists participated in scheduled lectures and symposia. In the mornings paired plenary lectures covered specific topics from two points of view; afternoons were devoted to two seriatim sets of three symposia, each set of

*This structuring was done primarily by Dr. Gardner C. Quarton, Program Director of NRP at that time.

three being held concurrently. Plenary lectures treated topics that could be defined sufficiently to be encapsulated in lecture form. The symposia, each occupying six afternoons, dealt with topics that are too broad or are burgeoning too rapidly to be covered adequately in lecture form. The chairman and one co-spokesman of each symposium reported its highlights to all participants at plenary sessions; the essays based on these lectures help integrate the several sections of this book.

The chairmen, besides being responsible for the organization, planning, and functioning of their symposia, also undertook the responsibility of the scientific editing of manuscripts of the plenary lectures that were related to their symposium themes. Any success that attends the publication of this volume is due in no small measure to the insight, tact, and hard work of these chairmen-editors whose names are cited on the title page.

The lecture-symposium format adopted in 1969 had the advantage of permitting in-depth examination and discussion of the four broad themes selected. As a result of formal and informal discussions at Boulder and of subsequent reconsideration by the authors, the majority of the chapters go beyond the manuscripts originally prepared and given at the ISP. The arrangement of this book follows the four thematic strands; plenary and symposium lectures are clustered in accordance with the subject matter. Ordering of the 88 chapters and section introductions is in an array of blocks that somewhat recapitulates the historical trend in the neurosciences—as, indeed, in all other sciences—toward studying ever-smaller entities and ever-faster phenomena.

The editorial and esthetic quality of the series has been maintained in the face of rising costs by the staff of The Rockefeller University Press, especially by its Director, William A. Bayless, Assistant Director Reynard Biemiller, and Editor Helene Jordan Waddell. The last-named once more began her arduous year-long editorial task by attending the entire program at Boulder and discussing manuscripts with authors after presentations. We thank the staff, as we thank the previous and the present Presidents of the Rockefeller University, Dr. Detlev W. Bronk and

Dr. Frederick Seitz, for permitting its nonprofit press to undertake this relatively costly venture.

At Boulder the opening words of welcome on behalf of the sponsoring institution were given by Dr. Howard W. Johnson, President of the Massachusetts Institute of Technology, who attested to the deep interest of M.I.T. and of himself in neuroscience and who attended the first week of lectures and symposia. His presence at the ISP and his hearty encouragement were greatly appreciated.

We also thank Dr. Irwin W. Sizer, Dean of the Graduate School at M.I.T., and Dr. John B. Goodenough, also of M.I.T., who participated throughout the meeting as counselors to Fellows and who, along with Dr. Dana L. Farnsworth and Dr. Seymour S. Kety, both of Harvard University, served on the NRP Fellowship Selection Committee.

We thank the morning chairmen who presided over the plenary lectures and steered the discussion along the most profitable lines. We also thank the administration and the staff of the University of Colorado at Boulder—particularly Mr. William E. Wright and his coworkers in the Bureau of Continuation Education—for their effective management, constant cooperation, and hospitality before, during, and after the three-week program. We thank the Federal and the philanthropic agencies acknowledged on page x for the generous support without which the ISP could not have been held or the book written and published. And finally we thank most heartily the NRP Associates and Staff for their untiring efforts to make ISP 1969 and this book maximally effective.

Francis O. Schmitt

*Brookline, Massachusetts*
*15 March 1970*

*Acknowledgment of Sponsorship and Support*

The Neurosciences Research Program is sponsored by the Massachusetts Institute of Technology and supported in part by U.S. Public Health Service, National Institutes of Health, Grant No. GM 10211; National Aeronautics and Space Administration, Grant No. NsG 462; Office of Naval Research Grant Nonr. (G)-00012-69; The Rogosin Foundation; and the Neurosciences Research Foundation, Inc. The 1969 Intensive Study Program in the Neurosciences was supported by National Institutes of Health, Grant No. GM 12568; North Atlantic Treaty Organization; and the Neurosciences Research Foundation, Inc. The publication of this book was partially subsidized by The Commonwealth Fund, John Deere Foundation, The Ruth H. and Warren A. Ellsworth Foundation, General Electric Company, The Grace Foundation, The Grant Foundation, Handy and Harman, The Charles A. Harrington Foundation, The Andrew W. Mellon Foundation, The Standard Oil Company of New Jersey, and The Teagle Foundation, Inc.

# CONTENTS

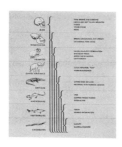

## EVOLUTION OF BRAIN AND BEHAVIOR

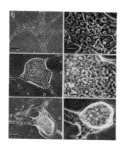

## DEVELOPMENT OF THE NERVOUS SYSTEM

## DETERMINANTS OF NEURAL
## AND BEHAVIORAL PLASTICITY

# COMPLEX PSYCHOLOGICAL FUNCTIONS

# NEURAL SUBSYSTEMS
# AND PHYSIOLOGICAL OPERATIONS

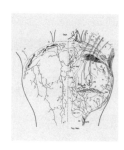

# COMMUNICATION AND CODING
# IN THE NERVOUS SYSTEM

# ASPECTS OF MOLECULAR NEUROBIOLOGY

# RECOGNITION AND CONTROL AT THE MOLECULAR LEVEL

# THE NEUROSCIENCES

## SECOND STUDY PROGRAM

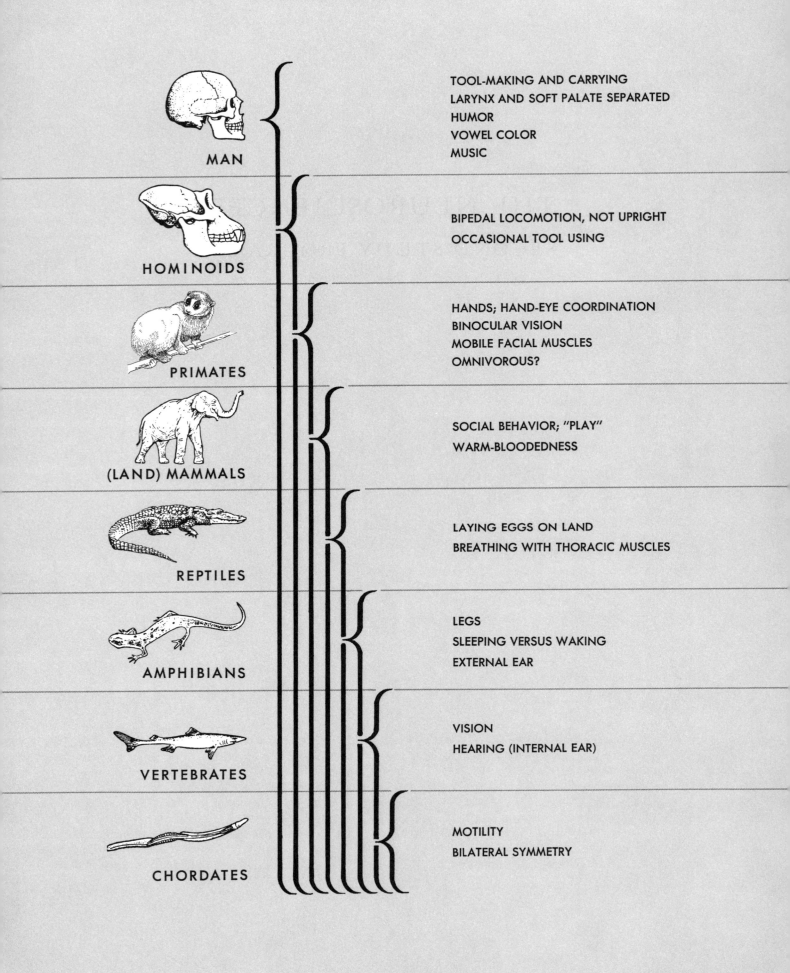

MAN

TOOL-MAKING AND CARRYING
LARYNX AND SOFT PALATE SEPARATED
HUMOR
VOWEL COLOR
MUSIC

HOMINOIDS

BIPEDAL LOCOMOTION, NOT UPRIGHT
OCCASIONAL TOOL USING

PRIMATES

HANDS; HAND-EYE COORDINATION
BINOCULAR VISION
MOBILE FACIAL MUSCLES
OMNIVOROUS?

(LAND) MAMMALS

SOCIAL BEHAVIOR; "PLAY"
WARM-BLOODEDNESS

REPTILES

LAYING EGGS ON LAND
BREATHING WITH THORACIC MUSCLES

AMPHIBIANS

LEGS
SLEEPING VERSUS WAKING
EXTERNAL EAR

VERTEBRATES

VISION
HEARING (INTERNAL EAR)

CHORDATES

MOTILITY
BILATERAL SYMMETRY

# EVOLUTION

# OF BRAIN AND

# BEHAVIOR

In this simple classification of chordates, the lowest
form of animal in each classification exhibits features
associated with communication behavior, as listed at
right. Brackets indicate that each group possesses
or has evolved beyond the characteristics exhibited by
all the groups below. (Adapted from "The Origin of
Speech" by Charles F. Hockett. Copyright © 1960 by
Scientific American, Inc. All rights reserved.)

# 1 Prefatory Comments on Evolution of Brain and Behavior

GARDNER C. QUARTON

THE LAST FIFTY YEARS have witnessed a dramatic upheaval in the study of evolution. Classically, scholars have attempted to demonstrate that evolution can account for the diversity of living things, to provide a systematic basis for categorization, and to describe the anatomy, physiology, and behavior of man's closest relatives in an attempt to find the key to man himself.

With the eruption of modern genetics, however, studies have diverged sharply from the traditional path. They have led toward using the new knowledge to compare different living species and to reconstruct the evolutionary continuum by analyzing gene expression at the levels of both molecular biology and population genetics.

Yet, the classical and the new approaches are not necessarily mutually exclusive. Because the structure and function of nervous systems help to determine the behavior of organisms and because behavior, in its turn, makes successful adaptation possible and, therefore, influences phylogenetic selective processes, the evolution of brains and of behavior should, ideally, be studied together. The reasons are similar to the rationale for studying the development of the nervous system and the behavior patterns which accompany that de-

GARDNER C. QUARTON Director, Mental Health Research Institute, The University of Michigan, Ann Arbor, Michigan

velopment. In both cases, investigation of the sequential parallel and interlocking steps that lead to highly differentiated structure and function will reveal information about the entire system—even one as complex as man—that would not be available if we dealt only with that system in its complete form. However, one obstacle has proved, so far, to be insurmountable. In the developmental sequence, all of the steps can, in principle, be studied longitudinally in real time, but a formidable road block lies in the difficulty of finding adequate ways to observe and separate *all* of the events in developmental biology that take place simultaneously in a very rapid sequence. In the biology of evolutionary change, an additional problem is posed. Most of the steps can never be observed in real time. Therefore, we must reconstruct the sequence by fragments of information, some acquired from comparative study of living animals and some from an analysis of fossils and other records of past events.

Without question, scientists tend to be uncomfortable when they speculate about how they might use information they do not have or perhaps can never have. At the same time, a clarification of the goals of comparative neurobiology and psychology, if such is possible, might substantially influence research strategies.

For instance, if we had relatively complete descriptions of the brains and behaviors of all of the animals that have ever lived, organized in a plausible evolutionary tree, we would be able to use the words "ancestor" and "descendant" without qualification. We could then examine change, and explore correlations, in the differentiation of brain structures, brain functions, and behavior. For instance, if a descendant possessed two cell types in a locus where the ancestor appeared to have only one, we could quite legitimately ask if we were correct in identifying only one type in the ancestor. If we agreed we were justified in our anatomical findings, we could explore changes in functions in the descendant associated with the differentiation of cell type. Similarly, we could examine the effect of changes in connectivity patterns between elements of the nervous system or in the function of the system.

At branch points in the evolutionary tree, it would be possible to study two descendants, or lines of descendants, each of which had evolved from a common ancestor and each of which had apparently made a successful adaptation to its environment. By comparing two different anatomical and physiological systems, identical in most respects but different in some one significant fashion, it might be possible to identify the biological basis for behavior differences or to identify two different biological mechanisms.

If the whole tree were in front of us, it also would be much easier to assess the part played by environmental conditions in producing parallel evolution. Animals with different ancestral lines but similar habitats could be compared systematically with animals with shared ancestral lines and different environments.

On the other hand, if we were able to identify the genes involved in the shift from ancestor to descendant and the intervening events in gene expressions, we should be able to clarify the mechanisms underlying the differences in behavior at a much more detailed level. This strategy is the basis for today's great interest in animals much simpler than vertebrates, in which it may at least be possible to construct reasonably complete gene maps and to identify the elements and connectivity patterns of the nervous systems.

If more knowledge were available concerning man's immediate ancestors, we might ask if some of the physiological and psychological traits that presumably were "inherited" phylogenetically were not more relevant for the adaptation of those ancestors than they are for living man. However, faced with the impossibility of reconstructing most ancestor-descendant relationships in detail and with the great complexity of brains and behavior of living animals, we must decide if any of these lines of research can be followed at all.

One happy consequence of current interest in comparative studies of brain and behavior is that it has forced upon us the inescapable conclusion that whereas our knowledge of neuroanatomy is often surprisingly incomplete, we have not let that deter us. The studies on the visual system of birds, reported in several essays in this volume, not only clarify visual mechanisms in vertebrates and stress the importance of two anatomical systems for vision. They also illuminate the fact that we now realize that for a long time we missed this vital second system. If such a significant system can be overlooked, isn't it possible that many other brain systems may have escaped detection by our anatomical methods? Fortunately, there appears to be an increased interest in more detailed anatomical investigation of the brains of many different species, and these studies can use the relatively new electron microscope, new staining methods that outline tracts with putative transmitters, new techniques that bring out small fibers and synapses, and more sophisticated combined anatomical and physiological investigations.

A complete review of comparative neurobiology and psychology would be out of place in this volume. The papers that appear in this section represent, instead, attempts to present overviews of selected problems.

Nauta and Karten summarize trends in the evolution of

the vertebrate brain, first by developing a schematized and generalized picture of that brain as a framework for the discussion of differences in structure and organization and, second, by analyzing some specialized systems in greater detail. Hodos has concerned himself with the assumptions we must make and the limitations that are necessary if we are to draw inferences from studies of living animals for the reconstruction of an evolutionary tree. Washburn and Harding have presented a picture of the evolution of primate behavior and have connected this sequence with what little is known of the evolution of primate brain.

All three papers deal in different ways with strategies for future research in comparative studies of nervous systems and of behavior.

# 2  A General Profile of the Vertebrate Brain, with Sidelights on the Ancestry of Cerebral Cortex

## WALLE J. H. NAUTA and HARVEY J. KARTEN

THE MOST elementary tenet of the theory of evolution is that animal specification followed a temporal sequence such that one order of species developed from another, and in time gave rise to one or more further orders. The reconstruction of the "tree of evolution," one of the most constantly pursued goals of biology, is attended by numerous difficulties, foremost among which is the circumstance that existing forms of life represent little more than "leaves on the ends of branches" of a tree, the trunk and limbs of which have long been extinct. Virtually all extant animals appear to be specialized forms that have diverged in greater or lesser degree from any of the identified or presumed mainlines of evolution. The identification of such "mainlines," furthermore, is often highly uncertain, the more so because several vertebrate classes appear to have evolved not from one, but from several ancestors. Modern amphibians, for example, are suspected of representing several developmental lines originating from various piscine forms. Similarly, monotremes, marsupials, and placental animals may represent parallel phyletic lines among mammals, each evolving from a different reptilian ancestor.

These and other constraints of evolutionary biology, reviewed more systematically elsewhere in this volume, apply to comparative neurology no less strictly than to other phylogenetic disciplines. Current knowledge of brain evolution is, however, limited by two additional circumstances. In the first place, neural tissues rapidly disintegrate after death, and there is thus no hope that any but indirect and relatively superficial information about the brain's architecture in extinct forms will ever be obtained. Second, and for the moment no less restricting, intensive systematic studies of brain organization so far have been limited very largely to mammalian species—a consequence, perhaps, of the fact that neuroanatomy received a major early impetus from the neuropsychiatric clinic. It would be unfair to say that nonmammalian brains have been neglected, but the enormous amount of work devoted to such brains during the past century has resulted almost exclusively in normal anatomical descriptions and, until quite recently, has included only sporadic studies by the more rigorous experimental methods that have been used so profitably in the study of mammalian brain organization.

As a result of this incongruity, our understanding of nonmammalian brains is really quite limited, too much so to permit far-reaching conclusions as to fundamental organizational differences among the brains of different classes. For example, the brain of a frog clearly differs from that of a lizard with respect to external features and, at a slightly less macroscopic level, with respect to the relative size and topography of nerve-cell groups and "fiber systems" (axon bundles). In the absence of more detailed knowledge of interneuronal relationships, however, it is impossible to determine whether such differences represent dissimilarities in

WALLE J. H. NAUTA and HARVEY J. KARTEN Department of Psychology, Massachusetts Institute of Technology, Cambridge, Massachusetts fundamental structural organization, or signify no more than variations in the relative degree of development of various central nervous subsystems common to both classes. Clearly, an enormous amount of experimental work—anatomical, physiological, biochemical, and behavioral—remains to be done before the fundamental steps of brain evolution and their functional corollaries can be fully recognized. It seems likely that future studies will lead to the identification of several crucial determinants of brain evolution. Such determinants may consist in the emergence somewhere in phylogeny of a novel neuronal element, or in a re-grouping of neurons or of synaptic junctions, possibly in response to the development of a new afferent relationship with one or another part of the brain, or in the first appearance of a distinctive chemical specification of a particular neuronal system, or even in a restructuring of glia-neuron relationships, leading to a significantly different partitioning of brain tissue.

Far from dealing with all these presumable aspects of brain evolution, the present chapter is focused on one of the central issues of comparative neurology, namely, the problem of *homology of neuronal systems*. This term needs the following explanation. The evolution of the vertebrate brain is characterized primarily not by a linear increase in the size of the central nervous system or in the number of its constituent neurons, but, instead, by more or less progressive modification of neuronal subsystems. Such tendencies are especially noticeable in the forebrain, where structural rearrangements can be profound enough to cause severe problems in determining which, if any, of a diversity of neuron groups found in a more recently evolved class of vertebrates corresponds to a particular cell group or system of interrelated cell groups in the ancestral class. A particularly precipitous recomposition of the vertebrate forebrain takes place at the transition between reptiles and mammals. Both birds and mammals have modified the ancestral reptilian forebrain structure, but, whereas in birds the resulting organization still appears relatively comparable to that of modern reptiles, the forebrain of modern mammals has deviated so much from the presumed precursory pattern that it can no longer be compared readily with the latter by routine cytoarchitectonic and fibroarchitectonic comparison. Instead, such identifications require detailed experimental studies of afferent and efferent relationships, as well as a search for common embryological derivations. The more refined, recent, histochemical techniques may provide an additional approach to the homology problem in brain evolution.

In the last part of this chapter, the search for interphyletic homologies in brain evolution is illustrated by some recent studies conducted by experimental-anatomical and histochemical methods. To place the results of these studies in proper perspective, however, it is necessary first to consider some general structural features of the vertebrate central nervous system. Needless to say, the following brief account can provide no more than a generalized survey of this highly complex subject. Its main purpose is to identify some of the major organizational features that distinguish the forebrains of mammals from those of nonmammalian forms. The scheme of the nonmammalian brain given below is based largely on recent experimental findings in birds and reptiles; in the absence of comparable data for other nonmammalian classes, it is impossible to say to what extent its fundamental aspects are valid for a generalization of "the nonmammalian vertebrate brain."

## General overview of the vertebrate brain

GROSS MORPHOLOGICAL FEATURES *The nonmammalian brain.* Some major general features of the central nervous system of nonmammalian vertebrates are represented schematically in Figure 1A. The central nervous system of all vertebrates can be subdivided into an unpaired, bilaterally symmetrical *neuraxis* and a paired subdivision, the *cerebral hemisphere,* or telencephalon. The neuraxial subdivision consists of (1) the spinal cord, (2) the rhombencephalon with the associated cerebellum, (3) the mesencephalon or midbrain, and (4) the diencephalon. In the diencephalon, three subdivisions are distinguished: *thalamus, subthalamus* (not shown in Figure 1), and *hypothalamus*. It represents the unpaired middle part of the prosencephalon, lateral parts of which become extruded in early embryonic development to form the cerebral hemisphere. The term *forebrain*, used repeatedly in the following account, is a literal translation of prosencephalon, and invariably refers to the complex formed by the diencephalon and cerebral hemisphere.

In the cerebral hemisphere, three major neural territories can be recognized:

1. The *olfactory system,* largely composed of primitive cortical formations *—the olfactory bulb, which receives the axons of the olfactory receptor cells, and a variety of other cortical regions that receive the projections from the olfac-

---

* The term *cortex* denotes variously complex neural organizations at the surface of the brain that are distinguished from subcortical (noncortical) cell organizations by the following combination of characteristics: (1) neuronal cell bodies are arranged in more or less clearly distinguishable layers, varying in number from one or two in the allocortex (see below) to five or six in the neocortex; (2) there is an additional, most superficial layer (plexiform or molecular layer), which essentially is a fiber stratum containing numerous axons and dendritic processes but few neuronal cell bodies; and (3) numerous neurons in at least the most superficial cell layer (in the neocortex, all cell layers) emit at least one major "apical" dendrite perpendicularly toward the brain surface; many such dendrites reach into the first (plexiform) layer even from the deepest cell layer of the neocortex.

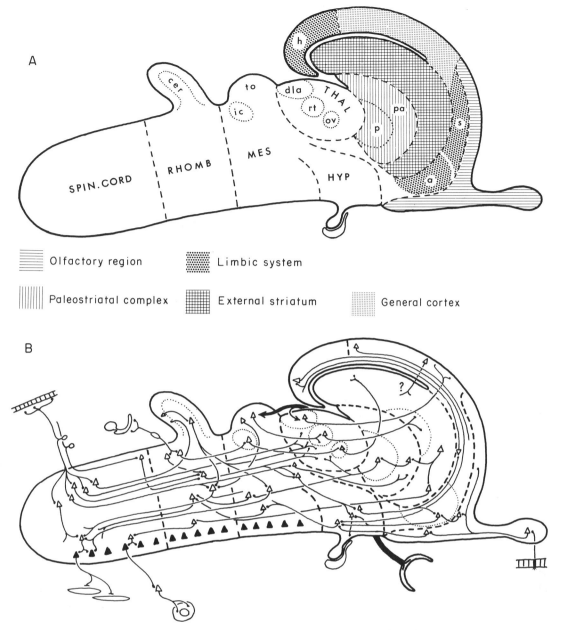

FIGURE 1  Diagramatic representation of the central nervous system of a nonmammalian vertebrate in longitudinal view. The diagrams are based on recent findings (explained in the text) with respect to connectional patterns in avian and reptilian brains in particular; it is uncertain to what extent they are valid for other nonmammalian vertebrate orders. A. General overview, showing the principal divisions of the neuraxis and, in various line and stipple patterns, the major territories of the cerebral hemisphere as itemized in the text. B. Schematic illustration of some major neural conduction pathways. For reasons of convenience, ascending fiber systems have been indicated in the dorsal half of the spinal cord and brainstem, descending pathways in the ventral half (in the forebrain this separation could not be maintained). Motor neurons are schematically indicated as a row of solid black triangles near the ventral periphery of spinal cord and brainstem. Only three primary sensory neurons are included in the diagram: one enters the spinal cord (left upper corner of diagram), a second one represents the auditory nerve (immediately behind the cerebellum), and a third one is an olfactory receptor cell (right lower extreme of drawing). All other neurons appearing in the diagram are indicated as open triangles and, strictly speaking, are intermediate between sensory and motor neurons. It should be kept in mind that the diagram represents an attempt to express some apparent principles rather than actual complexities of neuronal alignment. *Abbreviations:* a, amygdala; cer, cerebellum; dla, dorsolateralis anterior of thalamus; h, hippocampus; HYP, hypothalamus; ic, inferior colliculus; MES, mesencephalon; ov, nucleus ovoidalis thalami; p, paleostriatum primitivum; pa, paleostriatum augmentatum; RHOMB, rhombencephalon; rt, nucleus rotundus thalami; s, septum; SPIN. CORD, spinal cord; THAL, thalamus; to, optic tectum.

tory bulb and form a major basal part of the cerebral hemisphere, the piriform lobe.

2. The *corpus striatum,* which forms the central core of the cerebral hemisphere and (in reptiles and birds, at least) can be subdivided readily into two major territories, here to be named *internal striatum* (paleostriatum); and *external striatum,* each, in turn, composed of more or less clearly demarcated neuronal territories. In the avian brain, for example, the internal striatum consists of a magnocellular medial segment (the paleostriatum primitivum) and a larger lateral aggregate of smaller neurons (the paleostriatum augmentatum), and the external striatum is sharply subdivided into at least four segments—the ectostriatum, archistriatum, neostriatum, and hyperstriatum. Evidence mentioned in the final section of this paper suggests that the external striatum is a neuronal territory typical of nonmammalian phyla, and is absent as such from mammals.

3. *The limbic system.* This term, derived from mammalian neuroanatomy, is preferable to the more traditional name *rhinencephalon,* which too exclusively came to imply a specific relationship with the olfactory system. The term denotes a heterogeneous array of structures, some of which (the hippocampus and adjacent cortical regions) compose a considerable dorsomedial part of the pallium (cortical mantle of the hemisphere) separated from the more ventrally situated olfactory cortex by a smaller or larger expanse of corticoid tissue (dorsolateral or general cortex). Together with the olfactory cortex and the general cortex, the limbic cortexes are often classified as *allocortex,* a term used to distinguish these primordial forms of cortex, with only a small number of cell layers, from the multilayered *isocortex* or neocortex, which appears as an additional form of cortex only in the pallium of mammals. The limbic system, however, also includes noncortical formations such as the amygdaloid complex and septum.

Much like the olfactory apparatus, the limbic system is characterized by a fairly stable evolutionary history. Despite considerable phylogenetic modification of several of its components, the amygdala in particular, its character of a major forebrain complex connected with the hypothalamus remains preserved throughout vertebrate evolution.

*The mammalian brain.* Figure 2A epitomizes the gross morphological features of the mammalian brain. The diagram emphasizes only two of the major characteristics that distinguish the brain of the mammal from that of its phylogenetic prototypes, as well as from the avian brian, which represents a contemporary but far more conservative evolution from the ancestral reptilian model.

The most outstanding feature of the mammalian brain is the replacement of the "general cortex" by a large expanse of novel type, the multilayered *isocortex* (often called *neocortex*), a development that crowds the more primordial allocortical formations of the limbic system (hippocampus

and associated regions) relatively farther toward the edge (*limbus*) of the mantle. At the same time, the corpus striatum, although still massive, shows a marked reduction in relative size. This reduction, as is documented below, is attributable to an apparent loss of the external striatum, the structure accounting for the larger part of the striatal mass in reptiles and birds.

Even the most cursory survey of the external features of the vertebrate brain reveals a striking phylogenetic increase in the relative size of the forebrain (Figure 3). A small rostral prominence in the fish brain, the forebrain becomes progressively larger in the amphibian-reptilian sequence, and in birds and mammals (believed to have branched off from reptilian ancestors) forms the largest subdivision of the central nervous system. Progressive phylogenetic modification of the forebrain is, however, a matter not only of relative size but also of internal structure. Interphyletic differences in the configuration of the cerebral hemisphere, especially that of the corpus striatum and the pallium, pose the most challenging problems in attempts to trace the course of brain evolution. Such differences are paralleled by equally profound changes in the architecture of the diencephalon, especially that of the thalamus.

By comparison, more caudal parts of the brain, particularly the rhombencephalon, exhibit a much greater constancy of organization. Even though considerable interphyletic and species differences are encountered in comparative studies of these more caudal brain regions, such variations are only rarely so profound that homologies cannot be recognized readily. A functional corollary of this contrast in the structural evolution of forebrain and hindbrain suggests itself. Whereas the *intrinsic* mechanisms of the rhombencephalon and spinal cord appear to be involved largely with the maintenance of primordial stabilities, such as those of the internal milieu (internal homeostasis) and somatic posture (stability in space), the intrinsic function that most typifies the more rostral parts of the brain, and especially the forebrain, appears to lie in the perception of goals and goal priorities, as well as in the patterning of behavioral strategies serving the pursuit of these goals (stability in time, or stability of goal-directed behavior).

Regardless of class or species, it is likely that, in order to discharge this function, the forebrain has access to, and possesses mechanisms for, the integration of all available forms of sensory information concerning the environments, both internal and external, of the organism. It is unfortunate that comparative data on the subject of sensory afflux to the forebrain are at a premium; unfortunate because any concept of interphyletic homology in central nervous organization, if it is to be better than speculative, should be based first and foremost on reliable identification of similarities in the afferent and efferent connections of forebrain components.

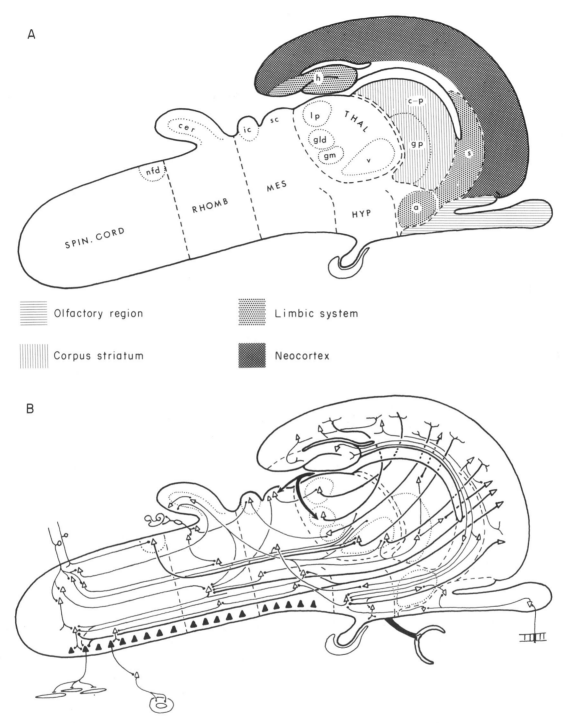

FIGURE 2  Schematic drawings representing the mammalian brain. In comparisons with the diagrams of the nonmammalian brain shown in Figure 1, the most pronounced differences appear in the composition of the pallial mantle in which the general cortex has become replaced by neocortex; the apparent absence of the nonmammalian external striatum; and the appearance of a circumscript somatic sensory nucleus (v) in the thalamus receiving, among other somatic sensory lemnisci, part of the spinothalamic tract and most of the medial lemniscus originating in the nuclei of the dorsal funiculus (nfd). Major conduction pathways afferent and efferent to the neocortex have been indicated in slightly bolder line. *Abbreviations:* a, amygdala; cer, cerebellum; c-p, caudoputamen; gld, lateral geniculate body; gm, medial geniculate body; gp, globus pallidus; h, hippocampus; HYP, hypothalamus; ic, inferior colliculus; lp, nucleus lateralis posterior of thalamus; MES, mesencephalon; nfd, nuclei of the dorsal funiculus; RHOMB, rhombencephalon; s, septum; sc, superior colliculus; SPIN. CORD, spinal cord; THAL, thalamus.

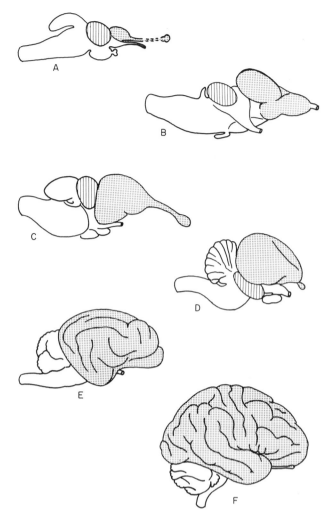

FIGURE 3 A series of side views of brains of various vertebrate orders, to illustrate the progressive increase in relative size of the cerebral hemisphere (stippled). A, codfish; B, frog; C, alligator: D, pigeon; E, cat; F, man. The optic tectum, indicated by vertical shading, is obscured from view in the two mammalian forms by the overlying cerebral hemisphere. The drawings, made on vastly different scales, are based in part on an illustration appearing in Strong and Elwyn, 1964.

INTERNAL STRUCTURE OF THE VERTEBRATE BRAIN *General considerations.* The central nervous system of all vertebrates, at its most elementary level, can be characterized as a *three-neuron nervous system.* Presumably evolved from one- and two-neuron stages of primitive Metazoa, the three-neuron organization is characterized by having a net of synaptically interconnected *intermediate* neurons interposed between the primary sensory nerve cells and the motor neurons. The appearance of this "great intermediate net" (Herrick, 1922) undoubtedly marks a fundamental step in

the evolution of the nervous system. It became virtually the sole recipient of sensory messages from the environments of the organism (it is by-passed only by the direct, i.e., mono-synaptic, reflex connections between primary sensory neurons and motor neurons, possibly a relatively late development of particular adaptive value for terrestrial vertebrates). By virtue of its intermediate position, it must have permitted a phenomenal amplification of the limited possibilities for sensory processing and integration afforded by the more primitive one- and two-neuron organizations.

It must be pointed out that the neuronal population of the vertebrate *central nervous system* (brain and spinal cord) consists exclusively of intermediate neurons and motor neurons: the cell bodies of the primary sensory neurons are outside the central organ, even though naturally their centrally directed axons must enter (and often extend far into) the brain and spinal cord to establish synaptic contacts.

Of modest size at the outset, the net of intermediate neurons rapidly expands in phylogeny so that in all vertebrates it forms the largest part by far of the central neuron population. In higher primates, the total number of brain cells is commonly estimated to be of the order of 10 billion ($10^{10}$), of which number no more than a few millions are motor neurons. These figures suggest that many mammalian brains contain at least as many as 2000 intermediate neurons to each motor neuron, a surprising ratio when it is realized that it is only by way of motor neurons that the activity of the central nervous system can be expressed in movement, whether in the form of a simple reflex or a complex, goal-directed behavior. A numerical relationship of this order suggests a very high degree of convergence of central neuronal conduction pathways toward the motor neurons, and emphasizes the appropriateness of Sherrington's characterization of the motor neuron as the "final common pathway" of the nervous system.

It is clear from these numerical data alone that the complexity of the brain is, in essence, the complexity of the intermediate net of neurons. Moreover, because the neurons composing the intermediate net account for something like 99.95 per cent of the neuron population of the mammalian brain and spinal cord, the term *intermediate net* becomes nearly synonymous with the central nervous system as a whole and, hence, loses much of its practical usefulness in discussions of neural organization. From a conceptual point of view, however, the term is nonetheless important, for it recalls the essential fact that an overwhelming majority of the neurons that make up the brain and spinal cord (in fact, all but the motor neurons) cannot be classified as either sensory or motor in the strict sense. This is true even though some intermediate neurons obviously are more directly involved in the receipt and processing of sensory information conveyed into the organization by true (i.e., primary) sensory neurons than are others that may, by contrast, have relatively direct (i.e., oligosynaptic) efferent connections with motor neurons. Such differences in relative position vis-à-vis the afferent and efferent portals of the central nervous system have made it customary to refer to certain groups of intermediate neurons as "sensory" and to others as "motor," but it should be emphasized that the terms in this application indicate relative rather than absolute characteristics of such cell groups.

Some basic organizational features of the nonmammalian central nervous system are illustrated in Figure 1B. For reasons of convenience, the motor neurons (solid black triangles) are shown aligned near the ventral surface of the spinal cord and brainstem. As shown in the diagram, motor neurons (strictly, neurons the axons of which leave the central nervous system and innervate peripheral effector tissues either directly, as in the case of skeletal musculature, or by the mediation of a second, peripheral motor-neuronal element, as in the case of smooth musculature and gland cells of the viscera) are limited to the spinal cord, rhombencephalon, and mesencephalon. Not shown in the diagram is a variant form of motor neuron that is limited to the diencephalon and is typical of the hypothalamic connection with the anterior lobe of the pituitary complex. The axon of this unique type of effector neuron does not leave the central nervous system, but terminates in close relationship to a superficial plexus of blood capillaries, the effluent venous vessels of which ("hypothalamo-pituitary portal system") enter the anterior lobe of the pituitary complex, in which they break up into a second capillary plexus. The effect of these hypothalamo-pituitary effector neurons on the gland is thought to be conveyed by means of neural transmitter substances ("releasing factors") released into the capillaries and reaching the anterior pituitary by way of the portal blood vessels.

Of the far more numerous intermediate neurons, only a small sample has been indicated, all in the form of open triangles. The sensory neurons are characterized by having cell bodies that are outside the central nervous system, usually in a peripheral sensory ganglion, but, in the unique case of the olfactory system, in the surface epithelium proper. The optic system represents a special case: the retina is a forwarded outpost of the brain itself, and impulses elicited by photic stimulation of the retinal receptor cells have passed at least two synapses in the retina before arriving at the midbrain and diencephalon by way of the optic tract (heavy arrow in Figure 1B immediately behind the cerebral hemisphere).

*Sensory Systems* SENSORY MECHANISMS OF SPINAL CORD AND RHOMBENCEPHALON Figure 1B diagramatically illustrates a common model of the possible neuronal sequences followed by impulses transmitted to the central nervous system over primary sensory neurons entering the spinal cord and rhombencephalon. At the far left of the diagram

the central axonal process of one primary sensory neuron forming part of a spinal segmental nerve is shown distributed to a group of intermediate neurons in the dorsal part of the gray matter of the spinal cord. Such cell groups of the intermediate net, in direct contact with primary sorysen neurons and hence "first in line" in terms of sensory processing mechanisms, are often termed *secondary sensory nuclei.*

The pool of secondary sensory neurons that come into contact with each sensory nerve shows considerable local differentiation, exemplified in the mammalian spinal cord by the different cellular characteristics of the substantia gelatinosa, nucleus proprius, and columna dorsalis Clarkei, respectively, and in the rhombencephalon by marked architectonic differences between various subdivisions of the cochlear nucleus. There are only scattered data with respect to the question whether such differentiations correspond to a selective processing of one particular component (submodality) of the incoming information by each subdivision of the secondary sensory complex or, alternatively, to a parallel processing of each submodality by various specialized secondary sensory cell groups. In the cochlear-nucleus complex, the latter of these alternatives appears to prevail.

The axons arising from second sensory nuclei form the so-called secondary sensory pathways, conduction channels which extend the spread of sensory information far beyond the first processing station. These secondary sensory channels can extend in any or all of the following directions: (1) the *local reflex channel* to motoneurons, either directly or by the intermediary of one or more internuncial neurons; (2) the *cerebellar channel* leading to the cortex of the cerebellum; (3) the *lemniscal channel.*

The term *lemniscus* refers somewhat loosely to fiber systems originating from secondary sensory cell groups and ascending toward the forebrain, in particular to the thalamus. The lemniscal pathway arising from the spinal cord in the nonmammalian animals largely corresponds to the so-called *spinothalamic tract* of mammalian forms. This pathway is sometimes referred to as the spinal paleolemniscus to distinguish it from a parallel spinal neolemniscus or *medial lemniscus,* originating from the so-called nuclei of the dorsal funiculus (nfd, in Figure 2A), a massive conduction system in mammals that appears to be either absent from or only weakly developed in nonmammalian forms. The spinal paleolemniscus is characterized by having a widespread distribution in the brainstem. Most of its fibers terminate in the reticular formation of the rhombencephalon and mesencephalon, and only a minority extend beyond these levels to the thalamus. The relatively small thalamic component of the paleolemniscus is probably augmented by tertiary fibers arising from neurons of the reticular formation, but, as these neurons are believed to receive neural afflux also from several sources other than the spinal cord, such tertiary links cannot be regarded as specific carriers of impulse patterns originating in the spinal cord or in any other source of secondary sensory pathways.

It is noteworthy that no well-defined thalamic cell group for the receipt and processing of spinal (i.e., somesthetic and visceroceptive) impulses has been identified in nonmammalian forms. Findings in experimental studies in birds and reptiles suggest that the relatively sparse spinothalamic fibers are distributed among a diffuse cell population in the caudal thalamus rather than in a characteristic, circumscript thalamic nucleus. This apparent absence of a well-defined homologue of the mammalian ventrobasal thalamic nucleus is the more remarkable as circumscript thalamic cell groups associated with the visual and auditory systems have recently been identified in birds and reptiles (see below).

With considerable individual variation, the central conduction pathways associated with sensory *cranial nerves* are patterned after the spinal cord model. With the exception of the optic and olfactory tracts—which, strictly speaking, neither are true cranial nerves nor are composed of primary sensory neurons—all the sensory cranial nerves have their central distribution in secondary sensory cell groups in the rhombencephalon. These first-order sensory processing stations include the complex of the *trigeminal nucleus* receiving somatic sensory afflux from the face region and the linings of the oral and nasal cavities, the *nucleus of the solitary tract* receiving impulses originating in taste receptors and in visceroreceptors (mechano-receptors and chemoreceptors in the walls of internal organs, particularly the cardiovascular system, the digestive, and the respiratory tracts), and the *vestibular and cochlear nuclei,* the impulse afflux of which originates in highly specialized receptor epithelia of the membranous labyrinth.

Each of these secondary sensory nuclei originates a local reflex channel, sometimes of great complexity, as in the cases of the vestibular nuclei and the nucleus of the solitary tract. The vestibular nuclei give rise to a prominent cerebellar channel, but cerebellar projections arise also from other secondary sensory nuclei of the rhombencephalon, with the possible exception of the nucleus of the solitary tract. As to lemniscal conduction channels, a variety of long, ascending fiber systems arising from rhombencephalic sensory nuclei has been identified, but in nonmammalian forms none of these is known to extend directly to the thalamus. In mammals, several fiber groups ascend from the trigeminal nucleus to the thalamus, at least one of which forms a component of the medial lemniscus, but similar pathways so far have not been demonstrated in nonmammalian forms.

It is not clear to what extent the avian *quinto-frontal tract* (Wallenberg) is comparable to any component of the trigeminal lemniscus of mammals. Originating from the principal nucleus of the secondary sensory trigeminal complex, this peculiar fiber system terminates in the so-called *nucleus basalis* near the ventral surface of the forebrain, a cell

group whose relationship to the thalamus, if any, is unknown.

THE AUDITORY SYSTEM  Of particular relevance, in view of recent findings mentioned below, is the thalamic representation of the auditory system. Neither in mammals nor in any nonmammalian form thus far examined have lemniscal fibers from the cochlear nuclei (*lateral lemniscus*) been traced directly to the thalamus; none appears to extend rostrally beyond the *inferior colliculus* of the mammalian midbrain tectum (ic, in Figure 2A), or, in nonmammalian forms, beyond the nucleus mesencephali lateralis dorsalis (ic, in Figure 1A), the apparent homologue of the mammalian inferior colliculus. From this mesencephalic structure, however, a massive fiber system, the brachium of the inferior colliculus, leads to the thalamus. In mammals, this last link in the auditory path ascending to the forebrain terminates in the *medial geniculate body* (gm, in Figure 2A), a major complex of thalamic cell groups from which originates a massive fiber radiation to the auditory region of the neocortex. Only recently has the circumscript *nucleus ovoidalis* (ov, in Figure 1A) in the thalamus of reptiles and birds been identified as the nonmammalian counterpart of the medial geniculate body, in the sense that it appears to be, like the mammalian nucleus, the major thalamic receiving station of the ascending auditory conduction system. It must be stressed, however, that the nucleus ovoidalis, unlike the mammalian medial geniculate body, does not appear to project to any part of the pallial mantle. Instead, in birds it has recently been found to have its major efferent relationship with the external striatum, more specifically the so-called "field L" (Rose, 1914) of the neostriatum (Figure 4B).

CENTRAL MECHANISMS OF VISION  Composed of the efferent fibers of the retina, the optic tract is the only major afferent system that is distributed from the sensory receptor organ immediately to the mesencephalon and diencephalon. In all vertebrate classes, a large number of retinal fibers (in nonmammalians the majority by far) terminate in the dorsal part of the midbrain, the tectum mesencephali. In nonmammalian vertebrates, the recipient region forms the *optic tectum* (to, in Figure 1A), a large, multilaminate structure in most classes, and especially highly differentiated in birds. In mammals, the optic tectum appears in modified and generally reduced form as the *superior colliculus* (sc, in Figure 2A).

The mammalian superior colliculus, as well as the nonmammalian optic tectum, emits a massive descending fiber system to the brainstem reticular formation (tecto-bulbar tract) and beyond it to at least the upper segments of the spinal cord (tecto-spinal tract). On the basis of physiological observations, it appears likely that, by these descending connections, the tectum is prominently involved in the conjugated eye- and body-axis (neck or whole body) movements subserving the tracking of moving objects.

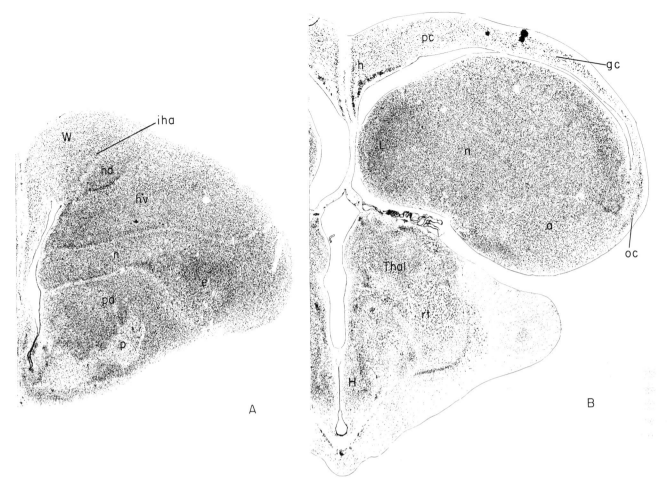

FIGURE 4 Two transverse sections of the pigeon brain, stained for cell bodies by the Nissl method. A, a rostral level of the cerebral hemisphere; B, a more caudal region. Also shown in B is the diencephalon (lower half of the figure). *Abbreviations*: a, archistriatum; e, ectostriatum; gc, general cortex; h, hippocampus; H, hypothalamus; hd, hyper- striatum dorsale; hv, hyperstriatum ventrale; iha, nucleus intercalatus of hyperstriatum accessorium; L, "field L" of neostriatum; n, neostriatum; oc, olfactory cortex; p, pale- ostriatum primitivum; pa, paleostriatum augmentatum; pc, parahippocampal cortex; rt, nucleus rotundus thalami; THAL, thalamus; W, Wulst.

Only recently has an important *ascending* connection of the optic tectum been identified. In mammals, the pathway in question extends from the superior colliculus to the thal- amus, where it terminates largely in the so-called *nucleus lateralis posterior* (lp, in Figure 2A), a cell group projecting to widespread (and thus far inadequately delineated) regions of the neocortex, probably including at least part of the "visual cortex," defined below. In birds and reptiles, the thalamic cell group receiving the ascending projection from the optic tectum is the *nucleus rotundus* (rt, in Figures 1A and 4B). The largest and most conspicuous thalamic nucleus in these non- mammalian forms, the nucleus rotundus has recently been found in birds to project to a well-defined subdivision of the external striatum, the *ectostriatum*.

Judged by its structural characteristics and efferent con- nections, the optic tectum appears certain to convey highly coded information from the retina, not only to effector mechanisms of the brainstem and spinal cord, but also to the forebrain. The path over the mesencephalic tectum, how- ever, is not the only route by which retinal impulses can be transmitted to the forebrain. In most (and possibly all) vertebrates, at least part of the fibers of the optic tract are distributed directly to the thalamus. Of modest volume in most nonmammalian forms, this direct retinothalamic con- nection in mammals becomes a major conduction route for retinal impulses. In most primates, for example, it has con- siderably outgrown the retinotectal pathway. The thalamic region involved in this connection in mammals is the *dorsal nucleus of the lateral geniculate body* (gld, in Figure 2A), a well- defined—and in primates and other advanced forms strikingly laminated—cell group from which originates the massive *optic radiation* distributed to a large occipital region of the neocortex, the so-called *visual cortex*.

In birds and reptiles, the thalamic cell region receiving direct retinal afflux has recently been identified as the *nucleus dorsolateralis anterior* (dla, in Figure 1A), a relatively indistinct territory in the pigeon, but a prominent, circumscript, and highly differentiated complex in the owl. There is convincing evidence, both anatomical and physiological, that this cell region of the avian thalamus projects selectively to a remarkable territory of the cerebral hemisphere, situated in the region of transition between the most dorsal zone of the external striatum (hyperstriatum) and the pallial mantle. The region in question occupies a substantial rostral part of the external prominence of the avian cerebral hemisphere known as the "Wulst." It is characterized by a sharply defined layer of small neurons (granule cells) quite reminiscent of the granular fourth layer of the mammalian visual cortex (Figures 4A and 7).

An important conclusion to be drawn from the foregoing accounting of central sensory systems in nonmammalian forms is that, contrary to earlier views, *the cerebral hemisphere of birds and reptiles receives well-defined projections from specific visual and auditory nuclei in the thalamus. Unlike their mammalian counterparts, however, these thalamo-cerebral afferents, with the exception of the retino-thalamo-cerebral pathway to the Wulst, do not (or at least not immediately) involve the pallial mantle; instead, they are each distributed to a circumscript region of the external striatum.* The question of the extent to which these systems of sensory conduction to the hemisphere are comparable to those of mammals is discussed further in a final section of this chapter.

Neuronal Systems of Nonspecific Afferentation
In the foregoing account of sensory conduction mechanisms believed to be relatively highly modality-specific, only occasional mention has been made of conduction systems of lesser specificity—in other words, systems composed of neurons receiving afferents from more than one single source. It seems likely that, in the neuraxis of all vertebrates, such multi-afferented neurons outnumber the more nearly unimodal nerve cells. Apparently all vertebrates, besides depending heavily for their orientation in the external environment on "labeled lines" conveying information of high spatial and temporal precision, also need for their survival other, more diffusely afferented, neuronal systems.

In vertebrates, neurons corresponding to this general description are found throughout the brainstem and spinal cord. In the rhombencephalon and mesencephalon they occupy major parts of the cross-section; the circumstance that a large part of their territory is traversed at these levels by numerous disseminated fascicles of myelinated fibers has led to the term "formatio reticularis alba et grisea," later

modified to *brainstem reticular formation*. The systemic connotation of the term reticular formation, however, cannot but arbitrarily be limited to regions that have such a "reticulated" gross appearance. On the basis of continuity of fiber connections and similarity of cytological characteristics, it would seem arbitrary not to extend the "brainstem reticular formation" forward beyond the midbrain into certain regions of the thalamus, and through the hypothalamus into the septal region, or down beyond the rhombencephalon into the spinal cord.

Such continuity, needless to say, does not imply functional equivalence throughout. The septal region may be comparable to the rhombencephalic tegmentum in terms of general cellular characteristics; it may be reciprocally connected with the latter by polysynaptic chains of neurons, but the fact remains that neurons of the rhombencephalic reticular formation receive direct afferents from the spinal cord, from the cerebellum, from such secondary sensory cell groups as the vestibular nuclei and the nucleus of the solitary tract, and from the sensorimotor cortex, but those of the septum, by contrast, receive a major afferent fiber system from the hippocampus. Clearly, the reticular formation, however continuous throughout the neuraxis, is by no means an undifferentiated neural continuum; witness also the considerable number of its cytoarchitectural subdivisions that have been recognized in studies by the Nissl method.

The reticular formation is quite generally characterized (1) by being composed of neurons with long, poorly ramified dendrites (a pecularity that has led Ramón-Moliner and Nauta [1966] to suggest the term "isodendritic core" as a substitute for "reticular formation") and (2) by a tendency of its conduction lines to be organized in the form of polysynaptic chains among which, however, some longer axonal lines are not completely lacking. Conduction systems of such predominantly polysynaptic composition can be described as "open systems" in the sense that each line at each synaptic interruption is open to at least several further afferent inputs. Convergence of various different afferent systems on a single neuron indeed appears to be among the most prominent characteristics of the brainstem reticular formation (Figures 1B and 2B include several examples of such heterogeneously afferented neurons).

Understandably, this mode of organization is often described as "nonspecific" or "diffuse." It deserves emphasis, however, that several major *effector manifestations* of the reticular formation cannot be considered diffuse. Among such functions are the maintenance of the respiratory rhythm and other homeostatic processes that require exquisitely specific rather than diffuse neural guidance. Furthermore, a majority of the "motor" systems descending from the forebrain converge on the pool of motor neurons by way of neurons of the brainstem reticular formation. Such reticular intermediaries can be interpreted as *interneurons* of the motor system in the same sense that adheres to this term as applied to groups of neurons near, and channeling impulses to, motoneuronal cell groups in the spinal cord and brainstem (see below). In this context, likewise, the reticular formation, despite its apparently diffuse and "open" organization, can hardly be considered diffuse functionally.

The same considerations apply to *ascending* conduction pathways involving the reticular formation. Although several sensory systems, notably the somatic (in mammals at least), visual, and auditory, have well-developed lemniscal pathways to the forebrain, others appear to lack such relatively "closed" ascending conduction lines, and it therefore must be assumed that information transmitted by such systems can reach the forebrain only by way of intermediary neurons in the rhombencephalic and mesencephalic reticular formation. Such extralemniscal pathways exist in all sensory systems, but they appear to be the only route to the forebrain available to, for example, the visceroceptive system represented by the nucleus of the solitary tract. The information conveyed by this afferent system must be of fundamental significance to the hypothalamic and limbic organizations governing visceral and endocrine effector mechanisms, and it would seem strange, at first glance, that its transmission to the forebrain should be less than direct.

In view of the foregoing considerations, a general interpretation of the reticular formation would seem to be based more appropriately on the effector characteristics of the organization than on the complexity and apparent diffuseness of its afferent relationships. Clearly, several major functional manifestations of the reticular formation must be classified as highly specific, and the "openness" of the structure would in these functions appear to subserve convergent integration, rather than diffusion, of multiple forms of information. Nonetheless, other functional aspects of the reticular formation, notably its well-documented central role in mechanisms determining the general activity state of the organism, could be considered "diffuse," at least from a phenomenological point of view. In this latter class of functioning, the reticular formation appears to have the significance of a general adjustment mechanism of the central nervous system itself, capable of modulating the processing of information in both afferent and efferent systems, and thereby the responsiveness of the organism to its external environment.

The great complexity and wide ramification of its neuronal concatenations have naturally interfered with a detailed anatomical analysis of the reticular formation. Polysynaptic, "short-neuron" systems are much more difficult to trace out by either anatomical or physiological methods than are the more nearly "closed" long-axon conduction pathways. Only relatively long links in the reticular organization have therefore been identified anatomically, and what is known about them has been derived almost exclusively from studies of mammalian species, the cat and rat in particular. Some of the longer conduction lines involving the reticular formation have been indicated schematically in Figures 1B and 2B, but it must be stressed that these diagrams emphasize the apparent principle rather than the details of the reticular organization. Interphyletic differences in the organization of this widespread neural apparatus are almost certain to exist, but comparative data on the subject are too scarce to supply a basis for phylogenetic conclusions.

THE MOTOR SYSTEM    This term is here used in its customary loose sense to indicate not only the motor neurons but also the neural systems descending upon these "final common pathway" elements in a generally convergent manner.

From quantitative anatomical impressions, it appears likely that most of the neural instructions issued to motor neurons in the spinal cord and brainstem come from a large population of nerve cells, the so-called *interneurons*. Neurons of this category are numerous in both the ventral horn and zona intermedia of the spinal cord, as well as in the dorsolateral (parvicellular) region of the rhombencephalic tegmentum. Together with the motor neurons, these generally smaller "pre-motoneuronal" elements compose a functional system that could be termed the *lower motor system,* an organization consisting in turn of local functional units here to be named *local motor apparatus,* for the reason that each appears to be organized in correspondence with the patterns of movement peculiar to the skeletomuscular mechanism that it controls (arm different from leg, facial musculature different from external eye muscles, and so forth).

The notion of a lower motor system composed of motor neurons and the associated interneurons is based, among other considerations, on the observation that the so-called motor pathways (with considerable variation) appear to terminate for the most part among the pools of interneurons, and to a lesser extent in direct synaptic contact with motor neurons. Such fiber systems converge on the lower motor apparatus from virtually all levels of the brain. In mammalian forms, the better-known of these systems arise from the rhombencephalon (reticular formation and vestibular nuclei), the mesencephalon (superior colliculus and red nucleus), and the sensorimotor region of the neocortex.

Several of these fiber systems, in particular the reticulospinal and reticulobulbar tracts, originate in cell territories converged on by afferents from diverse brain regions, and could thus be viewed as common last links of confluent encephalomotor systems. For example, the magnocellular medial region of the rhombencephalic reticular formation, which gives rise to a substantial reticulospinal pathway, is converged on by fibers from area 6, i.e., the trunk region of the sensorimotor cortex, and from medial, apparently also trunk-related components of the cerebellum. The red nucleus, originating in the rubrospinal and rubrobulbar tracts, likewise receives major fiber afflux from the sensorimotor cortex and the cerebellum, but the cortical and cerebellar regions from which these afferents arise appear to be involved in the motor mechanisms of face and extremities rather than trunk.

Other major effector systems appear to affect the lower motor apparatus only by way of more polysynaptic pathways of transreticular conduction. For example, the principal efferent path descending from the corpus striatum, the *ansa lenticularis* arising from the globus pallidus, does not extend directly beyond the mesencephalon, and it appears likely that impulses traveling over this pathway to the lower motor apparatus must pass at least one additional synapse in the rhombencephalic reticular formation. A similar polysynaptic organization appears to exist in the pathways leading from the forebrain to the lower motor apparatus of the *visceral motor system*. Such pathways arise largely from the hypothalamus, a region of the diencephalon receiving most of its afferent supply from the limbic and olfactory systems (in mammals, from the frontal neocortex as well) of the cerebral hemisphere, and from the mesencephalic reticular formation. No fibers of these descending hypothalamic tracts have been traced beyond the mesencephalon, and it therefore seems likely that the conduction of hypothalamic impulses to the lower visceromotor levels involves synaptic transmissions in the mesencephalic and rhombencephalic reticular formation. Judged by physiological observations (and contrary to the diagramatic representation in Figures 1B and 2B) the course followed by these transreticular visceromotor pathways is quite diffuse in the sense that they are spread over a large part of the cross-section of the brainstem.

Much less is known of the organization of the central motor system in nonmammalians. From normal anatomical descriptions as well as from sparse experimental studies, it appears likely that, at least in birds and reptiles, the descending fiber systems are largely comparable to those of mammals. Comprehensive comparisons, however, with the mammalian motor system must await more detailed experimental studies in a large sample of nonmammalian forms. If the highly variable modes of propulsion characterizing individual species be considered, fundamental variations in the organization of the central motor apparatus seem almost certain to exist, but it cannot be predicted whether such differences will be found to have their major anatomical expression at the level of the lower motor system, at higher levels, or in the disposition of descending conduction systems.

Despite this dearth of comparative data, there is one observation with respect to descending conduction pathways that might hold a clue applicable to the problem of forebrain homology. It is a remarkable fact that only one of the major fiber systems descending from the mammalian forebrain, namely the neocortical projection, has been found to extend directly (i.e., without synaptic interruption) beyond the mesencephalon. Neither the ansa lenticularis originating from the corpus striatum, nor the conduction pathways descending from the limbic system and hypothalamus, appear to exceed the caudal limits of the mesencephalic reticular formation (Figure 2B). By contrast, a considerable proportion of the corticofugal projection bypasses the mesencephalic level and distributes itself directly to the pontine nuclei—which, in turn, project to the cerebellum (this cortico-ponto-cerebellar system is not included in Figure 2B)—and to the rhombencephalic tegmentum and spinal cord. In most mammalian species, the neocortical efferents to the rhombencephalon and spinal cord form a compact, more or less prominent fiber bundle on the ventral surface of the medulla oblongata, the so-called *pyramidal tract*.

There is some recent evidence that in birds and reptiles, likewise, the fiber systems descending from the limbic system and hypothalamus, as well as the conduction pathways arising from those striatal subdivisions (internal striatum) that are comparable to the mammalian corpus striatum, do not extend directly below the mesencephalon. Naturally, as birds and reptiles do not have a neocortex, it is debatable whether any descending fiber system in such nonmammalian forms could be considered homologous with the mammalian pyramidal tract. Nonetheless, experimental studies in birds have produced evidence of a long fiber system descending from the forebrain and extending directly beyond the midbrain to the rhombencephalic tegmentum and to at least the upper segments of the spinal cord. It is remarkable that this avian cerebrobulbar and cerebrospinal projection system, in contrast to the mammalian pyramidal tract, does not originate in the pallial mantle but, instead, in a dorsal region of the so-called *archistriatum* (the archistriatum, labeled a in Figure 4B, is a ventral component of the "external striatum"; its ventrocaudal part appears to be homologous with the mammalian amygdala). In its course and distribution, this avian fiber system bears a marked resemblance to the corticobulbar fiber bundle described in ungulates by Bagley (1922) and Haartsen and Verhaart (1967).

## Structural and functional homology in the vertebrate forebrain

In the foregoing account, reference was made to a striking contrast between the evolution of the forebrain (diencephalon and cerebral hemisphere) and that of the mesencephalon, rhombencephalon, and spinal cord. The latter, more caudal, subdivisions of the central nervous system have evolved in a fairly stable manner that allows relatively easy interphyletic comparisons to be made for most of the motor and secondary sensory nuclei, and even for major subdivisions of the brainstem reticular formation. The evolution of the forebrain, by contrast, is characterized by considerable architectural modification, which culminates in the abrupt emergence of a neocortex in the pallial mantle of mammals, an event accompanied by seemingly profound changes in the composition of the thalamus and corpus striatum.

The following discussion is addressed largely to the radical departure in organization that distinguishes the mammalian brain from its probable ancestral prototypes.

*Histochemical observations* The principal features distinguishing the forebrain of mammals from that of nonmammalian forms are vividly illustrated by a comparison with the avian forebrain. The comparison is the more interesting, as birds (which share with mammals a reptilian ancestry) appear to have further elaborated, rather than modified, the basic plan of the reptilian forebrain.

Figure 4 shows two transverse sections of the forebrain of the pigeon. It will be noted that by far the largest part of the section is occupied by structures bearing designations ending with the suffix "striatum," whereas only a small amount of pallium appears. The large number of striatal subdivisions is striking when a bird's forebrain is compared with that of a rat (Figure 5): in this, as in all other mammalian forms, only two major striatal territories are evident, viz., the *caudoputamen* and the *globus pallidus*. It has been widely assumed in the past that the corpus striatum of birds is, in its totality, homologous with, even though far more differentiated than, that of mammals. This assumption was based exclusively on rather gross morphological criteria, such as the striated appearance of the gray matter, comparable positions of the structures in question with respect to the ventricle of the forebrain, and comparably subcortical (rather than cortical) arrangement of the component neurons. Furthermore, the absence of a structure in the pallium of reptiles and birds that could be compared with the mammalian neocortex was taken to imply a virtually *de novo* genesis in mammals of huge populations of specific sensory neurons in the telencephalon, without precedence in ancestral forms.

Reasonable as these assumptions appeared to be, recent

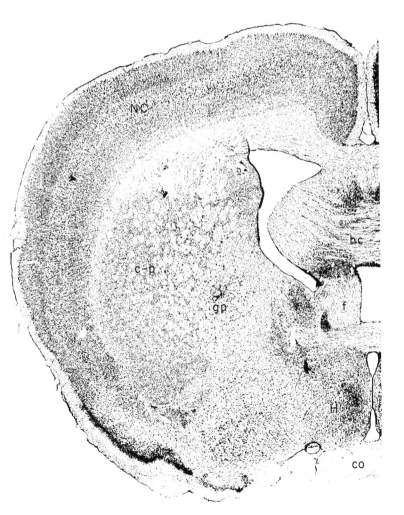

FIGURE 5 Transverse section through the rostral part of the forebrain of a small mammal (rat). Note that the mammalian corpus striatum consists of the caudoputamen (c-p) and globus pallidus (gp), and that it is adjoined laterally and dorsally by the neocortex (NC). The olfactory cortex, characterized by a dark band of pyramidal cells, can be seen at the left lower margin of the photograph. The hippocampus does not appear at this rostral level, but two of its major fiber systems, the hippocampal commissure (hc) and fornix (f) are visible in the section. *Other abbreviations:* co, optic chiasm; H, hypothalamus.

histochemical observations have made it questionable whether, indeed, the entire corpus striatum of birds is comparable to the mammalian corpus striatum. It is well-established that the mammalian caudoputamen contains exceedingly large amounts of both dopamine and acetylcholinesterase. Dopamine in the telencephalon is confined almost exclusively to the caudoputamen (c-p, in Figures 2A and 5), but acetylcholinesterase, although present throughout the telencephalon, is particularly highly concentrated in the same region. Recent studies of the pigeon (Juorio and Vogt, 1967; Karten and Iversen, unpublished) have yielded the unexpected observation that, in the avian telencephalon (Figure 6), such high concentrations of dopamine and acetylcholinesterase are characteristic of the *paleostriatum augmentatum* (pa, in Figure 1B). Furthermore, subsequent experimental studies (Karten, 1969) produced evidence that this particular subdivision of the avian striatum, much like the mammalian caudoputamen, has massive efferent connections with the large-celled, most internal striatal component, which is known in birds as the *paleostriatum primitivum* (p, in Figure 1B), in mammals as the globus pallidus (gp, in Figures 2A and 5). A homology of the avian paleostriatum primitivum with the mammalian globus pallidus is suggested by the observation that each in its respective class is characterized by having a selectively high content of iron compounds, and both project by way of comparable, well-defined fiber systems (the ansa lenticularis) to the mesencephalic tegmentum.

Considered together, these independent forms of evidence strongly indicate that the mammalian caudoputamen corresponds to the avian paleostriatum augmentatum, and the globus pallidus of the mammal to the paleostriatum primitivum of birds. Implied in this notion is the conclusion that the mammalian corpus striatum (caudoputamen and globus pallidus) corresponds to no more than the internal striatum (paleostriatal complex) of birds, and the question arises: What, then, is the nature of the huge and differentiated "external striatum" that overlies the avian paleostriatal complex?

That the external striatum of reptiles and birds appears to have no readily identifiable counterpart in the mammalian brain—identifiable in the gross-morphological sense of similarity in relative position with respect to the ventricle, for example—should not lead to the *a priori* conclusion that this vast neuronal population somehow became "lost" in the transition from the ancestral reptiles to their descendent mammalian forms. Before such a conclusion is drawn, an alternative possibility should be considered, namely, that the neurons composing the external striatum of birds and reptiles have come to occupy a radically different position in the mammalian brain. There is no direct evidence of such translocation, but circumstantial support for the notion would be provided if cell groups in the mammalian brain could be identified, the neural associations and physiological properties of which were comparable to those of the external striatum of nonmammalian forms. Such indirect evidence is indeed available.

All the recent anatomical findings point to the neocortex as the mammalian brain structure most comparable to the nonmammalian external striatum with respect to afferent and efferent relationships. Some of the principal findings supporting this comparison are mentioned in the foregoing overview of the vertebrate brain. In the following account, several further details of the neural connections in question are briefly reviewed.

*The representation of the visual system in the avian forebrain*
As mentioned earlier in this chapter, two separate circumscript pathways appear to be available in birds for the transmission of visual information to the forebrain. One of these leads from the retina to the optic tectum, and from there, via the nucleus rotundus of the thalamus (Karten and Revzin, 1966), to a circumscript subdivision of the external striatum, the ectostriatum (Revzin and Karten, 1966/67). In more detailed studies (Karten and Hodos, in press), the distribution of this tecto-thalamo-ectostriatal conduction pathway was found to be restricted to a cytologically

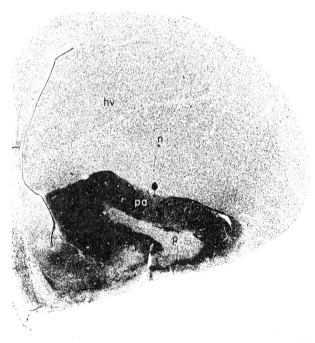

FIGURE 6 Transverse section of the pigeon forebrain stained for acetylcholinesterase by Koelle's method, and counterstained for cells by the Nissl technique. The histochemical reaction identifies the paleostriatum augmentatum (pa) as the forebrain structure containing by far the highest content of acetylcholinesterase. *Other abbreviations:* hv, hyperstriatum ventrale; n, neostriatum; p, paleostriatum primitivum.

distinct central core of the ectostriatum; a more peripheral "periectostriatal" zone, however, receives projections from this central core, as also do the adjoining neostriatum and a laminated population of cells on the dorsolateral surface of the hemisphere. That this tectofugal conduction system indeed plays a vital role in a bird's visual function has been demonstrated by the finding that destruction of the nucleus rotundus in the pigeon causes gross disturbances of visual discrimination, the exact nature of which is currently being investigated (Hodos and Karten, 1966; in press). Curiously, in a recent study Revzin (1967) found the neurons of the nucleus rotundus responsive to photic stimuli delivered at any point within an extremely large part of the visual field, a characteristic contrasting markedly with the generally more restricted receptive-field properties of neurons in other parts of the forebrain (see below).

The tecto-thalamo-ectostriatal conduction pathway of birds bears a resemblance to a conduction route in the mammalian brain leading from the superior colliculus to a relatively well-defined cell group in the caudal part of the thalamus, the nucleus lateralis posterior (Altman and Carpenter, 1961; Morest, 1965; Abplanalp, 1968; Schneider, 1969). This nucleus of the mammalian thalamus appears to project to widespread regions of the neocortex. These regions have not yet been adequately defined, but Snyder and Diamond (1968) have recently reported evidence that the nucleus projects in part to a peripheral region of the visual cortex, the so-called circumstriate belt (the term "striate," in this context, refers not to the corpus striatum but to the central region of the visual cortex, called striate cortex because of a particularly well-developed fiber plexus in its fourth layer, the stripe of Gennari).

The comparison between the respective representations of the visual system in the forebrains of birds and mammals cannot end at this point, for in both classes another important route is available for the conduction of visual impulses to the forebrain. As pointed out in the foregoing overview, in mammals this pathway leads from the retina over the lateral geniculate body of the thalamus to the visual region of the neocortex (striate cortex together with the surrounding "circumstriate belt"). Long considered a unique feature of the mammalian visual system, this massive pathway recently has been found to have an analogue in the avian brain, in the form of a conduction route from the retina to the thalamic nucleus dorsolateralis anterior (Cowan et al., 1961; Karten and Nauta, 1968), a cell group that, in turn, projects to a well-defined region of the "Wulst" of the cerebral hemisphere (Karten and Nauta, 1968). In reptiles, as well, a pathway from the retina to the dorsal thalamus has been identified (Knapp and Kang, 1968; Hall and Ebner, 1969), but in this class the corresponding projection on the cerebral hemisphere has not yet been determined.

The region of the avian Wulst that receives the retino-thalamo-cerebral projection is difficult to classify. Extremely well developed in the owl, this composite region is characterized by the presence of a distinct layer of closely spaced small neurons (granule cells), termed the *nucleus intercalatus of the hyperstriatum accessorium* (iha, in Figure 4A and Figure 7) and quite comparable in appearance to the conspicuous granular fourth layer of the mammalian striate cortex. The termination of the thalamic projection, however, is not entirely confined to this small-celled layer, for it also involves a deeper stratum of larger cells, the *hyperstriatum dorsale* (hd, in Figures 4 and 7). The thalamic projection to this remarkable cell region appears to have a precise ("point-for-point") topographical organization, much like the mammalian geniculo-cortical projection. The parallel between the visual region of the avian Wulst and the mammalian visual cortex is further emphasized by the recent observation (Revzin, 1969) that the neurons of the Wulst of pigeons have several of the physiological characteristics (small receptive fields, columnar organization of neurons with comparable stimulus requirements) that were earlier identified in the neuronal population of the mammalian visual cortex (Hubel and Wiesel, 1962).

The two routes for the conduction of visual impulses to

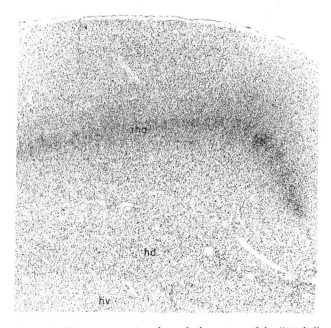

FIGURE 7  Transverse section through the region of the "Wulst" in the owl. Compare with the corresponding region of the pigeon, shown at lower magnification in Figure 4A, and note the great development of the granular layer iha ("nucleus intercalatus of the accessory hyperstriatum") in the owl. Also appearing in the photograph are the hyperstriatum dorsale (hd) and part of the hyperstriatum ventrale (hv).

the avian forebrain are illustrated in diagramatic form by Figure 8.

*Auditory structures in the avian forebrain* The auditory conduction pathway to the forebrain in birds is described briefly in an earlier section of this chapter and is illustrated in semidiagramatic form by Figure 9. As in mammals, a massive lateral lemniscus originating from the cochlear nuclei of the rhombencephalon (Am, Al, Av and Mm, Ml and Mvl) ascends to the mesencephalon, where it terminates in the nucleus mesencephali lateralis dorsalis (MLD), the apparent avian homologue of the mammalian inferior colliculus (Karten, 1967). From this mesencephalic cell group, a well-defined tractus ovoidalis, apparently homologous with the mammalian brachium of the inferior colliculus (BCI), extends forward to terminate in the nucleus ovoidalis (Ov) (Karten, 1967), a circumscript thalamic cell group, which, in turn, projects to field L, a medial region of the neostriatum characterized by a high density of its neuronal population (Karten, 1968; see also Figure 4B).

The problem of forebrain homology appears in sharp

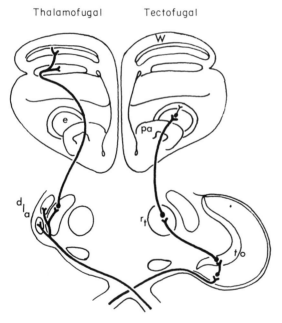

Thalamofugal          Tectofugal

FIGURE 8 Schematic drawing of the two principal pathways leading from the retina to the cerebral hemisphere in the owl. All retinal efferents, forming the optic tract, cross in the optic chiasm to the opposite side. The right half of the diagram shows the path over the optic tectum (to) and nucleus rotundus thalami (rt) to the ectostriatum (e), a ventral component of the external striatum. On the left, the conduction route over the nucleus dorsolateralis anterior thalami (dla) is shown; this pathway is distributed to various cell layers of the "Wulst" (W), a complex part of the pallial mantle. The drawing is oriented in the frontal plane. pa, paleostriatum augmentatum.

outline when the avian auditory and visual systems are compared with those of mammals. It appears likely enough that the avian nucleus ovoidalis thalami corresponds to the mammalian medial geniculate body, for each of these nuclei in its vertebrate class is the major thalamic recipient of the ascending auditory lemniscus. Major problems are, however, encountered in extending the suggested homology to the respective efferent relationships of the two thalamic cell groups: whereas the medial geniculate body projects primarily to the auditory area of the neocortex, the nucleus ovoidalis has its major efferent connection, not with any region of the avian pallial mantle, but with field L of the distinctly noncortical neostriatum. Many of the same problems arise in attempts to identify homologous structures in avian and mammalian visual systems. As in the auditory realm, homologies appear fairly self-evident up to the thalamic level: there is good reason to compare the avian nucleus rotundus to the mammalian nucleus lateralis posterior on the basis of common afferent relationships with the mesencephalic tectum, and the avian nucleus dorsolateralis anterior to the lateral geniculate body of mammals by the criterion that each in its class appears to be virtually the only thalamic cell group to receive afferents directly from the retina. Nonetheless, beyond the thalamus, comparison becomes questionable, for the mammalian nuclei project mainly to various components of the visual neocortex, whereas the avian nucleus rotundus projects to the noncortical ectostriatum, and the nucleus dorsolateralis anterior to a region that forms a transition between the hyperstriatum and the pallial mantle (the Wulst).

## Discussion and conclusions

The major conclusions to be drawn from the foregoing account can be stated as follows:

1. The "external striatum" of reptiles and birds is absent as such from mammals, in which, conversely, a "neocortex" appears in the pallial mantle.

2. The nonmammalian external striatum is comparable to the mammalian neocortex in the sense that, as is the latter, it is the major recipient of specific visual and auditory projections from the thalamus, and appears to be the only forebrain component to project directly to the rhombencephalon and the spinal cord.

Naturally, in view of these conclusions, it is tempting to suggest that the mutual exclusiveness of the two cerebral structures may be more than coincidental. In fact, it would seem reasonable to postulate that *the same neurons that in reptiles and birds compose the large external striatum, in mammals have come to occupy the pallial mantle and form a major proportion of the cell population of the neocortex.*

It is interesting that the same hypothesis arrived at here by histochemical evidence and comparisons of the respective

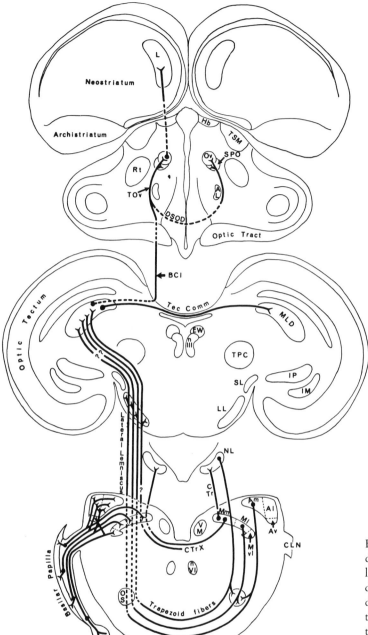

FIGURE 9 Diagramatic representation of the principal auditory conduction pathways of the pigeon. Shown on the left side of the lower drawing is the auditory nerve and its distribution to the cochlear nuclei. The remainder of the diagram illustrates the conduction pathway ascending from the cochlear nuclei to the nucleus ovoidalis thalami, and thence to field L of the neostriatum. (From Boord, 1969).

neural connections was postulated earlier by Källén (1962) on the basis of comparative-embryological findings. The external striatum is known to arise in embryonic development by cell proliferation in the so-called dorsal ventricular ridge. In nonmammalian forms, its neuroblasts mature *in situ*, that is to say, without radical migration away from their matrix. A dorsal ventricular ridge also appears as a zone of intensive cell proliferation in early mammalian embryos, but in later stages of embryonic development the structure gradually dwindles. Is it possible that, as sche-

matically indicated in Figure 10, the neuroblasts generated in this region in the mammalian embryo migrate around the lateral corner of telencephalic ventricle and invade the pallial mantle? Direct proof of such migration is not available and would be difficult to obtain, for it would require *selective* radioactive labeling of the dorsal ventricular ridge, followed at later developmental stages by the identification of labeled neuroblasts in the pallial mantle.

In the absence of such direct proof, the present hypothesis can be supported only by indirect evidence, in particular by

A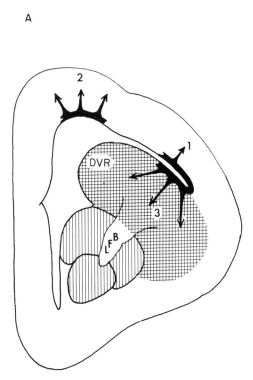

B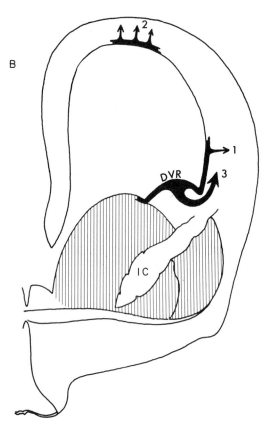

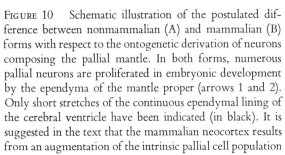

FIGURE 10    Schematic illustration of the postulated difference between nonmammalian (A) and mammalian (B) forms with respect to the ontogenetic derivation of neurons composing the pallial mantle. In both forms, numerous pallial neurons are proliferated in embryonic development by the ependyma of the mantle proper (arrows 1 and 2). Only short stretches of the continuous ependymal lining of the cerebral ventricle have been indicated (in black). It is suggested in the text that the mammalian neocortex results from an augmentation of the intrinsic pallial cell population by neuroblasts proliferated by the ependyma of the dorsal ventricular ridge (DVR, arrow 3 in B); the corresponding proliferation in nonmammalian forms produces the external striatum (arrows 3, in A). As in Figures 1 and 2, the paleostriatal complex of the nonmammalian brain and the caudoputamen-globus pallidus of the mammal are indicated by vertical shading; the nonmammalian external striatum is shown by cross-hatching. *Abbreviations:* DVR, dorsal ventricular ridge; IC, internal capsule; LFB, lateral forebrain bundle.

the observation that major neural connections of the nonmammalian external striatum are strikingly comparable to those of the mammalian neocortex. Obviously, the hypothesis is based on two assumptions: first, that the neurons originating in a particular matrix, or even in particular territories thereof, are somehow so specified as to be allowed synaptic connections only with particular neuronal systems elsewhere in the brain, and second, that such specifications remain, in principle, unchanged across class boundaries, in this case across the transition from reptiles to mammals. Some support for the first of these assumptions is supplied by a recent observation in the golden hamster suggesting that the fibers of the developing optic tract are subject to considerable territorial constraints in establishing their synaptic contacts (Schneider and Nauta, 1969). The second

of the aforementioned premises is at present only inferential and, although supported by the present data, cannot yet be fully evaluated.

It should be emphasized that the present notion does not imply that the whole of the mammalian neocortex is homologous with the nonmammalian external striatum. It is possible, and even likely, that the neocortex contains large numbers of neurons that were proliferated by the ependyma of the pallial mantle proper. It would therefore seem more appropriate to suggest that the neuronal population of the external striatum has become incorporated into the mammalian cortex, and thus to formulate the notion of *homologous neurons* rather than homologous composite structures. For example, it would seem conceivable that field L of the nonmammalian external striatum corresponds only to those neurons of the mammalian auditory cortex that stand in direct synaptic contact with the fibers of the auditory thalamocortical radiation.

A further point to be emphasized is the uncertainty as to whether all the sensory areas of the neocortex contain neuronal aggregates that have homologous counterparts in the nonmammalian external striatum. As stated earlier in this chapter, no circumscript cell territories specifically representing the somatic sensory system have been identified so far either in the thalamus or in the external striatum of nonmammalian forms. Future studies may yet show such cell groupings to exist, but until such time it must be considered possible that the somatic sensory cortex of the mammal has no functional correspondent in the nonmammalian external striatum.

For somewhat different reasons, the same doubt could be entertained with respect to the area striata of the mammalian visual cortex. As pointed out earlier, this principal recipient of the retino-geniculo-cortical conduction system has a parallel of sorts in the avian Wulst, a complex structure that straddles the transition between the external striatum and the pallium. Observations in the chick embryo (Kuhlenbeck, 1938; Northcutt, 1969) indicate that the Wulst develops from the ependyma of the pallium rather than that of the external striatum, a conclusion supported by the recent finding that in turtles (which, as do all reptiles, lack a structure readily comparable to the avian Wulst) the corresponding thalamic nucleus (dorsolateralis anterior) projects to the pallial mantle (personal communication from Dr. Ford F. Ebner). Taken together, these data suggest that the mammalian striate cortex has its ancestral prototype in the pallial mantle, rather than in the external striatum of nonmammalian forms.

The evidence reported and discussed in the foregoing parts of this paper suggests that the mammalian neocortex should not be regarded as a huge mass of neurons that had no precedence in ancestral forms. The comparison of mammalian ascending and descending systems with their nonmammalian counterparts leads to a quite different supposition, namely, that the neocortex resulted from a translocation of large masses of neurons, which in ancestral forms occupied subcortical stations, in particular the region of the external striatum. The indisputable uniqueness of the neocortex, according to this view, would reside not in any novel origin, but rather in the *alignment* of its neural components.

The question naturally arises as to what benefits might accrue to organisms in which subcortical cell groups are realigned so as to become integrated into a cortical organization. It could be argued that the characteristic "stacking" of neurons in layers, combined with the parallel, palisade-like orientation of apical dendrites, entails greater economy in the expenditure of axonal material, and thereby permits a greater number of neurons per unit volume to be connected with one another in cortical tissue than in subcortical gray matter. This point may have some relevance in a consideration of the respective evolutionary potentialities of cortical and noncortical neural tissues, but it remains to be seen whether the neocortical neurons of such mammals as the hedgehog or opossum are more numerous or more intricately interconnected than are the neurons of the formidable and differentiated external striatum of such avian forms as the owl. Moreover, it appears that whatever functional advantages may result from a cortical reorganization of subcortical gray matter are not necessarily manifested (as pointed out by Hodos elsewhere in this volume) either in a greater capacity for sensory discrimination or in greater refinements of movement. The essential functional characteristics of neocortical tissue as opposed to subcortical neuronal groupings have escaped definition thus far, but there is reason to hope that some pertinent answers may be provided by more detailed physiological comparison of neocortical areas with their apparent subcortical counterparts in nonmammalian forms. Obviously, comparative-anatomical studies, such as here discussed, cannot be expected to yield the answers. It can be said of neuroanatomical studies what was said of Coronado's children: It is perhaps not so important that they did not find the gold, as that they found a place in which to search for it.

## REFERENCES

ABPLANALP, P., 1968. An experimental neuroanatomical study of the visual system in tree shrews and squirrels. Massachusetts Institute of Technology, Cambridge, Massachusetts, doctoral thesis.

ALTMAN, J., and M. B. CARPENTER, 1961. Fiber projections of the superior colliculus in the cat. *J. Comp. Neurol.* 116: 157–177.

BAGLEY, C., JR., 1922. Cortical motor mechanism of the sheep brain. *Arch. Neurol. Psychiat.* 7: 417–453.

BOORD, R. L., 1969. The anatomy of the avian auditory system. *Ann. N.Y. Acad. Sci.* 167: 186–198.

COWAN, W. M., L. ADAMSON, and T. P. S. POWELL, 1961. An experimental study of the avian visual system. *J. Anat.* 95: 545–563.

HAARTSEN, A. B., and W. J. C. VERHAART, 1967. Cortical projections to brain stem and spinal cord in the goat by way of the pyramidal tract and the bundle of Bagley. *J. Comp. Neurol.* 129: 189–201.

HALL, W. C., and F. F. EBNER, 1969. Thalamotelencephalic projections in a turtle (*Pseudemys scripta*). *Anat. Rec.* 163: 193 (abstract).

HERRICK, C. J., 1922. Neurological Foundations of Animal Behavior. Henry Holt, New York.

HODOS, W., and H. J. KARTEN, 1966. Brightness and pattern discrimination deficits in the pigeon after lesions of nucleus rotundus. *Exp. Brain Res.* 2: 151–167.

HODOS, W., and H. J. KARTEN. Visual intensity and pattern discrimination deficits after lesions of ectostriatum in pigeons. *J. Comp. Neurol.* (in press).

HUBEL, D. H., and T. N. WIESEL, 1962. Receptive fields, binocular interaction and functional architecture in the cat's visual cortex. *J. Physiol.* (London) 160: 106–154.

JUORIO, A. V., and M. VOGT, 1967. Monoamines and their metabolites in the avian brain. *J. Physiol.* (London) 189: 489–518.

KÄLLÉN, B., 1962. Embryogenesis of brain nuclei in the chick telencephalon. *Ergebn. Anat. Entwickl.* 36: 62–82.

KARTEN, H. J., 1967. The organization of the ascending auditory pathway in the pigeon (*Columba livia*). I. Diencephalic projections of the inferior colliculus (nucleus mesencephali lateralis, pars dorsalis). *Brain Res.* 6: 409–427.

KARTEN, H. J., 1968. The ascending auditory pathway in the pigeon (*Columba livia*). II. Telencephalic projections of the nucleus ovoidalis thalami. *Brain Res.* 11: 134–153.

KARTEN, H. J., 1969. The organization of the avian telencephalon and some speculations on the phylogeny of the amniote telencephalon. *Ann. N.Y. Acad. Sci.* 167: 164–179.

KARTEN, H. J., and W. HODOS. Telencephalic projections of the nucleus rotundus in the pigeon (*Columba livia*). *J. Comp. Neurol.* (in press).

KARTEN, H. J., and W. J. H. NAUTA, 1968. Organization of the retinothalamic projections in the pigeon and owl. *Anat. Rec.* 160: 373 (abstract).

KARTEN, H. J., and A. M. REVZIN, 1966. The afferent connections of the nucleus rotundus in the pigeon. *Brain Res.* 2: 368–377.

KNAPP, H., and D. S. KANG, 1968. The retinal projections of the side-necked turtle (*Podocnemeis unifilis*) with some notes on the possible origin of the pars dorsalis of the lateral geniculate body. *Brain, Behav., Evol.* 1: 369–404.

KUHLENBECK, H., 1938. The ontogenetic development and phylogenetic significance of the cortex telencephali in the chick. *J. Comp. Neurol.* 69: 273–301.

MOREST, D. K., 1965. Identification of homologous neurons in the posterolateral thalamus of cat and Virginia opposum. *Anat. Rec.* 151: 390 (abstract).

NORTHCUTT, R. G., 1969. Discussion of the preceding paper (by H. J. Karten). *Ann. N.Y. Acad. Sci.* 167: 180–185.

RAMÓN-MOLINER, E., and W. J. H. NAUTA, 1966. The iso-dendritic core of the brain-stem. *J. Comp. Neurol.* 126: 311–335.

REVZIN, A. M., 1967. Unit responses to visual stimuli in the nucleus rotundus of the pigeon. *Fed. Proc.* 26: 656 (abstract).

REVZIN, A. M., 1969. A specific visual projection area in the hyperstriatum of the pigeon (*Columba livia*). *Brain Res.* 15: 246–249.

REVZIN, A. M., and H. J. KARTEN, 1966/67. Rostral projections of the optic tectum and the nucleus rotundus in the pigeon. *Brain Res.* 3: 264–276.

ROSE, M., 1914. Über die cytoarchitektonische Gliederung des Vorderhirns der Vögel. *J. Psychol. Neurol.* 21: 278–352.

SCHNEIDER, G. E., 1969. Two visual systems: Brain mechanisms for localization and discrimination are dissociated by tectal and cortical lesions. *Science* (Washington) 163: 895–902.

SCHNEIDER, G. E., and W. J. H. NAUTA, 1969. Formation of anomalous retinal projections after removal of the optic tectum in the neonate hamster. *Anat. Rec.* 163: 258 (abstract).

SNYDER, M., and I. T. DIAMOND, 1968. The organization and function of the visual cortex in the tree shrew. *Brain, Behav., Evol.* 1: 244–288.

STRONG, O. S., and A. ELWYN, 1964. Human Anatomy (R. C. Truex and M. B. Carpenter, editors), 5th edition, Williams and Wilkins, Baltimore, p. 8.

# 3 Evolutionary Interpretation of Neural and Behavioral Studies of Living Vertebrates

WILLIAM HODOS

THE EVOLUTION OF the nervous system and behavior has been a matter of enduring interest to scientists of many disciplines since the impact of Darwin's ideas was felt more than a century ago. In spite of the long-standing interest in

WILLIAM HODOS Department of Experimental Psychology, Walter Reed Army Institute of Research, Washington, D. C.

this area of research, relatively little progress has been made. The causes of such underdevelopment have been partly technical and partly conceptual. The ethologists, because of their close touch with systematic biology, have been keenly aware of the problems involved in evolutionary research. They have generally tended to avoid these problems by concentrating on comparisons at the generic and family levels. Comparative psychologists and comparative neurologists, on the other hand, either have been unaware of the problems or have pretended that they did not exist. However, if the evolutionary issues are faced realistically, means can be found of overcoming the difficulties in order to achieve a reconstruction of the historical sequences in the development of the brain and behavior.

The fundamental question in neural and behavioral evolutionary studies is: Can the study of animals that are alive today tell us anything about animals that lived tens of millions or hundreds of millions of years ago? I believe that the answer is a qualified "Yes," provided that one is willing to make certain assumptions and accept the limitations inherent in this type of research. My purpose here is to describe some of the assumptions and limitations.

THE PHYLOGENETIC SCALE  A principal impediment to progress in evolutionary research on the brain and behavior seems to be the widespread failure of workers to differentiate between two conceptual schemes. The first is the popular notion of a phylogenetic scale and the second is the evolutionist's concept of a family tree of organisms, i.e., a phylogenetic tree. The phylogenetic scale is a ranking of living things according to the pre-Darwinian zoological and theological view that all organisms occupy a natural rank in a continuous, unidimensional hierarchy, the *scala naturae* or Great Chain of Being (Wightman, 1950; Lovejoy, 1936). In its most scientifically based form, the *scala naturae* was only a taxonomic device and, according to Simpson (1961), not an attempt to depict evolution. A phylogenetic tree, on the other hand, is a geneology of living and extinct organisms based on the currently available data of paleontology, comparative morphology, physiology, behavior, and systematic biology. The basis of the confusion between the two organizational schemes seems to be that both arrange living things according to similarities; and similar organisms are often more closely related than are dissimilar organisms. The concept of a phylogenetic scale has no scientific status, but its superficial similarity to a phylogenetic tree has served to perpetuate its use in the scientific literature. The persistence of the phylogenetic-scale concept has had a particularly detrimental effect on the development of theory in comparative psychology (Hodos and Campbell, 1969).

The notion of "higher" and "lower" animals is an outgrowth of the phylogenetic-scale concept. Although the terms "higher" and "lower" have specific meanings to a taxonomist or to a paleontologist, as ordinarily used in categorizing animals, they are misleading and should be avoided, except in their specific technical sense. Other adjectives derived from the phylogenetic scale are "subhuman," "subprimate," and "submammalian." These terms should be avoided in favor of such designations as "nonhuman," "nonprimate," and "nonmammalian."

## Special problems of neural and behavioral evolution

In attempting to reconstruct the evolutionary history of the development of the nervous system and behavior, one encounters a number of problems that are not usually faced by students of the evolution of the skeleton, dentition, integument, and other hard morphological features. The foremost among these is that neither the nervous system nor behavior leaves a fossil record from which sequential developments can be inferred. In the case of the nervous system, some statements about gross morphology can be made by an examination of impressions within the cranial cavities of fossils (Romer and Edinger, 1942; Edinger, 1948; Radinsky, 1968), but these can tell us nothing about the details of connections between cell groups or physiology. Some behavioral inferences can be drawn from fossilized skeletons and the geological strata in which they were found. Such inferences would include feeding habits, adaptations to terrestrial, aquatic, arboreal, or aerial environments, ability to manipulate objects, and so forth. However, conclusions about courtship patterns, parental care, learning and reasoning ability, and other behavioral phenomena that are based exclusively on fossil finds would be extremely speculative. On the other hand, observations of living animals that are direct, relatively unchanged descendants of specific ancestral groups of animals can provide a firmer foundation for inferences about sequential developments within a particular lineage.

## Evolutionary history of vertebrates

Figures 1 to 3 are phylogenetic trees constructed by A. S. Romer, and the account of the evolutionary history of vertebrates presented here is based largely on Romer's (1966, 1968a, 1968b) interpretation of phylogenetic relationships. Other interpretations, differing in some details, may be found in the writings of Gregory (1951), Jarvik (1968), Colbert (1955), Le Gros Clark (1959), Ørvig (1968), Carter (1967), and Young (1962). All these contain extensive bibliographies of research reports and theoretical papers.

In discussing the evolutionary history of vertebrates, I follow Mayr's (1969) recommendation that the term "ancestral" be used instead of "primitive" and that "derived" character be used instead of "advanced" or "specialized." Mayr made these suggestions because "primitive" is too often confused with "simple," and "advanced" and "specialized" too often are taken to mean "complex." There are many examples of evolutionary trends in the direction of greater simplification (Simpson, 1967).

Figure 1 is a phylogenetic tree (Romer, 1968a) showing the pattern of evolutionary relationships among fishes. The primordial vertebrates (of which we have any record) were jawless fishes, the ostracoderms. They were most abundant in the Devonian period and are probably represented today by the cyclostomes (lampreys and hagfishes), which have in common with them a number of ancestral characteristics, the chief of which is the absence of jaws. Romer (1966) pointed out, however, that the living cyclostomes may not be good representatives of ancestral jawless fishes, because the latter were filter feeders, whereas the modern cyclostomes have evolved into a predacious, parasitic existence. Nevertheless, they possess many characters in common with ancestral ostracoderms, not only in terms of somatic morphology, but also in the gross anatomy of the central nervous system (Ariëns Kappers et al., 1960; Romer, 1966; Carter, 1967).

Another prominent group of the Devonian period were the placoderms, which were sharklike fishes that possessed a rudimentary jaw. They are probably represented by the modern chondrichthyes (cartilaginous, jawed fishes), such as

sharks and rays. Also attaining prominence in this age of fishes were the Osteichthyes, or bony fishes, which had branched into the actinopterygians (ray-finned fishes) and the sarcopterygians (fleshy-finned fishes). The fleshy fins of the sarcopterygians appear to have developed into tetrapod limbs.

Thus, in the Devonian period, the progenitors of the four major lineages of living vertebrates were present: the ostracoderms leading to the modern cyclostomes, the placoderms leading to the modern sharks and rays, the actinopterygian line leading to the modern bony fishes, and the sarcopterygian line leading to the modern tetrapods. Although each of these four lineages presumably arose from some common ancestral line, each has followed its own evolutionary course. They have evolved in *parallel,* and therefore their living descendants should not be used to represent an evolutionary *sequence.* Although the cyclostomes, cartilaginous fishes, and actinopterygians do not represent tetrapod ancestors, the study of these forms is nevertheless important for our understanding of tetrapod evolution; characters that are common to these animals and the sarcopterygians may be presumed also to have been present in their common ancestor. Moreover, within the actinopterygians the possibility exists for studying animals that could be taken as representatives of a sequence in fish evolution, for there are living descendants of the palaeoniscoid fishes, the holostean fishes, and the teleost forms. Even within the teleosts, much information exists about probable phylogenies, so that several quasi-evolutionary series might be used as a means of inferring trends in neural and behavioral evolution among these fishes.

Among the sarcopterygians are the Dipnoi, represented today by the modern lungfishes, and the crossopterygians (lobe-finned fishes), represented by the living coelacanths. Until recently, the latter were thought to be extinct. The coelacanths are extremely important, because, among living forms, they are the most closely related to the rhipidistians, which are believed to be the ancestors of tetrapods (Szarski, 1962). Unfortunately, their rarity makes them impractical for most types of laboratory investigations of the nervous system and behavior.

Figure 2 shows the radiation of tetrapods from the earliest amphibians, the labyrinthodonts. These had an alligator-like appearance and presumably formed the basic amphibian stock. Although the modern urodeles, or tailed amphibians (newts and salamanders), bear some resemblance to labyrinthodonts, they differ in a number of important ways. Even further removed from any resemblance to ancestral amphibians are the anurans (frogs and toads) and the legless Apoda. However, the relationship of the modern amphibians to either sarcopterygian ancestors or labyrinthodonts is unclear (Parsons and Williams, 1963; Thompson, 1968; Hecht, 1969; Bolt, 1969). In view of the uncertain relation-

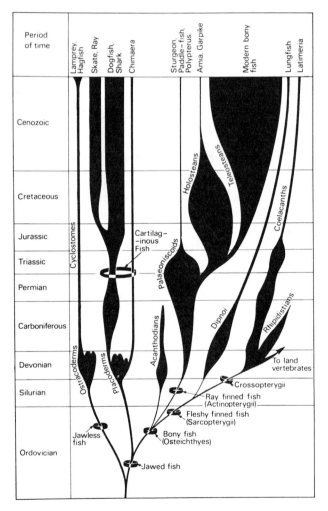

FIGURE 1    A presumed pattern of evolutionary relationships of the fishes (Romer, 1968a).

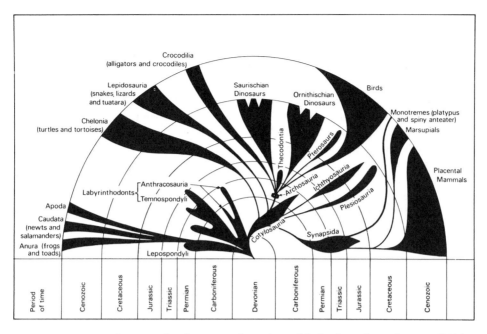

FIGURE 2   A presumed pattern of evolutionary relationships of the land vertebrates (Romer, 1968a).

ship of modern amphibians to the ancestral amphibians that gave rise to reptiles, as well as their structural differences, data from modern amphibians, which are used as representatives of the ancestors of reptiles, should be examined with great caution. This uncertain status should be carefully considered when one evaluates Herrick's (1948) attempt to use the nervous system of the tiger salamander to represent that of the common ancestor of all tetrapods. Although Herrick's monumental studies have cast considerable light on the organization of the central nervous system of this animal, we should seriously question whether the urodele nervous system should be regarded as a tetrapod prototype.

Somewhat similar to the labyrinthodonts were the cotylosaurs, the earliest reptiles. From this group a number of important reptilian forms appear to have radiated. The oldest surviving of these are the Chelonia (turtles and tortoises). The Lepidosauria (lizards and snakes) appear much later in the fossil record. Lizards appear in the Jurassic, and snakes first appear at about the same time as placental mammals. Although now extinct, the thecodonts of the Permian and Triassic periods formed the basic stock of the Archosauria and gave rise to a wide radiation of reptiles, including the dinosaurs and modern Crocodilia (alligators and crocodiles). Birds are also believed to have evolved from the archosaurian-stem reptiles. Romer (1968a, p. 221), in a description of the skeleton of an ancestral bird, stated that "the *Archaeopteryx* skeleton still preserves so many features of the ancestral archosaur pattern that, had feathers been absent, it might be argued that we were dealing with

some unusual type of very small dinosaur." Mammals are traceable to a quite different line of reptiles, the synapsids, which appear to have diverged from the cotylosaurs fairly early.

The choice of living reptiles as representatives of ancestral reptiles is difficult. Turtles are descendants of a very old reptilian lineage but have developed a number of specializations not found in the cotylosaurs. The tuatara (*Sphenodon*) probably descended from early lepidosaurs and retains many characteristics of the ancestral-stem reptiles (Romer, 1968b). It would, therefore, be an extremely useful animal for neural and behavioral investigations, but it is rare and protected, so its introduction into the laboratory on a wide scale seems unlikely. Alligators and crocodiles, as surviving archosaurs, are reasonable representatives of the reptilian ancestors of birds. Both the archosaurs and the lepidosaurs are, however, so far removed from the synapsid reptiles that their usefulness in investigations of mammalian evolution is somewhat diminished.

Mammals probably originated from an advanced group of synapsids, the therapsids, which were mammal-like reptiles. Many of the taxonomic characters that differentiate reptiles from mammals do not fossilize, so a clearcut distinction between the fossil of a reptile-like mammal and that of a mammal-like reptile may be beyond the means of current methodology. Our knowledge of the earliest evolutionary history of mammals is further marked by the uncertainty as to whether all mammals are modifications of a single line of mammal-like reptiles or whether the modern mammalian orders are descended from several different groups of mammal-like reptiles (Simpson, 1959; Reed, 1960). As far as experiments with living animals are concerned, the question is purely academic, as there are no known living mammal-like reptiles. On the other hand, the living monotremes (platypus and echidna) may provide some answers because they retain some reptilian characteristics (e.g., they lay eggs) and could be thought of as reptile-like mammals. The evolutionary relationship of the living marsupials to placental mammals is not entirely clear, but they seem to be more closely related to placentals than to monotremes. A study of marsupials, among which the opossums retain many ancestral characteristics, should provide us with some insights into the nervous system and behavior of the earliest mammals.

A presumed phylogenetic tree of placental mammals is shown in Figure 3. The stem placental mammals are generally believed to have been insectivores, of which the surviving members are the moles, shrews, and hedgehogs. The earliest primates, the prosimians, appear to be derived from insectivores, as were carnivores and probably rodents. Therefore, the often-made comparison of rats, cats, and monkeys as representing a phylogenetic series is erroneous from a historical point of view; primates, rodents, and carnivores each followed independent lines of evolution. Most of the orders of living mammals represent the results of such *parallel* evolution rather than *sequential* evolution. Each of these groups has followed its own course to its

present level of development. The only order of living mammals that can reasonably be regarded as representing the ancestors of primates are the insectivores.

I do not go into detail on the evolution of primates, as this is discussed by Washburn (this volume), but I wish to point out that, although primates have a closer affinity to one another than to members of other orders of mammals, not all primates are representative of man's ancestors. For example, the New World monkeys (squirrel monkeys, spider monkeys, and others) are descendants of New World prosimians; the Old World monkeys (macaques, mangabeys, baboons, and so on), the great apes, and man are all descendants of Old World prosimians. These two larger groups have had a long period of independent development. There is no more point in regarding New World monkeys as representatives of man's ancestors than there is in regarding rats as representing the ancestors of cats.

## Some illustrations of phylogenetic analysis

A Neuroanatomical Example    Figure 4, from a widely used textbook of neuroanatomy (Truex and Carpenter, 1964), shows the brains of representatives of a number of vertebrate classes. One could easily form the impression that these brains represent successive stages in the evolution of the human brain, but Figure 5 shows that such is not the case. Here, the brains shown in Figure 4 have been placed

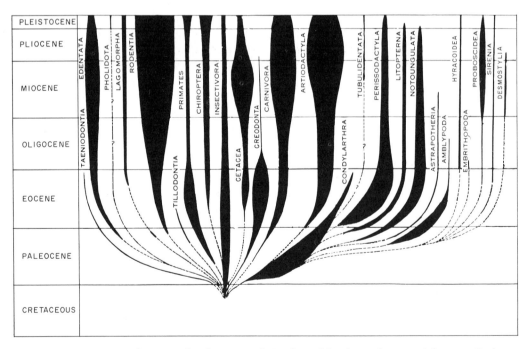

Figure 3    A presumed pattern of evolutionary relationships of the placental mammals (Romer, 1966).

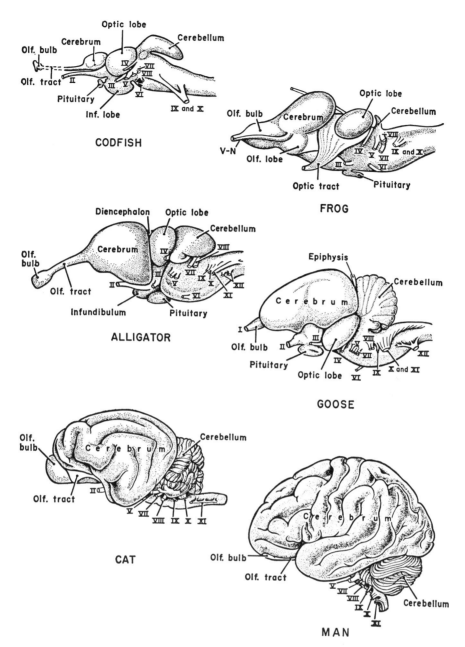

FIGURE 4   Representative vertebrate brains (Truex and Carpenter, 1964).

at the ends of branches of a phylogenetic tree. The codfish brain is irrelevant to a consideration of stages in the evolution of the human brain, because actinopterygians were not ancestral to tetrapods. The frog and alligator brains could serve as representatives of earlier stages in the evolution of the human brain, subject, of course, to the limitations inherent in the use of these animals as described earlier. The goose brain and cat brain are not directly relevant to the question of human brain evolution; neither birds nor carnivores were ancestral to primates. A frog-alligator-goose comparison could, however, represent stages in the

evolution of bird brains, and a frog-alligator-cat comparison could represent stages in the evolution of carnivore brains.

A BIOCHEMICAL EXAMPLE   Figure 6 presents data on evolutionary trends in the synthesis of L-ascorbic acid in birds (Chaudhuri and Chatterjee, 1969). The amount of ascorbic acid synthesized (μg/mg of protein) in liver, kidney, or both, of a large number of bird species has been plotted on a phylogenetic tree of birds. Birds close to the presumed stem of the avian phylogenetic tree generally synthesized ascorbic

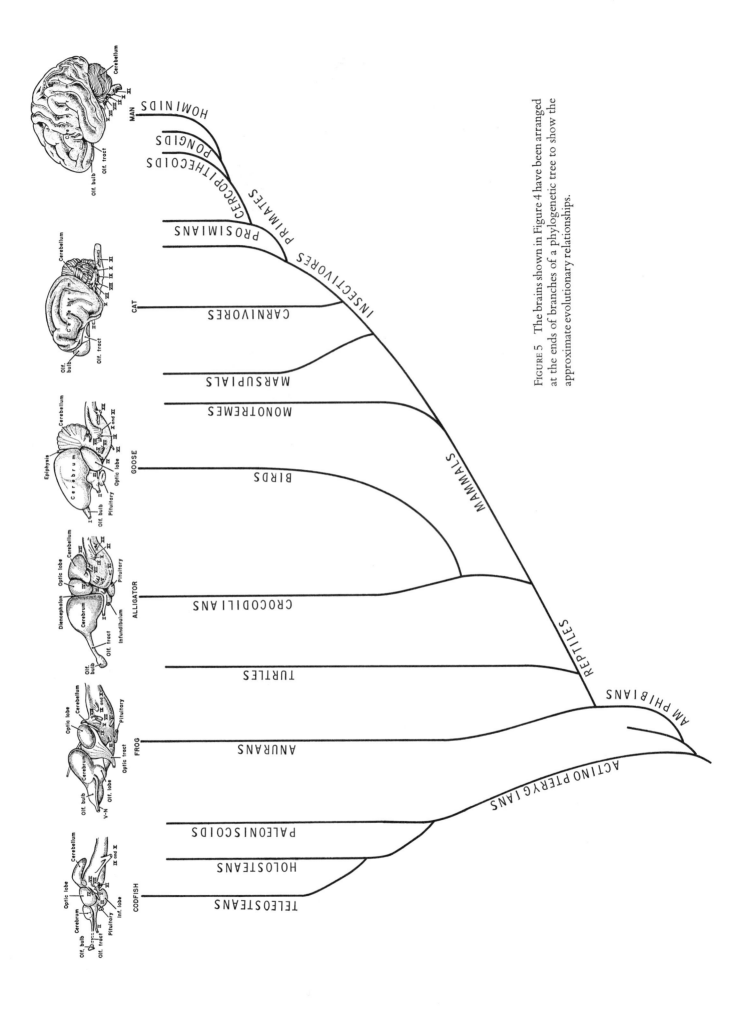

FIGURE 5   The brains shown in Figure 4 have been arranged at the ends of branches of a phylogenetic tree to show the approximate evolutionary relationships.

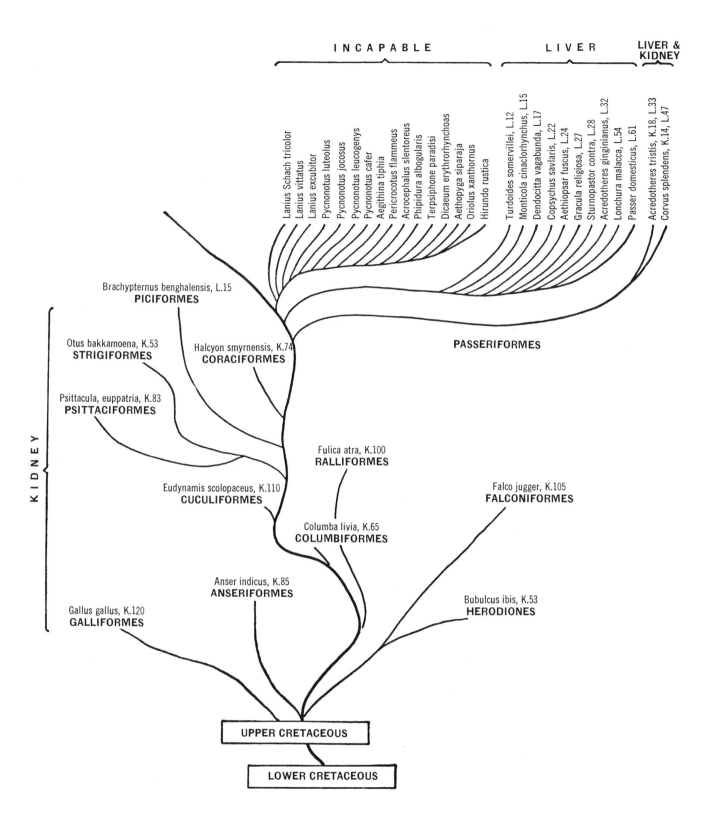

INCAPABLE    LIVER    LIVER & KIDNEY

Lanius Schach tricolor
Lanius vittatus
Lanius excubitor
Pycnonotus luteolus
Pycnonotus jocosus
Pycnonotus leucogenys
Pycnonotus cafer
Aegithina tiphia
Pericrocotus flammeus
Acrocephalus slentoreus
Phipidura albogularis
Terpsiphone paradisi
Dicaeum erythrorhynchoas
Aethopyga siparaja
Oriolus xanthornus
Hirundo rustica
Turdoides somervillei, L.12
Monticola cinaclorhynchus, L.15
Dendocitta vagabunda, L.17
Copsychus savlaris, L.22
Aethiopsar fuscus, L.24
Gracula religiosa, L.27
Sturnopastor contra, L.28
Acredotheres ginginianus, L.32
Lonchura malacca, L.54
Passer domesticus, L.61
Acredotheres tristis, K.18, L.33
Corvus splendens, K.14, L.47

Brachypternus benghalensis, L.15
**PICIFORMES**

PASSERIFORMES

Otus bakkamoena, K.53
**STRIGIFORMES**

Halcyon smyrnensis, K.74
**CORACIFORMES**

Psittacula, euppatria, K.83
**PSITTACIFORMES**

Fulica atra, K.100
**RALLIFORMES**

Falco jugger, K.105
**FALCONIFORMES**

Eudynamis scolopaceus, K.110
**CUCULIFORMES**

Columba livia, K.65
**COLUMBIFORMES**

KIDNEY

Anser indicus, K.85
**ANSERIFORMES**

Bubulcus ibis, K.53
**HERODIONES**

Gallus gallus, K.120
**GALLIFORMES**

UPPER CRETACEOUS

LOWER CRETACEOUS

FIGURE 6   Synthesis of L-ascorbic acid in the liver and kidney microsome fractions of different species of birds (Chaudhuri and Chatterjee, 1969).

acid in the kidney. Among the Passeriformes, synthesis takes place in the liver, in both liver and kidney, or not at all. These data suggest phylogenetic trends in the biosynthetic activity of birds.

A BEHAVIORAL EXAMPLE   As an example of how a behavioral phylogenetic analysis might be carried out and to illustrate the difficulties in interpretation of such analyses, I present some comparative behavioral data that I hope have a bearing on the evolution of behavioral plasticity. The data that I have chosen to present are from the formulation of learning sets, or "learning to learn." Although a number of other indicators of plasticity in animals have been suggested (for reviews of these, see Munn, 1965; Riopelle, 1967; Warren, 1965; Bitterman, 1965a, 1965b, 1969), the methods of animal testing and data presentation have been so variable as to render any comparison among them extremely difficult. On the other hand, learning set lends itself well to a phylogenetic analysis because of the relatively high degree of consistency in methods of testing and presentation of data. Undoubtedly learning-set performance is related to only one type of plasticity, and the failure of a particular group of animals to show progressive improvement should not be construed to mean that these animals are incapable of behavioral modification.

A typical learning-set experiment employs a series of discrimination problems, consisting of pairs of stimuli, usually visual. The subject is rewarded for choosing one stimulus of the pair but not the other. The position of the positive stimulus (i.e., the one associated with the reward) is varied from right to left randomly to insure that the subject is discriminating the stimuli and not merely the spatial location of the reward. A large number of problems are used, from several hundred to a thousand. In some experiments, the discrimination training is continued until the discrimination is learned to a criterion, such as 90 per cent correct or better. In other experiments, the subjects are tested for a fixed number of trials on each problem, irrespective of the per cent correct. The main dependent variable in these studies is the average performance of the subject on the *second* trial of a large block of problems, e.g., 50 or 100. A high proportion of correct second-trial responses in a block of problems indicates that the subject has acquired the learning set, or has learned to learn. Thus, the first trial of each problem serves as an information trial. It provides the subject with a cue about which stimulus is correct on that particular problem.

The subject that has acquired the learning set has not merely learned to discriminate stimuli; he has learned a strategy or principle of responding that will result in a greater frequency of reward. The principle is: correct on trial one, stay with that stimulus; incorrect on trial one, switch to the other stimulus. In terms of evolutionary significance, the ability to develop an abstract principle from past learning experiences would seem to have a higher survival value than the ability merely to learn about the motivational consequences of particular responses to particular stimuli. Learning set provides the organism with the means for acquiring new skills rapidly in new situations with minimal failure. Without a learning set each problem must be learned anew. An animal that can apply a principle to a new situation would seem to have considerable advantage over one that could not.

In order to carry out a phylogenetic analysis of learning set, curves of learning-set performance of as many different species as could be found were plotted on a common set of coordinates. Only those learning-set studies that reported data in the form of per cent correct on trial two were used. Additional studies that reported data only in other forms, such as trials to criterion, were omitted because they could not be directly compared. Likewise, data from immature animals were omitted, because learning-set performance varies ontogenetically (Harlow et al., 1968; Zimmerman and Torrey, 1965). An exception was the human data, which are from children; comparable adult data could not be found. Smooth curves were drawn through the empirical data points. In a number of cases, the data obtained from different species were so similar that they could be readily fitted to the same curve. The resulting curves are shown in Figure 7. In this figure, mean per cent correct on trial two is plotted as a function of the number of successive problems. Each curve is based on data points that represent the mean per cent correct during successive blocks of problems, e.g., every 50 or 100 problems. An important consideration in the interpretation of these curves is that, even when a learning-set curve shows little or no increase from problem to problem, the animals show progressive improvement *within* problems, i.e., they learn, although they do not acquire the learning set.

Devotees of the phylogenetic scale would be pleased, at first, to note in Figure 7 that squirrels and rats show little or no improvement after 1000 problems, cats improve at a modest rate, monkeys improve fairly rapidly, and the great apes and human beings, of course, acquire the learning set quite rapidly. A second look might be disconcerting; squirrel monkeys and marmosets do no better than cats, and ferrets and minks acquire the learning set at about the same rate as some primates. Perhaps most awkward for the phylogenetic scale is that the performance of pigeons is comparable with that of New World monkeys. As indicated above, however, the phylogenetic scale bears only the most superficial resemblance to a phylogenetic tree and leads to an erroneous view of the phylogenetic relationships among various groups of animals. Figure 8 presents a phylogenetic tree showing the presumed evolutionary relationships of the various animals listed in Figure 7. An index of the rate of

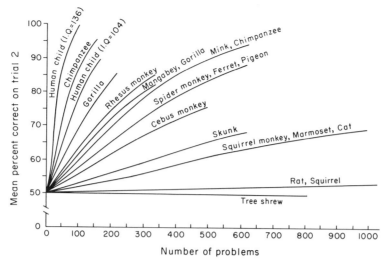

FIGURE 7 A family of "ideal" curves representing the development of learning set in various animals. The curves were obtained by the fitting of smooth curves to empirical data points. The data of the human children and chimpanzees were from Hayes et al. (1953); those of the gorillas, from Fischer (1962); those of the rhesus monkeys, from Harlow (1959); those of the mangabeys, from Behar (1962); those of the squirrel monkeys, from Miles (1957); those of the marmosets, from Miles and Meyer (1956); those of the cebus monkeys and spider monkeys, from Shell and Riopelle (1958); those of the squirrels and rats, from Rollin, cited in Warren (1965); those of the ferrets, minks, skunks, and cats, from Doty et al. (1967); those of the tree shrews, from Leonard et al. (1966); and those of the pigeons, from Ziegler (1961). As the human, gorilla, and chimpanzee in the development of learning set. The number next to the studies each reported data on only two subjects, the curves of each subject are shown separately.

acquisition of the learning set is indicated next to the common name of each animal at the top of the figure. This index is the mean per cent correct on trial two after 100 problems and is a rough approximation of the rate of learning-set formation. The index scale varies from 50 to 100. A score of 50 represents no improvement; a score of 100 represents maximum performance.

The only nonmammals about which we have learning-set data are pigeons. Because there appear to be no data on other birds or reptiles, not much can be said about the evolutionary significance of the performance index of pigeons. In regard to mammals, the only stem mammals about which we have data are the insectivores, which are represented by the tree shrews. These animals did not show the slightest indication of acquisition of the learning set, even after 800 problems.

Recently there has been considerable discussion as to whether tree shrews are arboreal insectivores or actually primitive prosimians (Campbell, 1966). Learning-set data of insectivores might add weight to one or the other side of the dispute, if any data on prosimians were available for comparison. Unfortunately, there seem to be none in the literature. Such lack of data is doubly unfortunate because the prosimians represent the stem of the primate lineage.

The data of Figure 8 seem to suggest that animals that are closely related tend to acquire the learning set at similar rates. The New World monkeys (Ceboidea) all acquire the set at low rates; the Old World monkeys (Cercopithecoidea) all acquire the set at intermediate rates; and the great apes and human beings (Hominoidea) acquire the set at the highest rates. Among the rodents, squirrels and rats acquire the set at about the same extremely low rates, which may not be effectively different from chance performance. The carnivores studied acquired the learning set at about the same rate as the New World monkeys. Particularly striking is the similarity in acquisition rates of the minks and ferrets as compared with the acquisition rate of the skunks. All three species belong to the family Mustelidae, but minks and ferrets are classed in the same genus, *Mustela*, whereas the common North American striped skunks belong to the genus *Mephitis*.

The data presented in Figure 8 can be regarded as only suggestive of evolutionary trends within the various orders of mammals indicated in the Figure. The data are far too sparse in numbers of species represented and numbers of animals within each species for any serious conclusions to be drawn about the phylogenetic development of learning-set ability. Furthermore, data are totally lacking for such critical taxonomic groups as the reptiles, monotremes, marsupials, and prosimians. Let us assume for the moment, however, that adequate representation had been given to all groups. Even so, a number of variables must be considered

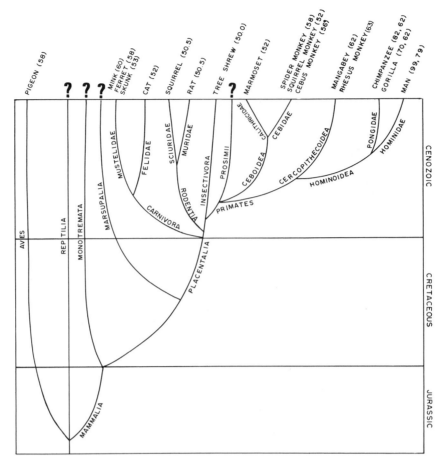

FIGURE 8   A phylogenetic tree to show evolutionary trends in the development of learning set. The number next to the name of each animal represents its rate of learning-set acquisition. The range of possible scores is 50 (no improvement) to 100 (maximum performance).

before one may conclude that the observed behavioral differences represent a phylogenetic effect. First among these is the problem of motivation and reward. We have no data to indicate whether a rat deprived of food for 24 hours is as hungry as a ferret deprived of food for the same period, or how much grain will be as rewarding to a pigeon as a raisin is to a monkey. Techniques are available (Hodos, 1961; Hodos and Kalman, 1963; Hodos and Trumbule, 1967) for scaling deprivation states and reward strengths, but, to my knowledge, these have not been applied to the matching of motivational conditions across species. A second problem is whether the animals studied were laboratory reared or reared in the wild. Several investigators have discussed the drastic effects that laboratory rearing can have on behavior (Lockard, 1968; Harlow and Harlow, 1965). In general, wild-reared animals seem to be preferable to laboratory- or zoo-reared animals. A related problem is the extent to which the testing situation is suited to the par-

ticular sensory and motor abilities of the animal under consideration. For example, all the stimuli in the experiments described above were visual. The relatively poor vision of the rats, compared with that of the other animals, may have contributed to their very low rate of learning-set acquisition. Their performance might have been far superior on a test of olfactory learning set. Indeed, many of the other animals with well-developed olfactory systems might have more rapidly acquired the learning set if the stimuli had combined visual and olfactory stimuli. Another point worth consideration is that the testing situation was designed by human beings and, as such, might have biased the results in favor of themselves and their closest primate relatives. In order to obtain the best possible performance from an animal, the testing situation should be designed to reflect its normal interaction with its natural environment.

Can these difficulties be circumvented? I believe that they can if we are willing to assume that each of the possibly confounding variables described above would always result in an underestimate of an animal's maximum capability, never in an overestimate. In other words, any one of a number of factors can cause an animal to perform at less than its best, but nothing can cause it to exceed its natural limitations. In practice, this would mean that the particular experimental technique that resulted in the highest performance level for an animal must be taken as giving the best estimate. All other techniques must be regarded as yielding underestimates. Just as the paleontologist would modify his phylogenetic trees to take new data into account, so the student of behavioral evolution would modify his conclusions about maximum levels of attainment as new data were acquired.

## Problems of phylogenetic analysis

One of the difficulties inherent in arranging data according to a paleontologist's or taxonomist's phylogenetic tree is the probabilistic nature of the tree. All phylogenetic trees are nothing more than working hypotheses about relationships among different organisms, living and extinct, based on their similarities and differences. Moreover, the full details about the evolution of vertebrates are not known. Therefore, unless we ourselves are prepared to do original research in paleontology and taxonomy, we must enlist the aid of experts in these fields. In many instances, the experts do not agree, and we have no recourse but to await new data that will support one hypothesis over the other.

Another problem in phylogenetic analysis is to avoid the tendency to regard a single living species as if it were ancestral to another living species. The nervous system and the behavior of ancestral animals cannot be inferred from the study of a single living species. Only by looking for common characteristics among related groups (including direct descendants of older lineages) can we hope to formulate an impression of what an ancestral form was like. Even then, we shall have a picture of only a hypothetical *archetype,* which may not bear a close resemblance to an actual common ancestor (Simpson, 1961; Mayr et al., 1953; Mayr, 1968). All too commonly, comparisons are made of the brains of "the reptile" and "the mammal" or the behavior of "the cat" and "the monkey," as if the particular reptile selected actually embodied the essence of reptilism or the particular monkey selected typified all monkeys, irrespective of their great degree of diversity.

Finally, not all comparative studies need be directed at the reconstruction of specific evolutionary sequences. An equally important purpose of comparative investigations is the analysis of adaptation and specializations. Indeed, most comparative studies fall into this category. The study of the relationship of structure to function and its implications for the survival of organisms in various taxonomic groups

is as necessary for an understanding of the evolutionary process as is the determination of phylogenies (Davis, 1954). Moreover, the student of the general mechanisms of adaptation and survival is not limited to the study of groups of animals that are descendants of a common lineage, as is the student of specific evolutionary sequences. In adaptation studies, the comparison of the visual systems of a honeybee, an octopus, a pigeon, and a monkey can provide the basis for meaningful generalizations, provided that there is no suggestion that these animals form a historical sequence.

In conclusion, neural and behavioral studies of living animals, if they are to be meaningful in terms of present-day interpretations of the evolutionary history of vertebrates, must be carried out with certain limitations in mind. First, the facts of vertebrate evolution are not fully known. Many of the facts that are known have varying degrees of uncertainty attached to them. Second, living animals in an evolutionary study should be descendants of a common lineage. Third, neural or behavioral characteristics found in a single living species cannot automatically be assumed to have been present also in the ancestral species that the living animal is intended to represent. Fourth, a single species should not be used to represent a larger taxon, just as a single individual should not be used to represent an entire species.

I believe that only by accepting the limitations imposed on us by paleontology and taxonomy, and operating within them, can we achieve the goal of a meaningful reconstruction of the historical sequences in the evolution of brain and behavior that have brought living animals to their present state and that may affect the course of their future development.

## Acknowledgments

I am grateful to Professor A. S. Romer, the University of Chicago Press, and the World Publishing Company for their permission to reproduce the phylogenetic trees shown in Figures 1 to 3; to Professor R. C. Truex and the Williams and Wilkins Publishing Company for permission to reproduce the brain drawings shown in Figure 4; and to Dr. I. B. Chatterjee and the American Association for the Advancement of Science for permission to reproduce the biochemical data shown in Figure 6.

I also wish to express my gratitude to Dr. Alice S. Powers for her assistance in surveying and interpreting the learning-set literature.

### REFERENCES

ARIËNS KAPPERS, C. U., G. C. HUBER, and E. C. CROSBY, 1960. The Comparative Anatomy of the Nervous System of Vertebrates, Including Man. Hafner, New York. (Reprint of 1936 edition.)

BEHAR, I., 1962. Evaluation of cues in learning set formation in mangabeys. *Psychol. Rep.* 11: 479–485.

BITTERMAN, M. E., 1965a. The evolution of intelligence. *Sci. Amer.* 212 (no. 1): 92–100.

BITTERMAN, M. E., 1965b. Phyletic differences in learning. *Amer. Psychol.* 20: 396–410.

BITTERMAN, M. E., 1969. Thorndike and the problem of animal intelligence. *Amer. Psychol.* 24: 444–453.

BOLT, J. R., 1969. Lissamphibian origins: Possible protolissamphibian from the Lower Permian of Oklahoma. *Science (Washington)* 166: 888–891.

CAMPBELL, C. B. G., 1966. The relationships of the tree shrews: The evidence of the nervous system. *Evolution* 20: 276–281.

CARTER, G. S., 1967. Structure and Habit in Vertebrate Evolution. University of Washington Press, Seattle.

CHAUDHURI, C. R., and I. B. CHATTERJEE, 1969. L-ascorbic aci: synthesis in birds: Phylogenetic trend. *Science (Washington)* 1964d 435–436.

CLARK, W. E. LE GROS, 1959. The Antecedents of Man. Edinburgh University Press, Edinburgh.

COLBERT, E. H., 1955. Evolution of the Vertebrates. John Wiley and Sons, New York.

DAVIS, D. D., 1954. Primate evolution from the viewpoint of comparative anatomy. *Hum. Biol.* 26: 211–129.

DOTY, B. A., C. N. JONES, and L. A. DOTY, 1967. Learning set formation by mink, ferrets, skunks and cats. *Science (Washington)* 155: 1579–1580.

EDINGER, T., 1948. Evolution of the horse brain. *Mem. Geol. Soc. Amer.* 25: 1–177.

FISCHER, G. F., 1962. The formation of learning sets in young gorillas. *J. Comp. Physiol. Psychol.* 55: 924–925.

GREGORY, W. K., 1951. Evolution Emerging. Macmillan, New York.

HARLOW, H. F., 1959. Learning set and error factor theory. *In* Psychology: A Study of a Science (S. Koch, editor). McGraw-Hill, New York, N.Y., pp. 492–537.

HARLOW, H. F., A. J. BLOMQUIST, C. I. THOMPSON, K. A. SCHILTZ, and M. K. HARLOW, 1968. Effects of induction age and size of frontal lobe lesions on learning in rhesus monkeys. *In* The Neuropsychology of Development (R. L. Isaacson, editor). John Wiley and Sons, New York, pp. 79–120.

HARLOW, H. F., and M. K. HARLOW, 1965. The affectional systems. *In* Behavior of Nonhuman Primates (A. M. Schrier, H. F. Harlow, and F. Stollnitz, editors). Academic Press, New York, vol. 2, pp. 287–334.

HAYES, K. J., R. THOMPSON, and C. HAYES, 1953. Discrimination learning set in chimpanzees. *J. Comp. Physiol. Psychol.* 46: 99–104.

HECHT, M. K., 1969. The living lower tetrapods: Their interrelationships and phylogenetic position. *Ann. N.Y. Acad. Sci.* 167: 74–79.

HERRICK, C. J., 1948. The Brain of the Tiger Salamander, *Ambystoma tigrinum.* The University of Chicago Press, Chicago.

HODOS, W., 1961. Progressive ratio as a measure of reward strength. *Science (Washington)* 134: 943–944.

HODOS, W., and C. B. G. CAMPBELL, 1969. Scala naturae: Why there is no theory in comparative psychology. *Psychol. Rev.* 76: 337–350.

HODOS, W., and G. KALMAN, 1963. Effects of increment size and reinforcer volume on progressive ratio performance. *J. Exp. Anal. Behav.* 6: 387–392.

HODOS, W., and G. H. TRUMBULE, 1967. Strategies of schedule preference in chimpanzees. *J. Exp. Anal. Behav.* 10: 503–514.

JARVIK, E., 1968. Aspects of vertebrate phylogeny. *In* Current Problems of Lower Vertebrate Phylogeny. Fourth Nobel Symposium (T. Ørvig, editor). Almqvist and Wiksel, Stockholm, pp. 497–527.

LE GROS CLARK, W. E. (See CLARK, W. E. LE GROS.)

LEONARD, C., G. E. SCHNEIDER, and C. G. GROSS, 1966. Performance on learning set and delayed-response tasks by tree shrews (*Tupaia glis*). *J. Comp. Physiol. Psychol.* 62: 501–504.

LOCKARD, R. B., 1968. The albino rat: A defensible choice or a bad habit? *Amer. Psychol.* 23: 734–742.

LOVEJOY, A. O., 1936. The Great Chain of Being. Harvard University Press, Cambridge, Massachusetts.

MAYR, E., 1968. The role of systematics in biology. *Science (Washington)* 159: 595–599.

MAYR, E., 1969. Principles of Systematic Zoology. McGraw-Hill, New York.

MAYR, E., E. G. LINSLEY, and R. L. USINGER, 1953. Methods and Principles of Systematic Zoology. McGraw-Hill, New York.

MILES, R. C., 1957. Learning-set formation in the squirrel monkey. *J. Comp. Physiol. Psychol.* 50: 356–357.

MILES, R. C., and D. R. MEYER, 1956. Learning sets in marmosets. *J. Comp. Physiol. Psychol.* 49: 219–222.

MUNN, N. L., 1965. The Evolution and Growth of Human Behavior. Second edition. Houghton-Mifflin, Boston.

ØRVIG, T. (editor), 1968. Current Problems of Lower Vertebrate Phylogeny. Fourth Nobel Symposium. Almqvist and Wiksel, Stockholm.

PARSONS, T. S., and E. E. WILLIAMS, 1963. The relationship of the modern Amphibia: A re-examination. *Quart. Rev. Biol.* 38: 26–53.

RADINSKY, L. B., 1968. Evolution of somatic sensory specialization in otter brains. *J. Comp. Neurol.* 134: 495–505.

REED, C. A., 1960. Polyphyletic or monophyletic ancestry of mammals, or: What is a class? *Evolution* 14: 314–322.

RIOPELLE, A. J. (editor), 1967. Animal Problem Solving. Penguin. Baltimore.

ROMER, A. S., 1966. Vertebrate Paleontology. Third edition. The University of Chicago Press, Chicago.

ROMER, A. S., 1968a. The Procession of Life. World, Cleveland.

ROMER, A. S., 1968b. Notes and Comments on Vertebrate Paleontology. The University of Chicago Press, Chicago.

ROMER, A. S., and T. EDINGER, 1942. Endocranial casts and brains of living and fossil Amphibia. *J. Comp. Neurol.* 77: 355–389.

SHELL, W. F., and A. J. RIOPELLE, 1958. Progressive discrimination learning in platyrrhine monkeys. *J. Comp. Physiol. Psychol.* 51: 467–470.

SIMPSON, G. G., 1959. Mesozoic mammals and the polyphyletic origin of mammals. *Evolution* 13: 405–414.

SIMPSON, G. G., 1961. Principles of Animal Taxonomy. Columbia University Press, New York.

SIMPSON, G. G., 1967. The Meaning of Evolution. Yale University Press, New Haven, Connecticut.

SZARSKI, H., 1962. The origin of the Amphibia. *Quart. Rev. Biol.* 37: 189–241.

THOMPSON, K. S., 1968. A critical review of the diphyletic theory of rhipidistian-amphibian relationships. *In* Current Problems of Lower Vertebrate Phylogeny. Fourth Nobel Symposium (T. Ørvig, editor). Almqvist and Wiksel, Stockholm, pp. 285–305.

TRUEX, R. C., and M. B. CARPENTER, 1964. Strong and Elwyn's Human Neuroanatomy. Fifth edition. Williams and Wilkins, Baltimore.

WARREN, J. M., 1965. Primate learning in comparative perspective. *In* Behavior of Nonhuman Primates (A. M. Schrier, H. F. Harlow, and F. Stollnitz, editors). Academic Press, New York, vol. 1, pp. 249–281.

WIGHTMAN, W. P. D., 1950. The Growth of Scientific Ideas. Oliver and Boyd, Edinburgh.

YOUNG, J. Z., 1962. The Life of Vertebrates. Second edition. Clarendon Press, Oxford.

ZEIGLER, H. P., 1961. Learning-set formation in pigeons. *J. Comp. Physiol. Psychol.* 54: 252–254.

ZIMMERMANN, R. R., and C. C. TORREY, 1965. The ontogeny of learning. *In* Behavior of Nonhuman Primates (A. M. Schrier, H. F. Harlow, and F. Stollnitz, editors). Academic Press, New York, vol. 2, pp. 405–447.

# 4 Evolution of Primate Behavior

S. L. WASHBURN and R. S. HARDING

The study of evolution provides an intellectual background, a setting, for the understanding of the contemporary forms of life. Knowledge of evolution should help in the selection of animals for experimental purposes, in the formulation of comparative problems, and in interpretation. No one thinks that evolutionary investigations should dominate biology as they did in the nineteenth century, but we believe that they should be a useful part of biological science.

## Theory

The present theory of evolution has been clearly stated by numerous authorities, especially Huxley (1964), Simpson (1949), Roe and Simpson (1958), Dobzhansky (1962), and Mayr (1963). The synthetic theory, which was formulated in the 1930s and accepted in the 1940s, has been greatly strengthened by the discovery of the nature of the genetic code and the new techniques which the discovery of DNA has made possible. The synthetic theory may be briefly stated as follows: Evolution is the result of changes in the gene frequencies in populations. Behaviors leading to reproductive success are favored by natural selection, and the genetic bases for these successful behaviors are incorporated into the gene pool of the population. There is a feedback relation between behavior and its biological base, so that behavior is both a cause of changing gene frequencies

and a consequence of changing biology. Evolution is adaptation over time, and there are no trends except those that result from continuing selection.

The most important conclusions that can be drawn from evolutionary studies stem from this basic theory, rather than from descriptive studies. For example, many persons have believed that large parts of the human brain (such as the frontal lobes or corpus callosum) had little functional importance, and surgical destruction of these and other parts of the brain has been defended on this basis. But, whatever particular theory of human origin one may believe, the brain has become larger over time, and, if selection has been for more and more frontal lobes, these parts must have important adaptive functions. Evolutionary gain or loss clearly indicates functional importance, even if the precise nature of the function is still debated by scientists. For example, the threefold to fourfold increase in the size of the human cerebellum in the last million years requires explanation in terms of new human adaptations.

Evolutionary theory demands a radical departure from the assumptions and methods of traditional comparative studies. Not only do a frog, a rat, and a man not constitute an evolutionary sequence, but the division of creatures into systems makes it almost impossible to understand adaptation. Obviously, as a temporary research strategy, it is useful to look at the skeleton, for example, but interpreting evolutionary changes in the skeleton necessarily involves other systems. Such an involvement is an inevitable consequence of the fact that evolution is the result of selection for the biological base of successful behaviors of popula-

S. L. WASHBURN and R. S. HARDING  Department of Anthropology, University of California, Berkeley, California

tions. Because the nervous system is a coordinating system, to view its evolution in isolation from the adapting animal of which it is a part is particularly misleading.

The theory of evolution requires that we consider both the fossils and the living forms, and that this information be considered in the light of genetics and experimental biology. Each kind of information is essential and supplements the other, and both are briefly considered before the problems of the primates and human evolution are discussed in more detail.

## The fossils

Science has revealed an incredible variety of life and an evolutionary process that has gone on for some three billion years. From the rocks comes evidence of multitudes of creatures which no longer exist and which have left no close living relatives. The variety of forms that are alive today are only a few end products of a few of the ancient lineages. For example, the 150 million years of dinosaur evolution could not be deduced on the basis of the study of contemporary reptiles, nor could the diversity and complexity of their adaptive radiations be appreciated. Comparative anatomy of the surviving few is a completely inadequate approach to understanding the diversity of life, and it is no accident that many of the scientists who have contributed most to the understanding of evolution have considered the fossil record, anatomy, and behavior (Romer, 1966; Simpson, 1953; Young, 1950; Roe and Simpson, 1958).

It is not only the faunas that are remote in time that are no longer represented, but without fossils there would be no evidence of the glyptodonts, giant sloths, camels, or saber-toothed cats of North America of only a few thousand years ago, and there would be no traces at all of the extinct faunas of South America (Simpson, 1965). It is clear that, even if a few forms survive with little change (such as the opossum), the succession of faunas can be appreciated only through a study of the fossils. There is no way to reconstruct the past without using evidence from paleontology.

## Living forms

The contemporary forms of life are the representatives of those few lineages that were successful, and all have been evolving. The point is simply that the study of both contemporary forms and the fossils gives a better understanding than studying either one alone (Jay, 1968). For example, fossil skulls show that the earliest primates had long faces, wide interorbital regions, and small brains (Clark, 1962). In the contemporary prosimians these same features can be seen, and part of the primitive structure is directly related to the sense of smell (e.g., turbinate bones, cribriform plate of ethmoid, large olfactory bulb). Study of the living forms

provides detailed information and permits reconstruction of the adaptive functions of the parts seen in the fossils. But, in addition, the living forms have a rhinarium, tactile hairs on the nose and elsewhere, scent glands, and a sense of smell that is far more important than it is in monkeys. These features were characteristic of primitive mammals in general, so it is reasonable to reconstruct far more anatomy and behavior than could possibly be determined from the fossil bone alone.

But the contemporary prosimians are highly diversified and have evolved away from the form seen in the fossils. There is a great diversity of locomotor patterns, from the slow-moving lorises to the rapid-jumping galagos and tarsiers. There is no suggestion of anything like an aye-aye or an indri in the ancestry of monkeys. Some have evolved far from proportions of the Eocene forms, but some have changed remarkably little. Examination of both the living animals and the fossils shows which have changed least, the nature of some of the changes, and gives a far fuller understanding of both.

No living form is *assumed* to be primitive or to represent an unmodified ancestral stage. No feature of a contemporary creature is *assumed* to be the same as that in ancestors of many millions of years ago. Studying both the living creatures and the fossils of their ancestors shows that the basic adaptation of the primates is in the hands and feet, that the sensory adaptation was that of primitive mammals, and that the adaptation of the special senses and brain to arboreal, diurnal life came millions of years after the appearance of other primate features. These general conclusions provide a setting for the understanding of the primate special senses, brain, and biology that is compatible with all that is known of evolution and in no way depends on the great chain of being or an assumption that any particular primate is an unmodified ancestor.

## Trends

When the adaptive radiations of any group of animals are investigated, trends can usually be discerned. As time passed, animals became larger, brains became larger, and teeth evolved. Such sequences were once attributed to orthogenesis, an inbuilt evolutionary momentum. Now it is believed that they are the result of selection, which, in the main, has been in one direction. Where data are abundant, the trends seem much less regular and much more complicated than was once thought, and, because the trend is the result of selection, there may be reversals. The existence of long-term adaptive trends brings some order into evolutionary comparisons and makes it easier to compare contemporary forms.

For example, the brains of the contemporary prosimians are much larger than those of their early Eocene ancestors.

Comparisons with insectivores and rodents show that there has been substantial adaptation to arboreal life, and much of this may have taken place after the separation of the lines leading to contemporary prosimians and monkeys. For this reason, differences in the intelligence of *Lemur catta* and a macaque, for example, may be less than between *Notharctus* and a macaque. Knowledge of the trends makes it possible to compensate for the biases which are introduced when the living forms are compared without reference to ancestral ones.

New World and Old World monkeys are probably descended independently from some group of prosimians (Omomyidae?). The apparent similarities are the result of parallel evolution, which should help in the interpretation of adaptive trends. Whether in the New World or the Old World, brains associated with stereoscopic color vision and reduction in the sense of smell (and related systems) have evolved to a greater degree than have prosimian brains. The New World monkey is important in research on the central nervous system mainly because its nervous system affords us the opportunity to discover the selection pressures that resulted in this parallel evolution. No Old World monkeys have prehensile tails, but in New World monkeys there is every variation from long, primitive, balancing tails to prehensile grasping tails, complete with hairless skin that is comparable with that of a hand and is associated with a large area of cortex.

This series offers a model of the evolution of the interrelations of ability and brain. The issue is not ancestry, but the understanding that comes from the differential adaptation in a series of related forms.

In summary, the existence of evolutionary trends, adaptive radiations, and diversified behaviors offers the opportunity to interpret the process of evolution, to plan experiments, and to understand biological differences or similarities.

## Time

Most of the research in comparative anatomy was done prior to the development of time scales based on the use of radioactive elements. Viewing contemporary forms as relatively unchanged is much more credible if the times involved are short, as they were thought to be when the methods of comparative anatomy were taking shape. For example, the age of the world was supposed to be something on the order of 30 to 40 million years in 1900 (Sollas, 1905). This is approximately half of the time in which the mammals have been dominant, a quarter of the age of the reptiles, and less than one per cent of present estimates of the age of the earth. When Lord Kelvin concluded that organic evolution was impossible because there was not enough time, he was entirely correct—it was only the estimate of time which he used that was wrong.

When Keith wrote that the ancestors of man and apes must have separated in the Oligocene, he thought that the time of the separation was approximately a million years ago. Middle Oligocene is now regarded as some 30 million years ago. Consider the difference of a separation between man and monkey a million years ago with subsequent evolution dominated by orthogenesis, and a separation 30 million years ago with evolution the result of selection. Old World monkeys have been evolving, adapting, and radiating for at least 30 million years after their lineage separated from that leading to man. And on the human side our ancestors were evolving, adapting, radiating for 30 million years; therefore when men and macaques are compared, it must be remembered that there are at least 60 million years of evolutionary events that separate the two.

It is useful, therefore, to look at the adaptations that have evolved in each group in light of this long biological separation. On the monkey side, locomotor patterns have remained much the same. On the ape-human side, quadrupeds evolved into brachiators, brachiators into knuckle-walkers, and knuckle-walkers into bipeds. On the monkey side, animals remained primarily vegetarian, mostly arboreal, and did not use objects. On the ape-human side, animals began to hunt, lived on the ground, and used objects. There is every indication that the contemporary apes have evolved farther from early Miocene ancestors than the monkeys have from theirs, and man has advanced structurally and behaviorally beyond the apes.

## Adaptation and comparative behavior

The problems which arise in comparing behaviors that result from adaptation over long periods of time may be illustrated by considering three of the most commonly compared animals: pigeons, rats, and men. Birds have had a distinct evolutionary history for at least 150 million years, which means that at least 300 million years of evolutionary events separate pigeon and rat. There are some parallel adaptations, such as a high metabolic rate, but, for the most part, adaptations to flight have been quite different from mammalian adaptations. The resulting behaviors, on the other hand, may seem similar at first glance. Two experiments typify the kind of behavioral adaptation in birds that appears understandable in terms of mammalian behavior, but is not. First, the courtship display of the male turkey before the female turkey seems quite comprehensible to the human observer. Yet Schein and Hale (1965) have shown that the male turkey will go through a complete display when confronted with nothing more than the stuffed head of a female; in other words, most of what the male turkey sees does not influence his behavior at all. Or, to take another example, a hen turkey shows what appears to be normal maternal behavior toward her chicks, and protects them vigorously. Yet a deafened turkey will kill her own chicks (Schleidt, 1960). Conversely, a turkey hen with normal hearing will attack a polecat or any other small predator, but will not attack a stuffed polecat that has been made to chirp like a turkey chick. In other words, the chirp of the turkey chick functions to keep the hen from attacking, and she will attack any small, moving object that cannot be heard to chirp. Clearly, then, avian behavior achieves the same ends as does comparable mammalian behavior, but it rests on a fundamentally different biological basis. The adaptation to flight has dominated the evolution of the brain of the bird, just as it has that of the skeleton and many other parts of the body. Complex behavior is highly adaptive, but flight puts very severe restrictions on the size of the brain. The avian solution is complex stereotyped behavior (dependence on vision and hearing) which is controlled by a highly simplified relation to the environment and a minimum of learning.

Now consider the social rat. When a strange rat approaches the nest, it is attacked (Barnett, 1963). The rats appear to recognize the members of their own group, but, if the stranger is rubbed with litter from the floor of the nest box, it is not attacked. And if a member of the group is removed and cleaned, it is then attacked. Recognition is by the sense of smell, and the rat makes a distinction that neither the pigeon nor man can make.

Obviously, pigeons and rats have different abilities which enable them to adapt to different ways of life. From the point of view of evolution, to ask which is more intelligent is meaningless. Pigeons cannot live like rats, nor can rats live like pigeons. Both represent successful ways of life, and the differences cannot be measured by any simple psychological test. Because man is visual and equates intelligence and learning, man presents pigeons and rats with learning sets based on visual clues, but rat intelligence is based on the sense of smell, touch, and proximity, rather than on vision and distance. It would be easy to devise tests in which the senses of rats were favored; the intelligence of pigeons and men would then be rated as low.

In summary, tests may be given to pigeons, rats, and men, and their performance on these tests may be compared. Performance will depend at least as much on the tests as on "intelligence." One may also be interested in the evolutionary process that has produced the very different adaptations of these highly diverse forms.

The rate of maturation in pigeons and man gives a useful measure of the fundamentally different adaptive nature of their nervous systems. Birds mature in less than one per cent of their lifetime span. Imprinting may take place in a few hours. Only a little practice is necessary for the maturation of skills. In man, on the other hand, critical periods (as in language acquisition) have ranges of many months or years. Adult skills (such as chipping a stone tool, throwing a spear) can take years to acquire. It has been adaptive for man to store a very large amount of information and to be able to learn a wide variety of behaviors. For a human to mature normally, there must be years of social interaction and learning.

Obviously, scientists may be interested in what a pigeon and a man have in common, and some generalizations about learning may apply to the most diverse vertebrates; however, the study of evolution attempts to give some historical understanding of how birds and man have adapted differently.

From an evolutionary point of view, the study of rats is very different from that of birds. The sense of smell and touch, and nocturnal activity, are as important to rats as they are to prosimians. Many of these features are basically mammalian, and many of the early primates had enlarged incisor teeth similar to those of the contemporary rodents. Study of the fossils, contemporary prosimians, and rodents helps in the understanding of an actual stage in human evolution.

## Human evolution

Human evolution is the result of the success of a series of ways of life. Each of these was based on abilities which have changed over time.

The general stages may be discerned in both the fossils and the most closely related living forms. The primates became separated from the insectivores as a result of their adaptation to climbing with grasping hands and feet. The replacement of claws by nails is anatomically exceedingly complicated, and this major locomotor adaptation appears to have taken place only in one group of mammals. The unity of the early primates is shown in features of the skull (Van Valen, 1965) and teeth (Szalay, 1968). Increase in the size of the brain followed (Hofer, 1962); the olfactory bulb remained large and the cerebrum and cerebellum small. Fossil forms were numerous and highly diversified (Russell et al., 1967), and the same is true of the surviving prosimians. The monkeys of both the New World and the Old World are probably descended from one family of prosimians, and the surviving prosimians have been evolving in separate lineages for some 50 million years. The existence of a wide variety of living prosimians and of fossils from many families allows considerable reconstruction of the way of life of our ancestors for some 30 million years, after the primates had separated from other mammals and before the evolution of the monkeys.

During the Oligocene period, some 40 to 30 million years ago, monkeys evolved. Again, with our understanding based on the living forms and the exceedingly fragmentary fossils, the principal changes were the evolution of stereoscopic color vision, together with a great increase in the size of the brain and a loss of the structures associated with the sense of smell. The major changes are clearly reflected in the anatomy of the skull, but the fossil record is so incomplete that there is no way of deciding how far the evolution may have progressed in the latter part of the Eocene. There are only fragmentary skulls from the critical period (Simons, 1959, 1967).

As shown by the limb bones, our ancestors were quadrupedal, arboreal forms for some 20 million years before the ape pattern of locomotion evolved. During this period, the general pattern of adaptation appears to have been very similar to that of the contemporary Old World monkeys. In spite of the variety of kinds of monkeys, many structures and behaviors are shared. All are quadrupedal, with many common features of hands, feet, limbs, and trunk. All sleep in a sitting position, which is reflected in the callosities and in the ischia. All are highly social, mature slowly, and bear single offspring. These monkeys have been highly successful, with the density of a single species rising to more than 200 per square mile in favorable localities.

Interest in human evolution leads to an emphasis on the behavior and structure of apes, but the fossil record shows that the monkeys have displaced all the small African apes. Small apes were numerous in the Miocene, but now only the large ones can compete in the African forests, and they are rare and most unsuccessful compared with the monkeys of the *Cercopithecus, Cercocebus,* and *Papio* groups. In Asia, where competition was with monkeys of the Colobinae only, small apes (gibbons) survive.

In short, study of the contemporary Old World monkeys does give many clues to the way our ancestors lived for millions of years—arboreal, visual, social, with learning important. Comparisons with parallel adaptations in New World forms allows controls on many of the interpretations.

## Apes (Pongidae)

Long after Old World monkeys (Cercopithecidae) and apes (Pongidae) could be distinguished on the basis of the dentition, a new way of climbing-feeding evolved in the apes. As shown in the contemporary apes, this pattern involves the independent use of individual arms and legs in climbing, hanging, and feeding. This new arboreal pattern is correlated with the structure of the trunk, the upper limbs, and the arrangement of the thoracic and abdominal viscera. It is clearly reflected in the bones forming the fingers, wrist (Lewis, 1969), elbow joint, and shoulder (Washburn, 1968a). Man shares this structural-functional

pattern with the contemporary apes, but in the fossils of some 20 million years ago only the shoulder shows the beginning of this locomotor specialization. Men and apes must have shared a common ancestry long after the kind of adaptation seen in the limb bones of *Proconsul* (Napier and Davis, 1959) or *Pliopithecus* (Zapfe, 1958). Unfortunately, there are no ape skeletons of Pliocene age.

The gibbons are the most extreme in their adaptations to hanging in the trees, and many lines of evidence suggest that their lineage diverged before that of the great apes and man, but after the locomotor adaptation described above. For example, gibbons sleep sitting and have the monkey kind of ischial callosities, whereas the great apes build nests, sleep in positions very similar to those of man, and have, at most, traces of callosities.

The great apes manipulate objects and use them in aggressive displays. Van Lawick-Goodall (1967, 1968) described chimpanzees throwing branches, grass, and stones in displays, and playing extensively with objects. Leaves are used for cleaning the body, and grass or sticks for obtaining termites. Lancaster (1968a) has put these activities into perspective by pointing out that, although they do not seem to be much from a human point of view, chimpanzees put objects to more different kinds of uses than all other mammals combined. Termiting involves a long period of learning for the young chimpanzee, and this activity, which requires much skill and concentration, may last for hours.

Chimpanzees hunt to a limited degree, and as many as four males may cooperate in extended hunting. Reynolds (1966) pointed to numerous features in which chimpanzee social behavior may afford the most useful model for the possible behavior patterns of our ancestors. But it must be stressed that chimpanzees have also evolved since their separation from the human lineage. No claim is made that chimpanzees are unmodified ancestors, but it is asserted that the rich data on chimpanzee behavior give a very different picture from what would be the case if the behavior of contemporary monkeys were used as the only standard. Further, any speculations may be checked against gorilla behavior, as described by Schaller (1963), and the whole understanding of ape behavior will soon be enriched by studies now in progress (Horr on orang; Fossey on gorilla). It should be emphasized that man is most similar to African ground-living, knuckle-walking apes, not to the forms that remained arboreal (Washburn, 1968b).

The brain of the apes has been traditionally described as advanced in a human direction over that of monkeys, and the behavioral data certainly support this interpretation. Certainly, slower maturation, object play, object use, cooperative hunting, some food sharing—all suggest a brain-behavior complex that is much more human than that of the monkeys.

## Men (Hominidae)

Isolated teeth show that the human lineage had separated from that leading to the apes by 3.7 million years ago, and the actual separation must have been more than four million years ago (F. C. Howell, personal communication). Many fossils dated on the order of two million years ago show that, by that time, our ancestors (*Australopithecus*) were bipedal. Apparently these creatures used stones and shaped some by knocking off a few chips. They lived in the savanna and hunted at least small game, and also possibly large animals. The teeth of these fossils were remarkably human. The foot was fully human, and the ilium was similar to that of man in most respects and very different from the comparable bone of any other primate.

The brains of these creatures were no larger than those of the contemporary great apes, representative capacities being from 400 to 550 cc. One hand has been well preserved, and the phalanges are halfway between that of a knuckle-walker and that of man. Judged from a terminal phalanx, the thumb was small compared with that of modern man, but large compared with an ape thumb (Napier, 1962). The molars erupted on the same delayed pattern as in man and very much more slowly than in the apes (Mann, 1968).

It is probable that this small-brained, bipedal stage in human evolution lasted for at least two million years and, possibly, twice as long. By some 600,000 years ago *Homo erectus* had replaced *Australopithecus,* and the brain had doubled in size. The large-brained men are associated with stone tools, which take great manual skill to make. Even when suitable material is supplied and careful instructions and demonstrations are given, it takes months to learn to copy some of the better tools of half a million years ago. These people used fire, killed large animals, and were human in the usual sense of this term. But, even so, there was very little change in the form of the stone tools. The same assemblages have been found over large geographic areas and over many thousands of years. It was not until 40,000 to 30,000 years ago, when men who were, in skeletal characters, virtually indistinguishable from ourselves appeared, that history starts to proceed at a rapid rate.

In the last 30,000 years before agriculture, the kinds of stone tools changed quickly, were locally highly diversified, and many technical advances were made, such as bows, boats, harpoons, ground stone, and so forth (Laughlin, 1968). The rate of change and degree of local cultural adaptation stand in marked contrast to earlier conditions in which the same kind of stone tool was made for hundreds of thousands of years.

If this view of human evolution is at all correct, then most of the distinctive evolution of the human brain was *long after* the separation of man and ape and was in response to uniquely human selection pressures. If the brain evolved in a feedback relation with hand skills, social skills, and, finally, language, the explanation of what is unique about the human brain lies in the evolutionary events of the last million years. In this sense, most of the human brain is a product of the success of the human way of life, and it cannot be understood by the study of brains that have not undergone that particular evolutionary history.

The point may be illustrated by a consideration of the hand. Among the contemporary primates, the human thumb is unique in its size, strength, and degree of opposition of the fingers. The hand of *Australopithecus* shows that its human features evolved long after the use of stone tools began. The human hand is an ape hand rebuilt by selection pressures for efficient tool use. Likewise, the large area of human brain related to hand skills is the result of the success of skillful object using. The brain is important, not only in determining the possibility of manual skills, but in making learning possible. Object use is fun for men, and they easily learn a wide variety of manual skills (Hamburg, 1963). Chimpanzees throw stones and may hit the intended target, but they do not make piles of stones and then practice throwing. Other chimpanzees do not applaud a hit, nor does an organized society help the individual to see the possibilities and develop his skills to the utmost. (A human brain may easily guide a chimpanzee to a level of performance that lies well beyond the normal behavior of the species.)

What has evolved in man is an ability, and one can see how remarkable this human ability is by an appreciation of the fact that, with all the families of primates equipped with grasping hands and feet, with dozens of genera and hundreds of species, there is only one instance of substantial object use. The study of evolution may be used either to see the common elements in hands and their actions or to see the unique elements and their special histories.

Numerous social skills are also unique to man. There is virtually no planned cooperation among the nonhuman primates. Estrous behavior is universal. Rage may be only slightly controlled. Dominance behavior is common. Sharing of food is almost nonexistent, except to a slight degree among chimpanzees. Put in evolutionary terms, the success of the human kind of social system has been so great that it has exerted a major influence on the evolution of the central nervous system. Man has evolved a brain that can mature normally only in a social setting; the advantage of slow maturation may be primarily that it allows the learning of social systems, including the knowledge and skills that make them possible. It must be remembered that, until very recently, there was little specialization in society, aside from the primary sexual division of labor. Each individual in a social system had to know all the skills, magic, religion, and so on, of the society. Only after the advent of agriculture, and especially after the scientific revolution, were there many individuals who had not mastered most of the skills of society. Division of labor changes the whole relation of the way of life to the evolution of the brain.

At the present time the greatest functional difference between man and the nonhuman primates is language. In the communication systems of the nonhuman primates, the information conveyed is concerned, for the most part, with the emotional states of the actors. Such systems are multimodal (Lancaster, 1968b), using gesture, postures, and sound to convey general meanings, but there is no way to express a name. In social situations, animals may convey their intentions (to be aggressive, to submit, to groom, to flee, and so forth), but no specific information is conveyed about the environment. For example, sound may mean "danger" (more likely fear on the part of the sender of the message); in some monkeys different sounds may signal ground predator or avian predator (Struhsaker, 1967), but there is no way to signal leopard or hyena, although this would seem to be of the greatest adaptive value. Language has been so important in human adaptation, and sounds are so common among nonhuman primates and mammals in general, that it is very difficult to understand why such a system of naming evolved only once. It is tempting to see this as originating in consequence of early tool using, which may have forced concentration on objects in a new manner. However the naming adaptation may have started, Lenneberg (1967) has suggested that all modern languages may have had a common origin on the order of 50,000 to 30,000 years ago, or about the time anatomically modern man appeared and the historical record rapidly became complex. Obviously, even if this speculation is correct, simpler antecedent forms of language may have existed. In any case, man is now so constituted that he can learn language, any language, and may learn several. This ability depends directly on an extensive biological base (Masland, 1968; Sperry and Gazzaniga, 1967) and indirectly creates a pressure for vastly more memory and the possibility of complex planning. Nowhere is the difference between man and ape more apparent. In spite of great efforts, chimpanzees cannot be taught to speak because they lack the biological base for this learning (Kellogg, 1968).

The brain has evolved in a feedback relation with the way of life, so the present structure of the brain reflects its evolutionary past. For example, the large area of brain associated with hand skills is an indication of the importance of this behavior in human evolution. Because the brain increased in size by approximately three times in less than a million years, the more that can be known about the new parts of the brain, the more we can understand the events of the last million years. Uncertainties of the fossil record and the gaps in our knowledge of the functions of the brain, however, make this correlation difficult. Penfield and Rasmussen's (1950) description of the function of the cortex corresponds with our view of this period in human evolution. What evolved were the hand skills and the technology they make possible, social skills with increased memory and planning and control, and linguistic skills, which are the basis of human communication. If Cobb (1958) was correct in his view that the newest part of the cerebellum is primarily concerned with learned hand skills, this would support the inferences we have drawn from Penfield and Rasmussen's interpretation of the function of the cortex. The threefold to fourfold increase in the size of the cerebellum in the last million years must be explained by the evolutionary importance of substantial new functions. Eccles et al. (1967) wrote that the cerebellum is important in learned skills and stated that "We have to envisage that the cerebellum plays a major role in the performance of all skilled actions. . . ." (p. 314).

As pointed out at the beginning of this chapter, those structures that have increased over the millennia as a result of selection must have important functions. The more fully the functional differences between man and ape can be understood, the more we can analyze the events that historically separate man and ape. The more complete the fossil record becomes, the more accurately the origin of the more recent behaviors can be dated, but it may well be that in the next decade the increase in understanding of the human brain will do more to illuminate the fossil record than the other way around.

In summary, human evolution must be seen as a succession of ways of life. The nature of each way of life may best be understood by a study of both the fossils and the living primates. Fortunately the wide variety of living primates provides a rich source of information. Unfortunately, the fossils, although numerous, are fragmentary, leaving wide areas for reasonable disagreement.

The ways of life are made possible by abilities, and each ability is made possible by inputs, internal coordination, and skilled performance. Over time, an ability may decrease (as has the sense of smell in the evolution of man or that of birds), or it may increase because of continued selection (as have hand skills in human evolution). Abilities may be determined almost entirely by heredity or be greatly influenced by learning. In the latter case, the biology determines the ease with which the behaviors can be initially learned or subsequently modified.

Evolution (adaptation over time) is a complex process in which behavior is in a feedback relation with the biology that makes it possible. This process has resulted in an almost infinite variety of forms of life, most of which have been long extinct. There is no way that the few extant forms can be arranged in any simple order of complexity or intelligence. Study of the fossils and of the living forms may give us deeper understanding of adaptation, of evolution, and of the processes and settings that have produced the existing varieties of life, including man.

NOTE: There have been many different theories of human origin and primate phylogeny. Recently, numerous studies of albumins, transferrins, hemoglobin, DNA, and chromosomes prove that men (Hominidae) are particularly close to the apes (Pongidae) and much less closely related to Old World monkeys (Cercopithecidae). The recent evidence supports the traditional conclusion that, among the apes, the order of relationship is: African apes, orangutan, gibbons. The times of separation of the various lineages is still under debate, but a relatively recent separation between man and ape seems the most probable. The evidence has been reviewed by Wilson and Sarich (1969) and by Goodman (1968).

## Acknowledgments

The research on which this paper is based was supported by U.S. Public Health Service Grant 08623. We wish to thank Dr. Phyllis C. Dolhinow for reading the manuscript and Mrs. Alice Davis for editorial assistance.

### REFERENCES

BARNETT, S. A., 1963. The Rat. A Study in Behaviour. Aldine, Chicago, p. 91.

CLARK, W. E. LE GROS, 1962. The Antecedents of Man. Second revised edition. Edinburgh University Press, Edinburgh.

COBB, S., 1958. Foundations of Neuropsychiatry. Sixth edition, revised and enlarged. Williams and Wilkins, Baltimore.

DOBZHANSKY, T. G., 1962. Mankind Evolving. Yale University Press, New Haven.

ECCLES, J. C., M. ITO, and J. SZENTÁGOTHAI, 1967. The Cerebellum as a Neuronal Machine. Springer-Verlag, New York.

GOODMAN, M., 1968. Phylogeny and taxonomy of the catarrhine primates from immunodiffusion data. I. A review of the major findings. In Taxonomy and Phylogeny of Old World Primates with References to the Origin of Man (B. Chiarelli, editor). Rosemberg and Sellier, Turin, Italy, pp. 95–107.

HAMBURG, D. A., 1963. Emotions in the perspective of human evolution. In Expression of the Emotions in Man (P. H. Knapp, editor). International Universities Press, New York, pp. 300–317.

HOFER, H., 1962. Über die Interpretation der ältesten fossilen Primatengehirne. Bibl. Primatol. 1: 1–31.

HUXLEY, J., 1964. Evolution: The Modern Synthesis. Reprint. John Wiley and Sons, New York.

JAY, P. C., 1968. Primate field studies and human evolution. In Primates: Studies in Adaptation and Variability (P. C. Jay, editor). Holt, Rinehart and Winston, New York, pp. 487–503.

KELLOGG, W. N., 1968. Communication and language in the home-raised chimpanzee. Science (Washington) 162: 423–427.

LANCASTER, J. B., 1968a. On the evolution of tool-using behavior. Amer. Anthropol. 70: 56–66.

LANCASTER, J. B., 1968b. Primate communication systems and the emergence of human language. In Primates: Studies in Adaptation and Variability (P. C. Jay, editor). Holt, Rinehart and Winston, New York, pp. 439–457.

LAUGHLIN, W. S., 1968. Hunting: An integrating biobehavior system and its evolutionary importance. In Man the Hunter (R. B. Lee and I. DeVore, editors). Aldine, Chicago, pp. 304–320.

LAWICK-GOODALL, J. VAN, 1967. My Friends the Wild Chimpanzees. National Geographic Society, Washington, D. C.

LAWICK-GOODALL, J. VAN, 1968. The behaviour of free-living chimpanzees in the Gombe Stream Reserve. Anim. Behav. Monogr. 1: 161–311.

LE GROS CLARK, W. E. (See CLARK, W. E. LE GROS).

LENNEBERG, E. H., 1967. Biological Foundations of Language. John Wiley and Sons, New York, p. 261.

LEWIS, O. J., 1969. The hominoid wrist joint. Amer. J. Phys. Anthropol. 30: 251–267.

MANN, A. E., 1968. The Paleodemography of Australopithecus. University of California, Berkeley, doctoral thesis.

MASLAND, R. L., 1968. Some neurological processes underlying language. Ann. Otol. Rhinol. Laryngol. 77: 787–804.

MAYR, E., 1963. Animal Species and Evolution. Belknap Press, Harvard University Press, Cambridge, Massachusetts.

NAPIER, J., 1962. The evolution of the hand. Sci. Amer. 207 (no. 6): 56–62.

NAPIER, J. R., and P. R. DAVIS, 1959. The fore-limb skeleton and associated remains of Proconsul africanus. In Fossil Mammals of Africa, No. 16. British Museum (Natural History), London.

PENFIELD, W., and T. RASMUSSEN, 1950. The Cerebral Cortex of Man. The Macmillan Co., New York.

REYNOLDS, V., 1966. Open groups in hominid evolution. Man 1: 441–452.

ROE, A., and G. G. SIMPSON (editors), 1958. Behavior and Evolution. Yale University Press, New Haven.

ROMER, A. S., 1966. Vertebrate Paleontology. Third edition. The University of Chicago Press, Chicago.

RUSSELL, D. E., P. LOUIS, and D. E. SAVAGE, 1967. Primates of the French early Eocene. Univ. Calif. Publ. Geol. Sci., Vol. 73.

SCHALLER, G. B., 1963. The Mountain Gorilla. The University of Chicago Press, Chicago.

SCHEIN, M. W., and E. B. HALE, 1965. The effect of early social experience on male sexual behavior of androgen injected turkeys. In Readings in Animal Behavior (T. E. McGill, editor). Holt, Rinehart and Winston, New York, pp. 314–329.

SCHLEIDT, W. M., M. SCHLEIDT, and M. MAGG, 1960. Störung der Mutter-Kind-Beziehung bei Truthühnern durch Gehörverlust. Behaviour 16: 254–260.

SIMONS, E. L., 1959. An anthropoid frontal bone from the Fayum Oligocene of Egypt: the oldest skull fragment of a higher primate. Amer. Mus. Novitates 1976.

SIMONS, E. L., 1967. The earliest apes. Sci. Amer. 217 (no. 6): 28–35.

SIMPSON, G. G., 1949. The Meaning of Evolution. Yale University Press, New Haven.

SIMPSON, G. G., 1953. The Major Features of Evolution. Columbia University Press, New York.

SIMPSON, G. G., 1965. The Geography of Evolution. Chilton, Philadelphia and New York.

SOLLAS, W. J., 1905. The Age of the Earth. T. F. Unwin, London.

SPERRY, R. W., and M. S. GAZZANIGA, 1967. Language following surgical disconnection of the hemispheres. *In* Brain Mechanisms Underlying Speech and Language (F. L. Darley and others, editors). Grune and Stratton, New York and London, pp. 108–121.

STRUHSAKER, T. T., 1967. Auditory communication among vervet monkeys. *In* Social Communication among Primates (S. A. Altmann, editor). The University of Chicago Press, Chicago and London, pp. 281–324.

SZALAY, F. S., 1968. The beginnings of primates. *Evolution* 22: 19–36.

VAN VALEN, L., 1965. Treeshrews, primates and fossils. *Evolution* 19: 137–151.

WASHBURN, S. L., 1968a. The Study of Human Evolution. Oregon State System of Higher Education, Eugene.

WASHBURN, S. L., 1968b. Speculations on the problem of man's coming to the ground. *In* Changing Perspectives on Man (B. Rothblatt, editor). The University of Chicago Press, Chicago, pp. 191–206.

WILSON, A. C., and V. M. SARICH, 1969. A molecular time scale for human evolution. *Proc. Nat. Acad. Sci. U. S. A.* 63: 1088–1093.

YOUNG, J. Z., 1950. The Life of Vertebrates. Clarendon Press, Oxford.

ZAPFE, H., 1958. The skeleton of *Pliopithecus* (*Epipliopithecus*) *vindobonensis* Zapfe and Hürzeler. *Amer. J. Phys. Anthropol.* 16: 441–457.

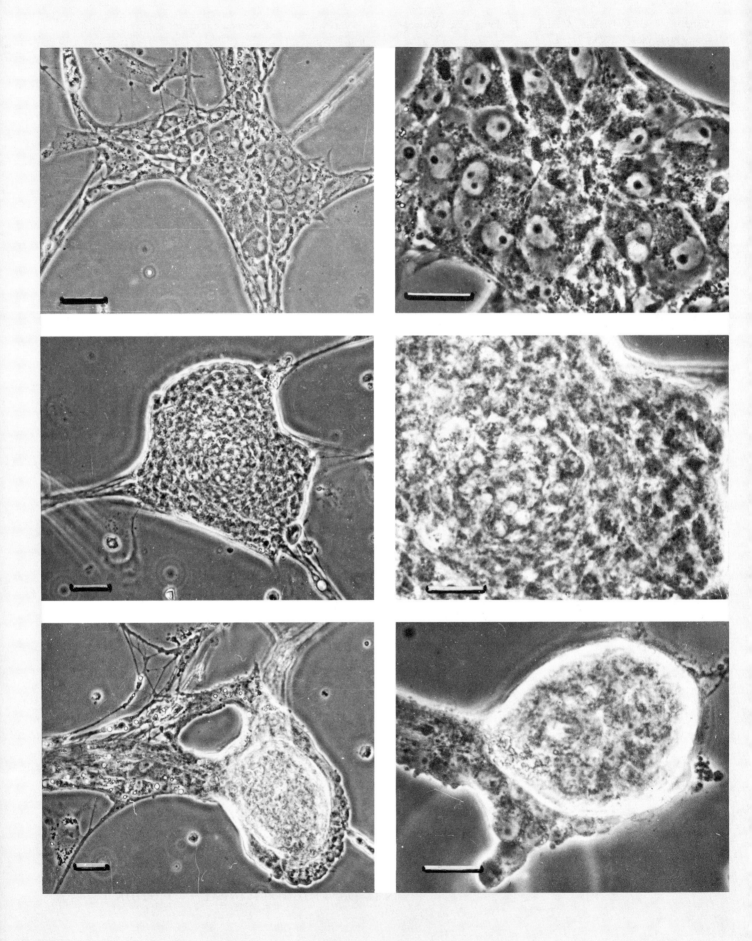

# DEVELOPMENT OF THE NERVOUS SYSTEM

*This is a photograph of dissociated sensory ganglia, in culture, of an eight-day-old chick embryo (*VARON, PAGE 90*), which serves to introduce the discussions of the "ontogeny of the nervous system after the appearance of its earliest rudiments at the beginning of embryonic life"* (EDDS, PAGE 51).

# 5 Prefatory Comments on Development of the Nervous System

M. V. EDDS, JR.

THE ONTOGENY OF the nervous system after the appearance of its earliest rudiments at the beginning of embryonic life has been of interest to relatively few investigators. To be sure, the roster of experimental neuroembryologists active since Harrison opened the field early in the century includes some of the most able developmental biologists. But one easily counts off their names without using up his fingers and, in retrospect, the field has had a disproportionately small influence on the neurosciences generally.

The reasons for this state of affairs are easily identified. Embryologists during the first half of the century were preoccupied with problems of early embryogenesis. Concern with the later phases of organogenesis and histogenesis was primarily descriptive and, until recently, not in the main stream of developmental studies. The emphasis on the mechanisms of initial cellular differentiation has provided some critical insight into early developmental processes at several levels from tissue rudiment to molecule. But even these insights, by and large, have not been made explicit for the nervous system, of which our understanding has remained frustratingly incomplete. Accordingly, the contributions of analytical neuroembryologists such as Harrison, Detwiler, Weiss, Hamburger, and Sperry were not just pioneering; they still stand out as lonely beacons in a research landscape given over to other matters.

Furthermore, during the years when neuroembryology

M. V. EDDS, JR. Division of Biological and Medical Sciences, Brown University, Providence, Rhode Island was emerging, both as a descriptive and as an experimental subject, most biological disciplines were being pursued in relative isolation from one another. Although embryological studies finally came to share in the postwar growth of interdisciplinary hybridization, they were more tardy than most. Even now, a deplorable tendency to accept a level of chemistry or pharmacology or neurophysiology from developmental biologists that is less rigorous than that practiced by colleagues working on similar problems in adult organisms is too slowly declining.

These facts, plus the admitted difficulties of analyzing neural development, make it understandable that most neuroscientists have accepted the nervous system as a finished product and have restricted themselves to such questions as: "What is it?" or "How does it work?"

But now we are witnessing a rapid increase of interest in various aspects of neuroembryology. The upsurge of the neurosciences in the past decade has made the relevance of developmental studies much more obvious. Even those neuroscientists with limited interest in embryological problems per se concede, or are at least willing to consider, that a developmental approach may shed important and otherwise unobtainable light on the logic of form and function of the nervous system. And, in their turn, developmental biologists are returning to the problems of neurogenesis in the recognition that a long period of neglect, coupled with the availability of new concepts and new tools of analysis, will now inevitably lead to exciting advances. Neuroscientists who now inquire, "How did it get that way?" are much more likely to get penetrating answers.

These considerations comprise the background for the section on the "Development of the Nervous System" that is reported in the following pages. The principal aim in organizing the symposium that produced these chapters was twofold: 1) to express in a selective but representative manner the current status of the field, and thereby to delineate the predominant themes of genetic neurology; and 2) to influence future strategies by exposing neglected areas and by calling attention to critical points from which optimal research leverage could be obtained.*

No preface could convey the diversity and the richness of the information set forth by the participants in this group; each author must speak for himself. But it may be helpful to start with a brief overview of the key topics and their interrelations.

The symposium opens with a paper by Paul Weiss, who takes a broad view of a field he did much to create; he thereby provides a base from which the rest of the symposium flows. Students of neurogenesis have long recognized that events at the cellular level are critical in the shaping of neural centers; hence, the first topic to be developed in depth is an analysis by Jay B. Angevine, Jr. of the role and the temporal gradients of patterned cell proliferations and migrations. His account is extended by Martin C. Prestige's treatment of cell differentiation and cell death, especially as these events depend quantitatively on interactions between developing nerve centers and the periphery. Studies of peripheral influences on developing neurons have been enriched by in vitro culture of neural tissues; the prospect of further exploiting this method, especially through cultures of nerve cells that have first been dissociated from one another, is outlined by Silvio Varon. Then additional data bearing particularly on the programed cellular activities of the primitive neural epithelium are presented in a summarizing commentary by Richard L. Sidman.

Once the requisite populations of young neurons have accumulated and positioned themselves appropriately, the next key event in neurogenesis is the establishment of connections, both among neurons and between the latter and peripheral tissues. The paper by Alfred J. Coulombre introduces the concept of neuronal specificity in the formation of these interconnections, particularly in emerging motor systems. Marcus Jacobson then elaborates the evidence for neuronal specificity, stressing examples from the cutaneous and visual systems.

All studies of the developing nervous system should, in the end, contribute to an understanding of the origins of behavior. But critical information about this aspect of the subject, despite repeated efforts, is still meager. David Bodian deals with fine-structural events during the development of synaptic connections as these relate to basic patterns of circuitry development and to the emergence of progressive motor capability. Finally, Viktor Hamburger describes the appearance in the embryo of patterns of motility. He assesses their relation both to morphological and physiological parameters and to the coordinated, purposeful movements that emerge later at hatching or birth.

The report of this section finishes with a commentary, both retrospective and prospective, on some of the issues related to neuronal specificity.

---

*Quite unplanned, but emerging as an important byproduct of informal discussion during the symposium, was a revision of the basic terminology of cells in the developing brain and spinal cord. Four new names were proposed for the fundamental zones of the differentiating neural tube, as defined by the form, behavior, and fate of their constituent cells: the *ventricular, subventricular, intermediate,* and *marginal* zones. Realizing that the proposal would receive greater attention if it came from the entire group, the participants in the neuroembryology symposium constituted themselves as a "Boulder Committee," and prepared a formal account for publication (Boulder Committee, 1970. Embryonic vertebrate central nervous system: Revised terminology. *Anat. Rec.* 166: 257-262). The new terms are used in the present symposium articles.

# 6 Neural Development in Biological Perspective

## PAUL A. WEISS

THIS THEMATIC INTRODUCTION to the problems of neurogenesis intends to indicate the place and role of neural development in the broad continuum of an integrated neuroscience. Addressing itself, therefore, not just to the neuroembryologist, but to students of neurobiology and behavioral science in general, its focus is on conceptual synthesis rather than on factual synopsis.

To understand the life of a city, one must know more about the city than its name. To understand the intricate structure and integrated operation of the mature nervous system, one must know the countless steps by which it has come up to its maturity from a plain plate of cells earmarked for that destination early in the developing germ. Descriptively, those steps have been roughly outlined in "stages" not unlike the series of anecdotal accounts and illustrations of the historical course of the development of a community; *static* morphology—macroscopic, microscopic, and, lately, ultramicroscopic—have predominated. Urban historians had to rely largely on overt criteria of political events, because a deeper penetration into the underlying dynamics of human habitations was handicapped by the rudimentary states of scientific psychology, ecology, sociology, economics, and demography. In the same way, neurogenesis—the development of the nervous system—has had to go for a long time without a penetrating study of the *dynamics,* i.e., the operation in space and time, of the forces and interactions of which the transitory morphological stages are but the manifest signs that happen to have surfaced into the range of visibility.

Toward the turn of this century, the needed shift of focus from a sheer *record* of serial changes of morphological features to the exploration of the *mechanisms* underlying those transformations became noticeable. Yet the resulting trend of analytical research on neurogenetic dynamics has been slow; it certainly has failed to gain momentum commensurate with its significance for an understanding of the developed nervous system, even though technical opportunities have been ample.

Not all the needed advances pertain to missing factual information. Some lie in the direction of better exploitation of valid data already available, the relevance of which is not properly taken into account for either the validation or the invalidation of existing hypotheses and theories. Moreover, research should be aimed at answering well-posed questions; the somewhat fuzzy and loose framework of concepts and terms in which we phrase our questions is in need of being sharpened and tightened, lest the expected answers suffer from equal indefiniteness.

As an example of one such conceptual bottleneck, it is still rare, except for writings in psychology, to find the nervous system treated as a true "system" in the full epistemological meaning of the term. The present occasion not being appropriate for an elaborate discussion of the point, let me just refer cursorily to two recent accounts (Weiss, 1969, 1970), in which I have set forth more fully the properties that distinguish a "system," especially an organic one, from a sheer assembly or complex of matter. The distinction lies 1) in the fact that the autonomy of the constituent parts of a system is rigorously restrained by the interdependency and interaction among them; and 2) in the peculiar *dynamics* of that interdependency, which is of such a nature as to preserve the cohesiveness, integrity, and entity of pattern of the whole dynamic complex, not by minutely stereotyped, microprecisely executed, and, hence, predictable courses of the individual components, but, on the contrary, in spite of the considerable range of unpredictable variability of the behavior of the parts in their individual courses. Applied to the nervous system, this means that two human brains are infinitely more similar in their standard, over-all configuration and operation as wholes than one could ever anticipate from the details of constellation and microcircuitry of the billions of constituent neurons, the micropatterns of which are unique in each individual case.

It is self-evident that one would be at a loss for how to explain that "systemic" integrative principle of the functioning mature nervous system unless its whole developmental course had already been subject to the same principle of systemic order. That is to say that, however much we might be able to learn about the ontogeny of a given, artificially isolated feature of the nervous system, that knowledge must remain fragmentary until we can supplement it by an understanding of how its own development is so coordinately linked with all the other emergent features that, despite the vagaries of their separate courses, the whole system ends up with the inner unity, consistency, and integration that is displayed by the finished product. In short,

PAUL A. WEISS  The Rockefeller University, New York, New York

the mere summation of even the most comprehensive accounts of isolated studies of singled-out components of the nervous system can yield no comprehension of the system as a whole unless we can establish how the component processes mesh.

That much of this program has remained a task for the future may be blamed in part on the widespread illusion that the course of development in general is lockstepped, microprogramed in precisely branching chains of reactions, precisely arranged in space, and precisely timed as to rates and sequential order. The faith of those who hold such a micromechanical concept is in direct proportion to their degree of remoteness from practical experience with developmental phenomena. The shorthand terms, to which development must of necessity be reduced for the second-hand information of persons who lack firsthand familiarity with developmental reality, cannot possibly reflect the immense variety of epigenetic detail by which the "genetic program" of a fertilized egg is executed by incessant interactions with the cytoplasmic, later cellular, and always extraorganismic, environments of the developing unit. These are environments, the fortuitous erratic fluctuations of which leave their marks on the detailed component steps of the whole course, without, within limits, throwing the total design of the program off course. Given this basic fact, the acceptance of an empirical principle of "*macrodeterminacy, not reducible to detailed microdeterminacy*" becomes logically cogent (see Weiss, 1968, Chapter 2; Weiss, 1969). Evidently, this principle rules out any interpretation of the function of the mature nervous system in terms of a micro-precisely predesigned machine; to explain its precision would be the charge of the embryologist as the student of the process through which it is supposed to have come about. His demonstration that neuronal development definitely does not proceed cliché-fashion thus renders untenable any theory of central nervous functions that is predicated on completely stereotyped connectivity.

Now, there are two possible objections to the validity of this argument. One might submit that evolutionary experience has endowed the germ with tolerances, that is, allowances for the variety of unforseeable, hence unpredictable, contingencies each specimen is bound to meet in its real and unique course of development. This contention is only partially correct. Evolution can operate only within and through the basic laws of dynamic nature; it cannot suspend, infract, or alter them and must submit to their rigorously set limitations. Evolution can fashion from the infinitely rich repertory of styles, in which the basic and unalterable dynamics of nature can find expression, mechanisms that have proved effective in meeting circumstances of *constant* repetitiveness. It can, for instance, establish predesigned control circuitry, comparable to a thermostatic device, such as the carotid-sinus control arc for maintaining normal blood pH or the myotatic reflex arc for modulating limb movement and posture. Yet evolution can hardly have anticipated and provided for the infinite number and variety of specific detailed *ad hoc* adjustments that would be necessary to insure relative invariance of the whole developmental pattern in the face of the capricious *inconstancy* in detail of the conditions to which the component developmental steps are exposed.

In short, structured neuronal feedback loops, such as those of the carotid and proprioceptive reflexes, molded by *past* evolutionary experience with major contingencies of rather consistent recurrence, cannot be put forward as rationale for claiming universal evolutionary prescience, as it were, of the myriad of minor and fortuitous *future* fluctuations of conditions during each individual and unique ontogeny. System dynamics provides for this primary general adaptiveness.

A second possible defense of microdeterminacy of neural detail stems from the remarkably stereotyped patterns of shape, size, number, localization, connectivity, and functional specialization that have been found to characterize the mosaic of neurons in the central ganglia of mollusks and crustaceans. Although little is known about the developmental history of those patterns, it obviously is tempting to put forth the constancy of their cellular mosaic as a fair model of the nervous system of the higher forms, with its incomparably larger and more intricate populations of units. Yet one cannot avoid the positive evidence that in the higher forms such a presumed invariance of microdetail just does not exist. Might, conceivably, the systemic functions of coordination, regulation, and integration, which in the higher forms are properties of large pools of neurons, be carried out in lower ones by *intracellular* dynamics within single units, comparable to the coordinated performances within the single cell of a ciliate protozoan? As we shall deal here exclusively with higher forms, we need not dwell on this unresolved problem.

At any rate, in the vertebrates there is definitely no precise microanatomical congruity between different individuals, even genetically identical ones ("identical twins" from single eggs), in either their developmental stages or their mature state. Despite that range of variance in detail, the developed organs turn out to be rather standard in organization and performance, raising to a prime problem of neurogenesis the *coordinated* and unified interaction among the component developmental steps. Without further resorting in this summary to the theory of system dynamics, I shall treat coordinated developmental team work analytically in terms of interactions among identifiable components.

Before entering on this task, let me dispel the antiquated notion that basic patterns of behavior are molded empirically within an undifferentiated matrix by associational trial-

and-error methods. Coordinated function develops in the embryo *autonomously,* without the benefit of tests of functional adequacy and, indeed, in experimental situations, in direct contravention of functional utility (Weiss, 1968, Chapter 22). In order to appreciate this fact, one need only remember that motor patterns develop quite normally in embryos deprived from the beginning of their sensory apparatus. For further reinforcement of this point, see Viktor Hamburger's presentation in this volume.

Our understanding of neurogenesis, therefore, hinges on the exploration of the *intrinsic* processes going on in the developing germ in an incessant series of interactions among the ordered units of genes, chromosomes, cell organelles and matrix, cells, ground substances, formed cell products, tissues, organs, the internal milieu, and, in a supporting measure, the external environment. Now, to be fit for analytical study, this general listing must be spelled out far more concretely and specifically.

Let me give an example. Figure 1 is a schematic diagram of the terminal organization of the neuronal fabric in the cerebellum. It represents levels or layers (symbolized by the lettered diamonds) populated by distinctly different species

of neurons of specifically different configurations, interconnected by sheaves or tangles of nerve fibers that are, not individually but statistically, rather predictable as to their terminations (indicated by arrows); the whole fabric is relevant to the adequate coordinated performance of that particular portion of the brain. Considering that the various cell and fiber groups interpenetrate and intertwine as profusely and haphazardly as plant roots in soil without obliterating the over-all spatial and functional order of the system, one becomes aware of the staggering number of specific questions that such a diagram raises in our minds. I quote from an earlier review (Weiss, 1955, p. 348):

How does the neural plate transform into primordia of brain, spinal cord, and ganglia? How does it grow? How do its cell groups specialize for their respective formative tasks, how early, in what places and what sequence? What makes them divide or cease to divide? What causes them to migrate and in what directions, and what to assemble in defined locations? What sets the numbers and quotas of the different neuron types, and adjusts them to the functional needs of the individual? How do they achieve selective interconnections on which their later functioning will depend? And which ones of these are really relevant to the specific patterns,

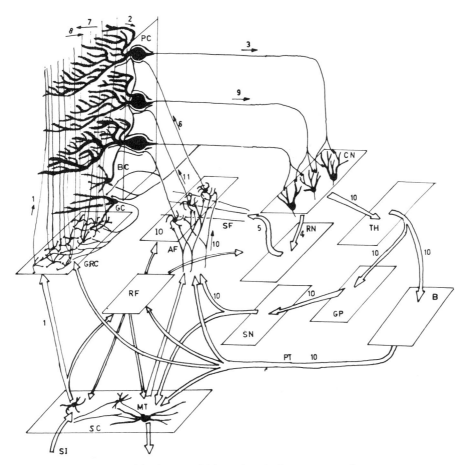

FIGURE 1   Model of neuronal fabric of cerebellum. (From Callatay, 1969.)

rather than just the general execution, of central functions? What provides the neuron population with the proper contingent of supportive, protective, and nutrient cells and structures of other origins, in varying combinations according to the local needs? And how much interdependence and interaction in growth and differentiation is there between different central regions before and after they have become segregated? Where does the axon arise from the neuroblast? What causes its elongation? What gives it its course? What determines the volume and pattern of the dendritic tree? Do the trunks and branches of the mature nerves reflect the orientation of early outgrowth? Is axonal outgrowth strictly oriented or is it haphazard, followed by selective abolition of unsuccessful connections? What determines deflections or other changes of course? What causes branches to form, and where? Are tissues flooded with nerve fibers, or is admission selective? If the latter, how is invasion held in check? And is penetration tantamount to functionally effective innervation? What causes the association of sheath cells and nerve fibers, and what is the mechanism of myelin formation? How do fibers group into bundles—by active aggregation or by the enveloping action of connective tissue? And what determines the places and proportions in which the various tissue elements combine to form nerves? How does it happen that fibers of similar function are often grouped together, and how do they each reach their appropriate destinations? Or do they? And if not, how could central functions fail to be confused? How does a nerve fiber gain in width, and what decides its final caliber? And how does it change with body growth? What controls the number of fibers available for a given area—size of the source, frequency of branching, overproduction followed by terminal screening, or all of these? Does exercise and practice have a constructive, or at least modifying, effect on central pathways and central size? Are fluctuating peripheral demands taken into account in the development of centers, and, if so, by what means? Can growing centers adjust to lesions or deformation, and how—by regeneration, compensatory growth, or substitutive functional corrections? And can the development of overt behavior be correlated with, or even explained by, the stepwise emergence of neural apparatuses?

Fundamental as these questions are to an understanding of the nervous system and of behavior, hardly any of them has thus far been answered adequately, partly for lack of conclusive observations and incisive experiments, but mostly because dynamic networks are, in general, refractory to analysis in isolated bits, pieces, and steps. By confining analytical studies to sufficiently small sectors of a network, one can identify reasonably self-contained "causal chains," but as soon as one abandons that confinement of vision for a more comprehensive view, one recognizes the fragmentary nature of those conclusions; for then one notes that those chains are intimately interlinked into a cohesive system of interactions by branches and anastomoses. Nonetheless, in tactical regards, the analytical procedure of singling out for study the simplest fragments of such relational networks first, and only afterward paying attention to their crosslinkages, remains unexcelled. One merely must bear in mind that such a procedure requires a deliberate act of abstraction

from "context," neglect of which bars true understanding of any system, and of the nervous system in particular.

Such, then, is the methodology that I have pursued in my own experimentation and tried to highlight in a review of neurogenesis in "Analysis of Development" (Willier et al., 1955), to which the reader must be referred here for further detail. Figure 2, reproduced from that book, gives a graphic representation of some of the identified routes and way stations of "neurogenesis," and of the net of their major interdependencies. The boxes denote sources and targets of relevant interactions (e.g., limb bud, neural crest, hormones), or phenomena and processes (e.g., elongation, recruitment, degeneration). The arrows represent sets of demonstrable pathways or operations by which the boxes are connected with one another. One notes that arrows bifurcate and that almost every box receives, as well as issues, more than one arrow. These features of the diagram express better than words the true systems character of the whole dynamic complex called "neurogenesis."

The function of this diagram is twofold. In the first place, in juxtaposition with Figure 1, it typifies the true relationship between the process and the product of development: every single diamond and single arrow of Figure 1, the finished brain, has passed through the whole coherent dynamic history of the network of Figure 2, and, conversely, every single box and arrow of Figure 2 is bound to have been involved in all the features of the total differentiated pattern of Figure 1. This simple mental exercise should uproot, once and for all, the naively preformistic notion of neurogenesis as simply the blind, piece-by-piece conversion of a minutely mapped-out cell mosaic into the configuration of the final product.

Although, as some of the other chapters in this volume describe, the epithelium of the primitive neural plate and tube does contain parcels of cells already blocked out for specific developmental tasks, the masses of cells arriving from further proliferation of those blocks turn into neurons that individually are by no means single-tracked. Their fates and paths are still indefinite in detail and gain their unique definitions only progressively through continuous interactions with the unpredictably varying constellations of their local environments. The latter, in turn, are functions of an equally varied past epigenetic history of the whole system. In this sense, neurogenesis illustrates faithfully the relationship between genetic determination and epigenetic execution of developmental patterns in general; indeed, Figures 1 and 2, set side by side, could stand more broadly as symbols of the dual epigenetic characterization of all development, according to which every gene affects potentially all characters of the developed organism, and every single feature of the latter, conversely, is influenced in some degree by all the genes.

The second and major function of the diagram of Figure

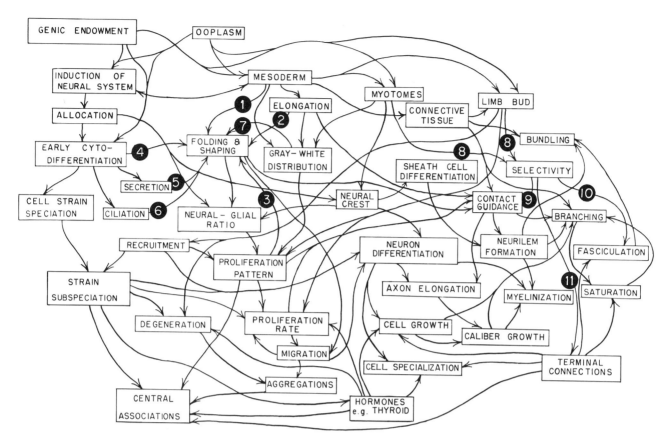

FIGURE 2  Diagram of some major "pathways" in the developmental dynamics of neurogenesis. (From Weiss, 1955.)

2, however, is: Experience has shown that the questions I spelled out above in the form in which a plain observer of the structure and function of the mature nervous system would logically ask them (in the legal terms of "Who is responsible" for a given act?) are scientifically unmanageable. They must be broken down into more specific and tangible issues to become answerable in analytical terms. The diagram offers some pointers on how to do so. Each arrow signifies a set of concrete developmental phenomena that have been recognized as causally (not just coincidentally) interrelated and have proved to be amenable to analytical exploration. In other words, the diagram is essentially a table of contents of a cohesive program for the disciplined study of a cohesive problem system.

Looking at it as one would look at the canvas for a huge painting in its early stages, and comparing it with the sparse crop of data on neurogenesis, one notes that the latter still correspond to widely separated daubs of paint, at times not even recognized as relevant to the whole design. Yet, notwithstanding their scarcity, the great majority can be fitted into the great over-all picture, mosaic-fashion, given a deliberate effort. In stressing this point, I wish to make clear that progress in developmental neurology hinges not only on promoting well-focused new research but, in equal measure, on exploiting more thoroughly the meager store of valid information that is already in our possession. This must be done by critical collation, correlation, and reconciliation of uncoordinated and unassimilated, but well-substantiated, data. The growing discrepancy, in our day, between the clamor for ever more research support, on the one hand, and the stringency of funds and wherewithal with which to meet the growing demands, on the other, might rekindle the search for treasures in the ashes of the past.

In order to lend some substance to this proposition, I use Figure 2 and select from it at random, for closer inspection, a few of the examples that for decades have failed to show a significant momentum of progress. As the first example, I choose the interactions marked in the diagram by numbers 1 to 7 (indicated in the text by brackets), which are part of the complex that is instrumental in the primitive *shaping of the central nervous system* (CNS). If one compares the significance attached in anatomy, physiology, and anthropology to the delineation of brain parts, folds, and fissures with our virtually complete lack of insight into the underlying formative dynamics, one might appreciate why I cite, in the following, an account I gave 15 years ago (Weiss, 1955, pp.

371–372) of what little we knew or conjectured then about brain morphodynamics. No substantial contribution to the reduction of the vacuum has been made since, except perhaps the recognition of the role of localized cell death (see the report by Prestige in this volume).

*Early Morphogenesis of Brain and Cord.* The gross shapes of the early brain and cord, respectively, are anticipated in the proportions of the neural plate, whose wide anterior part, upon folding upward, forms the large vault of a brain ventricle, while the narrower posterior part encloses the narrow lumen of the central canal of the spinal cord.

The shape of the canal varies with the details of the folding process. Uniform curling of the plate would leave a cylindrical lumen. This actually occurs in isolated pieces of plate in homogeneous surroundings (Holtfreter, 1934). The slit-shape of the normal tube has been shown to depend on the presence [1] of notochord (Lehmann, 1935). The effect may be credited to a vertical system of fibers, spanning the thickness of the plate along a median strip coextensive with the notochord and apparently attached to it, which holds the midline firmly anchored as a hinge about which the flanks of the plate fold up (Weiss, 1950). A similar fibrous plane seems to define the border between the alar and basal plate cell masses [4]; as the latter grow and bulge [3], it gives rise to the lateral sulcus. Because of their importance for the later regular distribution and grouping of cell columns [2, 7], such tangible traces of early subdivisions would merit more intensive study; at present, we have no more than vague hints as to their presumable role.

After the closure of the groove, the turgor of the fluid in the lumen assumes the morphogenetic role of firm support for the limp walls, which otherwise would collapse. The source of this turgor has been found in the secretion of fluid [5] from the cells of the inner lining of the early ventricles (Weiss, 1934; Holtzer, 1951). Furthermore, the ciliary beat [6] of the lining propels the fluid anteriorly, which in the normal embryo would help to maintain the distention of the brain cavity. . . . The shrinkage of the central canal by partial fusion of its walls (Hamburger, 1948), paralleled by the decline of mitotic activity [3], may reflect a reduction of turgor in the spinal portion.

With hydrostatic pressure on the inside and the confining skull capsule on the outside, continued enlargement of the brain wall by growth [3], cell migrations [4] and the deposition of white matter [7] must be expected to lead to deformations, which, depending on the local conditions, manifest themselves as cave-ins, outpocketings, fissures or folds. Practically nothing is known about the mechanics of these elementary shaping processes, although there are at least some indications that the fissures between major divisions of the cortex actually arise as cave-ins along lines of least resistance in the wall which tends to expand in confined space (Le Gros Clark, 1945; Källén, 1951). It must be emphasized, however, that the systematic pattern, according to which such mechanical events take place, is intrinsically prepared by the inequalities established previously by the locally differing processes of proliferation [3], migration, aggregation and differentiation [4] (see Bergquist and Källén, 1953); the gross mechanical factors do not create these differentials, but merely translate them into more conspicuous spatial configurations.

Accordingly, the attainment of normal brain configuration depends not only on the typical development of the brain wall, but also on the proper harmony between the latter and the growth of the skull capsule [1] (or in the case of the cord, the spine) on the outside, and the turgor of the cerebrospinal liquor [5] and its propulsion by forward-directed ciliary beat [6] on the inside. If this harmony is disturbed, either by a genetically determined imbalance between the component tissues or by later trauma or nutritional deficiencies, serious aberrations of the CNS will ensue. Genetically conditioned hypersecretion of central fluid, for instance, leads to hydrocephalus and brain herniation (Little and Bagg, 1924; Bonnevie, 1934); delayed closure of the folds past the onset of secretion, to various grades of spina bifida with draining fistulae (for an example of mechanical production of spina bifida, see Fowler, 1953); and retardation of skull growth in vitamin A deficiency, with unimpeded growth of the CNS, to brain compression and herniation (Wolbach and Bessey, 1942). The early cartilaginous capsule, at least in the spinal region, can accommodate its size to the actual dimensions of the enclosed CNS (Holtzer, 1952), but this adaptability is certainly greatly reduced in later stages.

Against the background of all the massive progress made in the recent past in molecular biology, enzymology, electron microscopy, membrane physiology, computer biology, immunology, and so forth, could one not expect at least a modicum of attention and work to be turned to the biomechanics of morphogenesis, exemplified by the shaping of the most precious human possession—the brain?

As a second example, I mention the problem of *selectivity.* Because a searching discussion of some instances of selectivity are presented in this volume by Coulombre and Jacobson, and because monologues, as well as reviews, dealing with neural specificity (see Weiss, 1966) are on record in abundance, I confine myself to a few supplementary comments on matters that seem to have escaped notice.

The term "selectivity" denotes the relational property that enables two bodies to communicate or to combine in pairs to the exclusion of other combinations. The dynamics and mechanisms involved are of so many kinds and forms, and the selective interactions vary so widely in discriminative sharpness, that it would be wholly unwarranted to postulate or search for any single master mechanism. The only common denominator to all of them seems to be a reciprocal matching of specific (nonrandom) spatial arrays or temporal sequences; matching steric configurations of molecules are as suitable for mutual recognition and conjugation as are complementary time patterns of impulses. One might question whether there is *any* relevant interaction in a living system that does not exhibit some trace of selectivity —from egg-sperm mating, enzyme-substrate interaction, hormone-target cell combination, to type-specific affinity among cells—but the development of the nervous system is certainly rich in examples; so much so that it would clutter up the diagram if every relation (arrow) that bears signs of specificity were marked. Accordingly, only those pertinent to the problem of selective structural coupling have been labeled; moreover, because problems of *molecular* specificity will be taken up in a later section of this book, I confine my comments to the specificity of *intercellular* relations, especially synaptic junctions.

Selectivity among tissue cells in general has been well documented by the active self-sorting according to types of indiscriminate cell mixtures. Pertinent relations to selectivity are numbered 8 to 11 in Figure 2, and again are bracketed in the text below. Significantly, cells treat as compatible not only members of their own tissue type, but also those of types with which they normally have symbiotic affiliations in the body (e.g., epithelia and corresponding stromas; lung epithelia and macrophages). Although direct observation, including cinemicrographic recording, of this process of association of like-to-like and segregation between like and unlike has been limited to non-neural cell types (summarized in Weiss, 1968, chapters 15, 17, and 19), the evidence of experiments in embryos extends the validity of the same principle to the neuronal cell population. Schwann cells apply themselves preferentially to nerve fibers (Abercrombie et al., 1949), but even more discriminative is the specific association between different classes of neurons and their appropriate non-neural destinations [10, 11]. Tips of pioneering nerve fibers, emigrating from motor and sensory centers, take distinctly different routes through the tissue matrix long before reaching muscles and skin, respectively (Hamburger, 1929; Taylor, 1944), which has led to the assumption that each is guided by corresponding chemical cues of the fibrous structures that serve as pathways ("selective contact guidance" [9]; "selective fasciculation" [10]; Chapter 21 in Weiss, 1968). Regardless of how they have arrived at the terminal tissues, however, motor fibers are accepted for intimate synaptic junction exclusively by muscle cells, and sensory fibers by sensory epithelia and end organs only [11]. Basically, this is just another instance of affinity between complementarily matching cell types, prematched for each other independently in the genetic program of the embryo, rather than by secondary epigenetic adaptation.

Primary *sub*specificities do not seem to exist among peripheral sensory neurons, although they are evidently instrumental in the establishment of intracentral connections between given sectors of the retina and corresponding sectors of the optic tectum (Sperry, 1951; Jacobson, this volume). The manner in which correct point-to-point projection of a given retinal spot upon the prematched tectal cells takes place despite arbitrary experimental dislocations of the eyeball relative to the midbrain prior to optic-fiber connections, and even later, after severance and regeneration of the optic nerve, is still unclear. It is even more obscure if one takes into consideration that the visual receptor neurons, oriented radially, are separated from the ganglionic cells of origin of the optic-nerve fibers by at least two intermediate

neurons, which are highly ramified, course horizontally, and connect with many of the radial processes. Such communication shunts would seem to rule out any possibility of isolated private structural channels, cell for cell, between given visual receptors and given optic fibers.

At any rate, in the *motor* field, the possibility of any primary subspecification of motor neurons for specific individual muscles can definitely be ruled out, because any motor fiber will accept transmissive junction with any uninnervated muscle at which it happens to arrive (for references, see Weiss, 1955). The further evidence, however, that eventually the various motor neurons do "know" the precise identity of the particular muscle with which each has become connected, and identifies it by an individual constitutional characteristic of that muscle, rather than by anatomical position or functional role (Weiss, 1968, Chapter 22), raises the crucial question of how that final discriminative subspecification has come about. My answer (summarized in the preceding quotation), based on the bare minimum of logical inferences that can be distilled from a comprehensive overview of the known data to be taken into account, is as follows: The primary subspecification must lie in the musculature, in the sense that each individual skeletal muscle would acquire, in the process of its ontogenetic differentiation, a discrete, individual, biochemical character that distinguishes it from other muscles, but is too subtle (as is "organ specificity" in general) to be demonstrated by present-day direct biochemical assays. These specific differentials would then *secondarily,* as if infectively, be passed on to the attached motor neurons [8], thus conferring upon them a mosaic pattern of subspecifications, which thereafter will represent faithfully the peripheral mosaic of muscles in the central system. I refrain from discussing what else goes on in the centers, as that would lead me into the broad problem field of the embryonic origin of the patterns of coordinated function, the phenomenology of which is treated in Hamburger's paper in this volume.

The postulated specific "modulation" (whatever this process may turn out to be) of motor neurons by their terminal effectors is no different in principle from the subspecification of sensory nerves by their peripheral receptor areas (proprioceptive, Verzar and Weiss, 1930; corneal, Weiss, 1942; cutaneous, Miner, 1956; Jacobson, this volume). Intracentral and peripheral neurons thus seem to differ in the mode of acquisition of their respective subspecificities: intracentral ones (e.g., retinal) becoming prematched by *primary* differentiation for later junction with correspondingly prespecified units; peripheral ones (e.g., somatic sensory and motor), which are destined for non-neural tissues, receiving their matching characteristics *secondarily* from those tissues after first establishing less specific junctions.

Alternative speculations about motor neurons being finely subspecified by primary differentiation so as to be enabled to find and mate with "their" appropriately pre-matched muscle partner are dealt with in Coulombre's review. Despite the force of factual arguments against them, I by-pass here the question of just how two elements become selectively related, rather than let its discussion distract attention from the fact that all the offered explanations have in common the explicit or implied acceptance of the *existence of discrete, qualitative differences among different (nonhomologous) muscles, among different sensory areas, and among the different neuron species connecting or connected with them, as well as among different neuron species engaged in intracentral communication.* The establishment of this qualitative subspecification as an incontrovertible fact, regardless of any predilections for its special explanation, has been the common tenor of virtually all the work that followed my first observations in 1922 on neuromuscular specification ("myotypic function"). Its crux lies in having disclosed a key element in the understanding of how orderly correspondences among the several billions of units in the nervous system can emerge, as they do in development, without the benefit of experience from empirical trial-and-error tests.

The modes by which cell-type subspecifications take place are still partly a matter of further research, partly of incisive logic. Our present ignorance about the processes underlying cell-strain diversification (subsumed under the conventional cover term "differentiation") is as profound in regard to the nervous system as it is in regard to the different muscles, skin territories, or any other tissue system or organ. The only certainty we have for all of them is that their inner diversity exceeds by far the range of distinctions fathomable by current optical or biochemical methods of detection; immunological approaches might come a little bit closer. The main conclusion, however, is that we must now definitely let the nervous system join the other tissues of the body in a highly diversified specificity of elements and rescue it from the former monotony of purely morphological criteria (shape, dimensions, structural connectivity) and electrical parameters, which have been the only recognizable variables to be admissible in the construction of models of neural integration and behavior. Evidently, the specific diversity of chemical constitution—the principle of "*specificity*"—has earned a new and major place in our constructs. Although we have focused here solely on its relevance to the establishment and maintenance of *developmental* order, it would be plausible to presume that it would continue to operate in the developed nervous system as an instrument of maintaining *functional* order, as well as of restoring order after structural or functional disturbances. Synaptic junctions, based on matching specificity between joined partners, could be assumed to disjoin if the members of a couple were to become chemically "alienated" by exposure to sufficiently disparate influences. Conversely, new synaptic relations could be entered into as a result of chemical convergence.

These prospects being largely for the future, I do not dwell on them. My purpose in bringing them up at all has been to indicate that, taking cognizance of the uninterrupted continuity of the life line from conception to death, the notion of a sharp discontinuity between a developmental and a postdevelopmental phase of an organism would be absurd, although certain aspects (e.g., native coordination) decline, while others (e.g., acquired skills, memory) rise in prominence. In other words, the diagram of Figure 2 is truncated, artificially lopped off at the bottom, and in order to be representative of the living nervous system, should be extended downward without break.

The lesson should be clear: whatever can and does happen in postembryonic behavior is rooted in the network of dynamic interactions, symbolized by the arrows of our diagram, recognized now as but an arbitrarily isolated portion of the neural time continuum. Postembryonic neural and behavioral history, as it continues from the bottom of the diagram throughout life, can only modify, but never escape, the frame of conditions which the epigenetic realities, the play of countless interactions beset by unpredictable variances, stabilized by an invariant genome at the core, have elaborated. This progressively evolving pattern, and not some fancied preordained and minutely predetailed genetic cast, forms the "subjective" background and matrix that personal experience can enlarge, mold, and adapt. It can be done within the limitations set by the interactions of genetic endowment and epigenetic dynamics, but also with the immense opportunities for further development they have left open.

The nervous system emerges from its embryonic phase well patterned and nothing could be more misleading than the impression that embryonic neurogenesis merely fabricates blank sheets on which experiential input from the outer world is then to inscribe operative patterns. To convey this message is perhaps the most important service that insight into neurogenesis can render in the search for a unified theory of the nervous system and of behavior.

## REFERENCES

ABERCROMBIE, M., M. L. JOHNSON, and G. A. THOMAS, 1949. The influence of nerve fibres on Schwann cell migration investigated in tissue culture. *Proc. Roy. Soc., ser. B, biol. sci.* 136: 448–460.

BONNEVIE, K., 1934. Embryological analysis of gene manifestation in Little and Bagg's abnormal mouse tribe. *J. Exp. Zool.* 67: 443–520.

BERGQUIST, H., and B. KÄLLÉN, 1953. On the development of neuromeres to migration areas in the vertebrate cerebral tube. *Acta Anat.* 18: 65–73.

CALLATAY, A. DE, 1969. Cerebellum and cerebrum model with periodic processing, neurotubules conduction hypothesis. *Curr. Mod. Biol.* 3: 45–61.

CLARK, W. E. LE GROS, 1945. Deformation patterns in the cerebral cortex. *In* Essays on Growth and Form (W. E. Le Gros Clark and P. B. Medawar, editors). Clarendon Press, Oxford, pp. 1–22.

FOWLER, I., 1953. Responses of the chick neural tube in mechanically produced spina bifida. *J. Exp. Zool.* 123: 115–151.

HAMBURGER, V., 1929. Experimentelle Beiträge zur Entwicklungsphysiologie der Nervenbahnen in der Froschextremität. *Wilhelm Roux' Arch. Entwicklungsmech. Organismen* 119: 47–99.

HAMBURGER, V., 1948. The mitotic patterns in the spinal cord of the chick embryo and their relation to histogenetic processes. *J. Comp. Neurol.* 88: 221–284.

HOLTFRETER, J., 1934. Formative Reize in der Embryonalentwicklung der Amphibien, dargestellt an Explantationsversuchen. *Arch. Exp. Zellforsch.* 15: 281–301.

HOLTZER, H., 1951. Reconstitution of the urodele spinal cord following unilateral ablation. Part I. Chronology of neuron regulation. *J. Exp. Zool.* 117: 523–557.

HOLTZER, H., 1952. An experimental analysis of the development of the spinal column. I. Response of pre-cartilage cells to size variation of the spinal cord. *J. Exp. Zool.* 121: 121–147.

KÄLLÉN, B., 1951. On the ontogeny of the reptilian forebrain. Nuclear structures and ventricular sulci. *J. Comp. Neurol.* 95: 307–347.

LE GROS CLARK, W. E. (See Clark, W. E. Le Gros.)

LEHMANN, F. E., 1935. Die Entwicklung von Rückenmark, Spinalganglien und Wirbelanlagen in chordalosen Körperregionen von Tritonlarven. *Rev. Suisse Zool.* 42: 405–415.

LITTLE, C. C., and H. J. BAGG, 1924. The occurrence of four inheritable morphological variations in mice and their possible relation to treatment with X-rays. *J. Exp. Zool.* 41: 45–91.

MINER, N., 1956. Integumental specification of sensory fibers in the development of cutaneous local sign. *J. Comp. Neurol.* 105: 161–170.

SPERRY, R. W., 1951. Mechanisms of neural maturation. *In* Handbook of Experimental Psychology (S. S. Stevens, editor). Wiley, New York, pp. 236–280.

TAYLOR, A. C., 1944. Selectivity of nerve fibers from the dorsal and ventral roots in the development of the frog limb. *J. Exp. Zool.* 96: 159–185.

VERZÁR, E., and P. WEISS, 1930. Untersuchungen über das Phänomen der identischen Bewegungsfunktion mehrfacher benachbarter Extremitäten. Zugleich: Direkte Vorführung von Eigenreflexen. *Pflügers Arch. Ges. Physiol.* 223: 671–684.

WEISS, P., 1934. Secretory activity of the inner layer of the embryonic mid-brain of the chick, as revealed by tissue culture. *Anat. Rec.* 58: 299–302.

WEISS, P., 1942. Lid-closure reflex from eyes transplanted to atypical locations in *Triturus torosus:* Evidence of a peripheral origin of sensory specificity. *J. Comp. Neurol.* 77: 131–169.

WEISS, P., 1950. Introduction to genetic neurology. *In* Genetic Neurology (P. Weiss, editor). The University of Chicago Press, Chicago, pp. 1–39.

WEISS, P., 1955. Nervous system (neurogenesis). *In* Analysis of Development (B. H. Willier, P. Weiss, and V. Hamburger, editors). W. B. Saunders Co., Philadelphia, pp. 346–401.

WEISS, P., 1966. Specificity in the neurosciences. Neurosciences Research Symposium Summaries (F. O. Schmitt and T. Melnechuk, editors). M. I. T. Press, Cambridge, Massachusetts, vol. 1, pp. 179–212.

WEISS, P., 1968. Dynamics of Development: Experiments and Inferences. Academic Press, New York.

WEISS, P., 1969. The living system: Determinism stratified. *Studium Gen.* 22: 361–400.

WEISS, P., 1970. Hierarchical Systems. Hafner Publishing Co., New York (in press).

WILLIER, B. H., P. A. WEISS, and V. HAMBURGER, 1955. Analysis of Development. W. B. Saunders Co., Philadelphia.

WOLBACH, S. B., and O. A. BESSEY, 1942. Tissue changes in vitamin deficiencies. *Physiol. Rev.* 22: 233–289.

# 7 Critical Cellular Events in the Shaping of Neural Centers

## JAY B. ANGEVINE, JR.

A MAJOR PROBLEM confronting the neural sciences is not to find *an* explanation for neuronal specificity but to find the right one. Edds (1967) stated it well. Most of us share his conviction that neurogenesis depends heavily on those individualized properties that enable neurons to enter selectively into functionally meaningful ensembles and interconnections. This specificity varies in degree and increases in steps (Jacobson, this volume). Its acceptance, nevertheless, as a mechanism for the attainment of neuronal connectivity provides a major but largely neglected key (Weiss, 1965) to greater understanding of the otherwise almost unbelievable story of brain development. In this regard, Ebert (1967) reminded us that the nervous system, its functional unity and cellular diversity notwithstanding, ontogenetically is not a privileged exception. Cells of all tissues acquire such properties during development. Moreover, elaboration of these properties occurs within a framework of four distinct but inseparable events (Hamburger and Levi-Montalcini, 1949) common to the genesis of all organ systems: cell proliferation, cell migration, cell differentiation, and cell death. Emphasis is placed here on the first, while the others are discussed in detail by Sidman and others (this volume).

In general, countless studies, utilizing classical and recent techniques of neuroembryology, affirm the orderliness and precise temporal sequence of these events. Indeed, the finished product certifies it in its blend of immense populations of neurons (more than $10^{12}$ in man) and astonishing individualities of constituent cells. Furthermore, ingenious experiments, selectively and vividly reviewed by Edds (1967), demonstrate that these events are partly determined by built-in factors, which permit some developmental autonomy, and partly influenced by immediate or remote surroundings, especially the periphery, which can affect dramatically the quality and quantity of neurons in certain centers (Prestige, this volume). In addition, hormone-like effects of diffusible protein agents such as the nerve growth factor on these events are well known (Varon, this volume).

JAY B. ANGEVINE, JR. Department of Anatomy, College of Medicine, The University of Arizona, Tucson, Arizona

## Cell proliferation

ORIGIN OF NEURAL PLATE AND FORMATION OF NEURAL TUBE The future nervous system first declares itself within the ectoderm of early vertebrate embryos as an elongated, oval, and thickened neural plate. Surrounding its margin is the presumptive epidermis. Beneath the plate, in the roof of the archenteron (primitive gut), lies the chordamesoderm which, preceded by the diffuse mesenchyme of the prechordal plate rostrally, becomes the discrete notochord and somites in caudal and middle regions of the embryo. These axial materials induce formation of the neural plate and confer regional specificity upon it (Saxén and Toivonen, 1962). The plan of the forebrain, induced by prechordal mesoderm, will differ strikingly from that of the midbrain and hindbrain, induced by the chordamesoderm. Morphogenetic effects of pure mesenchyme or of notochord and somites on the shape of the future neural tube (see Weiss, 1955) foretell these regional differences or similarities in neural organization.

In earliest stages, the prospective neural plate and epidermis are alike, but soon the plate thickens and its cells elongate. Then the edges of the plate, outlined by pigmented cells, elevate. By continued folding, the future neuraxis separates from the epidermis to form the neural tube. This process, neurulation, sweeps rostrally and caudally from the incipient cervical region. As neural folds unite, cells withdraw abruptly from the closure into the crevice between the tube and overlying epidermis, the margins of which also have united. This neural crest, originally a median flattened band, later becomes paired and segmented; it forms the craniospinal and sympathetic ganglia, adrenal medulla, and melanocytes. The Schwann cells, which form the neurilemmal and myelin sheaths of peripheral nerve fibers, are also derived from the neural crest, as shown by their absence following early removal of the crests and by autoradiography (Weston, 1963). Autoradiography also indicates, however, that some Schwann cells arise in the neural tube.

HISTORICAL PERSPECTIVES The study of histogenesis in the neural tube has been long and difficult. As one would imagine, many of the great figures in neurohistology participate; for every insight, a disagreement or uncertainty appears in an immense literature (see Sidman et al., 1959).

A key observation by Altmann in 1881 was that mitoses in all developing epithelia always occur at the surface farthest from the mesoderm. In 1889, Wilhelm His identified in the neural tube the large, round germinal cells, generally in mitosis, at the edge of the neurocoel (primitive ventricle). His concentric early neuroepithelial zones foreshadow the ependymal, gray and white layers of the mature CNS. The names persist, although their original significance has changed (Boulder Committee, 1970, see below), as do his Basle Nomina Anatomica and his designations for the brain subdivisions. The innermost columnar zone contained the slender inner processes of elongated, transformed epithelial cells, the spongioblasts. The processes intruded between the pale germinal cells, but the nuclei lay deeper in a nuclear zone. Columnar and nuclear zones together formed the *ependymal* zone of later stages and descriptions. The outer processes of the spongioblasts united syncytially in a surrounding *marginal* zone. Rapid proliferation of germinal cells produced neuroblasts but never spongioblasts. Neuroblasts migrated peripherally to accumulate between nuclear and marginal zones as a *mantle* zone.

In contrast, Schaper in 1897 believed that the primitive neuroepithelium should not be divided into zones. He thought that germinal cells could produce spongioblasts, thus accounting for the large number of the latter, and that the germinal cell and spongioblast were thus dividing and nondividing forms of a single, undifferentiated cell type. This idea was very close to the truth (see below). Proximity of mitoses to the neurocoel suggested lesser mechanical resistance and greater nourishment there.

Schaper further believed that germinal cells produced migratory indifferent cells which in the mantle zone could generate additional indifferent cells or differentiate into neurons or glia. Ramón y Cajal agreed that germinal cells could become either neuroblasts or spongioblasts but rejected wandering, indifferent cells. He concurred with His, who stated in 1901 that all neurons and glia originate directly from the primitive ependymal zone. As proof, Ramón y Cajal demonstrated there a series of apolar, bipolar, and unipolar forms transitional between the germinal cells and the maturing neuroblasts in the mantle zone.

Recognition that the early neural tube is, in fact, a pseudostratified columnar epithelium, in which cell nuclei migrate to the ventricular surface before mitosis and away afterward, was long in coming (F. C. Sauer, 1935a) and slow to win acceptance. Studies of metaphase arrest by colchicine and microphotometric determinations of DNA content after Feulgen staining provided some support, but only autoradiography (M. E. Sauer and Walker, 1959; Sidman et al., 1959) could furnish incontrovertible evidence for intermitotic migration of nuclei within the *ventricular zone* (see recommendations on terminology of the Boulder Committee, 1970). Furthermore, autoradiography showed that all *ventricular cells* divide and therefore constitute a homogeneous population. Hence, the spongioblast of His no longer exists; it was only a germinal cell in interphase.

HISTOLOGICAL FEATURES AND FINE STRUCTURE The concept of a syncytium is also a closed issue; electron microscopy proves the pseudostratified character of the ventricular zone and the separateness of its cells (Duncan, 1957; Tennyson and Pappas, 1962; Fujita and Fujita, 1963; and others). Moreover, terminal bars between cells in the neural plate or recently closed neural tube, another early observation by F. C. Sauer (1935b), are confirmed.

The fine structure of the ventricular cell is summarized here from Lyser's detailed study (1964) in the chick spinal cord. At 33 hours, prior to their visible and precocious differentiation in the ventrolateral region, the columnar cells extend throughout the thickness of the wall. The interphase cells are widest near the nucleus, which lies at various levels, and narrowest at the neurocoel. Plasma membranes completely separate adjacent cells. Junctional complexes, described in adult ependyma (Brightman and Palay, 1963), are not observed but may exist. Apically, however, lateral interdigitations and prominent membrane thickenings are seen. Palay (1967) discussed the significance of such contacts within epithelia as a communications system during development. Other features are many free ribosomes, a scant, granular endoplasmic reticulum, scattered mitochondria, a distinct Golgi complex above a large oval nucleus with two prominent nucleoli, a thin basal lamina, cilia extending into the neurocoel, and a pair of centrioles (see also Lyser, 1968). Microtubules (Weschsler, 1966; Herman and Kauffman, 1966) form the mitotic spindle as in other cells and hence do not signal cell differentiation.

GENERATION CYCLE OF THE VENTRICULAR CELL The cells of the ventricular zone generate all the neurons and macroglia of the CNS. Thanks to autoradiography, their rapid proliferative cycle is well known (for a thorough review, see Sidman, 1970). In early stages the total cycle is less than 10 hours and in 24-hour chick embryos (Fujita, 1962) can be as short as five. Slowing of the cycle, however, occurs with successive generations. For glial and Schwann cells, the precursors of which continue to divide for long periods of time, it can span 24 hours. Its four phases are S, the period of DNA synthesis (four to six hours); $G^2$, the premitotic growth interval (one to two hours); M, mitosis (one hour or less); and $G^1$, the postmitotic growth interval (which, as Sidman remarked, may last from almost no time to the entire life of the animal, as is true for neurons).

The position of the ventricular cell in each of these phases is also clear; the work of Fujita (1963, 1964, 1966, 1967) is eminent here, as in the kinetics. The nuclei lie deep in the ventricular zone in S phase and progressively approach the neurocoel in $G^2$. Mitosis (M) occurs only at the free surface, whereupon the daughter nuclei move again during $G^1$ to deeper positions.

Preliminary observations by Fujita (1966) show that ventricular cells exhibit changes in fine structure during the generation cycle. The major organelles are concentrated in the apical process in $G^2$ but in the basal process in $G^1$. Fujita also noted that the cilium and basal body disappear as the cell enters $G^2$ and reappear in late $G^1$; the rarity of cilia in such rapidly proliferating regions as the ventrolateral region of the neural tube could reflect the brief time available for ciliogenesis there. Lyser (1964, 1968), however, finds a cilium generally present on these cells (see above).

In M phase the spindle fibers are parallel to the inner surface of the ventricular zone (Fujita, 1962). Martin and Langman (1965) did note a few cells in which the spindle axis was perpendicular to the surface. Such orientation might free one daughter cell for migration to the *intermediate* (mantle) *zone* while the other remained anchored by terminal bars. But later Langman et al. (1966) counted only one perpendicular spindle among 134 metaphases; the rest had parallel spindles in side or polar view or oblique spindles. Spindles are visible for only 10 minutes, so that even an occasional perpendicular one, plus some of the oblique variety, could liberate many cells in 24 hours. Nevertheless, the mechanism of release of ventricular cells is not understood.

After the nuclei of daughter cells return to deep positions, the generation cycle reoccurs. In early stages, ventricular cells produce more ventricular cells. Later, they produce both ventricular cells and neuroblasts; the latter migrate rapidly into the intermediate zone. Gradually the ventricular cells cease dividing and differentiate into neuroglial and ependymal cells (Fujita, 1963). Observations vary on the timing of these three histogenetic stages (Källén and Valmin, 1963; Fujita, 1964; Martin and Langman, 1965) and overlap clearly occurs (Miale and Sidman, 1961; Sidman, 1970). Nonetheless, a program of origin for different cell types, both broadly as stated here and more explicitly for the neurons of given regions (see below), is certain, and there is general accord concerning its sequence.

REGIONAL ANALYSIS OF VENTRICULAR CELL PROLIFERATION   The single greatest application of the powerful new technique of autoradiography in the past decade has been the analysis of the generation cycle in cells exposed to tritiated thymidine (Sidman, 1970). Studies in the nervous system since the pioneering works of M. E. Sauer and Walker (1959) in the chick and Sidman et al. (1959) in the mammal have provided breakthroughs, as stressed above,

in our understanding of fundamentals of histogenesis and grasped previously inaccessible data on the kinetics of cell turnover. And, as data accumulate, controls upon these major cellular events (DeLong and Sidman, 1962; Sidman, 1970) become easier to identify and study.

The use of autoradiography for scrutinizing the development of regional neural organization is a new approach and greatly improves resolution over classical morphogenetic studies (Angevine, 1965). One can now trace individual neurons from proliferative through migratory states to final address in the adult brain. A label on a neuroblast is *permanent*; once the cell leaves the ventricular zone, it loses the capacity to divide. Moreover, rigorous schedules of neuron origin permit *selective* labeling by appropriate timing of radioactive thymidine injections. Thus, identity of neurons arising at a given developmental stage may be ascertained by noting their positions in the mature brain. Other neurons, originating earlier or later, remain nonradioactive.

Exploiting these principles, Angevine (1965), Taber Pierce (1966, 1967), and Hinds (1968a, 1968b) have studied adult mice injected at consecutive developmental stages and have constructed elaborate and hitherto unknown calendars of neuronal birth dates for the hippocampal region, brainstem, and olfactory bulb, respectively. Moreover, examination of serial embryonic stages after thymidine-$H_3$ had been injected allowed isolation and day-by-day surveillance of the paths of young neurons.

Regional autoradiographic studies are few compared with those of general features of neurogenesis. Exploration, however, has ranged widely over the cerebellum and its deep nuclei, the cerebral cortex and basal ganglia, the retina and superior colliculus, the spinal cord and ganglia—even to the subcommissural organ and human pulvinar (largest and most recent thalamic nucleus; see Angevine, 1970, for references).

The results of these studies are new and basic facts of neurogenesis, unattainable before autoradiography with tritium, the significance of which emerges when viewed in relation to other developmental features. For example, timetables of ventricular cell proliferation (Angevine, 1965) as intricate as those for neuroblast migration (Levi-Montalcini, 1964) characterize the origin of all brain regions. These timetables are fundamental schedules of preliminary events and may clarify how certain, although by no means all, long pathways arise. Moreover, an inside-out sequence in time of origin of neurons in the cerebrum (see below) foreshadows gradients in differentiation of cortical neurons (Pappas and Purpura, 1964; Stensaas, 1968; Åström, 1967; Caley and Maxwell, 1968) and in maturation of their patterns of electrical excitability (Purpura et al., 1964). Further search may uncover roles for times of origin in fashioning local connections within the neuropil.

PATTERNED CELL PROLIFERATION FOR SPECIFIC BRAIN COMPONENTS The orderly programing of neuronal origin is striking for three brain locales: the isocortical (six-layered) and allocortical (fewer-layered) regions of the cerebral cortex and the ventral thalamic, dorsal thalamic, and epithalamic zones of the diencephalon (betweenbrain). These regions are considered in turn.

Detailed studies of time of neuronal origin are not complete for the isocortex. General features, however, have been published (Angevine and Sidman, 1961, 1962; Sidman and Angevine, 1962; Hinds and Angevine, 1965). This region in the mouse is foretold on the eleventh day of gestation by proliferation of the future deepest neurons of the cortex, adjacent to the white matter (Figure 1, page 66). Some neurons arise even earlier, on day 10, but are not illustrated. More neurons originate subsequently; neurons arising on day 13 come to lie in the middle third of the mature cortex and on day 15 in the outermost third. A smaller population from day 17 becomes the most superficial cell layer. A few cells form on day 18, just before birth, but again are not shown. The more superficial positions of younger neurons to those of older cells attest to active migration of neuroblasts. Furthermore, they demonstrate that the younger neurons or at least their cell nuclei (Morest, 1968; personal communication) transpose with the older neurons or their nuclei. This fact hints at the future columnar plan of connectivity in the cortex (Colonnier, 1967; Hubel and Wiesel, 1968).

Time of neuronal origin is known in greater detail for the allocortex (Angevine, 1965). Its distinctive neuronal assemblies facilitate study and invite comparative analysis (Figure 2, page 68). The cytoarchitecture of the various cortical areas does not concern us here, and few anatomical terms are needed. Areal boundaries (indicated by dashed lines) are sharp and reflect well-documented regional peculiarities in neuronal morphology and connectivity.

Allocortical neurons begin to form on the tenth day of gestation, as in isocortex, but in fewer numbers at first. The oldest neurons, as before, are generally the deepest cells of the various subdivisions. Simultaneous origin of some neurons of more superficial layers, however, occurs here and there. Thereafter, neurons for all regions but one originate in the familiar inside-out sequence (indicated by arrows). The pyramidal cells and the various interneurons arise in the ventricular zone nearby and migrate during gestation into the cortex.

Neurons of the retrohippocampal formation stop forming on the sixteenth day; the last cells, as in other regions, are small neurons (of the presubiculum). In the adjacent hippocampal formation (subiculum, hippocampus, and area dentata), the subiculum achieves its neuronal complement even earlier, on day 15. The hippocampal sector CA2 displays a similar course, but the other two sectors, CA1 and

CA3, have longer periods, ending early on the eighteenth day. Interestingly, the latter sectors have a prominent fiber connection, not shared by CA2. The final neurons are at the outer margin (with reference to the white matter) or the very tip of the pyramidal cell band. These patterns coincide precisely with cytoarchitectonic divisions; CA2 and CA3 resemble each other in dye stains, but their calendars of neuronal birth dates contrast sharply, as do their unique connections noted with metallic methods.

Neurons of the deep and superficial layers of area dentata, as do those of the neighboring hippocampus, originate between days 10 and 15. But the small granule cells of the intervening layer are distinctive. They arise over a long period from the tenth day of gestation to the twentieth day after birth and perhaps longer. Embryos killed soon after thymidine-H³ labeling, before migration of postmitotic cells has taken place, show that the superficial granule cells formed prenatally arise in the ventricular zone. Those arising perinatally, in contrast, come from division of cells outside it, near the fimbria (fringe of the hippocampus). The labeled fusiform daughter cells migrate past the pyramidal cells of CA3, with which the future granule cells will connect, to deeper positions in the granular layer. Postnatally, proliferation of precursors in the granular layer itself produces the final granule cells, which migrate no farther but differentiate at their site of origin. Thus, time of origin of the granule cells follows an outside-in sequence, opposite to that seen elsewhere in allocortex and isocortex. Furthermore, these neurons stem from precursors which are not postmitotic at the time of migration, in contrast to the usual neuroblast. Precursors arising in the rhombic lip of the hindbrain (Miale and Sidman, 1961; Altman and Das, 1966; Taber Pierce, 1967) also proliferate persistently on the surface of the cerebellum and brainstem before differentiating into the small neurons of the cerebellar cortex and cochlear nuclei.

The most intensive regional study of patterned cell proliferation to date concerns the diencephalon (Angevine, 1970). This region, even in a mouse, is vast; its neurons aggregate in more or less discrete nuclei, rather than the extensive sheets found in cortex. Hence, adequate sampling necessitated analysis at three transverse levels (Figure 1). Unexpectedly, broad gradients in time of neuronal origin, which would have been missed at one level, are revealed. In fact, a more unified pattern emerges than for allocortex, where areal differences are perhaps distracting, although not dissonant, features.

Three conventionally recognized diencephalic zones were examined (Figure 3, page 70): ventral thalamus, dorsal thalamus, and epithalamus. Some neurons, as for the cortex, originate on the tenth day of gestation. These locate chiefly in the ventral thalamic zone and anticipate its well-known precocious differentiation. The period of neuronal origin for

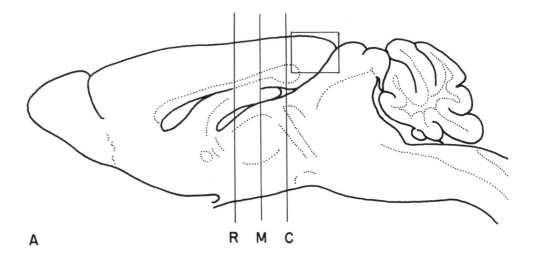

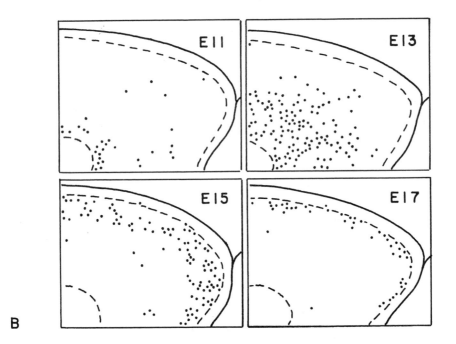

FIGURE 1 Paramedian sagittal section of mouse brain (A) illustrating region (rectangle) in four drawings (B) of occipital cortex to show inside-out sequence in time of origin of isocortical neurons (Angevine and Sidman, 1961). Each of four mice received *one* injection of thymidine-H³ on the eleventh, thirteenth, fifteenth, or seventeenth day of gestation (E11–E17); all were killed 10 days *after* birth when the radioactive neurons had reached final positions (dots). In B, outer broken line represents inner boundary of molecular layer of cerebral cortex; inner broken line, the boundary between cortex and white matter, chiefly corpus callosum. In A, lines R (rostral), M (middle), and C (caudal) show the three transverse levels in the autoradiographic study of time of neuronal origin in the diencephalon (Figure 3).

the ventral thalamus is brisk, and the final cells arise early on the thirteenth day at all three levels. These events foreshadow a *ventrodorsal* gradient of proliferation in the ventricular zone.

Other neurons arise on the tenth day, but only at the caudal level (lower picture) in the dorsal thalamus. These neurons subsequently form in massive numbers until early on the thirteenth day. These findings illustrate a *caudorostral* gradient in the dorsal thalamus; the middle and rostral levels (middle and upper pictures) lag one and two days, respectively, behind the caudal in the swift sweep of labeling phases.

A *lateromedial* gradient appears when dorsal thalamus and epithalamus are examined at the three levels. This sequence, opposing that for cortex, may reflect lateral displacement of neuroblasts or their cell nuclei as newer ones arise. The pattern may fit the nuclear organization of the diencephalon (Scheibel and Scheibel, 1966), just as the inside-out sequence may suit the columnar one of cortex.

Large neurons of the brain generally arise before smaller ones (Hinds 1968a), perhaps determining precedence of motor over sensory systems (Hamburger, this volume). This is striking in the epithalamus. Its principal structure, the habenula, has two nuclei, lateral (Hl) and medial (Hm). Neurons of the former ". . . are large, star shaped, and have long processes that make them look like most of the cells found in the other nuclei of the thalamus" (Ramón y Cajal, 1894—see Ramón y Cajal, 1966). But the latter ". . . is characterized by an extraordinary number of very small cells . . . nearly as small as the granule cells of the cerebellum." Hence, the habenula offers a model of the two major categories of diencephalic neurons, side-by-side and in pure culture. The large neurons arise in a short burst of proliferation beginning late on day 11 and ending early on day 13. The dwarf cells, however, form later and longer, from early on day 12 until late on day 16. Such persistent proliferation characterizes small neurons of many regions, notably cerebellum (Miale and Sidman, 1961), area dentata (Angevine, see above), and olfactory bulb (Hinds, 1968a, 1968b). It is not obvious in the diencephalon because few small neurons exist in the rodent thalamus (Scheibel and Scheibel, 1966, 1967). But in brain elaboration, in which increased numbers of short-axoned interneurons and "endless circuit reduplication" (Scheibel and Scheibel, 1966) are recognized trends, perhaps ". . . the logistics are easier to arrange if small cells form later . . . on surfaces capable of comfortable expansion" (Sidman, 1970). Origin of small cells is not triggered at or by birth (which occurs at a relatively early stage of forebrain development) or restricted to postnatal development (Altman and Das, 1967), although reciprocal influences between postnatal neurogenesis and environmental response may exist. Rather ". . . birth may be considered an arbitrary point in a con-

tinuous and relatively independent development of the brain" (Hinds, 1968a), as seen in the early but sustained origin of the small neurons for area dentata.

Thus, long periods of neuron origin may offer modulation of a different kind—a delicate mechanism for regulating the number of short-axoned cells. Indeed, Sidman (this volume) speculates on the possibly profound consequences of only minor changes in shutting off the rapid stream of small cells. Slight delay could mean major evolutionary increments, quantitated in cerebellum but equally obvious in area dentata. Conversely, cessation too soon could be a developmental disaster. Furthermore, the volume of the habenula ". . . seems to decrease in a relative manner as one ascends in phylogeny" (Ramón y Cajal, 1894—see Ramón y Cajal, 1966); this reduction might reflect shortening by only hours the term of neuronal origin. Although the design of the epithalamus is essentially constant in phylogeny, the habenular nuclei are not highly developed in mammals as compared with other vertebrates. Variations, when habenulae of different mammals are examined, involve both nuclei but are more striking in the medial, which in man is especially insignificant in size and cell population.

The habenula vividly demonstrates programed activity by the ventricular zone, which produces two neuronal types at successive times. Such programing is also clear in the brain stem (Taber Pierce, 1966), cerebellum (Fujita, 1967), and olfactory bulb (Hinds, 1968a), but the controls are unknown (Sidman, 1970). Furthermore, overlapping origin of large and small neurons and subsequent glial cells bespeaks multipotentiality of ventricular cells and progressive restriction with advancing development (Hinds, 1968a).

PROLIFERATIVE GRADIENTS IN THE VENTRICULAR ZONE
Several gradients in time of neuron origin have been mentioned: the ubiquitous inside-out gradient for cortex, the outside-in gradient for the granular layer of area dentata (plus evident progression along its curvature from suprapyramidal to infrapyramidal limbs), and the lateromedial gradient for dorsal thalamus and epithalamus. Much concerning possible relationships of these sequences to future connections and synaptology must here be left unsaid. But the ventrodorsal and caudorostral gradients in the diencephalon merit special attention. These intimate a sloping wave of origin (and hence cessation of DNA synthesis) of neuron precursors in the embryonic forebrain. Because cessation of DNA synthesis in ganglion cell neuroblasts correlates with specification of their future central connections in the amphibian brain (Jacobson, 1968), perhaps thalamic neuroblasts are so specified.

The ventrodorsal gradient suggests a wave of neuron generation that is originally rostrocaudal, foreshadowing

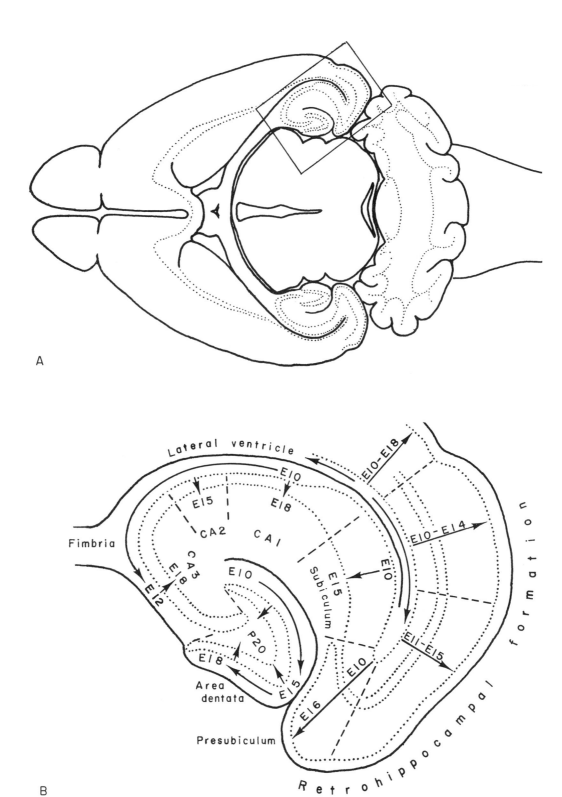

A

B

FIGURE 2   Horizontal section of mouse brain (A) illustrating location (rectangle) in drawing (B) of hippocampal region and its components to show inside-out sequence in time of origin of allocortical neurons and highly individual calendars of neuronal birth dates for each component (Angevine, 1965). A series of mice received *one* injection of thymidine-H³ at various days of gestation (E10–E18); all were killed *after* birth when radioactive neurons had reached final positions. Arrows show sequence and duration of neuronal origin for each discrete area (separated by dashed lines).

the craniocaudal progression of neural maturation (Hamburger, this volume), but revectored by reorientation of neuromeres I–III (transient early brain segments—Coggeshall, 1964; Angevine, 1970). Similarly, the caudorostral gradient in dorsal thalamus may be a redirected ventrodorsal one. Moreover, a known ontogenetic precocity of the basal telencephalon versus the convexity of the hemisphere and a caudorostral gradient in the isocortex (Angevine and Sidman, 1962) might be redirected rostrocaudal and ventrodorsal gradients, respectively, in neuromere I.

GLIAL CELL FORMATION—THE SUBVENTRICULAR ZONE
After a variable time, a *subventricular zone* of dividing cells appears between ventricular and intermediate zones in most regions of the CNS. It becomes prominent from late gestation to adulthood in the lateral ventricles (Smart, 1961). The cells incorporate thymidine-H³ but do not show intermitotic nuclear migrations. They come from ventricular cells, but probably have longer generation cycles (Sidman, 1970). They differentiate into neurons (Hinds, 1968b) and glia (Smart and Leblond, 1961; Altman, 1966), but it is not clear whether one subventricular cell can give rise to both. Finally, they migrate outward and continue to multiply; some derivatives may divide in cortex (Hommes and Leblond, 1967), but others are permanently postmitotic.

Glial cells are the last elements on the program of cell proliferation and their generation cycle is longer (see above), which explains their appearance chiefly postpartum. They may arise, however, almost as early as neuroblasts; some near the habenula arise on the thirteenth day of gestation (Angevine, unpublished observation). The early origin of glioblasts is obscured by the fact that ". . . the majority of glial cells in the embryonic brain continue to multiply, whereas few of the neuroblasts do so" (Sidman, 1970).

## Cell migration

BEHAVIOR OF NEUROBLASTS  A fundamental difference between neuroblasts and other formative cells bearing the *blast* suffix (erythroblast, myeloblast, osteoblast) is that the former no longer divide. Exceptions occur (see above) in the cerebellum, cochlear nuclei, and area dentata. Furthermore, neuroblasts, with few exceptions (area dentata, habenula), migrate from site of origin to another place to differentiate. Whether this migration is active or merely passive cell displacement has been an important issue. Evidence for true migration and against it is discussed by Sidman (this volume), together with the nature of the environment of migrating cells. "The orderly fashion of these movements and their time pattern . . . in all instances so rigorous as to permit prediction of when the first neuroblasts will start their long journey . . ." (Levi-Montalcini, 1964) derives, however, from patterned cell proliferation.

COMPLEXITY OF MIGRATIONS  Cell migration may be complex (Levi-Montalcini, 1964), and neuroblasts may collide with others formed earlier or simultaneously. The cerebellum offers the best example (Miale and Sidman, 1961). Purkinje and granule cells arise, respectively, from two opposed germinal matrices: the ventricular zone and a transient external granular layer derived from it. Daughter cells of the latter migrate inward past the Purkinje cell bodies to attain final positions. Migration follows proliferation closely, however, attested by advanced prenatal differentiation of many brain neurons.

## Cell differentiation

INDEPENDENCE OF NEUROGENETIC EVENTS  The interval between time of origin and other events may be more variable. The Purkinje cells just mentioned originate early and abruptly; the granule cells, later and persistently (Miale and Sidman, 1961). Yet elaboration of Purkinje dendrites is long delayed postnatally until development of parallel fibers from granule cells. Onset of function then ". . . occurs *pari passu* with the morphogenesis of axodendritic synaptic substrate" (Purpura, 1967).

An intriguing example (Morest, 1968) of independence of neurogenetic events is the growth of dendrites in elongated neuroblasts which retain attachments to the internal, or ex-

In B, outermost broken line represents inner boundary of molecular layer of cerebral cortex; innermost broken line, boundary between cortex and white matter, either corpus callosum or alveus of hippocampus. Note that allocortex has prominent outer and inner cell layers. Outer layer stops in presubiculum, as shown; its cells are tiny, and originate for a longer time than do those for adjacent regions. Inner layer continues from retrohippocampal formation into hippocampal formation, forming prominent band of pyramidal cells in subiculum and hippocampus (subdivided into CA1, CA2, and CA3). Note early cessation for CA2 and early origin, persistent proliferation (until 20 days after birth–P20), and contrasting outside-in sequence for small granule cells in intermediate layer of area dentata (surrounding tip of pyramidal cell band in hippocampus). Note also gradients: from suprapyramidal to infrapyramidal limbs of granule-cell layer in area dentata, from CA1 to CA3 in hippocampus, from lateral to medial regions in retrohippocampal formation, and from boundary with allocortex rostrally in isocortex.

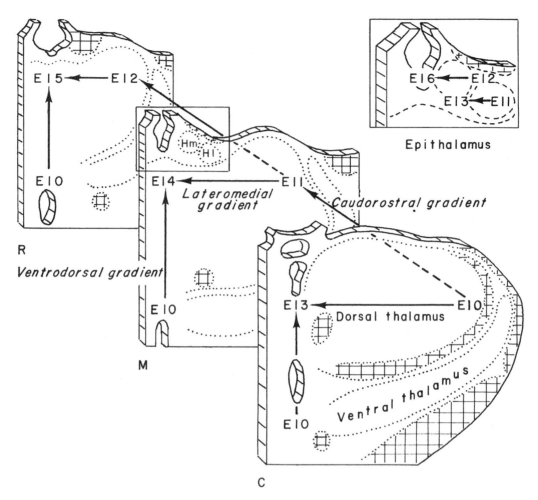

FIGURE 3 Three transverse levels (R, M, C; see Figure 1) of diencephalon of adult mouse to show time of neuronal origin for ventral thalamus, dorsal thalamus, and epithalamus (Angevine, 1970). Code numbers, arrows, and other details as described for Figures 1 and 2. Note early origin (on E10) of ventral thalamus at all three levels, foreshadowing a *ventrodorsal* gradient. Note also staggered onset and cessation of neuronal origin for dorsal thalamus at caudal level (C), middle level (M), and rostral level (R), expressing a *caudorostral* gradient. Finally, note lateral positions of first neurons in dorsal thalamus at all levels and in epithalamus (lateral habenular nucleus, Hl, and medial habenular nucleus, Hm) at middle level (inset), showing a *lateromedial* gradient.

Gradients and massive proliferation of neuroblasts overshadow nuclear differences and demonstrate developmental unity in dorsal thalamus. In epithalamus, persistent proliferation (until E16) of small granule cells for Hm follows pattern in other brain regions and illustrates programed activity by the ventricular zone, which produces large neurons for Hl earlier (until E13). Final habenular neurons differentiate near or at site of origin in ventricular zone of upper chamber of third ventricle.

ternal, limiting membranes of the ventricular zone, or both. Axon sprouting, an initial step in neuronal differentiation (Lyser, 1964), also occurs in these cells (Morest, personal communication), which resemble the spongioblasts of His (see above), but are indeed neuroblasts. In the optic tectum such primitive elements form axons, as well as dendritic buds, before loss of attachment. Thus, the growth of axon and dendrites, as well as other overt features of differentiation, is an independent variable. Furthermore, nuclear migration may not coincide with neuroblast retraction from the limiting membranes. In the prospective parietal cortex of the opossum, the nucleus migrates within the long ventricular cell from a position in the ventricular zone to one near the pia mater (vascular tunic). There, the cell loses attachment to the inner limiting membrane and "pulls up" its cytoplasm in "rope ladder" fashion. In other instances, Morest observed neuroblasts retracting inner attachments as they migrate.

Sidman (this volume) considers other aspects of cell differentiation: the fine structure of neuroblasts, development of histochemical properties, and attainment and modification of synaptic connections.

## Cell death

Programed cell degeneration, consistent in number, place, and time, is the most recently recognized (Glücksmann, 1951) cellular event. Neuron formation in excess was discussed by Hughes (1968) and by Prestige (this volume); in brief, this event is important for the shape, cell aggregation, and connectivity of the future CNS.

REFERENCES

ALTMAN, J., 1966. Autoradiographic and histological studies of postnatal neurogenesis. II. A longitudinal investigation of the kinetics, migration and transformation of cells incorporating tritiated thymidine in infant rats, with special reference to postnatal neurogenesis in some brain regions. *J. Comp. Neurol.* 128: 431–473.

ALTMAN, J., and G. D. DAS, 1966. Autoradiographic and histological studies of postnatal neurogenesis. I. A longitudinal investigation of the kinetics, migration and transformation of cells incorporating tritiated thymidine in neonate rats, with special reference to postnatal neurogenesis in some brain regions. *J. Comp. Neurol.* 126: 337–389.

ALTMAN, J., and G. D. DAS, 1967. Postnatal neurogenesis in the guinea-pig. *Nature (London)* 214: 1098-1101.

ANGEVINE, J. B., JR., 1965. Time of neuron origin in the hippocampal region. An autoradiographic study in the mouse. *Exp. Neurol.,* suppl. 2: 1–70.

ANGEVINE, J. B., JR., 1970. Time of neuron origin in the diencephalon of the mouse. An autoradiographic study. *J. Comp. Neurol.* (in press).

ANGEVINE, J. B., JR., and R. L. SIDMAN, 1961. Autoradiographic study of cell migration during histogenesis of cerebral cortex in the mouse. *Nature (London)* 192: 766–768.

ANGEVINE, J. B., JR., and R. L. SIDMAN, 1962. Autoradiographic study of histogenesis in the cerebral cortex of the mouse. *Anat. Rec.* 142: 210 (abstract).

ÅSTRÖM, K.-E., 1967. On the early development of the isocortex in fetal sheep. *Progr. Brain Res.* 26: 1–59.

BOULDER COMMITTEE, 1970. Embryonic vertebrate central nervous system: Revised terminology. *Anat. Rec.* 166: 257–261.

BRIGHTMAN, M. W., and S. L. PALAY, 1963. The fine structure of ependyma in the brain of the rat. *J. Cell Biol.* 19: 415–439.

CALEY, D. W., and D. S. Maxwell, 1968. An electron microscopic study of neurons during postnatal development of the rat cerebral cortex. *J. Comp. Neurol.* 133: 17–43.

COGGESHALL, R. E., 1964. A study of diencephalic development in the albino rat. *J. Comp. Neurol.* 122: 241–269.

COLONNIER, M., 1967. The fine structural arrangement of the cortex. *Arch. Neurol.* 16: 651–657.

DE LONG, G. R., and R. L. SIDMAN, 1962. Effects of eye removal at birth on histogenesis of the mouse superior colliculus: An autoradiographic analysis with tritiated thymidine. *J. Comp. Neurol.* 118: 205–223.

DUNCAN, D., 1957. Electron microscope study of the embryonic neural tube and notochord. *Tex. Rept. Biol. Med.* 15: 367–377.

EBERT, J. D., 1967. Gene expression. *Neurosci. Res. Program Bull.* 5 (no. 3): 223–306.

EDDS, M. V., JR., 1967. Neuronal specificity in neurogenesis. *In* The Neurosciences: A Study Program (G. C. Quarton, T Melnechuk, and F. O. Schmitt, editors). The Rockefeller University Press, New York, pp. 230–240.

FUJITA, H., and S. FUJITA, 1963. Electron microscopic studies on neuroblast differentiation in the central nervous system of domestic fowl. *Z. Zellforsch. Mikroskop. Anat.* 60: 463–478.

FUJITA, S., 1962. Kinetics of cellular proliferation. *Exp. Cell Res.* 28: 52–60.

FUJITA, S., 1963. The matrix cell and cytogenesis in the developing central nervous system. *J. Comp. Neurol.* 120; 37–42.

FUJITA, S., 1964. Analysis of neuron differentiation in the central nervous system by tritiated thymidine autoradiography. *J. Comp. Neurol.* 122: 311–327.

FUJITA, S., 1966. Applications of light and electron microscopic autoradiography to the study of cytogenesis of the forebrain. *In* Evolution of the Forebrain, Phylogenesis and Ontogenesis of the Forebrain (R. Hassler and H. Stephan, editors.) G. Thieme, Stuttgart, pp. 180–196.

FUJITA, S., 1967. Quantitative analysis of cell proliferation and differentiation in the cortex of the postnatal mouse cerebellum. *J. Cell Biol.* 32: 277–287.

GLÜCKSMANN, A., 1951. Cell deaths in normal vertebrate ontogeny. *Biol. Rev. (Cambridge)* 26: 59–86.

HAMBURGER, V., and R. LEVI-MONTALCINI, 1949. Proliferation, differentiation and degeneration in the spinal ganglia of the chick embryo under normal and experimental conditions. *J. Exp. Zool.* 111: 457–501.

HERMAN, L., and S. L. KAUFFMAN, 1966. The fine structure of the embryonic mouse neural tube with special reference to cytoplasmic microtubules. *Develop. Biol.* 13: 145–162.

HINDS, J. W., 1968a. Autoradiographic study of histogenesis in the mouse olfactory bulb. I. Time of origin of neurons and neuroglia. *J. Comp. Neurol.* 134: 287–304.

HINDS, J. W., 1968b. Autoradiographic study of histogenesis in the mouse olfactory bulb. II. Cell proliferation and migration. *J. Comp. Neurol.* 134: 305–321.

HINDS, J. W., and J. B. ANGEVINE, JR., 1965. Autoradiographic study of histogenesis in the area pyriformis and claustrum in the mouse. *Anat. Rec.* 151: 456–457 (abstract).

HOMMES, O. R., and C. P. LEBLOND, 1967. Mitotic division of neuroglia in the normal adult rat. *J. Comp. Neurol.* 129: 269–278.

HUBEL, D. H., and T. N. WIESEL, 1968. Receptive fields and functional architecture of monkey striate cortex. *J. Physiol. (London)* 195: 215–243.

HUGHES, A. F. W., 1968. Aspects of Neural Ontogeny. Academic Press, New York.

JACOBSON, M., 1968. Cessation of DNA synthesis in retinal ganglion cells correlated with the time of specification of their central connections. *Develop. Biol.* 17: 219–232.

KÄLLÉN, B., and K. VALMIN, 1963. DNA synthesis in the embryonic chick central nervous system. *Z. Zellforsch. Mikroskop. Anat.* 60: 491–496.

LANGMAN, J., R. L. GUERRANT, and B. G. FREEMAN, 1966. Behavior of neuro-epithelial cells during closure of the neural tube. *J. Comp. Neurol.* 127: 399–411.

LEVI-MONTALCINI, R., 1964. Events in the developing nervous system. *Progr. Brain Res.* 4: 1–29.

LYSER, K. M., 1964. Early differentiation of motor neuroblasts in the chick embryo as studied by electron microscopy. I. General aspects. *Develop. Biol.* 10: 433–466.

LYSER, K. M., 1968. An electron-microscope study of centrioles in differentiating motor neuroblasts. *J. Embryol. Exp. Morphol.* 20: 343–354.

MARTIN, A., and J. LANGMAN, 1965. The development of the spinal cord examined by autoradiography. *J. Embryol. Exp. Morphol.* 14: 25–35.

MIALE, I. L., and R. L. SIDMAN, 1961. An autoradiographic analysis of histogenesis in the mouse cerebellum. *Exp. Neurol.* 4: 277–296.

MOREST, D. K., 1968. Growth of cerebral dendrites and synapses. *Anat. Rec.* 160: 516 (abstract).

PALAY, S. L., 1967. Principles of cellular organization in the nervous system. *In* The Neurosciences: A Study Program (G. C. Quarton, T. Melnechuk, and F. O. Schmitt, editors). The Rockefeller University Press, New York, pp. 24–31.

PAPPAS, G. D., and D. P. PURPURA, 1964. Electron microscopy of immature human and feline neocortex. *Progr. Brain Res.* 4: 176–186.

PURPURA, D. P., 1967. Comparative physiology of dendrites. *In* The Neurosciences: A Study Program (G. C. Quarton, T. Melnechuk, and F. O. Schmitt, editors). The Rockefeller University Press, New York, pp. 372–393.

PURPURA, D. P., R. J. SHOFER, E. M. HOUSEPIAN, and C. R. NOBACK, 1964. Comparative ontogenesis of structure-function relations in cerebral and cerebellar cortex. *Progr. Brain Res.* 4: 187–221.

RAMÓN Y CAJAL, S., 1966. Studies on the Diencephalon (E. Ramón-Moliner, compilator and translator). Charles C Thomas, Springfield, Illinois, pp. 62–76.

SAUER, F. C., 1935a. Mitosis in the neural tube. *J. Comp. Neurol.* 62: 377–405.

SAUER, F. C., 1935b. The cellular structure of the neural tube. *J. Comp. Neurol.* 63: 13–23.

SAUER, M. E., and B. E. WALKER, 1959. Radioautographic study of interkinetic nuclear migration in the neural tube. *Proc. Soc. Exp. Biol. Med.* 101: 557–560.

SAXÉN, L., and S. TOIVONEN, 1962. Primary Embryonic Induction. Prentice-Hall, Englewood Cliffs, New Jersey.

SCHEIBEL, M. E., and A. B. SCHEIBEL, 1966. Patterns of organization in specific and nonspecific thalamic fields. *In* The Thalamus (D. P. Purpura and M. D. Yahr, editors). Columbia University Press, New York, pp. 13–46.

SCHEIBEL, M. E., and A. B. SCHEIBEL, 1967. Anatomical basis of attention mechanisms in vertebrate brains. *In* The Neurosciences: A Study Program (G. C. Quarton, T. Melnechuk, and F. O. Schmitt, editors). The Rockefeller University Press, New York, pp. 577–602.

SIDMAN, R. L., 1970. Autoradiographic methods and principles for study of the nervous system with thymidine-$H^3$. *In* Contemporary Research Techniques of Neuroanatomy (S. O. E. Ebbesson and W. J. H. Nauta, editors). Springer-Verlag, New York (in press).

SIDMAN, R. L., and J. B. ANGEVINE, JR., 1962. Autoradiographic analysis of time of origin of nuclear versus cortical components of mouse telencephalon. *Anat. Rec.* 142: 326–327 (abstract).

SIDMAN, R. L., I. L. MIALE, and N. FEDER, 1959. Cell proliferation and migration in the primitive ependymal zone; an autoradiographic study of histogenesis in the nervous system. *Exp. Neurol.* 1: 322–333.

SMART, I., 1961. The subependymal layer of the mouse brain and its cell production as shown by radioautography after thymidine-$H^3$ injection. *J. Comp. Neurol.* 116: 325–347.

SMART, I., and C. P. LEBLOND, 1961. Evidence for division and transformations of neuroglia cells in the mouse brain, as derived from radioautography after injection of thymidine-$H^3$. *J. Comp. Neurol.* 116: 349–367.

STENSAAS, L. J., 1968. The development of hippocampal and dorsolateral pallial regions of the cerebral hemisphere in fetal rabbits. VI. Ninety millimeter stage, cortical differentiation. *J. Comp. Neurol.* 132: 93–108.

TABER PIERCE, E., 1966. Histogenesis of the nuclei griseum pontis, corporis pontobulbaris and reticularis tegmenti pontis (Bechterew) in the mouse. An autoradiographic study. *J. Comp. Neurol.* 126: 219–239.

TABER PIERCE, E., 1967. Histogenesis of the dorsal and ventral cochlear nuclei in the mouse. An autoradiographic study. *J. Comp. Neurol.* 131: 27–53.

TENNYSON, V. M., and G. D. PAPPAS, 1962. An electron microscope study of ependymal cells of the fetal, early postnatal and adult rabbit. *Z. Zellforsch. Mikroskop. Anat.* 56: 595–618.

WATTERSON, R. L., 1965. Structure and mitotic behavior of the early neural tube. *In* Organogenesis (R. L. DeHaan and H. Ursprung, editors). Holt, Rinehart and Winston, New York, pp. 129–159.

WECHSLER, W., 1966. Die Feinstruktur des Neuralrohres und der Neuroektodermalen Matrixzellen am Zentralnervensystem von Hühnerembryonen. *Z. Zellforsch. Mikroskop. Anat.* 70: 240–268.

WEISS, P., 1955. Nervous system (neurogenesis). *In* Analysis of Development (B. H. Willier, P. A. Weiss, and V. Hamburger, editors). W. B. Saunders, Philadelphia, pp. 346–401.

WEISS, P. A., 1965. Specificity in the neurosciences. *Neurosci. Res. Program Bull.* 3 (no. 5): 1–64.

WESTON, J. A., 1963. A radioautographic analysis of the migration and localization of trunk neural crest cells in the chick. *Develop. Biol.* 6: 279–310.

# 8 Differentiation, Degeneration, and the Role of the Periphery: Quantitative Considerations

## M. C. PRESTIGE

IT WAS ESTABLISHED early in the history of experimental embryology that the size and state of the developing central nervous system (CNS) are not only intrinsically determined but are influenced by the size and nature of the peripheral tissue that it innervates. The size changes in the CNS may be in either direction, by changes in cell number (hypoplasia/hyperplasia) or by changes in cell size (atrophy/hypertrophy). They may involve a variety of adjustments of the center to the periphery (retrograde) in some qualitative or quantitative way; or the reverse (orthograde, transneuronal). There is still an argument as to whether orthograde actions on post-junctional tissue are caused by "use" (i.e., impulse traffic) or by "trophic factors" (i.e., release of chemicals from nerve terminals regardless of impulses in the presynaptic terminals). The nerve cell body itself undergoes a characteristic response to injury, such as axotomy, which histologists call chromatolysis. If prolonged, this leads to atrophy or even cell death (degeneration).

In the classic literature, the term "center" covers any nerve cell center, and "periphery" covers any tissue peripheral to the central nervous system. In experiments of similar design, however, it has been found that the same concepts appear to apply for nerve cells the axons of which lie entirely within the central nervous system, if the term "periphery" is held to mean the set of postsynaptic cells for the system under question. Unfortunately, in this sense, dorsal-root ganglion cells have two peripheries, whereas retinal ganglion cells are "central" to lateral geniculate cells. Even this terminology is stretched for cells such as those neurosecretory axons in insects (Scharrer, 1968) in which the release sites are strikingly similar to the presynaptic component of conventional interneuronal junctions, yet are completely without postsynaptic cells. The problem becomes urgent for cells in vitro, which can survive with only the medium as periphery, albeit always supplemented by serum or embryo extract. It may be that the terms "center" and "periphery" are useful in this context only as historical examples from a more general class of phenomena—the dependence of nerve cells on the environment in which their axons terminate.

Other papers in this section establish the relevance of considering neurogenesis at the cellular level, and explore the roles of cell division, differentiation, migration, and death. My task in this chapter is to consider the periphery as one of the key factors controlling these events, especially differentiation and death. In doing so, I stress the ways in which these systems are similar to other non-neural embryological contexts, in the belief that the particular conditions that we associate with the nervous system are but modifications of more general cellular properties.

The first question to be asked of the experiments that have been done is: "To what extent does the development of the nervous system display genetic pre-adaptation to the periphery that it will innervate, and to what extent does the periphery modify this subsequently?" Classically, this question has been approached by isolation and recombination experiments, usually in vivo. The second question is the one of mechanism; so far not much progress has been made, although I return to this topic at the end of the chapter.

To compile a list of ways in which developing nervous tissue is pre-adapted would be an impossible task. One can, however, attempt to find what properties are present before the center comes into contact with the periphery; an example is the determination of regional differences in the spinal cord (Wenger, 1951). In early chick embryos, transplantation of lengths of thoracic spinal cord to the brachial level does not induce the later formation of the motor enlargement that is characteristic of the spinal cord at limb levels, and vice versa. This differentiation is determined before there is any possibility of interaction between the limb bud and the spinal cord.

In this chapter, I examine the methods and evidence for two further examples concerning the control of cell number—the first, one that is pre-adapted, and the second, one in which the extent and time relations of the center/periphery relationship can be explored quantitatively.

## Cell balance sheet

To construct a complete cell balance sheet, we must examine the processes of proliferation, migration, and de-

M. C. PRESTIGE Department of Physiology, University of Edinburgh, United Kingdom

generation quantitatively by doing cell counts in both normal and experimental situations. The effect of the periphery is usually inferred from the results of isolating the center; such an experiment gives more reproducible results than the more complex one of increasing the periphery.

Ideally, evidence from cell counts should satisfy the following conditions: 1) consistent criteria for identification of cells to be counted both in position and appearance and, in the embryological context, also in staging; 2) reproducible data in absolute units; and 3) confirmation by an independent technique. Unfortunately, condition 1 is bedeviled by the considerable subjective element; condition 3 has so far usually proved elusive.

Before a cell population in vivo is discussed, it is instructive to consider the growth of a pure strain of cells in monolayer culture (Hahn et al., 1968). In this experiment (Figure 1), an accumulation phase of increasing cell number, during which time more than 75 per cent of the cells are capable of dividing, is, after several generations, succeeded

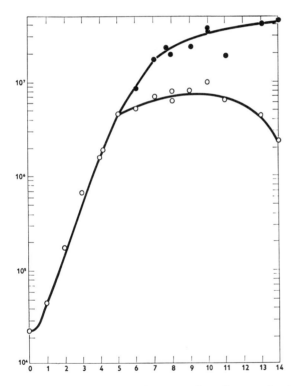

FIGURE 1 Growth of Chinese hamster cells. Cells were plated in replicate plastic petri dishes. Each day, cells from two (or four) dishes were trypsinized and counted on a Coulter counter. The plotted points represent averages of eight counter readings. The maximum cell number attained by the "fed" cultures (closed circles) is not exceeded if the feeding rate is doubled. Open circle: No medium renewal. Closed circle: Medium changed day 5 and daily thereafter. Abscissa, time after subculture; ordinate, cell number. (From Hahn et al., 1968.)

by a plateau in which the percentage of cells capable of division drops below 40 per cent. This plateau is not caused by stale medium, and seems to be the result of contact between the cells. During the exponential phase, there is no cell death, but, during the plateau, about 10 per cent of the cells are lost daily.

Many of these characteristics can be recognized in the CNS—with one important difference; nerve cells are not themselves capable of division. We must therefore identify a feeder population of cells, the ventricular zone, some cells of which differentiate to become nerve cells. Accumulation rates of the latter will depend not only on the size and generation time of this feeder population, but also on the proportion leaving to become nerve cells. It may be noted that the feeder population itself may be increasing or decreasing in size, depending on the proportion differentiating. Control mechanisms could act either on the generation-time parameter or on the proportion lost per cycle.

An example from the CNS is the development of the lateral ventral horn cells of *Xenopus*, the South African clawed toad. These cells, which innervate the limb and girdle muscles, lie in four thin columns (one to each limb) in the ventrolateral region of the gray matter. They are distinct from neighboring cells, both in position and appearance. At most, they number about 6000 per limb and can be enumerated without too much tedium in serial sections. Degenerating cells can be identified by the presence of a pycnotic nucleus. Counts made of dividing cells, either by autoradiography or mitoses, are difficult to interpret because there is no way of identifying which are the divisions of presumptive ventral-horn cells. It should be possible, although it has not yet been attempted, to estimate rates of migration by using autoradiography. The usual precautions must be taken to avoid sampling and counting errors and the inevitable individual variation.

In normal development, ventral-horn cells (Figure 2) can first be recognized and counted at stage 50 (Normal Table of Nieuwkoop and Faber, 1956; early limb bud). As do the cells in vitro, they accumulate up to a peak (stages 53–54, paddle and digit stages), and have a plateau phase, but thereafter, as they mature, they decline in number to about one quarter of the maximum (Beaudoin, 1955; Hughes, 1961; and unpublished observations). At the time of the peak, the cell bodies are closely packed together, with very little cytoplasm. As they differentiate, they become separated by glia. Cell degeneration reaches a peak at the time of most rapid decline (Hughes, 1961). Hughes also estimated the time that a cell takes to degenerate and thus calculated the total number of cells dying during this period. He found it was greater than the difference between the maximum and final number, and concluded that further recruitment of cells was taking place while other cells were dying (turnover). There is, therefore, the surprising observation that initially

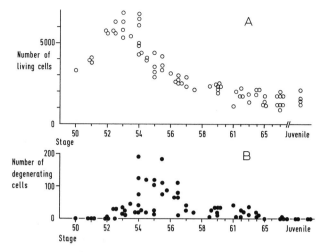

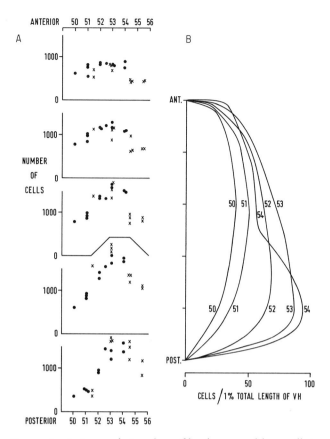

FIGURE 2   A. Counts of the number of living cells in the lumbar ventral horn of *Xenopus* during development. Each point represents one side of one animal. The peak in the number of cells is at stages 53–54 (limb bud transforming into leg). B. Counts of the number of degenerating cells from the same material. They are not seen before the decline in living cell number and are most frequent at stages 54–56 (first movements of the leg).

the nervous system produces many nerve cells that do not survive.

As the tadpole develops, it also grows. This means that, in order to use data from animals of different stages, comparable regions of the spinal cord must first be identified, and the effect of elongation must be allowed for. This can be done by using the points of entry of the dorsal roots as internal markers. These preserve the same relative positions to one another once they are established. In older animals (after stage 53), their relationship to the location of the motor neurons can be examined, and it can be seen that, although there may be some variation in the precise form of the plexus, roots 8, 9, and 10 invariably insert over the lumbar ventral horn; root 7 may, or root 11 may; rarely, both do. However, both sides of an individual are usually identical, and patterns in siblings are usually similar. In this way, using older individuals of the same batch, it is possible to identify the longitudinal extent of the presumptive ventral-horn regions at and before stages at which the measurable extent is indeterminate.

Using this technique, one can subdivide the column of ventral-horn cells into fifths, and these can be compared at different developmental stages, despite elongation. Figure 3A shows some counts of living cells from two batches of tadpoles during the accumulation and peak phase. The shape and position of the peaks in the counts indicate that there is a rough anteroposterior gradient of maturity, as in many other embryonic systems. The anterior peaks are

FIGURE 3   A. Accumulation phase of lumbar ventral-horn cells in developing *Xenopus*. The column of cells has been divided transversely into fifths, and the number of ventral-horn cells present in each has been counted. Stage 50, round limb bud; stage 53, paddle; stage 54, digits. Each point represents one animal. Data taken from two batches of eggs. The counts from one batch (denoted by ×) have been multiplied by 1.285 to match total numbers of cells. The counts demonstrate the characteristic heights of the peaks in each fifth and the occurrence of an anteroposterior gradient in maturity. B. Profiles of the distribution of ventral-horn cells along the longitudinal axis during the period covered by the data of A, plotted on a scale that removes the effect of elongation caused by growth. At stage 54, it can be seen that cells are being lost in the anterior half while still being gained posteriorly (positional turnover).

much flatter and the accumulation rates lower than the posterior ones. The maximum heights of the peaks vary, not according to any simple relation. Table I shows that this relative abundance of cells remains the same after metamorphosis.

The same data can be represented visually by plotting the distribution of cells against the total length of the ventral horn after compensating for elongation. Figure 3B shows that these profiles change in shape between stages 53 and 54, at which time the anterior portion of the ventral-horn column is decreasing in cell number while the posterior is still increasing. Considering the whole ventral horn without subdivision, this would appear as a population in turnover.

To test whether these early temporal relationships are preserved, one can investigate the further development by using the number of degenerating cells in each fifth as an indicator. In each (columns in Table II), the number of degenerating cells rises to a peak and then declines again, just as it does in counts of the whole ventral horn. Each fifth having a characteristic size relative to its fellows, the counts in each fifth have been expressed relative to the final number of cells that remain (row 3 of Table I). These transformed data are shown in Table III to give an index of the number of degenerating cells, seen at one time, per cell that finally matures, and enables comparison of the different-sized populations.

It is apparent from these Tables that the anteroposterior gradient of maturity that was present earlier is no longer absolute. The degenerating cell indexes agree with the primacy of the anterior two-fifths at the start, but the peak of degeneration in the anterior fifth is later than in fifth 2, 3, and 4, indicating slower development, which continues until at metamorphosis (stages 61–64) the anterior fifth is only as advanced as the most posterior portion.

These observations can be summarized as follows:

1. There is a characteristic relative abundance of cells at any level along the anteroposterior axis. This is conserved from the peak onward.

2. There is anteroposterior asynchrony of initial development, such that the anterior end starts first.

3. The anterior fifth develops more slowly than do the posterior four fifths.

It may be that this third observation reflects the markedly smaller number of cells, perhaps by exerting a decreased co-operative action. A similar relationship between the size of a population of developing nerve cells and its rate of growth in vivo can be found in dorsal-root ganglia. In *Xenopus,* ganglion 10 is more mature, larger at all times, and develops faster than ganglion 9. The same is even more true in a comparison of ganglia 10 and 8. An inverse relationship between cell number and their nutritional requirements in vitro is familiar in cell biology (e.g., Eagle and Levintow, 1965).

So too is the occurrence of an anteroposterior gradient in the development of the ventral horn. Hughes (1968) discussed this more fully, with its possible significance. It is a concept relevant to this chapter because, in the adult anuran, there is an approximate somatotopic relationship between the nerve cells along the anteroposterior axis of the ventral horn and the muscles of the limbs in the proximodistal axis; correspondingly, histogenesis and function in the limb progress proximodistally. These gradients are initially set up independently, for they also are found if isolated from one another; they are examples of pre-adaptation.

The characteristic relative abundance of cells along the anteroposterior axis provides a clue toward the solving of an old problem, that of what determines the size of a motor unit. At the most naive level, this will be determined in part by the number of motor neurons available for the muscle. The data presented show that the relative abundance of cells is set up largely, and perhaps totally, before any interaction.

## Two types of turnover and asynchrony

It appears that we should distinguish two types of turnover and two types of asynchrony. Positional asynchrony arises when separate parts of a cell population differentiate at dissimilar times or rates. Local asynchrony occurs when closely neighboring cells differ. Similarly, positional turnover can take place when one part of a tissue is gaining and another part is losing cells; this can be contrasted with local turnover.

It is important, therefore, not only to study a cell population as a whole, but also to examine its geometry. The ventral horn consists of a long, thin column that lies parallel to its feeder population. Metabolic interaction might be expected to be more complete transversely than longitudinally. The effectively interacting population would therefore be small; the accumulation phase can be short, and the cells can possess a high degree of local synchrony. A dorsal root ganglion, however, develops as a compact mass, in which there is every opportunity for interaction among any of the differentiated or dividing cells. The effectively interacting population is, therefore, probably large; the accumulation phase is much longer than that of the ventral horn (Prestige, 1965), and the cells do possess poor local synchrony.

In contrast, retinal cells lie in a thin plate that is surrounded at the rim by the feeder cells. Interaction between cells remote from one another must be difficult, so that they might possess only local synchrony without much positional synchrony.

## Effect of periphery on cell balance sheet

The ventral horn of *Xenopus* will continue to be used to illustrate these effects, although many of the results have been reported by earlier workers (notably Hamburger, Barron, Perri, Kollros, and Hughes) on other animals and in different parts of the nervous system. Typically, the experiments consist of removal of the hind leg on one side of each of a batch of similar larvae, and observation of the subsequent development of the lumbar ventral-horn cells. The opposite

### TABLE I

*Relative Abundance of Living Nerve Cells Along the*
*Length of the Lumbar Ventral Horn During Development*

| | Anterior $\longrightarrow$ | $\longrightarrow$ | $\longrightarrow$ | | Posterior |
|---|---|---|---|---|---|
| % number in each fifth at peak count (2 series) | 12 11 | 18 15 | 23 22 | 25 28 | 22 24 |
| % number in each fifth after metamorphosis (mean ± se. of 7 individuals) | 11.5 ± 1.05 | 18.5 ± 0.55 | 21.3 ± 0.92 | 27.2 ± 0.88 | 21.5 ± 1.38 |
| Number in each fifth after metamorphosis (mean of 7 individuals) | 170 | 272 | 315 | 400 | 310 |

### TABLE II

*Distribution of Degenerating Cells Along the*
*Lumbar Ventral Horn During Development*

| Stage | Mean number of degenerating cells in each fifth | | | | | Total | Number of individuals |
|---|---|---|---|---|---|---|---|
| | Anterior | $\longrightarrow$ | $\longrightarrow$ | $\longrightarrow$ | Posterior | | |
| 52/53 | 6.0 | 9.5 | 7.4 | 3.4 | 1.3 | 27.6 | 7 |
| 54 | 12.6 | 22.6 | 34.2 | 31.0 | 16.8 | 117.2 | 9 |
| 55/56 | 19.4 | 16.9 | 22.1 | 22.4 | 31.2 | 112.0 | 9 |
| 56/57 | 7.3 | 7.6 | 9.3 | 8.3 | 13.8 | 46.3 | 12 |
| 58/59/60 | 6.8 | 7.2 | 7.5 | 5.8 | 8.2 | 35.5 | 11 |
| 61/62/63/64 | 2.5 | 2.3 | 1.8 | 3.1 | 3.3 | 13.1 | 19 |

### TABLE III

*Relative Indices of the Rate of Cell Degeneration*
*Along the Lumbar Ventral Horn During Development*

Relative index for each fifth and stage =

$$\frac{\text{Mean number of degenerating cells seen}}{\text{Mean number of living cells present after metamorphosis}} \times 100$$

| Stage | Anterior $\longrightarrow$ | $\longrightarrow$ | $\longrightarrow$ | | Posterior |
|---|---|---|---|---|---|
| 52/53 | 3.5 | 3.5 | 2.3 | 0.8 | 0.4 |
| 54 | 7.4 | 8.3 | 10.8 | 7.8 | 5.4 |
| 55/56 | 11.4 | 6.2 | 7.0 | 5.5 | 10.0 |
| 56/57 | 4.3 | 2.8 | 2.9 | 2.8 | 4.5 |
| 58/59/60 | 4.0 | 2.6 | 2.4 | 1.5 | 2.6 |
| 61/62/63/64 | 1.5 | 0.8 | 0.6 | 0.8 | 1.1 |

side of the tadpole is not affected and can be used as a control. By doing this operation in different experiments on batches reared to different stages, one can follow the interaction of limb and spinal cord closely.

Ventral-horn cells react in one of three characteristic ways to this operation; the reaction depends mostly on the stage of the animal, but also on the cell's location in the anteroposterior axis.

*Phase I cells.* Small cells, with only a thin rim of cytoplasm. These are unaffected by amputation until the animal develops to the stage at which they become Phase II.

*Phase II cells.* Small, bipolar cells with a little basophilic cytoplasm. They disappear within two to three days after amputation, accompanied by excess pycnotic nuclei.

*Phase III cells.* Larger cells, with well-developed basophilic cytoplasm. All these cells probably have an axon in the ventral roots that is easily stained with silver. After limb amputation, these cells die only after a period that may extend for weeks or months; meanwhile, they undergo chromatolysis.

I shall now summarize the evidence. The amputation counts are presented diagramatically in Figure 4. For the purpose, a linear correction has been made for necessarily incomplete amputations. If the hind limb is removed at any time before stage 53 (paddle), there is no effect until stage 53 (Hughes and Tschumi, 1958). Then the cells apparently become sensitive to the absence of the leg and die off rapidly in the following stages. If the animal is prevented by hypophysectomy from achieving stage 53, the cells survive indefinitely (Race, 1961). Amputation at stages 53–56 causes immediate cell loss. At stage 57 (forelimb emergence) amputation causes only a small immediate cell loss, or none, but it stops the decline in cell number and the associated degenera-

tion, thus causing an apparent excess of cells over the control side. After a month or so, these cells then all die. At stage 61 (climax), the pattern is similar, and the cells take even longer to die. In two-month juveniles, decline of cell number has ceased on both sides at the time of the operation; cell death is delayed four to seven months. Between stages 55 and 56, cell loss follows amputation in two waves (Hughes, 1961): one early (which is comparable with that after amputation at stages 53–54) and one late (comparable with that after operations at stage 57 onward). This late loss seen after amputation at stages 55–56 onward must be the result of a delayed reaction; for if the cells had been dying because the general development of the animal had caused them to acquire acute sensitivity to the absence of the leg, then amputation at that later stage would have led to immediate cell loss. This did not happen. On the contrary, cell loss was then even more delayed. Therefore, the dying cells are showing a delayed reaction, rather than delayed sensitivity, a situation in contrast to amputation before stage 53.

It seems that these three ways of reacting to amputation are phases through which each ventral-horn cell must pass. One can make estimates of the number of cells showing any one type of reaction for each stage, and the sum of these curves is similar to the curve of the total number of cells that are normally within the ventral horn at the same stage (Prestige, 1967).

Amputation, however, is an unphysiological injury, and it could be asked if these results are not merely a nonspecific reflection of that. Evidence that cells die on leaving Phase I if nervous connection is not made with the limb can be obtained without any trauma to the nerve cells themselves (as in the 1958 slit experiment by Hughes and Tschumi). For Phase III cells, both the possibility of hyperplasia (changes opposite in sign to those seen after a decrease in the periphery can be induced by an increase) and the prolonged period between operation and cell death make it difficult to implicate trauma as the cause of hypoplasia. It is also difficult to see how trauma can reduce the rate of degeneration, as after amputation at stages 57 to 61. The possibility that cells develop directly from Phase I to Phase III, and that the reaction characteristic of Phase II is the result of trauma that causes degeneration in some or all of a homogeneous population of Phase III cells, is difficult to disprove. But the fact that Phase I cells do enter a period of acute dependence on contact with the leg suggests that this period could be the Phase II as defined in the acute experiments. Meanwhile, the populations are at least operationally distinct.

Race's (1961) observation that early hypophysectomy prevents cell death in the ventral horn despite amputation indicates that an endocrine action on the nerve cells must precede any interaction with or induction by the leg. It also suggests that the stimulus for Phase I cells to differentiate to Phase II is endocrine in nature. It cannot be caused by the leg

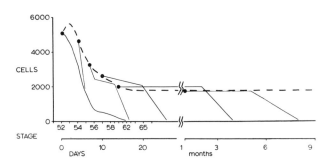

FIGURE 4 Composite diagram of the effects of amputation at differing stages on the subsequent development of the ventral horn. Dashed line, number of cells in normal development; continuous lines, number of cells remaining after amputation at the time of the filled circle. Data corrected for incomplete amputation. Note change of time scale. The data show that cells are lost after amputation either early (Phase II) or late (Phase III) or both.

itself, for cells will die at this stage if they have been previously prevented physically from innervating the leg by a slit barrier (Hughes and Tschumi, 1958). Phase II cells must thereafter have some contact with the leg, although it might not be necessary for each to have an axon if the cells are in metabolic cooperation. The transition from Phase II to Phase III must be at least partially controlled by the leg, and even in Phase III cells retain a dependence, although a diminishing one, on continued contact. The fact that the normal decline in cell number is arrested by amputation at stages 57 and 61 suggests that the stimulus for further differentiation in Phase III, with its associated cell loss, also comes from the peripheral tissue.

Phase I cells themselves are not simply undifferentiated cells but are already determined in their future course of development. They differ in position and morphology from other cells of the intermediate zone; they are incapable of mitosis. There must be a primary transition of ventricular cells to Phase I.

This formulation has one defect as a working hypothesis. Using *Rana pipiens,* Beaudoin (1955) and Race (1961) have reported that ablation of the early limb bud, although not affecting the initial assembly of cells, delayed the decline in cell number that normally accompanies differentiation. This latter observation is not predicted by the hypothesis. One possibility is that there may be peripheral control of the rate of cell migration into the ventral horn (the primary transition). The process might be similar to that suggested by Barron (1946, 1948). This topic is discussed more fully in a closely reasoned review by Kollros (1968).

It remains to ask which cells degenerate in the course of normal development. The most plausible answer is that, at first, they represent Phase II cells that have failed to make, or have lost, contact with the leg. Later, Phase III cells must also be dying. The evidence for this view is presented elsewhere (Prestige, 1967). Unfortunately, no attempt has yet been made to count myelinated and unmyelinated nerve fibers at each stage, so it is not yet known if functional contacts are made and withdrawn when cells die.

## Significance of cell degeneration

We owe to Glücksmann (1951) the recognition that cell death is a common feature of embryonic tissues. Cell death also occurs in culture and in many adult tissues along with continuous replacement. In this respect, the adult nervous system is atypical in that its cells are not replaced.

Glücksmann distinguished between morphogenetic and histogenetic degeneration, the former leading to changes in shape or structure of organs and the latter associated with the differentiation of cells within tissues. The formation of discrete ventral horns by the loss of cells at nonlimb levels in the developing chick (Levi-Montalcini, 1950) and mouse

(Harris, 1965) is an example of morphogenetic degeneration. These distinctions that arise between levels of the cord are, however, independent of the periphery (Wenger, 1951; Straznicky, 1963; Székely, 1963). The decline in cell number in anurans is related only to histogenesis and is profoundly modifiable by the periphery.

Glücksmann suggested that histogenetic degeneration represents one side of a balance mechanism that controls the size of the population. Cell death thus acts as a damper on excess production. The concept implies that a threshold number of cells is exceeded. Earlier, Hamburger and Levi-Montalcini (1949) had suggested that this mechanism was operative in trunk dorsal-root ganglia of the chick embryo and that the periphery was setting the threshold. It appears that this redundancy hypothesis obtains also in the ventral horn of *Xenopus.*

It should be pointed out that this hypothesis implies that the rate of degeneration can either be altered by a change in the rate of cell production or by a change in threshold; these could be independently manipulated.

Is there any other advantage to the animal? Because as cells mature they become less dependent on the limb, any loss of cells will be taking place among the most junior. Cells with large, stable, peripheral fields will be selected for. Hughes (1968) has suggested that selection and rejection may be even more precise, down to the level of specific nerve/muscle contacts. A third possibility is that differentiation of mature nerve cells requires the death of others or that the larval neuromotor system requires more cells than does the adult (smaller peripheral fields).

## Retrograde transfer of information

These experiments reveal some of the phenomena of peripheral control of cell populations in the CNS. The nature of the information passed from leg to ventral-horn cell remains unclear. I have suggested elsewhere (Prestige, 1967) that it takes place retrogradely along motor axons and is probably chemical. Two steps may be involved: (1) a nerve/peripheral tissue recognition factor, and (2) an axonal information transfer system. To call the former Maintenance Factor does little more than state that, in its absence, the nerve cell eventually dies. Unless the converse is also true, such a statement is of little interest. The evidence from hyperplasia would be striking corroboration if it were more repeatable. The literature on retrograde motor hyperplasia, however, is not unanimous (May, 1933; Detwiler, 1933; Hamburger, 1939; Bueker, 1945; Perri, 1957; Hughes, 1962). The disagreement is, perhaps, the result of differing techniques.

Since the demonstration by Weiss and Hiscoe (1948) that material passes in bulk down the axon, much work has been done on the nature and significance of this process. The retrograde transfer, however, has not attracted so much attention. That such a process does occur can be inferred from the following evidence:

1) Removal of the field to be innervated prior to the arrival of the developing nerve fibers causes the death of the presynaptic cells at the time when connection would normally have taken place, both in the peripheral (e.g., Hamburger, 1934) and in the central nervous system (e.g., Harkmark, 1956). This takes place without trauma to the presynaptic cells.

2) The characteristic maturation of motor-nerve fibers in postnatal life depends completely on the successful establishment of neuromuscular connections (Evans and Vizoso, 1951).

3) After section of a muscle nerve in adults, later reinnervation of the muscles can, to some extent, reverse the effects of the lesion on the injured nerve cell (e.g., Weiss et al., 1945; Aitken et al., 1947).

4) It is possible to produce hyperplasia or hypertrophy of nerve centers by permitting them, in suitable circumstances, to innervate an abnormally large periphery (e.g., Bueker, 1945). This also takes place without trauma to the nerve cells.

5) Changes in the condition of a muscle can lead to metabolic responses in the motor neurons. Watson (1969b) implanted the hypoglossal nerve into the sternomastoid muscle, and 70 days later cut the normal innervation (spinal accessory nerve). Hypoglossal motor neurons on that side reacted by an increased RNA synthesis. This also takes place without further trauma to the cells.

In cases 1, 3, and 5, it is possible to exclude the possibility that the necessary information is carried by afferent nerve fibers.

It is also known that labeled metabolites injected into muscle can be carried centripetally along the muscle nerve, both in vitro (Kerkut et al., 1967) and in vivo (Watson, 1968b).

Experiments of these types demonstrate that information is passed from axon tip to nerve-cell body, and that chemical substances also do so. They do not, however, imply that there is necessarily a separate mechanism or species of messenger for each message to be transferred. The coding and decoding specificities may conceivably arise from the specificity of the attachment site; in an analogous manner, the nerve impulse does not carry any idiosyncratic information.

## Epithelio-mesenchymal interactions

Motor neurons are representatives of the basal layer of cells in the stratified epithelium of the neural ectoderm, in that their axon tips are in contact with the basal lamina. All nerve cells have retained some epithelial characteristics in being polarized and in their relationships with neighboring cells. But in three respects the neuromuscular junction is not typical of central synapses, because 1) the former is between un-

like tissues of different germ layer origin, 2) the gap is much wider, and 3) the space has a thickened basement lamina within it.

Interactions between tissues of different origin have been extensively investigated in recent years by in vitro separation and recombination techniques. These inductions (review by Grobstein, 1967) can even take place across a filter into which cell processes may penetrate partly, but close or tight contacts are not made. It seems that extracellular materials produced by the inducing tissue may stabilize the differentiation of the reacting tissue (Hay, 1968). Dodson (1967), for example, showed that epidermis, which normally requires dermis for successful differentiation, could be grown on collagen gels, as a substitute for the collagenous basement lamina.

Interactions of this kind typically show a strong temporal pattern. For instance, whereas presumptive pancreatic endoderm will not differentiate in culture before the six-somite stage in the mouse embryo, it will subsequently, provided it is in contact with its mesoderm (Wessells and Cohen, 1967). Moreover, after the 30-somite stage, the pancreas rudiment can be grown on embryo extract without any mesoderm.

Inductions of epithelium by mesenchyme have different degrees of specificity. For example, thyroid epithelium from 16-day chick embryo was recombined, after disaggregation, with mesoderm from mesentery, ventricle, perichondrium, or its own thyroid capsule. Only with the last did normal lobules develop and the cells cytodifferentiate normally (Hilfer et al., 1967). On the other hand, pancreatic endoderm is much less demanding; it will differentiate successfully with any mesoderm so far tested (Golosow and Grobstein, 1962).

In these three respects, I view the amputation experiments as in vivo separations of neural ectoderm from homologous mesoderm. Interactions normally must take place at or across a wide junctional cleft filled with characteristic extracellular material; they show marked temporal constraints; and they possess specificity functions which vary from nonspecific (Weiss and Hoag, 1946; Guth, 1962) to highly specific selective reinnervation (Mark, 1965; Sperry and Arora, 1965) of muscles. The only radical recombination experiments with heterologous mesoderm are those of Bueker (1948), which led to the discovery of Nerve Growth Factor. However, Straznicky (1963) showed that homografted brachial spinal cord of the chick, if made to innervate a leg, differentiated normally and moved the leg as if it were a wing, suggesting that the information required for spinal cord from limb mesoderm is common to both wing and leg.

## Peripheral influences on cellular metabolism

It is desirable that concepts discussed here be translated into biochemical terms. Techniques are now available for quantitative analysis at the cellular level. As a result, we can now assess the significance of the "chromatolysis" that follows section of or injury to the axon. Changes in total nucleic acid and protein content have been measured in single regenerating nerve cells (Brattgård et al., 1957), and recently Watson (1968a) has measured rates of ribosomal RNA synthesis after nerve section in hypoglossal motoneurons. He found that this increases markedly at one to five days (depending on the amount of axon remaining), reaches a peak, and falls again to or below control level at three weeks. He showed that these changes cannot primarily be the result of loss of contact with muscle, because a similar response occurs after a second injury, when reinnervation has been prevented. Similarly, the drop in RNA synthesis after the peak cannot be due to reinnervation, because it still takes place when reinnervation is delayed. The cycle is triggered by the injury, rather than driven by loss of muscle contact. This form of chromatolysis, then, is not an example of interrupted peripheral induction.

In another study, using botulinum toxin, Watson (1969a) found a different reaction. The lesion in this case was at the site of nerve/muscle interaction without loss of axoplasm (Ambache, 1949; Burgen et al., 1949; Brooks, 1956). Changes in RNA synthesis and membrane characteristics occur in parallel both in nerve cell and in muscle fiber. It is noteworthy that the changes that occur as a result of failure of nerve/muscle interaction are similar to the changes observed in cells in vitro on the release of contact inhibition. Thus it is tempting to speculate that some of the developmentally significant changes that take place when nerve fiber reaches muscle should be regarded as examples of contact inhibition.

## Acknowledgments

I am very grateful to the [British] Medical Research Council for their support and to Miss Sheila Bourn for her invaluable help.

### REFERENCES

AITKEN, J. T., M. SHARMAN, and J. Z. YOUNG, 1947. Maturation of regenerating nerve fibres with various peripheral connexions. *J. Anat.* 81: 1–22.

AMBACHE, N., 1949. The peripheral action of *Cl. botulinum* toxin. *J. Physiol.* (*London*) 108: 127–141.

BARRON, D. H., 1946. Observations on the early differentiation of the motor neuroblasts in the spinal cord of the chick. *J. Comp. Neurol.* 85: 149–169.

BARRON, D. H., 1948. Some effects of amputation of the chick wing bud on the early differentiation of the motor neuroblasts in the associated segments of the spinal cord. *J. Comp. Neurol.* 88: 93–127.

BEAUDOIN, A. R., 1955. The development of lateral motor column cells in the lumbo-sacral cord in *Rana pipiens,* I. Normal development and development following unilateral limb ablation. *Anat. Rec.* 121: 81–95.

BRATTGÅRD, S-O., J.-E. EDSTRÖM, and H. HYDÉN, 1957. The chemical changes in regenerating neurons. *J. Neurochem.* 1: 316–325.

BROOKS, V. B., 1956. An intracellular study of the action of repetitive nerve volleys and of botulinum toxin on miniature end-plate potentials. *J. Physiol.* (*London*) 134: 264–277.

BUEKER, E. D., 1945. Hyperplastic changes in the nervous system of a frog (Rana) as associated with multiple functional limbs. *Anat. Rec.* 93: 323–331.

BUEKER, E. D., 1948. Implantation of tumors in the hind limb field of the embryonic chick and the developmental response of the lumbosacral nervous system. *Anat. Rec.* 102: 369–389.

BURGEN, A. S. V., F. DICKENS, and L. J. ZATMAN, 1949. The action of botulinum toxin on the neuro-muscular junction. *J. Physiol.* (*London*) 109: 10–24.

DETWILER, S. R., 1933. Experimental studies upon the development of the amphibian nervous system. *Biol. Rev.* (*Cambridge*) 8: 269–310.

DODSON, J. W., 1967. The differentiation of epidermis, I. The interrelationship of epidermis and dermis in embryonic chicken skin. *J. Embryol. Exp. Morphol.* 17: 83–105.

EAGLE, H., and L. LEVINTOW, 1965. Amino acid and protein metabolism, I. The metabolic characteristics of serially propagated cells. *In* Cells and Tissues in Culture, Vol. I (E. N. Willmer, editor). Academic Press, London, pp. 277–296.

EVANS, D. H. L., and A. D. VIZOSO, 1951. Observations on the mode of growth of motor nerve fibers in rabbits during postnatal development. *J. Comp. Neurol.* 95: 429–461.

GLÜCKSMANN, A., 1951. Cell deaths in normal vertebrate ontogeny. *Biol. Rev.* (*Cambridge*) 26: 59–86.

GOLOSOW, N., and C. GROBSTEIN, 1962. Epitheliomesenchymal interaction in pancreatic morphogenesis. *Develop. Biol.* 4: 242–255.

GROBSTEIN, C., 1967. Mechanisms of organogenetic tissue interaction. *Nat. Cancer Inst. Monogr.* 26: 279–299.

GUTH, L., 1962. Neuromuscular function after regeneration of interrupted nerve fibers into partially denervated muscle. *Exp. Neurol.* 6: 129–141.

HAHN, G. M., J. R. STEWART, S-J. YANG, and V. PARKER, 1968. Chinese hamster cell monolayer cultures, I. Changes in cell dynamics and modifications of the cell cycle with the period of growth. *Exp. Cell Res.* 49: 285–292.

HAMBURGER, V., 1934. The effects of wing bud extirpation on the development of the central nervous system in chick embryos. *J. Exp. Zool.* 68: 449–494.

HAMBURGER, V., 1939. Motor and sensory hyperplasia following limb-bud transplantations in chick embryos. *Physiol. Zool.* 12: 268–284.

HAMBURGER, V., and R. LEVI-MONTALCINI, 1949. Proliferation, differentiation and degeneration in the spinal ganglia of the chick embryo under normal and experimental conditions. *J. Exp. Zool.* 111: 457–501.

HARKMARK, W., 1956. The influence of the cerebellum on development and maintenance of the inferior olive and the pons. *J. Exp. Zool.* 131: 333–371.

HARRIS, A. E., 1965. Differentiation and degeneration in the motor horn of the foetal mouse. University of Cambridge, Cambridge, England, doctoral thesis.

HAY, E. D., 1968. Organization and fine structure of epithelium and mesenchyme in the developing chick embryo. *In* Epithelial-Mesenchymal Interactions (R. Fleischmajer and R. E. Billingham, editors). The Williams and Wilkins Co., Baltimore, pp. 31–55.

HILFER, S. R., E. K. HILFER, and L. B. ISZARD, 1967. The relationship between cytoplasmic organization and the epithelio-mesodermal interaction in the embryonic chick thyroid. *J. Morphol.* 123: 199–212.

HUGHES, A., 1961. Cell degeneration in the larval ventral horn of *Xenopus laevis* (Daudin). *J. Embryol. Exp. Morphol.* 9: 269–284.

HUGHES, A., 1962. An experimental study on the relationships between limb and spinal cord in the embryo of *Eleutherodactylus martinicensis. J. Embryol. Exp. Morphol.* 10: 575–601.

HUGHES, A., 1968. Development of limb innervation. *In* Growth of the Nervous System/A Ciba Foundation Symposium (G.E. W. Wolstenholme and M. O'Connor, editors). J. and A. Churchill, Ltd., London, pp. 110–117.

HUGHES, A., and P. A. TSCHUMI, 1958. The factors controlling the development of the dorsal root ganglia and ventral horn in *Xenopus laevis* (Daud). *J. Anat.* 92: 498–527.

KERKUT, G. A., A. SHAPIRA, and R. J. WALKER, 1967. The transport of $^{14}$C-labelled material from CNS ⇌ muscle along a nerve trunk. *Comp. Biochem. Physiol.* 23: 729–748.

KOLLROS, J. J., 1968. Order and control of neurogenesis (as exemplified by the lateral motor column). *Develop. Biol., Suppl.* 2: 274–305.

LEVI-MONTALCINI, R., 1950. The origin and development of the visceral system in the spinal cord of the chick embryo. *J. Morphol.* 86: 253–283.

MARK, R. F., 1965. Fin movement after regeneration of neuromuscular connections: An investigation of myotypic specificity. *Exp. Neurol.* 12: 292–302.

MAY, R. M., 1933. Réactions neurogéniques de la moelle à la greffe en surnombre, ou à l'ablation d'une ébauche de patte postérieure chez l'embryon de l'Anoure, *Discoglossus pictus,* Otth. *Bull. Biol. Fr. Belg.* 67: 327–349.

NIEUWKOOP, P. D., and J. FABER, 1956. Normal Table of *Xenopus laevis* (Daudin). North-Holland Publ. Co., Amsterdam.

PERRI, T., 1957. Sul trapianto di abbozzi di arti negli Anfibi anuri con particolare riguardo alle conseguenze sul sistema nervoso centrale e periferico. *Riv. Biol.* (*Perugia*) 49: 361–417.

PRESTIGE, M. C., 1965. Cell turnover in the spinal ganglia of *Xenopus laevis* tadpoles. *J. Embryol. Exp. Morphol.* 13: 63–72.

PRESTIGE, M. C., 1967. The control of cell number in the lumbar ventral horns during the development of *Xenopus laevis* tadpoles. *J. Embryol. Exp. Morphol.* 18: 359–387.

RACE, J., JR., 1961. Thyroid hormone control of development of lateral motor column cells in the lumbo-sacral cord in hypophysectomized *Rana pipiens. Gen. Comp. Endocrinol.* 1: 322–331.

SCHARRER, B., 1968. Neurosecretion. XIV. Ultrastructural study of sites of release of neurosecretory material in blattarian insects. *Z. Zellforsch. Mikroskop. Anat.* 89: 1–16.

SPERRY, R. W., and H. L. ARORA, 1965. Selectivity in regeneration of the oculomotor nerve in the cichlid fish, *Astronotus ocellatus. J. Embryol. Exp. Morphol.* 14: 307–317.

STRAZNICKY, K., 1963. Function of heterotopic spinal cord segments investigated in the chick. *Acta Biol. Acad. Sci. Hung.* 14: 143–155.

SZÉKELY, G., 1963. Functional specificity of spinal cord segments in the control of limb movements. *J. Embryol. Exp. Morphol.* 11: 431–444.

WATSON, W. E., 1968a. Observations on the nucleolar and total cell body nucleic acid of injured nerve cells. *J. Physiol.* (*London*) 196: 655–676.

WATSON, W. E., 1968b. Centripetal passage of labelled molecules along mammalian motor axons. *J. Physiol.* (*London*) 196: 122P–123P.

WATSON, W. E., 1969a. The response of motor neurones to intramuscular injection of botulinum toxin. *J. Physiol.* (*London*) 202: 611–630.

WATSON, W. E., 1969b. Some metabolic responses of motor neurones to axotomy and to botulinum toxin after nerve transplantation. *J. Physiol.* (*London*) 204: 138 P.

WEISS, P., M. V. EDDS, and M. CAVANAUGH, 1945. The effect of terminal connections on the caliber of nerve fibers. *Anat. Rec.* 92: 215–233.

WEISS, P., and H. B. HISCOE, 1948. Experiments on the mechanism of nerve growth. *J. Exp. Zool.* 107: 315–395.

WEISS, P., and A. HOAG, 1946. Competitive reinnervation of rat muscles by their own and foreign nerves. *J. Neurophysiol.* 9: 413–418.

WENGER, B. S., 1951. Determination of structural patterns in the spinal cord of the chick embryo studied by transplantations between brachial and adjacent levels. *J. Exp. Zool.* 116: 123–163.

WESSELS, N. K., and J. H. COHEN, 1967. Early pancreas organogenesis: Morphogenesis, tissue interactions, and mass effects. *Develop. Biol.* 15: 237–270.

# 9

# In Vitro Study of Developing Neural Tissue and Cells: Past and Prospective Contributions

SILVIO S. VARON

THE NEURAL TISSUE is extremely complex. Not only does it contain a far greater number of cell types than any other tissue, but the intercellular relationships that permit it to function as an integrated society must be of a highly critical and specific order. So, too, must be the dynamic processes that underlie the formation of even the simplest neural network capable of an elementary "behavioral" task. This fundamental complexity of neural tissue is also the greatest obstacle to its experimental analysis. One way to reduce it is to abstract a portion of neural tissue and study it in isolation under conditions in which it can be kept alive and functioning, in other words to establish a *neural culture*.

A neural culture would have a number of advantages. Cellular interactions would be restricted to those taking place within the cultured population, and the complex, usually obscure, influences of a living organism would be replaced by those of a limited environment (substrate and medium). Selected agents (hormones, drugs, ions) could be applied through the culture medium without concern as to whether, or to what extent, the agent is allowed to reach the tissue or whether its effects may have been mediated through some other tissue. Cells in the living stage would be directly accessible to visual observation and recording (e.g., phase contrast microphotography or time-lapse cinematography) and to electrophysiological and biochemical investigations. Finally, correlations pertaining to the same cells would become possible among the data obtained by different techniques, including the ultimate analysis of the fixed preparation by staining, autoradiographic, and electron-microscopic methods. On the other hand, a number of potential liabilities have raised great criticisms and generated even greater frustrations. One is the concept that cells, transferred from their natural habitat to the artificial environment of the culture, may permanently lose their specialized traits (de-differentiation) or fail to proceed in their normal differentiation (arrested development). A related problem is the difficulty encountered in selecting or establishing adequate parameters to assess the conditions of a culture at both cellu-

lar and supracellular levels. Finally, great variability may be expected from replicate cultures, reflecting biological variations in the source tissue, uncontrolled differences in the damage inflicted to it while establishing the cultures, and the general lack of adequate knowledge about its requirements and susceptibilities in culture.

Advantages and liabilities apply differently to tissue cultures and to cell cultures. A *tissue* (or *explant*) culture is set up by explanting in vitro a fragment of an organ or a tissue small enough to presumably allow adequate exchanges with the medium. The explant retains, at least initially, the original organotypic relationship among its component cells, but the individual cells within it cannot be directly visualized and selectively approached until some degree of resolution of the explanted mass is attained in culture. A *cell* culture, on the other hand, is started by seeding in vitro a cell suspension obtained by dissociation of the source tissue. This dissociation, whether achieved by mechanical, chemical, or enzymatic means, disrupts the original organization of the tissue and the numerical balance among its various cell types and also modifies the morphology and the state of the individual cells. It does, however, provide for an immediate visualization of, and direct access to, each element of the culture. Moreover, the cultured cells reorganize themselves into some "histiotypic" patterns that might mimic, or substitute for, the original "organotypic" relationships. In so doing, they could offer clues to normal histogenetic processes and to possible dependencies of certain cellular properties on specific social interactions. In cultures of both types, the cellular material is held in a semisolid medium, most commonly a plasma clot, or adheres to a solid surface, usually a collagen-coated coverslip. The physical and chemical nature of the substrate and the mode by which the cultured material is anchored are important in terms of both the accessibility and the behavior of the cells; this is particularly relevant in the case of dissociated cell cultures.

Defined media (balanced salt solutions enriched with glucose, amino acids, vitamins, and antibiotics) fail to support neural cultures unless supplemented by more complex, and much less defined, biological fluids, such as various fetal or adult sera, embryonic extracts, ascitic fluids, and so on. The role of these supplements may be to provide not only greater

SILVIO S. VARON Department of Biology and School of Medicine, University of California at San Diego, La Jolla, California nutritional support but also some specific agents necessary to the survival, differentiation or proliferation, or both, of the cultured cells. One classical example of such agents is the Nerve Growth Factor (NGF). NGF is a fascinating protein, or group of proteins, which specifically promotes growth and differentiation of peripheral sensory and sympathetic ganglia. Its sources, molecular properties, and biological activities have been amply reviewed in recent times (Levi-Montalcini, 1966; Levi-Montalcini and Angeletti, 1968; Shooter and Varon, 1970; Shooter, this volume). Through tissue-culture techniques NGF was demonstrated to act directly on the ganglionic targets and the only available assay of NGF activity has been provided.

Sixty-three years have passed since the first successful attempts to culture neural tissue in vitro (Harrison, 1907). With only a few, but impressive, exceptions, efforts to take advantage of the features specifically provided by the new approach were deferred in favor of attempts to achieve conditions under which the cultured neural tissue would mimic, at least morphologically, its normal counterpart. It took more than 50 years to obtain such a demonstration for explant cultures and, thus, to destroy the myth of an unavoidable de-differentiation of neural tissue in vitro. Neural cell cultures, on the other hand, have begun to receive some serious interest only in the past 10 years. Much of the explant culture work has been covered recently (Murray, 1965; Crain, 1966). In the present review, cell-culture studies are examined in some detail against the background of our knowledge of explant cultures, with respect first to peripheral and then to central neural tissues.

## Peripheral neural tissue

The peripheral neural tissue most extensively studied in both explant and cell cultures is the spinal sensory ganglion (dorsal-root ganglion). Its cellular composition is relatively simple. There are two populations of *neurons*, differing in size, intraganglionic distribution, developmental time course, and functional connections (Hamburger and Levi-Montalcini, 1949); only one population (the smaller, mediodorsally situated nerve cells) is reported to be affected by NGF, and its responsiveness is confined to a relatively short period of its embryonic life (in the chick, between six and 14 days in ovo). *Capsule cells* make up the periganglionic capsule and intraganglionic septs, are assumed to be of mesenchymal origin, and, in culture, are variably called fibroblasts, fibrocytes, or connective cells. The term *satellite cells* is used generically to cover nonneuronal intraganglionic elements of presumed neural-crest derivation; they comprise cells closely apposed to ganglionic neurons, or pericytes, and cells less specifically situated in the interneuronal spaces. *Schwann cells,* also assumed to be neural crest derivatives and possibly a specialized version of satellite cells, are found in specific

association with nerve fibers and are responsible for their myelination; in cultures, the term "spindle cells" often describes both fibroblast and Schwann cells, in view of their similarly elongated shape.

EXPLANT CULTURES   Gross examination of these cultures shows an early (10 to 15 hours) and progressive migration of spindle cells, but not of neurons, out of the explant, a slightly delayed (24 hours and onward) radial outgrowth of wavy and branching nerve fibers from the periphery of the explant, and a progressive degeneration (over the first week) of a number of neurons within the explant. As a result of spindle-cell migration and neuronal degeneration, the explant becomes progressively thinner; by the end of one week, cells within it can be visualized and cellular developments examined in detail over the next several weeks.

At the end of week 1, the soma of surviving neurons exhibits a "chromatolytic" pattern (nucleus situated eccentrically, basophilic material homogenous and confined to the periphery of the cell), which is typical of neurons subjected to anatomic or functional lesions. The following weeks show gradual enlargement of the somata, centralization of the nuclei, formation and centripetal dissemination of basophilic masses ("flakes") until Nissl patterns characteristic of mature neurons are achieved. Pericytes are tightly associated with the maturing perikarya. This *cytotypic* maturation does not proceed synchronously in all the surviving neurons; by the end of week 2 most of them are at some intermediate stage and by week 4 some mature neurons are present. Also by the end of week 1, axons cease to elongate and begin to get thicker. Schwann cells aligned along the nerve fibers become more firmly adherent and finally envelop them. A continuous fibrous capsule is formed around the neuronal soma with its attached pericytes (second to third week) and gradually extends distad into a sheath surrounding the axon with its adhering Schwann cells. As this *organotypic* maturation proceeds (weeks 3 and 4), myelin begins to appear in segments related to individual Schwann cells. The fine analysis of this orderly sequence of events in explant cultures of chick-embryo sensory ganglia (Peterson and Murray, 1960), later confirmed in detail with fetal rat ganglia (Bunge et al., 1967b), has led to the suggestion that both cytotypic and organotypic maturations depend on the interaction among different ganglionic elements, with specified relative positions and obligatory developmental steps.

The composition of the medium is also critical. Successful maturation was demonstrated in media heavily supplemented with chick-embryo extract. Additional supplementation with ultrafiltrate of human placental serum and bovine serum improved the longevity of cultures but did not appear essential to their maturation. On the other hand, totally unsupplemented medium allowed the survival of some neurons for only up to three weeks, with apparently normal fiber outgrowth and relatively advanced soma maturation, but no formation of fibrous capsule and sheath or of myelin. Addition of NGF to an otherwise unsupplemented medium produces characteristic, dose-dependent changes of the early behavior of the explant. Spindle-cell migration is inhibited until after two days in vitro. Fiber outgrowth is prominent by as early as 12 hours, and by 24 hours a dense halo of radially outgrowing, straight, and profusely branching fibers is established; it is the extent of this fiber halo that is used as the basis of current NGF bioassays. Finally, extensive degeneration of neurons within the explant has not been observed up to three days. No detailed, long-term study has yet been reported for these cultures, so that information on their cytotypic and organotypic maturation in vitro is still missing. Another, very intriguing, example of the influence of environmental agents on the evolution of neural cultures has been reported recently (Murray and Benitez, 1968). Incorporation of up to 25 per cent of heavy water ($D_2O$) in a supplemented medium markedly enhanced growth and proliferation of supportive cells and accelerated the growth and maturation of neuronal elements. The effects of $D_2O$, studied in detail with cultures of sympathetic ganglia (the other target tissue of NGF), have been also observed with sensory ganglia, hypothalamus, cerebellum, and cerebral cortex explants.

The demonstration of an adequate morphological maturation of sensory ganglia in explant cultures (Peterson and Murray, 1960) has been followed by the demonstration of the functional competence of their neurons (Crain, 1965). Resting potentials (−40 to −65 millivolts) and injury discharges, but no spontaneous activity, were recorded. Electrical stimulation elicited characteristic action potentials with spikes of 50 to 95 millivolts in amplitude and two to five milliseconds in duration, followed by a long hyperpolarization phase. Slow-rising prepotentials and subthreshold local responses were also observed. The bioelectric activity could be detected as early as seven days and as late as seven weeks in culture, lasted for up to two minutes and was often rather complex, apparently involving direct soma stimulation, or indirect soma activation by neurite-propagated spikes, or both. Many characters of the responses were comparable with those reported for various types of neurons *in situ* or shortly after isolation. Considerable difficulties were encountered in the direct visualization and the penetration of the cells because of relatively tough layers covering the entire culture and fibrous sheaths surrounding the individual neurons.

DISSOCIATED CELL CULTURES The pioneer work (Nakai, 1956) was carried out with eight- to 12-day sensory ganglia of chick embryo, dissociated enzymatically and cultured in a plasma clot and medium supplemented with chick-embryo extract and human placental serum. Time-lapse cinematography was used extensively for studying the cultures. The dissociated neurons were rounded and generally free of pericytes. Within 12 hours, surviving neurons began to show multiple cytoplasmic outpushings, most of which were rapidly withdrawn while one or more developed into growing processes; later, neurites frequently subdivided into two main branches, sprouted out a number of processes in a whiplike fashion, or sent recurrent collaterals back to the soma or the neuritic stem of origin. The neurons showed no mobility and only occasional passive displacements. Their somata exhibited typical chromatolytic patterns, with no signs of cytotypic maturation within the life span of the culture (up to 19 days). Fibrocytes were large, spindle-shaped, and highly mobile, and they often migrated across nerve fibers. Schwann cells, recognizable by their persistent and extreme slenderness, had rapid, ripple-like movements when isolated; they tended to align themselves permanently in series on the neurites, along which they exhibited only snail-like sliding excursions. No satellite cells were described in these cultures.

Earlier studies of the fiber outgrowth of explant cultures had suggested (Weiss, 1934) that growing fibers orient themselves along ultrastructural patterns in the colloidal substrate of the culture, and that such patterns are themselves influenced by the activity of the cultured cells (dehydration, proteolysis, and so on). Local changes in the liquidity of the substrate induce *fasciculation,* that is, the formation of bundles or plexuses, or both, among different nerve fibers, a phenomenon that mimics organotypic relationships among nerve processes in vivo. Extensive pinocytotic and other membrane activities had been described in the advancing tip, or *growth cone,* of growing fibers (Lumsden, 1951). With the new system of dissociated cell cultures, these studies were carried out in considerably greater detail (Nakai and Kawasaki, 1959; Nakai, 1960). Growth cones continually sprouted a number of thin "filopodia" feeling out the immediate surroundings, some retracting, others expanding into a new growth cone. This process of filopodial "palpation" was particularly evident whenever a growth cone approached within 20 $\mu$ of an obstacle. Filopodial contacts became consolidated into persisting connections with some obstacles (Schwann cells, own soma, or neuritic stem) but not with others (fibroblasts, macrophages), suggesting that selective adhesiveness may play a role in the interaction of nerve fibers with other cells or cellular structures (Nakajima, 1965). Fasciculation among fibers from different neurons, although favored by less-solid substrate gels, actually resulted from an active process of filopodial association between two growth cones, or a growth cone and a fiber stem, or even two fiber stems. Fasciculation was transitory, in that bundles would resolve again, or fan out into plexuses, or redistribute their component fibers into new fascicular patterns.

An important progress in the technique of dissociated cell cultures came with the replacement of the plasma clot with a collagen surface. The resulting· cultures exhibited nerve-fiber growth, Schwann cell alignment, and fasciculation entirely comparable with those of the previous system; the monolayered, rather than three-dimensional, distribution of their cells was, however, considerably more suitable for autoradiographic, electrophysiological, and other investigations. The use of collagen-supported ganglionic cell cultures (Utakoji and Hsu, 1965) provided new evidence in favor of another concept advanced by Weiss (Weiss and Hiscoe, 1948), namely, that elongation, thickening, and maintenance of a nerve fiber are sustained by a continuous, proximodistal flow of axoplasm produced exclusively in the neuronal soma. After short incubations with radio-uridine or radio-amino acids, cultured neurons became labeled only in their perikaryon. Longer exposures to the latter, but not the former, precursor resulted in progressively longer portions of the neurites also being labeled. The neuritic labeling proceeded in a proximodistal direction and was strictly proportional to the exposure times. Thus, it appeared that RNA and protein were both synthesized only in the soma, and that protein, but not RNA, was transported down the neurite at rates (10 mm per day) approximately 10 times faster than the rate of neurite elongation. The same cultures yielded another interesting piece of information, namely, that even neuronal somata of dissociated cells, having no recognizable association with pericytes, could proceed within two weeks to a full cytotypic maturation. The reverse conclusion had been suggested in a contemporary study (Shimizu, 1965), which, for the first time, described the behavior, in cell cultures, of epithelioid elements reasonably interpretable as satellite cells. It must be recognized, however, that different results may well derive from differences in the cellular composition of the dissociates (age of the source ganglia, dissociation technique) and the chemical make-up of the media used.

The importance of the medium composition was again impressively stressed by experiments with NGF (Levi-Montalcini and Angeletti, 1963). Neurons dissociated from eight- to 10-day sensory ganglia of chick embryo failed to grow fibers or to survive beyond the second day when cultured in a medium unsupplemented or supplemented only with horse serum. In contrast, with NGF (50 units per milliliter) added even to the unsupplemented medium, a large number of neurons put out processes by 12 hours. By 24 hours, an extensive fiber network had been produced, at the nodes of which nerve cells stood out, singly or in small clusters, on a background of supportive cells. It was suggested that NGF played an essential role in the survival and growth of these ganglionic neurons, and that the effectiveness of various biological fluids to support cultures of explanted, as well as dissociated, ganglia was related to their possible NGF content. Other investigators examined NGF-supported cell cultures over a longer period of time (Cohen et al., 1964), starting with a ganglionic harvest apparently nearly free of supportive cells. The small neuronal clusters, observed earlier at the nodes of the well-developed nerve-fiber network, became very dense by the end of one week. The neurons within them were no longer easily observable and had become more rounded. The fiber network had grown considerably thicker, and nerve bundles connected the clusters to one another. By three to four weeks, the entire neuronal population was aggregated in a few, large, ganglia-like masses surrounded by fiber networks and interconnected by fiber trunks. Shortly afterward, those interconnecting trunks became detached from the vessel's surface, the ganglia-like structure quickly degenerated, and non-neural cells, until then barely detectable, started to proliferate vigorously. Stained sections of the larger aggregates showed fully mature neuronal somata, long and poorly branched neurites and numerous finer fibers with a high affinity for silver, and a large number of smaller cells filling the interneuronal and interneuritic spaces. Ultrathin sections, examined by electron microscopy, exhibited a remarkable similarity to corresponding materials from intact, freshly dissected ganglia.

Although aggregation into histiotypic patterns was a well-described biological property of dissociated cells in suspension (Moscona, 1962), the report by Cohen et al. (1964) was the first description of similar events with neural cells cultured in a monolayer. The investigators suggested that a random migration of the initially dispersed neurons gave rise to the early clusters, which would eventually coalesce into the larger pseudogangliar masses. A large number of non-neuronal cells, however, were found in these later aggregates, and it is a general observation that cultured ganglionic neurons, whether isolated or at the margin of an explant, have little if any mobility. A somewhat different view was suggested by recent work (Varon and Raiborn, in preparation) using similar eight-day ganglionic dissociates, prepared relatively free of capsule cells, and a thin plasma layer (Varon and Raiborn, 1969) as the culture substrate. With no NGF in the unsupplemented medium, none of the neurons produced any fiber (Figure 1A). In the presence of NGF (7S species, 1–100 units per milliliter) neurons developed, within 24 hours, striking processes with all the features described by other workers, namely, extensive branching (Figure 1B–D), recurrent collaterals (Figure 1E), fasciculation (Figure 1F), and even the classical T-cell morphology of ganglionic sensory neurons in vivo (Figure 1C). With or without NGF, the non-neuronal population consisted almost exclusively of very slender spindle cells, presumably Schwann elements, typically linking with one another in long, linear or branched chains (Figure 2A–C) and forming, in one to two days, a loose cell network extended throughout the culture. It was at the nodal points of this cell network

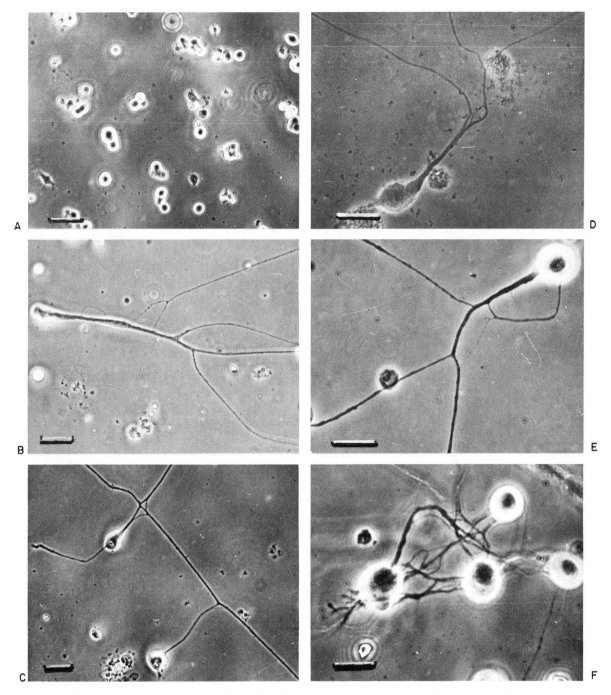

FIGURE 1  Dissociated sensory ganglia (8-day embryo) in culture: neurons. Selective dissociation by short trypsin treatment, yielding a harvest essentially free of capsule cells. Rose-chamber cultures, thin plasma-layer system. Time in culture: 1 day. Phase contrast. Bars = 20 $\mu$. A. In the absence of NGF. B to F. In the presence of 50 units/ml of NGF (7S).

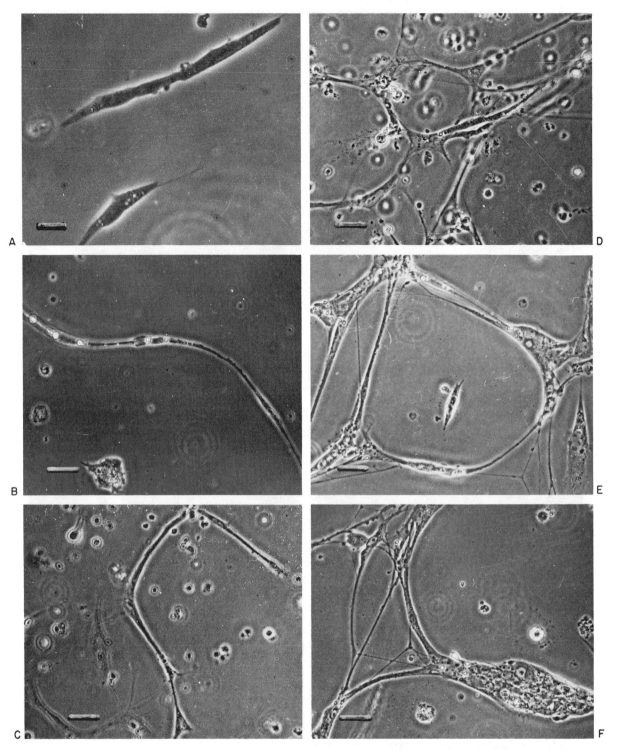

FIGURE 2 Dissociated sensory ganglia (8-day chick embryo) in culture: spindle cells and cell clusters. Materials as in Figure 1. Phase contrast. Bars = 20 μ. A, B, C. Chains of spindle cells: 1 day in culture. D, E, F. Nodal clusters of spindle and nerve cells: 2 days in culture with NGF.

that, by the second day, the NGF-supported cultures showed a number of cell aggregates ranging from tiny clusters to medium-sized patches (Figure 2D–F). These aggregates contained both neurons and spindle cells and were still essentially arranged in a monolayer. By 72 hours, one could observe, side by side within the same culture, large but still monolayered patches (Figure 3A, D), plurilayered but well-contoured aggregates (Figure 3B, E), and dense masses similar to the pseudoganglia described by Cohen et al. (Figure 3C, F). Silver-staining applied directly to the cultures showed the profuse intermingling of spindle cells, neurons, and nerve fibers both outside (Figure 4A, B) and within (Figure 4C) the monolayered patches. Aggregation in these cultures, therefore, appeared to have occurred through a special interaction between spindle and nerve cells, proceeding by progressive recruiting from two- to three-dimensional structures. The "organotypic" value of these "histiotypic" structures remains to be examined.

Recently, another type of intercellular organization has been described for dissociated ganglionic cell cultures, in which both cytotypic maturation and bioelectric competence were successfully demonstrated. Scott et al. (1969) cultured whole dissociates from 10-day-old ganglia, using collagen, a $CO_2$-controlled atmosphere and a complex but still defined medium moderately supplemented with fetal calf serum. After seeding, the rounded neurons were recognizable by their large size (30 $\mu$ in diameter) and bright birefringence. By 24 hours, a reticular background of stretched-out fibroblastic cells was evident, together with an extensive growth of nerve fibers. By three days, a confluent sheet had been achieved by fibroblasts arrayed in parallel to one another; the neurons were either still single or clumped in small nodes often interconnected by nerve fibers, an arrangement persisting for as late as five weeks. At five weeks, sections perpendicular to the culture plane showed the supportive cells to have grown to a plurilayered, connective-like tissue carpet, with cell-free regions in the middle strata. Most neurons had remained at the surface. Their somata showed intermediate or final stages of cytotypic maturation. Their processes were very long, with repeated bifurcations and extensive fasciculation, and formed a profuse network not only all over the surface but also within the intercellular spaces of the connective sheet. No myelinating fibers were observed. Some 300 neurons, five weeks in culture, were examined with intracellular microelectrodes under full visual control. More than 80 per cent of them had resting potentials of $-40$ to $-55$ millivolts, their unimodal size distribution suggesting a homogeneous population. Injury discharges appeared in all of them and subsided in some cells within one minute. Electrical stimulation was applied to 70 cells and in 44 of them gave rise to an action potential, with spike amplitudes of from 60 to 85 millivolts and durations of from 1.5 to 4 milliseconds, mostly followed by a long hyperpolarization phase. Subthreshold responses were obtained with lowered stimulus strength. In a few cases, stimulation could be applied peripherally to the nerve fiber; conduction velocities ranged from 0.1 to 0.6 meter per second.

In our laboratory (Marchiafava and Varon, in preparation), whole dissociates of 11-day embryonic ganglia yielded collagen-supported and $CO_2$-equilibrated cultures grossly, but not entirely, similar to those described by Scott et al. (1969). By five days, the neurons (singly or in small groups) were embedded in, but slightly protruding from, an almost confluent monolayer of fibroblasts. Most neurons had diameters of from 20 to 25 $\mu$, with pericytes adhering to only a few of the larger cells. Schwann cells were mainly in long chains within the fibroblastic sheet, possibly tracing the poorly visible nerve-fiber network. By 10 and up to 20 days, the fibroblastic sheet was fully confluent and very tightly packed (Figure 5A), with overt indications of plurilayered stacking. The neurons had retained their previous relationship to the sheet (Figure 5B), but only rarely exhibited a mature somal pattern. Attempts to record intracellularly from even the larger cells cultured for up to 20 days were unsuccessful; cultures older than 20 days have not yet been investigated. A marked contrast, however, was presented by cultures incubated in the presence of NGF (7S species, 100 units per milliliter). The fibroblastic sheet developed normally, but the neurons were strikingly more numerous (Figure 5C). Nerve fibers and Schwann cells lay clearly *over* the surface of the fibroblastic carpet (Figure 5D), while the neuronal perikarya, brightly glowing under phase (Figure 5C) and very plump (Figure 6C), almost ballooned on top of the culture (Figure 6A, B). Neuronal sizes had a bimodal distribution; the smaller cells (10 to 15 $\mu$ in diameter) often appeared in clusters (Figure 6D) and exhibited a chromatolytic somal pattern, whereas the larger cells (20 to 25 $\mu$ in diameter) had a sizable proportion of mature somata even as early as five days in culture. No pericytes could be observed around any of them. Strings of aligned Schwann cells etched portions of the extensive and conspicuous fiber network spread over the fibroblastic layer (Figure 6E, F). Bioelectric activity was consistently observed after about 10 days in culture and occasionally as early as the fifth day. About 100 neurons, 10 to 20 days in culture and with diameters ranging from 25 to 35 $\mu$, were tested intracellularly, with the use of one microelectrode for both stimulation and recording. Most of them exhibited a resting potential in the range of $-35$ to $-45$ millivolts (a value range probably lowered by the occurrence of junction potentials at the tip of the microelectrode) and injury discharges upon impalement that subsided to a stabilized resting level within two seconds in about half of the cells. Injections of rectangular pulses of outward current, 0.01 microampere in magnitude, in the recovered neurons induced initiation of spikes (Figure

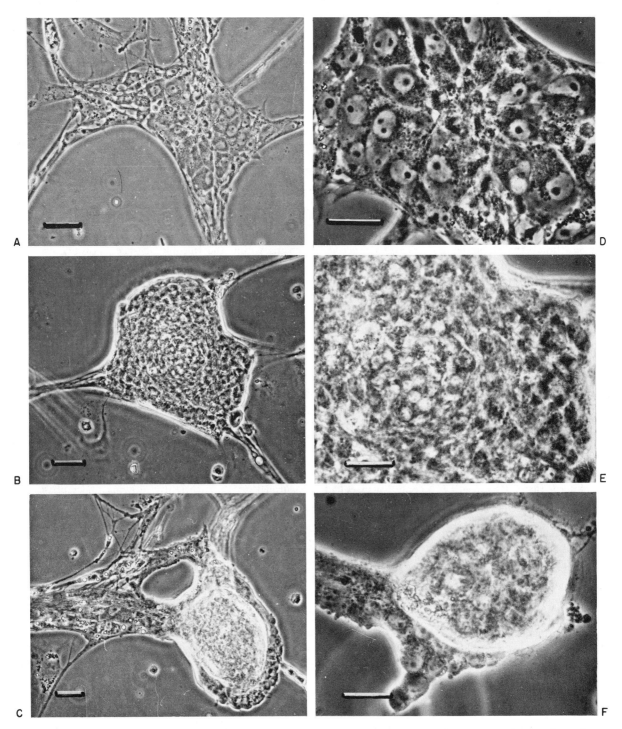

FIGURE 3 Dissociated sensory ganglia (8-day chick embryo) in culture: aggregates of spindle cells and neurons. Materials as in Figure 1 and 2. Time in culture: 3 days. Phase contrast. Bars = 20 μ.

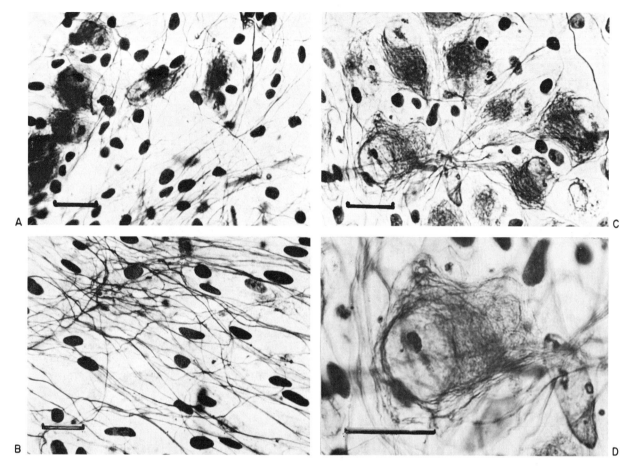

FIGURE 4 Dissociated sensory ganglia (8-day chick embryo) in culture: silver stain. Materials as in Figures 1, 2, 3.

Time in culture: 3 days. Holmes's silver nitrate. Bars = 20 μ.

7) from 40 to 60 millivolts in amplitude and about six milliseconds in duration, usually followed by a hyperpolarization phase. The action potentials were all-or-none responses, no potential being evoked with subthreshold stimulation. Repetitive stimulation (one pulse per second) elicited responses of similar amplitude and duration over a period of up to two minutes.

## Central nervous tissues

EXPLANT CULTURES Many studies have been reported over the years (Murray, 1965) on prenatal or early postnatal tissues from chick, mouse, rat, kitten, or man. The pioneer work of Harrison (1910), Russel and Bland (1933), Costero and Pomerat (1955), and many others was mainly on the morphology of normal and malignant neural tissue in vitro and the dynamic manifestations of their living cells. Notable examples of the latter are the rhythmic pulsatility of oligodendrocytes (Lumsden and Pomerat, 1951), later also observed in Schwann cells (Pomerat, 1959), the modulatory

capacity of glial cells (Pomerat, 1952), and the ciliary activity of ependymal cells (Hild, 1957a). Successful maturation in vitro, in terms of differentiation of neuronal and nonneuronal elements, tissue organization, and myelin development (Hild, 1964), was finally demonstrated for many central nervous tissues, in particular spinal cord (Peterson et al., 1965), brainstem (Hild, 1957b), cerebellum (Bornstein and Murray, 1958), and cerebrum (Bornstein, 1964). Suggestive examples have been provided of how neural tissue cultures can be applied to pathological (Fernandes and Pomerat, 1961; Bornstein, 1962), genetic (Sidman and Pearlstein, 1965), and biochemical (Lehrer and Bornstein, 1967) investigations.

Electrophysiological work on explant cultures of central nervous tissue, maintained in vitro for up to several months, has been extensive in the last decade (Crain, 1966). Cunningham (1962), using large electrodes embedded in cord, cerebellum, or cerebral tissue at explantation, described spontaneous activity in cultures up to two weeks that was enhanced by strychnine and blocked by anesthetics. Hild and

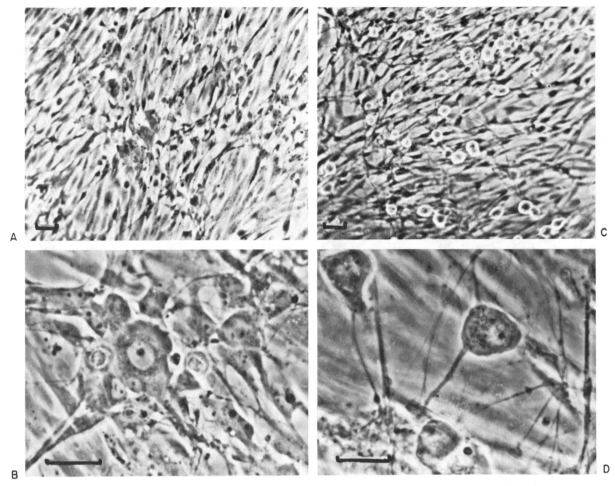

FIGURE 5 Dissociated sensory ganglia (11-day embryo) in culture: full dissociation by prolonged trypsin treatment. Falcon-dish cultures on collagen-coated coverslips, CO₂-equilibrated atmosphere. Time in culture: 10 days. Phase contrast. Bars = 20 μ. A, B. 2 ml of 1415 basal medium +5 per cent fetal calf serum and 600 mg per cent glucose. C, D. Same, plus 100 units/ml of NGF (7S). Note the number and phase glow of the nerve cells (C) and the position of nerve and Schwann cells above the confluent fibroblastic background (D).

Tasaki (1962) demonstrated spontaneous activity, excitability, and conduction in rat and kitten cerebellar cultures. Crain examined in detail the bioelectric activities of cultured spinal cord (Crain and Peterson, 1964) and cerebral cortex (Crain and Bornstein, 1964) and was able to show that synaptic function also remained a property of cultured neural tissue. Spontaneous activity often resembled, in long-term cord or cerebral cultures, some of the normal electro-encephalographic patterns of these tissues. In cultures maintained for a sufficiently long time, the asynchronous barrages of spike potentials evoked by electrical stimulation at an earlier stage gave way to more synchronized, oscillatory afterdischarges, indicative of the onset of special integrating mechanisms. The effects of drugs on these various activities were remarkably similar to those elicited in situ.

The next advance in the field came with the demonstration that synaptic function was not only preserved, but could actually be acquired in vitro by immature tissue—in other words, that organotypic maturation can be achieved in cultured neural tissue at both a morphological and a functional level. Embryonic rodent cord-myotome preparations continued to differentiate in vitro both morphologically (Bornstein and Breitbart, 1964) and functionally (Crain, 1964), with full development of competent neuromuscular junctions. In cultures of spinal cord still attached to its sensory ganglion (Crain, 1966), stimulation of the ganglion elicited a response from the dorsal portion of the cord explant after a longer time in culture than did the direct stimulation of the cord tissue, and a response from the ventral portion of the cord at an even later stage, both indications of

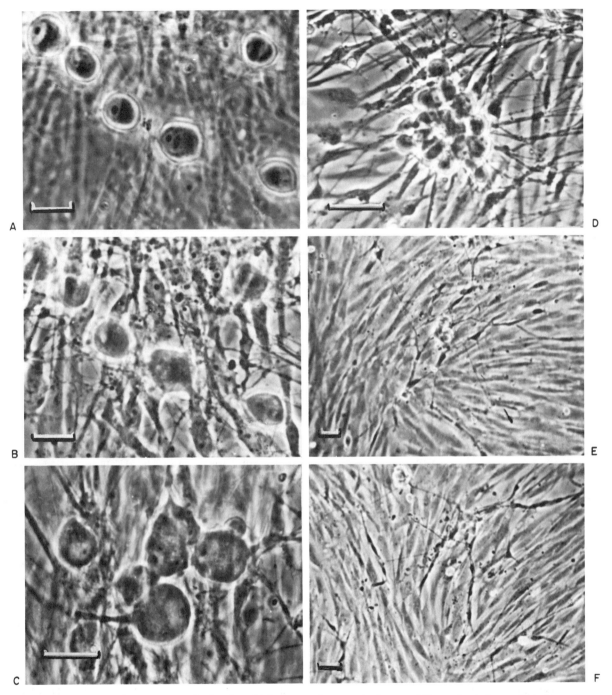

FIGURE 6  Dissociated sensory ganglia (11-day chick embryo) in culture: materials as in Figure 5. Time in culture: 10 days. Medium as in Figure 5, with 100 units/ml of NGF (7S). Phase contrast. Bars = 20 μ. A, B. Field of neurons, photographed at two different focal planes. C, D. Examples of the larger (C) and the smaller (D) neuronal populations. E, F. Network of nerve fibers and aligned Schwann cells overlying the fibroblastic carpet.

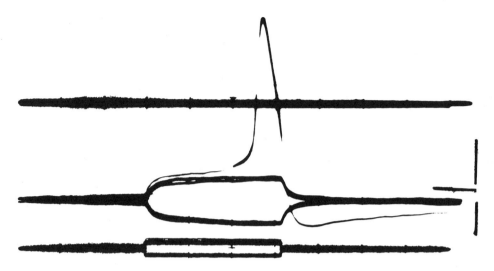

FIGURE 7 Action potential recorded from a sensory neuron in cell culture. Material and culture conditions as in Figure 6 (100 units/ml of NGF), 10 days in culture. A fine-tipped, glass micropipette filled with 3M KCl (resistance, ca. 60 megohms) was used as a single microelectrode for both stimulation and recording. Pulse of outward current, 0.01 μ amp in magnitude. Horizontal lines are: top, zero voltage reference; middle, resting potential level; lowest, step-current monitor. Upper vertical, lower vertical, and horizontal bars, at right lower corner, represent 10 mV, 0.1 μ amp, and 10 msec, respectively.

the orderly sequence in which bioelectric development was taking place in the cultured explant. A direct correlation between the appearance of synapses and of synaptic bioelectric activity was demonstrated with explant cultures of immature cord, studied in parallel at the electrophysiological (Crain and Peterson, 1967) and electron-microscope (Bunge et al., 1967a) levels.

Synaptic connections could also be established entirely *de novo*. When separate neural-tissue fragments were cultured side by side, neurites from one explant grew across the 1-mm gap and linked functionally to neurons in the other (Crain et al., 1968). Thus, stimulation of a sensory ganglion still connected to a spinal-cord explant induced complex discharges not only in the originally connected cord but also in neighboring, initially independent, explants of cord or brainstem tissue. The response in the newly coupled explant arose after long latencies, presumably through a gradual activation of extensive polysynaptic chains. Stimulation of the brainstem explant elicited bioelectrical activity in the newly coupled cord culture, but not in its originally connected sensory ganglion. Similar demonstrations were obtained with coupled cerebrum-medulla, cerebrum-cord, and, finally, cord-medulla-forebrain explants. The functional connection, in culture, of ordered arrays of central neural tissues is one of the most promising tools for neural investigation produced to date by the explant culture technique.

DISSOCIATED CELL CULTURES OF EMBRYONIC CEREBRAL TISSUE A culture system, intermediate between a tissue and a cell culture and of considerable potential interest, is that provided by dissociating a tissue, allowing the suspended cells to reaggregate, culturing the cell reaggregates in a liquid medium, and, finally, fixing them for histological examination (Moscona, 1962). Reaggregate cultures of chick-embryo eye-cup cells (Stefanelli et al., 1967) had provided electron-microscope evidence of differentiation in vitro of supportive, neuronal, and photoreceptor cells and of *de novo* formation of synaptic connections among retinal neurons in spite of the complete initial disruption of tissue organization. Very recently (Sidman, this volume), this approach has permitted the demonstration of specific histotypic reorganization patterns in reaggregates of mouse hippocampus and other cerebral tissues and the failure of such patterns to develop in corresponding reaggregate cultures from neurological mutant mice ("reeler").

Studies on dissociated cell cultures from central neural tissues are, as yet, extremely few. Cavanaugh (1955) reported that single or reaggregated nerve cells, dissociated from five- to seven-day-old chick-embryo spinal cord, did survive for at least four days and grew typical, naked processes. Long-term cultures of glial and endothelial elements from dissociated adult rabbit brain have been described (Varon et al., 1963). Recently, an extensive effort was started in our laboratory to dissociate, fractionate, and culture cells from central neural tissues. The techniques developed thus far (Varon and Raiborn, 1969) were directed mainly at 11-day chick-embryo cerebral tissue, but exploratory work has shown them to be applicable to similar tissue from seven- to 16-day-old chick embryos, newborn rat and, less effectively, newborn mouse. Attempts were also made with human bioptic material (McKhann et al., 1969).

Mechanical dissociation of 11-day chick-embryo cerebrum through a 200-mesh nylon cloth yields a suspension, or harvest, of fully dispersed, rounded cells among which two populations can be distinguished by size. The bigger cells (15 $\mu$ in average diameter) have a large, relatively clear nucleus and retain, in some cases, a thick process stump. The smaller cells (5 to 8 $\mu$ in diameter) are more dense and opaque and have a marked tendency, upon standing, to form small aggregates that rapidly coalesce into larger clumps. Seeding of harvest aliquots on coverslips coated with a thin layer of rooster plasma causes the plasma to clot and trap the cells at various depths. Within a few hours of incubation in a Rose chamber, three classes of cells can be grossly recognized in the culture. One class (A) comprises the large cells of the harvest; the A cell (Figure 8A), flattened out a little and with a grossly polygonal contour, grows multiple processes usually branching out at a short distance from the soma and assumes an over-all morphology typical of a nerve cell in culture. A second class (B) is made up of small cells which remain compact and rather dense, with an oval or pear-shaped profile and a poorly visible nucleus. B cells (Figure 8B) also appear to be neuronal elements but are almost exclusively unipolar or pseudo-unipolar, with one very long process branching at some distance from the soma, if at all. A and B cells do not appear to divide, have little detectable mobility, and will not attach directly to uncoated glass surfaces, all characters in keeping with their presumed neuronal identity. They develop an extensive fiber network, in which fibers fasciculate into bundles and plexuses and also make contact with perikarya of other A or B cells (Figure 8D, E). The third class (C) also originates from the smaller-sized harvest population. C cells (Figure 8C) quickly flatten out into thin, ragged-contoured, epithelioid elements, are very mobile, can readily attach directly to glass, and, under suitable conditions, can proliferate rapidly into an epithelial-like confluent monolayer, and can, in fact, be propagated in culture for at least several months. B, but not A, cells readily aggregate with C cells in suspension and, in the culture, will attach to individual C cells or sit on top of C-cell patches, much in the manner in which dissociated ganglionic neurons (Figure 5C,D) were shown to position themselves over a fibroblastic layer. Highly purified A and C cells can be fractionated out of a harvest suspension; it must be stressed, however, that all three classes probably encompass heterogeneous populations.

More recent work (Varon and Raiborn, in preparation) has been directed mainly toward exploring ways to improve the culture technique. Monolayer cultures were obtained by seeding the harvest on a collagen-coated coverslip sealed in a Rose chamber, allowing the cells to attach (two to three hours), and replacing the seeding medium with a small volume of medium supplemented with plasma; the added medium spreads over the attached cells in a thin film that rapidly clots and holds the cells sandwiched between it and the collagen floor. Except for their monolayer arrangements, cells cultured in this "sandwich" system were entirely comparable to those cultured within a plasma layer. Subsequent addition of a larger volume of medium causes the plasma film to peel off the cells, leaving them conveniently exposed for direct manipulations or for recoating with a fresh plasma film, a procedure that prolonged the average life span of the culture from four to 12 days. The behavior of collagen-supported cultures in a $CO_2$-controlled atmosphere is also under investigation; preliminary results suggest that the system strongly stimulates the proliferation of C cells, but might cause some loss of A cells.

Attempts to demonstrate bioelectric abilities in the A or the B cells have, thus far, failed; the cells are considerably smaller than the sensory ganglionic cells, and a refinement of present microelectrodes and impalement techniques appears necessary. On the other hand, a satisfactory procedure has been established (Hámori, Kruger, Miller, and Varon, in preparation) to prepare the cultures for electron microscopy in such a way that the relative position of the cells within the monolayer is maintained, and individual cells, identified by phase microscopy, can be followed up in the ultrathin sections. The preparations are extremely well preserved and promise to be of great future use for ultrastructural studies. Figure 9 shows electron micrographs of A-type and, presumably, B-type cells (Figure 9A, B, page 98) and one, not necessarily representative, of a C-type cell (Figure 9C).

## Conclusion and perspectives

The validity of approaching neurobiological problems by culture techniques has been well established by the impressive results of recent years. There are no longer any doubts that neural *explant cultures* permit the maturation, maintenance, and functional expression of specific neural characters at both the cytotypic and the organotypic levels. Some examples have been cited in the preceding sections, and many more can be found in other recent reviews (Murray, 1965; Crain, 1966; Crain et al., 1968) of the ways by which such cultures can and will contribute to the study of the nervous system.

The field of dissociated neural *cell cultures* is presently undergoing the same growth pains as did the field of explant cultures some decades ago. Their potential advantages lie in

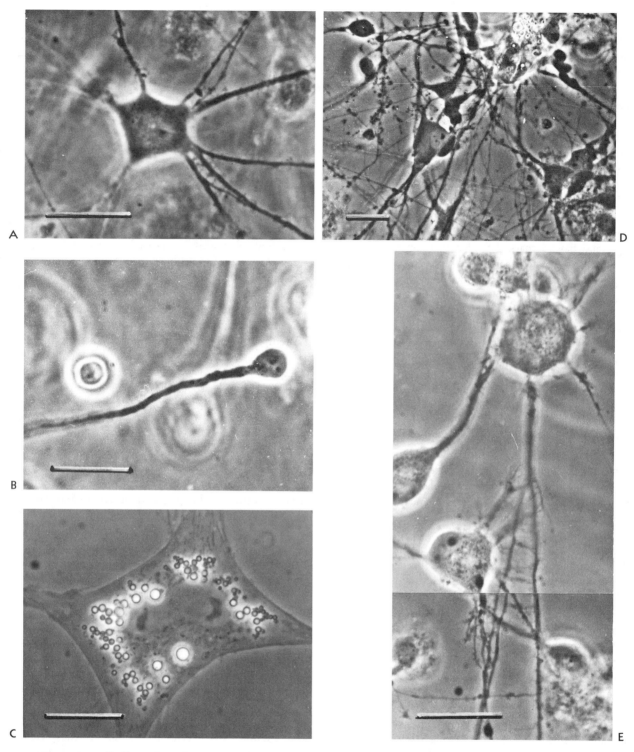

FIGURE 8  Cell cultures from dissociated cerebral tissue (11-day chick embryo). Tissue dissociated mechanically into Eagle's basal medium. Cultures in Rose chamber, thin plasma-layer system. Time in culture: 5 days. Phase contrast. Bars = 20 $\mu$. A, B, C. Individual cells of the A, B, and C type. respectively. D, E. Fields of the same culture.

that they uniquely offer both analytical and synthetic approaches to the study of neural systems. Analytical approaches are offered by the direct correlation that is possible, in these cultures, between visual and instrumental observations at the level of any individual cell, as well as by the possibility (Varon and Raiborn, 1969) of segregating different cell types from one another and studying their autonomous properties in the absence of heterotypic cellular interactions. Synthetic approaches are exemplified by the organotypic reorganization of cells in culture already observed for ganglionic (Cohen et al., 1964), retinal (Stefanelli et al., 1967), and cerebral (Sidman, this volume) cells and the possibility of analyzing which organization is necessary to, or conducive of, the resumption of cytotypic, organotypic, and functional properties characteristic of neural tissue. Although some of these perspectives are clearly not to become realities for some time, a number of points in their favor have already been firmly established. Dissociated nerve cells do survive in vitro and, at least in the case of ganglionic ones, can mature functionally as well as morphologically (Scott et al., 1969; Varon et al., in preparation). Their newly grown processes assume relationships with one another (fasciculation), with their own somata (recurrent collaterals), and with somata of other neurons (Varon and Raiborn, 1969) that grossly mimic those of normal neural tissue and may be revealed, by future ultrastructural studies, to have even greater organotypic significance. Neurites growing out of an explant (Crain et al., 1968) can connect functionally with other neurons that were not part of the same pre-laid tissue pattern. Finally, newly formed synaptic connections have been identified by electron microscopy (Stefanelli et al., 1967) in cultured reaggregates of neuroretinal cells. It remains to discover (I hope in the near future) under what conditions two initially isolated nerve cells will produce in culture a mature and functional synapse and, in so doing, generate the basic link of the simplest model for a neuronal network.

## Acknowledgment

The work described in this paper was supported by U. S. Public Health research Grant NB 07607 from the National Institute of Neurological Diseases and Stroke, and Research Grant E 513 from the American Cancer Institute.

### REFERENCES

BORNSTEIN, M. B., 1963. A tissue-culture approach to demyelinative disorders. *Nat. Cancer Inst. Monogr.* 11: 197–214.

BORNSTEIN, M. B., 1964. Morphological development of neonatal mouse cerebral cortex in tissue culture. *In* Neurological and Electroencephalographic Correlative Studies in Infancy (P. Kellaway and I. Petersén, editors). Grune and Stratton, New York, pp. 1–11.

BORNSTEIN, M. B., and L. M. BREITBART, 1964. Anatomical studies of mouse embryo spinal cord—skeletal muscle in long-term tissue culture. *Anat. Rec.* 148: 362 (abstract).

BORNSTEIN, M. B., and M. R. MURRAY, 1958. Serial observations on patterns of growth, myelin formation, maintenance and degeneration in cultures of new-born rat and kitten cerebellum. *J. Biophys. Biochem. Cytol.* 4: 499–504.

BUNGE, M. B., R. P. BUNGE, and E. R. PETERSON, 1967a. The onset of synapse formation in spinal cord cultures as studied by electron microscopy. *Brain Res.* 6: 728–749.

BUNGE, M. B., R. P. BUNGE, E. R. PETERSON, and M. R. MURRAY, 1967b. A light and electron microscope study of long-term organized cultures of rat dorsal root ganglia. *J. Cell Biol.* 32: 439–466.

CAVANAUGH, M. W., 1955. Neuron development from trypsin-dissociated cells of differentiated spinal cord of the chick embryo. *Exp. Cell Res.* 9: 42–48.

COHEN, A. I., E. C. NICOL, and W. RICHTER, 1964. Nerve growth factor requirement for development of dissociated embryonic sensory and sympathetic ganglia in culture. *Proc. Soc. Exp. Biol. Med.* 116: 784–789.

COSTERO, I., and C. M. POMERAT, 1955. Cellular prototypes of central gliomata. *Proc. 2nd Int. Congr. Neuropathol., London, 1955, part I.* Excerpta Med. Found. Amsterdam, pp. 273–277.

CRAIN, S. M., 1956. Resting and action potentials of cultured chick embryo spinal ganglion cells. *J. Comp. Neurol.* 104: 285–329.

CRAIN, S. M., 1964. Electrophysiological studies of cord-innervated skeletal muscle in long-term tissue cultures of mouse embryo myotomes. *Anat. Rec.* 148: 273–274 (abstract).

CRAIN, S. M., 1966. Development of "organotypic" bioelectric activities in central nervous tissues during maturation in culture. *Int. Rev. Neurobiol.* 9: 1–43.

CRAIN, S. M., and M. B. BORNSTEIN, 1964. Bioelectric activity of neonatal mouse cerebral cortex during growth and differentiation in tissue culture. *Exp. Neurol.* 10: 425–450.

CRAIN, S. M., and E. R. PETERSON, 1964. Complex bioelectric activity in organized tissue cultures of spinal cord (human, rat and chick). *J. Cell. Comp. Physiol.* 64: 1–14.

CRAIN, S. M., and E. R. PETERSON, 1967. Onset and development of functional interneuronal connections in explants of rat spinal cord-ganglia during maturation in culture. *Brain Res.* 6: 750–762.

CRAIN, S. M., E. R. PETERSON, and M. B. BORNSTEIN, 1968. Formation of functional interneuronal connections between explants of various mammalian central nervous tissues during development in vitro. *In* Growth of the Nervous System/A Ciba Foundation Symposium (G. E. W. Wolstenholme and M. O'Connor, editors). Little, Brown and Co., Boston, pp. 13–31.

CUNNINGHAM, A. W. B., 1962. Qualitative behavior of spontaneous potentials from explants of 15 day chick embryo telencephalon in vitro. *J. Gen. Physiol.* 45: 1065–1076.

FERNANDES, M. V., and C. M. POMERAT, 1961. Cytopathogenic effects of rabies virus on nervous tissue in vitro. *Z. Zellforsch. Mikroskop. Anat.* 53: 431–437.

HAMBURGER, V., and R. LEVI-MONTALCINI, 1949. Proliferation, differentiation and degeneration in the spinal ganglia of the chick embryo under normal and experimental conditions. *J. Exp. Zool.* 111: 457–501.

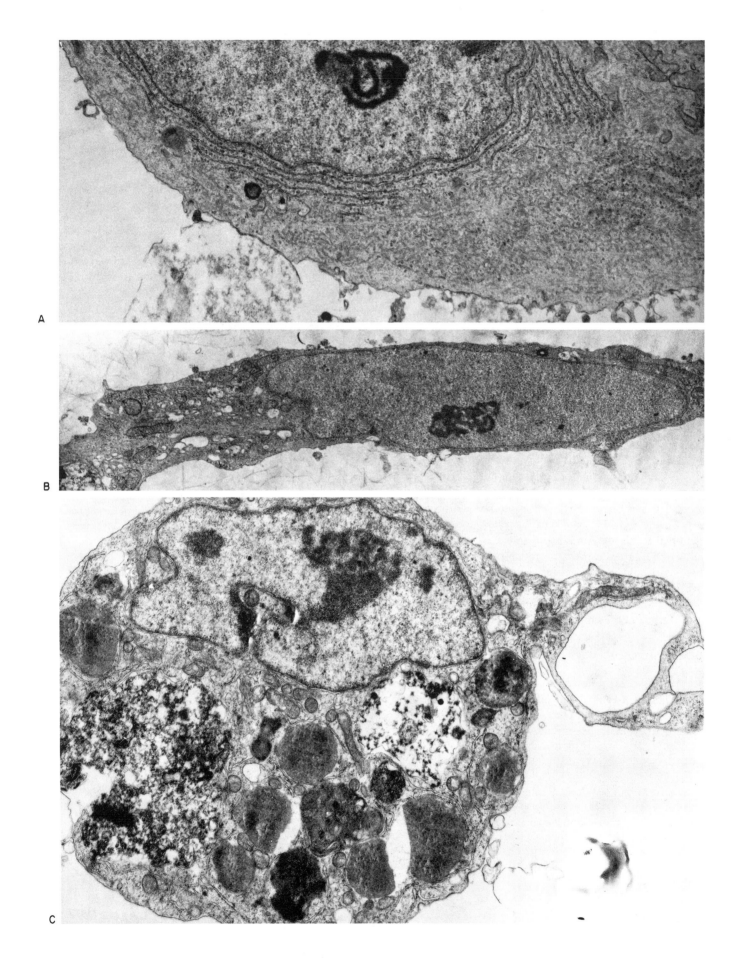

HARRISON, R. G., 1907. Observations on the living developing nerve fiber. *Proc. Soc. Exp. Biol. Med.* 4: 140–143.

HARRISON, R. G., 1910. The outgrowth of the nerve fiber as a mode of protoplasmic movement. *J. Exp. Zool.* 9: 787–847.

HILD, W., 1957a. Ependymal cells in tissue culture. *Z. Zellforsch. Mikroskop. Anat.* 46: 259–271.

HILD, W., 1957b. Observations on neurons and neuroglia from the area of the mesencephalic fifth nucleus of the cat in vitro. *Z. Zellforsch. Mikroskop. Anat.* 47: 127–146.

HILD, W., 1964. Myelinated neuronal perikarya in cultures of central nervous tissue. *Anat. Rec.* 148: 291 (abstract).

HILD, W., and I. TASAKI, 1962. Morphological and physiological properties of neurons and glial cells in tissue culture. *J. Neurophysiol.* 25: 277–304.

LEHRER, G. M., and M. B. BORNSTEIN, 1966. Carbohydrate metabolism of the developing brain *in vivo* and *in vitro*. *In* Variation in Chemical Composition of the Nervous System (G. B. Ansell, editor). Pergamon Press, New York, pp. 67–68 (abstract).

LEVI-MONTALCINI, R., 1966. The nerve growth factor: Its mode of action on sensory and sympathetic nerve cells. *Harvey Lect. (1964–65)* Ser. 60: 217–259.

LEVI-MONTALCINI, R., and P. U. ANGELETTI, 1963. Essential role of the nerve growth factor in the survival and maintenance of dissociated sensory and sympathetic embryonic nerve cells *in vitro*. *Develop. Biol.* 7: 653–659.

LEVI-MONTALCINI, R., and P. U. ANGELETTI, 1968. Nerve growth factor. *Physiol. Rev.* 48: 534–569.

LUMSDEN, C. E., 1951. Aspects of neurite outgrowth in tissue culture. *Anat. Rec.* 110: 145–179.

LUMSDEN, C. E., and C. M. POMERAT, 1951. Normal oligodendrocytes in tissue culture. A preliminary report on the pulsatile glial cells in tissue cultures from the corpus callosum of the normal adult rat brain. *Exp. Cell Res.* 2: 103–114.

MCKHANN, G. M., W. HO, C. RAIBORN, and S. VARON, 1969. The isolation of neurons from normal and abnormal human cerebral cortex. *Arch. Neurol.* 20: 542–547.

MOSCONA, A. A., 1962. Cellular interactions in experimental histogenesis. *Int. Rev. Exp. Pathol.* 1: 371–428.

MURRAY, M. R., 1965. Nervous tissues *in vitro*. *In* Cells and Tissues in Culture (E. N. Willmer, editor). Academic Press, New York, vol. 2, pp. 373–455.

MURRAY, M. R., and H. R. BENITEZ, 1968. Action of heavy water ($D_2O$) on growth and development of isolated nervous tissues. *In* Growth of the Nervous System/A Ciba Foundation Symposium (G. E. W. Wolstenholme and M. O'Connor, editors). Little, Brown and Co., Boston, pp. 148–174.

NAKAI, J., 1956. Dissociated dorsal root ganglia in tissue culture. *Amer. J. Anat.* 99: 81–129.

NAKAI, J., 1960. Studies on the mechanism determining the course of nerve fibers in tissue culture. II. The mechanism of fasciculation. *Z. Zellforsch. Mikroskop. Anat.* 52: 427–449.

NAKAI, J., and Y. KAWASAKI, 1959. Studies on the mechanism determining the course of nerve fibers in tissue culture. I. The reaction of the growth cone to various obstructions. *Z. Zellforsch. Mikroskop. Anat.* 51: 108–122.

NAKAJIMA, S., 1965. Selectivity in fasciculation of nerve fibers *in vitro*. *J. Comp. Neurol.* 125: 193–204.

PETERSON, E. R., S. M. CRAIN, and M. R. MURRAY, 1965. Differentiation and prolonged maintenance of bioelectrically active spinal cord cultures (rat, chick and human). *Z. Zellforsch. Mikroskop. Anat.* 66: 130–154.

PETERSON, E. R., and M. R. MURRAY, 1960. Modification of development in isolated dorsal root ganglia by nutritional and physical factors. *Develop. Biol.* 2: 461–476.

POMERAT, C. M., 1952. Dynamic neurogliology. *Tex. Rep. Biol. Med.* 10: 885–913.

POMERAT, C. M., 1959. Rhythmic contraction of Schwann cells. *Science (Washington)* 130: 1759–1760.

RUSSEL, D. S., and J. O. W. BLAND, 1933. A study of gliomas by the method of tissue culture. *J. Pathol. Bacteriol.* 36: 273–283.

SCOTT, B. S., V. E. ENGELBERT, and K. C. FISHER, 1969. Morphological and electrophysiological characteristics of dissociated chick embryonic spinal ganglion cells in culture. *Exp. Neurol.* 23: 230–248.

SHIMIZU, Y., 1965. The satellite cells in cultures of dissociated spinal ganglia. *Z. Zellforsch. Mikroskop. Anat.* 67: 185–195.

SHOOTER, E. M., and S. VARON, 1970. Macromolecular aspects of the nerve growth factor proteins. *In* Protein Metabolism of the Nervous System (A. Lajtha, editor). Plenum Press, New York, pp. 419–437.

SIDMAN, R. L., and R. PEARLSTEIN, 1965. Pink-eyed dilution (*p*) gene in rodents: Increased pigmentation in tissue culture. *Develop. Biol.* 12: 93–116.

STEFANELLI, A., A. M. ZACCHEI, S. CARAVITA, A. CATALDI, and L. A. IERADI, 1967. New-forming retinal synapses in vitro. *Experientia (Basel)* 23: 199–200.

UTAKOJI, T., and T. C. HSU, 1965. Nucleic acids and protein synthesis of isolated cells from chick embryonic spinal ganglia in culture. *J. Exp. Zool.* 158: 181–201.

VARON, S., and C. W. RAIBORN, JR., 1969. Dissociation, fractionation and culture of embryonic brain cells. *Brain Res.* 12: 180–199.

VARON, S. S., C. W. RAIBORN, JR., T. SETO, and C. M. POMERAT, 1963. A cell line from trypsinized adult rabbit brain tissue. *Z. Zellforsch. Mikroskop. Anat.* 59: 35–46.

WEISS, P., 1934. In vitro experiments on the factors determining the course of the outgrowing nerve fiber. *J. Exp. Zool.* 68: 393–447.

WEISS, P., and H. B. HISCOE, 1948. Experiments on the mechanism of nerve growth. *J. Exp. Zool.* 107: 315–395.

FIGURE 9 Electron micrographs of cerebral cell cultures. Material as in Figure 8, culture in Rose chamber and collagen-plasma film "sandwich" system. Time in culture: 2 days. Fixation in 0.5 per cent gluteraldehyde +0.5 per cent formaldehyde, postfixation in 1 per cent $OsO_4$. Embedding in Araldite. A ($\times$ 13,000), B ($\times$ 8000), and C ($\times$ 13,800): sections through A-type, B-type, and C-type cells, respectively.

# 10 Cell Proliferation, Migration, and Interaction in the Developing Mammalian Central Nervous System

## RICHARD L. SIDMAN

A FUNDAMENTAL THEME in this section of the book is the origin, character, and modifiability of intercellular relationships. The synapse has been the focus of attention, but the cells of the nervous system are in close contact with one another long before synapses develop, as the immature cells proliferate and migrate. A high degree of order characterizes these events. We plan to survey here what is known of that order and to indicate directions in current work of various laboratories that might illuminate how the patterns of cell proliferation and migration contribute to the organizing of the mature central nervous system (see also Fujita, 1966; Gracheva, 1967; Hicks and D'Amato, 1968; Källén, 1965; Langman, 1968; Levi-Montalcini, 1963; Watterson, 1965).

### Zones of the developing CNS

The first point to emphasize is that the developing central nervous system (CNS) is a different organ from its adult counterpart and has its own component parts and terminology (Angevine, this volume). The early nervous system is a sheet of columnar epithelial cells, each one stretching from ventricular to outside surface. The cells acquire this arrangement when the nervous system is a simple plate, and the pattern persists when the sheet of cells transforms into a tube (His, 1904). Cells are attached to one another side to side near the ventricular surface (Wechsler, 1966; Herman and Kauffman, 1966; Menkes et al., 1967) and round up when they divide (Stensaas and Stensaas, 1968), so that the cell nucleus is displaced to and fro during the generation cycle (Figure 1). Kinetic data obtained by autoradiography indicate that all the cells behave in the same way with respect to the cell generation cycle (Fujita, 1966; Kauffman, 1968). After an appropriate number of divisions (and we have no idea of the mechanisms controlling the number and

timing of the cycle), some cells quit the scene, move outward as postmitotic young neurons, and generate an axon (Ramón y Cajal, 1960). Until this time the cells are indistinguishable from one another by electron-microscope criteria (Lyser, 1968).

We must begin to assign some names and must face the problem that the terminologies in general use are inadequate. A suggested revision that developed from the discussions at ISP 1969 has been presented elsewhere (Boulder Committee, 1970); it is used here to help clarify the issue of where various cells of the CNS arise. As illustrated in Figure 1, the zone occupied by the moving nuclei of the original columnar cells is called the *ventricular* zone; it lines the entire ventricular system early in development. The distal parts of these columnar epithelial cells, outside the zone in which the nuclei move, constitute the *marginal* zone. A new zone that forms between the original two zones and is occupied initially by somas of postmitotic cells is named the *intermediate* zone. Another new population, not generally acknowledged by previous workers, then appears between the ventricular and intermediate zones to form the *subventricular* zone.

The cell types in these developmental zones are as follows: 1. The original columnar epithelial cells stretch through all zones during part of interphase; their nuclei are always confined to the ventricular zone, and during mitosis the entire cell is so confined (Stensaas and Stensaas, 1968). 2. Young neurons (postmitotic, usually designated by the confusing name neuroblasts) are the first occupants of the intermediate zone; later glial cells, vascular cells, and axon terminals of distant young neurons will join them. 3. Subventricular cells differ from young neurons in the intermediate zone in that they continue to divide, and they differ from the columnar epithelial cells of the ventricular zone in that their nuclei do not move to and fro during the generation cycle; subventricular cells give rise to several classes of neurons and macroglia. (We do not know if the ventricular or subventricular populations are divisible into

RICHARD L. SIDMAN Department of Neuropathology, Harvard Medical School, Boston, Massachusetts

clones of neuronal and glial precursors, or if one or the other, or both, still contain common stem cells.)

## Derivatives of the subventricular zone

The subventricular zone is present throughout the developing mammalian CNS at one time period or another, persisting for months in some areas, such as the cerebral hemispheres, and only for a few days in some others, such as the spinal cord. Most important, it adopts special configurations in particular areas and gives rise to some of the most interesting populations of neurons in the whole CNS. For example, a special focus of subventricular cells in the caudal part of the roof of the fourth ventricle gives rise to the external granule-cell population of the developing cerebellum, and eventually to the mature granule cells (Figure 2). A nearby group of similar cells migrates through the medulla and pons, forming the transient corpus pontobulbare (Figure 2) and eventually becoming inferior-olivary and pontine neurons.

These various patterns of proliferation provoke some

speculations about the control of brain size within and between species. There are developmental diseases in man in which brain size is either grossly reduced or moderately increased, with little alteration in morphogenesis (e.g., Cunningham and Telford-Smith, 1895; Ambler et al., 1969). A plausible explanation is that the developmental control of cell number went awry without major abnormality in the mechanisms that control cell arrangements, shapes, and perhaps connections. Might evolutionary differences likewise be accomplished simply by changing the rate or duration of cell division? Is the rat brain merely a mouse brain with an order of magnitude more cells, as a result of alteration in a rate-controlling gene? Probably not, was the consensus at the symposium, else there would be more frequent big jumps in evolution. Nauta and Karten (this volume) made the further point that the difference between guinea pig and rat appears to relate not so much to differences in cell number, but to the finesse with which interneurons are fitted in. Another argument is that, at least in some species, differences in organization of the CNS involve introduction of new developmental mechanisms.

A recently described example in man (Rakic and Sidman, 1969) concerns another specialized subventricular derivative (Figure 3), this one in the floor of the lateral ventricles and named the ganglionic eminence. This fetal component of

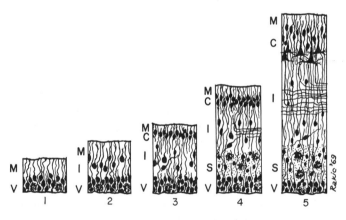

FIGURE 1 Schematic drawings to illustrate histogenesis in the cerebral isocortex. The ventricular surface is at the bottom and the external surface of the brain at the top. Picture 1 shows the stage at which the cerebral wall consists of only ventricular (V) and marginal (M) zones. In 2, the intermediate (I) zone is added. Picture 3 shows the first cellular constituents of the cortical plate (C). In picture 4, the subventricular (S) zone has appeared just external to the ventricular zone. It contains mitotic cells, as does the ventricular zone. The nuclei of additional cells are crossing the intermediate zone toward the cortical plate, and the intermediate zone is further complicated by the arrival of horizontally or obliquely disposed afferent axons, probably originating in the diencephalon. Picture 5 shows further elaboration of the intermediate zone, with afferent and efferent axons crossing it. The deep-lying pyramidal neurons of the cortical plate are differentiating, while later-forming cells are attaining final positions external to their predecessors. Further details are given in the text.

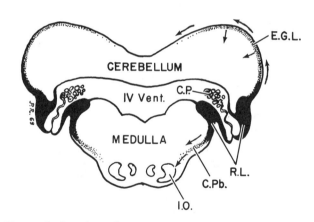

FIGURE 2 Drawing of a transverse section of cerebellum and medulla from a 12-week human fetus. The ventricular and subventricular zones lining the fourth ventricle (IV Vent.) are almost exhausted except at the posterior and lateral margins, where a thickened subventricular zone persists as the rhombic lip (R.L.) adjacent to the root of the choroid plexus (C.P.). From the rhombic lip, proliferating cells migrate around the external surface of the cerebellum rostrally and toward the midline (arrows) to form the external granular layer (E.G.L.). Postmitotic cells later will migrate inward (arrows). On the inferior side of the ventricle, postmitotic cells migrate from the inner fold of the rhombic lip through the lateral medulla (arrows), and form Essick's (1907) transient corpus pontobulbare (C.Pb.) while en route to the inferior olive (I.O.).

the brain has been recognized as the probable source of cells for the nearby basal ganglia, but it must have a broader significance, for it is a huge structure in the human fetus and is full of dividing cells for at least the final two thirds of the gestation period. We found that, if the fetal brain is cut in the horizontal plane (Figure 3), a migration path can be traced in Nissl and Golgi preparations from the telencephalon clear across into the pulvinar region of the diencephalon. From the fifth through the eighth months of gestation in man, cells stream across, passing en route through a previously undescribed structure which we named the *corpus gangliothalamicus* (by analogy with the corpus pontobulbare [Essick, 1907]). Thus, a brain region not represented significantly in simpler mammals is formed in man in part by a new mechanism. It remains to be learned whether other

parts of the brain are derived from the same extraordinary fountainhead, particularly some of the cortical association areas that are peculiar to primates and receive their inputs from the pulvinar.

## Control of cell number and pattern

The very tight control of cell number in ontogenesis should be emphasized. The cerebellar granule-cell population, for example, is enormous. In man it is estimated at one hundred billion neurons (Braitenberg and Atwood, 1958), more than all the neurons of the cerebrum combined. In the mouse, these cells begin to form on embryonic day 13 and increase at nearly a logarithmic rate for the next 21 days, reaching a crescendo late in the second postnatal week. More than 80 per cent of the postmitotic granule cells are formed between day 7 and day 12 (Figure 6 in Fujita, 1967). I need not belabor the point that a very slight alteration in rate or duration of cell genesis would make an enormous difference in the outcome.

Prestige (this volume) deals with another facet of this story by considering cell death as a normal developmental event. In *Xenopus* larvae, the number of ventral-horn cells in particular lumbar segments of the spinal cord reaches 6000, whereas the number at maturity is only about 1500. He makes some fairly plausible assumptions about rates of cell genesis and speed of removal of dead cells, and suggests that there is not only a loss of cells but actually a turnover of young neurons in the cord. By the term "turnover," he implies that there is further recruitment of cells even as others are dying. The loss of cells is programed, in that they die at characteristic stages in their maturation and do not die if their maturation is delayed by hypophysectomy. Several lines of evidence are presented in support of the idea that the periphery influences neuron number by passage of information in a retrograde direction along the axon.

There was speculative discussion during the symposium on which this section of the book is based, as to why cell death should be part of the mechanism for control of cell number. The major ideas bandied about were: (1) that the CNS might make cells in excess prior to the availability of the target organ, rather than make just the right number at the last moment; (2) that some kind of transfer of information might be effected from dying to neighboring cells; and (3) that a critical mass of cells might be needed for a major morphogenetic event, but not be needed thereafter. The major general point is that net cell number is determined by cell genesis *and* cell death; the latter has been given too little attention.

Another general point, emphasized by Angevine (this volume), is that there is a firm program, or timetable, of cell genesis, with reference both to the various cell types and to the neurons of given regions. This is a time-honored

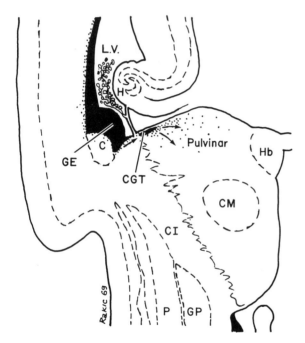

FIGURE 3  Drawing of the thalamus and part of the cerebrum on one side of a horizontal section through the brain of a 20-week human fetus. The posterior pole of the temporal lobe is off the picture at the upper left, and part of the insula is illustrated at the lower left. Internal to the insula are the putamen (P) and globus pallidus (GP), separated by the internal capsule (CI) from the thalamus, with orienting labels on its centrum medianum (CM), habenular nucleus (Hb), and pulvinar region. In the temporal lobe, the hippocampal formation (H) lies medial to the temporal horn of the lateral ventricle (L.V.), and laterally lies the tail of the caudate nucleus (C) separated from the ventricle by an enormous mass of subventricular cells known as the ganglionic eminence (GE). Postmitotic cells migrate (arrows) from the ganglionic eminence of the telencephalon and form the transient corpus gangliothalamicus (CGT) while en route to the pulvinar region of the diencephalon.

view, but autoradiography with triatiated thymidine has given numerical precision to the generalization (see also review by Sidman, 1970). Without repeating Angevine's detailed evidence in the hippocampal formation and elsewhere in the cortex, one may summarize simply with the statement that the pattern of times of origin, or "birthdays," of nerve cells can characterize given subregions of the CNS as precisely as the synaptic anatomy and may represent the more fundamental characteristic, for it takes precedence in time.

The birthdays of diencephalic neurons, in contrast to cortical ones, tell a slightly different story. Angevine emphasizes not so much the discrete patterns of cell origin for given thalamic nuclei, although there are such patterns in some instances, but rather three broad and continuous gradients that describe the thalamus as a whole. The gradients run, from early arising to late arising cells, in the ventral to dorsal, caudal to rostral, and lateral to medial axes. The segregation of individual clusters of neurons, which we refer to as the thalamic nuclei, must be established later, perhaps by superimposing factors resulting from thalamocortical interactions on the basic three-coordinate gradient system.

## Do cells migrate in the developing CNS?

These autoradiographic studies make another very general point, namely, that the sites of cell origin almost always differ from the sites where the mature cells will reside. This, in turn, raised the issue for general discussion as to whether cells actually do migrate in the developing brain. The answer is an unequivocal "yes" for many populations, but the autoradiographic method, with its focus on the labeled cell *nucleus* to the exclusion of other parts of the cell, gives only part of the story.

The clearest example of true migration is provided by the cerebellar granule cell, for which Golgi (Ramón y Cajal, 1960), autoradiographic (Miale and Sidman, 1961), and electron-microscope (Mugnaini and Forstrønen, 1967) data are in concordance. The details of the migration process are worth a description. When a round precursor cell in the external granule layer has divided for the final time, the daughter cells come to lie in the deeper parts of the external granular layer and take on a bipolar form, with horizontal processes aligned in the transverse plane (Figure 4). Thus, the orientation of the granule cell is already established at this time, probably before the cell makes contact with the Purkinje-cell dendrites, which are oriented 90 degrees differently in the sagittal plane. Then the cell moves into the incipient molecular layer and emits a cytoplasmic process inward. Next, the nucleus moves down that process, the segment behind the nucleus taking on axonal properties, while the segment ahead of the nucleus retains more general

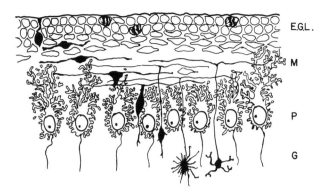

FIGURE 4  Schematic drawing of a transverse section through the developing cerebellar cortex, based on illustrations of Ramón y Cajal (1960). The external granular layer (E.G.L.) contains dividing cells in its superficial layers; its deeper layers contain postmitotic young neurons, some of which already possess bipolar horizontal processes oriented in the transverse plane. These cells are young granule-cell neurons, and their developmental sequence is illustrated in the series of blackened cells that extend diagonally from the upper left and reach the bottom of the picture to the right of center. They cross the developing molecular (M) layer, pass the immature Purkinje (P) cells with dendrites oriented in the sagittal plane, and reach the granular (G) layer as described further in the text.

cytoplasmic characteristics. When the cell body reaches the granular layer it develops a series of transient processes. Many afferent axons make contact with these and with the soma itself. The cell then remodels to its adult configuration, with a smooth-surfaced soma, a few dendrites, and only the tips of the dendrites in contact with axon terminals. The primary evidence for migration is the change in position of the postmitotic granule cell relative to a fixed landmark, the Purkinje-cell soma.

Another instructive example of migration has been described by Morest (1969a) in Golgi preparations of the opossum medial trapezoid nucleus, one of the auditory relay nuclei of the brainstem. Here some young nerve cells retain the ancestral columnar form, with nuclei in the ventricular or subventricular zones and cytoplasmic processes attached to ventricular and outer brainstem surfaces (Figure 5). Then the nucleus apparently moves outward the length of the cell, and the inner process may be discarded or resorbed so that the entire cell lies near the external surface of the brainstem. Next, a new inward-directed process forms, an axon branches outward from it, dendritic sprouts emerge from it directly congruent with the sprouting axon terminals of the afferent nerve supply, and, finally, the nucleus migrates inward to take up a position close to the point of emergence of the axon. There are several variations on this theme (Morest, 1969a, 1969b). The cell behavior is reminiscent of that described above for the developing granule-cell neuron of the cerebellum.

Morest (1969a, 1969b) emphasized the importance of the exquisitely detailed spatial correspondence between the developing dendrites and their input and the precise temporal relationships. He underscored the temporal factor by describing how the input arises. Neurons of the left cochlear nucleus are destined to innervate the right medial trapezoid nucleus. The axons grow first to the ipsilateral medial trapezoid nucleus but make no synaptic connections there; they then cross the midline, pass the corresponding axons coming from the other side, and engage developing dendrites on young neurons of the contralateral side. The growing axonal and dendrite surfaces may not yet be "ripe" at the time of the ipsilateral contact, or there may be other and more interesting mechanisms governing the attainment of synaptogenesis only on the second, contralateral, opportunity.

The same general mechanism may apply as well to the cerebral cortex. The migration of cells to the cortical plate was described (Angevine and Sidman, 1961) as an "inside-out" migration (Figure 1). Early-forming cells apparently went as far outward as they could go. Somas of successive waves of later cells attained more superficial final positions, as clearly established by the tritiated thymidine labeling patterns. The attainment of those patterns, however, might be achieved by more complex routes of migration than had been recognized—first all the way outward and then part way inward again (Figure 5, adapted from Morest, 1969b; see also the earlier report of Berry and Rogers, 1965).

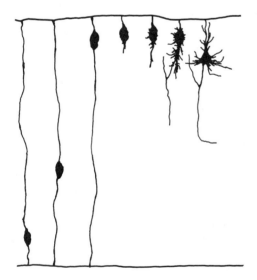

FIGURE 5 Schematic illustration of Morest's (1969a, 1969b) concept of the developmental displacement and modeling of some classes of neurons. See text for further details.

## Experimental analysis of cell shape and alignment

These elegant anatomical studies pose a special problem of research strategy. If the assembling of a nervous system were to depend on spatial and temporal events of such extraordinary precision that development would fail unless a myriad of axons that are constantly remodeling and often growing over considerable distances were to meet precisely corresponding dendrites and somas exactly on schedule, how could one hope to unravel mechanisms? Yet mechanisms are analyzable via grafting experiments, ablations, tissue culture, analysis of mutants, and combinations of these approaches.

The tissue-culture approach is reviewed by Varon (this volume). He and others are developing methods, still at an early stage, for maintenance of isolated neurons in vitro in order to evolve simple systems for studying neuron-neuron interactions (e.g., Goldstein, 1967; Shimada et al., 1969). The most interesting point to emerge so far is that Nerve Growth Factor (NGF) prolongs the survival of isolated neurons from sensory ganglia and promotes the development of their processes (e.g., Levi-Montalcini and Angeletti, 1963; Varon and Raiborn, 1969). Varon (this volume) also summarizes experiments demonstrating reaggregation and histotypic differentiation of chick-embryo sensory ganglion cells in the presence of NGF. One would anticipate that tissue-culture assays might be developed for other cell "maintenance" factors by the in vitro study of mutants in which particular classes of neurons degenerate selectively during postnatal life, or particular synthetic processes fail (e.g., Wolf and Holden, 1969).

Another example of the combined tissue culture-mutant approach brings the mechanism of cell patterning a bit closer within reach. The experiment involves the mutant mouse *reeler,* one of the most intriguing of the many available mutants (Sidman et al., 1965). The reeler brain develops normally except for the cerebellar cortex, cerebral isocortex, and hippocampal formation, which show disorganization of cell alignment and of intracortical synaptic connections (Hamburgh, 1963; Meier and Hoag, 1962; Sidman, 1968). These regions are the only parts of the CNS where late-forming cells and the earlier-forming cells with which they will make synaptic contact are known to become mutually transposed during normal development. Autoradiographic analysis indicates that, in reeler brain, cells arise in the normal places and at the normal times, migrate normally to the cortex, and then fall, all disoriented, into random positions instead of lining up correctly in relation to their neighbors.

The problem can be considered with reference to cell shape or cell position. Our impression is that the shapes of large neurons are controlled fundamentally by factors intrinsic to the cells themselves, although they are influenced secondarily by interactions with other cells. This point of view has been well expressed by Mugnaini (1969) on the basis of electron-microscope studies. Our evidence comes from analysis of Golgi preparations of CNS of reeler and other mutants (Sidman, 1968). Purkinje cells in the reeler cerebellum display somatic and dendritic forms recognizable as unique to that particular and special class of neurons, even though the cells are markedly abnormal in position, orientation, and synaptic connections. Likewise, in the cerebellum of the *weaver* mutant, the Purkinje cells acquire essentially their normal form even though very few external granule cells survive and migrate inward past them to become granule-cell neurons. In the normal human fetus, the timetable of cerebellar histogenesis is prolonged, and the Purkinje cells mature to a considerable extent before many granule cells have formed, again indicating the relative independence of the Purkinje cell (Rakic and Sidman, 1970). One must add the qualification that the volume and the fine details of dendritic architecture of the Purkinje cells are indeed abnormal when their synaptic input is reduced, as is the case in several other neuronal systems (e.g., Valverde, 1967; Ruiz-Marcos and Valverde, 1969).

Because reeler behaves genetically as a single locus mutation with presumably a single basic effect, can one pinpoint its action to some one aspect of development other than cell shape? Our working hypothesis has been that cell position is that critical parameter and that the reeler locus influences a specific recognition mechanism that allows patterned cell alignment in cortexes (Sidman, 1968).

There are two culture experiments on reeler that pertain to the testing of this hypothesis, one involving organotypic cultures and the other reaggregation cultures. Organotypic cultures of newborn cerebellum allow neurons to mature and make synaptic connections (Wolf, 1964; Crain et al., 1968). Wolf has now established further that the cultures attain a differentiated state with several features of the maturing cerebellar cortex. For example, granule cells migrate inward past Purkinje cells, as described earlier in vivo, make synaptic connections with Purkinje cells, and receive an input (source unknown) via their dendrite terminals in glomeruli of the newly formed granule layer (Wolf et al., 1967; Wolf and Dubois, 1970). In reeler cultures, as in vivo, granule and Purkinje neurons survive and differentiate, at least in part, but fail to attain a normal orientation relative to one another (Wolf, 1970). The reeler disorder thus is intrinsic to the cerebellar cortex and is expressed in vitro.

The reaggregation culture experiments do not involve much new differentiation in vitro, but do give insight into the properties that have already been acquired by cortical cells at the time of explantation. This culture experiment is as follows (DeLong, 1970). Particular regions of normal developing brain are treated with trypsin, Ca-free and Mg-free salt solutions, and mechanical agitation, to give a

suspension of isolated viable cells as described by Moscona for retina and other developing organs (Moscona, 1965). The cells are allowed to reaggregate in culture in shaking flasks that contain liquid medium. Within seven days in vitro, normal cells from cerebellar cortex, cerebral isocortex, hippocampal formation, or olfactory bulb reaggregate and form histotypic structures—not merely neural ones, as distinct from, say, cartilaginous or renal aggregates, but patterns representative of the particular region sampled. The most striking results are obtained with the hippocampal formation, which, in the best cases (Figure 6), forms a curved cell band of nearly constant thickness in which pyramidal-cell neurons are properly aligned relative to one another and are all oriented with normal polarity. Nerve-fiber layers form consistently on each side of the band of pyramidal cells. The position of the aligned cells bears no constant relationship to the surface of the aggregate. At one end, the pyramidal band in some cases widens into a "subiculum," different in cell pattern from hippocampus or from isocortex, and, in rare instances, something of a dentate gyrus appears at the other end of the pyramidal band.

The fidelity of the pattern implies that something more than an "aggregation factor" is at play (DeLong, 1970). There must be an alignment and orientation mechanism

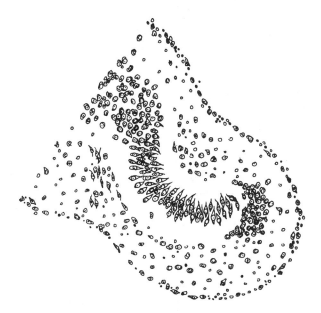

FIGURE 6 Drawing, after the photomicrographs of DeLong (1970), of a section through an aggregate of cells that had been dissociated from the hippocampal region of a normal 18.5-day mouse embryo and cultured for seven days in liquid medium on a rotating shaker at 36.5° C. Cells have not only reaggregated but have segregated, aligned, and oriented themselves in a pattern resembling the organization of the hippocampal formation in vivo at late fetal ages. See text for further details.

with reference to cells and their immediate neighbors. This mechanism must be differentially distributed over the cell surfaces (or must be mediated by an asymmetrically oriented molecule), for the cells always line up side to side, never end to end, and always show polar orientation. Also, the expression of the mechanism must be independent of afferent input or precise position of cell relative to the ependymal-pial coordinates.

The mechanism is closely time-dependent. Accurate alignment of cells, as just described in hippocampus, occurs only when embryos of 18.5 days are used. When the tissue is taken about half a day earlier, the alignment is incomplete. Half a day later, the cells of various sizes aggregate but show no tendency to sort out and align. The same general properties are found in the other regions of CNS that were tested, although the temporal requirements were not quite so exacting (DeLong, 1970).

Cultures of cerebral isocortex from reeler and control litter mates were compared (DeLong and Sidman, 1970). Cells of control fetuses regularly form a radial cortical pattern, but cells of reeler litter mates fail to develop a comparable architectural arrangement. Reeler cells aggregate and probably migrate within the aggregate, but do not align side to side in parallel, lack the normal polarity with outward-directed apical processes, and fail to form layers. These aggregating cultures thus show some of the same defects observed in intact fetal reeler cortex. Cells in cultures of cerebellar cortex behave similarly. Cultures of olfactory bulb, however, are the same in reeler and control, just as they are both normal in vivo, which argues against some nonspecific disorder of reeler cells in vitro.

The alignment mechanism controlled by the wild type of allele at the reeler locus must be (1) intrinsic to the dissociated cells, (2) operative during a restricted time period, and (3) dependent on a surface-bound agent, as inferred from the accuracy of the alignment process. The extraordinary precision of developing synaptic connections in vivo (as described by Bodian, Szentágothai, and others in this volume), with different types of synapses acquired at different times and yet coming to occupy adjacent territories on the surface of a given cell, also would seem to argue in favor of a surface-bound orienting mechanism rather than a diffusible agent.

The problem of cell alignment in the developing nervous system would seem to be ripe for the theoreticians, and almost ready for the chemists.

## REFERENCES

AMBLER, M., S. POGACAR, and R. SIDMAN, 1969. Lhermitte-Duclos disease (granule cell hypertrophy of the cerebellum). Pathological analysis of the first familial cases. *J. Neuropathol. Exp. Neurol.* 28: 622–647.

ANGEVINE, J. B., JR., and R. L. SIDMAN, 1961. Autoradiographic study of cell migration during histogenesis of cerebral cortex in the mouse. *Nature (London)* 192: 766–768.

BERRY, M., and A. W. ROGERS, 1965. The migration of neuroblasts in the developing cerebral cortex. *J. Anat.* 99: 691–709.

BOULDER COMMITTEE, 1970. Embryonic vertebrate central nervous system: Revised terminology. *Anat. Rec.* 166: 257–261.

BRAITENBERG, V., and R. P. ATWOOD, 1958. Morphological observations on the cerebellar cortex. *J. Comp. Neurol.* 109: 1–33.

CRAIN, S. M., E. R. PETERSON, and M. B. BORNSTEIN, 1968. Formation of functional interneuronal connexions between explants of various mammalian central nervous tissues during development *in vitro. In* Growth of the Nervous System/A Ciba Foundation Symposium (G. E. W. Wolstenholme and M. O'Connor, editors). J. and A. Churchill, Ltd., London, pp. 13–31.

CUNNINGHAM, D. J., and T. TELFORD-SMITH, 1895. The brain of the microcephalic idiot. *Sci. Trans. Roy. Dublin Soc., Ser. 2,* 5: 287–360.

DELONG, G. R., 1970. Histogenesis of fetal mouse isocortex and hippocampus in reaggregating cell cultures. *Develop. Biol.,* 22: 563–583.

DELONG, G. R., and R. L. SIDMAN, 1970. Alignment defect of reaggregating cells in cultures of developing brains of Reeler mutant mice. *Develop. Biol.,* vol. 22: 584–600.

ESSICK, C. R., 1907. The corpus ponto-bulbare—A hitherto undescribed nuclear mass in the human hind brain. *Amer. J. Anat.* 7: 119–135.

FUJITA, S., 1966. Applications of light and electron microscopic autoradiography to the study of cytogenesis of the forebrain. *In* Evolution of the Forebrain. Phylogenesis and Ontogenesis of the Forebrain (R. Hassler and H. Stephan, editors). G. Thieme Verlag, Stuttgart, pp. 180–196.

FUJITA, S., 1967. Quantitative analysis of cell proliferation and differentiation in the cortex of the postnatal mouse cerebellum. *J. Cell Biol.* 32: 277–287.

GOLDSTEIN, M. N., 1967. Incorporation and release of $H^3$-catecholamines by cultured fetal human sympathetic nerve cells and neuroblastoma cells. *Proc. Soc. Exp. Biol. Med.* 125: 993–996.

GRACHEVA, N. D., 1967. Autoradiography of Nucleic Acid and Protein Synthesis in the Nervous System. (In Russian.) Nauka Press, Leningrad.

HAMBURGH, M., 1963. Analysis of the postnatal developmental effects of "reeler," a neurological mutation in mice. A study in developmental genetics. *Develop. Biol.* 8: 165–185.

HERMAN, L., and S. L. KAUFFMAN, 1966. The fine structure of the embryonic mouse neural tube with special reference to cytoplasmic microtubules. *Develop. Biol.* 13: 145–162.

HICKS, S. P., and C. J. D'AMATO, 1968. Cell migration to the isocortex in the rat. *Anat. Rec.* 160: 619–634.

HIS, W., 1904. Die Entwickelung des menschlichen Gehirns während der ersten Monate. Hirzel, Leipzig.

KÄLLÉN, B., 1965. Early morphogenesis and pattern formation in the central nervous system. *In* Organogenesis (R. L. De Haan and H. Ursprung, editors). Holt, Rinehart and Winston, New York, pp. 107–128.

KAUFFMAN, S. L., 1968. Lengthening of the generation cycle during embryonic differentiation of the mouse neural tube. *Exp. Cell Res.* 49: 420–424.

LANGMAN, J., 1968. Histogenesis of the central nervous system. *In* The Structure and Function of Nervous Tissue (G. H. Bourne, editor). Academic Press, New York, Vol. 1, pp. 33–65.

LEVI-MONTALCINI, R., 1963. Growth and differentiation in the nervous system. *In* The Nature of Biological Diversity (J. M. Allen, editor). McGraw-Hill, New York, pp. 261–295.

LEVI-MONTALCINI, R., and P. U. ANGELETTI, 1963. Essential role of the nerve growth factor in the survival and maintenance of dissociated sensory and sympathetic embryonic nerve cells in vitro. *Develop. Biol.* 7: 653–659.

LYSER, K. M., 1968. An electron-microscope study of centrioles in differentiating motor neuroblasts. *J. Embryol. Exp. Morphol.* 20: 343–354.

MEIER, H., and W. G. HOAG, 1962. The neuropathology of "reeler," a neuro-muscular mutation in mice. *J. Neuropathol. Exp. Neurol.* 21: 649–654.

MENKES, B., C. ALEXANDRU, and O. TUDOSE, 1967. Investigation on pre- and postmitotic movement of the neuroblasts, in the embryonic neural tube. *Rev. Roum. Embryol. Cytol., Ser. Embryol.* 4: 29–33.

MIALE, I. L., and R. L. SIDMAN, 1961. An autoradiographic analysis of histogenesis in the mouse cerebellum. *Exp. Neurol.* 4: 277–296.

MOREST, D. K., 1969a. The differentiation of cerebral dendrites: A study of the post-migratory neuroblast in the medial nucleus of the trapezoid body. *Z. Anat. Entwicklungsgesch.* 128: 271–289.

MOREST, D. K., 1969b. The growth of dendrites in the mammalian brain. *Z. Anat. Entwicklungsgesch.* 128: 290–317.

MOSCONA, A. A., 1965. Recombination of dissociated cells and the development of cell aggregates. *In* Cells and Tissues in Culture, Vol. 1 (E. N. Willmer, editor). Academic Press, London, pp. 489–529.

MUGNAINI, E., 1969. Ultrastructural studies on the cerebellar histogenesis. II. Maturation of nerve cell populations and establishment of synaptic connections in the cerebellar cortex of the chick. *In* Neurobiology of Cerebellar Evolution and Development (R. Llinás, editor). American Medical Association, Education and Research Foundation, Chicago, pp. 749–782.

MUGNAINI, E., and P. F. FORSTRØNEN, 1967. Ultrastructural studies on the cerebellar histogenesis. I. Differentiation of granule cells and development of *glomeruli* in the chick embryo. *Z. Zellforsch. Mikroskop. Anat.* 77: 115–143.

RAKIC, P., and R. L. SIDMAN, 1969. Telencephalic origin of pulvinar neurons in the fetal human brain. *Z. Anat. Entwicklungsgesch.* 129: 53–82.

RAKIC, P., and R. L. SIDMAN, 1970. Histogenesis of cortical layers in the human cerebellum, particularly the lamina dissecans. *J. Comp. Neurol.*, 139: 473–500.

RAMÓN Y CAJAL, S., 1960. Studies on Vertebrate Neurogenesis (L. Guth, translator). Charles C Thomas, Springfield, Illinois.

RUIZ-MARCOS, A., and F. VALVERDE, 1969. The temporal evolution of the distribution of dendritic spines in the visual cortex of normal and dark raised mice. *Exp. Brain Res.* 8: 284–294.

SHIMADA, Y., D. A. FISCHMAN, and A. A. MOSCONA, 1969. Formation of neuromuscular junctions in embryonic cell cultures. *Proc. Nat. Acad. Sci. U. S. A.* 62: 715–721.

SIDMAN, R. L., 1968. Development of interneuronal connections in brains of mutant mice. *In* Physiological and Biochemical Aspects of Nervous Integration (F. D. Carlson, editor). Prentice-Hall, Englewood Cliffs, New Jersey, pp. 163–193.

SIDMAN, R. L., 1970. Autoradiographic methods and principles for study of the nervous system with thymidine-H[3]. *In* Contemporary Research Techniques of Neuroanatomy (S. O. E. Ebbesson and W. J. Nauta, editors). Springer-Verlag, New York (in press).

SIDMAN, R. L., M. C. GREEN, and S. H. APPEL, 1965. Catalog of the Neurological Mutants of the Mouse. Harvard University Press, Cambridge, Massachusetts.

STENSAAS, L. J., and S. S. STENSAAS, 1968. An electron microscope study of cells in the matrix and intermediate laminae of the cerebral hemisphere of the 45 mm rabbit embryo. *Z. Zellforsch. Mikroskop. Anat.* 91: 341–365.

VALVERDE, F., 1967. Apical dendritic spines of the visual cortex and light deprivation in the mouse. *Exp. Brain Res.* 3: 337–352.

VARON, S. S., and C. W. RAIBORN, JR., 1969. Dissociation, fractionation, and culture of embryonic brain cells. *Brain Res.* 12: 180–199.

WATTERSON, R. L., 1965. Structure and mitotic behavior of the early neural tube. *In* Organogenesis (R. L. De Haan and H. Ursprung, editors). Holt, Rinehart and Winston, New York, pp. 129–159.

WECHSLER, W., 1966. Die Feinstruktur des Neuralrohres und der neuroektodermalen Matrixzellen am Zentralnervensystem von Hühnerembryonen. *Z. Zellforsch. Mikroskop. Anat.* 70: 240–268.

WOLF, M. K., 1964. Differentiation of neuronal types and synapses in myelinating cultures of mouse cerebellum. *J. Cell Biol.* 22: 259–279.

WOLF, M. K., 1970. Anatomy of cultured mouse cerebellum II: A reappraisal of structures demonstrated by silver impregnation. *J. Comp. Neurol.* (in press).

WOLF, M. K., and M. DUBOIS, 1970. Anatomy of cultured mouse cerebellum I: Golgi and electron microscopic demonstration of granule cells, their afferent and efferent synapses. *J. Comp. Neurol.* (in press).

WOLF, M. K., and A. B. HOLDEN, 1969. Tissue culture analysis of the inherited defect of central nervous system myelination in jimpy mice. *J. Neuropathol. Exp. Neurol.* 28: 195–213.

WOLF, M. K., A. B. HOLDEN, and P. M. HARLAN, 1967. The fate of the granule cell precursors in myelinating cultures of mouse cerebellum. *Anat. Rec.* 157: 343 (abstract).

# 11 Development of the Vertebrate Motor System

## ALFRED J. COULOMBRE

### General considerations

THE DEVELOPMENT OF THE MOTOR SYSTEM of vertebrates is considered here from three points of view: the chronology of its component steps, the general principles that govern it, and some of the unanswered questions that relate to it. Because this essay is in no sense exhaustive, the several reviews on which it leans heavily are cited at appropriate points. General reviews of vertebrate neurogenesis have been published by Detwiler (1936), Edds (1967), Hamburger (1955), Harrison (1935), Hughes (1968), Kollros (1968), Piatt (1948), Ramón y Cajal (1960), Weiss (1950, 1955).

Several generalizations that relate to the nervous system as a whole must be considered before we proceed to an analysis of the developing motor system. During its development, the nervous system follows a sequence of steps which, although similar in different vertebrate species, take place in grossly different periods of time, ranging from a few weeks in some rodents and birds to many months in some mammals. Among poikilothermes, the rate of development of the nervous system has been accelerated or retarded even within the same species (Kollros, 1968). Such differences in the duration of neural development are not accompanied by any change in the order in which major developmental steps occur. Thus, the principal events in vertebrate neurogenesis appear to occur in an obligatory temporal sequence. Furthermore, synchrony of development is maintained between the nervous system and the periphery it innervates, regardless of the duration of development, which suggests that the synchrony is achieved by interactions between the center and the periphery as development proceeds.

Another generalization (Eccles, 1968) is that the nervous systems of vertebrates, the most complex organizations of matter known to us, develop during their early stages in the total absence of any sensory contact with the environment with which they must ultimately cope. Because the developing nervous system is thus removed from any pos-

sibility that sensory input can play a role in its early embryogenesis, the steps in its maturation must depend on factors that are, for the most part, intrinsic to the organism. This autonomy extends even to the relationships among parts of the nervous system. For example, the motor system, which always begins to differentiate before the sensory system (Ramón y Cajal, 1960; cf. Hamburger, this volume), completes most of its maturation without sensory input, and is even capable of initiating and sustaining some complex motor behaviors in the absence of that input (Weiss, 1941a).

### Origins of the motor system

EARLY DEVELOPMENT The vertebrate motor system is constructed of nervous elements derived from embryonic ectoderm and of effector elements derived principally from embryonic mesoderm. Early in development, the primitive neural plate and the neural crest that ajoins it arise from ectoderm under the influence of chorda mesoderm (reviewed by Saxén and Toivonen, 1962). At this stage, the earliest known topographic mapping takes place in the embryonic nervous system. A representation of the anteroposterior axis of the soma gradually becomes fixed, commencing at the anterior pole of the plate. The mediolateral axis is fixed later in development. An important generalization is that it is usual for the several axes defining a common topographic system to become fixed, each at a different point in developmental time. This observation limits the types of hypotheses we may entertain concerning the nature of the mechanisms that establish topographic axes.

The production of neural folds and their closure to form a neural tube are accompanied by a further segregation of the neuroepithelial cells into alar (or sensory) plate and basal (or motor) plate, separated by the lateral sulcus. This separation imposes yet another map on the central nervous system in providing a central representation of sensory systems, which are largely of ectodermal derivation, and of motor systems, which are almost exclusively of mesodermal derivation. We focus here on some aspects of the development of the posterior (spinal cord) portion of the basal plate and of the peripheral effectors with which it establishes connections.

ALFRED J. COULOMBRE Ophthalmology Branch, National Institute of Neurological Diseases and Stroke, National Institutes of Health, Bethesda, Maryland

STRATEGIES OF INTERMEDIATE DEVELOPMENT One of the central problems of neurogenesis is how specific central stations in the motor portion of the spinal cord become linked with appropriate effector structures. There are three types of developmental pattern, any one of which could result in the establishment of appropriate connections between the center and the periphery. Regional differences might develop first in the effector periphery, and secondarily determine the connections accepted by the motor neurons by means of centripetal influences transferred along the motor axons, a process called *myotypic specification* (Weiss 1941a).

Alternatively, a central representation of the peripheral topography might develop first and subsequently impose its pattern on an initially indifferent periphery by means of centrifugal influences through the motor axons. Both of these mechanisms would require that pathways be specified in the tissues intervening between successive levels in the basal plate and specific regions of premuscle mesenchyme in the periphery. A final possibility is that topographically congruous maps might arise independently at the center and at the periphery, and that the motor axons might subsequently grow out to interconnect corresponding portions of the two topographies. In this case, the motor neurons could reach the periphery along pathways that had been previously specified in the intervening tissues, or they might grow out initially at random, consolidating as specific nerve pathways only after some of them had made appropriate associations at the periphery. It remains to be determined which of these mechanisms obtains within each of the subsystems of the developing nervous system, whether there is a shift phylogenetically in the mechanism employed, and even whether different mechanisms operate for the several topographies and modalities that may be mapped within the same subsystem.

## Emergence of central topographies

MORPHOGENETIC CELL DEATH The central component of the motor system begins to differentiate earlier than does the effector system (Ramón y Cajal, 1908), and independently of it. The basal plate of the primitive spinal cord contains an undifferentiated column of motor neuroblasts along each side (Harris, 1965). In at least some urodeles this differentiates into a medially situated column of primary motor neurons that supply the axial musculature and a ventrolateral column of motor neurons (Woodburne, 1939). The tailed amphibians do not appear to progress beyond this level of motor organization and do not develop brachial (forelimb) or lumbar (hind-limb) enlargements of the spinal cord as do anurans, reptiles, birds, and mammals. These enlargements of the cord in the regions that supply the limbs are due to the development of the ventral horns

from cells belonging to the ventrolateral, motor-cell column. In the embryo of the domestic fowl (Levi-Montalcini, 1950; Hamburger, 1952), the limb enlargements are sculptured out of the ventrolateral motor column by extensive *morphogenetic* cell degeneration at the cervical level and by medial migration of cells to form the visceral columns at the thoracic and sacral levels. Extensive cell death also plays a role in the morphogenesis of the lateral motor column of the mouse (Harris, 1965). These early events in the differentiation of the motor column and the formation of the ventral horns proceed independently of the developing periphery (Harrison, 1904; Hamburger, 1928, 1958; Wenger, 1951; Hughes and Tschumi, 1958). Once the ventral horns have emerged, they continue to lengthen by addition of neuroblasts at their posterior ends, as was demonstrated in *Xenopus* by Hughes (1961, 1963).

HISTOGENETIC CELL DEATH Far more cells are recruited and differentiate in the ventrolateral motor column than will be used ultimately by the muscular periphery. The numbers are reduced to appropriate levels as development proceeds. This is accomplished by *histogenetic* cell degeneration (cf. Prestige, this volume). The number of ventral-horn cells that survive this late wave of cell death is under control of the developing limb, falling to lower-than-normal levels after the extirpation of limb rudiments and remaining larger than normal when supernumerary limbs are grafted on (Braus, 1906; Levi-Montalcini and Levi, 1942; Hamburger and Levi-Montalcini, 1949; Beaudoin, 1955; Kollros, 1956; Baird, 1957; Kollros and Race, 1960; Race, 1961; Hughes, 1962, 1963, 1965, 1968; Race and Terry, 1965; Prestige, 1967).

FORMATION OF FUNCTION-SPECIFIC GROUPINGS OF MOTOR CELLS The further development of the limb enlargements involves not only continuing cell recruitment and cell death to adjust cell number, but also cell differentiation, cell enlargement, and the remodeling of function-specific groupings of motor cells. As the ventral horns influence and are influenced by the developing limb musculature that they subtend, their cells form topographic arrays which represent the effector periphery. Partial descriptions of this architecture exist. Much less is known concerning the extrinsic and intrinsic forces that mold it.

There is evidence that the motor organization at limb levels of the cord is influenced, at least in part, and in at least some amphibians, by extrinsic influences originating in the limb musculature. In urodeles, a supernumerary limb grafted near the forelimb of a host will receive innervation from part of the adjacent brachial plexus and will move, joint for joint, in synchrony with the host limb (Weiss, 1922; Detwiler, 1925). This phenomenon was called by Weiss (1931, 1937) a *homologous response of homonymous muscles,* that is, muscles of similar name or function. In addition to its occurrence in urodeles, it has been demonstrated experimentally in fish (Sperry, 1950; Arora and Sperry, 1957; Sperry and Arora, 1965), in anurans (Weiss, 1941a; Sperry, 1947), and inferred indirectly in man from the behavior of supernumerary digits (Weiss, 1935; Weiss and Ruch, 1936). There seems to be no report of the phenomenon in reptiles and birds. Developmentally, homologous response can be elicited throughout life for some muscle systems in fishes and urodeles, during premetamorphic (larval) life in anurans (Weiss, 1941a; Sperry, 1947), and not at all in postnatal mammals, as judged by persistent dysfunction of muscles after nerves have been cross-sutured (Sperry, 1945). There are regional as well as temporal strictures on the phenomenon of homologous response. Limbs transplanted to the head region become innervated and subsequently move not in synchrony with the host limbs but usually with some of the nearby musculature of the head (Nicholas, 1929, 1930a, 1930b; Detwiler, 1930; Hibbard, 1965). Homologous responses are observed only when the donor limb is grafted close enough to that of the host to receive at least part of its innervation from the brachial plexus (Detwiler, 1933). After several earlier interpretative attempts, the phenomenon of homologous response gave rise to the concept of *neuronal modulation,* which embraced specification of the ventral motor neurons by the motor periphery (Weiss, 1941a). The concept suggests that regional specificities which develop peripherally are communicated to the ventral-horn cells centripetally along neuronal cell processes and determine the synaptic associations that will be formed with the central neurons. In the case of striated skeletal muscle, this dictation of central specificity by individual muscles (Weiss, 1931), or even myoneural units, has been called *myotypic specification* (Weiss, 1941a).

This influence of the periphery on the motor cells must be mediated via motor fibers, because homologous response will develop in the absence of sensory innervation (Weiss, 1937). The concept of myotypic specification would assure that each motor neuron would establish central synaptic associations appropriate to the muscle it supplied. Such connections are presumed to be made within the pattern of coordination among the muscles (e.g., the temporal sequence of stimulation or inhibition of the antagonists which move each limb joint). We shall see that such a functional pattern is, in fact, represented in the ventral cord and that it develops to some extent independently of peripheral influence.

As the ventral motor column matures, it becomes internally organized to represent the muscular topography of the limb it subtends. Early in development, the proximodistal axis of the limb is represented in anteroposterior sequence in the ventral horn. In the brachial ventral horn of the urodele, each muscle tends to become represented by several groups of motor neurons at different segmental levels (Székely and Czéh, 1967). These muscle-specific nests of neurons, although widely scattered, are not placed entirely at random, but rather are grouped into large fields representing functional groups of muscles (e.g., flexors, extensors, protractors, retractors) which operate on a given joint. The functional fields overlap and interpenetrate each other extensively. Székely (1968) has suggested that the widespread and multiple representation of limb muscles in the urodele ventral horn may explain why a nerve branch from any segment in this form can innervate a transplanted limb and cause it to move in synchrony with the host limb. On this basis, he has raised one of the rare questions concerning the explanation of the phenomenon of homologous responses.

Each muscle in the mammal tends to be mapped uniquely on the final, common, motor cells of the ventral horn. As development proceeds, the distal portions of the limb musculature become represented more dorsally in the ventrolateral motor column than does the proximal musculature, and extensor muscles more laterally than the flexor muscles. Finally, the topographic representations of individual muscles tend to become clustered in groups, each representing a joint (reviewed by Romanes 1953, 1964). Silver (1942) has demonstrated a somewhat similar arrangement in the frog. It is not yet known how the more primitive pattern of representation is restructured into the more complex patterns characteristic of the more mature animal. It remains to be clarified what roles are played in this process by cell migration, cell death, the retraction of older constellations of connections, and the establishment of new groupings. Beyond this we need information concerning how numerous morphological topographies and functional modalities are mapped in a common framework of time and space.

The integrated motor patterns that are represented by the functional grouping of motor neurons appear to develop independently of sensory input from the periphery. Deafferentation prevents neither the development (Weiss, 1937), nor the expression of coordinated movement (Weiss, 1936). These coordinated motor patterns not only develop autonomously, but are regionally specific. By transplanting different sections of larval or embryonic cord into juxtaposition with forelimbs or hind-limbs, or vice versa, Székely (1963) showed in salamanders, and Strasnicky (1963) in chickens, that the pattern of movement that ensued was dictated by the brachial and lumbar sections of the cord, respectively, and not by the limb (cf. Hamburger, this volume).

## Differentiation of the motor periphery

SUPPORT OF MUSCLE FIBER DIFFERENTIATION BY MOTOR FIBERS  Just as the central portions of the motor system show some autonomy in their early development, so the limb musculature at the periphery develops *to a point,* and persists *for a time* without innervation (Harrison, 1904; Hunt, 1932; Eastlick, 1943: Eastlick and Wortham, 1947; Hamburger and Levi-Montalcini, 1950; Singer, 1952). Myoblasts that have been explanted into tissue cultures with or without cloning develop cross-striations and contract (Capers, 1960; Nakai, 1965; Konigsberg, 1963), although their nuclei tend to remain centrally situated in the cell rather than assuming peripheral positions as they do in fully differentiated skeletal muscle fibers in vivo (Zelena, 1962). Muscle cells require innervation to retain their integrity and complete their differentiation. The muscle fibers of aneurogenic limbs tend to degenerate in time if they do not receive innervation (Hamburger, 1928, 1939). This ability of the motor-nerve fiber to stabilize the state of differentiation of muscle fibers is not shared by sensory fibers (Weiss and Edds, 1945; Gutmann, 1945).

Beyond assuring the mere survival of the muscle fibers, the axonal connection with the center supports a range of qualitative changes in the effector cells subtended by the spinal nerves (Tower, 1935; Buller et al., 1960a; Gutmann and Hník, 1962). Some of the changes occur in all innervated skeletal muscle fibers. These include: the formation of a sole plate in response to the tip of the axon of the motor neuron (Zelena and Szentágothai, 1957; Zelena, 1959); the exclusion, especially in higher vertebrates, of other axons from the formation of additional junctions on the same muscle fiber (Harrison, 1910; Elsberg, 1917; Aitken, 1950; Hoffman, 1951); a decrease in acetylcholine sensitivity on all surface areas not in or close to the myoneural junction (Ginetzinsky and Shamarina, 1942; Kuffler, 1943; Axelsson and Thesleff, 1959; Miledi, 1960); and an increase in cholinesterase activity near the myoneural junction (Mumenthaler and Engel, 1961) and a decrease elsewhere (Kovács et al., 1961).

NERVE-DEPENDENT DIFFERENTIATION OF "FAST" AND "SLOW" MUSCLE FIBERS Another group of changes that take place after innervation affects striated muscle fibers differentially. In mammals, muscle fibers normally begin to contract shortly after the nerve fibers touch them (Straus and Weddell, 1940; Diamond and Miledi, 1962). Initially, all muscle fibers have the relatively long contraction times characteristic of "slow" muscle (Denny-Brown, 1929; Buller et al., 1960a, 1960b; Buller and Lewis, 1963, 1964; Close, 1964, 1965). Later, some muscle units become "fast" (short contraction time), while others remain slow (Eccles et al., 1962; Ridge, 1967). "Fast" and "slow" muscles differ from each other quantitatively in a number of ways, e.g., speed of calcium transport (Mommaerts, 1968), density of capillary beds (Romanul and Pollack, 1969), myoglobin levels (McPherson and Tokunaga, 1967), and so forth.

Nerve fibers sustain these differences because muscles revert to the primitive "slow" state after denervation, and assume the fast or slow condition on reinnervation by their appropriate nerves. Each muscle fiber appears to be able to assume the "fast" or the "slow" state throughout life. The nerve fibers that innervate the muscle determine which state will be stabilized, as cross innervation of "fast" and "slow" muscles is followed by a reversal in their types (Buller et al., 1960a; Buller and Lewis, 1965; Close, 1965).

## Connections between center and periphery

ESTABLISHMENT OF SPECIFIC MOTOR PATHWAYS Whatever mechanisms assure the establishment and maintenance of appropriately linked central topographic maps and peripheral effector arrangements, the problem remains of how the motor axons find their ways along appropriate pathways from the basal plate derivatives across intervening tissues to specific motor regions in the limb. Harrison (1914) called attention to one ingredient of the solution of this problem when he demonstrated in vitro that fibers grew out of an explant only when a physical interface, such as that provided by an experimentally introduced spider web, was present along which they could grow. This phenomenon has been extensively investigated under the name "contact guidance" by Weiss (1928, 1934, 1941b, 1945, 1955). Yet an unanswered question remains of whether the motor axons, the earliest nerve cell processes to emerge from the nervous system, grow at random along the physical interfaces available to them or whether, at the outset, they enter pathways that are specific for their destinations in specific muscles. The early "pathfinder" fibers (as Harrison, 1910, called them) emerge before there is detectable differentiation of the limb musculature. The first fibers that have been stained with silver are parallel with one another and are directed toward the periphery. They very quickly form spinal nerves that correspond in number to the dorsal root ganglia. This segmental arrangement is secondarily imposed by the somatic segmentation (Detwiler, 1936). At limb levels, the motor and sensory fibers emerge from the spinal nerves in bundles that reflect the major nerve branches of the future limb (Harrison, 1910; Taylor, 1943; Nieuwkoop and Faber, 1956; Hughes, 1965). These cables form as younger fibers grow along the pathways laid down and consolidated by the pioneer fibers, a process called *fasciculation.* It is also not clear whether this early fasciculation is random or selective. It is clear that, later in development when the matter can be analyzed, selective associations of sensory or motor neurons, or both, have been parceled out, presumably as the result of a selective affinity among them earlier in development (Weiss, 1955; Hamburger, 1962). Because these early fascicles already reflect the nerve pattern of the adult limb, such patterns must be in part intrinsic to the central nervous system (Hughes, 1968, p. 53), although it seems plausible that these intrinsic patterns are expressed only in response to the cues provided by patterns in the otherwise undifferentiated mesenchyme of the limb.

In considering the forces that guide the motor axons to appropriate destinations within the limb, we appear to be dealing with both a generalized guidance (Detwiler, 1933, 1936, p. 80; Detwiler and Van Dyke, 1934; Wieman and Nussmann, 1929) to which neurons react indiscriminately, and a guidance system, which, at least in higher vertebrates, assures that specific groups of axons follow specific pathways to specific destinations. Although the limbs of adult vertebrates lie at relatively more posterior levels than do the cord segments that supply them, Miller and Detwiler (1936) showed that during the stage of its initial innervation the limb lies immediately adjacent to the involved segments of the spinal cord. By transplanting limb buds of embryos of *Amblystoma* to different anteroposterior levels at different stages of development, Detwiler (1933, 1936) demonstrated three phenomena relating to the innervation of the limb. First, the spinal-cord levels that innervate the limb are not completely determined by the level opposite which the limb happens to lie at the time of motor-axon outgrowth. Spinal nerves from the levels that normally innervate a limb will grow posteriorly *or* anteriorly, and make connection with a limb that has not been too far displaced from its usual site. It must be stressed that this generalized attraction is not necessarily limb-specific, as other organs, such as the eye, will become innervated by the limb plexus when they are transplanted to these sites (Detwiler and Van Dyke, 1934). Second, spinal nerves will thus "seek out" a displaced limb for only a limited "window" in developmental time. Third, a limb that has been transplanted so far from its normal site that it does not receive some input from those cord levels that normally supply it will not develop the coordinated movements that characterize a limb. Narayanan (1964) has confirmed these findings for the chick embryo. He has, in addition, shown that the detailed pattern of anastomosis and branching in the brachial plexus is reproduced even when limb primordia have been so far displaced that they are innervated by spinal levels that do not normally supply them. This observation could be interpreted to suggest that the pathways along which the nerve fibers grow into a limb are specified in the tissues of the limb prior to its invasion by nerve fibers. The results of other experiments, however, argue as forcibly that the spinal cord determines the geometry of the pathways that the nerve fibers follow into the limbs. Thus, Castro (1963) has shown that partial unilateral removal of the brachial cord of the chick embryo results in the formation of only that portion of the brachial innervation that the residual brachial cord would normally have formed. Piatt (1942, 1952, 1956, 1957) transplanted aneurogenic limbs of *Ambystoma* to different cord levels, or replaced the normal brachial segments by cord segments from different anteroposterior levels, and showed that a normal brachial nerve pattern resulted only when the forelimb became innervated by fibers arising from brachial levels of the cord. The data that indicate that the limb dictates those nerve channels the motor fibers will follow do not necessarily contradict the data that implicate the cord in determining how these axons assort into appropriate pathways. The connections of the motor-column cells with the peripheral muscles can be re-established by regeneration if the motor axons are severed. Later in the life of the organism, motor fibers of skeletal muscle re-establish endings rather indiscriminately with any type of skeletal muscle fiber that is not already pre-empted (Weiss and Hoag, 1946; Buller et al., 1960b; Bernstein and Guth, 1961; Guth, 1962; Eccles et al., 1962). Such lack of selectivity on the part of regenerating motor axons may however, vary in degree. For example, some preganglionic sympathetic fibers do "seek out" ganglion cells of appropriate modality (Langley, 1897) and will, in some cases, even displace inappropriate synapses that were established prior to reinnervation (Guth and Bernstein, 1961).

THE ROLE OF SCHWANN CELLS  The first motor axons emerge from the basal plate before any Schwann cells have reached the vicinity (Harrison, 1904, 1906, 1907, 1908, 1924; Ramón y Cajal, 1960, p. 19). Removal of the neural crest, which prevents development of the Schwann cells (reviewed by Hörstadius, 1950), retards but does not prevent the outgrowth of axons. Therefore, the Schwann cells do not influence the direction of growth of these earliest ventral-root fibers. During later development, however, as during regeneration of peripheral nerves (Ramón y Cajal, 1959), growing axons tend to follow pathways defined by Schwann cells.

The affinity of Schwann cells for axons at the proper stage of differentiation leads to their aggregation in chains (Speidel, 1933; Peterson and Murray, 1955). The Schwann-cell pathway produced by such aggregation may serve to guide the outgrowth of fibers that emerge later in development. The type of motor axon with which the Schwann cell becomes associated dictates whether myelin sheath is elaborated (Hillårp and Olivecrona, 1946). If the motor fiber is destined to be of the large-diameter, rapidly conducting variety, a myelin sheath will form in the manner reviewed by Fernández-Morán (1957). As soon as the motor axon has become invested by a myelin sheath interrupted by nodes of Ranvier, its rate of conduction increases appreciably (Carpenter and Bergland, 1957).

## REFERENCES

AITKEN, J. T., 1950. Growth of nerve implants in voluntary muscle. *J. Anat.* 84: 38–49.

ARORA, H. L., and R. W. SPERRY, 1957. Myotypic respecification of regenerated nerve-fibres in cichlid fishes. *J. Embryol. Exp. Morphol.* 5: 256–263.

AXELSSON, J., and S. THESLEFF, 1959. A study of supersensitivity in denervated mammalian skeletal muscle. *J. Physiol. (London)* 147: 178–193.

BAIRD, J. J., 1957. Normal development of motor nerve cells in the lumbosacral cord of anurans and development following partial limb ablation and subsequent regeneration. University of Iowa, Iowa City, doctoral thesis.

BEAUDOIN, A. R., 1955. The development of lateral motor column cells in the lumbo-sacral cord in *Rana pipiens*. I. Normal development and development following unilateral limb ablation. *Anat. Rec.* 121: 81–95.

BERNSTEIN, J. J., and L. GUTH, 1961. Nonselectivity in establishment of neuromuscular connections following nerve regeneration in the rat. *Exp. Neurol.* 4: 262–275.

BRAUS, H., 1906. Vordere Extremität und Operculum bei Bombinatorlarven. Ein Beitrag zur Kenntnis morphogener Correlation und Regulation. *Morphol. Jahrb.* 35: 509–590.

BULLER, A. J., J. C. ECCLES, and R. M. ECCLES, 1960a. Differentiation of fast and slow muscles in the cat hindlimb. *J. Physiol. (London)* 150: 399–416.

BULLER, A. J., J. C. ECCLES, and R. M. ECCLES, 1960b. Interactions between motoneurones and muscles in respect of the characteristic speeds of their responses. *J. Physiol. (London)* 150: 417–439.

BULLER, A. J., and D. M. LEWIS, 1963. Factors affecting the differentiation of mammalian fast and slow muscle fibers. *In* The Effect of Use and Disuse on Neuromuscular Functions (E. Gutmann and P. Hník, editors). Elsevier Publishing Co., Amsterdam, pp. 149–159.

BULLER, A. J., and D. M. LEWIS, 1964. Further observations on the differentiation of skeletal muscles in the kitten hind limb. *J. Physiol. (London)* 176: 355–370.

BULLER, A. J., and D. M. LEWIS, 1965. Some observations on the effects of tenotomy in the rabbit. *J. Physiol. (London)* 178: 326–342.

CAPERS, C. R., 1960. Multinucleation of skeletal muscle *in vitro*. *J. Biophys. Biochem. Cytol.* 7: 559–566.

CARPENTER, F. G., and BERGLAND, R. M., 1957. Excitation and conduction in immature nerve fibers of the developing chick. *Amer. J. Physiol.* 190: 371–376.

CASTRO, G. DE O., 1963. Effects of reduction of nerve centers on the development of residual ganglia and on nerve patterns in the wing of the chick embryo. *J. Exp. Zool.* 152: 279–293.

CLOSE, R., 1964. Dynamic properties of fast and slow skeletal muscles of the rat during development. *J. Physiol. (London)* 173: 74–95.

CLOSE, R., 1965. Effects of cross-union of motor nerves to fast and slow skeletal muscles. *Nature (London)* 206: 831–832.

DENNY-BROWN, D., 1929. On the nature of postural reflexes. *Proc. Roy. Soc., ser. B, biol. sci.* 104: 252–301.

DETWILER, S. R., 1925. Coordinated movements in supernumerary transplanted limbs. *J. Comp. Neurol.* 38: 461–493.

DETWILER, S. R., 1930. Obervations on the growth, function, and nerve supply of limbs when grafted to the head of salamander embryos. *J. Exp. Zool.* 55: 319–379.

DETWILER, S. R., 1933. Experimental studies upon the development of the amphibian nervous system. *Biol. Rev. (Cambridge)* 8: 269–310.

DETWILER, S. R., 1936. Neuroembryology: An Experimental Study. The Macmillan Co., New York.

DETWILER, S. R., and R. H. VANDYKE, 1934. The development and function of deafferented fore limbs in *Amblystoma J. Exp. Zool.* 68: 321–346.

DIAMOND, J., and R. MILEDI, 1962. A study of foetal and new-born rat muscle fibres. *J. Physiol. (London)* 162: 393–408.

EASTLICK, H. L., 1943. Studies on transplanted embryonic limbs of the chick. I. The development of muscle in nerveless and in innervated grafts. *J. Exp. Zool.* 93: 27–49.

EASTLICK, H. L., and R. A. WORTHAM, 1947. Studies on transplanted embryonic limbs of the chick. III. The replacement of muscle by "adipose tissue." *J. Morphol.* 80: 369–389.

ECCLES, J. C., 1968. Chairman's opening remarks. *In* Growth of the Nervous System / A Ciba Foundation Symposium (G. E. W. Wolstenholme and M. O'Connor, editors). Little, Brown and Co., Boston, pp. 1–2.

ECCLES, J. C., R. M. ECCLES, and W. KOZAK, 1962. Further investigations on the influence of motorneurones on the speed of muscle contraction. *J. Physiol. (London)* 163: 324–339.

EDDS, M. V., JR., 1967. Neuronal specificity in neurogenesis. *In* The Neurosciences: A Study Program (G. C. Quarton, T. Melnechuk, and F. O. Schmitt, editors). The Rockefeller University Press, New York, pp. 230–240.

ELSBERG, C. A., 1917. Experiments on motor nerve regeneration and the direct neurotization of paralyzed muscles by their own and by foreign nerves. *Science (Washington)* 45: 318–320.

FERNÁNDEZ-MORÁN, H., 1957. Electron microscopy of nervous tissue. *In* Metabolism of the Nervous System (D. Richter, editor). Pergamon Press, London and New York, pp. 1–34.

GINETZINSKY, A. G., and N. M. SHAMARINA, 1942. Tonomotor phenomenon in denervated muscle. *Usp. Sovrem. Biol.* 15: 283–294.

GUTH, L., 1962. Neuromuscular function after regeneration of interrupted nerve fibers into partially denervated muscle. *Exp. Neurol.* 6: 129–141.

GUTH, L., and J. BERNSTEIN, 1961. Selectivity in the re-establishment of synapses in the superior cervical sympathetic ganglion of the cat. *Exp. Neurol.* 4: 59–69.

GUTMANN, E., 1945. The reinnervation of muscle by sensory fibres. *J. Anat.* 79: 1–8.

GUTMANN, E., and P. HNÍK, 1962. Denervation studies in research of neurotrophic relationships. *In* The Denervated Muscle (E. Gutmann, editor). Czechoslovak Academy of Sciences, Prague, pp. 13–56.

HAMBURGER, V., 1928. Die Entwicklung experimentell erzeugter nervenloser und schwach innervierter Extremitäten von Anuren. *Arch. Entwmech. Org.* 114: 272–363.

HAMBURGER, V., 1939. Motor and sensory hyperplasia following limb-bud transplantations in chick embryos. *Physiol. Zool.* 12: 268–284.

HAMBURGER, V., 1952. Development of the nervous system. *Ann. N. Y. Acad. Sci.* 55: 117–132.

HAMBURGER, V., 1955. Trends in experimental neuroembryology. *In* Biochemistry of the Developing Nervous System (H. Waelsch, editor). Academic Press, New York, pp. 52–73.

HAMBURGER, V., 1958. Regression versus peripheral control of differentiation in motor hypoplasia. *Amer. J. Anat.* 102: 365–409.

HAMBURGER, V., 1962. Specificity in neurogenesis. *J. Cell. Comp. Physiol.* 60 (suppl): 81–92.

HAMBURGER, V., and R. LEVI-MONTALCINI, 1949. Proliferation, differentiation and degeneration in the spinal ganglia of the chick embryo under normal and experimental conditions. *J. Exp. Zool.* 111: 457–502.

HAMBURGER, V., and R. LEVI-MONTALCINI, 1950. Some aspects of neuroembryology. *In* Genetic Neurology (P. Weiss, editor). University of Chicago Press, Chicago, pp. 128–160.

HARRIS, A. E., 1965. Differentiation and degeneration in the motor horn of the foetal mouse. University of Cambridge, Cambridge, England, doctoral thesis.

HARRISON, R. G., 1904. An experimental study of the relation of the nervous system to the developing musculature in the embryo of the frog. *Amer. J. Anat.* 3: 197–220.

HARRISON, R. G., 1906. Further experiments on the development of peripheral nerves. *Amer. J. Anat.* 5: 121–131.

HARRISON, R. G., 1907. Experiments in transplanting limbs and their bearing upon the problems of the development of nerves. *J. Exp. Zool.* 4: 239–281.

HARRISON, R. G., 1908. Embryonic transplantation and development of the nervous system. *Anat. Rec.* 2: 385–410.

HARRISON, R. G., 1910. The outgrowth of the nerve fiber as a mode of protoplasmic movement. *J. Exp. Zool.* 9: 787–847.

HARRISON, R. G., 1914. The reaction of embryonic cells to solid structures. *J. Exp. Zool.* 17: 521–544.

HARRISON, R. G., 1924. Neuroblast versus sheath cell in the development of peripheral nerves. *J. Comp. Neurol.* 37: 123–205.

HARRISON, R. G., 1935. The Croonian Lecture. On the origin and development of the nervous system studied by the methods of experimental embryology. *Proc. Roy. Soc., ser. B, biol. sci.* 118: 155–196.

HIBBARD, E., 1965. Orientation and directed growth of Mauthner's cell axons from duplicated vestibular nerve roots. *Exp. Neurol.* 13: 289–301.

HILLARP, N., and H. OLIVECRONA, 1946. The role played by the axon and the Schwann cells in the degree of myelination of the peripheral nerve fibre. *Acta Anat.* 2: 17–32.

HOFFMAN, H., 1951. Fate of interrupted nerve fibres regenerating into partially denervated muscles. *Aust. J. Exp. Biol. Med. Sci.* 29: 211–219.

HÖRSTADIUS, S., 1950. The Neural Crest. Oxford University Press, London.

HUGHES, A., 1961. Cell degeneration in the larval ventral horn of *Xenopus laevis* (Daudin). *J. Embryol. Exp. Morphol.* 9: 269–284.

HUGHES, A., 1962. An experimental study on the relationships between limb and spinal cord in the embryo of *Eleutherodactylus martinicensis*. *J. Embryol. Exp. Morphol.* 10: 575–601.

HUGHES, A., 1963. On the labelling of larval neurones by melanin of ovarian origin in certain Anura. *J. Anat.* 97: 217–224.

HUGHES, A., 1965. A quantitative study of the development of the nerves in the hind-limbs of *Eleutherodactylus martinicensis*. *J. Embryol. Exp. Morphol.* 13: 9–34.

HUGHES, A., 1968. Aspects of Neural Ontogeny. Academic Press, New York.

HUGHES, A., and P. A. TSCHUMI, 1958. The factors controlling the development of the dorsal root ganglia and ventral horn in *Xenopus laevis* (Daudin). *J. Anat.* 92: 498–527.

HUNT, E. A., 1932. The differentiation of the chick limb bud in chorio-allantoic grafts, with special reference to the muscles. *J. Exp. Zool.* 62: 57–91.

KOLLROS, J. J., 1956. The further development of the spinal cord, ganglia and nerves. *In* Normal Table of *Xenopus laevis* (Daudin) (P. D. Nieuwkoop and J. Faber, editors). North-Holland Publishing Co., Amsterdam, pp. 67–73.

KOLLROS, J. J., 1968. Order and control of neurogenesis. *In* The Emergence of Order in Developing Systems (M. Locke, editor). Academic Press, New York, pp. 272–305.

KOLLROS, J. J., and J. RACE, JR., 1960. Hormonal control of development of the lateral motor column cells in the lumbosacral cord in *Rana pipiens* tadpoles. *Anat. Rec.* 136: 224.

KONIGSBERG, I. R., 1963. Clonal analysis of myogenesis. *Science* (*Washington*) 140: 1273–1284.

KOVÁCS, T., A. KÖVÉR, and G. BALOGH, 1961. Studies on the localization of cholinesterase in various types of muscle. *J. Cell. Comp. Physiol.* 57: 63–71.

KUFFLER, S. W., 1943. Specific excitability of the endplate region in normal and denervated muscle. *J. Neurophysiol.* 6: 99–110.

LANGLEY, J. N., 1897. On the regeneration of pre-ganglionic and of post-ganglionic visceral nerve fibres. *J. Physiol.* (*London*) 22: 215–230.

LEVI-MONTALCINI, R., 1950. The origin and development of the visceral system in the spinal cord of the chick embryo. *J. Morphol.* 86: 253–283.

LEVI-MONTALCINI, R., and G. LEVI, 1942. Les conséquences de la destruction d'un territoire d'innervation périphérique sur le développement des centres nerveux correspondants dans l'embryon de Poulet. *Arch. Biol.* 53: 537–545.

McPHERSON, A., and J. TOKUNAGA, 1967. The effects of cross-innervation on the myoglobin concentration of tonic and phasic muscles. *J. Physiol.* (*London*) 188: 121–129.

MILEDI, R., 1960. The acetylcholine sensitivity of frog muscle fibres after complete or partial denervation. *J. Physiol.* (*London*) 151: 1–23.

MILLER, R. A., and S. R. DETWILER, 1936. Comparative studies upon the origin and development of the brachial plexus. *Anat. Rec.* 65: 273–292.

MOMMAERTS, W. F. H. M., 1968. Muscle energetics: Biochemical differences between muscles as determined by the innervation. *Proc. Int. Union Physiol. Sci. XXIV Int. Congr.* 6: 116.

MUMENTHALER, E., and W. K. ENGEL, 1961. Cytological localization of cholinesterase in developing chick embryo skeletal muscle. *Acta Anat.* 47: 274–299.

NAKAI, J., 1965. Skeletal muscle in organ culture. *Exp. Cell Res.* 40: 307–315.

NARAYANAN, C. H., 1964. An experimental analysis of peripheral nerve pattern development in the chick. *J. Exp. Zool.* 156: 49–60.

NICHOLAS, J. S., 1929. Movements in transplanted limbs. *Proc. Soc. Exp. Biol. Med.* 26: 729–731.

NICHOLAS, J. S., 1930a. The effects of the separation of the medulla and spinal cord from the cerebral mechanism by the extirpation of the embryonic mesencephalon. *J. Exp. Zool.* 55: 1–22.

NICHOLAS, J. S., 1930b. Movements in transplanted limbs innervated by eye-muscle nerves. *Anat. Rec.* 45: 234 (abstract).

NIEUWKOOP, P. D., and J. FABER, 1956. Normal Table of *Xenopus laevis* (Daudin). North-Holland Publishing Co., Amsterdam.

PETERSON, E. R., and M. R. MURRAY, 1955. Myelin sheath formation in cultures of avian spinal ganglia. *Amer. J. Anat.* 96: 319–355.

PIATT, J., 1942. Transplantation of aneurogenic forelimbs in amblystoma punctatum. *J. Exp. Zool.* 91: 79–101.

PIATT, J., 1948. Form and causality in neurogenesis. *Biol. Rev. (Cambridge)* 23: 1–45.

PIATT, J., 1952. Transplantation of the aneurogenic forelimbs in place of the hindlimb in amblystoma. *J. Exp. Zool.* 120: 247–285.

PIATT, J. 1956. Studies on the problem of nerve pattern. I. Transplantation of the forelimb primordium to ectopic sites in *Amblystoma. J. Exp. Zool.* 131: 173–201.

PIATT, J., 1957. Studies on the problem of nerve pattern. II. Innervation of the intact forelimb by different parts of the central nervous system in amblystoma. *J. Exp. Zool.* 134: 103–125.

PRESTIGE, M. C., 1967. The control of cell number in the lumbar ventral horns during the development of *Xenopus laevis* tadpoles. *J. Embryol. Exp. Morphol.* 18: 359–387.

RACE, J., JR., 1961. Thyroid hormone control of development of lateral motor column cells in the lumbo-sacral cord in hypophysectomized *Rana pipiens. Gen. Comp. Endocrinol.* 1: 322–331.

RACE, J., JR., and R. J. TERRY, 1965. Further studies on the development of the lateral motor column in anuran larvae. I. Normal development in *Rana temporaria. Anat. Rec.* 152: 99–106.

RAMÓN Y CAJAL, S., 1908. Nouvelles observations sur l'évolution des neuroblastes, avec quelques remarques sur l'hypothèse neurogénétique de Hensen-Held. *Anat. Anz.* 32: 65–87.

RAMÓN Y CAJAL, S., 1959. Degeneration and Regeneration of the Nervous System (R. M. May, translator). Hafner, New York, 2 vols. (Translation of a 1928 edition.)

RAMÓN Y CAJAL, S., 1960. Studies on Vertebrate Neurogenesis. (Revised and translated by L. Guth.) Charles C Thomas, Springfield, Illinois. (Revision of a 1929 edition.)

RIDGE, R. M. A. P., 1967. The differentiation of conduction velocities of slow twitch and fast twitch muscle motor innervations in kittens and cats. *Quart. J. Exp. Physiol. Cog. Med. Sci.* 52: 293–304.

ROMANES, G. J., 1953. The motor cell groupings of the spinal cord. *In* The Spinal Cord/A Ciba Foundation Symposium (G. E. W. Wolstenholme, editor). Little, Brown and Co., Boston, pp. 24–38.

ROMANES, G. J., 1964. The motor pools of the spinal cord. *Progr. Brain Res.* 11: 93–116.

ROMANUL, F. C. A., and M. POLLACK, 1969. The parallelism of changes in oxidative metabolism and capillary supply of skeletal muscle fibers. *In* Modern Neurology: Papers in Tribute to Derek Denny-Brown (S. Locke, editor). Little, Brown and Co., Boston, pp. 203–213.

SAXÉN, L., and S. TOIVONEN, 1962. Primary Embryonic Induction. Prentice-Hall, Englewood Cliffs, New Jersey.

SILVER, M. L., 1942. The motoneurons of the spinal cord of the frog. *J. Comp. Neurol.* 77: 1–39.

SINGER, M., 1952. The influence of the nerve in regeneration of the amphibian extremity. *Quart. Rev. Biol.* 27: 169–200.

SPEIDEL, C. C., 1933. Studies of living nerves. II. Activities of ameboid growth cones, sheath cells, and myelin segments, as revealed by prolonged observation of individual nerve fibers in frog tadpoles. *Amer. J. Anat.* 52: 1–79.

SPERRY, R. W., 1945. The problem of central nervous re-organization after nerve regeneration and muscle transposition. *Quart. Rev. Biol.* 20: 311–369.

SPERRY, R. W., 1947. Nature of functional recovery following regeneration of the oculomotor nerve in amphibians. *Anat. Rec.* 97: 293–316.

SPERRY, R. W., 1950. Myotypic specificity in teleost motoneurons. *J. Comp. Neurol.* 93: 277–287.

SPERRY, R. W., and H. L. ARORA, 1965. Selectivity in regeneration of the oculomotor nerve in the cichlid fish, *Astronotus ocellatus. J. Embryol. Exp. Morphol.* 14: 307–317.

STRASNICKY, K., 1963. Function of heterotopic spinal cord segments investigated in the chick. *Acta Biol. Acad. Sci. Hung.* 14: 145–155.

STRAUS, W. L., JR., and G. WEDDELL, 1940. Nature of the first visible contractions of the forelimb musculature in rat fetuses. *J. Neurophysiol.* 3: 358–369.

SZÉKELY, G., 1963. Functional specificity of spinal cord segments in the control of limb movements. *J. Embryol. Exp. Morphol.* 11: 431–444.

SZÉKELY, G., 1968. Development of limb movements: Embryological, physiological, and model studies. *In* Growth of the Nervous System/A Ciba Foundation Symposium (G. E. W. Wolstenholme and M. O'Connor, editors). Little, Brown and Co., Boston, pp. 77–93.

SZÉKELY, G., and G. CZÉH, 1967. Localization of motoneurones in limb moving spinal cord segments of *Amblystoma. Acta Physiol. Acad. Sci. Hung.* 32: 3–17.

TAYLOR, A. C., 1943. Development of the innervation pattern in the limb bud of the frog. *Anat. Rec.* 87: 379–413.

TOWER, S. S., 1935. Atrophy and degeneration in skeletal muscle. *Amer. J. Anat.* 56: 1–43.

WEISS, P., 1922. Die Funktion transplantierter Amphibienextremitäten. *Anz. Akad. Wiss. Wien* 59: 199–201.

WEISS, P., 1928. Experimentelle Organisierung des Gewebewachstums *in vitro. Biol. Zentralbl.* 48: 551–566.

WEISS, P., 1931. Das Resonanzprinzip der Nerventätigkeit, dargestellt in Funktionsprüfungen an transplantierten überzähligen Muskeln. *Arch. Ges. Physiol.* 226: 660–658.

WEISS, P., 1934. In vitro experiments on the factors determining the course of the outgrowing nerve fiber. *J. Exp. Zool.* 68: 393–447.

WEISS, P., 1935. Homologous (resonance-like) function in supernumerary fingers in a human case. *Proc. Soc. Exp. Biol. Med.* 33: 426–430.

WEISS, P., 1936. A study of motor coordination and tonus in deafferented limbs of Amphibia. *Amer. J. Physiol.* 115: 461–475.

WEISS, P., 1937. Further experimental investigations on the phenomenon of homologous response in transplanted amphibian limbs. III. Homologous response in the absence of sensory innervation. *J. Comp. Neurol.* 66: 537–548.

WEISS, P., 1941a. Self-differentiation of the basic patterns of coordination. *Comp. Psychol. Monogr.* 17: 1–96.

WEISS, P., 1941b. Nerve patterns: The mechanics of nerve growth. *Growth* (suppl.) 5: 163–203.

WEISS, P. 1945. Experiments on cell and axon orientation *in vitro*: The role of colloidal exudates in tissue organization. *J. Exp. Zool.* 100: 353–386.

WEISS, P., editor, 1950. Genetic Neurology. University of Chicago Press, Chicago.

WEISS, P., 1955. Nervous system (neurogenesis). *In* Analysis of Development (B. H. Willier, P. A. Weiss, and V. Hamburger, editors). W. B. Saunders Co., Philadelphia, pp. 346–401.

WEISS, P., and M. V. EDDS, JR., 1945. Sensory-motor nerve crosses in the rat. *J. Neurophysiol.* 8: 173–193.

WEISS, P., and A. HOAG, 1946. Competitive reinnervation of rat muscles by their own and foreign nerves. *J. Neurophysiol.* 9: 413–418.

WEISS, P., and T. C. RUCH, 1936. Further observations on the functions of supernumerary fingers in man. *Proc. Soc. Exp. Biol. Med.* 34: 569–570.

WENGER, B. S., 1951. Determination of structural patterns in the spinal cord of the chick embryo studied by transplantations between brachial and adjacent levels. *J. Exp. Zool.* 116: 123–163.

WIEMAN, H. L., and T. C. NUSSMANN, 1929. Experimental modification of nerve development in amblystoma. *Physiol. Zool.* 2: 99–124.

WOODBURNE, R. T., 1939. Certain phylogenetic anatomical relations of localizing significance for the mammalian central nervous system. *J. Comp. Neurol.* 71: 215–257.

ZELENA, J. 1959. Effect of innervation on the development of skeletal muscle (in Czech with English and Russian summary) (as cited *in* Eccles, J. C., The Physiology of Synapses. 1964). (Babákova sbríka 12.) Státné Zdravotniché nakladatelství, Prague.

ZELENA, J., 1962. The effect of denervation on muscle development. *In* The Denervated Muscle (E. Gutmann, editor). Czechoslovak Academy of Sciences, Prague, pp. 103–126.

ZELENA, J., and J. SZENTÁGOTHAI, 1957. Verlagerung der Lokalisation spezifischer Cholinesterase während der Entwicklung der Muskelinnervation. *Acta Histochem.* (*Jena*) 3: 284–296.

# 12  Development, Specification, and Diversification of Neuronal Connections

## MARCUS JACOBSON

THE PROBLEM of specification can be regarded, simply, as the problem of how one neuron becomes different from others. It is not yet possible to define neuronal specificity in terms of cellular processes and products of synthesis that result in the development of specific differences between neurons. The term neuronal specificity is used to connote differences in the functions of neurons that are implicit in their individual anatomical characteristics, particularly in their con-

nectivity. In general, the parameters that appear to be of importance in the development of neuronal specificity are differences in the time of genesis of young neurons, differences in their cytodifferentiation, differences in their positions, contacts, and connections with other cells, and in the modifiability of these connections. Because there seem to be many different parameters by which the differences between neurons may be assessed, it is perhaps best to apply the principle of *one specificity at a time*. Here we are concerned mainly with the development of specific connections between neurons, and are attempting to account for the development of the great diversity of types of neurons and their interconnections that are found in the mature nervous

MARCUS JACOBSON  Thomas C. Jenkins Department of Biophysics, The Johns Hopkins University, Baltimore, Maryland system. Diversity may arise in a population of neurons in two different ways. Either the differences among various kinds of neurons may be genetically coded and therefore under direct genetic control, or they may arise because developing neurons have been freed from direct genetic control and therefore develop individual differences in size, shape, and connectivity.

Another way of approaching the problem of specification and diversification in the nervous system is to start with the postulate that the genome cannot contain all the information required to specify the detailed structure of the fully developed brain. This conclusion has been reached from estimates of the maximum information content of the genome (Elsasser, 1958, 1961, 1962) and its insufficiency to specify the structure of the mature brain (Bremermann, 1963; Jacobson, 1969), or the complexity of innate patterns of behavior (Thorpe, 1963). It is generally agreed that it is simpler and more economical to use genetic information to specify a developmental program than to specify the final structure in detail. A more parsimonious use of genetic information to specify a developmental program, including a sequence of changes in cellular adhesiveness, motility, and growth, could result in the development of neuronal circuits of a complexity that increases as the result of interactions within the system, without requiring detailed genetic specification of the fully developed structure.

Releasing the majority of neurons from genetic constraints would also result in an economy of genetic information that would be used to specify the connectivity of only a small percentage of the neurons. We may, therefore, speculate on the existence of a wide spectrum of different kinds of neurons, ranging from one extreme class of neurons that develop connections under direct genetic constraints to another extreme class of neurons that develop connections under indirect or lax genetic constraints. I merely wish to draw attention to the possibility of different degrees of neuronal specificity rather than to enunciate a hypothesis which may, at best, be only partly correct.

The following table is an attempt to list some of the characteristics of the two extreme classes of neurons, but it is assumed that a graded series of intermediate types also develop.

| *Class I* | *Class II* |
|---|---|
| 1. Severely constrained genetically | 1. Loosely constrained genetically |
| 2. Rigid specification | 2. Lax specification |
| 3. Generated early | 3. Continue to be generated late in ontogeny |
| 4. Mainly macroneurons (afferent and efferent neurons with long axons; connectivity often organized topographically) | 4. Mainly microneurons (interneurons with short axons) |
| 5. Invariant connectivity | 5. Variable connectivity |
| 6. Genetic specification sufficient | 6. Genetic specification not sufficient. Require functional validation |
| 7. Unmodifiable after specification that takes place early in development | 7. Modifiable until specification, which may be delayed until late in development |

A particular type of neuron may be assigned a place in this scheme based on its morphological and physiological characteristics. We may also observe whether its development is contingent on function or sensory experience, and whether modifications take place after surgical rearrangement or dislocation of its connections. The amount of functional modification can be measured behaviorally or by mapping the connections electrophysiologically. This strategy has been used in studies of the connections that develop between the eye and brain and between the skin and spinal cord.

## Specification of central connections of sensory neurons resulting from axonal contacts with peripheral organs

The sensory nerves of all vertebrates have been shown to project in somatotopic order onto the sensory centers in the spinal cord and brain. How does this point-to-point central representation of sense organs in the skin develop? Either it might, in principle, develop as a result of selective growth of nerves to sense organs, or the nerves might connect nonselectively in the skin, and then a compensatory adjustment of their central connections might occur. Several factors probably act in combination to control the development of peripheral innervation: mechanical guidance of the nerves by other tissues, chemotaxis, and chemoaffinity between nerves and sense organs have all been proposed, and I allude to these again. There is, however, experimental evidence that the reflex connections of some sensory nerves are specified by their peripheral contacts with the cornea in salamanders (Weiss, 1942; Kollros, 1943), the skin in frogs (Miner, 1956; Jacobson and Baker, 1968, 1969), and the muscle spindles in newborn cats (Eccles et al., 1962).

Evidence in favor of peripheral specification of cutaneous nerves in frogs was obtained by Miner (1956). She observed that back-to-belly inversion of a skin graft on the trunk of a frog tadpole resulted in the development of misdirected reflexes. Stimulation of a point on the skin graft elicited reflex limb movements aimed at the original position rather than at the grafted position of the stimulated skin. By dissection, Miner showed that the nerves had not grown back selectively to the grafted skin. She concluded that the nerves had connected nonselectively with the graft and had then switched their central connections to conform with their new location in the skin.

We have confirmed and extended Miner's observations on the misdirected reflexes which develop after inverted skin grafts have been performed on frog tadpoles (Figure 1). We have found that the results depend on the position and size of the skin graft and on the stage of development at which the graft is made. Misdirected reflexes develop in only some cases, and then only by replacing normal reflexes that have developed as usual, regardless of the position of the graft (Jacobson and Baker 1968, 1969).

If a large patch of skin, extending from the back to the belly on one side of a frog tadpole, is excised, all the cutaneous nerves are cut, and the skin is replaced in a back-to-belly inverted position, nerves regenerate into the skin graft and its sensitivity is restored. After metamorphosis, the cutaneous reflexes appear at the usual time and are at first normal: touching the graft results in a limb movement directed accurately to the point of stimulation. At this stage, the original location of the grafted skin has no effect in determining the reflex connections of its sensory nerves (Figure 2). After a few days, however, the normal reflexes are gradually replaced by misdirected reflex movements aimed at the original position of the stimulated skin. Stimulating the belly skin grafted on the back results in a movement of the leg aimed at the belly, and vice versa (Figure 3). These misdirected reflexes are clearly maladaptive, but they persist permanently. Experience cannot be responsible for the change from adaptive to maladaptive behavior. The change of behavior is the reverse of the change that would be expected as a result of learning, and might be called unlearning.

We have shown, by electrophysiological recording, that the cutaneous nerves connect nonselectively with the skin graft, back nerves to belly skin and vice versa, and restore normal cutaneous reflexes (Jacobson and Baker, 1968, 1969). No changes in the peripheral pattern of innervation take place while the reflexes change from normal to misdirected. Spread of the reflexogenic zone that gives rise to misdirected reflexes, as shown in Figure 3, is not the result of changes in the receptive fields in the skin but is assumed to be caused by changes in the reflex associations in the spinal cord. We infer that the central reflex associations of the cutaneous nerves change so that the reflex circuits are appropriate to the new connections of the nerves with the skin, even though the reflex movements are functionally inappropriate. Regulation or modification appears to have taken place in the neuronal circuits, and these changes could not have been related to function and experience. We postulate that each region of the skin produces a specific biochemical change in the nerve, resulting in the formation of specific central connections (Jacobson and Baker, 1968, 1969). The skin instructs the nerve to form central connections that are congruent with the position of the peripheral nerve in the skin. Switching of connections between neurons might be limited to adjacent neurons; the limiting distance apart is set by the length and branching pattern of their neurites.

Rearrangement of connections of sensory neurons in the frog's spinal cord might be easier than in other species because the central neurite of each sensory neuron has unusually extensive rostral and caudal branches. Each sensory neuron sends branches up and down the spinal cord to make connections at all levels, from the branchial to the lumbosacral, although the highest density of connections is made at the level at which the nerve enters the spinal cord (Liu and Chambers, personal communication).

At metamorphosis the skin loses its capacity to instruct the nerve to form the appropriate reflex connections, or the nerves lose their capacity to obey the instructions. We have found that back-to-belly inverted skin grafts made after larval stage XV, or in adult frogs, always give rise to normal reflexes, whereas grafts made before stage XV give rise to misdirected reflexes. The size of the graft is important: normal reflexes only are in every case elicited from back-to-belly inverted grafts less than 40 mm$^2$ in area (measured at the end of metamorphosis), whereas larger grafts exchanged between back and belly invariably give rise to some misdirected reflexes. The percentage of the total number of reflexes that are misdirected increases in direct proportion to the area of the graft, from zero for grafts less than 30 mm$^2$ in area to about 70 per cent for grafts with an area of more than 120 mm$^2$ (Baker and Jacobson, 1970). Normal reflexes are always obtained from grafts, of any size, that are merely translocated on the back or on the belly.

One of the prerequisites for the development of reflexes that are misdirected to the original position rather than to the translocated position of a skin graft is that nerves do not invade the graft from the surrounding skin. Electrophysiological mapping of the receptive fields shows that separate nerves supply the graft and the surrounding skin in all cases in which misdirected reflexes develop (Figure 4). In such cases, there is no overlap of receptive fields across the margins of the graft. Overlap occurs on the graft and on the surrounding skin but not between the graft and the surrounding skin. This is not the result of a mechanical barrier at the graft margin, because overlap of receptive fields takes place if back-to-belly grafts are below the critical size. Thus there seem to be unknown factors that prevent ingrowth of nerves from back-skin to belly-skin grafts and vice versa.

SPECIFICATION OF CONNECTION BETWEEN THE EYE AND BRAIN There is clear evidence that whole systems of neuronal circuits develop in the prefunctional stage of embryonic development before external stimuli evoke nervous activity. The development of the visual system affords the best evidence of prefunctional development of specific neuronal connections. In all vertebrates there is an orderly point-to-point projection of the retina on the visual centers. If the

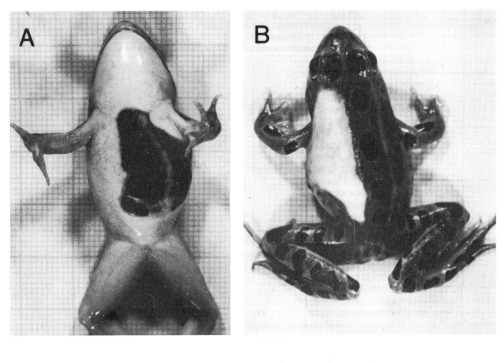

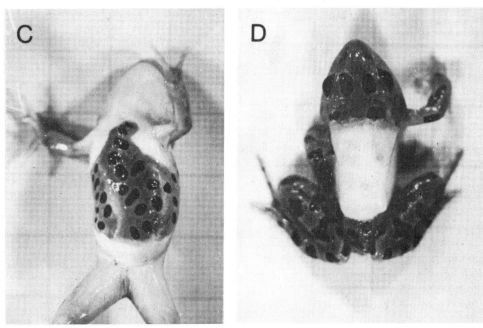

FIGURE 1  A. Ventral view of frog with a back-to-belly inverted skin graft.
B. Dorsal view of the frog shown in A.
C. Ventral view of a frog with 180-degree rotation of the skin on the trunk. The cutaneous reflexogenic zones and receptive fields of this frog are shown in Figure 3 and 4.
D. Dorsal view of the frog shown in C. (From Jacobson and Baker, 1969.)

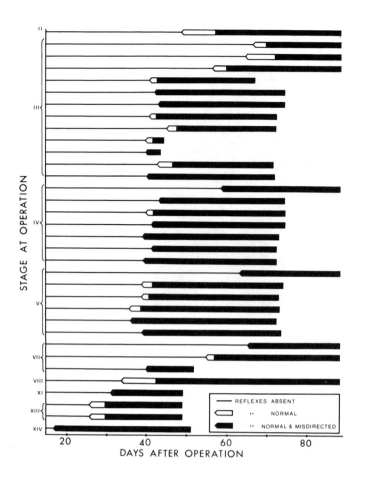

FIGURE 2  Time of onset of normal and misdirected reflexes in 34 frogs with back-to-belly inverted skin grafts made at larval stages shown on the ordinate. Reflexes were tested daily during the postoperative period shown on abscissa. (From Jacobson and Baker, 1969.)

optic nerve, which connects the retina to the midbrain tectum, is cut in fishes or frogs, the animal is totally blind at first but recovers normal vision within a few weeks as a result of optic-nerve regeneration. We have shown, by mapping the connections between the retina and tectum electrophysiologically, that, in adult frogs and goldfish (Gaze and Jacobson, 1963; Jacobson and Gaze, 1965), the normal point-to-point connections are completely restored after such regeneration. Sperry (1944, 1951) has proved that the restoration of vision is independent of function and experience, for, if the optic nerve is cut and the eye is inverted, the frog recovers with inverted visuomotor reflexes, which are never corrected by experience. These experiments show that the optic-nerve fibers regenerate to their proper places in the tectum regardless of the relative positions of the eye and brain and regardless of the functional effect. The regeneration of retinotectal connections, and presumably also their development, depend on growth processes that are not affected by experience.

Sperry (1963) has postulated that, during the development of the visual system, each retinal ganglion cell acquires a biochemical specificity uniquely related to its position in the retina. A matching map of biochemical specificities is assumed to develop in the tectal cells. Sperry proposed that

connections are then formed only between retinal and tectal cells with matching specificities. According to his theory, the organization of the nervous system is brought about by the development of specific chemoaffinity between neurons that become connected to one another. The development of these specificities occurs in the prefunctional stage of neurogenesis.

To discover the stage at which these specificities develop in the visual system, I inverted the eye of the toad *Xenopus* at various stages of development, before connections had formed between the retina and the tectum, and then mapped the connections after they had formed (Jacobson, 1967a, 1967b, 1968a, 1968b). These experiments showed that the ganglion cells of the retina are at first unspecified, and are pluripotent with respect to the connections they can form in the brain.

At a later stage, before any visual function is possible, each ganglion cell becomes uniquely specified. The experiments also showed that if the eye is inverted before the early tailbud stage (embryonic stage 29), its connections with the tectum develop normally and are the same as the connections of the other, normal eye (Figure 5). That is, at the time of eye rotation at stage 29, the retinal ganglion cells do not "know" what their central connections should be, and

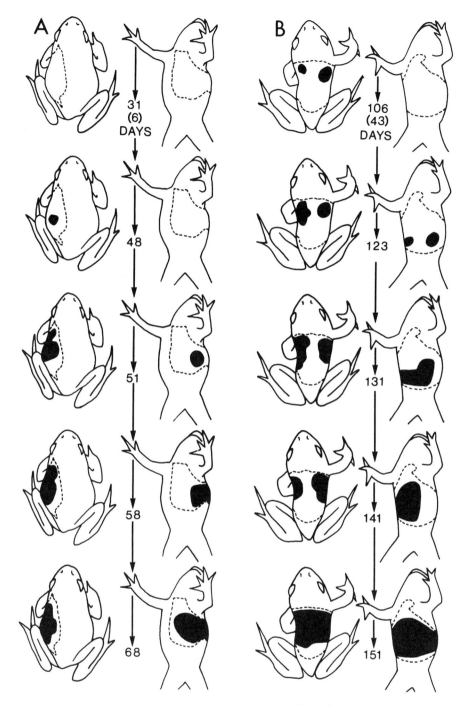

FIGURE 3 Spread of the area of skin graft from which mis-directed reflexes were evoked at progressively later post-operative times. The graft is outlined by a dashed line. Misdirected reflexes were evoked from the black area; normal reflexes, from the rest of the graft. A. Dorsal and ventral views of a frog with a back-to-belly inverted skin graft made at larval stage IV and tested 31 days postopera-tively (six days after metamorphosis) and 48, 51, 58, and 68 days postoperatively. B. Dorsal and ventral views of a frog with back-to-belly reversal of the skin of the trunk made at larval stage V, and tested 106 days postoperatively (43 days after metamorphosis) and 123, 131, 141, and 151 days postoperatively. Receptive field maps of these frogs are shown in Figure 4. (From Jacobson and Baker, 1969.)

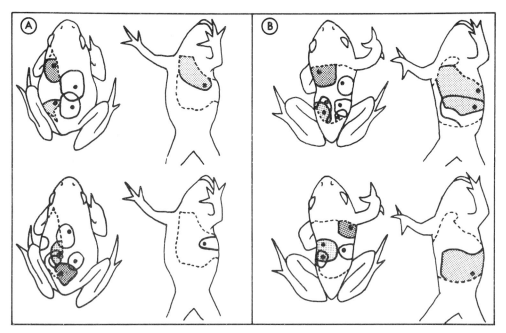

FIGURE 4 Receptive fields of cutaneous sensory nerves in two frogs with back-to-belly inverted skin grafts (same frogs as shown in Figure 3). A. The graft outlined with the dashed line was made at larval stage VI, and the receptive fields were mapped 68 days later. B. The graft outlined with the dashed line was made at larval stage V, and the receptive fields were mapped 151 days later. The area of skin within which action potentials were evoked by a stimulus in each cutaneous nerve is shown as an enclosed area, stippled in some cases. Each cutaneous nerve enters the skin at the position shown by a small circle in its receptive field. (From Jacobson and Baker, 1969.)

those connections are specified according to their new positions after rotation. The ganglion cells lose this capacity to regulate their connectivity within a period of about 10 hours during embryonic stages 29 to 30. If the eye is inverted during or after stage 30, the animal develops an inverted retinotectal projection and has permanently inverted vision. In stages at which the retinotectal connections are specified, the retinal receptors have not yet developed, so there is no question that these connections are specified in the prefunctional stages of neurogenesis.

Rotation of the eye at stage 30 results in the development of a retinotectal projection that is inverted in the anteroposterior (nasotemporal) axis of the retina but is normal in the dorsoventral axis (Figure 6). Such inversion means that specification of the ganglion cells in the anteroposterior axis of the retina takes place before rotation of the eye at stage 30, but dorsoventral specification occurs only later, in accordance with the position of the eye after inversion. Inversion of the eye at stage 31 and later, or in adults, results in total inversion of the retinotectal projection (Figure 7).

Our electron-microscope studies of the cytodifferentiation of the developing retina of *Xenopus* has shown that the first optic-nerve fibers start sprouting from the ganglion cells at stage 33. That happens many hours after the terminal connections of the optic fibers have been specified at stage 30 and about one week before the optic-nerve fibers grow into the tectum and begin forming connections there at stages 49 to 50 (Fisher and Jacobson, 1970).

Tritiated thymidine autoradiography of the developing eye of *Xenopus* (Jacobson, 1968b) has shown that before stage 28 the optic vesicle is composed of neuroepithelial germinal cells, which continue DNA synthesis and mitosis with a generation time of about 12 hours. At stage 29, a mantle layer of young neurons, which have ceased DNA synthesis, is formed external to the neuroepithelial cells of the optic vesicle. It has been shown that these young neurons differentiate into retinal ganglion cells, for if cumulative labeling is begun at stage 29 and continued until the retina is clearly differentiated at stages 38 to 50, only the ganglion cells are unlabeled. Therefore, the first neuroepithelial germinal cells to stop DNA synthesis differentiate into ganglion cells, and about 80 per cent of the ganglion cells in the center of the retina are formed at stages 29 to 30 (Figure 8). By stage 30, the final DNA synthesis has been completed in all the neuroepithelial cells that form retinal ganglion cells (except those formed at the periphery of the retina). The duration of stages 29 to 20 is approximately 10 hours. The generation time of the neuroepithelial cells

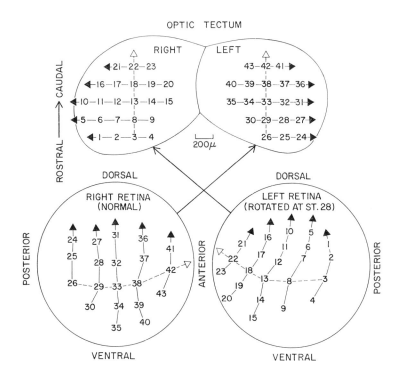

FIGURE 5  Map of the retinotectal projection in adult of *Xenopus* to the left tectum from the normal right eye, and to the right tectum from the left eye, which had been dorsoventrally and anteroposteriorly inverted (rotated 180 degrees) at larval stage 28. The projection from the inverted eye is normal. Each number on the tectum represents the position at which a microelectrode recorded action potentials in response to a small spot of light at the position shown by the same number on the retina. Figures 6 to 8 conform to the same conventions. (From Jacobson, 1968a.)

is about 12 hours. Therefore, all the neuroepithelial cells that give rise to the ganglion cells in the central part of the retina cease DNA synthesis within the period of one generation cycle.

## Determinative mechanisms of nerve-fiber growth and connections

The optic-nerve fibers appear to be destination-bound from the time they start growing from the retina. How does each optic fiber reach its correct destination in the tectum? Apart from the guidance provided by the optic stalk, we can rule out mechanical guidance as a factor in directing the 600,000 optic fibers that grow to their correct places in the contralateral optic tectum from each retina of the frog. There is no evidence that galvanotropism plays a part, either. The growth of optic-nerve fibers up a chemical concentration gradient is possible, but no evidence of chemotaxis in the nervous system has yet been produced. Some encouragement to look for it however, can be obtained from the evidence of chemotaxis in bacteria (Adler, 1966), leukocytes (McCutcheon, 1946; Harris, 1954), and the cellular slime molds (Bonner, 1947; Konijn et al., 1967, 1968).

We cannot lightly dismiss the conviction, expressed in many of Ramón y Cajal's writings, that chemotropism plays an important part in the formation of specific neuronal connections. His meticulous observations of the invariant patterns of growth of embryonic nerve fibers toward their targets led him to favor the theory of chemo-

tropism. According to him, the tips of the growing nerve fibers are destination-bound, probably being guided by chemical cues or, as he phrased it, "growth caused by the arrival at the axon of powerful currents of alluring substances" (Ramón y Cajal, 1960, p. 180). Many observations suggest that the growth of nerve fibers to specific places is determined by chemical rather than mechanical factors. Most interesting are the observations that nerve fibers can reach their correct destinations and form functionally adequate circuits after growing along abnormal routes. For example, in the goldfish, regenerating optic-nerve fibers that have been deflected surgically from their normal routes will grow back to their correct places in the optic tectum, bypassing all other tectal neurons on the way (Attardi and Sperry, 1963).

Optic-nerve fibers in *Xenopus* can connect selectively even after having been surgically deflected into the oculomotor nerve and after having grown along an aberrant pathway into the tectum (Hibbard, 1967). Correct synaptic connections in the central nervous system of the lobster may be formed by axons that have grown along aberrant pathways from heteromorph antennules that have regenerated in place of the eye (Maynard and Cohen, 1965). Selective central connections also are formed by sensory nerves that regenerate from anal cerci transplanted to the thorax of the cricket (Edwards and Sahota, 1967). These and many other observations that axons form appropriate connections after growing along abnormal routes suggest that selective nerve growth is not determined by mechanical or

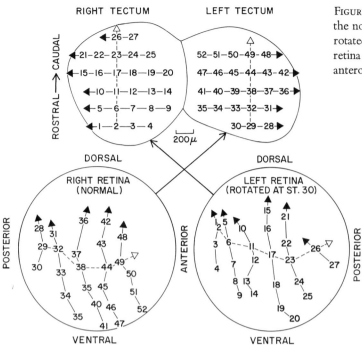

FIGURE 6  The retinotectal projection in adult of *Xenopus* from the normal right retina and from the left retina, which had been rotated 180 degrees at larval stage 30. The projection from the left retina is normal in the dorsoventral axis but is inverted in the anteroposterior axis. (From Jacobson, 1968a.)

electrical forces but by a highly refined chemotaxis or chemoaffinity of some kind (Sperry, 1955, 1963). Experiments showing that mechanical factors affect the growth of nerve fibers in tissue culture (Weiss, 1934, 1941, 1955) seem to me to have rather limited relevance to the selective formation of connections during development of the central nervous system. Conditions in tissue culture are very different from those in vivo, and mechanical forces alone lack the specificity required to guide nerve fibers to their correct targets.

It seems most probable that some kind of specific molecular interaction takes place between the growing nerve fibers

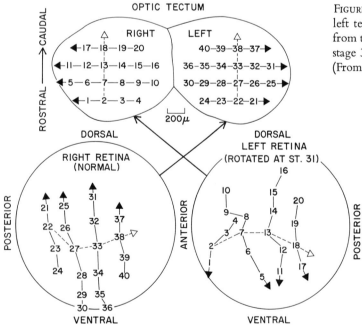

FIGURE 7  The retinotectal projection in adult of *Xenopus* to the left tectum from the normal right retina, and the right tectum from the left retina, which had been rotated 180 degrees at larval stage 31. The projection from the left retina is totally inverted. (From Jacobson, 1968a.)

and other neurons and glial cells with which they come into contact, as Weiss has suggested (Weiss, 1941, 1947). But the nature of the specificity is almost as elusive now as it was when Ramón y Cajal reflected on the problem long ago. He thought that the selection was due to some kind of chemotaxis, and argued that with a mechanical or electrical hypothesis "it becomes difficult to understand how, of the large nervous contingent arriving at the mammalian snout, some fibers travel without error toward the cutaneous muscle fibers, others toward the hair follicles, others to the epidermis, and, finally, some to the tactile apparatus of the dermis. A similar multiple specificity is found in the tongue, where the hypoglossal fibers invade the muscle, trigeminal fibers innervate the papillae, and facial (geniculate ganglion) and glossopharyngeal fibers go to the gustatory papillae" (Ramón y Cajal, 1960). He concluded: ". . . there is no doubt that, at first, many imperfect connections occur. But these incongruencies are progressively corrected . . . [by] a selection, due to the atrophy of certain collaterals and the progressive disappearance of disconnected or useless neurones" (Ramón y Cajal, 1960). A most attractive hypothesis is the possibility that specific connections are formed by the survival of the fittest as a result of natural selection among the nerve fibers. The excessive formation of collateral branches has been seen in developing and regenerating axons (Ramón y Cajal, 1959; Speidel, 1933, 1935, 1941, 1942; Edds, 1953). The chances that an axon will meet a favorable postsynaptic neuron are greatly increased by the restless activity of the growing axon terminal. The axon terminal is in constant motion, rapidly emitting filopodia, which do not usually extend more than a few hundred microns before being withdrawn or becoming the main line of axonal growth (Speidel, 1941, 1942; Pomerat, 1961).

It is highly probable that neuronal connections are formed as a result of random branching of axons which find their targets by trial and error. It appears that, during normal development of fiber tracts, determinative mechanisms result in the selection of correct pathways by nerve fibers en route to their terminal loci. The experiments that have already been cited, however, prove that nerve fibers can form correct connections after traversing aberrant pathways. What, then, is the basis of selection and of the persistence of the "correct" connections? Sperry (1963) has suggested that selection is made on the basis of chemoaffinity between axon terminals and the neurons with which they connect, the affinities being established by genetic and developmental mechanisms acting on the presynaptic and postsynaptic units before the connections are formed. This hypothesis has the merit of being consistent with, and appears to have been inspired by, the experimental evidence that sorting out of cells is the result of "tissue affinities" or of "cell specific differences in surface tension" (Holtfreter, 1939) or of "differential cellular adhesion" (Steinberg, 1962a, 1962b, 1963).

Townes and Holtfreter (1955) concluded that the sorting out of embryonic cells is the result of selective cell adhesion as well as chemotactic migration of cells along concentration gradients. Steinberg (1962a, 1962b) showed that the sorting out can be caused entirely by differential cell adhesion. According to the "differential adhesion hypothesis" (Steinberg, 1963, 1964), sorting out and morphogenetic movements of embryonic cells take place because (1) cells adhere to one another with adhesive strengths that are specific for the cell type; and (2) the cells within a tissue assume an equilibrium arrangement, determined by their intercellular adhesive strengths, such that the interfacial, or adhe-

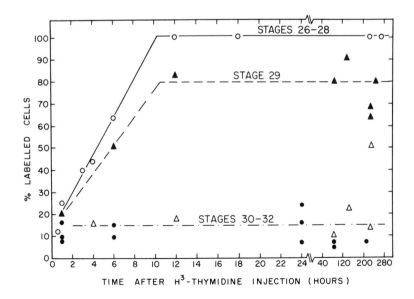

FIGURE 8 Percentages of labeled cell nuclei in the optic cup of *Xenopus* larvae after either one or several injections of tritiated thymidine. The stage at which the initial injection was given is indicated by the following symbols: open circles, stages 26 to 28; solid triangles, stage 29; open triangles, stage 30; solid circles, stages 31 and 32. (From Jacobson, 1968b.)

sive, free energy of the system is at a minimum. The "differential adhesion hypothesis" predicts that a mixture of two types of cells will eventually form an aggregate in which cells of the less cohesive type will envelop the cells with greater cohesion.

Little genetic information may be required to specify cellular adhesiveness, which differs only quantitatively but not qualitatively in different cells or in different parts of a single cell. As a consequence, the genetic control of morphogenesis may be relatively simple, and may consist of a programed sequence of changes in cellular adhesiveness and in cellular motility (Trinkaus, 1965; Gustafson and Wolpert, 1967; Curtis, 1967). Even such complex morphogenetic events as the migration of young neurons, the aggregation of neurons to form brain nuclei, and the association of axons to form nerve-fiber tracts may result from differential cellular adhesion.

The movements of masses of cells during fusion of fragments of tissues or during the sorting out of aggregates composed of different types of cells (Holtfreter, 1939; Townes and Holtfreter, 1955; Steinberg, 1962a, 1962b) closely resemble the movements of young neurons during morphogenesis of the nervous system. Many examples could be given to illustrate the resemblance, but only a few are mentioned here. The migration of cells from the neural crest is an obvious case of differential cell migration, which may depend, at least in part, on a program of changes in adhesiveness of the migrating cells (Weston, 1963; Weston and Butler, 1966). The migration of young neurons from the external granular layer at the surface of the cerebellum to form the internal granular layer (Ramón y Cajal, 1952–1955; Miale and Sidman, 1961), and the migration of young neurons from the superficial position they occupy in the rhombic lip of the medulla to the depths of the brainstem (Harkmark, 1954; Taber Pierce, 1966, 1967) may also be controlled by a programed sequence of changes of adhesiveness and of motility of the migrating cells.

Differential adhesiveness of nerve terminals may also provide a plausible explanation of the specificity of connections of nerves with various types of sense organs. Differential adhesiveness of nerve fibers may be a mechanism for the sorting out of the fibers in long-fiber tracts, with the more adhesive fibers taking up positions internal to those that are less adhesive. For example, the latitude and longitude of each retinal ganglion cell need only be translated into a specific adhesiveness of its optic-nerve fiber for the fibers to arrange themselves in retinotopic order in the optic nerve and tract. The final sorting out of nerve-fiber terminals, and the formation of specific connections between the presynaptic terminals and the postsynaptic neurons, may also be the result of differential cellular adhesiveness. Multiple branching of the presynaptic fibers takes place and great-

ly increases the chances that each fiber will find a postsynaptic cell with the same specific adhesiveness. Differential adhesiveness of cells may play a role in the sorting out of migrating young neurons or of growing axons in all these cases, but the roles of chemotaxis and of mechanical guidance of the migrating neurons, particularly by pre-existing fiber tracts, must also be taken into consideration. The relative importance of these various factors in the control of migration and in the formation of synaptic connections cannot be evaluated at present, because the facts are not known.

## Genetic determination vs. functional validation in the development of neuronal connections

These comments on the putative role of such relatively simple mechanisms as differential cellular adhesiveness in the morphogenesis of the nervous system serve to illustrate the principle that it is much simpler and more economical for the egg to contain instructions for a program of development than to contain all the information to specify the structure of the fully developed organism. They also serve to emphasize the importance of temporal and spatial order during neurogenesis. From the beginning, neurogenesis has a temporal and spatial pattern. It proceeds according to a timetable that is the same for all members of the species, and it is very similar in all classes of vertebrates. In the prefunctional stage, the timetable must be genetically determined. Presumably, the orderly development of the nervous system is the result of sequential activation of genes and a regulated pattern of synthesis and distribution of various species of molecules in the young neurons and glioblasts. The details of this developmental program are unknown. Although there is good evidence that organized connectivity develops in the prefunctional stage, there is no evidence that functional activity is necessary for the formation of any neuronal connections. I am not referring to the unknown functional changes, which we must assume are the basis of learning and memory, but to the possibility that synapses develop as the result of a specific kind of experience. There is no evidence of the latter. Rather, it seems that the circuitry of the nervous system is prefunctionally determined, and that subsequent functional validation of some connections may occur.

Hubel and Wiesel (1963, 1965) have shown that this process of functional validation operates in the maturation of binocular connections in the visual cortex of newborn kittens. The evidence that the binocular connections are formed by prefunctional mechanisms is that they are present before the kitten opens its eyes. The binocular connections break down if normal binocular vision does not take place during the fourth to the sixth postnatal weeks—for example, if one eye is occluded or squints. The binocular connections persist if one eye is occluded after the sixth week and if the kitten has enjoyed normal binocular vision for several weeks after opening its eyes. These connections are formed during the prefunctional stage, but they must be functionally validated during a critical period of development, which may be an example of the principle that functionally effective connections are selected in preference to those that are functionally ineffective. It also illustrates the importance of timing.

Another principle should be emphasized; there are great differences among classes of neurons with respect to their developmental timetables. The modifiability of some neurons is lost very early in ontogeny, but persists longer in others, and may take place only during a critical period in other types of neurons (Jacobson, 1969). During development, neurons undergo a progressive reduction in their capacity to form new connections and to modify existing ones. This occurs in different classes of neurons at different times during development. On the basis of their degree of specification, we can group neurons into two main classes, although there may also be transitional or intermediate types. Class I neurons are highly specified early in development and are unmodifiable thereafter. Class II neurons remain uncommitted and modifiable until late in ontogeny, and their connections may then become specified as a result of functional interactions among the components of the nervous system. How, then, can two classes of neurons be recognized? In the first instance, we must determine, by experimentation, if there are any functional modifications in connections after the connections are surgically rearranged or dislocated. The amount of functional modification can be determined behaviorally or by mapping the neuronal connections electrophysiologically. Both tests have shown that the connections of retinal ganglion cells in the optic tectum of frogs and fishes are completely and irreversibly determined very early in development (in the prefunctional stage) and are unmodifiable thereafter. Are there anatomical or biochemical differences by means of which the two classes of neurons can be distinguished? The modifiable neurons might be expected to have immature or embryonic features. It should be appreciated, however, that functional determination of the neuron need not be associated with any single cellular structure, so might not be recognizable by anatomical methods.

Therefore, we must regard the formation of neuronal circuits as caused initially by genetic mechanisms that operate only with forward reference to experience. The forward reference has developed as the result of the selection, during evolution, of the genetic mechanisms that control neurogenesis. These genetic mechanisms may be sufficient in the case of some kinds of neurons. In other types of neurons, genetic determination alone may not be sufficient, and functional validation may be necessary. The prefunctional period of neurogenesis is followed by a period in which functional modifiability of some neuronal circuits is possible. These modifications can take place only within the constraints imposed by the genetic endowment and developmental history of the neuron.

## Summary

Specificity and diversity may develop in different ways in different kinds of neurons. Some neurons may become different as a result of genetic control of their development, whereas others may diversify because they are released from genetic constraints. In other words, differentiation of the first class of neurons is deterministic; in the second class, it is probabilistic. Transitional and intermediate classes are also supposed to develop.

Neurons of Class I are highly specified early in development, after which their connectivity is tightly constrained and unmodifiable. The ganglion cells of the retina show all the characteristics of this class. They are generated early in ontogeny and are irreversibly specified at the time of their final mitosis before the growth of their processes commences and before connections develop. Other types of large neurons with long axons and invariant connectivity are believed to exhibit similar characteristics.

Neurons of Class II remain unspecified until late in ontogeny. It is postulated that these neurons are produced in excess and in great diversity. From this diverse population, only those neurons, or their connections, that are functionally most fitted to survive to maturity do so. Many types of interneurons are supposed to belong to Class II or have transitional characteristics. Their connections may be modified or may be maintained by function and experience of an appropriate type. One of the requirements for functional validation of connections of Class II neurons may be coincidence of patterns of impulses in convergent afferents. Evidence of this kind of requirement has been found in binocular connections in the visual cortex of newborn kittens. The cutaneous afferent neurons of the frog tadpole have characteristics intermediate between those of Class I and those of Class II, but after metamorphosis they clearly develop into neurons of Class I.

The examples discussed in this essay illustrate the principle that during development nerve cells undergo a progressive increase in their specificity and a corresponding decrease in their capacity to form new connections and to modify existing ones, although this change takes place at different times during the development of different types of neurons.

# REFERENCES

ADLER, J., 1966. Chemotaxis in bacteria. *Science* (*Washington*) 153: 708–716.

ATTARDI, D. G., and R. W. SPERRY, 1963. Preferential selection of central pathways by regenerating optic fibers. *Exp. Neurol.* 7: 46–64.

BAKER, R. E., and M. JACOBSON, 1970. Development of reflexes from skin grafts in *Rana pipiens:* Influence of size and position of grafts. *Develop. Biol.* (in press).

BONNER, J. T., 1947. Evidence for the formation of cell aggregates by chemotaxis in the development of the slime mold *Dictyostelium discoideum. J. Exp. Zool.* 106: 1–26.

BREMERMANN, H. J., 1963. Limits of genetic control. *Inst. Elec. Electron. Eng. Trans. Mil. Electron.* 7: 200–205.

CURTIS, A., 1967. The Cell Surface: Its Molecular Role in Morphogenesis. Logos Press, London.

ECCLES, J. C., R. M. ECCLES, C. N. SHEALY, and W. D. WILLIS, 1962. Experiments utilizing monosynaptic excitatory action on motoneurons for testing hypotheses relating to specificity of neuronal connections. *J. Neurophysiol.* 25: 559–580.

EDDS, M. V., JR., 1953. Collateral nerve regeneration. *Quart. Rev. Biol.* 28: 260–276.

EDWARDS, J. S., and T. S. SAHOTA, 1967. Regeneration of a sensory system: The formation of central connections by normal and transplanted cerci of the house cricket *Acheta domesticus. J. Exp. Zool.* 166: 387–395.

ELSASSER, W. M., 1958. The Physical Foundation of Biology. Pergamon Press, New York.

ELSASSER, W. M., 1961. Quanta and the concept of organismic law. *J. Theor. Biol.* 1: 27–58.

ELSASSER, W. M., 1962. Physical aspects of non-mechanistic biological theory. *J. Theor. Biol.* 3: 164–191.

FISHER, S., and M. JACOBSON, 1970. Ultrastructural changes during early development of retinal ganglion cells in *Xenopus. Z. Zellforsch. Mikroskop. Anat.* 104: 165–177.

GAZE, R. M., and M. JACOBSON, 1963. A study of the retinotectal projection during regeneration of the optic nerve in the frog. *Proc. Roy. Soc., ser. B, biol. sci.* 157: 420–448.

GUSTAFSON, T., and L. WOLPERT, 1967. Cellular movement and contact in sea urchin morphogenesis. *Biol. Rev.* (*Cambridge*) 42: 442–498.

HARKMARK, W., 1954. Cell migrations from the rhombic lip to the inferior olive, the nucleus raphe and the pons. A morphological and experimental investigation on chick embryos. *J. Comp. Neurol.* 100: 115–209.

HARRIS, H., 1954. Role of chemotaxis in inflammation. *Physiol. Rev.* 34: 529–562.

HIBBARD, E., 1967. Visual recovery following regeneration of the optic nerve through oculomotor nerve root in *Xenopus. Exp. Neurol.* 19: 350–356.

HOLTFRETER, J., 1939. Gewebeaffinität, ein Mittel der embryonalen Formbildung. *Arch. Exp. Zellforsch.* 23: 169–209.

HUBEL, D. H., and T. N. WIESEL, 1963. Receptive fields of cells in striate cortex of very young, visually inexperienced kittens. *J. Neurophysiol.* 26: 994–1002.

HUBEL, D. H., and T. N. WIESEL, 1965. Binocular interaction in striate cortex of kittens reared with artificial squint. *J. Neurophysiol.* 28: 1041–1059.

JACOBSON, M., 1967a. Starting points for research in the ontogeny of behavior. *In* Major Problems in Developmental Biology (M. Locke, editor.) Academic Press, New York, pp. 339–383.

JACOBSON, M., 1967b. Retinal ganglion cells: Specification of central connections in larval *Xenopus laevis. Science* (*Washington*) 155: 1106–1108.

JACOBSON, M., 1968a. Development of neuronal specificity in retinal ganglion cells of *Xenopus. Develop. Biol.* 17: 202–218.

JACOBSON, M., 1968b. Cessation of DNA synthesis in retinal ganglion cells correlated with the time of specification of their central connections. *Develop. Biol.* 17: 219–232.

JACOBSON, M., 1969. Development of specific neuronal connections. *Science* (*Washington*) 163: 543–547.

JACOBSON, M., and R. E. BAKER, 1968. Neuronal specification of cutaneous nerves through connections with skin grafts in the frog. *Science* (*Washington*) 160: 543–545.

JACOBSON, M., and R. E. BAKER, 1969. Development of neuronal connections with skin grafts in frogs: Behavioral and electrophysiological studies. *J. Comp. Neurol.* 137: 121–141.

JACOBSON, M., and R. M. GAZE, 1965. Selection of appropriate tectal connections by regenerating optic nerve fibers in adult goldfish. *Exp. Neurol.* 13: 418–430.

KOLLROS, J. J., 1943. Experimental studies on the development of the corneal reflex in Amphibia III. The influence of the periphery upon the reflex center. *J. Exp. Zool.* 92: 121–142.

KONIJN, T. M., D. S. BARKLEY, Y. Y. CHANG, and J. T. BONNER, 1968. Cyclic AMP: A naturally occurring acrasin in the cellular slime molds. *Amer. Natur.* 102: 225–233.

KONIJN, T. M., J. G. C. VAN DE MEENE, J. T. BONNER, and D. S. BARKLEY, 1967. The acrasin activity of adenosine-3', 5'-cyclic phosphate. *Proc. Nat. Acad. Sci. U. S. A.* 58: 1152–1154.

MCCUTCHEON, M., 1946. Chemotaxis in leukocytes. *Physiol. Rev.* 26: 319–336.

MAYNARD, D. M., and M. J. COHEN, 1965. The function of a heteromorph antennule in a spiny lobster, *Panulirus argus. J. Exp. Biol.* 43: 55–78.

MIALE, I. L., and R. L. SIDMAN, 1961. An autoradiographic analysis of histogenesis in the mouse cerebellum. *Exp. Neurol.* 4: 277–296.

MINER, N., 1956. Integumental specification of sensory fibers in the development of cutaneous local sign. *J. Comp. Neurol.* 105: 161–170.

POMERAT, C. M., 1961. Cinematology, indispensable tool for cytology. *Int. Rev. Cytol.* 11: 307–334.

RAMÓN Y CAJAL, S., 1952–1955. Histologie du Système Nerveux de l'Homme et des Vertébrés (translation by L. Azoulay). Instituto Ramón y Cajal del Consejo Superior de Investigaciones Científicas, Madrid, 2 vols. (Reprint of 1909–1911 issue, published in Paris.)

RAMÓN Y CAJAL, S., 1959. Degeneration and Regeneration of the Nervous System (translated by R. M. May). Hafner, New York. (Translation of an edition of 1928, Oxford University Press, London.)

RAMÓN Y CAJAL, S., 1960. Studies on Vertebrate Neurogenesis (revised and translated by L. Guth). Charles C Thomas, Springfield, Illinois. (Revision of a 1929 edition.)

Speidel, C. C., 1933. Studies of living nerves II. Activities of ameboid growth cones, sheath cells, and myelin segments, as revealed by prolonged observation of individual nerve fibers in frog tadpoles. *Amer. J. Anat.* 52: 1–79.

Speidel, C. C., 1935. Studies on living nerves III. Phenomena of nerve irritation and recovery, degeneration and repair. *J. Comp. Neurol.* 61: 1–80.

Speidel, C. C., 1941. Adjustments of nerve endings. *Harvey Lect.* (*1940–1941*), Ser. 36: 126–158.

Speidel, C. C., 1942. Studies of living nerves VII. Growth adjustments of cutaneous terminal arborizations. *J. Comp. Neurol.* 76: 57–73.

Sperry, R. W., 1944. Optic nerve regeneration with return of vision in anurans. *J. Neurophysiol.* 7: 57–69.

Sperry, R. W., 1951. Regulative factors in the orderly growth of neural circuits. *Growth, Suppl.* 15: 63–87.

Sperry, R. W., 1955. Problems in the biochemical specification of neurons. *In* Biochemistry of the Developing Nervous System (H. Waelsch, editor). Academic Press, New York, pp. 74–84.

Sperry, R. W., 1963. Chemoaffinity in the orderly growth of nerve fiber patterns and connections. *Proc. Nat. Acad. Sci. U. S. A.* 50: 703–710.

Steinberg, M. S., 1962a. On the mechanism of tissue reconstruction by dissociated cells, I. Population kinetics, differential adhesiveness and the absence of directed migration. *Proc. Nat. Acad. Sci. U. S. A.* 48: 1577–1582.

Steinberg, M. S., 1962b. On the mechanism of tissue reconstruction by dissociated cells, III. Free energy relations and the reorganization of fused, heteronomic tissue fragments. *Proc. Nat. Acad. Sci. U. S. A.* 48: 1769–1776.

Steinberg, M. S., 1963. Reconstruction of tissues by dissociated cells. *Science* (*Washington*) 141: 401–408.

Steinberg, M. S., 1964. The problem of adhesive selectivity in cellular interactions. *In* Cellular Membranes in Development (M. Locke, editor). Academic Press, New York, pp. 321–366.

Taber Pierce, E., 1966. Histogenesis of the nuclei griseum pontis, corporis pontobulbaris and reticularis tegmenti pontis (Bechterew) in the mouse. An autoradiographic study. *J. Comp. Neurol.* 126: 219–239.

Taber Pierce, E., 1967. Histogenesis of the dorsal and ventral cochlear nuclei in the mouse. An autoradiographic study. *J. Comp. Neurol.* 131: 27–53.

Thorpe, W. H., 1963. Ethology and the coding problem in germ cell and brain. *Z. Tierpsychol.* 20: 529–551.

Townes, P. L., and J. Holtfreter, 1955. Directed movements and selective adhesion of embryonic amphibian cells. *J. Exp. Zool.* 128: 53–120.

Trinkaus, J. P., 1965. Mechanisms of morphogenetic cell movements. *In* Organogenesis (R. L. DeHaan and H. Ursprung, editors). Holt, Rinehart and Winston, New York, pp. 55–104.

Weiss, P., 1934. In vitro experiments on the factors determining the course of the outgrowing nerve fiber. *J. Exp. Zool.* 68: 393–448.

Weiss, P., 1941. Nerve patterns: The mechanics of nerve growth. *Growth, Suppl.* 5: 163–203.

Weiss, P., 1942. Lid-closure reflex from eyes transplanted to atypical locations in *Triturus torosus:* Evidence of a peripheral origin of sensory specificity. *J. Comp. Neurol.* 77: 131–169.

Weiss, P., 1947. The problem of specificity in growth and development. *Yale J. Biol. Med.* 19: 235–278.

Weiss, P., 1955. Nervous system (neurogenesis). *In* Analysis of Development (B. H. Willier, P. A. Weiss, and V. Hamburger, editors). W. B. Saunders, Philadelphia, pp. 346–401.

Weston, J. A., 1963. A radioautographic analysis of the migration and localization of trunk neural crest cells in the chick. *Develop. Biol.* 6: 279–310.

Weston, J. A., and S. L. Butler, 1966. Temporal factors affecting localization of neural crest cells in the chicken embryo. *Develop. Biol.* 14: 246–266.

# 13    A Model of Synaptic and Behavioral Ontogeny

DAVID BODIAN

DAVID BODIAN   Department of Anatomy, The Johns Hopkins University School of Medicine, Baltimore, Maryland

Other chapters in this volume deal with important advances in our knowledge of the time course of proliferation of embryonic cells of the neural tube, of the origin and migration of cell classes to definitive locations, of the origin of neuron interdependence, and of the modulation of the size of functional neuronal populations through early cell death. Overlapping these events, and continuing beyond them, are the events of differentiation of cells, especially neurons, which are more directly linked to the establishment of functional systems of cells, and of their behavioral capabilities. In particular, the outgrowth of neuronal processes, axons, and dendrites, and the linkage of neuron to neuron through synaptic formation, laid the foundation for the development of functional "subsystems," such as the arm region of the spinal cord or·the retina, and of the linkage of subsystems into a total nervous system.

Pioneers of the theory of neuronal organization and of behavioral development, such as Ramón y Cajal (1960) and Coghill (1929), respectively, were well aware of the critical importance of synapse formation in the establishment of functional circuits. Only recently, however, with the coming of the electron microscope and with the development of refined electron-miscroscope techniques, has it been possible to explore with adequate tools the significant problems of synaptic development.

Although studies of synaptogenesis are in an early stage, it seems worthwhile to see whether our present knowledge of the morphology and sequence of synaptic development, and of the progressive development of behavior, can be correlated. If the two can be related, as Coghill proposed, is it possible to discern an adaptive "strategy" of construction of a functional subsystem, such as the spinal cord segment, in which progressive development of motor behavior and progressive development of circuitry and synaptic structure can be studied together? Moreover, if a discernible plan of construction emerges in the spinal cord, is it generalizable to other subsystems of the vertebrate nervous system, and can it serve for the planning of future research?

## Choice of subsystem

The studies on which this essay is based are part of an investigative program on the development of the fine structure in the spinal cord of macaque embryos (Bodian, 1966b, 1968). The monkey has a relatively long gestation period (160 days) as compared with that of other laboratory mammals. This was thought to be advantageous in respect to the timing of significant events of brief duration. Macaques, as do other primates, including man, possess a direct connection from cerebral cortex to spinal motoneurons, so that the range of input levels to motoneurons is thereby extended to the highest levels of control.

In the spinal model which we have chosen, it has become apparent that the critical event in the establishment of synaptic connections begins surprisingly early, as does the onset of reflexive muscular movements. Both reflex and spontaneous movements of the embryo were studied in the amniotic fluid, through the thin and transparent amniotic membrane, as well as after it was removed from the amniotic cavity. The placenta and umbilical cord were left intact, however, and only local anesthesia was used for the exposure of the maternal uterus. Pregnancies were accurately timed by brief exposure of females to the male. Table I indicates the available material. Cinematography of cutaneous reflex responses and of spontaneous movements assisted repeated reviews.

## Analysis of early stages of motility

Although we deal with the beginnings of behavior and are a long way from the goal of understanding the development of higher nervous functions, there is no reason to doubt that a study of the simplest behavioral subsystem in a vertebrate might be used as a model that could yield clues or principles which might strengthen the attack on a more complex subsystem, such as the primate and human cerebral cortex. Coghill used the urodele as a primitive model that might offer insight into the ontogeny of motility of all vertebrates, but we propose to use the ontogeny of the primate spinal segment as a model that might offer insight into the possibilities and difficulties in the study of synaptic ontogeny. The hopeful assumption is that we might at least discover certain basically similar principles of ontogenetic construction, as well as important particular differences between such subsystems as spinal cord, retina, olfactory bulb, cerebral cortex, and cerebellum. Needless to say, the cellular components of each subsystem in the central nervous system are basically similar, and each undergoes a more or less comparable sequence of maturational stages. The time course is, however, variable, so that development of the spinal cord, for example, is far in advance of that of cerebral cortex.

Of the two populations, the nerve cells and the neuroglial cells, the former, of course, play the primary role. In fact, neurons may complete the establishment of major synaptic circuitry in a particular center before neuroglial cells have multiplied to a significant degree or have differentiated for their major functions. Only ependymal and pial neuroglia (boundary neuroglia) are present in the early period of development of a particular "center." Thus, we may at first confine our attention solely to the nerve cells.

For the purpose of comparing the model we have chosen, the spinal-cord segment, with other subsystems of the central nervous system, the following terminology is employed to designate nerve-cell classes within the array of nerve cells of a particular subsystem:

Efferent nerve cells: origin of major output pathway—often have large perikarya.

Primary afferent nerve cells: develop concurrently with above. (Axons arise from neurons with distant perikarya, i.e., are extrinsic.)

Interneurons: small neurons, of both excitatory and inhibitory function, wholly within the particular "center" (intrinsic).

Secondary afferent nerve cells: extrinsic input neurons other than primary afferent ones.

TABLE I

TABLE I

*Onset of Motor Behavior in Relation to Early Fetal Growth in Macaques (from Bodian, 1968)*

| Stage | Monkey | Actual Age* | Estimated Ovulation Age (Days) | Crown–Rump Length (mm) | Body Weight (grams) | Brain Weight (grams) | Behavioral Stage |
|---|---|---|---|---|---|---|---|
| 1 | RB 6 | 41–47 | | 17 | 0.55 | | |
| | CB 43 | 42–45 | 42–45 | 20 | 0.52 | | Pre-reflex |
| | RB 5 | 37–42 | | 20 | 0.80 | | |
| | RB 2 | 42–47 | | 22 | 0.90 | | |
| 2 | CB 48 | | | 24 | 1.3 | | Early local spontaneous and re- |
| | CB 46 | 45–48 | 46–47 | 26 | 1.6 | | flex movements (trigeminal, |
| | CB 38 | 44–47 | | 28 | 1.7 | 0.30 | cervical, brachial) |
| 3A | RB 4 | 47–52 | | 32 | 1.8 | | Above, plus weak trunk and leg |
| | RB 3 | 43–48 | 48–49 | 33 | 2.6 | | responses to hand and perioral |
| | CB 50 | 48–49 | | 36 | 3.0 | 0.50 | stroking |
| 3B | CB 37 | 49–52 | | 41 | 3.6 | 0.52 | Active long intersegmental re- |
| | CB 42 | 48–51 | 50–51 | 41 | 3.6 | | flexes and onset of complex |
| | RB 1 | 48–52 | | 41 | 4.2 | 0.60 | arm reflexes |

*Based on period of exposure to male (mid-cycle).

## Stage I. Prior development of efferent and primary afferent neurons

One may define four general stages of development in the spinal cord of vertebrates; these are characterized by increments of synaptic development, as shown in Figure 1. The first is the stage of prior development of efferent and primary afferent neurons, but with no evidence of synaptic linkage on motor neurons, even within one or two days prior to expected local reflexes. Earlier studies have emphasized the precocity of differentiation of motor-column cells, and spinal ganglion neurons do not lag far behind. Ramón y Cajal's early Golgi study tended to create the impression that interneurons develop almost concurrently with primary afferent and efferent neurons (Ramón y Cajal, 1960). Most students of spinal-cord development, however, have commented on the relatively late development of most interneurons in the dorsal and intermediate cell columns. In Figure 2, showing Stage 1 in the monkey, primary afferent and efferent neurons have completed their link with peripheral end-organs, whereas the interneuron populations of the posterior horn are still in the replicative stage. In electron micrographs, no synaptic bulbs can be found in the primitive lateral motor column or in the adjacent marginal zone. Only growth cones are present.

During Stage I, the early development of efferent neurons is sometimes accompanied by spontaneous movements in some species, before the development of capability of a reflexly initiated response. Although other nervous centers have no way of demonstrating such a degree of functional "readiness" prior to effective synaptic transmission to the efferent neuron, the occurrence of such a phase in the spinal model suggests that electrical excitability of efferent neurons in general may precede that of interneurons, which is, of course, subject to experimental testing.

SIGNIFICANCE OF PRIOR DEVELOPMENT OF EFFERENT NEURONS AND PRIMARY AFFERENT NEURONS  Why does the last neuronal component of the spinal reflex circuit develop first, rather than last, in the linear sequence from primary afferent neuron to interneuron to efferent neuron? One may suppose, in a speculative way, that the prior development of efferent and primary afferent neurons must have a special adaptive value. Both the major input neurons and the output neurons of major functional centers are usually arranged to convey information that is of a topographic nature, or is expressed as a "projection," such as the projection of the body image. This major requirement of efferent and primary afferent neurons might mean that the population of such neurons in a center must be topographically specified by its axon terminations (motor periphery, in the instance of the spinal cord) before central interneuron complexity or some other neuropil complexity can interfere with the optimal array of output and input components. The bridge is built, as it were, by extension from both banks of the river toward the center.

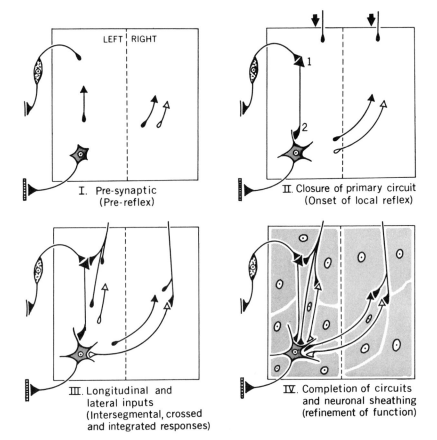

FIGURE 1 Schema indicating model of developmental stages of synaptic circuitry in a nervous center, based on analysis of the spinal motor system, in which the level of development may be correlated with level of behavioral capability.

Stage I. Initial differentiation of major efferent neurons, followed by major afferent neurons (extrinsic), and, finally, interneurons. Growth cones shown as oval tips of growing axons.

Stage II. Synaptic closure of circuit between primary afferent and primary efferent neurons, in linear sequence from afferent to efferent neuron. Synapses are numbered in order of linkage, and synaptic bulbs shown as triangular terminals of axons.

Stage III. Ingrowth of axons of extrinsic and intrinsic interneurons, longitudinal and lateral (including crossed interneurons in spinal cord). Integrated activity of system, with inhibitory (open neurons) as well as excitatory (shaded neurons) units.

Stage IV. Increased complexity of interneuron connections, and progressive refinement of function. Synaptic linkages completed, permitting neuroglial development to enable more efficient synaptic and axonal function without interference with linkage formation.

## Stage II. Closure of intrinsic circuits from afferent to efferent neurons

In the macaque, the earliest reflex responses that can be elicited are flexion (withdrawal) reflexes of the arm, which follow stroking of the hand with a soft bristle. The reflex is ipsilateral at first, and is contemporaneous with spontaneous movements of mouth and arms within the amniotic fluid. Unlike the human, the macaque at this stage does not exhibit jaw movements in response to cutaneous stimulation about the mouth; only axo-dendritic synapses are found on motor neurons (Figure 3). The precocious local reflex may take place either in the oral or in the brachial regions in different mammalian species, and perhaps it represents an unusual sensitivity to synaptic transmission at these levels at a stage when synaptic contacts with the efferent neuron are few and are limited to only one type of synaptic bulb on dendrites (Bodian, 1966b, 1968). Indeed, the early differentiation of dendrites of major efferent neurons and the direction of their growth in the path of interneuron axons appear to be designed to increase the probability of "correct" connectivity with the major afferent input. The primitive

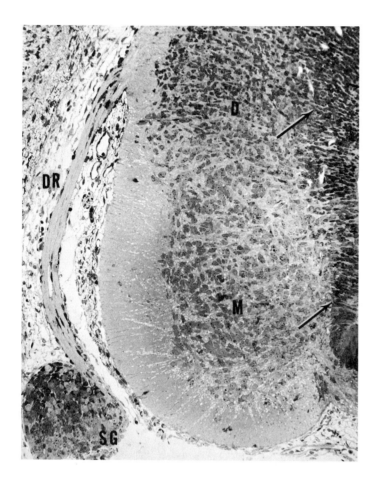

FIGURE 2 Light micrograph of cervical spinal cord of macaque, just prior to onset of earliest local reflexes. Replication of cells in the motor cell column (M) is completed, whereas proliferation and migration of neuroblasts of the dorsal cell region (D), where interneurons will differentiate, is actively in progress. Lower arrow shows margin of differentiated ependymal layer. Upper arrow shows a zone of active mitosis (matrix layer). Central canal is at right margin. DR, dorsal root; SG, spinal ganglion. Two-micron section, plastic imbedding, toluidin blue stain, × 150. Macaque CB 43, 20-mm stage.

synaptic bulbs are found mainly in the inner marginal zone, on dendrites only, and contain both spheroid synaptic vesicles and the larger vesicles that characterize growth cones (Bodian, 1966b, 1968).

The synaptic bulbs are remarkably few in number when the local reflex circuit is established. They appear to be capable of functioning very soon after linkage. Our recent studies have shown that this behavioral expression of closure of the local reflex circuit is a precocious characteristic of the brachial region in the macaque, whereas a comparable state of synaptic development is behaviorally silent in the lumbar region. Here, the capability of local reflex response awaits the input into the circuit of intersegmental synapses, which represent the substrate of Coghill's so-called "total reaction pattern," or the synchrony of local limb response with trunk response (Coghill, 1929).

SIGNIFICANCE OF STAGE OF CLOSURE OF PRIMARY CIRCUIT It now seems generally agreed that Coghill's concept of the "sovereignty" of the intersegmental integrating mechanism over the local response patterns requires reinterpretation (Bodian, 1968). Although intersegmental organi-

zation, as an example of progressive interneuron development, is of great importance in an early stage of spinal-reflex maturation, it is not necessarily the initial or primary behavioral or morphological component. Indeed, Coghill's picture of the "motor path" of the urodele (Coghill, 1929, Figure 10) is that of descending collaterals of motor neurons, rather than of interneurons. Evidence is lacking for such a "motor path" in other vertebrates. In the monkey, our electron-microscope studies show such an early appearance of two types of synaptic bulb, probably representing excitatory and inhibitory inputs to motor neurons, that their origin from interneurons seems more in accord with modern studies of motor-neuron reflex organization than with their possible origin from motor neuron collaterals. Coghill recognized the necessity of inhibitory as well as excitatory inputs to explain alternating swimming movements, but had no way of relating this to morphological differences in synaptic type. Today, the existence of excitatory and inhibitory spinal interneurons is well established (Florey, 1961).

Recognition of the likelihood that Coghill's whole body pattern of motor integration is an early and important, but not primary, aspect of development of organization within the spinal cord opens the way to interpret the significance of Stage II for nervous centers in general. *The closure of a primary afferent-efferent circuit as the first goal of neuronal development is consistent with the idea that the topographic relationship of input to output (as in sensory projection pathways) must be insured by early closure.* Once this is established, the elaboration of additional intrinsic and extrinsic synaptic inputs follows, and perhaps is not so rigidly determined, as is also suggested by Jacobson in this volume (also Jacobson, 1969).

Efferent neuronal cell bodies are prominent in any given center as the first to show differentiation of Nissl bodies and dendrites, so the impression has been created among some workers that a retrograde sequence of development takes place from efferent to afferent neurons. If the location of the major afferent neuron is well known, as in the spinal cord, it is clear that this neuron is not far behind the efferent neuron in its development. With respect to the closure of the first three-neuron circuits from afferent to efferent neurons, however, a linear sequence of synaptic linkage does appear to exist (Figure 1).

For example, in the spinal cord the simplest functional pathway, involving three neuron classes, becomes functional immediately upon closure of the circuit by means of synaptic contact with the efferent neuron (Bodian, 1968). Previous linkage of the first-order and second-order neurons must be assumed. In the spinal cord, the circuit closure may be represented by a simple, local, cutaneous reflex, or may be "silent."

*Stage III. Development of extrinsic secondary circuitry (linkage of intersegmental and lateral (crossed) synapses: axosomatic as well as axodendritic; inhibitory as well as excitatory).*

Our own studies of the phase of elaboration of interneuron and extrinsic synaptic linkages have only begun to explore the events of this immensely complex phase. Qualitative changes and quantitation of the rate of change in each variable are both difficult to deal with from the technical standpoint. It is clear, however, that synaptogenesis continues at an exponential rate between the stage of closure of the first circuit and the stage of glial differentiation. Onset of synaptogenesis can be recognized by the transformation of growth cones to synaptic bulbs of different types (attachment sites and synaptic vesicles), and later onset of myelination and astrocyte differentiation signals the near-completion of synaptic linkages.

During the intervening period, the entry and progressive differentiation of axonic growth cones of extrinsic neurons and of excitatory and inhibitory interneurons have certain new characteristics. Synaptic linkage with motor neuron cell bodies takes place, as contrasted with the purely axo-

dendritic synapses of Stage II (Figure 4). A second and possibly inhibitory synaptic type appears (F type, Figure 5) and increases to its maximal proportion in the total synaptic-bulb population in an astonishingly brief period—a few per cent of the total gestation period (Figure 6). However, the best estimate of the rate of total synaptic differentiation, as measured by the increase in volume of all synaptic bulbs in comparison with total neuropil volume, indicates a very gradual increase, approaching the adult proportion near term.

SIGNIFICANCE OF STAGE OF DEVELOPMENT OF EXTRINSIC SECONDARY CIRCUITRY  It is clear that, in the spinal cord, the appearance of synaptic bulbs on the somata of motor

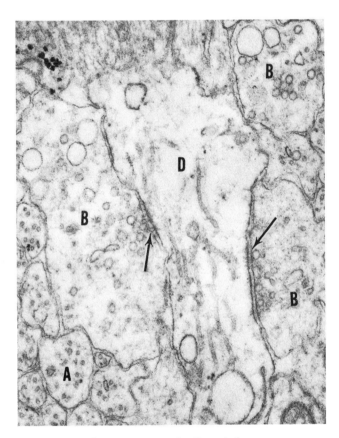

FIGURE 3  Electron micrograph of cervical motor-neuron neuropil of macaque embryo CB 46 (26 mm), at stage of onset of local cutaneous reflexes. Primitive synaptic bulbs (B) are first detectable on primitive dendrites (D) at this stage. Synaptic bulbs show spheroid microvesicles of 500 Å, as well as larger vesicles of 1000–2000 Å. The latter are not present in mature synaptic bulbs and are characteristic of growth cones. Thus, the primitive synaptic bulb is transitional between growth cones and synaptic bulbs. A. Primitive axon in cross section. × 42,000.

neurons and the appearance of significant numbers of a second type of synaptic bulb, containing flattened synaptic vesicles (Bodian, 1966a), suggest a causal relation to the appearance of intersegmental and crossed-reflex responses. These responses follow closely upon the stage of simple, local reflexes. Hamburger has shown (this volume) that, in the absence of primary afferent neurons, intersegmental neurons in the chick are capable of inducing the development of periodic limb movements. This indicates that the primary afferent synapses are not essential for the development of motor neuron activity, but does not exclude the primary afferent neurons from a major role in the development of normal motor patterns.

*At any rate, the significance of Stage 3 would seem to be the early integration of a series of related "subsystems."* In the motor system of the macaque, this appears to involve linkage of brachial spinal cord with a succession of caudal and rostral

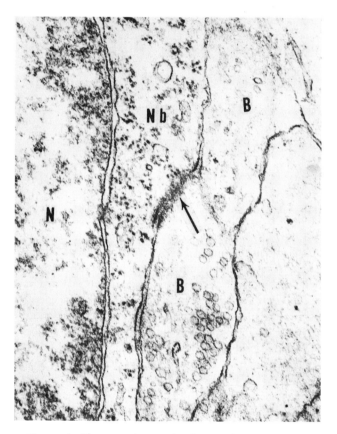

FIGURE 4 Electron micrograph of cervical motor-neuron neuropil of macaque embryo CB 42 (41 mm), showing neuroblast (Nb) with axosomatic synaptic bulb (B). These are first apparent at this stage of developing intersegmental reflexes. Synaptic bulbs with flattened synaptic vesicles also appear at this stage. Arrow indicates adhesion site of axosomatic synaptic bulb. N, nucleus of neuroblast. × 58,000.

"subsystems," concluding with the corticospinal neurons. The preceding stage may be behaviorally "silent," as in the lumbar spinal segments, but summation with intersegmental synapses of Stage III appears to insure reflex responses after stimulation of any segmental level.

### Stage IV. Completion of synaptic linkages and onset of neuronal sheathing (refinement of function)

Neuronal and axonal sheathing by neuroglial cells does not begin to an appreciable degree in the motor neuron neuropil of the monkey spinal cord until about the 65th day of gestation, or three weeks after first synaptic bulbs can be observed (Figure 7). During the next eight weeks, which represent one third of the gestation period, the number and variety of synaptic bulbs appear to approach the adult level, whereas neuroglial development continues to lag behind. Ramón y Cajal recognized astrocytes in the ventral horn of the chick spinal cord at a stage of nine to 10 days, when synaptic development is well advanced, and noted that they appear last in the dorsal horns (Ramón y Cajal, 1960). He did not comment on the important fact that the sequence of neuroglial differentiation follows that of neuronal and synaptic differentiation in each region.

Because synaptic linkages are well advanced soon after neuroglial differentiation and neuronal sheathing begin, it is obvious that the elaborate refinement of function, such as the voluntary movement of an adult animal, is not dependent merely on completion of circuits within the motor neuron neuropil. To a certain degree, maturation of the neurons themselves, as well as of other morphological components, such as neuroglia, plays a continuing role. Finally, functional maturation implies a progressive refinement of cerebral signals descending to the intrinsic spinal circuits, as their quality is developed in accord with forebrain differentiation. For example, the corticospinal linkages with the spinal circuits occur very early, before input to the corticospinal neurons is fully developed.

In early studies of myelin-sheath development it was established, first, that myelin sheathing of axons is not a necessary requirement for the functioning of embryonic axons, and, second, that the sequence of myelination in different regions of the nervous system, as well as the sequence of astrocyte development, roughly parallels the sequence of maturation. The delay in both myelin and neuropil sheathing by neuroglia has not, however, been adequately explained. Our studies of synaptic differentiation in the spinal cord indicate that the occurrence of appreciable neuroglial sheathing coincides with the terminal stages of synaptic linkages. The significance of axonal sheathing for making possible increased conduction velocity of myelinated axons, or of neuropil sheathing for restricting the extracellular diffusion of important cations, such as $K^+$ in synaptic regions,

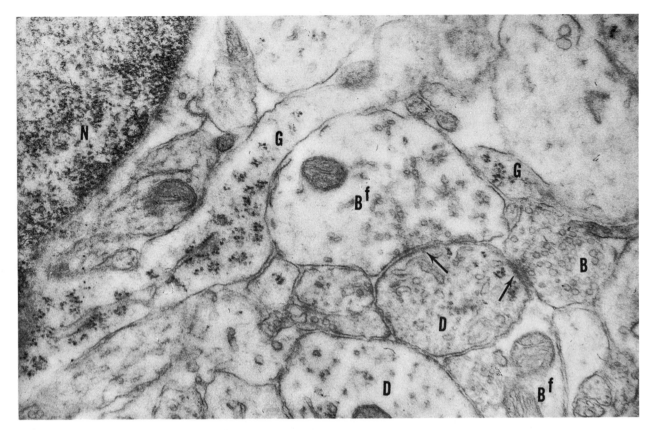

FIGURE 5 Electron micrograph of cervical motor neuronal neuropil of macaque embryo CB 52 (60 mm). This stage shows rapidly increasing density of synaptic bulbs (B), early intrusion of neuroglial processes (G) into the neuropil, and high proportion of synaptic bulbs with flattened synaptic vesicles (B$^f$). Arrows show adhesion sites of axodendritic synapses. D, dendrites; N, nucleus of primitive neuroglia cell. × 38,000.

is now supported by experimental data (Kuffler et al., 1966). The further refinement of behavior implied by the more rapid and efficient functioning of sheathed neurons is perhaps achieved, however, at the expense of restricting, but not necessarily eliminating, the further exploratory sprouting of axons and axon collaterals. *In these terms the delay in sheathing is understandable as an adaptive mechanism for insuring unhindered neuron-to-neuron recognition and linkage.* Figure 1 summarizes in diagramatic form the stages that have been discussed.

## Comparison of development of spinal cord with higher order integrating centers

We return now to the proposition that any one center, such as the spinal segment, may serve as a model for other centers in the nervous system. It seems unlikely at the outset that such a model could be quite general in its applicability. Of greater interest is the possibility that the model may be sufficiently general to be useful. A review of available data, in-

cluding Ramón y Cajal's pioneering work (1960) on spinal cord, retina, cerebellum, and cerebral cortex, (1) has convinced us that the model possesses certain general features, although only a limited number of centers have been studied to the extent that an adequate analysis is possible. Whenever we deal with a center that has a population of large efferent neurons, this neuron is the first to differentiate (spinal motor neuron, retinal ganglion cell, olfactory mitral cell, cerebellar Purkinje cell, and cortical pyramidal cell), as can be found in Ramón y Cajal's descriptions of the earliest stages of neuronal development in these centers. Contemporary workers have confirmed this fact, but seem to emphasize only that the earliest neurons to differentiate are large (macroneurons) and that successively smaller neurons are then differentiated (Altman, 1966; Fujita, 1967; Hinds, 1968). Angevine's data (this volume) on cortex are also in agreement. Little attention has been given to the fact that large neurons in a center are generally the major efferent neurons with a topographic projection, and that their early differentiation may be linked to the necessity for early cir-

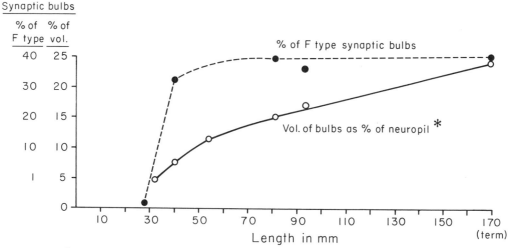

Synaptic bulbs

**% of F type synaptic bulbs**

*Vol. of bulbs as % of neuropil* \*

Length in mm

(term)

\*Exclusive of cell bodies and myelinated fibers

FIGURE 6 Chart showing the time of first occurrence and rate of development of populations of two major types of synaptic bulbs in the motor-neuron neuropil of macaque fetuses. Total volume of all synaptic bulbs increases gradu- ally as a fraction of the volume of the motor-neuron neu- ropil until birth, but the proportion of F type synaptic bulbs increases dramatically between 28 and 41 mm. Data ex- tended from Bodian, 1968.

cuit formation with primary afferent terminals in the center.

In the case of the retina, we have not only Ramón y Cajal's description of the precocious differentiation of gan- glion cells, followed by receptor cells, but also radioauto- graphic evidence for the sequence of completion of mitosis in conformity, as shown by Jacobson in this volume. Recent electron-microscope studies of mouse and rat retina also confirm this sequence. These studies show, in addition, that there is a reverse, or centripetal, sequence of synaptic link- ages from receptor cell to bipolar cell to ganglion cell (Olney, 1968; Weidman and Kuwabara, 1968; Nilsson and Crescitelli, 1969). A similar sequence appears to have been found in the developing olfactory bulb (Hinds, 1968). The similarity to spinal-cord development is striking.

It is well established that, in the cerebral cortex, the effer- ent pyramidal cells are the first neurons to differentiate. By using the Golgi method, Ramón y Cajal showed, in the four-day-old mouse, a very early entry of extrinsic axons into the superficial horizontal plexus in proximity to apical dendrites of pyramidal cells and to horizontal cells in layer I (Ramón y Cajal, 1960, Figures 1, 5, 8). Neurons with as- cending axons are present at this time, so that the possibility of an initial circuit from specific afferent endings in the deep cortical layers, relaying to pyramidal basal dendrites, must also be considered. Ramón y Cajal's figure of the cortex of the newborn mouse shows development of basal dendrites, for example, but no afferent fibers are shown. In more re- cent years, an active interest in evoked response patterns of

the developing cortex, and in morphological correlates of function, has emerged. Interpretation of early electrical activity, beyond the initial simple spike responses to electri- cal stimulation (Crain and Bornstein, 1964), has resulted in differing views as to the priority of the superficial or deep plexus as the site of closure of the first circuit connecting primary afferent and efferent neurons. Purpura et al. (1964) emphasized the early role of the superficial plexus and apical dendrites in the new-born cat, whereas Molliver (1967) and Åström (1967) have obtained electrophysiological and morphological data from sheep and dog fetuses that suggest a still earlier phase of evoked postsynaptic activity of pyramidal cells in the deeper layer of the cortex. This activity may involve the basal dendrites of the pyramidal neurons. The evidence was well reviewed by Molliver (1967). At any rate, the evidence at present points to an early role of pyramidal neurons and of an afferent system that can produce sharply localized evoked responses (Mol- liver and Van der Loos, 1970). Only later does the elabora- tion of interneuron circuits and of neuronal sheathing enter the picture. To this extent, the spinal-cord model and the cerebral-cortical analogue tend to coincide. Table II sum- marizes the available data, with insertions of a few specula- tive points in order to indicate the possible generality of the spinal model of nervous-tissue development.

Our interpretation of early embryonic data pertaining to the development of behavior offers some perspective on the problem of linkage of "subsystems" for integrated function,

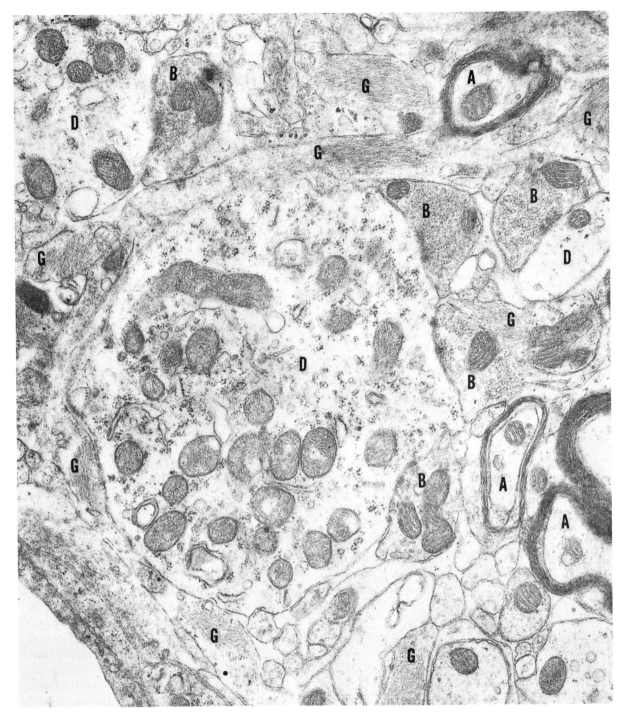

FIGURE 7 Electron micrograph of motor-neuron neuropil of macaque at term, showing the relatively large volume that is now occupied by processes of astrocyte neuroglia (G). These become significantly numerous midway through gestation. D, dendrites; B, synaptic bulbs; A, myelinated axons. × 30,000.

Table II

Comparison of Neuronal Classes in Spinal Model and in Other Centers,
in Presumed Sequence of Differentiation

| | Spinal Cord | Retina | Olfactory Bulb | Cerebellum | Cerebral Cortex |
|---|---|---|---|---|---|
| Efferent neuron | Motor neuron | Ganglion cell | Mitral cell | Purkinje cell | Pyramidal cell |
| Primary afferent neuron | Spinal ganglion cell | Receptor cells | Olfactory receptor neurons | Climbing fiber neurons (?) | Specific thalamic afferents |
| Intrinsic inter-neuron (major circuit) | Posterior horn cell | Bipolar cells | Granule cells | Granule cells (?) | Stellate cells |
| Intrinsic inter-neuron (lateral interactions) | Intersegmental interneurons Motor neuron collaterals | Horizontal and amacrine cells | Mitral cell collaterals | Basket, stellate cells, Purkinje collaterals | Horizontal cells Pyramidal collaterals |
| Secondary afferent neurons (extrinsic) | "Upper motor" neurons | Efferent optic nerve fibers | Efferent olfactory tract fibers | Mossy fibers (?) | Callosal and association neurons |

because the efferent neurons of many subsystems are the primary afferent neurons of other subsystems. In the motor realm, the succession of linkages of this type proceeds from spinal to more rostral levels. Afferent input to each level would seem to proceed concurrently, or at least with only a brief lag. If one were to speculate on the ultimate linkages of all subsystems into an integrated organism, one might begin by assuming that "functional readiness" of the entire motor apparatus is a pre-condition for sensorimotor behavior of the definitive organism, as it seems to be for the youngest embryo capable of movement. Interneuronal circuitry is completed long after the sensorimotor afferent and efferent neurons have been linked synaptically.

This point of view of embryonic development of behavior and circuitry appears to coincide with one aspect of MacKay's concept of perception and brain function (this volume). Perception, as distinct from reception, is viewed as the updating of the conditional organization of the brain for action. Extending this concept to the sensory component of experience, one might then regard perception as a function based on interneurons, concerned since early development with the updating of the conditional organization of action and of sensory awareness. The interneurons may establish their complex circuitry in each subsystem upon the previously developed framework of afferent-to-efferent primary circuit.

## Future prospects

Coghill believed that behavioral processes might be clarified through the study of development of structural and func-

tional organization, and vice versa. How far have we come in achieving his vision? Other papers in this volume deal with the considerable insight gained in recent times through analysis of neuroembryological processes, as contrasted with analysis of sequences of development. A high order of technical achievement has made possible new knowledge of events at the cellular or light-microscope level, such as cellular replicative patterns, cellular migration, cellular interdependence, and their genetic bases.

In the meantime, before we have progressed very much beyond the level of Coghill's perceptions, behavioral scientists have recognized that significant relationships between cellular organization and behavior in the realm of the forebrain have yet to be established. They have developed a whole new order of questions that press for as-yet-nonexistent anatomical information (Isaacson, 1968). It seems probable to me that the role of environmental stimulation in the development of cerebral competence, the role of hormones, including sex hormones, in the modulation of behavior, and the whole question of the nature of plasticity in behavior and its age limits (all of which will be considered in this volume) will not be helped toward clarification unless the synaptic level of structural and functional development is better understood. There is no obvious reason why, in time, the study of the ontogeny of higher nervous functions, and of its plasticity, cannot be made more rigorous through combined behavioral-structural studies.

As yet, however, we are only in the early stages of acquiring the kind of qualitative and quantitative understanding of synaptic development that is needed before experimental approaches to the synaptic level of behavioral development can be effectively pursued. Specific points of attack that seem promising are, first of all, further determination of the critical periods of differentiation of specific synaptic types, as their origin becomes known, making possible their experimental manipulation or deletion. The role of initial neuroglial sheathing of the neuropil components needs investigation in the light of newer knowledge of neuroglial functioning. Of great interest is the difficult problem of the time limits of synaptic outgrowth. Does such outgrowth continue, to at least some degree, beyond the period of neuroglial ensheathment—under the influence of environmental factors, perhaps—or is neuroglial ensheathment a final barrier to the stage of enrichment of connectivity? Is completion of circuitry hindered by premature sheathing, or, in contrast, is refinement of function affected by inadequacy of neuroglial sheathing? In other words, attention must be directed at both the determinants of dendritic and synaptic bulb development and linkage and the extent of the role of neuroglial sheathing in limiting or inhibiting synaptic development. Beginnings, at least, have been made in relation to the former. Although the ultimate goal may well be the analysis of the primate and human forebrain, the use of simpler model systems may make possible such basic information as the time of origin of excitatory and inhibitory synapses and of specific transmitter substances and related enzymes in major "subsystems."

## REFERENCES

ALTMAN, J., 1966. Autoradiographic and histological studies of postnatal neurogenesis. II. A longitudinal investigation of the kinetics, migration and transformation of cells incorporating tritiated thymidine in infant rats, with special reference to postnatal neurogenesis in some brain regions. *J. Comp. Neurol.* 128: 431–473.

ÅSTRÖM, K.-E., 1967. On the early development of the isocortex in fetal sheep. *Progr. Brain Res.* 26: 1–59.

BODIAN, D., 1966a. Synaptic types on spinal motoneurons: An electron microscopic study. *Bull. Johns Hopkins Hosp.* 119: 16–45.

BODIAN, D., 1966b. Development of fine structure of spinal cord in monkey fetuses. 1. The motoneuron neuropil at the time of onset of reflex activity. *Bull. Johns Hopkins Hosp.* 119: 129–149.

BODIAN, D., 1968. Development of fine structure of spinal cord in monkey fetuses. II. Pre-reflex period to period of long intersegmental reflexes. *J. Comp. Neurol.* 133: 113–165.

COGHILL, G. E., 1929. Anatomy and the Problem of Behaviour. Cambridge University Press, Cambridge, England. (Reprinted, 1964, by Hafner, New York.)

CRAIN, S. M., and M. B. BORNSTEIN, 1964. Bioelectric activity of neonatal mouse cerebral cortex during growth and differentiation in tissue culture. *Exp. Neurol.* 10: 425–450.

FLOREY, E. (editor), 1961. International Symposium on Nervous Inhibition, 2nd, Friday Harbor, Washington, 1960. Pergamon Press, Oxford.

FUJITA, S., 1967. Quantitative analysis of cell proliferation and differentiation in the cortex of the postnatal mouse cerebellum. *J. Cell Biol.* 32: 277–287.

HINDS, J. W., 1968. Autoradiographic study of histogenesis in the mouse olfactory bulb. I. Time of origin of neurons and neuroglia. *J. Comp. Neurol.* 134: 287–304.

ISAACSON, R. L. (editor), 1968. The Neuropsychology of Development. A Symposium. John Wiley and Sons, New York.

JACOBSON, M., 1969. Development of specific neuronal connections. *Science (Washington)* 163: 543–547.

KUFFLER, S. W., J. G. NICHOLLS, and R. K. ORKAND, 1966. Physiological properties of glial cells in the central nervous system of Amphibia. *J. Neurophysiol.* 29: 768–787.

MOLLIVER, M. E., 1967. An ontogenetic study of evoked somesthetic cortical responses in the sheep. *Progr. Brain Res.* 26: 78–91.

MOLLIVER, M. E., and H. VAN DER LOOS, 1970. The ontogenesis of cortical circuitry: the spatial distribution of synapses in somesthetic cortex of newborn dog. *Ergebn. Anat.* 42 (in press).

NILSSON, S. E. G., and F. CRESCITELLI, 1969. Changes in ultrastructure and electroretinogram of bullfrog retina during development. *J. Ultrastruct. Res.* 27: 45–62.

OLNEY, J. W., 1968. Centripetal sequence of appearance of receptor-bipolar synaptic structures in developing mouse retina. *Nature (London)* 218: 281–282.

PURPURA, D. P., R. J. SHOFER, E. M. HOUSEPIAN, and C. R. NOBACK, 1964. Comparative ontogenesis of structure-function relations in cerebral and cerebellar cortex. *Progr. Brain Res.* 4: 187–221.

RAMÓN Y CAJAL, S., 1960. Studies on Vertebrate Neurogenesis (revised and translated by Lloyd Guth). Charles C Thomas, Springfield, Illinois. (Revision of a 1929 edition.)

WEIDMAN, T. A., and T. KUWABARA, 1968. Postnatal development of the rat retina. *Arch. Ophthalmol.* 79: 470–484.

# 14 Embryonic Motility in Vertebrates

## VIKTOR HAMBURGER

IN THE 1920s and 1930s, the theoretical ideas concerning embryonic behavior were polarized in two schools of thought: Coghill, on the basis of his pioneer studies of the motility of the salamander, *Ambystoma,* considered behavior to be integrated from beginning to end, from the first movements of the head to swimming, walking, feeding and so forth. He generalized this concept to cover all vertebrates, including man. His ideas have been very influential, up to this day. The opposing school, including most of those working on mammalian fetuses, led by W. F. Windle, held the view that local reflexes were the building units of behavior. They were thought to be integrated secondarily into complex action systems. Both viewpoints seem now untenable as generalized theories of behavior development.

Our own investigations have led to a different polarization of ideas. The earlier work in the 1920s and 1930s was dominated completely by the reflex-response concept. Behavior was said to begin at the stage at which the embryo or fetus became responsive to stimuli. As a corollary, the role of "experience" through sensory channels, during embryonic and fetal development, was considered by many as an essential element in the structuring of postnatal action patterns. More recent studies, primarily dealing with motility in the chick embryo, have revealed the importance of nonreflexogenic spontaneous motility, up to advanced stages. At the same time, the role of sensory input in the performance of the embryo has been relegated to a minor position. Evidence is accumulating that this type of motility is basic also in other forms.

## Spontaneous motility—general

All vertebrate embryos perform movements when seemingly undisturbed and under adequate physiological conditions. In different forms, motility starts at different stages of development, and the movements exhibit changing frequencies and patterns in the course of development. What is the nature of these movements? In first approximation, we define "spontaneous" as nonreflexogenic. To establish spontaneous movements as a category distinct from

stimulated movements, it is necessary not only to exclude such obvious possible sources of stimulation as amnion and uterine contractions, but also possible hidden sources, such as self-stimulation by way of the proprioceptive system. The safest procedures are radical deafferentation experiments, some of which are described below. The first evidence for clearcut, nonreflexogenic motility, however, came to light not through experimentation but by the astute observation of the normal chick embryo by a great pioneer and innovator, the German physiologist, Wilhelm Preyer, who, almost singlehandedly, established the "Physiology of the Embryo" as a special branch of physiology. In his book *Specielle Physiologie des Embryo,* published in 1885, he reported his discovery that, although motility begins at about four days of incubation, responses to any kind of stimulation could not be elicited until after eight days (actually about seven days). He immediately recognized the importance of this prereflexogenic motility, and he also noticed the uncoordinated, aimless, seemingly nonadaptive nature of these movements, which he compared with the kicking and fidgeting movements of the infant. He called these spontaneous movements *"impulsive* movements," in distinction from *reflexive* and *instinctive* movements, two categories that are behaviorally adaptive and goal-directed. With uncanny premonition, he asserted that these impulsive movements are probably generated by some processes creating chemical energy in the motor cells that is then transformed into "actual energy," that is, motility, thus anticipating a more rigorous definition of "spontaneity."

Before I discuss spontaneous motility in detail, I should point out some limitations to an approach that relies on the unsolicited overt performance of the embryo. One never knows whether the embryo exhibits its full potential of motility. It is remarkable enough, and fortunate, that the salamander, chick, sheep, rhesus, and human embryos for which neurological data are available show spontaneous (or stimulated) motility very shortly after the necessary primitive neural connections are established; and one gets the impression that, in the chick embryo, new spontaneous movements of parts, such as limbs, beak, or eye, are added as soon as the prerequisite pathways and connections are established. But periods of silence have been reported during the middle period of gestation for such mammalian fetuses as those of cat and sheep, owing possibly either to the prevalence of inhibition from specific brain centers or to inadequate

VIKTOR HAMBURGER Department of Biology, Washington University, St. Louis, Missouri

permissive physiological conditions. Very few studies are devoted to a critical analysis of the relation of motility patterns to $O_2$ or $CO_2$ tension in the blood or to other physiological parameters, and asphyxiation as a source of error has bedeviled many of the earlier studies of mammalian and human fetal motility. A thorough analysis of the permissive conditions for spontaneous motility in different species probably would be rewarding.

## Analysis of spontaneous motility in the chick embryo

The spontaneous motility of the chick embryo has been described repeatedly (Orr and Windle, 1934; Hamburger et al., 1965; Hamburger, 1968b), and I confine myself to a brief characterization. Sawing a window in the shell directly above the embryo exposes it to direct observation; its position can be determined by means of candling. The movements consist of irregular, seemingly uncoordinated twistings of the trunk, jerky flexions, extensions, and kicking of the legs, gaping and later clapping of the beak with or without tongue movements, eye and eyelid movements, and occasional wing flapping in later stages. The movements are performed in unpredictable combinations. From their beginning at three and a half to four days up to about 13 days, a distinct periodicity is noticeable, activity phases alternating with inactivity phases (Figure 1). The activity phases become gradually longer and the inactivity phases shorter, so that between 14 and 17 days the rhythmicity becomes less clear. Even then, however, motility is not continuous but is frequently interrupted by short, quiet periods of a few seconds in duration. We have designated this type of motility as Type I, and occasional rapid, jerky, spasmodic movements passing through the whole body, or "startles," as Type II (Hamburger and Oppenheim, 1967). The total activity—that is, the time spent in activity during the standard observation period of 15 minutes—builds up gradually and attains a peak value of 75 to 80 per cent around day 13; this is maintained through day 17. Thereafter, Type-I motility decreases sharply to about 30 per cent at day 19. In these calculations, inactivity phase is defined as

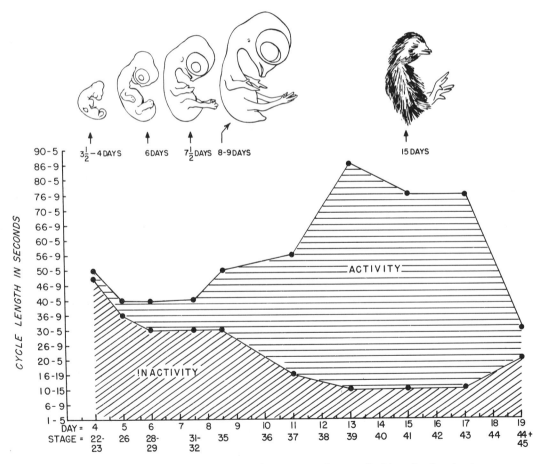

FIGURE 1 Mean duration of activity and inactivity phases and of length of cycles, in seconds, at different stages of the chick embryo. (From Hamburger et al., 1965, Figure 2.)

a period of quiescence, lasting 10 seconds or longer. If one includes shorter quiet periods in the calculation of inactivity phases, then peak total activity drops below 60 per cent.

Lack of integration of movements of different body parts and the cyclic nature of motility are characteristics indicative of nonreflexogenic activity, but rigorous deafferentation experiments are necessary to prove the point, at least for the embryonic period after seven days of incubation. As is pointed out above, the embryo is not sensitive to stimulation (Preyer, 1885), and the reflex circuits are not closed prior to that stage (Windle and Orr, 1934; Visintini and Levi-Montalcini, 1939). For later stages, deafferentation experiments are available. It is obvious that one cannot deafferent the whole embryo; one must do it piecemeal. The first such experiment was done on the legs (Hamburger et al., 1966). We made a gap several somites long in the thoracic spinal cord of two-day embryos to exclude sensory input from rostral levels. Simultaneously, we extirpated the dorsal part of the lumbosacral spinal cord, including the neural crest, thus eliminating the sensory ganglia and dorsal roots (Figure 2). These chronic preparations showed no response to exteroceptive or proprioceptive stimulation. The motor area developed normally up to 15 days (Figure 3), when progressive degeneration began. The motility of these operated embryos was compared not with that of normal embryos but with embryos in which only a thoracic gap had been made, because it had been shown that the separation of the spinal cord from the brain results in a reduction of body motility, the brain being a source of stimulation (Hamburger et al., 1965). In all stages, up to 15 days, spontaneous cyclic motility was comparable with that of control embryos with thoracic gap (Figure 4). The experiments demonstrated that sensory input is not necessary for the triggering and the maintenance of leg motility at the normal rate. It was suggested that the Type-I movements originated from spontaneous discharges of ventral internuncial or motor neurons.

In a similar operation, the deafferentation of all head structures supplied by the sensory trigeminal nerves was accomplished (Hamburger and Narayanan, 1969). The sensory nerves of the head emerge from the trigeminal ganglion, which is situated in front of the inner ear. The ganglion originates from two embryonic primordia, the neural crest of the preotic medulla and a local thickening of the epidermis, the so-called trigeminal placode. Both primordia were extirpated bilaterally in early embryonic stages. Stimulation tests of older embryos showed that tactile sensitivity was absent in all parts of the head skin in 31 of 35 operated embryos. In addition, the proprioceptive innervation of the jaw musculature was also greatly reduced or absent, in most cases, because the mesencephalic V nucleus which supplies the proprioceptive nerves was im-

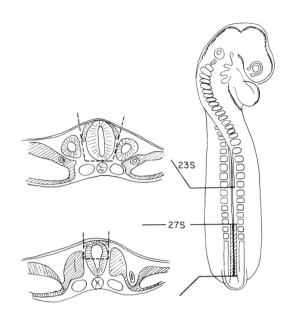

FIGURE 2  Schema of operation of deafferentation of the leg level in the two-day chick embryo. Upper left: Removal of the entire spinal cord along segments 23–27 inclusive. Lower left: Removal of the dorsal half of the spinal cord, including the neural crest, which gives rise to the sensory ganglia. (From Hamburger et al., 1966, Figure 1.)

paired, or absent, as a result of the operation. Nevertheless, recordings of Type-I motility showed that total activity, as well as the durations of activity and inactivity periods, was within normal range. Obviously, Type-I motility is independent of sensory input through the trigeminal system, up to 15 days. The decline of motility of the experimental embryos after that stage may be due to the absence of specific or nonspecific tonic sensory input. However, brain damage produced inadvertently in the majority of the embryos, and possible transneuronal degeneration, must be considered as alternative explanations. It should be noted that both the leg and the head deafferentation experiments also eliminate *self-stimulation* as a source of embryonic motility. In particular, the brushing of the legs against the head, which has been considered as a definite possibility for self-stimulation, actually does not serve this function. The same conclusion was reached in an experiment in which both leg buds were extirpated in four-day embryos. Again, total motility was not different from normal, up to 17 days (Helfenstein and Narayanan, 1970). Bilateral otocyst extirpation does not alter Type-I motility either, up to 17 days (Decker, in preparation). All these deafferentation experiments lead to the same conclusion: that Type-I motility is nonreflexogenic, up to 15 to 17 days of incubation. The obvious implication is that this type of motility is the result

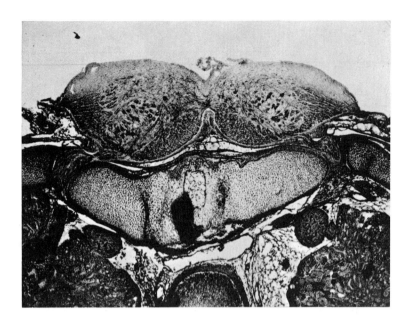

FIGURE 3    Cross section of lumbosacral spinal cord of 15-day embryo after removal of the dorsal half, as in Figure 2. (From Hamburger et al., 1966, Figure 9.)

of discharges that are generated in neurons of the CNS.

The above-mentioned experiment, in which the lumbosacral spinal cord was isolated, and similar experiments, in which a chronic gap was made in the cervical cord, show that the different sectors of the spinal cord are autonomous in the generation of discharges. In both instances, the time pattern of total activity—that is, its gradual rise to a peak at 13 days and its periodicity—is unchanged. The performance, however, is quantitatively at a lower level. We infer that the brain also generates discharges which are transmitted to the cord. No inhibitory brain effect was detected, up to 17 days (Hamburger et al., 1965). A more detailed analysis of the contributions of different brain regions (Decker and Hamburger, 1967) showed that we are not dealing with a simple mass effect of brain tissue; rather, different parts of brain have different effects at different developmental stages. For instance, a definite influence of the cerebellum is demonstrable from day 15 on. The situation changes after 17 days (see below).

It has been pointed out that, as sensory input is not the source of embryonic motility, the most plausible hypothesis is the assumption of spontaneous discharges in the embryonic nervous system. It would probably be difficult to pinpoint by histological and cytological techniques the neuron types in which the activity generates. The above-mentioned experiments, in which the isolated ventral half of the cord showed its capacity to generate overt motility, suggest an involvement of ventral internuncial neurons or motor neurons, or of both. It is at this point of the analysis that the exploration of the electrical activity of the embryonic cord became mandatory.

## Electrical activity of the embryonic spinal cord

In order to elucidate the neurophysiological basis of the motility patterns, electrophysiological investigations were started in 1967 by Dr. R. Oppenheim and Mr. R. Provine under the direction and with the generous aid of Dr. T. Sandel of the Psychobiology Laboratory of Washington University. The considerable technical difficulties in recording unit electrical activity from the spinal cord of chick embryos *in situ* were overcome eventually. In 1968, after the departure of Dr. Oppenheim, Dr. S. Sharma joined these efforts. In the following, I review briefly the results obtained by Dr. Sharma and Mr. Provine (see Provine et al., 1970; Sharma et al., 1970).

Numerous recordings have been made from normal embryos, particularly at the 17-day stage, to obtain information on the firing pattern within the lumbosacral spinal cord. Glass micropipettes, from 4 to $6\mu$ in diameter at their tips, were filled with 3 molar KCl agar solution and inserted in the spinal cord at the level of dorsal root 25, anterior to the glycogen body. (It should be remembered that the lumbosacral plexus is formed by nerves 23 to 30.) By probing the cord from dorsal to ventral, one can distinguish three regions of activity (Figure 5):

A. Approximately the upper third of the cord shows relatively continuous activity. The interspike intervals are rather regular but vary from unit to unit. Most units recorded in this region show long-lasting periods of activity, with relatively short quiet periods, or none at all. We call this the "sensory region." Anatomically it corresponds to the dorsal column.

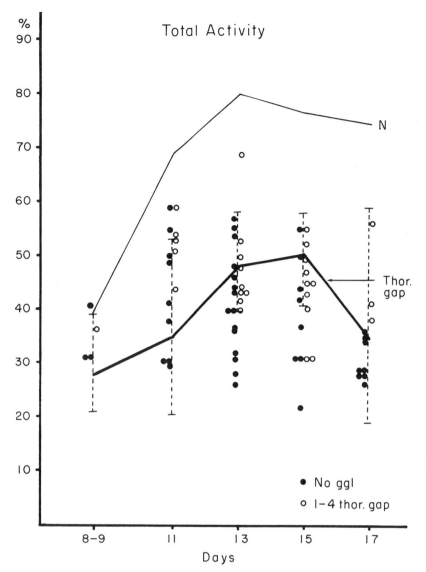

FIGURE 4 Per cent activity during 15-minute observation period. Upper solid line (N) means activity of normal embryos. Lower solid line (Thor. Gap) means activity of control embryos with thoracic gaps in the spinal cord. Each solid and open circle represents one recording from an experimental embryo. Dotted lines signify the range of controls. (From Hamburger et al., 1966, Figure 3.)

B. There follows a relatively silent region, which is about 100–200 μ in depth. This is the region immediately below the dorsal column.

C. Below this region, one obtains more discrete patterns of single-unit activity. Frequent "bursts" have been observed; these may result from the simultaneous discharge of several units. The bursts usually start abruptly and trail off into single units. Some units characteristically fire only in bursts. Others appear to fire continuously and may or may not fire synchronously during local burst activity deriving from other units. The deepest ventral region contains continuously firing units which possess relatively regular inter-spike intervals. All of region C contains the median and lateral motor columns and, in addition, a heterogeneous population of internuncial, commissural, and glial cells.

Thus, it is established for the first time that there exist in the spinal cord of the chick embryo patterns of single-unit electrical activity varying from intensive bursts to very low activity. Attempts are now being made to relate these patterns of neural activity to the behavioral periods of activity and inactivity which are discussed in earlier sections of this paper.

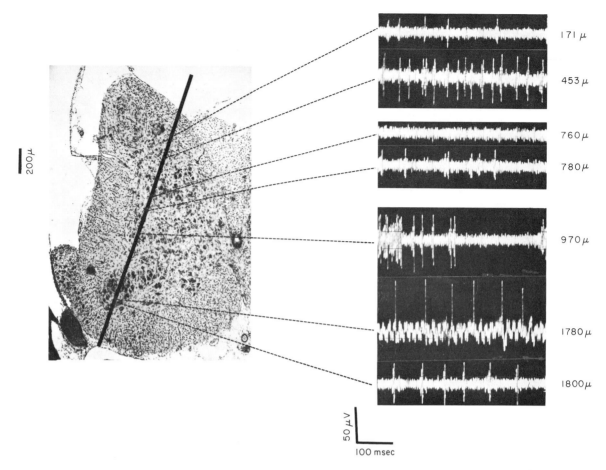

FIGURE 5 Electrode track plotted on the transverse section of the spinal cord of 17-day chick embryo at the level of dorsal root 25. Records at right show the activity picked up by the electrode tip at various points along the track. Upper two units represent the sensory region. Record at 760 μ represents zone B followed by the lower four units in motor column.

We discuss next some experimental results obtained by Sharma. They are relevant to the hypothesis formulated above—that we are dealing with spontaneous discharges of neurons, independently of sensory input.

1. Acute transection of the spinal cord was performed at the level of vertebrae one and two, in order to eliminate brain influence. A reduction of activity was observed at all levels of the cord. It is mentioned above that spinal transection reduced motility.

2. The same operation was performed and, in addition, dorsal roots 22 to 27 were transected on the ipsilateral side of the electrode. In this way, local sensory input to the recording site is blocked. As a result, the activity in sensory region A is lost, but activity in area C is not markedly affected. The contralateral side is unaffected.

3. After transection of the spinal cord, as under 1 above, 0.25 cc of 0.5 per cent xylocaine (Lidocaine-Ivenex) was injected into a thigh muscle on the ipsilateral side of the electrode. This drug blocks the afferent impulses by way of the dorsal roots without creating a trauma that might result from dorsal-root transection. All electrical activity is suspended within five minutes after injection, in region A, but recovery starts after 45 to 60 minutes. Activity in region C is not much affected, if at all. The contralateral side is not affected.

4. Acute transection of the spinal cord was performed immediately in front of the recording site (segment 21). In addition, all lumbosacral dorsal roots 23 to 30 were severed on both sides (Figure 6). This experiment tests directly our hypothesis that the ventral half of the spinal cord generates electrical activity in the absence of all sensory input. Some faint residual activity was recorded from the lower level of region A. Activity of units is present in region C throughout its depth (Figure 7).

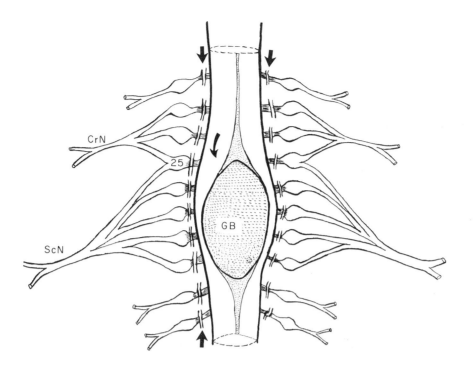

FIGURE 6 Diagram of the isolated lumbosacral spinal cord of 17-day chick embryo. Straight arrows point to the site of transection of the dorsal roots; curved arrow indicates the approximate electrode position. Spinal cord was cut at upper and lower extremities (marked by dots). CrN, cruralis nerve; GB, glycogen body; ScN, sciatic nerve.

It should be noted that, in all four experiments, the over-all level of spinal cord activity was reduced, as compared with that of normal embryos. It seems likely that this reduction can be attributed to the lack of input from higher centers. It is too early to speculate concerning the cause for the further reduction of electrical activity following deafferentation.

One point is of particular importance: in no experimental case was the electrical activity of the ventral region of the spinal cord completely silenced. One cannot escape the conclusion that, although higher centers and sensory input contribute to the over-all firing level of this region, the latter contains elements that continue to initiate discharges of nerve cells in the absence of these sources of input.

## Spontaneous motility in other vertebrate embryos

Spontaneous movements have been observed in all vertebrate embryos of which the behavior has been studied. Only an incomplete list can be given. Tracy (1926) found it in the teleost toadfish, *Opsanus tau,* in which it is cyclic, as in the chick embryo, although the inactivity phases are relatively longer. A prereflexogenic motility period of two and a half weeks precedes the stage at which the embryo becomes responsive to tactile stimulation. In the lizard *Lacerta vivipera*

(Hughes et al., 1967), spontaneous motility is also intermittent during a substantial part of the embryonic period. There is a sharp rise in total activity during a period of about 12 days, but, in contrast to the chick, it does not maintain its high level; rather, it drops sharply during the subsequent 12 days. Several days intervene between the onset of motility and the sensitivity to tactile stimulation. The situation is very similar in the turtle embryo, *Chelydra serpentina* (Decker, 1967); cyclic motility reaches a short peak-activity period (at 30 days) and immediately drops gradually to a very low level. In both lizard and turtle the maximal total activity is 40 per cent or less, that is, considerably lower than in the chick embryo. "Turtle embryos like lizard embryos become sensitive to exteroceptive stimulation a few days after the onset of motility" (Decker, 1967, p. 954). Observations on spontaneous motility in mammalian fetuses were reported for the cat (Windle et al., 1933), the rat (Angulo y González, 1932), the sheep (Barcroft and Barron, 1939), the rhesus monkey (Bodian, 1966; Bodian et al., 1968), man (Hooker, 1952), and others. In no instance is there a reference to periodicity, probably largely owing to the preoccupation of the early observers with stimulated activity. In the 1930s, spontaneous activity was considered an odd phenomenon and of no particular interest. In a reinvestigation of fetal behavior in the rat, we have found typical spontaneous motility (Narayanan, Fox, and Hamburger, unpublished). It begins during the second half of day 16. Total motility (i.e., time spent in activity during the standard observation period of 15 minutes) builds up to a peak, as in all other forms. This is reached at day 18 and maintained through day 20—a day before parturition. The activity of the fetus is lower than in other forms; the highest level is between 20 and 28 per cent. The motility is intermittent, but an analysis of the Poisson distribution shows that we are dealing with a random periodicity: the movements seem to start and stop at random intervals.

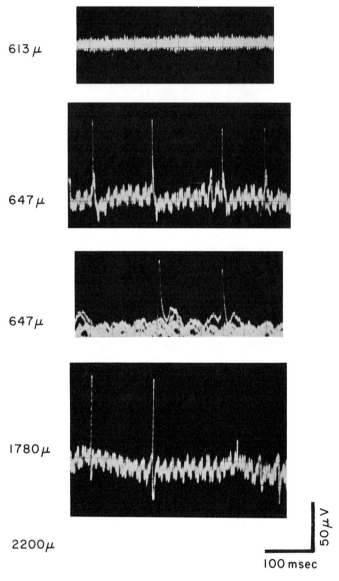

FIGURE 7 Electrical activity at various depths in the deafferentated lumbosacral spinal cord of 17-day chick embryo. The dorsal cord was almost inactive (up to 613 μ). Total depth in this probe was about 2200 μ.

Concerning the *nonreflexogenic* nature of spontaneous motility, deafferentation experiments of the kind reported above for the chick embryo are not available for any other form. A prereflexogenic period characteristic for the chick was found in the teleost, lizard, and turtle, but in none of the mammalian, fetuses. Hence, the only positive evidence we have for nonreflexogenic activity is for short periods in the development of motility of reptiles and for a longer period in the teleost, apart from the chick embryo. In this connection, it is of interest to note that amnion contractions, which are conspicuous in the chick embryo, are totally absent from the embryos of the turtle and of the rat. Hence, they are ruled out as stimulative agents in these two forms. They do occur in the lizard. The independence of embryonic motility from amnion contractions in the chick embryo was demonstrated by Oppenheim (1966). Uterine contractions occur in the rat, but they do not influence the fetal motility (Narayanan, unpublished).

The *patterns* and forms of behavior, that is, their qualitative aspect, have been described by different observers in different terms and from different viewpoints. It would be difficult and perhaps unrewarding to attempt to give a composite picture of the different types of head, trunk, and limb movements. The studies of many of the earlier investigators were guided by the question of whether behavior fitted in the scheme of Coghill, who, on the basis of his extensive work on the salamander, *Ambystoma,* had postulated that embryonic behavior is integrated from beginning to end. Although such may be true for urodeles and teleosts and perhaps other aquatic forms, which, for reasons of survival, must attain the capacity for swimming and other integrated performances as early as possible, the chick embryo presents an entirely different picture, as shown above. Its spontaneous movements are random, apparently based on massive discharges of large numbers of neurons that are not held in check or streamlined by higher inhibitory or integrating centers. As far as mammals are concerned, most investigators agree that, at least in the early phases, their movements are also mass movements, which involve all parts of the body, but that local movements of limbs, head, and so forth, make an early appearance. Our preliminary observations on rat fetuses give the impression, shared by others, that they are certainly not completely integrated but are more of the random type.

*In summary:* The contention of Preyer that spontaneous ("impulsive") motility is a special category of embryonic motility is amply confirmed; it is characteristic of most, or all, vertebrate embryos; it shows periodicity and a random pattern in many; and its nonreflexogenic nature is proved, at least for the chick embryo. I have pointed out elsewhere that its adaptive significance may be to insure the normal formation and maintenance of articulations and the intactness of the musculature. Our understanding of the neurophysiological basis of this type of motility is only beginning. Our hypothesis that we are dealing with discharges of motor neurons which fire indiscriminately and are probably driven by interneurons needs much further scrutiny. The periodicity cycles, which are on the scale of seconds or minutes, are, as yet, unexplained.

*Integrated movements*

The origins of integrated behavior have been reviewed on several occasions (Hamburger, 1963, 1968a, 1968b), and I limit myself here to a few general remarks. The pioneer work of Coghill on the salamander, *Ambystoma,* has clarified the issue for this form. In his "Correlated anatomical and physiological studies of the growth of the nervous system" (see Coghill, 1929), he has shown that behavior development in this urodele proceeds in a strictly programed way. The first bending of the head in the tail-bud stage is linked with the integrated swimming movements and with later walking and feeding action patterns by intermediate behavioral stages, which maintain the integrated involvement of all parts throughout behavior development. Local responses originate by a process of emancipation or "individuation" from the total pattern. The process is very similar in teleosts (Tracy, 1926). Coghill was successful in correlating the behavior sequence, step by step, with a corresponding sequence in the differentiation process of the central nervous system and with the formation of appropriate synaptic connections shortly before a new step in behavior development is attained.

Such an achievement was remarkable and provocative at the time, and an incentive for many other studies, but, although *Ambystoma* solves the problem of behavior integration in this simple and straightforward way, it was soon realized that the situation in amniotes is more complex and less transparent. As is stated above, the motility of the chick embryo, at least up to 17 days, is unintegrated in the sense that antecedents to organized posthatching behavior, such as walking, pecking, and so forth, cannot be clearly recognized. Wing flappings and alternate leg movements are rare. The same holds for reptilian embryos and, at least to a certain extent, for mammalian embryos. The first integrated activity in the chick embryo is the preparation for hatching and the hatching act itself, which occupies it from incubation days 17 to 20. The integrated head, trunk, and leg movements which are involved have been described in detail and designated as Type-III movements (Hamburger and Oppenheim, 1967; Hamburger, 1968b). It is not clear whether sensory input is necessary for this activity. Embryos with complete trigeminal or vestibular deafferentation fail to hatch. Yet, these experiments are inconclusive. The inability to perform these movements could be ascribed to the lack of specific sensory guidance or of tonic input. However, these chronic preparations show secondary transneuronal degeneration, and brain damage was produced inadvertently in most of the trigeminal operations. Hence, the behavioral deficiency could be the result of the impairment of central connectivity.

We have stated repeatedly that we do not find it possible to relate the Type-III motility to Type-I and Type-II motility. Not only are they different in pattern, but Type-I and occasionally Type-II movements continue during the prehatching period in the intervals between the episodes of integrated hatching movements. Nor do we see a direct continuity or relationship between any of the three embryonic motility types and the posthatching action patterns, such as walking, pecking, drinking, and so on. It seems that the neural apparatus for these activities is fully prepared during the embryonic period and that the action patterns are triggered by environmental or intrinsic signals.

We are not yet prepared to extend this notion to mammalian fetuses. It is conceivable that, particularly in forms that are very immature at birth and not immediately required to fend for themselves, the build-up of integrated patterns is more gradual. Alternating leg movements and trunk and leg movements that resemble righting (i.e., restoring the upright position) have been described, but that such antecedents are necessarily reflexogenic is not implied.

*Stimulated embryonic movements*

A wealth of information is available on responses to tactile and other stimulations in many forms, and particularly mammals, largely because research on embryonic motility was focused on this aspect in the 1930s and 1940s. The material on birds has been ably reviewed by Gottlieb (1968); the older literature, including that pertaining to mammals, is covered in the comprehensive review by Carmichael (1954). We limit ourselves to a few brief general remarks, mostly concerning birds and mammals.

The onset and progression of stimulated activity are as intimately tied to the differentiation of their neurological substrate as are those of spontaneous motility. A good example is the cutaneous sensitivity in the trigeminal area of the human fetus, which has been analyzed in detail by Humphrey (1964). The perioral region, which is the first reflexogenous zone, becomes sensitive at a menstrual age of seven and a half weeks, shortly after the cutaneous V fibers have reached the skin. The response, that is, contralateral neck flexion, coincides with the arrival of spinal tract V fibers at the level of the second and third cervical-cord segment. As soon as longitudinal fibers have grown caudally, the response is extended to include trunk and limb movements. It is of great interest that, for a considerable period (two weeks in the human fetus), local perioral stimulation elicits generalized total body movements. From the tenth week on, the generalized response subsides and gives way to local responses, such as mouth opening. In most vertebrate embryos, from amphibians to man, the trigeminal area is the first to become sensitive to tactile stimulation, but in the cat (Windle et al., 1933) and the rhesus monkey (Bodian, 1966; Bodian et al., 1968), the palmar surface of the forelimb becomes responsive at the same time. The sequence from generalized to local responses has been observed widely, but, in some forms, local reflexes occur simultaneously with, or even slightly earlier than, generalized movements. Obviously, the timetable for the differentiation of the central connections, and for the inhibitory mechanisms involved in restricted local movements, differs from form to form. Perhaps these differences have been overemphasized in the controversy over the primacy of total versus local responses as the basis of embryonic behavior.

Finally, I wish to raise the fundamental question of the relevance of these investigations of stimulated responses to the normal behavior development *in situ*. Windle, who has had extensive experience with the behavior of mammalian fetuses, has made the following comment: "It should not be assumed that all responses which can be induced occur spontaneously within the uterus of the normal intact individual. As a matter of fact, there is scanty evidence that any of them occur normally during the early part of the gestation period. . . . The fetus is adequately nourished and warmed in a medium lacking practically all the stimulating influences of the environment with which it will have to cope later. No significant excitation of the external receptors occurs" (Windle, 1940, pp. 164–165). The chick embryo is likewise sheltered in an environment that provides only a narrow range of stimulations—incomparably less than does the posthatching environment. Self-stimulation plays a minor role, if any, according to our experimental results (see above). Proprioceptive cues also have been excluded in several deafferentation experiments.

It follows that the stimulation experiments do not elucidate the actual overt performance of normal embryos *in situ*, at least up to fairly advanced stages. The stimulation experiments have revealed a great deal about the progressive differentiation of complex response patterns and the underlying neurological basis. These important results should have deserved a more extensive treatment in our presentation. However, if these reflexogenic response patterns hardly ever become overtly manifest during the major part of embryonic and fetal life, owing to the absence of appropriate stimuli, they cannot be considered as *the* elementary building stones in the genesis of behavior, as has

been claimed frequently in the past. On the other hand, it is just as difficult to relate the motility patterns which the normal, undisturbed embryo and fetus actually perform *in situ* (mostly spontaneous in nature) to postnatal behavior. In the chick embryo, at least, we were unable to obtain clues from prehatching behavior for an understanding of posthatching behavior. The situation in mammals requires much further study from this viewpoint. One point seems clear: most embryos are *active* before they manifest their capabilities to respond to environmental cues. The precedence of action over reaction in embryonic development of behavior has interesting evolutionary and theoretical implications that we cannot follow up at this point.

## *Summary*

Coghill's idea of a continuity of integration in behavior development from beginning to end cannot be generalized beyond his own material, the salamander, *Ambystoma,* and perhaps the teleosts. The antithesis: that local responses are the building blocks in behavior development is equally untenable as a general theory. We must admit frankly that the present state of our knowledge does not permit us to formulate broad generalizations. It is doubtful whether additional direct observations and simple stimulation procedures will carry us much farther. The neurological approach, particularly on the ultrastructural level; the neurophysiological approach combined with experimental procedures; and rigorously controlled behavior experiments would seem to give the best promise of further advances.

## *Acknowledgment*

All investigations from this laboratory were supported by NIH Grant 5R01-NB 05721 to the author.

REFERENCES

ANGULO Y GONZÁLEZ, A. W., 1932. The prenatal development of behavior in the albino rat. *J. Comp. Neurol.* 55: 395–442.

BARCROFT, J., and D. H. BARRON, 1939. The development of behavior in foetal sheep. *J. Comp. Neurol.* 70: 477–502.

BODIAN, D., 1966. Development of fine structure of spinal cord in monkey fetuses. I. The motoneuron neuropil at the time of onset of reflex activity. *Bull. Johns Hopkins Hosp.* 119: 129–149.

BODIAN, D., E. C. MELBY, and N. TAYLOR, 1968. Development of fine structure of spinal cord in monkey fetuses. II. Pre-reflex period to period of long intersegmental reflexes. *J. Comp. Neurol.* 133: 113–165.

CARMICHAEL, L., 1954. The onset and early development of behavior. *In* Manual of Child Psychology (L. Carmichael, editor). John Wiley and Sons, New York, pp. 160–185.

COGHILL, G. E., 1929. Anatomy and the Problem of Behaviour. Cambridge University Press, Cambridge, England.

DECKER, J. D., 1967. Motility of the turtle embryo, Chelydra serpentina (Linné). *Science* (*Washington*) 157: 952–954.

DECKER, J. D., and V. HAMBURGER, 1967. The influence of different brain regions on periodic motility of the chick embryo. *J. Exp. Zool.* 165: 371–384.

GOTTLIEB, G., 1968. Prenatal behavior of birds. *Quart. Rev. Biol.* 43: 148–174.

HAMBURGER, V., 1963. Some aspects of the embryology of behavior. *Quart. Rev. Biol.* 38: 342–365.

HAMBURGER, V., 1968a. Beginnings of co-ordinated movements in the chick embryo. *In* Growth of the Nervous System/A Ciba Foundation Symposium (G. E. W. Wolstenholme and M. O'Connor, editors). J. and A. Churchill, Ltd., London, pp. 99–105.

HAMBURGER, V., 1968b. Emergence of nervous coordination. Origins of integrated behavior. *In* The Emergence of Order in Developing Systems (M. Locke, editor). Academic Press, Inc., New York, pp. 251–271.

HAMBURGER, V., M. BALABAN, R. OPPENHEIM, and E. WENGER, 1965. Periodic motility of normal and spinal chick embryos between 8 and 17 days of incubation. *J. Exp. Zool.* 159: 1–14.

HAMBURGER, V., and C. H. NARAYANAN, 1969. Effects of the deafferentation of the trigeminal area on the motility of the chick embryo. *J. Exp. Zool.* 170: 411–426.

HAMBURGER, V., and R. OPPENHEIM, 1967. Prehatching motility and hatching behavior in the chick. *J. Exp. Zool.* 166: 171–204.

HAMBURGER, V., E. WENGER, and R. OPPENHEIM, 1966. Motility in the chick embryo in the absence of sensory input. *J. Exp. Zool.* 162: 133–160.

HELFENSTEIN, M., and C. H. NARAYANAN, 1970. Effects of bilateral limb bud extirpation on motility and prehatching behavior in chicks. *J. Exp. Zool.* 172: 233–244.

HOOKER, D., 1952. The Prenatal Origin of Behavior. University of Kansas Press, Lawrence.

HUGHES, A., S. V. BRYANT, and A. D'A. BELLAIRS, 1967. Embryonic behaviour in the lizard, *Lacerta vivipara*. *J. Zool.* (*London*) 153: 139–152.

HUMPHREY, T., 1964. Some correlations between the appearance of human fetal reflexes and the development of the nervous system. *Progr. Brain Res.* 4: 93–135.

OPPENHEIM, R., 1966. Amniotic contraction and embryonic motility in the chick embryo. *Science* (*Washington*) 152: 528–529.

ORR, D. W., and W. F. WINDLE, 1934. The development of behavior in chick embryos: The appearance of somatic movements. *J. Comp. Neurol.* 60: 271–285.

PREYER, W., 1885. Specielle Physiologie des Embryo. Grieben's Verlag, Leipzig.

PROVINE, R. R., S. C. SHARMA, T. T. SANDEL, and V. HAMBURGER, 1970. Electrical activity in the spinal cord of the chick embryo, in situ. *Proc. Nat. Acad. Sci. U. S. A.* 65: 508–515.

SHARMA, S. C., R. R. PROVINE, V. HAMBURGER, and T. SANDEL, 1970. Unit activity in the isolated spinal cord of the chick embryo, in situ. *Proc. Nat. Acad. Sci. U. S. A.* (in press).

TRACY, H. C., 1926. The development of motility and behavior reactions in the toadfish (Opsanus tau). *J. Comp. Neurol.* 40: 253–369.

VISINTINI, F., and R. LEVI-MONTALCINI, 1939. Relazione tra differenziazione strutturale e funzionale die centri e delle vie nervose nell'embrione di pollo. *Schweiz. Arch. Neurol. Psychiat.* 43: 381–393.

WINDLE, W. F., 1940. Physiology of the Fetus. Saunders, Philadelphia.

WINDLE, W. F., J. E. O'DONNELL, and E. E. GLASSHAGLE, 1933. The early development of spontaneous and reflex behavior in cat embryos and fetuses. *Physiol. Zool.* 6: 521–541.

WINDLE, W. F., and D. W. ORR, 1934. The development of behavior in chick embryos: spinal cord structure correlated with early somatic motility. *J. Comp. Neurol.* 60: 287–307.

# 15 Concluding Comments on Development of the Nervous System

## M. V. EDDS, JR.

We are at the beginning, and not near the end, of the process of understanding development. This is not the time to summarize, but a time to prospect and project into the future; and the best thing is to turn for guidance to the living object which teaches us the lessons and also teaches us the real problems to which we are to direct our questions. PAUL WEISS, 1958, p. 845.

DURING THE SYMPOSIUM on the "Development of the Nervous System," on which this section of the volume is based, many of the issues raised led to vigorous discussion. In most instances, these discussions are adequately reflected in the individual chapters. However, one issue (the ontogeny of neuronal specificity) promoted prolonged and intricate debate. To this observer, it seemed clear that the initiated saw the same evidence in different ways, while the uninitiated, despite evident interest, were often confused by what they heard. Accordingly, this summarizing commentary is restricted primarily to the topic of neuronal specificity; it aims at additional explication for those persons who are meeting the topic for the first time and at further clarification for those already immersed in it. These comments, in addition to their relevance to this section, logically stem from and extend earlier treatments of the same subject (Weiss, 1965; Edds, 1967) to which the reader is referred for further documentation.

## Neuronal Specification

Evidence set forth in the chapters by Coulombre and by Jacobson is usually regarded as demonstrating that primary afferent and somatic motor neurons receive their final specifications from the peripheral tissues they enter relatively late in their development. The term "specify," in the present context, is taken to mean the impression upon a population of young and relatively indifferent neurons of qualities related to the position and termination of each. These qualities, among the many others that gradually differentiate to set neurons apart from one another, correspond in some un-

known way to the topography of the periphery that induced them, e.g., individual muscles or small areas of skin. The specification of neurons, no matter how it may arise, occurs at different times in development, depending on the particular animal or neural structure involved.

In the meantime, matching specificities complementary to those of the peripheral neuron arise independently in central neurons. The specific properties of peripheral and central neurons are then expressed at the cellular level in rules of assembly that determine how they will become associated. Structural and functional interrelations are formed (or not formed) with varying degrees of precision. By selective matching of their respective properties, such cells become interwoven in the patterns that characterize the adult nervous system.

The evidence on which these generalizations are based, however, is drawn from only a few neuronal systems in a few different animals. It remains for the future to corroborate their general validity and to add essential points of detail—for example, the extent to which specificities differ in their mode of origin, their precision or their rigidity from one type of neuron to another, and the relation of any such variation to other properties of the same cells.

Although the specific quality of a neuron may be established only gradually during ontogeny, once set, it is not ordinarily revocable. In subsequent functional adjustments or artificially imposed regeneration, the neuron can no longer be modified by its periphery. If already determined neurons—either central or peripheral—regenerate after interruption of their axons and become reconnected, there are two possible outcomes: first, the regenerating nerve fibers may return selectively to their original end organs (or the equivalent in a supernumerary part) and re-establish the original function; second, the regenerating neurons may nonselectively form foreign connections and, in the absence of any respecification, produce some abnormal function.

In practice, it is often difficult to distinguish between these alternatives. Functional tests may give misleading or ambiguous results. Variability of response in behavioral tests may be an important clue or a distracting annoyance. And even if the functional observations are clear beyond challenge, the extent to which nerve regeneration was selective

M. V. EDDS, JR. Division of Biological and Medical Sciences, Brown University, Providence, Rhode Island

or nonselective is often surprisingly difficult to decide. If the original function is recovered, for example, the possibility must be entertained that the regenerated neurons were not irrevocably fixed after all and that they have, indeed, been respecified by new peripheral terminations. In the absence of independent evidence, the choice between that possibility and selective nerve regeneration may be impossible. Such ambiguities have been resolved in some cases but not in others, for which evidence is still incomplete or indirect. Hence, clear understanding of what is or is not established in particular cases is essential.

For these reasons, some further comment is in order concerning two principal cases that have been advanced in support of neuronal specification by peripheral end organs: the reversed wiping reflexes in frogs with rotated skin grafts, and the homologous motor responses of transplanted supernumerary limbs in the salamander. The question is: Considering the crucial relevance of these cases to theories of the origin of neuronal specificity, how secure are they? Can they be challenged persuasively?

## Rotated Skin Grafts

When Miner (1956) described the apparent respecification of anuran cutaneous neurons that had regenerated during larval life into dorsoventrally reversed skin grafts, she provided a relatively simple system for further analysis. Unlike the transplanted limb, with its many interacting muscles involved in a behavioral test, experiments with rotated skin grafts seemed to pose only the question: When the skin is stimulated, will the frog exhibit normal or reversed, misdirected wiping reflexes; will it wipe its back or its belly?

Miner's evidence for respecification seemed so clear and unexceptionable that the best part of a decade went by before anyone elected to repeat the experiment. Then, as Jacobson's account (this volume) makes clear, the results turned out to be more complicated, although in essence Miner's original conclusions have been confirmed. The key piece of new evidence adduced by Jacobson is the electrophysiological demonstration that the receptive fields of the individual cutaneous nerves are the same (both anatomically and in the pattern of impulses generated by stimulating them in some constant manner), whether or not the flank skin has been rotated. That is, the shape of a receptive field is a function of the nerve that supplies it. The new receptive fields that emerge after the reinnervation of rotated skin grafts have shapes that match those of adjacent receptive fields of normal skin. Thus, the grafts have become reinnervated by local nerve fibers that entered and terminated in the nearest available skin; selective reattachment of regenerating sensory nerve fibers to their original cutaneous terminals appears to be ruled out. The misdirected reflexes observed in the experimental frogs, then, depend on sensory

neurons that have been respecified by their new peripheral endings.

Of equal significance were Jacobson's observations that normal reflexes are elicited from skin pieces rotated later than mid-larval life or from grafts smaller than about 40 mm². No respecification occurs in such cases. The smaller grafts, no matter when made, become invaded by nerve fibers from surrounding normal skin, with consequent overlap of receptive fields. Grafts of whatever size rotated late in larval life are no different from those made in adult frogs; that is, they give rise to normal reflexes, indicating that the afferent neurons that enter them are specified beyond any possibility of change by a new periphery.

Finally, misdirected reflexes were often elicited from rotated skin only some days after functional connections had been achieved by the regenerating fibers. In about half of the animals followed closely, normal and misdirected reflexes developed simultaneously. But in the remaining animals, normal reflexes appeared first, followed within about a week by misdirected reflexes. For reasons still not clear, but presumably related to the progressiveness of respecification, the latter were at first elicited only from a small central region of the graft. Then this central reflexogenous region spread outward slowly at a rate of some 0.5 to 1.0 mm per day, finally covering most of the graft. Histological studies of the grafts showed no sign of changing populations of epidermal or other cells that might be postulated to explain these results. Thus, taken altogether, the body of evidence presented by Jacobson, although still partly indirect and, hence, subject to argument, speaks strongly in favor of specification of anuran cutaneous nerve cells by their end organs.

## Homologous Responses

What, then, of the motor side of the story? Is the phenomenon of homologous response, which is *the* evidence for the peripheral specification of developing motor neurons, equally secure? The phenomenon itself is secure beyond question; its interpretation is not. Although peripheral respecification was initially supported as the correct explanation of homologous response by a profuse and elaborate body of evidence (Weiss, 1936, 1965, for review), new lines of experimentation have reopened a possible explanation once thought to be ruled out, namely, selective regeneration of myoneural connections. Unfortunately, the new evidence is all indirect, and it thus provides not a crucial test, but rather a climate that promotes reappraisal of earlier evidence.

The phenomenon of homologous response, as once again reviewed in Coulombre's essay, has been taken to mean that the muscles of a transplanted larval salamander limb, having intercepted only a fraction of the brachial nerve fibers that normally supply them, somehow respecify those motor axons in accordance with their new endings. Judged from the behavior of larval limb transplants, this myotypic respecification occurs before the recovery of motor function; even the first fractional limb movements are already in phase with those of the host limb (Weiss, 1936). Thus, axons that once supplied a digital flexor muscle, say, would now supply an ulnar extensor with impulse patterns that precisely match those reaching the ulnar extensor of a nearby normal limb. Concomitant structural or functional changes in the relations between the respecified motoneurons and the interneurons of the spinal cord would, of course, have to occur if central impulses are to reach the correct muscles. Given these facts, the conclusion that respecification had taken place would be difficult to deny. If, however, the regenerating motor axons that entered the transplanted limb had merely returned selectively to their normal, or near-normal, terminals, the phenomenon of homologous response would require a different explanation.

It has long been recognized that the third, fourth, and fifth segments of the urodele spinal cord, the ones innervating the forelimb, comprise a functional subsystem. The pattern of movement displayed by the limb is, as it were, uniquely built into this part of the cord as a repertory of collective neuronal functions, intracentrally coordinated, arising independently of any connection with the periphery during development, and functioning thereafter independently of any sensory input. If a limb transplant is to function at all, to say nothing of functioning in phase with the normal host limb, it must become innervated by at least one of the three segments (for reviews, see Weiss, 1941, 1955, 1965; Hughes, 1968).

Furthermore, neurons supplying all the limb muscles are, at least in urodeles, distributed throughout the three cord segments. Especially pertinent evidence has recently been presented by Székely (1968): microelectrode stimulation of the brachial cord at various points lying in a three-dimensional, rectangular grid encompassing all three segments repeatedly elicits contraction of the same single limb muscles or parts of muscles. "Motor neurons supplying a number of different muscles are arranged in small groups alternating with other groups of motor neurons innervating different sets of muscles" (Székely, 1968, p. 83). The well-organized, somatotopic representation found in the spinal cord of mammals and birds is lacking in urodeles. In the latter, the topography of the muscles is not sharply localized in its central representation. Instead, there is a near-random representation with extensive overlap of individual muscles.

As would be expected, studies of the distribution of motor axons as they exit through the third, fourth, and fifth spinal nerves to enter the brachial nerve plexus demonstrate, with only occasional disagreement, that a given spinal nerve contains axons for essentially all limb muscles. Accordingly, a limb transplant innervated by a single spinal nerve does not have so qualitatively restricted a fraction of its normal inner-

vation as either gross or comparative anatomy might suggest.

The possible objections to the concept of myotypic respecification which these observations permit are rendered much less cogent, however, by the demonstration (Weiss, 1937) that even a distal limb nerve can provide all the requisite motor axons for a homologously responding transplant. The nerve in question is the inferior brachial which normally supplies only the flexor muscles of the wrist and hand. As we have argued before (Edds, 1967), this result must be explicitly contradicted or explained away if Weiss's general argument is to be successfully challenged. If, as seems likely, the inferior brachial nerve differs from a brachial spinal nerve in having only a qualitatively restricted portion of all the nerve fibers in a limb, the case for their myotypic respecification is persuasive.

Let us return to a consideration of instances of selective nerve regeneration in which axons clearly do get back to their original end organs. As Sperry (1965) has emphasized, "The matter of selectivity in nerve growth has always been rather controversial and has a long history of pros and cons." During the period when the phenomenon of homologous response was first being actively investigated, the prevailing view held that nerve regeneration is not selective, and that new nerve connections are formed indiscriminately. In the intervening years, however, a number of instances of selective nerve regeneration have been uncovered; only two are mentioned here (see also Sperry, 1965, and Székely, 1968).

One of them is elaborated in the chapter by Jacobson, namely, the regeneration of the severed optic nerve. Studies of this phenomenon, pursued incidentally on adult or near-adult teleosts and amphibians after neurons of the visual system are already definitively specified, show beyond dispute that the optic-nerve fibers end up in essentially the same parts of the optic tectum that they originally occupied, any uncertainties being based on the limited resolution obtainable with the physiological or anatomical measures employed (Attardi and Sperry, 1963; Jacobson and Gaze, 1965; cf. Gaze and Sharma, 1970). Despite efforts to detect it, there is no evidence from these studies that, during regeneration, the early-arriving axons spread out widely through the optic tectum and then later retract to form more restricted terminals. On the contrary, such evidence as we have (Attardi and Sperry, 1963; Sperry, 1965; DeLong and Coulombre, 1967) indicates that advancing nerve fibers follow specific clues all along the optic pathway back into the tectum, hence may be said to go, and not just to get back, to their original stations.

Another set of relevant observations is drawn from analyses of nerve regeneration to the fin musculature of a bony fish (Mark, 1965, 1969). Here, excellent recovery of normal fin movements follows regeneration of motor nerve fibers that have been interrupted in the brachial plexus. (This capacity, incidentally, is shared by amphibians, but not by higher vertebrates.) In this case, a clear decision is possible between the alternatives of selective regeneration and myotypic respecification. This is due, in part, to the relatively simple arrangement and function of the fin musculature, and in part to a fortunate anatomy of the nerves to adductor and abductor muscles; these nerves can be cross-anastomosed without fear that axons escaping from the nerve junctions will wander into their own distal nerve stumps. Transmissive neuromuscular connections are re-established by the axons thus forced into foreign muscles—an expected result, because motor axons readily make such connections with any available muscle fibers. However, the fin movements produced after nerve crossing remain uncoordinated and inappropriate. The inference seems inescapable that the recovery of normal fin movements after crushing the brachial plexus depends on the selective re-establishment of the original nerve–muscle relations.

These observations, and related arguments, have led Mark (1969) to challenge the concept of myotypic respecification of neurons, and to plead for renewed analysis of the phenomenon (cf. Sperry and Arora, 1965). He also supports his thesis with the results of a study (Mark et al., 1966) of normal nerve–muscle relations in the salamander limb, in which muscle fibers, unlike those of most higher vertebrates, but again like those of the fish, are usually innervated by more than one neuron. Such polyneuronal innervation patterns include large, multiple *en grappe* myoneural terminals. (For a discussion of the hazards of generalizing about vertebrate innervation patterns, see Bone, 1964.) Mark suggests that nerve fibers of diverse central origin regenerating into a salamander-limb muscle are confronted with a different kind of competition than would be the case in a mammal. In the latter, the first axon to reach a muscle fiber generally renders it refractory to supernumerary innervation—hence, competition is reduced or precluded almost from the start. But in muscles with polyneuronal innervation, the several neurons that attach to an individual muscle fiber might then compete, with the final advantage going to axons most resembling the original innervation.

Finally, the observations of Székely and colleagues should be recorded. The views of previous workers on the special anatomical and functional character of the limb-moving level of the urodele spinal cord—on its graded polysynaptic activity, on its parallel reticular nets—have already been mentioned. Székely et al. (1969) have proposed that, because of the relatively diffuse distribution of motor neurons in the salamander cord, the analysis of muscle actions from even the most refined visual observations of movements in transplanted limbs may lead to erroneous conclusions. The interaction of urodele limb muscles was examined by Weiss (1952) and again by Székely (1968) in theoretical myochronograms derived from cinematographic records of normal

and transplanted limbs. Székely et al. (1969) have now analyzed muscle interaction directly by electrical recordings from eight limb muscles obtained during free walking movements. These myochronograms reveal delicate patterns of interaction involving muscles in numerous co-contractions that differ from those that may be inferred from analyses of motion pictures.

The way is now open for comparable studies of muscle action both in transplanted limbs and in intact limbs that have been deprived of a fraction of their normal innervation by selective destruction of brachial spinal nerves. Székely's finding of the extensive co-contraction of muscles during walking permits the prediction that, even if nerve regeneration into transplanted limbs were to occur randomly and indiscriminately, many muscles might receive, if not their original innervation, a sufficiently close approximation to produce coordinated, homologous limb movements. This possibility is now being tested (Székely, personal communication).

Lastly, some note should be taken in passing of the fact, developed elsewhere in this volume by Cohen, that homologous response is also displayed in the cockroach. There, opportunities for further study are especially appealing because individual neurons can be identified and could be followed throughout the period when respecification is occurring. Some persons would contend that only by tracing the fate of individual neurons in such favorable preparations will the issues raised above ever be resolved.

## Coda

Although the various facts and arguments set forth in the preceding paragraphs render uncertain the mechanisms underlying the phenomenon of homologous response and indicate the need for more crucial experiments, they do nothing, of course, to shed any doubt on the phenomenon itself. Moreover, no matter which of the proposed explanations turns out to be correct, neuroscientists will be left with derivative puzzles of staggering proportions. If it were to be shown, for instance, that the motor neurons of the larval urodele spinal cord are indeed already specified definitively at the time when limb transplants are usually performed, the question would remain of when and how they were specified. What theory, for instance, would accommodate cutaneous sensory neurons that were peripherally specified in mid-larval life and motor neurons that were specified earlier in some other manner?

There are at least two directions in which studies of neuronal specificity must continue to move: 1) toward delineating much more carefully how the information derived from lower vertebrates relates to mammals; and 2) toward a better understanding of the nature of specificity.

Recall that all investigations bearing on neuronal specificity in mammals have been conducted postnatally, at a time when neurons are not subject to respecification (Sperry, 1965). The inference that they once were is just that, an inference. The limited ability of the mammalian central nervous system to regenerate places still another restriction on what we know or are likely to learn of intracentral specificities. However, nothing bars the way, except want of good ideas, to effective analysis of developing mammalian systems, as a recent study by Schneider and Nauta (1969) indicates. By destroying the superior colliculus in the neonatal hamster just before direct fibers from the retina reach it, these authors were able to show that the optic axons deprived of their normal end stations had preferentially entered instead a secondary nucleus in the optic pathway, namely, the lateroposterior nucleus of the thalamus.

This apparent reduction of normal specificity restraints, a kind of flexibility within limits, comes as no surprise to persons who are familiar with the dynamic nature of neuronal relations (cf., e.g., Barker and Ip, 1966; Liu and Chambers, 1958). Nor should one be surprised by Blinzinger and Kreutzberg's (1968) comparable demonstration that the synaptic endings on the facial motor neurons in the rat are at least temporarily displaced by adjacent microglial cells during nerve regeneration. These microglia take up ³H-thymidine two to four days after nerve crushing, and subsequently undergo mitosis, as Sjöstrand (1965) has shown. In a similar vein is Raisman's (1969a, 1969b) demonstration of synaptic lability in the septal nuclei of the mouse. The septal nuclei interconnect the hippocampus and the hypothalamus, normally sending axons from and to each. Lesions of the fimbria destroy the hippocampal input and lead to the vacating of synaptic sites (predominantly on dendrites). The latter are then taken over by fibers of hypothalamic origin (which usually terminate mainly on cell bodies). A similar result follows the reverse experiment of destroying the medial forebrain bundle, which eliminates fibers from the hypothalamus. Synaptic relations, then, are certainly not to be regarded as constant and unchangeable, even in the adult mammal. An exploration of the extent to which they also change in other animals in which respecification of nerves is apparently in process should be rewarding.

On the question of the nature of neuronal specificity, the members of this group could report no progress, other than to provide hints as to when, in particular cases, specification may occur in relation to DNA synthesis or to the determination of other parameters, such as axial gradients. As Jacobson's account indicates, moreover, the biological facts seem awesome. Somatotopic maps projected onto neural centers are relative rather than absolute (see Gaze and Sharma, 1968). Specificities probably occur in gradients, as Jacobson, and Sperry (1965), have both stressed, neurons close together being more alike than those farther apart. These gradients are plastic at some periods of development and will

re-form if disarranged; subsequently, they become fixed and can no longer adapt to experimental intervention.

But the question of mechanism is still wide open, and investigations from all possible points of vantage would be most welcome. As we have noted before (Edds, 1967), "The problem is not to find *an* explanation; it is to find the right one." Ideas for conceivable approaches may be found elsewhere in this volume, for instance, in contributions by Barondes (mucopolysaccharide complexes as interneuronal recognition substances), by Nomura (cell-surface recognition sites where key metabolic changes can be triggered), and by Edelman (postsynaptic substances that exercise feedback control on the synapse or on presynaptic events).

Are morphologically mature synapses also functionally mature? Can crucial functional changes occur without leaving a structural trace? How do events at the molecular level lead on to particular cellular behaviors? Which cellular attributes (motility, adhesiveness, and so forth) are crucial in determining how neurons will become associated? What are the rules of cellular assembly that achieve a desirable functional plasticity, yet avoid structural chaos? How directly or indirectly are these events mapped in the genome? These are questions for the future.

REFERENCES

ATTARDI, D. G., and R. W. SPERRY, 1963. Preferential selection of central pathways by regenerating optic fibers. *Exp. Neurol.* 7: 46–64.

BARKER, D., and M. C. IP, 1966. Sprouting and degeneration of mammalian motor axons in normal and de-afferented skeletal muscles. *Proc. Roy. Soc., ser. B, biol. sci.* 163: 538–554.

BLINZINGER, K., and G. KREUTZBERG, 1968. Displacement of synaptic terminals from regenerating motoneurons by microglial cells. *Z. Zellforsch. Mikroskop. Anat.* 85: 145–157.

BONE, Q., 1964. Patterns of muscular innervation in the lower chordates. *Int. Rev. Neurobiol.* 6: 99–147.

DeLONG, G. R., and A. J. COULOMBRE, 1967. The specificity of retinotectal connections studied by retinal grafts onto the optic tectum in chick embryos. *Develop. Biol.* 16: 513–531.

EDDS, M. V., JR., 1967. Neuronal specificity in neurogenesis. *In* The Neurosciences: A Study Program (G. C. Quarton, T. Melnechuk, and F. O. Schmitt, editors). The Rockefeller University Press, New York, pp. 230–247.

GAZE, R. M., and S. C. SHARMA, 1970. Axial differences in the reinnervation of the optic tectum by regenerating goldfish optic nerve fibres. *Exp. Brain Res.* 10: 171–181.

HUGHES, A. F. W., 1968. Aspects of Neural Ontogeny. Academic Press, New York.

JACOBSON, M., and R. M. GAZE, 1965. Selection of appropriate tectal connections by regenerating optic nerve fibers in adult goldfish. *Exp. Neurol.* 13: 418–430.

LIU, C. N., and W. W. CHAMBERS, 1958. Intraspinal sprouting of dorsal root axons. *A. M. A. Arch. Neurol. Psychiat.* 79: 46–61.

MARK, R. F., 1965. Fin movement after regeneration of neuromuscular connections: An investigation of myotypic specificity. *Exp. Neurol.* 12: 292–302.

MARK, R. F., 1969. Matching muscles and motoneurons. A review of some experiments on motor nerve regeneration. *Brain Res.* 14: 245–254.

MARK, R. F., G. VON CAMPENHAUSEN, and D. J. LISCHINSKY, 1966. Nerve-muscle relations in the salamander: Possible relevance to nerve regeneration and muscle specificity. *Exp. Neurol.* 16: 438–449.

MINER, N., 1956. Integumental specification of sensory fibers in the development of cutaneous local sign. *J. Comp. Neurol.* 105: 161–170.

RAISMAN, G., 1969a. Neuronal plasticity in the septal nuclei of the adult rat. *Brain Res.* 14: 25–48.

RAISMAN, G., 1969b. A comparison of the mode of termination of the hippocampal and hypothalamic afferents to the septal nuclei as revealed by electron microscopy of degeneration. *Exp. Brain Res.* 7: 317–343.

SCHNEIDER, G. E., and W. J. H. NAUTA, 1969. Formation of anomalous retinal projections after removal of the optic tectum in the neonate hamster. *Anat. Rec.* 163: 258 (abstract).

SJÖSTRAND, J., 1965. Proliferative changes in glial cells during nerve regeneration. *Z. Zellforsch. Mikroskop. Anat.* 68: 481–493.

SPERRY, R. W., 1965. Embryogenesis of behavioral nerve nets. *In* Organogenesis (R. L. DeHaan and H. Ursprung, editors). Holt, Rinehart and Winston, New York, pp. 161–186.

SPERRY, R. W., and H. L. ARORA, 1965. Selectivity in regeneration of the oculomotor nerve in the cichlid fish, *Astronotus ocelletus.*

*J. Embryol. Exp. Morphol.* 14: 307–317.

SZÉKELY, G., 1968. Development of limb movements: Embryological, physiological and model studies. *In* Growth of the Nervous System/A Ciba Foundation Symposium (G. E. W. Wolstenholme and M. O'Connor, editors). Little, Brown and Co., Boston, pp. 77–93.

SZÉKELY, G., G. CZÉH, and G. VÖRÖS, 1969. The activity pattern of limb muscles in freely moving normal and deafferented newts. *Exp. Brain Res.* 9: 53–62.

WEISS, P., 1936. Selectivity controlling the central-peripheral relations in the nervous system. *Biol. Rev. (Cambridge)* 11: 494–531.

WEISS, P., 1937. Further experimental investigations on the phenomenon of homologous response in transplanted amphibian limbs. II. Nerve regeneration and the innervation of transplanted limbs. *J. Comp. Neurol.* 66: 481–535.

WEISS, P., 1941. Self-differentiation of the basic patterns of coordination. *Comp. Psychol. Monogr.* 17: 1–96.

WEISS, P., 1952. Central versus peripheral factors in the development of coordination. *Res. Publ. Ass. Res. Nerv. Ment. Dis.* 30: 3–23.

WEISS, P., 1955. Nervous system: Neurogenesis. *In* Analysis of Development (B. H. Willier, P. A. Weiss, and V. Hamburger, editors). W. B. Saunders Co., Philadelphia, pp. 346–401.

WEISS, P., 1958. Summary and evaluation. *In* The Chemical Basis of Development (W. D. McElroy and B. Glass, editors). Johns Hopkins Press, Baltimore, pp. 843–854.

WEISS, P., 1965. Specificity in the neurosciences. *Neurosci. Res. Program Bull.* 3 (no. 5): 1–64.

# DETERMINANTS
# OF NEURAL
# AND BEHAVIORAL
# PLASTICITY

*Genes create a substrate that specifies the capacity of
behavior; environmental stimuli determine its specific
expression and content. "A major theme in the papers that
follow is the examination of factors that underlie such
postnatal differentiation of behavior," says Robert
Galambos in his prefatory remarks. This picture symbolizes
both the genetic endowment, as expressed in the faces and
attitudes of the children, and their environment, typified
here by the learning experience. (Courtesy of National
Education Association)*

# 16 Prefatory Remarks on Determinants of Neural and Behavioral Plasticity

ROBERT GALAMBOS

THE FOLLOWING PARAGRAPHS are intended to introduce the ideas that originally prompted the symposium on this topic and to orient the reader to the main themes it was intended to develop.

The papers in this section logically extend the concepts and problems raised in the preceding section on Development of the Nervous System. This follows from the idea that birth for mammals and hatching for chicks are merely episodes on their developmental continua. An infant nervous system, like the body that surrounds it, continues to grow and change, a phenomenon regulated in an important way by the genetic readout. This continuing development of the nervous system provides progressively new potentialities for behavioral response, and the environment in which the animal moves after birth largely determines whether these potentialities will be realized and what exact form they will take. Thus, as a human newborn brain weighing 380 grams changes into one weighing 1400 grams by puberty, the genes create a unique substrate that allows a language repertoire to expand from zero to that of a loquacious adolescent, and the environment determines which of

ROBERT GALAMBOS Neurosciences Department, University of California at San Diego, La Jolla, California

the thousands of known languages the child will actually speak. In this, as in most samples of behavior, the genetic readout specifies the capacity, while the environmental stimuli determine its specific expression and content. A major theme in the papers that follow is the examination of factors that underlie such postnatal differentiation of behavior.

The word "plasticity" in our section title requires some explanation. This term was chosen in order that we might cover more than is included under such conventional terms as learning and habituation (the alteration in response probability due to an interaction between organism and environment). It allows inclusion of the events responsible for maturation of a capacity to respond, as in the species-specific fixed action patterns of bees, fishes, and birds. It also admits switching events, such as those that make it possible to identify an object as an upholstered chair (a single response) regardless of whether it is seen, sat upon, or bumped into in the dark (different stimulus inputs), and those that allow a given stimulus to produce one or another response depending on context, as when a dog runs to retrieve a thrown stick and then cringes if threatened by it. Both plasticity and its opposite, stability, thus apply to a broad and heterogeneous class of neurobehavioral reorganizations. The phenomena of recovery of function after brain lesions, sex differentiation and expression induced by hormones, brain hypertrophy associated with enriched experience, heterosynaptic facilitation, and so forth, all represent plasticity, or the ways in which either the functional output of a neuronal aggregate or its structural status show alteration through time when tested by a fixed stimulus input. Stability, by contrast, is the word to be used to cover central events that make it impossible for animals to change their responses during life. Both terms imply that probably relatively few general principles exist for explaining structural stability and structural modifiability in the nervous system, and make welcome any and all experiments that aim to clarify the mechanisms underlying observed behavioral responses.

For psychologists interested in learning, a few additional comments are pertinent. Learning is a special case of plasticity, and the invariable performance of a learned response is a special case of stability. As the ethologists have long insisted, learning amounts to the fine tuning of a fully orchestrated behavioral program that is required for successful performance of the composer's original themes. The view of learning as an accumulation of independent units (habits), rather than as an integrated elaboration and differentiation of existing behavioral propensities, commits the error of reifying "differentials," which Weiss (1967) has eloquently deplored. The amazing fact that rats come equipped to run down mazes seems to have been ignored by theorists who define "learning" as the time it takes for the animals to do the running. All plastic events, including learning, can best

be understood as the dependent progeny of a genetically-directed embryological and developmental history that both provides a substrate for adaptive modification and restricts its extent and character. The puzzling problem of assigning relative weights to the two contributing factors (heredity, environment) was a major theme of the symposium, and pervades virtually all the papers in this section.

Current investigations of plastic phenomena proceed along many fronts, using single cells, cellular aggregates, and entire organisms as objects for study. The common goals are to identify what cellular areas or brain loci are involved, and to describe the mechanisms at work. Taken together, the papers that follow provide a fair sample of the techniques presently in use and of the conclusions such experiments permit. Each one reports measurements of behavioral, electrical, or chemical events, sometimes in some combination. Two of them, the studies on man (Williams, Ervin and Anders), illustrate the unique problems and opportunities this kind of investigation presents. Others examine the interplay of genetic and environmental factors (Delius, Goy, Valenstein); bring up to date the information electrophysiological studies can provide (Adey, Fox); or represent examples of current experiments upon single nerve cells or small groups of them (von Baumgarten, Hydén). The remaining papers deal with relatively new topics —the surprising plasticity of autonomic responses (DiCara) and the use of drugs that prevent protein synthesis, thus preventing memory consolidation (Barondes).

The reader searching for the generalizations threaded through these studies may wish to keep in mind four questions the members of the symposium held before themselves throughout the meetings.

1. What is the constant neural dimension that correlates with a fixed behavioral response? Much of the chemical, electrical, and other data collected during measurements on plastic and stable performance must be irrelevant; they amount to noise that confuses the issue. What event, or group of them, is the signal to be identified in that noise?

2. What mechanisms alter the genetically determined organization of a nervous system in such a way as to yield the responses determined by environmental stimuli? During acquisition of a behavioral response, the stimulus input triggers events within the substrate provided by the genome and thereby changes the brain; what are the processes by which this is accomplished?

3. A popular answer to the question just asked points to the endogenous electrical and chemical activities of the brain, both of which are impressively large, and speculates that the sequence followed is: sensory input $\longrightarrow$ altered electrical activity $\longrightarrow$ protein biosynthesis $\longrightarrow$ modified brain cells. Is this idea correct?

4. The several variables that seem to influence the amount of plasticity a brain displays include species, chronological

age, and environment (both external and internal). What brain states maximize stability or plasticity, and how can they be characterized and achieved?

The symposium contributed few direct answers to these questions, as the reader will discover. It did, however, cast some light upon them, as I will attempt to show in the final paper in this section.

REFERENCE

WEISS, P. A., 1967. $1 + 1 \neq 2$ (One plus one does not equal two). *In* The Neurosciences: A Study Program (G. C. Quarton, T. Melnechuk, and F. O. Schmitt, editors). The Rockefeller University Press, New York, pp. 801–821.

# 17 Normal and Pathological Memory: Data and a Conceptual Scheme

## FRANK R. ERVIN and TERRY R. ANDERS

MEMORY, APPARENTLY a general property of neural nets, is the basis for adaptive behavior of the individual organism. It has therefore attracted the attention of neurobiologists from Hartley through Ramón y Cajal to the present. In recent years, the power of the conceptual tools of molecular biology as an aid in understanding the mechanism of species "memory" (the genetic code) has led to the construction of tempting analogies between mechanisms for genetic storage and those for memory storage. The overview of experimental work in this mode, as reported in the first Boulder meeting (Quarton et al., 1967) by Agranoff, Altman, Chow, Eisenstein, Galambos, Hydén, John, Kandel, Miller, Nelson, and Sperry needs little modification today. The weakness in much of that approach to memory lies in the implicit confounding of *information storage*—in the sense in which it is used to refer to ferrite cores of magnetic tapes, a process that must include structural change in molecules or in their relationships—with "a memory" as we infer it from the behavior of the intact organism or understand it introspectively.

Another major experimental approach has been to produce in animals brain lesions similar to those found in humans who exhibit clinically defined deficits in memory. These experiments have been limited in the degree to which a behavioral change unambiguously related to "memory"

can be produced. The lesion studies seem paradoxical unless one assumes that no behavior tested to date in other animals is comparable to that from which we infer memory in man. Indeed, almost all definition and investigation of memory in man have been limited to verbal behavior or to behavior that is unusally mediated by "verbal," i.e., word-image, processes. (For example, maze learning need not be done verbally, but humans most efficiently memorize a maze as "...one right, two left, two right, etc.")

Finally, there is a small but tantalizing body of information gleaned from man during electrical stimulation of brain structures exposed by surgery. Detailed reporting of past experiences or perceptions elicited "on command" by stimulation have suggested a macro-organization of information in the brain comparable to a file that can be searched like a magnetic tape. This metaphor is an attractive one in a period of innovative computer technology and has most recently included a holographic model. Our own experience in human brain stimulation suggests a slightly different conception of the macro-organization.

The major effort in this essay is to summarize our concept of the prevailing evidence concerning the organization of memory in man. We believe that there is an orderly model and a methodology for examining memory that is all but neglected in studies of clinical pathology. We have also attempted to demonstrate the usefulness of such a schema for organizing existing information and future research. Judging from the literature, however, we believe it necessary to preface these remarks with a discussion of several central conceptual and methodological issues.

FRANK R. ERVIN AND TERRY R. ANDERS Stanley Cobb Laboratories for Psychiatry Research, Massachusetts General Hospital, Boston, Massachusetts

LEARNING AND MEMORY   Learning and memory are not separate functions. One cannot remember anything that has not been learned, nor is it possible to demonstrate learning without memory. Nevertheless, it is important to distinguish the operations employed to study these functions. As suggested by Melton (1963), a performance change from Trial *n* to *n* + 1 is considered to be a *learning* change when the variable of interest is the ordinal number of Trial *n*. A change from Trial *n* to *n* + 1 is viewed as a *memory* change when the variable of interest is the interval between trials. Performance changes over time, intervening activity, or both are the defining independent variables in the study of memory. For this reason, tests of memory *must* include at least two performance tests. The first, generally the last learning trial before the retention interval, provides an index of what was learned. The second, the memory test, measures the amount of forgetting that has taken place since the first test. Without an index of learning, the single-performance test fails to distinguish between that which has been forgotten and that which was not learned.

This definition alone does not provide a working model. Of additional value has been the division of information flow into the three functions of registration, retention, and retrieval. Registration, in this view, is simply the process of getting information from the outer world into storage. Retention is the storage of information over time, and retrieval makes stored information available for use. Retention and retrieval are the processes of major interest in memory research.

MEMORY AND TIME   The temporal parameters of retention have generally been divided into at least two parts. Retention over intervals measured in seconds or minutes is conventionally referred to as short-term memory (STM), and retention over intervals of hours, days, or months, as long-term memory (LTM). A demarcation between STM and LTM has been as difficult to specify behaviorally as neurologically (Hebb, 1949), and some have taken this difficulty as an indication that these "types" of memory may simply represent different points on a continuum (Peterson, 1966). Nevertheless, the arbitrary limit of five minutes, which has been suggested by Melton (1963) for the study of STM, seems useful.

A similar distinction between "recent" and "remote" memory is found in the clinical literature. There seems, however, to be some confusion regarding usage. Recent memory generally denotes the ability to retrieve information about the near past. For example, the retrograde amnesia following concussion is a marked loss of "recent" memory. Remote memory refers to the ability to retrieve the more distant past, again information acquired prior to the onset of a pathological condition. The loss of this function seems to occur only with widespread, severe brain disease. Loss of recent memory is also used to describe the underlying defect of anterograde amnesia, a defect in retention of information acquired after the pathological event. This use of "recent" memory differs considerably from that used in regard to retrograde amnesia. The latter loss includes a time span of seconds or minutes, comparable to STM. Evidence strongly suggests that the loss of recent memory associated with retrograde amnesia is a problem of retrieval, whereas the recent-memory defect of anterograde amnesia is a failure of registration. For these reasons, it seems best to avoid confusion by applying distinctive labels. We propose that the use of "recent memory" be restricted to clinical use in the description of retrograde amnesias and that STM be the preferred term to describe the process affected in anterograde amnesia.

CRITERIA FOR THE STUDY OF MEMORY   Several experimental criteria are important in attempts to explore the pathologies of memory. One cardinal rule is to use appropriate control groups. There are other safeguards, however, that are necessary to insure that only the variable of interest contributes to performance differences (Underwood, 1964). These rules apply to all memory comparisons but are particularly critical when one is working with clinical groups with disabilities that may influence test performance independent of memory ability. These criteria are as follows. 1. The level of learning must not be different for the control and for the test groups. Both groups must start with the same amount of information, because there is a high correlation between the level of learning and the rate of forgetting. In practice, the retention measure taken after the shortest delay interval is considered to be the valid index of the level of learning. 2. The performance measure of amount learned must be significantly below 100 per cent to insure that no overlearning has occurred and that the measure remains sensitive. 3. Only when these two conditions are met and different rates of forgetting occur may the effect be attributed to a memory difference alone.

## Human memory processes

In this model we present views that have grown out of contemporary memory research with normal individuals (Atkinson and Shiffrin, 1968; Bower, 1967; Broadbent, 1963; Sperling, 1963, 1967; Waugh and Norman, 1965; Wickelgren, 1969).

SENSORY MEMORY   One of the more interesting findings in the memory literature is the clear demonstration of sensory memory, the system that briefly holds the raw data of sensation available for attention, scanning, and further processing. Most of the available information has come from research in the visual mode, although some preliminary work has been reported with audition.

*Visual mode* It is an old observation that visual "images" persist for a short time after removal of the stimulus. The dramatic example of this effect is the so-called "afterimage." Only recently, however, have the functional properties of visual memory been explored under conditions representative of our everyday encounter with the visual world (Averbach and Coriell, 1961; Sperling, 1960, 1963).

The Averbach and Coriell (1961) report provides estimates of storage capacity and stimulus persistence derived from an experiment in which subjects were given a 50-millisecond glance at a 16-letter array of random letters. The subjects' task was to recall a single letter, which was indicated by a bar marker. The marker appeared simultaneously with the array or followed it by varying time intervals. The number of letters recalled under the simultaneous condition was considered to be an estimate of capacity. Results showed that over 70 per cent of the 16-letter array could be recalled immediately. This estimate, however, was considered to be somewhat lower than "true" capacity, owing to losses occurring during the time required to locate and "read" the letter. Estimates of storage time were derived from delayed recall performance, and these showed that forgetting started almost immediately after the array was removed and progressed steadily for 150 milliseconds. At this point, retention leveled off at a 25 to 35 per cent level. Ideally, this curve should have reached a zero level; the lower limit, however, remained at about four or five letters, suggesting that this number of items had been read into another, more permanent memory system. Adjustments were made on this assumption, and the upper limits of stimulus persistence in visual memory were then estimated to be about 250 milliseconds.

Haber and Standing (1969) have also provided estimates of stimulus persistence in visual memory that approximate 250 milliseconds. In their experiment, subjects were required to gauge the duration of fading between repetitions of a stimulus. The shortest cycle time in which complete fading was judged to occur was 250 milliseconds. An observation made in conjunction with this experiment is germane to the present discussion. It was found that estimates of duration were the same, whether the stimuli were presented to the same eye or to each eye in regular alternation, strongly suggesting a central locus for visual memory. In addition, Haber and Standing have pointed out some of the complexities involved in estimates of stimulus persistence by systematically varying background luminance, the adapting field, and length of exposure. These parametric changes do influence the duration of the "trace." One implication of this finding is that many of the variables, such as brightness and

contrast, which are known to improve the "perceptual quality" of a stimulus, do so simply by increasing the duration of the visual memory trace. In spite of this variability, the duration of visual memory is a few hundred milliseconds at best.

Averbach and Coriell (1961) also investigated the mechanism of forgetting in this system. Some sort of quick erasure mechanism seemed necessary, because normal perceptual abilities precluded the possibility that all forgetting occurred by simple decay. They tested this possibility for an active "erasure" mechanism by replacing the bar marker of the original experiment with a black circle around the letter. At intervals of less than 200 milliseconds there was "a quick substitution of the circle for the stored letter." At longer intervals there was no such effect. This erasure phenomenon requires that the new and old information occupy similar positions in the visual field. Monocular and binocular testing suggest that this effect, too, has a central locus.

*Auditory mode* Auditory experiments have been less successful in sharply defining the dimensions of a sensory memory. The work of a number of authors, however (Eriksen and Johnson, 1964; Fraisse, 1963; Guttman and Julesz, 1963), supports the suggestion of a memory store of less than a second's duration. Other modalities have not been investigated systematically.

To summarize from limited data, mostly in the visual mode, there seems to be a "sensory memory" entered automatically with sensation. Its capacity is not well measured, but may be limited only by the amount of information transmitted by the sense organ in a "perceptual moment." Its duration is a few hundred milliseconds, and information is removed from it both by decay over time and by active "erasure" in response to new information.

*Recoded memory* Successful transfer from this short-lived store requires that it be "attended to" and copied over into a more stable memory store. There seem to be at least two ways in which a successful transfer can be accomplished. One way is to recode the sensory data verbally. When asked to describe how they remember sensory information, subjects report scanning the sensory store, applying names to relevant items (Sperling, 1960, 1963), and preserving the verbal label. Little else is known about verbal recoding. Even less is known about sensory or nonverbal recoding, but evidence for a sensory recoder is persuasive. Children and animals profit from sensory experience without the aid of verbal skills. Further, language seems to be inadequate for describing many of the vivid images and recollections that we have tucked away.

The very process of recoding sensory data seems to copy it over into another, more stable memory system. Discussion of this more stable system follows an outline provided by the Waugh and Norman (1965) model of memory.

Three constructs are basic to their model: (1) primary memory (PM), (2) secondary memory (SM), and (3) a transfer mechanism. We have added to their model a suggested tertiary memory (TM).

PRIMARY MEMORY   Primary memory is a limited-capacity storage system that is the initial recipient of verbally recoded information. The actual capacity has not been established, although one is tempted to grasp at Miller's (1956) "magical number $7 \pm 2$" as a possibility. In terms of definable items such as letters or numbers, PM has a capacity smaller than sensory—at least visual—memory.

Information is organized in PM in temporal sequence. Forgetting occurs as new items enter and displace old ones. The duration of an item in PM is usually brief, because the organism is continuously processing new information. Nevertheless, information persists longer in PM than in sensory memory. Some idea of the duration of PM can be derived from the classic Peterson and Peterson (1959) experiment. They studied recall of small amounts of material that were well within the capacity of PM. Processing of this information for longer storage was prevented by interpolating a counting task during the retention interval. The material to be recalled was a simple nonsense syllable. The syllable and a three-digit number were read to the subject who repeated the number and counted backward until cued for recall of the syllable after intervals of from three to 18 seconds. The probability of recall was found to decrease exponentially to about 0.10 after 18 seconds. Based on these results, the duration of traces in PM seems to be several seconds rather than the fractions of a second found in sensory memory.

Sensory recoded information does not seem to be routed through PM as we have defined it. Evidence suggests that nonverbal information either has a separate memory buffer that serves a similar function, or that it enters the next more stable memory process (SM) direct from the sensory system. Posner and Konick (1966) demonstrated the critical difference between the retention of verbal and nonverbal information under conditions in which performance is considered to depend on PM. They studied the retention of a motor response, the measured excursion of a lever, following delays of 0, 10, 20, and 30 seconds. The subjects were either left to their own resources or were required to conduct various types of simple mental tasks during the delay intervals. Although some forgetting occurred, more important was the finding that test performances were comparable after all types of delay intervals. This is quite different from the usual finding in verbal learning, in which interpolated tasks result in considerable forgetting, and strongly suggests that PM does not play an integral part in the storage of nonverbal information. The best alternative to transfer through PM seems to be a direct link between the sensory system and SM, probably through some sort of sensory recoder.

There are three possible courses for information displaced in PM: (1) it may be dropped from the system and forgotten; (2) it may be actively retained by being recirculated through PM; or (3) it may be passively retained by being transferred to SM. The transfer mechanism from PM to SM is associated with rehearsal, a recycling of material through PM. Rehearsal serves the dual function of keeping an item temporarily in store and increasing the probability of its transfer to SM. The probability of successful transfer increases with the amount of time an item spends in PM. Because information flows through PM at a fairly constant rate, difficult material must be recycled to increase its time in PM and the likelihood of its transfer. The time required for this step varies with both subject and task variables—that is, more time for slow learners and for difficult material.

SECONDARY MEMORY   Secondary memory is characterized as a larger and more permanent storage system than PM. Only information that has been successfully stored in SM will be available for recall after extended periods of time. At present, it is impossible to give empirical estimates of capacity and duration. The problem with estimates of capacity has been the lack of a standard unit of information. Psychologists have generally relied on some physical segment of the external stimulus as a measuring unit. Miller (1956), however, has demonstrated the difficulties inherent in this approach. The adult human subject is rather facile in transforming stimulus "segments" into idiosyncratic units and combinations. Attempts to measure stimulus persistence in SM have also revealed that a large number of variables affect persistence. Indeed, a prominent characteristic of SM seems to be the variable duration of traces.

The process of forgetting in SM seems to be best described by the "interference theory" (Underwood, 1957). According to this theory, forgetting results from the "unlearning" of responses and from competition between responses at the time of recall. Therefore, both previous learning and learning that occurs during a retention interval are sources of information that compete with and extinguish task material. Interference from previously learned material is referred to as proactive inhibition; that from subsequent learning, retroactive inhibition. Proactive inhibition is usually assumed to be the most potent source of forgetting, because the individual brings a far greater store of information to an experimental situation than could possibly be accumulated during a 24-hour retention interval. In this view, we carry around with us the source of much of our forgetting in the form of prior learning.

The organization of SM also seems to differ from that of PM. Although much surely remains to be discovered, the present evidence is that semantic variables play a large role in the organization of SM. For example, a recent experiment by Baddeley and Dale (1966) has shown that PM and SM can be distinguished in terms of the types of errors that intrude during recall. In PM, most errors result from phonetic confusions. Items that sound alike (Bs and Vs, for example) are substituted for each other. Errors in SM, however, involve items that have similar meanings rather than similar sounds. Evidently, part of the recoding process in transferring information into SM is semantic, and semantics take precedence over phonetics. This is not to say that all organization in SM is semantic. Abundant evidence supports the existence of other types of relationships, such as temporal and spatial contiguity and syntactic structure.

Another characteristic of SM that distinguishes it from PM is the speed with which information can be retrieved. Retrieval seems to be much slower from SM (Waugh and Holstein, 1968). The longer retrieval latencies perhaps reflect the extra time needed for searching a larger storage system.

We have already suggested the possibility that nonverbal information is stored directly in SM. It is difficult to present supportive evidence for this contention, because psychologists have largely ignored the study of nonverbal memory. A major exception is the study of motor skills, although much of this literature emphasizes applied problems.

One issue of interest here involves the relevance of interference theory to the forgetting of nonverbal information. Adams (1967) reviewed those few studies that have attempted to demonstrate the effects of proactive and retroactive interference in motor retention (Duncan and Underwood, 1953; Lewis et al., 1951; Lewis and Shephard, 1950; and McAllister and Lewis, 1951). Retroactive interference seems to be a relatively stable phenomenon in motor retention as long as the interpolated learning involves the acquisition of antagonistic responses. The single attempt to investigate proactive interference (Duncan and Underwood, 1953) failed to find the usual decrement due to prior experience. To conclude from this negative finding that motor learning is impervious to proactive interference would be hasty, however, and introspection suggests that it is very sensitive indeed. It may be reasonable to assume for the present that the process of forgetting is similar for both verbal and motor memories, suggesting that they may reside in a common storage system.

Although the mechanism of forgetting may be the same, there is some question about the relative stability of the two types of traces. This question arises from the frequent finding of unusually high resistance to forgetting of certain types of motor skills. For example, Underwood (1957) has estimated that 15 to 25 per cent of a perfectly learned word list cannot be recalled after 24 hours. It may, however, take several years for this amount of forgetting to occur with

some types of motor learning. It is important to note that only certain types of motor skills show this high resistance (Adams, 1967). Highly stable traces are formed only during the acquisition of "continuous" motor acts, such as tracking tasks and others that involve a paced activity that requires continuous correction of errors (Adams, 1967, p. 218). By comparison, the traces laid down during the learning of a "discrete" motor task seldom have greater stability than those found in verbal learning.

The apparent reason for the relatively high rate of forgetting discrete motor skills is the dependence of such skills on verbal cues. As a result, it seems that only those conclusions drawn about continuous motor skills would apply to our understanding of memory for sensory recoded information. The extreme sophistication with which the normal individual can deal verbally with his environment must never be underestimated.

TERTIARY MEMORY  The prominent characteristic of TM is its storage durability. Once information has achieved storage at this level in the cognitive system, it seems to remain a permanent part of the individual. Evidence for the existence of memories with this quality may be found in the clinical literature. In most instances of brain damage or disease, intensive electroconvulsive therapy, old age, and other processes that erase less stable memories, there remain intact a number of highly overlearned items, which we define as the content of TM.

The capacity of TM must be very large, because it contains much information that we use frequently. The information required to walk and talk, to say and write our names, and to recognize the names and faces of old friends, are all characteristic of the contents of TM. In this sense, however, TM is the most difficult of the memory processes to get information into, because it takes months or years of overlearning to achieve such durable storage. A second characteristic of TM is the ready accessibility of the stored information. The same overlearned information that survives an extensive retrograde amnesia can be retrieved with little hesitation. For example, the time required to recall one's own name on command is miniscule in comparison to the time required to dredge up familiar but less frequently used names. In this sense, retrieval from TM is much faster than retrieval from SM.

Almost nothing is known about the organization of memories in TM, other than that it must be an impressive system to afford such ready access to so much information. It seems curious that such fascinating aspects of memory as those associated with TM have not attracted more research.

SUMMARY  In sequence, visual information is held in sensory memory many milliseconds, labeled, and thereby transferred to PM, where it displaces the oldest information. In

PM it is available for prompt recall, and it can be held by rehearsal indefinitely in the absence of new input. Without rehearsal, this information disappears after several seconds. Rehearsal (or other factors) allow for transfer from PM to SM, where the coding is reordered and the information is stored contextually rather than temporally. Retrieval is slower from SM than from PM, but the information is stable for hours, or days (experimentally), or years. If the information is used very frequently, it may again acquire short retrieval time and be even more permanently accessible as part of a postulated tertiary memory.

We have summarized most of the above discussion in Table I.

## Pathological memory

On the basis of the preceding outline of normal memory processes, let us attempt to analyze the nature of impairment in clinically defined memory derangements of anterograde and retrograde amnesia. A substantial literature exists on these topics, but typically the research has been unsystematic, has employed a variety of noncomparable methods, and bears little relationship to general memory theory. The tendency has been to infer memory processes from complex learning situations, while ignoring the techniques designed for the express purpose of investigating memory. For these reasons, we found it necessary to devise a set of criteria for selecting the material to be included here. 1. An attempt was made to be moderately critical of the scientific quality of the research included; each experiment or observation was evaluated according to the set of rules previously outlined regarding the semantics of memory research. 2. We placed a high premium on research conducted with patients whose cognitive defects were fairly well restricted to memory. 3. We also placed a premium on research with patients whose brain lesion or lesions were confirmed. The human literature is obviously limited in this regard.

ANTEROGRADE AMNESIA Anterograde amnesia is traditionally described as a failure of retention of newly acquired information. For example, when an amnesic patient is presented with new material, registration seems to occur normally, because his immediate responses are usually of an appropriate nature. If, however, the stimulus is removed and the patient is made to delay his response for as little as several seconds, he is now unable to produce the correct response. Unfortunately, the clinical picture of these patients is seldom so simple, and past study has included attempts to identify a constellation of symptoms defined as an "amnesic syndrome." Talland (1965b) has presented a

TABLE I

*Human Memory Processes*

| | Storage System | | | |
| | Sensory Memory | Primary Memory | Secondary Memory | Tertiary Memory |
|---|---|---|---|---|
| CAPACITY | Limited by amount transmitted by receptor (?) | The 7 $\pm$ 2 of the memory span | Very large (no adequate estimate) | Very large (no adequate estimate) |
| DURATION | Fractions of a second | Several seconds | Several minutes to several years | May be permanent |
| ENTRY INTO STORAGE | Automatic with perception | Verbal recoding | Rehearsal | Overlearning |
| ORGANIZATION | Reflects physical stimulus | Temporal sequence | Semantic and relational | ? |
| ACCESSIBILITY OF TRACES | Limited only by speed of read out | Very rapid access | Relatively slow | Very rapid access |
| TYPES OF INFORMATION | Sensory | Verbal (at least) | All | All |
| TYPES OF FORGETTING | Decay and erasure | New information replaces old | Interference: retroactive and proactive inhibition | May be none |

lucid review of the issues; because they exceed the scope and purpose of the present paper, they have not been included here but should be reviewed by the interested reader.

Reviews of various aspects of memory research on amnesic patients can be found in the literature (see Lewis, 1961; Ojemann, 1966; Piercy, 1964; Talland, 1965b; DeMorsier, 1967; and Milner et al., 1968). There is little controversy among these reviewers about the generality and severity of the memory impairment. The usual conclusion has been that severe anterograde amnesia involves a "virtual inability to acquire new information." There are, however, a number of observations of areas of unimpaired performance that have not previously been accounted for in theoretical explanations of the patient's memory deficit.

*Sensory memory* Sensory memory has not been extensively investigated in amnesic patients. This omission probably stems from the many reports of their good immediate registration of stimulus events. Scoville and Milner's (1957) now-famous patient H. M. is a good example of severe anterograde amnesia taking place in the absence of visual, attentional, or perceptual disorders. In fact, H. M. performs many nonmnemonic tasks with above-average ability, in accord with his superior intelligence (Milner et al., 1968). In what seems to be the only investigation of amnesic patients' sensory memory, Peter Schiller tested H. M.'s performance under conditions related to the visual masking task described earlier (Averbach and Coriell, 1961). He concluded from these observations that H. M.'s visual processes were equivalent to those of the normal controls. It would be interesting to know how this result will generalize to amnesic patients with known perceptual disorders, such as those demonstrated by the Korsakoff patients described by Talland (1958). Nevertheless, sensory memory may be intact in patients with severe anterograde amnesia.

*Primary memory* Primary memory also seems to function normally in these patients, as demonstrated by their normal digit span and by their ability to deal efficiently with subspan amounts of information (Wechsler, 1917; Zangwill, 1946; Talland, 1965b; Drachman and Arbit, 1966; Milner, 1966; Victor et al., 1959; Wickelgren, 1968). When the demands of the task exceed the capacity of PM, however, a performance deficit becomes apparent. One demonstration of the abrupt difference between the ability of these patients to deal with subspan and supraspan amounts of information has been reported by Drachman and Arbit (1966). They found that, although normal controls could more than double their recall by practicing, amnesic patients could add an average of only one item to their original span of seven after an equal number of repetitions. Other investigators have found comparable results in a variety of situations (Milner, 1966; Scoville and Milner, 1957; Penfield and Milner, 1958; Talland and Ekdahl, 1959; Talland et al., 1967).

The amnesic patient's deficit may also be demonstrated with subspan amounts of information simply by delaying recall. Typically, amnesic and normal subjects do equally well on the immediate recall of subspan lists. When recall is delayed, however, the amnesic patient loses the material much faster than does the normal subject (Talland, 1968; Talland et al., 1967). Some of the difficulty encountered by many amnesic patients is caused by their failure to use rehearsal (Talland et al., 1967), but even if they have been observed to use rehearsal as a memory aid, it did not seem to have its full effect on the stability of the learned material (Scoville and Milner, 1957; Penfield and Milner, 1958; and Sidman et al., 1968). Interrupting rehearsal, even after many minutes, was sufficient in most cases to erase completely any trace of the information being rehearsed. Thus, it seems that their mnemonic strategy reflects a complete dependence on primary memory. Rehearsal for these patients seems to be not so much a method of enhancing storage as a method of maintaining attention on the task at hand.

*Secondary memory* The many examples of the amnesic patient's inability to exceed the storage capacity of PM strongly implicate SM as the weak link in his cognitive processes. The specific difficulty may be one of encoding, storage, or retrieval. Our own work leads us to believe that it is encoding; that is, information never is stored properly in SM. The clarification of this question should be readily obtained by clearly designed experimental analysis of the model we have outlined.

One thing that is clear from the literature is that the amnesic patient may be amnesic mainly for verbal information. Our review revealed three kinds of learning tasks which these patients could perform much more proficiently than might be expected of an individual with a generalized memory deficit. These exceptions include the acquisition of certain types of motor skills, classical and operant conditioning, and "perceptual" learning. The common feature of these types of learning is that they involve the acquisition of nonverbal information.

MOTOR SKILLS There have been several demonstrations of the amnesic subjects' normal learning and retention of certain types of motor skills (i.e., Corkin, 1968; Milner, 1962; Talland, 1965b). The first was Milner's with the amnesic patient H. M. The patient was asked to trace a star-shaped pattern while viewing his progress only through a mirror. He was able to perform this task as well as normal control subjects, demonstrated comparable improvement during three days of practice, and showed no losses between days of testing. The intriguing aspect of the results was the report that H. M. was completely unaware of his improvement or even of his previous days of practice.

This normal retention of motor skills by amnesic patients applies only to a specific class of tasks, in much the same way that the high retention by normal subjects applies only to a specific class of motor skills. Only with a continuous-performance test, such as a tracking test, does the amnesic patient show normal improvement and retention, and there are numerous examples of the amnesic patient's failure to learn or retain discrete motor tasks (Milner, 1965; Corkin, 1965).

As previously mentioned, the apparent reason for the higher rate of forgetting with discrete motor tasks is the dependence of such tasks on verbal cues. Accordingly, the amnesic patients' ability to perform the continuous, but not the discrete, tasks supports the suggestion that their memory impairment is primarily for verbal material. If such is the case, certain similarities would be expected in their performance of verbal and discrete motor tasks. At least one important similarity does exist. This involves the relationship between successful performance and the amount of information involved in performing the task.

Milner et al. (1968) have investigated H. M.'s ability to learn mazes of short lengths. In one study they reduced Milner's (1965) visual maze from 28- to 10-choice points, and H. M. failed to learn this shortened version even after 125 trials. A further reduction of the path to eight-choice points enabled H. M. to learn the maze slowly, and he required 155 trials spread over two days. Retention of the correct path was tested after delays of one, two, three, and six days. Marked savings were evidenced after each delay. On the day of the last retention test for the eight-choice path, H. M. was given 25 more practice trials with the 10-choice version. Performance was as poor as on previous testings. Learning the eight-choice maze did not transfer, even though it was simply the first portion of the longer maze path.

The length of Corkin's (1965) tactual maze was also reduced from 10- to five-choice points for a second test of H. M.'s ability to learn short mazes, and H. M. did not reach the learning criterion within the prescribed 300 practice trials. He did, however, show fairly consistent improvement over the days of practice. This finding, together with his previous success, was taken as evidence of his ability to learn and retain limited maze sequences.

Milner et al. concluded by calling attention to the fact that H. M.'s successes were with mazes involving sequences within the limits of the memory span. Milner (1965) offered an account: "... the experience in the latter part of the maze interferes with the effective rehearsal of the first part of the path, so that a patient such as H. M., who seems to rely entirely upon verbal rehearsal to bridge a temporal gap, must start each new trial as if it were a fresh problem" (p. 332). This view is in agreement with our characterization of the amnesic patient as depending entirely on PM when dealing with verbal information. It seems, however, that with sufficient perseverance, small amounts of information that do not overload PM may eventually be processed into a more permanent storage. Several facets of this conclusion remain to be tested. 1. Will equal perseverance with a comparable verbal learning task, such as learning a list of words, eventually be learned, completing the analogy drawn between verbal and discrete motor learning? 2. Will such learning, if it does occur, be limited to subspan amounts of information? 3. Given that some learning does take place, will retrieval of this information have the characteristics of having been stored in SM or PM?

CLASSICAL CONDITIONING A number of observations and experiments in the literature deal with the classical conditioning of amnesic patients. Many of these observations are anecdotal and employ unorthodox procedures. The classic example is, of course, Claparede's (1907) insightful observation.

"I stuck the patient hard with a pin concealed in my hand. This tiny pain was forgotten as quickly as innocuous perceptions and, several instants after the prick, she didn't remember anything. However, when I brought my hand close to hers once more, she pulled back her hand in a reflex fashion without knowing why. In fact, if I asked her why she had done this, she answered, dumbfounded, 'Well, don't I have the right to pull my hand away?'" (page 101).

Gruenthal and Stoerring (1930) reported a similar observation, and Barbizet (1963) described a patient who became upset at the sight of the electroconvulsive shock apparatus even though he could not recall having seen it before.

The unconditioned stimulus (UCS) has not always been so obvious or of such a painful nature. For example, De-Morsier (1967) simply told his patient to lift his arm when he said a certain word. He then waited 15 seconds, said the word, and the patient lifted his arm. When asked why he had lifted it, the patient replied, "It's just an idea I had." Again, the performance was accomplished on command without recognition of the memory.

The more systematic attempts to condition amnesic patients have reported mixed results. Gantt and Muncie (1942) failed in an attempt to condition three alchoholic Korsakoff patients, and the reason may be similar to that given by Milner et al. (1968) for H. M.'s failure to demonstrate conditioning when an electric shock was used as the UCS. Their attempt was abandoned because H. M. did not show a galvanic skin response to the shock, "even at intensity-levels that normal control subjects found disagreeably painful" (p. 223). H. M. seemed to be aware of the shock, but he never complained or showed any of the signs of pain. It is interesting to note that a reduced responsiveness to normally noxious stimuli (p. 93) after bilateral lesions involving the amygdala, as is the case with H. M. and perhaps with Gantt's patients, is a frequent finding in lower animals (Goddard, 1964). As a result, conditioning may have failed because of the patients' abnormal reaction to electric shock. This view is further supported by reports of successful conditioning of amnesic patients using a different UCS (Talland, 1960; Linskii, 1954).

In summary, the existing evidence suggests that amnesic patients can be classically conditioned in the proper situation. The most dramatic examples are anecdotal.

OPERANT CONDITIONING   Sidman et al. (1968) trained H. M. to perform a discrimination task utilizing a series of successive approximations to the final task. The first approximation was simple and included only one choice. After short practice on this and several intermediate steps, H. M. was soon performing the eight-choice discrimination task without having been given any verbal instructions. This training was retained after several interruptions for counting pennies and for interviews. The interviews revealed an interesting discrepancy between the verbal and the nonverbal aspects of the task. H. M. appeared to be amnesic in his oral accounts, while actually performing the task with normal accuracy. Again, we see an example of the amnesic patient's ability to learn and remember without a "verbal awareness" of such occurrences. More important, however, is the potential function of this type of training procedure as a remedial step toward some alleviation of their severe handicap. Perhaps the success with which these training procedures have been used with certain other handicapped individuals, such as the mentally retarded, could be employed to improve the life circumstances of the amnesic patient. Further exploration of the use of operant conditioning procedures seems worthwhile for both theoretical and practical reasons.

PERCEPTUAL LEARNING   There remains one other class of tasks which the amnesic patient both learns and retains in a normal manner. It is difficult to know exactly what label to apply to them, because little is known about the abilities required for their successful performance. Their nature suggests that they represent a form of perceptual learning, in the sense that what is learned is recoded sensory information but is closer to the raw sensory data than a verbally recoded representation.

The clearest demonstration of the amnesic patient's ability to perform such tasks was reported by Warrington and Weiskrantz (1968). The task, originally used by Gollin (1960), included the recognition of incomplete drawings of simple objects. Each object was represented by five drawings at various stages of completion. The subject was first shown the most incomplete version and progressed through the series until he recognized the represented object. There were 10 objects; each was presented five times on each of three days. Initially, the amnesic patients required more steps than the normals to recognize the objects; how-

ever, their rate of improvement at each test session and their savings from one day to the next were equal to those of the normal controls. A second experiment with a graded series of incomplete words had similar results.

The memory traces established under these conditions seem to have considerable stability, comparable perhaps to those established during the acquisition of continuous motor tasks. This was demonstrated by Warrington and Weiskrantz in a retest of the temporal lobectomized amnesic patient after a three-month delay. The savings computed from the first five trials on each occasion was 20 per cent. Similarly, two groups of normal subjects were tested on the incomplete words test. One group was retested after a one-week delay; the other, after a four-week delay. Both groups evidenced considerable savings on relearning. The high retention of these materials by amnesic subjects is even more striking when compared with the results cited in a later report by Warrington and Weiskrantz (1968). In this study, the same stimulus words were used in more conventional tests of recall and recognition, which showed very poor retention after delays as short as five minutes.

In summary, we believe these exceptions to the generality of the amnesic patient's memory impairment support our contention that anterograde amnesia is a defect of verbal memory. In each of the above instances, amnesic patients demonstrated mnemonic skills well within the realm of those associated with SM. This has not been the case in tests of verbal memory.

RETROGRADE AMNESIA   Retrograde amnesia clinically refers to the loss of information acquired prior to the onset of some pathological state. Benson and Geschwind (1967) have summarized the usual findings in retrograde amnesia by distinguishing among three types of memories.

"There is a period of seconds to minutes in which a memory is fragile and may be permanently abolished. There is a longer period in which the memory is consolidated but in which retrieval may be impaired. Finally, many memories, especially old and overlearned ones, may be retrieved despite influences which abolish the most recently acquired memories and affect the retrieval of many later memories" (p. 542). These distinctions are reminiscent of our own PM, SM, and TM. The traces of PM are permanently lost. The traces of SM represent by far the greater part of the loss, which is seldom permanent. The traces of TM are those that are retained after even the most extensive retrograde amnesia. The loss from SM has correctly been summarized to be attributable to faulty retrieval, because in most instances the amnesia "shrinks" and the missing information again becomes available for recall without having been relearned (Benson and Geschwind, 1967; Russell and Nathan, 1946; Williams and Zangwill, 1952). In addition, several studies have shown that, under special circumstances, the "missing" information can be evoked; Russell and Nathan (1946) used barbiturate hypnosis and Talland (1965a) used prompting during an interview.

Another well-documented finding has been the observation that the borderland between lost and preserved information is a blurred area including "islands" of intact memory in a background of missing information (Williams and Zangwill, 1952). Alternating islands of lost and preserved memories seem to be present in both the early and the late stages of retrograde amnesia. The failure to find a distinct temporal break defining the retrograde loss has important theoretical implications, because it suggests that "consolidation" is not simply a function of time. Other factors must be involved, e.g., some mnemonic process that may selectively hasten storage. Affective significance of the information immediately comes to mind as such a process. Recent evidence has failed, however, to support the significance of affect as a mnemonic enhancer independent of more conventional devices, such as rehearsal (Weiner, 1966). It seems to provide the incentive for more vigorous rehearsal and review of the material, because affectively significant stimuli presented without an opportunity for rehearsal are remembered no better than neutral stimuli. Be that as it may, consolidation does not seem to occur simply as a function of time.

The similarities between the prognoses for retrograde and anterograde amnesia in individual cases have been noted by Benson and Geschwind (1967). Making the additional assumption that retrograde amnesia is a defect of retrieval and anterograde amnesia is a defect of registration, they concluded that registration and retrieval must be served by the same neurological substrate. This possibility is of major importance to our understanding of the neuropsychology of human memory. Evidence from the study of traumatic amnesia is fairly consistent with their hypothesis; but long-term follow-ups of the course of retrograde and anterograde amnesics in chronic amnesia cases do not consistently support it. For example, the 14-year reevaluation of H. M. by Milner et al. (1968) suggests some remission of his anterograde amnesia in the absence of any such recovery of the retrograde loss. Talland (1967) reported the complete recovery of the retrograde loss in the presence of a persisting anterograde amnesia for one of his amnesic patients; but in a similar case, both types of amnesias apparently persist unabated. As Milner et al. pointed out, however, an evaluation of the extent of the retrograde defect becomes very difficult after the passage of several years.

NEURAL MECHANISMS OF MEMORY The most important clues as to the neural structures in memory come from those human patients who suffer discrete lesions documented at autopsy or at the time of surgery. The distribution and varying etiologies of such lesions have been reviewed by several authors (Ojemann, 1966; Talland, 1965b). In summary, severe anterograde amnesia follows bilateral lesions of the hippocampus, the mammillary bodies, the anterior nucleus of the thalamus, and, possibly, the anterior columns of the fornix, and the dorsomedial nucleus of the thalamus. This interconnected system of structures, part of the classic circuit of Papez, seems to be the only place where discrete lesions can produce a clinical memory impairment, an anterograde amnesia, usually with a slight, persistent, retrograde amnesia. As we have pointed out, this impairment can be considered to be restricted to an inability to transfer properly coded information from primary to secondary memory; it is, perhaps further restricted to impairment of verbal memory.

This neuropsychological correlation is further supported by the instances of fugue states in temporal-lobe epilepsy. These states can be replicated in the laboratory in appropriate cases by bilateral electrical stimulation of the hippocampus, without spread of disruptive seizure discharge to the rest of the brain. In these instances, the individuals may carry out quite complex acts, travel distances, purchase objects, and read and respond to written material for many minutes or a few hours. That is to say, sensory and primary memory mechanisms function properly, and information in secondary and tertiary memory is retrieved and acted on appropriately. There is, however, total amnesia for the period, with no recovery under hypnosis or drugs or after time. Thus, during that period no information enters SM in a form that can be retrieved, implying that the transfer mechanism, the recoding mechanism, or both, are impaired.

Unfortunately, the conditions that produce relatively pure retrograde amnesia are all poorly understood pathologically and must be thought of as producing diffuse disturbance of the brain until they are further dissected experimentally. Concussion, electroshock therapy, apoplexy, and anesthesia are examples of such conditions.

One would like to think that meticulous examination of various clinical cases of memory pathology would reveal examples of relatively pure impairment of each of the conceptual steps in our model or serve to refine the model. That is, there should be at the very least, disturbances resulting from (a) limitations in PM; (b) impairment of transfer from PM to SM; (c) faulty storage or coding in SM; (d) faulty retention in SM; (e) faulty search; (f) recognition; and (g) retrieval from SM.

We have already suggested that the hippocampal lesion (or seizure) cases represent (b), and that the retrograde amnesia of postconcussion cases that fully recover represents (g). Certain limitations of the mental retardate seem related to (a) and to (c). There is, however, too little rigorously defined experimental data to allow us further speculation. The purpose of such a dissection of sharply defined mnemopathologies would, of course, be to relate them to neural processes that could be explored in the animal laboratory.

The attempts to prepare animal models for anterograde amnesia have been disappointing to date. In spite of the consistency of the neuropathological data for man, many careful lesion studies in animals of various species have failed to produce a behavioral deficit that all observers would define as memory impairment. Two explanations of this paradox are possible. One is that the lesions are not functionally equivalent because of a phylogenetic change in the organization of the hippocampal system. The other is that tests for memory impairment are not equivalent and that the appropriate animal behavior has not been isolated for measurement. We suggest strongly that the latter is true, with the reservation that verbal memory is most profoundly affected in man.

## Brain stimulation

Experience in stimulating chronically implanted electrodes placed stereotactically in a variety of subcortical structures has been no more revealing. In a review of more than 2000 such stimulations from our own experience, only 14 produced memory-like phenomena. These were of two kinds. From depth stimulation in the temporal lobe, in the region of the hippocampus and amygdala, we could evoke reminiscence-like states, which gave rise to reports of past experience. Such a report might be, "I have a feeling like a time when I was a boy and. . . ." Such a state might be reproducible, with similar mood images evoked and quite detailed accounts of experiences "long forgotten." On only one occasion was there a report of a detailed and out-of-context past experience. This experience could be evoked, but was sensitive to changes in stimulus parameters, and varied slightly on each repetition.

On two occasions, electrodes in the region of the posterior thalamus elicited complex, stereotyped motor acts and verbalizations, which were not remembered by the patient. On careful analysis, one turned out to be a childhood game, which had been a favorite of the patient's. In this instance, a long, unrehearsed, learned motor pattern was elicited in proper temporal sequence by activation of some kind of retrieval mechanism.

On the whole, however, the results of electrical stimulation of brain structures in waking man are anecdotally fascinating, but, as yet, they contribute little to our understanding of memory mechanisms. None of the data reported above, of course, give any support to the notion that there is a macroscopic "place" where memory is stored, or where any of the conceptual mechanisms described earlier are uniquely situated. The brain is not organized in such neat parcels. It is, however, not homogeneous mass. Its spatiotemporal patterning at the macroscopic level must be respected if it is to be understood.

## Summary

1. "Memory" is not a unitary function, nor do such divisions as "recent and remote" (clinically used) or "labile and fixed" (used in the animal laboratory) provide models of heuristic value.

2. In the last decade, a useful model of normal (verbal) memory has evolved, which permits rigorous testing of psychological hypotheses and should be respected in the elaboration of neurobiological theories.

3. Of particular importance in the present discussion is the distinction between "primary" and "secondary" memory and the transfer mechanism between them.

4. Applying this formulation to the two most broadly studied instances of memory pathology in man clarifies several points:

a) The anterograde amnesia typified by cases of Korsakoff's syndrome and the related purely amnesic disorders can best be understood as an impairment of transfer from primary to secondary memory.

b) Retrograde amnesia can be seen as a defect in the mechanisms of retrieval from secondary (but not primary) memory.

c) Although neither "transfer" nor "retrieval" mechanisms are unitary concepts, this dissection of the locus of psychopathology should aid in specifying the neural mechanisms involved in the mnemonic process.

For example, bilateral hippocampal ablation in man has little effect on primary memory processes, or on secondary memory or on retrieval mechanisms, but produces a profound disruption of transfer (at least of verbal information). Therefore, if one wishes to analyze the role of the hippocampus in "memory" in the experimental animal, one might most fruitfully utilize a behavioral model consistent with this formulation.

The lack of definitive information as to whether the model holds for nonverbal (or nonsymbolic) material makes it impossible to decide at present whether an animal paradigm is at all possible. A few simple experiments in man could settle this problem.

5. The last point above suggests at least a method for resolving what has seemed a painful paradox: No animal species has shown a clear impairment of what all observers would call "memory" after any set of lesions replicating those that produce a severe "amnesic syndrome" in man. Three logical possibilities now appear:

a) The memory deficit after bilateral Papez circuit lesions in man is limited to verbal (or verbally mediated) material, so would not be apparent in other animals.

b) Animal experimentation to date has not utilized sensitive enough tests of the early process of transfer from primary to secondary memory.

c) A qualitative change in neural organization from subhuman to human, one reflection of which is language ability, is also reflected in a changed role of the hippocampus (and related structures) in the memory mechanism. Possibilities A and B are subject to test.

6. The existence of a rigorous model has allowed for rapid progress in the description of normal human memory. Although this model will evolve (and perhaps change radically), it is important that observers of clinical-pathological states make observations and tests that can at least be interpreted, *inter alia,* in the framework of the model. By analogy with other pathological conditions, one would expect to find examples of clinical "memory disturbance" associated with each separate process in the model. Careful correlation of these with the etiological process might provide important hypotheses about the concomitant neural mechanisms. Such hypotheses could then be rigorously tested in the animal laboratory for precise delineation.

7. Stimulation of the exposed human brain leads to reports of "memory-like" phenomena in some subjects. These responses are most common in the temporal neocortex and, in modified form, in the temporal subcortical structures. The limited explorations of these phenomena have not yet clarified the neural structure of memory.

## Acknowledgments

Preparation of this chapter was supported in part by the National Institute of Mental Health (Grant MH-17110–01) to Frank R. Ervin, and by the National Institute of Child Health and Human Development (Grant MR-1-FO2-HD41527–01) to Terry R. Anders.

## REFERENCES

ADAMS, J. A., 1967. Human Memory. McGraw-Hill, New York.

AGRANOFF, B. W., 1967. Agents that block memory. *In* The Neurosciences: A Study Program (G. C. Quarton, T. Melnechuk, and F. O. Schmitt, editors). The Rockefeller University Press, New York, pp. 756–764.

ALTMAN, J., 1967. Postnatal growth and differentiation of the mammalian brain, with implication for a morphological theory of memory. *In* The Neurosciences: A Study Program (G. C. Quarton, T. Melnechuk, and F. O. Schmitt, editors). The Rockefeller University Press, New York, pp. 723–743.

ATKINSON, R. C., and R. M. SHIFFRIN, 1968. Human memory: A proposed system and its control processes. *In* The Psychology of Learning and Motivation: Advances in Research and Theory, Vol. 2 (K. W. Spence and J. T. Spence, editors). Academic Press, New York, pp. 89–195.

AVERBACH, E., and A. S. CORIELL, 1961. Short-term memory in vision. *Bell Syst. Tech. J.* 40: 309–328.

BADDELEY, A. D., and H. C. A. DALE, 1966. The effects of semantic similarity on retroactive interference in long- and short-term memory. *J. Verb. Learn. Verb. Behav.* 5: 417–420.

BARBIZET, J., 1963. Defect of memorizing of hyppocampal-mammillary origin: A review. *J. Neurol. Neurosurg. Psychiat.* 26: 127–135.

BENSON, D. F., and N. GESCHWIND, 1967. Shrinking retrograde amnesia. *J. Neurol. Neurosurg. Psychiat.* 30: 539–544.

BOWER, G. H., 1967. A multicomponent theory of the memory trace. *In* The Psychology of Learning and Motivation: Advances in Research and Theory, Vol. 1 (K. W. Spence and J. T. Spence, editors). Academic Press, New York, pp. 229–325.

BROADBENT, D. E., 1963. Flow of information within the organism. *J. Verb. Learn. Verb. Behav.* 2: 34–39.

CHOW, K. L., 1967. Effects of ablation. *In* The Neurosciences: A Study Program (G. C. Quarton, T. Melnechuk, and F. O. Schmitt, editors). The Rockefeller University Press, New York, pp. 705–713.

CLAPAREDE, E., 1907. Expériences sur la mémoire dans un cas de psychose de Korsakoff. *Rev. Med. Suisse Romande* 27: 301–303.

CORKIN, S., 1965. Tactually-guided maze learning in man: Effects of unilateral cortical excisions and bilateral hippocampal lesions. *Neuropsychologia* 3: 339–351.

CORKIN, S., 1968. Acquisition of motor skill after bilateral medial temporal-lobe excision. *Neuropsychologia* 6: 255–265.

DeMORSIER, G., 1967. Gliome hypothalamique avec syndrome amnésique de Korsakoff: Le problème de la mémoire. *Ann. Med. Psychol.* 1: 177–228.

DRACHMAN, D. A., and J. ARBIT, 1966. Memory and the hippocampal complex. *Arch. Neurol. Psychiat.* 15: 52–61.

DUNCAN, C. P., and B. J. UNDERWOOD, 1953. Retention of transfer in motor learning after twenty-four hours and after fourteen months. *J. Exp. Psychol.* 46: 445–452.

EISENSTEIN, E. M., 1967. The use of invertebrate systems for studies on the bases of learning and memory. *In* The Neurosciences: A Study Program (G. C. Quarton, T. Melnechuk, and F. O. Schmitt, editors). The Rockefeller University Press, New York, pp. 653–665.

ERIKSEN, C. W., and H. J. JOHNSON, 1964. Storage and decay characteristics of nonattended auditory stimuli. *J. Exp. Psychol.* 68: 28–36.

FRAISSE, P., 1963. The Psychology of Time. Harper and Row, New York.

GALAMBOS, R., 1967. Brain correlates of learning. *In* The Neurosciences: A Study Program (G. C. Quarton, T. Melnechuk, and F. O. Schmitt, editors). The Rockefeller University Press, New York, pp. 637–643.

GANTT, W. H., and W. MUNCIE, 1942. Analysis of the mental defect in chronic Korsakov's psychosis by means of the conditioned reflex method. *Johns Hopkins Hosp. Bull.* 70: 467–487.

GODDARD, G. V., 1964. Function of the amygdala. *Psychol. Bull.* 62: 89–109.

GOLLIN, E. S., 1960. Developmental studies of visual recognition of incomplete objects. *Percept. Mot. Skills* 11: 289–298.

GRUENTHAL, E., and G. STOERRING, 1930. Über das Verhalten bei unschriebener völliger Merkunfähigkeit. *Monatsschr. Psychiat. Neurol.* 74: 254–269.

GUTTMAN, N., and B. JULESZ, 1963. Lower limits of auditory periodicity analysis. *J. Acoust. Soc. Amer.* 35: 610.

HABER, R. N., and L. G. STANDING, 1969. Direct measures of short-term visual storage. *Quart. J. Exp. Psychol.* 21: 43–54.

HEBB, D. O., 1949. The Organization of Behavior. John Wiley and Sons, New York.

HYDÉN, H., 1967. Biochemical changes accompanying learning. *In* The Neurosciences: A Study Program (G. C. Quarton, T. Melnechuk, and F. O. Schmitt, editors). The Rockefeller University Press, New York, pp. 765–771.

JOHN, E. R., 1967. Electrophysiological studies of conditioning. *In* The Neurosciences: A Study Program (G. C. Quarton, T. Melnechuk, and F. O. Schmitt, editors). The Rockefeller University Press, New York, pp. 690–704.

KANDEL, E. R., 1967. Cellular studies of learning. *In* The Neurosciences: A Study Program (G. C. Quarton, T. Melnechuk, and F. O. Schmitt, editors). The Rockefeller University Press, New York, pp. 666–689.

LEWIS, A., 1961. Amnesic syndromes. *Proc. Roy. Soc. Med.* 54: 955–961.

LEWIS, D., D. E. McALLISTER, and J. A. ADAMS, 1951. Facilitation and interference in performance on the modified Mashburn apparatus: I. The effects of varying the amount of original learning. *J. Exp. Psychol.* 41: 247–260.

LEWIS, D., and A. H. SHEPHARD, 1950. Devices for studying associative interference in psychomotor performance: I. The modified Mashburn apparatus. *J. Psychol.* 29: 35–46.

LINSKII, V. P., 1954. Kvoprosu o vyrabotke vslovnykh refleksov u boliykh s korsakovskii sindromom. (On the formation of conditioned reflexes in patients exhibiting Korsakoff's syndrome.) *Zh. Vyssh. Nerv. Decatel.* 4: 791–798.

McALLISTER, D. E., and D. LEWIS, 1951. Facilitation and interference in performance on the modified Mashburn apparatus: II. The effects of varying the amount of interpolated learning. *J. Exp. Psychol.* 41: 356–363.

MELTON, A. W., 1963. Implications of short-term memory for a general theory of memory. *J. Verb. Learn. Verb. Behav.* 2: 1–21.

MILLER, G. A., 1956. The magical number seven, plus or minus two: Some limits on our capacity for processing information. *Psychol. Rev.* 63: 81–97.

MILLER, N. E., 1967. Certain facts of learning relevant to the search for its physical basis. *In* The Neurosciences: A Study Program (G. C. Quarton, T. Melnechuk, and F. O. Schmitt, editors). The Rockefeller University Press, New York, pp. 643–652.

MILNER, B., 1962. Les troubles de la mémoire accompagnant des lésions hippocampiques bilaterales. *In* Physiologie de l'Hippocampe. Colloques Internationaux No. 107, Paris, C.N.R.S.

MILNER, B., 1965. Visually-guided maze learning in man: Effects of bilateral hippocampal, bilateral frontal and unilateral cerebral lesions. *Neuropsychologia* 3: 317–318.

MILNER, B., 1966. Neuropsychological evidence for differing memory processes. Abstract of paper presented XVIII International Congress of Psychology, Moscow, August, 1966.

MILNER, B., S. CORKIN, and H.-L. TEUBER, 1968. Further analysis of the hippocampal amnesic syndrome: 14-year follow-up study of H. M. *Neuropsychologia* 6: 215–234.

NELSON, P. G., 1967. Brain mechanisms and memory. *In* The Neurosciences: A Study Program (G. C. Quarton, T. Melnechuk, and F. O. Schmitt, editors). The Rockefeller University Press, New York, pp. 772–775.

OJEMANN, R. G., 1966. Correlations between specific human brain lesions and memory changes: A critical survey of the literature. *Neurosci. Res. Program Bull.* 4 (suppl.): 1–70.

PENFIELD, W., and B. MILNER, 1958. Memory deficit produced by bilateral lesions in the hippocampal zone. *A.M.A. Arch. Neurol. Psychiat.* 79: 475–497.

PETERSON, L. R., 1966. Short-term verbal memory and learning. *Psychol. Rev.* 73: 193–207.

PETERSON, L. R., and M. J. PETERSON, 1959. Short-term retention of individual verbal items. *J. Exp. Psychol.* 58: 193–198.

PIERCY, M., 1964. The effects of cerebral lesions on intellectual function: A review of current research trends. *Brit. J. Psychiat.* 110: 310–352.

POSNER, M. I., and A. F. KONICK, 1966. Short-term retention of visual and kinesthetic information. *Organ. Behav. Hum. Perf.* 1: 71–86.

QUARTON, G. C., 1967. The enhancement of learning by drugs and the transfer of learning by macromolecules. *In* The Neurosciences: A Study Program (G. C. Quarton, T. Melnechuk, and F. O. Schmitt, editors). The Rockefeller University Press, New York, pp. 744–755.

RUSSELL, W. R., and P. W. NATHAN, 1946. Traumatic amnesia. *Brain* 69: 280–300.

SCOVILLE, W. B., and B. MILNER, 1957. Loss of recent memory after bilateral hippocampal lesions. *J. Neurol. Neurosurg. Psychiat.* 20: 11–21.

SIDMAN, M., L. T. STODDARD, and J. P. MOHR, 1968. Some additional quantitative observations of immediate memory in a patient with bilateral lesions. *Neuropsychologia* 6: 245–254.

SPERLING, G., 1960. The information available in brief visual presentation. *Psychol. Monogr.* 74: (Whole No. 498).

SPERLING, G., 1963. A model of visual memory tasks. *Hum. Factors* 5: 19–31.

SPERLING, G., 1967. Successive approximations to a model for short-term memory. *Acta Psychol.* 27: 285–292.

SPERRY, R. W., 1967. Split-brain approach to learning problems. *In* The Neurosciences: A Study Program (G. C. Quarton, T. Melnechuk, and F. O. Schmitt, editors). The Rockefeller University Press, New York, pp. 714–722.

TALLAND, G. A., 1958. Psychological studies of Korsakoff's psychosis: II. Perceptual functions. *J. Nerv. Ment. Dis.* 127: 197–219.

TALLAND, G. A., 1960. Psychological studies of Korsakoff's psychosis: VI. Memory and learning. *J. Nerv. Ment. Dis.* 130: 366–385.

TALLAND, G. A., 1965a. An amnesic patient's disavowal of his own recall performance, and its attribution to the interviewer. *Psychiat. Neurol. (Basel)* 149: 67–76.

TALLAND, G. A., 1965b. Deranged Memory. Academic Press, New York.

TALLAND, G. A., 1967. Amnesia: A world without continuity. *Psychol. Today* 1: 43–50.

TALLAND, G. A., 1968. Some observations on the psychological mechanisms impaired in the amnesic syndrome. *Int. J. Neurol.* 7: 21–30.

TALLAND, G. A., and M. EKDAHL, 1959. Psychological studies of Korsakoff's psychosis: IV. The rate and mode of forgetting narrative material. *J. Nerv. Ment. Dis.* 129: 391–404.

TALLAND, G. A., W. H. SWEET, and H. T. BALLANTINE, 1967. Amnesia syndrome with anterior communicating artery aneurysm. *J. Nerv. Ment. Dis.* 145: 179–192.

UNDERWOOD, B. J., 1957. Interference and forgetting. *Psychol. Rev.* 64: 49–60.

UNDERWOOD, B. J., 1964. Degree of learning and the measurement of forgetting. *J. Verb. Learn. Verb. Behav.* 3: 112–129.

VICTOR, M., G. A. TALLAND, and R. D. ADAMS, 1959. Psychological studies of Korsakoff's psychosis: I. General intellectual functions. *J. Nerv. Ment. Dis.* 128: 528–537.

WAUGH, N. C., and E. C. HOLSTEIN, 1968. Recall latencies of highly overlearned items. *Psychonomic Sci.* 11: 143.

WAUGH, N. C., and D. A. NORMAN, 1965. Primary memory. *Psychol. Rev.* 72: 89–104.

WARRINGTON, E. K., and L. WEISKRANTZ, 1968. New method of testing long-term retention with special reference to amnesic patients. *Nature* (*London*) 217: 972–974.

WECHSLER, D., 1917. A study of retention in Korsakoff's psychosis. *Psychiat. Bull. N. Y. State Hosp.* 2: 403–451.

WEINER, B., 1966. Effects of motivation on the availability and retrieval of memory traces. *Psychol. Bull.* 65: 24–37.

WICKELGREN, W. A., 1968. Sparing of short-term memory in an amnesic patient: Implications for strength theory of memory. *Neuropsychologia* 6: 235–244.

WICKELGREN, W. A., 1969. Coding, retrieval, and dynamics of multitrace associative memory. Paper prepared for the Fifth Annual Symposium on Cognition, Carnegie-Mellon University, April 3–4, 1969.

WILLIAMS, M., and O. L. ZANGWILL, 1952. Memory defects after head injury. *J. Neurol. Neurosurg. Psychiat.* 15: 54–58.

ZANGWILL, O. L., 1946. Some qualitative observations on verbal memory in cases of cerebral lesions. *Brit. J. Psychol.* 37: 8–19.

# 18 Aspects of Human Intelligence

## HAROLD L. WILLIAMS

IS THE POWER and sweep of man's thought attributable merely to a complex organization of elementary behavioral units such as reflexes, or are there emergent phenomena in his transactions, the understanding of which requires unique concepts and techniques? Optimists among neural and behavioral scientists are convinced that the answer is fundamentally simple:

The history of science suggests that the complexities of the child's real world are probably only multiple combinations of pure and simple processes. Hence the complexity of child behavior *in vivo* may result only from the large number of factors involved; yet each factor by itself may operate in a simple manner, readily and completely understandable in a laboratory setting (Bijou and Baer, 1960, p. 143).

Such psychologists in the tradition of American behaviorism are confident that complex behavioral phenomena can be analyzed into elementary behavioral units (e.g., reflexes), which are induced and modulated by more-or-less complex conditions of stimulation. The radical behaviorist stipulates that man inherits certain tendencies to act and certain adaptive characteristics, but his analyses of behavior focus on the ways in which a responsive and coherent environment gradually shapes behavior. The child is viewed as a reactive organism which acquires habits that are associatively linked to one another and to previously acquired habits. As a result, his behavioral repertoire is cumulative and continuously changing. Development is the quantitative accumulation of learned behavior, and the identification of stages of development is merely a shorthand device for labeling points on a continuous function.

Scientists of a more organismic persuasion, such as Werner, Piaget, and Goldstein, would agree that the development of intelligence depends on continuous and broad-ranging transactions with a richly varied environment, but their explanations of cognitive behavior emphasize the maturation of internalized cognitive structures, rather than environmental variables, as mediating mechanisms. The child is *active* rather than reactive. Even when well fed and rested, the infant creates stimulation by such behaviors as visual inspection of his environment. By acting upon the environment, and observing the results of his actions, he develops his knowledge of reality.

Thus, for the organicists, man's behavior is internally directed and teleological. The child's intelligence matures through a hierarchically arranged, maturationally deter-

HAROLD L. WILLIAMS Department of Psychiatry and Behavioral Sciences, University of Oklahoma Medical Center, Oklahoma City, Oklahoma

mined sequence of stages, each of which is a necessary result of the previous one, each containing emergent qualitative properties, and each defined by a unity of organization that characterizes all the individual's behavior at that stage. The tasks of the organicist are to describe the stages of cognitive development, to identify the processes that cause the developmental sequence, and to explicate the transformational functions that relate a given stage to the preceding one. For a normal child, the rate of progression through the sequence and the terminus of cognitive growth depend on training and a favorable environment, but the order of succession of cognitive stages is believed to be immutable. Thus, a question of fundamental significance for these models of development concerns the degree to which the sequence of cognitive stages can be altered by differences in culture or by the application of principles of behavioral modification. The relatively small set of experiments addressed to this problem has not settled the issue.

This paper presents some contributions from American learning theory, and examines the degree to which stimulus-response (S-R) reinforcement models can deal with such behavior as language, which appears to be hierarchically organized. There follows a brief survey of one organismic model of development, that of Piaget. Certain similarities between Piaget's view of cognitive processes and Lenneberg's analysis of language are then discussed, and compared with recent work by Soviet psychologists, whose ideas about cognitive development seem to be intermediate between the S-R theorists on the one hand and the organicists on the other. Finally, some implications of the three systems for educational practice are discussed.

## American learning theory

Current behavioristic conceptions tend to be problem-oriented rather than theory-oriented, but for problems such as the learning of complex discriminations—concepts, for example—one can distinguish at least two theoretical formulations. There are those that assume a direct association of physical stimuli and overt responses—the single-unit S-R theories (Spence, 1940)—and those that postulate implicit mediating responses with their accompanying response-produced cues—the sequential S-R theories (Lawrence, 1950; Kendler, 1963). American learning theorists often accord prime importance to the implicit verbal response.

Several experiments (Kendler, 1963; Kendler and D'Amato, 1955; Kendler and Kendler, 1956; Kendler et al., 1960) used a two-stage concept formation task, which, in its second stage, permits reversal shift (RS) or nonreversal shift (NRS). Children between three and four responded as do rats (Kelleher, 1956), performing best on NRS, for which mediation is not a necessary inference. But from about age

seven on, children showed superior performance on RS, for which mediation seems to be required. These results suggest that the behavior of very young children still primarily depends on environmental cues, with which relatively simple S-R connections are formed, whereas older children develop internal cues, or chains of responses, some links of which are covert and probably verbal. It is interesting that the transition occurs at an age when children characteristically begin to name relationships between stimuli. Thus, one aspect of the child's progression from animal-like to adult-like human patterns of performance is his ability to apply such terms as "larger-than" and "darker-than" to stimulus relations. It may be these naming systems that mediate his performance on concept-formation tasks. Zeaman and House (1963), however, have proposed and found evidence to support an alternative mediation theory, in which attention is the inferred construct. Before a subject can learn which stimuli to approach and which to avoid, he must learn to make observing responses, which then serve as mediators. Soviet psychologists have proposed similar theories about mediation; they are discussed in another section of this paper.

The simplest of the sequential S-R theories seems to say that performance consists of a chain of associated motor responses, such that a motor event becomes the stimulus for the next motor event, which determines a third, and so on. Can chained habits serve as adequate representations of complex behavior? Lashley (1951), in his famous lecture at the Hixon symposium, was one of the first American psychologists to see this problem in clear perspective. He found it inconceivable that motor activity could take place in orderly fashion without a hierarchical scheme of integration superimposed on a wide range of specific acts. By logical analysis he showed that chains of motor events cannot account for the sequential ordering of behavior. There are two major reasons for this: (1) motor events can happen in such rapid succession that there cannot be sufficient time for impulses to be sent from periphery to brain and back to another muscle; (2) a specific movement, say limb flexion, is part of not just one, but of many different, patterns.

Accurately controlled "whip-snapping" movements of the hand can be executed in less than the shortest times of reaction to tactile stimulation of the arm. Circular movements of the hand, involving coordinated and continuously changing contractions of shoulder, elbow, and wrist can occur in one-tenth of a second, and the finger-strokes of a pianist may reach 16 per second, in any order, too fast for the eye to follow. The various gaits of a horse, or of animals with more than one type of ambulation, involve different timing, different orders of footfalls, different styles of movement; yet the same flexions and extensions are part of each sequence. Lashley's examples suggest that even the simplest motor sequences require some sort of hierarchical plan for their execution.

What are the appropriate units for such behavioral analysis? It has always been possible to divide an action into a sequence of muscle twitches, but most behavioral psychologists have chosen for their datum a molar unit such as the operant response, because these larger response systems co-vary more systematically with changing conditions of stimulation. However, molar units are composed of movements of limbs, trunk, and other parts, which, in turn, are composed of precisely timed patterns of contraction and relaxation of groups of muscles. Surely, then, a complete description of behavior, and its neural substrates, will encompass all levels simultaneously.

That is to say, we are trying to describe a process that is organized on several different levels, and the pattern of units at one level can be indicated only by giving the units at the next higher, or more molar level of description. For example, the molar pattern of behavior X consists of two parts, A and B, in that order. Thus, X = AB. But A, in turn, consists of two parts, a and b; and B consists of three, c, d and e. Thus, X = AB = abcde, and we describe the same segment of behavior at any one of the three levels . . . the complete description must include all levels (Miller et al., 1960, p. 13).

The ethologists have attempted such complete analyses of behavior, and their strategies reveal its hierarchical structure, even in quite primitive animals. For example, Tinbergen (1951) found evidence for hierarchical organization in the reproductive behavior of the male stickleback. Hierarchical concepts have also proved useful for the analysis of behavior at higher phylogenic levels. For example, Lashley showed that, under certain conditions of experimentation, the rat must have organized a flexible, purposive, atemporal strategy for accomplishing a required task. In a famous experiment (1929) he found that rats that had learned a maze could still traverse it even though surgery prevented them from using their habitual motor movements. New motor movements could be substituted into the same general strategy for negotiating the maze. Apparently this freedom of function was accomplished by a shift from response learning to place learning—that is, the acquisition of a "cognitive map," a "plan" (Miller et al., 1960) permitting detours and substitutions.

Perhaps the most obvious representation of hierarchical organization of performance is found in human verbal behavior. Studies by Chomsky (1957), Brown (1965), Lenneberg (1967), and others have shown that sentences are not simply chains of verbal responses. First, a marvelously intricate, very high-speed set of movements of tongue, teeth, lips, oral cavities, thorax, and abdomen, involving something like 100 separate muscles, is programed to produce about 14 phonemes per second. The 15 to 85 phonemes used by the world's languages are programed by rule into morphemes, morphemes are organized by grammatical rules to form sentences, and so on.

The fact that morphemes are arranged in strings encouraged a number of psychologists to view the sentence-generating system as a fairly simple one, a Markov process. Unfortunately, such a simple left-to-right grammar will not do. First, all languages are open, in the sense that an infinite number of sentences can be generated under the syntactical rules. To generate only acceptable sentences, a machine model working from a left-to-right algorithm would require an infinite memory and unlimited experience. Second, sentences can be grammatically embedded in other sentences. Consider, for example, the following sentence, taken from Miller et al. (1960): "The man who said X is here." For X we can substitute "Because Y I am moving to Boulder" and we can let Y be the sentence "If occasionally Z I have allergies," and so on. Because English grammar would permit the construction of indefinitely long sentences composed of such nested dependencies, a left-to-right sentence generator is not adequate to the task.

The fact of nesting suggests a hierarchical organization of syntax. Chomsky (1957) has devised a strategy for hierarchical analysis that resembles the parsing procedure one learned in grammar school. This phrase-structure grammar consists of a set of forming rules for generating sentences such as "The boy hit the ball." Chomsky calls such rules "kernal strings." On top of this is a system of transformations that operates on the kernal strings to combine or permute them into more complex forms. He has shown that sentence generators of the hierarchical type can deal with nested dependencies, which embarrass the left-to-right grammar. It should be made clear that the adult speaker of a language must have an implicit understanding of the forming rules and the transformation rules for grammar if he is to generate understandable messages and recognize acceptable sentences.

Clearly, the simplest of the sequential S-R theories are not equipped to handle the complexity involved in hierarchical organizations of performance. However, there are concepts available to S-R theorists, such as Hull's (1943) habit-family hierarchy, which could probably be extended to handle such complex behavioral systems as language, and some S-R reinforcement theorists are moving in this direction. For example, the cautious, one-stage mediational hypothesis formulated by the Kendlers has been extended by Berlyne (1965) to an elaborate set of hierarchically organized networks of mediating responses that contain links and operations sufficient to account for the complex decision processes underlying language and thought.

The organicists see evidence of hierarchical organization both within and between the stages of cognitive development. Piaget, for example, uses the term "schema" to subsume sets of action sequences that form a recognizable and organized totality (e.g., the schema of sucking) and are organized by a central strategy. "The schema as it appeared to us, constitutes a sort of sensorimotor concept, or more broadly, the motor equivalent of a system of relations and classes" (Piaget, 1952, p. 385). In every cognitive stage, behavior is organized and controlled by interlocking the complexly organized schemata, and two or more schemata can coalesce into one supraordinate system. Thus, as the infant explores his environment, visual orientation becomes coordinated with sucking, hearing, touch, and prehension schemata in such a way that he constructs concepts of objects and relations that modify in a coherent way his further exploration of the world.

In organismic theories, cognitive growth is a process of continuous generalization and differentiation of these action systems through a sequence of stages which are themselves hierarchically organized. For example, in Piaget's system, the concrete operations which appear in the five-to-seven-year age group are performed upon the systems of sensori-motor operations which develop in the first three or four years. The stage of formal operations (the final phase of intellectual growth) "... involves cognitive activities which are performed upon the concrete operations elaborated in the stage just preceding. Concrete operations must precede formal operations in the temporal series, logically as well as psychologically, since the constitution of the former is absolutely necessary to the activation of the latter" (Flavell, 1963, p. 20).

## Piaget's view of cognitive processes

Piaget's astounding contribution to psychology cannot be summarized in a chapter-length paper. Since the 1920s, his publications, alone and with his colleagues at the Geneva Institute, comprise about a third of the world literature on the development of intelligence. Briefly, the Genevan studies usually begin with some aspect of human performance which adults can execute. Materials that pose a problem are provided for the child's manipulation, and by naturalistic observation, by rearranging objects, or by asking questions, the child's concepts, strategies, and techniques for mastering the problem are ascertained. For Piaget, normal adults at their best are eminently rational. At least implicitly, they understand causal relations, the invariance of properties in the presence of transformations, and the logic of classes. Above all, they are able to consider the barely possible—to turn the world upside down, as it were, with their words and their thoughts. To describe the thought of the child and the adult, Piaget has developed qualitative logico-algebraic models that combine certain properties of mathematical groups and lattices. The de-

velopment of intelligence is seen as a march toward these mathematical systems of thought (see Flavell, 1963).

The child is born with the organs and action systems necessary for interaction with his environment, and is endowed with the capacity to construct his own experience and knowledge of himself and his universe. His concepts of reality grow out of his inevitable, always differentiating, always integrating commerce with objects, space, temporal relations, and his community of adults and peers. He is impelled to move forward in this adaptive process because of lack of coherence between his own various action systems (his operative schemata) and new environmental input. Throughout childhood he is constantly engaged in two complementary and simultaneous forms of action: assimilation and accommodation (Piaget, 1950). The function of assimilation is to integrate new data from the environment into his current knowledge and operations. The function of accommodation is to alter his knowledge systems so that they will be consistent with reality. He moves toward a balanced view of himself and the environment by assimilation and accommodation, and, in the long run, his cognitive life will reach a kind of equilibrium. In the short run, however, his equilibrium is constantly perturbed by the results of his actions on his environment.

Three major stages can be distinguished in the cognitive development of the child, and, within each of these, several substages are recognized.

STAGE 1. SENSORIMOTOR OPERATIONS The first major stage culminates at about 18 months, when the child may first construct two-word sentences. Progressive adaptation begins with reflexive behaviors, which gradually develop into organized systems of movements (schemata) of the body and of objects. The child learns that movements made in one direction can be reversed, that a goal can be reached from different starting points or by different routes, that objects that disappear can be found, and that objects have permanence. During the six substages of sensorimotor development, the child generalizes relations between similar objects, discriminates between different objects, and acquires some primitive notions of causality.

STAGE 2. PREOPERATIONAL AND CONCRETE INTELLIGENCE Between two and four the child learns to name things, ask questions, and assert propositions, but he is governed by perception and lack of perspective. Given two identical balls of clay and asked to roll one of them into a long sausage, he asserts that there is now more clay because it is longer, or less clay because it is thinner. His geometry is non-Euclidian, in that space is not yet a stable container in which things occupy specific positions and have specific relations to one another. If he is seated facing a three-dimensional scene and asked to select a photograph representing the view of a doll facing the scene from a different perspective, he attributes to the doll, regardless of its position, his own point of view.

At about seven years, the child develops elementary but logical thought structures. No longer dominated by how things look, he begins to understand elementary principles of class relations, and he acquires the concepts of conservation (invariance) of quantity, weight, and volume.

STAGE 3. FORMAL OPERATIONAL INTELLIGENCE The child of 11 or 12 has the cognitive equipment·for scientific investigation with formal operations. He can work with combinations, permutations, and probability; he can test hypotheses and consider the barely possible. He can perform operations on operations, and understand relations of relations. Analysis of these thought processes convinced Piaget that they approximate formal mathematical structures, such as logical groups and lattices.

### Piaget's stages, cultural or biological?

How much of this sequence is cultural—a phenomenon of Western technology? How much is innate—the result of maturational readout? There is no simple answer to these questions. Although Piaget is very much aware of the problem, his theoretical writings do not specify in detail which aspects of the stage sequence he believes to be innate and which the results of experience. The sensorimotor systems operative at birth are said to be innate, but subject to immediate and continuous modification by experience. Inherited biological structures involved in sensation, perception, and motility condition what we can perceive, limit our actions, and influence the construction of our most fundamental concepts of reality; the rate of maturation of the CNS sets limits on the rate of maturation of intelligence. But the infant's biological endowment consists of more than a set of structures that limit intellectual progress. It includes the species-specific style of adaptation, the *modus operandi* represented by the concepts of assimilation and accommodation. These are the invariant adaptive processes found at birth and in every stage of development. Conflict between these processes is the source of drive for intellectual mastery, and cognitive development consists of their progressive equilibration (Piaget, 1967, pp. 102–114).

Progression through the sequence of stages also depends on broad-ranging transactions between the child and a richly varied environment. Furthermore, the stage of development finally achieved in adolescence is a function of schooling, technology, and language. Yet, Piaget believes that the sequential order of stages is the same across cultures, that a stage cannot be skipped, and that progression through the stages cannot be substantially accelerated by specific tuition (see, for example, Inhelder, 1962, pp. 19–40). The

case for these conclusions is based on the following kinds of facts:

1. Several European and North American replication studies found that, although there were shifts of a year or two in the age of transition, the same developmental sequence took place (Wallach, 1963).

2. Chinese children with no schooling (Goodnow, 1962), rural Mexicans (Maccoby and Modiano, 1966), and children of the African bush (Greenfield, 1966) apparently show the expected sequence at least up to the stage of concrete operations.

3. Socially deprived and mentally retarded children are slowed, but apparently progress through the same developmental sequence (Inhelder, 1944).

4. Vigorous efforts to move the child from one stage to the next have not been successful (e.g., Wohlwill and Lowe, 1962; Smedslund, 1960, 1961a, 1961b; Morf, 1959), and when the child is pushed to a new level of understanding in a given experiment, the effect usually does not show reliable generalization.

5. Deaf preschool children with little or no language apparently progress through the early stages of cognitive development at about the same rate as do normal children (Piaget, 1967, Chapter 3).

Efforts to accelerate Piageten cognitive processes include an excellent series by Smedslund (1960, 1961a, 1961b, and others) on conservation of weight, studies by Morf (1959) on class-inclusion relations, and those by Gréco (1959) on spatial order. None was very successful. But surely it is a paradox that, on the one hand, the course of development depends so much on environmental transactions, and, on the other, that it cannot be speeded up by training. Obviously, we do not know very much about how to potentiate intelligent behavior.

Bruner's (1964) studies suggest ways in which the developmental process might be modified. He agrees that preoperational children are dominated by perception—the iconic stage. He shows, however, that the language and even the logical organization necessary for solving the conservation problems may be available to the perceptually ruled child. When the potency of perceptual cues is reduced, the younger child may solve the problem by logical analysis. The results of experiments by Françoise Frank (Bruner, 1964, 1966) support this view.

The cross-cultural studies and training efforts cited above leave unanswered a number of fundamental questions. Does the relative invariance of the sequence of cognitive stages imply hereditary determination, or can it be explained by cross-cultural equivalence in the nature of spatiotemporal reality? Is there a biological clock that limits the rate at which experience can be converted into internalized logical operations, or could the process be speeded by more sophisticated application of principles of behavioral modification? What *are* the rules that govern the transition from stage to stage, and what specific behavioral criteria determine the onset and termination of a given stage?

Finally, recent analyses by such behavior theorists as Stevenson (1962) and Berlyne (1962) show that the differences between the developmental psychologies of Piaget and current American learning theory may be more apparent than real. Stevenson (1962, p. 114) pointed out that both are historical theories wherein behavior is progressively modified by experience. Both accept the empirical law of effect, the question being what is reinforcing and what is reinforced. Piaget's fundamental adaptive processes, assimilation and accommodation, are similar, conceptually and operationally, to Hull's (1943) three kinds of generalization between stimulus and response, i. e., stimulus generalization, response generalization, and stimulus-response generalization.

An important source of disagreement between the two theoretical systems concerns the action-reaction problem. The question is a fundamental one, because the answer to it determines the role attributed to the child in intelligent responding. It determines the kind of operating characteristics assigned to the child, the kind of model one holds for the child. The question of whether man is an active seeker of stimuli and a director of his own behaviors, or a responder, controlled by environmental stimuli, deserves close study.

## Language

Let us turn now to a brief examination of the development of language and to the relation of language to cognition. Is language an example of instrumental responding, entrained by a responsive environment, or is it a biologically determined, emergent phenomenon, found only in man?

Behavioristic formulations (e.g., Nissen, 1958) have usually assumed that language is a tool, gradually acquired through instrumental learning. Although admittedly complex, language, according to this view, does not introduce really new psychological processes. One can think of it as an instrumental technique that greatly increases the speed and efficiency of processes already present in nonverbalizing animals.

The case presented here for the biological origins of language and its specificity in man relies heavily on the monumental text by Lenneberg (1967) and on his recent article in *Science* (1969), in which he defended the view that language has specific biological foundations, that it develops parallel with brain maturation, and that it is a qualitatively new phenomenon found only in man.

Chomsky, Lenneberg, and Brown have shown that virtually every aspect of all known languages at all levels (phonology, syntax, and semantics) involves the expression of relations. In all languages of the world, words label sets of relational principles rather than specific objects. Lenneberg wrote (1969, p. 640), "Knowing a word is never a simple association between an object and an acoustic pattern, but the successful operation of those principles, or application of those rules that lead to using the word 'table' or 'house' for objects never before encountered." According to this view, communication systems found in other animals do not qualify as predecessors of the language of man. Language is not a pattern of noises or a group of warning signals. Lenneberg concluded that language is a species-specific phenomenon, derived from species-specific properties of man's nervous system.

The last sentence implies that language has strong biological foundations, and Lenneberg asserted such a thesis. He stated that language has the following six characteristics: (1) it is a form of behavior present in all cultures of the world; (2) in all cultures of the world its onset is age-correlated; (3) there is only one acquisition strategy—it is the same for all babies in the world; (4) it is based intrinsically on the same operating characteristics, whatever its outward form; (5) throughout man's recorded history these operating characteristics have been constant; and (6) it is a form of behavior that may be impaired specifically by circumscribed brain lesions, which leave other mental and motor skills relatively unaffected (1969, p. 635). Some of the support mustered for this view is the following:

1. The development of language proceeds through a series of stages, such that each stage is a necessary precursor for the succeeding stage.

2. These stages correlate with measures of physical growth, motor skill, and brain maturation better than with chronological age in both normal and retarded children.

3. Language development is relatively independent of the technological sophistication of a culture.

4. It is also relatively independent of specific forms of tuition. Both Lenneberg and Brown found that children who had not acquired the spontaneous use of certain grammatical rules could not be taught even to imitate sentences formed by such rules. If the language of a child of two years consists only of a small set of single words, he can be taught new single words but he can neither use nor imitate phrases or short sentences.

5. Rather retarded children growing up even in understaffed institutions may pick up an amazing degree of language skill.

6. Normal children of congenitally deaf parents develop language at the expected time and in the usual sequence, even though parents make different sounds from those of the children, and even though during infancy the children's vocalizations have no significant effect upon their parents' behavior.

Lenneberg wrote: "From these instances, we see that

language capacity follows its own natural history. The child can avail himself of this capacity if the environment provides a minimum of stimulation and opportunity. His engagement in language activity can be limited by his environmental circumstances, but the underlying capacity is not easily arrested" (1969, p. 637).

Clearly, the interpretation of these data is debatable. They raise the same sorts of questions about the acquisition of language that were raised about Piaget's view of cognitive development. Does the relative invariance of the sequence of stages across cultures, and across levels of ability and levels of sensory impairment, really imply hereditary determinants, or can they be explained by reference to constant environmental variables? Does the development of language competence in some retardates, whose brains are presumably abnormal, imply an innate basis for language or a constant perceptual environment? One could argue that, if the brain structure is altered, the course of language acquisition should be changed. Does the correlation between the development of language and the growth of physical and motor processes imply a causal connection between biological and linguistic variables? Not necessarily. The relation could mean that both attributes are equally sensitive to environmental influences. And again, what specific criteria determine the onset and end of a given stage, and what are the mechanisms by which a child is propelled from stage to stage? Is the progression of language from stage to stage simply a matter of maturational readout, or is it partly a function of experience?

It is fascinating that Piaget and Lenneberg have each examined similar complex phenomena with rather similar methods, evaluating them against similar criteria, organizing them into stages and hierarchical systems, but have arrived at drastically different etiological positions, with Piaget emphasizing transactions with the environment and Lenneberg stressing biological substrates as the determinants of development. Clearly, the data from cross-cultural comparisons, and from studies of the retarded or the brain-injured, have not sufficient depth or precision to delineate the contributions of innate and experiential factors to the development of either language or cognition. The apparent dichotomy between mechanisms for the acquisition of language and the acquisition of intelligence represented in the two theories may be a function of different theoretical assumptions, rather than different interactions between innate factors and experience.

## Language and cognition

Is the child's maturing intelligence a consequence, a cause, or simply a correlate of his maturing language skills?

When a child of two, who can construct two-word sentences, is compared with a baby of eight months, whose cognitive operations must be inferred from motor and perceptual action, it seems natural to assume that language has profoundly altered his intellectual apparatus by adding thinking to it. Thanks to language, the child can evoke past experiences and free his thinking and behavior from the compelling properties of the current perceptual field. Thanks to language, objects and events are experienced in a conceptual frame of reference, and the child understands something of their underlying properties and relations. In short, if the child's behavior prior to language is compared with his behavior after its onset, it is easy to assume that thought is the consequence of language.

Two approaches to this problem have been either to use the varying features of the world's natural languages as independent variables, examining their effects on aspects of cognition, or to use the presence or absence of language as the independent variable and evaluate intellectual development as a function of the acquisition of language.

For example, Lenneberg and Bastian found that Zuni and Navaho Indians, whose color vocabulary maps differently on the color continuum, did not differ in color discrimination (Lenneberg, 1967, p. 348). More recently, Lenneberg's group has shown that the semantic structure of a given language can have a biasing effect on color recognition, but only under special experimental conditions.

For the second type of analysis, both Lenneberg and Piaget have found that preschool deaf children (with very little language) can perform nonverbal cognitive tasks as well as normal-hearing children can. The extensive literature on performance of the deaf shows very little agreement on this problem, but the better-designed studies, such as those of Oléron (1957), Furth (1961), and Rosenstein (1961), agree with those of Lenneberg and Piaget.

The burden of Piaget's argument for relative independence of language and cognitive development rests on his observations of child performance prior to the acquisition of language. One example of a multitude of similar observations is the following:

Jacqueline, 19 months, watches me when I put a coin in my hand and then put my hand under a coverlet. I withdraw my hand closed; Jacqueline opens it, then searches under the coverlet until she finds the object (Piaget, 1967, p. 99).

Piaget asserts that for the child to perform the sequence he must understand, in action, a kind of transitivity of relations: i.e., the coin was in the hand; the hand was under the coverlet; therefore, the coin is under the coverlet. Such transitivity of actions is a functional equivalent of what, on the symbolic plane, will be transitivity of serial relations, and even class-inclusion relations.

How about concept formation? All languages label their concepts by means of words. Will the presence in the language of a relevant word make a given concept easy to attain? Carroll's (1964) survey of the literature shows that, in most instances, concepts that can be named or easily formulated in a language can be attained more easily than those that cannot. As Lenneberg (1967, p. 356) pointed out, however, in the language of experience there is considerable congruence among the languages of the world. Phenomena that have perceptual and cognitive salience in the environment of man are likely to be referenced in all languages.

What of the opposite view? Is intelligence a necessary and sufficient condition for the development of language? Longitudinal studies of Mongoloid children conducted by Lenneberg et al. (1964) showed that their sequence of language phases was the same as that for normal children. As with normals, developmental processes of language correlated better with measures of biological maturation than with chronological age or IQ. A certain "IQ threshold" was needed for the appearance of language, but, above that threshold, language development had little or no correlation with IQ.

Most cognitive theorists, including Piaget and Lenneberg, would acknowledge—even stress—the enormous importance a shared language has for the manifestations of symbolic thought, but the small research literature concerned with this problem suggests that the development of language is neither the cause nor the consequence of the development of intelligence. But what about their ultimate degree of interdependence? Piaget makes a convincing case that the developmental origins of thought are in the systematic sensorimotor transactions that characterize the child in the first months of life, and most theorists would agree that symbolic processes are possible in the absence of language. On the other hand, language is certainly a necessary condition for the propositions, the implications, the disjunctions that characterize formal operational thought, but is it a sufficient condition? As with the nature-nurture problem, the answer awaits more incisive questions and more powerful methods of analysis.

## Soviet views of cognitive processes

Russian scientists are vigorously engaged in the study of children. In fact, the Russian literature in child psychology accounts for about a third of the world's output (Berlyne, 1963). The brief summary to follow relies mainly on Berlyne's excellent review (1963).

The two major figures in the background of Russian child psychology are Pavlov (the second signal system) and Vygotsky (1962). Modern research apparently represents a coalescence of these two vastly different systems of thought. The second signal system, language, enables us to respond to words as signals of signals. The word "light" can stand for the conditioned stimulus, a light. But such verbal stimuli stand for whole classes of nonverbal stimuli that

have common properties. Thus, words are concepts, and they permit high-level abstraction. There are some important differences between the second and the first signal systems. First, at the verbal level, a powerful stimulus (the unconditioned stimulus) is not required for reinforcement, association being possible by contiguity alone. Second, associations on which the verbal system depends can be formed in a single trial, continuing reinforcement being unnecessary. Third, verbal concepts, despite their stability, can be altered instantly in the presence of new information. Thus, the second signal system is remarkably flexible.

Research on the second signal system examines the means by which behavior is brought under the control of verbal stimuli. Studies by Volkova, summarized by Berlyne, are especially interesting. In a salivary conditioning experiment, with school-age children as subjects, she used the word "ten" as a positive stimulus and the word "eight" as an inhibitory stimulus, and found that, through generalization, such verbal patterns as "5 + 5" or "80/8" became positive stimuli, whereas "4 × 2" and "4 + 4" became inhibitory. Another experiment showed that the words "correct" and "mistake" would generalize to such true or false sentences as "The doctor cures sick people" or "At night the sun shines."

Vygotsky was also concerned with the ways in which speech acquires control of behavior and in the relation between speech and thought. Speech passes through a number of stages before it reaches the status of "the main mechanism of conscious voluntary behavior" (Luria, 1961). At first it serves simply as a stimulus to action—a releasing function. In the young child, once the action has begun, it cannot be inhibited or changed by further speech signals. For example, Luria reported that by the end of the second year a child can respond to verbal stimuli, such as "put on your stockings." But if he is in the process of putting them on, verbal commands will not cause him to reverse the operation. Words can release behavior for which the child is already set, but they do not serve as semantic or discriminative stimuli. For example, if the child is ready to give the experimenter (E) a ball, but the E says "Give me the doll," the child gives E the ball.

Between four and five years, the child begins to respond to the semantic properties of speech. If told to press a bulb twice when a short signal appears and not to press for a long signal, he can perform correctly as long as he repeats the instructions aloud. Between five and six, he learns to carry out fairly complex discrimination tasks in accordance with instructions alone and without overt verbalization. The child's inner verbal operations become the major organizers of his behavior, and, having formulated verbal rules for behavior, he uses them to analyze and orient himself with respect to new incoming information.

Vygotsky's views (1962) of man's intelligence were similar in many ways to those of the organicists. His development of language represents a sudden discontinuity between man and other species. Language is characterized by the fact that its words have meaning; they are concepts. Verbal concepts are not merely conditioned associations. A concept is a complex act of thought involving deliberate attention, logical memory, and abstraction. However, language is not identical with thought, nor does thought grow out of speech, nor speech out of thought. The evolution of language and that of thought are parallel processes which constantly interact with and influence each other.

Vygotsky's theories have had a considerable effect on current Soviet psychology, particularly on the work of Zaporozhets and his colleagues (1965). They study implicit *observing* responses and their modulation by feedback as bases of behavioral mediation. In their view, Piaget's complex action systems and strategies of logical analysis are not sufficient. First, the child must learn to explore the environment, to ask relevant questions, to examine the problem—in short, to orient. With maturation, the increasing effectiveness of problem solving is partly a function of increasingly effective procedures for observing and analyzing the environment.

The child is born with relatively well-developed sensory systems, and within the first days of life he fixates interesting stimuli and traces their movements. By three or four months, he shows preferential orienting to complex forms or to unfamiliar stimuli. The transition to more complex systems of receptor (and manual) activity occurs gradually during the whole preschool period. Studies of the development of visual orienting behavior by Zinchenko and Ruzskaya (summarized by Zaporozhets) illustrate this kind of research. Irregular forms were projected on a screen through which eye movements could be filmed, and children were to look at the screen attentively so that they could recognize the form among other forms later.

In three- and four-year-olds, eye movements were infrequent and fixations were long, most being centered within the figure, and the child made no attempt to traverse its length and breadth or follow its outline. With this primitive technique of exploration, later recognition scores were low.

The eye movements of four- to five-year-olds suggested that they were oriented toward the length and width of the form. Although there were no movements that systematically followed the periphery, there were fixations at certain salient points of the object. These children were more successful at later recognition.

From age five, the child's eyes began tracing the outline of the figure, as if tracing or modeling its form. At first, the movements were halting and redundant, outlining only a portion of the display. Nevertheless, the five-year-olds later found the correct form with few errors. By age seven, the child moved his eyes in a systematic, orderly fashion around the periphery of the object. Having acquired a more efficient method of visual examination, the seven-year-olds could solve more difficult problems with the forms, such as drawing or constructing them.

In the Soviet view, the processes of perception are not developed in isolation. Instead, they develop in the course of the child's organized sensory and motor activity with objects. To build up a reliable image of the situation and of appropriate actions, he must learn to use feedback from the environment and from his own actions. The young child can be taught strategies of visual and manual exploration that considerably advance his problem-solving skills, and, as he acquires images of perceptual inputs and his own actions, he becomes capable of planning and reasoning. Thus, Zaporozhets related his work on observing behavior to the second signal system and thought.

It was suggested earlier that Soviet thinking about cognitive development is intermediate between American S-R theory on one side and organismic theories on the other. Consider, for example, the role assigned to the child in cognitive operations. Like the Kendlers, the Soviets find striking differences between younger and older children in their acquisition of complex discriminations. The child of three requires many trials to build up a discrimination, reversal training takes considerable time, and conditioning, once achieved, is unstable. The young child is essentially a reactive organism whose behavior is brought under the control of a responsive environment by the application of established principles of classical and instrumental conditioning. By age five or six, however, when the child has acquired the second signal system, he can instruct himself, carry out his own intentions, and verbalize what he intends to do and what he is doing. Thus he shifts from reflexive to self-directed behavioral programs, from reaction to action.

Like the organicists, the Soviets assert that some forms of human behavior—language, for example—and complex thought are qualitatively different from any behavior found in other species. Human actions can be voluntary and conscious, and investigation of the properties of conscious behavior is a legitimate scientific enterprise. This analysis begins with the identification of mediating processes. As in S-R theory, implicit verbal responses are considered important mediators, and, as in Piaget's system, internalized strategies for intelligent action can be acquired, but, as pointed out earlier, recent Soviet research has focused on observing behavior, i.e., the orienting response. "For Zaporozhets, the objective equivalent of a conscious attentive process . . . is the occurrence of an orientation reaction" (Berlyne, 1963, p. 178). Thus, while S-R theorists have generally been concerned with laws of response acquisition and control, and Piaget with the organization of cognitive processes (executive functions), the Soviets have concentrated on ways of exploring the stimulus and developing a plan of action for solving the problem. It seems likely that future theories of cognitive development will encompass all three of these points of view.

## Some implications for education

What implications do the studies of the Genevans, the Soviets, the linguists, and the learning theorists have for the educational process? The current views of some pedagogical theorists, many educators, and most middle-class parents seem remarkably optimistic. The radical environmentalist position that has been represented in the enormous Federal programs in education and mental health assumes that the slow learner and the retarded child got that way because they were deprived of optimal experiences and environments for learning. In this view, it would seem that intelligence consists of a system of technologies, problem-solving strategies, and procedures of analysis that can be taught by a responsive and rationally programed environment. The task of the pedagogical sciences is to discover optimal environments and programing technologies, and the task of teachers and parents is to apply these optimal systems to the entrainment of intelligent behavior. Listen to Bruner:

I shall take the view in what follows that the development of human intellectual functioning from infancy to such perfection as it may reach is shaped by a series of technological advances in the use of the mind. Growth depends upon the mastery of techniques and cannot be understood without reference to such mastery. These techniques are not, in the main, inventions of the individuals who are "growing up," they are, rather, skills transmitted with varying efficiency and success by the culture—language being a prime example (Bruner, 1964, p. 1).

Soviet behavioral scientists, although investigating somewhat different aspects of cognition than is Bruner, would find his philosophical position quite consistent with their own. For them, also, the problem is to understand (and train the child to) the strategies and tactics of intelligent behavior. The Soviet system of cognitive research is unique in the world, in that there is such highly systematic collaboration between the laboratory scientists and practical educators that reliable findings from the laboratory are briskly tested in the classroom, and the results of applied investigations are fed back briskly to the laboratory. Therefore, the laboratory scientist, although working from a strong theoretical base, tends to be problem-oriented, and the problems he studies have high social relevance.

The following study by Zankov (1962) is an example of

classroom research in which some of the findings of the Zaporozhets group were apparently replicated.

The inquiry concerned optimal means of combining oral and visual teaching in a botany lesson for a fifth-year class. The class was studying the cell structure of leaves. Questionnaires were administered before and after the lesson to evaluate the progress of pupils taught by two methods.

In method I, the teacher orally guided the pupil in the observations he made as he examined the structural properties of the leaf. The teacher asked questions to which the pupils replied. He said, for example, "Take a good look at the epidermis of the leaf; how are the cells arranged?" "Look at the pulp cells; are they as closely attached to one another as the cells of the epidermis?" and so on. The lesson was not limited merely to teaching pupils about the various leaf tissues. They also learned that light traverses the transparent epidermis of the leaf and reaches the pulp cells, and other aspects and relations.

Method II was a traditional laboratory-lecture procedure. The teacher described certain aspects of the leaf, its perceptible properties and relations, while the children examined their specimens. He might say, "The epidermis of the leaf is composed of closely adhering cells. The pulp cells are arranged in several layers, which are spaced out and separated from one another by intercellular spaces."

As the reader has already guessed, method I was the more successful. The postinstruction questionnaire showed that, with method I, the pupils had learned a good deal more about leaves than had the lecture group. Other subjects, such as geography and history, were also learned better under method I. One hopes that the obvious differences in level of language, and the apparent differences in the pacing of teaching, between the two methods were introduced by Zankov's translator and were not features of the studies.

Most good teachers would agree that a dialogue between teacher and pupil is more effective than a lecture and that it is useful to teach the pupil optimal tactics for examining a problem. Given reasonable motivation to learn, a reversal of the Zankov results would be startling.

But a question of deepest significance for both Soviet and American society is whether an enriched environment and improved educational technology can accelerate the growth of intelligence. Obviously, neither Piaget nor Lenneberg would expect any strategy of teaching to have much of a potentiating effect on the growth of intelligence or the acquisition of language. That is, the child must progress through each developmental stage before he can achieve its successor. If he has only two-word sentences he cannot be taught to program longer sentences until he is ready. If he is preoperational, he cannot be taught class-inclusion relations. His brain is not ready.

As mentioned earlier, several psychologists have used

recitation, reinforcement, and feedback procedures in an effort to accelerate the growth of understanding. Some have reported modest success, but the better-designed studies usually have not. Flavell stated the problem well.

Almost all the training methods reported impress one as sound and reasonable and well-suited to the educative job at hand. And yet most of them have had very little success in producing cognitive changes. It is not easy to convey the sense of disbelief that creeps over one in reading these experiments. It can be hard enough to believe that children systematically elect nonconservation in the first place; it is more difficult still to believe that trial after trial of carefully planned training is incapable of budging them from this aberrant position. Further, there is more than a suspicion from present evidence that when one does succeed in inducing some behavioral change through this or that training procedure, it may not cut very deep (Flavell, 1963, p. 377).

The preoperational cognitive structures apparently resist short-term training procedures, which seems to imply that there is considerable developmental reality about these structures. As Flavell pointed out, the relative failure of training procedures derived from American learning theory confers a kind of validity on Piaget's assertions.

Does Piaget mean that there is nothing to do but sit by and wait for intelligence to unfold? Certainly not. On scientific grounds, not just out of ethical considerations, he would not approve of deprived, impoverished environments. The child is born to learn, but stable knowledge of one's self and one's universe can develop only through frequent, widely varied, and meaningful transactions with the objects, spaces, perspectives, and relations of the environment. For a child who is permitted no commerce with the environment, growth beyond the earliest phases of the sensorimotor stage is impossible.

Since his first publications in the 1920s, Piaget has stressed the importance of peer transactions for intellectual growth. We become socialized, give up animistic notions, develop perspectives, and rationalize causality through the perturbations which occur in interaction with our peers. The preoperational child does not learn especially well from most adults, because adults usually cannot empathize with his cognitive situation. He learns best by comparing his ideas, his convictions, his morality, and his perspectives with those of others whose stage of development is close to his own.

In summary, the aspects of cognitive behavior with which Piaget and Lenneberg are concerned will probably not be accelerated by total-push education programs based on radical environmental models. There is something deeply developmental, even biological, about these maturing processes. Furthermore, education alone can never make a scientist of a child whose retardation has a biological basis. On the other hand, the development of biologically normal children can be slowed, even stopped, in sufficiently impoverished environments.

Perhaps it is worth saying that school performance and cognitive development are two entirely different matters. The poor learner may indeed be retarded, but, as Zigler and his colleagues' excellent studies (e.g., Zigler, 1963; Butterfield and Zigler, 1965; Gruen and Zigler, 1968; Zigler et al., 1968; Zigler and Butterfield, 1968) have shown, poor performance, even among the biologically retarded, is very often a form of underachievement. Poor performance may result from lack of motivation, lack of experience, lack of training, a history of failure, fear of adults, different goals, anxiety, physical illness, or hunger. Thus, therapeutic approaches based on environmentalist persuasions do have potential value in the management of school performance, but the reasons for poor performance are manifold, and they differ from child to child. Piaget would agree with those who argue that, for each child, the multiple blocks to optimal performance must be diagnosed and removed. Then, within his intellectual limits, the child can be expected to undertake with vigor the business of learning about his world.

## Acknowledgment

The writer is grateful to Dr. Kathryn L. West for her help in the preparation of the manuscript.

REFERENCES

BERLYNE, D. E., 1962. Comments on relations between Piaget's theory and S-R theory. In Thought in the Young Child (W. Kessen and C. Kuhlman, editors). Monographs of the Society for Research in Child Development, Serial No. 83, Vol. 27, No. 2, Antioch Press, Yellow Springs, Ohio.

BERLYNE, D. E., 1963. Soviet research on intellectual processes in children. In Basic Cognitive Processes in Children (J. C. Wright and J. Kagan, editors). Monographs of the Society for Research in Child Development, Serial No. 86, Vol. 28, No. 2. University of Chicago Press, Chicago, Illinois.

BERLYNE, D. E., 1965. Structure and Direction in Thinking. John Wiley and Sons, New York.

BIJOU, S. W., and D. M. BAER, 1960. The laboratory-experimental study of child behavior. In Handbook of Research Methods in Child Development (P. H. Mussen, editor). John Wiley and Sons, New York, pp. 140–197.

BROWN, R., 1965. Social Psychology. The Free Press, New York.

BRUNER, J. S., 1964. The course of cognitive growth. Amer. Psychol. 19: 1–15.

BRUNER, J. S., 1966. On the conservation of liquids. In Studies in Cognitive Growth (J. S. Bruner, R. R. Olver, P. M. Greenfield, et al., editors). John Wiley and Sons, New York, pp. 183–207.

BUTTERFIELD, E. C., and E. ZIGLER, 1965. The influence of differing institutional social climates on the effectiveness of social reinforcement in the mentally retarded. *Amer. J. Ment. Defic.* 70: 48–56.

CARROLL, J. B., 1964. Language and Thought. Prentice-Hall, Englewood Cliffs, New Jersey.

CHOMSKY, N., 1957. Syntactic Structures. Mouton, The Hague and Paris.

FLAVELL, J. H., 1963. The Developmental Psychology of Jean Piaget. Van Nostrand Company, Princeton, New Jersey.

FURTH, H. G., 1961. The influence of language on the development of concept formation in deaf children. *J. Abnorm. Soc. Psychol.* 63: 386–389.

GOODNOW, J. J., 1962. A test of milieu differences with some of Piaget's tasks. *Psychol. Monogr.* 76, no. 36 (whole no. 555).

GRÉCO, P., 1959. L'apprentissage dans une situation à structure opératoire concrète: Les inversions successives de l'ordre linéaire par des rotations de 180°. *In* Apprentissage et Connaissance. Études d'Épistémologie Génétique (P. Gréco and J. Piaget, editors). Presses Universités, Paris, France, vol. 7, pp. 68–182.

GREENFIELD, P. M., 1966. On culture and conservation. *In* Studies in Cognitive Growth (J. S. Bruner, R. R. Olver, P. M. Greenfield, et al., editors). John Wiley and Sons, New York, pp. 225–256.

GRUEN, G., and E. ZIGLER, 1968. Expectancy of success and the probability learning of middle-class, lower-class, and retarded children. *J. Abnorm. Psychol.* 73: 343–352.

HULL, C. L., 1943. Principles of Behavior. Appleton-Century-Crofts, New York.

INHELDER, B., 1944. Le diagnostic du raisonnement chez les débiles mentaux. Delachaux et Niestlé, Neuchâtel.

INHELDER, B., 1962. Some aspects of Piaget's genetic approach to cognition. *In* Thought in the Young Child (W. Kessen and C. Kuhlman, editors). Monographs of the Society for Research in Child Development, Serial No. 83, Vol. 27, No. 2. Antioch Press, Yellow Springs, Ohio.

KELLEHER, R. T., 1956. Discrimination learning as a function of reversal and nonreversal shifts. *J. Exp. Psychol.* 51: 379–384.

KENDLER, H. H., and M. F. D'AMATO, 1955. A comparison of reversal shifts and non-reversal shifts in human concept formation behavior. *J. Exp. Psychol.* 49: 165–174.

KENDLER, H. H., and T. S. KENDLER, 1956. Inferential behavior in preschool children. *J. Exp. Psychol.* 51: 311–314.

KENDLER, T. S., 1963. Development of mediating responses in children. *In* Basic Cognitive Processes in Children (J. C. Wright and J. Kagan, editors). Monographs of the Society for Research in Child Development, Serial No. 86, Vol. 28, No. 2. University of Chicago Press, Chicago, Illinois.

KENDLER, T. S., H. H. KENDLER, and D. WELLS, 1960. Reversal and nonreversal shifts in nursery school children. *J. Comp. Physiol. Psychol.* 53: 83–88.

LASHLEY, K., 1929. Brain Mechanisms and Intelligence. University of Chicago Press, Chicago, Illinois.

LASHLEY, K., 1951. The problem of serial order in behavior. *In* Cerebral Mechanisms in Behavior (L. A. Jeffress, editor). John Wiley and Sons, New York, pp. 112–136.

LAWRENCE, D. H., 1950. Acquired distinctiveness of cues: II.

Selective association in a constant stimulus situation. *J. Exp. Psychol.* 40: 175–188.

LENNEBERG, E. H., 1967. Biological Foundations of Language John Wiley and Sons, New York.

LENNEBERG, E. H., 1969. On explaining language. *Science (Washington)* 164: 635–643.

LENNEBERG, E. H., I. A. NICHOLS, and E. F. ROSENBERGER, 1964. Primitive stages of language development in mongolism. *In* Disorders of Communication (D. M. Rioch and E. A. Weinstein, editors). The Williams and Wilkins Company, Baltimore, Maryland, pp. 119–137.

LURIA, A. R., 1961. The Role of Speech in the Regulation of Normal and Abnormal Behavior. Liveright, New York.

MACCOBY, M., and N. MODIANO, 1966. On culture and equivalence: I. *In* Studies in Cognitive Growth (J. S. Bruner, R. R. Olver, P. M. Greenfield, et al., editors). John Wiley and Sons, New York, pp. 257–269.

MILLER, G. A., E. GALANTER, and K. H. PRIBRAM, 1960. Plans and the Structure of Behavior. Henry Holt and Co., New York.

MORF, A., 1959. Apprentissage d'une structure logique concrète (inclusion): Effets et limites. *In* L'Apprentissage des Structures Logiques. Études d'Épistémologie Génétique, Vol. 9 (A. Morf, J. Smedslund, Vinh-Bang, and J. F. Wohlwill, editors). Presses Universités, Paris, France, pp. 15–83.

NISSEN, H. W., 1958. Axes of behavioral comparison. *In* Behavior and Evolution (A. Roe and G. G. Simpson, editors). Yale University Press, New Haven, Connecticut, pp. 183–205.

OLÉRON, P., 1957. Studies of the Mental Development of Deaf-Mutes. Monograph, Centre National de la Recherche Scientifique.

PIAGET, J., 1950. Psychology of Intelligence. Harcourt, Brace and World, New York.

PIAGET, J., 1952. The Origins of Intelligence in Children. International Universities Press, New York.

PIAGET, J., 1967. Six Psychological Studies. Random House, New York.

ROSENSTEIN, J., 1961. Perception, cognition and language in deaf children (a critical analysis and review of the literature). *Except. Children* 27: 276–284.

SMEDSLUND, J., 1960. Transitivity of preference patterns as seen by pre-school children. *Scand. J. Psychol.* 1: 49–54.

SMEDSLUND, J., 1961a. The acquisition of conservation of substance and weight in children. I. Introduction. *Scand. J. Psychol.* 2: 11–20.

SMEDSLUND, J., 1961b. The acquisition of conservation of substance and weight in children. II. External reinforcement of conservation of weight and of the operations of addition and subtraction. *Scand. J. Psychol.* 2: 71–84.

SPENCE, K. W., 1940. Continuous versus non-continuous interpretations of discrimination learning. *Psychol. Rev.* 47: 271–288.

STEVENSON, H. W., 1962. Piaget, behavior theory, and intelligence. *In* Thought in the Young Child (W. Kessen and C. Kuhlman, editors). Monographs of the Society for Research in Child Development, Serial No. 83, Vol. 27, No. 2. Antioch Press, Yellow Springs, Ohio.

TINBERGEN, N. 1951. The Study of Instinct. Oxford University Press, Oxford.

VYGOTSKY, L. S., 1962. Thought and Language. (English translation.) M. I. T. Press, Cambridge, Massachusetts.

WALLACH, M. A., 1963. Research on children's thinking. *In* National Society for the Study of Education, 62nd Yearbook. University of Chicago Press, Chicago, Illinois, Part I, pp. 236–276.

WOHLWILL, J. F., and R. C. LOWE, 1962. Experimental analysis of the development of the conservation of number. *Child. Develop.* 33: 153–167.

ZANKOV, L. V., 1962. Enseignement oral et enseignement visuel. *Cited in* The Psychology of the Use of Audio-Visual Aids in Primary Education (G. Mialaret, editor). George G. Harrap and Company, London, 1966.

ZAPOROZHETS, A. V., 1965. The development of perception in the preschool child. *In* European Research in Cognitive Develop-ment (P. H. Mussen, editor). Monographs of the Society for Research in Child Development, Serial No. 100, Vol. 30, No. 2. University of Chicago Press, Chicago, Illinois.

ZEAMAN, D., and B. J. HOUSE, 1963. An attention theory of retardate discrimination learning. *In* Handbook in Mental Deficiency: Psychological Theory and Research (N. R. Ellis, editor). McGraw-Hill, New York, pp. 159–223.

ZIGLER, E., 1963. Social reinforcement, environmental conditions and the child. *Amer. J. Orthopsychiat.* 33: 614–623.

ZIGLER, E., D. BALLA, and E. C. BUTTERFIELD, 1968. A longitudinal investigation of the relationship between preinstitutional social deprivation and social motivation in institutionalized retardates. *J. Pers. Soc. Psychol.* 10: 437–445.

ZIGLER, E., and E. C. BUTTERFIELD, 1968. Motivational aspects of changes in IQ test performance of culturally deprived nursery school children. *Child Develop.* 39: 1–14.

# 19 The Ontogeny of Behavior

JUAN D. DELIUS

## Ethology

ETHOLOGY ORIGINATED with Konrad Lorenz's work in the 1930s, but did not have widespread impact until the early 1950s, and then mainly through the writings of N. Tinbergen, a student and collaborator of Lorenz's. In Tinbergen's *The Study of Instinct* (1951), ethology was revealed as a new, widely ranging theory of behavior, the cornerstone of which was the assertion, forcefully supported with evidence, that much of an animal's behavior is instinctive, innate, inherited. That proposition was in marked opposition to the then-prevalent view among American animal psychologists that most, if not all, adaptive behavior was acquired through conditioning. This difference led to a renewal of the hoary controversy of nature versus nurture. Much of that controversy seems to have been based on semantic misunderstandings (Lehrman, 1970); therefore a review is unnecessary. It must be emphasized, however, that the original ethological standpoint on the issue has undergone considerable modification and differentiation since *The Study of Instinct* was published, as have most other early theoretical views of ethology (Hinde, 1966; Marler and Hamilton, 1966). Nonspecialists often seem unaware of this change, but it is discernible even in the writings of the more conservative ethologists (Lorenz, 1965; Eibl-Eibesfeldt, 1967). There now seems to be a smooth cline of opinions, ranging from the original, extreme, nativistic ethological theory to the most radical behaviorism (Skinner, 1966). The present paper reflects an intermediate position, which can also be characterized, I believe, as highly pragmatic.

It is probably correct to say that ethology is no longer a theory of behavior so much as an attitude toward the study of behavior. It may be characterized as firmly biological, wherein ecology, evolution, genetics, ontogeny, and physiology are applied to the analysis of every class of behavior in all animal species, including man. In pursuing this broad approach, ethologists are not reserved about making use of whatever methods, facts, and theories they find useful, whether or not their origin is ethological. In their view, three fundamental questions can be asked about behavior; they concern its phylogeny, ontogeny, and physiology, and although the answers are often interrelated, ethologists have found them heuristically useful and are most insistent that the answers to all these questions are of equal importance for a true understanding of behavior. It is, perhaps, their insistence on this that most clearly sets them apart from other schools of behavioral studies.

JUAN D. DELIUS  Department of Neurosciences, University of California at San Diego, La Jolla, California. *Present Address:* Department of Psychology, University of Durham, Durham, England

## Behavior genetics

The concept of instinct, as originally proposed by ethologists, was complex. Instinctive behavior was considered to be innate, stereotyped, and species-specific, and a hypothetical, quasi-physiological mechanism was postulated as responsible for its occurrence. We now know that these attributes are independent of one another (Moltz, 1965). Only the first characteristic concerns us here. Innate has generally been equated with "inherited," although there is, perhaps, a case for distinguishing between them in the sense of their every-day meaning. Essentially, however, the question is whether the genetic material is capable of determining at least some patterns of behavior. Evidence that it is indeed so has accumulated rapidly over the last 10 to 15 years through studies on the genetics of behavioral traits (Fuller and Thompson, 1960; Hirsch, 1967). All sufficiently investigated traits have proved to be under some measure of genetic control. These studies have also highlighted the complexity of the gene-environment interactions that determine behavioral characters, to the extent that the inherited-learned dichotomy now appears irrelevant, as illustrated by the demonstration of Tryon and his followers that even learning performances of mice are affected by genetic factors.

From a population of laboratory mice, he chose good and poor maze learners and bred them separately, selecting from each successive generation the best and worst learners, respectively. After some eight generations of selection, the distribution of maze-performance scores of the two strains did not overlap, even when they were bred without further selection, i.e., when they bred true. Hybrids were intermediate (Tryon, 1942). Maze-learning scores, however, reflect a complex interaction of many behavioral parameters, such as locomotor activity, sensory capabilities, and so on, besides learning, *sensu stricto,* however that might be defined. Genetic factors may affect any of these. Detailed studies on the Tryon strains have shown that their brightness and dullness are the result of a number of such interdependent characteristics, some of which can even be demonstrated in nonlearning situations. It would be extremely difficult to make a statement of precisely what behavior is inherited.

Attempts at a genetic analysis have led to the conclusion that many gene loci are involved in controlling the maze-learning abilities of these strains of mice. Little, therefore, can be learned about the mode of action of the genes that give rise to these behavior syndromes (Fuller and Thompson, 1960).

More promising in this respect are studies on the learning capabilities of various, highly inbred, strains of mice that originally were bred for purposes other than learning experiments (Bovet et al., 1969). Using an active-avoidance paradigm in a shuttlebox, Bovet and his colleagues obtained learning curves for individuals of a normal, unselected breed of mice and of various inbred strains. Whereas the curves of the former scatter widely, those of the latter show little interindividual variance, but from strain to strain there are marked differences, some consistently giving rise to slow learners, others to fast ones, and so on. Clearly, the genetic homogeneity of the inbred strains and the genetic differences among them are responsible for the reduced variance and the differences in learning performance. Again, these differences may also result from the effect on performance by the genes through behavioral characteristics other than the learning process proper—for example, through different sensitivities to shock, differences in locomotion, and the like. This possibility is partially weakened by the finding that the learning performance of the various strains in the shuttlebox correlates highly with the performances in a maze-learning task.

A more detailed analysis of the learning differences among some of the strains is of interest. This involves a comparison of their performances in sessions of massed learning trials alternating with extended intersession intervals. The performance improvements within the sessions were comparable for both a poor and a good learning strain, but during the intervals the former showed drastic decrements and the latter showed continuing performance increments. This suggests, and other similar experiments support it, that the poor learners are deficient in memory consolidation although they are normal in terms of short-term memory. If a postsession convulsive shock was administered to the better-learning mice to prevent consolidation of short-term memory traces, the animals could be brought to show performances similar to those of the mice that were poor learners.

These studies suggest that the genetics of learning may be a powerful tool for investigating the neurophysiology and biochemistry of learning, particularly if it proves possible to develop strains that differ only in terms of genes that control specific components of the learning process.

Beyond doubt, virtually all behavioral characteristics are under some degree of genetic control, but we know little about the underlying causal processes, which is, perhaps, not surprising in view of the as-yet poorly understood mode of gene action in neurogenesis (Edds, concluding remarks, this volume). Studies on phenylketonuria go farthest in analyzing the biochemical pathways in the genetic determination of a behavioral trait. Patients with this abnormality have an extremely low intelligence, are tantrum-prone, and have an abnormally high urinary excretion of phenylpyruvic acid. Genetic analysis of family trees of phenylketonurics, recently facilitated by the possibility of recognizing heterozygotes, makes it virtually certain that phenylketonuria is caused by the homozygous presence of a single autosomal recessive gene. It is well established that the gene acts by causing a deficiency of phenylalanine hydroxylase, thereby blocking the conversion of phenylalanine into tyrosine, which in turn leads to the utilization of an alternative metabolic pathway, resulting in the excretion of phenylpyruvic acid. The behavioral effects, however, are said to be caused by the high level of phenylalanine, which interferes with the hydroxylation of tryptophan, thus causing the accumulation of abnormal levels of hydroxytryptophan. The further causal sequence is obscure (Hsia, 1967, 1969; Sourkes, 1962). At any rate, beyond some tendency for microcephaly, no consistent morphological correlates of phenylketonuria have as yet been found.

It is interesting to note that a diet lacking phenylalanine given to phenylketonurics from birth prevents the behavior defects. This indicates the risk of characterizing phenylketonuria as a purely inherited trait, for under certain environmental conditions it can fail to develop in spite of the corresponding gene constitution. Further, it has been suggested that the diet can be discontinued, once the neural development phase has terminated, without adversely affecting the patient's behavior. If so, it would be an example of the general principle that environmental conditions tend to have profound and irreversible behavioral consequences early in development, although they may have no, or only transitory, effects on the adult. This principle is further illustrated by the finding that genetically nonphenylketonurics born of phenylketonuric mothers are of subnormal intelligence, because in utero they develop under high levels of phenylalanine, even though they have normal levels postpartum.

The studies on the determination of sexual behavior, which Goy describes in this volume, are a further and striking example of the complex pathways of gene action on behavior and their interaction with environmental factors.

## Early environment

The action of environmental factors on behavioral development is often considered to be of two types: organic and behavioral (Anastasi, 1958). To the extent that genes depend on a physicochemical environment to be able to express themselves in neural structure and neurobiochemical dynamics, and that variations of that environment may modify their expression, we can talk of organic factors. To the extent that an individual's own behavior and that of the surrounding population are instrumental in modifying that individual's behavior on a long-term basis, we can speak of behavioral factors. In the final analysis, all behavioral factors must have organic consequences in the above sense; so, although convenient, this classification (as with many others in biology) cannot be applied too rigorously. In most instances, environmental influences on behavioral develop-

ment are, in any case, composite and just too complex to be encompassed by two adjectives, as some of the following examples illustrate.

Anoxia, toxins, and drugs affecting a pregnant mother are all acknowledged risks for the normal behavioral development of the embryo, even when they do not disrupt or only temporarily disrupt the mother's behavior. Differences in the intra-uterine position of members of monozygotic twins often lead to considerable intertwin differences in behavioral characteristics, notably in their intelligence, although they are genetically identical (Vandenberg, 1965). The reflexes of newborn babies are also modified by their previous uterine position. Babies born in the normal presentation, for example, have both a mild leg-extension reflex when a surface is placed gently against the sole of the foot and a mild leg-withdrawal reflex to a plantar pinprick. In babies in a breech delivery, the extension reflex is virtually absent, but the withdrawal reflex is strongly enhanced. Phenomena related to habituation and facilitation are probably at play (Prechtl and Knol, 1958).

Even more subtle differences in uterine environment can have permanent behavioral consequences. It has been demonstrated that rats subjected to stress early in life will give birth to young which, when adults, score high in a test devised to measure emotionality, even if their foster mothers were normal, in contrast to young born to, and fostered by, normal females. Probably the effect is due to hormonal differences between the two types of mothers. In the same experiment, it was shown that the postnatal environment provided by stressed foster mothers, as compared with normal foster mothers, does raise the emotionality of young born to normal mothers (Denenberg and Whimbey, 1963). Similar differences in adult emotionality can be brought about by the handling or not handling of baby rats, so it seems likely that differences in the behavior of stressed and nonstressed mothers are responsible, although one cannot exclude the possibility that hormonal factors are transmitted via the milk. At any rate, the handling experiments show that behavioral factors can, in some instances, have effects similar to organic factors (Joffe, 1969). Incidentally, handling of juveniles has been shown to have diverse effects in different genetic strains of mice. In some strains it leads to increased emotionality; in others handling decreases it. This warns against any generalization—across species, say— about the effect of environmental factors (Ressler, 1966; King and Eleftheriou, 1959).

An important type of influence of early environment on behavioral ontogeny is imprinting, the celebrated contribution of ethologists to the gamut of learning phenomena. A number of avian species and, in a less drastic way, some mammals become durably attached socially to whatever object they are exposed to within a limited period soon after birth. The attachment occurs rapidly and in the absence of obvious reinforcement. Outside the critical period, such attachment cannot easily be obtained. Once established, social preferences are difficult, if not impossible, to reverse. They last through the parental phase and influence the choice of sexual partners in later life, although here experience during adolescence may have modifying effects, at least in some species. Imprinting research is in a state of explosive development, and only some of the recent findings can be sketched here.

Some experiments indicate that the end of the critical period within which imprinting is possible may be due to a saturation-like phenomenon. A storage space set aside for the memorization of the attachment object is filled, during the imprinting period, with information about a suitable imprinting object or, if such an object is withheld, with information about incidental stimuli, such as the walls of the home cages. After that, the store is refractory to further information (Bateson, 1966).

It is also clear that imprinting does not surprise the newly born animal as a *tabula rasa*. It has been demonstrated (Gottlieb, 1966) that ducklings have strong preferences for the parental call of their species, which does not depend on prehatching experience with the call. Although it might be inviting to label this preference as innate, the vocalizations uttered by the embryos themselves during the few last days before hatching enhance the preference markedly. This cannot be labeled learning, because the preference enhanced is not toward the call they had experienced, but toward one they have never heard.

Similarly, a brief first exposure to an imprinting stimulus does not necessarily bind domestic chicks to it; in subsequent exposures they can express visual preferences for other stimuli with which they have had no previous experience, even though these preferences cannot be demonstrated before the first exposure. It is as if the first experience is necessary to awaken these latent preferences (Kilham et al., 1968). Can one call this preference innate if, in fact, it requires experience before it can become operant? Or can it be called learned if it refers to stimuli the animals have not experienced?

In dimorphic species of ducks, i.e., species in which the plumage of males and females is different, sexual imprintability is also different. Sexual choices of adult males are strongly affected by imprinting experiences in early life; those of females are far less so (Schutz, 1965). This suggests that perinatal sexual hormone levels gate the imprinting mechanism differentially. Whatever the mechanism, it would be clearly unadaptive if females became sexually imprinted on their (female) mothers. If females have accessible information, necessarily genetically coded, about the appearance of the males of their species without imprinting, why do the males not have similar information about the females? The answer, evolutionarily speaking, may lie in

the fact that males are distinctively colored, whereas the females (at least of several duck species) are inconspicuous and resemble one another. Much more subtle discriminations are necessary to distinguish them. Correspondingly, the mechanisms mediating these discriminations must be more complex; perhaps their specification through genetic information requires an inordinate amount of genetic space that cannot be afforded.

That early social experience has long-lasting effects on later social behavior, in ways other than those brought about through imprinting, has been impressively demonstrated in studies of the behavioral development of rhesus monkeys. Females raised in isolation with artificial "mothers" later become poor mothers themselves, as compared with females raised normally. They tend not to suckle or hug their young, but rather to react aggressively toward them. Other spheres of social behavior, notably sexual, are similarly disturbed by a lack of early social experience. Experience with peers, but not necessarily with parental animals, seems sufficient for a normal behavioral development (Harlow and Harlow, 1965; Mason, 1965).

A normal social environment also provides opportunities for the transmission of what is analogous to cultural traditions. Until the occasion on which a young female monkey learned by trial and error to wash sweet potatoes before eating them, all members of the semi-wild, isolated colony of Japanese macaques to which she belonged always ate them covered with sand. The washing habit spread rapidly to most of the colony members. Observations indicated that the other juvenile monkeys learned through observation and subsequent imitation. Other populations of the same monkeys never acquired this potato-washing culture (Kawai, 1965). Incidentally, at one time it was thought that observational learning did not take place in animals; it has now been shown to occur fairly regularly in several species under controlled laboratory conditions (John et al., 1968). Song learning in passerine birds, as discussed below, is an example, and local song traditions or "dialects" are well-documented phenomena.

The previous examples illustrate the variety of processes and phenomena involved in the ontogeny of behavior, but the development of song in birds provides the best evidence that essentially identical behavioral performances can arise through quite different ontogenetic processes in different species (Konishi and Nottebohm, 1969; Nottebohm, 1970).

Individuals of some nonpasserine species—chickens and doves, for example—can develop their species-typical vocalizations without ever having heard them. Furthermore, they develop normal adult calls even if they have been deafened shortly after hatching, that is, before they themselves have uttered any of the adult calls in question. In these species, then, genetic information specifies directly the development of motor mechanisms responsible for the generation of species-specific vocalizations.

In some passerine species, including chaffinches and white-crowned sparrows, normal song will develop only if the young have had the opportunity to hear their species song. Relatively few exposures to it are sufficient. Note that, while this information is being stored, the young birds are not yet capable of uttering the relevant vocalizations; only when fully grown will they give evidence of having learned. If these birds are deafened after exposure to the parental song, but before they have reached the age at which they themselves sing, their song develops in a highly abnormal way. If, however, deafening is performed after they have gone through the song-development phase and are singing the species-specific song, it has no effect whatsoever on the song. In brief, these birds, guided by auditory feedback, appear to learn to match an auditory template they have acquired early in life by hearing their parents. Once the acquired auditory template is transferred to a motor mechanism through the self-learning process, the latter is independent of auditory feedback.

A third, intermediate type of song development appears to occur in other passerines—the song sparrow, for example. Although the young never need to hear other members of their species, they must be able to hear themselves to attain a normal, species-specific song. This suggests that, although genetic information is capable of specifying the development of an auditory song expectation or template, the motor mechanism for the correct song can be arrived at only through a trial-and-error, self-learning process that requires auditory feedback.

There is suggestive evidence that the young, of those birds that must learn their song from adults, selectively and without any pre-training pick up the songs of their own species from a range of other songs they may hear. This indicates that they must have available genetic information leading to the development of mechanisms that insure a selective storage of the correct song. In a way, then, these birds are genetically programed as to what they will learn (Thorpe, 1961).

Bird vocalizations can be obtained by electrical stimulation of widely distributed telencephalic and hypothalamic sites, and also from or near auditory nuclei of the mesencephalon and the thalamus (Brown, 1969; Delius, 1970). Does this proximity of sensory and motor systems reflect the necessity of transferring a sensory template to a motor mechanism during development? The possibility that the co-extension is related to an auditory monitoring of motor output seems unlikely, because, once it has been established, species-specific song is accurately maintained for indefinite periods by deafened birds. It might be interesting to examine these auditory structures for changes in detailed histology,

unit physiology, and biochemistry, concurrent with the template acquisitions and transfers discussed above.

When I say that genetic information is capable of specifying a motor mechanism or a sensory template—or, rather, the neural network that represents them—I do not mean to imply that they do not themselves have a developmental history. Clearly they must, and further study of their ontogeny might well bring out unexpected interactions between genes and environment, possibly to the extent that, under certain environmental conditions, the genes are actually incapable of specifying the relevant mechanisms and learning processes may be required for implementing them.

Even under conditions in which genes do specify the development of such template mechanisms, learning is not necessarily excluded. One can imagine that, as a logical extension of the song self-learning just mentioned, a neural network could come to match a performance required by another through a trial-and-error process that reinforces successful matchings. In terms of conventional learning theory, such intraneural learning requires only postulating that reinforcement can take place as a consequence of neural activity patterns in the absence of overt behavior—a conception that seems reasonably realistic when one considers the development of human thinking (Williams, this volume).

Is it possible to assess the contribution of inheritance and environment on a single scale in spite of the variety of interactions that operate in behavior development? This question is like asking to what extent salt is the result of either the sodium or the chloride that constitutes it. Lorenz has attempted a solution by focusing on the adaptiveness of behavior, rather than on behavior itself, and on the sources of information that determine it, rather than on the molecular processes of behavioral ontogeny (Lorenz, 1965).

An individual's behavior must be precisely adjusted to deal with environmental conditions if it is to survive and reproduce. Only two sources of information are accessible to the individual to enable him to match his behavior in this way; one is that which he inherits with his genetic makeup; the other is that which he acquires through learning and related processes in the course of life. If we deny the animal access to the latter source by isolating him from aspects of the environment relevant to the development of a specific adaptive behavior—the so-called Casper Hauser experiment—and his behavior continues to be adapted to the normal environment, we can conclude that the relevant information must have come through his genes. There, in the genetic material, the information has, of course, accumulated during phylogeny through the evolutionary process. Incidentally, it has been pointed out that evolution closely parallels trial-and-error learning at the species level, mutations representing trials, and survival and reproduction being analogous to reinforcements (Pringle, 1951).

An example clarifies this argument. Newly hatched, inexperienced chicks prefer to peck at three-dimensional objects rather than at two-dimensional figures. The main cue they use to make this important discrimination is shadowing. In the chick's normal environment, objects are shadowed below because the sun shines from above. If one presents naive chicks with photographs of a three-dimensional object, they prefer one with a shadow beneath it rather than one that has been inverted 180 degrees so that the shadow falls above the object. Chicks raised in an environment in which the light comes from below, rather than from above, persist in showing the same preference. It follows that they have inherited information which indicates that shadowing below rather than above is a cue for three-dimensionality, because, even when given environmental information to the contrary, they continue to operate on the normally adaptive hypothesis (Dawkins, 1968). This does not imply that the development of the chick's pecking behavior is independent of environmental factors; for example, chicks reared in the dark tend to refuse to peck and show signs of disturbed visual perception (Kovach, 1969). Only a specific adaptiveness—an abstract property of the behavior—is inherited, not the behavior itself.

In many cases, the adaptivity of behavior is due to information from both sources, and in view of the as-yet-unavoidably conversational, rather than strictly theoretical, use of the term information, it is difficult to assess quantitatively the contributions of each source. Similarly, the concept of adaptiveness, although biologically well founded, can be elusive in specific instances. I do not want to belabor these operational difficulties, but it seems that Lorenz's ideas, although suggesting a theoretical solution, are not yet helpful to a reductionist approach.

## Learning

Once the growth phase of an organism has ended, the morphological and physiological characteristics of most organs remain reasonably stable. Not so in the case of the nervous systems of many species. There, even though conventional anatomical and physiological techniques fail to show obvious evidence of it, behavioral performance forcefully documents a continued and striking state of plasticity throughout the individual's life span. The main expression of this protracted plasticity is the capacity for learning, of storing information about the environment, so that subsequent behavioral responses to the same environment are enduringly altered.

Learning can be viewed as the building up of a noniso-

morphic model of the spatiotemporal contingencies of environmental events—more precisely, of the events about which the central nervous system receives neural information. This includes events produced by the animal itself, that is, its behavioral actions signaled either by the sensory re-afferents or by the efferent commands that initiate them. These latter signals, generated by the nervous system itself, make the model an active, responding, operational one, rather than just a passive, descriptive one.

The adaptive function of this model enables individuals to predict events and generate actions vital for successful survival and reproduction on the occurrence of neural events. A bird's chance of surviving is greatly enhanced if it is capable of inferring a striking predator and fleeing on sighting a strolling cat. Events critical for survival and reproduction are those that have either positive or negative reinforcing properties (Glickman and Schiff, 1967; Valenstein, this volume). The functional significance and, hence, rewarding quality of such events will often be conditional on the context within which they occur—that is, depend on the motivational state of the animal. For example, food ingestion is neither adaptive nor rewarding to a satiated animal.

The categorization of events as being aids to survival or reproduction may have taken place either during phylogeny, in which case they are innately primarily rewarding, or during ontogeny, through consistent contingency with primary reinforcers. Such dichotomous classification may, in many cases, be misleading because both processes contribute through complex interactions.

In recent years, a wider range of events than has previously been acknowledged has been shown to have reinforcing properties. For example, chaffinches will learn to seek a particular perch if it activates the replay of a tape-recorded chaffinch song, but not if it produces a burst of white noise (Stevenson, 1967). Fighting fish will swim through a loop more often if it leads to seeing another fighting fish (Thompson, 1963). Both events are critical, because they herald sexual competitors that must be sought out and warded off if reproduction is to be successful.

This is not to say that learning can be expected to occur only in conjunction with reinforcement. There is no doubt that mice bearing a genotype that makes them capable of remembering the whereabouts of food they found when they were sated, or a burrow they discovered when they were exploring, will be at an advantage, evolutionarily speaking, when they are hungry or are chased by a cat. Such latent learning can be demonstrated in rats which learn to run a maze for food faster if they explored it previously without reward than if they had not (Kimble, 1961).

Why is it (disregarding hairsplitting arguments) that the ontogeny of some behaviors is controlled predominantly by genetic information, while other behaviors are shaped mainly by learning? This diversity must have arisen during phylogeny. Behavior may be innate because there is a survival premium on having a response available without a preceding, time-consuming conditioning process. This is exemplified by such predator-avoidance behavior of young animals as the camouflage-enhancing, frozen crouch of just-hatched herring-gull chicks when they hear the parental, predator-released, alarm call (Goethe, 1955). Acquisition of this response through learning would clearly be inefficient, because chicks need be discovered only once by predators to be beyond any chance of further learning. Behavior patterns dealing with environmental situations that have remained stable throughout the phylogeny of a species are likely candidates for enhanced genetic control. The innate breathing behavior of most animals is an example (Spurway and Haldane, 1941). Responses that cope with situations too variable to permit evolutionary adaptation must necessarily arise under environmental control during the lifetime of the individual. An example is the sophisticated mnemotactical, spatial orientation behavior shown by digger wasps with respect to the surroundings of their nests (Tinbergen and Kruyt, 1939; Iersel and van den Assem, 1964), or, for that matter, the behavior of a rat in a laboratory maze.

Apart from these factors, system-inherent constraints may also play a role in determining whether the development of a given behavioral performance is mainly gene or environment controlled. For example, the chromosomal space required to encode ontogenetic specifications for a complex behavior pattern or, rather, for the mechanisms capable of generating it, may be prohibitive and must therefore be implemented by learning. This possibly applies to the hunting and feeding skills of animals, of which the oystercatcher is an example. The young birds take months to master the skills of either stabbing or chiseling the adductor muscle of mussels and the art of detecting annelid worms inches below the sand by probing with their bills. This learning pattern persists despite a high juvenile mortality rate that indicates a strong selective pressure for a more innate feeding behavior (Norton-Griffiths, 1969).

Compensatory behavioral adaptations soften the hardship of having to learn behavioral skills with which one is not born. Protracted parental care is one such compensation, and it can include behavior analogous to teaching. Cats, even primiparous ones, bring still-living prey to their kittens and, making sure that it does not escape, let the young play with it, thereby providing hunting practice (Ewer, 1969). The experimentally demonstrated inclination of kittens to imitate their mothers, but not other cats, may further facilitate the learning process (Chesler, 1969).

Innate propensities, although not completely specifying adequate behavior, may facilitate the acquisition of learning.

The preference, mentioned earlier, of naive chicks to peck at three-dimensional objects rather than two-dimensional specks certainly accelerates the development of adequate feeding behavior. So does fast conditioning in this context: usually chicks require only a single trial to learn the association between the color of a seed and its bad taste (Lee-Teng and Sherman, 1966).

The preceding comments reflect the ethologist's conviction that species-specific learning abilities are as much determined by selection pressures acting during phylogeny as are innate behavior patterns (Thorpe, 1963). The restricted capacity of rats for associating certain stimulus qualities with certain punishments is relevant. They learn to avoid food pellets of a certain size when their consumption is followed by electric shock, but they do not learn to avoid pellets of a particular taste when their ingestion is followed by a shock. On the other hand, they cannnot be taught to avoid X-ray illness by selecting a certain pellet size, although that is relatively easy to do if pellet taste is the cue (Garcia et al., 1968). In the rat's normal environment it is indeed unlikely that food-morsel size is a useful predictor of gastric upsets or whatever natural condition comes closest to radiation illness, whereas taste probably is. Conversely, taste is not likely to be serviceable for anticipating exteroreceptive pain, but food shape (as in the case of thorny berries) possibly could be.

The adaptedness of learning abilities also explains the lack of correlation between the phylogenetic status of a species and its learning capacities as measured by some contrived tests (Warren, 1965). Such attempts must necessarily fail on two counts. Learning abilities are determined not by phylogenetic status but by the requirements of the mode of life of each species, limited by whatever system-inherent constraints prevent the evolution of optimally adjusted learning abilities. Bees are capable of amazing learning feats, adjusted to their social way of life (von Frisch, 1967). Octopuses compare with cats on learning performances, a convergence possibly arising from their predatory habits (Young, 1964). Owing to the need for standardization, the measuring tests, ingenious as they may be, cannot pay full attention to the learning specializations of different species. Rats do poorly in learning-set tasks requiring visual discriminations (Hodos, this volume). But is it safe to extrapolate a similar deficiency if olfactory cues were used?

While I do not depreciate the results of comparative studies of learning, the implication is that it may be more profitable to relate species-specific learning characteristics to ecological needs—that is, to start with observations of animals in their natural settings. The conclusion that pleasurable field studies (in exotic places, of course) hold the key to the solution of a behavioral problem is a timely note on which to close an ethological review.

## Acknowledgment

I wish to thank Dr. S. A. Hillyard for criticizing, Mr. P. L. Appleton for editing, and my wife for typing, the drafts of this paper, and Prof. R. Galambos for encouragement and advice. While preparing this manuscript, the author was a temporary member of the Neurosciences Research Program, and his research was supported by the Science Research Council, London.

## REFERENCES

ANASTASI, A., 1958. Heredity, environment and the question "how?" *Psychol. Rev.* 65: 197–208.

BATESON, P. P. G., 1966. The characteristics and context of imprinting. *Biol. Rev.* 41: 177–220.

BOVET, D. F., F. BOVET-NITTI, and A. OLIVERIO, 1969. Genetic aspects of learning and memory in mice. *Science (Washington)* 163: 139–149.

BROWN, J. L., 1969. The control of avian vocalizations by the central nervous system. *In* Bird Vocalizations (R. A. Hinde, editor). Cambridge University Press, Cambridge, England, pp. 79–96.

CHESLER, P., 1969. Maternal influence in learning by observation in kittens. *Science (Washington)* 166: 901–903.

DAWKINS, R., 1968. The ontogeny of a pecking performance in domestic chicks. *Z. Tierpsychol.* 25: 170–186.

DELIUS, J. D., 1970. The neural substrate of vocalizations in gulls and pigeons. *Exp. Brain Res.* (in press).

DENENBERG, V. H., and A. E. WHIMBEY, 1963. Behavior of adult rats is modified by the experiences their mothers had as infants. *Science (Washington)* 142: 1192–1193.

EIBL-EIBESFELDT, I., 1967. Grundriss der vergleichenden Verhaltensforschung. Piper, Munich.

EWER, R. F., 1969. The "instinct to teach." *Nature (London)* 222: 698.

FRISCH, K. VON, 1967. The Dance Language and Orientation of Bees. The Belknap Press of Harvard University Press, Cambridge, Massachusetts.

FULLER, J. L., and W. R. THOMPSON, 1960. Behavior Genetics. John Wiley and Sons, New York.

GARCIA, J., B. K. McGOWAN, F. R. ERVIN, and R. A. KOELLING, 1968. Cues: Their relative effectiveness as a function of the reinforcer. *Science (Washington)* 160: 794–795.

GLICKMAN, S. E., and B. B. SCHIFF, 1967. A biological theory of reinforcement. *Psychol. Rev.* 74: 81–109.

GOETHE, F., 1955. Beobachtungen bei der Aufzucht junger Silbermöwen. *Z. Tierpsychol.* 12: 402–433.

GOTTLIEB, G., 1966. Species identification by avian neonates: Contributory effects of perinatal auditory stimulation. *Anim. Behav.* 14: 282–290.

HARLOW, H. F., and M. K. HARLOW, 1965. The affectional systems. *In* Behavior of Nonhuman Primates, Vol. 2 (A. M. Schrier, H. F. Harlow, and K. F. Stollnitz, editors). Academic Press, New York, pp. 287–334.

HINDE, R. A., 1966. Animal Behavior. McGraw-Hill, New York.

HIRSCH, J. (editor), 1967. Behavior Genetic Analysis. McGraw-Hill, New York.

HSIA, D. Y., 1967. The hereditary metabolic diseases. *In* Behavior Genetic Analysis (J. Hirsch, editor). McGraw-Hill, New York. pp. 176–193.

HSIA, D. Y., 1969. The role of phenylalanine in mental development. *In* The Future of Brain Sciences (S. Bogoch, editor). Plenum Press, New York, pp. 379–387.

IERSEL, J. J. A. VAN, and J. VAN DEN ASSEM, 1964. Aspects of orientation in the diggerwasp *Bembix rostrata*. *Anim. Behav. Suppl.* 1: 145–162.

JOFFE, J. M., 1969. Prenatal Determinants of Behavior. Pergamon Press, Oxford.

JOHN, E. R., P. CHESLER, F. BARTLETT, and I. VICTOR, 1968. Observation learning in cats. *Science (Washington)* 159: 1489–1491.

KAWAI, M., 1965. Newly-acquired pre-cultural behavior of the natural group of Japanese monkeys on Koshima Island. *Primates* 6: 1–30.

KILHAM, P., P. H. KLOPFER, and H. OELKE, 1968. Species identification and colour preferences in chicks. *Anim. Behav.* 16: 238–244.

KIMBLE, G. A., 1961. Hilgard and Marquis' Conditioning and Learning. Methuen, London; Appleton-Century-Crofts, New York.

KING, J. A., and B. L. ELEFTHERIOU, 1959. Effects of early handling upon adult behavior in two subspecies of deermice *Peromyscus maniculatus*. *J. Comp. Physiol. Psychol.* 52: 82–88.

KONISHI, M., and F. NOTTEBOHM, 1969. Experimental studies in the ontogeny of avian vocalizations. *In* Bird Vocalizations (R. A. Hinde, editor). Cambridge University Press, Cambridge, England, pp. 29–48.

KOVACH, J. K., 1969. Development of pecking behavior in chicks: Recovery after deprivation. *J. Comp. Physiol. Psychol.* 68: 516–532.

LEE-TENG, E., and S. M. SHERMAN, 1966. Memory consolidation of one-trial learning in chicks. *Proc. Nat. Acad. Sci. U. S. A.* 56: 926–931.

LEHRMAN, D. S., 1970. Semantic and conceptual issues in the nature-nurture problem (in press).

LORENZ, K., 1965. Evolution and Modification of Behaviour. Methuen, London, Chicago University Press, Chicago.

MARLER, P., and W. J. HAMILTON, III, 1966. Mechanisms of Animal Behavior. John Wiley and Sons, New York.

MASON, W. A., 1965. The social development of monkeys and apes. *In* Primate Behavior (I. DeVore, editor). Holt, Rinehart and Winston, New York, pp. 514–543.

MOLTZ, H., 1965. Contemporary instinct theory and the fixed action pattern. *Psychol. Rev.* 72: 27–47.

NORTON-GRIFFITHS, M., 1969. The organisation, control and development of parental feeding in the oystercatcher (*Haematopus ostralegus*). *Behaviour* 34: 55–114.

NOTTEBOHM, F., 1970. Ontogeny of bird song. *Science (Washington)* 167: 950–956.

PRECHTL, H. F. R., and A. P. KNOL, 1958. Fussohlenreflex beim neugeborenen Kind. *Arch. Psychiat. Z. Ger. Neurol.* 196: 542–553.

PRINGLE, J. W. S., 1951. On the parallel between learning and evolution. *Behaviour* 3: 174–215.

RESSLER, R. H., 1966. Inherited environmental influences on the operant behavior of mice. *J. Comp. Physiol. Psychol.* 61: 264–267.

Schutz, F., 1965. Sexuelle Prägung bei Anatiden. *Z. Tierpsychol.* 22: 50–103.

Skinner, B. F., 1966. The phylogeny and ontogeny of behavior. *Science (Washington)* 153: 1205–1213.

Sourkes, T. L., 1962. Biochemistry of Mental Disease. Harper and Row, New York.

Spurway, H., and J. B. S. Haldane, 1954. The comparative ethology of vertebrate breathing. I. Breathing in newts, with a general survey. *Behaviour* 6: 8–34.

Stevenson, J. G., 1967. Reinforcing effects of chaffinch song. *Anim. Behav.* 15: 427–432.

Thompson, T. I., 1963. Visual reinforcement in Siamese fighting fish. *Science (Washington)* 141: 55–57.

Thorpe, W. H., 1961. Bird–Song. Cambridge University Press, Cambridge, England.

Thorpe, W. H., 1963. Learning and Instinct in Animals. Methuen, London, Harvard University Press, Cambridge, Massachusetts.

Tinbergen, N., 1951. The Study of Instinct. Clarendon Press, Oxford, London.

Tinbergen, N., and W. Kruyt, 1939. Über die Orientiering des Bienenwolfes. *Z. Vergl. Physiol.* 25: 292–334.

Tryon, R. C., 1940. Genetic differences in maze-learning ability in rats. *In* 39th Yearbook, National Society for the Study of Education, Part I. Public School Publishers, Bloomington, Ill., pp. 111–119.

Vandenberg, S. (editor), 1965. Methods and Goals in Human Behavior Genetics. Academic Press, New York.

Warren, J. M., 1965. The comparative psychology of learning. *Annu. Rev. Psychol.* 16: 95–118.

Young, J. Z., 1964. A Model of the Brain. Clarendon Press, Oxford.

# 20 Early Hormonal Influences on the Development of Sexual and Sex-related Behavior

## ROBERT W. GOY

Under ideal conditions, the procedures for investigating the influences of early hormonal states on the development of behavior are simple and classical. The appropriate procedures involve removal or destruction of the endocrine gland (which is the only, or primary, source of particular hormones) so that the effects of development in the anhormonal state can be determined. In addition, replacement therapy with extracts of the gland or with pure synthetic forms of the hormone should be carried out to determine which of a variety of possible compounds most closely duplicates the effects of the glandular secretions. In actual practice, the problem of characterizing the hormonal influences on sexual development is more difficult to analyze. In part, the greater difficulty is attributable to the existence of two different endocrine glands (the ovary and the testis), each present in a distinctive genetic constitution. This situation poses a logical dilemma in the sense that we can never ask or answer the questions of what is the effect of ovariectomy in the genetic male or orchiectomy in the genetic female. Part of the difficulty also lies in the fact that our knowledge of the nature of the secretory products of these endocrine glands during early development is extremely limited. So limited, in fact, that it is questionable whether the principle of "replacement therapy" can be effectively applied to the analysis of the problem. To further complicate matters, the endocrine glands are not functional at all times in early development. Accordingly, experiments designed to test glandular influences will yield either negative or unparalleled results, unless they are carried out during those periods that are normally characterized by heightened levels of secretory activity.

These difficulties and complications do not pose any greater obstacle to the analysis of psychosexual differentiation than to the analysis of morphological differentiation. In the analysis of both problems, investigators have had to proceed by making certain assumptions and by using situations that are only approximations of the ideal case for endocrinological studies. What follows in this presentation is a review of the literature on psychosexual differentiation within the context of the ideal case, while at the same time pointing out the limitations of the data and the nature of the assumptions underlying specific interpretations.

ROBERT W. GOY Department of Reproductive Physiology and Behavior, Oregon Regional Primate Research Center, Beaverton, Oregon

## The effects of spontaneously occurring hormonal deficiencies during early development: the anhormonal case

In human development an event that prevents any gonad from forming rarely takes place. The condition is known clinically as Turner's Syndrome and characteristically it results in the development of a phenotypic female, who, because of the lack of gonads, fails to show the pubertal changes associated with gonadal activity at that time. Such individuals are not regarded as unfeminine at any time during their early development, and many cases are discovered only when they fail to manifest signs of puberty at the normal time. If treated with ovarian hormones during the period of adolescence, these individuals respond in a manner not measurably different from that of the normal female. Moreover, to the extent that the literature permits such a statement, the psychosexual orientation and libido of such individuals closely parallel those of the normal female. Individuals suffering from Turner's Syndrome, however, do not qualify as pure or ideal cases for demonstrating the effects of the development of the genetic female in the absence of an ovary. Such individuals are not genetic females (xx), nor are they genetic males (xy). The evidence available from karyotypes indicates that the majority of such cases possess only one sex chromosome, which, in the human, is invariably an x-chromosome. Presumably, karyotypes of the y-o variety fail to develop.

An animal case paralleling Turner's Syndrome was fortuitously discovered many years ago by F. A. Beach (1945). During a series of routine ovariectomies of female rats, Beach encountered one animal that contained no traces of ovarian or testicular tissue and that appeared phenotypically indistinguishable from normal females. When this subject was injected with suitable quantities of estrogen and progesterone, it responded by displaying the complete pattern of behavioral estrus characteristic of the normal female. The parallel with Turner's Syndrome is incomplete, however, as the genotype of this rat was never determined. Accordingly, we do not know whether the case demonstrates agonadal development of the genotypic male, female, or some other genotype.

From the literature on human beings, it is possible to approach the anhormonal case in a very indirect way. In another rare genetic anomaly, individuals incapable of responding to androgens have been identified. The genetic disorder in these instances is a disorder not of the sex chromosomes, but rather of the autosomes. As a result of this autosomal factor, the individual case may go undetected in the genetic female in whom physiological responses to androgens are relatively less important. Accordingly, the syndrome has been primarily associated with the genetic

male. The association of the condition with the male genotype has given rise to an unfortunate misnomer for the condition, and it is known clinically as Testicular Feminization. In line with current views, however, the testis of these individuals is not feminizing nor does it secrete hormones which in their types or pattern resemble the hormones of the ovary. At least in the adult state of these individuals, the testes secrete amounts of androgens and estrogens that are within the range of normal genetic males. Nevertheless, as the clinical term implies, genetic males possessing the abnormal genes are indistinguishable from phenotypic females in their external appearance as well as in their psychosexual orientation and libidinal interests. Interpretation of this clinical case as a demonstration of the effects of anhormonal development in the genotypic male requires specific assumptions. These are: (1) that the genetic disorder prevents or blocks responsiveness to androgens throughout all development, early as well as late; and (2) that the gonadal secretions during early development are essentially similar to those of the normal genetic male. Although no data exist which permit us to decide directly upon the validity of these assumptions, the normal-appearing testicular morphology and the absence of Müllerian-duct derivatives are indirectly supportive.

Cases of genetically determined disorders of sexual development in males occur spontaneously among lower mammals, but no studies have been carried out to assess the behavioral characteristics of afflicted individuals in adulthood. In a colony of rats maintained at Oklahoma City Medical Center, a high incidence of genital malformation and incomplete sexual differentiation occurs among genotypic males (Stanley and Gumbreck, 1964; Allison, 1965; Allison et al., 1965; Stanley et al., 1966). Experiments that would permit decisive interpretations of the nature of the disorder have not, however, been carried out. From the data available, decisions cannot be made as to whether the malformations are a result of insufficient androgen, insensitivity to normal amounts of androgen, or delayed formation of the gonad. In many cases, however, as in the human syndrome, individuals carrying the abnormal genes resemble females to varying degrees in their outward morphology.

The somatic manifestations of the abnormal gene actions are variable and present a continuum of effects. Allison et al. (1965) have described individuals completely devoid of any internal reproductive-tract structures. Externally they appear very feminine, with small vaginal orifices and nipples along the milk line. Much less severe examples of developmental aberration also are seen, in which the morphology is essentially masculine but the individuals are sterile.

Although the data summarized concerning spontaneous anomalies of sexual development are somewhat limited, nothing that has been described contradicts the view that the female form and psychosexual orientation can develop in the absence of hormonal influences from the gonads. Moreover, the development of the female pattern does not depend on a specific sexual genotype. Individuals that are genotypically XO or XY can and do develop (under specific circumstances) along lines as feminine as those characteristic of the normal female.

## Experimental approaches to the anhormonal case

Studies of behavioral development have been restricted to gonadectomy during relatively late stages, compared with studies in experimental embryology. In addition, most, if not all, of the definitive behavioral studies have been carried out on the rat. In this species, studies of early gonadal activity permit some estimate to be made of testicular secretions present during the early period of development. Noumura et al. (1966) have shown that the testis develops the capacity to synthesize androgens as early as the thirteenth day of fetal development. In our laboratory, biochemical assays demonstrate that testosterone can be found in the peripheral blood of newborn rats and continues to be present in measurable amounts until about the tenth day after birth (Resko et al., 1968) (Figure 1). These results mean that castration of the male rat on the day of birth deprives the individual of testicular androgens only for a little more than half of the period that these hormones are normally present. Despite this relatively crude approximation of the anhormonal case, experiments conducted by Grady et al. (1965) demonstrated that castration during the neonatal period had profound effects on the sexual development of genotypic males. Individuals deprived of their testes at either one or five days of age developed behavioral characteristics normally present only in genotypic females of this species. When these males reached adult ages and were injected with estrogen and progesterone, they displayed lordosis (the posture of the receptive female) in response to mounting by normal males. Males castrated

at 10 days of age or later showed little or no tendency to display lordosis under comparable conditions of testing (Table I). As illustrated in Table I, males castrated on the day of

TABLE I

*Means of the mean copulatory quotients of castrated male rats receiving estradiol benzoate and progesterone*

(Numbers in parentheses are the per cent of tests during which the experimental subjects were mounted.*)

| Groups | N, Ss | 3.3 μg. | N, Ss | 6.6 μg. |
|---|---|---|---|---|
| Spayed females | 9 | .452 ( 97) | 7 | .787 (100) |
| Day-1 males | 8 | .301 ( 88) | 7 | .572 (100) |
| Day-5 males | 8 | .253 ( 78) | 6 | .183 ( 96) |
| Day-10 males | 5 | .000 ( 75) | 6 | .028 ( 88) |
| Day-20 males | 7 | .003 ( 79) | 7 | .056 ( 89) |
| Day-30 males | 7 | .000 ( 71) | 7 | .085 ( 89) |
| Day-50 males | 8 | .000 ( 84) | 7 | .053 ( 96) |
| Day-90 males | 7 | .000 ( 78) | 7 | .038 (100) |

* Data adapted from Grady et al., 1965.

birth displayed lordosis which more closely resembled that displayed by normal females than did any other group.

When males castrated at these early ages are allowed to mature and are then injected with testosterone propionate, the pattern of sexual behavior is characterized by frequent mounting behavior, but intromission and ejaculation are either absent or infrequent (Beach and Holz, 1946; Grady et al., 1965). The investigators who have studied this problem most closely have repeatedly pointed to inadequacies of phallic development as one of the possible reasons for the general failure of early castrates to display intromission and ejaculation despite injection of high doses of testosterone in adulthood. The theoretical question with regard to whether the perinatal testis contributes to the development of male

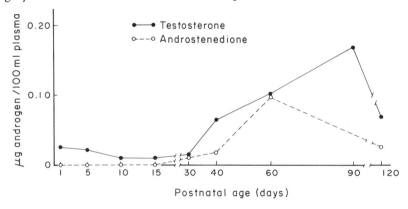

FIGURE 1 Concentration of testosterone in pooled samples of peripheral plasma obtained from independent groups of male rats at different postnatal ages. (From data published by Resko et al., 1968; Figure previously published in Goy, 1968.)

sexual behavior by altering central neural mechanisms or by altering the peripheral effector apparatus has not been answered experimentally. Since the initial study by Grady et al. (1965), the technical problems that stand in the way of a direct answer have not been overcome, and the neutrality of their position in regard to these alternatives has had to be maintained. In a very recent re-examining of the same problem, Beach et al. (1969) restated this position:

A choice between these alternatives (i.e. influences on central neural structures vs. peripheral effector structures, RWG) cannot be made on the basis of the present evidence, but they are not mutually exclusive and the most conservative hypothesis would seem to be that both types of effect are important.

The male hamster, as does the male rat, undergoes extensive psychosexual differentiation during the early postnatal period. Accordingly, it can provide comparative data bearing on the general question of the behavioral characteristics which develop in the absence of the testis during a period of development that is critical for psychosexual differentiation. Eaton (1969) has recently studied male hamsters castrated on the day of birth. As in the rat, such males develop an expression of the lordosis response closely resembling that of the normal female. When given estrogen and progesterone in adulthood, lordosis lasting for 267 out of a possible 300 seconds could be elicited readily by the sniffing behavior of a sexually active, normal male. The corresponding data for normal females was 290 out of 300 seconds.

Investigations of the genetic female are less satisfactory than those of the male from the point of view of providing evidence that early ovariectomy results in a state of hormonal deprivation. No direct biochemical measurements exist to support the notion that the ovary is active endocrinologically during fetal or neonatal life in the female rat or any other mammal. Although data have been collected showing that ovariectomy soon after birth leads to pituitary changes suggestive of a loss of negative feedback, the specific hormone involved in this feedback has never been identified.

If hormones secreted by the neonatal ovary contribute to female sexuality, as some investigators have postulated (A. A. Gerall, J. B. Dunlap, and C. N. Thomas, personal communication), they are not essential to the development of fairly complete sexual responsiveness. Adult female rats, ovariectomized on the day of birth, display vigorous lordosis patterns when the proper ovarian hormones are supplied by injection (Wilson and Young, 1941). Moreover, ovariectomy on the day of birth does not prevent the development of some behaviors that can be stimulated by testosterone in adulthood. When such females are injected with testosterone propionate, they display marked augmentation of mounting when paired with receptive female partners (Harris and Levine, 1965; Gerall and Ward, 1966; Whalen

et al., 1969). The ability of the genetic female rat to display increased frequencies of mounting behavior during testosterone treatment in adulthood does not depend, however, on removal of the ovary on the day of birth. A common finding among lower mammals is that mounting occurs spontaneously in many females and that testosterone will stimulate mounting activity in intact adult females or in females spayed as adults (Beach and Rasquin, 1942; Goy and Young, 1958; Phoenix et al., 1959; Gerall and Ward, 1966; Young, 1961).

The results obtained from these experimental studies are consistent with the interpretation of the spontaneous occurrences of gonadless development. Including the special case of Testicular Feminization, in which autosomal genetic factors prevent responsiveness to androgens, all cases of development in an early environment deficient either in gonadal hormones or in the physiological actions of the hormones are associated with the retention and elaboration of neural systems that mediate behavior characteristic of the normal genetic female. This generalization holds true whether we are considering mounting behavior of the normal female and its stimulation by a variety of hormones in adulthood, or are considering lordosis responses and their stimulation by estrogen and progesterone. Just as the consequences of development in the anhormonal environment are not limited to a particular form of behavior, so, too, they are not restricted to a particular sexual genotype. Both the genetic male and the genetic female develop along parallel, if not entirely identical, lines when deprived of their gonads. In short, both genetic sexes develop in a manner which so far has not been distinguished from that of a normal female, but which differs markedly from that of a normally differentiated male.

*Contributions of testicular hormones to the development of behavioral traits characteristic of the male*

The experiments reviewed so far show that testicular secretions contribute to the suppression or loss of the lordosis reflex, which otherwise develops in genetic male rats. Such an effect on the sexual behavior repertoire may be regarded as one of the contributions the early testis makes to the development of a male psychosexual orientation in this species. But this effect of testicular hormone early in development is not limited to the genotypic male. When female rats or guinea pigs are treated with testosterone propionate at the proper times in early life, the development of lordosis can be markedly interfered with, or in some cases prevented entirely (Phoenix et al., 1959; Barraclough and Gorski, 1962; Harris and Levine, 1965; Goy et al., 1962; Harris, 1964; Goy et al., 1964). The female guinea pig and rat differ markedly in the length of the gestational interval, which lasts for 68 days in the former and only 21 days in the latter. Corresponding to this difference in the length of prenatal life, the developmental period during which testosterone propionate is effective in suppressing the development of lordosis differs in the two species. For the genetic female guinea pig, testosterone is maximally effective when it is administered between the thirtieth and fortieth days of *prenatal life*. For the rat, in contrast, the maximally effective period for androgenic suppression is during the first five days of *postnatal life*.

Suppression of the development of lordosis can be effected by a wide variety of gonadal steroids including estradiol (Feder and Whalen, 1965; Levine and Mullins, 1966), androstenedione (Stern, 1969), and progesterone (Diamond and Wong, 1969). It is unlikely that all these hormones are involved in the normal testicular suppression of lordosis, for the quantity of each hormone that must be injected greatly exceeds that which could be expected to result from normal secretory activity. In contrast to the other hormones mentioned, estradiol is known to interfere with the development of male copulatory behavior (Whalen, 1964). Nevertheless, the experiments demonstrate that the developmental period when the testis is normally active can be viewed as a period of sensitivity to a broad spectrum of steroids, and each may make a different contribution to development.

Possibilities exist for the production of a wide variety of psychosexual abnormalities when the array of testicular hormones is considered singly and in various combinations. One recent example deserves special attention because of its possible relevance to the etiology of male homosexuality. During the 10 to 20 days of life immediately after birth, testosterone is the principal, if not the only, androgen present in the testis and peripheral blood of the male rat (Resko et al., 1968). There is substantial evidence supporting the interpretation that the presence of this hormone accounts for the suppression of such feminine characteristics as lordosis and the augmentation of such male characteristics as mounting, intromission, and ejaculation as the organism develops. When the testis is removed at birth, however, and androstenedione in critical dosages is the only androgen provided during this developmental period, a unique set of behavioral traits characterize the adult (Goldfoot, et al., 1969).

Such genetically male individuals develop normal male genitalia as well as mounting. intromission, and ejaculation behavior, but their over-all masculinity is incomplete. The development of lordosis is not suppressed by androstenedione in the amounts administered by Goldfoot et al., and accordingly the adult pattern includes the retention of behavioral traits that are normally characteristic of genetic females. The parallelism to homosexuality is illustrated not by the retention of lordosis per se, but by the implication that specific hormonal conditions permit the retention of broadly defined feminine characters while, at the same time, contributing to the formation of normal male genitalia. Inasmuch as androstenedione is a normal biosynthetic precursor of testosterone, any genetic disturbance that prevents the final formation of testosterone could contribute to the formation of morphologically phenotypic males with psychosexual abnormalities marked by the retention of feminine characters.

The genotypic female has been widely used as a subject to demonstrate the influence of early testosterone on the development of behaviors normally characteristic of the male. In the first demonstration of these effects to come from our laboratory, evidence was presented for an augmentation of mounting behavior in the female guinea pig that was independent of any hormonal stimulation in adulthood and that represented a permanent change in the behavior of these treated females. Corresponding effects have been reported for female hamsters (Crossley and Swanson, 1968) and some strains of female rats treated with testosterone propionate during the neonatal period (Harris and Levine, 1962; Gerall and Ward, 1966; Nadler, 1969) but not for others (Whalen et al., 1969). An additional aspect of the effects of early androgen was brought out in our first study by the demonstration that treated females showed a heightened sensitivity to testosterone propionate in adulthood. When both normal and treated females were injected with identical amounts of testosterone propionate, the treated females showed a more rapid rise in the rate of mounting displayed in standardized tests as well as a higher over-all frequency of mounting. Comparable results were obtained when prenatally treated and normal females were given daily testosterone from birth to 70 days of age. Again, the treated females developed mounting behavior more rapidly and to a higher level of expression than females not treated prior to birth (Goy et al., 1967). In more recent studies of the genetic female rat, Gerall and Ward (1966) and Ward (1969) have demonstrated that, if treatment of genetic females with testosterone propionate is begun before birth and continued for a short time into the neonatal period, the sexual behavior pattern can be almost completely reversed. Females treated in this way display the complete male copulatory pattern, including intromission and ejaculation, when given additional testosterone propionate in adulthood.

Throughout our work with the effects of testosterone given at early stages of development, we have believed that it is important to show that the behavioral outcomes were not attributable to toxic or other pharmacological effects of the hormone. For this purpose, the behavior of sibling males treated identically has formed a part of our studies (Phoenix et al., 1959). In no case has early testosterone in the doses we administered altered the behavior of treated males compared with normal males. Inasmuch as a damaging effect of excessive testosterone during early development has been reported in earlier work (Wilson and Wilson, 1943), behavioral measures demonstrating that treated males are behaviorally normal takes on a special importance.

## Masculine patterns of social behavior are influenced by early androgen

The effects of early testosterone propionate on the development of masculine patterns of responses are not limited to aspects of sexual behavior. Swanson (1967) has shown masculinizing effects of perinatal androgen on the open-field behavior of female hamsters, and Bronson and Desjardins (1968) have demonstrated comparable influences of androgen on the fighting and aggressive behaviors of mice. The results with fighting behavior in mice are entirely consistent with the view that early androgen has an organizing action on the nervous system in the sense that androgens in adulthood modify or augment aggressive behavior only in males, not in females (Beeman, 1947; Tollman and King, 1956).

So long as effects could be shown only for behaviors that form a part of the male and female sexual repertoire, or for behavioral traits requiring the actions of hormones in adulthood, a possible interpretation was that the primary effect of early androgen was an alteration of the sensitivity of the neural tissues to specific hormones. Such an interpretation is not inaccurate as it applies to sexual behaviors or to the fighting behavior of mice, but, for the past five years, C. H. Phoenix and I have been studying the development of behaviors that dictate the need for a broader view of the actions of early androgen. We have chosen to study the social behavior of young rhesus monkeys for a demonstration of these broad effects. The primary reason for this choice was that studies by Rosenblum (1961) and Harlow (1965) had demonstrated that patterns of infant and juvenile social interactions unrelated to sexual behavior were sexually dimorphic in this species. Studies we have conducted confirm these earlier reports for five different kinds of social interactions observed under our standardized conditions. For each of the five behaviors the distinction between the sexes is purely quantitative, rather than qualitative. That is, no behavior that is unique to one sex has been identified. In addition, all the behaviors we have studied are displayed much more frequently by males than by females.

In these experiments, genetic female monkeys were treated for varying periods of time beginning on day 39 of gestational age through the day 69 or day 105 of the average 168-day gestation period. In the monkey, which, like the guinea pig, is a long-gestation species, the critical period for psychosexual differentiation appears to be prior to birth rather than after birth, as it is in the rat and hamster.

Testosterone propionate dissolved in oil was injected intramuscularly into the mother and reached the fetuses via placental circulation. After weaning, usually at three months of age, the treated females were randomly assigned to groups containing normal males, castrated males, normal females, or combinations of these various subjects. The frequency with which these testosterone-treated females displayed patterns of threatening, play initiation, rough-and-tumble play, and chasing play are illustrated in Figures 2, 3, 4, and 5, respectively. The data are presented for four consecutive series of observations, which represent samples of behavior during the first, second, third, and fourth years of life. Comparable data for normal males and females are depicted in the same Figures. For all the behaviors illustrated, males differ significantly from females, and the treated females are intermediate.

The differences between the sexes in social behavior patterns are independent of gonadal secretions at the ages when these behaviors are being displayed. Although testicular secretory activity has been assayed biochemically and shown to be at low level during the infant and juvenile periods (Resko, 1967), low levels are not essential to the display of these behaviors in the genetic male. Conversely, low levels of ovarian activity play no measurable role in suppressing or diminishing the frequency with which these behaviors are displayed by the normal genetic female. The data presented in Table II demonstrate that castration of the male rhesus on the day of birth does not prevent the display of these social behaviors at frequencies characteristic of the normal intact male. Correspondingly, ovariectomy on the day of birth does not facilitate the display of these "masculine" behaviors by the genetic female.

The mounting of rhesus males is a complex behavior that develops gradually with maturation and depends in part on social experience (Mason, 1961; Harlow, 1961, 1965). During the infant, juvenile, and adolescent periods, the mounting of rhesus monekys shifts gradually from the infantile pattern of standing at the partner's rear or side to the mature pattern of clasping the partner's ankles or legs with the feet. The age at which the mature double-foot-clasp mount is first displayed by males under our conditions of rearing and observation is extremely variable, ranging from eight months to as much as three years of age. The possibility exists that, with more opportunity for social experience or more prolonged periods of social contact, less variability would be observed. Genetic female rhesus monkeys, treated prenatally with testosterone propionate, display a developmental pattern for mounting behavior which resembles that of the genetic male and differs from that of the genetic female (Tables III and IV).

The changes induced in female rhesus monkeys by prenatal testosterone have subtle features and include characteristics of the individual that cannot be tabulated as frequency

## Play Initiation

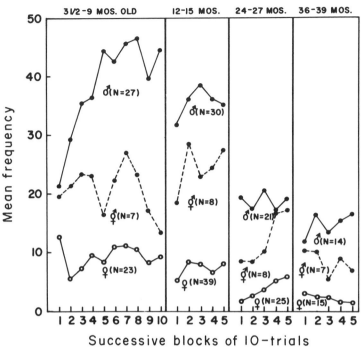

FIGURE 2  Average frequency of performance of play-initiation in standardized tests for social behavior by male (♂), female (♀), and prenatally testosteronized females (⚦) throughout the first 39 months of postnatal life.

## Threat

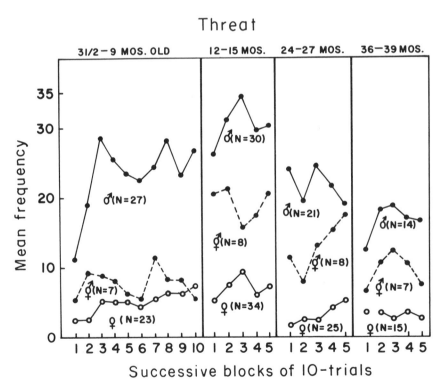

FIGURE 3  Average frequency of performance of stereotyped threat expressions in male (♂), female (♀), and prenatally testosteronized females (⚦) throughout the first 39 months of postnatal life.

# Rough and Tumble Play (Rhesus)

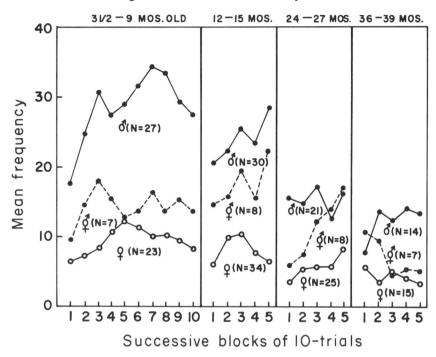

FIGURE 4  Average frequency of performance of rough-and-tumble play in male (♂), female (♀), and prenatally testosteronized females (♂) throughout the first 39 months of postnatal life.

# Pursuit Play

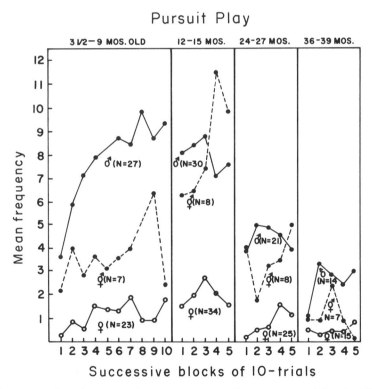

FIGURE 5  Average frequency of performance of pursuit play (chasing behavior) in male (♂), female (♀), and prenatally testosteronized females (♂) throughout the first 39 months of postnatal life.

## Table II

*Average frequency of performance of four kinds of social behavior by normal males and females and by male and female rhesus gonadectomized at birth*

| | | N | Threat | Play Initiation | Rough-and-Tumble Play | Pursuit Play |
|---|---|---|---|---|---|---|
| | | | *Average Frequency per Animal per Block of 10 Trials* | | | |
| Spayed | ♀ | 3 | 7.0 | 4.0 | 3.6 | .2 |
| Intact | ♀ | 23 | 5.6 | 9.2 | 9.0 | 1.1 |
| Castrated | ♂ | 2 | 35.0 | 49.2 | 39.3 | 5.3 |
| Intact | ♂ | 27 | 23.5 | 38.6 | 29.1 | 7.8 |

## Table III

*Changes with age in the frequency of performance of immature mounting postures by male, female, and prenatally testosteronized female rhesus*

| | 3½–9 mos old | | 12–15 mos old | | 24–27 mos old | | 36–39 mos old | |
|---|---|---|---|---|---|---|---|---|
| | N | Av. Freq. of Mts per Animal per Block of 10 Trials | N | Av. Freq. of Mts per Animal per Block of 10 Trials | N | Av. Freq. of Mts per Animal per Block of 10 Trials | N | Av. Freq. of Mts per Animal per Block of 10 Trials |
| ♂ | 27 | 2.8 | 30 | 2.1 | 21 | 1.5 | 14 | 0.9 |
| ♀ | 23 | 0.0 | 34 | 0.0 | 25 | 0.1 | 15 | 0.0 |
| ♀̣ | 7 | 1.0 | 8 | 0.5 | 8 | 0.4 | 7 | 0.3 |

## Table IV

*Changes with age in the frequency of performance of mature mounting posture by male, female, and prenatally testosteronized female rhesus*

| | 3½–9 mos old | | 12–15 mos old | | 24–27 mos old | | 36–39 mos old | |
|---|---|---|---|---|---|---|---|---|
| | N | Av. Freq. of Mts per Animal per Block of 10 Trials | N | Av. Freq. of Mts per Animal per Block of 10 Trials | N | Av. Freq. of Mts per Animal per Block of 10 Trials | N | Av. Freq. of Mts per Animal per Block of 10 Trials |
| ♂ | 27 | 0.2 | 30 | 1.2 | 21 | 3.7 | 14 | 2.7 |
| ♀ | 23 | 0.0 | 34 | 0.0 | 25 | 0.0 | 15 | 0.0 |
| ♀̣ | 7 | 0.1 | 8 | 1.0 | 8 | 1.1 | 7 | 0.9 |

data. In this species, the display of the mature double-foot-clasp mount requires the complete cooperation of the partner. The full manifestation of the behavior could be easily prevented at any time if the partner withdrew or sat down after the initial contact. Just how this cooperation is elicited in the partner is a subtlety that we have not been able to analyze. It does not appear to be a simple matter of social dominance, insofar as our criteria for social dominance are concerned. Some other set of factors, which, in the case of heterosexual mounting, may include sexual attractiveness, appears to be involved. Whatever characteristics are essential for the elicitation of cooperation in a partner by a mounter, females treated prenatally with testosterone seemed to possess these characteristics to a much greater extent than did normal females.

The experiments on rhesus monkeys demonstrate that the presence of testosterone during an embryonic or fetal stage contributes to the development of masculine patterns of social and mounting behavior, which are displayed only at later stages of life. Moreover, the changes induced in the behavior of treated individuals are not limited to behaviors controlled by later hormones, and the effects cannot in this case be interpreted solely as a permanent alteration in sensitivity of neural tissues to hormonal activation. Instead, it seems necessary to conclude that exogenous prenatal androgen alone is sufficient to cause the development in the genetic female of a variety of behaviors normally characteristic of the young developing male. Correspondingly, prenatal androgen alone (but from endogenous rather than exogenous sources) is sufficient to cause the development of these social behaviors in the genotypic male.

## Concluding remarks

A number of theoretical and experimental papers have adequately criticized the textbook view that sex is determined at the moment of fertilization (Chang and Witschi, 1956; Yamamoto et al., 1968; Turner, 1969). Such a point of view, although it may be adequate to describe the assemblage of genetic material at that moment in time, fails to provide insight into a wide variety of sexual phenomena known to biologists. For example, the occurrence of true functional hermaphroditism among many invertebrate forms clearly contradicts the concept that particular reproductive functions or capabilities are limited to a specific genetic endowment. Similarly, transformation from functional male to functional female with maturation not infrequently occurs among certain crustaceans (Charniaux-Cotton, 1962) and marine annelids. Among annelids of the genus *Ophryotrocha*, adult functional females can reconvert to males either when starved or when cut to pieces and allowed to regenerate. The genetic constitution very likely has not been altered during the sexual transformation in such individuals, any more than it has in a variety of decapod crustaceans, which live the first part of their lives as males and the later part as females.

Spontaneous sex reversal is more rare in vertebrate forms than among invertebrates. It is among vertebrates that the gonadal hormones as we know them (the estrogens and androgens) take on special importance in controlling the character of sexual development and, in some instances, are

capable of completely overriding the differences in genetic constitution. That is to say, among vertebrates as among invertebrates, a single genetic constitution is compatible with either sexual phenotype. Parallels to complete spontaneous sex reversals have been produced experimentally in two classes of vertebrates (fishes and amphibians) by hormonal treatment, thus demonstrating that even among these higher forms the sexual genotype is not solely determinative. Among birds and mammals, the genetic endowment appears to limit the hormonal modification of morphological sexual characters more strongly, but complete reversals of the behavioral phenotype are possible.

It is surprising that among mammals, which show relatively greater complexity of neural development than do fishes and amphibians, behavioral characteristics should be influenced by the fetal and larval hormones to a nearly comparable degree. Nevertheless, the behavioral studies we have reviewed permit us to regard the genetic constitution of mammals as a plastic, pluripotential matrix, highly susceptible to shaping and selection by environmental influences such as the hormones. For those mammalian species (guinea pig [Valenstein et al., 1955; Goy and Young, 1957]; rat [Hård and Larsson, 1968]; rhesus monkey [Mason, 1961; Harlow, 1965]; and human being [Money, 1963]) in which social experience also contributes to the development of sexual and sex-related behaviors, I propose the hypothesis that early hormonal influences predispose the individual to the acquisition of specific patterns of behavior.

The physiological basis for this predisposition to acquire sexually specific patterns of behavior may well be a unique organization of neural systems involved in the mediation of drive and reward. The experiments with rhesus monkeys can be interpreted broadly as suggesting that neural structures that have no dependence on the activational properties of hormones appear to be involved in the acquisition and maintenance of masculine social behavior. Speculatively, the hypothesis may be advanced that some of these neural structures are motivational in character and functionally related to the reinforcing events associated with, or derived from, the performance of such masculine activities as rough-and-tumble play, threatening, chasing, and prepubertal mounting. All these behaviors are displayed by infant and juvenile female, as well as male, monkeys. The distinction between the sexes is purely quantitative, and the reinforcing events involved in regulating the quantitative aspects of behavior are sexually specific.

### REFERENCES

ALLISON, J. E., 1965. Testicular feminization. *Okla. Med. Ass.* 58: 378–380.
ALLISON, J. E., A. J. STANLEY, and L. G. GUMBRECK, 1965. Sex chromatin and idiograms from rats exhibiting anomalies of the reproductive organs. *Anat. Rec.* 153: 85–91.

BARRACLOUGH, C. A., and R. A. GORSKI, 1962. Studies on mating behaviour in the androgen-sterilized female rat in relation to the hypothalamic regulation of sexual behaviour. *J. Endocrinol.* 25: 175–182.

BEACH, F. A., 1945. Hormonal induction of mating responses in a rat with congenital absence of gonadal tissue. *Anat. Rec.* 92: 289–292.

BEACH, F. A., and A. M. HOLZ, 1946. Mating behavior in male rats castrated at various ages and injected with androgen. *J. Exp. Zool.* 101: 91–142.

BEACH, F. A., R. G. NOBLE, and R. K. ORNDOFF, 1969. Effects of perinatal androgen treatment on responses of male rats to gonadal hormones in adulthood. *J. Comp. Physiol. Psychol.* 68: 490–497.

BEACH, F. A., and P. RASQUIN, 1942. Masculine copulatory behavior in intact and castrated female rats. *Endocrinology* 31: 393–409.

BEEMAN, E. A., 1947. The effect of male hormones on aggressive behavior in mice. *Physiol. Zool.* 20: 373–405.

BRONSON, F. H., and C. DESJARDINS, 1968. Aggression in adult mice: Modification by neonatal injections of gonadal hormone. *Science (Washington)* 161: 705–706.

CHANG, C.-Y., and E. WITSCHI, 1956. Genic control and hormonal reversal of sex differentiation in Xenopus. *Proc. Soc. Exp. Biol. Med.* 93: 140–144.

CHARNIAUX-COTTON, H., 1962. Androgenic gland of crustaceans. *Gen. Comp. Endocrinol. (Suppl.)* 1: 241–247.

CROSSLEY, D. A., and H. H. SWANSON, 1968. Modification of sexual behaviour of hamsters by neonatal administration of testosterone propionate. *J. Endocrinol.* 41: xiii–xiv (abstract).

DIAMOND, M., and C. L. WONG, 1969. Neonatal progesterone: Effect on reproductive functions in the female rat. *Anat. Rec.* 163: 178 (abstract).

EATON, G. G., 1969. Perinatal androgen's role in the ontogenesis of coital behavior in the male hamster (*Mesocricetus auratus*). University of California, Berkeley, doctoral thesis.

FEDER, H. H., and R. E. WHALEN, 1965. Feminine behavior in neonatally castrated and estrogen-treated male rats. *Science (Washington)* 147: 306–307.

GERALL, A. A., and I. L. WARD, 1966. Effects of prenatal exogenous androgen on the sexual behavior of the female albino rat. *J. Comp. Physiol. Psychol.* 62: 370–375.

GOLDFOOT, D. A., H. H. FEDER, and R. W. GOY, 1969. Development of bisexuality in the male rat treated neonatally with androstenedione. *J. Comp. Physiol. Psychol.* 67: 41–45.

GOY, R. W., 1968. Organizing effects of androgen on the behaviour of rhesus monkeys. *In* Endocrinology and Human Behaviour (R. P. Michael, editor). Oxford University Press, London, pp. 12–31.

GOY, R. W., W. E. BRIDSON, and W. C. YOUNG, 1964. Period of maximal susceptibility of the prenatal female guinea pig to masculinizing actions of testosterone propionate. *J. Comp. Physiol. Psychol.* 57: 166–174.

GOY, R. W., C. H. PHOENIX, and R. MEIDINGER, 1967. Postnatal development of sensitivity to estrogen and androgen in male, female and pseudohermaphroditic guinea pigs. *Anat. Rec.* 157: 87–96.

GOY, R. W., C. H. PHOENIX, and W. C. YOUNG, 1962. A critical period for the suppression of behavioral receptivity in adult female rats by early treatment with androgens. *Anat. Rec.* 142: 307 (abstract).

GOY, R. W., and W. C. YOUNG, 1957. Somatic basis of sexual behavior patterns in guinea pigs. *Psychosom. Med.* 19: 144–151.

GOY, R. W., and W. C. YOUNG, 1958. Responses of androgen-treated spayed female guinea pigs to estrogen and progesterone. *Anat. Rec.* 131: 560 (abstract).

GRADY, K. L., C. H. PHOENIX, and W. C. YOUNG, 1965. Role of the developing rat testis in differentiation of the neural tissues mediating mating behavior. *J. Comp. Physiol. Psychol.* 59: 176–182.

HÅRD, E., and K. LARSSON, 1968. Dependence of adult mating behavior in male rats on the presence of littermates in infancy. *Brain, Behav. Evolut.* 1: 405–419.

HARLOW, H. F., 1961. The development of affectional patterns in infant monkeys. *In* Determinants of Infant Behaviour (B. M. Foss, editor). John Wiley and Sons, New York, pp. 75–97.

HARLOW, H. F., 1965. Sexual behavior in the rhesus monkey. *In* Sex and Behavior (F. A. Beach, editor). John Wiley and Sons, New York, pp. 234–265.

HARRIS, G. W., 1964. Sex hormones, brain development and brain function. *Endocrinology* 75: 627–648.

HARRIS, G. W., and S. LEVINE, 1962. Sexual differentiation of the brain and its experimental control. *J. Physiol. (London)* 163: 42P–43P.

HARRIS, G. W., and S. LEVINE, 1965. Sexual differentiation of the brain and its experimental control. *J. Physiol. (London)* 181: 379–400.

LEVINE, S., and R. F. MULLINS, JR., 1966. Hormonal influences on brain organization in infant rats. *Science (Washington)* 152: 1585–1592.

MASON, W. A., 1961. The effects of social restriction on the behavior of rhesus monkeys: II. Tests of gregariousness. *J. Comp. Physiol. Psychol.* 54: 287–290.

MONEY, J., 1963. Psychosexual development in man. *In* Encyclopedia of Mental Health. Franklin Watts, Inc., New York, pp. 1678–1709.

NADLER, R. D., 1969. Differentiation of the capacity for male sexual behavior in the rat. *Hormones and Behav.* 1: 53–63.

NOUMURA, T., J. WEISZ, and C. W. LLOYD, 1966. *In vitro* conversion of 7-³H progesterone to androgen by the rat testis during the second half of fetal life. *Endocrinology* 78: 245–253.

PHOENIX, C. H., R. W. GOY, A. A. GERALL, and W. C. YOUNG, 1959. Organizing action of prenatally administered testosterone propionate on the tissues mediating mating behavior in the female guinea pig. *Endocrinology* 65: 369–382.

RESKO, J. A., 1967. Plasma androgen levels of the rhesus monkey: Effects of age and season. *Endocrinology* 81: 1203–1212.

RESKO, J. A., H. H. FEDER, and R. W. GOY, 1968. Androgen concentrations in plasma and testis of developing rats. *J. Endocrinol.* 40: 485–491.

ROSENBLUM, L. A., 1961. The development of social behavior in the rhesus monkey. University of Wisconsin, Madison, Wisconsin, doctoral thesis.

STANLEY, A. J., and L. G. GUMBRECK, 1964. Male pseudoher-linked recessive character. Program, 46th Meeting, The Endocrine Society, Abstract no. 36.

STANLEY, A. J., L. G. GUMBRECK, and R. B. EASLEY, 1966. FSH content of male pseudohermaphrodite rat pituitary glands together with the effects of androgen administration and castration on gland and organ weights. Program, 48th Meeting, The Endocrine Society, Abstract no. 235.

STERN, J. J., 1969. Neonatal castration, androstenedione, and the mating behavior of the male rat. *J. Comp. Physiol. Psychol.* 69: 608-612.

SWANSON, H. H., 1967. Alteration of sex-typical behaviour of hamsters in open field and emergence tests of neo-natal administration of androgen or oestrogen. *Anim. Behav.* 15: 209-216.

TOLLMAN, J., and J. A. KING, 1956. The effects of testosterone propionate on aggression in male and female C57BL/10 mice. *Brit. J. Anim. Behav.* 4: 147-149.

TURNER, C. D., 1969. Experimental reversal of germ cells. *Embryologia* 10: 206-230.

VALENSTEIN, E. S., W. RISS, and W. C. YOUNG, 1955. Experiential and genetic factors in the organization of sexual behavior in male guinea pigs. *J. Comp. Physiol. Psychol.* 48: 397-403.

WARD, I. L., 1969. Differential effect of pre- and postnatal androgen on the sexual behavior of intact and spayed female rats. *Hormones and Behav.* 1: 25-36.

WHALEN, R. E., 1964. Hormone-induced changes in the organization of sexual behavior in the male rat. *J. Comp. Physiol. Psychol.* 57: 175-182.

WHALEN, R. E., D. A. EDWARDS, W. G. LUTTGE, and R. T. ROBERTSON, 1969. Early androgen treatment and male sexual behavior in female rats. *Physiol. Behav.* 4: 33-39.

WILSON, J. G., and H. C. WILSON, 1943. Reproductive capacity in adult male rats treated prepuberally with androgenic hormone. *Endocrinology* 33: 353-360.

WILSON, J. G., and W. C. YOUNG. 1941. Sensitivity to estrogen studied by means of experimentally induced mating responses in the female guinea pig and rat. *Endocrinology* 29: 779-783.

YAMAMOTO, T., K. TAKEUCHI, and M. TAKAI, 1968. Male-inducing action of androstenedione and testosterone propionate upon XX zygotes in the medaka, *Oryzias latipes*. *Embryologia* 10: 116-125.

YOUNG, W. C., 1961. The hormones and mating behavior. *In* Sex and Internal Secretions, Vol. 2 (W. C. Young, editor). Williams and Wilkins, Baltimore, Maryland, pp. 1173-1239.

# 21 Stability and Plasticity of Motivation Systems

ELLIOT S. VALENSTEIN

THERE IS NO compelling reason to maintain that only one mechanism underlies the stability and plasticity of all behaving organisms. Also, it is not likely that behavioral stability and plasticity will be explained by a single principle even in a given species. It may be important to bear in mind while deriving molecular explanation from "simple nervous systems," therefore, that there may be more than one "code" and that some aspects of plasticity and stability may emerge only as properties of complex networks.

The satisfaction of biological and acquired needs requires both stable and plastic response systems. Where the environment provides a relatively constant supply of that which is needed, as, for example, oxygen, the response can be relatively stable; where the sources of supply are less dependable, the organism must behave in a manner that maximizes the possibility of making contact with the rel-evant goal objects (including trying patterns that have been successful in the past), and the consummatory behavior must be responsive to the various forms that the goal object might take. It is equally important to note, however, that plastic and stable responses may be interrelated (Galambos, 1967). Plasticity may result from a shifting between built-in, relatively stable responses ("fixed action patterns") or from a change in frequency and order of display of a series of such responses. A new behavior emerges from the "shaping" of a genetically specified nervous system, and even those persons who had been most optimistic about operant conditioning techniques have come to recognize that instrumental responses by animals trained in the laboratory often revert to patterns characteristic of the species ("instinctive drift," as illustrated by Breland and Breland, 1961).

To elaborate that point of view, this paper describes some recent studies of the shifting of stable, built-in response patterns, along with some speculations relating this work to the more general topic of plasticity and stability of

ELLIOT S. VALENSTEIN Department of Psychophysiology-Neurophysiology, Fels Research Institute, Yellow Springs, Ohio

motivational systems. Data collected by electrical stimulation of hypothalamic structures are emphasized, but a number of the conclusions bear on different techniques for activating neural structures and to brain areas other than the hypothalamus.

## Historical review

A few years after Ewald (as described by Talbert, 1900) explored the possibilities of employing chronic electrodes for cortical stimulation of dogs, Karplus and Kreidl initiated their classical studies of hypothalamic functioning. Electrical stimulation through deep electrodes produced pupillary dilation and constriction, as well as other indexes of automatic outflow (e.g., tearing, salivation). This work, which extended into the 1920s and stressed autonomic mechanisms and hypothalamus-brainstem interactions, seemed to confirm Sherrington's characterization of the hypothalamus as the "head ganglion of the autonomic nervous system." The utilization of electrical stimulation of the hypothalamus by Karplus and Kreidl was followed shortly by that of Ranson and his coworkers (Ranson, 1937; Ranson and Magoun, 1939), who generally used lightly anesthetized animals, and perhaps most dramatically by Hess and his associates (cf. summary of experimental work of W. R. Hess by Gloor, 1954), who studied fully awake, relatively unrestrained animals.

Whereas the early reports of the elicited behavior emphasized their isolated autonomic characteristics, there gradually evolved an appreciation that the responses were characteristic of such states as somnolence, rage, temperature regulation, and eating. In the meantime, such labels as "sham rage," which emphasized the motor aspects of the response and denied the presence of a motivational or affective component, were replaced—a trend particularly evident in Hess's later writings, in which he rejected the term "sham rage" and regarded Bard's phrase "angry behavior" as more appropriate to one of the states that could be produced by hypothalamic stimulation.

Hess's work in Zurich, which was relatively uninterrupted by World War II and continued into the 1950s, was still in progress when the report by Olds and Milner (1954) made it apparent that appropriate forebrain stimulation did indeed elicit strong motivational reactions. To deal with the fact that the frequency of emission of arbitrarily selected responses could be dramatically elevated when followed by stimulation of selected hypothalamic and limbic structures, such labels as "reinforcing brain area," "start or go system" and "positively motivating system" have been applied. During the past 15 years a great number of such studies examined either the various properties of these "reinforcing brain systems" (cf. reviews by Olds, 1962; Valenstein, 1966) or the biologically significant behaviors that could be

elicited. The behaviors described include eating (Coons, 1963; Coons et al., 1965; Delgado and Anand, 1953; Fantl and Schuckman, 1967; Hutchinson and Renfrew, 1966; Larsson, 1954; Morgane, 1961; Roberts et al., 1967; Steinbaum and Miller, 1965; Tenen and Miller, 1964), drinking (Andersson and McCann, 1955; Greer, 1955; Mendelson, 1967; Mogenson and Stevenson, 1967), gnawing (Roberts and Carey, 1965), hoarding (Herberg and Blundell, 1967), stalking-attack (Flynn, 1967; Hutchinson and Renfrew, 1966; Roberts and Kiess, 1964; Roberts et al., 1967), coprophagia (Mendelson, 1966), and male copulatory behavior (Caggiula, 1970; Caggiula and Hoebel, 1966; Roberts et al., 1967; Vaughn and Fisher, 1962). Because, in most cases, nondeprived animals exhibit these responses only during periods of stimulation, the term "stimulus-bound" behavior is applied to them.

Stimulus-bound behavior has generally been considered a motivated, rather than a stereotyped, motor act, because it occurs only when appropriate goal objects are present and because the stimulated animals both learn a task (Coons et al., 1965) and tolerate aversive stimulation, such as shock (Morgane, 1961) and quinine additives (Tenen and Miller, 1964), in order to perform it. It is widely accepted, furthermore, that stimulus-bound eating is elicited by activation of specific neural circuits underlying "hunger," drinking by activation of specific "thirst" circuits, and so on, with the other elicited behaviors studied. Distinguishable from the issue of specific neural circuits is the view, probably held by fewer investigators, that relatively discrete hypothalamic regions control behavior related to specific biological needs.

## Are specific hypothalamic circuits stimulated?

During the past year, I have undertaken a series of collaborative studies with Verne Cox and Jan Kakolewski on the modifiability of behavior elicited by hypothalamic stimulation and the nature of the motivation underlying this behavior. These studies suggest an alternative to the "specific circuit" view of the hypothalamus in the display of plastic and stable responses and provide the background for some theoretical speculation.

An examination of the relevant literature on stimulus-bound behaviors makes it apparent that an impressive overlap in the anatomical correlates of the different types exists. It therefore occurred to us that the particular behavior observed in a given study might simply reflect the experimenter's interests and his consequent limitation of the reinforcements provided in the testing situation. Consequently, we provided the animal with ample opportunity to express its "point of view." As Table I shows, the behavior elicited by hypothalamic stimulation could indeed change without any modification of stimulus parameters (Valenstein et al., 1968a; Valenstein et al., 1969b). If a rat initially eats in response to stimulation, removal of food is generally followed in time by drinking or wood-gnawing. This appearance of a second behavior superficially resembles the "displacement behavior" of the ethologists, but because the new, stimulus-bound behavior appears as frequently as the initial behavior when, in subsequent tests, all objects are again available, the two phenomena differ. The results depicted in Table 1 have been replicated in several hundred animals and suggest that no exclusive relationship exists between the hypothalamic site activated by electrical stimulation and the first behavior pattern evoked.

In a few instances, the initial behavior elicited by hypothalamic stimulation could not be "switched" to a second behavior; in such cases, there may have been an explanation for the failure. In many of these cases, the elicited behavior was either directed toward the animal's own body or consisted of locomotor activities and, as a result, it was difficult or impossible to prevent its expression by removing the goal object. (Such responses are not ordinarily scored as stimulus-bound, which may be only because they do not conform to an arbitrarily chosen classification system.) A

striking case illustrates the point. One animal picked up the end of its tail and moved it laterally in the mouth, seemingly as a form of preening, during stimulation. We could not "switch" this behavior until it occurred to us to fasten the tail to the animal's back with adhesive tape. When stimulated, the animal initially turned back toward where its tail should have been. After approximately 20 stimulations, the rat explored the other possibilities in its environment, and in time drank in response to stimulation. Drinking was later displayed as frequently as the "tail preening" after the tail was released. We are presently exploring the possibility that, when stimulation does not elicit a specific stimulus-bound behavior, the general "locomotor exploratory behavior" evoked may, like tail-preening, be responsible.

Our conviction that no specific relationship exists between a given hypothalamic site and a particular response is strengthened by other evidence. For instance, switches are not restricted to oral behavior such as eating, drinking, and gnawing; animals that initially shuffle food with the forepaws or dig in sand have been switched to eating or drinking in response to the same stimulation. Similarly, both

## TABLE I

*Eating (E), drinking (D), and gnawing (G) behavior elicited during hypothalamic stimulation. Each test had 20 stimulation periods. Maximum score for any one behavior is 20, but the animal could exhibit different behaviors during each period. The dash (—) in the second series of tests indicates which goal object had been removed. RP, retangular pulses; SW, sine wave.*

| Numbers of Animals | Behavior | First Series | | | Second Series | | Competition | Stimulus Parameters ($\mu a$) |
|---|---|---|---|---|---|---|---|---|
| | | 1 | 2 | 3 | 1 | 2 | | |
| 60 | E | 0 | 0 | 0 | 15 | 17 | 11 | RP, 80 |
| | D | 20 | 20 | 20 | — | — | 14 | RP, 80 |
| | G | 0 | 0 | 0 | 0 | 0 | 0 | RP, 80 |
| 61 | E | 0 | 0 | 0 | 20 | 20 | 15 | RP, 120 |
| | D | 20 | 20 | 20 | — | — | 12 | RP, 120 |
| | G | 0 | 0 | 0 | 0 | 0 | 0 | RP, 120 |
| 63 | E | 0 | 0 | 0 | 0 | 0 | 0 | RP, 500 |
| | D | 0 | 0 | 0 | 20 | 20 | 12 | RP, 500 |
| | G | 20 | 20 | 20 | — | — | 8 | RP, 500 |
| 74 | E | 0 | 0 | 0 | 20 | 20 | 12 | SW, 20 |
| | D | 20 | 20 | 20 | — | — | 13 | SW, 20 |
| | G | 0 | 0 | 0 | 0 | 0 | 0 | SW, 20 |
| 80 | E | 19 | 16 | 12 | — | — | 10 | RP, 120 |
| | D | 1 | 5 | 8 | 19 | 16 | 10 | RP, 120 |
| | G | 0 | 0 | 0 | 2 | 2 | 6 | RP, 120 |
| 89 | E | 0 | 0 | 0 | 18 | 20 | 16 | SW, 24 |
| | D | 19 | 19 | 20 | — | — | 4 | SW, 24 |
| | G | 0 | 0 | 0 | 0 | 0 | 0 | SW, 24 |

(After Valenstein et. al., 1968a)

Caggiula (1970) and Gallistel (1969) have recorded male sexual behavior and eating from the same electrode at the same current parameters. Also, von Holst and von Saint Paul (1963) showed that domestic fowls cluck, fluff the feathers, preen, or do nothing at all after shocks to the same electrode site, at different times, and without alteration in the stimulus parameters.

A controversy exists, however, over the interpretation of our findings. Wise (1968), for example, has confirmed our finding that in almost all cases a second elicited behavior may be obtained, but he concluded that his experiment explains our results in a way that "is consistent with the theory that separate, fixed neural circuits, functionally isolated from each other by biochemical specificity, mediate eating and drinking in this area of the brain" (p. 379). He reports that a second behavior can be elicited simply by increasing hypothalamic stimulation intensity to a sufficiently high level, and that the current intensity required to elicit a given behavior tends to decline with time. Thus, he claims, two neural circuits mediating different behaviors have different stimulation thresholds, and the second behavior emerges not because of environmental manipulation, but because of a gradual increase with time in the sensitivity of the neural system responsible.

As has been pointed out in a reply (Valenstein et al., 1969a), Wise has overlooked two important items in his experiment. Although a second behavior was elicited by increasing the stimulation intensity, it took place after the initially preferred object was removed. Furthermore, Wise did not state how long it took for the second behavior to emerge after current strength was increased and the preferred goal object was removed. Did the new behavior appear immediately (as would be expected if a functionally distinct neural circuit were activated) or gradually? Addressing these questions, we (Cox and Valenstein, 1969) demonstrated that, in many cases, high-intensity stimulation did not immediately elicit a new behavior in the presence of the initially preferred goal object. Furthermore, after its removal, stimulation only gradually acquired the ability to elicit a new behavior (Table II), whereas immediate display would be expected of a behavior mediated by a different neural substrate with a higher threshold.

It has also been suggested that stimulation through small electrodes might limit the response to one behavior alone. It turns out that the likelihood of obtaining a specific stimulus-bound behavior is radically decreased with small-diameter electrodes, but, if a response could be elicited, a second response could also be obtained by our procedure (Valenstein et al., 1969b). Dr. Bengt Andersson (personal communication, 1968) had noted earlier that, when he reduced the size of the electrodes implanted in goats, drinking could no longer be elicited. Miller (1965) reported similar results in the rat.

Evidence to support our conclusion comes also from inquiry into the motivation underlying the behavior elicited by hypothalamic stimulation. In several situations, animals displaying stimulus-bound eating or drinking did not behave as if hungry or thirsty. For example, animals eating in response to stimulation did not readily switch to another food when the first was removed from the test chamber; many switched to drinking even when the food was changed only in shape, as when pellets were ground (Valenstein et al., 1968b). A large percentage of rats reared from infancy on a liquid diet displayed, when mature, stimulus-bound eating of food pellets, but, if deprived of them, they did not eat the familiar liquid diet in response to stimulation (Valenstein and Phillips, 1970). These results are especially significant, because the rats generally preferred the liquid diet to the food pellets. As for drinking behavior, animals displaying stimulation-induced drinking did not switch to drinking from a different, but familiar, container. Also, their taste preferences differ from that of rats drinking when thirst has been induced by water deprivation: naturally thirsty animals prefer water to a glucose solution, but the converse is true of stimulus-bound drinking. Furthermore, animals exhibiting stimulus-bound drinking often continued to lap at an empty water tube, suggesting that the ingestion of water was not essential to the maintenance of the elicited lapping behavior (Valenstein et al. 1968). Prior to eating or drinking, animals have also engaged in stereotyped movement patterns that appeared to be as significant a component of the elicited response as the final ingestion. Finally, in several instances whether an animal ate or drank in response to stimulation seemed to depend on the location of the food or water. None of these observations is consistent with the view that stimulation activates a specific drive state, such as hunger or thirst.

Among the conclusions suggested by these results are the following: 1. *The relationship between a specific hypothalamic area and a behavior pattern is not fixed.* 2. *Animals displaying stimulus-bound eating or drinking do not exhibit response or stimulus generalization gradients similar to those of animals known to be hungry or thirsty.* If deprived of the opportunity to engage in the initial behavior, the selection of a new goal object is not dictated by the animal's capacity to satisfy the same presumed drive state. 3. *The behavior elicited by hypothalamic stimulation may be maintained by some other reinforcement than that derived from satisfying a simulated need state such as hunger or thirst.* 4. *Hypothalamic stimulation does not evoke specific drive states.*

These conclusions should make the report that a new behavior can become associated with the stimulus more credible and, in addition, eliminate the dilemma raised because animals will self-stimulate at electrode sites that are presumed to evoke such drives as hunger. The paradox of why an animal should stimulate itself and thereby produce a hunger-drive state dissolves if one does not assume that the animal is hungry. The assumption that an animal experiences the aversive consequences specifically associated with hunger, based primarily on the evidence of motivation to eat, has produced confusion. The awkwardness of this formulation becomes even more evident when one appreciates the necessity to postulate hypothalamic "centers" or even circuits for such drives as gnawing, food shuffling, tail preening, and so forth, and, as is described subsequently, object carrying. I shall offer an alternate view after I have reviewed some additional experimental work.

## Anatomical analysis

Figure 1A, B, and C present a portion of the anatomical results obtained from some 300 rats and 400 electrode placements, in which both the histological and behavioral data were judged to be reliable by a set of explicit criteria. These anatomical data will be completely analyzed elsewhere (Cox and Valenstein, 1970). They show that no precise correlations can be established between behavior elicited and site stimulated. Thus posterior hypothalamic points that produced eating, and so forth, were found in the region shown by others to elicit male copulatory behavior (Caggiula, 1970; Caggiula and Hoebel, 1966). Stimulus-bound eating and drinking have also been elicited from the medial portions of the ventral thalamic nucleus, a confirmation of other reports that extrahypothalamic sites elicit eating (cf. review by Morgane and Jacobs, 1969, pp. 153–171). The only hypothalamic regions that did not elicit eating, drinking, or gnawing were the anterior and ventromedial nuclei.

Furthermore, sites from which "nonspecific exploratory behavior" was evoked (cf. earlier comments on this arbitrary classification) overlap with sites eliciting one of the specific consummatory behaviors. Finally, the rate of self-stimulation, high versus low, was not correlated with the presence or absence of a specific elicited behavior.

It is possible that functionally discrete systems may be interdigitated, but the electrical-stimulation evidence does not provide strong support for such a view. Von Holst and von Saint Paul (1963), for example, have noted that, in the domestic fowl, the reaction that previously had followed stimulation could still be evoked after electrocoagulation of the tissue 0.3 to 0.5 mm around an electrode. One might also assume that changing stimulus frequency and wave shape would modify both the structures activated and the pattern of their excitation, but such changes do not seem to alter the elicited behavior pattern.

## Electrolytic lesions between hypothalamus and brainstem

In experiments still in the preliminary stages, Cox has shown that unilateral lesions in the ventral tegmental area of Tsai can significantly reduce stimulus-bound eating and drinking without interference with normal eating, drinking, growth, or general behavior. Furthermore, when each of two bilaterally placed electrodes was effective, a unilateral lesion disrupted the behavior produced by the ipsilateral electrode alone. Why do these lesioned animals not eat or drink if hypothalamic stimulation makes them hungry or thirsty? Either the stimulation produces "hunger" only in

TABLE II

*Stimuli Required Before Appearance of New Behavior After Removal of Initially Preferred Item*

| Subject Numbers | Current Intensity* | Number of Stimuli | | New Behavior** |
| | | First Appearance | Reliable Appearance† | |
|---|---|---|---|---|
| 10AA | 10 SW†† (high) | 6 | 174 | D |
| 21AA | 20 SW (high) | 18 | 184 | D |
| 97Z | 7 SW (low) | 19 | 318 | E |
| 49AA | 17 SW (low) | 3 | 39 | E |
| 53AA | 10 SW (low) | 52 | 52 | E |
| 56Y | 16 SW (high) | 541 | 541 | E |

*Low indicates the initial current intensity employed to elicit a response. High indicates the highest current that did not appear disruptive.

**D, drinking; E, eating.

†Eight responses out of 20 consecutive stimulations were used as criteria for reliable appearance of new behavior.

††SW = 60-cycle sine wave.

(After Cox and Valenstein, 1969)

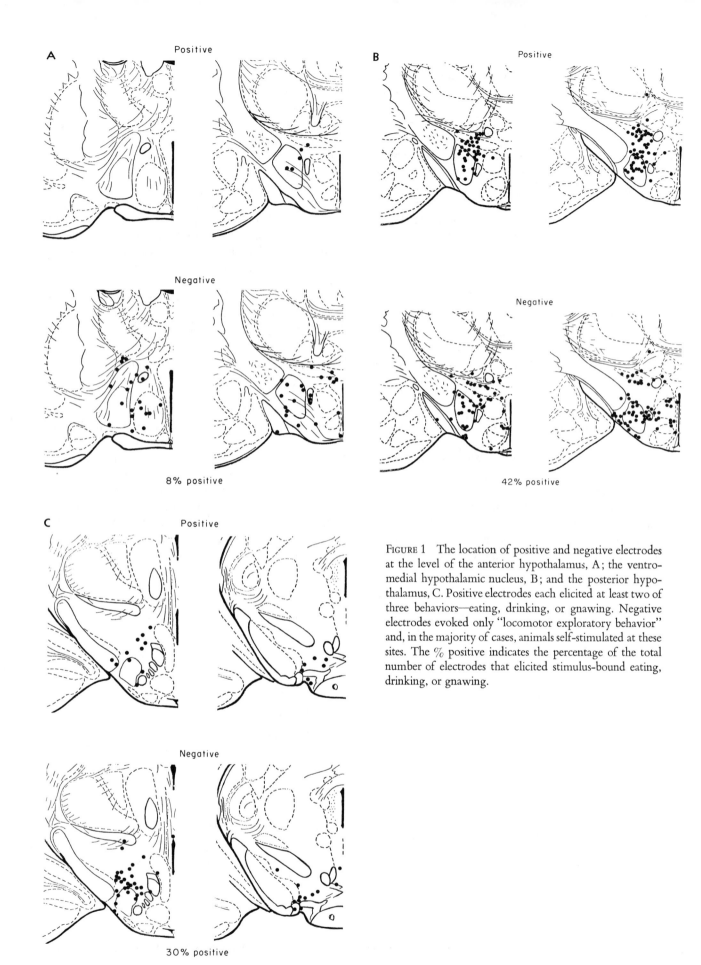

FIGURE 1 The location of positive and negative electrodes at the level of the anterior hypothalamus, A; the ventro-medial hypothalamic nucleus, B; and the posterior hypo-thalamus, C. Positive electrodes each elicited at least two of three behaviors—eating, drinking, or gnawing. Negative electrodes evoked only "locomotor exploratory behavior" and, in the majority of cases, animals self-stimulated at these sites. The % positive indicates the percentage of the total number of electrodes that elicited stimulus-bound eating, drinking, or gnawing.

concert with brainstem structures (in which case, brainstem structures may prove to be equally effective for producing these effects) or it achieves its result because it activates brainstem motor pathways. Obviously, I favor the latter view, if for no other reason than that the hypothalamic activation of a specific hunger drive has been rejected.

## Prepotent responses of individual animals

Although hypothalamic location does not provide a good indication of the behavior that will be elicited, an examination of the data obtained from a large number of animals does suggest a pertinent factor. We have observed that, when both of two hypothalamic electrodes in an animal elicit stimulus-bound behavior, it is highly probable that the behavior will be the same in quality, even though the *anatomical location of the electrodes may be widely disparate* (cf., exceptions listed under Anatomical Analysis). This was true of animals that initially exhibited only one behavior such as drinking, eating, or wood gnawing, as well as animals that exhibited two or more behaviors from the same electrode (Valenstein and Cox, 1970). On occasion, animals exhibited such uncommon stimulus-bound behavior as, for example, food-shuffling with the forepaws or tail-preening, and these relatively rare behaviors also tended to be displayed from both electrodes, even though their hypothalamic location was quite different. Apparently the response or responses that are prepotent in a given animal are those likely to be expressed in stimulus-bound behavior, which implies not that different electrodes in an animal always elicit the same behavior, but rather that some factor, which might be called an animal's "response hierarchy," must be considered in an evaluation of results. This point is particularly relevant to data collected from those multiple "strut" electrodes, in which several sites can be explored along a common anterior-posterior plane in a single animal. In such preparations, there is the danger that a number of sites, obtained from relatively few animals, can exaggerate the impression that a given behavior pattern is associated with a particular anterior-posterior plane in the hypothalamus.

As noted, we and others have observed instances in which two electrodes in a given animal may elicit different behavior (Valenstein et al., 1969b), but some of these cases may reflect the fact that selective experience has been provided. A study still in progress illustrates the possible relevance of special "training" (Valenstein, in preparation). Animals which initially exhibited only one stimulus-bound behavior have been compared after two different training procedures. Half the animals received a large number of additional tests, with only the one initially chosen goal object present; the other animals were provided with comparable time in the test chamber, but received no stimulation experience while there. Both groups were then stimulated in the absence of

their preferred goal object until a second stimulus-bound behavior pattern emerged. After a second behavior appeared, the animals were tested in a competitive situation, with both goal objects present. Those that had received the extra stimulation experience, with only the preferred goal object available, exhibited a stronger preference for the first behavior and displayed the second behavior relatively infrequently. This important result indicates that repetition of the response in the presence of the *eliciting* stimulation strengthens the bond between the two. It is evident that the significance of examples of different behavior elicited from two electrodes in the same animals cannot be determined unless details of the training procedure are provided. It is not, however, suggested that this is always the explanation for the elicitation of different behaviors in our animals.

## Prepotent responses characteristic of the species

The behavior elicited by hypothalamic stimulation does not consist of any arbitrary response that may be dominant at a given moment. The responses evoked are well-established motor responses that are especially significant to the species, as well as the individual animal. This has become even more evident as a result of some recent collaborative work with A. G. Phillips (Phillips et al., 1969).

The subjects were rats with electrodes implanted at different sites throughout the hypothalamus. These animals were tested in a Plexiglas chamber 60 cm in length in which photocells were placed close to each end. If the animal activated the photocells at one end (ON side), it received continuous hypothalamic stimulation until it interrupted the photocell beam at the other end (OFF side) to terminate the stimulus. In a matter of minutes, the animals learned to self-stimulate and to control the duration of stimulation by running back and forth. A variety of objects (food pellets, rubber erasers, wooden dowel sticks, and molding strips) was then distributed throughout the ON side of the chamber. All the animals that self-stimulated reliably (19 out of 22) began to pick up objects, carry them to the OFF side, and deposit them there as soon as the stimulus terminated. No animal carried objects from the OFF to the ON side, nor did any carry objects when not stimulated. The inedible as well as the edible objects were transported, and on a few occasions an animal even carried its own tail. In the number and kinds of objects carried, rats that had never exhibited specific stimulus-bound gnawing or eating behaved like those that had. Food-deprived animals self-stimulated at higher rates and consequently carried more objects per unit of time, but they did not show a preference for edible items.

In an extension of this study (Valenstein et al. 1970), the ON side of the chamber was divided in half, lengthwise, with a Plexiglas partition. After training, the animals were first tested without objects, and could self-stimulate by entering either half. Objects were placed in the least-preferred side; the data indicate a unanimous preference for stimulation with an opportunity to carry objects back over stimulation alone. These results resemble Mendelson's report (1966) showing that satiated rats prefer the combination of food and brain stimulation to brain stimulation alone. Apparently, brain stimulation does not fully activate all the neural circuits responsible for reinforcement. Execution of a stimulus-bound response provides reinforcement over and above that produced by hypothalamic stimulation. According to one view (which I do not share), this evidence that elicited object-carrying is reinforcing requires the addition of another specific drive system to the ever-growing hypothalamic list.

We also calculated the average duration of the stimulation and nonstimulation periods selected by the animals and programed the equipment to deliver these durations in a regular sequence without regard to the animal's behavior or location (Phillips et al., 1969). Initially, animals that happened by chance to be at the previous ON side when stimulated, picked up and carried objects to the opposite side; those stimulated on the previous OFF side seldom did so. Within several minutes, all carrying behavior stopped, and animals that previously had exhibited stimulus-bound eating or wood-gnawing reverted to these behaviors, but those that had previously displayed only general locomotor exploratory behavior did not orient to the objects. Object-carrying appears, therefore, only when the moment of stimulation bears some regular relationship to the animal's behavior, location in the environment, or both.

Two other facets of this study should be mentioned briefly before the implications of this research are discussed. A group of rats was reared to maturity in our laboratory on an exclusively liquid diet. Hypothalamic electrodes were then implanted, and the animals were tested with wooden objects in the Plexiglas chamber. These animals, which had had no contact with manipulable objects other than their own bodies and feces, carried objects as efficiently as did the normally reared animals. Similarly, Slonim (1968) has reported that the golden hamster (*Hemicricetus auratus*), reared with only liquid food available, stores food at the same age and with the same intensity as do normally reared animals. In contrast, guinea pigs reared under normal laboratory conditions, with food pellets available, self-stimulated by running back and forth in the test chamber, but they never carried objects, although they occasionally ate during stimulation.

It is well known that rats carry objects related to hoarding, nest building, and retrieving of young. They also carry objects that have no obvious utility at all. Under natural conditions, whether engaged in food hoarding or various aspects of maternal behavior, the transporting of objects is invariably from a more open and vulnerable location to one

that is relatively familiar and protected. In this context, it should be noted that Eibl-Eibesfeldt (1961) has stressed the fact that a home or sleeping site must be established in a new environment before carrying behavior is displayed.

Summarizing these experiments, the evidence suggests that object-carrying by the rat should be viewed as a prepotent, adaptive response that can be incorporated into more complicated behavior patterns that serve the purposes of specific motivational states related to hunger and maternal behavior. It is important, however, to recognize the independent existence of this basic response unit. The rats that are carrying objects in our studies are not hoarding food, building nests, or retrieving pups. Although object-carrying can be elicited from widely disparate hypothalamic sites, the rat does not carry objects unless the stimulation presentations are consistent with the establishment of a structure meaningful to its environment. Collectively, these results stress the importance of our viewing the elicited behavior in terms of prepotent, species-specific response patterns and the environmental conditions that normally elicit them in an aroused animal. The accumulation of ethological information relevant to these responses is likely to be more valuable than interpretations based only on presumed motivational states. It may also be necessary to explore ways to determine the prepotent responses characteristic of individual animals in order to understand the behavior elicited by stimulation (Valenstein, 1970).

## Conclusions and theoretical speculations

The main conclusions drawn for elicited object-carrying are applicable to other stimulus-bound behavior. 1. Stimulus-bound eating, drinking, or gnawing do not involve hunger, thirst, or gnawing drives as generally conceptualized, but seem to be derived from conditions that excite the neural substrate underlying existing prepotent species-specific responses ("fixed action patterns"). 2. Discharging this sensitized or excited substrate is reinforcing, and it can provide the motivation to engage in instrumental behavior, which is rewarded by the opportunity to make the response. The execution of the response is not reinforcing because it decreases some "hypothalamic drive state," but rather because of the discharge of the excited substrate underlying the response. When the presence of an external goal object such as food or water is essential to the execution of the response, the elicited behavior has been considered stimulus-bound. Evidence was provided, however, that many instances of elicited locomotion and bodily oriented behavior should be included in the same class of phenomena.

The consequences of hypothalamic activation at a particular locus are not limited to only one behavior pattern. The effect is more general than that, but it is not without direction or limits. It appears likely that the activational state that produces attack behavior (Hunsperger, 1965; Flynn, 1967) is different from that associated with eating and drinking. Stimulation of the ventromedial hypothalamic nucleus in the rat may evoke aggressive attack, but in our experience it does not produce eating or drinking. It is possible that under some conditions different states may be simultaneously activated, as Hutchinson and Renfrew (1966) reported that they could obtain both eating and attack behavior from stimulation of the same hypothalamic point in the cat, but this may be a common association only in predaceous animals. Von Holst and von Saint Paul (1963) referred to the general states elicited by electrical stimulation as "moods." Hypothalamic stimulation appears to be inducing a state that is functionally related to a group of responses. It is unlikely that cellular elements at the hypothalamic level are all irrevocably tied to only one response system. Under more natural conditions, there are neural, humoral, and systemic cues at numerous physiological levels that identify the state more specifically and are responsible for the sensitization of particular response patterns.

Figure 2 attempts to summarize some of the relationships that have emerged from the experimental results. Hypo-

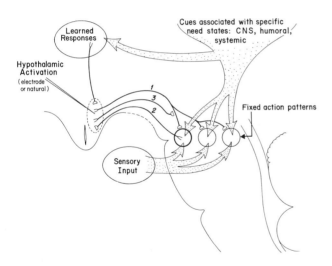

FIGURE 2 Reinforcement of motor patterns. Schema illustrates the relationships between the "reinforcing brain system" and the response system, as suggested by experimental results.

thalamic activation (1) has the capability of exciting the substrate of a number of "fixed action patterns." The motor patterns excited will consist of those responses that, in a given species, are related to the state induced by the particular hypothalamic activation. The specific response that is elicited will depend on the relative prepotency of the different responses. The prepotency may reflect the differential

sensitizing influence of the internal conditions associated with specific biological needs, but it has proved difficult to demonstrate that this has more than a temporary influence on the behavior elicited by stimulation. It is likely that "built-in" characteristics of individual animals, interacting with the modulating inputs from the environment, are the major determinants of the responses that are elicited. The opportunity to engage in the response is reinforcing, which has been schematized in Figure 2 by the depiction of a pathway (2) that has the capacity to activate the "reinforcing brain system." This sequence—that is, hypothalamic activation followed by the execution of the response—leads to a strengthening of the association, as illustrated by the connection (3). It may be important to note that this association is not strengthened by the existence of a temporal contiguity alone. If a response such as eating is displayed as a result of food deprivation, a bond will not be established with concurrent hypothalamic stimulation (Valenstein and Cox, 1970). Apparently, connection (3) in the diagram is effective in strengthening only those preceding responses just activated via a hypothalamic pathway. Lastly, the diagram indicates that instrumental responses may be elicited if they are followed by activation of the "reinforcing brain system," as in the self-stimulation experiments, or if they make it possible to obtain the reinforcement associated with the discharge of a sensitized response system. Under physiological conditions, when the cues associated with specific need states are present, the diagram indicates that there will be an increase in the excitation level of both the substrate for the "built in" appropriate consummatory responses and the particular learned responses that are capable of placing the animal in a position to exercise these consummatory responses.

The advantage to be gained from such a model is that the emphasis is shifted to the motor system and the reinforcement produced by the feedback from the response. It is commonly accepted that it is necessary to postulate an aversive component of a drive or need state. The animal's motivation, it is assumed, derives from the elimination of this aversive component. Conceivably, this is an unwarranted generalization from extreme states of bodily need. The concept of drive is justified as a means of denoting that an animal's response to a class of stimuli may vary as a function of time since last contact with such stimuli, but it would be interesting to entertain the notion that specific drives, such as hunger, gain their identity, at least in part, through an anticipation that the act of eating will be especially reinforcing at a particular time.

An earlier publication (Valenstein, 1966) stressed the "immediate reinforcement" produced by afferent neural patterns resulting from external stimulation. This "direct" activation of the reinforcing brain area was postulated as a mechanism to explain how some stimuli (e.g., food) could have motivational consequences that do not depend on feedback from the biological consequences of interaction with the stimulus (e.g., digestion and utilization). At that time, it was suggested that one of the main functions of "drives states" was to act as "gating mechanisms," which direct afferent neural impulses and thereby determine their ability to activate reinforcing neural areas. Figure 3 sum-

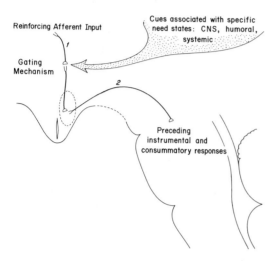

FIGURE 3 Role of sensory input. Schema illustrates the relationship between specific need states and the "immediate reinforcement" produced by sensory input.

marizes the stimulus aspect of the problem and illustrates one way of schematizing the process of reinforcement of responses that maintain the animal's contact with the reinforcing stimulus. Pathway 1 indicates that the capacity of sensory input to have reinforcing consequences depends on the "gating" accompanying specific need states. Pathway 2 is a schematic representation of the fact that the responses preceding the activation of the "reinforcing brain system" will be strengthened.

The present position, which stresses feedback accompanying the execution of responses, is meant to supplement rather than to replace the view that stresses the "immediate reinforcement" derived from sensory input. We speak of drives at those times when a configuration of stimuli and responses are most reinforcing. Normally, these configurations are meaningful in terms of some biological need, but that is only because this is a requirement for survival. The behavior of the individual animal at any given time may be guided by much more hedonistic principles.

## Acknowledgment

This work was supported by NIMH Research Grant M-4529, Research Scientist Award MH-4947, and NASA Research Grant NGL 36–005–001.

## REFERENCES

ANDERSSON, B., and S. M. McCANN, 1955. A further study of polydipsia evoked by hypothalamic stimulation in the goat. *Acta Physiol. Scand.* 33: 333–346.

BRELAND, K., and M. BRELAND, 1961. The misbehavior of organisms. *Amer. Psychol.* 16: 681–684.

CAGGIULA, A. R., 1970. Analysis of the copulation-reward properties of posterior hypothalamic stimulation in rats. *J. Comp. Physiol. Psychol.* 70: 399–412.

CAGGIULA, A. R., and B. G. HOEBEL, 1966. "Copulation-reward site" in the posterior hypothalamus. *Science (Washington)* 153: 1284–1285.

COONS, E. E., 1963. Motivational correlates of eating elicited by electrical stimulation in the hypothalamic feeding area. Yale University, doctoral thesis.

COONS, E. E., M. LEVAK, and N. E. MILLER, 1965. Lateral hypothalamus: Learning of food-seeking response motivated by electrical stimulation. *Science (Washington)* 150: 1320–1321.

COX, V. C., and E. S. VALENSTEIN, 1969. Effects of stimulation intensity on behavior elicited by hypothalamic stimulation. *J. Comp. Physiol. Psychol.* 69: 730–733.

COX, V. C., and E. S. VALENSTEIN., 1970. Distribution of hypothalamic sites yielding stimulus-bound behavior. *Brain, Behav., Evolut.* (in press).

DELGADO, J. M. R., and B. K. ANAND, 1953. Increase of food intake induced by electrical stimulation of the lateral hypothalamus. *Amer. J. Physiol.* 172: 162–168.

EIBL-EIBESFELDT, I., 1961. The interactions of unlearned behavior patterns and learning in mammals. *In* Brain Mechanisms and Learning (J. F. Delafresnaye, editor). Charles C Thomas, Springfield, Illinois, pp. 53–68.

FANTL, L., and H. SCHUCKMAN, 1967. Lateral hypothalamus and hunger: Responses to a secondary reinforcer with and without electrical stimulation. *Physiol. Behav.* 2: 355–357.

FLYNN, J. P., 1967. The neural basis of aggression in cats. *In* Neurophysiology and Emotion (D. C. Glass, editor). The Rockefeller University Press and Russell Sage Foundation, New York, pp. 40–60.

GALAMBOS, R., 1967. Introduction: Brain correlates of learning. *In* The Neurosciences: A Study Program (G. C. Quarton, T. Melnechuk, and F. O. Schmitt, editors). The Rockefeller University Press, New York, pp. 637–643.

GALLISTEL, C. R., 1969. Self-stimulation: Failure of pretrial stimulation to affect rats' electrode preference. *J. Comp. Physiol. Psychol.* 69: 722–729.

GLOOR, P., 1954. Autonomic functions of the diencephalon. A summary of the experimental work of Prof. W. R. Hess. *A. M. A. Arch. Neurol. Psychiat.* 71: 773–790.

GREER, M. A., 1955. Suggestive evidence of a primary "drinking center" in the hypothalamus of the rat. *Proc. Soc. Exp. Biol. Med.* 89: 59–62.

HERBERG, L. J., and J. E. BLUNDELL, 1967. Lateral hypothalamus: Hoarding behavior elicited by electrical stimulation. *Science (Washington)* 155: 349–350.

HOLST, E. VON, and U. VON SAINT PAUL, 1963. On the functional organisation of drives. *Anim. Behav.* 11: 1–20.

HUNSPERGER, R. W., 1956. Affektreaktionen auf elektrische

Reizung im Hirnstamm der Katze. *Helv. Physiol. Pharmacol. Acta* 14: 70–92.

HUTCHINSON, R. R., and J. W. RENFREW, 1966. Stalking attack and eating behaviors elicited from the same sites in the hypothalamus. *J. Comp. Physiol. Psychol.* 61: 360–367.

LARSSON, S., 1954. On the hypothalamic organisation of the nervous mechanism regulating food intake. *Acta Physiol. Scand.* 32(suppl. 115): 7–63.

MENDELSON, J., 1966. Role of hunger in T-maze learning for food by rats. *J. Comp. Physiol. Psychol.* 62: 341–349.

MENDELSON, J., 1967. Lateral hypothalamic stimulation in satiated rats: The rewarding effects of self-induced drinking. *Science* (*Washington*) 157: 1077–1079.

MILLER, N. E., 1965. Chemical coding of behavior in the brain. *Science* (*Washington*) 148: 328–338.

MOGENSON, G. J., and J. A. F. STEVENSON, 1967. Drinking induced by electrical stimulation of the lateral hypothalamus. *Exp. Neurol.* 17: 119–127.

MORGANE, P. J., 1961. Distinct "feeding" and "hunger motivating" systems in the lateral hypothalamus of the rat. *Science* (*Washington*) 133: 887–888.

MORGANE, P. J., and H. L. JACOBS, 1969. Hunger and satiety. *World Rev. Nut. Diet.* 10: 100–213.

OLDS, J., 1962. Hypothalamic substrates of reward. *Physiol. Rev.* 42: 554–604.

OLDS, J., and P. MILNER, 1954. Positive reinforcement produced by electrical stimulation of septal area and other regions of rat brain. *J. Comp. Physiol. Psychol.* 47: 419–427.

PHILLIPS, A. G., V. C. COX, J. W. KAKOLEWSKI, and E. S. VALENSTEIN, 1969. Object-carrying by rats: An approach to the behavior produced by brain stimulation. *Science* (*Washington*) 166: 903–905.

RANSON, S. W., 1937. Some functions of the hypothalamus. *Harvey Lect.* (1936–1937) Ser. 36: 92–121.

RANSON, S. W., and H. W. MAGOUN, 1939. The hypothalamus. *Ergeb. Physiol. Biol. Chem. Exp. Pharmakol.* 41: 56–163.

ROBERTS, W. W., and R. J. CAREY, 1965. Rewarding effect of performance of gnawing aroused by hypothalamic stimulation in the rat. *J. Comp. Physiol. Psychol.* 59: 317–324.

ROBERTS, W. W., and H. O. KIESS, 1964. Motivational properties of hypothalamic aggression in cats. *J. Comp. Physiol. Psychol.* 58: 187–193.

ROBERTS, W. W., M. L. STEINBERG, and L. W. MEANS, 1967. Hypothalamic mechanisms for sexual, aggressive and other motivational behaviors in the opossum, Didelphis virginiana. *J. Comp. Physiol. Psychol.* 64: 1–15.

SLONIM, A. D., 1968. Unconditional reflex as specific character and problem of studying instinct. *Progr. Brain Res.* 22: 506–517.

STEINBAUM, E. A., and N. E. MILLER, 1965. Obesity from eating elicited by daily stimulation of hypothalamus. *Amer. J. Physiol.* 208: 1–5.

TALBERT, G. A., 1900. Ueber Rindenreizung am freilaufenden Hunde nach J. R. EWALD. *Arch. Anat. Physiol.* (*Physiol. Abt.*) 1900: 195–208.

TENEN, S. S., and N. E. MILLER, 1964. Strength of electrical stimulation of lateral hypothalamus, food deprivation, and tolerance for quinine in food. *J. Comp. Physiol. Psychol.* 58: 55–62.

VALENSTEIN, E. S., 1966. The anatomical locus of reinforcement. *In* Progress in Physiological Psychology, Vol. 1 (E. Stellar and J. M. Sprague, editors). Academic Press, New York, pp. 149–190.

VALENSTEIN, E. S., 1970. Behavior elicited by hypothalamic stimulation: A prepotency hypothesis. *Brain, Behav., Evolut.* (in press).

VALENSTEIN, E. S., and V. C. COX, 1970. The influence of hunger, thirst, and previous experience in the test chamber on stimulus-bound eating and drinking. *J. Comp. Physiol. Psychol.* 70: 189–199.

VALENSTEIN, E. S., V. C. COX, and J. W. KAKOLEWSKI, 1968a. Modification of motivated behavior elicited by electrical stimulation of the hypothalamus. *Science* (*Washington*) 159: 1119–1121.

VALENSTEIN, E. S., V. C. COX, and J. W. KAKOLEWSKI, 1968b. The motivation underlying eating elicited by lateral hypothalamic stimulation. *Physiol. Behav.* 3: 969–971.

VALENSTEIN, E. S., V. C. COX, and J. W. KAKOLEWSKI, 1969a. Hypothalamic motivational systems: Fixed or plastic neural circuits? *Science* (*Washington*) 163: 1084.

VALENSTEIN, E. S., V. C. COX, and J. W. KAKOLEWSKI, 1969b. The hypothalamus and motivated behavior. *In* Reinforcement and Behaviour (J. T. Tapp, editor). Academic Press, New York, pp. 242–285.

VALENSTEIN, E. S., V. C. COX, and J. W. KAKOLEWSKI, 1970. A reexamination of the role of the hypothalamus in motivation. *Psychol. Rev.* 77: 16–31.

VALENSTEIN, E. S., J. W. KAKOLEWSKI, and V. C. COX, 1968. A comparison of stimulus-bound drinking and drinking induced by water deprivation. *Commun. Behav. Biol., Part A,* 2: 227–233.

VALENSTEIN, E. S., and A. G. PHILLIPS, 1970. Stimulus-bound eating and deprivation from prior contact with food pellets. *Physiol. Behav.* (in press).

VAUGHN, E., and A. E. FISHER, 1962. Male sexual behavior induced by intracranial electrical stimulation. *Science* (*Washington*) 137: 758–760.

WISE, R. A., 1968. Hypothalamic motivational systems: Fixed or plastic neural circuits? *Science* (*Washington*) 162: 377–379.

# Plasticity in the Autonomic Nervous System:

# Instrumental Learning of

# Visceral and Glandular Responses

LEO V. DiCARA

MANY BELIEVE that the autonomic nervous system, which mediates involuntary glandular and visceral responses, can be modified only by classical conditioning and not by the instrumental learning procedures employed in the training of voluntary skeletal responses. This belief has led to the notion that instrumental learning and classical conditioning are two basically different phenomena, with different neuro-anatomical and neurophysiological substrates. Establishing the truth or falsity of these traditional beliefs is of fundamental significance for theories of learning, for understanding the biological basis of learning, and for questions related to medicine. I would like to present the results of recent experiments which demonstrate that the control of visceral activities such as heart rate and the flow of urine can be learned in the same manner that skeletal responses can be learned.

Until recently, the experimental evidence on instrumental learning of visceral responses has been limited to exploratory or unpublished studies and vague references to the Russian literature (see Mowrer, 1938; Skinner, 1938). Recently, reports of success and of failure (Mandler et al., 1962; Harwood, 1962) have appeared. These experiments were reviewed and summarized by Kimmel (1967) and Katkin and Murray (1968).

A major problem encountered in research on the instrumental modification of visceral responses is that the majority of such responses are altered by voluntary responses such as tensing of specific muscles or changing the rate or pattern of breathing. One way to circumvent this problem, in animals at least, is to abolish skeletal activity by curare-like drugs such as d-tubocurarine; these interfere pharmacologically with the transmission of the nerve impulse to the skeletal muscle but do not affect the neural control of autonomically mediated responses.

Curarized subjects cannot breathe and must be artificially

LEO V. DiCARA    The Rockefeller University, New York, New York

respirated, and, because they cannot eat or drink, the possibilities for rewarding them in a training situation are somewhat limited. As is well known, however, training of instrumental skeletal responses can also be accomplished either by using direct electrical stimulation of rewarding areas of the hypothalamus, or by allowing escape from and/or avoidance of mildly noxious electric shock. Recent experiments in curarized animals with the use of these techniques have shown that either increases or decreases can be produced by instrumental procedures in visceral responses such as heart rate (Trowill, 1967; Miller and DiCara, 1967; Hothersall and Brener, 1969), intestinal motility (Miller and Banuazizi, 1968), blood pressure (DiCara and Miller, 1968a), vasomotor responses (DiCara and Miller, 1968b, 1968c), and urine formation (Miller and DiCara, 1968). Several investigators have reported similar instrumental learning of heart rate and blood pressure in man (Brener, 1966; Engel and Hansen, 1966; Engel and Chism, 1967; Shapiro et al., 1969) and have started to employ in the therapy of certain human cardiovascular disorders the powerful techniques used successfully in the animal experiments (Engel and Melmon, personal communication).

## Similarity between the learning of skeletal and that of visceral responses

In an experiment reported by Miller and DiCara (1967), we showed that curarized rats could be trained either to increase or to decrease their heart rates in order to obtain direct electrical stimulation of the medial forebrain bundle in the hypothalamus. By rewarding changes in the desired direction during the presentation of light and tone signals, we finally taught the rats to increase or decrease heart rate by approximately 20 per cent. In this experiment, we controlled the possibilities of experimenter bias by flipping a coin, after all the adjustments had been made in the respirator, to determine the direction of heart rate to be rewarded. Figure 1 shows the heart-rate changes induced over the first 90-minute training period, most of which were in

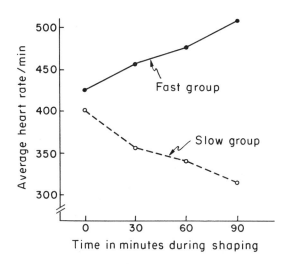

FIGURE 1 Learning curves for groups of curarized rats rewarded with intracranial brain stimulation for either increases or decreases in heart rate. (From Miller and DiCara, 1967.)

the form of an over-all increase or decrease. With prolonged training, however, the rats learned to respond more quickly and specifically. Thus the sample record from one rat in Figure 2 shows, at the beginning of discrimination training, no appreciable reduction in heart rate that is specific to the light and tone stimuli. Consequently, considerable time passed after the onset of these stimuli before the rat met the behavioral criterion and received the brain stimulation reward. By the end of the discrimination training, however, heart rate slowed, and the reward was delivered, almost im-

mediately after stimulus onset. These facts make it clear that instrumental heart-rate learning, as with instrumental skeletal learning, can be brought under the control of a discriminative stimulus.

In a different experiment, we demonstrated that heart-rate learning showed another of the important properties of the instrumental learning of skeletal responses—it could be remembered (DiCara and Miller, 1968d). Rats given a single training session for heart-rate increase (or decrease) exhibited reliable changes in the proper direction when retested under curare in the experimental situation for the first time three months later. Similar results have been reported by Hothersall and Brener (1969), who also provide evidence for progressive improvement as a function of daily training sessions.

In the above experiments, the reward was direct electrical stimulation of the brain. Miller and I have also conducted an experiment using the second of the two forms of thoroughly studied reward that can be conveniently used with curarized rats, namely, the escape from and/or avoidance of mild electric shock (DiCara and Miller, 1968e). We presented a shock signal to the rat for five seconds, after which brief pulses of mild electric shock were delivered to the rat's tail. During these five seconds, the rat could turn off the signal and avoid the shock by making the correct heart-rate responses. If it failed to do so, the shocks continued until the rat "escaped" by making the correct response, which immediately turned off both the shock and the shock signal. For half of the rats, the shock signal was a tone and the safe signal a flashing light; for the other half, these cues were reversed. During training, equal numbers

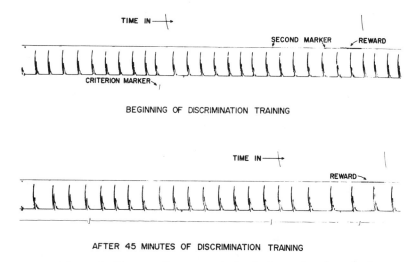

FIGURE 2 Electrocardiogram at the beginning and at the end of discrimination training of a curarized rat rewarded for slow heart rate. Slowing of heart rate is rewarded only during a "time-in" stimulus (light and tone). (From Miller and DiCara, 1967.)

of shock and safe signals were randomly interspersed with so-called "blank" trials—those without signals or shocks. The results of this study are shown in Figure 3. Rats re-

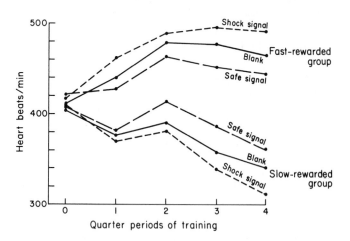

FIGURE 3 Changes in heart rate during avoidance training. (From DiCara and Miller, 1968e.)

warded by escape or avoidance for increasing their heart rate learned to increase it, whereas those identically rewarded for decreasing their heart rate learned to decrease it. The learning began with a general change in base line, as indicated by the blank trials. As training progressed, however, the shock signal came to elicit a progressively greater change in the rewarded direction and the safe signal a statistically reliable change in the opposite direction, toward the base line.

These results indicate that the instrumental learning of heart rate does not require a particular type of reward, such as electrical stimulation of the brain, but can be reinforced, as in other instrumental situations, by the withdrawal of a noxious stimulus. Banuazizi (1968) has reported similar discrimination learning of an intestinal response in curarized rats reinforced by the avoidance of mild electric shock.

In fact, all the phenomena tested thus far have been found to be similar for the instrumental learning of skeletal responses and that of visceral responses. Two additional examples can be cited. First, learned visceral responses, as do learned skeletal responses, progressively weaken, or experimentally extinguish, when a series of training trials is presented without reward (DiCara and Miller, 1968e). Second, an animal that has learned to use a particular cue as the positive stimulus for the performance of a given skeletal response and another as the negative stimulus readily learns to use these same cues in the same roles for a second discrimination that provides the same reward but requires a different skeletal response; similarly, rats that discriminate be-

tween particular positive and negative cues in a bar-pressing response exhibit the best discrimination when the same cues are used for the visceral response of increased or decreased heart rate (Miller and DiCara, 1967).

## Specificity of learning

The studies mentioned above demonstrate that the same reward can produce opposite visceral changes, which rules out at once the possibility that the heart-rate learning was produced by some unconditioned or innate effect of the reward. Furthermore, because the rats were completely paralyzed, as determined by electromyographic recordings, any obvious voluntary skeletal mediation is ruled out. What has not yet been eliminated is that the rat learned a general pattern of brain activation, or some particular initiation of impulses from the motor cortex that would have produced struggling skeletal movements had the motor endplates not been paralyzed by curare. As we pointed out in one of our first papers (Miller and DiCara, 1967), perhaps our training resulted in such central impulses, and the innate, or classically conditioned, effect of the central commands to struggle changed the heart rate. Three experiments that seem to have eliminated such a possibility can be described.

Miller and Banuazizi (1968) compared the instrumental learning of intestinal contractions with that of heart rate. Intestinal motility was recorded by means of a water-filled balloon inserted beyond the anal sphincter; changes in intestinal tone thus converted to changes in water pressure were measured by a pressure transducer connected to a tube attached to the balloon. The results are shown in Figures 4 and 5. The group rewarded for increases in intestinal contractions learned an increase, the group rewarded for decreases learned a decrease, but neither group showed an appreciable change in heart rate (Figure 4). Conversely (Figure 5), the group rewarded for increases in heart rate showed an increase, the group rewarded for decreases showed a decrease, but neither group showed a change in intestinal contractions. A statistically reliable negative correlation showed that the better the rewarded visceral response was learned, the less change occurred in the other, nonrewarded response. These results demonstrated that the instrumental learning of two visceral responses can occur independently of each other and can be made specific to the response rewarded; they also clearly rule out the possibility that the learning was mediated by some general reaction, such as arousal or a shift in sympathetic-parasympathetic balance.

We decided to test the unlikely possibility that the Miller and Banuazizi rats learned to direct toward skeletal muscles certain motor commands that affect intestinal contractions but not heart rate when intestinal changes are rewarded, and

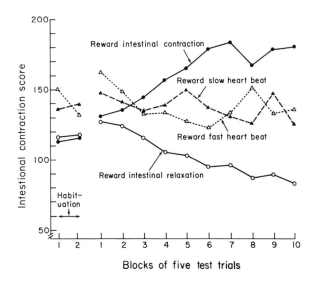

FIGURE 4 Graph showing that the intestinal contraction score is changed by rewarding either increases or decreases in intestinal contractions but is unaffected by rewarding changes in heart rate. (From Miller and Banuazizi, 1968.)

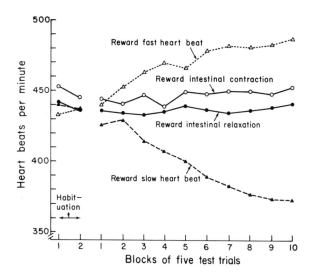

FIGURE 5 Graph showing that the heart rate is changed by rewarding changes in heart rate but is unaffected by rewarding changes in intestinal contractions. Comparison with Figure 4 demonstrates the specificity of visceral learning. (From Miller and Banuazizi, 1968.)

to direct toward the skeletal muscles other commands that affect heart rate but not intestinal contractions when heart rate changes are rewarded. We therefore trained rats under curare to alter heart rate and tested them later in the noncurarized state, reasoning that one might find muscle movements as well as heart-rate training in the noncurarized state (DiCara and Miller, 1969). The results indicated that both increases and decreases in heart rate of approximately 10 per cent, learned while under curare, transfer to the noncurarized state, but without correlated differences in respiration or other muscle activity.

Because of considerable evidence for normal interdependence of somatic motor and cardiovascular responses (Brener and Goesling, 1968; Black, 1967; Obrist et al., 1969), it is necessary to show that somatic-motor events do not mediate the response changes found in each experiment on visceral learning. In one of our more recent experiments, we showed that curarized rats not only make specific cutaneous vasomotor responses independently of changes in heart rate and blood pressure, but can even learn to produce these in the right ear and not the left (DiCara and Miller, 1968c). Another experiment on specificity showed that the p-p and p-r intervals of the rat electrocardiogram can be independently instrumentally conditioned (Fields, 1970.) This degree of specificity of visceral learning, which may surprise those persons who hold the traditional belief that the autonomic nervous system is less plastic than the cerebrospinal system, may be related to the evidence that various visceral responses have specific representation,

roughly analogous to that of skeletal responses, in the cerebral cortex (Chernigovskii, 1967).

*Behavioral consequences of instrumental heart-rate training*

Although the foregoing experiments demonstrate that visceral learning can be specific, under some circumstances the training procedure produces interesting generalized behavioral effects. For example, in handling the rats that had just recovered from curarization, I noticed that those that had been trained to avoid electric shock by increasing their heart rate were more likely to squirm, squeal, and defecate than were those that had been trained to reduce their heart rate. Could the instrumental learning of heart-rate increase have produced some unexpected additional effects, perhaps on the rat's level of activation or reactivity? To answer this question, Weiss and I used a modified shuttle-avoidance response to test the rats that had been subjected to curarized heart-rate training (DiCara and Weiss, 1969). In this apparatus, a danger signal warns the rat to run from one compartment to another before a specified time interval passes in order to avoid a foot shock. Previous work has shown that learning in this apparatus is an inverted, U-shaped function of the strength of the shocks; excessive shocks elicit emotional, instead of running, behavior (Theios et al., 1966). Weiss and I trained our rats in this apparatus, using an optimal shock level for normal animals (0.3 milliampere), and found that rats previously rewarded for decreasing their heart rates learned very well, but those rewarded for increasing their heart rate learned extremely poorly, as if their reactivity or emotionality had been greatly increased.

## Other examples of unconventional learning

ELECTRICAL ACTIVITY IN THE CENTRAL NERVOUS SYSTEM Elsewhere in this book, Fox summarizes experiments in which instrumental techniques have been used to alter the amplitude of various components of the electrical response evoked from the brains of animals (Fox and Rudell, 1968) and men (Rosenfeld et al., 1969). The spontaneous cortical waves (Carmona, 1967) and hippocampal theta waves (Black et al., 1970) have similarly been modified by instrumental techniques.

GLANDULAR RESPONSES The rate at which urine is formed by the kidney can be altered through learning (Miller and DiCara, 1968). Using rats, we placed catheters permanently in the bladder and counted, with an electronic device, the drops of urine formed each minute. In order to secure the high rate of urine formation required for prompt detection and rewarding of small changes in flow, the rats were loaded with water by infusion through another catheter in the jugular vein. To determine how the change in rate of urine formation was achieved, we measured glomerular filtration by the use of $^{14}$C-inulin, and rate of renal blood flow by the use of tritiated p-aminohippuric acid. The results show that curarized rats learn to increase or decrease the formation of urine in order to obtain electrical stimulation of the brain, and that they do so by changing renal blood flow. These blood-flow changes were not accompanied by changes in heart rate or blood pressure, so they must have resulted from vasomotor changes in the renal arteries. The vasomotor responses measured in the tail were the same in both the increase and decrease groups.

POSSIBLE ROLE OF INSTRUMENTAL VISCERAL LEARNING IN HOMEOSTASIS Because skeletal responses mediated by the cerebrospinal nervous system operate on the external environment, the ability to learn responses that bring rewards such as food, water, or escape from pain has survival value. To what extent does regulation of the internal visceral responses by the autonomic system have a similar adaptive value? Will an animal with experimentally altered internal homeostasis perform the instrumental tasks necessary to correct the imbalance? In a two-part experiment designed to test this question we (Miller, DiCara, and Wolf, 1968) injected antidiuretic hormone (ADH) into rats if they selected one arm of a T-maze, and isotonic saline if they selected the other. One group of normal animals, loaded with water by stomach tube in advance, learned to select the saline arm; choice of the ADH arm would have pre-

vented the excess-water secretion required for restoration of water homeostasis. By contrast, a group of rats with diabetes insipidus and stomach-tubed with hypertonic NaCl in advance regularly chose the ADH arm. For this group the homeostatic consequences of the injections are reversed: ADH causes concentration of urine, thus promoting excretion of the excess NaCl, but isotonic saline merely produces further homeostatic imbalance. The control rats received neither water nor NaCl and exhibited no preference in the maze.

These instrumental responses aimed at correcting internal homeostatic imbalance would appear to be no less adaptive than those learned skeletal responses that are directed toward altering the relationship of the animal to its external environment. Taken together with the other evidence that glandular and visceral responses can be instrumentally learned, these data raise the question of the degree to which animals normally learn the various autonomic responses required to maintain proper homeostatic levels. Whether such learning actually occurs may depend on whether the innate homeostatic mechanisms permit deviations large enough to function as a drive. Most innate controls may normally be so precise as to make learning impossible, but in abnormal circumstances or in pathological conditions where they falter, visceral learning rewarded by a return to homeostasis might be available as an alternative mechanism.

## Acknowledgments

The preparation of this manuscript and the research described were supported, in part, by United States Public Health Service Grants MH 13189 and MH 16569, by Career Development Award 1–K4–GM–34, 110, and by Research Grant 69–797 from the American Heart Association.

### REFERENCES

BANUAZIZI, A., 1968. Modification of an autonomic response by instrumental learning. Yale University, New Haven, Connecticut, doctoral thesis.

BLACK, A. H., 1967. Transfer following operant conditioning in the curarized dog. *Science (Washington)* 155: 201–203.

BLACK, A. H., G. A. YOUNG, and C. BATENCHUK, 1969. The avoidance training of hippocampal theta waves in flaxedilized dogs and its relation to skeletal movement. *J. Comp. Physiol. Psychol.* 70: 15–24.

BRENER, J. M., 1966. Heart rate as an avoidance response. *Psychol. Rec.* 16: 329–336.

BRENER, J., and W. J. GOESLING, 1968. Heart rate and conditioned activity. Paper presented at the meeting of the Society for Psychophysiological Research, Washington.

CARMONA, A., 1967. Trial and error learning of cortical EEG activity. Yale University, New Haven, Connecticut, doctoral thesis.

CHERNIGOVSKII, V. N., 1967. Interoceptors. State Publishing

House of Medical Literature, Medgiz, Moscow. English translation (D. Lindsley, editor). American Psychological Association, Washington.

DiCARA, L. V., and N. E. MILLER, 1968a. Instrumental learning of systolic blood pressure responses by curarized rats: Dissociation of cardiac and vascular changes. *Psychosom. Med.* 30: 489–494.

DiCARA, L. V., and N. E. MILLER, 1968b. Instrumental learning of peripheral vasomotor responses by the curarized rat. *Commun. Behav. Biol.* Part A, 1: 209–212.

DiCARA, L. V., and N. E. MILLER, 1968c. Instrumental learning of vasomotor responses by rats: Learning to respond differentially in the two ears. *Science (Washington)* 159: 1485–1486.

DiCARA, L. V., and N. E. MILLER, 1968d. Long term retention of instrumentally learned heart-rate changes in the curarized rat. *Commun. Behav. Biol., Part A*, 2: 19–23.

DiCARA, L. V., and N. E. MILLER, 1968e. Changes in heart rate instrumentally learned by curarized rats as avoidance responses. *J. Comp. Physiol. Psychol.* 65: 8–12.

DiCARA, L. V., and N. E. MILLER, 1969. Transfer of instrumentally learned heart-rate changes from curarized to noncurarized state: Implications for a mediational hypothesis. *J. Comp. Physiol. Psychol.* 68: 159–162.

DiCARA, L. V., and J. M. WEISS, 1969. Effect of heart-rate learning under curare and subsequent noncurarized avoidance learning. *J. Comp. Physiol. Psychol.* 69: 368–374.

ENGEL, B. T., and R. A. CHISM, 1967. Operant conditioning of heart rate speeding. *Psychophysiology* 3: 418–426.

ENGEL, B. T., and S. P. HANSEN, 1966. Operant conditioning of heart rate slowing. *Psychophysiology* 3: 176–187.

FIELDS, C., 1970. Instrumental conditioning of the rat cardiac control systems. *Proc. Nat. Acad. Sci. U. S. A.* 65: 293–299.

FOX, S. S., and A. P. RUDELL, 1968. Operant controlled neural event: Formal and systematic approach to electrical coding of behavior in brain. *Science (Washington)* 162: 1299–1302.

HARWOOD, C. W., 1962. Operant heart rate conditioning. *Psychol. Rec.* 12: 279–284.

HOTHERSALL, D., and J. BRENER, 1969. Operant conditioning of changes in heart rate in curarized rats. *J. Comp. Physiol. Psychol.* 68: 338–342.

KATKIN, E. S., and E. N. MURRAY, 1968. Instrumental conditioning of autonomically mediated behavior: Theoretical and methodological issues. *Psychol. Bull.* 70: 52–68.

KIMMEL, H. D., 1967. Instrumental conditioning of autonomically mediated behavior. *Psychol. Bull.* 67: 337–345.

MANDLER, G., D. W. PREVEN, and C. K. KUHLMAN, 1962. Effects of operant reinforcement on the GSR. *J. Exp. Anal. Behav.* 5: 317–321.

MILLER, N. E., and A. BANUAZIZI, 1968. Instrumental learning by curarized rats of a specific visceral response, intestinal or cardiac. *J. Comp. Physiol. Psychol.* 65: 1–7.

MILLER, N. E., and L. V. DiCARA, 1967. Instrumental learning of heart rate changes in curarized rats: Shaping, and specificity to discriminative stimulus. *J. Comp. Physiol. Psychol.* 63: 12–19.

MILLER, N. E., and L. V. DiCARA, 1968. Instrumental learning of urine formation by rats; changes in renal blood flow. *Amer. J. Physiol.* 215: 677–683.

MILLER, N. E., L. V. DiCARA, and G. WOLF, 1968. Homeostasis and reward: T-maze learning induced by manipulating antidiuretic hormone. *Amer. J. Physiol.* 215: 684–686.

MOWRER, O. H., 1938. Preparatory set (expectancy). A determinant in motivation and learning. *Psychol. Rev.* 45: 62–91.

OBRIST, P. A., R. A. WEBB, J. R. SUTTERER, and J. L. HOWARD, 1969. The cardiac-somatic relationship: Some reformulation. *Psychophysiology* (in press.)

ROSENFELD, J. P., A. P. RUDELL, and S. S. FOX, 1969. Operant control of neural events in humans. *Science (Washington)* 165: 821–823.

SHAPIRO, D., B. TURSKY, W. GERSHON, and M. STERN, 1969. Effects of feedback and reinforcement on the control of human systolic blood pressure. *Science (Washington)* 163: 588–590.

SKINNER, B. F., 1938. The Behavior of Organisms. Appleton-Century-Crofts, New York.

THEIOS, J., A. D. LYNCH, and W. F. LOWE, 1966. Differential effects of shock intensity on one way and shuttle avoidance conditioning. *J. Exp. Psychol.* 72: 294–299.

TROWILL, J. A., 1967. Instrumental conditioning of the heart rate in the curarized rat. *J. Comp. Physiol. Psychol.* 63: 7–11.

# 23 Spontaneous Electrical Brain Rhythms Accompanying Learned Responses

W. R. ADEY

THE HIGHER NERVOUS CENTERS of all vertebrates and many invertebrates are characterized by aggregates of nerve cells, collected in ganglia with distinctive morphological arrangements, and at the head end of the organism. The morphologist has studied their organization by dissection, at first grossly with the naked eye and, by ever-finer degrees, through microscopy of tissues. Sectioning of tissue in search of elemental aspects of its organization has been a pervasive concept not necessarily restricted to morphological studies. Just as a histological section may be regarded as a "time slice" that permanently isolates a tissue fragment at a particular moment in the vast ebb and flow of its metabolic exchanges, so too has the physiologist sought traditionally to isolate and simplify functional organization in cerebral ganglia. He has hoped thereby to reveal essential aspects of information transaction in these ganglia, stripped of electrical activity that continues in all their waking and sleeping states.

Through dissections suggested by his anatomist colleagues, or with the aid of anesthetic agents, the physiologist has minimized the intrusion of so-called "spontaneous" electrical activity, in order that he may better study single-cell discharges evoked by central or peripheral stimuli, or the characteristics of evoked potentials elicited on an essentially silent electrical background. It is inherent in such an approach that so-called spontaneous activity is viewed as noncontributory in processes of information transaction, or that it contribute in only minor or secondary ways. In this connection, the cortical EEG has been described as a mere noise in cerebral tissue and, even until quite recently, as arising in metabolic aspects of brain function.

In seeking a relationship between gross electrical rhythmic activity of central nervous structures and the transaction of information, it is, of course, necessary that the test stimuli be presented with strictly known time relations to the EEG record, and that these stimuli be effective in activating central sensory pathways. Yet it is important to avoid contamination of the spontaneous activity by recurrent, brief, evoked responses to transient stimuli. These can elicit an entrainment of the evoked response, with evocation of late components in the poststimulus period. They may be difficult to distinguish from altered background activity attributable to a changing "set" of attention, rather than specifically to the processing of information (Adey, 1969c; Walter et al., 1967a). It may further be emphasized that we do not live in a tachistoscopic world. If saccadic interruptions of sensory volleys do indeed play a role as a modulating process essential to information processing, then these normally occur centrally, or are at least under fine central control, as in eye movements essential to perception (Adey, 1969b; Ditchburn, 1963).

Our attempts to understand brain electrical rhythms demand that we look more closely at their origins. It is no longer seriously proposed that the EEG arises as the envelope of the firing of nerve cells, but we do not know with certainty any general laws that might relate the firing of a neuron to a particular phase of concurrent EEG waves in the same domain of tissue. Such relationships are more easily defined in relation to evoked potentials (Fox and O'Brien, 1965) and to specialized aspects of EEG activity, as in spindle waves of sleep (Creutzfeldt et al., 1964; Creutzfeldt et al., 1966) or in electrical activity of isolated cortex (Frost, 1968), than to the wave activity of the awake brain, as in hippocampal theta trains (Noda et al., 1969) or in cortex of unanesthetized pretrigeminal preparations (Frost and Elazar, 1968).

Our knowledge of the origins of the EEG now rests on considerable information from intracellular recording, which has revealed very large waves, many millivolts in amplitude, in many cortical regions (Fujita and Sato, 1964; Creutzfeldt et al., 1965; Jasper and Stefanis, 1965). These waves have a frequency spectrum similar to the much smaller EEG recorded simultaneously by gross electrodes at the same point, but are typically incoherent with this EEG (Elul, 1964, 1967a, 1967b, 1968). Even here our ability to extrapolate further is quite limited. Is the large intraneuronal wave totally attributable to postsynaptic potentials from impinging synaptic volleys, or is it in certain aspects an oscillation in membrane potential endogenously induced in

W. R. ADEY  Department of Anatomy and Physiology, Space Biology Laboratory, Brain Research Institute, University of California at Los Angeles, California

each neuron? How does the firing of a cerebral neuron relate at its threshold to the concurrent intracellular wave? Does the small extracellular field arising from volume conduction of large intracellular waves significantly modify excitability of neurons in the tissue which it pervades (Nelson, 1967)? Parenthetically, but by no means unimportantly, do non-neural cellular elements also play a part in the wave events so characteristic of unceasing cortical activity (Karahashi and Goldring, 1966)? The EEG appears to arise in major degree from intracellular waves generated by a population of cells. We shall pursue further the mathematical models that best describe relations between a population of contributing generators and the EEG, considered as a volume-conducted sum of these generators.

The following discussion will also be concerned with consistent patterns that have been detected in these volume-conducted integrals recorded at the brain surface or on the scalp, as concomitants of specific psychological performances. Their regular occurrence and reliable detection by pattern-recognition techniques (Hanley et al., 1968; Berkhout et al., 1969b) again direct our attention to this question of rigorous ordering of the electrical events in cerebral extracellular space, even when these extracellular processes are recorded as the summed activity of a neuronal population and are but a small fraction of the intraneuronal wave events from which they arise.

Clearly, detection of these ordered patterns does not imply that the volume-conducted EEG itself may directly modify excitability of neurons in its contributing population, but does indicate that the EEG can provide a useful window on cellular slow waves. Our studies have related retrieval of stored information to conductive changes in cerebral tissue and, thus, to altered membrane surface and intercellular conductance (Adey et al., 1966b). These correlated wave and conductance phenomena during information transaction and retrieval have led us to investigate macromolecular (mucopolysaccharide and mucoprotein) structures at neuronal surfaces during excitation (Wang et al., 1966; Wang et al., 1968; Adey et al., 1969; Tarby et al., 1968; Hafemann et al., 1969).

*The nature of spontaneous cerebral electrical rhythms: characterization of populations of neuronal wave generators*

The foregoing discussion has considered the genesis of the EEG on the basis of certain physiological phenomena: the occurrence of large intraneuronal waves, with spectral density distributions closely resembling those in the concomitant EEG (Elul 1964, 1967a, 1967b; 1968); a reduction in wave amplitude by a factor of about 200 between intracellular and extracellular recording sites, thus agreeing with

measurements of transmembrane conductance (Coombs et al., 1959) and of conductance in extracellular fluid (Cole, 1940) arising in a voltage-divider action; and, typically, a lack of coherence between the intraneuronal waves and the EEG in cross-spectral calculations. I documented these findings at the 1966 Intensive Study Program (Adey, 1967), and briefly summarize them here. It was concluded that the EEG arises as the sum of volume-conducted activity from a population of generators obeying Cramér's (1955) theorem, in being largely independent or nonlinearly related and in possessing a statistical mean and a finite standard deviation (Elul, 1968).

These studies have focused on the microcosm of the single neuronal wave generator in attempts to relate populations of such generators to the gross EEG. Such correlates as have been detected emphasize the potential for further studies to validate EEG recording as a window on phenomena at the neuronal level. There continue to be significant statistical dangers in spectral analyses attempted on very short data segments (Wiener, 1958; Meshalkin and Efremova, 1964). The natural desire of the physiologist to seek such correlates during a single, brief, behavioral event, or in relation to a single EEG episode (Gersch, 1968), such as an alpha burst or spindle train, must be tempered by the strong probability of falsely positive correlation and coherence functions (Blackman and Tukey, 1959; Efremova et al., 1964). On the other hand, there appears no reason to accept the view, so often espoused by those persons preoccupied with engineering aspects of pulse-coded nerve nets, that "the study of brain waves is equivalent to holding an oscilloscope probe six feet in diameter up to a computer and pronouncing from the resultant wave-form on the underlying structure and function" (Harmon, quoted by Perkel and Bullock, 1969).

In pursuing these elusive characteristics of a population of wave generators, it is indeed feasible to invert the process of sampling individual generators and extrapolating to the population as a whole. For example, we may study amplitude distributions in the EEG and make certain inferences about interrelations of the contributing generators within a single cortical domain.

AMPLITUDE DISTRIBUTION STUDIES OF NEURONAL WAVES AND OF THE EEG    For the past 30 years, quantitative analysis of the EEG has been based primarily on studies in the frequency domain. First by autocorrelation techniques and, more recently, by spectral analysis, peaks have been detected in the EEG spectral contours. By contrast, the implications of the similarity of the EEG and random noise have received only scant consideration (Elul, 1969). To the communications engineer, noise typically implies degradation of information transmission, or its complete absence, considered in the context of noisy components and noisy transmission paths (Adey, 1969b). Little is known about generation of noiselike activity in the brain, and it is conceivable that, unlike engineering systems, the brain may use noisy systems in ways essential for information processing. We may well fail to recognize the presence of information in such a system if we yield to preconceived notions that brain waves are merely noise.

If the EEG is treated as a stationary, or time-invariant, statistical process for brief epochs, its probability distribution can be constructed as a histogram of the instantaneous EEG amplitude values. As a statistical generalization, trends toward a normal, or Gaussian, amplitude distribution may be interpreted as indicating a tendency to independence of the contributing generators. Studies in our laboratory have examined these distributions in intraneuronal waves of the cat, and in the scalp EEG of man (Elul, 1967a; Adey et al., 1966a; Elul, 1969; Elul and Adey, in preparation).

In all cases so studied, the probability distribution of intracellular potential oscillations in cat cerebral cortex was non-Gaussian. By contrast, gross EEG activity recorded from an adjacent site on the surface of the suprasylvian cortex showed a strong tendency to follow a Gaussian distribution. We have considered above the likelihood that these slow neuronal waves contribute most significantly to gross EEG activity. If a large neuronal population were oscillating synchronously, the gross EEG would be a simple replica of concurrent waves in individual neurons. Because the probability distribution of amplitude changes for the neuronal wave is non-Gaussian, one might expect that the gross EEG would also be non-Gaussian. These observations on normal distribution support the view that individual contributions of unitary generators cannot be fully synchronized (Elul, 1969).

Although the simplest explanation would be based on application of the central-limit theorem of statistics (Cramér, 1955), and would postulate complete statistical independence of the unitary generators, such a view encounters anatomical and physiological difficulties. Synaptic potentials are undoubtedly major contributors to the intracellular waves, and arise in finite and invariate anatomical connections. Total independence of the generators is therefore unlikely. A more realistic physiological model would involve nonlinear relations between individual wave generators. Here, transitions from subliminal to supraliminal states of the generators in neuronal firing might provide the requisite nonlinearity (Elul and Adey, 1966). In either model, mutual asynchrony of unitary EEG generators is supported both by dissimilarities in neuronal and EEG amplitude-probability distributions, and by dissimilarities in the extracellular micro-EEG recorded in adjoining tissue domains only 30 microns apart (Elul, 1962).

This model of asynchronous unitary generators has been tested in a predictive frame from scalp-EEG data from man

in known behavioral states (Adey et al., 1966a; Elul, 1969; Elul and Adey, in preparation). The strongest trends to Gaussian distributions would be expected when unitary generators are desynchronized; when relations between neurons are stronger, as in sleep, the amplitude distributions would be shifted away from Gaussian. The EEG in the idle subject followed a Gaussian distribution in about two-thirds of the epochs analyzed. An EEG sample length of 2.0 seconds was used, because it was found that effects of inherent nonstationarity became increasingly serious with longer epochs and led to erroneously low estimates of goodness-of-fit (Campbell et al., 1967). In sleep, Gaussian trends were low in slow-wave and spindle sleep, and were closer to those in the awake state in rapid-eye-movement (REM) stages (Figure 1). During mental task performance, the non-Gaussian activity increased to as much as twice that in the idle state (Figure 2, page 228).

Thus, there appears to be increased cooperative activity of neuronal elements in a single cortical domain during performance of a mental task. A similar interpretation of data from different cortical regions was made by Gavrilova and Aslanov (1964), also using Gaussian trends in amplitude distributions. Not only do these data indicate that amplitude analysis of the EEG may provide significant information on mental functions; they also suggest that the EEG can be validly interpreted in terms of neuronal synchronization and desynchronization.

PHASE RELATIONS OF NEURONAL FIRING TO CONCURRENT EEG-WAVE ACTIVITY This problem is obviously of great importance to our understanding of information processing in the brain. Neuronal waves involve periodic depolarization or hyperpolarization of the cell membrane, and cell firing occurs on the depolarizing phase of the neuronal wave. At this point, however, further study fails to reveal any simple relationship between cell firing and simultaneous wave processes, whether the latter are viewed at the level of the neuronal wave or the gross EEG. It is only when cerebral organization is stripped of its essential complexity in the transactional processes of the waking state, as in anesthesia, epileptic discharges, certain sleep states, or in isolated cortical slabs, that simplistic patterns emerge, with neuronal firing highly correlated with cortical-wave activity.

Studies by von Euler and Green (1960a, 1960b) and Fujita and Sato (1964) indicated that some hippocampal pyramidal cells tend to fire at a particular phase of theta waves induced by nonspecific stimuli in anesthetized preparations, and that drugs such as eserine enhance this relationship during urethane anesthesia (Green et al., 1960).

We have examined this relationship in unanesthetized hippocampal neurons (Noda et al., 1969), using autocorrelograms and firing phase/theta wave histograms. In the

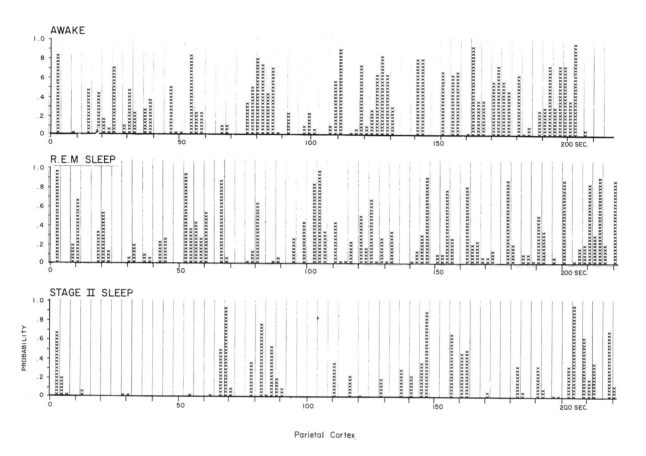

Parietal Cortex

FIGURE 1 Amplitude distributions in EEG records from parietal cortical screws in man in the awake state, during REM sleep, and in spindle sleep. Successive 2-second epochs were analyzed for probability of Gaussian amplitude distri-butions. The regular spindle train in Stage II sleep reduced probabilities of Gaussian distributions to a low level. (From Adey et al., 1966a.)

awake state, and in intermediate and deep sleep, the auto-correlograms showed an almost flat contour, indicating that each spike was independent of the preceding one. In the aroused state, and in rapid-eye-movement (REM) sleep, there was a slight periodicity with peaks spaced approx-imately equally at rates that might reflect the theta rhythm. Further studies with firing phase/theta wave histograms in REM sleep showed that although fixed phase relations between spikes and waves were occasionally seen they were rarely clear (Figure 3, page 229). Even in the same neuron, the phase relationship was not fixed. Phase relationships tended to change gradually with continued firing, or a con-stant relationship disappeared suddenly. When two neurons were observed simultaneously, the firing probability with respect to the theta waves usually differed from one cell to the other, in agreement with findings of Petsche and Stumpf (1962) for septal cells.

There is clearly a statistical sampling problem in deter-mining relationships between EEG and single-unit activity. Frost and Elazar (1968) found that a typical cell tends to fire near a negative peak in the EEG, and that the firing is sometimes followed by a positive EEG deflection. They emphasize that these relationships differ from cell to cell, but that the evidence strongly favors a nonrandom relation-ship between the EEG and the moment of cell firing. This is in agreement with Fox and Norman (1968), who de-scribed extremely high correlations between spike prob-ability and amplitude of the microelectrode-recorded EEG. High correlations with amplitude are interesting in view of the lack of coherence between micro-EEG records from electrodes spaced only 30 microns apart (Elul, 1962), and the difficulty in recording a fixed firing threshold with intra-cellular electrodes in cortical neurons. Firing certainly occurs close to the positive peak of the intraneuronal wave, but the wave frequently exceeds this level without initiation of firing, and successive spikes may arise at different levels of

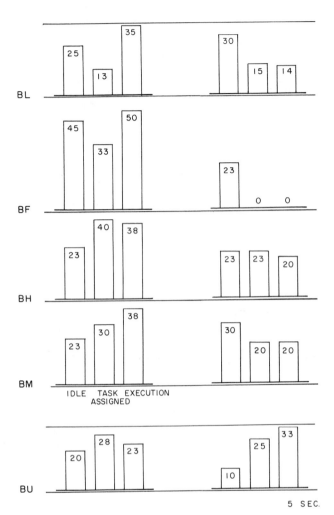

FIGURE 2  Probabilities of Gaussian distributions in five subjects in three states of alertness—sitting idle, after assignment of a mental task, and during execution of the task. The trends to Gaussian distributions in the top four subjects followed a similar pattern, increasing between idleness and task execution in frontofrontal records, and decreasing in fronto-occipital records. The latter finding is interpreted as an indication of increased neuronal "cooperativeness." Subject 5 (BU) participated unwillingly. (From Adey et al., 1966a.)

depolarization. Possible mechanisms for these nonlinear transforms of the neuronal wave to the spike train are discussed elsewhere (Elul and Adey, 1966).

We have considered the probable contribution of the neuronal wave to the gross EEG, and also its relationship to the initiation of the spike discharge. In neither respect does a clear picture of simple relations emerge. We should thus be

encouraged to accept the concept proposed above of an inherently noisy central processor, in which noise and uncertainty are essential aspects of information processing. It seems probable that the neuronal wave, by reason of its major origin in synaptic potentials, is an important transform of information in cerebral transactions, and that it ranks equivalently with spike generation, which would then be secondary to prior and concurrent wave phenomena within the individual neuronal generator.

## EEG patterns in cerebral systems: corticosubcortical interrelations in alerting, orienting, and discriminating

We have considered the EEG so far as an isolated phenomenon, within a single domain of tissue, and have reviewed the statistical models best characterizing its origins from cellular generators. Now, we may examine local EEG records from different cortical and subcortical regions known to be interrelated in physiological mechanisms subserving alerted behavior and discriminative processes. If we could discern consistent EEG relations between cerebral structures set considerable distances apart, we would have evidence supporting a much broader model in which the EEG relationships would indicate gross relations and would also be specific descriptions of shared processes at cellular levels between different populations of neuronal generators.

Much evidence has been presented elsewhere describing such corticocortical and corticosubcortical EEG patterns in animals and man (Adey, 1967a; Adey, 1969a; Grastyán et al., 1966; Caille and Bock, 1966; Remond, 1960; Storm van Leeuwen, 1965). Reliable detection of these patterns has depended increasingly on computed analyses. At first, these analyses were performed on data from single behavioral performances, but it soon became feasible to evaluate parameters common to repeated behavioral performances and to assess levels of variance within and between these parameters. Mathematical techniques employed have progressed from simple averaging of EEG wave trains (Adey and Walter, 1963) through correlation analysis (Adey et al., 1961) to sensitive and powerful methods of spectral analysis (Walter et al., 1967b, 1967c; Adey et al., 1967b).

At the same time, there has been a growing awareness that the psychological "set" of a particular behavioral state may be very short, lasting only a fraction of a second, or at most one or two seconds, in the course of decision making. It is therefore not appropriate to use mathematical techniques requiring constancy of EEG pattern (or mathematical stationarity) for much longer periods in order to insure statistically valid analysis. Nor is it justifiable to perform analyses with these methods in short EEG epochs and to avoid consideration of falsely positive correlates, likely in these circumstances. Cross-spectral calculations, for example

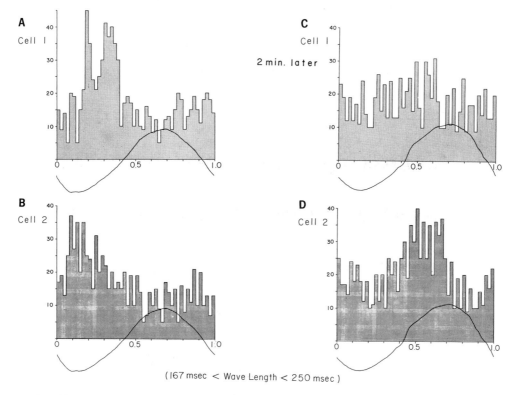

A Cell 1

C Cell 1
2 min. later

B Cell 2

D Cell 2

( 167 msec < Wave Length < 250 msec )

FIGURE 3 Phase/theta wave histograms, estimating the probability density function of neuronal discharges with respect to the phase of local slow waves. A and B are for one 30-second epoch in REM sleep, and C and D are from the same state two minutes later. The wave superimposed on each histogram represents summed waveforms recorded with the same microelectrode. In the first epoch (A and B), 54 waves with periods of between 167 and 250 milliseconds were recorded. The second epoch (C and D) included 58 waves. These data show that (1) the firing probability with respect to the phase of local theta waves is different in each neuron; (2) the phase relationship is not consistent, disappearing (from A to C), or shifting (from C to D); and (3) spike initiation can occur at different phases of local theta waves. (From Noda et al., 1969.)

(Walter, 1963), must consider data epoch length, data bandwidth, and data sampling rates in determining degrees of freedom that will ultimately determine the frequency resolution possible in a valid spectral analysis, and levels of coherence that are statistically significant. Such factors are apparently still not widely understood (Gersch, quoted by Perkel and Bullock, 1969; Creutzfeldt, 1969).

BEHAVIORAL STUDIES IN CAT: COMPOSITE SPECTRAL ANALYSIS TECHNIQUES FOR VERY SHORT EEG EPOCHS For a spectral resolution of 1.0 Hz over the band 0–30 Hz, a record length of 15 seconds is required. By summing together short EEG epochs lasting 0.5 to 2.0 seconds taken from repeated identical behavioral performances, it is possible to analyze in statistically valid fine detail the EEG spectral characteristics in brief performances associated with alerting, orienting, and visual discrimination (Elazar and Adey, 1967a, 1967b). This "evoked spectral analysis" was used with EEG segments as short as 1.6 seconds at a resolution of 1.0 Hz.

The technique was first applied in autospectral analysis of hippocampal EEG records (Elazar and Adey, 1967a; Figure 4). The alert cat's hippocampal EEG shows regular, high-amplitude waves at 4 to 7 Hz, the so-called "theta rhythms" (Grastyán et al., 1959; Adey et al., 1960; Radulovacki and Adey, 1965). Hippocampal autospectral analyses made during four behavioral states—after a warning stimulus, during performance of a light-dark discrimination and approach to food reward, and at the beginning and end of each trial—show that this theta rhythm shifts its frequency consistently during the successive behavioral states.

When performance is at the level of 80 per cent correct or more, the sharp peaks of spectral density move stepwise from 4.0 Hz, during the period before a warning tone, to 5.0 Hz with a narrow high peak in the warning epoch, and to 6.0 Hz at the beginning of the discrimination ap-

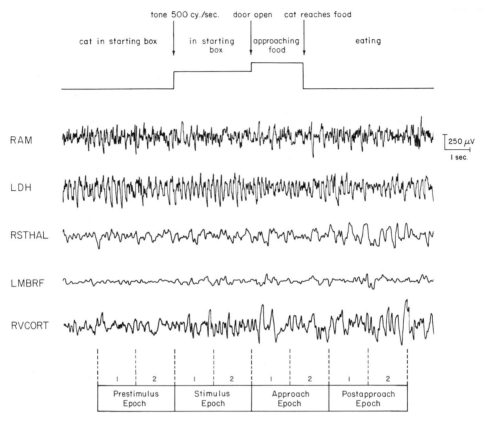

CAT NO. 4    RUN 50
CORRECT, PERFORMANCE 83%

tone 500 cy./sec.    door open    cat reaches food

cat in starting box | in starting box | approaching food | eating

RAM

LDH

RSTHAL

LMBRF

RVCORT

250 μV

1 sec.

1    2    1    2    1    2    1    2

Prestimulus Epoch | Stimulus Epoch | Approach Epoch | Postapproach Epoch

FIGURE 4 Typical EEG records from a cat in a single trial, with a sequence of an alerted state, an orienting response to a 500-Hz tone, and in approach to food with a light-dark discrimination. Event marker is in top trace, and at the bottom of the figure is the segmentation of the record for spectral analysis. For each behavioral epoch, two periods 1.6 seconds long were analyzed. Abbreviations: RAM, right amygdala; LDH, left dorsal hippocampus; RSTHAL, right subthalamus; LMBRF, left midbrain reticular formation; RVCORT, right visual cortex. (From Elazar and Adey, 1967a.)

proach epoch, at which point the peak contained 35 per cent of the total spectral power. As the goal is approached, the total EEG power drops, and the main spectral peak reverts to 5.0 Hz; thereafter, the spectrum trends toward that of the prewarning state, 4.0 Hz (Figure 5).

Distinct differences were found between correct and incorrect responses. "Attention sets" immediately preceding the incorrect performance were substantially different from those in correct responses, a phenomenon also described below for the monkey. There was no shift of theta peak activity to 6.0 Hz in the approach epoch in incorrect responses, and the spectral contour was broader than in correct responses. Incorrect responses showed very high and very regular 4.0 Hz activity (50 per cent of spectral power), by comparison with correct responses, and theta irregular-

ities beginning with the approach continued into the postapproach period.

Immediately before reaching the food reward, the EEG often appeared "desynchronized." Analysis showed an abrupt decrease in wave amplitude. This was considered a very tense state of the animal on confronting the food reward, and distinct from preceding decision-making behavior.

Later calculations of cross-spectra revealed components shared between cortical and subcortical records at each spectral frequency (Elazar and Adey, 1967b). In advanced training, there was increased 1.0–2.0 and 3.0 Hz activity in amygdala, subthalamus, midbrain reticular formation, and visual cortex, as well as in the hippocampus during the visual-discrimination task outlined above. All these struc-

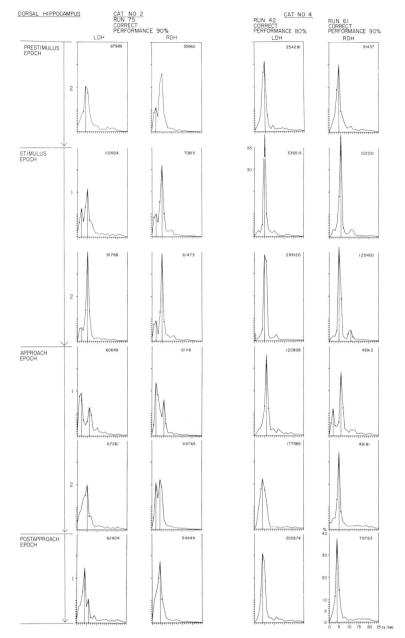

FIGURE 5   Selected plots of autospectral analyses of hippocampal EEG data epochs from two cats at advanced levels of training. There is a progressive shift from a peak at 4–5 Hz in the alerted state, to 5 Hz during orienting, to 6 Hz during discrimination, and to 4 Hz in the postapproach epoch. These analyses were based on compilations of separate 1.6-second epochs, in sufficient numbers to assure statistically valid 1-Hz resolution in the spectra (see text). (From Elazar and Adey, 1967a.)

tures showed theta spectral peaks at 5.0 and 6.0 Hz during the approach period, but these peaks were less pronounced and more variable than in the hippocampus (Figure 6).

Calculations of coherence (Walter, 1963) between pairs of EEG records from different structures revealed high degrees of linear interrelation. Coherences were consistently high between homotopic points in the two hippocampi at many frequencies. Significant coherences also showed in the theta range between hippocampus and subthalamus, less often between hippocampus and midbrain reticular formation, and between hippocampus and visual cortex, particularly during approach, and particularly in advanced training.

In summary, these studies of EEG interrelations between neocortical, subcortical, and limbic structures during the period of consolidation of learning have shown that the hippocampus plays a leading role in complex interacting neural circuits. They emphasize that, as pointed out by

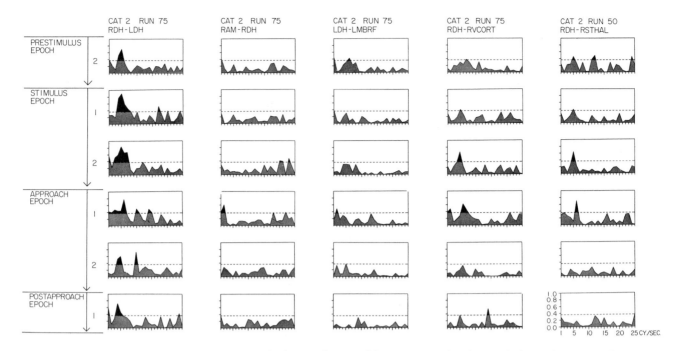

FIGURE 6   Plots of coherence between pairs of EEG records in a cat performing in the paradigm of Figure 4. Pairings were RDH-LDH, left and right hippocampus; RAM-RDH, amygdala-hippocampus; LDH-LMBRF, midbrain-hippocampus; RDH-RVCORT, hippocampus-visual cortex; RDH-RSTHAL, hippocampus-subthalamus. The sequence of plots should be read from top down, as indicated by arrows. Each column of plots is for a particular pairing of EEG records. Scales of coherence (ordinates) and spectral frequencies (abscissae) are shown on the bottom right graph. The horizontal dashed line on each graph indicates the significant level. (From Elazar and Adey, 1967b.)

Penfield (1958), it is in these interactions with other systems, particularly in caudal diencephalic and rostral mesencephalic areas, that the hippocampus appears to exert its functions in consolidation and recall of memory, rather than as an isolated and independent repository of memory traces.

Our window here on the EEG as an index of information transaction in cerebral systems has rested on estimates of linear interrelations. Although linear transfer functions may describe major aspects of the transaction of information within and between corticosubcortical systems in learning, we are unable to account for considerable EEG spectral power at cerebral sites that nevertheless have clear linear interrelations. We must thus devise new methods of evaluating nonlinear interrelations. This residual component would comprise nonlinearly related activity and a truly random component. Such techniques as complex demodulation may offer a modest but useful step, if the number of contributing generators is assumed to be restricted (Tukey, 1965; Walter and Adey, 1968). Because the EEG within a cortical domain bears a strikingly nonlinear relationship to its contributing cellular generators, it would be surprising if transactional mechanisms between cerebral loci did not also exhibit significantly nonlinear interrelations.

EEG PATTERNS IN MONKEY IN DELAYED-MATCHING-TO-SAMPLE TASKS   Similar studies of short EEG epochs were then made in a series of monkeys in preparation for a 30-day earth orbital space flight (Adey et al., 1967b), in which the greater behavioral repertoire of the primate was tested in a complex visual discrimination. The task involved a panel of five small, circular, display windows that also functioned as push-switches. Initially a central display was lit, presenting a circle, cross, square, or triangle in random sequence. The monkey was trained to press this display, extinguishing the light and initiating a 16-second delay, after which the peripheral displays were lit, presenting the four symbols simultaneously. A correct response required pressing the display matching the initial symbol.

Detailed analyses were made of EEG epochs recorded in both the initial single symbol displays and in the later four-symbol peripheral presentations (Berkhout et al., 1969a). The initial display epochs lasted 3 seconds and ended with the animal pressing the display-switch. The peripheral display epochs varied in duration. They began at the onset of the four-symbol display and ended at the moment of response. Autospectral and cross-spectral calculations of intensity and coherence for the band 0–32 Hz had a resolution of 1.0 Hz. Initial analysis of data from one monkey presented here has since been confirmed in generally similar patterns in nine others.

Most autospectral analyses of records from single structures did not clearly separate correct from incorrect responses. Only one hippocampal record showed this capability, based on decreased delta (1–3 Hz) and increased beta (14–32 Hz) activity in correct responses relative to incorrect. This difference characterized both the single-symbol and the later four-symbol presentation. On the other hand, coherence measurements showed clear differences between correct and incorrect responses, not only from analysis of EEG epochs during discrimination of the four-symbol display, but also from analysis of the initial single-symbol display 16 seconds before the success or failure actually occurred.

Using only initial display epochs for coherence calculations, the separation between later correct and incorrect responses could not be detected in the allocortical-neocortical (hippocampus to parietal cortex) coherence. On the other hand, coherence measures between pairs of deep structures (hippocampus to centrum medianum, or to midbrain reticular formation, or to opposite hippocampus) showed great differences that anticipated the correctness of the subsequent decision (Figure 7). A successful match was characterized by low coherence between these structures, whereas an erroneous outcome was characterized by a high

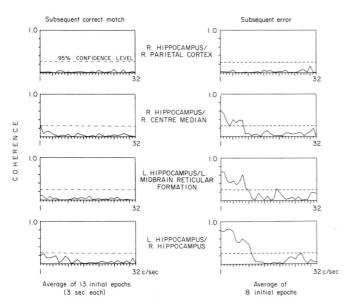

FIGURE 7 Coherence between EEG records from structures indicated in monkey performing a delayed matching task. Coherences were measured during initial symbol presentations according to success or failure of subsequent match 16 seconds later. Data were taken from a single day's performances. (From Berkhout et al., 1969a.)

coherence at frequencies below 12 Hz. These differences were also characterized by a technique of automated pattern recognition, using stepwise discriminant analysis, as described below. This automatic separation correctly classified more than 90 per cent of eventual successes or failures from examination of the initial single-symbol epochs.

Comparison of these findings in the monkey with those in cat cited above not only emphasize similarities of limbic and diencephalic systems participating in a visual choice, but also invite consideration of differences in EEG patterns that may relate to perception of form, as opposed to a simple brightness discrimination. Thus, although an "aroused" EEG pattern in the monkey during the initial symbol display was an apparent prerequisite for subsequent correct matching during the four-symbol display, it was not a prerequisite for correct response to the initial single stimulus. Speculatively, those initial responses leading to a correct delayed match involved correct perception of the symbolic or structural characteristics of the initial stimulus, whereas those initial responses followed by erroneous matches were based on the unstructured brightness aspects of the stimulus only. This simpler brightness discrimination apparently did not evoke an aroused EEG pattern in limbic and subcortical nuclei. Killam and Killam (1967) noted increased 40-Hz activity in cat during correct discrimination of form, but not during simple intensity discriminations. Our monkey data also showed increased intensity in the 20–30-Hz band only in those epochs in which perception of form must have occurred. High-intensity theta peaks characterized the cat hippocampal EEG during visual discrimination, and also occurred in temporo-occipital leads of man during difficult visual form discriminations (Walter et al., 1967b), but were not seen in these monkey data. This theta activity has been interpreted as relating to an anxiety component, as during impending launch of an astronaut (Adey et al., 1967a). These high theta coherences between hippocampal and subcortical leads in cat may also relate to perception of intensity, rather than form.

## Computed recognition of EEG patterns associated with decision making and psychological stress in animals and man

With increasing refinement of behavioral test procedures, particularly in repeatable temporal accuracy of test presentations, and with the realization of our goals to record scalp EEG data with minimal artifacts in performing man (Walter et al., 1967b; Kado and Adey, 1968), it has become clear that these records do indeed arise in changes in activity within the brain, and not, as has been suggested (Bickford, 1964), in indeterminate contributions from scalp, neck, and tongue musculature. There has been a comparable effort in analysis of neuroelectric data, but even with current analytic methods that can effectively detect only linear relations, there has been a plethora of computed parameters that may truly overwhelm the investigator, and compound, rather than resolve, his search for simplified aspects of general patterns. Our most acute difficulty lies in human perceptual weakness in dealing with computed outputs of large numbers of numerical variables. It profits us very little to deal with computer print-outs that exceed the bulk of original EEG records.

We have therefore entrusted this process of pattern detection increasingly to the computer, based on its initial calculation of comprehensive spectral parameters (Walter et al., 1967c), as described in the volume based on the last ISP (Adey, 1967). Briefly, the computer first selects a single parameter that best distinguishes a particular data set from all other sets, and then sequentially selects other parameters in order of decreasing capacity to achieve improved separation of data sets.

A cautionary attitude necessarily prevails in such applications. Are we using a method that achieves separations validly based on the physiology of cerebral systems, or is the computer at best an *idiot savant*, groping blindly for patterns that are little more than figments of its mechanical imagination? No categorical answer is yet possible, but the evidence increasingly supports the view that these classifications are significant on physiological grounds.

COMPUTER RECOGNITION OF PATTERNS IN CHIMPANZEE EEG DURING TASK PERFORMANCE We have used this pattern-recognition method for EEG data from stereotaxically implanted chimpanzees (Figure 8) playing an electronic version of the game of tick-tack-toe (noughts and crosses to the British reader). The animals may play against the trainer, or against each other, or against a computer (Hanley et al., 1968). This pattern recognition method was applied to two situations: (1) with the chimpanzee seated, not performing, but attending to the game, awaiting his opponent's move; and (2) performing with a response to the opponent's move. The technique successfully separated the two situations and also distinguished between correct and incorrect decisions.

A series of 156 EEG samples was analyzed for the two situations from three chimpanzees. Ten samples for right and wrong decisions were analyzed from a fourth animal. Electrodes were placed in amygdala, reticular formation, and nucleus centrum medianum. In the fourth animal, used in the right-versus-wrong experiment, electrodes were also placed in the caudate nucleus, and in paracentral and ventral anterior thalamic nuclei. Each EEG channel was separated in different frequency bands by digital filtering techniques. These preserve both power-density distributions at each frequency and the phase information necessary for comparison of wave trains from different locations.

In the examples shown, the frequency sub-bands were 1–3 Hz (Delta), 4–7 Hz (Theta), 8–12 Hz (Alpha), 13–18 Hz (Beta 1), 19–25 Hz (Beta 2), and 26–29 Hz (Beta 3). Four parameters were selected in each band. 1. Sum of spectral densities at each frequency. This is proportional to the mean square of the EEG amplitude, and is a measure of the power in that filter pass-band. 2. Mean frequency within the band. This is close to the dominant frequency, if one is present. 3. Bandwidth within the band. This expresses invariability of the dominant frequency. 4. Coherence, analogous to the correlation coefficient of classical statistics, was determined at each frequency or band of frequencies.

*Separation of performing and nonperforming states.* The results of these analyses are presented in both tabular and graphic form. The boundaries in the diagram enclose all the samples of the particular situation, and the asterisks indicate the position of the group means. In this display, a multidimensional plot has been reduced for purposes of reproduction to a bidimensional array. The "distance" between the means is obtained by calculating the sum of the products of the selected parameters, and their respective canonical coefficients. If a large number of competing parameters entered into the discrimination, only those contributing to the best discrimination were chosen for the display (Figure 9, page 236).

As discussed elsewhere (Hanley et al., 1968), the parameters distinguishing the two situations were not detectable by visual inspection of the EEG, nor did they arise in muscle-movement artifacts. Theta-band frequencies again figured prominently in these separations, both in their amplitude characteristics in amygdala and hippocampus, and in high coherence levels between hippocampus and midbrain reticular formation.

*Discrimination between correct and incorrect decisions.* A series of five incorrect responses was analyzed, the largest number of errors committed by one fully trained animal. They were successfully separated by analysis for both "before" and "after" epochs. Discriminant analysis of the decision-making ("before") epoch also involved consideration of competing successful variables. In this case, the discriminant analysis was continued beyond the first selection of a pair of parameters that gave 100 per cent separation. In this way, the successful pair first selected was thrown into competitive relation with second and subsequent successful selections. In all, three pairs were successful, but the third and last pair achieved the greatest separation. In the "before" epoch, one parameter, the theta bandwidth in the thalamic nucleus ventralis anterior, survived the final competition, but in the "after" epoch, none of the choices

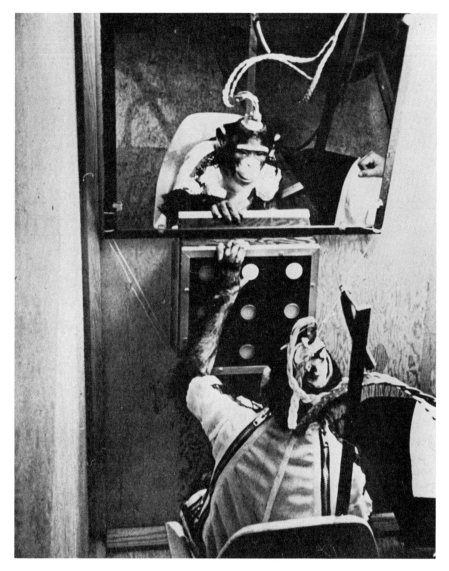

FIGURE 8 Chimpanzee playing tick-tack-toe by making linear configuration of lighted symbols. Game was played against moves by adversary (the trainer, a computer, or another chimpanzee). (From Hanley et al., 1968.)

survived. In the "before" epoch, all parameters decreased with correct decisions. This also occurred in all but one parameter in the "after" situation (Figure 10, page 236).

Although two parameters were necessary to achieve 100 per cent correct selection, only one parameter in one band from one structure was sufficient to classify with a high degree of success. For example, the sum of spectra in the theta band from nucleus ventralis anterior accurately assigned all correct decisions and misassigned only one incorrect response. In both "before" and "after" epochs, there was a striking incidence of parameters from thalamic ventralis anterior nucleus in all sets. As discussed elsewhere

(Hanley et al., 1968), this finding may relate to the part played by the nucleus ventralis anterior in the thalamo-striatal system described by Buchwald et al. (1961) as a "caudate loop," and because the game included a willed motor act, draws attention to the strong evidence favoring subcortical structures, including the striatum (rather than cortical structures), in the physiological basis of willed movements (Jung and Hassler, 1960).

These sensitive computer techniques have thus permitted a separation of visually indistinguishable EEG epochs that correlate with particular behavioral states as brief as those associated with decision making, and give further evidence

| Parameter in order of choice | Location | | Band | Direction with performance |
|---|---|---|---|---|
| 1. Sum of Spectra | L Hipp | | Delta | Value |
| 2. Mean Frequency | R F-T | | Alpha | Slowed |
| 3. Sum of Spectra | L Amyg | | Theta | Value |
| 4. Bandwidth | L Hipp | | Alpha | Narrowed |
| 5. Mean Frequency | R CM | | Alpha | Increased |

| Parameter in order of choice | Location | | Band | Comparison with correct and incorrect decision |
|---|---|---|---|---|
| Mean Frequency | Left ventral anterior thalamic nucleus | | Delta | Increased in correct decision |
| Coherence | Parcentral nucleus | LVA | Delta | Decreased in correct decision |
| Mean Frequency | Paracentral nucleus | | Beta-1 | Decreased in correct decision |
| Coherence | LVA | L Hipp | Alpha | Decreased in correct decision |
| Coherence | L Hipp | L Amyg | Beta-1 | Decreased in correct decision |
| Coherence | LVA | L Caudate | Beta-1 | Decreased in correct decision |

CASE 2

NOT PERFORMING          PERFORMING

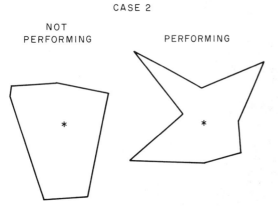

CASE 6

CORRECT                    INCORRECT

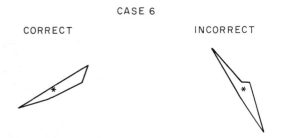

FIGURE 9  Pattern-recognition techniques applied to EEG data from chimpanzee performing tick-tack-toe task (see text). The boundaries enclose all samples of the particular situation; the asterisks indicate the position of the group means. Five steps were required to separate completely 26 samples of the two situations, not performing and performing. However, the first selection correctly classified them with more than 90 per cent accuracy. All samples were obtained the same day, and all were correct. (From Hanley et al., 1968.)

FIGURE 10  EEG patterns accompanying correct and incorrect decisions in the performing chimpanzee. Each set of two parameters classified the decision with 100 per cent accuracy; the sets are in order of increasing success.

of intimate and precise involvement of the EEG in cerebral transactions.

DETECTION OF EEG PATTERNS DURING PSYCHOLOGICAL STRESS IN MAN   More than 30 years ago, Hoagland et al. (1938) used an EEG delta-wave index in an experiment on emotion in humans involving hostile personal and sexual comments in a confrontation situation. The amount of EEG subalpha activity and its pharyngeal (hypothalamic) cortical phase lag were good indications of subjective annoyance and rage. Darrow and his colleagues (1950 and 1965) have described lead and lag changes in phase relations of dominant rhythms, measured between opposite sides and poles of the head, as functions of the emotional significance or information content of discrete symbolic or verbal stimuli. They were also sensitive to aspects of anticipation preceding a stressful stimulus, and thus resembled the long-lasting "contingent negative variation" described in these states by W. Grey Walter (1965). Our own studies have substantially confirmed the existence and direction of such transient change in both simulated and genuine stress (Adey, 1969a; Adey et al., 1967a; Walter et al., 1967a, 1967b).

The following study of the relationship of the EEG to verbal behavior was directed at two specific problems: (1) the extent to which the EEG is altered across the definable stages of a verbal exchange; and (2) the extent to which the EEG is altered by the specific information content of given verbal stimuli, or by the different alerting or stress value of a series of such stimuli (Berkhout et al., 1969b). The stages of this formal verbal exchange were described as perception, decision, and anticipation epochs, which were delimited by question duration, the question-answer interval, and the answer-next-question interval. From this arrangement of epochs, two analysis regimes evolved. For those subjects who generally delayed answering for more than 1.5 seconds past the termination of each question, the perception, decision, and anticipation epochs were separately analyzed. For those subjects whose answers followed questions very closely, only active and passive epochs were defined. The active epochs were defined as 5.0 seconds immediately following question onset and contained both question and answer. The passive epoch consisted of the subsequent 5.0 seconds of data without question or response (Figure 11).

Questions asked were classified as highly stressful, moderately stressful, or not stressful on the basis of the autonomic responses of a group of 65 subjects. The questions were separated into three autonomic response classes based on the criteria listed in Table I. The standard deviations referred to in that Table were obtained by pooling responses from all subjects across all questions. Our goal was to isolate homogeneous sets of verbal stimuli (rather than homogeneous sets of autonomic responses), so the high-stress and low-stress classes of stimuli were further divided into groups differing only in verbal content. This division was supported by analyses of concomitant EEG records. Spectral estimators used here resembled those in the chimpanzee study. To compensate for unequal bias of coherence measurements as a function of epoch length, they were defined as the difference between observed values and the calculated 95 per cent confidence level for nonrandom occurrences of that value. Interpretation of the discriminant

analysis output required an evaluation of the measurement of classification successes with the addition of each step. The possible physiological significance of a parameter chosen by the program at a given step was assessed from a detailed study of the distribution of its values for the initially defined groups and of changes in patterns at successive steps.

*EEG patterns in perception-decision-anticipation cycles of single subjects.* In four subjects, delays in question-answer sequences were long enough to permit valid estimates of spectra for all three epochs. The computation selected parameters most consistently identifying these epochs. These parameters were then combined into a multidimensional index. Cutoff points established a tentative identification of each of the total 60 epochs in this series. These subjects each had distinctive EEGs, and a discriminating canon for the whole group could not be found. On the other hand, considered separately, it was possible to achieve more than 90 per cent correct identification of the 60 epochs in each case.

The sequence of steps for one subject is described in detail (Table I). At the fifteenth step, 58 out of 60 were correctly identified. The initial parameter selected was a cross-spectral derivative, coherence between temporo-occipital ($T_5O_1$) and biooccipital leads ($O_1O_2$) at 6 Hz. Coherence was lowest during the perception epoch and highest during the decision epoch between question and answer. With this parameter alone, 50 per cent was correctly attributed. Correct attributions were primarily from perception and decision categories, which averaged the lowest and highest coherences,

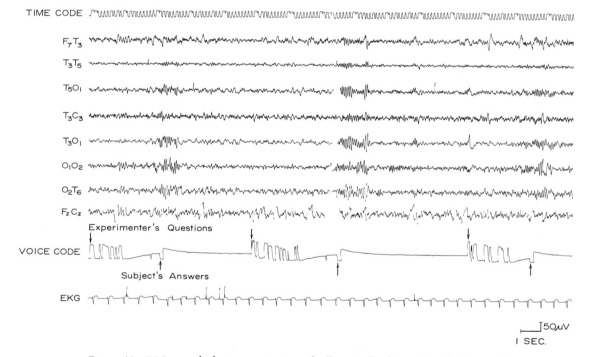

FIGURE 11   EEG records during questioning of a "reactive" subject. Note the bursts of alpha activity regularly following this subject's answers. (From Berkhout et al., 1969b).

TABLE I

*Summary of a stepwise discriminant analysis. Data for a single subject were evaluated by epoch type, 60 epochs contributing. Values of intensity are in $\log_{10} \mu V^2/Hz$. Phase was defined as the sine of the angle by which the first channel of each pair led or lagged the second in a given frequency band. Coherence was defined in nonphysical units as the difference between an observed value and the calculated 95 per cent confidence level for nonrandom occurrence of that value, based on epoch length and spectral resolution. (From Berkhout et al., 1969b.)*

| Step | Parameter (3-Hz Bandwidths) | Percent Correct after Step | Average Values over 20 Epochs | | |
|------|------|------|------|------|------|
| | | | Perception Epochs | Decision Epochs | Anticipation Epochs |
| **SUBJECT 8** | | | | | |
| 1 | $T_5O_1 \times O_1O_2$ Coherence, 6 Hz | 54 | −.34 | −.12 | −.23 |
| 2 | $T_5O_1$ Intensity, 12 Hz | 62 | 2.40 | 2.52 | 2.60 |
| 3 | $O_1O_2 \times O_2T_6$ Coherence, 6 Hz | 79 | −.23 | −.10 | −.33 |
| 4 | $O_1O_2 \times O_2T_6$ Coherence, 18 Hz | 83 | −.26 | −.18 | −.33 |

97 per cent correct attained at Step 15

respectively. This particular coherence parameter was the best discriminator among several coherence parameters similarly distributed over the three epochs; these included $T_5O_1$–$O_1O_2$ at 6 Hz, and $O_1O_2$–$O_2T_6$ at 12 Hz. In succeeding steps, the remaining parameters were corrected for variance common to items already chosen, so that co-varying families of parameters would be represented by their best number. Each increment of correctly classified epochs was usually drawn from the extreme tails of the distribution of the particular parameter added at each step.

Temporo-occipital ($T_5O_1$) intensity was selected as Step 2, and identified five additional epochs. All were anticipation epochs with high alpha intensity. The symmetrical channel ($O_2T_6$) had a similar alpha pattern, low during question presentation, slightly higher before answer, and highest while attending the subsequent question. No more than five of the 60 epochs were misclassified for each individual with 15 parameters contributing; with only five parameters contributing, no more than 15 of the 60 epochs were misclassified. For all subjects, perception epochs were characterized by an attenuation of parietal-occipital alpha, compatible with an alert state. Decision epochs showed increased coherence in theta and beta bands, compatible with a state of increased stress. Anticipation epochs were characterized by low coherences and generally high alpha intensities.

*EEG patterns in high and low stress questions for groups of subjects.* A limited account is presented of successful separations of EEG parameters characterizing a group of subjects during stressful questioning (Berkhout et al., 1969b). EEG changes associated with strings of alternating active and passive epochs (as defined above) were individual-specific. However, separate analysis of active epochs alone indicated similarities over many subjects simultaneously, within strict numerical limits. Moreover, these similarities were related to stress-values of particular questions, and with their verbal content.

For the purpose of this group assessment, EEG records were first classified as "reactive" or "nonreactive," according to whether alpha spindling or alpha attenuation accompanied the presentation of questions (Berkhout et al., 1969b, Figures 2 and 3).

Active epochs from five subjects with "nonreactive" EEGs were analyzed. All questions with high stress were combined in one set, and innocuous and low-stress questions in another. By restricting the discriminating procedure to these two classes, the parameters chosen as the basis of separation could be related to differential autonomic responses that defined the two sets. It should be emphasized that this did not imply a linear correlation of EEG and autonomic activity, as the criteria of stress in a question included contradictory patterns of autonomic arousal and widely varying baselines. Characteristics of high and low stress emerged from this analysis (Table II). More than 90 per cent of the epochs tested were correctly identified.

Tentatively, the parameters listed in Table II may be interpreted as an EEG index of stress. In high stress, temporal-parietal theta bandwidth and occipital alpha bandwidth were narrowed, with a 70 per cent success in frontal-

TABLE II

*Summary of a stepwise discriminant analysis for a group of five subjects, evaluated by stress level. One "active" epoch (see text) was included for each subject and question, providing 140 epochs for analysis. (From Berkhout et al., 1969b.)*

| | | | Average Values | |
| Step | Parameter | Percent Correct after Step | Stress Epochs | Non-stress Epochs |
| --- | --- | --- | --- | --- |
| GROUP A | NON-REACTIVE EEG | | | |
| 1 | $F_7T_3 \times T_5O_1$ Coherence 0–2 Hz | 67% | −.48 | −.38 |
| 2 | $T_3T_5$ Bandwidth 3–7 Hz | 68% | 2.2 Hz | 2.4 Hz |
| 3 | $O_1O_2 \times O_2T_6$ Coherence 8–12 Hz | 70% | +.01 | −.32 |
| 4 | $O_2T_6$ Bandwidth 8–12 Hz | 74% | 1.77 Hz | 1.9 Hz |
| 5 | $F_7T_3$ Intensity 20–30 Hz | 75% | 3.00 | 2.80 |
| 6 | $T_3T_5$ Phase 13–19 Hz | 84% | −.12 | .02 |

92 per cent correct at Step 12

temporal beta intensity. Cross-spectral discriminators included a slight but consistent decrease in delta-band coherence between left frontal and left temporal areas, and a major alpha-band coherence between right occipital and right parietal channels during the same epochs.

The EEG elements that appeared sensitive to verbal aspects of the presentation, in ways that generalized adequately across subjects, were almost completely restricted to such cross-spectral derivatives as coherence and phase. The distribution of incorrect attributions among subjects was uniform: no one subject generated more than three incorrect attributions for the 14 epochs evaluated per subject, and no one question generated more than two incorrect attributions for the five epochs evaluated per question.

Thus, EEG criteria can indeed be used to separate behavioral epochs differing only in the nature of a verbal exchange, and this separation is valid for a series of subjects. It is scarcely conceivable that such a separation, based on parameters regionally organized and specific to a functional scheme of brain systems, would occur with clearly identifiable similarities between individuals if the EEG did not bear a close relation to transaction of information in cerebral systems.

## Role of membrane polyanions in electrical rhythms and long-term states of cerebral tissue

This essay has rested its case on a sequence in phenomenology; waves generated by neurons are major contributors to the EEG, and patterns in the EEG are closely related to finer aspects of behavioral states, including levels of learned performance.

Do these wave processes also relate to storage and retrieval of information? We have shown that there is a progressive decrease in the scatter of phase relations in these wave trains at progressively higher levels of performance. In the probability bounds in wave processes at high levels of discriminative performances, we may discern elements of a "stochastic" or best-fit pattern, where wave relations in cerebral systems may closely resemble, but not necessarily be identical with, those present during initial laying down of the memory trace. Such wave phenomena would be associated with "write" and "read" phases of storage. Neuronal excitability would then be determined by previous experience of these particular patterns of waves (Adey and Walter, 1963).

Beyond this point, a search for causality between physiological phenomena and storage or retrieval of information in brain tissue must remain speculative in quite critical aspects. Even though a complete picture may not emerge with current investigative techniques, and although gaps in phenomenology are great, speculation may still be useful in suggesting broad lines of future research.

If we direct our attention to membranes of cerebral neurons, there is impressive evidence that their structural limits are not set by the double layers of the plasma membrane of classic electron microscopy. As a structural and functional unit, these membranes are now considered to include macromolecular coats of mucopolysaccharides and mucoproteins. These coats greatly increase membrane thickness and must be considered in any model of an excitable membrane (Schmitt and Davison, 1965; Pease, 1966; Rambourg and Leblond, 1967; Bondareff, 1967). These molecules may hold the key to essential steps in excitatory processes. They form highly hydrated networks in which the macromolecular lattice is extremely loose. Loss of water occurs with retraction of the network. Because they bear numerous anionic sites, they bind readily to cations, and particularly to divalent cations. Calcium ions exhibit a greater binding power than do other divalent cations (Katchalsky, 1964; Adey et al., 1969; Simpson, 1968; Wang and Adey, 1969), and in binding to calcium displacement of bound water occurs, with retraction of the network.

These substances have been characterized clearly by light and electron microscopy at the neuronal surface. It appears that they also occur as a less organized and more diffuse material in channels of the extracellular space of brain tissue. Our studies have suggested that changes are induced in them by environmental experiences, and perhaps by slow electrotonic wave processes that occur within cerebral neurons and relate to genesis of the EEG (Adey, 1966b; Adey et al., 1966b). It is conceived that a surface mosaic or patchwork of macromolecules would come to characterize each neuronal membrane, based on regional electrotonic events.

Susceptibility of these surface macromolecules to modification during neuronal experience is supported by our findings of regional modifications in electrical impedance of brain tissue during alerting, orienting, and discriminative responses (Adey et al., 1963; Adey et al., 1966b), and the more recent disclosure of impedance changes induced by altered calcium concentrations in brain tissue (Wang et al., 1966; Adey et al., 1969), which are abolished by prior enzymatic degradation of the mucopolysaccharides by hyaluronidase (Wang et al., 1968; Wang and Adey, 1969). Raised calcium concentrations consistently evoked seizure discharges, as did binding of calcium by chelating agents such as EDTA.

A recent model of nervous excitation, based on Wien dissociation effects, has proposed that excitation involves a change in binding sites for calcium and would be associated with local alkalosis (Bass and Moore, 1968). Affinity of anionic sites on the membrane for hydrogen ions is much greater than for other univalent cations, and usually greater than for divalent cations, such as calcium and magnesium. Competition between hydrogen and calcium ions would occur at different but interrelated anionic sites, with calcium ions attaching on the outer surface of the membrane and hydrogen ions on the inner (Tarby et al., 1968). Bass and Moore envisage either sequential or parallel interaction of calcium and hydrogen with polyanionic binding sites. Sequential attachment would only change the species of bound ions, but parallel independent action would lead to conformational changes in surface proteins.

It is in these conformation changes in protein that we may have the most definitive clue to lasting changes in structure that are the substrate of information storage. The suggested role for calcium in these phenomena leads logically to a consideration of a possible similar action in neurohumoral and neurohormonal processes (Simpson, 1968). In vitro experiments by Quarles and Folch-Pi (1965) showed that gangliosides in biphasic organic-aqueous systems can be induced by calcium to localize at the aqueous-organic interface when contaminated with protein.

Although we do not yet know what may be the normal triggers to lasting conformational changes in macromolecular structures at the neuronal surface, our model emphasizes the regional autonomy in membrane structure in the storage of information, rather than directly involving nuclear DNA and RNA mechanisms. It is encouraging to the neurobiologist that such schemes have been suggested by immunologists to explain antibody production by plasma cells, in response to antigens fixed on the external surface of the membrane (Nossal et al., 1967).

## Acknowledgments

Studies in our laboratory were supported by the United States Public Health Service (Grants NB-01883, MH-03708, NB-2501, and FR-7), the United States Air Force Office of Scientific Research (Contract AF-49-(638)-1387), the Office of Naval Research [Contract ONR 233 (91)], and the National Aeronautics and Space Administration (Contract NAS 237-62).

REFERENCES

ADEY, W. R., 1966. Neurophysiological correlates of information transaction and storage in brain tissue. *Progr. Physiol. Psychol.* 1: 1–43.

ADEY, W. R., 1967. Intrinsic organization of cerebral tissue in alerting, orienting, and discriminative responses. *In* The Neurosciences: A Study Program (G. C. Quarton, T. Melnechuk, and F. O. Schmitt, editors). The Rockefeller University Press, New York, pp. 615–633.

ADEY, W. R., 1969a. Spectral analysis of EEG data from animals and man during alerting, orienting and discriminating responses. *In* Conference on Attention in Neurophysiology (T. Mulholland and C. Evans, editors). Butterworth, London.

ADEY, W. R., 1969b. Neural information processing; windows without and the citadel within. *In* Biocybernetics of the Central Nervous System (L. D. Proctor, editor). Little, Brown and Co., Boston, pp. 1–27.

ADEY, W. R., 1969c. *In* evoked potentials as indicators of sensory information processing. *Neurosci. Res. Program Bull.* 7 (no. 3): 246–247.

ADEY, W. R., B. G. BYSTROM, A. COSTIN, R. T. KADO, and T. J.

TARBY, 1969. Divalent cations in cerebral impedance and cell membrane morphology. *Exp. Neurol.* 23: 29–50.

ADEY, W. R., C. W. DUNLOP, and C. E. HENDRIX, 1960. Hippocampal slow waves; distribution and phase relations in the course of approach learning. *Arch. Neurol.* 3: 74–90.

ADEY, W. R., R. T. KADO, J. DIDIO, and W. J. SCHINDLER, 1963. Impedance changes in cerebral tissue accompanying a learned discriminative performance in the cat. *Exp. Neurol.* 7: 259–281.

ADEY, W. R., R. ELUL, R. D. WALTER, and P. C. CRANDALL, 1966a. The cooperative behavior of neuronal populations during sleep and mental tasks. *Electroencephalogr. Clin. Neurophysiol.* 23: 88 (abstract).

ADEY, W. R., R. T. KADO, J. T. MCILWAIN, and D. O. WALTER, 1966b. The role of neuronal elements in regional cerebral impedance changes in alerting, orienting and discriminative responses. *Exp. Neurol.* 15: 490–510.

ADEY, W. R., R. T. KADO, and D. O. WALTER, 1967a. Computer analysis of EEG data from Gemini flight GT-7. *Aerospace Med.* 38: 345–359.

ADEY, W. R., R. T. KADO, and D. O. WALTER, 1967b. Analysis of brain wave records from Gemini flight GT-7 by computations to be used in a 30 day primate flight. *In* Life Sciences and Space Research. North-Holland Publishing Co., Amsterdam, pp. 65–93.

ADEY, W. R., and D. O. WALTER, 1963. Application of phase detection and averaging techniques in computer analysis of EEG records in the cat. *Exp. Neurol.* 7: 186–209.

ADEY, W. R., D. O. WALTER, and C. E. HENDRIX, 1961. Computer techniques in correlation and spectral analyses of cerebral slow waves during discriminative behavior. *Exp. Neurol.* 3: 501–524.

BASS, L., and W. J. MOORE, 1968. A model of nervous excitation based on the Wien dissociation effect. *In* Structural Chemistry and Molecular Biology (A. Rich and N. Davidson, editors). W. H. Freeman and Co., San Francisco, pp. 356–369.

BERKHOUT, J., W. R. ADEY, and E. CAMPEAU, 1969a. Simian EEG activity related to problem solving during a simulated space flight. *Brain Res.* 13: 140–145.

BERKHOUT, J., D. O. WALTER, and W. R. ADEY, 1969b. Alterations of the human electroencephalogram induced by stressful verbal activity. *Electroencephalogr. Clin. Neurophysiol.* 27: 457–469.

BICKFORD, R., 1964. Properties of the photomotor response system. *Electroencephalogr. Clin. Neurophysiol.* 17: 456.

BLACKMAN, R. B., and J. W. TUKEY, 1959. The Measurement of Power Spectra. Dover, New York.

BONDAREFF, W., 1967. An intercellular substance in rat cerebral cortex; submicroscopic distribution of ruthenium red. *Anat. Rec.* 157: 527–536.

BUCHWALD, N. A., E. J. WYERS, T. OKUMA, and G. HEUSER, 1961. The "caudatespindle" I. Electrophysiological properties. *Electroencephalogr. Clin. Neurophysiol.* 13: 509–518.

CAILLE, E. J. P., and P. BOCK, 1966. Apport de l'analyse des paramètres électrobiologiques dans l'objectivation des phases du sommeil. Ministère des Armées Françaises, Service de Psychologie Appliquée, Toulon, France, Study No. 17, 29 pp.

CAMPBELL, J., E. BOWER, S. J. DWYER, and G. V. LAGO, 1967. On sufficiency of auto correlation functions as EEG descriptors. *Inst. Elec. Electron. Eng. Trans. Bio-Med. Eng.* 14: 49–52.

COLE, K. S., 1940. Permeability and impermeability of cell membranes for ions. *Cold Spring Harbor Symp. Quant. Biol.* 8: 110–122.

COOMBS, J. S., D. R. CURTIS, and J. C. ECCLES, 1959. The electric constants of the motoneurone membrane. *J. Physiol.* (London) 145: 505–528.

CRAMÉR, H., 1955. The Elements of Probability Theory. John Wiley and Sons, New York, pp. 168–171.

CREUTZFELDT, O. D., 1969. The human alpha rhythm and its relation to "mental activity." *In* Conference on Attention in Neurophysiology (T. Mulholland and C. Evans, editors). Butterworth, London.

CREUTZFELDT, O. D., J. M. FUSTER, H. D. LUX, and A. C. NACIMIENTO, 1964. Experimenteller Nachweis von Beziehungen zwischen EEG-Wellen und Aktivität corticaler Nervenzellen. *Naturwissenschaften* 51: 166–167.

CREUTZFELDT, O. D., S. WATANABE, and H. D. LUX, 1966. Relations between EEG phenomena and potentials of single cortical cells. II. Spontaneous and convulsoid activity. *Electroencephalogr. Clin. Neurophysiol.* 20: 19–37.

DARROW, C. W., and M. CONVERSE, 1950. EEG phase and autonomic function. *Electroencephalogr. Clin. Neurophysiol.* 2: 225 (abstract).

DARROW, C. W., and R. G. HICKS, 1965. Interarea electroencephalographic phase relationships following sensory and ideational stimuli. *Psychophysiology* 1: 337–346.

DITCHBURN, R. W., 1963. Information and control in the visual system. *Nature (London)* 198: 630–632.

EFREMOVA, T. M., N. G. ZHEGALKINA, and L. D. MESHALKIN, 1968. Investigations of the dynamics of change of spectral composition of the bioelectrical activity in the cerebral cortex of the rabbit during rhythmic photic stimulation. *In* Mathematical Analysis of the Electrical Activity of the Brain (M. N. Livanov, V. S. Rusinov, and J. S. Barlow, editors). Harvard University Press, Cambridge, Massachusetts, pp. 45–52.

ELAZAR, Z., and W. R. ADEY, 1967a. Spectral analysis of low frequency components in the electrical activity of the hippocampus during learning. *Electroencephalogr. Clin. Neurophysiol.* 23: 225–240.

ELAZAR, Z., and W. R. ADEY, 1967b. Electroencephalographic correlates of learning in subcortical and cortical structures. *Electroencephalogr. Clin. Neurophysiol.* 23: 306–319.

ELUL, R., 1962. Dipoles of spontaneous activity in the cerebral cortex. *Exp. Neurol.* 6: 285–299.

ELUL, R., 1964. Specific site of generation of brain waves. *Physiologist* 7: 125.

ELUL, R., 1967a. Amplitude histograms of the EEG as an indicator of the cooperative behavior of neuron populations. *Electroencephalogr. Clin. Neurophysiol.* 23: 87 (abstract).

ELUL, R., 1967b. Statistical mechanisms in generation of the EEG. *In* Progress in Biomedical Engineering (L. J. Fogel and F. W. George, editors). Spartan Books, Washington, D. C., pp. 131–150.

ELUL, R., 1968. Brain waves: Intracellular recording and statistical analyses help clarify their physiological significance. *Data Acquis. Process. Biol. Med.* 5: 93–115.

ELUL, R., 1969. Gaussian behavior of the electroencephalogram: Changes during performance of mental task. *Science (Washington)* 164: 328–331.

ELUL, R., and W. R. ADEY, 1966. Instability of firing threshold and "remote" activation in cortical neurones. *Nature (London)* 212: 1424–1425.

EULER, C. von. and J. D. GREEN, 1960a. Activity in single hippocampal pyramids. *Acta Physiol. Scand.* 48: 95–109.

EULER, C. von, and J. D. GREEN, 1960b. Excitation, inhibition and rhythmical activity in hippocampal pyramidal cells in rabbit. *Acta Physiol. Scand.* 48: 110–125.

Fox, S. S., and R. J. NORMAN, 1968. Functional congruence: An index of neural homogeneity and a new measure of brain activity. *Science (Washingon)* 159: 1257–1259.

Fox, S. S., and J. H. O'BRIEN, 1965. Duplication of evoked potential waveform by curve of probability of firing of a single cell. *Science (Washington)* 147: 888–890.

FROST, J. D., 1968. EEG intracellular potential relationships in isolated cerebral cortex. *Electroencephalogr. Clin. Neurophysiol.* 24: 434–443.

FROST, J. D., and Z. ELAZAR, 1968. Three-dimensional selective amplitude histograms: A statistical approach to EEG-single neuron relationships. *Electroencephalogr. Clin. Neurophysiol.* 25: 499–503.

FUJITA, Y., and T. SATO, 1964. Intracellular records from hippocampal pyramidal cells in rabbit during theta rhythm activity. *J. Neurophysiol.* 27: 1012–1025.

GAVRILOVA, N. A., and A. S. ASLANOV, 1968. Application of electronic computing techniques to the analysis of clinical electroencephaloscopic data. *In* Mathematical Analysis of the Electrical Activity of the Brain (M. N. Livanov, V. S. Rusinov, and J. S. Barlow, editors). Harvard University Press, Cambridge, Massachusetts, pp. 53–64.

GERSCH, W., 1968. *In* Neural coding. *Neurosci. Res. Program Bull.* 6 (no. 3): 298.

GRASTYÁN, E., G. KARMOS, L. VERECZKEY, and L. KELLÉNYI, 1966. The hippocampal electrical correlates of the homeostatic regulation of motivation. *Electroencephalogr. Clin. Neurophysiol.* 21: 34–53.

GRASTYÁN, E., K. LISSÁK, I. MADARÁSZ, and H. DONFOFFER, 1959. Hippocampal electrical activity during the development of conditioned reflexes. *Electroencephalogr. Clin. Neurophysiol.* 11: 409–430.

GREEN, J. D., D. S. MAXWELL, W. J. SCHINDLER, and C. STUMPF, 1960. Rabbit EEG "theta" rhythm: Its anatomical source and relation to activity in single neurons. *J. Neurophysiol.* 23: 403–420.

HAFEMANN, D. R., A. COSTIN, and T. J. TARBY, 1969. Neurophysiological effects of tetrodotoxin in lateral geniculate body and dorsal hippocampus. *Brain Res.* 12: 363–373.

HANLEY, J., D. O. WALTER, J. M. RHODES, and W. R. ADEY, 1968. Chimpanzee performance: Computer analysis of electroencephalograms. *Nature (London)* 220: 879–881.

HOAGLAND, H., D. E. CAMERON, M. A. RUBIN, and J. J. TEGELBERG, 1938. Emotion in man as tested by the delta index of the EEG. II. Simultaneous records from cortex and from a region near the hypothalamus. *J. Gen. Psychol.* 19: 247–261.

JASPER, H. H., and C. STEFANIS, 1965. Intracellular oscillatory rhythms in pyramidal tract neurones in the cat. *Electroencephalogr. Clin. Neurophysiol.* 18: 541–553.

JUNG, R., and R. HASSLER, 1960. The extrapyramidal motor system. *In* Handbook of Physiology, Section I, Neurophysiology (J.

Field, H. W. Magoun, and V. E. Hall, editors). American Physiological Society, Washington, D. C., pp. 863–927.

KADO, R. T., and W. R. ADEY, 1968. Electrode problems in central nervous monitoring in performing subjects. *Ann. New York Acad. Sci.* 148: 263–278.

KARAHASHI, Y., and S. GOLDRING, 1966. Intracellular potentials from "idle" cells in cerebral cortex of cat. *Electroencephalogr. Clin. Neurophysiol.* 20: 600–607.

KATCHALSKY, A., 1964. Polyelectrolytes and their biological interaction. *In* Connective Tissue: Intercellular Macromolecules. New York Heart Association. Little, Brown and Co., Boston, pp. 9–41.

KILLAM, K. F., and E. K. KILLAM, 1967. Rhinencephalic activity during acquisition and performance of conditional behavior and its modification of pharmacological agents. *In* Structure and Function of the Limbic System (W. R. Adey and T. Tokizane, editors). *Progr. Brain Res.* 27: 388–399.

MESHALKIN, L. D., and T. M. EFREMOVA, 1968. Estimation of the spectra of physiological processes over short intervals of time. *In* Mathematical Analysis of the Electrical Activity of the Brain (M. N. Livanov, V. S. Rusinov, and J. S. Barlow, editors). Harvard University Press, Cambridge, Massachusetts, pp. 37–44.

NELSON, P. G., 1967. Brain mechanisms and memory. *In* The Neurosciences: A Study Program (G. C. Quarton, T. Melnechuk, and F. O. Schmitt, editors). The Rockefeller University Press, New York, pp. 772–775.

NODA, H., S. MANOHAR, and W. R. ADEY, 1969. Spontaneous activity of cat hippocampal neurons in sleep and wakefulness. *Exp. Neurol.* 24: 217–231.

NOSSAL, G. J. V., G. M. WILLIAMS, and C. M. AUSTIN, 1967. Antigens in immunity. XIII. The antigen content of single antibody-forming cells early in primary and secondary immune responses. *Aust. J. Exp. Biol. Med. Sci.* 45: 581–594.

PEASE, D. C., 1966. Polysaccharides associated with the exterior surface of epithelial cells: Kidney, intestine, brain, *J. Ultrastruct. Res.* 15: 555–588.

PENFIELD, W., 1958. Functional localization in temporal and deep sylvian areas. *Res. Publ. Ass. Res. Nerv. Ment. Dis.* 36: 210–226.

PERKEL, D. H., and T. H. BULLOCK, 1969. Neural coding. *Neurosci. Res. Program Bull.* 6 (no. 3): 221–348.

PETSCHE, H., and C. STUMPF, 1962. Hippocampal arousal and seizure activity in rabbits; toposcopical and microelectrode aspect. *In* Physiologie de l'Hippocampe (J. Cadilhac, editor). *Colloq. Int. CNRS* 107: pp. 121–134.

QUARLES, R., and J. FOLCH-PI, 1965. Some effects of physiological cations on the behaviour of gangliosides in a chloroform-methanol-water biphasic system. *J. Neurochem.* 12: 543–553.

RADULOVACKI, M., and W. R. ADEY, 1965. The hippocampus and the orienting reflex. *Exp. Neurol.* 12: 68–83.

RAMBOURG, A., and C. P. LEBLOND, 1967. Electron microscope observations on the carbohydrate-rich cell coat present at the surface of cells in the rat. *J. Cell Biol.* 32: 27–53.

REMOND, A., 1960. Recherche des renseignements significatifs dans les enregistrements électrophysiologiques et mechanisation possible. *In* Actualités Neurophysiologiques (A. M. Monnier, editor). Second Series. Masson, Paris, pp. 167–210.

SIMPSON, L. L., 1968. The role of calcium in neurohumoral and neurohormonal extrusion processes. *J. Pharm. Pharmacol.* 20: 889–910.

SCHMITT, F. O., and P. F. DAVISON, 1965. Role of protein in neural function. *Neurosci. Res. Program Bull.* 3 (no. 6): 55–76.

STORM VAN LEEUWEN, W., 1965. Relationships between the behavior of dogs and electrical activity of different brain regions. *In* Refleksi Golovnago Mozga (E. A. Asratyan, editor). Academy of Sciences of USSR, Moscow, pp. 82–95.

TARBY, T. J., A. COSTIN, and W. R. ADEY, 1968. Effects of tetrodotoxin on impedance in normal and asphyxiated cerebral tissue. *Exp. Neurol.* 22: 517–531.

TUKEY, J. W., 1965. Data analysis and the frontiers of geophysics. *Science* (*Washington*) 148: 1283–1289.

WALTER, D. O., 1963. Spectral analysis for electroencephalograms: Mathematical determination of physiological relationships from records of limited duration. *Exp. Neurol.* 8: 155–181.

WALTER, D. O., and W. R. ADEY, 1968. Is the brain linear? *In* Proceedings of the Symposium of International Federation of Automation and Computing. Yerevan, Armenia.

WALTER, D. O., J. BERKHOUT, and W. R. ADEY, 1967a. Patterns of EEG and autonomic reactivity to laboratory stress. *In* Progress in Biomedical Engineering (L. J. Fogel and F. W. George, editors). Spartan Books, New York, pp. 125–130.

WALTER, D. O., R. T. KADO, J. M. RHODES, and W. R. ADEY, 1967b. EEG baselines in astronaut candidates estimated by computation and pattern recognition techniques. *Aerospace Med.* 38: 371–379.

WALTER, D. O., J. M. RHODES, and W. R. ADEY, 1967c. Discriminating among states of consciousness by EEG measurements. A study of four subjects. *Electroencephalogr. Clin. Neurophysiol.* 22: 22–29.

WALTER, W. G., 1965. Brain responses to semantic stimuli. *J. Psychosom. Res.* 9: 51–61.

WANG, H. H., and W. R. ADEY, 1969. Effects of cations and hyaluronidase on cerebral electrical impedance. *Exp. Neurol.* 25: 70–84.

WANG, H. H., R. T. KADO, and W. R. ADEY, 1968. Calcium, mucopolysaccharides and cerebral impedance. *Fed. Proc.* 27: 749.

WANG, H. H., T. J. TARBY, R. T. KADO, and W. R. ADEY, 1966. Periventricular cerebral impedance after intraventricular injection of calcium. *Science* (*Washington*) 154: 1183–1184.

WIENER, N., 1958. Nonlinear Problems in Random Theory. Wiley and Sons, New York. M. I. T. Press, Cambridge, Massachusetts.

# 24    Evoked Potential, Coding, and Behavior

## STEPHEN S. FOX

DELINEATION OF coding of functional information by bioelectric processes in brain is essential to an understanding of the ways in which the organism undergoes change with experience. Advances in knowledge of how the evoked potential—that stable index of processes in large cell populations—represents changes in the nervous system permit investigation of the changes in neural state that parallel or underlie behavior modification. As parameters of slow activity and evoked potentials are delineated (Dill et al., 1968; Thomas et al., 1968; Verzeano et al., 1968; Norman and Fox, in preparation), it will become possible to utilize the easily recorded, and generally representative, spontaneous slow-activity and evoked potentials as indexes of underlying synaptic organizations. This development will have particular advantages in behavioral neurophysiology. Single-cell studies of learning (Jasper et al., 1960; Yoshii and Ogura, 1960; Kamikawa et al., 1964; Bureš and Burešová,

1965; Buchwald et al., 1966a; Hori et al., 1967; O'Brien and Fox, 1969a, 1969b), although very promising and reasonably indicative of associative processes, are still technically difficult to perform. They yield high variability in the outcome as a result of the particular eccentricities of the individual cells studied, the necessarily small samples, the common instability of many of the preparations, and the fact that an individual cell provides a very limited sample of the immense populations that are functional under normal behavioral conditions (O'Brien and Fox, 1969a, 1969b). A primary goal of this discussion is to evaluate the extent to which powerful analytical tools in the form of manipulations of evoked potentials are available and promising for an understanding of functional neural coding.

### Bioelectrical origins of the evoked response

Our knowledge has progressed from the early theoretical accounts of slow waves, which were dominated by the axonological framework originating in studies of nerve and which suggested that slow waves represented the envelope

STEPHEN S. FOX   Department of Psychology, University of Iowa, Iowa City, Iowa

of spike discharges from individual neurons firing with some dispersion (Adrian and Matthews, 1934). Doubt was cast on the concept of summation of action potentials in envelopes of integration, when appropriate physiological capacitances to serve the function of integration were found to be absent. Later, revised formulations suggesting that slow waves originated from the summation of afterdischarges (of the unit firings), which were of longer duration than the action potentials, were also laid to rest by the demonstrations (1) that in the asphyctic or anesthetized state EEG rhythms can be recorded in the absence of cortical-cell firing (Marshall et al., 1943; Li and Jasper, 1953) and (2) that the discharge, at least of pyramidal tract neurons, contributes negligibly to the cortical-surface response (Gorman and Silfvenius, 1967; Humphrey, 1968a, 1968b). Although these observations causally separated action potentials both from the slow potentials of the EEG and from the evoked potentials, they were taken also to indicate a poor or irregular relationship between the two events (Bishop and Clare, 1952; Li et al., 1952; Jung, 1953; Li et al., 1956; Adey, 1958; von Euler et al., 1958; Dunlop et al., 1959; Widen and Ajmone Marsan, 1960; Li, 1963; Gerstein and Kiang, 1964; Adey, 1966; Buchwald et al., 1966).

More recent conceptions of the source of slow potentials as the summation of excitatory and inhibitory postsynaptic potentials (Eccles, 1951; Purpura, 1959; Kandel and Spencer, 1961; Pollen, 1964; Stefanis and Jasper, 1964; Klee et al., 1965; Purpura and McMurtry, 1965; Creutzfeldt et al., 1966a, 1966b; Lux et al., 1966), carefully reviewed by Purpura (1967), Morrell (1967), and Landau (1967) in the earlier volume of this series, by Purpura in an extensive consideration (1959), and by Ajmone Marsan (1965), have made it increasingly clear that both the evoked potential and the variety of slow potentials included in the surface-recorded EEG are integrated reflections of synaptic potentials and local potentials from individual nerve cells and, possibly, in addition, from non-neural elements.

Our work has also consistently been an exception to the conclusion of a poor relationship of spike to wave. We were convinced that interpretations of spike-wave relationships, based on either single or multiple, superimposed, oscilloscope traces, yielding small and variable samples, could provide only impressions of the patterns of single-cell firing and make it difficult to reach conclusions regarding relations to the evoked potential. In a study from our laboratory (Fox and O'Brien, 1965), computer compilation of the probability of firing of single cells in cat cortex in response to physiological sensory stimulus (somatic or light flash) revealed that the frequency distribution (i.e., probability) of the firing for any one cell closely corresponded to the average waveform of the evoked potential recorded from the same microelectrode. We showed that

this high correlation between spike probability and waveform held throughout the poststimulus period by point-to-point cross-correlation (Fox et al., 1967). Moreover, if a single cell responded to more than one sensory stimulus, the probability of response to each stimulus was given by the respective waveform of the evoked potential.

We concluded from these data that no component (positive or negative, early or late) of the asynchronous evoked potential recorded in this way was uniquely related to responses of specific cell populations or to specific portions of cells, because, from each cell sampled, a probability curve could be obtained that closely resembled the waveform of the entire evoked response. This outcome suggested, therefore, that whatever potential sources contributed to the evoked potential, they were directly related to or reflected in the firing of single cells.

Thus, it was certain that, contrary to conclusions from other studies (Jung, 1953; Gerstein and Kiang, 1964), knowledge of the waveform of the evoked potential does, to a great extent, enable prediction of the response pattern of individual cortical cells. In other experiments, we recorded from the cerebellar vermis and studied single-cell spike responses to a click stimulus followed by hippocampal stimulation (Figure 1). As the probability of spikes in response to the click increased, the mean latency of response to the succeeding hippocampal stimulation increased until the spike response failed altogether. It should be noted that the latency of the hippocampal-stimulated spike is not a function of the number of spike responses to a click, but is only related probabilistically. We then measured the amplitude of the evoked-potential response to the click stimulus and related this measure to the latency of the single-cell responses to hippocampal stimulation (Figure 2). The shape of the amplitude-predicted curve was almost identical to that of the spike probability-predicted curve, and the failure point of the hippocampal response was almost exactly the same as when the latency was predicted by the probability of the spike response to click. Therefore, the evoked-potential amplitude in response to the first stimulus predicts the latency or probability of a spike in response to the next stimulus, as does spike probability.

The impressively high correlation between spike probability and the waveform and amplitude of the evoked potential can be extended to the moment-to-moment prediction of the probability of spike firing by the spontaneous microelectrode EEG in brain. We have established an extremely high correlation between the probability of single-cell firing and the amplitude of the microelectrode EEG in online, real-time computer experiments (Fox and Norman, 1968). By sampling the spontaneous EEG voltages in two ways—randomly each millisecond and at each moment the cell fired—we derived the conditional probability of spike firing for each voltage from the relationship of the number

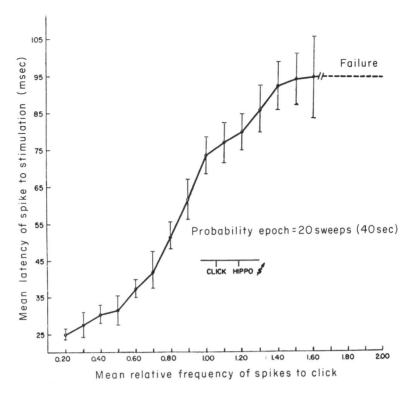

FIGURE 1 Prediction of spike latency by probability of spike to preceding stimulus. Latency of cell response to hippocampal stimulation increases proportionally with increased probability (mean relative frequency in 20 sweeps) of spike response to click. Note that mean failure occurs at approximately 1.60 (1.60 spikes per sweep in response to click). See text and Figure 2. (From Fox et al., 1967.)

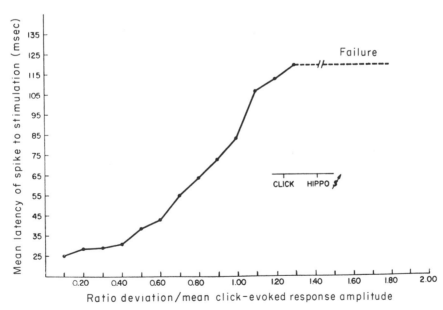

FIGURE 2 Prediction of spike latency by amplitude of evoked potential response to preceding stimulus. Latency of single cell response to hippocampal stimulation increases proportionally with increased amplitude of evoked response (relative to control) to preceding click. Note that mean failure of cell to fire occurs at approximately 1.40 (evoked response = 1.4 times control). See text and Figure 1. (From Fox et al., 1967.)

of times each voltage occurred to the number of times a spike occurred at each voltage (Figure 3). In these studies, 90 per cent of all cells observed showed significant correla-

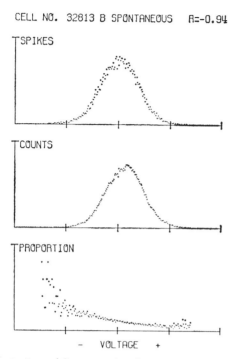

CELL NO. 32813 B SPONTANEOUS    R=-0.94

FIGURE 3  Record from a single cell in visual cortex of cat. Center curve: number of occurrences of each slow-wave voltage; top curve: number of occurrences of a cell spike at each voltage in center curve; bottom curve: conditional probability of spike firing at each voltage in center curve. The R value at top gives the Pearson correlation coefficient between slow-wave voltage and spike probability. Correlations in these and all following figures are based on the middle half of the spike and slow-wave values, and all are significant beyond the .001 level. Vertical axis is arbitrary and omitted. (From Fox and Norman, 1968.)

tions between the probability of firing and the slow-wave amplitude. This close correspondence of spike probability and spontaneous EEG amplitude is a general phenomenon in cat cortex.

These studies support and extend earlier conclusions, drawn from studies of the relation between spike probability and evoked-potential properties, that parameters of the general class of brain slow-wave activity, which includes evoked potentials and the EEG, are representative of the momentary excitability of the brain, and the excitability is indicated by the spike probability (Purpura, 1959). Such *congruence* between spike and wave has been confirmed independently in other studies (Frost and Elazar, 1968; Lass, 1968; Robertson, 1965), in which almost the same analytical procedure was often used. Perhaps more important, how-

ever, and contrary to some previous expectations, the correlation of spontaneous firing from a given cell with EEG voltage is not a biophysical constant, but varies from one replication to another—as much as 20 per cent from sample to sample. This variance may be more consistent with findings of cortical physiology than a constant relationship would be. If the field of the EEG activity were greater than unicellular, the slow activity might represent a mean excitability for a population of cells, and the excitability of the single cell might be expected to deviate from the mean excitability, depending on which of its afferents are active at a given moment. Congruence, then, was proposed as an expression of how well an observed cell is represented by the mean excitability and, as a dynamic and changing physiological measure, it should undergo orderly and biological variation under proper conditions.

We demonstrated that the spike-wave congruence could be varied experimentally for individual cells in visual cortex. When cells were driven with light, the spike-wave congruence increased significantly from its level during spontaneous activity in more than half of the observed (100) cortical cells. Congruence decreased in the remainder of the observed cells. This is understandable, because a small percentage of primary cortical sensory cells are not primarily responsive to the relevant modality, but either respond weakly or respond to stimuli in other modalities (Buser and Borenstein, 1957; Buser and Imbert, 1961; Jung, 1961; Fessard, 1961). Under driven conditions, cells related primarily to visual function are dominantly responsive to visual afferents that are not shared by such diffusely projecting or nonprimary afferent cells. In contrast, these two different cell populations may have more active afferents in common under spontaneous conditions.

Thus, the fact that not all cells are ever represented at a given moment or that not all single-cell probabilities correlate well with the amplitude of the wave may be a less important problem than is the issue of shifting probabilities. One may suggest, in fact, exactly the opposite of the proposition that all cell probabilities correlate well with the amplitude of the wave. The relevant problem may be to examine and partition the variance of the relation between unit firing and slow waves. This shifting relationship of spike probability to the individual wave parameters may elucidate the variability of the spike train.

Other studies extended spike-wave congruence based on amplitude (Fox and Norman, 1968) to studies based on slope, or first derivative properties, of the wave. For example, in a cell of which the spike probability is well related to both instantaneous amplitude and instantaneous slope of the spontaneous activity from the microelectrode under flash-driven conditions (Figure 4A), the spike probability may become dominantly slope-predicted, although amplitude-predicted congruence may continue to show a minor relationship (Figure 4B). Such observations of low amplitude-predicted congruence under driven conditions alone gives the common impression that the probability of this cell's firing is not well related to the wave. Similar changes have been seen in the opposite direction and within one slow-wave parameter. Thus, amplitude-prediction of the probability may be independent of the slope-prediction, and such independence may be experi-

mentally manipulated by the introduction of changes in the physiological state.

Slow waves, from the point of view of such experimental results, may carry more information about spike probability than does the spike itself. In the spike train taken alone, each spike is essentially an equal event, with the probability given by the just-preceding interval. Possibly, the wave carries information about two determinants of spike probability that may be mutually independent. If the shift of the probability of firing from amplitude-predicted to slope-predicted indicates that the determinants of spike probability involve a shift to a different set of afferents, it may be economical and reasonable to assume that the two wave categories are the result of two independent biogenic sources of spike probability, possibly on different locations on the cell, e.g., axosomatic and axodendritic postsynaptic sites. Rall (1967; Rall et al., 1967) has suggested in a series of mathematical papers that it is possible that different sources of bioelectrical activity may appear different to the cell body and therefore affect spike probability differentially. Furthermore, Rall's conception of different shape indexes for dendritic and somatic PSPs is consistent with the view that these amplitude and slope measures may be sensitive to the properties of different membranes.

The current question on the relationship between spike and wave may not be either of the following: (1) Why is a given cell not always well related to the wave? or (2) Why are some cells related and others not related to the wave? Instead, the question may be: What are the determinants of the shifting relationship of spike and wave?

Functional congruence can be experimentally useful as an index or continuous measure of the neural homogeneity of the tissue around the electrode. By the extent to which the activity of a single cell can be well predicted by the mean excitability, as represented by the evoked potential or spontaneous slow wave, the variance in the population around the electrode may be low and, therefore, the homogeneity of functioning high. Conversely, by the extent to which the activity of an individual cell is far from the mean, the variance in the population may be high and the homogeneity low, with considerable dispersion or asynchrony of function in the population. Such studies of functional congruence could be expanded with an aim toward understanding: (1) its time course; (2) its spatial extent; (3) its relation to behavior; and (4) its generality throughout a wide number of structures in brain.

Data from intracellular and extracellular studies continue to support the view that the macroelectrode EEG is the summation of membrane and local potentials from large numbers of individual cells situated in the various cortical and subcortical laminae (Fujita and Sato, 1964; Frost and Gol, 1966; Frost et al., 1966; Elul, 1968; Frost, 1967, 1968).

Together with our results, studies of unit activity and

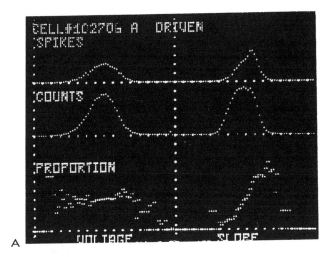

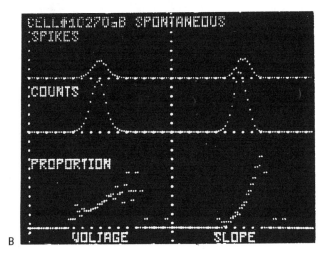

FIGURE 4 Simultaneous prediction of single-cell firing probability from slope (right) and from amplitude (left) of EEG recorded from microelectrode, under driven (A) and spontaneous (B) conditions. Slow wave is sampled in two ways: Top graphs in A and B show distribution of voltages in EEG as sampled at the moments of single-cell firing; middle graphs sampled each millisecond for the same time period. Bottom graphs, mean relative frequency (probability) of the cell firing at each voltage. In B, note that prediction by both slope and amplitude, contrasted with A, shifts to dominant slope-prediction. Zero voltage and slope is at center of each graph, positive to the right and negative to the left of center.

its relationship to evoked-potential and slow-wave phenomena indicate that the momentary fluctuation in the probability of a single cell's firing, as determined by the moment-to-moment fluctuation of its membrane potential, is stochastically represented by slow waves in brain. Thus, the continually oscillating slow-wave activity provides an accurate estimate of the sequential excitability on the average of single cells in the immediate vicinity of the electrode. Accurate estimates are not available, however, of the distances across which this mean represents the neuronal substrate. To the extent, therefore, that wave processes continue to be partitioned in terms of their individual parameters or components, and that the particular parameters, at least statistically, relate to specific aspects of the underlying synaptic organization, one may have increased confidence in this stable, general, representative, and reliable index of neural activity as an accurate reflection of functional details of the microanatomy. Therefore, one should be able to state with some precision, from this general index, the nature of the changes that take place in the neural tissue surrounding the electrode with changing behavioral conditions.

## Behavioral correlates: sources of instability

Relating the evoked potential to changes in experience or particular behaviors has not been without its difficulties. Individual correlative research efforts have clearly demonstrated a wide gamut of evoked-potential parameters that appear to vary, depending on the recording site in brain and on the particular behavior or experience involved, so that specificity or generality is yet to be forthcoming (John, 1961; Morrell, 1961). In attempts to relate changing aspects of the chronically recorded evoked potential to changes in behavior, a number of sources of instability have made it difficult to specify reliably the relevant or functional aspects of the behavior. A short review of such sources of instability may be useful in suggesting alternative strategies.

First, it may be mentioned that present methods for the measurement of behavior do not offer adequate knowledge or exact specification of the relevant aspects of the behavior being used as a correlate for neural events. By this I mean that, typically, only one reliably measurable, although arbitrary, aspect of the performance is followed, and very little is understood regarding the detailed *microtopography* of the behavior. Commonly assumed in such behavioral studies is that the specific microtopographical configuration of the behavior, e.g., muscular or receptor activity, is either similar from trial to trial, from animal to animal, and from early to late in learning, or is not relevant to the neural correlates. In addition, humoral, visceral, and sympathetic aspects of the physiology of behavior contribute to the total

state of an animal at any given moment or on any particular trial, but are also ignored.

Second, variability from animal to animal or subject to subject in the discrete microtopography or in response sequence in specific acquisition makes almost impossible, speaking neural-correlatively, the pooling of subject data from which the usual acquisition function emerges.

A third source of instability derives from inadequate or partial knowledge of the exact stimulus conditions. Even the assumption of a "neutral" stimulus may be brought into question. Prior to training, such a stimulus is neutral from the point of view of the behavior paradigm, but no stimulus is entirely neutral to any animal or human subject. Instead, it elicits specific sensory concomitants, orientations, and postural adjustments. Also, each subject differs, in its response to particular stimuli, along some generalization gradient based on prior experience, with stimuli varying along similar dimensions. Further, receptor orientation has been demonstrated to have dramatic effects on stimulus input. Finally, inherently variable spontaneous behavior additionally contributes to the bioelectrical response to a so-called "neutral" stimulus. Despite precise physical control of stimulus properties, knowledge of the "effective" stimulus is incomplete, at best.

A fourth source of considerable variability results from the arbitrary choice of a given behavior category in reference to a chosen brain location. Electrical responses to a single stimulus can be recorded from many widely separated areas of the brain, which has encouraged the placement of large arrays of electrodes to compensate, by sheer force of number, for the lack of specific information in locating brain responses that might be relevant to a specific behavioral paradigm. Our current state of knowledge regarding functional representation of complex behaviors in brain does not allow us to make assumptions regarding unique representation of behavior at any particular brain site. Nor does it allow us a logical rationale for combining a number of locations for recording purposes and expecting them to contribute conjointly in a substantial way to a given behavioral condition.

The representation of behavior in brain may be further complicated by the fact that the component nature of complex behavior may be both multiply located and sequentially represented in the brain. Continuation of the acquisition process may depend on the shifting of bioelectrical processes to new, relevant locations representing new configurations of microbehaviors that characterize new stages of behavioral acquisition. In this way, transient changes in bioelectrical activity may correlate with one or more portions of the acquisition function rather than duplicate the learning function as implicitly assumed. This contrasts with the conclusion, often reached in correlative studies, that a changed electrical activity in a given portion of brain is relevant, either early in learning but not late in learning, or vice versa.

Fifth, additional instability derives from the arbitrary definition of behavior, which often takes into account only one or a few endpoint responses in acquisition or performance, and not the infinitely complex and not known (possibly not parallel in time) set of collateral responses that are occurring and may be conditioned in the same training paradigm (Bykov, 1957; Cohen and Durkovic, 1966; Meurice et al., 1966; Gavalas, 1967; Miller, 1969). Such parallel conditioned responses, of course, may have unknown individual and conjoint influences on the bioelectrical responses, thereby making the possibility of unique coding even more unlikely.

A sixth, and perhaps most important, source of instability is the problem of common resolution of behavioral and bioelectrical measurements on a common time base with a common zero point. Correlation of momentary and discrete bioelectrical events with multiply determined molar behaviors may be in error by one or more orders of magnitude and probably by as many as seven orders (days compared to milliseconds). Perhaps preferable to such an approach is the millisecond resolution of analogue behaviors as they relate to the millisecond-to-millisecond resolution of analogue changes in brain.

Finally, the specification of parameters of the bioelectrical signal for evaluation in relation to behavior is often not done prior to experimental changes; it awaits the empirical outcome of an experiment. Therefore, parameter specification for correlation with molar behavior is necessarily arbitrary and may be unrelated to the major neural response system under conditioning control by the animal.

In summary, substantial variability of results in studies of bioelectrical correlates of behavior arises from lack of knowledge of the effective stimulus, of the total actual response under conditioning control, of the relevant recording site or combinations of sites, and of relevant parameters of the bioelectrically dependent variable, as well as a possibly unsuitable time base and resolution.

## Operant control of bioelectrical events

The above multiple sources of instability in correlative studies led us to devise a modified approach to the study of behaviorally significant bioelectrical events (Fox and Rudell, 1968; Rosenfeld et al., 1969; Fox and Rudell, 1970; Rudell, 1970; Rudell and Fox, in preparation). Our approach provides for the systematic study of bioelectrical response parameters that relate to or encode learned behavior. This approach is more a system of study than a collection of empirical findings. It is viewed as a way of refining the

neural-correlates problem and of redefining goals to be more consistent with realistic measurement of both behavior and bioelectrical events in brain.

One domain of relative stability and certainty is knowledge of the bioelectrical signal itself. One can understand the parameters (rise time, amplitude, and phase or latency) of the evoked potential in a specific and exact way; and, as suggested above, the parameters can be related to activity in the underlying neuronal population. Accordingly, our approach is the following. Instead of being made a dependent variable, the evoked potential is used as the criterion for reinforcement, or the independent variable. Prior to the experiment, we specify the temporal location and parameters of the evoked-response components to be reinforced. We allow the animals a free range of behavior, and they generate whatever behavior is available to bring about the specified bioelectrical response.

In general, an evoked-potential component of relatively low probability of occurrence is chosen from an animal's repertoire. The operant paradigm makes the assumption that bioelectrical events of low probability or of low amplitude are not, in fact, noise in the nervous system, but may be significant with respect to behavior. In addition, if the animal can be properly manipulated experimentally, such events will become more probable. With reinforcement, the probability of the occurrence would increase if it is relevant to behavior. As a food-deprived animal would be expected to maximize food-reinforced behaviors, an animal should generate a reinforced evoked-potential parameter with a higher probability than baseline only if the parameter is related to either behavioral or neural information.

In addition, operant control of neural events allows manipulation of components or parameters of the evoked potential to a biobehavioral steady state. Steady state does not mean a reduction in the variance of the potential component under experimental control, but rather a reliable, stable, and reproducible condition of the evoked potential, such that behavior may be observed repeatedly over a long period of time with a bioelectrical component in that given condition.

The experimental procedure for operant conditioning of aspects of the evoked potential in cats is as follows. Initially, measurements are taken of the mean and variance of an evoked response from an arbitrarily chosen brain location. A component criterion response is selected, and the probability of its occurrence without reinforcement is determined. In most experiments, this probability is determined anew each day before the training session. During training, reinforcement is presented immediately after the occurrence of a criterion response. We have used a variety of control techniques. Initially, we used yoked control animals, and the pattern of reinforcement generated by the experimental animal served as the basis for reinforcement for the yoked control animal. The control animal is, in every respect, treated as the experimental animal, except that the reinforcement is not contingent on its own response. Further control measures involve operant reversal of the initially trained polarity of a given component within the same animal; this procedure of reversal of the reinforcement criterion serves as a more effective control for non-reinforcement-related changes. The entire conditioning program is under the online real-time control of a computer system (PDP-8 or Interdata Model 4) (Fox and Rudell, 1968).

In the first study we successfully trained negativity by requiring that the mean voltage of a selected portion of the flash-evoked response (170–193 milliseconds after the stimulus) be 1.2 standard deviations more negative than the mean before conditioning. In a second study, we demonstrated that we could train the same voltage 1.2 standard deviations more positive than the mean. In these studies, criterion responses for the experimental animals increased to a high level under reinforcement (37.9 per cent), in contrast to the continued stable low performance (6.2 per cent) of the controls. The acquisition of such operant-controlled neural events was extremely rapid in some cases—as short as a single training session—depending on the complexity of the parameters and of the criteria. The most impressive result of these studies was the specificity of the change in the components. The definition of the criterion response was restricted to only the response amplitude between 170 and 193 milliseconds, and the animals generated highly specific responses that affected only this designated component of the evoked response.

Similar operant control of human scalp potentials, recorded vertex to mastoid in response to tones delivered through earphones, has also been demonstrated (Rosenfeld et al., 1969). The studies were conducted with and without visual feedback on the evoked potential, with and without the auditory signal, and under a variety of conditions controlling for muscle artifact, as well as for slow-potential shifts in the baseline. Subjects were reinforced with money for each successful criterion alteration of the evoked-potential component selected for reinforcement. They demonstrated at least a twofold increase in frequency of occurrence of the selected potential, an amplitude 1 standard deviation or more larger than that computed during baseline observation (Figure 5). The curves for the various replications are remarkably similar.

We have also raised the question of the number of independent functional "channels," or minimal information units, available in the time-locked evoked potential, and have investigated electrophysiological events that suggest partially independent behavioral coding by a late compon-

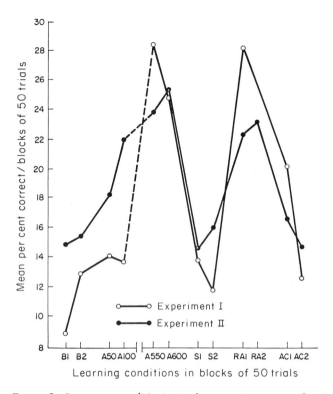

FIGURE 5  Late-wave conditioning in humans. Percentage of responses reaching criterion as a function of conditions. B1, 100 stimulus trials unreinforced; and B2, a second 100-trial block, unreinforced, constitute baseline measures. A50, trials on which criterion responses were reinforced, trials 1–50. A100, reinforced trials 50–100; A550, reinforced trials 500–550; A600, reinforced trials 550–600. $S_1$ and $S_2$ are blocks of 50 trials each, during which subjects suppressed criterion responding (lost money for each criterion response). RA1 and RA2 are each 50 trials of reacquisition trials during which reimbursement for criterion responses was restored. AC1 and AC2, artifact-control blocks of 50 trials each for which the tone was removed altogether in Experiment 1 and was substantially attenuated in Experiment 2. For Experiment 1, t = 2.55; d.f. = 4 and .025 < P < .05; for Experiment 2, t = 4.69, d.f. = 6 and .0005 < P < .005. In both cases, baseline was added to suppression scores and compared with acquisition plus reacquisition scores. (From Rosenfeld et al., 1969.)

ent (190–213 milliseconds) of the visual cortical evoked potential in cat brain (Fox and Rudell, 1970).

Mean waveshapes for three animals during baseline measurement and at the end of both negativity and positivity training revealed considerable individuality (1) in response style among the three animals, and (2) in the discrete nature of the change in waveform (Figure 6).

It is clear that bioelectrical criteria for reinforcement must be at least as critically and unambiguously defined as molar behavioral criteria. Even the discrete criteria employed have been sufficiently ambiguous to allow a variety of solutions.

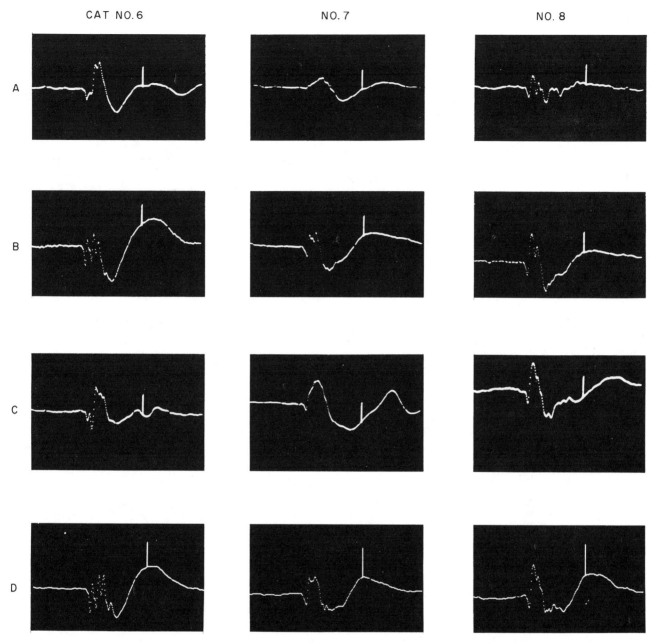

FIGURE 6  Averaged evoked potentials for three cats trained to modify a late visual cortical-response component.

    A: Evoked potentials before reinforcement baseline;
    B: Evoked responses after negativity training was completed;
    C: Responses after positivity training was completed.
    D: Responses as in B except criterion responses averaged only.

A, B and C averages include all trials, both criterion and noncriterion (successes and failures). Vertical bars mark the center of the five milliseconds averaged for the criterion point. Note discretely localized modifications for Cat 6C and Cat 8C and prolonged duration positivity for Cat 7C. Also note random variability of early waves. N = 1000 sweeps. Sweep duration = 500 msec. (From Fox and Rudell, 1970)

Of major interest are the changes in other portions of the evoked potential associated with changes in the reinforced component. Examination of the early or primary component revealed some changes from pretraining with positivity and negativity training, but these changes in the primary component were unsystematic with respect to changes in the reinforced component. More quantitative and detailed information concerning the independence of the evoked-potential components from one another and from the component modified by reinforcement comes from an additional animal. In this case, we reinforced for increased positivity (1 standard deviation from the pre-experimental mean) of a late component of the flash-evoked potential (188–192 milliseconds).

The progressive changes over days in amplitudes of four components of the evoked response were observed separately for comparison (Figure 7). Mean amplitudes for these four components of the evoked potential, each 5 milliseconds in duration, were collected during baseline, training, and extinction phases: point 1, the primary visual component, at 35 milliseconds after the flash; point 2, 100 milliseconds; point 3, the reinforced component, 190 milliseconds; and point 4, 265 milliseconds. Discrete changes in the mean daily amplitude of point 3, the reinforced point, took place with training, with little or no change in the amplitudes of the other components. Points 2 and 4 showed no trend. However, point 1, the primary visual component, did show some trend toward negativity during the last six days of training and a return to baseline during extinction, suggesting some parallel, classically conditioned process. The negative trend in the primary component is unrelated to the change in amplitude of the reinforced component. The shift in mean amplitude of point 3 is directly related to the shifting proportion of criterion over noncriterion responses during training (from 26 per cent to 68 per cent correct). The amplitude of point 1, in contrast, varies randomly with criterion and noncriterion point 3 responses, and its averaged value (both before and after training) for the two conditions is essentially superimposable (Figure 8). This is paradoxical in view of the progressive negative shift of point 1 during training (Figure 7). In general, computation of correlations of each of the four points with the rest of the wave support the preceding conclusions in showing relatively good functional independence of the evoked-potential components both during baseline and during substantial modification of one component.

These responses represented a specific and local change in direction of the reinforced component and not simply an alteration of over-all voltage. Therefore, the evoked potential in this case is not a single functional event, coding a single message throughout its extent. The late component that we manipulated experimentally may not be referred to legitimately as a "late negative" or a "late positive" com-

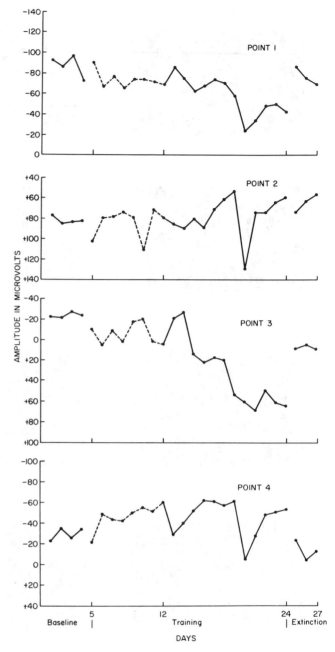

FIGURE 7 Daily mean amplitudes of four components of the flash-evoked potential. Point 1, average of 5 msec. beginning 35 msec. after flash; Point 2, 5 msec. starting 100 msec. after flash; Point 3, 5 msec. starting 190 msec. after flash; Point 4, 5 msec. starting 265 msec. after flash. Point 3 is the component reinforced for increased positivity. Absolute amplitudes are used in this figure. Note positive trends over days for Points 1 and 3. (From Fox and Rudell, 1970)

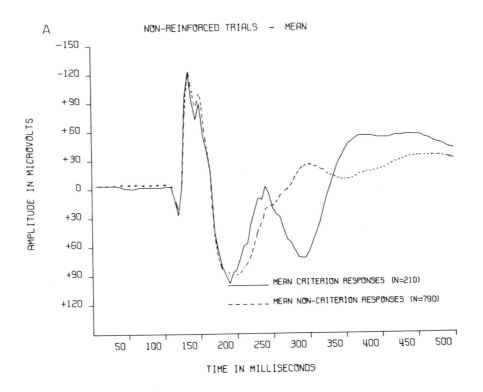

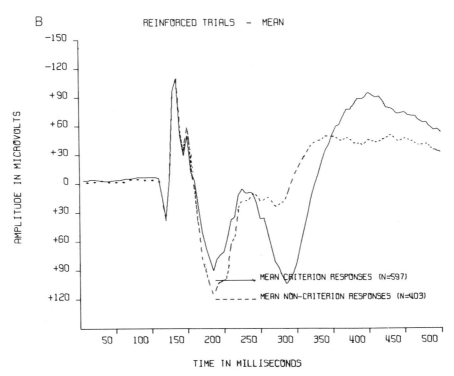

FIGURE 8 Averaged flash-evoked potentials A, before training, and B, after training for the same cat as in Figure 5. In A, solid line is average of responses which would have met reinforcement criterion; broken line is average of responses which would have failed. In B, after training, solid line is average of successful trials, broken line, of failures. Note the difference in proportion of successful to unsuccessful trials before and after training. Note also stability of early components. (From Fox and Rudell, 1970)

ponent, without specification of the exact concomitant behavioral state, because experimental manipulation of that state may be accompanied by either positivity or negativity. This is also true for the definition of components themselves. The conclusion that a certain number of components in a certain order in time best characterize the visual or auditory evoked potential is also a state-dependent conclusion for which the specific conditions must be described, because under reinforcement control animals have "eliminated" a component altogether. "No potential" of "no component" may simply be the neutral polarity state between two possible extremes. Widespread traditional agreement concerning the sequential components and their common polarities points only to the severely limited conditions under which evoked potentials have been studied.

Also important are the implications for the use of mean amplitudes in bioelectrical studies of behavior. The changes of the primary component in mean amplitude over days would, in traditional studies of neural correlates of behavior, be interpreted as positive evidence for a relation between the acquisition process and the amplitude of the early component. However, the changes were not correlated either with changes in the reinforced component or with the trial's success or lack of success. We believe that such an observation of a systematic, progressive, and statistically significant—but irrelevant—change in mean amplitude over trials sets serious limitations on the use of mean amplitudes in learning studies, except in combination with other, more specific measures that will partition the behavioral sources of variance that account for changes in the mean (see also John, 1967, p. 360).

Concern for the possible molar behaviors that accompany such changes in brain activity has led us to examine more closely some of the obvious possibilities. In the earlier studies (Fox and Rudell, 1968; Rosenfeld et al., 1969; Fox and Rudell, 1970), we were able to identify no gross motor behavior that could be related to the brain-potential performance. To the contrary, the cats sat still in a variety of positions in the recording box, and no differences between experimental and control animals were observed. Human subjects could report no gross motor behaviors associated with their successful modification of the brain wave, nor could the experimenters observe any.

Because modification of light intensity as controlled by receptor orientation and head position might contribute to the modification of the evoked potential, animals being trained to a positivity shift were equipped with a highly directional photocell fixed to the head-mount. The cell showed whether the animal was looking at the light. In the unrestrained state, the reinforced animal consistently looked away from the light toward the adjacent right wall. We found an extremely high correlation between percent of correct trials and percent of times the animal did not look

at the light during training and early extinction, although the correlation was not apparent during the initial baseline measurements or later extinction (Figure 9). However, the correlation proved to be spurious. When the animal's head was fixed in a restraining frame, oriented directly toward the light (that is, in a direction opposite from its spontaneous preference), it was immediately and consistently able to generate the required positivity for reinforcement (Figure 9C). When the animal was rotated to the left (Figure 9D) it also maintained its high level of positivity modification during reinforcement trials, generating waves very similar to those seen with the spontaneous choice (Figure 9B). Thus, the head-frame controls, which eliminate the effect of receptor orientation or body position, indicate that the correlations observed between brain events and behavior are in no way necessary or sufficient for the production of bioelectrical criterion responses. It should be noted that such a high correlation between brain and behavior can occur and yet be irrelevant, which raises another doubt regarding the suitability of a correlative approach.

## Evoked potentials and neural coding

Pulse codes have been suggested for the nervous system, derived from the orderliness of poststimulus-time histograms taken from single cells (Fox and O'Brien, 1965; Gerstein and Kiang, 1964; Kiang, 1965; Chow et al., 1968; John and Morgades, 1969; Gerstein and Perkel, 1969). The moment-to-moment variations in spike probability from such histograms may be viewed as sequential or temporal probabilities. We know that the instantaneous sequential voltages of the evoked potential mimic the shape of the single-unit firing pattern in response to a stimulus. Therefore, operant control of individual parameters or components, or both, of the evoked response may be viewed as directly producing changes in the sequential pulse code for neuronal populations larger than one cell.

Some reservation regarding ways to infer a "real code" from a "candidate code" is appropriate (Perkel and Bullock, 1968). The occurrence of ordered and patterned firing of a cell or group of cells following a stimulus does not indicate that such a response plays any role or represents any state that has behavioral relevance. Final validation has required demonstration of a "decoder." By physiological reinforcement on an operant schedule, however, functional validation of candidate codes is possible even without the demonstration of a decoder. When a parameter or component, or both, of a wave can be brought under operant control, they may be assumed to be behaviorally relevant and therefore a part of a functional code.

The number of independent, minimal, information-bearing units or durations, or independent information channels of the evoked response, have also been studied

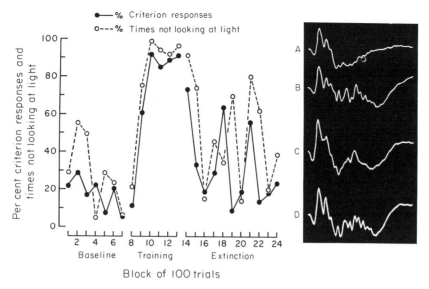

FIGURE 9 Graph, left: Percent criterion responses (solid lines) and percent of times cat was looking *away* from the light source. Measures taken on a single day. Seven blocks of baseline, 6 blocks of training, and 11 extinction blocks of 50 flashes each. Milk reward is followed by rapid increase in criterion responding and a parallel increase in tendency to turn away from stimulus source. Both measures return to baseline level in extinction. Correlation is high in training and early in extinction but absent during baseline measurements and later extinction. Note blocks 2, 19, and 22, in which cat looked away from the light more than 50 per cent of the time but in which only 28 per cent, 8 per cent, and 13 per cent, respectively, generated a criterion response. Traces, right: A, before training, in free situation; B, during training, in free situation; C, during training in holder, facing toward the light (nonpreferred direction); and D, in holder facing away from light (nonpreferred direction). For all training phases, the criterion notch appears on the negative late wave, despite high correlation of behavior measures N = 100 sweeps. Sweep duration = 250 msec.

(Rudell and Fox, 1970), as has conversion of large portions of the wave into a single channel by requiring longer duration-constant polarities. Demonstration of such prolonged constant polarity waves, however, does not necessarily indicate combination or elimination of independent channels but only experimentally controlled covariation. Identity of functional bioelectrical components, even in the steady state, does not indicate identity of related molar behaviors or states; the assumption of unique bioelectrical configurations for given states is not justified without direct evidence. We suggest at this time that it is more likely that a given molar state is related to a unique combination of spatial and temporal parameters, polarities, and components, with a given state of cortical excitability at a particular cortical location. Such a combination, we believe, plays a role in a large number of behavioral states, and the representation of a unique state also includes a particular combination of brain locations, each with a specific excitability pattern.

We believe that such a complex array of patterns of excitability and locations makes it unfruitful, therefore, to assign any unique significance to a correlation between any aspect of an evoked potential at any brain location and behavior. Rather, one may address oneself to the possible or probable combinations of codes available in the spatiotemporal array. Bioelectrical coding in brain may have its parallel in genetic coding, for which only the rules but not the specific combinational outcome may be specified. This suggestion is general, excepting the case of two matched analogue functions, one bioelectrical and one molar behavioral.

Space-related variations in voltage of the evoked response should be considered along with time-related changes. The relation between voltages in different areas of brain may also constitute an aspect of coding (Adey, 1967, and this volume), but anatomical connectivity does not assure functional connectivity. A surprising independence of lateral-geniculate activity from visual-cortex activity has been found when both locations were observed (Rudell, 1970; Rudell and Fox, in preparation); successes and failures based on cortical activity were not accounted for even grossly by simultaneous activity in the geniculate (Figure 10) or the brainstem. Space-related coding involves the *transfer functions* between two or more parts of brain. As a zero-transfer function has already been shown, more than one transfer function is possible and, therefore, the set of such transfer functions must be determined empirically (John et al., 1963; Ruchkin et al., 1964; John, 1967; John and Morgades,

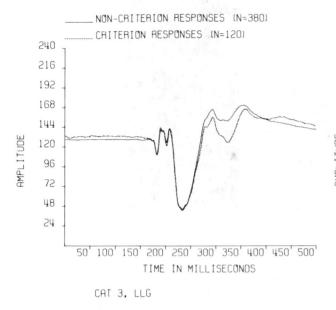

CAT 3. LV2 WITH LLG

_____ NON-CRITERION RESPONSES (N=380)
.......... CRITERION RESPONSES (N=120)

CAT 3. LLG

_____ DURING NON-CRITERION RESPONSES (N=380)
.......... DURING CRITERION RESPONSES (N=120)

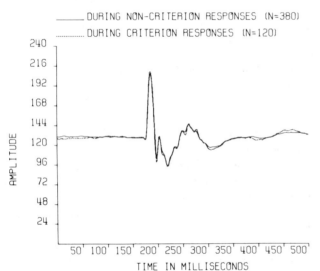

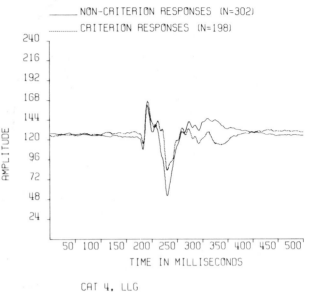

CAT 4. LV2 WITH LLG

_____ NON-CRITERION RESPONSES (N=302)
.......... CRITERION RESPONSES (N=198)

CAT 4. LLG

_____ DURING NON-CRITERION RESPONSES (N=302)
.......... DURING CRITERION RESPONSES (N=198)

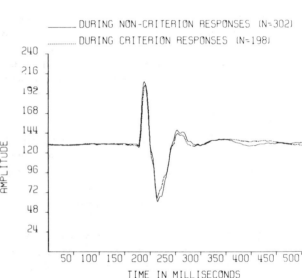

FIGURE 10 Averaged flash-evoked potentials in two cats. A, from left visual cortex II (LV2), and from B, left lateral geniculate (LLG), simultaneously. Solid line is average of noncriterion (failure) responses. Broken line is average of successes. Total = 500. Geniculate averages are based on cortical successes and failures. Geniculate does not differentiate cortical success and failure responses, suggesting independence of these structures for this state. Also note lack of difference in early cortical response with success and failure.

1969). Anatomical connectivity, therefore, does not assure spatial coding, and the function of such connections must be validated behaviorally. Functional anatomy may be limited by anatomical connection but may not be assumed from it. Direct, formal, and systematic validation by operant reinforcement is also useful in validating such candidate spatial codes or candidate transfer functions.

Finally, a more difficult aspect of bioelectrical behavioral coding relates to discrete coding under conditions in which there is no synchronous input and therefore no evoked potential. In such a case of spontaneous activity, the obvious orderliness of temporal coding disappears, and, although the wave may continue to represent the instantaneous pulse code, that code itself is more elusive. This is the "nonzero-time paradigm," because no simple stimulus-marked or time-locked zero is available. Analyses of general properties of spontaneous brain activity in terms of frequency or detailed power spectra are well known (see Adey, this volume). However, such methods, which average activity over arbitrary time epochs, lose the discrete sequential-firing probabilities represented by the moments of the wave. To preserve such representation of the pulse code, we consider individual spontaneous waves in the brain not as frequency (an extrapolation from slope) but as individual iterative events or discrete repetitions of particular coded statements. Thus, sequential neural events amenable to modification by operant control may be used as criteria to induce the neurally appropriate steady state of the animal (Kamiya, 1969; Sterman et al., 1970; Wyrwicka and Sterman, 1968; Olds, 1965; Mulholland, 1968; Fetz, 1969; Carmona, 1967; Dalton, 1969).

The operant-control approach to spontaneous activity of brain may perhaps be compared speculatively to formal structural and mathematical linguistic analysis, in which formal grammars may be derived, together with syntactic elements, without the need to refer to content. Such an analysis may permit delineation of the behavior-relevant parameter of bioelectrical coding without knowledge of the event being coded. The loss, however, of behavioral information may leave us no worse off than do some current ambiguities associated with the correlative design. To the extent that questions of coding, both temporal and spatial, may continue to be systematically asked in the absence of such knowledge, functional coding continues to be a viable and provocative area of research.

## REFERENCES

ADEY, W. R., 1958. Organization of the rhinencephalon. *In* Reticular Formation of the Brain (H. H. Jasper, L. D. Proctor, R. S. Knighton, W. C. Noshay, and R. T. Costello, editors). Little, Brown and Co., Boston, pp. 621–644.

ADEY, W. R., 1966. Neurophysiological correlates of information transaction and storage in brain tissue. *In* Progress in Physiological Psychology (E. Stellar and J. M. Sprague, editors). Academic Press, New York, pp. 1–43.

ADEY, W. R., 1967. Intrinsic organization of cerebral tissue in alerting, orienting, and discriminative responses. *In* The Neurosciences: A Study Program (G. C. Quarton, T. Melnechuk, and F. O. Schmitt, editors). The Rockefeller University Press, New York, pp. 615–633.

ADRIAN, E. D., and B. H. C. MATTHEWS, 1934. The interpretation of potential waves in the cortex. *J. Physiol.* (*London*) 81: 440–471.

AJMONE MARSAN, C., 1965. Electrical activity of the brain: Slow waves and neuronal activity. *Israel J. Med. Sci.* 1: 104–117.

BISHOP, G. H., and M. H. CLARE, 1952. Sites of origin of electrical potentials in striate cortex. *J. Neurophysiol.* 15: 201–220.

BUCHWALD, J. S., E. S. HALAS, and S. SCHRAMM, 1966a. Changes in cortical and subcortical unit activity during behavioral conditioning. *Physiol. Behav.* 1: 11–22.

BUCHWALD, J. S., E. S. HALAS, and S. SCHRAMM, 1966b. Relationships of neuronal spike populations and EEG activity in chronic cats. *Electroencephalogr. Clin. Neurophysiol.* 21: 227–238.

BUREŠ, J., and O. BUREŠOVÁ, 1965. Relationship between spontaneous and evoked unit activity in the inferior colliculus of rats. *J. Neurophysiol.* 28: 641–654.

BUSER, P., and P. BORENSTEIN, 1957. Réponses corticules "secondaires" à la stimulation sensorielle chez le chat aurarisé non-anesthésié. *Electroencephalogr. Clin. Neurophysiol.* 6 (suppl.): 89–108.

BUSER, P., and M. IMBERT., 1961. Sensory projections to the motor cortex in cats: A microelectrode study. *In* Sensory Communication (W. A. Rosenblith, editor). The M.I.T. Press, Cambridge, Massachusetts, pp. 607–626.

BYKOV, K., 1957. The Cerebral Cortex and the Internal Organs. Chemical Publishing Co., New York.

CARMONA, A. B., 1967. Trial and error learning of the voltage of the cortical EEG activity. *Diss. Abstr.* 28: 1157B–1158B.

CHOW, K. L., D. F. LINDSLEY, and M. GOLLENDER, 1968. Modification of response patterns of lateral geniculate neurons after paired stimulation of contralateral and ipsilateral eyes. *J. Neurophysiol.* 31: 729–739.

COHEN, D. H., and R. G. DURKOVIC, 1966. Cardiac and respiratory conditioning differentiation and extinction in the pigeon. *J. Exp. Anal. Behav.* 9: 681–688.

CREUTZFELDT, O. D., S. Watanabe, and H. D. Lux, 1966a. Relations between EEG phenomena and potentials of single cortical cells. I. Evoked responses after thalamic and epicortical stimulation. *Electroencephalogr. Clin. Neurophysiol.* 20: 1–18.

CREUTZFELDT, O. D., S. WATANABE, and H. D. LUX, 1966b. Relations between EEG phenomena and potentials of single cortical cells. II. Spontaneous and convulsoid activity. *Electroencephalogr. Clin. Neurophysiol.* 20: 19–37.

DALTON, A. J., 1969. Discriminative conditioning of hippocampal electrical activity in curarized dogs. *Commun. Behav. Biol.*, Part A, 3:283–287.

DILL, R. C., E. VALLECALLE, and M. VERZEANO, 1968. Evoked potentials, neuronal activity and stimulus intensity in the visual system. *Physiol. Behav.* 3: 797–801.

DUNLOP, C. W., K. F. KILLAM, M. BRAZIER, and W. R. ADEY, 1959. Effects of gamma-aminobutyric acid and thiosemicarbazide on cerebellar activity in the cat. *Fed. Proc.* 18: 38 (abstract).

ECCLES, J. C., 1951. Interpretation of action potentials evoked in the cerebral cortex. *Electroencephalogr. Clin. Neurophysiol.* 3: 449–464.

ELUL, R., 1968. Brain waves: Intra-cellular recording and statistical analysis help clarify their physiological significance. *In* Data Acquisition and Processing in Biology and Medicine. Proceedings, Rochester Conference, 1966, Vol. 5 (K. Enslein, editor). Pergamon Press, Oxford.

EULER, C. VON, J. D. GREEN, and G. RICCI, 1958. The role of hippocampal dendrites in evoked responses and afterdischarges. *Acta Physiol. Scand.* 42: 87–111.

FESSARD, A., 1961. The role of neuronal networks in sensory communication within the brain. *In* Sensory Communication (W.

A. Rosenblith, editor). The M. I. T. Press, Cambridge, Massachusetts, pp. 585–606.

FETZ, E. E., 1969. Operant conditioning of cortical unit activity. *Science* (*Washington*) 163: 955–957.

FOX, S. S., J. C. LIEBESKIND, J. H. O'BRIEN, and R. H. HUGH DINGLE, 1967. Mechanisms for limbic modification of cerebellar and cortical afferent information. *Progr. Brain Res.* 27: 254–280.

FOX, S. S., and R. J. NORMAN, 1968. Functional congruence: An index of neural homogeneity and a new measure of brain activity. *Science* (*Washington*) 159: 1257–1259.

FOX, S. S., and J. H. O'BRIEN, 1965. Duplication of evoked potential waveform by curve of probability of firing of a single cell. *Science* (*Washington*) 147: 888–890.

FOX, S. S., and A. P. RUDELL, 1968. Operant controlled neural event: Formal and systematic approach to electrical coding of behavior in brain. *Science* (*Washington*) 162: 1299–1302.

FOX, S. S., and A. P. RUDELL, 1970. The operant controlled neural event: Functional independence in behavioral coding by early and late components of the visual cortical evoked response in cats. *J. Neurophysiol.* 33: 548–561.

FROST, J. D., JR., 1967. An averaging technique for detection of EEG-intracellular potential relationships. *Electroencephalogr. Clin. Neurophysiol.* 23: 179–181.

FROST, J. D., JR., 1968. EEG-intracellular potential relationships in isolated cerebral cortex. *Electroencephalogr. Clin. Neurophysiol.* 24: 434–443.

FROST, J. D., JR., and Z. ELAZAR, 1968. Three-dimensional selective amplitude histograms: A statistical approach to EEG-single neuron relationships. *Electroencephalogr. Clin. Neurophysiol.* 25: 499–503.

FROST, J. D., JR., and A. GOL, 1966. Computer determination of relationships between EEG activity and single unit discharges in isolated cerebral cortex. *Exp. Neurol.* 14: 506–519.

FROST, J. D., JR., P. KELLAWAY, and A. GOL, 1966. Single-unit discharges in isolated cerebral cortex. *Exp. Neurol.* 14: 305–316.

FUJITA, Y., and T. SATO, 1964. Intracellular records from hippocampal pyramidal cells in rabbit during theta rhythm activity. *J. Neurophysiol.* 27: 1011–1025.

GAVALAS, R. J., 1967. Operant reinforcement of an autonomic response: two studies. *J. Exp. Anal. Behav.* 10: 119–130.

GERSTEIN, G., and N. Y.-S. KIANG, 1964. Responses of single units in the auditory cortex. *Exp. Neurol.* 10: 1–18.

GERSTEIN, G. L., and D. H. PERKEL, 1969. Simultaneously recorded trains of action potentials: Analysis and functional interpretation. *Science* (*Washington*) 164: 828–830.

GORMAN, A. L. F., and H. SILFVENIUS, 1967. The effects of local cooling of the cortical surface on the motor cortex response following stimulation of the pyramidal tract. *Electroencephalogr. Clin. Neurophysiol.* 23: 360–370.

HORI, Y., I. TOYOHARA, and N. YOSHII, 1967. Conditioning of unitary activity by intracerebral stimulation in cats. *Physiol. Behav.* 2: 255–259.

HUMPHREY, D. R., 1968a. Re-analysis of the antidromic cortical response. I. Potentials evoked by stimulation of the isolated pyramidal tract. *Electroencephalogr. Clin. Neurophysiol.* 24: 116–129.

HUMPHREY, D. R., 1968b. Re-analysis of the antidromic cortical response. II. On the contribution of cell discharge and PSPs to the evoked potentials. *Electroencephalogr. Clin. Neurophysiol.* 25: 421–442.

JASPER, H. H., G. RICCI, and B. DOANE, 1960. Microelectrode analysis of cortical cell discharge during avoidance conditioning in the monkey. *Electroencephalogr. Clin. Neurophysiol.* 13 (suppl.): 137–155.

JOHN, E. R., 1961. High nervous system functions: Brain functions and learning. *Annu. Rev. Physiol.* 23: 451–484.

JOHN, E. R., 1967. Mechanisms of Memory. Academic Press, New York.

JOHN, E. R., and P. P. MORGADES, 1969. Neural correlates of conditioned responses studied with multiple chronically implanted moving microelectrodes. *Exp. Neurol.* 23: 412–425.

JOHN, E. R., D. S. RUCHKIN, and J. VILLEGAS, 1963. Signal analysis of evoked potentials recorded from cats during conditioning. *Science* (*Washington*) 141: 429–431.

JUNG, R., 1953. Neuronal discharge. *Electroencephalogr. Clin. Neurophysiol.* 4 (suppl.): 57–71.

JUNG, R., 1961. Neuronal integration in the visual cortex and its significance for visual information. *In* Sensory Communication (W. A. Rosenblith, editor). The M.I.T. Press, Cambridge, Massachusetts, pp. 627–674.

KAMIKAWA, K., J. T. MCILWAIN, and W. R. ADEY, 1964. Response patterns of thalamic neurons during classical conditioning. *Electroencephalogr. Clin. Neurophysiol.* 17: 485–496.

KAMIYA, J., 1969. Operant control of the EEG alpha rhythm and some of its reported effects on consciousness. *In* Altered States of Consciousness (C. Tart, editor). John Wiley and Sons, New York, pp. 507–517.

KANDEL, E. R., and W. A. SPENCER, 1961. Electrophysiological properties of an archicortical neuron. *Ann. N. Y. Acad. Sci.* 94: 570–603.

KIANG, N., 1965. Discharge patterns of single fibers in the cat's auditory nerve. *M. I. T. Res. Monogr.* 35.

KLEE, M. R., K. OFFENLOCH, and J. TIGGES, 1965. Cross-correlation analysis of electroencephalographic potentials and slow membrane transients. *Science* (*Washington*) 147: 519–521.

LANDAU, W. M., 1967. Evoked potentials. *In* The Neurosciences: A Study Program (G. C. Quarton, T. Melnechuk, and F. O. Schmitt, editor). The Rockefeller University Press, New York, pp. 469–482.

LASS, Y., 1968. A quantitative approach to the correlation of slow wave and unit electrical activity in the cerebral cortex of the cat. *Electroencephalogr. Clin. Neurophysiol.* 25: 503–506.

LI, C.-L., 1963. Cortical intracellular synaptic potentials in response to thalamic stimulation. *J. Cell. Comp. Physiol.* 61: 165–179.

LI, C.-L., C. CULLEN, and H. H. JASPER, 1956. Laminar microelectrode analysis of cortical unspecific recruiting responses and spontaneous rhythms. *J. Neurophysiol.* 19: 131–143.

LI, C.-L., and H. H. JASPER, 1953. Microelectrode studies of the electrical activity of the cerebral cortex in the cat. *J. Physiol.* (*London*) 121: 117–140.

LI, C.-L., H. MCLENNAN, and H. H. JASPER, 1952. Brain waves and unit discharge in cerebral cortex. *Science* (*Washington*) 116: 656–657.

Lux, H. D., J. M. Fuster, A. Nacimiento, and O. D. Creutzfeldt, 1966. Relationship between changes of the membrane potential of cortical neurons and EEG phenomena. *Electroencephalogr. Clin. Neurophysiol.* 20: 99 (abstract).

Marshall, W. H., S. A. Talbot, and H. W. Ades, 1943. Cortical response of the anesthetized cat to gross photic and electrical afferent stimulation. *J. Neurophysiol.* 6: 1–15.

Meurice, E., H. Weiner, and W. Sloboda, 1966. Operant behavior and the galvanic skin potential under DRL schedule. *J. Exp. Anal. Behav.* 9: 121–129.

Miller, N. E., 1969. Learning of visceral and glandular responses. *Science (Washington)* 163: 434–444.

Morrell, F., 1961. Electrophysiological contributions to the neural basis of learning. *Physiol. Rev.* 41: 443–494.

Morrell, F., 1967. Electrical signs of sensory coding. *In* The Neurosciences: A Study Program (G. C. Quarton, T. Melnechuk, and F. O. Schmitt, editors). The Rockefeller University Press, New York, pp. 452–469.

Mulholland, T., 1968. Feedback electroencephalography. *Activ. Nerv. Sup. (Prague)* 10 (no. 4): 410–438.

O'Brien, J. H., and S. S. Fox, 1969a. Single-cell activity in cat motor cortex. I. Modifications during classical conditioning procedures. *J. Neurophysiol.* 32: 267–284.

O'Brien, J. H., and S. S. Fox, 1969b. Single-cell activity in cat motor cortex. II. Functional characteristics of the cell related to conditioning changes. *J. Neurophysiol.* 32: 285–296.

Olds, J., 1965. Operant conditioning of single unit responses. *Proc. 23rd Int. Congr. Physiol. Sci. Tokyo* 4: 372–380.

Perkel, D. H., and T. H. Bullock, 1968. Neural Coding. *Neurosci. Res. Program Bull.* 6 (no. 3): 221–348.

Pollen, D. A., 1964. Intracellular studies of cortical neurons during thalamic induced wave and spike. *Electroencephalogr. Clin. Neurophysiol.* 17: 398–404.

Purpura, D. P., 1959. Nature of electrocortical potentials and synaptic organizations in cerebral and cerebellar cortex. *Int. Rev. Neurobiol.* 1: 47–163.

Purpura, D. P., 1967. Comparative physiology of dendrites. *In* The Neurosciences: A Study Program (G. C. Quarton, T. Melnechuk, and F. O. Schmitt, editors). The Rockefeller University Press, New York, pp. 372–393.

Purpura, D. P., and J. G. McMurtry, 1965. Intracellular activities and evoked potential changes during polarization of motor cortex. *J. Neurophysiol.* 28: 166–185.

Rall, W., 1967. Distinguishing theoretical synaptic potentials computed for different soma-dendritic distributions of synaptic input. *J. Neurophysiol.* 30: 1138–1168.

Rall, W., R. E. Burke, T. G. Smith, P. G. Nelson, and K. Frank, 1967. Dendritic location of synapses and possible mechanisms for the monosynaptic EPSP in motoneurons. *J. Neurophysiol.* 30: 1169–1193.

Robertson, A. D. J., 1965. Correlation between unit activity and slow potential changes in the unanesthetised cerebral cortex of the cat. *Nature (London)* 208: 757–758.

Rosenfeld, J. P., A. P. Rudell, and S. S. Fox, 1969. Operant control of neural events in humans. *Science (Washington)* 165: 821–823.

Ruchkin, D. S., J. Villegas, and E. R. John, 1964. An analysis of average evoked potentials making use of least mean square techniques. *Ann. N. Y. Acad. Sci.* 115: 799–826.

Rudell, A. P., 1970. The operant conditioning of primary and secondary components in the visual evoked potential with measurement of collateral neural and behavioral activity. University of Iowa, Iowa City, doctoral thesis.

Rudell, A. P., and S. S. Fox, 1970. The operant conditioning of early and late components in the visual evoked potential. *Fed. Proc.* 29: 590 (abstract).

Stefanis, C., and H. Jasper, 1964. Intracellular microelectrode studies of antidromic responses in cortical pyramidal tract neurons. *J. Neurophysiol.* 27: 828–854.

Sterman, M. B., R. C. Howe, and L. P. MacDonald, 1970. Facilitation of spindle-burst sleep by conditioning and electroencephalographic activity while awake. *Science (Washington)* 167: 1146–1148.

Thomas, J., P. Groves, and M. Verzeano, 1968. The activity of neurons in the lateral geniculate body during wakefulness and natural sleep. *Experientia (Basel)* 24: 360–362.

Verzeano, M., R. C. Dill, E. Vallecalle, P. Groves, and J. Thomas, 1968. Evoked responses and neuronal activity in the lateral geniculate. *Experientia (Basel)* 24: 696–698.

Widen, L., and C. Ajmone Marsan, 1960. Unitary analysis of the response elicited in the visual cortex of cat. *Arch. Ital. Biol.* 98: 248–274.

Wyrwicka, W., and M. B. Sterman, 1968. Instrumental conditioning of sensorimotor cortex EEG spindles in the walking cat. *Physiol. Behav.* 3: 703–707.

Yoshii, N., and H. Ogura, 1960. Studies on the unit discharge of brainstem reticular formation in the cat. I. Changes of reticular unit discharge following conditioning procedure. *Med. J. Osaka Univ.* 11: 1–17.

# 25 Plasticity in the Nervous System at the Unitary Level

## RUDOLF J. VON BAUMGARTEN

### The problem of complexity

Learned behavior in animals and conscious memory in man are both highly complex. Therefore, the neuronal apparatus that helps in acquiring, storing, and reading-out these effects must also be complex. Light- and electron-microscope studies reveal that the functional complexity is indeed matched in the neuropil by nearly countless bifurcations and synapses, which resemble a "wiring diagram." This diagram can, perhaps, never be untangled, understood, or even described adequately.

It seems reasonable to assume, however, that complicated behavioral and mental changes are based on elementary events occurring at the neuronal, synaptic, or glial levels, that is, on processes that take place in the immediate environment of the neuronal membrane. Such changes, even though they play a key role, could be very small, and neuronal networks could amplify them at the output into larger, more effective changes by means of divergent cascade circuitry and reciprocal innervation. If such were the case, behavior and memory would depend on a large number of unitary changes, which the complex neuronal interconnections integrate into the compound picture, or *Gestalt*.

The search for cellular mechanisms involved in classical conditioning should have priority over those affecting instrumental conditioning or other, more complex, forms of learning that might be based on or cannot be separated from classical conditioning (Hilgard and Marquis, 1961).

No known physiological mechanism exists by which information in the nervous system is durably stored at the unitary level. In "simple systems" the few plastic changes so far measured within individual neurons seem related only to short-term information storage, and it is not clear that they have anything to do with "real" or long-term memory. Nevertheless, it appears useful to explore and discuss them in this context.

Most so-called simple systems are invertebrate and offer

definite advantages to the investigator of memory mechanisms.

1. The number of neurons and synapses is limited, which improves the chance of detecting plastic changes in a single neuron without interference from other, unwanted, neurons.

2. Some simple systems contain extraordinarily large nerve cells that can easily be identified and named.

3. The activity of an individual neuron can be studied in vitro for a period of up to several days; the input and output are electrically monitored and the chemical environment is controlled.

On the other hand, simple systems present certain disadvantages.

1. Changes observed after training may originate in cells other than the one examined, in which case the cell studied is merely an indicator, or follower, cell.

2. All simple systems under study possess a complex neuropil containing thousands of axo-axonal synapses; it is difficult to establish which of these synapses may have learning capabilities.

3. Presumably, relatively few neurons can "learn"; the majority show only integrating and coding functions on the input or output side of the system. It is possible that the ideal "learning cell," if there is one, has not yet been identified in *Aplysia* or in any other available system. It may even be that the animal ideally suited for these studies has so far escaped our attention.

### Possible sites of information storage at the neuron

From the standpoint of economy, it would be practical to store a memory near the synapses at which the information arrives and is read out. The cell membrane is, therefore, a more likely candidate, at least for short-term information storage, than is the nucleus, which interacts with the membranes only by the relatively slow pathway of protoplasmic transport. The changes might take place, of course, either in glial membranes adjacent to the neuronal membranes (Galambos, 1961; Hydén, 1967) or in the endoplasmic reticulum immediately beneath the cell surfaces. It appears possible, moreover, that permanent membrane changes at a

RUDOLF J. VON BAUMGARTEN   Mental Health Research Institute, University of Michigan, Ann Arbor, Michigan

later stage involve RNA-dependent mechanisms of the nucleus (Hydén and Egyházi, 1962).

The neuronal membrane itself appears not to be uniform, but to present a mosaic of morphologically, biochemically, and functionally different structures (Grundfest, 1957). This article considers two of these as possible sites of information storage—the synaptic regions and those parts of the membrane that perform pacemaker activities.

## Information storage by synapses

CAN USE AND DISUSE OF SYNAPSES EXPLAIN MEMORY? One of the most common memory hypotheses assumes that the frequent use of a synapse establishes a more stable connection between two neurons (Eccles, 1961). Extending this idea to a complex polysynaptic network, if a particular pathway between input and output were used over and over again, it would "consolidate," and information would pass preferentially through this, rather than some other, chain of neurons. This hypothesis seems to be supported by a number of psychological, physiological, and morphological data: learning curves rise with the number of repetitions; a burst of impulses in a presynaptic fiber enhances the amount of transmitter subsequently released per impulse (a phenomenon termed post-tetanic potentiation or PTP [Lloyd, 1949; Hughes, 1958]); and the number of spines decreases on cortical dendrites after dissection of the presynaptic afferent fibers (Mathieu and Colonnier, 1968) and after functional reduction of the synaptic input (Valverde, 1967).

Much of this evidence, it may be pointed out, concerns only a single input channel; data more applicable to the memory formation problem would deal with the convergence and interaction of two different input channels on one output channel (Olds, 1961). Since such evidence does not exist, it can be argued that the use-disuse hypothesis is inapplicable to the memory problem.

In Pavlov's original experiment on the conditioned salivation reflex, for instance, the dogs were frequently exposed to a ringing bell, but they did not salivate unless the sound was paired with food presentation during a training period. The bell alone must have activated any preformed synapses that connect the acoustic input channel to the efferent neurons of the salivation channel. The absence of salivation under such a condition proves that these are not the ones that "improve" to cause the postsynaptic action potentials responsible for conditioned salivation. In fact, an established conditioned salivation response gradually disappears when the acoustic input channel is the only one activated.

We can therefore conclude that the mere use and disuse of only one synaptic channel could not rechannel information in the sense of memory. The effect of use and disuse, however, could help to explain some forms of short-term sensitization or pseudoconditioning.

INTERACTION OF TWO DIFFERENT INPUT CHANNELS AT THE SYNAPSE If it is true that memory resulting from classical conditioning procedures depends on the formation of new synaptic connections or on the improvement of pre-existing ones, this process could take place under the very special geometric and temporal conditions shown in Figure 1. Two neurons, A and B, converge at the same neuron, C, and are excited in close temporal sequence, often at an interval not exceeding 500 milliseconds. As a result of this converging and properly timed activity, a plastic change takes place that makes neuron C fire even if activated only by previously subthreshold synaptic impulses from neuron A.

It is obviously of the utmost importance, if we are to understand the basic mechanisms of this kind of memory, to pinpoint the site of this synaptic interaction. Possible changes in the synaptic region include a change in size of presynaptic and postsynaptic areas; altered uptake, availability, and release of transmitter; changes in receptor number, size, and sensitivity; and plastic structural changes in the synaptic cleft that enhance or decrease diffusion of the neural transmitter from the presynaptic to the postsynaptic side.

It is interesting to consider, in addition, the idea that a structural change in the synaptic membranes may require a neurosecretion other than, and in addition to, the transmitter release (Figure 2). Under full depolarization, the presynaptic terminal would release this additional "substance A," which would then interact with a "substance B" released by depolarization of the postsynaptic membrane. Their combination would, in a catalytic or in a slower immunological process, eventually lead to a long-lasting structural change of the interface between two neurons. The process would be analogous to the hardening of epoxy glue after the components are mixed together. An attractive feature of this hypothesis is that improvements of synaptic connectivity would take place only when the postsynaptic membrane spike closely follows the EPSP, that is, before substance A has been dissipated. In behavioral terms, such improvements would take place only when the unconditioned stimulus (UCS) immediately follows the conditioned stimulus (CS), as is the case in classical conditioning.

This hypothesis, however, does not explain why a reversed sequence of the CS and the UCS (backward conditioning) is ineffective, unless we make additional assumptions; for instance, that substance B can be released from the postsynaptic membrane only after substance A has come in contact with the membrane. Another difficulty of the hypothesis is that the postsynaptic membrane might be electrically inexcitable (Grundfest, 1957) and that dendrites

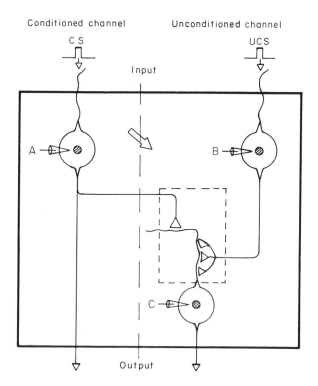

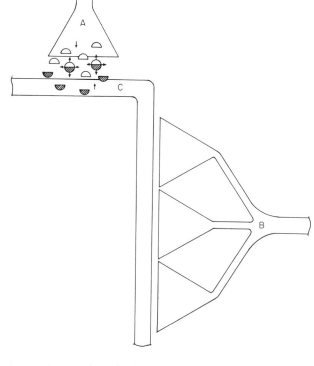

FIGURE 1 Schematic diagram of a conditioned reflex at the unitary level. A signal CS at the input of the conditioned channel crosses over and eventually appears at the output of the unconditioned channel. This rechanneling of information takes place only when neuron B is excited shortly after excitation of neuron A during the conditioning period.

The microrecordings at the bottom of the Figure indicate the membrane activity of three neurons before, during, and after conditioning. Neuron A fires action potentials after each conditioned stimulus. Neuron C responds with an EPSP when neuron A is fired. After the conditioning period, neuron C would respond to the firing of neuron A with an action potential. This change in the responsiveness of neuron C to impulses in neuron A could depend either on increased synaptic efficacy (a) or on changes in the pacemaker apparatus of neuron C that lead to "self re-excitation" (b).

FIGURE 2 Hypothetical scheme of synaptic consolidation after neurosecretion from two neurons. (This Figure and Figure 3 correspond to the rectangular area outlined in Figure 1.) During the excitation of synaptic terminal A, a presynaptic substance (open half-circles) is released into the synaptic cleft. An impulse fired in neuron C shortly afterward releases a "postsynaptic substance" (hatched half-circles). The reaction of the two substances would form a third substance (full circles), which induces morphological changes of the synaptic cleft or the adjacent membranes, thereby increasing synaptic efficacy.

might not participate in the spike generation (Grundfest and Purpura, 1956). In such cases, we could assume that a strongly passive, rather than an active, depolarization is causing the release of substance B.

Scilard (1964) proposed a hypothesis resembling that just outlined, in which proteins contained in the presynaptic terminal diffuse into the postsynaptic membrane whenever both membranes are simultaneously depolarized during a learning situation. The two complementary proteins then dimerize in an immunological reaction. The new dimer could enhance the transmitter efficacy of the synapse by binding substances like acetylcholinesterase.

At the present time, little evidence is available to prove or disprove either hypothesis. On the other hand, it is hard to imagine how two neurons, connected merely by a chemical synapse, could so interact as to change synaptic efficacy except through the mutual release and interaction of specific substances. More electron optical and histochemical data on the synapses before, during, and after functional or behavioral changes might further clarify the problem.

INTERACTION OF CHANNELS AT EPISYNAPTIC SITES   There is good evidence that synapses can be gated quantitatively by synaptic terminals sitting on synaptic terminals. Such a gating effect, as in presynaptic facilitation and inhibition (Eccles, 1964; Tauc, 1965), can be mediated by synaptic hyperpolarization or depolarization of the terminal. It ceases as soon as the action potentials in the episynaptic terminal stop. Therefore, presynaptic facilitation and inhibition, as we understand them today, cannot account for any long-term changes in connectivity, unless as-yet-unknown structural and functional changes take place at the terminal. If such is the case, the problem remains the same, but its site of origin shifts from synaptic junctions to episynaptic junctions (Figure 3). To prove that such a mechanism exists, it would

be necessary to make simultaneous intracellular recordings of the postsynaptic neuron, the presynaptic neuron, and the episynaptic neuron as close as possible to the junction. This task has not yet been accomplished; it seems to be just beyond the limits of current techniques.

Some indirect support, however, exists for the view that episynaptic events could play a role in heterosynaptic facilitation (HSF) in *Aplysia* neurons (Kandel and Tauc, 1965a). Aside from post-tetanic potentiation (PTP), HSF is the only instance of changed synaptic effectiveness measured over a time span of minutes with intracellular microelectrodes. It therefore seems justifiable to bring Kandel's brief account (1967) of HSF up to date.

HSF describes the increase in amplitude of a test EPSP produced by activity in one nerve if a strong prior shock (priming stimulus) is applied to a different nerve. HSF does not take place if, instead of the priming shock, a postsynaptic action potential is initiated by artificial depolarization of the recorded cell. Hence, the interesting events in the recorded cell seem to occur presynaptically and not postsynaptically. It is not clear, however, if this facilitation necessarily involves electrical events at an episynaptic junction or diffusion of a facilitating substance released by excited "priming neurons" situated some distance away.

HSF was shown to be of different origin from post-tetanic potentiation (Tauc and Epstein, 1967). The repetitive discharges of the test nerve after strong priming stimuli suggest, however, that, under this condition, HSF might be contaminated to some degree by simultaneous PTP (von Baumgarten and Jahan-Parvar, 1967).

After threshold stimulation of the test nerve, HSF can be observed in EPSPs that have a short and constant latency and follow the all-or-none rule when the stimulation strength is altered. But that such unitary EPSPs in *Aplysia* originate from only one synaptic terminal is still unproved. Recent scanning electron micrographs show that axonal terminals in *Aplysia* bifurcate and terminate in a large number of synaptic endings (Lewis et al., 1969). If this evidence holds true, EPSPs that show an all-or-none behavior when the stimulation strength at the test nerve is changed, could show variations in the size of the EPSP when changes take place in the number of simultaneously discharging end knobs. In this case, how the priming stimulation can switch additional end knobs of multiple synaptic structure on or off must still be explained.

The value of HSF as a possible mechanism of memory depends heavily on its degree of specificity. But HSF is usually composed of a large nonspecific component and a very small specific one. In the majority of cases, HSF can be elicited in several test synapses by one strong, unpaired shock on the priming nerve. In this respect, the nonspecific part of HSF does not meet the requirement for true con-

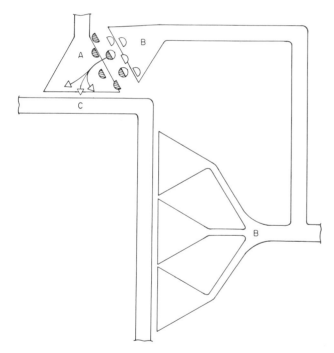

FIGURE 3   Hypothetical scheme of synaptic consolidation following episynaptic interaction. When an impulse in neuron A is followed closely by an impulse in neuron B, it is assumed that there is neurosecretion from both sides into the cleft between terminals A and B. The combination of both substances could induce morphological changes (indicated by arrows) at the membrane either inside or outside terminal A, and would result in increased synaptic efficacy between A and C.

ditioning as outlined above, that is, that two different channels should interact only when excited in close temporal sequence. Kandel and Tauc (1965b), however, reported three isolated cases in which HSF did not appear unless the test shock was paired closely with the priming shock.

Furthermore, it has been shown that, even in cells that react nonspecifically, paired stimulation caused a stronger facilitation than did priming alone (Jahan-Parvar and von Baumgarten, 1967). Short intervals of about 350 milliseconds between the test shock and the priming shock caused more HSF than did longer or shorter intervals (von Baumgarten and Hukuhara, 1969). The optimal interstimulus interval, as measured in that study, compared closely with the optimal interstimulus interval in behavioral conditioning. The specific component of HSF that depends on paired stimulation could be related to conditioning, whereas the nonspecific component could create a general sensitization of the background on which the true conditioning takes place.

The mechanism of heterosynaptic facilitation is still essentially unexplained. Because the presynaptic neuron has never been recorded intracellularly, we do not know if it is hyperpolarized during heterosynaptic facilitation, as we would expect in presynaptic facilitation. Presynaptic facilitation would require the continuous firing of an interneuron during the whole period of HSF—up to 45 minutes. Such an interneuron has not been found. We recently observed that the minimal latency period after the priming shock and before HSF appears often exceeds several seconds (Haigler and von Baumgarten, 1970). Such a long delay is hard to explain in terms of presynaptic facilitation unless slow diffusion or metabolic processes take place at the test terminal. HSF lasts from a few seconds up to an hour; its consolidation into a more permanent form of facilitation has not yet been observed, but cannot be excluded on the basis of the few cell types thus far investigated.

## Nonsynaptic possibilities for information storage in single neurons

No major breakthrough in the synaptic connectivity hypothesis of memory has been achieved, despite extensive work during the past 10 years, which justifies our focusing attention on sites and possibilities other than the synapses in which information may be stored.

Temporal parameters play a major role in memory, which could indicate that "biological clocks" participate in the process. Not only is the sequence of CS and UCS stored, but the time interval elapsed between them is also. In a delayed conditioned reflex (CR), even a long interval is stored and read out with astonishing precision; in temporal conditioning, the time interval itself replaces the CS (Lock-

hart, 1966). In the last ISP volume, several contributors stressed that learning is related to recognition, storage, and read-out of *temporal relations* and *time intervals* (Eisenstein, 1967; Livingston, 1967; John, 1967b). It has been shown that the time intervals of flickering light can be stored and will show on an electroencephalogram (EEG) after the stimulation period (Livanov and Poliakov, 1945; John, 1967a). "A central problem is whether this phenomenon [assimilation of rhythms] depends on the establishment of a resonance-like tuning of a network, or pacemaker properties in the discharging neuron" (John, 1967b).

Synapses do not change their delay time markedly and probably cannot store information about time intervals. The existence of polysynaptic delay times is not probable from an economic standpoint, because a single synaptic delay is in the order of only one millisecond. As a result, an enormously large number of neurons in a closed chain would be needed for an explanation of behavioral delays of seconds, minutes, or even hours. The existence of reverberating circuits can be excluded, as demonstrated by cooling experiments and by electroconvulsive shock. We must, therefore, consider John's second alternative—that information storage is related to neuronal pacemaker potentials. In the following discussion, some observations are assembled to support this view.

"Vestibular Memory" It is commonly known that sailors and passengers who land from a boat trip on a rough sea are subject to unpleasant after-effects (Livingston, 1967). We tried to explore this observation experimentally by subjecting rabbits to rhythmic rolling, pitching, or yawing movements on motor-driven platforms. Their eye movements were recorded with the electronystagmographic method before, during, and after prolonged exposure to the movements of the rocking platform. When, after a "training period" of two to three weeks, the rocking was suddenly stopped, the eye movements tended to continue for some minutes at the same approximate frequency as during rocking (Figure 4). Even after the eye movements ceased, periods of nystagmus reappeared at intervals. This phenomenon could best be explained by the assumption that some neurons in the vestibular or oculomotor nuclei took on rhythmic pacemaker properties and discharged at the approximate frequency of the rocker.

Optomotoric nystagmus also is stored and can be reproduced (Mackensen and Wiegmann, 1959). Afternystagmus appears after several minutes of exposure to the sight of a rotating drum, which has alternating black and white perpendicular stripes on its surface. Afternystagmus can be recorded continuously in monkeys for periods of up to 420 seconds (Krieger and Bender, 1956). Even when the afternystagmus had disappeared, it could be re-established by a nonspecific stimulation, such as touching the animal. It is

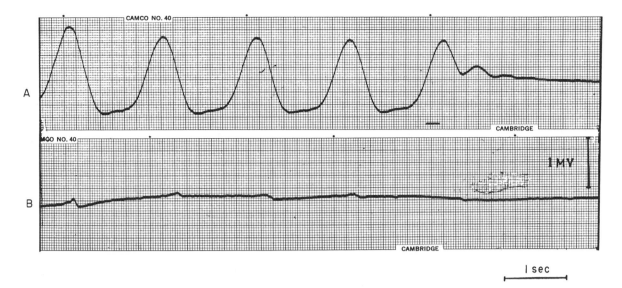

CAMCO NO. 40

A

CAMBRIDGE

CO NO. 40

1 MV

B

CAMBRIDGE

1 sec

FIGURE 4 Rhythmic eye movements of a rabbit after prolonged exposure on a rocking platform. A. Electronystagmographic recording, made in the dark, of the vertical, counter-rolling eye movements during exposure on a rocking platform. The rocking device tilted the rabbit 45 degrees to each side. The rocking was stopped suddenly after three weeks of continuous movement. B. Twenty seconds after the platform was stopped, a spontaneous vertical nystagmus appeared with the approximate frequency of the former rocking movements.

important to note that the frequency and initial direction of the afternystagmus resemble the frequency and direction of the stimulation. Different stimulation frequencies in the same subject are followed by corresponding changes in the frequency of the afternystagmus.

We performed experiments on human subjects, who were asked to follow with their eyes a sweeping light point on a cathode-ray oscilloscope. The scope was then switched off, and the subjects sat with open eyes in an absolutely dark room. Afternystagmus appeared, lasted several minutes, and in some cases even exceeded in amplitude and regularity the eye movements that took place during the original stimulation period (Figure 5). Afternystagmus has been conditioned by pairing a conditioned stimulus with the rotation of a striped drum (Guinsberger and Zikmund, 1956). The presentation of the conditioned stimulus alone then elicited rhythmic eye movements.

When Bechterew, at the end of the last century, studied the effects on eye movements of the destruction of the labyrinth, he observed an interesting phenomenon: bilateral labyrinthectomy causes nystagmus only when the two labyrinths are destroyed in two sessions with an interval of several days or weeks between (Bechterew, 1909). Unilateral labyrinthectomy always causes a nystagmus, which has been explained as peripheral "imbalance" of the vestibular system. Because this nystagmus disappears again after several days, it was concluded that a *plastic compensating mechanism* exists; it is activated by the initial imbalance and

tends to counteract the disturbing rhythmic eye movements. When the second labyrinth is destroyed, the rhythmic eye movements reappear, this time toward the other side, indicating the intrinsic rhythmic activity of the compensatory mechanisms. To obtain more insight into the nature of the plastic changes of their pacemaker systems, it would be highly interesting to record neurons of the vestibular or oculomotor nuclei with intracellular microelectrodes, both during the Bechterew nystagmus and the optokinetic or vestibular afternystagmus.

SHORT-TERM PLASTIC CHANGES IN RESPIRATORY CENTERS The respiratory movements of all vertebrates and many invertebrates are the consequence of rhythmic bursts of action potentials in neurons of the respiratory centers in the lower brainstem. These centers offer a unique opportunity to investigate problems of spontaneous rhythms in neurons of the central nervous system, with the use of techniques of extracellular and intracellular microrecording. Respiratory neurons continue to discharge in rhythmic bursts, even when the respiratory muscles have been paralyzed by injection of succinylcholine or curarine (von Baumgarten, 1956) or the respiratory centers have been isolated from the rest of the brain stem (Hukuhara and Okada, 1956). Intracellular recordings of respiratory neurons reveal slow oscillations of the resting potential (von Baumgarten et al., 1959) such as those in the Purkinje fibers of the pacemaker areas of the heart (Trautwein and Kas-

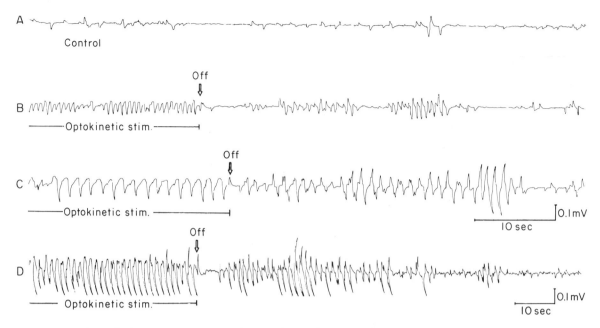

FIGURE 5 Prolonged afternystagmus in man after optokinetic stimulation at different frequencies. The optokinetic stimulus was a point of light sweeping across the screen of a cathode-ray oscilloscope. At the signal, the scope was switched off, and the subject was recorded in absolute darkness. A. Control in the dark before stimulation. B. Optokinetic stimulation at a frequency of 16 per second. When the stimulation was switched off after five minutes, periods of afternystagmus appeared at approximately the same frequency as during the stimulation. C. Same experiment, same subject, but lower frequency of stimulation (seven per second). The afternystagmus again appeared at approximately the same frequency as during the stimulation period. D. Another example of optokinetic afternystagmus. Note the changes in amplitude.

sebaum, 1961) and in pacemaker cells of *Aplysia* (Arvanitaki and Chalazonitis, 1965).

We recently performed experiments in which inflation of the lungs (Hering and Breuer reflex) was used to impose artificial rhythms on the pacemaker cells of the respiratory centers. In each trial, a young rat was anesthetized with pentobarbital, and its nose was connected to a respiratory pump. The rat shown in Figure 6 had a respiration rate of 120 per minute, but the pump rate was adjusted to only 60 per minute. When the pump was switched on, the rat continued to breathe for some time at its original spontaneous rate. It managed to do so by breathing with the pump, but it intercalated additional shallow inspirations between the large strokes. These intercalated inspirations soon faded, and the respiratory centers locked their own rhythm with that of the pump. When the pump was stopped, the rat continued to breathe at the low frequency of the pump for some time, indicating that the respiratory centers indeed had assimilated the new rhythm; but soon this new rhythm was replaced by the original frequency of 120 per minute.

STORAGE OF INTERVALS IN *Aplysia* CELLS In our laboratory, Chun-Fan Chen, a graduate student, has made intracellular microrecordings in the hope of observing the "assimilation of rhythm" phenomenon in single *Aplysia* pacemaker cells. Burster cells—those that discharge in regular bursts (Strumwasser, 1967; Kandel, 1967)—and those cells that discharge with single spikes at regular intervals without observable synaptic input (Arvanitaki and Chalazonitis, 1961a) were both subjected to intracellular stimulation by means of a transmembrane current. The stimuli consisted of rhythmic periods of a depolarizing current, mediated by a bridge circuit to the recording microelectrode. The frequency of stimulation was adjusted to be either slightly (but significantly) higher, or lower than the naturally occurring bursts.

When such stimulation was continued over a period of from several minutes to an hour and then stopped suddenly, the recorded nerve cell responded in a significant number of cases by delivering the subsequent spontaneous bursts at the imposed new intervals. After the spontaneous activity locked with the stimulus, the first spontaneous burst appeared at the approximate time of the anticipated stimulus, which was not actually delivered (Figure 7, page 268). Pacemaker neurons, which discharge in single spikes, also could be influenced this way (Figure 8, page 268). During the

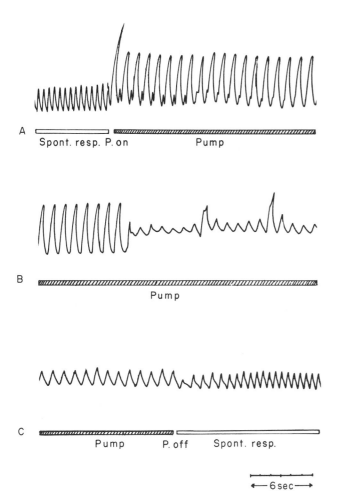

A. Spont. resp. P. on     Pump

B. Pump

C. Pump     P. off     Spont. resp.

←— 6 sec —→

FIGURE 6 Assimilation of rhythms in the respiratory centers. A. The rat breathes spontaneously 120 times per minute. After artificial respiration was started at half this rate, the old spontaneous respiratory rhythm was maintained for some time by intercalating an additional breath between the strokes of the pump. Eventually the respiratory movements locked with the pump rhythm. B. The stroke volume of the pump was reduced to a minimum in order to avoid hyperventilation. Note the two spontaneous gasps. C. After the pump was stopped suddenly, the rat continued to respirate spontaneously for several breaths at the pump frequency before falling back to its own rhythm.

stimulation, some neurons that displayed rhythmic, oscillating, pacemaker potentials showed progressive adjustments of the pacemaker potential from its original wavelength to that imposed by the stimulus. The slope of the pacemaker potential became steeper when the stimulation frequency exceeded the spontaneous oscillation rate, or flatter when the stimulation frequency was less than the spontaneous rate. After the pacemaker reached the firing level of the neuron at the approximate time of the next stimulus, the stimulus could be switched off and the next spontaneous discharges came at the right time.

The induced changes of endogenous pacemaker activity were usually of only short duration, and the neuron returned to its original discharge frequency after one or a few cycles. In selected cases, however, the interstimulus interval could be "recovered" after periods of up to 20 minutes by means of single intracellular stimuli. When such a stimulus was given, long after the training period, the cell responded with the interval imposed during the previous training period (Figure 9, page 269).

In similar experiments with regularly discharging *Aplysia* neurons, a single intracellular depolarizing shock, which was below the threshold for an action potential, was paired with a stronger, suprathreshold depolarization that followed it by six seconds (Figure 10, page 269). After repetition of such paired stimulation in a regularly discharging neuron, the subthreshold stimulation alone led to a delayed action potential, the latency of which was approximately equal to the interval that had existed formerly between the subthreshold and superthreshold stimuli. The first depolarizing intracellular stimulus in this case was subthreshold for an action potential, so any feedback from the synaptic network of the neuropil can be excluded. The change, in all probability, took place in the recorded neuron alone.

CAN CLASSICAL CONDITIONING BE EXPLAINED BY "SELF RE-EXCITATION" OF NEURONS AFTER SYNAPTIC POTENTIALS? In a neuron on which the CS and the UCS converge synaptically, as in Figure 1, the EPSP caused by the UCS could, after a training period of paired stimulation, trigger a slow depolarizing potential, which, at its peak, would lead to an action potential or at least increase the probability of spike-firing at the anticipated time (see Figure 1, bottom trace, Cb). In other words, the CS would gain the effect of the UCS as in classical conditioning.

This hypothesis of learning by self-re-excitation is based on two major assumptions: (1) that the UCS can trigger a slow depolarizing potential, or at least can reset an ongoing pacemaker oscillation; or (2) that the wavelength and amplitude (slope) of such a slow depolarization are not constant but undergo plastic changes, depending on the interstimulus interval during conditioning.

Support for the first assumption comes from experiments that proved that synaptic activity can trigger relatively long-lasting changes of the membrane potential (Arvanitaki and Chalazonitis, 1961b; Nishi and Koketsu, 1968; Pinsker and Kandel, 1969). In this volume, Bloom proposes a possible biochemical link between the synaptic and the pacemaker activities. In his opinion, norepinephrine not only represents a synaptic transmitter (in that it causes the postsynaptic membrane to produce a synaptic potential) but also reacts with cyclic adenosine monophosphate (AMP) to alter the activity of electrogenic pumps in the membrane.

The second assumption is still vague, but is supported by

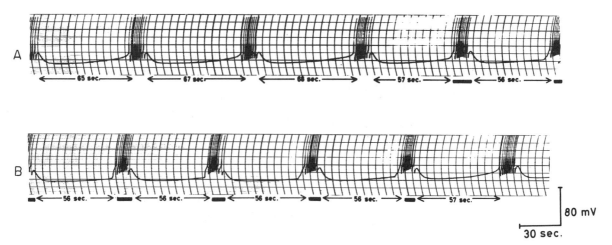

80 mV

30 sec.

FIGURE 7 Effect of rhythmic intracellular stimulation of a burster cell (R15) of *Aplysia*. The average spontaneous burst interval is 67 seconds. Rhythmic intracellular depolarization is applied (black bars) six times, with an interstimulus inter. val of 56 seconds. After the end of the stimulation period, the first spontaneous burst appears after the entrained inter-stimulus interval. Maximal spontaneous variability of the mean interburst interval, 3.5 per cent. Induced change, 15 per cent.

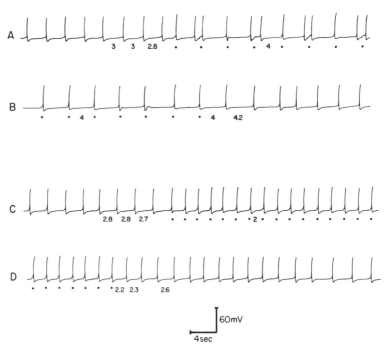

60mV

4sec

FIGURE 8 Lengthening and shortening the interspike inter-val by rhythmically applied depolarizing pulses to an un-identified cell in the left side of the *Aplysia* abdominal gan-glion. The intracellular depolarizing pulses (black dots) are just at threshold for a spike response. A and B (continuous records). Spontaneous action potentials at an interval be-tween 2.5 and 3 seconds. When rhythmic intracellular stim-ulation at a longer interval (four seconds) is switched on, the cell at first discharges irregularly but eventually locks its dis-charges with the stimulus. After the stimulus is switched off, the cell discharges three times at about the "entrained" lengthened interval. C and D. Same neuron. Stimulus inter-val shorter than the average spontaneous interval. The dis-charges again lock with the stimulus during the training period. After the stimulus is switched off, the cell spontane-ously discharges several times with a shortened interval.

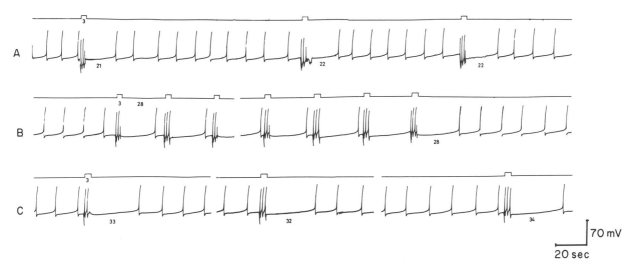

FIGURE 9  Effect of single depolarizing impulses before and after a rhythmic stimulation period in unidentified cell on right side of the abdominal ganglion of *Aplysia*. A. Control. Single intracellular depolarizing pulses cause short bursts followed by silent periods of irregular length. B. Fifteen-minute "training period" with depolarizing pulses of four seconds each, spaced at 28-second intervals. The last stimulus is followed by the "entrained" interval of 28 seconds. C. Single pulses, if applied after "training," cause longer pauses than in the control, A.

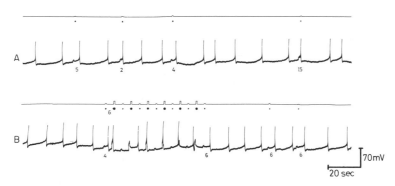

FIGURE 10  Effect of subthreshold depolarization before and after "conditioning" of an unidentified, spontaneously discharging cell in the abdominal ganglion. A. Subthreshold stimuli are followed by action potentials at irregular intervals. B. Each subthreshold depolarizing pulse is paired with a strong intracellular depolarization, which causes an action potential. The interstimulus interval is six seconds. After this "conditioning period," the interval between a single subthreshold depolarizing stimulus and the next "spontaneous" action potential is six seconds in three subsequent cases.

the observations that (1) spontaneous, rhythmic, action potentials lock with the rhythm of imposed stimulation (Perkel et al., 1964); that (2) such changes of spontaneous cellular rhythms can outlast the stimulation period; and that (3) rhythmic neuronal oscillations, as expressed in the instances of respiratory rhythms, nystagmus, and EEG waves "assimilate" externally imposed rhythms.

The endogenous pacemaker oscillations seem to be based on metabolic processes energizing electrogenic pumps rather than on permeability changes of the membrane. Unfortunately, we are still far from understanding the con-trolling forces behind the slow metabolic processes of the membrane, and it therefore appears premature to speculate about biochemical mechanisms responsible for plastic changes within the pacemaker apparatus of the cell.

Anatomical connectivity, by means of fiber connections and synapses, is only one of the several preconditions of functional connectivity. Probably many more channels than are actually used are anatomically provided for in the network of the neuropil. The transition of a channel from the functionally inactive to the functioning state could be interpreted as the elementary basis of learning. But it would be shortsighted to regard only the synapses as responsible for such changes.

Traditionally, neurophysiologists describe nervous integration mainly in terms of EPSPs, IPSPs, and action potentials. It might be an underestimation of the function of a nerve cell to say that it only passively summates EPSPs and IPSPs at the axon hillock. A nerve cell might very well have an individual "history" and contribute its own vote to the "democratic" decision of whether a spike is fired. It may do so by following an EPSP with activation of a slowly depolarizing ionic pump, initiation of rhythmic pacemaker activity after synaptic bombardment, or a change in firing level. In any case, if we assume that the the postsynaptic neuron can undergo plastic changes, there is no need to make the less practical (from an engineering standpoint) assumption that the incoming information to the neuron is altered (i.e., "falsified") at the synapse before it even reaches the neuron.

## Summary

Our present-day neurophysiological knowledge about plasticity at the unitary level raises more questions than it can answer. All observed changes in synaptic efficacy, as well as pacemaker properties of neurons, are short-lived and cannot account for permanent memory. It could be argued, however, that most of the cell types used for these studies were selected for technical reasons only (i.e., large size, easy access, limited tissue pulsation). Obviously, the central nervous system contains follower cells that do not learn as well as do other cells with a higher level of independence. The cells that are specialized in learning are not necessarily large enough to be recorded easily with intracellular microelectrodes. They might be as small as the granule cells in the cerebral cortex. Future research should be directed toward the problem of finding the animal and the cell types that are ideally suited for studies on neuronal plasticity.

In conclusion, I wish to stress again that the elementary mechanism of memory at the neuronal level remains unknown. All presently available data are insufficient and must be supplemented by hypotheses and speculations. We should broaden our view to include new hypotheses and experimental materials that deserve more detailed study.

In the course of evolution, learning and memory were important mechanisms of adaptation to the environment and consequently had high survival value. It therefore appears possible that several redundant mechanisms developed to secure this capability. For this reason, various hypotheses about the elementary mechanism of learning do not necessarily exclude one another. At the present stage in our understanding, the best strategy seems to be a multilateral approach to the problem, including physiological, morpho-

logical, and biochemical methods to permit the pursuit of all promising leads.

## Acknowledgment

This work was supported in part by NIH Grant No. 1R01–07753–01.

## REFERENCES

ARVANITAKI, A., and N. CHALAZONITIS, 1961a. Slow waves and associated spiking in nerve cells of *Aplysia*. *Bull. Inst. Oceanogr.* 581: 1224.

ARVANITAKI, A., and N. CHALAZONITIS, 1961b. Phases d'inhibition prolongées consécutives aux potentiels postsynaptiques d'excitation ou d'inhibition (neurones d'*Aplysia*). *J. Physiol.* (*Paris*) 53: 253–254.

ARVANITAKI, A., and N. CHALAZONITIS, 1965. Les oscillations de basse fréquence du potentiel de membrane somatique (neurone d'*Aplysia*). *Compt. Rend. Lecanes Soc. Biol.* 159: 1179–1191.

BAUMGARTEN, R. von. 1956. Koordinationsformen einzelner Ganglienzellen der rhombencephalen Atemzentren. *Pflügers Arch.* 262: 573–594.

BAUMGARTEN, R. von, K. BALTHASAR, and H. P. KOEPCHEN, 1959. Über ein Substrat atmungsrhythmischer Erregungsbildung im Rautenhirn der Katze. *Pflügers Arch.* 270: 504–528.

BAUMGARTEN, R. von, and T. HUKUHARA, 1969. The role of interstimulus interval in heterosynaptic facilitation in *Aplysia californica*. *Brain Res.* 16: 369–381.

BAUMGARTEN, R. von, and B. JAHAN-PARVAR, 1967. Beitrag zum Problem der heterosynaptischen Facilitation in *Aplysia californica*. *Pflügers Arch.* 295: 238–346.

BECHTEREW, W. von, 1909. Die Functionen der Nervencentra, Vol. 2. Gustav Fischer, Jena.

ECCLES, J. C., 1961. The effects of use and disuse on synaptic function. *In* Brain Mechanisms and Learning. A Symposium (J. F. Delafresnaye, editor). Blackwell Scientific Publications, Oxford, pp. 335–348.

ECCLES, J. C., 1964. The Physiology of Synapses. Springer-Verlag, Berlin, Göttingen, Heidelberg.

EISENSTEIN, E. M., 1967. The use of invertebrate systems for studies on the bases of learning and memory. *In* The Neurosciences: A Study Program (G. C. Quarton, T. Melnechuk, and F. O. Schmitt, editors). The Rockefeller University Press, New York, pp. 653–655.

GALAMBOS, R., 1961. A glia-neural theory of brain function. *Proc. Nat. Acad. Sci.U.S.A.* 47: 129–136.

GRUNDFEST, H., 1957. Electrical inexcitability of synapses and some consequences in the central nervous system. *Physiol. Rev.* 37: 337–361.

GRUNDFEST, H., and D. P. PURPURA, 1956. Inexcitability of cortical dendrites to electrical stimuli. *Nature* (*London*) 178: 416–417.

GUINSBERGER, E., and V. ZIKMUND, 1956. Conditioned optokinetic nystagmus. *Physiol. Bohem.* 5: 368–375.

HAIGLER, H. J., and R. J. von BAUMGARTEN, 1970. The minimum

latency of heterosynaptic facilitation in *Aplysia*. *Fed. Proc.* 29: 589 (abstract).

HILGARD, E. R., and D. G. MARQUIS, 1961. Classical and instrumental conditioning compared. *In* Conditioning and Learning. Second Edition, Revised by G. A. Kimble. New York, Appleton-Century-Crofts, Inc., pp. 78–108.

HUGHES, J. R., 1958. Post-tetanic potential. *Physiol. Rev.* 38: 91–113.

HUKUHARA, T., and H. OKADA, 1956. On the automaticity of the respiratory centers of the catfish and crucian carp. *Jap. J. Physiol.* 6: 313–320.

HYDÉN, H., 1967. RNA in brain cells. *In* The Neurosciences: A Study Program (G. C. Quarton, T. Melnechuk, and F. O. Schmitt, editors). The Rockefeller University Press, New York, pp. 248–266.

HYDÉN, H., and E. EGYHÁZI, 1962. Nuclear RNA changes of nerve cells during a learning experiment in rats. *Proc. Nat. Acad. Sci. U.S.A.* 48: 1366–1373.

JAHAN-PARVAR, B., and R. J. VON BAUMGARTEN, 1967. Untersuchungen zur Spezifitaetsfrage der heterosynaptischen Facilitation bei Aplysia californica. *Pflügers Arch.* 295: 347–360.

JOHN, E. R., 1967a. Mechanisms of Memory. Academic Press, New York.

JOHN, E. R., 1967b. Electrophysiological studies of conditioning. *In* The Neurosciences: A Study Program (G. C. Quarton, T. Melnechuk, and F. O. Schmitt, editors). The Rockefeller University Press, New York, pp. 690–704.

KANDEL, E. R., 1967. Cellular studies of learning. *In* The Neurosciences: A Study Program (G. C. Quarton, T. Melnechuk, and F. O. Schmitt, editors). The Rockefeller University Press, New York, pp. 666–689.

KANDEL, E. R., and L. TAUC, 1965a. Heterosynaptic facilitation in neurones of the abdominal ganglion of *Aplysia depilans*. *J. Physiol.* (*London*) 181: 1–27.

KANDEL, E. R., and L. TAUC, 1965b. Mechanism of heterosynaptic facilitation in the giant cell of the abdominal ganglion of *Aplysia depilans*. *J. Physiol.* (*London*) 181: 28–47.

KRIEGER, H. P., and M. B. BENDER, 1956. Optokinetic afternystagmus in the monkey. *Electroencephalogr. Clin. Neurophysiol:* 8: 97–106.

LEWIS, E. R., T. EVERHART, and Y. Y. ZEEVI, 1969. Studying neural organization in *Aplysia* with the scanning electron microscope. *Science* (*Washington*) 165: 1140–1143.

LIVANOV, M. N., and K. L. POLIAKOV, 1945. The electrical reactions of the cerebral cortex of a rabbit during the formation of a conditioned defense reflex by means of rhythmic stimulation. *Izv. Akad. Nauk. USSR, Ser. Biol.,* 3: 286.

LIVINGSTON, R. B., 1967. Brain circuitry relating to complex behavior. *In* The Neurosciences: A Study Program (G. C. Quarton, T. Melnechuk, and F. O. Schmitt, editors). The Rockefeller University Press, New York, pp. 499–515.

LLOYD, D. P. C., 1949. Post-tetanic potentiation of response in monosynaptic reflex pathways of the spinal cord. *J. Gen. Physiol.* 33: 147–170.

LOCKHART, R. A., 1966. Temporal conditioning of BSRR. *J. Exp. Psychol.* 71: 438–440.

MACKENSEN, G., and O. WIEGMANN, 1959. Untersuchungen zur Physiologie des optokinetischen Nachnystagmus. *Graefes Arch. Ophthalmol.* 16: 497–509.

MATHIEU, A.-M., and M. COLONNIER, 1968. Electron microscopic observations in the molecular layer of the cat cerebellar cortex after section of the parallel fibers. *Anat. Rec.* 160: 391 (abstract).

NISHI, S., and K. KOKETSU, 1968. Analysis of slow inhibitory postsynaptic potential of bullfrog sympathetic ganglion. *J. Neurophysiol.* 31: 717–727.

OLDS, J., 1961. Discussion of the paper by J. C. Eccles. *In* Brain Mechanisms and Learning. A Symposium (J. F. Delafresnaye, editor). Charles C Thomas, Springfield, Illinois, p. 350.

PERKEL, D., J. H. SHULMAN, T. H. BULLOCK, G. P. MOORE, and J. P. SEGUNDO, 1964. Pacemaker neurons: Effects of regularly spaced synaptic input. *Science* (*Washington*) 145: 61–63.

PINSKER, H., and E. R. KANDEL, 1969. Synaptic activation of an electrogenic sodium pump. *Science* (*Washington*) 163: 931–935.

SCILARD, L., 1964. On memory and recall. *Proc. Nat. Acad. Sci. U. S. A.* 51: 1092–1099.

STRUMWASSER, F., 1967. Neurophysiological aspects of rhythms. *In* The Neurosciences: A Study Program (G. C. Quarton, T. Melnechuk, and F. O. Schmitt, editors). The Rockefeller University Press, New York, pp. 516–528.

TAUC, L., 1965. Presynaptic inhibition in the abdominal ganglion of *Aplysia*. *J. Physiol.* (*London*) 181: 282–307.

TAUC, L., and R. EPSTEIN, 1967. Heterosynaptic facilitation as a distinct mechanism in Aplysia. *Nature* (*London*) 214: 724–725.

TRAUTWEIN, W., and D. G. KASSEBAUM, 1961. On the mechanism of spontaneous impulse generation in the pacemaker of the heart. *J. Gen. Physiol.* 45: 317–330.

VALVERDE, F., 1967. Apical dendritic spines of the visual cortex and light deprivation in the mouse. *Exp. Brain Res.* 3: 337–352.

# 26 Multiple Steps in the Biology of Memory

SAMUEL H. BARONDES

BECAUSE SOME form of memory is established immediately upon learning, there is a tendency to assume that "memory" is a simple and immediate consequence of learning. Yet many experiments indicate that the storage of memory goes on for hours and even days after training. The point of the present brief report is to emphasize that a biological analysis of memory must consider mechanisms by which new, functional, interneuronal relationships are created over a relatively long period of time.

Formation of memory may resemble the prolonged sequence of reactions that end in organ differentiation rather than the relatively simple and immediate reactions characteristic of induction of a specific enzyme in response to a change in cellular environment. This need not mean that the search for a single limiting factor in memory storage is unjustified, for a complex and prolonged sequence of reactions is not necessarily irreducible. The full development of an antibody response may take a period of time equal to that required to deposit a permanent memory in the brain, yet this complex biological process can be viewed in terms of one limiting reaction and its manifold secondary consequences.

The biology of memory was considered in detail in the book that resulted from the 1966 Intensive Study Program. The papers by Galambos (1967), Miller (1967), Eisenstein (1967), Kandel (1967), Quarton (1967), Agranoff (1967), and Hydén (1967) provide a background for the present brief report. In addition, reviews by McGaugh and Herz (1970), Glassman (1969), and Barondes (1969) may be consulted for details of more recent work in this area. The last critically considers the evidence that protein synthesis in the cerebrum is required for long-term memory storage. Although many of the experiments discussed in this paper employ inhibitors of protein synthesis in the cerebrum, the means by which these drugs influence memory storage is not of primary concern here.

## Short-term and long-term memory

MEMORY THAT DIMINISHES OVER HOURS Studies in which mice are trained while 90–95 per cent of their protein synthesis in the cerebrum is inhibited by acetoxycycloheximide suggest the existence of a short-term memory process, which diminishes over hours, and a long-term memory process, which is much more stable. Such mice acquire a simple position discrimination or light-dark discrimination in a one-choice T-maze in the same number of trials as do controls and show identical learning curves (Cohen and Barondes, 1968a). Furthermore, retention (measured by the number of trials required to reach criterion on retesting) is normal two or three hours after training (Figure 1). However, retention at six hours, one day, seven days (Figure 1), or six weeks (Barondes and Cohen, 1967)

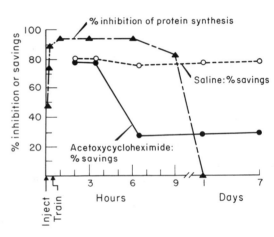

FIGURE 1 Effect of acteoxycycloheximide on protein synthesis in the cerebrum and memory. Mice were injected subcutaneously with 240 μg of acetoxycycloheximide 30 minutes before training to escape shock by choosing the lighted limb of a T-maze to a criterion of five out of six consecutive correct responses. Different groups were tested for retention (per cent savings) at each of the indicated times. (For details, see Barondes and Cohen, 1968a.)

after training is markedly impaired. This suggests that a "short-term" memory-storage process is established during training and persists for more than three hours thereafter, despite marked inhibition of protein synthesis in the cerebrum during training, but that a "long-term" storage process does not develop normally in mice after they have been treated with acetoxycycloheximide.

The long-term process blocked by acetoxycycloheximide is normally initiated during training, or within minutes

SAMUEL H. BARONDES Departments of Psychiatry and Molecular Biology, Albert Einstein College of Medicine, Yeshiva University, Bronx, N. Y. Present Address: Department of Psychiatry, School of Medicine, University of California at San Diego, La Jolla, California

thereafter, or both, in the maze tasks that have been studied. To impair this process markedly, the inhibitor of protein synthesis in the cerebrum must be administered before training. If acetoxycycloheximide is injected subcutaneously five minutes before training, so that marked inhibition is established during training, there is marked impairment of retention when measured seven days later (Figure 2). In

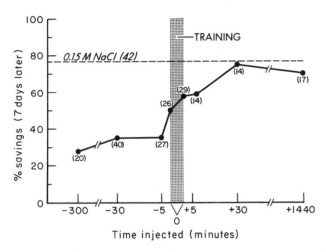

FIGURE 2 Effect on memory of subcutaneous administration of acetoxycycloheximide at times before or after training. All groups were tested for retention seven days later. (For details, see Barondes and Cohen, 1968a.)

contrast, administration of the drug immediately after training has only a slight effect, and, if delayed for 30 minutes, it has none (Figure 2). Such findings suggest that the critical protein synthesis in the cerebrum, which is apparently required for long-term memory storage, is already under way during training and is insusceptible to inhibition within minutes.

Although the long-term process is normally initiated during training, or within minutes thereafter in the maze task, under certain circumstances it may be initiated as long as three hours after training, as is suggested from studies of mice trained 30 minutes after they had received cyclo-heximide injections, which markedly inhibited protein synthesis in the cerebrum for only several hours (Barondes and Cohen, 1968b). Such mice did not normally develop long-term memory, although their capacity for protein synthesis in the cerebrum was restored while the short-term process was still demonstrable. However, if they were treated three hours after injection with foot shock, amphetamine, or corticosteroids, agents that are interpreted as producing a state of "arousal," they subsequently demonstrated substantial long-term memory (Barondes and Cohen, 1968b). Identical treatments given six hours after training, a time when the short-term process was no longer

measurable, did not establish long-term memory (Barondes and Cohen, 1968b). Furthermore, reinstatement of inhibition of protein synthesis in the cerebrum by injection of acetoxycycloheximide prior to introduction of amphetamine blocked the effect of the latter (Figure 3).

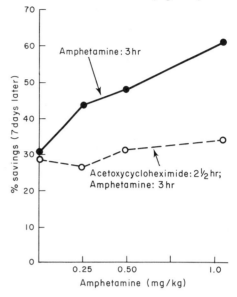

FIGURE 3 Effect of amphetamine on memory and its antagonism by acetoxycycloheximide. All mice were injected with 0.12 gm/kg of cycloheximide 30 minutes before training. To reinstate marked inhibition of protein synthesis in the cerebrum, one group received acetoxycycloheximide two and a half hours after training. Both groups received amphetamine three hours after training, and retention was measured seven days later (For details, see Barondes and Cohen, 1968b).

These experiments suggest that the information acquired from training can be converted into the long-term form only if intact protein-synthesizing capacity, an appropriate "state of arousal," and the availability of the information in the short-term store all coexist. They do coexist during normal learning. When cycloheximide is administered before training, however, protein-synthesizing capacity in the cerebrum is inhibited during the arousal that normally accompanies training; and when the protein-synthesizing capacity returns, the required arousal has subsided. Livingston (1967) has speculated on the role of "arousal" in directing the formation of long-term memory storage.

RETROGRADE AMNESIA Evidence for a change in memory during some period after training was first obtained from studies in which electroconvulsive shock was administered after training was completed (see reviews by McGaugh, 1966; Jarvik, 1968; Weissman, 1967). Convulsions induced immediately after training produce marked impairment of retention, but if delayed they are ineffective. The period during which electroshock convulsions induce retrograde amnesia varies in different experiments. In some it is measured in seconds, whereas in others some retrograde amnesic effect is detectable if electroconvulsive shock is given up to six hours after training (Kopp et al., 1966). Production of spreading depression in the cortex by injections of potassium chloride into the brain (Bureš and Burešova, 1963) and administration of other agents which affect the nervous system (Weissman, 1967) may also produce retrograde amnesic effects. In some situations, even acetoxycycloheximide or cycloheximide effectively impair long-term memory if given as long as 30 minutes after training, as found in goldfish by Agranoff et al. (1966) or in mice (Figure 4). Studies of this type have been taken to mean that the memory-storage process "consolidates" or is strengthened over some period of time after training in such a way that it may be influenced by these treatments very shortly after training but not thereafter. The enhancement of memory by administration of drugs within a short period after training has also been interpreted in this way (McGaugh, 1969).

Experiments in which disruptive manipulations are made after training have also provided some evidence consistent with separate short-term and long-term processes in memory storage. Albert (1966) has shown that cathodal polarization shortly after training produces retrograde amnesia, but that retention remains normal for several hours and then becomes markedly impaired. Geller and Jarvik (1968)

have shown that the retrograde amnesic effects of electroconvulsive shock given immediately after passive avoidance training may not be apparent for hours after the convulsions are produced. These studies resemble those with cycloheximide and acetoxycycloheximide, in that retention remains for hours after training before diminishing markedly.

## More delayed effects

Flexner, Flexner, and Stellar (1963) found what appeared to be retrograde amnesic effects of puromycin if the drug was injected into the temporal regions of the brain up to several days after training. Although puromycin is a protein-synthesis inhibitor, subsequent studies suggest that its retrograde amnesic effect need not be attributed to inhibition of protein synthesis in the cerebrum for memory storage but rather to production of cerebral electrical abnormalities (Cohen et al., 1966) and the production of peptidyl-puromycin (Flexner and Flexner, 1968b; Gambetti et al., 1968), which in some manner impairs brain function. Furthermore, these retrograde effects are reversible if saline is injected into the original injection sites days after the puromycin is administered (Flexner and Flexner, 1968a), unlike the effects produced by administration of acetoxycycloheximide before training (Rosenbaum et al., 1968).

A major finding in these studies is that the puromycin injections in the temporal regions of the brain are effective

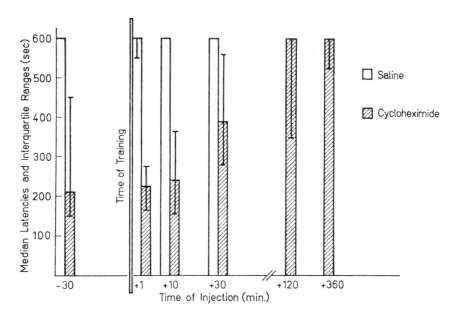

FIGURE 4   Effect of cycloheximide injections before, or at several times after, training. Mice were injected subcutaneously with cycloheximide before, or at several times after, receiving one-trial "passive-avoidance" training and tested for retention seven days later. Increased latency of entry into the box in which shock had been received is a measure of memory. (For details, see Geller et al., 1969).

only if given up to several days after training but not if given one week or more after training (Flexner, Flexner, and Stellar, 1963). Apparently something is changing during this long period. Complex changes in performance were subsequently found by Deutsch et al. (1966) after intracerebral injection of diisopropylfluorophosphate (which inhibits a number of enzymes, including cholinesterase) at various times in the days after training in a maze. Furthermore, injections of actinomycin-D in doses that do not inhibit over-all cerebral RNA synthesis extensively but that produce local neuronal necrosis and electrical abnormalities impair memory if given one day but not seven days after training (Figure 5). This series of studies suggests that some

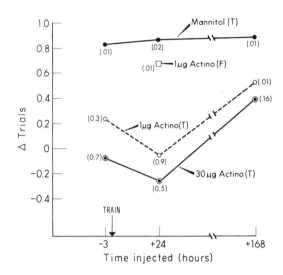

FIGURE 5 Effect of intracerebral injections of actinomycin-D at times before or after training on retention measured 24 hours after injection. Mice were trained in a position-discrimination until they had made two correct responses. Significant retention after such brief training is demonstrated by the decrement in average trials required to reach criterion on retesting, which is indicated by $\Delta$ trials. The $P$ values (shown in parentheses) represent the probability that there is significant retention. Both the small and large doses of actinomycin-D produce local neuronal necrosis and disturbances in hippocampal electrical activity. The small dose has little effect on over-all cerebral RNA synthesis. Injections in the temporal regions (T) up to one day, but not seven days, after training impair performance measured one day later. Injections in the frontal regions (F) are ineffective even at one day. (For details, see Squire and Barondes, 1970).

aspect of the memory-storage process is in flux for days after training is completed. From the studies that show that the amnesic effect of puromycin injection given one day after training can be *reversed* by subsequent saline injections (Flexner and Flexner, 1968a, 1968b; Rosenbaum et al., 1968), it appears that this drug may interfere with *retrieval* of stored information rather than with the storage process itself. The same may prove true of the other treatments that are effective long after training. Although the long-term storage process may be established during training or within hours thereafter, normal "retrievability" may require intact temporal-lobe function for days after training. The nature of the change in some aspect of the over-all memory-storage system for at least one day after training is one of the most challenging aspects of this whole problem.

## Multiple processes in memory

The memory storage process is being examined largely through behavioral techniques rather than by direct study of relevant neuronal events, so inferences about the underlying processes can be made only with considerable caution. The available evidence has been interpreted as being consistent with one or more dependent or independent processes. McGaugh has diagrammed these possibilities (Figure 6). In the "single-trace" hypothesis, one process is established by training and is progressively strengthened for some period after training. Only one critical change effected by training is gradually augmented. In the "dual-trace" hypothesis there are separate mechanisms for short-term and long-term memory. They make use of different biological mechanisms, and the long-term process can be directed by the short-term process or directly by training. The "multiple-trace" hypothesis is a further elaboration of the "dual trace."

At present, we cannot definitively choose among these possibilities. The retrograde amnesia studies are consistent with a "single-trace" hypothesis. One can readily imagine a single-memory process the progressive development of which after training decreases its susceptibility to disruption. On the other hand, the studies with acetoxycycloheximide that show transient intact memory are consistent with a "dual-trace" hypothesis but do not conclusively prove it. One can readily imagine a short-term process not involving protein synthesis in the cerebrum that lasts from three to six hours, and another long-term process that does depend on such synthesis. These experiments however, are also consistent with a single memory-storage process based on protein synthesis in the cerebrum. In this view, the small amount of residual protein-synthesizing capacity not inhibited by the drug may mediate memory for from three to six hours after training. Because the amount of protein made would be only a small percentage of that made normally, the memory might diminish rapidly if the protein had a short half-life. One's willingness to accept the idea that memory storage is mediated by a protein that turns over so rapidly will determine his opinion of the plausibility of such a hypothesis.

The data on further changes in memory for days after training suggest only that integration of new information

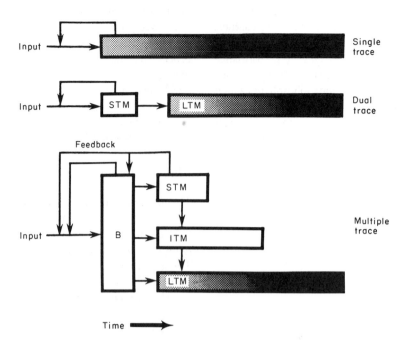

FIGURE 6 Alternative mechanisms of memory storage, as diagramed by McGaugh (1969). Abbreviations: B, buffer memory; ITM, intermediate-term memory; LTM, long-term memory; STM, short-term memory. The short-term memory, postulated by McGaugh on the basis of studies with human beings, lasts for seconds, whereas the memory that lasts for several hours and is called "short-term" in the present report corresponds to his intermediate-term memory. (For details, see text; McGaugh, 1969.)

into the system may continue for such a long period. The fundamental mechanism for storing and information may or may not be changing during this period. Therefore, despite extensive study, all we can say is that the memory-storage process is in flux for a considerable period after learning.

## Needed: simplifications

The studies on memory reported here are all concerned with the effects of manipulations of brain functions on the behavior of the whole organism. Although the ultimate cellular and molecular basis of memory storage may prove to be extremely simple, the system that processes the behavioral information and effects this change appears to be extremely complex. A classical response to a complex process is a reductionistic one—find a simpler system or study components of the system. Discovery of a relatively simple nervous system which displays learning and memory would clearly be a major advance. Attempts to identify such systems have been summarized recently (Bullock, 1967) and are discussed elsewhere in the present book (von Baumgarten). Portions of invertebrate nervous systems with relatively small numbers of neurons which demonstrate memory for many hours or days ("long-term" mem-

ory) have not yet been identified. However, if the "single-trace" hypothesis proves to be correct, studies of the mechanism of relatively short-lived memory phenomena in simple systems may prove relevant to the critical problem of long-term information storage.

Reductionism also dictates more general studies of regulation of the development of functional neuronal relationships. Our understanding of the molecular regulation of neuronal excitability is rudimentary. Likewise, the mechanism of development of neuronal pathways during the differentiation of the nervous system (see Jacobson, this volume; Barondes, this volume) may be crucial for work on the memory problem, although the "refined" differentiation characteristic of memory may be based on different processes.

Work on the effect of gross manipulations on memory in intact organisms has identified some pharmacological complexities of the problem. The most striking of these is that storage and retrievability of a memory may develop over a considerable period of time. Further work of this type will lead to a deeper understanding of these complexities at the level of the whole behaving animal. Knowledge of the factors that regulate interneuronal communication limits research with this approach. Further behavioral studies must be based on such information.

## Summary

Evidence that development of memory storage continues for minutes or hours after the learning experience is ended, and that the over-all memory system, including retrievability, is modified for one day or more after training, is considered briefly. A view of memory storage is considered in which there is: (1) a "short-term" process, independent of protein synthesis in the cerebrum, that persists for hours after training; (2) a "long-term" process, normally initiated during training, that also depends on such protein synthesis; and (3) a process that can last for more than a day after training and influences the retrievability of the stored information. The difficulty of drawing conclusions about such biological processes from behavioral studies is recognized and bemoaned.

## Acknowledgments

Supported by Career Development Award MH 18,232 and Research Grant MH 12773 from the National Institute of Mental Health.

### REFERENCES

AGRANOFF, B. W., 1967. Agents that block memory. *In* The Neurosciences: A Study Program (G. C. Quarton, T. Melnechuk, and F. O. Schmitt, editors). The Rockefeller University Press, New York, pp. 756–764.

AGRANOFF, B. W., R. E. DAVIS, and J. J. BRINK, 1966. Chemical studies on memory fixation in goldfish. *Brain Res.* 1: 303–309.

ALBERT, D. J., 1966. The effects of polarizing currents on the consolidation of learning. *Neuropsychologia* 4: 65–77.

BARONDES, S. H., 1969. Cerebral protein synthesis inhibitors block long-term memory. *Int. Rev. Neurobiol.* 12: 177–205.

BARONDES, S. H., and H. D. COHEN, 1967. Delayed and sustained effect of acetoxycycloheximide on memory in mice. *Proc. Nat. Acad. Sci. U. S. A.* 58: 157–164.

BARONDES, S. H., and H. D. COHEN, 1968a. Memory impairment after subcutaneous injection of acetoxycycloheximide. *Science* (*Washington*) 160: 556–557.

BARONDES, S. H., and H. D. COHEN, 1968b. Arousal and the conversion of "short-term" to "long-term" memory. *Proc. Nat. Acad. Sci. U. S. A.* 61: 923–929.

BULLOCK, T. H., and G. C. QUARTON, 1967. Simple systems for the study of learning mechanisms. *In* Neurosciences Research Symposium Summaries, vol. 2 (F. O. Schmitt, T. Melnechuk, G. C. Quarton, and G. Adelman, editors). M. I. T. Press, Cambridge, Massachusetts, pp. 203–327.

BUREŠ, J., and O. BUREŠOVÁ, 1963. Cortical spreading depression as a memory disturbing factor. *J. Comp. Physiol. Psychol.* 56: 268–272.

COHEN, H. D., and S. H. BARONDES, 1967. Puromycin effect on memory may be due to occult seizures. *Science* (*Washington*) 157: 333–334.

COHEN, H. D., and S. H. BARONDES, 1968. Effect of acetoxycyclo-heximide on learning and memory of a light-dark discrimination. *Nature* (*London*) 218: 271–273.

COHEN, H. D., F. ERVIN, and S. H. BARONDES, 1966. Puromycin and cycloheximide: Different effects on hippocampal electrical activity. *Science* (*Washington*) 154: 1557–1558.

DEUTSCH, J. A., M. D. HAMBURG, and H. DAHL, 1966. Anticholinesterase-induced amnesia and its temporal aspects. *Science* (*Washington*) 151: 221–223.

EISENSTEIN, E. M., 1967. The use of invertebrate systems for studies on the bases of learning and memory. *In* The Neurosciences: A Study Program (G. C. Quarton, T. Melnechuk, and F. O. Schmitt, editors). The Rockefeller University Press, New York, pp. 653–665.

FLEXNER, L. B., and J. B. FLEXNER, 1968a. Intracerebral saline: Effect on memory of trained mice treated with puromycin. *Science* (*Washington*) 159: 330–331.

FLEXNER, L. B., and J. B. FLEXNER, 1968b. Studies on memory: The long survival of peptidyl-puromycin in mouse brain. *Proc. Nat. Acad. Sci. U. S. A.* 60: 923–927.

FLEXNER, J. B., L. B. FLEXNER, and E. STELLAR, 1963. Memory in mice as affected by intracerebral puromycin. *Science* (*Washington*) 141: 57–59.

GALAMBOS, R., 1967. Brain correlates of learning. *In* The Neurosciences: A Study Program (G. C. Quarton, T. Melnechuk, and F. O. Schmitt, editors). The Rockefeller University Press, New York, pp. 637–643.

GAMBETTI, P., N. K. GONATAS, and L. B. FLEXNER, 1968. The fine structure of puromycin-induced changes in mouse entorhinal cortex. *J. Cell Biol.* 36: 379–390.

GELLER, A., and M. E. JARVIK, 1968. The time relations of ECS induced amnesia. *Psychonomic. Sci.* 12: 169–170.

GELLER, A., F. ROBUSTELLI, S. H. BARONDES, H. D. COHEN, and M. E. JARVIK, 1969. Impaired performance by post-trial injections of cycloheximide in a passive avoidance task. *Psychopharmacologia* 14: 371–376.

GLASSMAN, E., 1969. The biochemistry of learning: An evaluation of the role of RNA and protein. *Annu. Rev. Biochem.* 38: 605–646.

HYDÉN, H., 1967. RNA in brain cells. *In* The Neurosciences: A Study Program (G. C. Quarton, T. Melnechuk, and F. O. Schmitt, editors). The Rockefeller University Press, New York, pp. 248–266.

JARVIK, M. E., 1968. Consolidation of memory. *In* Psychopharmacology: A Review of Progress (D. H. Efron et al., editors). U. S. Government Printing Office, Washington, PHS Publ. 1839, pp. 885–889.

KANDEL, E. R., 1967. Cellular studies of learning. *In* The Neurosciences: A Study Program (G. C. Quarton, T. Melnechuk, and F. O. Schmitt, editors). The Rockefeller University Press, New York, pp. 666–689.

KOPP, R., Z. BOHDANECKY, and M. E. JARVIK, 1966. Long temporal gradient of retrograde amnesia for a well-discriminated stimulus. *Science* (*Washington*) 153: 1547–1549.

LIVINGSTON, R. B., 1967. Brain circuitry relating to complex behavior. *In* The Neurosciences: A Study Program (G. C. Quarton, T. Melnechuk, and F. O. Schmitt, editors). The Rockefeller University Press, New York, pp. 499–515.

McGaugh, J. L., 1966. Time-dependent processes in memory storage. *Science (Washington)* 153: 1351–1358.

McGaugh, J. L., 1969. Facilitation of memory storage processes. *In* The Future of the Brain Sciences (S. Bogoch, editor). Plenum Press, New York, pp. 355–370.

McGaugh, J. L., and M. J. Herz, 1970. Controversial issues in consolidation of the memory trace. *In* Memory Consolidation: Progress and Prospects (J. L. McGaugh and M. J. Herz, editors). Albion Press, San Rafael, California (in press).

Miller, N. E., 1967. Certain facts of learning relevant to the search for its physical basis. *In* The Neurosciences: A Study Program (G. C. Quarton, T. Melnechuk, and F. O. Schmitt, editors). The Rockefeller University Press, New York, pp. 643–652.

Quarton, G. C., 1967. The enhancement of learning by drugs and the transfer of learning by macromolecules. *In* The Neurosciences: A Study Program (G. C. Quarton, T. Melnechuk, and F. O. Schmitt, editors). The Rockefeller University Press, New York, pp. 744–755.

Rosenbaum, M., H. D. Cohen, and S. H. Barondes, 1968. Effect of intracerebral saline on amnesia produced by inhibitors of cerebral protein synthesis. *Commun. Behav. Biol., Part A,* 2: 47–50.

Squire, L., and S. H. Barondes, 1970. Actinomycin-D: Effect on memory at different times after training. *Nature (London)* (in press).

Weissman, A., 1967. Drugs and retrograde amnesia. *Int. Rev. Neurobiol.* 10: 167–198.

# 27 Protein Changes in Nerve Cells Related to Learning and Conditioning

## HOLGER HYDÉN and PAUL W. LANGE

This paper deals mainly with the biochemical changes observed in neurons during three different learning experiments. First, however, some experiments are described that show differences in the protein composition of nerve and glial cells. Of the learning experiments, the first is on instrumental learning in rats, in which changes took place in the synthesis of three acidic neuronal proteins and in the base composition of neuronal RNA. We present the arguments that these changes are specifically related to training and are an expression of increased gene activation. In the next study, we describe the protein changes observed in brain cells during simple sensory conditioning in rats, and argue that these are caused by an increased level of attention rather than to learning per se. Finally, we report some RNA data from neurons in monkeys performing a visual discrimination test.

### Neuronal and glial proteins

Five years ago, Moore (1965) and Moore and MacGregor (1965) described a brain-specific protein called S100, so

HOLGER HYDÉN and PAUL W. LANGE Institute of Neurobiology, Faculty of Medicine, University of Göteborg, Göteborg, Sweden

named because it is soluble in saturated ammonium sulfate. It is an acidic protein, has a molecular weight of about 20,000, constitutes 0.1 per cent of the brain proteins, and moves close to the anodal front in electrophoresis. In the rat, it develops 12 days after birth, and is present only in nervous tissue. Thirty per 100 moles of its amino acids are acidic. It contains 30 per cent glutamic acid and no tryptophan. S100 can be further separated into at least three fractions, of which two have a high turnover and react immunologically with antiserum against S100 (McEwen and Hydén, 1966). The S100 protein is not linked to carbohydrates (Figure 1).

Hydén and McEwen (1966) have shown by antiserum-precipitation reactions supported by the Coons' technique (1957) that S100 is mainly a glial protein, which in nerve cells is found only in the nuclei. Recently, Benda and collaborators (1968) confirmed its presence in glia and showed its tenfold growth in a clonal strain of glial tumors. Perez and Moore (1968) have also presented evidence that S100 is mainly a glial protein. Moore and Perez (1966, 1968) have described another brain-specific protein, which seems to be localized exclusively in the nerve cells and has been named the 14-3-2 protein.

There is evidence for the existence of still other brain-specific, soluble proteins. MacPherson and Liakopolou (1965) have described one in the β-globulin range; Kosinski and Grabar (1967) have described five soluble proteins; and Warecka and Bauer (1966, 1967) recently described an α-glycoprotein rich in neuraminic acid, which develops three months after birth in man and is probably derived from glia. Bennett and Edelman (1968) have purified and characterized still another acidic brain-specific protein.

## An immunological study of Deiters' nucleus

We have examined the properties of antibodies prepared against neurons and glia (Mihailović and Hydén, 1969) obtained from Deiters' nucleus in a continuing attempt to identify brain-specific proteins in them. The antigens in brain cells presumably number in the order of hundreds; Huneeus-Cox and his coworkers (Huneeus-Cox, 1964; Hunneeus-Cox et al., 1966), for instance, successfully prepared antisera against 11 antigens in preparation of squid axoplasm that did not include the external membranes. In our study, antigen consisted of glial material dissected from the Deiters' nucleus of the rabbit by the freehand technique we have already described (Hydén, 1959). The dissection was carried out at 4° C, with careful removal of capillaries and nerve-cell bodies and processes; in this way, 3.2 milligrams of Deiters' nucleus glia were collected from 40 rabbits. The other antigen consisted of 1.3 grams of whole Deiters' nucleus, containing both neurons and glia, dissected from 100 rabbits.

Each of these antigens was homogenized and mixed with both complete and incomplete Freund's adjuvant. A group of six rhesus monkeys, each weighing 3 to 3.5 kilograms, was injected intramuscularly with 0.6 milliliter of one or the other emulsion once a week for four weeks. None ever showed neurological symptoms or signs of tuberculosis. The animals were bled after one week. On day 44, each monkey received a booster injection of 0.2 milliliter of its antigen emulsion precipitated with $Al_2(SO_4)_3$, and was bled one week later. These sera were tested on Ouchterlony plates against extracts of glia and of Deiters' nucleus, and their precipitation activities against sucrose-Triton X-100 extracts of both glia and of Deiters' nucleus material were also evaluated. In addition, we used the micromethod for double diffusion in one dimension in glass capillaries previously described (Hydén and McEwen, 1966) as an assay system. We also used the Coons' (1957) multiple-layer, indirect method for immunofluorescence applied to cryostat sections through the Deiters' nucleus, with evaluation of the specific fluorescence appearing in the nerve and glial cells. Some samples of the antisera were absorbed in two or three steps with sucrose-Triton X-100 homogenates of glia and of rabbit spleen; others were twice absorbed with rabbit spleen and then absorbed with glia.

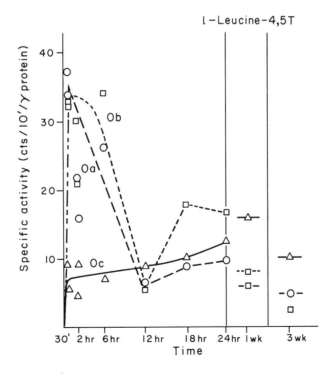

FIGURE 1 Specific radioactivity of bands Oa, Ob, and Oc separated on 11.2 per cent polyacrylamide gels as a function of time between injection of the isotope and sacrifice. Radio-activity was determined after combustion of slices of the polyacrylamide gels by liquid scintillation counting. Isotope: 1-leucine-4,5 T.

TABLE I

*A. Gel precipitation reactions (+) between anti-Deiters' nucleus antiserum (1:512) and a homogenate of Deiters' nucleus.*

| Antigen | Reaction | Antigen | Reaction | Antigen | Reaction | Antigen | Reaction |
|---------|----------|---------|----------|---------|----------|---------|----------|
| ($\mu$g/$\mu$l) | | | | | | | |
| 8.20 | − | 0.80 | + | 0.30 | + | 0.05 | − |
| 4.10 | − | 0.50 | + | 0.20 | + | 0.02 | − |
| 2.10 | − | 0.40 | + | 0.10 | + | 0.01 | − |
| 1.00 | − | | | | | | |

*B. Gel precipitation reactions (+) between anti-Deiters' glia antiserum (1:512) and an antigen homogenate of Deiters' glia.*

| Antigen | Reaction | Antigen | Reaction |
|---------|----------|---------|----------|
| ($\mu$g/$\mu$l) | | | |
| 0.67 | + | 0.08 | − |
| 0.60 | + | 0.04 | − |
| 0.16 | + | 0.02 | − |

TABLE II

*Gel precipitation reactions (+) between anti-Deiters' glia antiserum and 0.9 $\mu$g of protein extracted from nerve and glia cells dissected from Deiters' nucleus. Normal serum controls negative in each case.*

| Antiserum Dilution | Protein from | |
|--------------------|--------------|--------------|
| | Nerve cell | Glia cells |
| 1:64 | − | + |
| 1:128 | − | + |
| 1:256 | − | + |
| 1:512 | − | + |
| 1:1024 | − | − |
| 1:2048 | − | − |

Tables I to IV summarize some results of these studies. Both the anti-Deiters' nucleus and the antiglia sera formed well-defined precipitates with microgram per microliter amounts of their respective antigens (Table I). Table II shows that the antiglia serum formed precipitates with the glia but not with nerve cells obtained from Deiters' nucleus, and that no precipitates formed when normal rabbit serum was used against these antigens.

Table III shows the results of an antigen dilution study: homogenates of isolated nerve cells and of the same volumes of glial cells were tested against the antiglia antiserum in the dilution ratio of 1 to 512. Even when 300 isolated nerve cells were used, no precipitation was obtained, but glial homogenates gave well-defined precipitates.

TABLE III

*Precipitation between anti-Deiters' glia antiserum (1:512) and diluted homogenates of Deiters' nerve cells and of glia. Method: double diffusion in glass capillary. Neuronal protein estimate based on 12,000 $\mu\mu$g of protein per cell. Glial protein per unit volume estimated at 50 per cent neuronal.*

| Deiters' Neurons | | | Deiters' Glia (same volumes as nerve cells) | |
|------------------|------------------|------------------|------------------|------------------|
| Number of nerve cells | Calculated protein in $10^{-6}$g | Precipitates obtained | Calculated protein in $10^{-6}$g | Precipitates obtained |
| 300 | 3.6 | 0 | 1.8 | 2 |
| 150 | 1.8 | 0 | 0.9 | 2 |
| 70 | 0.9 | 0 | 0.45 | 2 |
| 60 | 0.72 | 0 | 0.36 | 1 |
| 30 | 0.36 | 0 | 0.18 | 1 |
| 15 | 0.18 | 0 | 0.09 | 0 |
| 6 | 0.09 | 0 | 0.045 | 0 |
| 3 | 0.045 | 0 | 0 022 | 0 |

Precipitates were obtained when the anti-Deiters' glia antiserum was tested against glia dissected from other parts of the brain, e.g., from the hypoglossal nucleus and from the spinal cord and cerebral cortex; but none appeared against homogenates of motor neurons, pyramidal nerve cells of the hippocampus, and granular cells from the cerebellum, all containing from 3.5 to 0.01 micrograms of protein per microliter.

Antiserum against the whole Deiters' nucleus gave two

TABLE IV

*Number of precipitation lines after absorption of anti-Deiters' nucleus antiserum. Antigen: homogenates from 120 isolated nerve cells and corresponding amount of glia containing 1.6 μg used in each case. All dilutions tested (1:2, 1:4, 1:8, 1:16) gave the same result.*

|  | Protein from | |
|---|---|---|
|  | Nerve Cells | Glia Cells |
| Unabsorbed | 2 | 2 |
| Absorbed with glia | 1 | 0 |
| Absorbed with spleen | 1 | 0 |

precipitation lines with both glia and nerve cells as antigens. However, when this antiserum was absorbed with glia or with spleen, only the nerve-cell homogenates gave precipitates (Table IV).

The results with the fluorescence technique matched those obtained with the immunodiffusion technique, as summarized in these Tables. Experiments were carried out according to the multiple-layer method of Coons (1957, 1958). Cryostat sections 5μ thick, cut through the lateral vestibular nucleus, were first dried (in some cases left overnight in the refrigerator at +4° C) and subsequently fixed in cold acetone for 30 seconds. After being washed for five minutes in the buffered saline, the sections were covered for 30 minutes with the antiserum to be investigated. After thorough washing (three to five minutes) in a cold pH 7.1, phosphate-buffered saline, a goat-antimonkey globulin-gamma-globulin conjugated with fluorescein isothiocyanate (Difco product) was applied to the sections for 30 minutes, and the excess was removed by repeated washing (again three to five minutes) in the buffer. Control sections treated only with normal monkey serum and conjugated gamma globulin were used regularly with each experimental series. The sections were finally mounted in a small drop of buffered glycerol (nine parts glycerol, one part buffered saline) under a coverslip and immediately observed in a Zeiss fluorescence microscope. Photographs were taken with high-speed Ektachrome film, using exposures varying from one to five seconds. The sections then were restained with Ehrlich's hematoxylin eosin. The fields previously photographed were identified under the light microscope and rephotographed in black and white, thus enabling comparisons to be made of conventional microscopical appearances of structural details with fluorescent pictures.

Antiserum to whole Deiters' nucleus when absorbed with glia, with spleen, or with both gave no fluorescence in glial cells, but did so in nerve cells; this fluorescence was localized in the outer rim of the cell body and in the dendritic processes, which could be traced through the section by their brilliant fluorescence, suggesting that the antigens were localized in the plasma membranes. Furthermore, the reaction was positive in the nerve-cell nucleus, but not at the site of the nucleolus.

From these observations, the following conclusions can be made. Neurons and glia differ with respect to antigen composition. This finding is interesting, in that both types of cells develop from the type of ectodermal stem cell. The question, then, is whether the antigens are specific for the type of cell in which they occur. Judged by the absorption experiments, the neuronal antigens seem to be specific for that type of cell. It should be noted that the neuron-specific antigens were concentrated in the processes, in the outermost part of the cell, and in the nucleus—especially the nuclear membrane.

On the other hand, it seemed clear that the antigens in the glia were not glia-specific. They were localized throughout the cell body, but not in the nucleus. Any immunological organ specificity seems, therefore, to be due to the presence of antigens in the neurons.

Conversely, glial cells possess the acidic S100 protein that is confined only to the nerve tissue and is shared with neurons. The presence of this antigen cannot be demonstrated by the method used in this study for preparing immune sera (Levine and Moore, 1965).

## Altered protein synthesis during training

EXPERIMENT 1. HANDEDNESS TRANSFER *Incorporation of* $^3$*H-leucine into the acidic protein fractions 4 and 5.* Most rats are left-handed or right-handed when permitted to perform complicated paw movements, and they can be induced to transfer this handedness in the retrieval of food (Hydén and Egyházi, 1964). Thus, when given access to a glass tube partly filled with small protein pellets, they reach down into the tube to retrieve the pellets, one by one; when first tested, clear preference for the left or right forepaw is found in 92 per cent of the animals. When the glass tube is placed next to a wall to prevent the use of the preferred paw (e.g., wall to the right for a right-handed rat), the animal begins to retrieve the food pellets with the nonpreferred paw. When given two training sessions of 25 minutes per day in this situation, the number of pellets obtained per day increases linearly up to day 8. In the experiment to be described next, such performance curves were obtained on all rats; they were similar to the curve shown in Figure 2. Once learned, this new behavior is retained for a long time. No stress (surgical, mechanical, or shock) is applied to induce the new behavior, so this procedure has distinct advantages over other behavioral experiments with rats.

To trace protein synthesis during this learning, the rats, under halothane anesthesia, received 60 μCi of $^3$H-leucine

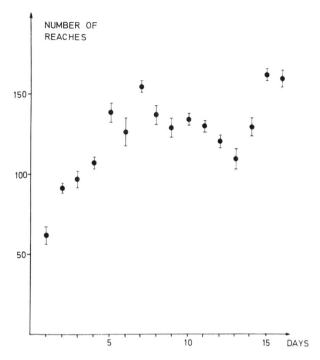

FIGURE 2  Performance of reversal of handedness as the average number of successful reaches for 10 rats trained twice daily, 25 minutes per day, for 16 days.

in 60 μl intraventricularly in each hemisphere half an hour before the final training period. Hippocampal nerve-cell samples were then taken for analysis 15 minutes after the last training period. Nerve cells of the hippocampus were selected because (1) several clinical and behavioral studies have shown the importance of this structure for the formation of long-term memory (see, e.g., Meissner, 1966; Ojemann, 1966; Penfield, 1952; Burešová et al., 1962; Adey et al., 1963); (2) its bilateral destruction results in severe defects in learning and formation of memory (Meissner, 1966; Ojemann, 1966); (3) during attentative learning, impedance changes occur in the hippocampus (Adey et al., 1963); and (4) no memory is formed if protein synthesis in the hippocampus is inhibited by 90 per cent (Barondes and Cohen, 1967; Flexner and Flexner, 1968).

The micromethod used for protein analysis was as follows. About 300 pyramidal cells from the CA3 region of the hippocampus, separately dissected out freehand on a cooling table, were analyzed for protein by a technique already described (Hydén and Lange, 1968a). (One may ask why it seems necessary to struggle with such minute amounts of material and with the dissection of such small areas within the brain. As an answer, we advocate the view that altered synthesis, if any, is more likely to be found in a uniform cell population from an area that is clearly involved functionally. In a mixed cell population from a whole brain such changes disappear easily in the background noise.) An outline of this procedure is given in Figure 3. The left side of the scheme gives the various steps leading to the value of the specific activities per amount of protein in each protein microfraction.

These specific activity values vary because of variation in the local concentration of ³H-leucine, so the correction procedure shown on the right side of Figure 3 was applied in order to allow a comparison of values from identical parts of both hemispheres or from different animals. This was accomplished in a separate experiment, in which the relation between the uncorrected specific activities and the concentration of the free ³H-leucine in the hippocampal nerve cells was determined and found to be linear. Dividing the specific activity values obtained by the values of the ³H-leucine concentration determined locally permitted all specific activities to be compared at a uniform free ³H-leucine concentration.

In an earlier study (Hydén and Lange, 1968b), the incorporation of ³H-leucine in the CA3 nerve-cell protein fractions 4 and 5 (Figure 4) was evaluated on the fifth day of training, i.e., on the linear, increasing part of the performance curve. The specific activities of these protein fractions were significantly greater in trained rats than in control rats of the same age (P < 0.005), and there was some evidence for higher incorporation in the hippocampus contralateral to the training paw.

Presumably, protein fractions 4 and 5 each contain several species of proteins and there is no reason, as yet, to believe that the qualitative characteristics of the protein formed during training is specific for the process, as no data exist on the composition of these proteins. Nevertheless, it is pertinent to ask whether the increased synthesis of fractions 4 and 5 is specific for the training.

This we attempted to do in the present study by measuring fractions 4 and 5 in rats given five, seven, and 10 training sessions according to the following schedule. A group of 24 rats was given five days of training. Five of these (Group 1) received ³H-leucine prior to the last training session, and the CA3 hippocampal nerve-cell material was taken for analysis as described above. The remaining animals were placed in cages, and given food and water ad libitum. After 14 days, they were all subjected to two training periods of 25 minutes each; five of these animals (Group II) were given ³H-leucine, and the CA3 nerve-cell material was taken for analysis. The remaining rats (Group III) were returned to their cages for 14 additional days, then trained for three days with two training periods per day (each of 25 minutes), and, after a ³H-leucine injection, their hippocampal brain cells were taken for analysis. The controls were untrained rats

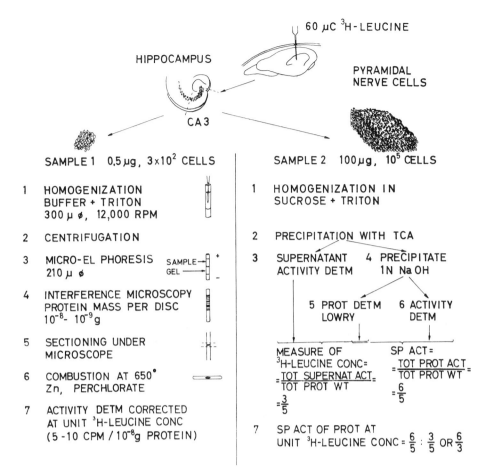

FIGURE 3  Outline of the microdiscelectrophoresis procedure for separation of $10^{-7}$ to $10^{-9}$ g of protein, and evaluation of incorporation of radioactive amino acid into the individual fractions. Volume of sample is proportional to weight of total protein in the sample. *Abbreviations: activity detm, activity determination; Prot detm, protein determination; Sp act, specific activity; sp act of prot at, specific activity of protein at unit a.s.o.; tot supernat act, total supernatant activity.*

of the same age; 50 per cent of them were litter mates of the experimental animals.

The performance of the rats in the three groups is shown in Table V. Table VI demonstrates that the specific activities of protein fractions 4 and 5 were significantly increased after five and seven training days but *not* after 10. We have shown (Hydén and Lange, 1968b) that the corrected specific activities (counts per minute per microgram) of protein fractions 4 and 5 differ from the corresponding values, as a result of a more refined separation technique, which allowed a better separation of smaller amounts of the protein sample. The values found for the unseparated protein were, of course, not affected. The unseparated protein of the CA3 pyramidal nerve cells behaved like that of fractions 4 and 5 in showing higher incorporation values in Groups I and II, but not in Group III (Table VI).

After a chance observation, we made a study of the incorporation of $^3$H-leucine by two rats that were subjected to half of the initial training time allowed the other rats. Whereas Group I had 10 training sessions (two of 25 minutes each for five days), these two rats had five (two of 25 minutes each on days 1 and 2; one of 25 minutes on day 3), and on day 14 they were given only a single 25-minute training session before being killed for analysis. On training-day 3 they had made 120 reaches; in the final session, 100 reaches. These animals (Group IIA in Table VI) gave a greater protein-synthesis response than did those receiving the longer training.

Because the incorporation of $^3$H-leucine into protein fractions 4 and 5 increased significantly during the training, it became important to know the relation of these fractions to the remainder of the nerve-cell protein in terms of $^3$H-

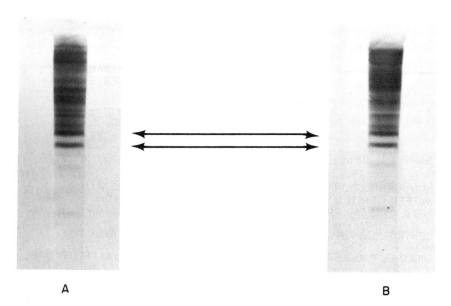

A                                        B

FIGURE 4 Protein of pyramidal nerve cells of the hippo-campus, CA3 region, separated on polyacrylamide gels, 400 $\mu$ in diameter, and stained with amido black. Fractions 4 and 5 from the anodal front are indicated by arrows.

TABLE V

*Average number of pellets obtained per day by rats using the nonpreferred paw.*

|  | Number of rats | Pellets |
|---|---|---|
| Group I (performance on day 5) | 24 | 100 |
| Group II (performance on day 14) | 19 | 90 |
| Group III (performance on day 30) | 14 | 90 |

TABLE VI

*Corrected specific activities of hippocampal CA3 nerve-cell proteins, both the unseparated and fractions 4 and 5. Corrected specific activity refers to counts per minute per microgram $\pm$ standard error of the mean.*

|  | Fractions 4 and 5 | | | Unseparated protein | |
|---|---|---|---|---|---|
|  | No. of rats | No. of gels | Corrected specific activity | No. of samples | Corrected specific activity |
| *Group I* (training 5 days) | 5 | 10 | 3.3 ± 0.40 | 10 | 14.20 ± 1.90 |
| *Group II* (resumed training day 14) | 5 | 10 | 3.9 ± 0.48 | 10 | 15.50 ± 1.90 |
| *Group IIA* (half training time) | 2 | 5 | 13.0 ± 0.60 | | |
| *Group III* (resumed training day 30) | 14 | 35 | 1.8 ± 0.17 | 28 | 5.10 ± 0.58 |
| *Control* | 10 | 24 | 1.5 ± 0.16 | 20 | 6.00 ± 0 92 |

leucine incorporation. Protein of the CA3 nerve cells from Group I rats (trained for five days) was, therefore, separated on polyacrylamide gels, divided in four parts, and the radioactivity was determined in each part; as can be seen in Figure 5, the radioactivity of protein fractions 4 and 5 is relatively high.

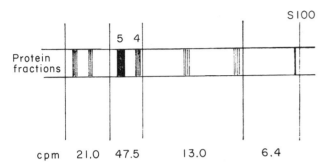

FIGURE 5  Gels 400 $\mu$ in diameter, containing separated $^3$H-labeled pyramidal nerve-cell protein from the CA3 region, were cut in four pieces, as indicated, and the radioactivity (numbers below) was determined as counts per minute after combustion. Note that the radioactivity in protein fractions 4 and 5 is relatively great.

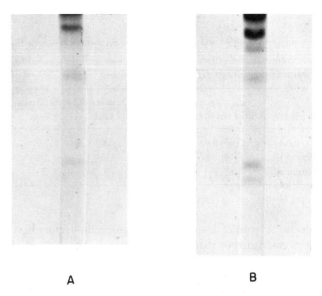

FIGURE 6  Photographs of nerve-cell protein of the hippocampus, CA3 region, separated on polyacrylamide gels 400 $\mu$ in diameter and stained with amido black. A. From control rat using one forepaw; B. From rat on day 5 of training using the nonpreferred paw. The acidic proteins migrate toward the bottom of the gel.

INCREASED SYNTHESIS OF S100 PROTEIN  Both the electrophoretic pattern of the soluble CA3 nerve-cell protein isolated in the experiment just described and microdensitometer recordings (Figure 7) made of 75 protein separations stained with brilliant blue showed two protein bands at the front in the trained rats compared with only one in the controls (Figures 6 and 7, Table VII). This protein fraction of the controls gave a positive immunological reaction when treated with antiserum against the S100 protein. Figure 7 shows that the amount of protein contained in the two anodal bands from trained rats was greater than the amount of protein contained in the one band from the controls. Furthermore, when gel cylinders (from experimental rats) with two anodal front bands were immersed in saturated $(NH_4)_2SO_4$ solution for 20 minutes, the band closest to the anode disappeared, identifying it as S100. These facts—the electrophoretic localization of the new protein fraction, its disappearance in saturated $(NH_4)_2SO_4$, and the increased amount of anodal protein in the trained rats—suggest that brain-specific S100 protein increased in amount during training. This S100 protein was presumably localized in the nuclei of the hippocampal neurons.

Our interpretation of the result given above is that the increase of the S100 protein during reversal of handedness specifically relates the S100 protein to the learning processes. As we pointed out above, however, training involves several factors not related to learning per se. In experiments

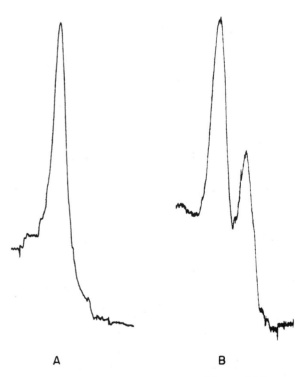

FIGURE 7  Microdensitometric recordings of the anodal front protein fractions as shown in Figure 6. A. Control. B. Trained animal.

## TABLE VII

*Frequency of single- and double-front anodal protein bands in the electrophoretic pattern of 75 polyacrylamide gels from 23 rats; Group II and Group III as in Tables V and VI.*

|  | Number of rats | Number of gels tested | Number of Protein Bands | |
|---|---|---|---|---|
|  |  |  | One | Two |
| Control | 7 | 19 | 20 | 0 |
| Group II | 4 | 15 | 5 | 10 |
| Group III | 12 | 41 | 20 | 20 |

on the reversal of handedness, the unspecific factors have been eliminated or reduced to a minimum. The motor and sensory activity, attention, motivation, and reward are equated among the experimental and control animals, and the stress involved in reversal of handedness is minimal. In view of these considerations, we performed the following experiments, which specifically relate the S100 protein during reversal of handedness to learning per se.

A group of eight rats was trained during two 25-minute periods per day for three days. Between the first and second training session of the fourth day, half of the rats were injected intraventricularly on both sides with $2 \times 25$ µg of antiserum against S100 in $2 \times 25$ µl. The other four were similarly injected with the same amount and volume of antiserum against rat γ-globulin. During injection, the rats were lightly anesthetized with halothane. The animals were trained in a second session on the fourth day 45 minutes after the injections. After this treatment, the rats were trained for three additional days in two 25-minute sessions per day. The results are presented in Figure 8. The number of reaches is plotted against the number of training days. Before injection of antisera, all rats followed an identical performance curve. After the injection of antiserum against rat γ-globulin, these rats followed a performance curve that was an extrapolation of the performance curve before the injection.

By contrast, the performance of rats injected with antiserum against S100 protein did not increase, i.e., the number of reaches made per day remained the same as those immediately before the injection. As is seen from the curves in Figure 8, the difference in number of reaches between the two groups of injected rats is clearly significant. Thus, the S100 protein is specifically correlated to learning processes.

At this point, it seems appropriate to comment on the protein changes in the rats undergoing intermittent training (Groups I-II, Table V). These rats performed well on day 5, day 14, and day 30, i.e., when they had received training for five, seven, and 10 days. If the increased synthesis in the hippocampal nerve cells had been an expression of increased and sustained neural function, then the values found on the

fifth and fourteenth days would presumably also be found when the rats received three additional days of training at 30 days. The fact that the incorporation values at 30 days do not differ from the controls is a strong indication that the observed increase in synthesis is to be correlated to learning processes occurring during the training. We suggest the interpretation that, when the novelty of the task has passed, the hippocampal nerve cells cease to respond with increased synthesis of this type of protein (S100) which is brain-specific and thus can be expected to mediate specific brain functions. It is an open question whether a specific synthetic response takes place in other parts of the brain at this time.

EXPERIMENT 2. SENSORY CONDITIONING TEST IN RATS
*Incorporation of ³H-leucine into neuronal protein.* This experiment was designed to limit participation of such factors as motivation, motor activity, and reinforcement in an experiment involving change of behavior (Hydén, Lange, and Wood, in preparation). For that purpose we measured protein synthesis in rats subjected to paired and unpaired tone and light stimuli, and to light stimulus alone. A total of 80 rats were used in pilot tests, behavioral checks, and for final experiments. A rat was placed in a cage with a wired floor in a sound-absorbing, dimly lit room for 20 minutes prior to the experiment. One group of rats received an acoustical signal of 1000 Hz (conditioning stimulus), followed by a visual (unconditioned) stimulus. The tone and light stimuli both lasted 0.2 second and were presented automatically at a frequency of six per minute. Another group of rats was used as behavioral controls. In this

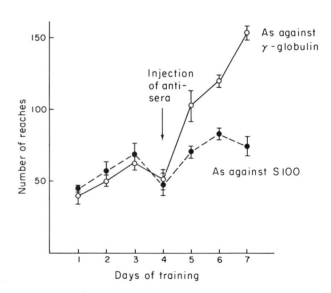

FIGURE 8 Performance curves of two groups of rats, four in each. One group injected with antiserum against S100; the other, with antiserum against rat γ-globulin on the fourth day of training.

286    DETERMINANTS OF NEURAL AND BEHAVIORAL PLASTICITY

group, the tone and light stimuli were followed by an electric shock; in 10 trials the rats learned to jump up to a shelf to escape the electric shock (criterion, eight of 10) when the tone-light stimuli were presented. This test, a type of sensory conditioning described and discussed by Morrell (1967), demonstrated that a linkage had been formed between the two sensory areas. A third group of rats received tone and light stimuli distributed at random. A fourth group received only light stimuli at a frequency of six per minute.

The experimental rats were first injected bilaterally and intraventricularly with $^3$H-leucine during light fluothane anesthesia, then given sound-light or sound-light-at-random stimulation for 15 minutes. The time lapse from the last injection to the sampling of brain material was 40 minutes.

Both types of stimulation *increased* the incorporation of $^3$H-leucine into the neuronal protein of the hippocampus but *decreased* it in the visual cortex (Table VIII). Control

TABLE VIII

*Tone-light conditioning in rats. Incorporation of $^3$H-leucine into protein isolated from hippocampal nerve cell and visual cortex of rats. Data in cpm total protein per cpm total supernate.*

| Stimuli | Number of samples | Incorporation of $^3$H-leucine | |
|---|---|---|---|
| | | Visual cortex | Hippocampus |
| Paired | 8 | 3.49 ± 0.23 | 3.43 ± 0.41 |
| Unpaired | 8 | 3.50 ± 0.23 | 4.45 ± 0.56 |
| Light only | 12 | 3.48 ± 0.18 | — |
| Control | 8 | 4.46 ± 0.41 | 2 86 ± 0.41 |

animals given a light stimulus alone showed decreased incorporation in the visual cortex.

Two findings in this sensory conditioning experiment seem to exclude the possibility that the protein changes were correlated with learning processes during the conditioning. The first is that a light stimulus alone gave the same incorporation values for the visual cortex as did the paired and unpaired tone-light stimuli. The second finding is that the tone-light stimuli distributed at random gave the highest incorporation of $^3$H-leucine into the protein of the hippocampal nerve cells. Therefore, the conclusion is that the protein changes observed during the conditioning presumably are expressions of increased attention or orientation reflexes. In all three types of sensory experiments, the finding that the incorporation values for the cells of the visual cortex were lower than those of the controls agrees with electrophysiological observations.

## Discussion

The aim of the studies reported here has been to correlate protein changes in nerve cells (and glial cells) in specific parts of the brain with learning processes that occur during training. It seems evident that mapping the areas that respond with defined changes in protein fractions during behavioral experiments is a prerequisite for a comprehensive theory on the mechanisms relating macromolecules in brain cells to storage and retrieval of information. The observations relating behavioral responses to synthesis and composition of RNA in nerve and glial cells (Hydén and Egyházi, 1962, 1963, 1964; Hydén and Lange, 1965; Shashoua, 1968), taken in conjunction with the observations on protein reported here, may be considered as a beginning that may eventually form the basis of such a theory.

It is interesting that the immunological study reported here brought out such clear differences in the antigen composition of neurons and glia. This finding brings into question the matter of transfer of RNA from glia to neuron, for which view there exists some evidence (Hydén and Lange, 1966). Such transfer could still take place, even if the neuronal protein programed by the glial RNA were not antigenic for rabbits challenged by our technique. S100 protein, for instance, is not antigenic unless injected under special circumstances.

This S100 protein seems, however, to be definitely linked to learning, as is demonstrated especially well in the experiment in which antiserum against S100 impaired learning, but that against γ-globulin did not (Figure 8). All factors, including stress, were identical for the control and experimental rats in this study. Before and after the antisera injections, all were subjected to the same training program, and the injections into the brain ventricles were carried out under identical conditions. Additional food was supplied to rats receiving S100 protein antiserum to compensate for the different amounts of reinforcement obtained. The result showing that only the S100 protein antiserum inhibited further learning seems clearly to link this brain-specific protein to learning processes that take place during training.

The nerve-cell proteins 4 and 5 are acidic, even though their composition is still unknown. Their response during intermittent training, which was spread over one month, seems significant and is pertinent for the interpretation that the synthetic response was linked to learning processes within the training. The hippocampal nerve cells did not respond with increased synthesis of these proteins after the last training sessions (a month after the initial training), which excludes the possibility that the increased protein synthesis during the two previous training sessions was merely an expression of increased motor activity, sensory activity, attention, or change in age.

The protein changes in the transfer of handedness experiment, which was instrumental learning, can be compared with those in the sensory-sensory experiments, which was classical conditioning. The instrumental learning is a complicated type involving, among other things, motor-sensory activities, motivation, and attention. The acidic (including S100) protein in hippocampal nerve cells rises during acquisition of behavior in this case, but during sensory conditioning no systematic change related to learning can be seen in either the hippocampal or the cortical cells. The protein-synthetic response of hippocampal cells in this type of classical conditioning cannot, therefore, be equated with that taking place during instrumental learning. In sensory conditioning, the direction and magnitude of protein synthesis changes in the hippocampus seem only to follow the response of the cells to the sensory input and to have no relation to the learning factors involved.

Thus, both the light-tone and the at-random stimuli gave high incorporation values of $^3$H-leucine into the hippocampal nerve-cell protein. In visual cortex, the incorporation values were equal and lower, compared with the controls for all conditions of stimulation. It may, therefore, be tentatively concluded that, in hippocampal cells, the biochemical activities taking place during instrumental learning differ from those during classical sensory conditioning.

## Acknowledgments

We thank Dr. Robert Galambos for his generous help and advice with the preparation of the manuscript. Antiserum against the S100 protein was obtained from Dr. Lawrence Levine of Brandeis University, Waltham, Massachusetts. The studies reported in this paper were supported by Grant K68-11X-86 from the Swedish Medical Research Council and Riksbankens Jubileumsfond.

### REFERENCES

ADEY, W. R., R. T. KADO, J. DIDIO, and W. J. SCHINDLER, 1963. Impedance changes in cerebral tissue accompanying a learned discriminative performance in the cat. *Exp. Neurol.* 7: 259–281.

BARONDES, S. H., and H. D. COHEN, 1967. Delayed and sustained effect of acetoxycycloheximide on memory in mice. *Proc. Nat. Acad. Sci. U. S. A.* 58: 157–164.

BENDA, P., J. LIGHTBODY, G. SATO, L. LEVINE, and W. SWEET, 1968. Differentiated rat glial cell strain in tissue culture. *Science (Washington)* 161: 370.

BENNETT, G. S., and G. M. EDELMAN, 1968. Isolation of an acidic protein from rat brain. *J. Biol. Chem.* 243: 6234–6241.

BUREŠOVÁ, P., J. BUREŠ, E. FIFKOVÁ, O. VINOGRADOVA, and T. WEISS, 1962. Functional significance of corticohippocampal connections. *Exp. Neurol.* 6: 161–172.

COONS, A. H., 1957. The application of fluorescent antibodies to the study of naturally occurring antibodies. *Ann. N. Y. Acad. Sci.* 69: 658–662.

COONS, A. H., 1958. Fluorescent antibody methods. *In* General Cytochemical Methods, Vol. 1 (J. F. Danielli, editor). Academic Press, New York, pp. 400–422.

FLEXNER, L. B., and J. B. FLEXNER, 1968. Studies on memory: The long survival of peptidyl-puromycin in mouse brain. *Proc. Nat. Acad. Sci. U. S. A.* 60: 923–927.

HUNEEUS-COX, F., 1964. Electrophoretic and immunological studies of squid axoplasmic proteins. *Science (Washington)* 143: 1036–1037.

HUNEEUS-COX, F., H. L. FERNANDEZ, and B. H. SMITH, 1966. Effects of redox and sulfhydryl reagents on the bioelectric properties of the giant axon of the squid. *Biophys. J.* 6: 675–689.

HYDÉN, H., 1959. Quantitative assay of compounds in isolated, fresh nerve cells and glial cells from control and stimulated animals. *Nature (London)* 184: 433–435.

HYDÉN, H., and E. EGYHÁZI, 1962. Nuclear RNA changes of nerve cells during a learning experiment in rats. *Proc. Nat. Acad. Sci. U. S. A.* 48: 1366–1373.

HYDÉN, H., and E. EGYHÁZI, 1963. Glial RNA changes during a learning experiment in rats. *Proc. Nat. Acad. Sci. U. S. A.* 49: 618–624.

HYDÉN, H., and E. EGYHÁZI, 1964. Changes in RNA content and base composition in cortical neurons of rats in a learning experiment involving transfer of handedness. *Proc. Nat. Acad. Sci. U. S. A.* 52: 1030–1035.

HYDÉN, H., and P. W. LANGE, 1965. A differentiation in RNA response in neurons early and late during learning. *Proc. Nat. Acad. Sci. U. S. A.* 53: 946–952.

HYDÉN, H., and P. W. LANGE, 1966. A genetic stimulation with production of adenic-uracil rich RNA in neurons and glia in learning. *Naturwissenschaften* 53: 64–70.

HYDÉN, H., and P. W. LANGE, 1968a. Micro-electrophoretic determination of protein and protein synthesis in the $10^{-9}$ to $10^{-7}$ gram range. *J. Chromatogr.* 35: 336–351.

HYDÉN, H., and P. W. LANGE, 1968b. Protein synthesis in the hippocampal pyramidal cells of rats during a behavioral test. *Science (Washington)* 159: 1370–1373.

HYDÉN, H., and B. S. MCEWEN, 1966. A glial protein specific for the nervous system. *Proc. Nat. Acad. Sci. U. S. A.* 55: 354–358.

KOSINSKI, E., and P. GRABAR, 1967. Immunochemical studies of rat brain. *J. Neurochem.* 14: 273–281.

LEVINE, L., and B. W. MOORE, 1965. Structural relatedness of a vertebrate brain acidic protein as measured immuno-chemically. *Neurosci. Res. Program Bull.* 3 (no. 1): 18–22.

MCEWEN, B. S., and H. HYDÉN, 1966. A study of specific brain proteins on the semi-micro scale. *J. Neurochem.* 13: 823–833.

MACPHERSON, C. F. C., and A. LIAKOPOLOU, 1965. Water soluble antigens of brain. *Fed. Proc.* 24: 176 (abstract).

MEISSNER, W. W., 1966. Hippocampal functions in learning. *J. Psychiat. Res.* 4: 235–304.

MIHAILOVIĆ, L., and H. HYDÉN, 1969. On antigenic differences between nerve cells and glia. *Brain Res.* 16: 243–256.

MOORE, B. W., 1965. A soluble protein characteristic of the nervous system. *Biochem. Biophys. Res. Commun.* 19: 739–744.

MOORE, B. W., and D. MACGREGOR, 1965. Chromatographic and electrophoretic fractionation of soluble proteins of brain and liver. *J. Biol. Chem.* 240: 1647–1653.

Moore, B. W., and V. J. Perez, 1966. Complement fixation for antigens on a picogram level. *J. Immunol.* 96: 1000–1005.

Moore, B. W., and V. J. Perez, 1968. Specific acidic proteins of the nervous system. *In* Physiological and Biochemical Aspects of Nervous Integration (F. D. Carlson, editor). Prentice-Hall, Englewood Cliffs, New Jersey, pp. 343–359.

Morrell, F., 1967. Electrical signs of sensory coding. *In* The Neurosciences: A Study Program (G. C. Quarton, T. Melnechuk, and F. O. Schmitt, editors). The Rockefeller University Press, New York, pp. 452–469.

Ojemann, R. G., 1966. Correlations between specific human brain lesions and memory changes. *Neurosci. Res. Program Bull.* 4 (suppl.): 1–71.

Penfield, W., 1952. Epileptic automatism and the centrencephalic integrating system. *Res. Publ. Ass. Res. Nerv. Ment. Dis.* 30: 513–528.

Perez, V. J., and B. W. Moore, 1968. Wallerian degeneration in rabbit tibial nerve: Changes in amounts of the S-100 protein. *J. Neurochem.* 15: 971–977.

Shashoua, V. E., 1968. RNA changes in goldfish brain during learning. *Nature (London)* 217: 238–240.

Warecka, K., and H. Bauer, 1966. Studies of brain proteins: Immunochemical studies of the water-soluble fractions. *Deut. Z. Nervenheilk.* 189: 53–66.

Warecka, K., and B. Bauer, 1967. Studies on "brain-specific" proteins in aqueous extracts of brain tissue. *J. Neurochem.* 14: 783–787.

# 28 Determinants of Neural and Behavioral Plasticity: An Overview

ROBERT GALAMBOS and STEVEN A. HILLYARD

## Heredity and environment

As I suggested in my introduction to this section, an explanation of behavioral plasticity and stability in terms of transactions between the genetic readout and the environment has its roots in the neuro-embryological problems to which the preceding section addressed itself. An orderly sequence of exchanges between the genome and the milieu characterizes both the embryonic sculpturing of the nervous system and the subsequent emergence and transformation of behavior. We believe that a detailed examination of this continuity between embryogenesis and behavioral ontogeny can shed some light on the heredity-environment interaction and on the meaning of such developmental concepts as maturation and critical period.

First, we should point out that some mechanisms of embryogenesis, including cellular growth, differentiation, and motility, genetic "switching," and myelination, continue to operate during times of behavior modification. As Altman has described (1967), the postnatal mammalian brain exhibits, at least for a time, a continuation of histogenesis, neuronal and glial mitosis, cell migration, and the growth of axonal and dendritic processes. Moreover, the impressive body of evidence showing that macromolecular synthesis and speciation accompany stable alterations of behavior (Glassman, 1969) suggests (subject to further proof) that the orderly switching along gene-directed biosynthetic pathways that characterizes embryonic metamorphosis (Bonner, 1965) also participates in the postnatal restructuring of the central nervous system.

The questions of *how* these postnatal "maturational" events are related to the concomitant emergence of behavior and how vulnerable they are to environmental variation are the central issues. How, for example, does one relate the well-known efficacy of "early experience" in altering the course of behavioral development (Fuller and Walker, 1962; Delius, this volume) to the histogenesis and cellular growth still in progress? Does the incompleteness of neuronal and/or glial differentiation determine a heightened receptivity to certain inputs, which, when supplied, create both an enduring change and a new substrate for subsequent differentiations? An embryological analogue of such an ef-

ROBERT GALAMBOS AND STEVEN A. HILLYARD University of California at San Diego, La Jolla, California

fect was described earlier in this volume by Jacobson: the cells of amphibian retinae retain their receptivity to chemical gradients only as long as neurogenesis continues.

Conceiving the neonatal nervous system to be highly vulnerable to the environment may seem paradoxical, because the neuromaturational processes under way are supposed to be mainly genetically determined and species specific. Some neurological "maturation" does indeed seem relatively impervious to experimental factors (such as that underlying locomotor development), but other types of "maturation" seem to establish critical periods, or sequential stimulus sensitivities, that depend upon and respond to specific schedules of experience. Thus, certain aspects of postnatal behavioral and neurological development can be accelerated by enriched environments (Diamond et al., 1966; Rosenzweig et al., 1969) or depressed by sensory or nutritive deprivation (Wiesel and Hubel, 1965; Himwich, 1969; Dobbing, 1964). Visual experience also evidently alters the regional concentrations of synapses (Cragg, 1967; Bloom, this volume) as well as promoting cellular hypertrophy. It is not known, however, whether the consequences of synaptogenesis and cytogenesis are different, with, say, the former encoding specific experiential parameters and the latter permitting a general increase in level of behavioral function. Perhaps both these aspects of differentiation in the central nervous system are equally involved in representing specific patterned behaviors and stimulus-response contingencies. Hence, possibilities of relating stable species-specific behavior to patterns of neurogenesis and of relating behavioral plasticity and learning to intermittent bursts of renewed gene-directed morphogenesis must await more detailed knowledge of how stimulus-response sequences are represented in the substrate.

EMBRYOLOGICAL MODELS OF BEHAVIORAL DEVELOPMENT Embryological principles and concepts have considerable currency (and varying success) in explaining the ontogeny of behavior in man and animals. The fixed temporal sequence in which behavior develops in an embryo (see Hamburger, this volume) has prompted psychologists to segment the regularities of human intellectual development into an analogous series of behaviorally defined "stages." When it comes to specifying the etiology of this sequence of stages, however, two divergent schools have arisen, one propounding a "maturational unfolding" of genetically predetermined capabilities and the other a historical epigenetic interplay between genetic readout and experience.

The former tradition is exemplified by Gesell, who asserted in his *Embryology of Behavior* (1945), that the "successions of behavior patterns are determined by the innate process of growth called maturation." The role of the milieu in sustaining or modulating the genetically fixed patterns of both morphogenesis and behavioral development was denigrated. The Gestaltist theorists have taxed the genome even further, demanding the encoding of the intricate mental structures underlying insight and abstraction which emerged full blown without practice. Although the strict maturationalist viewpoint has few adherents today, the key issue remains unanswered: How specific are the "innate ideas" represented in the genome?

As Williams related (Chapter 18), the course of intellectual development of the child has been conceived quite differently by Piaget, even though he also adopted an explicit embryological metaphor (the morphogenesis of the mollusk) to explain the process (Piaget, 1952). He firmly rejects the idea that the evolving infantile modes of thought are encoded *a priori* within the maturing biological substrate. Like the shell of the clam, the child's intellectual apparatus is continuously sensitive to the constituency of the environment, progresses stepwise in complexity by means of active interactions with it, and by this process achieves more and more veridical representations of his worldly encounters. In this epigenetic theory, each level of problem-solving capacity emerges by way of a long historical series of interactions of the earlier stages with the environment.

The crux of Piaget's interactionist position can be understood by referring to a current version of the epigenetic embryological model. Most developmental biologists accept that each characterizable stage of embryonic (or cellular) differentiation is the product of its immediate predecessor's lawful interaction with an appropriate environment; no stage springs *de novo* or exclusively from a pattern of genetic activity. In terms of molecular biology, a specific gene-enzyme-structure pattern is active at each stage, making the embryo receptive to a restricted range of environmental substrates. Any such substrates present are incorporated into their waiting molecular niches, changing the morphological structure and initiating a new pattern of gene-enzyme activity, thereby creating the next stage with new substrate preferences. And so forth. The course of development is thus governed by the sequential throwing of genetic switches, the exact pattern of switching being a cofunction of the extant structure and the external impingement. The problem of how a newly activated gene expresses itself as a patterned structural embellishment, however, remains enigmatic.

To translate this scheme into a model of psychological development, it is necessary to replace the "morphological structures" with "psychological structures," which in this case incorporate their environmental assimilanda from sensory inputs rather than from the embryonic broth. A psychological structure is simply a construct designed to predict and interrelate the stage-specific behavioral input-output phenomena; it may be as simple as the arrow in the

S-R equation (Galambos, 1967) or as complex and abstract as one of Piaget's mathematical lattice structures.

According to Piaget, only the most elementary psychological structures—a handful of sensory-motor reflex propensities—are specified by the genetic endowment. Progression to loftier levels of competence depends upon the child's innate tendency to expand and differentiate his structures in accordance with environmental contingencies. This occurs by the active "assimilation" of all newly encountered stimulus-response relationships of a level of complexity appropriate to the receptive categories within the cognitive structure; or, if the observed stimulus configuration taxes the structure sufficiently, it may "accommodate" (differentiate) itself in order to encode this new relationship. Cognitive development, like embryonic development, becomes a leapfrogging of ever more highly differentiated structures, each with its range of stimulus preferences.

The relevance of the epigenetic model of developmental biology to language learning was forcefully documented by Lenneberg (1969, this volume; see also Williams, Chapter 18). His main argument is that the capacity for language depends upon the unique biological organization of the developing human brain and differs fundamentally from the strings of conditioned habits that might be encoded with lesser efficiency into the brain of lower species. Although Lenneberg's model is explicitly epigenetic, with the child becoming "sensitive to successively different aspects of the language environment," he seems to place more explanatory weight upon genetic contributions to the formation of language structures than Piaget places on cognitive structures. In making this point, Lenneberg cites the correlation of language development with neuromuscular maturation, the existence of "universal" language structures, and the broad range of training procedures that suffice for normal development. Even when the learning environment is abnormal, as in schools for the deaf, "a number of aspects of language are automatically absorbed," implying that certain receptive niches had been determined by heredity factors. His position is far from that of the heredity predetermination school, however, in emphasizing that specific, essential ingredients (as yet ill-defined) must be supplied by the environment at each stage if normal development is to ensue.

In explaining the correspondence between brain maturation and language development, Lenneberg assumes that it is the neural maturation, presumably also generated by epigenesis, that "sets the pace" for the emergence of language. Because the brain must encode language, it is not surprising that the developmental indexes of both are correlated. But might not there be some interaction in the other direction as well, with language experience altering the maturation and differentiation of its neural basis?

An epigenetic approach to behavioral development can also be applied to the more prosaic learning situations in animals. In order to reach the stage of maze-running-for-reward, a rat must first have developed, through active practice, the capacity for orienting his body towards visual targets (Held and Hein, 1963) and for discriminating along the stimulus dimensions of the choice points, not to mention the earlier phases of locomotor and sensory development. After experience with a number of mazes, the problem-solving strategies will be generalized and accommodated into a stage of higher order, the formation of a learning set. It is thus more accurate, and perhaps more useful, to consider that the particular learned response under study is one stage in a historical continuum rather than an isolated habit.

Certain of Valenstein's new observations on the plasticity of appetitive behaviors elicited by electrical stimulation (Chapter 21) would also seem susceptible to interpretation in this developmental framework. They challenge the commonly held view that the rodent hypothalamus is equipped at birth with a fixed, separate, neural circuitry for satisfying each of the homeostatic needs (food, water, mating, and so forth), and that these "drive" circuits generate oriented behavior when appropriately activated by bodily need, releasing stimuli, or direct electrical stimulation. Valenstein contends that hypothalamically stimulated rats do not strive for a specific goal object or endstate, but are content to perform one of a class of stereotyped motor acts that may or may not result in a tangible reward. They may either gnaw, carry, chew, or drink upon stimulation at a particular point, depending on the environment and the recent history of responding. Furthermore, the same act can be evoked from diverse stimulus sites, suggesting that large portions of the hypothalamus act together in determining the plasticity of responding.

These findings certainly suggest the existence of a flexible response-selection system in the hypothalamus which can facilitate various behavioral subsystems according to the current organismic and environmental milieu. Perhaps understanding of these phenomena is promoted by recognizing that drinking, eating, gnawing, and so on, are members of an "appetitive" repertoire that evolves with experience, not the outputs of "drive circuits" built in at birth. At any point in development, the coordination of, say, gnawing with other appetitive behaviors and with environmental signals reflects a lifetime of experience with objects to be gnawed. The electrical stimulation, in this formulation, prompts the rat to seek an interaction with the environment, and the opportunities provided by both present and past environments determine which act in the repertoire will be executed. Any of a large class of acts may become satisfying or "reinforcing," some apparently inappropriate and maladaptive at the time. Perhaps Valenstein has thus shown, experimentally, how to create the brain state that brings the developing organism to the environment in search of the interactions by which it changes itself.

IMPLICATIONS OF THE EPIGENETIC MODEL   An essential attribute of epigenetic theories is their prediction that a precise end product will result despite the genetic specification of only the initial structure and at least some of the laws of transition between stages. Given a relatively consistent environment, members of a species then will inevitably grow through a stereotyped and highly predictable parade of stages. Piaget is not too explicit about how interstage transitions are initiated and guided, but presumably each cognitive structure has, in association, a set of laws governing how it can differentiate in response to each of a host of environmental provocations. These transition rules may not be exclusively specified by the genome, but rather by the entire organismic history of heredity-environment interchanges responsible for that structure. Piaget's model does not necessarily require that portions of the genome be triggered at crucial junctures to guide behavioral adaptation, as Hydén's evidence (1969; this volume) and the embryological analogy both suggest, and he would dispute that anything so fully formed as an "innate idea" is supplied by such switching of genes.

The epigenetic framework implies that life is a succession of "critical periods" during which specific experiences are requisites for the development of specific behavioral capacities. It explains why a newly hatched chick, if prevented from pecking for two weeks, can starve to death beside a pile of grain, a fact revealed many years ago (Padilla, 1935); in this case, when the emerging capacity to act was not practiced during a critical period, it was lost. A critical period occurs whenever the psychological structure has reached an appropriate stage of receptivity; if the sought-for configuration fails to appear at that time, the structure will differentiate to a higher stage (by accommodating to the available inputs), perhaps losing forever any receptivity at that level. Depending upon the species and the trait, such periods critical for behavior modification are known to last a matter of hours (imprinting in ducks), years (language learning without accent), or a lifetime (human adaptation to visual distortion.) The duration of a critical period probably depends on the rate at which the underlying structure differentiates, which would again be a co-function of genetic readout rates and relevant stimulation.

Critical periods during the acquisition of language skills are considered in some detail by Williams (this volume). Termination of the critical period for learning a foreign language without accent is generally attributed to a loss of "plasticity" due to maturation and fixation of the linguistic neuronal assemblies that "lock[s] certain functions into place at adolescence" (Lenneberg, 1969). An alternative that may

be entertained in the present context, however, is that at this time the language facility becomes so fully differentiated in the mode of the native tongue that there is no latitude for acceptance of foreign phrases. The issue is how much of this plasticity is lost because of saturation with the native language form and how much because of a maturational timetable, independent of experience. An experimental approach to this problem would be to study an adolescent with a normal brain but little language experience. A similar question can be asked about the progressive inability, with age, for the nondominant cerebral hemisphere to acquire control of the language function after the dominant hemisphere has been damaged. Rather than postulating an age-dependent rigidity of neuronal structure, it is possible for us to consider that the minor hemisphere is being increasingly differentiated for its own tasks of visual-spatial orientation (Levy-Agresti and Sperry, 1968) and hence passes beyond the original state of receptivity to linguistic material. Further research is needed to test the generality of proposals such as that the onset and termination of behavioral plasticity result from the level of functional differentiation rather than from a time-locked, genetically-determined neuronal differentiation.

The model of development as a historical sequence of stages, each building on the last, also predicts the empirical finding that the detailed steps in both cognitive and language development always emerge in an immutable, irreversible order (Williams, this volume). It is apparently not possible to impose an isolated, advanced functional capacity upon the child out of context. The higher human intellectual functions can be attained only by going through the entire species-specific sequence, which permits the genome and the environment to be properly timed and coordinated in shaping the subtleties of intellect.

Such concepts are also being exploited in educational and psychotherapeutic contexts. Recently, Delacato (1966) argued that many learning disorders result from the wrong kind of early experience, which caused perturbation in all subsequent stages of cognitive development. His epigenetic brand of therapy may consist of having the patients re-enact an entire orthodox developmental history, from the crawling stage onward. Freudian psychotherapy is also based on the idea that an extreme experience at an early stage of (affective) development can distort and dominate subsequent emotional attachments. Again, the treatment is to cause the patient to "regress" to a level of functioning characteristic of that primitive formative stage, so that the offending idea can be expunged from its source.

Some rather self-evident educational strategies also follow from the epigenetic position. When presenting a lesson, the teacher should discuss concurrently the subject matter at several levels of complexity, so that each pupil has available an input appropriate for his level of sophistication. In Piagetian terms, the most efficient learning takes place when the material in each new lesson is only slightly more complex than can be assimilated into the existing structure, resulting in the stepwise accommodations known as learning. Such considerations can make one wonder, as do some childhood educators, whether sufficiently complex material is actually being presented to small children, whether accelerated programs contain material that is too complex, and how learning failure can be rectified by remedial techniques. They raise, in short, the question of how to maximize the plastic capacities of organisms. While Piaget has pioneered the detailed definition and description of the intellectual stages of children, the core experiences that optimize each level of transition have yet to be understood.

MATURATION   It is generally accepted that the neuronal circuitry underlying species-specific "behavior organs" (such as the knee-jerk) is assembled by a process called "maturation." Maturation is normally defined as growth differentiation, regulated primarily in rate and form by the genetic readout, although such extreme environments as malnutrition or hormone treatments may alter even the most natavistic trait. Such maturation, however, always takes place within a broth of metabolites that contribute structural elements to the nervous tissues. In the embryo, moreover, feedback from the intracellular and extracellular milieu triggers and controls successive genetic activations in such a way that the rate of neurological development is regulated by an oscillatory interplay between the genetic readout and the biochemical milieu rather than by a clock confined to the genome. The differential growth rates of various parts of the nervous system and the phenomenon of embryonic induction of growth by adjacent tissues suggest that biochemical oscillations between separate nervous structures also contribute to the timing of maturation. In an analogous fashion, the imperturbable rate of progression through the cognitive and language states (Williams, Chapter 18) may be determined as much by the temporal characteristics of oscillations between the growing psychological structure and experience (i.e., between assimilation and accommodation) as by a neurological "readiness" due solely to a time-locked brain maturation. What, then, is the proper definition of maturation? Is it genetically regulated neuropsychological development taking place within a constant environment, or development in which environmental variation makes little or no difference? In either case, maturation cannot be divorced from the environmental constituency.

The difficulty of defining maturation relates to the problem of specifying the nature and extent of the genetic contribution to a behavioral trait on a molecular level. The genetic contribution derives from the sum total of the successive patterns of genetic activity during embryology, plus any subsequent genetic switching during behavioral adaptation: at each stage of development, however, the genetic influence may be modified by the environmental constituency at the level of translation, transcription, enzyme activities, and so forth. As stated earlier, some segments of any developmental sequence will be relatively sensitive to environmental interventions, while in others a clocked genetic readout may ensue as a function of prior historical events. An environmental provocation, then, elicits a response from the organismic structure that may tend towards either "instruction" (the passive mirroring of impressed forms) or "selection" (releasing of genetically predetermined patterns). Both types of response-altering mechanisms may be considered "plastic."

From this, it is clearly pointless to ask *how much* of an individual's behavioral repertoire is determined by each of the sources, as environmental elements have been assimilated into the central nervous system since conception, and genetic expression independent of an environmental context is impossible. An appropriate quantitative question, however, might be this: Under what range of environmental fluctuations will the behavior in question remain invariant or, alternatively, what are the minimum essential environmental ingredients needed to permit the behavior to emerge? One might seek to assign meaningful numbers, for instance, to the knee-jerk response, stable under a vast range of environments, and hence more "genetically determined" by this index than are eating after food deprivation or stopping at a red light.

In conclusion, the qualitative question of *how* the behavior-generating structures are shaped by hereditary and environmental sources (Anastasi, 1958) transcends the trivial quantitative matter of how much each contributes. One experimental approach to the problem would require identifying and analyzing the key transitional junctures at which increments in form are added. Studying the temporal and structural stability of a given transition under the influence of different classes of environments would permit conclusions about which qualitative aspects of behavior are added by heredity and which by experience. This approach is being used by Nottebohm (1970) in the analysis of bird song ontogeny by deafening chaffinches at each stage of a long, stereotyped history. In general, if a stage N + 1 inevitably follows the precursor stage N, despite wide environmental manipulations, it can be concluded that an "innate hint" of that stage is being contributed from the genes in conjunction with the existing structure. If a behavioral analysis does not proceed in such a painstaking, stage-by-stage manner, the historical interwining of heredity and experience will make separation of the factors virtually impossible.

HORMONES   A discussion of Goy's analysis (Chapter 20) of hormonal influences on behavior may shed further light upon the obscure interchange that organizes the neural substratum into a patterned response generator. Androgens act on the central nervous system in two ways, first by permanently channeling the maturational readout of the neuronal genome and, later, by acting as a phasic "environmental" stimulant of stable cell assemblies. Thus, androgen administered perinatally results in permanent differentiation of a male phenotype (anatomically, physiologically, and behaviorally), irrespective of the chromosomal sex specification. Subsequent to this critical period, its role becomes "activational" rather than "organizational" (Young, 1962), because the sex hormone then provokes the species-specific mating patterns that had been validated by the prenatal hormonal milieu.

The early administration of androgen quite possibly achieves its specification of nervous structure by engaging the genetic machinery of the target neurons. Hormones are products of gene-directed biosynthesis, are distributed throughout tissues as gradients, and induce growth and differentiation in those cells that are in receptive or "plastic" stages. In some cases, this induction is accompanied by enhanced mRNA and protein synthesis (Cohen, 1970; Korner, 1969), suggesting the occurrence of orderly genetic "switching," a prevalent concept in modern theories of maturation. Thyroxine has a similar effect upon the infantile nervous system, accelerating the maturation of certain neurobehavioral functions, whereas cortisol has a retarding influence (Shapiro et al., 1970).

The foreing suggests that hormones might be functionally equivalent to genetic factors in the patterning of behavioral maturation. If this is so, hormonal action may serve as an experimental model of the gene-directed neurological differentiation underlying species-specific behavior. While many actual genes are also susceptible to experimental control by interference with their enzymatic expression, most do not have well-defined behavioral syndromes as products. The analysis of CNS structural development in different behavioral genotypes is another approach to this problem, but hormones have the advantage of discrete application, both temporally and structurally.

The nature of the enduring modification in the hormone-treated brain could be sought with a variety of methodologies. A biochemical analysis, perhaps in vitro, could ascertain if the hormone does indeed function as a gene derepressor and, if so, could determine the products of the activated genome. Autoradiographic incorporation studies might identify not only where the receptive cells are located, but also which metabolites are involved.

In mature animals, on the other hand, sex hormones do not engender any enduring modifications of the CNS, but act as co-releasers of patterned sexual behavior that can be elicited just as well by electrical stimulation of certain sites in the limbic system. At this stage, the hormone induces a *short-term* reversible plasticity, modifying a pattern of stimulus-response probabilities to effect precisely coordinated behavior. The hormonal influence on sensory transmission, specific motor thresholds, and their integration by the activated "patterning" mechanism (Flynn, 1967) can also be explored by electrophysiological stimulation and recording techniques.

It is important to determine whether this response facilitation is accomplished by the same neurochemical mechanism that initiated the perinatal sexual differentiation, and whether the pathways activated in adulthood are identical to those established during the early critical period. The hormones may have a secondary role as neurohumors or neurotransmitters in modulating the electrical excitability of organized classes of "receptive" neurons. An understanding of the cytological basis of this responsiveness may be relevant to other types of plasticity in which early experience with a stimulus increases the later effectiveness of that stimulus. Goy has demonstrated, however, that neonatal androgenization has an impact upon behavioral characteristics that is more widespread than is the simple sensitization to subsequent hormone exposures.

This raises the important question of the extent of functional overlap between neurotransmitters, neurohumors, and hormones. It is conceivable that hormones may at times act as neurohumors and that some transmitters or neurohumors are able to organize neural tissues or to differentiate neurons via an activation of the genome. A recent NRP Work Session (Guth, 1969), in fact, has documented the "trophic effects" exerted by neurons in modifying the molecular and metabolic structure of adjacent tissues. If a recognized transmitter or other product of neuroelectrical activity were found to perform such hormonal functions, it could constitute a direct route for producing enduring modifications of neuronal organization through environmental stimulation. Such a hormone-like agent is a candidate for mediating the neuroelectrical-biochemical transduction, the topic to which we can now turn.

## The link between electrical and chemical events in the brain

A human brain comprises two to three per cent of the body in weight, yet it commands up to 50 per cent of the resting energy consumption and oxygen utilization. What fraction of this impressive brain metabolism is devoted to the synthetic and catabolic activities associated with plastic phenomena? Certainly some of it is, for both Hydén (Chapter 27) and Barondes (Chapetr 26) present evidence that increased synthesis of protein is essential for the formation of stable memories. Barondes tested the possibility that this synthesis by an mRNA pool takes place when mice learn how to escape shocks in a maze. Using cycloheximide, a specific and powerful inhibitor of protein synthesis, he has shown that short-term (three to six hours) plastic events (learning, retention) take place in the presence of the drug, while the long-term stable ones (memory) do not. His discussion judiciously places limits on the conclusions to be drawn from these results, but he, like Glassman (1969), interprets them as evidence that protein synthesis is an obligatory event in the laying down of permanent memories.

Hydén elected to measure protein synthesis in hippocampal neurons during acquisition of a motor skill. Several methods show the synthesis of brain-specific protein to be increased, especially during the early training stage. Of these, the S100 fraction would appear to have unusually interesting properties, as injection of an antibody against this protein into the brain blocks further acquisition. This remarkable experiment would appear to be the first direct evidence that a specific protein is essential for the production of a plastic change (learning) in the brain.

Neither Barondes nor Hydén specifies, however, how much new protein is synthesized, where it goes, or what it does there. What fraction of the total metabolism this activity consumes, therefore, remains unknown, as does any additional energy that may be required for altered readout from the genome, excess transmitter activity, and related events. Obtaining accurate estimates of these quantities would seem to be an important task for future research.

A human brain is also an impressive generator of electrical currents. It continuously maintains standing potentials of several millivolts between its various regions upon which slow oscillations of similar dimensions—the EEG and the evoked potential—are superimposed; furthermore, the constituent cells exhibit reversibly collapsible potentials in the range of tens of millivolts and at the rate of billions of instances per second. These electrical events have, of course, all been examined for years in a search for clues to be correlated with plastic events.

Adey (Chapter 23), Fox (Chapter 24), and von Baumgarten (Chapter 25) provide new and useful information. Adey presents the results of sophisticated computer analyses of the EEGs generated by animals and men performing a variety of discrimination tasks. By comparing such quantities as phase information and the power-density distributions at each frequency, meaningful comparisons of the wave trains from different brain locations become possible. In one such study, correctness and incorrectness of motor acts by a trained chimpanzee were accurately predicted by analyzing the waves appearing in only one of several brain regions examined; this nucleus, the ventralis anterior of the thalamus, can also be suspected from its connections to be a place in which neural events required for switching from one motor act to another might take place.

Adey also reports increased cooperative activity of the neuronal discharges in a single cortical area during per-

formance of a mental task. Fox, recording from the cortex of cats and men, showed that the shape of the potential evoked by a stimulus can be altered by simple reinforcement techniques, a surprising demonstration which implies that organisms can control in fine detail the way in which their neurons fire. These results are consonant with those of John and Morgades (1969), who have examined simultaneously the waves and the neuronal spikes evoked by visual stimuli in trained cats. Their evidence shows that domains of brain cells measuring a cubic millimeter or larger produce virtually the same average slow wave and spike activity during performance of a learned act. Furthermore, the same cellular domain produces a different average slow wave and spike response when a different learned act is performed (John, 1967). To varying degrees, all these studies support the conclusion that the crucial event taking place during training in the interaction between a neural substrate and the environment consists of assembling brain cells into groups that have common response properties. As Hebb suggested long ago (1949), each learned task seems to assemble the same units in a different manner, and the ability to perform one task or another depends upon which response pattern the stimulus generates within the brain.

This increase in cooperative activity within a brain region during training—an example of plasticity—must somehow be linked to the biosynthetic activities, including protein synthesis already discussed. The ongoing electrical patterns, altered by the environmental stimuli, may be assumed to trigger the macromolecular events, which, in turn, alter the response propensities of the cells in order to create the assembly. What could be the mechanisms by which this interaction sequence—electrical-chemical-electrical—is accomplished?

The first step, triggering new biosynthetic activity by electrical activity, would appear to require some specific event in addition to the synaptic depolarizations continuously present in the waking brain. Perhaps the hormone that aids in learning consolidation, as hypothesized by Kety (this volume), must be applied to the synaptic regions to induce the printing of information held in temporary storage. If so, the brain regions initiating the necessary activity may well be outlined by Ervin and Anders (Chapter 17), and the possibility that cyclic AMP mediates the synthetic activities in these cells (as it seems to do in other hormone-mediated increases of synthesis) would require exploration. Brain regions so activated, it should be pointed out, are likely to be widely distributed and even to include synaptic areas in the medulla, for Hydén showed years ago that the neurons and glia in the vestibular nuclei engage in new synthetic activity during learning of a balancing task (Hydén, 1967).

The fate of the newly synthesized material is an important unanswered question. One such component, the S100 protein identified by Hydén and Lange (Chapter 27), is normally present in the glia and neurons that have been studied so far, but its function in these cells is unknown. Whether other and even more interesting products are synthesized in addition to the acidic proteins isolated to date must be left open. The critical products of the new synthesis, however, might act in different sites to create the structural change that forms the basis of the engram—at the synapses by altering transmitter efficiency or site sensitivity, in other membrane regions of the cell, or at the level of cellular metabolism to facilitate firing to the changed input. Of these three possibilities, the last two received direct attention in the Symposium; von Baumgarten (Chapter 25) developed the idea that nerve cells may not be mere slaves to their synaptic drive because, by altering either their metabolic machinery or their membranes at nonsynaptic sites, they could change the consequences of a fixed synaptic bombardment.

Neither the fate nor the function of products newly synthesized during plastic processes is understood, so little can be said about the way in which the cells in a brain region become assembled into units that act coherently. If our model is correct, however, the end result of the biosynthesis, which, Barondes shows, requires many hours to accomplish, is to create a unique assemblage of brain cells that act as a unit and whose output is a particular behavioral act triggered by a particular stimulus input. Whether such stable configurations of cells actually do represent the end result of plastic interactions between the neural substrate and the environment will undoubtedly be the subject of many future symposia.

## REFERENCES

ALTMAN, J., 1967. Postnatal growth and differentiation of the mammalian brain, with implications for a morphological theory of memory. *In* The Neurosciences: A Study Program (G. C. Quarton, T. Melnechuk, and F. O. Schmitt, editors). The Rockefeller University Press, New York, pp. 723–743.

ANASTASI, A., 1958. Heredity, environment, and the question "how?" *Psychol. Rev.* 65: 197–208.

BONNER, J. F., 1965. The Molecular Biology of Development. Oxford University Press, New York.

COHEN, P. P., 1970. Biochemical differentiation during amphibian metamorphosis. *Science* (*Washington*) 168: 533–534.

CRAGG, B. G., 1967. Changes in visual cortex on first exposure of rats to light. *Nature* (*London*) 215: 251–253.

DELACATO, C. H., 1966. Neurological Organization and Reading. C. C Thomas, Springfield, Ill.

DIAMOND, M. C., F. LAW, H. RHODES, B. LINDNER, M. R. ROSENZWEIG, D. KRECH, and E. L. BENNETT, 1966. Increases in cortical depth and glia numbers in rats subjected to enriched environment. *J. Comp. Neurol.* 128: 117–125.

DOBBING, J., 1964. The influence of early nutrition on the development and myelination of the brain. *Proc. Roy. Soc., ser. B, biol. sci.* 159: 503–509.

FLYNN, J. P., 1967. The neural basis of aggression in cats. *In* Neurophysiology and Emotion (D. C. Glass, editor). The Rockefeller University Press, New York, pp. 40–60.

FULLER, J. L., and M. B. WALKER, 1962. Is early experience different? *In* Roots of Behavior (E. L. Bliss, editor). Harper Bros., New York, pp. 235–245.

GALAMBOS, R., 1967. Introduction: Brain correlates of learning. *In* The Neurosciences: A Study Program (G. C. Quarton, T. Melnechuk, and F. O. Schmitt, editors). The Rockefeller University Press, New York, pp. 637–642.

GESELL, A., 1945. The Embryology of Behavior: The Beginnings of the Human Mind. Harper Bros., New York.

GLASSMAN, E., 1969. The biochemistry of learning: an evaluation of the role of RNA and protein. *Annu. Rev. Biochem.* 38: 605–646.

GUTH, L., editor., 1969. "Trophic" effects of vertebrate neurons. *Neurosci. Res. Program Bull.* 7 (no. 1): 1–73.

HEBB, D. O., 1949. Organization of Behavior. John Wiley and Sons, New York.

HELD, R., and A. HEIN, 1963. Movement-produced stimulation in the development of visually guided behavior. *J. Comp. Physiol. Psychol.* 56: 872–876.

HIMWICH, W. A., 1969. The effect of environment on the developing brain. *In* The Future of the Brain Sciences (S. Bogoch, editor). Plenum Press, New York, pp. 237–252.

HYDÉN, H., 1967. Biochemical changes accompanying learning. *In* The Neurosciences: A Study Program (G. C. Quarton, T. Melnechuk, and F. O. Schmitt, editors). The Rockefeller University Press, New York, pp. 765–771.

HYDÉN, H., 1969. Trends in brain research on learning and memory. *In* The Future of the Brain Sciences (S. Bogoch, editor). Plenum Press, New York, pp. 265–279.

JOHN, E. R., 1967. Mechanisms of Memory. Academic Press, New York.

JOHN, E. R., and P. P. MORGADES, 1969. Neural correlates of conditioned responses studied with multiple chronically implanted moving microelectrodes. *Exp. Neurol.* 23: 412–425.

KORNER, A., 1969. The hormonal control of protein synthesis. *Biochem. J.* 115: 30P–31P.

LENNEBERG, E. H., 1969. On explaining language. *Science* (*Washington*) 164: 635–643.

LEVY-AGRESTI, J., and R. W. SPERRY, 1968. Differential perceptual capacities in major and minor hemispheres. *Proc. Nat. Acad. Sci. U.S.A.* 61: 1151 (abstract).

NOTTEBOHM, F., 1970. Ontogeny of bird song. *Science* (*Washington*) 167: 950–956.

PADILLA, S. G., 1935. Further studies on the delayed pecking of chicks. *J. Comp. Psychol.* 20: 413–443.

PIAGET, J., 1952. The Origins of Intelligence in Children (Margaret Cook, translator). International Universities Press, New York.

ROSENZWEIG, M. R., E. L. BENNETT, M. C. DIAMOND, S.-Y. WU, R. W. SLAGLE, and E. SAFFRAN, 1969. Influences of environmental complexity and visual stimulation on development of occipital cortex in rat. *Brain Res.* 14: 427–445.

SHAPIRO, S., M. SALAS, K. VUKOVICH, 1970. Hormonal effects on ontogeny of swimming ability in the rat: Assessment of central nervous system development. *Science (Washington)* 168: 147–151.

WIESEL, T. N., and D. H. HUBEL, 1965. Comparison of the effects of unilateral and bilateral eye closure on cortical unit responses in kittens. *J. Neurophysiol.* 28: 1029–1040.

YOUNG, W. C., 1962. Patterning of sexual behavior. *In* Roots of Behavior (E. L. Bliss, editor). Harper Bros., New York, pp. 115–122.

# COMPLEX
# PSYCHOLOGICAL
# FUNCTIONS

*"For centuries, men have speculated about perception,*
*learning, language, thought processes, emotions, and*
*other mental activities," says Gardner C. Quarton in*
*his prefatory remarks to this section. Illustrating one such*
*complex psychological function is this group of patterns.*
*It is used by MacKay in his chapter, beginning on page*
*303, in which he discusses the challenges offered the*
*conjunction of physiological and psychological techniques*
*in the study of perception.*

# 29 Prefatory Remarks on Complex Psychological Functions

GARDNER C. QUARTON

FOR CENTURIES, men have speculated about perception, learning, language, thought processes, emotions, and other mental activities. Yet the systematic study of the brain really began barely 100 years ago with the investigation of the pathologies accompanying neurological and psychiatric illnesses. It was clear that if the brain was damaged by a blow on the head or in some other manner, mental function was interfered with in a variety of ways. Gradually the investigation of brain function that accompanies psychological function became better organized but, as it did so, serious scientists came to realize that these complex functions are too difficult to explain fully in the light of present knowledge, and that investigations of sensory systems and motor function have, in some sense, a kind of priority. Much of a current neurophysiology text is, therefore, based on studies of sensory or motor systems or on even more fundamental processes. At the present time, we do not have an adequate explanation, in terms of events in the brain, for any of those complex mental processes that seem so real subjectively and are discussed so much psychologically.

The research strategies employed by investigators of brain and behavior have changed dramatically in the last century. Even fifty years ago most scientists were primarily

GARDNER C. QUARTON Director, Mental Health Research Institute, The University of Michigan, Ann Arbor, Michigan

interested in accounting for subjective experience. With the search for more objectivity in psychology, emphasis has shifted toward the investigation of the determinants of measurable behavior and of changes in that behavior. Partly for this reason, partly because of the great complexity of human behavior, and partly because ethical considerations restrict experiments on man, physiological psychology has now become devoted largely to the search for brain mechanisms that correlate with observable behavior in laboratory animals.

Our curiosity about man and his behavior remains, however, and it is reasonable to try to pull together what is known about the brain mechanisms that provide the basis for those aspects of man's behavior that interest us most.

In this volume, the major emphasis is on areas of research that seem critical to the development of the neurosciences. As a result, most of the papers included are concerned with a better understanding of the operation of brain components and small cellular systems. We have, however, selected for consideration four topics concerned with complex psychological functions. Learning and memory are discussed in some detail in the preceding section. In this section, we have included six papers on three topics—perception, emotions and mood, and communication and language. (We could equally well have attempted to clarify levels of awareness, attention, or even logical thinking.) The topics were selected because, taken together, they illustrate the problems posed by the search for brain mechanisms that might explain complex psychological functions.

There are two essays on each topic. To begin, two authors examine studies of perception. MacKay approaches the subject of perception and brain function by considering perceptual states as clues to neural processes that produce a repertoire of action. Held reviews the logic of modern theoretical and experimental approaches to the studies of perception. He also discusses the organization of perceptual systems and methods of processing sensory input.

Evidence points to the significant role played by the limbic system in states of emotion and selective attention. An elucidation of these mechanisms requires an understanding not only of anatomical connections but also of the relationships of biochemical and humoral reactions. Kety evaluates the role of norepinephrine and other putative neurotransmitters and discusses the possible linkages of biogenic amines and affective states to reinforcement mechanisms and the consolidation of learning. Speculation on the possible adaptive functions of such mechanisms, as well as their implications for the evaluation of behavior, leads to the paper by MacLean, who discusses brain correlates of emotion examined in the light of evolutionary changes in the brain. The paleomammalian limbic system receives special attention because of its importance in emotional function.

The third topic, communication and language, is approached first by Ploog, who investigates social communication among animals, particularly primates. Social signals —whether they are approach and avoidance or partner interaction for providing reproductive isolation in species— have their brain correlates, which developed from primitive to highly complex neuronal machinery. Social communication seems to be a basic need for all primates, and Lenneberg discusses brain correlates of perhaps the most complex of all methods of social communication—human language. Its uniqueness has produced many controversies over the years as to its origin, structure, and function, and Lenneberg reviews the evidence that has led to present-day theories on anatomical and physical correlates of man's capacity for verbal behavior.

The plan represented by these papers was adopted to provide better coverage of the material. The two authors on each topic present different aspects of their area of interest and, concomitantly, at least two points of view on research strategy.

# 30 Perception and Brain Function

## DONALD M. MACKAY

### *The challenge of perceptual phenomena to the neurosciences*

At first blush, it might seem obvious that the subjective content of our perceptual experience should be a rich source of clues to the brain processes that mediate perception. Granted the working assumption that all features of the world as currently perceived are represented in some way in the current state of the brain, it would indeed follow that someone who knew the "representational code"—who knew which brain state (or alternative states) corresponded to each possible state of perception—should be able to tell from his subjective experience much of what a physiologist should find going on in his sensory system. It would not at all follow, however, that someone who did not know the "code," or the principles on which sensory information was processed in the brain, should be able to discover anything about these matters by the introspective study of perception. On the contrary, if all perceptual experience were self-consistent, so that we never found any conflict between different samples of sensory evidence, I can see no way in which introspection alone could lead to any worthwhile neuroscientific predictions. The most we could predict is that for any *change* in perceptual experience there should be a correlated *change* in brain state, but not vice versa.

We might be tempted, on these grounds, to mistrust and ignore all perceptual phenomena as irrelevant to brain science, but to do so I think would be to err in the opposite direction. The relation of the two is indeed not obvious, and many pitfalls beset the would-be integrator of psychological and physiological data, as we shall see. But I hope to show that integration has already proved to be both possible and profitable, and that it promises a more rapid advance in our understanding of brain mechanisms than we could hope to achieve by physiological methods in isolation. Its key is the existence of perceptual anomalies and illusions—indications, usually, that some particular function of the physiological machinery is under abnormal strain. From the nature of the strains thus revealed, we can discover unsuspected categories in terms of which sensory information is analyzed and processed, and from the pattern of physiological change, as

each strain is applied and removed, we may hope to identify the neural basis of the function concerned.

### *Pitfalls of perceptual psychophysiology*

A good illustration of the difficulty in making neurophysiological inferences from perceptual data comes from the classical field of sensory psychophysics. When the physical intensity of a sound is increased, it is perceived as "louder." What should the physiologist expect to find happening in the brain as the correlate of this experience? We may expect that some change must occur in the firing pattern of some neural activity, but data reviewed by Segundo elsewhere in this volume indicate the bewildering variety of "candidate codes" that must be considered. The frequency in some fibers, or the number of fibers that are active, may increase with perceived loudness; they could equally well decrease. The effect of increased intensity might be only to increase the regularity or coherence of firing while leaving the mean frequency unchanged; and so on. It may be plausible to expect the perception of an increase in intensity to correlate with an increase in some over-all measure of neural activity, but it is far from necessary.

With some relief, we learn from physiology that the over-all firing rate in peripheral sensory neurons does increase with intensity. Can we, then, treat the subjective data as a kind of internal measurement of this firing rate? If so, we might expect the loudness that is psychophysically estimated to vary with physical intensity according to the same law as the average neural firing rate. In the days when psychologists accepted Fechner's (1860) method of estimating subjective intensity in terms of the sum of "just noticeable increments," the psychophysical law was considered to be logarithmic: "perceived intensity" ($\psi_F$) varied as the logarithm of physical stimulus intensity (I) or

$$\psi_F = k \log I \tag{1}$$

When physiologists began to find roughly logarithmic transfer functions between stimulus-intensity and neural-firing rates, they believed that Fechner's law was corroborated.

The work of S. S. Stevens (1961) and his school, however, has disturbed this harmony with the finding that perceived intensity ($\psi_S$), as estimated by more direct methods, is not at all proportional to the sum of just noticeable increments

DONALD M. MACKAY  Department of Communication, University of Keele, Staffordshire, England

($\psi_F$). In fact, over a wide range of modalities, the psychophysical relation found by Stevens is a power law,

$$\psi_S = a\,I^\beta \qquad (2).$$

Stevens (1961) and others took it to mean that the physiological transfer functions previously thought to be logarithmic should turn out, on closer examination, to be power laws.

I have elsewhere (1963) questioned the logic of this conclusion and shown that, on at least one plausible model, the data of both Stevens and Fechner could be expected, even on the assumption that the receptor transfer functions were logarithmic. I mention it here to illustrate the dangers of what I take to be a wrong way of seeking to integrate perceptual and physiological data—namely, by treating the first as an "internal description" of the form of the second. The two (we may assume) are *correlates* of each other in the sense used earlier in this chapter, but it is quite unsafe to assume *a priori* that they are *analogues* (i.e., similar in form) and, still more so, to assume that they must be numerically equivalent in the way that two independent measurements of the same variables should be. To appreciate the weakness of such assumptions we need only to consider the counter example of a digital computer. Here there is a one-to-one correlation between the physical states of the machine and the numbers represented in it, but this leads to no simple proportionality between the numbers and any physical measure of activity, such as current strength, impulse frequency, or the like.

## Neural information processing and perceptual anomalies

A similar argument applies to the prediction of perceptual phenomena on the basis of the physiological information-processing operations performed on sensory signals. If lateral inhibition between neighboring visual channels causes contrast enhancement, for example, it may seem obvious that the human subject should perceive contrasts as exaggerated; but then the assumption would be, again, that the content of perception is analogous to (rather than simply correlated with) the configuration of processed signals from the sensory system.

This leads us to ask under what circumstances a perceptual sign of sensory processing might be expected. A general answer is difficult to find; but in one large class of cases the processing machinery is betrayed only when it is overloaded or underloaded with particular input features to which it is sensitive. When the sensory input is rich enough, or poor enough, in features to which some processing sub-system is specifically sensitive, then, if the subsystem concerned is plastic or self-adjusting, we may expect its operating characteristics to change in such a way that a normally balanced

input will be processed anomalously for some time afterward (MacKay, 1957a, 1961a). Thus, after the visual system is exposed to a bright light, a test light appears abnormally dim; after it is exposed to red light, a neutral gray appears green; after exposure of a moving pattern, a stationary field seems to move in the opposite direction (see below), and so on.

A second group of perceptual anomalies results from the interaction of simultaneously stimulated regions of a sensory field. Here, again, there is no basis for assuming that all such interactions must give rise to perceptual anomalies. On the contrary, the majority of these (lateral inhibition included) seem to operate on normal, everyday inputs in such a way as to maintain more or less veridical perception. Anomalies result only when the statistical properties of the field are sufficiently abnormal; once again, the circumstances giving rise to these anomalies become informative to the neuroscientist hungry for clues to the mechanics of sensory processing (MacKay, 1961a).

For example, perceptual signs of lateral inhibition can be detected (such as Mach bands; see Ratliff, 1965), but only with particular simplified stimuli. Moreover, such phenomena are often complicated by features that defy simple explanation in the same terms. In Figure 1, for example, the contrast in subjective brightness between left and right halves of the horizontal band is much enhanced after a boundary has been introduced between them (see legend). An analogous enhancement of color contrast between neighboring areas gives rise to the well-known Land effect (Land, 1959a, 1959b), whereby a surprisingly wide range of subjective hues can be generated by different admixtures of, say, red and white light. Here again, however, certain boundary conditions must be met for the full effect to be observed (Rushton, 1961).

On the auditory side, similar evidence of contrast enhancement is well known; specific sensitivity to frequency gradients and the like has been discovered by Suga (1964) and Evans and Whitfield (1964), and a range of corresponding perceptual anomalies are being found.

The argument of the present paper is that, for the most rapid progress in this area, the study of neural sensory mechanisms and of perceptual anomalies must go hand in hand; that the challenge of perceptual phenomena, although tantalizing, is not unconstructive. Each anomaly observed can suggest a worthwhile experiment in sensory neurophysiology, or at least can help to narrow the almost infinite range of possible experiments competing for priority. Each neural information-processing mechanism discovered can, in turn, suggest some corresponding experiment in perceptual psychophysics, the outcome of which may lead to more refined physiological hypotheses and experiments; and so on, in hypothetico-deductive fashion.

What has been said applies *a fortiori* to correlations be-

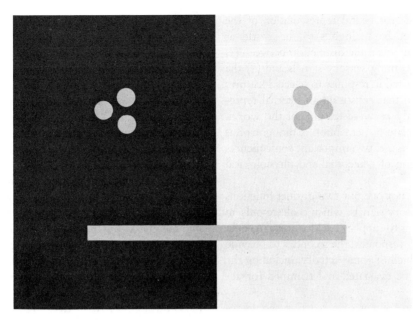

FIGURE 1 Anomalies of simultaneous contrast. Although lateral inhibition can explain in general terms the enhanced subjective brightness of areas surrounded by regions of lower objective luminance, it leaves certain complications unexplained. For example, if this figure is bisected by a black thread parallel to its vertical edge, contrast between the two halves of the horizontal grey bar is enhanced; and the region of enhanced contrast can be "drawn" to left or right by moving the thread away from the midline.

tween anomalies of perception and lesions or ablations of parts of the nervous system. The pioneering work of Head (1920) and Holmes (1918) and of Teuber and his collaborators (1960) with cases of cerebral gunshot wounds, for example, has been of immense importance in establishing a rough correlation between the location of cerebral lesions and specific defects in perception. Perceptual studies by Sperry (1966) and Gazzaniga et al. (1965) in "split brain" patients (in whom the corpus callosum and anterior commissure had been cut for clinical reasons) have added further dimensions to the problem of identifying the cerebral correlate of conscious perception. Conversely, Brindley and Lewin (1968) have recently been successful in inducing spatially localized phosphenes by means of an array of stimulating electrodes placed on the occipital cortex of a blind subject. Their work has demonstrated a correlation between the topography of cortical activity and that of the corresponding visual space—a correlation that had begun to be doubted on the basis of recent discoveries of cortical "feature detectors" (Hubel and Wiesel, 1959; Lettvin et al., 1959).

For the remainder of this paper, however, I confine myself to what can be learned from the study of perception under normal physiological conditions and to the problems of relating this fruitfully to sensory neurophysiology. Some of the lessons of perceptual adaptation to deprivation, distortion, or redistribution of the sensory input, and of

surgical intervention, are dealt with by Held in another chapter.

## A functional approach to perception

If the content of perception cannot safely be assumed to be simply analogous to the neural activity pattern, what alternatives are open to us? At what level can we expect to find any illuminating isomorphism between the two? For a constructive answer we must move farther back to consider the *functional consequences* of perceiving. An organism can pursue goals in a structured field of action only if its "conditional repertoire" of possible sequences of action is set up to take proper account of the current structure of the field. In a world of changing structure, and in the face of the fading and corruption of internally stored information, fresh information from sense organs is continually needed to update the form of the "conditional repertoire" or "state of conditional readiness" for all possible internal or external action. The success of an action (as evaluated by current internal goal criteria) depends on its being adapted to the current state of the world. The updating of the conditional repertoire to match the changing structure of the field of action, in response to sensory signals, suggests itself in these terms as the functional correlate of perception (MacKay, 1952, 1956, 1967).

This functional approach leaves unspecified (thus far) the

extent to which the internal neural representation of the world may or may not be analogous to the world as perceived. What it highlights is the distinction between (1) the configuration of incoming sensory signals and (2) the updating operations (elicited in response to selected features of those sensory signals) that constitute an internal representation of the currently *perceived* features of the world. Once the physical correlate of perception is thought of as an internal matching response, two important consequences follow for the integration of perceptual and physiological data.

First, we clearly have not one but two distinct functions to be performed by sensory signals, which coalesce only in extreme cases. The first is to supply "feed forward" to help elicit the right up-dating responses; the second is to provide feedback in terms of which ongoing activity, including the internal responses, can be evaluated and trimmed for accuracy.

Second, the subjective content of perception should reflect more directly the generative process giving rise to the matching responses than the sensory signals evoking them. In particular, if the matching response is some transform ($T_M$) of the generative activity (G), and if the processed sensory signals are some other transform ($T_S$) of their source distribution (S), then a match between them entails some comparison of $T_S(S)$ with $T_M(G)$, or, in general terms, a process of "evaluation" (Figure 2). In the simplest case of direct comparison, it is evident that any processing operation ($T_S$) could in principle be "balanced out" by the introduction of a *similar* operation ($T_M$) in the feedback path, such that no sign of $T_S$ would be reflected in G (and hence in perceptual experience). An operation in the feedback path here could have the same effect as the *inverse* operation applied to the input signal.

When the matching criteria are more complex, the effect of $T_M$ is less simply specifiable, but the general principle

holds that an operator in a feedback path has an over-all effect roughly inverse to its effect in the input path. In particular (to revert to Fechner's and Stevens's laws), if the stimulus (S) has a physical intensity (I), and if both $T_S$ and $T_M$ were logarithmic, then, for a simple proportional match, we would have $k_1 \log I = k_2 \log G$, or $G = I^\beta$, where $\beta = k_1/k_2$. Thus the strength of matching generative activity (G) would, in this case (i.e., even with logarithmic transducers), be related to stimulus intensity I by a power law (MacKay, 1963).

Another particular implication is that it is not necessary to assume that the production of a smooth and stable percept from discontinuous sensory samples requires some smoothing integrator on the input side. The evaluative process could "steer" a continuous matching response as readily on the basis of discontinuous as of continuous samples. (Consider, for example, the way in which one can maintain a stable percept of the visual world as one walks past and looks through a row of trees.) Finally, it implies that perception of the field of action could, in principle, occur without the subject's having any clear awareness of the sensory modality responsible (see *Multiple Levels of Perception*, below).

## "Feature detectors"

As mentioned above, a sensory input abnormally rich or poor in features to which a processing subsystem is sensitive might induce the latter to betray its existence by shifting the "neutral point" on its operating characteristics. The resulting perceptual illusion can help to show which pairs of qualities are functionally complementary for the subsystem concerned. We have already noticed the examples of brightness, color, and motion in this connection. A somewhat more surprising example turns out to be *contour direction*. If a field of near-parallel lines is superimposed on a neutral test field of randomly changing dot patterns ("visual noise"), the test field is perceived as organized into shadowy, wavy lines (the "complementary image") roughly at right angles to the stimulus lines (MacKay, 1957a, 1957b). This peculiar adaptation is specific to the region of retinal field that is stimulated and can be seen also as a fleeting aftereffect (see Figure 3). From the information-engineering standpoint, it seemed natural to interpret this as suggesting the existence of a population of neural subsystems that were differentially sensitive to contour direction, and that treated directions at right angles as complementary, in much the same sense as are red and green (MacKay, 1957a, 1961a). Contrast enhancement between simultaneously stimulated members of this population could also account for certain geometrical illusions, especially those involving the overestimation of angles (MacKay, 1957a).

As noted by Creutzfeldt (this volume), single units (so-

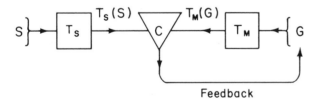

FIGURE 2 Essentials of a "matching response" mechanism. Internal generative activity (G) is forced to follow changes in the activity of a stimulus source (S) by feedback from an evaluator, here represented by the comparator (C). The comparator receives and compares transforms $T_S(S)$ and $T_M(G)$ of the activities S and G, respectively. The response of G would normally also be aided by feed forward from suitable filtrates of S, but this path has been omitted for simplicity.

called "edge detectors") sensitive to the direction of optical contours on the retina were discovered soon afterward (Hubel and Wiesel, 1959; Lettvin et al., 1959); these seemed naturally fitted to fill the role suggested by the perceptual findings. Whether the two are directly connected, however, remains to be demonstrated. No conclusive evidence has yet been reported of contrast enhancement between contour detectors with neighboring receptive fields, although units sensitive to more complex features of the optical stimulus pattern have been found in embarrassing abundance (see Bishop, this volume).

A further complication has turned up from the perceptual side. It has long been known that the colored fringes seen along the edges of contours tend to disappear when prism spectacles are worn (Kohler, 1951, 1962). McCollough (1965), when investigating this effect, discovered that after prolonged exposure to red lines in one direction alternating with green lines orthogonal to them, a test field of parallel

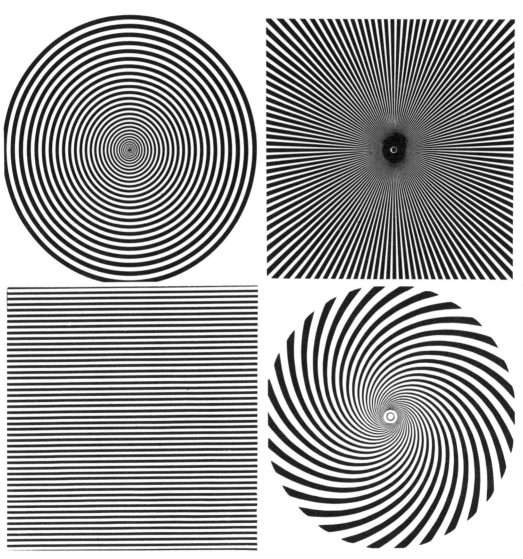

FIGURE 3   Some patterns that evoke the perception of complementary images, either as fleeting aftereffects (lasting one second or less) or when moving "visual noise" is superimposed on the figure. A suitable source of visual noise is the "snowstorm" seen on the screen of a television receiver tuned to an empty channel. If a flat sheet of glass is held at 45 degrees to the surface of this page, in such a position that the image of the television screen is optically superimposed on the figure, the speckled field will be seen apparently to stream in directions roughly at right angles to the lines of each figure.

black-and-white lines was perceived as colored according to the direction of the lines, the hue being complementary to that which was previously associated with each direction. She interpreted this as perceptual evidence of color adaptation in visual contour detectors—a conclusion recently questioned by Harris and Gibson (1968) but suggestive of some interesting physiological experiments. Recently I have made some observations (unpublished) that may point in the same direction. If a field of red parallel lines is superposed on a field of green parallel lines oriented in a different direction, a state of perceptual rivalry is set up, even with monocular viewing, each field alternately suppressing the other in a manner not observed with similar black-and-white patterns, or to the same extent with superposed irregular patterns in red and green. The simplest interpretation might be that orthogonal directions of complementary colored lines are represented by mutually exclusive states of the same neural system (perhaps resulting from mutual inhibition of two antagonistic subsystems); but, in view of what is said above about perception as a matching response, it is evident that the "rivalry" could arise either on the input or the response side, and no firm inference can be drawn without further data.

A more abstract category of feature analysis is suggested by the adaptation seen on prolonged exposure to a nonuniformly textured field such as is shown in Figure 4 (MacKay, 1961b). Perceived texture becomes more and more uniform, the coarser features disappearing to leave a fine-grained (sometimes periodic) mottled pattern. This suggests the existence of a family of feature filters sensitive to texture density or "space frequency," with a shorter time constant of adaptation at lower frequencies. The work of Campbell and Robson (1968) on the visual space-frequency response to parallel-line fields provides further data on "texture processing" of this sort, which is obviously of biological importance in estimating the velocity of approach to textured surfaces (Gibson, 1950).

## "Motion detectors"

Reference has already been made to the long-known negative aftereffect of exposure to a steadily moving field (the "waterfall effect"). The sensation characteristic of this aftereffect can be described as one of "motion without displacement"—a combination of properties impossible in an analogue. In information-engineering terms, this suggests the existence of a subsystem of elements sensitive to the "brushing" of the retinal array by a moving image; the elements signal *motion* as distinct from change of position, and adapt to unidirectional motion with a shift of "zero-level," so giving rise to a negative motion signal when stimulus motion ceases (MacKay, 1961a). Physiological evidence of just such processes has been found by Barlow

and Hill (1963) in the eye of the rabbit, encouraging expectations that similar processes will explain the waterfall effect in human subjects.

Once again there are intriguing subsidiary phenomena on the perceptual side to suggest further physiological questions. For example, we have found that the *subjective* velocity of a population of large and small spots moving across the retina with the same *objective* velocity depends on the size of the spots; the subjective velocity is less for larger spots. The waterfall aftereffect can be reduced or suppressed by stroboscopic illumination of the stationary test field (Anstis et al., 1963), although the effect returns when the stroboscope is shut off. We have also found that stroboscopic illumination superposed on a well-lit moving stimulus pattern reduces the perceived velocity (as distinct, of course, from the displacement observed in a given time!) for the duration of the flash train, with a dramatic "recovery" as soon as the train is discontinued. Both of these phenomena suggest that a reversible reduction of the response of velocity-sensitive units by stroboscopic illumination might be worth looking for—although once again the interaction might as readily take place on the matching-response side as in the sensory input processors.

An indication that perception of motion depends on two distinct mechanisms with different sensitivities comes from the illusion of instability seen when a stroboscope illuminates a field containing small, self-luminous objects (MacKay, 1958, 1961a). For very small displacements of the retinal image, the self-luminous objects can be seen to move, but the flash-lit background does not. Although the displacement of their respective images over the retina is the same, the "brushing" of the retina by the self-luminous image gives rise to signals not generated by the discontinuous displacement of the flash-lit image.

Further quantitative evidence in the same direction is reported by Körner and Dichgans (1967), who found the perceived velocity of a moving stimulus to be nearly doubled when the retina was held stationary (causing it to be "brushed" by the image) rather than allowed to fixate and follow the moving pattern. In the latter case, the perception of motion depends, of course, not on retinal image displacement (for there is none) but directly on the generation of the matching oculomotor response. In corroboration, it is well known that when the extraocular muscles are paralyzed, any effort to move the eyes is accompanied by an illusion of motion of the visual field, the image of which naturally remains stationary on the retina, which the subject thinks he is moving.

## The stability of the perceived world

This leads us to the vexed question of the stability of the perceived world during voluntary movements, particularly ex-

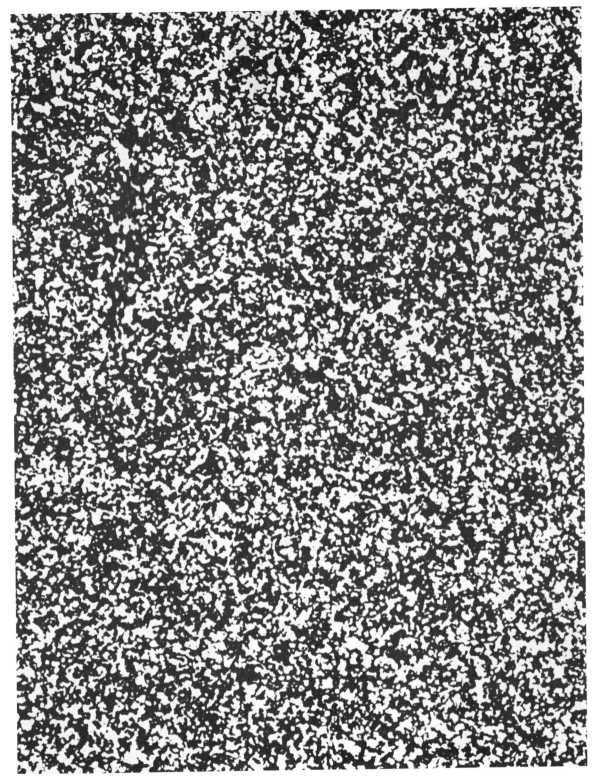

FIGURE 4  Adaptation to inhomogeneity of texture. The appearance of this static visual noise field changes considerably after half a minute's steady fixation on any particular point. The texture perceived becomes much more uniform, and in some cases periodic, as if the pattern in the region of fixation were replicated over the whole field. Moving the fixation point "reactivates" the original percept.

ploratory movements of the eye. Only a brief discussion is possible here. The chief question of interest to the neuroscientist is whether the maintenance of stability requires the visual system to subtract or suppress from the input signals the changes resulting from voluntary movement before these signals reach "higher centers." If the content of perception were supposed to be analogous to the output of the sensory system, then indeed some such "cancellation" process would be required, as proposed by von Holst (1957) and illustrated herein in Figure 5. An *Efferenzkopie,* or "corollary discharge" (Teuber, 1960), from the oculomotor innervation is here supposed to perform the required subtraction before the signals "reach perception."

It seems undeniable that some form of interaction is required between the generator of exploratory activity and the sensory input, if the sensory world is not to be misperceived as moving when exploration occurs. From our present standpoint, however, there is no good reason to assume that removing the signs of exploratory activity from the sensory input is needed. Indeed, the counter example of tactile exploration, as when we use the arm muscles to move the palm over a stationary surface, shows that we can perceive our sensory world as at rest not *in spite of* but even *because of* the sensory changes (the displacement of the tactile image, and so forth) resulting from our explorations.

What is needed, in terms of information engineering, is that the *criteria of evaluation* of the sensory input should depend on the internally planned and generated movement. In other words, the question for the internal information system is whether the current sensory input contains any information demanding change in the internal representation of the world. Is there any mismatch between the internal state of conditional readiness and the current signals from the sensory system? The normal sensory changes consequent on exploratory movement, so far from constituting such a mismatch, are part of the evidence that the explora-

tory program (organized on the basis of the current internal representation) is succeeding according to plan. They confirm, rather than disturb, the current representation, whether or not they themselves are perceived.

If, then, as neuroscientists we are encouraged to look in sensory areas of the brain for signals from oculomotor and other generators of bodily movement, we may do well not to jump too quickly to the conclusion that any signals we find must have the function of eliminating sensory signs of such movement. The foregoing brief discussion, which abbreviates arguments presented in more detail elsewhere (MacKay, 1957c, 1962, 1970) suggests that there are alternative functions worth considering for such "corollary discharges" and that cancellation of motion signals may even be the last thing the system needs for perceptual stability.

## Perception of form

Perhaps the most challenging of all perceptual problems to the neuroscientist is the perception of form. As the familiar Frazer "spiral" shows (Figure 6), this is clearly not a matter of "template matching" on the basis of the topography of the neural image, for only a circular template would match the "twisted cords" of the figure, and these are not seen as circles. Instead, it would seem that the integration of evidence from local areas proceeds on a logical, rather than a topological, basis—more as the integration of sample descriptions to form a specification than as pieces of a jigsaw puzzle to form a picture.

It is natural to think nowadays in terms of a hierarchy of analyzing operations, of which the signaling of brightness contrast and edge or line orientation are low-level members. The "hypercomplex cells" of Hubel and Wiesel (see Bishop, this volume) may be involved in higher-level operations of this family. Jeffreys (1969), in our laboratory at Keele, has found evidence that visual-pattern features of a higher order than the over-all contour length or density can have a powerful effect on evoked cortical response as monitored by the occipital evoked potential. As Figure 7 shows, the early component of evoked response to a parallel-line field can be much enhanced by breaking up the lines, and still more by introducing contrast between the directions of different segments. A uniform grid pattern can be made to induce a still bigger evoked response if some of the lines are removed to leave squares that are not surrounded by other squares, and so on. These findings are consistent with the idea that contrast enhancement may operate between neighboring feature-sensitive elements up to fairly abstract levels of pattern analysis.

The level of the sensory system at which such "descriptive analysis" takes place can also be investigated perceptually. For example, we can recognize the form of a tree

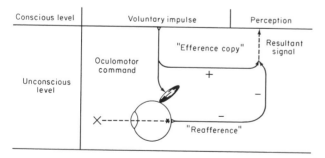

FIGURE 5 The "cancellation" model proposed by von Holst (1957) to account for the stability of the perceived world during voluntary eye movement. Motion signals from the retina were presumed to be canceled by the subtraction of equal and opposite signals generated in the oculomotor system.

FIGURE 6   The Frazer "spiral." The twisted cords all lie on concentric circles.

at the edge of a forest with the help of binocular vision even when the tree is indistinguishable from its background to either eye alone. This capacity shows that much of our pattern-analyzing machinery must be at a level at which signals from both eyes have been integrated to extract depth information—the basis of well-known stereoscopic methods of unmasking camouflaged buildings by aerial reconnaissance. (More elaborate ways of demonstrating these and other facts of binocular vision have been developed by Julesz [1960], using computer-generated patterns.)

Again, there is evidence of a close link between the perception of pattern and of motion in the perceptual distortions that result when a moiré figure is set in motion in such a way that symmetrically placed moiré contours move in different directions (MacKay, 1964b). Figure 8A shows one example of a pattern that results from the superposition of Figure 8B on the "ray pattern" of Figure 3; displacement of the two superposed figures causes a marked perceptual asymmetry that is seen to "collapse" over a period of a second or so after relative motion ceases, suggesting that integrated velocity signals play a part in the perceptual location of contours.

A question currently debated is whether the internal representation of a percept is simply a compresence of descriptive signals from "feature filters," each attributing one feature to the object perceived, or whether we should look for something quite different. Subjective experience is not a

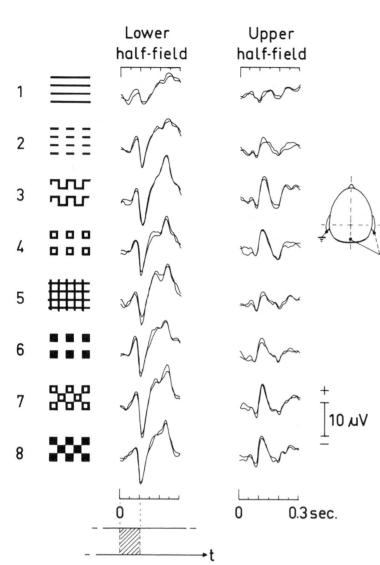

FIGURE 7 Sensitivity of evoked potentials to high-order contrast in optical patterns. The amplitude of the human occipital evoked potential can in many cases be increased by removing portions of a stimulus pattern. The evidence suggests that it is sensitive to the degree of contrast between neighboring areas not only in respect of luminance, but also of higher-order features, such as the presence and direction of line elements. Comparison of the responses with patterns 1 and 2 or 4 and 5 makes it clear that the total length of contour is not the main determinant of response amplitude. Note the contrast in form and even in polarity between the scalp responses to stimulation of upper and lower half-fields (right and left columns). This is presumed to be due to the different anatomical orientation of the corresponding cortical areas in the neighborhood of the calcarine fissure.

good guide here, but it can be questioned whether the *unity* of objects as we perceive them could have an adequate correlate in terms of a bundle of discrete descriptive signals.

Moreover, we must keep in mind that feature filters could subserve many other functions that are not directly connected with the subjective recognition of pattern. These include the provision of suitable feedback signals for the servo-systems that maintain the focus and the fixation and convergence of the eyes, and the extraction of information about the motion of the body through the visual environment from such global features of the retinal image as the gradient of texture density, the divergence of local image velocity, and the like (Gibson, 1950). In any case, the functional approach we have been taking suggests that to look for the correlate of perceiving in the activity of the feature filters, as such, would be to stop short, and that perceiving is achieved only in and through the matching response of

internal *adjustment to reckon with* the source of the sensory signals, either proximal or distal. In this view, the unity of the perceived object has its counterpart in the unity of the matching organizer of conditional readiness, set up, perhaps, in response to a bundle of discrete demands from sensory feature filters, but itself naturally seamless and unitary unless the sensory information content requires its fragmentation. The parceling of sensory samples into discrete internal channels does not of itself constitute any such requirement.

Here once again we must contemplate two distinguishable processes, in either of which illusions and other anomalies of perception might have their origin: (1) the processing of sensory information into "feed-forward" channels with suitable connections to evoke an appropriate internal matching-response configuration; and (2) the generation and evaluation of that matching response itself. In the alternating perception of ambiguous figures (Figure 9), for

A                                                                        B

FIGURE 8 Dynamic distortion of perceived form. (A) A moiré figure resulting from the superposition of a transparency of (B) on the "ray" pattern of Figure 3. If the transparency is moved gently to and fro, asymmetries appear in the moiré figure. These disappear each time the pattern is allowed to come to rest.

example, it seems economical to suppose that the oscillation we see is not in the sensory feed-forward system, but in the matching response generator, two of whose states are equally compatible with the constraints of the input, so succeed each other as do the two possible states of a multivibrator. In the Frazer illusion (Figure 6) the balance of economy of hypothesis is harder to judge. In the case of illusions of brightness contrast, the evidence points to the sensory processing side as mainly responsible, although some complications may be attributable to the response generator (see legend of Figure 1).

But whether these particular judgments are right is not important. I mention them only to make clearer the two classes of possibility that must always be kept in mind as we search for physiological correlates of perceptual phenomena.

## Multiple levels of perception

So far, we have tended to speak as if one internal matching response were necessary and sufficient for any given stimulus configuration, but there are many cases in which this is clearly not so. The contents of this page may be perceived equally accurately as ink on paper, as a succession of letters, or as a succession of English sentences. In each case, the internal matching response is characteristically distinguished by the *class* of conditional readinesses associated with it. In our terms, "seeing. . . as" implies "being ready to reckon with. . . as." When we are perceiving the contents of a page

as an argument with which we disagree, it comes as something of a shock to be asked about the color or granularity of the ink or the type font of the letters. We may even need to look again; our current matching response has not been at a level that made us ready to reckon with these particular aspects of what lay before us. Our internal representation, although complete in itself, was in other terms—not inconsistent with, but complementary to, the terms required to do justice to the aspects now under discussion.

In terms of neural organization, this points to a *hierarchic* structure in the generator of perceptual matching responses. A response that matches a given visual or auditory input, for example, may be initiated at different levels in this hierarchy, and the criteria of "match" or "mismatch" may be set in correspondingly different terms. In principle, there is no reason why matching responses should not take place simultaneously at different levels, but the evidence of subjective experience is that this is normally possible only at the expense of over-all speed or accuracy. The more closely we try to follow changes at one level, the more the changes become exclusive of perception of details at other levels. The correlate in information-engineering terms would seem to be that the central evaluator of mismatch has a limited channel capacity, which must be shared between different levels, if more than one is to be "attended to" at one time. The quality of perception is determined by the kind of internal activity called into being to match the sensory demands, rather than by the configuration of sensory signals *per se*.

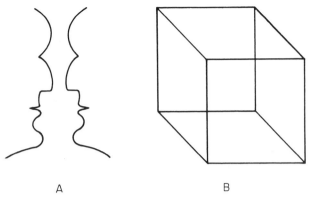

FIGURE 9 Ambiguous figures. A. Faces or candlestick? B. The Necker cube.

This theme is dramatically illustrated by cases of ambiguity of perceived modality—in which the subject is uncertain or mistaken as to which of his senses provides him with a particular percept. Totally blind people, for example, can often detect obstacles by what they call "facial vision." This feels to them as if it depended on the skin of their faces; but experiment shows that they are relying on acoustical echoes of ambient noise, in the absence of which they become "facially blind" (Cotzin and Dallenbach, 1950; Wilson, 1966). Only by making special efforts can they perceive the acoustical phenomena as such. Bach-y-Rita (1969) reports a similar ambiguity of perceived modality in subjects presented with optical information through a tactile display on the skin of the back, derived from the output of a forward-facing television camera. After a few hours of active use, the subject spontaneously senses the objects televised as in front of him and likens this experience to visual perception.

A particular case of some interest is the perception of speech. In an early paper on automata (MacKay, 1951), I suggested that analysis of complex patterns might be carried out advantageously by a process of "internal replication"—the structure of commands to the internal generator then serving as a compact specification of the pattern replicated. This would be particularly useful if the replicating procedure was unaffected by transformations (such as translation or magnification of an optical figure) that left the pattern invariant.

In the field of speech perception, this idea (termed "analysis by synthesis") has been espoused by a number of workers (e.g., Halle and Stevens, 1962). It has also been criticized by others (e.g., Lane 1965) on experimental grounds, such as that subjects unable to speak a given language can identify sounds in that language just as well as those who do speak it. In terms of our present discussion, the question to ask is *at what level* (if any) of the generative hierarchy an "internal replication" of speech would be advantageous. I have argued elsewhere (MacKay, 1967) that, although identification

of speech sounds as *acoustical* events might be negligibly facilitated by analysis by synthesis, the perception of the *communication intended by their originator*, i.e., of what he is trying to do by emitting them, might indeed be best achieved by the imitative running of the corresponding high-level organizers in the listener's own system. These could be activated under feedback and feed forward, in such a way as to keep up a dynamic match with the generation of the speaker's utterance. In subjective terms, I perceive what a speaker is up to when I find myself "internally shadowing" not the words as such, but the goal-directed strategy that (in me) would be expressed by the words he is using. To revert to the formalism discussed above (*A Functional Approach to Perception*), the matching transformation ($T_M$) is now from goal strategy (G) to word sequence $[T_M(G)]$. If the sensory signal is a corresponding transform $[T_S(S)]$ of the speaker's goal strategy (S), the matching goal strategy (G) in the listener becomes an internal model of that of the speaker (S). He perceives the speaker as a *purposeful user* of speech, rather than as merely an acoustical source.

In short, the notion of a matching response is much more general than that of simple imitation; but when it would make sense for the perceiver to use part of his own generative machinery as a model or analogue (rather than just a correlate) of the source of sensory signals, internal imitation or analysis by synthesis would be worth looking for in the neural machinery. This point is particularly relevant to interpersonal perception and communication, including failure of communication (MacKay, 1964a).

## Conclusion

This paper has not been designed as a systematic review of the literature of perceptual psychophysiology. Excellent surveys of this kind are available (Teuber, 1960; Vastola, 1968). The question we have been considering is, rather, a methodological one. On what principles, and with what kinds of questions in mind, should we as neuroscientists hope to use perceptual phenomena to guide or discipline our theorizing and experimentation? How may we expect to relate the content of subjective experience with the contents and activity of the brain?

We have rejected attempts to read the subjective description of a percept as a description of a neural image, noting some of the pitfalls that have beset brain science when such attempts have been made. Instead, we have suggested that the two be regarded merely as correlates, which only in special cases might turn out to be analogous.

If the activity of the brain includes the correlate of the activity of perceiving, we must expect to find corresponding features in the two patterns of activity, once we know how to look for them; but normal perception gives few clues to any possible correspondence. Hence the importance of perceptual singularities and anomalies. When a stimulus becomes just detectable, or shows a just-noticeable perceptual change, it makes sense to ask which feature of what brain process changes significantly. With a perceptual illusion, we can ask what sensory processing elements show corresponding signs of bias and the like. When two concurrent stimuli interfere perceptually, we can look for concomitant lateral interactions between their presumed neural counterparts. With a brain lesion, we can at least seek hints to normal function from the nature of perceptual deficits, although this method, too, has many pitfalls (Teuber et al., 1960).

From an information-engineering analysis of the function of a sensory system, we have seen the importance of keeping distinct the two correlated brain activities presumably involved in perceiving: (1) the elaboration of the sensory demand, and (2) the resulting organization of the conditional readiness to reckon with the state of affairs perceived. Our suggestion has been that the brain activity correlated with perceiving (as a conscious experience) be thought of as the *updating* activity required on the part of the internal evaluator of the match or mismatch between demand and conditional readiness. Biases, changes of operating characteristics, anomalous interactions, and the like on either side of this matching process may be expected to give rise to distortions of perception. On this basis, the neural correlate of conscious experience could be identified with the constantly shifting frontier of evaluation and adjustment of the balance between demand and response. The latter would include the running readjustment of goal priorities that we call self-determination (MacKay, 1966).

Space limitations do not permit a discussion of experimental evidence bearing on the neural location of this shifting frontier, which has been reviewed in the symposium *Brain and Conscious Experience* (Eccles, 1966). Some recent work by Libet (1965) and Libet et al. (1967) is of particular interest in showing how elaborate and extensive the pattern of evoked activity may be in sensory cortex without giving rise to conscious sensation. Clear occipital evoked potentials may also be recorded in response to modulated light under conditions in which a subject is unable to detect any flicker (Van der Tweel and Verduyn Lunel, 1965; Regan, 1968). Corresponding experiments in dogs with implanted electrodes have shown a strong correlation between the behavioral threshold for perception of flicker and the amplitude of second harmonic signals at the lateral geniculate, rather than at the cortical level (Lopes da Silva, 1970).

In short, it is clear that much cortical activity can occur in response to sensory stimulation without eliciting conscious experience, and it may even be doubted whether the traditional view of the cerebral cortex as the "essential physiological substrate of consciousness" has much to justify it. We may be tempted to take the other extreme, on the ground that, if all parts of the sensorimotor causal chain are active in ways correlated with what is perceived, we might as well regard the whole chain as the "essential substrate." This, however, is too simple; for it is already possible to break into the chain at various points, activating one part while the rest remains undisturbed, and in this way narrowing down the necessary and sufficient physiological conditions for conscious experience. Sooner or later, no doubt, we shall find ourselves at a point where we are "peeling the onion," and further analysis would become self-defeating. But that time is not yet; further important discoveries may be expected to result from the contemporary conjunction of psychological and physiological methods in the study of perception.

## REFERENCES

ANSTIS, S. M., R. L. GREGORY, N. DE M. RUDOLF, and D. M. MAC-KAY, 1963. Influence of stroboscopic illumination on the aftereffect of seen movement. *Nature* (*London*) 199: 99–100.

BACH-Y-RITA, P., 1969. A tactile vision substitution system based on sensory plasticity. Proc. Second Conf. on Visual Prosthesis, Chicago (in press).

BARLOW, H. B., and R. M. HILL, 1963. Selective sensitivity to direction of movement in ganglion cells of the rabbit retina. *Science* (*Washington*) 139: 412–414.

BRINDLEY, G. S., and W. S. LEWIN, 1968. The sensations produced by electrical stimulation of the visual cortex. *J. Physiol.* (*London*) 196: 479–493.

CAMPBELL, F. W., and J. G. ROBSON, 1968. Application of Fourier analysis to the visibility of gratings. *J. Physiol.* (*London*) 197: 551–566.

COTZIN, M., and K. M. DALLENBACH, 1950. "Facial vision": The role of pitch and loudness in the perception of obstacles by the blind. *Amer. J. Psychol.* 63: 485–515.

ECCLES, J. C. (editor), 1966. Brain and Conscious Experience. Springer-Verlag, New York.

EVANS, E. F., and I. C. WHITFIELD, 1964. Classification of unit responses in the auditory cortex of the unanesthetized and unrestrained cat. *J. Physiol.* (*London*) 171: 476–493.

FECHNER, G. T., 1860. Elemente der Psychophysik, Vol. 2. Breitkopf and Härtel, Leipzig, pp. 548–560.

GAZZANIGA, M. S., J. E. BOGEN, and R. W. SPERRY, 1965. Observations on visual perception after disconnexion of the cerebral hemispheres in man. *Brain* 88: 221–236.

GIBSON, J. J., 1950. The Perception of the Visual World. Houghton Mifflin, Boston.

HALLE, M., and K. N. STEVENS, 1962. Speech recognition: A model and a program for research. *Inst. Radio Eng. Trans. Inform. Theory* 8: 155–159.

HARRIS, C. S., and A. R. GIBSON, 1968. Is orientation-specific color adaptation in human vision due to edge detectors, afterimages, or "dipoles"? *Science* (*Washington*) 162: 1506–1507.

HEAD, H., 1920. Studies in Neurology. Oxford Medical Publications, London, 2 vols.

HOLMES, G., 1918. Disturbances of visual orientation. *Brit. J. Ophthalmol.* 2: 353–384.

HOLST, E. VON., 1957. Aktive Leistungen der menschlichen Gesichtswahrnehmung. *Studium Gen.* 10: 231–243.

HUBEL, D. H., and T. N. WIESEL, 1959. Receptive fields of single neurones in the cat's striate cortex. *J. Physiol.* (*London*) 148: 574–591.

JEFFREYS, D. A., 1969. *In* Evoked brain potentials as indicators of sensory information processing (D. M. MacKay, editor). *Neurosci. Res. Program Bull.* 7 (no. 3): 217.

JULESZ, B., 1960. Binocular depth perception of computer-generated patterns. *Bell Syst. Tech. J.* 39: 1125–1162.

KÖRNER, F., and J. DICHGANS, 1967. Bewegungswahrnehmung, optokinetischer Nystagmus und retinale Bildwanderung. *Albrecht von Graefes Arch. Klin. Exp. Ophthalmol.* 174: 34–48.

KOHLER, I., 1951. Über Aufbau und Wandlungen der Wahrnehmungswelt. *Sitzungsber. Oster. Akad. Wiss.* (*Phil.-Hist. Kl.*) 227 (no. 1): 1–118.

KOHLER, I., 1962. Experiments with goggles. *Sci. Amer.* 206 (no. 5): 62–72.

LAND, E. H., 1959a. Color vision and the natural image. Part I. *Proc. Nat. Acad. Sci. U. S. A.* 45: 115–129.

LAND, E. H., 1959b. Color vision and the natural image. Part II. *Proc. Nat. Acad. Sci. U. S. A.* 45: 636–644.

LANE, H., 1965. The motor theory of speech perception: A critical review. *Psychol. Rev.* 72: 275–309.

LETTVIN, J. Y., H. R. MATURANA, W. S. MCCULLOCH, and W. H. PITTS, 1959. What the frog's eye tells the frog's brain. *Proc. Inst. Radio Eng.* 47: 1940–1951.

LIBET, B., 1965. Cortical activation in conscious and unconscious experience. *Perspect. Biol. Med.* 9: 77–86.

LIBET, B., W. W. ALBERTS, E. W. WRIGHT, JR., and B. FEINSTEIN, 1967. Responses of human somatosensory cortex to stimuli below threshold for conscious sensation. *Science* (*Washington*) 158: 1597–1600.

LOPES DA SILVA, F. H., 1970. Dynamic Characteristics of Visual Evoked Potentials. Grafisch Bedrijf Schotanus and Jens, Utrecht.

MCCOLLOUGH, C., 1965. The conditioning of color-perception. *Amer. J. Psychol.* 78: 362–378.

MACKAY, D. M., 1951. Mind-like behaviour in artefacts. *Brit. J. Phil. Sci.* 2: 105–121.

MACKAY, D. M., 1952. In search of basic symbols. *In* Transactions of the 8th Conference on Cybernetics, 1951, New York (H. von Foerster, editor). Josiah Macy, Jr., Foundation, New York, pp. 181–221.

MACKAY, D. M., 1956. Towards an information-flow model of human behaviour. *Brit. J. Psychol.* 47: 30–43.

MACKAY, D. M., 1957a. Moving visual images produced by regular stationary patterns. *Nature* (*London*) 180: 849–850.

MACKAY, D. M., 1957b. Some further visual phenomena associated with regular patterned stimulation. *Nature* (*London*) 180: 1145–1146.

MACKAY, D. M., 1957c. The stabilization of perception during voluntary activity. *In* Proceedings of the 15th International Congress of Psychology, North-Holland Publishing Co., Amsterdam, pp. 284–285.

MACKAY, D. M., 1958. Perceptual stability of a stroboscopically lit visual field containing self-luminous objects. *Nature* (*London*) 181: 507–508.

MACKAY, D. M., 1961a. Interactive processes in visual perception.

*In* Sensory Communication (W. A. Rosenblith, editor). M. I. T. Press, Cambridge, Massachusetts, pp. 339–355.

MACKAY, D. M., 1961b. The visual effects of non-redundant stimulation. *Nature* (*London*) 192: 739–740.

MACKAY, D. M., 1962. Theoretical models of space perception. *In* Aspects of the Theory of Artificial Intelligence (C. A. Muses, editor). Plenum Press, New York, pp. 83–104.

MACKAY, D. M., 1963. Psychophysics of perceived intensity: A theoretical basis for Fechner's and Stevens' laws. *Science* (*Washington*) 139: 1213–1216.

MACKAY, D. M., 1964a. Communication and meaning—a functional approach. *In* Cross-Cultural Understanding: Epistemology in Anthropology (F. S. C. Northrop and H. Livingston, editors). Harper and Row, New York, pp. 162–179.

MACKAY, D. M., 1964b. Dynamic distortions of perceived form. *Nature* (*London*) 203: 1097.

MACKAY, D. M., 1966. Cerebral organization and the conscious control of action. *In* Brain and Conscious Experience (J. C. Eccles, editor). Springer-Verlag, New York, pp. 422–445.

MACKAY, D. M., 1967. Ways of looking at perception. *In* Models for the Perception of Speech and Visual Form (W. Wathen-Dunn, editor). M. I. T. Press, Cambridge, Massachusetts, pp. 25–43.

MACKAY, D. M., 1970. Visual stability and voluntary eye movement. *In* Handbook of Sensory Physiology (R. Jung, editor). Springer-Verlag, New York (in press).

RATLIFF, F., 1965. Mach Bands: Quantitative Studies on Neural Networks in the Retina. Holden-Day, San Francisco.

REGAN, D., 1968. A high frequency mechanism which underlies visual evoked potentials. *Electroencephalogr. Clin. Neurophysiol.* 25: 231–237.

RUSHTON, W. A. H., 1961. The eye, the brain and Land's two-colour projections. *Nature* (*London*) 189: 440–442.

SPERRY, R. W., 1966. Brain bisection and mechanism of consciousness. *In* Brain and Conscious Experience (J. C. Eccles, editor). Springer-Verlag, New York, pp. 298–313.

STEVENS, S. S., 1961. The psychophysics of sensory function. *In* Sensory Communication (W. A. Rosenblith, editor). M. I. T. Press, Cambridge, Massachusetts, pp. 1–33.

SUGA, N., 1964. Recovery cycles and responses to frequency modulated tone pulses in auditory neurones of echo-locating bats. *J. Physiol.* (*London*) 175: 50–80.

TEUBER, H.-L., 1960. Perception. *In* Handbook of Physiology (J. Field, editor). American Physiological Society, Washington, D. C., Section 1: Neurophysiology, vol. 3, pp. 1595–1668.

TEUBER, H.-L., W. S. BATTERSBY, and M. B. BENDER, 1960. Visual Field Defects After Penetrating Missile Wounds of the Brain. Harvard University Press, for the Commonwealth Fund, Cambridge, Massachusetts.

VAN DER TWEEL, L. H., and H. F. E. VERDUYN LUNEL, 1965. Human visual responses to sinusoidally modulated light. *Electroencephalogr. Clin. Neurophysiol.* 18: 587–598.

VASTOLA, E. F., 1968. Localization of visual function in the mammalian brain—A review. *Brain, Behav. Evolut.* 1: 420–471.

WILSON, J. P., 1966. Psychoacoustics of obstacle detection using ambient or self-generated noise. *In* Les Systèmes Sonars Animaux, Biologie et Bionique, Vol. 1 (R. G. Busnel, editor). INRA-CNRZ, Jouy-en Josas, France, pp. 89–114.

# 31 Two Modes of Processing Spatially Distributed Visual Stimulation

## RICHARD HELD

IN THE LAST DECADE, investigators in the field of perception have been enormously and, I believe, justifiably impressed with the discoveries made with microelectrodes applied to sensory (especially visual) nervous systems. The recorded responses of single units to complex and highly specific stimuli are of great interest to perceptionists, as attested by an ever-increasing number of references made in their writings. Why have these discoveries been so impressive? The answer is implicit in the meaning of the word "perception" —the apprehension of objects and events in the world. Classically, perception has been contrasted with sensation, which was often assumed to be a direct registration of properties provided by peripheral receptors that transduce stimuli affecting them (Boring, 1942). Many problems have been posed by this opposition between sensation and perception. How are the objects of perception derived from the discontinuous set of sensory elements provided by the receptor mosaic? The visual world consists of global entities— segregated objects with continuous contours. How are their neural correlates organized out of the set of elements provided by the receptor mosaic? The objects of visual perception retain their identity despite various perspectival distortions and displacements of their images over the sensory mosaic produced by changing conditions of view. What accounts for their invariance of shape under these transformations? Clearly, further processing above the level of the receptors has been indicated, but physiological evidence for such processing has not been available. This ignorance has been breached only with the discoveries during the last few years of many types of receptive fields, feature detectors, and selective filters in sensory nervous systems. These new discoveries are regarded by commentators as the beginning of an understanding of how the brain performs the processing underlying perception. Indeed, Hubel and Wiesel, who are responsible for major advances in the area, have referred to the response specificity shown by cells in the visual cortex as revealing the ". . . building blocks of perception . . ." (1965).

In order to make the opposition between classical and modern approaches more concrete and to raise the issue under discussion, let us consider two models of the visual nervous system (Figures 1 and 2). Both are concerned with the neural processing of spatially distributed stimulation on the retinae. In the seventeenth century, Descartes proposed his schema with the intent of solving a problem posed by the laws of image formation, then recently discovered by Kepler (Descartes, 1664). The troublesome inversions and doubling of the retinal images produced by the lens system of the eye were eliminated by an inverse transformation imputed to the nervous system. The point-to-point projection of the retinal images onto the ventricle surfaces and thence to the pineal body preserved information about the loci of these points of origin on the retinae. Preservation of their ordering maintained the shape of the object. Descartes wisely left the problems of perception to the action of the soul. His model of point-to-point projection retained its importance at least through the nineteenth century, and shades of it still remain, as is witnessed by a recently published statement of the late Kenneth Ogle (1969). "We must attribute the discrimination of differences in visual direction, first, to the discrete character of the organization of the retinal receptors, an organization that consists of a mosaic of separated and effectively insulated light-sensitive elements and, second, to the continued topographical identity of those elements in the organization of nerve fibers connecting them to the terminal areas in the occipital cortex of the brain."

Another type of model (Figure 2), derived from consideration of the responses of single cells in visual cortex, was recently proposed by Hubel and Wiesel (1962). This schema envisages repeated convergence of the outputs of lower-order cells upon higher-order cells in the visual nervous system to yield the complex receptive fields that these authors have observed. For each cell of its type, there are myriads of similar ones with slightly differing receptive-field properties. From this array, the extraction of more complex shapes by further combination seems plausible. As the size of the receptive field increases, however, specificity of locus of origin on the retina is reduced. The spatial resolution provided by such cells cannot be better than the size of

RICHARD HELD  Massachusetts Institute of Technology, Cambridge, Massachusetts

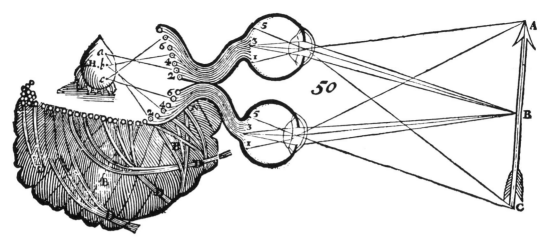

FIGURE 1   Descartes' diagram of the formation of images on the retinae of the eyes and the paths of transmission in the visual nervous system.

their receptive fields, which typically measure several degrees of visual angle. Better resolution can, of course, be obtained from the aggregate of these cells by further combinations also envisaged in principle by Hubel and Wiesel (1965). But such recombinations for the purpose of recovering information available at lower levels in the system, prior to initial convergence, seem uneconomical at best.

In the Cartesian model, how figural properties are extracted from the aggregate of centrally projected points, which preserve locus information, is a mystery. In the modern model, central cells respond to interesting global features of stimulation at the cost of a loss of locus-specific information. The difference of circuitry in these models suggests that processing for figural, as opposed to locus, information may be performed by different modes of analysis, and that economy might be served by separation of the mechanisms performing them.

The problem of locus becomes interesting when we are concerned with metrical properties of perceived space; direction, distance, and other magnitudes. It is most interesting in connection with the control of movement. Sensorially guided movements are always directed with respect to target loci in space. Their control must be guided by information about locus that is at least as accurate and precise as that of the terminus of the movement. Consequently, one should expect that locus-specific information in the visual nervous system would be closely associated with the system for control of orienting and localizing movements.

I shall review several experiments done in my laboratory and several from other laboratories; some strictly behavioral and some combining behavioral studies with either

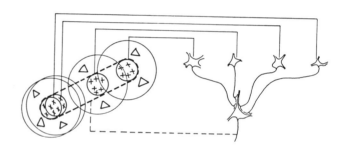

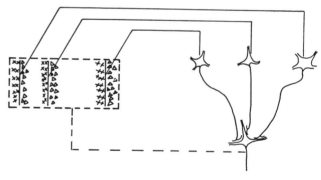

FIGURE 2   Hubel and Wiesel model of convergence of the outputs of lower-order on higher-order neurons shown on the right side of diagrams. By this means, more complex and larger receptive fields (dashed lines) are built from simpler ones (solid lines), as shown on the left side of the diagrams. The upper diagram shows how cells in the lateral geniculate may connect with cells in striate cortex; the lower diagram shows how cells in visual cortex may connect with other cells in the same region to form combined receptive fields.

neurosurgery or electrophysiological recording. The review has a dual purpose: first, to convince the reader that it is useful to make the distinction between two modes of analysis of spatially distributed stimulation; second, to present samples of several types of research, done by neuropsychologists, which can be brought to bear on a particular systematic issue.

## Experiments on behavior

A PERCEPTUAL ILLUSION  In the 1860s, Helmholtz observed that when a checkerboard is placed quite close to the eye, with the gaze centered on the board (Figure 3), the contours will appear straight only if they are bowed toward the point of fixation (1962). He believed that the correct curvature was hyperbolic. Dr. Mary Parlee and I became interested in this illusion, for reasons that will become apparent, and made the following measurements (Parlee, 1969). While fixating a point (gazing at it steadily), observers viewed a section of a hyperbola flashed on a screen for 100 milliseconds. The width of the line was 15 minutes of angle, and its length subtended 32 degrees of visual angle. Other dimensions are shown in Figure 4A. The hyperbolic section was varied in curvature from concave to the right, through straight, to concave left, and back in successive presentations. The observer was instructed to choose that curvature which appeared "straight." The curve was presented either 10 degrees to the right of fixation, or 10

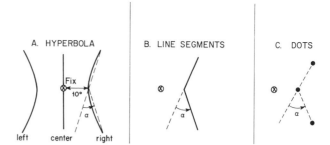

FIGURE 4 A–C  Flashed stimuli in which the angle, *a*, was varied in a psychophysical procedure designed to assess the observer's judgment of their straightness.

degrees to the left, or centered through the fixation point. The same procedure was used to present two straight-line segments rotated about their joint so as to vary the angle $\alpha$ (Figure 4B). In addition, observers were asked to align three dots by the same technique (Figure 4C). The results showed that the settings to apparent straightness for both hyperbolas and line segments, presented 10 degrees removed from fixation, yielded $\alpha$ values that average fractions of a degree. Angle $\alpha$ for the dots, however, approached 10 degrees. The results suggested that the dot stimuli were processed in a manner that differed from the line stimuli—a manner that introduced a 10-degree bias into setting to the subjective criterion of straightness.

We suspected that the dots were, of necessity, processed by a locus-specific mode of analysis, because they were a very sparse stimulus for contour analysis. Following the logic above, we decided to tap the system for guided control of movement by having our observer actually point at the central position and the two extreme positions of the stimuli without seeing his hand. As shown in Figure 5, the observer was given a pen with which he could repeatedly mark the apparent locations of the middle and end positions of the target stimuli, which were viewed fully reflected in a mirror. The results may seem surprising to those who think of vision as a unitary process: they implied that a bias of $\alpha$ equal to many degrees will appear irrespective of whether the subject points at either lines or dots set in objective alignment. When the three dots were, in fact, set so as to appear to form a straight line, markings of them formed an angle $\alpha$ not significantly different from zero. The same bias was found when the dots were presented one at a time to be marked, thereby indicating the independence of these responses from global properties of the stimulus. On the other hand, markings of the line-segment stimuli, set at apparent straightness, did not form a straight line.

This experiment, and others like it, can be interpreted as revealing the operation of two modes of analysis, in accord with my previous discussion. We conclude that the locus-

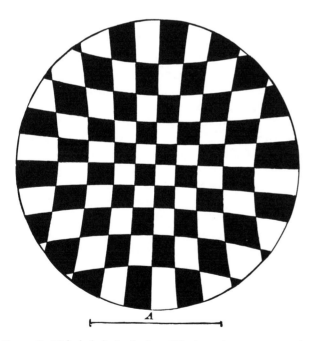

FIGURE 3  Helmholtz's checkerboard illusion. Lines appear straight when viewed at distance A drawn to the same scale as the figure.

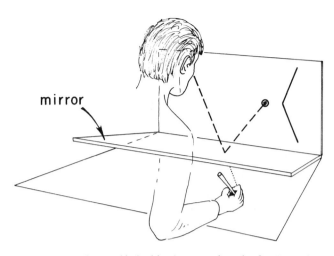

FIGURE 5 With gaze (dashed line) centered on the fixation point, the observer marked with a pen three positions (middle and two ends) of the flashed stimulus viewed in a fully reflecting mirror. The apparent locations of the three positions coincided with the marking surface under the observer's hand.

specific mode of analysis is invoked either by impoverishing the stimulus figure—by eliminating extended contour in the case of the dots—or by the requirement of the motor task to point at stimulus loci irrespective of the contouring of the field. These results are reminiscent of findings from another type of experimentation, namely, study of the adaptive consequences of visual rearrangement.

PERTURBING THE SYSTEM BY REARRANGEMENT Donald MacKay has given examples (this volume) of perceptual effects attributed to fatiguing of feature filters. If we want to perturb the visuomotor system, we can do so by means of an optical device that rotates, translates, or distorts the retinal image. The initial errors of orientation and localization, as well as the changed appearances of objects, are predictable. If the subject who suffers the rearranged vision is, however, allowed to perform routine activities in his normal environment while wearing the optical device, all the visuomotor errors diminish toward zero with length of time of exposure, while the nonlinear visual distortions that adapt merely become lessened to some asymptote representing a fraction of the induced distortion. The relative degree of adaptation in overt localizing responses as compared with figural judgments makes a revealing comparison. Of the large number of rearrangement experiments reported in the literature (Howard and Templeton, 1966), a few are relevant to the present question.

The wedge prism has been the most often used optical-rearranging device, at least since the time of Helmholtz (1962). I shall discuss two of the effects of the prism on

vision. First, it produces a gross displacement of the apparent direction of the object in the apexward direction that may be called the linear component. Second, there is a nonlinear component of refraction (variation in the displacement with angle of incidence of rays) that causes straight lines, perpendicular to the base-apex axis of the prism, to appear curved (Ogle, 1952). If one measures the orientation of head and body to a small visible target before and after wearing prisms fitted in goggle frames, full and exact compensation can take place, provided the rearranged subject wears the goggles in his normal environment for a sufficient time (Held and Bossom, 1961). The subject eventually localizes the target correctly despite the optical displacement produced by the prism. We can regard this change as a recalibration of the linkage between locus-specific analyzers and the generator of orienting responses.

If we consider only the linear component of prism displacement, no distortion of forms is seen by the subject. As mentioned above, however, there is also the nonlinear component, which makes straight lines appear curved. We can test for adaptation to the curvature distortion by using a psychophysical procedure. The subject views a grating through a prism of variable power that he controls and, by this means, nulls out the apparent curvature. When this is done, one finds a rather limited amount of adaptation that asymptotes at a small fraction of the prism-induced curvature (Kohler, 1964).

It is revealing to compare the full compensation shown possible for linear components of rearrangement with the fractional compensation shown for nonlinear rearrangement, even after prolonged exposure. Only the latter type effects a change in the apparent shapes of objects. Returning to the consideration of two modes of processing, these results may be interpreted in terms of a constraint that the system for figure identification exerts on the orienting system (Held, 1968). Compensation for the nonlinear distortion would be realized as a set of changes in the apparent direction of positions in the field of vision, the intersection of which in a frontal plane in space would define a changed shape. The system for figure identification, if operating independently and not subject to the same kind of adaptation, would have, however, an unchanged criterion for the psychophysical judgment of shape. The latter could constrain the amount of compensation revealed.

The operation of these two systems has been dramatically shown by Kohler (1964), who used a prism that covered half of the field of view. The discontinuity introduced into the field of vision is adapted to, in the sense that orienting movements to targets seen both above and below the discontinuity are reported to become accurate. The appearance of the discontinuity is never corrected, however, leading to the paradox that the two separated parts of a vertical line, seen above and below the discontinuity in the field, although reported to lie in the same direction, nevertheless appear discontinuous (Figure 6).

Two further observations on a special kind of adaptation to rearrangement are relevant. If only the hand is viewed through the prism, as was originally done by Helmholtz, then the direction of reach shifts during exposure to compensate for the optical displacement. Moreover, the shifts have been found to be specific to the exposed hand and do not transfer to the opposite hand, which was not previously viewed through the displacing prism. The recalibration of the linkage between locus-specific analyzers and oriented responses may be made specific to one or the other limb. This specificity contrasts sharply with the lack of specificity in responses contingent on the figure-analyzing system. It suggests that the locus-specific analysis occurs close to the origin of motor commands. Clearly, if a perceiver learns either to identify a particular shape, or otherwise respond to it, we expect that he should be able to do so by any of a variety of response indicators. Indeed, one of the defining properties of perceptual activity is its generalization over types of response. Finally, we have performed these experiments with the dark-adapted as well as the light-adapted eye and have found comparable effects (Graybiel and Held, in press). Scotopic vision is adequate for adaptation, which indicates that the high-acuity apparatus which serves for detailed pattern vision is not necessary for recalibration of the orienting response.

DEPRIVATION EXPERIMENTS   Observations of the modifiability of visual-motor coordination in both human and animal subjects led us to question the role of this modifiability in the development of the neonate (Held and Bossom, 1961). A traditional approach to questions of early development is the procedure of deprivation. If one suspects that some form of exposure to the environment is crucial for development of function, then simply eliminate that experience until the time in the life cycle when a normal animal would show the behavior of interest. If the animal is deficient in the predicted manner, the absent experience is

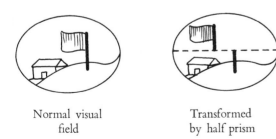

Normal visual
field

Transformed
by half prism

FIGURE 6   Appearance of objects without and with half-prism spectacles in place over the eyes.

implicated, although the mechanism involved may remain in question.

A series of experiments, performed in several laboratories, has demonstrated that early deprivation of patterned stimulation to both eyes produces deficits in visual-motor coordination. Contrary to older accounts (Riesen, 1958), a recent report has suggested that form discrimination does not suffer greatly from this type of deprivation (Meyers and McCleary, 1964). Dr. Alan Hein and I found very similar visual-motor deficits when we used the milder deprivation procedure of allowing pattern vision to kittens during development but preventing visual feedback from self-produced movements. One such procedure involved a comparison of active with passive movement (Held and Hein, 1963). The active animals achieved guided movements; their passively transported litter mates did not. More important for the present question, we also demonstrated that the two eyes may be dissociated in their capacity to control movement if one eye is closed only during active movement, the other only during passive movement (Figure 7). No such dissociation between eyes performing figure discriminations was found in testing procedures that preclude the use of visually guided responses (Meyers and McCleary, 1964).

Along similar lines, Alan Hein and I recently reared kittens which during their early life, wore ruffs around their necks when in light. These obscured the view of their limbs and bodies but reduced mobility only minimally (Hein and Held, 1967). In order to test their visually guided limb movements, we used a variant of the classic visual-placing response. The animals were brought down toward a set of horizontal prongs, which triggered extension of the forelimbs. In a normal animal, this limb extension is guided so that the paws almost always strike the prongs and rarely fall between them. Our deprived animals, however, which had never viewed their forelimbs, could not guide them accurately, according to the frequency with which their paws extended into the interstices between the prongs. Similar results have been obtained with infant monkeys reared without sight of their limbs (Held and Bauer, 1967).

Relevant to the question of the two modes of analysis was the outcome of a procedure designed to rear kittens in such a way that one eye viewed the limbs but the other eye did not (Figure 8). The eye that did not view the limbs failed to guide their movements on the prong test, but the contralateral eye did so. Consequently, it has been shown possible to dissociate the two eyes with respect to control of the limbs (Hein et al., in press). To repeat, this finding is important in light of the evidence that the two eyes of a normal animal cannot be dissociated with regard to figure identification. I should also point out that Alan Hein has shown that it is possible to expose selectively one limb of a kitten and thereby to produce an animal that can accurately

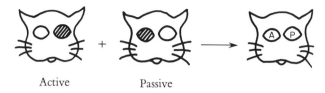

Active      Passive

FIGURE 7 When in an illuminated environment, each kitten received an equivalent amount of visual stimulation during active movement when one eye was occluded (shading) and during passive transport when the other eye was occluded.

guide one limb from only one eye, the other limb from both eyes (Hein, in press).

To sum up the results of the three types of experimentation so far discussed, dissociation of contour-specific from locus-specific modes of analysis apparently can be inferred from the results of behavioral studies of illusions and of adaptation to rearrangement. The constraints exerted by figural properties on directed responses indicate an interaction between these modes which is inadequately understood at this time. When the locus-specific system is tapped by early deprivation, it appears that, unlike the contour-specific system, both receptor organs (the two eyes) and effectors (the two forelimbs) may be dissociated.

## Behavioral effects of intervention in the nervous system

SPLIT-BRAIN EXPERIMENTS The effects of surgical intervention on certain behaviors bear on our problem. In recent

Opaque ruff        Transparent ruff
3 hours daily       3 hours daily

FIGURE 8 During each day, when free to move about in an illuminated environment, each kitten wore an opaque ruff for three hours while one eye was occluded and a transparent ruff for another three hours while the other eye was occluded.

years, a long series of experiments has been performed on monkeys and cats with surgical transections along the midline of the brain. Most of this research has been done under the impetus provided by Roger Sperry and his collaborators (Sperry, 1961). The experiments have shown that, in animals with optic chiasm and forebrain commissures split, discriminations of visual form, established by the training of only one eye, are not performed when tested with the contralateral eye. Comparison between figures presented to opposite eyes, and hence to opposite hemispheres, is difficult if not impossible. It appears, then, that these forebrain commissures are, in general, required for the transfer of figural information from one side of the brain to the other. Contrary to this finding, Bossom and Hamilton (1963) and Hamilton (1967) found no deficit in the use of any eye-hand pair in these animals, which implies that locus-specific information is available to both sides of the brain in the absence of forebrain commissures. Moreover, studies of prism adaptation, in which only one eye of the split-brain animal was exposed, show complete transfer to the other eye, as in normal animals. The authors of these experiments have suggested that the organization of the visual-motor system, as contrasted with that for figure discrimination, entails midbrain and even brain-stem mechanisms, which allow subcerebral cross-communication between hemispheres.

The results with split-brain animals suggest that the locus-specific information is available at subcortical levels and, hence, can cross to the opposite hemisphere through subcortical commissures. Figure-specific information, on the contrary, appears to be available only in the cerebrum, as judged by its ready transfer only through the cerebral commissures. This inference is at least partially borne out by the results of lesions limited either to the forebrain or midbrain structures concerned in vision.

THE DE-STRIATE MONKEY Contrary to previous accounts, Humphrey and Weiskrantz (1967) and Humphrey (in press) report that monkeys suffering removal of striate cortex are capable of performing visually guided behaviors. Such animals have now been shown to orient eyes and head to an object that is isolated on an otherwise homogeneous background, and then to reach for it with reasonable accuracy in direction but not in depth. They cannot do so for objects presented in an otherwise configured field of vision. They cannot discriminate form in the manner in which the normal monkey is so proficient. They appear to be incapable of identifying visible objects by any of the gross criteria of recognition. De-striation appears, then, to have crippled the apparatus for figure-specific analysis. Visual centers in either extra-striate cortex or in subcortical regions must, then, mediate the visual control of movement in these animals.

THE DE-STRIATE AND THE COLLICULECTOMIZED HAMSTER The role of a midbrain visual center in visual-motor control has been demonstrated clearly in at least one species of mammal. In an ingenious study of the behavior of the Syrian golden hamster, Gerald Schneider compared the effects of ablating either visual cortex or the superior colliculus of his animal of choice. The results showed a striking double-dissociation of function (Schneider, 1969). The animals deprived of visual cortex showed a loss of ability to discriminate between simple forms, although they were quite capable of orienting their heads toward visually presented objects (usually sunflower seeds). On the other hand, the colliculectomized animals were quite incapable of orienting to visible objects or of going from starting position to goal under visual guidance. Yet they showed themselves capable of making a form discrimination when allowed to guide themselves to the stimulus figure (entrance to goal box) by tactile means. In this species, at least, there is a highly selective effect of the ablation of one or the other structure.

MONOCULAR PATTERN DEPRIVATION A nonsurgical technique for producing central dysfunction in the visual system was discovered by Wiesel and Hubel (1963). They have shown that prolonged monocular occlusion in kittens, if begun at a very young age, causes a marked loss in responsivity of cells in the visual cortex that would normally be driven by stimulation of the formerly closed eye. In a follow-up study, Ganz and Fitch (1968) also came to the conclusion that there are profound losses after monocular deprivation, both in perceptual-motor coordination and in the discrimination of visual form by the affected eye. After exposure to a normal environment, however, much of the missing visual-motor coordination develops, although pattern vision remains deficient, and this recovery takes place despite the absence of cortical units with what Ganz et al. (1968) called highly selective receptive fields. We might refer to these cats as partially de-striated with respect to the deprived eye. Yet, they are quite capable of visual control of at least gross orienting movements, given the exposure to the environment required for perfecting these capabilities. To summarize our interpretation of these experiments, we should like to believe that the animal reared with monocular occlusion is one in which the occluded eye not merely fails to drive cortical units responsible for figural analysis, but also fails to guide orienting behavior. Orienting, however, is at least partially recoverable. The two eyes have been temporarily dissociated with regard to control of visually guided behavior but permanently dissociated with regard to figural analysis.

The results of Ganz and his colleagues are a clear example of the double-edged-sword aspect of the deprivation technique. Any significant deviation from the normal conditions under which an animal is reared can have two or more consequences for the growth of the nervous system and the behaviors it controls.

## Conclusions

Some of the grounds for distinguishing between two modes of analysis, and a number of examples of research on vision that justify the distinction, have been reviewed. Further discussions along these lines can be found in a symposium entitled "Locating and Identifying: Two Modes of Visual Processing" by Held (1968), Ingle (1967), Schneider (1967), and Trevarthen (1968). Many other observations make sense when viewed from the perspective of the two modes of analysis, and relevant new observations are being made under its guidance.

## Acknowledgments

Preparation of this report and research in the author's laboratories have been supported by NIMH Grant MH-07642 and NASA Grant NGR 22–009–308.

REFERENCES

BORING, E. G., 1942. Sensation and Perception in the History of Experimental Psychology. Appleton-Century-Crofts, New York.

BOSSOM, J., and C. R. HAMILTON, 1963. Interocular transfer of prism-altered coordinations in split-brain monkeys. *J. Comp. Physiol. Psychol.* 56: 769–774.

DESCARTES, R., 1664. *In* Traité de l'Homme.

GANZ, L., and M. FITCH, 1968. The effect of visual deprivation on perceptual behavior. *Exp. Neurol.* 22: 638–660.

GANZ, L., M. FITCH, and J. A. SATTERBERG, 1968. The selective effect of visual deprivation on receptive field shape determined neurophysiologically. *Exp. Neurol.* 22: 614–637.

GRAYBIEL, A. M., and R. HELD, 1970. Prismatic adaptation under scotopic and photopic conditions. *J. Exp. Psychol.* 85:16–22.

HAMILTON, C. R., 1967. Effects of brain bisection on eye-hand coordination in monkeys wearing prisms. *J. Comp. Physiol. Psychol.* 64: 434–443.

HEIN, A., 1970. Visual-motor development of the kitten. *Opt. Weekly* (in press).

HEIN, A., and R. HELD, 1967. Dissociation of the visual placing response into elicited and guided components. *Science (Washington)* 158: 390–392.

HEIN, A., R. HELD, and E. C. GOWER, 1970. Development and segmentation of visually-controlled movement by selective exposure during rearing. *J. Comp. Physiol. Psychol.* (in press).

HELD, R., 1968. Dissociation of visual functions by deprivation and rearrangement. *Psychol. Forsch.* 31: 338–348.

HELD, R., and J. A. BAUER, JR., 1967. Visually guided reaching in infant monkeys after restricted rearing. *Science (Washington)* 155: 718–720.

HELD, R., and J. BOSSOM, 1961. Neonatal deprivation and adult rearrangement: Complementary techniques for analyzing plastic sensory-motor coordinations. *J. Comp. Physiol. Psychol.* 54: 33–37.

HELD, R., and A. HEIN, 1963. Movement-produced stimulation in the development of visually guided behavior. *J. Comp. Physiol. Psychol.* 56: 872–876.

HELMHOLTZ, H. VON, 1962. Helmholtz's Treatise on Physiological Optics (translated and edited by J. P. C. Southall), Vol. 3. Dover Publishing, New York. (Originally published in German as Handbuch der Physiologischen Optik, III, 1866.)

HOWARD, I. P., and W. B. TEMPLETON, 1966. Human Spatial Orientation. Wiley, New York, Chapter 15.

HUBEL, D. H., and T. N. WIESEL, 1962. Receptive fields, binocular interaction and functional architecture in the cat's visual cortex. *J. Physiol.* (*London*) 160: 106–154.

HUBEL, D. H., and T. N. WIESEL, 1965. Receptive fields and functional architecture in two non-striate visual areas (18 and 19) of the cat. *J. Neurophysiol.* 28: 229–289.

HUMPHREY, N. K., 1970. What the frog's eye tells the monkey's brain. *Brain, Behav., Evolut.* (in press).

HUMPHREY, N. K., and L. WEISKRANTZ, 1967. Vision in monkeys after removal of the striate cortex. *Nature* (*London*) 215: 595–597.

INGLE, D., 1967. Two visual mechanisms underlying the behavior of fish. *Psychol. Forsch.* 31: 44–51.

KOHLER, I., 1964. The formation and transformation of the perceptual world. *Psychol. Issues* 3 (Monogr. 12).

MEYERS, B., and R. A. MCCLEARY, 1964. Interocular transfer of a pattern discrimination in pattern deprived cats. *J. Comp. Physiol. Psychol.* 57: 16–21.

OGLE, K. N., 1952. Distortion of the image by ophthalmic prisms. *A. M. A. Arch. Ophthalmol.* 47: 121–131.

OGLE, K. N., 1969. Visual acuity. *In* The Retina: Morphology, Function, and Clinical Characteristics (B. R. Straatsma, M. O. Hall, R. A. Allen and F. Crescitelli, editors). UCLA Forum in Medical Sciences No. 8, University of California Press, Berkeley and Los Angeles, California, pp. 443–483.

PARLEE, M. B., 1969. Continuous and discontinuous contours: Their roles in localization vs. perception of straightness. M.I.T., Cambridge, Massachusetts, doctoral thesis.

RIESEN, A. H., 1958. Plasticity of behavior: Psychological aspects. *In* Biological and Biochemical Bases of Behavior (H. F. Harlow and C. N. Woolsey, editors). University of Wisconsin Press, Madison, pp. 425–450.

SCHNEIDER, G. E., 1967. Contrasting visuomotor functions of tectum and cortex in the golden hamster. *Psychol. Forsch.* 31: 52–62.

SCHNEIDER, G. E., 1969. Two visual systems. *Science* (*Washington*) 163: 895–902.

SPERRY, R. W., 1961. Cerebral organization and behavior. *Science* (*Washington*) 133: 1749–1757.

TREVARTHEN, C. B., 1968. Two mechanisms of vision in primates. *Psychol. Forsch.* 31: 299–337.

WIESEL, T. N., and D. H. HUBEL, 1963. Single-cell responses in striate cortex of kittens deprived of vision in one eye. *J. Neurophysiol.* 26: 1003–1017.

# 32 The Biogenic Amines in the Central Nervous System: Their Possible Roles in Arousal, Emotion, and Learning

SEYMOUR S. KETY, M.D.

EMOTIONS MAY BE seen as adaptive states of a generalized nature involving the brain as well as the autonomic and endocrine systems in response to a significant change in the environment, and serving to bring about or to facilitate appropriate behavior. Some basic emotions are arousal, pleasure, fear, and rage, and their counterpart behaviors, to attend, approach, avoid, or attack. A body of evidence suggests that some of the biogenic amines play crucial roles at various points in bringing about these states.

*Catecholamine action in the periphery*

The evidence is clear for the involvement of two catecholamines, epinephrine and norepinephrine, in a number of

SEYMOUR S. KETY, M.D.  Department of Psychiatry, Harvard Medical School; Psychiatric Research Laboratories, Massachusetts General Hospital, Boston, Massachusetts

affective states (Kety, 1966). The peripheral adrenergic synapse is one of the few synapses at which a neurotransmitter has been identified and its role established beyond serious doubt (von Euler, 1956). The hypotheses that attribute transmitter functions to biogenic amines in the central nervous system have, in fact, been suggested by analogy with the peripheral adrenergic synapse. Its presynaptic ending has in it all the enzymes that are necessary for the synthesis of norepinephrine (Goodall and Kirshner, 1958; Stjärne and Lishajko, 1967). It has recently been demonstrated by Geffen and Rush (1968) that, in the splenic nerve, all but a negligible amount of the norepinephrine released by stimulation is synthesized at the nerve endings. Much earlier it was shown that norepinephrine is released by stimulation of the sympathetic nerves to an organ. The postsynaptic effects of such nerve stimulation can be duplicated by the application of norepinephrine to that region and can be blocked by the same drugs that block the effects of sympathetic stimulation. There is also reason to believe that norepinephrine is inactivated enzymatically within the nerve ending by monoamine oxidase, and postsynaptically by catechol 0-methyl transferase (Axelrod, 1959). The bulk of the amine, however, is removed from the synaptic cleft by a process of reuptake into the presynaptic terminal (Iversen, 1967).

In recent years, some additional information has appeared about the peripheral adrenergic synapse, and is, I think, relevant to our further discussion. A number of investigations (Alousi and Weiner, 1966; Roth et al., 1966; Sedvall et al., 1968) have shown that the synthesis of norepinephrine in sympathetic nerve endings is firmly coupled to its release. Table I represents some data from Sedvall and his co-authors, who studied the formation of $^{14}$C-norepinephrine from $^{14}$C-tyrosine in the rat submaxillary gland. They examined the gland under three conditions: during the normal resting state, while decentralized, i.e., removed from central impulses, and during stimulation. Because the amine is largely confined to the adrenergic endings, the norepinephrine content of the gland represents the norepinephrine in the endings. The total norepinephrine did not show very significant changes, but the amount of $^{14}$C-norepinephrine that was synthesized under each of these conditions showed a marked variation. From these data it was possible to calculate the minimum rate of synthesis of norepinephrine in the nerve endings of the gland. This shows a good correlation with the amount of nervous activity, falling to less than one-half in the decentralized state, and rising more than two-fold in the stimulated state (Table I).

Much evidence indicates that tyrosine hydroxylase is the rate-limiting enzyme in the synthesis of norepinephrine (Nagatsu et al., 1964; Levitt et al., 1965), and that this amine blocks tyrosine hydroxylase by a process of end-product inhibition (Udenfriend et al., 1965; Weiner and Rabadjija,

1968a). The increased synthesis brought about by stimulation is thought (Weiner and Rabadjija, 1968a) to result at first from a reduction in norepinephrine at the ending and therefore, presumably, a release of end-product inhibition. But, with more chronic stimulation, evidence has been adduced suggesting an induction of tyrosine hydroxylase, as puromycin prevents much of the increased synthesis accompanying chronic stimulation (Weiner and Rabadjija, 1968b).

There is little doubt that norepinephrine, like acetylcholine, is stored in specialized vesicles found in abundance in the presynaptic axon terminals (Bondareff and Gordon, 1966; Hökfelt, 1968). The prevailing assumption has been that the transmitter is synthesized in these vesicles, and released from them by a process of exocytosis (Iversen, this volume).

Although it is compatible with much of the existing data, that model does not readily or parsimoniously explain some observations. The vesicles appear not to decrease in number with fairly severe stimulation, and there is some difficulty in accounting for the formation and disposal of vesicles in sufficient number at the synapse to account for the quantities of transmitter released during moderately high activity. More important, perhaps, release of norepinephrine sequestered in vesicles would not affect the norepinephrine concentration in the vicinity of tyrosine hydroxylase and thus signal increased synthesis. The possibility of refillable vesicles has been suggested in order to take these discrepancies into account, which it does to a considerable extent. Fairly recently, results have been reported by Kopin et al. (1968) that are difficult to explain by models in which both storage and release are accomplished by the same structures.

In one of their experiments (Figure 1) the spleen was perfused with $^{14}$C-tyrosine at a constant rate while the splenic nerve was stimulated continuously. The specific activity of

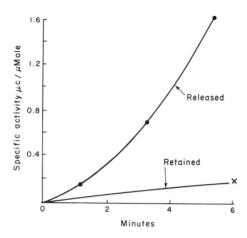

FIGURE 1 Specific activity of NE recovered from perfusate or remaining in the tissue at various times during perfusion of spleen with $^{14}$C-tyrosine under constant stimulation of the splenic nerve. (From Kopin et al., 1968)

the norepinephrine released was measured and compared with that remaining in the spleen, presumably in the adrenergic nerve endings. The specific activity of the released norepinephrine rose rapidly to a level several times that which was retained. Thus, the existence of at least two norepinephrine pools was confirmed, but, more important, this experiment suggests that the bulk of the amine released by stimulation was newly synthesized. These pools could, of course, represent different populations of axons, some with rapid and others with slow turnover, except that those workers who have examined large populations of such neurons by histofluorescent techniques have not observed markedly different rates of depletion or replenishment in adjacent neurons. It is more likely, but not established, that these pools with different rates of turnover are subcellular rather than supracellular. More recently, the preferential release of newly synthesized dopamine has been reported from axon terminals in the isolated striatum (Besson et al., 1969).

One can adduce, as Kopin has in fact done, the existence of two types of vesicles, one primarily for storage and the other for release, with different turnover rates. A more parsimonious model, in which the vesicles serve simply to store the amine, seems capable of accounting for these and many of the other observations. In that model (Figure 2), the newly synthesized amine would be outside the vesicles, perhaps cytoplasmic, and in equilibrium with at least two stores. Some may be concentrated in the presynaptic membrane, from which it could be released into the synaptic cleft during depolarization. This membrane is rich in gangliosides, which have a high affinity for amines. Through dynamic equilibrium, some may be taken up by the vesicles when the cytoplasmic levels are high, and released when the levels are low. An alternative to simple

physical storage and release at the presynaptic membrane would be some active transport process coupled to sodium and calcium (Bogdanski and Brodie, 1969). Monoamine oxidase in the mitochondria and feedback inhibition would tend to keep the cytoplasmic level from rising indefinitely, but a fall in the level would immediately call forth an increased activity of the cytoplasmic tyrosine hydroxylase by release of feedback inhibition.

One can imagine the norepinephrine shuttling back and forth between presynaptic and postsynaptic membranes, driven by depolarization first of one and then the other, with an active re-uptake process to conserve the undegraded amine, and synthesis replacing net loss. Although some of the assumptions of this concept seem worth testing, it is unlikely that it will account for all synaptic release of norepinephrine, because approximately one-third of the amine released during stimulation appears to be derived from a storage pool (Kopin et al., 1968). The demonstration of small amounts of chromogranin, a specific vesicular protein, in the efflux of certain adrenergic endings after stimulation, implies some release directly from vesicles (de Potter et al., 1969), but the quantities that have been recovered would account for only a few percent of the transmitter released.

## Biogenic amines in the central nervous system

Although there is much to be learned regarding the mechanisms of release and action of neurotransmitters at peripheral synapses, a transmitter function appears well es-

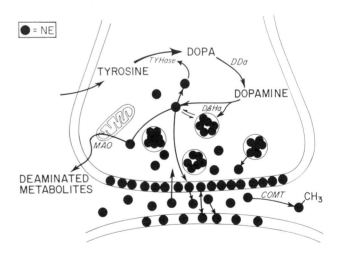

FIGURE 2 A model of the adrenergic synapse, in which vesicles are responsible only for storage, and release occurs from the cytoplasmic pool via some mechanism of accumulation in and efflux from the presynaptic membrane. Abbreviations: TYHase, tyrosine hydroxylase; DDa, dopa decarboxylase; DβHa, dopamine β hydroxylase; COMT, catechol-O-methyl-transferase; MAO, monoamine oxidase.

tablished for norepinephrine, as is its peripheral involvement, along with epinephrine, in emotional expression. In the brain, however, no biogenic amine has been shown conclusively to be a transmitter, or even to be specifically associated with a particular behavioral or affective state. The evidence is indirect and inferential, but, nevertheless, sufficiently extensive and consistent to permit the formulation of hypotheses that central as well as peripheral actions of biogenic amines mediate arousal and emotion.

By means of the characteristic fluorescence of their condensation products after reaction with formalin vapor (Eränkö, 1956), serotonin and catecholamines have been demonstrated throughout the brain and spinal cord, concentrated in the presynaptic varicosities along the terminal axons of particular neurons (Hillarp et al., 1966). The neuronal cell-bodies themselves appear to be clustered in the brainstem, the serotonin-containing neurons in the midline raphé nuclei and those containing catecholamines more laterally disposed with a high density in the locus ceruleus. Axons from those amine-containing neurons pass downward into the spinal cord and upward by way of the medial forebrain bundle to innervate most of the brain (Fuxe, 1965). There are especially high densities of their terminals in the various nuclei of the hypothalamus, but the entire cerebral cortex and even the cerebellar cortex contain many very fine fibers with the characteristic serotonin or catecholamine fluorescence, which is intensified after treatment with monoamine oxidase inhibitors and diminishes following reserpine (Dahlström and Fuxe, 1965) or lesions of the medial forebrain bundle. Although dopamine and norepinephrine cannot be distinguished by histofluorescence, complementary chemical studies indicate that most of the catecholamine in the telencephalon is norepinephrine, except for the important dopamine-containing nigro-striatal tract (Andén et al., 1966). Sufficient attention has been given to norepinephrine in the brain and its possible relationship to mood to permit using that amine as an example of the group without suggesting that it has a predominant or exclusive role.

The early studies relied on the total content of norepinephrine in the brain or its concentration in particular regions. Thus, Barchas and Freedman (1963) showed a decrease in this amine along with an increase in serotonin in brains of rats forced to swim to exhaustion. Maynert and Levi (1964) found a 40 per cent decrease in norepinephrine in the brainstem after the administration of 30 minutes of foot shocks. Reis and Gunne (1965) induced rage in cats by stimulating the amygdala, and demonstrated a highly significant decrease in the norepinephrine levels of the telencephalon. More recent studies (Reis and Fuxe, 1969) have confirmed the relationship between induced rage and norepinephrine levels in the brain and have adduced further evidence for the crucial involvement of this amine in that

form of behavior by augmenting and inhibiting the manifestations of rage by drugs that respectively potentiate and block the pharmacological actions of norepinephrine at receptors in the periphery.

Closer to the dynamic relationships involved in synthesis and release is the examination of the turnover of norepinephrine in the brain. If synthesis is coupled to release in the central nervous system, as it appears to be in the periphery, simple levels of the endogenous amine would reflect only disparities between the two processes and not the magnitude of either. On the other hand, the disappearance of labeled amine from a pool into which it had been introduced would be affected by alterations in its rate of release, even though these were completely compensated by corresponding changes in synthesis. Although norepinephrine does not cross the blood:brain barrier (Weil-Malherbe et al., 1961), the tritium-labeled amine is rapidly distributed to various parts of the brain after intraventricular (Glowinski and Axelrod, 1966) or intracisternal (Schanberg et al., 1967) injection and concentrated at presynaptic endings, largely of norepinephrine-containing neurons (Descarries and Droz, 1969). The disappearance of the labeled amine from the brain does not follow a mono-exponential curve, suggesting the presence of two or more compartments with different rates of turnover. Its initial slope, however, should approximate the average turnover rate, and the decrements over standard periods of time have been used for qualitative comparison of turnover rates between experimental and control groups of animals (Thierry et al., 1968). An alternative method of examining turnover rate is afforded by following the curve of disappearance of the endogenous amine after blocking its synthesis by means of $\alpha$-methyl tyrosine (Costa and Neff, 1966). Although this technique does not require the use of exogenous material and obviates questions regarding the specificity of the material the turnover of which is being examined, it does require the assumption that turnover rate is independent of endogenous levels and would not be expected to reflect changes in rate of turnover brought about by changes in synthesis, because the latter process is blocked.

In addition to yielding valuable information on the effects of drugs on norepinephrine metabolism (Glowinski and Axelrod, 1966; Schanberg et al., 1967), the rate of turnover has also been examined in some behavioral states, in which it has been found to be more sensitive and informative than study of endogenous levels alone. In foot shock of milder form than that used previously, it was possible for Thierry et al. (1968) to demonstrate a significant increase in turnover throughout the brain and spinal cord with no systematic change in endogenous levels of the amine. This suggested an augmented synthesis coupled with release, which has been demonstrated more directly in the periphery (Alousi and Weiner, 1966; Sedvall et al., 1968).

Repeated exposure to such stress over a period of three days was associated with a significant elevation of endogenous levels of the amine, further supporting the hypothesis that synthesis was stimulated, and compatible with an induction of the rate-limiting enzyme.

The turnover of norepinephrine was found (Kety et al., 1967) to be substantially increased after one week of a regimen of twice-daily electroconvulsive shocks and 24 hours after the last shock, at a time when the behavior of the animals was quite normal (Figure 3). This, coupled with a significant elevation in endogenous norepinephrine, implied a persistent increase in synthesis of the amine. An increase in tyrosine hydroxylase levels in the brain was demonstrated in animals 24 hours after the same regimen of electroconvulsive shocks (Musacchio et al., 1969). The drugs effective in relieving clinical depression (amphetamine, monoamine oxidase inhibitors, imipramine) all affect norepinephrine turnover or metabolism in the brain in ways that would be expected to increase the activity of that amine at central synapses, whereas drugs causing depression have an opposite effect (Kety, 1967). The finding that electroconvulsive shock, which is probably the most effective treatment for depression, could also increase the availability of norepinephrine at synapses in the brain (in this instance by increasing its synthesis) is also compatible with the possibility that a deficiency of that amine may exist in the brain in states of depression.

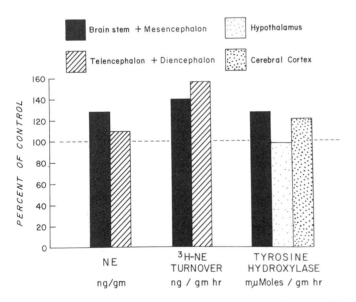

FIGURE 3 Changes in endogenous NE, turnover of ³H-NE and tyrosine hydroxylase activity in regions of the rat brain 24 hours after the last of a series of electroconvulsive shocks administered twice daily for one week. (From Kety et al., 1967, and Musacchio et al., 1969.)

Pharmacological tools in conjunction with a form of appetitive behavior have been used to adduce evidence that norepinephrine may be involved in the "reward" system in the brain. Stein (1964) has examined the effects of various drugs on the self-stimulating behavior in rats described by Olds and Milner (1954) and has found that imipramine or amphetamine will significantly increase such activity, but reserpine will suppress it. The effects of amphetamine are greatly diminished following reserpine (Stein, 1964) or α-methyl tyrosine (Crow, 1969). The most parsimonious explanation of all these observations is that this form of appetitive behavior requires the intervention of one of the catecholamines. The conditioned avoidance response is another form of behavior that can be blocked by drugs (Rech et al., 1966) which deplete catecholamine stores in the brain or by lesions in the posterolateral hypothalamic midbrain junction or medial forebrain bundle (Sheard et al., 1967), resulting in a loss of serotonin and norepinephrine from much of the telencephalon. The effects of catecholamine-depleting drugs are rather specific, because they can be demonstrated while escape behavior is unaffected. Depletion of serotonin alone is apparently insufficient to affect conditioned avoidance (Tenen, 1967).

Correlations of turnover rates of biogenic amines with behavioral states or the ability of drugs that affect amines in the brain to alter behavior offer evidence that is compatible with hypotheses that one or another amine is involved in a particular form of behavior. Such evidence hardly constitutes proof, however, because alternative explanations of the various findings must usually be entertained. Better evidence would be the demonstration of the release of a particular putative transmitter in the brain in constant association with a specific type of behavior, as has been achieved for acetylcholine, norepinephrine, and gamma-aminobutyric acid at particular peripheral synapses. Brain slices (Baldessarini and Kopin, 1967) have been shown to release norepinephrine or other amines (Katz et al., 1968) when stimulated with electric current or potassium ion, or when exposed to low concentrations of certain drugs. Lithium ion administered in vivo or added in vitro appears to block the stimulated release of norepinephrine or serotonin from brain slices, suggesting that lithium ion, rather specifically effective in treating mania, may act by inhibiting the release of biogenic amines.

Although a release of norepinephrine from the brain in vivo has been reported in association with appetitive stimulation (Stein and Wise, 1967), the specificity of the release has not been established; in other experiments, neuronal activation has been found to release not only norepinephrine but also urea and inulin (Chase and Kopin, 1968). On the other hand, d-amphetamine has been found to release norepinephrine and its methylated metabolite, but not inulin, from the brain (Carr and Moore, 1969).

The ability of acetylcholine or norepinephrine applied locally on "receptor" regions of a postsynaptic structure to elicit a response that closely mimics the effect of neural stimulation, functionally, electrically, and pharmacologically, constitutes what is probably the best evidence that these substances are neurotransmitters at certain peripheral synapses. For that reason, the effects of putative transmitters applied locally in the brain have been of considerable interest. When norepinephrine is injected into the ventricles, or intravenously in animals with a poorly developed blood:brain barrier, the effect produced is not arousal, but, in virtually every instance, some form of somnolence (Mandell and Spooner, 1968). The microinjection of this amine in the region of individual cells while their electrical activity is being recorded (Salmoiraghi and Bloom, 1964) usually produces an inhibition of spontaneous activity. These observations do not necessarily argue against an important involvement of norepinephrine in arousal, because neither the dose nor the site of application is controlled in the first type of experiment; an inhibition of random and spontaneous activity throughout the brain may consititute a characteristic feature of arousal with a facilitation of only small, sharply focused, and specifically activated regions, which may have gone undetected in the recording of unit activity.

On the other hand, the elicitation of particular forms of behavior after more specific administration of the amine would suggest its involvement in similar types of natural activity. A number of such observations have been reported. Wise and Stein (1969) suppressed self-stimulation (through electrodes in the medial forebrain bundle of rats) by the administration of diethyldithiocarbamate or disulfiram, which block dopamine $\beta$-hydroxylase and deplete the brain of norepinephrine, but not of dopamine. In such animals, the appetitive behavior could be restored by intraventricular injection of l-norepinephrine (5 micrograms) but not by the dextro-isomer, or by dopamine or serotonin.

Slangen and Miller (1969) have carried out a series of well-designed experiments that strongly suggest the involvement of norepinephrine in a type of feeding behavior. By implanting fine cannulae into the perifornical region at the posterior portion of the anterior hypothalamus, they were able to test the effects of various substances on a region at which electrical stimulation is known to induce eating in a previously satiated rat. Small doses of l-norepinephrine (20 micromillimoles) were found promptly to elicit the same type of behavior, although serotonin had no effect. Dopamine induced no immediate change in behavior but a delayed and weak eating response, compatible with its conversion to norepinephrine. This region responded to other pharmacological agents as do norepinephrine alpha receptors in the periphery. Phentolamine, which blocks alpha receptors, antagonized the norepinephrine response, and a beta-receptor stimulant (isoproterenol) or antagonist

(propranolol) had no effects per se or on the norepinephrine response. The norepinephrine-induced behavior was potentiated eightfold by previous treatment of the animal with desipramine. An effect was also obtained from tetrabenazine, which releases norepinephrine and other amines from storage depots, provided monoamine oxidase had previously been blocked by nialimide. It is difficult to avoid the interpretation that this behavior is mediated by the release of norepinephrine acting on specific adrenergic receptors in this region.

Even the intraventricular administration of norepinephrine need not always lead to generalized sedation. Segal and Mandell (1970) have observed activation and improved performance on a continuous avoidance task in animals receiving a constant infusion of low concentrations of norepinephrine; the activated behavior gave way to sedation with higher concentration.

In summary, the evidence appears to be good, but hardly conclusive, that norepinephrine and other amines play important roles in the mediation of various emotional and behavioral states. Norepinephrine especially appears to be implicated in arousal, aggression, and certain appetitive behaviors, although some observations (Seiden and Peterson, 1968; Creveling et al., 1968) are more compatible with an important role for dopamine. The best evidence for the involvement of serotonin appears to be in the production of sleep (Jouvet, 1969). No simplistic hypothesis, however, is compatible with all of the observations, and it appears futile to attempt to account for a particular emotional state in terms of the activity of one or more biogenic amines. It seems more likely that these amines may function separately or in concert at crucial nodes of the complex neuronal networks that underlie emotional states. Although this interplay may represent some of the common features and primitive qualities of various affects, the special characteristics of each of these states are probably derived from those extensions of the networks that represent apperceptive and cognitive factors based on the experience of the individual. A possibility which deserves some exposition and exploration is that an important adaptive role of the biogenic amines is to favor the elaboration of such networks in the learning process and the association of cognitive with appropriate affective elements.

## Possible role of biogenic amines in memory and learning

The peripheral autonomic and humoral components of affective states have well-recognized functions in the anticipation, facilitation, and maintenance of a variety of adaptive responses. Neither the central components of these states nor their functions have been so well defined, but it is possible that they subserve even more important adaptations—reinforcing significant inputs, suppressing irrelevant ones, evoking or facilitating responses, which, in the experience of the species or the individual, have the greatest survival value, and thus influencing the neuronal processes involved in memory to permit the development, reinforcement, and maintenance of the most appropriate responses.

Most of the earlier hypotheses concerning the neural mechanisms involved in memory emphasized repeated activation of a synapse as a necessary antecedent to some sustained alteration in its function, but failed to consider what may be the crucial importance of contingency with affective states in inducing such persistent changes. In 1963, Young emphasized the importance of the outcome of an act for memory in the octopus and designated the anatomical pathways and physiological processes whereby such interaction could occur. Szilard (1964) proposed a contingency model of learning in which an interaction between specific complementary proteins between synaptic membranes was the basis of alterations in synaptic efficacy. The interaction of amines with proteins, facilitated by cyclic AMP, to form antigenic complexes specific for each engram was suggested by Hechter and Halkerston (1964).

In his hypothesis, Roberts (1966) linked arousal and memory with the hypothalamic-pituitary neuroendocrine system. A concept that stressed the adaptive association between affect and memory emerged from the 1966 Intensive Study Program (Livingston, 1967). The dependence of learning on attention and on reward or punishment appears well established at the behavioral level, and there is obvious adaptive advantage in a mechanism that consolidates not all experience equally but only those experiences that are significant for survival (Kety, 1965).

It is not difficult to see how, as a result of selective pressure, some rudimentary adaptive responses to certain prevalent exogenous stimuli could become genetically endowed —crude, aversive behavior to noxious stimuli (loud noise, pain, extremes of heat and cold) or appetitive behavior (approach, sucking, swallowing) to stimuli associated with suckling (warmth, nipple in proximity of the mouth, milk in the pharynx). Each response would include a primitive motor pattern, appropriate autonomic, endocrine, and metabolic changes, and a state of arousal common to them all.

Now it is useful to make some assumptions, which, although plausible, are far from established on the basis of existing evidence: (1) that the aroused state induced by novel stimuli, or by stimuli genetically recognized as significant, is pervasive and affects synapses throughout the central nervous system, suppressing most, but permitting or even accentuating activity in those that are transmitting the novel or significant stimuli; (2) that this state, through one or more of its components, favors the development of persistent facilitatory changes in all synapses that are currently in a state of excitation or have recently been active; (3) that

there is a fairly random network of synapses with complexity sufficient enough for pathways to exist and be reinforced between many other neurons and the relatively few that are involved in mediating the primitive and genetically endowed adaptive responses.

A nervous system so constructed would have the remarkable capability not only of responding on the basis of a genetically determined input and response code, but of developing a much more elaborate and adaptive neuronal network or state between input and response, now on the basis of its idiosyncratic experiences.

Thus, the animal so equipped might wander into a new territory and experience some form of pain, responding with reflexive avoidance movements and an appropriate affective state. In addition, however, the animal's state of arousal, which would have risen to a higher level during its exploration of unfamiliar surroundings, would have tended to accentuate all the novel sensory inputs from a number of modalities associated with the new experiences. The powerful affective state and intense arousal induced by the pain, in addition to its classical autonomic and endocrine concomitants in the periphery, may also act centrally to induce some persistent chemical change in all recently active synapses, especially in those associated with the novel sensory experiences of the unfamiliar territory, the activity of which the state of arousal had accentuated. With many repetitions of that sequence and by a process of algebraic summation, such as that employed in the computer of average transients, those inputs and activated pathways—in short, that pattern and state of neuronal excitation which preceded and was repeatedly associated with the pain— would be translated into a chemical change of greater magnitude, and therefore longer duration, than that of random activity. In that way, subsequent presentation of some of the sensory stimuli peculiar to that particular territory could eventually evoke the same affective state, the heightened arousal, the peripheral autonomic and metabolic responses, and the aversive behavior without the intervention of the painful stimulus. Thus, a conditioned avoidance and emotional response would have been established, and what was originally an inborn response only to pain would have been cognitively elaborated on the basis of idiosyncratic experience, to permit the anticipating, preparing for, or avoiding the noxious stimulus in a remarkable new type of adaptation.

It is possible to suggest certain anatomical pathways and neurochemical mechanisms that may satisfy some of the requirements of this hypothetical process or that seem to merit further investigation from such a point of view. If protein synthesis is crucially involved in the consolidation of the memory trace, as much recent work appears strongly to suggest, then the hypothesis outlined above would imply that some chemical components of arousal should be able to facilitate synaptic protein synthesis, probably at synapses, contingent on neuronal activity wherever it occurs throughout the brain.

At the 1966 Intensive Study Program, I presented a highly speculative model of a neuronal process whereby this could be achieved (Figure 4). This assumed a population of neurons containing and releasing norepinephrine or another amine in affective states, in this instance arousal and pain, with widely distributed axons, not only to the neurons which mediate the peripheral autonomic and endocrine concomitants, but also feeding back onto central synapses generally, and thus including those the activity of which was coincident with or closely antecedent to the pain. It was speculated that the amine released might affect protein synthesis so as to facilitate consolidation at synapses recently activated and still reverberating or otherwise sustaining some differentiating change. Kandel and Spencer (1968) have recently reviewed the evidence for various persistent neuronal changes after activation.

Since that time, a number of observations have been made which lend some credence to that hypothesis and suggest one or more sites at which such amine-stimulated consolidation could occur. The hypothetical consolidating feedback, by playing upon synapses indiscriminately and affecting those recently active, would reinforce patterns not because they were adaptive, but simply because they were consistently associated with a particular affective state. It is well known that nonadaptive or irrelevant responses, e.g., specified changes in the cortical evoked potential in man (Rosenfeld et al., 1969) can be induced merely by rewarding them consistently. One seeks, therefore, a widely distributed neural system that may be activated in affective states and capable of releasing a trophic substance indiscriminately in the region of a great number of synapses. Scheibel and Scheibel (1967) have described multibranched axons of cells in the brainstem that make synapses with thousands of neurons up and down the neuraxis. They have also described in greater detail the architectonics of the "unspecific afferents" to the cerebral cortex with evidence that these long climbing axons of cells, some of which are in the brainstem reticular system, weave about the apical dendrites of pyramidal cells with an extremely loose axodendritic association, in contrast to the vast number of well-defined synapses established by the terminals of the "specific afferents" (Figure 5). In 1968, Fuxe, Hamberger, and Hökfelt described the terminations in the cortex of norepinephrine-containing axons of brainstem neurons, pointing out the similarities in their distribution to that of the unspecific afferents of Scheibel and Scheibel.

If some of the unspecific afferents are indeed "adrenergic" or "aminergic" terminals invading the millions of sensory-sensory and sensory-motor synapses of the cortex, they would provide a remarkably effective mechanism whereby amines released in arousal could affect a crucial population of synapses throughout the brain. The hippocampal and cerebellar cortex are also characterized by "climbing fibers," some of which are norepinephrine-containing (Andén et al., 1967; Blackstad et al., 1967), and one group has adduced evidence for a transmitter role for that amine between the terminals of certain climbing fibers and Purkinje cells (Siggins et al., 1969). There is even some evidence that axons from the same adrenergic neurons in the brainstem may be distributed to cerebral, hippocampal, and cerebellar cortex, as well as to hypothalamus and other areas. Marr (1969) has proposed a novel theory of the cerebellar cortex that implies a "learning" of patterns of motor activity by that structure, based on a conjunction of excitation in mossy and climbing fibers. It is possible that, through similar and simultaneous processes in these three cortexes, the state of arousal by means of adrenergic input to each may serve concurrently to reinforce and to consolidate the significant sensory patterns, the affective associations and the motor programs necessary in the learning of a new adaptive response. It is of interest that Phillips and Olds (1969) have described single units in the midbrain that fire not in response to a stimulus, but to its significance in terms of the previous experience of the animal.

Just as the peripheral expressions of arousal and affect are mediated by both neurogenic and endocrine processes, it is possible that the central components employ humoral, as well as neural, modes. There is a great tendency for these apical dendrites and their afferents in a remarkably similar fashion in the cerebrum, hippocampus, and cerebellum to seek the cortical surface; indeed, the cortical convolutions that increase that surface by several times must have some adaptive function. Developmental history and the geometrical requirements of the circuitry have been invoked to explain this phenomenon, but it is also possible that the position of these structures in close proximity to the cerebrospinal fluid serves another adaptive function. The constant flow of this medium from the ventricles over the whole cortical surface on its way to arachnoid villi offers a means of superfusing the cortex with substances derived from the blood stream at the choroid plexus or intracerebrally secreted by various stations along its path. It is noteworthy that $^3$H-norepinephrine injected into one ventricle or into the cisterna magna rapidly penetrates the superficial layers of the brain; the amines do not easily pass the blood:brain barrier and would not readily be removed by the capillaries, so such a mechanism would further assure the widespread distribution of any that may be released from endings near the cortical surface. Acetylcholine (Collier and Mitchell, 1967) and prostaglandins (Ramwell and Shaw, 1966) have been shown to be released at the cortex by neuronal activity, and it is likely that other substances are released and may be broadcast by this process.

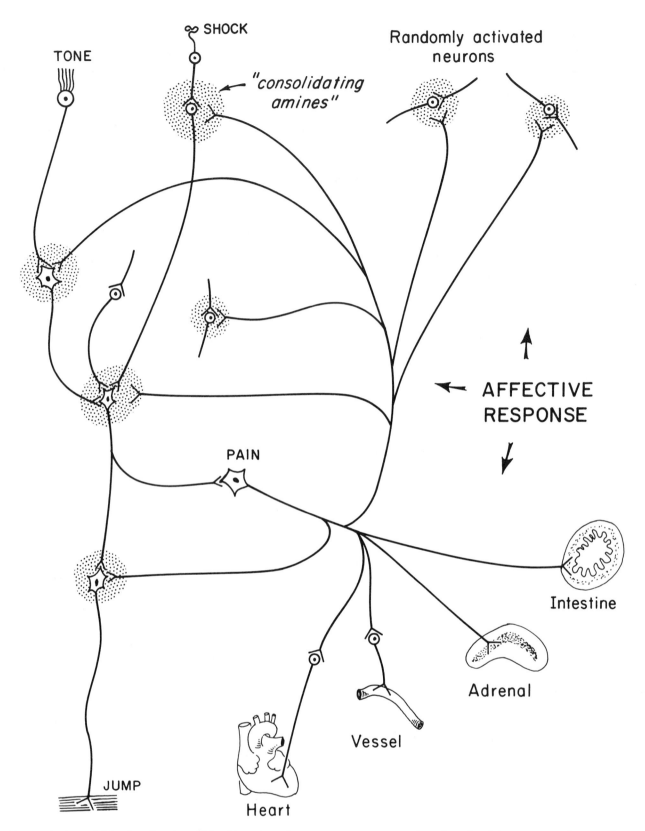

FIGURE 4  A model for learning, in which a biogenic amine released generally throughout the brain during an affective state (arousal, punishment, reward) is assumed to facilitate consolidation at synapses recently activated and, by a process of summation, to reinforce those pathways that regularly preceded or accompanied the affective state.

Secretions of the hypothalamus, trophic hormones of the pituitary, and the steroid hormones of the adrenal cortex, some of which are regularly secreted in states of arousal and stress, may thus have additional access to this rich population of synapses. Many of these substances have, in one system or another, displayed a capacity to stimulate the synthesis of RNA or a protein. The steroid hormones and ACTH clearly affect conditioning (Levine, 1968), and, in addition to their well-established abilities to induce enzymes in other tissues, one of them has recently been found to restore tryptophan hydroxylase activity in the midbrain of the adrenalectomized rat (Azmitia and McEwen, 1969). It is tempting to speculate upon the possible trophic actions of such stress-related hormones on cortical synapses.

A substantial number of recent observations are compatible with this model, and in some cases the hypothesis helps to explain some inconsistencies. RNA and protein synthesis have both been found to be stimulated at sites of increased neuronal activity (Glassman, 1969; Berry, 1969), but inconsistently so, perhaps because the crucial contingent factor of arousal has not been held constant (Altman and Das, 1966). Conversely, Roberts and Flexner (1969) have pointed out that the increased protein synthesis demonstrated by some experiments during learning is considerably greater than would be required by the newly established neuronal patterns, unless a generalized facilitation of synthesis of new protein took place and then decayed except for that in the repeatedly reinforced patterns.

The hypothesis would predict that drugs which release or enhance norepinephrine in the brain or exert its neuronal effects would favor consolidation and facilitate memory. Amphetamine and caffeine appear capable of producing such effects (Bignami et al., 1965; Oliverio, 1968), and recently a more specific ability of amphetamine or foot-shock to counteract the suppression of consolidation brought about by cycloheximide has been reported (Barondes and Cohen, 1968).

Conversely, lesions or drugs that deplete or block norepinephrine in the brain should retard consolidation and prevent acquisition. Although there are many reports on the ability of such drugs (reserpine, $\alpha$-methyl tyrosine) to block the established conditioned-avoidance response, there is little information on their effects during acquisition. Recently, however, W. B. Essman (personal communication) has found a significant impairment in acquisition, after $\alpha$-methyl tyrosine, which was most severe when the brain levels of norepinephrine were lowest. Lesions of the medial forebrain bundle, which would be expected to deplete the telencephalon of norepinephrine and serotonin, appeared in one study (Sheard et al., 1967) to depress acquisition. Conversely, stimulation of this region has been used as a means of positive reinforcement in conditioning curarized animals (Trowill, 1967). In a recent clinical study, Lester et al. (1969)

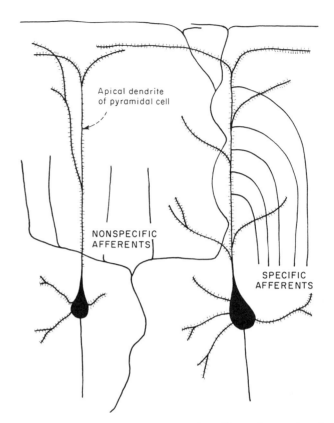

FIGURE 5 Diagrammatic representation (from the work of Scheibel and Scheibel, 1967) of the associations between specific and nonspecific afferents on the apical dendrites of pyramidal cells.

reported apathy and an apparent inability to consolidate recent memory in patients during treatment with a low phenylalanine-tyrosine diet. They have also observed (personal communication) that these effects were alleviated by the administration of l-dopa.

The suggestion that the release of norepinephrine may favor consolidation of learning by stimulating protein synthesis is made more tenable by recently acquired information on the possible action of cyclic adenosine monophosphate (AMP) in the brain. This substance, present in surprisingly high concentration in the central nervous system (as is adenyl cyclase, the enzyme that brings about its synthesis) is crucially involved in enzyme induction and protein synthesis in a wide variety of bacterial and mammalian cells, and appears to increase the activity of a protein kinase in brain (Miyamoto et al., 1969). Evidence is accumulating that cyclic AMP may mediate the effects of norepinephrine on central neurons (Siggins et al., 1969), as it is believed to do for actions of catecholamines and other hormones on liver, muscle, and other peripheral tissues. It is interesting that the stimulation of protein kinase by cyclic AMP can be potentiated markedly by magnesium ions (Miyamoto et al.,

1969) or potassium ions (Shimizu et al., 1970) and inhibited by calcium, which suggests means whereby an effect of adrenergic stimulation could be differentially exerted on recently active, as opposed to inactive, synapses.

Earlier hypotheses, which assigned an important role to central norepinephrine and other biogenic amines in mediating emotional states, have stimulated considerable research. Possibly these speculations regarding their possible function in the biochemical processes that underlie memory and learning may be of some heuristic value.

## REFERENCES

ALOUSI, A., and N. WEINER, 1966. The regulation of norepinephrine synthesis in sympathetic nerves: Effect of nerve stimulation, cocaine, and catecholamine-releasing agents. *Proc. Nat. Acad. Sci. U. S. A.* 56: 1491–1496.

ALTMAN, J., and G. D. DAS, 1966. Behavioral manipulations and protein metabolism of the brain: Effects of motor exercise on the utilization of leucine-H³. *Physiol. Behav.* 1: 105–108.

ANDÉN, N.-E., K. FUXE, B. HAMBERGER, and T. HÖKFELT, 1966. A quantitative study on the nigro-neostriatal dopamine neuron system in rat. *Acta Physiol. Scand.* 67: 306–312.

ANDÉN, N.-E., K. FUXE, and U. UNGERSTEDT, 1967. Monoamine pathways to the cerebellum and cerebral cortex. *Experientia* (*Basel*) 23: 838–839.

AXELROD, J., 1959. Metabolism of epinephrine and other sympathomimetic amines. *Physiol. Rev.* 39: 751–776.

AZMITIA, E. C., JR., and B. S. McEWEN, 1969. Corticosterone regulation of tryptophan hydroxylase in midbrain of the rat. *Science* (*Washington*) 166: 1274–1276.

BALDESSARINI, R. J., and I. J. KOPIN, 1967. The effect of drugs on the release of norepinephrine-H³ from central nervous system tissues by electrical stimulation *in vitro*. *J. Pharmacol. Exp. Ther.* 156: 31–38.

BARCHAS, J. D., and D. FREEDMAN, 1963. Brain amines: Response to physiological stress. *Biochem. Pharmacol.* 12: 1232–1235.

BARONDES, S. H., and H. D. COHEN, 1968. Arousal and the conversion of "short-term" to "long-term" memory. *Proc. Nat. Acad. Sci. U. S. A.* 61: 923–929.

BERRY, R. W., 1969. Ribonucleic acid metabolism of a single neuron: Correlation with electrical activity. *Science* (*Washington*) 166: 1021–1023.

BESSON, M. J., A. CHERAMY, P. FELTZ, and J. GLOWINSKI, 1969. Release of newly synthesized dopamine from dopamine-containing terminals in the striatum of the rat. *Proc. Nat. Acad. Sci. U. S. A.* 62: 741–748.

BIGNAMI, G., F. ROBUSTELLI, I. JANKŮ, and D. BOVET, 1965. Psychopharmacologie. *Compt. Rend. Acad. Sci.* 260: 4273–4278.

BLACKSTAD, T. W., K. FUXE, and T. HÖKFELT, 1967. Noradrenaline nerve terminals in the hippocampal region of the rat and the guinea pig. *Z. Zellforsch. Mikroskop. Anat.* 78: 463–473.

BOGDANSKI, D. F., and B. B. BRODIE, 1969. The effects of inorganic ions on the storage and uptake of H³-norepinephrine by rat heart slices. *J. Pharmacol. Exp. Ther.* 165: 181–189.

BONDAREFF, W., and B. GORDON, 1966. Submicroscopic localization of norepinephrine in sympathetic nerves of rat pineal. *J. Pharmacol. Exp. Ther.* 153: 42–47.

CARR, L. A., and K. E. MOORE, 1969. Norepinephrine: Release from brain by d-amphetamine in vivo. *Science* (*Washington*) 164: 322–323.

CHASE, T. N., and I. J. KOPIN, 1968. Stimulus-induced release of substances from olfactory bulb using the push-pull cannula. *Nature* (*London*) 217: 466–467.

COLLIER, B., and J. F. MITCHELL, 1967. The central release of acetylcholine during consciousness and after brain lesions. *J. Physiol.* (*London*) 188: 83–98.

COSTA, E., and N. H. NEFF, 1966. Isotopic and non-isotopic measurements of the rate of catecholamine biosynthesis. *In* Biochemistry and Pharmacology of the Basal Ganglia (E. Costa, L. J. Côté, and M. D. Yahr, editors). Raven Press, New York, pp. 141–156.

CREVELING, C. R., J. DALY, T. TOKUYAMA, and B. WITKOP, 1968. The combined use of α-methyltyrosine and threo-dihydroxyphenylserine in selective reduction of dopamine levels in the central nervous system. *Biochem. Pharmacol.* 17: 65–70.

CROW, T. J., 1969. Mode of enhancement of self stimulation in rats by methamphetamine. *Nature* (*London*) 224: 709–710.

DAHLSTRÖM, A., and K. FUXE, 1965. Evidence for the existence of monoamine neurons in the central nervous system. II. Experimentally induced changes in the intraneuronal amine levels of bulbospinal neuron systems. *Acta Physiol. Scand.* 64 (suppl. 247): 5–36.

DESCARRIES, L., and B. DROZ, 1970. Intraneuronal distribution of exogenous norepinephrine in the central nervous system of the rat. *J. Cell Biol.* 44: 385–389.

ERÄNKÖ, O., 1956. Histochemical demonstration of noradrenaline in the adrenal medulla of the hamster. *J. Histochem. Cytochem.* 4: 11–13.

EULER, U. S. VON, 1956. Noradrenaline. Charles C Thomas, Springfield, Illinois.

FUXE, K., 1965. Evidence for the existence of monoamine neurons in the central nervous system. IV. The distribution of monoamine terminals in the central nervous system. *Acta Physiol. Scand.* 64 (suppl. 247): 37–85.

FUXE, K., B. HAMBERGER, and T. HÖKFELT, 1968. Distribution of noradrenaline nerve terminals in cortical areas of the rat. *Brain Res.* 8: 125–131.

GEFFEN, L. B., and R. A. RUSH, 1968. Transport of noradrenaline in sympathetic nerves and the effect of nerve impulses on its contribution to transmitter stores. *J. Neurochem.* 15: 925–930.

GLASSMAN, E., 1969. The biochemistry of learning: An evaluation of the role of RNA and protein. *Annu. Rev. Biochem.* 38: 605–646.

GLOWINSKI, J., and J. AXELROD, 1966. Effects of drugs on the disposition of H³-norepinephrine in the rat brain. *Pharmacol. Rev.* 18: 775–785.

GOODALL, McC., and N. KIRSHNER, 1958. Biosyntheses of epinephrine and norepinephrine by sympathetic nerves and ganglia. *Circulation* 17: 366–371.

HECHTER, O., and I. D. HALKERSTON, 1964. On the nature of macromolecular coding in neuronal memory. *Perspect. Biol. Med.* 7: 183–198.

HILLARP, N.-Å., K. FUXE, and A. DAHLSTRÖM, 1966. Demonstration and mapping of central neurons containing dopamine, noradrenaline, and 5-hydroxytryptamine and their reactions to psychopharmaca. *Pharmacol. Rev.* 18: 727–741.

HÖKFELT, T., 1968. *In vitro* studies on central and peripheral monoamine neurons at the ultrastructural level. *Z. Zellforsch. Mikroskop. Anat.* 91: 1–74.

IVERSEN, L. L., 1967. The Uptake and Storage of Noradrenaline in Sympathetic Nerves. Cambridge University Press, London.

JOUVET, M., 1969. Biogenic amines and the states of sleep. *Science (Washington)* 163: 32–41.

KANDEL, E. R., and W. A. SPENCER, 1968. Cellular neurophysiological approaches in the study of learning. *Physiol. Rev.* 48: 65–134.

KATZ, R. I., T. N. CHASE, and I. J. KOPIN, 1968. Evoked release of norepinephrine and serotonin from brain slices: Inhibition by lithium. *Science (Washington)* 162: 466–467.

KETY, S. S., 1965. The incorporation of experience into the central nervous system. *In* Désafférentation expérimentale et clinique (J. de Ajuriaguerra, editor). Georg et Cie, Geneva, pp. 251–256.

KETY, S. S., 1966. Catecholamines in neuropsychiatric states. *Pharmacol. Rev.* 18: 787–798.

KETY, S. S., 1967. The central physiological and pharmacological effects of the biogenic amines and their correlations with behavior. *In* The Neurosciences: A Study Program (G. C. Quarton, T. Melnechuk, and F. O. Schmitt, editors). The Rockefeller University Press, New York, pp. 444–451.

KETY, S. S., F. JAVOY, A.-M. THIERRY, L. JULOU, and J. GLOWINSKI, 1967. A sustained effect of electroconvulsive shock on the turnover of norepinephrine in the central nervous system of the rat. *Proc. Nat. Acad. Sci. U. S. A.* 58: 1249–1254.

KOPIN, I. J., G. R. BREESE, K. R. KRAUSS, and V. K. WEISE, 1968. Selective release of newly synthesized norepinephrine from the cat spleen during sympathetic nerve stimulation. *J. Pharmacol. Exp. Ther.* 161: 271–278.

LESTER, B. K., R. E. CHANES, and P. T. CONDIT, 1969. A clinical syndrome and EEG-sleep changes associated with amino acid deprivation. *Am. J. Psychiat.* 126: 185–190.

LEVINE, S., 1968. Hormones and conditioning. *In* Nebraska Symposium on Motivation, 1968 (W. J. Arnold, editor). University of Nebraska Press, Lincoln, pp. 85–101.

LEVITT, M., S. SPECTOR, A. SJOERDSMA, and S. UDENFRIEND, 1965. Elucidation of the rate-limiting step in norepinephrine biosynthesis in the perfused guinea-pig heart. *J. Pharmacol. Exp. Ther.* 148: 1–8.

LIVINGSTON, R. B., 1967. Reinforcement. *In* The Neurosciences: A Study Program (G. C. Quarton, T. Melnechuk, and F. O. Schmitt, editors). The Rockefeller University Press, New York, pp. 568–577.

MANDELL, A. J., and C. E. SPOONER, 1968. Psychochemical research studies in man. *Science (Washington)* 162: 1442–1453.

MARR, D., 1969. A theory of cerebellar cortex. *J. Physiol. (London)* 202: 437–470.

MAYNERT, E. W., and R. LEVI, 1964. Stress-induced release of brain norepinephrine and its inhibition by drugs. *J. Pharmacol. Exp. Ther.* 143: 90–95.

MIYAMOTO, E., J. F. KUO, and P. GREENGARD, 1969. Cyclic nucleotide-dependent protein kinases. III. Purification and properties of adenosine 3′5′-monophosphate-dependent protein kinase

from bovine brain. *J. Biol. Chem.* 244: 6395–6402.

MUSACCHIO, J. M., L. JULOU, S. S. KETY, and J. GLOWINSKI, 1969. Increase in rat brain tyrosine hydroxylase activity produced by electroconvulsive shock. *Proc. Nat. Acad. Sci. U. S. A.* 63: 1117–1119.

NAGATSU, T., M. LEVITT, and S. UDENFRIEND, 1964. Tyrosine hydroxylase: The initial step in norepinephrine biosynthesis. *J. Biol. Chem.* 239: 2910–2917.

OLDS, J., and P. MILNER, 1954. Positive reinforcement produced by electrical stimulation of septal area and other regions of the rat brain. *J. Comp. Physiol. Psychol.* 47: 419–427.

OLIVERIO, A., 1968. Neurohumoral systems and learning. *In* Psychopharmacology; A Review of Progress 1957–1967. U. S. Public Health Service Publication No. 1836. U. S. Govt. Printing Office, Washington, pp. 867–878.

PHILLIPS, M. I., and J. OLDS, 1969. Unit activity: Motivation-dependent responses from midbrain neurons. *Science (Washington)* 165: 1269–1271.

POTTER, W. P. DE, A. F. DE SCHAEPDRYVER, E. J. MOERMAN, and A. D. SMITH, 1969. Evidence for the release of vesicle-proteins together with noradrenaline upon stimulation of the splenic nerve. *J. Physiol. (London)* 204: 102P–104P.

RAMWELL, P. W., and J. E. SHAW, 1966. Spontaneous and evoked release of prostaglandins from cerebral cortex of anesthetized cats. *Amer. J. Physiol.* 211: 125–134.

RECH, R. H., H. K. BORYS, and K. E. MOORE, 1966. Alterations in behavior and brain catecholamine levels in rats treated with α-methyltyrosine. *J. Pharmacol. Exp. Ther.* 153: 412–419.

REIS, D. J., and K. FUXE, 1969. Brain norepinephrine: Evidence that neuronal release is essential for sham rage behavior following brainstem transection in cat. *Proc. Nat. Acad. Sci. U. S. A.* 64: 108–112.

REIS, D. J., and L.-M. GUNNE, 1965. Brain catecholamines: Relation to the defense reaction evoked by amygdaloid stimulation in cat. *Science (Washington)* 149: 450–451.

ROBERTS, E., 1966. Models for correlative thinking about brain, behavior, and biochemistry. *Brain Res.* 2: 109–144.

ROBERTS, R. B., and L. B. FLEXNER, 1969. The biochemical basis of long-term memory. *Quart. Rev. Biophys.* 2: 135–173.

ROSENFELD, J. P., A. P. RUDELL, and S. S. FOX, 1969. Operant control of neural events in humans. *Science (Washington)* 165: 821–823.

ROTH, R. H., L. STJÄRNE, and U. S. VON EULER, 1966. Acceleration of noradrenaline biosynthesis by nerve stimulation. *Life Sci.* 5: 1071–1075.

SALMOIRAGHI, G. C., and F. E. BLOOM, 1964. Pharmacology of individual neurons. *Science (Washington)* 144: 493–499.

SCHANBERG, S. M., J. J. SCHILDKRAUT, and I. J. KOPIN, 1967. The effects of psychoactive drugs on norepinephrine-³H metabolism in brain. *Biochem. Pharmacol.* 16: 393–399.

SCHEIBEL, M. E., and A. B. SCHEIBEL, 1967. Structural organization of nonspecific thalamic nuclei and their projection toward cortex. *Brain Res.* 6: 60–94.

SEDVALL, G. C., V. K. WEISE, and I. J. KOPIN, 1968. The rate of norepinephrine synthesis measured *in vivo* during short intervals; influence of adrenergic nerve impulse activity. *J. Pharmacol. Exp. Ther.* 159: 274–282.

SEGAL, D. S., and A. J. MANDELL, 1970. Behavioral activation of rats during intraventricular infusion of norepinephrine. *Proc. Nat. Acad. Sci. U. S. A.* 66: 289–293.

SEIDEN, L. S., and D. D. PETERSON, 1968. Reversal of the reserpine-induced suppression of the conditional avoidance response by L-dopa: Correlation of behavioral and biochemical differences in two strains of mice. *J. Pharmacol. Exp. Ther.* 159: 422–428.

SHEARD, M. H., J. B. APPEL, and D. X. FREEDMAN, 1967. The effect of central nervous system lesions on brain monoamines and behavior. *J. Psychiat. Res.* 5: 237–242.

SHIMIZU, H., C. R. CREVELING, and J. DALY, 1970. Factors affecting cyclic AMP formation in brain: Use of slices pre-labelled by incubation with adenine-$^{14}$C. *Trans. Amer. Soc. Neurochem.* 1: 67.

SIGGINS, G. R., B. J. HOFFER, and F. E. BLOOM, 1969. Cyclic adenosine monophosphate: Possible mediator for norepinephrine effects of cerebellar Purkinje cells. *Science (Washington)* 165: 1018–1020.

SLANGEN, J. L., and N. E. MILLER, 1969. Pharmacological tests for the function of hypothalamic norepinephrine in eating behavior. *Physiol. Behav.* 4: 543–552.

STEIN, L., 1964. Self-stimulation of the brain and the central stimulant action of amphetamine. *Fed. Proc.* 23: 836–850.

STEIN, L., and G. D. WISE, 1967. Release of hypothalamic norepinephrine by rewarding electrical stimulation or amphetamine in the unanesthetized rat. *Fed. Proc.* 26: 651 (abstract).

STJÄRNE, L., and F. LISHAJKO, 1967. Localization of different steps in noradrenaline synthesis to different fractions of a bovine splenic nerve homogenate. *Biochem. Pharmacol.* 16: 1719–1728.

SZILARD, L., 1964. On memory and recall. *Proc. Nat. Acad. Sci. U. S. A.* 51: 1092–1099.

TENEN, S. S., 1967. The effects of p-chlorophenylalanine, a serotonin depletor, on avoidance acquisition, pain sensitivity and related behavior in the rat. *Psychopharmacologia* 10: 204–219.

THIERRY, A.-M., F. JAVOY, J. GLOWINSKI, and S. S. KETY, 1968. Effects of stress on the metabolism of norepinephrine, dopamine and serotonin in the central nervous system of the rat. I. Modifications of norepinephrine turnover. *J. Pharmacol. Exp. Ther.* 163: 163–171.

TROWILL, J. A., 1967. Instrumental conditioning of the heart rate in the curarized rat. *J. Comp. Physiol. Psychol.* 63: 7–11.

UDENFRIEND, S., P. ZALTZMAN-NIRENBERG, and T. NAGATSU, 1965. Inhibitors of purified beef adrenal tyrosine hydroxylase. *Biochem. Pharmacol.* 14: 837–845.

WEIL-MALHERBE, H., L. G. WHITBY, and J. AXELROD, 1961. The blood-brain barrier for catecholamines in different regions of the brain. *In* Regional Neurochemistry (S. S. Kety and J. Elkes, editors). Pergamon Press, Oxford, pp. 284–292.

WEINER, N., and M. RABADJIJA, 1968a. The effect of nerve stimulation on the synthesis and metabolism of norepinephrine in the isolated guinea-pig hypogastric nerve-vas deferens preparation. *J. Pharmacol. Exp. Ther.* 160: 61–71.

WEINER, N., and M. RABADJIJA, 1968b. The regulation of norepinephrine synthesis. Effect of puromycin on the accelerated synthesis of norepinephrine associated with nerve stimulation. *J. Pharmacol. Exp. Ther.* 164: 103–114.

WISE, C. D., and L. STEIN, 1969. Facilitation of brain self-stimulation by central administration of norepinephrine. *Science (Washington)* 163: 299–301.

YOUNG, J. Z., 1963. Some essentials of neural memory systems. Paired centres that regulate and address the signals of the results of action. *Nature (London)* 198: 626–630.

# 33 The Triune Brain, Emotion, and Scientific Bias

## PAUL D. MacLEAN

IT IS TRADITIONAL to regard the exact sciences as completely objective. The cultivation of this attitude is illustrated by a statement of Einstein quoted by C. P. Snow (1967): "Perception of this world by thought, leaving out everything subjective became . . . my supreme aim." Bertrand Russell

PAUL D. MACLEAN Section on Limbic Integration and Behavior, Laboratory of Neurophysiology, National Institute of Mental Health, Bethesda, Maryland

(1921) maintained that introspective data are scientifically inappropriate for investigation because they do not obey physical laws. Early in this century, the behaviorist school, with Watson as its leading exponent, sought to revive the spirit of the Helmholtz tradition and to establish psychology as an exact science on an equal footing with physics and the other natural sciences (cf. Shakow and Rapaport, 1964). In the study of animals and man, the approach was to be completely objective; consciousness, subjectivity, and introspection were to be treated as though nonexistent (Watson,

1924). The irony of all such attitudes is that every behavior selected for study, every observation and interpretation, requires subjective processing by an introspective observer. Logically, there is no way of circumventing this or the more disturbing conclusion that the cold, hard facts of science, like the firm pavement underfoot, are all derivatives of a soft brain. No measurement or computation obtained by the hardware of the exact sciences enters our comprehension without undergoing subjective transformation by the software of the brain. The implication of Spencer's statement (1896) that objective psychology owes its origin to subjective psychology could apply to the whole realm of science.

For such reasons it is scientifically important to consider how a fifth dimension, the subjective brain, affects man's relative view of the world. In discussing this question, I do not intend to deal with the familiar Cartesian topic of perceptual illusions, striking examples of which are demonstrated in Donald MacKay's chapter in this volume. Rather, I focus attention on mechanisms underlying emotional cerebration, to show how primitive systems of the brain have the capacity to generate affective feelings of what is real, true, and important and thereby to influence scientific attitudes.

In my own research on the brain I have been concerned with the question of what accounts for the difference between rational and emotional cerebration. In simplest terms, cerebration refers to mental actions of the brain. In logic and mathematics the steps of mentation can be such as to lead to indisputable conclusions, but in emotional cerebration there is no prediction of what the outcome will be. Emotional cerebration appears to have the paradoxical capacity to find equal support for opposite sides of any question. It is particularly curious that in scientific discourse, as in politics, the emotions seem capable of standing on any platform. Different groups of reputable scientists, for example, often find themselves in altercation because of diametrically opposed views of what is true. Although seldom commented on, it is equally bewildering that the world order of science is able to live comfortably for years, and sometimes centuries, with beliefs that a new generation discovers to be false. How is it possible that we are able to build higher and higher on the foundations of such beliefs without fear of their sudden collapse? In addition, there is the paradox that the emotional investment of some scientists is such that they remain convinced of the truth of a theory long after it has been proved to be false. As E. G. Boring (1964) emphasized, in paraphrasing a comment of Max Planck: "Important theories, marked for death by the discovery of contradictory evidence, seldom die before their authors."

DEFINITIONS Before delving into brain mechanisms, I find it necessary to elaborate on the meaning of three key words—"emotion," "affect," and "fact." We commonly speak of the subjective and expressive aspects of emotion. As the subjective aspect is purely private, it must be distinguished by some such word as "affect." Only we as individuals can experience affects. The public communication of affects requires expression through some form of verbal or other behavior. Rosenblueth et al. (1943) defined behavior as any change of an entity with respect to its environment. The behavioral expression of affect is appropriately denoted by Descartes' meaning of the word "emotion."

Scientifically, it is important to recognize that whether we are dealing with affect or emotion, each is manifest to us as observers only as information. What this signifies is expressed concisely by paraphrasing Berkeley and Hume in Wiener's words: "Information is information, not matter or energy" (Wiener, 1948). Subjectively, we gain the impression that there never can be communication of information without the intermediary of things recognized as behaving entities, no matter how small or how detected. Behaving entities, in other words, are unconditional correlates of communication. In a forthcoming book (MacLean, 1966b), I have developed the reasoning by which we may recognize as facts only those things that can be agreed upon publicly as entities behaving in a certain way. The term "validity" does not apply to the facts themselves, which are neither true nor false per se, but rather, to what is agreed upon as true by subjective individuals after a public assessment of the facts. What is agreed upon as true or false by one group may be quite contrary to the conclusions of another group.

The attribute that most clearly distinguishes psychological from other functions of the brain is the by-product of subjectivity (MacLean, 1960). All conventionally recognized forms of psychological information—awareness, sensations, perceptions, compulsions, affects, thoughts—are characterized by a co-existing state of subjectivity.

Affects differ from other forms of psychic information insofar as they are subjectively qualified in a physical sense as being either agreeable or disagreeable (Figure 1). There are no neutral affects because, emotionally speaking, it is impossible to feel unemotionally (MacLean, 1966b, 1969b).

The affects can be further subdivided (Figure 1) into three main types which for purposes of discussion I refer to as basic, specific, and general. The basic affects are informative of basic bodily needs that we subjectively recognize as hunger, thirst, and the various urges to breathe, defecate, urinate, have sexual outlet, and so forth. The specific affects are those occurring with activation of specific sensory systems, as illustrated by the sense of disgust in smelling a foul odor or the feeling of pain after a noxious stimulus (MacLean, 1966b, 1969b).

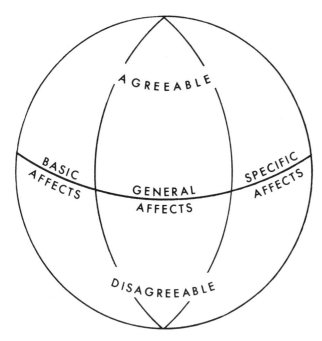

FIGURE 1   A scheme for viewing the world of affects.

The traditionally regarded emotions such as love, anger and so forth belong to what I call general affects. I call them "general" because they are feelings that may pertain to situations, individuals, or groups. Unlike the other forms of affect, they may persist or recur after the inciting circumstances. All the general affects may be considered in the light of self-preservation or the preservation of the species (MacLean, 1960). Affects giving information of threats to the self or species fall into the category of disagreeable affects. In the opposite category are affects informative of the removal of threats and the gratification of needs.

If one excludes verbal behavior, one can identify in animals and man six types of behavior that are inferred to be guided by the general affects. These six behaviors are recognized as (1) searching, (2) aggressive, (3) protective, (4) dejected, (5) gratulant, and (6) caressive. Verbally, they may be respectively characterized by such words as (1) desire, (2) anger, (3) fear, (4) sorrow, (5) joy, and (6) affection. Symbolic language makes it possible to identify many variations of these affects, but, in working with animals, one must base inferences about emotional states largely on these six general types of behavior (MacLean, 1966b, 1969b).

EVOLUTIONARY RELICS   We hear much protest these days against the Establishment. But people generally fail to realize that the model for the offending Establishment is built into man's nervous system. The on-going struggle in the outside world is, so to speak, but a reflection of the constant strife within ourselves. In the investigation of brain correlates of emotion, it is helpful to consider the evolution of the hierarchical organization of the brain. Perhaps the most revealing thing about the study of man's brain is that he has inherited the structure and pattern of organization of three basic types, which, for simplifying discussion, I refer to as reptilian, paleomammalian, and neomammalian (MacLean, 1962, 1964, 1966b, 1967, 1968a, 1968b, 1969b, 1969c). Although these fundamental brain types show great differences in structure and chemistry, all three must intermesh and function together as a triune brain.

The hierarchy of the three brains is schematized in Figure 2. The oldest heritage of man's brain is basically reptilian. Represented in black, it forms the matrix of the upper brainstem and comprises much of the reticular system, midbrain, and basal ganglia. The reptilian brain is characterized by greatly enlarged basal ganglia that resemble the corpus striatum of mammals (Papez, 1929), but, in contrast to mammals, there is only a rudimentary cortex.

The paleomammalian brain is distinguished by a marked outgrowth of primitive cortex, which, as is explained later, is synonymous with the limbic cortex. Finally, there mush-

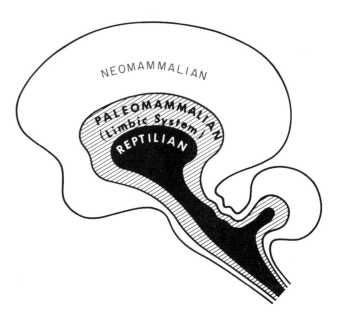

FIGURE 2   Diagramatic representation of hierarchical organization of three basic brain types, which in evolution of the mammalian brain become part of man's inheritance. For purposes of discussion, they are labeled in ascending order as reptilian, paleomammalian, and neomammalian. The paleomammalian counterpart of man's brain corresponds to the limbic system which is believed to play a special role in emotional functions. (From MacLean, 1967.)

rooms late in evolution a more highly differentiated form of cortex called neocortex, which is "the hallmark of the brains of higher mammals and which culminates in man to become the brain of reading, writing, and arithmetic" (MacLean, 1968b).

In popular terms, the three basic brains might be regarded as biological computers, each with its own special form of subjectivity and intelligence, its own sense of time and space, and its own memory, motor, and other functions (MacLean, 1966b, 1968a, 1968b, 1969b). The comparative observations of ethologists would lead one to infer that the reptilian brain programs stereotyped behaviors according to instructions based on ancestral learning and ancestral memories. In our new field laboratory at Poolesville, Maryland, we plan to test the hypothesis that the counterpart of the reptilian brain in mammals plays a crucial role in such genetically constituted forms of behavior as establishing territory, finding shelter, hunting, homing, mating, breeding, imprinting, forming social hierarchies, selecting leaders, and the like.

The reptilian brain appears to be a slave to precedent. This has obvious survival value because, for example, if animals once have found a safe watering hole they avoid the risk of drinking elsewhere. It would be interesting to know to what extent the reptilian counterpart of man's brain contributes to his superstitions and obeisance to precedent in ceremonial rituals, religious convictions, legal actions, and political persuasions.

Obeisance to precedent is conducive to obsessive compulsive behavior, an archaic form of which is illustrated by the sea turtle's returning to the same place year after year to lay its eggs. Recently it has been observed that a number of mammals, such as seals and sheep, have a tendency to return to home grounds (see Harper, in press, for review). Elsewhere I have touched on this topic in relation to man and have also cited other possible relics of reptilian patterns of behavior in human activities, including a predisposition to imitation in social situations (MacLean, 1969b, 1969c).

In summary, and metaphorically speaking, it was as though the reptilian brain were neurosis-bound by an ancestral superego (MacLean, 1964). At the same time, it appears to have inadequate neural machinery for learning to cope with new situations.

The evolutionary development of a respectable cortex in lower mammals would seem to represent nature's attempt to provide the reptilian brain with a "thinking cap" and to emancipate it from stereotyped behavior (MacLean, 1968a, 1968b). The primitive cortex might be imagined as comparable to a crude television screen, giving the animal a better picture of its internal and external environment for adapting to new situations (cf. MacLean, 1958). Memories of current experiences begin to override ancestral memories in guiding behavior. Most of the primitive cortex in all mammals is found in a large convolution that Broca, in 1878, called the great limbic lobe because it surrounds the brainstem. Limbic means "forming a border around." From the standpoint of behavioral implications, it should be emphasized that this lobe, as illustrated in Figure 3, is found as a common denominator in the brains of all mammals. The evolutionary stability of the limbic cortex contrasts with the mushrooming neocortex, shown in white, which culminates in man and gives him a large screen on which a picture can be portrayed by a written and spoken language.

Because of its apparent relationship to olfactory structures, the limbic lobe was formerly believed to subserve purely olfactory functions, and in many textbooks was included as part of the rhinencephalon (see MacLean, 1955, p. 355; 1968a, p. vi). The classic paper of Papez in 1937 struck a mortal blow to this line of thinking. Since then, clinical and experimental work has shown that, in addition to olfactory functions, the limbic cortex is involved in emotional behavior and associated endocrine and viscerosomatic activities. In 1952, I suggested the term "limbic system" as a suitable designation for the counterpart of the paleomammalian brain. This system includes the limbic cortex and structures of the brainstem with which it has primary connections (MacLean, 1952).

## The limbic system and emotional functions

From a large accumulation of clinical and experimental findings, I have selected for emphasis what I consider to be the best evidence that the limbic system derives information in terms of emotional feelings that guide behavior required for self-preservation and the preservation of the species. Finally, in bringing this subject close to home, I suggest how this ancient part of the brain may influence scientific belief and fashions in research. In particular, I call attention to limbic mechanisms underlying affective feelings of individuality and reality which are so fundamental to a sense of what is true and important. This entails a brief summary of our microelectrode findings on interoceptive and exteroceptive inputs to the limbic cortex.

BACKGROUND CONSIDERATIONS As background for these considerations, it will be helpful to refer to a simplified anatomical diagram of the limbic system and to indicate briefly functions of three of its main subdivisions. In Figure 4 the ring of limbic cortex is shown in stipple. It should be emphasized that the limbic cortex has similar features in all mammals and is structurally primitive compared with the new cortex, suggesting that it continues to function at an animalistic level in man as in animals. Also, as opposed to the new cortex, it has strong connections with the hypothalamus and other structures of the brainstem that play a basic role in integrating emotional expression. The Figure

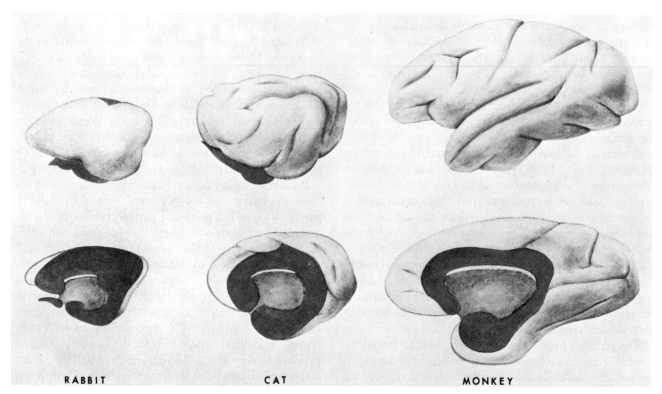

RABBIT                  CAT                  MONKEY

FIGURE 3 Lateral and medial views of brains of rabbit, cat, and monkey drawn roughly to scale. This Figure illustrates that the limbic lobe (dark shading in medial view) is found as a common denominator of the cerebrum throughout the mammalian series. Surrounding the brainstem, the limbic lobe contains most of the cortex corresponding to that of the paleomammalian brain. The greater part of the neocortex, which mushrooms late in evolution, occupies the lateral surface. (After MacLean, 1954b.)

focuses on three pathways that link the hypothalamus to three main subdivisions of the limbic system. You will note that the two upper pathways branching from the medial forebrain bundle meet with descending fibers from the olfactory apparatus and feed into the upper and lower parts of the ring through the amygdala and septum at the points marked no. 1 and no. 2. Clinical and experimental findings indicate that the lower part of the ring fed by the amygdala is primarily concerned with emotional feelings and behavior that insure self-preservation (MacLean, 1958). Its circuits are, so to speak, kept busy with the selfish demands of feeding, fighting, and self-protection (MacLean, 1968b, p. 29).

There is evidence, on the other hand, that the structures associated with the septum in the upper part of the ring are involved in expressive and feeling states that are conducive to sociability and the procreation and preservation of the species (see MacLean, 1962, for review).

Presumably because of the intimate role of the olfactory sense in both feeding and mating, oral and genital functions are brought into close relationship in the amygdala and septal regions. As opposed to this situation, you will observe that the third pathway by-passes the olfactory apparatus; it leads to the anteromedial thalamus and then into cortex in the upper part of the ring. In the phylogeny of the primate brain, it is notable that the septal region remains relatively undeveloped, whereas the structures connected by the third pathway increase in size and become most prominent in man. Our brain and behavioral studies suggest that this condition reflects a shifting of emphasis from olfactory to visual influences in sociosexual behavior (MacLean, 1962, 1966b, 1968b).

There are two more background considerations. First, histochemical studies have shown that both adrenergic (Fuxe, 1965) and "cholinergic" (Lewis and Shute, 1967) fiber systems supply the limbic cortex and that serotonin is found in high concentration in some structures in the lower part of the ring (Paasonen et al., 1957). These findings have obvious relevance to the psychological action of drugs that affect the metabolism of biogenic amines and acetylcholine. Some of these agents, such as reserpine, induce distinctive EEG changes in the hippocampal formation (MacLean et al., 1955/1956).

Second, let me emphasize that a seizure discharge, if induced in the hippocampus, has the tendency to spread throughout and be confined to the limbic system. No experiment shows so convincingly that the limbic system is a functionally, as well as an anatomically, integrated system. At the same time it provides a striking demonstration of the dichotomy of function—or what I have called a "schizophysiology" (MacLean, 1954a, 1958)—of limbic and neocortical systems.

CLINICAL EVIDENCE  The most convincing evidence that the limbic system is involved in emotional functions is derived from the study of patients with psychomotor epilepsy. In such cases, the epileptogenic focus is commonly in or near the limbic cortex in the lower part of the ring and may be the result of scarring after head injury, infection, or

circulatory insufficiency. Malamud (1966, p. 194) has arrived at the conclusion that sclerosis of the hippocampus is the "common denominator" in these conditions. Penfield and Jasper (1954) have shown that electrical stimulation in the involved region may elicit the same kind of symptoms as those occurring in a spontaneous seizure. During the initial epileptic discharge, patients typically experience one or more of a wide variety of vivid affects. The basic and specific affects include feelings of hunger, thirst, nausea, suffocation, choking, cold, warmth, and the need to defecate or urinate. Among the general affects are feelings of terror, fear, sadness, depression, foreboding, familiarity or strangeness, reality or unreality, wanting to be alone, paranoid feelings, and anger. Sometimes a patient will experience an alternation of opposite feelings, suggesting that

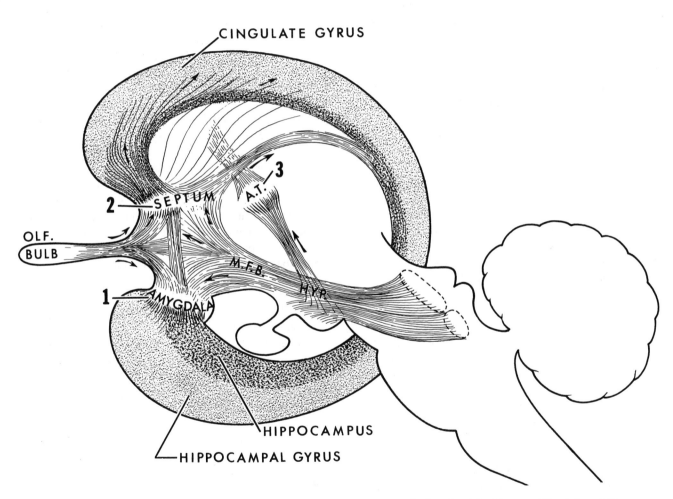

FIGURE 4  The limbic system comprises the limbic cortex and structures of the brainstem with which it has primary connections. This diagram shows the ring of limbic cortex in light and dark stipple and focuses on three pathways (1, 2, and 3) that link three main subdivisions of the limbic system. See text for functional significance. *Abbreviations:* A.T., anterior thalamic nuclei; HYP, hypothalamus; M.F.B., medial forebrain bundle; OLF, olfactory. (From MacLean, 1958, 1967)

there may be a reciprocal innervation for feeling states comparable to the reciprocal innervation of muscles (MacLean, 1955, p. 363).

In proceeding to consider neural mechanisms that underlie an affective feeling of conviction of the reality of ourselves and our environment, I want to call attention particularly to the type of aura experienced by Dostoyevsky and which he described in *The Idiot* as "a sensation of existence in the most intense degree" (1962). This heightened sense of reality is illustrated by the words of a patient (R.A.) known to me who had an epileptogenic focus in the left medial temporal region. Describing his aura, he said, "Each time this happens, thoughts occur very clear and bright to me . . . as if this is what the world is all about . . . [this is] the absolute truth." Here is evidence that a primitive system of our brain that represents an inheritance from lower mammals is able to generate, all out of context, a feeling of what is real, true, and important. More commonly, patients experience a *déjà vu*, which carries with it a sense of conviction that one has already seen and experienced what is happening. In other words, the feeling has the important affective quality responsible for a sense of familiarity and personal identification with what is experienced visually. This, of course, is basic to the memory of visual experiences.

In other cases, the seizure may generate a feeling of depersonalization, conveying the impression that one is looking at oneself from a distance. Or there may be alterations of perception involving any of the sensory systems: objects may seem unusually large or small, near or far; sounds may seem loud or faint; one's tongue, lips, or extremities may seem swollen to large proportions. Time may appear to speed up or slow down. Significantly, these symptoms occur in the endogenous and toxic psychoses and are familiar to many users of psychedelic drugs.

QUESTION OF INDIVIDUALITY   The question next arises: What is it that psychologically distinguishes each one of us as individuals? An appeal to introspection reveals that the condition most crucial for our feeling of individuality is our dual source of information from the external public world and our own private internal world. Signals to the brain from the world within are entirely private, being self-contained, whereas those of the outside world can be publicly experienced and lend themselves to comparison among individuals. A sense of individuality, therefore, would seem to depend on a privacy owing to the inaccessibility of internal signals to anyone but the individual himself (MacLean, 1966b, 1969a).

Kubie (1953) claimed that memory depends on a similar duality of experience. One might say that the union of internal and external experience is as important for memory as the combination of antigen and antibody in developing an enduring immunity. There is evidence from studies of limbic epilepsy that, with epileptic interference of somato-visceral integrative functions of the limbic system, experiences cannot be remembered (cf. MacLean, 1954a). At the same time, it should be emphasized that during the post-ictal disruption of limbic function some individuals are capable of complex motor and intellectual performance that presumably depends on a functioning neocortex. This is illustrated by a classical case (Case Z) described by Hughlings Jackson (1889) and Jackson and Colman (1898). The patient, a young doctor, suffered from epilepsy as the result of a small lesion in the region of the uncus. During one of his seizures he examined a patient, made a correct diagnosis, and wrote an appropriate prescription, but had no recollection of it afterward. In a sense, such individuals function temporarily as if they were disembodied spirits.

NEURAL MECHANISMS   For many years I have had a special interest in the question of forebrain mechanisms that may account for the fusion of internal and external experience and the feeling of individuality. For the reasons already mentioned, one might suspect that the limbic cortex is fundamental to our sense of individuality and the reality of things.

In the "visceral brain" paper of 1949, I showed a diagram suggesting that all the sensory systems (both interoceptive and exteroceptive) feed into the limbic cortex of the hippocampal formation. The implication was that emotional feelings represent a "mentational" blend of internal and external experience. At that time it was known that the olfactory apparatus was indirectly connected with the hippocampal formation, but there was no experimental evidence of a representation of the other senses. Since then, physiological findings (Green and Arduini, 1954), supported by anatomical studies (Daitz and Powell, 1954), have shown that the septum provides a connecting link between the hypothalamus and the hippocampal formation. Here would be a source of interoceptions.

I myself have devoted particular attention to the question of visual, auditory, and somatic inputs. For their demonstration we have tested large populations of units in experiments on awake, sitting, squirrel monkeys prepared with a chronically fixed stereotaxic device that provides a closed system for cerebral exploration with either metal or glass microelectrodes. Investigating first the visual question, we found that units in areas of the posterior parahippocampal cortex (Figure 5) are activated by photic stimulation (MacLean, 1966a; MacLean et al., 1968). As illustrated in Figure 6, a large percentage of responding units in the posterior hippocampal gyrus gave sustained on-responses during illumination of the eye, suggesting that such neurons may signal changes in background illumination and thus possibly play some role in wakefulness, alerting, light-depen-

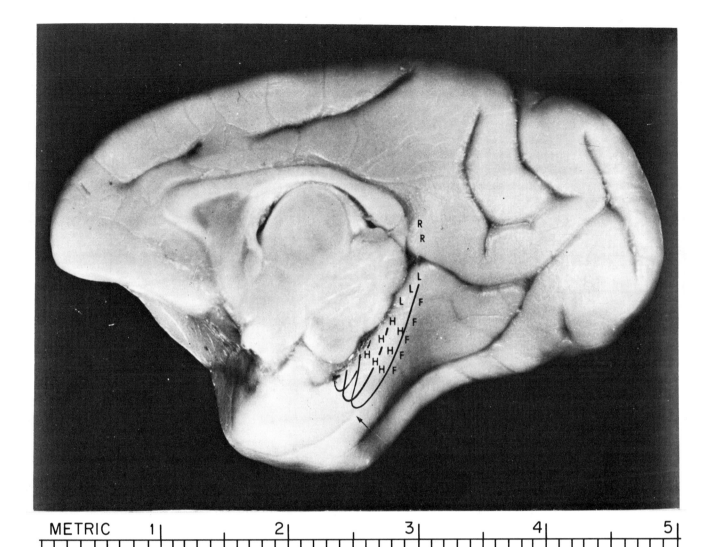

FIGURE 5 Medial view of squirrel monkey's brain, showing approximate location of the photically responsive areas in the medial temporo-occipital cortex. The responsive limbic areas are the posterior hippocampal gyrus (H), parahippocampal portion of the lingual gyrus (L), and retrosplenial cortex (R). The responsive part of the fusiform gyrus is labeled F. The hippocampus lies folded underneath the hippocampal gyrus. Curved black lines schematize the temporal detour of that part of the optic radiations known as Flechsig's knee, or Meyer's loop. See Figure 7 for actual anatomy. The loop does not extend forward into the entorhinal area in the anterior hippocampal gyrus. Arrow points to caudal extremity of the rhinal fissure, which marks the posterior limit of the entorhinal area. (From MacLean, 1969a.)

dent neuroendocrine changes or in all three. In each case, the locus of the recording electrode was in or near the granular layer. Such tonic on-units were not encountered in other limbic areas or in the classical visual cortex.

How do visual impulses reach the parahippocampal cortex? We have found that after a lesion in the ventrolateral part of the lateral geniculate body (the main nucleus for transmitting visual impulses) a continuous band of degeneration extends into the core of the posterior part of the hippocampal gyrus (Figure 7) and that some fibers enter the cortex here and in the neighboring areas (MacLean, 1966a; MacLean and Creswell, 1970). The ventral part of this degenerating band corresponds to the temporal loop in man. It has always been a mystery why some of the optic radiations make this long temporal detour, but, on the basis of the anatomical and microelectrode findings, it would now appear that they travel this roundabout way in order to distribute fibers to the posterior limbic cortex. The

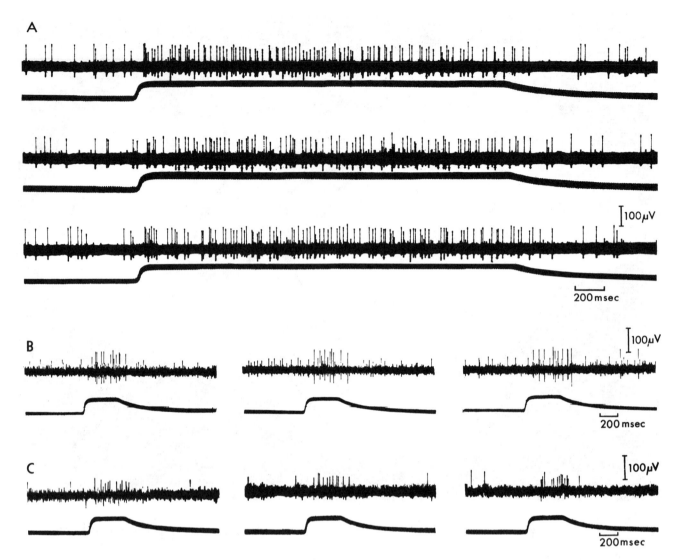

FIGURE 6  Series of sustained on-responses to ocular illumination recorded from units in the posterior hippocampal gyrus (labeled H in Figure 5) of three squirrel monkeys. Such tonic on-units, which possibly signal changes in background illumination, were not found in the other areas labeled in Figure 5 or in the classic visual areas. Duration of photic stimulation is shown in accompanying records by response of a photo cell. (From MacLean et al., 1968.)

inferior pulvinar, regarded as a visual association nucleus, also contributes fibers by projections lying just lateral to the optic radiations (MacLean and Creswell, 1970).

The parahippocampal cortex transmits impulses to the hippocampus, which in turn projects to the hypothalamus and other structures of the brainstem involved in emotional, endocrine, and somatovisceral functions. I have suggested that herein lie possible mechanisms by which "the brain transforms the cold light with which we see into the warm light which we feel" (MacLean, 1966b, 1969b).

Recently, we explored the insular cortex overlying the claustrum for evidence of inputs from the somatic and auditory systems (Reeves et al., 1968). This cortex is limbic by definition, because it comprises part of the phylogenetically old cortex bordering the brainstem. As indicated in Figure 8, it projects to the hippocampal formation (cf. Pribram and MacLean, 1953). Auditory and somatic stimulation evokes responses of a significant number of units in the claustral insula. We found two main types of auditory units, one of which responded with latencies as short as 10 milliseconds. (Parenthetically, there was one unit in the overlying frontal opercular cortex that responded only to the vocalization of another monkey.) Somatic units were activated by pressure alone or by pressure and light touch.

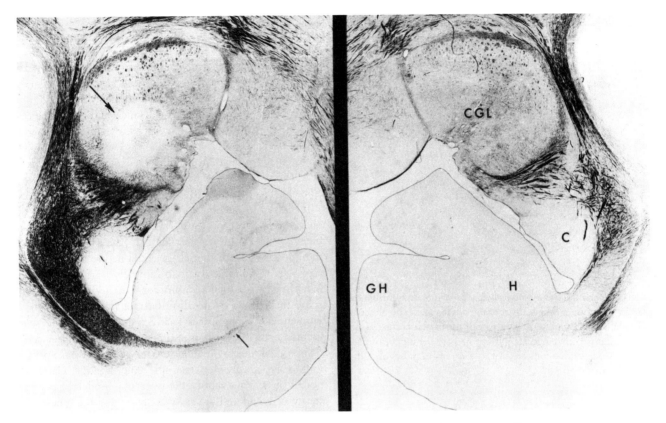

FIGURE 7 A lesion in the ventrolateral part of the lateral geniculate body (large arrow) results in degeneration that funnels into the temporal loop and its continuation into the core of the posterior hippocampal gyrus (small arrow). Some degenerating fibers can be traced into the posterior hippocampal gyrus and adjacent areas labeled in Figure 5.

Section on right shows corresponding structures on unoperated side where only normal fibers are stained. *Abbreviations:* C, nucleus caudatus; CGL, corpus geniculatum laterale; GH, gyrus hippocampi; H, hippocampus. (From MacLean and Creswell, 1970).

The receptive fields were usually large and bilateral. No units responded to more than one modality.

As the insular cortex overlying the claustrum projects to the hippocampal formation, these findings indicate at least one corticofugal pathway by which impulses of auditory and somatic origin may reach the hypothalamus and influence emotional and vegetative functions.

CONCLUDING COMMENTS One might infer that the neocortex performs many of its nice discriminations unhindered by signals and noise generated in the internal world. It is probable on anatomical grounds, however, that the limbic cortex is constantly bombarded by impulses from the internal environment. In neuronal terms, how might this distinction between the limbic and neocortex shed light on feelings of individuality and reality?

As a concluding illustration, I describe briefly microelectrode experiments in awake, sitting, squirrel monkeys,

in which we recorded from both outside and inside hippocampal cells while applying shocks to the midline septum and to the olfactory bulb (Yokota et al., 1967; 1970). The salient anatomy pertaining to this study is diagrammed in Figure 9. The pathways from the septum and bulb may be considered representative of internal and external inputs to the hippocampus. Through the septum, the hippocampus receives impulses from the hypothalamus, which plays an important role in aversive, appetitive, visceral, and humoral reactions of an *unconditional* nature. Our anatomical findings (Gergen and MacLean, 1964), supported by observations of others (Raisman, 1966; Lewis and Shute, 1967), indicate that the septum projects predominantly to *stratum oriens,* which contains the basal dendrites of hippocampal pyramids as well as a number of interneurons. On the contrary, there is evidence that olfactory impulses follow the perforant pathway that terminates on the distal portions of the apical dendrites (Black-

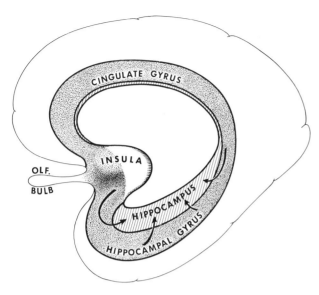

FIGURE 8 The insular cortex overlying the claustrum is limbic by definition, because it forms a part of the phylogenetically old cortex bordering the brainstem. This part of the insular cortex, as do the hippocampal and cingulate gyri, projects to the hippocampus. Microelectrode studies indicate that it transmits impulses of auditory and somatic origin.

stad, 1958; Cragg, 1960). As illustrated in Figure 10, it was found that septal stimulation is highly effective in eliciting excitatory postsynaptic potentials (EPSPs) and neuronal discharge, whereas olfactory volleys generate only EPSPs without neuronal discharge. The anatomical considerations suggest that septal is more effective than olfactory stimulation because it results in depolarization nearer the soma of the cells.

These experiments indicate that, in at least one major subdivision of the limbic cortex, the opportunity exists for the interplay of signals from the internal and external environment, with the internal signals having an overriding influence on neuronal discharge. In addition to the posed question of individuality, the findings have interesting implications in regard to *conditioning*, which is a first step to learning and memory. In terms of classical conditioning, interoceptive impulses transmitted by the septum would be analogous to unconditional stimuli as they are capable by themselves of causing cellular discharge, whereas olfactory and other exteroceptive impulses conducted by the perforant pathway would be comparable to conditional stimuli (Gergen and MacLean, 1964).

The hippocampal formation occupies a central position within the limbic system. Hence, the evidence that signals from the internal world have a pre-emptory role in the function of hippocampal neurons provides some basis for the suggestion that the mentation of the limbic brain is

referentially anchored in the self. This will recall the emphasis that was given earlier to the clinical evidence that the limbic brain has the capacity to generate vivid, affective feelings of what is real, true, and important. Under ordinary circumstances, how might such feelings become attached to scientific propositions and other products of thought? Little knowledge exists about anatomical and functional relationships of limbic and neocortical systems. One might conjecture that reciprocal mechanisms exist in the upper brainstem by which affect could facilitate, distort, or paralyze thought and by which thought could generate or control a state of emotion (MacLean, 1958).

In addition to lending an affective sense of conviction of what is real, true, and important, how else might limbic function bias scientific attitudes? I have wondered, for example, how much the self-oriented nature of affect may have been responsible for man's initially learning to count on his fingers and then kept him for so many centuries from seeing the "sunya"—the all-important nothing, or zero—represented by the spaces between his fingers. The self-oriented nature of affect is also relevant to man's proclivity to choose familiar, workaday models for scientific explanations. One recalls, for example, the bygone Cartesian models of nervous action; the hydraulic models of psychic energy of Freud and the early analysts; and the nineteenth century concept of heat as a flowing substance. As we browse in the attic of history and find ourselves smiling at such discarded models, it is a little disquieting to be reminded that we, too, in our vaunted scientific position, have a similar tendency to project ourselves into our

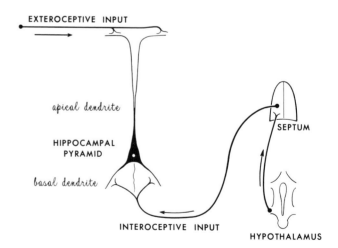

FIGURE 9 Sketch of essential anatomical details outlined in text, providing a possible explanation of microelectrode findings of a differential action of representative interoceptive and exteroceptive inputs on hippocampal neurons. (From MacLean, 1969a.)

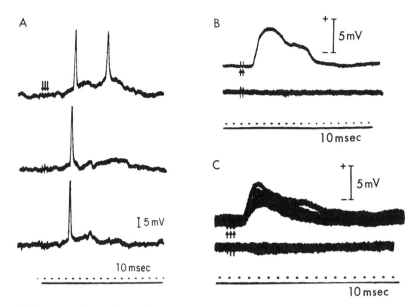

FIGURE 10 Intracellular recordings from hippocampal neurons in awake, sitting, squirrel monkeys. A illustrates that septal stimulation is highly effective in eliciting excitatory postsynaptic potentials (EPSPs) and neuronal discharge. In B and C, olfactory volleys generate only EPSPs without spikes. A shows three successive responses. In C several sweeps are superimposed. Lower tracings in B and C are recordings just outside cell. Resting membrane potentials in every case were stable and measured between −40 and −50 mV. (Excerpts of figures from Yokota et al., 1970.)

scientific concepts. In molecular and genetic biology, for example, do we not refer to messengers, operators, promoters, master genes, slave genes, and the like, somewhat as though the behaving entities of our discourse were like page boys, postmen, or other members of a Lilliputian society? It is important to keep asking how such anthropomorphic modeling may interfere with detached thinking and thereby limit our intellectual horizons. It is also pertinent to consider how reptilian imitation and limbic feelings of what is real and important may contribute to diversionary fads and fashions in science.

Elsewhere, I have discussed the possibility that the psychic functions of the limbic system are such as to predispose us to apprehend ourselves and the world as an aesthetic continuum. On the contrary, the neocortex, with its nice differentiating ability, appears to have the propensity to subdivide things into smaller and smaller entities. Is it possible that these opposing tendencies underlie our difficulties in resolving such time-worn questions as those concerning the-one-and-the-many, finite and infinite space, and the simultaneous existence of the wave and quantum? In what direction does reality lie? With the soft brain always standing between us and what we observe, can we ever tell? The search for an answer presents a challenge far surpassing that of the recent exploit of landing men on the moon, because the cranial vault into which we must launch our probing rockets contains the most complicated constellation of behaving entities in the known universe.

## Summary

The subject of brain correlates of emotion has important relevance to questions of epistemology and the shaping of scientific attitudes. In the present analysis, the emotions are viewed from a new perspective and considered in the light of evolutionary changes of the brain. In evolution, the brain of man retains the pattern of organization of three basic types which, for purposes of discussion, are referred to as reptilian, paleomammalian, and neomammalian. The paleomammalian counterpart is represented by the limbic system, which is discussed primarily because of evidence of its central role in generating affective feelings, including those important for a sense of reality of oneself and the environment and a conviction of what is true and important. Some microelectrode findings in awake, sitting monkeys are described in considering neural mechanisms that possibly underlie the affective feelings in question. Finally, the subject is brought close to home by suggesting how limbic affective influences are involved in biasing scientific thought.

# REFERENCES

BLACKSTAD, T. W., 1958. On the termination of some afferents to the hippocampus and fascia dentata. *Acta Anat.* 35: 202–214.

BORING, E. G., 1964. Cognitive dissonance: Its use in science. *Science (Washington)* 145: 680–685.

BROCA, P., 1878. Anatomie comparée des circonvolutions cérébrales. Le grand lobe limbique et la scissure limbique dans la série des mammifères. *Rev. Anthrop.* 1: 385–498.

CRAGG, B. G., 1960. Responses of the hippocampus to stimulation of the olfactory bulb and of various afferent nerves in five mammals. *Exp. Neurol.* 2: 547–571.

DAITZ, H. M., and T. P. S. POWELL, 1954. Studies of the connexions of the fornix system. *J. Neurol. Neurosurg. Psychiat.* 17: 75–82.

DOSTOYEVSKY, F., 1962. The Idiot (translated by C. Garnett). The Modern Library, New York, p. 214.

FUXE, K., 1965. Evidence for the existence of monoamine neurons in the central nervous system. IV. The distribution of monoamine nerve terminals in the central nervous system. *Acta Physiol. Scand.* 64 (suppl. 247): 37–84.

GERGEN, J. A., and P. D. MACLEAN, 1964. The limbic system: Photic activation of limbic cortical areas in the squirrel monkey. *Ann. N. Y. Acad. Sci.* 117: 69–87.

GREEN, J. D., and A. A. ARDUINI, 1954. Hippocampal electrical activity in arousal. *J. Neurophysiol.* 17: 533–557.

HARPER, L. V., 1970. Ontogenetic and phylogenetic functions of the parent-offspring relationship in mammals. *In* Advances in the Study of Behavior, Vol. 3 (D. S. Lehrman, R. A. Hinde, and E. Shaw, editors). Academic Press, New York (in press).

JACKSON, J. H., 1889. On a particular variety of epilepsy ("Intellectual aura"), one case with symptoms of organic brain disease. *Brain* 11: 179–207.

JACKSON, J. H. and W. S. COLMAN, 1898. Case of epilepsy with tasting movements and "dreamy state"—Very small patch of softening in the left uncinate gyrus. *Brain* 21: 580–590.

KUBIE, L. S., 1953. Some implications for psychoanalysis of modern concepts of the organization of the brain. *Psychoanal. Quart.* 22: 21–68.

LEWIS, P. R., and C. C. D. SHUTE, 1967. The cholinergic limbic system: Projections to hippocampal formation, medial cortex, nuclei of the ascending cholinergic reticular system, and the subfornical organ and supra-optic crest. *Brain* 90: 521–540.

MACLEAN, P. D., 1949. Psychosomatic disease and the "visceral brain." Recent developments bearing on the Papez theory of emotion. *Psychosom. Med.* 11: 338–353.

MACLEAN, P. D., 1952. Some psychiatric implications of physiological studies of frontotemporal portion of limbic system (visceral brain). *Electroencephalogr. Clin. Neurophysiol.* 4: 407–418.

MACLEAN, P. D., 1954a. The limbic system and its hippocampal formation. Studies in animals and their possible application to man. *J. Neurosurg.* 11: 29–44.

MACLEAN, P. D., 1954b. Studies on the limbic system ("visceral brain") and their bearing on psychosomatic problems. *In* Recent Developments in Psychosomatic Medicine (E. Wittkower and R. Cleghorn, editors). Pitman, London, pp. 101–125.

MACLEAN, P. D., 1955. The limbic system ("visceral brain") in relation to central gray and reticulum of the brain stem. Evidence of interdependence in emotional processes. *Psychosom. Med.* 17: 355–366.

MACLEAN, P. D., 1958. Contrasting functions of limbic and neocortical systems of the brain and their relevance to psychophysiological aspects of medicine. *Amer. J. Med.* 25: 611–626.

MACLEAN, P. D., 1960. Psychosomatics. *In* Handbook of Physiology, Section I: Neurophysiology (J. Field, H. W. Magoun, and V. E. Hall, editors). American Physiological Society, Washington, D. C., vol. 3, pp. 1723–1744.

MACLEAN, P. D., 1962. New findings relevant to the evolution of psychosexual functions of the brain. *J. Nerv. Ment. Dis.* 135: 289–301.

MACLEAN, P. D., 1964. Man and his animal brains. *Mod. Med.* 32: 95–106.

MACLEAN, P. D., 1966a. The limbic and visual cortex in phylogeny: Further insights from anatomic and microelectrode studies. *In* Evolution of the Forebrain (R. Hassler and H. Stephan, editors). Plenum Press, New York, and Georg Thieme, Stuttgart, pp. 443–453.

MACLEAN, P. D., 1966b. Brain and Vision in the Evolution of Emotional and Sexual Behavior. Thomas William Salmon Lectures, New York Academy of Medicine.

MACLEAN, P. D., 1967. The brain in relation to empathy and medical education. *J. Nerv. Ment. Dis.* 144: 374–382.

MACLEAN, P. D., 1968a. Ammon's Horn: A continuing dilemma. *Foreword in* S. Ramón y Cajal, The Structure of Ammon's Horn (translated from the Spanish by L. Kraft). Charles C Thomas, Springfield, Illinois.

MACLEAN, P. D., 1968b. Alternative neural pathways to violence. *In* Alternatives to Violence (L. Ng, editor). Time-Life Books, New York, pp. 24–34.

MACLEAN, P. D., 1969a. The internal-external bonds of the memory process. *J. Nerv. Ment. Dis.* 149: 40–47.

MACLEAN, P. D., 1969b. The paranoid streak in man. *In* Beyond Reductionism (A. Koestler and J. R. Smythies, editors). Hutchinson and Co., London, pp. 258–278.

MACLEAN, P. D., 1969c. A Triune Concept of the Brain and Behavior: Lecture I. Man's Reptilian and Limbic Inheritance; Lecture II. Man's Limbic Brain and the Psychoses; Lecture III. New Trends in Man's Evolution. Clarence Hincks Memorial Lectures, Queen's University, Kingston, Ontario, Toronto University Press (in press).

MACLEAN, P. D. and G. CRESWELL, 1970. Anatomical connections of visual system with limbic cortex of monkey. *J. Comp. Neurol.* 138: 265–278.

MACLEAN, P. D., S. FLANIGAN, J. P. FLYNN, C. KIM, and J. R. STEVENS, 1955–1956. Hippocampal function: Tentative correlations of conditioning, EEG, drug, and radioautographic studies. *Yale J. Biol. Med.* 28: 380–395.

MACLEAN, P. D., T. YOKOTA, and M. A. KINNARD, 1968. Photically sustained on-responses of units in posterior hippocampal gyrus of awake monkey. *J. Neurophysiol.* 31: 870–883.

MALAMUD, N., 1966. The epileptogenic focus in temporal lobe epilepsy from a pathologoical standpoint. *Arch. Neurol.* 14: 190–195.

PAASONEN, M. K., P. D. MACLEAN, and N. J. GIARMAN, 1957. 5-Hydroxytryptamine (serotonin, enteramine) content of structures of the limbic system. *J. Neurochem.* 1: 326–333.

PAPEZ, J. W., 1929. Comparative Neurology. T. Y. Crowell, New York.

PAPEZ, J. W., 1937. A proposed mechanism of emotion. *Arch. Neurol. Psychiat.* 38: 725–743.

PENFIELD, W., and H. JASPER, 1954. Epilepsy and the Functional Anatomy of the Human Brain. Little, Brown and Co., Boston.

PRIBRAM, K. H., and P. D. MACLEAN, 1953. Neuronographic analysis of medial and basal cerebral cortex. II. Monkey. *J. Neurophysiol.* 16: 324–340.

RAISMAN, G., 1966. The connexions of the septum. *Brain* 89: 317–348.

REEVES, A. G., K. SUDAKOV, and P. D. MACLEAN, 1968. Exploratory unit analysis of exteroceptive inputs to the insular cortex in awake, sitting, squirrel monkeys. *Fed. Proc.* 27: 388 (abstract).

ROSENBLUETH, A., N. WIENER, and J. BIGELOW, 1943. Behavior, purpose and teleology. *Phil. Sci.* 10: 18–24.

RUSSELL, B., 1921. The Analysis of Mind. George Allen Unwin, London, and the Macmillan Co., New York.

SHAKOW, D., and D. RAPAPORT, 1964. The influence of Freud on American psychology. *Psychol. Issues* 4 (Monogr. 13).

SNOW, C. P., 1967. Variety of Men. Charles Scribner's Sons, New York, p. 90.

SPENCER, H., 1896. Principles of Psychology. D. Appleton and Co., New York, 2 vols. in 3.

WATSON, J. B., 1924. Behaviorism. The People's Institute Publishing Co., Inc., New York.

WIENER, N., 1948. Cybernetics; or, Control and Communication in the Animal and the Machine. John Wiley and Sons, New York, p. 155.

YOKOTA, T., A. G. REEVES, and P. D. MACLEAN, 1967. Intracellular olfactory response of hippocampal neurons in awake, sitting squirrel monkeys. *Science* (*Washington*) 157: 1072–1074.

YOKOTA, T., A. G. REEVES, and P. D. MACLEAN, 1970. Differential effects of septal and olfactory volleys on intracellular responses of hippocampal neurons in awake, sitting monkeys. *J. Neurophysiol.* 33: 96–107.

# 34 Social Communication Among Animals

DETLEV PLOOG

## Phylogenetic aspects of communication processes

CLASSES OF SOCIAL BEHAVIOR   There is good evidence to show that the evolution of the nervous system is closely paralleled by a corresponding evolution of behavior. The phylogenetic process has produced not only increasingly effective sense organs, faster action, more acute discrimination, and many species-specific characteristics, but also a variety of more and more elaborate social systems, mediated by an increasing refinement of communication processes. I shall not go so far as to speculate about the most primitive forms of social behavior. Even for the vertebrates, it would be extremely difficult to be precise as to when, during evolution, social behavior began.

There are four classes of behavior that may be called social. First, the interactions of the male and female for the purpose of reproduction; second, the actions and interactions of a male or a female, or both, for the care and upbringing of the young as well as interactions of the young with parents and siblings; third, all forms of cohesive interactions in herds, groups, troops, bands, and the like; fourth, all kinds of agonistic interactions among members of the same species, such as attacking, defending, threatening, avoiding, withdrawing, fleeing, and so forth.

RITUALIZATION   If we take an extreme reductionistic view of behavior patterning, we might postulate that, in the beginning, the simplest modes of social behavior were nothing but approach and avoidance among members of the same species. The outcome of such an encounter was always unpredictable and depended on each partner's actions and reactions. These, in turn, depended not only on external stimuli but also on the internal state of the individual. This internal state is sometimes discussed in terms of concepts such as motivation or disposition or readiness for action. Because communication between partners was desirable, social signals evolved, as the theory goes, to permit a more

DETLEV PLOOG   Max-Planck Institute for Psychiatry, Munich, Germany flexible way of encounter and a greater degree of information about the outcome.

Whether we consider the vertebrates or other phyla, there seems to be a basic rule in the development of social signals, which was first described by Julian Huxley: if it is of advantage to a sender for a recipient to perceive the disposition or the state of the sender by a specific behavior, that pattern is likely to be modified until it becomes conspicuous. This transformation of a behavior into a signal is called ritualization. Animal ritualization constitutes an adaptive canalization of expressive behavior, produced by natural selection. It has a built-in genetic basis, although learning may also play a part (Huxley, 1923). These transformed behavioral patterns are commonly derived from a basic repertoire not originally employed for communication purposes.

From old-timers of evolution—the so-called living fossils, such as the tiny reptile tuatara (*Sphenodon punctatus*), which is found on some of the small southern islands of New Zealand—we are able to learn something about the fixed pattern of social signals, which in this case possibly remained unchanged for about 150 million years. This reptile shows head-nodding as one of several social signals that can also be seen in recent species of lizards. Figure 1 shows a South American male lizard, *Iguana iguana,* at two phases of its chief display. The extended dewlap makes the signal especially conspicuous. The signal consists of a sequence of head-noddings that are highly stereotyped in amplitude, frequency, and time course, which are shown in the line above the head. This has been revealed by motion-picture analysis of many events in various animals.

Communication processes of this kind, although more flexible, are obviously still present in mammals, including nonhuman primates (Altmann, 1967; Ploog and Melnechuk, 1969; Sebeok, 1968). Figure 2 shows two slightly different kinds of back-rolling in the squirrel monkey. At the left, the back-rolling is the typical pattern that serves to clean the fur, especially when it is wet. It has no signal function. At the right the back-rolling is more refined, a kind of "short-hand," and it appears largely in conjunction with penile erection. This serves as a social signal. The signal content is appeasement. A dominant intruder employs it to stop or diminish the fleeing or avoidance behavior of the excited group. It facilitates mutual approaches of the intruder and of members of the group (Castell and Ploog, 1967; Castell et al., 1969).

In addition to movements of parts of the body with or without participation of such autonomic responses as penile erection, dilation of pupils, pilo-erection, changes in color, and so on, the vertebrates very early employed vocalizations as social signals. These include the mating croaks of frogs, the hissing of turtles, the vocalization of tuatara, the songs of birds, and, eventually, the vocal repertoire of monkeys, which I discuss later in greater detail.

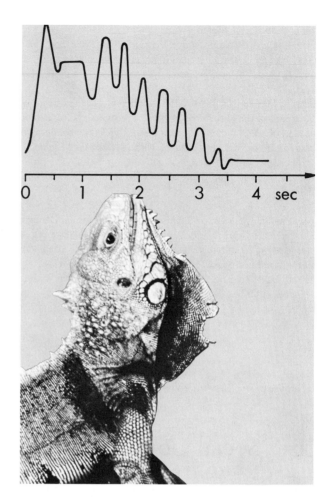

FIGURE 1 The head-nodding display of *Iguana iguana,* a South American lizard. Onset (top) and termination (bottom) of the conspicuous social signal. Amplitude, frequency, and time course of the nodding shown in the line above the head.

FIGURE 2 Back-rolling of the squirrel monkey, *Saimiri sciureus*. Drawings from motion-picture sequences. Left: Back-rolling in its noncommunicative function, i.e., cleaning the wet fur. Right: Back-rolling, ritualized as a social signal of appeasement, in conjunction with penile erection. (From Castell et al., 1969.)

Approach and avoidance, with their extremes of attack and flight, and partner interaction for mediating communication within a species and for providing a means of reproductive isolation between the species—these basic mechanisms of behavior have brain correlates that developed from very primitive to highly complex neuronal machinery. The older parts were preserved, although modified, while newer parts came into existence. Some aspects of the neuronal basis for social behavior are discussed by Paul MacLean in this volume.

## Communication processes and brain functions

GENITAL DISPLAY AS A SOCIAL SIGNAL  I have mentioned the increasing refinement and differentiation of communication processes and now give but one example. In order to learn more about the social behavior of the squirrel monkey, I observed a colony of these animals in MacLean's laboratory and found, among other social signals, one that is now referred to as genital display (Ploog et al., 1963; Ploog and MacLean, 1963). This signal is used in various types of agonistic, dominance, and courtship behaviors. From sociometric data, I have concluded that genital display is a social sign stimulus that contributes decisively to the formation of group structure (Ploog, 1963, 1967). Figure 3 shows two variations of genital display. Both consist of several components: lateral positioning of the leg with the hip and knee

bent; marked supination of the foot and abduction of the big toe; as well as erection of the penis. In females, enlargement of the clitoris replaces erection (Maurus et al., 1965). The display is frequently accompanied by specific vocalizations and occasionally by a few spurts of urine. It may last for a second or so to many minutes. A fraction of this display, the penile erection, can be elicited by electrical stimulation of subdivisions of the limbic system (MacLean and Ploog, 1962).

Here, again, we may refer to the process of ritualization. Genital organs, when functioning as part of copulatory behavior, serve to maintain the species. These same organs, when used in genital display, become a ritual that provides the individual with a means of intraspecific communication that enables him to assert himself in a communal situation. We do not know precisely which brain structures are involved in mediating this signal, but we are sure that at least some of the areas that have yielded genital responses, on electrical stimulation, contribute to its generation. In this connection, we have been interested in the developmental aspects of brain functions and communication processes.

ONTOGENETIC ASPECTS OF SOCIAL SIGNALING  Various features of neonate behavior in nonhuman primates suggest the built-in nature of certain behavior patterns and their sequential order of occurrence. For instance, grasping the fur and rooting, searching for, and sucking the nipples take place as soon as the neonate has emerged from the birth canal (Bowden et al., 1967). We were surprised to find that

FIGURE 3 Genital display, a social signal of the squirrel monkey, used in agonistic and courtship behavior by both sexes. Two of several variations; top; displaying at distance; bottom, counterdisplay at proximity. (From Ploog, 1967.)

such a complex behavior as genital display can also be observed in the neonate. Motion-picture analysis of the behavior of an infant a day and a half old revealed it displaying to a cage-mate. The signal seems to be a highly integrated pattern that includes visual orientation and recognition, directed partner relations, and coordination of motor and autonomic functions. It indicates that brain structures involved in this response are functionally mature at birth. The innate nature of genital display is also suggested by observations of a hand-raised, male squirrel monkey, which directed this signal against its substitute mother, a stuffed woolen sock (Hopf, 1970; Ploog, 1969).

This conclusion is supported by experiments with other primates. For instance, Sackett (1966) observed changes of the stimulus value of facial expressions in rhesus monkeys during ontogenesis. He found that infants raised in strict isolation did not respond differentially to the picture of an adult male's threatening face until they were three and a half months old. At that age, however, the infants suddenly began to develop avoidance behavior to this social signal.

With increasing age there is a change in the function of genital display in the squirrel monkey. Up to a certain age, the signal, although used with increasing frequency, does not affect the adult partner's behavior. At the end of the first and the beginning of the second year, which is also the time of final weaning, the situation changes. I take the example of the dominant male of a group and a young animal. Until the infant is about nine months old, the "boss" of the group, the alpha male, threatens only mildly and infrequently, even if the young male displays toward him and makes use of this opportunity time and again. In fact, the boss and the mother are the ones that are most frequently addressed by genital display. When the young animal grows older, however, relations with the alpha male change radically. The latter threatens severely and may attack the young male when it displays toward him. There is an amusing transitional phase, during which the young male wishes not to give up its behavior and turns away while displaying but nevertheless glances over its shoulder at the boss. At the beginning of the second year, the display behavior decreases markedly, and the young male establishes a new relationship with the dominant male. Before this time, contact between them is rare. Afterward, contacts are more frequent and consist mainly of huddling and playing. The play includes a characteristic pattern that is, no doubt, schooling for fighting—a playful but well-controlled training by the alpha animal (Ploog et al., 1967).

The example of genital display as a social signal shows some features that may be important for all primates, including man. On the one hand, it is a built-in, species-specific behavior pattern. On the other hand, as the infant matures, both the infant and the group change their interaction patterns with respect to a given signal. First, the infant uses the signal over and over again with little or no consequence for either sender or recipient. Despite this lack of reinforcement, it persists up to a predetermined age. The signal is then followed by severe consequences, and the young monkey learns, within a short period of time, the context in which it may or may not use the signal. This protracted maturation, which coincides with other biological data, such as second dentition and final weaning, seems to depend on the species, in one form or another, and is characteristic of the social development of primates. This process of socialization is genetically determined by maturational processes in which innate and learned components are interlocked.

VOCALIZATIONS AS SOCIAL SIGNALS  So far, we have considered communication processes that are guided by visual perception. From the other senses that can mediate social communication, such as audition, olfaction, and touch, I choose audition, because it plays a fundamental role in human communication. The generation of sounds for auditory communication is, as I have already mentioned, an old capacity in vertebrates. Some classes, chiefly the birds, have developed a highly elaborate form of sound production which has never been achieved by mammals. Auditory communication however, plays an important role among the nonhuman primates. The Old World monkeys and the great apes have only a few, but variable, calls. The New World monkeys have a greater repertoire of rather discrete sounds, which carry specific information.

The adult squirrel monkey has a very elaborate repertoire of 26 to 30 calls, the physical characteristics of which are distinguishable by means of sound spectrography. Figure 4 is a schematic presentation of the vocal repertoire. This is divided into five groups. The sixth group is comprised of calls formed by elements of the five groups. Although the groups were formed primarily on the basis of physical characteristics, there is much evidence that a common denominator exists in each group for the information content of the signals.

We used four methods to investigate the meaning of the calls. First, the calls were, whenever possible, elicited repeatedly by well-defined visual stimuli. Second, vocal and motor behavior patterns were elicited by play-backs of tape recordings of animals' calls. Third, an animal's motivational state was varied, and the accompanying vocal behavior was observed. Fourth, types of calls were correlated with the social role and status of the emitters and receivers of calls within the group (Winter et al., 1966).

At this point, I give only an example of varying the motivational state (Figure 5). In Figure 5A and B, the degree of motivation was changed by food deprivation for 24 and 30 hours, respectively. Figure 5A represents a distribution of frequencies of calls typical during feeding and charac-

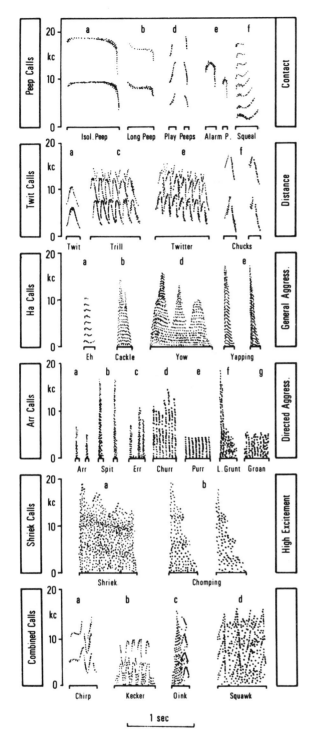

FIGURE 4 Vocal repertoire of the squirrel monkey, schematic presentation according to spectrographic analysis. Letters refer to the original sound spectrograms. *Abbreviations:* aggress., aggression; alarm p., alarm peep; l. grunt, labor grunt. For division into six groups and for further explanation, see text. (From Winter et al., 1966.)

terized by a predominance of twittering and trilling calls. Only about 10 per cent of the total vocalization consisted of cackling, which has an aggressive connotation. After 30 hours (Figure 5B), the distribution was considerably altered. The amount of twittering and cackling changed as compared with that in A. In addition, keckering and shrieking were recorded. These are calls of high excitement and may trigger aggressive interactions of groups. When a strange male was introduced (Figure 5C), twittering and trilling calls ceased. The percentage of cackles decreased in favor of churrs and purrs. This took place together with an increase in directed aggressive actions, such as genital display or persecuting, biting, and fighting.

VOCALIZATIONS ELICITED BY ELECTRICAL BRAIN STIMULATION  The discreteness of the vocal signals makes this animal particularly suitable for electrical brain-stimulation studies. A pilot study revealed that the majority of the naturally occurring calls could be elicited reliably and repeatedly in specific brain sites by the use of implanted electrodes (Jürgens et al., 1967). The results of an elaborate study (Jürgens and Ploog, 1970), which has just been finished, are shown in a schematic view (Figure 6).

Two major groups of vocalizations, the cackling and growling calls, which correspond to the third and fourth groups in Figure 4, are elicitable within a continuous system of fibers. Cackling expresses general excitement, with an aggressive motivation. Growling calls are directed against specific animals in an aggressive context. The system for cackling leads from the caudal end of the periaqueductal gray through the periventricular gray of the diencephalon. At the level of the inferior thalamic peduncle the system branches off in three components: the first follows the inferior peduncle dorsally toward the anteromedial thalamic nucleus; the second follows the inferior peduncle ventrolaterally into the amygdala and farther through the external capsule and the uncinate fasciculus to the rostroventral temporal cortex; the third follows the anterior thalamic radiation along the ventromedial border of the internal capsule into the ventromedial orbital cortex and the precallosal cingulate gyrus.

The system for growling calls partly overlaps the cackling system, but also has its own extensions, for instance, into the area ventralis tegmenti of Tsai and via the medial forebrain bundle into the lateral hypothalamus. At the level of the inferior thalamic peduncle, another important deviation from the cackling system leads into the preoptic region just anterior to the anterior commissure, turns to the stria terminalis, and follows these fibers into the amygdala. The distribution of the growling calls is essentially identical with the hissing and growling of cats, with the amygdala, stria terminalis, perifornical hypothalamus, periaqueductal gray, and tegmentum as mediating structures.

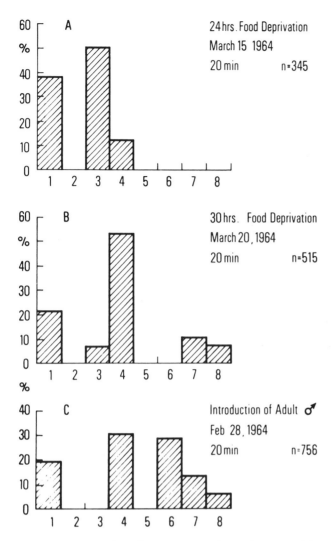

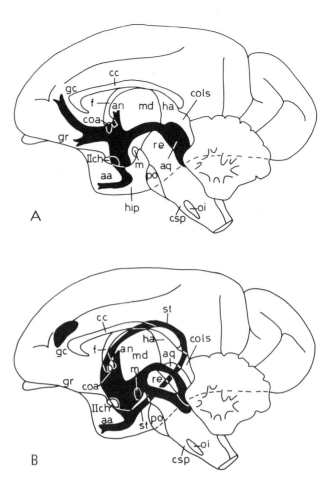

FIGURE 5 Change of vocalization pattern due to alteration of the motivational state. A, B. Two sexually mature squirrel monkeys in one cage. C. Introduction of an unfamiliar adult male into the cage after normal feeding of all animals concerned. 1, peeps and chirps; 2, play peeping; 3, twittering/trilling; 4, cackling; 5, yapping; 6, churr/purr; 7, kecker; 8, shrieking. (From Winter et al., 1966.)

FIGURE 6 Vocalizations elicited by electrical stimulation of the squirrel monkey's brain. General view of the cerebral system (in black). A, yielding cackling calls; B, yielding growling calls. *Abbreviations:* aa, area anterior amygdalae; an, nucleus anterior; aq, substantia grisea centralis; cc, corpus callosum; coa, commissura anterior; cols, colliculus superior; csp, traetus corticospinalis; f, fornix; gc, gyrus cinguli; gr, gyrus rectus; ha, nucleus habenularis; hip, hippocampus; m, corpus mamillare; md, nucleus medialis dorsalis thalami; oi, nucleus olivaris inferior; po, griseum pontis; re, formatio reticularis tegmenti; st, stria terminalis; Ilch, chiasma nervorum opticorum. (From Jürgens and Ploog, 1970.)

Other groups of calls, such as peeps, trills, and shrieks, are represented not in continuous systems but in circumscribed areas only. For example, the peep calls, which serve for contact and group cohesion, are elicitable in the subcallosal gyrus, the nucleus accumbens of the septal area, the medioventral head of the caudate, the rostral hippocampus, the midline thalamus, and the spinothalamic tract. Trill calls, often associated with feeding and location, were found along the precommissural fornix between Zuckerkandl's bundle and the genu of the corpus callosum.

I do not intend, here, to deal with anatomical details and

considerations, but it is necessary to demonstrate the specificity of the vocal system and its correlation with certain brain structures and systems. On the other hand, it is noteworthy that the brain structures that mediate these calls, especially the amygdala, the hypothalamus, the stria terminalis, and the midbrain periventricular gray, are also associated with attack and defense behavior. Furthermore, almost all structures that are shown to mediate vocalizations are known to yield other behavioral or autonomic responses, including genital responses. The latter is of special interest in social signaling, because erection during genital display is frequently accompanied by two types of vocalization—calls of directed aggression in an aggressive context and peep calls in a nonaggressive context. These alternative types of vocalization and the genital response can be elicited simultaneously by brain stimulation.

TELESTIMULATION AS MEANS OF SIGNAL ANALYSIS So far we have dealt with motor and vocal signals as such. We have shown that a great variety of these signals can be elicited from distinct brain structures. Therefore, one should be able to control and to manipulate the communication processes within a group to permit a precise analysis of the signal function and thereby of the determinants of group dynamics.

Such control and manipulation can be accomplished by radio-stimulation of electrodes chronically implanted in the brains of free-moving animals (Delgado, 1963; Robinson and Warner, 1967; Maurus and Ploog, 1969). By this means, both the electrically elicited response or sequences of this response and the reaction or sequences of reactions of the group can be studied. This latter part of the communication process, i.e., the reaction of the recipient to a social signal, is indicative of the signal function. Telestimulation in such a quasi-natural group situation reveals most interesting results.

First, one and the same electrical stimulus applied at the same brain site in the same animal at different times does not necessarily elicit the same response. Rather, the animal's response depends on the composition and disposition of the group, and on the rank, role, and sex of the stimulated and the interacting animal. For example, in our first study of this kind, dominance gestures of squirrel monkeys in a number of different situations were elicited more frequently in females than in the subordinate male, and very rarely in the dominant male. This finding cannot be attributed to possible variations in stimulated brain structures.

Second, an electrically elicited behavior pattern with social relevance can lead to different social interactions, which again depend on the actual situation, and on the rank, role, and sex of the participants. For instance, lower-ranking animals never (without any exception in 475 events) responded to certain signals, such as dominance gestures, with any of the known social signals.

Third, sequences of behavioral events may be composed of events that carry information and those that have no communicative significance. The observer may be unable to differentiate the information-carrying signals from all other activities. Furthermore, the animals may use means of communication that we do not yet recognize as social signals because they are imperceptible to the human observer, even when careful motion-picture analysis is employed (Maurus and Ploog, in preparation).

Automation of the behavioral control by use of refined techniques of signal recognition would open up a new era of quantitative analysis of communication processes. For the time being, without adequate techniques, one must rely on motion pictures and observation.

## Communication in the light of information theory

THE STOCHASTIC ANALYSIS OF A TWO-PARTNER SITUATION Mathematical description of the stochastics and information flow in a real-life social situation can be made provided the variables of the situation are considerably reduced (Altmann, 1965, 1967). The bidirectional communication theory of Marko and Neuburger (1967), an expansion of Shannon's basic theory, provides a means of analyzing a multi-information-generating system, in which each information-generating system is simultaneously a sender and a receiver and in which information may be flowing in two or more directions at any given time. It would not be appropriate to attempt to explain here the mathematics of the system, other than to note that it is based on the assumption of two stochastic processes whereby the generated signals of each of the systems depend on the preceding signals of that system and also on the signals received from the second system.

A general illustration of the two-system situation (Figure 7) will serve to introduce the basic terminology. Each system generates a certain amount of free entropy, $F_1$ or $F_2$ (information arising entirely from its own internal states). This free entropy combines with the transinformation, $T_{12}$ or $T_{21}$ (information received from the other system), and the two quantities together form the system entropy, $H_1$ or $H_2$ (the total information transmitted by each system). Of this entropy, not all is actually received by the other

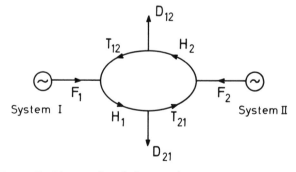

FIGURE 7 Diagram for a bidirectional communication situation. System I and system II: information-generating system, i.e., animals. $F_1$ and $F_2$: free entropy, i.e., spontaneously generated information of each system. $T_{12}$ or $T_{21}$: transinformation, i.e., information received from the other system. $H_1$ or $H_2$: total information transmitted by each system, i.e., the combination of the system entropy of $T_{12}$ or $T_{21}$ and $F_{12}$ or $F_{21}$. $D_{12}$ or $D_{21}$: discrepancy, i.e., the lost amount of the total information. (From Marko, 1966.)

system. The fraction received is called the transinformation, mentioned above, and the amount lost is known as the discrepancy, $D_{12}$ or $D_{21}$.

The following example shows how such an analysis can be applied to a real social situation. Five adult male squirrel monkeys, who had lived together for several years, were separated overnight and caged singly in different rooms. In the experimental situation on the following day, two animals were brought together at random in one cage and observed for 15 minutes per session for a total of 217 observation periods. The reunion was always accompanied by a greeting ceremony, which can be described by a set of five behavior patterns, plus a sixth category that arbitrarily includes all other behavioral events (Mayer, 1969).

Figure 8 illustrates the results. In the upper third of the figure, the dominant monkey, Tell, at the beginning of the observation period, generated as much free entropy or spontaneous information (0.59) as did his subordinate part-

ner, Fan. But he received a much larger amount of information from Fan (0.21) than Fan did from him (0.03). The respective discrepancies and system entropies reflect this difference in transinformation. It is evident, therefore, that the dominant animal reacts to its partner by showing a much greater diversity of behavior than does the subordinate.

The situation is different in the middle third of the Figure. Fighting and a severe bite resulted in a dramatic shift of dominance. Then Tell was subordinate. But the strategy of Fan's dominance was different. He generated almost twice as much free entropy as did Tell, i.e., 0.70 versus 0.38. But the transinformation of 0.22 from the now-dominant Fan to the now-subordinate Tell was almost equal to the previous situation. Fan dominated Tell but, as judged by Tell's strategy, was doing too much; that is, Fan employed his repertoire of dominance behavior much more frequently than did Tell when he was dominant.

After six weeks, the new dominance situation stabilized, and some interesting changes took place in the information flow. Fan had learned how to be dominant efficiently, i.e., without rattling off his aggressive repertoire. The upper third and the lower third of the Figure are almost mirror images. The partners generated nearly the same amount of free entropy. The dominant partner received a large transinformation from the subordinate partner (0.21), but the subordinate partner received very little transinformation from the dominant partner (0.02). This result, although surprising at first, fits precisely with the evaluation of the aforementioned telestimulation results, in which low-ranking animals do not respond to dominance gestures by high-ranking animals.

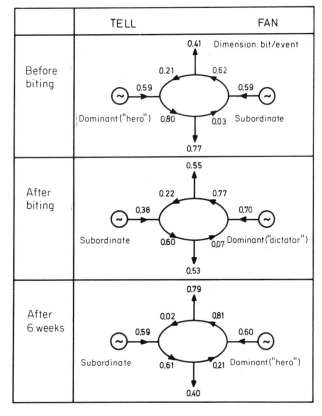

FIGURE 8 Communication processes between partners—Tell and Fan, two squirrel monkeys—in an experimental situation. The figures are calculated from observed behavioral events during 217 observation periods. The dimension for the calculus is bit per event. The positions of the figures correspond to the letters F, T, H, and D of Figure 7. For further explanation, see legend to Figure 7 and text. (From Mayer, 1969.)

THE "HERO" AND THE "DICTATOR": TWO STRATEGIES OF CONTROL  This example of Fan and Tell demonstrates two strategies of being dominant. In a socially stabilized situation, the dominant animal reacts to the behavior of the subordinate and takes minimal but effective measures to maintain dominance. To the observer of such a group, the subordinate animal does not appear to be ruled by the dominant. With a humorous connotation and with reference to analogous findings in social psychology (Hofstätter, 1957), such a dominant individual can be called the "hero." The strategy employed in a socially unstable situation seems to be quite different, and an animal dominant in such a situation behaves as a "dictator"; it establishes dominance by intimidation. The spontaneous behavior of the subordinate animal is greatly reduced, and the ruler does not pay attention to it.

This example illustrates two points: First, information theory is applicable to rather complex, although limited, behavioral situations. In this case, it was shown mathematically that the behavioral sequences correspond to Markov chains of the fourth order (Mayer, 1969). Second, an important conclusion must be drawn from the fact that a single event, in this case a severe fight, can shift a whole system and thereby cause various behavioral patterns and social signals in the repertoire to carry different weights, which are often unknown and difficult to measure. The influence of a signal from the partner and the consequences of the partner's behavior are different before and after the fight. One might say, therefore, "a bite is worth a thousand glares."

## Nonverbal communication in man

COMMENTS ON THE VOCAL SYSTEM AND THE HUMAN BRAIN  The communication system in nonhuman primates has reached a very high degree of complexity. The sender of signals can employ posture, motion, and vocalization. He can address a whole group with certain vocal signals and a single individual either vocally, visually, or by combined signals, without eliciting response from the whole group. The recipient of these signals must discriminate visual, auditory, and combined signals in a serial order and respond appropriately. Furthermore, all the partners communicate in an ever-changing context of events, in which the motivational state and the social rank or role of each individual must be taken into account at any given moment.

A primate requires many years to attain not only physical maturity but also the social maturity typical of adult social behavior. The struggle for food and other needs seems, under normal conditions at least, less frequent than the struggle for partner relations and social status. Social interactions seem to be a basic need of all primates, including man. Nonhuman primates handle their affairs, as we have seen, by means of vocal and other nonverbal communication. In man, a similar communication system, with representation in limbic and other subcortical structures, still plays a role and is seen clearly in all vocal, facial, postural, and other motor expressions of emotions, in songs without words, and in the cooing and babbling of small infants. In fact, this system is involved in every human conversation and in every direct partner interaction, but is dominated by human language, the most effective and unique communication system that Nature has produced. It constitutes the chief trait of human social behavior and is the basis for the cultural evolution of mankind.

In this context, it is important to note that speech depends not only on cortical but also on subcortical structures, including certain thalamic nuclei, which are connected with the cortical speech areas, the head of the caudate nucleus, parts of the diencephalon, and the periaqueductal gray matter. Although there are not sufficient comparative data on man, apes, and monkeys, there is evidence for the homology of the subcortical vocal system between man and the rest of the primate community.

Because of the importance of speech, interest in human communication was, for a long time, focused on language problems. During the last decade, and increasingly over the last few years, human nonverbal communication has become a new research area. Smiling as a visual, and crying as an auditory, social signal have been investigated (Koehler, 1954; Ambrose, 1961; Wolff, 1963, 1969; Ploog, 1964; and others), and the gaze as a means of human communication has attracted much attention (Argyle, 1967).

THE GAZE AS A SOCIAL SIGNAL  From the many different types of nonverbal communication, I choose only the example of the direction of gaze as a social signal (Gibson and Pick, 1963; Cranach, 1970). In this analysis of signal functions I have often made the assumption that the behavioral response to the signal discloses its function. At the same time, I have noted that, as in telestimulation, there is much signaling but no response. In animal experiments, it is always difficult to find out what simulus is equivalent to a response and what detail of a known signal is really perceived (Klüver, 1936, 1965); in a social setting, it often becomes impossible to discover which details are necessary to produce the appropriate signals.

In man, the situation is different. He can report what he perceives. The human gaze is considered an important social signal, so one wishes to know more about the recipient's perceptive properties. Cranach and his coworkers tried to approach this problem by a series of experiments of the following type (Figure 9).

The sender looked in random order at target points on lines that crossed the receiver's eye region horizontally and vertically; some of these points were within and some were outside his face area. The receiver announced his opinion of whether the sender was looking into his face, and an impartial observer was asked to do the same. In these experiments, gaze direction, gaze movement, head position, and head movement were investigated as cues for the perception of gaze direction. The over-all results are striking (Figure 10).

The receiver responds differently to the different stimulus configurations; unnoticed by the receiver, head position, gaze movement and head movement influence his perception of gaze direction. In general, a significant number of times the recipient's judgment proved wrong; the sender's gaze actually was directed to target points outside the face. It seems that each person has a slight disposition for a paranoid response.

Whether this tendency has a built-in component is an open question. For an explanation of the genesis of the receiver's reaction, we must consider that, in early childhood, eyes and ocula have the function of releasers (Kaila, 1932; Spitz and Wolf, 1946; Ahrens, 1954). This results in a very marked attention to the ocular component of orientation in further development. The disproportionately high sub-

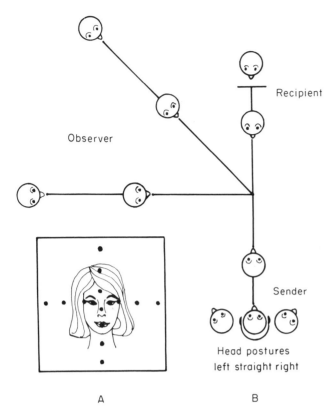

FIGURE 9  The human gaze as a social signal. A. The face of the experimental subject who receives gazes from another subject who sends the signals. The dots are target points distributed on and around the face of the recipient. B. Sender and recipient at two different distances from each other. The sender wears earphones to receive commands for the target point he must look at. He employs three different head postures for glancing at the recipient, who gives his opinion of which target point the sender selected. An impartial observer, at two different distances and angles, is asked to do the same. (From Krüger and Hückstedt, 1969.)

jective certainty in the judgment of gaze direction might originate from the innate basis for this reaction. Visual orientation serves important functions in interaction, so attention to it is positively reinforced in the sense of a conditioning process (Cranach).

The results also show that the discrimination of observers was generally worse than that of receivers. Eye contact cannot be judged reliably by an observer, which makes it difficult to use the gaze between partners as a measure for social interaction. Cranach stated that looking behavior is part of a total system of orienting reactions, but very little is known about its function in human social interaction. Further research should find methods for recognizing and assessing motor behavior as part of visual social signals in man.

## Determinants of social communication processes: summary and evaluation

I discussed four major areas of psychobiological significance in which investigation is necessary for analysis of the determinants of social communication processes. These are (1) characteristics of visual and auditory perception; (2) the motivational state; (3) the social status; and (4) the stage of maturation of a communicating animal. One can employ various methods and techniques of analysis for each of these areas. But the research strategy in this field, as in others, should provide as much as possible for measuring the interactions of the animals under a set of definable variables. This measurement is admittedly difficult, because the more restricted the behavioral situation, the more the communicative repertoire of the species under investigation will shrink and alter.

In the field of visual perception, it has been demonstrated that many of the social signals are conspicuous, such as the nodding of the iguana or the genital display of the squirrel monkey. These signals lend themselves particularly well to the analysis of signal function or, we might say, to the analysis of the information content. On the one hand, the onset, the time course, and other possible stimulus properties of the signal are observable and thereby usually measurable; on the other hand, the consequences of the signal, in optimal cases, may also be directly observable as, for instance, turning away after a treat or a counter-display after a genital display.

But now a new difficulty arises. Although the signal may be definable, the outcome may be variable; that is to say, the animal has a certain number of responses from which to choose. How large this number is depends on the size of the behavioral repertoire of the animal. An iguana certainly has fewer response choices than does a primate. The fact that there are a number of possible responses to a given social stimulus is the empirical basis for the stochastic treatment of communication processes.

The reverse situation obtains in visual social signals. That is, the signal is imperceptible to the human observer but the consequences of the animal's behavior are clearly observable and even stereotyped. This poses severe problems for the analysis of social communication and immediately raises the question of whether there may also be a third prototype of interaction in which neither the signal nor the immediate response is perceptible by the human observer. There are some hints that such may be the case in nonhuman primates, especially in well-adapted and stabilized colonies.

The examples of auditory signals demonstrated that the response-to-signal properties are different from those of visual ones. They are perceptible for the observer as long as they are in the range of the human auditory capacity. They are easier to characterize by sound spectrography, and the

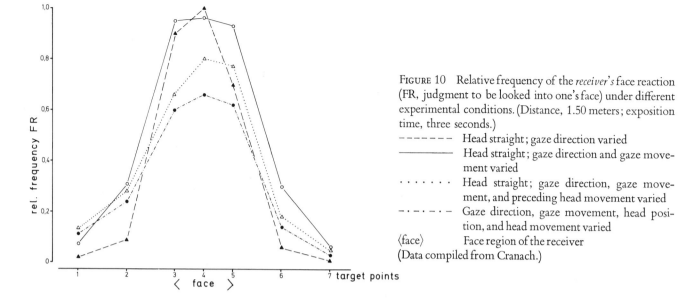

FIGURE 10 Relative frequency of the *receiver's* face reaction (FR, judgment to be looked into one's face) under different experimental conditions. (Distance, 1.50 meters; exposition time, three seconds.)

------ Head straight; gaze direction varied

———— Head straight; gaze direction and gaze movement varied

· · · · · · Head straight; gaze direction, gaze movement, and preceding head movement varied

-·-·-·- Gaze direction, gaze movement, head position, and head movement varied

⟨face⟩ Face region of the receiver

(Data compiled from Cranach.)

response of the animal to an auditory signal seems to be more stereotyped than that to visual signals. This holds at least for the communication system of the squirrel monkey and various other primate species. The spectrum of the vocal repertoire that is used in certain communal situations expresses the motivational state of an animal, and it has been shown that the alteration of the motivational state modifies the distributional pattern of the calls as well as the disposition for such action patterns as cohesive or agonistic behavior.

The vocal repertoire of the squirrel monkey is also particularly suitable for the demonstration of the correlation of brain mechanisms and social communication. It was possible to elicit almost all the approximately 30 distinguishable calls by electrical brain stimulation. These artificially elicited calls are in most cases indistinguishable from the natural ones. The description of the brain structures mediating these vocal signals gave rise to considerations of the vocal communication system in reptiles, mammals, and particularly in subhuman primates on the one hand, and the vocal system and its relation to human language on the other. It is stated that the natural evolution of the subcortical vocal system leads to a species-specific determinant of social communication and that it is one of the prerequisites for the evolution of human language.

Another aspect of social communication can be related to the evolution of human social behavior. This is the hierarchical social organization of many vertebrates, in particular of the primate family. It was demonstrated on one hand that responses to social signals are based on innate behavioral components, but on the other that these responses to social signals are specified by learning according to the role of the communicating participant. A particular reaction evoked by a particular signal depends not only on the specific information content but also on the social roles and ranks of the partners.

Such a conclusion is brought out very clearly by the various studies of the maturation of social behavior in which the social roles of the newborns, the infants, the juveniles, and the adults are well defined and closely paralleled by their biological developmental state. Growth and maturation are long-term modifiers of the communication patterns and provide for stage-specific interactions with group mates.

In closing, I revert to the beginning of the paper and the discussion of the natural history of social communication, which, I believe, can improve our understanding of the present-day behavior of man. It would be erroneous to suppose that what appear to be the most recent acquisitions of the primate family—logic, arithmetic, the language arts, and extension of the learning period—can completely suppress our earlier heritage and govern our lives exclusively. These newly acquired facilities are cognitive. Our social communication, however, although certainly containing cognitive elements, appears to be more closely related to our primitive heritage and not necessarily governed by logic and reasoning. To control the progress of the cultural evolution that has taken place so much more rapidly than our natural evolution, and to provide for the survival of mankind, it seems of utmost importance that we study the nonlogical rules and principles of social communication in much greater detail and with less bias, and that these rules and principles be utilized more effectively in the future.

# REFERENCES

AHRENS, R., 1954. Beitrag zur Entwicklung des Physiognomie- und Mimikerkennens. *Z. Exp. Angew. Psychol.* 2: 412–454 (pt. 1), 599–633 (pt. 2).

ALTMANN, S. A., 1965. Sociobiology of rhesus monkeys. II: Stochastics of social communication. *J. Theor. Biol.* 8: 490–522.

ALTMANN, S. A. (editor), 1967. Social Communication Among Primates. The University of Chicago Press, Chicago.

AMBROSE, J. A., 1961. The development of the smiling response in early infancy. *In* Determinants of Infant Behaviour (B. M. Foss, editor). John Wiley and Sons, New York, vol. 1, pp. 179–201.

ARGYLE, M., 1967. The Psychology of Interpersonal Behavior. Penguin Books Ltd., Harmondsworth, Middlesex.

BOWDEN, D., P. WINTER, and D. PLOOG, 1967. Pregnancy and delivery behavior in the squirrel monkey (*Saimiri sciureus*) and other primates. *Folia Primatol.* 5: 1–42.

CASTELL, R., and D. PLOOG, 1967. Zum Sozialverhalten der Totenkopf-Affen (*Saimiri sciureus*): Auseinandersetzung zwischen zwei Kolonien. *Z. Tierpsychol.* 24: 625–641.

CASTELL, R., H. KROHN, and D. PLOOG, 1969. Rückenwälzen bei Totenkopfaffen (*Saimiri sciureus*): Körperpflege und soziale Funktion. *Z. Tierpsychol.* 26: 488–497.

CRANACH, M. VON, 1970. The role of orienting behavior in human interaction. *In* The Use of Space by Animals and Men (A. H. Esser, editor). Indiana University Press, Bloomington (in press).

DELGADO, J. M. R., 1963. Telemetry and telestimulation of the brain. *In* Bio-Telemetry (L. E. Slater, editor). Pergamon Press, New York, pp. 231–249.

GIBSON, J. J., and A. D. PICK, 1963. Perception of another person's looking behavior. *Amer. J. Psychol.* 76: 386–394.

HOFSTÄTTER, P., 1957. Gruppendynamik. Die Kritik der Massenpsychologie (E. Grassi, editor). [Tenth edition.] *In* Rowohlts Deutsche Enzyklopädie, Vol. 38. Rowohlt Verlag, Hamburg.

HOPF, S., 1970. Report on a hand-reared squirrel monkey (*Saimiri sciureus*). *Z. Tierpsychol.* (in press).

HUXLEY, J. S., 1923. Courtship activities in the red-throated diver (*Colymbus stellatus* Pontopp); together with a discussion of the evolution of courtship in birds. *J. Linn. Soc. London Zool.* 35: 253–292.

JÜRGENS, U., and D. PLOOG, 1970. Cerebral representation of vocalization in the squirrel monkey. *Exp. Brain Res.* 10: 532–554.

JÜRGENS, U., M. MAURUS, D. PLOOG, and P. WINTER, 1967. Vocalization in the squirrel monkey (*Saimiri sciureus*) elicited by brain stimulation. *Exp. Brain Res.* 4: 114–117.

KAILA, E., 1932. Die Reaktionen des Säuglings auf das menschliche Gesicht. *Ann. Univ. Aboensis, Ser. B.* 17: 114.

KLÜVER, H., 1936. The study of personality and the method of equivalent and non-equivalent stimuli. *Character Personality* 5: 91–112.

KLÜVER, H., 1965. Neurobiology of normal and abnormal perception. *In* Psychopathology of Perception (P. H. Hoch and J. Zubin, editors). Grune and Stratton, New York, pp. 1–40.

KOEHLER, O., 1954. Das Lächeln als angeborene Ausdrucksbewegung. *Z. Menschl. Vererb. Konstitutionslehre* 32: 390–398.

KRÜGER, K., and B. HÜCKSTEDT, 1969. Die Beurteilung von Blickrichtungen. *Z. Exp. Angew. Psychol.* 16: 452–472.

MACLEAN, P. D., and D. W. PLOOG, 1962. Cerebral representation of penile erection. *J. Neurophysiol.* 25: 29–55.

MARKO, H., 1966. Die Theorie der bidirektionalen Kommunikation und ihre Anwendung auf die Nachrichtenübermittlung zwischen Menschen (Subjektive Information). *Kybernetik* 3: 128–136.

MARKO, H., and E. NEUBURGER, 1967. Über gerichtete Grössen in der Informationstheorie. Untersuchungen zur Theorie der bidirektionalen Kommunikation. *Arch. Elek. Uebertrag.* 21: 61–69.

MAURUS, M., J. MITRA, and D. PLOOG, 1965. Cerebral representation of the clitoris in ovariectomized squirrel monkeys. *Exp. Neurol.* 13: 283–288.

MAURUS, M., and D. PLOOG, 1969. Motor and vocal interactions in groups of squirrel monkeys, elicited by remote-controlled electrical brain stimulation. *In* Proceedings of the 2nd International Congress of Primatology, Atlanta, Georgia, 1968, Vol. 3 (H. O. Hofer, editor). Karger, Basel and New York, pp. 59–63.

MAYER, W., 1969. Beschreibung des Gruppenverhaltens von Totenkopfaffen unter besonderer Berücksichtigung der Kommunikationstheorie. Technical University, Munich, Doctoral Thesis.

PLOOG, D., 1963. Vergleichend quantitative Verhaltensstudien an zwei Totenkopfaffen-Kolonien. *Z. Morphol. Anthropol.* 53: 92–108.

PLOOG, D., 1964. Verhaltensforschung und Psychiatrie. *In* Psychiatrie der Gegenwart (H. W. Gruhle et al., editors). Springer Verlag, Berlin, Göttingen, Heidelberg, vol. 1, pt. 1B, pp. 291–443.

PLOOG, D., 1967. The behavior of squirrel monkeys (*Saimiri sciureus*) as revealed by sociometry, bioacoustics, and brain stimulation. *In* Social Communication among Primates (S. A. Altmann, editor). The University of Chicago Press, Chicago, pp. 149–184.

PLOOG, D., 1969. Early communication processes in squirrel monkeys. *In* Brain and Early Behavior: Development in the Fetus and Infant (R. J. Robinson, editor). Academic Press, New York, pp. 269–298.

PLOOG, D., J. BLITZ, and F. PLOOG, 1963. Studies on social and sexual behavior of the squirrel monkey (*Saimiri sciureus*). *Folia Primatol.* 1: 29–66.

PLOOG, D., S. HOPF, and P. WINTER, 1967. Ontogenese des Verhaltens von Totenkopf-Affen (*Saimiri sciureus*). *Psychol. Forsch.* 31: 1–41.

PLOOG, D., and P. D. MACLEAN, 1963. Display of penile erection in squirrel monkey (*Saimiri sciureus*). *Anim. Behav.* 11: 32–39.

PLOOG, D., and T. MELNECHUK, 1969. Primate communication. *Neurosci. Res. Program Bull.* 7 (no. 5): 419–506.

ROBINSON, B. W., and H. WARNER, 1967. Telestimulation of the primate brain. *Arch. Phys. Med. Rehabil.* 48: 467–473.

SACKETT, G. P., 1966. Monkeys reared in isolation with pictures as visual input: Evidence for an innate releasing mechanism. *Science* (*Washington*) 154: 1468–1473.

SEBEOK, T. A. (editor), 1968. Animal Communication. Indiana University Press, Bloomington.

SPITZ, R. A., and K. M. WOLF, 1946. The smiling response: A

contribution to the ontogenesis of social relations. *Genet. Psychol. Monogr.* 34: 57–125.

WINTER, P., D. PLOOG, and J. LATTA, 1966. Vocal repertoire of the squirrel monkey (*Saimiri sciureus*), its analysis and significance. *Exp. Brain Res.* 1: 359–384.

WOLFF, P. H., 1963. Observations on the early development of smiling. *In* Determinants of Infant Behaviour (B. M. Foss, editor). John Wiley and Sons, New York, vol. 2, pp. 113–138.

WOLFF, P. H., 1969. The natural history of crying and other vocalizations in early infancy. *In* Determinants of Infant Behaviour, Vol. 4 (B. M. Foss, editor). John Wiley and Sons, New York, and Methuen and Co. Ltd., London, pp. 81–109.

# 35    Brain Correlates of Language

## ERIC H. LENNEBERG

*Neuroanatomy alone cannot*
*explain language capacities*

PROBLEMS OF COMPARATIVE ANATOMY   It would be satisfying if we could explain the mechanism of human language by certain unique structures of the human brain. Unfortunately, this is not possible. The difficulty is largely because of a general discrepancy between our knowledge of neuroanatomy and our knowledge of behavior. Neuroanatomy does not tell us, for instance, why a cat and a dog have different types of vocal behavior with rather different ethological functions. A further difficulty arises from the problem of identifying homologies between certain parts of primate brains; this, in turn, makes it difficult to decide just how unique certain aspects of our brain really are. Consider, for example, the left convexity of the cortexes of rhesus, chimpanzee, and man, with their peculiar folding patterns. The arrangement of gyri and convolutions is, of course, quite different in each species, but this may have no relevance whatever to the behavior patterns typical of each animal. After all, the folding is merely a consequence of mechanical factors that impinge on the embryonic development of these structures.

Attempts have been made to search for homologies of histological fields (von Bonin and Bailey, 1961; Kreht, 1936), and here we could argue, on the one hand, that each histological type of man's cortex has homologues, at least

in chimpanzee cortex (which would minimize the uniqueness of human features), or, on the other hand, we could stress that the specific configuration of each field is somewhat different in man (which would underline uniqueness). So far, there are few objective facts that could help us make a decision in dilemmas of this kind. Geschwind (1965) has argued that man has a peculiar pattern of transcortical fiber connections (which is undoubtedly correct), and that this is the anatomic substrate of his ability to speak. But the difficulty here is that we do not yet know the physiological function of any particular transcortical connection; further, we shall see presently that the nature of language capacity is such that one may question the relevance of connections between certain areas in the cortex to the basic skill of acquiring language-knowledge.

SIZE AND VOLUME   The size of the human brain does not govern language capacity (see Table I). Human beings with dwarfed brains are often capable of perfect language. Schultz (1962) has reported on gorilla skulls from the capacity of which we can infer brain sizes of more than 700 cc, no smaller than some human brains that have language capacity. The question is frequently raised whether the number of cellular units is of critical importance, rather than the brain weight. The issue is difficult. On the one hand, large amounts of cortical tissue in the critical areas of the human left hemisphere can be removed at an early age without hindering the emergence of language. On the other hand, the size of a lesion in the adult is an important factor for the prognostication of language deficits. Apparently size is the reason that lesions even in the area of

---

ERIC H. LENNEBERG   Professor of Psychology and Neurobiology, Cornell University, Ithaca, New York

TABLE I

*Language capacity and its relation to brain and body weight*

|  | Age | Speech Faculty | Body Weight (Kg) | Brain Weight (Kg) | Ratio |
|---|---|---|---|---|---|
| Man (m) | 2 ½ | Beginning | 13 ½ | 1.100 | 12.3 |
| Man (m) | 13 ½ | Yes | 45 | 1.350 | 34 |
| Man (m) | 18 | Yes | 64 | 1.350 | 47 |
| Man (dwarf) | 12 | Yes | 13 ½ | 0.400 | 34 |
| Chimp. (m) | 3 | No | 13 ½ | 0.400 | 34 |
| Chimp. (f) | Adult | No | 47 | 0.450 | 104 |
| Rhesus | Adult | No | 3 ½ | 0.090 | 40 |

(From Lenneberg, 1967)

Broca, in Heschl's gyrus, or in the angular gyrus may, under certain circumstances, spare language functions. Further, the importance of size may well vary with cortical location. A small lesion is perhaps better tolerated in certain parietal areas than in the more classical speech regions (Russell and Espir, 1961; Conrad, 1954; Penfield and Roberts, 1959), which suggests that, in the adult brain, where topographic language specificity is fairly well developed, there is, nevertheless, redundancy of elements with threshold values for the tolerance of tissue destruction. On the other hand, microscopic examination of the brains of nanocephalic dwarfs has not revealed abnormal cytoarchitecture (Seckel, 1960). It is, therefore, possible (although far from certain) that these brains have markedly fewer cells than normal brains, which suggests that developmental factors also play an important role in the specification of cells, including the functions they shall have in language processes.

Geschwind and Levitsky (1968) have examined 100 normal adult brains and have found marked anatomical asymmetries between the upper surfaces of the right and left temporal lobes. The planum temporale (the area behind Heschl's gyrus) was larger on the left in 65 per cent of the brains and larger on the right in 11 per cent. We cannot, however, be certain that language is responsible for or depends on this relative hypertrophy.

On balance, it is clear that to attribute the capacity for language simply to quantitative parameters such as size or number of cells is a gross oversimplification.

TOPOGRAPHY There are several regions in the brain from which language may be interfered with in rather characteristic ways. The most dramatic regional differentiation is the lateral asymmetry. In more than 95 per cent of right-handed individuals, aphasia ensues from left hemisphere lesions, whereas in about 50 per cent of clearly left-handed

individuals, aphasia results from lesions on the right (see Table II). There is also anteroposterior differentiation in the cortex of the left hemisphere, with the precentral medioventral frontal regions being involved in productive aspects of language, and the postcentral parietal areas appearing to be involved primarily in more cognitive aspects. Lesions in the dorsomedial region of the left temporal lobe produce language deficits in which the receptive component dominates.

There is a temptation to give cortical topography a functional, especially psychological, interpretation and to guess at the "meaning" of the size of the so-called "association areas" and their peculiar positions between sensory projection areas. This guesswork is based on specific psychological theories for which there is yet but little physiological or even biological justification. Large areas are said to be "uncommitted cortex," which means merely that we have not been able to discern the true physiological function of this tissue. Although fairly specific symptoms result from various regional lesions, it is generally agreed that such are not evidence for the existence of "language centers," as had been assumed by earlier neurologists. Unfortunately, we still have but scanty evidence that would enable us to attribute unfailingly specific cortical locations to specific clinical symptomatology. Penfield and Roberts (1959) have published the results of cortical stimulation and have attempted to provide new maps of stimulation points according to behavioral disturbances. When one compares all these maps, one cannot escape the notion that almost any kind of speech disturbance may result from stimulating almost any point within a fairly wide region. Facts of this sort should not serve to revive the stale controversy between holists and localizationists. I merely wish to point out that we still do not know the functions of much cortical tissue or its subcortical connections.

There is also agreement that subcortical lesions may produce fairly specific interference with either language or

TABLE II

*Incidence of cerebral dominance with respect to handedness*

|  | Per cent* |
|---|---|
| Clearly right-handed; clearly left cerebral dominance | 82 |
| Clearly right-handed; cerebral dominance not clearly left (i.e., mixed or right dominance) | 3 |
| Handedness not clearly established | 9 |
| Clearly left-handed; clearly left cerebral dominance | 3 |
| Clearly left-handed; clearly right cerebral dominance | 3 |

*Percentages vary with age, and incidence of handedness varies slightly from country to country.

speech. The best example is destruction of the internal capsule; with the increase in surgical correction of Parkinsonism, some structures in the basal ganglia and the diencephalon have also become suspect as being involved in speech mechanisms. Lesions in the lateroventral nucleus of the thalamus, particularly, have a relatively large incidence of untoward sequelae of this sort; everything from mild dysarthria to complete anarthria has been observed. Also, palidectomies may compromise the elaboration of speech. As a rule, these disturbances are limited to motoric events, with perception or language-knowledge remaining intact.

There is considerable evidence (see Ploog, elsewhere in this volume) that in carnivores, and probably also in lower primates, the peri-aqueductal gray in the mesencephalon subserves motor coordination in vocalizations of various kinds. Speech calls for highly specialized and well-integrated coordination of some 100 muscles. This coordination can be disturbed to the point of complete anarthria, and there are some suggestions (Lenneberg, 1962) that lesions (particularly when sustained during embryogenesis or at the time of birth) in the peri-aqueductal gray are responsible for this disorder.

PERIPHERAL ANATOMY    The vocal tract in man shows a number of structural changes that make the production of speech sounds uniquely possible (Lenneberg, 1967; Lieberman et al., 1969). On the other hand, these modifications are not the essential prerequisite for language capacity. Children with greatly deformed fauces can learn to understand English, even though their own speech is unintelligible. Further, a careful study of the congenitally deaf shows that these individuals learn to communicate via a natural language, such as English, even though they cannot hear and usually cannot speak (their medium of communication may either be sign language, which is a direct derivative of English, or simply reading and writing). Also, congenital blindness is no impedance to language acquisition, which demonstrates that the capacity for language is not simply a capacity for associating auditorially perceived patterns with visually perceived patterns. Moreover, children with fixed mesencephalic lesions that interfere with muscular coordination for speech (congenital anarthria) may nevertheless acquire knowledge of a natural language if the rest of their brain functions normally. (An objective test to demonstrate this finding is described below.)

## The neurological correlates of language are processes, not information confined to structures

Neither the knowledge of words nor the knowledge of any particular rule for the formation of sentences appears to be lodged in any particular tissue, which suggests that the capacity for language is based on peculiar modes of inter-cellular activities instead of on the presence of special cells that respond only to peculiar speech stimuli. In the remainder of this paper, I argue against the hypothesis that language capacities can be explained by an assembly of analyzing or detecting units and in favor of the idea that the neural correlates of language capacities are a propensity for certain activity patterns in the human brain. The activity patterns that I am thinking of would have to be conceived of as a hierarchy. At the highest level we consider only the state of the entire brain. This state, however, must be analyzed as the composite of the activities of all the suborgans of the brain. The suborgans, in turn, are a composite of different activities in their various regions and nuclei, those of their cells, and those of their organelles.

CLINICAL SYMPTOMS OF APHASIA    Many typologies for aphasic symptoms have been proposed, but none has found universal acceptance. The least controversial classification is a threefold system: (1) *productive*—saying words is either impossible or very difficult, but language-knowledge is intact (patient understands everything and can usually read and often write a little with left hand); (2) *cognitive*—production of words is possible, but patient has difficulty understanding what is being said to him and his own utterances are either incoherent or incomprehensible; (3) *amnestic*—patient can produce words and can usually understand spoken and written material fairly well; he experiences, however, enormous difficulties in finding the words he needs for any given utterance.

GENERALITY OF DEFICITS    It is possible to make one generalization about language disorders that holds for any of the three types: no lesion can abolish selected linguistic units. Deletion of a few items from an otherwise intact repertoire does not occur. A patient cannot lose three of his 36 phonemes, or two of his 12 prepositions, or seven phrase-structure rules and 20 transformational rules à la Chomsky's generative grammar (Chomsky, 1965). Disease alters the modes of normal function (interferes with the way of speaking, understanding, doing) but does not wipe out accomplishments item by item. The alterations of the underlying neurophysiological processes interfere with the smooth operation of speech and language, but these interferences do not follow either the linguist's or the psychologist's theoretical construct. This fact tends to be obscured because patients are so frequently examined by some standard test constructed by linguists or psychologists who simply look for features suggested by their own theories. When individual patients are studied exhaustively and without the constraints of a test that merely allows the recording of a few fixed items, discordance between the patients' deficits and traditional theoretical models becomes more obvious.

DISSOCIABILITY OF SELECTED LANGUAGE AND SPEECH SKILLS    The clinician who attempts to examine his aphasic patients as an unbiased descriptivist (or naturalist) will at first be puzzled by an apparently endless variety of symptoms. Every patient will show a slightly different pattern in what he can no longer do and in what he can still do. Faced with such a sea of unordered facts, one will wish to do two things: (1) try to discern correlations in order to reduce the mass of data to a smaller number of underlying factors; (2) see whether the factors also reflect some independent ("real") brain correlate. Step 1 is too obvious to deserve further comment here. Step 2 should also be an obvious endeavor, but it has been badly neglected in aphasiology, although it is of the greatest heuristic value for the neuropsychologist. It suggests what questions are to be asked.

The logic implied by Step 2 is familiar to us from other types of biological investigations, such as genetics. We observe all the variations in the morphology or behavior of an organism, say the fruit fly. Symptoms that are strictly correlated are attributed to a single factor, say a gene or a chromosomal locus. This suggests criteria for systematic breeding experiments and for a further search for the genetic mechanism involved.

There is a simple way of ordering our clinical observations so as to show up the lacunae in our knowledge; it tells us at once whether the classification system adopted is relevant to neurophysiological inquiry. Consider, for instance, the question of whether it makes neurophysiological sense to treat the power to speak (let us call it globally *verbal communication* and not worry, for the moment, about the different aspects of this complex) as distinct from other cognitive capacities (let us call them collectively *nonverbal cognitive processes*). Does *verbal communication* have brain correlates that are independent of the correlates underlying such capacities as making inferences, developing new learning sets, recognizing common denominators in physically different objects and situations? We shall say this dichotomy promises to be of heuristic value for the neurosciences if the two types of processes can be affected independently by disease—if double dissociation can be demonstrated; there must be some patients who lose the power of verbal communication without suffering appreciable loss in their nonverbal cognitive processes, and vice versa.

Conveniently, one sets up a matrix such as that shown in Table III; the entries in the cells record the clinical occurrences. All available cells must have positive entries, i.e., the particular combination of symptoms must be observable. If not, we are probably deceiving ourselves by giving separate labels to behavioral disorders that do not have independent and specific brain correlates and are therefore of little interest to the student of brain functions. As long as we are dealing with such gross classes as are shown in Table

TABLE III

*Dissociability of verbal communication from general cognition*

| | | INTACT | |
|---|---|---|---|
| | | Verbal Communication | Nonverbal Cognitive Processes |
| Lost | Verbal Communication | | Commonly seen in traumatic aphasia |
| Lost | Nonverbal Cognitive Processes | Seen in senility; in congenital feeble-mindedness; sometimes in certain drug intoxications | |

III, the answer is fairly simple. In this case, we may assert confidently that verbal communication in general is dissociable from nonverbal cognitive processes. We must add, however, that even here we are dealing with questions of degree. The patient whose intellect is affected by disease, who has difficulty with subtractions, with understanding proverbs, with recognition of the use of objects, and so forth, would clearly not be saying exactly the same sorts of things as he used to say before he fell ill. When we look at language behavior in detail, we often have difficulty in keeping language skills separate from cognitive skills as manifest in what a person is saying. In general terms, however, the statement is correct; speaking is clinically dissociable from other cognitive functions.

If we make a similar matrix out of the threefold classification (productive, cognitive, amnestic) mentioned in the previous section, as in Table IV, we begin to see that the more specific language skills, as suggested by clinical ex-

TABLE IV

*The dissociability of aphasic symptoms*

| | | INTACT | | |
|---|---|---|---|---|
| | | Speech-praxia | Language-knowledge | Word-presence |
| Lost | Speech-praxia | | Broca's aphasia | No data |
| Lost | Language-knowledge | Sensory or Wernicke's aphasia | | No data |
| Lost | Word-presence | Amnestic type of aphasia | No data | |

perience, do not present so clear a situation. By *speech-praxia* I mean the capacity to say whatever one wants to say, i.e., marshaling the muscles of the vocal tract and coordinating their action for speech. *Language-knowledge* is the general capacity to understand what is being said (of course, on a level of interest, complexity, and education appropriate to the patient's background) and to communicate coherently through the verbal medium, by either speaking or writing. *Word-presence* is the capacity to think promptly of the right word at the right time.

The construction of a matrix such as that shown in Table IV reveals our lack of knowledge. At present we do not have sufficient evidence to assert that there is complete dissociation among the three symptom complexes. We see that speech-praxia and language-knowledge are clearly independent, i.e., they may be affected selectively, but it is not certain if the memory disorder underlying word-presence is entirely dissociable from language-knowledge. The Table does not give any information on whether either language-knowledge or word-presence is dissociable from other nonverbal cognitive processes, but the clinician who took the trouble to set up the appropriate matrix would soon convince himself that there is, in fact, no evidence yet that complete dissociability is possible. Probably when language is affected in these more particular ways, certain specific nonverbal capacities are also affected.

If we now take the complex previously called language-knowledge and try to split this further, for instance into *speaking coherently* and *understanding sentences* (Table V), we gather from the matrix that double-dissociation is not possible. Perhaps understanding may be relatively more intact than speaking, but, to my knowledge, the reverse does not occur. Thus, the relationship between the two is one of a particular within a general.

We are led to the conclusion that the only clearly independent skill is what we have called speech-praxia. This skill may be abolished while apparently all other intellectual

TABLE V

*The dissociability of certain cognitive aspects of language*

|  |  | INTACT | |
|---|---|---|---|
|  |  | Speaking Coherently | Understanding Sentences |
| Lost | Speaking Coherently |  | May occur but degree or extent of dissociability uncertain |
|  | Understanding Sentences | Complete dissociability does not seem to occur |  |

functions remain intact. It is also interesting that this skill has the least controversial anatomical representation. On the other hand, although the oral transmission of language is a biologically constant and universal feature of modern man, we have also argued that it is not the necessary and sufficient basis for verbal communication. This, too, is illustrated by our patients. Although they cannot speak, they can receive and send coherent verbal messages (in contrast, for example, to a chimpanzee). But language-knowledge, the kernel of language capacity, is physiologically more closely tied to more general cognitive processes, a point to be expanded presently.

INERADICABILITY OF LANGUAGE CAPACITY  Although disease may impede the capacity for verbal communication, circumscribed lesions cannot neatly wipe out *all* aspects of language (without, at the same time, compromising severely the patient's general cognitive function). In other words, a patient to whom we can no longer talk or who can no longer talk to us is usually not devoid of any and all language capacities (he is not like an infant who has never learned to speak), but his language capacities are so distorted as to be of no practical value for communication. As we have seen before, the patient who cannot say anything frequently continues to understand, and the patient who cannot understand often speaks very fluently (although incoherently). In fact, he may speak too fluently (logorrhea), reducing further his avenues of communication. To abolish language completely would require nothing short of a subtotal left hemispherectomy, and even then we could hardly expect the patient to be comparable to a normal human being who has simply not acquired language (as seen in congenital deafness before formal training).

Moreover, in early childhood, the hemispheres are not yet specialized for language, and the left-to-right polarization resulting in dominance is easily blocked by different types of pathologies (Basser, 1962; Lenneberg, 1967). One may also find, although rarely, adults with poorly developed dominance, so that language processes seem more diffuse than is normal. These observations argue against the existence of anatomically definable language centers and in favor of *activity patterns* as the neurological correlates of language. Lesions alter the activity patterns by making certain types of modulation impossible, but they cannot suppress just one type of activity pattern, allowing others to persist entirely unimpaired. Overt language behavior and its underlying neurophysiological process are restructured by disease but in a maladaptive fashion.

NONSPECIFIC ETIOLOGY  Language disorders are the result not only of structural lesions. They may be caused by a great variety of factors, including high fever, exhaustion, pharmacological agents, metabolic disorders, and intoxications. It is true that the clinical picture in all these conditions varies considerably, and that one could not mistake an aphasia caused by a vascular accident for a language disorder connected with alcoholic intoxication—merely because each of these conditions has its own side-effects. If we compare only the language symptoms and if we disregard the course of the condition, differences can be minimal. If such profoundly different pathogenic events can have such similar effects, we can conclude only that the effect is due to a very delicate mechanism and probably also a very general one; almost anything that is bad for brain function may be bad for language. Again, the theoretical constructs of processes and activity patterns seem to be more germane to the facts than any construct that implies the acquisition of individual, independent items lodged in specific cells or tissues.

## The recognition process as a homologue of the language process

It is sometimes possible to investigate one biological phenomenon by studying it simultaneously with another, closely related phenomenon. I wish to suggest here that the recognition process has many interesting parallels to the language process; in fact, I wish to treat the two as biological homologues. I am implying not that they are the product of a single physiological mechanism, but that they have distinct neurological correlates, which are based on similar, general principles of operation. I believe that each has its peculiarities, but that their common ontogenetic origins are still discernible in their analogous modes of functioning. The two sections that follow are presented in the hope that the juxtaposition has heuristic value; in going back and forth from language processes to recognition processes, we may be able to use the discoveries in one area as stepping stones or pointers toward discoveries in the other.

CLINICAL SIMILARITIES BETWEEN DISORDERS OF RECOGNITION AND OF LANGUAGE  Some of the most striking similarities between recognition, especially in the realm of vision, and language are brought out under conditions of disease. Virtually every aspect of aphasia has its parallel in disordered perceptual processes.

Take the phenomenon of incomprehension in so-called sensory aphasia. The patient cannot make any sense of what is being said to him, although his hearing is intact. Something very similar may occur in recognition as a consequence of a variety of cerebral insults. The patient may experience a visually perceived object as entirely strange and in some sense incomprehensible. He does not seem to know what to do with it or what it is for, although under certain conditions he may even know the name of the object; or he may recognize the importance and use of the object but feel

that he has never encountered it before—the *jamais vu* phenomenon. In other instances, the environment suddenly and unpleasantly appears to be changing its shape, curling, bending, receding, or approaching. Movements may be perceived as abnormally fast or slow, or parallax motion may appear as irregular. Also, individual objects that are being regarded may change their shape. In these instances, the sense of the meaning of the objects and the sense of continuity, constancy, and nexus with the subject's past experience are inextricably intertwined. We find the same thing in sensory aphasia.

Semantic difficulties, i.e., failure to understand the meaning of words, as a rule are mixed with difficulties concerning the meaning of sentences, that is, the relationship between words. The patient picks one or two words out of the examiner's question and complains that he does not know what is being said about these objects. There are further parallels in the realm of recognition—for instance, the phenomenon of simultanagnosia, in which the patient is able to understand the meaning, that is, to recognize one object that is placed before him, but gives evidence of agnosia when he is asked to deal with a second object or to relate various objects in his environment one to the other.

The productive end of language disorders also has its counterparts in recognition. Consider the uncontrollable sequence of visions caused by certain drug intoxications (notably mescal and LSD) or the disconnected emergence of meaningless patterns that may be caused by structural lesions. These symptoms may range from well-formed and "syntactically integrated" hallucinations, when the patient cannot tell his own fabrications from what he is experiencing, to disconnected and entirely meaningless flashes or patterns of sensations (phosphenes) that are experienced as if they were visually perceived phenomena. The close relationship between the more productive and the more receptive sides is illustrated by the rare experience of superimposed sights. The patient who is watching a person getting up and approaching him experiences a number of simultaneous, coexisting images; he sees the person still sitting while he also sees him standing and walking (Klüver, 1966; Teuber et al., 1960; Milner and Teuber, 1968).

Even the memory defect in aphasia, the extraordinary difficulty in finding words at the right moment, has its counterpart in recognition. The patient may have the sensation of being just about to see something, of almost having a visual image; he strives toward achieving that image but does not reach it—the *presque vu* phenomenon.

It appears to me that there is a general common denominator underlying this entire range of symptoms, both in language and in the recognition processes. It has to do with a general loss of nexus, a difficulty in integrating over time, of relating things from one moment of experience to the next, as well as of relating the present to the past. Some of the distortions may have to do with the distortion of certain temporal parameters in the processes underlying the experience. Local relative slowing or speeding of processes that contribute to the total activity pattern may cause a disintegration or distortion in the interaction of functions.

In the light of this interpretation, the topographic anatomy of language functions assumes new meaning. Presumably the topographic differentiation is the peculiar anatomical substrate prerequisite for the specific temporal modulation of the physiological activities responsible for language. This point of view would also explain the difficulty of totally wiping out language capacities by circumscribed lesions. Only certain temporal aspects of neural integration for speech depend on given loci; language still remains a mode of functioning that apparently involves much more of the brain than any of the older center doctrines implied.

We have not said anything about etiologies of the perceptual disorders. As a matter of fact, many different etiologies can cause symptoms such as those described. This point has been stressed by Klüver (1966), who cites insulin hypoglycemia, high-grade fever from a variety of infectious diseases, thyrotoxicosis, hashish and cocaine poisoning, schizophrenia, and neurosyphilis as possible causes. Milner and Teuber (1968) have reviewed similar conditions caused primarily by trauma and epilepsy. We may add to this list delirium and other mental derangements, especially in connection with chronic thiamine deficiency. Note that, as in the case of aphasia, entirely different conditions, with totally different pathological processes, can cause similar symptoms. There is only one possible interpretation; some very general aspect of brain function must be involved, so that almost anything can affect this general function. Parameters of temporal integration strike me as a general enough notion to be a good candidate here. Of course, speculation on the meaning is hazardous, but it seems reasonable to assume that the cognitive component of visual perception is a dynamic process, just as in the case of language.

Clinical disorders in recognition-function resemble language disorders in many other ways. In both cases, the experiences or the behavior resulting from the disease are rearrangements rather then obliterations; capacities are restructured but in a maladaptive fashion. A similar explanation was first proposed by the German neurologist V. von Weizsäcker (1950), who called it *Funktionswandel*. Although it is possible to produce both experiences and behavioral derangements through electrical stimulation, it is not clear whether destruction of the stimulated loci would establish the same disturbance permanently. Further, just as in language, it seems as if no lesion could wipe out visual recognition without at the same time introducing other dimensions of cognitive disorders. (Such seems to be the case with man, although some reports of parallel animal experiments give but scanty support for this assertion. We must remember, however, that information on behavioral deficits in animals is never so detailed as in human patients. In fact, the testing of lesioned animals is always and necessarily restricted to a very few areas of behavior, and in most instances the behaviors tested are ethologically artificial for the animal.)

We have stressed that language disorders are never deletions of specific items from a repertoire. The same may be said of recognition disturbances. Patients do not encounter difficulty with any one item or a fixed list of them; whatever the nature of their disturbance it applies to percepts in general. Only whole modes of functioning become distorted in disease, not one or another association or learned act.

SIMILARITIES BETWEEN THE NORMAL PROCESSES OF RECOGNITION AND LANGUAGE  To the layman it may appear that language differs from visual recognition in two fundamental ways: he may believe that the former is arbitrary and sequential whereas the latter is supposedly "natural" and simultaneous. But this view is incorrect with respect to recognition. Recognition also involves a certain degree of arbitrariness—it has a learned component, in that the organism must associate many arbitrarily paired features. Animals (at least most mammals) and man must interact with the environment during the developmental period for the recognition process to be well adapted to the exigencies of adult life. (This similarity between language and recognition had already impressed the behaviorists who discussed language some 20 years ago. It led them to think that there was virtually nothing special to be said about language and that any animal has the basic capacity for language acquisition. The latter conclusion, however, is not warranted; although language is not different from recognition with respect to arbitrariness, it is different in certain other important ways, to which we shall return.)

In order to understand language, it is, of course, necessary to process temporal patterns and to discern sequential relationships. Visual recognition, however, requires very similar tasks. It is well recognized today that all aspects of perception have a temporal component and that the brain must establish connections between what is seen from one moment to the next and between the present and the past. Nor is there a fundamental difference between language and recognition simply because one appears to be productive, whereas the other appears to be only passive and receptive. The comprehension of the visually perceived environment requires an active process. This has been the main subject of Gestalt psychology, which has provided an endless variety of examples that show that physical stimuli may become imbued with different meanings, depending on a variety of subjective factors, including disposition and previous experience. A similar point has been made by phenomenologists and, before them, by exponents of various schools of philosophy, particularly the German rationalists. MacKay (elsewhere in this volume) gives further demonstrations of the subject's active contributions to perception.

Similarities between language and recognition (or even cognition in general) become rather striking when attempts are made to characterize the two in some formal metalanguage—say, an algebra. Such an attempt has been made recently for language by Chomsky (1965) and his school, and some hundred years earlier similar attempts were made for the laws of thought and knowledge (or, as we would say today, cognition) by Boole (1854). That Chomsky's generative grammar has the essential features of a Boolean algebra is not simply because Chomsky arbitrarily chose that metalanguage, but because the two processes are quite objectively classificatory in nature.

One may discover the rules that govern the formation of sets in visual recognition as well as in language. When this is done for visual recognition, we have an attempt to discover equivalence classes of stimuli—to use Klüver's (1965) terminology—for a particular animal. In other words, we ask what kinds of patterns are processed by the animal as if they were similar; what kinds of distortions and transformations preserve or destroy the sense of similarity for the animal; and the problem essentially remains the same, no matter whether the animal is man, dolphin, or honeybee. When questions of this sort are pursued, and the results for different species are compared, we discover (at least in many instances) that the "sense of similarity" is often species-specific with respect to certain stimulus phenomena. Cats, for example, do not seem to be able to recognize objects when they are presented as line drawings. It is not even certain whether they recognize a scene that is shown on a TV screen. In contrast, there is evidence that chimpanzees recognize photographs and certain other graphic representations. It is not yet known, however, whether the conditions or criteria for similarity, if tested systematically, would be identical for chimpanzee and man. It would not be surprising if certain discrepancies were found. After all, their brains are structurally not identical; might we not also expect differences in the mode of function?

This discussion is relevant to language because in language we see the extreme form of classificatory behavior, and in many instances we can quite easily state what the rules are that govern the formation of language classes. The specific rules of the grammar of language are, of course, different from the specific rules of a "grammar of recognition" (where *rule of grammar* must be understood as the symbolic representation of operations or computation-like processes). It can be shown (I am completing a monograph on the subject), however, that the *general mode* of calculating with classes is remarkably similar for man's recognition behavior

and for his language behavior. This discovery has strengthened my belief that the similarity between language and cognition is based on a true and biologically based affinity. The difficulties encountered by any other animal in acquiring language (as characterized in the next section) must be the result of the species-specificities in types of cognition.

## Some characteristics of human language capacity

The basic units in verbal communication are neither speech sounds nor words but sentences, or, as Hughlings Jackson preferred, propositions. "The food is bad." "I want milk." "Tomorrow the sun will shine." We are so bound by the propositional nature of language that it is impossible for us *not* to impute a proposition even to single-word utterances such as those spoken by an infant, for instance. When he says "milk" or "daddy," we at once believe he uttered a wish or commented on a situation—"I want milk," "I see daddy." In the discourse of adults, every single word utterance is invariably understood as a sentence in which the self-understood components have been omitted.

Sentencehood is a class, in fact the most characteristic class of all languages. It is an open, or intensive, class; that is, membership is defined by a general principle—a relational principle called predication or attribution. To discover whether a sequence of words is or is not a member of this class, an operation must be performed in which it is decided whether the peculiar relationship, *predication,* does or does not hold. What we have said so far is familiar to every fourth grader: a sentence consists of a subject and a predicate. But what is not so obvious is that the predicate relationship is of a rather abstract and very general nature. It turns out to be the basis of virtually all concatenations of words. If we characterize language in terms of a formal system, predication has the status of a binary operation. The meaning of this operation could roughly be translated as: "There are two elements, a and b, and the relationship is that b is a specification of a." In the phrase "old man," *old* is the b element, which specifies *man,* the a element, producing "the man is old," from which "old man" is derived (by another set of operations technically called *transformations*).

A further and strictly formal characteristic of all languages is that the basic binary operation may be applied iteratively. Consider an imperative: "Put the shoe on the bed." This might be expressed in any language, and it is simple enough so that any child who is just beginning to learn to speak will understand it. But notice how the semantic or meaning-bearing elements are syntactically related to one another. The subject is *you,* the a element; the b element that specifies it is the activity *putting.* This nuclear a-b pair is bracketed (see Figure 1), forming a new a element that is specified by a b, in this case the object *shoe,* itself

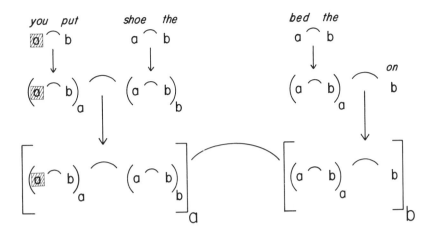

FIGURE 1   Diagramatic representation of all predicate relationships (and the relation between these relations) of the imperative: "Put the shoe on the bed." The covert subject *you* is shaded; words appear disordered because of the arbitrary convention of letting the *b* element, the modifier, follow its subject, the *a* element.

predicated by the definite article. Once more, the entire result is bracketed to form a new *a,* specified by the *b* element, a prepositional phrase, which is a syntactic class also made up of a composition of predications. In this case, the phrase *on the bed* is made up of a double iteration: *bed* is specified by the article *the,* and this phrase is specified by the preposition *on.* We now have the logical structure of the concatenations. To make a well-phrased sentence out of this net of predicate relationships, further operations of the transformational type must be brought to bear on the words.

Much detail has been left out of this characterization, but it serves to show that even very simple sentences are based on very abstract relationships.

So far, we have confined ourselves to the syntactic aspect of language in very general terms. If we say that the sentence "Put the shoe on the bed" has been understood, we imply that the subject has decomposed the sentence into the relational pattern diagramed in Figure 1. Syntax is an indispensable characteristic of human language. Any communication that lacks the type of relationship just sketched out is simply not comparable to human language.

The semantic side of language is even more complicated and still more dependent on abstract relational principles. Words are never simply the tags of individual objects. Words also stand for classes, and the classes are specified by families of relational principles. The child who learns to speak and to give names to objects learns the particular principles that define a class. Thus, *table* is not the name of a thing in the same way as *Napoleon* is the name of a certain man; *table* is the name of a set of relations that have to do with physical appearance as well as with function and use.

Because the child does *not* learn an association between a sound pattern and a light pattern, but instead learns what the principle of naming is, he can give names to objects he has never seen before and be quite creative in this process (for instance, by calling a rock a table if he wishes to use it as such). Switching for a moment to the psychologists' terminology, we may say that names label conceptualization processes.

The complexity of the semantic relationships becomes most obvious if we imagine a computer-like device endowed with pattern-perception capacities equaling those of a child. If it is to understand such a sentence as "Did I put the shoe on the bed?" it must have computational facilities that allow it to recognize physical patterns, such as shoes and beds, and also to ascertain whether the relationship *putting* or *on* holds, in this case, and whether either the shoe or the bed is the object of the preposition *on.* Further, if the computer misunderstands the sentence and construes it analogously to the question "Did I put the shoe on the child?" it would program itself to do quite a different computation on the optic arrays. It would return an answer "No," based on the observation that I did not dress the bed. We see, then, that, even if naming objects were a simple operation, the semantics of a sentence go far beyond a simple pattern matching and involve the computation of relationships, first in the sentence and then upon the environment.

Processing of the sound patterns of a language is also based on a complicated computational procedure of which we now know that it cannot be accomplished by simple tracking of acoustical components or by template matching. This point has been dealt with in detail elsewhere (Lenneberg, 1967) and cannot be reviewed here. It is also somewhat irrelevant to the basic problems of language, because children, for instance, may acquire language-knowledge through reading and writing only and in the absence of the special skills needed for phonological analysis.

Clearly, language cannot be defined exhaustively in a few sentences. But, in lieu of a definition, I give below an imaginary primitive language that contains some fundamental features of all human natural languages. Any animal that can understand sentences in this language, i.e., execute commands or answer questions by signaling yes or no, may be said to have at least some of the basic capacities for language. Here is a typical lexicon:

7 object words: cup, box, ball, pin, bed, table, window
3 action words: take, put, point
4 attribute words: small/big, roundish/squarish
7 relational words: on/under, out/in, to/from, and
3 syntactic markers: affirmation, negation, question
—
24 total vocabulary

The syntax of this language consists of only two types of sentences, imperatives and questions. Delivery of the sentence may be in any form whatever—spoken words, hand-signs, written symbols, or any other code—and the subject to be tested must be able to signal only two words, *yes* and *no*. On the semantic side, however, the sentences produced in this language should follow the conventions of some natural language, e.g., English. Language capacity in a subject, say an autistic child, a normal infant at age 12 months, or a chimpanzee, is demonstrated if he correctly executes commands in this language and also answers questions appropriately. Notice that our concern is only with *understanding sentences*. We do not require the subject to say or signal anything except affirmation or negation—*yes* or *no*. Furthermore, we do not care about the method used to prepare the subject for this task; it might be anything from just speaking to the subject to training him in a formal conditioning paradigm. The test is designed merely to see whether a basic capacity for language-specific computations is present.

Sometimes it is difficult, even in normal children, to adduce rigorous proof of language-knowledge. In fact, many a doting mother imputes language comprehension to her infant long before the child is likely to be able to have passive language-knowledge, such as tested for above. The dog that heels on command, the baby who waves bye-bye when asked to, the autistic child who comes to the dinner table when he hears the words "dinner is ready," or the chimpanzee that makes a number of hand signals to which observers ascribe meanings by using their own imagination, have not really demonstrated a capacity for language-specific computations. The subjects have merely shown that a given stimulus may come to elicit a fixed behavioral se-

quence or that they can emit spontaneously a volley of signals. All this is still far too general to qualify even as a precursor of the operations on which human language is based. The subjects have yet to give evidence that they are capable of extracting the specific relationships (and relations between relationships) that are encoded in simple sentences, of analyzing the environment in terms of these relationships, and of matching up these two sets of operations.

## Conclusions

As a working hypothesis, I propose that language is a specialization of the human type of cognitive processes. These, in turn, are a special form of the more general primate type of cognition; this of a mammalian type, and so forth. A biological view of language induces me to postulate a common descent of all existing natural languages. Language always has its roots in the physiological processes of cognition.

The characterization of language is slightly different, depending on whether the linguist, the mathematician, or the physiologist speaks. In linguistic terms, language consists of classes and of principles or rules. These same concepts may be translated into the terminology of a Boolean algebra, and now we are dealing with sets, operations, and relations, each implying the other. The physiologist must look at language as a set of activities and processes. Notice that each approach forces on us a dynamic point of view. The fact that language also has a repertoire such as a vocabulary is not its most outstanding feature. In fact, it shares this with many other accomplishments of animals and men.

Although the importance of language for social cohesion in man cannot be overestimated, and its dependence on interaction between individuals is established, the communicative function of language can also mislead the investigator. An overemphasis on the communicative aspects engenders the notion that language has been purposely devised, like a tool, to serve as a vehicle for the transfer of information. This attitude glosses over the real problems: what is the *nature* of the sender-receiver, and how is the range of messages limited or even determined by the structure and function of the devices that do the communicating, i.e., the brains involved? Just what does it mean, in biological terms, to have knowledge of a particular language, and, more generally, what is the capacity for acquiring such knowledge?

The basic proposition stressed is that language-knowledge is an activity (not a static storehouse of information), namely, the extraction of peculiar relationships from the environment and the interrelating of these relationships. This activity is a special form of a general family of activities, collectively called cognition; the point is illustrated by showing that there are important parallels between the cognitive aspects of perception and those of language. The neurological clinic provides further support for the view that language function is patterned in ways similar to cognitive (especially perceptual) function. A comparison of acquired language disturbances with other acquired disorders of cognition, such as perceptual recognition, emphasizes the similarity in symptomatology, as well as in the natural history of development and of loss due to disease. A scrutiny of the clinical facts suggests that disease of any form causes merely a disturbance in the integration of activities and that this pathological restructuring of activities—the physiological imbalance—manifests itself behaviorally in various incapacities. The incapacities, however, must never be viewed as the simple deletion of given, circumscribed items from a repertoire of skills or as the elimination of information from an otherwise intact collection. Instead, a new configuration of behavioral propensity—mostly of an ill-adaptive nature —ensues from disease.

These considerations lead to new directions in research on brain correlates of language. Because language-knowledge is seen as the result of an interaction between activities, i.e., between physiological processes, aphasic symptoms, as an example, should be investigated for their temporal parameters, such as changes in rate, rhythm, sequence, continuity, capacity for preserving a sense of nexus, modes of transforming spatial patterns into temporal ones, and so on. Further and systematic comparison of aphasic symptoms with other intellectual disorders is also apt to lead to new hypotheses concerning the general nature of the physiological processes underlying mentation in man.

## REFERENCES

BASSER, L. S., 1962. Hemiplegia of early onset and the faculty of speech with special reference to the effects of hemispherectomy. *Brain* 85: 427–460.

BONIN, G. VON, and P. BAILEY, 1961. Pattern of the cerebral isocortex. *In* Primatologia: Handbook of Primatology (H. Hofer, A. H. Schultz, and D. Starck, editors). Karger, Basel, vol. 2, pt. 2, fasc. 10.

BOOLE, G., 1854. An Investigation of the Laws of Thought. Walton and Maberly, London. (Reprinted by Dover, New York, 1951.)

CHOMSKY, N., 1965. Aspects of the Theory of Syntax. M. I. T. Press, Cambridge, Massachusetts.

CONRAD, K., 1954. New problems of aphasia. *Brain* 77: 491–509.

GESCHWIND, N., 1965a. Disconnexion syndromes in animals and man, Part I. *Brain* 88: 237–294.

GESCHWIND, N., 1965b. Disconnexion syndromes in animals and man, Part II. *Brain* 88: 585–644.

GESCHWIND, N., and W. LEVITSKY, 1968. Human brain: Left-right asymmetries in temporal speech region. *Science* (*Washington*) 161: 186–187.

KLÜVER, H., 1965. Neurobiology of normal and abnormal perception. *In* Psychopathology of Perception (P. J. Hoch and J. Zubin, editors). Grune and Stratton, New York, pp. 1–40.

KLÜVER, H., 1966. Mescal and Mechanisms of Hallucinations. The University of Chicago Press, Chicago.

KREHT, H., 1936. Cytoarchitektonik und motorisches Sprachzentrum. *Z. Mikroskop. Anat. Forsch.* 39: 331–354.

LENNEBERG, E. H., 1962. Understanding language without ability to speak. *J. Abnorm. Soc. Psychol.* 65: 419–425.

LENNEBERG, E. H., 1967. Biological Foundations of Language. Wiley, New York.

LIEBERMAN, P. H., D. H. KLATT, and W. H. WILSON, 1969. Vocal tract limitations on the vowel repertoires of rhesus monkey and other nonhuman primates. *Science* (*Washington*) 164: 1185–1187.

MILNER, B., and H.-L. TEUBER, 1968. Alteration of perception and memory in man: Reflections on methods. *In* Analysis of Behavioral Change (L. Weiskrantz, editor). Harper and Row, New York, pp. 268–375.

PENFIELD, W., and L. ROBERTS, 1959. Speech and Brain Mechanisms. Princeton University Press, Princeton, New Jersey.

RUSSELL, W. R., and M. L. E. ESPIR, 1961. Traumatic Aphasia. Oxford University Press, London.

SCHULTZ, A. H., 1962. Die Schädelkapazität männlicher Gorillas und ihr Höchstwert. *Anthropol. Anz.* 25: 197–203.

SECKEL, H. P. G., 1960. Bird-headed Dwarfs. Charles C Thomas, Springfield, Illinois.

TEUBER, H.-L., W. S. BATTERSBY, and M. B. BENDER, 1960. Visual Field Defects After Penetrating Missile Wounds of the Brain. Harvard University Press, Cambridge, Massachusetts, for the Commonwealth Fund.

WEIZSÄCKER, V. VON, 1950. Der Gestaltkreis. Thieme, Stuttgart.

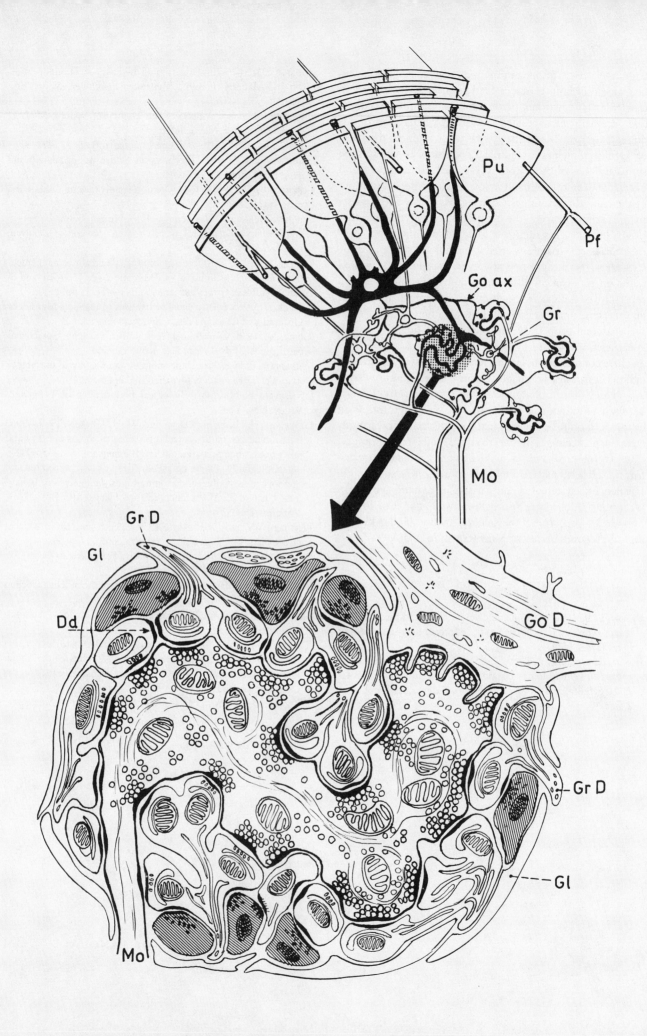

# NEURAL SUBSYSTEMS AND PHYSIOLOGICAL OPERATIONS

*"The scope [of this section] is defined by the levels of integration between higher nervous functions and basic cellular mechanisms."* BULLOCK, PAGE 375. *This diagram, which illustrates the position of the glomerulus in the neuronal network of the cerebellar cortex (above) and an idealized version of its ultrastructure (below), is from* SZENTÁGOTHAI, PAGE 430.

# 36 Operations Analysis of Nervous Functions

## THEODORE HOLMES BULLOCK

WE ARE IN midcourse in the rough progression from large-scale to smaller-scale problems. The focus is therefore shifted, for several chapters, away from the prime concern with over-all consequence in behavior and onto the organizational aspects of how the nervous system operates. It is not yet centered, as in the later blocks, on the hardware or componentry; that is, on just which molecules and configurational changes account for the cellular actions. Rather, given the membrane machinery for generating potentials and responding to signals, this block of chapters deals with such questions as: What is going on in a brain center? What events, interactions, and distribution of influences comprise the workings of the communication device or information machine we call the brain?

In brief, the scope of the present block is defined by the levels of integration between higher nervous functions and basic cellular mechanisms. We may call this broad, intermediate area that of *neural subsystems*. By this term, we mean any functionally related subsets of the set of all nerve cells and associated cells considered from a systems viewpoint.

There are many subsystems. They embrace larger and smaller sets of elements; they overlap and are delimited arbitrarily, in some cases by function, in others by structure. Examples are the visual, olfactory, somesthetic, limbic, reticular activating, extrapyramidal, temperature-regulating, startle response, and hypophyseal control systems—any fraction we wish to define. Its role in the whole system

THEODORE HOLMES BULLOCK   University of California at San Diego, La Jolla, California

may or may not be clear, but the assumption in each is that there is a particularly significant *functional* relationship defining the subsystem. (Classically, many of these are "systems," which is quite appropriate when considered by themselves; in the context of the whole nervous system, they are subsystems.)

From the great array of subsystems we can deal with only a small sample, chosen for the diversity of approaches, lessons, and challenges they represent. One of the aims is to bring out transferable principles, both of brain organization and of research strategy. This essay deals with the latter, and is highly subjective. It is not a synthesis and is only in part a sermon; the main effort is to be suggestive. This chapter relates to the other chapters in this section by a vital thread which is explained as the argument unfolds; in brief they are selected examples of the approach to understanding the nervous system via disclosure of the operations of the subsystems.

NEED FOR REVOLUTIONS    T. S. Kuhn, in *The Structure of Scientific Revolutions* (1962), pointed out that "normal science" aims not to upset but to consolidate, document, and quantitate the accepted general framework or "paradigm" of a field. But the real advances are made in revolutions.

I want to disturb, to incite, to disinhibit our imaginative thinking. The state of neurophysiology, rich as it is in excitement of new approaches, euphoric with its wealth of weapons, inundated with its accelerating flood of findings, yet needs an infusion. Our rate of progress toward understanding "what is going on here?", i.e., the main principles of operation of large numbers of cells in organized arrays, is unsatisfactory. We ought, therefore, to encourage the dreaming up of candidate principles, to be looked for and then invalidated or permitted. We should be better prepared to recognize the emergent properties that must inhere in a system so complex. The hard-won framework of our present understanding of the nervous system is surely going to prove wrong in major ways. As in the history of classical physics and classical genetics, the order we have created out of ancient chaos will be short-lived and displaced by revolutions and, one hopes, a new order. This means we should provide positive encouragement to break out of the framework that we are relatively satisfied with today and, while continuing to hammer out the best work we know how to do, constantly seek saltations.

Let me emphasize, lest these statements be mistaken as license for dilettantism, that a solid grounding in the accumulated body of knowledge remains a prerequisite for sophisticated evolution of ideas. But it is futile to use the reins of caution on a horse that has no spirit, and one cannot strike out in new directions on a tried and true trail.

Whereas the revolutions in physics, genetics, and some other fields can be traced to accumulated concern over dis-

crepancies in highly quantified data with mathematically derived, precise predictions, this is not the only genesis of revolution. In neuroscience there is not yet a quantified framework, except in special branches. In most of the field, the rigor permitting real discrepancies between expectation and finding is too far off to wait for. Yet we do have a "paradigm" in Kuhn's usage—the guiding set of generally accepted propositions. A glance at our own history supports the call for saltations, for imaginative ideas, to break out into new paradigms rather than waiting for specific discrepancies.

Let us recall some unexpected discoveries, made by looking, not by reinterpreting, doubtless aided by a thorough familiarity with the literature, by asking "Is that all there is?", "What for?", "What if?", "How about?", and "What is going on?"

HISTORIC SALTATIONS    What do I mean by saltation? To go back only a few decades, remember when local potentials came in? They were resisted, disbelieved, and explained away, but they eventually put us on a new plane of understanding. Remember when direct inhibition came in? When the McCulloch-Pitts neuron was quite an accurate model of the best we knew? Then dendrites were rediscovered, and we had symposia about their differences from axons. Remember when electrical transmission was discovered? And then electrical inhibitory and electrical excitatory, chemical inhibitory and chemical excitatory, one-way and two-way synapses? We soon became blasé about those, only to be jolted by presynaptic synapses. These were at first a negligible curiosity in some no-account invertebrates, but then the wave of new instances swept up the mammalian cord and brain. Another great leap began when direct, low-resistance electrotonic connections were found between neurons in the cardiac ganglion of lobsters, a discovery that soon spread like wildfire through other invertebrates, lower vertebrates, and, only recently, mammals.

INCIPIENT SALTATIONS    It is a little too early to say, but we may witness a major saltation if it turns out, as it now seems, that there are several distinct codes in nerve impulse trains, not merely the code of mean frequency. Another vast change in our conceptual model seems sure to come when we begin to assess the numbers of neurons that perform their normal roles without spiking, as the majority of neurons in the retina are now known to do. An old idea will become a revolutionary one if it can be shown that diffuse electrical fields in nervous tissue are influential as causes and not only as inevitable consequences of nervous activity.

Other major saltations wait in the wings for new evidence. We heard at the ISP the tentative suggestion that a large fraction of the synaptic vesicles in invertebrate axons are not waiting to be released as synaptic transmitters but may be en route to release as neurosecretory messengers. S. A. Raymond has just completed a doctoral thesis at M. I. T. that gives new life to the idea that axonal branch points may be systematic, complex filters of temporal patterns of spike trains, and that the filter properties depend on the recent history of the axon as well as on activity in specific classes of neighboring fibers. The possibility of labile growth of fibers and endings recurs. The two-way traffic in axons, involving both fast and slow transport of substances to and from the cell body, gives ample basis for specific molecular information flow. The suggestive evidence at hand that terminals can enlarge or shrink, even withdraw or multiply, is unpopular today but has not been explained away; if there is something to these suggestions, the electron micrographs of fixed material are poor representations of functionally available connectivity, and the probability of a critical significance of such structural lability seems high enough to be taken seriously.

## Examples of recent saltations at the levels of multineuronal subsystems

All the preceding instances of past or potential saltations concern essentially the neuronal level. Let me remind you of a few at higher levels.

DIFFUSE, NONSPECIFIC ACTIVATING SYSTEM   What a tremendous change in our thinking, our whole framework of simplifications about the mammalian nervous system, took place with the discovery of the diffuse, nonspecific activating system. Here, in parallel with the classical specific pathways ascending and descending for sensory and motor functions, was found another whole system in the core of the neuraxis. Although interconnected with the others, its neurons lose local sign and modality. This nonspecific activating system has a new kind of significance in generalized arousal, and provides a platform state without which the specific pathways cannot elicit behavior. No doubt this picture will be corrected and refined in turn.

PARALLEL SENSORY SYSTEMS   Much newer, and not yet assessed in many species, is the equivalently interesting distinction between two parallel visual pathways (Held et al., 1967, 1968; Held, this volume). To oversimplify, the proposition is that the pathway via geniculate to cerebral cortex has the function of determining the "what," the pathway to mid-brain tectum that of determining the "when and where," of visual stimuli. The cortex permits pattern recognition, but the tectum is necessary for useful behavioral orientation toward a desirable pattern.

Mountcastle (1961) has clearly pointed to a basic distinction between two parallel somesthetic systems, the lemnis-

cal-ventrobasal and the spinothalamic-posterior thalamic group pathways. The lemniscal elements end in cortical units that are exquisitely discriminating in locus, intensity, and dynamics, and the spinothalamic elements end in cortical units that are vastly less specific in excitatory receptive field, inhibited by large skin areas, and not poised for action in rapid cadence. Katsuki et al. (1954) pointed out a similar distinction between thick and thin fibers of some sensory systems, and Bullock (1953) noted a number of examples. Something similar is emerging between dorsal and ventral cochlear nuclei (Evans and Nelson, 1966, 1968).

Julian Bigelow (personal communication) points out that biologists have been slow to recognize that such separation or factorization of different aspects of input into distinct processing channels is necessary in any system, living or not, which is adapted to adjust to a varying environment; the principles of such factorization are, he asserts, the really important generalizations to be made with regard to the evolution of higher forms.

THE LIMBIC SYSTEM   Paul MacLean, elsewhere in this volume, speaks of the vast change in our general view that came with the recognition of the limbic system as a mediator of affective state, instinct, and mood. Against the background of classical localization of functions, with cortical areas for vision, audition, skin senses, movements, speech, and the like, and deep areas for temperature regulation, respiration, autonomic functions, and the like, it was a great advance to learn that such a vague, diffuse, and variegated role as affective state could be localized. The memorable image that represents this dénouement is of Professor Delgado of Yale University facing a fighting bull and holding a transmitter control that could send stimulating pulses through electrodes implanted in the bull's amygdaloid nuclei—limbic structures in the temporal lobes. Whereas, a moment before, the beast had been charging aggressively toward the unarmed scientist, it stopped in its tracks within seconds of the onset of transmission, not paralyzed, insensible, or incapacitated in any of the classical specific systems, but just plain tame. A great body of experiments has firmly established that stimulation or lesions in the limbic structures can cause tameness or aggressiveness, hypersexuality, various signs of feeding and drinking, of emotional expression, and of recent memory impairment, and disturbance of normal sequencing of behavior. A single common denominator is not yet clear, but a short while ago the idea that species- and individual-preserving mechanisms in the brain should be segregated was revolutionary. Factoring out the basic parameters, independent of modality, that nature has really played upon in evolving this organization is a prime challenge.

We neuroscientists, who should recognize a revolution in thinking, are as guilty as anyone of becoming blasé too quickly; we become used to it, and too soon cease to be impressed by the wonder of functional localization of rewarding, of punishing, of instinctive mood switching, of appetitive searching. Who would have thought a local stimulus would make a goat drink himself bloated, or a hungry cat spurn food? If we did not accommodate so quickly, the creative processes of germinating more saltations might be facilitated.

It seems to me we have been for years on the verge of showing, in the limbic system, a neurophysiological basis of the old ethological concepts of action-specific potential and the hierarchical organization of instincts. These ideas were criticized by behaviorists, because they lacked a neurophysiological basis. Actually, they are not only physiologically reasonable but are strongly indicated by the fragmentary evidence. The clear and satisfactory demonstration has not been made, however, and in the meantime the ethologists have nearly abandoned these terms. The obstacle has been not the ethologist's evolving terminology and improved precision but a too-rapid assimilation of a new paradigm or acceptance of a new view (the limbic system and its role) without applying to it thought as fresh as its significance deserves.

## The challenge of study of operations at the level of changes in meaning of messages in neurons

Another class of advances that has opened a whole new era in brain research is the characterization of the differences in meaning of signals in neurons, both in parallel and in series. By meaning I mean *what* information, relevant to the system, is encoded in the neuronal discharge, therefore the message content, which includes the label on the line that must be known to the system. As we follow either afferent pathways or motor systems from lower to higher levels, the message content of the neurons must change. Beyond the question of coding (what parameters of neuronal activity carry the information), lies a realm of questions, such as those outlined above, that require insights into the natural history of the creatures (types of neurons) encountered with our collecting devices (electrodes) in the jungle of the brain, based on the aspects of input having interest for the organism, not on arbitrary measures chosen for the convenience of our arithmetic manipulation. The change of meaning of impulse discharge from one level to the next is, in terms of units, the final useful result of the convergence and connectivity—the processing and integration that constitute a behaviorally defined subsystem. It is the problem of the labels on the lines and calls for a taxonomy in the best sense.

Let me recount one example, a type of cell in the frog tectum postsynaptic to the optic nerve fibers, which Lettvin et al. (1961) called "sameness" neurons. These lie deep in the tectum and have receptive fields almost as large as the visual field of the frog. I quote the original description:

. . . It is a bit embarrassing to present the following description so batrachomorphically, but at least it reflects what we have found so far. Every such cell, in fact, acts so complexly that we can hardly describe its response save in terms ordinarily reserved for animal behavior.

Let us begin with an empty gray hemisphere for the visual field. There is usually no response of the cell to turning on and off the illumination. It is silent. We bring in a small dark object, say 1 to 2 degrees in diameter, and at a certain point in its travel, almost anywhere in the field, the cell suddenly "notices" it. Thereafter, wherever that object is moved it is tracked by the cell. Every time it moves, with even the faintest jerk, there is a burst of impulses that dies down to a mutter that continues as long as the object is visible. If the object is kept moving, the bursts signal discontinuities in the movement, such as the turning of corners, reversals, and so forth, and these bursts occur against a continuous background mutter that tells us the object is visible to the cell.

When the target is removed, the discharge dies down. If the target is kept absolutely stationary for about two minutes, the mutter also disappears. Then one can sneak the target around a bit, slowly, and produce no response, until the cell "notices" it again and locks on. Thereafter, no small or slow movement remains unsignaled. There is also a place in the visual field, different for different cells, that is a sort of Coventry to which a target can retire and escape notice except for sharp movements. This Coventry, or null patch, is difficult to map. The memory that a cell has for a stationary target that has been brought to its attention by movement can be abolished by a transient darkness. These cells prefer small targets, that is, they respond best to targets of about 3 degrees.

There is also (we put this matter very hesitantly) an odd discrimination in these cells, which, though we would not be surprised to find it in the whole animal, is somewhat startling in single units so early behind the retina. Not all "sameness" cells have this property. Suppose we have two similar targets. We bring in target A and move it back and forth along a fixed path in a regular way. The cell sees it and responds, signaling the reversals of movement by bursts. Now we bring in target B and move it about erratically. After a short while, we hear bursts from the cell signaling the corners, reversals, and other discontinuities in the travel of B. Now we stop B. The cell goes into its mutter, indicating that what it has been attending to has stopped. It does not signal the reversals of target A, which is moving back and forth very regularly all the time, until after a reasonable time, several seconds. It seems to attend one or the other, A or B; its output is not a simple combination of the responses to both.

These descriptions are provisional and may be too naturalistic in character. However, we have examined well over a hundred cells and suspect that what they do will not seem any simpler or less startling with further study. There are several types, of which the two mentioned are extremes. Of course if one were to perform the standard gestures, such as flashing a light at the eye, probably the cells could be classified and described more easily. However, it seems a shame for such sophisticated units to be handled that way

—roughly the equivalent of classifying people's intelligence by the startle response.

Here is a cell that is extracting features of a pretty high degree of complexity, and one that is history-dependent in a quite special way. It is only about a fourth-order cell, by a minimum count. Two main lessons stand out. First, the cell would not have been discovered, characterized, and labeled without a considerable use of unconventional, imaginative stimulus regimes and the devotion of a lot of time. Very likely our present understanding of this cell type is incomplete (Fite, 1969) and will be altered by further experiments ethologically designed.

Second, there is no reason to underestimate the degree of sophistication achievable by successive stages of convergence and change of meaning of discharge. The argument is sometimes heard that recognition of naturally interesting stimuli, such as other individuals of the species, could not be in such units because they have not been found, despite all attempts to find them. But in fact there has been much less effort to find subtle units than to find simpler types, whereas it will doubtless require much more effort. Moreover, the number of units for any given subtle specification must be far fewer, and the number of types to be distinguished possibly much higher, than for simpler classes of units. In the realm of visual shape recognition, we have almost no psychophysics for those species commonly used in neurophysiology and, even if we did, no *a priori* assurance that the natural hierarchy of recognition units can be usefully described or sorted in terms of the parameters we choose for convenient definition of stimuli. Indeed, we have little basis for the usual tacit assumption of unapproachable complexity in our higher discriminations. In principle, it is possible to discriminate each of thousands of faces, of words, of manual operations using a dozen or so either-or allelomorphs. To test for such a mechanism, a search for units would have to be done in the spirit of a naturalist rather than limited to preconceived parameters.

I am trying to say that we should not give up and conclude that the jungle has no anteaters; we have hardly begun to look. In fact, the types of units already known are rather numerous and some are pretty complex, taking together the specialized auditory and the visual, the frog higher tectal, the cat cortical, and the insect and crustacean higher-order neurons. In this volume, as a splendid example of a fresh kind, Bishop adds a new dimension to the visual cortical units of the cat.

## The challenge of study of operations at the level of major functionally segregated cell masses

I started by calling for new ideas, for breaking out onto new planes. The presupposition is that our best present knowl-

edge will prove significantly wrong and that our rate of progress in generating saltations is unsatisfactory.

Let me underline that point. I just said that the types of units known are numerous and that some are complex. But our knowledge is so primitive that we cannot yet say whether these so-called types are significantly plastic, so that readiness to fire is subject to the level of hormone, of odor, of hunger or mood, or whether the set of criteria for firing can be acquired by learning.

These recognition units are, in effect, decision-making. When we know the set of input criteria for the "command cells," discussed herein by Wilson—cells able to trigger complex normal movements and sequences employing widespread muscles—I have no doubt that we will also find these cells to be decision-making recognition units, the narrow funnel of converging input and diverging output networks.

This mechanism is one available candidate, but there are several other plausible possibilities (Bullock, 1961b; MacKay, this volume). Curiously, we do not know whether these in fact operate, although they are at least alternatives and are more popular among neurophysiologists today than are recognition neurons.

Turning to a grosser level, with respect to clinical and macroscopic observations on structures such as the cerebellum, caudate nucleus, and pulvinar, on which we have many decades of contributions and thousands of papers, one might expect a reasonably satisfactory agreement on functions. But on the cerebellum, we have, as Llinás points out in his chapter, only statements like "something to do with motor coordination and tone." Even this least common denominator of textbook treatments of the cerebellum may be wrong as a basic characterization. The cerebellum, with its relatively stable histological organization, evolved as an exteroceptive (lateral line) analyzing device, hence of course also with indirect influence on motor performance. But it functions actively even in the absence of movement, as in electric fish, in which it contributes to on-going analysis of electroreceptor input while the fish is quiescent (Bullock, unpublished). The point is that fresh thought and candidate functions should be canvassed; the functional operations being performed are not known.

Even the geniculates and the so-called relay nuclei of the thalamus are doing things far beyond our present ken, to judge from the anatomy (Szentágothai, and Scheibel and Scheibel, this volume) and the physiology (Levick et al., 1969; Purpura, this volume). We really do not know what is happening in most of the parietal, the occipital, or the temporal, let alone the frontal cortex. Saying so does not mean we are asking for the circuit diagram. This may come, fractionally, but, as Lewis shows (this volume), it will be of extremely limited help unless we have at the same time all the dynamics of each junction, a knowledge of the normal input, and *a priori* understanding of the purpose. It is the last that I am speaking of here—the functional role, at the intermediate level.

## A suggested approach by analogy with operations research

The methods in use, then, have not gone far enough. New approaches are needed. One that may help is a kind of operations research with an ethological base; that is, based on the normal repertoire of actions of the species and the normally occurring stimuli that are meaningful to it. If this knowledge were combined with all the tools of experimental neurology, it might put us ahead.

One generally thinks of operations research as a term for maximizing some desired output in human affairs. But it particularly refers to research on interacting processes in complex systems, when we do not know all the components or influences and cannot hold all but one of the variables constant or the system at will. An important element is discernment of operations, by imaginative insight or guessing. I have the word of Jacob Bronowski, one of the pioneers in operations research in World War II, that the term is appropriate for our use, which is here simply to help develop inferences about candidate internal transformations in a language suitable for carrying over lessons from one subsystem to another.

To attempt a specific proposition, let us develop the analogy of the study of a complex social institution, such as a university (although one could also pick a corporation or a city). We can recognize operations at several levels of analysis; arbitrarily I have chosen four (Table I).

At the lowest level, all the units (which are people) perform, whatever their role, certain functions in common: listening, reading, speaking, moving fingers this way or that way, walking, thinking, even some fairly advanced operations such as interpreting and remembering. Although the process is the same in each unit, in some units it has high significance for the whole university; in others, less.

At the highest level, we recognize functions defined by the external relations of the university, by its output and the factors in its success or failure. These functions are often organized into departments and may be geographically segregated, but it is important to note that they may be diffuse (planning, recruiting).

The crux is that we have not studied how the system works until we have looked between the first and last levels to see what kinds of operations and operators are there and how they are related. The two intermediate columns show some examples. They do not show the essential relations among them, the web of communication and influence, the code or language, or languages, they use, the organizational hierarchy, specific and nonspecific effects, lability, redundancy, and similar properties. But, as does a naturalist examining a new ecosystem, a first approximation must be made of some of the main kinds of creatures and processes there are. These are not self-evident, but need to be ferreted out. Only then can one begin to study relationships.

In the nervous system, the first list represents some of the operations common to neurons. The last list represents the achievements at the level of behavior and external relations. These two lists are relatively better established. The intermediate levels are those concerned with how the system works, therefore with the central nervous physiology. It is remarkable and worth some thought that we do not have more than a few items corresponding to lists II and III in the accepted corpus of physiology today. As for the naturalist in the field, the entities that belong here are not self-evident. They cannot be listed easily and do not belong merely because they seem plausible.

That is one of the main points of this essay. We cannot trust our intuition at these levels. But we *are* dependent on the use of our imagination to invent candidate items, else they may long go overlooked. Only after justifying a reasonable list can we begin to establish relationships, rigor, and subjects worth representing mathematically. The lists given are crude drafts to open discussion and doubtless grossly neglect important functions; perhaps they also artificially include inappropriate categories. As long as this is true, our understanding of the web of interrelations that comprise the organization of the system is primitive.

It is striking that most current textbooks of neurophysiology deal with main headings, above list I, that are all either anatomical (the physiology of the spinal cord or of the cerebellum) or come from the fourth list, which are behaviorally and ecologically defined accomplishments. I am suggesting that it is time to add chapter headings from the truly physiological how-it-works lists, which will encourage and compel us to study the evidence for the naturalness and adequacy of these items, and then to work out their interrelations. Until we do, the only physiological principles common to many special applications, such as vision and hearing, for example, are elementary neuronal properties from list I.

I am sure this challenge will be difficult and that the subject will be fluid. We shall make mistakes of commission and omission. The attempt to systematize intermediate-level integration that I made a short time ago (Bullock and Horridge, 1965) seems inadequate today in many respects.

Filtering or its equivalent should be a broad category, it seems clear, and should probably be subdivided into spatial and temporal filtering or other subclasses (Ratliff et al., 1969). Another major heading, and class of operations, is the generation of patterned discharge in time and hence in space and intensity. The generation may be initiated either spontaneously (as by rhythmic internal clocks or by non-

TABLE I

*Lists of sample operations and operators in a social institution and in a nervous system, at four levels of complexity. Items below broken lines need only be found in relatively derived or advanced instances; such items seem to be more numerous in levels to the right.*

## THE UNIVERSITY

| I | II | III | IV |
|---|---|---|---|
| Listening | Phoning | Technician | Anatomy |
| Reading | Typing | Bookkeeper | Anthropology |
| Speaking | Filing | Scholar | Counseling |
| Moving fingers | Selecting |   Beginning | Purchasing |
| Walking | Drawing |   Advanced | Mailing |
| Thinking | Ordering | Librarian | Planning |
| Interpreting | Arranging | Janitor | Extension service |
| Remembering | Composing | Supervisor | Building and grounds |
| | Programing | Analyst | Educational policy |
| | Grading | Physician | Recruiting |
| | Lecturing | ----- | Publishing |
| | | Coach | ----- |
| | | Radiation safety officer | Parking |
| | | Grants manager | Lobbying |
| | | | Museums |
| | | | Galleries |
| | | | Endowments |

## THE NERVOUS SYSTEM

| | II | III | IV |
|---|---|---|---|
| Transducing | Filtering | Orienting | Vision |
| Encoding | Gating | Expecting | Hearing |
| Conduction | Tuning | Recognizing | Pain |
| Impeding/spreading current | Modulating |   Affective situation | Posture |
| Transmitter production | Contrasting |   Rival, offspring . . . | Locomotion |
| Synaptic response | Pattern generating | Deciding | Reproduction |
| Pacemaking | Comparing | Arousing | Sleep |
| Neurosecretion | Erasing | Focusing attention | Temperature regulation |
| Ion pumping | Feature extracting | Controlling input | Respiration |
| Facilitation | Synchronizing; desynchronizing | Set point adjustment | Eating |
| Accommodation | Storing | Habituating | Emotion |
| | Switching | Consolidating | Mood switching |
| | Multiplexing | ----- | ----- |
| | Spatial transformation | Associating | Insight learning |
| | | Awareness | Eugnosia |
| | | Symbol use | Planning |
| | | | Language |
| | | | Altruism |

rhythmic accumulation of "action-specific potential") or triggered by adequate input. Some time ago I proposed (Bullock, 1961a) that each of five plausible permutations of purely central and purely reflex generation of pattern actually occurs, in different cases. Wilson* (this volume) brings this subject up to date. It is important beyond the motor sphere, which is where the available evidence is concentrated. As with each of the items proper to lists II and III,

*We were shocked to learn, as this book was going to press, that Donald Wilson was drowned on June 23, 1970, in a boating accident. We are all poorer for the loss of this brilliant young scientist.

the same principles and problems must recur in many places in the nervous system.

Switching, recognizing, decision making, like some other candidate operations, both overlap and are subdivisible. But some such formulation is surely a major heading in physiology, transcending both the particular sense modality in which our present knowledge is concentrated and the efferent systems in which similar operations occur. The formulation also embraces the equivalent processes in the great intervening domains that are neither afferent nor efferent.

Some other candidates seem attractive or plausible, but, as far as I am aware, do not yet find support in physiological evidence, although they may be suggested by psychophysical data. These might include sequential scanning, sampling, multiplying, comparing, generalizing, and others. Part of the reason for proposing such lists is to heat up the debate over each candidate.

Closely similar are storing and its various subdivisions, such as temporary holding, printing, filing in context, duplicating (as on the other side of the brain), retrieving, and the like. These have physiological and even anatomical support, but are heavily influenced by our knowledge of level-IV accomplishments.

The higher we go the more chance there is of unexpected principles of organization. Probably the lists lengthen; certainly they become more difficult. Here is where one searches for the right terms for functions performed by the so-called association areas of the cortex and thalamus. For example, Pribram (1960) obtained evidence that led him to speak of "problem solving" as a function of the intrinsic forebrain mechanisms and to distinguish two aspects characteristic, respectively, of the posterior and the frontal portions. He called these "differentiative" and "intentional." Lesions of the former affect the "delineation of the problem"; lesions of the frontal, intentional system affect the "economic solution of a problem." Such terminology is not easy, but is a healthy sign of effort to define more adequately the reality we face in the brain; it deserves both emulation and criticism toward a refined taxonomy.

There are other proposals for the functions of regions of association cortex that have such an unconventional quality (outside the "accepted paradigm") that they are usually ignored and not debated. But I believe we should give them a hearing if only to loosen up our thinking and encourage the proposal of alternative candidacy. Let me illustrate with a little-thought-of proposition of Nielsen (1958). Certain dominant hemisphere lesions in a few patients suggested to him that categorizing objects in the world related to the symbolization inferred from aphasias is localized. Nielsen thought the deficit in these patients was a blindness to a category of objects that seemed to share the characteristic of "animateness." When asked to name objects visible from a window or on a tray, they overlooked horses, people, flowers—and a doll and false teeth. I mention this, not as a consistent clinical entity or confirmed finding, but as the kind of unexpected clue we, as good naturalists, should be prepared to notice and evaluate.

There is no attempt here to list the candidate operations in cortex and thalamus. The chapters in this block by Purpura and Scheibel and Scheibel re-examine the hard evidence in search of general, transferable processes, as do the other chapters for other examples of central masses. The purpose here is simply to emphasize the need for new thinking of candidate operations and processes, especially unconventional or unfamiliar formulations, in order to escape our unconscious assumption that the brain is organized, with respect to higher functions, along lines that we expect from our systems of logic and social organization.

I believe you will agree that, despite a huge literature, we still have ahead of us the main elucidations, namings, and discoveries, in a sophisticated sense, of actual operations and processes. A part of our ignorance is due to failure to think of terms for candidate processes, limited as we are in thought by our language or, perhaps more, by constraints in our use of the reservoir of language. These lead to a tragic waste of the opportunity that presents itself in clinical material. We are like the blind men studying the elephant (Bullock, 1965), each probing with a long electrode that filters in unknown ways, its data filtered again through our more or less prepared minds.

The suggestion I have been making, for attempting to discern and propose candidate "operations," is actually an eclectic way of encouraging different approaches with certain common elements. "Operations" may be both means (mechanisms) and ends (performances); the mechanisms of one level of analysis are phenomenological performances of the next level below.

Francis Bacon, in his *Novum Organum,* worried about the distortions of objectivity in the scientist's mind by the "idols of the tribe, the idols of the den, the idols of the market, the idols of the theatre." We live in a different era and may well worry instead about the limitations of our subjectivity, our vision, or our creative insight in interpreting evidence and designing tests at this highly complex level of brain function.

Our vision is limited by our successes. We open up rich veins and dig for ore with the tunnel vision of a miner.

Our vision is limited by our nerve. Forgetting that the goal of science is induction of new and larger principles, we act as though the ultimate sin is speculation. The complexity of the brain surely hides a vast reservoir of novel principles yet to be discovered, but there may be no field in which it is harder to publish speculation. That is our own fault because referees are representatives of the prevailing attitude.

Our vision is limited by orthodoxy. It is limited by our specialization and ignorance of the literature beyond our own subsystem. It is limited by our separatism—our habit of working alone and shunning real collusion. It is limited by our budgets. We are guilty of lowering our sights to the horizon of the research grant we think is obtainable. I live on a hill, from which I look out every day at Mt. Palomar. I am reminded that we shall not have a tool like the Palomar reflecting telescope until we ask for it. And we cannot ask for it until we conceive it.

## Acknowledgments

Original research reported herein was aided by grants from the National Institutes of Health, the National Science Foundation, the U.S. Air Force Office of Scientific Research, and the Office of Naval Research.

### REFERENCES

BULLOCK, T. H., 1953. Comparative aspects of some biological transducers. *Fed. Proc.* 12: 666–672.

BULLOCK, T. H., 1961a. The origins of patterned nervous discharge. *Behaviour* 17: 48–59.

BULLOCK, T. H., 1961b. The problem of recognition in an analyzer made of neurons. *In* Sensory Communication (W. A. Rosenblith, editor). M. I. T. Press, Cambridge, Massachusetts, pp. 717–724.

BULLOCK, T. H., 1965. Strategies for blind physiologists with elephantine problems. *Soc. Exp. Biol. Symp.* 20: 1–11.

BULLOCK, T. H., and G. A. HORRIDGE, 1965. Structure and Function in the Nervous Systems of Invertebrates. W. H. Freeman and Company, San Francisco, 2 vols.

EVANS, E. F., and P. G. NELSON, 1966. Behaviour of neurones in cochlear nucleus under steady and modulated tonal stimulation. *Fed. Proc.* 25: 463 (abstract).

EVANS, E. F., and P. G. NELSON, 1968. An intranuclear pathway to the dorsal division of the cochlear nucleus of the cat. *J. Physiol.* (*London*) 196: 76P–78P.

FITE, K. V., 1969. Single-unit analysis of binocular neurons in the frog optic tectum. *Exp. Neurol.* 24: 475–486.

HELD, R., D. INGLE, G. E. SCHNEIDER, and C. B. TREVARTHEN, 1967–1968. Locating and identifying: Two modes of visual processing. A symposium. *Psychol. Forsch.* 31: 42–62, 299–348.

KATSUKI, Y., J. CHEN, and H. TAKEDA, 1954. Fundamental neural mechanism of the sense organ. *Bull. Tokyo Med. Dent. Univ.* 1: 21–31.

KUHN, T. S., 1962. The Structure of Scientific Revolutions. University of Chicago Press, Chicago.

LETTVIN, J. Y., H. R. MATURANA, W. H. PITTS, and W. S. McCULLOCH, 1961. Two remarks on the visual system of the frog. *In* Sensory Communication (W. A. Rosenblith, editor). M. I. T. Press, Cambridge, Massachusetts, pp. 757–776.

LEVICK, W. R., C. W. OYSTER, and E. TAKAHASHI, 1969. Rabbit lateral geniculate nucleus: Sharpener of directional information. *Science* (*Washington*) 165: 712–714.

MOUNTCASTLE, V. B., 1961. Some functional properties of the somatic afferent system. *In* Sensory Communication (W. A. Rosenblith, editor). M. I. T. Press, Cambridge, Massachusetts, pp. 403–436.

NIELSEN, J. M., 1958. Memory and Amnesia. San Lucas Press, Los Angeles.

PRIBRAM, K. H., 1960. The intrinsic systems of the forebrain. *In* Handbook of Physiology. Section I: Neurophysiology (J. Field, H. W. Magoun, and V. E. Hall, editors). American Physiological Society, Washington, D. C., vol. 2, pp. 1323–1344.

RATLIFF, F., B. W. KNIGHT, and N. GRAHAM, 1969. On tuning and amplification by lateral inhibition. *Proc. Nat. Acad. Sci. U. S. A.* 62: 733–740.

RAYMOND, S. A., 1969. Physiological influences on axonal conduction and distribution of nerve impulses. M. I. T., Cambridge, Massachusetts, doctoral thesis.

# 37 Neural Subsystems: Goals, Concepts, and Tools

EDWIN R. LEWIS

## Interaction and some of its consequences

INTERACTION DUE TO INTRINSIC PROPERTIES OF ELEMENTS AND PROCESSES  Figure 1 shows a formal system element. Few natural objects (or manufactured components), however, fit this scheme. Most objects lack specified input and output terminals, and the behavior of most objects is not uniquely describable within the context of

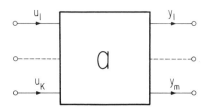

FIGURE 1   Formal abstract system element. It has specified inputs (u), specified outputs (y), and describable relationships among them (Zadeh and Desoer, 1963).

Figure 1. Consider an electrical resistor (Figure 2). Although it has two terminals and well-defined relationships among its variables (Ohm's law), it cannot be represented by a single abstract system element. The behavior of the resistor (from the point of view of formal system theory) depends on what is connected to the resistor. In other words, it depends on the structural context of the resistor; the resistor exhibits *structural contextuality*. The behavior of the resistor also depends on which variable (current or voltage) is considered input (cause) and which is considered output (effect). In other words, the behavior of the resistor (from a system-theory point of view) depends on its *orientation* with respect to the direction of causality; and the resistor exhibits no *a priori* orientation.

Although general system theory purports to treat non-oriented objects and objects exhibiting structural contextuality, most of the techniques for handling such objects are left to circuit-theory textbooks. System-theory texts

EDWIN R. LEWIS  Department of Electrical Engineering and Computer Sciences, and Electronics Research Laboratory, University of California, Berkeley, California

treat oriented representations, free from structural context. Thus we have either *systems* of oriented, context-free objects or *circuits* of objects that are nonoriented, structurally contextual, or both. Based on this scheme, resistors are circuit elements, not system elements; so are almost all natural objects.

Whether they are linear or nonlinear, time-invariant or time-varying, deterministic or stochastic, anticipatory or nonanticipatory, circuits of nonoriented, structurally contextual objects have certain inescapable characteristics:

1. The properties of its individual parts are obscured completely in the operation of the circuit.
   a. Circuit input-output descriptions are infinitely ambiguous with respect to underlying mechanism and organization.
   b. Unambiguous determination of underlying mechanism and organization requires complete decomposition of the circuit.
2. The whole circuit is nonoriented and structually contextual.
3. Until the orientation and structural context of the circuit have been determined, the circuit itself is its own most economical description.

Although quite general, these points are demonstrated most easily with linear processes such as those illustrated in Figure 3. The most important parameter of a linear process is its rate constant. Figure 3A illustrates the first point and its corollaries by showing how individual rate constants are masked increasingly in whole-circuit operation as structural context becomes progressively more complex. Figure 3B illustrates the second point by showing how strongly whole-circuit input-output relationships depend on orientation. Figure 3C illustrates the final point by demonstrating that the minimal-state equations are no less complicated than the circuit itself, nor do they convey more information.

For a perturber-observer, these inescapable characteristics generate a dilemma. On the one hand, circuits are irreducible entities: any measured circuit property is determined by the entire circuit and is not attributable to any of its individual elements; if a single element is isolated, its properties will have little obvious relevance to its operation in the intact circuit. On the other hand, unambiguous determination of the relationship between circuit operation and underlying mechanism and organization requires effective isolation of every element and every interaction and determination of the relevance in each case.

This dilemma exists not only in passive, linear circuits, in which rate constants are the important properties, but in all circuits. In active circuits, for example, stability measures are important properties. As are rate constants, stability properties of individual elements are obscured completely in the actions of the whole circuit; and the stability of the whole circuit is not attributable to any individual element, but is a property of the circuit. In nonlinear circuits, state trajectories are important measures; these, too, are whole-circuit properties, not attributable to individual elements. In stochastic circuits, probabilities are important. Once again, the probabilities associated with individual processes are obscured in the complex joint probabilities affecting whole-circuit operation.

INTERACTION DUE TO FEEDBACK   Some neurons may be exceptions to the generalization that natural objects are non-oriented and structurally contextual. Once triggered, the spike apparently does not depend further on the events leading to its generation. Axons thus tend to isolate, send-

ing ends of neurons from receiving ends; cause and effect can be separated. Furthermore, the nature of the chemical signal at a synapse may be determined solely by the state of the sending neuron and be independent of loading by the receiving neuron. Thus, synaptic transmission may provide freedom from structural context.

Unfortunately, neurons commonly are connected reciprocally as well as serially, which brings about an interaction via feedback. When interaction becomes profuse, as apparently it often does, cause and effect become obscure and structural contextuality becomes strong. These effects of feedback are illustrated in Figure 4. Here is a process with a rate constant $\alpha$; but it is embedded in a feedback loop. Consequently, $\alpha$ is obscured completely in the whole-system response. If $\alpha$ were negative, the process in the oval would be unstable; yet if K were larger than $\alpha$, it would never exhibit that instability while the loop remained intact. Furthermore, evaluation of $\alpha$ requires breaking of the loop. Thus, if feedback is present, systems of oriented, structurally independent objects generate the same dilemma as circuits of nonoriented, structurally contextual objects. They are irreducible entities, yet they must be decomposed

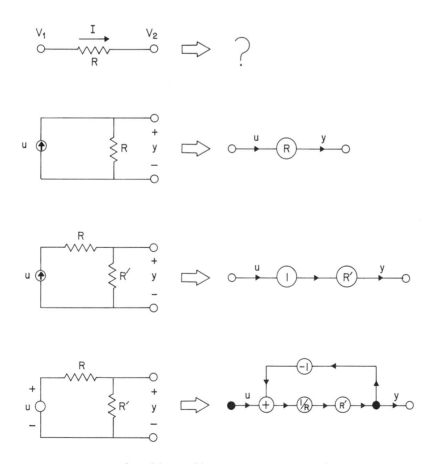

FIGURE 2   A few of the possible system representations of a resistor.

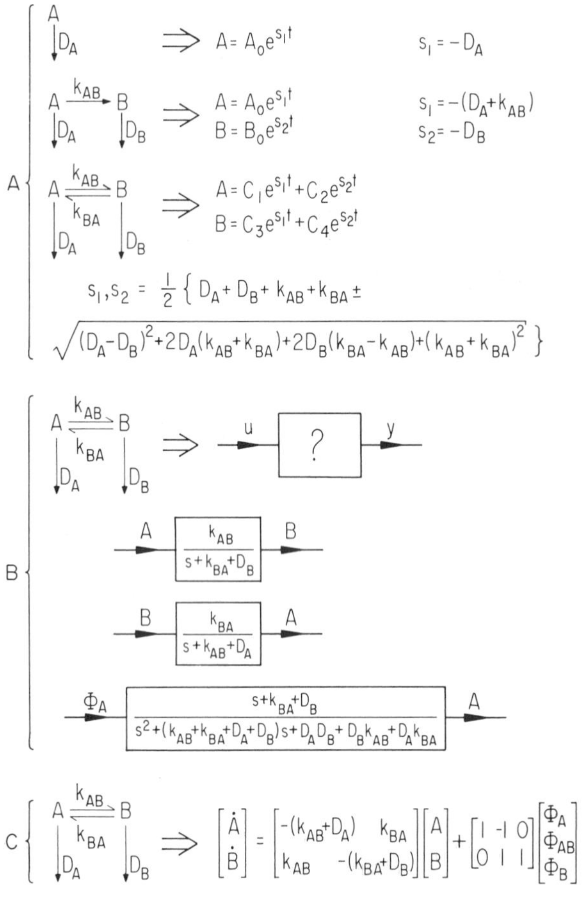

$$A \begin{cases} \begin{array}{l} A \\ \big\downarrow D_A \end{array} \Rightarrow A = A_0 e^{s_1 t} \qquad\qquad s_1 = -D_A \\[1.5em]
\begin{array}{l} A \xrightarrow{k_{AB}} B \\ \big\downarrow D_A \quad \big\downarrow D_B \end{array} \Rightarrow \begin{array}{l} A = A_0 e^{s_1 t} \\ B = B_0 e^{s_2 t} \end{array} \qquad \begin{array}{l} s_1 = -(D_A + k_{AB}) \\ s_2 = -D_B \end{array} \\[1.5em]
\begin{array}{l} A \underset{k_{BA}}{\overset{k_{AB}}{\rightleftharpoons}} B \\ \big\downarrow D_A \quad \big\downarrow D_B \end{array} \Rightarrow \begin{array}{l} A = C_1 e^{s_1 t} + C_2 e^{s_2 t} \\ B = C_3 e^{s_1 t} + C_4 e^{s_2 t} \end{array} \\[1.5em]
s_1, s_2 = \tfrac{1}{2}\left\{ D_A + D_B + k_{AB} + k_{BA} \pm \right. \\
\left. \sqrt{(D_A - D_B)^2 + 2D_A(k_{AB} + k_{BA}) + 2D_B(k_{BA} - k_{AB}) + (k_{AB} + k_{BA})^2} \right\}
\end{cases}$$

$$B \begin{cases}
\begin{array}{l} A \underset{k_{BA}}{\overset{k_{AB}}{\rightleftharpoons}} B \\ \big\downarrow D_A \quad \big\downarrow D_B \end{array} \Rightarrow \quad u \to \boxed{\ ?\ } \to y \\[1.5em]
A \to \boxed{\dfrac{k_{AB}}{s + k_{BA} + D_B}} \to B \\[1.5em]
B \to \boxed{\dfrac{k_{BA}}{s + k_{AB} + D_A}} \to A \\[1.5em]
\Phi_A \to \boxed{\dfrac{s + k_{BA} + D_B}{s^2 + (k_{AB} + k_{BA} + D_A + D_B)s + D_A D_B + D_B k_{AB} + D_A k_{BA}}} \to A
\end{cases}$$

$$C \begin{cases}
\begin{array}{l} A \underset{k_{BA}}{\overset{k_{AB}}{\rightleftharpoons}} B \\ \big\downarrow D_A \quad \big\downarrow D_B \end{array} \Rightarrow
\begin{bmatrix} \dot{A} \\ \dot{B} \end{bmatrix} = \begin{bmatrix} -(k_{AB}+D_A) & k_{BA} \\ k_{AB} & -(k_{BA}+D_B) \end{bmatrix}\begin{bmatrix} A \\ B \end{bmatrix} + \begin{bmatrix} 1 & -1 & 0 \\ 0 & 1 & 1 \end{bmatrix}\begin{bmatrix} \Phi_A \\ \Phi_{AB} \\ \Phi_B \end{bmatrix}
\end{cases}$$

FIGURE 3 Some effects of orientation and structural context. A. As structural context becomes increasingly complex in a simple chemical circuit, the basic responses (second column) become more complicated and the diffusion constants (D) and rate constants (k) are obscured. B. The input-output relationship of a simple chemical circuit depends on which concentration (A, B) or flux (Φ) is input and which is output; s denotes the complex variable of the Laplace transform. C. The state equation for the curcuit. $\dot{A}$ and $\dot{B}$ denote time derivatives; Φ denotes flux, e.g., the rate at which A is supplied, $\Phi_A$, or converted to B, $\Phi_{AB}$.

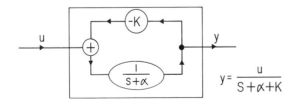

FIGURE 4 A first-order process (in the oval) embedded in a feedback loop: s denotes the complex variable of the Laplace transform.

completely if the properties of individual elements are to be observed.

## Circuit questions and system questions

SUBSYSTEM If profuse interaction occurs without interruption throughout an entire collection of neurons, then no neuron or group of neurons in that collection will act as a true individual. No measured property of the collection will be attributable to a particular neuron or group of neurons. The properties of the whole collection will be infinitely ambiguous with respect to underlying mechanism and organization. Because no group of neurons within the collection acts as an individual, although any such group may be selected arbitrarily as a subsystem, it seems futile to consider it as a separate entity.

However, considerable evidence supports the view that nervous systems do exhibit some localization of function, some isolation of cause and effect within them. Many vertebrate retinas, for example, are believed to exhibit only afferent fibers and thus must act very much as independent entities. One might reasonably suspect that *any* nervous-system segment exhibiting well-defined and separate input and output tracts not only would be oriented, but also would be somewhat free of structural context. Such segments, or subsystems, might very profitably be considered separately, not merely because they meet the criteria of Figure 1, but because their input and output variables are, in principle, identifiable, and they very likely operate with independence and ascribable functions. Because of their potential independence, it seems reasonable to consider them independently.

CIRCUIT QUESTIONS Two distinct classes of questions exist concerning such nervous subsystems. One class comprises inwardly directed questions, questions about the internal mechanisms and organization of the subsystem and how they relate to its operation. Because unambiguous answers to such questions can be achieved only through complete decomposition of the subsystem, they are essentially circuit questions. The other class comprises system questions, those that treat the subsystem as an entity and do not demand its decomposition.

With careful and complete isolation and observation of each process, one could, in principle, determine *what* mechanisms and organzation are present in a circuit; but this would lead merely to a collection of facts. Mechanistic understanding demands that one also know how those mechanisms and that organization underlie circuit operation. Given the component processes, one should be able to predict theoretically the operation of the circuit. Thus, mechanistic analysis has two crucial aspects, *identification* of the parts, and theoretical *synthesis* of the circuit from those parts. Circuit operation is infinitely ambiguous with respect to underlying organization, which implies that an infinite number of possible theoretical syntheses exist for every operation. As an example, Figure 5 shows just a few of the fundamentally different ways one could synthesize a one-Hertz (Hz) sinusoidal oscillator, and demonstrates that considerable identification should precede any attempts at synthesis.

Circuits such as nervous subsystems clearly are capable of operating in modes that never occur naturally. A perturber-observer may employ signals that evoke mechanisms such as metabolic processes or hysteretic-state changes that normally never affect operation of the subsystem. In view of the obvious inherent difficulties of mechanistic analysis, it seems the height of folly to find explanations for behavior that is not natural. Purely from an economic viewpoint, therefore, one should have considerable knowledge of the

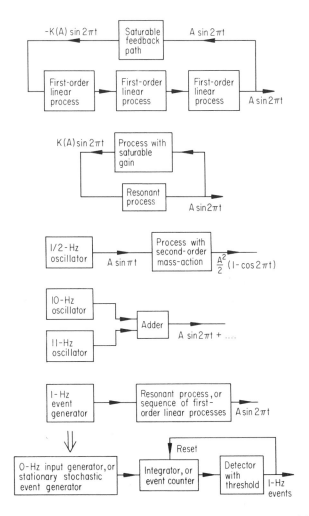

FIGURE 5  Five of the many ways to realize a one-Hertz sinusoidal oscillator.

nature of signals normally impinging on a subsystem before attempting *any* mechanistic analysis.

SYSTEM QUESTIONS  A neural circuit, if it is oriented and its structural context is determined, can be considered as a subsystem and treated as an entity. One can begin to consider its position and ponder its functional significance with respect to the systems of which it is part and, ultimately, with respect to the homeostasis or the ecological or evolutionary status of the entire organism or species. One can attempt to identify its inputs and outputs, to develop generalized descriptions of its input-output characteristics, and from these attempt to determine its roles or predict its behavior in novel situations or determine the qualitative natures of its internal operations or the constraints it imposes on whole-system operation. The major theme of such questions is not "How does a nervous subsystem work?" but "What does it normally do, and why?" System questions are, as a result, inherently somewhat teleological (Mesarović, 1968).

Although the answers to most system questions are obtained with the help of generalized input-output descriptions, such descriptions by themselves invariably are insufficient. In the case of the linear processes of Figure 3B, for example, if it were determined that A was input and B was output, then the input-output description would have the form $1/(s + \alpha)$. This describes, in an abstract way, how the subsystem will respond to input disturbances. It does not reveal the nature of the disturbances that actually will be imposed upon the subsystem when it is embedded in a larger system, nor does it reveal the aspects of the responses of the subsystem that have any effect on the rest of the system. Interpretation of this subsystem's role will depend not only on a general description of its input-output relationships, but also on the nature of signals normally impinging upon it and on the significance of those signals. This signal-dependence of interpretation is illustrated in Figure 6.

*Techniques for circuit questions*

DECOMPOSITION  The techniques of neural-circuit decomposition are well known and need no elaboration here. Unfortunately, they have not been completely successful. Consider, for example, the saga of the lobster cardiac ganglion (for references, see Hartline and Cooke, 1969; Mayeri, 1969).

Situated on the dorsal wall of the heart, the nine-neuron cardiac ganglion produces periodic bursts of spikes which drive the heart into contraction. Although it receives regulatory fibers from the central nervous system, the ganglion is an independent oscillator, and is an excellent example of a separable subsystem. Furthermore, it is very probably the most thoroughly studied neural circuit

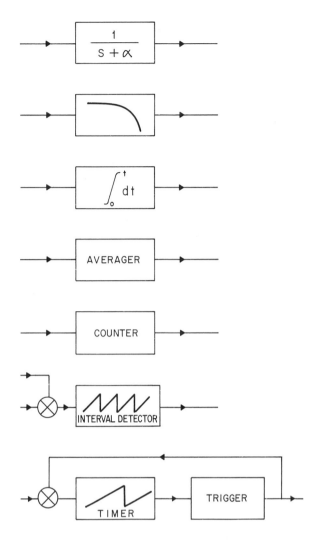

FIGURE 6 A few of the many things a first-order process can accomplish, depending on the input and output signals and their significance. It can be a "filter" passing low frequencies, a "damper" preventing rapid transients, or an "inherent limitation." It can be an "integrator" of high-frequency signals, a "demodulator," an "averager" of continuous signals, or a "counter" of discrete signals. If pulses can repeatedly reset it to the same state, it can be an "interval detector" or a "timer." It even can be approximately one-third of a sinusoidal oscillator (see Figure 5).

(neuron for neuron). Investigators have gathered extensive information about whole-ganglion operation and about the effects on this operation of many parameters, including tension of the heart muscle, temperature, ionic concentrations, and levels of extrinsic inhibition and excitation.

In addition to being studied *in situ*, the ganglion has been studied after removal from the heart and during systematic removal of its various parts—all in attempts to identify the seat of burst formation. Weak DC currents have been imposed across various parts of the ganglion. Extraneous sig-

nals have been introduced into specific axons and axon bundles; spike trains in nerves and in single axons have been observed in minute detail; and both spatial and temporal spike correlations have been examined thoroughly. Microelectrodes have been used to observe the fine details of electrical activity within the somata, to impose voltage and current steps across soma membranes, and to observe the effects in the same cell and in others. The correlations between extracellularly recorded spikes and intracellularly recorded synaptic potentials have been studied thoroughly.

Considerable pharmacological evidence has been gathered, including the effects of drugs such as acetylcholine, epinephrine, gamma amino butyric acid, ouabain, and dinitrophenol and of anesthetics such as procaine, which was used for selective anesthetization of ganglion parts in attempts to locate the burst mechanism. Histological studies have been performed in attempts to determine both ganglion organization and placement of electrodes after electrophysiological measurements.

The lobster cardiac ganglion has been subjected to thorough ablation studies, thorough gross electrophysiological studies, thorough microelectrophysiological studies, thorough paramacological studies, extremely thorough statistical studies, and thorough voltage-clamp and current-clamp studies. In short, the repertoire of neurophysiological circuit identification techniques has been very nearly exhausted; in fact, several new techniques have been invented and have been employed simultaneously in most of their reasonable combinations—all for the study of this relatively simple group of nine neurons.

The results of the identification procedures have been impressive. In addition to being very probably the most thoroughly studied neural circuit (neuron for neuron), the lobster cardiac ganglion also very probably is the best-understood group of neurons with respect to the details of circuit interactions. Thus, although many of the cardiac-ganglion investigators were interested in the operations of single neurons rather than in the operation of the entire circuit, the cardiac ganglion nevertheless provides a test case. With all of the available information, can one specify the mechanisms and organization underlying the major cardiac-ganglion circuit operation—namely, its oscillation (burst formation)? The answer is *no*. We still do not know how bursts are initiated or how they are terminated, nor do we know whether bursts are the results of endogenous oscillation in a single neuron or of circuit interactions among several neurons. The circuit has not been decomposed sufficiently.

What is required for sufficient decomposition? Diligent application of available techniques certainly should add to our knowledge of the ganglion. However, sufficient decomposition requires that one be able to determine organization. Unfortunately, current techniques apparently are inadequate. Light microscopy has not provided sufficient resolution to map invertebrate neuropil, and mapping it with 500 Å serial sections in a transmission electron microscope has proved to be an enormous task (Frazier et al., 1967). Furthermore, even the most thorough statistical analyses of spike trains could not reveal organization. Clearly, *new tools* are needed for circuit mapping. In addition, sufficient decomposition requires observation of all important circuit variables. In invertebrates, many of these occur in neuropil, far from somata and often not measurably reflected by them (Kennedy et al., 1969). Presently, no tools are available for observing details of interactions in neuropil.

## New tools emerging

Although the saga of the cardiac ganglion can be made to appear gloomy, two facts are encouraging. First, our general knowledge of neural circuitry has increased enormously since the cardiac-ganglion studies began. Second, many new and potentially powerful tools for identification are emerging.

Fluorescent dyes are enhancing considerably the possibility of coupling electrophysiology and histology. After recording from the soma of a neuron, one can inject electrophoretically a dye such as procion yellow; it will spread throughout the neuron's processes, apparently with very little toxic effect (Stretton and Kravitz, 1968; Kennedy et al., 1969). The dye can be fixed within the neuron and its processes, which then can be traced through serial sections. These dyes are adding considerably to our knowledge of the geometry of the neuropil.

Ultimately, one still will be faced with the resolution limitations of light microscopy. Not having this limitation, yet allowing one to view large pieces of tissue rather than ultrathin sections, the *scanning* electron microscope bridges the gap between the visual world of light microscopy and the largely inferential world of conventional electron microscopy (Hayes and Pease, 1969). With its sensitivity to surfaces, its high resolution over extraordinary depths of field, and its ability to provide micrographs with three-dimensional perspective, the scanning electron microscope also promises to be an excellent tool for mapping. Preliminary studies in a retina (see Figure 7A) and in an invertebrate ganglion (Figure 7B) have proved quite successful (Lewis et al., 1969a; 1969b.)

Although not so imminent as the histological tools, new physiological tools for neural-circuit observations also seem to be emerging. Microelectronics technology, for example, has progressed enormously during the past decade, and the most advanced techniques of this field now are being applied to the development of multiple microelectrodes for extracellular recording (Kalisch and Angell,

1968). Up to six gold electrodes with 10-$\mu$m spacing have been mounted on an 80-$\mu$m silicon probe, and the array successfully employed to record from single neurons in cat cortex. The workers now are attempting to integrate microcircuit amplifiers into the probe.

Although too weak to be useful with available sensors and signal-analysis devices, optical and infrared concomitants of neuroelectric activity have been found which offer the hope that some day we shall be able to observe visually the interactions of neuronal circuits, and with considerably greater spatial resolution than we can anticipate now with arrays of microelectrodes (Cohen et al., 1968; Tasaki et al., 1968).

New and promising techniques for inference of neural circuit interactions from spike-train statistics also are being developed (Perkel, this volume; Gerstein, this volume). By these techniques we may eventually be able to examine correlations among spike trains and infer from them the connections among the corresponding neurons, thus avoiding, in some cases, the necessity of observing the intricacies of signal flow in neuropil. The cardiac ganglion makes one point clear. When axons bifurcate so that we can observe a spike *after* we see its effect on another cell, we often will need spatial correlations as well as temporal correlations in order to determine which cell was driving which.

## Tools of synthesis and analysis

Because irreducible neural subsystems exhibit the characteristics of circuits, their theoretical synthesis and analysis is similar to formal circuit synthesis and analysis. When the interactions in a neural network are found to be linear, some of the well-developed and powerful techniques of *linear* circuit theory may be applicable. The static effects of lateral inhibition in the lateral eye of *Limulus,* for example, can be analyzed expeditiously with a circuit-theory technique for ladder networks. One begins at the extremity of the circuit (i.e., the edge of the eye) and works back toward the input, developing a continued fraction (Figure 8). The simple, iterative expression resulting from the process can be manipulated very easily with a digital computer, once the K values have been established.

When active, dynamic, nonlinear aspects of neural networks are considered, the well-developed tools of formal circuit analysis become useless, and synthesis generally is achieved by means of simulation (see review by Harmon and Lewis, 1968). With respect to mechanistic analysis, however, synthesis by simulation often leaves much to be desired (Fein, 1966). Simulations generally are complicated, and their operations are often difficult to comprehend. Consequently, published simulation studies often have produced desired neural-network operations without revealing *how* those operations were generated. In fact one often cannot

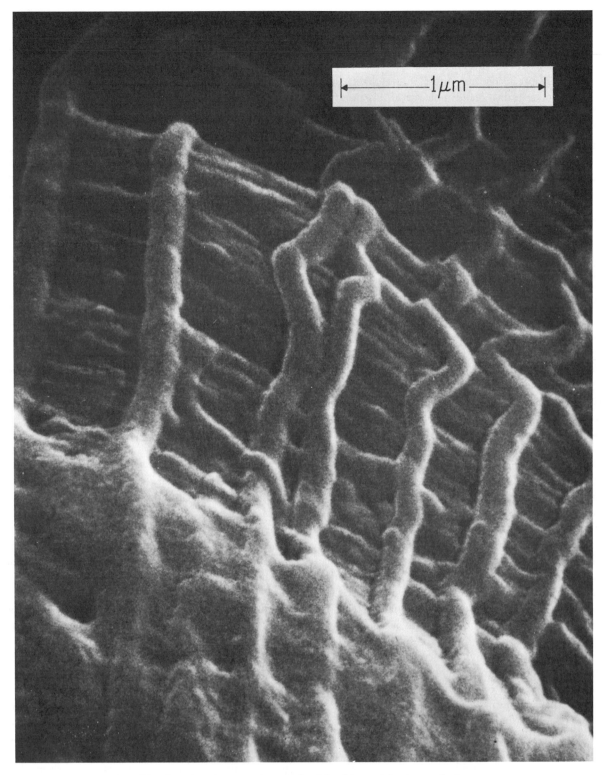

FIGURE 7A  Scanning electron micrograph of dendrites originating at the inner segment (bottom) and coursing over the outer segment of a cone in the retina of *Necturus*.

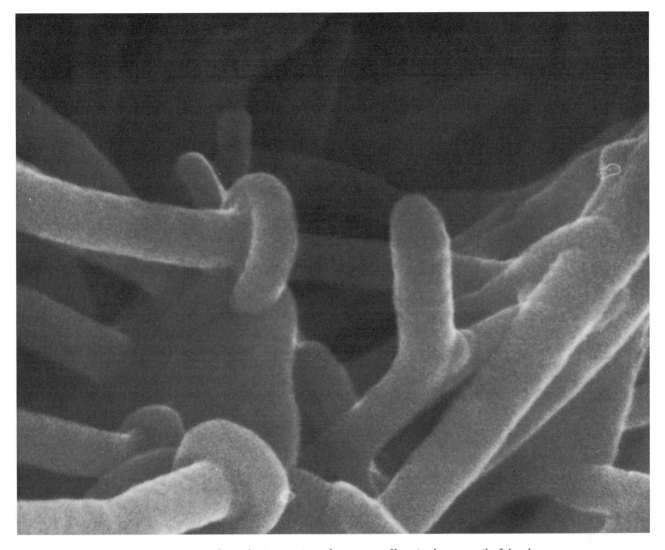

FIGURE 7B    Knobs at the intersection of two nerve fibers in the neuropil of the abdominal ganglion of *Aplysia californica*. Magnification: approximately 35,000.

be sure whether the operations resulted from simulated interacting biological variables, or from accidental properties indigenous to the model employed in the simulation.

Kalman (1968) pointed out that part of this problem can be eliminated by requiring that models be *nothing more* than summaries of experimental data (i.e., models should be MINIMAL), so that "repeating an experiment on the model should yield exactly the same data as was assumed in constructing the model." Unfortunately, he also pointed out that theories of building minimal models are available only for linear, time-invariant systems or circuits. Nonetheless, he proposed two theorems which he considered to be general:

Theorem 1. There exists always a model; there exists always a minimal model.

(A minimal model is one which is simplest, has the minimal number of parameters, etc.; a specific definition must be given in each case.)

Theorem 2. Properties of the model are always either (a) inherent in experimental data used to construct the model, or (b) completely arbitrary. The second possibility is ruled out if the model is minimal.

Even with linear, time-invariant circuits, perturber-observers can face considerable difficulty in attempting to apply the notion of minimal model. Visual transduction in the *Limulus* lateral eye, for example, is linear over small ranges of light intensity. The best available data indicate that transduction comprises a sequence of eight or more linear processes (Fuortes and Hodgkin, 1964; Pinter, 1966). If the processes are completely irreversible and identical, eight will suffice to reproduce the data. If they are not completely irreversible and identical, more than eight generally

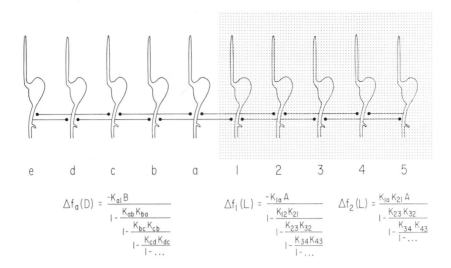

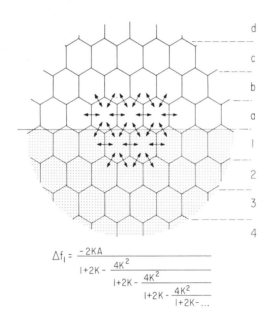

FIGURE 8  Light-shadow transitions on two models of the compound eye of *Limulus;* each ommatidium inhibits its nearest neighbors, as indicated by the knobs and arrows. The expressions of Ratliff et al. (1966) have been made linear by considering incremental eccentric-cell spike frequencies ($\Delta f$) rather than absolute frequencies. $\Delta f(L)$ denotes the incremental response of a shadowed unit due to the presence of the lighted region [$\Delta f(L) \equiv 0$ when the entire eye is shadowed]. $K_{ij}$ relates the frequency of ommatidium j to its inhibitory effect on ommatidium i:

$$(\Delta f_i (L) = - \sum_{j=1}^{n} K_{ij} \, \Delta f_j (L).$$

In the two-dimensional model, the Ks are assumed to be equal.

will be required. In fact, increasingly more stages will be required as the processes become increasingly different from one another or increasingly reversible. Finally, if the number of processes becomes infinite, the model represents a diffusion process, and it still can reproduce the data. These three models are illustrated in Figure 9. With respect to state-space dimension (the usual criterion), the model with eight identical, isolated processes is minimal, but this fact does not seem to make it a more reasonable model than the other two. Indeed, one reasonably might question the probability of finding in nature a sequence of eight identical and completely irreversible reactions.

For neural networks, the notion of a minimal model must be tempered by neurophysiological constraints, embryological constraints, pharmacological constraints, and probably many others. For example, Reiss (1962) demonstrated by means of simulation that two reciprocally inhibiting neurons were sufficient to generate the alternating bursts of spikes required to drive antagonistic muscles. The only real examples of such systems found to date are the Mauthner-cell pairs in the teleosts and in tailed Amphibia.

According to the best available data, these are embedded in networks considerably more complex than Reiss's "minimal model" (Figure 10). Thus the designer is in constant danger of having his designs considerably modified or even nullified by the perturber-observer. Nonetheless, synthesis (i.e., a theoretical putting-together of the parts) well founded in circuit identification will be crucial if we are ever to achieve complete analysis.

## Techniques for system questions

INPUT-OUTPUT DESCRIPTION  Description of input-output characteristics (often called "system identification") usually involves two steps—determination of the number of state variables, followed by determination of the parameters associated with each variable. As the example of linear visual transduction in *Limulus* illustrates, even the first step can be extremely difficult to achieve. Nonlinear systems offer more difficulties. Troelstra (1969) has shown, for ex-

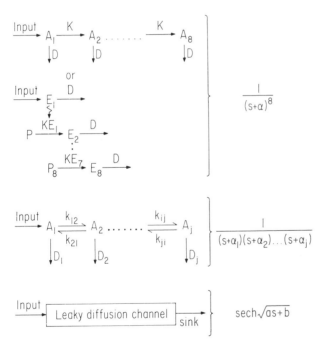

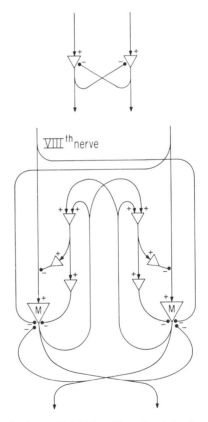

FIGURE 9  Three models that can reproduce the data from visual transduction in the *Limulus* compound eye. Top: Two irreversible processes: a sequence of irreversible chemical reactions with diffusion, and a sequence of proenzyme-enzyme reactions with diffusion. Middle: a sequence of reversible chemical reactions with diffusion. Bottom: a diffusion along a leaky channel (e.g., electrotonic conduction along a dendrite or diffusion of a photochemical product). The letter s is the complex variable of the Laplace transform.

FIGURE 10  Reciprocal inhibition. Top: A minimal model for alternated driving of antagonistic muscles. (From Reiss, 1962.) Bottom: Estimated circuit for alternated driving of the tail of a startled fish. Diamonds denote somata; knobs denote inhibitory synaptic endings; arrows denote excitatory endings; M denotes a Mauthner neuron. (From Horridge, 1968.)

ample, that with a sequence of two or more first-order, linear processes connected to a threshold element, it is extremely difficult to determine from the response of the threshold element just how many processes are present.

System identification characteristically is an elaboration of output capabilities. It comprises highly quantitative measurement and analysis of output dynamics in response to extremely simple input signals, preferably signals from which all other inputs can be constructed easily. Such elaboration of output dynamics seems reasonable for cases such as motor subsystems in which the natural outputs are expected to be elaborate and the inputs are expected to be rather simple. Sensory subsystems, however, may generate simple "yes" or "no" responses to extremely complicated spatial and temporal input patterns. Hubel and Wiesel (1962), Lettvin et al. (1959), Frishkopf et al. (1968), and others have shown that conventional system identification with its elaboration of the output is not at all useful in such cases. In fact, one often needs to consider only one aspect of the output—its maximum response amplitude. The inputs producing those responses, however, deserve careful analysis and elaborate description. A similar approach (i.e., an "inverted" system identification) should be useful for any subsystem suspected of being a decision maker.

In several cases, carefully quantified conventional system identification has led to interesting new insights about neural subsystems (see reviews by Stark, 1968; Harmon and Lewis, 1968). In each case, success depended on considerable knowledge of the signals normally impinging upon the subsystem and considerable knowledge of the roles of the subsystem. In light of the examples of Figures 5 and 6, these conditions are not surprising.

TELEOLOGICAL APPROACHES Although we may not know the electrical properties of the axon of the Mauthner cell, how its axon loads its soma, precisely how signals interact in its dendrites, or much else about the mechanisms or internal organization underlying its behavior, we nonetheless have a relatively clear, unambiguous picture of the Mauthner cell as a nervous subsystem (Horridge, 1968; Furukawa and Furshpan, 1963). This is so because (1) it has been identified as an oriented object, and its orientation has been determined (it receives inputs from both eighth nerves and from the other Mauthner cell and sends outputs to the other Mauthner cell and to a giant motor fiber in the spinal cord); and (2) its role is reasonably well established (it is part of a system that decides on the basis of eighth-nerve vibratory input whether to elicit tail-flips and how those tail-flips should be directed). Many neurons (e.g., those in the cerebellum) have inhibitory inputs, yet receive little notice for it. In light of the decision role of the Mauthner cell, however, one finds rich meaning in the fact that its ipsilateral inputs are excitatory, its contralateral inputs are in-

hibitory, and its output feeds back to inhibit the other Mauthner cell.

Similarly, lateral inhibition by itself commands little attention. Purkinje cells of the cerebellum, for example, apparently are laterally inhibiting, but such a fact receives little notice because the role of the cerebellum is unknown. In peripheral sensory systems, however, lateral inhibition can provide contrast enhancement and increased acuity (von Békésy, 1967). Its presence in visual systems explains a well-known psychophysical phenomenon, Mach bands (Ratliff, 1965). One finds rich meaning, therefore, in the presence of lateral inhibition in peripheral sensory systems.

Responses of the frog eye to minute spots of light, the excitatory centers with inhibitory surrounds, elaborate dynamics resulting in "on cells," "off cells," and "on-off cells," all provide some information about possible circuit interactions within the retina; but they did not provide much insight about the normally operating frog retina as a subsystem. Minute spots of light are not behaviorally significant visual patterns for frogs. When the role of the retina in natural frog behavior finally was considered, however, and input signals were chosen accordingly, the retina was found to be an extremely refined decision-making machine (Lettvin et al., 1959).

In the bullfrog ear, the fact that the basilar papilla exhibits a response peak at 1300 Hz and the amphibian papilla exhibits a peak at 200–400 Hz and is inhibited by sound energy between 500 Hz and 1000 Hz is interesting. It assumed considerably greater significance when behavioral studies showed that the mating call of the adult male bullfrog has spectral peaks at 200 Hz and 1300 Hz and very little energy between 500 Hz and 1000 Hz, while sexually immature male bullfrogs call with a spectral peak between 500 Hz and 1000 Hz; only the call of a mature bullfrog will evoke calling in other males and the establishment of a chorus (Frishkopf et al., 1968).

Many other examples can be cited: studies of mammalian reflexes (Sherrington, 1961), studies of the cockroach cercal system (Roeder, 1963), studies of the noctuid auditory system (Roeder and Treat, 1961), studies of insect flight motor systems (Wilson, 1966), and many more. All these led to important new insights about nervous-system operation; each of them treated an oriented *nervous subsystem* as an entity, and each of them asked system questions rather than circuit questions.

Similarly, the lobster cardiac ganglion (1) has been identified as an oriented object, and much of its orientation has been determined (it receives inputs from three pairs of extrinsic fibers, two accelerator and one inhibitor, and from stretch receptors in the heart, and it sends motor fibers to the heart); and (2) much of its role has been determined (it controls the rhythm of the lobster heart).

If one were interested in system questions rather than

circuit questions, one might ask about the ultimate sources of accelerator and inhibitor inputs to the ganglion. Do they mediate regulation of some homeostatic variable such as oxygen tension in the central nervous system? Do they respond directly to motor activity, anticipating oxygen-tension decline? Are they connected to baroceptors? If feedback regulation is involved, what is the open-loop gain of the system? Is it stable? These all are legitimate nervous-subsystem questions; answers to them may be even more revealing and important than the answers to all the circuit questions that have been posed. Even more important, answers to them may be achievable with present tools and present techniques.

## Summary: some priorities

A computer specialist recently stated that if we consider a nervous system to be an information processor, then in order really to understand it we must know as much about it as we need to know about information processors of our own design. He went on to say that we must have *complete knowledge* of every aspect of it, from the physics underlying the operations of its individual elements to its over-all structural and functional design. We must be able to design it, redesign it, maintain it, and repair it.

This certainly appears to be a reasonable definition of understanding. On the other hand, most information processors of our own design are composed of semiconductor devices and we do not have *complete knowledge* of the physics underlying semiconductor-device operation. In fact, given the geometry and doping profile of a semiconductor device, we still cannot predict the performance of that device. Nonetheless, we understand our information processors so well that we *can* design them and build them. Needless to say, we do not design them in terms of basic device physics.

When it comes to nervous systems, on the other hand, we are not designers and builders. For the present, at least, we must be content to be perturbers and observers. What we need are some reasonable goals and some reasonable priorities for perturbers and observers. Many of the system and circuit concepts discussed in this paper are strongly relevant to such goals and priorities.

STRUCTURAL CONTEXT  The existence of structural contextuality implies that input-output descriptions are enormously ambiguous with respect to underlying mechanism and organization. It implies that neuronal circuits will be something other than the sum of their parts and that if a neuronal circuit element (e.g., a neuron or group of neurons) is isolated by surgery or anesthesia or any other means, its properties may have little relevance to its operation in the intact circuit. However, complete mechanistic analysis requires that every part somehow be isolated for observa-

tion, and presently we do not have sufficient tools for either isolation or observation. Recent work, such as that on retina (Werblin and Dowling, 1969) and that on *Tritonia* (Willows, 1968) proves that much remains to be learned by diligent application of available tools, but thorough mechanistic analysis will require new tools, and their development should be given high priority. On the other hand, for mechanistic analysis, the study of input-output relationships should be given low priority.

SIGNAL CONTEXT  With respect to analysis of functions of a neural subsystem, input-output descriptions may be useful, but they can be interpreted only in the light of their natural signal context (i.e., the physical attributes of normal input and output signals of the subsystem). Furthermore, if one wishes to discover and analyze normal mechanisms, he must do it by employing normal signals. Thus, knowledge of natural signal context should be given a high priority, both for functional analysis and for mechanistic analysis.

FUNCTIONAL CONTEXT  Finally, if one wishes to understand the operation of a complete system, he normally must do so through functional analysis, in terms of the roles of its subsystems. An engineer would understand a computer, for example, not as a circuit of resistors, diodes, capacitors, magnetic cores, et cetera. He would understand it as a system of clocks, scalers, storage bins, et cetera. If a neurobiologist wishes to attain similar understanding of a nervous system, he certainly may do it in the same way, in terms of the functional significance of each system element, or subsystem. Therefore, if understanding of whole nervous-system operation is the ultimate goal, the highest priority should be given to determination of roles. The final implication of system theory is that neurobiologists should assume a somewhat teleological approach (Mesarović, 1968; Frishkopf et al., 1968).

## Acknowledgment

The work of E. R. Lewis is supported by the National Science Foundation under Grant NSF-GK-3845.

REFERENCES

BÉKÉSY, G. VON, 1967. Sensory Inhibition. Princeton University Press, Princeton, New Jersey.

COHEN, L. B., R. D. KEYNES, and B. HILLE, 1968. Light scattering and birefringence changes during nerve activity. *Nature (London)* 218: 438–441.

FEIN, L., 1966. Biological investigations by information processing simulations. *In* Natural Automata and Useful Simulations (E. A. Edelsack, L. Fein, A. B. Callahan, and H. H. Patee, editors). Spartan, Washington, pp. 181–202.

FRAZIER, W. T., E. R. KANDEL, I. KUPFERMANN, R. WAZIRI, and R. E. COGGESHALL, 1967. Morphological and functional properties of identified neurons in the abdominal ganglion of *Aplysia californica*. *J. Neurophysiol.* 30: 1288–1351.

FRISHKOPF, L. S., R. R. CAPRANICA,-and M. H. GOLDSTEIN, JR., 1968. Neural coding in the bullfrog's auditory system a teleological approach. *Proc. Inst. Elec. Electron. Eng.* 56: 969–980.

FUORTES, M. G. F., and A. L. HODGKIN, 1964. Changes in time scale and sensitivity in the ommatidia of *Limulus*. *J. Physiol.* (*London*) 172: 239–263.

FURUKAWA, T., and E. J. FURSHPAN, 1963. Two inhibitory mechanisms in the Mauthner neurons of goldfish. *J. Neurophysiol.* 26: 140–176.

HARMON, L. D., and E. R. LEWIS, 1968. Neural modeling. *Advan. Biomed. Eng. Med. Phys.* 1: 119–241.

HARTLINE, D. K., and I. M. COOKE, 1969. Postsynaptic membrane response predicted from presynaptic input pattern in lobster cardiac ganglion. *Science* (*Washington*) 164: 1080–1082.

HAYES, T. L., and R. F. W. PEASE, 1969. The scanning electron microscope: A nonfocused, multi-informational image. *Ann. N. Y. Acad. Sci.* 157: 497–509.

HORRIDGE, G. A., 1968. Interneurons. W. H. Freeman and Co., San Francisco.

HUBEL, D. H., and T. N. WIESEL, 1962. Receptive fields, binocular interaction and functional architecture in the cat's visual cortex. *J. Physiol.* (*London*) 160: 106–154.

KALISCH, R. B., and J. B. ANGELL, 1968. Microprobes for medical electronics. *U. S. Air Force Office Aerospace Res., Res. Rev.* 7: 7.

KALMAN, R. E., 1968. New developments in systems theory relevant to biology. *In* Systems Theory and Biology (M. D. Mesarović, editor). Springer-Verlag, New York, pp. 222–232.

KENNEDY, D., A. I. SELVERSTON, and M. P. REMLER, 1969. Analysis of restricted neural networks. *Science* (*Washington*) 164: 1488–1496.

LETTVIN, J. Y., H. R. MATURANA, W. S. McCULLOCH, and W. H. PITTS, 1959. What the frog's eye tells the frog's brain. *Proc. Inst. Radio Eng.* 47: 1940–1951.

LEWIS, E. R., T. E. EVERHART, and Y. Y. ZEEVI, 1969a. Studying neural organization in *Aplysia* with the scanning electron microscope. *Science* (*Washington*) 165: 1140–1143.

LEWIS, E. R., Y. Y. ZEEVI, and F. S. WERBLIN, 1969b. Scanning electron microscopy of vertebrate visual receptors. *Brain Res.* 15: 559–562.

MAYERI, E. M., 1969. Integration in the lobster cardiac ganglion. Dissertation, Biophysics Group, University of California, Berkeley.

MESAROVIĆ, M. D., 1968. Systems theory and biology—view of a theoretician. *In* Systems Theory and Biology (M. D. Mesarović, editor). Springer-Verlag, New York, pp. 59–87.

PINTER, R. B., 1966. Sinusoidal and delta function responses of visual cells in the *Limulus* eye. *J. Gen. Physiol.* 49: 565–593.

RATLIFF, F., 1965. Mach Bands: Quantitative Studies on Neural Networks in the Retina. Holden-Day, San Francisco.

RATLIFF, F., H. K. HARTLINE, and D. LANGE, 1966. The dynamics of lateral inhibition in the compound eye of *Limulus*. I. *In* The Functional Organization of the Compound Eye (C. G. Bernhard, editor). Pergamon Press, New York, pp. 399–424.

REISS, R. F., 1962. A theory and simulation of rhythmic behavior due to reciprocal inhibition in small nerve nets. *Joint Computer Conf.* 21: 171–194.

ROEDER, K. D., 1963. Nerve Cells and Insect Behavior. Harvard University Press, Cambridge, Massachusetts.

ROEDER, K. D., and A. E. TREAT, 1961. The detection and evasion of bats by moths. *Amer. Sci.* 49: 135–148.

SHERRINGTON, C. S., 1961. The Integrative Action of the Nervous System. Yale University Press, New Haven, Connecticut. (Reissue of 1906 edition, C. Scribner's Sons, New York.)

STARK, L., 1968. Neurological Control Systems. Plenum Press, New York.

STRETTON, A. O. W., and E. A. KRAVITZ, 1968. Neuronal geometry: Determination with a technique of intracellular dye injection. *Science* (*Washington*) 162: 132–134.

TASAKI, I., A. WATANABE, R. SANDLIN, and L. CARNAY, 1968. Changes in fluorescence, turbidity, and birefringence associated with nerve excitation. *Proc. Nat. Acad. Sci. U. S. A.* 61: 883–888.

TROELSTRA, A., 1969. System identification with threshold measurements. *Inst. Elec. Electron. Eng. Trans. Syst. Sci. Cybernetics* 5: 313–321.

WERBLIN, F. S., and J. E. DOWLING, 1969. Organization of the retina of the mudpuppy, *Necturus maculosus*. II. Intracellular recording. *J. Neurophysiol.* 32: 339–355.

WILLOWS, A. O. D., 1968. Behavioral acts elicited by stimulation of single identifiable nerve cells. *In* Physiological and Biochemical Aspects of Nervous Integration (F. D. Carlson, editor). Prentice-Hall, Englewood Cliffs, New Jersey, pp. 217–243.

WILSON, D. M., 1966. Central nervous mechanisms for the generation of rhythmic behaviour in arthropods. *Symp. Soc. Exp. Biol.* 20: 199–228.

ZADEH, L. A., and C. A. DESOER, 1963. Linear System Theory. McGraw-Hill, New York.

# 38 Neural Operations in Arthropod Ganglia

## DONALD M. WILSON

ALTHOUGH IT IS difficult to prove the inheritance of behavior for any particular example, it is now generally conceded that many kinds of behavior, from simple to complex, are instinctive, or genetically programed, and that learning, if present in these cases, accomplishes a modification of hereditary instructions. Many adaptive changes in behavior in animals may, in fact, be merely examples of postembryonic development or maturation. Arthropods, especially insects that have dramatic metamorphoses of form and function of the nervous system in later life stages, would seem especially suitable for studies of development of behavior, but remarkably little attention has been given to this significant problem by arthropodologists. On the other hand, the neural networks that run whole-animal activities in the adult arthropod are probably closer to being completely understood than are networks in any other taxa. Hence, in this paper I deal with some of those presumably genetically programed subsystems that are fully developed and operating. I attempt to find in them some of the elementary operations that networks of neurons may be performing and that permute and combine to make up the functioning system.

Information, either genetic or learned, may be stored in the subcellular structures or network structures within the central nervous system. Because the morphology of sense organs determines modality and submodality of sensitivity, it also represents information storage. Peripheral interconnections between receptor and neural elements in the sense organs begin those filtering and abstracting functions that are necessary before the whole world of input can result in decisions appropriate to regulating output. Less often noted is the fact that all body form, even non-nervous aspects, is relevant to behavior, most obviously in the case of the muscular and skeletal systems. A motor command is effective and adaptive only when it is based on some knowledge of body form, just as a sensory input is correctly interpreted only if the central nervous system knows the labels of its inputs. To a large degree, at least, it appears that the central nervous system and the body periphery, including sense organs and motor machinery, follow parallel but independent courses of development. Through this development much of the information for behavior control is fixed in the

DONALD M. WILSON was Professor of Biology at Stanford University, Stanford, California, until his untimely death on June 23, 1970.

structure of either central nervous system or body periphery, or both.

Genetically encoded behavioral information can be read out in three functional types of nervous structure at the neural network level. One is the *sensory filter,* which I discuss only briefly. Networks, such as lateral inhibition networks in visual systems, can abstract special qualities of a stimulus pattern, such as motion or edge detection. Information stored in the network makes it selective of the pattern of input that can get through, but the network itself does not generate that pattern. *Pattern generators,* which are more characteristic of motor systems, may accept any input (or even be spontaneous) and produce a relatively stereotyped output, the pattern of which is not related to the input pattern. The same general network morphology can work either as a filter or as a pattern generator, depending on the parameters of the interactions between elements. Finally, closed-loop *reflexes* can incorporate aspects of peripheral body structure into preprogramed behavior control. The present paper deals principally with central pattern generation and reflex modulation of central commands.

### Central pattern generators

Coordinated motor output that controls behavioral acts can occur in the absence of specific sensory input. So many examples are known now that the list seems interminable. An emerging generality is that most simple behavioral acts, such as breathing, walking, flying, swimming, heartbeat control, singing in insects and birds, copulation, and so on, do not necessarily rely on special sensory input or sensory feedback for the patterning of their motor output. One of the first such cases to be demonstrated really clearly was that of flight control in grasshoppers (Wilson, 1961). In these animals 80 motor neurons drive the muscles of the four wings. Each neuron discharges in a pattern of brief bursts of impulses (zero-to-four) on each wingstroke. The 80 comprise four subsets in which nearly synchronous volleys occur. The subsets have different relative timing according to whether they drive upstroke or downstroke muscles of forewings or hind wings. The whole pattern of output with correct relative phasing can occur when the nerve cord is isolated from sensation and stimulated randomly (Figure 1).

In spite of intensive efforts to determine the coordinating mechanisms of this numerically simple system, we have as yet no idea how the output pattern is generated. We do

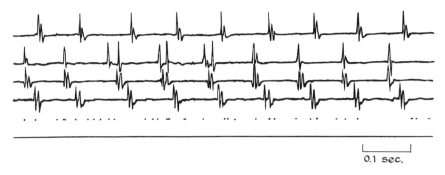

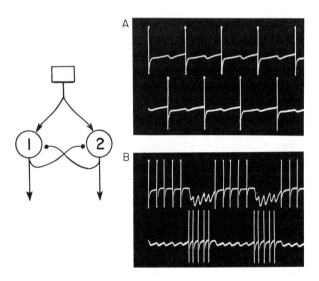

FIGURE 1 Random stimulation of the isolated grasshopper nerve cord elicits coordinated motor output such as that during flight. The CNS was totally deafferented and a Poisson train of the stimuli (bottom line) was delivered to the nerve cord. The four motor units from which recordings were made behave as in flight, except that the burst repetition rate is low. (Redrawn from Wilson and Wyman, 1965.)

know that there is not a single pacemaking center that merely drives all the motor neurons. The system can be surgically reduced to at least four subsystems, each of which is capable of producing the alternating pattern required by one wing. On the other hand, we believe that the known motor-neuron interactions are too weak to account for the pattern by themselves. Hence, some type of coupled system of multiple pacemakers is implicated.

For some even simpler examples we can provide models that are probably correct. The first suggested model for the production of alternating output between two cells or two cell populations was the *reciprocal inhibition network*. This model, which has a long history, is presented in Figure 2 and

its legend. Only one proved real-life example of this model is known. The two Mauthner neurons of fishes excite many motor neurons in the longitudinal body muscles, and the alternating sequence of Mauthner axon impulses produces vigorous swimming. Intracentral reciprocal inhibition could also produce the alternation of bursts of impulses in vertebrate respiratory neurons, in motor neurons used in vertebrate locomotion, in the grasshopper flight-control system, and so on, but as yet the mechanism is not proved for any of these.

Another model for the coordination of motor output is the *positive feedback network*. Groups of cells in the brain of the nudibranch *Tritonia* (Willows, 1967) have mutually excitatory synaptic coupling. Another such group of cells is found in the stomatogastric ganglion of decapod crustaceans (Maynard, 1966). Activity in any cell in one of these networks causes activation of others, and excitatory feedback results in a runaway burst of impulses. Fatigue or other self-limiting processes terminate the burst. After recovery, a new burst may begin, and the network may oscillate. Synergistic vertebrate respiratory neurons are thought to excite one another synaptically and, by that mechanism, produce the burst cycle during respiration, but the evidence for that is not strong.

*Single cells may have inherent impulse-burst capabilities* (Strumwasser, 1967), and network properties can then serve merely to coordinate the activities of individual elements. The heart ganglion of the mantis shrimp *Squilla* (Watanabe et al., 1967) contains several cells, each of which is capable of the normal output rhythm. They are tightly coupled electrotonically and, hence, always fire synchronously. The lobster cardiac ganglion contains four small driving cells and five follower motor neurons. There is mutual excitation among at least some of these cells, and it has been considered possible that the output bursts are caused by the positive feedback. Accumulating evidence suggests that the small cells may have endogenous burst capability, however, and

FIGURE 2 Reciprocal inhibition networks can achieve simple one-to-one alternation (A), as in the fish Mauthner cells, or may produce alternating bursts (B). These records are from electronic neuromimes. (Redrawn from Wilson and Waldron, 1968.)

their excitatory coupling may be as important for synchronization as for pattern production within the unit. Such multiple synergistic mechanisms could be common, and their analysis is especially difficult (Wilson and Waldron, 1968).

The last network mechanism for pattern production that I mention is that of the *lateral* or *recurrent inhibition network*. As is the case with the reciprocal inhibition mechanism, it has been demonstrated for only a single case of pattern generation—the coordination of motor-neuron output in certain flies. Each dorsal longitudinal flight muscle of flies is innervated by five or six motor neurons. In some calliphorids and drosophilids these neurons discharge in a phase-locked, ordered pattern at identical frequencies (Figure 3). Occasionally the pattern shifts abruptly to a new order (Wyman, 1966). These patterns were mimicked in computer-simulation experiments in which each model neuron inhibited every other one (Wilson, 1966). Subsequent physiological experimentation involving antidromic stimulation of single motor neurons has shown that each unit in the small pool does indeed inhibit every other one strongly (Mulloney, 1969a). Variations in the strength of the lateral inhibition account for some variations in output pattern in various insect species and even mutant races in *Drosophila* (Wilson, 1968a; Mulloney, 1969b; Levine, personal communication).

This pattern-generating, lateral-inhibition network is formally similar to the filtering network in the *Limulus* eye. It differs in that the inhibitory strength is much greater between the few units controlling one muscle. Weaker inhibition probably spreads to the motor neurons of other muscles, on which it influences the frequency but not the exact timing of discharge. The known recurrent inhibition between synergistic motor neurons in vertebrates (Thomas and Wilson, 1967) probably has only this weaker effect.

## Arthropod neurons

The best-known arthropod neurons are certain motor neurons and the giant fibers of crayfish. Recent work by Selverston and Kennedy (1969) and others using the procion yellow injection technique has provided an exact description of the fine anatomy of these cells, including all the dendritic branches visible in the light microscopice. An example of a motor neuron is shown in Figure 4A. The motor neurons that have been studied share several features, and incomplete studies suggest that these features are common to insect motor neurons as well. The cell bodies lie in the outer rind of the ganglion and lack synaptic contacts. The neurons are monopolar, sending a single process, often of rather small diameter, into the neuropil. There the process swells and connects to two or more large branches with unequal diameters. One of these leaves the ganglion either ipsilater-ally or contralaterally as the motor axon. Shorter branches either enter the fine neuropil, or *Punctsubstanz,* or are in contact with the large, intersegmental fibers of the longitudinal tracts, often including the giant fibers in crayfish. The cell bodies not only lack synaptic input, but most are also electrically passive; impulses often do not invade the soma. Furthermore, in smaller motor neurons in crayfish and in all those studied in insects, the cell body is electrically so distant from the large processes of the neuropil that no activity can be recorded from it, even when the axon is known to be active. Hence, our knowledge of electrical activities in these neurons derives mainly from recordings in the larger processes in the neuropil. There one may record a usual variety of excitatory or inhibitory synaptic potentials. Impulses vary in amplitude, suggesting that they arise from spike-initiating zones that are multiple or spatially labile, and that impulses do not necessarily even invade all regions of the main process. Hoyle (1970) has proposed that most of the intraganglionic portion of insect motor neurons is electrically inexcitable. Such is certainly not always true in crayfishes. Some synapses between giant fibers and motor neurons are electrical, with a large safety factor producing one-to-one transmission at moderate frequencies.

The lateral giant fibers of crayfish are more interestingly complex. Each has inputs and outputs in nearly every segmental ganglion. Each is in contact with the other via tight electrotonic junctions in each ganglion, and each has a cell body in every ganglion. In fact, each represents a sequential series of cells with a large axon coursing one segment anteriorly, where it makes electrotonic contact with the next (Figure 4B). By virtue of these connections the series of axons act as a unit.

The most complex neurons I know of in arthropods are intersegmental sensory interneurons found in the crayfish cord (Kennedy and Mellon, 1964). These fibers also receive inputs in several segments and can conduct impulses either anteriorly or posteriorly, depending on which sensory field is stimulated. If impulses arise nearly simultaneously from spatially widespread inputs, impulses may clash and cancel. Sequential, temporal stimulation of segmentally arrayed inputs can give rise to several impulses. Conduction delays and time and space features of the stimulus sequence determine the temporal spacing of the impulses recorded at either end of the fiber (Figure 5). An anterior-going stimulus sequence produces a shorter interspike-interval burst at the anterior end than at the posterior end, and vice versa. These fibers do complex integrative chores; however, until anatomical studies that may prove otherwise are completed, it remains possible that they, like the lateral giant fibers, really are series of tightly coupled intersegmental neurons and not single cells.

I wish to add one important comment about arthropod neurons. Only the larger cells have been studied in detail—

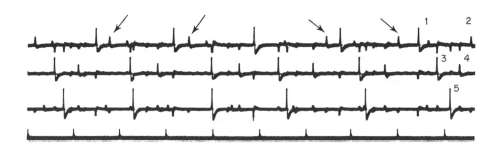

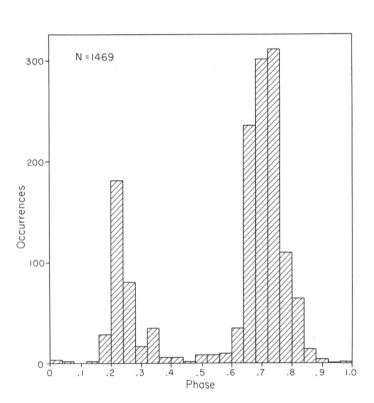

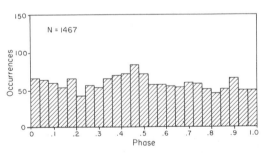

FIGURE 3  Multistable firing patterns in flight motor neurons of the fly. In the electromyogram, the upper two traces are from two electrodes in the same muscle. Several different motor units can be discerned in each trace. The four numbered units in the top traces could be identified throughout a long record. Phase relationships between selected units are illustrated in the top two histograms. The phases of unit 2 in the interval of unit 1 are bimodally distributed (left histogram). (Interval is the time between two impulses in the same channel. Phase is defined as the amount of time from an impulse in one channel to an impulse in a second channel, divided by the time from the same impulse in the first channel to the next impulse in that same channel, i.e., its interval.) In the short section of record in the upper trace, the two quasi-stable patterns are shown with a sudden transition between. The phases of unit 4 in unit 2 intervals have four modes (upper right histogram). Unit 5 (third trace), which is in another muscle, shows no preferred phase with respect to unit 1 (lower right histogram) or any other unit in that muscle. (Lower trace, 10 per second time marker.) (Redrawn from Wyman, 1966.)

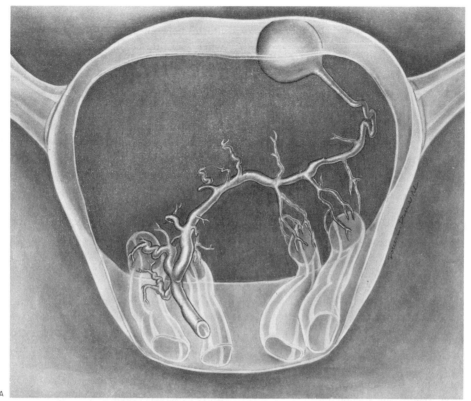

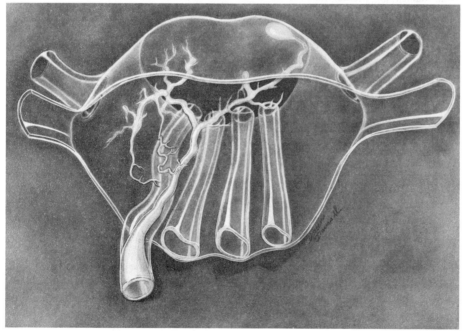

FIGURE 4 Reconstructions from procion yellow injections of the two crayfish neurons. A. A motor neuron, the giant motor fiber of the abdominal flexors, has a contralateral cell body, "dendrites" that contact each longitudinal giant fiber and others ending in other parts of the neuropil, and an axon which exits from the ganglion (not shown). B. Each lateral giant fiber is really a segmental series of neurons with a broad zone of contact in each segment. The soma is contralateral; most of the "dendrites" are presumed to be input processes. There are electrotonic tight junctions in the serial overlap zone, and also between the two lateral homologues in each segment at the point of decussation. (From Kennedy, Selverston, and Remler, 1969.)

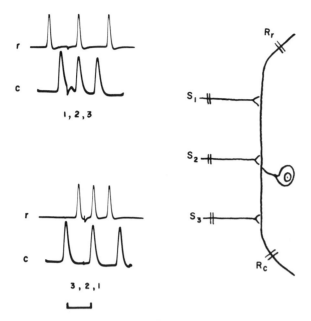

FIGURE 5 A complexly integrating crayfish interneuron with inputs in several segments. Recording sites at the rostral end (Rr) or caudal end (Rc) of the fiber reveal different apparent rates of activity, depending on whether the stimulus sequence was from front to back, $S_1$ $S_2$ $S_3$, or vice versa. (Redrawn from Kennedy and Mellon, 1964.)

the giant fibers, motor neurons, and large intersegmental interneurons. Certainly these account for less than 10 per cent of the neurons of even the lower segmental ganglia. The many small cells, most of which are probably short-axoned, may be comparable with vertebrate neurons such as Renshaw cells, granule cells, and so on. We have little physiological information on them. On the basis of studies of the large cells, we cannot construct a good model for any arthropod behavioral control systems as complex as those for walking or flight, so I expect that the little cells are necessarily involved. Unfortunately, at present there is little effort directed toward study of them. Yet, if we can extrapolate from vertebrate studies, such as those of Rall and Shepherd (1968) on the olfactory bulb, we can expect the small cells to be involved in such a basic subsystem operation as providing a mechanism for network oscillation and, possibly, doing that without even producing impulses themselves.

## Command fibers

Given that there are pattern-generating subsystems in the ganglia of arthropods, we may go on to ask, What turns them on and off, and how are they modulated for the purposes of orientation or steering? The answer will be more satisfactory than the preceding discussion. A few years ago, Wiersma and Ikeda (1964) used the term *command fiber* to

designate certain interneurons in crayfish that activate many motor neurons in complexly coordinated concert, even when the stimulation of the single fiber is a simple train of evenly spaced electrical shocks without phasic or patterning relationship to the output. Years earlier, Wiersma (1952) had first described such a fiber. When stimulated in the circumesophageal connective, this single fiber elicited the "defense posture," which involves probably more than half of the musculature of the whole crayfish body. The abdomen is arched and the legs are extended; the body rests on the tail and the last two or three pairs of legs; the chelipeds are outstretched and the claws opened—all as a result of the stimulating of one interneuron! Analogous cell types are known in insects and mollusks. For example, Willows (1967) found a neuron in *Tritonia* that elicited a long sequence of swimming movements involving much of the musculature of the animal when only a single impulse occurred in the stimulated cell.

The literature on command fibers is now extensive; these fibers may be divided roughly into two categories—those controlling posture and those driving oscillatory systems. Interactions between the two types and intermediate types undoubtedly occur, but I discuss them in these separate contexts. Rather than attribute each observation, I suggest the more general recent articles by Atwood and Wiersma (1967) and Evoy and Kennedy (1967).

The beat of the swimmerets in such decapod Crustacea as crayfish consists of a forward-running, metachronous rhythm of alternating protractions and retractions. It can occur in the absence of all sensory feedback and sometimes arises spontaneously. In quiescent preparations, stimulation of any one of a few long interneurons can activate the whole system and the output pattern is normal. There may be a long latency between the start of stimulation and the onset of output. This latency is partly a function of input frequency. Above the minimal frequency needed to activate the system, increases in input frequency have relatively little effect on the output, but there is a correlation between input and output frequencies. Several of the command fibers controlling the beat of the swimmerets appear to have identical effects on output; perhaps they are truly redundant. Others are differentiable in that they may give rise to a greater power of beat on one side or in different segments. It could be that the apparently redundant command fibers have undiscovered individual effects.

Each of the several ganglia that control swimmeret beat contains oscillatory drive mechanisms. Isolated ganglia can still work, but independently of one another. Because the command fibers do not set the phase of even one ganglionic oscillator, they cannot mediate the coordination of the several in the metachronal rhythm. Recently, Paul Stein (personal communication) has found a small set of fibers only a few segments long that normally fire in a phasic pattern related to that of an isolated ganglion oscillation. Stimulation of a bundle of these in an appropriate rhythm can set the phase of oscillation in a ganglion one or two segments away.

The command fibers controlling crayfish abdominal posture are highly diverse and no two appear to be identical. Some affect large sets of muscles (some up to 300 motor neurons); others, relatively few. Some effects are differential with respect to side or segment. Cinematic analysis during stimulation of single fibers shows cases in which abdominal extension may be principally in the anterior segments, be uniformly spread, or be principally posterior, and shows that rate of development of the commanded posture may vary with different identifiable fibers (Figure 6). Recordings from motor neurons, even within the small group that drives a single large muscle, indicate differentiation in connectivity of single interneurons of quite similar behavioral function. Each of these command fibers codes a particular posture by virtue of the set of its output connections, not its temporal pattern of activity. It is a labeled line in the same sense that sensory fibers are. Postural command fibers working together with those driving rhythmic systems may possibly steer locomotory activities.

One should, perhaps, also classify interneuronal giant fibers under the title command fibers. They often evoke motor acts involving much of the musculature. They are involved in well-known startle or escape reactions, e.g., the sudden "jump" of aquarium fishes when the glass is tapped, is mediated by their Mauthner axons. The giant fibers differ from the sorts of command fibers already mentioned in that there is a strong phasic relationship between impulses in the giant fibers and the resulting output. Wiersma's use of the term referred to cases of the type in which the effects of long trains of command-fiber impulses are summated in the next stage of neural integration and detailed phasic information in the command is, by and large, lost or filtered out.

Command fibers can be thought of as connecting neural links between sensory filters and pattern generators. Some higher-order sensory interneurons affect motor output (Kennedy, personal communication), but, in general, the question What turns on a command fiber? is still unanswered.

## Hierarchical organization in arthropod ganglia

Without presenting the relevant evidence, I assert that it is unlikely, in most cases, that interactions between motor neurons provide the basis for their patterned oscillatory behavior. Something like a hierarchically higher-level pacemaker or a pattern generator is called for. Hierarchical structure also is indicated by the fact that command fibers do not generally excite motor neurons monosynaptically, and some "sensory interneurons," when stimulated singly, produce no effects at all in motor neurons. We think of the interneurons of the lower arthropod nerve cord as being secondary or tertiary sensory neurons, pre-motor neurons, or intermediate between sensory and pre-motor. Whether there are definable levels in the intermediate category has not even been discussed.

Relevant to a discussion of the notion of structural hierarchy are comments made earlier on the lobster cardiac ganglion. The five large cells of that ganglion are motor neurons incapable of producing bursts of impulses by themselves. The four small cells are pre-motor interneurons, which, in concert or possibly singly, can produce impulse bursts. The latter drive the former. But the large motor neurons also excite the small pre-motor cells. This excitatory feedback is not very strong; hence, it is still possible to describe a principal order through the two-layered system, but the notion of structural hierarchy in nerve networks is made more complex by the possibility of feedback from lower to higher levels. Another complexity is introduced by the fact that inputs may bypass levels. For example, not all inputs to the crayfish abdominal motor neurons must work through

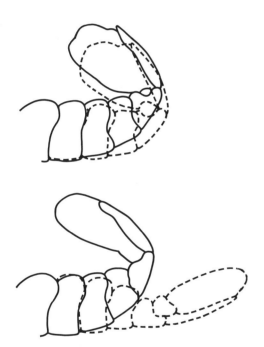

FIGURE 6 Command fibers for abdominal postural control in crayfishes are differentiated with respect to both rate of change of posture and final position. Depicted here are initial (solid) and final (dotted) positions of the abdomen, traced from cinematographic records at the beginning and at the end of stimulation of two different extension command fibers at 100 pulses per second. Top: A relatively rapid movement involving primarily the second and third joints (the more caudal joint angles change only a little). Below: A slower extension, involving all the caudal segments, produced by a second command fiber in the same preparation. (From Kennedy et al., 1967.)

the same number of levels of network hierarchy. The giant fibers make synaptic contact directly upon motor neurons.

Aside from the glaring exception that we have no knowledge of the basis of the oscillatory mechanism, observations like those above, complete in detail, are probably enough to explain the nonreflex behavior of the crayfish abdomen. We must also know what turns on the command fibers, of course. But, at the lower level, the present approach may be sufficient for a complete understanding of central mechanisms. A striking feature of these analyses is the degree of uniqueness or lack of redundancy at the cellular level. Motor neurons are constant in number, position, and fields of innervation. Particular interneurons, such as command fibers and intersegmental phasing fibers, can be located repeatedly in different animals in the same region of the cord cross section. In crayfish, the only demonstrated cases of nonconstancy at the whole-neuron level are a few "sports" identified by Florey (1966). In these individual animals one or two extra bilateral pairs of cells appear. Florey suggested that the supernumerary cells, plus the normal ones, are daughter cells of neuroblasts that have undergone one more division than usual. His electrophysiological studies show that the anomalous cells either share functions with the normal ones or, in one case, result in peripheral neuromuscular inhibition of a muscle that does not normally receive inhibition. However, again I must warn that all these studies are on larger neurons of the lower centers. There is some evidence that in invertebrates there may be natural neuronal death or cell addition in late stages of the life history. However, we do not know if changes in cell number take place within the behavioral control systems that are being studied. If they do, the optimism I expressed above—that present techniques may allow us to achieve complete understanding—may not be justified.

## Proprioceptive reflexes

The central pattern generators of arthropod segmental ganglia are turned on and off and steered by command-fiber inputs from higher centers. They are also influenced by proprioceptive feedback resulting from their own output actions. Proprioceptive feedbacks have diverse functions in different cases. One reflex regulates the power of the lobster swimmeret beat, but alone cannot determine its phase or frequency (Davis, 1969). Another affects intersegmental phase in insect locomotory activities (Weevers, 1966). I discuss in detail only one, perhaps rather special, reflex because it well illustrates some of the issues and operations that are the subject of Part VI of this volume.

In the hinge of each of the four wings of a grasshopper is a single-unit stretch receptor that fires one to a few impulses toward the end of each upstroke. If the four sensillae are destroyed, the central oscillator runs at about half normal speed, and flight power is so drastically reduced that the ani-

mal cannot actually fly. If three are damaged, the oscillator runs at about 75 per cent of normal frequency. With two stretch receptors intact, the frequency is within 10 per cent of normal, and with three it is in the normal range. In addition to the effect on oscillator frequency, the stretch receptors influence the output amplitude measured as the number of motor impulses per unit per wingstroke cycle. Ablation of stretch receptors affects power doubly through both oscillator frequency and amplitude. It does not matter which two or three sensillae are ablated. The effects on output are equal no matter what spatial combination is chosen; the two receptors of one side have no more effect on that side than on the other, and cutting those of one segment affects both segments equally. I recently examined these preparations with an eye to detecting even small amplitude differences associated with particular receptors, and could find none. The four stretch receptors have different peripheral addresses, but their central effects are pooled nonspecifically, at least in that their effects on output are not differentiable. Although their output effects are nonunique, however, they are not redundant. Two receptors are not so good as four; the system cannot adapt to the loss of more than two of the stretch receptors. The four stretch receptors contribute nonspecifically and nonlinearly to a tonic excitatory state of the flight-control system.

Another point (relevant to this section of the book) with respect to this reflex is that each stretch-receptor axon carries considerable information about wing position and velocity, but that information is not used by the CNS. Recording from them during flight, the experimenter can estimate wingbeat frequency, phase, and amplitude (Figure 7). This information is carried in the number of impulses per burst, the time of onset of bursts, and the bursting frequency. Each of the four carries similar information. But this detailed information is apparently not used by the grasshopper. If the stretch receptors are cut and stimulated electrically, according to other than normal patterns, the oscillator frequency and amplitude are normal as long as the frequency of input is normal, averaged over several output cycles. The ganglion does not analyze for phasic information in the input pattern but only sums total input from the four receptors over averaging times that obliterate fine temporal detail. A fair physical analogy is a long, time-constant, resistance-capacitance network that integrates input pulses into a more or less ripple-free steady voltage, which is a function of over-all input frequency. The filtering that occurs in this grasshopper reflex discards much information about the wingbeat that we might find interesting and passes only that which is necessary to maintain a proper degree of excitatory state in the oscillating central system. The stretch receptors have the nature of *Stimulationsorganen*, a term coined by German authors to designate receptors that contribute to central tone rather than provide specific information. This reflex pro-

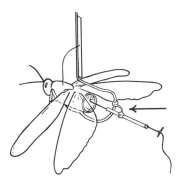

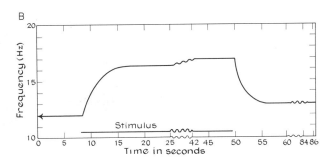

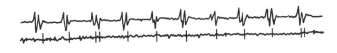

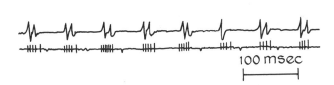

100 msec

FIGURE 7 Stretch receptor of the grasshopper wing hinge. A. Firing of a mesothoracic stretch receptor during tethered flight. The upper trace in each pair is a record of activity of the first basalar muscle; the lower trace, a stretch receptor. At lower wingbeat frequency the stretch receptor fires more impulses per cycle. (From Wilson and Gettrup, 1963.) B. Output frequency of a deafferented grasshopper preparation before, during, and after a long bout of stimulation of two stretch receptors. (Redrawn from Wilson and Wyman, 1965.)

vides a good example for stressing the point that in assigning a code or in assessing information-content in a signal, one should (if he is concerned about biological relevance) pay attention to the natural reader of that code.

## Uniqueness, but nonnecessity

If one motor unit, one muscle, or even the whole set of muscles that drive one wing are incapacitated, a grasshopper can still fly. Such ability may not surprise a naturalist, but, given that the flight machinery is driven by a central pattern generator that has no need for special patterns of input, it is not obvious that it could accommodate for changes in the efficacy of its output. Indeed, if a grasshopper is flown when tethered (fixed in space) it does not change its pattern of motor output when muscles are cut or wings ablated. Removal of an entire hind wing, which should give rise to a threefold to fourfold power asymmetry between the two sides, does not result in adaptive changes in motor output as long as the animal is fixed so that it cannot roll or yaw. Yet, when the same animal is thrown in the air, it can fly straight and level. Exteroceptive rather than proprioceptive feedbacks are utilized in regulating flight path and stability. In the tethered animal, these are ineffective because the animal cannot turn, and there is no error signal even though the preprogramed central pattern generator is not appropriate to the deranged peripheral anatomy. Two kinds of exteroception are used in free flight; one is visual and the other is a wind-direction sense mediated by hair sensillae on the head. If an animal that has had asymmetrical muscle or wing abla-

tion is flown in the dark (infrared light techniques allow observation), with the head hairs cemented over, it flies in a curved path, the radius of which is influenced by the degree of surgical damage. Proprioception is insufficient to correct the turn. If either vision or wind-direction sense is intact, the animal can adapt completely to major damage.

The role of the visual input alone can be studied in a special tethering apparatus that allows roll only (Figure 8). Slow rolling in this apparatus does not give rise to a wind-direction error signal because the body axis remains constantly aligned with the direction of the wind stream. Electrodes can be placed in the muscles of the animal in this apparatus and the motor-output pattern recorded during rolling behavior. In the dark, animals do not orient relative to the roll axis. They may remain stationary in any position, in which case the motor output to bilaterally homologous muscles is symmetrical; or they may roll continuously, owing to asymmetrical output. With lights on, especially if there is a simulated horizon, they fly stably with the dorsum toward the light or the dorsoventral axis at right angles with respect to the plane of the horizon. If muscles or wings are cut, all animals roll continuously in the dark in the expected direction, because there is no change in motor output. There are drastic and appropriate motor output changes in the light. The animal may assume a completely normal orientation if the surgical damage is not too severe, or hold a stable position which deviates, in the dark, in the direction of the error; or, when the damage is excessive, roll continuously, but unevenly, with a nearly stable point in the cycle. The visual control-of-roll reflex has the properties of a *proportional error*

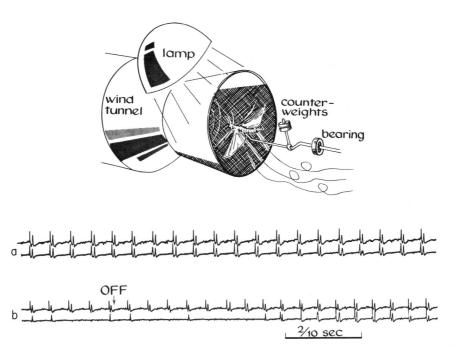

FIGURE 8 Apparatus for recording motor output during rolling flight shows that the natural output may be symmetrical in the dark even when extensive peripheral damage exists (a). In (b) the light had been on and the output suitably modified to prevent rolling until the point marked *off*, when the output again became symmetrical and the animal began to roll again. (From Wilson, 1968b.)

*servomechanism*. This servomechanism could not be discovered with the older recording techniques, in which muscle potentials were monitored but which did not allow an overt roll error during flight.

Grasshopper flight control, like many other invertebrate behavioral control systems, consists of a central pattern generator that can be turned on by unpatterned input from command channels. The output pattern is complete in the absence of feedbacks, but proprioceptive feedbacks modify the pattern to control wingbeat frequency (and pitch), and exteroceptive feedbacks produce motor-output asymmetries which can correct surgically-induced turning tendencies.

During the course of study of these exteroceptive controls, I found that many grasshoppers have long-term, naturally asymmetrical output and always roll in the dark. The central pattern generator is imperfect, and feedback modulation is necessary in order to permit them to fly well at all. The feedback mechanisms may have evolved not only to correct damage to peripheral body structures, but also to correct inherent central error. Such errors might arise through developmental accidents. Current studies on *Drosophila, Oncopeltus,* and *Tribollium* suggest that the turning tendency "error" may also in some cases be genetic, because stocks of animals with different degrees of bias may be bred. These selected animals are not incapacitated in a lighted environment, but in the dark may circle compulsively with a radius of only a few body lengths.

Reflex mechanisms by themselves do not explain simple arthropod behavior; central programs, by themselves, are also inadequate. The interaction of the two types of control system produces the coupling between CNS and body periphery and environment that is necessary for adaptation to genetic or developmental error, accidental damage, varying load, or unpredictable environment. A result of the superstructure of correcting feedbacks that modulate central programs is that the whole behavioral control systems have *fail-safe* characteristics. One may remove a muscle or a wing, a stretch receptor, the eyes, the wind receptors, or any one of the other known input sources of the grasshopper flight system without preventing stable flight. Only if too many parts of the system are damaged at once is the behavior impaired. Although every part is unique and nonredundant and is repeated from animal to animal, no one part is necessary.

## Muscle control and the need for proprioception

Another inadequacy of purely central programing of motor output has come to my attention through recent studies on the neural control of muscle. If a skeletal muscle of a vertebrate or an arthropod is excited by a sinusoidally frequency-modulated pulse train, its length under a variety of load conditions varies hysteretically (Figure 9). During the decreasing-frequency half of the modulation cycle, the muscle holds the maximum position achieved until the stimulus frequency drops to relatively low values. The degree of hysteresis is large even when the modulation period is many times the relaxation time for the muscle. If, during the decreasing-frequency half of the cycle, extra load is applied briefly to the muscle, it will reset to a longer length, which it then holds until it re-enters the normal loop (Figure 9). During stimulus trains of decreasing frequency, the muscle can hold any length within the area bounded by the hysteresis loop, depending on load variation, which means that a central motor-command system cannot be precalibrated unless all variations in load can be predicted. In order to command the length of a single muscle, the CNS must have available devices that monitor the effects of its own output if unpredictable variations in load are to be accommodated.

Other complications in the relationship between motor output and movement have been described by Hoyle (1964). He recorded motor output to the leg muscles of freely locomoting grasshoppers and found that, even during apparently identical series of leg steps, the output patterns were highly diverse. He could not find a stereotyped output pattern associated with the stereotyped motion. Sometimes flexors and extensors alternated, as expected, but at other times one muscle remained tonically active while the other fired in bursts. The leg oscillated as the phasically activated muscle pulled against the other as if it were a spring. Per-

haps several or many output patterns can produce the same behavior. We can think about the possibility that centrally patterned motor outputs command temporal sequences of muscle *lengths,* or limb positions, and that load variations cause proprioceptive feedbacks which indicate mismatch between command and achievement. If such a comparison of goal and realization occurs, subsequent motor output may have more the nature of an error signal than of primary length command.

## Central control of the sequencing of complex behavior

Finally, I wish to discuss one part of a study by Loher and Huber (1966) on the central programing of courtship behavior in the grasshopper *Gomphocerus rufus.* The importance of this example is that it provides a possible way of thinking about how neural subsystems are coupled into more complex programs of behavior. The mature male grasshopper orients in front of a female and performs a sequence of movements. These consist of three phases or subroutines. The first consists of three to four seconds of head-shaking and small-amplitude hind-leg movements, starting at one to two cycles per second, and increasing in frequency to four to six per second at the end of this phase. During the second phase, the antennae are thrown back and then brought forward again and the hind legs kick once or twice,

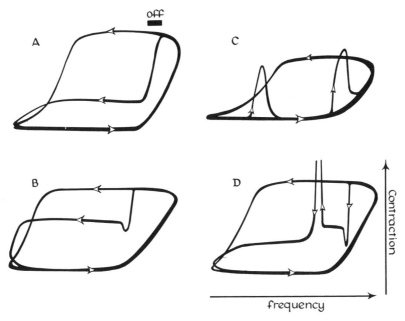

FIGURE 9   Hysteresis loops for isotonic contractions of an insect muscle stimulated by an impulse train modulated from 10 to 100 pps. at 0.1 Hz. (The stimulus frequency axis [abscissa] was driven by the stimulus-modulating sine wave.) A. A brief interruption of the stimulus train results in rapid relaxation until stimulation resumes. B. A similar curve results when the muscle is lengthened by an external force during continuous stimulation. C. An externally imposed shortening of the muscle does not cause resetting of its length. D. A length reset caused by passive elongation cannot be followed by a reset to shorter length, although the muscle could hold a shorter length under the identical stimulus history. (From Wilson and Larimer, 1968.)

producing sharp, stridulatory sounds. Phase two lasts one second. Phase three begins with one-half second of silence followed by four to five seconds of prolonged stridulatory movements. The whole sequence may be repeated as many as 50 times. Visual cues (sight of female) or auditory ones (female song) can elicit the male behavior. Either is sufficient. A blind or deaf male can be stimulated to court. In fact, sexually deprived males may court *in vacuo*. Input may trigger the behavior, but it does not pattern behavior further, nor is special input even necessary. Once triggered, each sequence runs to its end in spite of interference with output or feedback. If the hind legs are fixed in any position, or the nerve cord is cut in front of the ganglion that innervates them, stridulation is impossible. The male courts silently, but according to the time schedule of an intact animal. Phase one begins every eight to nine seconds, even though there is no output and, hence, no feedback, during phase three. It is as if the nervous system contains a tape-recorded program, which when turned on plays over and over, calling up subroutines, but not monitoring output and, therefore, ignoring output malfunction. This example is not analyzed in cellular detail, and it may not even be very rigorously interpreted, but it suggests what may be an important principle in arthropod behavior control, namely, that local subsystems drive the movements of individual segments and limbs and that complex sequences are programed by higher-level timers which activate the local systems in sequence.

## Summary

Neural subsystems controlling arthropod behavior include, at the nerve network level, both *sensory filters* and *pattern-generating networks*.

These may be connected by *command interneurons*.

Filters, interneuronal links, and pattern generators are arranged in a *hierarchy* of an unknown number of levels. Probably signals can bypass some levels of that hierarchy.

By and large, each arthropod neuron that has been studied is unique, but is repeated from animal to animal. There seems to be little or no *redundancy*. However, in general, no single cell is necessary, because feedback controls assure adaptive function even when central control mechanisms are damaged or inherently deficient.

*Proprioceptive* and *exteroceptive feedback loops* are utilized to provide the central mechanisms with information about the effect of their own output as well as about body periphery or environment.

## Acknowledgment

Recent work by the author was supported by Grant NB 07631 from the U.S.P.H.S.

## REFERENCES

ATWOOD, H. L., and C. A. G. WIERSMA, 1967. Command interneurons in the crayfish central nervous system. *J. Exp. Biol.* 46: 249–261.

DAVIS, W. J., 1969. Reflex organization in the swimmeret system of the lobster. I. Intrasegmental reflexes. *J. Exp. Biol.* (in press).

EVOY, W. H., and D. KENNEDY, 1967. The central nervous organization underlying control of antagonistic muscles in the crayfish. I. Types of command fibers. *J. Exp. Zool.* 165: 223–238.

FLOREY, E., 1966. Anomalous cardioregulator innervation in certain individuals of the crayfish *Pacifastacus leniusculus*. *J. Exp. Zool.* 163: 93–97.

HOYLE, G., 1964. Exploration of neuronal mechanisms underlying behavior in insects. *In* Neural Theory and Modeling (R. F. Reiss, editor). Stanford University Press, Stanford, California, pp. 346–376.

HOYLE, G., 1970. Cellular mechanisms underlying ethology. Neuroethology. *Advan Insect Physiol.* (in press).

KENNEDY, D., W. H. EVOY, B. DANE, and J. T. HANAWALT, 1967. The central nervous organization underlying control of antagonistic muscles in the crayfish. II. Coding of position by command fibers. *J. Exp. Zool.* 165: 239–248.

KENNEDY, D., and DeF. MELLON, JR., 1964. Receptive-field organization and response patterns in neurons with spatially distributed input. *In* Neural Theory and Modeling (R. F. Reiss, editor). Stanford University Press, Stanford, California, pp. 400–413.

KENNEDY, D., A. I. SELVERSTON, and M. P. REMLER, 1969. Analysis of restricted neural networks. *Science* (*Washington*) 164: 1488–1496.

LOHER, W., and F. HUBER, 1966. Nervous and endocrine control of sexual behaviour in a grasshopper (*Gomphocerus rufus* L., Acridinae). *Symp. Soc. Exp. Biol.* 20: 381–400.

MAYNARD, D. M., 1966. Integration in crustacean ganglia. *Symp. Soc. Exp. Biol.* 20: 111–149.

MULLONEY, B. C., 1969a. Organization of the flight motor neurons of Diptera. *J. Neurophysiol.* 33: 86–95.

MULLONEY, B. C., 1969b. Impulse patterns in the flight motor neurons of *Bombus californicus* and *Oncopeltus fasciatus*. *J. Exp. Biol.* (in press).

RALL, W., and G. M. SHEPHERD, 1968. Theoretical reconstruction of field potentials and dendrodendritic synaptic interactions in olfactory bulb. *J. Neurophysiol.* 31: 884–915.

SELVERSTON, A. I., and D. KENNEDY, 1969. Structure and function of identified nerve cells in the crayfish. *Endeavour* 28: 107–113.

STRUMWASSER, F., 1967. Types of information stored in single neurons. *In* Invertebrate Nervous Systems (C. A. G. Wiersma, editor). The University of Chicago Press, Chicago, pp. 291–319.

THOMAS, R. C., and V. J. WILSON, 1967. Recurrent interactions between motoneurons of known location in the cervical cord of the cat. *J. Neurophysiol.* 30: 661–674.

WATANABE, A., S. OBARA, and T. AKIYAMA, 1967. Pacemaker potentials for the periodic burst discharge in the heart ganglion of a stomatopod, *Squilla oratoria*. *J. Gen. Physiol.* 50: 839–862.

WEEVERS, R. DE G., 1966. The physiology of a lepidopteran muscle receptor. III. The stretch reflex. *J. Exp. Biol.* 45: 229–249.

WIERSMA, C. A. G., 1952. Neurons of arthropods. *Cold Spring Harbor Symp. Quant. Biol.* 17: 155–163.

Wiersma, C. A. G., and K. Ikeda, 1964. Interneurons commanding swimmeret movements in the crayfish, *Procambarus clarki* (Girard). *Comp. Biochem. Physiol.* 12: 509–525.

Willows, A. O. D., 1967. Behavioral acts elicited by stimulation of single, identifiable brain cells. *Science* (*Washington*) 157: 570–574.

Wilson, D. M., 1961. The central nervous control of flight in a locust. *J. Exp. Biol.* 38: 471–490.

Wilson, D. M., 1966. Central nervous mechanisms for the generation of rhythmic behaviour in arthropods. *Symp. Soc. Exp. Biol.* 20: 199–228.

Wilson, D. M., 1968a. The nervous control of insect flight and related behavior. *Advan. Insect Physiol.* 5: 289–338.

Wilson, D. M., 1968b. Inherent asymmetry and reflex modulation of the locust flight motor pattern. *J. Exp. Biol.* 48: 631–641.

Wilson, D. M., and E. Gettrup, 1963. A stretch reflex controlling wingbeat frequency in grasshoppers. *J. Exp. Biol.* 40: 171–185.

Wilson, D. M., and J. L. Larimer, 1968. The catch property of ordinary muscle. *Proc. Nat. Acad. Sci. U. S. A.* 61: 909–916.

Wilson, D. M., and I. Waldron, 1968. Models for the generation of the motor output pattern in flying locusts. *Proc. Inst. Elec. Electron. Engrs.* 56: 1058–1064.

Wilson, D. M., and R. J. Wyman, 1965. Motor output patterns during random and rhythmic stimulation of locust thoracic ganglia. *Biophys. J.* 5: 121–143.

Wyman, R. J., 1966. Multistable firing patterns among several neurons. *J. Neurophysiol.* 29: 807–833.

# 39 Neuronal Operations in Cerebellar Transactions

## R. LLINÁS

Any serious attempt to discuss general aspects of cerebellar function must make clear at the outset that, although the phylogeny, ontogeny, neuroanatomy, and electrophysiology of the cerebellar cortex are better understood than any other region of the central nervous system, no clear formulation has as yet been attained regarding its over-all function. Before entering into the actual complexities of the neuronal interactions in the cerebellar cortex, it is appropriate to describe briefly some of the morphological and electrophysiological details that we shall consider throughout this paper.

### Morphological organization

The "Basic Cerebellar Circuit," a Typical Simple Circuit Organization  Following the study of the morphological and functional characteristics of the cerebellar cortex in different vertebrates, a working hypothesis was advanced regarding the existence of a "common denominator" for the neuronal organization of the cerebellar cortex throughout phylogeny. The hypothesis assumes that the cerebellar cortex of all vertebrates is basically organized around Purkinje cells (the only output system from the

Rodolfo Llinás  Institute for Biomedical Research, Chicago, Illinois. *Present Address:* Division of Neurobiology, Department of Physiology and Biophysics, University of Iowa, Oakdale Campus, Iowa City, Iowa

cerebellar cortex), which receive two afferent systems—the climbing-fiber and the mossy-fiber input. The superimposed interneuronal systems then represent a specialization of this basic pattern of organization (Llinás, 1969b; Llinás et al., 1969b). (The term "interneuron" is used here to refer to "those nerve cells whose axons are distributed in the vicinity of their place of origin.")

*Climbing fibers*  The climbing-fiber system, as first described by Ramón y Cajal (1888b), is a monosynaptic input which terminates in direct contact with the soma and dendrites of the Purkinje cells. The actual junction takes place via small spines which protrude from the main dendrite of the cell (Larramendi and Victor, 1967; Uchizono, 1967, 1969; Hillman, 1969a, 1969b; Sotelo, 1969) (Figure 1). These fibers, which branch profusely and twine in creeper-like fashion around the main dendrite of the Purkinje cell, establish a large number of synaptic contacts with the dendritic spines. It has been calculated that in the frog there are approximately 300 contacts between a climbing fiber and a Purkinje cell (Llinás et al., 1969a), and in the alligator the number is approximately 170 (Hillman, 1969b). Such an order of magnitude seems to be the norm for this input in different vertebrates (Hillman, 1969b). Light- and electron-microscope characteristics of the synapses are quite similar in different terrestrial vertebrates (Hillman, 1969b; Llinás and Hillman, 1969) and in fish (Houser, 1901; Kaiserman-Abramof and Palay, 1969; Nicholson et al., 1969; Nieuwenhuys and Nicholson, 1969; Schnitzlein and Faucette,

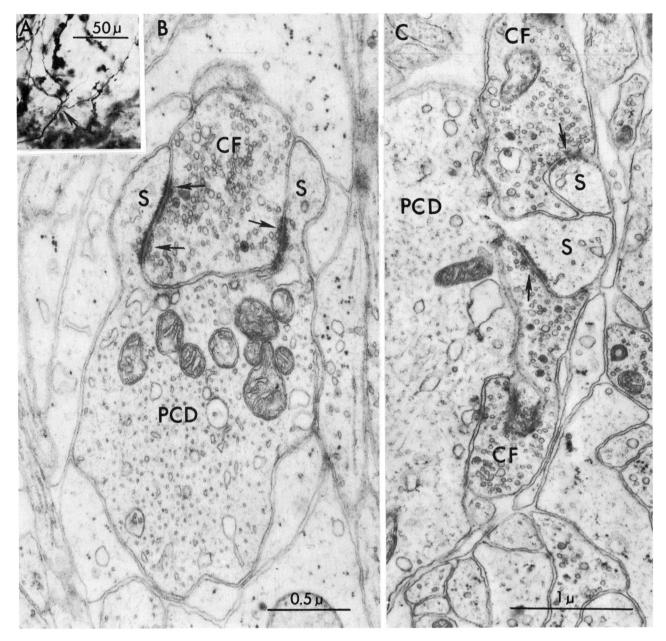

FIGURE 1 Light and electron micrographs illustrating the relation of the climbing fiber and the Purkinje cell in the frog *Rana catesbeiana*. A. Golgi stain of a climbing fiber in the molecular layer of the cerebellar cortex. The fiber trifurcates (arrow) and has the same orientation as main dendrites of Purkinje cells. B and C. Electron micrographs of the synapse of the climbing fiber and the Purkinje cell. In B, the Purkinje-cell dendrite (PCD) has been sectioned transversely to show the close relationship between the ascending climbing fiber (CF) and the dendrite of the Purkinje cell. Note that the "synaptic contacts" (arrows) take place at dendritic spines (S). In C, a longitudinal section of a Purkinje-cell dendrite (PCD). A climbing fiber (CF) is seen to follow the dendritic tree longitudinally. As in B, the synaptic contact occurs at the dendritic spines (arrow). (Unpublished results, courtesy of D. E. Hillman.)

1969). Therefore it can be stated that the climbing-fiber input seems to have been modified little throughout the evolution of the central nervous system. Ontogenetically, the climbing fiber is the first input to be in contact with the Purkinje cell (Ramón y Cajal, 1904; Scheibel and Scheibel, 1964; Larramendi, 1965, 1969; Mugnaini, 1966, 1969; Kornguth et al., 1968). This first contact occurs at the level of the Purkinje-cell soma where the well-known "capu-

"chon" arrangement described by Ramón y Cajal (1904) is the typical form of junction between the climbing fiber and its Purkinje cell.

*Mossy fibers* The mossy-fiber input is also present throughout phylogeny (Ramón y Cajal, 1904; Ariëns Kappers et al., 1936; Mugnaini, 1969; Llinás et al., 1969b). In all cases these fibers terminate in contact with the dendrites of the granule cells in the granular layer of the cerebellar cortex. The character of this synapse is glomerular, i.e., a single afferent mossy fiber produces a saclike expansion, with digits that enter in contact with a large number of granule-cell dendrites (Hámori and Szentágothai, 1966; Fox et al., 1967) (Figure 2). This type of contact is complicated in higher vertebrates but is simple in lower vertebrates, as can be seen in frogs (Hillman, 1969a, 1969b; Sotelo, 1969), alligators (Hillman, 1969b), or fish (Nieuwenhuys and Nicholson, 1969b; Kaiserman-Abramof and Palay, 1969; Nicholson et al., 1969).

The granule cell, the second link in the mossy-fiber input to the Purkinje cells, produces throughout phylogeny an ascending axon, which divides in a T fashion (Figure 3A) and establishes contact with the dendrites of Purkinje cells by means of small spines situated in the so-called spiny branchlets (Ramón y Cajal, 1904) (Figure 3B). The electron-microscopical characteristics of this "crossing-over" synapse (Hámori and Szentágothai, 1964; Fox et al., 1964; Uchizono, 1967) are once again quite similar throughout cerebellar phylogeny (Llinás et al., 1969b). We have, therefore, a system that seems to be basically similar in most vertebrates, as shown at both the light- and the electron-microscope levels.

As for the development of the mossy-fiber system, these fibers reach the granule layer and establish contact with the dendrites of the granule cells once those cells have migrated down the molecular layer to the granular layer. Prior to this migration, the granule neuroblast emits two axonic prolongations at opposite poles of its soma. These form the parallel fibers as the soma descends to the granular layer, generating the ascending portion of the granule-cell axon (Ramón y Cajal, 1904).

The chronological development of synaptic junctions in this system is such that the parallel fiber-Purkinje cell synapse is the first to be formed, whereas the mossy-fiber and granule-cell junction develop later. It appears, therefore, that the ontogenetic organization of the synaptic links along this pathway has a reverse direction from that of nerve propagation in the adult form (Larramendi, 1969).

*Intrinsic neurons of the cerebellar cortex* Superimposed on this "basic cerebellar circuit," and with varying degrees of complexity as one goes up the phylogenetic scale, are the cerebellar interneurons; the two main groups are the basket and stellate cells of the molecular layer and the Golgi cells of the granular layer. The synaptic action exerted by these neurons on their target cells has been shown to be inhibitory and is discussed in detail below.

Morphologically the interneurons of the molecular layer (the stellate and basket cells) are basically a homogeneous group, with the basket and small stellate cells representing the limits of the continuum. Characteristically, these interneurons receive their most voluminous input from parallel-fiber synapses in their dendrites and soma, and their axons terminate, for the most part, on the dendrites and soma of the Purkinje cells. Those neurons whose axons terminate around the lower part of the Purkinje-cell soma and initial segment are the so-called "basket cells" (Ramón y Cajal, 1888a), while those terminating mostly on the dendrites are the "stellate cells" (Ramón y Cajal, 1904). It must be clear, however, that both types of cells, basket and stellate, have a large number of terminals on Purkinje-cell dendrites. A second important point regarding their morphology is that their axons run transversely at right angles to the parallel fibers (Ramón y Cajal, 1904; Hámori and Szentágothai, 1965; Fox et al., 1967).

In addition to parallel fibers, these cells are also known to receive terminals from other interneurons and from collaterals of the climbing fibers (Scheibel and Scheibel, 1954; Lemkey-Johnston and Larramendi, 1968a, 1968b) and the lower basket cells also receive synaptic inputs from axon collaterals of Purkinje cells (Hámori and Szentágothai, 1968; Lemkey-Johnston and Larramendi, 1968b).

The Golgi cells seem to receive inputs from both parallel fibers (Ramón y Cajal, 1904; Hámori and Szentágothai, 1966; Fox et al., 1967) and mossy fibers (Hámori and Szentágothai, 1966; Uchizono, 1967; Hillman, 1969b; Mugnaini, 1969; Nicholson et al., 1969). From the phylogenetic point of view, it seems likely that the Golgi cells were primarily granular-layer interneurons, the main input of which was the mossy fibers (Llinás and Hillman, 1969), and that only later did the dendritic arbor reach the level of the molecular layer. The terminals of the Golgi cell spread throughout the granular layer and terminate on the dendritic digits of the granule cells, forming a ringlike arrangement which surrounds the glomerulus (Ramón y Cajal, 1904). Other afferents terminating in contact with the bodies of the Golgi cells are the recurrent collaterals of Purkinje cells (Ramón y Cajal, 1912; Hámori and Szentágothai, 1966; Fox et al., 1967; Lemkey-Johnston and Larramendi, 1968b), as well as the collaterals of the climbing fibers (Scheibel and Scheibel, 1954; Hámori and Szentágothai, 1966; Lemkey-Johnston and Larramendi, 1968b).

The ontogeny of the interneuronal system is now fairly clear. As first hypothesized by Ramón y Cajal (1904), the Golgi cells arise from the mesencephalic rhombic lip and migrate to the cerebellar anlagen very much as do the Purkinje cells, although two or three days later in time (Fujita et al., 1966; Fujita, 1969). The basket and large and small

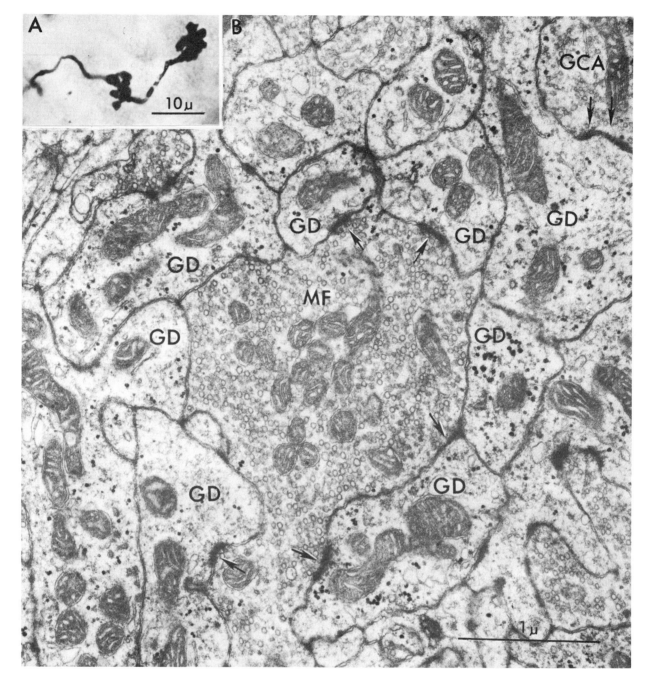

FIGURE 2 Light and electron micrographs of mossy fibers and ultrastructural organization of cerebellar glomerulus in *Caiman sclerops*. A: Golgi stain of a mossy fiber rosette at the cerebellar granular layer. B: Electron micrograph of a similar area, showing the mossy-fiber terminal (MF). The mossy fiber sac is seen to establish multiple synaptic junctions (arrows) with the dendrites of the granule cells (GD), which surround the mossy-fiber terminal. Other smaller areas of synaptic contact between mossy fiber and granule-cell dendrites are shown in the upper left and lower right corners of the micrograph. The upper right corner shows a synaptic relationship between a Golgi-cell axon (GCA) and a granule-cell dendrite (two arrows). Note the difference in size and shape between the synaptic vesicles in the mossy fiber and in the Golgi cell terminals. (Unpublished results, courtesy of D. E. Hillman.)

stellate cells, on the other hand, seem to arise from the external granular layer, which is formed by cell migration from the posterior tip of the roof of the fourth ventricle (Schaper, 1894). The synapses that these cells establish with the Purkinje cells indicate that the dendrites of granule cells are the last to form in the development of the cerebellar cortex.

The "basic cerebellar circuit" that is described above was postulated because the interneurons of the cerebellar cortex appear to be the most variable elements in this system (cf. Nieuwenhuys and Nicholson, 1969b) and because there are fewer of them as one goes down the phylogenetic scale. The rather tardy organization of the interneuronal system in ontogeny strengthens this viewpoint (Larramendi, 1969).

## Functional organization

FUNCTIONAL PROPERTIES OF THE CLIMBING FIBER-PURKINJE CELL SYSTEM   The climbing fiber was first shown to be an excitatory input to the Purkinje cell in the cerebellum of the cat (Eccles et al., 1966a). In all animals we have so far studied (Llinás et al., 1969b; Llinás and Hillman, 1969), climbing-fiber activation is always characterized by its ability to generate burstlike responses on Purkinje cells (Figure 4). These responses, which can be recorded intracellularly or extracellularly, are generated by a large, all-or-none, excitatory postsynaptic potential (EPSP) which reflects the all-or-none nature of the action potential in the climbing fiber itself (Eccles et al., 1966a; Llinás and Nichol-

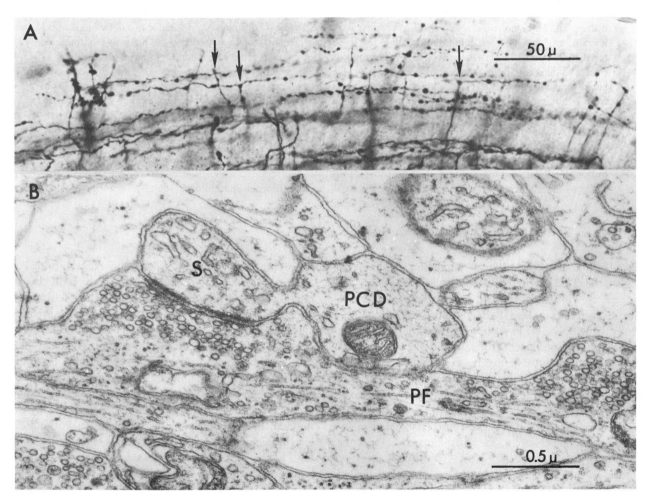

FIGURE 3   Light and electron micrographs of parallel fibers in the cerebellar molecular layer of *Rana catesbeiana*. A: Golgi stain of parallel fibers in the molecular layer. The parallel fibers are formed by a T-shaped bifurcation of the ascending axons of the granule cells (arrows), the parallel fibers having a beaded appearance. As shown in B, the beads are the sites of synaptic contact between the parallel fibers (PF) and the spines (S) of Purkinje-cell dendrites (PCD). This type of junction has been referred to as a "crossing-over synapse" (Hámori and Szentágothai, 1964). (Unpublished observations, courtesy of D. E. Hillman.)

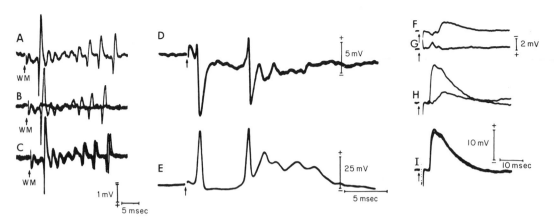

FIGURE 4 Intracellular and extracellular recordings from Purkinje cells after climbing fiber activation (*Rana catesbeiana*). A, B, and C. Extracellular recordings illustrating the burst-type activation of the Purkinje cell evoked by electrical stimulation of the underlying cerebellar white matter. The all-or-none character of this spike burst is illustrated in B, in which the stimulus is straddling threshold for the climbing-fiber activation. C. Three superimposed sweeps to demonstrate the stereotyped nature of the spike burst. D and E. Intracellular and extracellular recordings from a Purkinje cell after white-matter stimulation. In D, the electrode is situated extracellularly in the immediate vicinity of a Purkinje cell. White-matter stimulation (arrow) evokes an antidromic action potential followed by the climbing-fiber burst response. In E, the electrode is moved to an intracellular position. The first action potential is produced by the antidromic invasion, and the second spike potential, which is followed by a prolonged depolarization, is generated by the stimulation of the climbing fiber in synaptic contact with the impaled Purkinje cell. F to I. Intracellular Purkinje-cell records to illustrate the synaptic potential evoked by climbing-fiber stimulation. F. Graded excitatory postsynaptic potential evoked by mossy-fiber activation at white matter level. Stimulus was subthreshold for climbing-fiber activation. G. Extracellular field potentials recorded in the vicinity of the Purkinje cell by a stimulus, as in F. In H, the white-matter stimulus is increased to threshold value for the climbing-fiber activation. A large, all-or-none, excitatory postsynaptic potential is recorded. In I, the white-matter stimulus is increased to evoke a climbing-fiber EPSP every time. Note the regularity in amplitude and time course of the synaptic potential evoked by this repeated stimulation of the climbing fiber. (From Llinás, Bloedel, and Hillman, 1969.)

son, 1969; Llinás et al., 1969a; Nicholson et al., 1969) (Figure 4H, I). This electrophysiological finding is in total agreement with the anatomical concept of one climbing fiber to each Purkinje cell (Ramón y Cajal, 1904). In the cat cerebellum, however, two climbing fiber responses were seen in a Purkinje cell on two occasions (Eccles et al., 1966a).

The unique one-to-one massive excitatory action on the Purkinje cell by the climbing fiber has produced a series of working hypotheses regarding the functional significance of that input. We deal here with five of these hypotheses: the read-out system, the phasic-control system, the intended-movement system, the learning system, and the "clearing" system.

*Read-out system* This hypothesis was first promulgated by Eccles et al. (1966a). It held that the climbing fiber, because of its one-to-one relation with the Purkinje cell, can activate a burst of action potentials that is characteristic of every pair of climbing fiber and Purkinje cell. The climbing fiber could function, so to speak, as a testing probe by which to measure the excitability level of the Purkinje cell

at any time. Given that the changes in excitability of Purkinje cells could be generated synaptically only by the activation of the mossy-fiber, granule-cell system and its associated interneuronal complex, it was suggested that the climbing fibers would "read out" the result of mossy-fiber activity. The number of action potentials evoked by a climbing fiber at any particular time would thus be a function not only of the magnitude of the depolarization of the climbing fibers but also of the over-all state of excitability of a Purkinje cell. As it has been found experimentally that the number of action potentials evoked in a Purkinje cell by climbing-fiber activation is related to the excitability level of the Purkinje cell, this working hypothesis is very attractive. Furthermore, to establish this hypothesis as correct, it would have to be demonstrated that the climbing fiber activates the Purkinje cells in a stereotyped and reproducible manner, and that modulation of this burst of spikes can, in fact, occur. These prerequisites have been found in the cat (Eccles et al., 1966a) as well as in other vertebrates (Llinás et al., 1969b). On the other hand little has been said regarding the possible functional meaning of such a "read-out" system, and some objections to the hypothesis have been raised on the basis of comparative physiology (Llinás et al., 1969b).

*Phasic-control system* A second hypothesis, put forward more recently, views the climbing-fiber afferents as a phasic control system, that is, as a system with an abrupt and short-lasting activation of the type produced by derivative function operators. This phasic control system would utilize the Purkinje-cell mantle concomitantly with the mossy-fiber afferents in a time-sharing fashion. Owing to the lack of complicated neuronal connectivities at the Purkinje-cell level, rapidly changing patterns of Purkinje-cell activation could be superimposed by the climbing-fiber system on the more tonic type of activity evoked by the mossy-fiber afferents (Llinás, 1969b; Llinás et al., 1969a). The type of phasic-control system suggested would be especially relevant in situations in which a rather strong and well-localized inhibitory pattern needs to be imposed on the target cells of the Purkinje neurons, e.g., when a step-function type of correction is required during a particular movement. At the end of this paper, I discuss this point further in connection with the concept of ballistic movements. Implicit in the hypothesis of a phasic-control system is the assumption that climbing-fiber and mossy-fiber inputs function independently and that, owing to the action of the mossy-fiber, parallel-fiber system, the rather high safety factor of the climbing fiber is present as a hierarchical device to override any ongoing activity in the Purkinje cell.

*Intended-movement system* Somewhat similar to the preceding is the hypothesis recently advanced by Oscarsson (1969) that, because of the functional organization of the spinal inputs to the cerebellum, the climbing-fiber system is particularly involved in supplying to the cerebellar cortex information about ongoing central activity in the motor pathways. The system would thus generate a temporospatial analogue of the internal patterns of motor regulation prior to motor execution, while the mossy fiber could convey to the cerebellum information regarding performed movement. The cerebellum is then visualized as an error-correcting device in which intended and performed movements are organized and compared. According to this hypothesis, as opposed to the previous one, mossy- and climbing-fiber systems perform in a strictly conjoint manner.

*Learning system* A fourth hypothesis, lately put forth by Marr (1969), suggests that the cerebellum is the seat for learning motor skills. The climbing-fiber system, he suggests, has as its main function the modification of the synaptic efficiency between the parallel fiber and the Purkinje cell. This system would operate in such a way that, after a climbing-fiber activation, all parallel fibers acting on the Purkinje cell up to 100 milliseconds later would increase their synaptic efficiency permanently. That is to say, every

time a climbing fiber is activated, the parallel fibers, which are activated simultaneously, are reinforced, and in this manner a Purkinje cell "learns" to "recognize" patterns of activation of parallel fibers, which would then be related to particular movements. The validation of this hypothesis awaits new and revolutionary findings, such as the existence of heterosynaptic modification of synaptic action.

*Clearing system* Harmon, Kado, and Lewis (Harmon et al., in press) have suggested that, inasmuch as the climbing fiber generates a large excitatory input to a Purkinje cell, which is, in fact, able to reach "depolarization inactivation" in many cases, the climbing fiber may be regarded as a "clearing system" utilized to erase the ongoing activity of the Purkinje cell at a given instant. This hypothesis can be considered only if it can be shown that, after climbing-fiber activation, the Purkinje-cell axon is inactivated before it can generate a burst of action potentials.

FUNCTIONAL PROPERTIES OF THE MOSSY-FIBER, GRANULE-CELL, PURKINJE-CELL SYSTEM This second input, which has been demonstrated to be excitatory in cats (Eccles et al., 1966c) and other vertebrates (Llinás et al., 1969a; Llinás and Nicholson, 1969), has, as opposed to the climbing fiber and Purkinje cell relationship, an enormous convergence with maximum divergence at any Purkinje cell (Fox and Barnard, 1957). In fact, it has been postulated that the typical spatial organization of the dendrites of the Purkinje cell (i.e., the fact that the dendritic tree tends to be isoplanar and oriented at right angles to the axis of the parallel fiber) has as its raison d'être the insurance of maximum convergence with maximum divergence (cf. Fox and Barnard, 1957).

From the functional point of view, the mossy fiber-parallel fiber system has been viewed in at least four different ways: (a) as a timing-device system; (b) as a coincidence-detector system; (c) as a tonic activating system; and (d) as a pattern-generating system (codon).

*Timing-device system* This particular view was first expressed by Braitenberg (Braitenberg and Onesto, 1962; Braitenberg, 1967) and has recently been re-examined by Freeman (1969). The basic assumption has been that, as parallel fibers are relatively uniform in diameter, traverse the molecular layer at right angles to the Purkinje-cell dendrites, and enter into contact with many of these cells along the path, the conduction time of a synchronous parallel-fiber volley would generate a strict sequential activation of the Purkinje cells as the action potentials are conducted along the length of the parallel fibers. According to this hypothesis, the minimum fraction of time ($\delta t$) that the cerebellum could compute would be the conduction time between the arbors of two adjacent Purkinje cells along the direction of the parallel-fiber spread. This has been calculated to be in the order of a fraction of a millisecond (Braitenberg, 1967). Given the length of the parallel fibers,

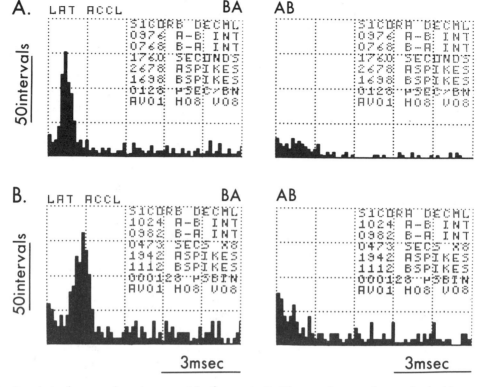

FIGURE 5  Correlation between the action potentials of two Purkinje cells recorded simultaneously during natural (rotatory) stimulation (*Rana catesbeiana*). A and B. Serial correlation histograms of two different pairs of Purkinje cells, computed during horizontal angular acceleration of the animal. The cells lie 0.15 mm apart in A and 0.3 mm apart in B. The prominent peaks seen in the histograms on the left, but not in those on the right, indicate a marked correlation in the times of firing between the members of a pair; cell B of each pair tended to fire at a preferred time delay after cell A, but cell A tended to fire randomly after cell B. (From Freeman, 1969.)

the cerebellum would have a maximum time range of about five milliseconds.

Direct evidence for sequential activation by parallel fibers has been recently obtained by Freeman (1969), who has shown that, after a frog has been physiologically stimulated by rotation, a definite, unidirectional, time correlation is found between adjacent Purkinje cells under a particular parallel-fiber beam (Figure 5).

This hypothesis has recently attained further impetus through anatomical and physiological findings in some teleosts. It has been observed (Kaiserman-Abramof and Palay, 1969; Nieuwenhuys and Nicholson, 1969b) that in mormyrid fish the granule cells are situated in the cerebellar regions of the valvula in such a fashion that the Purkinje cells are activated in an exclusively sequential manner. This is because the granule cells have only an ascending axon without the T-like bifurcations found in other vertebrates. These axons are so organized that they are in contact with the dendrites of the Purkinje cells sequentially, given that the cells are arranged in an orderly stack fashion at right angles to the direction of the ascending granule cells (see Figure 6).

The granule cells are restricted to a layer and are more or less equidistant from the Purkinje cells, so any large, mossy-

FIGURE 6  Schematic diagram of relationships of neural elements within the ridge of mormyrid valvula cerebellaris. Broken lines indicate probable axonic connections. Mossy fibers (mf) enter granule-cell layer beneath ridge and synapse with granule cell (gr) dendrites and, probably, with descending dendrite of central cells (cc). Axons of granule cells enter molecular layer at base or ridge and become nonbifurcated parallel fibers (pf). Parallel fibers synapse with dendrites of basal cells (bc), central cells (cc), vertical cells (vc), stellate cells (sc), Purkinje cells (P), and Golgi cells (Gc). Basal, central and vertical cells send their axons into the efferent basal bundle (bb); stellate-cell axons probably synapse on Purkinje cells, which, in turn, probably send their axons to the somata of basal cells. The Golgi-cell axons return to the dendrites of the granule cells. The circuit at the top left of the figure diagramatically illustrates the connectivities between the neural elements within one half of a ridge. Arrows indicate direction of transmission of neural activity. (From Nieuwenhuys and Nicholson, 1969b.)

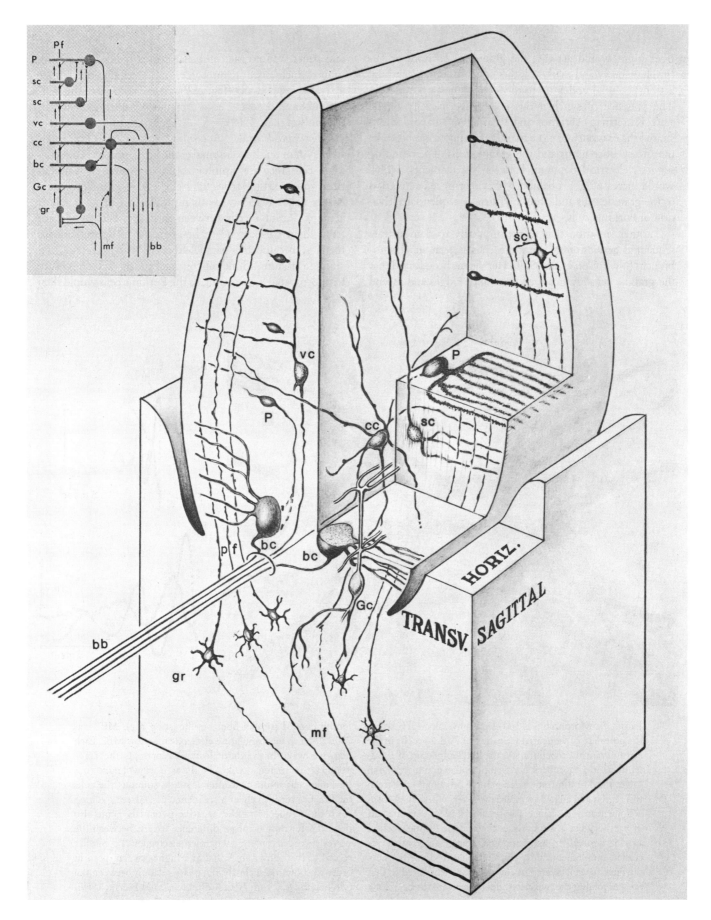

fiber input would activate the granule cells more or less simultaneously and evoke a rather synchronous, ascending, action-potential volley up to the level of the Purkinje cells. The presence of synchronous volleys is especially likely, since Kaiserman-Abramof and Palay (1969) have demonstrated the existence of close membrane apposition between the mossy-fiber input and the granule cells. Their finding suggests electrical coupling between the elements, which would increase the probabilities of synchronized activation of the granule cells and, thus, of a strictly sequential activation of Purkinje cells.

Another situation that seems implicitly to demonstrate sequential activation of Purkinje cells is seen in elasmobranchs (Nicholson et al., 1969). Here it has been found that the granule cells are grouped in midline ridges and ascend

and penetrate into the cerebellar cortex from only one region, which once again forces the Purkinje cells to be activated sequentially (Figure 7). Actual records of this activation have, in fact, been obtained (Figure 7A and B) (Nicholson et al., 1969).

*Coincidence detection* According to this view, the parallel-fiber system tends to emphasize the activation of those Purkinje cells that are simultaneously activated by parallel-fiber volleys arising from different regions of the cerebellar cortex. That is to say, with multiple loci of granule-cell activity, Purkinje cells between these loci and equidistant from them would tend to be activated in a simultaneous manner, because the parallel-fiber volleys are conducted along a folium. Simultaneous activation would ensure temporospatial summation, so the Purkinje cells would then

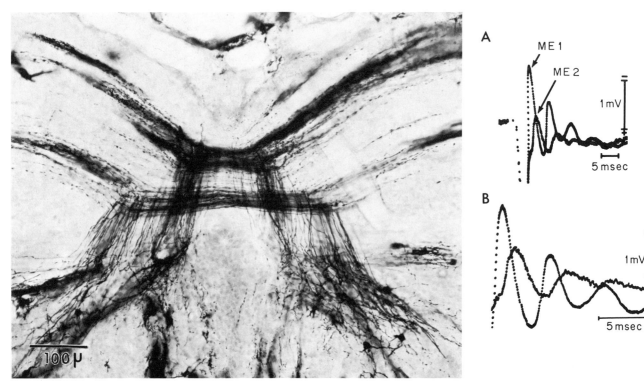

FIGURE 7 Morphological and electrophysiological basis for the postulate of sequential activation of Purkinje cells in the elasmobranch cerebellum. On the left, Golgi stain of parallel fibers in a thornback ray. Prominentiae granulares lie in lower third of illustration and granule cell axons arise from this region and ascend to molecular layer, where they form a T junction and, thence, parallel fibers. Note that one half of parallel fiber always crosses the midline (center vertical axis of figure). A: Superimposed averaged potentials recorded simultaneously by two microelectrodes (ME) near the Purkinje cell layer. The electrodes were oriented so that the axis joining the two electrodes lay beneath the beam of a locally evoked parallel-fiber bundle (sting ray). ME1 was $250\mu$ closer to the stimulating electrode than was ME2. Each negative wave (upwards deflection) represents the compound action currents evoked by the activation of numerous Purkinje cells. Note that after a single stimulus the cells tend to fire repetitively at a frequency of 200/sec, judging from the frequency of the negative potentials. Note also that a fairly constant phase difference is seen between the waves recorded on the two microelectrodes. B: Similar record to that shown in A, but at a higher sweep speed in order to show more clearly the phase relation between the two potentials. (From Nicholson, Llinás, and Precht, 1969.)

function as a coincidence detector system, signaling the simultaneous arrival of parallel-fiber volleys at a particular region of the molecular layer. This idea is especially intriguing when it is borne in mind that the cerebellum is one of the few regions of the central nervous system that is continuous throughout the midline, and so can integrate coincidence between activation of the left and right halves of the central nervous system.

*Tonic activation* Somewhat simpler is the hypothesis that the parallel-fiber system serves as a tonic, widespread activator of a large number of Purkinje cells (Llinás, 1969b). According to this view, the rather widespread ramification of the mossy fibers and the spatial organization of the parallel fibers themselves would tend to convert the input of the mossy fibers into more or less homogeneous activation of large groups of Purkinje cells. By such means, these cells would exercise a tonic control of cerebellar and midbrain nuclear masses. In agreement with this hypothesis, the impulse activity of the granule cell in the unanesthetized animal is rather high and continuous. Likewise, the spontaneous activity of the Purkinje cell demonstrates a more or less continuous firing pattern. The mossy-fiber input would, therefore, tend to modulate the activity of the Purkinje cell en masse.

Experimentally, the physiological stimulation of the cerebellum by means of rotation (Precht and Llinás, 1969b) has demonstrated that spike activity generated in the Purkinje cell by the vestibular input has a tonic firing frequency very closely related to the amplitude of the stimulus.

*Codon system* Marr (1969) has suggested that the mossy-fiber, granule-cell, parallel-fiber systems function to formulate specific patterns of Purkinje-cell activation. These patterns, or "codons," represent specific functional states in the motor system and are defined as "subsets" of a collection of active mossy fibers. This codon can thus be regarded as a "language of small subsets," of which the information is stored by the activation of the climbing fibers. Although this hypothesis is alleged to be testable (Marr, 1969), some of its assumptions lack a proper physiological basis, e.g., "long-term storage" evoked by synaptic activation.

INTERNUNCIAL NEURONS OF THE CEREBELLAR CORTEX Also related to the mossy-fiber, climbing-fiber Purkinje-cell system are the two interneuronal systems, the basket cells and the stellate cells of the molecular layer and the Golgi cells of the granular layer. As stated above, these two systems have been shown to be inhibitory in the different cerebellums so far studied.

*Lateral inhibition* Basket-cell and stellate-cell inhibition on the soma and dendrites of the Purkinje cell has been hypothesized to serve as a lateral inhibitory system (Eccles et al., 1966b, 1966d) restricting the activation of the Purkinje cells to those situated immediately under a "beam" of

activated parallel fibers. This point of view is strongly supported by the highly specialized spatial organization of the axons of the basket cells and large stellate cells that run strictly at right angles to the parallel fibers (Eccles et al., 1967).

*Overflow-preventing system* A different view of the functional meaning of this inhibitory system can, however, be stated here. Because synchronous activation of parallel fibers after physiological stimulation might not be the common form of parallel-fiber function, the basket cells and the stellate cells could serve more as general "overflow-preventing systems," tending to minimize the possibility of inactivation by depolarization of the Purkinje cells. They would function then as overdampening systems that are able to reduce nonlinearities caused by depolarization inactivation.

This latter view seems to be in accordance with the rather high level of inhibitory unitary potentials observed in Purkinje cells under spontaneous conditioning, especially after intracellular Cl⁻ injection (Eccles et al., 1966d).

## Granule-cell interneurons—the Golgi cells

*Simple feedback system* The Golgi-cell system has been hypothesized to be a simple, recurrent, inhibitory system (Eccles et al., 1966c). This view suggests that the cells would serve as a straightforward negative feedback system by means of their direct inhibitory action upon the granule-cell dendrites (Eccles et al., 1966c). The actual anatomical arrangement would be such that, when a given group of granule cells is activated, the excited Golgi cells would have a return inhibitory action upon the granule cells immediately beneath the activated beam of parallel fibers.

*Gating device* A variation of this theme has been postulated by Precht and Llinás (1969a). These investigators found that the Golgi cell, possibly owing to direct input by the mossy fiber, may serve not only as a feedback overflow-preventing system but also as a "gating" device, which, by means of a feed-forward mechanism, would prevent sequential activation through a particular mossy-fiber channel, but would not prevent the activation of the same granule-cell population by a mossy fiber of different origin.

INTEGRATIVE PROPERTIES OF PURKINJE CELLS Although the particular morphology that characterizes Purkinje cells has been known since the turn of the century, little information is yet available as to the functional meaning of the complex spatial organization of this neuron. The fact that the dendritic tree is close to isoplanar has been known since the time of Henle (1879), and it now seems evident that a certain degree of isoplanarity is characteristic of Purkinje cells throughout phylogeny. Equally constant within phylogeny is the complexity of the dendritic ramification in a Purkinje cell. As noted above, several features in the morphological organization of the tree seem to be omnipresent in this neuron.

The first feature is the presence of a widely branching dendritic tree, which may arise from one or two or, in particular cases, several somatic stems. The ramification of the dendritic tree seems, however, to be well organized throughout. In the Reptilia and Amphibia so far examined, the dendritic tree arises from a single-stem dendrite that bifurcates at an almost constant distance from the soma (Hillman, 1969b). This particular morphological character seems to imply a specific functional organization. Among the most strictly organized Purkinje cells are those found in the cerebella of certain teleosts, such as the mormyrid, in which the dendrites of the Purkinje cells are known to stem from basal dendrites situated near the level of the Purkinje-cell somas. The branches of these main dendrites ascend in an almost perfectly perpendicular course to the surface of the cerebellar cortex. This astonishing regularity, which seems comparable in degree with that of the retina, suggests once again a specific functional correlate.

For the most part, physiological investigations on Purkinje cells have placed little emphasis on the functional properties of the dendritic tree. Most investigators have assumed that the dendritic arbor of the Purkinje cell serves, as in the case of a motor neuron, as a simple integrating device which electrotonically transmits a synaptic depolarization generated either by the parallel-fiber, Purkinje-cell synapse or by the climbing-fiber, Purkinje-cell synapse, as well as by the interneurons of the molecular layer. More recently, however, studies on the alligator cerebellum (Llinás et al., 1968, 1969c) have strongly suggested that the dendritic tree of the Purkinje cell is not simply an apparatus for electrotonic summation of excitatory and inhibitory potentials. Rather, it appears that it must also be regarded as a site of complex integratory mechanisms involving the generation of dendritic local responses and action potentials which can propagate unidirectionally and bidirectionally. This possibility once again opens the question of the functional meaning of the dendritic arbor and suggests the need for a re-evaluation of the functional properties of dendrites in other neurons.

*Dendritic spike initiation in Purkinje cells*   The studies in electrical activity of Purkinje-cell dendrites in alligators have suggested the presence of dendritic action potentials which could be initiated at any level of the Purkinje-cell tree from the quite peripheral ramification of the dendritic tree. The presence of such a "boosting" mechanism would bring into question the simple view that the orthodromic activation of the Purkinje-cell soma must require the electrotronic spread of the synaptic potential from the tips of the dendrite to the Purkinje-cell soma. Although it is difficult to calculate the electrotonic distance between a small ramification of a spiny branchlet of a Purkinje cell and its initial segment, it has appeared reasonable, on the basis of our results, to suggest that such a boosting mechanism would aid enormously the communication between different afferent systems and the axon of the Purkinje cell.

Once the presence of active responses at the dendritic level was demonstrated, two other electrophysiological properties of the Purkinje-cell dendrites became apparent: (a) *functional independence,* i.e., the ability of dendrites to fire independently of one another, and (b) *unidirectionality,* i.e., the tendency of dendritic spikes to be conducted in a somatopetal direction (Llinás et al., 1969c). As shown in Figure 8, the current density analysis of the field potentials evoked by parallel-fiber activation in the alligator *Caiman sclerops* strongly suggests that dendritic spikes generated in the peripheral dendritic tree are conducted towards the soma (C and E), but not antidromically to other dendritic branches (D and F). This unidirectional tendency of dendritic spikes seems to be the result of a decreasing rate of excitability of dendrites with respect to distance from the soma (Llinás and Nicholson, 1969).

*Vertical integrative properties of dendrites*   As the Purkinje-cell dendritic tree can be envisaged as having two main directions of spread, one vertical with respect to the surface of the cerebellum and one horizontal (parallel to the surface), the integrative properties of dendrites can be postulated to have a vertical and transverse vectorial component (Figure 9).

Vertical integration would imply algebraic summation of excitatory and inhibitory inputs in one particular vertical segment of a dendrite. Given that the parallel fibers in the molecular layer are stratified in such a way that deeply situated granule cells project to deep regions in the molecular layer and superficial granule cells project to superficial areas of the molecular layer, a vertical integration in the Purkinje cell would necessitate the simultaneous activation of deep and superficial strata within the granular layer.

Szentágothai demonstrated some years ago (Eccles et al., 1967) that the mossy-fiber projection to the granular layer was not random, but that the input arising from the spinal cord projected mainly to the deeper regions of the granular layer of the cerebellar cortex, whereas the superficial regions of the granular layer activated by its mossy fibers originated in the brain stem (reticular formation, etc.). This particular columnar stratification of the afferents would be adequate in evoking vertical integration in the Purkinje-cell dendritic tree through spatial and temporal summation of the actions of the two different input categories.

In this situation, once a dendritic segment reached a firing level, the dendrite would generate action potentials conducted in a noncontinuous manner up to the level of the Purkinje-cell soma.

*Transverse integrative properties of dendrites*   Transverse in-

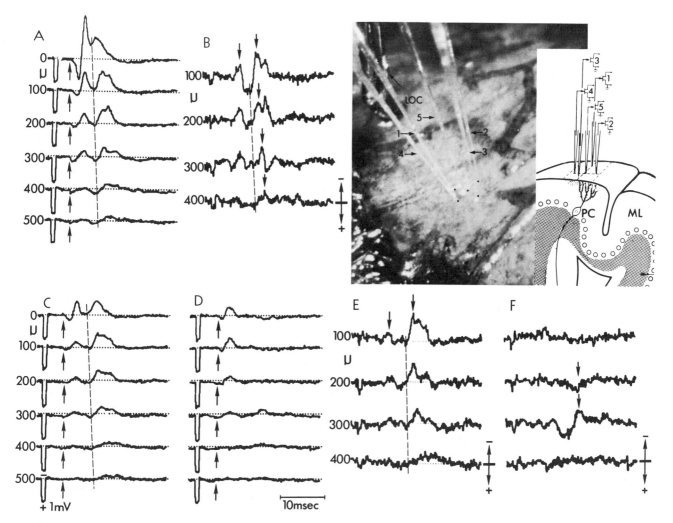

FIGURE 8 Current sources and sinks calculated from the field potentials evoked by local stimulation of the surface of the cerebellar cortex in the alligator (*Caiman sclerops*). The field potentials were recorded simultaneously by an array of five microelectrodes (see photograph and diagram to the right). A shows a computer average of the fields produced by a set of eight successive local stimuli, each set administered at a different depth and recorded from electrode #5. Upward arrows represent stimulus artifacts. In B, current densities calculated from potentials recorded with the five microelectrodes at depths indicated to the left of each record. First downgoing arrow indicates current generated by the parallel-fiber volley; second arrow the current generated by the activation of parallel fiber and Purkinje cell. Upgoing transients indicate current sinks; downgoing transients indicate current sources. Second downgoing arrows in B demonstrate that, after local stimulation, a sink of current is generated and moves downward to the level of the soma. Note increase in latency of the current sink with depth. Time calibration as in D. In C and D, a series of field potentials recorded by electrode #5 after a weak stimulus to the cerebellar cortex. C was recorded immediately in line, and D was recorded 200 μ out of beam. E and F show the current densities associated with the potentials shown in C and D. Note that the weak local stimulus is able to generate a dendritic sink which moves downward in time and which seems to move transversely so that in F a source-sink relation is generated which does not, however, invade antidromically but instead produces a dendritic current source at higher levels (see arrow at 200 μ depth. Abbreviations: PC, Purkinje cell; ML, molecular layer. (From Llinás and Nicholson, 1969.)

tegration would be present, on the other hand, when a localized activation of a particular region of the granule-cell layer generates a more or less simultaneous barrage of parallel-fiber spikes at a given level within the transverse spread of the Purkinje-cell dendritic tree. This situation would initiate a transverse integration that would lead to different functional situations. For instance, summation of electrotonic subthreshold depolarization from different den-

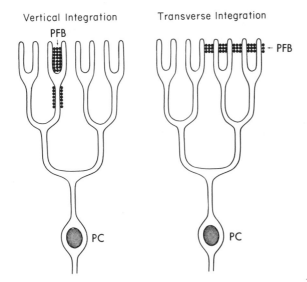

FIGURE 9 Schematic drawing of proposed vertical and transverse integrative configurations in Purkinje-cell dendritic tree. PFB, parallel fiber beam.

dritic segments, at the level of the transverse dendritic linkage, could generate the continuous activation of the cell by spatial integration. If, however, the synaptic barrage were to increase to the point at which action potentials would be generated simultaneously at many sites in the peripheral dendritic tree, the transverse system would function as a self-regulating system because of collision and refractoriness, so that only the early dendritic spikes would reach the soma actively. In this respect, therefore, transverse integration differs from vertical integration. In the latter, the only rate-

limiting function for the decoder would be the refractoriness of the dendritic-spike system, because each dendritic branch would function as an independent unit, with a "direct line" to the soma as opposed to the "party line" organization of the transverse integratory conformation. These two situations are illustrated in Figure 9.

It must, of course, be understood that vertical and transverse integrative properties are only highly idealized simplifications of the integratory ability of Purkinje cells. A more realistic approach to Purkinje-cell function should take into consideration the whole gamut of temporospatial patterns of parallel-fiber activity. When the parallel fibers crossing the dendritic tree of a particular Purkinje cell tend to be activated in a horizontal, sheetlike fashion, the Purkinje cell approximates the transverse integrative mode. Such form of action would be the best to achieve tonic firing in a Purkinje cell. The firing frequency should be related to an electrotonic summation of dendritic depolarization at the transverse dendritic system, so this form of integration would approximate, at midrange, a direct relation to the amplitude of the input. The system would tend to be nonlinear at high levels of afferent pressure as a result of collision. The postulation of vertical and transverse integration modes of Purkinje cells would be completely hollow if all that were available for speculation were the Purkinje cells of higher vertebrates. However, the study of the spatial organization of dendrites of Purkinje cells in different vertebrates suggests that these two forms of integration do indeed occur. Thus, the Purkinje cell of a mormyrid would be, for the most part, a vertical, integrating Purkinje cell, given that the transverse length of the dendritic arbor is minimal (Figure 10A). This Purkinje cell would be most

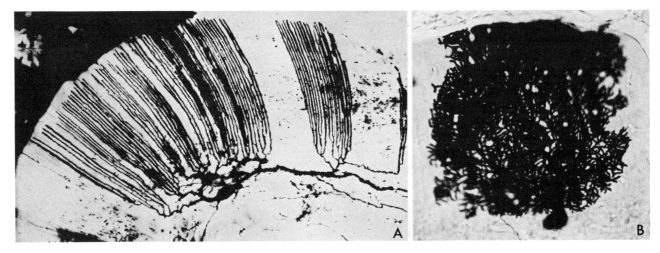

FIGURE 10 A. Composite photomicrograph of a Purkinje cell in a sagittal section through C2 area of cerebellum in mormyrid fish. Golgi method, ×280. (From Nieuwenhuys and Nicholson, 1969b.) B. Golgi stain of a Purkinje-cell dendritic tree in cat. ×225. (From Llinás and Hillman, 1969.)

responsive to vertically oriented sheets of parallel-fiber activation. When the rather small diameters of the Purkinje-cell dendrites are considered, it is difficult to imagine that a horizontal beam of activated parallel fibers impinging on their tips would be of much functional consequence to the Purkinje cell unless action potentials were generated at each branch, because transverse integration would necessarily be minimal. On the other hand, in the Purkinje cells of such higher vertebrates (Figure 10B) as mammals (which are characterized by the huge size of the dendritic-cell arbor and the more transverse character of the dendritic organization), vertical or horizontal sheets of parallel fibers would tend to have a mixture of both vertical and transverse integrative properties.

Consistent with the idea of vertical integratory properties of Purkinje cells is the presence of rather widespread inhibitory terminals in the dendritic tree. It is obvious that the dendritic inhibitory action on a Purkinje cell would be most effective when directed to a vertical integrating conformation, inasmuch as an inhibitory input in a dendrite should tend to produce a functional amputation of the dendritic branchlets distal to the synaptic site. On the other hand, to produce a complete inhibition of the transverse integrating system, a large number of dendritic trees would have to be inhibited more or less simultaneously; alternatively, a somatic or even axonic inhibitory synapse (such as the basket cell) would be needed to regulate the transverse integratory firing mode. This latter type of inhibition, however, would tend to lump all regulatory action at the axonic level and thus would tend to be much grosser than the dendritic inhibition.

DISFACILITATION AS A MECHANISM OF PURKINJE-CELL ACTION  Even prior to the elegant demonstration by Ito and Yoshida (1964) of the inhibitory nature of the Purkinje cell, it had been shown that cerebellar inhibition was mediated by disfacilitation (Terzuolo, 1959; Llinás, 1964). This form of inhibition, as opposed to that produced by an inhibitory synapse, is brought about by the removal of a background synaptic depolarization. It is the most common mechanism for the spread of an inhibitory action in the central nervous system, because, for the most part, inhibitory neurons have short axons. When studying the evolution of cerebellar function in different vertebrates, the conclusion is quickly reached that disfacilitation, as a mechanism for cerebellar control, must have been a secondary specialization of the primordial function of Purkinje cells. As one goes down the phylogenetic scale, it is evident that the presence of direct Purkinje-cell connectivities with mesencephalic centers such as the oculomotor apparatus (Kidokoro, 1969) and even such peripheral receptor systems as the vestibular organ (Llinás and Precht 1969) seem to be the rule, but in higher vertebrates the number of Purkinje-

cell axons that actually leave the cerebellum becomes rather small. This shift of direct inhibition toward disfacilitation, which the cerebellum undergoes during phylogeny, seems to emphasize the functional role of the cerebellar nuclei in higher vertebrates as opposed to those in lower forms (Tsukahara, 1969). It appears urgent for us to understand the functional role of the cerebellar nuclei of higher vertebrates and to elaborate further our ideas regarding direct inhibitory action via Purkinje cells.

## General comments on cerebellar function

As stated at the beginning of this paper, we find ourselves, in regard to general aspects of cerebellar function, in the seemingly paradoxical situation of knowing an enormous amount of detail about the organization of neural elements in the cortex and, at the same time, being in almost complete ignorance about its over-all function. It is, of course, quite possible that, given the rather far-removed position of the cerebellar machinery from the primary sensory motor mechanisms (at least in higher vertebrates), its function will ultimately become apparent when we understand better the sensorimotor levels of interaction that lie functionally and anatomically below the cerebellar mantle. On the other hand, for the last 100 years or so (Flourens, 1842) it has been alleged that the cerebellar cortex is implicated in the coordination of movement, because cerebellar lesion is known to produce well-defined and easily recognizable neurological syndromes. In discussing the general concepts of possible cerebellar function, one must take into consideration its possible role in the regulation of movement. For many years, however, physiologists (Orbeli, 1940; Moruzzi, 1950; Snider and Stowell, 1944) have suggested that the cerebellar cortex might also be implicated in functions other than motor control. The reason the cerebellum has been said to have a strictly motor function is probably related to the fact that motor disorders are easier to detect than is more subtle sensory loss.

Very much in line with a more general concept of cerebellar function are the incongruities found, throughout phylogeny, between the size of the cerebellum in different species and their locomotor abilities. Indeed, this particular question was discussed at a recent meeting by Bennett, Bullock, Llinás, and Nicholson (Llinás, 1969a, pp. 242, 536–537.) Again, direct sensory control by the cerebellum has, in fact, been demonstrated in the frog, as stated above. It is tempting to suppose that the cerebellum may have a type of function that is related to the temporospatial organization of body image from sensory and motor information that is then utilized as an error-correcting servomechanism. That such may be true is suggested by the enormous development of specific areas of the cerebellar cortex in animals with specialized sensory systems—in particular, those systems related to the central formulation of the animal's surrounding, both in the temporal and in the spatial domain. Examples can be found in elasmobranch fish, which are known to have an elaborate electroreceptor system and a large cerebellum. A more dramatic case, however, is found in some species of electrosensitive teleosts, especially mormyrid fish, in which the cerebellum reaches veritably giant dimensions (Kaiserman-Abramof and Palay, 1969; Nieuwenhuys and Nicholson, 1969a). For this reason, it is almost imperative that our concept of cerebellar function be enlarged from that of a motor-regulating system to one involving more general attributes. To quote Bullock, "The point here is that we have evidence that the fish is doing a complex job of analysis with the cerebellum *without moving*; it is not necessarily a movement-coordinating organ at all" (personal communication).

Regarding the specific, over-all type of functions performed by cerebellar circuits, one can imagine that the cerebellum may be functioning either as an open-loop or as a closed-loop system, and that its main function may, in some aspects, be that of an error-correcting device. A typical example of a closed-loop system is that which continuously integrates the ongoing activity of the nervous system as a movement is being carried out. Such a hypothesis has recently been stated in a new light by Eccles (1967) as *the evolving movement theory*. Here the cerebellum is seen as a purely closed-loop system, the error-correcting abilities of which would be operant only after the movement had been initiated and then only as a response to the *actual* production of a particular error, which would signal back through a proprioceptor system to the cerebellum.

A different approach is to view the cerebellum as a *ballistic error-correcting device* that would function in an open-loop manner. From this conceptual stand, one could state that, as a particular pattern of motor action is organized in the central nervous system and sent toward the final common path, the cerebellum could, almost simultaneously, be correcting supraspinal errors that had not yet been integrated into motion but that were already included in the descending motor-triggering signals. Such a *ballistic* or open-loop system would have to be employed in all motor behavior involving rapid movements, because the closed-loop turnover time, due to transmission delays, would not allow the control system to exercise its correction at a sufficient speed. This "hit-or-miss" type of error-correcting system can be recognized in the types of activities performed by musicians or athletes, in which the ballistic properties of motor action are at a very high premium. Even in an everyday activity, such as handwriting, it is now known that movements are of the open-loop ballistic type. For instance, in cursive writing, speeds of 50 mm per second are common, so an accuracy of 0.5 mm corresponds to about 10 milliseconds, which does not allow time for a closed-loop

correcting device to operate (Denier van der Gon and Thuring, 1965; Denier van der Gon and Wiencke, 1969). In any event, a realistic way of looking at the cerebellum leaves both alternatives available, that is, both an open-loop and a closed-loop type of function. In this respect, our statements regarding the possible tonic and phasic functions of mossy fibers and climbing fibers, respectively, might be related to the open- and closed-system view of motor control.

## REFERENCES

Ariëns Kappers, C. U., G. C. Huber, and E. C. Crosby, 1936. The Comparative Anatomy of the Nervous System of Vertebrates, Including Man. Macmillan, New York.

Braitenberg, V., 1967. Is the cerebellar cortex a biological clock in the millisecond range? *Progr. Brain Res.* 25: 334–346.

Braitenberg, V., and N. Onesto, 1962. The cerebellar cortex as a timing organ. Discussion of a hypothesis. Atti 1° Congr. Int. Med. Cibernetica. Giannini, Naples.

Denier van der Gon, J. J., and J. Ph. Thuring, 1965. The guiding of human writing movements. *Kybernetik* 2: 145–148.

Denier van der Gon, J. J., and G. H. Wieneke, 1969. The concept of feedback in motorics against that of preprogramming. *In* Biocybernetics of the Central Nervous System (L. D. Proctor, editor). Little, Brown and Company, Boston, pp. 287–304.

Eccles, J. C., 1967. Circuits in the cerebellar control of movement. *Proc. Nat. Acad. Sci. U. S. A.* 58: 336–343.

Eccles, J. C., M. Ito, and J. Szentágothai, 1967. The Cerebellum as a Neuronal Machine. Springer-Verlag, Berlin, Heidelberg, New York.

Eccles, J. C., R. Llinás, and K. Sasaki, 1966a. The excitatory synaptic action of climbing fibres on the Purkinje cells of the cerebellum. *J. Physiol.* (*London*) 182: 268–296.

Eccles, J. C., R. Llinás, and K. Sasaki, 1966b. Parallel fibre stimulation and the responses induced thereby in the Purkinje cells of the cerebellum. *Exp. Brain Res.* 1: 17–39.

Eccles, J. C., R. Llinás, and K. Sasaki, 1966c. The mossy fibre-granule cell relay of the cerebellum and its inhibitory control by Golgi cells. *Exp. Brain Res.* 1: 82–101.

Eccles, J. C., R. Llinás, and K. Sasaki, 1966d. Intracellularly recorded responses of the cerebellar Purkinje cells. *Exp. Brain Res.* 1: 161–183.

Flourens, P., 1842. Recherches Expérimentales sur les Propriétés et les Fonctions du Système Nerveux dans les Animaux Vertébrés. Édition 2. Baillière, Paris.

Fox, C. A., and J. W. Barnard, 1957. A quantitative study of the Purkinje cell dendritic branchlets and their relationship to afferent fibres. *J. Anat.* 91: 299–313.

Fox, C. A., D. E. Hillman, K. A. Siegesmund, and C. R. Dutta, 1967. The primate cerebellar cortex: a Golgi and electron microscopic study. *Progr. Brain Res.* 25: 174–225.

Fox, C. A., K. A. Siegesmund, and C. R. Dutta, 1964. The Purkinje cell dendritic branchlets and their relation with the parallel fibers: Light and electron microscopic observations. In Morphological and Biochemical Correlates of Neural Activity

(M. M. Cohen and R. S. Snider, editors). Harper and Row, New York, pp. 112–141.

FREEMAN, J. A., 1969. The cerebellum as a timing device: An experimental study in the frog. *In* Neurobiology of Cerebellar Evolution and Development (R. Llinás, editor). American Medical Association, Education and Research Foundation, Chicago, pp. 397–420.

FUJITA, S., 1969. Autoradiographic studies on histogenesis of the cerebellar cortex. *In* Neurobiology of Cerebellar Evolution and Development (R. Llinás, editor). American Medical Association, Education and Research Foundation, Chicago, pp. 743–747.

FUJITA, S., M. SHIMADA, and T. NAKAMURA, 1966. H$^3$-thymidine autoradiographic studies on the cell proliferation and differentiation in the external and the internal granular layers of the mouse cerebellum. *J. Comp. Neurol.* 128: 191–208.

HÁMORI, J., and J. SZENTÁGOTHAI, 1964. The "crossing over" synapse: An electron microscope study of the molecular layer in the cerebellar cortex. *Acta Biol. Acad. Sci. Hung.* 15: 95–117.

HÁMORI, J., and J. SZENTÁGOTHAI, 1965. The Purkinje cell baskets: Ultrastructure of an inhibitory synapse. *Acta Biol. Acad. Sci. Hung.* 15: 465–479.

HÁMORI, J., and J. SZENTÁGOTHAI, 1966. Participation of Golgi neurone processes in the cerebellar glomeruli: An electron microscope study. *Exp. Brain Res.* 2: 35–48.

HÁMORI, J., and J. SZENTÁGOTHAI, 1968. Identification of synapses formed in the cerebellar cortex by Purkinje axon collaterals: An electron microscope study. *Exp. Brain Res.* 5: 118–128.

HARMON, L. D., R. KADO, and E. R. LEWIS. Cerebellar modeling problems. *In* To Understand Brains (L. D. Harmon, editor). Prentice-Hall, Englewood Cliffs, New Jersey (in press).

HENLE, J., 1879. Handbuch der Nervenlehre des Menschen. Brunswick, Germany.

HILLMAN, D. E., 1969a. Morphological organization of frog cerebellar cortex: A light and electron microscopic study. *J. Neurophysiol.* 32: 818–846.

HILLMAN, D. E., 1969b. Neuronal organization of the cerebellar cortex in Amphibia and Reptilia. *In* Neurobiology of Cerebellar Evolution and Development (R. Llinás, editor). American Medical Association, Education and Research Foundation, Chicago, pp. 279–325.

HOUSER, G. L., 1901. The neurones and supporting elements of the brain of a selachian. *J. Comp. Neurol.* 11: 65–175.

ITO, M., and M. YOSHIDA, 1964. The cerebellar-evoked monosynaptic inhibition of Deiters' neurones. *Experientia* (*Basel*) 20: 515–516.

KAISERMAN-ABRAMOF, I. R., and S. L. PALAY, 1969. Fine structural studies of the cerebellar cortex in mormyrid fish. *In* Neurobiology of Cerebellar Evolution and Development (R. Llinás, editor). American Medical Association, Education and Research Foundation, Chicago, pp. 171–205.

KIDOKORO, Y., 1969. Cerebellar and vestibular control of fish oculomotor neurones. *In* Neurobiology of Cerebellar Evolution and Development (R. Llinás, editor). American Medical Association, Education and Research Foundation, Chicago, pp. 257–276.

KORNGUTH, S. E., J. W. ANDERSON, and G. SCOTT, 1968. The de-velopment of synaptic contacts in the cerebellum of Macaca mulatta. *J. Comp. Neurol.* 132: 531–546.

LARRAMENDI, L. M. H., 1965. Purkinje axo-somatic synapses at seven and 14 postnatal days in the mouse. An electron microscopic study. *Anat. Rec.* 151: 460 (abstract).

LARRAMENDI, L. M. H., 1967. Synaptogenic period of mossy terminals in mouse cerebellum. An electron microscopic study. *Anat. Rec.* 157: 275 (abstract).

LARRAMENDI, L. M. H., 1969. Analysis of synaptogenesis in the cerebellum of the mouse. *In* Neurobiology of Cerebellar Evolution and Development (R. Llinás, editor). American Medical Association, Education and Research Foundation, Chicago, pp. 803–843.

LARRAMENDI, L. M. H., and T. VICTOR, 1967. Synapses on the Purkinje cell spines in the mouse. An electron microscopic study. *Brain Res.* 5: 15–30.

LEMKEY-JOHNSTON, N., and L. M. H. LARRAMENDI, 1968a. Morphological characteristics of mouse stellate and basket cells and their neuroglial envelope: An electron microscopic study. *J. Comp. Neurol.* 134: 39–72.

LEMKEY-JOHNSTON, N., and L. M. H. LARRAMENDI, 1968b. Types and distribution of synapses upon basket and stellate cells of the mouse cerebellum. An electron microscopic study. *J. Comp. Neurol.* 134: 73–112.

LLINÁS, R., 1964. Mechanisms of supraspinal actions upon spinal cord activities. Differences between reticular and cerebellar inhibitory actions upon alpha extensor motoneurons. *J. Neurophysiol.* 27: 1117–1126.

LLINÁS, R. (editor). 1969a. Neurobiology of Cerebellar Evolution and Development. American Medical Association, Education and Research Foundation, Chicago.

LLINÁS, R., 1969b. Functional aspects of interneuronal evolution in the cerebellar cortex. *In* The Interneuron (M. A. B. Brazier, editor), UCLA Forum in Medical Sciences, no. 11, University of California Press, Berkeley and Los Angeles, pp. 329–348.

LLINÁS, R., J. R. BLOEDEL, and D. E. HILLMAN, 1969a. Functional characterization of the neuronal circuitry of the frog cerebellar cortex. *J. Neurophysiol.* 32: 847–870.

LLINÁS, R., and D. E. HILLMAN, 1969. Physiological and morphological organization of the cerebellar circuits in various vertebrates. *In* Neurobiology of Cerebellar Evolution and Development (R. Llinás, editor). American Medical Association, Education and Research Foundation, Chicago, pp. 43–73.

LLINÁS, R., D. E. HILLMAN, and W. PRECHT, 1969b. Functional aspects of cerebellar evolution. *In* The Cerebellum in Health and Disease (W. S. Fields and W. D. Willis, Jr.). Warren H. Green, St. Louis, pp. 269–291.

LLINÁS, R., and C. NICHOLSON, 1969. Electrophysiological analysis of alligator cerebellum: A study on dendritic spikes. *In* Neurobiology of Cerebellar Evolution and Development (R. Llinás, editor). American Medical Association, Education and Research Foundation, Chicago, pp. 431–465.

LLINÁS, R., C. NICHOLSON, J. A. FREEMAN, and D. E. HILLMAN, 1968. Dendritic spikes and their inhibition in alligator Purkinje cells. *Science* (*Washington*) 160: 1132–1135.

LLINÁS, R., C. NICHOLSON, and W. PRECHT, 1969c. Preferred centripetal conduction of dendritic spikes in alligator Purkinje cells. *Science* (*Washington*) 163: 184–187.

LLINÁS, R., and W. Precht, 1969. The inhibitory vestibular efferent system and its relation to the cerebellum in the frog. *Exp. Brain Res.* 9: 16–29.

MARR, D., 1969. A theory of cerebellar cortex. *J. Physiol.* (*London*) 202: 437–470.

MORUZZI, G., 1950. Problems in Cerebellar Physiology. Charles C Thomas, Springfield, Illinois.

MUGNAINI, E., 1966. Ultrastructural aspects of cerebellar morphology in the chick embryo. *Anat. Rec.* 154: 391 (abstract).

MUGNAINI, E., 1969. Ultrastructural studies on the cerebellar histogenesis. II. Maturation of nerve cell populations and establishment of synaptic connections in the cerebellar cortex of the chick. *In* Neurobiology of Cerebellar Evolution and Development (R. Llinás, editor). American Medical Association, Education and Research Foundation, Chicago, pp. 749–782.

NICHOLSON, C., R. LLINÁS, and W. PRECHT, 1969. Neural elements of the cerebellum in elasmobranch fishes: Structural and functional characteristics. *In* Neurobiology of Cerebellar Evolution and Development (R. Llinás, editor). American Medical Association, Education and Research Foundation, Chicago, pp. 215–243.

NIEUWENHUYS, R., and C. NICHOLSON, 1969a. A survey of the general morphology, the fiber connections, and the possible functional significance of the gigantocerebellum of mormyrid fishes. *In* Neurobiology of Cerebellar Evolution and Development (R. Llinás, editor). American Medical Association, Education and Research Foundation, Chicago, pp. 107–134.

NIEUWENHUYS, R., and C. NICHOLSON, 1969b. Aspects of the histology of the cerebellum of mormyrid fishes. *In* Neurobiology of Cerebellar Evolution and Development (R. Llinás, editor). American Medical Association, Education and Research Foundation, Chicago, pp. 135–169.

ORBELI, L. A., 1940. New notions on cerebellar functions. *Usp. Sovrem. Biol.* 13: 207–220.

OSCARSSON, O., 1969. The sagittal organization of the cerebellar anterior lobe as revealed by the projection patterns of the climbing fiber system. *In* Neurobiology of Cerebellar Evolution and Development (R. Llinás, editor). American Medical Association, Education and Research Foundation, Chicago, pp. 525–537.

PRECHT, W., and R. LLINÁS, 1969a. Functional organization of the vestibular afferents to the cerebellar cortex of frog and cat. *Exp. Brain Res.* 9: 30–52.

PRECHT, W., and R. LLINÁS, 1969b. Comparative aspects of the vestibular input to the cerebellum. *In* Neurobiology of Cerebellar Evolution and Development (R. Llinás, editor). American

Medical Association, Education and Research Foundation, Chicago, pp. 677–702.

RAMÓN Y CAJAL, S., 1888a. Estructura de los centros nerviosos de las aves. *Rev. Trim. Histol. Norm. Patol.* 1: 305–315.

RAMÓN Y CAJAL, S., 1888b. Sobre las fibras de la capa molecular del cerebelo. *Rev. Trim. Histol. Norm. Patol.* 2: 343–353.

RAMÓN Y CAJAL, S., 1904. La Textura del Sistema Nervioso del Hombre y los Vertebrados. Moya, Madrid.

RAMÓN Y CAJAL, S., 1912. Sobre ciertos plexos pericelularos en la capa de los granos del cerebelo. *Trab. Lab. Invest. Biol.* 10.

SCHAPER, A., 1894. Die morphologische und histologische Entwicklung des Kleinhirns der Teleostier. *Anat. Anz.* 9: 489–501.

SCHEIBEL, M. E., and SCHEIBEL, A. B., 1954. Observations on the intracortical relations of the climbing fibers of the cerebellum. *J. Comp. Neurol.* 101: 733–760.

SCHEIBEL, M. E., and SCHEIBEL, A. B., 1964. Some structural and functional substrates of development in young cats. *Progr. Brain Res.* 9: 6–25.

SCHNITZLEIN, H. N., and J. R. FAUCETTE, 1969. General morphology of the fish cerebellum. *In* Neurobiology of Cerebellar Evolution and Development (R. Llinás, editor). American Medical Association, Education and Research Foundation, Chicago, pp. 77–106.

SNIDER, R. S., and A. STOWELL, 1944. Receiving areas of the tactile, auditory, and visual systems in the cerebellum. *J. Neurophysiol.* 7: 331–357.

SOTELO, C., 1969. Ultrastructural aspects of the cerebellar cortex of the frog. *In* Neurobiology of Cerebellar Evolution and Development (R. Llinás, editor). American Medical Association, Education and Research Foundation, Chicago, pp. 327–371.

TERZUOLO, C. A., 1959. Cerebellar inhibitory and excitatory actions upon spinal extensor motoneurons. *Arch. Ital. Biol.* 97: 316–339.

TSUKAHARA, N., 1969. Electrophysiological study of cerebellar nucleus neurones in the dogfish *Mustelis canis*. *In* Neurobiology of Cerebellar Evolution and Development (R. Llinás, editor). American Medical Association, Education and Research Foundation, Chicago, pp. 251–256.

UCHIZONO, K., 1967. Synaptic organization of the Purkinje cells in the cerebellum of the cat. *Exp. Brain Res.* 4: 97–113.

UCHIZONO, K., 1969. Synaptic organization of the mammalian cerebellum. *In* Neurobiology of Cerebellar Evolution and Development (R. Llinás, editor). American Medical Association, Education and Research Foundation, Chicago, pp. 549–583.

# 40 Glomerular Synapses, Complex Synaptic Arrangements, and Their Operational Significance

## JOHN SZENTÁGOTHAI

TOPOGRAPHIC ARRANGEMENT between presynaptic axon terminals and the postsynaptic neuron is generally envisaged as essentially a side-by-side location of a number of axon terminals, scattered in varying densities on various parts of the soma-dendritic surface. It is not known if the respective synaptic actions (either excitatory or inhibitory) in neighboring axon terminals influence one another or, if they do, to what degree. Conversely, it is not known if the net result of simultaneous synaptic actions of the terminals in contact with the whole or with a large part of the neuron (for example, the soma of one of the major dendrites) is computed by the perikaryon (or, if the dendrites have propagated impulses, it is computed separately by the major dendrites). There are synaptic regions in which the structural arrangement itself suggests the first possibility. These are the so-called *synaptic glomeruli* or *synaptic complexes,* meaning a definite and circumscribed structural arrangement between a number of specific axon terminals and one (or several) dendritic end (or ends). If such an arrangement is surrounded and separated from the environment by a capsule of glial processes, it may be called a glomerular synapse. If the separation from the surrounding neuropil is not sufficiently clear, or if the shape and outlines of the arrangement are irregular, it is more appropriate to call it a synaptic complex or a synaptic cluster.

There is reason to assume that glomerular synaptic arrangements are highly effective and versatile tools of information processing. In addition, their presence in various parts of the nervous system makes it worth-while to take a closer view of various types of these structures. Although a number of unorthodox, functional interpretations of glomerular synapses could be presented, I do not propose, for the time being, to venture beyond what follows directly from established concepts of modern, unit-level neurophysiology. The conclusions that must be drawn are sufficiently exciting (and even baffling) if this more moderate course is adopted. This does not, however, imply that the

J. SZENTÁGOTHAI Anatomy Department, University Medical School, Budapest, Hungary

correct explanation of glomerular synapses is the one given here.

## Cerebellar glomeruli

The cerebellar glomeruli, or islands, are the archetype of glomerular synapses that were recognized and correctly interpreted structurally in Ramón y Cajal's classic description (1911). Electron microscope (EM) studies (Gray, 1961; Kirsche et al., 1965) have substantiated that cerebellar islands are the sites of synaptic articulation between the mossy fiber and the granule-cell dendrites (Figure 1). Further EM analysis, using degeneration methods (Hámori, 1965; Szentágothai, 1965; Hámori and Szentágothai, 1966; Fox et al., 1967), revealed that the cerebellar glomeruli have a standard structure containing (as correctly surmised by Ramón y Cajal) two kinds of axon terminals and two kinds of dendrites.

The structure of the cerebellar glomeruli and their position in the neuronal network of the cerebellar cortex are explained diagrammatically by Figure 2. As seen in the lower part of the diagram, the center of the glomerulus is occupied by the large, sinusoid terminal of the mossy fiber, the so-called mossy rosette. This is in synaptic contact with a considerable number of dendritic terminals (or digits) of the granule neurons and often has a larger contact with a Golgi-cell dendrite. In the outer part of the glomerulus, the terminals of Golgi-cell axons establish synaptic contacts either with the dendritic digits of the granule-cell neurons opposite their contacts with the mossy rosette, or somewhat "downstream" on the necks of the dendrite terminals.

The dendritic terminals, or "digits," of the granule-cell neurons are bulbous enlargements of the thin granule-cell dendrites that invariably contain one mitochondrion in the center. An endoplasmic vacuole is generally wrapped halfway around the mitochondrion. A row of postsynaptic granules is often encountered beneath the subsynaptic cell membrane. Desmosomoid dendrodendritic attachments between neighboring dendritic digits are frequent. In many cases (although by no means all) small, irregular, spinelike processes of the dendritic terminals can be seen to protrude

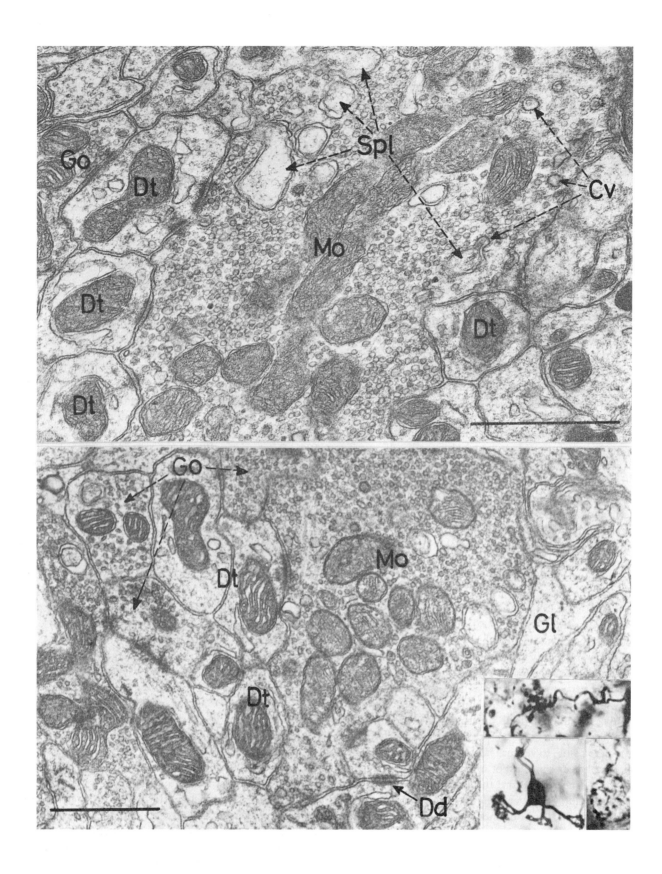

deeply into invaginations of the mossy rosette, and large so-called "coated vesicles" often appear to emerge (Figure 1) from the invaginations of the mossy-rosette membrane (Szentágothai, 1965; Eccles et al., 1967). Similar minute spines or spinules occur regularly at the contact surfaces between the Golgi dendrites and the mossy rosettes (Hámori and Szentágothai, 1966). Strangely, in neither case are there specifically differentiated synaptic contact regions on the spinules themselves.

The axon terminals of the Golgi cells differ from the mossy rosettes characteristically in structure, not only by being smaller and situated at the periphery, but by containing smaller and flattened (Figure 1) synaptic vesicles (Uchizono, 1967). This corresponds to their assumed inhibitory nature.

There appears to be a fair correlation between the recent physiological findings of Eccles and coworkers (see Eccles et al., 1967, Chapter VII) and these structural features. The mossy rosettes excite both granule-cell and Golgi-cell dendrites. The granule-cell dendrites are inhibited by Golgi-cell axons. The Golgi cells can therefore be considered to be negative feedback devices that operate over the parallel-fiber system to limit the mossy-fiber input to the Purkinje cells. Because of the large span of the typical Golgi-cell dendrites, the notion has been put forward (Eccles et al., 1967, p. 211) that the device is particularly constructed to limit inputs that would cause simultaneous activities in longitudinal strips of cerebellar folia, which are significantly broader than the dendritic tree of a Purkinje cell.

However it may be, the cerebellar glomerulus is undoubtedly a synapse with remarkably complex integrative capacities. This integration consists of an excitatory input (the mossy rosette) that has a very high divergence of 1:100 in the mossy granule and none in the relation between the mossy fiber and the Golgi-cell dendrite. (The number of granule-cell dendrites in contact with a mossy rosette once was calculated to be in the order of 15 to 20 [Fox et al.,

1967; Eccles et al., 1967]. A detailed analysis of quantitative neuronal relations of the cerebellar cortex, however, now in progress in our laboratory [Palkovits et al., in preparation], has shown that the earlier calculations based on silver-stained material were grossly misleading. The real number of granule-cell dendrites entering the average glomerulus may be close to a hundred.) Arborization of the mossy-fiber afferent within the folium is considerable. Recent calculations based on the number of mossy afferents entering smaller folia (to be published in detail elsewhere) show that the arborization is much more abundant than hitherto suspected. According to this calculation, the average mossy-fiber afferents entering a smaller folium produce about 50 to 60 rosettes that appear to be fairly evenly distributed in the granule layer of the folium. Convergence upon the individual granule-cell neuron is four to one, as the average granule cell has four (varying between two and seven) dendrites that usually are involved with as many glomeruli, the rosettes of which belong to different mossy-fiber afferents. This can be deduced from the fact that the rosettes of any given mossy afferent are arranged at greater distances from one another than the maximal spans of two dendrites of granule neurons (Szentágothai, 1968a). Convergence upon the dendrites of the Golgi-cell neuron is probably much larger.

In the same apparatus is an inbuilt, inhibitory gate control for the transmission of impulses in one of the two possible directions from mossy rosette to granule cell. It is probably unitary with respect to the individual glomerulus, as only one Golgi-axon branch enters each cerebellar island. Conversely, the divergence is considerable; each Golgi axonal arborization takes up considerable space, within which it supplies all the glomeruli. There is little, if any, overlap between the dendritic trees and the axonal arborizations of the Golgi-cell neurons (Ramón y Cajal, 1911; Jakob, 1928; Eccles et al., 1967).

---

FIGURE 1 Ultrastructural details of cerebellar glomeruli. The mossy axon terminal or so called mossy rosette (Mo) occupies the central position. This huge axon terminal is surrounded by numerous terminals of the granule dendrites (Dt), each typically with one mitochondrion in their center and an endoplasmic sac wrapped halfway around the mitochondrion. Synaptic contacts of usual structure are established between the mossy terminal and the granule dendrite terminals. Neighboring granule dendrite terminals frequently have dendro-dendritic attachments (Dd) of desmosomoid character. Small, spine-like processes of the granule dendritic terminals (so-called spinules, Spl) intrude into deep invaginations of the mossy terminal. At such sites so-called "coated vesicles" (Cv) seem to detach themselves from the mossy rosette membrane. In the outer part of the glomeruli smaller axon profiles can be seen. These contain somewhat smaller and flattened synaptic vesicles, and can be identified as the endings of Golgi-cell axons (Go). Part of the glial capsule (Gl) of the glomerulus can be seen in the lower electron micrographs. Scale, 1μ. Inset: Relevant Golgi structures as seen with the light microscope. Top: A typical mossy rosette surrounded by the shadow of the spherical glomerulus; bottom left, a typical granule cell with the characteristic dendrites that form the dendritic terminals inside the glomeruli. Bottom right: The arborization of Golgi-axon branch in the outer zone of the spherical glomerulus.

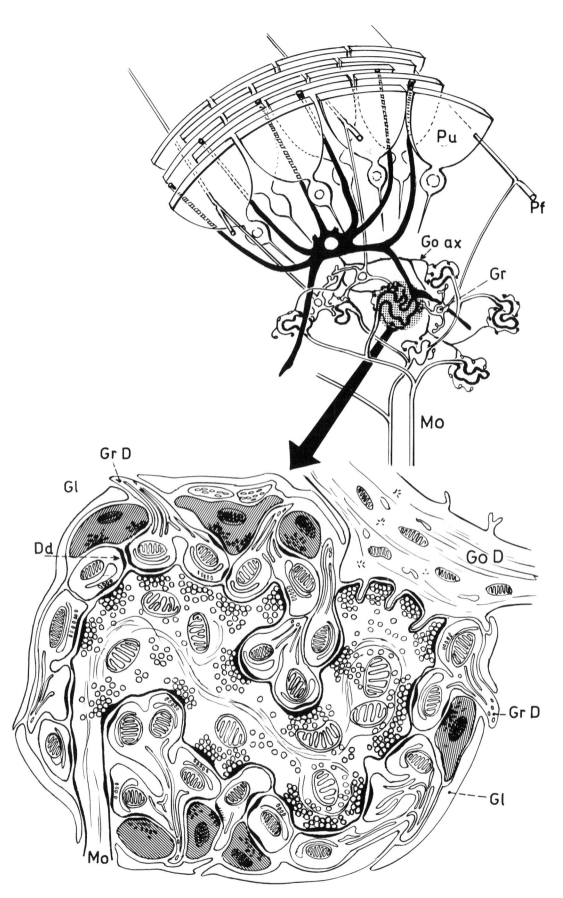

## The glomerular synapses in the lateral geniculate body

Glomerular synapses of characteristic structure, in which synaptic coupling of neurons is arranged in a highly specific manner, have been recognized in the lateral geniculate body (LGB) relatively recently (Szentágothai, 1962, 1963; Colonnier and Guillery, 1964; Peters and Palay, 1966; Karlsson, 1967; Pecci Saavedra and Vaccarezza, 1968). Apart from small species differences, their structure seems to be remarkably uniform in most laboratory animals, and their neuronal linkage has been unraveled by the combined study of Golgi pictures and degeneration, on both the light-microscope and the EM level (Szentágothai et al., 1966). The structure of the LGB glomeruli is illustrated by an electron micrograph (Figure 3) and a diagram showing both the gross neuronal arrangement and the details of synaptic articulation within the glomerulus (Figure 4). The axon terminals of retinal afferents occupy the central position in the glomeruli. In contrast to the cerebellar glomeruli, in which there is only one mossy rosette, several club-shaped, optic-fiber terminals may occur in an LGB glomerulus. It is not known if these are terminals of the same optic afferent or if they originate from different afferent fibers. Fortunately, the optic afferents can be recognized in most laboratory mammals, because their mitochondria tend to be swollen (Figure 3). This swelling is probably an artifact caused by imperfect fixation, but it is also a favorable artifact which makes it easy to identify the optic axon terminal in the normal material. The arborization of the retinal afferents is rich and occupies a space of considerable size (roughly 300 × 500 × 100 microns). It would be difficult, however, to make specific statements on the number of glomeruli that a retinal afferent might enter. Several tens, as an order of magnitude, might be a fair guess.

No terminals of retinal afferents are encountered outside the glomerular synapses. At least two other kinds of axon terminals can be found in the LGB glomeruli. Both are relatively small in comparison to the optic terminals, but they differ in density of plasma structure; one is rather light and the other is dark. We have labeled the light ones type 2 and the darker ones type 3; the optic terminals are type 1 (Szentágothai et al., 1966).

On evidence based on Golgi pictures and on degeneration observations, the type-2 terminals can be identified with the axon terminals of the Golgi second-type cells, which are found in abundance in the LGB. Golgi second-type cells have extended dendritic trees and a profuse arborization of their axons, the branches of which, typically, do not reach beyond the limits of the dendritic tree. The axonal ramification can thus reach many glomeruli, but only those that are within the reach of the dendrites of the same cell. An extensive overlap between dendrites, as well as between axons of neighboring cells, certainly does occur, although no more specific (numerical) statements are as yet possible.

As seen in Figure 3, the synaptic vesicles of the type-2 axon terminals are smaller and appear to be flattened, as compared with the large and spherical vesicles of the optic (type-1) and type-3 terminals. The type-2 axon terminals thus belong to the inhibitory type of synapse, according to the concept of Uchizono (1965). At least some of the type-3 terminals could be traced back by degeneration to descending corticogeniculate pathways that arise primarily from the occipital lobe. There are many other kinds of synapses in the LGB in addition to those of the glomeruli, but they are not considered here. Probably the majority of the synapses of the descending corticogeniculate pathways, for example, are not in the glomeruli (Szentágothai et al., 1966) but in the outer, general neuropil.

The dendritic elements participating in the glomeruli of the lateral geniculate body are themselves rarely, if ever,

---

FIGURE 2 Diagram illustrating the position of the glomerulus in the neuron network of the cerebellar cortex (above) and an idealized version of its ultrastructure (below). The upper diagram gives a quasi-stereoscopic view of a small part of a cerebellar folium, into which the mossy afferents (Mo) enter; their synaptic expansions are the rosettes. Rosettes are connected mainly by the short, claw-shaped dendrites of the granule cells (Gr) and by the descending dendrites of the large Golgi neurons (black). The ascending axons of the granule cells give rise to the parallel fibers (Pf) which while running in the longitudinal axis of the folium pierce the flattened dendritic trees of the Purkinje neurons (Pu, represented here as spade-shaped boxes). The axon branches of the Golgi neurons (Go ax) enter the glomeruli and give rise to a plexus of small beaded terminals. The diagram below shows the synaptic relations of these elements within the glomerulus as seen in the electron microscope (see also Figure 1). Mo, mossy afferent; GrD, granule dendrites entering through the glial capsule (Gl) of the glomerulus and terminating in the characteristic bulbous terminals or digits; Dd, desmosomoid dendrodendritic contacts; GoD, descending dendrite of Golgi-cell neuron with characteristic small spines, which makes broad contact with the mossy rosette. The Golgi axon terminals situated in the periphery of the glomerulus (hatched) establish synaptic contacts exclusively with granule cell dendrites.

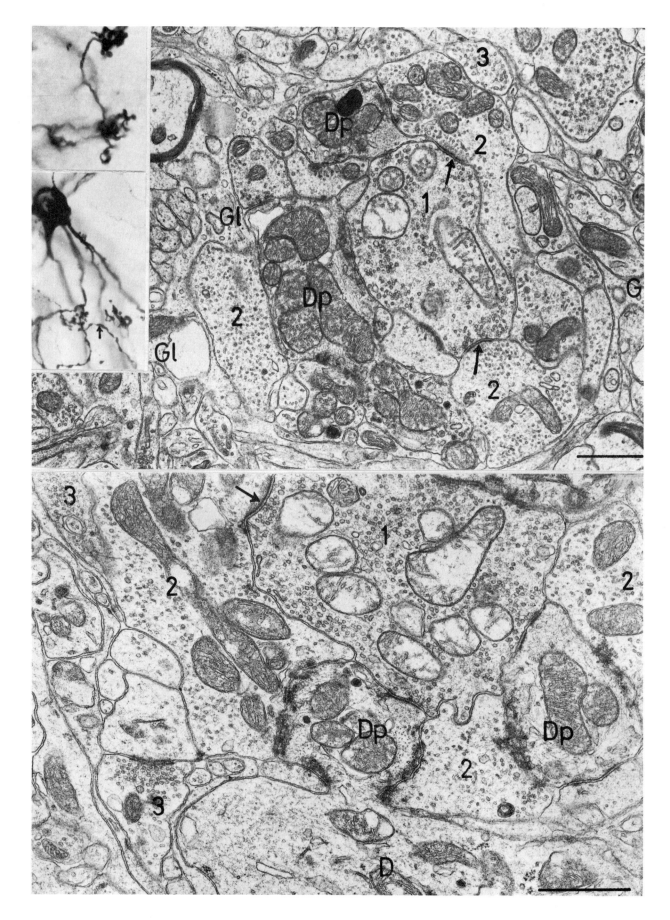

dendrites. Instead, they are short, blunt protrusions of the dendrites (Figure 3) that can be found in abundance on the principal dendrites of the geniculocortical relay cells, at the sites of which they break up into secondary branches. These dendritic protrusions contain two to three densely packed and densely cristated mitochondria, and have a strongly filamentous structure. The filaments are interwoven in a particularly dense feltwork immediately beneath the postsynaptic surface. Interdendritic attachment plaques are usually found, and all three kinds of axon terminals make synaptic contacts with the dendritic protrusions.

A most remarkable feature of the glomeruli is the frequent occurrence of axo-axonic synapses. These have been interpreted as the possible structural basis of presynaptic inhibition, which has been described (Angel et al., 1965; Iwama et al., 1965) as taking place in the LGB. It was soon recognized however, that the structural polarity of the axo-axonic synapses is in the wrong direction, if a depolarization of the optic terminals is required. The afferents of retinal origin (type 1) are invariably presynaptic, by structural standards, mainly to type-2 axon terminals and less often to those of type 3. Type 3 is invariably presynaptic to type 2.

If the structural arrangement were taken as the basis of interpretation, there could be one possible explanation: optic afferents would, by depolarization of the Golgi terminals, presynaptically inhibit (or, conversely, enhance?) an inhibitory influence exercised upon the geniculate-body relay cells by the Golgi second-type neurons. This possibility was tentatively mentioned by Pecci-Saavedra and Vaccarezza (1968), based on their observation that the vesicles in the "postsynaptic" axon terminals are flattened. I elaborated on the same concept independently in more detail, on the basis of the observation that the type-2 axon terminals are from Golgi second-type neurons, which are often inhibitory in nature (Szentágothai, 1968b). The concept of a presynaptic disinhibition exercised by the retinal afferent is more attractive in view of physiological observations that much less activity is experienced in LGB neurons

if the visual field is uniformly light than if there is some sharp contrast projected upon the retina. One could explain this by assuming that the relatively large number of Golgi neurons (a ratio of about 1:2 or 1:3 Golgi to geniculate relay neurons in the LGB; Tömböl et al., 1969), if uniformly excited by optic afferents, would keep most glomeruli under tonic inhibition. Whenever the excitation that arrives from the retina is patterned, i.e., if some retinal afferents are excited and others are not, the optic terminals that are in a strategic position could suppress the Golgi inhibition by presynaptic disinhibition; then the retinal impulses might get through to the relay cells. Theoretically, it could as well be the other way around. If the action of retinal afferents on Golgi endings were excitatory, the effect would be an enhancement of the Golgi inhibition. It is, however, difficult to imagine the physiological use of such a mechanism.

## Other thalamic relay nuclei

Similar glomerular synapses have most recently been described in the ventrobasal nucleus of the thalamus (ventralis posterolateralis, VPL) by Jones and Powell (1969b). The structure of the glomeruli agrees in many respects with that of the lateral geniculate-body glomeruli; the lemniscal afferent terminates in the large central axon terminals, and various smaller axon terminals probably are Golgi second-type endings and descending corticothalamic fibers. Both have axodendritic contacts with the VPL neurons and axo-axonic contacts with the lemniscal fibers. In the axo-axonic contacts, the lemniscal fiber is also invariably presynaptic.

A less regular but otherwise similar arrangement was described recently by Majorossy and Réthelyi (1968) in the ventral (acoustic) portion of the medial geniculate body (MGB). Although lacking the nice regularity of the LBG glomeruli, the structural and neuronal arrangement in the MGB "synaptic clusters" is, in all essentials, exactly the same as in the LGB and VPL glomeruli.

One can only conclude that this kind of synaptic arrange-

FIGURE 3 Ultrastructure of the glomeruli in the lateral geniculate body. The glomerulus in the upper electron micrograph is well surrounded by a capsule of glial elements (Gl). The center is generally occupied by the terminals of optic fibers (1) recognizable in normal material by the tendency of the mitochondria to swell. Pale axon terminals containing much smaller, and often somewhat flattened vesicles (2) are interpreted as Golgi second-type neuron axons. They often have axo-axonic contacts with the optic terminals (arrows) in which the optic terminal appears always to be presynaptic, whereas the subsynaptic web is always seen on the side of terminal (2). There is also a third kind (3) of axon terminal, with larger and more spherical vesicles. The filamentous dendritic profiles are not dendrites, but, as seen in the lower electron micrograph, are spheroid protrusions (Dp) of larger dendrites (D) that remain outside the glomerulus. (Scale 1 $\mu$) Insets, upper left: Light-microscope photographs show the Golgi picture of optic-fiber terminal arborization (top), and a Golgi second-type neuron (bottom), a dendrite of which becomes entangled in glomerular fashion with axon of another Golgi cell (arrow).

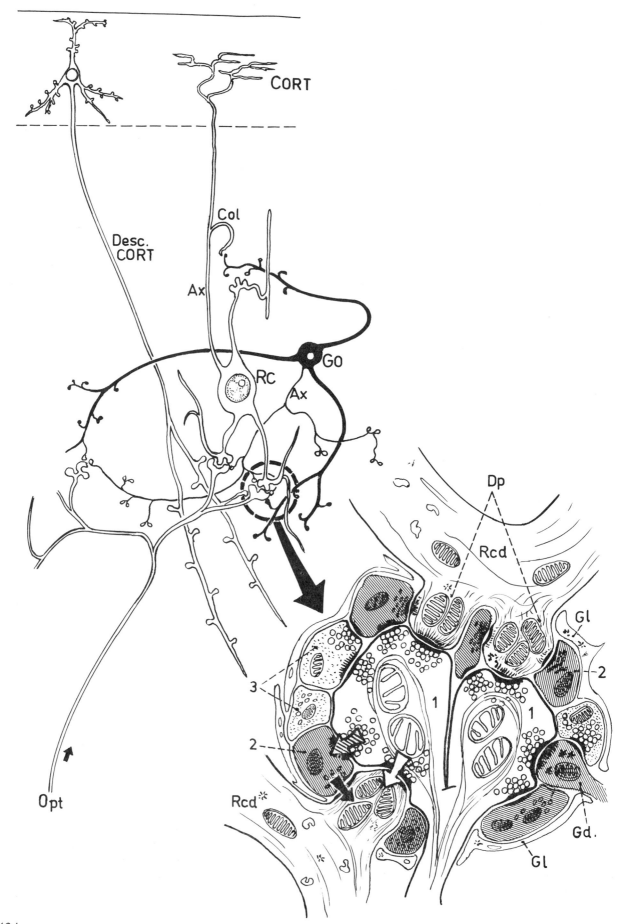

ment is a tool for the type of pre-processing of the information pattern, taken up by various combinations of receptors, that is essential before that pattern can be transmitted to the primary sensory cortical fields. The similarity in the essential features of the neuronal and synaptic arrangements of the main subcortical relay nuclei for such different senses may seem surprising. Similarity in the principle of neuron coupling, however, does not in itself mean similarity in the mode of processing, which depends decisively on the actual connectivity, i.e., both the geometry and topology of the branching of axons and dendrites. In spite of certain similarities, modes of branching are very different, particularly in the LGB and the MGB. This leaves room for considerable differences in the kinds of transforms to which the impulse patterns are subjected in the various relay nuclei, despite the essential similarities of the tools by which the "switching" is performed.

## Other kinds of thalamic glomeruli

Various kinds of glomerular synapses have been described in other parts of the thalamus, mainly in specific nonsensory nuclei (Pappas et al., 1966). I mention only one because of its inverse arrangement: a single dendritic club is its center and various kinds of axon terminals are packed closely in the outer zone (Figure 5). This type of glomerulus is found in the pulvinar and the lateralis posterior nuclear group (LP) of the thalamus (Majorossy et al., 1965). Majorossy and his colleagues have, to some extent, analyzed the neuronal composition of this type of glomerulus by using the light microscope for indirect approach. A direct EM analysis has not, so far, been attempted. The structure and the supposed neuron connectivity of the pulvinar glomeruli are shown in Figure 6. This type of glomerulus shows an even greater regularity than that of the cerebellar glomeruli, and their separation from the environment by a

glial capsule is more obvious than in any other glomerulus hitherto known. The center of the glomerulus is invariably occupied by a single side branch of a principal dendrite or of one of its secondary branches. This is relatively large, bluntly terminating, and often club-shaped.

The diagram in Figure 6 is still highly conjectural, and more detailed degeneration studies on the EM level may require that it be altered considerably. For the time being, the most reasonable explanation of the glomeruli (Szentágothai, 1966) is that they are sites of convergence from corticothalamic fibers of different origin. Light-microscope degeneration studies have shown that descending fibers from various regions (with the possible exception of the frontal region) converge on the same cells and probably even on the same glomerular synapse of the pulvinar. In the absence of direct, unit-level, physiological information, nothing can be said at present about the possible ways in which the incoming impulse patterns are processed in these associative or elaborative thalamic nuclei. Yet the glomeruli, with their unusual wealth of axons thoroughly interlocked and impinging on a single postsynaptic site, are most impressive in themselves.

I cannot resist the temptation to speculate on how such a processing device might work. It would be difficult to imagine that each of the axons, many of which may not even come into direct contact with the central dendrites, has an independent synaptic action. One might conceive, instead, a mechanism of mutual interaction of various axon terminals, both excitatory and inhibitory, the net balance (or outcome) of which would be the synaptic action exercised on the central dendrite. If one now considers that a single neuron may send dendrites to 20 or more glomeruli, one must also envisage a two-step integration in each neuron. One step is the interaction in each glomerulus caused by many converging synaptic influences; the other step is a final computation, at the cellular level, of the net

FIGURE 4 Diagram showing the neuronal arrangement in the lateral geniculate body and the synaptic articulation in the geniculate glomeruli. Optic afferent (Opt) from the retina effects synapses mostly in the glomeruli and generally with the spheroid protrusions (Dp) of the relay cell (Rc) dendrites (Rcd). The protrusions are characteristically localized at sites at which the dendrites make sharp turns or at sites close to their branchings. Delicate spine-like branchlets of Golgi second-type neuron (Go) dendrites (Gd) also participate in the glomeruli (Figure 3 inset), although they cannot be identified with certainty in the electron microscope picture. The axons of Golgi second-type neurons enter the glomeruli and establish synapses with the relay cell dendritic protrusions by means of terminals (2) that contain smaller and often flattened synaptic vesicles. As the Golgi second-type neurons are supposed to be of inhibitory nature, they are drawn in black or are hatched (in the lower right part of the diagram). A third type (3) of axon terminals belongs partly to descending fibers of cortical origin; the origin of the other type-3 endings is uncertain. Axoaxonic contacts between the several axon terminals of the glomeruli are frequent, particularly between the optic (1) and the Golgi axon (2) terminals. In this case, the optic terminal always appears to be structurally presynaptic. The probable synaptic actions at various synaptic sites are indicated in the lower diagram: white arrow indicates excitatory synaptic action from optic terminal to relay dendrite; black arrow the inhibitory action of Golgi axon upon relay cell dendrite; hatched arrow the presynaptic inhibition of inhibitory terminal by optic terminal. Gl, glial capsule.

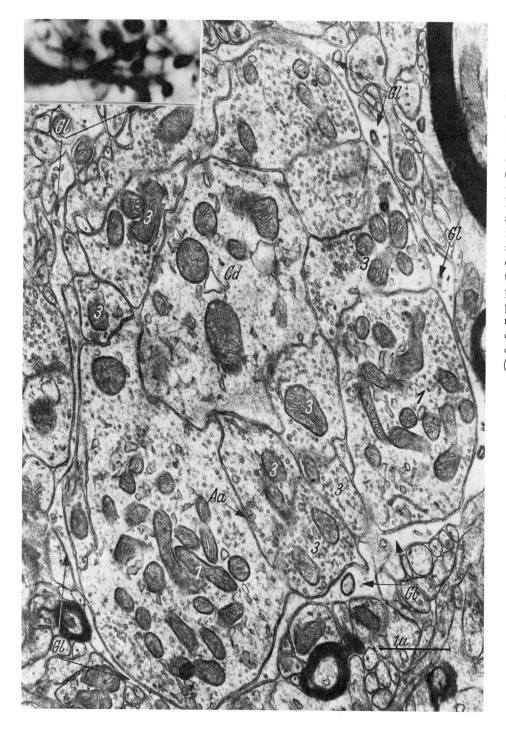

FIGURE 5 Pulvinar glomerulus of the cat. The glomerulus is surrounded by a clear glial capsule (Gl). Its center is occupied by a large, club-shaped central dendrite (Cd), which is surrounded by numerous closely interlocked axon terminals with both ordinary synaptic contacts with the central dendrite and axo-axonic contacts (Aa) with each other. Identification of various kinds of axon terminals is still highly tentative: a large, poorly vesiculated axon terminal with an irregular endoplasmic tubular system has been labeled 1; another has been labeled 3 on the basis of certain similarities to the third type of terminals in the geniculate glomeruli. Upper left inset: light photomicrograph shows principal dendrite of pulvinar neuron with numerous bulbous side branches which correspond in size to the central dendrites of glomeruli. Reumont neurofibrillar stain. (From Majorossy et al., 1965)

FIGURE 6 Tentative explanatory diagram of neuronal connections and synaptic arrangement in the pulvinar. Cortico-thalamic fibers from various areas (sensory motor, sens-mot; parietal, pariet; occipital, occip; temporal, temp) converge upon the same club-shaped branch of the pulvinar relay cell (drawn in outline). Probably axons (Ax) of Golgi second-type neurons (Go) enter the same glomeruli. Synaptic arrangement shown at lower left indicates that, besides usual kinds of synaptic actions (excitatory and inhibitory), various other kinds of axo-axonic (Aa) interactions might play a significant role in these glomeruli. Col, collaterals; D, principal dendrite of pulvinar neuron; Gl, glial capsule; Nps, synapse of the nonglomerular neuropil; Pva, large, poorly vesiculated axon terminal with endoplasmic tubular system.

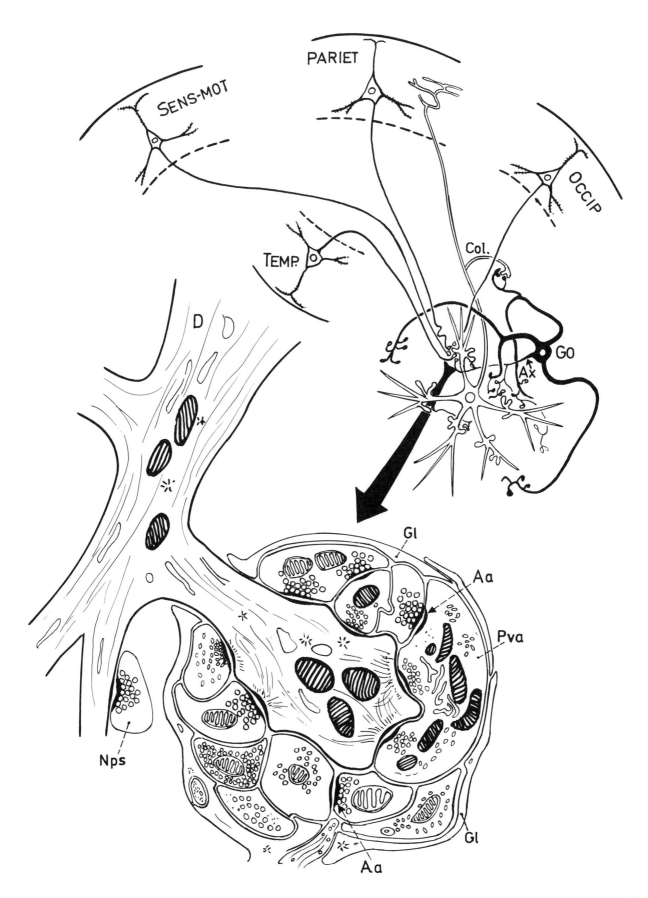

results from synaptic influences at glomerular and non-glomerular synaptic sites (Szentágothai, 1965).

## The synaptic complexes in the substantia gelatinosa

Synaptic complexes very similar to glomeruli have been described in the substantia gelatinosa of the spinal cord (Ralston, 1965; Réthelyi and Szentágothai, 1965). More details are now available as a result of a recent analysis in which we used Golgi, EM, and degeneration information in combination (Réthelyi and Szentágothai, 1969). As seen in Figure 7, a large and sinuous axon terminal occupies a central position, with almost alternating axo-axonic and axodendritic synaptic contacts placed on the surface and protruding into the depression of the surface of the large central axon. The large axon is invariably presynaptic, by structural standards, to the smaller axon terminals. Although the large central axon has axodendritic contacts, these contacts apparently are far fewer than those of the axo-axonic ones. A very regular arrangement is composed of three elements in which the large axon is presynaptic to a small axon profile, which, in turn, is presynaptic to a dendritic profile. The dendritic elements of these complexes undoubtedly belong to the small neurons of the substantia gelatinosa (see Figure 7), the axons of which are supposed to come into contact with the large relay neurons of the dorsal horn (Szentágothai, 1964).

No functional interpretation of the synaptic complexes was possible until we had some information on the origin of the large central axons. As can be recognized in the Golgi picture (Réthelyi and Szentágothai, 1969), the large axons are of local intraspinal origin and derive from pyramid-shaped neurons of the dorsal horn. These neurons receive—as can be deduced from the distribution of their dendrites (Figure 7)—synaptic contacts from both the axonal feltwork of the substantia gelatinosa and the terminal axonal plexus of descending pathways. Electron microscope degeneration studies have shown that at least part of the smaller axons that participate in the synaptic complexes derive from primary sensory neurons, so we have tentatively explained the neuronal arrangement diagrammatically, as shown in the network diagram in Figure 7 (Réthelyi and Szentágothai, 1969). This kind of neuronal network might serve as a structural basis for the primary afferent depolarization, for which some mechanism of this kind was postulated on physiological grounds by Schmidt (1968). Contacts established by descending fibers arising from the somatosensory area $S_I$ (Nyberg-Hansen and Brodal, 1963) on the ventral dendrites of the pyramid-shaped neurons might give a simple anatomical explanation of the observation by Andersen et al. (1964b) of depolarization of primary spinal afferents induced by stimulation of the same area. In addition, such a neuronal circuit might also be the tool for certain gating mechanisms, envisaged by Melzack and Wall (1965), by which the forward conduction of impulses arriving through the dorsal roots could be modified both by descending impulses and by simultaneously active, primary afferent channels. Many observations, particularly on mechanisms for pain perception, point toward the existence of neuronal mechanisms that could modify afferent input, even at the level of the first synapses. Structural and functional information are both, perhaps, too inadequate, for the time being, to allow us to make specific suggestions on the neuronal structure and mode of operation of such a gating neuronal circuit. But we still might venture to say that synaptic complexes with many axo-axonic synapses, probably working on the principle of presynaptic inhibition, may be the decisive tools by which such mechanisms are engineered.

## General comments

These examples may suffice to illustrate the structural principle of the glomerular synapse or the synaptic complex. The assembly of various kinds of axonal and dendritic terminals in a circumscribed structural unit, which has a definite structure and which corresponds to a specific con-

FIGURE 7   Neuronal network diagram of the substantia gelatinosa of spinal cord (top) and synaptic articulation in its "synaptic complexes" (bottom). Primary afferents send axon collaterals to substantia gelatinosa (S gel) where they establish synapses with the dendrites of the small substantia gelatinosa neurons (Sgn). Axons (Ax) of these neurons can both secure forward conduction through the large ascending relay neurons (ARC) of lamina IV of Rexed (hatched) and/or stimulate pyramidal neurons (Pyr and dotted) with coarse, sinusoid ascending axons. The minute synaptic relations are shown in part below. Primary sensory axons (Paa) have contact with both the substantia gelatinosa cell dendrites and with the large sinusoid axon terminals (Pyr ax) of the pyramidal cells. Relatively few synaptic contacts are established between the pyramidal cell axons and the substantia gelatinosa cell dendrites. The synapses between primary sensory axons and gelatinosa cell dendrites appear structurally to be of the usual excitatory type; the vesicles of the pyramidal cell axons are unusually large. The most likely explanation of the various synaptic actions is indicated by the arrows: white arrows signify excitation and hatched ones presynaptic inhibition and depolarization of the primary afferents.

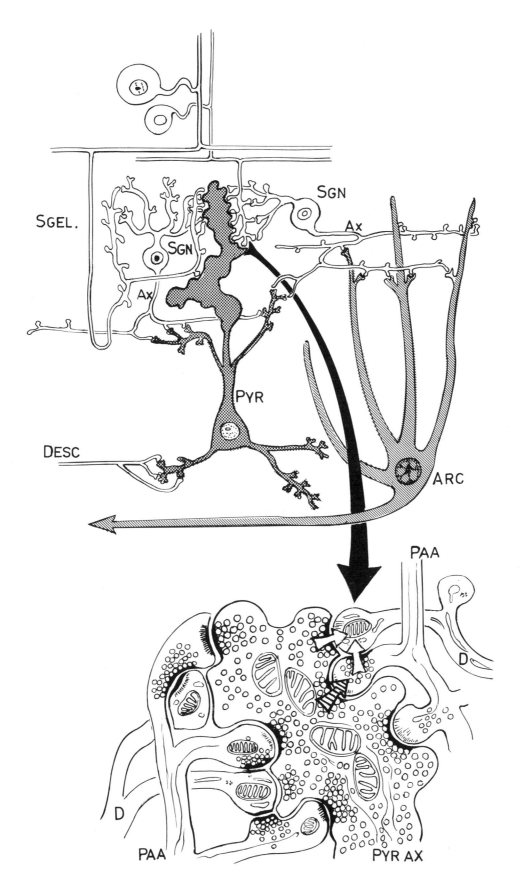

nectivity pattern of neurons, suggests that the glomeruli are units of integrative functions. This integrative function may be relatively simple, as in the cerebellar glomeruli, in which there is a conventional excitatory action from the mossy rosettes to granule dendrites and to Golgi dendrites. In addition, a postsynaptic inhibitory action of Golgi axons is exercised on the granule-cell dendrites.

The task that presents itself in the cerebellum could have been solved in a more usual way—for example, by having the mossy-granule synapse at the dendrites and the Golgi-granule synapse at the cell bodies. If, however, the extremely large number of granule cells is considered (about $6 \times 10^6/\text{mm}^3$ of granular-layer tissue), the simplification and economy in the use of living matter achieved by this arrangement become apparent. One mossy rosette in the center can transmit excitation to a hundred granule dendrites. Such transmission, however, can be inhibited by a single preterminal branch of a Golgi-cell axon that distributes its inhibitory terminals in the outer zone of the glomerulus (see Figure 1, inset at bottom right). It is obvious that such use of living matter is highly economical, when compared with a more conventional kind of connectivity—for example, if the mossy afferents and the Golgi-cell axon terminals had to approach each granule cell separately. Such an arrangement would require roughly a hundred times more branches of the mossy afferents and of the Golgi-cell axons to reach each individual granule cell.

Apart from the obvious advantages in structural economy, the cerebellar glomeruli also may be tools that increase the safety factor in transforming the information patterns they receive. Each granule-cell neuron picks up excitation from four different mossy afferents from among a population that is probably highly mixed in origin. Golgi inhibition is probably unitary in all the glomeruli with which any given Golgi neuron has contact, because of their territorial (nonoverlap) organization. The granule neurons, in turn, secure a further generalization of the received excitation pattern by transmitting through their parallel fibers to 200 to 300 Purkinje cells situated in a longitudinal row but with a convergence of about $2 \times 10^5$:1 (i.e., so many parallel fibers to one Purkinje cell).

The situation is relatively simple, however, in the cerebellar glomeruli, because there is no known interaction between the two kinds of axon terminals (the mossy rosettes and the Golgi endings), and there are no axo-axonic contacts. The glomerular structure becomes more meaningful and of specific functional significance when, as is the case in all other glomeruli, the structure makes it almost certain that such an interaction does occur. The concept of presynaptic inhibition (Eccles et al., 1961) postulates specific sites of interaction, mainly between axon terminals. The primary (or other sensory) afferent depolarization, which

has been analyzed so thoroughly in various parts of the CNS, requires axo-axonic contacts between the axons of central neurons and the terminals of the afferents, in which the latter are postsynaptic. These conditions are clearly fulfilled in the substantia gelatinosa complexes, where the large, presynaptic, sinuous axon terminals belong to local interneurons that could well be activated by primary afferents over one or two interneurons, exactly as required by physiological observations (Schmidt, 1968). Because the structural polarity of axo-axonic contacts is the reverse of what would be required, the glomeruli of the geniculate body and the VPL do not lend themselves to similar explanations.

The role of the glomerular synapses is probably much more specific in the sensory relay nuclei. The very fact that glomeruli are always found in these nuclei and in other central structures that have a decisive function in the analysis of sensory patterns (for example, in the optic tectum of lower vertebrates, Sétáló and Székely, 1967), points to some specific function that analyzes spatial information patterns. An obvious possibility is the mechanism suggested for the LGB. A large body of accumulated evidence favors the assumption that Golgi second-type neurons in these subcortical nuclei are inhibitory (e.g., Andersen et al., 1964a). Additionally, the terminals of their axons are of the inhibitory type described by Uchizono (1965). The structure of the glomeruli in the subcortical sensory-relay nuclei suggests, therefore, that they are sites for the transform of information patterns that use either presynaptic disinhibition, or, possibly, the reverse—enhancement of inhibition. The glomeruli may, of course, be sites of external descending cortical or reticular interaction as well, interfering with the transmission of sensory information to the cortex.

Because we lack any direct physiological information, next to nothing can be said about the various kinds of glomerular synapses in other nonsensory parts of the thalamus. The mutual entanglement of axons of various origins becomes prevalent above the axodendritic contacts, which may indicate that integrative functions of hitherto completely unknown character must be considered.

It is probably premature to make far-reaching generalizations about the possible functional significance of the glomeruli. Structural economy, as shown in the example of the cerebellar glomeruli, may be a decisive factor in other glomeruli as well. All the possible functions attributed to the glomeruli of the sensory-relay nuclei could be achieved by conventional types of synaptic articulations, but this would require infinitely more neuronal connecting material and separate contacts, i.e., all kinds of structural complexities. As a crude analogy to electronics, one might consider the glomeruli as "integrated circuits" of the nervous system. It remains to be seen, of course, whether such analogies can help in our understanding of the functional significance of the glomerular synapses. For the time being, there is little, if any, direct physiological evidence (except for the cerebellar glomeruli) about their mode of functioning.

In many glomeruli, the regular occurrence of dendro-dendritic attachments with a specialized desmosomoid structure may also have some specific significance. Ralston and Herman (1969) have interpreted some of the glomerular neural processes that have synaptic contacts, including the typical accumulations of synaptic vesicles as dendrites, on the basis that they contain tubules and some ribosomes. It remains to be seen whether this new concept of dendritic synapses, which would introduce additional intricacies, will stand the trial of continued studies. The spinelike processes (spinules) of the dendritic ends that are frequently met with in the cerebellar glomeruli (Figure 1) and the more ordinary spines that are found, for example, in the LGB glomeruli of certain species (especially the dog), are of great interest, as they appear to depend on functional circumstances. It has been tentatively suggested (Eccles et al., 1967, Chapter VII) that they indicate lasting changes induced by functional loading. In the LGB glomeruli of the dog, the glomerular spines that invade the optic terminals exclusively do not develop if the animal is deprived of patterned vision (Szentágothai, 1968c; Szentágothai and Hámori (1969). All this might indicate that, functionally, glomerular synapses may have more complexity than economically built switches for the interaction of more than two elements.

## Summary

Glomerular synapses or synaptic complexes are specific contact arrangements between different but determined kinds of axon terminals and a single type (or sometimes several) of dendritic terminals (or dendritic appendages). In order to appraise the similarities or differences, or both, of glomerular synapses, four characteristic examples have been compared: (1) the glomeruli in the cerebellar cortex; (2) glomerular synapses of the subcortical thalamic and geniculate sensory relay nuclei, of which the LGB glomeruli are treated in more detail as the most characteristic and best known of this kind; (3) the glomerular synapses of the pulvinar; and (4) the synaptic complexes of the substantia gelatinosa.

A remarkable similarity exists in the essential features of glomerular synapses in all subcortical sensory relay nuclei (VPL, LGB, MGB). The main synapse in each complex is invariably the contact between the specific afferent and the dendrites of the corticopetal relay cells. But there is also invariably an inbuilt synaptic apparatus of local (probably inhibitory) interneurons. The interneuronal axon terminals are clearly presynaptic to the dendritic terminals (or appendages) of the relay cells. A third standard contact arrangement in these glomerular synapses is the axo-axonic contact between the sensory afferents and the local (inhibitory) interneuron terminals, in which the former invariably appears to be presynaptic. Various speculations can be made on the functional significance of this arrangement. A fourth common feature of the glomeruli in the subcortical sensory-relay nuclei are small terminals of descending cortical fibers that originate from the respective sensory cortical region. They resemble excitatory terminals, but their involvement in the glomerular synapses is less unequivocal, so that their functional significance may be not so much an inbuilt direct action on the transmission from primary afferents to relay cells as an indirect action on the interneurons of the relay nuclei.

The similarities of the glomerular synapses in the subcortical sensory relay nuclei might indicate that they are essential tools for information processing in that particular type of neural subsystem. In spite of obvious differences in the requirements of subcortical processing in the somatosensory, optical, and acoustical systems, the operations needed for projecting the sensory pattern to the respective sensory cortexes—which show such clear similarities among themselves—can very well be imagined as depending on subcortical subsystems built on essentially similar principles.

The synaptic complexes of the substantia gelatinosa are those closest in principal arrangement to the glomerular synapses of the sensory-relay subsystems. They differ from the latter, nevertheless, in that they appear to be devices for depolarization of primary afferents and probably for presynaptic inhibition. In this respect, the functional interpretation of the substantia gelatinosa complexes seems to be more straightforward and more realistic.

The glomerular synapses of the cerebellar cortex are essentially different and relatively easy to interpret functionally. They are devices to insure a very high divergence from a single mossy terminal to about 100 granule neurons and have an inbuilt, inhibitory control of this transmission. With respect to the enormous number of granule-cell neurons, the highly economic solution of this neuronal coupling is significant.

The glomerular synapses of the pulvinar give another example that is fundamentally different from all other kinds. Even if the functional explanations that could be given today are highly tentative, at best, all these examples show that various neural subsystems have peculiar and specific kinds of synaptic arrangements. The speculations offered for the functional significance of various kinds of synaptic glomeruli may prove inadequate and will probably soon need major revisions, but even in the preliminary stage they might contribute to our knowledge of the structural substrate of some subsystem operations.

ANDERSEN, P., C. McC. BROOKS, J. C. ECCLES, and T. A. SEARS, 1964a. The ventro-basal nucleus of the thalamus: Potential fields, synaptic transmission and excitability of both presynaptic and post-synaptic components. *J. Physiol.* (*London*) 174: 348–369.

ANDERSEN, P., J. C. ECCLES, and T. A. SEARS, 1964b. Cortically evoked depolarization of primary afferent fibers in the spinal cord. *J. Neurophysiol.* 27: 63–77.

ANGEL, A., F. MAGNI, and P. STRATA, 1965. Evidence for presynaptic inhibition in the lateral geniculate body. *Nature* (*London*) 208: 495–496.

COLONNIER, M., and R. W. GUILLERY. 1964. Synaptic organization in the lateral geniculate nucleus of the monkey. *Z. Zellforsch. Mikroskop. Anat.* 62: 333–355.

ECCLES, J. C., R. M. ECCLES, and F. MAGNI, 1961. Central inhibitory action attributable to presynaptic depolarization produced by muscle afferent volleys. *J. Physiol.* (*London*) 159: 147–166.

ECCLES, J. C., M. ITO, and J. SZENTÁGOTHAI, 1967. The Cerebellum as a Neuronal Machine. Springer Verlag, New York.

FOX, C. A., D. E. HILLMAN, K. A. SIEGESMUND, and C. R. DUTTA, 1967. The primate cerebellar cortex: A Golgi and electron microscope study. *Progr. Brain Res.* 25: 174–225.

GRAY, E. G., 1961. The granule cells, mossy synapses and Purkinje spine synapses of the cerebellum: Light and electron microscope observations. *J. Anat.* 95: 345–356.

HÁMORI, J., 1965. Identification in the cerebellar isles of Golgi II axon endings by aid of experimental degeneration. *In* Proceedings of the Third European Regional Conference on Electron Microscopy, Prague, 1964 (M. Titlbach, editor). Publishing House of the Czechoslovak Academy of Sciences, Prague, vol. B, pp. 291–292.

HÁMORI, J., and J. SZENTÁGOTHAI, 1966. Participation of Golgi neuron processes in the cerebellar glomeruli: An electron microscope study. *Exp. Brain Res.* 2: 35–48.

IWAMA, K., H. SAKAKURA, and T. KASAMATSU, 1965. Presynaptic inhibition in the lateral geniculate body induced by stimulation of the cerebral cortex. *Jap. J. Physiol.* 15: 310–322.

JAKOB, A., 1928. Das Kleinhirn. *In* Handbuch der mikroskopischen Anatomie des Menschen (W. v. Möllendorff and W. Bargmann, editors). Band 4, Teil 1, Nervensystem. Springer Verlag, Berlin, pp. 674–916.

JONES, E. G., and T. P. S. POWELL, 1969a. Electron microscopy of synaptic glomeruli in the thalamic relay nuclei of the cat. *Proc. Roy. Soc., ser. B, biol. sci.* 172: 153–171.

JONES, E. G., and T. P. S. POWELL, 1969b. An electron microscopic study of the mode of termination of cortico-thalamic fibres within the sensory relay nuclei of the thalamus. *Proc. Roy. Soc., ser. B, biol. sci.* 172: 173–185.

KARLSSON, U., 1967. Three-dimensional studies of neurons in the lateral geniculate nucleus of the rat. III. Specialized neuronal contacts in the neuropil. *J. Ultrastruct. Res.* 17: 137–157.

KIRSCHE, W., H. DAVID, E. WINKELMANN, and I. MARX, 1965. Elektronenmikroskopische Untersuchungen an synaptischen Formationen im Cortex cerebelli von Rattus rattus norvegicus, Berkenhoot. *Z. Mikroskop. Anat. Forsch.* 72: 49–80.

MAJOROSSY, K., and M. RÉTHELYI, 1968. Synaptic architecture in the medial geniculate body (ventral division). *Exp. Brain Res.* 6: 306–323.

MAJOROSSY, K., M. RÉTHELYI, and J. SZENTÁGOTHAI, 1965. The large glomerular synapse of the pulvinar. *J. Hirnforsch.* 7: 415–432.

MELZACK, R., and P. D. WALL, 1965. Pain mechanisms: A new theory. *Science* (*Washington*) 150: 971–979.

NYBERG-HANSEN, R., and A. BRODAL, 1963. Sites of termination of corticospinal fibers in the cat. An experimental study with silver impregnation methods. *J. Comp. Neurol.* 120: 369–391.

PAPPAS, G. D., E. B. COHEN, and D. P. PURPURA, 1966. Fine structure of synaptic and nonsynaptic neuronal relations in the thalamus of the cat. *In* The Thalamus (D. P. Purpura and M. D. Yahr, editors). Columbia University Press, New York, pp. 47–75.

PECCI SAAVEDRA, J., and O. L. VACCAREZZA, 1968. Synaptic organization of the glomerular complexes in the lateral geniculate nucleus of the cebus monkey. *Brain Res.* 8: 389–393.

PETERS, A., and S. L. PALAY, 1966. The morphology of laminae A and $A_1$ of the dorsal nucleus of the lateral geniculate body of the cat. *J. Anat.* 100: 451–486.

RALSTON, H. J., III, 1965. The organization of the substantia gelatinosa Rolandi in the cat lumbosacral spinal cord. *Z. Zellforsch. Mikroskop. Anat.* 67: 1–23.

RALSTON, H. J., III, and M. M. HERMAN, 1969. The fine structure of neurons and synapses in the ventrobasal thalamus of the cat. *Brain Res.* 14: 77–97.

RAMÓN Y CAJAL, S., 1911. Histologie du Système Nerveux de l'Homme et des Vertébrés. Vol. 2, Maloine, Paris.

RÉTHELYI, M., and J. SZENTÁGOTHAI, 1965. On a peculiar type of synaptic arrangement in the substantia gelatinosa of Rolando. *In* 8th International Congress of Anatomists. (Abstracts of papers, scientific demonstrations, and films), Wiesbaden, 1965. Georg Thieme Verlag, Stuttgart, p. 99.

RÉTHELYI, M., and J. SZENTÁGOTHAI, 1969. The large synaptic complexes of the substantia gelatinosa. *Exp. Brain Res.* 7: 258–274.

SCHMIDT, R. F., 1968. The functional organization of presynaptic inhibition of mechanoreceptor afferents. *In* Structure and Function of Inhibitory Neuronal Mechanisms (C. von Euler, S. Skoglund, and U. Söderberg, editors). Pergamon Press, Oxford and New York, pp. 227–233.

SÉTÁLÓ, G., and G. SZÉKELY, 1967. The presence of membrane specializations indicative of somato-dendritic synaptic junctions in the optic tectum of the frog. *Exp. Brain Res.* 4: 237–242.

SZENTÁGOTHAI, J., 1962. Anatomical aspects of junctional transformation. *In* Information Processing in the Nervous System. Vol. 3. Proceedings of the International Union of Physiological Sciences. XXII. International Congress Leiden, 1962 (R. W. Gerard and J. W. Duyff, editors). International Congress Sr. No. 49, Excerpta Medica Foundation, Amsterdam, pp. 119–136.

SZENTÁGOTHAI, J., 1963. The structure of the synapse in the lateral geniculate body. *Acta Anat.* 55: 166–185.

SZENTÁGOTHAI, J., 1964. Neuronal and synaptic arrangement in the substantia gelatinosa Rolandi. *J. Comp. Neurol.* 122: 219–239.

SZENTÁGOTHAI, J., 1965. Complex synapses. *In* Aus der Werkstatt des Anatomen (W. Bargmann, editor). Georg Thieme Verlag, Stuttgart, pp. 147–167.

SZENTÁGOTHAI, J., 1966. Some general structural principles of thalamo-cortical and metathalamo-cortical connections. *In* Cortico-Subcortical Relationship in Sensory Regulation (D. Gonzales Martin, editor). Academy of Sciences, Havana, Cuba, pp. 327–342.

SZENTÁGOTHAI, J., 1968a. Structure-functional considerations of the cerebellar neuron network. *Proc. Inst. Elec. Electron. Engrs.* 56: 960–968.

SZENTÁGOTHAI, J., 1968b. Neuronhálózatok és neuronhálózati modellek (Neuron networks and network models). Inaugural lecture. *Magyar Tud. Akad. Biol. Oszt. Közl.* 11: 61–71. (In Hungarian.)

SZENTÁGOTHAI, J., 1968c. Growth of the nervous system: An introductory survey. *In* Growth of the Nervous System/A Ciba Foundation Symposium (G. E. W. Wolstenholme and M. O'Connor, editors). Little, Brown and Co., Boston, pp. 3–12.

SZENTÁGOTHAI, J., and J. HÁMORI, 1969. Growth and differentiation of synaptic structures under circumstances of functional deprivation and of lack in distant connexions. *In* Symposium of the International Society for Cell Biology, Vol. 8. (S. H. Barondes, editor). Academic Press, New York, pp. 301–320.

SZENTÁGOTHAI, J., J. HÁMORI, and TÖMBÖL, 1966. Degeneration and electron microscope analysis of the synaptic glomeruli in the lateral geniculate body. *Exp. Brain Res.* 2: 283–301.

TÖMBÖL, T., G. UNGVÁRY, F. HAJDU, M. MADARÁSZ, and G. SOMOGYI, 1969. Quantitative aspects of neuron arrangement in the specific thalamic nuclei. *Acta Morph. Acad. Sci. Hung.* 17: 229–313.

UCHIZONO, J., 1965. Characteristics of excitatory and inhibitory synapses in the central nervous system of the cat. *Nature (London)* 207: 642–643.

UCHIZONO, K., 1967. Synaptic organization of the Purkinje cells in the cerebellum of the cat. *Exp. Brain Res.* 4: 97–113.

# 41 Elementary Processes in Selected Thalamic and Cortical Subsystems—the Structural Substrates

MADGE E. SCHEIBEL and ARNOLD B. SCHEIBEL

*My hypothesis then is that thought models, or parallels reality—that its essential feature is not 'the mind', 'the self', 'sense-data', nor propositions but symbolism, and that this symbolism is largely of the same kind as that which is familiar to us in mechanical devices which aid thought and calculation.* CRAIK, 1943, The Nature of Explanation.

THE INTRINSIC ORGANIZATION of the thalamus, especially in the carnivore and primate, attains a level of complexity which sets it apart from all other regions of the brainstem. Even the cerebral cortex, for which it serves as portal, can afford no greater challenge in terms of field organization or anatomical diversity. The range of its operational modes (see Purpura, this volume) documents the opulence of its structural variety and marks it as an especially fertile field for attempts at structuro-functional correlation and operations analysis.

MADGE E. SCHEIBEL and ARNOLD B. SCHEIBEL Departments of Anatomy and Psychiatry and Brain Research Institute, U.C.L.A. Medical Center, Los Angeles, California

## Historical introduction

The word thalamus comes down to us from the writings of Galen, who thought it supplied animal spirits to the optic nerve. Because the word is translatable as "anteroom," it appropriately reflects our contemporary notion of its portal function to the hemispheres. The first coherent description of the intrinsic nuclei was contributed by Luys (1865), who also called attention to the probable sensory function of large portions of the thalamic fields. In a series of important experiments involving selected cortical ablations, Gudden (1870) documented the projection of a number of thalamic subsystems upon cortex and, a quarter of a century later, Ramón y Cajal (1900) described the intricacy of its neuropil. The classical work on thalamic descriptive morphology and projection systems remains that of Walker (1938); its functional counterpart, emphasizing the importance of two-way circulation between thalamus and cortex is probably to be found in the work of Dusser de Barenne and McCulloch (1938, 1941). The rapidly expanding roster of recent functional studies can be sampled in the bibliography that follows the chapter by Purpura (this volume).

The purpose of this review is to evaluate certain features of telodiencephalic morphology, not as an end in itself, but on the basis of their probable substrate importance to a group of operations generally considered significant in the processing of information. It follows that not all aspects of thalamic structure will be reviewed. Rather, we limit consideration to those systems that appear capable of providing a biological base for a selected group of elementary operations. These are discussed in their functional context by Purpura (this volume). In some cases, it will seem indicated to discuss neural assemblies whose functional correlates are not considered elsewhere. Here, again, our conviction that such system substrates play a significant role in information processing, whether the relevant operation has been clearly identified or not, motivated our choice.

It has been customary to epitomize anatomical data as a series of circuits that presumably represent those lines of communication, cell sequences, and synapses over which information is most likely to flow. Such an approach suggests, by implication, a similarity in design, and possibly in process, to fairly complex yet ultimately knowable electronic circuits. Although this approach is not without its power, the fact remains that nervous systems are far more than complex electronic circuits, and multidimensional approaches are ineluctably necessary. We have only to look at the profusion of neurochemically discrete subsystems that have recently been identified in the brainstem reticular core (Dahlström and Fuxe, 1965) to document such a notion. Our own studies of neural substrate are based primarily on modifications of the silver impregnation methods of Golgi, which are, so far as we know, silent on the chemical nature of the neural ground. Nevertheless, we intend to avoid the usual linear circuit diagrams as much as possible in favor of an approach that we first used in our study of the inferior olive, 15 years ago (Scheibel and Scheibel, 1955). Then, as now, we believed that *field analysis* might ultimately prove more significant than the traditional line-dot-line drawings. As did Herrick, we conceive of the field as "an organized living structure in action characterized both by an architecture and by a continuing pattern of fluctuation in the excitatory state" (Herrick, 1948).

The range of structural entities that we consider, i.e., the grain of our inquiry, obviously depends on the methodology involved. Most of these data depend on Golgi impregnations examined under the light microscope, so it is clear that the vast trove of information available at the submicroscopic level is not considered here, save when we specifically utilize some datum from a colleague's work. Nevertheless, we believe that, at the level of resolution to which we are committed, several meaningful statements about the structure and interrelationship of field elements are possible. This should serve as an adequate preface, not only to the elegant microelectrode analyses of Purpura (this volume) but also to the more detailed ultramicroscopic studies that are just becoming available.

## Modality and topographical specificity

One of the operational characteristics of the thalamic somatosensory relay appears to be the considerable degree of synaptic security that marks transmission through the ventrobasal (Vb) nuclear complex. Synaptic effectiveness combined with maintenance of mode and locus specificity suggests the presence of structural substrates uniquely adapted to this type of *high-fidelity transmission*. The presence of bushy terminals capping the thalamopetal relay from more caudal sensory relay structures (the gracile and cuneate nuclei) undoubtedly provide such a structural motif. These arbors are generated by repeated branching of the terminal, unmyelinated segments of each parent axon, with the consequent production of a roughly cone-shaped neuropil plexus. Each resultant structure is characterized in turn by a geometrical increase in the amount of axonal surface area relative to that of the parent fiber and the presence of large numbers (between $10^3$ and $10^4$) of branch points (Figure 1). Additionally, in the adult preparation, it appears that such fibers develop clusters of large boutons on at least some of their terminal elements, a secondary specialization that may be related to a proportion of the synaptic transactions effected by these elements.

Immersed in these terminal arbors are the postsynaptic receptive elements. These include the thalamocortical relay cells, several populations of local circuit (short-axoned cell) elements the relative densities of which are species-dependent, and restricted numbers of large, sparsely branched, dendrite-bearing neurons which we have called "integrator elements" (Figure 2).

The thalamocortical relay cells are most notable for the peculiar structure of the dendritic domain. Each of the four to 10 principal dendrite shafts leaving the cell body (soma), which is 20 to 30 $\mu$ in size, quickly breaks up into an intensively reduplicated shaft system that produces a tufted or bushy appearance (*Buschzellen* or *cellules en buisson* of Ramón y Cajal, 1911). It would appear that this structural characteristic maximizes postsynaptic surfaces within a relatively limited area, i.e., the area subtended by one or several of the presynaptic arbors. The net result of these presynaptic and postsynaptic structural specializations offers optimal conditions for axodendritic interaction between ascending sensory relay (lemniscal) fibers and thalamocortical relay cells. Szentágothai (1967) believes that the late-developing, grapelike clusters of boutons on the bushy terminal elements represent an added specialization ensuring *high-priority transmission* in this system. Although such an

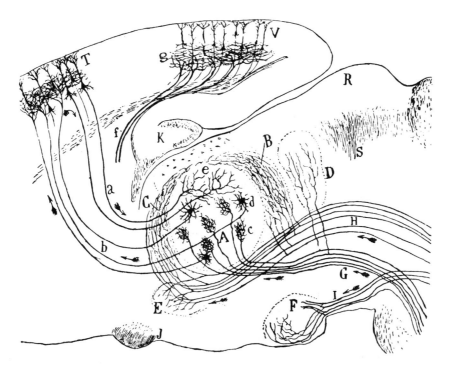

FIGURE 1 Summarizing illustration by Ramón y Cajal which shows some of the ascending thalamopetal pathways and the main axonal elements interconnecting thalamus and cortex. A, principal sensory nucleus; B, C, accessory sensory or trigeminal nuclei (*sic*); D, posterior thalamic nucleus; E, nucleus of the zona incerta; F, lateral mammillary nucleus; G, medial lemniscus; H, central path for the trigeminal nuclei and other systems; I, mammillary peduncle; J, optic chiasm; K, hippocampus; R, superior colliculus; S, elements of posterior commissure; T, motor cortex; V, visual cortex; a, corticothalamic fibers; b, thalamocortical fibers; c, bushy arbor generated by terminating medial lemniscus fiber; d, thalamocortical projection cell; e, terminal plexus of corticothalamic fiber; f, fibers of the optic thalamocortical radiation; g, terminal afferent plexus of the optic radiation. (From Ramón y Cajal, 1911, Figure 323; reproduced by permission of Consejo Superior de Investigaciones Cientificas, Madrid.)

FIGURE 2 Summary of cell types found in the ventrobasal nuclear complex. A, thalamocortical projection cell; B, local circuit (Golgi type II) cell with very restricted axonal path; C, local circuit cell with more extensive axon system and clawlike terminals on some dendrites; D, similar, presumably local-circuit cell, but without demonstrable axon; E, local circuit cell with flattened axonal and dendritic domain. This type of cell is found in the ventral anterior nucleus and also in the ventrobasal nucleus, shaped against the periphery of the nuclear field. F, large reticular-like cell with bifurcating axon (integrator cell); G, protoplasmic astrocyte; H, oligodendrocyte variant with multiple swirled processes; a, clawlike structures at dendrite tips of some local circuit cells. Drawn from many sections of 20- to 70-day cat thalamus stained by Golgi variant. × 600.

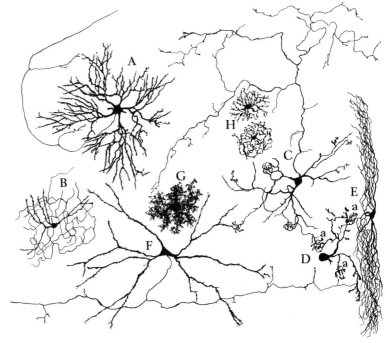

idea is attractive and appears to gain some degree of support from electron-microscope analysis (Tömböl, 1967; Ralston and Herman, 1969), we have also found that such bouton clusters appear to interdigitate with clawlike structures crowning the tips of some local circuit cells (Figure 2). These structures are reminiscent of the granule-cell claws in the cerebellum and underline our conviction that present knowledge of the synaptic anatomy of the thalamus is still far from complete.

No matter what further structural analyses of this system have to offer, we can be reasonably certain that sequences of thalamopetal sensory afferents generate ensembles of bushy arbors of relatively restricted diameter (250 to 400 $\mu$, depending on species), partially overlapping one another, and engulfing limited numbers of thalamocortical relay cells (as few as three to 10 in rodents, and 20 to 50 in monkeys). The resultant anatomical pattern appears to insure spatially localized, intensively reduplicated, axodendritic interactions, a notion that receives powerful support from physiological data (Purpura, this volume).

Because of the importance we attach to the bushy arbor in insuring high-fidelity transmission, further reflection on this unique structure seems appropriate. This structural paradigm, marked by spatially localized and intensively reduplicated bifurcations of terminating sensory fibers, has been recognized in primitive vertebrates (Herrick, 1948) and in such invertebrate forms as the Cephalopoda (Ramón y Cajal, 1917) and the Arthropoda (Ramón y Cajal and Sanchez, 1915). These studies point to a relationship between sensory representational systems, featuring some degree of precision, and the bushy terminal pattern. We have previously suggested that this structural idiom ". . . may well represent a neurologically archaic mode of generating a relatively intense, localized disturbance of appreciable duration among a group of postsynaptic components following a stimulus with some localizing signature" (Scheibel and Scheibel, 1966c).

Mountcastle (1961) has called attention to another functional characteristic of lemniscal neurons, whether at lower brainstem, thalamic, or cortical levels. This is the tendency of postsynaptic elements in these systems to respond with brief repetitive trains of impulses, even when the input over the first-order afferents is ". . . made as synchronous as possible by electrical stimulation of the skin." He contrasted this type of postsynaptic volley with the individual impulse that characterizes the output of the motor neuron, even when activated by trains of presynaptic stimuli. Intracellular analyses show that such output trains occur on the summits of prolonged excitatory postsynaptic potentials (EPSPs) (Andersen et al., 1964). Eccles (1964) has suggested that this reflects prolonged activity of synaptic transmitter substance, a notion that receives some degree of support from the discovery of glomeruli (Szentágothai, 1963; Pappas et

al., 1966) or synaptic complexes (Ralston and Herman, 1969) in many thalamic stations. These structures appear to be characterized by the presence of a cluster of synaptic structures of varying morphology, usually surrounded by a protective sheath of glial lamellae. Their possible significance as an anatomically economic means of information processing is reviewed by Szentágothai in this volume, but it is the presence of the glial capsule which is assumed to provide the means for prolongation of the postsynaptic potential by delaying dissipation of the transmitter. Although this notion is attractive, it clearly depends on the efficacy of several thicknesses of glial membrane to dam transmitter molecules, which are ordinarily rather mobile. It also presupposes that such sheaths encapsulate the synaptic complex spherically. So far, we are not familiar with any three-dimensional reconstructions of such systems that would support this concept. For the moment, then, the "lingering transmitter" explanation of prolonged thalamic EPSPs and the accompanying repetitive unit discharge patterns must await both functional confirmation and a proved anatomical substrate.

In the meantime, it seems worthwhile to mention an alternative substrate for the prolonged action of the postsynaptic elements that we have discussed previously (Scheibel and Scheibel, 1966c). The development of each bushy arbor involves increasing numbers of branch points, gradual attenuation in the diameter of the parent axon as its branching continues, and a geometric increase in axonal surface area. Each of these conditions must tax the spike-propagative abilities of the system, and it seems reasonable to expect that, in many cases, the original afferent spike sequence might die at branch points or at many stations in the arbor as branches become finer and more numerous and conduction safety factors are exceeded (McCulloch et al., 1952). The result could well be the generation of an intense, relatively long-lasting depolarization, maintained within the confines of the arbor and presumably capable of influencing the contained postsynaptic elements for an appreciable period of time. We conceive of the "field effect" as intensifying the powerful axodendritic articulations effected by synaptic structures on terminal arboreal fibers, thereby adding to the synaptic effectiveness and specificity of this lemniscal link.

We have paid particular attention to terminal patterns generated by the medial lemniscus, because it represents the main source of new information from body surface and underlying muscle masses. This may accordingly be considered the principal means for *updating information* in the ventrobasal-cell matrix and in the thalamocortical complex for which it serves as portal. The individual terminal arbors appear (in rodents at least) to be arranged in concentric lamellae, which can be seen in both sagittal and frontal sections (Figure 3). We assume that these represent part of the anatomical substrate for mode and locus specificity maintained through this field. Equally specific, if differently patterned, arrangements can, however, be mapped out for axon terminals from other sources, each involving unique groupings and sequences of Vb neurons from the total pool (Figures 3 and 4). The resulting convergence of afferent systems and fractionation of cell domains generated by each afferent must lead to essentially unique combinations of afferent loading on virtually every cell of the Vb pool (Figure 10). The possible significance of this complex

patterning to *information readout and retrieval* is worth noting and is commented on below.

We have made no mention of the possible role of collateral or "surround" inhibition as a means of enhancing the modal and topographical specificity maintained through the ventrobasal thalamocortical relay. As is indicated in a subsequent section on synchronization, the usually subsumed substrate of recurrent collaterals and a population of local-circuit (short-axoned) interneurons cannot be dem-

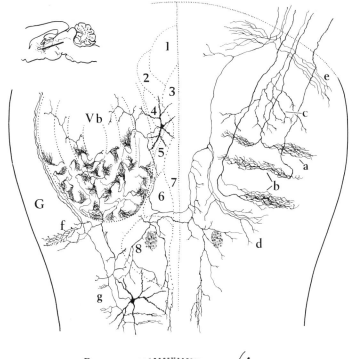

FIGURE 3 Neuropil patterns characteristic of thalamic fields as seen in horizontal section. Vb, Ventrobasal complex; G, Lateral geniculate; a, corticothalamic discoid afferent terminal developing symmetrically from parent fiber; b, similar discoid afferents developing asymmetrically from parent fibers; c, corticothalamic fiber generating broadly branching terminal pattern; d, part of the terminal system of corticothalamic fiber to medial nonspecific nuclear system composed in part of: 1, parataenial; 2, anterior ventral; 3, interanteromedial; 4, anterior medial; 5, central medial; 6, central lateral; 7, interventricular; 8, centre median-parafascicular complex. Also, e, part of plexus generated by centrifugal fibers in area of n. reticularis thalami; f, afferent terminals in lateral geniculate nucleus; and g, in tegmentum. Drawn from several sections of 10- to 20-day-old rat brain stained by rapid Golgi variant. × 200.

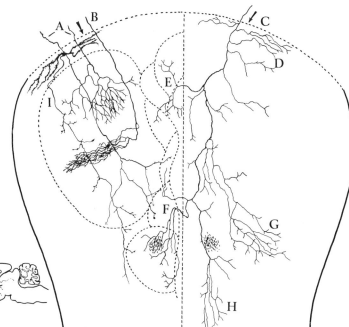

FIGURE 4 Thalamic afferent systems descending from more rostral stations as seen in horizontal section. Corticothalamic axons (A) generate widely branched terminal patterns spreading over appreciable portions of the ventrobasal nucleus. Corticothalamic axon (B) terminates in a dense, chiplike terminal field at right angles to its course. Corticothalamic axon (C) produces a complex system of terminals in multiple stations of the thalamic nonspecific system, including the medial half of the ventral anterior nucleus (D); anterior medial nuclei (E); posterior components of the medial nuclei (F); parts of the posterior thalamic cell mass (G); and the medial part of the tegmentothalamic interface (H). Axon of neuron in nucleus reticularis thalami forming part of the intrathalamic feedback system (I). Drawn from several sections of 10- to 30-day-old rat and cat brain stained by rapid Golgi variants. × 200.

onstrated with the degree of regularity which we would expect necessary to fulfill a function of this type. Furthermore, there has been no convincing physiological demonstration that an inhibitory surround characterizes the operation of this relay.

No such problem faces us at the level of the sensory receptive cortex, where an elaborate architecture of columnar modules linked by feed-forward and feed-back inhibitory lines serves to intensify modal and topographic specificity and to enhance contrasts. Although operations analysis of thalamic substrates remains our primary focus, cortical architecture presents sufficiently compelling examples of operation-tagged subsystems to warrant brief consideration.

## Modularization

Cortical elements are so organized in repetitively recognizable sequences as to allow their categorization in groups, each of which appears structurally determined and functionally significant.

1. The unique structure of the individual cortical pyramid, and the topographical precision with which various presynaptic systems are applied along the soma-dendrite surface, suggest that each pyramidal neuron should be considered a modular entity capable of *data sampling* over a wide range of inputs. Using fifth-layer pyramids from primary visual cortex of the rabbit as an example, recent studies (Globus and Scheibel, 1967a) show that corticipetal volleys from the lateral geniculate nucleus are applied densely along the middle third of the apical shafts by terminals of the thalamocortical relay (Figure 5). Cortical projections from medial thalamic (nonspecific) fields and probably from the upper portion of the brainstem reticular core terminate in long, climbing, axodendritic sequences along the entire length of the apical shaft of each pyramid, probably including portions of the apical arches (Scheibel and Scheibel, unpublished). Callosal afferents connecting mirror-image sectors of the two hemispheres make synaptic connections entirely upon oblique branches of the apical shafts (Globus and Scheibel, 1967b), but recurrent collaterals of adjacent pyramidal cell axons appear to be limited to basilar dendrites and to the tips of the apical arches (Globus and Scheibel, unpublished). It is clear from these data that information of extracortical origin reaches only the vertical component of the individual pyramidal cell module, and intracortically derived information is brought exclusively to horizontal or obliquely oriented elements. It has been found that a cylindrical figure can be inscribed around each pyramid, the height being determined by the length of the apical shaft, whereas the diameter depends on the total spread of apical arches and basilar dendrites. Because of the striking quantitative similarity in horizontal

dendritic spread for any pyramid in a specific cortical area (within 5 to 7 per cent), it follows that the diameter of the cylindrical module for any one cortical area may be considered a constant (Globus and Scheibel, 1967c). The length of the apical shaft, however, and therefore the height of the module, may vary from approximately 200 $\mu$ to somewhat more than 2000 $\mu$. If we assume a rough relationship between the number of synaptic termini and the available length of synaptic surface, we conclude that all pyramidal cell modules of any one cortical area receive approximately equal numbers of intracortically derived terminal elements but vastly different numbers of extracortically derived

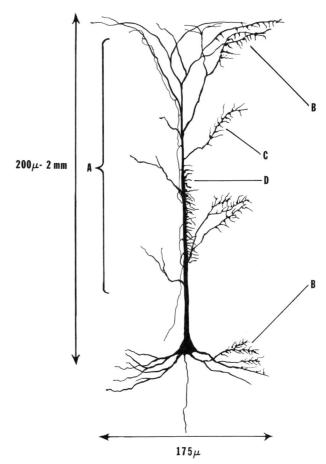

FIGURE 5 The individual cortical pyramid conceived as the simplest modular element of cortex, showing the location of major presynaptic terminal ensembles. A, diffuse distribution of nonspecific (brainstem reticular and intralaminar) afferents along the entire vertical dendrite system; B, distribution of recurrent collaterals on basilar dendrites and apical arches; C, distribution of callosal (contralateral) afferents on oblique branches; D, distribution of specific afferent projection on central third of apical shafts (of fifth-layer pyramids). Approximate vertical and horizontal measurements of typical pyramidal module are indicated. Drawn on basis of Golgi impregnations at various magnifications.

synaptic terminals. It is therefore conceivable that depth of the individual pyramidal cell in the cortex determines, in part, the nature of its *correlation function*.

2. That group of cortical pyramids and associated local circuit (stellate, short-axoned, Golgi type II) cells which are included within the field of arborization of a single terminating thalamocortical sensory fiber can be considered to constitute a module. All such elements are exposed to approximately the same input, although the degree of contamination by inputs from immediately adjacent components varies because of the partially shifted overlap displayed by successive terminal fields. Because the diameter of an individual sensory terminal field is of the order of 300 to 500 $\mu$, depending on cortical area and species, it seems likely that the functional columnarization of sensory cortex (Mountcastle, 1961; Werner, this volume) is significantly determined by this substrate. It is worth noting that, although arrays of neurons with specific responsive signatures are distributed in columnar arrays of the order of 500 $\mu$ in diameter, cells toward the edges of these modules have lower response rates than those more central in position, owing probably to decreasing synaptic densities from branches of the parent module fiber (Figure 6). Similarly, such peripheral elements are driven with increasing effectiveness by afferents to adjacent cortical-cell columns. On the basis of this structural paradigm, one might predict a series of maxima and minima for each mode and locus with relatively smooth interface gradients and fairly low resolution, save for the presence of three further structural components, each of which adds another dimension of processing to the cortical arrays.

3. The nonspecific systems of brainstem and thalamus reach cortex via several routes, the details of which are not particularly relevant at this point. Once at the grey-white interface, each of these axons appears to break up into a series of branches that ascend vertically toward the cortical surface, establishing sequences of axodendritic contacts with the spines of apical shafts and terminal arches. This array of ascending axon components appears to generate somewhat asymmetric modules of from 1 to 3 mm in diameter. The resulting roughly cylindrical modules are characterized by relatively low synaptic densities compared with those produced by specific afferents. There is also considerable mutual interpenetration of axonal domains, so that within the confines of a single 2- to 3-mm module, components of several dozen adjacent elements may be represented. Most of the several tens of thousands of pyramidal-cell apical shafts in the area of this domain will carry at least one such "climbing fiber" of reticular or medial thalamic origin. Some dendrites may carry more than one such component. In any case, the low density of synaptic endings from this source along the shaft and the presence of such termini as far from the somal trigger area as the apical arches suggest that these elements are not likely to exert significant degrees of phasic control over the output of the pyramid. As suggested by the model proposed by Rall (1964), this pattern of synaptic distribution is more likely to result in low levels of tonic control over the trigger zone. The resultant *biasing*

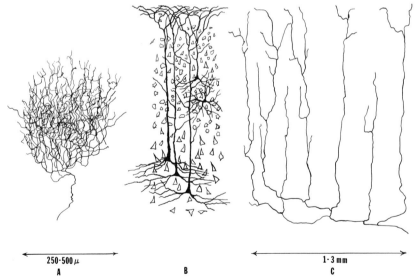

250-500 $\mu$

A                    B                    C

FIGURE 6   B: Fragment of the cortical cell matrix and two of the afferent elements which impinge upon it. A: Terminal axonal plexus generated by specific afferent sensory fiber. C: Diffuse terminal domain established by nonspecific afferent fiber from brainstem reticular formation of thalamic intralaminar system. Drawn from a number of sections of 10- to 100-day-old rat and cat brain stained with rapid Golgi variants. $\times$ 200.

of pyramidal-cell output may not only introduce a significant component of vertical organization into cortex, but may also insure a plastic element in this organization. For it is quite conceivable that as the relationship of the organism to its environment changes, resulting in qualitative and quantitative differences in output of the ascending reticular activating system and medial thalamic core, consequent changes of biasing activity at the cortical level may alter sequence or emphasis in the columnar sensory arrays. This reversible type of regulation, or *focusing,* in a system with a high degree of modal and topographical specificity offers a number of advantages in sophisticated information processing under changing input conditions.

4. Each pyramidal-cell axon releases three to 10 collaterals before entering the subgriseal white matter. Most of these collaterals run for hundreds or thousands of micra within the cortex, ascending obliquely toward the surface, or running parallel to the cortical surface at varying depths. Figure 7 shows most, although probably not all, of the collaterals generated by three pyramidal axons in one plane 250 $\mu$ in thickness. There is now little doubt that the role of these collaterals is largely inhibitory, reducing by a fairly constant fraction the output of a large number of pyramids in the surrounding area (Kameda et al., 1969) and serving thereby as a source of low-level feedback control. Each pyramid or cluster of pyramids thus constructs for itself an inhibitory surround of moderate effectiveness, extending up to 3 mm around the core element.

5. A more effective, if less broadcast, type of inhibition (of the feed-forward type) is apparently generated by axons of some local circuit cells in an activated column. As first shown by Ramón y Cajal (1911), many of these cells send their axons laterally for short to moderate distances (100 to 1000 $\mu$), terminating in baskets upon small numbers of pyramids in adjacent modules. The presumed effectiveness of the soma-initial segment locale as the site for inhibition (Eccles et al., 1964) marks this structural element as a powerful factor in enhancing contrast through afferent inhibition.

We have assumed that a plastic and effective contrast enhancement mechanism is maintained in cortex through a combination of nonspecific system biasing, low-level feedback inhibition via recurrent collaterals, and more intensive, if selective, feed-forward inhibition via local-circuit "basket-type" cells. It appears reasonable to assume that these mechanisms represent some of those necessary to maintain the locus and mode-specificity characteristic of the primary receptive areas. These areas represent, however, only a fraction of the total cortical mass. So far as we can tell from histological analysis of "motor" and "association" areas, essentially similar structural paradigms are found. It will be of considerable interest to determine whether they serve as substrate to dissimilar mechanisms in nonreceptive cortical areas.

### The thalamic synchronization process

The characteristic alternation of excitatory and inhibitory postsynaptic potentials occurring widely throughout the thalamus synchronously with long-latency, evoked cortical potentials now constitutes a well-documented aspect of thalamic physiology (e.g., Adrian, 1941, 1951; Dempsey and Morison, 1942; Purpura and Cohen, 1962; Andersen and Andersson, 1968). Of the several hypotheses advanced to account for this impressive degree of thalamic synchro-

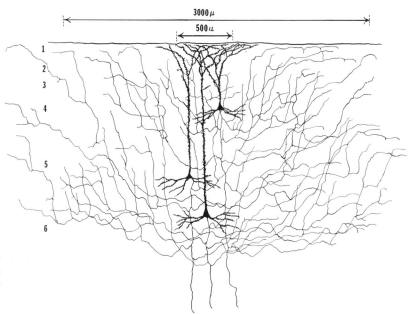

Figure 7 Ensemble of recurrent collaterals which surround three pyramidal cells in cortex. For any small group of core neuronal elements this may be considered the domain of recurrent inhibition. Drawn from several sections of 60-day cat cortex stained by a rapid Golgi variant. ×200.

nization, the one proposed by Andersen and Eccles (1962) and elaborated by Andersen and Andersson (1968) is probably best known. These workers have invoked a local, recurrent, inhibitory circuit depending on short-axoned cells of Renshaw type (Renshaw, 1941; Eccles et al., 1954) to provide the necessary inhibitory feedback modulated by shorter periods of postanodal exaltation. Although it is assumed that this model would, in itself, adequately account for the rhythmically occurring sequences, an added factor of innate rhythmicity peculiar to thalamic neurons has been proposed to provide intrinsic pace-making properties of a facultative nature.

To develop a reasonable anatomical model to account for the rapid development of widespread thalamic neuronal synchronization, and particularly of the prolonged IPSPs which suppress spontaneous and evoked thalamic discharge, several factors must be considered. 1. Although recurrent collaterals are a frequent feature of thalamocortical axons, they are by no means regularly present, nor are they always found at locations appropriately distant from the parent soma for initiating a recurrent loop. Even Eccles has observed that ". . . the large [thalamo-cortical] cells are reported to have fewer axon collaterals [Ramón y Cajal, 1911] than would be expected according to our postulate of a powerful recurrent inhibitory action" (Eccles, 1966). 2. The short-axoned cell, which is considered necessary to the recurrent loop to change the sign of excitation to inhibition (thereby fulfilling the same role as the hypothesized Renshaw cell of the spinal cord), is not an invariable feature of the ventrobasal field. Our own studies indicate that there are very few, if any, of these local circuit elements in the thalami of small rodents, and their number becomes appreciable only in carnivores and primates.

If we assume that mechanisms as basic to vertebrate brain systems as thalamic synchrony characterize the operations of rodent thalami (Burke and Sefton, 1966), it is unreasonable to attribute the effect to a structure (the short-axoned cell) which may not even be present in these species.

Another problem inherent in this proposed model is the rapid development of the synchronous process throughout large areas of the thalamus (Andersen and Andersson, 1968). If its spread depended on participation of the recurrent collateral-interneuron system as proposed by Andersen and Eccles (1962), very appreciable latencies would be involved as progressive sequences of synapses were crossed by the burgeoning process. In suggesting an alternative model, we draw on structural data concerning the nucleus reticularis thalami, which we have reported previously (Scheibel and Scheibel, 1966b, 1967a). This sheetlike nuclear complex surrounding the lateral and anterior borders of the thalamus has, until recently, been considered a final common pathway on the route from nonspecific systems in the thalamus to cortex (Hanbery et al., 1954). However, Golgi prepara-

tions clearly show that the vast majority of reticularis-cell axons project caudally upon thalamus and upper brainstem rather than rostrally upon cortex (Figure 4). Virtually all specific and nonspecific thalamocortical systems perforate the nucleus reticularis on their rostral path, generating a complex neuropil of collaterals and terminals as they pass through. An ideal substrate is thereby provided for generating feedback control over thalamic operations, activated by each thalamofugal volley. Furthermore, the vast majority of thalamic neurons are only one synapse away from the postulated source of inhibition, thereby assuring rapid development and spread of synchrony. Physiological support is supplied for this notion by the recent studies of Massion (1968).

## Thalamically induced desynchronization

Thalamocortical desynchronization can be achieved by stimulation of the medial thalamic nuclear complex at relatively high frequencies (Moruzzi and Magoun, 1949). The process appears characterized by suppression of synchronized high-amplitude, low-frequency rhythms in conjunction with augmentation of excitatory postsynaptic drive and inhibition of prolonged synchronizing IPSPs (Purpura, this volume). With regard to the underlying circuit paradigm, Schlag and Chaillet (1963) observed that, when the mesencephalic tegmentum directly behind the thalamus was destroyed, cortical desynchronization could no longer be obtained after high-frequency stimulation of the medial thalamus. The authors concluded that "activation" apparently depends exclusively on the mediation of the ascending reticular formation of the brain stem.

We believe it is possible to account for this *frequency-selection* effect by referring to certain structural features which we have mentioned above and elsewhere (Scheibel and Scheibel, 1966a, 1966b, 1967a, 1967b). It has already been noted that virtually all thalamocortical transactions, including those of the nonspecific system, are carried out by thalamofugal elements ascending through the domain of the nucleus reticularis thalami. Rostral-coursing axons of the brainstem reticular core, however, follow a different path. Beyond the crucial branch point, which occurs just caudal to the centre median-parafascicular complex (Nauta and Kuypers, 1958; Scheibel and Scheibel, 1958), the shorter dorsal lamella is lost in a group of posterior and medial thalamic nuclei. The ventral component ascends through zona incerta and hypothalamus and continues forward into the fields of the preoptic and basal forebrain region. Approximately 10 per cent of these elements may reach cortical stations without synaptic interruption (Scheibel and Scheibel, 1958; Magni and Willis, 1963), but the majority reach these levels following one or more synapses. Significantly, the ascending reticular fibers are situated ventral to the nucleus reticularis thalami and therefore do not penetrate its domain on their rostral course.

If the nucleus reticularis is conceived of as a *frequency-sensitive gate,* open to low frequencies and closed to high, the meaning of the Schlag-Chaillet observation becomes clear. Low-frequency stimulation of the medial thalamus is immediately effective in synchronizing electrical activity over huge areas of thalamus and cortex, because this roughly matches the frequency range established by the reticularis gate. Higher-frequency stimulation is blocked once the transmitting elements enter the field of the nucleus reticularis. Caudally directed components of the medial cell masses (Figure 8), however, establish abundant synaptic links with elements of the ascending reticular formation at mesencephalic levels and thus, more indirectly, can produce changes at cortical levels that include flattening and frequency enhancement of ongoing cortical electrical activity.

The existence of *frequency-specific* dual pathways from axial portions of the brainstem to cortex recalls the experiments of Rossi (1963) and Jouvet (1967), who have shown that appropriately placed lesions can destroy the capability of the organism to experience slow-wave, spindle-rich

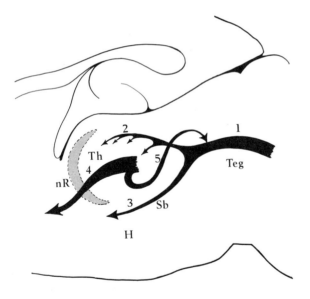

FIGURE 8   Schematic representation of the projection paths of the two nonspecific systems. The brainstem reticular core (1) ascends through the mesencephalic tegmentum (Teg) and divides into a smaller dorsal lamina (2) which projects into several nuclei of the thalamus, Th, and a larger ventral lamina (3) which projects into subthalamus (Sb) and hypothalamus (H). The thalamic nonspecific (intralaminar) system (4) sends a major projection rostrally through the nucleus reticularis thalami, nR, and a secondary projection (5) caudally into tegmentum. (From Scheibel and Scheibel, 1967a, with permission of the publisher.)

sleep on the one hand, or "activated," rhombencephalic, REM (low-voltage, fast wave) sleep on the other.

## Multiplexing and parallel processing

The thalamocortical relay-cell axon exemplifies the type of tagged line that projects specifically to a receptive site without major branching along its course. The only exceptions here are the collaterals that may develop in the first few hundred microns of its trajectory through the mass of the ventrobasal field and, possibly, during its penetration of the nucleus reticularis thalami. In contrast, the axons of most cells making up the complex of medial-thalamic nonspecific and intralaminar nuclei are characterized by complex trajectories, which include both rostral and caudal branches of bifurcation and elaborate lateral branching (Figure 9). As in the case of neurons of the brainstem reticular core (Scheibel and Scheibel, 1958, 1967b), these axons of cells in the thalamic nonspecific system appear to penetrate a wide variety of nuclei of highly diverse functions. As an example, the axon of a single cell in the anterior third of the thalamic nonspecific system may project major branches anteriorly upon orbitofrontal cortex and posteriorly into the mesencephalic tegmentum. Physiological support for these structural observations are available in the literature (Skinner and Lindsley, 1967; Starzl and Whitlock, 1952). Secondary branches may simultaneously innervate multiple nuclei of the nonspecific thalamic complex on both sides of the midline while penetrating obliquely across one or more of the adjacent specific thalamic fields. Many branches have been followed into the densest neuropil fields of ventrolateral and ventral anterior nuclei, where frequency-specific control exerted by these nonspecific systems over projection neurons has been demonstrated (Purpura, this volume). Furthermore, branches have been followed in various sagittal planes of section into hypothalamus, preoptic areas, and basal ganglia, thereby rounding out a picture characterized by extensive *multiplexing* of thalamic nonspecific output (Scheibel and Scheibel, 1967a). It might also be noted that thalamopetal afferents from more rostral, presumably cortical, centers projecting upon the medialthalamic cell complex show similar patterns of extensive branching and widespread terminal patterns throughout the length of thalamus and in some cases even farther caudally (Figures 3 and 4).

One of the significant problems associated with this type of multiplexed distribution lies in the nature of the output at the various synaptic stations that have been established. There is no reason to doubt that each of the presynaptic terminals can be functionally active, but there are insufficient data for a definition of the nature and sequence of these activities. Purpura (this volume) has shown that both EPSPs and IPSPs, of varying duration and intensity, may be

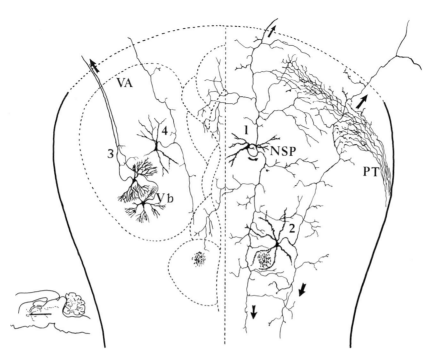

FIGURE 9  Contrast in axonal trajectories of specific and nonspecific thalamic cell systems. 1 and 2: Bifurcating, richly branched axons of thalamic nonspecific neurons (NSP). 3: Relatively simple and almost unbranched axons of two thalamic ventrobasal (Vb) cells. The bifurcating axon of an integrator cell, 4, projects both caudally and rostrally through the ventral anterior (VA) nucleus. Terminals of the pallidothalamic bundle (PT) cross the VA nucleus and provide substrate for synaptic interface with the nonspecific system. Drawn from several sections of 10- to 50-day rat and cat cut in horizontal plane. Rapid Golgi variant. × 200.

produced in a number of different neural subcenters supplied by the medial thalamic structures. But it is still to be determined whether this range of effects can be produced by branches of the individual nonspecific axon, and whether the complete roster of terminals from one such fiber is always active. With regard to the first question, a number of investigators (Tauc and Gerschenfeld, 1961; Kandel et al., 1967) have now shown that individual terminals of a single axon in *Aplysia* can produce differing postsynaptic effects, owing presumably to specificity of receptor sites along postsynaptic surfaces. Somewhat more indirect evidence from vertebrates may be drawn from our own structural studies in spinal cord (Scheibel and Scheibel, 1969), which indicate that at least 30 per cent of primary afferent collaterals directly supply both flexor and extensor motor neurons without intercalated interneurons. The presumption here is that both facilitatory and inhibitory effects can be produced appropriately at postsynaptic neuronal sites of antagonistic function. Clearly, the critical experiment has yet to be done in the mammalian vertebrate nervous system.

In the same way, we are at present equally without a solution to the problem of obligatory versus facultative activity at all the presynaptic loci on the various branches of a multi-plexed, nonspecific axon. We first faced this problem with similar elements in the brainstem reticular core (Scheibel and Scheibel, 1958, 1967b) and are attempting to approach it through the use of multiple microelectrode techniques. As methodologically difficult as the problem still appears, whether the individual axon can gate its output through multiple branch points or whether all branches passively transmit all volleys that are carried over the main axonal lines is certainly of more than theoretical interest.

*The corticothalamic reflux*

One of the characteristics of interaction between neural subsystems is reciprocity of connections, such that if A → B, then B → A. The highly ordered projection of the ventrobasal nucleus upon somesthetic receptive (sensory) cortex is matched by complex corticothalamic fiber systems supplying the ventrobasal fields. At least two general configurations of terminal neuropil have been identified, although the specific intracortical source of each has not been precisely determined. The first type of fiber (type 1), originally described by Ramón y Cajal (1911), ramifies widely but sparsely over large portions of the sensory thalamic field, generating a repetitively branching axonal field of roughly cone-shaped configuration. The branches of a single fiber may encompass as much as 20 to 30 per cent of the entire Vb nucleus. The density of terminal structures that it generates in any one locale is accordingly low. The second type (type 2) appears characterized by a somewhat thicker parent fiber and a unique terminal-axonal domain. Each fiber generates, at right angles to its trajectory through the thalamus, one or more highly reduplicated, yet essentially two-dimensional, neuropil plates. The individual plate is composed of innumerable small branches, each bearing clusters of knoblike terminal structures. The maximal dimension of each plexus is realized in the transverse, or coronal, plane, where it may cover two-thirds or more of the ventrobasal field. In the rostrocaudal dimension, however, its extension is minimal (Figures 3 and 4). As a result, each Vb field appears to be transsected by hundreds (in rodents) or thousands (in carnivores) of these chiplike terminal modules stacked one after another along the entire rostrocaudal axis of the nucleus.

These two corticothalamic elements form a dramatic contrast in terms of the terminal patterns that they generate, and presumably serve different intrathalamic roles. Unfortunately, there are, as yet, insufficient data to document the functional status of either of these elements, so assignment of roles to either must be considered purely hypothetical.

It is difficult to conceive of a specific *naming or addressing* function for the more diffusely patterned terminal, if it is conceived as operating individually. The extent of its tridimensional penetration of the Vb area and the apparent low density of terminal synaptic structures from any one element argue for a highly general type of function appropriate to broad neuronal fields, and presumably independent of bin position or domain locale. Drawing from our experience with broadly distributed, nonspecific, projection fibers to cortex, we suggest some type of across-the-board *threshold manipulation* or *bias adjust* function for these elements.

The terminal organization of the second element appears to fit it for a function involving rather precise degrees of locus definition, in this case a chiplike sector across the height and width of the Vb nucleus. As these terminal neuropil fields are stacked in rostrocaudal sequence, the entire array can be thought of as constituting a series of trans-domain samples from front to rear. In this case, the read-out system would apparently seek information integrated across the successive tiers of concentric lamellae, which characterize the terminal distribution of medial lemniscal fibers (Figure 3).

A somewhat different hypothesis as to the role of both corticothalamic fibers *operating in concert* can be entertained, based very freely on certain core-selection techniques in current use in computer circuitry. Although the type-1

afferents spread widely as they run caudally into the Vb, effecting low-density innervation over wide neuronal fields, structural evidence suggests that they are stacked rather tightly in rostrocaudal sequence, with their apices anterior. In aggregate, they have the opportunity of contributing to the synaptic ensemble of virtually all Vb cells. The type-2 discoid afferents are also tightly stacked and cut perpendicularly across the axis of this sequence of conical domains, so each Vb cell may be considered to receive double cortical innervation. Imposition of an X, Y grid upon a spatially regular matrix of discrete elements is an effective way to provide read-out capability for any one element of the matrix. Such arrangements are, in fact, used in the *coincident-current selection technique* for reading out an individual core from a computer memory array. In this case, each member of the array can be individually selected by convergence of half-strength current pulses in two sets of selection lines perpendicular to each other. At this point, we can do no more than speculate that these two categories of corticothalamic fibers may also represent the substrate for a *cortical read-out system* playing upon thalamus. Implicit in such a suggestion is a role for the ventrobasal nucleus far richer than the "relay" function all too frequently assigned to it.

## Conclusion

In the preceding pages we have described the structural organization of several thalamic and cortical subsystems to the extent that they can be revealed by light-microscope techniques. On the basis of these descriptions, we have suggested putative relationships to a group of operations significant in information processing. The extreme complexity of the neurohistological matrix and the virtual exclusion from these studies of data from electron microscopy and molecular biology assure the inadequacy of our analogies. Nevertheless, system analysis must begin somewhere, and we have selected a level of resolution at which patterns are clearly etched. The beauty and consistency of these structural arrangements throughout much of the vertebrate line offer eloquent testimony to the meaningfulness of their form. Appropriate interpretation of their patterns clearly awaits more powerful research methods and more penetrating insights.

## Summary

Thalamus and cortex have been scanned for systems in which structural and functional data are conducive to some degree of operations analysis. In so doing, we have sought possible relations between neural operations and those generally considered significant in information processing.

1. The lemniscal relay through the ventrobasal nucleus of

the thalamus is examined as an example of a high-priority transmission system. Structural patterns generated by presynaptic and postsynaptic elements appear to insure spatial localization and temporal extension of ascending thalamopetal volleys. The operational role of the system is conceived as updating the thalamic sensory neuron banks.

2. Cortical operations may be considered within the framework of a hierarchy of interacting modules of increasing size and complexity.

A. The soma-dendrite system of the individual pyramidal cell constitutes the simplest module. Its horizontal and vertical components receive spatially ordered arrays of intracortically and extracortically derived afferents, respectively, providing substrate for data sampling and correlation analysis.

B. The cell ensemble within the domain of a single, terminating, specific, thalamocortical fiber constitutes a module the elements of which are, to varying degrees, under the synaptic drive generated by a single thalamic element. This module probably provides the structural substrate for the physiologically demonstrated cortical sensory column, although peripheral limits of the two are not necessarily congruent.

C. Cells within the terminal field generated by a corticipetal, nonspecific, afferent fiber constitute a modular entity under the bias adjustment range of one nonspecific (brainstem or thalamic) neuron.

D. The ensemble of recurrent collaterals generated by a core group of pyramidal neurons encompasses a cortical domain of considerable size. Virtually all the cortical pyramids in this area can be considered as constituting a module exposed to low-level feedback control by the central core elements.

E. Horizontally oriented modules of smaller extent result

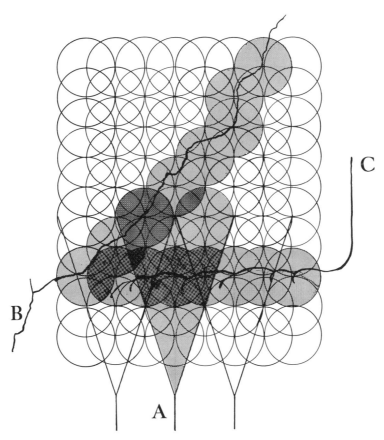

FIGURE 10  Highly simplified schematic drawing suggesting the interpenetration and overlapping of presynaptic and postsynaptic domains in the ventrobasal nucleus. The ensemble of overlapping circles represents the dendritic domains of a number of thalamocortical cells. The partially overlapped conical figures (A) indicate the fields generated by presynaptic terminals of the lemniscal afferents. Thalamic nonspecific afferents (B) move obliquely across the Vb field matrix, making synaptic contact with irregular linear ensembles of neurons, and the corticothalamic discoid afferents (C) establish rather precise synaptic domains across the Vb nucleus. The different intensities of the gray screen give some idea of the convergence of afferents and the consequent fractionation of domains.

from the selective impingement of local circuit (Golgi type II) cell terminals on limited numbers of cortical cells. Neurons in these domains apparently receive a relatively high level of inhibition of feed-forward type.

3. The development of thalamic synchronization is usually considered to depend on a substrate of recurrent collaterals from thalamic relay neurons playing back on local, short-axoned elements, which, in turn, modulate the parent relay elements. Problems inherent in this model are considered and an alternative is suggested, involving the nucleus reticularis thalami as a major short-latency feedback that controls and synchronizes thalamic activity.

4. Thalamically induced desynchronization after high-frequency stimulation of nonspecific thalamic nuclei has been shown to depend on caudal projections upon the mesencephalic tegmentum. It is suggested that the rostrally directed tegmental reticular relay escapes the frequency-sensitive gating action of the nucleus reticularis thalami by running ventral to it through the hypothalamus and the preoptic region. The resultant operation of frequency-specific dual pathways from axial portions of the brainstem to cerebral cortex can also be demonstrated by selective abolition of activated or slow-wave sleep by appropriately placed brainstem lesions.

5. The output of cells in the thalamic nonspecific system is characterized by richly branching axons, in contradistinction to the almost unbranched projection of neurons in the thalamic specific system. The significance of multiplexed outputs is considered in the light of local gating control of the individual branches.

6. The corticothalamic reflux upon somatic-sensory thalamus (Vb) appears characterized by at least two major types of fibers, one with extensive divaricating branches covering large areas of the field, and a second with dense discoid arbors lined up along the rostrocaudal axis like a row of chips. One possible functional interpretation of this structural substrate is modeled on the coincident current selection technique for retrieval of items from a computer memory core. The possibility of a cortical read-out system for the thalamic field argues against a purely relay function for this nucleus.

Although neural circuit organization as depicted by the Golgi methods appears to offer a reasonable starting point for operations analysis, the restricted range of our data and the consequently tentative nature of our hypotheses are emphasized.

## Acknowledgment

The work summarized here has been supported by the United States Public Health Service through Grants NB 1063 and HD 00972.

## REFERENCES

ADRIAN, E. D., 1941. Afferent discharges to the cerebral cortex from peripheral sense organs. *J. Physiol.* (*London*) 100: 159–191.

ADRIAN, E. D., 1951. Rhythmic discharges from the thalamus. *J. Physiol.* (*London*) 113: 9P–10P.

ANDERSEN, P., and S. A. ANDERSSON, 1968. Physiological Basis of the Alpha Rhythm. Appleton-Century-Crofts, New York, pp. 15–30.

ANDERSEN, P., and J. C. ECCLES, 1962. Inhibitory phasing of neuronal discharge. *Nature* (*London*) 196: 645–647.

ANDERSEN, P., J. C. ECCLES, and T. A. SEARS, 1964. The ventrobasal complex of the thalamus: Types of cells, their responses and their functional organization. *J. Physiol.* (*London*) 174: 370–399.

BURKE, W., and A. J. SEFTON, 1966. Inhibitory mechanisms in lateral geniculate nucleus of rat. *J. Physiol.* (*London*) 187: 231–246.

DAHLSTRÖM, A., and K. FUXE, 1965. Evidence for the existence of monoamine neurons in the central nervous system. II. Experimentally induced changes in the intraneuronal amine levels of bulbospinal neuron systems. *Acta Physiol. Scand.* 64 (suppl. 247): 5–36.

DEMPSEY, E. W., and R. S. MORISON, 1942. The production of rhythmically recurrent cortical potentials after localized thalamic stimulation. *Amer. J. Physiol.* 135: 293–300.

DUSSER DE BARENNE, J. G., and W. S. McCULLOCH, 1938. The direct functional interrelation of sensory cortex and optic thalamus. *J. Neurophysiol.* 1: 176–186.

DUSSER DE BARENNE, J. G., and W. S. McCULLOCH, 1941. Functional interdependence of sensory cortex and thalamus. *J. Neurophysiol.* 4: 304–310.

ECCLES, J. C., 1964. The Physiology of Synapses. Springer-Verlag, Berlin, pp. 42–46.

ECCLES, J. C., 1966. Properties and functional organization of cells in the ventrobasal complex of the thalamus. *In* The Thalamus (D. P. Purpura and M. D. Yahr, editors). Columbia University Press, New York, pp. 129–138.

ECCLES, J. C., P. FATT, and K. KOKETSU, 1954. Cholinergic and inhibitory synapses in a pathway from motor-axon collaterals to motoneurones. *J. Physiol.* (*London*) 126: 524–562.

GLOBUS, A., and A. B. SCHEIBEL, 1967a. Synaptic loci on visual cortical neurons of the rabbit. The specific afferent radiation. *Exp. Neurol.* 18: 116–131.

GLOBUS, A., and A. B. SCHEIBEL, 1967b. Synaptic loci on parietal cortical neurons: Terminations of corpus callosum fibers. *Science* (*Washington*) 156: 1127–1129.

GLOBUS, A., and A. B. SCHEIBEL, 1967c. Pattern and field in cortical structure: The rabbit. *J. Comp. Neurol.* 131: 155–172.

GUDDEN, B., 1870. Experimentaluntersuchungen über das peripherische und zentrale Nervensystem. *Arch. Psychiat. Nervenkrankh.* 2: 693–723.

HANBERY, J., C. AJMONE-MARSAN, and M. DILWORTH, 1954. Pathways of non-specific thalamo-cortical projection system. *Electroencephalogr. Clin. Neurophysiol.* 6: 103–118.

HERRICK, C. J., 1948. The Brain of the Tiger Salamander. The University of Chicago Press, Chicago, p. 89.

JOUVET, M., 1967. Neurophysiology of the states of sleep. In The

Neurosciences: A Study Program (G. C. Quarton, T. Melnechuk, and F. O. Schmitt, editors). The Rockefeller University Press, New York, pp. 529–544.

KAMEDA, K., R. NAGEL, and V. B. BROOKS, 1969. Some quantitative aspects of pyramidal collateral inhibition. *J. Neurophysiol.* 32: 540–553.

KANDEL, E. R., W. T. FRAZIER, and R. E. COGGESHALL, 1967. Opposite synaptic actions mediated by different branches of an identifiable interneuron in Aplysia. *Science (Washington)* 155: 346–349.

LUYS, J., 1865. Recherches sur le système nerveux cérébro-spinal: Sa structure, ses fonctions et ses maladies. J. B. Baillière et Fils, Paris, 2 vols. (Quoted from J. F. Fulton, 1943, Physiology of the Nervous System, edition 2, Oxford University Press, New York, p. 252.)

McCULLOCH, W. S., J. Y. LETTVIN, W. H. PITTS, and P. C. DELL, 1952. An electrical hypothesis of central inhibition and facilitation. *Res. Publ. Ass. Res. Nerv. Mental. Dis.* 30: 87–97.

MAGNI, F., and W. D. WILLIS, 1963. Identification of reticular formation neurons by intracellular recording. *Arch. Ital. Biol.* 101: 681–702.

MASSION, M. J., 1968. Étude d'une structure motrice thalamique. Le noyau ventrolatéral et de sa régulation par les afférences sensorielles. *Thèse de doctorat d'état es-sciences naturelles,* pp. 22–24.

MORUZZI, G., and H. W. MAGOUN, 1949. Brain stem reticular formation and activation of the EEG. *Electroencephalogr. Clin. Neurophysiol.* 1: 455–473.

MOUNTCASTLE, V. B., 1961. Some functional properties of the somatic afferent system. *In* Sensory Communication (W. A. Rosenblith, editor). M. I. T. Press, Cambridge, Massachusetts, pp. 403–436.

NAUTA, W. J. H., and H. G. J. M. KUYPERS, 1958. Some ascending pathways in the brain stem reticular formation. *In* Reticular Formation of the Brain (H. H. Jasper et al., editors). Little Brown and Co., Boston, pp. 3–30.

PAPPAS, G. D., E. B. COHEN, and D. P. PURPURA, 1966. Fine structure of synaptic and nonsynaptic neuronal relations in the thalamus of the cat. *In* The Thalamus (D. P. Purpura and M. D. Yahr, editors). Columbia University Press, New York, pp. 47–71.

PURPURA, D. P., and B. COHEN, 1962. Intracellular recording from thalamic neurons during recruiting responses. *J. Neurophysiol.* 25: 621–635.

RALL, W., 1964. Theoretical significance of dendritic trees for neuronal input-output relations. *In* Neural Theory and Modeling (R. F. Reiss, editor). Stanford University Press, Stanford, California, pp. 73–97.

RALSTON, H. J., 1969. The synaptic organization of lemniscal projections to the ventrobasal thalamus of the cat. *Brain Res.* 14: 99–115.

RALSTON, H. J., and M. M. HERMAN, 1969. The fine structure of neurons and synapses in the ventrobasal thalamus of the cat. *Brain Res.* 14: 77–97.

RAMÓN Y CAJAL, S., 1900. Contribución al estudio de la via sensitiva central y de la estructura del tálamo óptico. *Rev. Trim. Microgr.* 5: 185–198.

RAMÓN Y CAJAL, S., 1911. Histologie du Système Nerveux de l'Homme et des Vertébrés. A. Maloine, Paris, vol. 2, pp. 381–503.

RAMÓN Y CAJAL, S., 1917. Contribución al conocimiento de la retina y centros ópticos de los cefalópodos. *Trab. Lab. Invest. Biol. (Univ. Madrid)* 15: 1–82.

RAMÓN Y CAJAL, S., and D. SANCHEZ, 1915. Contribución al conocimiento de los centros nerviosos de los insectos. *Trab. Lab. Invest. Biol. (Univ. Madrid)* 13: 1–164.

RENSHAW, B., 1941. Influence of discharge of motoneurons upon excitation of neighboring motoneurons. *J. Neurophysiol.* 4: 167–183.

ROSSI, G. F., 1963. Sleep inducing mechanisms in the brain stem. *Electroencephalogr. Clin. Neurophysiol., suppl.* 24: 113–132.

SCHEIBEL, M. E., and A. B. SCHEIBEL, 1955. The inferior olive. A Golgi study. *J. Comp. Neurol.* 102: 77–131.

SCHEIBEL, M. E., and A. B. SCHEIBEL, 1958. Structural substrates for integrative patterns in the brain stem reticular core. *In* Reticular Formation of the Brain (H. H. Jasper et al., editors). Little Brown and Co., Boston, pp. 31–55.

SCHEIBEL, M. E., and A. B. SCHEIBEL, 1966a. The organization of the ventral anterior nucleus of the thalamus. A Golgi study. *Brain Res.* 1: 250–268.

SCHEIBEL, M. E., and A. B. SCHEIBEL, 1966b. The organization of the nucleus reticularis thalami: A Golgi study. *Brain Res.* 1: 43–62.

SCHEIBEL, M. A., and A. B. SCHEIBEL, 1966c. Patterns of organization in specific and nonspecific thalamic fields. *In* The Thalamus (D. P. Purpura and M. D. Yahr, editors). Columbia University Press, New York, pp. 13–46.

SCHEIBEL, M. E., and A. B. SCHEIBEL, 1967a. Structural organization of nonspecific thalamic nuclei and their projection toward cortex. *Brain Res.* 6: 60–94.

SCHEIBEL, M. E., and A. B. SCHEIBEL, 1967b. Anatomical basis of attention mechanisms in vertebrate brains. *In* The Neurosciences: A Study Program. (G. C. Quarton, T. Melnechuk, and F. O. Schmitt, editors). The Rockefeller University Press, New York, pp. 577–602.

SCHEIBEL, M. E., and A. B. SCHEIBEL, 1969. Terminal patterns in cat spinal cord. III. Primary afferent collaterals. *Brain Res.* 13: 417–443.

SCHLAG, J. D., and F. CHAILLET, 1963. Thalamic mechanisms involved in cortical desynchronization and recruiting responses. *Electroencephalogr. Clin. Neurophysiol.* 15: 39–62.

SKINNER, J. E., and D. B. LINDSLEY, 1967. Electrophysiological and behavioral effects of blockade of the nonspecific thalamo-cortical system. *Brain Res.* 6: 95–118.

STARZL, T. E., and D. C. WHITLOCK, 1952. Diffuse thalamic projection system in monkey. *J. Neurophysiol.* 15: 449–468.

SZENTÁGOTHAI, J., 1963. The structure of the synapse in the lateral geniculate body. *Acta Anat.* 55: 166–185.

SZENTÁGOTHAI, J., 1967. Models of specific neuron arrays in thalamic relay nuclei. *Acta Morphol. Acad. Sci. Hung.* 15: 113–124.

TAUC, L., and H. M. GERSCHENFELD, 1961. Cholinergic transmission mechanisms for both excitation and inhibition in molluscan central synapses. *Nature (London)* 192: 366–367.

TÖMBÖL, T., 1967. Short neurons and their synaptic relations in the specific thalamic nuclei. *Brain Res.* 3: 307–326.

WALKER, A. E., 1938. The Primate Thalamus. The University of Chicago Press, Chicago, p. 321.

# 42 Operations and Processes in Thalamic and Synaptically Related Neural Subsystems

## DOMINICK P. PURPURA

### Scope of the inquiry

THE THALAMUS IS A major target of exteroceptive, intra-cerebral, and special sensory projection systems, the chief source of input to the cerebral cortex and the basal ganglia, and an essential link in the limbic-midbrain circuit. The strategic role of the thalamus in virtually all forebrain trans-actions compels examination of those processes and opera-tions that are likely to be consequences of special features of thalamic synaptic organizations. In such a survey, emphasis is to be placed on internuclear interactions that not only illustrate key operations of thalamic integrative mechanisms but also provide an understanding of thalamic influences on synaptically related neuronal subsystems.

The analysis of intrathalamic and thalamocortical synaptic organizations that began with the pioneering studies of Dempsey and Morison (1942) and Morison and Dempsey (1942) has been considerably advanced as a result of the in-tensive application of intracellular recording techniques. Al-though many details of functional relationships have been elucidated, only rarely has it been possible to define the morphological basis for these relationships (cf. Scheibel and Scheibel, this part of the volume). The reason is to be sought, in part, in the nature of the rich internuclear syn-aptic pathways which link fundamentally different types of neuronal organizations. These pathways are of the utmost complexity, representing as they do the highest degree of elaboration of central internuncial systems. The synaptic mechanisms underlying thalamic operations are generated in this interneuronal matrix, the functional organization of which has been sketched only in broad brush strokes. Nevertheless, some of the impressions gleaned from in-tracellular studies should serve as a useful guide to future analyses of thalamic operations.

An attempt is made here to describe elementary processes in these operations and to indicate, whenever feasible, the probable synaptic substrate required for specific processes.

This approach aims to define principles that can be ex-pressed in relatively simple operational terms. Operations demonstrable in thalamic and related neuronal subsystems include high-fidelity transmission, input selection, output tuning, synchronization, desynchronization, filtering, gat-ing, parallel processing, storing, alternation, attenuation, and step-down transformation. Suffice it to say that, in some obscure fashion, all these operations and many more are smoothly integrated in thalamic neuronal organizations to provide a suitable substrate for the most complex func-tions of the mammalian brain.

### Features ensuring modality and topographical specificity

Requirements for maintaining segregation and specificity of input in relay nuclei that have prominent and direct pro-jections to cortex are set largely by operations of first-order elements in which are effected major transformations of sensory information into spatiotemporal impulse codes (Mountcastle et al., 1969; Werner, elsewhere in this vol-ume). The secure synaptic relations that obtain at succes-sively higher relays in these projection systems involve postsynaptic elements in the thalamus that are topographi-cally and functionally linked to labeled lines, with minimal or no cross-modality dispersion. The consequences are well illustrated in fine-grained modality and place-specific maps of the thalamic ventrobasal complex, as displayed in text-books of neurophysiology (cf. Mountcastle and Darian-Smith, 1968).

Major features of specific relay activity are evident in dis-crete input and high-fidelity transmission with minimal distortion of output signal, at least in unanesthetized prep-arations (Mountcastle et al., 1963; Poggio and Mount-castle, 1963; Mountcastle et al., 1969). Synaptic security is achieved despite penalties imposed by predominant dis-tribution of second-order (lemniscal) afferents to dendrites of ventrobasal (VB) neurons (Ralston, 1969; Scheibel and Scheibel, 1966, and elsewhere in this volume). Similar axodendritic relations obtain between brachium con-junctivum afferents and ventrolateral (VL) neurons that

DOMINICK P. PURPURA   Department of Anatomy, Albert Ein-stein College of Medicine, New York, New York

give rise to primary afferent projections to motor cortex. Compensation for signal attenuation is achieved in these elements by "all-or-none" excitatory postsynaptic potentials (EPSPs) and by partial and full spike initiation in dendrites (Maekawa and Purpura, 1967a; Purpura, 1967) (Figure 1). Intrinsic properties of specific relay neurons, such as their minimal adaptation to injected depolarizing currents and absence of significant threshold differences between initial segment and soma membrane, also undoubtedly contribute to the maintenance of dynamic relations in input-output characteristics (Maekawa and Purpura, 1967a).

Entirely different input-output characteristics are observed in VB cells in barbiturate-anesthetized animals (Mountcastle, 1961). Presynaptic inhibition at prethalamic and thalamic levels may be greatly enhanced by barbiturate anesthesia, which may also facilitate the operation of recurrent inhibition of VB neurons (Andersen et al., 1964a; 1964b). Both processes probably serve to suppress dynamic aspects of input-output relations without influencing static responsiveness of VB elements. The development of synchronized thalamocortical rhythms during barbiturate

narcosis also has important consequences for specific relay transmission, as discussed below.

Modality and topographically specific information transferred through thalamic relays is deposited in cortical regions representing primary projection target areas of specific thalamocortical afferents. Specifically responsive neurons are distributed in columnar arrays, in which constituent elements exhibit limited cross-modal interaction (Werner, elsewhere in this volume). The remarkable similarity in discharge pattern of cortical neurons at different depths in a particular radial column that is activated by specific thalamocortical afferents is seen in an examination of intracellular records obtained during a single penetration of cortex (Figure 2). The generation of similar patterns of excitatory and inhibitory postsynaptic potentials (EPSPs-IPSPs) in these elements requires activation, by primary afferents, of elements distributed in a rigorously specified connectivity pattern. As in the case of thalamic relays, specific thalamocortical afferents powerfully excite or inhibit cortical neurons (Purpura and Shofer, 1963). In fact, it is rare indeed to record subliminal EPSPs in cortical neurons in response to peripheral or specific thalamic stimulation that is capable of eliciting a primary evoked potential.

Columnar organizations of neurons in somatosensory cortex are replicated in motorsensory cortex in relation to corticofugal projection systems (Welt et al., 1967). Still to

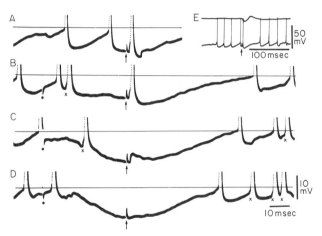

FIGURE 1 Characteristics of "all-or-none" excitatory postsynaptic potentials (EPSPs) and variations in firing level of a ventrobasal (VB) thalamic neuron activated by medial lemniscal stimulation (at arrows). A-D. Photographically enlarged segments of records from continuous recording. Spikes are truncated for display purposes. E. Example to show full spike amplitude and primary response (upper channel) at cortical surface to lemniscal stimulation. A. Lemniscal stimulation elicits a fast-rising EPSP which initiates a spike discharge at a higher level of membrane potential than do preceding spontaneous spikes. B–C. Lemniscal stimulation is preceded by internal capsule stimulation (at dots), which evokes long-latency, inhibitory postsynaptic potentials (IPSPs) in the VB relay neuron. In C, spike generation is suppressed by the IPSP, whereas in D, the EPSP fails in an "all-or-none" fashion. Variations in firing level are indicated by x, which also identifies spikes preceded by fast depolarizing prepotentials. (Data from Maekawa and Purpura, 1967a.)

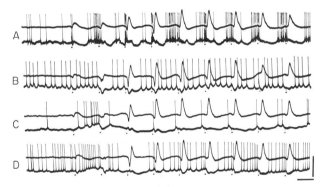

FIGURE 2 PSP patterns evoked during augmenting responses in four nonpyramidal-tract cells impaled during a single penetration of motor cortex. Approximately 10 to 20 minutes elapsed between the recording of each series of responses. A. Cell situated at a depth of 0.8 mm. exhibits high-frequency repetitive discharges during surface positivity, and late IPSPs during surface negativity. Discharges reappear prior to stimulation. B. Cell at a depth of 0.85 mm. Short-latency, complex IPSPs observed throughout period of stimulation. C. Cell at 0.95 mm. shows IPSPs during augmentation. D. Cell at 1.00 mm. exhibits IPSPs during all phases of cortical-surface response. Note similar PSP patterns observed in B-C during third response of each series. Calibrations 0.1 second; 50 mV. (From Purpura, Shofer, and Musgrave, 1964.)

be determined is the extent to which intercolumnar labeling (sensory to motor, and vice versa) also operates to achieve selectivity and specificity in sensorimotor performance (Asanuma et al., 1968). In both types of columnar organization, afferent inhibition (Mountcastle, 1961) and recurrent inhibition are demonstrable ancillary mechanisms for enhancing contrast (Brooks and Asanuma, 1965; Phillips, 1959; Stefanis and Jasper, 1964). These are reinforced by localized, corticothalamic, excitatory and inhibitory inputs to thalamic relays, which fulfill a similar role in input selection within a specific thalamocortical projection system (Andersen et al., 1964b; Maekawa and Purpura, 1967b).

## Thalamic synchronization process

Neurons in widespread parts of the thalamus, including many functionally different nuclear groups, are capable of exhibiting synchronized discharges that are succeeded by "silent periods" lasting 80 to 100 milliseconds (Verzeano and Negishi, 1960). Coarse electrodes applied to the surface or inserted into different parts of the thalamus may record six- to 12-per-second rhythms during these sequences of neuronal discharge-synchronization and silence. A similar rhythm, recordable from the cortical surface or scalp, represents the typical alpha rhythm of the EEG (Brazier, 1968). EEG-synchronization in the intact animal and in man is, in part, a consequence of the operation of thalamocortical synaptic mechanisms that are predominantly, but not uniquely (Andersen and Andersson, 1968), related to the activity of nonspecific thalamic nuclei and their interactions with specific projection nuclei. Intracellular studies of the thalamic (Purpura and Cohen, 1962) and cortical (Purpura et al., 1964) stages of the synchronization process provide a clear illustration of the operation of thalamic interneuronal circuits involved in one of the most impressive electrographic phenomena associated with the behavioral state of drowsiness or light sleep.

The thalamic synchronization process is best demonstrated during low-frequency (six- to 12-per-second) stimulation of medial components of the thalamic reticular system (Jasper, 1949). Such stimulation induces the classical recruiting response (Dempsey and Morison, 1942; Morison and Dempsey, 1942)—a progressive build-up of long-latency evoked potentials which characteristically "wax and wane" in amplitude during continued stimulation, much in the fashion of spontaneous EEG waves of the drowsy animal or an animal lightly narcotized with barbiturates (Brazier, 1968).

Intracellular recording from thalamic neurons during evoked EEG synchronization reveals that widespread synchronization of neuronal discharges is effected by excitatory and inhibitory postsynaptic potentials (EPSP-IPSP sequences) with an over-all duration of 100 to 150 milli-

seconds (Purpura and Cohen, 1962) (Figure 3). The major feature of the synchronization process is a prolonged IPSP that effectively suppresses spontaneous and evoked discharges. Two features of the synchronization process are noteworthy: (1) the latencies of evoked PSPs and (2), the synchronous timing of these PSPs in neurons in fundamentally different types of thalamic nuclear organizations (Purpura and Shofer, 1963; Purpura et al., 1966a) (Figure 4). Although EPSP latencies of 8 to 10 milliseconds are common, it is not unusual for succeeding IPSPs to develop with a latency of 20 to 40 milliseconds. Thus, the overt simplicity of the synaptic mechanisms that underlie the synchronization process belies the extraordinary complexity of the interneuronal networks which generate the EPSP-IPSP responses. Similar PSP responses are observed in thalamic neurons during spontaneous six- to 12-per-second EEG waves in unanesthetized (Maekawa and Purpura, 1967a, 1967b) and barbiturized animals (Andersen et al., 1964b).

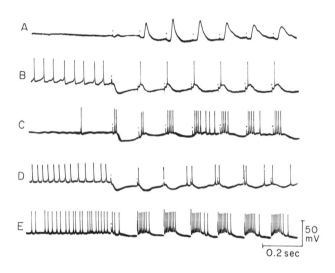

FIGURE 3 Patterns of intracellularly recorded activities of thalamic neurons during evoked EEG synchronization (recruiting responses) elicited by low-frequency (seven per second), medial-thalamic stimulation. A. Characteristics of surface-negative recruiting responses (motor cortex) elicited throughout the experiment from which the intracellular records (B–E) were obtained. B. Neuron in ventral anterior region of thalamus exhibiting prolonged IPSP after first stimulus, then EPSP-IPSP sequences with successive stimuli. C. Relatively quiescent ventrolateral neuron develops double discharge with first stimulus. The ensuing IPSP is succeeded by another evoked EPSP and cell discharge. Note alternation of IPSP. D. Neuron with discharge characteristics similar to that shown in B. E. Neuron in intralaminar region exhibiting an initial, prolonged IPSP that interrupts spontaneous discharges. The second and all successive stimuli evoked prolonged EPSPs with repetitive discharges that are terminated by IPSPs. (From Purpura and Shofer, 1963.)

Two mechanisms have been proposed to account for the synchronization of thalamic neuronal discharge. The view championed here implicates complex organizations of excitatory and inhibitory interneurons in the synchronization process. The hypothesis originally proposed by Andersen and Eccles (1962) and elaborated by Andersen and Andersson (1968) relies on additional processes of recurrent inhibition and postanodal excitation in the mechanism of rhythmically recurring IPSP sequences in some thalamic neurons. It has also been proposed by the latter workers that thalamic neurons have intrinsically unstable excitability states, which endow them with unique properties as "facultative pacemaker" elements in generating localized thalamocortical spindle bursts (Andersen and Andersson, 1968). The data supporting this view would appear to be applicable under conditions of barbiturate anesthesia and partial destruction of nonspecific thalamic nuclei, which, under ordinary circumstances, are the major source of synchronizing input to specific thalamic nuclei.

GATING AND FILTERING OPERATIONS OF THALAMIC SYNCHRONIZATION That specific thalamic nuclei participate in the evoked EEG-synchronization process that is initiated by low-frequency stimulation of medial thalamic regions provides a useful illustration of gating and filtering operations, and has important consequences for behavior—

clearly revealed in studies of ventrolateral (VL) relay neurons in the cerebello-thalamocortical projection pathway (Cohen et al., 1962). As noted above, stimulation of the brachium conjunctivum elicits monosynaptic, "all-or-none" EPSPs, which invariably initiate spike discharges in VL-relay cells (Purpura et al., 1965).

In contrast, medial thalamic stimulation induces prolonged synchronizing EPSP-IPSP sequences in VL neurons, as noted above. The combination of brachium conjunctivum and medial thalamic stimulation will result in complete inhibition of relay discharges when the brachium stimulus is applied at any time during the IPSP of 100-millisecond duration that is evoked by medial thalamic stimulation (Figure 5). Conversely, relay activity will be facilitated during the brief EPSPs induced from the medial thalamus. Inasmuch as VL cells are under active inhibition for 70 to 80 per cent of the time during which EEG

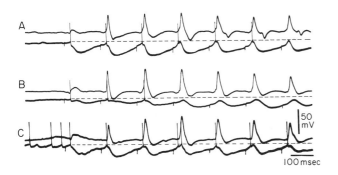

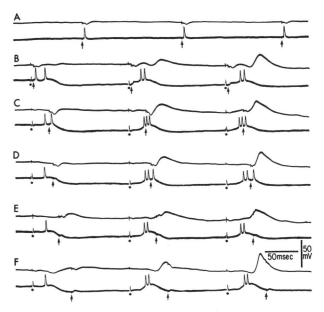

FIGURE 4 Temporal relations of EPSP-IPSP sequences in neurons situated in different thalamic regions during motor-cortex recruiting responses (upper channel records) evoked by seven-per-second stimulation of medial-thalamic, nonspecific nuclei. The three unidentified neurons were impaled at different depths from the dorsolateral surface of the exposed thalamus during a single penetration of the microelectrode. Depths as follows: A, 2.0 mm.; B, 4.5 mm.; C, 6.5 mm. Timing of the first three spike discharges elicited by EPSPs is synchronized in the three elements. IPSP latencies and duration are similar except during late phases in B. Note IPSP summation leading to progressive hyperpolarizing shift of membrane potential during continued stimulation. Dashed lines are drawn through "firing levels" determined by first synaptically evoked response of each series. (From Purpura et al., 1966a.)

FIGURE 5 Gating effect of synchronizing EPSP-IPSP sequences in a ventrolateral (VL) relay neuron. A-F. Segments taken from a continuous recording in a VL cell during interaction of PSPs evoked by stimulation of the brachium conjunctivum and medial thalamus. Upper channel, primary and recruiting responses recorded from motor cortex. Lower channel, intracellular records obtained from a VL neuron monosynaptically activated by brachium conjunctivum stimulation (at arrows). A. Brachium conjunctivum stimulation alone. B-F. Variations in temporal relations of applied stimuli. Note in C that brachium conjunctivum stimulation is effective during the EPSP phase of the EPSP-IPSP sequence induced by medial-thalamic stimulation. Prior to and after this, EPSPs evoked from the brachium conjunctivum fail to initiate spikes because of the prolonged summating action of the IPSPs evoked from the medial thalamus.

synchronization is elicited, it is not unlikely that this partial "functional deafferentation" of motor-cortex elements may play a major role in effecting the alteration in discharge patterns of corticospinal neurons that has been observed in freely moving monkeys during the transition from wakefulness to drowsiness or light sleep (Evarts, 1964).

It has already been recognized that the suppression of VL relay activity during EEG-synchronization and behavioral sleep (Steriade et al., 1969) is explicable on the basis of the gating operation of synchronizing EPSP-IPSP sequences in VL cells. Gating effects of synchronizing PSPs initiated by medial-thalamic stimulation are also observed in VB neurons, but to a lesser extent than that observed in VL cells (Maekawa and Purpura, 1967b). Nevertheless, it is clear that spontaneous thalamocortical synchronizing activities may be associated with similar EPSP-IPSP sequences, which effectively modulate VB relay activity. It is not unlikely that the internuclear interactions effected between nonspecific thalamic nuclei and VB relay cells and interneurons may be involved in the periodic behavior of spontaneous VB cell discharges (Poggio and Viernstein, 1964).

## Desynchronization as a thalamic operation

The transition from light sleep to wakefulness is accompanied by a dramatic alteration in the EEG, consisting of suppression of synchronized low-frequency, high-amplitude rhythms and the production of low-amplitude, high-frequency activity (Brazier, 1968). The discovery of the brain-stem reticular activating system (Moruzzi and Magoun, 1949) and subsequent studies of the effects high-frequency, medial-thalamic stimulation have on the initiation of EEG desynchronization have provided adequate models for examining the role of thalamic synaptic organizations in the desynchronizing process (Purpura and Shofer, 1963).

It will be recalled that the evoked EEG-synchronization process initiated by medial, nonspecific, thalamic stimulation is frequency specific, requiring stimuli in the range of six to 12 per second for optimum development of EPSP-IPSP sequences in thalamic neurons. Increases in stimulus frequency beyond this range effectively curtail and then suppress the prolonged IPSPs that are characteristic of the synchronizing process. One result of this input tuning is seen in the summation of EPSPs and an augmentation of cell discharge (Figure 6). Intracellular studies that replicate the original experimental paradigm of Moruzzi and Magoun (1949) illustrate the dramatic attenuation of synchronizing IPSPs produced by high-frequency, brain-stem, reticular stimulation delivered on a background of EEG synchronization that is elicited by low-frequency, medial-thalamic stimulation (Purpura et al., 1966b) (Figure 7). It is apparent from Figures 6 and 7 that the basic thalamic

operations underlying the transition from synchronization to desynchronization of neuronal activity involve an attenuation or inhibition of prolonged synchronizing IPSPs and an augmentation of excitatory synaptic drives. The consequences of this alteration in the synaptic pattern of nonspecific input to specific relay cells are evident in the disinhibition and facilitation of the VL relay discharges that accompany reticulocortical activation (Figure 7). The operations illustrated in Figures 6 and 7 adequately account for the facilitation of specific evoked potentials recorded from VL during the transition from sleep to wakefulness in chronically implanted animals (Steriade et al., 1969).

The analysis of synchronizing and desynchronizing processes in thalamic neuronal networks has revealed additional operations of these circuits apart from their gating, filtering, and modulating effects on relay and other elements. The PSP patterns that underlie these fundamentally different processes are elicited by frequency specific mechanisms, which suggests the operation of reciprocally acting, inhibitory pathways with different dynamic ranges of input selection. The analysis also discloses postactivation facilitation of augmented excitatory drives and suppression of inhibitory (synchronizing) pathways as a consequence of

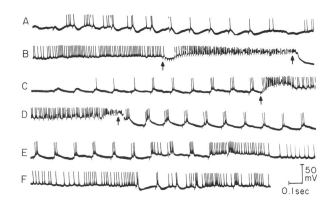

FIGURE 6 Effects of repeated, high-frequency, medial-thalamic stimulation on a ventromedial cell that exhibited a synchronization pattern characterized by short-latency EPSPs and prolonged IPSPs prior to activation. A–E. Continuous record. A. Low-frequency, medial-thalamic stimulation succeeded by a phase of hyperexcitability (B). At first arrow in B, a prolonged IPSP is initiated by the first stimulus of the high-frequency (60 per second) repetitive train. Successive stimuli after the IPSP evoke summating EPSPs associated with high-frequency spike attenuation. D. Second period of low-frequency, medial-thalamic stimulation after repolarization initiates only prolonged, slowly augmenting EPSPs. Changes in stimulus frequency between arrows, in C and D, induce high-frequency, repetitive discharges superimposed on depolarization, the magnitude of which is related to stimulus frequency. F. Several seconds later. Note reappearance of IPSPs during low-frequency, medial-thalamic stimulation. (From Purpura and Shofer, 1963.)

brief periods of high-frequency stimulation of the thalamic and brain-stem reticular system. "Storage" of these effects in interneuronal circuits is also evident from intracellular observations of persisting excitatory synaptic drives after brief periods of stimulation (Figure 6).

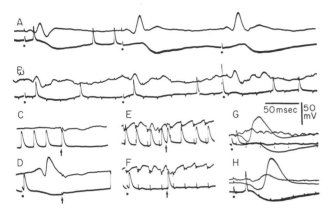

FIGURE 7 Intrathalamic synaptic events associated with desynchronizing effect of pontine-reticular stimulation. Upper channel records obtained from motor cortex. A, B. From an unidentified ventrolateral thalamic neuron. A. EPSP-IPSP sequence induced by low-frequency, medial-thalamic stimulation that elicits a prominent polyphasic recruiting response in cortex. B. Simultaneous low-frequency, medial-thalamic and high-frequency, pontine-reticular stimulation. Note marked attenuation of IPSP during reticular stimulation. (From Purpura et al., 1966a.)

C–E. Disinhibiting effect of reticular activation on VL relay neuron inhibited during low-frequency, medial-thalamic stimulation. C. Evoked response of VL neuron to brachium conjunctivum stimulation (at arrow) alone. D. Low-frequency, medial-thalamic stimulation (dot) precedes brachium stimulation by 50 milliseconds. A small EPSP is revealed in isolation when the brachium stimulus is applied during the prolonged IPSP. E. Effects of reticular stimulation on excitability of VL neuron. F. Same as in D, but during concomitant reticular stimulation that attenuates the IPSP evoked by medial thalamic stimulation. (From Purpura et al., 1966b.)

G. Details of the blocking effect of high-frequency, pontine-reticular stimulation on a prolonged IPSP evoked in an unidentified thalamic neuron by low-frequency, medial-thalamic stimulation. Superimposed traces show cortical surface and intracellular records during response to seven-per-second, medial-thalamic stimulation and during continued low-frequency, medial-thalamic and high-frequency, reticular stimulation. Reticular stimulation virtually eliminates the cortical potential evoked by medial thalamic stimulation. Note short-latency, reticulocortical responses at cortical level and small EPSPs in thalamic neuron. Low-frequency, medial-thalamic stimulation is unable to elicit IPSPs during reticulocortical activation.

H. Same experimental design as in G, but from another preparation. Reticular activation suppresses recruiting response but does not affect IPSP. This was observed in a small proportion of neurons. (Modified from Purpura et al., 1966a.)

## Reciprocity in specific-nonspecific internuclear interactions

The foregoing internuclear operations in the thalamus have emphasized various effects of medial, nonspecific, thalamic stimulation on elements in specific relay nuclei. Recent studies have sought the existence of reciprocal internuclear pathways in experimental designs, in which stimulation of a specific VL nucleus was carried out during intracellular recording from neurons of the medial, nonspecific thalamus (Desiraju et al., 1969). The salient results of this study are summarized in Figure 8 by recordings taken from four different preparations. Stimulation in VL, which elicits typical primary evoked responses in motor cortex, evokes short-latency (1.5 to 4.0 milliseconds), prolonged IPSPs in a large proportion of medial, nonspecific, thalamic neurons. Similar effects are observed during low-frequency repetitive stimulation in VL, which elicits typical augmenting responses. Although a somewhat larger proportion of medial thalamic neurons exhibits short-latency EPSPs that precede IPSPs during primary evoked responses, summation of IPSPs with repetitive VL stimulation effectively suppresses the discharge capabilities of these EPSPs.

The observations in Figure 8 not only point to the operation of powerful synaptic pathways linking specific and nonspecific nuclei; they also indicate that transit time in the specific-nonspecific pathways is extremely rapid in contrast to nonspecific-specific interactions. Suffice it to say that the inhibition in medial thalamic neurons subsequent to VL stimulation represents the shortest latency inhibitory effect

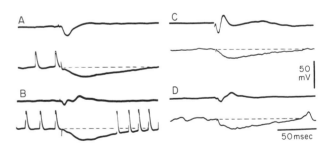

FIGURE 8 Inhibitory interactions between specific and nonspecific thalamic nuclei. Examples of intracellular recordings (lower channel records) obtained from medial-thalamic neurons in four different preparations. Specific responses to VL stimulation recorded monopolarly from motor cortex are shown in upper channel records (negative upward). A, B. Cells exhibiting 30- to 40-mV spike potentials prior to VL stimulation and short-latency IPSPs immediately after stimulation. C, D. Cells partially depolarized after impalements. In these elements, as well as in A and B, IPSPs are not preceded by EPSPs. Dashed horizontal lines are drawn through baseline membrane potential level. (From Desiraju et al., 1969.)

observed to date in intracellular studies of thalamic neurons.

Elucidation of the inhibitory action of VL stimulation on medial, nonspecific, thalamic elements provides an adequate explanation of the short-lasting suppression produced by prior VL stimulation on medial-thalamic, evoked EPSPs in pyramidal-tract neurons (Figure 9). Without knowledge of the internuclear synaptic transactions taking place at the thalamic level, it might have been inferred from the findings of Figure 9 that the transient suppression of medial-thalamic, evoked EPSPs was an example of "presynaptic inhibition" in cortical pathways. The designation of this as "very remote" inhibition is probably far more appropriate, inasmuch as the EPSP suppression is a consequence of the persisting inhibition of thalamic nonspecific pathways that is caused by VL stimulation.

Several examples of thalamocortical operations that enhance channel selection have been noted previously. The demonstration that VL stimulation inhibits nonspecific thalamic neurons provides a powerful mechanism for enhancing the effectiveness of a particular VL input pathway to motor cortex by suppression of irrelevant spontaneous activity in nonspecific thalamocortical projection systems.

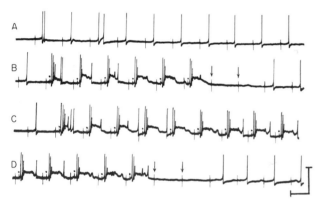

FIGURE 9 Interaction of specific and nonspecific evoked postsynaptic activities in a pyramidal-tract (PT) neuron of motor cortex. A–D. From a continuous recording. A. Low-frequency stimulation in medial thalamus elicits a single or, occasionally, a double discharge of the PT neuron. The discharge is initiated by a small EPSP of 20- to 30-millisecond latency. B. Concomitant stimulation of the ventrolateral thalamic nucleus (at dots) induces powerful and prolonged EPSPs, associated spike discharges, and spike inactivation. Arrows in B identify the two medial-thalamic stimuli immediately after cessation of VL stimulation. Note that the medial-thalamic evoked EPSP is either not detectable or is subthreshold for several hundred milliseconds after the last VL stimulus. C. Onset of a second period of VL stimulation after recovery of nonspecific evoked responses. D. Same as in B, to show suppression of medial-thalamic evoked EPSPs after the last VL stimulus. As explained in the text, the data reflect at the cortical level the operation of an inhibitory internuclear pathway in the thalamus that is activated by VL stimulation. Calibrations: 50 mV, 100 milliseconds.

## Parallel processing of nonspecific thalamic output

Examination of the synaptic effects observed in a wide variety of forebrain structures activated by stimulation of medial, nonspecific, thalamic nuclei discloses a feature of thalamic reticular-system operations that is not encountered in specific projection pathways. Stimulation of specific thalamic nuclei, such as VL or VB, elicits responses localized to relatively restricted areas of neocortex.

In contrast, stimulation of medial-thalamic nuclei activates synaptic organizations in the neocortex, the corpus striatum (Shimamoto and Verzeano, 1954), the amygdaloid complex, and the mesencephalic reticular regions, in addition to producing the powerful intrathalamic synaptic effects noted above. One can infer from this that once the thalamic reticular system is stimulated by appropriate inputs it distributes activity in many parallel projection systems. The synaptic transformations that occur in different structures receiving nonspecific thalamic input depend on intrinsic differences in the organization of excitatory and inhibitory neurons in these structures, as well as on differences in properties of responding elements (Purpura, 1969).

Experimental methods similar to those described above have been utilized to characterize the different patterns of synaptic activity initiated in forebrain structures during cortical-surface recruiting responses, the electrographic "signature" of adequate stimulation of medial and intralaminar, nonspecific, thalamic nuclei (Figure 10). At the level of the neocortex, nonspecific thalamic stimulation elicits, in pyramidal-tract neurons, slowly rising EPSPs of long latency (Purpura et al., 1964) (Figure 10A). Such EPSPs are generally uninfluenced by hyperpolarizing currents that are applied intracellularly, which suggests that they are generated predominantly at axodendritic synapses (Creutzfeldt et al., 1966; Purpura and Shofer, 1964). These EPSPs in pyramidal-tract neurons induce variable spike discharges in contrast to the powerful EPSPs initiated by specific thalamic stimulation (Figure 9). Nonpyramidal-tract neurons are readily driven by nonspecific thalamic stimulation, and may exhibit prominent EPSP-IPSP sequences not unlike those observed in thalamic neurons during induced EEG synchronization (Purpura et al., 1964).

Different components of the basal ganglia exhibit fundamentally different PSP patterns during nonspecific thalamic stimulation. A large proportion of caudate neurons develop prolonged (80- to 100-millisecond) EPSPs of 10- to 20-millisecond latency; these bear a remarkable resemblance to EPSPs observed in pyramidal neurons in the motor cortex (Purpura and Malliani, 1967). Caudate neurons show little or no spontaneous discharges and exhibit one, or at the most a few, spike discharges in association with rather prominent EPSPs (Figure 10B). Thus, despite their different phylogenetic history and markedly different morphology, caudate neurons and corticospinal neurons appear to be similarly influenced by nonspecific thalamic stimulation.

It has been suggested elsewhere that, with the phylogenetic elaboration of the neocortex, synaptic relations established between nonspecific thalamic nuclei and neostriatum were functionally replicated in the development of a parallel projection system to neocortex (Purpura, 1969). However, in response to medial-thalamic stimulation, not all elements of the neostriatum exhibit PSP patterns that are typically encountered in neocortical pyramidal-tract neurons. For

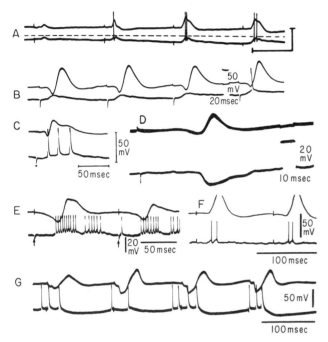

FIGURE 10. Parallel processing of unspecific thalamic output in synaptically related neuronal subsystems. Upper channel records, obtained from motor cortex, display the characteristics of recruiting responses elicited in different preparations during medial-thalamic stimulation. A. Typical responsiveness of a pyramidal-tract neuron in motor cortex. Dashed line drawn through firing level. Nonspecific thalamic stimulation elicits a progressively incrementing EPSP with a slow rise time and a prolonged decay phase. Calibrations: 50 mV, 100 milliseconds. (From Purpura et al., 1964.) B. Intracellular synaptic events recorded in a caudate neuron during nonspecific thalamic stimulation. Note remarkable similarity in EPSP characteristics in the caudate and PT neuron. (From Purpura and Malliani, 1967.) C. Pattern of EPSP and spike discharges generally observed in putamen neurons. D. Long-latency IPSP detectable in a neuron of the nucleus entopeduncularis during medial-thalamic stimulation. (C and D, from Malliani and Purpura, 1967). E, F. Characteristics of discharge patterns of neurons in mesencephalic reticular formation during medial thalamic stimulation (From Maekawa and Purpura, unpublished). G. Responsiveness of a neuron in the amygdaloid complex during stimulation of intralaminar nuclei of the nonspecific projection system. (From Santini and Purpura, in preparation.)

example, neurons in ventral parts of the caudate show PSP patterns similar to those observed in ventrolateral and ventroanterior nuclei of the thalamus (Purpura and Malliani, 1967), whereas putamen neurons characteristically respond with more potent EPSPs and associated spike discharges (Malliani and Purpura, 1967) (Figure 10C). Cells of the globus pallidus and entopeduncular nucleus generally exhibit IPSPs with very long latencies after medial-thalamic stimulation (Figure 10D). It is to be noted that none of the synaptic effects observed in different components of the corpus striatum after nonspecific thalamic stimulation depend on indirect relays involving neocortex. But it is not to be inferred from this brief survey that the physiological data indicating widespread activation (and inhibition) of neuronal organizations in the corpus striatum are indicative of the operation of direct projection systems from medial and intralaminar thalamic nuclei, despite neuroanatomical evidence in favor of such connections (Powell and Cowan, 1954; 1956). In point of fact, there can be little doubt that virtually all the synaptic effects initiated by nonspecific thalamic stimulation in the corpus striatum, the amygdaloid complex (Santini and Purpura, 1969) (Figure 10G), and the mesencephalic reticular regions (Maekawa and Purpura, unpublished) (Figure 10E and F) involve polysynaptic and small-fiber projection systems (Verhaart, 1950) that describe complex trajectories through the basal ganglia and brain stem (Frigyesi and Purpura, 1967).

The functional significance of parallel processing of nonspecific thalamic output is not difficult to envision, because this provides a mechanism whereby activities in diverse structures can be influenced from central, nonspecific, modulator systems that receive convergent input from different sources (Albe-Fessard and Fessard, 1963). The notion that nonspecific influences are predominantly modulatory, rather than pacemaker, in operation is consistent with the data at hand, inasmuch as true pacemaker or autorhythmic activities have not yet been observed in intracellular recordings taken from a large number of medial and intralaminar thalamic neurons.

## Operations of a thalamic interneuronal "interface"

The ansa lenticularis is generally regarded as the major outflow pathway of the corpus striatum (Wilson, 1914). In the cat, this projection system arises from the globus pallidus and entopeduncular nucleus, the latter representing the homologue of the medial pallidal segment of primates (Fox et al., 1966; Nauta and Mehler, 1966). Lenticular projections to the thalamus are prominent in ventroanterior-ventrolateral (VA-VL) nuclei as well as in medial or paramedial nuclei, such as the nucleus centrum medianum (Mehler, 1966). The most impressive afferents to VL and, to a lesser extent, VA arise from lateral cerebellar nuclei and project to the thalamus via the brachium conjunctivum. The VL nucleus of the thalamus, as noted above, is the major source of afferents to the cerebral motor cortex. From the standpoint of its relationships to the cerebellum, corpus striatum, and motor cortex, VL may be viewed as a unique "interface" between motor-control systems of the basal ganglia and cerebellum on the one hand and the neocortex on the other. It will be recalled at this juncture that VL neurons exhibit prominent EPSP-IPSP sequences during EEG synchronization induced by medial, nonspecific, thalamic stimulation (Purpura et al., 1965). Therefore, particular importance attaches to the analysis of convergent operations of VL relay cells and interneurons activated by pallidofugal, cerebellofugal, and nonspecific-specific internuclear pathways in the assessing of the integrative activities of interneuronal synaptic organizations of the thalamus.

Recent intracellular studies have established that there is a population of VL relay neurons that receives convergent monosynaptic excitation from the cerebellum and outflow pathways of the corpus striatum (Desiraju and Purpura, 1969). Convergent monosynaptic EPSPs evoked in VL relay neurons by stimulation of the brachium conjunctivum and the ansa lenticularis have basically similar characteristics but differ in latency by 0.5 to 0.8 millisecond, the time required for impulse conduction in the longer cerebellofugal pathway (Figure 11A–D). Monosynaptic EPSPs evoked in

VL cells by stimulation of the brachium conjunctivum are frequently succeeded by IPSPs. Although similar EPSPs are elicited by stimulation of the ansa lenticularis, these are generally not followed by IPSPs (Figure 11E and F). The inference here is that collaterals of brachium conjunctivum afferents to VL relay neurons engage a parallel inhibitory circuit that is not available to projections of the ansa lenticularis that distribute to the same VL relay cells.

Disynaptic and polysynaptic PSPs are also observed in VL neurons after stimulation of the ansa lenticularis and the brachium conjunctivum. In contrast to the exclusively excitatory monosynaptic actions of these inputs to relay cells, however, synaptic effects observed in nonrelay elements (interneurons) may be reciprocal in nature (Figure 11G and H). Latencies of such reciprocal PSPs are in the range of four to six milliseconds, indicating considerable dispersion of cerebellofugal and striatofugal projection activity in VL interneuronal organizations. Longer (>10 milliseconds) latency PSPs have also been noted in VL neurons in response to stimulation of the ansa lenticularis, but these are rarely encountered with stimulation of the brachium conjunctivum. Such VL neurons commonly exhibit convergence of long-latency responses evoked by stimulation of lenticular and thalamic nonspecific nuclei (Figure 11 I–L).

Remarkably, the PSP patterns induced in these interneurons by pallidal-entopeduncular stimulation resemble in all

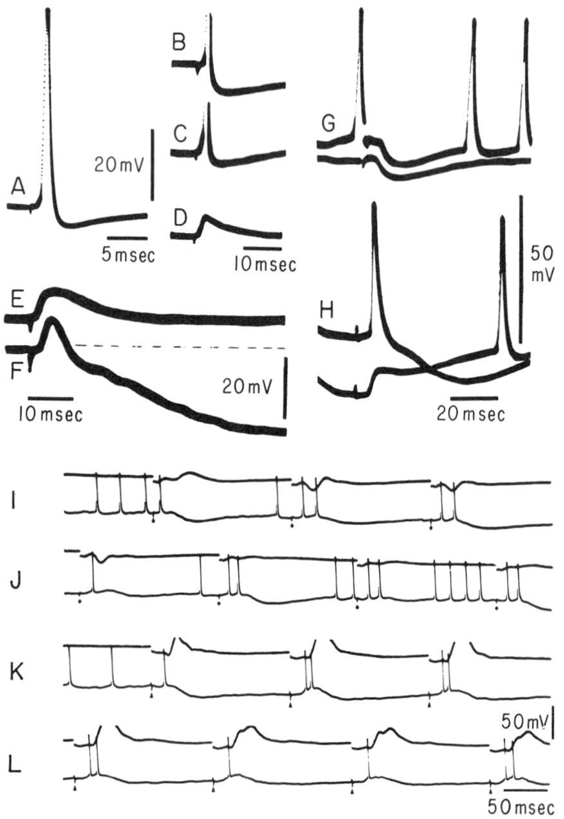

FIGURE 11 Intracellular recording of convergent monosynaptic excitation of a VL neuron by stimulation of the ansa lenticularis (A, C, and D) and the brachium conjunctivum (B). Spikes in B and C truncated for display purposes. Note minimal latency differences of EPSPs in B and C. EPSP evoked by ansa lenticularis stimulation is shown in isolation in D. E, F. Records obtained from a different VL neuron after stimulation of ansa lenticularis (E) and brachium conjunctivum (F). Only the brachium-evoked EPSP is succeeded by a prolonged IPSP, the early phase of which exhibits low-amplitude oscillations. G, H. Example of convergent but reciprocal synaptic effects observed in a VL neuron following ansa lenticularis (G) and brachium conjunctivum (H) stimulation. Responses in each case were elicited at two levels of membrane polarization. G. Upper record obtained during spontaneous discharges; lower record, during a phase of increased membrane polarization, during which spontaneous discharges were eliminated. In each instance, ansa lenticularis stimulation evokes a four- to six-millisecond IPSP. H. Lower record of the pair obtained during a spontaneous long-duration IPSP. Brachium conjunctivum stimulation elicits a four- to six-millisecond latency EPSP and spike discharge. The EPSP is revealed in isolation during the spontaneous IPSP. I–L. Example of similar long-latency EPSP-IPSP sequences elicited in a VL neuron by repetitive stimulation in the region of nucleus entopeduncularis (I and J, continuous recording) and stimulation of medial thalamus (K and L, continuous recording). Upper channel records obtained from motor cortex. Note prominent long-latency, surface-negative, recruiting response evoked by medial-thalamic stimulation (From Desiraju and Purpura, 1969.)

respects those evoked by stimulation of nonspecific-specific internuclear projections, possibly as a consequence of the fact that ansa lenticularis stimulation is also capable of evoking PSP patterns of variable latency in medial and intralaminar neurons (Desiraju and Purpura, unpublished observations).

The data illustrated in Figure 11 indicate a wide range of complex interrelationships of lenticulofugal, cerebellofugal, and nonspecific evoked activities in VL neuronal organizations. At one end of the temporal spectrum, and "spatially" in respect to VL relay elements, monosynaptic excitatory effects are induced by convergent projection of the brachium conjunctivum and the ansa lenticularis. Polysynaptic convergence may also take place within interneuronal organizations, but this is more likely to be reflected in reciprocal synaptic actions. Finally, at the other end of the temporal spectrum, long-latency PSPs are observed with ansa lenticularis and nonspecific thalamic stimulation but not with brachium conjunctivum stimulation. The difference suggests that VL neurons are organized in functional domains related to inputs that define particular synaptic territories within the domains.

Whereas cerebellar inputs are relatively restricted in distribution to relay cells and interneurons closely linked to relay neurons, ansa lenticularis projections are distributed to elements that gain access to virtually all components of the interneuronal network. Polysynaptically activated VL neurons beyond the "synaptic range" of cerebellofugal projections appear to be readily accessible to nonspecific-specific internuclear pathways and ansa lenticularis projections. When viewed in this fashion, it is clear that the important transactions which take place in programing afferent volleys to motor cortex from VA to VL are consequences of interactions carried out in interneuronal domains related to cerebellar, basal ganglia, and nonspecific inputs (Frigyesi and Purpura, 1964). Within these domains, signal selection by one input may be replicated, facilitated, inhibited, or uninfluenced by another input. The operations suggested by intracellular studies of the VA-VL interneuronal "interface" between nonspecific thalamus, cerebellum, and basal ganglia on the one hand, and the motor cortex on the other, are undoubtedly capable of generating a wide range of discharge patterns in corticospinal neurons. Little wonder, then, that, in the search for appropriate "high-priority" sites for detecting tremorogenic neuronal activities in patients with various types of abnormal movement disorders, the interface here defined as the VA-VL neuronal organization (cf. Jung and Hassler, 1960) has been a most favorable target (Albe-Fessard et al., 1966; Gaze et al., 1964; Jasper and Bertrand, 1966).

I have emphasized some of the transactions carried out in the VL interface by using examples of internuclear interactions that illustrate gating, output tuning, and filtering of

VL relay activity. Another particularly interesting operation has been demonstrated recently in studies of the effects on responses of some VL neurons of prolonged, repetitive stimulation of pallidal-entopeduncular regions. The essential features of this operation are illustrated in Figure 12. In this study, low-frequency stimulation in the entopeduncular nucleus initially elicited little or no synaptic activity in a VL neuron. Continued stimulation then resulted in progressive enhancement of a long-latency EPSP, which evoked one or more spike discharges. When this EPSP became well established, continuing stimulation resulted in a pattern of responsiveness in which only *alternate* stimuli elicited the EPSP and associated spikes. Thus, despite a steady input, the output burst from the VL neuron was at half of the frequency of the input stimulus.

The results shown in Figure 12 had not been observed previously in intracellular studies of thalamic neurons or, for that matter, in studies of other vertebrate neurons. It is of some significance that the phenomenon of EPSP alternation

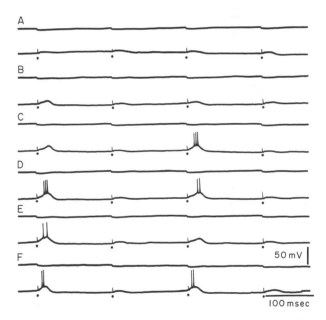

FIGURE 12 Example of step-down transformation of striopallidothalamic input to VL neuronal organizations. Upper channel records obtained from motor cortex to illustrate absence of evoked response during continued six-per-second stimulation in region of nucleus entopeduncularis-ansa lenticularis. A–F. Continuous recording. Stimuli marked by dots. A, B. Initial period of stimulation elicits a slowly incrementing EPSP in a VL neuron. C. With continued stimulation, a delayed EPSP of considerable magnitude makes its appearance with alternate stimuli of the six-per-second train. When the alternation pattern is well established, as in D–F, discharges are superimposed on the long-latency EPSP. These occur in bursts at half the input frequency as a consequence of the alternating EPSP. (From Desiraju and Purpura, in preparation.)

is detectable in those basal ganglia-thalamic synaptic organizations that have been implicated in tremorogenic processes encountered in human dyskinesias (Yahr and Purpura, 1967). This raises the intriguing possibility that, under conditions of prolonged and excessive excitatory synaptic drive from pallidal entopeduncular elements, VA-VL neuronal organizations may be capable of converting a high-frequency input to a low-frequency output via their intrinsic excitatory and inhibitory elements. The alternation of the EPSP observed under these conditions is reminiscent of the IPSP alternation that characterizes the effect of low-frequency stimulation of medial, nonspecific, thalamic nuclei on many thalamic neurons (Purpura and Cohen, 1962). Both alternation processes or step-down transformations must surely result from complex and prolonged excitatory and inhibitory interneuronal interactions with different temporal characteristics. The reader will be spared details of the many possible permutations and combinations of circuit elements required for the observed effects.

## Some consequences of conjoint thalamic operations

The foregoing analyses of several obvious operations and processes, identifiable from intracellular studies, might be considered little more than an amusing treatment of experimental data were it not that, when taken together, these operations provide important clues to integrative functions of the thalamus. For integration implies the cooperative activity of neurons in many and diverse functional organizations intrinsic both to the thalamus and to other structures reciprocally related to these neuronal organizations. Perhaps there has been too much emphasis on detailed studies of "specific" relay nuclei on the one hand and of "nonspecific" nuclei on the other, each nuclear system being considered in isolation, without concern for their interactions and interrelations. Whenever possible, attempts have been made to focus on aspects of internuclear relations, in the expectation that a more meaningful view of thalamic integrative function than has been available heretofore will emerge from the analysis of conjoint operations.

The complex but orderly functional organization of thalamic interneuronal systems is clearly exemplified in the processes that underlie synchronization and desynchronization of thalamic neuronal activity. Both events have operational consequences that are secondary derivatives of alterations in thalamic discharge patterns. Synchronizing and desynchronizing processes gate transmission effectively through specific relays with prominent and reciprocal cortical projections, and induce a variety of synaptic effects in structures activated in parallel with cortex. The consequences of parallel processing are to be seen in dramatic positive and negative feedback actions on thalamic neurons,

which, in turn, will exert feed-forward influences on other subsystems.

Additional operations, including storing, filtering, input-selection attenuation, alternation, and output tuning, are also demonstrable as elements of the functional integration of intrathalamic and extrathalamic synaptic systems. These examples of thalamic operations gain in importance when it is appreciated that they reflect the functional properties of a neural substrate consisting of VA-VL and synaptically related structures (motor cortex, cerebellum, basal ganglia, and nonspecific nuclei), generally considered essential for the elaboration of sensorimotor functions of the utmost complexity. It is not difficult to envision the results of abnormal perturbations in any of these cortical operations concerned with information processing in cerebello-thalamic, thalamocortical, thalamo-striatal, and intrathalamic subsystems. Even relatively minor disturbances in the latency, timing, and proportion of excitatory and inhibitory drives in a particular organization will exert widespread effects on related subsystems and lead to marked alterations in discharge patterns in pyramidal and extrapyramidal outflow pathways of the forebrain. A major task that remains is to specify the nature and origin of the functional disturbances resulting from pathophysiological processes involving one or more of the subsystem operations illustrated in this report. Such an endeavor should permit formulation of hypotheses of dynamic systems operations that are capable of embracing the wealth of isolated and seemingly unrelated data on the role of neocortex, cerebellum, thalamus, and basal ganglia in sensorimotor integrative processes.

## Summary

Physiological processes that underlie a variety of operations of thalamic neuronal subsystems are considered in terms of their elementary synaptic mechanisms and organizational substrate. Operations demonstrable in thalamic and related neuronal subsystems include high-fidelity transmission, input selection, output tuning, synchronization, desynchronization, filtering, gating, parallel processing, storing, signal attenuation, and step-down transformation. These operations are illustrated in intracellular studies of neurons in specific and nonspecific nuclei of the thalamus, cerebral cortex, and corpus striatum.

Such studies emphasize the functional significance of reciprocal, internuclear, synaptic pathways linking specific and nonspecific nuclei and the mode of activation of neocortical and striatal elements by thalamic projection. Lenticulofugal and cerebellofugal projections to the thalamus are shown to engage elements that constitute an interneuronal interface between corpus striatum and cerebellum on the one hand and motor cortex on the other. Physiological dissection of this interneuronal interface in VA-VL permits definition of synaptic domains described by convergent inputs from medial nonspecific thalamus, corpus striatum, and cerebellum. Operations and processes effected in the interneuronal interface are considered of fundamental importance for the generation of different temperospatial patterns in input pathways to motor cortex. The intracellular data illustrated in relation to identifiable thalamic processes provide an understanding of synaptic mechanisms in neuronal subsystems implicated in sensorimotor integrative processes and their disturbances.

## REFERENCES

ALBE-FESSARD, D., and A. FESSARD, 1963. Thalamic integrations and their consequences at the telencephalic level. *Progr. Brain Res.* 1: 115–148.

ALBE-FESSARD, D., G. GUIOT, Y. LAMARRE, and G. ARFEL, 1966. Activation of thalamocortical projections related to tremorogenic processes. *In* the Thalamus (D. P. Purpura and M. D. Yahr, editors). Columbia University Press, New York, pp. 237–249.

ANDERSEN, P., and S. A. ANDERSSON, 1968. Physiological Basis of the Alpha Rhythm. Appleton-Century-Crofts, New York.

ANDERSEN, P., C. McC. BROOKS, J. C. ECCLES, and T. A. SEARS, 1964a. The ventro-basal nucleus of the thalamus: Potential fields, synaptic transmission and excitability of both presynaptic and post-synaptic components. *J. Physiol.* (*London*) 174: 348–369.

ANDERSEN, P., and J. C. ECCLES, 1962. Inhibitory phasing of neuronal discharge. *Nature* (*London*) 196: 645–647.

ANDERSEN, P., J. C. ECCLES, and T. A. SEARS, 1964b. The ventro-basal complex of the thalamus: Types of cells, their responses and their functional organization. *J. Physiol.* (*London*) 174: 370–399.

ASANUMA, H., S. D. STONEY, JR., and C. ABZUG, 1968. Relationship between afferent input and motor outflow in cat motorsensory cortex. *J. Neurophysiol.* 31: 670–681.

BRAZIER, M. A. B., 1968. The Electrical Activity of the Nervous System. Edition 3. Williams and Wilkins, Baltimore, Maryland.

BROOKS, V. B., and H. ASANUMA, 1965. Recurrent cortical effects following stimulation of medullary pyramid. *Arch. Ital. Biol.* 103: 247–278.

COHEN, B., E. M. HOUSEPIAN, and D. P. PURPURA, 1962. Intrathalamic regulation of activity in a cerebellocortical projection pathway. *Exp. Neurol.* 6: 492–506.

CREUTZFELDT, O. D., H. D. LUX, and S. WATANABE, 1966. Electrophysiology of cortical nerve cells. *In* The Thalamus (D. P. Purpura and M. D. Yahr, editors). Columbia University Press, New York, pp. 209–230.

DEMPSEY, E. W., and R. S. MORISON, 1942. The production of rhythmically recurrent cortical potentials after localized thalamic stimulation. *Amer. J. Physiol.* 135: 293–300.

DESIRAJU, T., G. BROGGI, S. PRELEVIC, M. SANTINI, and D. P. PURPURA, 1969. Inhibitory synaptic pathways linking specific and nonspecific thalamic nuclei. *Brain Res.* 15: 542–543.

DESIRAJU, T., and D. P. PURPURA, 1969. Synaptic convergence of cerebellar and lenticular projections to thalamus. *Brain Res.* 15: 544–547.

EVARTS, E. V., 1964. Temporal patterns of discharge of pyramidal tract neurons during sleep and waking in the monkey. *J. Neurophysiol.* 27: 152–171.

FOX, C. A., D. E. HILLMAN, K. A. SIEGESMUND, and L. A. SETHER, 1966. The primate globus pallidus and its feline and avian homologues: A Golgi and electron microscopic study. *In* Evolution of the Forebrain (R. Hassler and H. Stephan, editors). Thieme, Stuttgart, pp. 237–248.

FRIGYESI, T. L., and D. P. PURPURA, 1964. Functional properties of synaptic pathways influencing transmission in the specific cerebello-thalamocortical projection system. *Exp. Neurol.* 10: 305–324.

FRIGYESI, T. L., and D. P. PURPURA, 1967. Electrophysiological analysis of reciprocal caudato-nigral relations. *Brain Res.* 6: 440–456.

GAZE, R. M., F. J. GILLINGHAM, S. KALYANARAMAN, R. W. PORTER, A. A. DONALDSON, and I. M. L. DONALDSON, 1964. Microelectrode recordings from the human thalamus. *Brain* 87: 691–706.

JASPER, H., 1949. Diffuse projection systems: The integrative action of the thalamic reticular system. *Electroencephalogr. Clin. Neurophysiol.* 1: 405–420.

JASPER, H. H., and G. BERTRAND, 1966. Thalamic units involved in somatic sensation and voluntary and involuntary movements in man. *In* The Thalamus (D. P. Purpura and M. D. Yahr, editors). Columbia University Press, New York, pp. 365–384.

JUNG, R., and R. HASSLER, 1960. The extrapyramidal motor system. *In* Handbook of Physiology. Section I: Neurophysiology (J. Field, H. W. Magoun, and V. E. Hall, editors). American Physiological Society, Washington, D. C., vol. 2, pp. 863–927.

MAEKAWA, K., and D. P. PURPURA, 1967a. Properties of spontaneous and evoked synaptic activities of thalamic ventrobasal neurons. *J. Neurophysiol.* 30: 360–381.

MAEKAWA, K., and D. P. PURPURA, 1967b. Intracellular study of lemniscal and non-specific synaptic interactions in thalamic ventrobasal neurons. *Brain Res.* 4: 308–323.

MALLIANI, A., and D. P. PURPURA, 1967. Intracellular studies of the corpus striatum. II. Patterns of synaptic activities in lenticular and entopeduncular neurons. *Brain Res.* 6: 341–354.

MEHLER, W. R., 1966. Further notes on the center median nucleus of Luys. *In* The Thalamus (D. P. Purpura and M. D. Yahr, editors). Columbia University Press, New York, pp. 109–127.

MORISON, R. S., and E. W. DEMPSEY, 1942. A study of thalamocortical relations. *Amer. J. Physiol.* 135: 281–292.

MORUZZI, G., and H. W. MAGOUN, 1949. Brain stem reticular formation and activation of the EEG. *Electroencephalogr. Clin. Neurophysiol.* 1: 455–473.

MOUNTCASTLE, V. B., 1961. Some functional properties of the somatic afferent system. *In* Sensory Communication (W. A. Rosenblith, editor). M. I. T. Press, Cambridge, Massachusetts, pp. 403–436.

MOUNTCASTLE, V. B., and I. DARIAN-SMITH, 1968. Neural mechanisms in somesthesia. *In* Medical Physiology, vol. II (V. B. Mountcastle, editor). Edition 12. Mosby, St. Louis, Missouri, pp. 1372–1423.

MOUNTCASTLE, V. B., G. F. POGGIO, and G. WERNER, 1963. The relation of thalamic cell response to peripheral stimuli varied over an intensive continuum. *J. Neurophysiol.* 26: 807–834.

MOUNTCASTLE, V. B., W. H. TALBOT, H. SAKATA, and J. HYVÄRINEN, 1969. Cortical neuronal mechanisms in flutter-vibration studied in unanesthetized monkeys. Neuronal periodicity and frequency discrimination. *J. Neurophysiol.* 32: 452–484.

NAUTA, W. J. H., and W. R. MEHLER, 1966. Projections of the lentiform nucleus in the monkey. *Brain Res.* 1: 3–42.

PHILLIPS, C. G., 1959. Actions of antidromic pyramidal volleys on single Betz cells in the cat. *Quart. J. Exp. Physiol. Cog. Med. Sci.* 44: 1–25.

POGGIO, G. F., and V. B. MOUNTCASTLE, 1963. The functional properties of ventrobasal thalamic neurons studied in unanesthetized monkeys. *J. Neurophysiol.* 26: 775–806.

POGGIO, G. F., and L. J. VIERNSTEIN, 1964. Time series analysis of impulse sequences of thalamic somatic sensory neurons. *J. Neurophysiol.* 27: 517–545.

POWELL, T. P. S., and W. M. COWAN, 1954. The connexions of the midline and intralaminar nuclei of the thalamus of the rat. *J. Anat.* 88: 307–319.

POWELL, T. P. S., and W. M. COWAN, 1956. A study of thalamo-striate relations in the monkey. *Brain* 79: 364–390.

PURPURA, D. P., 1967. Comparative physiology of dendrites. *In* The Neurosciences: A Study Program (G. C. Quarton, T. Melnechuck, and F. O. Schmitt, editors). The Rockefeller University Press, New York, pp. 372–393.

PURPURA, D. P., 1969. Interneuronal mechanisms in thalamically induced synchronizing and desynchronizing activities. *In* The Interneuron (M. A. B. Brazier, editor). UCLA Forum in Medical Sciences, no. 11, University of California Press, Berkeley and Los Angeles, pp. 467–496.

PURPURA, D. P., and B. COHEN, 1962. Intracellular recording from thalamic neurons during recruiting responses. *J. Neurophysiol.* 25: 621–635.

PURPURA, D. P., T. L. FRIGYESI, J. G. McMURTRY, and T. SCARFF, 1966a. Synaptic mechanisms in thalamic regulation of cerebello-cortical projection activity. *In* The Thalamus (D. P. Purpura and M. D. Yahr, editors). Columbia University Press, New York, pp. 153–172.

PURPURA, D. P., J. G. McMURTRY, and K. MAEKAWA, 1966b. Synaptic events in ventrolateral thalamic neurons during suppression of recruiting responses by brain stem reticular stimulation. *Brain Res.* 1: 63–76.

PURPURA, D. P., and A. MALLIANI, 1967. Intracellular studies of the corpus striatum. I. Synaptic potentials and discharge characteristics of caudate neurons activated by thalamic stimulation. *Brain Res.* 6: 325–340.

PURPURA, D. P., T. SCARFF, and J. G. McMURTRY, 1965. Intracellular study of internuclear inhibition in ventrolateral thalamic neurons. *J. Neurophysiol.* 28: 487–496.

PURPURA, D. P., and R. J. SHOFER, 1963. Intracellular recording from thalamic neurons during reticulocortical activation. *J. Neurophysiol.* 26: 494–505.

PURPURA, D. P., and R. J. SHOFER, 1964. Cortical intracellular potentials during augmenting and recruiting responses. I. Effects of injected hyperpolarizing currents on evoked membrane potential changes. *J. Neurophysiol.* 27: 117–132.

PURPURA, D. P., R. J. SHOFER, and F. S. MUSGRAVE, 1964. Cortical intracellular potentials during augmenting and recruiting responses. II. Patterns of synaptic activities in pyramidal and nonpyramidal tract neurons. *J. Neurophysiol.* 27: 133–151.

RALSTON, H. J., 1969. The synaptic organization of lemniscal projections to the ventrobasal thalamus of the cat. *Brain Res.* 14: 99–115.

SANTINI, M., and D. P. PURPURA, 1969. Thalamo-striatal control of amygdaloid neuron activity. *Anat. Rec.* 163: 255–256 (abstract).

SCHEIBEL, M., and A. B. SCHEIBEL, 1966. Patterns of organization in specific and nonspecific thalamic fields. *In* The Thalamus (D. P. Purpura and M. D. Yahr, editors). Columbia University Press, New York, pp. 13–46.

SHIMAMOTO, T., and M. VERZEANO, 1954. Relation between caudate and diffusely projecting thalamic nuclei. *J. Neurophysiol.* 17: 278–288.

STEFANIS, C., and H. JASPER, 1964. Recurrent collateral inhibition in pyramidal tract neurons. *J. Neurophysiol.* 27: 855–877.

STERIADE, M., G. IOSIF, and V. APOSTOL, 1969. Responsiveness of thalamic and cortical motor relays during arousal and various stages of sleep. *J. Neurophysiol.* 32: 251–265.

VERHAART, W. J. C., 1950. Fiber analysis of the basal ganglia. *J. Comp. Neurol.* 93: 425–440.

VERZEANO, M., and K. NEGISHI, 1960. Neuronal activity in cortical and thalamic networks. *J. Gen. Physiol.* 43 (no. 6, pt. 2): 177–195.

WELT, C., J. C. ASCHOFF, K. KAMEDA, and V. B. BROOKS, 1967. Intracortical organization of cat's motorsensory neurons. *In* Neurophysiological Basis of Normal and Abnormal Motor Activities. (M. D. Yahr and D. P. Purpura, editors). Raven Press, Hewlett, New York, pp. 255–288.

WILSON, S. A. K., 1914. An experimental research into the anatomy and physiology of the corpus striatum. *Brain* 36: 427–492.

YAHR, M. D., and D. P. PURPURA (editors), 1967. Neurophysiological Basis of Normal and Abnormal Motor Activities. Raven Press, Hewlett, New York.

# 43 Beginning of Form Vision and Binocular Depth Discrimination in Cortex

P. O. BISHOP

IT IS PERHAPS NATURAL to assume that seeing with two eyes is more complex than seeing with one, and to believe that the solution of the problems of binocular vision must wait on an understanding of monocular vision. Because it is not easy to appreciate any change in the appearance of objects when we close one eye, the intuitive conclusion—hardly surprising—is that form is first elaborated for each eye separately and that stereopsis is added at a subsequent stage by the conjunction of uniocular forms. This view has been widely held: Sherrington (1906), for example, concluded "that during binocular regard . . . each uniocular mechanism develops independently a sensual image of considerable completeness. The singleness of binocular perception results from union of these elaborated uniocular sensations." Paradoxically, the reverse is probably more nearly the case. I believe that estimates of binocular depth are made on the fairly raw data before form analysis has proceeded very far. If so, it opens the possibility of providing a neural mechanism for stereopsis without confronting the much more formidable problem of form perception. A strong hint in this direction is provided by the fact that, although a well-developed body of knowledge exists in relation to the psychophysics of binocular depth perception, the same can hardly be said about form and pattern recognition. Neurophysiological advance is certainly handicapped by the absence of a satisfactory psychophysics of form. It is perhaps to be expected, then, that this paper will be concerned largely with binocular single vision and depth discrimination and only to a relatively minor extent with form vision.

There are many ways in which both monocular and binocular visual depth may be estimated. In this paper only binocular depth perception, or stereopsis, is considered. Stereoscopic perception is based on the fact that each eye sees the world from a slightly different vantage point. As a result, the retinal images in the two eyes are slightly different, and this horizontal, retinal-image disparity provides the cue for depth discrimination.

Although stereoscopic techniques have been used to nullify attempts at camouflage, the theoretical implications of these earlier outdoor observations have only recently been appreciated as a result of Julesz' (1964) use of random dot patterns as stereo pairs (Figure 1). When viewed mo-

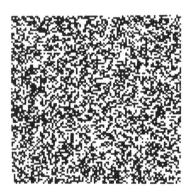

FIGURE 1   Random dot stereo pair. When viewed monocularly, the two fields appear as random sets of dots, without recognizable features. Viewed stereoscopically, there is a vivid depth impression, with a central square floating in front of its surround, an effect caused by the displacement of the central square region in one field laterally in respect to the same region in the other. (From Julesz, 1964.) (Copyright 1964 by AAAS.)

P. O. BISHOP   Department of Physiology, John Curtin School of Medical Research, Australian National University, Canberra, Australia

nocularly, the two fields appear as random sets of dots, without recognizable features. Viewed stereoscopically, however, there is a vivid depth impression, with a central square floating out in front of its surround. This effect is caused by the displacement of the central square region in one field laterally with respect to the same region in the other field. A number of important suggestions can be made as a result of Julesz' work:

1. As stated above, the neural mechanisms for binocular depth discrimination are likely to be, in a sense, less complex than those required for form recognition and also antecedent to them in the brain.

2. The analyses by which depth discriminations are made have a mosaic point-by-point or feature-by-feature basis. Line contours of any appreciable length are not required.

3. That form perception can arise binocularly by the viewing of a pattern of dots that are merely random to monocular inspection suggests that the mechanisms concerned are an integral part of the chain of events leading to pattern recognition.

The partial chiasmal decussation in the carnivores and primates, whereby the fibers from the temporal retina of the ipsilateral eye and the nasal fibers from the contralateral eye come together in an optic tract, provides the kind of anatomical substratum necessary for binocular mechanisms underlying stereopsis. The pathways from the two eyes pass through the lateral geniculate nucleus with only minimal interaction (Sanderson et al., 1969) and come together effectively, for the first time, immediately on reaching the striate cortex. The receptive field organization of geniculate neurons resembles that of retinal ganglion cells, which suggests that the analyses underlying form recognition are deferred for the cerebral cortex and take place after the two pathways have come together in Layer IV. Furthermore, the retinocerebral fibers from corresponding retinal areas come together in the striate cortex with considerable precision (see below; Nikara et al., 1968). All these observations are in keeping with the suggestions above, based on Julesz' work

Any mechanism for depth discrimination based on precise retinocerebral connections from corresponding retinal areas clearly requires that the extraocular muscles maintain a level of binocular fixation of a very high order. The observation that the movements of the two eyes are precisely correlated has long been known as Hering's law of equal innervation. More recently, work with binocularly stabilized retinal images has shown that the retinal disparity due to imperfect correlation between the motions of the two eyes is of the same order as, or somewhat less than, the extent of Panum's fusional area (D. Fender, personal communication).

From the above considerations, it is possible to suggest, in fairly general terms, some of the features of the neural mechanisms underlying human stereoscopic perception. Stereopsis is likely to be based, not on a higher-order disparity estimate between fully elaborated uniocular images,

but rather on a precise "bundle-by-bundle" conjunction of the retinocerebral pathways from very small but closely corresponding areas in the two retinas, the conjunction taking place at or soon after the arrival of the pathways at the level of the striate cortex. Such an anatomical arrangement would be difficult to achieve without a partial decussation at the optic chiasma, and its operation requires an accurate synergy between the two sets of extraocular muscles whereby corresponding retinal areas are kept in fairly precise visual register at all times. The nature of binocular depth-discrimination mechanisms in animals is still largely unknown, although the development of Julesz' random dot patterns now makes it possible to test for stereopsis. As vertebrates below the Mammalia have complete decussation at the chiasma, it is likely that any mechanisms for binocular depth discrimination to be found among them will be radically different from those for human stereopsis. Even among the Mammalia it is probable that stereoscopic mechanisms are, for the most part, fairly rudimentary, because only the carnivores and primates have a well-developed partial decussation at the chiasma. However, the quality of the binocular control of eye movements among the carnivores, subhuman primates, and the mammals generally has not been investigated.

Having put forward some general ideas, I shall now show how recent studies of the receptive fields of single neurons in the striate cortex have made it possible to develop these ideas in considerable detail, particularly in respect to binocular depth discrimination.

*Receptive field organization: some basic concepts*

Over the past 10 years or so, a number of concepts have gradually emerged, based on the idea of a receptive field; these have a general relevance for the organization of the visual system. Each neuron in the visual pathway, at least up to the level of the occipital cortex, is thought of as concerned with the analysis of features of objects or patterns present in its receptive field. The particular stimulus pattern that brings out the strongest response from the neuron is sometimes referred to as its trigger feature (Barlow et al., 1964). A growing number of classes of units at all levels of the visual pathway has been distinguished on the basis of trigger features. The directionally selective ganglion cells in the rabbit retina form the class that has been analyzed in the greatest detail (Barlow and Levick, 1965). The cells respond to small movements anywhere in the receptive field, provided they are in the preferred direction. There is no response in the opposite, or null, direction. Barlow and Levick (1965) suggested that the stimulus specificity for direction of movement is achieved by a local inhibitory action that prevents discharge in the null direction. The proposed mechanism for stimulus generalization involves the pooling on the ganglion cells of the outputs of a series of like elements—the bipolar cells.

A second important concept concerns the sequential processing of visual information at successive levels in the visual system. Hubel and Wiesel (1962, 1965a) have developed the idea of a hierarchical order of visual neurons that permits cells with simple stimulus requirements to discharge onto higher-level cells with more complex requirements. They have distinguished a number of functional classes of neurons in the visual cortex that have receptive fields of ascending complexity, namely, simple, complex, lower-order hypercomplex, and higher-order hypercomplex. These cells distinguish contour features of increasing complexity, from simple lines or edges with a specific location defined by the receptive field of a simple cell up to specific angles or corners situated anywhere within the receptive field of a higher-order hypercomplex cell. The transformations that are assumed to take place over the four classes of cell involve the repetition of two basic steps: increased stimulus specificity at one level, followed by the generalization of the same specific stimulus at the next level.

Many cortical neurons have receptive fields for both eyes, so that they have binocular stimulus specificities in addition to those relating to one eye. It is to be expected that, when the eyes converge, the binocular receptive-field pairs will superimpose in the plane of the object of regard. The distribution of the receptive fields of binocularly activated neurons is, however, such that only a small proportion can correspond exactly with one another at any one time. The remainder are said to show receptive-field disparity (Nikara et al., 1968). This concept has been developed to provide a basis for binocular single vision and depth discrimination (Barlow et al., 1967; Bishop, 1970).

## Classification of striate neurons

Based on the organization of their receptive fields and the nature of the stimuli to which they respond, the cells in the striate cortex have been classified as simple, complex, hypercomplex, and nonoriented (Hubel and Wiesel, 1962, 1968). All four types occur in both cat and monkey. As mentioned above, a valuable working hypothesis is to think of these cells as belonging to a hierarchical order of increasing complexity, whereby simple cells feed onto complex, complex onto hypercomplex, and so on. Nevertheless, there are difficulties with the above classification at the lowest and the highest levels of organization. Current accounts of the organization of the striate cortex place the simple neurons as the first link in the chain. However, it is now clear that there are striate cells with nonoriented receptive fields that do not have directional selectivity, but

have properties that are, in many ways, intermediate between those of lateral geniculate neurons and simple cortical cells (Baumgartner et al., 1965; Denney et al., 1968; Joshua and Bishop, 1970; Henry, Bishop, and Coombs, unpublished observations). The recognition of them as cortical neurons as opposed to geniculate axons seems clear whenever the units are binocularly discharged and their spikes have a cell type of waveform (Bishop et al., 1962a). The distinction, however, may be difficult to make when the units are discharged only monocularly, particularly as units with nonoriented receptive fields are most commonly found at about Layer IV, the layer of termination of geniculostriate axons (Hubel and Wiesel, 1968). Cortical units of this type have, as yet, been studied relatively little, and it is by no means clear whether, in the sequence of cortical activation, they are prior to or belong at the same level as simple cells. At the other extreme of the hierarchical order, cells with hypercomplex receptive fields are being discovered in increasing numbers in the striate cortex (Hubel and Wiesel, 1968), and the classification will doubtless eventually require upward extension.

The account of the cortical beginning of visual form and binocular depth discrimination given below is confined largely to a consideration of the relevant properties of simple cells in the striate cortex. Because their responses are relatively simple and the units concerned are found most commonly in Layer IV of the striate cortex, a reasonable hypothesis is to identify units of the simple type as stellate cells. Probably the geniculostriate axons terminate on the stellate cells in this layer.

## Simple cells and inhibitory phenomena

We also have found it convenient to classify striate units as simple and complex (Pettigrew et al., 1968a), but our interpretation of the receptive-field organization of simple cells differs in important respects from that of Hubel and Wiesel (Henry et al., 1969, and unpublished observations). The difference arises largely because, up to the present, inhibitory influences have been generally neglected or misinterpreted. Hubel and Wiesel (1959, 1962) distinguished units as simple because, using flashing but stationary spots or slits of light, they were able to subdivide the receptive fields into regions that responded by discharging either at light ON or at light OFF. These regions were then regarded as excitatory and inhibitory, respectively, excitation being equated with an ON discharge and inhibition with an OFF discharge. In their descriptions, however, cortical cells were said to differ from those in the retina and lateral geniculate nucleus by reason of the elongated side-by-side arrangement of the "excitatory" and "inhibitory" regions.

It is important to recognize that no necessary, or even

usual, relationship exists between ON discharges and excitation on the one hand and OFF discharges and inhibition on the other. Both ON and OFF stimuli can lead to inhibition, as well as to excitation. Because cortical cells, and particularly simple units, may have a relatively low or absent maintained discharge, true inhibitory effects (i.e., a reduction in firing as opposed to an OFF discharge) have been generally neglected. We have recently developed a method for demonstrating inhibition that has a general application throughout the visual system (Henry et al., 1969). A brief description of the method is given below. By its use, we have been able to demonstrate both the inhibitory and the excitatory regions in the receptive fields of striate neurons. In particular, we have shown that the receptive fields of simple cells are largely inhibitory in nature and that the excitatory zone is confined to a relatively small region situated toward the center of the field. For simple cells, this excitatory zone probably corresponds to the receptive fields as mapped by Hubel and Wiesel (1959, 1962).

There are also receptive fields which are apparently only inhibitory. In addition, our method has revealed a new component in the organization of receptive fields, namely, zones of subliminal excitation. We believe that these zones in the receptive field of the nondominant eye supply an important gating effect for striate neurons in relation to binocular depth discrimination (see below). Our method has other important applications in the study of binocular vision, because the excitatory and inhibitory effects of binocular receptive-field interactions can be determined much more easily and in much greater detail than by the prism-shift method introduced by Pettigrew et al. (1968b). It is to be expected that special procedures would be necessary to demonstrate inhibition because, in their study of binocular vision, Burns and Pritchard (1968) could find no evidence of mutual inhibition between the two retinal inputs.

## Simple receptive fields: monocular stimulus specificities

The responses of simple cells used to illustrate this paper have all been recorded from the striate cortex of the cat, the cells being distinguished as a class in the following way (cf. Pettigrew et al., 1968a). They have little or, commonly, no spontaneous or maintained discharge. For this reason, they are easily missed and must be sought by the continual presentation of a stimulus, which itself is being changed continually. One hopes that, in this way, the stimulus complex specific for each cell will be tried as the cell comes within the recording range of the advancing microelectrode. The excitatory zone in the receptive field is usually very small (frequently 0.5 degree or less across), particularly when it is close to the center of gaze. The units

respond best to a straight edge or contrast border with a particular angle of orientation, and moved slowly across the receptive field in a specific or preferred direction (Hubel and Wiesel, 1959, 1962). If the direction of movement of the slit departs from the optimum by about 20 degrees, the firing is markedly reduced or abolished (Figure 3F) (Campbell et al., 1968). Most frequently they respond in one direction of movement and little (Figure 3A) or not at all (Figure 2) in the opposite direction. As a stimulus, stationary spots of light flashed on and off are not so effective as a moving slit, and may be completely ineffective. The average response histogram obtained by means of a multichannel scaler is probably the simplest quantitative measure of the extent and sensitivity of the excitatory zone.

Our general experimental method may be described briefly (Pettigrew et al., 1968a; Joshua and Bishop, 1970). A cat, anesthetized ($N_2O/O_2$) and completely paralyzed, faces the rear-projection tangent screen, which is 2 meters in front of the animal. The striate unit is stimulated by a slit of light projected onto the back of the translucent screen by reflection from a system of two mirrors, each mounted on the coil of a moving coil galvanometer (see Figure 6B). In this way, the slit of light can be positioned readily anywhere on the screen. The slit may be varied in length, width, and orientation, and the speed and amplitude of movement are controlled precisely by applying the output of a function generator to the galvanometer coils. A pulse from the function generator at the start of a cycle sets the multichannel scaler advancing from channel to channel in step with the steadily advancing slit. As the slit passes over the excitatory zone of the receptive field, counts of the single-unit discharges are added to the appropriate bins in the scaler. The slit then passes back over the receptive field before it begins another cycle.

Figure 2 shows typical, sharply-defined, unimodal average-response histograms from a simple unit. They were produced by an optimally-oriented long narrow slit of light moved broadside forward and backward across the receptive field. The unit was highly selective in that it responded in only one direction of movement and then only to the trailing edge of the slit. The single peak progressively shifted to a later and later position on the back sweep of the histogram as the slit was widened from 9 minutes of arc to about 6 degrees (Figure 2), while the starting position of the trailing edge on the forward sweep was kept constant.

Figure 3 was prepared from another simple cell. Histogram A is the response to the movement of a slit 17 minutes of arc in width. The cell is almost completely direction selective, and the response is confined largely to a single peak on the back sweep. This peak was due to the summed effect of the responses to the leading (positive) and trailing (negative) edges of the slit as is shown by the fact that when the slit width was increased from 1 cm to 12 cm (Figure 3B), the responses to the leading (L) and trailing (T) edges were clearly separated from one another by an amount closely related to the width of the slit. As the starting position of the trailing edge was kept constant, it was easy to assign the peaks in the histogram to the appropriate edge. The nature of the organization of the receptive field center may, however, lead the relationship between slit width and response separation to be nonlinear for very narrow widths, and this effect will then persist as a constant small difference between the parameters at larger slit widths (Figure 3A and B; Henry, Bishop, and Coombs, unpublished observations).

Although still only responding to edges, various units show a range of different response patterns. Some units respond only to one edge, either positive or negative, and then only in one direction. Others respond to one edge but in both directions of movement. Still others respond to both edges in the forward direction and only one edge in the backward sweep, and so on. Furthermore, whether a response occurs to a particular edge depends on additional factors, such as the presence or absence of other edges and their spatial relation to the edge in question.

Recently we have shown that the excitatory zones in the receptive fields of simple units are always partly enclosed by a large inhibitory surround, which is particularly intense immediately adjacent to the excitatory zone (Figure 3C and D; Henry et al., 1969, and unpublished observations). We demonstrated the inhibitory regions by using two independently controlled slits of light, moving asynchronously with respect to each other, but both at about the same optimal

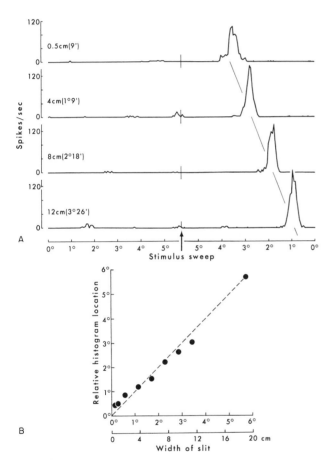

FIGURE 2   Average response histograms from a simple striate unit to movements of a slit of light forward and backward across its receptive field. The arrow indicates the turnaround point. The unit responds only to movement in the backward direction and then only to the trailing edge of the slit. Successive responses are to increasing slit widths as indicated, the starting position of the trailing edge for forward movement being kept constant throughout. The graph (B) plots slit width against position of the response peak in the histogram. (Henry, Bishop, and Coombs, unpublished observations.)

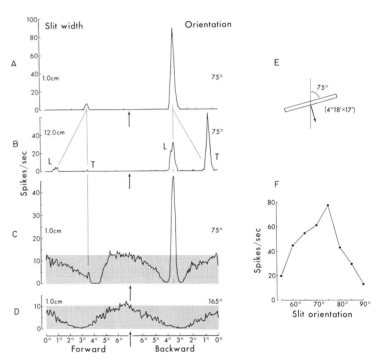

FIGURE 3   A, B. Average response histograms from a simple unit to movements of a slit of light forward and backward across its receptive field. Responses to both leading and trailing edges of the slit on the backsweep are indicated by the use of slits of different width. C, D. Average response histograms from the same unit, showing how inhibitory regions in the receptive field are revealed by the use of the testing slit of light against the background of an artificially induced discharge. Dotted areas show the level of the induced discharge in the absence of the testing stimulus. The histograms are for two directions of movement of the testing stimulus: C, in the preferred direction; D, at right angles to the preferred direction. (Henry, Bishop, and Coombs, unpublished observations.)

speed. The movement of one slit (conditioning) was of small amplitude and confined to the excitatory zone of the receptive field, and the other (testing) had a large sweep that moved right across the whole of the receptive field. Whenever the conditioning stimulus moved in the preferred direction it produced a relatively high spike discharge, which, in the absence of the testing stimulus, filled the bins in the multichannel scaler fairly uniformly (Figure 3, dotted areas). The filling was uniform because the spike discharge caused by the conditioning stimulus was random with respect to the cycling of the multichannel scaler. The recycle pulse for the scaler always came from the function generator used to drive the testing stimulus, and the function generator for the conditioning stimulus operated asynchronously to it. Inhibitory regions in the receptive field are readily demonstrated against this artificially-induced maintained discharge by having the two slits, conditioning and testing, operating at the same time.

All the histograms in Figure 3 were obtained from one simple cell. The responses A and B give no indication of the presence of inhibitory regions in the receptive field. The latter were, however, clearly shown when the receptive field was mapped by the above method. The two examples, C and D, are for movements of the testing stimulus in two directions, one in the preferred direction (C) and the other at right angles to it. In sharp contrast to the excitatory zone, the intensity of the inhibitory surround is relatively independent of the direction of stimulus movement. By confining the testing stimulus to the excitatory zone, it can be shown that the surround inhibition is not an essential component in the directional selectivity for excitation (Henry, Bishop, and Coombs, unpublished observations). The stimulus movement is always least effective for excitation at right angles to the preferred direction. The excitatory zone, which has an elongated shape like the simple receptive field plots described by Hubel and Wiesel (1959, 1962), is always much smaller than the inhibitory surround (Figure 4; center 0.6 degree across, surround about 6 degrees across). In the axis of the preferred movement, the inhibitory regions are not a surround but are more in the nature of bands on each side of the excitatory center. It must be stressed that no response takes place in the surround either at light ON or light OFF; it is purely inhibitory, whatever the stimulus.

## Beginnings of form discrimination

Simple units require highly specific stimulus features for their discharge. They monitor small parts of the visual field for light-dark contours or those segments of contours that fall within their receptive fields. A contour is analyzed in terms of its orientation at the particular locality, the speed

and direction of its movement, and the polarity of its contrast (i.e., light-dark or dark-light). In addition, the analysis takes into consideration the presence of more than one contour within the receptive field, as when there are slits of light or dark bars. Both facilitatory and inhibitory mechanisms contribute to this specialization. It is clear that inhibitory mechanisms suppress the activity of a cell whenever stimulus features inappropriate to its specialization are presented in the receptive field of the unit. Furthermore, at any one time, a given small part of the visual field is unlikely to provide appropriate stimuli for more than a small proportion of the cells with receptive fields in the area, so that the activity of most of them will be suppressed. A common observation is that the level of the maintained discharge of cortical units is much less than for units lower down on the visual pathway (Herz et al., 1964). Undoubtedly the inhibitory effects of inappropriate stimuli play an important part in reducing the maintained activity of cells, particularly those of the simple type.

By contrast, facilitatory effects are highly specific and must be relatively uncommon occurrences. Undoubtedly, certain combinations of appropriate stimuli, such as additional contours, boost the responses of the cell, although the operation of these facilitatory mechanisms is not so clear as is that for inhibition. The general picture that emerges is one of highly selective facilitation against a background of widespread general inhibition. The dominant theme, however, is inhibitory.

## Simple receptive fields: binocular stimulus specificities

So far we have considered the properties of the receptive field for one eye. Almost without exception, however, simple units have receptive fields associated with both eyes (Henry et al., 1969), and these binocular receptive-field pairs provide the essential basis for binocular depth discrimination. Although nearly all the cells in the lateral geniculate have a binocular influence of an inhibitory nature (Sanderson et al., 1969), there is apparently no excitatory interaction between the two retinal inputs until the visual pathways come together on simple cells in the striate cortex (see Figure 6A). A very important feature of the binocular receptive fields of striate neurons is that the highly specific stimulus complex required for their discharge is always closely similar for the two eyes. The two receptive fields have the same size and arrangement, so that whatever stimulus is the most effective for one eye (in form and orientation and in direction and rate of movement) is also the most effective for the other (Hubel and Wiesel, 1962, 1965a; Barlow et al., 1967; Pettigrew et al., 1968b). Almost without exception, the only difference between the two receptive fields, as judged by hand plotting, relates to the phenomenon of eye dominance which has been studied in detail by Hubel and Wiesel (1962, 1963, 1965a, 1965b; Wiesel and Hubel, 1963, 1965a, 1965b).

The use of binocular parallax as a basis for depth discrimination clearly requires that the cortical neurons make estimates of the parallax produced by viewing *the same feature* in the visual field from the separate viewpoints of the two eyes. The essential first step in binocular depth discrimination is, therefore, the selection (from the retinal images in each eye) of those separate parts that are images of the same feature in the visual field. This first step is provided by the fact that each striate neuron is discharged only when its two receptive fields, one for each eye, are presented with the same highly specific stimulus complex, so each separate part of the retinal image in one eye is accurately paired with its fellow image in the other eye. Because of eye drifts and flicks, the pairings are being continually lost and remade, as first one pair of receptive fields and then another pick up the same object feature. The next step is the assessment of the magnitude and direction of the binocular parallax between the corresponding parts of the two retinal images. This step involves the phenomenon of receptive-field disparity.

## Receptive-field disparity

The concept of receptive-field disparity, which has been developed in detail elsewhere (Barlow et al., 1967; Nikara et al., 1968; Joshua and Bishop, 1970; Bishop, 1970), is illustrated in Figure 4. The receptive fields are conventionally represented as variously-shaped rectangles (cf. Bishop, 1970). Paralysis of the extraocular muscles in the cat preparation causes the eyes to assume a fixed position of slight divergence. For this reason, the two receptive fields of a binocularly activated unit are horizontally separated on the tangent screen. Let us now consider a group of binocular receptive-field pairs chosen so that all the members for the right eye have the same, but arbitrary, visual direction (Figure 4). Because they have precisely the same direction, all the right-eye fields lie accurately one on top of the other on the tangent screen. If we now look at the left-eye members of the receptive-field pairs, we can see that they are not in register but form a two-dimensional scatter to which the term "receptive field disparity" has been applied (Nikara et al., 1968). The upper part of Figure 4 is purely diagrammatic, but the histogram at A, taken from the experiments of Nikara et al. (1968), shows the actual distribution of the horizontal receptive-field disparities for a group of receptive fields within about 5 degrees of the visual axis. When the vertical dimension is taken into account, the scattergram of the receptive-field disparities for the same group is shown

in B. The distribution is approximately Gaussian, with a standard deviation of about 0.5 degree in both horizontal and vertical directions. The mean receptive-field position for the left-eye distribution may be taken as exactly corresponding to the center of the superimposed receptive fields for the right eye. Under experimental conditions, however, it has not been possible to plot a sufficient number of binocular receptive-field pairs having all the members for one eye with the same visual direction. The distribution in Figure 4B was thus obtained by moving the receptive fields for one eye to a common position and adjusting the members for the other eye by a similar amount in each case. In this way, all the receptive-field scatter has been transferred to the group for the second eye.

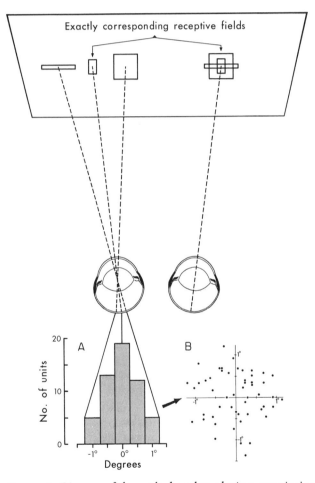

FIGURE 4 Diagram of the method used to obtain a quantitative measure of receptive-field disparities. A. Histogram showing actual distribution of horizontal receptive-field disparities of 54 binocular receptive-field pairs within about 5 degrees of the visual axis. B. Scattergram of the horizontal and vertical receptive-field disparities of the same units as in A. (Bishop, 1970; data for A and B from Nikara et al., 1968.)

Figure 5 has been developed from Figure 4 by a converging of the eyes, as in life, so that they fixate a point F in the plane of the tangent screen. Under these circumstances, the corresponding points for central vision in the two eyes will coincide at F, and the receptive field at the center of the left-eye distribution will fall precisely over its partner for the right eye. It can be seen, however, that the receptive fields at the periphery of the left-eye distribution fail to superimpose their partners for the right eye and do so only in planes closer to or farther away from that of the tangent screen (Figure 5, dotted receptive fields). In order to use receptive-field disparities for binocular depth discriminations, the cortical mechanisms must be able to take account not only of the amounts by which particular receptive-field pairs depart from the condition of exact correspondence, but also of the direction of the disparity.

In embryological development there is presumably a limit to the precision with which the fibers from corresponding parts of the two retinas can come together on

neurons in the striate cortex. Possibly this initially random element in the fiber connections is subsequently developed into a "known" pattern of receptive-field disparities and used as a basis for binocular depth discriminations. It is shown below that the condition of exact correspondence of receptive fields is signaled in the brain by the marked facilitation of the striate-neuron discharge that occurs when the receptive fields are precisely superimposed in the plane of the stimulus. A slight departure from exact correspondence leads to mutual inhibition by the two receptive fields and to the suppression of firing by the striate neuron.

## Binocular mechanisms of depth discrimination

Before describing the binocular mechanisms by which depth discriminations are made, I must refer again to our experimental methods (Pettigrew et al., 1968a, 1968b; Henry et al., 1969). Figure 6A shows the pathways from the two eyes passing independently through the lateral geniculate nucleus

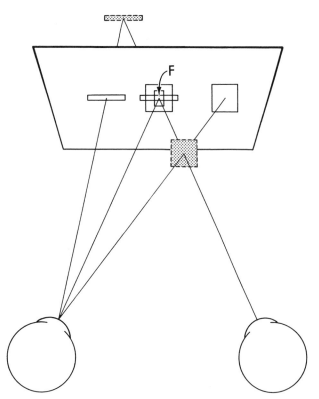

FIGURE 5 Diagram of a rearrangement of the upper part of Figure 4 to indicate the way in which the phenomenon of receptive-field disparity might form the basis of a neural mechanism for binocular depth discrimination. The exactly corresponding receptive fields have been superimposed in the plane of the tangent screen. The dotted receptive fields show the planes at which the other receptive fields superimpose. F is the fixation point at which the eyes converge.

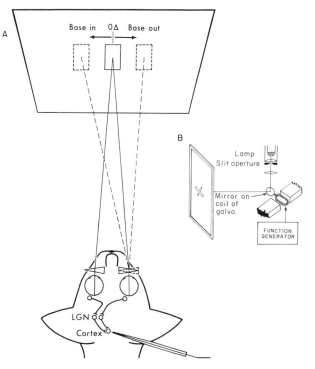

FIGURE 6 General experimental arrangement for recording from single, binocularly activated, striate neurons in cat. The animal, anesthetized ($N_2O/O_2$) and completely paralyzed, faces a rear-projection tangent screen 2 meters in front of it. A. Prisms before each eye serve to position the receptive fields on the translucent tangent screen. B. A slit of light projected onto the back of the screen by reflection from a mirror mounted on the coil of a moving coil galvanometer. LGN, lateral geniculate nucleus; 0Δ, zero prism diopters.

before finally coming together on neurons in the striate cortex. The divergence of the eyes due to the paralysis is overcome by means of prisms in front of each eye. In this way, the two receptive fields of a striate neuron can be moved over the tangent screen so that they come into exact correspondence, designated 0Δ (zero prism diopters) for descriptive purposes. By keeping the position of the left-eye receptive field constant and moving the right-eye field to one or other side of this position, we can study the nature of the binocular response when the receptive fields are out of correspondence by varying amounts. When the right-eye prism is placed base out, the receptive field is moved to the right of the left-eye field and, in like manner, to the left for the base in position. Shifting the receptive fields in this way is equivalent to moving the plane of the stimulus closer to and farther away from the animal, while keeping the relative positions of the receptive fields constant, which is, of course, the situation that obtains under natural conditions.

As described above, the organization of the receptive fields of simple striate neurons includes a long, narrow, directionally selective, excitatory center and a roughly circular, nondirectionally selective surround. The term "receptive axis" has been used for the line of sight passing through the center of a receptive field of a visual neuron (Bishop et al., 1962b). By definition, therefore, the two receptive fields of a binocularly-activated neuron will superimpose in the frontoparallel plane where their receptive axes cross. The diagrams in Figure 7 show the essential features of the interaction between the receptive fields of a

simple striate neuron at successive frontoparallel planes along the two receptive axes. By way of simplification, the excitatory centers of the receptive fields in Figure 7A have been given a vertical orientation, the cell responding preferentially to a horizontally moving stimulus. In this case, the centers extend over the greater part of the vertical dimension of the receptive fields, the inhibitory surrounds or side bands, drawn conventionally as ellipses, lying mainly to the left and right.

Figure 7B is a highly simplified diagram showing the average response histograms to be expected at the different frontoparallel planes as a result of the interactions between the two receptive fields. In the plane of exact correspondence, the receptive axes cross and the excitatory centers superimpose. Under these circumstances, the binocular response is markedly facilitated. Just in front of and behind this position, the receptive fields have moved out of register so that the excitatory center of each receptive field lies in the inhibitory surround of the other field. The binocular response is then greatly depressed or abolished. Still farther in front and behind, the receptive fields have become sufficiently separated for each to produce a response independently of the other.

It must be emphasized that the diagrams in Figure 7 show the interactions of the two receptive fields of a single striate neuron. The extrapolation from these ideas to concepts such as Panum's fusional area and regions for diplopia must be made on the basis of populations of striate neurons with partially shifted, reciprocally overlapping distributions of receptive-field disparities. For the one eye there are, of

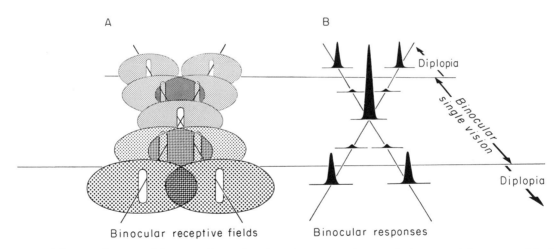

Binocular receptive fields          Binocular responses

FIGURE 7 Schematic diagram of the essential neural mechanisms underlying binocular single vision and stereopsis. A. Superimposition of the binocular receptive fields of a striate neuron at successive frontoparallel planes in the vicinity of the intersection of the two receptive axes. The receptive axis is the line of sight passing through the center of a receptive field. B. Schematic representation of average response

histograms, showing facilitation and depression of the binocular responses to stimuli situated at the same frontoparallel planes as the receptive fields in A. Marked facilitation occurs at the intersection of the receptive axes, and depression at positions immediately in front of and behind the intersection.

course, a large number of sets of striate neurons, such that the members of each set have receptive fields with the same visual direction, but with the visual direction varying from set to set. For the other eye, each set will have associated with it a two-dimensional distribution of receptive fields, similar to that shown in Figure 4. The separate distributions for the latter eye will be spatially offset with respect to one another in a way that closely corresponds to the spatial arrangement of the sets of receptive fields for the former eye. The above concepts have been developed more fully elsewhere (Bishop, 1970).

A rather more complicated pattern of binocular responses is usually obtained under experimental conditions even when the stimulus is a single light-dark border. The added complexities arise mainly from asymmetries both in the spatial arrangement and in the intensity of the excitatory and inhibitory areas of the two receptive fields as well as from the presence in the fields of regions of subliminal excitation. Furthermore, when slits or bars are used as stimuli, responses may occur to both leading and trailing edges.

Figure 8 shows the experimental results obtained by interacting a binocular receptive-field pair in a manner analogous to that shown in Figure 7 (Henry, Bishop, and Coombs, unpublished observations). The unit was selected for the relative simplicity of its responses, and the stimulus was a single, vertically orientated, light-dark border moved horizontally from left to right across the receptive field and then back again. Before recording began, the two receptive fields were brought into exact correspondence by means of prisms (see Figure 6A-0Δ). Then the monocular responses were obtained first by the occlusion of one eye and then of the other (Figure 8, Right eye, Left eye). The two monocular average-response histograms show that the unit responded about equally from the two eyes and that the discharges in each case were also about equal for the two directions of stimulus movement.

The interaction of the two receptive fields was then studied by keeping the position of the left-eye receptive field constant on the tangent screen and moving the right-eye receptive field from the superimposed position (0Δ)

first to the left (4Δ base in) and then to the right (6Δ base out). Without changing the stimulus parameters used for the monocular histograms, we recorded a series of binocular responses by sweeping the light-dark border over the two receptive fields at successive one-diopter (1Δ = 0.57 degree) steps of receptive-field separation. In order to simplify the illustration, only the portions of the histograms for the backward sweep are shown, although those for the forward sweep are quite similar. During the backward sweep, the light-dark border moved from right to left on the tangent screen. Hence the response from the right eye appears toward the end of the histogram when the location of its receptive field is to the left of the central position (i.e., prism base in).

When the receptive fields were exactly superimposed (0Δ), the binocular response was greatly facilitated, being 84 per cent greater than the sum of the monocular responses. For positions on either side of exact correspondence the responses from the two eyes are virtually abolished. Farther again to one or the other side, the responses recover, but, although the prism steps covered a total range of 4 degrees, this was, unfortunately, still not sufficient to show the full extent of the receptive-field interaction. At the two extremes of the range, the response from one eye was still fairly depressed, while that from the other had passed through a stage of slight facilitation corresponding to peripherally situated zones of subliminal excitation in the receptive field. Despite such complication, the general pattern of binocular interaction closely resembles the schematic representation in Figure 7.

The quality of stereoscopic acuity in man suggests that the sensitivity of striate neurons to a change in binocular receptive-field alignment must be of a very high order. A striking and characteristic feature of the binocular responses of simple cells is the extremely sharp transition from facilitation to inhibition for very small shifts in binocular receptive-field alignment. The binocular responses in Figure 7 changed from maximal facilitation to almost total inhibition for a change in alignment of only about 0.5 degree (0Δ to 1Δ base in). The cell may well have revealed a greater sensitivity had finer prism settings been used. Other units have

---

FIGURE 8   Upper right: Monocular average-response histograms of a simple striate neuron to horizontal movements of a vertically orientated light-dark border passing across each receptive field from left to right and back again. The arrow indicates the turnaround point in each case. The responses are about equal for the two eyes and for the two directions of movement in one eye. Left column: Binocular average-response histograms for the same unit with the use of the same stimulus. Only the second halves (backward sweep) of the histograms are shown. Binocular responses were recorded over a range of right-eye receptive-field positions to one (6Δ base out) and the other (4Δ base in) side of the left-eye receptive field which was held constant throughout. The full extent of the facilitated response when the fields were exactly superimposed (0Δ) is shown at the right of the column. (Henry, Bishop, and Smith, unpublished observations.)

been studied with changes in prism power as small as 0.1Δ. In one instance, the transition from the peak of facilitation (167 per cent) to complete inhibition occurred for a change in receptive-field alignment of less than 10 minutes of arc (Henry et al., 1969). The determination of the limits of the sensitivity of striate neurons for binocular depth discriminations is set by present techniques for eliminating or obviating residual eye movements.

## Binocular gate neurons

Many of the cells in the striate cortex, particularly those of the simple type, are reported to be exclusively monocular and without influence from the other eye (Hubel and Wiesel, 1962, 1968). Recently, however, Henry et al. (1969) have shown, in the cat, at least, that many, if not most, of these monocularly discharged simple cells have powerful inhibitory and subliminal excitatory receptive fields for the nondominant eye. This new kind of receptive field has been revealed against the background of an artificially induced maintained discharge, with the use of a technique similar to that described above for demonstrating inhibitory regions in the receptive field of the dominant eye. Although discharged from only one eye, these simple cells, it has been suggested, act as binocular gate neurons (Bishop, 1970). In the binocular situation they allow the

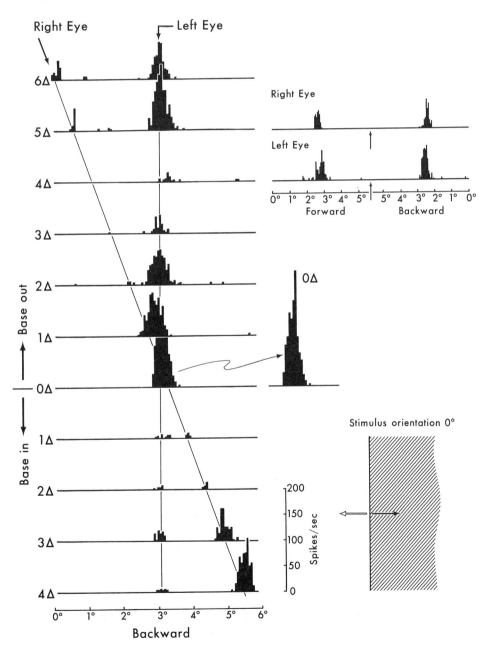

discharge of the cell only when the specific stimulus is at the precise depth in space at which the subliminal excitatory region in the field of the nondominant eye corresponds with the excitatory region for the other eye (Figure 9). The binocular gate mechanism is shown diagrammatically in Figure 9.

A simplified explanation can now be given in neurophysiological terms. A more adequate statement calls for a detailed extrapolation from single units to neuron populations (cf. Bishop, 1970). A striate neuron will be discharged when the specific stimulus is presented at the precise depth in space where the two receptive axes cross. Binocular

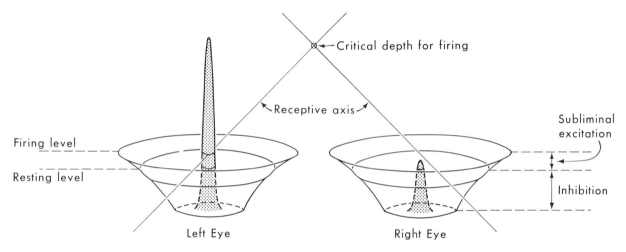

BINOCULAR GATE NEURON

FIGURE 9 Diagrammatic representation of the interaction between the receptive fields of a binocular gate neuron. With monocular stimulation, the unit is discharged only from the left (dominant) eye, but, when the optimal stimulus is applied monocularly to the small central region of the receptive field of the nondominant eye, the excitatory state of the cell is raised above the resting, but below the firing, level. In the binocular situation, the critical depth for firing occurs where the receptive axes cross, because the liminal and subliminal excitatory regions of the two receptive fields coincide at this point. Small misalignments of the receptive fields, either in front of or behind the critical depth, lead to mutual inhibition. Under monocular stimulating conditions, the discharge from the dominant eye would not, of course, be subject to this inhibitory influence, and the cell would fire much as it does when stimulated binocularly at the critical depth.

## Binocular single vision and stereopsis

The study of receptive-field disparities in the paracentral visual field has provided a definition of the horopter in neurophysiological terms and has led to the development of a construction analogous to Panum's fusional area (Joshua and Bishop, 1970). The detailed application of this new knowledge to the problems of binocular single vision and stereopsis has been considered elsewhere (Bishop, 1970). Figures 7 and 8, however, provide the opportunity for referring briefly to the solution of one of the hitherto unresolved key problems. There have long been difficulties associated with any simple concept of corresponding retinal points as a basis for binocular single vision and stereopsis. Binocular depth perception entails the stimulation of horizontally disparate retinal elements by a single object point, yet the ultimate percept is that of a single object point. How can single vision arise from the stimulation both of corresponding retinal points and of noncorresponding retinal points?

single vision results from the simultaneous stimulation of the two receptive fields, and the depth assessment that is made will depend on the receptive-field disparity of the neuron concerned (Figure 4). Double vision results from the successive stimulation of the two receptive fields, which is to be expected when the stimulus is presented in a plane either nearer to the animal or farther from the site where the receptive axes cross. Double vision in the immediate vicinity of the crossing point is prevented, however, by the mutual inhibition exerted by the two receptive fields. This effect is doubtless related to the phenomenon of obligatory single vision that obtains within Panum's fusional area. Figure 7 shows that diplopia will nevertheless occur when the plane of the stimulus departs still farther from the specific depth for binocular facilitation. The two receptive fields then become sufficiently separate for their successive stimulation to escape mutual inhibition. In man, when single objects are observed outside Panum's fusional area, diplopia occurs, although a measure of stereoscopic experience still persists for a short space both in front of and behind the area. It is postulated that diplopia results when the majority of binocular receptive-field pairs are stimulated successively, although some depth discrimination may still be possible in the face of the diplopia if a significant population of binocular receptive fields remains with sufficient disparity to be stimulated simultaneously (Bishop, 1970).

## Higher-level discrimination

It is beyond the scope of the present paper to consider what is known of the subsequent stages in the discrimination of form by the higher-order cells, both in the striate cortex itself and in the parastriate area (cf. Hubel and Wiesel, 1965a), but brief mention is made of the role that complex cells may play at a higher level by generalizing the binocular depth specificities of simple cells (Pettigrew et al., 1968b).

## Complex units: specificities and generalization

Complex receptive fields have been distinguished in the following way (Pettigrew et al., 1968a). The neurons concerned generally have a fairly brisk spontaneous activity, and their receptive fields are relatively large—generally

more than 3 degrees across. The response to moving contours is highly directionally selective, and the discharge is of relatively high frequency and well maintained. The average response histograms are complex and multimodal in form. An important characteristic of this type of receptive field is that very small amplitude movement, much smaller than the field dimensions, of a slit or line stimulus will produce a response, provided the orientation and direction of movement are not too far from optimal. The direction of movement and orientation of the slit are important, but the exact position within the receptive field is not. In other words, the unit has specificity for movement in a particular direction, but has generalization for position (i.e., visual direction), and commonly this generalization also extends to the sign of the contrast edge (i.e., whether positive or negative).

Hubel and Wiesel (1962) have suggested that the properties of complex units are derived from pooling the outputs of a number of simple units, all of which have the same directional selectivity for movement, and our work suggests that there may, in turn, be directionally selective subunits even within the small fields of these simple-cell components (Pettigrew et al., 1968a).

Figure 10 shows the binocular responses of a complex unit to a precisely oriented slit moved in the preferred

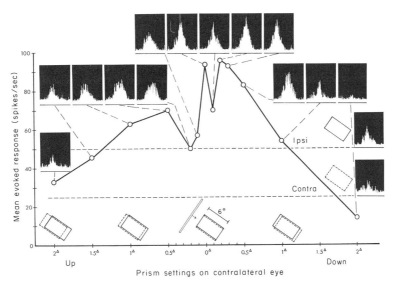

FIGURE 10   Binocular interaction on a complex unit with its receptive fields in varying degrees of noncorrespondence to show the narrow range of facilitation despite large receptive-field size. The receptive fields are represented by rectangles in the lower portion of the graph, and the relative positions of the IPSI (unbroken outlines) and CONTRA (broken outlines) receptive fields are shown at contralateral prism settings of 2 Δ up, 1 Δ up, 0 Δ and 1 Δ down. The receptive fields were quite large, measuring 6 degrees across. The histograms show responses to movement of a slit stimulus (3 degrees × 0.2 degree) in the preferred direction across the monocular (lower right: IPSI and CONTRA) and across both receptive fields at different prism settings on the contralateral eye. Horizontal broken lines indicate the amplitudes of the separate monocular responses; the two monocular average-response histograms are shown at the extreme right. The graph shows mean evoked response as a function of prism setting. The ipsilateral eye was dominant. (From Pettigrew et al., 1968b.)

direction across the length of the two receptive fields, both when they were in exact correspondence and also when in varying degrees of nonalignment (Pettigrew et al., 1968b).

When the receptive fields were out of correspondence by little more than 1 degree ($2\Delta = 1.1$ degree), the binocular response fell close to or below the level for the nondominant eye, and would doubtless have fallen still farther had tests been made at greater receptive-field separations. In other words, the two receptive fields had to be moved out of alignment by only about one-sixth of their length to produce marked depression of the binocular response. They would, of course, have been able to discriminate disparities that were much smaller than 1 degree. The specificity of some complex units for a narrow range of binocular disparities may therefore be comparable with that found in the case of simple units.

If we again consider the complex unit having as input a number of simple cells, all with the same directional selectivity, we must now suppose that the input cells may also have receptive fields with the same (or very nearly the same) binocular correspondence. Thus, we may suppose that there has been a pooling of units, all of which have the same directional selectivity and the same binocular discrimination with respect to depth. Under natural conditions, such a unit would be optimally activated by a line stimulus over a relatively wide range of visual directions (6 degrees for the unit in Figure 10) but only over a much narrower range of depth in space. Depth discrimination would be generalized for visual direction and probably also for contrast. The selectivity for depth discrimination is again achieved by binocular facilitation for the optimal stimulus when it is presented at the depth in space at which the receptive fields correspond and by mutual inhibition for stimuli at other than the preferred depth.

## Conclusion

The beginnings I have described have not taken us very far toward an understanding of the brain mechanisms by which we discriminate the visual forms of everyday life. Even the most complex cells so far studied in the higher visual areas still seem to be concerned with the elements of form rather than coherent wholes. Cells have yet to be discovered that respond only when the simplest of forms, such as a rectangle or a circle, is presented in the visual field. Of course, it may well be that the simplest forms for the particular animal are not simply geometrical. Unfortunately, we still have only a very rudimentary understanding of form discrimination in animals, using behavioral methods, and a satisfactory psychophysics of form perception has yet to be developed for man. The discrimination of form may always be the property of groups of neurons, perhaps always of large assemblies of neurons and never in any sense the

trigger feature for one neuron. If such were the case, it would explain why the stimulus requirements attributed to the higher-order hypercomplex cells in Area 19 are still surprisingly simple in relation to what might have been expected from the enormous complexity of the three visual areas. It also means that the concept of a trigger feature for single neurons, as applied in the more peripheral parts of the visual pathway, may no longer be relevant in the cerebral cortex or may at least become increasingly less relevant.

## Summary

The concept of a receptive field is the central idea in visual neurophysiology at the present time. The receptive field of a visual neuron is the small part of the total visual field in which the specific stimulus must be applied in order for the cell to be discharged or for that discharge to be inhibited. Recent studies of the receptive fields of single neurons in the visual cortex, particularly in the cat, have made it possible to present a detailed neurophysiological theory of binocular single vision and stereopsis for the first time. By contrast, much less progress has been made toward an understanding of the neural basis of form perception, reflecting the much greater difficulties inherent in such a task. Relative progress in these two fields depends in large measure on corresponding developments in psychophysics. The neurophysiologist studying visual mechanisms looks to psychophysics to define the nature of the stimulus and to give an account of the perceptual process. The psychophysics of binocular vision has been well developed since the nineteenth century, and important advances continue to be made, but a psychophysics of form barely exists today.

The present account of the cortical beginning of binocular depth discrimination and form vision is largely limited to a consideration of the receptive fields of simple cells in the striate cortex. Simple cells are presumed to be the stellate cells situated principally in Layer IV where they receive geniculocortical axon terminals. These cells require highly specific stimulus features for their discharge and monitor small parts of the visual field for light-dark contours or those segments of contours that fall within their receptive fields. A contour is analyzed in terms of its orientation at a particular locality, the speed and direction of its movement, and the polarity of the contrast. Both facilitatory and inhibitory mechanisms contribute to this specialization, and stimuli that are inappropriate for the particular cell lead to the inhibition of its discharge. The picture that emerges is one of highly selective facilitation against a background of widespread general inhibition. Although it may be said that simple cells are concerned with the elements of form rather than coherent wholes, such also appears to be true of the most complex cells yet studied in the higher visual areas.

Many cortical neurons have receptive fields for both eyes,

each field having the same highly specific stimulus requirements. There are, in addition, binocular stimulus specificities. At the crossing point of the lines of sight through their centers, the two receptive fields are accurately in register, one over the other. When such is the case, the stimulus that is optimal for one receptive field must also be so for the other, and the firing of the cell is greatly facilitated. At other depths, the receptive fields are out of alignment, and mutual inhibition occurs. The spatial distribution of the binocular receptive-field pairs is such that only a small proportion can exactly correspond in the frontoparallel plane which contains the fixation point. The remainder are said to show receptive-field disparity. This concept has been developed to provide a neural mechanism for binocular single vision and stereopsis, and resolves the long-standing difficulty posed by the perception of a single object from the stimulation of disparate retinal elements.

## Acknowledgments

I wish to thank my colleagues Mr. G. H. Henry and Mr. J. S. Coombs for allowing me to use unpublished data. My thanks are also due to Mrs. Eva Elekessy and Miss Isabel Sheaffe for their assistance in the preparation of this paper.

## REFERENCES

BARLOW, H. B., C. BLAKEMORE, and J. D. PETTIGREW, 1967. The neural mechanism of binocular depth discrimination. *J. Physiol.* (*London*) 193: 327–342.

BARLOW, H. B., R. M. HILL, and W. R. LEVICK, 1964. Retinal ganglion cells responding selectively to direction and speed of image motion in the rabbit. *J. Physiol.* (*London*) 173: 377–407.

BARLOW, H. B., and W. R. LEVICK, 1965. The mechanism of directionally selective units in rabbit's retina. *J. Physiol.* (*London*) 178: 477–504.

BAUMGARTNER, G., J. L. BROWN, and A. SCHULZ, 1965. Responses of single units of the cat visual system to rectangular stimulus patterns. *J. Neurophysiol.* 28: 1–18.

BISHOP, P. O., 1970. Neurophysiology of binocular single vision and stereopsis. *In* Handbook of Sensory Physiology, Vol. 7 (R. Jung, editor). Springer-Verlag, Berlin (in press).

BISHOP, P. O., W. BURKE, and R. DAVIS, 1962a. The interpretation of the extracellular response of single lateral geniculate cells. *J. Physiol.* (*London*) 162: 451–472.

BISHOP, P. O., W. KOZAK, and G. J. VAKKUR, 1962b. Some quantitative aspects of the cat's eye: Axis and plane of reference, visual field co-ordinates and optics. *J. Physiol.* (*London*) 163: 466–502.

BURNS, D. B., and R. PRITCHARD, 1968. Cortical conditions for fused binocular vision. *J. Physiol.* (*London*) 197: 149–171.

CAMPBELL, F. W., B. G. CLELAND, G. F. COOPER, and C. ENROTH-CUGELL, 1968. The angular selectivity of visual cortical cells to moving gratings. *J. Physiol.* (*London*) 198: 237–250.

DENNEY, D., G. BAUMGARTNER, and C. ADORJANI, 1968. Responses of cortical neurones to stimulation of the visual afferent radiations. *Exp. Brain Res.* 6: 265–272.

HENRY, G. H., P. O. BISHOP, and J. S. COOMBS, 1969. Inhibitory and sub-liminal excitatory receptive fields of simple units in cat striate cortex. *Vision Res.* 9: 1289–1296.

HERZ, A., O. CREUTZFELDT, and J. FUSTER, 1964. Statistische Eigenshaften der Neuronaktivität im ascendierenden visuellen System. *Kybernetik* 2: 61–71.

HUBEL, D. H., and T. N. WIESEL, 1959. Receptive fields of single neurones in the cat's striate cortex. *J. Physiol.* (*London*) 148: 574–591.

HUBEL, D. H., and T. N. WIESEL, 1962. Receptive fields, binocular interaction and functional architecture in the cat's visual cortex. *J. Physiol.* (*London*) 160: 106–154.

HUBEL, D. H., and T. N. WIESEL, 1963. Receptive fields of cells in striate cortex of very young, visually inexperienced kittens. *J. Neurophysiol.* 26: 994–1002.

HUBEL, D. H., and T. N. WIESEL, 1965a. Receptive fields and functional architecture in two nonstriate visual areas (18 and 19) of the cat. *J. Neurophysiol.* 28: 229–289.

HUBEL, D. H., and T. N. WIESEL, 1965b. Binocular interaction in striate cortex of kittens reared with artificial squint. *J. Neurophysiol.* 28: 1041–1059.

HUBEL, D. H., and T. N. WIESEL, 1968. Receptive fields and functional architecture of monkey striate cortex. *J. Physiol.* (*London*) 195: 215–243.

JOSHUA, D. E., and P. O. BISHOP, 1970. Binocular single vision and depth discrimination. Receptive field disparities for central and peripheral vision and binocular interaction on peripheral single units in cat striate cortex. *Exp. Brain Res.* (in press).

JULESZ, B., 1964. Binocular depth perception without familiarity cues. *Science* (*Washington*) 146: 356–362.

NIKARA, T., P. O. BISHOP, and J. D. PETTIGREW, 1968. Analysis of retinal correspondence by studying receptive fields of binocular single units in cat striate cortex. *Exp. Brain Res.* 6: 353–372.

PETTIGREW, J. D., T. NIKARA, and P. O. BISHOP, 1968a. Responses to moving slits by single units in cat striate cortex. *Exp. Brain Res.* 6: 373–390.

PETTIGREW, J. D., T. NIKARA, and P. O. BISHOP, 1968b. Binocular interaction on single units in cat striate cortex: simultaneous stimulation by single moving slit with receptive fields in correspondence. *Exp. Brain Res.* 6: 391–410.

SANDERSON, K. J., I. DARIAN-SMITH, and P. O. BISHOP, 1969. Binocular corresponding receptive fields of single units in the cat dorsal lateral geniculate nucleus. *Vision Res.* 9: 1297–1303.

SHERRINGTON, C. S., 1906. Integrative Action of the Nervous System. Yale University Press, New Haven, Connecticut.

WIESEL, T. N., and D. H. HUBEL, 1963. Single-cell responses in striate cortex of kittens deprived of vision in one eye. *J. Neurophysiol.* 26: 1003–1017.

WIESEL, T. N., and D. H. HUBEL, 1965a. Comparison of the effects of unilateral and bilateral eye closure on cortical unit responses in kittens. *J. Neurophysiol.* 28: 1029–1040.

WIESEL, T. N., and D. H. HUBEL, 1965b. Extent of recovery from the effects of visual deprivation in kittens. *J. Neurophysiol.* 28: 1060–1072.

# 44 Neural Subsystems: An Interpretive Summary

LEON D. HARMON

AMBROSE BIERCE, in his *The Devil's Dictionary,* defined mind as follows:

Mind, n.—A mysterious form of matter secreted by the brain. Its chief activity consists in the endeavor to ascertain its own nature, the futility of the attempt being due to the fact that it has nothing but itself to know itself with.

Perhaps Bierce's pessimism is as overemphasized as it is oversimplified, but in all truth we are obliged to face up to some staggering problems now that neurophysiology is emerging from its naive infancy. Because the problems range from molecular to behavioral, we may find it useful to ask a rather curious question: What exactly do we mean by saying we wish to "understand" the brain? There are several different answers, descriptions, and models, each capable of being true, but each profoundly different.

Consider an analogy couched in computer terms. I shall give you a desert island and on it place a high-speed digital computer, complete with terminal equipment. I shall keep it in good repair indefinitely, through any trauma you impose. Your task is to "understand" that system. My conditions are that you may import any number of people of any discipline, any knowledge, except that they must know nothing *a priori* about computer theory, structure, or function.

You are to obtain "understanding" in three different ways. First, one of your teams must find out how to operate the computer; they must discern its gross input/output functions. They will convince me of their "understanding" when they are able to write a complete programing manual for the system.

Another group must seek understanding at a quite different level. They are required to find out, for instance, what a shift register is and how an adder works. This group will convince me of their understanding when they are able to design a similar machine—not identical, but using the same principles of subsystem organization, operation, and interaction.

A still different level of "understanding" must be provided by the third group. It is required that they find out

how the elemental components operate. They must, for example, discern how a transistor works, elucidate the properties of electronic tubes, and find out about magnetic storage. They will convince me of their "understanding" when they are able to produce complete equivalent-circuit diagrams of all components, both active and passive.

Now, the conclusion at each of these three levels of understanding will be perfectly true and yet very different. That is, there are several distinct yet equally valid kinds of understanding, depending on one's domain of discourse. How well can relations be seen among these various conclusions? To begin with, it is difficult to see how understanding at one level helps that at another. For example, those persons who wish to understand at the programing-manual level could not care less whether the system uses tubes or transistors. Those who are trying to understand at the subsystem circuit level (adders and shift registers, for instance) need know nothing about programing languages or pattern-recognition algorithms. And those who wish to understand how a transistor functions are not helped either by adder theory or by Fortran IV.

Of course, there are logical and causal relationships among all three levels, but in this computer analogy, at any event, there seems to be little useful extrapolation from one level to another. At the very least, the ease of extrapolation is markedly direction-sensitive. For instance, it appears unlikely that any knowledge of the structure or function of an adder can lead to an understanding of a pattern-recognition operation, because the adder is the same whether the computer is calculating $\pi$, is recognizing a geometric figure, or is composing a fugue. Similarly, in neurophysiology, it seems equally hopeless to accept the standard argument that one must start at the lowest levels (e.g., molecular, subcellular), opening *all* black boxes before going on to greater system complexity.

Extrapolation in the other direction may, however, be somewhat easier. Knowledge at a higher level may help one develop strategy for investigation at a lower level. For example, if one knows something about the process involved in a computer's pattern-recognition algorithm, there is *some* chance of deducing the role of an adder (which may be required to execute that process).

The difference is primarily one of hierarchical levels of

LEON D. HARMON   Bell Telephone Laboratories, Incorporated, Murray Hill, New Jersey code and language. That is, the descriptions of information flow and process are entirely different at these various levels; the simpler (lower level) signal structures do not yield a prediction of those at higher levels. Here is an example of wholes being not only greater than but quite different from the parts that constitute them. Furthermore, there is a matter of nonuniqueness. An adder may play a role in an infinite number of higher-order processes, whereas a size-invariant, triangle-recognition processor may have only a limited kind and number of underlying logical and arithmetical operators. Extrapolation from low level to high level diverges, but that from high to low converges.

Another kind of difficulty in coming to an understanding of nervous systems is that we may be conditioned into thinking about them in ways that are more constrained than we like to admit or are sufficiently aware of. Such concepts as serial computation, discrete subsystems, Aristotelian logic, and coding theory may well be forcing us to view, say, neocortical function in lights that are wholly inappropriate.

## Subsystems

The positions taken by the contributors to this section of this book are closest to level two of the computer analogy. Neural subsystems, i.e., small nets of neurons, can be likened to shift registers and adders. We wish to understand the functions of a relatively complicated cooperative ensemble (network) of working devices (neurons), in which the signal-processing properties of the ensemble can be clearly and specifically determined. Furthermore, we wish to understand the operationally significant relationships among such subsystems, i.e., the functional organization of the entire system. Thus, the emphasis here is on *function* rather than on *structure* or on *signal codes*.

There are many kinds and numbers of subsystems in a human being; for example, there are several hundred thousand different kinds of molecules with their replicating mechanisms, thousands of metabolic subsystems, and perhaps two or three thousand neural subsystems. We shall be concerned only with several of the latter (both in man and in lower animals), as they are involved in particular processing of input (sensory), central, and output (motor) signals.

It is necessary to point out that in nervous-system analysis, the idea of "subsystems" may be misleading or even fallacious. Neural systems may be holistic to the extent that subsystem subdivision is arbitrary and useless; despite observed specificities of some gross tracts or particular nets, perhaps the mammalian CNS operates as an entity according to information-processing principles that we do not even suspect at present.

The two distinctions between *circuits* and *systems* that Lewis (this volume) has drawn have considerable impor-

tance, both conceptually and practically, for neurophysiological research. Because their implications bear strongly on the ensuing discussion, I first restate them in slightly different form.

Lewis defines a circuit as a collection of interacting elements, no subgroup of which can readily be analyzed to disclose separation of cause (input) and effect (output). This is a matter of orientation; in general, circuits are not unambiguously oriented; input and output terminals are ambiguous with respect to underlying organization. In contrast, systems (entire organisms) usually contain subgroups which have inputs that can be differentiated from outputs; this permits specification of cause and effect and hence facilitates analysis. The latter subgroups are called *subsystems*—the subject of this section of the book.

The second distinction is that of structural context. Lewis raises the troublesome question of whether neurons (or possibly even subsystems) can, in general, be considered in isolation. Owing to profuse interconnection and widespread reciprocal interaction, many of the measured properties of a system are likely to be determined ultimately by the entire system and are not uniquely and unequivocally assignable to *individual components* (just as epistatic interaction in chromosomes demonstrates intergene influences). Consequently, in the worst case, a perturbation anywhere in the system will be reflected in reaction throughout that system. Conversely, the state and the response of an individual element are not context-free—they must ultimately reflect the states and responses of all other elements.

The elements that comprise circuits generally are embedded in a structure that has strong interactive influence on those elements, so a clear separation of element function and, hence, circuit analysis is impossible unless complete structural context is taken into account. Properties of a circuit cannot be attributed to individual, particular elements. On the other hand, subsystems are by definition structurally (hence functionally) isolable; although they may be surrounded by other subsystems, that context need not be taken into account.

In what follows we define element, circuit, subsystem, and system by example and in terms of orientation and context. Elements (such as neurons) are interconnected into circuits (such as neural nets); one or more circuits forms a subsystem (such as a ganglion); one or more subsystems forms a system—perhaps an entire organism. Thus we might view a retinal bipolar cell as an element, a bipolar-amacrine-ganglion cell network as a circuit, and an entire retina as a subsystem.

The critical distinction between circuit and subsystem should be viewed in terms of a bounded domain which has clearly observable input and output terminals. Within that domain we have one or more circuits; interactions are profuse, and cause and effect are obscure. Thus, with respect to input and output terminals the domain is irreducible. Outside that domain the signal flows are discernible; context can be assessed, and cause and effect can be traced. The bounded domain is a *subsystem,* is viewed as a black box, and consists of at least one *circuit.* We refer to the domain either as a subsystem or as a circuit, depending on whether it is being viewed from outside or inside.

In the light of Lewis's observations on the immense difficulty of coming to an understanding of such a relatively small circuit as the nine-cell cardiac ganglion of the lobster, I wonder whether it is profitable to continue to seek the full *internal* (circuit) analysis of such a subsystem. Again the question arises—How far down in level must black boxes be opened in order to comprehend a system? One class of answer to that question, although addressed to a higher level of organization than our subsystem level, uses control-system analytic techniques. Such systems as eye tracking (Stark et al., 1962; Fender, 1964), pupillary servomechanisms (Stark and Sherman, 1957), and optokinetic orienting (Reichardt, 1961) have been elucidated by the identification of larger units of organization that might correspond to collections of subsystems. In this work, the black boxes are control-system functional operators such as multipliers, coincidence gates, function generators, and the like. Each is taken to subsume many neurons in networks which intrinsically are of no concern to the analysis. These are formal equivalents that are sufficiently representative *and* accurate for precise predictions of response in the living system to be obtained. This work has led to important and revealing results at a level of inquiry including organism behavior itself.

Because circuit analysis implies total commitment to analysis of *all* components and their interactions, the commitment may be as futile as it is irrelevant. However, if the goal is to understand *circuits* and *components* as such, and there is no hope of elucidating systems, then I would argue that such effort is justified and of great interest, simply in the investigation of nature. But if one's goal is to understand *systems* as such, then in all likelihood approaches that lie at levels between irreducible subsystems (such as the lobster cardiac ganglion) and reducible control systems (such as optokinetic supersystems) hold the most hope for meaningful neurophysiological research.

## Teleology

Teleology, taken here *solely* to mean mechanical determinism, plays an important part in the analysis of living systems. Inference about role is essential, not merely tolerable, for both circuit analysis and subsystem analysis. Consider, for instance, an electronic circuit of which the orientation has somehow been established and for which even the transfer function is obtained. Simply possessing the transfer

function alone is virtually meaningless without knowledge of the circuit's role *and* of normal input signals, as an infinite number of forcing functions (and therefore of input-output pairs) are possible. Such a situation is equivalent to having a differential equation without having any boundary conditions. Even the best of electrical-circuit engineers, given a schematic of modest proportions and no clue as to intended function, will be unable to assign a meaningful role to that circuit. He will remain in essential ignorance of its functional significance unless he knows what was in the designer's mind. Induction of role from form is, in general, impossible.

It simply is not enough to be able to describe a circuit and its component action, however completely, in order to comprehend its role. Suppose, for example, we assume that genetic instructions completely and unequivocally specify some particular neural system, structure, codes, and all. We shall assume a completely causal chain—all principles and mechanisms of the subsystem are absolutely contained in the genetic structure. I suspect that you may examine, test, describe, and catalogue everything about that genetic system, yet you will *not* be able to extrapolate to the implicit subsystem *unless* you also know the role and principles underlying the genetic system and its transforms.

A teleological approach in neurophysiology seems essential and should have high priority. Because of the impossibility of inferring function from form, some *a priori* prejudice as to role is mandatory. Furthermore, systems analysis may reveal $n$ disjoint parameters. Because there are $2^n$ possible subsets, complete assessment is virtually impossible if $n$ is large. One must apply criteria such as teleological constraints to narrow the possibilities. Consequently, the most revealing and significant experiments may be expected to be those in which some prior notion of operational detail (e.g., gating, filtering, and so on) is imposed.

## Principles of function

An implicit assumption, distributed throughout the foregoing discussion, is that principles of neural action do or must resemble those of our machines. Consider, for example, the prevalence of contemporary opinion that nervous systems, being information processors, can be viewed as computer-like devices. It is clear that many aspects of our engineering technology are found in physiological measurements. Such properties as triggering, gating, switching, and synchronizing have been found. Amplitude discrimination, filtering, amplification, linear-to-logarithmic signal transformation, and waveform generation have been seen. The basic arithmetical operations have been documented in nervous-system responses, as have integration, differentiation, sign inversion, and correlation. In addition, there is evidence for counting, coincidence detection, delay, phase shift, and duration measurement.

Although these properties can be found experimentally, it has yet to be demonstrated that systems use them in going about their business. An especially striking example is found in Wilson's analysis (this volume) of wing-muscle stretch receptors in the grasshopper. Those units precisely encode wing-beat frequency, phase, and amplitude, yet the axon-impulse time, rate, and number that code that information are subsequently smeared out to give the CNS only a *gross* command relating to over-all average wing-beat rate and power. Here is a simple example of the danger of assessing information, code, or notion of relevant processing until and unless one knows something about, as Wilson neatly puts it, "the natural reader of that code." To correlate is not necessarily to elucidate.

Despite such seemingly negative comment, there is no reason at present not to consider neural subsystems in engineering terms. In fact, theoretical studies guiding experimental probes for subsystem operation may contribute entirely new concepts in information processing. It is perhaps not too wild a speculation that actual principles of nervous-system operation may be different, perhaps far different, from those of our present computer technology. Certainly such a difference obtains for associative memory, for example. To suspect, to look for, and perhaps to find such new principles constitute one of the most exciting problem areas of living-system analysis.

## Special-purpose vs. general-purpose systems

In considering identification, measurement, and analysis of subsystems, we run into an old problem that takes on new significance for understanding at this level. It has to do with network plasticity—whether we are dealing with a network that is rigidly fixed or one that is in some manner adaptable.

Another way to speak of plasticity is to ask whether a system is "special purpose" or "general purpose," in the sense of computers. Is a particular animal special-purpose in that it is inflexible, more or less completely preprogramed, and therefore constructed with a great *a priori* response specificity, or is it general purpose in being capable of considerable modification of internal structure and external behavior? At least in the domain of computers, a special-purpose machine that has fixed structure (wiring) and fixed function may be much easier to analyze than is a general-purpose machine the very "structure" (signal-flow paths) of which is infinitely modifiable by programs and by states of its memory cells. Because of such distinctions, the ways of thinking about the two kinds of systems differ considerably, as it must also be for living systems. Grasshopper and man would appear to lie at extremes of a spectrum ranging from special-purpose to general-purpose systems, and the re-

quired conceptual approaches to system analysis might differ as much as do the analyses required for understanding a typewriter and a computer.

Of course, to what extent man is a general-purpose system may be questioned. Clearly, many of his subsystems are special purpose, genetically determined, homeostatic, and quite inflexible. To the extent that there is fixed wiring and fixed processing we can proceed with greater ease and, indeed, even afford to take teleological approaches. Experiment and analysis guided by *a priori* consideration of what an animal does and needs to know have paid off handsomely (Maturana et al., 1960; Hubel and Wiesel, 1962; Nikara et al., 1968).

## Command and control

A topic closely related to that of special-purpose machinery is the complex neuromuscular sequence that is activated by simple "command" stimuli. Wilson (this volume) cites a considerable body of evidence in invertebrates, particularly of genetically preprogramed structures and functions in arthropod ganglia. He holds that there are three principal classes: sensory filters, closed-loop reflexes, and pattern generators.

Such preprogramed subsystems epitomize the idea of special-purpose neural machinery. Here are structures specifically arranged to execute particular functions; presumably they are immutable, and any possible overlay of learning apparently is minimal or nonexistent (at least for simple behavior, like locomotion).

Because the ultimate business of virtually all CNS processing is to generate motor-command signals, questions pertaining to behavioral control via neuromuscular activating systems are of fundamental importance. William James, in 1897, had a remarkably modern view of this point regarding the significance of motor output when he observed: "The structural unit of the nervous system is in fact a triad, neither of whose elements has any independent existence. The sensory impression exists only for the sake of awaking the central process of reflection, and the central process of reflection exists only for the sake of calling forth the final act. All action is thus *re*-action upon the outer world; and the middle stage of consideration or contemplation or thinking is only a place of transit, the bottom of a loop, both of whose ends have their point of application in the outer world. If it should ever have no roots in the outer world, if it should ever happen that it led to no active measures, it would fail of its essential function, and would have to be considered either pathological or abortive. The current of life which runs in at our eyes or ears is meant to run out at our hands, feet or lips. The only use of the thoughts it occasions while inside is to determine its direction to whichever of these organs shall, on the whole, under the circumstances actually present, act in the way most propitious to our welfare" (James, 1897, *in* Gibson, 1962).

One of the rather striking generalities of pattern generation in insect motor-control systems is that stereotyped signal sequences appear to be produced from unpatterned stimuli (e.g., random pulse trains), or they may even be spontaneously generated. Wilson's description of the behavior of a central, pattern-generated, flight-control subsystem in the grasshopper illustrates the point. In that example, four subsets of phased, synchronized, motor-neuron volleys arise from relatively simple, unpatterned, central control. Although the known motor-neuron interactions appear to be too weak to account for the synchrony, the preservation of precise coordination among the four sets of wing motor neurons that have an imprecise, common, central-control input seems to argue for a mechanism (as yet unknown) whereby undifferentiated CNS signals set in motion highly precise, peripheral, temporal-pattern sequences.

The ability of a subsystem to generate intricate patterned outputs or complex motor sequences from relatively simple central commands constitutes an intriguing and fairly common class of subsystem operations in invertebrates and, perhaps, in vertebrates as well. Elaborate temporal sequences of behavior, involving many sets of muscles, can be initiated by single pulses, unpatterned pulse trains, or even random stimuli. It is as though a simple command to "go!" triggers an entire subsystem into one particular behavior pattern, as is seen in "subroutines" of swimming, walking, flying, posturing, copulating, and singing.

The numbers of circuit elements in the subsystems triggered by command fibers are not known with precision, but some reasonable guesses follow: 80 motor neurons plus a probably greater number of interneurons and oscillation producers in grasshopper-flight and cricket-song subsystems; 200–300 motor neurons plus some indefinite number of interneurons for crayfish posture and swimmeret subsystems and for cockroach walking; possibly 1000 to 10,000 motor neurons and interneurons in the only known vertebrate "control fiber," the Mauthner neurons.

At least one simple but potent circuit arrangement is known to produce temporal spike-patterns—it is recurrent inhibition. One of Wilson's examples shows that phase-locked patterned discharge occurs in simple neural networks that are laterally inhibited. In such cases, the input is a relatively indifferent common stimulus. Modeling studies of reciprocally inhibiting units have demonstrated a great variety of pattern production from simple, regular stimuli (Reiss, 1962; Harmon, 1964; Harmon and Lewis, 1966).

It appears that the production of stable, intricate, temporal-output patterns by cross-inhibited single units may be of fundamental importance to "simple" neural subsystems. Interestingly, as Wilson points out, the sensory corollary

(e.g., lateral recurrent inhibition in *Limulus*) also demonstrates a potent network-organization principle, this time for sensory analysis rather than for motor synthesis. (It should be noted, however, that reciprocally inhibiting *pairs* generally are associated with temporal processing such as switching or spike-pattern generation, whereas laterally inhibiting two-dimensional *networks* usually are associated with spatial image processing such as contrast sharpening or detail filtering.)

That stimulation of a *single* command fiber may elicit coordinated activation of half of the musculature of the entire body (as in the crayfish defense posture) seems to argue for a kind of single-neuron importance virtually unknown in vertebrate physiology. Curiously, however, there are some vertebrate experiments which at least suggest similar action. Gross electrode stimulus was observed to cause a falcon to peer at a particular corner of a room despite the animal's being turned to various different orientations (Strumwasser and Cade, 1957). The simple stimulation of the brain of a pocket-mouse initiated a prolonged sequence of rapid two-pawed stuffing of hallucinated seeds into its cheek pouches (Strumwasser and Cade, 1957). Gross square-wave shocks to the temporal cortex of man elicited complete reruns of memory sequences (Penfield and Perot, 1963).

The foregoing examples, for both invertebrate command fibers and vertebrate simple, gross stimulation, relate to involuntary stimulus/response situations. Of course, the enormously potent *ad hoc* arrangements of neural operation available to a consciously attending human permits an even more striking unleashing of complex sequences from simple stimuli. For example, the impingement of a few photons of light on a retina in a jungle at night can trigger the shouldering, aiming, and firing of a rifle—a sequence of neuromuscular events that is more complicated by orders of magnitude than the evocation of a crayfish defense posture.

No artificial excitation of precommand-fiber structures has yet stimulated a command fiber to action, probably because the required stimulus pattern is complex. For instance, the full array of inputs required for a crayfish to rear up in defense posture probably includes visual patterns, movement, exposed position (not in cave), time of day, and so on. All these must funnel down to the command unit. The subsystem coding-decoding precursors appear to be remarkably selective and fail-safe. Presumably, interposed loops or parallel, multiple processes link the visual subsystem with the command subsystem.

It is interesting to speculate that command neurons in invertebrates have a homologue in vertebrate "decision units" (if, indeed, such exist). This would be some sort of final funneling point in a high-order animal, just before divergence to effectors. Consider, for example, a vastly oversimplified example as is schematized in Figure 1.

Suppose that we ask a man to raise his right arm if a particular photograph is a picture of an individual, X, who is known to him. Let me represent the situation as one in which, if, and *only if,* the appropriate sensory input occurs, some central command-and-control subsystem initiates the neuromuscular, sequence-activation signals needed to produce the required response.

In schematic form, at least, such a situation is homologous to the invertebrate sensory filter → command fiber → pattern-generator. The principles of subsystem coupling and of signal-coding transformations may be qualitatively similar, at least as far as we now know. Thus, in disclosing principles of nervous-system organization and hierarchical processing, studies of relatively simple nervous systems may include some helpful clues to analysis of the more complex systems. The potential "payoff" is great, because the command-and-control box of Figure 1 encloses the great unknown turn-around zone between sensory and motor systems, i.e., generalization, decision making, and, at least in some actions and animals, consciousness.

## Functional inhibition

Bishop (this volume) places particular stress on the definition and role of inhibition, especially as some controversy over interpretation of receptive-field organization stems from it. Conventionally, the idea is that an ON discharge is equivalent to an excitatory influence, and an OFF discharge reflects an inhibitory one. Bishop points out that the relationships of ON/excitatory and OFF/inhibitory are neither necessary nor common.

In one modeling study of ON, OFF, and ON/OFF responses, *both* excitation and inhibition are played off in combination against each other in just one unit to obtain these temporal responses (Harmon, 1966). In this case, excitatory postsynaptic potential (EPSP) and inhibitory postsynaptic potential (IPSP) voltage time-courses and asymptotic levels were used as variables and clearly demonstrated the nonnecessity of relationships between circuit inhibition and over-all subsystem or functional inhibition.

Bishop's point is important, particularly because inhibition in membrane action is often confused with inhibition in system action; it is essential to distinguish between circuit inhibition and functional inhibition. A notable example is seen in the commonly held view that cerebellar Purkinje cells are inhibitory, a circuit view based on the evidence of their direct output. However, from a subsystem point of view, if Purkinje axons happen to terminate on inhibitor interneurons which disinhibit particular subsystems, the effect of Purkinje activation is excitatory. Perhaps *Limulus* retinal action is the most familiar example of the importance of analysis in terms of functional inhibition rather than in terms of EPSPs and IPSPs.

By viewing visual receptive-field responses in terms of inhibition and excitation rather than in terms of ON and OFF, Bishop shows that receptive fields of "simple" cells are largely inhibitory (having a small, central, excitatory zone), and some are completely inhibitory. It remains to be demonstrated how much the delineation and analysis of subsystem function will improve, given this "new look," but indications are that there is increased utility. For example, Bishop shows that in binocular depth discrimination, the excitatory and inhibitory effects of stimulating a receptive field in one eye have an important gating influence on the response to stimulus of a corresponding zone in the other eye, as read by a striate neuron.

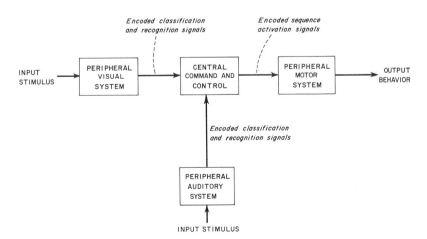

FIGURE 1 Hypothetical system representing a human requested to respond manually when he sees an appropriate visual pattern. Auditory instructions, recognized and encoded, prime central command-and-control system. Subsequent visual-input stimulus, recognized as sufficient, releases primed central system to evoke sequence of neuromuscular activation signals appropriate to response requested.

A particularly interesting result of testing receptive fields of simple units for inhibitory effects came from the use of two simultaneous stimuli (e.g., slits of light). A conventional, preferred-direction response is obtained in a typical receptive field, while stimulation of a *direction-independent* inhibitory surround simultaneously influences that response, producing no ON or OFF transient, only inhibition.

Just as inhibition is seen to play a dominant role in visual systems, providing highly selective discrimination, so too the functional role of inhibition also should be expected to be profound in all sorts of neural subsystems, if for no other reason than to avoid continuous cataclysmic discharge. As neural circuits of reciprocal and lateral inhibition are more and more ubiquitously seen, the great potency of opposed excitatory/inhibitory functions becomes increasingly apparent (Florey, 1961; von Euler et al., 1966). It would not be surprising if an extensive theory of neural subsystem signal filters, generators, and processors is evolved in the near future, the basic development of which would rest on relatively regular structures that use simple combinations of excitation, inhibition, and their time courses.

## Hierarchical processing and decision making

The concept of hierarchical processing and control appears to be a natural, efficient way to deal *in extenso* with otherwise unwieldy systems. Such is equally true in rockets, governments, and nervous systems. The discovery over the last decade of hierarchical levels of visual-data processing in many animals from insects to vertebrates has provided one of the most illuminating new chapters in neurophysiology.

In cat and monkey, for example, as one traces the effects of patterned optical signals from retina to lateral geniculate to striate cortex to higher visual cortex, the ascending hierarchy of data processing presents an orderly and impressive progression (Hubel and Wiesel, 1962; Barlow and Levick, 1965; Barlow et al., 1967; Bishop, 1969). The most striking aspect of this level-by-level processing is feature extraction. Successive hierarchies of neural actions analyze two-dimensional retinal space to represent simple aspects of shape and temporal changes of illumination, and, through spatially discrete representations of those changes, motion is also represented.

Despite such gains in our knowledge, Bishop noted that "cells have yet to be discovered which respond only when the simplest of forms, such as a rectangle or a circle, is present in the visual field." This observation leads immediately to what are perhaps the most intriguing, yet intractable, questions in all neurophysiology—those that relate to decision making, perception, and consciousness. What, for instance, is the system (subsystem) that observes lower-order subsystems and is "aware" of a complex pattern? What is the nature of the "little man" who sits on top

of the hierarchical processors, observing? Although these are not precise, well-formed, or even operational queries, it is relatively easy to formulate one that is closely related and that is of central importance to the ultimate analysis of, say, the primate visual system. It is as follows:

Given that states of single neurons in striate and peristriate cortex represent primitive features of visual scenes (lines, edges, angles) and, further, that those representations are, for some cells, invariant over retinal position but are constrained with respect to motional aspects (direction, velocity), where, how, and by what are higher-order classification responses made? An enormous amount of visual brain has so far been required to discern these relatively simple features. Where does one look next—and how? After area 19, where?

There seem to be two major possibilities. One is that other, higher structures combine the computational results of the striate/peristriate subsystem to produce these sophisticated operations we seek. The infratemporal cortex might seem to be a reasonable candidate; however, in that case, one is faced with the curious possibility that a relatively small processor (compared with everything prior, from retina to area 19) does most of the incredibly complex integrative work. The other alternative is that the *modus operandi* changes; single units no longer are used to represent feature states. Instead, simultaneous, combined (not statistically smeared but highly articulated) states of very large numbers of cells in striate and peristriate cortex conspire to represent, say, a recognizable face. That is, the representation is now multicellular, involving perhaps hundreds of thousands (or millions) of simultaneously signaling units, which, when read individually, are those discussed above— the feature detectors. The states of the parts taken separately do not represent the whole; rather, it is the relations among them that are important, just as individual silver halide grains do not "represent" the photograph of a face, but the ensemble does. Similarly, flecks of chalk on a blackboard are individually meaningless but in relationship may represent a triangle.

Although a multistate response of a neural pool may represent a complex event of recognition, it may or may not actually constitute the conscious awareness of that recognition response. The ultimate conscious perception, attending to one thing at a time, may reside elsewhere, its neural substrate receiving as input the response of the recognition network. The question is open as to whether a single "pontifical" neuron or a large ensemble (as suggested above) is used to represent a final recognition state. We tend to be seduced into thinking in terms of "final common paths," i.e., ultimate convergence of many processes to one channel. But even a binary decision, such as the hand-raising example shown in Figure 1 or the vocal signaling "yes" in response to a recognized face, may involve huge numbers of neurons just to do the responsive signaling. Similarly, the binary recognition-decision itself may very well be represented by the multistate condition of a vast pool of units. This issue, discussed in detail by Bullock (Bullock and Horridge, 1965, p. 281 ff.), is one of the most tantalizing and difficult in all neurophysiology.

One final article of faith on this speculative point: I am not surprised to find grasshoppers absolutely dependent on the integrity of one or a few neurons, but I find it difficult to believe, especially in the face of cortical ablation and natural-attrition evidence, that in man there can be much dependence on unique cells, small groups, or local circuits.

It is tempting to conjecture that arrangements exist in which both individual units *and* the pools they comprise compute and represent significant information. For instance, single-unit feature detectors and even pontifical decision-makers might be part of a pool in which wave formation and propagation also play an essential role. Such subsystem design could reflect completely new and potent principles of functional organization. If this notion turns out to be at all valid, a rather different concept of "subsystem" must be developed. For here is the idea of a self-nested system, layered like an onion, a thing that is both parts and whole, a multilayered, hierarchical, signal processor, which, rather than being sequentially laminated and connected, is an everything-overlaid-on-everything holistic entity.

## Summary

This chapter is, in part, a summary of some issues discussed in the section of this volume entitled "Neural Subsystems and Physiological Operations," and, in part, some personal responses to those and related issues. The central idea is that a "subsystem" point of view is especially useful in the analysis of nervous-system function.

The particular issues I have responded to range rather widely: subsystems, teleology, principles of function, special-purpose versus general-purpose systems, command and control, functional inhibition, and hierarchical processing and decision-making. Although these topics are more or less disparate, they are discussed with one common theme in mind: using a particular engineering point of view and level of inquiry to seek an understanding of neural function.

I mentioned above that many of these problems are of staggering complexity, both experimentally and theoretically. Let me suggest still one more difficulty—one that harks back to Bierce's worry that the mind has only itself to understand itself with. Conceivably, the complexity of the brain (its principles of organization and operation) may exceed the ability of its output portion (its conscious attention, language, logics of discourse, and analysis) to deal with it. In other words, the mental functions produced by the brain may well be very much simpler than the principles

of operation of the underlying mechanisms giving rise to those functions.

Whether such worries are justified, and whether our circuit and our subsystem problems ultimately prove intractable in many cases, ultimate limitations still appear remote. With the continuing development of new technologies and new theory, the prospect for further and deeper understanding seems excellent.

## *Acknowledgments*

I appreciate the feedback, both positive and negative, from the readers of an early draft of the manuscript. They include Theodore Bullock, chairman of this section of the book, and those participants of ISP 1969 whose ideas I maligned, and four relatively detached colleagues, Newman Guttman, Bela Julesz, R. F. Reiss, and Eric Wolman.

### REFERENCES

BARLOW, H. B., C. BLAKEMORE, and J. D. PETTIGREW, 1967. The neural mechanism of binocular depth discrimination. *J. Physiol. (London)* 193: 327–342.

BARLOW, H. B., and W. R. LEVICK, 1965. The mechanism of directionally selective units in rabbit's retina. *J. Physiol. (London)* 178: 477–504.

BISHOP, P. O., 1970. Neurophysiology of binocular single vision and stereopsis. *In* Handbook of Sensory Physiology, Vol. 7 (R. Jung, editor). Springer-Verlag, Berlin (in press).

BULLOCK, T. H., and G. A. HORRIDGE, 1965. Structure and Function in the Nervous Systems of Invertebrates, Vol. I. W. H. Freeman and Co., San Francisco.

EULER, C. VON, S. SKOGLUND, and U. SÖDERBERG, 1966. Structure and Function of Inhibitory Neuronal Mechanisms. Pergamon Press, Oxford and New York.

FENDER, D. H., 1964. The eye-movement control system: Evolution of a model. *In* Neural Theory and Modeling (R. F. Reiss, editor). Stanford University Press, Stanford, California, pp. 306–324.

FLOREY, E. (editor), 1961. Nervous Inhibition. Pergamon Press, New York.

HARMON, L. D., 1964. Neuromimes: Action of a reciprocally inhibitory pair. *Science (Washington)* 146: 1323–1325.

HARMON, L. D., 1968. Modeling studies of neural inhibition. *In* Structure and Function of Inhibitory Neuronal Mechanisms (C. von Euler, S. Skoglund, and U. Söderberg, editors). Pergamon Press, Oxford and New York, pp. 537–563.

HARMON, L. D., and E. R. LEWIS, 1966. Neural modeling. *Physiol. Rev.* 46: 513–591.

HUBEL, D. H., and T. N. WIESEL, 1962. Receptive fields, binocular interaction, and functional architecture in the cat's visual cortex. *J. Physiol. (London)* 160: 106–154.

JAMES, W., 1897. Reflex action and theism (from "The Will to Believe and other Essays"). *In* The Limits of Language (Walker Gibson, editor). Colonial Press, Clinton, Massachusetts (1962).

MATURANA, H. R., J. Y. LETTVIN, W. S. McCULLOCH, and W. H.

PITTS, 1960. Anatomy and physiology of vision in the frog (*Rana pipiens*). *J. Gen. Physiol.* 43 (no. 6, pt. 2): 129–176.

NIKARA, T., P. O. BISHOP, and J. D. PETTIGREW, 1968. Analysis of retinal correspondence by studying receptive fields of binocular single circuits in cat striate cortex. *Exp. Brain Res.* 6: 353–372.

PENFIELD, W., and P. PEROT, 1963. The brain's record of auditory and visual experience. *Brain* 86: 595–696.

REICHARDT, W., 1961. Autocorrelation, a principle for the evaluation of sensory information by the central nervous system. *In* Sensory Communication (W. A. Rosenblith, editor). M. I. T. Press, Cambridge, Massachusetts, pp. 303–317.

REISS, R. F., 1962. A theory and simulation of rhythmic behavior due to reciprocal inhibition in small nerve nets. *Joint Computer Conf.* 21: 171–194.

STARK, L., and P. M. SHERMAN, 1957. A servoanalytic study of the consensual pupil reflex to light. *J. Neurophysiol.* 20: 17–26.

STARK, L., G. VOSSIUS, and L. R. YOUNG, 1962. Predictive control of eye tracking movements. *Inst. Radio Eng. Trans. Human Factors Electron.* 3: 52–57.

STRUMWASSER, F., and T. J. CADE, 1957. Behavior elicited by brain stimulation in freely moving vertebrates. *Anat. Rec.* 128: 630–631 (abstract).

# 45 The Insect Eye as a Model for Analysis of Uptake, Transduction, and Processing of Optical Data in the Nervous System

## WERNER E. REICHARDT

AT PRESENT, investigations in the field of sensory physiology are based on the application of three methods: anatomical and histological studies; electrophysiological microprobing; and evaluation of evoked behavioral responses.

The first method allows us to draw conclusions from a group of functional operations, which, in principle, could take place in the circuitry of the nerve-nets under consideration. Moreover, it is sometimes possible to exclude certain operations—a procedure that may result in omitting classes of possible perceptual processes.

By use of the second method, one may trace the information flux, especially in those parts of the nervous system that are, in one or two dimensions, isotropic in structure and in function, i.e., the interactions of which depend on the distance between two or more interacting neurons and not on their position. The technological difficulties of this method become formidable whenever the subsystem under consideration is nonisotropic.

The third method can be applied to the analysis of functional principles of input-output relations in behavioral experiments, regardless of the structural complexities of the physical parameters of components in the system. This method, however, does not give us information about how these functional principles are realized physically and which components of the nervous structure actually are involved.

The account presented here concerns the application of the three methods to the analysis of the structure and function of the optical system of flies (*Drosophila, Musca*). Their nervous systems consist of about $10^5$ to $10^6$ neurons, which, in part, make up definable classes of elements that are replicated many times in an orderly way, so the principal circuitry can be worked out without identifying specific elements.

### Representation of the optical surround in the photoreceptor layer of the compound eye

The compound eyes of a fly consist of individual ommatidia that are optically screened from one another by pigments. The kind of representation of the visual environment in the

WERNER E. REICHARDT Max-Planck-Institut für biologische Kybernetik, Tübingen, Germany.

layer of the photoreceptors (rhabdomeres) is determined by (1) the optics of the individual ommatidium, (2) the number and arrangement of rhabdomeres in every ommatidium, and (3) the geometry of the eye.

Figure 1 is an electron micrograph of a cross section of an individual ommatidium. It shows seven rhabdomeres arranged in a typical asymmetric pattern. It should be noted that the central rhabdomere (small cross section) is com-

pound, consisting of two from individual retinular cells. They are aligned in such a way that the distal part belongs to one cell and the proximal part to another. The rhabdomeres are separated from one another and remain separated down the whole length of the ommatidium. Consequently, their optical axes differ, so that each rhabdomere receives light from different directions in the optical environment (Autrum and Wiedemann, 1962; Wiedemann, 1965).

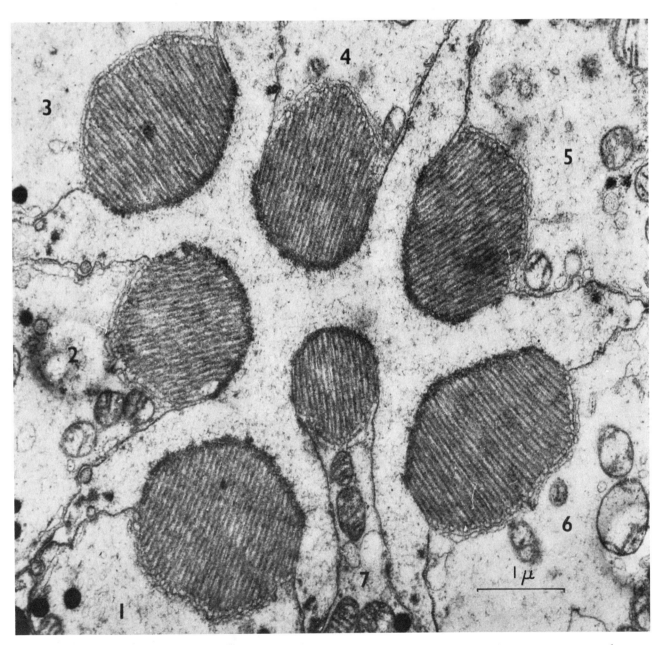

FIGURE 1   Cross section through an ommatidium of *Musca*. Seven sensory cells with their attached rhabdomeric structures are shown. The diameter of each of the peripheral rhabdomeres (1–6) is about twice that of the central one (7). Rhabdomere 8 cannot be seen in the Figure, as it is situated proximal to 7. Note the different orientations of the microvilli in the different rhabdomeres. Fixation: glutaraldehye-dosmium. (From C. B. Boschek, unpublished.)

Kirschfeld (1967) determined the angular distribution of the optical axes of the rhabdomeres, and found that these axes mark points in the optical surround that constitute a pattern similar to that of the distal endings of the rhabdomeres in an ommatidium, but inverted by 180°. It follows that each ommatidium acts as a small lens eye in which the image generated by the lens can be resolved by the seven (eight) rhabdomeres into seven points.

A striking feature of the angular separation of the optical axes of the rhabdomeres is that they match the ommatidial divergence angles. A point source of light in the optical surround may always be aligned with respect to the compound eye in such a way that seven (eight) rhabdomeres of seven different ommatidia receive light from a single source, as indicated in Figure 2. Consequently, as Kirschfeld (1967)

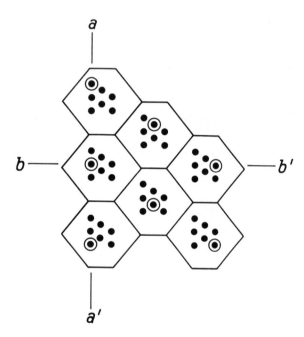

FIGURE 2 Drawing of a cross section of seven facets (*Musca*). The rhabdomeres, indicated by black dots, are shown in their typical arrangement in the upper eye region. Rhabdomeres surrounded by a circle receive light from one and the same point of the optical environment. (From Kirschfeld, 1967.)

has shown, seven (eight) rhabdomeres, each from a different ommatidium, are "looking" at one and the same point in the optical environment. Therefore, between individual points in the optical environment and individual photoreceptors of the fly's eye, a one-to-seven, not a one-to-one, relationship exists. As there are seven (eight) separate rhabdomeres per ommatidium, optical resolution of the eye should be determined by the angle $\Delta\varphi$ between the optical axes of adjacent ommatidia, not by the angles between optical axes of individual rhabdomeres.

The alignment of the optical axes of rhabdomeres in different ommatidia has been demonstrated in the intact animal by the method of "antidromic" illumination (Kirschfeld and Franceschini, 1968). Light is shone into the brain of the fly through a small hole made in the head. Part of the light reaches the rhabdomeres, where it is guided to their distal endings and finally leaves the ommatidia through their dioptric systems. By this method, the rhabdomeres act as secondary radiants, the distal endings of which are imaged on the optical surround. Figure 3 shows how the radiation leaving the compound eye superimposes at different distances from the eye. At a large distance (greater than $1500\mu$), seven images from seven (eight) different rhabdomeres of seven different ommatidia join as described.

## Angular sensitivity distributions of rhabdomeres

The fundamental optical properties of a compound eye are characterized by two parameters: 1. The divergence angle $\Delta\varphi$ between the axes of adjacent ommatidia. This angle sets the limit of optical resolution. In *Drosophila*, the mean of $\Delta\varphi$ amounts to $\overline{\Delta\varphi} = 4.6°$; in *Musca*, to $\overline{\Delta\varphi} = 2.0°$. 2. The angular sensitivity distributions of the photoreceptors (rhabdomeres), expressed by $\Delta\rho$, the half-width of the distribution. These distributions determine the contrast transmission from the optical environment onto the receptor layer.

Recent electrophysiological probing by Scholes (1969) on the dark-adapted eye of *Musca* have led to values of about $\Delta\rho = 2.5°$. The data were taken in the dorsofrontal quadrant, where facet dimensions are largest and divergence angles are smallest. There is good evidence that the value of $\Delta\rho$ reported here relates to one of the peripheral retinular cells (numbers 1 to 6) and not to the central ones (numbers 7, 8). Because the sensitivity distribution of a rhabdomere is determined by the convolution of the lenslet Airy disk with the cross section of the rhabdomere ending, it is likely that the acceptance angles of cells 7 and 8 are narrower than those of cells 1 to 6 (Kirschfeld and Franceschini, 1970). Their $\Delta\rho$ is estimated to be in the order of 1.5° in *Musca*.

## Spectral sensitivities of rhabdomeres

The first direct proof that photolabile pigments are present in the fly's rhabdomeres stems from microspectrophotometric measurements and their interpretations by Langer and Thorell (1966) and Langer (1966, 1967). They found two types of extinction spectra from individual rhabdomeres of *Calliphora*: one has a maximum of about 515 millimicrons (green receptor); the maximum of the other is about 470 millimicrons (blue receptor). The absorption with its maximum in the blue is associated with the central rhabdomere (number 7), whereas the absorption with its

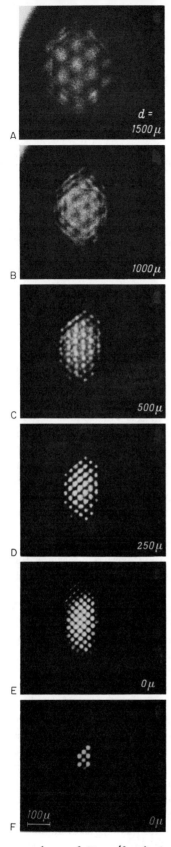

maximum in the green is related to rhabdomeres 1 to 6. It appears that only two photopigments are present in the eyes of flies and, consequently, that their color-vision system, if it exists, is dichromatic.

## Dichroic absorption properties

Langer's microspectrophotometric measurements on single rhabdomeres of the *Calliphora* eye (1965) revealed dichroic absorption effects, mainly in the spectral range of from 500 to 520 millimicrons, to linearly polarized light. As the dichroism disappears if the photopigment present in the rhabdomeres is bleached, the oriented arrangement of the chromophores of pigment molecules probably causes the observed effect.

Kirschfeld (1969) succeeded in demonstrating dichroic effects directly by microscopic inspection of *Musca* eye preparations. He found that maximal extinction in the different rhabdomeres (at 510 millimicrons for 1 to 6 and at 480 for 7) of an ommatidium takes place at different angular orientations of a linearly polarized light beam. The directions of polarization are, for rhabdomeres 1 to 6, oriented parallel to the axes of their microvilli, whereas, for rhabdomere 7, maximal extinction is found if the vector of polarization is oriented vertically to the microvilli axes. This finding is in accord with Langer's results (1965) for rhabdomeres 1 to 6 in the *Calliphora* eye and also with those obtained with different methods by Giulio (1963) and Eguchi and Waterman (1968).

Moody and Parriss (1961) have tried to explain the dichroic absorption properties of rhabdomeres on the assumption that the chromophores of the pigment molecules are arranged randomly within the microvilli membranes. If this model is adapted, more light is absorbed if the E-vector of the stimulating light is parallel and less if it is oriented perpendicularly to the long axes of the microvilli of the rhabdomeres. This agrees with the findings from rhabdomeres 1 to 6 in *Musca*. Furthermore, the model leads to a 0.5 ratio of the extinction coefficients for minimal and maximal absorption. Taking into account the experimental data of Burkhardt and Wendler (1960), Kuwabara and Naka (1959), and Scholes (1969), Kirschfeld (1969) determined this ratio for rhabdomeres 1 to 6 as about 0.43, a result that is, at least, not in conflict with the hypothesis suggested by Moody and Parriss (1961). The experimental results also lead to the conclusion that the concentration of photopigments should amount to about $10^4$ molecules per $\mu^2$ of microvillar membrane. A rhabdomere (1 to 6) 200 $\mu$ in length therefore contains some $2 \times 10^8$ pigment molecules, which absorb about 65 per cent of unpolarized light at a wavelength of 516 m$\mu$.

Although the Moody-Parriss model is not in conflict with the experimental findings from rhabdomeres 1 to 6 in

FIGURE 3 Compound eye of *Musca* (female, intact animal) inspected microscopically under antidromic illumination. The focal plane of the microscope was adjusted at the following distances from the eye surface. Distance: d = 1500$\mu$ (A); d = 1000$\mu$ (B); d = 500$\mu$ (C); d = 250$\mu$ (D); d = O$\mu$ (E). (F) shows the so-called pseudopupil at the level of the cornea. Aperture A of the microscope objective is 0.11 (u = 6.3°) in A to E, and 0.03 (u = 1.7°) in F. (From Kirschfeld and Franceschini, 1968.)

497

*Musca,* it fails to explain that extinction in rhabdomere 7 (*Musca*) is maximal if the E-vector of the stimulating light is oriented perpendicularly to the long axes of the microvilli. If it is assumed that dichroism of rhabdomeres is caused by the orientation of chromophores situated in parallel to the microvillar membranes, a higher degree of orientation than that assumed by the Moody-Parriss model should therefore be expected in central rhabdomere 7.

## Single-quantum effects

Light impinging on the photopigment molecules in the rhabdomeric structures of the compound eye elicits elementary photochemical reactions. These reactions are the first link in the chain of excitation processes in the receptor cells. We may ask whether a single quantum of light is sufficient, or the coincidence of several quanta is necessary, to trigger these elementary reactions (Reichardt, 1965; Reichardt et al., 1968).

In order to simplify the theoretical considerations, it is assumed here that a number of photopigment molecules, $N$, of a single rhabdomere are situated in an area $F$. Every molecule with its molecular cross section, $q$, takes up the fraction $q/F$ of $F$. If a light flux, $j$, of the number of quanta, $n$, per unit of time impinges on $F$, the pigment molecules may absorb quanta. If we make the general assumption that it takes $z$ or more quanta per time, $\tau$, to convert an inactive molecule into an activated one, the mean rate, $\bar{r}$, of photochemical reactions per rhabdomere and flux $j$ is then given by the expression

$$\bar{r}/j = N \times q[j \times (q \times \tau/F)]^{z-1}/z! \times F . \qquad (1)$$

In the case of a one-quantum process ($z = 1$), equation (1) reduces to $\bar{r}/j = (N \times q)/F$. The quantum yield does not depend on the flux, $j$. The molecular cross section, $q$, is a constant. In the case of a two-quantum reaction ($z = 2$), we have $\bar{r}/j = N \times Q/F$, in which $Q = j \times (q \times \tau/2F)$. The quantum yield depends linearly on $j$, which is to say that the effective molecular cross section, $Q$, changes proportionally with the quantum flux, $j$.

The following two light programs were selected (Reichardt, 1965) to test whether the optomotor response depends on a single-quantum or on a multiquantum reaction. Program 1 consisted of a light stimulus, the flux of which was constant in time and equal to $j_1$. In program 2, the stimulus consisted of a sequence of short light pulses: duration, $\Delta t$; peak amplitude, $j_2$; pulse frequency, $f = 1/T$; and average flux, $j_1$. The average number of quanta applied per time unit is evidently the same in both programs.

Both programs should evoke the same average rate of photochemical reactions when applied to a one-quantum receptor. If, however, these programs are applied to a multiquantum receptor, program 2 may trigger a higher photochemical reaction rate than does program 1. The maximum factor by which the average reaction rates may differ is easily derived for the two-quantum case. As $j_2$ is greater than $j_1$ by a factor $T/\Delta t$, the effective molecular cross section, $Q$, is enlarged by this factor. Consequently, under program 2 the two-quantum receptor is $T/\Delta t$ times more efficient than under program 1 and, conversely, the average rate of photochemical reactions is enlarged by the same factor. The effect predicted for the two-quantum receptor depends on the magnitudes of $\tau$, $\Delta t$, and $T$. One expects to find the predicted effect if $\tau < \Delta t$; if, however, $\Delta t < \tau < T$, the individual light pulse is integrated over the time interval, and therefore the expected increase in the average rate of photochemical reactions is smaller than $T/\Delta t$. The increase factor declines to one when $\tau \geq T$, because in this case two or more pulses are integrated in time and, consequently, the effect does not appear. Under these conditions, multiquantum receptors behave as one-quantum receptors do, and do not respond differently to the two different light programs.

It should be pointed out that our considerations are valid only if $j(q/F) \times \tau \ll 1$. This condition contains the requirements for extremely low quantum rates per receptor. Furthermore, the concentrations of the inactive photopigments in light should be essentially the same as those for experiments carried out in darkness. Only under these conditions can one expect that the kinetics of the photochemical reactions depend exclusively on the quantum rates.

The test experiments were carried out with females of *Musca.* The torque exerted by *Musca* during fixed flight was utilized as the quantitative measure of the optomotor reactions of the insect to light programs 1 and 2. A test fly was suspended by a torque compensater at the axis of a rotating cylinder that carried an illuminated periodic pattern of 45° in spatial wavelength. In Figure 4, a typical experimental result is presented.

These results are also typical for a number of experiments carried out with different light-pulse frequencies down to $10^{-1}$ Hz. The tests were made at a brightness level of $2 \times 10^{-2}$ Apostilb, or about 200 absorbed quanta per second in one of rhabdomeres 1 to 6. As the concentration of pigment molecules per rhabdomere amounts to about $2 \times 10^8$ for the quantum rates in question, an individual pigment molecule is hit, on the average, every $10^6$ seconds by a single quantum of light. If we assume that regeneration of the bleached pigment, as measured for rod and cone pigments by Rushton (1964), takes about 100 seconds, only 0.01 per cent of the pigment molecules in a rhabdomere are bleached. Therefore, it is unlikely that changes in the concentrations of the inactive photopigments influenced our experimental results and their interpretations.

The observations summarized in this section accord

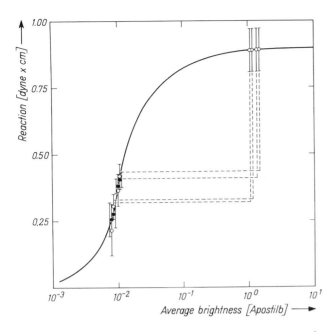

FIGURE 4  Test of single-quantum or multiquantum reaction in the eye of *Musca*. Solid curve: Optomotor reaction versus average brightness (given in Apostilb units; 1 Apostilb = $10^{-4}$ Lambert) of patterned cylinder. Illumination with light program 1. Spatial wavelength of patterned cylinder, $\lambda = 45°$; pattern contrast, m = 21 per cent; angular speed of pattern, w = 49.25°/sec. Response curve represents the averages obtained from five individual flies. These averages and standard errors are not shown in the Figure. Average standard error is $\pm 0.15$ dyne $\times$ cm. Closed circles and open circles: closed circles represent averages obtained from four individual flies with light program 1. The standard error, $\pm\sigma$, of each average, indicated by vertical lines, is calculated from five measurements taken from one individual fly. The open circles near the abscissa value $10^{-2}$ Apostilb represent response averages obtained from the same flies under light program 2. Pairs of closed and open circles at the *same* abscissa value were obtained from one individual fly. Light program 2 consisted of pulsed light with the same average quantum flux as in program 1. Pulse frequency, 500 c/sec; pulse duration, $1.5 \times 10^{-5}$ sec; peak intensity of pulse to average intensity, $j_2/j_1 = 133$. Open circles near abscissa value $10^0$ Apostilb demonstrate strengths of optomotor responses to be expected under light program 2 for a two-quantum receptor if condition $\tau \le 1.5 \times 10^{-5}$ sec is fulfilled. The abscissa values of these open circles are derived from the abscissa values of open circles actually measured and multiplied by the factor $j_2/j_1 = 133$. This factor determines the increase in effectiveness during program 2, because the effective molecular cross section of the two-quantum receptor changes by this factor. Horizontal broken lines indicate these shifts of open circles; vertical broken lines give significant limits for one- versus two-quantum receptors. Spatial wavelength of patterned cylinder, pattern contrast, and pattern speed is the same as described above. (From Reichardt, 1965.)

with the hypothesis that one single quantum of light is sufficient to elicit an elementary photochemical reaction, which, in turn, triggers an elementary photoreceptor response, detectable by the mechanism of movement perception.

## Signal-to-noise ratio

In the preceding section, I have shown that optomotor responses of flies can be elicited under low quantum rates. At the absolute response threshold, these rates may be as low as about one quantum absorbed per second and per rhabdomere. The Poisson statistics of the light quanta should therefore result in a significant level of noise in the light signals received by the photoreceptors.

We (Fermi and Reichardt, 1963) investigated the way in which the contrast required of a rotating stimulus pattern depends on the mean brightness, $\bar{I}$, of the pattern if a just-measurable response is elicited. A typical result of this dependence is plotted in Figure 5. Below quantum rates of about 200 per second, in one of the rhabdomeres 1 to 6, pattern contrast, m, and mean brightness, $\bar{I}$, are empirically related by

$$\log m + \tfrac{1}{2}\log \bar{I} = \text{const. or } m \sim \bar{I}^{-1/2}. \qquad (2)$$

One can easily calculate that this equation is consistent with the hypothesis that the noise resulting from the statistics of the light quanta absorbed by the photoreceptors is the principal cause of the breakdown of the optomotor response. During a fixed time interval, the mean number,

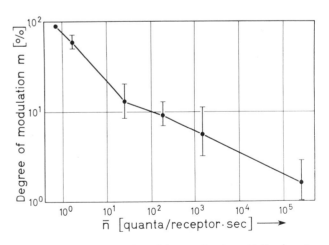

FIGURE 5  The degree of modulation of a sinusoidally changing light flux received by the receptors of the compound eyes of *Musca* as a function of the mean number, $\bar{n}$, of absorbed quanta per second, and receptor (1–6) at the optomotor response threshold. Dots represent means of individual measurements, and vertical lines are standard errors of the means. (From Eckert, 1970.)

ñ, of quanta absorbed by a photoreceptor is proportional to the mean brightness $\bar{I}$. The actual number of quanta absorbed will vary randomly in accord with a Poisson distribution with mean ñ. The fluctuations of n about its mean, ñ, represent a "quantum noise" that sets a limit to the brightness differences that can be distinguished by the receptor. A measure of the size of these fluctuations is given by the standard deviation of the distribution of n. Because n has a Poisson distribution, its standard deviation is $(\bar{n})^{\frac{1}{2}}$. It follows that if $\Delta \bar{n}$ is the additional mean number of quanta that are absorbed when the brightness is raised from $\bar{I}$ to $\bar{I} + \Delta \bar{I}$, then $\Delta \bar{n}$ must be greater than $(\bar{n})^{\frac{1}{2}}$ if the two brightnesses are distinguishable. Hence, the relationships $\Delta \bar{I} \sim \Delta \bar{n} \sim (n)^{\frac{1}{2}} \sim \bar{I}^{\frac{1}{2}}$ would hold at the threshold of brightness discrimination. This relation is equivalent to $\Delta \bar{I}/\bar{I} \sim m \sim \bar{I}^{-\frac{1}{2}}$, which represents the measured dependence of threshold contrast on background brightness.

The measured relationship is different from that which would be expected if the threshold were determined by the thermal noise in the photoreceptors, the neuronal noise in the nervous system, or both. If the breakdown of the reaction were determined by these "internal" noises, the signal-to-noise ratio should have a constant, critical value at the threshold of the reaction. The amplitude of the internal noise would, at low quantum rates, not be expected to depend on the light signals received by the photoreceptors, so the value of the signal-to-internal-noise ratio should depend only on the signal amplitude. The amplitude of the light signal received by each photoreceptor is proportional to the product $m\bar{I}$ of the contrast and mean brightness of the rotating stimulus pattern. Hence, the relationship $m\bar{I} =$ const. or, equivalently,

$$\log m + \log \bar{I} = \text{const.} \tag{3}$$

should hold at the threshold of the reaction. It may be seen in Figure 5 that this relationship would represent a poor fit to the measured dependence of threshold contrast on mean pattern brightness at very low quantum rates.

These considerations therefore indicate that internal noise obviously does not represent a principal factor determining the threshold of the optomotor response.

### *The representation of the optical environment in the nerve-cell layer of the first optical ganglion*

I have pointed out that the projection of the optical environment onto the retina is determined by the structure and by the geometry of the compound eye, as well as by the properties of the dioptric system. Light from a point source mounted in the optical environment is received by seven (eight) different rhabdomeres in adjoining ommatidia. The correspondence between individual points in the optical environment and the individual receptors of the retina is,

therefore, a one-to-seven and not a one-to-one relationship.

The projection from the retina on the first optical ganglion lamina merely depends on the nerve-fiber connections between these two levels (Vigier, 1907, 1908, 1909; Trujillo-Cenóz, 1965; Trujillo-Cenóz and Melamed, 1966; Braitenberg, 1967; Strausfeld, 1970a). The fibers leading from the sensory cells of individual ommatidia twist 180° and distribute themselves at the level of the lamina, which has a periodic structure of elements called optic cartridges (Ramón y Cajal and Sanchez, 1915) or neurommatidia (Villianes, 1892), arranged in a hexagonal array. The total number of cartridges matches that of the ommatidia. Every cartridge does not have two interneurons associated with it, as was once thought. Rather, it has four (Braitenberg and Strausfeld, 1970; Strausfeld, in preparation) and receives fibers from a group of six adjoining ommatidia—those that contain the six retinular cells of the group that collects optical information from one point of the environment only. Consequently, the projection between the retina and the lamina is based on a six-to-one, and not on a one-to-one, correspondence between retinular cells and cartridges. Hence, the representation of the optical environment in the nerve-cell layer of the lamina is such that an individual point of the environment is represented in an individual cartridge. Cells 7 and 8 in each ommatidium (the long visual fibers) project as pairs (Strausfeld and Blest, 1970) through the lamina and terminate in the medulla, after crossing the first optic chiasm.

Figure 6A and B are drawings from Strausfeld of the laminar region. Detailed explanations of these drawings are given in the figure captions. Most important in this connection are the findings of Braitenberg (1969), Strausfeld (1970b), and Braitenberg and Strausfeld (1970), who claim that there are at least two, and possibly four, types of lateral connections present in the lamina. Collaterals (Figure 6A) from one of the cartridge neurons (L4) extend vertically to two adjacent cartridges, where they possibly make synaptic contacts. Tangential fibers (Figure 6B), originating from cell bodies beneath the lamina, spread laterally over an area of about three to four cartridges in both vertical and horizontal planes. Information from these areas could be relayed together by single collaterals to the medulla (Strausfeld, 1970a).

An important consequence of these results derived from *Musca*, as well as other species of Diptera, is that an individual cartridge does not receive information from two distinct positions of the environment via receptor elements, a necessary condition for any mechanism that must perceive movements by the correlation of data. Nevertheless, the cartridges can no longer be excluded as the possible sites for movement detection, because lateral interconnections are present between neighboring cartridges. Further electrophysiological and behavioral work seems necessary before the functional roles of the synaptic links between neighboring cartridge interneurons can be explained fully.

At this point we may come back to the surprising finding I reported earlier in this chapter—that internal noise does not influence the threshold of the optomotor response. This result is understandable for the receptors, because their activation energies are about a factor 80 above the kT level, but it is not easy to conceive that the neuronal pathways for the next synaptic stages involved in the information transfer —as, for instance, the lamina—operate at the theoretical limit, particularly because nerve spikes, as information carriers, have not yet been recorded either from the receptor cells or from the laminar region. The question arises whether there are coincidence-anticoincidence devices present in the circuitry of the laminar region that may be responsible for a drastic increase in signal-to-noise ratio. The structural complexity of the lamina and its connection with the second synaptic region (the medulla) does not rule out such a possibility, but at present only a highly speculative statement could be put forward to support the suggestion.

## Evidence for representation from electrophysiological data

I have reported that nervous responses should be evoked in the "output" of a laminar cartridge by light from a single point in the environment, received by the associated group of six ommatidia.

Scholes (1969) succeeded in recording graded potentials to short flashes of light, provided by a point source, from a laminar component in the upper frontal quadrant of the eye of *Musca*. The time course of these graded laminar responses turned out to be very similar to those he recorded from retinular cells 1 to 6. The responses were also found when light rays were directed on a single ommatidium, rather than on the whole eye. This was achieved by shielding the corneal surface with a mask that contained a hole 10 to 15 $\mu$ in diameter. The mask was moved over the corneal surface, some 75–100 $\mu$ distant from it, and contour lines were drawn round the points on the corneal surface at which light was equally effective in eliciting a cartridge response. The result of the experiment, shown in Figure 7, shows that the contours correspond convincingly to the pattern of ommatidia that project onto the single cartridge. Sensitivity to illumination of rhabdomeres 7 and 8 is very low, a finding that accords with the histological observation that the long visual fibers pass the lamina without making synaptic contacts with monopolar interneurons.

Scholes (1969) was also able to show that the graded responses found at the laminar output to light stimulating the whole eye are the sum of the partial responses from the six individual receptors that contribute to the laminar response.

This result was to be expected, as the optomotor responses to pulsed light (see section on Single-Quantum Effects, above) did not reveal any coincidence of signals either at the receptor level or at those stages peripheral to the process of perception.

It should also be mentioned that angular sensitivity distributions, measured at laminar outputs, amounted to about the same half-width as in distributions measured from retinular cells. This result suggests that the optical axes of the associated group of six rhabdomeres are nicely aligned in parallel.

## Movement perception

The localization of the movement-perception system in the insect optical ganglia, the central brain, or both, is still an unsolved problem. An analysis of the system, carried out on the beetle *Chlorophanus* and the flies *Drosophila* and *Musca*, was originally based on input and output response relations (Hassenstein, 1951; Reichardt, 1957; Hassenstein, 1958; Reichardt and Varjú, 1959; Varjú, 1959; Reichardt, 1961; Götz, 1964; McCann and McGinitie, 1965; Varjú and Reichardt, 1967). More recently, electrophysiological investigations have been undertaken with the aim of localizing the movement perception system in the higher optical ganglia (McCann et al., 1966; Bishop and Keehn, 1967; Bishop et al., 1968).

Behavioral studies have led to the formulation of a mathematical model by means of which responses to given pattern stimuli can be predicted (Reichardt, 1957, 1961). An essential feature of the model is that it must take into account only the transformations of signals from adjacent sensory inputs and their interactions during the process of perception. This is because the strength of the optomotor response elicited by the stimulation of the compound eyes is the sum of all the partial responses evoked by the interactions of data from neighboring sensory inputs. This finding can be expressed by saying that, with respect to movement perception, the compound eye is isotropic in function.

The model of movement perception envisages two cross-connected information input channels (A and B) and a common output channel to the motor system (see Reichardt, 1961). The inputs A and B represent two adjacent cartridge outputs, fiber terminals, or both, that lead from sensory cells 7 and 8; A and B obviously act as input elements that provide the minimum detector requirement for the optomotor response. The two inputs transform the space and time coordinates of the stimulus into the time functions $L_A$ and $L_B$. The time functions can be considered to be sums of an average light flux, C, and of a fluctuating light flux, G(t), the average value of which is zero. That is

$$L_A = C + G(t), \quad L_B = C + G(t - \Delta t), \qquad (4)$$

A

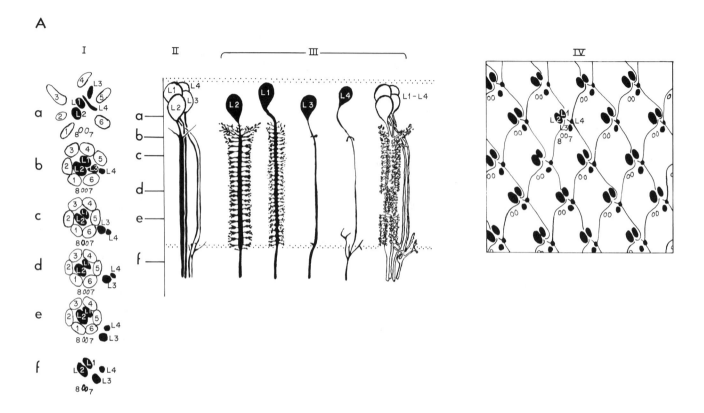

FIGURE 6.   A. Centripetal interneurons of optic cartridges (ending in the second synaptic region, the medulla). I. Cross-sections of an optic cartridge at six levels (a to f) through the lamina. Open circles represent receptor elements. Closed circles represent interneuron axis fibers. Retinular-cell endings (1–6) form a crown around two axial monopolar cells (L1 and L2). Long visual fibers (7 and 8) remain outside the crown. The midget monopolar cell (L3) lies within the crown at level b. Sometimes two side branches can be seen at levels b and e, pointing back toward the crown. The tripartite cell (L4) runs parallel to L3. II. Reduced silver stains show the outline of the relative positions of L1, L2, L3 and L4 of each optic cartridge. III. Silhouettes of monopolar cells as seen in Golgi preparations. L2 has two distinct sets of processes, one of which, at levels a and b, could extend as far as the processes of L2 in the two cartridges adjacent to it. However, their synaptic relationships are, as yet, unknown. The processes of L2 deeper in the lamina at levels c, d, and e are arranged radially from the axis fiber and extend between the retinular cells of a crown (Straus-feld, 1970b). L1 has a single set of processes, also disposed radially from the axis fiber, which are restricted to within the crown of retinular-cell endings. Characteristically, L3 has a kinked appearance and usually has two tiny unilateral side branches. L4, like L3, is invariably situated posterior to the crown of the retinular cell. It has one or two side branches at level b, and three branches at level f. The shortest of these extends into the adjacent crown, anteriorly. The other two extend dorsally and ventrally in a characteristic pattern, as shown in IV. L1–L4, in III, shows a reconstruction of the geometrical relationships between the four monopolar cells. L2 has been partly cut away to reveal the narrower spread of L1. IV. A schematic drawing, derived from reduced silver preparations, showing a portion of the lamina, cut tangentially, at level f. The exceedingly regular network of the L4 side branches could provide lateral interaction between triplets of optic cartridges.

B. The periodic arrangements of elements in the lamina. *Abbreviations:* R, retinular cell ending; LV, long visual fibers (7 and 8 in Figure 6AI); F, fenestration layer of the lamina (in which retinular cell decussation is localized); P, external plexiform layer (synaptic layer) of the lamina; 10C, the outer 15–20 mµ of the first optic chiasma. I. Crowns of retinular cells (indicated in horizontal section) are repetitively arranged in the lamina, as are the pairs of long visual fibers that project through the lamina and end, finally, in the medulla. II. The repetitive arrangement of four monopolar cells (L1–L4). Tangential processes (T) are regularly arranged, with respect to cartridges, there being a system of ascendent branches spaced one to about every five cartridges. Subsequent processes invest the outer face of the plexiform layer and invade it between crowns of the retinular cell. These tangentials have connections to the medulla (Strausfeld, 1970a) and could possibly relay information from groups of cartridges to the second synaptic region, as well as provide an anatomical basis for intercartridge interaction in the lamina. III. The repetitive arrangement of morphologically "centrifugal" endings derived from elements in the medulla. Of the three forms, one (be) invests the outside of a crown of retinular-cell endings and the other two (c) are situated between the long, visual-fiber pair and the L3 and L4 pair outside the crown. All three types cross over the first optic chiasma parallel to the fibers of centripetal elements from the lamina to the medulla. Lateral interaction could also be mediated by amacrine cells (Ramón y Cajal and Sanchez, 1915), which may have a periodic arrangement in the lamina. They are not included in this diagram. (Diagrams are from Strausfeld, 1970a, and Strausfeld, in preparation. The orientation of elements in AI is from Braitenberg, 1970.)

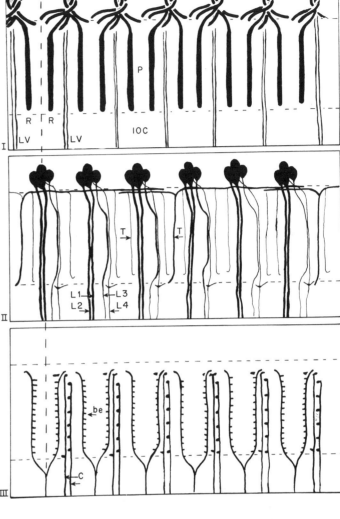

B

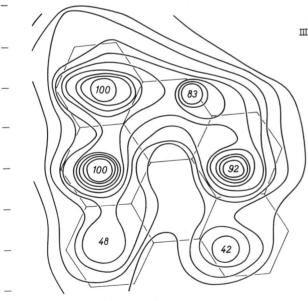

FIGURE 7 A contour map showing the distribution of sensitivity over the surface of the cornea in the lamina of a *Musca* compound eye. The sampling points, each of which was tested two to four times during the experiment, are indicated by the intersections of the bars in the margins. The dimensions of the facets in the eye regions concerned, measured after the experiments, are sketched behind the contour lines. The experiment was made on the upper frontal quadrant. (From Scholes, 1969.)

in which t is the time and $\Delta t$ the time interval between the reception of a stimulus by A and its reception by B.

The model envisages further that the time functions $L_A$ and $L_B$ are transformed by a succession of directly connected and cross-connected components that carry out linear transformations (filters D, F, and H), multiplication, and time averaging.

The strength of the response, R, predicted by the model, to a moving contrasted pattern is given by the expression

$$R = \int_0^{+\infty} W_{DF}(\eta) \int_0^{+\infty} W_{DH}(\xi)\Phi_{GG}(\eta - \xi - \Delta t)d\xi d\eta$$

$$- \int_0^{+\infty} W_{DH}(\eta) \int_0^{+\infty} W_{DF}(\xi)\Phi_{GG}(\eta - \xi - \Delta t)d\xi d\eta , \quad (5)$$

in which $W_{DF}$ and $W_{DH}$ describe the transformations in the directly connected and cross-connected channels in terms of filter-weighting functions.

$$\Phi_{GG}(\eta - \xi - \Delta t) = \frac{1}{2T} \int_{-T}^{+T} G(t - \eta)G(t - \xi - \Delta t)dt$$

$$\lim T \to \infty \quad (6)$$

designates the autocorrelation function of the light flux fluctuation G(t) and $\eta$ and $\xi$ integration variables.

The response, R, depends on two parameters of the stimulus: the speed $w \sim 1/\Delta t$ and the structure of the moving pattern, which determines the autocorrelation function $\Phi_{GG}$. The special character of the response also depends on the transformations of the components D, F, and H, which reflect functional properties of the movement-perception system.

Most important in this connection is the dependence of R on $\Phi_{GG}$ and not on G(t) itself. If G(t) is decomposed into Fourier components, any changes in the phase relations of these components do not influence R, as they belong to the class of transformations to which $\Phi_{GG}$ and, consequently, R, are invariant. This prediction of the theory was tested experimentally and confirmed.

We may conclude from these considerations that in the process of movement perception, as in any perceptual process, sensory information is destroyed—in the present case, in order to abstract motion from the displacement of individual points in the optical surround relative to the eye of the insect.

## Determination of resolution and contrast transmission

The fundamental optical properties of the fly's compound eyes, expressed by $\Delta\varphi$ and $\Delta\rho$, can be determined by optomotor response measurements. Before turning to the experimental results, we shall derive from equation (5), which describes the input-output relations of the model for movement perception, how the strength of reaction R depends on the parameters $\Delta\varphi$ and $\Delta\rho$ under elementary stimulus conditions. Such conditions are realized by a test pattern, the contrast of which changes sinusoidally. If one designates $\lambda$ the spatial angular wavelength of the pattern, m the contrast (expressed by the degree of modulation), and w the angular velocity of the pattern, it can be calculated from equation (5) that R is given by the expression

$$R \sim m^2 \times f\left(\frac{w}{\lambda}\right) \times \sin\left(\frac{2\pi \times \Delta\varphi}{\lambda}\right). \quad (7)$$

Apparently R depends on three terms: (1) the square of the pattern contrast m; (2) the function $f(w/\lambda)$, which reflects transfer properties, and (3) the so-called interference term $\sin(2\pi \times \Delta\varphi/\lambda)$, which contains $\lambda$ and $\Delta\varphi$, but not w.

So far we have established only the predicted dependence of R on $\Delta\varphi$, under the special stimulus conditions selected here, but not yet the dependence on $\Delta\rho$. The reason is that, in setting up the mathematical model for movement perception, we have tacitly assumed that the model inputs A and B receive their information from only two distinct points in the optical surround but not from areas of finite size. On this mathematical assumption, the pattern contrast, even for extremely small spatial wavelengths ($\lambda$), is transmitted to A and B without loss. In reality, however, such is not the case, as the size of the visual fields of the receptors is finite. Consequently, for decreasing spatial wavelengths of the stimulating pattern, the contrast received by the rhabdomeres is diminishing. As has been stated before, no distinct boundaries of these visual fields exist, but a region within the light sensitivity decreases continuously, somewhat in accordance with a Gaussian distribution function with half-width $\Delta\rho$.

When a Gaussian-like, visual-field distribution is taken into account, and $m_1$ is designated as the contrast of the sinusoidal pattern and $m_2$ as the contrast transmitted to the receptor level (Fermi and Reichardt, 1963; Götz, 1964, 1965), $m_1$ and $m_2$ are related by the expression

$$m_2 = m_1 \times h(\lambda; \Delta\rho) , \quad (8)$$

with the contrast transfer function $h(\lambda; \Delta\rho) = \exp[-(\pi^2/4\ln2) \times (\Delta\rho/\lambda)^2]$ and $\Delta\rho \ll \pi$. Replacing m in equation (7) by $m_2$ from equation (8), for the strength of reaction R under the given stimulus conditions, we arrive at the expression

$$R \sim m_1{}^2 \times h^2(\lambda; \Delta\rho) \times f\left(\frac{w}{\lambda}\right) \times \sin\left(\frac{2\pi \times \Delta\rho}{\lambda}\right), \quad (9)$$

which consequently depends on both parameters $\Delta\varphi$ and $\Delta\rho$.

Experiments on *Drosophila* and *Musca*, carried out with sinusoidally changing contrast patterns, have shown that, if $\lambda$ is kept constant but w—and, therefore, also $w/\lambda$—are changed, the response R throughout the whole range of w does not change sign. Consequently, $f(w/\lambda)$ does not change sign under a variation of $w/\lambda$. From equation (9) it follows that the sign and, therefore, the direction of the optomotor response, depend exclusively on the interference term

$$\sin\left(2\pi \times \Delta\varphi/\lambda\right).$$

Within the wavelength region $2\Delta\varphi \le \lambda < +\infty$, the moving sinusoidal pattern is resolved, as the number of samples received per period $\lambda$ at any time is greater than or equals two. If, however, $\lambda$ enters the region $0 < \lambda < 2\Delta\varphi$, optical resolution of the periodic pattern breaks down, because fewer than two samples per wavelength are received under this condition, and, as a result, the information content of the pattern is transmitted incompletely. The resolution limit of the model with $\Delta\varphi$ spacing between the adjacent inputs is therefore determined by $\lambda = 2\Delta\varphi$, the first zero crossing of the interference term $\sin(2\pi \times \Delta\varphi/\lambda)$.

These theoretical considerations constitute the necessary background for our understanding experimental results from *Drosophila* (Götz, 1964, 1965) and *Musca* (Eckert, 1970) undertaken with the aim of determining the parameters $\Delta\varphi$ and $\Delta\rho$.

The strength of optomotor reaction was measured as a function of $\lambda$, whereas $w/\lambda$ and consequently $f(w/\lambda)$ were kept constant. In these experiments, the contrast of the sinusoidal patterns amounted to $m_1 = 1$. The first zero crossing of the response was found for $\lambda_0 = 9.2°$ in *Drosophila* and for $\lambda_0 = 4.0°$ in *Musca*. From these results one derives an average divergence angle of $\overline{\Delta\varphi} = 4.6°$ in *Drosophila* and $\overline{\Delta\varphi} = 2.0°$ in *Musca*. The angles coincide with values measured between the axes of adjacent ommatidia in horizontal cross sections of eye preparations (Braitenberg, 1967).

As an important consequence of these behavioral tests, we can now state that optical resolution is determined by the divergence angle between the optical axes of adjacent ommatidia, and not by the smaller angles between the optical axes of neighboring rhabdomeres. This result is in full agreement with the established one-to-seven correspondence between an individual point in the optical environment and its representation in seven different rhabdomeres in seven different ommatidia.

If patterns with spatial wave lengths, $\lambda$, below the optical resolution limit are applied, the strength of the optomotor reaction decreases, which is the result of the increasing contrast reduction by the sensitivity distributions, $h(\lambda; \Delta\rho)$, of the receptors. From a pair of experiments, in which $f(w/\lambda)$ was kept constant but $\lambda$ and $m_1$ were varied, the values of $\overline{\Delta\rho}$ were determined. It turned out that $\overline{\Delta\rho}$ is about 3.5° for *Drosophila* and 1.7° for *Musca*.

The determination of the two parameters $\Delta\varphi$ and $\Delta\rho$ lead to the question of whether their values are related to each other in such a way that they form an optimal pair. That is to say, the optical resolution limit of the eye is, as we have seen, determined by the divergence angle $\Delta\varphi$ between the optical axes of adjacent ommatidia. The smaller $\Delta\varphi$ is, the better the resolution and, therefore, the finer the details of the optical surround that can be resolved by the eye. A corresponding statement holds for $\Delta\rho$. The smaller $\Delta\rho$ is, the better the contrast transmission from the optical surround to the receptor and the cartridge layer of the eye. From a purely theoretical point of view, the best resolution and contrast transfer could be reached if $(\Delta\varphi; \Delta\rho) \to 0$ and, consequently, the number of ommatidia per eye equals $N \to +\infty$. If these limits are approached, the light flux $\Phi$ into an individual ommatidium or a receptor would also approach zero as $\Phi \sim (\Delta\varphi \times \Delta\rho)^2$. Therefore, the signal-to-quantum noise ratio would reach intolerable values and the optical information received by the receptors of the eye would be destroyed, so that the advantages of perfect resolution and contrast transfer would be lost. Hence, the product $\Delta\varphi \times \Delta\rho$ must exceed a certain limit, a condition that does not determine the numerical value of the ratio $\Delta\rho/\Delta\varphi$. Again, from a theoretical point of view, it can be seen easily that, for a fixed value of the product $\Delta\varphi \times \Delta\rho$, there must exist a ratio $\Delta\rho/\Delta\varphi$, which leads to an optimal combination of both resolution and contrast transfer. For the sake of argument, let us assume that $\Delta\varphi$ is large and consequently $\Delta\rho$ is small. Under these conditions, contrast transfer would markedly exceed the limit of optical resolution. If, on the other hand, $\Delta\varphi$ is small and $\Delta\rho$ is large, optical resolution could not be utilized to its limit, as contrast transfer at this limit would be very poor. A mathematical treatment of this problem by Götz (1965) has led to the result that under the condition $\Delta\varphi \times \Delta\rho = \text{const.}$, optimal resolution *and* contrast transfer are established if $\Delta\rho/\Delta\varphi = 0.88$. The experimental values derived from *Drosophila* and *Musca* are 0.76 and 0.85. The value of the optical parameters $\Delta\varphi$ and $\Delta\rho$ of the compound eyes of *Drosophila* and *Musca* therefore nearly meet the requirements for optimal resolution and contrast transmission.

### Visual detection and fixation of objects

In the past, the analysis of movement perception in insects has been carried out by measuring the torque response of fixed flying insects to the movements of patterns. In these experiments, the pattern velocities relative to the test fly were always determined and kept under control by the experimenter ("open loop" conditions).

Investigations of the behavior of flies (*Musca*) were recently undertaken by Reichardt and Wenking (1969) with the aim of analyzing the processes of visual detection and fixation of objects. For this purpose, a servo-system was developed that enables a test fly to operate under "closed-loop" conditions; that is to say, a fly controls the angular velocity of its surround by its own torque response. The principal components of the mechano-electronic device are given in Figure 8.

If an object (a single, black, vertically oriented stripe) is mounted on the inner surface of the cylinder, no significant indication of any reaction by the fly to the object, irrespective of its angular position, is found in the average torque response under open-loop conditions. If, however, the servo-system driving the cylinder is coupled to the torque response (closed-loop condition) of the fly, the object is moved into the direction of flight, where it reaches a stable fixation position.

In Figure 9, a typical experimental result is presented, giving the probability of object positioning versus the angular position of the object during the phase of object fixation for different coupling-in conditions. At present,

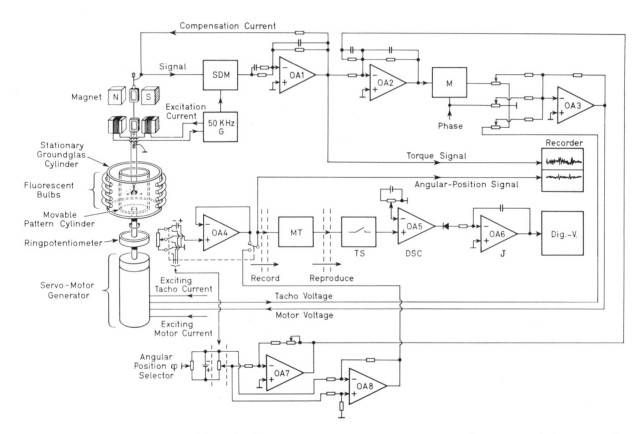

FIGURE 8 Mechano-electronical device for the quantitative study of detection, fixation, and discrimination of visual objects. The essential components of the device consist of a torque compensator with linear response characteristics and a cut-off frequency near 1 kHz; a servomotor with a driving shaft that carries a patterned ground-glass cylinder and a ring potentiometer that signals the angular position of the patterned cylinder; an electronic device that controls the motor speed and direction by the torque signal of the fixed flying test fly suspended from the torque compensator. The cut-off frequency of the control device is limited to 50 Hz. The instrument can be used in two different modes of operation: 1. Free running mode. A constant torque signal generates a constant angular *velocity* of the patterned cylinder. The direction of rotation is determined by the direction of the torque. 2. Stabilized mode. The pattern center is elastically stabilized at a selected angular position, $\varphi$. In this mode, a constant torque signal generates a constant angular *displacement* of the patterned cylinder. The direction of the displacement is determined by the direction of the torque.

*Abbreviations:* Dig. V., digital voltmeter; DSC, discriminator; G, generator; J, integrator; M, signal multiplier; MT, magnetic-tape recorder; OA1–OA8, operational amplifiers; SDM, signal demodulator; TS, time switch.

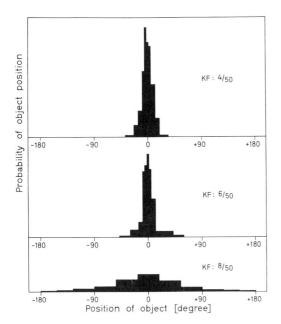

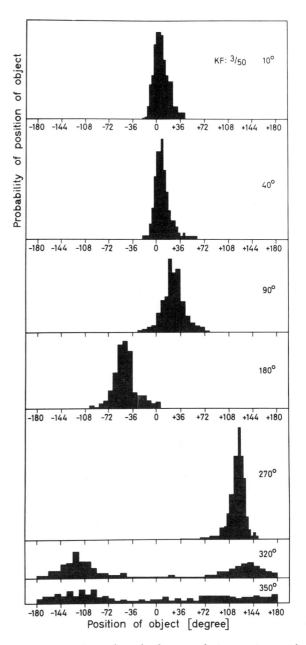

FIGURE 9 Experimental results obtained from the instrument and described as mode 1 in legend of Figure 8. The object consists of a single, black, vertically oriented, stripe 6° in width. The Figure gives the probability of object positioning versus the angular position of the object during the phase of object fixation for different coupling-in conditions, KF. Calibration: 1 dyn × cm torque signal generates 3.4 revolutions per second of the patterned cylinder for KF = 2/10. Fixation of object is very pronounced for KF = 4/50 and less pronounced under the condition of stronger coupling, KF = 8/50. Direction of flight: 0 degree. (Fly was fixated and pointed in a direction for which the angular coordinate was set at zero.) The data for an individual distribution were taken during a time interval of 180 seconds. Number of data per distribution is about 6000.

the correct simulation of the natural coupling-in conditions is not known; consequently it is an open question whether the instability during fixation at high coupling-in values is introduced by the fly, by the motor drive system, or by both.

In another sequence of experiments, presented in Figure 10, the angular width of the black stripe was varied. When the stripe is as much as about 40° in angular width, the flies fixate at its center; between 60° and 320° they fixate at a portion of the stripe near the edges. When a 350° black stripe is presented, the fly is confronted with the reversed-contrast condition with respect to the 10° stripe experiment shown in the Figure. That is to say, a black-on-white pattern is replaced by a white-on-black pattern, both 10° in angular width. Whereas the black-on-white pattern is clearly fixated, the white-on-black pattern essentially is not, which follows from the nearly flat probability distribution in Figure 10.

FIGURE 10 Experimental results from mode-1 operation, with single, black, vertically oriented stripes of different angular widths (10°, 40°, 90°, 180°, 270°, 320°, and 350°). Coupling-in condition is kept constant and amounts to KF = 3/50. The probability of object positioning is plotted versus the angular position of the object during the phase of object fixation. For stripe widths up to 40°, the test fly fixates the center of the stripe. For increasing stripe widths, up to 320°, the test fly fixates a portion of the object near one of the two edges. At a stripe width of 350° fixation disappears. Data for an individual distribution were taken during 180 seconds.

The experimental findings reported so far suggest that the resulting movement of the object, during the phase of object transfer into the fixation position, results from an asymmetry in the strength of the response of the fly to progressive and regressive motions of the object relative to its compound eyes. This suggestion was tested and confirmed with the aid of an additional feedback loop inserted into the control device (see Figure 8), by means of which the object is elastically stabilized at a selected angular position. If, under these conditions, the torque response of the test fly is coupled to the cylinder-control device (closed-loop condition), the fly generates a probability distribution of object positioning, the maximum of which is displaced to an angular position situated between the direction of flight and the stabilized open-loop position.

Figure 11 shows that the angular displacement, which is

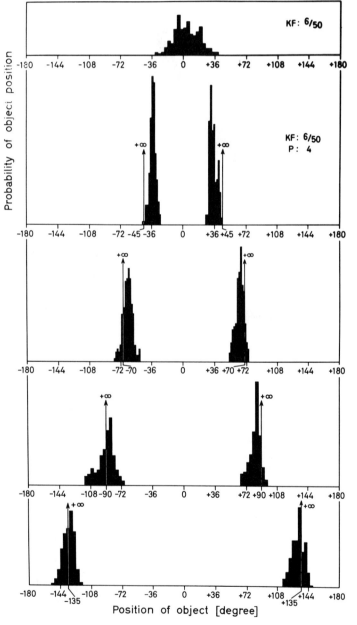

strongest in the frontal part of the two compound eyes, reflects the asymmetry of the response and indicates, for every position, the resulting average torque response that is responsible for the observed phenomenon of object transfer in the direction of flight. The upper distribution in Figure 11 represents a control experiment of the same test fly under normal (nonstabilized) fixation conditions.

The method described seems to be appropriate and will be applied, in both modes of operations, to the analysis of pattern discrimination by insects.

## Summary

During the past years, the visual system of insects, particularly of flies, has been used increasingly as a model for an analysis of uptake, transduction, and processing of data in the central nervous system. The total number of neurons in the eye of a fly is about $5 \times 10^5$. These neurons are composed of definable classes of elements that are replicated many times in an orderly way. Electrophysiological microprobing of the retina and the ganglia is possible and has been applied with the aim of tracing the information flux at different levels. Flies possess a rich repertoire of behavior. *Drosophila*, especially, has been investigated genetically in great detail.

The account I present here concerns the application of three methods—histological, behavioral, and electrophysiological—to the analysis of structure and function of the optical system of flies (*Drosophila* and *Musca*).

The discussion of the optical properties summarizes the way in which the optical environment is represented in the photoreceptor layer of the compound eye. An individual point in the optical surround is imaged onto seven different receptor structures (rhabdomeres) in seven different ommatidia. Therefore, the correspondence between individual

FIGURE 11 Experimental results from mode-1 and mode-2 operations (see Figure 8). The object consists of a single black, vertically oriented stripe 6° in angular width. The coupling-in condition throughout the various experiments amounts to KF = 6/50. The probability of object positioning is plotted versus the angular position of the object during the phase of object fixation. The upper distribution was recorded in mode-1 operation (Figure 8). The other distributions were recorded in mode-2 operation. The object was here elastically stabilized at the following angular positions: ±45°, ±70°, ±90°, and ±135°. Calibration: KF = 6/50 in connection with an elastic stabilization of strength P = 4 results in a 12° angular displacement of the object for 1 dyn × cm torque signal. Each distribution generated by the test fly was recorded during 180 seconds. The distributions are shifted toward the direction of flight. The shift is most pronounced at ±45° and decreases with the increase in angular position of the stabilized object.

points in the surround and the receptor mosaic is one to seven, and not one to one. Consequently, the limit of optical resolution of a compound eye is by a factor seven below the limit one would expect from the spatial separation of rhabdomeres.

Two parameters determine the optical properties of a compound eye: (1) the divergence angle, $\Delta\varphi$, between the optical axes of adjacent ommatidia which sets the limit of optical resolution; and (2) the angular sensitivity distributions of individual rhabdomeres, expressed by $\Delta\rho$, the half-width of the distribution. These distributions determine contrast transmission from the optical environment onto the receptor layer. Their quantitative values have been determined and are reported in the section on Determination of Resolution and Contrast Transmission. The product $\Delta\varphi \times \Delta\rho$ determines the quantum flux into an individual rhabdomere, whereas the degree of adjustment between optical resolution and contrast transfer is expressed by the ratio $\Delta\rho/\Delta\varphi$. Also in the section mentioned, it is shown that $\Delta\rho$ and $\Delta\varphi$ form an optimal pair, that is to say, optical resolution meets the requirements of optical contrast transfer.

Two different spectral sensitivities of rhabdomeres are described. It seems that only two photopigments are present in the eyes of flies. The arrangement of the chromophores of the pigment molecules, situated in the rhabdomeres, is such that a strong dichroic absorption effect is observed.

I emphasize the problem of whether an elementary photochemical reaction in the photopigments can be elicited by a single quantum of light or whether this process requires the coincidence of absorption of two or more quanta. The evidence is based on quantitative behavioral data, which accord with the hypothesis that a single quantum of light is sufficient to elicit an elementary photochemical reaction. The reaction, in turn, triggers an elementary photoreceptor response.

If weak optomotor responses are elicited in flies, with very dim illumination, pattern contrast must be set at a high level. Behavioral experiments show that the optomotor response threshold is invariant under a change of pattern contrast, m, and average pattern brightness, $\bar{I}$, if the product $m\bar{I}^{\frac{1}{2}}$ is kept constant. This finding suggests that the breakdown of the response is caused by the fluctuation of quanta (quantum noise) and not by an internal noise.

I devote the section entitled Movement Perception to summaries of former experiments that led to the formulation of a mathematical theory of the optomotor response. The theory states that, in the formation of movement perception, optical data are processed in the central nervous system in accordance with a first-order correlation process. The important consequence of this finding is that 50 per cent of the information present in the optical environment is selectively destroyed in the process of movement perception. During the last few years, considerable anatomical and histological work has been done in an attempt to localize correlation processes in the structures of the optical ganglia. These investigations have not been successful so far, but it has been shown that the synaptic junctions that connect the fibers leading from the sensory cells in the retina with the interneurons of the first optical ganglion (lamina) are not responsible for correlating optical data. The evidence is discussed in the sections preceding the one on Movement Perception. The interneurons of the lamina are lumped together in identical groups (cartridges), equal in number to those of the ommatidia. Each group receives synaptic inputs from sensory fibers, which in turn collect optical information from a single point in the optical environment. This finding rules out the possibility of correlating data from two distinct points of the optical environment, a necessary condition for the perception of motion.

Recently, optomotor-response investigations have been extended toward a quantitative analysis of visual detection, fixation, and discrimination of patterns. Some of these experiments are reported. Under fixed flight conditions, flies respond only if an object or pattern is moved relative to its eyes. Progressive motion of an object (progressive with respect to one of the two eyes) results in an optomotor response stronger than regressive motion. This asymmetry of the response is the cause for pattern fixation in the overlap region of the visual fields of the two compound eyes. The center of a vertically oriented black stripe on a white background is fixated if the angular width of the stripe is no greater than about 40°. For a stripe wider than 40°, the test flies fixate portions of the black stripe near the edges. Fixation seems to be limited to black stripes. A white stripe on a black background is not fixated if the stripe width is less than about 10°. The experimental results reported so far suggest that flies possess a nervous mechanism for pattern discrimination.

## REFERENCES

AUTRUM, H., and J. WIEDEMANN, 1962. Versuche über den Strahlengang im Insektenauge (Appositionsauge). *Z. Naturforsch., B,* 17: 480–482.

BISHOP, L. G., and D. G. KEEHN, 1967. Neural correlates of the optomotor response in the fly. *Kybernetik* 3: 288–295.

BISHOP, L. G., D. G. KEEHN, and G. D. McCANN, 1968. Motion detection by interneurons of optic lobes and brain of the flies *Calliphora phaenicia* and *Musca domestica. J. Neurophysiol.* 31: 509–525.

BRAITENBERG, V., 1967. Patterns of projection in the visual system of the fly. I. Retina-lamina projections. *Exp. Brain Res.* 3: 271–298.

BRAITENBERG, V., 1969. The anatomical substratum of visual perception in flies. A sketch of the visual ganglia. *In* Processing of Optical Data by Organisms and by Machines Proceedings of the International School of Physics "Enrico Fermi," Course 43

(W. Reichardt, editor). Academic Press, New York and London, pp. 328–340.

BRAITENBERG, V., and N. J. STRAUSFELD, 1970. The elementary mosaic organisation of the visual system's neuropil in *Musca domestica* (L). *In* Handbook of Sensory Physiology, Vol. II. Springer-Verlag, Heidelberg, Germany (in press).

BURKHARDT, D., and L. WENDLER, 1960. Ein direkter Beweis für die Fähigkeit einzelner Sehzellen des Insektenauges, die Schwingungsrichtung polarisierten Lichtes zu analysieren. *Z. Vergl. Physiol.* 43: 687–692.

ECKERT, H., 1970. Verhaltensphysiologische Untersuchungen am visuellen System der Stubenfliege Musca domestica. Max-Planck-Institut für biologische Kybernetik. Doctoral thesis.

EGUCHI, E., and T. H. WATERMAN, 1968. Cellular basis orf polarized light perception in the spider crab, *Libinia*. *Z. Zellforsch. Mikroskop. Anat.* 84: 87–101.

FERMI, G., and W. REICHARDT, 1963. Optomotorische Reaktionen der Fliege *Musca domestica*. *Kybernetik* 2: 15–28.

GIULIO, L., 1963. Elektroretinographische Beweisführung dichroitischer Eigenschaften des Komplexauges bei Zweiflüglern. *Z. Vergl. Physiol.* 46: 491–495.

GÖTZ, K. G., 1964. Optomotorische Untersuchung des visuellen Systems einiger Augenmutanten der Fruchtfliege *Drosophila*. *Kybernetik* 2: 77–92.

GÖTZ, K. G., 1965. Die optischen Übertragungseigenschaften der Komplexaugen von *Drosophila*. *Kybernetik* 2: 215–221.

HASSENSTEIN, B., 1951. Ommatidienraster und afferente Bewegungsintegration. *Z. Vergl. Physiol.* 33: 301–326.

HASSENSTEIN, B., 1958. Die Stärke von optokinetischen Reaktionen auf verschiedene Mustergeschwindigkeiten. *Z. Naturforsch., B,* 13: 1–6.

KIRSCHFELD, K., 1967. Die Projektion der optischen Umwelt auf das Raster der Rhabdomere im Komplexauge von Musca. *Exp. Brain Res.* 3: 248–270.

KIRSCHFELD, K., 1969. Absorption properties of photopigments in single rods, cones and rhabdomeres. *In* Processing of Optical Data by Organisms and by Machines (W. Reichardt, editor). Proceedings of the International School of Physics "Enrico Fermi," Course 43. Academic Press, New York and London, pp. 116–136.

KIRSCHFELD, K., and N. FRANCESCHINI, 1968. Optische Eigenschaften der Ommatidien im Komplexauge von *Musca*. *Kybernetik* 5: 47–52.

KIRSCHFELD, K., and N. FRANCESCHINI, 1970. Etude optique in vivo des éléments photorecepteurs dans l'oeil composé de Drosophila. *Kybernetik* (in press).

KUWABARA, M., and K. NAKA, 1959. Response of a single retinula cell to polarized light. *Nature* (*London*) 184: 455–456.

LANGER, H., 1965. Nachweis dichroitischer Absorption des Sehfarbstoffes in den Rhabdomeren des Insektenauges. *Z. Vergl. Physiol.* 51: 258–263.

LANGER, H., 1966. Spektrometrische Untersuchung der Absorptionseigenschaften einzelner Rhabdomere im Facettenauge. *Zool. Anz.* 29 (Suppl.): 329–338.

LANGER, H., 1967. Grundlagen der Wahrnehmung von Wellenlänge und Schwingungsebene des Lichtes. *Zool. Anz.* 30 (Suppl.): 195–233.

LANGER, H., and B. THORELL, 1966. Microspectrophotometry of single rhabdomeres in the insect eye. *Exp. Cell Res.* 41: 673–677.

MCCANN, G. D., and G. F. MACGINITIE, 1965. Optomotor response studies of insect vision. *Proc. Roy. Soc., ser. B, biol. sci.* 163: 369–401.

MCCANN, G. D., Y. SASAKI, and M. C. BIEDEBACH, 1966. Correlated studies of insect visual nervous systems. Proceedings of the International Symposium on Functional Organization of the Compound Eye. Pergamon Press, Oxford, pp. 559–583.

MOODY, M. F., and J. R. PARRISS, 1961. The discrimination of polarized light by *Octopus*: A behavioural and morphological study. *Z. Vergl. Physiol.* 44: 268–291.

RAMÓN Y CAJAL, S., and D. SANCHEZ, 1915. Contribución al conocimiento de los centros nerviosos de los insectos. *Trab. Lab. Invest. Biol.* (*Univ. Madrid*) 13: 1–164.

REICHARDT, W., 1957. Autokorrelations-Auswertung als Funktionsprinzip des Zentralnervensystems. *Z. Naturforsch., B,* 12: 448–457.

REICHARDT, W., 1961. Autocorrelation, a principle for the evaluation of sensory information by the central nervous system. *In* Sensory Communication (W. A. Rosenblith, editor). M.I.T. Press, Cambridge, Massachusetts, pp. 303–317.

REICHARDT, W., 1965. Quantum sensitivity of light receptors in the compound eye of the fly *Musca*. *Cold Spring Harbor Symp. Quant. Biol.* 30: 505–515.

REICHARDT, W., V. BRAITENBERG, and G. WEIDEL, 1968. Auslösung von Elementarprozessen durch einzelne Lichtquanten im Fliegenauge. *Kybernetik* 5: 148–169.

REICHARDT, W., and D. VARJÚ, 1959. Übertragungseigenschaften im Auswertesystem für das Bewegungssehen. *Z. Naturforsch., B,* 14: 674–689.

REICHARDT, W., and H. WENKING, 1969. Optical detection and fixation of objects by fixed flying flies. (*Musca domestica*). *Naturwissenschaften* 56: 424–425.

RUSHTON, W. A. H., 1964. Flash photolysis in human cones. *Photochem. Photobiol.* 3: 561–577.

SCHOLES, J., 1969. The electrical responses of the retinal receptors and the lamina in the visual system of the fly *Musca*. *Kybernetik* 6: 149–162.

STRAUSFELD, N. J., 1970a. Golgi studies on insects. I. The optic lobes of Diptera. *Phil. Trans. Roy. Soc.* (*London*), *ser. B,* 258: 135–223.

STRAUSFELD, N. J., 1970b. Cell types in the lamina of Diptera. Monopolar Cells. Zellforsch Medizin. Anat. (in press).

STRAUSFELD, N. J., and A. D. BLEST, 1970. Golgi studies on insects. II. The optic lobes of Lepidoptera. *Phil. Trans. Roy. Soc.* (*London*), *ser. B,* 258: 81–134.

TRUJILLO-CENÓZ, O., 1965. Some aspects of the structural organization of the intermediate retina of dipterans. *J. Ultrastruct. Res.* 13: 1–33.

TRUJILLO-CENÓZ, O., and J. MELAMED, 1966. Compound eye of dipterans: Anatomical basis for integration—An electron microscope study. *J. Ultrastruct. Res.* 16: 395–398.

VARJÚ, D., 1959. Optomotorische Reaktionen auf die Bewegung periodischer Helligkeitsmuster. *Z. Naturforsch., B,* 14: 724–735.

VARJÚ, D., and W. REICHARDT, 1967. Übertragungseigenschaften im Auswertesystem für das Bewegungssehen II. *Z. Naturforsch.*, B, 22: 1343–1351.

VIGIER, P., 1907. Sur les terminaisons photoréceptrices dans les yeux composés des *Muscides. Compt. Rend. Acad. Sci.* 145: 532–536.

VIGIER, P., 1908. Sur l'existence réelle et le rôle des appendices piriformes des neurones. Le neurone périoptique des Diptères. *Compt. Rend. Soc. Biol.* 64: 959–961.

VIGIER, P., 1909. Mécanisme de la synthèse des impressions lumineuses recueilliés par les yeux composés des Diptères. *Compt. Rend. Acad. Sci.* 148: 1221–1223.

VILLIANES, J., 1892. Contribution a l'histologie du système nerveux des invertebrés. La lame ganglionaire de la langouste. *Ann. Sci. Nat.*, 7e *Serie*, 13: 385–398.

WIEDEMANN, J., 1965. Versuche über den Strahlengang im Insektenauge (Appositionsauge). *Z. Vergl. Physiol.* 49: 526–542.

# 46 Olfactory Receptors for the Sexual Attractant (Bombykol) of the Silk Moth

## DIETRICH SCHNEIDER

### Chemoreception and olfaction

RESPONSES to environmental influences or stimuli are fundamental phenomena of living matter. Because organisms are complex chemical systems, it is obvious that chemical stimulus-response functions are found on all levels of organization, from the microbes to the highest metazoans such as insects and vertebrates. In the latter groups we find well-defined chemoreceptor systems: olfaction and taste (Schneider, 1969). These two are distinct modalities; one finds anatomically different input channels responding to certain groups of (adequate) chemical stimuli.

Olfactory organs are highly developed detector systems which respond with short latencies to small numbers of stimulating molecules. They also analyze gas mixtures quite efficiently, as we know from our sense of smell. Our present problem is the detection of one, and not the discrimination of many, different odorous molecules. I describe a specialized odor receptor which transforms a chemical signal of great biological importance into a nervous message. It is necessary to analyze the catching and transport of the odor molecules in the outer structures of the receptor and, subsequently, to look for ways to approach the transducer function, which is the transformation of the stimulus into the bioelectrical response. Finally, I discuss the possible fate of

DIETRICH SCHNEIDER Max-Planck-Institut für Verhaltensphysiologie, Seewiesen, Germany

the signal molecule after it has delivered its message, and of other such molecules which did not reach their destination but were adsorbed on inexcitable structures of the animal.

In our laboratory we have studied intensively the activity of insect olfactory receptors in general and sexual attractant-receptor responses in particular, because insects offer a number of opportunities for the experimental approach to olfaction:

1. The receptor system is situated on an accessible extremity, the antenna.
2. The primary sensory (receptor) cells belong to the epidermal cell layer and, with their dendrites, are in intimate contact with specialized cuticle differentiations.
3. All the individual receptor-cell axons run directly to the deutocerebrum without axon fusion. Receptor interaction by collaterals and synapses, or by ephapses, is unknown (Steinbrecht, 1969a).
4. Insects not only are able to differentiate among many odors (as do vertebrates), but they also have large numbers of highly specialized receptors for biologically important odors (unknown in vertebrates).
5. Odor receptors for such key substances (as, for instance, the excretory messenger substances, or pheromones) are maximally sensitive and may comprise as much as 50 per cent of the olfactory cells of the animal.
6. The chemical structure of some pheromones has been analyzed, which facilitates the quantitative physiological analysis.
7. Electrophysiological recordings can be made with single

receptor cells and whole sets of receptor organs (antennae).

8. Receptor responses can be compared with the behavior of the animal.

## The attractant and the response of the antennal receptors

The striking lure effect of female moths on their mates was known to biologists of the nineteenth century (see Schneider, 1957). At the turn of the century, it was found that the female's lure emanates only from the abdominal epidermal lure glands, but the functional principle of the attraction phenomenon was doubted by some workers in the field. The main reason for these doubts was that none of the glandular products had been isolated and, consequently, had not been tested as a potential olfactory stimulus. Furthermore, the luring stimulus seemed, in some cases, to carry for such long distances that it could not be explained by an odorous message transfer. Last, but not least, the situation was and still is complicated for the experimenter because most of the sexual attractants that have thus far been identified are not odorous to human beings.

The first successful isolation of a sexual-lure pheromone of a moth was accomplished by Butenandt and his coworkers (1959). These investigators chose the domesticated silk moth, *Bombyx mori,* as their experimental animal, because large numbers of glands are available and bioassays are comparatively easy. After many years of chemical work it was found that the lure principle of the female silk moth is an alcohol, which was named "bombykol" (Butenandt et al., 1959):

$$CH_3 - (CH_2)_2 - \overset{H}{\underset{H}{C}} = \overset{H}{C} - C = \overset{H}{C} - (CH_2)_8 - CH_2OH$$

Hexadeca-10-trans, 12-cis-dien-1-ol = bombykol

This substance, as well as its geometrical isomers, was subsequently synthesized and became a useful tool for the physiologist (Butenandt and Hecker, 1961; Butenandt et al., 1962; Hecker, 1960). Recently bombykol has again been synthesized and labeled with tritium (Kasang, 1968). Since the early 1950s we have been working with the *Bombyx* lure system, using behavioral, electrophysiological, and histological techniques. Because the subject has previously been reviewed (Schneider, 1957, 1963a, 1963b, 1966, 1969; Boeckh et al., 1965), a brief summary of the status of published results will suffice here. Then I shall report in more detail the most recent advances we have made in studies of the ultrastructure of the receptor, the absolute threshold with the single receptor cell, the behavior of the animal, and finally the biochemistry of bombykol after it is adsorbed on the antenna.

Isolated male antennae, or antennae of mechanically immobilized male moths, respond with a slow electrical potential (electroantennogram = EAG) when stimulated with a puff of air containing bombykol. The stimulus normally induces a temporary negativity of the electrode situated at the distal end of the antenna. The EAG is assumed to be composed of a large number of receptor potentials of simultaneously stimulated bombykol receptors, which are arranged in the manner of serial electrical batteries.

Female antennae of *Bombyx,* or of any other moth so far tested, do not respond to bombykol or the lure gland. Female and male moth antennae, however, show EAGs when stimulated with other odorants.

Stimulus-response characteristics of the antennal receptor system were determined by a series of odor sources of rising strength. We loaded pieces of filter paper with different amounts of bombykol and put them in short glass tubes (cartridges). EAGs were elicited by blowing air through the tubes onto the antenna. The EAG-threshold concentration of bombykol administered in this manner was between $10^{-3}$ and $10^{-2}$ $\mu$g. The response curve rises in proportion to the logarithm of bombykol concentration on the paper, and the functional range of the whole antenna covers a concentration ratio of $1:10^5$ or more (Schneider and Hecker, 1956; Schneider, 1957, 1962, 1963a, 1963b, 1966; Boeckh et al., 1965; Schneider et al., 1967; Schneider, 1967, 1969; Kaissling, 1969). Filter papers containing 100 $\mu$g or more of bombykol are "overloaded" because of their limited surface. These strong stimuli induce a marked adaptation of the responding system. Amplitudes of EAGs elicited with fresh female glands are equal to EAG responses of between 1 and 10 $\mu$g elicited by bombykol on filter paper (Steinbrecht, 1964a).

Stimulus-response curves to the bombykol isomers (10-cis, 12-trans; 10-cis, 12-cis; 10-trans, 12-trans) are shifted to odor concentrations between 100 and 1000 times higher. In addition, these curves level off at submaximal amplitudes, thus showing "saturation" with strong stimuli (Schneider et al., 1967).

Earlier estimates of the absolute bombykol threshold led to values of approximately $10^3$ and $10^7$ bombykol molecules per cm³ of air for the behavioral and EAG responses, respectively (Boeckh et al., 1965; Schneider, 1967). The accuracy of our estimates was improved by the availability of tritiated bombykol and was later calculated as $10^4$ molecules per cm³ of air with 50 per cent of the animals responding (Schneider et al., 1968).

The sensory organs that are responsible for the reception of bombykol are long olfactory hairs (sensilla trichodea), according to microelectrode recordings (E. Priesner, unpublished; see also Boeckh et al., 1965; Schneider and Steinbrecht, 1968; Kaissling and Priesner, 1970). This finding corresponds to our observation with the wild silk moth,

*Antheraea pernyi*, in which sensilla of this morphological type also responded to the attractant of the female (Schneider et al., 1964).

## Structure of the Bombykol receptor

In earlier publications we described the histological structures of the *Bombyx* antenna as seen with the light microscope (Schneider and Kaissling, 1956, 1957, 1959). After the sensilla trichodea were identified as the bombykol receptor organs (Priesner, unpublished), their ultrastructure was studied with the electron microscope (Schneider and Steinbrecht, 1968; Steinbrecht, unpublished). The general scheme of such a sensillum is shown in Figure 1. Sensilla

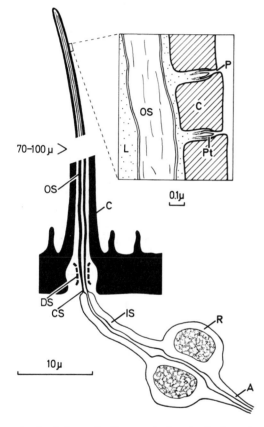

FIGURE 1 Semischematic diagram in longitudinal section of a sensillium trichodeum of *Bombyx mori*. Only the base and the tip of the thick-walled sense hair are shown (approximately 25 per cent of total hair length). The hair is innervated by two sensory nerve cells which respond to the sexual attractant bombykol. Accessory cells and epidermis cells are omitted. The hair lumen is filled with a fluid, the "sensillum liquor," which bathes the dendrites. Inset: Enlarged area of the hair wall as seen with the electron microscope. *Abbreviations:* A, axon; C, cuticle; CS, ciliary zone of the outer dendrite; DS, extracellular sheath of the dendrites; IS, inner segment of the dendrite; L, sensillum liquor; OS, outer segment of the dendrite; P, pore; Pt, pore tubule; R, soma of the receptor cell. (Courtesy of Dr. R. A. Steinbrecht.)

trichodea are olfactory insect sensilla of the pore-tubule type. The cuticle of these organs is perforated by a complex system of outer pores and inner tubules which reach into the lumen of the hair. In the best-known sensillum of this type (the sensillum basiconicum of the carrion beetle: Ernst, 1969) it was shown that the pores and tubules are open to the outside, and that the inner end of the tubules is clogged by some electron-dense material. Despite claims of several investigators, these hollow structures are certainly not "neurofilaments" or fine protrusions of the dendrite that permit a direct cell contact with the air (see Ernst, 1969).

The dendrites of the sensillum trichodeum receptor cells are found in the lumen of the hair and are bathed in a liquid, the "sensillum liquor" (Ernst, 1969). Whether the tubules of the hair cuticle reach all the way to the cell membrane of the dendrite or end somewhere in the liquor is still an open question (Schneider and Steinbrecht, 1968; Steinbrecht, 1969b). Bombykol receptor hairs are between 45 and 140 $\mu$ in length, are innervated by one or two sensory nerve cells, and have 1000–3000 pores between 100 Å and 150 Å in diameter, with as many as 20,000 tubules approximately 200 Å in width. The receptor cell is of the ciliary type common to many insect receptors. The distal parts of the dendrites contain only small vesicles and neurotubules and contain no mitochondria or other cell organelles.

On the *Bombyx* antenna we found an average of 16,000 sensilla trichodea, which are supplied by one receptor cell or, at the most, two, amounting to a maximum of 32,000 bombykol receptor cells in one antenna, or 64,000 in both (Steinbrecht, unpublished). Eighty per cent, or 25,000, of these cells in each antenna are of the sensitive type, as described below (Kaissling and Priesner, 1970).

Counts of the 0.2-$\mu$-diameter axons in the antennal nerve revealed that each axon is connected directly with the deutocerebrum. Axon fusion, which was assumed to be the rule in afferent insect nerves, was not apparent in the *Bombyx* antenna and may be rare or nonexistent in insects (Steinbrecht, 1969a).

Preliminary histological inspection of the *Bombyx* deutocerebrum indicates that the number of neurons that are potentially the "secondary olfactory elements" is at least 100 times smaller than the number of odor receptor cells. This, together with the occurrence of glomeruli, which seem to be areas where the telodendrion of the receptor axons is in synaptic contact with the secondary cells, is strikingly analogous to the bulbus olfactorius of vertebrates.

## Bombykol diffusion on and in the sensillum

The structural peculiarities of insect olfactory sensilla suggest that the stimulating molecules do not, as other investigators have thought (Slifer, 1961; Boeckh et al., 1965), pass directly from the gas phase to the extended receptor surface. Instead, we now assume that the odor molecules are first adsorbed on the hair cuticle and thereafter diffuse through the pores and tubules into the lumen of the hair (Schneider and Steinbrecht, 1968; Adam and Delbrück, 1968; Ernst, 1969). Possibly they then continue to diffuse through the liquor before they eventually make contact with the dendritic membrane to start transduction by interaction with a hypothetical acceptor molecule. In principle, this system is analogous to the vertebrate nose (air passage, mucus, receptor hairs) and many other animal receptors, for instance, the vertebrate ear (pinna, outer canal, tympanum, middle ear, fluid, hair cells). In all these we are dealing with stimulus-conducting systems with potential filter qualities which may distort, select, and retard the stimulus before it reaches the transductor, the receptor cell.

This model of the olfactory sensillum is still hypothetical to a large extent, but for some of the critical steps we have direct or indicative evidence, which stems mainly from our studies with radioactive bombykol.

*Adsorption* on the cuticle of the hair is apparent when $^3$H-bombykol is blown on the antenna, where it is adsorbed readily and also quantitatively under physiological conditions. These measurements prove the correctness of earlier physical calculations (Adam and Delbrück, 1968) which predicted that the antenna, with its hairs arranged strategically, could filter the pheromone quantitatively out of the air stream. The assumption is inescapable that large parts of the hair surface are a catching area, because the threshold must be 1000 times higher if only the molecules that hit the pores directly are effective (Kaissling, 1969). The direct evidence of adsorption and the low threshold make obsolete our earlier considerations on a possible pore-hitting chance after molecule reflection or adsorption (see Boeckh et al., 1965).

*Molecule transfer* The next question is: How are the adsorbed molecules transferred to the dendrite? One possibility is a direct penetration of the cuticle, analogous to processes in the bombykol-producing gland of the female moth, which does not have pores, tubules, or any kind of canals (Steinbrecht, 1964b). A much more plausible assumption is that the bombykol molecules make use of the pore-tubule "gate-system" to enter the hair lumen. The structural geometry of this system strongly suggests such a process, which is also physically possible with a diffusion time which is short in relation to the reaction time of the cell (Adam and Delbrück, 1968; Kaissling and Priesner, 1970). Some of our experiments also indicate that a molecule-transfer system exists on the hair, the capacity of which may be a critical limiting factor. EAGs elicited by strong odor stimuli show long-lasting aftereffects which depend on the absolute stimulus strength and not on the stimulatory ef-

ficiency of a molecule, possibly owing to an overloaded sensillum surface, which slowly empties the adsorbed molecules through the tubules to the dendrite (Kaissling, 1969). So far, we still lack direct evidence that bombykol enters the pores and tubules, let alone that it diffuses through the liquor to the dendrite.

Another question is the *location of the acceptor,* which is generally supposed to be the sensitive and selective receiving structure, designed to "read" the message of the odor molecule (Kaissling, 1969). Again, the likely structure is the receptor membrane, the dendrite, which is, in all probability, also the locus of transduction. But we have no direct information on either the nature of the acceptor or its location. Could this critical deciphering system be situated somewhere between the hair cuticle and the membrane of the dendrite? Such a position is possible, if we assume, for instance, that the tubules are selective filters that are fully open for bombykol, partially open for the isomers, and closed for any molecule that is not eliciting a cell response. In that case, it would not be necessary for the membrane to be specialized for bombykol, but the membrane should be able to respond equally well to the isomers and some other molecules, because the tubules have already accomplished the selection. This somewhat provocative displacement of the specificity (away from the receptor cell membrane to an extracellular structure) is difficult to understand in view of the differentiating processes during morphogenesis, because the tubules are not the product of the receptor cell. Extracellular critical selection is also hardly understandable with all those insect olfactory sensilla, in which two or more cells with their dendrites share a common liquor space and have different reaction spectra for odorants (Schneider and Boeckh, 1962; Schneider, 1963a; Schneider et al., 1964; Lacher, 1964, 1967; Kaissling and Renner, 1968).

Finally, as with most other receptors, transduction in any olfactory cell is not understood. As with visual, auditory, and vibration receptors of extreme sensitivity, we must explain the function of a highly triggered system. A single odor molecule or very few can apparently start an explosive spreading of a permeability change or, in other words, induce the *receptor-generator potential.*

## Threshold for the behavior and single-cell response

We have recently conducted new large-scale behavior tests and numerous single-cell recordings to improve the accuracy of our earlier estimates of the minimum number of molecules necessary to elicit a response (Kaissling and Priesner, 1970). Now, the prerequisite for an evaluation of these new experiments was the availability of radioactive bombykol. This compound was resynthesized, with one out of four molecules labeled with tritium in one of the cis-positions (Kasang, 1968). We have now been able to measure the output of the bombykol sources, the cartridges, and to determine the amount of bombykol adsorbed on the antenna and on the hairs (Figure 2). As in our earlier work

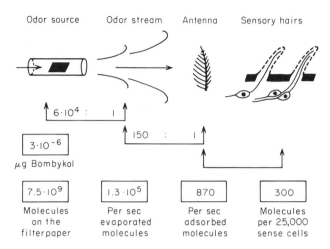

FIGURE 2   Radiometric determination of molecule numbers on the antenna of *Bombyx*. Only the number of molecules on the filter paper ($3 \times 10^{-6}$ μg bombykol) can be measured directly at the behavior threshold. All other values are extrapolations from measurements at higher intensities. See text for further explanations. (Courtesy of Dr. K. E. Kaissling.)

(Boeckh et al., 1965; Schneider et al., 1967), a direct measurement of the bombykol output was not possible with threshold stimuli, because the sensitivity of the receptor system of the animal is much greater than the sensitivity of the scintillation counter. The minimum amount of bombykol detected by the counter was the 10-second output of a cartridge holding $10^{-2}$ μg. A significant number of animals (18 per cent beyond a control response of 5 per cent, with $P = 0.002$), however, reacted with wing fluttering if stimulated for less than one second by an air stream passing through a cartridge holding $3 \times 10^{-6}$ μg of bombykol, which can be called the behavior threshold (Figure 3, left curve).

For the extrapolation from the measured output down to the threshold output, we could rely on the observation that bombykol sources from 100 μg down to $10^{-2}$ μg always gave outputs that were proportional to the load, and assume such a proportionality to be true also with the even smaller charges. With the extrapolation in time, we had to account for an initial output peak that bombykol sources show during the first 100 milliseconds after the onset of the stimulus.

After correcting for all these factors, including the molecules adsorbing on insensitive antennal structures, we found

that, during a threshold stimulus, 300 bombykol molecules are available for the 25,000 receptor cells of one antenna. To learn the chances for hits of one, two, or more molecules on any of the cells, we applied the Poisson statistics. The result is that only two of the cells encounter two hits, and 300 encounter single hits (Kaissling and Priesner, 1970).

These findings have been decisively supplemented by single-cell recordings. The resting signal frequency of the receptor cells is low but not negligible: 25,000 cells send approximately 1600 impulses per second to the brain. One can predict that a minimum of 120 additional "bombykol-impulses" would be necessary to overcome this background noise.

What is the actual signal threshold of the single cells? Figure 3 illustrates the answer: with a $10^{-5}$ μg bombykol cartridge, few cells react with one impulse; a stimulus 10 times higher elicits one or more impulses in 30 per cent of the cells; two or more impulses in only a few cells; and so on. Because of the many measurements that led to these values, it can be said that the one- and two-impulse percentages are significantly positioned on Poisson curves that have been drawn through these points. Interestingly, the 10-impulse curve does not correspond to the 10-hit curve.

The close correspondence of the theoretical Poisson

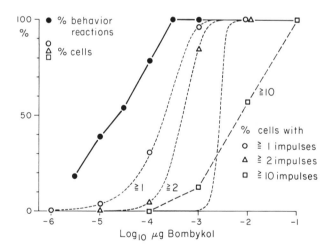

FIGURE 3   *Bombyx mori* male. Behavior reactions and nerve impulses of single receptor cells. Stimulus: bombykol in micrograms on the odor source (glass cartridge with filter paper, see Figure 2). Behavior reactions (left curve) counted during two seconds. Eighteen per cent of the animals (exceeding a control reaction by 5 per cent) react at $3 \times 10^{-6}$ μg of bombykol ($P = 0.002$). The percentage of cells firing $\geqq 1$ and $\geqq 2$ impulses (hatched curves) fits the theoretical expectation of the corresponding Poisson distribution. The 10-impulse curves do not fit. Each impulse value is the mean of approximately 800 measurements. See text for additional explanations. (Courtesy of Drs. K. E. Kaissling and E. Priesner.)

curves with the increase of the number of responding receptor cells means that the average rate of impulses is, at these low stimulus intensities, strictly random and proportional to the stimulus strength. Impulse measurements necessarily failed to show a significant signal increase at the behavior threshold of $3 \times 10^{-6}$ $\mu$g of bombykol, because of the extremely reduced hit chance for a given sensillum from which one is recording. The precise correspondence of the impulse curves with the Poisson curves allows, however, an extrapolation to the range of the behavior threshold. Here we find that 200 impulses are elicited per 25,000 cells, a figure that, in the range of experimental error, is identical with the value of 300 cells as hit by one molecule each at the behavior threshold. This latter value stems from $^3$H-bombykol measurements, which also led to the conclusion that only two of all the cells of one antenna received two molecule hits each under these conditions. This then also means that only two of the cells gave two impulses each.

Bombykol adsorption in relation to structural geometry, cell number, behavior- and impulse-responses led to the conclusion that *single* bombykol *molecules elicit* at low stimulus intensities *single impulses* in the receptor cells. In order to reach the same response in 200 cells with double molecule hits, one would need a stimulus concentration 10 times higher. Therefore we may say that the most sensitive type of bombykol receptor cells "counts" the molecules at these low stimulus intensities.

## Transduction and the fate of the bombykol molecule

According to neurophysiological knowledge, transduction in a receptor is understood as the transformation of the stimulus energy into the excitatory state of the cell. To analyze this process or the chain of processes in the case of the bombykol receptor, we need detailed information on the following:

1. The ionic composition of the dendritic cytoplasm and of the sensillum liquor.
2. The molecular composition and architecture of the membrane elements that maintain the membrane potential.
3. The chemical nature of the acceptor molecules that presumably interact with the bombykol molecules.
4. The mode of interaction between the acceptor and the bombykol.
5. The principle of the trigger which works on a quantal basis at low stimulus intensities: one molecule elicits one impulse, two elicit two impulses.
6. The fate of the bombykol molecule before and after the interaction.

Our extracellular electrophysiological recordings led to

the assumption that the receptor-generator potential is of dendritic origin, because the microelectrode which is placed at the hair base shows a relative negativity with excitatory stimuli. The polarity of the simultaneously recorded impulses, however, is usually positive, indicating that the impulses are a depolarization of the axon membrane near the basement membrane where the reference electrode is placed in the hemolymph (see Schneider et al., 1964; Boeckh et al., 1965). For a detailed insight into the effects of one, two, and more molecule hits, recording intracellularly is desirable, but at present this is hardly possible because of the small size of the cells. For the same reason, the chemistry of the dendrite and its extracellular and intracellular media is still unknown.

Recently Kaissling (1969) gave an interesting formalistic approach to the transducer function of the bombykol receptor. His calculations are based on the amplitudes and time constants of the rise and decay of the EAG, which is assumed to be a sufficiently reliable image of the receptor potential of the bombykol-transducing sensillum. In principle, the bombykol "bonding," which is thought to be responsible for the EAG, could be either an adsorptive or an enzymatic interaction. The calculation leads to an estimate of the number of bombykol acceptors, which is given as $10^8$ per antenna. Surprisingly, this figure is almost equal to the total number of tubules in all the bombykol receptor hairs. So far, we do not understand the meaning of this correspondence, but it indicates that the tubules may be responsible for the kinetics of the slow potentials.

The whole receptor system, with its huge branched antenna and long receptor hairs, is obviously designed for an optimal molecule-catching capacity. In contrast, the number of bombykol acceptors seems to be uneconomically low. If one speculates that the dendritic membranes are composed of medium-sized protein molecules, which are also the acceptors, one antenna would have between $10^{10}$ and $10^{12}$ acceptors. One way to overcome this dilemma would be to assume that the tubules lead the molecules directly to the acceptor loci on the membrane, thus permitting the molecule to hit the right acceptor without delay. Unfortunately, micromorphological studies of these structures are extremely difficult, and one cannot as yet determine whether the tubules end directly on the membrane or somewhere in the liquor (see Figure 1).

Recently we have looked for the bombykol after exposing many antennae to tritiated bombykol for different periods of time (Kasang, unpublished). So far, the results indicate that bombykol is first adsorbed on the antennal surface. After several minutes, the bulk of the radioactivity has traveled into the antenna, whence it can now be eluated with strong organic solvents (chloroform-methanol). Interestingly, however, the radiochromatograms now show three fractions. The alcohol fraction, which presumably still is bombykol, is progressively metabolized into acid and ester.

There is no direct indication as yet that this process is related to transduction, because it is observed with male and female antennae and also occurs on other parts of the insect body (but not on cotton fibers).

This comparatively slow bombykol degradation, however, may very well be an important biological mechanism. The pheromone, aside from the question of whether it is used in transduction or only adsorbed on inexcitable structures, must become inactivated to prevent subsequent, and erroneous, receptor stimulation that may come from a later bombykol desorption. This consideration would, of course, also apply to the female. Female moths have been reported to be attractive only when their lure glands are expanded (Brady and Smithwick, 1968; Götz, 1951; Jacobson, 1965; Shorey and Gaston, 1967). Metabolic pheromone degradation would explain this phenomenon. Any attractant that also contaminates the scaly body of the calling female is progressively destroyed.

## Summary

The sex-attractant (bombykol) system of the silk moth, *Bombyx mori,* consists of: (1) the female pheromone-emanating glands; (2) the transport medium (air); and (3) the receptor organs on the male antennae. The dendrite of the olfactory receptor cell is in a fluid-filled hair, the cuticular wall of which is penetrated by pores and microtubules. Bombykol molecules are adsorbed on the hair surface and presumably diffuse to and through the pores and tubules into the fluid, where they hit the dendritic membrane and elicit the cell response. Behavioral and electrophysiological threshold studies reveal that the receptor cell responds to single-molecule hits, whereas the animal needs the activation of approximately 200 cells in each antenna to start its motor response. The close correspondence of the impulses as elicited in the receptor cell, with the Poisson distribution of molecule hits, indicates that one molecule elicits one impulse and two molecules elicit two impulses. The 10-molecule hit curve, however, does not fit the 10-impulse curve. The mechanism of transduction is still unknown but can be treated in terms of a mathematical formalism which is based on the electrical responses. Biochemical studies indicate that bombykol is progressively, but only partially, metabolized into acid and ester on male and female antennae and on other parts of the body of the animals. Any connection between this process and the transducer function is so far unknown. The possible biological meaning of the bombykol degradation is discussed.

## Acknowledgments

The observations, calculations, and speculations that are presented in this review are the results of the closely correlated teamwork of a group of scientists and technicians. The goal of our work is to analyze insect olfaction in general and bombykol reception in particular. Recently the latter complex of problems has been worked on successfully by Drs. K. E. Kaissling (physiology), G. Kasang (chemistry), E. Priesner (physiology), and R. A. Steinbrecht (histology). We are all indebted to our technical staff members (last but not least) for their skillful assistance.

### REFERENCES

ADAM, G., and M. DELBRÜCK, 1968. Reduction of dimensionality in biological diffusion processes. *In* Structural Chemistry and Molecular Biology (A. Rich and N. Davidson, editors). W. H. Freeman and Co., San Francisco and London, pp. 198–215.

BOECKH, J., K.-E. KAISSLING, and D. SCHNEIDER, 1965. Insect olfactory receptors. *Cold Spring Harbor Symp. Quant. Biol.* 30: 263–280.

BRADY, U. E., and E. B. SMITHWICK, 1968. Production and release of sex attractant by the female Indian meal moth, *Plodia interpunctella. Ann. Entomol. Soc. Amer.* 61: 1260–1265.

BUTENANDT, A., R. BECKMANN, D. STAMM, and E. HECKER, 1959. Über den Sexual-Lockstoff des Seidenspinners *Bombyx mori.* Reindarstellung und Konstitution. *Z. Naturforsch. B* 14: 283–284.

BUTENANDT, A., and E. HECKER, 1961. Synthese des Bombykols, des Sexual-Lockstoffes des Seidenspinners, und seiner geometrischen Isomeren. *Angew. Chem.* 73: 349–353.

BUTENANDT, A., E. HECKER, M. HOPP, and W. KOCH, 1962. Über den Sexuallockstoff des Seidenspinners, IV. Die Synthese des Bombykols und der *cis-trans*-Isomeren Hexadecadien-(10.12)-ole-(1). *Justus Liebigs Ann. Chem.* 658: 39–64.

ERNST, K.-D., 1969. Die Feinstruktur von Riechsensillen auf der Antenne des Aaskäfers *Necrophorus* (Coleoptera). *Z. Zellforsch. Mikroskop. Anat.* 94: 72–102.

GÖTZ, B., 1951. Die Sexualduftstoffe an Lepidopteren. *Experientia* (*Basel*) 7: 406–418.

HECKER, E., 1960. Chemie und Biochemie des Sexuallockstoffes des Seidenspinners (*Bombyx mori* L.). *XI Int. Kong. Entomol.* (*Vienna*), B 3: 69–72.

JACOBSON, M., 1965. Insect Attractants. Interscience Publishers, John Wiley and Sons, New York.

KAISSLING, K.-E., 1969. Kinetics of olfactory receptor potentials. *In* Olfaction and Taste III (C. Pfaffmann, editor). The Rockefeller University Press, New York, pp. 52–70.

KAISSLING, K.-E., and E. PRIESNER, 1970. Die Riechschwelle des Seidenspinners. *Naturwissenschaften* 57: 23–28.

KAISSLING, K.-E., and M. RENNER, 1968. Antennale Rezeptoren für Queen Substance und Sterzelduft bei der Honigbiene. *Z. Vergl. Physiol.* 59: 357–361.

KASANG, G., 1968. Tritium—Markierung des Sexuallockstoffes Bombykol. *Z. Naturforsch., B,* 23: 1331-1335.

LACHER, V., 1964. Elektrophysiologische Untersuchungen an einzelnen Rezeptoren für Geruch, Kohlendioxyd, Luftfeuchtigkeit und Temperatur auf den Antennen der Arbeitsbiene und der Drohne (*Apis mellifica* L.). *Z. Vergl. Physiol.* 48: 587–623.

LACHER, V., 1967. Elektrophysiologische Untersuchungen an einzelnen Geruchsrezeptoren auf den Antennen weiblicher Moskitos (*Aëdes aegypti* L.). *J. Insect Physiol.* 13: 1461–1470.

SCHNEIDER, D., 1957. Elektrophysiologische Untersuchungen von Chemo- und Mechanorezeptoren der Antenne des Seidenspinners *Bombyx mori* L. *Z. Vergl. Physiol.* 40: 8–41.

SCHNEIDER, D., 1962. Electrophysiological investigation on the olfactory specificity of sexual attracting substances in different species of moths. *J. Insect Physiol.* 8: 15–30.

SCHNEIDER, D., 1963a. Electrophysiological investigation of insect olfaction. *In* Olfaction and Taste I (Y. Zotterman, editor). Pergamon Press, Oxford and New York, pp. 85–103.

SCHNEIDER, D., 1963b. Vergleichende Rezeptorphysiologie am Beispiel der Riechorgane von Insekten. *Jahrb. Max-Planck Ges. Göttingen* 149–177.

SCHNEIDER, D., 1966. Chemical sense communication in insects. *Symp. Soc. Exp. Biol.* 20: 273–297.

SCHNEIDER, D., 1967. Wie arbeitet der Geruchssinn bei Mensch und Tier? *Naturwiss. Rundsch.* 20: 319–326.

SCHNEIDER, D., 1969. Insect olfaction: Deciphering system for chemical messages. *Science* (*Washington*) 163: 1031–0137.

SCHNEIDER, D., B. C. BLOCK, J. BOECKH, and E. PRIESNER, 1967. Die Reaktion der männlichen Seidenspinner auf Bombykol und seine Isomeren: Elektroantennogramm und Verhalten. *Z. Vergl. Physiol.* 54: 192–209.

SCHNEIDER, D., and J. BOECKH, 1962. Rezeptorpotential und Nervenimpulse einzelner olfactorischer Sensillen der Insektenantenne. *Z. Vergl. Physiol.* 45: 405–412.

SCHNEIDER, D., and E. HECKER, 1956. Zur Elektrophysiologie der Antenne des Seidenspinners *Bombyx mori* bei Reizung mit angereicherten Extrakten des Sexuallockstoffes. *Z. Naturforsch., B,* 11: 121–124.

SCHNEIDER, D., and K.-E. KAISSLING, 1956. Der Bau der Antenne des Seidenspinners *Bombyx mori* L. I. Architektur und Bewegungsapparat der Antenne sowie Struktur der Cuticula. *Zool. Jahrb., Abt. Anat. Ontog. Tiere* 75: 287–310.

SCHNEIDER, D., and K.-E. KAISSLING, 1957. Der Bau der Antenne des Seidenspinners *Bombyx mori* L. II. Sensillen, cuticulare Bildungen und innerer Bau. *Zool. Jahrb., Abt. Anat. Ontog. Tiere* 76: 223–250.

SCHNEIDER, D., and K.-E. KAISSLING, 1959. Der Bau der Antenne des Seidenspinners *Bombyx mori* L. III. Das Bindegewebe und das Blutgefäss. *Zool. Jahrb., Abt. Anat. Ontog. Tiere* 77: 111–132.

SCHNEIDER, D., G. KASANG, and K.-E. KAISSLING, 1968. Bestimmung der Riechschwelle von *Bombyx mori* mit Tritium-markiertem Bombykol. *Naturwissenschaften* 55: 395.

SCHNEIDER, D., V. LACHER, and K.-E. KAISSLING, 1964. Die Reaktionsweise und das Reaktionsspektrum von Riechzellen bei *Antheraea pernyi* (Lepidoptera, Saturniidae). *Z. Vergl. Physiol.* 48: 632–662.

SCHNEIDER, D., and R. A. STEINBRECHT, 1968. Checklist of insect olfactory sensilla. *Symp. Zool. Soc. London* 23: 279–297.

SHOREY, H. H., and L. K. GASTON, 1967. Pheromones. *In* Pest Control (W. W. Kilgore and R. L. Doutt, editors). Academic Press Inc., New York, pp. 241–265.

SLIFER, E. H., 1961. The fine structure of insect sense organs. *Int. Rev. Cytol.* 11: 125–159.

STEINBRECHT, R. A., 1964a. Die Abhängigkeit der Lockwirkung des Sexualduftorgans weiblicher Seidenspinner (*Bombyx mori*) von Alter und Kopulation. *Z. Vergl. Physiol.* 48: 341–356.

STEINBRECHT, R. A., 1964b. Feinstruktur und Histochemie der Sexualduftdrüse des Seidenspinners *Bombyx mori* L. *Z. Zellforsch. Mikroskop. Anat.* 64: 227–261.

STEINBRECHT, R. A., 1969a. On the question of nervous syncytia: Lack of axon fusion in two insect sensory nerves. *J. Cell Sci.* 4: 39–53.

STEINBRECHT, R. A., 1969b. Comparative morphology of olfactory receptors. *In* Olfaction and Taste III (C. Pfaffmann, editor). The Rockefeller University Press, New York, pp. 3–21.

# 47 General Principles of Neuroendocrine Communication

## BERTA SCHARRER

ONE OF THE important insights gained in the neurobiological sciences during the past decade is recognition of the high degree of functional interdependence between nervous and endocrine centers. Their effectiveness in coordinating the body's regulatory activities requires pathways of communication between these two systems of integration (Figure 1). Because each of them functions in its own characteristic way, it cannot be assumed *a priori* that the "language" of one is suitable for the other. By definition, signals from endocrine organs to the nervous system are carried by blood-borne chemical messengers (hormones). In the framework of the present discussion, nothing more needs to be said about these afferent stimuli than that certain nerve cells are capable of recording endocrine signals and converting them into two types of neural activities. One leads to changes in the organism's behavior; the other elicits endocrine responses and thus becomes part of the ever-changing sequence of regulatory events that constitute neuroendocrine integration.

The nature of the efferent pathways, linking the neural with the endocrine apparatus, turns out to be rather complex. Its general analysis, which is the central topic of this chapter, is based on correlative structural and functional studies in a large variety of animal forms, both vertebrate and invertebrate. The literature in this field has become so voluminous that only a small selection of representative references can be given. The examination in depth of a few specific cases is the subject of the chapter that follows.

In principle, there are two ways in which the switchover from nervous to endocrine systems can be accomplished. One is by conventional innervation; the other, by the activity of specialized neurosecretory centers. The two modes of communication have certain features in common, but differ in others. In both instances, the neuron, in response to afferent stimuli, transmits information by means of a special neurochemical mediator to an effector cell that is endowed with an appropriate receptor site. Major differences in the nature of the various available mechanisms concern the duration of the signal and the spatial relation between the site of release of the messenger substance and the receptor (effector cell).

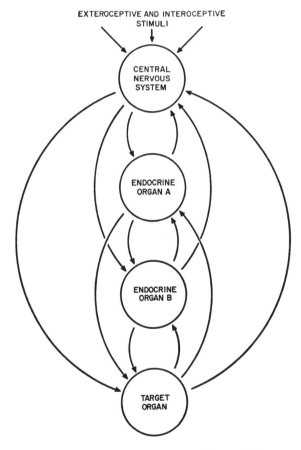

FIGURE 1 Diagram illustrating theoretically possible neuroendocrine interactions. The arrows symbolize both nervous and hormonal connections. In any given case only some of the pathways are actually used. (From Scharrer and Scharrer, 1963.)

BERTA SCHARRER Department of Anatomy, Albert Einstein College of Medicine, Bronx, New York

## Control of endocrine cells by conventional innervation

Direct nervous control of endocrine cells by conventional synaptic transmission is feasible, as the recipient cell is geared to chemical mediation. However, the strictly localized nature of the contact areas involved, and the exceedingly short duration of the signals conveyed by standard neurotransmitter substances (e.g., acetylcholine, noradrenalin), impose certain restrictions on the efficacy of such a control mechanism. Optimally, every gland cell would have to be endowed with its own secretomotor junction. In order to result in appropriate responses, simultaneous and continuous signals would have to be directed, via these synaptic complexes, to the endocrine effector cells over periods of time that by far exceed those characteristic of regular interneuronal or neuromuscular communication. Therefore, the question is whether such conventional neural control, provided it exists in a given endocrine organ, plays a major or only a minor role.

An obvious feature to look for in organs suspected of being under direct neural control is morphological evidence of innervation. In the case of endocrine cells, secretomotor junctions have long been thought to be absent. Proof to the contrary was impossible under the restrictions of light microscopy. Even today, after much painstaking scrutiny of a variety of tissues under the electron microscope, neuroeffector junctions with endocrine cells have not been recorded with sufficient frequency to warrant the conclusion that each of these cells receives its own nerve supply, or at least the majority do. On this basis alone, therefore, it seems unlikely that all the existing neuronal directives, be they stimulatory or inhibitory, are conveyed to endocrine centers by conventional synaptic transmission.

Nevertheless, reports on ultrastructural evidence for innervation of endocrine cells are no longer a rarity, and cover a variety of organs. Among these, as might be expected, are effector cells of neural derivation, viz., those of the adrenal medulla (von Euler, 1967) and of the pineal gland (Wurtman et al., 1968; Machado et al., 1968). Neuroeffector junctions that establish contact with non-neural endocrine cells tend to lack some of the attributes of typical synapses, especially membrane specializations (see below). Such contact sites occur in beta and alpha cells of mammalian pancreatic islets (Legg, 1967; Esterhuizen et al., 1968; Watari, 1968; Shorr and Bloom, 1970); in all three layers of the adrenal cortex (Unsicker, 1969); in Leydig cells of the avian testis (Baumgarten and Holstein, 1968); and in the ultimobranchial body of amphibians (Robertson, 1967) and birds (Stoeckel and Porte, 1967), as well as in the parafollicular cells of a mammal (Young and Harrison, 1969). Control of the renin-producing juxtaglomerular cells by adrenergic fibers is suggested by fluorescence-histochemical and electron-microscopic information (Barajas, 1964; Gomba et al., 1969). Of particular interest are data on the innervation of adenohypophyseal elements.

For example, in the pars intermedia of the cat (Figure 2, Figure 3A), Bargmann et al. (1967) observed one type of synaptic specialization, presumably adrenergic, with small, dense-core vesicles interspersed among electron-lucent ones (see also Streefkerk, 1967; Howe and Maxwell, 1968). Another, with only electron-lucent presynaptic vesicles, was tentatively interpreted as a cholinergic synapse but, like other secretomotor junctions in the endocrine system (Doerr-Schott and Follenius, 1969), these, too, might in reality be aminergic, even though they do not happen to display "typical" dense-core vesicles. The identification of adrenergic junctional complexes in the meta-adenohypophysis ("pars intermedia") of teleosts was accomplished by means of ultraradioautography after the administration of tritiated noradrenalin (Follenius, 1968). Several additional morphological data are available, for instance, for amphibians (Dent and Gupta, 1967; Jørgensen and Larsen, 1967; Enemar et al., 1967; Pehlemann, 1967; Saland, 1968; Goos, 1969).

Indirect experimental evidence for the existence of neural, as contrasted with hormonal, input was obtained from animals in which unilateral interruption of nerve supply affected merely one side of the neuroendocrine axis. This was true for the alteration of nuclear size in the zona fasciculata deprived of its connection with the ventromedial nucleus of the hypothalamus (Smollich and Döcke, 1969). In a comparable situation, among cephalopods, one of the existing pair of optic glands, the hormonal product of which stimulates gonadal maturation, was released from inhibitory neural control by unilateral severence of the connection with the central nervous system (Wells and Wells, 1969). Again, the result speaks for direct nervous control over endocrine tissue, even though the ultrastructural identification of regular secretomotor junctions is still missing (Bern, personal communication).

The question is what specific role can be assigned to synaptically transmitted information of this sort. In frogs, for example, aminergic light-dependent neurons in contact with pars intermedia cells seem to be responsible for a tonic type of inhibitory control over the release of the chromatophorotropin MSH (Oshima and Gorbman, 1969; see also Iturriza, 1969). Control over hormone release was also postulated for the adrenal cortex (Unsicker, 1969). On the whole, however, this question is still largely unanswered, because, in the majority of cases discussed in the preceding paragraphs, conventional synaptic signals are not the only controlling factors in operation. In fact, it would be difficult to state with certainty if, in any of the existing glands of internal secretion, conventional innervation would suffice as the sole means for control. As is discussed in the next section, the efficient operation of the neuroendocrine axis depends on the availability of additional mechanisms, which overcome the limitations inherent in synaptic control.

## Control of endocrine cells by neurosecretory neurons

Neurochemical communication with endocrine structures in ways other than synaptic is accomplished by a special class of nerve cells called "neurosecretory neurons." An appreciation of their significance in neuroendocrine mediation requires a brief general characterization of the phenomenon of neurosecretion.

THE CLASSICAL CONCEPT OF NEUROSECRETION The concept that certain nerve centers specialize in "neurosecretory activity" has gradually evolved from a combination of cytological and physiological investigations. (For reviews, see Scharrer and Scharrer, 1963; E. Scharrer, 1965, 1966; B. Scharrer, 1967a, 1969a; Bern, 1966; Bern and Knowles, 1966; Picard and Stahl, 1966; Hagadorn, 1967a, 1967b; Bargmann, 1968; Hofer, 1968; Sachs, 1969.) Neurosecretory neurons are found widely in the animal kingdom, from primitive metazoans to mammals (Gabe, 1966; Lentz, 1968), which bespeaks their basic functional importance. In essence, the unique property of these neurons is that they produce chemical messengers in quantities that permit dissemination by circulatory channels. These substances, being more stable than regular neurotransmitters, are capable of eliciting sustained effects in multiple and diverse "target cells" that are situated at some distance from the site of release of the active principle. Many, but by no means all, of the responding cells are endocrine.

Because these blood-borne neurochemical messengers have a large sphere of operation, and because re-use of the mediator, as seems to occur in adrenergic and possibly other kinds of synaptic transmission by a "shuttle-service" type of operation (see Iversen, 1967), is unknown in neurosecretory substances, the latter must be synthesized in larger quantities than the conventional agents operating in "chemical synaptic transmission." In short, substances with the specifications outlined resemble regular hormones in every respect except that their cells of origin are neurons. The term "neurohormone" commonly applied to them is, therefore, indeed appropriate.

In cells manufacturing neurohormones, secretory activity takes precedence over other neuronal attributes. Thus, they are neither ordinary nervous elements nor glandular cells, but a combination of both (Figure 4). This structural and functional specialization sets them sufficiently apart from conventional neurons to warrant the designation of neurosecretory neurons (see B. Scharrer, 1967a, 1969a, 1969c).

Typical neurons of this kind are easily recognized at the light- and electron-microscopic levels by their pronounced glandular features and the fact that, instead of establishing synaptic contact with contiguous (neuronal or non-neuronal) effector cells, they terminate in close proximity to vascular channels. The proteinaceous nature of the secretory product (active polypeptides bound to carrier proteins called neurophysins; see Berde, 1968) is responsible for its identification under the light and electron microscopes. The presence of characteristic, more or less electron-dense, membrane-bounded granules in several size categories has been demonstrated in numerous ultrastructural studies and need not be elaborated on here.

Groups of terminals laden with neurosecretory storage material, ready to enter the circulation, often form "neurohemal organs." These are exemplified by the neurohypophysis of vertebrates, the corpus cardiacum of insects, and the sinus gland of crustaceans. Together with the neural centers in which the neurohormones are manufactured and the axonal tracts in which they are conveyed to the terminals and stored, the neurohemal organs form neurosecretory systems.

The most widely studied among these organ complexes are the hypothalamic-neurohypophyseal systems of mammals and other vertebrates. The neurohormones released from them fall into two categories. 1. The so-called posterior-lobe hormones, derived from neurosecretory neurons in the supraoptic and paraventricular hypothalamic nuclei of higher vertebrates, reach nonendocrine "terminal targets" (such as kidney tubules and uterine muscles) by way of the general circulation. Their activities are examples of first-order neuroendocrine mechanisms, a term proposed by Rothballer (1957) for neurohormonal activities that control "terminal targets" directly. Because they do not involve endocrine way stations, they lie outside the topic with which the present text is concerned. 2. Neurohormones originating in other hypothalamic centers, such as the arcuate nucleus, and released into the hypothalamic-adenohypophyseal portal system at the level of the median eminence are responsible for second- or third-order neuroendocrine mechanisms. These intrahypophyseal functions are of direct concern for processes of neuroendocrine interaction, and are discussed in somewhat more detail below.

In analogous organ systems among invertebrates, the central role of the neurosecretory neuron as an intermediary is essentially the same. It rests on its capacity to receive and process a variety of signals from neural centers and to pass it on to centers of internal secretion, thus exerting control over their multiple and intricate functions (E. Scharrer, 1952).

MODULATIONS OF THE CLASSICAL VIEWS ON NEUROSECRETION The preceding paragraphs have summarized the classical interpretation of the phenomenon of neuro-

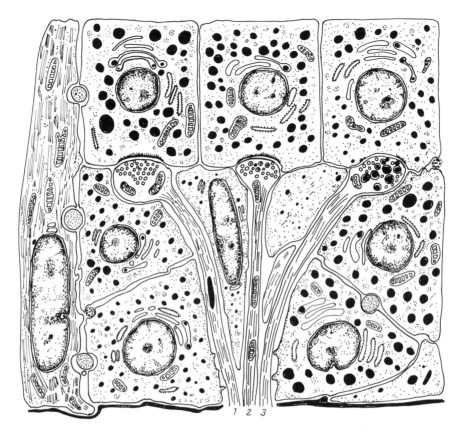

FIGURE 2 Diagram showing innervation of endocrine cells of pars intermedia of cat pituitary. Note presumed cholinergic (1), adrenergic (2), and "peptidergic" (3) fibers and terminals. (From Bargmann et al., 1967.)

secretion. In recent years, important contributions to this core of knowledge have yielded new insights on two fronts, as a result of which the dividing line between conventional and neurosecretory neurons is no longer so sharp as it appeared in the past. Some of the existing borderline cases more or less defy classification, and the validity of criteria for the classification of neurosecretory neurons has changed.

1. In addition to the prototypes, i.e., the "peptidergic" neurosecretory A fibers, there are others in which the material to be discharged into the circulation or elsewhere is of nonpeptide character. Such B fibers (which should not be confused with the conventional neuron type for which the same letter is used in the neurophysiological literature) furnish active principles that resemble chemically those operating as transmitters in ordinary aminergic neurons, but do so in greater quantity and apparently with different results (see, for example, Knowles, 1965a, 1965b; Björklund et al., 1968; B. Scharrer and Weitzman, 1970). These elements are further discussed in the section on hypophysiotropic factors. The source of another kind of nonpeptide neurohormone may exist in the insect brain (Williams, 1967).

2. Not all the A and B fibers in existence terminate at or near vascular channels. They may establish direct contact with an effector cell which may be neuronal, endocrine, or neither. They may also terminate close to several putative effector cells, i.e., at a distance of no more than several thousand Ångstrom units. An extracellular stromal material that occupies the intervening space may perhaps serve as a tem-

FIGURE 3 Examples of electron micrographs showing nerve terminals in synaptic or synaptoid contact with endocrine cells. A. Fiber (arrow) terminating on parenchymal cell of pars intermedia of cat pituitary. ×24,000. (From Bargmann et al., 1967.) B. Nerve terminal on cell of adrenal cortex of Syrian hamster. ×42,000. (From Unsicker, 1969.) C. Terminal with numerous agranular synaptic vesicles penetrating beta cell of pancreatic islet of cat. ×48,000. (From Legg, 1967.) D. Ovoid profile of terminal on alpha cell (at left) of pancreatic islet of brown bat. ×24,000. (From Watari, 1968.)

porary reservoir and as a vehicle for the relatively slow propagation of the mediator substance (see Barer, 1967; B. Scharrer, 1969a).

Sites of direct contact ("neurosecretomotor junctions," Bern, 1966) and terminations close to endocrine cells can be found in the adenohypophysis of vertebrates and in analo-

gous organs among invertebrates, such as the corpus allatum and the prothoracic gland of insects (B. Scharrer, 1964a, 1964b), and perhaps the infracerebral gland of nereids (Golding et al., 1968). Because, in such special instances, the dispatch of neurosecretory signals does not involve the circulatory system, and the action presumably occurs in the

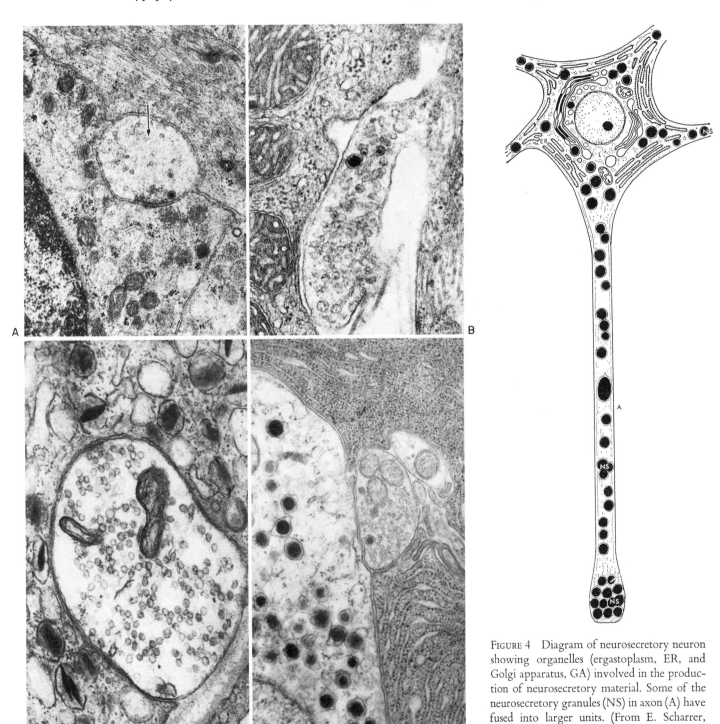

FIGURE 4   Diagram of neurosecretory neuron showing organelles (ergastoplasm, ER, and Golgi apparatus, GA) involved in the production of neurosecretory material. Some of the neurosecretory granules (NS) in axon (A) have fused into larger units. (From E. Scharrer, 1966.)

direct vicinity of the site of release of the messenger substance, the latter should not be designated as a neurohormone. On the other hand, the proteinaceous nature of the mediator sets it apart from conventional neurotransmitters, and virtually nothing is known about the mode of operation of peptidergic neurosecretomotor junctions. Therefore, these cannot be ranked among the standard types of synaptic contacts, but rather among the "synaptoid" configurations (see below). The interpretation of this class of ultrastructural specializations has much to gain from the correlation of electron-microscopic and physiological information. This applies, for example, to the analysis of the events in-

volved in the discharge of neurosecretory material from the axon, a problem that has been given much attention in recent years.

RELEASE MECHANISM FOR NEUROSECRETORY SUBSTANCES Irrespective of the nature of their site of termination, neurosecretory neurons display ultrastructurally distinctive areas at which the messenger substances are released (Figure 5). These subcellular specializations, which are rather widespread in the terminal part of the axon, mimic presynaptic parts of conventional interneuronal junctions (clustered "synaptic-type" vesicles, electron-dense deposits

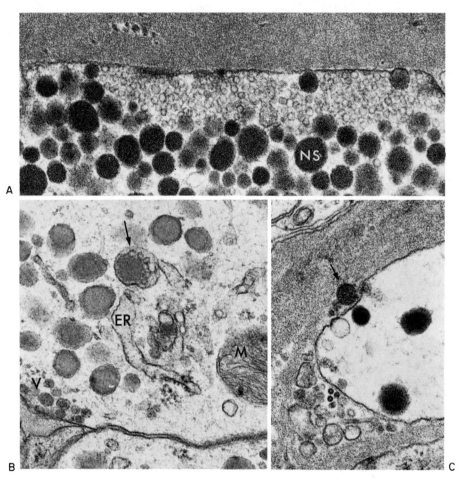

FIGURE 5 Electron micrographs illustrating sites of release of neurosecretory material in the corpus cardiacum of insects. A. Synaptoid configuration in neurosecretory axon facing extracellular stroma (top) which separates neuron from hemolymph (*Leucophaea maderae*). ×48000,. Note A-type neurosecretory granules (NS) and numerous small, agranular, and few dense-core vesicles plus diffuse opaque material pressing against axolemma. B. Profile of neurosecretory axon with A-type granules, one of which

buds off small electron-lucent vesicles (arrow). Also note small dense-core vesicles (V) and free dense particles clustered near stromal area (presumed site of release) (*Periplaneta americana*). ×42,500. ER, endoplasmic reticulum; M, mitochondrion. C. Part of neurosecretory axon with three electron-dense granules, one of which faces a similar granule in extracellular location (arrow) (*Byrsotria fumigata*). ×42,000. (A and B, from B. Scharrer, 1968a; C, original.)

at the junctional membranes, and so forth) and have, therefore, been called "synaptoid" configurations (see Knowles, 1967; Lederis, 1967; B. Scharrer, 1967b, 1968a).

At release sites where postsynaptic cells are missing (Figure 5), regular "chemical transmission" is out of the question. Therefore, the small electron-lucent vesicles in synaptoid areas need not be considered as receptacles for acetylcholine. Instead, at least some of them seem to be related to the mechanism by which the active material is released from the neurosecretory vesicle (see Knowles, 1965b; Bern, 1966; Herlant, 1967; Lederis, 1967; Streefkerk, 1967; Giller and Schwartz, 1968; B. Scharrer, 1968a).

More or less intact neurosecretory granules in extra-axonal location, i.e., having been discharged by exocytosis (Figure 5C) have been described in some species, particularly arthropods (see Weitzman, 1969; Normann, 1969). More often, however, the ultrastructural details observed speak for the fragmentation of the larger neurosecretory vesicles before their products are extruded from the neuron. In either case, the membranes bounding small synaptoid vesicles seem to be derived largely from those that formerly enclosed the neurosecretory granule (Figure 5B). It remains to be seen whether the more or less electron-lucent content of some of these small vesicles is active neurosecretory material that is no longer bound to its protein carrier and is ready to be discharged. Equally uncertain, therefore, is the chemical composition of the diffuse, electron-dense material that gives the impression of having been spilled from the neurosecretory vesicles and forms deposits close to the plasma membrane, ready to enter the extracellular compartment. Is it protein carrier (neurophysin) devoid of, or still affiliated with, active neurosecretory messenger substances? Further exploration of their mode of discharge and dissemination, the clarification of which is so important for the functional interpretation of these neurochemical mediators, should be based on the premise that one of several possible patterns may be prevalent, although not exclusive, in one animal species and absent or of minor importance in another.

The view that the synaptoid configurations discussed are indeed release sites of neurosecretory material is supported by experimental evidence. Their numbers were found to be increased after electrical stimulation in vitro of nerves, causing neurohormone release from the corpus cardiacum of insects (Scharrer and Kater, 1969; see also Normann, 1969). Because these more or less transient specializations of neurosecretory axons are functionally not equivalent to conventional synaptic elements, their presence is, in itself, not suggestive of impulse conduction. Nevertheless, the same mechanism, involving depolarization of the plasma membrane and calcium dependency, seems to be responsible for the transfer to the extracellular compartment of neuro-

secretory products, as well as other neurochemical mediators (Douglas and Poisner, 1964; Douglas, 1968).

HYPOPHYSIOTROPIC FACTORS Because of its dominant role in the control of other endocrine as well as "terminal" effector sites, the adenohypophysis and its dependence on neural input occupy the center of interest within the framework of neuroendocrinology (see Scharrer and Scharrer, 1963; Bajusz and Jasmin, 1964; Weitzman, 1964–1970; E. Scharrer, 1965, 1966; B. Scharrer, 1967a, 1968b, 1969a; Martini and Ganong, 1966–1967; Bajusz, 1969). Extrinsic and intrinsic afferent stimuli, transformed into uniform efferent commands within the hypothalamus, are relayed to the various glandular elements of the pars anterior and pars intermedia by neurochemical mediators, many of which qualify as neurohormones. These substances all have in common that they alter the release and apparently also the rate of synthesis of specific adenohypophyseal hormones. Instead of reaching their destinations by way of the general circulation, the hypothalamic factors involved make use of special pathways, in particular the primary plexus of the hypophyseal portal system.

Because a number of them elicit the discharge of adenohypophyseal tropic hormones, such as the follicle-stimulating and the adrenocorticotropic hormones, the former have been given the designation of "releasing factors." However, because this control over the pituitary by the central nervous sytem includes additional, e.g., inhibitory, activities, the more general terms of "hypophysiotropic factors" or "regulating factors" are preferable.

There is neither space nor need for a detailed discussion of the hypophysiotropic factors, which are referred to by abbreviations such as FSHRF (or FRF, follicle-stimulating hormone releasing, or regulating, factor). These have been extensively discussed and reviewed recently by several authors (see Schally et al., 1967–1968; Brodish, 1968; McCann and Porter, 1969).

The chemical nature of these relatively small molecules has not yet been entirely elucidated in mammals, and much less so in lower vertebrates (Jørgensen and Larsen, 1967). Most of them seem to be of peptide nature or, perhaps, of a more complex structure with peptide linkages (McCann and Porter, 1969). The existence of a carrier protein comparable to neurophysins of posterior-lobe material is suggested by the appearance of a histologically demonstrable substance that parallels an increase in CRF (corticotropin releasing factor) in the median eminence of the adrenalectomized mouse (Bock and aus der Mühlen, 1968).

There is, however, evidence for the presence of additional, i.e., nonpeptide mediators that are also released into the hypophyseal portal system of the median eminence. For example, in the rat, catecholamine-containing neurons

(B fibers of Knowles) which form the tubero-infundibular tract, have been shown to participate in the regulation of gonadotropin secretion in a manner that has not, as yet, been fully clarified. There is some evidence that the time course characteristic for the operation of B fibers is intermediate between that of neurosecretory A fibers and conventional neurons. Perhaps such B fibers play an auxiliary role by monitoring the activity of peptidergic hypophysiotropic factors (Kamberi et al., 1969), or they may function as regulating factors in their own right.

The relevant principle involved in the control of rhythmic reproductive phenomena in the rat does not appear to be noradrenalin (Sandler, 1968) but dopamine, accumulations of which are present at terminal sites where release into the capillary plexus takes place. This class of materials can be demonstrated either histochemically by the highly specific fluorescence technique of Hillarp and his coworkers (see Falck and Owman, 1965), or electron microscopically by the presence of numerous small (about 500 Å), dense-core vesicles after appropriate treatment (incubation with α-methylnoradrenalin, fixation in potassium permanganate).

The extensive analysis of Hökfelt (1967, 1968) deserves much of the credit for elucidating the ultrastructural criteria for aminergic release sites in this and other parts of the nervous system. His results are in line with those of Devine and Simpson (1968), Machado et al. (1968), Budd and Salpeter (1969), and others.

In mammals, the only known pathway for the neurochemical control of anterior pituitary cells, irrespective of the chemistry involved, is via the portal system just described.

The pars intermedia, on the other hand, shows peptidergic and aminergic terminals in direct contact with endocrine elements (Figure 2), as shown for the cat by Bargmann et al. (1967). Such junctional areas are also found on anterior pituitary cells of some fishes; in others, somewhat intermediate situations have been observed in which release sites of neurosecretory axons abut on the extracellular stroma discussed earlier (see Vollrath, 1967; B. Scharrer, 1968a, 1969b).

It is of particular interest that the mammalian pars intermedia, in addition to peptidergic neurosecretomotor contacts, also contains junctional complexes that possess the characteristics of conventional synapses (Bargmann et al., 1967). These are dealt with accordingly in the preceding section.

## General conclusions and outlook

The general impression gained from this survey of neuroendocrine communication is that of a remarkable diversity by which neurons, both conventional and neurosecretory, address themselves to hypophyseal and other endocrine centers. Depending on the kind of animal and the glandular effector site under investigation, cellular conduction, diffusion, or vascular transport of the neuromediator may predominate. In other words, the transfer of information may occur at close contact or over a wide range of extracellular distances, and thus both neurohumoral and neurohormonal messengers are involved.

For the dispatch of localized signals, conventional "chemical synaptic" and special "peptidergic" junctions are available. The latter may provide relatively sustained "private" messages and thus overcome the temporal limitations of regular synaptic control. Yet, both forms of local signaling seem to have their place in special situations and, as they involve more than one type of chemical messenger, they can be said to enrich the "vocabulary" of neuroendocrine communication.

The fact remains, however, that the gap between the nervous and the endocrine systems could not be bridged effectively by conventional innervation alone. The major load in this important process is handled to best advantage by neurohormones, provided by neurosecretory neurons.

The realization that this distinctive class of neurons itself makes use of more than one chemical and functional type of neurochemical mediator has further widened the scope of the study of neuroendocrine mediation. As a result, an intermediate category of neuroendocrine interactions must now be recognized in which neither neurohormonal nor regular neurotransmitter-type messengers are in operation. These interactions, about which knowledge is still fragmentary, require much further illumination. A search along evolutionary lines should yield valuable information on the gradual evolvement of different levels of specialization and their functional significance. This phylogenetic process concerns not only a progressively restricted distribution of specific mechanisms for neurochemical signaling, but a concomitant increase in discrimination on the part of the effector cells. The biochemical background of this specialization is still in the realm of speculation. Although the common denominator in the different chains of cellular events elicited by various kinds of neurochemical mediators may be cyclic AMP (adenosine monophosphate) (Jutisz and Paloma de la Llosa, 1969), the specificity of the responding cell may have evolved in step with that of the enzyme adenyl cyclase, as suggested by Bonner (1969).

Irrespective of the existing spectrum of neuroendocrine mediation, the distinction between neurohormones and neurohumors should be retained. Similarly, the concept of neurosecretion should not be abandoned, in spite of recent modulations in its scope. On the other hand, the term "neurosecretory" would become meaningless if it were extended to include conventional neurons as well, even though they have the biosynthetic equipment for secreting small amounts of specific substances. This well-established term should continue to denote only that distinctive class of neuronal elements in which the synthesis, storage, and release of glandular products have become the dominant feature.

Hypothalamic neurosecretory neurons, together with the mammalian pineal and adrenal medulla, have been classified as "neuroendocrine transducers" by Wurtman et al. (1968), because they "act to convert a neural input into an endocrine output." In this respect, they are said to differ from "endocrine transducers (thyroid, ovary, adrenal cortex) whose physiological input comes only from the circulation." After careful assessment of current information, the question arises as to whether this classification is accurate and conceptually useful.

At first glance, the separate status of these three neuroendocrine transducers would seem to be justified, because it is concordant with the fact that, because of their neural derivation, they are singularly suited for neuroendocrine communication.

But what about endocrine glands of non-neural ancestry that display previously unsuspected signs of innervation? As discussed above, examples of these are found among various vertebrates in the pars intermedia and pars anterior of the pituitary, and adrenal cortex, the pancreatic islets, and in calcitonin-producing tissue; among invertebrates, an example is the corpus allatum of insects. Or, what about endocrine organs whose input of signals does not come from the circulation alone, but via additional pathways, such as extracellular stromal channels in the adenohypophysis of some fishes? Obviously, these and other cases do not fit into either of the two categories of transducers proposed by Wurtman.

Furthermore, is an endocrine effector cell that receives neurochemical instruction from a neurosecretory neuron an endocrine transducer if the mediator is a neurohormone? Does its status change to that of a neuroendocrine transducer if the message is conveyed by a representative of the same class of neurosecretory neurons, but in this case by way of a neurosecretomotor junction? In the latter instance, the neurosecretory neuron would be one step removed from the position of neuroendocrine transducer and would become the penultimate link in the chain of commands constituting neuroendocrine integration. Clearly, the multiplicity of existing and potential modes of operation does not lend itself to rigid categorization.

As it turns out, neuroendocrine "transduction" is not confined to the activity of modified neural elements capable of hormone production, but makes use of a whole spectrum of possibilities. The recognition of this versatility appears to be the salient feature in the present analysis of the subject of neuroendocrine communication.

## REFERENCES

BAJUSZ, E. (editor), 1969. Physiology and Pathology of Adaptation Mechanisms: Neural-Neuroendocrine-Humoral. International Series of Monographs in Pure and Applied Biology, Modern Trends in Physiological Sciences, Vol. 27. Pergamon Press, Oxford.

BAJUSZ, E., and G. JASMIN (editors), 1964. Major Problems in Neuroendocrinology. Karger, Basel and New York.

BARAJAS, L., 1964. The innervation of the juxtaglomerular apparatus. An electron microscopic study of the innervation of the glomerular arterioles. Lab. Invest. 13: 916–929.

BARER, R., 1967. Speculations on the storage and release of hormones and transmitter substances. Bibl. Anat. 8: 72–75.

BARGMANN, W., 1968. Neurohypophysis, structure and function. In Handbuch der Experimentellen Pharmakologie, N. S. Vol. 23 (B. Berde, editor). Springer-Verlag, Berlin, Heidelberg, New York, pp. 1–39.

BARGMANN, W., E. LINDNER, and K. H. ANDRES, 1967. Über Synapsen an endokrinen Epithelzellen und die Definition sekretorischer Neurone. Untersuchungen am Zwischenlappen der Katzenhypophyse. Z. Zellforsch. Mikroskop. Anat. 77: 282–298.

BAUMGARTEN, H. G., and A.-F. HOLSTEIN, 1968. Adrenerge Innervation im Hoden und Nebenhoden vom Schwan (Cygnus olor). Z. Zellforsch. Mikroskop. Anat. 91: 402–410.

BERDE, B. (editor), 1968. Neurohypophysial hormones and similar polypeptides. In Handbuch der Experimentellen Pharmakologie, N. S. Vol. 23. Springer-Verlag, Berlin, Heidelberg, New York.

BERN, H. A., 1966. On the production of hormones by neurones and the role of neurosecretion in neuroendocrine mechanisms. Symp. Soc. Exp. Biol. 20: 325–344.

BERN, H. A., and F. G. W. KNOWLES, 1966. Neurosecretion. In Neuroendocrinology, Vol. 1 (L. Martini and W. F. Ganong, editors). Academic Press, New York, pp. 139–186.

BJÖRKLUND, A., A. ENEMAR, and B. FALCK, 1968. Monoamines in the hypothalamo-hypophyseal system of the mouse with special reference to the ontogenetic aspects. Z. Zellforsch. Mikroskop. Anat. 89: 590–607.

BOCK, R., and K. AUS DER MÜHLEN, 1968. Beiträge zur funktionellen Morphologie der Neurohypophyse. I. Über eine "gomoripositive" Substanz in der Zona externa infundibuli beidseitig adrenalektomierter weisser Mäuse. Z. Zellforsch. Mikroskop. Anat. 92: 130–148.

BONNER, J. T., 1969. Hormones in social amoebae and mammals. Sci. Amer. 220 (no. 6): 78–91.

BRODISH, A., 1968. A review of neuroendocrinology, 1966–67. Yale J. Biol. Med. 41: 143–198.

BUDD, G. C., and M. M. SALPETER, 1969. The distribution of labeled norepinephrine within sympathetic nerve terminals studied with electron microscope radioautography. J. Cell Biol. 41: 21–32.

DENT, J. N., and B. L. GUPTA, 1967. Ultrastructural observations on the developmental cytology of the pituitary gland in the spotted newt. Gen. Comp. Endocrinol. 8: 273–288.

DEVINE, C. E., and F. O. SIMPSON, 1968. Localization of tritiated norepinephrine in vascular sympathetic axons of the rat intestine and mesentery by electron microscope radioautography. *J. Cell Biol.* 38: 184–192.

DOERR-SCHOTT, J., and E. FOLLENIUS, 1969. Localisation des fibres aminergiques dans l'hypophyse de *Rana esculenta*. Étude auto-radiographique au microscope électronique. *Compt. Rend. Acad. Sci., sér. D* 269: 737–740.

DOUGLAS, W. W., 1968. Stimulus-secretion coupling: The concept and clues from chromaffin and other cells. *Brit. J. Pharmacol.* 34: 451–474.

DOUGLAS, W. W., and A. M. POISNER, 1964. Stimulus-secretion coupling in a neurosecretory organ: The role of calcium in the release of vasopressin from the neurohypophysis. *J. Physiol.* (*London*) 172: 1–18.

ENEMAR, A., B. FALCK, and F. C. ITURRIZA, 1967. Adrenergic nerves in the pars intermedia of the pituitary in the toad, *Bufo arenarum. Z. Zellforsch. Mikroskop. Anat.* 77: 325–330.

ESTERHUIZEN, A. C., T. L. B. SPRIGGS, and J. D. LEVER, 1968. Nature of islet-cell innervation in the cat pancreas. *Diabetes* 17: 33–36.

EULER, U. S. VON, 1967. Adrenal medullary secretion and its neural control. *In* Neuroendocrinology, Vol. 2 (L. Martini and W. F. Ganong, editors). Academic Press, New York, pp. 283–333.

FALCK, B., and C. OWMAN, 1965. A detailed methodological description of the fluorescence method for the cellular demonstration of biogenic monoamines. *Acta Univ. Lund, Sect. II,* 7: 1–23.

FOLLENIUS, E., 1968. Innervation adrénergique de la méta-adéno-hypophyse de l'Épinoche (*Gasterosteus aculeatus* L.). Mise en évidence par autoradiographie au microscope électronique. *Compt. Rend. Acad. Sci., sér. D,* 267: 1208–1211.

GABE, M., 1966. Neurosecretion. Pergamon Press, Oxford, London, New York.

GILLER, E., JR., and J. H. SCHWARTZ, 1968. Choline acetyl-transferase: Regional distribution in the abdominal ganglion of *Aplysia. Science* (*Washington*) 161: 908–911.

GOLDING, D. W., D. G. BASKIN, and H. A. BERN, 1968. The infracerebral gland—A possible neuroendocrine complex in *Nereis. J. Morphol.* 124: 187–215.

GOMBA, S., W. BOSTELMANN, V. SZOKOLY, and M. B. SOLTÉSZ, 1969. Histochemische Untersuchung der adrenergen Innervation des juxtaglomerulären Apparates. *Acta Biol. Med. Ger.* 22: 387–392.

GOOS, H. J. T., 1969. Hypothalamic control of the pars intermedia in *Xenopus laevis* tadpoles. *Z. Zellforsch. Mikroskop. Anat.* 97: 118–124.

HAGADORN, I. R., 1967a. Neurosecretory mechanisms. *In* Invertebrate Nervous Systems—Their Significance for Mammalian Neurophysiology (C. A. G. Wiersma, editor). University of Chicago Press, Chicago, pp. 115–124.

HAGADORN, I. R., 1967b. Neuroendocrine mechanisms in invertebrates. *In* Neuroendocrinology, Vol. 2 (L. Martini and W. F. Ganong, editors). Academic Press, New York, pp. 439–484.

HERLANT, M., 1967. Mode de libération des produits de neuro-sécrétion. *In* Proceedings of the 4th International Symposium on Neurosecretion (F. Stutinsky, editor). Springer-Verlag, Berlin, pp. 20–35.

HOFER, H. O., 1968. The phenomenon of neurosecretion. *In* The Structure and Function of Nervous Tissue, Vol. 1 (G. H. Bourne, editor). Academic Press, New York and London, pp. 461–517.

HÖKFELT, T., 1967. The possible ultrastructural identification of tubero-infundibular dopamine-containing nerve endings in the median eminence of the rat. *Brain Res.* 5: 121–123.

HÖKFELT, T., 1968. *In vitro* studies on central and peripheral monoamine neurons at the ultrastructural level. *Z. Zellforsch. Mikroskop. Anat.* 91: 1–74.

HOWE, A., and D. S. MAXWELL, 1968. Electron microscopy of the pars intermedia of the pituitary gland in the rat. *Gen. Comp. Endocrinol.* 11: 169–185.

ITURRIZA, F. C., 1969. Further evidences for the blocking effect of catecholamines on the secretion of melanocyte-stimulating hormone in toads. *Gen. Comp. Endocrinol.* 12: 417–426.

IVERSEN, L. L., 1967. The Uptake and Storage of Noradrenaline in Sympathetic Nerves. Cambridge University Press, London.

JØRGENSEN, C. B., and L. O. LARSEN, 1967. Neuroendocrine mechanisms in lower vertebrates. *In* Neuroendocrinology, Vol. 2 (L. Martini and W. F. Ganong, editors). Academic Press, New York, pp. 485–528.

JUTISZ, M., and M. PALOMA DE LA LLOSA, 1969. L'adénosine-3′, 5′-monophosphate cyclique, un intermédiaire probable de l'action de l'hormone hypothalamique FRF. *Compt. Rend. Acad. Sci., sér. D* 268: 1636–1639.

KAMBERI, I. A., R. S. MICAL, and J. C. PORTER, 1969. Luteinizing hormone-releasing activity in hypophysial stalk blood and elevation by dopamine. *Science* (*Washington*) 166: 388–390.

KNOWLES, F., 1965a. Neuroendocrine correlations at the level of ultrastructure. *Arch. Anat. Microscop. Morphol. Exp.* 54: 343–357.

KNOWLES, F., 1965b. Evidence for a dual control, by neurosecretion, of hormone synthesis and hormone release in the pituitary of the dogfish *Scylliorhinus stellaris. Phil. Trans. Roy. Soc.* (*London*) *ser. B, biol. sci.* 249: 435–455.

KNOWLES, F., 1967. Neuronal properties of neurosecretory cells. *In* Proceedings of the 4th International Symposium on Neurosecretion (F. Stutinsky, editor). Springer-Verlag, Berlin, pp. 8–19.

LEDERIS, K., 1967. Ultrastructural and biological evidence for the presence and likely functions of acetylcholine in the hypothalamo-neurohypophysial system. *In* Proceedings of the 4th International Symposium on Neurosecretion (F. Stutinsky, editor). Springer-Verlag, Berlin, pp. 155–164.

LEGG, P. G., 1967. The fine structure and innervation of the beta and delta cells in the islet of Langerhans of the cat. *Z. Zellforsch. Mikroskop. Anat.* 80: 307–321.

LENTZ, T. L., 1968. Primitive Nervous Systems. Yale University Press, New Haven, Connecticut.

McCANN, S. M., and J. C. PORTER, 1969. Hypothalamic pituitary stimulating and inhibiting hormones. *Physiol. Rev.* 49: 240–284.

MACHADO, A. B. M., C. R. S. MACHADO, and L. E. WRAGG, 1968. Catecholamines and granular vesicles in adrenergic axons of the developing pineal body of the rat. *Experientia* (*Basel*) 24: 464–465.

MARTINI, L., and W. F. GANONG (editors), 1966–1967. Neuroendocrinology. Academic Press, New York, vols. 1, 2.

NORMANN, T. C., 1969. Experimentally induced exocytosis of neurosecretory granules. *Exp. Cell Res.* 55: 285–287.

OSHIMA, K., and A. GORBMAN, 1969. Pars intermedia: Unitary

electrical activity regulated by light. *Science (Washington)* 163: 195–197.

PEHLEMANN, F. W., 1967. Ultrastructure and innervation of the pars intermedia of the pituitary of *Xenopus laevis*. Proc. Fourth Confer. Europ. Comp. Endocr. Carlsbad, Czechoslovakia, (abstract).

PICARD, D., and A. STAHL., 1966. La cellule neurosécrétrice chez les vertébrés. *Compt. Rend. Ass. Anat.* 51: 1–75.

ROBERTSON, D. R., 1967. The ultimobranchial body in *Rana pipiens*. III. Sympathetic innervation of the secretory parenchyma. *Z. Zellforsch. Mikroskop. Anat.* 78: 328–340.

ROTHBALLER, A. B., 1957. Neuroendocrinology. *Excerpta Med. Sect. III*, 11: iii–xii.

SACHS, H., 1969. Neurosecretion. *Advan. Enzymol.* 32: 327–372.

SALAND, L. C., 1968. Ultrastructure of the frog pars intermedia in relation to hypothalamic control of hormone release. *Neuroendocrinology* 3: 72–88.

SANDLER, R., 1968. Concentration of norepinephrine in the hypothalamus of the rat in relation to the estrous cycle. *Endocrinology* 83: 1383–1386.

SCHALLY, A. V., W. LOCKE, A. J. KASTIN, and C. Y. BOWERS, 1967–1968. Some aspects of neuroendocrinology. *In* Year Book of Endocrinology (T. B. Schwartz, editor). Year Book Medical Publishers, Chicago, pp. 5–29.

SCHARRER, B., 1964a. Histophysiological studies on the corpus allatum of *Leucophaea maderae*. IV. Ultrastructure during normal activity cycle. *Z. Zellforsch. Mikroskop. Anat.* 62: 125–148.

SCHARRER, B., 1964b. The fine structure of the blattarian prothoracic glands. *Z. Zellforsch. Mikroskop. Anat.* 64: 301–326.

SCHARRER, B., 1967a. The neurosecretory neuron in neuroendocrine regulatory mechanisms. *Amer. Zool.* 7: 161–169.

SCHARRER, B., 1967b. Ultrastructural specializations of neurosecretory terminals in the corpus cardiacum of cockroaches. *Amer. Zool.* 7: 721–722 (abstract).

SCHARRER, B., 1968a. Neurosecretion. XIV. Ultrastructural study of sites of release of neurosecretory material in blattarian insects. *Z. Zellforsch. Mikroskop. Anat.* 89: 1–16.

SCHARRER, B., 1968b. Neuroendocrine factors in the control of reproduction. *In* Perspectives in Reproduction and Sexual Behavior (M. Diamond, editor). Indiana University Press, Bloomington, pp. 145–149.

SCHARRER, B., 1969a. Neurohumors and neurohormones: Definitions and terminology. *J. Neuro-viscer. Relat. Suppl.* 9: 1–20.

SCHARRER, B., 1969b. Comparative aspects of neurosecretory phenomena. *In* Progress in Endocrinology. Proceedings of the IIIrd International Congress of Endocrinology, Mexico, 1968 (C. Gual and F. J. G. Ebling, editors). Excerpta Medica Foundation, Amsterdam, pp. 365–367.

SCHARRER, B., 1969c. Current concepts in the field of neurochemical mediation. *Med. Coll. Virginia Quart.* 5: 27–31.

SCHARRER, B., and S. B. KATER, 1969. Neurosecretion. XV. An electron microscopic study of the corpora cardiaca of *Periplaneta americana* after experimentally induced hormone release. *Z. Zellforsch. Mikroskop. Anat.* 95: 177–186.

SCHARRER, B., and M. WEITZMAN, 1970. Current problems in invertebrate neurosecretion. *In* Aspects of Neuroendocrinology (W. Bargmann and B. Scharrer, editors). Springer-Verlag, Berlin (in press).

SCHARRER, E., 1952. The general significance of the neurosecretory cell. *Scientia* 87: 176–182.

SCHARRER, E., 1965. The final common path in neuroendocrine integration. *Arch. Anat. Microscop. Morphol. Exp.* 54: 359–370.

SCHARRER, E., 1966. Principles of neuroendocrine integration. *In* Endocrines and the Central Nervous System. *Res. Publ. Ass. Res. Nerv. Ment. Dis.* 43: 1–35.

SCHARRER, E., and B. SCHARRER, 1963. Neuroendocrinology. Columbia University Press, New York.

SHORR, S. S., and F. E. BLOOM, 1970. Fine structure of islet-cell innervation in the pancreas of normal and alloxan-treated rats. *Z. Zellforsch. Mikroskop. Anat.* 103: 12–25.

SMOLLICH A., and F. DÖCKE, 1969. Nervale Einflussnahme des Hypothalamus auf die Funktion der Nebennierenrinde. *J. Neuroviscer. Relat.* 31: 128–135.

STOECKEL, M.-E., and A. PORTE, 1967. Sur l'ultrastructure des corps ultimobranchiaux du Poussin. *Compt. Rend. Acad. Sci., sér. D*, 265: 2051–2053.

STREEFKERK, J. G., 1967. Functional changes in the morphological appearance of the hypothalamo-hypophyseal neurosecretory and catecholaminergic neural system, and in the adenohypophysis of the rat. A light, fluorescence and electron microscopic study. Drukkerij Wed. G. Van Soest N. V., Amsterdam, doctoral thesis.

UNSICKER, K., 1969. Zur Innervation der Nebennierenrinde vom Goldhamster. Eine fluoreszenz- und elektronenmikroskopische Studie. *Z. Zellforsch. Mikroskop. Anat.* 95: 608–619.

VOLLRATH, L., 1967. Über die neurosekretorische Innervation der Adenohypophyse von Teleostiern, insbesondere von *Hippocampus cuda* und *Tinca tinca*. *Z. Zellforsch. Mikroskop. Anat.* 78: 234–260.

WATARI, N., 1968. Fine structure of nervous elements in the pancreas of some vertebrates. *Z. Zellforsch. Mikroskop. Anat.* 85: 291–314.

WEITZMAN, M. (editor)., 1964–1970. Bibliographia Neuroendocrinologica. Albert Einstein College of Medicine, New York, vols. 1–7.

WEITZMAN, M., 1969. Ultrastructural study on the release of neurosecretory material from the sinus gland of the land crab, *Gecarcinus lateralis*. *Z. Zellforsch. Mikroskop. Anat.* 94: 147–154.

WELLS, M. J., and J. WELLS, 1969. Pituitary analogue in the octopus. *Nature* (London) 222: 293–294.

WILLIAMS, C. M., 1967. The present status of the brain hormone. *In* Insects and Physiology (J. W. L. Beament and J. E. Treherne, editors). Oliver-Boyd, Edinburgh and London, pp. 133–139.

WURTMAN, R. J., J. AXELROD, and D. E. KELLY, 1968. The Pineal. Academic Press, New York.

YOUNG, B. A., and R. J. HARRISON, 1969. Ultrastructure of light cells in the dolphin thyroid. *Z. Zellforsch. Mikroskop. Anat.* 96: 222–228.

# 48 Neuroendocrine Transducer Cells in Mammals

## RICHARD J. WURTMAN

IN GENERAL, the brain sends signals to other organs by utilizing chains of neurons connected by synapses. Peripheral nerves constitute the last links in these chains; their axons terminate directly on skeletal muscle, adipose tissue, and exocrine glands, and transmit instructions to these cells by releasing acetylcholine or norepinephrine. Among the many tasks the brain must perform, at least two cannot be mediated solely through neuronal channels, because the organs with which the brain must communicate do not respond to neurotransmitters, which are released in their immediate vicinity, but to other chemical signals, which are delivered to them by the blood stream. These tasks include regulation of concentrations of certain substances in the extracellular fluid, and control of the reproductive organs.

For example, one of the effector mechanisms activated by the brain when plasma glucose concentrations fall below an allowable minimum operates by accelerating the rate of glycogenolysis in the liver. The immediate signal that instructs the hepatic cells to initiate the depolymerization of glycogen is a hormone, epinephrine, which reaches the hepatocytes not by diffusing across a synapse, but by being taken up from the circulation. Similarly, the brain of the female mammal instructs the ovary to mature, and subsequently directs it to ovulate, not by sending nerve impulses to individual ovarian cells but by causing the pituitary gland to secrete specific hormones, the gonadotropins. These substances reach the ovary via the blood stream, and activate biochemical processes leading to the maturation of the ovum, ovulation, and the synthesis of ovarian hormones.

The output signals emitted by neurons differ in important ways from those emitted by cells that communicate with their target tissues via the general circulation, i.e., endocrine cells. Hence, the ability of the mammalian brain to regulate substances in the extracellular fluid and to control gonadal function requires that the brain have access to a specialized group of communications cells capable of transducing neuronal inputs to hormonal outputs. These cells may be termed "neuroendocrine transducer cells." Their input signals resemble the input (or output) of a neuron, in that they respond to a neurotransmitter, which releases them by dif-

fusing across a short distance, typically at a synapse. Their output signal is similar to the output (or input) of a true endocrine cell, in that they emit coded chemical messages that are delivered to all cells in the body, but provide information to only a relatively small population of receptor cells (Figure 1).

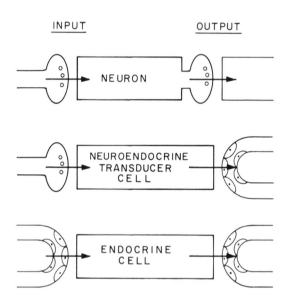

INPUT                                    OUTPUT

FIGURE 1 Schematic diagram showing inputs and outputs associated with neurons, neuroendocrine transducer cells, and endocrine cells.

The present paper compares some of the communications properties of neurons, endocrine cells, and neuroendocrine transducer cells, and describes two specific examples of the latter—pinealocytes and adrenomedullary chromaffin cells.

## Some Differences Between Neural and Endocrine Communications

The transmission of signals from one neuron to another (or from a neuron to the cells that it innervates) is mediated by a familiar process: the neurotransmitter, a specific substance stored within characteristic subcellular vesicles, is liberated near the receptor cell. Most commonly, this release occurs at an identifiable anatomical locus, the synapse. The neuro-

RICHARD J. WURTMAN   Massachusetts Institute of Technology, Cambridge, Massachusetts

transmitter then diffuses across a short distance to reach a specialized zone (the postsynaptic membrane) on the receptor cell, where it alters the flux of specific ions. This causes a change in electrical potential within the postsynaptic cell and, if the cell is a neuron, alters the probability that an action potential will be generated. Nearly all the compounds thought to function as neurotransmitters have similar chemical characteristics, i.e., low-molecular-weight, water-soluble amines and, possibly, amino acids. Moreover, they are rapidly inactivated by physical and chemical processes such as reuptake into their cell of origin or enzymatic transformation. Their concentrations in the blood tend to be very low.

The transmission of signals between organs via the blood stream utilizes an array of chemicals far broader than the current list of putative neurotransmitters; furthermore, the hormones seem to lack common chemical characteristics. Thus, insulin is water-soluble, but progesterone is highly nonpolar; thyroxine is a low-molecular-weight amino acid, but thyroid-stimulating hormone (TSH) appears to be a large glycoprotein; epinephrine is rapidly cleared from the circulation (by enzymatic transformation or uptake into sympathetic nerve endings), but cortisol persists in the blood for relatively long periods. The specific anatomic locus on the receptor cell at which a hormone acts has yet to be identified, and almost certainly lacks the well-defined structural features of the postsynaptic membrane. Similarly, no characteristic electrical response has been identified in hormone-responsive cells analogous to the ion fluxes and potential changes observed in the postsynaptic neuron or at the neuromuscular junction. Within seconds one can usually tell whether a given neuron has received and responded to a neurotransmitter; considerably more time is required to determine whether a thyroid cell has responded to circulating TSH.

Perhaps the most characteristic difference between the transmission of signals by neurotransmitters and that by hormones lies in the techniques used by these communications systems to attain "privacy." Nervous systems obtain privacy by anatomical means: a given neuron apparently transmits signals only to the small number of cells with which it makes synapses, or to cells lying within a few hundred Ångstroms of its terminal boutons. Thus, even though the particular chemical signal (e.g., acetylcholine) emitted when a particular neuron fires might be capable of stimulating $10^8$ other neurons within the brain, only $10^2$ or $10^3$ cells respond, because only this smaller number of neurons actually receives quanta of the neurotransmitter.

Communications systems that utilize the circulation to transmit signals obtain privacy by biochemical means: a given signal may be distributed by the blood to every cell in the body; the signal is, however, coded, and only a relatively small number of cells are able to perform the necessary decoding operation to obtain information. The high degree of

specificity attainable by hormonal communications systems is well illustrated by the physiological regulation of the thyroid gland. TSH, the input to this organ, is carried by the circulation to every organ in the body; thyroxine, its output, is distributed in essentially the same volume. However, only the thyroid gland appears capable of responding to the information content in circulating TSH levels, while the heart, the liver, and most other organs show biochemical responses to circulating thyroxine.

Table I describes some characteristic differences in the ways signals are transmitted by neurons and glandular cells.

TABLE I

*Information Transfer Via Neurons and Endocrine Cells*

|  | Neurons | Endocrine Cells |
|---|---|---|
| Primary input | Neurotransmitter at synapse | A hormone or other blood-borne substance (no characteristic locus of input) |
| Responses | Fast: Release of neurotransmitter | Fast: Release of hormone |
|  | Slow: Synthesis of specific neuronal proteins | Slow: Effects on glandular structure |
| Output: | | |
| Language | Few words: compounds of low molecular weight, high polarity, similar structure | Large vocabulary; wide range of molecular species |
| Distribution | Local | General |
| Privacy | By anatomy | By biochemical coding |
| Extinction | Rapid | Slower |

## Neuroendocrine Transducer Cells

The conversion of neural to hormonal signals is accomplished by a third group of communications cells, the neuroendocrine transducers (Wurtman and Axelrod, 1965). As do neurons and endocrine cells, the neuroendocrine cells transmit messages from one locus within the body (i.e., the cells that provide their input) to specific groups of receptor cells. Data available on the input signals to several specific neuroendocrine transducers suggest that these cells respond to typical neurotransmitter substances (acetylcholine and norepinephrine), which reach them by diffusing across synapses or from nearby boutons. Their output signals exhibit all the variety typical of hormones: epinephrine is water-soluble, but melatonin is nonpolar; renin is a high-molecular-weight protein, but epinephrine and melatonin are low-molecular-weight derivatives of single amino acids. The output signals (the hypophysiotropic hormones) emitted by hypothalamic transducer cells, which mediate the neural control of the anterior pituitary, apparently act only on this single target organ. In contrast, oxytocin, a hormonal signal emitted by the paraventricular nucleus, carries instructions to both the uterus and the myoepithelium of the mammary glands.

The demonstration that a given cell functions as a neuroendocrine transducer requires two types of evidence: (1) it must be shown by anatomical methods that the cell receives a direct innervation; and (2) it must be demonstrated that the ability of the cell to secrete its hormone under appropriate physiological conditions is impaired on interruption of this innervation. With these criteria, at least five groups of cells can now be termed neuroendocrine transducers. They are:

1. The chromaffin cells of the adrenal medulla, which respond to a sympathetic cholinergic input by releasing the hormone epinephrine (Wurtman, 1966).
2. The parenchymal cells of the mammalian pineal organ, which respond to a sympathetic noradrenergic input by synthesizing and releasing the hormone melatonin (Wurtman et al., 1968).
3. The cells of the supraoptic and paraventricular hypothalamic nuclei, which respond to noradrenergic or cholinergic inputs, or both, by releasing the hormones vasopressin and oxytocin (Scharrer and Scharrer, 1963).
4. Hypothalamic cells which may reside within the arcuate nuclei and which appear to secrete "releasing factors" or hypophysiotropic hormones into the pituitary portal circulation, possibly in response to a noradrenergic or dopaminergic input (Martini and Ganong, 1967).
5. The juxtaglomerular cells of the mammalian kidney, which respond to a sympathetic noradrenergic input by releasing renin into the blood stream (Bunag et al., 1966).

It seems likely that this list will continue to expand. Whenever it can be demonstrated that the brain influences the secretion of a hormone from a peripheral organ (e.g., insulin from the pancreas), a *prima facie* case has been made for the participation of a neuroendocrine transducer in the secretory process.

Neuroendocrine transducer cells in the adrenal medulla and the supraoptic nucleus contain characteristic granular vesicles visible by electron microscopy, segregated by ultracentrifugation, and shown by chemical or biological assays to store the secretory products of their respective cells. On the other hand, pineal parenchymal cells do not seem to store their secretion in a characteristic organelle. Hence, the tendency for the secretion of a cell to be localized within a visible particle cannot be used for the identification of neuroendocrine transducer cells. This tendency seems to be correlated with the chemical nature of the secretion and not with the communications properties of its cells of origin. Water-soluble hormones are often present in relatively high concentrations within their cells of origin. Their partial confinement within vesicles allows these concentrations to exist without unduly raising the cytoplasmic solute loads. Moreover, the binding of epinephrine and insulin to subcellular particles probably protects these hormones from catabolic intracellular enzymes (i.e., mitochondrial monamine oxidase and lysosomal proteases), which would destroy them if they floated freely in the cytoplasm. Lipid-soluble hormones tend not to be localized within vesicles or similar subcellular particles. For example, estrogens and thyroxine are stored in reservoirs outside the cell, that is, in the follicular fluid of the ovary or the colloid of the thyroid, whereas corticosterone is probably synthesized *de novo* on demand.

## The Chromaffin Cells of the Mammalian Adrenal Medulla as Neuroendocrine Transducers

The adrenal medulla, situated outside the cranial cavity, is considerably more accessible for study than the neuroendocrine transducers of the pineal and the hypothalamus. The adrenal vein can easily be cannulated, and the epinephrine and norepinephrine in venous blood can be measured in response to altered physiological states or to the stimulation of the splanchnic nerve that innervates the adrenal.

The adrenal medulla is probably the sole source of circulating epinephrine in mammals (Wurtman, 1966). This unique ability to synthesize and secrete the catecholamine derives from the fact that the adrenal chromaffin cells contain almost all the enzyme phenylethanolamine-N-methyl transferase (PNMT) that is present in adult mammals (Figure 2). (Small amounts of PNMT activity can be demonstrated in parts of the brain [Pohorecky et al., 1969]; however, it is unlikely that significant quantities of unchanged epinephrine enter the blood stream from this organ.) PNMT catalyzes the transfer of a methyl group from S-

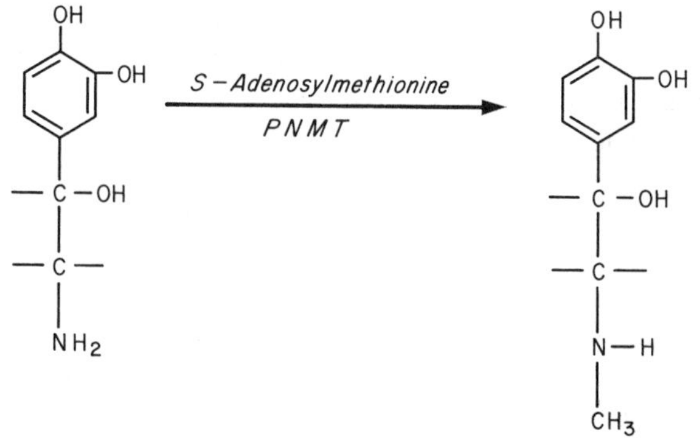

FIGURE 2  Biosynthesis of epinephrine in mammalian adrenal medulla. The N-methylation of norepinephrine is catalyzed by the enzyme phenylethanolamine-N-methyl transferase (PNMT).

adenosylmethionine to the amine nitrogen of norepinephrine (Axelrod, 1962).

Specific mechanisms exist for controlling both the chemical nature of the material secreted by the adrenal medulla and the net rate of secretion:

1. At any given time, the percentage of epinephrine in the total amount of catecholamine secreted depends on a hormonal factor, i.e., the delivery of glucocorticoid hormones secreted by the adrenal cortex to the medullary chromaffin cells (Figure 3). The steroid hormones elevate PNMT activity, probably by accelerating the synthesis of the enzyme protein (Wurtman and Axelrod, 1966). If adrenocortical secretion is suppressed by the removal of the pituitary gland, both PNMT activity and the ratio of epinephrine to total adrenal catecholamines fall. This ratio in the adrenal venous effluent also declines (Wurtman et al., 1968c), inasmuch as the chromaffin cells appear to secrete epinephrine and norepinephrine in roughly the same proportions as they store them. The administration of small amounts of adrenocorticotrophic hormone (ACTH), or of very large doses of natural or synthetic glucocorticoids, restores PNMT activity (and the in vivo synthesis of epinephrine) in hypophysectomized animals (Table II). This high glucocorticoid requirement

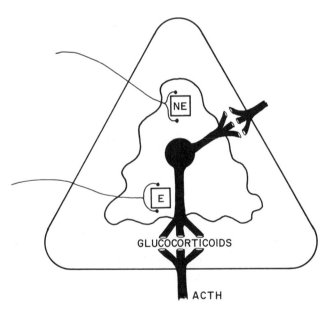

FIGURE 3 Control of the synthesis and secretion of epinephrine in the mammalian adrenal medulla. The synthesis of epinephrine (E) from norepinephrine (NE) is catalyzed by the enzyme phenylethanolamine-N-methyl transferase (PNMT), which is induced by glucocorticoid hormones secreted from the adrenal cortex (e.g., hydrocortisone, corticosterone). These compounds are preferentially delivered to the medullary chromaffin cells by an intra-adrenal portal venous system; their secretion is stimulated by a hormone (adrenocorticotrophic hormone, ACTH) from the pituitary gland. The secretion of epinephrine is stimulated by preganglionic cholinergic neurons, the terminal boutons of which form synapses with the adrenal chromaffin cells.

TABLE II

*Effect of Hypophysectomy and ACTH on Phenylethanolamine-N-Methyl Transferase Activity*

| Treatment | PNMT Activity mμM/gland | mμM/gm |
|---|---|---|
| Normal | 123.1* | 122.3* |
| Hypophysectomized | 28.8 | 43.1 |
| Hypophysectomized + ACTH | 190.0* | 147.3† |

*$P < 0.001$ differs from hypophysectomized.
†$P < 0.01$ differs from hypophysectomized.

is not surprising, for it is known that the adrenal medulla is normally perfused with glucocorticoid concentrations 100 to 1000 times greater than those present in peripheral blood. The adrenal chromaffin tissue of mammals is surrounded by an envelope of adrenal cortex, and preferentially receives the undiluted secretions of this layer via a portal vascular system (Coupland, 1965). Treatment of hypophysectomized animals with low doses of ACTH restores the preferential delivery of adrenocortical steroids to the medulla. Doses of glucocorticoids that restore the levels of hydrocortisone in peripheral blood to normal fail, however, to replace the hormone levels normally available to the medulla.

It is interesting to note that in the frog—an animal in which PNMT activity and epinephrine are found in many organs and in which epinephrine may function as a neurotransmitter—the adrenal gland lacks a well-defined cortex and medulla and PNMT activity is not modified by hypophysectomy, nor is the enzyme protein induced by treatment with glucocorticoids (Wurtman et al., 1968b). The adrenal medulla as a neuroendocrine transducer is probably a phylogenetically new organ in mammals.

2. The release of epinephrine (or norepinephrine) from the adrenal medulla is stimulated by nerve impulses delivered to the chromaffin cells by the cholinergic terminals of the splanchnic sympathetic nerves (Figure 3). Physiological stimuli, such as hypoglycemia, which normally accelerate epinephrine secretion, fail to do so if the medulla is denervated, or if the animal is treated with cholinergic blocking agents.

The hormonal signal represented by epinephrine can be decoded by a relatively large number of target cells. For example, epinephrine release accelerates hepatic glycogenolysis, stimulates the rate and force of cardiac contraction, and suppresses the insulin secretion from the beta cells of the pancreas.

## Mammalian Pinealocytes as Neuroendocrine Transducers

NEURAL AND PHOTIC REGULATION OF MELATONIN SYNTHESIS In mammals, the parenchymal cells of the pineal organ are unique in their ability to synthesize melatonin and other methoxyindoles (Figure 4). This characteristic results

described below indicate that in the pineal, as elsewhere, the catecholamine is the true sympathetic neurotransmitter; the presence of serotonin within the nerve terminals probably results from the extremely high concentrations of the indoleamine in the neighboring pineal parenchymal cells.

Nerve impulses generated in rats when they are exposed to light apparently *decrease* the flow of action potentials to the pineal gland; hence, less of the sympathetic neurotransmitter is liberated. Within the parenchymal cells, HIOMT activity falls, and the rate of melatonin synthesis declines. If

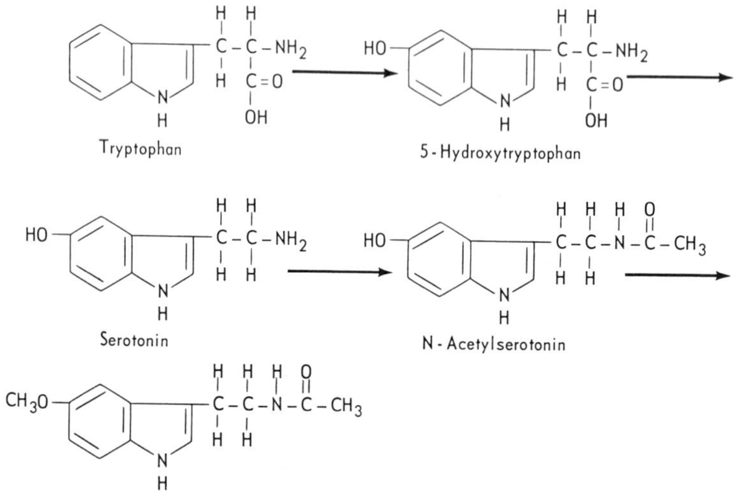

FIGURE 4 Biosynthesis of serotonin and melatonin in the mammalian pineal organ. The transfer of a methyl group from S-adenosylmethionine to the 5-hydroxy position of N-acetylserotonin is catalyzed by the enzyme hydroxyindole-O-methyl transferase (HIOMT).

from the highly specific localization of the enzyme hydroxyindole-O-methyl transferase (HIOMT) to these cells (Axelrod et al., 1961). HIOMT catalyzes the transfer of a methyl group from S-adenosylmethionine to hydroxyindoles such as N-acetylserotonin (Figure 4). Environmental lighting indirectly controls its activity. In rats, exposure to light generates nerve impulses that travel to the pineal organ by a complex pathway, including the retina, the optic nerves and chiasm, the inferior accessory optic tract, the medial forebrain bundle, descending neurons that mediate the central control of sympathetic nervous function, preganglionic fibers from the cervical spinal cord to the superior cervical ganglia, and postganglionic fibers that terminate in the vicinity of, or directly upon, pineal parenchymal cells (Figures 5A, B, and C, pages 536 and 537; Table II) (Wurtman et al., 1968a). The terminal boutons of pineal sympathetic nerves are unique in that their storage vesicles contain serotonin as well as norepinephrine. In vitro studies

one blinds the rats, or blocks the photic input to the pineal by transecting the inferior accessory optic tracts or by removing the superior cervical ganglia, HIOMT activity within the gland no longer responds to changes in the photic environment. This dependence on an indirect, nervous pathway for its photic input constitutes a major difference between mammalian pineal organs and the photoreceptive pineals of such lower vertebrates as the frog (*Rana pipiens*). Pineal cells in this species contain characteristic outer segments; these transduce light energy of specific wavelengths to nerve impulses, which are then transmitted directly to the brain. In mammals, pineal organs lack photoreceptive elements and cannot be shown electrophysiologically to respond directly to light. Moreover, they neither send axons to the brain nor receive a central innervation. The functions of melatonin also appear to have changed with evolution. In the frog, HIOMT activity is demonstrable in the pineal, but the enzyme is also present in the brain and elsewhere.

The distribution of HIOMT is compatible with the hypothesis that its product, melatonin, functions as a neurotransmitter rather than as a hormone. In mammals, the pineal gland is the sole site of melatonin synthesis; the indole is synthesized and probably released as a hormone in response to sympathetic nervous inputs.

THE EFFECTS OF NOREPINEPHRINE ON MELATONIN SYNTHESIS The regulation of pineal function by suggested neurotransmitters can be studied in vitro: If individual organs are incubated with $^{14}$C-labeled tryptophan, they convert the amino acid to $^{14}$C-serotonin, $^{14}$C-melatonin, and $^{14}$C-5-hydroxyindole acetic acid (5-HIAA), which are released into the incubation medium; they also synthesize and store $^{14}$C-labeled proteins. These processes proceed linearly for at least 48 hours and show surprisingly little variation from one pineal organ to another. Compounds thought to be candidates for the sympathetic neurotransmitter substance in the pineal can be added to the medium in various concentrations, and their effects on indole and protein synthesis can be examined.

The addition of L-norepinephrine and related catecholamines to the incubation medium causes a marked increase in the amounts of $^{14}$C-melatonin and $^{14}$C-serotonin synthesized and released into the medium (Axelrod et al., 1969) and smaller increases in the amounts of $^{14}$C-protein and $^{14}$C-tryptophan accumulated within the pineal organ (Wurtman et al., 1969). The increase in $^{14}$C-protein content probably depends on a pool effect, i.e., norepinephrine increases the uptake of labeled tryptophan from the medium, causing its specific activity within the cell to rise. The much greater effect of the catecholamine on $^{14}$C-melatonin synthesis almost certainly results from an action on the biosynthetic pathway connecting serotonin and melatonin. This explanation is supported by the observations that the addition of norepinephrine to the medium also causes increased HIOMT activity assayed in vitro. The addition of serotonin to the incubation medium has no effect on $^{14}$C-melatonin synthesis, nor does this amine modify the responses of the culture to added norepinephrine. These data support the hypothesis that norepinephrine, not serotonin, is the neurotransmitter released from pineal sympathetic neurons.

The role of the adenyl cyclase system in the effects of norepinephrine of the pineal has recently been examined by using dibutyryl-3'-5'-adenosine monophosphate (DAMP), a synthetic derivative of cyclic AMP (Shein and Wurtman, 1969). The addition of relatively small amounts of this substance to cultured pineal organs mimics the stimulation of $^{14}$C-melatonin and $^{14}$C-serotonin synthesis produced by norepinephrine (Table III). However, dibutyryl-AMP does not increase the accumulation of $^{14}$C-trytophan or of $^{14}$C-protein within pineal cells (Wurtman et al., 1969). Because norepinephrine has indeed been shown to enhance the ac-

tivity of adenyl cyclase (a membrane enzyme) in pineal homogenates, this dissociation of indole synthesis from tryptophan uptake is compatible with the general hypothesis that cyclic AMP mediates *some* of the biochemical effects of norepinephrine, e.g., those affecting processes within the cytoplasm of the cell, but does not mediate the direct effects of the neurotransmitter on cellular membranes, e.g., the stimulation of $^{14}$C-tryptophan uptake.

The complete demonstration that mammalian pineal cells function as neuroendocrine transducers would require evidence that the sympathetic nervous input to these cells accelerates both the synthesis and the secretion of melatonin. Unfortunately, it has not yet been possible to study the rate of melatonin release in vivo. It is clear that melatonin *is* released from the pineal, for the indole is present in the urine and in tissues such as peripheral nerve, which lack HIOMT activity. It is, however, undetermined whether the indole is secreted into the general circulation as are most other hormones, or released directly into the cerebrospinal fluid. Teleological considerations, if not experimental evidence, support the latter possibility. Melatonin apparently produces its biological effects in mammals by acting on the brain. Isotopic studies have shown that the likelihood that a given molecule of melatonin can gain entry to the brain is almost 100-fold greater if it is placed in the cerebrospinal fluid than if injected into the blood stream. Methods must be developed for collecting the output of the pineal and assaying this material for melatonin, before it can be claimed with certainty that pineal cells are true neuroendocrine transducers.

## Summary

The three kinds of mammalian cells specialized to transmit signals from one organ to another are neurons, neuroendocrine transducer cells, and endocrine cells. Neurons and neuroendocrine transducer cells respond to neurotransmitter inputs, which usually reach them at synapses; the signals emitted by neurons are also neurotransmitters. The hormonal signals released by neuroendocrine transducer cells and

TABLE III

*Effects of Dibutyryl-AMP on Pineal Tryptophan Metabolism*

|  | Medium | | | Pineal | |
|---|---|---|---|---|---|
|  | $^{14}$C Melatonin (cpm) | $^{14}$C Serotonin (cpm) | $^{14}$C 5-HIAA (cpm) | $^{14}$C Protein (cpm) | $^{14}$C Tryptophan (cpm) |
| Control | 374 | 348 | 87 | 222 | 449 |
| D-AMP | 1464* | 870* | 63 | 230 | 473 |

*$P < 0.001$ differs from control.

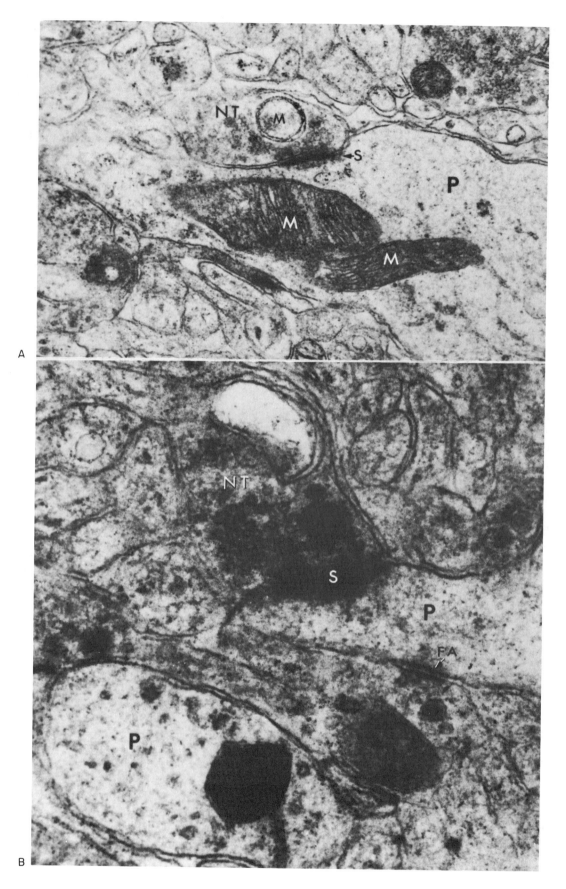

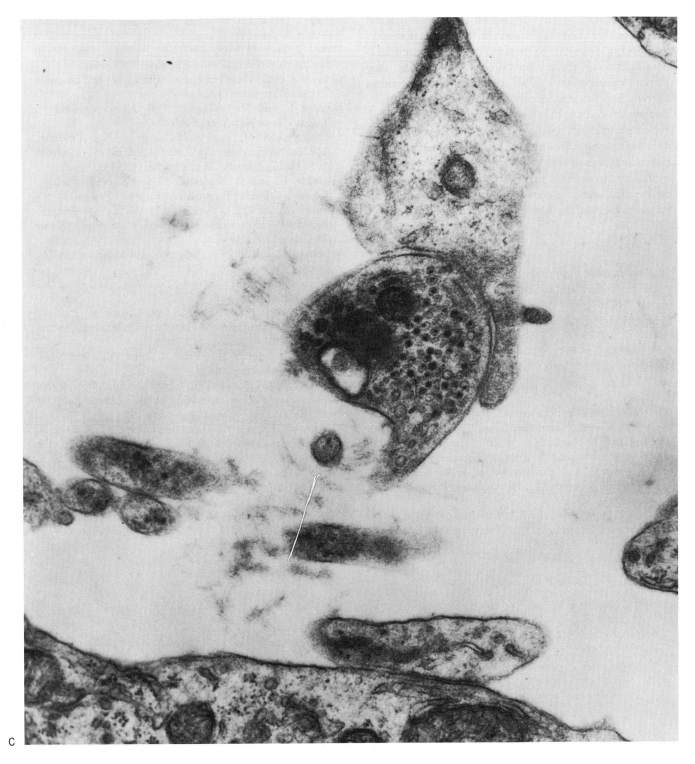

C

FIGURE 5 (*left*)  Electron micrographs of rat pineal organs, showing sympathetic nerve terminals near pineal parenchymal cells. S in A and B and FA in B show presynaptic and postsynaptic thickening. C (*above*) is the same as A and B, at a lower magnification. (Courtesy of Drs. Johannes Ariëns Kappers and David Wolfe.) *Abbreviations:* M, mitochondrion; NT, nerve terminal; P, pinealocyte.

endocrine cells are coded and complex chemicals distributed throughout the body by the general circulation; the inputs to endocrine cells are also hormones. Neural communications links attain privacy by anatomical means, i.e., they transmit signals only to cells that lie close to their terminal boutons. Endocrine communications systems attain privacy by chemical coding: their output is distributed universally, but only a relatively small population of cells is able to decode their message. Examples of neuroendocrine transducer cells include mammalian pinealocytes, mammalian adrenomedullary chromaffin cells, the "hypophysiotropic" cells of the hypothalamus, the cells of the supraoptic and paraventricular nuclei, and the juxtaglomerular cells of the kidney.

*Acknowledgment*

Studies described in this report were partially supported by grants from the National Aeronautics and Space Administration (NGR-22-009-272) and the National Institutes of Health (AM-11237 and AM-11709).

## REFERENCES

AXELROD, J., 1962. Purification and properties of phenylethanolamine-N-methyl transferase. *J. Biol. Chem.* 237: 1657–1660.

AXELROD, J., P. D. MacLean, R. W. Albers, and H. Weissbach, 1961. Regional distribution of methyl transferase enzymes in the nervous system and glandular tissues. *In* Regional Neurochemistry (S. S. Kety and J. Elkes, editors). Pergamon Press, Oxford, pp. 307–311.

AXELROD, J., H. M. Shein, and R. J. Wurtman, 1969. Stimulation of C¹⁴-melatonin synthesis from C¹⁴-tryptophan by noradrenaline in rat pineal in organ culture. *Proc. Nat. Acad. Sci. U. S. A.* 62: 544–549.

BUNAG, R. D., I. H. Page, and J. W. McCubbin, 1966. Neural stimulation of release of renin. *Circ. Res.* 19: 851–858.

COUPLAND, R. E., 1965. The Natural History of the Chromaffin Cell. Longmans Green, London.

MARTINI, L., and W. F. Ganong, 1967. Neuroendocrinology. Academic Press, New York, 2 vols.

POHORECKY, L. A., M. Zigmond, H. Karten, and R. J. Wurtman, 1969. Enzymatic conversion of norepinephrine to epinephrine by the brain. *J. Pharmacol. Exp. Ther.* 165: 190–195.

SCHARRER, E., and B. Scharrer, 1963. Neuroendocrinology. Columbia University Press, New York.

SHEIN, H. M., and R. J. Wurtman, 1969. Cyclic AMP: Stimulation of C¹⁴ indole synthesis by cultured rat pineals. *Science (Washington)* 166: 519–520.

WURTMAN, R. J., 1966. Catecholamines. Little, Brown and Co., Boston.

WURTMAN, R. J., and J. Axelrod, 1965. The pineal gland. *Sci. Amer.* 213 (no. 1): 50–60.

WURTMAN, R. J., and J. Axelrod, 1966. Control of enzymatic synthesis of adrenaline in the adrenal medulla by adrenal cortical steroids. *J. Biol. Chem.* 241: 2301–2305.

WURTMAN, R. J., J. Axelrod, and D. E. Kelly, 1968a. The Pineal. Academic Press, New York.

WURTMAN, R. J., J. Axelrod, E. Vessell, and G. T. Ross, 1968b. Species differences in inducibility of phenylethanolamine-N-methyltransferase. *Endocrinology* 82: 584–590.

WURTMAN, R. J., A. Casper, L. A. Pohorecky, and F. C. Bartter, 1968c. Impaired secretion of epinephrine in response to insulin among hypophysectomized dogs. *Proc. Nat. Acad. Sci. U. S. A.* 61: 522–528.

WURTMAN, R. J., H. M. Shein, J. Axelrod, and F. Laren, 1969. Incorporation of ¹⁴C-tryptophan into ¹⁴C-protein by cultured rat pineals: stimulation by *l*-norepinephrine. *Proc. Nat. Acad. Sci. U. S. A.* 62: 749–755.

# 49 The Olfactory Bulb as a Simple Cortical System: Experimental Analysis and Functional Implications

GORDON M. SHEPHERD

NEUROBIOLOGISTS ARE INTERESTED in the mammalian olfactory bulb for two main reasons. One is its obvious importance for olfactory discrimination. Less obvious is its relevance to the study of what may be called "cortical systems." It is, however, from this latter point of view that I wish to describe the experimental analysis of neuronal interactions in the olfactory bulb. This work has suggested a general scheme for the organization of neuronal systems within the bulb (Phillips et al., 1963; Shepherd, 1963b), and it has provided the physiological data for the theoretical analysis, described by Rall elsewhere in this volume, which predicted a dendrodendritic synaptic pathway for recurrent inhibition through the amacrine-type granule cells of the bulb (Rall et al., 1966; Rall and Shepherd, 1968). My aim is to describe what we now know about the neuronal organization of the bulb, to outline where we stand with regard to experimental evidence for the functional properties of granule cells, and to indicate some implications for our understanding of neuronal integration in the bulb and in other cortical systems in the central nervous system.

We may begin by defining a "cortical system" as a region of the central nervous system that has the following characteristics. The neurons are differentiated into several distinct types, and their cell bodies and cell processes are organized into several nonrepeating layers. One type, which may be designated as the output neuron, is large; its processes span most of the layers, and it sends its axon to other parts of the CNS. There is also usually a free surface. This definition includes regions as diverse as neocortex, cerebellum, hippocampus, prepyriform cortex, olfactory bulb, and, as I show below, retina; it generally excludes spinal, brain stem, and thalamic relay nuclei. By this it is not meant to imply that many of the neuronal mechanisms in cortical regions are not similar, in principle, to those in the relay nuclei; what is implied is that, in the course of vertebrate evolution, particular demands have arisen for the processing and retention of information, perhaps within certain spatial

GORDON M. SHEPHERD  Department of Physiology, Yale University School of Medicine, New Haven, Connecticut

constraints placed on the brain, and these demands have been met by a "cortical" organization.

If it be accepted that there are common principles of neuronal organization for the processing of information in parts of the vertebrate brain that have a cortical organization, then attention is naturally directed to cortical regions with the simplest construction, providing thereby the best hope for unambiguous physiological results. Among cortical regions, as above defined, the olfactory bulb is noteworthy for its simple and symmetrical organization. Moreover, it is one of the first cortical structures to emerge in the course of vertebrate evolution, and certain features of its organization are clearly recognizable throughout the vertebrate series. Within the bulb are several distinct neuronal types, the dendrites and axon collaterals of which are restricted to an unusual degree to different layers. These features, and others that are pointed out below, suggest that the olfactory bulb may be viewed as a "reduced" or simplified cortical system, an *E. coli* for cortical physiology, one might say, and it is from that point of view that our experiments have been undertaken.

## Anatomy of the bulb

The anatomical features of interest to the experimenter are illustrated in Figure 1. At the top is a drawing of the head of a rabbit; the olfactory bulb is situated in front of the forebrain, and is easily exposed for the purpose of recording with micropipettes simply by removing the overlying layer of bone. The main input to the bulb is by way of the bundles of olfactory nerve fibers from the nose; the main output is by way of the lateral olfactory tract on the surface of the prepyriform cortex of the brain.

The bottom diagram in Figure 1 represents Golgi-stained neurons from the layers in the dorsal part of the bulb. The neuron of first importance is the mitral cell; it sends out a primary dendrite to receive the input from the olfactory nerves coming from the surface of the bulb, and its axon is directed deeply and emerges posteriorly to form the lateral olfactory tract. Our basic experimental procedure (Phillips et al., 1963) was to place a fine stimulating wire on the ex-

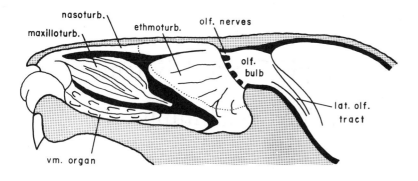

FIGURE 1    Top: Diagram of head of rabbit, showing positions of nasal cavity, olfactory bulb, and brain. Extent of olfactory mucosa on ethmoturbinate is indicated by dotted line. *Abbreviations:* ethmoturb., ethmoturbinate; lat. olf. tract, lateral olfactory tract; maxilloturb., maxilloturbinate; nasoturb., nasoturbinate; olf. bulb, olfactory bulb; olf. nerves, olfactory nerves; vm. organ, vomeronasal organ. (Modified from Allison, 1953.) Bottom: Golgi-stained neurons in bulb. Olfactory nerve fibers enter from dorsal surface, above. *Cell types:* G, granule; M, mitral; T, tufted. (Modified from Ramón y Cajal, 1911.)

posed tract and send single volleys of impulses into the bulb; there the volleys invade the mitral cells directly and synchronously in the antidromic direction. Another wire was placed on a bundle of olfactory nerve fibers, so that single orthodromic volleys could be set up (Shepherd, 1963a), and these caused synchronous synaptic activation of the mitral cells through the mitral primary dendrite.

It should be appreciated that the bulb can be activated over input and output pathways that are completely separated from each other, so that there is no danger of affecting the one when stimulating the other. In addition, the

pathways form discrete bundles and are therefore easily stimulated, and they lie fully exposed on the brain surface so that the stimulation is always under direct visual control. The separation of input and output pathways for the mitral cell is almost as complete as for the motoneuron in the spinal cord; the only qualification is the possibility of centrifugal fibers in or near the lateral olfactory tract at the point of stimulation (Ramón y Cajal, 1911; Heimer, 1968; Price, 1968).

These experimental advantages are unusual for a cortical structure in the CNS, and they are matched by some unusual characteristics of the neurons themselves within the bulb. The neurons are differentiated into several distinct types, and, most importantly, their processes are restricted to clearly defined layers. Thus, the olfactory nerve fibers all terminate in a single layer of rounded structures called glomeruli, a feature of all vertebrate olfactory bulbs from the lamprey to man. The mitral cell bodies all lie deeper, in a separate layer. The consequence of this arrangement is that the only way the olfactory nerve input can set up impulses in the mitral axons is through the mitral primary dendrite. Thus, as Ramón y Cajal (1911) stressed, the mitral primary dendrite is unique proof, on anatomical grounds alone, that dendrites in the CNS can play an essential role in transmitting nerve signals, rather than being limited to serving long-term or metabolic functions. The primary dendrite has a radial orientation and spans the entire external plexiform layer; in the young rabbits that we used its minimal length was 400–500 $\mu$ and its average diameter was of the order of 6 $\mu$.

The mitral cell also gives rise to several secondary basal dendrites which run laterally in the external plexiform layer. Their branches are entirely confined within this layer, which means that they are functionally isolated from the glomerular input and form a second, separate, dendritic field for the mitral cell. This is one of the clearest instances of what may be termed "fractionation of dendritic field" (cf. Sprague, 1958) in the CNS, the significance of which is mentioned below. The mitral axon gives off two kinds of collateral: deep ones directed laterally, and recurrent ones which run radially and terminate in the external plexiform layer. Ramón y Cajal suggested that the recurrent collaterals would provide for "avalanche conduction" by connections with neighboring mitral cells; our evidence for their action is discussed below.

Although the mitral cell is the primary output neuron of the bulb, the most numerous neuron is the granule cell. These have small cell bodies which lie in clusters below the mitral cell body layer. They give off a short central dendrite and a long peripheral dendrite which branches and terminates in the external plexiform layer. The dendritic branches are covered with gemmules (spines), resembling thereby the apical dendrites of cortical pyramidal cells; in contrast, the mitral dendrites are smooth, like motor-neuron dendrites. In specimens fixed in Bouin's fluid and stained with cresyl violet (Shepherd, 1966), the mitral dendrites are clearly apparent, but the granule dendrites (like cortical pyramidal dendrites) are not clearly seen. Thus, two distinctive types of dendrite seem to be intermingled in the external plexiform layer.

The classical Golgi studies were in agreement that the granule cell lacked an axon and was analogous in this respect to the amacrine cell in the retina. Ramón y Cajal supposed the peripheral process to carry activity in a cellulifugal di-

rection; although he could draw no further functional implication, he was impressed by the "absolute constancy of the connections between the granule cell processes and the mitral secondary dendrites in the external plexiform layer" (Ramón y Cajal, 1955). The evidence for dendrodendritic synapses gives a new meaning to that acute observation by Ramón y Cajal. Later I return to the possible functional analogies between granule cells in the bulb and amacrine cells in the retina. A note about terminology: although "amacrine" (*a*, neg. + *makros*, long + *inos*, fiber; having no long processes: the American Illustrated Medical Dictionary, 1951) may be an etymologically appropriate term for both these cell types, I retain the term "granule" in this report because of its established use for these cells in the olfactory bulb.

The other neuronal types in the bulb include, briefly, tufted cells that are scattered throughout the external plexiform layer. Their dendritic patterns are similar to those of mitral cells, but their axonal projections are different and, in fact, a matter of debate. Around the glomeruli are numerous cells with short axons (not shown in Figure 1) which connect neighboring glomeruli. Finally, there are some short-axon cells scattered in the granule cell layer. Although I concentrate on the mitral and granule cells in this report, our microelectrode studies have yielded information on the functional connections of these other cell types, too, and they are included in the schema for intrabulbar pathways in Figure 7.

## Excitation and inhibition of mitral cells

A shock to the lateral olfactory tract sets up a synchronous volley of impulses in the mitral-cell axons. The impulses travel antidromically in the axons into the olfactory bulb, where they invade the mitral cells. The mitral cells in turn activate other cells in the bulb to which they are synaptically connected. This activity generates extracellular current around the activated cells, mitral and others, and a microelectrode inserted into the bulb records the potential changes caused by the summed current flow generated by the populations of active cells.

A typical sequence of recordings from the surface to the depth of an olfactory bulb is shown on the right in Figure 2. On the left is a tracing of a histological section containing the electrode path over which these records were obtained. The records are shown connected to the depths at which they were recorded. Each trace begins with a short base line, followed by a gap indicating the artifact of the tract shock, and then a sequence of positive- or negative-potential deflections. The analysis of these summed potential patterns is described by Rall. Here it may be noted that the patterns are distinctive for different layers of the bulb and can therefore be used at the time of recording of unit potentials to

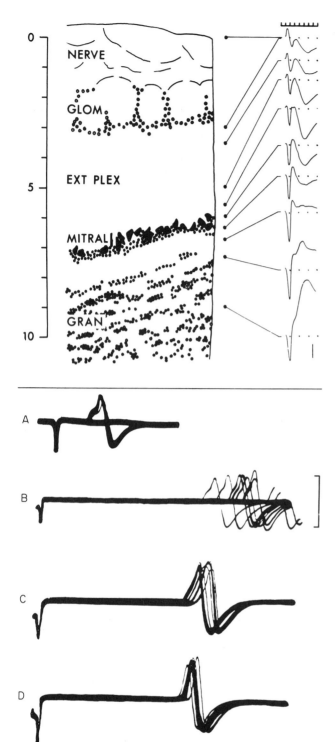

FIGURE 2 Correlation between extracellular potential wave responses to lateral olfactory tract volleys, and histological layers of rabbit's olfactory bulb. At left is tracing of Nissl-stained section containing track of recording pipette; at right, each response is connected to dot showing position along electrode track from which recording was made. Time scale above in milliseconds; small dots show recording base line at 2-millisecond intervals. Vertical bar at lower right is 1 millivolt. Depth scale at left in 100-$\mu$ divisions. *Histological layers:* olfactory nerve, glomeruli, external plexiform, mitral cell body, granule cell. (From Rall and Shepherd, 1968.)

Positivity of pipette recording tip is upward in this recording and in those that follow.

FIGURE 3 Impulse generation in single mitral cell. Extracellular recording with micropipette; bar at right indicates 30 millivolts. Ten sweeps are superimposed for each record. A. Antidromic impulse after stimulation of lateral olfactory tract. B-D. Orthodromic impulses after stimulation of olfactory nerve bundles at increasing intensities. Time in milliseconds. (From Shepherd, 1963a.)

find the depth of the micropipette tip (Phillips et al., 1961; 1963). Recordings similar to those in Figure 2 have been reported by Baumgarten et al. (1962) and Ochi (1963).

The first question of physiological interest is how impulses are generated in the mitral cell. Figure 3 shows micropipette recordings from a unit at the layer of mitral-cell bodies, as judged by the potential wave pattern. A single shock to the lateral olfactory tract elicited a single, all-or-nothing, positive-negative spike in successive trials. The recording tip was extracellular; the spike is of the order of 25 millivolts peak-to-peak, and is therefore a so-called "giant spike," like those that have been recorded from many other neurons in the CNS (cf. Granit and Phillips, 1956; Phillips, 1959). Record A in Figure 3 shows direct antidromic invasion at a fixed latency of about 1.0 millisecond and permits this unit to be positively identified as a mitral cell.

In Figure 3B a single shock to an olfactory nerve bundle also elicited a single spike. The latency was much longer because of the slow conduction velocity of the unmyelinated olfactory nerves, and the latency was variable because activation was across the synapse in the glomerulus, and the volley was near threshold. With stronger shocks (C and D), the latency decreased and became less variable. There was a prominent hesitation on the spike during antidromic invasion and during threshold orthodromic activation; this was reduced by the stronger volleys, and the spike duration also decreased. It appears that there is an A-B sequence of impulse generation, from, presumably, axon hillock to cell body, and that there is tighter coupling with increasing synaptic depolarization. Thus, the mitral cell generates spikes in a sequence like that in motor neurons (cf. Fuortes et al., 1957; Eccles, 1957) and many other neurons, despite the particular morphology of its dendrites. That the spikes in mitral cells are brief is important in regard to the recurrent input from the granule cells, as is discussed below.

Prolonged "dendritic" spikes (Tasaki et al., 1954) were never seen, except during injury discharges.

Do these volleys produce any other effect on the excitability of the mitral cell? This was studied by testing the ability of a mitral cell to respond with an antidromic impulse after a conditioning antidromic or orthodromic volley. Some mitral cells could respond at any time except during the first few milliseconds, corresponding to the absolute refractory period; in such cases, the refractory period was similar after either an antidromic or an orthodromic impulse.

In many mitral cells, however, there was a long-lasting blockage of impulse generation, as illustrated in Figure 4.

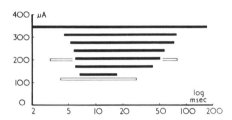

FIGURE 4 Relation between strength of conditioning shocks to lateral olfactory tract (ordinate) and duration of blockage of impulse generation (abscissa) in two mitral cells. Axon thresholds: closed bars, 330 microamperes; open bars, 750 microamperes. Bars indicate periods of blocking of testing antidromic spikes. (From Phillips et al., 1963.)

The closed bars indicate the periods during which antidromic impulse invasion in one mitral cell was blocked after a conditioning antidromic volley. It can be seen that the onset, duration, and recovery of blockage were graded with the strength of the conditioning shock. All the conditioning shocks, except the strongest (340 microamperes), were below threshold for the axon of this mitral cell. It may be presumed that the shocks were above threshold for other mitral axons in the lateral olfactory tract, and that connections from these other mitral cells caused the blockage of impulse generation in the cell. Therefore, the blockage can be ascribed to true inhibition, as distinct from depression following a spike in this cell. When a conditioning spike did occur (340 microamperes in Figure 4), the blockage lasted much longer, and it could be shown that in many cells the duration was graded with the strength of the suprathreshold conditioning shocks. In such cases, the blockage appears to be due to a combination of refractoriness after the spike, self-inhibition of the cell, and inhibition caused by other mitral cells.

The latency of onset of the inhibition became shorter as the shocks were made stronger, presumably by activation of more mitral axons in the tract. In Figure 4 (closed bars)

the minimum latency with the shock just subthreshold was 4.8 milliseconds; with the shock just suprathreshold the refractoriness in the aftermath of the spike merged with the period expected for inhibition. In some mitral cells these two periods were not continuous, and a second antidromic spike could be interjected between the two periods. Detailed study showed that the minimum delay for the onset of the inhibition was about 3.0 milliseconds, with a range of from 3 to 5 milliseconds in our population of mitral cells. In some cells the inhibition had a very long latency, up to tens of milliseconds; in other cells it could be elicited only by repetitive stimulation of the tract. All these characteristics suggest that the inhibition is mediated by a pathway that contains more than one synaptic relay. The maximum duration of the inhibition was of the order of several hundred milliseconds.

Closely similar results were obtained when an antidromic impulse was interjected in the aftermath of a conditioning shock to the olfactory nerves. In some cells the antidromic impulse was blocked only in the brief refractory period after the orthodromic impulse. In many cells, however, blockage was long-lasting and resembled the inhibition after a tract shock. Thus, it occurred after a subthreshold conditioning shock; its timing of onset and recovery was graded with conditioning shock strength; the minimum latency of onset was several milliseconds after the arrival of the orthodromic volley in the bulb. It was concluded that an orthodromic volley activated the same pathways within the bulb for mitral inhibition, in roughly the same temporal sequence, as did an antidromic volley. This similarity extended to the finding that an antidromic spike could also be interjected between the periods of refractoriness and inhibition after a spike that was elicited orthodromically. The results from orthodromic conditioning were therefore important in establishing that mitral-cell inhibition could be produced by activation of natural orthodromic pathways.

The analysis of mitral-cell inhibition has been carried out largely with extracellular unit recordings, because intracellular recordings are exceedingly difficult to obtain from the relatively small mitral cell and, when they are obtained, the membrane potential is low and deteriorates rapidly. The few successful recordings by Yamamoto et al (1963) and ourselves (Phillips et al., 1963), recently confirmed by Nicoll (1969), show clearly, as in Figure 5A, that a hyperpolarization of the mitral-cell membrane occurs after a volley in the tract or olfactory nerves, and that the timing of onset and recovery, and grading with shock strength, are similar to those of the spike inhibition recorded extracellularly. In the recording in Figure 5A, hyperpolarization began at about a 5-millisecond latency and lasted for about 30 milliseconds; the peak hyperpolarization was about 7 millivolts, which is not very meaningful because the resting potential was low and deteriorating. The recordings per-

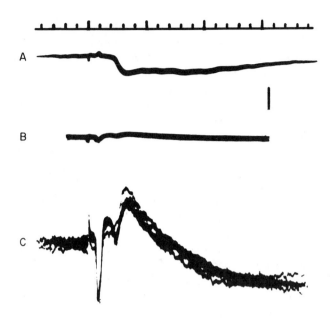

FIGURE 5  A. Intracellular recording from a mitral cell, showing hyperpolarization following lateral olfactory tract shock. Vertical bar indicates 10 millivolts. Extracellular recording at same (B) and higher amplification (C) after loss of cell. Time in 2-millisecond intervals.

mit, however, the tentative conclusion that this is an inhibitory postsynaptic potential (IPSP), and we presume, in the absence of any other evidence, that it arises from ionic-conductance changes in the mitral-cell membrane similar to those responsible for IPSPs in motor neurons and other CNS neurons (cf. Eccles, 1964).

## Neuronal systems in the bulb

The relatively long latency of the mitral-cell inhibition raises the likelihood of an interneuron in the inhibitory pathway. It will be recalled from the description of bulbar anatomy that the secondary dendrites of the mitral cells are intermingled with the radial processes of the granule (amacrine) cells in the external plexiform layer. We therefore suggested, on the basis of our unitary studies (Phillips et al., 1963; Shepherd, 1963b), that the granule cell is the interneuron which mediates inhibition of the mitral cells, through inhibitory synapses on the mitral-cell secondary dendrites.

What, then, is the pathway for activation of the granule cells? One possibility is that deep collaterals from mitral axons have excitatory synapses on deep dendrites of granule cells; the activated granule cell would then inhibit the mitral cells. This would be analogous to the pathway for Renshaw inhibition in spinal motor neurons (Eccles et al.,

1954). Evidence relating to such a pathway was gained from unit recordings in the granule layers. As shown in Figure 6A, a threshold shock to the tract was followed by the brief field potential associated with mitral-body invasion and then a single spike. With stronger shocks—two, three, and four times threshold—there was a repetitive discharge of two, three, and four spikes. In successive trials at ×1 threshold, the spike latency varied somewhat; with increasing shock strength the latency of the first spike decreased. These are properties of synaptic activation, and are in contrast to antidromic spikes in mitral cells, which always had fixed latencies.

It is of interest that a different type of repetitive response is found in units situated superficially, near the olfactory glomeruli. The unit in Figure 6B responded to an olfactory nerve volley at threshold with a train of seven spikes, lasting about 30 milliseconds. The shock to the nerves was increased from (a) to (e); at three times threshold (c) the discharge increased only from seven to nine spikes, with the same train duration, and at 12 times threshold (e) the discharge still consisted of nine spikes, and the train duration was actually shorter. It can be seen that the spikes in the middle of the train were markedly reduced in amplitude. This appears to be the kind of "inactivation" found in Purkinje cells (Granit and Phillips, 1956) and hippocampal pyramidal cells (Kandel and Spencer, 1961) when they are excessively depolarized. Intracellular recordings by Yamamoto et al. (1963) have revealed the appropriate depolarizing response of cells in this region to olfactory nerve volleys. An interesting feature of these cells is that the impulse discharge is a very stereotyped affair, independent to a certain extent of the amount of depolarization impressed on the cell by the afferent volley.

A summary of the relation between the physiological findings and the anatomy of the bulb is provided by the schematic diagram in Figure 7. The anatomy is based on the studies of Ramón y Cajal (1911), as in Figure 1; the excitatory (E) and inhibitory (I) synaptic connections are based primarily on the results of our unit recordings.

To summarize briefly, an antidromic impulse in one mitral axon invades the cell and subsequently causes inhibition of that cell and neighboring cells through inhibitory synapses from the granule cell onto the mitral secondary dendrites. The repetitive discharges of units in the granule layer seemed to provide evidence for activation of the granule cells through the deep collaterals of the mitral axons, in analogy with the Renshaw pathway from motoneuron axon collaterals to interneurons in the spinal cord (Eccles et al., 1954). We were puzzled, however, at the time (Phillips et al., 1963; Shepherd, 1963b) by the fact that the granule layer units were found only infrequently, in contrast to the widespread nature of the mitral inhibition and the large numbers of granule cells. Furthermore, the repeti-

A         B

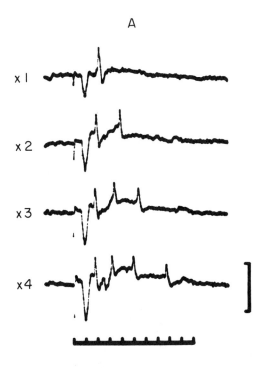
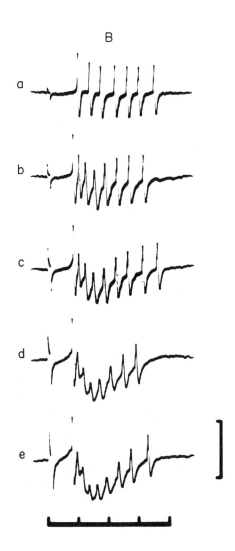

FIGURE 6 Two types of repetitive response found in bulbar units. A. Unit recorded in granule layer; synaptic responses to tract volley at threshold (×1) and at two, three, and four times threshold. Time in milliseconds; vertical bar, 2 millivolts. (From Shepherd, 1963b.) B. Unit recorded in outer zone of external plexiform layer; synaptic responses to olfactory nerve volley at threshold (a), and at two (b), three (c), six (d) and 12 (e) times threshold. Time, 10 milliseconds; vertical bar, 5 millivolts.

tive discharges were very brief, so that one had to postulate some special property, such as prolonged transmitter action, to account for the prolonged mitral inhibition.

Recent work has shown (Rall et al., 1966; Rall and Shepherd, 1968) that there is another pathway for activating the granule cells, through the same mitral dendrites that are then inhibited. This new type of functional pathway was suggested by the analysis, described by Rall in this volume, of the extracellular field potentials in the bulb (cf. Figure 2). The synaptic connections that would provide for such a pathway have been demonstrated in several electron-microscope studies (Hirata, 1964; Andres, 1965; Rall et al., 1966). The electron micrograph in Figure 8, from the work of

Renshaw type of pathway from mitral-axon collaterals would be basically the same: activation of the granule cell followed by inhibition of the mitral cell (cf. Figure 9).

It remains to be determined whether the repetitively firing units in the granule layer are, in fact, granule cells, or are short-axon cells playing a similar or perhaps different functional role. In contrast, units in the external plexiform layer respond to a tract volley with a single variable latency spike. This type of response suggests synaptic excitation, most probably via the recurrent collaterals of the mitral cells, followed by synaptic inhibition via the granule cells. Tract volleys rarely activate glomerular cells, but sometimes cause weak inhibition, suggesting that granule cells may

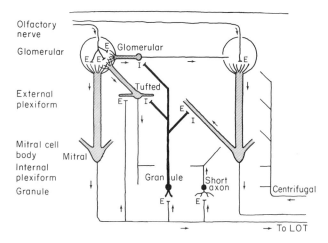

FIGURE 7 Schematic diagram of olfactory bulb pathways and connections between different neuron types, based on Ramón y Cajal (1911) and recent findings. Presumed excitatory and inhibitory connections are labeled E and I, respectively. Small arrows indicate direction of activity. Histological layers of the bulb are indicated at left. For explanation, see text. (Modified from Shepherd, 1963b.)

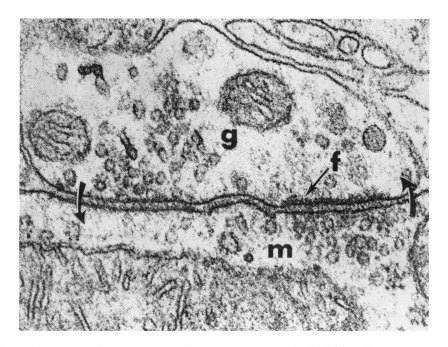

FIGURE 8 Characteristic synapses between mitral-cell dendrite (m) and gemmule of a granule-cell dendrite (g). There are two synaptic contacts with opposite polarities (determined by cluster of vesicles), as indicated by the arrows. These synapses provide for mitral-to-granule excitation and granule-to-mitral inhibition (see text). In mitral-to-granule synapse, (f) identifies dense filamentous material attached to postsynaptic membrane. Lead citrate. ×100,000 (From Rall et al., 1966.)

Reese and Brightman (Rall et al., 1966), shows clearly the side-by-side synapses of opposite orientation which would provide for the sequence of mitral-to-granule excitation followed by granule-to-mitral inhibition that is postulated by the physiological analysis. The synapses connect mitral secondary dendrites with the peripheral dendrites of the granule cells, and we have therefore termed this a *dendro-dendritic synaptic pathway for recurrent inhibition* of the mitral cells. The effect of either the dendrodendritic pathway or a

also have inhibitory synapses on them. Thus, the unit recordings lead to the concept of the granule cell as the common inhibitory interneuron in the olfactory bulb, mediating inhibition of the tufted and glomerular cells as well as of the mitral cells.

To turn to the pathways for orthodromic activation—a volley in the olfactory nerves synaptically excites all the cells connected to the glomeruli. The glomerular cell tends to respond with repetitive firing, the tufted and mitral cells with a single impulse followed by inhibition. An important point is that the inhibition is brought about over the same pathways through the granule cells as in the antidromic case. The synaptic actions of glomerular and tufted cells are still a matter for conjecture. The *glomerular cell* has horizontal connections which could provide for a spread of activity to mitral-cell dendrites in neighboring glomeruli; this spread would be opposed by inhibition from the granule cells. Thus, inhibition from the granule cell could be pitted against excitation from the afferent nerves and glomerular cells, thereby providing a mechanism for modulating excitability at the level of the afferent input to the mitral dendritic tuft in the glomerulus. The *tufted-cell* axon, according to Ramón y Cajal, gives off many collaterals in the internal plexiform layer just below the mitral-cell bodies. These could provide for synaptic activation of granule cells, with the consequent recurrent inhibition, or they could provide for a synaptic modulation of granule-cell excitability, to control the amount of inhibition delivered by the granule cells when activated by the mitral cells. The diagram in Figure 7 emphasizes that the bulb actually contains several neuronal systems related to the glomerular, tufted, and mitral cells. To a degree, these systems operate independently; to a degree, they compete for control of the excitability of the mitral and granule cells.

## Functional implications of neuronal interactions in the bulb

I suggested in the introduction that the bulb could be viewed as a simple cortical system. The test of such a view is the extent to which the experimental results have implications beyond the bulb. Specifically, do the neuronal interactions as summarized in Figure 7 have relevance only to the peculiar demands of discriminating olfactory stimuli, or do they, in addition, provide insights into the functional organization of other "cortical" regions? The results from the unit studies that I have described, and from the theoretical studies of summed extracellular potentials that Rall describes, do appear to have functional implications of general interest. These may be summarized as follows; some of the points are discussed here, and some are treated in more detail by Rall in his chapter.

1. KEY INTEGRATIVE FUNCTION FOR AXONLESS CELL The unit analysis, potential wave analysis, and electron-microscope studies provide clear evidence that a cell that lacks an axon can, nonetheless, play a key role in nervous integration. In the granule cell, this role appears to be an inhibitory one. The dendrodendritic synapses between granule and mitral-cell dendrites provide for close control of the mitral-cell output. It appears that the granule cell also inhibits tufted and glomerular cells; thus, an interneuron

may have inhibitory synapses on more than one neuron type.

2. THE DEFINITION OF DENDRITE The finding of synaptic connections between the dendrites of mitral and granule cells raises questions about our present terminology for the different parts of a neuron. The neuron doctrine has never been satisfactory in this regard; the problem has been well reviewed by Bodian (1962), who offered the following definitions consistent, at the time, with both morphological and physiological evidence. Dendrites, he suggested, are those processes that "receive synaptic endings of other neurons"; axons "conduct nervous impulses away from the dendrites," and axon telodendria (terminals) provide for "synaptic transmission to the dendrites of other neurons."

These definitions, as they stand, must now be modified to account for the findings in the olfactory bulb. The peripheral process of the granule cell not only receives synaptic excitation *from* mitral cells, but also provides synaptic inhibition *to* the mitral cells. It might be argued that, because the granule cell lacks a morphologically identifiable axon, the peripheral dendritic processes are functionally analogous to axonal telodendria. Such an explanation cannot apply to the mitral cell, however, which clearly does have an axon; the mitral-cell dendrites are presynaptic as well as postsynaptic to the peripheral dendrites of the granule cells. Thus, both the mitral and granule dendrites perform functions as synaptic effectors as well as synaptic receptors. In addition, the effector function of one kind of dendrite (mitral) is excitatory, and the other kind (granule) is inhibitory.

Several suggestions can be offered toward resolving this terminological dilemma. One is that the criteria of Bodian may still be valid if they are not exclusive, that is, if dendrites may transmit as well as receive, and axonal terminals may receive as well as transmit. The latter concept has already been introduced by work on presynaptic inhibition in peripheral and central neurons (cf. Eccles, 1964). The designation of a neuronal process as a "dendrite" might then be based on its *pre-axonal position* in the overall, normal flow of activity in a neuron, rather than on an exclusively receptor function. It would, of course, be convenient if one could identify a dendrite by specific ultrastructural criteria, or by a functional property such as the generating of graded potentials in contradistinction to propagating impulses, as in axons. Evidence for such criteria that would apply to all vertebrate neurons, however, is insufficient at present.

If a process is to be identified as a dendrite on the basis of its pre-axonal position, how then to identify the peripheral process of the granule cell, which lacks a morphologically identifiable axon? Here one can begin with the common-sense observations that what is not axon must be dendrite, that with Golgi stains the peripheral process can be seen to have branches and spines which resemble those of the apical dendrites of pyramidal cells in neocortex and hippocampus, and that a close resemblance of these dendrites is also found at the ultrastructural level (cf. Rall et al., 1966; Price, 1968). With regard to the pre-axonal position of this process, it may be suggested that it must be seen in the context of the synaptic connections with the mitral-cell dendrites. In this context, the peripheral process of the granule cell is a dendrite with respect to the axon of the mitral cell.

This view can be spelled out in more detail by reference to the diagram in Figure 10. To anticipate the discussion (point 7, below), this diagram shows the similar positions of granule and amacrine cells in the organization of the bulb and retina. In the normal flow of activity through the bulb, the granule cell, as do the mitral-cell dendrites, provides for the spread of activity (excitatory or inhibitory) prior to integration at the mitral-axon hillock and conversion there into the frequency-coded impulse discharge of the mitral axon. Seen in this context, the granule cell is "dendritic" vis-à-vis the mitral-axon hillock in both its receptor and its effector synaptic functions. Similar considerations apply to the retinal amacrine cell. In addition, the same argument can be applied to the bipolar cells in the retina. Figure 10 shows that the bipolar cell is analogous to the primary dendrite of the mitral cell, in that both of these processes pass on activity from the receptor level to the axon hillock of the output neuron. The entire bipolar cell functions as a dendrite with respect to the axon hillock site (on the ganglion cell) where its activity is integrated with that of the horizontal (amacrine) pathways.

It may be noted that this extension of the basis for neuron terminology to include the functional context is in line with Bodian's (1962) statement that "the basic role of the neuron is that of its relation to other cells." This view may now be seen to be completely general, in the sense that instances may be found in which any of the conventionally designated parts of a neuron (dendrites, soma, axons, telodendria) may have receptive (postsynaptic) or transmissive (presynaptic) sites.

3. DIFFERENT TYPES OF DENDRITE  In Golgi and electron-micrograph sections the mitral dendrites are smooth, as are motor neuronal dendrites, whereas the granule dendrites are invested with spines, as are apical dendrites of pyramidal cells. The two types of dendrite stain differentially in Bouin's fixed material. Are these two basic types of dendrite, and, if so, what is their functional significance? Is it related to their excitatory and inhibitory synaptic connections? Or is it related to rapid impulse generation in the smooth-dendrite cell and slow, partial, or prolonged regenerative activity in the spinous-dendrite cell? In this view

the smooth-dendrite cell might provide for instantaneous integration of synaptic inputs, while the spinous-dendrite cell might provide for long-term integration and possibly retention of information.

4. FULL ACTION POTENTIAL NOT REQUIRED FOR DENDRITIC FUNCTION  Neither mitral nor granule dendrites appear to need a propagating action potential to carry out their different functional roles. In the mitral cell, the primary dendrite provides for coupling between the synaptic input in the glomerulus and impulse generation in the mitral-axon hillock, and the secondary dendrite provides coupling between the hillock and the sites of dendrodendritic synaptic action. The electrotonic lengths of both types of dendrite appear to be adequate for effective spread of impulse depolarization, even if the dendritic membrane were passive. There is reason to belive that regenerative activity in the granule cell may be partial or decrementing; spread of depolarization to neighboring gemmules could occur because of the relatively short distances in the granule dendritic tree. It is clear, however, that the granule-cell response to the mitral dendritic input does not involve the entire granule cell. Regenerative activity, if present, must be largely restricted to the peripheral dendritic branches in the external plexiform layer.

5. GENERALIZES RECURRENT INHIBITION  Figure 9 shows a comparison between the new dendrodendritic pathway for recurrent inhibition and the classical Renshaw pathway through an axon collateral. In both diagrams it is assumed

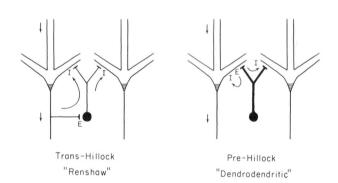

Trans-Hillock
"Renshaw"

Pre-Hillock
"Dendrodendritic"

FIGURE 9  Two types of pathway for recurrent inhibition. Left: Classical Renshaw inhibitory pathway (Eccles et al., 1964) from axon collateral through interneuron. Right: New dendrodendritic inhibitory pathway (Rall et al., 1966; Rall and Shepherd, 1968) from dendrite through interneuron. Direction of normal flow of activity is shown by arrows. Excitatory and inhibitory synapses are labeled E and I, respectively. Site of impulse generation at axon hillock is indicated by shading.

that a synaptic input from above generates an impulse in the hillock region. In Renshaw inhibition, the impulse on the axonal side of the hillock, through synaptic excitation of the interneuron, delivers synaptic inhibition to the same cell and to neighboring cells. In dendrodendritic inhibition, the impulse spreading back into the dendrites excites the interneuron synaptically, which, in turn, inhibits the same cell or neighboring cells.

We have suggested that the term "recurrent inhibition" should be generalized to cover each type of pathway, because both provide for inhibitory feedback to the same or neighboring cells. As the Renshaw circuit loops back from axon to dendrite, it can be viewed as "trans-hillock." In comparison, the dendrodendritic circuit is seen to be "pre-hillock." Further work is required on the functional significance of the inhibitory control that is possible over these pathways in different regions of the CNS. It is of interest that both pathways appear to be present in the granule cell.

It may be noted that the pre-hillock pathway appears to require a "fractionation" of the dendritic field (cf. Sprague, 1958), so that the initial synaptic excitation of the cell does not "shunt out" the dendrodendritic synapses. The synapses are activated by the impulse spreading from the hillock, and, as mentioned above, the impulse is brief, so that shunting between the synapses does not occur.

6. RHYTHMIC POTENTIALS  The electrical activity of the bulb is characterized by potential oscillations ("EEG waves") of large amplitude. We have noted that the dendrodendritic synaptic interactions are well suited for the development of such rhythmic activity. The mechanism can be understood with reference to Figure 10 (bulb). Impulse discharge in mitral cells results in synaptic excitation of many granule cells; these granule cells then deliver graded inhibition to many mitral cells. This inhibition cuts off the source of synaptic excitatory input to the granule cells. As the granule-cell activity subsides, the amount of inhibition delivered to the mitral cells is reduced, which permits the mitral cells to respond again to the excitatory input from the glomeruli. In this way a sustained excitatory input to the mitral cells would be converted into a rhythmic sequence of impulse followed by inhibition, locked in timing to a rhythmic activation of the granule cell pool.

7. MECHANISM FOR SELF AND LATERAL INHIBITION  It appears probable that, during natural activity, a mitral cell exerts on its neighbors the kind of lateral inhibition that is a common feature of many sensory systems. This mechanism might operate in the bulb to sharpen contrast, or it might be related to the adaptation that is such a marked feature of olfactory sensations. The lateral spread of the inhibition would be promoted by the long lengths of the mitral

secondary dendrites, the numerous reciprocal synapses on granule dendrites, and the spread of depolarization in the excited granule dendrites. Within this framework, the inhibition may have a selective action because of the local character of the feedback. This kind of selective action is not possible with the Renshaw pathway, because the impulse in the axon of the interneuron presumably invades all the axonal terminals (cf. Figure 9).

8. SIMILARITIES IN ORGANIZATION OF OLFACTORY BULB AND RETINA  Because the granule cell and the retinal amacrine cell are the two outstanding examples in the CNS of neurons without axons, it is of interest to see what further comparisons can be made between the two. These lead to the recognition of some basic similarities in the organization of the bulb and the retina, which are discussed in relation to the two schematic diagrams in Figure 10.

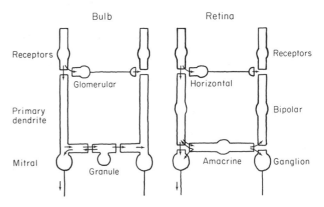

FIGURE 10  Diagram showing similarities in organization of olfactory bulb and retina. Neurons are shown schematically; note especially the compression of receptor and granule cells in bulb, to emphasize their primary functional connections. Normal flow of activity is indicated by arrows. For presumed synaptic actions and full discussion, see text.

The bulb appears to be organized along the following principles. There are input cells (the olfactory receptors) and output neurons (the mitral cells). There are vertical pathways for direct activation of the bulbar output neurons and horizontal pathways for integrating the vertical paths. The horizontal connections are organized into two tiers, the first apparently for controlling the bulbar input, the second for controlling the bulbar output. It can be seen that the granule cells are at the level of output control; this control is carried by the reciprocal synaptic connections with mitral secondary dendrites, which can be viewed as sidearms on the vertical pathways. The control exerted by the granule cells on the mitral-cell output appears to be exclusively inhibitory. The synaptic actions of the glomerular cells, at the input level, are not known. Further modification of the mitral-cell output would arise from activity in pathways related to tufted cells, mitral-axon collaterals, and short-axon cells (cf. Figure 7).

The diagram of the retina on the right of Figure 10 is based on the Golgi studies of Ramón y Cajal (1911) and the recent electron-microscope studies of Dowling and Boycott (1966). It shows that the retina appears to be organized on principles similar to those of the olfactory bulb. There are input cells (the visual receptors) and output neurons (the ganglion cells). There are vertical pathways from receptors through bipolar cells to ganglion cells, and horizontal pathways connecting the vertical paths. The horizontal connections are also organized in two distinct tiers, the first related to control of receptor input to the bipolar cells, and the second related to control of retinal output by the ganglion cells. It can be seen that the amacrine cell, as is the granule cell in the bulb, is at the level of output control; this control is through connection with the vertical pathway by reciprocal synapses (Kidd, 1962; Dowling and Boycott, 1966) between the amacrine cells and the bipolar terminals. In the primate retina, the bipolar terminals form synapses with both ganglion cells and amacrine cells, as depicted in the diagram. In the more complex retina of the frog, one or more amacrine processes are interposed between the bipolar terminals and the ganglion cells. This does not alter the basic plan shown in the diagram, except for the removal of most of the direct bipolar-to-ganglion synapses, and the substitution of amacrine-to-amacrine relays where necessary.

On morphological grounds, therefore, granule cells and amacrine cells exhibit striking similarities. Both lack an axon, and both take part in reciprocal synaptic connections. The granule cell is both postsynaptic and presynaptic to the mitral cells, and the amacrine cell is both postsynaptic and presynaptic to the bipolar cells. On the basis of these similarities we suggested (Rall et al., 1966; Rall and Shepherd, 1968) that the retinal amacrine cells might provide a type of dendrodendritic inhibitory pathway similar to that demonstrated for the granule cell in the bulb. A further similarity, brought out by the diagrams in Figure 10, is that the inhibition by the amacrine cells would also be at the level of output control, by monitoring the bipolar input to the ganglion cells.

This hypothesis must, of course, be tested by work on the retina. Recent studies by Werblin and Dowling (1969), A. Kaneko (personal communication), and D. A. Baylor (personal communication), combining unit recordings and histological identification, are therefore of great interest. These workers are in general agreement that, in the sequence from receptors to ganglion cells, center-surround antagonism is first found in bipolar cells. This is consistent with the hypothesis that amacrine cells are inhibitory to

bipolar terminals, although it does not rule out an inhibitory input from horizontal cells to bipolars. Inhibition from amacrine cells might also be related to adaptation or to directional sensitivity; these possibilities are under investigation (Barlow and Levick, 1965; Werblin and Dowling, 1969).

In view of the similarities in the organization of retina and bulb, work on the retina takes on added significance for the bulb. Little is known of the bulbar mechanisms underlying olfactory discrimination, mainly because the odorous stimulus cannot be adequately controlled. By contrast, precise control of light stimuli has enabled much to be learned about the functional properties of the retina. If the retina and bulb are similar in their organization, the kinds of information and the ways in which the information is handled in the bulb might be similar to those in the retina. As an example, lateral inhibition in the retina provides a mechanism for heightening contrast at a light-dark boundary. In view of the pathway for lateral inhibition demonstrated in the bulb, does this imply a mechanism to heighten contrast at heretofore unknown "boundaries" in the olfactory input to the bulb? As another example, the analytical capacity of ganglion cells of different species appears to be related to the complexity of the synaptic connections between bipolar, amacrine, and ganglion cells (cf. Michael, 1969). These synapses provide the ganglion cells with additional degrees of freedom from vertical and horizontal control, in contrast to the direct control in the case of the mitral cell. Does this imply that the mitral cell is relatively limited in the amount of analysis that it can carry out on the olfactory input? Such questions may give insight into the nature of the olfactory stimulus and suggest new experimental approaches to the difficult problem of olfactory discrimination.

## Summary

The mammalian olfactory bulb can be viewed as a simple cortical system. Because of the separation of afferent and efferent pathways, selective orthodromic and antidromic activation of the neuron populations within the bulb can be achieved. Unit recordings have permitted analysis of spike generation in mitral cells and of pathways for synaptic excitation and inhibition of other types of neurons. Biophysical reconstruction of field potentials evoked in the bulb led to the postulation of dendrodendritic synaptic interactions between mitral and granule cells. Electronmicrographs have revealed the postulated synapses between mitral and granule-cell dendrites. These synapses provide a pathway for recurrent inhibition of the mitral cells; we have suggested that the term "recurrent inhibition" be generalized to cover this dendrodendritic synaptic pathway as well as the classical Renshaw pathway through an axon collateral. Although it lacks an axon, the granule cell nevertheless plays a key integrative role by virtue of its inhibitory control of mitral cells.

The results thus far in the olfactory bulb have implications of general interest to neurobiologists, particularly with regard to the nature of recurrent inhibition, the integrative properties of neuronal dendrites, and the organization of the retina and of other cortical structures. An important aspect of this work has been the extent to which several methodologies and techniques—histological, ultrastructural, physiological, biophysical—have been brought to bear on single problems, e.g., the synaptic pathway for recurrent inhibition of mitral cells. This may be taken to reflect the advantages of working on a simple or "reduced" cortical system like the bulb. As Rall shows in his chapter in this volume, the simple features of the bulb have also permitted realistic biophysical models to be constructed for both the mitral and granule-cell populations. Thus far our interest has been in gaining understanding of the functional organization of the neuronal populations within the bulb; it is hoped that this will lay a foundation for future work on the neuronal mechanisms underlying the sensory discrimination of odors.

## Acknowledgments

A number of my colleagues have kindly read the manuscript and given valuable advice. Among them I am particularly indebted to Drs. W. Rall, C. G. Phillips, T. S. Reese, D. Ottoson, C. M. Michael, and D. Bodian.

## REFERENCES

ALLISON, A. C., 1953. The morphology of the olfactory system in the vertebrates. *Biol. Rev.* (*Cambridge*) 28: 195–244.

ANDRES, K. H., 1965. Der Feinbau des Bulbus olfactorius der Ratte unter besonderer Berücksichtigung der synaptischen Verbindungen. *Z. Zellforsch. Mikroskop. Anat.* 65: 530–561.

BARLOW, H. B., and W. R. LEVICK, 1965. The mechanism of directionally selective units in rabbit's retina. *J. Physiol.* (*London*) 178: 477–504.

BAUMGARTEN, R. VON, J. D. GREEN, and M. MANCIA, 1962. Slow waves in the olfactory bulb and their relation to unitary discharges. *Electroencephalogr. Clin. Neurophysiol.* 14: 621–634.

BODIAN, D., 1962. The generalized vertebrate neuron. *Science* (*Washington*) 137: 323–326.

DOWLING, J. E., and B. B. BOYCOTT, 1966. Organization of the primate retina: Electron microscopy. *Proc. Roy. Soc., ser. B, biol. sci.* 166: 80–111.

ECCLES, J. C. 1957. The Physiology of Nerve Cells. Johns Hopkins Press, Baltimore.

ECCLES, J. C., 1964. The Physiology of Synapses. Springer-Verlag, Berlin.

ECCLES, J. C., P. FATT, and K. KOKETSU, 1954. Cholinergic and inhibitory synapses in a pathway from motor-axon collaterals to motoneurons. *J. Physiol.* (*London*) 126: 524–562.

FUORTES, M. G. F., K. FRANK, and M. C. BECKER, 1957. Steps in the production of motoneuron spikes. *J. Gen. Physiol.* 40: 735–752.

GRANIT, R., and C. G. PHILLIPS, 1956. Excitatory and inhibitory processes acting upon individual Purkinje cells of the cerebellum in cats. *J. Physiol.* (*London*) 133: 520–547.

HEIMER, L., 1968. Synaptic distribution of centripetal and centrifugal nerve fibres in the olfactory system of the rat. An experimental anatomical study. *J. Anat.* 103: 413–432.

HIRATA, Y., 1964. Some observations on the fine structure of the synapses in the olfactory bulb of the mouse, with particular reference to the atypical synaptic configurations. *Arch. Histol. Jap.* 24: 293–302.

KANDEL, E. R., and W. A. SPENCER, 1961. Electrophysiology of hippocampal neurons. II. After-potentials and repetitive firing. *J. Neurophysiol.* 24: 243–259.

KIDD, M., 1962. Electron microscopy of the inner plexiform layer of the retina in the cat and pigeon. *J. Anat.* 96: 179–187.

MICHAEL, C. R., 1969. Retinal processing of visual images. *Sci. Amer.* 220 (no. 5): 104–114.

NICOLL, R. A., 1969. Inhibitory mechanisms in the rabbit olfactory bulb: Dendrodendritic mechanisms. *Brain Res.* 14: 157–172.

OCHI, J., 1963. Olfactory bulb response to antidromic olfactory tract stimulation in the rabbit. *Jap. J. Physiol.* 13: 113–128.

PHILLIPS, C. G., 1959. Actions of antidromic pyramidal volleys on single Betz cells in the cat. *Quart. J. Exp. Physiol. Cog. Med. Sci.* 44: 1–25.

PHILLIPS, C. G., T. P. S. POWELL, and G. M. SHEPHERD, 1961. The mitral cells of the rabbit's olfactory bulb. *J. Physiol.* (*London*) 156: 26P–27P.

PHILLIPS, C. G., T. S. P. POWELL, and G. M. SHEPHERD, 1963. Responses of mitral cells to stimulation of the lateral olfactory tract in the rabbit. *J. Physiol.* (*London*) 168: 65–88.

PRICE, J. L., 1968. The termination of centrifugal fibres in the olfactory bulb. *Brain Res.* 7: 483–486.

RALL, W., and G. M. SHEPHERD, 1968. Theoretical reconstruction of field potentials and dendrodendritic synaptic interactions in olfactory bulb. *J. Neurophysiol.* 31: 884–915.

RALL, W., G. M. SHEPHERD, T. S. REESE, and M. W. BRIGHTMAN, 1966. Dendrodendritic synaptic pathway for inhibition in the olfactory bulb. *Exp. Neurol.* 14: 44–56.

RAMÓN Y CAJAL, S., 1911. Histologie du Système Nerveux de l'Homme et des Vertébrés. Maloine, Paris.

RAMÓN Y CAJAL, S., 1955. Studies on the Cerebral Cortex (L. M. Kraft, translator). Year Book Publishers, Chicago.

SHEPHERD, G. M., 1963a. Responses of mitral cells to olfactory nerve volleys in the rabbit. *J. Physiol* (*London*) 168: 89–100.

SHEPHERD, G. M., 1963b. Neuronal systems controlling mitral cell excitability. *J. Physiol.* (*London*) 168: 101–117.

SHEPHERD, G. M., 1966. The orientation of mitral cell dendrites. *Exp. Neurol.* 14: 390–395.

SPRAGUE, J. M., 1958. The distribution of dorsal root fibres on motor cells in the lumbosacral spinal cord of the cat, and the site of excitatory and inhibitory terminals in monosynaptic pathways. *Proc. Roy. Soc., ser. B, biol. sci.* 149: 534–556.

TASAKI, I., E. H. POLLEY, and F. ORREGO, 1954. Action potentials from individual elements in cat geniculate and striate cortex. *J. Neurophysiol.* 17: 454–474.

WERBLIN, F. S., and J. E. DOWLING, 1969. Organization of the retina of the mudpuppy, *Necturus maculosus*. II. Intracellular recording. *J. Neurophysiol.* 32: 339–355.

YAMAMOTO, C., T. YAMAMOTO, and K. IWAMA, 1963. The inhibitory systems in the olfactory bulb studied by intracellular recording. *J. Neurophysiol.* 26: 403–415.

# 50  Dendritic Neuron Theory and Dendrodendritic Synapses in a Simple Cortical System

## WILFRID RALL

GORDON SHEPHERD, in his presentation, focused attention on the experimental evidence that provided the basis for our theoretical work on the olfactory bulb. He also discussed several wider implications of these experimental and theoretical results. Here, I wish to focus attention on theoretical aspects of this study. First, I describe a general theoretical approach that can be applied to different neurons or neuron populations and to different parts of nervous systems. Then I consider the specific application to the olfactory bulb and focus attention on particular points that proved to be most important for our interpretations. A comprehensive report of our analysis, results, and interpretations has already been published (Rall and Shepherd, 1968).

### Theoretical Method

Briefly, the theoretical method combines several mathematical models. One group of models is based on the biophysical models of nerve membrane; the simplest represents passive nerve membrane; another provides for synaptic excitation and inhibition; still others provide for the generation and propagation of active nerve impulses. Another group of models is concerned with the abstraction and simplification of dendritic branching. Still another group is based on considerations of geometric symmetry in a cortical

arrangement of neurons, especially when these neurons are activated synchronously. Such mathematical models were combined in a computer program, which was then used to predict spatiotemporal distributions of extracellular potential; these distributions were compared with those previously recorded experimentally in the olfactory bulb.

DENDRITIC BRANCHING    That neuronal dendritic branching can be very extensive has been well known for many years and has been widely illustrated. The details of such branching often appear to be impossibly complicated. The purpose of Figure 1 is to illustrate how dendritic branching can be abstracted, idealized, and simplified. The dendritic tree (Figure 1A) preserves branching but idealizes it to a symmetric branching pattern; the equivalent cylinder (Figure 1B) and the two chains of compartments (Figure 1C and D) reduce the spatial aspect of the dendritic tree to a single dimension, namely, the distance from the neuronal soma out into the dendritic tree. For those problems that involve either disturbances that spread from the neuronal soma outward into all of the branches or those that arise at corresponding locations in all members of a set of corresponding branches, consideration of the symmetry provides a justification for such simplification to a single space dimension.

It has been demonstrated mathematically (Rall, 1962a) that there is a class of dendritic trees which can be transformed into an equivalent cylinder. Briefly, this demonstration involved a consideration of the partial differential equation (cable equation) for spatiotemporal distributions of

WILFRID RALL    Mathematical Research Branch, National Institute of Arthritis and Metabolic Diseases, Bethesda, Maryland membrane potential in passive membrane cylinders, together with a consideration of the boundary conditions to be satisfied at every point of branching and of branch termination. In addition to providing a constraint upon the relative diameters of parent and daughter branches at every branch point, this analysis demonstrated the advantage of expressing the actual length, $\ell$, of each dendritic branch, as an *electrotonic length,* L. This is defined as $\ell/\lambda$, where $\lambda$, known as the *characteristic length,* is a basic parameter in the original cable equation. The value of $\lambda$ depends on the resistivities of the materials and on the square root of the diameter of each membrane cylinder. The equivalent cylinder (Figure 1B) is characterized by a single value of $\lambda$, whereas the dendritic tree (Figure 1A) is characterized by many different values of $\lambda$ for the different values of branch diameter.

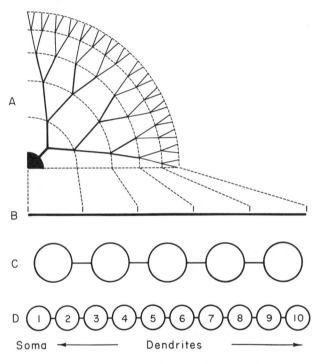

FIGURE 1    Schematic diagrams of simplified models of a neuron. A. Dendritic tree with an idealized, symmetrical, branching pattern; its trunk arises from a portion of the neuronal soma. B. Membrane cylinder that can be regarded as equivalent (electrotonically) to the entire dendritic tree (including its portion of the soma) when certain conditions are satisfied (Rall, 1962a). C. Chain of five equal compartments. Each compartment corresponds to a lumping of one-fifth of the length of the cylinder (B), or one-fifth of the surface area of the dendritic tree (A), as indicated by the dashed lines in (A). D. Chain of 10 equal compartments. Each compartment corresponds to a lumping of one-tenth of the equivalent membrane cylinder (B); compartment 1 represents the soma, and compartments 2 through 10 represent successive increments of electrotonic distance from the soma out to the dendritic terminals (Rall, 1964).

ELECTROTONIC LENGTH    For the equivalent cylinder, the electrotonic length is a simple concept; its importance results from the fact that this L also represents the electrotonic length of the entire dendritic tree (Figure 1A), including its portion of the neuronal soma. For some neurons it is useful to combine all dendritic trees and to characterize the entire soma-dendritic complex as an equivalent cylinder with some particular value of L. For other neurons, and for some patterns of neuronal activation, it is necessary to consider several equivalent cylinders coupled to a common soma, and in some cases to consider also subdivisions of dendritic trees. Our theoretical computations of activity in the olfactory bulb required us to assume values of L for mitral-cell dendrites and for granule cells. For motoneurons of cat spinal cord, the significance of the value of L for an understanding of synaptic potentials and their properties has recently been demonstrated in some detail (Rall et al., 1967). Also, a theoretical basis for estimating the value of L from an analysis of electrophysiological transients has recently been provided (Rall, 1969); this approach may prove to be useful with neurons of many different types.

COMPARTMENTAL MODELS    As indicated by Figure 1C and 1D, a chain of equal compartments can be used to approximate the equivalent cylinder. Each compartment represents a lumping of dendritic membrane. The essential, simplifying assumption is that nonuniformity of membrane potential is neglected within each compartmental region; nonuniformity is represented only by differences between compartmental regions. The chain of 10 compartments (Figure 1D) represents the one dimension, electrotonic distance, extending from the soma (compartment 1) outward into all the dendritic branches in nine lumped steps that correspond to equal increments of electrotonic distance and of membrane surface area.

An advantage of compartmental models is that they are not restricted to the special case of the equivalent cylinder. Compartments of different sizes are used whenever needed to represent significant departures from the idealized equivalent cylinder. Also, branching systems of compartments are used whenever it is necessary to distinguish among several dendritic trees of a single neuron, or among several divisions of a single dendritic tree. Furthermore, different synaptic inputs and different membrane properties can be specified for different compartments when required.

The mathematical model implied by such a compartmental diagram is a system of ordinary differential equations (see Rall, 1964, for details and references). These equations are linear and of first order for those compartments in which active nerve membrane properties are not present. The system of equations becomes complicated by additional nonlinear differential equations when one or more compartments must exhibit active membrane properties.

MEMBRANE MODELS The diagram in Figure 2 indicates the lumped membrane parameters for a single compartment. This equivalent electrical circuit is related to the papers of Hodgkin and Katz (1949), Fatt and Katz (1951 and 1953), Hodgkin and Huxley (1952), and Coombs et al. (1955); see also Cole (1968). In a completely passive membrane, the excitatory and inhibitory conductances, $G_\varepsilon$ and $G_j$, are both zero, and there remains the resting conductance, $G_r$, in parallel with the membrane capacity, $C_m$. For synaptic regions, the synaptic input intensity is represented by the variables $\varepsilon = G_\varepsilon/G_r$ and $\jmath = G_j/G_r$, which are assumed to be independent of changes in membrane potential. The dependence of membrane potential on $\varepsilon$ and $\jmath$ follows from standard circuit theory; the equations have been presented and discussed elsewhere (Rall, 1964; Rall and Shepherd, 1968). The 1968 reference also provides the specific pair of coupled nonlinear differential equations that was used to generate transient changes of $\varepsilon$ and $\jmath$ in active membrane. Such computed transients of $\varepsilon$ and $\jmath$ may be thought of as corresponding roughly to the transients of sodium and potassium conductance in the Hodgkin and Huxley (1952) model for active nerve membrane; they proved capable of generating and propagating well-shaped action potentials.

AXON, SOMA, AND DENDRITES Figure 3 shows three diagrams that preserve axonal, somatic, and dendritic regions of a neuron. Although these diagrams have been used to represent a mitral cell of olfactory bulb, they obviously have more general applicability. When the several dendrites (Figure 3A) are combined into an approximately equivalent cylinder, this system is reduced to one distance dimension (Figure 3B), but there is an obvious change in diameter from axon, through soma, to the (dendritic) equivalent cylinder. The compartmental abstraction (Figure 3C) indicates not only changes in diameter, but also a difference in membrane properties. The blackened compartments (axon and soma) are assumed to have active membrane properties; the dendritic compartments have passive membrane properties in some computations and active membrane properties in others.

SIMULATION OF ANTIDROMIC ACTIVATION The purpose in assembling these various component models was to generate (with a digital computer) a fairly realistic sequence of

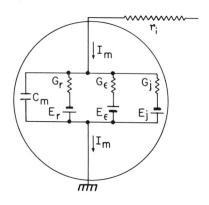

FIGURE 2 Diagram of the electrical equivalent circuit used to represent the lumped membrane parameters of a single compartment, in the compartmental models of Figures 1 and 3. The membrane current, $I_m$, flows through four parallel pathways provided by three conductances, $G_r$, $G_\varepsilon$, and $G_j$, and by the capacity, $C_m$. $G_r$ represents the resting membrane conductance and lies in series with the resting EMF, or battery, $E_r$. $G_\varepsilon$ represents an excitatory conductance which lies in series with an excitatory EMF, or battery, $E_\varepsilon$. Also, $G_j$ represents an inhibitory, or quenching, conductance which lies in series with its corresponding EMF, or battery, $E_j$ The resistor, $r_i$, represents the intracellular (core) resistance that joins one compartment to the next. (See Rall, 1964, and other references cited in the text.)

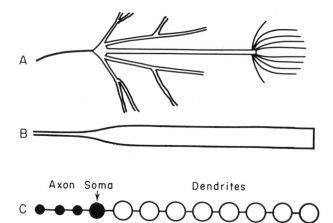

FIGURE 3 Schematic diagrams of successive abstractions of a mitral cell, or of another cell of this general type. A. Axon, soma, primary dendrite, and several (truncated) secondary dendrites of such a cell. B. The dendrites of A are represented as an approximately equivalent cylinder; the soma appears as the transition from the small-diameter axon cylinder to the large-diameter combined dendritic cylinder. C. This compartmental model corresponds to A and B above; the three axonal compartments and the soma (all shown in black) are assumed to have active membrane properties; the dendritic compartments are assumed to have passive membrane properties in some computations and active membrane properties in others.

the membrane states that occur during antidromic activation of a single representative mitral cell. A suprathreshold depolarization of the distal axonal compartment results in the generation of an action potential that propagates from one axonal compartment to the next. It took special care to insure that this propagating impulse did not fail to invade the soma compartment. This is because the rate of depolarization is much slower in the soma compartment than in the axonal compartments, for two reasons: (1) the soma compartment has a larger electrical capacity, but, even more important, (2) the spread of current from the soma into the dendritic compartments provides heavy damping to the soma. We could reduce this damping by reducing the size of the dendritic compartments or by providing the dendritic compartments with a small background level of depolarization or of synaptic excitation, or by doing both. When the soma does fire an action potential, membrane depolarization spreads out into the dendritic compartments, either rapidly and without decrement in active dendritic membrane, or more slowly and with some electrotonic decrement in passive dendritic membrane. We computed many variations of such antidromic activation and used them as the basis for calculations of the associated extracellular potential distributions.

Synchrony and Symmetry of Population   The key to the computation of the extracellular field potentials is provided by a consideration of the synchrony of activation in the neuron population and the geometric symmetry of the cortical anatomy. Figure 4A indicates the arrangement of the mitral-cell population in the olfactory bulb of rabbit. A single strong shock to the lateral olfactory tract initiates essentially synchronous antidromic activation of the mitral-cell population. The somas of these mitral cells lie in an almost spherical shell, and their primary dendrites are directed radially outward. It is useful to consider the implications that follow when this approximate spherical symmetry is idealized to perfect spherical symmetry, as indicated diagrammatically at the upper right (Figure 4B). Given complete symmetry and synchrony, with equally spaced mitral cells generating identical amounts of extracellular current, it follows that the current generated by each of these mitral cells will be confined to a small (conical) volume element, such as the one indicated in Figure 4B. Because the number of such mitral cells is very large, these conical volume elements are thin slivers, and the extracellular current flow has an essentially radial orientation in the sphere. Because of the symmetry, no current can flow outside the sphere from one region to another, which is an example of the "closed field" characterized by Lorente de Nó (1947); see also Rall (1962b) and Rall and Shepherd (1968) for additional details on this point.

Punctured Symmetry   That the actual olfactory bulb is not a completely closed sphere is significant because of the even more important fact that the experimentally observed field of extracellular potentials is not closed. Specifically, a microelectrode at the bulb surface detects a significant transient of potential, relative to a distant reference electrode, during synchronous mitral-cell activation. This means that some of the extracellular current generated by the mitral cells must flow away from the bulb surface; it can return through the bulb puncture. This secondary extracellular current is indicated in the diagram at lower left (Figure 4C); the primary extracellular current is confined within each conical volume element. The equivalent circuit, at lower right (Figure 4D), shows both the primary extracellular current (PEC) and the secondary extracellular current (SEC) produced by a generator of extracellular current (GEC), which is shown as a compartmental neuron model.

Potential Divider Effect   The secondary extracellular current path has a finite resistance, and the distant reference electrode is effectively situated somewhere along this resistance; this reference electrode divides the over-all potential difference into two parts. For this particular application to the olfactory bulb, the resulting potential divider ratio of outer resistance to puncture resistance is about 1:4. Consider, for example, the moment at which synchronous mitral activity generates extracellular current such that the extracellular potential at the depth of the mitral-cell bodies is 2.5 mV negative relative to the potential at the dendritic terminals; the 1:4 potential divider ratio then implies 0.5 mV across the outer resistance, and 2.0 mV across the puncture resistance. Therefore, when recording relative to the reference electrode, a microelectrode placed near the dendritic terminals should be 0.5 mV positive, and a microelectrode placed at the depth of the mitral cell bodies should be 2.0 mV negative. Such values are observed experimentally, and this potential divider effect represents an essential part of the computational model. The diagram at lower right (Figure 4D), by a combination of the compartmental model of Figure 3C with the punctured symmetry considerations of Figure 4C, provides a compact representation of the several theoretical models that have been combined in the computational model for the olfactory bulb field potentials. My reason for presenting these details is based on a belief that this combined model (with different potential divider ratios) can be applied to synchronous activation in other cortical structures; a recent example is an application to the cerebellum by Zucker (1969).

Experimental Potentials to be Interpreted   Figure 5 shows how the experimental recordings obtained by Phil-

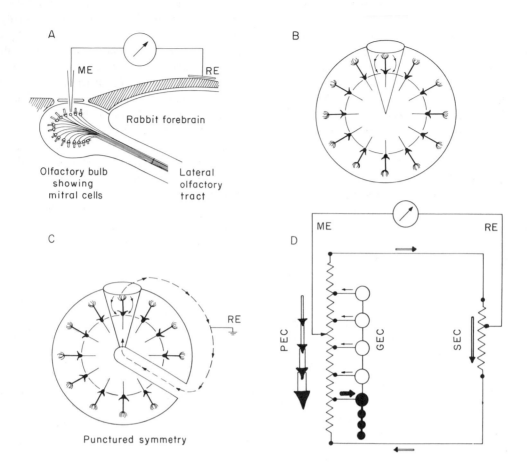

FIGURE 4 Diagrams related to cortical symmetry and to synchronous activation of the mitral-cell population. A. Schematic diagram of the experimental recording situation. The microelectrode (ME) penetrates the olfactory bulb; the reference electrode (RE) is distant. The mitral cells are arranged in an almost spherical cortical shell; their axons all project into the lateral olfactory tract. Single-shock stimulation to the lateral olfactory tract results in synchronous antidromic activation of the mitral-cell population. Circle with slanted arrow, here and in D, represents the recording system. B. Diagram of complete spherical symmetry of a cortical arrangement of mitral cells. The cone indicates a volume element associated with one mitral cell; the arrows indicate extracellular current generated by this mitral cell; the current is confined within its cone when activation is synchronous for the population. C. Diagram of punctured spherical symmetry. The arrows inside the cone represent the primary extracellular current generated per mitral cell; the dashed line (with arrows) represents the secondary extracellular current per mitral cell. The location of the reference electrode (RE) along the resistance of this secondary pathway serves as a potential divider. D. Diagram combines the potential divider aspect (C) with a compartmental model (from Figure 3C). The relations of both the microelectrode (ME) and the reference electrode (RE) to the primary extracellular current (PEC) and the secondary extracellular current (SEC) are shown. The generator of extracellular current (GEC) is a compartmental model representing the synchronously active mitral-cell population. The solid arrows next to the compartmental model represent the direction of membrane current flow at the moment of active, inward, soma-membrane current (heavy black arrow); the dendritic membrane current is outward. The open arrows represent the direction of extracellular current flow (both PEC and SEC) at this same moment.

lips, Powell, and Shepherd (1961, 1963) were related to four depths in the cortical structure. These field potentials, recorded in response to synchronous antidromic activation, were found to be highly reproducible; thus we regard this potential as a function of two variables, depth and time. Because Shepherd has already presented olfactory bulb anatomy, I simply point out the four depths: the glomerular layer (GL), the external plexiform layer (EPL), the mitral body layer (MBL), and the granular layer (GRL). At first we concentrated on time periods I and II, to see if we could predict the same dependence on depth and time from our computational model for synchronous antidromic activation of the mitral-cell population. Later we turned to a consideration of time period III.

COMPUTED RESULTS FOR MITRAL POPULATION    Figure 6 summarizes the results of a computation which simulated the synchronous antidromic activation of the mitral-cell population. This computation was based on the concepts already summarized in Figure 4C. This particular computation assumed passive dendritic membrane. An action potential was initiated in the most distal axonal compartment and propagated to the soma (shown as compartment 4 in Figure 6). The transient at lower left shows the transmembrane action potential (intracellular potential relative to extracellular potential) at the soma compartment. Above this are shown the transmembrane potential transients that spread into the dendritic compartments—an example of passive electrotonic spread from an active soma into passive dendrites. During period I, the soma membrane is more depolarized than the dendritic membrane, and the intracellular current flows from soma to dendrites, which means that the extracellular current must flow from dendrites to soma. This provides an intuitive explanation of the computed extracellular potentials during period I, shown in the right half of Figure 6. During period II, the active repolarization

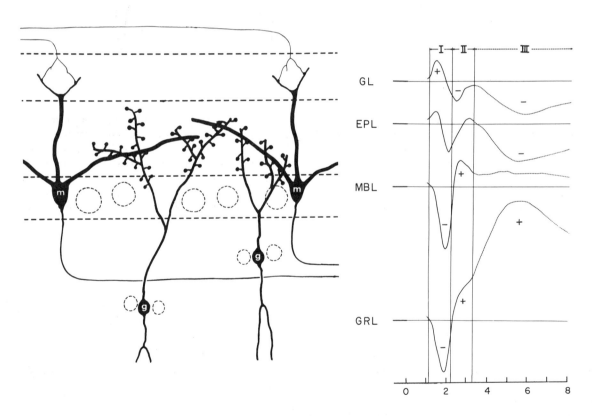

FIGURE 5    Experimental field potentials shown in relation to histological layers. A voltage transient is shown at right for each of the four histological layers at left. The schematic Golgi histology at left shows two mitral cells (m) and two granule cells (g). GL designates the glomerular layer; EPL, the external plexiform layer; MBL, the mitral-body layer; and GRL, the granular layer. I, II, and III designate three time periods used to facilitate the analysis and discussion of the voltage transients; the time scale at the lower right is in milliseconds. The + and − signs indicate polarity recorded at the microelectrode, relative to the reference electrode.

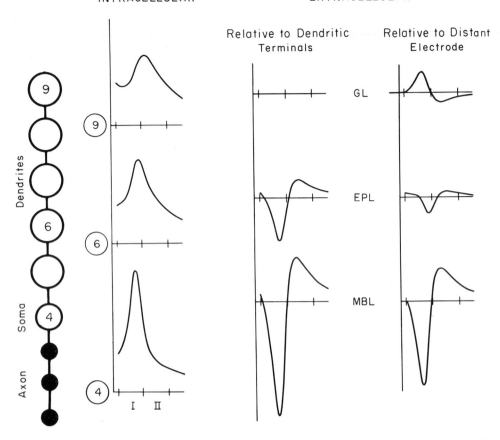

INTRACELLULAR                    EXTRACELLULAR

Relative to Dendritic          Relative to Distant
Terminals                      Electrode

GL

EPL

MBL

9

6

4

Dendrites

Soma

Axon

I    II

FIGURE 6 Computed voltage transients obtained from equations corresponding to the diagrammatic combined model of Figure 4C. Nine compartments, shown on the left, were used in the computations, but results are illustrated for only three of these: compartment 4 as soma, compartment 6 as mid-dendritic, and compartment 9 as dendritic terminals. The three transients on the left represent passive electrotonic spread of intracellular potential from a soma action potential into passive dendritic compartments. The two sets of tran-sients on the right represent the corresponding extracellular potential transients; the middle column expresses extracellular potential relative to the terminal (GL) level; the column on the right shows the same extracellular potential differences re-expressed relative to the distant reference electrode (potential divider effect). Time scale designates periods I and II; GL, EPL, and MBL designate the same histological layers as in Figure 5.

of the soma membrane makes the soma less depolarized than the dendritic membrane, and the intracellular current flows from dendrites to soma, which means that the extracellular current also becomes reversed and flows from soma to dendrites. This provides an intuitive explanation of the computed extracellular potentials during period II. The far right column of Figure 6, based on a potential divider ratio of 1:4, can be seen to provide a first approximation to periods I and II of the experimental transients shown in Figure 5.

SIGNIFICANCE OF DIFFERENT DEPTH DISTRIBUTIONS AND RELATIVE RESISTANCES    The depth distribution of poten-tial during time period III differs significantly from that of periods I and II. The schematic diagrams in the right half of Figure 7 were designed to illustrate the argument that activity in the mitral-cell population can account for the extracellular potential gradients during most of periods I and II, but that such activity cannot account for the extracellular gradients of potential during period III. Furthermore, the depth distribution of potential during period III could result from appropriate activity in the population of granule cells.

During periods I and II, large gradients of extracellular potential are present in the EPL, from GL to MBL, but not deeper in the granular layer, GRL. This experimental find-

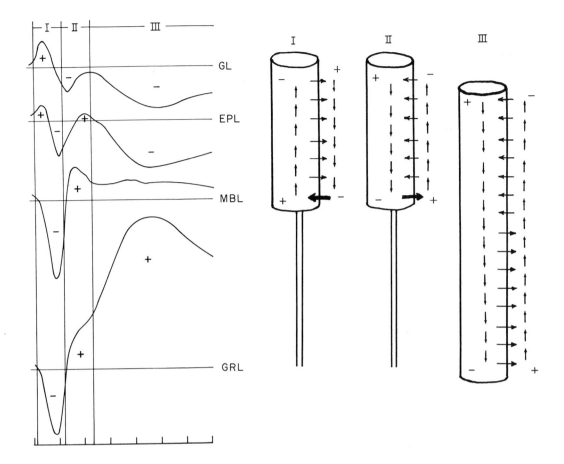

FIGURE 7 Depth distribution of potential, and equivalent cylinders corresponding to different neuron populations. The field potentials on the left are the same experimental records as in Figure 5. The three diagrams on the right show equivalent cylinders and the direction of electric-current flow at three different times—period I, period II, and period III. The arrows show intracellular current, membrane current, and primary extracellular current; secondary extracellular current is not shown. GL, EPL, MBL, and GRL designate the same histological layers as in Figure 5. In both I and II, the large cylinder extending from GL to MBL corresponds to the combined dendrites of the mitral-cell population, and the thin cylinder extending from MBL into GRL corresponds to the mitral axons. In period III, the long cylinder extending from EPL far down into GRL corresponds to the large population of granule cells.

ing was predicted by our computations, the essential reason being that the mitral axon has a much smaller area of cross section (and hence a much larger intracellular resistance, $r_1$, per unit length) than does the equivalent mitral dendritic cylinder. The ratio is estimated to lie in the range 1:25 to 1:50. As indicated in Figure 7 (middle two diagrams), this means that significant mitral-cell extracellular current is generated only between GL and MBL, and negligible mitral-cell extracellular current is generated in GRL. This quantitative agreement between theory and experiment was found not only for the peak gradients of periods I and II, but also in the negligible contribution to the early portion of the GRL transient that was made by the propagating axonal action potential.

It was the quantitative agreement between theory and experiment during periods I and II that convinced us of the impossibility of accounting for period III in terms of current generated by mitral cells. During period III, the gradient of extracellular potential reaches far down into the granular layer; it has its maximum value near MBL. The kind of equivalent cylinder that could generate such current is shown at the far right in Figure 7; it must provide an effective intracellular conductance down into the granular layer. Although individual granule cells are small, the granule-cell population is very large. The granule cells have the required depth distribution; no other sufficiently numerous type of cell has this depth distribution. Thus we were forced to conclude that the major generator of extracellular current during period III must be the large population of granule cells.

GRANULE-CELL ACTIVITY To generate the current flow shown in III in Figure 7, the granule-cell population could either undergo membrane depolarization in the EPL, membrane hyperpolarization in the GRL, or both. We were led to postulate synaptic excitation of the granule dendrites in the EPL as the primary activity leading to the generation of period III. Some synaptic inhibition could also be present at the deep granule-cell processes in GRL, but this was ruled out as the primary activity for two reasons: (1) such synaptic inhibition would not fit well with subsequent mitral-cell inhibition by granule cells, for which Shepherd (1963) already had experimental evidence; (2) such synaptic inhibition seemed less likely to generate a sufficient magnitude of current to account for period III. It should be added, however, that such synaptic inhibition of deep granule-cell processes could play an important secondary role of preventing action potential propagation in those granule-cell computations that include active membrane properties for the granule cells. One of the interesting findings of our granule-cell computations was that even with a "cool" version of active membrane kinetics, the resulting propagation (in the absence of synaptic inhibition) of active membrane depolarization was incompatible with the monophasic, spatially stationary character of the field potential observed during period III. The simplest successful granule-cell computations were obtained by assuming passive granule-cell membrane.

TWO-WAY DENDRODENDRITIC SYNAPTIC POSTULATE Thus, our quantitative fitting of theory to experiment convinced us that the granule-cell population must receive massive synchronous excitation delivered to the granule dendrites in the EPL. Phillips, Powell, and Shepherd postulated that granule cells inhibit mitral cells, but they implicitly assumed that the granule cells would be conventional in receiving their synaptic input at one region, propagating an impulse to another region, and then initiating synaptic inhibition there. By ruling out impulse propagation as predominant during period III, we required that synaptic excitation be delivered directly to the granule dendrites in the EPL. Although such could conceivably be done by recurrent axon collaterals of mitral cells (and we have not ruled them out as contributory), we were struck by the remarkable possibility, provided both by timing and by contiguity, that mitral secondary dendrites could deliver the needed synaptic excitation directly to the granule dendrites. That is, the depolarization of the mitral dendrites during period II is well timed to initiate the required synaptic excitation; also, Golgi preparations had demonstrated that the EPL is primarily a neuropil composed of these two kinds of dendrites.

Thus, we formed the postulate of two-way, dendroden-

dritic, synaptic interactions before we learned any of the details that were provided by electron-microscope studies of others. We postulated that membrane depolarization of the mitral secondary dendrites would initiate synaptic excitation, delivered from mitral to granule dendrites, and that the resulting membrane depolarization of the granule dendrites would then initiate synaptic inhibition, delivered from granule dendrites to mitral dendrites and somas.

ELECTRON-MICROSCOPE EVIDENCE It would be disingenuous to pretend that we were not pleased and excited when this two-way, dendrodendritic, synaptic postulate was provided with confirmatory histological support and elaboration by the electron-microscope findings of Reese and Brightman, who had been conducting an independent study of the olfactory system. These histological findings and our theoretical postulate complemented each other so well that we agreed to present them jointly (Rall et al., 1966). While preparing that joint paper, we learned of the similar histological findings of Hirata (1964), Andres (1965), and, later, of Hama (personal communication).

The schematic Golgi diagram (Figure 8A) shows only one mitral cell and one granule cell and indicates the intermingling of their dendrites in the external plexiform layer (EPL). A large number of additional granule cells and mitral cells (not shown in this diagram) essentially fill the EPL with a dense neuropil composed primarily of these two kinds of dendrites. In the Golgi diagram (Figure 8A), the open circle designates, at low magnification, the structures shown at higher magnification in Figure 8B, in which a granule-cell gemmule (GG) arises from a granule-cell dendrite (GD) and makes contact with a mitral secondary dendrite (MSD). This drawing (Figure 8B) was prepared from electron-microscope serial sections; the two exploded views reveal two different synaptic contacts between the gemmule (GG) and the mitral dendrite (MSD). The contact on the left has vesicles clustered close to the mitral side of its contact region; by the established criterion (Palay, 1958; Gray, 1959), this contact is assigned the polarity mitral-to-granule as indicated by an arrow. The other contact has vesicles clustered close to the granule side of its contact region; by the same criterion, this contact is assigned the polarity granule-to-mitral as indicated also by an arrow. In a quantitative study, based on many reconstructions from serial sections, all taken from one small region of the EPL, Reese (1966, 1970) found that most of the gemmules make such pairs of synaptic contacts of opposite polarity.

Our postulate that the mitral-to-granule synaptic contacts are excitatory, and that the granule-to-mitral synaptic contacts are inhibitory, was based not on morphological criteria, but on the theoretical considerations outlined in the preceding section. Nevertheless, it is interesting that our as-

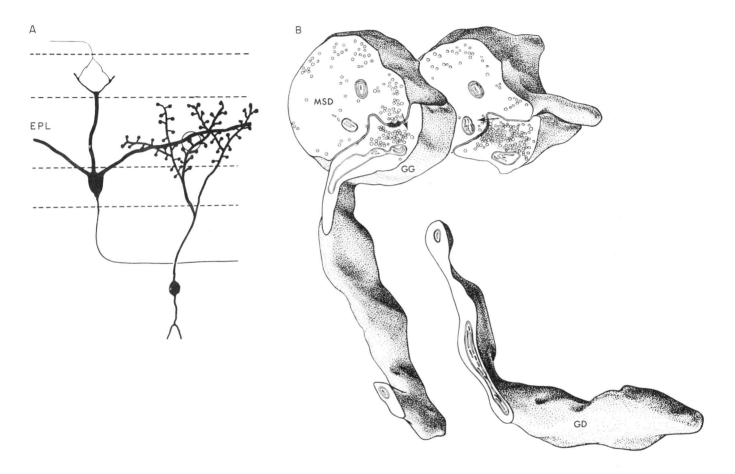

FIGURE 8  Dendrodendritic synaptic contacts at two magnifications. A. Schematic Golgi histology, similar to that shown in Figure 5 but with emphasis on the external plexiform layer (EPL). The circle designates an example of a mitral secondary dendrite touching the gemmule of a granule dendrite, corresponding to the higher magnification shown in B. B. Drawing based on electron microscopy. The mitral secondary dendrite (MSD) touches a granule cell gemmule (GG) the stem of which is a process of a granule-cell dendrite (GD). This drawing by Mrs. G. Turner represents a reconstruction from 23 consecutive serial sections less than 0.1 micron in thickness, prepared and assembled by T. S. Reese. Two exploded views serve to reveal two synaptic contacts of opposite polarity, as indicated by vesicles and by arrows. Further details in Rall et al., 1966; Reese, 1970.

signment of synaptic excitatory and synaptic inhibitory function is consistent with the two tentative morphological criteria that are currently being developed and explored. One criterion depends on the presence or absence of postsynaptic dense material. The other depends on a distinction between spherical and flattened vesicles; (See Price, 1968, for evidence on the vesicles and for references on both tentative criteria.)

## Discussion

TWO-WAY, DENDRODENDRITIC SEQUENCE    The reaction of several neuroscientists, on first sight, was to regard such oppositely oriented synapses adjacent to each other as a short-circuit which could have no useful function. Such an objection disappears, however, when temporal sequence is taken into account. Our interpretation can be summarized with the help of Figure 9.

Whenever a mitral-cell body fires an impulse, whether it has been activated by the antidromic or the orthodromic route (AD or OD in Figure 9D), it results in a spread of membrane depolarization into the mitral-cell secondary dendrites. This depolarization takes place whether one assumes active or passive dendritic membrane properties. As can be verified in Figure 6, such dendritic membrane depolarization begins in the later portion of period I and extends through period II. The schematic sequence (Figure 9A, B, C) summarizes the postulated sequence of synaptic interactions. Depolarization of mitral-cell dendritic membrane activates mitral-to-granule synaptic excitatory contacts. The resulting membrane depolarization of the granule gemmules activates granule-to-mitral synaptic inhibitory contacts. Such synaptic activity inhibits the mitral cell and causes membrane hyperpolarization of the mitral-cell dendrites and soma. It is important to supplement Figure 9 with the reminder that every mitral cell is in contact with the gemmules of many granule cells, and every granule cell is in contact with many mitral cells.

SELF INHIBITION AND LATERAL INHIBITION    It is evident that such two-way, dendrodendritic, synaptic interactions provide a negative feedback (recurrent inhibition) both to the mitral cells that have just fired impulses, and to any neighboring mitral cells that may not have fired. The self-inhibition feature, and the result that the granule-cell population can exert a damping effect on the entire mitral-cell population, could account (at least in part) for two phenomena: (1) an adaptive adjustment of the system to a wide range of input intensity; and (2) a tendency for rhythmic activity in the olfactory bulb. Also, because secondary dendrites of the mitral-cell extend laterally for long distances, it is likely that this mechanism can provide significant spatial

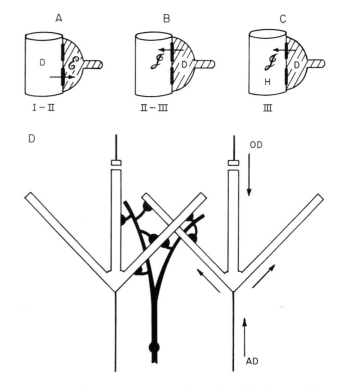

FIGURE 9   Schematic summary of postulated dendrodendritic synaptic interactions. A, B, and C show a single granule-cell gemmule (hatched) in contact with a mitral secondary dendrite (cylinder); a pair of synaptic contacts is indicated by thickenings in black. A. Membrane depolarization (D) of the mitral-cell dendrite during late period I and period II activates the mitral-to-granule synaptic contact (arrow) to deliver synaptic excitation ($\mathcal{E}$) to the gemmule during period II. B. The resulting depolarization (D) of the gemmule then activates the granule-to-mitral synaptic contact (arrow) to deliver synaptic inhibition ($\mathcal{I}$) to the mitral dendrite. C. During period III, the synaptic inhibition ($\mathcal{I}$) can cause both hyperpolarization (H) and inhibition of the mitral dendrite and also of the mitral soma. D. Several gemmules of a granule cell are shown in contact with the dendrites of two mitral cells. A mitral cell can be activated either antidromically (AD) or orthodromically (OD). When the mitral soma generates an action potential, membrane depolarization spreads out into the mitral secondary dendrites, either passively or actively. This membrane depolarization of the mitral dendrite causes synaptic excitation to all granule-cell gemmules in contact with this mitral cell. The synaptically excited granule dendrites then deliver synaptic inhibition to all the mitral cells with which they are in contact, including mitral cells that did not contribute to this activity.

gradients of lateral inhibition (see Ratliff, 1965), which might contribute to olfactory discrimination.

AMACRINE CELLS  By definition, an amacrine cell is a neuron that possesses no axon. The possible function of such neurons has troubled neuroscientists in the past. Our interpretation of granule-cell function in the olfactory bulb is therefore of some general interest. The function we envisage requires neither an axon nor an action potential in such cells; their synaptic interactions are transacted locally and probably in a graded, nonimpulsive manner. Axons and all-or-none action potentials are needed when information must be transmitted over a long distance, from one portion of the nervous system to another. Action potentials are not, however, needed in dendrites if their electrotonic length (L = $\ell/\lambda$) values are less than about 2.

Two further points deserve emphasis. To function, the dendrites of such cells must both send and receive information. Also, the lack of an impulse provides for slow, graded activity which could serve as a rudimentary short-term memory, i.e., a weighted spatiotemporal integral of previous and surrounding levels of input, in which the weights of contributions decrease both with distance and with time.

OTHER AMACRINE CELLS  The analogies between the olfactory bulb and the retina are not elaborated here; they are presented and discussed in Shepherd's chapter (this volume). The amacrine cells of octopus brain have been studied by Gray and Young (1964), who have postulated that these cells function as inhibitory interneurons and that they are important to the memory system of octopus. Quite recently, Best and Noel (1969) have observed reciprocal synaptic arrangements in the neuropil of planarian brain and have noted their possible significance.

COMMENT ON GRADED SYNAPTIC ACTIVATION  Questions are often asked about our assumption that these dendrodendritic synapses can be activated in a graded manner. Many persons have been under the impression that only a presynaptic action potential is capable of triggering synaptic activity. Our reply is in two parts. First, our experimental and theoretical study has strongly suggested that the granule dendrites do not generate action potentials in the EPL during period III, and that passive spread of depolarization into the mitral secondary dendrites can be significant in magnitude. Second, and probably more convincing to others, is that graded synaptic activation has been produced experimentally by means of graded membrane depolarization, both at the neuromuscular junction (Castillo and Katz, 1954; Liley, 1956; Katz and Miledi, 1965 and 1967) and at squid synapses (Katz and Miledi, 1966; Bloedel et al., 1966). Given such experimental "existence proofs," we believe

that our assumption of graded activation of the dendrodendritic synaptic contacts by nonimpulsive membrane depolarization can be regarded as tenable.

COMMENT ON NEURAL CIRCUITS  Because neural circuits are often presented in diagrams in a way that suggests simple one-to-one-to-one sequences of neuronal somas connected by axons, it seems desirable to comment on this. Most neurons have extensively branched dendritic trees. At least, such dendrites provide the possibility of receiving from many different input sources. At most, some of these dendrites may also *send* information, and some may be involved in graded, long-lasting, two-way interactions with other dendrites. In other words, wherever there are dendrites, one must consider the possibility of spatiotemporal integration, differentiation, and even more complicated operations or computations. Most diagrams of neural circuits provide no hint of the richness of such possibilities.

## Summary

As stated in the introduction, this contribution describes a general theoretical method which can be applied to different neurons and to various populations of neurons. It also describes the specific application of this method to the olfactory bulb, and explains how this led us to postulate two-way, dendrodendritic synaptic interactions in the external plexiform layer, between the dendrites of mitral cells and granule cells.

No mathematical details are presented here; these are available elsewhere. Diagrams and descriptive discussion are used to explain how considerations of dendritic branching, of electrotonic length, and of both passive and active membrane properties can be incorporated into a compartmental model representing the axon, soma, and dendrites of a neuron. Such a model was used for computing a spatiotemporal sequence of membrane states corresponding to antidromic activation of a representative mitral cell. Diagrams and descriptive discussion are also used to explain how considerations of cortical symmetry, and of synchronous activation in the mitral-cell population, together with a consideration of the reference electrode (location on a potential divider), can be incorporated into a model for computation of the extracellular field potentials.

The computed results obtained for synchronous antidromic activation of the mitral-cell population are shown to agree well with the early part (periods I and II) of the experimental records; such agreement provides useful insights for the interpretation of the experimental results. Special emphasis is given to the contrast in the depth distribution of the experimental field potentials when period III is compared with periods I and II; it is argued that mitral-cell activity could not account for period III; and it is concluded that activity in the granule-cell population must be the major generator of extracellular current during period III. Furthermore, it is explained that the polarity, the slow time course, and the spatial stationarity of the potentials during period III are best accounted for by massive synaptic excitation of the granule-cell dendrites in the external plexiform layer and by absence of action-potential propagation in these dendrites.

These theoretical results, when considered together with the previously known electrophysiology and Golgi histology, led us to postulate dendrodendritic mitral-to-granule synaptic excitation followed by dendrodendritic granule-to-mitral synaptic inhibition. Electron microscopy, done independently by others, has revealed synaptic contacts that could account for such two-way, dendrodendritic, synaptic interactions, and the latest electron-microscope studies further strengthen the plausibility of this interpretation.

The discussion restates the dendrodendritic sequence of membrane depolarizations and synaptic activations that we assume, and points out that the resulting mitral-cell inhibition could contribute to damping and rhythmic effects as well as to lateral inhibition in the olfactory bulb. Implications for amacrine cells include dendrites that send, as well as receive, synaptic information, and synapses that are presumably activated in a graded, nonimpulsive manner. In neural circuits and in interacting populations of neurons, the presence of dendrites opens possibilities of spatiotemporal integration, differentiation, and even more complicated operations or computations.

## REFERENCES

ANDRES, K. H., 1965. Der Feinbau des Bulbus olfactorius der Ratte unter besonderer Berücksichtigung der synaptischen Verbindungen. Z. Zellforsch. Mikroskop. Anat. 65: 530–561.

BEST, J. B., and J. NOEL, 1969. Complex synaptic configurations in planarian brain. Science (Washington) 164: 1070–1071.

BLOEDEL, J., P. W. GAGE, R. LLINÁS, and D. M. J. QUASTEL, 1966. Transmitter release at the squid giant synapse in the presence of tetrodotoxin. Nature (London) 212: 49–50.

CASTILLO, J. DEL, and B. KATZ, 1954. Changes in end-plate activity produced by pre-synaptic polarization. J. Physiol. (London) 124: 586–604.

COLE, K. S., 1968. Membranes, Ions and Impulses. University of California Press, Berkeley.

COOMBS, J. S., J. C. ECCLES, and P. FATT, 1955. The inhibitory suppression of reflex discharges from motoneurones. J. Physiol. (London) 130: 396–413.

FATT, P., and B. KATZ, 1951. An analysis of the end-plate potential recorded with an intracellular electrode. J. Physiol. (London) 115: 320–370.

FATT, P., and B. KATZ, 1953. The effect of inhibitory nerve impulses on a crustacean muscle fibre. J. Physiol. (London) 121: 374–389.

GRAY, E. G., 1959. Axo-somatic and axo-dendritic synapses of the cerebral cortex: An electron microscopic study. J. Anat. 93: 420–433.

GRAY, E. G., and J. Z. YOUNG, 1964. Electron microscopy of synaptic structure of Octopus brain. J. Cell Biol. 21: 87–103.

HIRATA, Y., 1964. Some observations on the fine structure of the synapses in the olfactory bulb of the mouse, with particular reference to the atypical synaptic configurations. Arch. Histol. Jap. 24: 293–302.

HODGKIN, A. L., and A. F. HUXLEY, 1952. A quantitative description of membrane current and its application to conduction and excitation in nerve. J. Physiol. (London) 117: 500–544.

HODGKIN, A. L., and B. KATZ, 1949. The effect of sodium ions on the electrical activity of the giant axon of the squid. J. Physiol. (London) 108: 37–77.

KATZ, B., and R. MILEDI, 1965. Propagation of electric activity in motor nerve terminals. Proc. Roy. Soc., ser. B, biol. sci. 161: 453–482.

KATZ, B., and R. MILEDI, 1966. Input-output relation of a single synapse. Nature (London) 212: 1242–1245.

KATZ, B., and R. MILEDI, 1967. The release of acetylcholine from nerve endings by graded electric pulses. Proc. Roy. Soc., ser. B, biol. sci. 167: 23–38.

LILEY, A. W., 1956. The effects of presynaptic polarization on the spontaneous activity at the mammalian neuromuscular junction. J. Physiol. (London) 134: 427–443.

LORENTE DE NÓ, R., 1947. Action potential of the motoneurons of the hypoglossus nucleus. J. Cell. Comp. Physiol. 29: 207–287.

PALAY, S. L., 1958. The morphology of synapses in the central nervous system. Exp. Cell. Res., suppl. 5: 275–293.

PHILLIPS, C. G., T. P. S. POWELL, and G. M. SHEPHERD, 1961. The mitral cell of the rabbit's olfactory bulb. J. Physiol. (London) 156: 26P–27P.

PHILLIPS, C. G., T. P. S. POWELL, and G. M. SHEPHERD, 1963. Responses of mitral cells to stimulation of the lateral olfactory tract in the rabbit. J. Physiol. (London) 168: 65–88.

PRICE, J. L., 1968. The synaptic vesicles of the reciprocal synapse of the olfactory bulb. Brain Res. 11: 697–700.

RALL, W., 1962a. Theory of physiological properties of dendrites. Ann. N. Y. Acad. Sci., 96: 1071–1092.

RALL, W., 1962b. Electrophysiology of a dendritic neuron model. Biophys. J. 2 (no. 2, pt. 2): 145–167.

RALL, W., 1964. Theoretical significance of dendritic trees for neuronal input-output relations. In Neural Theory and Modeling (R. F. Reiss, editor). Stanford University Press, Stanford, California, pp. 73–97.

RALL, W., 1969. Time constants and electrotonic length of membrane cylinders and neurons. Biophys. J. 9: 1483–1508.

RALL, W., R. E. BURKE, T. G. SMITH, P. G. NELSON, and K. FRANK, 1967. Dendritic location of synapses and possible mechanisms for the monosynaptic EPSP in motoneurons. J. Neurophysiol. 30: 1169–1193.

RALL, W., and G. M. SHEPHERD, 1968. Theoretical reconstruction of field potentials and dendrodendritic synaptic interactions in olfactory bulb. J. Neurophysiol. 31: 884–915.

Rall, W., G. M. Shepherd, T. S. Reese, and M. W. Brightman, 1966. Dendro-dendritic synaptic pathway for inhibition in the olfactory bulb. *Exp. Neurol.* 14: 44–56.

Ratliff, F., 1965. Mach Bands: Quantitative Studies on Neural Networks in the Retina. Holden-Day, San Francisco.

Reese, T. S., 1966. Further studies on dendro-dendritic synapses in the olfactory bulb. *Anat. Rec.* 154: 408 (abstract).

Reese, T. S., 1970. Discussion. *In* Mechanisms of Taste and Smell in Vertebrates/A Ciba Foundation Symposium (O. E. Loewenstein, editor), J. and A. Churchill, London (in press).

Shepherd, G. M., 1963. Neuronal systems controlling mitral cell excitability. *J. Physiol.* (*London*) 168: 101–117.

Zucker, R. S., 1969. Field potentials generated by dendritic spikes and synaptic potentials. *Science* (*Washington*) 165: 409–413.

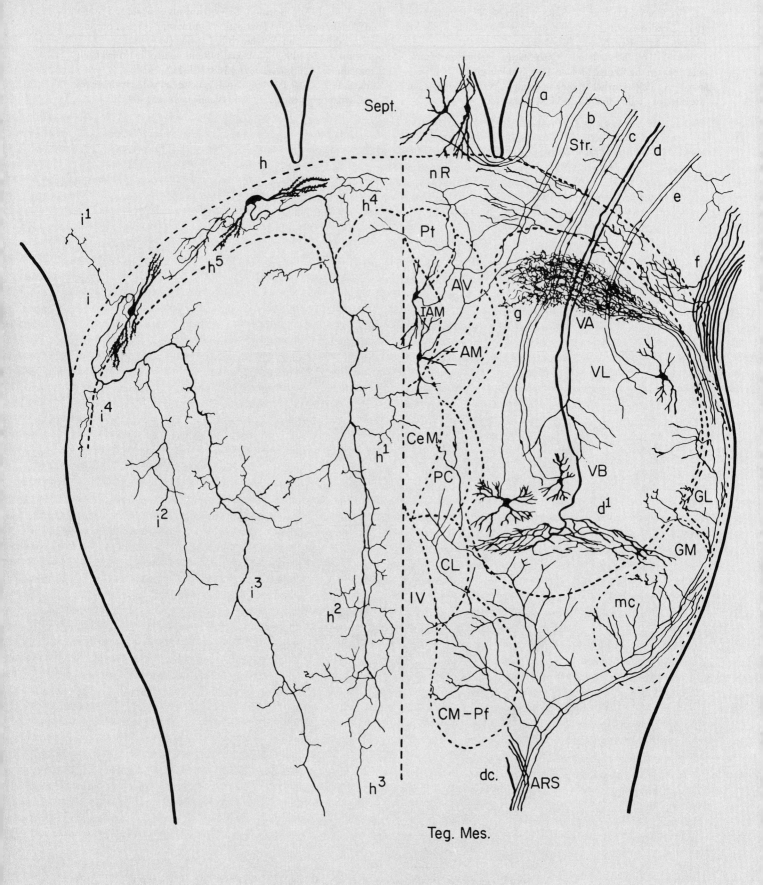

# COMMUNICATION AND CODING IN THE NERVOUS SYSTEM

*The complexity of neuronal circuitry is seen in this drawing, provided courtesy of M. E. and A. B. Scheibel and of Brain Research. It shows, in horizontal section, the intricate afferent, efferent, and intrinsic organization of a part of the mouse thalamus. Dr. José Segundo, contributing editor for this section, feels it is characterized by words of J. L. Borges: ". . . a growing, dizzying net of divergent, convergent, and parallel times."*

# 51 Communication and Coding by Nerve Cells

## J. P. SEGUNDO

AN EXPERIMENT in basic neurophysiology often consists of the study of the relations between two activities. These, which for purposes of this discussion are designated M and N, are selected because they either reflect links in a causal chain or, being influenced by common factors, serve as indicators of each other.

It is possible to conceive of a general relation, designated as R, which matches all M and N events (Figure 1). It is possible also to imagine that each event in R is described fully. Different subsets of M, N, and R may be considered. Some of these, $M_L$, $N_L$, and $r_L$, respectively, take place naturally in real life. Others, $M_E$, $N_E$, and $r_E$, respectively, are observed experimentally. The experimental events are selected by the investigator because they reproduce natural occurrences or because they may provide information about mechanisms, and usually are summarized by representative parameters (if spike trains, for example, by mean rates).

The relation R between the two activities can be approached prospectively or retrospectively (Figure 2). The prospective way is to observe in M a particular event, m, and then look for possible associations in N. It may be desirable to assign to each N event a probability of occurrence conditional for the given m; this value will depend on the mechanisms that underlie R. For instance, if the skin is in-

J. P. SEGUNDO Department of Anatomy and Brain Research Institute, University of California, Los Angeles, California

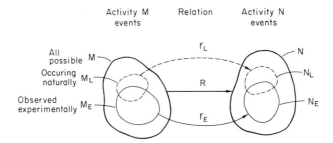

Activity M      Relation      Activity N
events                       events

FIGURE 1 Schematic representation of a typical neurophysiological experiment. Heavy continous lines represent the complete relation (R), broken lines that which occurs in real life ($r_L$), and light continuous lines that observed experimentally ($r_E$). M, N pairs may be: skin indentation-cutaneous afferent spike activity; postsynaptic spike activity; presynaptic spike activity-postsynaptic membrane potential; blood $CO_2$ concentration-EEG activity; state in the diurnal cycle-eye movements; and so on. The relation $r_E$ is often described probabilistically, because $M_E$ and $N_E$ do not correspond on a one-to-one basis. Respective sizes and positions of M, $M_L$, $M_E$, N, $N_L$, and $N_E$ illustrate the possibility that events may not occur either naturally or experimentally, that events observed experimentally do not necessarily take place naturally, and vice versa.

dented by 200 microns, can 30 afferent spikes be observed and, if so, with what probability? (Figure 2B) (Werner and Mountcastle, 1965, 1968). This viewpoint applied to $r_E$ is implicit in the current experimental strategy in which the investigator imposes certain conditions and then observes their effects.

The second, or "restrospective" approach is to recognize a particular N event, n, and then look for each of its possible associations in M. It may be desirable to assign to each M event a probability of occurrence conditional for the given n. This value will depend on R, as well as on the laws that govern M events and, perhaps, other issues. For instance, if 40 spikes were observed, could the delivered skin stimulus have been (among others *a priori* equally probable) an indentation of 400 microns and, if so, with what probability? This viewpoint applied to $r_L$ reflects how a subject must proceed when using his sensory input to identify some stimulus condition (Fujita et al., 1968). The distinction between the two approaches is especially useful when M and N reflect links in a causal chain. When M and N are influenced by common factors and serve as indexes of each other, the two approaches are essentially similar. The corresponding probabilities will depend on the relations between the common influences and each of the M and N activities, as well as on the laws that govern the occurrences of the common-influence events.

FIGURE 2 A. Different approaches to the relations between two activities. Prospective (R): Given a particular M event, m, the observer must predict its possible N event consequences and their respective probabilities. Retrospective ($R^{-1}$): Given a particular N event, n, the observer must infer its possible M event causes and their respective probabilities. B. Relations between skin indentation stimuli and number of impulses in a monkey mechanoreceptive fiber. Stimulus-response matrix of tallies. Each column corresponds to a separate stimulus category: stimuli were allocated to 30 equal categories on the basis of probe displacement. Each row corresponds to a separate response category: responses were allocated to 23 equal categories on the basis of the number of the impulses within a specified period. Total number of stimuli and that of responses in each category are shown across the bottom row or down the rightmost column, respectively. The proportion of cases in each entry relative to the total in the corresponding column (or row) estimates the prospective probability of that response given that applied stimulus (or the retrospective probability of that stimulus given that observed response). (From Werner and Mountcastle, 1965.) C. Relations between head roll in space and interspike interval mean in a cat lateral vestibular nucleus cell. The relations are represented by a probability density contour diagram over the position-mean plane; darker shadings correspond to higher densities. The densities along the constant abscissa line AA' corresponding to 5° SD (side down) are plotted as a histogram on the upper right; they estimate the prospective probabilities of the different mean intervals given that the head is rolled 5° SD. The densities along the constant ordinate lines BB' (30 msec mean interval, corresponding to a rate of 33.3/sec) and CC' (16 msec mean interval, corresponding to 62.5/sec) are plotted as clear and shaded histograms, respectively, in the lower right; they estimate the retrospective probabilities of the different roll positions given means of 30 msec and 16 msec, respectively. (From Fujita et al., 1968.)

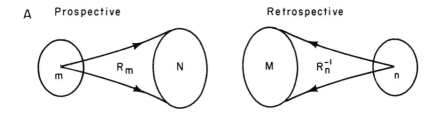

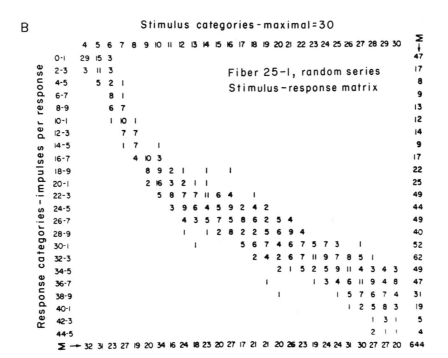

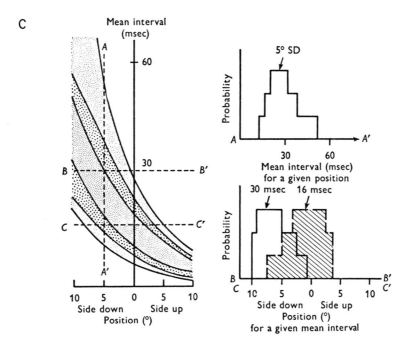

## Communication between nerve cells

The general concepts of communication are applicable to all situations in which one entity influences or relates to another (Shannon and Weaver, 1963). We envision the nervous system as a set of interacting components. Therefore, the formalisms of communication theory are appropriate for its study, because they comply with the goal desirable in any scientific endeavor—of having explicit critical evaluations of the assumptions, logic, and conclusions that are implicit in the thinking in the field.

Important entities in any communication scheme are the source that generates signals and the channel that transmits them (Khinchin, 1957; Shannon and Weaver, 1963). When the source and the channel-input alphabets differ, it is necessary to transform or "code" every sequence of letters from the former into a sequence of letters from the latter. The communication scheme involving a channel exclusively appears the most reasonable in many neurophysiological contexts: from the activity M, the activity N, and the relation R, the investigator extracts the channel-input alphabet, the output alphabet, and the corresponding probability measure, respectively. If he is also interested in a probability distribution over M, he may connect to the channel a source the alphabet of which is identical to that at the channel input and of which the probability measure is the desired one. Because relations between physiological variables can be composed or decomposed, and new sources and channels arise when source-channel connections are made, several schemes may be adjusted to essentially the same biological system.

All communication studies pose problems at technical, semantic, and effectiveness levels (Shannon and Weaver, 1963). At the technical level, the investigator is concerned with quantification of symbol transmission. This is achieved by information theory, a specialized branch of applied probability theory, in which the amount of information conveyed by an experiment is equated with the amount of uncertainty or entropy its outcome removes (e.g., Khinchin, 1957; Shannon and Weaver, 1963). Basic theorems often apply only when the source uncertainty is less than the channel capacity and the message is sufficiently long; when these conditions are not met, as could happen in nerve cells, the corresponding propositions are neither demonstrated nor intuitively obvious (Khinchin, 1957). At the semantic level, the investigator asks how precisely the transmitted symbols convey the "desired meaning"; at the effectiveness level, how successfully the received message elicits the "desired conduct."

Identification of a desired meaning or conduct in situations involving nerve cells has varying degrees of difficulty because of the anthropomorphic sense of the word "desired" and because of our limited knowledge of nervous-system physiology. The desired meaning is clear for first-order sensory fibers, the most likely function of which is providing information about the environment; so is the desired conduct for a motorneuron, the most likely function of which is muscle control. Such identifications are, on the other hand, unclear in areas in which functional role is uncertain. In such cases, the semantic and effectiveness analysis, although equally important, may be highly conjectural because of, and increasing with, our ignorance of the role of the structure in question. As more complex functions are involved, difficulties become more marked, perhaps reflecting the inherent shortcomings of our own understanding.

## Uses of information theory

In spite of these semantic and effectiveness qualifications, it is desirable, in some situations, to go beyond the tacit recognition that some communication scheme applies, and it is meaningful actually to measure information. Particularly warranted are calculations performed with the aim of quantifying the capacity of neural systems to inform as to the stimuli they normally receive (Stein, 1967b). Useful examples are, among others, MacKay and McCulloch's (1952) demonstration that a synapse is capable of transmitting as much information in interval-coded as in binary-coded form, although it probably actually transmits less, and several calculations in the visual system (Fitzhugh, 1957; Grüsser et al., 1962). Stein (1967b) derived valuable approximate equations for neurons that reflect the intensity of a constant stimulus lasting t seconds by the number of spikes in t; their information-transmitting capacity relates to the interspike interval distributions and, for large t's, takes into account the interval serial correlations, the number of cells in parallel, and the response reproducibility. In this model, the transmitted information is maximized by a probability of stimulus-intensity occurrence that varies inversely with the standard deviation of the response to each, and that often is close to uniform over its range.

Naturally and justifiably, spike trains play a major part in this section of the book (see also Segundo, 1970). Pioneering statistical studies by Hagiwara (1949) and Katsuki et al. (1950) were followed by others on methods (e.g., Bayly, 1968; Cox and Lewis, 1966; Gerstein and Kiang, 1960; Kogan et al., 1966; Perkel et al., 1967a, 1967b; Perkel, this volume). Each parameter measurement must be clarified explicitly to avoid apparent discrepancies. The mean rate, for example, can be measured over a specified period or over a specified number of spikes. Commonly it is measured over the entire record, but more restricted periods or sets can be chosen for physiological reasons (Segundo et al., 1966; Segundo and Perkel, 1969), to approximate an average (Walløe, 1968), or arbitrarily. The "window," involving specified periods or intervals, can be opened at prescribed times (e.g., at every second or spike), or can be moved continuously (Segundo et al., 1966; Segundo and Perkel, 1969). All spikes can be equivalent (e.g., Werner and Mountcastle, 1965; Walløe, 1968), or each can be weighted. A rate weighted with a decreasing function of the time since its occurrence (Segundo et al., 1966; Segundo et al., 1968) reflects jointly the nonweighted rate that relates to the inter-spike-interval mean and the pattern that relates to other statistics. In each case, parameters have certain defining conditions and domain parts that physiologically are more relevant than others (Segundo and Perkel, 1969).

Afferents from touch spots and in the dorsal spinocerebellar tract (DSCT) inform about skin indentation and muscle length, respectively (Jansen and Walløe, this volume; Walløe, 1968; Werner and Mountcastle, 1965, 1968). Contingent on the degree to which the neural response depends on the stimulus, the information provided by observing only the neural response approaches more or less that which would be obtained by observing the stimulus. Stimulus response matrixes (Figure 2B) were constructed on the reasonable assumption of the significance of some nonweighted rate under steady-state conditions. (The informative value also of other one-cell and two-cell parameters, especially under continually varying conditions, is discussed below.) The information transmitted, or transinformation, increased in a bounded manner when there were increases either in the number of equally probable-stimulus categories (which thus tended to a continuous, uniform distribution), in the number of observed-rate categories (which also tended to a continuous distribution), or in the observation times (Figure 3A and B). Figure 3A also illustrates the relation of the experimentally found transinformations with the calculated maximum values (Stein, 1967b). The natural order of the IA afferent intervals offered advantages over randomly shuffled sequences (Figure 3A and B). In

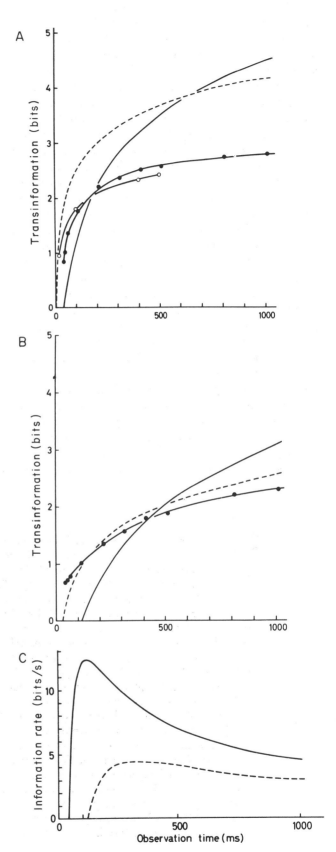

FIGURE 3 A and B. *Transinformations.* Values are from DSCT cells (continuous line with black circles) and from mechanoreceptive fibers (continuous line with open circles) (Werner and Mountcastle, 1965), and were calculated by a stimulus-response matrix method. The maximum possible values (broken line) were predicted by Stein (1967b). The stimulus-response matrix method probably overestimates and underestimates transinformations for short and long observation times, respectively. Transinformations are greater when they correspond to DSCT spike trains with interspike intervals in the order in which they occur in nature (A) than when they correspond to hypothetical trains with the same intervals after they have been shuffled (B). The solid lines reflect calculations based upon other methods. C. Rate of information transfer. The rate exhibits a maximum. It is consistently greater for intervals in natural order (continuous line) than for shuffled ones (broken line). All abscissae are observation times. (From Walløe, 1968.)

DSCT cells, in which variability, convergence, divergence, and interdependence participate, loss of information about muscle length is small (Walløe, 1968). Transinformations (Figure 3C) are moderate for brief observations; they then increase steeply and finally level off. With DSCT cells, the mean-rate variability when the same stimulus is applied several times is comparable to that during each trial. Contrastingly, with utricular afferents, rates change from one trial to another (Vidal et al., in preparation); hence, accurate subjective estimates should be based on several statistics of each cell, on several distinguishable cells, or on both.

The single powerful excitatory junction is appropriate for deducing formulas that relate uncertainties to mean rates and patterns (Segundo and Perkel, 1969; Segundo et al., 1966). Calculations are simplified when the statistics of the presynaptic discharge, i.e., the generating probabilities, are known (Figure 4A). In the prospective situation, the observer knows the timing of recent influential EPSPs (defined below), but, when predicting, is uncertain whether the postsynaptic cell will fire. He can base a prediction on prospective probabilities that are estimated experimentally (Figure 4B). In the retrospective situation, the observer knows the recent postsynaptic behavior, i.e., whether a postsynaptic spike has occurred, but is uncertain when inferring the presynaptic timings immediately preceding that event. He can base an inference on retrospective probabilities that are estimated experimentally (Figure 4C). Prospective and retrospective uncertainties are minimized when they are expressed as joint functions of rate and pattern.

Hence, parameters defined in information theory and estimated experimentally can be used to quantify conclusions with physiological interest. Particularly, they permit investigators to measure the uncertainties that exist in certain situations, with and without knowledge of relevant physiological indexes. Moreover, they enable them to analyze precisely the influence of several issues: at an excitatory junction, for example (Segundo and Perkel, 1969; Segundo et al., 1966), the generating probabilities reflect factors that act on presynaptic spike production and conduction; the prospective probabilities reflect such junctional and postsynaptic aspects as PSP characteristics (size, shape, summation rules, and so on) and spike consequences (afterpotentials, refractory cycles, and so on); and the retrospective probabilities reflect all these properties.

## Coding in the nervous system: Natural codes

Neurophysiologists apply the word "code" frequently, but less in the mathematical sense defined above than in one "borrowed from the common discourse" (Perkel and Bullock, 1968) and in use before the rigorous meaning was coined. Perkel and Bullock (1968) have defined it as the representation and transformation of information. More precisely, the relation R transforms, "codes," or "encodes" the M events into the N events (Figure 1), and the rules of this mapping constitute the code. The mapping "used" by the animal in the course of its normal life, i.e., the "natural code," is that implicit in the restricted relation $r_L$. The mapping detected by the investigator, i.e., the "experimental code," is that implicit in the restricted relation $r_E$; it is

FIGURE 4 Generating, prospective, and retrospective probabilities at a powerful excitatory junction. Whether the postsynaptic cell will fire after a particular EPSP (or presynaptic spike) is related to the timing of only a limited number of recent "influential" EPSPs (or presynaptic spikes), at the most. For a sequence of $j + 1$ brief events, *timing* is the ordered set of j first-order intervals $i_1, \ldots, i_j$; *span* T is their sum $i_1 + \ldots + i_j$; *pattern* is the ordered set of $j - 1$ ratios $i_1/T, \ldots, i_j - 1/T$; and *mean rate* is the ratio $j/T$, (or $j + 1/T$, as mentioned in Segundo et al., 1966). In order to plot on a plane, it is assumed that only three influential EPSPs have timing that involves only two intervals used as abscissa and ordinate (inset on the lower right). Higher probabilities are represented by darker shadings in the contour diagrams. Gradients in general, along lines through the origin, and along lines with $-1$ slopes reflect probability dependence on timings, rates, and patterns, respectively. (Computer-simulated cases.) A: "Generating" probability, i.e., probability that a set of influential EPSPs will exhibit a particular timing. This example corresponds to EPSPs driven by a Geiger counter (From Segundo et al., 1966.) B. "Prospective" probability, i.e., probability that a set of influential EPSPs with a particular timing will trigger a postsynaptic spike. C. "Retrospective" probability, i.e., probability that an observed postsynaptic spike was preceded by a set of influential EPSPs with a particular timing. Each of the three lower rows involves a different form of EPSP summation: linear (1), with EPSP reduction (2), and with EPSP augmentation (3). The reduction of closely placed EPSPs in B makes the timings with the shortest intervals (i.e., with the highest rates) prospectively (B) less effective in triggering possible spikes than those with intermediate intervals, and therefore, retrospectively (C) less likely to have triggered an observed spike. The highest prospective probabilities 1 and 3 are not found close to the origin, because if the interval between the early EPSPs in the influential set (i.e., the abscissa) is short, it is likely that there will be a postsynaptic spike early in the set and that, therefore, refractoriness will prevent the cell from firing again after the last EPSP in the set. (Also compare plots in Figure 164 of Segundo and Perkel, 1969.)

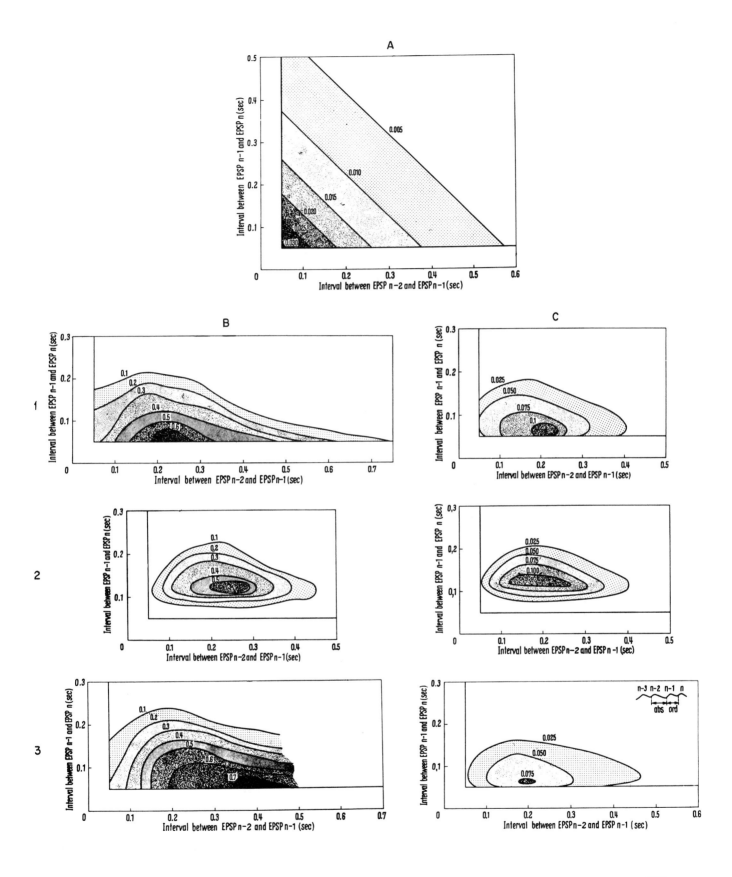

frequently limited to particular features (e.g., to a relation between rates) of the general code. This, we believe, is the sense in which it is said that a muscle length is coded as a mean rate (Jansen and Walløe, this volume), that a shift in presynaptic timing is coded as a change in postsynaptic firing (Segundo et al., 1966), and so on. The expression is particularly meaningful functionally when R involves a causal relation. The words "decoding" and "decoder," applied in communication (Shannon and Weaver, 1963) to the reconstruction of the original message from the signal and to the entity that performs it, may be misleading in a neurophysiological context if they imply that the resulting message is similar to the input (Perkel and Bullock, 1968); "reading" and "read-out mechanisms," respectively, would perhaps be clearer (Segundo and Perkel, 1969).

As the pertinent conditions vary over a certain space, which may well be continuous and nondenumerable, there

are corresponding alterations in neuronal activity. (The possibly probabilistic nature of the relation between conditions and changes should not be forgotten.) Any neural change, for example, that which results from a shift in conditions from $C_1$ to $C_2$ (Figure 5A), is characterized by spatial or temporal features, or both. The spatial features (Figure 5B) arise when the different conditions $C_1$ and $C_2$ are reflected by different sets, $S_1$ and $S_2$, of nerve cells, the activities of which are outside pre-established control ranges. They have been discussed extensively for most sensory modalities (e.g., Creutzfeldt, this volume; Mountcastle, 1967). As pointed out by Werner (this volume), the geometrical properties of central mappings differ from those of the stimulus space and thus preclude simple correspondences. In somatosensory area I, small shifts in stimulus location lead to marked changes in the responding sets, because receptive fields of individual neurons are highly

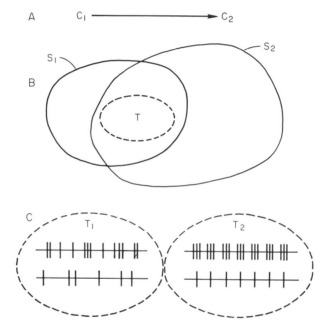

FIGURE 5 Schematic representation of the spatial and temporal components of the neuronal response to a naturally occurring change. The natural change from $C_1$ to $C_2$ (A) may occur in some sensory situation, be the passage from sleep to wakefulness or from one endocrine state to another, and so on. The resulting neuronal response (B and C) is also natural, thereby complying with the first necessary requirement for being a useful part of a code in a living communication scheme. The other requirement is that there be elements (nerve cells or effectors) that respond to that change (see text). The spatial component of the response (B) involves the shift from the set $S_1$ to the set $S_2$ of "re-

sponding" neurons. The temporal component (C) involves changes in the activities of and in the correlations between, the cells in T, which is a part of the intersection of $S_1$ and $S_2$. The discharges of two cells in T are represented below: on passing from conditions C, with activities $T_1$, to conditions $C_2$, with activities $T_2$, each cell becomes faster and more regular, and each of the evenly spaced triplets in one (upper record) coincides in time with a single spike in the other (lower record). This Venn diagram carries no implications as to the actual topographical distributions in the nervous system of $S_1$, $S_2$, and T.

restricted (e.g., Werner, this volume); contrastingly, in reticular nuclei, larger shifts are necessary to obtain significant changes, because fields are often less restricted (e.g., Segundo et al., 1967b). The temporal features (Fig. 5C) of the neural change arise when the conditions $C_1$ and $C_2$ are reflected by different activities or correlations, or both, in at least some cells responding to both. These also have been discussed.

Recognition that a particular neural change is, in some sense, a useful part of the natural codes in a neuronal communication scheme implies two necessary, but perhaps not sufficient, requirements (Segundo and Perkel, 1969). The first is to recognize that it occurs in normal life. This requirement is met in practice when the change from $C_1$ to $C_2$ is a natural one, i.e., occurs within a certain $M_L$ subset, and the neural change, taken here as an $N_L$-activity shift, is observed experimentally in a preparation as physiological as possible (e.g., in a freely moving and unanesthetized animal). The second necessary requirement is that there must be neural or effector elements, the activity of which is modified by it (Perkel and Bullock, 1968; Segundo and Perkel, 1969; Segundo et al., 1968; Terzuolo, this volume; Uttal and Krissoff, 1968; Walløe, 1968). If none exists, the change constitutes a dead end, inconsequential as far as subsequent processing is concerned. This requirement implies that the same change, taken now as an M event, must modify the activity of certain neuronal or effector elements, i.e., must elicit a change in some N space. It will be met commonly for the spatial features, because dissimilar sets usually distribute their influences differently; in fact, a neuron remains a "labeled line" (Bullock and Horridge, 1965; Perkel and Bullock, 1968) to the extent that its distribution and effects are unique. The requirement will be met for the temporal features, too, if certain cells respond differently to dissimilar presynaptic timings or correlations in a particular set of presynaptic terminals. Identification of temporal changes which are "read" by postsynaptic cells is thus an important physiological problem (Segundo and Perkel, 1969).

## The neuron as an analyzer of spike trains

The presynaptic terminals that make contact with a particular postsynaptic cell, produce PSPs therein and influence a certain spike-trigger locus, can be classified in, or between, two extreme categories based on the size of their PSPs and on the manner of their influence (Segundo and Perkel, 1969; Segundo et al., 1968). In the first extreme category are terminals that elicit large PSPs, i.e., "powerful" terminals. The terms "large" and "small" are relative to the average voltage between membrane potential and threshold and apply to PSPs recorded at that spike-trigger zone (Segundo et al., 1963). This category has been analyzed extensively (Bittner, in preparation; Bullock and Horridge, 1965;

Gillary and Kennedy, 1969; Jansen and Walløe, this volume; Moore et al., 1963; Perkel et al., 1964; Ripley and Wiersma, 1953; Segundo et al., 1963; Segundo et al., 1966; Segundo et al., 1968; Segundo and Perkel, 1969; Walløe, 1968; Wiersma and Adams, 1950). These terminals influence the postsynaptic discharge by way of their interspike interval mean (and, therefore, mean firing rate), standard deviation, histogram, serial correlations, and so forth. The postsynaptic discharge, then, depends on both the presynaptic mean rate and its pattern, measured over either the entire train or through specified windows (see above). The general trend is for faster excitatory and slower inhibitory presynaptic rates to produce faster postsynaptic discharges (Figure 6). The terms "fast" and "slow" as applied to the input relate to such junctional and postsynaptic features as time constants and aftereffects of PSPs and spikes (Segundo et al., 1963; Segundo and Perkel, 1969; Segundo et al., 1966; Terzuolo, this volume). Figure 6 represents findings in many pacemaker nerve cells either computer-simulated or living, such as those in molluscan ganglia and in crustacean stretch-receptor organs (but not necessarily in all of them, Terzuolo and Bayly, 1968). Irregular inputs produce monotonic curves. Regular inputs lead to zigzag curves with oppositely sloped segments that imply such special results as "paradoxical" rate changes and disproportionate effects (Moore et al., 1963; Perkel et al., 1964; Schulman, 1969; Segundo and Perkel, 1969). Changes around extreme rate values are usually insignificant, because it is not possible to go beyond stopping the cell or driving it maximally.

The pattern is influential at intermediate rates (Figure 7, top, page 579) (Moore et al., 1963; Perkel et al., 1964; Segundo et al., 1963; Segundo and Perkel, 1969; Segundo et al., 1966; Segundo et al., 1968). The separation of the curves in Figure 6 illustrates a sensitivity to variability, and the time structures of input trains at given mean rates influence the output of simulated DSCT and granule cells (Jansen and Walløe, this volume; Walløe, 1968). Greater presynaptic and postsynaptic irregularities commonly, but not always, go together (Hegstad, Segundo, and Perkel, in preparation; Segundo and Perkel, 1969; Segundo et al., 1968). Certain timings exert the maximum effect under each set of restrictions (Segundo and Perkel, 1969).

The input-output relation can be explored in detail when a single strong terminal excites a nonpacemaker cell (Segundo and Perkel, 1969; Segundo et al., 1966). At each instant, the latter bases the decision of firing on an evaluation of the timing, i.e., of the mean rate and pattern, of the EPSPs within a bounded period, and, therefore, of a restricted number of them. (The decision to fire also is biased by recent previous postsynaptic discharges.) The boundedness of the "integration" period reflects those of the postsynaptic time constant and of the presynaptic terminal excitability cycles; the restriction on the number of "influential" PSPs

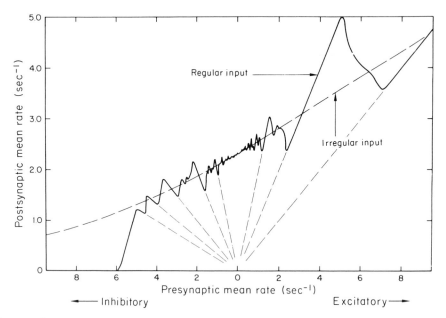

FIGURE 6 Influence of presynaptic mean rates. Computer simulation of a single powerful presynaptic terminal acting on a postsynaptic pacemaker cell (the spontaneous rate of which is about 2 per sec). On the abscissa, presynaptic mean rates conventionally affected by a positive or negative sign if corresponding to EPSPs or IPSPs, respectively. On the ordinate, postsynaptic mean rate. Broken lines correspond to irregular, and solid lines to regular, presynaptic train patterns. Graphs were drawn by hand on the basis of more than 150 points. The general trend is for "higher" presynaptic rates, i.e., slower IPSPs or faster EPSPs, to produce faster postsynaptic discharges. The irregular patterns produce a monotonic curve; the regular patterns produce a zigzag curve with oppositely sloped segments. This leads to special results, such as a sensitivity to variability, "paradoxical" rate changes, and some disproportionate effects (see also Segundo and Perkel, 1969). Comparable displays can be constructed for other variables (e.g., presynaptic rate-postsynaptic standard deviation). (From Perkel et al., 1964.)

reflects the facts of a bounded integration period and of presynaptic refractoriness. A mean rate weighted exponentially according to physiologically suggested parameters is, in this case, a functionally meaningful description (Segundo and Perkel, 1969).

The order in which terminals are activated may be influential, even beyond the obvious example that large IPSPs are more effective when they precede EPSPs (Hegstad et al., in preparation). Indeed, EPSP summation may be noncommutative (Gerstein, this volume; Segundo et al., 1963), and invertebrate neurons with segregated input zones separating spiking regions may "sense" stimulus direction and velocity (Kennedy and Mellon, 1964).

In the other extreme category of presynaptic terminals are those that elicit small PSPs, i.e., "weak" terminals (Segundo and Perkel, 1969; Segundo et al., 1968). Their contribution is demonstrated easily only if they are numerous. The degree of interterminal correlation is a critical issue under these conditions. The discharges of cells A and B are said to be "correlated" when, as estimated by the cross-correlation histogram, there is, within a certain time from an A spike, a different-from-average probability that B will fire (Perkel et al., 1967b). When, on the one hand, the weak terminals are independent, their influence arises exclusively from their summed mean rates and not from other discharge parameters (Figure 7, center). When, on the other hand, the weak terminals are correlated, their influence reflects other statistics in addition to mean rates (Figure 7, bottom) (Hegstad et al., in preparation; Segundo and Perkel, 1969; Segundo et al., 1968). This holds even if the dependence is of physiological degree, involves only some terminals, or occurs within separate groups.

Correlation between converging terminals is, therefore, most important physiologically (Figure 8, page 580). It arises because of shared influences (which may depend on common sensory drives, e.g., Gerstein, this volume; Terzuolo, this volume), or synaptic connections (Aladzhalova et al., 1969; Moore et al., 1970; Person, 1965; Vasilevsky, 1968), or both. Each anatomical-functional arrangement (in terms of cell properties, synaptic effects, and correlations with other cells) determines a characteristic cross-correlation histogram, but the converse is not true (Moore et al., 1970). Parallel correlated fibers with similar influences on the same post-synaptic cells (Figure 9A, page 581) may allow the

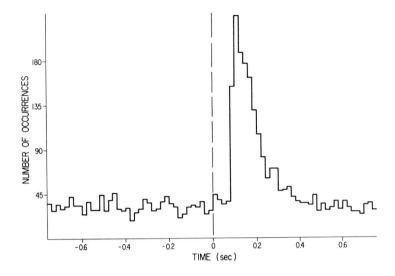

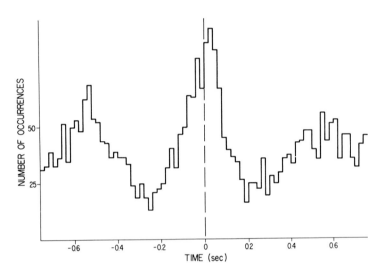

FIGURE 7  Influence of presynaptic patterns at similar mean rates. Interspike interval histograms of postsynaptic activities in computer simulation experiments in which the same pacemaker cell was submitted to inhibitory presynaptic influences that are either "irregular" (left column) or "regular" (right column). Single powerful presynaptic terminal (upper row), or 64 weak terminals that either are independent (center row) or have physiological degrees of correlation (lower row). All terminals have approximately the same mean firing rates; the powerful terminal elicits a large IPSP, and all 64 weak terminals elicit small IPSPs of about the same sizes. The postsynaptic activities corresponding to regular and irregular inputs differ with a single powerful terminal (upper row) and with 64 weak correlated terminals (lower row), but do not differ with 64 weak independent ones (center row). The irregular presynaptic terminals (lower left) have a stronger correlation than the regular ones (lower right). These cases were chosen because they illustrate to the naked eye the difference in histogram shapes. However, weaker and more comparable degrees of correlation could still produce outputs that were significantly different. The effect of presynaptic correlation was clearer when the weak terminals excited a nonpacemaker cell, as discussed in Segundo et al. (1968), than when they inhibited a pacemaker cell as in this Figure. In these comparisons, statements as to similarities or differences were based on the application to the observed cumulative interval histograms of 90 per cent confidence-level tests of the hypothesis that both samples were from populations with the same distributions (Dixon and Massey, 1969). Unfortunately, and as illustrated by most references, tests for statistical significance of, for example, differences and goodness-of-fit have not been used extensively in this field. N, number of cases; $\mu$, mean; $\sigma$, standard deviation; CU, coefficient of variation. (From Hegstad et al., in preparation.)

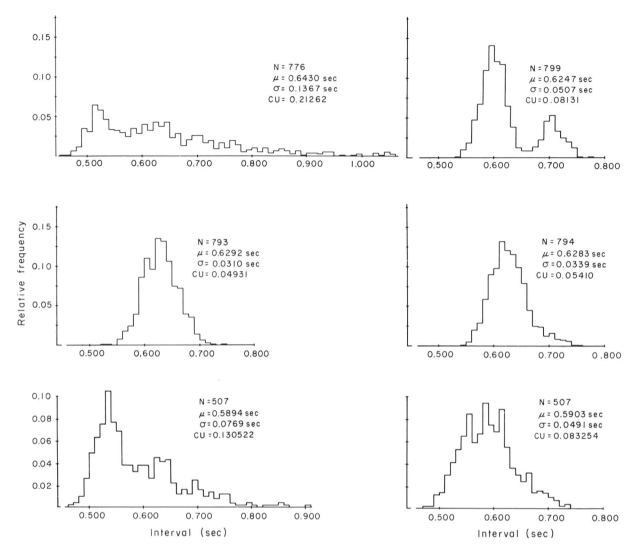

Figure 8 Correlated spike discharges. A. Cross-correlation histogram between the source of excitation, a Geiger-counter-driven shock train to a connective trunk, and the spike train of an excited nerve cell. In this case, the evoked EPSP closely paralleled the higher-than-average histogram profile to the right of zero. B. Cross-correlation histogram between two pacemaker cells sharing an inhibitory influence elicited by a Geiger-counter-driven stimulation of a connective trunk. As does shared excitation (but in a somewhat lesser degree), shared inhibition can produce a higher-than-average profile around the origin. The cross-correlation histogram between two activities reflects the individual features of the cells, the anatomical-functional characteristics of their connection, and correlations with other neurons. Empirical confidence bands are obtained by shuffling the intervals at random. Unidentified cells in an isolated *Aplysia* ganglion. (From Moore et al., 1970.)

nervous system to use spatial averaging, thus abbreviating temporal processing (Fitzhugh, 1957; Gerstein, this volume; Maffei, 1968; Perkel and Bullock, 1968; Segundo et al., 1963; Stein, 1967a, and this volume; Terzuolo, this volume).

The preceding considerations suggest that, when sufficient knowledge about the anatomical and physiological properties of a junction is available (i.e., when PSP sign and summation rules, postsynaptic firing characteristics, synergistic inputs, and so on, are known), it will be possible to antici- pate with acceptable certainty the correspondence between presynaptic and postsynaptic events, i.e., it will be possible to predict the corresponding code. The most general post-synaptic cell is that influenced by few powerful terminals and by larger numbers of intermediate and weak ones, with varying degrees of correlation (Segundo and Perkel, 1969).

The presynaptic mean rate is highly influential in neuron-al input-output relations. This holds for the nonweighted rate over complete records or restricted sets (e.g., Figures

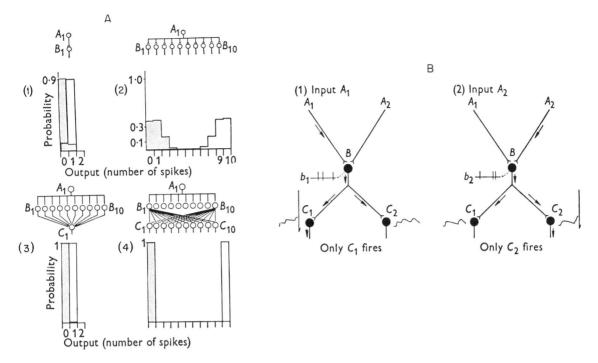

FIGURE 9 A. Advantages of parallel processing. The anatomical arrangements 1, 2, 3, and 4 exhibit increasingly complex excitatory connections (in 4, the paths from B cells to C cells are shown only for $B_1$ and $B_{10}$ so as to make the Figure less intricate). Beneath each diagram is a histogram of the possible outputs in response to timings $u$ (shaded histograms) and $f$ (unshaded histograms) fired by neuron $A_1$. 1. The prospective probability of each B cell firing is 0.09 for $u$ and 0.90 for $f$. 2. With 10 similar and noninteracting cells in parallel, $u$ usually will elicit about one spike and $f$ will elicit close to 10 spikes. 3 and 4. A third order set of C cells is such that each is influenced by all B cells and will fire only if at least any five of them discharge almost simultaneously; this leads to the virtually certain outcomes of "no spike" for $u$ and "10 spikes" for $f$. B. Switching" role of a neuron with different kinds of output junctions. If different inputs, $A_1$ and $A_2$, trigger B into producing different timings, $B_1$ and $B_2$, respectively, and each acts preferentially on a different set of postsynaptic cells, $C_1$ and $C_2$, respectively, the common neuronal link B may channel selectively different arriving influences down separate outflow channels. (From Segundo et al., 1963.)

158, 167, and 170 in Segundo and Perkel, 1969). In certain cases, for example, with many independent terminals at similar rates, the nonweighted, over-all rate is the only influential statistic. In other cases, such as those with few powerful terminals or numerous correlated weak ones, other single-cell or two-cell intervals or count statistics (such as standard deviations, histograms, serial coefficients, autospectra, cross-correlation histograms, phase relations, and so on) become important, together with the non-weighted rate. Then the ongoing modulation of the presynaptic intervals (or instantaneous rates) around their average determines corresponding variations of the postsynaptic intervals. (The modulation itself reflects the upstream generation of large PSPs, of bursts of correlated small ones, special threshold characteristics, and so on; Segundo and Perkel, 1969). Certain aspects of these modulations are quantified by the cross-spectral densities of the presynaptic and postsynaptic spike counts (Bayly, 1968; Cox and Lewis, 1966; Perkel, this volume; Terzuolo, this volume). The gain and phase shift of the transynaptic transfer of each periodic component depends on issues such as the relative lengths of its period and of junctional cycles and time constants. High gains and small phase shifts may not always be useful functionally. Linearity, even if defined satisfactorily, may, in this context, have a range that is no more than a small part of that of physiological variations. Nevertheless, a systems-analysis approach may serve to describe some of the time-invariant properties of the junction within linear limits (Perkel, this volume; Terzuolo, this volume).

Interspike-interval variability may or may not benefit the functional role of a particular cell. In sensory systems, for example, it may hinder information transmission about steady-states, but favor that about variable situations (Stein, 1967b, and this volume); thus, it can be undesirable or desirable, respectively. Shifts from constant to varying conditions often imply changes in the discharge of individual cells and, in addition, increased correlations: this happens in Ia muscle afferents (Jansen and Walløe, this volume;

Terzuolo, this volume) and possibly explains the discharge independence of steadily illuminated, photosensitive crayfish neurons (Stark et al., 1969).

Valuable general suggestions have arisen from sensory neurophysiology in the form of evoked changes that reasonably should be "read" by the nervous system (e.g., Creutzfeldt, this volume; Mountcastle, 1967; Uttal and Krissoff, 1968). For instance, the fact (Mountcastle et al., 1969) that a vibratory stimulation to a specified region of the skin is coded jointly in a particular set of correlated cortical neurons as a rate and as an interval order that preserves periodicity, is strongly suggestive evidence for the significance of several statistics in sensory processes (Figure 10). Other possibly influential issues were reviewed by Perkel and Bullock (1968).

## Mechanisms

Important aspects of the processes that determine the neuronal input-output relation affect the transmembrane potential value (Bittner, in preparation; Calvin, 1968; Gillary and Kennedy, 1969; Hartline and Cooke, 1969; Jansen and Walløe, this volume; Kogan et al., 1966; Levitan

et al., 1968; Moore et al., 1963; Perkel, this volume; Perkel et al., 1964; Schulman, 1969; Segundo et al., 1966; Segundo et al., 1968; Terzuolo, this volume; Walløe, 1968). When large PSPs are elicited by few terminals, the postsynaptic response to changes in presynaptic timing is explained partly by straightforward arguments, which include shapes, as well as linear and nonlinear summation rules, about postsynaptic, pacemaker, and after-potentials (Segundo et al., 1963; Segundo and Perkel, 1969). The relation of the prospective probability and other probabilities with separate physiological issues is discussed above. If several terminals are independent, their combined sequence is almost Poisson, and therefore relates little to individual patterns (Segundo et al., 1968). There are no known examples of this with powerful terminals, the DSCT case probably differing because of "beating" between several inputs, which, although independent in the steady state, are not quite enough to compensate for their regularity (Jansen and Walløe, this volume; Perkel, 1965; Walløe, 1968). When the independent terminals are weak, their effect approximates that of a DC bias, under constraints involving average intervals, postsynaptic time constants, and so on (Hegstad et al., in

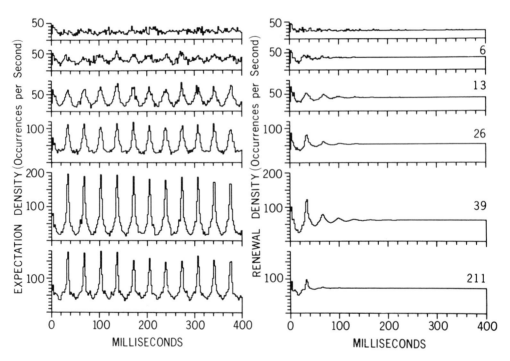

FIGURE 10 Coding of a vibratory stimulation as a periodic discharge timing within a specific set of correlated nerve cells. Autocorrelation histograms of the spike discharges of a neuron in the somatic sensory cortex of a monkey. Thirty-Hz sinusoidal stimulations were applied to the skin at amplitudes indicated for each row at the extreme right. In the left column, the intervals are taken in their natural order. In the right column, the same intervals are shuffled. (From Mountcastle et al., 1969.)

preparation; Segundo and Perkel, 1969; Segundo et al., 1968; Walløe, 1968). When excitatory, they will therefore elicit regular firing unless issues such as accommodation are strong. If, however, they are not independent and fire synchronously, they will elicit larger oscillations, and therefore resemble powerful terminals in terms of their influence (Segundo et al., 1967a).

The membrane potential can be described by such measures as means, histograms, waiting times to reach threshold, and frequency spectra, among others (Calvin, 1968; Levitan et al., 1968; Segundo and Perkel, 1969; Walløe, 1968), the values of which depend on the sign (excitatory, inhibitory), size, and statistics (rate, regularity, and so forth) of the arriving impulses, as well as on such postsynaptic features as time constants, pacemaker and after-potentials, and other related phenomena. Greater potential fluctuations around similar averages are seen with irregular or serially correlated large EPSPs than with regular or uncorrelated ones at the same mean rates. One or the other pattern will be more effective, depending on the threshold (Figure 11A, B, C, page 584). Average potential values increase linearly with arrival rates in simulated granular cells, but the variability is less with naturally ordered than with shuffled inputs (Figure 11D). There is also a linear relation in *Aplysia* between the mean rate of the influential EPSPs and the probability of the postsynaptic membrane depolarizing by a certain amount (Figure 170 in Segundo and Perkel, 1969). The distribution of the membrane potential might play a role in normal operation by relating to the proportion of cells that fire to test EPSPs (Levitan et al., 1968). A membrane-potential distribution may be estimated for each time after a presynaptic spike or stimulus, and there may be clear parallelism between the median value profile, the evoked PSP wave, and the corresponding cross-correlation histogram (Segundo et al., 1967a).

The important roles in the junctional input-output relation of the postsynaptic spike-generating and related processes also have been analyzed (Jansen and Walløe, this volume; Segundo et al., 1963; Segundo and Perkel, 1969; Segundo et al., 1968; Terzuolo, this volume; Walløe, 1968). Postsynaptic refractoriness is responsible for certain multimodal interval histograms (Segundo et al., 1963; Segundo et al., 1968), for the shift of the maximum prospective probability from the lowest abscissae in the "passage through a certain level" plots to intermediate values in the "spike" plots, for certain unexpected absences of EPSP acceleration effects (see Figures 160 and 164, respectively, in Segundo and Perkel, 1969), and so on.

A single nerve cell may have more than one kind of terminal (Bittner, 1968; Jansen and Walløe, this volume; Kandel et al., 1967; Roeder, 1966). Hence, to a particular activity change in the same presynaptic cell there may be a corresponding, different activity shift in each of several postsynaptic elements, even when (as may not always be the case) the trains in the different terminals remain identical (Segundo et al., 1963; Segundo et al., 1966). In other words, each junction type of the same parent axon may have its own particular code and prospective and retrospective probability mappings. This widens the functional range of single cells, and may enable them to play a switching role, whereby at certain timings the cell may influence some postsynaptic units and channel its message down one pathway and at other timings it may affect a different postsynaptic set and path (Figure 9B) (Bittner, 1968; Bittner, in preparation; Segundo et al., 1963; Segundo and Perkel, 1969; Segundo et al., 1966).

Part of the role of neurons as devices that communicate and code is explained by a simplified conception that lumps the entire synaptic area and a unique spike-trigger zone at a single somatodendritic point, and is defined by simple potential and threshold dynamics. This conception, implicit in much current thinking, is also explicit in several models and simulations. Their operations on the input have been studied extensively (e.g., Geisler and Goldberg, 1966; Harvey, 1967; Jansen and Walløe, this volume; Perkel, 1965; Perkel et al., 1964; Segundo et al., 1966; Segundo et al., 1968; Stein, 1967a; Walløe, 1968). Some of their stages are reminiscent of low-pass filters (Terzuolo, this volume). There are, however, additional subtleties and issues, some well demonstrated, some hypothetical, as summarized by, for example, Bullock (1967), Bullock and Horridge (1965), Eccles (1964), and Segundo (1970). Although they play a minor role in certain cells, for which the simplified models are reasonable approximations, they are highly developed in others, and the corresponding behavior is far more elaborate. Hence, the simplified conception cannot be considered a complete description of all nerve cells. In fact, the large numbers and varieties of nerve cells, of operational modes, and of influences present in any nervous system, even at the simplest level, should remind the neuroscientist of the assertion by Teilhard de Chardin (1962) that, whereas some disciplines are concerned with what is very large and others with what is very small, biology in general and, I may add, the neurosciences in particular, explore the special dimension of that which is immensely complex.

## Acknowledgments

The author's research was supported by a Research Career Program and a grant from the United States Public Health Service, and assisted by the University of California at Los Angeles Health Sciences Computing Facility and the Brain Research Institute Data Processing Laboratory.

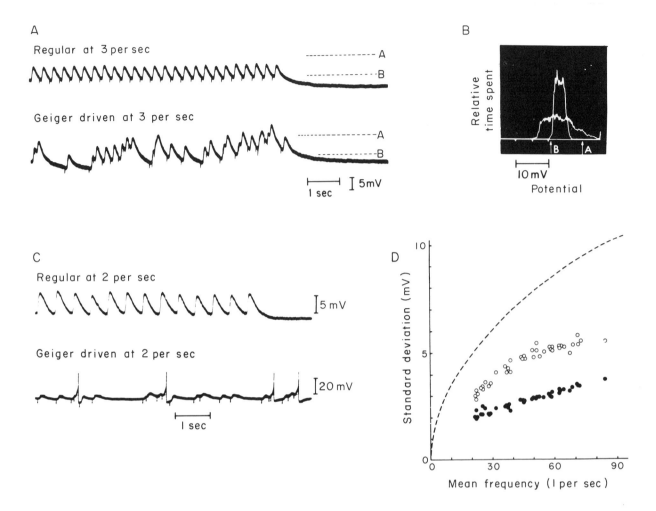

A

Regular at 3 per sec

Geiger driven at 3 per sec

1 sec    5mV

B

Relative time spent

10 mV

Potential

C

Regular at 2 per sec

5 mV

Geiger driven at 2 per sec

20 mV

1 sec

D

Standard deviation (mV)

Mean frequency (1 per sec)

FIGURE 11 Regular and irregular EPSPs at identical mean rates: Comparison of their effects on the membrane potential and their effectivenesses. A, B, C. Unidentified cell in *Aplysia* visceral ganglion. (Stimuli at 3 per sec.) The membrane potential oscillations during regular stimuli (A, upper record) had an average value of 10.7 mV (measured with respect to the resting value), a standard deviation of 2 mV, and a narrow, high, and relatively uniform histogram (B). The oscillations during irregular stimuli (A, lower record) had the same average of 10.7 mV, a standard deviation of 4.5 mV, and a broad, flat, and skewed histogram (B). The potential levels indicated as A and B on the records correspond to the values so labeled on the histograms. C. (Stimuli at 2 per sec.) In the records shown, where the threshold is high, e.g., about A, no spikes are evoked by a regular excitation, but several occur during irregular stimuli. If, on the other hand, the threshold were low, e.g., around B, the regular input might keep the cell firing continually, but the irregular input would allow for pauses. (From Levitan et al., 1968.) D. Computer simulation of a single, powerful, excitatory DSCT terminal acting on a nonpacemaker cerebellar granule cell. Standard deviation of the value of the postsynaptic membrane potential as a function of the corresponding presynaptic mean rate. The membrane fluctuations are smaller when the sequence of the DSCT interspike interval reproduces that which occurs naturally (black circles) than when the same intervals are shuffled randomly (open circles). (From Walløe, 1968.)

## REFERENCES

ALADZHALOVA, N. A., M. E. SAKSON, and G. P. POTYLITSIN, 1969. A statistical analysis of the interrelationship of two spike trains in a small volume of the cat cerebral cortex. *Dokl. Akad. Nauk. SSSR* 184: 735–738.

BAYLY, E. J., 1968. Spectral analysis of pulse frequency modulation in the nervous system. *Inst. Elec. Electron. Eng. Trans. Bio-Med. Eng.* 4: 257–265.

BITTNER, G. D., 1968. Differentiation of nerve terminals in the crayfish opener muscle and its functional significance. *J. Gen. Physiol.* 51: 731–758.

BULLOCK, T. H., 1967. Signals and neuronal coding. *In* The Neurosciences: A Study Program (G. C. Quarton, T. Melnechuk, and F. O. Schmitt, editors). The Rockefeller University Press, New York, pp. 347–452.

BULLOCK, T. H., and G. A. HORRIDGE, 1965. Structure and Function in the Nervous Systems of Invertebrates. W. H. Freeman, San Francisco.

CALVIN, W. H., 1968. Evaluating membrane potential and spike patterns by experimenter-controlled computer displays. *Exp. Neurol.* 21: 512–534.

COX, D. R., and P. A. W. LEWIS, 1966. The Statistical Analysis of Series of Events. Methuen, London; John Wiley and Sons, New York.

DIXON, W. P., and F. J. MASSEY, JR., 1969. Introduction to Statistical Analysis. McGraw-Hill, New York.

ECCLES, J. C., 1964. The Physiology of Synapses. Academic Press, New York.

FITZHUGH, R., 1957. The statistical detection of threshold signals in the retina. *J Gen. Physiol.* 40: 925–948.

FUJITA, Y., J. ROSENBERG, and J. P. SEGUNDO, 1968. Activity of cells in the lateral vestibular nucleus as a function of head position. *J. Physiol. (London)* 196: 1–18.

GEISLER, C. D., and J. M. GOLDBERG, 1966. A stochastic model of the repetitive activity of neurons. *Biophys. J.* 6: 53–69.

GERSTEIN, G. L., and N. Y. S. KIANG, 1960. An approach to the quantitative analysis of electrophysiological data from single neurons. *Biophys. J.* 1: 15–28.

GILLARY, H. L., and D. KENNEDY, 1969. Neuromuscular effects of impulse pattern in a crustacean motoneuron. *J. Neurophysiol.* 32: 607–612.

GRÜSSER, O.-J., K. A. HELLNER, and V. GRÜSSER-CORNEHLS, 1962. Die Informationsübertragung im afferenten visuellen Systems. *Kybernetik* 1: 175–192.

HAGIWARA, S., 1949. On the fluctuation of the interval of the rhythmic excitation. I. The efferent impulse of the human motor unit during voluntary contraction. *Rep. Physiograph. Sci. Inst. Tokyo Univ.* 3: 19–24.

HARTLINE, D. K., and I. M. COOKE, 1969. Postsynaptic membrane response predicted from presynaptic input pattern in lobster cardiac ganglion. *Science (Washington)* 164: 1080–1082.

HARVEY, R. J., 1967. The Spino-olivo-cerebellar Pathway of the Cat. Australian National University, Canberra, doctoral thesis.

KANDEL, E. R., W. T. FRAZIER, and R. E. COGGESHALL, 1967. Opposite synaptic actions mediated by different branches of an identifiable interneuron in *Aplysia*. *Science (Washington)* 155: 346–349.

KATSUKI, Y., S. YOSHINO, and J. CHEN, 1950. Action currents of the single lateral-line nerve fiber of fish. 1. On the spontaneous discharge. *Jap. J. Physiol.* 1: 87–99.

KENNEDY, D., and DE F. MELLON, JR., 1964. Receptive field organization and response patterns in neurons with spatially distributed input. *In* Neural Theory and Modeling, Proceedings of the 1962 Ojai Symposium (R. F. Reiss, editor). Stanford University Press, Stanford, California, pp. 400–413.

KHINCHIN, A. I., 1957. Mathematical Foundations of Information Theory. Dover Publications, Inc., New York.

KOGAN, A. B., Y. I. PETUNIN, and O. G. CHORAYAN, 1966. Investigation of the impulse activity of neurones by the methods of the theory of random processes. *Biofizika* 11: 1020–1027.

LEVITAN, H., J. P. SEGUNDO, G. P. MOORE, and D. H. PERKEL, 1968. Statistical analysis of membrane potential fluctuations. Relation with presynaptic spike train. *Biophys. J.* 8: 1256–1274.

MACKAY, D. M., and W. S. MC CULLOCH, 1952. The limiting information capacity of a neuronal link. *Bull. Math. Biophys.* 14: 127–135.

MAFFEI, L., 1968. Spatial and temporal averages in retinal channels. *J. Neurophysiol.* 31: 283–287.

MOORE, G. P., D. H. PERKEL, and J. P. SEGUNDO, 1963. Stability patterns in interneuronal pacemaker regulation. *In* Proceedings of the San Diego Symposium for Biomedical Engineering, La Jolla, California, pp. 184–194.

MOORE, G. P., J. P. SEGUNDO, D. H. PERKEL, and H. LEVITAN, 1970. Statistical signs of synaptic interactions in neurons. *Biophys. J.* (in press).

MOORE, G. P., D. H. PERKEL, and J. P. SEGUNDO, 1966. Statistical analysis and functional interpretation of neuronal spike trains. *Annu. Rev. Physiol.* 28: 493–522.

MOUNTCASTLE, V. B., 1967. The problem of sensing and the neural coding of sensory events. *In* The Neurosciences: A Study Program (G. C. Quarton, T. Melnechuk, and F. O. Schmitt, editors). The Rockefeller University Press, New York, pp. 393–408.

MOUNTCASTLE, V. B., W. H. TALBOT, H. SAKATA, and J. HYVÄRINEN, 1969. Cortical neuronal mechanisms in flutter-vibration studied in unanesthetized monkeys. Neuronal periodicity and frequency discrimination. *J. Neurophysiol.* 32: 452–484.

PERKEL, D. H., 1965. Applications of a digital computer simulation of a neural network. *In* Biophysics and Cybernetic Systems (M. Maxfield, A. Callahan, and L. J. Fogel, editors). Spartan Books, Washington, D. C., pp. 37–51.

PERKEL, D. H., and T. H. BULLOCK, 1968. Neural coding. *Neurosciences Res. Program Bull.* 6 (no. 3): 221–348.

PERKEL, D. H., G. GERSTEIN, and G. P. MOORE, 1967a. Neuronal spike trains and stochastic point processes. I. The single spike train. *Biophys. J.* 7: 391–418.

PERKEL, D. H., G. GERSTEIN, and G. P. MOORE, 1967b. Neuronal spike trains and stochastic point processes. II. Simultaneous spike trains. *Biophys. J.* 7: 419–440.

PERKEL, D. H., J. SCHULMAN, T. H. BULLOCK, G. P. MOORE, and J. P. SEGUNDO, 1964. Pacemaker neurons: Effects of regularly spaced synaptic input. *Science (Washington)* 145: 61–63.

PERSON, R. S., 1965. Investigation analysis of time relations between discharges from motoneurons of antagonistic muscles in humans (by means of crosscorrelation analysis of EMG). *Sechenov J. Physiol. USSR* 51: 73–75.

RIPLEY, S. H., and C. A. G. WIERSMA, 1953. The effects of spaced stimulation of excitatory and inhibitory axons of the crayfish. *Physiol. Comp. Oecol.* 3: 1–17.

ROEDER, K. D., 1966. Auditory system of noctuid moths. *Science (Washington)* 154: 1515–1521.

SCHULMAN, J., 1969. Information Transfer Across an Inhibitor to Pacemaker Synapse at the Crayfish Stretch Receptor. University of California, Los Angeles, doctoral thesis.

SEGUNDO, J. P., 1970. Functional possibilities of nerve cells for communication and for coding. *Acta Neurol. Latinoamer.* 14: 340–344.

SEGUNDO, J. P., G. P. MOORE, J. STENSAAS, and T. H. BULLOCK, 1963. Sensitivity of neurones in Aplysia to temporal pattern of arriving impulses. *J. Exp. Biol.* 40: 643–667.

SEGUNDO, J. P., and D. H. PERKEL, 1969. The nerve cell as an analyzer of spike trains. *In* UCLA Forum in Medical Sciences No. 11, The Interneuron (M. A. B. Brazier, editor). University of California Press, Berkeley and Los Angeles, California, pp. 349–390.

SEGUNDO, J. P., D. H. PERKEL, and G. P. MOORE, 1966. Spike probability in neurons: Influence of temporal structure in the train of synaptic events. *Kybernetik* 3: 67–82.

SEGUNDO, J. P., D. H. PERKEL, H. WYMAN, H. HEGSTAD, and G. P. MOORE, 1968. Input-output relations in computer-simulated nerve cells. Influence of the statistical properties, strength, number of interdependence of excitatory presynaptic terminals. *Kybernetik* 4: 157–171.

SEGUNDO, J. P., T. TAKENAKA, and H. ENCABO, 1967a. Electrophysiology of bulbar reticular neurons. *J. Neurophysiol.* 30: 1194–1220.

SEGUNDO, J. P., T. TAKENAKA, and H. ENCABO, 1967b. Somatic sensory properties of bulbar reticular neurons. *J. Neurophysiol.* 30: 1221–1238.

SHANNON, C. E., and W. WEAVER, 1963. The Mathematical Theory of Communication. University of Illinois Press, Urbana, Illinois.

STARK, L., J. NEGRETE MARTINEZ, G. YANKELEVICH, and G. THEODORIDIS, 1969. Experiments on information coding in nerve impulse trains. *Math. Biosci.* 4: 451–485.

STEIN, R. B., 1967a. Some models of neuronal variability. *Biophys. J.* 7: 37–68.

STEIN, R. B., 1967b. The information capacity of nerve cells using a frequency code. *Biophys. J.* 7: 797–826.

TEILHARD DE CHARDIN, P., 1962. La Place de l'Homme dans la Nature. Le Groupe Zoologique Humain. Union Générale d'Editions, Paris.

TERZUOLO, C. A., and E. J. BAYLY, 1968. Data transmission between neurons. *Kybernetik* 5: 83–85.

UTTAL, W. R., and M. KRISSOFF, 1968. Response of the somesthetic system to patterned trains of electrical stimuli. An approach to the problem of sensory coding. *In* The Skin Senses (D. R. Kenshalo, editor). Charles C Thomas, Springfield, Illinois, pp. 262–303.

VASILEVSKY, N. N., 1968. Statistical analysis of some parameters of the background activity of cortical neurons. *Sechenov Physiol. J. USSR* 54: 389–397.

WALLØE, L., 1968. Transfer of Signals Through a Second Order Sensory Neuron. Institute of Physiology. University of Oslo, Universitetsforlagets Trykningssentral, doctoral thesis.

WERNER, G., and V. B. MOUNTCASTLE, 1965. Neural activity in mechanoreceptive cutaneous afferents: Stimulus-response relations, Weber functions, and information transmission. *J. Neurophysiol.* 28: 359–397.

WERNER, G., and V. B. MOUNTCASTLE, 1968. Quantitative relations between mechanical stimuli to the skin and neural responses evoked by them. *In* The Skin Senses (D. R. Kenshalo, editor). Charles C Thomas, Springfield, Illinois, pp. 112–137.

WIERSMA, C. A. G., and R. T. ADAMS, 1950. The influence of nerve impulse sequence on the contractions of different crustacean muscles. *Physiol. Comp. Oecol.* 2: 20–33.

# 52 Spike Trains as Carriers of Information

### DONALD H. PERKEL

NERVE IMPULSES or "spikes" are a nearly ubiquitous accompaniment to neural functioning. Rejecting the idea that they are solely artifactual or epiphenomenal, we accept, rather, the notion that trains of impulses embody or carry information. Major questions arise immediately: (1) How do they carry information? (2) What kinds of information do they carry? (3) Where do they carry it from, and to? (4) What interprets the delivered information? (5) How well is the information carried? (6) What is the role of spike trains vis-à-vis other media of neural communication?

These questions are, of course, by no means independent. Moreover, the answers depend very much on which part of what nervous system is under consideration, as well as on the particulars of the experimental arrangement, especially with regard to surgical procedures, anesthesia, and the stimulus regimen.

We first draw a primary distinction between two kinds of information carried by spike trains (Segundo and Perkel, 1969): the information potentially available to the human investigator is of a different order and calls for different procedures of interpretation (e.g., statistical analysis) from the information normally extracted from a spike train by the next-order neural or effector element in the system. Thus, the experimenter can record and analyze very long trains of impulses, whereas the capacity of a neuron for temporally integrating its synaptic input is stringently limited. On the other hand, a neuron has access to all the synaptic input impinging on it, whereas with current techniques the electrophysiologist can record at most a handful of simultaneously detected spike trains, while having at best a rudimentary knowledge of the anatomical relationships among the corresponding neurons.

We may therefore investigate on the one hand the information extracted from spike trains in an ongoing fashion by a nerve cell that is "integrating" that information or "making decisions" based at least partially on that information; we may call this approach the "neuron-centered" study of neural coding in spike trains. The alternative approach emphasizes the information extractable by the experimenter. This "observer-centered" investigation addresses the prob-lems of identifying the functional connections among the neurons of which the spike trains are observed, and of characterizing their corresponding input-output relations.

The proposed kinds of neural codes have been reviewed, enumerated, and summarized recently (Perkel and Bullock, 1968), and the criteria—formal and biological—for adequately describing their properties were discussed in terms of establishing that a candidate scheme of neural coding is, in fact, functionally valid in a particular nervous system. In this presentation, the relationship between several spike-train codes is stressed in an effort to show that the particular operative information-encoding scheme depends on the context of information input to the system, and in particular on its characteristic modulation times relative to the interspike intervals of the observed neuron.

We retain here the working definition of neural coding as "the representation and transformation of information in the nervous system" (Perkel and Bullock, 1968), using the term "information" in its colloquial sense rather than with the technical meaning of Shannon.

Much of this chapter, therefore, is devoted to methods of extracting information from spike sequences. The statistical techniques are summarized, examples are given of their use in elucidating neuronal relationships, and, finally, new directions are indicated for extending and strengthening these techniques.

## The variety of spike-train codes

The peculiarity of spike-train codes is that the information is carried by means of impulses: brief, discrete, virtually indistinguishable events, occurring most usually at irregular intervals. Hence, they differ from continuous "signals" such as the electroencephalogram, and from trains of pulses occurring at regular intervals or at multiples of a fixed interval, such as those flowing in a digital computer. It follows that, as bearers of information in nervous systems, spike trains have special characteristics, capabilities, and limitations.

The two "classic" spike-train codes have been called "labeled-line" coding and rate coding. The concept of labeled lines means, quite simply, that an essential aspect of the information conveyed by a nerve impulse is embodied in the particular fiber it traverses (Perkel and Bullock, 1968). This notion goes back to the doctrine of "specific nerve energies" of Müller (1838 and 1842; Adrian, 1928), which

DONALD H. PERKEL The RAND Corporation, Santa Monica, California and Department of Anatomy, University of California at Los Angeles, California. *Present Address:* Department of Biological Sciences, Stanford University, Stanford, California

states that the subjective quality of a sensation—the "meaning" of an afferent electrical event—depends primarily on the central connection of the nerve fiber.

Such a view of neural functioning as an analogue of a telephone system is not nearly so trivial as it might first appear. The elaboration of the labeled-line principle, as seen in the complex schemata of several neuronal organizations (Werner, this volume; Hubel and Wiesel, 1962), illustrates the pervasiveness and functional complexity of codes of this sort. Moreover, increasing evidence for the homology of neurons identifiable from individual to individual (Bullock, 1970) bespeaks a more basic role for "wired-in" neural connections than has been attributed to them.

The second kind of neural code that might be considered classical is the rate (or frequency-of-occurrence) code. It may be stated most broadly as the representation of a scalar quantity—such as stimulus intensity—by means of the rate of nerve impulses in a particular fiber or group of fibers. First clearly annunciated by Adrian (1928), it embodies a fundamental principle of neural functioning: the nerve impulses that give rise to greater and lesser sensations of intensity of a stimulus—or to greater and lesser degrees of muscle contraction—are not in themselves different; the difference lies in the number of essentially identical impulses arriving per unit of time. The operating characteristics of rate codes have been described quantitatively in terms of information capacity by Stein (1967).

In both of these "textbook" codes, little emphasis is placed on the temporal structure of the impulse train. The labeled line might carry any type of signal, and the fact that the signals are impulses is irrelevant to this concept. For a rate code, the discrete nature of the impulses appears as rather a nuisance, and fluctuations in interval lengths give rise to fluctuations in the transmitted signal, which may have to be smoothed out through time-averaging or through summation of many parallel, redundant channels. If one were to design an artificial system using the rate-code principle exclusively, it would be most efficient and reliable if, at any given signal level, impulses occurred at regular intervals (Stein, 1967).

A number of "nonclassical" codes have been proposed, however, in which other aspects of the temporal structure of the impulse train are assigned a role in the transmittal of information (Perkel and Bullock, 1968); these aspects include the standard deviation and the coefficient of variation of intervals, serial correlations of intervals, and burst characteristics, such as number of intervals, duration, and temporal pattern (e.g., accelerando or ritardando). It has, in fact, been shown that some cells are measurably sensitive to variations in several temporal characteristics, implying that such codes are indeed capable of being "read" by neurons (e.g., Wiersma and Adams, 1949; for a review, see Segundo and Perkel, 1969).

The extreme example of codes that depend on temporal structure is that in which the length of each individual interspike interval carries information. The information capacity of these codes has been investigated theoretically (MacKay and McCulloch, 1952); they can handle far more bits per second than can frequency codes.

Interesting cases of "temporal-structure" spike codes are those involving more than a single channel. For reliable and rapid "reading-out" of rate codes, it is usually necessary to invoke parallel channels, but other kinds of temporal codes require several channels in a more fundamental way. Most proposed multiple-channel codes involve some elaboration of the idea of coincidence detection. In its simplest terms, a neuron may receive synaptic input from two excitatory sources, each of which alone produces a subthreshold response; if, however, the two impulses arrive within a short time interval, the neuron will be sufficiently depolarized to produce a spike. If more than two input channels are operative, and if some of the synapses are inhibitory rather than excitatory, extremely individualized "logic circuits" can be devised; these have exquisite sensitivity to slight temporal differences in arrivals.

Multiple-channel spike codes are obviously more difficult to investigate experimentally than are single-channel codes. Successful efforts in this direction have been concerned with binaural sound localization by neurons in the inferior colliculus (e.g., Rose et al., 1966). The temporal relationships between spikes in multiple, simultaneously recorded spike trains have been used for elucidating underlying functional relationships among the corresponding neurons (e.g., Gerstein, this volume), as discussed later in this article.

First, however, it is necessary to indicate that the several types of spike-train codes are not clearly separable and distinct, but rather that one tends to merge into another as characteristics of the signals or of the neurons are changed.

## Rate and interval codes

Let us examine the workings of rate codes in somewhat more detail. Suppose that there is an external signal, s(t), to be represented by the firing rate of a neuron. We may think of s(t) as representing the intensity of a stimulus as it varies with time; we assume that it always has a positive value, in order to avoid irrelevant notational complications.

Suppose that a spike occurs at a time $T_i$; the next spike, occurring at time $T_{i+1}$, defines the interspike interval, $\tau_i = T_{i+1} - T_i$, as well as the corresponding *instantaneous rate*, $\rho_i = 1/\tau_i$. A *rate code* is operative if the instantaneous rate, $\rho_i$, or its probability distribution, is determined by the history of the signal s(t), for $t \leq T_{i+1}$.

For example, the encoding unit—a receptor, perhaps—may operate as an integrating device, producing a spike when the integral of the signal reaches a threshold, $\theta$; i.e., the time of the next spike, $T_{i+1}$, is determined by

$$\theta = \int_{T_i}^{T_{i+1}} s(t)dt \,,$$

which implies that the integrator is reset to zero after each spike.

As another example, the interval $\tau_i$ can be governed by a time-dependent Poisson process, i.e., $\tau_i$ is drawn from an exponential distribution with rate parameter $\lambda(T_i)$. The rate parameter itself may depend simply on the instantaneous signal at the time of the previous spike, e.g.,

$$\lambda(T_i) = a\, s(T_i) + b \,.$$

Alternatively, the rate parameter may represent a weighted average of the signal over some previous time period:

$$\lambda(T_i) = \int_{T_i - W}^{T_i} w(t - T_i + W)s(t)dt \,,$$

or over a number, k, of preceding interspike intervals:

$$\lambda(T_i) = \int_{T_{i-k}}^{T_i} w(t -- T_{i-k})s(t)dt \,.$$

If the signal varies slowly with respect to the typical times between spikes, then the operation of any of these rate codes is exemplified by Figure 1A. The regularity of the spikes will, of course, depend on the particular coding scheme operative (Stein, 1967); in a large class of cases, the time-averaged rate of spike occurrence will closely reproduce the original signal.

Now let us suppose that the signal varies rapidly with respect to the interspike intervals, but that the same encoding mechanisms are at work. Then the spike train cannot follow or reflect the rapid changes in the signal, but rather will represent a smoothed or sampled value of the signal, as seen in Figure 1B. Even if the encoder is an integrating device, the spikes will, in general, occur at irregular intervals. If a reasonable choice of weighting function, w, and of a sufficiently small value of W or k is assumed, the various types of Poisson encoders, as does the integrating encoder, will produce trains in which each interval carries information about the "local" signal level, and may be nearly or totally independent of the adjacent intervals. Here we may speak more properly of an *interval code* than of a rate code, because every interval has a meaning of its own.

The essential difference, therefore, between a rate code and an interval code does not lie in the encoding mechanism; as we have seen, any of several mechanisms may subserve either type of code. Rather, it is the relationship between the typical times of "interesting" signal variations and the interspike interval times that prescribes the nature of the code.

It follows, then, that in experimental investigations of neural codes it is important that the variations in input signal must correspond, in magnitude and time course, with those encountered under physiological conditions. The same neuron may operate with a rate code, or an interval code, or a code of intermediate type, depending entirely on

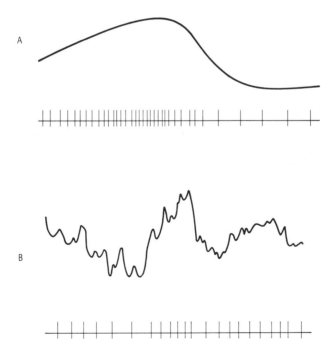

FIGURE 1    Representation of a continuous signal by a train of impulses (schematic). A: The signal varies slowly with respect to the mean interval between impulses; the representation is a "mean rate" code. B: The signal varies rapidly with respect to interpulse interval; the representation is an "interval code." See text.

the nature of the input signal imposed. Furthermore, any quantitative estimate of information-handling capacity will depend on the input signal; a slowly varying input will yield modest information rates, as in the studies of Stein (1967), whereas a rapidly varying input to the same neuron can approach the higher information rates characteristic of an interval code (MacKay and McCulloch, 1952). We emphasize, then, that it is meaningful to talk about information capacity of a neuron or a neural pathway only within the context of a particular stimulus environment.

## Statistical measures for single spike trains

The use of statistical techniques to describe the properties of spike trains has been outlined by Perkel et al. (1967a), and their application to neurophysiological problems has been reviewed by Moore et al. (1966). Here we confine ourselves to a description of only those basic statistical measures that are most useful for drawing inferences about underlying mechanisms and functional relationships.

The interval histogram is an important and widely used statistical measure of the temporal structure of a spike train. It represents an estimate of an underlying probability density function of interspike intervals. If we choose a large bin width for the histogram, the over-all profile is relatively smooth, but structural detail may be lost; if the bin width is small, detail is preserved, but individual bin tallies exhibit greater sampling fluctuations.

The interval histogram is an order-independent statistic; i.e., if intervals are randomly shuffled, the interval histogram of the reconstructed train will be identical with that of the unshuffled train.

A regularly firing neuron produces a narrow interval histogram; a "randomly" firing neuron (i.e., a Poisson process) gives rise to an interval histogram that has the form of a descending exponential function; a neuron that has exceptionally long silent periods between spikes produces an interval histogram with a long "tail"; patterned discharges, including bursts, may give rise to multimodal interval histograms.

The *autocorrelation histogram* may be thought of as the sum of histograms of intervals of all orders, i.e., intervals between adjacent spikes, between every other spike, and so on. It is also equivalent to an estimate of the likelihood (per unit time) of encountering a spike (not necessarily the nearest one) as a function of the time elapsed since the occurrence of an actual spike. The *autocorrelation density* (known also under several other names, such as renewal density and expectation density), which is estimated by the autocorrelation histogram, is defined formally as

$$h(\tau) = \lim_{\Delta t \to 0} [(1/\Delta t) \text{ Prob } \{\text{spike in } (t_0 + \tau,$$
$$t_0 + \tau + \Delta t)|\text{spike at } t_0\}] .$$

For a Poisson process, the autocorrelation density is flat. Neurons that fire at fairly regular intervals, such as pacemaker cells, give rise to an autocorrelation histogram that displays regularly spaced peaks and valleys, which flatten out progressively. For essentially all neurons, the autocorrelation density approaches a constant asymptotic value for sufficiently large values of its argument. This reflects the fact that knowing the time of occurrence of one spike enables us to predict the future likelihood of encountering other spikes in that train, but that this predictive power does not extend indefinitely into the future (barring the theoretical case of a "perfect" clock). Eventually, the time of occurrence of the original spike is "forgotten," and the likelihood of a spike is determined only by the over-all mean firing rate. For a Poisson process—completely "random" firing—the forgetting takes place immediately.

All neurons, because of refractory effects, display an initial zero autocorrelation, followed by a steeply rising portion. In certain neurons, this refractory effect, shown in the early bins of the autocorrelation histogram, is all that distinguishes the spike train from a true Poisson process.

The autocorrelation histogram reflects, in a complex fashion, the presence of serial dependencies among interspike intervals. If the intervals are shuffled at random, the autocorrelation histogram of the reconstructed train will remain essentially the same only if successive interval lengths are independent. If they are not, the two autocorrelation histograms will differ; the peaks may become sharper or broader, depending on the nature of the serial dependencies.

A more sensitive statistical measure of the serial dependencies of interval lengths is the *serial correlogram*. This is the set of serial correlation coefficients $\gamma_1, \gamma_2, \gamma_3, \cdots$, defined by

$$\gamma_j = \sigma^{-2} E(T_i - \mu)(T_{i+j} - \mu) ,$$

in which $\mu$ and $\sigma$ are the mean and standard deviation of interval length, respectively. In other words, the serial correlation coefficient of order j is the expected value of the product of the (normalized) interval length with that of the j'th following interval length. For example, the first-order serial correlation coefficient, $\gamma_1$, measures the serial dependence between each interval and its immediately following one. A value of $\gamma_1$ close to unity means that relatively short intervals are nearly always followed by short ones, and long ones by long ones. A value of $\gamma_1$ close to $-1$ indicates that short intervals are highly likely to be followed by long ones, and vice versa. A value close to zero indicates that successive interval lengths are uncorrelated (although it does not necessarily imply that they are independent).

A usual contribution to nonzero serial correlation coefficients is that of trends or relatively slow rate changes in the experimental record. Either a long-term slowing-down or speeding-up of the firing rate during the observation will produce positive serial correlations for high as well as low orders. More rapidly changing rates produce positive serial correlations of the lower orders, which then typically become negative before returning to zero. Bursts and patterned discharges typically show more rapid oscillations in the earlier portions of the serial correlogram.

Shuffling the intervals in the spike train destroys all serial dependence. The serial correlogram of the shuffled train represents a control case, the fluctuations of which around zero reflect solely sampling variations.

The serial correlation coefficient was used in one of the earliest examples of statistical analysis of single spike trains (Hagiwara, 1950). Hagiwara observed that the first-order serial correlation coefficient (in the muscle-spindle afferent fiber) was equal to $-0.5$, and that the serial correlation coefficients of higher order were all very close to zero. It is easily shown (McGill, 1962) that, if the impulses arising from a regular clock are subjected to independent random delays (with the provision that the variations are not so great as to destroy the serial order of the impulses), the train of delayed impulses will have serial correlation coefficients equal to $-0.5, 0, 0, \cdots$. Hagiwara interpreted this serial correlogram as being consistent with an extremely regular source of impulses transmitted over pathways that produced random delays before the impulses reached the recording site. In this case, the existence and certain properties of an unobserved unit were inferred from statistics of a spike train observed in another unit.

Other statistical analyses of single trains have yielded insights concerning intracellular mechanisms. For example, the analysis of interval fluctuations in pacemaker neurons of *Aplysia* by Junge and Moore (1966), aided by computer simulations, pinpointed the most likely source of the fluctuations as the asymptotic level of the pacemaker potential, in turn attributable to irregular shifts of the spike-initiating locus. Further examples of the interpretation of single-train statistical analyses were given by Moore et al. (1966) and by Harmon and Lewis (1966).

It is in the analysis of several spike trains observed simultaneously, however, that the strongest inferences can be made about underlying mechanisms.

## Comparison of simultaneously observed spike trains

The principal statistical tool for the comparison of two simultaneously observed spike trains is the cross-correlation histogram (Perkel et al., 1967b). It is defined and used in a fashion analogous to that of the autocorrelation histogram, and represents an estimate of the density

$$k_{AB}(\tau) = \lim_{\Delta t \to 0} [(1/\Delta t) \text{ Prob } \{\text{spike in B in } (t_0 + \tau,$$
$$t_0 + \tau + \Delta t) | \text{spike in A at } t_0 \}].$$

If two neurons are functionally *independent*, the cross-

correlation histogram is flat, to within sampling fluctuations only. We have proposed that the flatness of the cross-correlation histogram be used as a criterion of functional independence between two neurons during the period of observation (Moore et al., 1966).

Various types of functional relationships between the monitored neurons give rise to characteristic "signatures" in the cross-correlation histogram, as described in detail by Moore et al. (1970). The material presented in this section is drawn from that paper. By observing features in the cross-correlation histogram, one can, in some cases, infer the underlying functional relationships and estimate some of the relevant parameters; in other, less favorable cases, it is still possible to narrow the choice of admissible hypotheses about the underlying neurons.

If spike trains A and B are recorded from presynaptic and postsynaptic members of a pair of neurons, respectively, several features are observable in the cross-correlation histogram (Figure 2). The *primary effect* is a peak or valley, situated to the right of, and close to, the origin. A peak is observed if the synapse is excitatory; a valley, if it is inhibitory. If the inhibition is strong enough, firing may be suppressed completely for a time, as indicated by a drop in the cross-correlation histogram to zero. When not obscured by secondary effects (described below), and when well revealed by a sufficiently long sample and a fine enough bin width in the histogram, the primary peak reflects the time course of the typical postsynaptic potential induced in neuron B. Although the mapping is in general not precisely linear, it nevertheless enables the making of useful estimates of such parameters as the rise and fall times of the postsynaptic potential. The displacement of the primary peak or valley to the right of the origin reflects the over-all delay between the recording of the presynaptic impulse by the first electrode and the postsynaptic impulse by the second.

In addition to the primary effect, a number of *secondary effects* arise, owing to rhythmicities and refractory effects in the two neurons, i.e., departures from flatness in the respective autocorrelation histograms.

The *presynaptic* autocorrelation is mapped bilaterally and symmetrically around the peak or valley representing the primary effect. It has the same sign as the primary effect, so that if the synapse is excitatory, peaks in the presynaptic autocorrelation give rise to peaks in the cross correlation; if the synapse is inhibitory, peaks in the presynaptic autocorrelation give rise to valleys in the cross correlation. The peaks and valleys are broadened to an extent that corresponds to the effective width of the postsynaptic potential, or to that of the primary peak or valley.

*Postsynaptic* rhythmicities and other features of the autocorrelation of train B are mapped unilaterally, to the right of the primary peak or valley. Again, the sign of the primary peak is preserved: peaks in the postsynaptic autocorrelation correspond to peaks in the cross correlation if the synapse is excitatory; and to valleys, if it is inhibitory. The intensity with which these rhythmicities are mapped depends on the extent to which the presynaptic impulses reset the rhythm of the postsynaptic cell. If the latter is not an autonomous pacemaker, but rather is driven by a third, independent pacemaker neuron, the rhythmicity shown in the oscillations of the autocorrelation histogram of cell B will not be reflected in the AB cross-correlation histogram. From such an observation, it is possible to infer the existence of the unobserved driver neuron.

Combinations of these effects can result in a complex structure of the cross correlation, even in this simplest case of a trans-synaptic pair. Unraveling the several effects appears to be most readily accomplished through the techniques of spectral analysis, described in the concluding section of this article.

If cell B is the presynaptic element and cell A the postsynaptic member of the pair, the preceding effects are still present, *mutatis mutandi,* e.g., the primary peak and the postsynaptic secondary effects will appear to the left of the origin in the cross-correlation histogram.

Another basic configuration is that in which the two monitored neurons, A and B, have no synaptic connections between them, but rather receive synaptic input from a common source, C, either directly or through interneurons. The observable features in the cross-correlation histogram between A and B are rather more complex than the trans-synaptic case outlined above (Figure 3): the *primary effect* occurs in general on both sides of the origin; it may be dis-

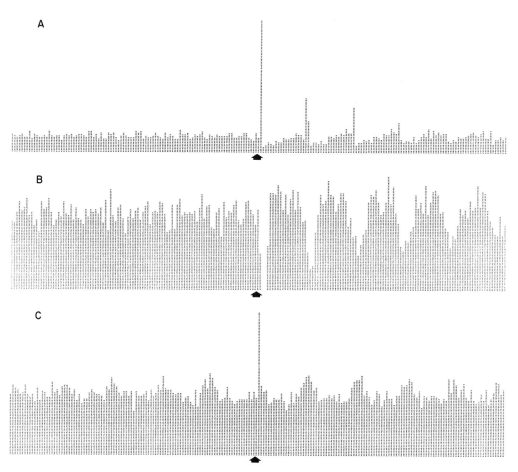

FIGURE 2 Cross-correlation histograms between presynaptic and postsynaptic neurons; computer-simulated data, arbitrary time scale. Arrow indicates origin of time coordinate. A: A Poisson (arrhythmic) train excites a pacemaker neuron. Resetting of the pacemaker's rhythmicity induces a succession of peaks. B: A Poisson train inhibits a pacemaker.

C: One slightly irregularly firing pacemaker excites a more irregularly firing pacemaker. The secondary peaks to the left of the arrow reflect the autocorrelation of the presynaptic cell; those to the right of the arrow include features of the autocorrelations of both cells. See text.

placed to the right if the conduction time from C to A is shorter than that from C to B, or to the left if the converse is true. It will be centered about the origin if the conduction times are about equal.

If the synapses on A and B are both excitatory or both inhibitory, the primary effect consists of a rather broad, symmetrical peak. If one synapse is excitatory and the other inhibitory, the shape will involve a valley and a peak; the precise configuration depends on the shapes and amplitudes of the two postsynaptic potentials, as well as on the relative times of arrival. A cross-correlation histogram of Rodieck (1967, Figure 10B) appears to be consistent with this class of underlying configuration; for a fuller discussion, see Moore et al. (1970).

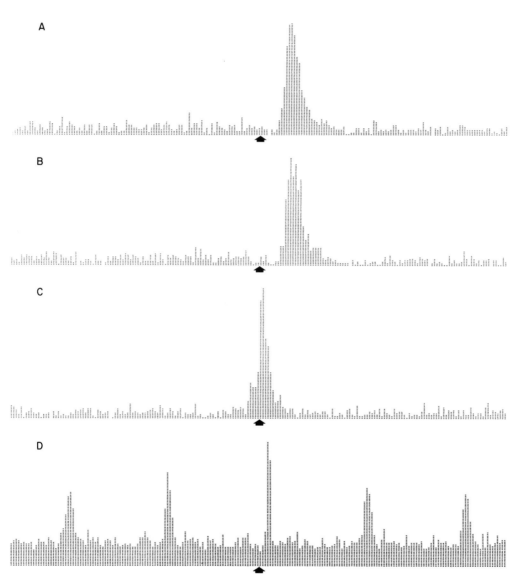

FIGURE 3 The effects of common excitation. A Poisson train of impulses derived from a Geiger counter excites two simultaneously monitored ganglion neurons in *Aplysia* (unpublished data of H. Levitan and J. P. Segundo). A: Cross-correlation histogram between stimulus train and one neuron. Abscissa extends from −1 sec to +1 sec; arrow marks time origin. B: Cross-correlation histogram between stimulus train and second neuron. C: Cross-correlation histogram between the two neurons. Peak is more symmetrical and centrally placed than in upper examples. D: Computer-simulated neurons. Cross-correlation histogram between two neurons excited by a common rhythmic "driver"; arbitrary time scale. See text; for further examples see Moore et al. (1970).

The *secondary effects* include the autocorrelations of all three neurons. That of the common source, C, is reflected symmetrically and bilaterally around the peak (or other feature) representing the primary effect. The autocorrelation of cell A is mapped unilaterally to the left of the primary peak; that of cell B is mapped unilaterally to the right of the primary peak. In addition, a complicated admixture of the autocorrelations of A and B will be mapped on both sides of the primary peak. The more conspicuous of these effects are illustrated in Figure 3.

In addition to synaptic effects, shared rate changes can elevate or depress the cross-correlation histogram (Perkel et al., 1967b). Typically, the time scales of shared rate changes are much longer than those of synaptic effects, so that the two can readily be distinguished. The cross-correlation histograms of Holmes and Houchin (1966) between widely separated cortical units appear to arise from a shared rhythmicity attributable to a common activating mechanism.

Recent examples of cross-correlation histograms, interpreted in terms of the underlying functional anatomy, appeared in the papers of Noda et al. (1969) and of Bell and Grimm (1969).

The enormous tangle of primary, secondary, and perhaps also tertiary effects in even the simplest anatomical configurations makes it difficult to infer neuronal connections directly from correlational analysis of spike trains. However, two additional approaches have genuine promise in this area. The first approach is to apply a periodic stimulus to the preparation, while monitoring two or more spike trains. The statistical analysis and interpretation of the results of experiments of this kind are discussed by Gerstein (this volume).

The other approach is the use of spectral analysis, according to the notions briefly sketched in the remainder of this article.

## Spectral analysis of spike trains

In place of the autocorrelation and cross-correlation histograms discussed above, which are in the time domain, the spectral analysis of spike trains makes use of the autospectrogram and cross-spectrogram, in the frequency domain. The autospectrogram is an estimate, from an experimental sample, of an autospectrum, or autospectral density, which is defined as the Fourier transform of the autocovariance of the spike train. Similarly, the cross-spectrum, or cross-spectral density, is the Fourier transform of the cross covariance.

The autocovariance is related simply to the autocorrelation density (or renewal density) of a spike train: one adds a delta function at the origin, subtracts the mean rate, and multiplies by the mean rate, to yield the *complete autocovariance*:

$$\zeta'(\tau) = \rho[h(\tau) + \delta(\tau) - \rho].$$

Eliminating the delta function gives the *incomplete autocovariance*:

$$\zeta(\tau) = \zeta'(\tau) - \rho\delta(\tau) = \rho[h(\tau) - \rho].$$

The delta function is introduced to reflect the fact that every spike is in coincidence with itself.

A *right-hand autocovariance* is defined by

$$\eta(\tau) = \begin{cases} \zeta(\tau) & \tau > 0 \\ 0 & \tau \le 0. \end{cases}$$

The cross covariance is obtained from the cross-correlation density simply by subtracting the mean rate of the second train, and then multiplying by the mean rate of the first train:

$$\xi_{ij}(\tau) = \rho_i[k_{ij}(\tau) - \rho_j].$$

The cross covariance between a train and itself is simply the complete autocovariance:

$$\xi_{ii}(\tau) = \zeta_i'(\tau) = \rho_i\delta(\tau) + \zeta_i(\tau).$$

The autocovariance is an even function:

$$\zeta(-\tau) = \zeta(\tau),$$

and the cross covariance between two trains taken in reverse order is simply the mirror image of the original cross covariance:

$$\xi_{ji}(\tau) = \xi_{ij}(-\tau).$$

The Fourier transform of any function, $f(\tau)$, is defined as follows:

$$f^*(\omega) = T[f(\tau)] = \int_{-\infty}^{\infty} e^{-i\omega t} f(t)\, dt$$

$$= \int_{-\infty}^{\infty} [\cos \omega t - i \sin \omega t] f(t)\, dt.$$

If the function $f(\tau)$ in the time domain is even, its Fourier transform is real; otherwise, it is complex. The Fourier transform may be inverted to yield the original time-domain function:

$$f(\tau) = \frac{1}{2\pi} \int_{-\infty}^{\infty} e^{i\omega\tau} f^*(\omega)\, d\omega.$$

The Fourier transform is a linear operator, and has the very useful convolution property which states that, if

$$h(t) = \int_{-\infty}^{\infty} f(t - z)g(z)\, dz,$$

then

$$h^*(\omega) = f^*(\omega)g^*(\omega);$$

that is, the Fourier transform of the convolution integral of two functions is the product of their individual transforms.

The properties of Fourier transforms in general are given in many standard texts, e.g., Brown (1965) and Jenkins and Watts (1968). The spectral analysis of point events was described by Bartlett (1963) and by Cox and Lewis (1966).

The fundamental application of the Fourier-transform approach to spike-train analysis lies in the following situation: we may express the cross covariance measured between a presynaptic and a postsynaptic location by means of a convolution equation between the presynaptic autocovariance and a function, $\epsilon(\tau)$, which we call the *synaptic response:*

$$\xi_{jk}(\tau) = \int_{-\infty}^{\infty} \xi_{jj}(\tau - z)\,\epsilon(z)\,dz .$$

Upon taking Fourier transforms, we express the cross spectrum as the product of the presynaptic autospectrum and the transform of the synaptic response:

$$\xi_{jk}^*(\omega) = \xi_{jj}^*(\omega)\epsilon^*(\omega) .$$

The foregoing equations define the function $\epsilon(\tau)$; it always exists in any steady-state situation, and can be estimated by dividing

$$\epsilon^*(\omega) = \xi_{jk}^*(\omega)/\xi_{jj}^*(\omega)$$

and then inverting the Fourier transform. When a linear system is dealt with, the function analogous to $\epsilon^*(\omega)$ is known as the transfer function, and it does not change when the input is changed. In the (nonlinear) system under discussion (presynaptic and postsynaptic spike trains) the question of the dependence of the function $\epsilon(\tau)$ on the nature of the input train is an empirical one, which is under investigation in both living and simulated preparations.

The convolution equations above describe the presynaptic secondary effect on the cross-correlation histogram: the bilateral mapping of the presynaptic autocorrelation. The primary effect and the postsynaptic secondary effect arise on further analysis of the synaptic response function, as follows:

$$\epsilon(\tau) = \sigma(\tau) + \kappa \int_0^{\infty} \eta(\tau - z)\sigma(z)\,dz$$

$$\epsilon^*(\omega) = \sigma^*(\omega)\,[1 + \kappa\,\eta^*(\omega)]$$

in which the first term, $\sigma(\tau)$, may be designated the *primary response,* and the term containing the convolution integral may be designated the *secondary response.* The function $\eta(x)$ is the one-sided autocovariance (of the postsynaptic discharge), as defined previously, and $\kappa$ is a scalar coefficient; it reflects the extent to which the postsynaptic cell's rhythmicity is reset by presynaptic impulses.

The primary response, $\sigma(\tau)$, describes the mapping of the postsynaptic potential. It can be recovered, once $\epsilon^*(\omega)$ has

been estimated, by assuming several values for $\kappa$, dividing, and inverting.

Thus, the spectral formalism not only describes quantitatively all the qualitative features of the trans-synaptic cross correlation, but also furnishes a means whereby the primary synaptic response can be disentangled from confusing secondary effects. That is, numerical processing of spike-train data can yield an estimate of the time course of the effect of the postsynaptic potential on the firing probability of the postsynaptic neuron.

We can systematically apply the same techniques to the case of shared synaptic input, using some of the symmetry properties of the Fourier transform, to yield a final expression for the cross spectrum as follows:

$$\xi_{AB}^*(\omega) = \xi_{A3}^*(\omega)\epsilon_3^*(\omega) = \xi_{A1}^*(\omega)\alpha_{13}^*(\omega)\epsilon_3^*(\omega)$$

$$= [\rho_C + \zeta_C^*(\omega)]\alpha_{12}^*(-\omega)\alpha_{13}^*(\omega)\epsilon_2^*(-\omega)\epsilon_3^*(\omega)$$

in which the expressions $\alpha^*(\omega)$ refer to the Fourier transforms of the conduction delays (constant or randomly fluctuating) between the points indicated by the subscripts.

By expanding the last two factors in the above expression, we can show how it accounts for all the qualitative properties we have described previously:

$$\epsilon_2^*(-\omega)\epsilon_3^*(\omega)$$
$$= \sigma_A^*(-\omega)[1 + \kappa_A\eta_A^*(-\omega)]\sigma_B^*(\omega)[1 + \kappa_B\eta_B(\omega)]$$
$$= \sigma_A^*(-\omega)\sigma_B^*(\omega)[1 + \kappa_A\eta_A^*(-\omega) + \kappa_B\eta_B^*(\omega)$$
$$\qquad + \kappa_A\kappa_B\eta_A^*(-\omega)\eta_B^*(\omega)] .$$

The first term in this expression represents the "primary response" in the shared-input case. Note that if both synapses are excitatory, or if both are inhibitory, the first term will be positive, as will be the corresponding feature in the cross covariance. This is in agreement with the observed finding that both shared excitation and shared inhibition give rise to a central peak in the cross-correlation histogram, whereas a central valley and peak of more complicated form arise from a common source that is excitatory to one observed neuron and inhibitory to the other.

The second term reflects a unilateral mapping of the autocovariance of neuron A to the *left* of the central primary response; the third term reflects the corresponding unilateral mapping of the autocovariance of neuron B to the *right* of the central response. The fourth term represents a more diffuse mixture (i.e., a difference convolution) of the two unilateral covariances, reflected in the cross covariance on both sides of the central peak or valley.

Note also that if the previous equation held, and if we knew or could satisfactorily estimate the coupling coefficients $\kappa_A$ and $\kappa_B$, we could solve for

$$[\rho_C + \zeta^*_C(\omega)]\alpha^*_{12}(-\omega)\alpha^*_{13}(\omega)\sigma^*_A(-\omega)\sigma^*_B(\omega) ,$$

which gives the autospectrum of the unobserved common source, convolved with the difference of the conduction times and with the two primary synaptic responses with appropriate sign.

It can be shown, in addition, that the technique illustrated above can be extended to arbitrary networks of neurons, so that we can prescribe a generally applicable set of rules for expressing the cross spectrum between any two points in a loopfree network in terms of the delays and synaptic properties of the intervening pathways, together with the rhythmicities of injected sources of impulses. Loops in the network yield homogeneous equations, leading to the hope that statistical criteria can be developed for detecting and characterizing loops on the basis of spike data.

With the exception of the work of Terzuolo and his group (this volume), the technique of spectral analysis of series of point events (Cox and Lewis, 1966) has seldom been applied to neuronal spike data. Its application to simultaneously observed spike trains is under active investigation at the time of writing (Perkel and Segundo, 1969). The technique holds the promise of being able to provide quantitative answers to certain problems raised by the time-domain statistics of spike trains, and thereby to provide additional information about underlying structures and about the ways nerve cells communicate with one another.

*Acknowledgment*

Supported in part by Research Grant NB 07325–03 from the National Institutes of Health.

## REFERENCES

ADRIAN, E. D., 1928. The Basis of Sensation. W. W. Norton and Co., New York.

BARTLETT, M. S., 1963. The spectral analysis of point processes. *J. Roy. Statist. Soc., B* 25: 264–287.

BELL, C. C., and R. J. GRIMM, 1969. Discharge properties of Purkinje cells recorded on single and double microelectrodes. *J. Neurophysiol.* 32: 1044–1055.

BROWN, B. M., 1965. The Mathematical Theory of Linear Systems. Chapman and Hall, London.

BULLOCK, T. H., 1970. The Reliability of Neurons. *In* Jacques Loeb Memorial Lectures, Woods Hole Oceanographic Institution and Marine Biological Laboratory, Woods Hole, Massachusetts.

COX, D. R., and P. A. W. LEWIS, 1966. The Statistical Analysis of Series of Events. Methuen, London.

HAGIWARA, S., 1950. On the fluctuation of the interval of the rhythmic excitation. II. The afferent impulse from the tension receptor of the skeletal muscle. *Rep. Physiogr. Sci. Inst. Tokyo Univ.* 4: 28–35.

HARMON, L. D., and E. R. LEWIS, 1966. Neural modeling. *Physiol. Rev.* 46: 513–591.

HOLMES, O., and J. HOUCHIN, 1966. Units in the cerebral cortex of the anaesthetized rat and the correlations between their discharges. *J. Physiol. (London)* 187: 651–671.

HUBEL, D. H., and T. N. WIESEL, 1962. Receptive fields, binocular interaction and functional architecture in the cat's visual cortex. *J. Physiol. (London)* 160: 106–154.

JENKINS, G. M., and D. G. WATTS, 1968. Spectral Analysis and its Applications. Holden-Day, San Francisco.

JUNGE, D., and G. P. MOORE, 1966. Interspike-interval fluctuations in *Aplysia* pacemaker neurons. *Biophys. J.* 6: 411–434.

McGILL, W. J., 1962. Random fluctuations of response rate. *Psychometrika* 27: 3–17.

MACKAY, D. M., and W. S. McCULLOCH, 1952. The limiting information capacity of a neuronal link. *Bull. Math. Biophys.* 14: 127–135.

MOORE, G. P., D. H. PERKEL, and J. P. SEGUNDO, 1966. Statistical analysis and functional interpretation of neuronal spike data. *Annu. Rev. Physiol.* 28: 493–522.

MOORE, G. P., J. P. SEGUNDO, H. LEVITAN, and D. H. PERKEL, 1970. Statistical signs of synaptic interaction in neurons. *Biophys. J.* (in press).

MÜLLER, J., 1833. Handbuch der Physiologie des Menschen. Vol. 1. Holscher, Coblenz. (English translation by William Baly, Vol. 1, 1838).

MÜLLER, J., 1840. Handbuch der Physiologie des Menschen. Vol. 2. Holscher, Coblenz. (English translation by William Baly, Vol. 2, 1842).

NODA, H., S. MANOHAR, and W. R. ADEY, 1969. Correlated firing of hippocampal neuron pairs in sleep and wakefulness. *Exp. Neurol.* 24: 232–247.

PERKEL, D. H., and T. H. BULLOCK, 1968. Neural coding. *Neurosci. Res. Program Bull.* 6 (no. 3): 221–348.

PERKEL, D. H., G. L. GERSTEIN, and G. P. MOORE, 1967a. Neuronal spike trains and stochastic point processes. I. The single spike train. *Biophys. J.* 7: 391–418.

PERKEL, D. H., G. L. GERSTEIN, and G. P. MOORE, 1967b. Neuronal spike trains and stochastic point processes. II. Simultaneous spike trains. *Biophys. J.* 7: 419–440.

PERKEL, D. H., and J. P. SEGUNDO, 1969. Statistical techniques for inferring functional relationships in a set of neurons producing spikes. *In* Abstracts, Third International Biophysics Congress of the International Union for Pure and Applied Biophysics, Massachusetts Institute of Technology, Cambridge, Massachusetts, p. 4.

RODIECK, R. W., 1967. Maintained activity of cat retinal ganglion cells. *J. Neurophysiol.* 30: 1043–1071.

ROSE, J. E., N. B. GROSS, C. D. GEISLER, and J. E. HIND, 1966. Some neural mechanisms in the inferior colliculus of the cat which may be relevant to localization of a sound source. *J. Neurophysiol.* 29: 288–314.

SEGUNDO, J. P., and D. H. PERKEL, 1969. The nerve cell as an analyzer of spike trains. *In* UCLA Forum in Medical Sciences No. 11, The Interneuron (M. A. B. Brazier, editor). University of California Press, Berkeley, pp. 349–390.

STEIN, R. B., 1967. The information capacity of nerve cells using a frequency code. *Biophys. J.* 7: 797–826.

WIERSMA, C. A. G., and R. T. ADAMS, 1949. The influence of nerve impulse sequence on the contractions of different crustacean muscles. *Physiol. Comp. Oecol.* 2: 20–33.

# 53 The Role of Spike Trains in Transmitting and Distorting Sensory Signals

## R. B. STEIN

IN RECENT YEARS there has been a tremendous growth of statistical techniques for analyzing the patterns of nerve impulses generated by single cells. This has led to greater insight into the detailed mechanisms of neuronal activity, but to what extent has it increased our understanding of the function of nerve cells in transmitting information? Can the ability of nerve cells to convey information from one part of the nervous system to another by means of spike trains be assessed accurately? If so, what parameters of the train convey information and how well do they convey it? Conversely, what types of signals will be distorted and how badly? Before considering these questions, let us examine an underlying assumption—that information is coded in the form of trains of all-or-none nerve impulses, rather than analogue parameters of the membrane potential.

Rushton (1961) argued strongly in favor of a code employing impulse trains by pointing out that the space constant of nerve fibers is short. The space constant λ is the distance over which the voltage produced by a steady point source of current applied to a linear, cable-like nerve fiber will decay to $1/e$ of its value at the point of application. Hodgkin and Rushton (1946) showed that

$$\lambda = \sqrt{g_a/g_m}, \tag{1}$$

in which $g_m$ is the membrane conductance and $g_a$ the axoplasmic conductance of a *unit length of cable*. Because the axoplasm must inevitably contain a dilute salt solution, $g_a$ must be low and therefore λ must be small. In a long length of nerve, the signal must be boosted repeatedly, and in myelinated nerve, the nodes of Ranvier can be thought of as just such boosters. Because there may be hundreds of nodes between a sense organ and the central nervous system, accurate transmission of an analogue signal would be extremely difficult. The boosters would have to be made of very high-tolerance components if the analogue signal at the end is to bear any relation at all to that at the beginning. Experimental evidence for sufficiently high tolerance components was lacking, so Rushton concluded that little information could be transmitted in this way.

R. B. STEIN Department of Physiology, University of Alberta. Edmonton, Alberta, Canada

Rushton's argument is certainly true for much of the nervous system, but it has important limitations. It applies only to fiber tracts that are long enough to contain a number of nodes of Ranvier. Evidence is accumulating that cells in the central nervous system (Shepherd, this volume; Rall, this volume), and in the retina (Werblin and Dowling, 1969) that have only short processes may use analogue signals in addition to or in place of nerve impulses. Rushton's argument also applies with less force to large axons. The values of $g_m$ and $g_a$ are conductances per unit length of cable, and with a cable of diameter $d$, these quantities can be quickly converted to more basic ones which vary much less from axon to axon. Let $G_m$ be the membrane conductance per *unit area of membrane* and $G_a$ the axoplasmic conductivity. Then geometric considerations indicate that $g_m = \Pi d\, G_m$, and $g_a = \Pi d^2\, G_a/4$, so the length constant

$$\lambda = \sqrt{\frac{d\, G_a}{4\, G_m}} \tag{2}$$

increases as the square root of fiber diameter. Corresponding arguments for myelinated nerve (Rushton, 1951; Noble and Stein, 1966) indicate that the characteristic length of myelinated fibers increases linearly with diameter, so the advantages of size are more marked.

In agreement with the idea that nerve impulses may be less important in large fibers, recent experimental work indicates that some large invertebrate sensory axons do not use nerve impulses at all. Bush and Roberts (1968) studied muscle receptors in the walking legs of the crab. Those receptors supply two sensory fibers, each of which is about 50 microns in diameter and several millimeters in length. During stretch of the muscles, intracellular and extracellular records show only graded responses. The fibers were not unduly damaged by the insertion of microelectrodes, because simultaneous records from motor axons indicated that the graded responses were capable of producing reflex effects. Furthermore, electrical stimulation also produced only graded responses, but nonetheless had reflex effects. Thus, some sensory fibers may use analogue signals exclusively, but their limitations in transmitting information should be recognized. Not only must the fibers be of large diameter and fairly short; their frequency response would probably be very limited if the effects were purely passive. The transfer function $L(s)$ for linear cable with time constant $\tau$ and space constant $\lambda$ is (Stein, 1967, equation 2.33)

$$L(s) \propto \exp(-\sqrt{1+s\tau}\ x/\lambda)/\sqrt{1+s\tau}. \qquad (3)$$

Thus, the decline as a function of distance for signals with a frequency component $s$ is much sharper than that of a steady signal. From equation (3) the steady-state response, $e^{-x/\lambda}$, is raised to the $\sqrt{1+s\tau}$ power and divided by $\sqrt{1+s\tau}$ for signals with a frequency component $s$. This very sharp frequency cutoff means that only slowly changing signals can be transmitted considerable distances in a linear cable. Rall (1962) has presented similar arguments based on the particular geometry of dendritic structures, indicating that distal dendrites probably act to produce a nearly steady bias level at the cell body of a motor neuron. *Thus, information transfer, particularly about rapidly varying signals over considerable distances, must take place mainly by means of all-or-none nerve impulses.*

Even if nerve impulses were strictly all-or-none, there is no insurance that each nerve impulse in a train will have identical effects synaptically. The effect of one impulse in potentiating or depressing the release of transmitter by a subsequent impulse is well known, and presumably very important in the adaptive behavior of the nervous system. Such a situation is basically different, however, from that of a signal based on analogue parameters of the membrane potential, and merely indicates that the device receiving the signal weights more or less strongly nerve impulses that take place at certain intervals. This weighting clearly affects input-output relationships, but the response of the next cell still depends entirely on the times of occurrence of nerve impulses. In other words, the discharge pattern of a nerve cell represents a statistical point process, and the information conveyed by the nerve cell is in terms of parameters of this process. Let us now consider what parameters are important in such a process.

The number of nerve impulses over a period of time or impulse frequency is clearly an important parameter, as Adrian and Zotterman (1926) first noted from their records of single sensory fibers. Increasing the sensory stimulus increased the frequency of nerve impulses smoothly, and this property is often referred to as frequency coding or frequency modulation of a sensory stimulus. If the steady discharge produced by a constant stimulus is either completely regular or completely random (Poisson process), then the mean is the only parameter available to convey information. In intermediate examples, or with time-varying stimuli, other parameters may convey information, but first let us analyze the amount of information that can be conveyed by mean rate alone. This requires some ideas from the mathematical theory of communication, which I have developed in detail elsewhere (Stein, 1967b).

## Information transmission by constant signals

Consider a discrete series of sensory stimuli, each lasting a time $t$ and repeated at a suitable repetition rate. Figure 1 shows schematically a single stimulus trial applied to a nerve cell discharging with a steady rate $\nu_{min}$ prior to stimulation. The stimulus of duration $t$ increases the frequency and produces $x$ impulses *on average*. This variable $x$ is often considered as the neuronal response, but is treated here as the input to a communications channel, because the average frequency or mean number of nerve impulses may be the only quantity wholly determined by the stimulus. The response variable $y$ is the number of nerve impulses actually discharged on a single trial. It will vary from trial to trial because of random fluctuations in the part of the cycle at which the stimulus begins, because of fluctuations in the duration of successive interspike intervals, and a number of other factors that can be lumped together under the heading of the "state" of the neuron.

These sources of variability limit the information that the nervous system (or the experimenter) can obtain about the stimulus from knowing the number of nerve impulses discharged on a single trial. The amount of information depends on three probability density functions: $p(x)$, the probability that the stimulus on a particular trial is $x$; $p(y)$, the probability that the response is $y$; and, most important of all, $p(y/x)$, the conditional probability that the response is $y$, given that the stimulus is $x$. Only two of these three functions are independent. If the stimulus probability function $p(x)$ and the conditional probability function $p(y/x)$ are known for all $x$, the response probability function $p(y)$ can be directly calculated. In fact, the function $p(y/x)$ alone determines the properties of a neuron as a communi-

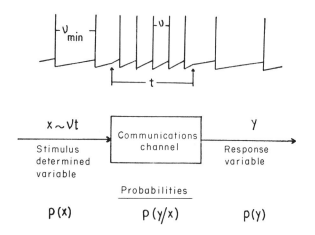

FIGURE 1   An idealized spike train and its analysis in terms of the probabilities of variables associated with a communications channel. Further explanation in text.

cations channel. From this function, the maximum amount of information a neuron could transmit about steady signals can be calculated, together with the stimulus distribution necessary to achieve this maximum. The amount of information, $I$, can be thought of as the logarithm of a number $M$ which represents the maximum number of discriminable categories with the use of an optimal stimulus distribution. Logarithms to the base 2 are normally used, so

$$I = \log_2 M,$$

and the units of $I$ are then denoted by the term binary bits of information. In general, to obtain $I$ directly is tedious, but some simplifications result when there are substantial numbers of nerve impulses, either from a single neuron or a population. Then, the amount of information, $I$, available from a single stimulus trial can be calculated by means of the equation (Stein, 1967b, equation 2.9)

$$I \sim \log \left[ \int_{x_{min}}^{x_{max}} \frac{dx}{\sqrt{2\Pi e}\, s(x)} \right]. \qquad (4)$$

I shall not attempt to derive this equation here, but simply note that the information and, hence, the number of discriminable categories $M$ depends on two quantities—the range in the mean number of nerve impulses that can be produced by stimulation, $x_{max} - x_{min}$, and $s(x)$, which is the standard deviation in the number of impulses produced from trial to trial when the mean number is $x$. If the discharge pattern behaves as a renewal process (i.e., each interspike interval is independent of every other one), with a mean interval $\mu$ and a standard deviation $\sigma$, then the mean number of impulses in time $t$ approaches $t/\mu$ and the standard deviation in number approaches

$$s(x) \sim \sigma \sqrt{t}/\mu^{3/2}. \qquad (5)$$

Equation (5) is the product of two factors, the coefficient of variation $\sigma/\mu$ for a single interval and the square root of the mean number $\sqrt{t/\mu}$. If equation (5) be substituted into equation (4), it follows that

$$I \sim \log \left[ \sqrt{\frac{t}{2\Pi e}} \int_{\mu_0}^{\mu_1} \frac{d\mu}{\sigma \sqrt{\mu}} \right], \qquad (6)$$

in which $\mu_0$ and $\mu_1$ are the minimum and maximum mean interspike intervals that a sensory stimulus of duration $t$ can produce. The standard deviation, $\sigma$, of the interval distribution is considered as a function of the mean, $\mu$. Equations can also be derived (Stein, 1967b) for single nerve cells, the discharge of which shows serial dependencies, and this analysis can be extended to a functionally similar group of cells whose individual properties are known. The parameters required are all readily measurable with present techniques, and the maximum number of discriminable stimulus categories can be estimated from, for example, a discharge of one second. Values obtained either by using the equations above or by direct measurement of the probability functions required are now available for a number of sensory receptors and have been collected in Table I.

These values represent the maximum amount of information that can be obtained about a constant stimulus, with trials lasting about one second. Grüsser (1962) suggested much higher values (nearly 100 bits) but neglected neuronal variability, which greatly restricts transmission of information. Theoretically, an even greater capacity for information

TABLE I

*Ability of some sensory fibers in the cat to transmit information about steady signals*

| Fibers | Preparation | Reference | Maximum Information (bits) | No. of Discriminable Categories |
|---|---|---|---|---|
| Slowly adapting cutaneous afferents | Anesthetized | Werner and Mountcastle, 1965 | 2.7 | 6–7 |
| Facial cutaneous afferents | Decerebrate | Darian-Smith et al., 1968 | 2.4 | 5 |
| Trigemino-thalamic fibers: | | | | |
| N. oralis | Decerebrate | Darian-Smith et al., 1968 | 2.1 | 4–5 |
| N. caudalis | | | 1.1 | 2 |
| Arterial chemoreceptors | Anesthetized | Stein, 1968 | 2.0 | 4 |
| Primary muscle spindle afferents | Decerebrate | Matthews and Stein, 1969b | 2.7 | 6–7 |
| Secondary muscle spindle afferents | Decerebrate | Matthews and Stein, 1969b | 4.7 | 25 |
| Group Ia dorsal spino-cerebellar fibers | Anesthetized | Walløe, 1968 | 2.6 | 6 |

transmission in one second would be possible by use of time-varying stimuli and a form of binary coding or interval coding (Mackay and McCulloch, 1952; Rapoport and Horvath, 1960; Färber, 1968), although these ideas have not been tested experimentally. In a binary code, time is divided into discrete periods, and in each period a pulse is either present or absent (one bit of information). A single nerve cell may discharge as many as 1000 impulses a second, so nearly 1000 bits per second would be transmitted in this way. Adjacent fibers might further increase the rate by transmitting independent information in much the way a paper tape or card reader utilizes a number of channels in parallel. By continuously grading the interspike intervals, rather than using discrete intervals of time, one could obtain even higher rates.

Figure 2 shows a comparison of information rates with the use of these various codes. They diverge widely for long times, because with steady inputs the information transmitted by a frequency code increases as the logarithm of time, rather than linearly. Mackay and McCulloch (1952) did not assume that a nerve cell used either a pulse or an interval code, although later workers (Rapoport and Horvath, 1960; Färber, 1968) were less explicit on this point. Nonetheless, for steady signals, these higher estimates are clearly unrealistic, as no neural mechanisms known could encode a steady signal as a series of binary coded digits or coded intervals. Nor are mechanisms known for decoding such signals centrally. Finally, the presence of neuronal variability would severely limit the operation of such a system, which depends either on a high degree of reliability or on elaborate error-checking procedures. Even the estimates for information as a function of time with the use of a frequency code may be too high. The amount of information available about a steady signal will ultimately be limited by slow background changes, which affect the mean rate of discharge of the signal neuron or group of neurons considered.

The values in Table I are in general agreement with psychological measurements, which indicate that only about seven categories in a single stimulus dimension can be discriminated by human subjects (Miller, 1956) and that the maximal rate of information transmission is less than 100 bits per second (Quastler, 1956). Further verification of this analysis must await comparison of behavioral and neurophysiological experiments carried out at several levels of the nervous system under closely controlled conditions (see, for example, Talbot et al., 1968).

The analysis up to this point has a number of weaknesses that should be explored further:

1. Only maintained stimuli that produce constant responses have been considered, whereas the nervous system appears to be more interested in change (Barlow, 1961).

2. Only the nerve impulses in a clearly defined period have been counted. The response of a real nerve cell will not be limited to any period in time, but will be an ongoing process that reflects recent presynaptic impulses more strongly. An exponential rather than a rectangular weighting of impulses would be more appropriate.

3. Parameters other than the mean number of spikes in a given period have been neglected. In particular, the role of neuronal variability remains obscure.

## Time-varying signals

A different approach has been used to overcome these problems. Consider now the response of a nerve cell to a smoothly varying sinusoidal stimulus, again assuming that the number of impulses per unit time is the appropriate output variable. If this assumption is valid, the response of the nerve cell can be measured with the use of a post-stimulus time histogram, in which each sweep of the histogram is triggered at a particular phase of the sine wave. Then each bin contains the number of impulses occurring at some phase of the sine wave within a time determined by the bin width and the number of sweeps. The units can be easily converted to impulses per second, and the values in each bin give a measure of the *expected density of nerve impulses* at a particular time after the start of the sine wave. The impulse density may be measured over a period that is short compared with single interspike intervals, and, although the units are the same, it should be distinguished from an average frequency measured over a long period of time. This technique has been applied to a variety of sensory

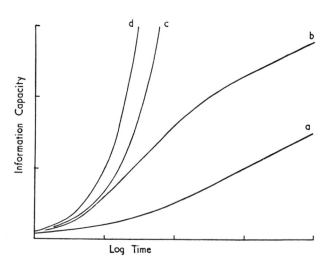

FIGURE 2 Information transmitted as a function of stimulus duration for an idealized neuron (A) discharging randomly and using a frequency code, (B) discharging fairly regularly and using a frequency code, (C) using a binary pulse code, and (D) using an interval code.

systems (Kiang, 1965; Talbot et al., 1968; Matthews and Stein, 1969a) and has recently been simulated by use of electronic neural analogues (French and Stein, 1970). These studies indicate that neuronal variability, far from limiting information transmission with time-varying stimuli, may actually permit more accurate transmission of the details of such stimuli.

Figures 3 and 4 illustrate the results using the neural analogue. Nerve cells that discharge regularly often become phase locked to a sinusoidal stimulus. Figure 3 shows simulation of this behavior, in which a 67-Hz sine wave was applied to a neural analogue that was producing 100 pulses per second in the absence of stimulation. This Figure represents the superposition of several sweeps, so the response is strictly locked to the stimulus, with three pulses occurring every two cycles. Pulses take place only at those three times in the cycle, so the poststimulus time histogram (Figure 4A) has three sharp peaks and is obviously a very distorted version of the sine wave stimulus. Increasing amounts of white Gaussian noise, however, added as an additional input to the analogue, will disrupt the strict timing of the responses (Figure 4B), and eventually (Figure 4C, D) the impulse density shows a smooth sinusoidal variation. The dots give the best-fitting sinusoidal curve to the data in the sense of minimum-square deviation from the data points. The amount of noise required to disrupt these phase-locked patterns clearly depends on the amplitude and frequency of the applied sinusoid. Figure 5 shows the frequency response of the neural analogue at a number of noise levels. The frequency in the absence of stimulation was again 100 pulses per second, and the amplitude of the stimulus was kept constant at a value which produced a modulation at low cyclic frequencies of just under 20 pulses per second. The amplitude of the response in the

absence of noise (Figure 5A) has obvious peaks at 50, 100, 200 Hz, and so on. These peaks and the corresponding effects on the phase of the response (B) represent distortions caused by the phase locking of the response to sine waves of these frequencies. These distortions are progressively reduced with increasing noise levels (Figure 5C, D), although at high noise levels the frequency response of the system begins to decline (Figure 5E, F). Also, the mean-square deviation of the best-fitting curve derived from the data obtained by averaging a given number of sweeps eventually begins to increase again at high noise levels. *For a given range of stimulus frequencies and amplitudes, an optimal level of neuronal variability will exist that permits efficient transmission of information with only a small amount of distortion.* In general, this optimal level will be greater the larger the amplitude of the stimuli and the closer the stimulus frequencies are to the discharge rate of the cells in the absence of stimulation or small multiples of this rate.

This argument suggests that cells responding to rapidly changing stimuli would require a higher level of variability to operate efficiently than those responding to slowly changing or static stimuli. Consideration of experimental data from various mechanoreceptors in the cat tends to confirm this idea. Those that respond to the most rapidly changing stimuli, e.g., cochlear neurons (Kiang, 1965) and fast-adapting receptors in the foot pad (Armett et al., 1962), discharge with extreme variability, whereas those that adapt slowly and respond to maintained stimuli, e.g., muscle spindle afferents (Matthews and Stein, 1969b) and slowly adapting skin afferents (Werner and Mountcastle, 1965), discharge with great regularity. Even between primary and secondary muscle-spindle afferent endings (Matthews and Stein, 1969b) and Type I and Type II slowly adapting cutaneous afferents (Chambers and Iggo, 1967), the more slowly adapting types—in the sense that they have a relatively smaller dynamic sensitivity—also discharge more regularly under a variety of conditions.

## Spatial averaging

There is still a weakness in this analysis, because it is based on long-time averages from single nerve cells or neural analogues. The response of higher-order neurons is presumably based on an average over the large number of fibers that make synaptic contact with it in a limited period of time. However, there is a class of statistical processes known as *ergodic* processes, in which an average over $n$ records of duration $t$ is equivalent to a single longer record of length $nt$. To prove that a population of nerve cells represents a single ergodic process, it must be shown that (1) the parameters necessary to specify the discharge of each cell are identical; (2) the discharge of the cells during maintained stimulation are uncorrelated; and (3) the time $t$ is

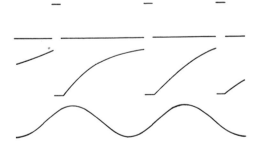

FIGURE 3 A neural analogue discharging at 100 pulses per second was stimulated by a 67-Hz sine wave as shown in the bottom trace. The resultant voltage changes in a "leaky" integrator with a time constant of 10 milliseconds are shown in the middle trace. The output pulses shown in the top trace are produced each time a threshold level (1V) is exceeded and they reset the integrator. This Figure represents several sweeps which superimposed exactly, with three pulses occurring for every two cycles of the applied sinusoid.

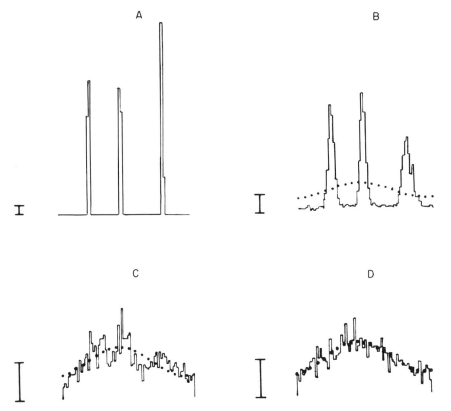

FIGURE 4 Poststimulus time histograms for the response of a neural analogue to the sinusoid shown in Figure 3, with increasing amounts of noise added (A-D). The extent of the histogram represents exactly one full cycle of the applied sinusoid, and each sweep was begun so that the peak of the applied sine wave came in the middle of the histogram. The ordinate has been converted to pulses per second by divid- ing the number of pulses in each bin by the bin width and the number of sweeps. The vertical bars indicate a rate of 100 pulses per second, and the dots represent the best- fitting sine wave for the response. Note the increasingly sinusoidal character of the response produced by adding larger amounts of noise with little change in the best- fitting sine wave.

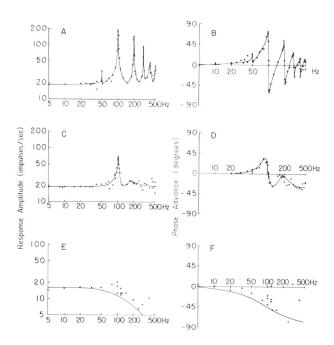

FIGURE 5 The amplitude and phase of the best-fitting sine wave computed at various frequencies with (A-B) small, (C-D) medium, and (E-F) large amounts of added noise. Further explanation in text.

A

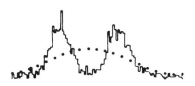

B

C

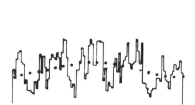

D

FIGURE 6   Poststimulus time histograms as in Figure 4 for neural analogues discharging at approximately (A) two, (B) three, and (C) four spikes per sinusoidal cycle and the sum (D) of these three histograms. The dots give the best-fitting sine curve to each histogram. The amplitude of the sine curve fitted to the sum of the histograms (D) 78.1 pulses per second is very close to the sum of the individual amplitudes (A) 35.9, (B) 22.8, and (C) 19.4 pulses per second. The phase of the sum is also very close to that of the vector sum of the results of the three individual histograms. However, the mean square deviation of the data from the curve fitted to the sum of the histograms is only about 50 per cent higher than any of the individual histograms. Thus, spatial averaging over a number of nerve fibers should result in a very considerable increase in the signal-to-noise ratio when individual fibers are phase locked to some extent.

long enough so that end effects caused by starting or stopping the stimulus can be neglected.

Gerstein (this volume) has shown that the discharges of many central neurons are correlated. Moreover, the mean rate and other parameters of single nerve cells in a population often vary considerably, and the times of interest are often not long enough to make the third criterion above valid. Thus, the requirements for ergodicity are not fulfilled, although spatial averaging over a number of cells, as opposed to temporal averaging, has considerable advantages. Figure 6 illustrates the advantages of spatial summation by showing the response to a 20-Hz sine wave of neural analogues discharging with some variability at 40, 60, and 80 pulses per second. At all three frequencies, there is considerable phase locking. The sum of the three histograms is also shown (Figure 6D), and is relatively smoother than any of the three individual histograms. The best fit to the sum is very close to the sum of the best fits for individual

histograms, so averaging three units produces a signal three times as large. The root mean-square deviation from the fitted curve is only 1.5 times as great. Thus, a very considerable reduction in distortion results from spatial averaging over even three units, which would not be achieved by further temporal averaging of single units that show phase locking to the stimulus.

## Summary

Four main conclusions result from the material presented here:

1. Nerve fibers can transmit very little information, particularly about rapidly varying stimuli, over any considerable distance with the use of a code based on analogue parameters of membrane potential.

2. The ability of nerve cells to transmit information about maintained stimuli by use of only the mean number of im-

pulses discharged in a period of time (frequency coding) can be calculated and compared with behavioral estimates. Neuronal variability severely limits this capacity.

3. With cyclic stimuli, the presence of neuronal variability prevents distortions caused by the phase locking of the response to the stimulus. Parameters such as the variance of intervals in a spike train may be adjusted for efficient information transmission, with little distortion over a given range of sensory stimuli, rather than themselves conveying information.

4. Use of a number of channels in parallel (spatial summation) further reduces distortion and can permit the details of a time-varying stimulus to be accurately transmitted.

## REFERENCES

ADRIAN, E. D., and Y. ZOTTERMAN, 1926. The impulses produced by sensory nerve-endings. Part 2. The response of a single end-organ. *J. Physiol.* (*London*) 61: 151–171.

ARMETT, C. J., J. A. B. GRAY, R. W. HUNSPERGER, and S. LAL, 1962. The transmission of information in primary receptor neurones and second order neurones of phasic system. *J. Physiol.* (*London*) 164: 395–421.

BARLOW, H. B., 1961. Possible principles underlying the transformations of sensory messages. *In* Sensory Communication (W. A. Rosenblith, editor). M. I. T. Press, Cambridge, Massachusetts, pp. 217–234.

BUSH, B. M. H., and A. ROBERTS, 1968. Resistance reflexes from a crab muscle receptor without impulses. *Nature* (*London*) 218: 1171–1173.

CHAMBERS, M. R., and A. IGGO, 1967. Slowly-adapting cutaneous mechanoreceptors. *J. Physiol.* (*London*) 192: 26P–27P.

DARIAN-SMITH, I., M. J. ROWE, and B. J. SESSLE, 1968. "Tactile" stimulus intensity: Information transmission by relay neurons in different trigeminal nuclei. *Science* (*Washington*) 160: 791–794.

FÄRBER, G., 1968. Berchnung und Messung des Informationsflusses der Nervenfaser. *Kybernetik* 5: 17–29.

FRENCH, A. S., and R. B. STEIN, 1970. A flexible neural analog using integrated circuits. *Inst. Elec. Electron. Eng. Trans. Bio-Med. Eng.* (in press).

GRÜSSER, O.-J., 1962. Die Informationskapazität einzelner Nervenzellen für die Signalübermittlung im Zentralnervensystem. *Kybernetik* 1: 209–211.

HODGKIN, A. L., and W. A. H. RUSHTON, 1946. The electrical constants of a crustacean nerve fibre. *Proc. Roy. Soc., ser. B, biol. sci.* 133: 444–4479.

KIANG, N. Y. S., 1965. Discharge Patterns of Single Fibers in the Cat's Auditory Nerve. M. I. T. Press, Cambridge, Massachusetts.

MACKAY, D. M., and W. S. McCULLOCH, 1952. The limiting information capacity of a neuronal link. *Bull. Math. Biophys.* 14: 127–135.

MATTHEWS, P. B. C., and R. B. STEIN, 1969a. The sensitivity of muscle spindle afferents to small sinusoidal changes of length. *J. Physiol.* (*London*) 200: 723–743.

MATTHEWS, P. B. C., and R. B. STEIN, 1969b. The regularity of primary and secondary muscle spindle afferent discharges. *J. Physiol.* (*London*). 202: 59–82.

MILLER, G. A., 1956. The magical number seven, plus or minus two: Some limits on our capacity for processing information. *Psychol. Rev.* 63: 81–97.

NOBLE, D., and R. B. STEIN, 1966. The threshold conditions for initiation of action potentials by excitable cells. *J. Physiol.* (*London*) 187: 129–162.

QUASTLER, H., 1956. Studies on human channel capacity. *In* Information Theory: Third London Symposium (C. Cherry, editor). Butterworths, London, and Academic Press, New York, pp. 361–370.

RALL, W., 1962. Electrophysiology of a dendritic neuron model. *Biophys. J.* 2 (suppl.): 145–167.

RAPOPORT, A., and W. J. HORVATH, 1960. The theoretical channel capacity of a single neuron as determined by various coding systems. *Inform. Contr.* 3: 335–350.

RUSHTON, W. A. H., 1951. A theory of the effects of fibre size in medullated nerve. *J. Physiol.* (*London*) 115: 101–122.

RUSHTON, W. A. H., 1961. Peripheral coding in the nervous system. *In* Sensory Communication (W. A. Rosenblith, editor). M. I. T. Press, Cambridge, Massachusetts, pp. 169–181.

STEIN, R. B., 1967a. Some models of neuronal variability. *Biophys. J.* 7: 37–68.

STEIN, R. B., 1967b. The information capacity of nerve cells using a frequency code. *Biophys. J.* 7: 796–826.

STEIN, R. B., 1968. Some implications of the variability in chemoreceptor discharge. *In* Arterial Chemoreceptors (R. W. Torrance, editor). Blackwell Scientific Publications, Oxford, pp. 205–210.

TALBOT, W. H., I. DARIAN-SMITH, H. H. KORNHUBER, and V. B. MOUNTCASTLE, 1968. The sense of flutter vibration: Comparison of the human capacity with response patterns of mechanoreceptive afferents from the monkey hand. *J. Neurophysiol.* 31: 301–334.

WALLØE, L., 1968. Transfer of Signals Through a Second Order Sensory Neuron. Institute of Physiology, Oslo, Norway, doctoral thesis.

WERBLIN, F. S., and J. E. DOWLING, 1969. Organization of the retina of the mudpuppy, *Necturus maculosus*. II. Intracellular recording. *J. Neurophysiol.* 32: 339–355.

WERNER, G., and V. B. MOUNTCASTLE, 1965. Neural activity in in mechanoreceptive cutaneous afferents: Stimulus-response relations, Weber functions and information transmission. *J. Neurophysiol.* 28: 359–397.

# 54    The Topology of the Body Representation in the Somatic Afferent Pathway

## GERHARD WERNER

IN THE COURSE OF the past 30 years, experimental studies and clinical observations firmly established the existence of a fixed relation between regions on the body and areas on both the sensory and the motor cortex. This suggested the idea of an orderly and systematic mapping of peripheral events in the brain (cf. Adrian, 1947). Specifically, neuroanatomical and neurophysiological studies determined that the spatial orderliness of the representation of the body at the cortical receiving areas and at subcortical relay stations follows the segmental (i.e., dermatomal) innervation pattern of the body periphery, at least in the medial lemniscal system and as far as overlap between consecutive dermatomes permitted one to recognize (Woolsey et al., 1942; Rose and Mountcastle, 1959). Equally abundant is the evidence that the cortical and subcortical projections in the visual system are also topographically organized, in the sense that fiber tracts and neurons preserve the general spatial arrangement that is to be found in the retina itself (Talbot and Marshall, 1941; Bishop et al., 1962; Garey and Powell, 1968; Whitteridge, 1965).

Prior to these discoveries, the notion of topographic orderliness of sensory projections was part of some of the psychophysiological theories of the nineteenth century, notably those of nativistic inclination. These theories, and in particular the theory of Bernstein (1868), postulated that perceived space is correlated with spatial relations established in the nervous system. At least in some degree, the neurophysiologists of the 1940s and 1950s seemed to supply the evidence needed to validate these older views, as well as Köhler's pronouncement of 1929 that "all experienced order in space is a true representation of a corresponding order in the underlying dynamical context of physiological processes."

Objections to applying the results of neurophysiological mapping studies as a basis for psychophysiological theories were, however, inevitable. In 1958, Doty produced evidence which led him to conclude that "the topographical arrange-ment of the retino-cortical projection is in itself of minor or no importance in the visual analysis of geometric patterns" (see also Doty, 1961). Nevertheless, the spatial orderliness of sensory mapping recurs with considerable persistence in diverse neural structures, the complete topographical representation of the cochlea in each of the 13 divisions of the human cochlear nucleus being, perhaps, the most illustrative example (Lorente de Nó, 1933). Therefore, reasons remain for inquiring into finer details of the mapping process that underlies the regularity of correspondence between an array of neural elements in the central nervous system and the peripheral sites they represent. Specifically, the question is whether some form of "coding" for the place of a sensation can be attributed to this mapping process, perhaps similar to that envisaged by Head (1920) to subserve, as a permanent physiological disposition, the appreciation of position and extension of a stimulus in a field of spatial relations (see also Oldfield and Zangwill, 1942).

For an investigation of the basic principles of mapping in the central nervous system, and of its possible implications, the component of the somatosensory system commonly designated as the dorsal column-medial lemniscal system offers features that promise to be particularly revealing. This system represents not only the cutaneous body surface, but, in addition, the receptors in joint capsules and periost situated in the space enclosed by the body surface. Thus, the question arises as to how this neural system accomplishes the task of mapping both the surface and the interior of the three-dimensional body periphery.

### The tactile-kinesthetic system of primates

In the primate organism, the dorsal column-medial lemniscal component of the somatic sensory nervous system subserves the modalities of touch and kinesthesis with exquisite specificity and synaptic security (Rose and Mountcastle, 1959; Mountcastle, 1961). Its anatomical constituents are the afferent nerve fibers which originate from peripheral receptors, ascend in the posterior funiculi of the spinal cord, form synaptic relays in the dorsal-column nuclei and the ventrobasal nuclei of the thalamus, and finally project to a

GERHARD WERNER Department of Pharmacology, University of Pittsburgh School of Medicine, Pittsburgh, Pennsylvania

map-like representation of the body in the postcentral gyrus of the cerebral cortex, i.e., somatic sensory area I (S-I) (Woolsey et al., 1942).

In the projection of afferents from the body to the cortex, the neurons receiving sensory signals from tactile (cutaneous) and kinesthetic receptors are arranged in separate cell columns, oriented perpendicular to the cortical surface. For the neurons composing each "modality pure" cell column, Powell and Mountcastle (1959) ascertained that the peripheral receptive field (RF) of a neuron situated at one depth of a cortical cell column is representative for the approximate RF location of all neurons of that same cell column. Therefore, the neurons of each cell column contribute but one body area to the cortical body representation, and the cell columns are the elementary units of which the cortical map is composed. The cell columns of the two submodalities (i.e., the tactile and the "deep" modality) intermingle in a mosaic in which cell columns with RFs interior to the body surface and columns representing the overlying skin area are in proximity to one another. Although the cortex has extension in depth, the cortical map is equivalent to a two-dimensional sheet to which each cortical cell column as a whole contributes only one receptive field. This circumstance enables one now to state the problem of the preceding section in more specific terms: one can view the projection to somatic sensory area I as a solution of the geometric problem of mapping the surface as well as the interior of a three-dimensional body onto a single two-dimensional plane.

## The map of the hind limb
## in the somatosensory area I

Because our primary interest was to determine with fine resolution what aspects of the body geometry are preserved in the body's cortical representation, we performed microelectrode penetrations that would cut across many neighboring cell columns of the cerebral cortex of the macaque. The penetrations were oriented in various directions with regard to the mediolateral and anteroposterior boundaries of somatosensory area I. Accordingly, we directed the microelectrodes in such a manner that they would traverse several millimeters of cortical tissue as nearly parallel to the cortical surface as possible, in view of its curvature and gyration. The data from these electrode traverses were supplemented by others obtained in penetrations perpendicular to the cortical surface, or with small deviations from the perpendicular (Werner and Whitsel, 1968).

Figure 1 illustrates, as an example, the data from one particular microelectrode penetration, which, in addition to the orientation from lateral to medial in the coronal plane, also subtended a small angle from anterior to posterior in the sagittal plane. The composition of the data,

such as those illustrated in Figure 1 and obtained in 43 histologically verified penetrations, enabled the delineation of the hind-limb representation in S-I, which is displayed in Figure 2 (Werner and Whitsel, 1968).

The general plan of the map in the anteroposterior direction is that the receptive fields of the neurons progress, essentially, in bands around the limb, much as did the laces of a Roman soldier's footwear. Medially on the cortex, the postaxial leg representation changes continuously into a preaxial representation at the anterior and the posterior borders of somatosensory area I; the converse occurs laterally on the cortex where postaxial RFs at the anterior and posterior borders of the cortical map adjoin the preaxial leg representation. These "fringe" zones of the map tend in both cases toward the respective midlines of the limb (Figure 2).

Another characteristic feature of the composite map shown in Figure 2 is the representation of the dermatomes in their serial order, as indicated by the brackets to the right in the figure. The sacral dermatomes project medially to the cortex, and their preaxial fringes occupy the rostral and the caudal borders of their representation. The lumbar dermatomes project laterally on the cortex, and their postaxial fringes occupy the rostral and caudal borders of the map. There is a particular sequence of RFs within each dermatomal band in the cortical map as the microelectrode traverses that band from medial to lateral; in the cortical region of the sacral dermatomes the RFs progress from proximal to distal on the leg; conversely, in the cortical dermatomal bands of segments L-6 and L-5, the RFs progress from distal to proximal on the leg.

The position of the foot in the map between postaxial and preaxial dermatomal projection is in accord with its principal dermatomal affiliation to dermatome L-7. The RFs of the neurons of this latter dermatomal band move from lateral to medial on the foot as one traverses the cortical map from medial to lateral. As a result, the sum total of all RFs represented in any mediolateral traverse of the cortical map describes a continuous spiral path around the limb.

Such a way of looking at the relation between the body and its cortical image is the reverse of the experimental approach: instead of determining the projection from the body to the cortex, we propose to examine the relation between linear arrays of cortical neurons and the peripheral patterns made up from the sum total of the RFs of these same arrays of cells. The schematic displays of Figure 3 are predictions of this relation, based on the experimentally determined landmarks of the cortical map. The mediolateral cortical traverse in Figure 3 (left) corresponds to a hairpin-like peripheral path, crossing the sole of the foot from lateral to medial; another cortical traverse is seen to involve RFs from the sole and dorsum of the foot, in that sequence

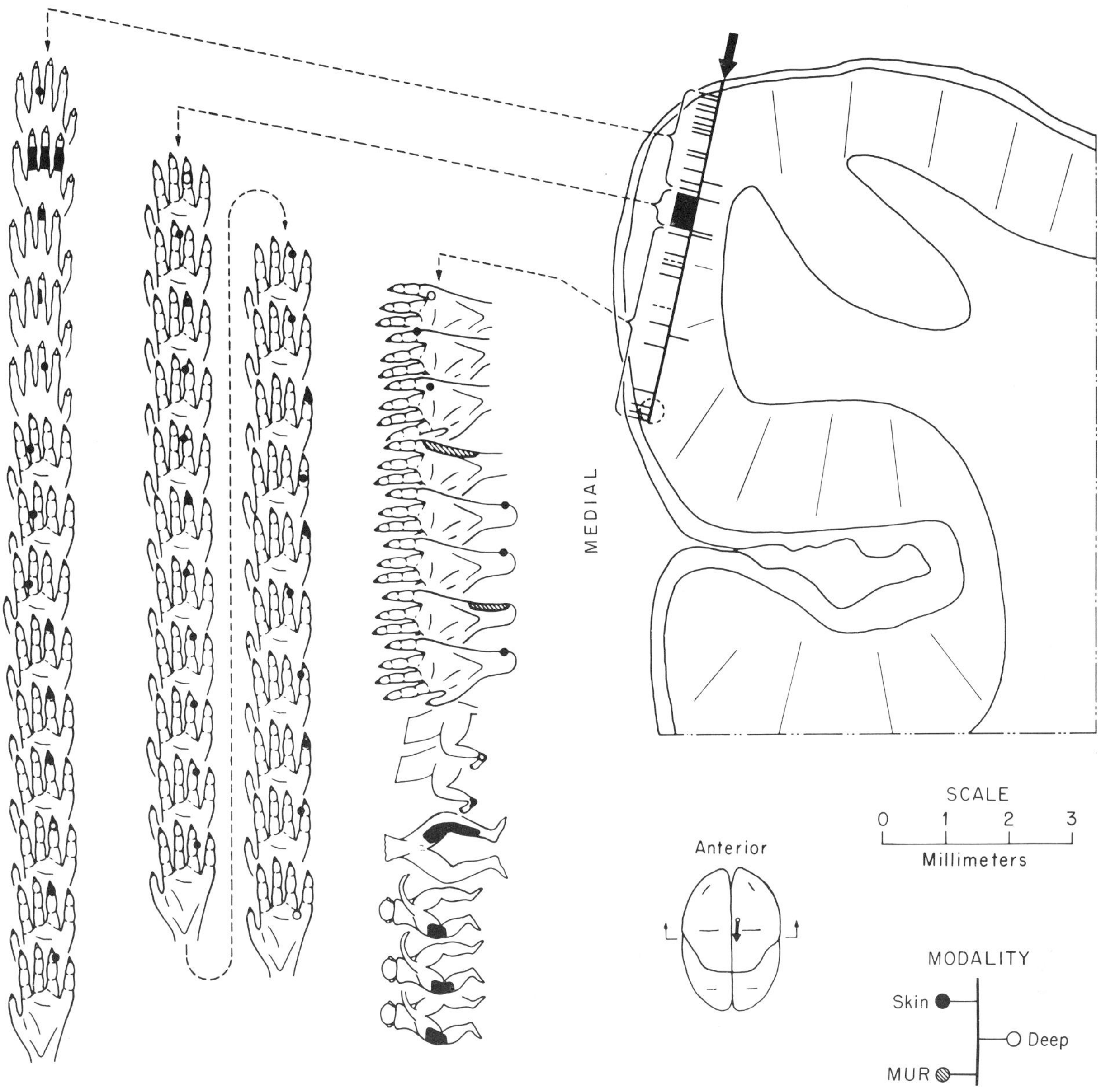

FIGURE 1    A microelectrode penetration through somato-sensory area I of the macaque. The electrode advanced from lateral to medial in the coronal plane and, to a lesser degree, from anterior to posterior in the sagittal plane. Diagram of the cortical surface is in the lower portion of the figure. Right-angled arrows indicate the plane of the coronal section containing the electrolytic lesion at the end point of the penetration. Descending arrow shows the electrode track as viewed from the cortical surface. Projection drawing of the coronal section containing the lesion shows the entire electrode track as well as the approximate depth of each isolated and identified cortical neuron. Neurons responding to skin stimuli point to the left, and neurons driven from joints and periost to the right, of the track. Lines crossing cortex indicate the orientation of the cell columns. The modality and receptive-field center of each unit are displayed by codes on the appropriate region of a figurine (see key in lower right corner for an explanation of codes; MUR, multiunit record). Figurines are arranged in the sequence in which the neurons were encountered in the microelectrode penetration. (By permission of the American Physiological Society.)

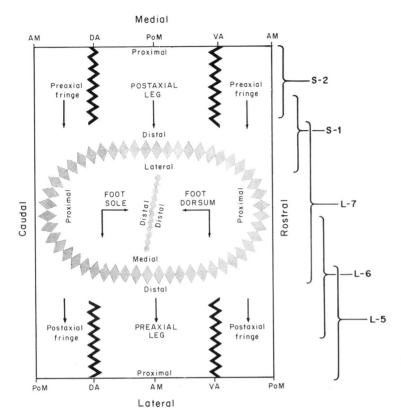

FIGURE 2 Schematic map of the hind-limb representation in somatosensory area I. The labels medial, rostral, lateral, and caudal indicate orientation of this map on the cerebral cortex. Abbreviations on the medial and lateral boundaries of map refer to positions on the body: AM, anterior midline; DA, dorsoaxial line; PoM, posterior midline; VA, ventroaxial line. Arrows in the map point in the direction of the dermatomal trajectories. Within the area of the foot projection, the labels proximal, distal, medial, and lateral refer to positions on the foot. Brackets to the right of the figure label the dermatomal bands of the cortical map. The zig-zag shape of the axial lines indicates transition zones between preaxial and postaxial leg representation in the map. Shaded band that encloses the foot representation stands for neurons the RFs of which link dorsum and sole of foot, or the foot with preaxial or postaxial calf, respectively. Diagonal band separating sole and dorsum represents the neurons with RFs on the tip of the toes. (By permission of the American Physiological Society.)

(right, Figure 3). These and all other peripheral paths which correspond to mediolateral cortical traverses progress on the leg from proximal to distal, traverse the foot from lateral to medial, and ascend the limb from distal to proximal.

Furthermore, the cortical map of Figure 3 demonstrates that the geometrical properties of the cortical body representation appear quite different from those of the body itself, for several reasons. First, unlike the relation in the body periphery, the projection from the skin does not form a continuous boundary in the cortical map, enclosing the projection from the "deep sensors." A second reason for the difference between the body and its map is that relations of proximity and distance between points on the body do not consistently remain preserved in the cortical map; for instance, RFs on the heel of the foot and the dorsum of the ankle are closely adjacent or contiguous on the body, but they map to the far ends of the anterior and posterior edges, respectively, of the cortical projection of dermatome L-7.

Most of the penetrations were directed through the cortical area of the hind-limb and trunk representation, and the discussion focuses on the projection of this body area. However, additional data on the forelimb representation are in accord with the idea that the general principles of the conclusions apply also to the body projection as a whole.

## The topology of somatic sensory area I

These apparent discrepancies between the geometrical properties of the body and its map preclude a simple correspondence between the two, at least in the framework of ordinary spatial intuition which is based on the metric structure of space, i.e., on relations of proximities and distances of points. For the purpose of characterizing how one space maps to another, however, it is possible to conceive of spaces that possess a different and more general structure than is inherent in a metric structure. The requirement is that such spaces can be described by a list of sets of elements, and that the *unions* and *intersections* of these sets must themselves be members of the list. Then it is said that the list of sets, their union and intersections, and the null set form a topology. The mapping between the two spaces is said to be a homeomorphism if every set of the topology in one space corresponds uniquely to a set in the other space, and vice versa. Figure 4 illustrates these definitions for the two simple topologies T and T*. If the mapping between the two spaces can be shown to be a homeomorphism, it follows that any topological property of one space is also a topological property of the other (see, for example, Hu, 1966; Lipschutz, 1965).

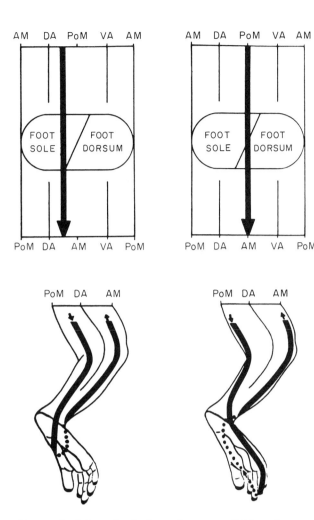

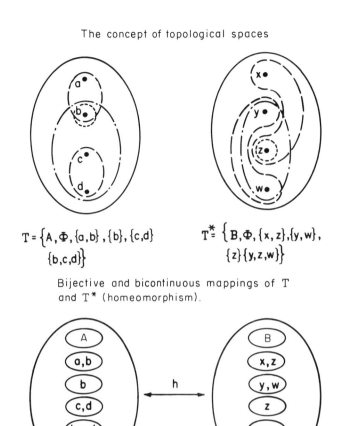

The concept of topological spaces

$$T = \left\{ A, \Phi, \{a,b\}, \{b\}, \{c,d\} \atop \{b,c,d\} \right\}$$

$$T^* = \left\{ B, \Phi, \{x,z\}, \{y,w\}, \atop \{z\}\{y,z,w\} \right\}$$

Bijective and bicontinuous mappings of T
and T* (homeomorphism).

FIGURE 3 Schematic display of two medial-lateral traverses through the hind-limb position of somatosensory area I. Figures at the bottom display the paths on the body, consisting of the totality of RFs that one would encounter in microelectrode penetrations that cross the cortical projection in the manner indicated in the diagrams at the top. The dotted portion of the path traverses the dorsum of the foot, which cannot be seen in this view. (By permission of the American Physiological Society.)

FIGURE 4 The upper portion illustrates two topological spaces, T and T*. In T, the array of elements {a,b,c,d} forms the sets as shown. T is a topology because all unions and intersections of the sets {a,b}; {b}; {c,d}; and {b,c,d} are themselves members of the list of sets; e.g., {c,d} ∩ {b,c,d} = {c,d}, which is a member of the list. Similarly, the sets {x,z}; {y,w}; {z}; and {y,z,w} have been chosen from the array = {x,y,x,w} to form a topology. The lower portion illustrates the idea of a homeomorphism, h, for it shows a mapping h which uniquely relates sets in T to sets in T*, and vice versa. An example for h is: {a,b} ↔ {x,z}; {b} ↔ {z}; {c,d} ↔ {y,w}; {b,c,d} ↔ {y,z,w}.

To apply these concepts to the characterization of the somatosensory projection, it is necessary to determine properties of the body and the cortical map which can be identified with topologies, in the sense of the preceding definition. The paragraphs that follow describe the interpretation of certain patterns of RFs on the body and of neurons in the cortex as topologies, and prove that the mapping between the body and its cortical image is a homeomorphism.

The clue of this analysis is the observation that the afferent fibers at the entry zone of each dorsal root and in the corresponding fiber laminae of the posterior funiculi in the

spinal cord are arranged in a certain order (Figure 5) (Werner and Whitsel, 1967). In each dermatome, a family of linear arrays of afferent fibers can be discerned; the orientation of these linear arrays in the dorsal roots and in the corresponding portions of the posterior funiculi is shown in Figure 6B. The distinguishing property of the fiber arrays with their orientation, indicated by the arrows, is that they represent peripheral RFs which, if taken together, combine in each segment to an essentially continuous path in the periphery (Figure 6A). We designated each path that can be generated in this fashion as a dermatomal trajectory (see Figure 5). Within each dermatome, all trajectories have a common direction on the body. The trajectories of consecutive dermatomes, although partially overlapping, form a loop around the hind limb, as shown schematically in Figure 6A. The afferent fibers from deep and cutaneous receptors in a dermatomal trajectory intermingle, the sole criterion of fiber order being the sequence of the RFs along the trajectory.

To emphasize the characteristic features of the body representation in the posterior funiculi of the spinal cord, it can be examined from the point of view of a hypothetical observer capable of "reading" the RF labels carried by individual afferent fibers. As this observer moves along from dorsomedial to ventrolateral in the spinal cord cross-sectional plane (Figure 6B), he would traverse a path on the body which begins postaxially at the sacrum and hip, descends to the lateral edge of the foot, crosses the foot from lateral to medial, and ascends the leg preaxially toward groin and hip (Figure 6A). On this path, the observer would "see" receptive fields on the surface of the leg intermingled with others deep in the leg, but at no point would he cross a continuous boundary which would enable him to tell whether he is at the "inside" of or "on" the body surface.

In the projection to the cortex, the dermatomal trajectories retain a characteristic role. This is illustrated in Figure 6C. In the cortical region of the sacral dermatomes, the RFs progress from proximal to distal on the leg. Conversely, in the cortical dermatomal bands of segments L-6

and L-5, the RFs progress from distal to proximal on the leg. The projections of the foot fall between segments S-1 and L-6; the RFs of the neurons in this cortical band progress sequentially from lateral to medial on the foot as one traverses its cortical map. As a result, the sum total of all RFs represented in any mediolateral traverse of the cortical map describes a continuous loop around the limb (see also Figure 3).

There is one striking difference between the appearance of this cortical path and the path traversed by the hypothetical observer in the posterior funiculi: the path would not be smooth; rather, the observer would see sequences of RFs which progress continuously along the body, but at each transition to the next dermatome he would backtrack part of the sequence just completed. Thus, his course would vacillate as he proceeded to complete the loop. In contrast, the cortical observer would find that to each cortical traverse in the direction shown in Figure 6C, there corresponds a continuous progression of RFs around the limb, as a result of dermatomal overlap in the cortical projection. It appears that the afferents from the dermatomal trajectories in each loop around the limb assemble according to the body region of origin.

In the projection from the periphery to the cortex, the body is first segmented into a series of overlapping dermatomes whereby the afferents from each dermatome retain a specific order; namely, that of the dermatomal trajectories. The trajectories align and partially superimpose to the effect that they map a continuous path on the body. Therefore, it appears that the body and the cortical map can be considered a composite of dermatomal trajectories, and that

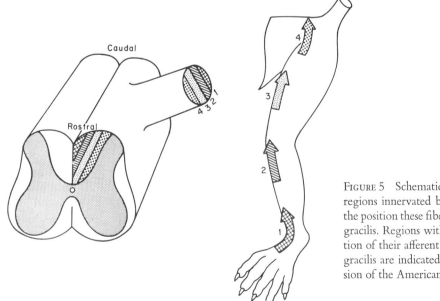

FIGURE 5  Schematic representation of the peripheral body regions innervated by fibers entering dorsal root L-6, and the position these fibers occupy in the root and the fasciculus gracilis. Regions within the dermatome (1–4) and the location of their afferent fibers in the dorsal root and fasciculus gracilis are indicated by corresponding codes. (By permission of the American Physiological Society.)

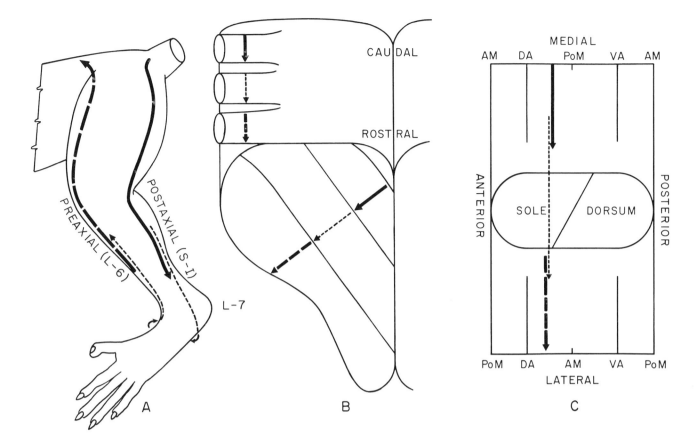

FIGURE 6  The role of the dermatomal trajectories in the somatosensory projection. A. A view of the hind limb with a hairpin-like path on its surface that is shown to consist of afferents in three dermatomes. The full line is the trajectory in S–I; it overlaps at its distal end with a trajectory in L–7; on the preaxial side, the loop is completed by a trajectory in L–6. The symbols of the respective trajectories are the same in A through C. Where the lines of the trajectories run in parallel, there is overlap between adjacent dermatomes. B. The position and direction of the dermatomal trajectories with the peripheral paths shown in A, as they appear in the dorsal roots and the fiber laminae of the posterior funiculi of the spinal cord. In the latter, the arrows of the trajectories point from dorsomedial to ventrolateral, as the cord is positioned in the body. C. The projection of the hind limb to somatic sensory area I, with the dermatomal trajectories that run side-by-side in the cortical map, as they do in the body periphery (see A). The landmarks of the map in C are as follows: AM, anterior midline; DA, dorsoaxial line; PoM, posterior midlimb; VA, ventroaxial line. Note that the hairpin-like path in the periphery appears on the cortex as a straight and continuous traverse of the map.

each dermatomal trajectory is treated in the projection as a unit, and it maps as a whole. This type of projection enables one, in the language of topology, to interpret the body as an aggregate of RFs that combine to ordered sets in the form of the dermatomal trajectories. Similarly, the cortical map appears in the projection as an aggregate of neurons that also combine to ordered sets, each set receiving afferents from one dermatomal trajectory. The totality of all ordered sets in the periphery is a topology in the body space, as are the sets of neurons in the cortical map. In the mapping, each set of elements in the periphery comes to correspond to one set of elements in the cortex, and vice versa; consequently, the mapping is a homeomorphism. Accordingly,

the body and its cortical map are topologically equivalent in this sense: two adjacent peripheral RFs that form part of one trajectory remain neighbors in the cortical projection. Neighboring RFs that belong to different trajectories need not, however, always be neighbors in the cortical map; they are not if their afferents enter the spinal cord at different segmental levels.

## The spatial characteristics of the cortical map

We are now in a position to develop a rigorous argument that the body representation in the cortical map has the property of a nonorientable surface. The argument rests on the principle that a surface can be specified by the orientation of its edges and their sequence, as one moves around the perimeter of that surface in a predetermined direction. An edge is marked with the exponent $(-1)$ if its orientation is opposite to the direction around the perimeter. The entire sequence is set equal to 1 as a notational device to show that the complete perimeter of the polygon has been given (see, for example, Blackett, 1967; Alexandroff, 1961).

This procedure is applied to the schematic hind limb of Figure 7 which defines the cortical map unambiguously as the surface of a Klein bottle (see legend to Figure 7).

This topological characterization of the cortical map implies also a set of rules for approximating a model of the cortical map in three-dimensional space. The individual steps of this construction are depicted in Figure 8. To construct a three-dimensional model of the Klein bottle, one would first roll the plane of the cortical map into a cylinder so that the interior edges of Am-PoM of the left map in Figure 7 come to superimpose; next, one would bend the cylinder, rotate the bottom plane by 180 degrees, and bring the bottom opening into superposition with the top opening. The last step would require that the surface of the cylinder intersect itself. The individual steps of this construction are depicted in Figure 8. The important characteristic of the Klein bottle and, therefore, of the cortical map as well is that it is nonorientable. The model of Figure 8 makes this intuitively clear, for there is no distinction possible between inside and outside, as one imagines a point to move along the surface of this model; nowhere would this point encounter a boundary.

The process of self-intersection which was part of the set of procedures used to aid in the visualization of the Klein bottle is physically not realizable in three-dimensional space; therefore, the model of Figure 8 is merely a stylized approximation. The principle we employed in this procedure is the ability to describe a space Y (i.e., the Klein bottle) as obtained by identification from a simpler space X (i.e., the cortical map). The description may be adequate to show the properties of Y, even though Y cannot be properly visualized.

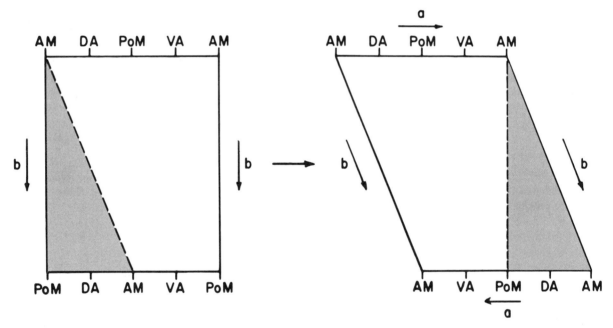

FIGURE 7 A schematic picture of the hind-limb portion of the cortical map. The anatomical body landmarks are indicated to permit identification of edge orientation and correspondence of edges. The abbreviations for these landmarks are the same as those in Figure 6D. The two lateral edges, AM-PoM, represent identical zones on the body periphery: they are interior edges of the map, and their orientation is that of the RF progression of the dermatomal trajectories, indicated by the arrows labeled b. To establish the orientation of the upper and the lower edges in the map, we transform the polygon of the cortical map, as shown by the shaded areas in the figure. The figures to the right and to the left are topologically equivalent. We merely remove the shaded triangle of the left side of the polygon and join it to the right side of the polygon, making sure that the proper landmarks superimpose. This transformation proves that the upper and the lower edges of the original polygon have an opposite orientation. We can now assemble the complete edge equation: a.b.a.b.$^{-1}$ = 1. This equation defines the cortical map unambiguously as the surface of a Klein bottle.

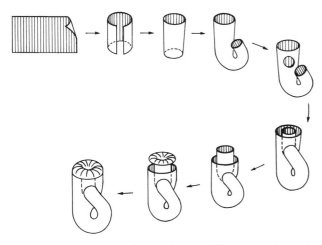

FIGURE 8  The sequence of operations to fold the cortical map in somatosensory area I to a three-dimensional model of a Klein bottle. (This figure was designed by T. Melnechuk of NRP.)

## Implications for the haptic sense

Animals and man possess the mode of behavior and experience known as haptics (Gibson, 1966; Révéz, 1950). The latter goes beyond the functional capabilities of the individual modalities of touch and kinesthesis in isolation; rather, in certain invariant combinations, the two sense modalities are thought to specify jointly the stimulus information of the form of objects in an environmental space. One might argue that this presumes a functional unit between sensory events signaling the position of external objects with respect to the body, and sensory events signaling the position of body parts relative to one another such that "skin space and bone space are all one piece" (Gibson, 1966). The suggestion is that the tacile-kinesthetic representation in one common map is the mechanism by which the central nervous system combines contactual information about objects in space external to the body, and of geometric information from within the body into *one* common space, suited to reflect the haptic sense.

The general implications of this conceptualization are that the sensory representation in the cerebral cortex can possess properties that are not inherent in the raw data originating from the peripheral sense organs themselves, and that the nature of new properties can be a consequence of the characteristics of the neural mapping process which links body periphery and central sensory representation.

## Somatosensory spaces and abstract mapping

The definition of a topology on the body periphery and in the cortical map was based on the concept of the dermatomal trajectory. With respect to those peripheral stimulus configurations that are part of the dermatomal trajectory, the relation of a homeomorphism between body periphery and the cortical image was ascertained in the preceding sections of this report. We shall now discuss the mapping of any arbitrary path, not necessarily coinciding with a dermatomal trajectory. The most striking characteristic of the experimentally determined geometric relation between the periphery and its cortical map in somatosensory area I is that any linear array of neurons in S-I (such as is traversed by a surface-parallel microelectrode penetration) traces on the body a continuous path of RFs (see, for example, Figure 1). But the converse is not generally true: to illustrate this, we select the peripheral path $1 \rightarrow 10$ on the body as shown in Figure 9. This path is composed of several dermatomal trajectories. Accordingly, the cortical image that can be predicted from the experimentally determined landmarks and other properties of the cortical map will not have the simple appearance of the mapping of a single dermatomal trajectory as illustrated in Figure 3. Rather, as shown in Figure 9, it will consist of fragments of the cortical images of those dermatomal trajectories that compose the peripheral path $1 \rightarrow 10$: for instance, the portion $1 \rightarrow 5$, which on the body is a composite of a dermatomal trajectory in L-6 and S-1, will map the cortical images $1'' \rightarrow 5''$ and $1' \rightarrow 5'$, respectively. Similarly, the portion $6 \rightarrow 7$ on the body will appear in a cortical region of dermatome L-7 in the form of a U, reflecting the reversal of direction as the peripheral path progresses from its preaxial to its postaxial position.

If one now applies the instructions of the previous section to embed the planar cortical map according to its topological characteristics in three-dimensional space, it becomes apparent that the branches of the cortical image $1' \rightarrow 10'$ and $1'' \rightarrow 10''$ in the planar map can combine on the surface of the Klein bottle in three-dimensional space to *one* continuous image of the peripheral path $1 \rightarrow 10$.

One way to conceptualize some possible implications of these relations between the body and its map in S-I is to employ the idea of the *path-lifting property* of mappings as an interpretative device (Hurewicz, 1955; Fadell, 1959; Hu, 1959). The heuristic power of this line of thought was demonstrated by Greene (1964) in an attempt to formalize the relation between spaces of various degrees of abstractness.

The essentials of this line of thought are schematically illustrated in Figure 10. Let B represent the metric space of the body in which P′ designates an arbitrary stimulus pattern (e.g., a path, with a beginning in O′); let C stand for the cortical map in S-I, which, as experiments established, projects by p continuously onto B (see above). Furthermore, let Y represent a structure in the central nervous system that contains a repertoire of stimulus specifications, $O'_i P'_i$, of which $E_i$ is a specific instance. To

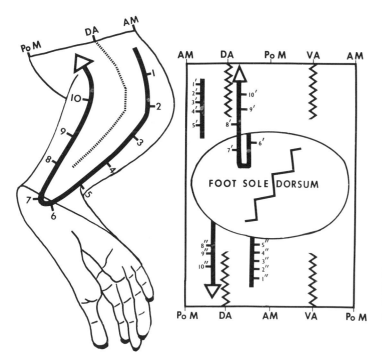

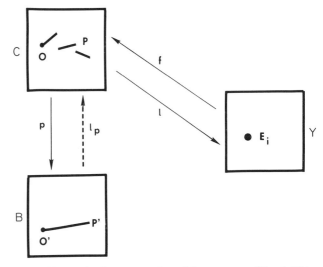

FIGURE 9 Diagrammatic representation of an arbitrary peripheral stimulus configuration (shown as path 1 → 10 on the limb) in the schematic cortical map of somatosensory area I in the postcentral gyrus.

give an example: an element in this repertoire could be the concatenation of a path, P′, with a starting point, O′; this example is chosen because of its obvious analogy to the type of concatenations known to occur in the visual system as the characteristic of hypercomplex cells (Hubel and Wiesel, 1962, 1968).

If there exists in this system a mapping, f: Y → C, and an inverse mapping, $l$, such that the sequence of mappings, $l \times f$, projects $E_i$ back onto itself, then p: C → B has the *path-lifting property* and f is said to be a *lifting function;* moreover, if f exists, it follows that a mapping, $l_p$, can be found that determines the image of O′P′ in C.

In more general terms, given these prerequisites, if the repertoire of central representation in Y contains signal specifications that correspond to a particular peripheral stimulus event, O′P′, then a procedure exists for finding a mapping of this event in C. In that sense, C may be viewed as possessing a "model" of B with respect to the specifications contained in Y. Similarly, C can depict classes of states of B that are "well formed" and admissible with respect to rules laid down in Y.

The formal resemblance to Sutherland's (1968, 1969) theory of visual pattern recognition is apparent: the mappings f and $l$ reflect the interactions between a "processor," C, and a "store," Y. The latter fulfills the role of a device that evaluates projection patterns from the periphery to C, in terms of their conformity with "schemata" stored in Y. That part of the total projection pattern in C that is relevant for the "recognition" by Y can be considered a mapping of

O′P′ to C under the mapping function $l_p$. Accordingly, the matching process between OP and $E_i$ is equivalent to the computation of a mapping function, $l_p$, such that $l_p$: O′P′ → OP. The advantage of expressing these relations in the framework of lifting functions is that the latter implies the existence of a solution, $l_p$, relative to Y in all instances, provided the existence of the functions $l$, f, and p, and with B being a metric space.

FIGURE 10 A visual representation of the concepts of "path-lifting property" and "lifting function."

The existence of p is experimentally established: it is that continuous mapping function that depicts any linear array of neurons on the cortex as a continuous path of RFs on the body (see above). The mappings *l* and *f* can be equated with neural connections between specific cortical receiving areas and "intrinsic" cortex lacking specific thalamocortical projections (Rose and Woolsey, 1949). Neurological and neurobehavioral studies suggested that the intrinsic cortical sector operates by efferents on the events occurring in the extrinsic cortical mechanisms and, thereby, accomplishes the partition of events occurring in the extrinsic system into elements of increasing complexity (Pribram, 1959). In this concept, the intrinsic sectors are conceived as programing mechanisms that function to partition events initiated by the sensory input into complex categories.

The general manner in which cortical receiving areas (exemplified by C in Figure 10) and a repertoire of interpretative rules, Y, are proposed to interact in the model with path-lifting property formally mimics the process of parsing a sentence and is, therefore, in line with the idea that linguistic analysis provides a model for description of *all* kinds of behavior (Miller et al., 1960).

## Is there a common principle in somatosensory mapping?

The experimental results reviewed in the preceding sections ascribed a specific connotation to the general idea of "topographic orderliness" of cortical representation (see beginning of this paper). In the first place, this orderliness reflects in S-I the fact that the net shift of RFs along linear arrays of cortical cell columns traces a continuous path on the body. Second, it is as if, for the purpose of mapping, the body periphery were decomposed into more-or-less narrow bands, and as if the afferents from these same bands assembled again in the cortical receiving areas to terminate on corresponding linear arrays of cell columns. It is now established that the projection to the second somatic area (Woolsey, 1943; Woolsey and Fairman, 1946) follows the same rule of mapping, except for the deletion of the "deep" submodality and for the introduction of the bilateral symmetry of the cutaneous representation (Whitsel et al., 1969).

In the projection to the first and second somatic areas, the fragmentation of the body into bands follows a specific order: the first-order afferent fibers with receptive fields on the body enter the spinal cord along its rostrocaudal extension in a characteristic sequence, namely, that of the "dermatomal trajectories" (Werner and Whitsel, 1967, 1968). The mapping of the body to the cortex represents the family of the trajectories rather than the body topography as such.

To view the cortical mapping in somatosensory areas I and II in this manner has, perhaps, some bearing on the understanding of the developmental factors that lead to these maps, for it permits one to see a rank order in the first-order afferent fibers, given by the sequence of entry along the rostrocaudal spinal axis. In the mapping, these afferents sort themselves to restore this peripheral rank order "like soldiers, on the next man" (Horridge, 1968). The important point is that, whatever carries in each nerve the label of rank order, it is preserved in the projection to the cortex, even if the afferents travel different central pathways, for in S-II the afferents arrange in the same rank order as they do in S-I, although the projection to the former is composed of distinctly different ascending pathways, which merge into one cortical representation pattern (cf. Bowsher, 1965).

This view establishes a connection to some current ideas which conceptualize morphogenesis in terms of computer programs and automata theory (Ulam, 1962; Arbib, 1967; Stahl, 1966; Apter and Wolpert, 1965; Bonner, 1965). The implication is that it is more meaningful to discuss a complex pattern (such as a cortical map) in terms of a set of instructions that can generate it, than in terms of the information contained in the final product. Accordingly, in a program that would generate the geometry of central projections, the sequence of entry of first-order afferent nerve fibers into the neuraxis would be a distinguishing attribute, capable of imparting an essentially identical topology to all central representations of these afferents. One is then led to conjecture that the multiplicity of central representations of a given set of peripheral receptors, each topologically in register with all others and each abstracting a different stimulus feature (cf. Maturana et al., 1960; Hubel and Wiesel, 1962, 1968; Werner and Whitsel, 1970), enables the parallel processing of sensory input similar to that applied in the design of pattern-recognition machines (Selfridge and Neisser, 1960).

## Acknowledgments

The author gratefully acknowledges the support of his research by the National Institute of Mental Health (Research Grant MH 11682); by the United States Army Medical Research and Development Command (contract DADA 17-67-C-7032); and by the Office of the Schizophrenia Research Program, The Supreme Council 33° A. A. Scottish Rite Northern Masonic Jurisdiction.

REFERENCES

ADRIAN, E. D., 1947. The Physical Background of Perception. Clarendon Press, Oxford.

ALEXANDROFF, P., 1961. Elementary Concepts of Topology (A. E. Farley, translator). Dover Publishing Co., New York.

Apter, M. J., and L. Wolpert, 1965. Cybernetics and development, I. Information theory. *J. Theor. Biol.* 8: 244–257.

Arbib, M. A., 1967. Automata theory and development: Part I. *J. Theor. Biol.* 14: 131–156.

Bernstein, B., 1868. Zur Theorie des Fechnerschen Gesetzes der Empfindung. *Arch. Anat. Physiol. Wiss. Med.*, pp. 388–393.

Bishop, P. O., W. Kozak, W. R. Levick, and G. J. Vakkur, 1962. The determination of the projection of the visual field on to the lateral geniculate nucleus in the cat. *J. Physiol.* (*London*) 163: 503–539.

Blackett, D. W., 1967. Elementary Topology. Academic Press, New York.

Bonner, J. F., 1965. The Molecular Biology of Development. Oxford University Press, New York.

Bowsher, D., 1965. The anatomicophysiological basis of somatosensory discrimination. *Int. Rev. Neurobiol.* 8: 35–75.

Doty, R. W., 1958. Potentials evoked in cat cerebral cortex by diffuse and by punctiform photic stimuli. *J. Neurophysiol.* 21: 437–464.

Doty, R. W., 1961. Functional significance of the topographical aspects of the retino-cortical projection. *In* The Visual System: Neurophysiology and Psychophysics (R. Jung, editor). Springer-Verlag, Berlin, pp. 228–245.

Fadell, E., 1959. On fiber spaces. *Trans. Amer. Math. Soc.* 90: 1–14.

Garey, L. J., and T. P. S. Powell, 1968. The projection of the retina in the cat. *J. Anat.* 102: 189–222.

Gibson, J. J., 1966. The Senses Considered as Perceptual Systems. Houghton Mifflin Co., Boston.

Greene, P. H., 1964. New problems in adaptive control. *In* Computer and Information Sciences (J. T. Tou and R. H. Wilcox, editors). Spartan Books, Washington, D. C.

Head, H., 1920. Studies in Neurology, Vol. 2. Hodder and Stoughton, London.

Horridge, G. A., 1968. Interneurons, Their Origin, Action Specificity, Growth and Plasticity. W. H. Freeman Co., San Francisco.

Hu, Sze-Ten, 1959. Homotopy Theory. Academic Press, New York.

Hu, Sze-Ten, 1966. Introduction to General Topology. Holden-Day, San Francisco.

Hubel, D. H., and T. N. Wiesel, 1962. Receptive fields, binocular interaction and functional architecture in the cat's visual cortex. *J. Physiol.* (*London*) 160: 106–154.

Hubel, D. H., and T. N. Wiesel, 1968. Receptive fields and functional architecture of monkey striate cortex. *J. Physiol.* (*London*) 195: 215–243.

Hurewicz, W., 1955. On the concept of fiber spaces. *Proc. Nat. Acad. Sci. U. S. A.* 41: 956–961.

Köhler, W., 1929. Gestalt Psychology. Liveright Publishing Corp., New York.

Lipschutz, S., 1965. General Topology. Schaum Publishing Co., New York.

Lorente de Nó, R., 1933. Anatomy of the eight nerve; the central projection of the nerve endings of the internal ear. *Laryngoscope* 42: 1–38.

Maturana, H. R., J. Y. Lettvin, W. S. McCulloch, and W. H. Pitts, 1960. Anatomy and physiology of vision in the frog (*Rana pipiens*). *J. Gen. Physiol.* 43 (pt. 2): 129–175.

Miller, G. A., E. Galanter, and K. H. Pribram, 1960. Plans and the Structure of Behavior. Holt, Rinehart and Winston, New York.

Mountcastle, V. B., 1961. Some functional properties of the somatic afferent system. *In* Sensory Communication (W. A. Rosenblith, editor). M. I. T. Press, Cambridge, Massachusetts, pp. 403–436.

Oldfield, R. C., and O. L. Zangwill, 1942. Head's concept of the schema and its application in contemporary British psychology. *Brit. J. Psychol.* 32: 267–286.

Powell, T. P. S., and V. B. Mountcastle, 1959. Some aspects of the functional organization of the cortex of the postcentral gyrus of the monkey: A correlation of findings obtained in a single unit analysis with cytoarchitecture. *Bull. Johns Hopkins Hosp.* 105: 133–162.

Pribram, K., 1959. On the neurology of thinking. *Behav. Sci.* 4: 265–287.

Révéz, G., 1950. Psychology and Art of the Blind (H. A. Wolff, translator). Green and Co., London.

Rose, J. E., and V. B. Mountcastle, 1959. Touch and kinesthesis. *In* Handbook of Physiology, Neurophysiology, Sect. 1, Vol. 1. American Physiological Society, Washington, D. C., pp. 387–429.

Rose, J. E., and C. N. Woolsey, 1949. Organization of the mammalian thalamus and its relationship to the cerebral cortex. *Electroencephalogr. Clin. Neurophysiol.* 1: 391–404.

Selfridge, O. G., and U. Neisser, 1960. Pattern recognition by machines. *Sci. Amer.* 20 (no. 2): 60–68.

Stahl, W. R., 1966. A model of self-reproducing automata based on string processing finite automata. *In* Natural Automata and Useful Simulations (H. H. Pattee, E. A. Edelsack, L. Fein, and A. B. Callahan, editors). Spartan Books, Washington, D. C., pp. 43–72.

Sutherland, N. S., 1968. Outlines of a theory of visual pattern recognition in animals and man. *Proc. Roy. Soc., ser. B, biol. sci.* 171: 297–317.

Sutherland, N. S., 1969. Outlines of a theory of pattern recognition in animals and man. *In* Animal Discrimination Learning (R. M. Gilbert and N. S. Sutherland, editors). Academic Press, New York, pp. 385–411.

Talbot, S. A., and W. H. Marshall, 1941. Physiological studies on neural mechanisms of localization and discrimination. *Amer. J. Ophthalmol.* 24: 1255–1264.

Ulam, S., 1962. On some mathematical problems connected with patterns of growth of figures. *In* Mathematical Problems in the Biological Sciences (R. E. Bellman, editor). American Mathematical Society, Providence, Rhode Island, pp. 215–222.

Werner, G., and B. L. Whitsel, 1967. The topology of the dermatomal projection in the medial lemniscal system. *J. Physiol.* (*London*) 192: 123–144.

Werner, G., and B. L. Whitsel, 1968. The topology of the body representation in somatosensory area I of primates. *J. Neurophysiol.* 31: 856–869.

WERNER, G., and B. L. WHITSEL, 1970. Stimulus feature detection by neurons in the somatosensory areas I and II of primates. *Inst. Elec. Electron. Eng. Trans. Man-Mach. Syst.* 11: 36–38.

WHITSEL, B. L., L. M. PETRUCELLI, and G. WERNER, 1969. Symmetry and connectivity in the map of the body surface in somatosensory area II of primates. *J. Neurophysiol.* 32: 170–183.

WHITTERIDGE, D., 1965. Geometrical relations between the retina and the visual cortex. *In* Mathematics and Computer Science in Biology. Medical Research Council, H.M.S.O., London, pp. 269–276.

WOOLSEY, C. N., 1943. "Second" somatic receiving areas in the cerebral cortex of cat, dog and monkey. *Fed. Proc.* 2: 55–56 (abstract).

WOOLSEY, C. N., and D. FAIRMAN, 1946. Contralateral, ipsilateral and bilateral representation of cutaneous receptors in somatic areas I and II of the cerebral cortex of pig, sheep and other mammals. *Surgery* 19: 684–702.

WOOLSEY, C. N., W. H. MARSHALL, and P. BARD, 1942. Representation of cutaneous tactile sensibility in the cerebral cortex of the monkey as indicated by evoked potentials. *Bull. Johns Hopkins Hosp.* 70: 399–441.

# 55 Signal Transmission Between Successive Neurons in the Dorsal Spinocerebellar Pathway

## JAN K. S. JANSEN and LARS WALLØE

FORMULATION of an adequate model for studying the efficiency of signal transmission in central neurons requires, initially, a quantitative determination of the entire neuronal input and output. The neuronal output is ordinarily simple to record as somatic or axonal spikes. Usually the input is much more difficult to determine. The present paper is, therefore, a discussion of the neuronal properties that are most important for the input-output transformation and of the extent to which the input activity of a neuron can be inferred from its output behavior.

The study of sensory pathways offers obvious advantages for such a purpose. The most important of these is the possibility of input control by adequate stimuli to the relevant receptors. However, in many sensory systems, there are complex neuronal interconnections at even early stages of signal transfer, and these represent a major obstacle for the interpretation of the transfer under physiological conditions. The complexity is presumably due to the complex content of the relevant signals. For instance, in general somatic, auditory, and visual pathways, space parameters, as well as intensity parameters, of the stimulus are of obvious physiological significance.

Therefore, in our attempt to study the transfer of signals in a nerve cell we have selected the second-order neurons of the dorsal spinocerebellar tract (DSCT) which offer several advantages. 1. The synaptic coupling of the system is comparatively well known (Lundberg, 1964; Oscarsson, 1965). 2. The properties of the relevant receptors (muscle spindles and tendon organs) have been studied extensively, and they give slowly adapting signals to steady-state inputs (Matthews, 1964; Jansen and Rudjord, 1964; Houk and Simon, 1967; Alnaes, 1967). 3. The content of the first-order signals can be described as functions of muscle length or tension. 4. The second-order neurons are localized in Clarke's column of the spinal cord; their perikarya, as well as their axons in the dorsolateral funicle of the cord, can be investigated with existing microelectrode techniques.

The present paper reviews some of our own work on the DSCT in cats and discusses its relevance to other groups of nerve cells.

## The specificity of synaptic coupling

A high degree of specificity characterizes several aspects of the synaptic coupling of the DSCT. Each group of second-order neurons is activated by the stretch receptors of a given muscle and is usually not excited by the stretch of other muscles, not even close synergists. A second type of

JAN K. S. JANSEN and LARS WALLØE   Institute of Physiology, University of Oslo, Norway specificity concerns the connections from the different types of stretch receptors. Each second-order neuron appears to be exclusively, or at least predominantly, activated from primary endings, Golgi tendon organs, or secondary endings of a given muscle. This has provided the basis for the identification of the different types of second-order neurons of the pathway (Jansen and Rudjord, 1965).

A final type of specificity is in the inhibitory connections within the DSCT. Although far from fully explored as yet, the pattern appears to be as specific as that of the excitatory connections (Jansen et al., 1967b). Thus, the tendon-organ afferents from the triceps muscle would inhibit the excitatory signal from pretibial flexor muscles, whereas the afferents from spindle primary endings inhibited the signal from the muscle of the great toe. These specificities of the synaptic couplings have enabled us to generate homogeneous excitatory or inhibitory inputs to the second-order neurons and thereby facilitate the interpretation of the neuronal behavior.

*The linear input-output relationship*

The second-order DSCT neurons give a maintained discharge to a maintained excitatory input. Two types of input have been studied systematically. The simplest is an excitation of the cell with a depolarizing transmembrane current step (Figure 1A). After an initial overshoot, the firing rate settled at a new steady level, at which it remained apparently indefinitely. The rate of firing increased linearly with increasing current intensity over a surprisingly wide range, at least up to 200 impulses per second (Figure 1C). The slope of the frequency current curves varied between 5 and 9 imp/sec/nA, with a mean of 6.7 (Eide et al., 1969a). The DSCT neurons appear to be unique among neurons so far studied in their wide range of linearity.

A comparable type of response is found in a maintained synaptic excitation elicited by muscle stretch (Figure 1B). Again the cell firing continued as long as the stimulus was maintained, and there was no indication of adaptation or fatigue in the synaptic transmission. As illustrated in Figure 1D, the firing frequency increased linearly with the degree of muscle extension. As it has been established that the rate of primary endings in muscle spindle increases linearly with muscle extension (Matthews, 1964), we may conclude that this synaptic transfer is linear as well. It is equally significant that the frequency range and the slope of the frequency-extension curve of the second-order neurons (3–5 imp/sec/mm) are very much the same as those of the corresponding receptors (Jansen et al., 1966).

The importance of the linear input-output relations (Figure 1) is emphasized by additional observations on summation of synaptic and injected transmembrane currents. According to its direction, a constant transmembrane current adds or subtracts a given number of imp/sec independently of the level of synaptic activation of the neuron. Similarly, a given synaptic input caused a constant increase in rate at all physiological firing levels, as a result of injected transmembrane current (Eide et al., 1969a). This shows that the synaptic currents and the injected currents act on a common spike-generating site in these DSCT neurons, as is also common in other nerve cells (Granit et al., 1963; Brown and Stein, 1966). It also permits an estimate of the average synaptic current generated by a 1-mm muscle stretch. With an average frequency-current slope of 7 imp/sec/nA (Eide et al., 1969a) and a frequency-length slope of 3.5 imp/sec/mm (Jansen et al., 1966), the average synaptic current is 0.5 nA per millimeter of muscle extension. We shall return to this when discussing the synaptic effect of each presynaptic impulse.

The input-output relation also has been investigated with sinusoidal stretch at different frequencies (Jansen et al., 1967c). Over the range studied and within the linear range of the stretch receptors, the frequency responses of the first- and second-order elements were indistinguishable. Accordingly, the transfer function of the synaptic transmission is frequency-independent over the range examined.

From comparisons between receptor response and psychophysical estimates of stimulus-response relations in the general somatic system, Mountcastle (1967) has suggested that the signal transformations that take place in these pathways are entirely linear. The nonlinearities that may be present in the over-all response appear to be due to the properties of the receptors. This important generalization and astonishing simplicity are supported at the first-order to second-order level by our observations on the DSCT.

This high degree of linearity is at present unexplained in terms of neuronal membrane properties. Stein (1967a) has shown that the Hodgkin-Huxley membrane equations generate a neuronal output that is essentially nonlinear. The rate of increase of firing frequency decreases rapidly with increasing excitation. Such behavior is seen, for instance, in the slowly adapting receptor of the crayfish (Enger et al., 1969). Motor neurons, on the other hand, exhibit a different type of nonlinearity (Kernell, 1965), in which the slope of the frequency current relationship is increased at higher levels of excitation. Kernell (1968) has given evidence that this may be caused by the prolonged afterhyperpolarization and the associated increase in membrane conductance. In DSCT neurons, the afterhyperpolarization is smaller than in motor neurons and apparently is caused by a different mechanism, because there is little concomitant increase in membrane conductance (Eide et al., 1969a). Also, when it is considered that their dendrites have largely unknown but probably nonlinear properties, it appears reasonable to suggest that the wide range of linearity exhibited by the DSCT neurons is the

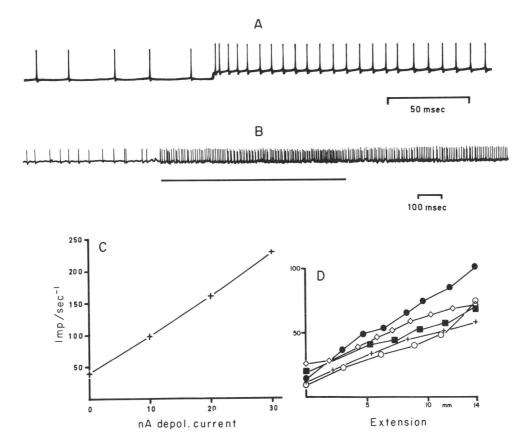

FIGURE 1 Excitation of DSCT cells. A. Activation of neuron by outward transmembrane current step of 10 nA starting at the break of the base line and lasting throughout the record. Intracellular record, spike amplitude about 70 mV. (From Eide et al., 1969a.) B. Activation of neuron by muscle stretch. Linear stretch of triceps muscle up to full physiological extension at a rate of 19 mm/sec in period indicated by line. Full extension maintained throughout rest of record. Extracellular axonal spikes recorded from dorsolateral funicle. (From Jansen et al., 1966.) C. Steady-state firing frequency (ordinate) against intensity of outward transmembrane current (abscissa). Same cell as in A. Observations from experiments reported by Eide et al., 1969a. D. Steady-state firing frequencies (ordinate) against muscle extension (abscissa) for five different second-order neurons, each indicated by separate symbol. All observations from de-efferented preparations.

result of mutual canceling of several nonlinear relationships.

An additional feature of the output of DSCT neurons requires comment. DSCT cells commonly exhibit a highly regular background activity at a rate of about 10 imp/sec even without any specific primary afferent input (Holmqvist et al., 1956; Jansen et al., 1966). The mode of generation of the background discharge is at present unknown. Holmqvist et al., (1956) gave some evidence that it might be due to a low-level synaptic input from spinal interneurons, but, as discussed below, our simulation experiments make this appear doubtful.

## The firing pattern

There is a striking difference between the regular firing pattern of DSCT cells activated by injected current and the irregular firing pattern observed during activation by muscle stretch (Figure 1A, B). Such irregularity of firing is common in central neurons during steady-state activation (Hilali and Whitfield, 1953; Gerstein and Kiang, 1960; Werner and Mountcastle, 1963; Goldberg and Greenwood, 1966). Two hypotheses have been advanced to explain this irregularity of firing. One postulates a noise generator intrinsic to the neuron, such as a random fluctuation in threshold or membrane potential (Geisler and Goldberg, 1966). The second ascribes the irregularity of firing to the discontinous nature of synaptic transmission (Stein, 1965). For the DSCT neurons there is direct evidence in favor of the latter hypothesis. The background firing can be highly regular, and the characteristic irregularity appears immediately when activated synaptically. Accordingly, it appears reasonable that the fine structure of the firing pattern contains information about the synaptic input. A systematic description of the firing pattern is required to determine it. The initial part of this description is the probability of occurrence of intervals of different duration as estimated by the interval histogram. These are usually unimodal and fairly symmetrical (Figure 2A). Their shape, independent of the degree of activation of the cell, is different from that of many other neurons in the absence of an approximately exponential right tail. The degree of irregularity, as measured by the coefficient of variation (CV = SD/mean interval), is independent of the frequency of firing of the cell. This appears from the plot of the standard deviation of the distribution against its mean interval (Figure 2C). The observations are adequately described by a straight line, the slope constant of which gives the CV. The CV varies between approximately 0.3 to 0.6 for DSCT cells activated from the primary ending of muscle spindles. The background firing, on the other hand, may have a CV as low as 0.05 (Jansen et al., 1966).

Another important feature of the firing pattern of the DSCT neurons is the remarkable degree of serial dependency. The value of the serial correlation coefficient (R) is commonly as high as −0.6 to −0.8 for neighboring intervals. This means that there is a tendency for short and long intervals to appear alternatively. We shall return to the possible significance of the serial dependency. The value of R is independent of the level of excitation once the cell is firing above some 20 imp/sec (Jansen et al., 1966).

*Synaptic mechanisms*

Intracellular records provided direct information about the synaptic organization of the DSCT neurons. The first experiments showed that the synaptic transmission to the DSCT cells was qualitatively similar to that of motor neurons (Curtis et al., 1958; Eccles et al., 1961). The high efficiency of transmission, as evidenced by the large amplitude of the compound excitatory postsynaptic potential

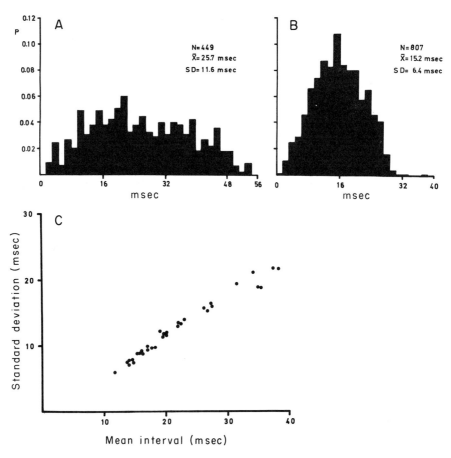

FIGURE 2  Distributions of firing intervals of second-order neuron activated by 5 mm (A) and 14 mm (B) maintained the extension of the anterior tibial muscle. C. Relationship between standard deviation and mean interval of firing at various degrees of maintained synaptic activation. All data from same cell.

(EPSP) to peripheral nerve stimulation, was also a feature of the early observations. In the following, we concentrate on two points of crucial importance for the interpretation of the firing pattern of the cells. These are the synaptic effect of each presynaptic impulse and the number of primary afferent fibers converging onto each second-order neuron.

With a small to moderate load on the appropriate muscle, characteristic short-lasting depolarizations appear in the membrane potential of DSCT neurons (Figure 3). Their rate of appearance increases with increasing load on the muscle, and from their time course they can be safely interpreted as EPSPs evoked by single presynaptic impulses (unitary [u] EPSPs) (Eide et al., 1969b). In favorable situations the u EPSP of a particular presynaptic fiber can be identified and followed by its amplitude and regularity of appearance. The regularity is due to their origin in muscle-stretch receptors, which fire regularly during steady-state activation.

There is a considerable variation in the amplitudes of the u EPSPs evoked from the different primary afferent fibers converging onto the DSCT neurons. The remarkable observation is, however, their large size. The largest u EPSP measured as much as 5 mV, and the average size of all the u EPSPs of Ia afferents to one neuron was 2.7 mV (Eide et al., 1969b). A second important finding was the relative constancy of the time course of the different u EPSPs. Because the period of transmitter liberation appears to be short in these cells (Kuno and Miyahara, 1968), the time course of potential change is determined by the position of the synapses in relation to the recording site, presumably in the cell body. Their constant time course, therefore, suggests that they are due to activity at synaptic terminations which are electronically equidistant from the cell body. The rapid time course and the large amplitudes of the EPSPs are

in agreement with the anatomical observation that primary afferent fibers end as "giant synapses" on the proximal dendrites of DSCT neurons (Szentágothai and Albert, 1955; Rèthelyi, 1968).

At higher levels of stretch, additional units are recruited, at which point the picture of the synaptic activity becomes complicated and the individual units cannot be identified separately (Figure 3). Therefore, the degree of convergence cannot be established in this way. Instead, it was determined from the ratio between the amplitude of the compound EPSP elicited by synchronous activation of all Ia afferents in the muscle nerve and the average size of the u EPSPs in the same neuron. Taking the nonlinear summation of EPSPs into account, we found, in a limited number of cells, 12 to 16 primary afferent fibers converging onto each second-order neuron (Eide et al., 1969b), which suggests that something like 30 per cent of the primary endings of the muscle employed project to each neuron of this type. Although the range of variation is not established, additional data to be discussed suggest that the estimate is reasonable, and we can get a crude check on the observations. The average synaptic current per millimeter of stretch was 0.5 nA (see above). The average increase in firing frequency for each primary ending of the muscle employed is 3.6 imp/sec (Jansen and Matthews, 1962). The total input to a second-order neuron is therefore about 50 imp/sec/mm, and the average amount of charge removed by each presynaptic impulse becomes $0.5 \times 10^{-9}/50 = 10^{-11}$ coulomb. With a membrane capacity of $3 \times 10^{-9}$ farad (Eide et al., 1969a), the average amplitude of the u EPSPs should be approximately 3 mV.

The large u EPSPs of DSCT neurons invite some comments on the mode of termination of the primary afferent fibers. The collaterals of the same Ia afferents terminate on motor neurons as well as on DSCT neurons, which il-

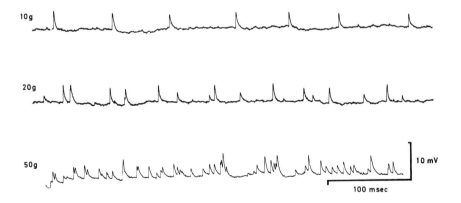

FIGURE 3 Unitary EPSPs of second-order neuron. Intracellular records. Increasing degrees of synaptic activity produced by loads of 10, 20, and 50 gram weight on the tendon of the soleus muscle. (From experiments by Eide et al., 1969b.)

lustrates instructively the differences in synaptic efficiency and coupling on two groups of second-order neurons of entirely different function. The Ia afferents from a particular muscle appear to establish monosynaptic connections with all the homonymous motor neurons (Mendell and Henneman, 1968). The amplitude of their u EPSP is small (about 0.1 mV), because each presynaptic impulse liberates only 2 to 3 quanta of transmitter on the average (Kuno and Miyahara, 1969a). The monosynaptic excitation of heteronymous motor neurons is less efficient on account of a smaller degree of convergence on these. The mean quantal content of the heteronymous u EPSPs is, however, the same as that of the homonymous ones (Kuno and Miyahara, 1969b). On the DSCT neurons, on the other hand, approximately 30 per cent of the Ia afferents appear to converge onto each neuron, and the average quantal content of their u EPSPs appears to be at least 10 times that of the motor neuronal u EPSPs (Eide et al., 1969b). These differences have their morphological correlates in the extent of synaptic contacts at the two types of cells (Szentágothai and Albert, 1955; Haggar and Barr, 1950). The DSCT synapses appear to be constructed for signal transmission, and the motor-neuronal synapses, for signal integration.

From the point of view of the output of the second-order neuron, the amplitude of the u EPSP in relation to the firing threshold of the cell, rather than the absolute amplitude of the u EPSP (Stein, 1965), is the important parameter. The firing threshold of the DSCT cells proved difficult to determine, partly because of the background firing of the cells and partly because the cells are easily injured by microelectrode penetration (Eide et al., 1969b). Even so, observations were obtained suggesting that the simultaneous occurrence of as few as two to three u EPSPs might be sufficient to fire the second-order neuron.

### A model of the second-order neuron

The problem discussed in this section is whether the observed firing pattern of the DSCT neurons can be explained quantitatively from the observed properties of the cells and their input. Various simplified models of nerve cells have been formulated, and their behavior has been explored by simulation on a digital computer (Walløe, 1968; Walløe et al., 1969). With the evidence at hand indicating that the irregularity of firing appeared to be associated with synaptic activation of the cells, the models have all been elaborations of the "quantal excitation" model orginally proposed by Stein (1965). Similar models have also been studied by Segundo et al. (1968).

In outline, the models describe a nerve cell subjected to input impulses, each of which generates a reduction in the membrane potential of the cell with a time course like a

u EPSP. The background activity is introduced by a progressive depolarization of the cell. Background depolarizations and EPSPs are summated linearly. Whenever the depolarization exceeds a certain limit, the nerve cell fires an impulse, and the membrane potential is reset to its original value. Certain additional complexities could be introduced in the models, such as a time-dependent recovery of excitability after each neuronal firing. For a complete description of the models and their behavior, the reader is referred to our original paper (Walløe et al., 1969).

Such models exhibit behavior that also appears to be relevant with respect to the firing pattern of the DSCT neurons. Some features of the behavior of the nerve cells were also commonly found for the models within the range explored, such as an approximately linear increase in firing frequency with increasing input frequency and a linear increase in standard deviation of output intervals with increasing mean interval. Introduction of a time-dependent recovery of excitability or a variation in the amplitude of the EPSPs over a range considered reasonable in relation to the DSCT neurons produced only minor and trivial changes in the behavior of the models. On the other hand, to generate interval distributions and serial dependencies comparable to those of the DSCT required specific models of considerable interest. The critical feature turned out to be the time structure of the input process. With a number of afferent fibers converging onto a nerve cell, the distribution in time of the combined input activity will approximate that of a Poisson process, if independent activity in each of the input fibers is assumed. However, with a Poisson distribution of the input activity it was impossible to generate interval distributions similar to those of the DSCT neurons. The distributions obtained always contained exponential right tails. Negative serial dependencies were introduced in such models by making the speed of recovery of excitability after a neuronal firing an increasing function of the duration of the preceding interval. The effect of this mechanism was explored over a range that exceeded the range observed experimentally (Eide et al., 1969a), but it did not influence the shape of the interval distributions appreciably. Furthermore, the degree of negative serial dependency obtained in this way was much less than that found in the discharge pattern of the DSCT cells ($R_{12} \approx -0.2$ compared with $-0.6$ obtained experimentally). Small degrees of negative serial dependency exist, however, in the discharge pattern of several other types of neurons (Kuffler et al., 1957; Poggio and Viernstein, 1964; Goldberg and Greenwood, 1966; Firth, 1966), which may well be explained by such an accumulation of subnormality (Geisler and Goldberg, 1966).

A more realistic model, in which the input activity was generated by a number of independent "elements" with properties similar to those of the primary endings of muscle spindles, exhibited a strikingly different behavior. The important characteristics of these "muscle spindles" were that their firing rate increased linearly with simulated "muscle length" (Matthews, 1964), and that their firing pattern at any given length was highly regular (CV = 0.04; Stein and Matthews, 1965). Even though the total input generated by a number of such "spindles" had an exponential distribution and a CV $\approx$ 1, and thus was distinguishable from a Poisson process in these respects, the second-order neuron now generated output distributions closely similar to those of the DSCT neurons (Figure 4). Furthermore, with such "spindle" input there were strong negative serial dependencies in the output of the model. First-order serial correlation coefficients as high as −0.5 were found even in models with a time-independent recovery of excitability. Thus, two qualitative conclusions are reached. The particular shapes of the interval distributions and a large part of the serial dependency of the DSCT neurons appear to be due to a particular time structure of the input of these cells. The serial dependencies were generally slightly smaller in models with a realistic "spindle" input than in the DSCT neuronal discharge. With the additional mechanism of accumulation of subnormality as outlined above, the $R_{12}$ increased to values observed in the DSCT neurons. Because summation of afterhyperpolarization and subnormality takes place in DSCT neurons (Eide et al., 1969a), it appears reasonable to conclude that this latter factor also contributes to their behavior.

Certain additional aspects of the behavior of the models were of some interest. As expected intuitively, the CV of the output intervals increased with increasing amplitude of the input EPSPs. In order to obtain a variability similar to that of the DSCT neurons (CV = 0.40) the EPSPs had to be approximately 50 per cent of the firing threshold of the cell. Increasing the amplitude of the EPSP also increased the slope of the input-output relationship of the model. As mentioned above, the frequency-length slopes of the DSCT neurons are approximately the same as those of the corresponding muscle-spindle afferents. Accordingly, the steady state "gain factor" can be expressed as 1/N, in which N is the number of input fibers converging onto a DSCT cell. Now, with the restriction on the size of the EPSPs required to generate the observed variability, one obtains an estimate of N required to obtain output frequencies and slopes in the observed range.

Admittedly this type of estimate is relatively crude. It was found, however, that models with 10 to 20 input "spindles" generated output frequencies similar to those of the DSCT cells. The degree of serial dependency also depended on the size of the input EPSPs and the number of input fibers. Here, again, it appeared that the models with 10 to 20 input fibers and EPSPs of half-threshold amplitude produced output intervals with serial dependencies like those of the DSCT neurons. Accordingly, the simulated models were able to generate an output activity closely approximating that of the DSCT cells with respect to interval distributions, frequency ranges, and serial dependencies, with parameters (EPSP amplitudes, number of input fibers, and recovery functions) reasonably similar to those determined experimentally.

If we accept the model, some considerations on the mode of generation of the background activity of the DSCT neurons may be justified. As we have mentioned, the background activity is usually highly regular in the absence of a primary afferent input to the cells (CV as low as 0.04). Originally, it was suggested that the background activity might be due to an excitatory synaptic input from interneurons of the spinal cord (Holmqvist et al., 1956). To account for the low rate of the background activity (10 imp/sec), the postulated EPSPs must have a correspond-

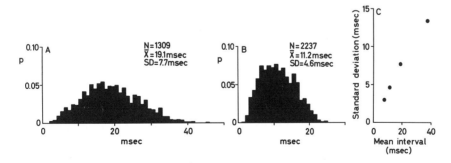

FIGURE 4 Behavior of neuronal model described in text. A and B. Distribution of output intervals at two different levels of excitation. C. Relationship between standard deviation and mean interval in the same experiment. Model consists of 15 independent "muscle spindles." Amplitude of unitary EPSPs, 3/5 of threshold; time constant of decay, 3 milliseconds. Exponential time course of recovery after firing with time constant determined by duration of the preceding interval. (From Walløe, 1968).

ingly slow time course, but this might be explained by the assumption of a peripheral dendritic position for the inter-neuronal synapses. However, considered within the framework of the present model for the behavior of the DSCT neurons (Walløe et al., 1969), it was found that something between 20,000 and 60,000 imp/sec, each producing EPSPs of correspondingly small amplitude, were required to generate the background rate of firing with its high degree of regularity. This must be a highly inefficient mechanism for producing a 10 imp/sec firing rate. Alternatively, it was suggested (Walløe et al., 1969) that the background firing of the DSCT cells might be caused by a genuine pacemaker mechanism intrinsic to the cell itself.

A second prediction based on the behavior of the model concerns a group of units activated from the secondary endings of a muscle and with ascending axons among the DSCT fibers in the dorsolateral funicle of the cord. Their behavior is similar to that of the DSCT neurons activated from primary endings, except for a more regular pattern of firing (Jansen et al., 1967a). The CV of the interval distributions of their steady-state activity may be as low as 0.2. Our data on such neurons are more limited, but if it is assumed that they are activated as suggested above, the model behavior suggests that they are excited by 30 to 40 fibers generating unitary EPSPs with mean amplitudes slightly less than one-fifth of the threshold of firing.

To conclude, the study of model behavior has permitted the formulation of a reasonable hypothesis of the generation of the firing pattern of DSCT neurons, and it has given a basis for predictions of important parameters in simple neuronal chains based on their observed output behavior.

Evidence from other types of neurons indicates that their irregularity and pattern of firing may be due to mechanisms similar to those of the DSCT neurons. The observations on the eccentric cell of the *Limulus* eye are most instructive (Ratliff et al., 1968). In the dark-adapted state, this cell fires with a high degree of irregularity, and there are prominent random fluctuations in membrane potential as a result of the quantal nature of the light stimulus. In the light-adapted state, these fluctuations are greatly reduced, and the firing pattern of the cell is much more regular. The cell firing to a steady transmembrane outward current is characterized by a high degree of regularity (Ratliff et al., 1968). Calvin and Stevens (1968) have shown that the firing pattern of motor neurons is explained by the synaptic noise recorded in the same cells. But even though it appears that simple models of the synaptic excitation, such as the model proposed by Stein (1965), may be sufficient to explain the behavior of many simple neurons, more complicated situations can easily be visualized. For instance, in a nerve cell with multiple spike-generating regions the mode of functioning might be entirely different.

## An assessment of transfer efficiency

Considerable importance is attached to the development of a quantitative measure of the efficiency of synaptic transmission (Perkel and Bullock, 1968). Information theory as developed by Shannon (1948) provides a possible approach, and its application to nerve signals has been explored by Stein (1967b). The application of information theory requires a measure of the degree of correspondence between input and output signals. Frequency coding is assumed and appears reasonable (Figure 1D). At the second-order level of the DSCT, the input is the mean frequency of the spike trains of the primary afferent fibers. The mean frequency, measured over a short period of time, of the second-order neuron represents the output. The problem then is to determine the distribution of output signals for any possible input signal. Different methods have been employed on the signals of DSCT neurons (Walløe, 1968). Because it has not been possible to measure the input and output simultaneously, all assume that the firing irregularity of the second-order neuron to a given steady input signal is representative of the uncertainty of the output signal to this input. Therefore, the problem is reduced to how accurately the mean frequency of firing of the second-order neuron can be determined. Because of the irregularity of firing, this accuracy obviously increases with increasing "observation time."

One method was similar in form to that of Werner and Mountcastle (1965), and was based on the estimation of the transinformation in an input-output matrix of contingent probabilities. The second method depends on the increasing accuracy of the estimate of mean frequency with increasing observation time. For the details of the methods employed, the reader is referred to the original publication (Walløe, 1968).

Both methods gave transinformations increasing logarithmically with observation time, and qualitatively similar results were obtained by both. Some quantitative differences can probably be accounted for by the various approximations involved.

Certain findings were of particular interest. It appeared that the transinformation decreased significantly (from 1 to 1.5 bits at any observation time) if the discharge trains were read in a randomly permutated sequence instead of in their original order. The DSCT transmission accordingly provides an example of the possible importance of the detailed time structure of spike trains for the efficiency of signal transmission, even when the signal is read as a frequency code.

The further question of a possible physiological significance of this property of the discharge pattern requires, first, that the receiver of the signal reads the activity of only one such nerve fiber. If several fibers with independent activity converged onto the third-order neuron, the time structure of each fiber's signal would largely be lost. Therefore, it appears significant that the DSCT axons terminate in the rather unique synaptic structures called cerebellar glomeruli. Each glomerulus consists of the processes of several third-order neurons, all ending in relation to the end bulb of only one mossy fiber (Ramón y Cajal, 1911).

For the DSCT transmission, input as well as output signals are reasonably well known. It is, therefore, possible to make rough estimates of the amount of information that is lost during the synaptic transfer. Stein (1967b) has derived expressions that are convenient for this comparison, and his formulas give results that are reasonably close to those of the methods outlined above. With an observation time of 200 milliseconds, each primary afferent fiber transmits about 4.6 bits if one assumes a frequency range of 10 to 80 imp/sec, like that experimentally observed. The main reason for the relatively high information content is the great regularity of firing of these primary afferent fibers (CV = 0.04; Stein and Matthews, 1965). About 15 primary afferent fibers converge onto each second-order neuron. They all carry the same kind of information (i.e., muscle length), but are firing independently of one another, so the second-order neuron will receive a quantity of information equivalent to that of a 15-times-longer record of the signal from one fiber. This is estimated to about 6.6 bits in 200 milliseconds. The signal of the second-order neuron contains about 3.0 bits over a frequency range equal to that of each input fiber. Accordingly, about half of the information is lost during the synaptic transfer.

The information content of the second-order signal is even smaller than that of each primary afferent fiber. Two points are worth mentioning in this connection: first, that some 30 per cent of what is transmitted is caused by the particular time structure of the second-order signal and, second, that the greater part of the loss is caused by the convergence of several primary afferent fibers. The estimation that the total content of the input is equal to a 15-times-longer record of one primary afferent assumes that the regularity of firing is preserved in each fiber. Alternatively, if the input signals are regarded as randomly distributed and covering a frequency range of 15 times that of each input fiber, the information content of the total input is approximately the same as that of the second-order neuron.

It is interesting that this high cost of synaptic transmission is found in a highly effective system like the DSCT. The justification for the introduction of synapses at this high cost must be the possible control of the transfer process, for instance by inhibition (see below).

The present estimates consider the transmission to each second-order neuron only. There is some justification for this from the evidence that the second-order signals are read individually (see above). When the system is considered as a whole, however, the information lost during transmission from first- to second-order neurons is less, because each primary afferent fiber presumably projects to several second-order neurons. The degree of divergence at the first- to second-order level is unknown at present.

By comparing psychophysical estimates of stimulus intensity and the signals of primary afferent fibers from cutaneous receptors, Werner and Mountcastle (1968) found that there is sufficient information in the signal of a single afferent fiber to account for the magnitude estimates. In view of the appreciable loss of information even in the powerful DSCT synapses, it can safely be suggested that human behavior can be explained only if one assumes the participation of a considerable number of neurons at all stages of the relevant pathway. The quantitative correspondence between subjective stimulus estimates and the signal of a single primary afferent (Werner and Mountcastle, 1968) may therefore be largely fortuitous.

## The effects of inhibition

A further illustration of the usefulness of the model appears when the effects of inhibitory inputs on the firing pattern of DSCT neurons are analyzed. Inhibition of neurons relaying signals from the anterior tibial muscle and the flexor digitorum muscle was obtained regularly by repetitive electrical stimulation of group I afferent fibers in the gastrocnemius-soleus nerve (Jansen et al., 1969). An example is shown in Figure 5A. After an initial transient, the inhibition is maintained for the entire duration of the inhibitory input, which permitted a study of the effect of "steady-state" inhibition on the firing rates and discharge pattern of the DSCT neurons.

Two contrasting effects were observed in different neurons. For some, the inhibitory effects measured as the reduction in firing frequency were independent of the degree of excitation of the cells (Figure 5B). These neurons showed no appreciable change in firing pattern, as evidenced by their interval histograms, CVs (Figure 6C), and serial dependencies (Jansen et al., 1969). In other neurons, a constant inhibitory input caused increasing reduction in firing frequency at higher levels of excitatory drive (Figure 5C). This type of behavior was associated with an increased regularity of firing (Figure 6D) and a reduced serial dependency of the discharge. However, all degrees of transitions were found between these two extremes of behavior. Thus, there was nothing in the experimental data to suggest that basically different mechanisms might be involved.

For the discussion of possible synaptic mechanisms underlying the inhibitory effects, our model of the DSCT neuron is accepted tentatively. Postsynaptic (Eccles et al., 1961; Hongo and Okada, 1967) as well as presynaptic

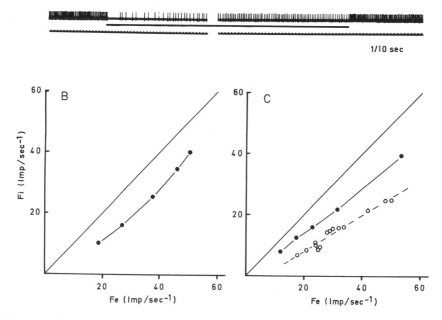

FIGURE 5 Inhibition of transmission in second-order neurons by stimulation of peripheral nerves. A. Sample record (7 seconds cut out from middle of continuous record). Neuron activated by steady extension of anterior tibial muscle. Period of inhibitory repetitive stimulation of triceps nerve indicated by line. B. Firing frequencies during maintained inhibition, Fi (ordinate), plotted against firing frequencies in control period, Fe (abscissa). Unit activated by varying degrees of maintained stretch of flexor digitorum longus muscle and inhibition elicited by 100-per-second stimulation of triceps nerve. C. As B, another neuron. Closed circle: data obtained during 50-per-second stimulation of triceps nerve. Open circle: data obtained during 100-per-second stimulation of triceps nerve. (A, B, and C from experiments reported by Jansen et al., 1969.)

(Eccles et al., 1963) inhibition of DSCT neurons after activation of spinal nerves have been described. To consider the simplest situation first, presynaptic inhibition should, by definition, reduce the size of the EPSPs without altering the properties of the postsynaptic neuron. In the model, the result would be an increased regularity of firing and a linear reduction in the slope of the input-output relation of the neuron. Accordingly, a presynaptic mechanism cannot explain the inhibition with an unchanged slope-constant and firing pattern. Furthermore, the inhibitory input regularly terminated the background firing of the DSCT neurons (Jansen et al., 1969), and this effect cannot be caused by presynaptic inhibition, as the background firing appears to reflect inherent pacemaker activity of the second-order neurons (see above).

Postsynaptic inhibitory mechanisms can, on the other hand, explain the observed effects. A postsynaptic inhibition, which acts predominantly by hyperpolarizing the cell, should not change the size of the EPSPs appreciably (Eide et al., 1969a). Accordingly, the firing pattern and the slope of the input-output relation should remain unchanged. If, however, the inhibition is associated with an increased conductance that reduces the input impedance of the cell, the amplitudes of the EPSPs will be reduced in proportion to the reduction in input impedance. Consequently, input-output slopes will be reduced, and the firing pattern will be changed in the observed direction. Thus, the simplest explanation for the observed inhibitory effects is that they are due to a postsynaptic inhibition causing various degrees of inhibitory "shunting" in the different DSCT neurons. If the axons of the inhibitory interneurons involved terminated more or less peripherally on the dendrites of DSCT neurons, such effects would be expected, which is in line with Rèthelyi's recent anatomical description of Clarke's column synaptology (1968). The postsynaptic nature of the inhibition is also suggested by the similar effects of the inhibition of transmission evoked from the somatosensory cortex (Jansen et al., 1969). This inhibition has been shown to be entirely postsynaptic (Hongo and Okada, 1967).

The observations on inhibition of signal transmission in the DSCT illustrate two important inhibitory effects. The

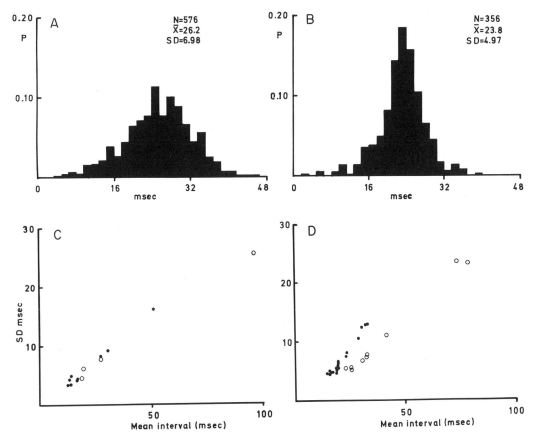

FIGURE 6 Firing patterns during maintained inhibition. A and B. Interval distributions of neuron, showing more regular firing during inhibition. A. Distribution during maintained activation of neuron by 50-gram-weight load on tibialis anterior tendon. B. Same cell as A. Excited by a 150-gram load on muscle tendon and inhibited simultaneously by repetitive nerve stimulation. Mean frequency of firing is approximately the same as in A. Inhibition caused a reduction in firing frequency of 29 imp/sec from preceding control period. C, D. Data from two different units. Scatter diagrams of standard deviation against mean intervals during control periods (closed circle) and during inhibitory nerve stimulation (open circle). (Data from Jansen et al., 1969.)

reduction of input-output slope during inhibition represents a reduction of the "gain" of transmission. Inhibition with maintained input-output slope, on the other hand, does not affect the gain. Comparable effects have been obtained for both the peripheral inhibition of crayfish receptors and lateral inhibition in *Limulus* eyes (Terzuolo et al., 1968). The behavior of motor neurons is also of interest in this connection. Postsynaptic inhibitory inputs always reduce the frequency of firing of these cells, usually without any change of slope of the input-output relation (Granit and Renkin, 1961; Granit et al., 1966), even though the inhibitory input commonly reduces the input resistance of the cell. According to Kernell (1969), the explanation is that the afterhyperpolarization is the dominating mechanism that controls the firing frequency of motor neurons.

## Conclusions

The signal transmission of the DSCT exhibits several properties that may also be relevant for other central nervous pathways. The most important is the wide range of linearity of transmission from first- to second-order neurons. This width of range is due to the linear current-frequency relation of the DSCT neurons and to the linear summation of their excitatory synaptic input. These properties and the time structure of presynaptic activity determine the firing pattern of the postsynaptic neuron. The particular firing pattern generated may significantly improve the efficiency of information transfer. Postsynaptic inhibitory mechanisms, which partly hyperpolarize and partly reduce the input impedance of the second-order neurons, explain inhibitory effects on the signal transmission. With its relatively simple organization and experimental accessibility, the DSCT appears to be a suitable system for further analysis of signal transmission in the central nervous system.

## REFERENCES

ALNAES, E., 1967. Static and dynamic properties of Golgi tendon organs in the anterior tibial and soleus muscles of the cat. *Acta Physiol. Scand.* 70: 176–187.

BROWN, M. C., and R. B. STEIN, 1966. Quantitative studies on the slowly adapting stretch receptor of the crayfish. *Kybernetik* 3: 175–185.

CALVIN, W. H., and C. F. STEVENS, 1968. Synaptic noise and other sources of randomness in motoneuron interspike intervals. *J. Neurophysiol.* 31: 574–587.

CURTIS, D. R., J. C. ECCLES, and A. LUNDBERG, 1958. Intracellular recording from cells in Clarke's column. *Acta Physiol. Scand.* 43: 303–314.

ECCLES, J. C., O. OSCARSSON, and W. D. WILLIS, 1961. Synaptic action of group I and II afferent fibres of muscle on the cells of the dorsal spinocerebellar tract. *J. Physiol. (London)* 158: 517–543.

ECCLES, J. C., R. F. SCHMIDT, and W. D. WILLIS, 1963. Inhibition of discharges into the dorsal and ventral spinocerebellar tracts. *J. Neurophysiol.* 26: 635–645.

EIDE, E., L. FEDINA, J. JANSEN, A. LUNDBERG, and L. VYKLICKÝ, 1969a. Properties of Clarke's column neurones. *Acta Physiol. Scand.* 77: 125–144.

EIDE, E., L. FEDINA, J. K. S. JANSEN, A. LUNDBERG, and L. VYKLICKÝ, 1969b. Unitary components in the activation of Clarke's column cells. *Acta Physiol. Scand.* 77: 145–158.

ENGER, P. S., J. K. S. JANSEN, and L. WALLØE, 1969. A biological model of the excitation of a second order sensory neurone. *Kybernetik* 6: 141–145.

FIRTH, D. R., 1966. Interspike interval fluctuations in the crayfish stretch receptor. *Biophys. J.* 6: 201–215.

GEISLER, C. D., and J. M. GOLDBERG, 1966. A stochastic model of the repetitive activity of neurons. *Biophys. J.* 6: 53–69.

GERSTEIN, G. L., and N. Y.-S. KIANG, 1960. An approach to the quantitative analysis of electrophysiological data from single neurons. *Biophys. J.* 1: 15–28.

GOLDBERG, J. M., and D. D. GREENWOOD, 1966. Response of neurons of the dorsal and posteroventral cochlear nuclei of the cat to acoustic stimuli of long duration. *J. Neurophysiol.* 29: 72–93.

GRANIT, R., D. KERNELL, and Y. LAMARRE, 1966. Algebraical summation in synaptic activation of motoneurones firing within the "primary range" to injected currents. *J. Physiol. (London)* 187: 379–399.

GRANIT, R., D. KERNELL, and G. K. SHORTESS, 1963. The behaviour of mammalian motoneurones during long-lasting orthodromic, antidromic and trans-membrane stimulation. *J. Physiol. (London)* 169: 743–754.

GRANIT, R., and B. RENKIN, 1961. Net depolarization and discharge rate of motoneurones, as measured by recurrent inhibition. *J. Physiol. (London)* 158: 461–475.

HAGGAR, R. A., and M. L. BARR, 1950. Quantitative data on the size of synaptic end-bulbs in the cat's spinal cord. *J. Comp. Neurol.* 93: 17–35.

HILALI, S., and I. C. WHITFIELD, 1953. Responses of the trapezoid body to acoustic stimulation with pure tones. *J. Physiol. (London)* 122: 158–171.

HOLMQVIST, B., A. LUNDBERG, and O. OSCARSSON, 1956. Functional organization of the dorsal spinocerebellar tract in the cat. V. Further experiments on convergence of excitatory and inhibitory actions. *Acta Physiol. Scand.* 38: 76–90.

HONGO, T., and Y. OKADA, 1967. Cortically evoked pre- and postsynaptic inhibition of impulse transmission to the dorsal spinocerebellar tract. *Exp. Brain Res.* 3: 163–177.

HOUK, J., and W. SIMON, 1967. Responses of Golgi tendon organs to forces applied to muscle tendon. *J. Neurophysiol.* 30: 1466–1481.

JANSEN, J. K. S., and P. B. C. MATTHEWS, 1962. The effects of fusimotor activity on the static responsiveness of primary and secondary endings of muscle spindles in the decerebrate cat. *Acta Physiol. Scand.* 55: 376–386.

JANSEN, J. K. S., K. NICOLAYSEN, and T. RUDJORD, 1966. Discharge pattern of neurons of the dorsal spinocerebellar tract activated by static extension of the primary endings of muscle spindles. *J. Neurophysiol.* 29: 1061–1086.

JANSEN, J. K. S., K. NICOLAYSEN, and T. RUDJORD, 1967a. On the firing pattern of spinal neurones activated from the secondary endings of muscle spindles. *Acta Physiol. Scand.* 70: 188–193.

JANSEN, J. K. S., K. NICOLAYSEN, and L. WALLØE, 1967b. On the inhibition of transmission to the dorsal spinocerebellar tract by stretch of various ankle muscles of the cat. *Acta Physiol. Scand.* 70: 362–368.

JANSEN, J. K. S., K. NICOLAYSEN, and L. WALLØE, 1969. The firing pattern of dorsal spinocerebellar tract neurones during inhibition. *Acta Physiol. Scand.* 77: 68–84.

JANSEN, J. K. S., R. E. POPPELE, and C. A. TERZUOLO, 1967c. Transmission of proprioceptive information via the dorsal spinocerebellar tract. *Brain Res.* 6: 382–384.

JANSEN, J. K. S., and T. RUDJORD, 1964. On the silent period and Golgi tendon organs of the soleus muscle of the cat. *Acta Physiol. Scand.* 62: 364–379.

JANSEN, J. K. S., and T. RUDJORD, 1965. Dorsal spinocerebellar tract: Response pattern of fibers to muscle stretch. *Science (Washington)* 149: 1109–1111.

KERNELL, D., 1965. High-frequency repetitive firing of cat lumbosacral motoneurones stimulated by long-lasting injected currents. *Acta Physiol. Scand.* 65: 74–86.

KERNELL, D., 1968. The repetitive impulse discharge of a simple neurone model compared to that of spinal motoneurones. *Brain Res.* 11: 685–687.

KERNELL, D., 1969. Synaptic conductance changes and the repetitive impulse discharge of spinal motoneurones. *Brain Res.* 15: 291–294.

KUFFLER, S. W., R. FITZHUGH, and H. B. BARLOW, 1957. Maintained activity in the cat's retina in light and darkness. *J. Gen. Physiol.* 40: 683–702.

KUNO, M., and J. T. MIYAHARA, 1968. Factors responsible for multiple discharge of neurones in Clarke's column. *J. Neurophysiol.* 31: 624–638.

KUNO, M., and J. T. MIYAHARA, 1969a. Non-linear summation of unit synaptic potentials in spinal motoneurones of the cat. *J. Physiol. (London)* 201: 465–477.

KUNO, M., and J. T. MIYAHARA, 1969b. Analysis of synaptic efficacy in spinal motoneurones from 'quantum' aspects. *J. Physiol. (London)* 201: 479–493.

LUNDBERG, A., 1964. Ascending spinal hindlimb pathways in the cat. *Progr. Brain Res.* 12: 135–163.

MATTHEWS, P. B. C., 1964. Muscle spindles and their motor control. *Physiol. Rev.* 44: 219–288.

MENDELL, L. M., and E. HENNEMAN, 1968. Terminals of single Ia fibers: Distribution with a pool of 300 homonymous motor neurons. *Science (Washington)* 160: 96–98.

MOUNTCASTLE, V. B., 1967. The problem of sensing and the neural coding of sensory events. *In* The Neurosciences: A Study Program (G. C. Quarton, T. Melnechuk, and F. O. Schmitt, editors). The Rockefeller University Press, New York, pp. 393–408.

OSCARSSON, O., 1965. Functional organization of the spino- and cuneocerebellar tracts. *Physiol. Rev.* 45: 495–522.

PERKEL, D. H., and T. H. BULLOCK, 1968. Neural coding. *Neurosci. Res. Program Bull.* 6 (no. 3): 221–348.

POGGIO, G. F., and L. J. VIERNSTEIN, 1964. Time series analysis of impulse sequences of thalamic somatic sensory neurons. *J. Neurophysiol.* 27: 517–545.

RAMÓN Y CAJAL, S., 1911. Histologie du Système Nerveux de l'Homme et des Vertébrés. A. Maloine, Paris, vol. 2, 993 pp.

RATLIFF, F., H. K. HARTLINE, and D. LANGE, 1968. Variability of interspike intervals in optic nerve fibers of *Limulus*: Effect of light and dark adaptation. *Proc. Nat. Acad. Sci. U. S. A.* 60: 464–469.

RÈTHELYI, M., 1968. The Golgi architecture of Clarke's column. *Acta Morphol. Acad. Sci. Hung.* 16: 311–330.

SEGUNDO, J. P., D. H. PERKEL, H. WYMAN, H. HEGSTAD, and G. P. MOORE, 1968. Input-output relations in computer-simulated nerve cells. *Kybernetik* 4: 157–171.

SHANNON, C. E., 1948. A mathematical theory of communication. *Bell Syst. Tech. J.* 27: 379–423.

STEIN, R. B., 1965. A theoretical analysis of neuronal variability. *Biophys. J.* 5: 173–194.

STEIN, R. B., 1967a. The frequency of nerve action potentials generated by applied currents. *Proc. Roy. Soc., ser. B, biol. sci.* 167: 64–86.

STEIN, R. B., 1967b. The information capacity of nerve cells using a frequency code. *Biophys. J.* 7: 797–826.

STEIN, R. B., and P. B. C. MATTHEWS, 1965. Differences in variability of discharge frequency between primary and secondary muscle spindle afferent endings of the cat. *Nature (London)* 208: 1217–1218.

SZENTÁGOTHAI, J., and A. ALBERT, 1955. The synaptology of Clarke's column. *Acta Morphol. Acad. Sci. Hung.* 5: 43–51.

TERZUOLO, C. A., R. L. PURPLE, E. BAYLY, and E. HANDELMAN, 1968. Postsynaptic inhibition—Its action upon the transducer and encoder systems of neurons. *In* Structure and Function of Inhibitory Neuronal Mechanisms (C. von Euler, S. Skoglund, and U. Söderberg, editors). Pergamon Press, Oxford and New York, pp. 261–275.

WALLØE, L., 1968. Transfer of signals through a second order sensory neuron. Institute of Physiology, University of Oslo, Oslo, doctoral thesis.

WALLØE, L., J. K. S. JANSEN, and K. NYGAARD, 1969. A computer simulated model of a second order sensory neuron. *Kybernetik* 6: 130–141.

WERNER, G., and V. B. MOUNTCASTLE, 1963. The variability of central neural activity in a sensory system, and its implications for the central reflection of sensory events. *J. Neurophysiol.* 26: 958–977.

WERNER, G., and V. B. MOUNTCASTLE, 1965. Neural activity in mechanoreceptive cutaneous afferents: Stimulus-response relations, Weber functions, and information transmission. *J. Neurophysiol.* 28: 359–397.

WERNER, G., and V. B. MOUNTCASTLE, 1968. Quantitative relations between mechanical stimuli to the skin and neural responses evoked by them. *In* The Skin Senses (D. R. Kenshalo, editor). Charles C Thomas, Springfield, Illinois, pp. 112–137.

# 56 Some Principles of Synaptic Organization in the Visual System

## OTTO D. CREUTZFELDT

Nobody would attempt to describe and define within any practical amount of space the general concept of analogy which dominates our interpretation of vision. There is no basis for saying whether such an enterprise would require thousands or millions or altogether impractical numbers of volumes. Now it is perfectly possible that the simplest and only practical way actually to say what constitutes a visual analogy consists in giving a description of the connections of the visual brain. . . . It is . . . not at all unlikely that it is futile to look for a precise logical concept, that is, for a precise verbal description, of "visual analogy." It is possible that the connection pattern of the visual brain itself is the simplest logical expression or definition of this principle.

JOHN VON NEUMANN

THE PRESENTATION OF the following material and the discussion of some of its implications may be considered out of place within the context of "Communication and Coding in the Nervous System," but I selected this material with the notion that principles of such coding are best demonstrated by analyzing the principles of connectivity in the nervous system. I believe that the thoughts of John von Neumann, quoted above, express very well the approach I favor in this context. One of the main aspects of coding within the nervous system is the spatial arrangement of its elements, the relations between them, and the combination of signals arriving at any given ganglion cell and at groups of ganglion cells from the different input sources of the periphery. This combination of inputs is the reason for the apparently increasing complexity of response characteristics and stimulus parameters of neurons at the higher levels of the nervous system, and emphasizes the intimate link between structure and function. A further reason for this presentation is that one of our aims should be to discover principles of input combinations based on mechanisms that force some types of neurons into excitatory or inhibitory contacts with certain neurons but not with others. My aim is to point to some of these principles within the visual system, although I realize that such an

analysis of the "functional Bauplan" of the nervous system is only at its starting point. Therefore, this aspect may be taken mainly as a stimulus for further research.

One of the basic electrophysiological principles of information transmission is that the output discharge frequency of a ganglion cell is linearly related to the current passing across the soma membrane (Hodgkin, 1948; Jansen and Walløe, this volume.) This statement is limited, however, because the excitability of a ganglion cell may change during a longer-lasting suprathreshold excitation (adaptation: Creutzfeldt et al., 1964; Granit et al., 1963). Furthermore, nonlinearities in the neuronal input-output relation may result from nonlinear properties of postsynaptic potential (PSP) summation and variable anatomical relationships between the input area (dendrite) and the trigger zone.

## Methods

The experiments were done on cat retina, lateral geniculate body, and visual cortex (area 17). In all cases the animals were fixed in a head frame and faced a tangent screen placed 1 meter from the eye. The pupils were dilated with synephrine-atropine and the corneas were covered with contact lenses of 0.00, +1.0 or +1.5 diopter. We conducted the retinal experiments under a continuous $N_2O$-anesthesia, most of the geniculate experiments after a single Nembutal injection of 30 mg/kg, and the cortical experiments after ether anesthesia. In all experiments we anesthetized the fixation points and wounds locally. A Flaxedil-curare mixture was applied continuously through an intravenous cannula. Body temperature and artificial respiration were controlled carefully.

Retinal activity was recorded from single optic-tract fibers with stereotactically introduced steel or tungsten electrodes. Geniculate and cortical units were recorded with glass micropipettes in an intracellular or "quasi-intracellular" electrode position. In all cases, only the skull and the dura were removed, at the place of electrode introduction, and during recording the hole in the skull was covered around the electrode.

Light stimuli of varying form, intensity, and exposure

OTTO D. CREUTZFELDT Department of Neurophysiology, Max-Planck-Institute for Psychiatry, Munich, Germany

frequency were projected on the tangent screen. Such projection could be done by computer-controlled stimulators and with on-line computer analyses of the responses during the experiments, or with a separate programing device and off-line analysis via tape recording. On-line analysis was done with an IBM 1130 system developed in our laboratory (Färber and Probst, 1968); off-line analysis, either with the IBM 1130 or with a CAT 1000. Photographic recordings could be taken from the tape.

## Results

FUNCTIONAL ORGANIZATION AT THE RETINAL LEVEL
The retina performs linear and nonlinear operations. I describe some of these and, by using appropriate experiments as examples, try to localize them within the different layers. Figure 1 is a schematized drawing of the different levels of the retina. The structure, as revealed by anatomical work (Polyak, 1941; Dowling and Boycott, 1966; Boycott and Dowling, 1969; Brown, 1965; Brown and Major, 1966; Leicester and Stone, 1967), suggests that the activity of many receptors is integrated by one bipolar cell, and that

several bipolar cells converge on one ganglion cell. Only relatively few bipolar and ganglion cells have an anatomical arrangement compatible with the one-to-one connection between a single receptor and a single ganglion cell as found in the primate retina (Boycott and Dowling, 1969). We have never recorded from cat ganglion cells, the properties of which suggested such a one-to-one relationship with only one receptor. Hence, all ganglion cells whose properties are discussed sum activity from many receptors. In what follows, I am assuming that more than one bipolar cell is in contact with each ganglion cell.

The summation area of a ganglion cell is called its receptive field center (RFC). It is surrounded by an area (the receptive field surround [RFS]) the stimulation of which leads to an effect that is the reverse of the center-stimulus effect. Excitation of receptors within the RFC may lead to an excitation of the underlying ganglion cell (on-center cells) or to an inhibition (off-center cells). In the latter case, the darkness after the light stimulus leads to an acceleration.

Let us assume that the size and shape of a ganglion cell RFC are determined by the distribution of its dendrites. Anatomical studies show that the distribution is in one plane

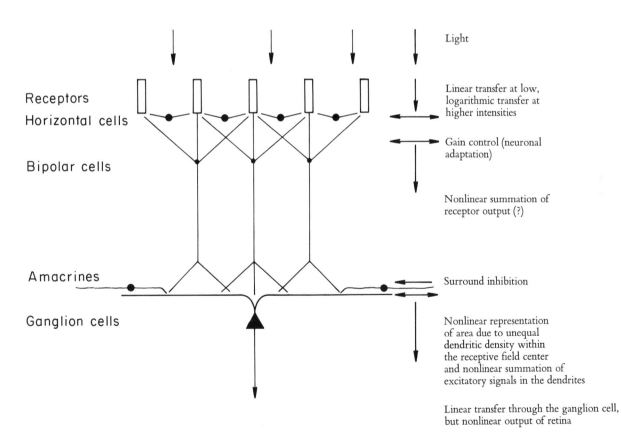

FIGURE 1   Schema of the retina. The integrative and transfer functions of the different levels are described at right. Horizontal arrows: Spatial integration, interaction, or both. Vertical arrows: Input into and transfer through the various elements. (See text.)

(Figure 2A) and that the dendritic "domains" are almost round (Brown, 1965; Brown and Major, 1966; Leicester and Stone, 1967), with diameters of between 80μ and 800μ, as determined from Golgi impregnations. These correspond to visual fields of about 0.5 degree to 4 degrees in diameter (Leicester and Stone, 1967).

Let us further assume that each receptor that lies above a dendritic branch of an on-center ganglion cell has an excitatory contact with the latter (through a bipolar cell, of course), and that synapses are distributed nearly uniformly along the dendrites. The sensitivity of that ganglion cell to a small stimulus within its RFC should be determined not by the total number of receptors illuminated but by the number of illuminated receptors that are in contact with the cell (Figure 2B–D). The spatial sensitivity of retinal ganglion cells should then be a function of the spatial density of their dendrites.

If one measures the area density of ganglion-cell dendrites within small areas at variable distances from the soma, one gets curves that are similar to, although somewhat more irregular than, the sensitivity distribution as determined in the physiological experiment (Figure 3). The differences can be corrected either by introducing a factor which takes into account the electrotonic attenuation of excitatory post-synaptic potentials (EPSPs) along the dendrites, or by assuming that each bipolar cell makes several contacts with one dendritic tree within the area of the domain of bipolar branching (20μ–40μ in diameter, Leicester and Stone, 1967), or by both. At the periphery of the domain of a dendritic tree, only few or one contact with the bipolar cells situated there can be expected. The effect of electrotonic attenuation can be estimated by the weighting factor $g(x) = \frac{1}{\cosh x/\lambda}$ in which x is the dendritic length and λ the length constant (Creutzfeldt et al., 1968). For a dendritic length constant of 100μ–150μ, the curves expected theoretically, based on histology, are virtually identical to experimental sensitivity curves (Figure 3, page 634).

This model can be tested. If a logarithmic summation of light quanta in the receptor-bipolar cell complex and a linear summation of excitation at the ganglion cell are assumed, the response (R) of the ganglion cell is related to the intensity (I) of a stimulus at different locations within the receptive field by $R = c \cdot f(1/r) \cdot \log I$. The factor $f(1/r)$ depends on the dendritic density and the electrotonic attenuation as a function of the stimulus distance (r) from the RFC middle. In a semilogarithmic plot, the slope of the response-intensity curves should then vary according to the factor $f(1/r)$. Figure 4 shows that within a distance of 1 degree from the middle of the receptive field the slopes are identical, indicating only small changes of $f(1/r)$ in this part of the RFC. At greater distances the curves are more horizontal, indicating a decrease of $f(1/r)$.

As suggested by the PST histograms and as discussed else-where, this effect can hardly be due to increased lateral inhibition from the surround. It agrees with the model and indicates fewer excitatory contacts at the periphery of the RFC than within its center. The plateau in the middle of the RFC (which is also found in the threshold experiments of Cleland and Enroth-Cugell, 1968) supports the view that here the overlap of bipolar branches compensates for electrotonic attenuation. The dendritic density in this central part of the RFC does, in fact, change relatively little, as shown in Figures 2 and 3.

If the ganglion cell should sum excitatory inputs linearly, one would expect that the response to round stimuli concentric to the RFC would be identical to the sum of responses to small spots of light within the larger stimulus area. They should then be proportional to the spatial sensitivity distribution along the radius. Such is, in fact, the case for stimuli that are close to threshold (Cleland and Enroth-Cugell, 1968) and for weak suprathreshold stimuli, as shown in Figure 5 (circles in A). The plateau of the sensitivity curve of most neurons near the center of the RFC explains that, for a restricted central area of the receptive field, Ricco's law (I × A = constant) is applicable. But if stronger stimuli are used, the ganglion-cell responses are considerably smaller than those calculated (Figure 5, page 636, triangles in A), indicating a nonlinear summation of excitation signals within the ganglion cell. One can investigate this result further by recording the responses to stimuli of variable diameter and intensity.

If the excitatory signals arriving at the ganglion cell were proportional to log I and were summed up linearly, the responses (R) should be of the form $R = f(A) \times \log I$. As in the first experiment, the slope of the intensity-response curves in the log plot should change considerably with the stimulus area (A). But such is not the case, and the slopes rise only slightly with increasing area (Figure 6, page 637). A completely parallel course of the intensity-response curves would suggest a logarithmic summation of incoming signals in the ganglion cell, whether they result from increase of area or increase of intensity. The ganglion-cell response would correspond to $R = c \times \log(I \times A) = c \times (\log I + \log A)$, which would imply a linear signal transmission in the receptors. The light flux I × A for evoking equal responses at different intensities and stimulus areas should then also be equal. Consequently, Ricco's law would also be applicable to suprathreshold stimuli.

Such a possibility is also excluded. The equal-response lines (broken lines in Figure 6), which would be parallel to the 45-degree lines (that correspond to log I + log A = constant), become more horizontal as the responses become larger, which indicates a higher efficiency of a low-intensity stimulus which covers a large area, than that of a high-intensity stimulus which covers a small area. Hence, the receptor-bipolar complex is more sensitive at low than at

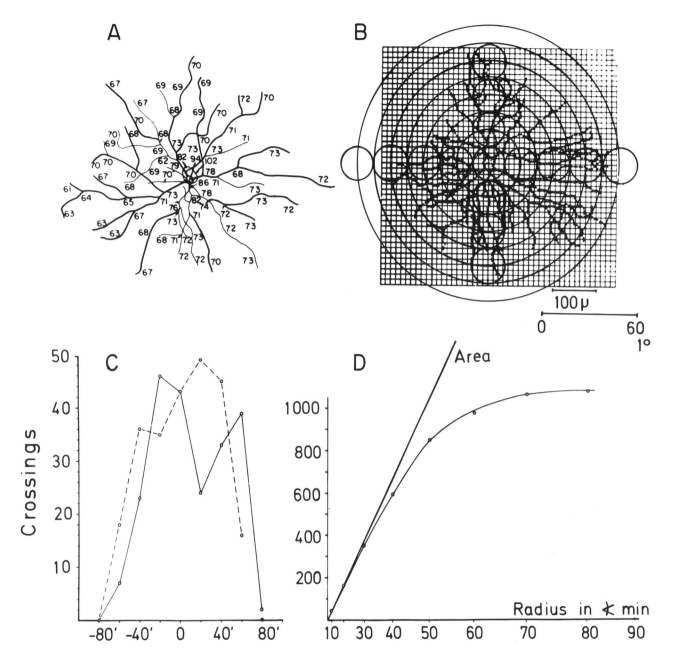

FIGURE 2 Dendritic density of retinal ganglion cells. A. The dendritic field of a rat retinal-ganglion cell, seen from above. The numbers on the dendrites indicate the depth in microns from the inner retinal surface. Golgi-Cox preparation (Brown, 1965). B. Dendritic field of cat retinal-ganglion cell (Brown and Major, 1966). A grid is projected over the dendritic tree. Each square corresponds to the area of about four receptors. Whenever a dendrite passes through a quarter of one square (receptor), a dot is made. The small and large circles indicate the retinal areas covered by light spots in the following experiments. C. Number of "dendrite-receptor crossings" (ordinate, see B) which are covered by the small circles (simulated light stimuli) at different distances from the center of the receptive field (abscissa). Solid line: Simulated "stimuli" along the horizontal axis. Broken line: Simulated stimuli along the vertical axis. D. Cumulative number of dendrite-receptor crossings (ordinate) covered by disks (stimuli) of increasing radius (abscissa). The straight line represents the increase in stimulus area. (From Creutzfeldt et al., 1968.)

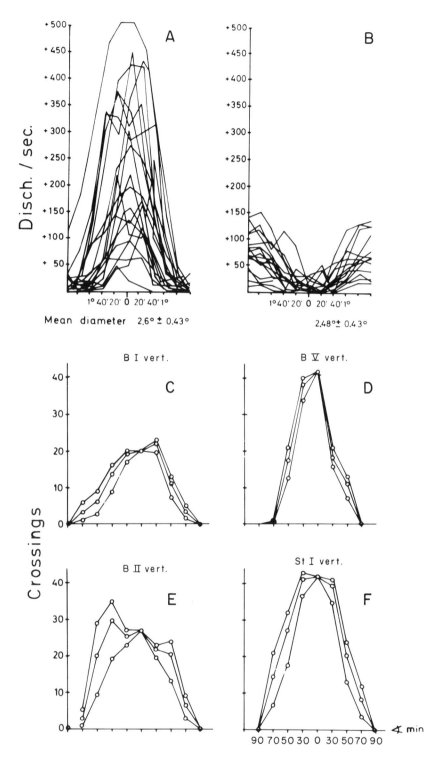

FIGURE 3 Response profile (A, B) and dendritic-density distribution across the receptive field of retinal ganglion cells. A, B. Responses of retinal ganglion cells to small suprathreshold spots of light (diameter 10 min ⋊) shone into different parts of the receptive field along one axis of the receptive field. Ordinate: Discharges per second during the first 120 milliseconds of the response. A. On-center neurons. B. Off-center neurons. Below each set of curves are indicated the mean RFC diameters of 19 retinal on-center and 13 retinal off-center neurons, as determined from the sensitivity curves. C–F. The dendritic-density distribution of four retinal ganglion cells as determined by the method explained in Figure 2B and C. The diameter of the simulated stimuli corresponds to that of the stimuli used in the experiments of A and B. Distance from middle of RFC is on the abscissa (values in F). The counts were done on Golgi preparations of retinal-ganglion cells (C–E), published by Brown and Major (1966), and (F) Leicester and Stone (1967). The upper curves in each diagram represent the original counts of dendritic receptor crossings; the two inner curves are the result of weighting the counts according to dendritic-length constants ($\lambda$) of 200 and 100 $\mu$ (see text). (From Creutzfeldt et al., 1968.)

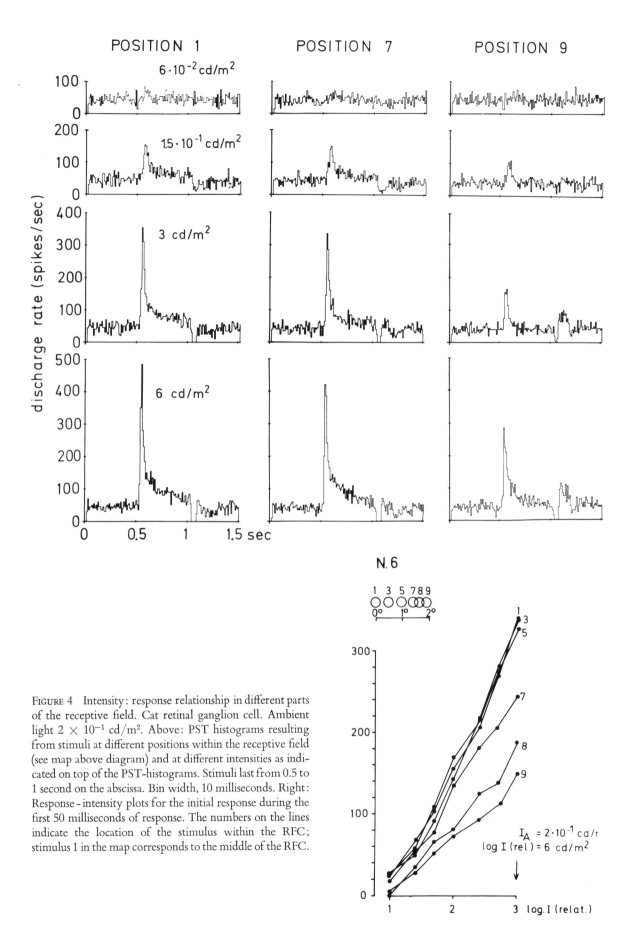

FIGURE 4  Intensity: response relationship in different parts of the receptive field. Cat retinal ganglion cell. Ambient light $2 \times 10^{-1}$ cd/m². Above: PST histograms resulting from stimuli at different positions within the receptive field (see map above diagram) and at different intensities as indicated on top of the PST-histograms. Stimuli last from 0.5 to 1 second on the abscissa. Bin width, 10 milliseconds. Right: Response-intensity plots for the initial response during the first 50 milliseconds of response. The numbers on the lines indicate the location of the stimulus within the RFC; stimulus 1 in the map corresponds to the middle of the RFC.

high light intensities, and thus does not transmit the intensities linearly. At the largest stimulus diameter, the equal-response lines turn upward toward higher light intensities, because the peripheral parts of these large stimuli already cover receptors that are not in excitatory contact with this ganglion cell.

The spatial sensitivity distribution and the area summation experiments suggest more than one mechanism for nonlinear signal summation within the RFC: (1) nonlinear signal summation in the receptors; (2) nonlinear sensitivity distribution of ganglion cells because of their dendritic geometry; and (3) nonlinear summation of excitatory signals in the ganglion cell. If our earlier statement on linear transmission of ganglion cells is also valid for retinal ganglion cells, we would have to locate the nonlinearity in the dendrites: simultaneous excitation of many neighboring synapses might lead to conductivity changes of the den-

dritic membrane and thus to nonlinear input-output summation properties. Another possibility of nonlinear transmission within retinal ganglion cells was discussed by Stone and Fabian (1968), who assumed that the relative refractoriness of the initial segment increases the threshold for as long as 20 milliseconds. As yet we have had no indication that an active inhibitory process peripheral to the ganglion cell is responsible for the nonlinear signal transmission in the RFC, as was suggested by Büttner and Grüsser (1968).

So far, I have discussed retinal nonlinearities caused by retinal-cell geometry and by nonlinear summation of excitatory signals in receptors and ganglion-cell dendrites. A further nonlinearity of spatial and intensity transfer is brought about by lateral inhibition from the receptive-field surround. I shall not go into the problem of lateral inhibition and contrast perception, as this is discussed elsewhere (Reichardt, this volume; for further information, see

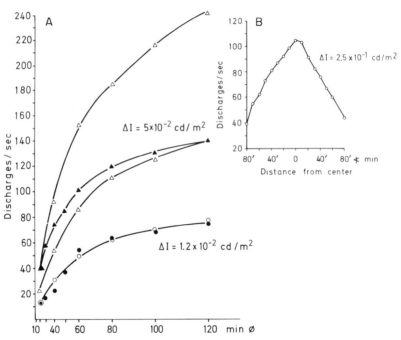

FIGURE 5 Experimental responses of a cat retinal-ganglion cell to stimuli of variable diameter (closed symbols) compared with expected responses (open symbols) calculated from the sensitivity distribution in different parts of the receptive field (figure on right). A. Responses (discharge rate during first 100 milliseconds of response, ordinate) as function of stimulus diameter (in minutes of arc, ordinate) at low intensity (closed circles) and high intensity (closed triangles). The two intensities ΔI are indicated on the side. Open circles: Expected responses for low intensity as calculated from the sensitivity distribution shown in B and from the response to the smallest spot. Open triangles: Expected

responses to high-intensity stimuli. The response to the largest stimulus is taken as the basis for calculation of the lower curve, and the response to the smallest spot is taken as the basis for the upper curve. B. Sensitivity distribution as determined by responses (ordinate, discharge rate during first 100 milliseconds of response) to small spots (10 minutes in diameter) shone into different parts of the receptive field (abscissa, distance from middle of RFC). The diagram shows the average of four stimulus series along different axes through the receptive field. This explains the regular "pyramidal" shape of the sensitivity distribution.

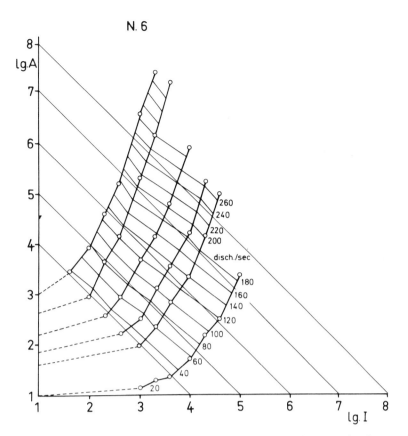

FIGURE 6   Effect of increase of stimulus intensity and size on response of a retinal on-center ganglion cell of cat. These response-intensity plots are for spots of light with various stimulus diameters concentric to the RFC. Each open circle is the average from 10 stimulus applications (discharge rate during first 50 milliseconds of response). Abscissa is the log of the intensity, and the discharge rates are plotted against variable linear ordinates the zero points of which, for each stimulus size, are at the intersection of the broken lines with the vertical log-A axis. This intersection also indicates the logarithm of the surface (log area) of the stimulus with which the connected stimulus-response curve was made. The numbers 20 to 260 are discharge rates. The thin lines connecting equal abscissa and ordinate values represent the $\log I + \log A = $ constant, and therefore respond to Ricco's law ($I \times A = $ constant). The thicker lines, which have negative slopes and go from 40 to 260 per second, connect equal responses of the different stimulus series. These lines run roughly parallel to $\log I + \log A = $ constant for only small responses (e.g., 40 and 60), but not for larger ones (e.g., 160 and 180). Those, being more horizontal, indicate that at larger diameters less intensity is needed to produce a given response than at small diameters. Only at the largest diameter (top curve) do the equal response curves turn up again because the stimulus is larger than the RFC.

Ratliff, 1965). Another effect of lateral inhibition in the retina is that it controls the background activity of on-center neurons at various light intensities. If diffuse light is shone on the retina, the discharge rate of on-center units increases and that of off-center units decreases only when relatively low intensities ($10^{-4}$ to $10^{-5}$ cd/m²) are used, i.e., only in the scotopic and lower mesopic range. At higher intensities, many on-center cells are slowed, whereas only a few are accelerated up to values of $10^{-3}$ to $10^{-2}$ cd/m² (candella per square meter) and then level off. This variable behavior results from the strength of the inhibitory surround, which, if strong, will slow the neuron discharge. Because on-center neurons increase their maintained discharge rates only up to intensities of $10^{-2}$ cd/m² and decrease at greater intensities and because most off-center units are completely inhibited above a light intensity of $10^{-3}$ to $10^{-4}$ cd/m², a decrease of the mass activity in the optic tract results, as Arduini and Pinneo (1962) observed by using large electrodes. This makes one doubt that the perception of ambient light is transmitted by the increased firing frequency of a large neuronal population.

Finally, if retinal transfer functions are discussed, the regulation of retinal sensitivity due to adaptation must be included. In the photopic range, the adaptation is mainly photochemical and is proportional to the amount of bleached photopigment. In the scotopic and mesopic range, a neuronal adaptation mechanism has been proposed (Rushton, 1965). It is assumed that the excitation of the receptors is pooled and changes the gain of the receptor-bipolar signal transmission. Recent measurements of this gain control showed that it is not proportional to the intensity I of the adapting light, but to $I^n$ (with a mean value of n = 0.6; Sakmann and Creutzfeldt, 1969). In the scotopic-mesopic range, the sensitivity of the retina does not, therefore, decrease in proportion to I as implied in the Weber-Fechner law ($\Delta I/I$ = constant), but to $I^{0.6}$. The Weber-Fechner relation is found only in the photopic range with photochemical adaptation.

The localization of the neuronal adaptation mechanism can only be approximated: if one plots the response of a ganglion cell to small light stimuli shone into the center of the receptive field against the logarithm of intensity, the result is a straight line over an intensity range of about 1.5 log units. It levels off at the lower and upper limits because of the retinal threshold sensitivity and the maximum discharge rate that a ganglion cell can maintain for a certain length of time (initial 50–100 milliseconds of the response). The curves obtained at different (scotopic and mesopic) adaptation levels run parallel, but the intensity range of the ganglion-cell response from threshold to the upper limit (1.5 log units) is independent of adaptation (Figure 7). The response (R) versus intensity (I) curves are described by R = a × log I; a determines the slope of the curves and is independent of the adaptation level. Therefore, the gain-control mechanism must be situated peripherally from—or in the same—elements in which the logarithmic summation of light intensity takes place. If the log transformation of intensity takes place in the receptors, adaptation must directly influence the gain of the receptors. It could be localized either within the receptors themselves or in such a way that it could change receptor sensitivity by a feedback control through the horizontal cells. But if the nonlinear summation of the retina is largely a function of the ganglion cells, the gain-control mechanism may be localized anywhere between the receptor and the input site of the ganglion cell.

Figure 7 also demonstrates the improvement of relative sensitivity at higher adaptation levels. Two lines are drawn through the responses at equal $\Delta I/I$ fractions. If the sensitivity $\Delta I/I$ would not change, the lines should be parallel to the abscissa, but, in fact, they show an inclination.

From the experiments reported, it is clear that the retina is a nonlinear signal transmitter, although under certain conditions and within certain limitations, linear behavior can be demonstrated (Hughes and Maffei, 1966). I have tried to demonstrate how the different components of these non-linearities might be analyzed, and even tentatively be local-

ized, in the different levels of the retina, as shown in Figure 1. As yet, we cannot localize some mechanisms satisfactorily. Whether this is because of an incorrect model or a combination of similar mechanisms at different levels must be determined by further experiments.

## Synaptic organization of transmission in the lateral geniculate body (LGB)

Synaptic organization in the LGB differs from that in the retina. Here, the retinocortical flow is interrupted by relay cells, and retinocortical impulses come under the influence of activity from other brain structures, as well as under some inhibitory influence from the contralateral retina. I do not discuss this latter aspect here (for further references, see Creutzfeldt and Sakmann, 1969); we shall deal mainly with the organization of the retinocortical relay.

Receptive fields of geniculate cells differ little from those of the retina (Hubel, 1963), especially the receptive-field centers, which, in our experience, are the same size as those of retinal ganglion cells. If one records intracellularly or quasi-intracellularly from geniculate relay cells, one can isolate single EPSPs (McIlwain and Creutzfeldt, 1967). These EPSPs are of large amplitude (up to 5–10 mV) and of equal shape. Their amplitude variation is within the limits

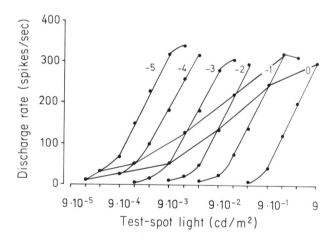

FIGURE 7 Intensity-response plots of a retinal on-center ganglion cell at different adaptation levels. The response (ordinate) is the discharge rate during the first 100 milliseconds. The intensity of the test spot is shown on the abscissa. The spot (10 minutes in diameter) is shone into the middle of the RFC. The numbers on the response-intensity curves indicate the exponent of the brightness of the adapting light (−5 means adapting light = $10^{-5}$ cd/m²). The two lines through the plots are drawn through points of equal $\Delta I/I$. If $\Delta I/I$ results in a constant response (Weber's law), the lines should be horizontal. See text for further details. (After Sakmann and Creutzfeldt, 1969).

that can be attributed to the fluctuations of the membrane potential (see Figure 8). Their frequency during spontaneous and evoked activity is equal to that found in single retinal ganglion cells under the same conditions. From these and other findings, we concluded that one geniculate relay cell receives its specific retinal excitatory input from only one optic-tract fiber (Creutzfeldt, 1968a; Singer and Creutzfeldt, 1970). Because optic-tract fibers divide into several branches before they terminate in brushlike endings (Guillery, 1966), the implication is that—at least in the foveal and parafoveal area—one tract fiber must innervate several relay cells and that the number of relay cells should be greater than the number of optic-tract fibers (for further discussion, see Singer and Creutzfeldt, 1970).

Electrical stimulation of the optic tract reveals that on-center as well as off-center fibers primarily excite LGB relay cells (Fuster et al., 1965). This excitatory contact must be very powerful, as suggested by the large amplitude of the EPSPs. The anatomical basis of such synaptic connections has been described by Szentágothai et al. (1966). The significance of such a synaptic organization for information

transmission is obvious. The duration of one EPSP is about 20 milliseconds (with an early linear decay), and the threshold of the cell is about twice the amplitude of an EPSP. Therefore the probability of a threshold depolarization depends on the EPSP rate and pattern, as is shown schematically in Figure 9, in which a random discharge rate of the afferent fiber (input) that elicites the EPSPs is assumed. Such a distribution is found in the optic tract (Herz et al., 1964). The input-output relationship depends on the mean discharge rate of the afferent fiber and improves as the latter increases (Figure 9A–C). Also, the transmission latency decreases with the discharge rate (Figure 9D). Above a maximal input rate, the output frequency of the cell is limited by its own maximal firing rate. Therefore, the transinformation of such a transmitting system is optimal at an intermediate discharge rate of the afferent input. The synaptic "noise" at a lower frequency is not transmitted.

The receptive field of geniculate relay cells has a powerful inhibitory surround, both because of the relative lack of excitation during illumination of the surround area of the retinal afferent fiber and because of an active inhibition

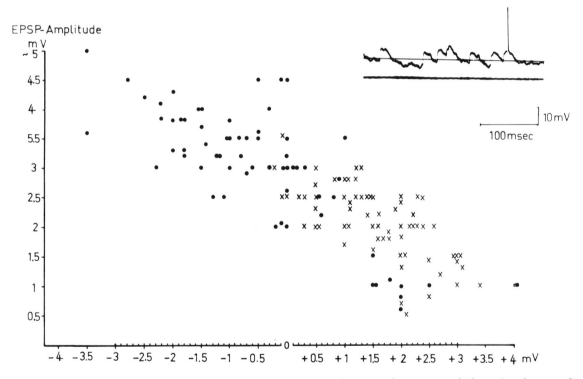

FIGURE 8  EPSPs from one optic-tract fiber in a lateral geniculate relay cell (on-center cell). Cat, Nembutal. Inset shows original record from a quasi-intracellular recording. The horizontal line through the record indicates the resting potential (membrane potential when no PSP activity is present). Diagram: Amplitude of EPSPs (ordinate) plotted against relative membrane potential (abscissa) at the start of the EPSP. The membrane potential was measured relative to the resting level as shown in the record. Closed circles, EPSPs during darkness; crosses, EPSPs during light stimulation. (Courtesy of Dr. W. Singer.)

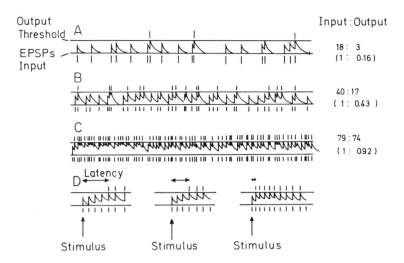

FIGURE 9 Schematic drawing of input-output relation of a cell that receives its excitatory input from one fiber. It is assumed that the fiber discharges at irregular intervals and that the duration of the EPSP is short in relation to the average interval when the discharge rate of the fiber is low (A), but long compared with the average interval at high-discharge rate (C). D. Input:output latency as a function of discharge rate of the fiber response. (From Creutzfeldt, 1968b.)

(McIlwain and Creutzfeldt, 1967). An analysis of this inhibitory input showed that LGB on-center cells are inhibited by off-center neurons and off-center cells by on-center neurons (Singer and Creutzfeldt, 1970). This conclusion is based mainly on the amplitudes and latencies of the inhibitory and excitatory components, seen after focal stimulation in different parts of the receptive field. This analysis suggested the model of LGB synaptic organization that is shown in Figure 10. It is not clear whether the inhibitory input comes directly from retinal on-center or off-center fibers or from geniculate relay cells via recurrent collaterals. Electrical stimulation of the optic radiation shows that recurrent inhibition exists in the geniculate, and the computer simulation of the geniculate responses gave the best results if recurrent inhibition was assumed. But evidence is not yet sufficient for hypothesizing exclusively a recurrent mechanism. A simulation is shown in Figure 11.

The model suggests an important principle of synaptic organization in the LGB. On-center cells of the LGB receive excitatory input only from on-center fibers of the optic tract and inhibitory input only from the off-center fibers, which implies a high degree of specificity in the selection of afferent connections. The same is true of off-center cells, which receive excitation exclusively from off-center fibers and inhibition from on-center fibers. This specificity suggests that the contact of a geniculate relay cell with one type of afferent fiber (on or off) may determine the cell type and therefore render it selective for further contacts with fibers (interneurons or recurrent collaterals) of another type. A chemical determination of this selectivity could be an attractive approach for verification.

It is not obvious how this synaptic organization of the LGB aids in information transmission. Some functions may be discussed. The active inhibition guarantees a cut-off of nonspecific, and perhaps corticofugal, excitatory

FIGURE 10 Schematic drawing of the synaptic organization in the LGB. A. On-center cell. B. Off-center cell. Top: Sketch of the assumed synaptic connection. The on-center cell is shown receiving excitation from one retinal on-center ganglion cell, and inhibition from a number of off-center ganglion cells. Possible inhibitory interneurons were omitted for simplicity. *Mutatis mutandis* for the off-center LGB cell in B. The black spots in the receptive field of the LGB cells indicate the positions of light stimuli shown below. Lower part: Responses of the assumed input fibers. A. Excitation of the on-center fiber leads to EPSPs; excitation of off-center fibers leads to IPSPs. The approximate sum of EPSPs and IPSPs at the different positions of the stimulus is shown in the smooth curves below the semischematic drawings of fiber activity. B. Off-center cell. Activation of the retinal off-center fiber leads to EPSPs; activation of on-center fibers to IPSPs. Bottom: Light stimulus. The vertical lines show the latencies of the various responses (short-latency excitation at light on in A and at light off in B; short latency inhibition at light off in A and at light on in B; and so on). (From Singer and Creutzfeldt, 1970).

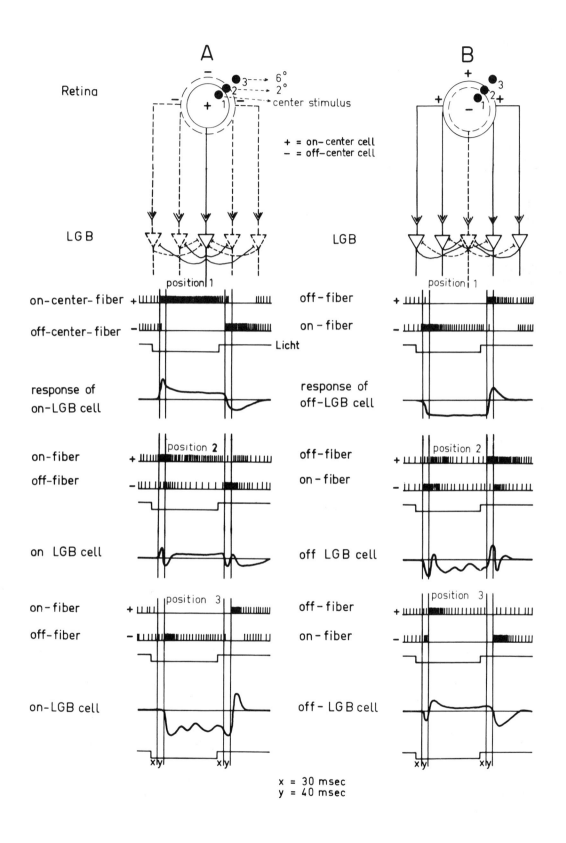

A

Retina

$+$ = on-center cell
$-$ = off-center cell

LGB

position 1

on-center-fiber $+$
off-center-fiber $-$
Licht

response of
on-LGB cell

position 2

on-fiber $+$
off-fiber $-$

on LGB cell

position 3

on-fiber $+$
off-fiber $-$

on-LGB cell

B

LGB

position 1

$+$
$-$

off-fiber
on-fiber

response of
off-LGB cell

position 2

off-fiber $+$
on-fiber $-$

off LGB cell

position 3

off-fiber $+$
on-fiber $-$

off-LGB cell

x|y|  x|y|

x = 30 msec
y = 40 msec

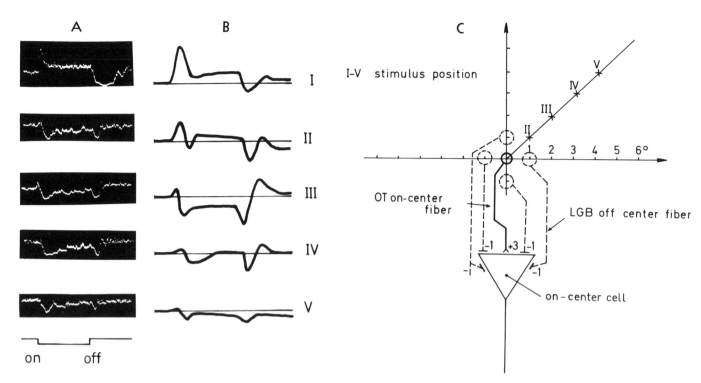

FIGURE 11 Simulation of the response of an on-center cell of the LGB to light spots shone into different parts of its receptive field. A. Original averaged responses recorded with a quasi-intracellular electrode. The stimulus positions I–V (top to bottom) correspond to the sketch in C. The responses are averaged from 10 identical stimuli. B. Simulated responses based on the model shown in Figure 10 and drawn schematically in C. The excitatory input is the activity of one retinal on-center ganglion cell. The inhibitory input is the weighted sum of four geniculate off-center cells (broken lines in C). Same time scale in A and B. (From Singer and Creutzfeldt, 1970).

influences when light is not shone into the RFC. Furthermore, an improvement in the discrimination of successive brightness contrasts could result from the diminution of inhibition simultaneously with the increase of excitatory input (disinhibition). No significant improvement of spatial-contrast discrimination, however, is brought about by the proposed mechanism.

*Cortical synaptic organization*

In the visual cortex we find yet another principle of synaptic organization. Again one must distinguish between the primary afferents (geniculocortical pathway or optic radiation) and those from such other origins as other cortical areas, other visual structures, and nuclei not primarily concerned with vision ("nonspecific" systems). Only the primary specific input is considered here.

Cortical EPSPs are much smaller than geniculate relay cell EPSPs, which are about half of the threshold. The ongoing synaptic bombardment, present without stimulation, keeps the membrane potential fluctuating between 5 and 10 mV below threshold and only rarely reaches the latter

(Creutzfeldt and Ito, 1968). For this reason, the "spontaneous" rate of most cortical cells is low. If an excitatory afferent fiber is activated by light, the temporal summation to reach threshold takes 30–60 milliseconds, even if the fiber discharges as fast as 150 to 200 times per second. Thus, the response latency to a light stimulus does not say much about the number of synapses lying between the afferent fiber and the cortical cell. Because various afferents with different functional properties converge on single cortical cells, stimuli in different parts of the receptive field may result in excitation, inhibition, or disinhibition. The time course of this disinhibition is even slower than that of direct excitation. The question of linear or nonlinear transmission of cortical cells is of little significance in the context of the organization of their synaptic input. A linear relation between membrane potential and discharge frequency is also of minor importance for transfer properties of cortical nerve cells, because nonlinear interaction of EPSPs and IPSPs must be expected on the dendrites in a manner that so far has defied a simple experimental or theoretical approach (Creutzfeldt et al., 1969).

The many receptive-field shapes of neurons in the primary visual cortex (Hubel and Wiesel, 1962, 1968) are caused by the numerous combinations of afferent fibers impinging upon them. Intracellular recordings show that excitation or inhibition of cortical neurons may originate from geniculate on-center or off-center fibers (Creutzfeldt and Ito, 1968). The question of whether the inputs are monosynaptic or polysynaptic remains open. The input combinations of excitatory and inhibitory on-center and off-center fibers varies from cell to cell, but nearby cells often have a similar or an identical input.

Figures 12 and 13 show examples of the intracellular analysis of the receptive fields of two cortical cells. In the cell in Figure 12, the central excitatory field (stimuli 3, 4, 8, 14) is flanked by two inhibitory fields (stimuli 1, 2, 6, and 10, 11). The excitatory field is caused by the excitatory input from an on-center fiber, the inhibitory fields by the inhibitory inputs from two on-center fibers with different receptive-field centers. The responses of this cell could be simulated in a computer model (Figure 12D) when only these three input fibers with the field centers a, b, and c, as shown in the stimulus map, are assumed. This cell would respond maximally to a horizontal bar of light that does not touch the inhibitory flanks.

Figure 13 shows the more complicated field of a neuron that was directionally sensitive to movement. The excitatory responses were maximal to stimuli 12, 13, 6, and 7. Inhibitory-on responses were elicited from areas 10 and 11, and a less powerful excitatory-on response, interrupted by a strong inhibition, from area 9. The functional behavior of this neuron, in spite of its complexity, could be simulated when the convergence of two excitatory on-center fibers with centers between stimulus 12 and 13, an inhibitory on-center fiber with a center between 10 and 11, and an excitatory off-center fiber with its center in the lower left quadrant of the stimulus map are assumed. This neuron responded best to a stimulus moving from right to left (Figure 14, page 646). When the stimulus was moved from left to right, inhibitory potentials (arrow in C) from the inhibitory-on area near 11 interrupted the excitation brought about by the excitatory field between 12 and 13. In many cells, however, the directional movement sensitivity was the result of excitatory connections with off-center fibers; the sensitivity was maximal when a dark stimulus entered the field and when a light stimulus left the field of this off-center input.

The most interesting conclusion of our intracellular analysis was that the responses of most simple visual cortex cells (Hubel and Wiesel, 1962) could be explained by the convergence of only a few afferent fibers with relatively simple response patterns, i.e., probably of direct geniculate origin. Often the resulting fields were such that lines of light in a certain direction were maximally effective (as most neurons in the studies of Hubel and Wiesel, 1962 and

1968). The exact anatomical basis of such an organization is still uncertain, as we cannot say from which cell type (pyramidal or Golgi cells) the records were taken. In any case, it indicates that a few afferent fibers provide the major synaptic input and, thus, the basis for the functional behavior of cortical cells. On the other hand, the similarity of certain aspects of some central receptive fields with regard to stimulation of both eyes (especially the responses to moving stimuli, as shown by Bishop, this volume) suggests that the selection of a limited number of afferents may not be the only determinant of the functional organization of cortical cells, because it would then be necessary to make the improbable assumption that the same types of afferents from both eyes are selected by the same cells. The "modular shape" of the dendritic tree of these neurons may also play a role. Further experimental data are necessary before such questions can be clarified.

Intracellular analysis of visual cortex cells has led us to the following conclusions (Creutzfeldt and Ito, 1968; Creutzfeldt, 1968b): The combination of inputs to a cell explains the existence of certain "optimal" stimulus shapes. The number of stimuli influencing each cortical cell and the variability of receptive fields in different cells make unlikely the hypothesis of an organization in which one particular cell "recognizes" a specific pattern, i.e., acts as a line, movement, or directionality detector, and conveys this information to the next group of neurons. It is true that, at first glance, this organization suggests a decoding mechanism with separate, well-defined functions for different cell groups. However, the within-cell and between-cell variety of functional combinations is difficult to force into such a scheme, and if the living cortex actually operated in this manner the available information "seen" by the eye would be reduced considerably. In fact, computers with pattern-recognition programs based on such concepts "recognize" very little. In short, we are unable to interpret the functional organization of the visual cortex in terms of behaviorally relevant information analysis. We know only that the cortex is necessary for visual-pattern analysis, but we lack sufficient knowledge about the behaviorally important signals extracted and transmitted by the visual system and about what actually "reaches" the animal.

## Summary

This paper discusses some principles of synaptic organization in the retina, the lateral geniculate body (LGB), and the visual cortex and their implications for the transfer properties of the various parts of the visual system.

The supported view is that the receptive-field center of a retinal on-center ganglion cell is determined by the distribution of its dendrites. The sensitivity distribution within the receptive-field center corresponds to the distribution of

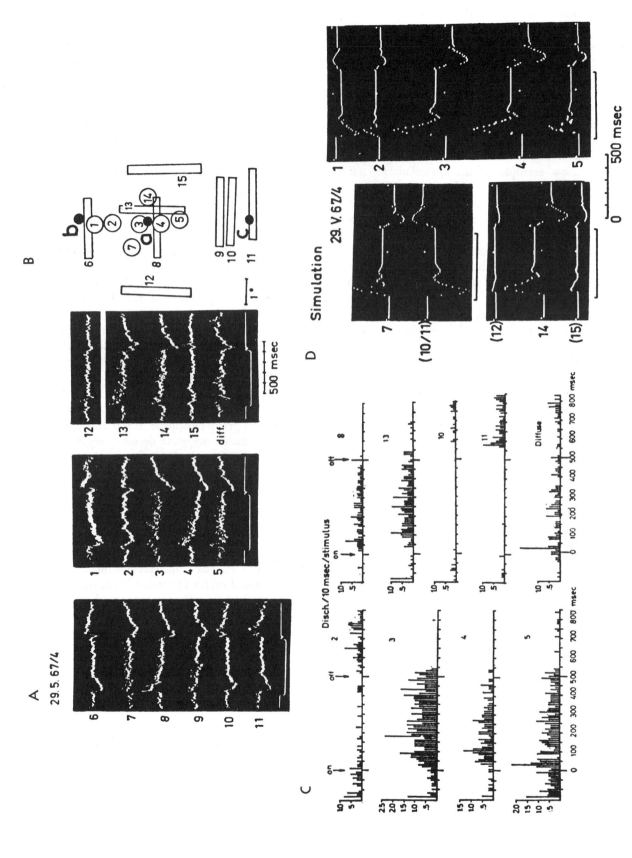

FIGURE 12 Response of a cortical cell (area 17) to spots of light shone into different parts of its receptive fields. A. Averaged slow potential responses (sum of EPSPs and IPSPs) from 10 successive stimuli shone into different parts of the receptive field. The numbers on the records correspond to the stimulus positions on the stimulus map in B. C. PST histograms of some stimuli. D. Computer simulation of the cell responses, based on the assumption that the excitatory input comes from one geniculate on-center fiber situated at A on the B diagram, and the inhibitory input from two geniculate on-center fibers situated in B and C. (After Creutzfeldt and Ito, 1968.)

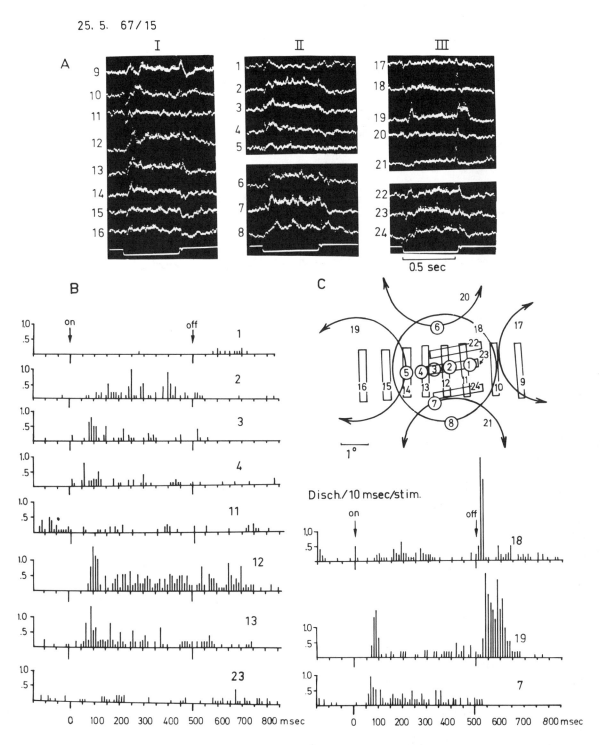

25. 5.  67 / 15

FIGURE 13   Responses of a cortical cell (area 17) to spots and bars of light shone into different parts of the receptive field. The cell is movement-sensitive with directional preference (see Figure 14). A. Averaged slow-potential responses (sum of IPSPs and EPSPs) to 10 stimuli. The numbers on the responses correspond to the stimulus numbers in the map (C). B. PST histograms of some stimuli. (From Creutzfeldt and Ito, 1968.)

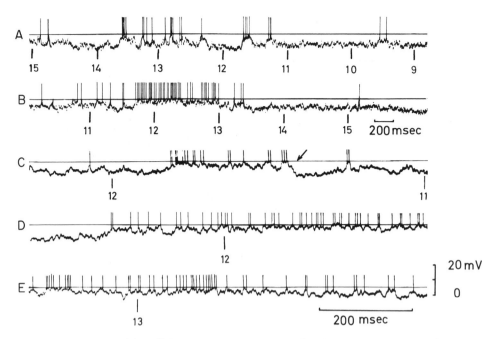

FIGURE 14 Movement response of the cell shown in Figure 13. In A (slow recording speed) and C (fast record), the stimulus moves from left to right (the numbers below the record correspond to the stimulus positions in the stimulus map of Figure 13). In B (slow record) and D (fast record) the stimulus moves from right to left. The response is stronger for these movements than during movement in the other direction (A and C), because IPSPs (arrow in C) interrupt the excitatory response. Note different time scale in A and B from that in C and D. (From Creutzfeldt and Ito, 1968.)

dendritic density. It is suggested that synaptic contacts between bipolar and ganglion cells are made by chance, i.e., whenever a bipolar-cell "axon" meets a ganglion-cell dendrite. A morphological model based on this assumption is suggested, and simulation of responses to light stimuli are demonstrated and compared with experimental data. Excitation is not summated linearly within the receptive-field center. Possible mechanisms for this are discussed, and experiments are described. Nonlinear signal summation takes place in the receptor-bipolar complex and, because of electrotonic spread and interaction of excitatory processes along the dendrites, at the ganglion-cell input level. At higher excitation levels, the maximal firing rate of retinal ganglion cells limits their response. The result of these different nonlinearities is that individual ganglion cells give incomplete information on stimulus brightness and spatial extent. Gain control, caused by neuronal adaptation at the receptor site, is another mechanism of nonlinear retinal transmission.

In the lateral geniculate body (LGB), the synaptic organization is such that the branches of optic-tract fibers have essentially one-to-one excitatory contacts with the relay cells. Each LGB relay cell thus receives a direct excitatory input from only one optic-tract fiber. Cells in the LGB that have an on-center fiber input are inhibited by off-center cells; off-center cells are inhibited by on-center cells. The receptive fields of these mutually inhibitory on-center and off-center cells are near each other. This arrangement does not improve the spatial dissolving power of the visual system. Only the sensitivity to successive contrasts in brightness may be slightly improved because of simultaneous excitation of the excitatory input and inhibition of the inhibitory input (disinhibition), or vice versa. The main function of the LGB may therefore be seen in the combination of the retinal and nonretinal inputs into the same relay cells. The information-transmission properties of a monosynaptic one-to-one relay are discussed, and it is shown how a nonlinear transfer is brought about by summation of irregularly timed EPSPs.

Visual-cortex cells receive a variety of inputs. In area 17, the main retinogeniculate input into one cortical cell consists of two to five radiation fibers. They may be excitatory or inhibitory and may come from geniculate on-center or off-center cells. These receptive-field centers overlap and have center distances of up to 2–3 degrees. Optimal stimuli are those that simultaneously inhibit inhibitory and activate excitatory input fibers. Such optimal stimuli are often bars of light or darkness. Moving stimuli are usually more efficient than stationary stimuli, because of the successive activation of several excitatory inputs. Direction-specific movement responses result from the successive invasion of the receptive fields by excitatory off-center and on-center input fibers or of excitatory and inhibitory on-center input fibers. Many possible input combinations were found. Cortical neurons are part of a pattern-recognition mechanism, but whether the observed optimal stimulus shapes are significant elements of its algorithm is questioned.

## Acknowledgment

This paper is based on experimental material which was collected in collaboration with Drs. M. Ito, B. Sakmann, H. Scheich, and W. Singer, and which has been, or will be, published in extenso elsewhere.

### REFERENCES

ARDUINI, A., and L. R. PINNEO, 1962. Properties of the retina in response to steady illumination. *Arch. Ital. Biol.* 100: 425–448.

BOYCOTT, B. B., and J. E. DOWLING, 1969. Organization of the primate retina: Golgi study. *Phil. Trans. Roy. Soc. (London) ser. B, biol. sci.* 255: 109–176.

BROWN, J. E., 1956. Dendritic fields of retinal ganglion cells of the rat. *J. Neurophysiol.* 28: 1091–1100.

BROWN, J. E., and D. MAJOR, 1966. Cat retinal ganglion cell dendritic fields. *Exp. Neurol.* 15: 70–78.

BÜTTNER, U., and O.-J. GRÜSSER, 1968. Quantitative Untersuchungen der räumlichen Erregungssummation im rezeptiven Feld retinaler Neurone der Katze. I. Reizung mit 2 synchronen Lichtpunkten. *Kybernetik* 4: 81–94.

CLELAND, B. G., and C. ENROTH-CUGELL, 1968. Quantitative aspects of sensitivity and summation in the cat retina. *J. Physiol. (London)* 198: 17–38.

CREUTZFELDT, O., 1968a. Functional synaptic organization in the lateral geniculate body and its implication for information transmission. *In* Structure and Function of Inhibitory Neuronal Mechanisms (C. von Euler, S. Skoglund, and U. Söderberg, editors). Pergamon Press, Oxford and New York, pp. 117–122.

CREUTZFELDT, O. D., 1968b. Physiologie der Hirnrinde. *Jahrb. Max-Planck Ges. 1968,* pp. 61–89.

CREUTZFELDT, O. D., and M. ITO, 1968. Functional synaptic organization of primary visual cortex neurones in the cat. *Exp. Brain Res.* 6: 324–352.

CREUTZFELDT, O. D., H. D. LUX, and A. C. NACIMIENTO, 1964. Intracelluläre Reizung corticaler Nervenzellen. *Pflügers Arch. Ges. Physiol.* 281: 129–151.

CREUTZFELDT, O. D., K. MAEKAWA, and L. HÖSLI, 1969. Forms of spontaneous and evoked postsynaptic potentials of cortical nerve cells. *Progr. Brain Res.* 31: 265–273.

CREUTZFELDT, O. D., and B. SAKMANN, 1969. Neurophysiology of vision. *Annu. Rev. Physiol.* 31: 499–544.

CREUTZFELDT, O. D., B. SAKMANN, and H. SCHEICH, 1968. Zusammenhang zwischen Struktur und Funktion der Retina. *In* Kybernetik 1968. (H. Marko and G. Färber, editors). Oldenbourg-Verlag, Munich, pp. 239–262.

DOWLING, J. E., and B. B. BOYCOTT, 1966. Organization of the primate retina: electron microscopy. *Proc. Roy. Soc., ser. B, biol. sci.* 166: 80–111.

FÄRBER, G., and W. PROBST, 1968. Automatische Versuchssteuerung und Versuchsauswertung mit dem IBM System 1130. *IBM Nachr.* 188: 126–131.

FUSTER, J. M., O. D. CREUTZFELDT, and M. STRASCHILL, 1965. Intracellular recording of neuronal activity in the visual system. *Z. Vergleich. Physiol.* 49: 605–622.

GRANIT, R., D. KERNELL, and G. K. SHORTESS, 1963. Quantitative aspects of repetitive firing of mammalian motoneurons, caused by injected currents. *J. Physiol. (London)* 168: 911–931.

GUILLERY, R. W., 1966. A study of Golgi preparations from the dorsal lateral geniculate nucleus of the adult cat. *J. Comp. Neurol.* 128: 21–49.

HERZ, A., O. D. CREUTZFELDT, and J. M. FUSTER, 1964, Statistische Eigenschaften der Neuronaktivität im ascendierenden visuellen System. *Kybernetik* 2: 61–71.

HODGKIN, A. L., 1948. The local electric changes associated with repetitive action in a non-medullated axon. *J. Physiol. (London)* 107: 165.

HUBEL, D. H., 1963. Integrative processes in central visual pathways of the cat. *J. Opt. Soc. Amer.* 53: 58–66.

HUBEL, D. H., and T. N. WIESEL, 1962. Receptive fields, binocular interaction and functional architecture in the cat's visual cortex. *J. Physiol. (London)* 160: 106–154.

HUBEL, D. H., and T. N. WIESEL, 1968. Receptive fields and functional architecture of monkey striate cortex. *J. Physiol. (London)* 195: 215–243.

HUGHES, G.W., and L. MAFFEI, 1966. Retinal ganglion cell response to sinusoidal light stimulation. *J. Neurophysiol.* 29: 333–352.

LEICESTER, J., and J. STONE, 1967. Ganglion, amacrine and horizontal cells of the cat's retina. *Vision Res.* 7: 695–705.

McILWAIN, J. T., and O. D. CREUTZFELDT, 1967. Microelectrode study of synaptic excitation and inhibition in the lateral geniculate nucleus of the cat. *J. Neurophysiol.* 30: 1–21.

NEUMANN, J. VON, 1967. The general and logical theory of automata. *In* Cerebral Mechanisms in Behavior (L. A. Jeffress, editor). Hafner Publishing Co., New York, London, pp. 1–41. (First edition, 1951.)

POLYAK, S., 1941. The Retina. The University of Chicago Press, Chicago.

RATLIFF, F., 1965. Mach Bands: Quantitative Studies on Neuronal Networks in the Retina. Holden-Day, Inc., San Francisco, California.

RUSHTON, W. A. H., 1965. Visual adaptation. *Proc. Roy. Soc., ser. B, biol. sci.* 162: 20–46.

SAKMAN, B., and O. D. CREUTZFELDT, 1969. Scotopic and mesopic light-adaptation in the cat's retina. *Pflügers Arch. Ges. Physiol.* 313: 168–185.

SINGER, W., and O. D. CREUTZFELDT, 1970. Reciprocal lateral inhibition of on- and off-center neurones in the lateral geniculate body of the cat. *Exp. Brain Res.* 10: 311–330.

STONE, J., and M. FABIAN, 1968. Summing properties of the cat's retinal ganglion cell. *Vision Res.* 8: 1023–1040.

SZENTÁGOTHAI, J., J. HÁMORI, and T. TÖMBÖL, 1966. Degeneration and electron microscope analysis of the synaptic glomeruli in the lateral geniculate body. *Exp. Brain Res.* 2: 283–301.

# 57 Functional Association of Neurons:

# Detection and Interpretation

## GEORGE L. GERSTEIN

DURING THE PAST 15 YEARS, the functions of many portions of the nervous system have been examined through the study of single neurons, each individually observed for some period of time. In sensory systems, in spite of the prevalence of spontaneous neural firing, it has been possible to determine the probabilistic relations between the presentation of an appropriate stimulus and the resulting change in firing pattern. Both in terms of stimulus detection and in terms of stimulus coding, it is apparent that the firing pattern of a single neuron does not carry enough information to account for the behavioral capabilities of an entire organism. To detect a change in firing pattern caused by a stimulus, for example, it is often necessary to sum the responses to repeated stimulus presentations, even though the stimulus intensity is far above the psychophysical detection level. Even if the response pattern of a single neuron is well defined, most measurable characteristics of the response pattern (such as latency, amplitude, shape of envelope, and so forth) turn out to be multivalued functions of various stimulus parameters. The characteristics of the firing pattern can therefore not be used uniquely to decide what the stimulus parameters were. Fortunately, the multivalued functions relating response pattern and stimulus parameter differ from neuron to neuron, so that the ambiguity can be resolved by combining information from several neurons. The ability of an organism to detect and determine stimulus parameters in a single presentation overwhelmingly suggests that such ensembles of neurons must be involved in the processing of sensory information. It would seem probable that most complex integrative functions in the nervous system also involve simultaneous activity in groups of neurons. Such functional groupings may well have boundaries and internal relationships that change in time or with the nature of the task.

The number of neurons comprising a functional asssembly might be in the hundreds for mammalian nervous systems; smaller groupings are likely in simpler nervous systems. Although the technology for examining such large groupings does not yet exist, methods have become available for the recording and interpretation of the simultaneous activity of small numbers of neurons. Two general types of experiment can yield such information. With intracellular recording, a group of functionally related neurons is available for study, if it be assumed that individual postsynaptic potentials are indentifiable. Little is known, however, under these conditions about the relative spatial arrangement of such neurons. In contrast, with extracellular recording, a group of neurons that lie within a small volume may be studied, but little is known *a priori* about their interconnections. The two types of experiment record, respectively, a functional group or a spatial group of neurons. The purpose of the present paper is to illustrate the detection and interpretation of neuronal interactions within such a group.

## Recording methodology

The simultaneous recording of electrical activity from several neurons can be accomplished by a variety of methods. In order to study neuronal interactions by the analytic methods reviewed in this paper, it is necessary to sort such compound data so that the activity of each observed neuron as a function of time is known individually. Various solutions for this problem have appeared in the literature; choice among the alternatives depends on the circumstances of the experiment and the detailed nature of the data.

INTRACELLULAR RECORDINGS   The action potential of the impaled neuron is easily separated from all other electrical activity by an amplitude discriminator. Postsynaptic potentials (PSPs) from synaptic connections incident on the impaled neuron can be resolved individually only if they arrive at a rate sufficiently low so that they overlap rarely in time, a condition that is more often realized in invertebrate recordings. PSPs from synapses at different locations on the impaled neuron should have different amplitudes and shapes because of the electrical properties of the dendritic membrane (Rall, 1967). These PSP shape differences can be used by a computer program to identify similar shapes in the recording, and thus to mark trains of particular synaptic events. An elegant program for this purpose was described

GEORGE L. GERSTEIN  Departments of Physiology and Biophysics, University of Pennsylvania, Philadelphia, Pennsylvania

by Hiltz (1965); the program will sort PSP shape even if there is partial overlap or a rapidly changing baseline potential. A much simpler program using a library of PSP shapes that are chosen by the operator has been described by W. Simon (personal communication).

EXTRACELLULAR RECORDINGS  If multiple electrodes are used, each of which isolates a single action potential, the experimental objectives are immediately realized. With multiple microelectrodes, however, there is difficulty in controlling the relative positions of the tips, and there is an unknown amount of damage to the tissue. A number of workers have chosen to use single microelectrodes with enough tip area exposed to allow the recording of several different neurons. The relative geometry of electrode tip and each neuron is different, so that, in general, each neuron produces a different action-potential shape and amplitude in the composite recording. Individual spike trains can then be determined by sorting the data according to action-potential shape. Methods for doing this have been devised with computer programs (Gerstein and Clark, 1964; Simon, 1965; Keehn, 1966) and with hardware devices (Glaser and Marks, 1968). Each of these methods of sorting of spike shape is subject to error; a careful evaluation of the reliability and sources of error for these methods can be found in the review by Glaser (1970).

An illustration of spike shape-sorting methods taken from our own work (Gerstein and Clark, 1964) is shown in Figure 1. Short sections of the raw data are shown at the top of the Figure, and include action potentials of several different amplitudes. More careful examination with the oscilloscope, as in the time-exposure at the top right of the Figure, shows that several different action-potential shapes can indeed be identified. Several stages of the sorting process are illustrated by the histograms in the lower part of the Figure. A "standard" waveform, labeled "0," is chosen arbitrarily from the data, and all other spike shapes in the data are compared with it. The comparison is made by a mean-square difference across the 32 data points that comprise each action potential. The weighting of different points in this calculation can be adjusted at will. The histogram labeled "Stage 0" shows the distribution of the population of spikes according to increasing dissimilarity of shape from the standard waveform. The population falls into a multimodal distribution; the farthest left peak can be defined as the portion of the population of action-potential shapes that is most similar to the standard waveform. This portion of the population is identified as shape class 0 and is removed from further consideration. A new standard waveform is chosen from the residue, and the process iterated in order to identify shape class 1. The process is sensitive: shape class 1 and shape class 2 action potentials have similar amplitudes and differ mainly by the shape of the afterpotential. The long tail of the distri-

bution in the Stage 3 histogram represents the various waveform overlap events. Such superpositions cannot be handled by this technique, and represent an unavoidable gap in the data.

The mean-square method of shape comparison used here can be explained in terms of a geometric model. Consider a 32-dimensional property space; the 32 voltage values that determine a given action potential define a single point in this space. We seek to determine whether there are clusters of points that correspond to populations of action potentials that are discrete in shape. The choice of a standard waveform corresponds to the choice of a single point in the property space. We then calculate the distance to all other data points in the space, and plot these distances as a histogram. If the original choice of origin for the distance measurement fell within a well-defined cluster, the histogram would be multimodal. Only the histogram peak corresponding to the smallest distance is well defined; two or more point clusters far from the chosen origin might occur at the same distance, and hence would be indistinguishable in the histogram. Other methods of shape comparison can also usefully be viewed in terms of a multidimensional property space.

### Analytic methodology

In the following discussions we assume that the basic data have been sorted into separate spike (or PSP) trains. We assume that (identical) stimuli occurred at instants $S_i$, while action potentials from spike train A occurred at instants $A_j$, and action potentials from spike train B occurred at instants $B_k$.

We may apply all the usual single-unit measures to such data in order to make a statistical description of each spike train (interval histograms, autocorrelograms, and so forth). It is also possible to examine the relationship between each firing pattern and the stimulus by computing the peri-stimulus-time histogram (PST histogram). A more detailed discussion of the mathematical meaning of the various computations that are appropriate for the characterization of a single train of action potentials can be found in Perkel et al. (1967a). In the present context of multiple, simultaneously recorded spike trains, these measures may be used to assess the similarity of response properties within the group of neurons that is being sampled. If data are gathered with a single extracellular electrode, all the observed neurons lie within a small volume, which might be several hundred microns in diameter. We may examine whether there is a relationship between such spatial contiguity and the response pattern to an appropriate stimulus. An example of such a measurement is shown in Figure 2. This small volume of cochlear nucleus contained a heterogeneous population of active neurons, although other, similar experiments have shown more homogeneous response properties.

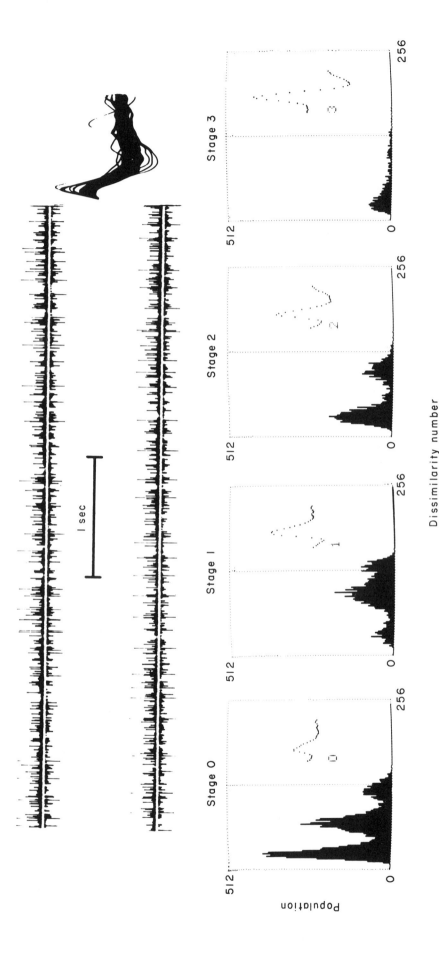

FIGURE 1 Sorting of spike shapes in a multiple neuron recording. Raw data at top; inset (top right) shows several distinct spike shapes in time exposure on a triggered oscilloscope. Four stages of the sorting process are below. In each histogram, below, all spike waveforms in the data are compared with the "standard" waveforms shown. In each case, the peak at the left represents spike waveforms that are most similar to the standard waveform. The identified waveforms are removed from the data at each stage of the separation process, so that each successive stage is performed on a smaller population of waveforms and waveform classes. Dissimilarity number represents the difference in waveform.

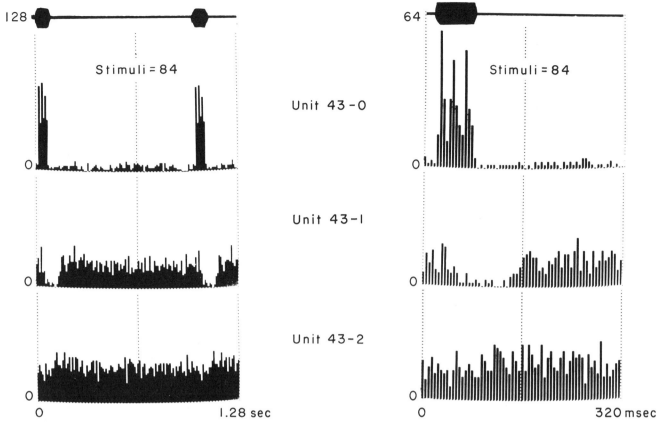

Figure 2 PST histograms from three simultaneously recorded neurons in cochlear nucleus of an anesthetized cat. A special tungsten electrode was used, and neurons were probably within a 200-micron sphere. Tone-burst stimulus is indicated at the top. PST histograms are shown with high time resolution in the right column to show the detailed firing patterns during stimulus presentation. Neuron shown at top was excited by stimulus, neuron in middle was inhibited, and neuron at bottom was unaffected.

When data recorded simultaneously from several neurons are available, it is interesting to make various interspike-train measurements. These allow the detection and assessment of direct interaction between the neurons. In addition, it is possible to evaluate the effect of the stimulus on such interactions and to estimate the degree of common input from sources other than the stimulus. These measurements can be used to derive a "logical wiring diagram" for the neurons under observation. Furthermore, together with the appropriate experimental design, they can be used to search for parallel-line coding in the transfer of sensory information. A detailed discussion of computations appropriate for the characterization of simultaneously recorded spike trains can be found in Perkel et al. (1967b). More recent developments are in Gerstein and Perkel (1969).

Several different interspike-train measurements have proved appropriate for various types of problems, as described below. For each of these measurements, it is important to have available a control calculation that predicts the appearance of the measurement if the two (or more) spike trains are completely unrelated. Such control calculations are made necessary by the statistical variability of the spike trains and by the need to sort out effects of stimulation from effects of direct interaction between, or of common input to, the observed neurons. The three measurements described here are variously related to cross correlation of two (or more) spike trains.

CROSS-INTERVAL HISTOGRAM Time intervals are measured from a given spike of train A to the nearest preceding and succeeding spikes of train B. This is repeated, using each spike of train A as origin, and a histogram is compiled. It should be noted that, unlike the ordinary cross correlation, this cross-interval histogram is in no way related to the histogram that would be obtained if the roles of the two spike trains in the measurement were interchanged. In general, it is necessary to compute the cross-interval histogram with both the A and the B spike trains as origins. A satisfactory control for this measurement is known from renewal theory (Cox, 1962) and is calculated from the interval histograms of the two spike trains. The cross-interval histogram is particularly useful for the detection of relationships that have a short time course of the same general duration as the mean interspike interval of the two spike trains. Relationships with a longer duration are lost in this measurement, because they may involve a time considerably longer than that to the next spike.

CROSS-CORRELATION HISTOGRAM   From a given spike of train A, the time intervals are measured to all preceding and succeding spikes of train B that fall within a selected time duration. This is repeated, using each spike of train A as origin, and a histogram is compiled. Unlike the cross-interval histogram, interchange of the roles of the two spike trains simply produces the same cross-correlation histogram with a reversal of the time scale. If the two spike trains are completely uncorrelated, and if their statistical properties are constant in time, the cross-correlation histogram will be flat (to within statistical uncertainty). If, however, data were taken during appropriate stimulation, both spike trains could show rate variations that are related to the stimulus, and a more complex procedure would be needed to obtain a control measurement. One possible control is obtained by convolution of the PST histograms from each spike train. A more readily calculated control measurement is applicable if the stimuli are periodic. Either spike train A or spike train B is segmented at the stimulation instants, and these segments are shuffled. A cross-correlation histogram is then constructed between the intact spike train and the shuffled spike train. The effect of this shuffling procedure is to destroy detailed correlation between the two spike trains while all effects caused by the periodic stimulation are left intact. The cross-correlation histogram and its controls are particularly useful for the detection of relationships that have a time course much longer than typical interspike intervals, but shorter than typical interstimulus intervals. Such relationships will appear in the cross-correlation histogram as deflections from the shape of the control histogram.

JOINT PERI-STIMULUS–TIME SCATTER DIAGRAM   This consists of a scatter diagram of the joint occurrences of spikes in trains A and B relative to the times of stimulation. The ordinate of each point plotted corresponds to the time between a stimulus event and a spike in train A (for example, $A_j–S_i$), and the abscissa corresponds to the time between the same stimulus event and a spike in train B (for example, $B_k–S_i$). For each stimulus event, a point is plotted for each combination of S–A and S–B time intervals, both of which fall into a specified range. A spike in train A will give rise, in the scatter diagram, to the same number of points as there are spikes in train B within the specified time range about the stimulus (and vice versa). This scatter diagram is a generalization of the PST histograms for the two spike trains and

of the cross-correlation histogram, and contains information beyond that available in those histograms. The diagram is particularly useful in sorting out the portions of relationships between spike trains that result from direct interaction between the neurons and those that result from common, or stimulus-related, input to the two neurons. Point densities in the scatter diagram are differently affected by each of these relationships: (1) direct stimulus effects on each neuron are expressed as bands of altered point density that are parallel to the coordinate axes; (2) direct interactions between the neurons are expressed as bands of altered point density that are parallel to the principal diagonal; (3) stimulus-related changes in the direct interactions are expressed as changes in the band structure along the direction of the principal diagonal; and (4) common input to the two neurons other than from the stimulus is also expressed as diagonal band structure, but is frequently more diffuse than the diagonal bands produced by direct interactions between the neurons.

In effect, the joint PST scatter diagram permits examination of the time structure of correlation between two spike trains. It should be noted that the scatter diagram is easily extended to three (or more) simultaneously recorded spike trains (Figures 10 and 11). Thus it allows detection and measurement of functional grouping of more than two neurons at a time.

An appropriate control measurement can be obtained for the scatter diagram by the same technique that has been described for the cross-correlation histogram. One spike train is segmented at the stimulus events and shuffled. The scatter diagram is again constructed between the intact spike train and the shuffled spike train. The principal effect of this procedure is to destroy all detailed correlation between the two spike trains, and consequently to eliminate all diagonal structure from the scatter diagram, although directly stimulus-related structure and the general "background" point densities are preserved.

*Categories of functional interaction*

The several measures described above can be used individually or in combinations to study multiple, simultaneously recorded spike trains. The initial objective of such study for extracellularly recorded spike trains is to define an equivalent "wiring diagram" for the connections that may exist between the observed neurons. This step, of course, is not always needed if the data originated from intracellular recording with PSP identification, because the functional relationships between the presynaptic and postsynaptic neurons are already defined. If, however, relationships between different identifiable PSPs are to be examined, the problem is logically equivalent to the case of multiple extracellular recording.

In subsequent portions of this paper, I discuss connections

between the observed neurons and connections from inferred (but not observed) neurons in terms of the stylized diagram shown in Figure 3. The two directly observed neurons are labeled A and B; known stimulus sources are represented by S. Inferred interneurons are labeled I, and inferred sources of common input are represented by D. Arrows show the connections between the various elements and represent either excitatory or inhibitory effects. Several of the pathways are redundant, even in this simple diagram. For example, the functions of the explicit pathways drawn from S to A, S to B, D to A, D to B, and S to D could be accomplished within the interneuron box I. The logical functions indicated in Figure 3 could also, of course, be accomplished by a hierarchy of more complex circuits, because chains of additional interneurons could be interposed between any two elements without altering function. Unfortunately, the sensitivity of our analytic methods is not sufficient to allow choice between such alternatives, so that the description of the observed interrelationships can be left in the simplest possible terms. It should be noted that in most data there is evidence for only a few of the possible interconnections shown in Figure 3.

Once the connections of the observed neurons have been established in the rather oversimplified terms of Figure 3, the stimulus conditions of the experiment may be varied. Such strategy allows detection of changes in the neural connections as a function of the stimulus conditions, and simultaneously allows detection of possible parallel-line coding schemes.

NEURONAL INTERACTION NOT AFFECTED BY THE STIMULUS Cross-interval histograms for two neurons in cochlear nucleus of an anesthetized cat during spontaneous and stimulated conditions are shown in Figures 4 and 5. The histograms in Figure 4 were taken with spikes from neuron 22–0 used as origin; in Figure 5 the roles of the two spike trains were interchanged, and spikes from neuron 22–1 were used as origin. Superimposed on each histogram is the appropriate control calculation corresponding to the assumption that there was no correlation between spike trains.

The general shape of these cross-interval histograms is different, depending on whether the stimulus was presented and on which of the two spike trains was used as origin. Despite the statistical variability of the histograms, there is clear agreement between the over-all histogram shapes and the control calculation. (The control curve is much smoother than the histograms in each case, because it is calculated from a probability distribution function; the cross-interval histograms themselves are estimators of probability densities.) Near the origin of each histogram, however, is a single peak that deviates from the control calculation by an amount that is large in comparison to the typical statistical variability. These peaks are at −3 milliseconds in the histograms of

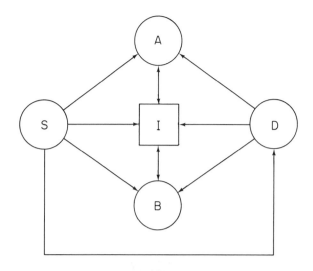

FIGURE 3 A schematic diagram useful for considering pathways to and between the two observed neurons, A and B. S represents stimulus sources; I represents interneurons; D represents other common inputs. Connections can be excitatory or inhibitory.

Figure 4 and at +3 milliseconds in the histograms of Figure 5. An interpretation that is consistent with all four histograms is that neuron 22–1 has a far greater-than-chance probability of firing 3 milliseconds *before* neuron 22–0, independent of the stimulus conditions; conversely, neuron 22–0 has a far greater-than-chance probability of firing 3 milliseconds *after* neuron 22–1. Both neurons fire frequently at other relative time intervals, although such cross intervals have a chance distribution. In physiological terms, the preferred relative time interval may represent either a direct synaptic connection between the two observed neurons with a conduction time of 3 milliseconds, or it may represent a common input to the two observed neurons. If the latter is true, there is a 3-millisecond difference in the conduction time between the common source and the two observed neurons. Both pathways are indicated in Figure 3. Choice between the alternatives is difficult without additional information. Experience with other data and with computer stimulation, however, suggests that very narrow peaks, as in the cross-interval histograms of Figures 4 and 5, are caused by direct connection between the two observed neurons. Histogram peaks that are the result of common input to the two observed neurons tend to be considerably wider than those shown in Figures 4 and 5.

NEURONAL INTERACTION AFFECTED BY THE STIMULUS Cross-interval histograms for two neurons in visual cortex of an unanesthetized, chronically prepared cat are shown in Figure 6, page 656. Neither of these units showed much spontaneous activity, but both responded actively to a moving parallel-bar pattern.

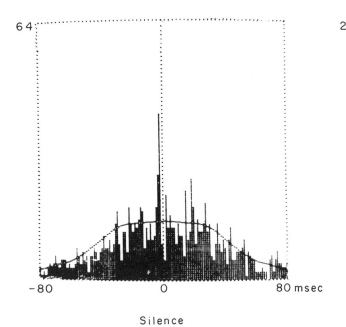

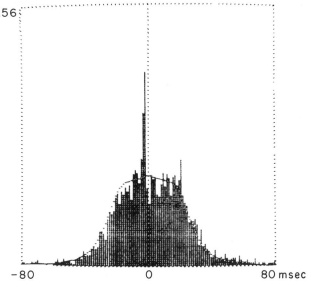

Silence

Tone bursts 1/sec, 800 msec

FIGURE 4 Cross-interval histograms of firings from two neurons in cochlear nucleus of an anesthetized cat. The left histogram was taken during spontaneous conditions; the right histogram was taken during stimulation by tone bursts of 800 msec, presented at 1 per second, containing a frequency that was within the auditory response area of both neurons. Control calculation for the case of no interaction between the observed neurons is indicated by the dotted curve.

There is excellent agreement between the control calculation and the left cross-interval histogram that corresponds to vertical motion of the bar pattern. The missing bars of the histogram near the origin are an artifact resulting from the inability of the spike shape-sorting system used here to deal with superposition of action-potential waveforms when the two neurons fire almost synchronously. This means that, during vertical motion of the bar pattern, the two spike trains showed only chance time relationships. In the second histogram of Figure 6, which corresponds to diagonal motion of the bar pattern, agreement between the control calculation and the cross-interval histogram is good only to the right of the origin. To the left of the origin, the histogram shows a series of peaks situated at approximately 3, 11, and, possibly, 22 milliseconds. This means that neuron 50–2 tends to follow the firings of neuron 50–3 by these preferred intervals, although no reciprocal relationship is evident. This time structure is unlikely to be related directly to the passage of the stimulus bars over the receptive fields, because such passage occurred only every 600 milliseconds. In physiological terms, the preferred relative time intervals may represent direct synaptic connection between the two observed

neurons via interneuronal chains of three different lengths. Alternatively, the preferred time intervals may represent a common source of input to the two observed neurons through similar interneuronal chains. Choice between these alternative explanations cannot be made uniquely, although the narrow peaks would favor direct connection as the more likely situation. In either case, it should be noted that the occurrence of a preferred timing relationship between the two spike trains depended on presentation of a "favored" stimulus. In terms of Figure 3, these data imply a pathway S to I that is able to modulate the pathway A to I to B.

The data shown in Figure 6 represent a situation in which the two stimulus conditions could not be differentiated clearly by examining the peri-stimulus-time (PST) histograms of the two observed neurons. The individual spike trains showed little difference during the two stimulus conditions. The correlation structure between the two spike trains did, however, show a dependence on the stimulus condition. Potentially, such a correlation structure between spike trains could, if it is stimulus selective, be used as a coding mechanism. Whether such codes are actually used in the nervous system must be determined in each case by identify-

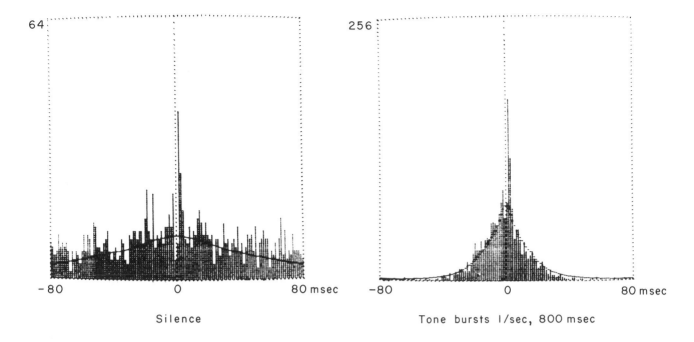

FIGURE 5   Cross-interval histograms of the same data as in Figure 4, but with the roles of the two spike trains interchanged. Control calculation as in Figure 4.

ing an appropriate decoding mechanism at the neural level upon which the correlated neurons project.

COMMON INPUT CAUSED BY STIMULUS   Figure 7 shows cross-correlation histograms under different stimulus conditions for the same two neurons in cochlear nucleus as are seen in Figures 4 and 5. These cross-correlation histograms were calculated by using the spike train of unit 22-0 as origin; it is not necessary to make the reciprocal calculation. The time scale was chosen to encompass the time between stimuli. The short-duration interaction effects discussed with the aid of cross-interval histograms in Figures 4 and 5 are barely resolved by the long time scale in Figure 7.

The cross-correlation histogram corresponding to the spontaneous condition is flat to within the statistical variability. This suggests that, during such spontaneous conditions, there is no correlation between the two spike trains. Both cross-correlation histograms in the stimulated conditions show, however, considerable deviations from flatness. These deviations appear to be periodic, with the same period as the stimulus presentations, and signify that the two spike trains are correlated in a time-varying way that is related to the stimulus. It is now necessary to determine whether such correlation structure is the result of direct interaction be-

tween the two observed neurons, or whether it simply represents the simultaneous changes in firing probability that are produced in each neuron by the presentation of the stimulus. Such determination is most easily achieved by comparing the cross-correlation histograms with the appropriate control calculations. Such controls correspond to spike trains with the identical time structure with respect to the stimulus, but with no direct correlation (see section on Analytic Methodology). The desired comparison for the middle cross-correlation histogram in Figure 7 is shown in Figure 8. The black dots represent the control calculation and are superimposed on a gray histogram, which replicates the one in Figure 7. Unlike the control calculation used for cross-interval histograms, the present calculation does not show a reduced statistical variability. Nevertheless, it seems clear that there is good agreement between the two measurements shown in Figure 8. We conclude that the correlation structure that is implied by the histograms of Figure 7 is entirely the result of stimulus-related firing by the two observed neurons, and that no direct interactions or connections between them need be invoked (except for those discussed in conjunction with Figures 4 and 5). In terms of the diagram of Figure 3, these data imply only pathways from S to A and from S to B.

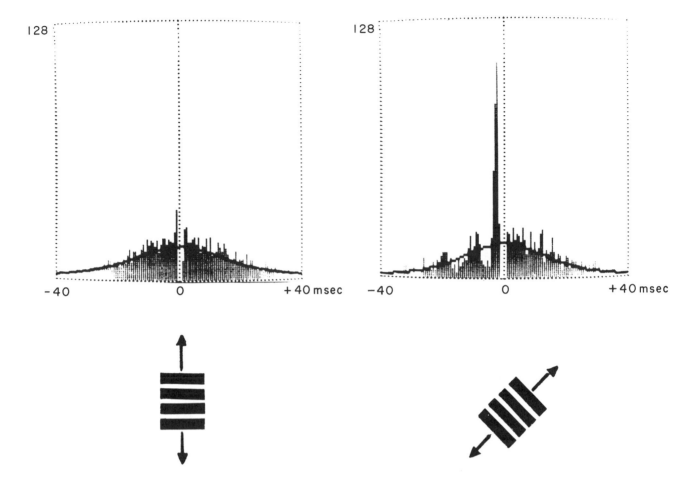

FIGURE 6    Cross-interval histograms of firings from two neurons in visual cortex of a chronically prepared cat. Each stimulus bar subtended 2° from the cat's eye and the entire pattern was moved sinusoidally through 20° once every 6 seconds. The direction of movement is indicated below each histogram. As on Figures 4 and 5, the control calculation is indicated by the dotted curve.

COMMON INPUT ALTERED BY STIMULUS    Cross-correlation histograms for two neurons in cochlear nucleus of an anesthetized cat are shown in Figure 9. The left histogram shows the cross correlation during the spontaneous condition. The two histograms on the right represent the cross correlation and its control calculation for data obtained during acoustic stimulation by 50-millisecond tone bursts of an appropriate frequency.

The cross-correlation histogram corresponding to the spontaneous condition shows a small, broad peak placed symmetrically around the origin. This means that the observed neurons tend to fire within approximately 100 milliseconds or less of each other, although on the average neither neuron leads the other. The most likely physiological explanation of the histogram is that these neurons have a com-

mon source of synaptic input. Other inputs to each neuron exist also, including input from stimulus sources (as discussed under Common Input Caused by Stimulus and below). These other inputs would insure that neither neuron tends to lead in response to input from the common source.

When the cross correlation and control calculation corresponding to the stimulated condition are compared, there appears to be a good fit in regions to the left of the origin and in regions to the right of +0.2 second. In the first 0.2 second to the right of the origin the correlation histogram is somewhat higher than the control calculation. The correlation in excess of that attributable to stimulus effects (as determined by the control calculation) is therefore not symmetrically arranged with respect to the origin. In this case, presentation of an appropriate stimulus has changed a cross

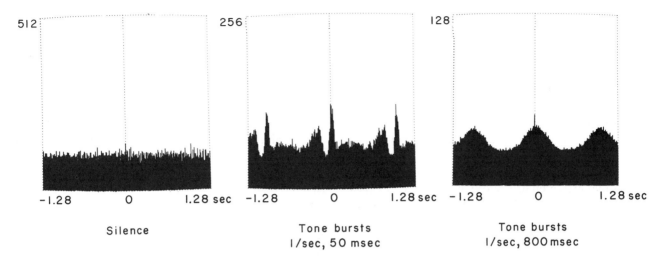

Silence · Tone bursts 1/sec, 50 msec · Tone bursts 1/sec, 800 msec

FIGURE 7 Cross-correlation histograms for two neurons in cochlear nucleus of an anesthetized cat. Left histogram for data taken during spontaneous conditions; middle histogram for data taken during stimulation by 50-msec tone bursts of "best" frequency, 1 per second; right histogram for data taken during stimulation by 800-msec tone bursts of same frequency, 1 per second.

Unit 22-1 from unit 22-0

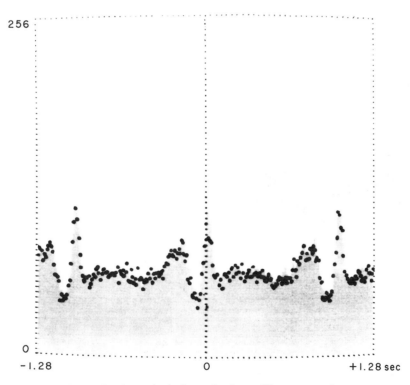

FIGURE 8 Control calculation for the middle cross-correlation histogram shown in Figure 7. The dots are the control; the original histogram is shown in gray bars.

FUNCTIONAL ASSOCIATION OF NEURONS 657

correlation that is symmetrical around the origin into one that is asymmetrical around the origin. In physiological terms this suggests that the nature of the common input to the observed neurons has been altered in a nonspecific way by the presentation of the stimulus. In terms of Figure 3, these data suggest pathways directly from S to A and S to B, pathways from D to A and D to B, and finally from S to D. The last pathway presumably has tonic effects on D, the common source.

TEMPORAL STRUCTURE OF NEURONAL INTERACTION    The methods we have examined above for determining the correlation between two spike trains all treat the entire available stretch of data uniformly. It may be desirable to know whether the correlation structure itself is time varying, particularly with respect to the instants of stimulus presentation. It also may be desirable to extend the analysis to three or more simultaneously recorded spike trains. Both objectives are met by plotting joint PST scatter diagrams; these are shown in Figure 10 for three neurons from the pleural ganglion of *Aplysia*, and in Figure 11 for three neurons from auditory cortex of a chronically prepared cat. Both Figures are presented as stereo pairs and are best viewed if merged. The origin for each cube is at the center rear, and the rear edges are the coordinate axes along which are

plotted, respectively the S to A, S to B, and S to C time intervals, in which A, B, and C are the three spike trains. The two rear faces and the bottom of each cube are planes which contain the two-dimensional, joint PST-scatter diagrams for the neuron pairs A–C, B–C, and A–B, respectively.

Various clusters of points are visible in both Figures, although with greater clarity in Figure 10, because the long-time scale reduces the apparent variability of firing. Each such cluster is readily identifiable in terms of particular aspects of each spike train or in terms of correlation between them. The lines of high density near and parallel to each of the coordinate axes represent the direct, stimulus-related firing of each of the three neurons. Some caution is needed in assigning these regions of altered density, because there is multiple representation. The line of high density in the A–B plane near the B axis represents the stimulus-related increase of firing in neuron A; conversely, the two lines of high density in the A–B plane near the A axis represent the stimulus-related increase of firing in neuron B.

Clearly visible in each coordinate plane are diagonal lines of high density. These represent highly correlated firing of the neurons in pairs. It should be noted that the point density along any of these diagonals is not uniform. There is a region of higher point density near the origin. This means that each pair of neurons has a high probability of firing

Spontaneous

Tone bursts 1/sec

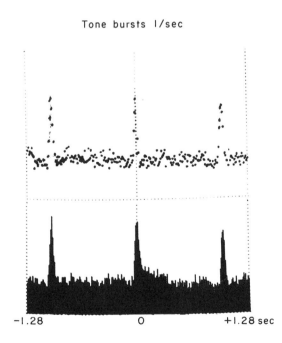

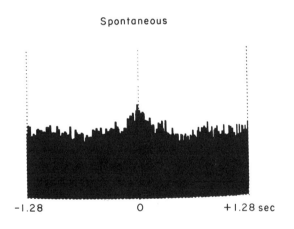

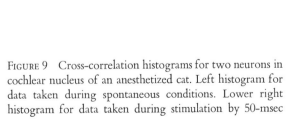

Units 32-3, 32-2

-1.28          0          +1.28 sec

FIGURE 9   Cross-correlation histograms for two neurons in cochlear nucleus of an anesthetized cat. Left histogram for data taken during spontaneous conditions. Lower right histogram for data taken during stimulation by 50-msec

tone bursts of "best" frequency, 1 per second. Upper right histogram (dots) is control calculation for the cross-correlation histogram immediately below.

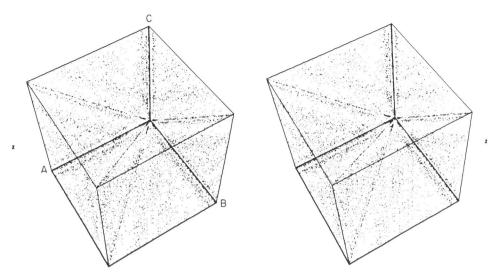

FIGURE 10   Joint PST scatter diagram for spike trains from three simultaneously recorded neurons in pleural ganglion of *Aplysia*. Stimulus was an electric shock to one of the nerves (W. Kristan, personal communication). Origin is at the center rear corner of the cube, and represents the instant of stimulus delivery. Activity of each neuron is plotted, respectively, along axes labeled A, B, C. Each edge of the cube is 10 sec long. This and Figure 11 are arranged for stereoscopic viewing. The left image is intended for the left eye. Thus, the eyes should be converged as if looking at a distant object, although focused on the page.

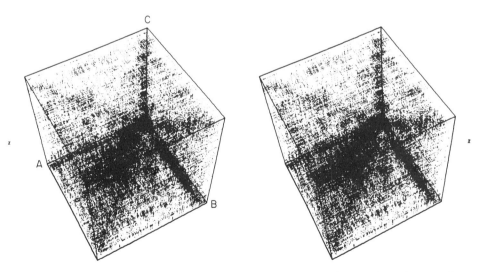

FIGURE 11   Joint PST scatter diagram for spike trains from three simultaneously recorded neurons in auditory cortex of a chronically prepared cat. Stimuli were 50-msec tone bursts of "best" frequency, 2 per second. Each edge of the cube is 0.5 second long.

with a particular small relative time interval; this probability is itself related to the instants of time that stimuli are presented.

In addition to the diagonal lines of high density in the coordinate planes, a line of high but varying density extends along the major diagonal of the cube in Figure 10. This represents the near-coincident firing of all three neurons. An interpretation consistent with the above observations is that all three neurons represented in Figure 10 had input from the stimulus and from some other common source. The latter input was itself modulated by the stimulus. The identification of the diagonal densities in Figure 10 as resulting from common input rather than direct connections (for example, from A to B to C) was possible, because all three neurons were recorded intracellularly and showed coincident PSP input.

The analysis of three cat neurons shown in Figure 11 has more apparent variability because of the shorter time scale. Nevertheless, most of the same features are visible; bands of higher density fall along the coordinate axes, on the diagonal of each coordinate plane, and along the principal diagonal of the cube. The interpretation can be that all three neurons had input from the stimulus and from some other common source, which was itself modulated by the stimulus. The choice between the possibility of direct connection and common source for these data is based on the width of the diagonal structures in Figure 11. Far narrower diagonal structure is found in some cases (Gerstein and Perkel, 1969) and is more likely to be the sign of direct interconnection between the observed neurons.

## Conclusion

As shown in the examples above, various inferences may be drawn from the analysis of multiple, simultaneously recorded, spike trains. The initial result is establishment of an equivalent "wiring diagram" for the connections that may exist to and between the observed neurons. When the appropriate control calculation is made, the effects of stimulus on the observed neurons can be assessed. Residual correlation suggests the presence of direct interaction between, or common input to, the observed neurons.

If interactions between the observed neurons exist, they may vary in time or in a manner related to the presentation of the stimulus. By appropriate experimental design, it is possible to seek the existence of stimulus coding in such form. If it is stimulus selective, correlation in the firings of two or more neurons may itself serve as a method of coding sensory (or other) information in the nervous system. In searching for this type of coding, one must identify correlation between spike trains that exceeds that resulting from the stimulus-related variations in each spike train. It is also necessary to show that a decoding mechanism for spike-train correlation exists among the neurons upon which the experimentally observed neurons synapse. We do not have such information for the particular example of "candidate code" described above. Both spatial and temporal sequence of incoming synaptic activation have, however, been shown to be important in determining the firing probability of a neuron. Experimental evidence for sensitivity to synaptic sequence can, in part, be found in Hall (1965), Moushegian et al. (1964), and Rose et al. (1966). Theoretical calculations of the effect of synaptic sequence on the resulting PSPs have been carried out in part by Rall (1964), MacGregor (1968), and Fernald (1970).

Perhaps the most interesting application of these approaches to the study of neuronal grouping is in experiments with plasticity of the nervous system. The basic experimental design examines the correlation structure of spike trains as a function of a conditioning paradigm. Thus, it is possible to measure whether the strength or the nature of the interconnections between the observed neurons is varied by the conditioning procedure. Experiments at our laboratory have shown that such changes of interactions between neurons do take place (W. Kristan, personal communication). As such studies progress, it may well turn out that the functional grouping of neurons is a dynamic property and that association of neurons depends on past usage as well as present task. Just as a picture of the nervous system in terms of single-neuron function has emerged, we may hope that the next few years will bring an understanding of neural groupings as functional elements of the nervous system.

## Acknowledgment

This research was supported by NIH Grants NB-05606 and FR-15.

### REFERENCES

Cox, D. R., 1962. Renewal Theory. Methuen, London, and John Wiley and Sons, New York.

Fernald, R., 1970. A neuron model with spatially distributed synaptic input. *Biophys. J.* (in press).

Gerstein, G. L., and W. A. Clark, 1964. Simultaneous studies of firing patterns in several neurons. *Science (Washington)* 143: 1325–1327.

Gerstein, G. L., and D. H. Perkel, 1969. Simultaneously recorded trains of action potentials: Analysis and functional interpretation. *Science (Washington)* 164: 828–830.

Glaser, E. M., 1970. Recognition and separation of interleaved neuronal activity. *In* Advances in Biological and Medical Engineering, Vol. 1 (S. Fine and R. M. Kenedi, editors). Academic Press, N. Y. and London (in press).

GLASER, E. M., and W. E. MARKS, 1968. On line separation of interleaved neuronal pulse sequences. *In* Data Acquisition and Processing in Biology and Medicine, Vol. 5 (K. Enslein, editor). Pergamon Press, New York.

HALL, J. L., 1965. Binaural interaction in the accessory superior-olivary nucleus of the cat. *J. Acoust. Soc. Amer.* 37: 814–823.

HILTZ, F. F., 1965. A method for computer recognition of intracellularly recorded neuronal events. *Inst. Elec. Electron. Eng. Trans. Bio-Med. Eng.* 12: 63–72.

KEEHN, D. G., 1966. An iterative spike separation technique. *Inst. Elec. Electron. Eng. Trans. Bio-Med. Eng.* 13: 19–28.

MacGREGOR, R. J., 1968. A model for responses to activation by axodendritic synapses. *Biophys. J.* 8: 305–318.

MOUSHEGIAN, G., A. RUPERT, and M. A. WHITCOMB, 1964. Brainstem neuronal response patterns to monaural and binaural tones. *J. Neurophysiol.* 27: 1174–1191.

PERKEL, D. H., G. L. GERSTEIN, and G. P. MOORE, 1967a. Neuronal spike trains and stochastic point processes. I. The single spike train. *Biophys. J.* 7: 391–418.

PERKEL, D. G., G. L. GERSTEIN, and G. P. MOORE, 1967b. Neuronal spike trains and stochastic point processes. II. Simultaneous spike trains. *Biophys. J.* 7: 419–440.

RALL, W., 1964. Theoretical significance of dendritic trees for neuronal input-output relations. *In* Neural Theory and Modeling (R. F. Reiss, editor). Stanford University Press, Stanford, California, pp. 73–97.

RALL, W., 1967. Distinguishing theoretical synaptic potentials computed for different soma-dendritic distributions of synaptic input. *J. Neurophysiol.* 30: 1138–1168.

ROSE, J. E., N. B. GROSS, C. D. GEISLER, and J. E. HIND, 1966. Some neural mechanisms in the inferior colliculus of the cat which may be relevant to localization of a sound source. *J. Neurophysiol.* 29: 288–314.

SIMON, W., 1965. The real-time sorting of neuro-electric action potentials in multiple unit studies. *Electroencephalogr. Clin. Neurophysiol.* 18: 192–195.

# 58 Data Transmission by Spike Trains

## C. A. TERZUOLO

IN THIS PAPER I summarize some of the data and the main conclusions on the subject of data transmission in the nervous system as gleaned from the work of Drs. Bayly, Knox, Poppele, and myself. First, I provide an outline of the approach to be used. Then I touch upon the following topics:

1) The information-carrying parameter of nerve-impulse trains.
2) Demodulation of the information by biological systems (nerve cell and effector muscle).
3) The implication of "regular" and "irregular" carrier rates for data transmission in single and parallel channels.

### Systems-analysis approach

In selecting the methodological approach for determining the time-dependent characteristics of neuronal systems a few considerations are relevant. First, it can be contended

C. A. TERZUOLO Laboratory of Neurophysiology, University of Minnesota Medical School, Minneapolis, Minnesota

that communication between the organism and its environment occurs essentially by way of continuous signals. In receptor organs, for example, the potential changes known as generator potentials are the first neuronal correlate of the input signal. The subsequent transformation of this continuous voltage into a discontinuous process, namely impulse activity, does not preclude the possibility of using the techniques of systems analysis that were developed for deterministic and continuous linear systems. To this end, deterministic input functions can be used and the output can then be derived by measuring a certain parameter or parameters of the spike train or by demodulating the information by means of procedures similar to those utilized by the organism in transforming impulse activity into continuous signals. With either approach, sample averaging is often required for the purpose of extracting the deterministic properties of the system under study. Indeed, the variables affecting the parameters responsible for neuronal activities combine, in most instances, to produce outputs that are essentially probabilistic in nature. This is particularly true when the occurrence time of individual action potentials is considered. However, the underlying deterministic properties of the system are still definable, by using systems analysis, if the distribution of output states is such that no distortions attributable to nonlinearities are present at the output. Sample averaging removes the effects of variables that are either intrinsic to the system, or result from changes in inputs that cannot be controlled by the experimenter. These effects are, therefore, equated with "noise."

To be sure, the approach will not permit, when spike trains are considered, to predict the time of occurrence of individual spikes if "noise" is present. However, when considering the integrative aspects of nerve functions in vertebrates it would seem that: 1) the time over which a spike can occur, in any given neuron, is only a small fraction of the integration time of the subsequent neuron; and 2) a large number of input channels usually converge upon central neurons, whereby only the density of spike occurrence within these converging channels is significant.

The experimental situation, when dealing with spike trains, can therefore be summarized as follows:

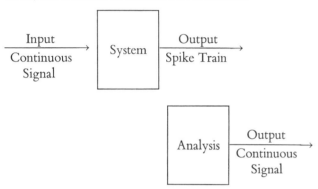

When the input is deterministic and continuous, as for a sinusoidal modulation, the analysis is designed to extract the modulation component present in the spike train. By measuring the gain and phase relations between this modulation component, which is contained in certain parameters of the spike train (see below), the time-dependent characteristics of the underlying deterministic properties of the system can be rigorously defined within the limits of its linear behavior. These results can then be used either to develop a model or to test the validity of previously proposed models.

## The information-carrying parameter of nerve-impulse trains

When confronted with a device whose output is in the form of a train of pulses, a choice must be made about the parameter to be quantitated for the purpose of recovering the information applied at the input. Indeed, several related but nonequivalent parameters can be described for a train of pulses. When the device considered is a biological system, the choice of the output parameter should take into account

the fact that the train of nerve impulses is going to be the input to other biological systems. Therefore, the parameter to be measured cannot be chosen arbitrarily, provided that we are interested in describing the way in which signals are transformed and operated upon in the nervous system. This restriction imposes the necessity of defining the biologically relevant, information-carrying parameter or parameters of nerve-impulse trains, namely those used in the process of data transmission between neurons and between neurons and effector organs. A review of the work by other authors on this subject has been recently provided by Perkel and Bullock (1968).

For our purpose, it will be sufficient to consider impulse rate (defined as the number of events for a given unit of time), intervals between impulses, and instantaneous rate. The latter is defined as the reciprocal of interval, and it is therefore equivalent to impulse rate only in the special case of an unmodulated pulse train. In a recent paper (McKean et al., 1970) data were presented that define the difference between the information carried by the parameters in question (see also Knight, 1969). Briefly, if a voltage-to-frequency converter is used, the relation between the sinusoidal input voltage to this device and the output pulse train are different, depending on whether impulse rate or intervals are considered. In the case of impulse rate (Figure 1, dashed line) the gain is constant and there is no phase shift, regardless of the input frequency.

$$\text{Gain is defined as } 20 \log_{10} \frac{\text{output}}{\text{input}} K,$$

where K is a constant. On the contrary, when instantaneous rate is measured at the output, the gain and phase relations as a function of modulation frequency become dependent on the average pulse rate, as shown in the solid curves of Figure 1.

Starting from this observation, two experiments that are formally equivalent can be designed. It is possible to measure either the transmembrane potential changes resulting from synaptic activity in a neuron receiving a frequency-modulated impulse train, or the tension changes of a muscle activated by applying a frequency-modulated pulse train to its nerve can be measured. The second experiment has the advantage that the output, muscle tension, is a behaviorally meaningful parameter. By using different input carriers (carrier being defined as the unmodulated pulse rate), one can establish if the phase between the input sinusoid to the voltage-to-frequency converter and the tension output is or is not dependent on the carrier rate. By stimulating all channels in unison, the experimental design makes a multiple-channel system equivalent to a one-channel system, therefore avoiding the effect to be observed in a population of voltage-to-frequency converters as described by Knight (1969).

Figure 2 shows the results of one such experiment. Four

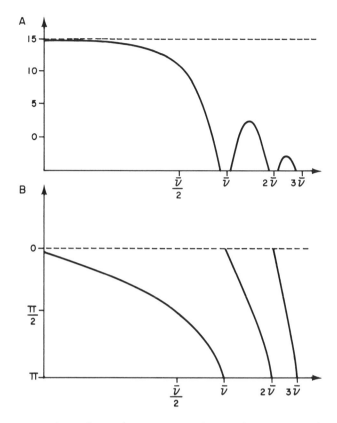

FIGURE 1 Difference between rate and interval parameters. The Bode plots describe the relations of the gain (A), in decibels, and the phase (B), in radians, between a sinusoidally modulated voltage applied to a voltage-to-frequency converter and the pulse rate (dashed line) or the instantaneous rate (solid curve) parameters. $\bar{\nu}$ is the average carrier rate. Fully explained in the text.

carrier rates were used. There is no appreciable difference in the phase of the output tension for all the carrier rates. Therefore, impulse rate or the average change around the carrier rate, rather than the interval between impulses, is the biologically relevant parameter.

## Demodulation of impulse frequency by biological systems

The above experiment also provides, within the limits of linear behavior of the systems involved, the transfer function between impulse rate and behavioral output (changes in muscle tension). The phase plot of Figure 2 is characteristic of a process of low-pass filtering. The same behavior has also been observed at the output of medial geniculate neurons (Maffei and Rizzolatti, 1967) and in the isolated sensory neuron of the crustacean stretch receptor (Terzuolo and Bayly, 1968). In the latter case, the low-pass filtering properties could be ascribed to the synaptic processes combined with the cable properties, because the dynamic characteristics of the encoder system were known and therefore could be subtracted from the Bode plot describing the input-output relation of the cell (see below).

Examination of the harmonic distortions and of the power spectrum in the nerve-muscle experiment proves that the recovered information can be free of distortion if the frequency of the input signal, or information, is low with respect to the carrier rate. This agrees well with the view expressed above that the demodulation process is equivalent to low-pass filtering. In fact, under the above conditions, the input information is simply a variation in the average, or DC value, of the pulse train and stands isolated

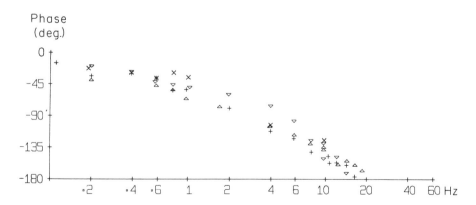

FIGURE 2 Absence of dependency between the phase angle of the sinusoidally modulated muscle tension and the average impulse frequency of the motor nerve in decerebrate cat. A sinusoidally modulated pulse train was used to stimulate the distal end of the cut gastrocnemius nerves. Stimulus carrier rates were 14.5 (×), 19.4 (pyramid), 29 (+), and 37 (inverted pyramid) pulses per second.

in the spectrum (Bayly, 1968). Provided that the filter time constant is short with respect to the modulating frequency and long with respect to the period of the carrier, recovery of information by low-pass filtering removes sideband frequency components that occur at each multiple of the carrier rate (Figure 3). The demodulated information is therefore free from distortion if the modulation depth and frequency are suitably restricted. As the frequency of the input information is increased with respect to that of the carrier rate, sample averaging becomes necessary if a single input channel is used. The results obtained by this procedure, however, are similar to those obtained, without need for averaging, when parallel input channels are used, or the mean carrier rate is increased (see next section of this chapter). Also, the process of demodulation is formally equivalent whether one considers the transfer of information between nerve cells or between a neuron and an effector muscle (McKean et al., 1970). In the first case, the filtering process results from characteristics of the postsynaptic potential; in the nerve-muscle preparation, the filtering process results from the characteristics of the contraction and relaxation processes. In both instances, the output (transmembrane potential change or muscle tension) is a function of the number of input impulses occurring in a given unit of time (the input-output relation at the neuromuscular junction being one-to-one). The averaging characteristics are determined, as in a low-pass filter, by the impulse response of the filter, which, in biological systems, is given either by the form of the postsynaptic potential or by the time course of the muscle twitch.

The operation of a low-pass filter can be demonstrated by an electrical analogue (resistance-capacitance [RC] network). Figure 4 shows the response of a filter when a sinusoidally modulated train of pulses is applied at the input. Note the similarity of this response to the muscle-tension response in Figure 2, and that this, too, is independent of carrier rate.

This conceptual model of low-pass filtering for the process of demodulation in the nervous system (cf. also Stevens, 1968) is equally applicable to both chemical and electrical synapses. Models that utilize an RC filter to mimic the time of decay of the postsynaptic potential have been used by several authors to investigate input-output relations under several conditions (cf. Segundo et al., 1968).

When the dynamic characteristics of the low-pass filtering process are specifically considered, additional implications become obvious. First of all, the averaging time of the filter must be short with respect to the frequency content of the input signal if this must be recovered unchanged (Bayly, 1968; Terzuolo and Bayly, 1968). Second, there will always be a restriction on the dynamic range of the information transfer (upper-modulation frequency), no matter how high the input and output carrier rates. Moreover, the gain of the demodulated signal will depend on both input and output carrier rates.

Although these points have been discussed previously (Terzuolo and Bayly, 1968), it is more relevant in the present context to stress that, given certain low-pass characteristics of the demodulation processes, a continuous function can be obtained at the output even when the input pulse train is modulated by more than 100 per cent, at least for certain frequencies of modulation (McKean et al., 1970).

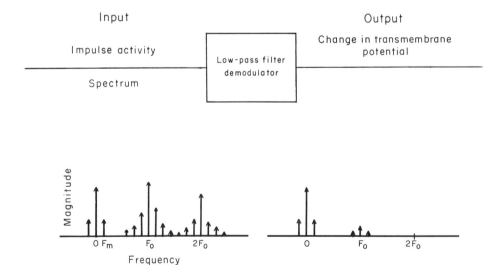

FIGURE 3 Demodulation of a "regular" impulse train by low-pass filtering. $F_m$: modulation frequency, $F_0$: carrier rate. Note that the reduction of the magnitude at frequencies above the modulation frequency depends upon the characteristics of the low-pass filter. (Modified from Bayly, 1967.)

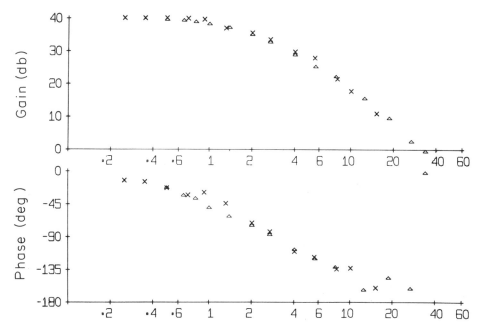

FIGURE 4 Low-pass filter behavior. The Bode plots describe the relation between the signal applied to a voltage-to-frequency converter (sinusoidal voltage) and the output of an RC filter, which demodulated the frequency-modulated pulse train produced by the voltage-to-frequency converter. Gain, in decibels, is shown in the upper plot, while the phase angle, in degrees, is seen in the lower plot. Carrier rates of 10 (×) and 20 (pyramid) pulses per second were used. Note that the gain and phase of the filter output are independent of the carrier rate (as in Figure 2).

Figure 5 illustrates this point. A filter with characteristics similar to those of the nerve-muscle preparation was used. In A, the input to the filter was a sinusoidally modulated pulse train. Separate bursts of pulses do lead, however, to similar results (only the gain being different), even if such bursts are substantially shorter than the period of the input sinusoid and if the interval between pulses, within the burst, is equal to that of A (Figure 5B). This fact demands the use of much restraint when considering the possible significance of patterns of unitary impulse activity. Also, several parameters of the pulse train become significant, under these conditions, for the recovery of the signal at the output of the low-pass filter. These parameters (duration of the burst, interval between bursts, phase relation between the burst and the input signal) are significant in the present context only when related to the frequency content of the input signal and the dynamic characteristics of the demodulation process (receiver). They cannot be considered as separate "codes" (see Perkel and Bullock, 1968) when they are found, under physiological conditions, in the same channel (the case of spinal motor neurons is a well-demonstrated instance, cf. Rosenthal et al., 1969). Indeed, as emphasized above, a knowledge of the dynamic characteristics of the receiver is a necessary condition for defining the signal-carrying parameter or parameters of impulse trains.

One more point. The model is consistent with the obser-

vation that input patterns may be significant in determining the output of the cell *when* the voltage changes produced by each presynaptic impulse are a significantly large portion of the voltage difference between resting level and threshold. This has indeed been shown under certain experimental conditions (Segundo et al., 1963). Also, presynaptic and postsynaptic changes are known to be capable of altering the effectiveness of synaptic inputs (see Eccles, 1964, for a review). This condition may also occur when more than one input is activated at the same time, owing to the fact that changes in membrane properties caused by the activation of an input can affect the postsynaptic potential caused by another input (Rall, 1964; Terzuolo and Llinás, 1966). These situations are still compatible with the low-pass filter model, as only the filter properties, or the contribution by a given input, is eventually different at any one time. However, in some of these cases, specific parameters of impulse trains can acquire an overriding biological significance (cf. Perkel and Bullock, 1968).

Of more general interest, although seldom considered, are the changes that the demodulated information may undergo during the encoding process as a consequence of the dynamic characteristics of this encoder system (spike-producing processes). Neurons, in general, can be conceptually subdivided into two systems connected by a cable, as in Figure 6A. The adequacy of this conceptualization has

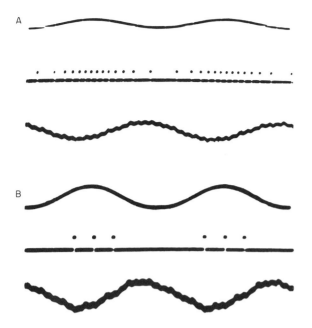

FIGURE 5 Behavior of a low-pass filter to a sinusoidally modulated and to a discontinuous pulse train. Upper trace in both A and B is the input voltage to the voltage-to-frequency converter; the second trace is the resulting train of pulses that is actually fed to the RC filter; the bottom trace is the output of the filter. In A, the train is sinusoidally modulated. In B, only a burst of equally spaced pulses was used. The modulation frequency was 10 Hz. The filter characteristics are similar to those of the nerve-muscle preparation whose frequency response is shown in Figure 2. Fully explained in the text.

been verified experimentally in the isolated neuron of the crustacean stretch receptor, in which the dynamic properties of the over-all neuron, as well as those of each of its component systems, were first studied (see Terzuolo and Knox, 1970), and in the eccentric cells of the *Limulus* eye (Dodge et al., 1968). By applying a sinusoidal current through a microelectrode placed inside the soma, the transfer function described by the Bode plot of Figure 6B was obtained for the encoder of the crustacean neuron. The gain and phase are not flat through the range of input frequencies (Terzuolo et al., 1968). Moreover, the behavior is dependent on the carrier rate even if impulse frequency, rather than instantaneous frequency, is the parameter measured.

Similar carrier-dependent behavior is also observed in the secondary endings of muscle spindles but not in the primary endings (Poppele and Bowman, 1970). Moreover, the differences in the response of different neurons to a current pulse, although in most cases not expressed in the form needed to define the frequency response characteristics of their encoder systems, certainly indicate that these characteristics vary from cell to cell (cf. Eccles, 1964). Finally, the dynamic properties of an encoder model have been shown to

be influenced by such factors as: 1) the amount of membrane that is antidromically invaded; 2) changes in time constant and space constant; 3) amount of local response; and 4) accommodation (Bayly, 1967). It is our view, therefore, that the properties of this system are a most significant factor in determining functional differences between neurons.

### Implications of "regular" and "irregular" carrier rates for data transmission in single and parallel channels

The presence of a "regular" rate of firing, when no time-variant function is applied at the input, was considered above as the equivalent of the carrier rate in a man-made communication system. It is obvious, however, that the average rate of the unmodulated impulse train, that is, its DC component, can be utilized by the nervous system to sense or impose levels of activity. This is certainly the case for those sensory systems subserved by receptors whose impulse frequency is proportional to the applied stimulus under steady-state conditions. However, distributions of impulse intervals vary in different systems and even within the same system under different conditions. While in some cases it is well known that the interval distribution is Gaussian, in other instances Gamma or Poisson distributions are present (as examples, see Gestri et al., 1966, and Knox, 1969, 1970).

One may therefore ask what influence the different types of carriers have on the signal that is transmitted by the different types of channels and what the consequences might be for the subsequent demodulation of the information by low-pass filtering.

To answer these questions, a common representation for all types of spike trains is desirable, and preferably in a form amenable to signal analysis in the frequency domain. The power-spectral density satisfies these requirements, because only this representation can be used for "irregular" carriers (the properties of these spike trains being definable only probabilistically [Perkel et al., 1967a, 1967b; Moore et al., 1968]).

For "regular" carrier rates, the spectral density shows that the DC component is accompanied by a single component at the modulation frequency, without multiples or harmonics. In addition, the spectrum contains components at all harmonics of the carrier rate, each accompanied by a set of adjacent sideband components that are displaced from these harmonics by multiples of the modulation frequency (Bayly, 1968), as shown in Figure 3. This situation places restrictions, when a single input channel is used, on the possibility of demodulating high-frequency information by low-pass filtering. Indeed, as the frequency of the input information approaches the carrier rate, distortions will arise from the

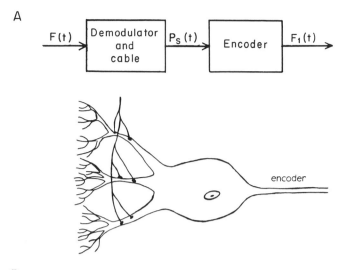

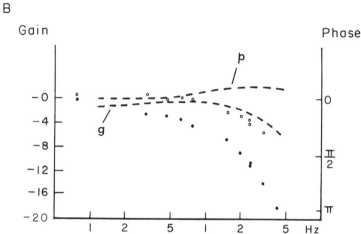

FIGURE 6  Data transmission between neurons. A: Schematic diagram of the process of data transmission between neurons. The presynaptic impulse train F(t) is demodulated by a process of low-pass filtering that leads to the transmembrane potential change $P_s(t)$. This is the input to the second system which encodes $P_s(t)$ in a frequency modulated output train $F_1(t)$. B: Dynamic properties of the two systems described in A. The neuron is the sensory cell of the slowly adapting stretch receptor organ of Crustacea. The input of the cell was a sinusoidally modulated impulse train, produced by stimulating the inhibitory axon of the cell. The output measured was instantaneous frequency. Gain, in decibels (open circles), and phase, in radians (closed circles), are plotted versus the modulation frequency, the carrier rates of the input-and output being the same (10 impulses per second). The broken lines describe the gain (g) and the phase (p) of the encoder system alone, as obtained by applying a sinusoidally modulated current through a microelectrode placed inside the soma of a crustacean neuron. These values must be subtracted from the experimental points to determine the transfer function of the demodulation process. (Modified from Terzuolo and Bayly, 1968.)

lower sidebands of the carrier component, because these first enter the filter passband (Bayly, 1968). Time and sample averaging must therefore be used experimentally to recover the applied information. In the central nervous system, however, parallel channels are available. The convergence of these input channels upon the same cell results in an effective increase in the average carrier rate. Therefore, while the information conveyed by any given channel may be ambiguous and distorted for a given combination of input frequency and carrier rate, the convergence of these inputs permits the cell to perform a sample average (Bayly, 1968;

Terzuolo and Bayly, 1968). The information can therefore be immediately available to the postsynaptic cell with little or no distortion (Poppele and Terzuolo, 1968; Maffei, 1968).

An example of the effect of increasing the carrier rate in a single channel is shown in Figure 7, and Figure 8 shows the effect of parallel channels converging upon a single motor neuron. Actually, an estimate can be made of the improvement in the signal-to-noise ratio to be expected as a function of the number of input channels.

If the carrier rates of the individual channels differ, then an improvement in the signal-to-noise power ratio, S/N, occurs because of the dispersion in frequency of the spectral components representing the distortion. In this case, and if the modulation depth is low, it may be shown that S/N increases as the number of channels increase (Bayly, 1968). If the carrier rates are identical and the channels are independent, improvement in signal-to-noise power ratio may still be realized. For a two channel system, S/N is given approximately by

$$S/N = \frac{\pi^2(1-\alpha)^2}{\text{Sin}^2(\pi-\pi\alpha)},$$

if $\alpha$, the percentage modulation, is low.

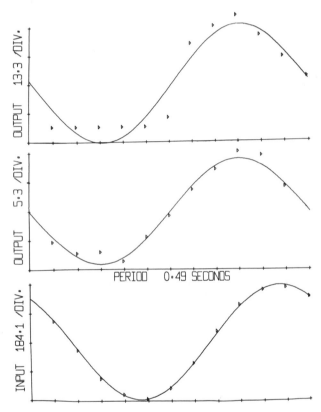

FIGURE 7  Linearization of the spindle output by the increase of the average carrier rate. The output from a primary ending of a gastrocnemius muscle spindle was modulated by stretching the muscle sinusoidally in the presence and in the absence of gamma activity. Upper plot: Spindle response in the absence of gamma stimulation. Note the absence of activity throughout one-half of the cycle. Middle plot: Spindle response in the presence of gamma stimulation of 28 pulses per second. Ordinate: changes in spindle discharge frequency in pulses per second. Lower plot: averaged data points of the input, with the fundamental frequency fitted to these points. Ordinate: microns of displacement. (From Rosenthal et al., 1969.)

When discussing multiple channels, it should be realized that, irrespective of the nature of the carrier in any given channel and its rate, the combined activity quickly tends to resemble that of a random Poisson process (Cox and Lewis, 1966), if sufficiently large numbers of independent channels are utilized in the absence of an applied input. Therefore, this type of carrier may be taken as being more general than the preceding one. Not surprisingly, the power spectrum of the random spike train shows that there are no components attributable to the carrier rate, $\lambda_0$, but only a large DC component whose power is proportional to $\lambda_0^2$ and a continuum of noise components whose power is proportional to $\lambda_0$ (Knox, 1969a, 1969b). In the case of sinusoidal modulation of the mean rate of impulse occurrence, a single component appears in the spectrum, the power of which is proportional to the modulation amplitude, $\lambda_m$, squared (Figure 9). Demodulation of the random spike train is therefore possible by low-pass filtering. However, there will always be distortion present at the output of the filter because of the continuum of the noise components. As discussed by Knox (1969, 1970), if the filter passband is $f_B$, the signal-to-noise power ratio is given by

$$S/N = \alpha^2\lambda_0/4f_B$$

for modulation frequencies much less than the reciprocal of the pulse width, and where $\alpha = \lambda_m/\lambda_0$. Notice that this expression applies equally to single random spike trains and to

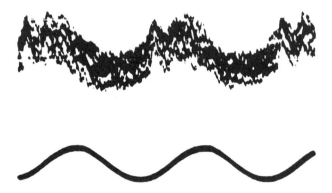

FIGURE 8  Sample average by a postsynaptic neuron resulting from the convergence of parallel input channels. Upper trace is an intracellular recording of transmembrane potential of a gastrocnemius motor neuron during sinusoidal muscle stretch (lower trace). Stretch frequency was 4 Hz. About ten sweeps are superimposed to show that the input information (sinusoidal stretch) becomes immediately available to the postsynaptic cell because of the spatial averaging provided by the convergence of parallel channels. This, in spite of the fact that the information contained in each channel is only partial and ambiguous (upper plot of Figure 7). Fully explained in the text. (Poppele and Terzuolo, unpublished.)

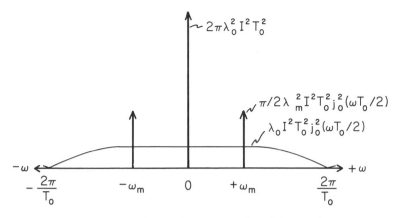

FIGURE 9 Power spectral density of the random pulse train. Sinusoidal modulation of the mean rate of occurrence, $\lambda(t) = \lambda_0 + \lambda_m \cos \omega_m t$, impulses per second, introduces a single component at the modulation frequency. In the ab- sence of modulation, the spectrum contains the DC and band limited noise components. (Modified from Knox, 1969.)

the combined activity of multiple, independent channels, because, if we suppose for simplicity that $\lambda$ is constant, then $\lambda_0$ can represent the sum of the individual channel carrier rates. The relationship shows the trade-offs available between number of channels, carrier rates, and filter passbands for any given signal-to-noise ratio. As an example, for $\alpha = 0.5$, $f_B = 16$ Hz and S/N = 10, $\lambda_0 = 2300$ impulses per second. This illustrates the disadvantages of using a single input channel with a random carrier.

However, the rate shown above is well within the physiological limits of systems that utilize parallel channels. For instance, all the primary endings from muscle spindles are said to distribute information to the entire motor neuron population belonging to the same muscle (Mendell and Henneman, 1968). In the gastrocnemius muscle of the cat, as an example, there are roughly 140 spindles (Swett and Eldred, 1960), so an average carrier rate of 20 impulses per second could satisfy the above requirement. This rate is well below that which can be attended under gamma activity. Moreover, it is likely that randomness, even in the presence of a forcing function affecting all spindles alike (e.g., stretch), may be insured by the random action of gamma motor neurons upon the individual spindles.

## Conclusions

In closing this brief review of some of our findings, I would like to stress that the study of the dynamic properties of single neurons alone, although a necessary step in the quest for the operational principles utilized by the nervous system, might not be sufficient for the formulation of models capable of accounting for relatively complex functions. The properties of populations of nerve cells could indeed exceed those of the individual elements of the population. Confronted with the task of "explaining" the input-output relations of the nervous system in terms suitable to the formulation of quantitative models, the choice of the experimental approach and data-processing techniques often acquires an overriding importance.

In several physiological functions the use of systems-analysis techniques and the approximations to linear analysis, for the purpose of defining the time-dependent characteristics of component systems, is now warranted over a behaviorally meaningful portion of the operational range of the systems involved. Only by defining such dynamic characteristics in relation to behaviorally meaningful outputs can we eventually acquire an appreciation of the significance, or insignificance, of parameters that are presently utilized only because they can be measured by available techniques. It is likely that the emphasis placed on some of these parameters might not be justifiable in every case. For instance, we have presented here data showing that certain details of given patterns of impulse activity may have no functional significance, owing to the dynamic characteristics of the system that receives such patterns. (For a review on this subject, in which those data supporting a different viewpoint are also considered, see Perkel and Bullock, 1968.) Conversely, certain properties that have not yet been adequately explored may be present.

For instance, the possible role of neurons as interface devices, instead of only summing integrators, is rarely considered. The dynamic properties of the encoder alone could provide for this possibility, as it is not inconceivable that the encoder could act to transmit information selectively within a certain frequency range, thereby transforming the output of one system into a form compatible with the function of another. Indeed, it seems unlikely that the output of one neuron is in a form appropriate for all those systems to which it distributes. The existence of such neurons may be now suspected, given the observation that only pyramidal-tract fibers apparently impinge upon certain spinal cord interneurons (Kostyuk et al., 1968).

If confirmed and extended to other cells, this observation can be highly significant. One would expect, in these cases, a highly different behavior from that observed in integrative neurons, such as the alpha motor neuron, in which uniform populations display a linear input-output relation, in terms of phase and relative gain, within a large frequency range (Poppele and Terzuolo, 1968; Rosenthal et al., 1969). Similar frequency independency of the input-output relations was also observed in neurons of the dorsal spinocerebellar tract (Jansen et al., 1967) over a modest frequency range. This behavior can easily be accounted for on the basis of the arguments developed above concerning the signal-carrying parameter of nerve-impulse trains and the demodulation processes by low-pass filtering. This simplicity is also found by comparing psychophysical measurements to unit responses in the somatic system, and has led Mountcastle (1967) to suggest that signal transformation in this system is entirely linear. In this context, one may notice that when "regular" carrier rates are present, a condition that appears to be particularly advantageous in a few input channels, the presence of a cascade of suitable low-pass filters (successive synapses) may provide a means of removing those distortions of the input signal that are the result of the presence of the carrier (sidebands).

## Acknowledgments

It is a pleasure to thank Drs. Bayly, Knox, McKean, Poppele, and Rosenthal for reading the manuscript critically and suggesting several improvements. The research program is supported by PHS Grant NB 2567, and by the Air Force Office of Scientific Research Grant AF-AFOSR-1221-67.

### REFERENCES

BAYLY, E. J., 1967. Measuring and modeling the dynamic properties of a nerve cell. University of Minnesota, Minneapolis, doctoral thesis.

BAYLY, E. J., 1968. Spectral analysis of pulse frequency modulation in the nervous systems. *Inst. Elec. Electron. Eng. Trans. Bio-Med. Eng.* 15: 257–265.

COX, D. R., and P. A. W. LEWIS, 1966. The Statistical Analysis of Series of Events. John Wiley and Sons, New York.

DODGE, F. A., JR., B. W. KNIGHT, and J. TOYODA, 1968. How the horseshoe crab eye processes optical data. IBM Research Report RC 2248.

ECCLES, J. C., 1964. The Physiology of Synapses. Springer-Verlag, Berlin.

GESTRI, G., L. MAFFEI, and D. PETRACCHI, 1966. Spatial and temporal organization in retinal units. *Kybernetik* 3: 196–202.

JANSEN, J. K. S., R. E. POPPELE, and C. A. TERZUOLO, 1967. Transmission of proprioceptive information via the dorsal spinocerebellar tract. *Brain Res.* 6: 382–384.

KNIGHT, W., 1969. Frequency response for sampling integrator and for voltage to frequency converter. *In* Systems Analysis Approach to Neurophysiological Problems (C. A. Terzuolo, editor). University of Minnesota, Minneapolis, pp. 61–72.

KNOX, C. K., 1969. An analysis of the discharges of crustacean statocyst receptors and their responses to sinusoidal rotations. University of Minnesota, Minneapolis, doctoral thesis.

KNOX, C. K., 1970. Signal transmission in random spike trains. *Kybernetik* (in press).

KOSTYUK, P. G., D. A. VASILENKO, and A. G. ZADOROZHNY, 1968. Pyramidal activation of propriospinal neurons. *Proceedings of the International Union of Physiological Sciences* 7: 245 (abstract).

McKEAN, T. A., R. E. POPPELE, N. P. ROSENTHAL, and C. A. TERZUOLO, 1970. The biologically relevant parameter in nerve impulse trains. *Kybernetik* 6: 168–170.

MAFFEI, L., 1968. Spatial and temporal averages in retinal channels. *J. Neurophysiol.* 31: 283–287.

MAFFEI, L., and G. RIZZOLATTI, 1967. Transfer properties of the lateral geniculate body. *J. Neurophysiol.* 30: 333–340.

MENDELL, L. M., and E. HENNEMAN, 1968. Terminals of single Ia fibers: Distribution within a pool of 300 homonymous motor neurons. *Science (Washington)* 160: 96–98.

MOORE, G. P., D. H. PERKEL, and J. P. SEGUNDO. Applications of the theory of stochastic point processes in the detection and analysis of neuronal interaction. Proceedings of the International Federation of Automatic Control Symposium on Technical and Biological Problems of Control 1968, Yerevan, Armenia (in press).

MOUNTCASTLE, V. B., 1967. The problem of sensing and the neural coding of sensory events. *In* The Neurosciences: A Study Program (G. C. Quarton, T. Melnechuk, and F. O. Schmitt, editors). The Rockefeller University Press, New York, pp. 393–407.

PERKEL, D. H., D. L. GERSTEIN, and G. P. MOORE, 1967a. Neuronal spike trains and stochastic point processes. I. The single spike train. *Biophys. J.* 7: 391–418.

PERKEL, D. H., D. L. GERSTEIN, and G. P. MOORE, 1967b. Neuronal spike trains and stochastic point processes. II: Simultaneous spike trains. *Biophys. J.* 7: 419–440.

PERKEL, D. H., and T. H. BULLOCK, 1968. Neuronal coding. *Neurosciences Res. Program Bull.* 6 (no. 3): 221–348.

POPPELE, R. E., and C. A. TERZUOLO, 1968. Myotatic reflex: Its input-output relation. *Science (Washington)* 159: 743–745.

POPPELE, R. E., and R. J. BOWMAN, 1970. Quantitative description of linear behavior of mammalian muscle spindles. *J. Neurophysiol.* 33: 59–72.

RALL, W., 1964. Theoretical significance of dendritic trees for neuronal input-output relations. *In* Neuronal Theory and Modeling (R. F. Reiss, editor). Stanford University Press, Stanford, California, pp. 73–97.

Rosenthal, N. P., T. A. McKean, W. J. Roberts, and C. A. Terzuolo, 1969. Linear analysis of the myotatic reflex and component systems. *In* Systems Analysis Approach to Neurophysiological Problems (C. A. Terzuolo, editor). University of Minnesota, Minneapolis, pp. 141–156.

Segundo, J. P., G. P. Moore, L. Stensaas, and T. H. Bullock, 1963. Sensitivity of neurones in *Aplysia* to temporal pattern of arriving impulses. *J. Exp. Biol.* 40: 643–667.

Segundo, J. P., D. H. Perkel, H. Wyman, H. Hegstad, and G. P. Moore, 1968. Input-output relations in computer-simulated nerve cells. Influence of the statistical properties, strength, number and inter-dependence of excitatory pre-synaptic terminals. *Kybernetik* 4: 157–171.

Stevens, C. F., 1968. Synaptic physiology. *Proc. Inst. Elec. Electron. Eng.* 56: 916–930.

Swett, J., and E. Eldred, 1960. Distribution and numbers of stretch receptors in medial gastrocnemius and soleus muscles of the cat. *Anat. Rec.* 137: 453–460.

Terzuolo, C. A., and R. Llinás, 1966. Distribution of synaptic inputs in the spinal motoneurone and its functional significance. *In* Nobel Symposium I—Muscle Afferents and Motor Control (R. Granit, editor). Almqvist and Wiksell, Stockholm, pp. 373–384.

Terzuolo, C. A., and E. J. Bayly, 1968. Data transmission between neurons. *Kybernetik* 5: 83–85.

Terzuolo, C. A., R. L. Purple, E. Bayly, and E. Handelman, 1968. Postsynaptic inhibition—its action upon the transducer and encoder systems of neurons. *In* Structure and Function of Inhibitory Neuronal Mechanisms (C. von Euler, S. Skoglund, and U. Söderberg, editors). Pergamon Press, Oxford and New York, pp. 261–275.

Terzuolo, C. A., and C. K. Knox, 1970. Static and dynamic behavior of the slowly adapting stretch receptor organ of Crustacea. *In* Handbook of Sensory Physiology (W. Lowenstein, editor). Vol. 1, Springer-Verlag, Berlin (in press).

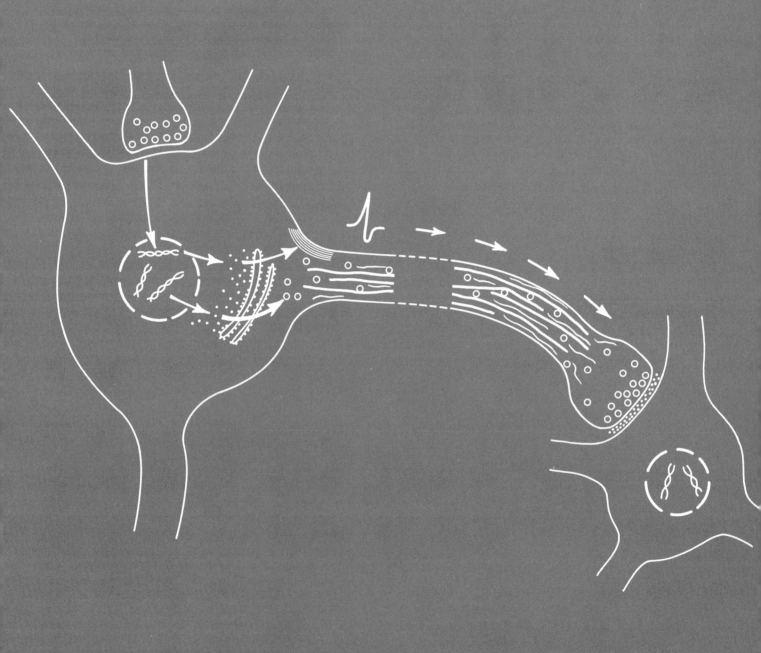

# ASPECTS OF
# MOLECULAR
# NEUROBIOLOGY

*Neuronal dynamics:* Synaptic release, *in a complex cleft, of transmitter, trophic, and macromolecular materials that modify the electric, metabolic, and genetic expression of the postsynaptic neuron;* linkage *between membrane componentry and genetic machinery that results in induction of enzyme synthesis and in changes of initial segment excitability; and* production *of vesicular organelles and enzyme cargo, translocated down the axon possibly by chemomechanical energy-coupling on microtubules.*

# 59 Prefatory Remarks on Aspects of Molecular Neurobiology

## FRANCIS O. SCHMITT

THE TERM "molecular biology" was initially popularized by William T. Astbury, the first professor of biomolecular structure, to denote that branch of life science which attempts to determine the molecular parameters of cell structure and function (Hess, 1970). Before electron microscopy, molecular biology was based primarily on X-ray crystallography and polarization optics; these and other physical methods, used in what was then called "colloid" science, helped deduce molecular parameters, particularly with regard to protein effectors of cells. Two decades later, DNA and RNA were characterized chemically and structurally, thereby ushering in the era of molecular genetics, with its concepts of transcription and translation in biosynthesis. With this breakthrough, the term "genetics" was broadened beyond the confines of problems of heredity as such. The implications of molecular genetics grew so large that it became widely identified with the parent field of molecular biology.

In the Second Intensive Study Program of the Neurosci-

FRANCIS O. SCHMITT Chairman, Neurosciences Research Program, Massachusetts Institute of Technology, Brookline, Massachusetts

ences, molecular neurobiology was considered from the broad purview of the molecular basis of the structure and function of brain cells and their organization into the systems and subsystems of the central nervous system. The coverage was not meant to be systematic or encyclopedic; for example, it included relatively little systematic neurochemistry, particularly that of the qualitative and the quantitative "What's-there-and-how-much?" kind. The aim was to be selective: it was not merely to deal with the brain as another tissue—like liver or kidney, which have special patterns of metabolism—but to be concerned especially with cellular processes that are unique to the brain; with phenomena that underlie the development of the brain in all its intricate circuitry; with the molecular basis of bioelectrical events, e.g., impulse propagation in axons and in neuronal nets; and, finally, with processes that confer plasticity, such as might provide a basis for learning and memory.

The lectures on molecular neurobiology articulated the current status of selected topics, suggested fruitful strategies of research and experimentation, critically examined current dogmas, however venerable and established, and asked questions that might open new avenues of research, especially to investigators who prepared for this inflectional point in the history of science by becoming proficient in both physical biochemistry and neurobiology.

Contributions to neural development at the molecular level were made by various speakers and participants in this symposium; therefore, there was substantial integration with the symposium on Developmental Neurobiology, organized and chaired by M. V. Edds, Jr.

In addition to the most obvious themes of molecular neurobiology already mentioned, the overriding problem of the molecular basis of learning, storage, delocalization and retrieval of memory, and other "higher" nervous functions were considered in a lower key. No formal lectures dealt with subjects such as the biochemical inhibition or stimulation of memory processing or the attempted transfer of memory or savings by administration of material obtained from trained animals. To some extent these topics were treated in the symposium on Determinants of Neural and Behavioral Plasticity, organized and chaired by R. Galambos.

The plenary and symposium lectures are conveniently grouped under four major headings: membranes; neuronal junctions; neurobiosynthesis; and neuroplasmic transport.

MEMBRANES AND INTERCELLULAR INTERACTION  Important ideas about the molecular organization of neuronal membranes are presented by M. Delbrück, who discusses new advances in specific ion transport in model membranes. M. Eigen provides a penetrating and highly original analysis of the mechanism by which ions as similar physically as

Na$^+$ and K$^+$ may be distinguished in model membrane systems and, presumably, also in natural membranes; this hypothesis may well prove to be the answer to this central problem that has long eluded biologists, biochemists, and biophysicists. Application of physical methods, particularly electron spin labeling, are presented by H. M. McConnell. Molecular mechanisms underlying impulse propagation, as deduced from optical studies (birefringence, light scattering, fluorescent probes) conducted chiefly in his own laboratory, are discussed by R. D. Keynes.

NEURONAL JUNCTIONS  The molecular biology of the synapse is portrayed at the electron-microscope level by J. D. Robertson and F. E. Bloom. Synaptic membranes are presumably sites of gene-directed molecular recognition that determines the ordering of neuronal circuitry in the developing central nervous system S. H. Barondes presents the view that glycoproteins and glycolipids, especially carbohydrate terminal residues, are responsible for this recognition. The biochemical aspects of neuronal junctions are given by V. P. Whittaker, with special reference to studies of isolated synaptosomes. Interwoven with the concepts of the synapse are current views about the synthesis, release, and mode of action of transmitters, presented by L. L. Iversen.

Glial membranes and glial-neuronal interaction are considered by R. P. Bunge, who, after dealing with the general phenomenon of neuronal ensheathment by glia and with permeability and bioelectrical parameters, concludes that, although not indispensable to neurons (under specialized or experimental conditions), glia are important regulators of neuronal activity.

BIOSYNTHESIS AND GENE EXPRESSION IN BRAIN CELLS  Comparative studies suggest that the portrayal of the prototypical vertebrate neuron given in textbooks is not a good model for the neurons of invertebrates, which represent some 95 per cent of the known animal species. M. J. Cohen points out that the soma is not so involved biochemically and physiologically in the neurons of invertebrates (at least in insects) as in those of vertebrates, which may explain why the former show more stereotyped behavior than do the latter.

E. M. Shooter discusses brain-cell organelles in the framework of the synthesis of products, chiefly proteins and enzymes, of gene expression. He also considers the effect on the regulation of gene expression by internal stimulation, e.g., by hormones and by the nerve growth factor, which has been the object of intensive study in his laboratory.

A. L. Lehninger authoritatively characterizes mitochondrial function, especially as manifested in brain cells, and offers cogent reasons supporting the view that mitochondrial membranes, readily available for experimentation, may represent a fruitful model of the neuronal membrane.

NEUROPLASMIC FLOW AND FAST TRANSPORT   The subject of slow neuroplasmic bulk flow and fast specific transport is portrayed in its broad neurobiological setting by P. A. Weiss, whose work over the last three decades has revolutionized our concepts of neuronal dynamics. Following his overview, P. F. Davison analyzes the possible role in these phenomena of the fibrous proteins, microtubules, and neurofilaments.

*Correlation and summational essays*

P. D. Nelson assesses the impact of the discussions on concerns at the higher levels of neurophysiology and psychology.

As organizer and chairman of the symposium on Molecular Neurobiology, I provide a summational essay.

### REFERENCE

Hess, E. L., 1970. Origins of molecular biology. *Science* (*Washington*) 168: 664–669.

# Lipid Bilayers as Models of Biological Membranes

M. DELBRÜCK

## Real membranes

WHEN WE TALK about biological membranes, we must recognize, first, their unity in some general features and, second, their enormous diversity in many specific features.

The most obvious aspects of unity are their thickness, around 70 Å; their composition, protein and lipid; the barrier they present to permeation; and the property of fragments of membranes to round up into vesicles and of holes to seal.

The most obvious features of membrane diversity are variations in thickness; great variations in the protein-to-lipid ratio; great variations in the kinds of lipid and kinds of protein; great variations in permeability; and infinite variations in function.

SEVERAL KINDS OF NATURAL MEMBRANES   Glancing briefly at the varieties of function, we have, first, a membrane enclosing every cell; this is the *cytomembrane* or plasmalemma, which exerts chemostatic controls on the interior and transmits numerous signals. It is characterized by the presence of glycolipids (Rouser et al., 1968). There

is the *double membrane surrounding every nucleus*. This membrane has, in many cases, been shown to have "holes," seemingly violating the principle that the real membranes minimize their perimeter. The contradiction is, however, only apparent, because the perimeters of these holes are not perimeters of the membrane; they are places at which the nuclear membrane folds over (Figure 1).

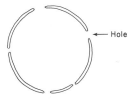

FIGURE 1   Nuclear double membrane with "holes." The perimeters of the holes are not perimeters of the membranes, but are places at which the membrane folds over.

There is the *myelin sheath,* an elaboration of the cytomembrane of the glial cells associated with nerve fibers. Perhaps this membrane is the one of least sophistication and the simplest function—that of electrical insulation. Its asymmetry is more conspicuous than that of most other mem-

M. DELBRÜCK   Division of Biology, California Institute of Technology, Pasadena, California

branes. Because it is multilayered in a very regular way, it has been a favorite subject for physical studies by means of low-angle X-ray diffraction.

There is the *mitochondrion,* with its interesting construction involving two totally different kinds of membrane, the inner one and the outer one. The *inner one* forms the cristae and is perhaps the most complex of all membranes. It is the carrier of the complicated machinery coupling the energy derived from respiration to the synthesis of adenosine triphosphate (ATP). It is, perhaps, unfortunate that such a disproportionately large share of the efforts of the biochemical and biophysical fraternities is spent on this superelaborate structure (and on the analogous chloroplasts), rather than on those membranes that show simpler specializations. The mitochondrion as a whole, as Lehninger explains in his chapter in this volume, is thought to have been derived in a very distant way from originally free-living bacteria, which, in the course of evolving, have become stripped of a great number of functions. The principal argument in favor of this hypothesis derives from some similarities of this inner membrane with the cytomembrane of bacteria. These similarities consist of the following: (1) both are the locus of oxidative phosphorylation; (2) both contain a specific phospholipid, diphosphatidyl glycerol (DPG); and (3) both lack sterols. As to DPG, it is an interesting lipid, which carries two negative charges in close proximity and, therefore, is prone to conformational changes when binding doubly-charged positive ions, such as $Ca^{++}$ (Shah and Schulman, 1965). The absence of sterols is a characteristic that appears to be limited to the inner membrane of mitochondria and, possibly, to the lamellar membrane of chloroplasts (Mercer and Treharne, 1965). All other membranes of plants and animals contain ergosterol or cholesterol, or some closely related compounds.

The *outer membrane of the mitochondrion* is perhaps a close relative and an elaboration of the *endoplasmic reticulum,* the locus of many synthetic activities in the cell. Both the outer membrane of the mitochondrion and the endoplasmic reticulum are characterized by the absence of glycolipids.

In a different direction of sophistication, we should mention the excitable membranes of the *nerve axon,* of the *muscle fiber,* and of their *synaptic specializations.* Here the most conspicuous aspect is the high degree of modifiability, in an all-or-none fashion, suggesting a very high degree of cooperativity between the elements of the membrane. It seems probable to me that the changes in state observable in these types of membrane are analogous to phase transitions of thermodynamics. These phase transitions, here occurring in two-dimensional objects, would be governed by somewhat different rules from those of ordinary thermodynamics. The rules have, as yet, been incompletely worked out. An important specific model for the functioning of nerve excitation has been presented (Adam, 1970).

Finally, we should mention the most "clever" membranes, which can be found in the *specific portions of sensory cells,* where they are attuned to the transduction of various kinds of stimuli. These systems can be triggered to change their state and thus give a macroscopic response by absolutely minimal inputs. The input may be a quantized one—a single molecule or a photochemical change in a single molecule, in olfaction and vision, respectively—or it may involve minimal changes in some continuous parameter, as in mechanical, electrical, or thermal stimuli.

PERMEATION   Let us stick to the simpler aspects and concentrate on permeation. Here we owe to the work of physiologists (Stein, 1967) a general classification:

*Simple permeation,* largely determined by the partition coefficients of the solutes between aqueous and lipid phases, and by diffusion within the lipid phase.

*Facilitated permeation,* involving carriers or pores or, perhaps, as we shall see, in some cases "carried pores." This kind of permeation is characterized by the phenomena of saturation and competition.

Next in sophistication we have *coupled carrier transport,* involving cotransport or countertransport of two species owing to the fact that the mobility of the carrier depends on more than one ligand. This kind of transport can cause motion of one component against its electrochemical gradient, by coupling it to the transport of another component that moves with its gradient.

We must and do have, in addition, *primary active transport* coupled to energy consumption in a manner that is not completely understood in any one case. Perhaps the closest to a complete understanding has been achieved in sugar transport in bacteria, in which it turns out that the transported species is a transiently phosphorylated one (Kaback, 1968).

## The approach through the study of lipid bilayers

NAKED MEMBRANES   It will take many years and many lines of attack for us to understand the structure and mode of function of all these membranes in their molecular details. The approach that I discuss here is that which starts out with lipid bilayers between aqueous phases and advances by modifying these bilayers in various ways, leading to simulation of some of the functions mentioned. This approach was pioneered by Mueller and Rudin, and to them we owe also the majority of the very interesting modifications of the properties of simple bilayers that can be obtained with the help of a variety of additives.

The simple bilayer is a miraculous structure in itself. It can be formed by smearing a lipid mixture across a hole in a partition separating two aqueous solutions. Under appropriate conditions, the nature of which is not clearly understood, these lipid mixtures thin out to "black membranes," truly bimolecular structures in which the polar heads of the phospholipids face the water phases, and the hydrocarbon chains form a somewhat disordered and liquid layer about 50 Å thick. The disorderliness is an essential feature. It is generally assured, at body temperature, by there being a sufficient number of kinks, cis-double bonds, in the hydrocarbon chains. Also necessary for good membrane formation is the presence of cholesterol or tocopherol or tetradecane or some such neutral compound, for reasons not clearly understood. The membrane part is surrounded by a rim of lipid in the "bulk phase." The structure of this bulk phase has never been characterized. To which of the several possible lipid-water phases (Luzzati, 1968) does it correspond? Or does it retain enough solvent to remain a disordered liquid?

This form of lipid bilayer is under tension, and is mechanically in a metastable state. When the membrane tears, the lipid is swallowed up by the bulk material along the rim of the hole across which the membrane is formed. It does not easily break because a relatively large activation energy is needed to start a tear. The liquidity of the hydrocarbon middle part is essential for giving the membrane plasticity. If the membrane is too stiff, it becomes mechanically fragile. One should appreciate the astonishing disproportions of such a membrane: a membrane 1 mm in diameter and 70 Å in thickness, if scaled up, would correspond to a thin piece of paper 30 feet in diameter.

The so-called surface tension (a few dyne cm$^{-1}$) of such double layers, measured by bowing them out with overpressuring of the liquid on one side, is not an ordinary interfacial tension. It is a measure of the relative energy of the double-layer versus the bulk phase.

Torn membrane joins the bulk phase, which is distinct from natural membranes, which do not ball up on being torn but retain their thickness. They reduce their contour length to zero by forming vesicles. Natural membranes have a *contour tension* rather than a surface tension. By contour tension I mean an analogue to the surface tension of liquids. Liquid droplets minimize their surface area, at constant volume, by rounding up to form a sphere. Liquids do so because each molecule tries to maximize its interaction with neighboring molecules of the liquid, and a molecule in the interior of the liquid has more neighbors, and thus more interaction, than one on the surface. Therefore, the liquid tries to have as few molecules on the surface as possible, i.e., it minimizes its surface area and forms a sphere. Analogously, any molecule in a natural membrane has more neighbors if it is in the interior of the membrane *area* than if it is at the *perimeter*. Therefore, a natural membrane, if it is constrained to be a membrane of constant thickness, will try to minimize its perimeter. It can do so in three-dimensional space by forming vesicles, thus re-

ducing the perimeter to zero. This is a general feature of fragments of real cellular membranes.

Besides the Mueller and Rudin technique, there are several other methods for forming lipid bilayers. Pagano and Thompson (1967) have described a procedure for forming bilayers in a manner analogous to the formation of soap bubbles: the lipid solution is put into the tip of a pipette, overlayered with an aqueous phase, and blown out under water. Large bubbles can thus be formed that have a bilayer around most of their surface and a cap of bulk material at the top (Figure 2). Such bubbles can be kept in suspension

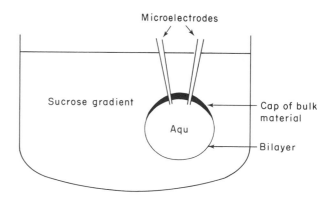

FIGURE 2  Lipid bilayer bubbles (Pagano and Thompson, 1968). The bubble is suspended in a sucrose gradient and carries a cap of bulk lipid solution. The cap can be penetrated by microelectrodes, and the bubble as a whole can be transferred from one aqueous environment to another.

by being blown into a sucrose gradient. The bubbles can be picked up by being sucked into a pipette and transferred to a different environment, and they can be impaled with microelectrodes. In this way, the bubbles can be used for measuring the permeation of tagged ions ($^{36}$Cl or $^{22}$Na), and they can be used for measuring the transference of ions during the passage of electrical current (Pagano and Thompson, 1968; Price and Thompson, 1969).

Another interesting procedure has been described by Tsofina et al. (1966). These authors found it difficult to incorporate proteins into lipid bilayers of the type produced by Mueller and Rudin. On the other hand, it has been known from earlier work that *monolayers* of lipids formed at the interface of air and water or at the interface of an organic solvent and water can be strongly modified by the presence of proteins in the aqueous phase. It has been widely believed that this modification is due to a true incorporation of protein into the monolayer. Therefore, Tsofina et al. proceeded to construct bilayers from two lipid–protein monolayers in the arrangement sketched in Figure 3. Bilayers were indeed formed, but it was not actually

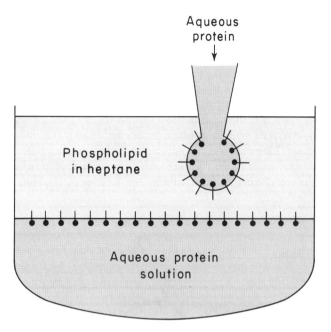

Aqueous
protein

Phospholipid
in heptane

Aqueous protein
solution

FIGURE 3  Bilayer constructed from two lipid-protein monolayers (Tsofina et al., 1966). One monolayer is formed at the interface of the lower aqueous protein solution (Aqu) and the benzene layer containing the phospholipid. A second monolayer is formed at the interface of the benzene phase, and the drop of aqueous protein solution is extruded from the pipette by reaching in from the top. These two monolayers are moved toward each other by appropriate mechanical procedures and, at the place of contact, form a true bilayer.

established whether the bilayer did contain protein. The method is a very interesting one, because it offers the hope of constructing bilayers containing *different proteins* in the two monolayers.

Finally, I wish to mention a procedure invented by Träuble and Grell (1970) in Eigen's laboratory (Figure 4). Here, too, the bilayers are formed stepwise. In the first step, an emulsion of water droplets is formed in a bulk phase of benzene to which a small amount of phosphatidyl choline has been added. The emulsion is formed by intense ultrasonic irradiation. The droplets are of fairly uniform size, about 400 Å in diameter, and are coated with a *monolayer* of the phospholipid. This emulsion is then layered on top of a different aqueous phase. Some more phospholipid is added to the benzene to form an interfacial layer between the bulk aqueous phase and the benzene, and the droplets of the emulsion (the prevesicles) are forced from the benzene phase into the water phase by high-speed centrifugation. In their passage through the interface, the prevesicles cover themselves with a second layer of phospholipid and now constitute vesicles surrounded by a true lipid bilayer. Here the possibility exists of making the bilayer asymmetric, with respect to the polar lipid, and making the inside aqueous phase different from the outside one. Moreover, that these vesicles can be prepared with a total surface area per milliliter vastly exceeding that obtainable by any other method opens the way to a great number of physical studies not accessible with the other procedures. Vesicles of phospholipids can also be formed in a single step, by sonication of a suspension of lipid in water (Huang, 1969).

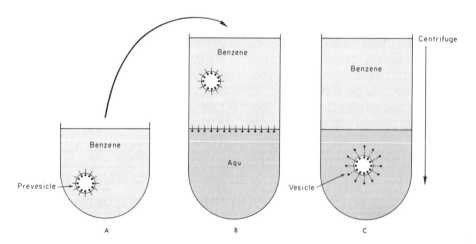

FIGURE 4  Production of masses of microvesicles by a two-step procedure (Träuble and Grell, 1970). Step 1. Formation of a water emulsion in a benzene phase containing small amounts of phospholipid. Prevesicles are formed on intense sonication. Step 2. Another aqueous phase is overlaid with the emulsion, and an interfacial monolayer of phospholipid is formed. Step 3. The prevesicles of the emulsion are forced through the interfacial monolayer (high-speed centrifugation). Vesicles bounded by a lipid bilayer are formed.

Let us return to the flat bilayers introduced by Mueller and Rudin. These are reasonably permeable to substances soluble in both water and lipid. One of these substances is water itself, whose solubility in hydrocarbon solvents is actually surprisingly high, around $10^{-3}$ molar. The nature of water dissolved in hydrocarbon has not been studied by modern methods. It is not known to what extent it is aggregated and, if it is, in what kinds of aggregates. Such a study would certainly be feasible and highly desirable.

These membranes are impermeable to substances of low solubility in hydrocarbon solvents or, more precisely, to substances for which the partition coefficient between water and the hydrocarbon phase is unfavorable for the latter. Foremost among these are anions and cations. As a result, the electrical conductance of such membranes is fantastically low, in most cases of the order of $10^{-8}$ mho cm$^{-2}$.

Läuger et al. (1967) made the interesting discovery that such membranes are relatively permeable to the iodide ion and much more so in the presence of the $I_2$ molecule. Probably this case is also a simple one of partitioning between water and lipid; the larger the volume over which the charge of the ion is distributed, the less energy is needed to transfer such an ion from its aqueous environment with a dielectric constant around 80 to the lipid phase with a dielectric constant around 2. The iodine molecule enhances the partition of the iodide ion in favor of the lipid phase. The mechanism of this enhancement has not yet been clarified unambiguously. It is possible that $I_3^-$ or $I_5^-$ ions are formed, thus spreading the charge over a larger volume and reducing the electrostatic energy needed for transfer to the lipid phase (Finkelstein and Cass, 1968). However, more indirect effects of $I_2$ on the lipid phase also can be conceived.

The unmodified lipid bilayer membrane imitates the thickness, electric capacitance ($\sim 1\mu f$ cm$^{-2}$), and the water permeability of real membranes. It fails in its electrical conductance in not being constrained to constant thickness when torn, and in all higher functions.

PROTON CARRIERS    Next in sophistication we may consider the remarkable effects of *uncouplers,* i.e., substances that, in mitochondria, uncouple oxidation from phosphorylation. Dinitrophenol is the classic example, and related compounds have been introduced recently. When such compounds are added to mitochondria, the electron transport system of respiration continues to operate at a high rate, but the energy supplied by this transport is not used for the synthesis of ATP. The theories concerning the mechanism of this uncoupling naturally depend on our theoretical understanding of the coupling of oxidation and phosphorylation in the first place. One class of such theories derives from ideas of Peter Mitchell (1966; see also Lehninger, this volume). Mitchell noted that membranes, be-

cause they are invariably contour free, must necessarily divide space topologically into an inside and an outside. Thus, one can conceive of a biochemistry, situated in the membrane, which both creates and depends on the differences in the electrochemical potentials of solutes on the inside versus those on the outside. One can also conceive that the coupling between electron transport and phosphorylation might be mediated by such gradients. Specifically, Mitchell conceived that the electron transport creates a pH differential between inside and outside, and that the pH differential drives the synthesis of ATP. Uncoupling, from this point of view, would occur if the pH gradient fails to be established, for instance by making the membrane highly permeable to protons. It is still uncertain if this idea is correct with respect to the uncoupling of oxidative phosphorylation in mitochondria, but it has turned out that the uncouplers are indeed good proton carriers in lipid bilayers (Hopfer et al., 1968; Liberman and Topaly, 1968). At least that is what they seem to be at first sight. They increase the conductivity of the membrane in a manner that would be expected for a mechanism that allows both the neutral uncoupler and the anion to shuttle back and forth. The dependence of the conductance on the concentration of uncouplers and on the pH support the idea, but whether this is the whole story is still not clear. Certain experimental findings on the saturation of this mechanism point to complications. These have been pursued actively by L. J. Bruner (Bruner, 1970).

DEVIATION FROM OHMIC BEHAVIOR    A striking feature of the conductance both of simple lipid bilayers and of those made more conductive by uncouplers is a symmetric non-linearity in the I versus the V curve: with increasing applied voltage the current rises faster than proportionally to the voltage. The deviation from linearity becomes obvious around 30 mV and reaches a factor of about 5 near 100 mV. Läuger and his associates have made a thorough theoretical study of this nonlinearity (Walz et al., 1969; Neumcke and Läuger, 1969), and point out three possible causes, each of which is analogous to certain mechanisms well known in solid-state physics.

The first possible cause is referred to as "ion injection." When an insulating layer is interpolated between two layers containing a large number of mobile charge carriers, an applied voltage will effectively be limited to the insulating layer, caused by the accumulation of charge carriers of opposite sign at the two interfaces. Because of the interplay of thermal diffusion and electrical fields, however, the charge layers at the interfaces will not be infinitely thin. They will extend some distance into the conducting phases. The extent of this spread of the space charge and of the voltage is the greater the lower the carrier concentration. The net effect of this space-charge distribution is a dependence on the applied voltage of the concentration of the majority carrier at the lipid-water interface: the greater the applied voltage the higher the majority-carrier concentration and, therefore, the higher the ion injection. Walz et al. (1969) showed that for lipid bilayers this effect is appreciable only at ionic strengths of about $10^{-3}$ molar or lower, and that it should rapidly decrease with increasing ionic strength. Both these predictions are contrary to the actual findings, and therefore this cause cannot account for the bulk of the observed effect.

The second cause studied in detail (Neumcke and Läuger, 1969) considers the effects of the so-called image forces to which the ions in the lipid phase are subjected. These image forces are an expression of the polarization of the dielectric environments of the ions. The forces are situated very close to a lipid-water interface and lead to a strong attraction of any ion dissolved in the lipid phase toward the interface and thus to a lowering of its energy. In lipid bilayers, the effect of these forces is to increase the partition of the ions in favor of the lipid phase, and the increase is greater as the applied voltage is greater. In this case, quantitative analysis leads to results much closer to reality. One obtains an I curve versus a V curve, which is independent of the ionic strength and gives deviations from linearity of the right sign and the right order of magnitude, but they still are not quite large enough to account for the observations.

A third effect has, therefore, been considered and calculated in detail. This is the field dissociation effect first discovered by Wien many decades ago and accounted for quantitatively by Onsager in a classic paper (Onsager, 1934). Onsager showed that the applied field affects the association-dissociation equilibrium of dissolved salts, leading to a net increase in dissociation in the presence of high applied field strength and, therefore, in the number of charge carriers. The theory of Onsager has been modified (Neumcke et al., 1970) to the situation of thin bilayers and has been shown to account satisfactorily for the observed nonlinearity, when combined with the contribution of image forces mentioned above.

ION CARRIERS This nonlinearity does not seem to be present and is probably swamped out when lipid bilayers are modified by the addition of ion carriers, substances that increase the conductance of lipid bilayers still more than do the uncouplers. The history of this discovery is somewhat as follows. Among the vast number of antibiotics against bacteria discovered during the last decades are some that, when applied to mitochondria, lead to a strange accumulation of potassium ions. The mechanism of this $K^+$ accumulation is still controversial, but again the application of these agents to lipid bilayers has led to a beautifully simple result: the bilayers become highly permeable to $K^+$ and highly selectively so. Mueller and Rudin (1967b), who discovered this phenomenon, thought at first that these substances,

which are ring molecules, accommodate the hydrated potassium ion in the center of the ring and were tailor-made to fit this hydrated ion. They conceived of these ring molecules as inserting themselves in the membrane-like washers, forming pores through which the ions might move. The truth turned out to be even more startling. When the first X-ray diffraction structure of one of these molecules (nonactin) came out (Kilbourn et al., 1967), it showed that the carrier wraps itself completely around the *naked* ion, suggesting that this complex could act as a "carrier pore" for shuttling particular ions across the membrane. A similar structure has recently been established for another $K^+$ carrier, valinomycin (Pinkerton et al., 1969; Ivanov et al., 1969). Eigen, elsewhere in this volume, deals with the features that permit these molecules to carry out such a very specialized function.

HOLE PUNCHERS The next class of additives might be called "*hole punchers.*" These are cyclic antibiotics containing a polyene chain as part of the molecule, but other portions of the ring are highly hydrophilic. Their effects on real membranes and on bilayers are contingent on the presence of cholesterol or some other sterol in the membrane. Nystatin and amphotericin B are the ones that have been studied most carefully (Finkelstein and Cass, 1968). These additives also increase the conductance of bilayers, but the effects that they produce are very different from those of ion carriers.

1. The additive must be present on both sides of the membrane to give reproducible results.

2. The conductance increase depends on a very high power of the drug concentration.

3. The conductance is based on an increased permeability to *anions* and is coupled with an increased permeability to water.

4. Anions are discriminated according to size.

5. The conductance has a very *high negative temperature coefficient* in contrast to that produced by the ion carriers. This permits the experimenter the neat trick of switching the conductance from one due to potassium to one due to anions by using bilayers to which both a $K^+$ carrier (valinomycin) and a hole puncher (nystatin) have been added. Simply by a change of a few degrees in temperature, the dominant conductance mechanism can be switched, and with it the membrane potential, if the solutions on the two sides are chosen appropriately.

GATEABLE ADDITIVES The last class of additives I wish to discuss are the gateable ones: EIM (excitation-inducing material, Mueller and Rudin, 1967a) and alamethicin (Mueller and Rudin, 1968). The first of these materials is a protein produced by a strain of *Enterobacter cloacae;* the second, a cyclic peptide containing 19 amino acids. Both substances are still very imperfectly characterized chemically. Both substances confer on lipid bilayers a conductance which varies over several orders of magnitude, depending on the applied voltage. Conductance is cationic but can be changed to anionic by the addition of such polycations as protamine, polylysine, or spermine. By playing with the applied voltage, salt gradients, or the polycationic additives, one can simulate many striking electrical phenomena characteristic of nerve-axon membranes, including resting potentials, action potentials, and rhythmic discharges. The molecular mechanisms here involved are still obscure. Especially, the basic question still unanswered is whether the sudden transitions from one state of conductivity to another bespeak a cooperative phenomenon on a large scale, such as a phase transition of the whole membrane, or a microcooperativity involving local conformational changes of oligomers. Perhaps the strongest argument in favor of the latter view comes from the observations of Bean et al. (1969), which showed that, on the addition of EIM, quantized increases in conductance can be seen, each quantum amounting to a conductance change of about $4 \times 10^{-10}$ mho. It seems plausible that these increases in conductance correspond to the opening of individual gates controlled by one or a few molecules of EIM. If such an interpretation is correct, then the excitations of the lipid bilayers produced with the help of these additives are not likely to be close relatives of true nerve excitations, because, for the latter, it seems probable to me that we are dealing with a true phase transition on a macroscopic scale (Adam, 1970).

NEW TECHNIQUES COMING OR HOPED FOR    There is much need for more basic studies on lipid bilayers, naked or with additives. The lipids used should be chemically defined, as in the studies of the Läuger group; otherwise we cannot hope to obtain data interpretable in terms of specific mechanisms. Electrical measurements must be supplemented by other physical measurement, by the attachment of spin labels or fluorescent labels, or the development of technologies for absorption and reflection measurements adapted to these very thin layers.

A most promising approach is the study of the bulk phases of lipid-protein-water mixtures (Gulik-Krzywicky et al., 1969). Such mixtures give rise to a variety of phases which can be studied by X-ray diffraction techniques. The clarification of the structures possible in laminar phases, especially, should go a long way in helping us to assess the relative roles of lipid and protein in contributing to the basic structure of real membranes.

Another approach of great value for the understanding of the interaction of lipids and proteins in membranes is that pursued by Colacicco (1969). Here a monolayer of phospholipid is first formed at an air-water interphase. Various proteins are then introduced into the aqueous subphase, and the penetration of the protein into the monolayer is assessed by its effect on the surface pressure of the monolayer.

## Perspective

It would seem that Nature, when she had invented the principle of membranes, found that she had caught on to a good thing, and proceeded to exploit this principle with the passion of the true inventor. As yet we have as little understanding of this general principle as we had of chromosomes, say, 30 years ago. Studies of physiologists, biochemists, and electron-microscopists have yielded some guidance in the characterization of special membranes, but the gap between this characterization and the light shed by the study of model membranes is still enormous.

Considering the relevance of the study of membranes to our over-all progress in biology and psychobiology, I do not think that we shall learn from such studies how any central nervous system functions or even be led directly to the solution of the great puzzles of neuroembryology. I do have a strong feeling, however, that radical progress in any of these fields will not come until the gap has been bridged that now exists between our understanding of the simplest model membranes and, say, the transducer membranes of sensory physiology. Here I am sure discoveries are still to be made that will rank with the greatest we have had in molecular biology.

REFERENCES

ADAM, G., 1970. Theory of nerve excitation as a cooperative cation exchange in a two-dimensional lattice. In Physical Principles of Biological Membranes (F. Snell, J. Wolken, G. Iverson, and J. Lam, editors). Gordon and Breach, Science Publishers, New York, pp. 35–64.

BEAN, R. C., W. C. SHEPHERD, H. CHAN, and J. EICHNER, 1969. Discrete conductance fluctuations in lipid bilayer protein membranes. J. Gen. Physiol. 53: 741–757.

BRUNER, L. J., 1970. Blocking phenomena and charge transport through membranes. Biophysik 6: 241–256.

COLACICCO, G., 1969. Applications of monolayer techniques to biological systems: Symptoms of specific lipid-protein interactions. J. Colloid Interface Sci. 29: 345–364.

FINKELSTEIN, A., and A. CASS, 1968. Permeability and electrical properties of thin lipid membranes. J. Gen. Physiol. 52 (no. 1, pt. 2): 145s–172s.

GULIK-KRZYWICKY, T., E. SHECHTER, M. FAURE, and V. LUZZATI, 1969. Interactions of proteins and lipids. Structure and polymorphism of protein-lipid-water phases. Nature (London) 223: 1116–1121.

HOPFER, U., A. L. LEHNINGER, and T. E. THOMPSON, 1968. Protonic conductance across phospholipid bilayer membranes induced by uncoupling agents for oxidative phosphorylation. Proc. Nat. Acad. Sci. U. S. A. 59: 484–490.

HUANG, C., 1969. Studies on phosphatidylcholine vesicles. Formation and physical characteristics. Biochemistry 8: 344–352.

IVANOV, V. T., I. A. LAINE, N. D. ABDULAEV, L. B. SENYAVINA, E. M. POPOV, Y. A. OVCHINNIKOV, and M. M. SHEMYAKIN, 1969. The physicochemical basis of the functioning of biological membranes: The conformation of valinomycin and its K⁺ complex in solution. *Biochem. Biophys. Res. Commun.* 34: 803–811.

KABACK, H. R., 1968. The role of the phosphoenolpyruvate-phosphotransferase system in the transport of sugars by isolated membrane preparations of *Escherichia coli. J. Biol. Chem.* 243: 3711–3724.

KILBOURN, B. T., J. D. DUNITZ, L. A. R. PIODA, and W. SIMON, 1967. Structure of the K⁺ complex of nonactin, a macrotetrolide antibiotic possessing highly specific K⁺ transport properties. *J. Mol. Biol.* 30: 559–563.

LÄUGER, P., W. LESSLAUER, E. MARTI, and J. RICHTER, 1967. Electrical properties of bimolecular phospholipid membranes. *Biochim. Biophys. Acta* 135: 20–32.

LIBERMAN, E. A., and V. P. TOPALY, 1968. Selective transport of ions through bimolecular phospholipid membranes. *Biochim. Biophys. Acta* 163: 125–136.

LUZZATI, V., 1968. X-ray diffraction studies of lipid-water systems. *In* Biological Membranes. Physical Fact and Function (D. Chapman, editor). Academic Press, London and New York, pp. 71–123.

MERCER, E. I., and K. J. TREHARNE, 1965. Occurrence of sterols in chloroplasts. *In* Biochemistry of Chloroplasts (T. W. Goodwin, editor). Academic Press, London and New York, pp. 181–186.

MITCHELL, P., 1966. Chemiosmotic coupling in oxidative and photosynthetic phosphorylation. *Biol. Rev. (Cambridge)* 41: 445–502.

MUELLER, P., and D. O. RUDIN, 1967a. Action potential phenomena in experimental bimolecular lipid membranes. *Nature (London)* 213: 603–604.

MUELLER, P., and D. O. RUDIN, 1967b. Development of K⁺-Na⁺ discrimination in experimental bimolecular lipid membranes by macrocyclic antibiotics. *Biochem. Biophys. Res. Commun.* 26: 398–404.

MUELLER, P., and D. O. RUDIN, 1968. Action potentials induced in bimolecular lipid membranes. *Nature (London)* 217: 713–719.

NEUMCKE, B., and P. LÄUGER, 1969. Nonlinear electrical effects in lipid bilayer membranes. II. Integration of the generalized Nernst-Planck equations. *Biophys. J.* 9: 1160–1170.

NEUMCKE, B., D. WALZ, and P. LÄUGER, 1970. Nonlinear electrical effects in lipid bilayer membranes. III. The dissociation field effect. *Biophys. J.* (in press).

ONSAGER, L., 1934. Deviations from Ohm's law in weak electrolytes. *J. Chem. Phys.* 2: 599–615.

PAGANO, R., and T. E. THOMPSON, 1967. Spherical lipid bilayer membranes. *Biochim. Biophys. Acta* 144: 666–669.

PAGANO, R., and T. E. THOMPSON, 1968. Spherical lipid bilayer membranes: Electrical and isotopic studies of ion permeability. *J. Mol. Biol.* 38: 41–57.

PINKERTON, M., L. K. STEINRAUF, and P. DAWKINS, 1969. The molecular structure and some transport properties of valinomycin. *Biochem. Biophys. Res. Commun.* 35: 512–518.

PRICE, H. D., and T. E. THOMPSON, 1969. Properties of lipid bilayer membranes separating two aqueous phases: Temperature dependence of water permeability. *J. Mol. Biol.* 41: 443–457.

ROUSER, G., G. J. NELSON, S. FLEISCHER, and G. SIMON, 1968. Lipid composition of animal cell membranes, organelles and organs. *In* Biological Membranes. Physical Fact and Function (D. Chapman, editor). Academic Press, London and New York, pp. 5–69.

SHAH, D. O., and J. H. SCHULMAN, 1965. Binding of metal ions to monolayers of lecithins, plasmalogen, cardiolipin, and dicetyl phosphate. *J. Lipid Res.* 6: 341–349.

STEIN, W. D., 1967. The Movement of Molecules Across Cell Membranes. Academic Press, New York.

TRÄUBLE, H., and E. GRELL, 1970. Formation of asymmetrical spherical lecithin vesicles. *Neurosci. Res. Program Bull.* (in press).

TSOFINA, L. E., E. A. LIBERMAN, and A. V. BABAKOV, 1966. Production of bimolecular protein-lipid membranes in aqueous solution. *Nature (London)* 212: 681–683.

WALZ, D., E. BAMBERG, and P. LÄUGER, 1969. Nonlinear electrical effects in lipid bilayer membranes. I. Ion injection. *Biophys. J.* 9: 1150–1159.

# 61 Alkali-ion Carriers: Dynamics and Selectivity

MANFRED EIGEN and RUTHILD WINKLER

## Alkali-ion carriers; concepts and facts

THE EXCITATION of the nerve membrane is intimately coupled to a fast exchange of sodium and potassium ions. Processes of this kind have been studied for a long time, and various mechanisms involving "pumps" and "carriers" have been proposed. Whatever those "demons" are, they must have a size-specific recognition site by which they distinguish selectively a given alkali ion from its (smaller and larger) homologues.

Actually, the hypothetical concept of the carrier reflects largely the difficulty in explaining selectivity in terms of simple ionic interactions. On the other hand, how the interaction of alkali ions with such carriers could be explained in chemical terms remained an unsolved problem as well.

Moore and Pressman (1964), Shemyakin et al. (1967), and Graven et al. (1966), have recently found that certain antibiotics bind potassium ions selectively and transport them through natural and artificial (lipid) membranes. Whether these antibiotics are, in fact, used as "carriers" has not yet been found out. Because they possess all the qualities once ascribed to a specific "carrier," however, we may well use them as model compounds in order to see how nature can account for such properties.

Table I contains a compilation of antibiotics which selectively either facilitate or interfere with specific alkali-ion transport through lipid membranes. The composition and spatial structure of these compounds have been studied in detail in various laboratories (for references, see Table I). Essentially, two groups of compounds can be distinguished.

1. Neutral macrocyclic compounds: All members of this group are cyclic molecules, and at neutral pH are uncharged (in the absence of the alkali ion). The carrier complex will therefore respond to an electric field via the charge of the alkali ion. The molecular weights are of medium magnitude, and the molecules are just large enough to surround the alkali ion with a hydrophobic coat. The composition varies widely. Thus, we find oligopeptides, such as gramicidin-S*; oligodepsipeptides (i.e., sequences of alternating amino acids and oxyacids), such as valinomycin and the

---

*The gramicidins A, B, and C are similar, but acyclic compounds.

---

MANFRED EIGEN AND RUTHILD WINKLER   Max Planck Institute for Physical Chemistry, Göttingen, Germany

enniatins; or other cyclic structures, e.g., simple polyethers or the macrotetrolides, which are discussed below in more detail. It is obvious that similar structures could be found also in protein molecules. Actually, some enzymes are known to bind sodium or potassium quite selectively. A feature of the above-mentioned cyclic compounds is that their molecular weights are considerably lower than those of even small protein molecules.

2. Charged open-chain compounds: These linear structures of various compositions all contain a carboxylic group. Thus, in the neutral pH range the uncomplexed molecules are negatively charged. The carboxylic group, however, is not necessarily involved in the coordinative binding with the metal ion. Rather, it is believed to form H-bonds to hydroxyl groups at the opposite end of the linear structure, thereby forming a cyclic conformation. The molecular weights are of a similar order of magnitude as those for the members of the first group.

All carriers possess a number of polar ligands, which are used to form a multidentate complex with the alkali ion. Large parts of the molecule, however, must be hydrophobic so that the carrier may easily penetrate into lipid membranes. As a consequence, the solubility in water is very poor. It will be seen that the particular architecture of the carrier is responsible for its unique functional properties.

It is quite obvious that a selective carrier action must be a consequence of a selective binding capacity for the different alkali ions. Thus, direct measurements of complex stability

### TABLE I
*Ion-selective antibiotics*

| Valinomycin Group | Nigericin Group |
|---|---|
| Valinomycin[1,2,3a,3b] | Nigericin[11] |
| Gramicidin S, Tyrocidin A[4] | Monensin[12] |
| Enniatins[5,6a,6b] | Dianemycin[13] |
| Macrotetrolides[7,8] | |
| Monazomycin[9,10] | |

1. Brockman et al., 1963
2. Shemyakin et al., 1963
3a. Pinkerton et al., 1969
3b. Ivanov et al., 1969
4. Schwyzer and Sieber, 1957
5. Plattner et al., 1963
6a. Quitt et al., 1963
6b. Quitt et al., 1964
7. Gerlach and Prelog, 1963
8. Kilbourn et al., 1967
9. Akasaki et al., 1963
10. Lardy et al., 1967
11. Steinrauf et al., 1968
12. Agtarap et al., 1967
13. Lardy et al., 1958

constants by the use of different nonaqueous solvents, have been performed in several laboratories. Table II contains some examples. They generally confirm the above assumption, although the correlation between complex stability and membrane permeability is not always straightforward, indicating that other factors (kinetic properties, such as lifetime, loading rate, and so forth) are involved in the dynamics of ion transport across membranes. Nevertheless, the most striking result is the unique size dependence of the complex stability showing a pronounced maximum for a given ion, disfavoring larger as well as smaller sizes. This behavior is quite different from any size dependence due to simple electrostatic interaction.

## Complexes of main group metal ions: stability and dynamics

Before starting a more detailed discussion of experimental data, we may ask for possible explanations for such a size specificity of complex stability. For alkali ions, specificity in binding is certainly not a consequence of a peculiar electron configuration, as it is for transition metal ions, such as $Ni^{2+}$, $Cu^{2+}$, and so forth. We are dealing here with ions of "noble gas"-like electron configuration, and we must therefore try to base our model on factors such as charge and size of the metal ion and peculiarities of structure of the complexing agent only.

The complex formation involves the substitution of one or several solvent molecules from the inner coordination sphere of the metal ion. If the incoming ligands can be arranged as freely as the solvent molecules (in the coordination shell), then—at least for a symmetrical, e.g., octahedral, arrangement—we should expect one of the two following monotonic size dependences:

1. The incoming ligand is more tightly bound than the solvent molecule to be substituted. The stability of the com-

plex will decrease with increasing radius of the metal ion. The smaller the metal ion, the larger will be the gain in binding energy for each ligand.

2. The incoming ligand is less tightly bound than the solvent molecule. Here the size dependence of complex stability will be inversed, because the smallest metal ion will most strongly prefer the solvent molecules.

Such behavior is depicted in Figure 1. It is very important that in these cases—with decreasing metal-ion size—the cavity formed by the ligands can contract just as freely as does the solvent coordination shell. This may not always be possible, e.g., if the ligands are interconnected in a multi-

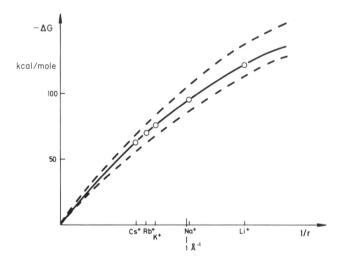

FIGURE 1 Free energy of hypothetical ligand binding (dotted lines) and solvation (solid lines, experimental values) as function of the reciprocal radius of alkali ions. (The ligand-binding curve is related to a fixed ligand concentration.)

TABLE II
*Complex formation constants of some antibiotics*

| Compound | Solvent | Anion | K [1/mole] | | $\dfrac{K_{K+}}{K_{Na+}}$ |
| | | | $Na^+$ | $K^+$ | |
|---|---|---|---|---|---|
| Nonactin | MeOH | SCN⁻ | $(1.6 \pm 0.3) \times 10^2$ | $(6.3 \pm 0.9) \times 10^3$ | 38 |
| | MeOH | SCN⁻ | | $(3.9 \pm 0.7) \times 10^3$ | |
| Monactin | MeOH | SCN⁻ | $(1.4 \pm 0.1) \times 10^3$ | $(3.2 \pm 1.3) \times 10^5$ | 230 |
| Enniatin B | EtOH *) | SCN⁻ | $1.3 \times 10^3$ | $3.7 \times 10^3$ | 2.8 |
| | MeOH | I⁻ | $(2.4 \pm 0.5) \times 10^2$ | $(8.4 \pm 1.0) \times 10^2$ | 3.5 |
| "Enniatin A" | MeOH | I⁻ | | $(1.2 \pm 0.1) \times 10^3$ | |
| Valinomycin | EtOH *) | SCN⁻ | not detectable | $5.2 \times 10^3$ | $>10^3$ |
| | MeOH | I⁻ | $(1.2 \pm 1.7) \times 10^1$ | $>8 \times 10^3$ | $>670$ |

\* Cf Pioda et al., 1967; Shemyakin et al., 1967

dentate chelate complex. Here the different ligands usually have some freedom of motion only as long as the cavity formed around the metal ion exceeds a certain size. Below this size the ligands may "freeze" into fixed positions, because steric hindrances, as well as ligand-ligand repulsion, will prevent further contraction of the cavity. Thus, the cavity will show an optimal fit for a metal ion of a certain size. For all metal ions larger than this critical size, the complex stability will increase with decreasing metal-ion radius, supposing that the ligands are more tightly bound than the solvent molecules. Below the critical size, however, this behavior is inversed, because there will be little or no gain in ligand-binding energy with decreasing metal-ion radius, and hence there will be no compensation for the increasing solvation energy. Note that this optimal fit will produce a maximum in the thermodynamic stability constant, but that this does not at all mean that the optimally fitted complex is the most stable complex in "absolute" terms.

The behavior described above is demonstrated in Figure 2. The lower curve describes the size dependence of the free

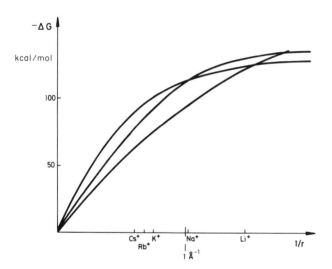

FIGURE 2 Free energy of solvation (lower curve) and chelation (upper curve for two hypothetical cases) as function of the reciprocal radius of alkali ions.

energy of solvation (r being the metal-ion radius). The upper curves represent two different size dependences for the free energy of ligand binding (including rest solvation). These curves are monotonic, i.e., the smallest ion shows the largest values, but, because the free energy of ligand binding bends sooner toward saturation (brought about by steric fixation of the ligands), the difference of the ligand binding and solvation curve may well show a maximum. Just such a difference determines the stability constant ($K_{stab}$) of the

metal complex. Note that the increase in free energy of ligand binding will be monotonic, regardless of whether $K_{stab}$ increases or decreases. Note furthermore that the maximum of $K_{stab}$ could be found at any position, depending on how soon the ligand-binding energy reaches saturation, but that this position does not necessarily coincide within the $1/r$ value at which saturation actually is reached. (The picture of optimal adaptation of a metal ion to the minimal size of the ligand cavity is oversimplified here in order to demonstrate the principle.) We may briefly summarize the main consequences of the above paragraph. A nonmonotonic size dependence, i.e., size specificity, is *not* a consequence of a special force, but is rather the result of the superposition of two influences—solvation and chelation. Both size dependences are monotonic per se but, owing to the special architecture of the complexing agent, level off with decreasing metal-ion size at different positions, so that the difference is no longer monotonic. If we remember that the solvation energies of $Na^+$ and $K^+$ differ by about 20 kcal/mole and that at room temperature 1.4 kcal/mole correspond to a difference of one order of magnitude in binding constants, we see easily that this "source" is sufficient to explain the observed binding specificities of carriers.

We now return to a discussion of experimental data. Some stability constants for various alkali-ion complexes are shown in Table II. A nonmonotonic size dependence—as striking as it appears to be for some of these complexes—actually has been observed in many other chelate complexes of main group metal ions. Some data for alkaline earth complexes are compiled in Table III. Note that the solvation energy for divalent ions is much larger than for monovalent ions, and thus differences (e.g., for $Mg^{2+}$ and $Ca^{2+}$) may be much more pronounced. The above interpretation is also supported by the rates of complex formation, which appear to be much more uniform than complex stabilities. Table IV shows some rate constants of complex formation for $Mg^{2+}$ and $Ca^{2+}$. The differences of the second-order rate constants within one group (e.g., $Mg^{2+}$) are due mainly to differences in the charges of the complexing ligands, because the rates involve ion pair formation. If one accounts for this effect, i.e., if one relates the rate constant with the rate-limiting step of substitution of a water molecule from the inner coordination sphere of the metal ion, a quite uniform value of about $10^5$ $sec^{-1}$ for $Mg^{2+}$ and $5 \times 10^8$ $sec^{-1}$ for $Ca^{2+}$ is obtained throughout (Eigen and Wilkins, 1965).

Almost all this information about rates of ligand substitution in metal complexes has been collected during the last decade, after the introduction of modern techniques for the study of rapid reactions, such as nuclear magnetic resonance, electron paramagnetic resonance, and relaxation spectrometry (Eigen, 1963). A uniformity of complex formation rates, as shown with $Mg^{2+}$ and $Ca^{2+}$, was found for almost all metal ions studied so far. It is therefore possible to present

TABLE III

*Stability constants of metal complexes*

($\log K_1$ in aq. sol., $\mu \to 0$, *$\mu = 0.1$; T $= 18 - 25°$)

| | $SO_4^{2-}$ | $F^-$ | $Ac^-$ | $IDA^{2-}$ | $NTA^{3-}$ | $EDTA^{4-}$ | $DGITA^{4-}$ | Metal-Phthal.$^{6-}$ | Erio R$^{3-}$ | Tiron$^{4-}$ | Oxine$^-$ |
|---|---|---|---|---|---|---|---|---|---|---|---|
| $Mg^{2+}$ | 2.3 | 1.8 | 0.8 | 3.7 | 7.0 | 9.1 | 5.2 * | 8.9 * | 7.6 * | 6.9 * | 4.7 |
| $Ca^{2+}$ | 2.3 | <1.0 | 0.8 | 3.4 | 8.2 | 11.0 | 11.0 * | 7.8 * | 5.4 * | 5.8 * | 3.3 |
| $Sr^{2+}$ | | | 0.4 | | 6.7 | 8.8 | 8.5 * | | | 4.6 * | 2.6 |
| $Ba^{2+}$ | | <0.5 | 0.4 | 1.7 | 6.4 | 7.8 | 8.4 * | 6.2 * | | 4.1 * | 2.1 |

| | | | | |
|---|---|---|---|---|
| $Ac^-$ | = acetate | | DGITA | = 2, 2 -ethylenedioxybis (ethyliminodi [acetic acid]) |
| IDA | = iminodiacetic-acid | | Metal-Phthal. | = metalphthaleine |
| NTA | = nitrilotriacetic-acid | | Erio-R | = eriochrom-black |
| EDTA | = ethylene-diaminotetraacetic-acid | | | |

a condensed survey on characteristic rates. A more detailed discussion of such a "periodic table of rate constants" for metal ions in a given solvent (Figure 3) can be found elsewhere (Diebler et al., 1970). Here, we want to emphasize only two points:

1. Deviations from a monotonic size dependence are found only with the transition metal ions, in which special influences such as ligand field stabilization and Jahn-Teller effect are involved. On the other hand, main-group metal ions show a uniform behavior throughout. If the rate can be associated with an elementary step, it reflects the monotonic size dependence, as expected for simple electrostatic interactions.

2. Despite the high solvation energies, the substitution of a single solvent molecule (or ligand) from a "complete" solvation sphere appears to be rapid. For the alkali ions this substitution occurs within $10^{-9}$ seconds. This rate is found for any single ligand belonging to a "complete" solvation sphere before any further substitution occurs. The loss of solvation energy is compensated for by the ligand binding. It would be more difficult to remove a solvent molecule from an incomplete solvation sphere, i.e., without compensation for the loss of solvation energy.

The high speed of single-ligand substitution is important for an understanding of the dynamics of carrier action (see below).

TABLE IV

*Rate constants for complex formation reactions of $Mg^{2+}$ and $Ca^{2+}$*

| | $k_R$ | |
|---|---|---|
| | $Mg^{2+}$ | $Ca^{2+}$ |
| $SO_4^{2-}$ | $1 \times 10^5$ * | |
| $S_2O_3^{2-}$ | $1 \times 10^5$ * | |
| $CrO_4^{2-}$ | $1 \times 10^5$ * | $\geq 5 \times 10^7$ * |
| $F^-$ | $3.7 \times 10^4$ | |
| HF | $\sim 4 \times 10^4$ | |
| $ATP^{4-}$ | $1.3 \times 10^7$ | $\geq 1 \times 10^9$ |
| $ATPH^{3-}$ | $\sim 3 \times 10^6$ | |
| $ADP^{3-}$ | $3 \times 10^6$ | $\geq 3 \times 10^8$ |
| $ADPH^{2-}$ | $1 \times 10^6$ | |
| Metal Phthal. | $(\sim 2 \times 10^6)$ | $(\sim 7 \times 10^8)$ |
| $IDA^{2-}$ | $9 \times 10^5$ | $2.5 \times 10^8$ * |
| | | $Sr^{2+}$: $3.5 \times 10^8$ * |
| | | $Ba^{2+}$: $7 \times 10^8$ * |
| Glycine$^-$ | | $4 \times 10^8$ * |
| Oxine$^-$ | $3.8 \times 10^5$ | $\geq 1 \times 10^8$ |
| Oxine H | $\sim 1 \times 10^4$ | |

* first-order rate constants

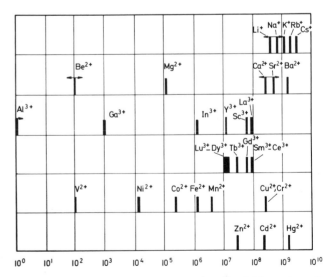

FIGURE 3  Characteristic rate constants ($sec^{-1}$) of substitution of water molecules from the inner coordination sphere of metal ions by various ligands.

## The study of alkali-ion complexes: solvents and indicators

The complexing mechanism of alkali-ion carriers have been studied in homogeneous solution. Methanol offers several advantages as a solvent.

1. Biological alkali-ion carriers are usually insoluble in water, owing to their lipophilic character, but are fairly soluble in methanol.

2. Because of the lower dielectric constant of methanol, complex stabilities, higher than those in water, are found, especially if charged ligands are involved. Thus, a more precise determination of stability and rate constants of otherwise quite weak complexes is possible.

3. The solvation of the cation in water and methanol is quite similar, as was shown by Strehlow (1952). There are only small changes of $\Delta G$, $\Delta H$, and $\Delta S$ if an alkali ion is transferred from water to methanol. The substitution behavior (mechanism and rate) might be quite similar for both solvents, so that studies in methanol can be considered representative for water as well.

4. Methanol can be used as a solvent for temperature-jump (T-jump) studies by both the Joule-heating and the microwave-heating methods. It even shows some advantages for field-effect and sound-absorption studies. As a consequence of the lower dielectric constant, the field effect for charged ligands is larger than in water. For uncharged ligands (e.g., the carriers described below under "The Na⁺-Monactin Complex"), relatively large $\Delta H$ values are to be expected. Thus, the equilibrium will be very strongly temperature-dependent. Sound waves in methanol involve a pronounced temperature wave, which, in water of about room temperature, is nearly absent (owing to the density maximum at 4° C).

In order to observe the equilibration of alkali-ion complexes and to follow rapidly any change in equilibrium, it was necessary to find a suitable indicator for the alkali ions in methanol. No such indicator has been reported in the literature so far. However, Schwarzenbach and Gysling (1949) described murexide as a specific indicator for Ca²⁺ ions in aqueous solution.

Murexide is the ammonium salt of purpuric acid and has the following chemical structure:

The anion has a strong absorption maximum around $\lambda = 527$ mµ ($\epsilon = 11700$ in methanol), which is shifted on complexing with Ca²⁺ to shorter wave lengths. Most likely, complexing occurs between two of the oxygens next to the N-bridge.

The stability constant for the Ca complex in aqueous solution, as reported by Schwarzenbach, is not large (pK 2. 7). Mg²⁺ was found to form no complex to any appreciable extent, whereas Sr²⁺ and Ba²⁺, according to Geier (1967), showed some tendency for complexing.

Although several other indicators have been reported in the literature as having much higher stability constants for the alkaline-earth ions, murexide seemed to be the best candidate for the indication of alkali ions in methanol for several reasons.

1. Murexide has a negative charge (distributed among the four oxygens next to the N-bridge), which is not protonated in the neutral pH range. Actually, the pK in aqueous solution is about zero and in methanol about 4 to 5. Alkali ions generally are quite weak in forming complexes, so it is very important that protons do not compete too strongly for the complexing site.

2. The spectral properties of murexide are peculiar. When complex formation occurs, the blue shift of the absorption maximum is probably due to some change in the orientation of the two rings relative to each other. Such an effect should be sensitive for weakly complexing ions, which otherwise would have only little influence on the electronic structure of the dye molecule. Figure 4 shows the spectral shift of the absorption maximum on titration with Na⁺. The well-defined isosbestic point indicates a simple complexing behavior. Those titrations were carried out for both the alkali and alkaline-earth series.

In the alkali series, the highest pK values were found for Na⁺ in the alkaline-earth series corresponding to Ca²⁺, as

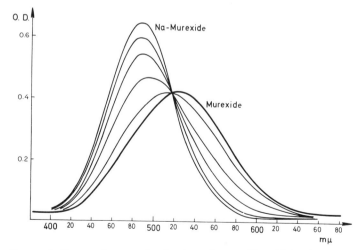

FIGURE 4  Spectrophotometric titration of murexide with Na⁺. (25°C; $C_{Mur} = 4 \times 10^{-5}$M).

was to be expected from Schwarzenbach's results for aqueous solutions. Owing to the lower dielectric constant of the solvent, the stability constants in methanol are higher than in water by orders of magnitude, and show much more clearly the nonmonotonic size dependence as described in the preceding section of this paper. One may consider murexide as a "carrier" for $Na^+$ and $Ca^{2+}$. Here again, the maxima in the stability constants result from the balance between ligand binding and solvation energy, whereas the "absolute" stabilities of the complexes should parallel monotonically the sequences of ionic sizes. This is clearly indicated by the amount of wavelength shift and the increase in $\epsilon$. Both $\lambda_{max}$ and $\epsilon$ are largest for the smallest ion and decrease monotonically with increasing size, as seen in Figure 5.

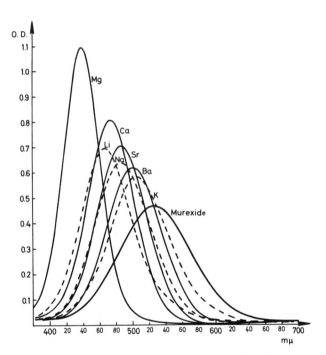

FIGURE 5 Absorption curves for various alkali- and alkaline-earth complexes with murexide.

For the alkali complexes, the stability constants can easily be determined from the titration curves (see Table V). Here $K_{diss}$ is always much larger than the indicator concentration, and therefore can immediately be obtained from plots of $(E - E_0)/(E_\infty - E_0)$ versus $1/C_M$, in which the extinctions $E$, $E_0$, and $E_\infty$ refer to the metal-ion concentrations $C_M$, 0, and $\infty$ (saturation), respectively.

A suitable indicator for studies of dynamic properties should have not only pronounced spectral shifts but also a sufficiently rapid performance. Only then can it be used for

the observation of other complexing reactions. Actually, the indicator may still be used if it reacts with a speed comparable to that to be followed. In that case, however, one must know precisely the rate constants of the indicator for the evaluation of the relaxation spectrum of the coupled reaction system. Therefore, it was necessary to carry out rate measurements with the indicator–metal-ion system.

For all the alkali ions, these reactions turned out to be too fast to be resolved by the temperature-jump technique, the half times being below the microsecond range. An electric-field traveling wave technique with spectrophotometric observation—developed by Ilgenfritz (1966)— could, however, be used for these studies. In this technique the perturbation of equilibrium is brought about by a strong electric field utilizing the dissociation field effect. The method is applicable to any equilibrium between charged reaction partners. The present time-resolution of this method is 30 nanoseconds; thus the method is just suitable for detecting the relaxation effects for the reactions of $Li^+$ and $Na^+$ with murexide. Figure 6 shows a typical oscillogram. $K^+$ (requiring a higher concentration because of its lower stability constant) was already beyond the resolution of this method. However, a lower limit for the rate constant could be derived; that constant almost coincides with the upper limit for a diffusion-controlled process. Even the measured value for $Na^+$ is very close to this limiting value of about $2 \times 10^{10}$ $M^{-1}$ $sec^{-1}$. (For charged species, the rate constant of a diffusion-controlled recombination is somewhat higher in methanol than in water, as a result of the lower dielectric constant of methanol, which favors the electrostatic attraction of the reactants.) The present results suggest that the reactions for $K^+$, $Rb^+$, and $Cs^+$ are diffusion controlled. The reaction of $Na^+$ is almost diffusion controlled; only $Li^+$ shows a somewhat lower value (see Table V).

We may conclude that, from both a static and a dynamic point of view, murexide appears to be an ideal indicator for alkali ions in methanol. The spectral shifts are characteristic and easily detectable. The stability constants are high enough and in a convenient range (most convenient for an investigation of biological carriers; see below). The rates are as high as they could be, i.e., they are diffusion controlled. Mu-

TABLE V

Rate and stability constants of alkali-ion-murexide complexes in methanol

| 25° C | $k_R[M^{-1} \times sec^{-1}]$ | $k_D[sec^{-1}]$ | $K[M^{-1}]$ |
|---|---|---|---|
| $Li^+$ | $5.5 \times 10^9$ | $7.7 \times 10^6$ | $7.1 \times 10^9$ |
| $Na^+$ | $1.5 \times 10^{10}$ | $5.9 \times 10^6$ | $2.55 \times 10^3$ |
| $K^+$ | $\sim 2 \times 10^{10}$ * | $\geq 10^7$ | $1.1 \times 10^3$ |

* diffusion controlled

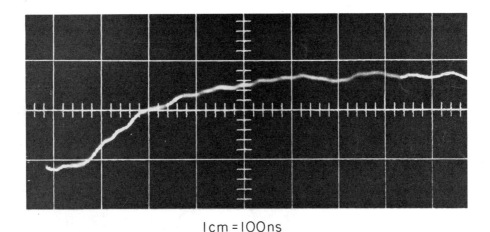

1cm = 100ns

FIGURE 6   Oscillogram of field effect relaxation for Na⁺-murexide complex in methanol.

rexide is thus suitable for indicating quickly any other complex reaction. (For details, see Winkler, 1969.)

## The Na⁺-monactin complex: structure and dynamics

Several carrier systems for alkali ions are described in the first section. We have chosen the macrotetrolide "monactin" as a model compound for a more detailed study of the equilibria and dynamics of complex formation.

Monactin belongs to a group of antibiotics that can be isolated from such microorganisms as actinomycetes. The chemical structure, as shown in Figure 7, was determined by Gerlach and Prelog (1963). X-ray analysis carried out by Kilbourn et al. (1967) indicates that the metal ion is surrounded by eight oxygens in a quasi-cubic arrangement, whereby the cyclic molecule is wrapped up like the seam of a tennis ball (see Figure 8A). All polar groups are "inside" the cyclic molecule, leaving a cavity for the metal ion that is optimally adapted to the size of K⁺ as reflected by the stability constants (see Table II). The values in Table II were determined by Pioda et al. (1967), using the technique of vapor-phase osmometry. As is seen, the Na⁺ complex of monactin is weaker than the K⁺ complex by more than two orders of magnitude, Li⁺ does not show detectable complex formation, and Rb⁺ and Cs⁺ complex measurably but more weakly than K⁺. The complex formation can be followed spectrophotometrically, with murexide used as indicator. Figure 9 shows the decrease of the Na⁺ murexide complex absorption on the addition of monactin. This method was used to follow the reaction behavior of monactin "instantaneously."

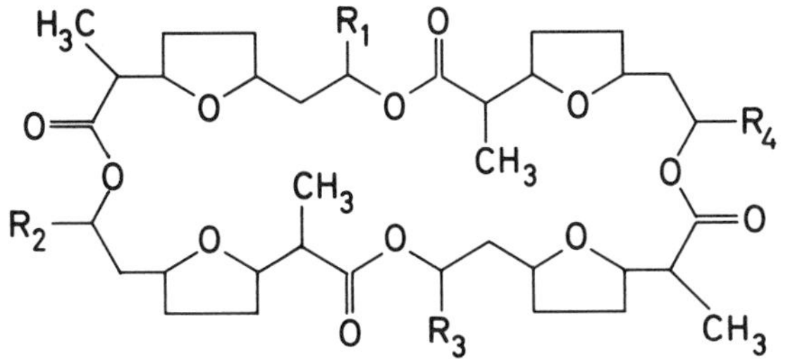

A few words may be added with regard to equilibrium measurements. Simon was the first to detect some equilibrium constants by using vapor-phase osmometry. With murexide as indicator, one may now also carry out spectrophotometric titrations. Moreover, temperature-jump measurements used for a study of the dynamics of equilibration also provide information on both equilibrium constants and enthalpies of reaction. Actually, this latter method has scarcely been used for equilibrium studies so far, but it turns out that it is more sensitive than most other methods known. The reason is that the T-jump amplitude (with spectrophotometric observation) is given by:

$$\delta E = \frac{\partial E}{\partial \ln K} \frac{\Delta H}{RT} \frac{\delta T}{T},$$

$\delta E$ being the change in extinction due to the T-jump $\delta T$, K the equilibrium constant, and $\Delta H$ the enthalpy of reaction.

In $\partial E/\partial \ln K$, all terms not containing $\ln K$ drop out so that this quantity is a unique function of K. For comparison,

R₁ = R₂ = R₃ = R₄ = CH₃                          Nonactin

R₁ = R₂ = R₃ = CH₃          R₄ = C₂H₅          Monactin

R₁ = R₃ = CH₃          R₂ = R₄ = C₂H₅          Dinactin

R₁ = CH₃          R₂ = R₃ = R₄ = C₂H₅          Trinactin

FIGURE 7   Chemical structure of macrotetrolides.

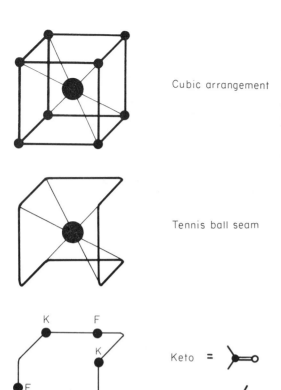

FIGURE 8   Schematic representation of the spatial structure of alkali-ion macrotetrolide complexes, left. Molecular model of nonactin as compared with solvated K⁺ ion, above.

Cubic arrangement

Tennis ball seam

Keto =

Furane =

titration curves essentially yield $\partial E/\partial \ln C_M$, $C_M$ being the concentration of the titration agent, such as the metal ions, but this quantity might become quite insensitive to $\ln K$ under certain conditions.

Another advantage of the relaxation amplitude method is that it yields direct information about the enthalpy of reaction. A special method has been worked out in which the

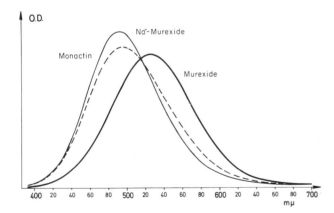

FIGURE 9   Decrease of Na⁺-murexide absorption upon addition of monactin (dotted line).

T-jump amplitudes of the metal-ion indicator are compared with the same system in the presence and absence of the carrier. The $\Delta H$ values obtained by this method are accurate by about $\pm 0.1$ kcal/mole, although the total temperature variation may not amount to more than 3° C (T-jump). Such a small temperature variation, of course, is of great advantage in the study of biological systems. The method is described in detail elsewhere (Winkler, 1969). For monactin with sodium a $\Delta H$ value of $-6.0$ kcal/mole was obtained, indicating an appreciable entropy term of $R\ln K$.

The equilibrium data obtained show clearly that specificity or selectivity is a property of the carrier molecule itself which is present as well in the absence of a membrane, i.e., in homogeneous solution. From the analysis of solvation energies and from the known structures, we also know how a carrier must be constructed in order to exhibit such a selectivity, but it is difficult to understand how such a system can be dynamically active in competing with any other type of transport.

Let us consider the carrier molecule as depicted in Figure 8B. The cavity is just large enough to accommodate the unsolvated metal ion. How does the metal ion get in? To strip off the whole solvation sphere would cost considerable activation energy and would slow the process. Thus, after we knew the architecture of the carrier molecule, an inescapable question was: How fast can such a carrier be loaded (and unloaded)? We knew from equilibrium studies that equilibration is complete within seconds. Thus, we tried first to measure the rates with the help of the temperature-jump technique. Because the sample cell represents the resistance of the discharge circuit, we could not reach very high time resolution. (Any cation concentration must be kept small so as not to interfere with the complex formation to be observed.)

Typical oscillograms are shown in Figure 10. The amplitude of the solution containing both the indicator and the carrier is more than twice as large as for the reference solution (containing only the indicator). That tells us that equilibration was complete within the time of resolution (i.e., about 1 millisecond). Actually, these very precise amplitude values (signal/noise $> 10^4$) were used to determine K and $\Delta$H of the carrier complex, as shown above.

A considerably higher time resolution can be reached by the electric-field pulse technique, which was used for the study of the metal indicator equilibrium. However, no field effect can be detected for the carrier complex, because monactin is a neutral molecule. On the other hand, enough heat is produced during the field pulse so that a (linear) shift of the extinction occurs, while the electric field is switched on (see Figure 11). Again, this shift was more than twice as large in the presence of the carrier, indicating that the equilibrium is established even within about 5 microseconds.

The only method with sufficient time resolution, therefore, that could be tried was sound absorption. With the help of a resonance method recently developed by Eggers (1967/1968), a pronounced rise of absorption toward frequencies of some hundred kilocycles per second was indeed observed. This absorption curve was evaluated with the help of known equilibrium data and yielded the rate constant for the complex formation of $Na^+$ with monactin.

The surprising fact is that this rate constant is as high as about $3 \times 10^8$ $M^{-1}$ $sec^{-1}$. It is just one order of magnitude lower than the expected value for a diffusion-controlled recombination. (Note that monactin is a neutral molecule, so that the limiting value is considerably lower than for the negatively charged indicator ion.)

## Alkali-ion carrier complexes: general behavior

How can such a process be envisaged? Apparently it is impossible for a completely or appreciably desolvated metal ion to slip into the carrier cavity. Removal of several solvent molecules in one step would cost too much free energy of activation. On the other hand, in the discussion of substitution rates we have seen that replacement of a single solvent molecule from an otherwise complete coordination sphere of any alkali ion can occur within $10^{-9}$ seconds. Thus, we must conclude that carrier-complex formation is a stepwise "redressing" process, in which every solvent molecule is substituted stepwise, so that a balance of solvation and

ΔOD = 0.0015/cm

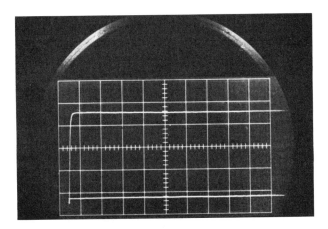

4 x 10$^{-5}$ M murexide

1 x 10$^{-3}$ M monactin

1 x 10$^{-3}$ M Na$^+$

ΔOD = 0.00072/cm

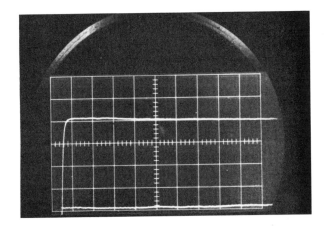

4 x 10$^{-5}$ M murexide

6 x 10$^{-4}$ M Na$^+$

FIGURE 10 Oscillograms of T-jump experiments with $Na^+$-murexide and monactin. The amplitudes of the relaxation effects for sample and reference solution differ by a factor of $\sim$2 (cf. sensitivity).

5 $\mu$ sec/cm

$\Delta$ OD = 0.0052/cm

$\lambda$ = 546 m$\mu$

E = 70 kV/cm

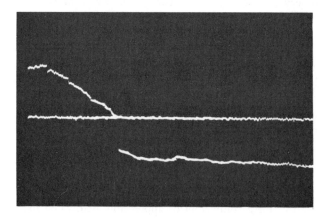

$4 \times 10^{-5}$ M murexide

$1 \times 10^{-3}$ M monactin

$1 \times 10^{-3}$ M Na$_0^+$

$(6 \times 10^{-4}$ M Na$^+)$

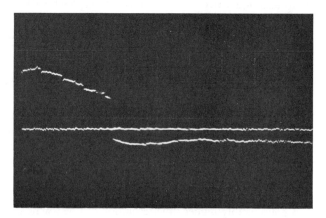

$4 \times 10^{-5}$ M murexide

$6 \times 10^{-4}$ M Na$^+$

FIGURE 11   Amplitudes of field effect of Na$^+$-murexide and monactin solutions. Due to heating by the electric field, the amplitudes shift linearly with time. As in Figure 10, this shift is about twice as large for the solutions containing monactin. (Note also the constant shift due to temperature change after the field is switched off.)

ligand-binding energy is always maintained. We may write the mechanism as:

$$M_N + X_0 \rightleftharpoons M_N X_0 \rightleftharpoons M_{N-1}X_1 \rightleftharpoons \cdots\cdot M_1 X_{N-1} \rightleftharpoons M_0 X_N ,$$

in which the index in M denotes the number of solvent molecules in the coordination sphere of the metal ion and the index in X the number of ligands bound to the metal ion. Only the initial and final states are appreciably populated; thus we may consider the intermediates to be in a steady state. Then we obtain an expression for the over-all rate constant that tells us that the probability for the reaction to proceed at any intermediate position in either direction must be not far from 50 per cent and that the rate constants for the single steps of substitution are of the order of magnitude of $10^9$ sec$^{-1}$, i.e., of the same order of magnitude as was found for the substitution of a single $H_2O$ molecule from the coordination sphere of alkali ions in aqueous media (see Figure 3). If any of the steps of substitution were appreciably slower, the value of $3 \times 10^8$ M$^{-1}$ sec$^{-1}$ for the over-all rate constant (which is only about one order of magnitude below the limiting value for diffusion control) could not be explained. In order to allow such a stepwise substitution, the carrier molecule must unfold the flexible "tennis-ball seam" to an almost annular conformation. This unfolding in the absence of the metal ion is favored by solvation of the polar groups inside the carrier molecule, which otherwise would repel one another considerably, and this solvation energy may balance the stress occurring in the "annular" conformation.

In conclusion, we may now formulate some rules for a design of an efficient metal-ion carrier:

1. The carrier molecule should possess electrophilic groups that can compete with the solvent molecules for metal-ion binding. These groups should be inside an otherwise lipophilic structure, which easily dissolves in membranes.

2. As many solvent molecules of the inner coordination sphere as possible should be replaced by the coordinating sites of the carrier molecule. For two ions of different size,

the reference state may then involve as much as the total difference of free energy of solvation.

3. The ligand should form a cavity adapted to the size of the metal ion. "Optimal fit" is related to an arrangement in which the difference of the free energies of ligand binding and solvation is maximal. Often it almost coincides with "fittest" geometrical arrangement. Cavity formation involves ligand-ligand repulsion as well as steric fixation of the chelate.

4. The carrier molecule should possess sufficient flexibility to allow for a stepwise substitution of the solvent molecules. Otherwise, if complete or substantial desolvation were required for the ion to slip into the cavity, the activation barrier would be high and the reaction slow.

Rule 1 fulfills the biological requirement, i.e., to gate the ion through a (lipid) membrane. Rules 2 and 3 take care of a high selectivity, whereas Rule 4 allows fast loading and unloading of the carrier. Not many structures of low molecular weight are known that would allow a simultaneous observation of all four rules. Almost all the classical complexing agents are poor in one respect or another.

So far, this paper has not dealt with any biological process. Actually, the problem we have been concerned with looks more like a problem of inorganic chemistry than one of neurobiology. This is certainly true, but before we could find out about the mechanism of sodium-potassium exchange in the nerve membrane we had to replace the demons "carrier" and "pump" by well-understood molecular processes.

We see that nature has designed very special molecules that are both dynamically active and highly specific, two properties that often exclude each other. For instance, the rate constant of dissociation $k_D$ is related to the binding constant by:

$$k_D = \frac{k_R}{K_{stab}}.$$

If $k_R$ reaches a limiting value (e.g., diffusion control), $k_D$ decreases if $K_{stab}$ increases. The rate constant $k_D$ determines the rate of unloading. Thus, as long as the over-all process is determined by the transport rate, one may gain an advantage by increasing $K_{stab}$, because this constant determines the specificity of binding. If, however, the rate of unloading becomes so slow that it determines the over-all rate, all further gain in specificity is lost by slowing down the unloading rate.

For $K^+$ and monactin, $K_{stab}$ is $2.5 \times 10^5$ $M^{-1}$. Thus, in order to keep $k_D$ above $10^3$ $sec^{-1}$ (millisecond range), $k_R$ had to be as high as $2.5 \times 10^8$ $M^{-1}$ $sec^{-1}$. Such a $K^+$ carrier would then be optimally adapted to the millisecond time range. The molecule solves the problem of quick loading by a conformation change, i.e., from an annular to a spherelike

conformation—a principle that is used in a similar way by protein molecules.

## Problems to be solved

We may now proceed to try to solve the problems that are more closely associated with a biological system. As a next step, we must answer the following questions:

1. How fast is the rate of loading and unloading if the carrier is sitting in the membrane and the whole process takes place in the interface?

2. What is the rate-limiting step of transport in artificial and natural membranes?

3. What are the carriers used for alkali-ion transport in nerve membranes?

4. How can the mechanism be switched on and off by an electric field and how is it coupled to a pump-facilitating active transport?

To answer the first and second questions, measurements with artificial and natural membranes are required. Relaxation spectrometry would be the adequate method to find out about the rate-limiting step, because the relaxation spectrum contains the time constants of all detectable intermediate steps involved in the over-all mechanism. For those studies, small membrane vesicles would be required. Max Delbrück (this volume) mentioned the various methods for the preparation of those vesicles. Actually, the work of Huang (1969), as well as that of H. Träuble and E. Grell (private communication), was started for this purpose. Relaxation studies on single bilayers using optical techniques with high time resolution may be considerably more difficult. On the other hand, the vesicles should also include protein-lipid systems.

The third question, of course, is a key question. The carriers studied here are antibiotics; it is not known whether they transport alkali ions in bacterial and, possibly, in mammalian membranes. Studies with $Tl^+$ ions (which have been found to replace specifically $K^+$ and $Rb^+$ in certain enzymes [Kayne, private communication]) are being carried out by Müldner to answer this important question.

The fourth question finally will bring us closer to processes related to the nerve membrane. The answer to it (and the corresponding model studies) will depend on the nature of the carrier. The carrier may be a small molecule like the antibiotics discussed in this paper. Such a molecule could be linked with an "allosteric" protein involved in the pump mechanism. The transport may just as well be facilitated by a membrane protein directly, or by a special membrane structure. In all these cases, specificity will depend on similar factors, as discussed in this paper (a balance of solvation and ligand binding by either of the mentioned structures), but the switching mechanism may be quite different for the various models.

Note: For a more detailed description of experiments and methods, see Winkler (1969). The aspect of metal complexing is treated in more detail in Eigen (1963) and Diebler et al. (1970). The broad subject of specific ion carriers was reported in an NRP work session (Eigen and De Maeyer, in press).

## REFERENCES

AGTARAP, A., J. W. CHAMBERLIN, M. PINKERTON, and L. STEINRAUF, 1967. The structure of monensic acid, a new biologically active compound. *J. Amer. Chem. Soc.* 89: 5737–5739.

AKASAKI, K., K. KARASAWA, M. WATANABE, H. YONEHARA, and H. UMEZAWA, 1963. Monazomycin, a new antibiotic produced by a streptomyces. *J. Antibiot. (Tokyo), Ser. A* 16: 127–131.

BROCKMANN, H. M., SPRINGORUM, G. TRÄXLER, and I. HÖFER, 1963. Molekulargewicht des Valinomycins. *Naturwissenschaften* 50: 689.

DIEBLER, H., M. EIGEN, G. ILGENFRITZ, G. MAASS, and R. WINKLER, 1969. Kinetics and mechanism of reactions of main group metal ions with biological carriers. *Pure Appl. Chem.* 20: 93–115.

EGGERS, F., 1967/1968. Eine Resonatormethode zur Bestimmung von Schall-Geschwindigkeit und Dämpfung an geringen Flüssigkeitsmengen. *Acustica* 19: 323–329.

EIGEN, M., 1963. Fast elementary steps in chemical reaction mechanisms. *Pure Appl. Chem.* 6: 97–115.

EIGEN, M., and L. C. M. DE MAEYER, 1970. Carriers and specificity in membranes. *Neurosci. Res. Program Bull.* (in press).

EIGEN, M., and R. G. WILKINS, 1965. The kinetics and mechanism of formation of metal complexes. *In* Summer Symposium on Mechanisms of Inorganic Reactions (J. Kleinberg, Symposium Chairman). Advances in Chemistry Series No. 49. American Chemical Society, Washington, D. C., pp. 55–80.

GEIER, G., 1967. Die Koordinationstendenz des Murexid-Ions. *Helv. Chim. Acta* 50: 1879–1884.

GERLACH, H., and V. PRELOG, 1963. Über die Konfiguration der Nonactinsäure. *Justus Liebigs Ann. Chem.* 669: 121–135.

GRAVEN, S. N., H. A. LARDY, D. JOHNSON, and A. RUTTER, 1966. Antibiotics as tools for metabolic studies. V. Effect of nonactin, monactin, dinactin, and trinactin on oxidative phosphorylation and adenosine triphosphatase induction. *Biochemistry* 5: 1729–1735.

HUANG, C., 1969. Studies on phosphatidylcholine vesicles. Formation and physical characteristics. *Biochemistry* 8: 344–352.

ILGENFRITZ, G., 1966. Chemische Relaxation in starken elektrischen Feldern. *In* Fourth Colloqium of the Johnson Research Foundation (in press).

IVANOV, V. T., I. A. LAINE, N. D. ABDULAEV, L. B. SENYAVINA, E. M. POPOV, YU. A. OVCHINNIKOV, and M. M. SHEMYAKIN, 1969. The physicochemical basis of the functioning of biological membranes: The conformation of valinomycin and its K+ complex in solution. *Biochem. Biophys. Res. Commun.* 34: 803–811.

KILBOURN, B. T., J. D. DUNITZ, L. A. R. PIODA, and W. SIMON, 1967. Structure of the K+ complex with nonactin, a macro-

tetrolide antibiotic possessing highly specific K+ transport properties. *J. Mol. Biol.* 30: 559–563.

LARDY, H. A., S. N. GRAVEN, and S. ESTRADA-O, 1967. Specific induction and inhibition of cation and anion transport in mitochondria. *Fed. Proc.* 26: 1355–1360.

LARDY, H. A., D. JOHNSON, and W. C. MCMURRAY, 1958. Antibiotics as tools for metabolic studies. I. A survey of toxic antibiotics in respiratory, phosphorylative and glycolytic systems. *Arch. Biochem. Biophys.* 78: 587–597.

MOORE, C., and B. C. PRESSMAN, 1964. Mechanism of action of valinomycin on mitochondria. *Biochem. Biophys. Res. Commun.* 15: 562–567.

PINKERTON, M., L. K. STEINRAUF, and P. DAWKINS, 1969. The molecular structure and some transport properties of valinomycin. *Biochem. Biophys. Res. Commun.* 35: 512–518.

PIODA, L. A. R., H. A. WACHTER, R. E. DOHNER, and W. SIMON, 1967. Komplexe von Nonactin und Monactin mit Natrium-, Kalium- und Ammonium-Ionen. *Helv. Chim. Acta* 50: 1373–1376.

PLATTNER, P. A., K. VOGLER, R. O. STUDER, P. QUITT, and W. KELLER-SCHIERLEIN, 1963. Synthesen in der Depsipeptid-Reihe. 1. Mitteilung. Synthese von Enniatin B. *Helv. Chim. Acta* 46: 927–936.

QUITT, P., R. O. STUDER, and K. VOGLER, 1963. Synthesen in der Depsipeptid-Reihe. 2. Mitteilung. Synthese von Enniatin A. *Helv. Chim. Acta* 46: 1715–1720.

QUITT, P., R. O. STUDER, and K. VOGLER, 1964. Synthesen in der Depsipeptidreihe. 3. Mitteilung. Anwendung der N-Nitroso-Schutzgruppe zur Synthese von Depsipeptiden. Eine weitere Synthese von Enniatin A. *Helv. Chim. Acta* 47: 166–173.

SCHWARZENBACH, G., and H. GYSLING, 1949. Metallindikatoren. I. Murexid als Indikator auf Calcium- und andere Metall-Ionen. Komplexbildung und Lichtabsorption. *Helv. Chim. Acta* 32: 1314–1325.

SCHWYZER, R., and P. SIEBER, 1957. Die Synthese von Gramicidin S. *Helv. Chim. Acta* 40: 624–639.

SHEMYAKIN, M. M., N. A. ALDANOVA, E. I. VINOGRADOVA, and M. YU. FEIGINA, 1963. The structure and total synthesis of valinomycin. *Tetrahedron Lett.* 28: 1921–1925.

SHEMYAKIN, M. M., YU. A. OVCHINNIKOV, V. T. IVANOV, V. K. ANTONOV, A. M. SHKROB, I. I. MIKHALEVA, A. V. EVSTRATOV, and G. G. MALENKOV, 1967. The physicochemical basis of the functioning of biological membranes: Conformational specificity of the interaction of cyclodepsipeptides with membranes and of their complexation with alkali metal ions. *Biochem. Biophys. Res. Commun.* 29: 834–841.

STEINRAUF, L. K., M. PINKERTON, and J. W. CHAMBERLIN, 1968. The structure of nigericin. *Biochem. Biophys. Res. Commun.* 33: 29–31.

STREHLOW, H., 1952. Zum Problem des Einzelelektrodenpotentials. *Z. Elektrochem.* 56: 119–129.

WINKLER, R., 1969. Kinetik und Mechanismus der Alkali und Erdalkalimetallkomplexbildung in Methanol. Doctoral thesis, Vienna.

# 62  Molecular Motion in Biological Membranes

## HARDEN M. MC CONNELL

BIOLOGICAL MEMBRANES offer many unsolved and challenging problems in biochemistry and structural chemistry. Our present discussion is concerned with *molecular motion* in membranes. We first describe in some quantitative detail the motion of a selected number of paramagnetic probes ("spin labels") in various biological membranes. Our analysis of the resonance spectra leads us to the conclusion that some membranes contain highly fluid hydrophobic regions, doubtless arising from the flexible hydrocarbon chains of the lipids. We then discuss the molecular basis of membrane fluidity. Finally, we speculate on the biological functions of membrane fluidity, particularly in relation to transport through membranes.

### Motions of spin-labels in membranes

During the past two years W. Hubbell, A. Horwitz, R. Kornberg, and I have undertaken a study of membranes using spin labels.

It will be recalled that a spin label is a synthetic organic free radical that can be incorporated in a biological system in order to yield kinetic and structural information through the paramagnetic resonance spectrum (Hamilton and Mc-Connell, 1968; Griffith and Waggoner, 1969). Significant kinetic and structural information can also be obtained from the nuclear resonance spectra of biochemical systems containing spin labels (Weiner, 1969; Mildvan and Weiner, 1969a, 1969b). The technique has a number of interesting features. For example, the paramagnetic resonance signals are essentially free from interference, because an extremely small proportion of the molecules in a biological system is paramagnetic, and only a small fraction of these gives detectable resonance signals under physiological conditions. On the other hand, the spin-label molecule or group always represents some perturbation on the system under study, and the significance of the structural and biochemical actions of the labeling molecules must always be carefully assessed.

The spin labels we have used in our work are nitroxide free radicals having the general formula

HARDEN M. MC CONNELL   Stauffer Laboratory for Physical Chemistry, Stanford, California

In this formula, R is usually designed so that the label is directed to some particular site or group of sites in a biological macromolecule. When the label is bound to a biological system so that it has a fixed orientation in space (static case), information on this orientation can be obtained from the resonance spectrum (McConnell, 1967; Hamilton and McConnell, 1968). When the molecule is tumbling, isotropically or anisotropically, information on this motion can be obtained when the correlation times are in the range of $10^{-7}$ to $10^{-9}$ seconds (Stone et al., 1965; Hubbell and McConnell, 1969a, 1969b). Finally, under favorable conditions, the resonance spectra in biological systems are sensitive to the polar or hydrophobic character of the environment (Hubbell and McConnell, 1968).

One of our earliest and most striking results was the observation that certain (intact) membranes contain hydrophobic regions of high fluidity (low viscosity) (Hubbell and McConnell, 1968). Figure 1 shows the paramagnetic resonance of the spin-label 2, 2, 6, 6-tetramethylpiperidine-1-oxyl (TEMPO) that has been added to a Ringer's solution containing an active vagus-nerve fiber of the rabbit.

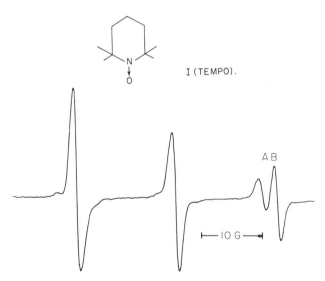

FIGURE 1   Paramagnetic resonance spectrum of the spin label TEMPO in the fluid hydrophobic region of the rabbit vagus-nerve fiber (A) and in the surrounding aqueous solution (B).

In Figure 1, signal A arises from TEMPO dissolved in a liquid-like hydrophobic region of the nerve fiber, and signal B arises from TEMPO dissolved in the surrounding aqueous solution. The widths of the individual lines are of the same order of magnitude ($\sim$2 gauss), showing that in both cases TEMPO is tumbling very rapidly and nearly isotropically, with a correlation time of the order of magnitude $10^{-9}$ seconds. Essentially the same spectrum has been seen for TEMPO in the presence of the walking-leg nerve fiber of the Maine lobster, *Homarus americanus* (Hubbell and McConnell, 1968).

In order to study further the hydrophobic regions of membranes, Hubbell has prepared a number of other spin labels, including the following molecules (Hubbell and McConnell, 1969a, 1969b):

II.

III (m,n).

Spin label II is the **N**-oxyl-4′, 4′-dimethyloxazolidine derivative of 5$\alpha$-androstan-3-one-17$\beta$-ol, and spin labels III (m, n) are the **N**-oxyl-4′, 4′-dimethyloxazolidine derivatives of the n + 2 keto acids having m + n + 3 carbon atoms. Models of spin labels I, II, and III (12, 3) and III (5, 10) are shown in Figures 2, 3, and 4. (Label III [5, 10] was first used by Keith et al. (1968) for biosynthetic incorporation into phospholipids of *Neurospora crassa*.)

Amphiphilic labels of types II and III (m, n) are particularly useful for studying membranes, because they bind to a great variety of membranes, their resonance spectra are often easily analyzed, and in favorable cases their orientation relative to membrane surfaces can be determined from the anisotropy of the resonance spectra. As examples, Figures 5 and 6 show the resonance spectra of amphiphilic labels in the nerve-fiber membranes, and in erythrocytes (oriented by hydrodynamic shear) as a function of the orientation of the anisotropic membrane distribution relative to the applied magnetic field. In the case of the erythrocytes, the preferred orientation of the long amphiphilic axis is parallel to the cylinder axis of the erythrocyte, and, in the case of the nerve fiber, the preferred orientation of the long amphiphilic axis is perpendicular to the direction of the common axonal cylinder axis (Hubbell and McConnell, 1969b). In the case of the steroid label II, the long amphiphilic axis is $\mu$. For labels III (12, 3) and III (17, 3), the preferred conformation of the label in the membranes is evidently one in which the oxazolidine ring is parallel to the local membrane surface.

FIGURE 2 Pauling-Corey-Koltun model of spin label I, 2, 2, 6, 6-tetramethylpiperidine-1-oxyl (TEMPO). Note the *approximately* spherical shape of the molecule.

The *motions* of labels II and III (12, 3), III (17, 3), and III (5, 10) in various membranes are particularly interesting. In brief, these motions are deduced from the observed paramagnetic resonance spectra, as follows. In the absence of any motion, the paramagnetic resonance of the nitroxide group in a spin label can be represented by the spin Hamiltonian, $\mathcal{H}$.

$$\mathcal{H} = |\beta|\mathbf{S} \cdot \mathbf{g} \cdot \mathbf{H}_o + h\mathbf{S} \cdot \mathbf{T}' \cdot \mathbf{I} - g_N\beta_N\mathbf{I} \cdot \mathbf{H}_o . \quad (1)$$

In equation (1), $\mathbf{S}$ and $\mathbf{I}$ represent the odd-electron and $N^{14}$ spin angular momentum operators, and $\mathbf{g}$ and $\mathbf{T}$ are the g-factor and nuclear hyperfine tensors. $\beta$ and $\beta_N$ are the electron and nuclear Bohr magnetons, h is Planck's constant, and $\mathbf{H}_o$ is the applied field vector. In the presence of molecular motion, this Hamiltonian becomes effectively time-dependent, because the tensors $\mathbf{g}$ and $\mathbf{T}$ are molecule fixed, and the field vector $\mathbf{H}_o$ is laboratory fixed. Under suitable conditions the solution of the Schrodinger equation, or density matrix equation,

$$ih \frac{\partial}{\partial t} \rho = \mathcal{H}\rho - \rho\mathcal{H} , \quad (2)$$

leads to an effective or average Hamiltonian, $\mathcal{H}'$,

$$\mathcal{H}' = |\beta|\mathbf{S} \cdot \mathbf{g}' \cdot \mathbf{H}_o + h\mathbf{S} \cdot \mathbf{T}' \cdot \mathbf{I} - g_N\beta_N\mathbf{I} \cdot \mathbf{H}_o , \quad (3)$$

of which the tensors $\mathbf{g}'$ and $\mathbf{T}'$ are suitable averages of the elements of $\mathbf{g}$ and $\mathbf{T}$ (Hubbell and McConnell, 1969a, 1969b). The paramagnetic resonance spectra of II and III (12, 3) and III (17, 3) in the nerve-fiber membrane can be accounted for by an anisotropic rotational diffusion of these molecules about their long amphiphilic axes. This means that the rotational frequencies are large compared with the corresponding anisotropies in the spin Hamiltonian, that is, large compared with frequencies in the range of $10^7$ to $10^8$ seconds$^{-1}$. In our experimental studies, the elements for $\mathbf{g}$

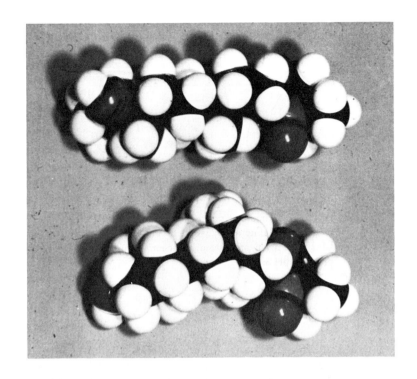

FIGURE 3 Pauling-Corey-Koltun model of spin label II (upper), the **N**-oxyl-4', 4'-dimethyloxazolidine derivative of 5-α-androstan-3-one-17-β-ol. Note the approximate cylindrical symmetry of this amphiphilic molecule. For comparison, another isomer (below) is shown, the **N**-oxyl-4', 4'-dimethyl-oxazolidine derivative of 5-β-androstan-3-one-17-β-ol.

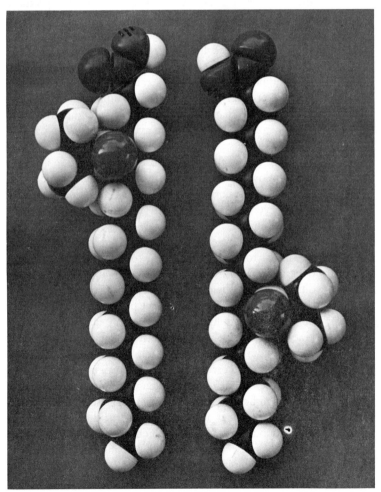

FIGURE 4 Pauling-Corey-Koltun models of the spin labels I (12, 3) (left) and I (5, 10) (right). These are the **N**-oxyl-4', 4'-dimethyl-oxazolidine derivatives of 5-ketostearic acid and 12-ketostearic acid, respectively.

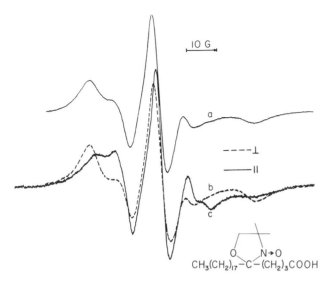

FIGURE 5 Paramagnetic resonance of the **N**-oxyl-4', 4'-dimethyl-oxazolidine derivative of sodium 5-ketotricosanoate (spin label I [17, 3]) in the walking-leg nerve fiber of *Homarus americanus*. (a) minced nerve; (b) applied field perpendicular to cylinder axis of nerve fiber; (c) applied field parallel to cylinder axis of nerve fiber. The g-factor for the center of the outer (inner) hyperfine extrema is 2.0031 ± 0.001 (2.0064 ± 0.001).

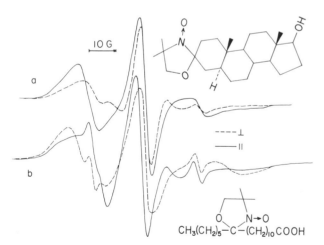

FIGURE 6 Paramagnetic resonance of the **N**-oxyl-4', 4'-dimethyl-oxazolidine derivatives of (a) 5-α-androstan-3-one-17-β-ol (spin label II) and (b) sodium 12-ketostearate (spin label I [5, 10]) in erythrocytes oriented by hydrodynamic shear, with the applied field perpendicular and parallel to the shear plane. For II, the center point g-factor is close to the isotropic value for both orientations, 2.0057. For I (5, 10) the outer (inner) hyperfine extrema center at g = 2.0027 ± 0.001 (2.0057 ± 0.001). The sharp signals at the isotropic splitting are due to I (5, 10) in solution in equilibrium with the bound label.

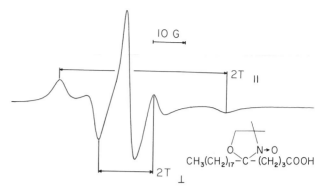

FIGURE 7 Paramagnetic resonance of the **N**-oxyl-4', 4'-dimethyloxazolidine derivative of sodium 5-ketotricosanoate (spin label I [17, 3]) in a sonicated dispersion of purified soybean phosphatides, pH 8.0.

and **T** are determined from an analysis of the resonance spectra of spin labels in single crystals of a suitable host, and the elements of **g**′ and **T**′ are deduced from the spectra of the labels in the membrane of interest.

As a specific example, the paramagnetic resonance of III (17, 3) in sonicated lipid dispersions is shown in Figure 7. The quantitative analysis of this spectrum shows that the oxazolidine ring undergoes a rapid anisotropic rotational motion (probably around the long amphiphilic axis of the molecule) and that the root mean square angular deviation of the oxazolidine ring from its average orientation is ∼26°. The resonance spectrum of III (17, 3) in the nerve fiber is very similar, but the anisotropic rotational motion is clearly slightly slower, and the corresponding mean angular deviation is 24°. The corresponding angular deviations are 16° in erythrocytes and 0° when III (17, 3) is bound to bovine serum albumin.

Although the resonance spectra of III (17, 3) and III (12, 3) are similar to one another in the sonicated lipid and nerve-fiber membranes, the spectra of III (5, 10) are remarkably different. Here the nitroxide group shows rapid and effectively isotropic motion. Thus, the polar head-group regions of the sonicated lipid dispersions and nerve-fiber membranes are relatively rigid; the hydrocarbon tail region is correspondingly more fluid.

As mentioned earlier, label II takes up a preferred orientation in membranes, with the long amphiphilic axis doubtless perpendicular to the local membrane surface. An analysis of the spectrum of II also shows a rapid anisotropic motion around the long amphiphilic axis **u**, this time with a large frequency compared to $10^8$ seconds$^{-1}$.

The anisotropic motions of labels II, III (17, 3), and III (12, 3) are associated with the facts that the nerve membrane binding sites are fluid, and one end of these molecules is "anchored" at a polar surface. What would happen to the motion of II if the 17-OH group were suddenly removed? We have found that the corresponding label, the **N**-oxyl-4′, 4′-dimethyloxazolidine derivative of 5α-androstan-3-one, undergoes rapid and effectively isotropic motion in the nerve-fiber membrane. *This result certainly supports the plausibility of models of perpendicular transport involving the rotations of relatively rigid proteins in the fluid hydrophobic region.* Note that the time required for rotation for the rigid steroid molecule (without the OH group) is $\sim 10^{-8}$ seconds, even though the molecular weight is relatively large ($\sim 376$). If we make the reasonable assumption that the correlation time for rotation is roughly proportional to molecular weight, a protein of molecular weight 37, 600 could still undergo a rotation in $\sim 10^{-6}$ seconds, if substrate binding produced a conformation change resulting in a disappearance of all exposed polar groups on the surface of the protein.

In our exploratory studies of membranes using spin labels, we have used labels I, II, and III (m, n) to study many different types of membranes (nerve fibers, myelin, bacteria, mycoplasma, synaptosomes, the sarcoplasmic system of muscle, erythrocytes, mitochondria, and *Nitella*). Unfortunately, these experiments are far from complete. Not all the above labels have been used with each membrane, quantitative binding studies have not been carried out, and many more labels must be synthesized and tested. Even so, we shall attempt to make some preliminary generalizations from our results.

The sarcoplasmic system of rabbit muscle shows the highest fluidity of any membrane we have yet examined. It is possible to isolate closed membrane vesicles from rabbit muscle which actively accumulate calcium ion in the presence of adenosine triphosphate (ATP) (Hasselbach, 1964; Martonosi, 1968). These vesicles show this high fluidity. The study has been carried out in this laboratory by R. Kornberg and W. Hubbell. The same degree of fluidity in these membranes is observed in the intact muscle system (Hubbell and McConnell, 1968).

A relatively high degree of fluidity has also been observed in the rabbit vagus-nerve fiber and in the walking-leg nerve fiber of the Maine lobster. We think it likely that the high fluidity of these nerve-fiber membranes is due to the axolemma, rather than to the surrounding Schwann cells. Thus, the axolemma in the squid axon is more strongly perturbed by low concentrations of detergents than is the surrounding Schwann cell membrane (Villegas and Villegas, 1968), which would be expected if the axolemma were the more fluid. Also, the relatively low proportion of myelin in the rabbit vagus is not expected to contribute significantly to the observed fluidity in the vagus-nerve fiber, because preliminary experiments indicate that myelin is not highly fluid (Hubbell and McConnell, unpublished). Other than

myelin, membranes of relatively high viscosity include the membranes of erythrocytes, mitochondria, and *E. coli*. The limiting membrane of the *Halobacterium* is the most rigid of any lipid-containing membrane we have yet examined (McConnell, Hubbell, and Stoeckenius, unpublished).

What is the molecular structure of the highly fluid regions we have detected in the neural and sarcoplasmic membranes? The spin-label spectra indicate very strongly that it must be a lipid bilayer. Our first reason for this assertion is the similarity between the resonance spectra of labels I, II, III (17, 3) and III (5, 10) in the nerve fiber and in sonicated phospholipid dispersions. For example, the resonance spectrum of III (17, 3) in sonicated phospholipids is shown in Figure 7, and may be compared with the resonance spectrum of this label in the nerve fiber as seen in Figure 5. Also, the spectral change observed in going from label III (12, 3) to label III (5, 10) is quite similar in both the nerve fiber and in sonicated phospholipids, and this change demonstrates the high fluidity of the hydrocarbon tail region relative to the polar head-group region. Our second reason for believing that these labels bind to fluid bilayers is the observed preferred orientation of the amphiphilic labels, which is perpendicular to the membrane surface.

In concluding this section on the motion of spin labels in membranes, it should be noted that other investigators have used other physical techniques to study membranes and phospholipid suspensions. Studies particularly relevant to the question of molecular motion in the hydrophobic regions of membranes have been summarized by Luzzati (1968), Chapman (1968), and Branton (1969). Insofar as there is direct overlap, our spin-label results are in excellent accord with these other studies. We think, however, that the spin-label spectra provide the most *direct* evidence for molecular *motion* in the hydrophobic regions of *intact* membranes.

## Molecular origin of membrane fluidity

The fluid hydrophobic regions revealed by the above spin-label studies are doubtless associated with the long hydrocarbon chains of the membrane lipids. It is well known that long-chain fatty acids as well as diacylphosphatides show one or more endothermic phase transitions associated with a stepwise melting of the hydrocarbon chains. An excellent review has been given by Chapman (1968). This investigator has shown by nuclear resonance studies that the hydrocarbon chains of certain pure lipids (e.g., 1, 2-dimyristoylphosphatidylethanolamine) show gradually increasing motion with increasing temperature until the chains "melt" at the temperature of a marked endothermic phase transition.

In phospholipids (e.g., diacylglycerolphosphatides) as well as in the fatty acids themselves, the endothermic transition temperatures are reduced when the fatty-acid chains are shortened, or when one or more carbon-carbon double bonds are introduced into the chains, or both. The introduction of *cis* double bonds lowers the transition temperatures more than does the introduction of *trans* double bonds, but the chains containing *trans* double bonds do melt at temperatures below those of the corresponding saturated chains. The literature on this subject has been reviewed by Chapman (1968). Here we only summarize briefly what we believe are the crucial structural principles involved in determining the melting points, and fluidities, of hydrocarbon chains.

In comparing different hydrocarbon chains, we assume them to be in a completely hydrophobic environment, and so neglect possible second-order interactions between the hydrocarbon chains and polar groups, or molecules, or both.

The preferred low-temperature conformation of a saturated hydrocarbon is extended, all *trans*, as shown in Figure 8 for octadecanoic acid, $C_{17}H_{35}(COOH)$. Salem (1962) has shown that the van der Waals dispersion interaction between extended (all *trans*) saturated hydrocarbon chains is of the form

$$W_{disp} = A \frac{3\pi}{8\ell} \frac{N}{D^5} ,$$

in which $\ell = 1.26$ Å (the projection of the carbon-carbon vector along the chain direction), D is the perpendicular distance between the chains, N is the number of $CH_2$ groups in the chains, and A is a constant ($-1340$ kcal/Å$^6$ mole). At a typical distance of 5 Å, this formula yields an interaction energy of 0.4 kcal/mole per $CH_2$ group. Two stearic acid molecules (N = 18) at a separation of 4.8 Å interact with an attractive dispersion energy of $-8.4$ kcal/mole. The strong dependence of the attractive intermolecular dispersion energy on intermolecular distance signifies that any modification of the chains that reduces the closeness of packing should increase the fluidity of the hydrocarbon chains, or reduce their melting points if they are not already fluid.

Unsaturated fatty-acid chains in bacterial and mammalian phospholipids have *cis* carbon-carbon double bonds. A model of the unsaturated *cis* fatty acid, 9-octadecenoic acid, is shown in Figure 8, in a lowest-energy conformation (see Flory, 1969). It is clear that the "kink" in the chain must reduce the closeness with which such hydrocarbon chains can be packed one against the other. This should be particularly significant in the usual situation in which the lipids contain *mixtures* of fatty acids differing in the degree of unsaturation.

The introduction of a *trans* carbon-carbon double bond into a saturated fatty-acid chain also produces a kink in an otherwise straight hydrocarbon chain. Figure 8 shows a model of one low-energy conformation of 9-octadecenoic acid with a *trans* carbon-carbon double bond between carbon atoms 9 and 10. Here, again, we expect a reduction in the ease with which hydrocarbon chains can be packed one

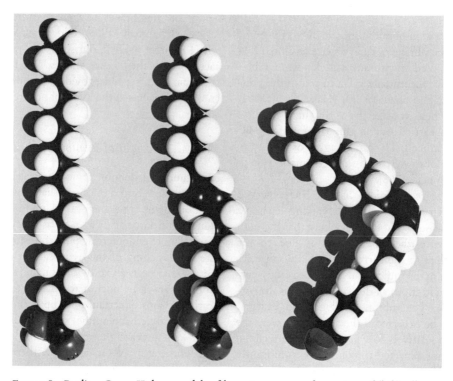

FIGURE 8  Pauling-Corey-Koltun models of lowest energy conformation s of (left), all *trans* octadecanoic acid; (center), *trans*-9-octadecenoic acid; and (right), *cis*-9-octadecenoic acid.

against the other, especially in typical lipid systems in which a mixture of saturated and various unsaturated chains is present.

The presence of carbon-carbon double bonds in fatty-acid chains must also play another significant role in contributing to the fluidity of the hydrophobic regions. That is, in highly fluid regions of membranes it is clear that the long hydrocarbon chains must be flexing and twisting so as to require internal rotations around the carbon-carbon single bonds. The introduction of carbon-carbon double bonds *reduces* the barriers to internal rotation for adjacent carbon-carbon single bonds. Thus, in the following sequence of carbon-carbon bonds,

$$C^a - C^b - C^c$$

the barrier to rotation about the b-c single bond is of the order of 3 kcal/mole, when a-b is a single bond (Flory, 1969). On the other hand, when carbon atoms a and b are joined together by a double bond,

$$C^a = C^b - C^c$$

the barrier to internal rotation about the b-c single bond is only 2 kcal/mole (Flory, 1969).

Thus, the introduction of carbon-carbon double bonds into long hydrocarbon chains enhances fluidity by two distinct mechanisms: (1) the double bonds introduce "kinks" that decrease the efficiency of close molecular packing of the hydrocarbon chains; and (2) the double bonds *facilitate* internal rotations around adjacent carbon-carbon single bonds.

We may now consider factors that can be expected to reduce the molecular fluidity of the hydrocarbon chains in a bilayer structure, for a fixed hydrocarbon chain composition. It is known from both nuclear resonance and spin-label studies that the fluidity of phospholipid dispersions is reduced by the incorporation of cholesterol (Chapman, 1968; Hubbell and McConnell, 1968). We interpret this as signifying no highly specific interactions between the fatty-acid hydrocarbon chains and the cholesterol molecule, but rather the fact that the cholesterol molecule itself has a rigid structure that inhibits the bending motions of neighboring hydrocarbon chains. If this interpretation is correct, then the introduction of any relatively large and rigid hydrophobic structure into an otherwise highly fluid chain region should reduce the molecular motion. (The effect of the antibiotic gramicidin S on phospholipid fluidity has been noted by Hubbell and McConnell [1968].) On the other hand, the introduction of cholesterol into an otherwise relatively rigid hydrocarbon chain region (say, straight-chain saturated hydrocarbons) may very well increase the molecular motion (Chapman, 1968).

If we make the very reasonable assumption that proteins "inside" the hydrophobic region of membranes have hydrophobic exteriors, and are just as "rigid" as are the water-soluble proteins familiar to us from X-ray diffraction studies, we may conclude that such proteins will tend to reduce the fluidity of an otherwise fluid ensemble of lipids. The great differences we have seen in the fluidity of different membrane systems may be due in part to this effect of proteins with hydrophobic surfaces.

### Biological role of membrane fluidity

The high membrane fluidity we have observed in our experiments is in sharp contrast to the relatively rigid structures that are so familiar to us from X-ray studies of crystalline proteins and nucleic acids. Indeed, this high hydrophobic fluidity is so unique relative to other assemblages of biological macromolecules that one may assume that this fluidity has a special function in biological processes. Of course the hydrophobic character of the fluid region provides a permeability barrier for water-soluble substances. Certainly pure protein membranes could likewise provide excellent permeability barriers, judged from the (lipid-free) protein gas vacuole membrane of *Halobacterium*. Thus, a significant role of lipids in membranes must be to provide a *fluid* permeability barrier. Regions of membranes having (protein)-(lipid bilayer)-(protein) structures may have both rigid protein and fluid lipid permeability barriers.

A fluid permeability barrier must have special biological functions. For example, it surely offers a self-sealing capacity in the event a membrane is injured, as well as a degree of flexibility in the biosynthetic self-assembly of the membrane. Of perhaps even greater interest is the likelihood that this fluidity permits transport both perpendicular and parallel to the membrane surface. We shall not attempt to discuss parallel transport, except to mention that it may well be involved in lipid movement from the synthetic point of origin to the membrane target site. (The insect sex attractant bombykol [Schnieder and Kaissling, 1956; Butenandt and Hecker, 1961] may also move parallel to the surfaces of membrane-like structures [Adam and Delbrück, 1968].)

We suggest that a crucial role of membrane fluidity in connection with perpendicular transport is the maintenance of a permeability barrier near proteins that are "moving" in the hydrophobic region, to provide transport of ions, or molecules, or both through the membrane. For example, consider the (facilitated) transport of the polar, water-soluble molecule lactose,

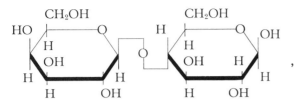

through a membrane with a typical thickness of the order of 100 Å (e.g., the limiting membrane of *E. coli*). This large polar molecule must be in contact with polar groups as it moves from one side of the membrane to the other; the polar environment is doubtless provided by stereospecific binding site (or sites) on one or more membrane proteins. Thus, not only does a large molecule traverse the membrane, but an even larger protein binding site moves through the membrane. The protein itself is certainly not fluid, so a very large rigid structure must move from one side of the membrane to the other, without great frictional resistance, and without opening up to other water-soluble ions and molecules, holes, cracks, or other breaks in the permeability barrier. These conditions surely require that the transporting protein be contained in a fluid hydrophobic environment. A speculative model of the membrane with a transport protein is shown in Figure 9. In this Figure it is assumed that binding of the lactose produces the loss of an otherwise exposed polar region of the protein, which could be accomplished if the lactose were so bound in a crevice in the protein as to be almost "buried" in the protein. The lactose binding should eliminate all exposed or "bare" polar groups

on the carrier protein, so that it can rotate or move in the lipid region without loss of hydrophilic solvation energy.

Not all the membrane systems we have studied show a high degree of fluidity. The membranes that appear to be most fluid (sarcoplasmic vesicles and axonal membranes) probably have particularly highly concentrated transport sites. It appears reasonable to conjecture at the present time that the highly fluid regions of membranes may be associated with sites for perpendicular transport, or at least certain types of perpendicular transport. (Chapman, 1968, has also suggested that fluidity may be necessary for diffusional processes in tissues.) Our conjecture has a corollary: high fluidity must also have one or more disadvantages; otherwise all membranes would be highly fluid. We suggest that one major disadvantage is that high fluidity is also associated with leakiness. In membranes in which little transport is required, or leakiness is disadvantageous, the membrane lipids may be less fluid.

There is evidence that fluidity in membranes may be regulated in response to environmental variables that would otherwise tend to change the fluidity. For example, as Chapman (1968) has pointed out, the proportions of unsaturated fatty acids in bacterial membranes, as well as in the membranes of the goldfish brain, increase when the external temperature is lowered, which may well be an adjustment to maintain membrane fluidity as the temperature is lowered.

Recent experiments by Fox (1969) also suggest a close connection between hydrocarbon chain unsaturation and perpendicular transport. This investigator studied the induction of lactose transport in an unsaturated fatty-acid auxotroph of *E. coli*. This bacterium requires an exogenous

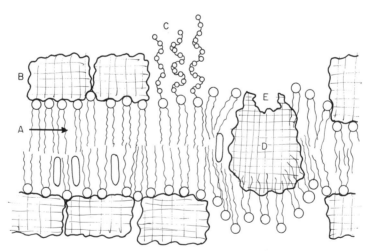

FIGURE 9   Schematic and speculative diagram of a membrane structure, showing lipophilic protein D with polar lactose binding site E, to provide facilitated transport of lactose through the fluid hydrophobic region. A. Phospho-lipids arranged in a bilayer. B. Protein. C. Lipopolysaccharides. D. Permease protein. E. Binding site. Note that lipopolysaccharides may not be present in the same bacterial membrane that contains the permease.

source of unsaturated fatty acid for growth (Silbert and Vagelos, 1967). Fox showed that, in the absence of exogenous unsaturated fatty acid, the lactose permease (M-protein) could be induced, but that this protein did not bind to the membrane in a functional form. On the other hand, in the presence of exogenous unsaturated fatty acid, the M-protein could be induced and was functionally incorporated into the membrane. As a first crude approximation, we may therefore picture the unsaturated fatty acids as being incorporated in the membrane and forming a fluid pool in which the M-protein is dissolved.

Our conjectures regarding the localization of fluid hydrophobic regions in the vicinity of sites for perpendicular transport suggest a number of experiments. Three examples are enumerated below.

1. In bacteria in which transport sites in membranes can be induced in relatively large numbers, it may be possible to demonstrate a quantitative increase of the extent of fluid membrane regions associated with induction of transport with the use of a gratuitous inducer.

2. It may be possible to visualize perpendicular transport sites in membranes, in the electron microscope, by gentle treatment of the membrane tissue with detergents, lipid soluble materials, or both. Under these conditions it can be anticipated that such substances will bind preferentially to the most fluid regions of the membrane, producing recognizable distortions or imperfections in the vicinity of the transport sites.

3. It may be possible to obtain a direct demonstration of the relation between fluidity and perpendicular transport if one can obtain a "temperature sensitive" transport system that shows an abrupt change in transport versus temperature curve. In this case, spin labels localized in the transport region should show a correspondingly abrupt change in spectrum at the same temperature.

In conclusion, we may note that in our discussion we have assigned a very important but relatively passive role to membrane fluidity: the fluid hydrophobic region of a membrane makes it possible for a relatively rigid protein with a hydrophobic surface to rotate, be translated (or both), within the membrane. It is quite plausible that fluidity may also serve as a regulator or even a switch for biochemical and biophysical events in membranes. For example, an effector, a hormone, or both, may bind to a hydrophobic region, modify the fluidity, and increase or decrease the rate of membrane transport. Although we have no direct evidence for this effect at present, it is interesting to note that we have observed that DDT binds to phospholipids in a way that makes them less fluid, and this substance also *prolongs* the action potential of axonal membranes (Hubbell and McConnell, unpublished).

The insect sex attractant bombykol (hexadeca-10-*trans*-12-*cis*-diene) has a chemical formula,

$$CH_3 - (CH_2)_2 - C^H = C^H \longrightarrow C_H$$
$$= C^H \longrightarrow (CH_2)_8 - CH_2OH,$$

that suggests it may fluidize an otherwise nonfluid bilayer region of a membrane. This, in turn, could "free" a protein sodium carrier, leading to neural excitation, and appropriate response of the insect.

Photoisomerization of the visual pigment retinaldehyde could likewise change membrane fluidity, leading to membrane permeability changes, and excitation.

Although our discussion of the possible biological roles of molecular motion and fluidity in membranes is clearly somewhat speculative at the present moment, it is also clear that it suggests a number of definitive experiments whereby these speculations can be tested.

## Acknowledgments

I am greatly indebted to my graduate students, Messrs. W. Hubbell, A. Horwitz, and R. Kornberg, for many helpful and critical discussions of the contents of this manuscript. Most of the experiments cited here have been carried out by Mr. Hubbell, and many of the ideas have been developed in discussions with him. Mr. A. Horwitz first brought the important works of F. Fox and of Silbert and Vagelos to my attention. The spin-label studies of the sarcoplasmic vesicle system by Mr. Kornberg and Mr. Hubbell have been a most significant contribution to our conviction that high fluidity in membranes is associated with perpendicular transport. Spin label III (5, 10) was first synthesized in my laboratory by Dr. Joachim Seelig, and I am greatly indebted to him for the sample used in the work described here. I am also indebted to Professor Paul Flory for a helpful discussion concerning barriers to internal rotation in hydrocarbon chains.

This work has been sponsored by the National Science Foundation under Grant GB7559, by the National Institutes of Health under Grant NB-08058-01, and by the Office of Naval Research under Contract No. 225(88), and it has benefited from facilities made available to Stanford University by the Advanced Research Projects Agency through the Center for Materials Research.

### REFERENCES

ADAM, G., and M. DELBRÜCK, 1968. Reduction of dimensionality in biological diffusion processes. *In* Structural Chemistry and Molecular Biology (A. Rich and N. Davidson, editors). W. H. Freeman and Co., San Francisco, pp. 198–215.

BRANTON, D., 1969. Membrane structure. *Annu. Rev. Plant Physiol.* 20: 209–238.

BUTENANDT, A., and E. HECKER, 1961. Synthese des Bombykols, des Sexual-Lockstoffes des Seidenspinners, und seiner geometrischen Isomeren. *Angew. Chem.* 73: 349–353.

CHAPMAN, D. (with a contribution by D. F. H. WALLACH), 1968. Recent physical studies of phospholipids and natural membranes. *In* Biological Membranes (D. Chapman, editor). Academic Press, New York, pp. 125–202.

FLORY, P. J., 1969. Statistical Mechanics of Chain Molecules. Interscience Publishers, New York, pp. 192–196.

FOX, F., 1969. A lipid requirement for induction of lactose transport in *Escherichia coli*. *Proc. Nat. Acad. Sci. U. S. A.* 63: 850–855.

GRIFFITH, O. H., and A. S. WAGGONER, 1969. Nitroxide free radicals: Spin labels for probing biomolecular structure. *Accounts Chem. Res.* 2: 17–24.

HAMILTON, C. L., and H. M. MCCONNELL, 1968. Spin labels. *In* Structural Chemistry and Molecular Biology (A. Rich and N. Davidson, editors). W. H. Freeman and Co., San Francisco, pp. 115–149.

HASSELBACH, W., 1964. Relaxing factor and the relaxation of muscle. *Progr. Biophys. Mol. Biol.* 14: 167–222.

HUBBELL, W. L., and H. M. MCCONNELL, 1968. Spin-label studies of the excitable membranes of nerve and muscle. *Proc. Nat. Acad. Sci. U. S. A.* 61: 12–16.

HUBBELL, W. L., and H. M. MCCONNELL, 1969a. Motion of steroid spin labels in membranes. *Proc. Nat. Acad. Sci. U. S. A.* 63: 16–22.

HUBBELL, W. L., and H. M. MCCONNELL, 1969b. Orientation and motion of amphiphilic spin labels in membranes. *Proc. Nat. Acad. Sci. U. S. A.* 64: 20–27.

KEITH, A. D., A. S. WAGGONER, and O. H. GRIFFITH, 1968. Spin-labeled mitochondrial lipids in *Neurospora crassa*. *Proc. Nat. Acad. Sci. U. S. A.* 61: 819–826.

LUZZATI, V., 1968. X-ray diffraction studies of lipid-water systems. *In* Biological Membranes (D. Chapman, editor). Academic Press, New York, pp. 71–123.

MCCONNELL, H. M., 1967. Spin-labeled protein crystals. *In* Magnetic Resonance in Biological Systems (A. Ehrenberg, B. G. Malmström, and T. Vännegård, editors). Pergamon Press, New York, pp. 313–323.

MARTONOSI, A., 1968. Sarcoplasmic reticulum. IV. Solubilization of microsomal adenosine triphosphatase. *J. Biol. Chem.* 243: 71–81.

MILDVAN, A. S., and H. WEINER, 1969a. Interaction of a spin-labeled analog of nicotinamide-adenine dinucleotide with alcohol dehydrogenase. II. Proton relaxation rate and electron paramagnetic resonance studies of binary and ternary complexes. *Biochemistry* 8: 552–562.

MILDVAN, A. S., and H. WEINER, 1969b. Interaction of a spin-labeled analogue of nicotinamide adenine dinucleotide with alcohol dehydrogenase. III. Thermodynamic, kinetic, and structural properties of ternary complexes as determined by nuclear magnetic resonance. *J. Biol. Chem.* 244: 2465–2475.

SALEM, L., 1962. The role of long-range forces in the cohesion of lipoproteins. *Can. J. Biochem. Physiol.* 40: 1287–1298.

SCHNEIDER, D., and K.-E. KAISSLING, 1956. Der Bau der Antenne des Seidenspinners *Bombyx mori* L. I. Architektur und Bewegungsapparat der Antenne sowie Struktur der Cuticula. *Zool. Jahrb., Abt. Anat. Ontog. Tiere* 75: 287–310.

SILBERT, D. F., and P. R. VAGELOS, 1967. Fatty acid mutant of *E. coli* lacking a $\beta$-hydroxydecanoyl thioester dehydrase. *Proc. Nat. Acad. Sci. U. S. A.* 58: 1579–1586.

STONE, T. J., T. BUCKMAN, P. L. NORDIO, and H. M. MCCONNELL, 1965. Spin-labeled biomolecules. *Proc. Nat. Acad. Sci. U. S. A.* 54: 1010–1017.

VILLEGAS, G. M., and R. VILLEGAS, 1968. Ultrastructural studies of the squid nerve fibers. *J. Gen. Physiol.* 51 (no. 5, pt. 2): 44s–60s.

WEINER, H., 1969. Interaction of a spin-labeled analog of nicotinamide-adenine dinucleotide with alcohol dehydrogenase. I. Synthesis, kinetics, and electron paramagnetic resonance studies. *Biochemistry* 8: 526–533.

# 63 Evidence for Structural Changes During Nerve Activity and Their Relation to the Conduction Mechanism

R. D. KEYNES

THE MAIN PURPOSE of this article is to describe some recent work on the thermal and optical properties of stimulated nerves which was undertaken in the hope that it would throw some light from a fresh direction on the nature of the molecular events underlying conduction of the nervous impulse. But before this new experimental evidence is considered, it is necessary to summarize the two opposed views on the mechanism of excitation that currently hold the field.

## The two stable-states theory

The idea has been proposed by Tasaki (1968) that the nerve membrane is to be regarded essentially as a fixed charge structure that can be switched between two different stable states. As I understand it, the genesis for this concept was the fact that, under certain conditions, the nerve membrane can generate greatly prolonged action potentials, from the plateau of which it can be triggered to revert abruptly and in all-or-none fashion back to the resting state in the same way as it does when the spike is initiated. Figure 1 shows an extreme example of this type of behavior. It was observed when a squid giant axon was first injected with tetraethyl-

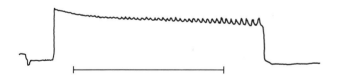

FIGURE 1   The action potential recorded with an internal voltage wire in a squid giant axon containing 15 mM TEA and immersed in isethionate artificial sea water. Amplitude at peak was 95 mV. Time calibration bar = 0.5 sec. Temperature 12° C. (Cohen and Keynes, unpublished experiment.)

R. D. KEYNES   Agricultural Research Council, Institute of Animal Physiology, Babraham, Cambridge, England

ammonium (TEA) ions to obtain the roughly tenfold prolongation of the spike described by Tasaki and Hagiwara (1957), and was then immersed in a solution in which nonpenetrating isethionate ions (Koechlin, 1955) had been substituted for the normal chloride. The further prolongation of the spike, by about two orders of magnitude, which resulted from this exchange of external anions does not seem to have been anticipated by Tasaki (1968), but it is readily understood from the alternative viewpoint as being the inevitable consequence of removing the principal source of ionic current for repolarizing the membrane. This source is an inward transfer of chloride in an axon in which the outward potassium current has already been blocked by TEA. It is not entirely clear whether the standard equivalent circuit is able, in its basic form, to account for the gradual buildup of oscillations during the plateau and the final regenerative repolarization seen under these abnormal conditions. However, there would be no difficulty in incorporating additional elements to explain this kind of behavior, such as a slowly inactivated or anomalous potassium conductance change.

It is impossible, in the space at my disposal, to examine the two stable-states theory in any detail, or to explain fully why I find it somewhat unsatisfactory. An essential feature of the theory is the role attributed to calcium ions, the displacement of which from association with the fixed negative charges in the membrane by potassium ions moving outward during the initial depolarization is supposed to characterize the "active" stable state, and the binding of which once more to the membrane then restores the "resting" stable state. Therefore, I would draw attention to the absence from the published accounts of the theory of any quantitative or even qualitative explanation for several generally accepted features of the action of calcium on excitable membranes. Although the theory might be expected to predict the observed rise in threshold of a nerve fiber with external $[Ca^{2+}]_o$, it would surely predict also that the duration of the active state, and hence of the spike, should decrease with rising $[Ca^{2+}]_o$. In actual fact, the spike duration is normally independent of external calcium, and it is only under severely unphysiological conditions that the spike can be shortened by raising the calcium level. Again, the well-documented finding (Frankenhaeuser and Hodgkin, 1957) that, under a voltage clamp, changes in $[Ca^{2+}]_o$ shift the curves relating sodium and potassium conductance to membrane potential along the voltage axis without altering either their shape or the peak values of the conductances, has been ignored by Tasaki (1968). He also makes no attempt to account for the difference in magnitude between the amounts of sodium and potassium that are transferred across the membrane during a single impulse (of the order of 4 pmole/cm$^2$) and the corresponding figure for calcium (shown by Hodgkin and Keynes, 1957, to be only about 0.006 pmole/cm$^2$).

## The ionic theory

The most widely accepted hypothesis to account for the properties of excitable membranes is that due to Hodgkin and Huxley, according to which the membrane can be regarded as possessing a capacitative element (Figure 2), in

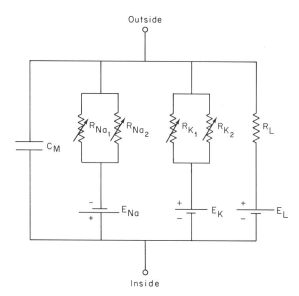

FIGURE 2 Equivalent circuit for an excitable membrane. In order to account for such phenomena as pacemaker activity and anomalous rectification, $R_{Na}$ and $R_K$ have each been shown as two variable resistors in parallel, which might behave as different functions of potential and time and be differently affected by drugs.

parallel with which there are batteries representing the Nernst equilibrium potentials $E_{Na}$ and $E_K$ for sodium and potassium ions, each in series with variable resistances. An additional battery and a fixed "leak" resistance $R_L$ are also included to cover the contribution to the membrane current, which is relatively small, from Cl$^-$ and other uniden-

tified ions. In order to account for the electrical events during excitation, it is proposed (for a general review of the theory, see Hodgkin, 1964) that $R_{Na_1}$ and $R_{K_1}$ go through a sequence of voltage-dependent and time-dependent changes. It is not necessary to describe in detail here the mathematical formulation of the hypothesis in terms of equations governing the sodium and potassium conductances that incorporate three coefficients that are continuous functions of membrane potential and time. It will suffice to note that the theory has successfully survived the tests of predicting from voltage-clamp data the behavior of the propagated action potential under a wide variety of conditions and of predicting from purely electrical measurements a correct value for the chemically determined ionic movements during the impulse.

It must be emphasized at this point that a description of the electrical properties of the nerve membrane, however complete and accurate it may be, does not provide any direct information about the nature, at the molecular level, of the events involved in the changes of ionic permeability. Certain features of the Hodgkin-Huxley equations, however, do have implications for the underlying molecular mechanism that should not be overlooked. Possibly the most important of these is the steepness of the conductance-voltage relationships in the neighborhood of the normal resting potential, an e-fold increase in conductance corresponding to a depolarization of only 4 or 5 millivolts (see Hodgkin, 1964). This means that the permeability must be controlled either by the displacement or reorientation of a single particle carrying as many as six negative charges, or by a mechanism involving the simultaneous redistribution of an equal number of singly charged groups or a correspondingly greater number of groups whose effective charge is less than one. The former possibility seems unlikely, but has not yet been ruled out experimentally (see Hille, 1970).

## Na and K channels: common or separate?

Two questions that are highly relevant to any consideration of possible conformational changes in excitable membranes are, first, whether the sodium and potassium ions pass through the same or separate channels and, second, how many channels there are. Both problems have recently been discussed by Hille (1970), to whom reference should be made for a full presentation of the arguments. There seems little doubt that the sodium and potassium channels are distinct entities. Thus, treatment of nerve membranes with the Japanese puffer-fish poison tetrodotoxin (TTX) totally blocks the sodium channels without affecting in any way the behavior of the potassium channels, and, conversely, TEA strikingly reduces the potassium conductance without altering the sodium conductance.

Several other approaches, such as a detailed consideration of the ionic selectivity of the membrane, of the action of calcium, of the degree of interaction between individual ions traversing the membrane, and of the relative numbers of channels, all lead to the same conclusion.

## The number of channels

From measurements of the amount of TTX removed from a small volume of a dilute TTX solution by passing several lobster-leg nerve trunks through it, Moore et al. (1967) concluded that not more than 13 TTX molecules were taken up by 1 $\mu m^2$ of membrane. If, as seems reasonable, this figure is taken to represent the number of sodium channels, it may appear surprisingly small. But it is large compared with that for sodium-pump sites in cells like erythrocytes, in which there is only about one ouabain-binding site per $\mu m^2$ (see, for example, Ellory and Keynes, 1969); and Hille (1970) has argued that it is not incompatible with the evidence from other directions. Taking the peak sodium conductance of a lobster axon as 0.53 mho/cm$^2$ (Narahashi et al., 1964), the conductance of a single channel comes out as 0.4 nmho. It may be no more than a coincidence, but this is precisely the value obtained by Bean et al. (1969) for the quantal increase in conductance when the so-called "excitability inducing material" is incorporated into phospholipid bilayers. In the squid giant axon, the peak sodium conductance is slightly smaller than it is in the lobster axon (0.12 mho/cm$^2$), so that the packing density of sodium channels is only of the order of 3 per $\mu m^2$. The number of potassium channels is rather uncertain, but their conductance is probably much lower than that of the sodium channels, and there are correspondingly more of them, perhaps by a factor of 50 (Hille, 1970). The importance of these numbers in the present context lies in a comparison with the number of phospholipid molecules in the membrane, which, if the area of a single molecule is 50 Å$^2$, is no fewer than $2 \times 10^6$ per $\mu m^2$ of each surface. One cannot avoid the conclusion that, because the current-carrying channels are evidently so sparsely distributed in the membrane, it will be much more difficult to detect conductance-linked conformational changes taking place only at these special places than potential-linked effects that might involve all, or at any rate a majority, of the molecules present.

## Entropy changes during nervous activity

Recent measurements of the temperature changes during the conduction of impulses along the nonmyelinated fibers of the cervical vagus nerve of rabbit have shown that, as illustrated in Figure 3, the upstroke of the action potential is accompanied by a liberation of heat, most of which is then reabsorbed during the downstroke (Howarth et al., 1968).

After allowing as well as possible for the effects of overlap between the downstroke of the spike in the fibers that conduct fastest and the upstroke in the slower ones, we estimated that the quantity of heat released during the upstroke of each impulse was 24.5 $\mu$cal/g. The "negative" heat during the recovery phase was 22.2 $\mu$cal/g, leaving a net initial positive heat of 2.3 $\mu$cal/g. The question then arose of explaining these results in terms of the events known to be taking place during the conduction process.

The first possibility to be examined was whether the heat of exchange of sodium and potassium could account for the initial heat. This idea could be quickly dismissed, because for 0.15 M solutions the heat of ionic mixing is only of the order of 10 cal/mole of Na or K, so that the observed exchange of 12 n-mole/g impulse (which was determined with $^{42}$K) would not produce more than about 0.12 $\mu$cal/g impulse. A more promising candidate would be the heat of displacement by the K$^+$ ions of Ca$^{2+}$ ions bound to membrane-fixed charges; that heat is likely to be positive and of the order of several thousand calories per mole (Coleman, 1952). One could account completely for the initial heat by an exchange of this kind only if *all* the K$^+$ ions lost from the nerve during the impulse were involved in it. For reasons already discussed, this is not an acceptable proposition. However, it is not impossible that a K$^+$–Ca$^{2+}$ exchange does make an appreciable contribution to the initial heat.

Another obvious source of heat is the liberation by circulating currents of the electrical energy ($= \frac{1}{2}CV^2$) stored in the membrane capacity (C) when the resting potential (V) is discharged. The membrane is then repolarized at the expense of the thermal energy of some of the ions in the surrounding solutions, so a corresponding quantity of heat would be reabsorbed during the recovery phase of the spike. But this "condenser theory," although it fits the facts extremely well as far as the timing of the positive and negative initial heats is concerned, does not properly explain

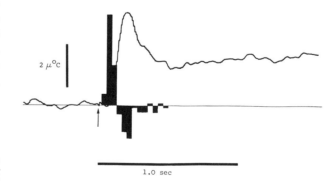

FIGURE 3 The temperature changes during the passage of an impulse along a nonmyelinated nerve trunk. The superimposed blocks show the computed evolution and subsequent reabsorption of heat. (From Howarth et al., 1968).

their magnitudes. There is some doubt as to what values of C and V should be used for the calculation, but, even on the most generous assumptions, $\frac{1}{2}CV^2$ cannot account for more than 11 $\mu$cal/g impulse.

We believe that the discrepancy is to be explained by the occurrence of an appreciable *decrease* in the entropy of the membrane when the potential across it is *reduced*. This is an effect in the opposite direction to that which would obtain if the dielectric behaved like a simple liquid, but the rather meager evidence on the effect of temperature on the capacity of nerve membranes and phospholipid bilayers supports the view that in such membranes the entropy change would be of the right order of size and in the right direction to fit the experimental observations. This argument does not, unfortunately, enable us to identify the type of conformational change that is presumably involved, and all that can be said is that it is unlikely to be confined to the sodium channels, because a positive heat of 25 $\mu$cal/g liberated at 13 $\times$ 10$^8$ channels/cm$^2$ in a nerve containing 6000 cm$^2$ membrane/g would correspond to 3 $\times$ 10$^{-18}$ cal/channel, or 2 $\times$ 10$^6$ cal/mole if there was one transport molecule at each channel. On the other hand, the same positive heat distributed among all the phospholipid molecules would give only 6 cal/mole, which can hardly be regarded as an improbable figure for an oriented molecule subjected to a voltage gradient of the order of 40,000 V/cm.

## Birefringence changes in excitable membranes

Optically detectable changes in excitable tissues have recently been observed by studying (1) birefringence, (2) light scattering, and (3) fluorescence emission from nerves stained with fluorescent dyes (Cohen and Keynes, 1968, 1969; Cohen et al., 1968, 1969; Tasaki et al., 1968; Carnay and Barry, 1969). The most straightforward of these approaches is the measurement of birefringence. The results obtained when nerves and slices of electric organ are mounted between a crossed polarizer and analyzer at 45 degrees to the plane of polarization, and illuminated brightly with white light, are described first.

Figure 4 shows a typical record of the change in light

FIGURE 4   Birefringence (thick line) and the action potential (thin line) in a squid giant axon. The arrow represents a light increase of 5 $\times$ 10$^{-6}$/impulse. Temperature, 7° C. (From Cohen et al., 1968.)

intensity during the passage of an impulse along a squid giant axon mounted on the stage of a polarizing microscope. The optical change is seen to follow very closely the membrane potential recorded with an intracellular electrode at the center of the illuminated region. A similar correspondence between the optical and electrical signals was observed in the small, nonmyelinated fibers of crab-leg nerve, the abdominal vagus nerve of rabbit, and in quite a different type of excitable tissue, the electric organ of the electric eel (Figure 5). An important point to establish at the outset was

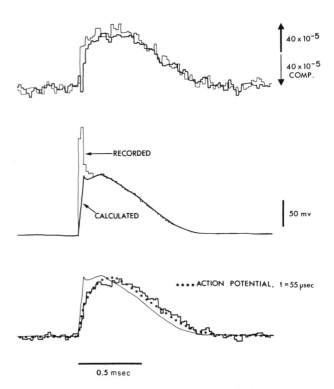

FIGURE 5   Birefringence changes during the discharge of *Electrophorus* electroplates. Upper records: Birefringence with and without optical compensation. Middle records: The action potential as recorded, and corrected for the stimulus artifact. Bottom records: Evidence that the birefringence follows the membrane potential change with a time constant of about 55 $\mu$sec. (From Cohen et al., 1969.)

that the change in light intensity did represent a change in birefringence and not a change in absorption or scattering of light. That such was indeed the case is proved by the top records in Figure 5, which show that, when optical compensation was introduced to subtract 150 nm of retardation from the 75 nm resting retardation of the tissue, the signal was the exact inverse of that without compensation. The squid giant axon gave precisely the same result when an appropriate compensator was inserted in the light path.

Except under favorable conditions, the changes in light intensity were too small to be distinguished properly from the random fluctuations in the resting intensity during a single sweep, and signal averaging devices were therefore used in order to improve the signal to noise ratio. In the squid axon the light intensity decreased by about one part in $2 \times 10^5$ at the peak of the spike, and several thousand sweeps were averaged. In the electric organ the light intensity increased by about one part in $2 \times 10^3$, and it was necessary to average only 50 to 100 sweeps.

The localization of the birefringence change in the membrane or its immediate vicinity was established by two kinds of experiment. About half of the resting birefringence of the squid axon arises from longitudinally oriented filaments in the axoplasm; but, when most of the axoplasm was removed by the "garden roller" technique (Baker et al., 1961) and the axon was perfused with a buffered solution of potassium fluoride, the change in birefringence during the spike was unimpaired. On several occasions we used a light stop at the image plane of the microscope to examine only a selected region of the axon, and established that a strip including 100 $\mu$M of an edge of a 1000-$\mu$M axon gave at least 20 times as large a fractional change in intensity as a strip covering the central 500 $\mu$M of the same axon. It follows that the birefringence change has a radial optic axis and arises from sources disposed in a cylindrical region at the outer edge of the axon or in the sheath (including the Schwann cell). We tentatively assume that, in fact, the change takes place either in the phospholipids of the membrane itself or in proteins or other molecules adsorbed to its surface. Although the Schwann cell certainly contributes substantially to the resting birefringence, it seems unlikely that it is responsible for the changes during activity, partly because the voltage drop across it is much smaller than that across the membrane, and partly because large birefringence changes are observed in the electric organ and frog muscle fibers in which there are no Schwann cells. By using the data of Schmitt and Bear (1939) for the absolute magnitude of the birefringence of cell membranes, it can be calculated (Cohen et al., 1968) that at the peak of the spike the intrinsic radial birefringence of the squid axon membrane is *increased* by about 0.15–0.5 per cent. In the electric organ the retardation is decreased by about the same amount, but there is not necessarily a contradiction in the direction of the change, for, in this case, there is considerable uncertainty as to the predominant orientation of the membrane, because the innervated faces of the electroplates are crowded with so many papillae and fine tubules.

The observations described so far suggest that the birefringence change was primarily a voltage-dependent phenomenon, but in order to place this conclusion on a firm basis it was essential to make observations under voltage-clamp conditions. Figure 6 shows that a square hyperpolar-

izing voltage pulse gave a roughly square birefringence response, and a depolarizing pulse gave a somewhat less square response. The departure from rectangularity of the response to depolarization was almost certainly due to the employment of a clamping circuit that did not allow for the series resistance of the Schwann cell. In all later work, electrically "compensated" feedback was used, and it became clear that, if there was a component of the birefringence change linked with ionic current flow or conductance, it was relatively small in comparison with the voltage-dependent effect. Experiments of this kind made it possible to study the quantitative relationship between birefringence and membrane potential, and it was established that, as expected, the relationship was a good approximation to a square law. However, the birefringence was not, as might have been predicted on the simplest basis, symmetrical about zero membrane potential, but about a potential displaced some 100 mV beyond zero in the depolarizing direction, i.e., inside positive. There is evidently some kind of electrical or structural asymmetry in the membrane. This aspect of the work seems one of the most promising for further study.

Another important parameter of the birefringence change that could be studied under a voltage clamp was its time constant or relaxation time. At 15° C, the relaxation time was about 30 $\mu$sec, and, although the absolute size of the birefringence change varied very little with ambient temperature, the relaxation time had a large temperature coefficient and was considerably greater at 5° C. The electric organ appeared to have a very similar relaxation time. Time constants of this order of size would require the orienting molecules or side chains to be improbably large if they were moving freely in water, and their environment must therefore be relatively viscous.

One further observation that must be mentioned is that in the presence of 60–150 nм tetrodotoxin or 300 nм

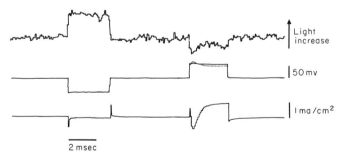

FIGURE 6  Changes in birefringence of the squid giant axon under a voltage clamp. Upper trace: Birefringence, the arrow representing an increase in light intensity of $1 \times 10^{-5}$. Middle trace: Membrane potential. Lower trace: Membrane current. Uncompensated feedback was used to clamp a 10-mm length of axon. Temperature, 17° C. (From Cohen et al., 1968.)

saxitoxin, the changes in birefringence for voltage-clamp pulses were doubled or quadrupled in size, while at the same time the relaxation time lengthened to an early component of 100–200 μsec followed by a late one of 2–3 msec. This striking and quite unexpected effect was not associated directly with the blocking action of the toxins because sometimes it did not manifest itself until several minutes after the sodium current had fallen to zero; conversely, when the toxin was washed away, the effect persisted for some time after the sodium current had recovered. As has already been seen, the numbers of toxin molecules taken up by the membrane at these low concentrations are apparently very few, and the effect of TTX on birefringence is all the more surprising. The effect of these and other molecules when adsorbed to the membrane will clearly repay further detailed examination.

## Changes in light scattering during nervous activity

The starting point for our recent work on optical changes in excitable tissues was an attempt to follow up the observations on light scattering in crab nerves made by Hill (1950) and Tobias (1952). The occurrence of relatively long-lasting light-scattering effects, apparently associated with a transient and still not properly explained swelling of the axons, was confirmed (Figure 7), but we were soon diverted from examining these slow changes by the discovery that there was also a fast change in light scattering, the time course of which was not unlike that of the action potential. The effect was a small one, the light scattered at 90 degrees increasing by

about one part in 20,000 for a single impulse, but by averaging 25 to 200 sweeps with a CAT computer, an acceptable signal-to-noise ratio could be attained. Apart from verifying that the change was reversed in direction when the light scattered forward instead of at right angles was measured, and showing that its size (expressed as $\Delta I/I$, in which I is the resting intensity) varied approximately as the reciprocal of (wave length)$^2$, it was not possible to make much progress in analyzing the light-scattering effect without studying it under voltage-clamp conditions. However, it turned out that similar effects could be observed in the squid giant axon, which enabled us to learn a little more about their origin.

As can be seen in Figure 8, the time course of the early changes in light scattering during propagation of an impulse along a squid axon bears some relation to that of the spike, but it is clear that several components are involved, and the record obtained at 90 degrees is not exactly the inverse of that at 35 degrees. One component appears to be voltage-dependent, which is most easily seen when a hyperpolarizing voltage-clamp pulse is applied and there are no large ionic currents flowing through the membrane (Figure 9). It seems to differ in origin from the voltage-dependent

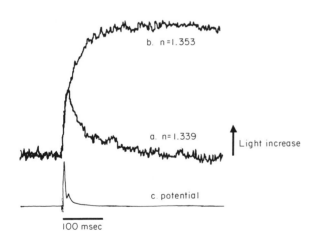

FIGURE 7 Light scattered at 90 degrees by a crab-leg nerve stimulated 15 mm from the illuminated region. The vertical arrow represents an increase of $2 \times 10^{-5}$/impulse: a, in artificial sea water with refractive index of 1.339; b, dextran added to give a refractive index of 1.353; c, the compound action potential recorded 10 mm beyond the illuminated region. Temperature, 18° C. (From Cohen et al., 1968.)

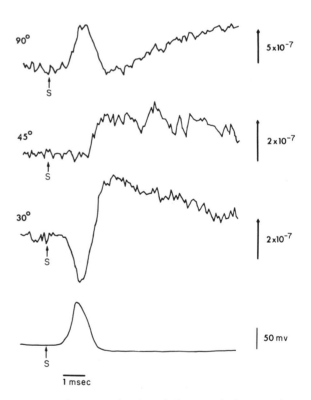

FIGURE 8 Light scattered at three different angles by a squid giant axon during a propagated impulse. Temperature, 13° C. (Cohen and Keynes, unpublished experiment.)

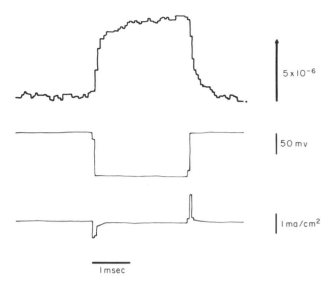

$5 \times 10^{-6}$

$50 \, mv$

$1 \, ma/cm^2$

1 msec

FIGURE 9 The voltage-dependent component of the light-scattering change in a squid giant axon, recorded at 35 degrees to the incident illumination. Compensated feedback was used, and the voltage pulse was in the hyperpolarizing direction. Temperature, 12° C. (Cohen and Keynes, unpublished experiment.)

birefringence change, because its relaxation time is appreciably greater (around 100 $\mu$sec) and it is unaffected by TTX. Then there are current-dependent components the sizes of which at the end of a voltage-clamp pulse are roughly proportional to the integrals of inward sodium current and outward potassium current. These can be abolished or reduced in size by treating the axon with TTX or TEA. There are also slower changes that persist after the passage of an impulse or the end of a voltage-clamp pulse, which have not yet been studied in any detail. Disappointingly, we have so far been unable to identify any light-scattering change of which the time course would fit with either the activation or the inactivation of the sodium and potassium conductances.

It seems probable that the current-dependent components and the slower effects arise from alterations in the refractive indexes and physical dimensions of various parts of the complex system that we are studying. Thus the concentrations of sodium and potassium in the space between the axon membrane and the Schwann cell are temporarily changed when current flows, with consequent movements of water and swelling or shrinking of the space. Similar light-scattering effects have been observed in the electric organ (Cohen et al., 1969), with one component that could plausibly be attributed to comparable effects involving the microtubules that invaginate the membrane. Attempts to correlate the time constant for recovery of the light scattering after a pulse of current with the time constant identified by Frankenhaeuser and Hodgkin (1956) as that for diffusion

of potassium out of the Schwann cell space, have not been successful. By studying the dependence of all these changes on the scattering angle, and by altering such factors as the refractive index of the external medium, we should eventually find it possible to determine their precise location and cause, but the task is not an easy one.

## Changes in fluorescence emission

It has been reported by Tasaki et al., (1968, 1969) that nerves stained with the fluorescent dyes 8-anilinonaphthalene-1-sulphonic acid (ANS) and acridine orange (AO) exhibit a small increase in fluorescence emission during the passage of an impulse. Carnay and Barry (1969) have since found that frog-muscle fibers stained with pyronine B give a decreased emission during the spike itself, followed by a somewhat larger increase in emission during the after-potential and period of latency relaxation. The fact that in nerve the change in light intensity was in the same direction for fluorescence emission as it was for 90-degree light scattering and had a similar time course (Tasaki et al., 1968) might suggest that the fluorescence effect was some kind of scattering artifact. But in muscle, both the directions and the time courses of the two effects were different (Carnay and Barry, 1969), which seems to dispose of this possibility.

The records published by Tasaki et al. (1969) for squid axons, and the similarity between the time courses of the birefringence and fluorescence emission changes observed in muscle by Carnay and Barry (1969), suggest that membrane potential is the main factor affecting fluorescence emission. Fluorescence must be studied under voltage-clamp conditions, however, if the precise origin of the changes in emission is to be determined. It is also clear that the nature of the dye with which the tissue is stained is of cardinal importance in any interpretation of the results of fluorescence measurements, because pyronine B gives a decrease in emission whereas ANS and AO give an increase. When the dye molecule can be tailor-made so that it is incorporated into the membrane structure in a predictable fashion, with the fluorescent group located in a specific region, studies of fluorescence emission will clearly become a valuable tool for probing the structural changes associated with nerve activity.

## Conclusions

Two distinctly different types of structural change have been detected in excitable tissues by optical means. In the first place, there are potential-dependent changes in birefringence and light scattering that can be seen in untreated tissues, and changes in fluorescence emission, probably similar in origin, that are seen after staining with appropriate dyes. These effects are presumably confined to the membrane, because only in its immediate neighborhood are the potential gradients really large, but there is little indication as to which part of the membrane is involved— the phospholipids of the membrane interior, or the proteins and other macromolecules of the membrane surface. Nor is it clear whether the potential-dependent changes arise primarily from reorientation of charged side groups in the electric field, as in the Kerr effect, or from a compression of the membrane by the attractive force exerted between the charges on its opposite sides. Second, there are current-dependent changes in light scattering and possibly in fluorescence emission as well. In all probability, these do not involve the membrane itself, but rather any external compartment from which the movement of ions is diffusion-limited, such as an endoplasmic reticulum communicating with the surface only through narrow channels, or the Schwann-cell space in the squid giant axon. In addition, there are relatively slow light-scattering changes, which appear to involve water movements into and out of the fibers, the origin of which has not yet been explored in detail.

Although we have done our best to find them, no optical effects have yet been observed that can be correlated definitely with alterations in membrane conductance. Perhaps this is not really surprising, in view of the enormous numerical disparity between the special sodium and potassium channels and the phospholipid and other molecules of which the general membrane structure is built. At these special sites there must surely be large conformational changes accompanying or causing the sequence of ionic conductance changes that take place during excitation and the propagation of an impulse, but they will be very hard to detect optically against a noisy background, and will produce a much smaller signal than any change affecting the whole membrane. It must therefore be concluded with reluctance that this approach is unlikely to yield direct evidence about the nature of the conformational changes at the active sites. However, it should give valuable information about the way in which the components of a cell membrane are affected by alterations of the electric field, and such knowledge is essential for the formulation of hypotheses concerned with the molecular basis of the mechanism of excitation.

## REFERENCES

BAKER, P. F., A. L. HODGKIN, and T. I. SHAW, 1961. Replacement of the protoplasm of a giant nerve fibre with artificial solutions. *Nature* (*London*) 190: 885–887.

BEAN, R. C., W. C. SHEPHERD, H. CHAN, and J. EICHNER, 1969. Discrete conductance fluctuations in lipid bilayer protein membranes. *J. Gen. Physiol.* 53: 741–757.

CARNAY, L. D., and W. H. BARRY, 1969. Turbidity, birefringence, and fluorescence changes in skeletal muscle coincident with the action potential. *Science* (*Washington*) 165: 608–609.

COHEN, L. B., B. HILLE, and R. D. KEYNES, 1969. Light scattering and birefringence changes during activity in the electric organ of *Electrophorus electricus*. *J. Physiol.* (*London*) 203: 489–509.

COHEN, L. B., and R. D. KEYNES, 1968. Evidence for structural changes during the action potential in nerves from the walking legs of *Maia squinado*. *J. Physiol.* (*London*) 194: 85P–86P.

COHEN, L. B., and R. D. KEYNES, 1969. Optical changes in the voltage-clamped squid axon. *J. Physiol.* (*London*) 204: 100P–101P.

COHEN, L. B., R. D. KEYNES, and B. HILLE, 1968. Light scattering and birefringence changes during nerve activity. *Nature* (*London*) 218: 438–441.

COLEMAN, N. T., 1952. A thermochemical approach to the study of ion exchange. *Soil Sci.* 74: 115–125.

ELLORY, J. C., and R. D. KEYNES, 1969. Binding of tritiated digoxin to human red cell ghosts. *Nature* (*London*) 221: 776.

FRANKENHAEUSER, B., and A. L. HODGKIN, 1956. The after-effects of impulses in the giant nerve fibres of *Loligo*. *J. Physiol.* (*London*) 131: 341–376.

FRANKENHAEUSER, B., and A. L. HODGKIN, 1957. The action of calcium on the electrical properties of squid axon. *J. Physiol.* (*London*) 137: 218–244.

HILL, D. K., 1950. The effect of stimulation on the opacity of a crustacean nerve trunk and its relation to fibre diameter. *J. Physiol.* (*London*) 111: 283–303.

HILLE, B., 1970. Ionic channels in nerve membranes. *Progr. Biophys. Mol. Biol.* 21: 1–32.

HODGKIN, A. L., 1964. The Conduction of the Nervous Impulse. Charles C Thomas, Springfield, Illinois.

HODGKIN, A. L., and R. D. KEYNES, 1957. Movements of labelled calcium in squid giant axons. *J. Physiol.* (*London*) 138: 253–281.

HOWARTH, J. V., R. D. KEYNES, and J. M. RITCHIE, 1968. The origin of the initial heat associated with a single impulse in mammalian non-myelinated nerve fibres. *J. Physiol.* (*London*) 194: 745–793.

KOECHLIN, B. A., 1955. On the chemical composition of the axoplasm of squid giant nerve fibers with particular reference to its ion pattern. *J. Biophys. Biochem. Cytol.* 1: 511–529.

MOORE, J. W., T. NARAHASHI, and T. I. SHAW, 1967. An upper limit to the number of sodium channels in nerve membrane? *J. Physiol.* (*London*) 188: 99–105.

NARAHASHI, T., J. W. MOORE, and W. R. SCOTT, 1964. Tetrodotoxin blockage of sodium conductance increase in lobster giant axons. *J. Gen. Physiol.* 47: 965–974.

SCHMITT, F. O., and R. S. BEAR, 1939. The ultrastructure of the nerve axon sheath. *Biol. Rev.* (*Cambridge*) 14: 27–50.

TASAKI, I., 1968. Nerve Excitation. Charles C Thomas, Springfield, Illinois.

TASAKI, I., L. CARNAY, R. SANDLIN, and A. WATANABE, 1969. Fluorescence changes during conduction in nerves stained with acridine orange. *Science* (*Washington*) 163: 683–685.

TASAKI, I., and S. HAGIWARA, 1957. Demonstration of two stable potential states in the squid giant axon under tetraethylammonium chloride. *J. Gen. Physiol.* 40: 859–885.

TASAKI, I., A. WATANABE, R. SANDLIN, and L. CARNAY, 1968. Changes in fluorescence, turbidity, and birefringence associated with nerve excitation. *Proc. Nat. Acad. Sci. U. S. A.* 61: 883–888.

TOBIAS, J. M., 1952. Some optically detectable consequences of activity in nerve. *Cold Spring Harbor Symp. Quant. Biol.* 17: 15–25.

# 64  The Ultrastructure of Synapses

## J. DAVID ROBERTSON

THE SUBJECT OF synaptic ultrastructure has been reviewed several times in recent years (Gray, 1966; Gray and Guillery, 1966; Couteaux, 1963; De Robertis, 1959; Palay, 1956). This review, because of space limitation, is highly selective and does not attempt to cover the field. It aims at a description of the ultrastructure of selected synapses in both vertebrate and invertebrate nervous systems, with emphasis on structural features that can be directly related to function.

### Structures related to chemical transmission

The first electron micrographs of motor end plates were reported in abstract form in 1954 (Palade and Palay, 1954; Reger, 1954; Robertson, 1954). The complete reports of two of these papers soon followed (Reger, 1955; Robertson, 1956b). The major features of the electron micrographs are indicated diagrammatically in Figure 1. The terminal axon is filled by about 300- to 500-Å, membrane-limited, synaptic vesicles. A gap of about 500 Å separates the presynaptic and postsynaptic membranes, and the gap is filled by dense material continuous with the basement lamina of the muscle fiber and that surrounding the terminal Schwann cell of the nerve. The basement lamina also extends into the junctional folds. Here the enfolded unit membranes (Robertson 1956b, 1960, 1964) are brought into apposition with a gap approximately 500 Å wide that is expanded slightly in the bottom of the fold. The junctional folds in rat and lizard endings are very complicated and extend in branching arrays all over the nerve terminals.

In a study of the vertebrate central nervous system in 1956, Palay clearly identified synaptic boutons and established that they contained accumulations of mitochondria and also the characteristic synaptic vesicles reported in motor end plates (Palade and Palay, 1954; Robertson, 1956b) and other synapses (De Robertis and Bennett, 1954, 1955; Fernández-Morán, 1955). Palay recognized the vesicles as likely bearers of packets of neurotransmitters, the presence of which in motor end plates had been suggested by Fatt (1954). Castillo and Katz (1954, 1956) had shown that characteristic, sudden changes in membrane potential occurred randomly at motor end plates at a frequency of a few milliseconds. The term miniature end-plate potential (MEPP) was applied to these spikes because they resembled the postsynaptic end-plate potential produced by discharge of the nerve to the ending, although they are much smaller. Their amplitude was approximately 0.1 to 0.5 millivolt, and their duration was only a few tenths of a millisecond. It was found, by the use of external microelectrodes, that the application of acetylcholine (ACh) to the surface of the presynaptic muscle fiber near the ending produced a potential change exactly equivalent to that of the MEPP, and a much larger amount produced a propagated action-potential spike. ACh was assumed to be released by the terminal nerve fiber as a transmitter, and it was also assumed that it was being continuously released in small packets, each producing one MEPP. The action potential was supposed to be the result of a massive synchronous release of these quanta.

The concept of quantal release of a neurotransmitter originated with this work. It seemed to be a reasonable analysis, but I did not find it an entirely satisfactory one for several reasons. For instance, it was apparent that, as its ACh was released, the membrane of the synaptic vesicles would, of necessity, have to become incorporated into the axonal membrane, which, as its surface area increased, would rapidly be converted into a membrane identical with that surrounding the vesicles. This seemed improbable, and it was subsequently proved incorrect as a general notion by Whittaker (1966), who showed that the membrane of the synaptosome is chemically distinct from the membrane of the synaptic vesicle.

In reviewing the above events, I have touched on the groundwork for much of current synaptology at the ultrastructural level. Thus, the synaptic vesicle, concentrated presynaptically, has become an almost necessary feature of a synapse; the notion of quantal release of transmitter by fusion of vesicles with the synaptic membrane is widely accepted, and the association of chemical transmission and a fairly wide synaptic gap is now known to be a characteristic feature of many kinds of putative chemical synapses. I turn now to a different kind of structural differentiation that is not associated with chemical transmission.

### Morphological changes in membranes resulting from electrical stimulation

In recent work, Dr. Camillo Peracchia (1969), in my laboratory, has observed a hexagonal pattern in *en face* views of

J. DAVID ROBERTSON  Department of Anatomy, Duke University Medical Center, Durham, North Carolina

THE ULTRASTRUCTURE OF SYNAPSES    715

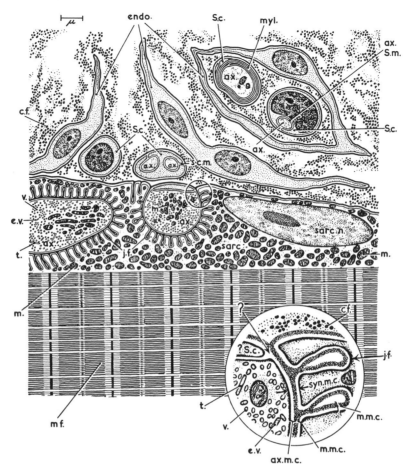

FIGURE 1 Interpretative diagram of a motor end plate in *Anolis carolinensis*. The surface membrane complexes (SMCs) of the Schwann cells (S.c.), muscle fiber (m.f.), and endoneural sheath cells (endo.) are shown. The muscle surface is thrown into junctional folds to make the subneural apparatus of Couteaux. The five layers of the SMC that separate terminal axoplasm from sarcoplasm are shown. The main features of terminal axoplasm and sarcoplasm are indicated. The region marked by the circle is enlarged to show more detail. The continuities between the membrane structures at the region marked "?" was somewhat interpretative, as was also the identification of Schwann cytoplasm at the time this diagram was constructed, but subsequent work has established the correctness of the interpretation. Magnification, × 60,000.

the synaptic membrane complexes (SMCs) of the lateral giant septal synapses of the crayfish, as indicated in Figure 2. The center-to-center spacing of the hexagonal facets is about 250 Å. In the median-giant-to-motor synapse (MGM), he has found another feature that may be related to this hexagonal pattern. Occasionally a peculiar scalloping of the synaptic membranes is seen in these synapses after certain special procedures. Here the membranes are alternately slightly separated and in very close contact. The repeat period of this scallop is, again, about 250 Å. It should also be noted that Peracchia has isolated certain membrane fragments from the crayfish ganglia by a special freeze-shatter method devised by Dr. Juan Vergara. He has found in these fragments a very regular hexagonal pattern in which the facets show a center-to-center spacing of about 150 Å. One of these micrographs is shown in Figure 3. It is too early to say where these fragments may have originated and what relationship, if any, they may have to the hexagonal pattern seen in the SMCs.

Peracchia has recently conducted some experiments designed to see whether structural alterations, demonstrable by the electron microscope, take place in the MGM synapses with activity. He has succeeded in eliciting and identifying a specific morphological response in electrically stimulated fibers. It appears frequently with stimulation and never in control animals. The abdominal nerve cord is per-

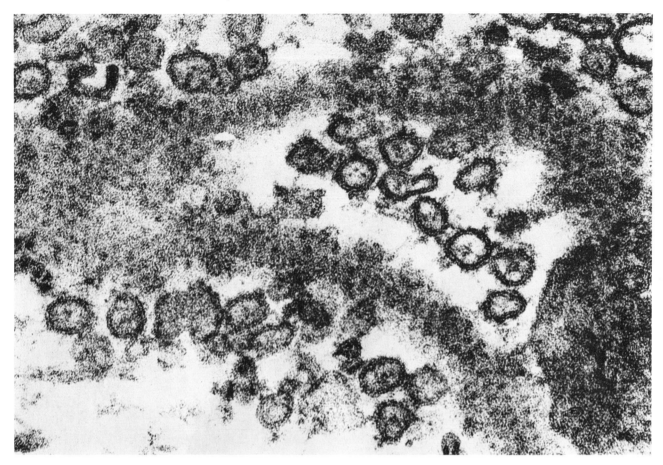

FIGURE 2 (*Above*) Section of lateral giant septal synapse in crayfish nerve cord, including synaptic membrane complex in an *en face* view. Note the hexagonally arranged spots spaced at a center-to-center distance of about 250 Å. Magnification, × 170,000.

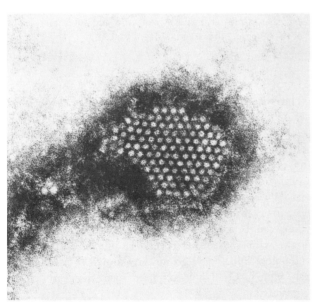

FIGURE 3 (*Left*) Hexagonal pattern in a membrane fragment isolated from crayfish nerve cord by C. Peracchia. Center-to-center spacing of the hexagons is about 90 Å. PTA-negative stain preparation. Magnification, × 170,000.

fused with glutaraldehyde during stimulation of one of the circumesophageal commissures at a frequency of 30 to 40 per second and a strength of approximately 10 volts. After some hours in glutaraldehyde, the nerve cord is post-fixed with $OsO_4$, according to the usual routine procedures. A distinct morphological change appears selectively in the stimulated cords.

Figure 4C shows both a stimulated fiber (the giant fiber,

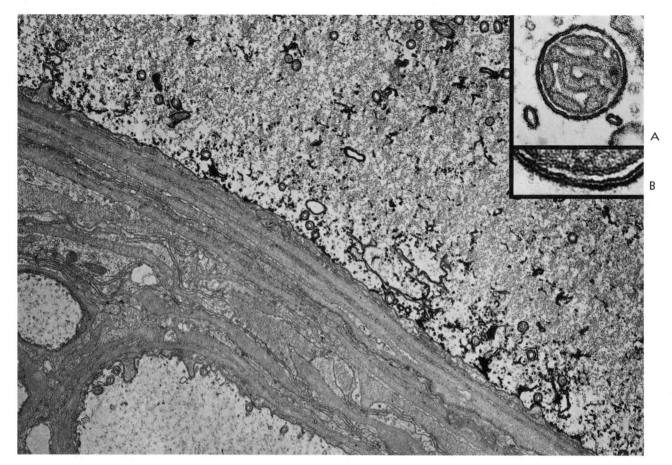

FIGURE 4  Stimulated crayfish giant nerve fiber (above), with two unstimulated small fibers (below). Fixed with glutaraldehyde-OsO₄ and stained with lead citrate. Note the increased density of the membranes of the giant fiber. The insets (A and B) of a mitochondrion show the dense material that increases the thickness of inner and outer leaflets of the outer membrane only. Magnifications, × 10,000; inset A, × 90,000; inset B, × 180,000.

above) and two smaller unstimulated fibers (below). An obvious, clearcut difference between the two is apparent. All the membranes of the stimulated fibers appear to be very dense. This phenomenon occurs if the perfusion of the stimulated nerve cord is begun after as few as 30 to 40 propagated impulses have been set up. The densification occurs in all the unit membranes bounded by axoplasm, including the surface membrane, endoplasmic reticulum and vesicle membranes, and the outer mitochondrial membranes. It is caused by the appearance of a dense amorphous material that effectively increases the thickness of the inner and outer dense leaflet (inset, Figure 4A, B). The inner one is increased by approximately 300 to 500 per cent; the outer, about 50 to 100 per cent. The material added to the membrane leaflets is not seen after glutaraldehyde fixation alone. If sections of the ganglion are treated with peroxide after osmium fixation, the dense material disappears. Apparently it represents a deposition of reduced osmic reaction products.

The reaction can be regarded as indicative of an increased osmiophilia of the membranes. At first it was thought that the dense material might be calcium (Peracchia and Robertson, 1968; Peracchia, 1969), but it did not show up after glutaraldehyde fixation alone, which seemed to rule out calcium. It is known from the work of Huneeus-Cox et al. (1966) that there is a protein in giant invertebrate nerve fibers that contains sulfhydryl groups, and that these groups can be reversibly bound by certain organomercurials, with consequent reversible blocking of impulse conduction. Sulfhydryl groups are known to react vigorously with OsO₄, but in oxidized form these groups are nonreactive.

The possibility then arose that a protein high in disulfide groups was present mainly in or on the membrane leaflets and was affected by impulse propagation in such a way that the disulfide groups were reduced. To test whether the osmiophilia might be due to such a protein, Peracchia treated unstimulated glutaraldehyde-fixed nerve fibers with thioglycollate in an effort to reduce disulfide groups before osmication. It was expected that an increased osmiophilia would appear as a result of the thioglycollate. Such proved to be the case, as indicated in Figure 5.

To carry the analysis a step further, Peracchia followed the glutaraldehyde-thioglycollate treatment with N-ethyl maleimide (NEM) before $OsO_4$. NEM binds sulfhydryl groups and makes them inaccessible to reaction with $OsO_4$. After this procedure, the osmiophilia presumably conferred by the thioglycollate is lost, as indicated in Figure 6. It thus appears reasonable to conclude that, during action-potential propagation, disulfide groups in a protein lying either on or within the cytoplasmic leaflet of the axon-unit membranes, as well as, to a lesser extent, on or within the outer leaflet, undergo a change whereby they are converted to sulfhydryl groups. It should be emphasized that this reaction never occurs in Schwann-cell membranes and can be elicited in axon membranes only by active stimulation during perfusion with glutaraldehyde or by treatment with thioglycollate. It thus appears that a meaningful morphological change is frequently demonstrable in electrically stimulated nerve membranes and that this change could conceivably be used as a morphological tag of active neurons in a neuropil, if it can be shown to be related to activity irrespective of electrical stimulation.

That the effect observed by Peracchia involves all the membranes bounded by axoplasm is interesting. One explanation may be simply that all these membranes are neuronal, and, although doubtless chemically specific, they may have at least one property in common, i.e., the presence of the sulphur-rich protein in question. Another way to explain the phenomenon is to assume that this compound is in solution in axoplasm and that all membranes in contact with axoplasm tend to adsorb some of it.

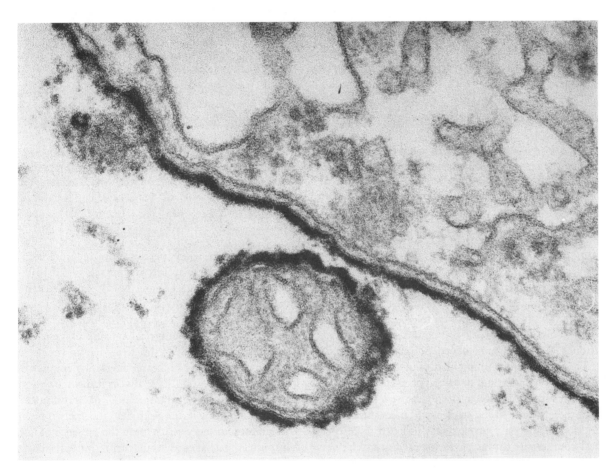

FIGURE 5  Crayfish nerve fiber treated with thioglycollate after fixation with glutaraldehyde but before postfixation with $OsO_4$ and lead staining. Note the dense amorphous material causing thickening of the axon membrane and the outer mitochondrial membrane. Compare this with the stimulated nerve fiber in Figure 4 and note the similarity in appearance, even though the thickness of the dense precipitate here is exaggerated. Magnification, $\times$ 140,000.

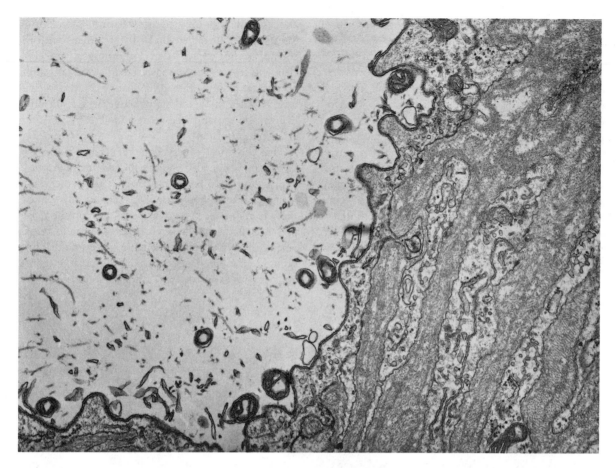

FIGURE 6   Crayfish nerve fiber treated with thioglycollate after fixation with glutaraldehyde and then with NEM before fixation with $OsO_4$ and lead staining. Note that this fiber resembles the control unstimulated nerve fibers shown in Figure 4. It is believed that the thioglycollate first reduced disulfide groups to produce sulfhydril groups that were then bound up by NEM, making them inaccessible to reaction with $OsO_4$. Magnification, $\times$ 21,000.

## Electrical synaptic transmission in invertebrates

The giant fiber synapses in the ventral nerve cord of the crayfish (Johnson, 1924) were the object of one of the earliest studies of synapses by electron microscopy (Robertson, 1953). This work has been reviewed fairly recently (Robertson 1961, 1964), and hence is treated here only briefly. The main point is simply that in these synapses the gap between presynaptic and postsynaptic membranes was occluded. Furshpan and Potter (1959) showed that transmission at these synapses is by an electrical mechanism, and the occlusion of the synaptic cleft was believed to offer a structural counterpart to this physiological feature. It was also noted that synaptic vesicles were not a prominent feature of these electrical synapses, although some clear vesicles about 300 to 500 Å in diameter were present. Aggregates of mitochondria were noted in postsynaptic axoplasm and rows of vesicles were sometimes seen close to both presynaptic and postsynaptic membranes. These aligned vesicles were spaced at intervals of 0.1 to 0.2$\mu$. In the case of the lateral giant septal synapses that were physiologically unpolarized, the synaptic cleft is not completely occluded; there is a gap approximately 20 Å wide.

The presence in an electrical synapse of vesicles superficially indistinguishable from the synaptic vesicle seen in neurohumoral synapses poses an interesting enigma. This is, in fact, another reason I have always had some reservations about the idea of quantal release of transmitter from vesicles. Certainly it suggests that the mere presence of vesicles does not necessarily mean the presence of chemical transmitters packaged within them. Indeed, their exact function cannot be specified on present evidence. One may speculate, perhaps, by saying that unit membrane vesicles are universal components of axoplasm and that they tend to concentrate at synaptic membrane complexes (SMCs) because there is something special about the electric field changes in such loci caused by the close apposition of two nerve membranes conducting action potentials. It is only in this region that *two* nerve membranes are in close apposition. Elul (1966, 1967) has recently shown that nerve membranes move toward a microelectrode tip if positive current is flowing from it. The currents involved are comparable to those that flow during action-potential propagation. The observed membrane movement was explained as being caused by their possessing a negative fixed charge. If one assumes that both the inside of the nerve membrane and the cytoplasmic side of the vesicle membrane have a fixed negative charge, vesicles ordinarily would not approach the membrane closely except during the current flows associated with action-potential propagation. Because this is ordinarily diphasic, one might expect little net effect on vesicles except, perhaps, at synapses, where one might imagine that special features of current flow may exert a net attractive electrophoretic effect leading to the vesicles becoming concentrated passively in this region. If this is so, the vesicles might—at least in some cases—have little or nothing to do with synaptic transmission. Another obvious, nonspecific way to explain the presence of vesicles in boutons is to note that these are nerve *terminals*. Axoplasmic flow might thus propel and trap them there. However, this would not explain their presence near the SMC of the crayfish MGM synapses.

In MGM synapses, another type of synaptic bouton is present on the synaptic processes themselves. They are packed with typical synaptic vesicles and in some cases contain glycogen particles, but there is never an occlusion of the presynaptic and postsynaptic membrane cleft, which is 100 to 150 Å wide, regardless of fixation method. The origin of these boutons is unknown. At one time they were thought to come from the adjacent motor fiber (Robertson, 1961), but serial sections failed to confirm this. The synapse shows neither dense material in the presynaptic or postsynaptic axoplasm near the synaptic membranes nor a condensation in the synaptic cleft. These might be inhibitory neurohumoral synapses, but this is complete speculation.

## Electrical synaptic transmission in vertebrates

I now review briefly a few relevant points in some work that has been reported elsewhere (Furshpan and Furukawa, 1962; Robertson, 1963; Robertson et al., 1963) on the club endings on the Mauthner cell lateral dendrite. Furshpan and Furukawa (1962) showed that the club endings transmit by an electrical mechanism, and we have produced morphological evidence that is consistent with this finding. These endings are characterized by SMCs that show disklike regions of complete closure of the synaptic cleft. Such a synaptic disk is seen diagrammatically in Figure 7A-E. In A and B, the disk is shown sectioned perpendicularly. In C it

is slightly tilted. Here we see lines that repeat at a period of about 90 Å. As the disk is tilted still farther, as in D, one begins to see rows of dots between these lines. In the synaptic disk tilted completely on its side and thus viewed *en face*, as in E, one sees a regular honeycomb arrangement of subunits made up of small facets in hexagonal array, with a lattice spacing of ~90 Å. The main point for this paper is that this synapse, which transmits electrically, is characterized by numerous close contacts between presynaptic and postsynaptic membranes in which the gap between the membranes is obliterated. It should be noted that other synapses specialized for electrical transmission show similar synaptic-cleft occlusions. Martin and Pilar (1963) showed that the calyciform endings on certain nerve cells in the chick ciliary ganglion transmit by an electrical mechanism, and Takahashi and Hama (1965) found the expected regions of synaptic-cleft occlusion. Pappas and Bennett (1966) again found a clearcut association of tight junctions with electrotonic coupling between giant pacemaker neurons in certain fish brains.

In the early days of application of permanganate fixation to cells and tissues, I noted tight junctions between adjacent mouse intestinal epithelial cells, similar to those seen in the electrical synapses, as well as close contacts between smooth

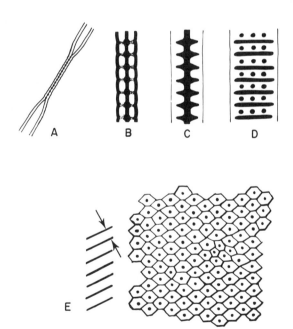

FIGURE 7 (*A-E*) Diagrams of synaptic disks from the club endings of the lateral dendrite of the Mauthner cell of the goldfish brain. A. The synaptic disk with the adjacent SMC segments at medium power in vertical section. B. A higher-power view with more details. C. The disk tilted slightly. D. A greater degree of tilt. E. The disk in frontal view. The main repeat period of about 95 Å is designated by the aligned arrows.

muscle cells (Robertson, 1959). Furthermore, Sjöstrand and Andersson (1954) noted extremely close contacts between adjacent membranes in cardiac muscle. Farquhar and Palade 1964) conducted an extensive survey of epithelial cell junctions and found that close contacts, usually called 'tight junctions," occurred extensively in epithelial tissues. Subsequently, Revel and Karnovsky (1967) studied such tight junctions by using a variant of the lanthanum staining method originated by Doggenweiler and Frenk (1965), and found the same kind of hexagonal array of subunits that had been demonstrated in the synaptic disk of the club endings. Even the repeat period of the subunits in their material was the same. Loewenstein (1966) demonstrated that many kinds of epithelia are characterized by electrotonic coupling between adjacent cells associated either with tight junctions or septate desmosomes.

As a result of all these lines of evidence, it is accepted that the tight junction is a region in which electrical coupling might be mediated. However, it also seemed likely that electrotonic coupling could also be mediated by septate desmosomes. The general concept has arisen that an essential result of electrotonic coupling is the prevention of lateral leakage of current. This could be accomplished by making the membranes come very close together or by interposing some element, such as the dense material of the septate desmosomes, to impede lateral current leakage. It is interesting to note here that nodes of Ranvier apparently use both of these methods in achieving saltatory conduction, because in some nodes there is a complete occlusion of the cleft between the Schwann cell membrane and the axonal membrane in the juxtaterminal myelinated region (Robertson, 1954), but in other nodes (Doggenweiler and Heuser, 1967; Hirano and Dembitzer, 1967) septate desmosomes that repetitively encircle the axons apparently serve the same purpose of limiting longitudinal leakage of current along the cleft between the axon and the Schwann cell from the internodal regions to the nodes. This kind of mechanism appears to be important in insuring current flow through the nodal membrane to satisfy the conditions for saltatory conduction.

Before leaving the subject of electrotonic coupling via tight junctions, I should mention that electrophysiological evidence of electrotonic coupling between cardiac muscle cells via tight junctions was soon forthcoming (Dewey and Barr, 1962, 1964). Recently McNutt and Weinstein (1969), Kreutziger (1968a, 1968b), Somer and Steere (1969), and Somer and Johnson (1970) have studied tight junctions in cardiac muscle tissue by means of the freeze-fracture technique. They have produced evidence that there is, in the central plane of the nexus, a regular hexagonal array of subunits with a center-to-center spacing of about 90 to 100 Å. McNutt and Weinstein regarded each subunit as a "contact cylinder" about 60 Å high by 60 Å in diameter.

The cylindrical particles are separated from one another by 20 to 30 Å of open space and have a center-to-center spacing of 90 Å. There is a 40-Å-deep central depression in each subunit. Opposite the contact cylinders on the cytoplasmic surface of the membrane are seen globular subunits in hexagonal array, with each subunit ~40 Å high and spaced with a center-to-center distance of 90 Å. These authors also believe that each of the subunits has a central pore about 20 to 25 Å in diameter and that these juxtacytoplasmic subunits each lie opposite a central cylinder. There is also evidence (Kreutziger, 1968b) of a similar subunit array in tight junctions or nexuses in mouse liver. Outside the region of tight junctions in cardiac muscle, Somer and Johnson (1970) have evidence that there is a smaller hexagonal array with a spacing of ~60 Å within the membranes, but it is not yet clear whether it is in the inner leaflet or the interior. However, their findings are compatible with the notion that some sort of special apparatus is added to the underlying unit membrane in regions of tight junctions that are designed for electrotonic coupling.

GENERAL SYNAPTOLOGY    In 1959, Gray applied a vigorous staining method ($OsO_4$ followed by ethanolic PTA) to vertebrate cerebral-cortical synapses, with resultant new findings. The method produced very heavy staining of certain elements, which permitted him to define two types of synapses. Both show presynaptic clusters of vesicles and membrane specializations manifested by dense material built up on both the presynaptic and postsynaptic membranes. In Type 1 there is a widening of the synaptic cleft to about 300 Å and there is a dense band about 100 Å thick in the cleft, but placed asymmetrically, being closer to the postsynaptic than to the presynaptic membrane. Gray observed, immediately beneath the postsynaptic membrane, a dense band of amorphous material, several hundred angstrom units wide, which appeared presynaptically as clumps spaced at intervals of a few hundred angstrom units. The synapse was found to occupy most of the area of synaptic contact. Gray regarded this particular morphological configuration as being the actual site of synaptic transmission. Type 1 appears characteristically in dendritic spines, whereas Type 2 is found more often on cell bodies. Here the regions of membrane specialization are less extensive, occupying only about 0.25 $\mu$. The presynaptic and postsynaptic specializations, as well as the clusters of synaptic vesicles, are, however, somewhat similar.

Whittaker and Gray (1962) conducted a combined electron microscope and biochemical study of certain fractions obtained from brain homogenates and showed that they contained predominantly synaptic boutons. Whittaker (1959) and, independently, De Robertis et al. (1961) showed that the boutons contain synaptic vesicles, which in turn contain ACh, thus lending support to the notion of quantal release by a mechanism of vesicle fusion with the presynaptic membrane, as mentioned above. Evidence in recent work reported by Whittaker (this volume) suggests that the acetylcholine in the nerve endings exists in two pools. One of these is in the vesicles, and the other is in the axoplasm of the ending. Present evidence seems to suggest that the ACh released during activity is from the free pool rather than from the vesicles. This finding could be interpreted to indicate that the synaptic vesicles are a repository for reserve ACh, but the ACh secreted as a neurotransmitter is that which is newly synthesized and found in the axoplasm of the ending. This topic is extremely controversial at present, and the above conclusion is in direct contradiction to the currently accepted quantal-release hypothesis, whereby the ACh is released from the vesicles by a mechanism of exocytosis.

The synapse of the retinal-rod receptor cell and the bipolar cell in vertebrate retina was one of the earliest synapses studied by electron microscopy (Sjöstrand, 1961), but it is not dealt with here beyond mention of certain special features of retinal synapses noted by Kidd (1962), who found many of the kinds of synaptic specialization described by Gray. In addition, Kidd frequently found in the pigeon retina what appeared to be three terminal axons in synaptic relationship with one another, as indicated in the diagram of Figure 8. A synaptic membrane specialization resembling the Gray Type 1 is seen between $P_1$ and $P_2$; $P_2$, in turn, contains a similar specialization in relationship to $P_3$. Kidd postulated that this relationship provided a morphological basis for presynaptic inhibition. An impulse arriving in $P_1$ would activate $P_2$ so as to diminish its response to an incoming impulse and hence decrease its effect on $P_3$. Normally, $P_2$ would fire $P_3$, but, because of the influence of $P_1$, it might not. Eccles (1964) predicted that

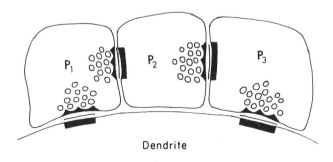

FIGURE 8 Diagram adapted from Kidd (1962) showing three ($P_1$–$P_3$) boutons, each containing clusters of synaptic vesicles related by two membrane specializations like the one seen in Gray Type 1 synapses. $P_3$ is related to a dendrite with a similar specialization.

there should be such structures, particularly in spinal cord, and Gray (1962), as well as Kidd (1962), studied the matter further and indeed found them. Gray made the point that they have been found in every place where physiological evidence indicates presynaptic inhibition should occur. However, it is interesting that, although Kidd found large numbers of these structures in the pigeon retina, he found them infrequently—only four times, in fact—in analogous parts of the retina of the cat. In any case, this particular relationship of synaptic specializations has been taken by many as indicative of presynaptic inhibition.

## Excitatory and inhibitory synapses

The recognition of excitatory and inhibitory synaptic endings by specific morphological features is obviously of great importance. The first clearcut claim for the morphological differentiation of excitatory and inhibitory endings was made by Uchizono (1965). He claimed that endings on Purkinje cells, which Eccles had designated physiologically as inhibitory, characteristically contained flattened and relatively small vesicles, and he noted that other types of endings thought to be excitatory had relatively large spherical vesicles. He associated the latter with the Gray Type 1 ending and the former with the Gray Type 2 ending.

In 1966, Bodian published micrographs of spinal motor neurons of the monkey that showed two distinct types of nerve endings; one of these contained spherical vesicles ~200 to 400 Å in diameter ("S" type); another contained vesicles of ~100 to 200 by 300 to 600 Å ("F" type). He interpreted the latter as representing flattened vesicles. He did not claim to be able to distinguish if one or the other of these types of endings might be excitatory or inhibitory, but concluded that he might be seeing a morphological feature that would serve to differentiate excitatory and inhibitory endings.

In 1967, Uchizono reported a study of the crayfish stretch receptor in which excitatory and inhibitory endings could be clearly differentiated. He obtained micrographs of the inhibitory endings on the dendrites of the receptor cell and found that they contained vesicles of the small, flattened type. He found that the excitatory endings on the muscle fibers contained vesicles of the rather large spherical type. Uchizono also studied certain crab muscles which have a double innervation, one excitatory and one inhibitory, and found that some of these endings contained flattened vesicles and some spherical vesicles, again consistent with his hypothesis.

In 1967, Larramendi et al. conducted an extensive statistical study of glutaraldehyde-OsO$_4$-fixed mouse cerebellum in which they attempted to relate vesicle size and shape to age. Their findings supported Uchizono's conclusions and added the interesting point that synaptic vesicles tend to decrease in size with age.

It is too early, at this time, to state with certainty that a definite morphological basis for the differentiation of excitatory and inhibitory endings has been established. It appears reasonably certain, however, that different types of endings may be distinguished by the appearance of the synaptic vesicles after glutaraldehyde-OsO₄ fixation. This is purely an empirical matter, and already some contradictory elements are beginning to appear. For instance, it is now recognized that some of the synaptic vesicles are extremely labile. Thus it has been found that some vesicles may become flattened after glutaraldehyde fixation if they are placed in a buffer for as short a time as 30 minutes. If the buffer wash is omitted, they may appear largely spherical. It is thus necessary to define very accurately the conditions of preparation, and to the extent that these have not yet been clearly worked out, the differentiation of excitatory and inhibitory endings on the basis of vesicular size and shape remains somewhat uncertain. It is to be hoped that, within the near future, it will be possible to make clearcut distinctions of excitatory and inhibitory endings based on such distinctive morphological features as the shape and size of synaptic vesicles. Mitochondrial characteristics may also play a part. Until more work has been done, it seems fair to say that the differentiation of excitatory and inhibitory endings remains uncertain.

## Juxtamembranal specializations

In the above discussion, several kinds of juxtamembranal specializations that take place in synaptic regions have been mentioned. I shall now turn to some more extreme specializations that have been found in synaptic endings. These are closely related either to the presynaptic membrane or to the postsynaptic membrane. For the most part, these specializations are simply amorphous dense bodies of various shapes, sizes, positions, and aggregations.

One of the first such specializations described was presented by Taxi in 1965 in his studies of superior cervical sympathetic-ganglion synapses. He described several irregular rows of dense bodies 200 to 300 Å in diameter, beginning about 250 to 600 Å away from the postsynaptic membrane, and of decreasing length proceeding successively away from the postsynaptic membrane. In 1966, Milhaud and Pappas, in a study of the habenula of the cat, noted, in about one third of the synapses they studied, peculiar arrays of subjunctional dense bodies rather like those seen by Taxi but in a regular hexagonal array.

In 1967, Akert et al. described the fine structure of a peculiar kind of synapse found in the subfornical organ of the cat. This has subjunctional dense bodies about 200 to 300 Å in diameter, about 500 Å apart in a hexagonal array,

placed in a row about 500 Å from the postsynaptic membrane. These synapses are characterized by presynaptic densities and clusters of vesicles much like those described by Gray.

Recently Akert and Sandri (1968) applied a zinc-iodide osmic method to a study of the crest synapses, with some interesting new results. This procedure impregnates certain synaptic vesicles with dense material. At first it was thought that this might be specific for cholinergic endings, because the reaction was prominent in motor end plates, but it was soon found that the impregnation was not specific. Nevertheless, it is certainly striking that some synaptic vesicles are impregnated and some in the same block are not. However, Akert and Sandri (1968) and Pfenninger et al. (1969) used this technique profitably with a considerable clarification of the nature of presynaptic and postsynaptic dense bodies. Figure 9 is a diagram taken from Pfenninger et al. (1969) summarizing their findings on the presynaptic densities.

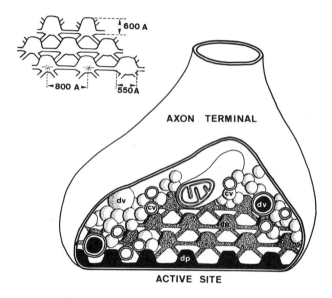

FIGURE 9  Diagram of axon terminal taken from Pfenninger et al. (1969). *Abbreviations:* cv, clear vesicle; dv, dense core vesicle; dp, dense projection. See text for further details.

They now visualize the presynaptic dense projections as being each about 550 Å in diameter, 600 Å high, and 800 Å apart in a regular hexagonal array. They believe the dense projections are interconnected by strands of dense material, as indicated in the diagram. They note that there is approximately enough space to fit one synaptic vesicle into the open region between the dense bodies. These workers have also described a considerable enhancement of the postsynaptic dense material after the zinc-iodide procedure, and also an enhancement of the dense material in the cleft between the presynaptic and postsynaptic membranes. They have observed that this material is split into two thin, dense layers in some preparations. They favor the concept that the presynaptic dense bodies in some way guide the presynaptic vesicles to the appropriate place on the presynaptic membrane, where their contents are released. Gray (1966) made a similar proposal, but regarded the dense bodies as sites of attachment.

Moor et al. (1969) have recently begun studies on freeze-etched preparations of brain tissue, and are elucidating the structure of synapses by the use of this method. In boutons they saw irregularly spaced dimples that might very well represent caveolae. Interestingly, these appear to be distributed more-or-less randomly over the whole surface of the boutons rather than concentrated in the region of membrane specialization, as might be expected if the caveolae do, indeed, represent regions in which synaptic vesicles are caught and where they empty their contents after having fused with the presynaptic membrane, as would be called for by the quantal-release exocytosis hypothesis. Of particular interest in these micrographs is the demonstration that the synaptic vesicles are discrete isolated bodies, as commonly believed. This provides a powerful new piece of evidence attesting to the nature of the vesicles.

In conclusion, I wish to discuss speculatively whether the vesicle really is related to quantal release and "exocytosis," because I believe this to be a fundamental problem. As mentioned above, I regarded this concept as suspect from the very beginning. The fact that the vesicle membrane and the synaptosome membrane are chemically distinct makes the simple exocytosis concept difficult to accept. Further, the evidence that ACh exists free in the cytoplasm of the synaptosome and that this is the material that is first released casts further doubt on the concept. On present evidence, it would appear at least possible that the vesicles may, in fact, represent a storage site for ACh and that the release occurs by some other mechanism. It is not at all difficult to imagine some sort of molecular gate, distributed in numbers over the presynaptic membrane, that, in response to an electrical depolarization, would open just enough to let a reasonably constant number of acetylcholine molecules pass through to the synaptic cleft before closing. Such a mechanism would account for quantal release and for the miniature end-plate potentials if the gates were assumed to open randomly with a certain low frequency without arrival of an impulse. The ACh in the vesicles would then be called upon if the immediate store in the axoplasm were exhausted. This viewpoint represents a radical departure from the apparently well-established central dogma of synaptology, but it has the virtue of getting around some of the difficulties of release by exocytosis.

If it is necessary to retain the concept of quantal release from the vesicles by some variant of the original idea, I wish to propose a tentative mechanism that could work without excessive destruction of membrane or intermixing of vesicular and axonal membrane substance. It may possibly be true that the synaptic vesicles are indeed derived from an interconnected tubular system that is branched many times with only a limited number of connecting points with the presynaptic membrane. Arrival of a membrane depolarization might cause this system to contract suddenly, perhaps by means of a spread of a protein conformation change over the membrane surface, causing it to expel the globular aggregates of acetylcholine in packets, as required by the quantal-release theory. (For a contractile model of transmitter release, see Poisner, 1970.) The necks between vesicles and the neck connecting to the synaptic membrane would remain patent, according to this view, between discharges and, as the tubules became recharged, one might expect ACh to become aggregated into pools by a microperistaltic action produced by membrane contraction. This would be further expected to lead to occasional random expulsion of the aggregated packets at the regions where the system connects with the presynaptic membrane. There are two ways that one could rationalize such a system with the observed facts. The simplest is to assume that the system is highly unstable, requires energy to be maintained, and breaks down into isolated vesicles at the slightest insult.

This is not an unreasonable assumption, because many such membrane systems now are known to break up during fixation into vesicles. The myelin sheath of the prawn is a good case in point; the slightest variation in fixation techniques from the optimal ones results in sheaths made of aggregates of vesicles (Heuser and Doggenweiler, 1966). The "T-system" of the skeletal muscle fiber is another example (Franzini-Armstrong and Porter, 1964). In 1956 I postulated that tubular membranous elements reported in lizard muscle (Robertson, 1956a) constituted a transverse membranous system connecting with the surface of the muscle fiber and explaining Hill's (1949) paradox. These connections were, however, not observed for many years, simply because fixation methods good enough to preserve them had not been developed. Other cases could be cited in which the introduction of new methods has resulted in the demonstration of tubular membranous continuity, whereas only vesicular systems were seen with older methods. The failure to demonstrate such connections initially in freeze-fracture preparations does cause some pause in this argument. However, the technique is plagued by its own artifacts and interpretative problems, and its results should not be taken too literally. Nevertheless, there does seem to be some hope that the freeze-fracture technique may settle this long-standing problem.

There is one additional way in which a mechanism may be proposed that would account for all the observed facts.

To be sure, it has the drawback of being complex. I base it on some studies by Schneider (1960) on contractile vacuoles in certain Protozoa. It has been found that the expulsion of water from contractile vacuoles occurs from the general kind of interconnected tubular system that is postulated above. If fixed at the proper stage in formation and filling before discharge, the interconnected tubular system can be seen clearly. However, if the system is fixed immediately after discharge, only a single channel surrounded by numerous isolated short tubules can be seen. We might draw an analogy with this system, and say that, in preparation for discharge, synaptic vesicles line up in rows and fuse with one another to make an effectively interconnected tubular system, which, in turn, is fused with the surface membrane; in this state, the system would then be, so to speak, "cocked" and ready to fire. The arrival of the nerve impulse would cause an immediate discharge of the contents of the system, which would be followed by an immediate breakup of the system into isolated vesicles. Figure 10 is designed to illustrate this idea, the advantage of which is that no vesicular membrane would become a permanent part of the presynaptic membrane. It would also account both for the

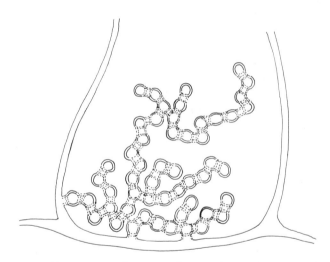

FIGURE 10  Diagram of a synaptic bouton, showing how synaptic vesicles might originate by breakdown of an interconnected tubular system. The dotted lines represent the two states of the system.

presence of isolated vesicles in the presynaptic terminals most of the time and for the findings in freeze-etch preparations. It would only be during the cocking of the system that the vesicles would be interconnected, although the external connections might be somewhat more persistent. It may be assumed that the cocking is a very fast event, taking place with the arrival of the nerve impulse by some kind of electrophoretic effect and lasting only as long as the

membrane depolarization persisted. Thus, successful display of the cocked system would be difficult. This mechanism is admittedly complex, but really no more so than the exocytosis mechanism. All it really adds is a plausible way to reverse the membrane flow quickly. On balance, even though it does not explain the presence of caveolae in the presynaptic membrane, I would say that the membrane gating mechanism in which the vesicle is only a storage vehicle is more likely.

## REFERENCES

AKERT, K., K. PFENNINGER, and C. SANDRI, 1967. Crest synapses with subjunctional bodies in the subfornical organ. *Brain Res.* 5: 118–121.

AKERT, K., and C. SANDRI, 1968. An electron-microscopic study of zinc iodide-osmium impregnation of neurons. I. Staining of synaptic vesicles at cholinergic junctions. *Brain Res.* 7: 286–295.

BODIAN, D., 1966. Electron microscopy: Two major synaptic types on spinal motoneurons. *Science* (*Washington*) 151: 1093–1094.

DEL CASTILLO, J., and B. KATZ, 1954. Quantal components of the end-plate potential. *J. Physiol.* (*London*) 124: 560–573.

DEL CASTILLO, J., and B. KATZ, 1956. Biophysical aspects of neuromuscular transmission. *Progr. Biophys. Biophys. Chem.* 6: 121–170.

COUTEAUX, R., 1963. The differentiation of synaptic areas. *Proc. Roy. Soc., ser. B, biol. sci.* 158: 457–480.

DE ROBERTIS, E. D. P., 1959. Submicroscopic morphology of the synapse. *Int. Rev. Cytol.* 8: 61–96.

DE ROBERTIS, E. D. P., and H. S. BENNETT, 1954. Submicroscopic vesicular component in the synapse. *Fed. Proc.* 13: 35 (abstract).

DE ROBERTIS, E. D. P., and H. S. BENNETT, 1955. Some features of the submicroscopic morphology of synapses in frog and earthworm. *J. Biophys. Biochem. Cytol.* 1: 47–58.

DE ROBERTIS, E. D. P., A. PELLEGRINO DE IRALDI, G. RODRIGUEZ, and C. J. GOMEZ, 1961. On the isolation of nerve endings and synaptic vesicles. *J. Biophys. Biochem. Cytol.* 9: 229–235.

DEWEY, M. M., and L. BARR, 1962. Intercellular connection between smooth muscle cells: The nexus. *Science* (*Washington*) 137: 670–672.

DEWEY, M. M., and L. BARR, 1964. A study of the structure and distribution of the nexus. *J. Cell Biol.* 23: 553–585.

DOGGENWEILER, C. F., and S. FRENK, 1965. Staining properties of lanthanum on cell membranes. *Proc. Nat. Acad. Sci. U.S.A.* 53: 425–430.

DOGGENWEILER, C. F., and J. E. HEUSER, 1967. Ultrastructure of the prawn nerve sheaths. Role of fixative and osmotic pressure in vesiculation of thin cytoplasmic laminae. *J. Cell Biol.* 34: 407–420.

ECCLES, J. C., 1964. The Physiology of Synapses. Springer-Verlag, Berlin.

ELUL, R., 1966. Dependence of synaptic transmission on protein metabolism of nerve cells: A possible electrokinetic mechanism of learning. *Nature* (*London*) 210: 1127–1131.

ELUL, R., 1967. Fixed charge in the cell membrane. *J. Physiol.* (*London*) 189: 351–365.

FARQUHAR, M. G., and G. E. PALADE, 1964. Functional organization of amphibian skin. *Proc. Nat. Acad. Sci. U.S.A.* 51: 569–577.

FATT, P. 1954. Biophysics of junctional transmission. *Physiol. Rev.* 34: 674–710.

FERNÁNDEZ-MORÁN, H., 1955. Studies on the submicroscopic organization of the thalamus. *In* Proceedings of the Sixth Latin American Congress on Neurosurgery. VI. Montevideo, pp. 599–753.

FRANZINI-ARMSTRONG, C., and K. R. PORTER, 1964. Sarcolemmal invaginations constituting the T system in fish muscle fibers. *J. Cell Biol.* 22: 675–696.

FURSHPAN, E. J., and T. FURUKAWA, 1962. Intracellular and extracellular responses of the several regions of the Mauthner cell of the goldfish. *J. Neurophysiol.* 25: 732–771.

FURSHPAN, E. J., and D. D. POTTER, 1959. Transmission at the giant motor synapses of the crayfish. *J. Physiol.* (*London*) 145: 289–325.

GRAY, E. G., 1959. Axo-somatic and axo-dendritic synapses of the cerebral cortex: An electron microscope study. *J. Anat.* 93: 420–433.

GRAY, E. G., 1962. A morphological basis for pre-synaptic inhibition? *Nature* (*London*) 193: 82–83.

GRAY, E. G., 1966. Problems of interpreting the fine structure of vertebrate and invertebrate synapses. *Int. Rev. Gen. Exp. Zool.* 2: 139–170.

GRAY, E. G., and R. W. GUILLERY, 1966. Synaptic morphology in the normal and degenerating nervous system. *Int. Rev. Cytol.* 19: 111–182.

HEUSER, J. E., and C. F. DOGGENWEILER, 1966. The fine structural organization of nerve fibers, sheaths, and glial cells in the prawn, *Palaemonetes vulgaris. J. Cell Biol.* 30: 381–403.

HILL, A. V., 1949. The abrupt transition from rest to activity in muscle. *Proc. Roy. Soc., ser. B, biol. sci.* 136: 399–420.

HIRANO, A., and H. M. DEMBITZER, 1967. A structural analysis of the myelin sheath in the central nervous system. *J. Cell Biol.* 34: 555–567.

HUNEEUS-COX, F., H. L. FERNANDEZ, and B. H. SMITH, 1966. Effects of redox and sulfhydryl reagents on the bioelectric properties of the giant axon of the squid. *Biophys. J.* 6: 675–689.

JOHNSON, G. E., 1924. Giant nerve fibers in crustaceans with special reference to Cambarus and Palaemonetes. *J. Comp. Neurol.* 36: 323–373.

KIDD, M., 1962. Electron microscopy of the inner plexiform layer of the retina in the cat and the pigeon. *J. Anat.* 96: 179–187.

KREUTZIGER, G. O., 1968a. Freeze-etching of intercellular junctions of mouse liver. *In* Proceedings of the 26th Annual Meeting of the Electron Microscope Society of America (C. J. Arceneaux, editor). Claitor's Publishing Division, Baton Rouge, Louisiana, pp. 234–235.

KREUTZIGER, G. O., 1968b. Specimen surface contamination and the loss of structural detail in freeze-etch preparations. *In* Proceedings of the 26th Annual Meeting of the Electron Microscope Society of America (C. J. Arceneaux, editor). Claitor's Publishing Division, Baton Rouge, Louisiana, pp. 138–139.

LARRAMENDI, L. M. H., L. FICKENSCHER, and N. LEMKEY-JOHNSTON, 1967. Synaptic vesicles of inhibitory and excitatory terminals in the cerebellum. *Science* (*Washington*) 156: 967–969.

LOEWENSTEIN, W. R., 1966. Permeability of membrane junctions. *Ann. N. Y. Acad. Sci.* 137: 441–472.

McNUTT, N. S., and R. S. WEINSTEIN, 1969. Interlocking subunit arrays forming nexus membranes. *In* Proceedings of the 27th Annual Meeting of the Electron Microscope Society of America (C. J. Arceneaux, editor). Claitor's Publishing Division, Baton Rouge, Louisiana, pp. 330–331.

MARTIN, A. R., and G. PILAR, 1963. Dual mode of synaptic transmission in the avian ciliary ganglion. *J. Physiol.* (*London*) 168: 443–463.

MILHAUD, M., and G. D. PAPPAS, 1966. The fine structure of neurons and synapses of the habenula of the cat with special reference to subjunctional bodies. *Brain Res.* 3: 158–173.

MOOR, H., K. PFENNINGER, and K. AKERT, 1969. Synaptic vesicles in electron micrographs of freeze-etched nerve terminals. *Science* (*Washington*) 164: 1405–1407.

PALADE, G. E., and S. L. PALAY, 1954. Electron microscope observations of interneuronal and neuromuscular synapses. *Anat. Rec.* 118: 335–336 (abstract).

PALAY, S. L., 1956. Synapses in the central nervous system. *J. Biophys. Biochem. Cytol.* 2 (suppl., no. 4, pt. 2): 193–201.

PAPPAS, G. D., and M. V. L. BENNETT, 1966. Specialized junctions involved in electrical transmission between neurons. *Ann. N. Y. Acad. Sci.* 137: 495–508.

PERACCHIA, C., 1969. Unmasking of SH groups in certain axonal membranes as a result of electrical stimulation. *J. Cell Biol.* 43 (no. 2, pt. 2): 102a (abstract).

PERACCHIA, C., and J. D. ROBERTSON, 1968. Changes in unit membranes of nerve fibers resulting from electrical stimulation. *J. Cell Biol.* 39 (no. 2, pt. 2): 103a (abstract).

PFENNINGER, K., C. SANDRI, K. AKERT, and C. H. EUGSTER, 1969. Contribution to the problem of structural organization of the presynaptic area. *Brain Res.* 12: 10–18.

POISNER, A. M., 1970. Release of transmitters from storage: A contractile model. *In* Second International Meeting of International Society for Neurochemistry, Milan, Italy, September 1–5, 1969 (in press).

REGER, J. F., 1954. Electron microscopy of the motor end-plate in intercostal muscle of the rat. *Anat. Rec.* 118: 344 (abstract).

REGER, J. F., 1955. Electron microscopy of the motor end-plate in rat intercostal muscle. *Anat. Rec.* 122: 1–16.

REVEL, J. P., and M. J. KARNOVSKY, 1967. Hexagonal array of subunits in intercellular junctions of the mouse heart and liver. *J. Cell Biol.* 33: C7–C12.

ROBERTSON, J. D., 1953. Ultrastructure of two invertebrate synapses. *Proc. Soc. Exp. Biol. Med.* 82: 219–223.

ROBERTSON, J. D., 1954. Electron microscope observations on a reptilian myoneural junction. *Anat. Rec.* 118: 346 (abstract).

ROBERTSON, J. D., 1956a. Some features of the ultrastructure of reptilian skeletal muscle. *J. Biophys. Biochem. Cytol.* 2: 369–380.

ROBERTSON, J. D., 1956b. The ultrastructure of a reptilian myoneural junction. *J. Biophys. Biochem. Cytol.* 2: 381–394.

ROBERTSON, J. D., 1959. Preliminary observations on the ultrastructure of nodes of Ranvier. *Z. Zellforsch. Mikroskop. Anat.* 50: 553–560.

ROBERTSON, J. D., 1960. The molecular structure and contact relationships of cell membranes. *Progr. Biophys. Biophys. Chem.* 10: 343–418.

ROBERTSON, J. D., 1961. Ultrastructure of excitable membranes and the crayfish median giant synapse. *Ann. N. Y. Acad. Sci.* 94: 339–389.

ROBERTSON, J. D., 1963. The occurrence of a subunit pattern in the unit membranes of club endings in Mauthner cell synapses in goldfish brains. *J. Cell Biol.* 19: 201–221.

ROBERTSON, J. D., 1964. Unit membranes: A review with recent new studies of experimental alterations and a new subunit structure in synaptic membranes. *In* Cellular Membranes in Development (M. Locke, editor). Academic Press, New York, pp. 1–81.

ROBERTSON, J. D., T. S. BODENHEIMER, and D. E. STAGE, 1963. The ultrastructure of Mauthner cell synapses and nodes in goldfish brains. *J. Cell Biol.* 19: 159–199.

SCHNEIDER, L., 1960. Elektronenmikroskopische Untersuchungen über das Nephridialsystem von *Paramaecium*. *J. Protozool.* 7: 75–90.

SÖJSTRAND, F. S., 1961. Electron microscopy of the retina. *In* The Structure of the Eye (G. K. Smelser, editor). Academic Press, New York, pp. 1–28.

SJÖSTRAND, F. S., and E. ANDERSSON, 1954. Electron microscopy of the intercalated discs of cardiac muscle tissue. *Experientia* (*Basel*) 10: 369–370.

SOMER, J. R., and E. A. JOHNSON, 1970. Comparative ultrastructure of cardiac cell membrane specializations. *Amer. J. Cardiol.* 25: 184–194.

SOMER, J. R., and R. L. STEERE, 1969. The nexus freeze-etched. *J. Cell Biol.* 35: 136 (abstract).

TAKAHASHI, K., and K. HAMA, 1965. Some observations on the fine structure of the synaptic area in the ciliary ganglion of the chick. *Z. Zellforsch. Mikroskop. Anat.* 67: 174–184.

TAXI, J., 1965. Contribution a l'étude des connexions des neurones moteurs du système nerveux autonome. *Ann. Sci. Natur., ser.* 12, *zool. biol. anim.* 7: 413–674.

UCHIZONO, K., 1965. Characteristics of excitatory and inhibitory synapses in the central nervous system of the cat. *Nature* (*London*) 207: 642–643.

UCHIZONO, K., 1967. Inhibitory synapses on the stretch receptor neurone of the crayfish. *Nature* (*London*) 214: 833–834.

WHITTAKER, V. P., 1959. The isolation and characterization of acetylcholine-containing particles from brain. *Biochem. J.* 72: 694–706.

WHITTAKER, V. P., 1966. Some properties of synaptic membranes isolated from the central nervous system. *Ann. N. Y. Acad. Sci.* 137: 982–998.

WHITTAKER, V. P., and E. G. GRAY, 1962. The synapse: Biology and morphology. *Brit. Med. Bull.* 18: 223–228.

# 65 Correlating Structure and Function of Synaptic Ultrastructure

## F. E. BLOOM

THE BASIC DETAILS of synaptic fine structure have been known for more than 15 years (Palay, 1956, 1958).Although scores of major and minor variations on the basic theme have been described, the functional importance of the relatively few cellular organelles apparent at the synapse has scarcely been elucidated. Many recent chemical studies (Appel et al., 1969; Mahler, 1969; and Barondes, 1968, an earlier paper in this volume) have indicated that isolated synaptic endings may be able to restructure some of their macromolecular protein constituents. Consequently, those persons interested in testing the "code of life" dogma (DNA to RNA to protein to function) as a mechanism for the learning or memory trace have again turned to fine structure, where the seductively high resolution of the electron microscope offers the promise of equating structure with function. In this essay, I attempt to explore the factual constraints of synaptic fine structure, re-emphasizing some of the topics described by Dr. Robertson in the previous chapter. From these basic constraints, I have tried to indicate the most obvious types of microstructural changes compatible with induced subcellular neuroplasticity, emphasizing proposals that can best be subjected to experimental testing. From these constraints and proposals, we can each judge the width of the "synapse gap" as one of the weak links in molecular neurobiology. Even the freshest novitiate to the neurosciences will realize, however, that the resolution of the mechanics and molecular potentialities of the synapse is unlikely to come from fine-structural experiments alone.

THE BASIC STRUCTURAL CONSTRAINTS  The synapse is truly a "mixed bag." To the physiologists, the synapse is the functional transducer that converts the electrical activity of the all-or-none action potential of the axon into chemical secretion of a transmitter, thereby producing electrical activity in the postsynaptic cell. Most electron microscopists would apply the term synapse when two morphological components are present: (1) a specialized interneuronal

contact zone between nerve terminals and dendrites, cell bodies, or other postsynaptic elements, that appears morphologically different from all other intermembranous conjunctions; and (2) the presynaptic accumulation of synaptic vesicles with partial or complete exclusion of axonal microtubules (Figure 1). The visualization of both these essential structural synaptic features depends on the type of fixation used for tissue preparation.

### Synaptic vesicles

In general, when such fixatives as osmium or permanganate are used to heighten the opacity of the membranous components, the most striking feature of the synapse is the large number of synaptic vesicles. The first question to be asked, then, is whether the size, shape, electron opacity, and intraaxonal distribution of vesicles can be related to synaptic functioning. Let us note, in passing, that, although the shape and configuration of synaptic vesicles seen in electron micrographs may not represent their native conformation in the brain prior to examination, no methods for examining native vesicles in situ have yet been devised.

Furthermore, some basic clarification is needed as to the origin of synaptic vesicles within the nerve terminal. Akert and his group have recently published freeze-etch electron micrographs (Akert et al., 1969) in which they have observed what appear to be pinocytotic invaginations of the axonal membrane. These invaginations do not occur at the area of the synaptic specialization. If synaptic vesicles were to form by budding from the axonal membrane (Westrum, 1965), their lipid composition must also change in the process, because Whittaker (1966) has shown that the lipid composition of synaptic vesicles is different from that of synaptosomal membrane. Where else, then, might the synaptic vesicles arise? The only other known source of "unit plasma membrane" in the terminal is axonal mitochondria. There is no evidence that vesicles bud from mitochondria, so might they be assembled at the mitochondrion from mobile proteolipid subunits? Or, despite the rarity of their sighting within axons, did all these vesicles pass to the terminals only by axoplasmic flow from the soma? We simply do not know!

F. E. BLOOM  Laboratory of Neuropharmacology, Division of Special Mental Health Research, National Institute of Mental Health, Saint Elizabeth's Hospital, Washington, D. C.

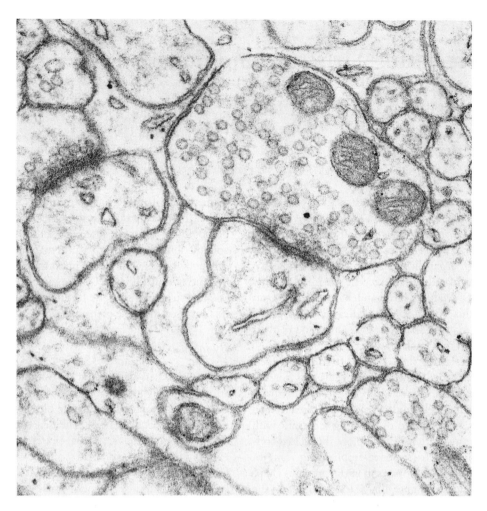

FIGURE 1 Neuropil in the molecular layer of the rat cerebellar cortex, illustrating two axodendritic contacts that exhibit presynaptic spherical, electron-lucent synaptic vesicles, and distinctly specialized zones of contact. An axodendritic contact with presynaptic vesicles, but no specialized zone of contact, is seen at the lower right. Note that, for the contact at upper left, the entire contacting surfaces between the axon and dendrite are specialized; for the larger contact at upper right center, only a portion of the apposing surfaces are morphologically specialized. Magnification, ×65,000.

Perhaps the best connection between synaptic vesicles and synaptic function is derived from brain homogenate studies, which have shown that the nerve endings (or the vesicle fractions subsequently derived from nerve endings) are the richest sources of suspected transmitter substances (see both Whittaker and Iversen, this volume). Because the peripheral autonomic transmitters, acetylcholine (ACh) and norepinephrine (NE), can be related to particular nerve endings within the autonomic nervous system (see review by Eränkö, 1967), analogous cytochemical methods have been applied to the central nervous system for purposes of identifying nerve endings on the basis of their neurotransmitters. These studies can best be reviewed by classifying vesicles on the basis of their electron opacity in routinely prepared electron micrographs (Figure 2).

ELECTRON-LUCENT SYNAPTIC VESICLES  The majority of synaptic vesicles seen in central nerve endings are electron-lucent (that is, of low electron density) and measure between 200 and 600 Å in diameter. This type of synaptic vesicle abounds in the classically cholinergic neuromuscular junction; it also occurs in isolated nerve-ending particle fractions of brain homogenates rich in ACh (see Whittaker, this volume). It is not clear, however, whether this basic vesicle morphology is indicative of nerve endings containing ACh or is simply the basic model of synaptic vesicles. Electron-lucent synaptic vesicles have also been described in nerve endings found in homogenate fractions not containing large amounts of ACh, and presumably, then, they must also serve to contain such transmitters as amino acids and the large population of as-yet-unspecified synaptic

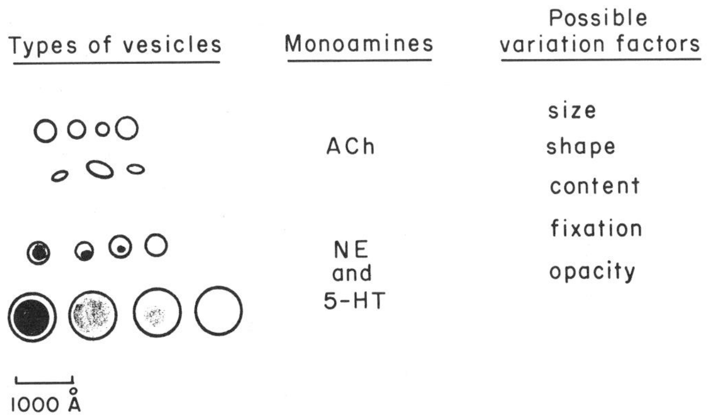

Types of vesicles        Monoamines        Possible
                                          variation factors

                          ACh              size
                                           shape
                                           content
                          NE               fixation
                          and              opacity
                          5-HT

1000 Å

FIGURE 2 Illustration of the variations among sizes, shapes, and electron-opacity of contents for the various central and peripheral synaptic vesicles and two of the transmitters to which they may be related. The upper two rows indicate the electron-lucent vesicles, most commonly related to acetylcholine content (ACh), as at the neuromuscular junction. The lower two rows illustrate the varia- tions among the electron-opaque synaptic vesicles related to con- tent of monoamines, such as norepinephrine (NE) and serotonin (5-HT). The reference mark indicates 1000 Å. At right are noted some of the factors that may relate to the variations in vesicle mor- phology.

transmitter agents. We may presume that only a portion of the electron-lucent synaptic vesicles will be shown to con- tain ACh, for which no cytochemical test has yet been con- clusively demonstrated. The zinc-iodide-osmium method of Akert (Akert and Sandri, 1968) and the metallic impreg- nation studies of Howard (personal communication) sug- gest the emergence of such a method. In its stead, enzyme histochemical localizations for cholinesterase, acetylcholin- esterase, and cholineacetyltransferase have been used to specify cholinergic content at the vesicle level (see Koelle, 1969; Karczmar, 1969).

Electron-lucent synaptic vesicles have also been sub- classified on the basis of their shape. Although the majority are spherical, certain nerve terminals, particularly when fixed by formalin perfusion techniques (Uchizono, 1965; Bodian, 1966a, 1966b), contain smaller, pleomorphic vesi- cles. The latter, typically flat synaptic vesicles, were func- tionally related to "inhibition" on the basis of their appear- ance in certain cerebellar-cortex nerve terminals, the func- tional characteristics of which are specifiable (Eccles et al., 1967), and on the basis of a comparison of the development of inhibition with the appearance of flat vesicles in their nerve terminals (Bodian, 1966b).

This area is one of the few in electron microscopy of brain in which quantitative measurements have been made on a series of material large enough to provide certain sta- tistically significant correlations. Lenn and Reese (1966) analyzed the ventral cochlear nucleus of several rodents and interpreted their vesicle measurements to indicate that the nerve endings containing smaller synaptic vesicles were associated with the inhibitory afferent terminals of higher auditory centers. Larramendi et al. (1967) have made large numbers of measurements on synaptic vesicles in rodent

cerebellar cortex to evaluate the conclusion of Uchizono (1965) that inhibitory nerve terminals, particularly those basket-cell terminals ending upon Purkinje cell bodies, ex- hibited small, oval, synaptic vesicles, whereas excitatory endings, such as those of parallel fiber and mossy fiber, have vesicles that are larger and more spherical in shape (Figure 1). The interesting conclusion reached by Larramendi and his group was that the size of the vesicle tended to decrease with development, and that those nerve terminals whose general functional classification is inhibitory were more likely to have smaller synaptic vesicles than were those found in cerebellar cortical nerve terminals thought to be excitatory (Eccles et al., 1967). The difference in vesicle size could be related to the rate at which the vesicles are utilized (metabolically consumed or emptied for refilling): large synaptic vesicles in less mature nerve terminals might in- dicate a faster utilization than do the smaller vesicles, which have survived longer (Larramendi et al., 1967).

The functional classification of electron-lucent vesicles based on criteria of shape or size has not yet been chemically related to specific transmitter agents. In my own experi- ence with autonomic nerve endings of the rat vas deferens, which is generally considered to be exclusively adrenergic, I have observed that vesicles that are typically spherical when they contain demonstrable electron-granular cores can become flat when the catecholamine content has been pharmacologically depleted. Central nerve endings thought to contain NE have also been observed to contain either round or pleomorphic vesicles (Descarries, personal com- munication). Thus, the size and shape of the vesicle could be related to: (1) function; (2) types of transmitter content; (3) amount or state of transmitter content; or (4) age of the vesicle within the nerve terminal. These underlying factors could predispose certain vesicles to alter their shape during fixation (Walberg, 1966), although the pleomorphic vesicles can be observed with several different methods of fixation (see Fukami, 1969).

ELECTRON-OPAQUE SYNAPTIC VESICLES    There are two populations of electron-opaque synaptic vesicles, one 400–600 Å in diameter, the other 800–1200 Å (Figure 2; Grillo and Palay, 1962). To the eyes of some, these vesicles represent diminutive declensions of the adrenal chromaffin granule, which, in turn, belongs to the general line of secretory particles such as those containing zymogens (Palade et al., 1961) or antidiuretic hormones (ADH) (Bargmann, 1966). Fine-structural analyses of the peripheral sympathetic nervous system have provided many data on the identification of the smaller synaptic vesicles that exhibit electron-granu-

lar content (Figure 3A) as the storage organelles of catecholamines. This identification is based on four types of data. First, the small granular vesicles have been observed in virtually every sympathetically innervated structure examined, provided the fixation has been adjusted to the tissue (see reviews by Grillo, 1966; Bloom and Giarman, 1968a; Tranzer et al., 1969; Jaim-Etcheverry and Zieher, 1969). The vesicles become even more opaque after the nerves have taken up substituted catecholamines such as 5-OH dopamine (Tranzer and Thoenen, 1967) and alpha-methyl NE (Bondareff, 1966; Hökfelt, 1968). Second, the granular contents of these granular vesicles 400–600 Å in diameter can be correlated with the intensity of the fluorescence-histochemical or biochemical estimation of catecholamines after pharmacological depletion or enhancement (Van Orden et al., 1966, 1967a). Third, by the use of

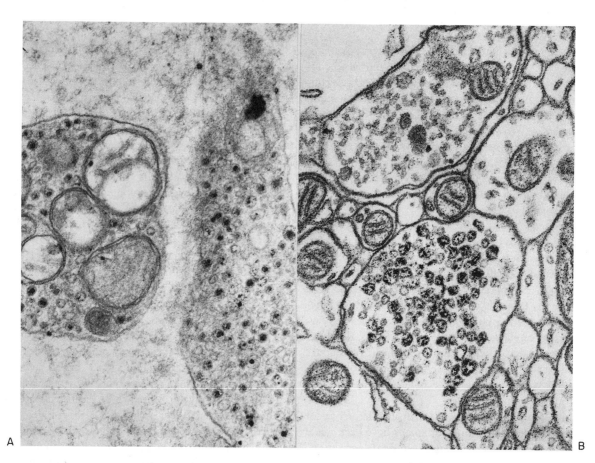

FIGURE 3    A. Two perivascular nerve terminals from the pineal of the rat, after fixation with glutaraldehyde and standard exposure to osmium tetroxide. These nerves are filled with vesicles of the small granular type (SGV), which in this normal animal constitute about 60–80 per cent of the vesicles seen. Modified from Bloom and Giarman, 1969. Magnification, ×50,000. B. Two nerve terminals from the paraventricular hypothalamus of the rat after glutaraldehyde perfusion and exposure to osmium tetroxide at 60°C for 30 minutes. The spherical synaptic vesicles in the lower ending are filled with electron-opaque osmiophilic precipitates, while the smaller, pleomorphic vesicles in the upper terminal are electronlucent. (Modified from Bloom and Aghajanian, 1968c.) Magnification, ×50,000.

electron-microscope autoradiography, radioactivity can be shown to concentrate in nerve endings that exhibit these small granular vesicles (Wolfe et al., 1962). Fourth, the nerves that exhibit the small granular vesicles degenerate on destruction of the postganglionic sympathetic nerve trunks (Pellegrino de Iraldi et al., 1965; Van Orden et al., 1967b). A more recent approach to the localization and understanding of granular vesicles has been the use of 6-hydroxydopamine (6-HDM) as a specific toxin causing local nerve-ending degeneration among various types of catecholamine-containing nerve endings in the periphery (Tranzer and Thoenen, 1968). However, the specificity and precise pathogenesis of the drug effects on fine structure still require additional data.

With this rather compatible story at hand for the periph-

eral nervous system, it was natural to apply the methods to the central nervous system (CNS) in an attempt to identify the catecholamine-containing nerve endings there. Until recently, however, small granular vesicles were not visible within the CNS, and attention was focused on the large granular vesicles (LGV) (Figure 4) (see Bloom and Aghajanian, 1968a; Bloom and Giarman, 1968a). The LGV could be seen to exhibit variable amounts of electron-granular content which was similar, although not identical, to the granularity seen in the small granular vesicles (see Grillo and Palay, 1962). The granularity of the small vesicles is typically solid and homogeneous, although the relative filling of the vesicle interior varies widely. Possibly this variation is related to species, tissues, and relative amine content, but it appears most closely related to fixation. The

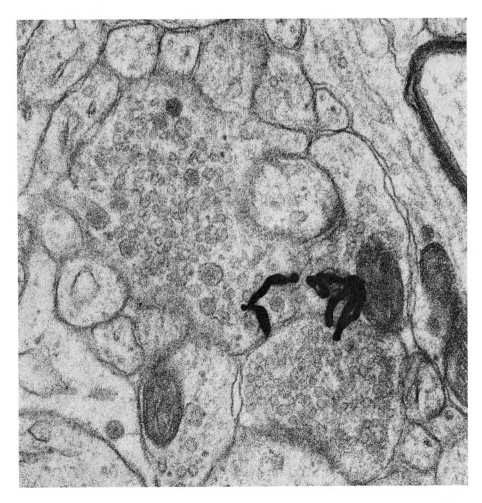

FIGURE 4 Autoradiograph of nerve terminals in the pontine raphé nucleus of the rat brainstem. Tissue was prepared two hours after intracisternal labeling with 0.8 micrograms of $^3$H-serotonin, and otherwise routinely prepared by the methods of Aghajanian and Bloom (1967a, 1967b). The nerve terminal with two large silver grains is making an axodendritic contact with a small dendritic spine; this terminal has many large granular vesicles and even more numerous electronlucent vesicles. (From Bloom and Couch, unpublished.) Magnification, ×65,000.

granularity of the large vesicles varies mainly in electron-opacity, the interior appearing either fibrillar or wispy, but usually filling most of the vesicle interior. The LGV were found in brain-homogenate fractions rich in NE nerve endings (de Robertis et al., 1965), and early evaluation of the effects of catecholamine-depleting drugs suggested that the granular content could be depleted (Pellegrino de Iraldi et al., 1963; Bak, 1965, 1967; Hashimoto et al., 1965; Ishii et al., 1965; Matsuoka et al., 1965; Shimizu and Ishii, 1964). The LGV have been observed in the majority of nerve processes over which autoradiographic grains are found after labeling endogenous brain stores by intra-ventricularly injected $^3$H norepinephrine (Aghajanian and Bloom, 1966, 1967b; Lenn, 1967; Descarries and Droz, 1968) or $^3$H serotonin (5-HT) (Figure 4; Aghajanian and Bloom, 1967a). In studies utilizing lesions to cause degeneration of specific catecholamine- or 5-HT-containing nerve pathways, those nerve endings that degenerate can also be shown to exhibit LGV (Aghajanian et al., 1969; Raisman, 1969a). Furthermore, the LGV which have been seen in electron micrographs specifically correlate with selective fluorescence-histochemical examination of the hypothalamus (Eneström and Svalander, 1967).

However, other combined light and electron-microscope surveys (Lenn, 1965; Fuxe et al., 1965, 1966) indicated that the LGV could not be the only storage organelles of brain NE. When the large granular vesicles were quantified for several regions of rat brain and the values were compared with the reported regional brain content of catecholamine and 5-HT by either biochemical or fluorescence-histochemical measurement, a correlation was found between the total percentage of nerve endings containing large granular vesicles and the combined content of NE and 5-HT (Bloom and Aghajanian, 1968a). The electron-opacity of the content of the LGV, however, did not vary in magnitude with the expected change in the monoamine level when brain monoamine content was depleted or enhanced pharmacologically.

All the above data on LGV in the brain could be interpreted to mean that such vesicles are but one subcellular signature that indicate nerve endings able to store monoamines. They do not indicate that the LGV are the storage site of the monoamine; in fact, in virtually all the nerve terminals that exhibit LGV, large numbers of spherical electron-lucent synaptic vesicles can also be seen. Furthermore, LGV are also seen in autonomic terminals regarded as classically "cholinergic" (Taxi, 1961; Grillo, 1966; Bloom and Barrnett, 1966), although they differ with respect to loading by 5-OH dopamine (Tranzer et al., 1969).

Although many aspects of both central and peripheral LGV remain to be solved, their significance to transmitter storage has become more remote since the successful application of permanganate fixation. When potassium per-manganate is used as the fixative on either fresh tissue blocks (Richardson, 1966) or thin slices of brain tissue incubated in NE, 5-HT, or alphamethyl NE (Hökfelt, 1967, 1968), synaptic vesicles in those regions of the brain rich in monoamines can be shown to contain small granular synaptic vesicles (SGV). Hökfelt (1967, 1968; Hökfelt and Jonsson, 1968) has shown, in an extensive series of experiments, that brain slices of caudate nucleus incubated in vitro in solutions of NE (1–10 micrograms per milliliter) exhibit the highest frequency of SGV in the nerve terminals. The same SGV can be seen in unincubated sections of hypothalamus and locus coeruleus, although the frequency is less. If the slices are taken from the brain of animals pretreated with reserpine to block storage of amines, no SGV are seen after subsequent incubation and fixation.

Depletion experiments have been shown to decrease the frequency with which small granular vesicles can be seen in brain structures after permanganate fixation (Hökfelt, 1968), but the experiment is difficult to perform because the fixation is variable and large sampling is not always possible; furthermore, the reproducibility of the fixation appears to vary from poor to adequate preservation. No autoradiographic studies have yet been performed to confirm that the permanganate granular vesicles are equivalent to catecholamine; those instances in which the experiments have been attempted (Hökfelt and Jonsson, 1968; Devine and Laverty, 1968; Bloom and Giarman, 1970) suggest that permanganate is not a superior preservative of radioactive catecholamine content. Furthermore, permanganate-fixed tissues, despite their equivocal retention of radioactive norepinephrine, fail to exhibit autoradiographic activity (Bloom and Giarman, 1970; Taxi, personal communication; Descarries and Droz, personal communication). Whether the failure is due to some oxidative breakdown of the NE isotope (Harrison et al., 1968) or to some chemographic reaction between the tissue and the emulsion, the results of this critical experiment are needed for complete interpretation of the small granular vesicle revealed in the brain with permanganate fixation.

Similar SGV can also be made to appear in electron micrographs by exposing glutaraldehyde-fixed brain to osmium tetroxide at elevated temperatures (Figure 3B). The question then arises as to whether the electron opacity of these granular vesicles represents catecholamine content either qualitatively or quantitatively. Attempts to correlate the "hot osmium" SGV with catecholamine content have so far been unsuccessful (Bloom and Aghajanian, 1968c), because both drug-depletion studies and autoradiographic studies fail to confirm that the granularities are associated with norepinephrine content. If the granular content of the small vesicles seen in the brain after permanganate or hot osmium could be associated with norepinephrine storage by autoradiographic or other independent cytochemical techniques, it would then be possible to utilize the granular material as a quantitative marker for functional experiments on monoamine-containing terminals. Methods will still be required to distinguish serotonin from NE-containing central nerve endings (see Jaim-Etcheverry and Zieher, 1968, 1970) and to establish a basic biological explanation for what distinguishes the reactivity of central small granular vesicles from the SGV of peripheral nerve endings.

On the other hand, the granular content of the large and small synaptic vesicles could be related to catecholamines indirectly, as a reflection of the chemical reactivity or avidity of the "matrix" within synaptic vesicles. What little evidence exists on the nature of adrenergic synaptic-vesicle matrixes suggests that they may contain a metallic cation, adenine nucleotides, and presumably some sort of phospholipoprotein (Colburn and Maas, 1965).

## Synaptic specializations

The fine-structural specializations repeatedly observed at sites of interneuronal contacts represent the second major morphological feature of the synapse. These points of specialized contact have either been described in terms of their morphology (e.g., membranous thickening, intermembranous bridges, desmosomoid differentiation, synaptic membrane complex, open junction, and so forth) or simply named for their functional connotation as synapses. In osmium-fixed tissues (see Palay, 1956, 1967), this specialization is characterized by sites at which the plasma membranes of the apposing processes are parallel to each other and accentuated to variable degrees by electron-dense material that occurs in the adjacent presynaptic and postsynaptic cytoplasm and within the intersynaptic space (Figures 1, 2, 4, 5).

When properly sectioned, the intersynaptic cleft appears to be somewhat wider than other intercellular spaces (Kelly, 1967; Lenn and Reese, 1966). It must be remembered, however, that the intercellular spaces seen in commonly prepared electron micrographs of brain do not take into account the swelling and shrinkages that have occurred through the steps of fixation and dehydration (Van Harrevald et al., 1965).

When tissue is fixed with dilute permanganate (Figure 5A), the cytoplasmic electron densities at the specialized contact zone are poorly seen, and the zone is, therefore, characterized only by parallel membranes of the presynaptic and postsynaptic elements (see Gray and Guillery, 1966). When osmium-fixed material is stained with phosphotungstic acid (PTA), the electron-dense material is particularly prominent (Figures 5B–D, 6A–C). Aghajanian and I (1966, 1967a, 1968b) have modified the PTA-staining procedure used by Gray (1959) on osmium-fixed blocks to study the nature of the paramembranous material found at specialized

interneuronal contacts. Our version of the staining procedure utilized glutaraldehyde-fixed nervous tissue stained in the block with ethanolic phosphotungstic acid (E-PTA). This procedure does not make membranes electron-opaque, but has the advantage that material stained with the electron-dense E-PTA is now selectively revealed (Figure 6A, B) in the same configuration as that described by Gray (Figure 5). This staining result clearly demonstrates that the electron-dense material is independent of the plasma membrane of the contacting neural processes. We have referred to the material that stains with E-PTA as "synaptic material," implying without proof that the sites at which it occurs are functional synapses. No other interneuronal surfaces, however, appear to be so specialized.

The organization of the synaptic material stained with E-PTA in different mammalian and vertebrate species and in different regions of the central and peripheral nervous system is, in general, similar. Four separate accumulations of material reactive with E-PTA can be consistently observed in thin sections that give a normal view of the contact site (Figure 6C). At the preterminal edge of the axoplasm of the terminal bouton, two to eight aggregates of moderately dense electron-opaque material project into the nerve terminal; these dense projections measure 300–500 Å in widest diameter and are separated from one another by 160–500 Å. Similar structures seen by Gray, who used PTA on osmium-fixed material (Gray, 1959), and by Akert and coworkers, who used bismuth iodide (BI) impregnation of glutaraldehyde-fixed material (Figure 5E) (Akert et al., 1969; Pfenninger et al., 1969), suggest that the dense projections occur in hexagonal arrays when viewed in frontal sections. These dense projections are not readily apparent in the neuropil of brains exposed to osmium tetroxide, where the material adjacent to the presynaptic membrane is seen only as dense granular patches varying in both size and shape (Figure 6B, C). Furthermore, their outline in osmium-fixed material is vague, irregular, and frequently amorphous, in part owing to the superimposition of the osmium-stained membrane of the synaptic vesicles.

A second element of the organized, dense, synaptic material is the highly dense electron-opaque band, 150–200 Å in thickness, situated just within the cytoplasm of the postsynaptic process. This element appears either as a continuous or as a discontinuous element equal in length to the opposing row of presynaptic dense projections.

The third element of synaptic material is an electron-opaque line, 50–150 Å in width extending through what appears to be the center of the intercleft space. In our material fixed with glutaraldehyde and stained with E-PTA, the intercleft opaque line occasionally appeared to be perforated by small electron-lucent gaps. Material stained with BI (Akert et al., 1969; Pfenninger et al., 1969) exhibits an electron-lucent line down the center of the intercleft line,

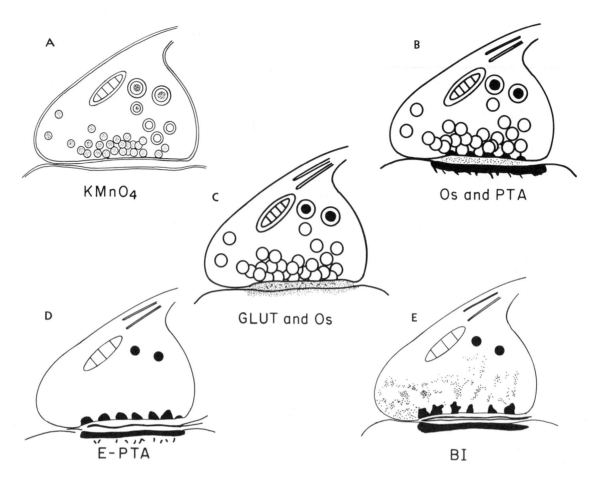

A  KMnO₄

B  Os and PTA

C  GLUT and Os

D  E-PTA

E  BI

FIGURE 5  Illustration of the variation in the electron-microscopic appearance of specialized interneuronal contacts after various types of fixative procedures (see text).

but this is not seen with the E-PTA preparation. This intra-synaptic element of the organized synaptic material can be compared with the thin fibrous residua occasionally seen as narrow lines or filaments within the synaptic cleft of osmium-fixed tissue.

Occasionally, a fourth element of synaptic specialization can be seen as fine wisps of electron-opaque material extending from the postsynaptic band into the cytoplasm of the postsynaptic neuron. The various synaptic staining procedures are summarized by the drawings in Figure 5.

Material reactive with E-PTA, and seemingly identical with the synaptic material, can also be visualized at myoneural junctions in several species. In this junction, presynaptic dense projections and continuous postjunctional bands lining the primary and secondary clefts of the junction are heavily stained. The "dendrodendritic junctions" (see both Rall and Shepherd in this volume) of the olfactory bulb of the rat also exhibit the polarized appearance of typical synaptic material. When the CNS of more primitive

invertebrate nervous systems is examined, we have difficulty in finding distinct and clear accumulations of synaptic material that is equally specialized. The presumed osmium equivalent, however, can be seen in the CNS of squid (Castejón and Villegas, 1964) and octopus (Jones, 1969), insects (Smith, 1965; Boadle and Bloom, unpublished results), and, possibly, planaria (Best and Noel, 1969).

We have found that certain tissue elements, other than the synaptic material, exhibit affinity for the E-PTA stain (Bloom and Aghajanian, 1968b). In tissues of both nervous and non-nervous origin, the nucleoli and nuclei were stained, as were collagen, erythrocytes z bands in striated muscle, and the large granular synaptic vesicles (Bloom and Aghajanian, 1968a, 1968b; Jaim-Etcheverry and Zieher, in press). Sheridan and Barrnett (1969) have recently investigated the staining of nuclei, and their cytochemical investigations support our suggestion that the staining with E-PTA is the result of epsilon amino groups of basic amino acids present in macromolecular proteins or

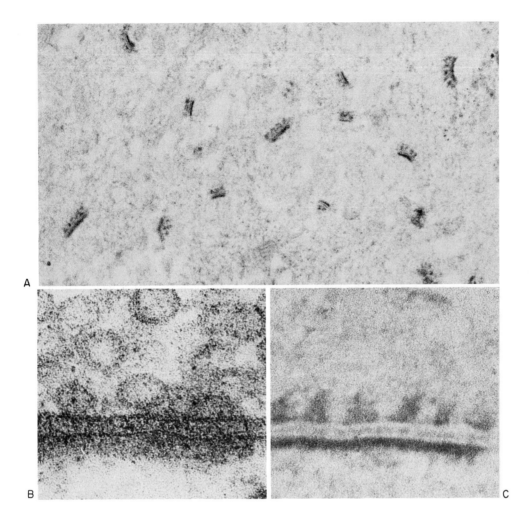

FIGURE 6   A. Low-power electron micrograph of molecular layer of rat cerebellar cortex after glutaraldehyde fixation and staining with ethanolic phosphotungstic acid (E-PTA). Most prominently stained are the numerous specialized interneuronal junctions, characterized by the continuous postsynaptic density and the two to eight presynaptic dense projections. Vague paramembranous outlines, microtubules, and mitochondrial matrixes can be seen faintly in the background. (Modified from Bloom and Aghajanian, 1968b.) Magnification, ×30,000. B, C. High-magnification comparison between specialized synaptic contact zones as seen after standard preparative techniques and heavy metal staining of the thin section (B) and after E-PTA staining (C). Although little, if any, distinct material can be seen presynaptically in B, except for synaptic vesicles, the intersynaptic cleft and the immediate postsynaptic cytoplasm are filled with a fibrillar electron-opaque material. In C, the presynaptic dense projections, a somewhat faint intersynaptic line, and a prominent postsynaptic band are seen; the electron-opaque lines occupied by the unstained presynaptic and postsynaptic plasma membranes can be directly related to their osmium-opaque structures in B. Both B and C are from adjacent blocks of rat cerebellar cortex, perfused 14 days postnatally. Magnification, ×210,000.

glycoproteins. Essentially, these data are based on the fact that proteolytic enzymes removed the material that stains with E-PTA (Bloom and Aghajanian, 1968b), whereas we could find no other cytochemical digestion methods that could eliminate the E-PTA staining property. E-PTA staining could also be prevented by acetylation of amino groups. In quantitative cytochemical reactions, Silverman

and Glick (1969) have recently provided data that confirm this explanation for PTA staining.

Similarly, proteolytic enzymes prevent the BI impregnation of specialized contacts, and in vitro tests indicate high affinity of BI for polylysine and polyarginine (Akert and Pfenninger, 1969).

The protein interpretation of the E-PTA staining differs

from that of Pease (1966), who has proposed that aqueous PTA, when used to stain sections of unfixed tissues dehydrated with various glycols, stains "polysaccharides." This polysaccharide staining is also reproduced by a modification of the periodic acid-Schiff stain, with silver as an electron-opaque marker for electron microscopy. With the use of this technique, polysaccharides can be stained in cell coats of the CNS, and this material appears to be somewhat thickened in areas identified as synaptic clefts (Rambourg and Leblond, 1967). In none of the studies by Rambourg and Leblond, however, do the presynaptic or postsynaptic components of the synaptic material stain, although they are regularly stained with E-PTA.

We may therefore assume, on the basis of these data, that a protein comprises the synaptic specialization and accounts for the electron-dense material in the cytoplasm of presynaptic and postsynaptic elements, but differs from the composition of the more general types of cell coat. This synaptic protein is less apparent in osmium-fixed material, possibly because osmium has a tendency to deaminate and break up proteins (Hake, 1965) as it fixes the tissue. The macromolecular protein matrix may be superimposed with carbohydrates, particularly within the synaptic cleft (Pappas and Purpura, 1966; Kelly, 1967; Bondareff, 1967) (Figure 7). Material that reacts with PTA can also be seen at other types of intercellular junctions, particularly those defined by Palade (Farquhar and Palade, 1963) as the junctional complexes, such as desmosomes or maculae adhaerens. This general type of junctional complex, seen also between certain elements of the nervous system (especially between dendrites or cell bodies, and between glia and vessels [Brightman and Reese, 1969]), differs from the specialized synaptic material in that the sides of the complexes are identical; that is, there are neither dense projections nor structural polarization of the general junctional material.

Does synaptic material assist in the process of synaptic transmission or only maintain adhesion between contacting elements? This material is difficult to find in the immature cortex, although the rate and time of its development can be correlated with the onsets of organized electrical activity and neuronal maturation (Figure 8; Aghajanian and Bloom, 1967c). The onset of neuronal reactivity within tissue cultures can also be temporally related to the formation of specialized interneuronal contacts (Bunge et al., 1965), a process that takes place without requiring electrical activity (Crain et al., 1968). These data suggest that the synaptic material may in some way be associated with the synaptic process, but do not specify what the participation is likely to involve.

It is the unique presynaptic configuration of the dense projections which distinguishes morphologically defined synapses from all other forms of junctional complexes. As to the functions of these projections, there appear to be two schools of thought: in one (Gray, 1966) the projections are

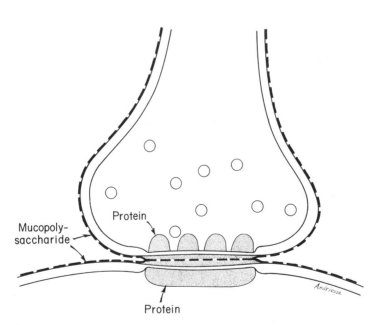

FIGURE 7 Drawing of idealized "synapse" showing the morphological features brought out by a variety of cytochemical techniques, including the specialized elements of the synaptic connection believed to be composed of proteins, and the cell-coat materials, which may be interspersed with proteins of the synaptic cleft. (Drawn by Paul Andriesse for NRP.)

sites for vesicle attachment (? quantal release); in the other (Akert et al., 1969), they serve to guide vesicles to the uncluttered "synaptic" membrane between the dense projections.

## Classification of mammalian synaptic types

Let us attempt to summarize the above data into a classification of the types of synapses seen and reported in a wide variety of fixations, regions, and species. This classification seems most easily broken into three major sets of criteria, based on the presynaptic, the postsynaptic, or the intersynaptic microstructure.

PRESYNAPTIC CLASSIFICATIONS  The first major presynaptic classification would not require electron microscopy at all, but can be confirmed with Golgi or Glees staining of

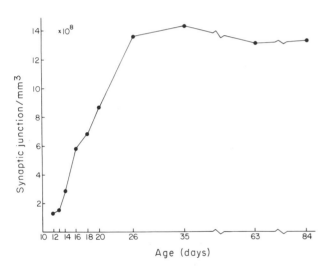

FIGURE 8  The number of synaptic junctions per cubic millimeter in the molecular layer of rat parietal cortex at various ages during development. Each point is mathematically derived from crude counts of profiles of synaptic junctions within thin sections. The number of synaptic profiles in a sample of 10 electron micrographs was counted for a single animal at each age. An area of 90 square microns was covered in each micrograph. The density of synaptic-junction "centers" was then estimated by first applying the Abercrombie correction, using the measured mean lengths of synaptic profiles. The mean lengths of synaptic profiles at each age were as follows: 12 days, 2466 Å; 13 days, 2359 Å; 14 days, 2242 Å; 16 days, 2152 Å; 18 days, 2242 Å; 20 days, 2271 Å; 35 days, 2310 Å; 63 days, 2516 Å; 84 days, 2289 Å. The thin sections (silver interference color) were assumed to be 700 Å thick. The corrected values for density were then doubled, because profiles were estimated to represent only about half of the total synaptic-junction centers within the sections. (From Aghajanian and Bloom, 1967c.)

neuroterminals, which distinguish them on the basis of their shape; that is, bouton versus club versus calyx-shaped endings of various sizes (see Bodian, 1967). A second criterion for presynaptic classification is whether the terminal is the actual termination of the axon or is one of a series of axonal enlargements that bear synaptic vesicle accumulations and make specialized contacts with dendrites or cell bodies: so-called en passant terminals. In this case, the word terminal really implies a functional point of secretion rather than a termination (Richardson, 1962, 1964); however, no experimental data exist on the assumed functional equivalence of seriatim "secretory" enlargements. The presynaptic element can also be classified structurally by vesicle shape, size, and electron opacity, or other cytochemical data on the transmitter. The appearance of the vesicle in the electron microscope may reflect, in part, the type of transmitter, the type of macromolecule binding the transmitter within the vesicle, and the protein synthetic properties of the membranes and axoplasm surrounding various types of vesicles (see above).

INTERSYNAPTIC COMPLEX  Synapses may also be classified by the morphology of the specialized contact zone. Gray, on the basis of his material, classified synapses into two types, depending on the relative predominance of the specialized postsynaptic element stained with PTA (Gray, 1959, 1969; Palay, 1967). Those junctions that exhibited little or no postsynaptic specialization appeared to be more symmetrical than did others in which the postsynaptic specialization was its full 250 Å in thickness. The recent work of Colonnier (1968) has confirmed these gradations of synaptic specialization in osmium-stained material. On the basis of our E-PTA staining of glutaraldehyde-fixed material, examined without intervening osmium exposure, Aghajanian and I (Aghajanian and Bloom, 1967c) classified the morphology of synaptic junctions into three groups, which were distinguished primarily on the basis of the size and shape of the presynaptic dense projection. It was our belief that the variations were related to the relative maturation of the synaptic complex, although it may well be that the variations are actually caused by two distinct types of synaptic contacts, one with relatively little postsynaptic material and one with the full complement of such material.

POSTSYNAPTIC ELEMENTS  Other specialized structures found within the postsynaptic elements also serve as a basis for discriminating types of synapses. These are not detailed here (see Milhaud and Pappas, 1966; and Robertson, this volume), but it may be well to point out that specialized forms of smooth endoplasmic reticulum and various configurations of postsynaptic electron-opaque material have been described without explanation.

## Are synaptic variations significant?

These factual constraints, listed with, I hope, a minimum of interpretation, indicate the basic components of synaptic ultrastructure. Clearly, there is a high degree of variability, particularly in relation to the synaptic vesicles, to the synaptic specialization, and to the size, site, and shape of the contact. Let us then inquire whether these variations in synaptic form and composition are indications of differences in function. That is, do they point to: (1) inhibition or excitation; (2) the chemical transmitter secreted (such as ACh, NE, serotonin, dopamine, amino acid, or, possibly, some combination of these); (3) the location of the synapse, for example, axo-somatic, axo-dendritic (on either primary, secondary, or tertiary dendrites or on dendritic spines), axo-axonic on the hillock, or on a distal nerve terminal, or "dendrodendritic" contacts; or (4) a property of the two cells forming the contact. That is, do certain types of pre-synaptic cell tend to form particular types of junctions with special classes of postsynaptic cells? Clearly, that they do so would seem to be demonstrated for cerebellar cortex (Larramendi et al., 1967); cerebral cortex (Colonnier, 1968) and septum (Raisman, 1969a, 1969b). This concept indicates not only the possibility that growing nerve terminals have sought out certain cellular recognition phenomena in order to establish the proper kinds of contact, but, further, that the type of presynaptic innervation can modify the functional response of the postsynaptic cell (see Guth, 1969).

In terms of our original quest for information permitting us to do experiments at the structural level on possible changes involved in plasticity and learning, it would seem essential to eliminate variability and morphology caused by intrinsic geometry, transmitter type, location of synapse on cell (positioning), and the particular elements in the circuit. If these variables could be quantitated for particular defined junctions, it should be easier to design experiments in which changes in synaptic morphology could be related to the activity state of the joined neuronal elements. Without knowing such variables, certain investigators have attempted experiments of this nature in the retina (de Robertis and Franchi, 1956; Mountford, 1963), the superior cervical ganglion (Green, 1966), the neuromuscular junction (see Hubbard and Kwanbunbumpen, 1968), and the adrenal medulla (de Robertis and Vaz Ferreira, 1957). At best, the results indicate that there is some correlation between the distribution and number of synaptic vesicles and activity, but much further experimental confirmation is needed.

## Proposals for synaptic plasticity

Kandel and Spencer (1968), in their review on cellular neurophysiological mechanisms of learning, pointed out that there are two major schools of thought on the way in which interneuronal connections could be modulated by experience. One suggested basic mechanism is the formation of new connections between nerve cells; a second is modulation (that is, changes in efficiency of communication) across existing synaptic connections.

NEW CONNECTIONS   If learning or memory requires the formation of new connections between neurons, the E-PTA staining method would seem to be ideal for experimental testing, because it simplifies the quantitation of synaptic contacts. With this in mind, we have performed several types of experiments attempting to quantitate the numbers of synaptic contacts in the molecular layer of the cerebral cortex of rats after a variety of behavioral and functional experiments (Aghajanian and Bloom, unpublished). We have compared the rate of formation of synaptic contacts in normal rats with those made hypothyroid by surgery. We have also tried to quantify synaptic contacts in the cortical molecular layer of animals subjected to "enrichment" (Krech et al., 1960). We have further compared the quantity of synaptic contacts in the visual cortex and motor cortex of rats raised in the dark with animals raised under normal lighting conditions, and have performed similar experiments on rabbits with one or both eyes sewn shut at birth, prior to the known perception of visual information (Foote, Aghajanian, and Bloom, unpublished). Such experiments are much easier to conceive and execute than to interpret. Thus far, the data have shown that little, if any, reproducible change takes place with these rather gross variations in functional loads. At best, hypothyroidism seems to delay, rather than to diminish, the total formation of synaptic contacts. The brain-enrichment procedures have had one intriguing result: although the enriched animals are different from the normal animals in numbers of contacts per unit volume of molecular-layer cortex, they are as likely to have more synapses as fewer.

The most reproducible changes we have observed have been in relation to visual deprivation, in which both the visual and the motor cortex of rats reared in the dark appear to have more synaptic contacts per unit area of molecular-layer cortex than do animals reared under normal lighting. One explanation could be that, because the rat is a nocturnal animal, being in the dark for extended periods of time caused its brain to be more active, hence forming more contacts. On the other hand, the increased number per unit area could simply be an artifact of variation in the size of nerve dendrites or terminals. Light-microscope analysis of Golgi-stained material (Globus and Scheibel, 1967; Valverde, 1967) shows that visual deprivation prevents the formation of dendritic spines in layer four of the visual cortex. Although this layer is different from that analyzed in our synapse-quantitating studies, it is possible that the apical dendrites of these cells could also have been less well developed than under normal conditions. A concomitant decrease in the size of the dendrites would make the nerve endings appear to be closer together and, hence, result in a slightly higher content of synaptic contacts per unit area. This experiment has also been performed for electron microscopy with osmium-stained material rather than PTA-stained material (Cragg, 1967, 1969). Cragg interpreted his results on both visual cortex and lateral geniculate to indicate that there were fewer nerve contacts in animals deprived of visual information. His data are based on the assumption that every process that contains synaptic vesicles is a synapse. Unless these nerve endings are followed to specialized contacts, and unless one can verify the relative diameters of the nerve endings, the assumption may not be valid. Nevertheless, Cragg (1969) reported that, under the conditions of his experiments, geniculate terminals of light-deprived rats are about 10 per cent wider, but 30 per cent less frequent. No corresponding dendrite measurements were made.

The most appropriate experiments for structural analysis of neural plasticity would include measurements of contact numbers and size, nerve terminals and their volume, and dendrites and their volume. These measurements would help significantly to eliminate apparent changes in number caused by changes in size of contacts or of contacting elements, which is the essence of the "new-versus-better-synapses" choice. Still, the primary visual relay may not be the appropriate place to seek functional plasticity, for the physiological effects of disuse are small (Burke and Hayhow, 1968). Despite these possibly significant variations resulting from visual deprivation, we have been most impressed with the relative constancy of the E-PTA contacts from animal to animal of the same species. Without proving it, this constancy suggests that the genetic composition of the neuron may well have programed the number and types of connections it is to make. If so, it might be more profitable to propose experiments in which changes of plasticity would be analyzed in terms of the modulation of existing connections.

SYNAPTIC SPROUTING OR TRANSPOSITION   We might also consider a compromise between forming new connections and modulating old ones. For instance, the genetic components of a neuron might program the basic number of axonal contacts with specific postsynaptic cells. With increases in activity, the original axon might sprout additional axon collaterals to the same postsynaptic cell. These secondary axon collaterals might even be permitted to be in contact with the cell closer to the axon hillock, where the influence of the presynaptic secretion on postsynaptic potentials would be relatively more important in terms of the discharge frequency or firing patterns of the

postsynaptic cell. One might suppose that the same situation could arise if it were possible for nerve endings that originally come in contact with neurons on distal portions of the dendrite to migrate in their position of contact to regions closer to the cell body or axon hillock. The existence of sprouting can hardly be doubted; many light-microscope studies have attested to it (Liu and Chambers, 1958; Goodman and Horel, 1966). The methods for demonstrating it most effectively, however, have required that the area investigated first be partially de-afferented surgically, allowing weeks or months for the "sprouting" to take place from the residual nerves, which are then sectioned and again stained with silver techniques. This type of experiment has recently been analyzed fine-structurally (Raisman, 1969b) in the medial and lateral septal nuclei of the rat. The results here strongly support the existence of nerve-terminal plasticity, in that certain systems of axons appear able to reinnervate synaptic sites vacated by degenerated lesioned axons of a separate system. As Raisman pointed out, if this potentiality of certain systems for terminal plasticity in any way reflects the changes associated with learning, it suggests that our concepts of the relative importance of specific connections may require re-evaluation.

BETTER SYNAPSES   With such a possible exception, then, I propose that changes in synaptic plasticity be examined in terms of changes that could account for increased or decreased efficiency of transmission across existing terminals. We might think, offhand, of several cellular mechanisms by which synaptic efficiency could be affected. We have already alluded to the possibility that synaptic activity could modulate the size of the presynaptic nerve terminal (Bazanova et al., 1966) or the postsynaptic dendrite and, as a consequence, modify the size or amount of specialized synaptic material. One could then examine nerve endings for the ratio between the size of the apposed nerve membranes and the amount of E-PTA or BI staining material. Changes in the glial configurations could, depending on what cellular functions are eventually attributed to the glia, change the diffusion rate of the transmitter or its catabolism at the site, and contribute to transmission efficiency.

There are at least three other possible modulating circumstances we could consider, although their proof depends not only on fine-structure but, perhaps more importantly, on combined pharmacological and electrophysiological experimentation. One circumstance would be that activity within a given nerve circuit could alter the quantal efficiency by which transmitter is released for a given depolarization of the nerve terminal. It is not known, for example, whether "quanta" (see Katz, 1966; also Iversen, this volume) are equal among nerve terminals of different presynaptic cells or in their effects on different postsynaptic cells.

Although the physical explanation of the quantal effect is almost universally linked conceptually to the presence of synaptic vesicles (see Katz, 1966), this relationship is unproved, and the quantal nature of spontaneous transmitter release could depend on any quantalized aspect of excitation-secretion coupling (see Douglas, 1968). Furthermore, when the nerve terminal is depolarized, the amount of transmitter released to act on the postsynaptic cell is roughly proportional to the resting transmembrane potential at that time (Bloedel et al., 1966). The biological properties that account for the transfer constant relating change in nerve-terminal membrane potential to amount of transmitter released could be a testable opportunity for plasticity in synaptic efficiency. The modulation of this constant of all-or-none spike activity might be approachable from a structural point of view, if the relationship of the dense projections to the excitation-secretion coupling process could be clarified. In an earlier proposal, Gray (1966) suggested that synaptic vesicles tend to cluster around the dense projections, and that the dense projections might be a form of attachment of the synaptic vesicles. If such were the case, one might expect that the proteinaceous matrix of the dense projection could have some enzymatic function that causes the release of transmitter or facilitates access to the covert receptor site. In neuromuscular junctions incubated in vitro to induce increases in quantal release, increased numbers of presumed dense projections (actually seen as presynaptic membrane thickenings) were reported (Hubbard and Kwanbunbumpen, 1968).

Synaptic efficiency could also be modulated if the receptiveness of the postsynaptic element could be altered by changes in activity of the presynaptic element. In experiments on the postsynaptic receptivity of regenerating and reinnervating neuromuscular junctions (see Guth, 1969), the presence of the presynaptic terminal appears to decrease the receptive area of the postsynaptic element. Recently, we found that, in the immature cerebellar cortex of the rat (Woodward, Hoffer, Siggins, and Bloom, in preparation), the Purkinje cells are responsive to several of the potential transmitters before any synaptic contacts have appeared. Such data could indicate that the postsynaptic cells begin by being overly receptive and subsequently localize their responsiveness with nerve activity, as occurs with neuromuscular junctions. The factors responsible for such changes in size and quality of receptivity would also be suitable for plasticity.

A third, and equally unsubstantiated, proposal for synaptic modulation is that the transmitter substance may be able to do more than simply generate typical rapid postsynaptic ionic potentials. In addition to the more commonly accepted type of postsynaptic potentials that last 10 to 100 milliseconds (see Eccles, 1964), long-lasting postsynaptic potentials of seconds have been described recently from many portions of the autonomic nervous system of frogs and certain mammals (Tosaka et al., 1968; Koketsu et al., 1968), as well as in the molluscan nervous system (Pinsker and Kandel, 1969; Kerkut et al., 1969). These long-lasting postsynaptic potentials appear in some ways to be metabolic, in that they can be influenced by metabolic poisons and by temperature, but not by ions. Most importantly, they do not exhibit concurrent conductance changes to ions (Kobayashi and Libet, 1968). One such prolonged postsynaptic potential may be the result of the influence of the transmitter substance (in this case, dopamine) in stimulating an electrogenic pump, which results in a prolonged hyperpolarizing postsynaptic potential (Kerkut et al., 1969). From our studies on the cerebellar cortex (Siggins et al., 1969), we have proposed that one explanation for norepinephrine-induced depression of cerebellar Purkinje-cell activity could be by way of intermediation with cyclic adenosine monophosphate (C-AMP). The C-AMP reproduces the changes in spontaneous firing patterns seen with NE, although the onset and offset of the response are much faster with C-AMP. Furthermore, the response to NE can be blocked by prostaglandins, which are known to block the ability of NE to stimulate adenyl cyclase, the enzyme that synthesizes C-AMP. Additionally, the response to NE is potentiated by such inhibitors of phosphodiesterase as theophylline and aminophylline. By putting these bits of information together, we could postulate that the transmitter substance could somehow generate a prolonged postsynaptic potential in addition to the immediate postsynaptic ionic potential; the prolonged potential shift could arise by stimulating a metabolic electrogenic pump to hyperpolarize or depolarize the membrane over long periods of time. Such prolonged shifts in membrane potential could "modulate" the postsynaptic element and make it more-or-less receptive to its incoming signals. This proposal could account for the concept of neuromodulation, often postulated without precise, testable cellular mechanisms.

### The synapse gap: its causes and cures

The prolonged descriptive history of electron microscopy sometimes frustrates the reader anxious to get on with experimental, fine-structural neurosciences. The preceding proposals require us to clarify certain of the descriptive variables among nerve terminals before we can hope to analyze for those changes that might take place with synaptic activity. If we were to propose the most ideal circumstances, we would require a system in which we could identify, with the electron microscope, the distinct type of nerve ending on a special type of postsynaptic cell (see Hendrickson, 1969, for a new method) the output of which is behaviorally important. We would hope to identify the transmitter or transmitters by which cell A can influence the activity of cell B, and to show that the connectivity of cell A to cell B could be modified by use. We would need to verify the existence of the transmitter by some cytochemical technique within these nerve terminals, as well as to satisfy the neurophysiological and neuropharmacological criteria of transmitter identification that are described in detail elsewhere (for reviews, see Bloom and Giarman, 1968a, 1968b).

But would such data establish the feasibility of ultrastructural analysis of neural plasticity? From our combinations of electrophysiology and fine structure, we believe that the responsiveness to NE and C-AMP that has been observed in Purkinje cells in the cerebellar cortex of the rat suggests a noradrenergic synapse. By using electron-microscopic techniques we can identify these nerve terminals studied by the electrophysiological approach. Yet, would these data on the potential noradrenergic innervation to the cerebellar cortex satisfy the criteria we have proposed here for a model pathway in which to examine the effects of synaptic plasticity? The answer to this question must be no, because, even though we can identify the site of the nerve ending on the postsynaptic cell and the cell bodies most likely to give rise to the synapse, the fact still remains that we do not know that Purkinje cells can "learn" anything in response to activity through the norepinephrine synapse. According to some (Young, 1966), the large cells, which are the ones most commonly recorded by electrophysiological methods, are the least likely to indicate modulating changes with experience or memory.

We may find, therefore, at the end of an infinitely difficult period of describing, organizing, and classifying nerve-ending patterns and particular synaptic connections, that those cells on which experiential modulation of synaptic activity takes place are the cells that are most difficult, if not impossible, to record with conventional electrophysiological methods. Thus, the main cause of the "synapse gap" in a molecular explanation of neural plasticity is that we do not yet know where to look nor what to look for. I hope that the factors pointed out here will at least serve to stimulate clarification. I believe the synaptic E-PTA material is protein; its implications are protean. Whether it will fill the "synapse gap" remains to be shown.

## Conclusions

At the cellular level of inquiry, the synaptic connections of neurons present a specifiable restrictive site at which to analyze possible plastic changes induced by activity in a particular nerve circuit. Fine-structural descriptions provide a number of morphological indexes against which the activity-linked changes might be measured, such as the size and shape of the terminals, their relative position on the postsynaptic cell, the number, shape, and size of the pre-terminal vesicles, and the relative size and morphology of the synaptic specialization. To this purely morphologic base we can add both the histochemical dimensions, by attempting to identify the transmitter stored in the vesicles, and the physiologic dimensions, by observing the response of the postsynaptic cell to this transmitter administered by microelectrophoresis. Furthermore, we can approach the molecular level of inquiry by isolating the macromolecular proteins constituting the synaptic specialization; in the dynamic concept of the synapse, these macromolecules may influence the efficiency of transjunctional events. To close the synapse gap in our understanding of the link between behavioral and cellular events, however, the primary requirement is a bona fide synaptic connection in the brain at which physiologically induced activity can reveal verified plasticity in function.

## REFERENCES

AGHAJANIAN, G. K., and F. E. BLOOM, 1966. Electron microscopic autoradiography of rat hypothalamus after intraventricular H³-norepinephrine. *Science* (*Washington*) 153: 308–310.

AGHAJANIAN, G. K., and F. E. BLOOM, 1967a. Localization of tritiated serotonin in rat brain by electron-microscopic autoradiography. *J. Pharmacol. Exp. Ther.* 156: 23–30.

AGHAJANIAN, G. K., and F. E. BLOOM, 1967b. Electron-microscopic localization of tritiated norepinephrine in rat brain: Effect of drugs. *J. Pharmacol. Exp. Ther.* 156: 407–416.

AGHAJANIAN, G. K., and F. E. BLOOM, 1967c. The formation of synaptic junctions in developing rat brain: A quantitative electron microscopic study. *Brain Res.* 6: 716–727.

AGHAJANIAN, G. K., F. E. BLOOM, and M. H. SHEARD, 1969. Electron microscopy of degeneration within the serotonin pathway of rat brain. *Brain Res.* 13: 266–273.

AKERT, K., and K. PFENNINGER, 1969. Synaptic fine structure and neural dynamics. *Symp. Int. Soc. Cell Biol.* 8: 245–260.

AKERT, K., H. MOOR, K. PFENNINGER, and C. SANDRI, 1969. Contributions of new impregnation methods and freeze etching to the problems of synaptic fine structure. *Progr. Brain Res.* 31: 223–240.

AKERT, K., and C. SANDRI, 1968. An electron-microscopic study of zinc iodide-osmium impregnation of neurons. I. Staining of synaptic vesicles at cholinergic junctions. *Brain Res.* 7: 286–295.

APPEL, S. H., L. AUTILIO, B. W. FESTOFF, and A. V. ESCUETA, 1969. Biochemical studies of synapses in vitro. III. Ionic activation of protein synthesis. *J. Biol. Chem.* 244: 3166–3172.

BAK, I. J., 1965. Electron microscopic observations in the substantia nigra of mouse during reserpine administration. *Experientia* (*Basel*) 21: 568–570.

BAK, I. J., 1967. The ultrastructure of the substantia nigra and caudate nucleus of the mouse and the cellular localization of catecholamines. *Exp. Brain Res.* 3: 40–57.

BARGMANN, W., 1966. Neurosecretion. *Int. Rev. Cytol.* 19: 183–201.

BARONDES, S. H., 1968. Further studies of the transport of protein to nerve endings. *J. Neurochem.* 15: 343–350.

BAZANOVA, I. S., S. A. EVDOKIMOV, V. N. MAIOROV, O. S. MERKULOVA, and V. N. CHERNIGOVSKII, 1966. Morphological and electrical changes in interneuronal synapses during passage of rhythmic impulses. *Fed. Proc.* 25: T187–T190.

BEST, J. B., and J. NOEL, 1969. Complex synaptic configurations in planarian brain. *Science* (*Washington*) 164: 1070–1071.

BLOEDEL, J., P. W. GAGE, R. LLINÁS, and D. M. J. QUASTEL, 1966. Transmitter release at the squid giant synapse in the presence of tetrodotoxin. *Nature* (*London*) 212: 49–50.

BLOOM, F. E., and G. K. AGHAJANIAN, 1966. Cytochemistry of synapses: Selective staining for electron microscopy. *Science* (*Washington*) 154: 1575–1577.

BLOOM, F. E., and G. K. AGHAJANIAN, 1968a. An electron microscopic analysis of large granular synaptic vesicles of the brain in relation to monoamine content. *J. Pharmacol. Exp. Ther.* 159: 261–273.

BLOOM, F. E., and G. K. AGHAJANIAN, 1968b. Fine structural and cytochemical analysis of the staining of synaptic junctions with phosphotungstic acid. *J. Ultrastruct. Res.* 22: 361–375.

BLOOM, F. E., and G. K. AGHAJANIAN, 1968c. An osmiophilic substance in brain synaptic vesicles not associated with catecholamine content. *Experientia* (*Basel*) 24: 1225–1227.

BLOOM, F. E., and R. J. BARRNETT, 1966. Fine structural localization of noradrenaline in vesicles of autonomic nerve endings. *Nature* (*London*) 210: 599–601.

BLOOM, F. E., and N. J. GIARMAN, 1968a. Physiologic and pharmacologic considerations of biogenic amines in the nervous system. *Annu. Rev. Pharmacol.* 8: 229–258.

BLOOM, F. E., and N. J. GIARMAN, 1968b. Current status of neurotransmitters. *Annu. Rep. Med. Chem.* 1967: 264–278.

BLOOM, F. E., and N. J. GIARMAN, 1970. The effects of *p*-Cl-phenylalanine on the content and cellular distribution of 5-HT in the rat pineal gland: Combined biochemical and electron microscopic observations. *Biochem. Pharmacol.* 19: 1213–1219.

BODIAN, D., 1966a. Electron microscopy: Two major synaptic types on spinal motoneurons. *Science* (*Washington*) 151: 1093–1094.

BODIAN, D., 1966b. Synaptic types on spinal motoneurons: An electron microscopic study. *Bull. Johns Hopkins Hosp.* 119: 16–45.

BODIAN, D., 1967. Neurons, circuits, and neuroglia. *In* The Neurosciences: A Study Program (G. C. Quarton, T. Melnechuk, and F. O. Schmitt, editors). The Rockefeller University Press, New York, pp. 6–24.

BONDAREFF, W., 1966. Localization of α-methylnorepinephrine in sympathetic nerve fibers of the pineal body. *Exp. Neurol.* 16: 131–135.

BONDAREFF, W., 1967. An intercellular substance in rat cerebral cortex: Submicroscopic distribution of ruthenium red. *Anat. Rec.* 157: 527–535.

BRIGHTMAN, M. W., and T. S. REESE, 1969. Junctions between intimately apposed cell membranes in the vertebrate brain. *J. Cell Biol.* 40: 648–677.

BUNGE, R. P., M. B. BUNGE, and E. R. PETERSON, 1965. An electron microscope study of cultured rat spinal cord. *J. Cell Biol.* 24: 163–191.

BURKE, W., and W. R. HAYHOW, 1968. Disuse in the lateral geniculate nucleus of the cat. *J. Physiol.* (*London*) 194: 495–519.

CASTEJÓN, O. J., and G. M. VILLEGAS, 1964. Fine structure of the synaptic contacts in the stellate ganglion of the squid. *J. Ultrastruct. Res.* 10: 585–598.

COLBURN, R. W., and J. W. MAAS, 1965. Adenosine triphosphate-metal-norepinephrine ternary complexes and catecholamine binding. *Nature* (*London*) 208: 37–41.

COLONNIER, M., 1968. Synaptic patterns on different cell types in the different laminae of the cat visual cortex. An electron microscope study. *Brain Res.* 9: 268–287.

CRAGG, B. G., 1967. Changes in the visual cortex on first exposure of rats to light. *Nature* (*London*) 215: 251–253.

CRAGG, B. G., 1969. The effects of vision and dark-rearing on the size and density of synapses in the lateral geniculate nucleus measured by electron microscopy. *Brain Res.* 13: 53–67.

CRAIN, S. M., M. B. BORNSTEIN, and E. R. PETERSON, 1968. Maturation of cultured embryonic CNS tissues during chronic exposure to agents which prevent bioelectric activity. *Brain Res.* 8: 363–372.

DEROBERTIS, E., and A. VAZ FERREIRA, 1957. Submicroscopic changes of the nerve endings in the adrenal medulla after stimulation of the splanchnic nerve. *J. Biophys. Biochem. Cytol.* 3: 611–614.

DEROBERTIS, E., and C. M. FRANCHI, 1956. Electron microscope observations on synaptic vesicles in synapses of the retinal rods and cones. *J. Biophys. Biochem. Cytol.* 2: 307–318.

DEROBERTIS, E., A. PELLEGRINO DE IRALDI, G. RODRÍGUEZ DE LORES ARNAIZ, and L. M. ZIEHER, 1965. Synaptic vesicles from the rat hypothalamus. Isolation and norepinephrine content. *Life Sci.* 4: 193–201.

DESCARRIES, L., and B. DROZ, 1968. Incorporation de noradrenaline-³H(NA-³H) dans le système nerveux central du rat adulte. Étude radio-autographique en microscopie électronique. *Compt. Rend. Acad. Sci., Ser. D* 266: 2480–2482.

DEVINE, C. E., and LAVERTY, R., 1968. Fixation for electron microscopy and the retention of ³H-noradrenaline by tissues. *Experientia* (*Basel*) 24: 1156–1157.

DOUGLAS, W. W., 1968. Stimulus-secretion coupling: The concept and clues from chromaffin and other cells. *Brit. J. Pharmacol.* 34: 451–474.

ECCLES, J. C., 1964. The Physiology of Synapses. Springer-Verlag, Berlin.

ECCLES, J. C., M. ITO, and J. SZENTHÁGOTHAI, 1967. The Cerebellum as a Neuronal Machine. Springer-Verlag, New York.

ENESTRÖM, S., and C. SVALANDER, 1967. Liquid formaldehyde in catecholamine studies. A new approach to the morphological localization of monoamines in the adrenal medulla and the supraoptic nucleus of the rat. *Histochemie* 8: 155–163.

ERÄNKÖ, O., 1967. Histochemistry of nervous tissues: Catecholamines and cholinesterases. *Annu. Rev. Pharmacol.* 7: 203–222.

FARQUHAR, M. G., and G. E. PALADE, 1963. Junctional complexes in various epithelia. *J. Cell Biol.* 17: 375–412.

FUKAMI, Y., 1969. Two types of synaptic bulb in snake and frog spinal cord: The effect of fixation. *Brain Res.* 14: 137–145.

FUXE, K., T. HÖKFELT, and O. NILSSON, 1965. A fluorescence and electronmicroscopic study on certain brain regions rich in monoamine terminals. *Amer. J. Anat.* 117: 33–45.

FUXE, K., T. HÖKFELT, O. NILSSON, and S. REINIUS, 1966. A fluorescence and electron microscopic study on central monoamine nerve cells. *Anat. Rec.* 155: 33–40.

GLOBUS, A., and A. B. SCHEIBEL, 1967. The effect of visual deprivation on cortical neurons: A Golgi study. *Exp. Neurol.* 19: 331–345.

GOODMAN, D. C., and J. A. HOREL, 1966. Sprouting of optic tract projections in the brain stem of the rat. *J. Comp. Neurol.* 127: 71–88.

GRAY, E. G., 1959. Axo-somatic and axo-dendritic synapses of the cerebral cortex: An electron microscope study. *J. Anat.* 93: 420–432.

GRAY, E. G., 1966. Problems of interpreting the fine structure of vertebrate and invertebrate synapses. *Int. Rev. Gen. Exp. Zool.* 2: 139–170.

GRAY, E. G., 1969. Electron microscopy of excitatory and inhibitory synapses: A brief review. *Progr. Brain Res.* 31: 141–155.

GRAY, E. G., and R. W. GUILLERY, 1966. Synaptic morphology in the normal and degenerating nervous system. *Int. Rev. Cytol.* 19: 111–182.

GREEN, K., 1966. Electron microscopic observations on the relationship between synthesis of synaptic vesicles and acetylcholine. *Anat. Rec.* 154: 351 (abstract).

GRILLO, M. A., 1966. Electron microscopy of sympathetic tissues. *Pharmacol. Rev.* 18: 387–399.

GRILLO, M. A., and S. L. PALAY, 1962. Granule-containing vesicles in the autonomic nervous system. *In* Electron Microscopy, 5th International Congress for Electron Microscopy, Philadelphia, 1962 (S. S. Breese, Jr., editor). Academic Press, New York, vol. 2, p. U-1.

GUTH, L., 1969. "Trophic" effects of vertebrate neurons. *Neurosci. Res. Program Bull.* 7 (no. 1).

HAKE, T., 1965. Studies on the reactions of $OSO_4$ and $KMnO_4$ with amino acids, peptides, and proteins. *Lab. Invest.* 14: 1208–1212.

HARRISON, W. H., W. W. WHISLER, and B. J. HILL, 1968. Catecholamine oxidation and ionization properties indicated from the $H^+$ release, tritium exchange, and spectral changes which occur during ferricyanide oxidation. *Biochemistry* 7: 3089–3094.

HASHIMOTO, Y., S. ISHII, Y. OHI, N. SHIMIZU, and R. IMAIZUMI, 1965. Effect of DOPA on the norepinephrine and dopamine contents and on the granulated vesicles of the hypothalamus of reserpinized rats. *Jap. J. Pharmacol.* 15: 395–400.

HENDRICKSON, A., 1969. Electron microscopic radioautography: Identification of origin of synaptic terminals in normal nervous tissue. *Science (Washington)* 165: 194–196.

HÖKFELT, T., 1967. On the ultrastructural localization of noradrenaline in the central nervous system of the rat. *Z. Zellforsch. Mikroskop. Anat.* 79: 110–117.

HÖKFELT, T., 1968. *In vitro* studies on central and peripheral monoamine neurons at the ultrastructural level. *Z. Zellforsch. Mikroskop. Anat.* 91: 1–74.

HÖKFELT, T., and G. JONSSON. 1968. Studies on reaction and binding of monoamines after fixation and processing for electron microscopy with special reference to fixation with potassium permanganate. *Histochemie* 16: 45–67.

HUBBARD, J. I., and S. KWANBUNBUMPEN, 1968. Evidence for the vesicle hypothesis. *J. Physiol. (London)* 194: 407–420.

ISHII, S., N. SHIMIZU, M. MATSUOKA, and R. IMAIZUMI, 1965. Correlation between catecholamine content and numbers of granulated vesicles in rabbit hypothalamus. *Biochem. Pharmacol.* 14: 183–184.

JAIM-ETCHEVERRY, G., and L. M. ZIEHER, 1968. Cytochemistry of 5-hydroxytryptamine at the electron microscope level. II. Localization in the autonomic nerves of the rat pineal gland. *Z. Zellforsch. Mikroskop. Anat.* 86: 393–400.

JAIM-ETCHEVERRY, G., and L. M. ZIEHER, 1969. Selective demonstration of a type of synaptic vesicle with phosphotungstic acid staining. *J. Cell Biol.* 42: 855–859.

JAIM-ETCHEVERRY, G., and L. M. ZIEHER, 1970. Ultrastructual aspects of neurotransmitter storage in adrenergic nerves. *In* Proceedings of the 1st International Symposium on Cell Biology and Cytopharmacology (in press).

JONES, D. G., 1969. The morphology of the contact region of vertebrate synaptosomes. *Z. Zellforsch. Mikroskop. Anat.* 95: 263–279.

KANDEL, E. R., and W. A. SPENCER, 1968. Cellular neurophysiological approaches in the study of learning. *Physiol. Rev.* 48: 65–134.

KARCZMAR, A. G., 1969. Is the central cholinergic nervous system overexploited? *Fed. Proc.* 28: 147–157.

KATZ, B., 1966. Nerve, Muscle, and Synapse. McGraw-Hill, New York, pp. 133–151.

KELLY, D. E., 1967. Fine structure of cell contact and the synapse. *Anesthesiology* 28: 6–30.

KERKUT, G. A., L. C. BROWN, and R. J. WALKER, 1969. Post synaptic stimulation of the electrogenic sodium pump. *Life Sci. Pt. I* 8: 297–300.

KOBAYASHI, H., and B. LIBET, 1968. Generation of slow postsynaptic potentials without increases in ionic conductance. *Proc. Nat. Acad. Sci. U. S. A.* 60: 1304–1311.

KOELLE, G. B., 1969. Significance of acetylcholinesterase in central synaptic transmission. *Fed. Proc.* 28: 95–100.

KOKETSU, K., S. NISHI, and H. SOEDA, 1968. Acetylcholine-potential of sympathetic ganglion cell membrane. *Life. Sci. Pt. I* 7: 741–749.

KRECH, D., M. R. ROSENZWEIG, and E. L. BENNETT, 1960. Effects of environmental complexity and training on brain chemistry. *J. Comp. Physiol. Psychol.* 53: 509–519.

LARRAMENDI, L. M. H., L. FICKENSCHER, and N. LEMKEY-JOHNSTON, 1967. Synaptic vesicles of inhibitory and excitatory terminals in the cerebellum. *Science (Washington)* 156: 967–969.

LENN, N. J., 1965. Electron microscopic observations on monoamine-containing brain stem neurons in normal and drug-treated rats. *Anat. Rec.* 153: 399–406.

LENN, N. J., 1967. Localization of uptake of tritiated norepinephrine by rat brain in vivo and in vitro using electron microscopic autoradiography. *Amer. J. Anat.* 120: 377–389.

LENN, N. J., and T. S. REESE, 1966. The fine structure of nerve endings in the nucleus of the trapezoid body and the ventral cochlear nucleus. *Amer. J. Anat.* 118: 375–389.

LIU, C. N., and W. W. CHAMBERS, 1958. Intraspinal sprouting of dorsal root axons. *A. M. A. Arch. Neurol. Psychiat.* 79: 46–61.

MAHLER, H. R., 1969. Protein turnover and synthesis: Relation to synaptic function. *Advan. Biochem. Psychopharmacol.* 1: 49–70.

MATSUOKA, M., S. ISHII, N. SHIMIZU, and R. IMAIZUMI, 1965. Effect of Win 18501-2 on the content of catecholamines and the number of catecholamine-containing granules in the rabbit hypothalamus. *Experientia* (*Basel*) 21: 1–5.

MILHAUD, M., and G. D. PAPPAS, 1966. The fine structure of neurons and synapses of the habenula of the cat with special reference to subjunctional bodies. *Brain Res.* 3: 158–173.

MOUNTFORD, S., 1963. Effects of light and dark adaptation on the vesicle populations of receptor-bipolar synapses. *J. Ultrastruct. Res.* 9: 403–418.

PALADE, G. E., P. SIEKEVITZ, and L. G. CARO, 1961. Structure, chemistry and function of the pancreatic exocrine cell. *In* Exocrine Pancreas/A Ciba Foundation Symposium (A. V. S. de Reuck and M. P. Cameron, editors). Little, Brown, Boston, pp. 23–49.

PALAY, S. L., 1956. Synapses in the central nervous system. *J. Biophys. Biochem. Cytol.* 2 (suppl., no. 4, pt. 2): 193–202.

PALAY, S. L., 1958. The morphology of synapses in the central nervous system. *Exp. Cell Res., Suppl.* 5: 275–293.

PALAY, S. L., 1967. Principles of cellular organization in the nervous system. *In* The Neurosciences: A Study Program (G. C. Quarton, T. Melnechuk, and F. O. Schmitt, editors). The Rockefeller University Press, New York, pp. 24–31.

PAPPAS, G. D., and PURPURA, D. P., 1966. Distribution of colloidal particles in extracellular space and synaptic cleft substance of mammalian cerebral cortex. *Nature* (*London*) 210: 1391–1392.

PEASE, D. C., 1966. Polysaccharides associated with the exterior surface of epithelial cells: Kidney, intestine, brain. *J. Ultrastruct. Res.* 15: 555–588.

PELLEGRINO DE IRALDI, A., H. FARINI DUGGAN, and E. DE ROBERTIS, 1963. Adrenergic synaptic vesicles in the anterior hypothalamus of the rat. *Anat. Rec.* 145: 521–531.

PELLEGRINO DE IRALDI, A., L. M. ZIEHER, and E. DE ROBERTIS, 1965. Ultrastructure and pharmacological studies of nerve endings in the pineal organ. *Progr. Brain Res.* 10: 389–422.

PFENNINGER, K., C. SANDRI, K. AKERT, and C. H. EUGSTER, 1969. Contribution to the problem of structural organization of the presynaptic area. *Brain Res.* 12: 10–18.

PINSKER, H., and E. R. KANDEL, 1969. Synaptic activation of an electrogenic sodium pump. *Science* (*Washington*) 163: 931–935.

RAISMAN, G., 1969a. A comparison of the mode of termination of the hippocampal and hypothalamic afferents to the septal nuclei as revealed by electron microscopy of degeneration. *Exp. Brain Res.* 7: 317–343.

RAISMAN, G., 1969b. Neuronal plasticity in the septal nuclei of the adult rat. *Brain Res.* 14: 25–48.

RAMBOURG, A., and C. P. LEBLOND, 1967. Electron microscope observations on the carbohydrate-rich cell coat present at the surface of cells in the rat. *J. Cell Biol.* 32: 27–53.

RICHARDSON, K. C., 1962. The fine structure of autonomic nerve endings in smooth muscle of the rat vas deferens. *J. Anat.* 96: 427–442.

RICHARDSON, K. C., 1964. The fine structure of the albino rabbit iris with special reference to the identification of adrenergic and cholinergic nerves and nerve endings in its intrinsic muscles. *Amer. J. Anat.* 114: 173–205.

RICHARDSON, K. C., 1966. Electron microscopic identification of autonomic nerve endings. *Nature* (*London*) 210: 756.

SHERIDAN, W. F., and R. J. BARRNETT, 1969. Cytochemical studies on chromosome ultrastructure. *J. Ultrastruct. Res.* 27: 216–229.

SHIMIZU, U., and S. ISHII, 1964. Electron microscopic observation of catecholamine-containing granules in the hypothalamus and area postrema and their changes following reserpine injection. *Arch. Histol. Jap.* 24: 489–497.

SIGGINS, G. R., B. J. HOFFER, and F. E. BLOOM, 1969. Cyclic adenosine monophosphate: Possible mediator for norepinephrine effects on cerebellar Purkinje cells. *Science* (*Washington*) 165: 1018–1020.

SILVERMAN, L., and D. GLICK, 1969. The reactivity and staining of tissue proteins with phosphotungstic acid. *J. Cell Biol.* 40: 761–767.

SMITH, D. S., 1965. Synapses in the insect nervous system. *In* The Physiology of the Insect Central Nervous System (J. E. Treherne and J. W. L. Beament, editors). Academic Press, New York, pp. 39–58.

TAXI, J., 1961. Étude au microscope électronique de ganglions sympathetique de grenouille. *Compt. Rend. Ass. Anat.* 47: 786–797.

TOSAKA, T., S. CHICHIBU, and B. LIBET, 1968. Intracellular analysis of slow inhibitory and excitatory postsynaptic potentials in sympathetic ganglia of the frog. *J. Neurophysiol.* 31: 396–409.

TRANZER, J. P., and H. THOENEN, 1967. Electron microscopic localization of 5-hydroxydopamine (3,4,5-trihydroxy-phenyl-ethyl-amine), a new "false" sympathetic transmitter. *Experientia* (*Basel*) 23: 743–745.

TRANZER, J. P., and H. THOENEN, 1968. An electron microscopic study of selective acute degeneration of sympathetic nerve terminals after administration of 6-hydroxydopamine. *Experientia* (*Basel*) 24: 155–156.

TRANZER, J. P., H. THOENEN, R. L. SNIPES, and J. G. RICHARDS, 1969. Recent developments on the ultrastructural aspects of adrenergic nerve endings in various experimental conditions. *Progr. Brain Res.* 31: 33–46.

UCHIZONO, K., 1965. Characteristics of excitatory and inhibitory synapses in the central nervous system of the cat. *Nature* (*London*) 207: 642–643.

VALVERDE, F., 1967. Apical dendritic spines of the visual cortex and light deprivation in the mouse. *Exp. Brain Res.* 3: 337–352.

VAN HARREVALD, A., J. CROWELL, and S. K. MALHOTRA, 1965. A study of extracellular space in central nervous tissue by freeze-substitution. *J. Cell Biol.* 25: 117–137.

VAN ORDEN, L. S., III, K. G. BENSCH, and N. J. GIARMAN, 1967a. Histochemical and functional relationships of catecholamines in adrenergic nerve endings. II. Extravesicular norepinephrine. *J. Pharmacol. Exp. Ther.* 155: 428–439.

VAN ORDEN, L. S., III, K. G. BENSCH, S. Z. LANGER, and U. TRENDELENBURG, 1967b. Histochemical and fine structural aspects of the onset of denervation supersensitivity in the nictitating membrane of the spinal cat. *J. Pharmacol. Exp. Ther.* 157: 274–283.

VAN ORDEN, L. S., III, F. E. BLOOM, R. J. BARRNETT, and N. J. GIARMAN, 1966. Histochemical and functional relationships of catecholamines in adrenergic nerve endings. I. Participation of granular vesicles. *J. Pharmacol. Exp. Ther.* 154: 185–199.

WALBERG, F., 1966. Elongated vesicles in terminal boutons of the central nervous system, a result of aldehyde fixation. *Acta Anat.* 65: 224–235.

WESTRUM, L. E., 1965. On the origin of synaptic vesicles in cerebral cortex. *J. Physiol.* (*London*) 179: 4P–6P.

WHITTAKER, V. P., 1966. Some properties of synaptic membranes isolated from the central nervous system. *Ann. N. Y. Acad. Sci.* 137: 982–998.

WOLFE, D. E., L. T. POTTER, K. C. RICHARDSON, and J. AXELROD, 1962. Localizing tritiated norepinephrine in sympathetic axons by electron microscopic autoradiography. *Science* (*Washington*) 138: 440–442.

YOUNG, J. Z., 1966. The Memory System of the Brain. University of California Press, Berkeley, p. 128.

# 66 Brain Glycomacromolecules and Interneuronal Recognition

## SAMUEL H. BARONDES

THIS PAPER is concerned with one major hypothesis: that polysaccharides on neuronal surfaces play a crucial role in defining and regulating interneuronal synaptic relationships. This hypothesis has not wanted for supporters. For several decades, students of embryogenesis (e.g., Holtfreter, 1939; Weiss, 1941; Moscona, 1963, 1968; Steinberg, 1963; Grobstein, 1967; Heinmets, 1968) have suggested that intercellular recognition is determined by substances on the surface of cells. Because glycoproteins and glycolipids (referred to here collectively as either glycomacromolecules or heterosaccharides) are characteristic constituents of cell surfaces, they have frequently been proposed as candidates for this function (Gesner and Ginsburg, 1964; Dische, 1966; Kalckar, 1965; Brunngraber, 1969). Bogoch (1965, 1968) has emphasized their possible importance in the brain.

At present, the evidence in support of this hypothesis is, at best, indirect. It is being considered largely to stimulate investigation of the mechanism of interneuronal recognition. This problem, although generally identified as one of the most critical in neurobiology, has not received the experimental attention it deserves because of the anticipated technical difficulties. The purpose of this paper is to emphasize that glycomacromolecules may provide the key to the solution of this problem and that relevant biochemical experiments are possible.

The fact that polysaccharides, like those that determine blood-group specification, may be linked covalently either to lipid or protein (Watkins, 1967), underscores the importance of considering both these classes of glycomacromolecules in a discussion of cell surfaces. Because glycolipids of the nervous system are already being studied extensively for other reasons, I emphasize the glycoproteins, but to do so in no way disregards the possible role of glycolipids. Indeed, gangliosides, which are concentrated in neurons (Lowden and Wolfe, 1964)—particularly at nerve endings (Wiegandt, 1967)—merit special consideration in a discussion of the neuronal surface. Yet glycoproteins may prove to be of even greater significance in determining the properties of synaptic surfaces, because they appear to possess greater structural variability and because of the enormous volume occupied by their long, extensively hydrated polysaccharide chains (Schubert, 1964). When the properties of a membrane surface (e.g., Figure 1) are considered, the polysaccharide side chains of glycoproteins deserve particular notice not only because they may stick out fairly far but also because they may occupy enormous volumes caused by hydration. Adjacent cells would be expected to interact at this most superficial level.

Before I discuss what is known about glycoproteins in the nervous system, it is necessary to review general studies of their structure, metabolism, cellular localization and function. For more detailed summaries, the reader is directed to recent volumes edited by Rossi and Stoll (1968), Davis and Warren (1967), and Manson (1968); to recent reviews by Ginsburg and Neufeld (1969) and Cook (1968); and to a Neurosciences Research Program Bulletin on *Brain Cell Microenvironment* (Schmitt and Samson, 1969).

SAMUEL H. BARONDES Departments of Psychiatry and Molecular Biology, Albert Einstein College of Medicine, Bronx. New York. *Present Address:* Department of Psychiatry, University of California at San Diego, La Jolla, California

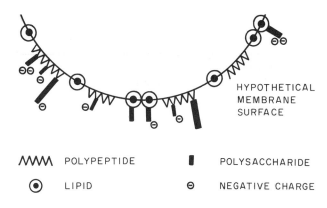

ᴧᴧᴧ POLYPEPTIDE ▮ POLYSACCHARIDE

◉ LIPID ⊖ NEGATIVE CHARGE

FIGURE 1 Hypothetical neuronal surface. Protruding carbohydrate residues from glycolipids and glycoproteins are shown, as envisioned by Lehninger (1968). Because of their greater length, the polysaccharide chains of glycoproteins may protrude farthest and may constitute the outermost portion of the cell surface.

## General studies of glycoproteins

STRUCTURE OF GLYCOPROTEINS Confusing terminologies of the past have been superseded by the definition of glycoproteins as all molecules that contain carbohydrate residues covalently linked to protein (Jeanloz, 1968). A special class of these substances, the acidic mucopolysaccharides, consists of a long chain of two repeating sugars covalently linked to protein. The other glycoproteins have less regular structures. Because the component sugars exist in anomeric forms and have multiple potential sites for linear and branched linkages, the potentialities for structurally unique glycoproteins are enormous (Table I), although the polysaccharides of mammalian glycoproteins are generally constructed from only six sugars, some of which are found in characteristic positions (Table I).

The characterized glycoproteins differ markedly in structure (Table II). Some contain a large number of separate disaccharides per molecule of protein (ovine submaxillary glycoprotein), a single longer chain of carbohydrates per molecule of protein (hen ovalbumin), several identical, fairly long chains of carbohydrates per molecule of protein (fetuin), or a number of chains of two types per molecule of protein (thyroglobulin). The similar carbohydrate units may not be precisely identical either within the same molecule or within different molecules from the same tissue. This slight heterogeneity may be due to variability during synthesis or to variable degrees of degradation either in vivo or as an artifact of the isolation procedure. The function of this heterogeneity is not known.

METABOLISM Knowledge about the site of synthesis, mechanism of synthesis, and site of degradation of glyco-

proteins is only now being accumulated. It is believed that carbohydrates are not added to glycoproteins until after the completion of the entire polypeptide chain. Although some polypeptides released from polysomes by puromycin have been found to contain glucosamine residues (Lawford and Schacter, 1966; Molnar and Sy, 1967), there is general agreement that carbohydrate incorporation is not dictated by ribonucleic acid templates. Rather, the addition of sugars is determined by the structure of the protein and the presence of appropriate glycosyl transferases in the cell. These enzymes are believed to be concentrated in the smooth endoplasmic reticulum (Hagopian et al., 1968).

The generally accepted position is that, on release of polypeptides from polysomes, appropriate sugars are sequentially incorporated from nucleotide sugars by specific glycosyl transferases. The sugars are presumably added one at a time, so that incomplete glycoproteins become substrates for the next enzyme in the sequence. Uridine diphosphate (UDP)-trisaccharides have, however, been found (Jourdian and Roseman, 1963) that could be intermediates in transfer of the trisaccharides to appropriate acceptors.

Polypeptide chains that will become glycoproteins are probably recognized by glycosyl transferases in a highly specific manner. The enzyme that transfers N-acetylgalactosamine to the hydroxyamino acid residues of ovine submaxillary glycoprotein will not transfer this sugar to a large number of other potential acceptors (McGuire and Roseman, 1967). Specificity is also shown in genetic studies correlating specific blood group substances with the presence or absence of specific glycosyl transferases. These studies indicate that addition of the same sugar to different incomplete polysaccharide chains may be dictated by different transferases. Thus, there are specific fucosyl transferases (Shen et al., 1968) responsible for the addition of each of the two fucose residues that determine the Lewis blood group:

$$\text{Fuc-}\alpha\text{-}(1{\to}2)\text{-Gal-}\beta\text{-}(1{\to}3)\text{-GluNac—R} \atop \text{Fuc-}\alpha\text{-}(1{\to}4) \diagup$$

Regulation of glycoprotein synthesis may involve other factors, in addition to the availability of appropriate acceptors, transferases, and nucleotide sugars. In bacterial systems, there is evidence for the transfer of sugars from nucleotides to lipid intermediates prior to their incorporation in polysaccharide chains (Weiner et al., 1965). An "acceptor complex" that contains lipids in addition to the sugar acceptors has been described by Rothfield and Horecker (1964). The entire complex interacts in some manner with the transferase. The existence of such complexes may be of considerable importance in preventing incorporation of erroneous sugars. In systems treated in vitro with detergents, "artificial glycoproteins" may be produced by the incorporation of sugars they do not contain normally (Bosmann et al., 1968).

TABLE I

*Potentialities for and Constraints on Structural Complexity of Glycoproteins*

I. *Potentialities for Structural Complexity*
1. Sequence of sugars
2. Anomeric form of sugar ($\alpha$ or $\beta$)
3. Position of attachment to neighboring sugar (e.g., $1 \rightarrow 3$; $1 \rightarrow 4$)
4. Branching
5. Length of polysaccharide units
6. Position on polypeptide to which polysaccharide attached
7. Number of polysaccharides per polypeptide
8. Substitution of other groups (e.g., sulfate)

II. *Some Constraints on Structural Complexity*
1. Limited number of sugars:
    N-acetylglucosamine, N-acetylgalactosamine, galactose, mannose, fucose, sialic acids (also: glucose, xylose, glucuronic acid)
2. All in D-configuration, except L-fucose
3. All in pyranose ring form
4. Innermost linkages:
    N-acetylglucosamine-amide N of asparagine (commonest)
    N-acetylgalactosamine-OH of serine or threonine
    also: galactose-OH of hydroxylysine (collagen)
        xylose-OH of serine (mucopolysaccharides)
5. Fucose and sialic acids always terminal

(Based on reviews by Roseman, 1968, and Ginsburg and Neufeld, 1969.)

TABLE II

*Carbohydrate Units of Glycoproteins*

| Glycoprotein | galactose | mannose | hexosamine | sialic acid | fucose | Mol. Wt. of Carbohyd. Unit | No. of Units per Molecule Protein |
|---|---|---|---|---|---|---|---|
| Ovalbumin (hen) | | 5 | 3 | | | 1420 | 1 |
| Submaxillary glycoprot (ovine) | | | 1 | 1 | | 494 | 800 |
| Fetuin | 4 | 3 | 6 | 4 | | 3520 | 3 |
| Thyroglobulin (calf) | | | | | | | |
| Unit A | | 10 | 2 | | | 2030 | 4–5 |
| Unit B | 4 | 3 | 5 | 2 | 1 | 2880 | 14 |

(Adapted from Spiro, 1968.)

A general principle of glycoprotein metabolism, which has been emphasized by Eylar (1966), is that addition of carbohydrates to proteins is a prelude to *secretion*. The completed glycoprotein is secreted either onto the cell surface or into the extracellular space (including the blood stream). Recent studies (Swenson and Kern, 1968) have shown that the addition of sialic acid to gamma globulin immediately precedes its secretion from plasma cells. The addition of glycomacromolecules to the cell surface apparently occurs in an orderly manner by spread from one region of the membrane rather than by random appearance at multiple sites over the cell surface, as was shown in studies of the pattern of reappearance of sialic acid residues removed from cultured cells by neuraminidase (Marcus and Schwartz, 1968).

Little is known about the mechanism of degradation of the glycoproteins. Although they are typically extracellular, the enzymes which apparently degrade them are present in lysosomes within the cell (Aronson and de Duve, 1968). That these enzymes normally function in their degradation has been shown by the lysosomal accumulation of glyco-peptides and glycolipids in patients with a genetically determined deficiency in one of these degradative enzymes. Degradation is believed to occur in lysosomes, so there may be a mechanism for entry into cells of surface glycoproteins and circulating glycoproteins.

LOCALIZATION ON CELL SURFACES  Evidence for the localization of glycomacromolecules on cell surfaces has been derived from histological studies and from enzymatic degradation of the surface of living cell suspensions. The histochemical techniques are based on reactions believed to be specific for polysaccharides or on reactions based on the presence of negatively charged groups often inferred to be sialic acid residues. Because both glycoproteins and glyco-lipids are probably retained on the cell despite fixation and histological processing, these techniques do not distinguish between them. Many studies show the presence of thick or thin polysaccharide coats on a variety of cells. Much of this work has been reviewed by Revel and Ito (1967). Particularly noteworthy are electron-microscope studies by Rambourg and Leblond (1967) showing glycomacromolecules on cell surfaces that have been treated first with periodic acid and then with silver methenamine, and the studies of Pease (1966), who has used phosphotungstic acid as a stain for what he believes are polysaccharide chains.

In the enzymatic studies, suspension cultures of cells are washed, incubated with a proteolytic enzyme or with neur-aminidase, and the released sugars and residual sugars in the living cells are analyzed. Wallach and Eylar (1961) found that similar amounts of sialic acid were released by neur-aminidase treatment of intact or ruptured ascites carcinoma cells, suggesting that sialic acid-containing molecules are abundant on the cell surface. Shen and Ginsburg (1968) found extensive release of both sialic acid and a variety of other sugars, characteristically present in glycoproteins, when HeLa cells in suspension culture were treated with trypsin. The combination of the histochemical and enzymatic studies leaves little doubt that the surface of the cell is rich in bound carbohydrates and that a substantial portion of the total cellular content of these substances is localized on the surface.

ROLE IN INTERCELLULAR RECOGNITION  That carbohydrate substances on cell surfaces have unique and recognizable structures has been well known since the demonstration that capsular polysaccharides of pneumococcus and blood-group substances on human erythrocytes may be recognized by specific antibodies. Whether there is such high specificity in recognition of one cell surface by another is unknown. Two types of experiments suggest that these substances play some role in intercellular recognition—isolation of substances that promote cellular interactions and studies of the effects of enzymatic removal of sugars from cell surfaces. The first method has been pursued most extensively in sponge (reviewed by Humphreys, 1967; Moscona, 1968). A glyco-protein-rich material, which is released from the surface of sponge cells in calcium-free media, has been shown to play a role in aggregating specific species of sponge cells. The material from one species aggregates only dissociated cells from the same species and not from a different species, which is taken as evidence of considerable specificity. Likewise, glycoproteins from yeast have been implicated as being responsible for sexual recognition (Crandall and Brock, 1968).

Other studies have been less direct and concern themselves primarily with the role of surface neuraminic acid on various recognition functions of cells. The binding of viruses to cells has been inhibited by pretreatment of the cells with neuraminidase (Gottschalk, 1960), and the action of serotonin on smooth-muscle cells has been interfered with by treatment of the cells with that enzyme (Woolley and Gommi, 1965). Treating lymphocytes with neuraminidase (Woodruff and Gesner, 1967) or with a mixture of glyco-sidases (Gesner and Ginsburg, 1964) has interfered with their accumulation in the spleen. Treating rabbit ova with the same enzyme has been shown to inhibit sperm penetration (Soupart and Clewe, 1965). These studies suggest, in general, that neuraminic acid on cell surfaces, which might be either in glycoproteins or glycolipids, confers some recognition function on these surfaces. On the basis of available evidence the recognition shown could be very crude. It remains to be shown that more highly specialized recognition functions are mediated by sugar-rich substances. Presently this is supported only by the teleological notions that the sugar-rich structures present on cell surfaces are suitable for coding functions (Bogoch, 1968), and that such functions apparently exist.

## Glycoproteins in brain

AMOUNT, DISTRIBUTION, AND COMPOSITION  Brunn-graber (1969) has estimated that between 3 per cent and 10 per cent of all brain proteins are glycoproteins. This calculation is based on the finding that there is approximately 1 milligram of carbohydrate incorporated in 100 milligrams of brain protein, and that the percentage of carbohydrate in glycoproteins is generally between 10 per cent and 33 per cent. More than 50 per cent of total brain hexosamine and approximately 20 per cent of brain sialic acid are present in nondialyzable glycoprotein fragments obtained by papain digestion of brain protein after removal of glycolipids (Brunngraber, 1969).

The distribution of brain glycoproteins has been studied by histochemical techniques, with the use of light and electron microscopy. Periodic acid-Schiff staining material is concentrated in the neuropil (Hess, 1955, 1958), and several studies with the electron microscope (Pease, 1966; Bondareff, 1967; Rambourg and Leblond, 1967) show that it is localized on cell surfaces and particularly at synaptic junctions (Figure 2). Regenerating axon tips are rich in polysaccharides (Young and Abood, 1960). Nerve-ending fractions isolated by sucrose gradient centrifugation are richer than other brain fractions in both glycoproteins and gangliosides (Brunngraber et al., 1967). All these studies point to a special role for sugar-containing proteins and lipids at synapses.

Relatively little is known about the composition of brain glycoproteins. Mucopolysaccharides of the types found in many other tissues have been identified in brain (Margolis, 1967). A sialic acid containing glycoprotein was found in the $a_2$-globulin fraction of soluble extracts of brain (Warecka and Bauer, 1967). Immunoelectrophoretic studies suggest that this glycoprotein is present only in brain. A glycoprotein with a molecular weight estimated at 30,000 has been identified in 90 per cent acetone extracts of brain and is also believed to be found uniquely in this tissue (Kuhn and Müldner, 1964; Gielen, 1966). The major glycoprotein detectable by periodic acid-Schiff staining in soluble extracts of five-day-old mouse brain studied by electrophoresis has an apparent molecular weight in the range of 250,000 (Dutton and Barondes, 1970), whereas particulate fractions from brain contain a wide range of glycoproteins.

"Sialomucopolysaccharides," obtained by papain digestion of defatted brain, have been studied extensively (Brunngraber, 1969). Although these substances can be fractionated in a variety of ways, Brunngraber believes that they all possess the same repeating unit, which consists of hexose and hexosamine, but differ in the relative number of attached sialic acid and fucose residues.

METABOLIC STUDIES Direct studies of glycosyl transferases in brain have been made primarily with sialyl transferases involved in ganglioside synthesis (Roseman, 1968). It is not clear whether similar or different sialyl transferases are involved in glycoprotein and glycolipid synthesis in brain. Den and Kaufman (1968) have reported that glycosyl transferases are abundant in nerve-ending particles prepared from embryonic chick brain. A fucosyl transferase has been found in brain microsomal fractions that transfers fucose from guanosine diphosphate (GDP)-fucose to a large number of incomplete endogenous glycoproteins in the preparation (Zatz and Barondes, unpublished). A mannosyl transferase, which transfers mannose from GDP-mannose to an unidentified glycolipid, has also been identified in brain (Zatz and Barondes, 1969). Caccam et al. (1969) have speculated that a similar enzyme might create a glycolipid intermediate, which is involved in glycoprotein synthesis.

In vivo glycoprotein metabolism in mouse brain has been examined somewhat more extensively. A unique aspect of brain glycoprotein metabolism appears to be the marked delay between synthesis of some polypeptide acceptors in the nerve-cell body and the addition of carbohydrates to these polypeptides only after they have been transported to nerve endings. This has been studied by observing the time course of incorporation of radioactive leucine (Barondes, 1964, 1968a) and glucosamine (Barondes, 1968b; Barondes and Dutton, 1969) into subcellular fractions of mouse brain at various times after intracerebral injection of these precursors. Radioactive leucine was incorporated rapidly into the protein of all subcellular fractions of brain, with the exception of the soluble component of the nerve-ending fraction (Table III). The soluble component of the nerve-ending fraction is obtained by lysing the fraction with water. This component has been a primary object of study because it is less contaminated by other brain constituents than is the particulate component of the nerve-ending fraction. The specific activity of the protein of the soluble component of nerve endings rose progressively for days after administration of the radioactive amino acid (Barondes, 1964, 1968a), presumably owing to transport of protein from the perikaryon to the nerve ending. In contrast, radioactive glucosamine was incorporated without delay into all subcellular fractions of brain, including the soluble protein of nerve endings (Table III), suggesting that some polypeptide chains formed in the nerve-cell body were modified by the addition of carbohydrates on arrival at the nerve ending.

This conclusion was confirmed by studies of the effect of acetoxycycloheximide, a potent inhibitor of cerebral pro-

TABLE III

*Incorporation of ³H-leucine and ¹⁴C-glucosamine into macromolecules of subcellular fractions of brain 3 hours and 6 days after injection. Mice were injected intracerebrally with a solution containing L-leucine-4, 5-³H and D-glucosamine-1-¹⁴C. Groups of mice were sacrificed either 3 hours or 6 days after injection; subcellular fractions were obtained, precipitated, washed and counted. (For details see Barondes, 1968b.)*

| | Counts/min/mg protein | | | |
| --- | --- | --- | --- | --- |
| | ³H-leucine | | ¹⁴C-glucosamine | |
| | 3 hrs. | 6 days | 3 hrs. | 6 days |
| Soluble | 1560 | 431 | 330 | 150 |
| Soluble of nerve-ending fraction | 142 | 412 | 295 | 127 |
| Microsomes | 1441 | 546 | 987 | 530 |
| Mitochondria | 549 | 339 | 242 | 130 |
| Myelin | 540 | 392 | 220 | 189 |
| Particulate of nerve-ending fraction | 376 | 397 | 547 | 386 |

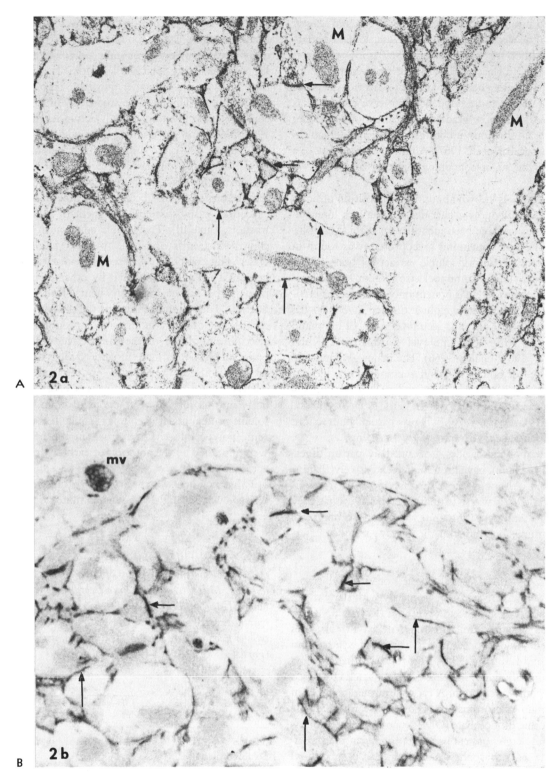

FIGURE 2 Glycomacromolecules at cell surfaces and synapses. Rat cerebral cortex was prepared (A) by the periodic acid-silver methionine technique (Rambourg and Leblond, 1967) and (B) by the chromic acid-phosphotungstic acid technique (Rambourg et al., 1969), both of which stain glycomacromolecules. Neuronal and glial processes are separated from each other by stained material (vertical arrows), but staining of the intercellular space is markedly increased in the region of the synaptic cleft (horizontal arrows). Mitochondria (M) and multivesicular bodies (mv) are also stained. Magnification × 20,000. The electron micrographs were kindly provided by Dr. A. Rambourg.

tein synthesis, on incorporation of glucosamine into glyco-proteins (Barondes and Dutton, 1969). Acetoxycyclo-heximide interferes with such incorporation by inhibiting the synthesis of protein acceptors for sugars. Within two hours after injection of acetoxycycloheximide, marked inhibition of glucosamine incorporation into brain glyco-proteins was observed. There was, however, relatively little inhibition of incorporation into the soluble glycoproteins of the nerve-ending fraction (Table IV), suggesting that the polypeptide chains synthesized before the administration of acetoxycycloheximide continued to flow to the nerve ending, where carbohydrates may be added. Rahmann (1968) has recently shown, by autoradiographic studies with the light microscope, that tritiated glucose is incorporated into what appear to be macromolecules in the neuropil. Although the resolution that can be achieved is limited and the products Rahmann is studying have not been clearly identified, these results are consistent with those arrived at by subcellular fractionation.

Studies with the subcellular fractionation technique have also indicated that turnover of soluble glucosamine-labeled glycoproteins is relatively rapid at nerve endings. This has been demonstrated in studies in which acetoxycyclohexi-mide was used to diminish the amount of labeled perikaryal glycoprotein arriving at the nerve ending (Figure 3) and in the absence of this inhibitor. Although fucose is incorporated relatively poorly into all soluble brain glycopro-teins, there was a relatively rapid turnover of fucose-labeled

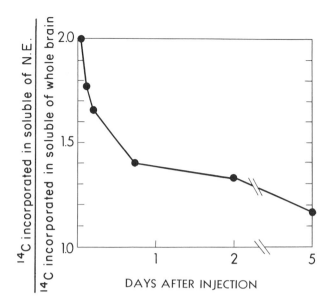

FIGURE 3 Relatively rapid turnover of some soluble glycopro-teins of nerve endings (N.E.). Mice were injected with $^{14}$C-glucos-amine two hours after subcutaneous administration of acetoxycy-cloheximide. They were sacrificed at the indicated times after in-jection of glucosamine. (For details, see Barondes and Dutton, 1969).

TABLE IV

*Effect of acetoxycycloheximide on incorporation of radioactive glucosamine into brain glycoproteins. Mice were injected sub-cutaneously with 240 micrograms of acetoxycycloheximide or with saline 2 hours before intracerebral injection of $^{14}$C-gluco-samine. They were killed 2 hours after glucosamine administra-tion, and their cerebral hemispheres were fractionated and assayed. (For details see Barondes and Dutton, 1969.)*

| Fraction | Counts/min/mg protein | | |
| --- | --- | --- | --- |
| | Control | Acetoxycycloheximide | % Inhibition |
| Whole brain | 1923 | 524 | 73 |
| Soluble of whole brain | 1299 | 376 | 71 |
| Soluble of nerve ending fraction | 841 | 782 | 7 |
| Particulate of nerve ending fraction | 1020 | 560 | 48 |
| Microsomes | 3220 | 641 | 80 |
| Myelin | 548 | 142 | 74 |
| Mitochondria | 1070 | 334 | 69 |

soluble glycoproteins which were associated with the nerve-ending fraction (Zatz and Barondes, 1970).

Some aspects of neuronal glycoprotein metabolism sug-gested by these studies are shown in Figure 4. Some, possibly all, polypeptide chains that will become nerve-ending glycoproteins are made in the perikaryon. At this site, some may receive only a fraction of the sugars they will ultimately contain, or none at all. Incomplete glycoproteins are transported down the axon to the nerve ending. At the nerve ending, glycosyl transferases add further carbohy-drates. The completed glycoproteins then "turn over" relatively rapidly, either because of degradation or, more likely, because of secretion from the nerve terminal. The secreted material may coat the nerve-terminal membrane or may migrate into the intersynaptic gap substance or some other site.

DEVELOPMENTAL STUDIES    Studies of the accumulation of periodic acid-Schiff reacting material in the neuropil as mouse brains matured were reported by Hess (1955). He found that this material progressively accumulates between the fifth and fourteenth postnatal days. Warecka and Müller (1969) and Bogoch (1968) have reported maturational changes in human brain glycoproteins. Accumulation of gangliosides in rat brain occurs between five and 20 days after birth, and there is maximal synthesis at about days 10 to 12 (Kishimoto et al., 1965; Suzuki, 1967).

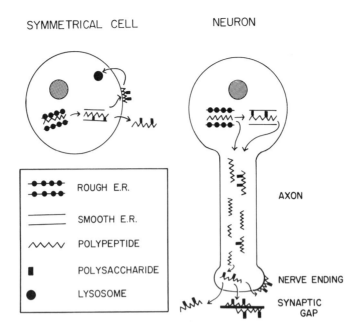

SYMMETRICAL CELL    NEURON

ROUGH E.R.

SMOOTH E.R.

POLYPEPTIDE

POLYSACCHARIDE

LYSOSOME

AXON

NERVE ENDING

SYNAPTIC GAP

FIGURE 4    Transport of glycoprotein precursors and glyco-proteins in usual mammalian cells and in neurons. In usual cells, sugars are believed to be added to glycoproteins in the smooth endoplasmic reticulum (E.R.) before they are se-creted onto the cell surface or into the intercellular space. In neurons, polypeptides that were transported in axons may receive sugars at nerve terminals prior to secretion.

Because of the possible role of glycoprotein synthesis in synapse formation, we have begun studies on synthesis of glycoproteins in developing mouse brain (Dutton and Barondes, 1970). Studies of incorporation into both soluble and particulate glycoproteins have been made with the polyacrylamide gel techniques developed by Maizel (1966). Incorporation of either radioactive glucosamine or radio-active fucose into glycoproteins of developing brain was studied. Although the results with both sugars are very similar, we have preferred to use the latter, because fucose, in contrast with glucosamine, is not a known constituent of the mucopolysaccharide group of glycoproteins, and is neither converted to other sugars nor incorporated into glycolipids in brain (Zatz and Barondes, 1970).

The major finding of these studies is that, during brain development, there is extensive synthesis of glycoproteins, which migrate slowly in an electrophoretic system in which migration is correlated with molecular weight (Shapiro et al., 1967). Synthesis of these substances diminishes markedly between 10 and 15 days after birth, and is at a very low level thereafter. When one-day-old mice were injected intra-cerebrally with radioactive fucose, and the soluble brain glycoproteins, obtained 24 hours later, were electro-phoresed, one major, slowly migrating peak (peak I) of radioactive glycoprotein was found (Figure 5). The soluble proteins obtained when fucose injections were made in five-day-old mice showed two prominent slowly migrating

peaks (I and II, Figure 5) and a smaller third peak, which migrated slightly farther. Synthesis of the glycoproteins in peak I is prominent in one-day-old mice, whereas peak II becomes prominent about three days after birth. The glycoproteins in peaks I and II are extensively synthesized until about 11 days after birth, but their synthesis has diminished markedly at 15 days after birth, and is not prominent in the adult (Figure 5).

Particulate brain glycoproteins synthesized by mice in-jected with labeled fucose at five days of age are distributed over a wide range (Figure 6). In contrast with adult brains particulate glycoproteins that, relatively, migrate more slowly, are synthesized in five-day-old brains (Figure 6). Upon calculation of the ratio of five-day-old to adult radioactive glycoprotein at each point on the gel, in order to factor out the adult particulate glycoproteins that are being synthesized in the five-day-old brain, the presence of peaks I and II becomes apparent in the particulate prepara-tions of the five-day-old group (Figure 7). This was true when the mice were sacrified either three or 24 hours after the injection of radioactive fucose (Figure 7). In contrast with brain, the patterns of glycoproteins synthesized in five-day-old and adult kidney do not show this difference (Figure 7).

The major glycoprotein detected by staining gels with the periodic acid-Schiff procedure, after electrophoresing the soluble fraction from five-day-old mice, corresponds

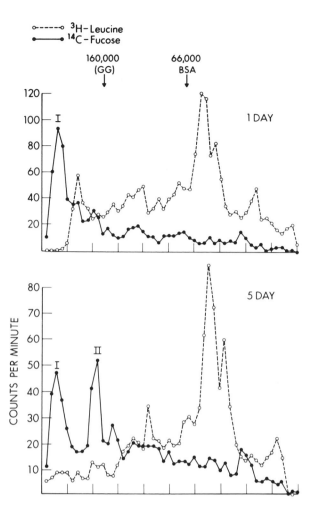

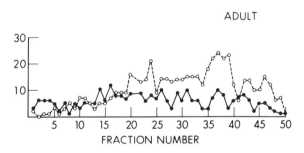

FIGURE 5 Polyacrylamide gel electrophoresis of soluble brain proteins. Adult, 5-day-old and 1-day-old mice were injected intercerebrally with ³H-leucine and ¹⁴C-fucose. They were sacrificed one day later, and soluble proteins were obtained by centrifugation of homogenates for one hour at 100,000 × gravity. The labeled proteins were treated with sodium dodecyl sulfate (SDS) and urea, reduced, alkylated, applied to polyacrylamide gel columns at 0, and electrophoresed in 0.1 per cent SDS by the method of Maizel (1966). The migration of proteins in this system is a function of molecular weight (Shapiro et al., 1967), but glycoproteins might behave anomalously. The migration of two standards, gamma globulin and albumin, is shown. The large peak of leucine-labeled protein has been shown to be microtubular protein (Dutton and Barondes, 1969). For details, see Dutton and Barondes, 1970.

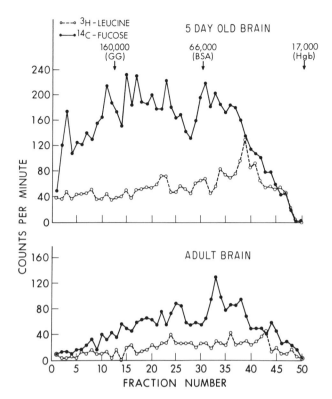

FIGURE 6 Polyacrylamide gel electrophoresis of particulate brain proteins. The particulate proteins were obtained, treated as in Figure 5, and electrophoresed. The sample from the 5-day-old mice contained a wide range of high-molecular-weight, fucose-labeled proteins. (From Dutton and Barondes, 1970.)

with peak I. Studies of turnover after injection of radioactive fucose in five-day-old mice indicate that the fucose-labeled glycoproteins of peak II have a half-life in the range of one week, which is similar to that of total fucose-labeled soluble brain glycoproteins. In contrast, the turnover of the fucose-labeled glycoproteins in peak I is very much slower (Dutton and Barondes, 1970).

We are presently attempting to analyze chemically the glycoproteins in peaks I and II and to localize them by radioautography. Preliminary studies demonstrate marked incorporation of radioactivity in the neuropil of five-day-old mice. In view of the developmental studies of Hess (1955), it is possible that these substances are related to the PAS-stained material in neuropil. The striking differences in the metabolism of the glycoproteins of peaks I and II suggest that they may play different roles in brain development.

## A model for interneuronal recognition

In this section, I attempt to demonstrate the plausibility of the notion that interneuronal recognition could be mediated

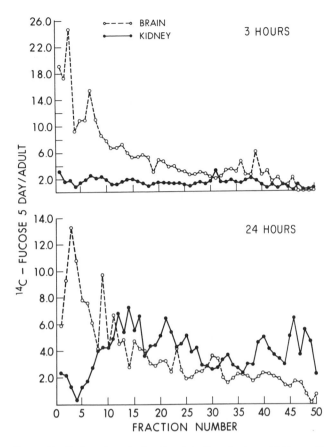

FIGURE 7 Ratios of particulate fractions from brain and kidney of 5-day-old-mice to those of adults. Adult and 5-day-old mice were injected with ¹⁴C-fucose and sacrificed 3 or 24 hours later, as indicated. Particulate proteins from brain and kidney were electrophoresed as described in Figure 5. By plotting the indicated ratio, the presence of peak I and peak II in particulate fractions of 5-day-old brain is apparent. This is not seen in the samples from kidney.

by surface glycomacromolecules by proposing, for the sake of argument, means whereby this could be brought about.

The best description of the specificity of interneuronal connections and the laws it apparently follows is in the retinotectal system, the development of which is considered in this volume by Marcus Jacobson. In essence, the work of Sperry (1963), Jacobson and Gaze (1965), De Long and Coulombre (1965), and Jacobson (1968) has demonstrated that the connections between regions of the retina and the tectum are rigidly specified at a precise time in development, long before the axons of retinal neurons have grown out to meet and make synapses with tectal neurons. Beyond this point in development, specification is so fixed that, when the eye is rotated, retinal cells still synapse with those tectal cells they would have sought out had the eye remained in its normal position. Jacobson (1968) has presented

evidence that specification occurs first in one direction and then perpendicularly.

From these studies it seems likely that both the retina and the tectum are independently specified. Models based only on retinal cells' maintaining positions relative to one another, without specification of tectal cells, could also explain the findings in the normal animal. The evidence for the re-establishment of normal connections, however, after optic-nerve section and *ablation* of a portion of the retinal or tectal cells (Attardi and Sperry, 1963; Jacobson and Gaze, 1965), is more suggestive of mutual recognition by the cells that connect.

If specific groups of retinal cells do indeed recognize specific groups of tectal cells, this could be based on "identity" reactions (like recognizes like), as suggested by Roberts and Flexner (1966), or on "complementary" reactions (opposites attract). The experiments show what may be considered a "complementarity" of connections, in that dorsal cells in the left retina make contact with ventral cells in the right tectum and anterior cells in the left retina make synapses with posterior cells in the right tectum (Figure 8). As will be shown, this is consistent with recognition based on complementary cell surfaces.

For the purpose of this discussion, let us consider that both retinal cells and tectal cells are specified by "gradients of inducers." In this view, one inducer, localized in the midline, spreads laterally and progressively induces changes in responsive tissues. The anterior (nasal) retinal and tectal cells, which are closer to the midline, would be exposed to a higher concentration of this inducer than would the posterior (temporal cells). A second inducer might spread dorsoventrally, thereby inducing changes primarily in the dorsal tectal or retinal cells. Were this view correct, retinal cells could be considered as making synapses with tectal cells which had received complementary quantities of inducer—that is, the most dorsal and anterior retinal cells would have been strongly influenced by both the midline and the dorsoventral gradients and would make contact with those most posterior and ventral tectal cells that had been least exposed to these inducers.

The nature of the cellular changes produced by the putative inducers is not clear. Presumably a *permanent* alteration in gene expression is produced by some mechanism common to all differentiative processes. For the purpose of this discussion, it is assumed that this change leads ultimately to a systematic alteration in the surface heterosaccharides of retinal and tectal cells. If such an assumption be accepted, a scheme will be proposed to illustrate how surface heterosaccharides could mediate the observed phenomena. Let us assume the following: (1) there is a large number of protein and lipid molecules on the membranes of all neurons to which polysaccharide chains may be linked covalently; (2) the addition of sugars to these surface

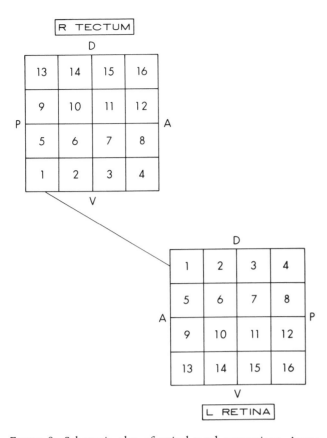

FIGURE 8 Schematic plan of retinal-tectal connections. Axons from retinal ganglion cells grow out and synapse with tectal cells in the regions indicated by the same number. Symbols used are: A, anterior or nasal; D, dorsal; P, posterior or temporal; V, ventral. The positions shown are those during the period of specification and were confirmed in discussions with C. Coulombre and M. Jacobson.

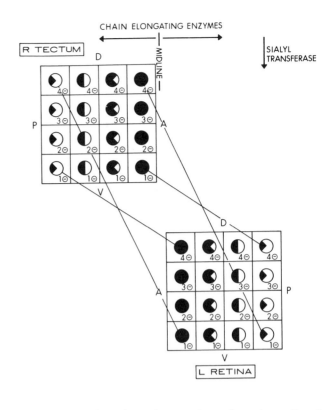

FIGURE 9 Hypothetical specification of retinal ganglion cells and tectal cells by the same gradients of inducers. The putative inducer, which spreads from the midline in a nasotemporal direction, influences primarily the nasal retinal and tectal cells, which, in most terminologies, are referred to as "anterior." The putative inducer that spreads in a dorsal-ventral direction influences primarily the dorsal retinal and tectal cells. Connections are ultimately made between cells that have complementary surfaces. Symbols used are the same as in Figure 8.

proteins and lipids is linearly related to the level of the various glycosyl transferases in the cell over a wide range; (3) the polysaccharide chains that are formed have a far greater binding affinity for membrane proteins than for other polysaccharide chains.

Given these assumptions, let us now postulate that the putative inducer in one direction induces the synthesis of several glycosyl transferases, perhaps a hexosyl transferase and a hexosaminyl transferase, the presence of which in the cells will lead to the synthesis of long-repeating polysaccharide chains. Let us also postulate that an inducer in a perpendicular direction leads to the synthesis of graded amounts of a sialyl transferase. This enzyme may transfer charged sialic-acid residues to polysaccharide chains, which are present in relatively large quantities on the surface of all the cells in question, so that the quantity of the sialyl transferase is strictly limiting over a wide range. Because graded quantities of these enzymes are present in the cells

in question, the pattern of carbohydrates on the entire surface of neurons in both the retina and the tectum could be schematized as shown in Figure 9. Cells close to the inducer of hexosyl and hexosaminyl transferases have surfaces with many long polysaccharide chains composed of repeating units and are designated as completely black in the Figure. Those most distant from the focus have chains that are relatively fewer and shorter; these are shown as mostly white. The negatively charged sialic acid residues added to these polysaccharide chains would be distributed as indicated in the Figure.

Retinal cells abundant in polysaccharides would tend to make contact with tectal cells, which are poor in polysaccharides, because the latter would have many free surface proteins, which are not already binding polysaccharides. Interneuronal interaction would also be regulated by surface charge, with strongly negative surfaces repelling one another. The net result would be that the black negative cells would tend to come in contact with the primarily white and uncharged tectal cells as shown in the Figure. Through this comparatively simple scheme the observed retinotectal relationships could be established.

I must emphasize that there is no evidence at present for such a scheme. It is based on a large number of assumptions, any of which could be incorrect. It deliberately avoids assumptions about specific interactions between polysaccharides (other than mutual repulsion of negative charges), because interactions between these compounds is poorly understood at present. Rather, it emphasizes an attraction between polysaccharides on the one hand and proteins on the other. Although these could be highly specific, as shown by the existence of highly specific antibodies (proteins) for given polysaccharide chains, much less specific interaction could be operative (Steinberg, 1963).

The assumptions about gradients of inducers and about complementarity of the retina and tectum are consistent with present information but are not necessary to explain what is already known. Therefore, the primary value of this model is that it illustrates how, without very detailed information, protuberant glycomacromolecules on neuronal surfaces might dictate interneuronal recognition in the absence of rigid templates for highly deterministic structural specificity. Although the model is consistent with known facts, it does not deal with everything that is known. In particular, no attempt is made to explain how the effect of the "gradients of inducers," presumed to be present only in embryogenesis, is maintained through the life of the organism.

## Summary

This paper considers the hypothesis that glycomacromolecules (glycolipids and glycoproteins) on neuronal surfaces mediate specific interneuronal recognition. The structure, metabolism, and localization of glycoproteins in cells are briefly reviewed. The capacity of nerve terminals to add carbohydrate residues to polypeptides synthesized in the neuronal perikaryon is shown, and the synthesis of several high-molecular-weight glycoproteins during a specific period in the development of the nervous system is described. A model for interneuronal recognition based on complementarity of cell surfaces in the retinotectal system is proposed, primarily to provoke biochemical investigation of this crucial problem.

## Acknowledgment

The work reported in this paper was supported by Career Development Award K3-MH-18232.

## REFERENCES

ARONSON, N. N., and DE DUVE, C., 1968. Digestive activity of lysosomes. II. The digestion of macromolecular carbohydrates by extracts of rat liver lysosomes. *J. Biol. Chem.* 243: 4564–4573.

ATTARDI, D. G., and R. W. SPERRY, 1963. Preferential selection of central pathways by regenerating optic fibers. *Exp. Neurol.* 7: 46–64.

BARONDES, S. H., 1964. Delayed appearance of labeled protein in isolated nerve endings and axoplasmic flow. *Science* (*Washington*) 146: 779–781.

BARONDES, S. H., 1968a. Further studies of the transport of proteins to nerve endings. *J. Neurochem.* 15: 343–350.

BARONDES, S. H., 1968b. Incorporation of radioactive glucosamine into macromolecules at nerve endings. *J. Neurochem.* 15: 699–706.

BARONDES, S. H., and G. R. DUTTON, 1969. Acetoxycycloheximide effect on synthesis and metabolism of glucosamine-containing macromolecules in brain and in nerve endings. *J. Neurobiol.* 1: 99–110.

BOGOCH, S., 1965. Brain proteins in learning. *Neurosci. Res. Program Bull.* 3 (no. 6): 38–41.

BOGOCH, S., 1968. The Biochemistry of Memory with an Inquiry Into the Function of the Brain Mucoids. Oxford University Press, New York.

BONDAREFF, W., 1967. Demonstration of an intercellular substance in mouse cerebral cortex. *Z. Zellforsch. Mikroskop. Anat.* 81: 366–373.

BOSMANN, H. B., A. HAGOPIAN, and E. H. EYLAR, 1968. Glycoprotein biosynthesis: The characterization of two glycoprotein: fucosyl transferases in HeLa cells. *Arch. Biochem. Biophys.* 128: 470–481.

BRUNNGRABER, E. G., 1969. Glycoproteins in the nervous system. *In* Handbook of Neurochemistry, Vol. 1 (A. Lajtha, editor). Plenum Press, New York, pp. 223–244.

BRUNNGRABER, E. G., H. DEKIRMENJIAN, and B. D. BROWN, 1967. The distribution of protein-bound N-acetylneuraminic acid in subcellular fractions of rat brain. *Biochem. J.* 103: 73–78.

CACCAM, J. F., J. J. JACKSON, and E. H. EYLAR, 1969. The biosynthesis of mannose-containing glycoproteins: A possible lipid intermediate. *Biochem. Biophys. Res. Commun.* 35: 505–511.

COOK, G. M. W., 1968. Glycoproteins in membranes. *Biol. Rev.* (*Cambridge*) 43: 363–391.

CRANDALL, M. A., and T. D. BROCK, 1968. Molecular aspects of specific cell contact. *Science* (*Washington*) 161: 473–475.

DAVIS, B. D., and L. WARREN (editors), 1967. The Specificity of Cell Surfaces. Prentice-Hall, Englewood Cliffs, New Jersey.

DE LONG, G. R., and A. J. COULOMBRE, 1965. Development of the retinotectal topographic projection in the chick embryo. *Exp. Neurol.* 13: 351–363.

DEN, H., and B. KAUFMAN, 1968. Ganglioside and glycoprotein glycosyltransferases in synaptosomes. *Fed. Proc.* 27: 346 (abstract).

DISCHE, Z., 1966. The informational potentials of conjugated proteins. *In* Protides of the Biological Fluids, 13th Colloquium, Bruges, 1965 (H. Peeters, editor). Elsevier Publishing Co., Amsterdam, pp. 1–20.

DUTTON, G. R., and S. H. BARONDES, 1969. Microtubular protein: Synthesis and metabolism in developing mouse brain. *Science (Washington)* 166: 1637–1638.

DUTTON, G. R., and S. H. BARONDES, 1970. Glycoprotein metabolism in developing mouse brain. *J. Neurochem.* (in press).

EYLAR, E. H., 1966. On the biological role of glycoproteins. *J. Theor. Biol.* 10: 89–113.

GESNER, B. M., and V. GINSBURG, 1964. Effect of glycosidases on the fate of transfused lymphocytes. *Proc. Nat. Acad. Sci. U.S.A.* 52: 750–755.

GIELEN, W., 1966. Über ein hirnspezifisches Glykoprotein. *Naturwissenschaften* 53: 504–505.

GINSBURG, V., and E. F. NEUFELD, 1969. Complex heterosaccharides of animals. *Annu. Rev. Biochem.* 38: 371–388.

GOTTSCHALK, A., 1960. The Chemistry and Biology of Sialic Acid and Related Substances. Cambridge University Press, London.

GROBSTEIN, C., 1967. Mechanisms of organogenetic tissue interaction. *Nat. Cancer Inst. Monogr.* 26: 279–299.

HAGOPIAN, A., H. B. BOSMANN, and E. H. EYLAR, 1968. Glycoprotein biosynthesis: The localization of polypeptidyl:N-acetyl-galactosaminyl, collagen:glucosyl, and glycoprotein:galactosyl transferases in HeLa cell membrane fractions. *Arch. Biochem. Biophys.* 128: 387–396.

HEINMETS, F., 1968. Cell-cell recognition and interaction. *Curr. Mod. Biol.* 1: 299–313.

HESS, A., 1955. Blood-brain barrier and ground substance of central nervous system. *A.M.A. Arch. Neurol. Psychiat.* 73: 380–386.

HESS, A., 1958. Further histochemical studies on the presence and nature of the ground substance of the central nervous system. *J. Anat.* 92: 298–303.

HOLTFRETER, J., 1939. Gewebeaffinität, ein Mittel der embryonalen Formbildung. *Arch. Exp. Zellforsch.* 23: 169–209.

HUMPHREYS, T., 1967. The cell surface and specific cell aggregation. *In* The Specificity of Cell Surfaces (B. D. Davis and L. Warren, editors). Prentice-Hall, Englewood Cliffs, New Jersey, pp. 195–210.

JACOBSON, M., 1968. Specification of neuronal connections during development. *In* Physiological and Biochemical Aspects of Nervous Integration (F. D. Carlson, editor). Prentice-Hall, Englewood Cliffs, New Jersey, pp. 195–214.

JACOBSON, M., and R. M. GAZE, 1965. Selection of appropriate tectal connections by regenerating optic nerve fibers in adult goldfish. *Exp. Neurol.* 13: 418–340.

JEANLOZ, R. W., 1968. Structure of the oligosaccharide side-chains of glycoproteins. *In* Biochemistry of Glycoproteins and Related Substances (E. Rossi and E. Stoll, editors). S. Karger, New York and Basel, pp. 94–107.

JOURDIAN, G. W., and S. ROSEMAN, 1963. Intermediary metabolism of the sialic acids. *Ann. N.Y. Acad. Sci.* 106: 202–217.

KALCKAR, H. M., 1965. Galactose metabolism and cell "sociology." *Science (Washington)* 150: 305–313.

KISHIMOTO, Y., W. E. DAVIES, and N. S. RADIN, 1965. Developing rat brain: Changes in cholesterol, galactolipids, and the individual fatty acids of gangliosides and glycerophosphatides. *J. Lipid Res.* 6: 532–536.

KUHN, R., and H. MÜLDNER, 1964. Über Glyko-lipo-sialo-proteide des Gehirns. *Naturwissenschaften* 51: 635–636.

LAWFORD, G. R., and H. SCHACTER, 1966. Biosynthesis of glycoprotein by liver. *J. Biol. Chem.* 241: 5408–5418.

LEHNINGER, A. L. 1968. The neuronal membrane. *Proc. Nat. Acad. Sci. U. S. A.* 60: 1069–1080.

LOWDEN, J. A., and L. S. WOLFE, 1964. Studies on brain gangliosides. III. Evidence for the location of gangliosides specifically in neurones. *Can. J. Biochem.* 42: 1587–1594.

McGUIRE, E. J., and S. ROSEMAN, 1967. Enzymatic synthesis of the protein-hexosamine linkage in sheep submaxillary mucin. *J. Biol. Chem.* 242: 3745–3747.

MAIZEL, J. V. JR., 1966. Acrylamide-gel electrophorograms by mechanical fractionation: Radioactive adenovirus proteins. *Science (Washington)* 151: 988–990.

MANSON, L. A. (editor), 1968. Biological Properties of the Mammalian Surface Membrane. Wistar Institute Press, Philadelphia.

MARCUS, P. I., and V. G. SCHWARTZ, 1968. Monitoring molecules of the plasma membrane: Renewal of sialic acid-terminating receptors. *In* Biological Properties of the Mammalian Surface Membrane (L. A. Manson, editor). Wistar Institute Press, Philadelphia, pp. 143–147.

MARGOLIS, R. U., 1967. Acid mucopolysaccharides and proteins of bovine whole brain, white matter and myelin. *Biochim. Biophys. Acta* 141: 91–102.

MOLNAR, J., and D. SY, 1967. Attachment of glucosamine to protein at the ribosomal site of rat liver. *Biochemistry* 6: 1941–1947.

MOSCONA, A. A., 1963. Studies on cell aggregation: Demonstration of materials with selective cell-binding activity. *Proc. Nat. Acad. Sci. U. S. A.* 49: 742–747.

MOSCONA, A. A., 1968. Cell aggregation: Properties of specific cell-ligands and their role in the formation of multicellular systems. *Develop. Biol.* 18: 250–277.

PEASE, D. C., 1966. Polysaccharides associated with the exterior surface of epithelial cells: Kidney, intestine, brain. *J. Ultrastruct. Res.* 15: 555–588.

RAHMANN, H., 1968. Transport route of ³H-glucose and site of synthesis of polysaccharides in the central nervous system of teleosts. *Exp. Brain Res.* 6: 32–48.

RAMBOURG, A., W. HERNANDEZ, and C. P. LEBLOND, 1969. Detection of complex carbohydrates in the Golgi apparatus of rat cells. *J. Cell Biol.* 40: 395–414.

RAMBOURG, A., and C. P. LEBLOND, 1967. Electron microscope observations on the carbohydrate-rich cell coat present at the surface of cells in the rat. *J. Cell Biol.* 32: 27–53.

REVEL, J.-P., and S. ITO, 1967. The surface components of cells. *In* The Specificity of Cell Surfaces (B. D. Davis and L. Warren, editors). Prentice-Hall, Englewood Cliffs, New Jersey, pp. 211–234.

ROBERTS, R. B., and L. B. FLEXNER, 1966. A model for the development of retina-cortex connections. *Amer. Sci.* 54: 174–183.

ROSEMAN, S., 1968. Biosynthesis of glycoproteins, gangliosides, and related sybstances. *In* Biochemistry of Glycoproteins and Related Substances (E. Rossi and E. Stoll, editors), S. Karger, New York and Basel, pp. 244–269.

ROSSI, E., and E. STOLL (editors), 1968. Biochemistry of Glycoproteins and Related Substances: Cystic Fibrosis, Part II. S. Karger, New York and Basel.

ROTHFIELD, L., and B. L. HORECKER, 1964. The role of cell-wall lipid in the biosynthesis of bacterial lipopolysaccharide. *Proc. Nat. Acad. Sci. U. S. A.* 52: 939–946.

SCHMITT, F. O., and F. E. SAMSON (editors), 1969. Brain cell microenvironment. *Neurosci. Res. Program Bull.* 7 (no. 4): 277–411.

SCHUBERT, M., 1964. Intercellular macromolecules containing polysaccharides. *Biophys. J.* (*Suppl.*) 4 (no. 1, pt. 2): 119–138.

SHAPIRO, A. L., E. VIÑUELA, and J. V. MAIZEL, 1967. Molecular weight estimation of polypeptide chains by electrophoresis in SDS-polyacrylamide gels. *Biochem. Biophys. Res. Commun.* 28: 815–820.

SHEN, L., and V. GINSBURG, 1968. Release of sugars from HeLa cells by trypsin. *In* Biological Properties of the Mammalian Surface Membrane (L. A. Manson, editor). Wistar Institute Press, Philadelphia, pp. 67–71.

SHEN, L., E. F. GROLLMAN, and V. GINSBURG, 1968. An enzymatic basis for secretor status and blood group substance specificity in humans. *Proc. Nat. Acad. Sci. U. S. A.* 59: 224–230.

SOUPART, P., and T. H. CLEWE, 1965. Sperm penetration of rabbit zona pellucida inhibited by treatment of ova with neuraminidase. *Fert. Steril.* 16: 677–689.

SPERRY, R. W., 1963. Chemoaffinity in the orderly growth of nerve fiber patterns and connections. *Proc. Nat. Acad. Sci. U. S.A.* 50: 703–710.

SPIRO, R. G., 1968. Carbohydrate units in glycoproteins. *In* Biochemistry of Glycoproteins and Related Substances (E. Rossi and E. Stoll, editors). S. Karger, New York and Basel, pp. 59–78.

STEINBERG, M. S., 1963. Reconstruction of tissues by dissociated cells. *Science* (*Washington*) 141: 401–408.

SUZUKI, K., 1967. Formation and turnover of the major brain gangliosides during development. *J. Neurochem.* 14: 917–925.

SWENSON, R. M., and M. KERN, 1968. The synthesis and secretion of γ-globulin by lymph node cells, III. The slow acquisition of the carbohydrate moiety of γ-globulin and its relationship to secretion. *Proc. Nat. Acad. Sci. U. S. A.* 59: 546–553.

WALLACH, D. F. H., and E. H. EYLAR, 1961. Sialic acid in the cellular membranes of Ehrlich ascites-carcinoma cells. *Biochim. Biophys. Acta* 52: 594–596.

WARECKA, K., and H. J. BAUER, 1967. Studies on "brain-specific" proteins in aqueous extracts of brain tissue. *J. Neurochem.* 14: 783–787.

WARECKA, K., and D. MÜLLER, 1969. The appearance of human "brain-specific" glycoprotein in ontogenesis. *J. Neurol. Sci.* 8: 329–345.

WATKINS, W. M., 1967. Blood group substances. *In* The Specificity of Cell Surfaces (B. D. Davis and L. Warren, editors). Prentice-Hall, Englewood Cliffs, New Jersey, pp. 257–279.

WEINER, I. M., T. HIGUCHI, L. ROTHFIELD, M. SALTMARSCH-ANDREW, M. J. OSBORN, and B. L. HORECKER, 1965. Biosynthesis of bacterial lipopolysaccharide, V. Lipid-linked intermediates in the biosynthesis of the O-antigen groups of *Salmonella typhimurium*. *Proc. Nat. Acad. Sci. U. S. A.* 54: 228–235.

WEISS, P. 1941. Nerve patterns: The mechanics of nerve growth. *Growth* (*Suppl*) 5: 163–203.

WIEGANDT, H., 1967. The subcellular localization of ganglioside in the brain. *J. Neurochem.* 14: 671–674.

WOODRUFF, J. J., and B. M. GESNER, 1967. The effect of neuraminidase on the fate of transfused lymphocytes. *J. Clin. Invest.* 46: 1134–1135.

WOOLLEY, D. W., and B. W. GOMMI, 1965. Serotonin receptors, VII. Activities of various pure gangliosides as the receptors. *Proc. Nat. Acad. Sci. U. S. A.* 53: 959–963.

YOUNG, I. J., and L. G. ABOOD, 1960. Histological demonstration of hyaluronic acid in the central nervous system. *J. Neurochem.* 6: 89–94.

ZATZ, M., and S. H. BARONDES, 1969. Incorporation of mannose into mouse brain lipid. *Biochem. Biophys. Res. Commun.* 36: 511–517.

ZATZ, M., and S. H. BARONDES, 1970. Fucose incorporation into glycoproteins of mouse brain. *J. Neurochem.* 17: 157–163.

# 67 The Investigation of Synaptic Function by Means of Subcellular Fractionation Techniques

## V. P. WHITTAKER

TRADITIONAL BIOCHEMICAL techniques have given much information about the chemical composition of the brain; indeed, many of the ubiquitous constituents of animal cells (e.g., the phospholipids) were first isolated from brain tissue.

However, such information is not sufficient for an understanding of how the molecular constituents of nervous tissue are organized at the cellular level. Histochemical and autoradiographic techniques can make valuable contributions in particular instances, but a more general approach is provided by the technique of subcellular fractionation, in which cell structure is broken down by the controlled application of liquid shear forces. The cell fragments and organelles so produced are identified and characterized by a combination of morphological and chemical analysis.

Brain tissue, with its multiplicity of cell types, would be an unpromising subject for this kind of examination, if it were not for an unexpected property of presynaptic nerve terminals: their resistance to breakdown by liquid shear forces is much greater than is that of glial cells and other portions of neurons (Figure 1). This resistance permits the isolation of presynaptic nerve terminals (Gray and Whittaker, 1962) as sealed structures, to which we have given the name synaptosomes (Whittaker et al., 1964). Other structures that can be isolated by subcellular fractionation include fragments of myelinated axons, glial fragments, nuclei, mitochondria, and lengths of preterminal axon.

Synaptosomes may be used for a variety of purposes, of which the most important are as follows. Most obviously, they can be used for studying the mechanisms of synthesis, storage, release, and ultimate destruction of transmitter substances, and the effects of drugs and toxins on these processes.

By subfractionating synaptosomes, one can study the molecular organization of the presynaptic terminal—the composition of the external membranes, synaptic vesicles, soluble cytoplasm, and intraterminal mitochondria (Eichberg et al., 1964; Whittaker, 1966b).

A certain proportion of synaptosomes have postsynaptic adhesions (Figure 1); as a result, they can, in principle, be used as a source of cleft material, of postsynaptic membranes, and of postsynaptic receptors, although caution is needed (Marchbanks and Whittaker, 1969; Whittaker, 1969a) in evaluating work done so far (Azcurra and De Robertis, 1967) on these lines.

Synaptosomes are formed, as far as we know, from any type of ending; the mossy-fiber endings in the glomerular layer of the cerebellar cortex pinch off to form synaptosomes that are sedimented at an unusually low speed, on account of their large size (Israël and Whittaker, 1965). Synaptosome preparations are, accordingly, source material for the identification of new transmitter substances, and extracts can be tested on neurons from regions of the brain from which the synaptosomes have been isolated (Krnjević and Whittaker, 1965).

Synaptosomes can be used to provide samples of terminal axoplasm in studies of axoplasmic flow (Barondes, 1966).

The degree of metabolic autonomy of the terminal region of the neuron may be gauged by comparing the incorporation of radioactive precursors into synaptosomes in vitro with that of nerve terminals isolated from more organized preparations (e.g., slices) after exposure of the latter to the precursors.

Synaptosomes are sealed, organized cell fragments with an intact external membrane, cytoplasm containing the full complement of glycolytic enzymes, and, usually, one or more small mitochondria. They can thus be regarded as miniature non-nucleated neurons or even as models of cells in general (Whittaker, 1968b). When warmed in a saline medium suitably fortified with metabolites and co-factors, they synthesize high-energy phosphate (Marchbanks and Whittaker, 1967; Bradford, 1969), extrude potassium (Bradford, 1969), take up sodium (Ling and Abdel-Latif, 1968), synthesize protein (Morgan and Austin, 1968), glycoprotein (Barondes, 1968), and glycolipids (S. Roseman, personal communication), and display sodium-dependent, carrier-mediated uptake of various substances: transmitters, including norepinephrine (Colburn et al., 1968; Bogdanski et al., 1968), and, under special conditions, acetylcholine (Marchbanks, 1969), amino acids (D. G. Grahame-Smith, L. Austin, personal communications), and choline (March-

V. P. WHITTAKER Department of Biochemistry, University of Cambridge, Cambridge, England, and Department of Neurochemistry, Institute for Basic Research in Mental Retardation, Staten Island, New York

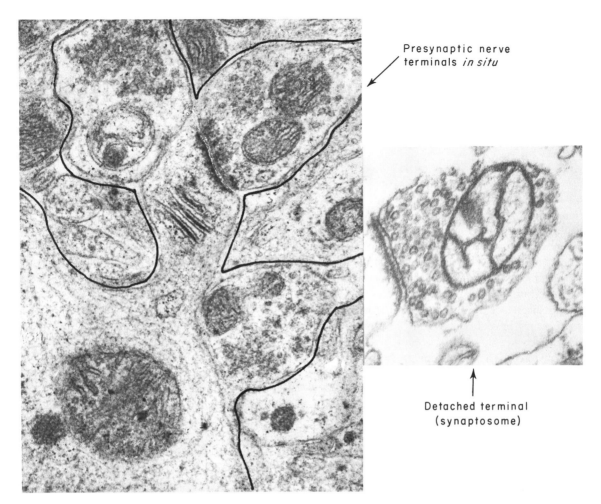

Presynaptic nerve
terminals *in situ*

Detached terminal
(synaptosome)

FIGURE 1    Left. Portion of cortical tissue showing dendrite spine and three presynaptic nerve terminals (outlined). Right. Detached presynaptic nerve terminal (synaptosome) showing similarity of structure to terminal *in situ* and postsynaptic adhesion.

banks, 1968a; Potter, 1968; Diamond and Kennedy, 1969); as a result, they should be useful source material for the isolation of carriers. For studies of membrane function, they are a useful "halfway house" between whole cells and phospholipid spherulites.

Synaptosomes may be used in various ways to enlarge our knowledge of the morphology of the terminal region. High-resolution techniques, such as negative staining, which cannot be used with intact tissue, can be applied to the synaptic region by utilizing synaptosomes and their constituent organelles (Horne and Whittaker, 1962). Portions of postsynaptic membranes often remain adherent to the periphery of synaptosomes, which shows that the cleft contains an effective adhesive binding the presynaptic and postsynaptic membranes together. Other forms of controlled damage to the terminal region give additional information. Thus on hypo-osmotic lysis, synaptosomes do not all disintegrate into their component parts; some hang together, appearing as clumps of synaptic vesicles and intraterminal mitochon-

dria surrounded by broken external membranes but devoid of soluble cytoplasmic constituents (Johnson and Whittaker, 1963; Whittaker et al., 1964), suggesting that the cytoplasm contains a gel-like material. In negative staining, this material resembles a polysaccharide such as Ficoll (observations of V. P. Whittaker reported in part by Hosie, 1965). Quantitative morphological studies can be carried out with synaptosomes (Clementi et al., 1966; Whittaker and Sheridan, 1965) more easily than with whole tissue; synaptosome fractions provide randomized samples of nerve terminals that could overcome sampling problems in evaluating the degree of synaptic connection in normal and pathological development.

## Limitations of the synaptosome technique

The main limitation of the synaptosome technique is that a synaptosome preparation is a mixed population of nerve endings utilizing different transmitters. In special cases, however, specific types of terminal can be separated, e.g., the mossy-fiber and parallel-fiber terminals from the cerebellar cortex (Israël and Whittaker, 1965; Johnston and Larramendi, 1968; Del Cerro et al., 1969). There is also the possibility of using peripheral tissue with a single type of innervation, such as electric organ, with its purely cholinergic innervation (Sheridan et al., 1966; Israël et al., 1968). Unfortunately, in peripheral tissues, the mechanical factors favoring the "pinching-off" of nerve terminals are not present, and the best that can be done so far is to isolate vesicles. This limits the possibility of studying the organization of the whole terminal region. In mammalian peripheral tissues, an additional problem is the small proportion of the whole tissue represented by nerve endings and the consequent low concentration of transmitter.

Another problem is the amount of material that can be processed. Terminals represent only 2 to 3 per cent of brain tissue; synaptic vesicles, only about 0.1 per cent (Whittaker, 1968a). Thus, when one is working with normal centrifuge equipment and the 1 to 10 grams of brain tissue that can be processed in a single run, the yield of material is too small for many purposes, especially if quite homogeneous preparations are required. There is, however, no difficulty in the scaling up of preparations with the zonal rotors now commercially available (Figure 2). At the other end of the scale, fractionation with 1 to 10 milligrams of tissue is feasible (V. P. Whittaker, unpublished observations; Giacobini, 1969), provided sufficiently sensitive analytical methods are available. Such scaled-down operations will be useful in studies of axonal flow and for work on small, relatively homogeneous samples of nervous tissue, such as sympathetic ganglia.

Yet another problem is the sensitivity of synaptosomes to autolysis and mechanical damage. They are much less stable than mitochondria, a fact that may partially explain why they remained undiscovered for so long. If one works at all times under conditions of good refrigeration (preferably in a cold laboratory), however, synaptosomes survive well, and short-term metabolic experiments at 25 to 37° C can be performed on them. There is some evidence that synaptosomes separated in Ficoll density gradients are more stable than those prepared in hyperosmotic sucrose, although they are more contaminated by other structures because of the absence of differential osmotic dehydration. Certain polymers (Ficoll, dextran, polyvinylpyrollidine) are thought to exert a stabilizing effect on cell membranes. However, critical comparisons between synaptosomes prepared in sucrose and Ficoll have not been made, and the latter are metabolically competent. Synaptosomes are considerably more sensitive to damage during fixation, staining, and embedding for electron microscopy than are other subcellular structures. Certain histochemical techniques (e.g., cholinesterase staining) do not give good results with synaptosomes, presumably because of the greater opportunities for the diffusion of products in a pellet compared with that in a whole-tissue block (L. L. Ross, personal communication), but others (e.g., fluorescence histochemistry of monoamines) work well (Masuoka, 1965).

The sensitivity of synaptosomes to autolysis may be the consequence of a local concentration of proteolytic enzymes in the terminal region concerned with the disassembly of proteins arriving there as a result of axoplasmic flow.

A comprehensive account of work on synaptosomes would be out of place here; for further information, the

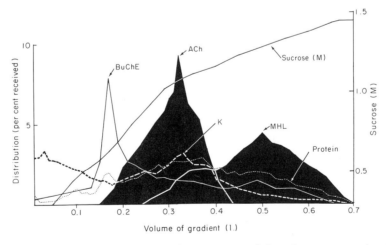

FIGURE 2  Separation of mitochondria (black peak at right) and synaptosomes (central black peak) from myelin and glial fragments (peak at left) from guinea-pig cortex in a 660 ml zonal rotor. Abbreviations: ACh, acetylcholine, a synaptosomal marker; BuChE, butyrylcholinesterase, a putative glial marker; MHL, malate hydrolyase, a mitochondrial marker. Note coincidence of potassium peak (K) and shoulder of mitochondrial marker with synaptosomes. Amount of brain processed was about 20 g.

reader is referred to recent reviews and printed lectures (Whittaker, 1965; Whittaker, 1969a, 1969b, 1969c, 1969d; Marchbanks and Whittaker, 1969). Attention is confined here to current work in my laboratory on the organization of the cholinergic terminal.

## Organization of the cholinergic terminal

LOCALIZATION OF ACETYLCHOLINE It has long been known that only about 25 per cent of the acetylcholine of forebrain can be extracted when the tissue is homogenized or comminuted in iso-osmotic aqueous media (e.g., 0.32 M sucrose) containing a cholinesterase inhibitor (for references, see Hebb and Whittaker, 1958). This acetylcholine is pharmacologically detectable, without further treatment, in the homogenate or in a high-speed supernatant fraction derived from it and is readily hydrolyzed by cholinesterases; the presence of a cholinesterase inhibitor in the suspension medium is necessary to stabilize it.

The remainder of the tissue acetylcholine is particle-bound, resistant to attack by cholinesterases, and undetectable pharmacologically unless first liberated from the particulate material by a variety of treatments that have in common their ability to destroy the integrity of lipoprotein membrane structures. A series of papers (Hebb and Whittaker, 1958; Whittaker, 1959, 1961; Gray and Whittaker, 1960, 1962) established that the "particle" containing the bound acetylcholine is the synaptosome. Bound acetylcholine represents the presynaptic store of acetylcholine in the tissue and owes its existence in homogenates and fractions derived therefrom to the ease with which presynaptic nerve terminals are detached to form synaptosomes. The free acetylcholine must originate either from synaptosomes disrupted during the initial homogenization or from cholinergic axons and cell bodies. The latter origin is more likely, because nerve terminals appear to be converted in high yield to synaptosomes (Clementi et al., 1966), and there are very few vesicles in forebrain homogenates.

Other work (Hebb and Whittaker, 1958; Whittaker, 1959; Whittaker et al., 1964; Whittaker and Sheridan, 1965) established that bound (i.e., synaptosomal) acetylcholine consists of more than one fraction. Disrupting of synaptosomes by suspension in hypo-osmotic solutions causes the release of 40 to 60 per cent of the bound acetylcholine (the "labile-bound" fraction); a variety of evidence (Marchbanks, 1968b) suggests that this is acetylcholine sequestered within the soluble cytoplasm of the synaptosome and bound only in the sense that the external membrane of the synaptosome constitutes a barrier to its free diffusion into the surrounding medium.

Fractionation of disrupted synaptosomes on a sucrose density gradient shows that the osmotically resistant ("stable-bound") fraction of synaptosomal acetylcholine is associated with vesicles; the observed amount per cholinergic

vesicle could be accounted for if the vesicle core is assumed to be filled with an approximately iso-osmotic solution of acetylcholine. This acetylcholine is released by osmotic stresses greater than the stress originally applied to the synaptosomes, suggesting that the acetylcholine is not simply adsorbed to the surface of the vesicle. The actual state of acetylcholine within the vesicle is, however, unknown. It may well be bound to a macromolecule analogous to chromogranin, the protein-binding norepinephrine in adrenergic nerve terminals and chromaffin granules. If so, the greater stability of synaptic vesicles to osmotic shock, compared to that of synaptosomes, might be simply a function of the greater ratio of surface area to volume of the vesicles, also compared to synaptosomes. This greater ratio would permit more rapid equilibration with media of lowered osmotic pressure.

The amount of acetylcholine per cholinergic vesicle (estimated to be about 2000; Whittaker and Sheridan, 1965; Whittaker, 1966a) is consistent with the requirements of the vesicle hypothesis. However, as we shall see, the vesicular pool of acetylcholine turns over more slowly than can account for the specific radioactivity of acetylcholine released from cholinergic terminals labeled with radioactive choline, suggesting an involvement of nonvesicular acetylcholine in transmitter release.

Attempts have been made recently (Chakrin and Whittaker, 1969; Richter and Marchbanks, 1969) to find out by means of radioactive precursors, whether the three fractions of brain acetylcholine simply originate from the redistribution of acetylcholine from a single pool during homogenization or pre-exist in the original tissue. Radioactive choline is incorporated into brain-cortex acetylcholine by intracortical injection in vivo or by the incubation of cortical tissue blocks in a physiological saline medium containing the precursor; the tissue is then homogenized, and synaptosomes and synaptic vesicles are isolated and extracted in the usual way. The extracts contain other radioactive products in addition to acetylcholine (mainly choline and phosphorylcholine). These are removed by submitting the extracts to column or thin-layer chromatography. A liquid ion-exchange extraction method developed by Fonnum (1969) has proved useful for separating choline derivatives from other constituents in the tissue extract. The identity of the radioactive acetylcholine that is isolated is checked by treating a sample of extract with acetylcholinesterase; any cholinesterase-resistant counts appearing in the acetylcholine region are deducted.

The results show (Table I) that vesicular acetylcholine is less labeled, at one hour, than synaptosomal cytoplasmic acetylcholine, and that free acetylcholine is even less labeled. This indicates that the various pools are turning over at different rates and do not result from the redistribution of a single pool during homogenization.

RELATION OF SUBCELLULAR ACETYLCHOLINE TO RELEASED ACETYLCHOLINE    Both types of preparation release acetylcholine spontaneously and, at a greatly increased rate, on stimulation. In the cortical preparation, the released, radioactive acetylcholine is collected by sealing a small sleeve or cup onto the surface of the cortex (Mitchell, 1966); acetylcholine released from the subjacent tissue diffuses into eserinized Ringer solution that has been placed in the cup. Cholinergic afferents can be stimulated by suitably placed electrodes. In the experiments with incubated blocks of cortical tissue, stimulation is brought about by 33 millimolar potassium.

Both types of preparation show a phenomenon that is difficult to explain on the basis of current ideas about transmitter release. The specific radioactivity of the acetylcholine released during stimulation and, still more, that of the acetylcholine released under resting conditions, are considerably higher than that of vesicular acetylcholine. The specific radioactivity of the acetylcholine released under resting conditions is, in fact, higher than *any* subcellular fraction isolated by current techniques. The fall in specific activity that takes place on stimulation may result from the augmentation of the resting release by acetylcholine derived from less radioactive stores, e.g., vesicles.

The identity of the rapidly labeled pool is obscure. It may represent a population of metabolically active but labile vesicles from which acetylcholine is preferentially released but which does not survive isolation. Or it may represent acetylcholine synthesized from choline and taken up from the medium before this has had time to mix completely with endogenous choline; Kety's concept (this volume) of a preferentially released pool of transmitter adsorbed onto the external membrane may be relevant here.

LOCALIZATION OF CHOLINE ACETYLTRANSFERASE    Acetylcholine is synthesized by an enzyme, choline acetyltransferase, that catalyzes the transfer on an acetyl group

from acetylcoenzyme A to choline. The subcellular distribution of this enzyme closely resembles acetylcholine itself, as far as the primary fractions are concerned. As is acetylcholine, about 25 per cent of the homogenate enzyme is in a free, fully active, nonparticulate form; the remainder is synaptosomal. Within the synaptosome, there is probably only one form—cytoplasmic. Evidence to the contrary (McCaman et al., 1965) overlooked the fact (Fonnum, 1967) that at the low ionic strengths of hypo-osmotic synaptosome suspensions, the enzyme tends to be adsorbed onto external membrane fragments, and that such fragments heavily contaminate the vesicle fraction used by these authors. The degree of binding of the enzyme to the external membrane as a function of ionic strength varies from one species to another, increasing from pigeon to guinea pig to rat to rabbit. The cat and sheep enzymes are also relatively strongly adsorbed. At the presumed ionic concentration of the synaptosome cytoplasm, the enzyme, in all species, is largely in the soluble form. However, a small amount of adsorption onto the external membrane in vivo cannot be excluded, and this, rather than extracellular enzyme, might be responsible for the pool of acetylcholine that turns over rapidly and is involved in resting transmitter release.

The mechanism whereby acetylcholine synthesized in the cytoplasm is transferred to the vesicles remains obscure. Possibly the system resembles that of the $Ca^{2+}$-accumulating vesicles derived from the pinching-off of the sarcotubular system of muscle (Makinose and Hasselbach, 1963), but all attempts to detect an acetylcholine-stimulated adenosine triphosphatase or ATP-stimulated radioactive acetylcholine exchange with vesicle preparations have so far failed.

LOCALIZATION OF ACETYLCHOLINESTERASE    The third characteristic component of cholinergic neurons, acetylcholinesterase, has a more complex subcellular distribution than either acetylcholine or choline acetyltransferase. In this case, fortunately, a specific histochemical method exists; applied at the electron-microscope level, it greatly assists in clarifying the otherwise somewhat confusing subcellular distribution data.

The fraction with the highest *specific* activity is the microsomal fraction. Electron-microscope studies (Lewis and Shute, 1966) show that the lumen of the endoplasmic reticulum of cholinergic neurons stained strongly for reaction products; fragments of this will be recovered in the microsome fraction. On the other hand, there are appreciable amounts of the enzyme in the myelin and synaptosome fractions; on subfractionating synaptosomes, the enzyme is localized in the external membrane fraction (Whittaker et al., 1964). Properly prepared vesicles (Figure 3) free from external membrane fragments (but not those prepared by the method of McCaman et al., 1965) have negligible

TABLE I

*Specific radioactivity of acetylcholine of subcellular fractions from guinea-pig brain after intracortical injection of [N-Me-³H] choline given one hour previously*

| Fraction | Specific Radioactivity of Acetylcholine in Fraction (Mean ± S.D.) | |
| --- | --- | --- |
| | Disintegration per Minute per Picomole | As % of Synaptosomes |
| Synaptosomes | 7.1 ± 1.8 | 100 |
| Vesicles | 4.6 ± 0.3 | 66 ± 5 |
| High-speed supernatant* | 2.5 ± 0.1 | 36 ± 1 |

* Eserine present in homogenate.

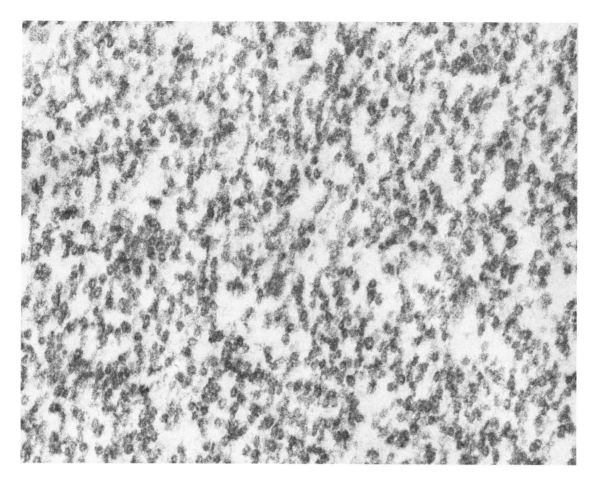

FIGURE 3 Isolated synaptic vesicles prepared by density-gradient centrifuging. Note absence of contaminating membranes.

amounts of esterase activity. These observations agree with the morphological findings: the enzyme is localized in the plasma membrane underneath the myelin sheath and, in high concentration, on the external membrane of the cholinergic nerve terminal.

The coexistence of cytoplasmic acetylcholine and acetylcholinesterase in the synaptosome seems less paradoxical if it is assumed that the functional groups of the enzyme face outward and are therefore not accessible to endogenous acetylcholine. There is some evidence for this: on standing, endogenous acetylcholine declines at a rate determined by diffusion through the external membrane, and far more slowly than a similar concentration of acetylcholine applied externally.

In conclusion, the discovery of synaptosomes has opened up a useful, new, and more direct approach to many problems of synaptic function. Much work must be done, however, before the mechanisms of storage and release of chemical transmitters are fully understood.

REFERENCES

AZCURRA, J. M., and E. DE ROBERTIS, 1967. Binding of dimethyl-$C^{14}$-$d$-tubocurarine, methyl-$C^{14}$-hexamethonium, and $H^3$-alloferine by isolated synaptic membranes of brain cortex. *Int. J. Neuropharmacol.* 6: 15–26.

BARONDES, S. H., 1966. On the site of synthesis of the mitochondrial protein of nerve endings. *J. Neurochem.* 13: 721–727.

BARONDES, S. H., 1968. Incorporation of radioactive glucosamine into macromolecules at nerve endings. *J. Neurochem.* 15: 699–706.

BOGDANSKI, D. F., A. TISSARI, and B. B. BRODIE, 1968. Role of sodium, potassium, ouabain and reserpine in uptake storage and metabolism of biogenic amines in synaptosomes. *Life Sci., Pt. I* 7: 419–428.

BRADFORD, H. F., 1969. Respiration *in vitro* of synaptosomes from mammalian cerebral cortex. *J. Neurochem.* 16: 675–684.

CHAKRIN, L. W., and V. P. WHITTAKER, 1969. The subcellular distribution of [*N-Me*-$^3$H] acetylcholine synthesized by brain *in vivo. Biochem. J.* 113: 97–107.

CLEMENTI, F., V. P. WHITTAKER, and M. N. SHERIDAN, 1966. The yield of synaptosomes from the cerebral cortex of guinea pigs estimated by a polystyrene bead "tagging" procedure. *Z. Zellforsch. Mikroskop. Anat.* 72: 126–138.

COLBURN, R. W., F. K. GOODWIN, D. L. MURPHY, W. E. BUNNEY, JR., and J. M. DAVIS, 1968. Quantitative studies of norepinephrine uptake by synaptosomes. *Biochem. Pharmacol.* 17: 957–964.

DEL CERRO, M. P., R. S. SNIDER, and M. L. OSTER, 1969. Subcellular fractions of adult and developing rat cerebellum. *Exp. Brain Res.* 8: 311–320.

DIAMOND, I., and E. P. KENNEDY, 1969. Carrier-mediated transport of choline into synaptic nerve endings. *J. Biol. Chem.* 244: 3258–3263.

EICHBERG, J., JR., V. P. WHITTAKER, and R. M. C. DAWSON, 1964. Distribution of lipids in subcellular particles of guinea-pig brain. *Biochem. J.* 92: 91–100.

FONNUM, F., 1967. The "compartmentation" of choline acetyltransferase within the synaptosome. *Biochem. J.* 103: 262–270.

FONNUM, F., 1969. Isolation of choline esters from aqueous solutions by extraction with sodium tetraphenylboron in organic solvents. *Biochem. J.* 113: 291–298.

GIACOBINI, E., 1969. Molecular basis of synaptic plasticity. *In* Proceedings of the 1st International Symposium on Cell Biology and Cytopharmacology, Venice (in press).

GRAY, E. G., and V. P. WHITTAKER, 1960. The isolation of synaptic vesicles from the central nervous system. *J. Physiol. (London)* 153: 35P–37P.

GRAY, E. G., and V. P. WHITTAKER, 1962. The isolation of nerve endings from brain: An electron-microscopic study of cell fragments derived by homogenization and centrifugation. *J. Anat.* 96: 79–88.

HEBB, C. O., and V. P. WHITTAKER, 1958. Intracellular distributions of acetylcholine and choline acetylase. *J. Physiol. (London)* 142: 187–196.

HORNE, R. W., and V. P. WHITTAKER, 1962. The use of the negative staining method for the electron-microscopic study of subcellular particles from animal tissues. *Z. Zellforsch. Mikroskop. Anat.* 58: 1–16.

HOSIE, R. J. A., 1965. The localization of adenosine triphosphatases in morphologically characterized subcellular fractions of guinea-pig brain. *Biochem. J.* 96: 404–412.

ISRAËL, M., J. GAUTRON, and B. LESBATS, 1968. Isolement des vésicules synaptiques de l'organe électrique de la Torpille et localisation de l'acétylcholine à leur niveau. *Compt. Rend. Acad. Sci., ser. D* 266: 273–275.

ISRAËL, M., and V. P. WHITTAKER, 1965. The isolation of mossy fibre endings from the granular layer of the cerebellar cortex. *Experientia (Basel)* 21: 325–326.

JOHNSON, M. K., and V. P. WHITTAKER, 1963. Lactate dehydrogenase as a cytoplasmic marker in brain. *Biochem. J.* 88: 404–409.

JOHNSTON, N. L., and L. M. H. LARRAMENDI, 1968. The separation and identification of fractions of nonmyelinated axons from the cerebellum of the cat. *Exp. Brain Res.* 3: 326–340.

KRNJEVIĆ, K., and V. P. WHITTAKER, 1965. Excitation and depression of cortical neurones by brain fractions released from micropipettes. *J. Physiol. (London)* 179: 298–322.

LEWIS, P. R., and C. C. D. SHUTE, 1966. The distribution of cholinesterase in cholinergic neurons demonstrated with the electron microscope. *J. Cell Sci.* 1: 381–390.

LING, C.-M., and A. A. ABDEL-LATIF, 1968. Studies on sodium transport in rat brain nerve-ending particles. *J. Neurochem.* 15: 721–729.

McCAMAN, R. E., G. RODRÍGUEZ DE LORES ARNAIZ, and E. DE ROBERTIS, 1965. Species differences in subcellular distribution of choline acetylase in the CNS. A study of choline acetylase, acetylcholinesterase, 5-hydroxytryptophan decarboxylase, and monoamine oxidase in four species. *J. Neurochem.* 12: 927–935.

MAKINOSE, M., and W. HASSELBACH, 1963. Die Regulation der freien Calcium-Konzentration in den Muskelfasern durch die Erschlaffungsvesikel. *Pflügers Arch. Ges. Physiol.* 278: 9–10.

MARCHBANKS, R. M., 1968a. The uptake of [$^{14}$C] choline into synaptosomes *in vitro. Biochem. J.* 110: 533–541.

MARCHBANKS, R. M., 1968b. Exchangeability of radioactive acetylcholine with the bound acetylcholine of synaptosomes and synaptic vesicles. *Biochem. J.* 106: 87–95.

MARCHBANKS, R. M., 1969. Biochemical organization of cholinergic nerve terminals in the cerebral cortex. *In* Cellular Dynamics of the Neurone (S. H. Barondes, editor). Academic Press, New York, pp. 115–135.

MARCHBANKS, R. M., and V. P. WHITTAKER, 1967. Some properties of the limiting membranes of synaptosomes and synaptic vesicles. *In* Abstracts of the 1st International Meeting of the International Society for Neurochemistry, Strasbourg, p. 147.

MARCHBANKS, R. M., and V. P. WHITTAKER, 1969. The biochemistry of synaptosomes. *In* The Biological Basis of Medicine, Vol. 5 (E. E. Bittar and N. Bittar, editors). Academic Press, New York, pp. 39–76.

MASUOKA, D., 1965. Monoamines in isolated nerve ending particles. *Biochem. Pharmacol.* 14: 1688–1689.

MITCHELL, J. F., 1966. Acetylcholine release from the brain. *In* Mechanisms of Release of Biogenic Amines (U. S. von Euler, S. Rosell, and B. Uvnäs, editors). Pergamon Press, Oxford, pp. 425–437.

MORGAN, I. G., and L. AUSTIN, 1968. Synaptosomal protein synthesis in a cell-free system. *J. Neurochem.* 15: 41–51.

POTTER, L. T., 1968. Uptake of choline by nerve endings isolated from the rat cerebral cortex. *In* The Interaction of Drugs and Subcellular Components in Animal Cells (P. N. Campbell, editor). J. and A. Churchill Ltd., London, pp. 293–304.

RICHTER, J. A., and R. M. MARCHBANKS, 1969. Effects of potassium and levorphanol in brain slices on the release of acetylcholine and its synthesis from radioactive choline. *In* Abstracts of the 2nd International Meeting of the International Society for Neurochemistry, Milan, pp. 338–339.

SHERIDAN, M. N., V. P. WHITTAKER, and M. ISRAËL, 1966. The subcellular fractionation of the electric organ of *Torpedo. Z. Zellforsch. Mikroskop. Anat.* 74: 291–307.

WHITTAKER, V. P., 1959. The isolation and characterization of acetylcholine-containing particles from brain. *Biochem. J.* 72: 694–706.

WHITTAKER, V. P., 1961. The binding of neurohormones by subcellular particles of brain tissue. Proceedings of the IVth International Neurochemical Symposium, Varenna, Italy, 1960. *In* Regional Neurochemistry: The Regional Chemistry,

Physiology and Pharmacology of the Nervous System (S. Kety and J. Elkes, editors). Pergamon Press, Oxford, pp. 259–263.

WHITTAKER, V. P., 1965. The application of subcellular fractionation techniques to the study of brain function. *Progr. Biophys. Mol. Biol.* 15: 39–96.

WHITTAKER, V. P., 1966a. The binding of acetylcholine by brain particles *in vitro*. *In* Mechanisms of Release of Biogenic Amines (U. S. von Euler, S. Rosell, and B. Uvnäs, editors). Pergamon Press, Oxford, pp. 147–163.

WHITTAKER, V. P., 1966b. Some properties of synaptic membranes isolated from the central nervous system. *Ann. N. Y. Acad. Sci.* 137: 982–998.

WHITTAKER, V. P., 1968a. The subcellular distribution of amino acids in brain and its relation to a possible transmitter function for these compounds. *In* Structure and Function of Inhibitory Neuronal Mechanisms (C. von Euler, S. Skoglund, and U. Söderberg, editors). Pergamon Press, Oxford and New York, pp. 487–504.

WHITTAKER, V. P., 1968b. The storage of transmitters in the central nervous system. *Biochem. J.* 109: 20P–21P.

WHITTAKER, V. P., 1969a. Subcellular localization of neurotransmitters. *In* Proceedings of the 1st International Symposium on Cell Biology and Cytopharmacology, Venice (in press).

WHITTAKER, V. P., 1969b. The nature of the acetylcholine pools in brain tissue. *Progr. Brain Res.* 31: 211–222.

WHITTAKER, V. P., 1969c. The subcellular fractionation of nervous tissue. *In* Structure and Function of the Nervous System, Vol. 3 (G. Bourne, editor). Academic Press, New York, pp. 1–24.

WHITTAKER, V. P., 1969d. The synaptosome. *In* Handbook of Neurochemistry, Vol. 2 (A. Lajtha, editor). Plenum Press, New York, pp. 327–364.

WHITTAKER, V. P., I. A. MICHAELSON, and R. J. A. KIRKLAND, 1964. The separation of synaptic vesicles from nerve-ending particles ("synaptosomes"). *Biochem. J.* 90: 293–303.

WHITTAKER, V. P., and M. N. SHERIDAN, 1965. The morphology and acetylcholine content of isolated cerebral cortical synaptic vesicles. *J. Neurochem.* 12: 363–372.

# 68  Neurotransmitters, Neurohormones, and Other Small Molecules in Neurons

## LESLIE L. IVERSEN

NEURONS CONTAIN a very large number of substances of low molecular weight, but most of these, such as the sugars, amino acids, fatty acids, nucleotides, and coenzymes involved in intermediary energy metabolism and other metabolic pathways, are universal components of living tissues and have no special relevance for neuronal function. I have chosen to discuss a few aspects of certain substances that are intimately involved in the functional properties of neurons. After a brief description of a group of organic anions that play an important role in maintaining the electrical polarization of the neuronal membrane, the rest of this chapter is devoted to a description of neurotransmitters and other neurosecretory products. All neurons are secretory cells, in the sense that they release a neurotransmitter or a neurohormone. Different types of neuron often have obvious,

morphologically distinguishing features, but they may be differentiated equally characteristically according to the particular neurotransmitter or neurohormone they manufacture and secrete. This chemical individuality of neurons can be exploited by appropriate histochemical techniques as a powerful tool for mapping the anatomical distribution of a particular neuronal type in the heterogeneous population of the nervous system.

This chapter necessarily suffers from major omissions; for instance, space does not allow the inclusion of any description of important groups of substances, such as the carbohydrates involved in neuronal glycoproteins and gangliosides (Barondes, this volume), or the neuronal lipids.

### Organic anions

The presence of organic anions in the axoplasm of neurons is an important feature in the ionic balance leading to the establishment of the resting membrane potential across

LESLIE L. IVERSEN   Department of Pharmacology, University of Cambridge, Cambridge, England

the neuronal membrane. From the studies of Schmitt and his coworkers on the ionic composition of the axoplasm of squid giant nerve fibers and lobster nerves, it was apparent that the concentration of inorganic anions, such as chloride, in the axoplasm was considerably lower than the total concentration of inorganic cation, indicating that a considerable "anion deficit" existed in axoplasm (Bear and Schmitt, 1939). This deficit has since been accounted for by various organic anions that are present in high concentrations in axoplasm. Koechlin (1955), for instance, showed that, in the axoplasm of squid giant fibers, the major organic anion is isethionic acid (2-hydroxyethane sulphonic acid), which is present in very high concentrations, accounting for some 20 per cent of the dry weight of axoplasm (Table I). In other species, various other organic anions may serve a similar role—for example, aspartic acid, glutamic acid, and N-acetyl aspartic acid (Figure 1). N-acetyl aspartic acid is thought to be particularly important in this respect in mammalian brain, in which it occurs in some abundance (Tallan, 1957).

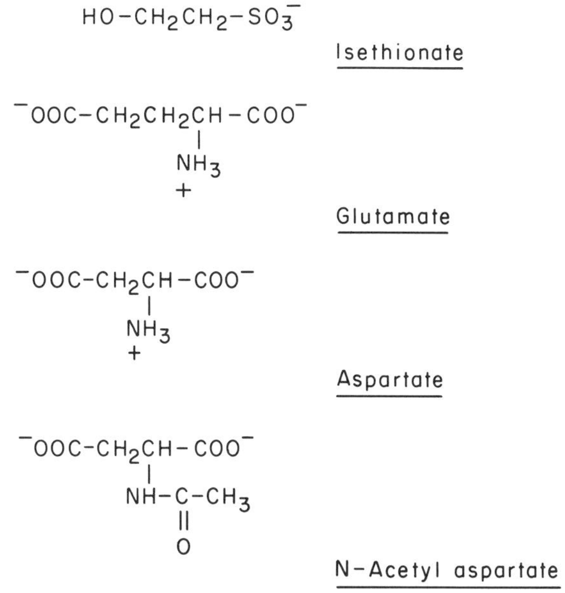

FIGURE 1 Structures of some organic anions found in axoplasm.

## TABLE I

*Acid-base balance in squid nerve axoplasm microequivalents per gram of axoplasm*

| ANIONS | | CATIONS | |
|---|---|---|---|
| Chloride | 140 | Potassium | 344 |
| Phosphates | 24 | Sodium | 65 |
| Aspartic acid | 65 | Calcium | 7 |
| Glutamic acid | 10 | Magnesium | 20 |
| Carboxylic acids | | Organic base | 84 |
| (succinate and fumarate) | 15 | | |
| Unidentified sulphate | 35 | | |
| Isethionate | 220 | | |
| TOTAL | 509 | TOTAL | 520 |

(From Koechlin, 1955.)

## Neurotransmitter substances

IDENTITY AND OCCURRENCE IN VARIOUS TISSUES  Table II lists those compounds known to act as neurotransmitters and the other compounds that probably fill this role. The Table describes, for each compound, the particular synapses for which the most extensive evidence is available. A discussion of the criteria used to establish the identity of neurotransmitter substances, or the detailed evidence available for each of the compounds in the Table are not possible here; these points are covered in the references cited. Although it seems likely that at least 10 different neurotransmitters exist, only three substances have been conclusively established as

transmitters to date: acetylcholine (ACh), norepinephrine (NE), and γ-aminobutyric acid (GABA). Apart from these three compounds, glutamate is a probable candidate as the excitatory transmitter released at insect neuromuscular junctions (Usherwood et al., 1968), and 5-hydroxytryptamine (5-HT, or serotonin) as the cardiac excitatory transmitter in molluscan heart (Welsh, 1957). GABA, glycine, and dopamine are possible inhibitory transmitters in the mammalian CNS (see Table II). The list of neurotransmitters and probable candidates for this role includes a diversity of chemical structures (Figure 2), but all the substances are small diffusible molecules (molecular weight < 300), and there is a preponderance of amines.

Many of the compounds listed in Table II are found throughout the animal kingdom. In particular, the amines (NE, dopamine, 5-HT, and ACh) occur in considerable abundance in many invertebrate nervous systems (Cottrell and Laverack, 1968), but are found only in relatively low concentrations in the mammalian CNS, perhaps reflecting the prominence of other, more recently evolved, transmitter systems in higher animals. Few generalizations can be made about the physiological functions of the various transmitters. ACh may serve both as an excitatory and as an inhibitory transmitter at different synapses in the same organism, and the same applies to NE. Furthermore, the same physiological function may be served by different transmitters in different species, for instance, the excitatory innervation of vertebrate skeletal muscle is exclusively cholinergic, but this function may be served by L-glutamate in crustaceans or by 5-HT in annelids. Some compounds appear to be almost exclusively inhibitory or excitatory in their trans-

| | Transmitter | Synapse Concerned | Effects Excitatory or Inhibitory | Antagonist Drug | Reference |
|---|---|---|---|---|---|
| **ESTABLISHED** | Acetylcholine (ACh) | Vertebrate neuromuscular junctions | + | d-Tubocurarine | Kravitz, 1967 |
| | Acetylcholine | Vertebrate autonomic ganglia | + | Hexamethonium | |
| | Acetylcholine | Vertebrate parasymp. fibers/smooth muscle | + or − | Atropine | |
| | Norepinephrine (NE) | Vertebrate sympathetic fibers/smooth muscle, etc. | + or − | Phenoxybenzamine ($\alpha$) propranalol ($\beta$) | von Euler, 1951 |
| | Adrenalin | Amphibian sympathetic fibers/smooth muscle, etc. | + or − | (As above) | |
| | γ-Aminobutyric acid (GABA) | Crustacean inhibitory neuromuscular junctions | − | Picrotoxin | Kravitz, 1967 |
| **PROBABLE** | 5-hydroxytryptamine | Molluscan cardiac excitatory nerves | + | LSD | Welsh, 1957 |
| | Glutamate | Insect excitatory neuromuscular junctions | + | Not known | Usherwood et al., 1968 |
| | Glycine | Mammalian spinal cord interneurons | − | Strychnine | Werman and Aprison, 1968 |
| | Dopamine | Mammalian nigrostriatal neurons | − | Not known | Hornykiewicz, 1966 |
| | Acetylcholine Dopamine 5-HT GABA Glutamate | Molluscan ganglion | + and − | Numerous | Tauc, 1967 Gerschenfeld, 1965 Kerkut and Walker, 1962 Stefani and Gerschenfeld, 1969 |

mitter role; thus, GABA and dopamine are usually found to produce inhibitory effects, and glutamate to produce excitatory effects. Excitatory effects for GABA and dopamine and inhibitory effects for glutamate have, however, been reported in some cells of molluscan ganglia (Tauc, 1967; Gerschenfeld and Lasansky, 1964; Kerkut and Walker, 1962).

It is clear that the nature of the effect produced by a particular transmitter substance is determined largely by the type of postsynaptic receptor with which the transmitter interacts. Several different types of receptor probably exist for most transmitters; for instance, at least three receptor types are known for acetylcholine in mammalian systems (Table II), and further types of ACh receptors have been described in invertebrates (Tauc, 1967). Two cell types, termed D and H, have been described in molluscan ganglia. Both are depolarized or hyperpolarized respectively by

ACh. Although the receptors involved in these two cell types are distinguishable by pharmacological criteria (hexamethonium blocks ACh effects on D cells but not on H cells), the effect of ACh on both types of cell appears to be an increased permeability to chloride ions (Tauc, 1967). The factor that determines the nature of the response appears to be the intracellular chloride-ion concentration of the two types of cell—this has been measured as $11.2 \pm 0.6$ millimolar for H cells and $24.7 \pm 0.8$ millimolar for D cells. In the D cells, therefore, the chloride equilibrium potential is less negative than the resting potential, so that ACh produces a depolarization, whereas the converse holds for the H cells (Kerkut and Meech, 1966). Even the same transmitter acting on the same receptor type may produce excitatory or inhibitory effects in different tissues. Thus, NE acting on "$\beta$-adrenergic" and "$\alpha$-adrenergic" receptors produces excitatory effects on cardiac and many smooth muscles, but

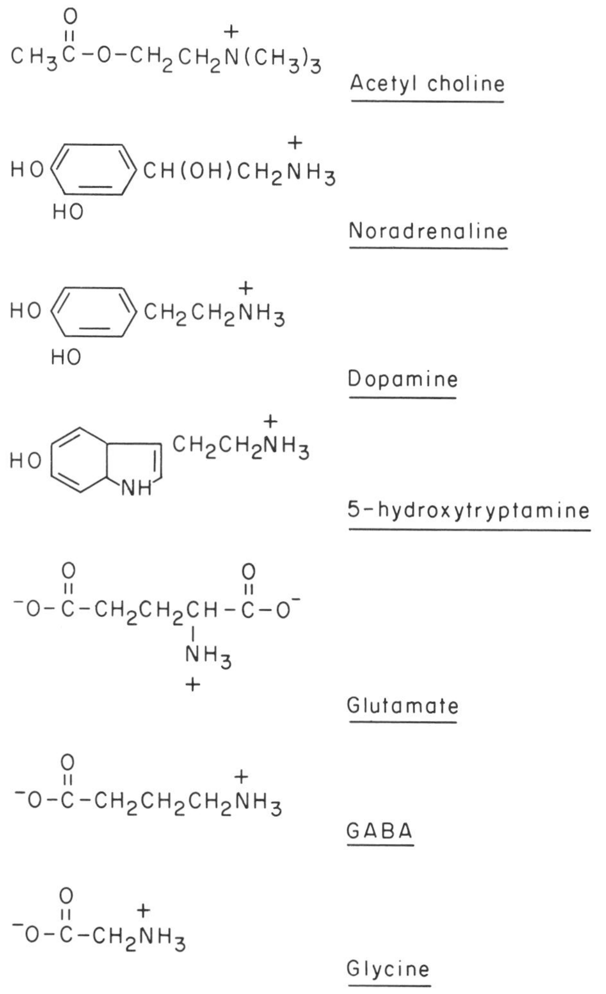

$CH_3\overset{\displaystyle O}{\overset{\|}{C}}-O-CH_2CH_2\overset{+}{N}(CH_3)_3$

Acetyl choline

$HO-\langle\ \rangle-CH(OH)CH_2\overset{+}{N}H_3$ (with HO)

Noradrenaline

$HO-\langle\ \rangle-CH_2CH_2\overset{+}{N}H_3$ (with HO)

Dopamine

$HO-\langle\ \rangle-CH_2CH_2\overset{+}{N}H_3$ (indole, NH)

5-hydroxytryptamine

$^-O-\overset{\displaystyle O}{\overset{\|}{C}}-CH_2CH_2\underset{\underset{+}{NH_3}}{CH}-\overset{\displaystyle O}{\overset{\|}{C}}-O^-$

Glutamate

$^-O-\overset{\displaystyle O}{\overset{\|}{C}}-CH_2CH_2CH_2\overset{+}{N}H_3$

GABA

$^-O-\overset{\displaystyle O}{\overset{\|}{C}}-CH_2\overset{+}{N}H_3$

Glycine

FIGURE 2 Structures of some neurotransmitters and putative neurotransmitters.

produces inhibitory effects on intestinal smooth muscle (Bülbring and Tomita, 1969).

As the number of chemically transmitting synapses with identified transmitters is increased, it may be necessary to consider some formal system of nomenclature to describe the many types of synapse that exist. The currently accepted terms "cholinergic" and "adrenergic" (Dale, 1934) are now seen to be somewhat lacking in precision. The term "cholinergic," for example, does not distinguish between the various types of synapse in which ACh is the transmitter. The term "adrenergic" does not allow a distinction to be made between the various types of synapse in which NE, dopamine, or adrenalin is a transmitter. For a complete description of a particular synapse, a formal system of nomenclature should perhaps describe (1) the transmitter released from the presynaptic nerve terminal; (2) the nature of the effect produced in the postsynaptic cell (i.e., excita-tory or inhibitory); and (3) the type of receptor involved (usually characterized by a specific antagonist drug). The vertebrate neuromuscular junction would thus be described as "acetylcholine, excitatory, d-tubocurarine," the synapses in autonomic ganglia as "acetylcholine, excitatory, hexamethonium," and so on. These might be abbreviated, for example, as ACh-E-C and ACh-E-H, respectively, and so on.

The points outlined above emphasize that the postsynaptic receptor sites with which transmitters interact are of great importance in determining the biological activity of these substances. This is further emphasized when it is realized that the neurotransmitter substances are by no means unique components of nervous tissues. ACh, for example, is found in certain non-neuronal tissues, notably in large amounts in mammalian placenta—which appears to lack any cholinergic innervation (Hebb, 1957). The catecholamines are found in glandular chromaffin cells in the adrenal medulla and in other mammalian tissues (Coupland, 1965); dopamine is present in a specialized form of mast cell in ruminant mammals (Bertler et al., 1959). Serotonin has a particularly ubiquitous distribution; large amounts of this substance are present in enterochromaffin cells of the vertebrate intestinal mucosa, blood platelets, mast cells, certain exocrine and endocrine glandular cells, pinealocytes, frog skin, fish spermatophores, and many insect and molluscan venoms (Erspamer, 1966). The catecholamines and 5-HT are even found in substantial amounts in certain microorganisms and in some plant tissues (Erspamer, 1966). The *reductio ad absurdum* for the notion that neurotransmitters are uniquely neuronal components is seen with the putative transmitters glycine and L-glutamate, which are found in the free amino-acid pool of every living cell as precursors for protein synthesis. The universal occurrence of glycine and glutamate might appear to argue against the view that these substances could act as neurotransmitters, but although such an argument has a certain emotional impact, it is clear that *no* transmitter substance has been shown to have a uniquely neuronal localization. The necessary specificity for chemical transmission is determined not by the occurrence of unique chemicals, but rather by the existence within a given neuron of specialized mechanisms for the manufacture, storage, and release of a particular substance, and by the occurrence of specific receptor sites on the postsynaptic cell. Those sites recognize and respond to the released chemical in a characteristic manner. On this basis, almost any small diffusible substance could act as a transmitter, which does not in any way preclude that substance from serving various other physiological or metabolic functions.

THE BIOSYNTHESIS OF NEUROTRANSMITTERS    The biosynthetic pathways for ACh, the catecholamines, 5-HT, and GABA are outlined in Figure 3 and are not described in

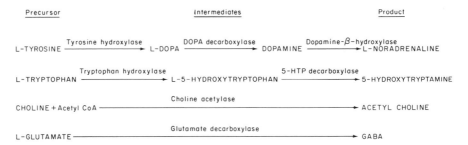

| Precursor | Intermediates | Product |
|---|---|---|

FIGURE 3   Biosynthetic pathways for some neurotransmitters.

detail here. Certain features of transmitter biosynthesis, however, appear to be common to all these systems.

1. Transmitter biosynthesis takes place within the neuron that secretes that transmitter, and the precursor for this biosynthesis is a substance found in some abundance in plasma or other extracellular fluids (choline, L-tyrosine, L-tryptophan, or L-glutamate). Each type of neuron probably possesses a specific transport mechanism to mediate the uptake of the appropriate precursor from the extracellular fluid into the neuron.

2. The enzymes responsible for the biosynthesis of a particular transmitter are found throughout all regions of the neurons that contain that transmitter, i.e., in cell body, preterminal axon, and nerve terminals. The major portion of the biosynthetic enzymes, however, is found in the region of the nerve terminals. Thus, in brain homogenates, most of the choline acetylase, tyrosine hydroxylase, tryptophan hydroxylase, and glutamic decarboxylase activities are recovered in nerve-ending particles or "synaptosomes" (Whittaker, 1965, and this volume). The concentration of the transmitter substance itself in different regions of the neuron often shows a similar distribution, with small amounts occurring in all regions of the cell, but with the highest concentrations in the nerve terminals (Figure 4).

3. In several cases, there is evidence that the rate of transmitter biosynthesis may be controlled by a negative-feedback inhibition exerted by the end product on an early step in the biosynthetic pathway. Thus, in vitro studies have shown that acetylcholine inhibits choline acetylase (Potter et al., 1968), norepinephrine and dopamine inhibit tyrosine hydroxylase (Udenfriend, 1968), and, in crustacean inhibitory neurons, γ-aminobutyric acid inhibits glutamic decarboxylase (Kravitz et al., 1965). These may constitute simple mechanisms by which the rate of transmitter biosynthesis is geared to the rate of utilization. It has been shown that the rate of transmitter synthesis in adrenergic and cholinergic nerves is increased during nerve activity, although the magnitude of this increase seems to be somewhat greater in cholinergic neurons, in which the rate of ACh synthesis may increase by more than seven-fold during nerve stimulation, than in adrenergic nerves, in which in-

creases of threefold in the rate of NE synthesis have been found (Birks and MacIntosh, 1961; Udenfriend, 1968). Feedback inhibition may not be the only mechanism involved in regulating the rate of transmitter synthesis; the synthesis of GABA in mammalian brain, for example, is not subject to feedback inhibition by the product, but is sensitive to inhibition by such anions as chloride (Roberts and Kuriyama, 1968). The synthesis of ACh may also be influenced by the ionic composition of the axoplasm, and, as the synthetic reaction is reversible, may also be governed by simple mass-action effects (Potter et al., 1968).

4. It seems likely that transmitter biosynthesis takes place most rapidly in nerve terminals, which contain the bulk of the biosynthetic enzymes. Indeed, it is difficult to imagine how any mechanism involving a remote synthesis of transmitter could meet the rapid demands for newly synthesized transmitter at the nerve terminal during sustained nerve activity. Estimates of the amount of ACh or NE released by a single nerve impulse in both cholinergic and adrenergic nerves indicate that the total reserves of transmitter stored in the terminals of either type of neuron are sufficient for only about 10,000 impulses, and, because only a portion of the total store is available for release, the functional reserves are sufficient for only a few minutes of sustained activity (Birks and MacIntosh, 1961; Stjärne et al., 1969). In the cholinergic nerves to the cat superior cervical ganglion, for example, the amount of ACh released during a 60-minute period of stimulation at a frequency of 20 impulses per second is equivalent to more than five times the total content of ACh in the ganglion (Birks and MacIntosh, 1961). Furthermore, normal rates of NE biosynthesis can be maintained for long periods in isolated organs, such as the heart or spleen, even though only the adrenergic terminal innervation is preserved in such preparations (Levitt et al., 1965). On the other hand, the manufacture of the enzymes themselves probably occurs exclusively in the perikaryal region of the neuron, which contains the nucleus and most of the mechanisms involved in protein synthesis; the possibility of enzyme synthesis by mitochondria at endings has been considered, but its role is not clear (Lehninger, this volume). There is, furthermore, a continuous flow of newly synthe-

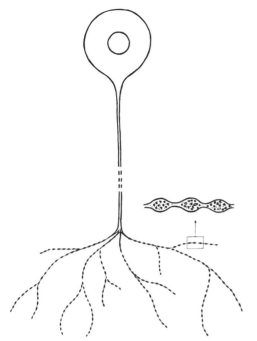

Cell body:

Total content ca. 0.4 pg NE
    (Dahlström and Häggendal, 1966)

    Conc. ca. 10–100 µg/g wet w.
    (Norberg and Hamberger, 1964)

Turn-over of granules ca. 1–5 h.


Varicosities:

Content per varicosity ca. $5 \cdot 10^{-3}$ pg NE

Conc. ca. 1000–3000 µg/g wet w.

Approx. number of amine storage
    granules ca. 1500 per varicosity
    (Dahlström, Häggendal, and Hökfelt,
    1966)


Terminals:

Total content per neuron ca. 0.1–0.4 ng NE;
    Total length per neuron ca. 10–30
    cm (Dahlström and Häggendal 1966)

FIGURE 4    Schematic illustration of the intraneuronal distribution of norepine-phrine in a sympathetic adrenergic neuron of the rat or cat. (Reproduced with permission from Dahlström, 1967; Dahlström and Häggendal, 1966; Dahlström et al., 1966; Norberg and Hamberger, 1964.)

sized macromolecular components from the cell body along the axon toward the terminal regions of the neuron (Weiss, this volume; Davison, this volume), so that the biosynthetic machinery at the nerve terminal is continuously renewed. The biosynthetic enzymes are made in the cell body region of the neuron and must pass down the axon to the terminal regions, which probably explains why the enzymes are found in all regions of the neuron, and may also account for the presence of the transmitter substance in the cell body and axon, although only that in the terminals is functional.

Substances and structures involved in transmitter storage, such as the macromolecular components of synaptic vesicles, are also continuously replenished by perikaryal synthesis and axoplasmic flow. Studies of the rate of transport of adrenergic vesicles (detected by virtue of their NE content) down sympathetic nerve trunks (Dahlström and Häggendal, 1966, 1967) have indicated that the average lifetime of an adrenergic vesicle in the terminals of such neurons is between 35 and 70 days (Figure 4). This finding is of considerable interest, for the value of 35 to 70 days is much longer than the average lifetime (approximately 12 hours) of a norepinephrine molecule in the adrenergic nerve terminals. It follows that the vesicle stores of NE must be constantly replenished by a biosynthesis of NE in the nerve terminals,

and that the amounts of NE which travel down the axon in association with preformed adrenergic vesicles cannot make any substantial contribution to the functional pool of NE in the terminal stores. This conclusion was also reached from the studies of Geffen and Rush (1968) on the rate of transport of NE down sympathetic nerve fibers. The concept of synaptic vesicles as re-usable storage packages is also in accord with current views concerning the mechanisms involved in transmitter release, which are discussed below.

A question of great interest is whether the rate of synthesis and axonal transport of the biosynthetic enzymes and storage-vesicle components is controlled by nerve activity. Although very little information is available on this point, recent reports indicate that the amount of tyrosine hydroxylase activity in adrenergic nerves or in adrenal medullary cells increases when these cells are subjected to stimulation (Viveros et al., 1969; Weiner and Rabadjija, 1968; Mueller et al., 1969). These findings may, perhaps, open new vistas in one of the most exciting directions for neurobiology, namely, the description of mechanisms by which neuronal activity can modify synaptic function.

TRANSMITTER RELEASE    When an action potential reaches the terminal regions of a neuron, the accompanying depolarization is rapidly followed by a release of transmitter substance into the synaptic cleft. Transmitter release takes place with a finite delay which has been measured as between 0.3 and 0.5 milliseconds for the release of ACh at the vertebrate neuromuscular junction (Katz and Miledi, 1965). The precise mechanisms involved in transmitter release are not yet known, but research on this problem and the general area of stimulus-secretion coupling is now proceeding rapidly. It seems likely that transmitter release at all chemically transmitting synapses may, as in the case of the neuromuscular junction, be characteristically of a quantal nature, i.e., the amount of transmitter released from the presynaptic nerve terminal is not continuously graded, but occurs as multiples of a basic quantum amount (Katz, 1966). In the absence of nerve activity, random spontaneous "miniature endplate potentials" (m.e.p.p.s.) were recorded from the innervated muscle fibers when recording electrodes were inserted into the muscle at a point close to the motor endplate. The m.e.p.p.s. appeared to be due to a spontaneous release of ACh from the nerve terminals, as they were blocked by anticholinergic drugs and prolonged by anticholinesterases. Furthermore, the amplitude of the m.e.p.p.s. was remarkably constant, indicating the quantal nature of the release. It could, furthermore, be shown that the release of ACh evoked by nerve stimulation was composed of a burst of some 300 quanta per nerve impulse. This was most convincingly demonstrated in experiments in which stimulation was applied to preparations in a low-calcium or high-magnesium medium, in which the amount of ACh release was much less than that normally found; under such conditions, the amplitude of the evoked potential recorded from the muscle cells during repeated stimulations showed a multimodal frequency distribution, with peaks occurring at one, two, three, and four times the mean amplitude of the spontaneous m.e.p.p.s. (Figure 5) (Boyd and Martin, 1956). Spontaneous miniature synaptic potentials have subsequently been recorded from other noncholinergic synapses: from adrenergic synapses on smooth muscle (Burnstock and Holman, 1966), from crayfish and insect neuromuscular junctions (Dudel and Orkand, 1960; Usherwood, 1963), and from squid giant fiber synapses (Miledi, 1966). It is now generally accepted, although not definitively proved, that the quantal release of transmitter substances is related to the storage of the transmitters in discrete packages in the form of synaptic vesicles, 300 to 500 Å in diameter, which feature prominently in the morphological appearance of all chemically transmitting synapses. For ACh, NE, dopamine, and 5-HT there is evidence that, within the nerve terminals containing these substances, the transmitters are present in high concentrations in the vesicle storage sites (Whittaker, 1965, and this volume). The possibility that they may be present in appreciable amounts in cytoplasmic (nonvesicle) solution is shown by Whittaker (this volume) for ACh and by Kety

(this volume), who postulates substantial storage in the presynaptic membrane. These newer data may have an important influence on our theories of quantal release, storage, and release mechanisms.

If the concept of quantal release is accepted, the problem of the mechanism of release becomes one of explaining how the transmitter substance is transferred from the storage site within a membrane-bound vesicle to the extracellular fluid outside the axonal membrane. As yet, there is very little detailed evidence concerning the mechanisms. Transmitter release does not involve the membrane systems which are responsible for the ionic permeability changes associated with the conduction of the nerve action potential; m.e.p.p.s. and evoked transmitter release can occur in the presence of tetrodotoxin, which specifically blocks the action potential mechanisms (Katz and Miledi, 1965).

The transmitter release process seems likely to prove similar to that in other secretory systems, such as those in the secretion of hormones or other products from glandular tissues. The secretory process in nerves and glandular tissues shows an absolute dependence on the presence of calcium ions in the external fluid surrounding the secretory tissue; glandular secretion is, furthermore, known to involve an inward movement of calcium into the secretory cell. Both neurotransmission and glandular secretion can also be suppressed by elevated magnesium concentrations. The secretory products of glandular tissues are stored in packets, in the form of secretory granules (1000 to 3000 Å in diameter). A glandular secretory system which has been extensively studied is that involved in the release of catecholamines (NE and adrenalin) from the chromaffin cells of the adrenal medulla in response to splanchnic-nerve stimulation or to applied ACh (for reviews, see Douglas, 1968; Smith, 1968). The secretory cells of the adrenal medulla are homologous with adrenergic neurons, and the two types of cell resemble each other in many ways. The release of catecholamines from adrenergic nerves and from the medulla is stimulated by similar secretogogues (ACh, nicotine, DMPP [1, 1-dimethyl-4-phenylpiperazinium iodide], barium); both processes are calcium-dependent. In the medullary cells, the catecholamines are stored in dense-core particles that are similar to, but larger (800 to 1200 Å) than, adrenergic synaptic vesicles.

In both types of storage particle, the amines appear to be bound by a mechanism that causes the formation of a complex with adenosine triphosphate (ATP) or other adenine nucleotide; ATP is present in each case in a stoichiometric ratio of 1 mole of ATP for each 4 moles of catecholamine (Hillarp and Thieme, 1959; Stjärne, 1964). In addition, both medullary particles and adrenergic vesicles contain the enzyme dopamine-$\beta$-hydroxylase, which catalyzes the terminal step in the NE biosynthetic pathway (Figure 3); in medullary particles, the enzyme is partly bound to the

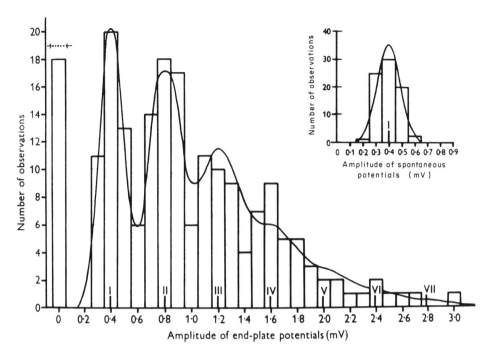

FIGURE 5 Histograms of endplate potentials (e.p.p.) and spontaneous amplitude distribution in a cat muscle fiber in which neuromuscular transmission was blocked by increasing the magnesium concentration of the Krebs solution to 12.5 mM. Peaks in e.p.p. amplitude distribution occur at one, two, three, and four times the mean amplitude of the spontaneous miniature potentials. Gaussian curve is fitted to spontaneous potential distribution and used for calculating theoretical distribution of e.p.p. (continous curve). Arrows indicate expected number of failures of response to nerve stimuli. (From Boyd and Martin, 1956.)

membranes of the storage particles and is partly in free solution within the particles (Viveros et al., 1969). A group of soluble proteins, known collectively as the chromogranins, has also been described in the purified storage particles from the adrenal medulla. The major component of the chromogranins has been isolated and purified (Smith and Winkler, 1967) and a sensitive immunological assay has been developed (Sage et al., 1967). This component, chromogranin A, is a protein of 77,000 molecular weight and is unusally rich in acidic amino-acid residues.

When the adrenal medulla is stimulated to secrete catecholamines, all the soluble contents of the storage particles appear to be discharged to the exterior of the cell. Chromogranin A, and the other chromogranins, ATP, and dopamine-$\beta$-hydroxylase have all been detected in the perfusate from stimulated adrenal glands (see Kirshner et al., 1967; Smith, 1968; Douglas, 1968; Viveros et al., 1969). Furthermore, the ratios of ATP and chromogranin A to catecholamine in the released material are similar to the ratios found for these components in the intact storage particles (Table III). On the other hand, insoluble components associated with the membranes of the medullary storage particles (lysolecithin, bound dopamine-$\beta$-hydroxylase) are not released, nor is there any detectable release of soluble components of the cytoplasm of the medullary cells (lactic dehydrogenase, phenylethanolamine N-methyl transferase). These results thus favor a mechanism of release in which all the soluble contents of the particles are discharged to the exterior, by a process of reverse pinocytosis or "exocytosis" (Figure 6). Such a mechanism was also suggested on the basis of morphological studies of adrenal medullary tissue after intense secretory activity, in which empty storage

TABLE III

*Release of protein and adenosine triphosphate (ATP) from the adrenal gland during stimulation*

| Released Material | Ratio of Catecholamine to Released Material | |
|---|---|---|
| | In Perfusate of Gland | In Intact Storage Particles |
| Storage particle Protein (chromogranin A) | $1.7 \pm 0.19$ | $1.6 \pm 0.19$ |
| Adenine nucleotides (ATP) | $4.22 \pm 0.07$ | Approx. 4.5 |

(From Kirshner et al., 1967; Douglas and Poisner, 1966.)

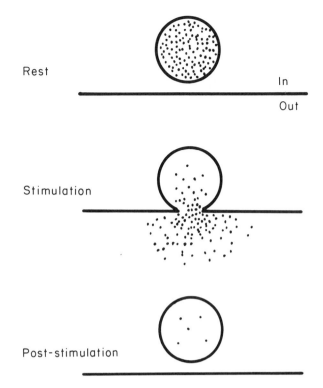

Rest

In

Out

Stimulation

Post-stimulation

FIGURE 6   Diagrammatic illustration of the principle of secretion by exocytosis.

particles were described (de Robertis, 1964). Recently, Poisner and Trifaró (1967) proposed a detailed model of exocytosis in the adrenal medulla that incorporates much of the curently available experimental evidence.

Far less experimental evidence is available concerning the mechanisms of neurotransmitter release, largely because of the technical difficulties involved in measuring the much smaller amounts of material released from nerve terminals as opposed to glandular tissues. Recent reports indicate, however, that a protein immunologically similar to chromogranin A is found in adrenergic vesicles (Banks et al., 1969), and that this protein is present in sympathetic ganglion cells and is transported down sympathetic axons (Livett et al., 1969). That a protein similar to chromogranin A may be implicated in the storage of NE in adrenergic synaptic vesicles and may be released along with NE from adrenergic nerve terminals is an attractive idea. It does, however, raise the question of how such a component could be replaced at the nerve terminals. As discussed above, it seems that most of the transmitter released from nerve terminals is synthesized locally at the terminals—this seems unlikely for a protein such as chromogranin A. If materials such as this are released from nerve terminals, it may be that special rapid-transit systems exist to convey them from the neuronal cell body to the terminal regions.

The probable application of the findings in the adrenal medulla to neurotransmitter release mechanisms raises many interesting questions. It is by no means universally agreed that transmitter release takes place via a mechanism of exocytosis (see, for example, Robertson, this volume). If exocytosis, however, is the universal mechanism of secretion, the implication is that soluble components of synaptic vesicles other than the neurotransmitters themselves may be released along with the transmitter from nerve terminals. Indeed, in the adrenal medulla the amount of protein and ATP secreted is greater (on a weight-for-weight basis) than the amount of catecholamine released. Hitherto, attention has focused entirely on the release of transmitter substances from neurons, but the concomitant release of other substances—both low molecular and macromolecular—may also play an important role in the over-all relationship between the presynaptic neuron and the innervated tissue. For instance, there is good evidence that nerves exert long-term trophic influences on the cells they innervate, a phenomenon that has been most extensively studied in the relationship between nerve and muscle (Guth, 1968, 1969). In vertebrate striated muscle, the presence of the motor innervation influences the sensitivity of the muscle fibers to ACh, their cholinesterase content, and even the type of proteins that are synthesized by the muscle (Guth, 1968). These trophic influences depend on the presence of an intact functional innervation and are abolished if the innervation degenerates or if transmitter release is blocked with botulinum toxin. Although it has been suggested that ACh itself might be the "trophic substance" released from the nerve (Drachman, 1967), there are reasons for believing that some substance other than ACh may be responsible (Miledi, 1960). It seems possible, therefore, that the trophic substance may represent some soluble component of the synaptic vesicles that is released along with ACh, although the existence or chemical identity of such a substance is, of course, still unknown.

TRANSMITTER CATABOLISM AND INACTIVATION   After the neurotransmitter has been released into the synaptic cleft and has interacted in some still-mysterious manner with the postsynaptic receptor sites, it is important that effective concentrations of transmitter should not persist in the synaptic cleft, because the information-transmitting properties of the synapse depend on the recognition by the postsynaptic cell of different firing frequencies of the presynaptic terminal. Although diffusion of the transmitter substance away from the synaptic cleft region may be sufficiently rapid in certain cases to account for the disappearance of transmitter from this region (Gerschenfeld, 1966), some specialized mechanism is thought to exist at most synapses to facilitate the rapid removal of the transmitter after its release. The classical example of such an inactivation system is the presence of acetylcholinesterase at cholinergic synapses to catalyze the rapid hydrolysis of ACh after its release from the presynaptic nerve terminals (see Kravitz, 1967). Although catabolic enzyme systems exist, however, for most of the other transmitter substances, it may be that an enzymic destruction of the transmitter substance will prove to be the exception rather than the rule. At many types of chemically transmitting synapse, the actions of the released transmitter may be terminated not by enzymic degradation but by a physical removal of the transmitter from its extracellular site of action, a process that is mediated by some specific transport system (for review, see Iversen, 1967, 1970). Such transport systems are known to exist for many of the putative transmitters (Table IV); in all cases, the systems have the properties of active-transport mechanisms (saturable, energy-dependent, sodium-dependent). The transmitter uptake systems have exceptionally high affinities for their substrates (affinity constants 0.3–30.0 micromolar) when compared with transport systems for sugars and amino acids in non-neuronal tissues (affinity constants 0.5–10.0 millimolar). There is evidence that the uptake system for catecholamines is situated in the neuronal membrane of adrenergic neurons (Iversen, 1967; this paper, Figure 7). It appears that at adrenergic synapses the transmitter substance is recaptured, in part, after its release from the nerve terminals, and that the same molecules of NE may subsequently be re-used in a further cycle of release and recapture.

In addition to transport systems, which seem likely to be important in terminating the actions of released transmitters, a catabolic enzyme, or enzymes, appropriate for the stored transmitter are known to exist in most neurons. Thus, acetylcholinesterase is not only present at the region of the synapse, but is found in all parts of the cholinergic neuron. In adrenergic neurons, the enzyme monoamine oxidase (MAO) is present, although it does not play an obvious role in adrenergic neurotransmission. Similarly, GABA-glutamate transaminase, the enzyme responsible for the degradation of GABA, is found in the axoplasm of crustacean inhibitory neurons, but does not play an obvious part in inhibitory transmission there (Kravitz et al., 1965; Kravitz, 1967). The presence of such catabolic enzymes may, however, be of some importance in determining the steady-state level of transmitter stored in these neurons. Inhibition of MAO or GABA-glutamate transaminase leads to a pronounced rise in the storage levels of catecholamines or GABA, suggesting that the catabolic enzymes may limit the normal storage levels of these transmitters.

## Neurosecretion

In addition to secreting neurotransmitters, which act directly on the cells in close synaptic contact with the nerve terminals, neurons may also secrete products that act on more remote target cells. The term "neurosecretion" is now commonly used to describe the latter process (for review, see Bern and Knowles, 1966). The axon terminals of neurosecretory neurons do not form synaptic contacts, but are situated in a specialized area of the nervous system termed a "neurohemal organ," in which the materials secreted by the terminals are released into a vascular system. The secretory products ("neurohormones") of these neurons may act on remote peripheral organs through the systemic circulation (color changes in crustaceans, tanning of the integument of insects, water balance in vertebrates, growth and regeneration in annelids). Alternatively, neurohormones may act on other endocrine glands, either through the systemic circulation (ecdysiotropic brain hormone in insects, molt-inhibitory hormone of crustacean eyestalk), or through a specialized system of portal blood vessels (hypothalamic releasing factors in vertebrates). These relationships are illustrated in Figure 8. Neurosecretory systems appear to be particularly common in invertebrate nervous systems, and only a limited number are known to exist in vertebrates. Of these, the hypothalamo-neurohypophyseal system is the most important and the most thoroughly studied. This system (Figure 9) has two principal compo-

TABLE IV

*Properties of some transport systems for neurotransmitters*

| Substance | Tissue | Affinity Constant ($\mu$M) | Reference |
|---|---|---|---|
| l-Norepinephrine | Sympathetic fibers, rat heart | 0.28 | Iversen, 1967 |
| l-Norepinephrine | Rat brain slices | 0.40 | Snyder and Coyle, 1969 |
| Dopamine | Caudate nucleus rat brain | 0.40 | |
| 5-HT | Rat brain slices | 0.56 | Blackburn et al., 1967 |
| GABA | Rat brain slices | 22.0 | Iversen and Neal, 1968 |
| Glycine | Rat spinal cord | 31.0 | Neal and Pickles, 1969 |

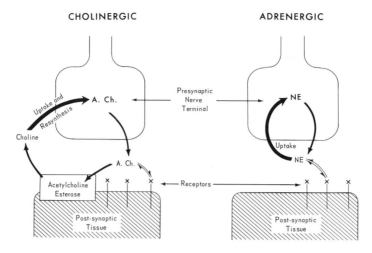

FIGURE 7   Comparison of transmitter inactivation mechanisms at cholinergic and adrenergic synapses.

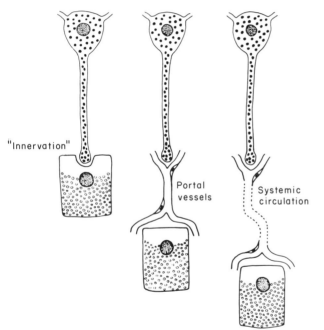

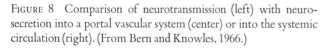

FIGURE 8   Comparison of neurotransmission (left) with neurosecretion into a portal vascular system (center) or into the systemic circulation (right). (From Bern and Knowles, 1966.)

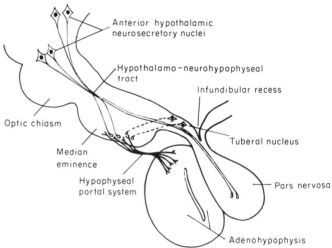

FIGURE 9   Neurosecretory systems in the vertebrate hypothalamoneurohypophyseal region. (From Bern and Knowles, 1966.)

nents. First, several as-yet-unidentified neuronal systems in the hypothalamus secrete "releasing factors" into the specialized portal capillary system leading from the region of the median eminence to the anterior lobe of the pituitary gland. There are at least six such "releasing factors," each of which controls the rate of secretion of one of the endo-crine hormones of the anterior pituitary (McCann and Dhariwal, 1966). Second, two groups of neurons in the anterior hypothalamus and in the supra-optic nucleus send axons to the posterior lobe of the pituitary gland. These neurons manufacture and secrete into the systemic circulation the hormones vasopressin and oxytocin. The hypothalamo-hypophyseal neurosecretory systems are of great importance in vertebrates, particularly in controlling through the anterior pituitary many of the major endocrine glands of the vertebrate body. The hypothalmic neurosecretory neurons thus act as the final common pathway in conveying information from the nervous system to the endocrine system.

The exact chemical identity of the neurohormones released by most neurosecretory neurons is unknown. In almost all cases about which information is available, however, the secretory products appear to be small polypeptides. The structure of oxytocin and vasopressin, basic cystine-containing octapeptides, are illustrated in Figure 10. Within neurosecretory neurons the neurohormones are stored in neurosecretory granules, which are usually larger than synaptic vesicles—1000 to 3000 Å in diameter. Unlike neurotransmitters, neurohormones appear to be synthesized exclusively in the cell-body region of the neuron and are then transported by axoplasmic flow to the nerve terminals. There is evidence (Douglas and Poisner, 1964) that neurosecretion involves mechanisms similar to those described for other secretory processes. Thus, the release of vasopressin from the posterior pituitary shows an absolute requirement for calcium.

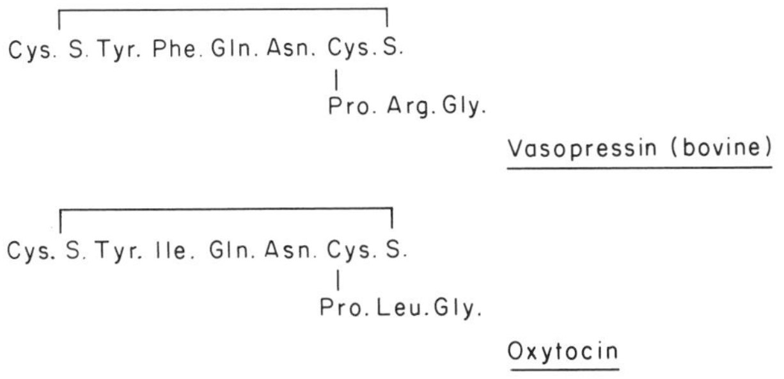

```
 ┌─────────────────────────────────────┐
Cys. S. Tyr. Phe. Gln. Asn. Cys. S.
                                 |
                           Pro. Arg. Gly.
                                        Vasopressin (bovine)

 ┌─────────────────────────────────────┐
Cys. S. Tyr. Ile. Gln. Asn. Cys. S.
                                 |
                           Pro. Leu. Gly.
                                        Oxytocin
```

FIGURE 10   Structures of the neurohormones oxytocin and vasopressin.

The parallel between neurosecretion and glandular secretion has recently been further extended by the discovery that purified neurosecretory granules from the posterior lobe of the pituitary contain two proteins—neurophysin I and neurophysin II. Each of these has a molecular weight of approximately 20,000, and they bind the neurohormones oxytocin and vasopressin. Both neurophysins form complexes with the neurohormones, which can be crystallized; there are three polypeptide binding sites per molecule of neurophysin (Hope and Dean, 1968; Hollenberg and Hope, 1968). There is also evidence that the neurophysins are released from the posterior lobe of the pituitary during secretory activity (Fawcett et al., 1968).

## Conclusion—Dale's law and its implications

Dale's law (Dale, 1935) states that a given neuron secretes the same transmitter substance from all its synaptic terminals. This does not imply that the postsynaptic effects induced by activity in a single neuron are necessarily purely inhibitory or excitatory at all its terminal branches. Indeed, the elegant neurophysiological studies of Kandel et al. (1967) show that in the *Aplysia* ganglion the terminal branches of a single cholinergic interneuron produce both inhibitory and excitatory effects on different postsynaptic cells. This emphasizes again that the actions of the transmitter are determined primarily by the nature of the receptor sites with which it interacts: in this case, the different synaptic actions of ACh are caused by the existence of at least two types of postsynaptic receptor.

Dale's principle, however, is in accord with modern knowledge of transmitter biochemistry. If a single neuron secreted different transmitters from its various branches, the existence of different sets of macromolecular mechanisms for the biosynthesis and storage of such transmitters in the different branches would be indicated. We believe that such macromolecular components are synthesized in the perikaryal region of the neuron and are subsequently transported to the terminal branches; if so, the existence of more than one set of transmitter mechanisms in a single neuron seems unlikely. Thus, the implication of Dale's law is that, during development, some process of differentiation determines which particular secretory product a given neuron will manufacture, store, and release. In biochemical terms, the difference between one neuronal type and another may not be very great. For example, studies of the axoplasmic components of lobster inhibitory and excitatory motor nerve fibers suggest that the major difference between the two may be the presence of a significantly higher activity of the enzyme glutamic decarboxylase in the inhibitory axons; that could account for the accumulation of the inhibitory transmitter GABA in these neurons (Kravitz et al., 1965). It has been suggested that the intracellular chloride-ion concentration in the vertebrate CNS might determine whether a given neuron will accumulate GABA (Roberts and Kuriyama, 1968).

## Summary

In this chapter, I have tried to summarize in broad terms some of the knowledge now available concerning the neuronal mechanisms associated with neurotransmission and neurosecretion. A great deal remains to be learned of such mechanisms. In particular, we must complete the list of transmitter substances, learn more about the importance of substances that may be released along with the transmitter at the presynaptic nerve terminal, determine the role of perikaryal protein synthesis in neurotransmission, and elaborate the pharmacological armory of drugs with specific effects on particular transmitter systems.

Perhaps most importantly we might hope to begin to understand the molecular mechanisms involved in the interaction of the released transmitter with the postsynaptic receptor molecules.

## REFERENCES

BANKS, P., K. B. HELLE, and D. MAYOR, 1969. Evidence for the presence of a chromogranin-like protein in bovine splenic nerve granules. *Mol. Pharmacol.* 5: 210–212.

BEAR, R. S., and F. O. SCHMITT, 1939. Electrolytes in the axoplasm of the giant nerve fibers of the squid. *J. Cell. Comp. Physiol.* 14: 205–215.

BERN, H. A., and F. G. W. KNOWLES, 1966. Neurosecretion. *In* Neuroendocrinology (L. Martini and W. F. Ganong, editors), Vol. 1. Academic Press, New York, pp. 139–186.

BERTLER, Å., B. FALCK, N.-Å. HILLARP, E. ROSENGREN, and A. TORP, 1959. Dopamine and chromaffin cells. *Acta Physiol. Scand.* 47: 251–258.

BIRKS, R., and F. C. MACINTOSH, 1961. Acetylcholine metabolism of a sympathetic ganglion. *Can. J. Biochem. Physiol.* 39: 787–827.

BLACKBURN, K. J., P. C. FRENCH, and R. J. MERRILLS, 1967. 5-hydroxytryptamine uptake by rat brain *in vitro*. *Life Sci.* 6: 1653–1663.

BOYD, I. A., and A. R. MARTIN, 1956. The end-plate potential in mammalian muscle. *J. Physiol.* (*London*) 132: 74–91.

BÜLBRING, E., and T. TOMITA, 1969. Suppression of spontaneous spike generation by catecholamines in the smooth muscle of the guinea pig taenia coli. *Proc. Roy. Soc., ser. B, biol. sci.* 172: 103–119.

BURNSTOCK, G., and M. E. HOLMAN, 1966. Junction potentials at adrenergic synapses. *Pharmacol. Rev.* 18: 481–493.

COTTRELL, G. A., and M. S. LAVERACK, 1968. Invertebrate pharmacology. *Annu. Rev. Pharmacol.* 8: 273–298.

COUPLAND, R. E., 1965. The Natural History of the Chromaffin Cell. Longmans, London.

DAHLSTRÖM, A., 1967. The intraneuronal distribution of noradrenaline and the transport and life-span of amine storage granules in the sympathetic adrenergic neuron. *Arch. Exp. Pathol. Pharmakol.* 257: 93–115.

DAHLSTRÖM, A., and J. HÄGGENDAL, 1966. Studies on the transport and life-span of amine storage granules in a peripheral adrenergic neuron system. *Acta Physiol. Scand.* 67: 278–288.

DAHLSTRÖM, A., and J. HÄGGENDAL, 1967. Studies on the transport and life-span of amine storage granules in the adrenergic neuron system of the rabbit sciatic nerve. *Acta Physiol. Scand.* 69: 153–157.

DAHLSTRÖM, A., J. HÄGGENDAL, and T. HÖKFELT, 1966. The noradrenaline content of the varicosities of sympathetic adrenergic nerve terminals in the rat. *Acta Physiol. Scand.* 67: 289–294.

DALE, H. H., 1934. Nomenclature of fibres in the autonomic system and their effects. *J. Physiol.* (*London*) 80: 10P–11P.

DALE, H. H., 1935. Pharmacology and nerve endings. *Proc. Roy. Soc. Med.* 28: 319–332.

DEROBERTIS, E. D. P., 1964. Histophysiology of Synapses and Neurosecretion. Pergamon Press, Oxford.

DOUGLAS, W. W., 1968. Stimulus-secretion coupling: The concept and clues from chromaffin and other cells. *Brit. J. Pharmacol.* 34: 451–474.

DOUGLAS, W. W., and A. M. POISNER, 1964. Stimulus-secretion coupling in a neurosecretory organ: The role of calcium in the release of vasopressin from the neurohypophysis. *J. Physiol.* (*London*) 172: 1–18.

DOUGLAS, W. W., and A. M. POISNER, 1966. Evidence that the secreting adrenal chromaffin cell releases catecholamines directly from ATP-rich granules. *J. Physiol.* (*London*) 183: 236–248.

DRACHMAN, D. B., 1967. Is acetylcholine the trophic neuromuscular transmitter? *Arch. Neurol.* 17: 206–218.

DUDEL, J., and R. K. ORKAND, 1960. Spontaneous potential changes at crayfish neuromuscular junctions. *Nature* (*London*) 186: 476–477.

ERSPAMER, V., 1966. Occurrence of indolealkylamines in nature. *In* Handbuch der Experimentellen Pharmakologie, Vol. 19, 5-Hydroxytryptamine and Related Indolealkylamines (V. Erspamer, editor). Springer-Verlag, Berlin, pp. 132–181.

EULER, U. S. VON, 1951. The nature of adrenergic nerve mediators. *Pharmacol. Rev.* 3: 247–277.

FAWCETT, P., A. POWELL, and H. SACHS, 1968. Biosynthesis and release of neurophysin. *Endocrinol.* 83: 1299–1310.

GEFFEN, L. B., and R. A. RUSH, 1968. Transport of noradrenaline in sympathetic nerves and the effect of nerve impulses on its contribution to transmitter stores. *J. Neurochem.* 15: 925–930.

GERSCHENFELD, H. M., 1966. Chemical transmitters in invertebrate nervous systems. *Symp. Soc. Exp. Biol.* 20: 299–323.

GERSCHENFELD, H. M., and A. LASANSKY, 1964. Action of glutamic acid and other naturally occurring amino acids on snail central neurons. *Int. J. Neuropharmacol.* 3: 301–314.

GUTH, L., 1968. "Trophic" influences of nerve on muscle. *Physiol. Rev.* 48: 645–687.

GUTH, L., 1969. "Trophic" effects of vertebrate neurons. *Neurosci. Res. Program Bull.* 7 (no. 1): 1–73.

HEBB, C. O., 1957. Biochemical evidence for the neural function of acetylcholine. *Physiol. Rev.* 37: 196–220.

HILLARP, N.-Å., and G. THIEME, 1959. Nucleotides in the catechol amine granules of the adrenal medulla. *Acta Physiol. Scand.* 45: 328–338.

HOLLENBERG, M. D., and D. B. HOPE, 1968. The isolation of the native hormone-binding proteins from bovine pituitary posterior lobes. Crystallization of neurophysin-I and -II as complexes with [8-arginine]-vasopressin. *Biochem. J.* 106: 557–564.

HOPE, D. B., and C. R. DEAN, 1968. Neurosecretory granules of the posterior pituitary. *In* The Interaction of Drugs and Subcellular Components in Animal Cells (P. N. Campbell, editor). J. and A. Churchill Ltd., London, pp. 305–334.

HORNYKIEWICZ, O., 1966. Dopamine (3-hydroxtryamine) and brain function. *Pharmacol. Rev.* 18: 925–964.

IVERSEN, L. L., 1967. The Uptake and Storage of Noradrenaline in Sympathetic Nerves. Cambridge University Press, London.

IVERSEN, L. L., 1970. Neuronal uptake processes for amines and amino acids. *In* Biochemistry of Simple Neuronal Models (E. Costa and E. Giacobini, editors). Raven Press, New York (in press).

IVERSEN, L. L., and M. J. NEAL, 1968. The uptake of ${}^3$H-GABA by slices of rat cerebral cortex. *J. Neurochem.* 15: 1141–1149.

KANDEL, E. R., W. T. FRAZIER, and R. E. COGGESHALL, 1967. Opposite synaptic actions mediated by different branches of an identifiable interneuron on *Aplysia*. *Science (Washington)* 155: 346–349.

KATZ, B., 1966. Nerve, Muscle, and Synapse. McGraw-Hill, New York.

KATZ, B., and R. MILEDI, 1965. The measurement of synaptic delay, and the time course of acetylcholine release at the neuromuscular junction. *Proc. Roy. Soc., ser. B, biol. sci.* 161: 483–495.

KERKUT, G. A., and R. W. MEECH, 1966. The internal chloride concentration of H and D cells in the snail brain. *Comp. Biochem. Physiol.* 19: 819–832.

KERKUT, G. A., and R. J. WALKER, 1962. The specific chemical sensitivity of *Helix* nerve cells. *Comp. Biochem. Physiol.* 7: 277–288.

KIRSHNER, N., H. J. SAGE, and W. J. SMITH, 1967. Mechanism of secretion from the adrenal medulla. II. Release of catecholamines and storage vesicle protein in response to chemical stimulation. *Mol. Pharmacol.* 3: 254–265.

KOECHLIN, B. A., 1955. On the chemical composition of the axoplasm of squid giant nerve fibers with particular reference to its ion pattern. *J. Biophys. Biochem. Cytol.* 1: 511–529.

KRAVITZ, E. A., 1967. Acetylcholine, γ-aminobutyric acid, and glutamic acid: Physiological and chemical studies related to their roles as neurotransmitter agents. *In* The Neurosciences: A Study Program (G. C. Quarton, T. Melnechuk, and F. O. Schmitt, editors). The Rockefeller University Press, New York, pp. 433–444.

KRAVITZ, E. A., P. B. MOLINOFF, and Z. W. HALL, 1965. A comparison of the enzymes and substrates of gamma-aminobutyric acid metabolism in lobster excitatory and inhibitory axons. *Proc. Nat. Acad. Sci. U. S. A.* 54: 778–782.

LEVITT, M., S. SPECTOR, A. SJOERDSMA, and S. UDENFRIEND, 1965. Elucidation of the rate-limiting step in norepinephrine biosynthesis in the perfused guinea-pig heart. *J. Pharmacol. Exp. Ther.* 148: 1–8.

LIVETT, B. C., L. B. GEFFEN, and R. A. RUSH, 1969. Immunohistochemical evidence for the transport of dopamine-β-hydroxylase and a catecholamine binding protein in sympathetic nerves. *Biochem. Pharmacol.* 18: 923–924.

McCANN, S. M., and A. P. S. DHARIWAL, 1966. Hypothalamic releasing factors and the neurovascular link between the brain and the anterior pituitary. *In* Neuroendocrinology (L. Martini and W. F. Ganong, editors), Vol. 1. Academic Press, New York, pp. 261–296.

MILEDI, R., 1960. The acetylcholine sensitivity of frog muscle fibres after complete or partial denervation. *J. Physiol. (London)* 151: 1–23.

MILEDI, R., 1966. Miniature synaptic potentials in squid nerve cells. *Nature (London)* 212: 1240–1242.

MUELLER, R. A., H. THOENEN, and J. AXELROD, 1969. Adrenal tyrosine hydroxylase: Compensatory increase in activity after chemical sympathectomy. *Science (Washington)* 163: 468–469.

NEAL, M. J., and H. G. PICKLES, 1969. Uptake of $^{14}C$-glycine by spinal cord. *Nature (London)* 222: 679–680.

NORBERG, K. A., and B. HAMBERGER, 1964. The sympathetic adrenergic neuron. Some characteristics revealed by histochemical studies on the intraneuronal distribution of the transmitter. *Acta Physiol. Scand.* 63 (suppl. 238): 1–42.

POISNER, A. M., and J. M. TRIFARÓ, 1967. The role of ATP and ATPase in the release of catecholamines from the adrenal medulla. I. ATP-evoked release of catecholamines, ATP, and protein from isolated chromaffin granules. *Mol. Pharmacol.* 3: 561–571.

POTTER, L. T., V. A. S. GLOVER, and J. K. SAELENS, 1968. Choline acetyltransferase from rat brain. *J. Biol. Chem.* 243: 3864–3870.

ROBERTS, E., and K. KURIYAMA, 1968. Biochemical-physiological correlations in studies of the γ-aminobutyric acid system. *Brain Res.* 8: 1–35.

SAGE, H. J., W. J. SMITH, and N. KIRSHNER, 1967. Mechanism of secretion from the adrenal medulla. I. A microquantitative immunologic assay for bovine adrenal catecholamine storage vesicle protein and its application to studies of the secretory process. *Mol. Pharmacol.* 3: 81–89.

SMITH, A. D., 1968. Biochemistry of adrenal chromaffin granules. *In* The Interaction of Drugs and Subcellular Components in Animal Cells (P. N. Campbell, editor). J. and A. Churchill Ltd., London, pp. 239–292.

SMITH, A. D., and H. WINKLER, 1967. Purification and properties of an acidic protein from chromaffin granules of bovine adrenal medulla. *Biochem. J.* 103: 483–492.

SNYDER, S. H., and J. T. COYLE, 1969. Regional differences in $H^3$-norepinephrine and $H^3$-dopamine uptake into rat brain homogenates. *J. Pharmacol. Exp. Ther.* 165: 78–86.

STEFANI, E., and H. M. GERSCHENFELD, 1969. Comparative study of acetylcholine and 5-hydroxytryptamine receptors on single snail neurons. *J. Neurophysiol.* 32: 64–74.

STJÄRNE, L., 1964. Studies of catecholamine uptake storage and release mechanisms. *Acta Physiol. Scand.* 62 (suppl. 228): 1–97.

STJÄRNE, L., P. HEDQUIST, and S. BYGDEMAN, 1969. Neurotransmitter quantum released from sympathetic nerves in cat's skeletal muscle. *Life Sci.* 8 (pt. 1): 189–196.

TALLAN, H. H., 1957. Studies on the distribution of N-acetyl-l-aspartic acid in brain. *J. Biol. Chem.* 224: 41–45.

TAUC, L., 1967. Transmission in invertebrate and vertebrate ganglia. *Physiol. Rev.* 47: 521–593.

UDENFRIEND, S., 1968. Physiological regulation of noradrenaline biosynthesis. *In* Study Group on Adrenergic Neurotransmission, London, 1968. Ciba Foundation Study Group. J. and A. Churchill, London, pp. 3–11.

USHERWOOD, P. N. R., 1963. Spontaneous miniature potentials from insect muscle fibres. *J. Physiol (London)* 169: 149–160.

USHERWOOD, P. N. R., P. MACHILI, and G. LEAF, 1968. L-glutamate at insect excitatory nerve-muscle synapses. *Nature (London)* 219: 1169–1172.

VIVEROS, O. H., L. ARQUEROS, R. J. CONNETT, and N. KIRSHNER, 1969. Mechanism of secretion from the adrenal medulla. III. Studies of dopamine β-hydroxylase as a marker for catecholamine storage vesicle membranes in rabbit adrenal glands. IV. The fate of the storage vesicles following insulin and reserpine administration. *Mol. Pharmacol.* 5: 60–68, 69–82.

WEINER, N., and M. RABADJIJA, 1968. The regulation of norepinephrine synthesis. Effect of puromycin on the accelerated synthesis of norepinephrine associated with nerve stimulation. *J. Pharmacol. Exp. Ther.* 164: 103–114.

WELSH, J. H., 1957. Serotonin as a possible neurohumoral agent:

Evidence obtained in lower animals. *Ann. N. Y. Acad. Sci.* 66: 618–630.

WERMAN, R., and M. H. APRISON, 1968. Glycine: The search for a spinal cord inhibitory transmitter. *In* Structure and Function of Inhibitory Neuronal Mechanisms (C. von Euler, S. Skoglund, and U. Söderberg, editors). Pergamon Press, Oxford and New York, pp. 473–486.

WHITTAKER, V. P., 1965. The application of subcellular fractionation techniques to the study of brain function. *Progr. Biophys. Mol. Biol.* 15: 39–96.

# 69 Structure and Function of Neuroglia: Some Recent Observations

## RICHARD P. BUNGE

IF WE WERE GIVEN a fleeting glance of the image of an infant cradled in his mother's arms, we might immediately presume the mother was feeding the child. Motherhood implies sustenance—it may not occur to us that the mother was, in fact, seeking to protect the child, or attempting to change a diaper. It seems correct to say that the neurobiologist, viewing the image of the nerve cell cradled in the arms of the neuroglia, has generally preferred the assumption that the glia were nourishing the neuron; this assumption has considerably influenced our thinking regarding neuroglial form and function.

Undoubtedly, this view was strengthened by the early recognition (e.g., Clayton, 1932) that some form of neuroglia was found in virtually every type of nervous system studied, both vertebrate and invertebrate (discussed in Bullock and Horridge, 1965; Bunge, 1968). It also was recognized that neuroglia could be found throughout all parts of the nervous system,* both central and peripheral, and that in many regions their numbers far exceeded the number of neurons. It is not difficult to understand how the concept of

---

* In this discussion, the term neuroglia is used for satellite cells and sheath cells of both the central and peripheral nervous system, and therefore will include peripheral ensheathing cells, commonly called Schwann cells.

---

RICHARD P. BUNGE Department of Anatomy, College of Physicians and Surgeons, Columbia University, New York. *Present Address:* Department of Anatomy, Washington University School of Medicine, St. Louis, Missouri

the glial cell nourishing the neuron was gradually extended in the minds of many neurobiologists until the neuron was considered dependent on glial presence.

It is now clear, however, that some situations demonstrate that certain neurons are not dependent on the presence of glia for their day-to-day existence. There are lower forms, e.g., certain coelenterates (Horridge and Mackay, 1962), in which nerve cells are not admixed with glia, and neuronal processes course among a variety of tissue types without any apparent cellular ensheathment (Figure 1A). In the development of higher nervous systems, a large part of early neuron development, including some synapse formation, occurs before substantial glial development has taken place (Bodian, this volume). It has also been observed that some genetic defects of glial development, e.g., the defective myelination of jimpy mice (Wolf and Holden, 1969), do not lead to discernible differences in neuronal maturation. Furthermore, in some regions that normally contain glial cells, neurons and their processes have survived after removal of the glial component. The cytoplasm of the massive packet glial cell surrounding the large neurons in leech ganglia can be removed mechanically; the neurons thus denuded incorporate glucose (Wolfe and Nicholls, 1967) and maintain their resting potential and ability to generate action potentials (Kuffler and Potter, 1964). A substantial number of the satellite cells and Schwann cells in mammalian sensory ganglion cultures can be destroyed by high doses of X-rays, thus denuding large areas of the neuronal soma and leaving many of the unmyelinated axons unsheathed. These neuronal elements survive for several

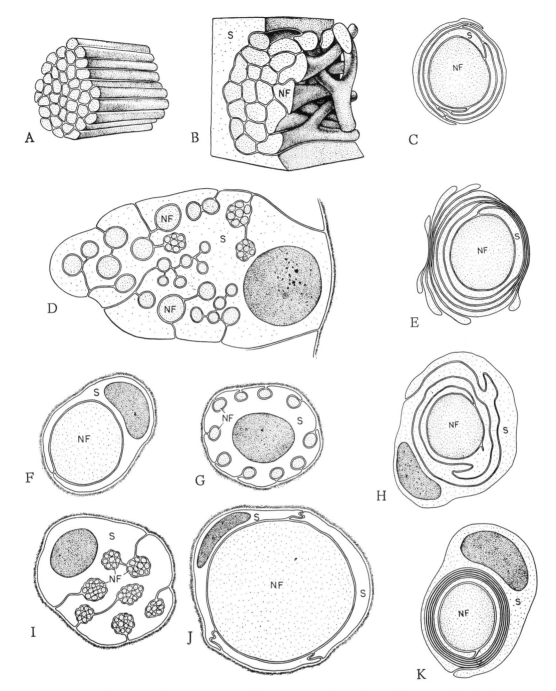

FIGURE 1 Schematic diagrams of various types of glial ensheathment. NF, nerve fiber; S, sheath cell. The glial component of the sheath is emphasized rather than connective tissue elements. A. In some forms nerve fibers have no sheaths at all (as in coelenterates). B. In many forms nerve fibers (or neurons) are ensheathed as a group, with a glial sheath disposed between them and their source of nutrient (as in vertebrate CNS). C. Large nerve fibers (or neurons) may be surrounded by multiple layers of glial processes (as in insects). D. In some forms, large glial cells ensheath axons singly or in groups (as in the leech). E. Multiple layers of glial processes may have some compaction between them (as in the earthworm). F. Large nerve fibers may be sur- rounded by a single Schwann cell (as in insect peripheral nervous system [PNS]). G. Small unmyelinated nerve fibers may be enclosed in individual troughs of glial membrane (as in mammalian PNS). H. In some cases, loosely applied glial processes are irregularly spiraled around the axon (as in insect PNS). I. Very fine fibers may, in some cases, be enclosed as a group in glial membrane troughs (as in vertebrate olfactory nerve). J. Large axons may, in some instances, have a mosaic of Schwann cells applied to their surface (as in the squid). K. In vertebrate PNS (and CNS) systematic spiralization and compaction lead to the forma- tion of myelin. (From Bunge, 1968.)

weeks despite having themselves been damaged by the radiation (Masurovsky et al., 1967). It has also been observed that neurons of the autonomic nervous system can be maintained for substantial periods in culture under conditions in which few satellite cells or fibroblasts develop (Levi-Montalcini and Angeletti, 1963; Burdman, 1968). Finally, it should be noted that, in many areas of the nervous system of both invertebrates and vertebrates, neurons lie in direct contact with one another, and axons course together in bundles directly adjacent to one another without intervening sheaths (Figure 1) (for discussion, see Bunge, 1968).

## The nature of "loose" ensheathment

In addition to the observations outlined above, indicating a degree of metabolic independence for the neuron, we now have a much clearer picture of the anatomical nature of the neuroglial-neuronal relationship. Some of the patterns of glial disposition that have been observed are shown schematically in Figure 1. The generalized relationship between neuronal and glial elements in many of these forms of ensheathment is depicted in Figure 2A. This Figure represents the more basic, unspecialized pattern that might be termed "loose" ensheathment. The neuronal element is harbored in an encircling glial process (or group of processes). The membrane of the glial process is brought into close apposition with the neuronal element, but generally makes no specialized contact with it. This point must be emphasized—in this simple and ubiquitous form of ensheathment there are no specialized contacts between glia and neuron, such as the regions of close membrane apposition that, it has been demonstrated, allow the direct passage of materials between cells in other systems (reviewed in Furshpan and Potter, 1968). Furthermore, the cleft between the neuron and glial cell, represented by shading in the various parts of Figure 2, is known to be relatively open and in continuity with extracellular space. In situations typical of the central nervous system of higher vertebrates, the glial cell is often interposed between blood vessels and neuronal elements, as is shown schematically in Figure 2C. Here, too, the clefts between neuron, glia, and blood vessel constitute spaces open to the movement of a variety of solutes.

Evidence for the relative looseness of this form of glial investment now is available from a variety of experiments. Observations on tissues fixed after the extracellular application of materials that are visible (or can be made visible) in electron micrographs indicate: (1) that the protein ferritin can pass into the 150-Å clefts between satellite cells and neurons in spinal ganglia of the toad (Rosenbluth and Wissig, 1964); (2) that the systems of clefts on the surface of squid nerve fibers (the mesaxon gaps, the Schwann layer channels, and the axolemma-Schwann cell spaces) allow the

penetration of particles as large as those of thorium dioxide (Villegas and Villegas, 1968) (Figure 3); and (3) that ferritin molecules travel quite freely in the clefts between glia and neurons in the vertebrate central nervous system (Brightman, 1965), as do also the marker molecules horseradish peroxidase (Figure 4) (Brightman, 1968; Brightman and Reese, 1969), and saccharated iron oxide (Pappas and Purpura, 1966).

In addition to these anatomical observations, some observations on physiological aspects of neuroglial cells provide direct information regarding intercellular spaces in nervous tissue. Work on the nervous system of the leech (in which the large size of the glial cells facilitates intracellular recordings) and on the optic nerve of the mud puppy and the frog has recently been summarized (Kuffler and Nicholls, 1966; Kuffler, 1967) as follows. It has been established that the neuroglia in these species are surrounded by a high-resistance membrane and have a rather high resting potential. In the mud puppy the resting potential is about 90 millivolts, which is at the equilibrium potential for $K^+$ in Ringer fluid (3 mEq per liter). These glial cells, like most cells, have a high internal $K^+$ concentration; unlike neurons, they do not generate propagated impulses. They are frequently linked to one another, but not to neurons, by special low-resistance junctions. These aspects of glial physiology having been established, it was then possible to use the membrane of glial cells and of neurons as indicators of the external ionic environment, and to establish that, in the preparations studied, the intercellular clefts in the nervous system serve as a rapid and effective pathway for the diffusion of such ions as $Na^+$ and $K^+$ and for such molecules as sucrose, insulin, dextran, or choline. The movement of these substances is thought to be around the glial cells rather than through the glial cytoplasm and can be quite rapid; the time for 75 per cent exchange of $Na^+$ with sucrose is about 10 to 20 seconds in the *Necturus* optic nerve (Kuffler et al., 1966).

Similar types of observations have also been made on peripheral nervous tissue of the squid, in which it is possible to impale axon and Schwann cell sheath cells separately for electrical recording. Here, too, the conclusion is that the narrow, intercellular clefts between Schwann cells surrounding the axon are no barrier for the diffusion of test molecules; the barrier is the axolemma itself (Figure 3) (Villegas and Villegas, 1968). Using this same preparation, these workers have also observed that axonal electrical activity is not accompanied by electrical change in the Schwann cells (Villegas et al., 1962a, 1962b, 1963).

How do we reconcile these properties of "loose" ensheathment with the known presence of firm barriers surrounding nervous tissues in many higher forms? In higher animals, there appears to be provision for a separate barrier

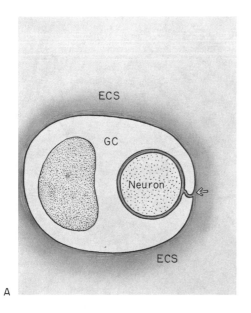

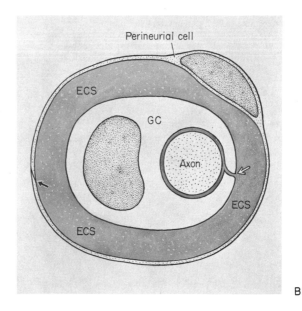

A

B

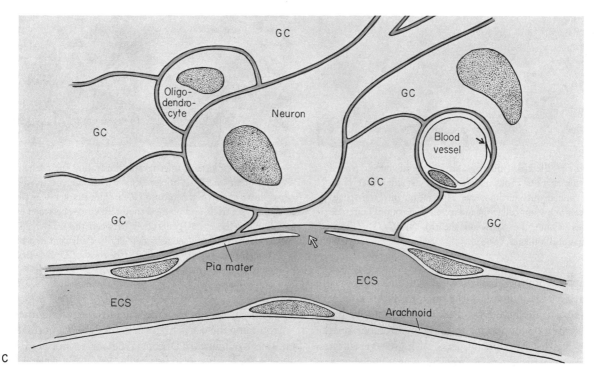

C

FIGURE 2  GC, glial cell; ECS, extracellular space. A. A generalized scheme of neuron-glial relationships is shown to emphasize the continuity (at the arrow) between the spaces around the axon and the general extracellular space. B. Manner in which ECS in certain vertebrate peripheral nerves is separated from general body fluids by a cellular sleeve, the perineurium. The close junction between overlapping perineurial cells (as at the black arrow in B) appear to form the major physical barrier to the general body environment. As shown in C, the extracellular spaces be-tween neurons and glia in mammalian CNS, although small, are believed to be relatively open. The barrier for the passage of protein out of or into CNS blood vessels is known to be at the endothelium of the blood vessels (black arrow). The narrow extracellular spaces of the brain are considered in general continuity with both ventricular and extraventricular cerebrospinal fluid (labeled ECS). A small nonensheathing oligodendrocyte is shown in C. (See text for details.)

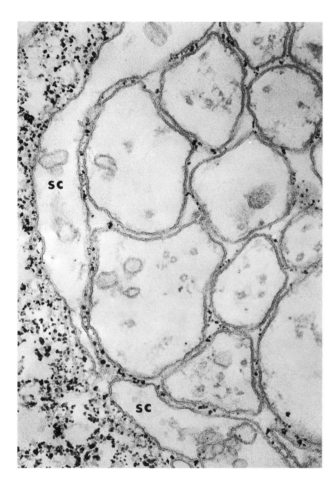

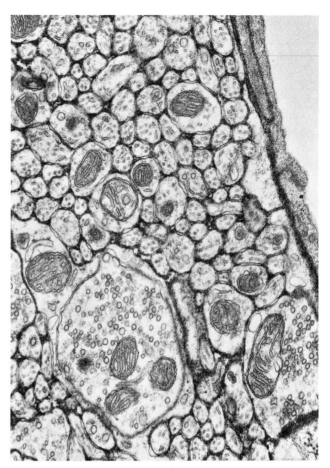

FIGURE 3 Electron micrograph of squid nerve fibers, showing marker molecules (thorium dioxide) that have been applied outside the nerve penetrating into the spaces between Schwann cell (SC) and axons (the roughly circular profiles within the Schwann sheath). × 87,000. (From Villegas and Villegas, 1968.)

FIGURE 4 Electron micrograph illustrating the presence of a protein marker, horseradish peroxidase, in the extracellular clefts of the mouse CNS. The marker was placed in the ventricle and entered the brain spaces between ependymal cells; it has penetrated between the numerous cell processes and between astrocyte endfeet adjacent to a blood vessel (upper right). × 34,000. (Courtesy of Dr. M. Brightman; see Brightman and Reese, 1969.)

system apart from the glial ensheathment under discussion. This barrier system involves the anatomical arrangement and function of the blood vessel endothelium on one hand, and the provision of special encasing membranes on the other. In the distal peripheral nerve of higher vertebrates, as shown schematically in Figure 2B, an encasing sleeve of cells, termed the perineurium, surrounds individual nerve fascicles (see review by Shantha and Bourne, 1968). The inner layer of this encasing sleeve is complete, like a squamous epithelium, and the cells comprising the layer are joined to one another by regions of close membrane apposition. These regions form the barrier that prevents entry of externally applied horseradish peroxidase and lanthanum hydroxide (Reese and Olsson, 1970) and ferritin (Waggener et al., 1965). Epineurial blood vessels, which course within the distal nerve outside the perineurium, appear to be permeable to albumin; when these vessels course internal

to the perineurium, into the spaces and tissues immediately surrounding the nerve fiber itself, they are not normally permeable to albumin (Olsson, 1968) or to horseradish peroxidase or lanthanum hydroxide (Reese and Olsson, 1969). If vascular permeability of vessels entering this space is abnormal, proteins may leak into the region inside the perineurium and perfuse the tissues of the endoneurium. There is some evidence that proteins in this space can gain access to the clefts between Schwann cells and axon (Kaye et al., 1963; Waggener et al., 1965). Thus, the direct barrier for large molecules in these regions of peripheral nerve appears to be not the glial cell but a "tight" vascular system and a special encasing membrane, the perineurium.

In the central nervous system, the arrangement of barriers is much too complex to be discussed in detail here, but there are certain parallels. In Figure 2C, a portion of the central nervous system is shown schematically to emphasize the point that the extracellular fluids in the clefts between cells are apparently part of a compartment that is continuous with the cerebrospinal fluid and is isolated from the blood plasma. The anatomical evidence for this diagram is substantial and has recently been reviewed (Brightman, 1968; Brightman and Reese, 1969; among others). Available evidence indicates that large molecules within the blood stream do not normally enter the brain parenchyma because they are blocked by close junctions between the endothelial cells of blood vessels (Reese and Karnovsky, 1967). Large molecules introduced into the cerebrospinal fluid (CSF) are not restrained by ependyma or pia, and enter the clefts of the nervous system readily; in a similar way, large molecules within the clefts of the CNS parenchyma move readily into the CSF compartment (discussed in Brightman and Reese, 1969). Work on the ionic environment of neurons and glial cells in the brain of an amphibian have indicated that cells within the optic nerve are surrounded by an ionic environment that corresponds more closely to that of the CSF than to that of blood plasma (Cohen et al., 1968). The intercellular clefts of this CNS preparation were found to be open, and ions diffused freely into them from the CSF. This compartment, which includes the CSF and the brain spaces, has special homeostatic mechanisms keeping the ion concentrations around neurons and glia within a narrow range and relatively independent of large changes in the blood plasma (see also Katzman et al., 1968).

It should be emphasized that this cursory discussion of barriers and spaces applies to higher vertebrates; such discrete anatomical barriers may be absent from certain invertebrate nervous systems (e.g., Eldefrawi et al., 1968). The generalization seems justified that where firm barriers exist between nervous tissue and general body fluids they are generally provided by special, nonglial membranes, rather than by glia. The clefts between neuron and glia are open; the glial cell does not comprise the extracellular space for the neuron. (The site of the blood-brain barrier for protein in the shark brain may be an exception to this generalization, according to Brightman et al., 1970.)

## Specializations of glial ensheathment: the myelin sheath

In contrast to the relative "openness" of the clefts between glia and neurons discussed above, in a variety of instances a compaction of glial cell processes around the neuron clearly results in a direct regulation of the environment in the vicinity of the neuron and thus alters its functional properties. One such instance is the myelin sheath. Normally, the ionic currents that are generated during the action-potential flow in the spaces between sheath cell and axon,

as well as in the clefts between sheath-cell processes (refer to Figures 1 and 2). If the lips of the ensheathing process were brought together and "sealed" against one another, as in a close junction (Figure 5), the ionic currents generated by axolemmal conduction would be constrained; this form of sheath would be expected to have a high resistance to radial current flow. If the capacitance of the ensheathing membrane was low, so that little current would be used to "charge up" the sheath, the ionic currents would tend to move along the core of the axon until a site of lower resistance for radial current flow was encountered. These are, of course, the properties of the myelin sheath. Myelin ensheathment differs from that discussed previously in that the ensheathing membranes are systematically compacted together in a form of close junction that is known, in other systems, to provide barriers to extracellular movement of many forms of materials, including ions (e.g., Farquhar and Palade, 1965). Periodically, this compacted form of ensheathment must be interrupted and the constraints on the periaxonal space lifted to allow regeneration of the action potential; these interruptions are the nodes of Ranvier. Here the axolemma is exposed to the usual (or an increased) amount of extracellular space and appears specialized (reviewed by Peters, 1968). The periaxonal space under the myelin would be open to the general extracellular space in the nodal area, but special forms of contacts are also provided in this region (Figure 5). Whether these special nodal glial-axon contacts are needed to aid in the "sealing" of the internodes against ionic currents or whether they provide necessary structural support is not known. Details of peripheral and central myelin and nodal structure have been recently reviewed (Peters, 1968; Elfvin, 1968).

Although it is known that the myelin sheath does provide a high resistance and a low capacitance (Hodgkin, 1964), the exact "tightness" of the seal it provides for the axon is not known. Lanthanum has been observed to penetrate the space between the myelin sheath and the axon of the CNS (Brightman and Reese, 1969), and recently Hirano et al. (1970) have observed that externally applied lanthanum hydroxide marks the region between the special contacts made by the glial cell and axon at the region adjacent to the node. It has also been recently observed that myelin lamellae may present a potential cleft in the interperiod line (Napolitano and Scallen, 1969; Revel and Hamilton, 1969). The myelin "seal" may be only relative, but it is sufficient to provide the control of current flow necessary for saltatory conduction.

## Metabolic properties of myelin

The prevailing view for much of the past decade has been that the components of myelin were relatively stable and, hence, that myelin was relatively inert. Myelin was some-

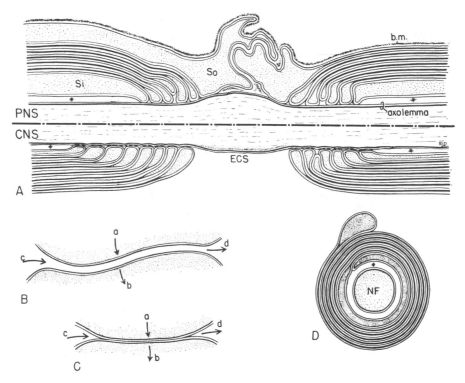

FIGURE 5   A. A nodal region from vertebrate PNS (above) and CNS (below). In the PNS, the Schwann cell provides an internal collar (Si) and an outer collar (So) of cytoplasm in relation to the compact myelin. Terminating loops of the compact myelin come into close apposition to the axolemma in regions near the node in both PNS and CNS; in PNS, Schwann cell processes are coated externally by basement membrane (b.m.) and are loosely interdigitated at the node. D illustrates the confinement of the periaxonal space (marked *) by the compaction of the spiraled membrane of the myelin sheath. In part B a normal intercellular con-figuration is shown; in part C a region of close membrane apposition. When cells are in contact, as in B, materials move quite freely from c to d, but not from a to b. In cells closely apposed, as in C, the movement of small molecules from a to b often is facilitated, and the movement of material from c to d restrained. If the close contacts between myelin terminal loops and axolemma and between compacted myelin lamellae function in this manner, the periaxonal space is relatively isolated from the general extra-cellular space. For details, see text. (From Bunge, 1968.)

times viewed simply as a physical insulator, without reference to its derivation from, and dependence on, a living cell.

During the past few years this picture has been changing rapidly. The mechanism by which the peripheral myelin sheath is derived from the spiral disposition of Schwann-cell membrane has been clarified, and a similar mechanism has been observed during the formation of central myelin from the oligodendrocyte (see review by Bunge, 1968). It has been directly observed that in the CNS more than one segment of myelin may derive from one glial cell; it has recently been estimated that as many as 30 segments may be related to a single oligodendrocyte in the optic nerve of the rodent (Peters and Proskauer, 1969). Our current concept of the anatomy of the central myelin sheath and the myelin-related oligodendrocyte (reviewed by Hirano and Dembitzer, 1967; Bunge, 1968; Peters, 1968) is shown schematically in the three parts of Figure 6. If the myelin segments believed to be related to one oligodendrocyte are unrolled from their usual spiral disposition around the axon, the cell would appear as shown in Figure 6C. The cytoplasm related to central myelin is thus seen to be in tenuous (but probably permanent) connection with the cell body of an oligodendrocyte (Figure 7). At first glance, the configuration of this glial cell seems incredible. When depicted as in Figure 6, however, the cell reveals remarkable similarities to certain neurons that also have long, slender extensions (as axons and dendrites) from the cell body. Is it surprising that a single oligodendrocyte supports several dozen segments of myelin if a single neuron can support a similar number of distally located neuromuscular junctions? It is also interesting to note that the chief cellular organelle present in the narrow extension of oligodendrocyte cytoplasm bordering the compacted regions of myelin is the microtubule (inset, Figure 7). These microtubules (so much

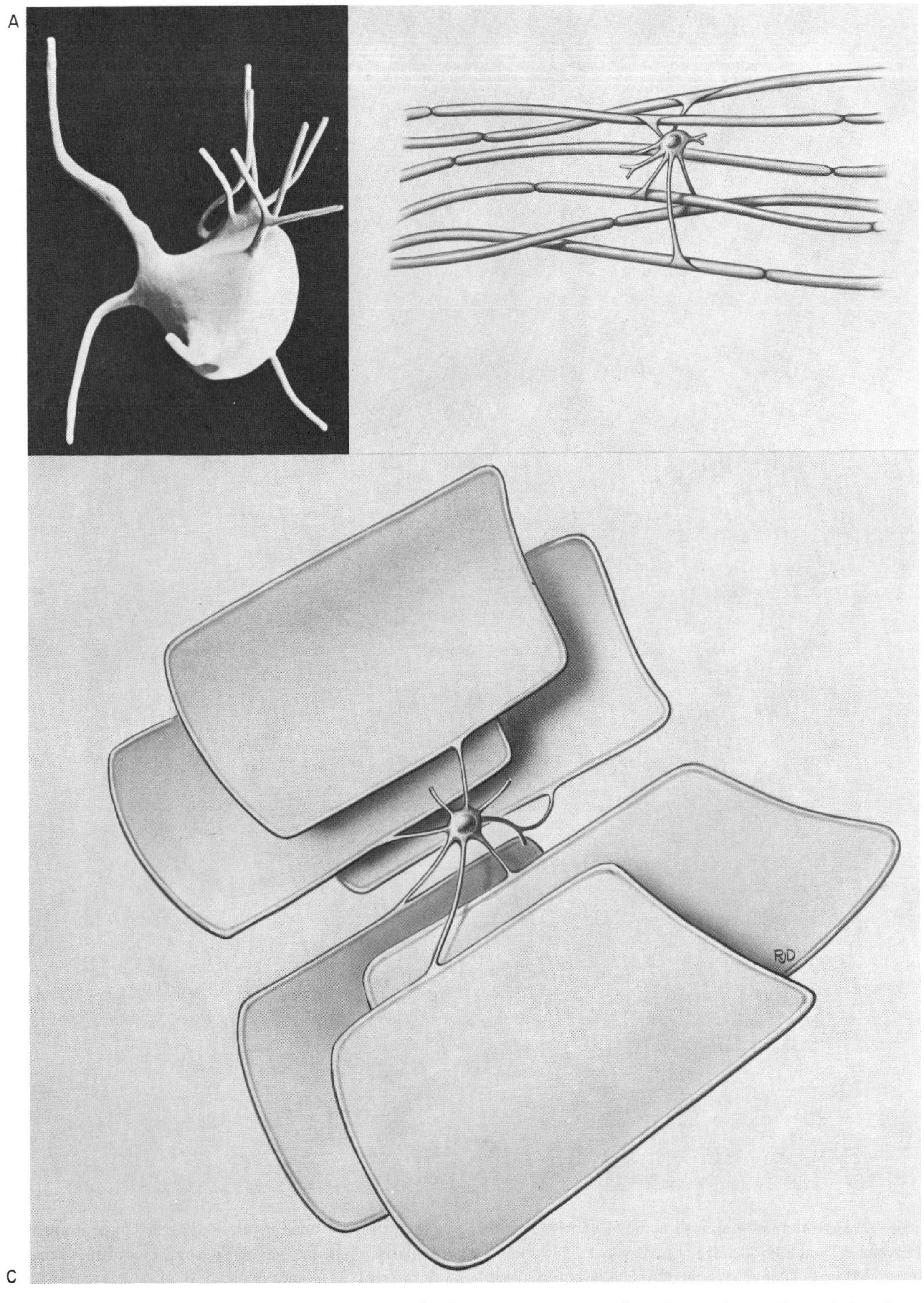

FIGURE 6   Drawings to show the CNS myelin-related cell (the oligodendrocyte) and its postulated relation to segments of myelin. A. The configuration of the cell body as reconstructed from serial sections. (From Stensaas and Stensaas, 1968.) B. The oligodendrocyte in continuity with many segments of myelin in the usual spiraled and compacted configuration. C. The myelin segments in an unrolled configuration. The cytoplasmic extensions of the oligodendrocyte related to myelin have certain similarities to the axonal extensions of neurons.

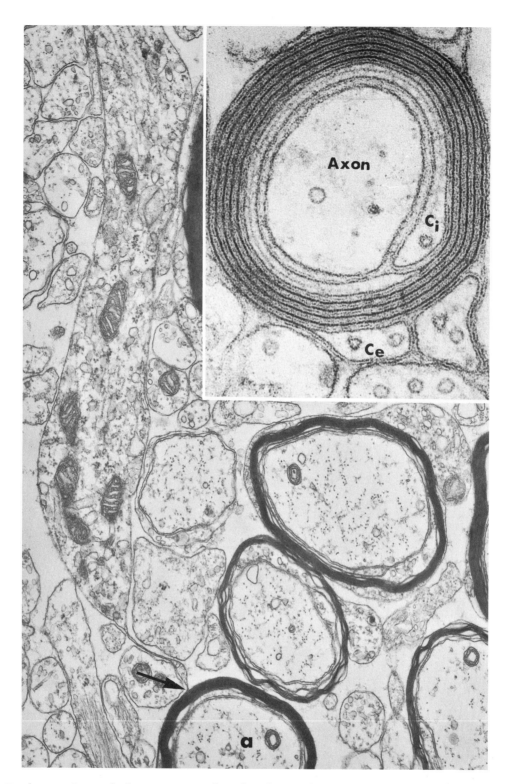

FIGURE 7 Electron micrographs showing CNS myelin and the myelin-related oligodendrocyte. Large E. M. shows the extended process of an oligodendrocyte in kitten spinal cord in tenuous connection (at the arrow) with the membrane of the myelin sheath forming around the axon marked a. The process contains mitochondria, ribosomes, numerous elements of the endoplasmic reticulum, and microtubules. Other myelin sheaths are in various stages of formation. The inset shows a small myelinated axon and demonstrates particularly well the myelin-related cytoplasm situated both internal ($C_i$) and external ($C_e$) to the compacted myelin layers. This cytoplasm contains profiles of microtubules cut in cross section. × 28,000; inset, × 166,000. (Figure courtesy of M. B. Bunge; inset, from Hirano and Dembitzer, 1967.)

studied in relation to axon transport systems) may, in the case of the myelin-related cell, function in facilitating the passage of materials from the cell body into the remote regions of the myelin internode. Indeed, agents such as vinblastine, which affect the organization of microtubules in neurons and axons, similarly affect microtubules in the myelin-related cytoplasm of oligodendrocytes (Schochet and Lampert, 1969). We thus have the picture of the central myelin-related cell with its major cellular machinery concentrated in a perikaryal area but devoted to maintenance of an extensive membrane system some distance away. The same pattern is seen in the neuron—a cell body devoted in large part to the maintenance of its distant extensions and synaptic endings.

Along with this advance in anatomical thinking regarding myelin, there has been increasing evidence that certain chemical components of myelin are turning over at considerably faster rates than would have been expected from early work (discussed in Hendelman and Bunge, 1969). Recent experiments (Horrocks, 1969) indicate that the ethanolamine phosphoglycerides of the myelin fraction of central nervous tissue have an apparent half-life of less than three days (see also Ansell and Spanner, 1967). A half-life of two and a half days has been reported for the sulfatides of the myelin fraction from mammalian brain (Davison and Gregson, 1966). Although these components may be only a portion of the total myelin lipids, the results do point to the necessity for a supporting cell as an active site of the synthetic activity necessary for the maintenance of an intact myelin sheath. In addition, there is the intriguing observation that diphtheria toxin, the potent and specific demyelinating agent (see McDonald and Sears, 1969), acts in certain tissues as a powerful inhibitor of protein synthesis (discussed by Gill et al., 1969). These observations point to specific roles for the myelin-related cell in maintenance as well as in myelin formation, and offer new hope of understanding whether specific demyelinating diseases may result from metabolic deficiencies rather than from the more commonly considered external attack.

Additional data that are changing our concept of the myelin sheath as a static structure are the observations that myelin may undergo a most dramatic splitting and swelling when poisoned with triethyltin (e.g., Hirano et al., 1968) and the recent report (Lampert and Schochet, 1968) that this reaction is similar to that observed in edema in the cerebellar white matter of ducklings treated with isonicotinic acid hydrazide (INH). It seems likely that, when the pharmacology of triethyltin action is better understood and a possible relation to INH action is explored, we may gain new insight into the agencies stabilizing the myelin sheath.

Recently it has been suggested that the myelin sheath, in addition to its effects on impulse conduction, may function to nourish the axon directly (Fleischhauer and Wartenberg, 1967; Singer, 1968). This suggestion derives in part from the observation that radioactivity from labeled amino acids (and uridine) injected systemically can be demonstrated in the myelin sheath (and myelin-related Schwann cell) earlier than it is observed in the axon; this same pattern can be seen if the axon has been severed from its parent cell. There can be many explanations of this observation, taking into consideration the possibility of metabolism directly related to myelin maintenance, as well as the possibility of synthetic mechanisms in the axon itself (Edström and Sjöstrand, 1969). Because it is known that axons can survive for extensive periods (McDonald, 1967) without myelin, and because there is evidence for incorporation of materials directly into myelin (and its related cell) as well as into the axon (Edström and Sjöstrand, 1969), a clear demonstration that materials move between myelin and axon seems most difficult.

## Forms of ensheathment related to myelin

It should be pointed out that certain invertebrates apparently achieve a form of saltatory conduction of the nerve impulse without the systematic spiralization and compaction of glial processes that characterize vertebrate myelin. The prawn has nerve fibers surrounded by partially compacted glial processes that are disposed concentrically around the axon (Heuser and Doggenweiler, 1966) and interrupted periodically by nodes. Without providing systematic compaction of the glial processes, this system allows conduction velocities of up to 20 meters per second. In a giant nerve fiber in the ventral nerve cord of certain shrimps, a highly compacted sheath apparently constrains ionic currents to an extracellular space between the axon and the sheath (Kusano, 1966). This sheath extends without interruption from one ganglion to the next, a distance of about 5 mm, and allows conduction velocities of up to 210 meters per second. Recently Hess et al. (1969) have established a relationship between the appearance of the myelin that ensheaths the synaptic apparatus of the avian ciliary ganglion and the appearance and persistence of electrical coupling in this region.

## Functional aspects of loose ensheathment

The "tight" type of glial sheath discussed above dramatically influences the properties of the periaxonal region, and thus alters the functional properties of the neuron. Do the "loose" forms of ensheathment discussed earlier influence the physiological properties of the underlying neuronal membrane? There is evidence that this type of glial sheath functions in the regulation of the neuronal environment; this is the theme of the ensuing discussion.

We know, first, that the very narrowness of the intercellular gap between neuron and glia can influence axolemmal responses. Experiments on the squid giant axon

(Frankenhaeuser and Hodgkin, 1956) support the view that the increment of K$^+$ released during the passage of an action potential is concentrated briefly in the narrow clefts between the axon and the adjacent glial sheath (see Figure 1). This momentary K$^+$ accumulation accounts for the negative after-potential recorded across the axonal membrane. Orkand et al. (1966) have demonstrated that nerve impulses in the optic nerve of the mud puppy and frog consistently lead to a decrease in the resting potential of glial cells. They further demonstrated that this effect was not mediated electrically, but was caused by K$^+$ release from unmyelinated nerve fibers during activity. In fact, prolonged maximum stimulation at 10 per second or more could lead to a complete block of conduction, the glial cells being depolarized as much as 48 millivolts in some experiments. This depolarization could definitely be attributed to K$^+$ accumulation in the extracellular clefts, and could also be demonstrated after natural stimulation of the retina. Nicholls and Kuffler (1964) showed that the glial-cell resting potentials, in another preparation, are much more sensitive to depolarization by increasing K$^+$ concentrations than are neuronal resting potentials. In discussing their observations, these authors suggest that glia surrounding especially active axons might serve to distribute the K$^+$ released from these areas to adjacent regions via an intracellular pathway (continuous from glia to glia via low-resistance junctions). This movement of current would then contribute to surface potentials recorded with extracellular electrodes. The contribution of the electrical properties of glia to slow-potential changes in nervous tissue has been frequently discussed (Kuffler and Nicholls, 1966; MacKay, 1969; see also Karahashi and Goldring, 1966; Grossman and Hampton, 1968).

Kuffler and Nicholls (1966) suggested that this signal, mediated by K$^+$ and traveling from neurons to glia, could serve as a trigger for release of metabolites from glial cells for use by neurons. The large, ensheathing glia under discussion commonly contain large amounts of glycogen, especially in invertebrates (Wigglesworth, 1960). It is tempting to suggest that these reserves can be made available to the neuron, but direct evidence for such an exchange is lacking (see Wolfe and Nicholls, 1967).

There is, in fact, no direct evidence for the transfer of material from glia to neurons. The search for such an exchange has generally focused on metabolites useful to the neuron in energy production or anabolic processes. Recently, however, there has been considerable discussion of how the glia may assist the neuron in catabolic and protective processes. Palay (1966) suggested that the disposition of a glial mantle around synaptic groupings in certain regions of the CNS may protect these regions from being influenced by adjacent synaptic activities, presumably by somehow limiting the spread of neurotransmitters (Figure 8). Sal-

peter (1969) has demonstrated that a portion of the acetylcholinesterase activity at the neuromuscular junction resides in the glial cap overlying the nerve terminal. Observations made by Smith and Treherne (1965) on cockroach ganglia may be related. Acetylcholine has been considered as a probable neurotransmitter in the insect nervous system, but, paradoxically, large amounts of topically applied acetylcholine have little effect on electrical activity in the cockroach ganglion. Smith and Treherne (1965) found that $^{14}$C-labeled acetylcholine not only penetrates rapidly into the CNS of the insect, encountering no apparent diffusion barrier, but also is rapidly metabolized on entry. In the presence of the anticholinesterase, eserine, the penetration of acetylcholine into the nervous system was not impaired, but its hydrolysis was inhibited. They concluded that in explaining the relative insensitivity of the insect's central nervous system to acetylcholine it was not necessary to postulate an exclusion of this material from CNS tissue; the phenomenon could be accounted for by the activity of the cholinesterase system in the ganglion. Their histochemical studies indicated that cholinesterase activity was particularly well marked in the glial sheath surrounding axons and neurons.

In a situation somewhat parallel to the one above, Hoskin et al. (1966) applied a potent cholinesterase inhibitor, the insecticide di-isopropylphosphorofluoridate (DPF) to preparations of squid giant axons. When $^{14}$C-labeled DPF is applied externally, radioactivity rapidly accumulates in the interior of the axon. The compound appearing in the axon is not DPF, however, but its inactive hydrolysis product, di-isopropyl phosphoric acid. Both the axoplasm and, more importantly (to use their words), the axonal envelope, including the Schwann cell and associated connective tissue, contain sufficient DPF-hydrolyzing enzyme to account for the relatively high concentration of DPF required to affect electrical activity in this preparation. Another observation that supports the general concept that glia regulate the content of extracellular fluids is that glial cells are active in the uptake of proteins present in abnormally high amounts in the intercellular clefts (Blakemore, 1969; Brightman, 1968).

These examples provide the background for what has been termed the gantlet concept (Bunge, 1968) of glial function. The neuron exposes extremely sensitive membranes to the general extracellular environment of the nervous system. Yet it cannot seal off the neuronal plasma membrane to protect it against straying neurotransmitters, neurohumors, toxins, and the like. By forcing substances of this type to run the narrow gantlet of the ensheathing extracellular space in which enzymatic breakdown or uptake can take place, a mechanism may be provided for protection of the axolemma without providing an actual seal. Recent observations (Orkand and Kravitz, in press) on the uptake of gamma aminobutyric acid (GABA) by lobster peripheral

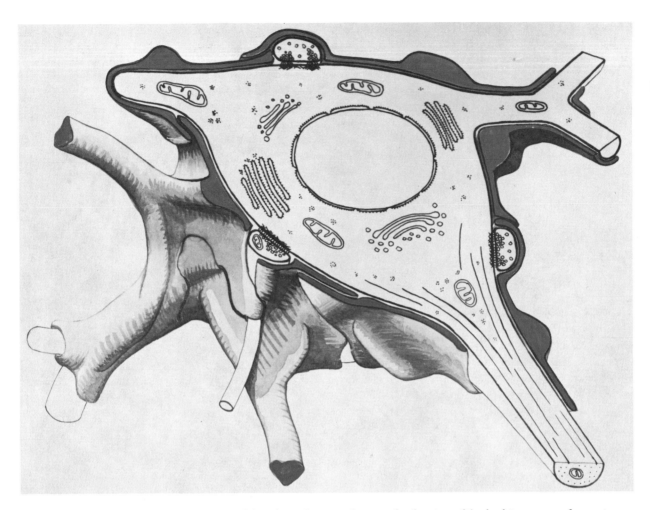

FIGURE 8    Diagrammatic representation of the relationship between a neuron and astrocytic processes. The surface of the neuronal soma is covered with thin cytoplasmic extensions of astrocytes. These also ensheath synaptic regions. In the more distal regions of the dendrite, groups of synaptic terminals are often surrounded by a common astrocytic mantle. (From Wolff, 1968.)

nervous tissue have indicated that this agent (a known transmitter at lobster neuromuscular junctions), when applied externally, is concentrated in the glial and connective tissue sheath cells of the nerve, rather than in the neuronal element (Figure 9). If this uptake is a mechanism for terminating the action of GABA released at synaptic clefts, the system provides a specific example of the contribution of nonneuronal cells to the control of the immediate milieu of the neuron.

A related view regarding the functioning of the narrow extracellular clefts of the brain is discussed in detail elsewhere in this volume. Focusing on the properties of the mucoproteins and mucopolysaccharides that are known to coat cell membranes, including neuronal and glial cell membranes (discussed in Schmitt and Samson, 1969), various workers have suggested that important properties of the extracellular space could be expected to vary as the physical state of the components varied (e.g., Adey et al., 1969). Factors such as hydration and divalent-cation binding by these extracellular components could influence the dimensions and the conductance properties of the intercellular spaces and could significantly alter neuronal function, possibly for extended periods. This view comes into conflict with the discussion above only if these large-molecular-weight intercellular materials are considered to be a complete "filler" material for the regions between nervous system cells. They cannot effectively fill these regions, as the nonselective extracellular movement of materials as large as protein has been clearly demonstrated (as discussed above). The matter is further complicated by observations which suggest that the amount of space between cells in nervous tissue may change in response to alterations in the permeability of adjacent cell membranes (e.g., Van Harreveld et al., 1965). We now know there are extracellular spaces in

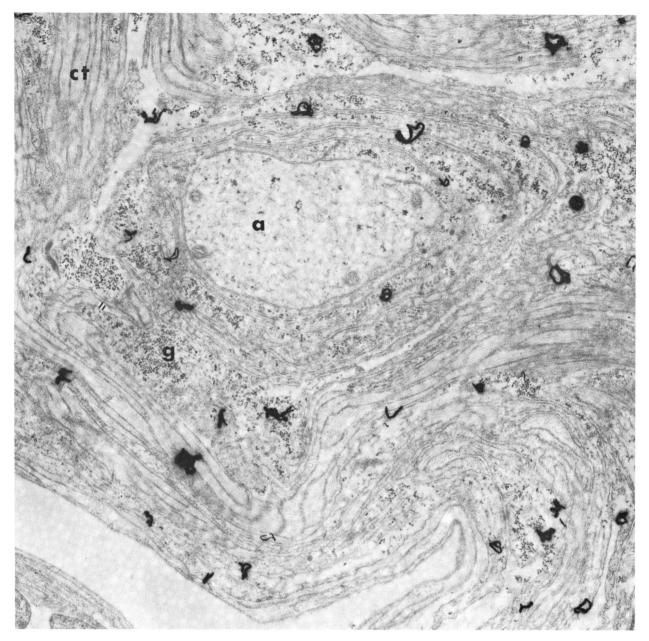

FIGURE 9  Electron-microscope autoradiogram illustrating the sites of GABA binding in lobster peripheral nerve. Labeled GABA was applied externally and, after uptake, was bound to the tissue with a fixative containing glutaraldehyde. The autoradiogram shows label over glial and connective-tissue components surrounding the axon (a). The ensheathing cells contain glycogen-like granules (g) and are interspersed with connective tissue fibers (ct). × 18,000. (From Orkand and Kravitz, in press.)

nervous tissue, but we need to know much more about their exact dimensions and their specific properties.

## Other glial-neuronal interactions

The discussion above has concerned primarily those glial cells that function as ensheathing cells in the central and peripheral nervous systems. There are, especially in higher nervous systems, small cells (the perineuronal oligodendrocytes) that are often disposed as neuron satellites (Figure 2c). Functional aspects of these cells have been discussed in detail by Hydén (1967), with special emphasis on metabolic exchange between neurons and glia in a variety of physiological states. These aspects cannot be considered in detail here, but it should be noted that the type of preparation used in the experiments has evoked considerable recent discussion. A clean separation of glial material from neurons has been attempted, but the purity of the glial preparation has been questioned by several investigators (Kuffler and Nicholls, 1966; Sotelo and Palay, 1968). Similar reservations apply to the recent observations by Hertz (1966) on alterations in "glial" respiration induced by altered Na$^+$ and K$^+$ levels; we have no information regarding the cellular homogeneity of the glial material. Some of the problems of glial isolation from mature nervous tissue have been discussed recently by Cremer et al. (1968).

## What do glial cells do?

The general theme I have presented above is that glia assist in regulating the neuronal environment. When this is done in a dramatic way, such as by the deposition of myelin, the mechanism of glial function can be discerned relatively clearly. When more subtle regulation is involved, the exact function of the glia becomes less clear. The brisk glial responses that occur in the region of the neuronal cell body after an injury to a distant portion of the axon (e.g., Friede and Johnstone, 1967; Blinzinger and Kreutzberg, 1968) suggest that there is, in fact, considerable metabolic "crosstalk" between neurons and glial cells. The earlier allusion to the mother-child relationship again seems appropriate. There is a multitude of overt and subtle interchanges between mother and child; after centuries of observation new aspects of this relationship are still being defined. It is perhaps not surprising that so few precise functions of neuroglia are known at present.

## Summary

Several recent observations suggest that neurons may not depend so directly on neuroglia as has generally been believed. Also, recent evidence indicates that the general form of neuroglial investment (such as that provided by astrocytes and by Schwann cells related to unmyelinated fibers) is a "loose" ensheathment, permitting considerable movement of solutes in the extracellular spaces. Myelin is presented as as an exception to this generalization, representing a systematic compaction of glial processes which are arranged to influence directly the ionic current movement in the vicinity of the axon. The anatomical disposition of the myelin-related cell in the CNS (the oligodendrocyte) is considered, and the mechanisms by which the metabolic demands of myelin may be met are discussed.

Observations indicating that neuroglia generally have a comparatively high resting-membrane potential, and a high internal K$^+$ level, are reviewed. The K$^+$-mediated glial depolarization that follows neuronal activity is discussed.

Several other instances are cited in which glia appear to contribute to the control of the immediate environment of the neuron. The general picture presented is that of neuroglia functioning primarily as regulating agents rather than as indispensable partners of the neuron.

## Acknowledgments

Work in the author's laboratory is supported by Grant NB-04235 from the National Institutes of Health, and Grant 428 from the National Multiple Sclerosis Society. I am grateful to Drs. G. Villegas, J. Wolff, A. Hirano, L. Stensaas, and M. Brightman for providing illustrative material from published work, and particularly thankful to Drs. Orkand and Kravitz for providing details of their work and the electron-microscope autoradiogram in Figure 9, prior to their publication. Mr. Robert Demarest provided the drawings in Figures 1, 2, 5, 6B, and 6C.

### REFERENCES

ADEY, W. R., B. G. BYSTROM, A. COSTIN, R. T. KADO, and T. J. TARBY, 1969. Divalent cations in cerebral impedance and cell membrane morphology. *Exp. Neurol.* 23: 29–50.

ANSELL, G. B., and S. SPANNER, 1967. The metabolism of labelled ethanolamine in the brain of the rat *in vivo*. *J. Neurochem.* 14: 873–885.

BLAKEMORE, W. F., 1969. The fate of escaped plasma protein after thermal necrosis of the rat brain: An electron microscope study. *J. Neuropathol. Exp. Neurol.* 28: 139–152.

BLINZINGER, K., and G. KREUTZBERG, 1968. Displacement of synaptic terminals from regenerating motoneurons by microglial cells. *Z. Zellforsch. Mikroskop. Anat.* 85: 145–157.

BRIGHTMAN, M. W., 1965. The distribution within the brain of ferritin injected into cerebrospinal fluid compartments. II. Parenchymal distribution. *Amer. J. Anat.* 117: 193–220.

BRIGHTMAN, M. W., 1968. The intracerebral movement of proteins injected into blood and cerebrospinal fluid of mice. *Progr. Brain Res.* 29: 19–37.

BRIGHTMAN, M. W., and T. S. REESE, 1969. Junctions between intimately apposed cell membranes in the vertebrate brain. *J. Cell Biol.* 40: 648–677.

BRIGHTMAN, M. W., T. S. REESE, Y. OLSSON, and I. KLATZO, 1970. Morphological restrictions to the passage of exogenous protein within shark brains. *J. Neuropathol. Exp. Neurol.* 29: 123–124.

BULLOCK, T. H., and G. A. HORRIDGE, 1965. Structure and Function in the Nervous Systems of Invertebrates. Freeman and Co., San Francisco.

BUNGE, R. P., 1968. Glial cells and the central myelin sheath. *Physiol. Rev.* 48: 197–251.

BURDMAN, J. A., 1968. Uptake of [$^3$H] catecholamines by chick embryo sympathetic ganglia in tissue culture. *J. Neurochem.* 15: 1321–1323.

CLAYTON, D. E., 1932. A comparative study of the non-nervous elements in the nervous system of invertebrates. *J. Entomol. Zool.* 24: 3–22.

COHEN, M. W., H. M. GERSCHENFELD, and S. W. KUFFLER, 1968. Ionic environment of neurones and glial cells in the brain of an amphibian. *J. Physiol.* (*London*) 197: 363–380.

CREMER, J. E., P. V. JOHNSTON, B. I. ROOTS, and A. J. TREVOR, 1968. Heterogeneity of brain fractions containing neuronal and glial cells. *J. Neurochem.* 15: 1361–1370.

DAVIDSON, A. N., and N. A. GREGSON, 1966. Metabolism of cellular membrane sulpholipids in the rat brain. *Biochem. J.* 98: 915–922.

EDSTRÖM, A., and J. SJÖSTRAND, 1969. Protein synthesis in the isolated Mauthner nerve fibre of goldfish. *J. Neurochem.* 16: 67–81.

ELDEFRAWI, M. E., A. TOPPOZADA, M. M. SALPETER, and R. D. O'BRIEN, 1968. The location of penetration barriers in the ganglia of the American cockroach, *Periplaneta americana* (L.). *J. Exp. Biol.* 48: 325–338.

ELFVIN, L.-G., 1968. The structure and composition of motor, sensory and autonomic nerve fibers. *In* The Structure and Function of the Nervous System, Vol. 1 (G. H. Bourne, editor). Academic Press, New York, pp. 325–377.

FARQUHAR, M. G., and G. E. PALADE, 1965. Cell junctions in amphibian skin. *J. Cell Biol.* 26: 263–291.

FLEISCHHAUER, K., and H. WARTENBERG, 1967. Elektronenmikroskopische Untersuchungen über das Wachstum der Nervenfasern und über das Auftreten von Markscheiden in Corpus callosum der Katze. *Z. Zellforsch. Mikroskop. Anat.* 83: 568–581.

FRANKENHAEUSER, B., and A. L. HODGKIN, 1956. The after effects of impulses in the giant nerve fibres of *Loligo*. *J. Physiol.* (*London*) 131: 341–376.

FRIEDE, R. L., and M. A. JOHNSTONE, 1967. Response of thymidine labeling of nuclei in gray matter and nerve following sciatic transection. *Acta Neuropathol.* 7: 218–231.

FURSHPAN, E. J., and D. D. POTTER, 1968. Low-resistance junctions between cells in embryos and tissue culture. *Curr. Top. Develop. Biol.* 3: 95–127.

GILL, D. M., A. M. PAPPENHEIMER, JR., R. BROWN, and J. T. KURNICK, 1969. Studies on the mode of action of diphtheria toxin VII. Toxin-stimulated hydrolysis of nicotinamide adenine dinucleotide in mammalian cell extracts. *J. Exp. Med.* 129: 1–21.

GROSSMAN, R. G., and T. HAMPTON, 1968. Depolarization of cortical glial cells during electrocortical activity. *Brain Res.* 11: 316–324.

HENDELMAN, W. J., and R. P. BUNGE, 1969. Radioautographic studies of choline incorporation into peripheral nerve myelin. *J. Cell Biol.* 40: 190–208.

HERTZ, L., 1966. Neuroglial localization of potassium and sodium effects on respiration in brain. *J. Neurochem.* 13: 1373–1387.

HESS, A., G. PILAR, and J. N. WEAKLY, 1969. Correlation between transmission and structure in avian ciliary ganglion synapses. *J. Physiol.* (*London*) 202: 339–354.

HEUSER, J. E., and C. F. DOGGENWEILER, 1966. The fine structural organization of nerve fibers, sheaths and glial cells in the prawn, *Palaemonetes vulgaris. J. Cell Biol.* 30: 381–403.

HIRANO, A., and H. M. DEMBITZER, 1967. A structural analysis of the myelin sheath in the central nervous system. *J. Cell Biol.* 34: 555–567.

HIRANO, A., H. M. DEMBITZER, and H. M. ZIMMERMAN, 1970. Observations and reflections on the periaxonal space of the central myelinated axon. *J. Neuropathol. Exp. Neurol.* 29: 124.

HIRANO, A., H. M. ZIMMERMAN, and S. LEVINE, 1968. Intramyelinic and extracellular spaces in triethyltin intoxication. *J. Neuropathol. Exp. Neurol.* 27: 571–580.

HODGKIN, A. L., 1964. The Conduction of the Nerve Impulse. Charles C Thomas, Springfield, Illinois.

HORRIDGE, G. A., and B. MACKAY, 1962. Naked axons and symmetrical synapses in coelenterates. *Quart. J. Microscop. Sci.* 103: 531–541.

HORROCKS, L. A., 1969. Metabolism of the ethanolamine phosphoglycerides of mouse brain myelin and microsomes. *J. Neurochem.* 16: 13–18.

HOSKIN, F. C. G., P. ROSENBERG, and M. BRZIN, 1966. Reexamination of the effect of DFP on electrical and cholinesterase activity of squid giant axon. *Proc. Nat. Acad. Sci. U. S. A.* 55: 1231–1235.

HYDÉN, H., 1967. RNA in brain cells. *In* The Neurosciences: A Study Program (G. C. Quarton, T. Melnechuk, and F. O. Schmitt, editors). The Rockefeller University Press, New York, pp. 248–266.

KARAHASHI, Y., and S. GOLDRING, 1966. Intracellular potentials from "idle" cells in cerebral cortex of cat. *Electroencephalogr. Clin. Neurophysiol.* 20: 600–607.

KATZMAN, R., L. GRAZIANI, and S. GINSBURG, 1968. Cation exchange in blood, brain and CSF. *Progr. Brain Res.* 29: 283–296.

KAYE, G. I., S. DONAHUE, and G. D. PAPPAS, 1963. Electron microscopical evidence for the uptake of colloidal particles by Schwann cells in situ. *J. Microscop.* 2: 605–612.

KUFFLER, S. W., 1967. Neuroglial cells: Physiological properties and a potassium mediated effect of neuronal activity on the glial membrane potential. *Proc. Roy Soc., ser. B, biol. sci.* 168: 1–21.

KUFFLER, S. W., and J. G. NICHOLLS, 1966. The physiology of neuroglial cells. *Ergeb. Physiol. Biol. Chem. Exp. Pharmakol.* 57: 1–90.

KUFFLER, S. W., J. G. NICHOLLS, and R. K. ORKAND, 1966. Physiological properties of glial cells in the central nervous system of Amphibia. *J. Neurophysiol.* 29: 768–787.

KUFFLER, S. W., and D. D. POTTER, 1964. Glia in the leech central nervous system: Physiological properties and neuron-glia relationship. *J. Neurophysiol.* 27: 290–320.

KUSANO, K., 1966. Electrical activity and structural correlates of giant nerve fibers in *Kuruma* shrimp (*Penaeus japonicus*). *J. Cell Physiol.* 68: 361–384.

LAMPERT, P. W., and S. S. SCHOCHET, 1968. Electron microscopic observations on experimental spongy degeneration of the cerebellar white matter. *J. Neuropathol. Exp. Neurol.* 27: 210–220.

LEVI-MONTALCINI, R., and P. U. ANGELETTI, 1963. Essential role of the nerve growth factor in the survival and maintenance of dissociated sensory and sympathetic embryonic nerve cells *in vitro*. *Develop. Biol.* 7: 653–659.

MCDONALD, W. I., 1967. Structural and functional changes in human and experimental neuropathy. *Mod. Trends Neurol.* 4: 145–164.

McDonald, W. I., and T. A. Sears, 1969. Effect of demyelination on conduction in the central nervous system. *Nature (London)* 221: 182–183.

MacKay, D. M., 1969. Evoked brain potentials as indicators of sensory information processing. *Neurosci. Res. Program Bull.* 7 (no. 3): 181–273.

Masurovsky, E. B., M. B. Bunge, and R. P. Bunge, 1967. Cytological studies of organotypic cultures of rat dorsal root ganglia following X-irradiation in vitro. I. Changes in neurons and satellite cells. II. Changes in Schwann cells, myelin sheaths, and nerve fibers. *J. Cell Biol.* 32: 467–518.

Napolitano, L. M., and T. J. Scallen, 1969. Observations on the fine structure of peripheral nerve myelin. *Anat. Rec.* 163: 1–6.

Nicholls, J. G., and S. W. Kuffler, 1964. Extracellular space as a pathway for exchange between blood and neurons in the central nervous system of the leech: The ionic composition of glial cells and neurons. *J. Neurophysiol.* 27: 645–671.

Olsson, Y., 1968. Topographical differences in the vascular permeability of the peripheral nervous system. *Acta Neuropathol.* 10: 26–33.

Orkand, P., and E. Kravitz, 1970. Localization of gamma aminobutyric acid in lobster nerve preparations. *J. Cell Biol.* (in press).

Orkand, R. K., J. G. Nicholls, and S. W. Kuffler, 1966. Effect of nerve impulses on the membrane potential of glial cells in the nervous system of Amphibia. *J. Neurophysiol.* 29: 788–806.

Palay, S. L., 1966. The role of neuroglia in the organization of the central nervous system. *In* Nerve as a Tissue (K. Rodahl and B. Issekutz, Jr., editors). Harper and Row, New York, pp. 3–10.

Pappas, G. D., and D. P. Purpura, 1966. Distribution of colloidal particles in extracellular space and synaptic cleft substance of mammalian cerebral cortex. *Nature (London)* 210: 1391–1392.

Peters, A., 1968. The morphology of axons of the central nervous system. *In* The Structure and Function of the Nervous System, Vol. 1 (G. H. Bourne, editor). Academic Press, New York, pp. 142–186.

Peters, A., and C. C. Proskauer, 1969. The ratio between myelin segments and oligodendrocytes in the optic nerve of the adult rat. *Anat. Rec.* 163: 243 (abstract).

Reese, T. S., and M. J. Karnovsky, 1967. Fine structural localization of a blood-brain barrier to exogenous peroxidase. *J. Cell Biol.* 34: 207–217.

Reese, T. S., and Y. Olsson, 1970. Fine structural localization of a blood-nerve barrier in the mouse. *J. Neuropathol. Exp. Neurol.* 29: 123.

Revel, J.-P., and D. W. Hamilton, 1969. The double nature of the intermediate dense line in peripheral nerve myelin. *Anat. Rec.* 163: 7–16.

Rosenbluth, J., and S. L. Wissig, 1964. The distribution of exogenous ferritin in toad spinal ganglia and the mechanism of its uptake by neurons. *J. Cell Biol.* 23: 307–325.

Salpeter, M. M., 1969. Electron microscope radioautography as a quantitative tool in enzyme cytochemistry. II. The distribution of DFP-reactive sites at motor endplates of a vertebrate twitch muscle. *J. Cell Biol.* 42: 122–134.

Schmitt, F. O., and F. E. Samson, Jr., 1969. Brain cell microenvironment. *Neurosci. Res. Program Bull.* 7 (no. 4): 277–411.

Schochet, S. S., Jr., P. W. Lampert, and K. M. Earle, 1969. Oligodendroglial changes induced by intrathecal vincristine sulfate. *Exp. Neurol.* 23: 113–119.

Shantha, T. R., and G. H. Bourne, 1968. The perineural epithelium—A new concept. *In* Structure and Function of the Nervous System, Vol. 1 (G. H. Bourne, editor). Academic Press, New York, pp. 380–459.

Singer, M., 1968. Penetration of labelled amino acids into the peripheral nerve fibre from surrounding body fluids. *In* Growth of the Nervous System/A Ciba Foundation Symposium (G. E. W. Wolstenholme and M. O'Connor, editors). Little, Brown and Co., Boston, pp. 200–215.

Smith, D. S., and J. E. Treherne, 1965. The electron microscopic localization of cholinesterase activity in the central nervous system of an insect, *Periplaneta americana* L. *J. Cell Biol.* 26: 445–465.

Sotelo, C., and S. L. Palay, 1968. The fine structure of the lateral vestibular nucleus in the rat. I. Neurons and neuroglial cells. *J. Cell Biol.* 36: 151–179.

Stensaas, L., and S. Stensaas, 1968. Astrocytic neuroglial cells, oligodendrocytes and microgliacytes in the spinal cord of the toad. II. Electron microscopy. *Z. Zellforsch. Mikroskop. Anat.* 86: 184–213.

Van Harreveld, A., J. Crowell, and S. K. Malhotra, 1965. A study of extracellular space in central nervous tissue by freeze-substitution. *J. Cell Biol.* 25: 117–137.

Villegas, G. M., and R. Villegas, 1968. Ultrastructural studies of the squid nerve fibers. *J. Gen. Physiol.* 51 (no. 5, pt. 2): 44s–60s.

Villegas, R., C. Caputo, and L. Villegas, 1962a. Diffusion barriers in the squid nerve fiber. The axolemma and the Schwann layer. *J. Gen. Physiol.* 46: 245–255.

Villegas, R., M. Giménez, and L. Villegas, 1962b. The Schwann-cell electrical potential in the squid nerve. *Biochim. Biophys. Acta* 62: 610–612.

Villegas, R., L. Villegas, M. Giménez, and G. M. Villegas, 1963. Schwann cell and axon electrical potential differences. Squid nerve structure and excitable membrane location. *J. Gen. Physiol.* 46: 1047–1064.

Waggener, J. D., S. M. Bunn, and J. Beggs, 1965. The diffusion of ferritin within the peripheral nerve sheath: An electron microscopy study. *J. Neuropathol. Exp. Neurol.* 24: 430–443.

Wigglesworth, V. B., 1960. The nutrition of the central nervous system in the cockroach *Periplaneta americana* L. *J. Exp. Biol.* 37: 500–512.

Wolf, M. K., and A. B. Holden, 1969. Tissue culture analysis of the inherited defect of central nervous system myelination in jimpy mice. *J. Neuropathol. Exp. Neurol.* 28: 195–213.

Wolfe, D. E., and J. G. Nicholls, 1967. Uptake of radioactive glucose and its conversion to glycogen by neurons and glial cells in the leech central nervous system. *J. Neurophysiol.* 30: 1593–1609.

Wolff, J., 1968. Die Astroglia im Gewebsverband des Gehirns. *Acta Neuropathol. (suppl.)* 4: 33–39.

# 70 A Comparison of Invertebrate and Vertebrate Central Neurons

## MELVIN J. COHEN

THE PAST 30 YEARS of intensive study on conductile processes within single units of the nervous system have emphasized that neurons throughout the animal kingdom have many features in common (Bullock and Horridge, 1965). Information from the squid axon, crustacean nerve fibers, or neuromuscular junctions in the crab and the frog have been used for formulating models of nerve conduction and transmission that appear to apply to neurons in general, including the central nervous system (CNS) of man. Similar generalities have been arrived at in the sensory realm through a comparative approach; the eye of *Limulus* (Hartline et al., 1952) and the muscle receptor organ of a crustacean (Kuffler, 1954) provide classic examples of studies in sensory function that are of broad significance.

It may, however, be imprudent to assume that such generality of function also applies to questions involving interaction between members of a population of central neurons. Such questions deal with the geometry of connections underlying a particular act of behavior, how these connectivity patterns develop, and how they are stabilized or modified during the lifetime of the organism. Here we must look at the entire neuron and expand our consideration beyond those facets concerned with the propagation and transmission of electrochemical signals. In this context, the relationship between the bioelectric activity of the neuron and the control of macromolecular synthesis in the nerve cell body may be of particular importance. The central neurons of invertebrates differ from those of vertebrates in the degree to which the specialized synaptic and conducting components of the neuron are separated from the genetic apparatus of the soma. This isolation of the central nerve-cell body from the synaptic field in invertebrates may be related to the predominance of genetically determined stereotyped behavior found within the invertebrates in contrast to the marked labile behavioral potential of the vertebrates. To these differences between the central neurons of vertebrates and invertebrates, and their possible relationship to behavior, this essay is addressed.

MELVIN J. COHEN Department of Biology, Yale University, New Haven, Connecticut

## Morphological comparisons between invertebrate and vertebrate central nervous systems

Figure 1 is a photograph of a living central ganglion in the larva of an invertebrate, the phantom midge (*Chaoborus*). Notice that the soma of the two large neurons are pushed to the periphery of the ganglion. The central portion of the ganglion is essentially devoid of cell bodies and consists of neurites, which form a dense neuropil. This is the area in which transmission and propagation of signals take place. The separation of neuronal somata from neuropil is further demonstrated in Figure 2, which is a transverse section through a thoracic ganglion of the cockroach *Periplaneta americana* (Cohen and Jacklet, 1967). It is stained with the basic dye pyronine-malachite green, to show the distribution of nerve-cell bodies. The nerve-cell bodies are confined to a peripheral rind about the ganglion; the central core, containing neuronal processes (neurites), is seen to be devoid of cell bodies. Compare this peripheral distribution of neuronal somata with that seen in sections through the vertebrate central nervous system, as illustrated in the chapter by Nauta and Karten in this volume. The cell bodies in the vertebrate material are seen to be distributed throughout the entire depth of the tissue, in marked contrast to their peripheral location in the invertebrate ganglia.

Figure 3 shows a neuron from the ganglion of the larva of the dragonfly *Aeschna* (Zawarzin, 1924). The soma is smooth and gives rise to only one process, which runs centrally into the neuropil. There, the single neurite may branch in complex patterns, and it is these fine processes that form the postsynaptic complex that is analogous to the dendritic tree of the vertebrate neuron. The cell body itself is devoid of processes, and the central neuron has no known synapses on its surface. It is characteristic that the central neuronal soma in invertebrates is physically removed from the area of synaptic transmission and impulse initiation.

A summary of types of neurons in vertebrates is shown in Figure 4 (Bodian, 1967). A primary feature of the interneurons and effector neurons in the vertebrate central nervous system is the complex branching of the dendritic tree from the soma. The dendritic tree is involved in the transmission of electrochemical signals from one unit to another.

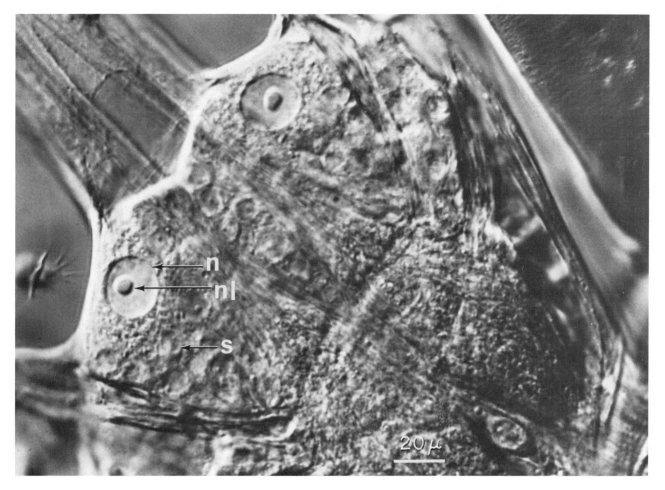

FIGURE 1 A ventral nerve cord ganglion of the phantom midge *Chaoborus crystallinus* seen with Nomarsky interference optics through the transparent cuticle of the living intact animal. The nucleus (n) and nucleolus (nl) are clearly visible in each of two large nerve cell bodies at the posterior edge of the ganglion. Small cells (s) are seen around the perimeter of the large cells. (From Cohen, 1969.)

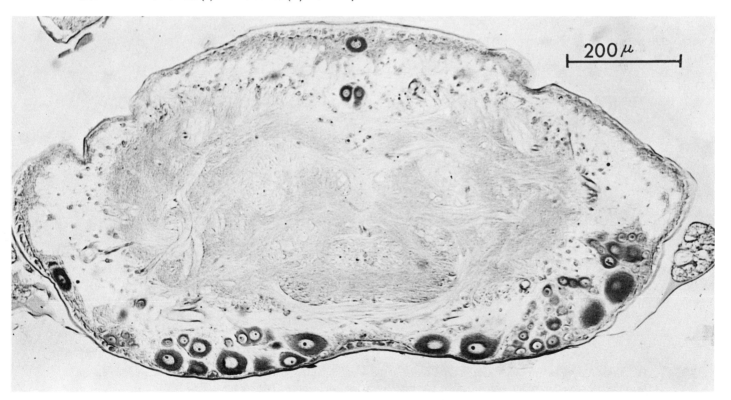

FIGURE 2 Transverse section through the anterior region of the metathoracic ganglion of the cockroach *Periplaneta americana*. The stain is pyronine-malachite green to show primarily the nerve cell bodies. Note the peripheral distribution of nerve cell bodies and their absence in the central neuropil region. (From Cohen and Jacklet, 1967.)

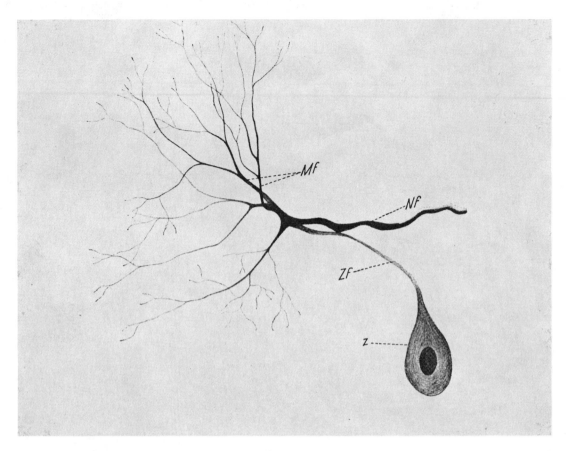

FIGURE 3 Motor neuron from the thoracic ganglion of the dragonfly larva *Aeschna* drawn from a methylene blue preparation. The thin link segment (Zf) joins the cell body (z) to an expanded area from which the dendrites (Mf) and the axon (Nf) originate. (From Zawarzin, 1924.)

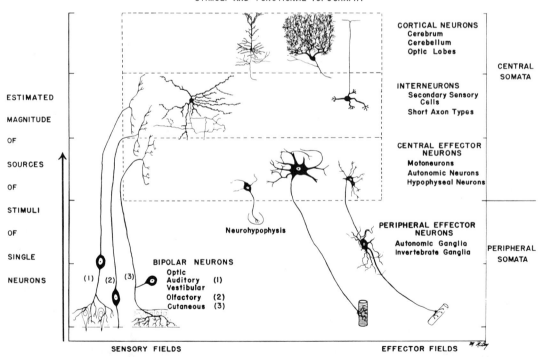

FIGURE 4 A summary of major neuron types found in the mammalian central nervous system. Note the extensive dendritic expansions of the central somata as contrasted to the smooth soma of the typical invertebrate central neuron seen in Figure 3. (From Bodian, 1967.)

Thus the soma, together with its genetic apparatus for the control of macromolecular synthesis, is thrust directly into the path of ionic and chemical transmitter fluxes related to synaptic transmission and impulse initiation. This is in contrast to the monopolar neuron, typical of the invertebrate CNS, the soma of which is often far removed from the sites of synaptic transmission and impulse initiation. Indeed, the greatest resemblance of neuron types in the two groups is between the invertebrate central neuron and the vertebrate dorsal-root ganglion cell, pictured in Figure 4. This vertebrate cell type also has its soma removed from the path of impulse propagation. The role this sensory cell plays in providing reliable information about the environment is also most effectively fulfilled by minimizing the variability of its response to a given stimulus throughout the lifetime of the organism.

## The identification of individual central neurons

CELL BODY MAPS An examination of Figure 2 gives the impression of relatively few nerve-cell bodies present throughout the metathoracic ganglion of the cockroach. Many of the cells range from $40\mu$ to $80\mu$ in diameter. Cell counts provide estimates of approximately 3500 nerve-cell bodies for the entire ganglion (Cohen and Jacklet, 1967). Pringle (1939) indicated that each leg muscle in the cockroach (*Periplaneta*) receives a small number of motor axons, and it has been determined electrophysiologically that no leg muscle receives more than five axons (Pearson and Bergman, 1969). Such a relatively simple system appears open to an identification of the actual neural elements involved in a particular behavioral act. This led us to attempt to map individual motor nerve-cell bodies that drive particular muscles in the metathoracic leg of the cockroach.

If the axon of a cockroach motor neuron is cut in the leg nerve, a dense ring of RNA appears within 12 hours after injury in the cytoplasm surrounding the nuclear membrane, and reaches a peak density at about 48 hours, as seen in Figure 5 (Cohen and Jacklet, 1965; Cohen, 1967b). The perinuclear RNA ring disappears in approximately one week, and at this time the injured neuron may begin to regenerate its axon (Bodenstein, 1957; Guthrie, 1962). The cytoplasmic response to axon injury bears some resemblance to that seen in vertebrate neurons and classified under the general term of chromatolysis (Nissl, 1892; Bodian, 1947). We have used the perinuclear RNA ring evoked by injury as an anatomical marker to construct cell maps of the cockroach metathoracic ganglion. These maps indicate which motor nerve-cell bodies send their axons out a particular peripheral nerve trunk (Cohen and Jacklet, 1967). As seen in Figure 6 the distribution of motor nerve-cell bodies is bilaterally symmetrical within the ganglion. Comparison of cell maps constructed from several animals indicates that the location of individual cell bodies within the ganglion is similar from one animal to the next. Such neuronal maps, illustrating the constancy of cell number and location, have been constructed for a number of invertebrates, including the leech (Nicholls and Baylor, 1968) and the mollusks *Aplysia* (Hughes and Tauc, 1962; Kandel et al., 1967) and *Tritonia* (Dorsett, 1967; Willows, 1967). Among the arthropods, the work of Otsuka et al. (1967) on the lobster locates individual central neurons and correlates the presence of presumed specific chemical transmitters with the determined function of these cells. The studies of Kennedy et al. (1969) have provided detailed maps correlating cell-body localization with a precise analysis of motor function in the crayfish central nervous system. These studies all point to a remarkable consistency of cell location and function within central ganglia in a wide variety of invertebrates.

THE DISTRIBUTION OF NEURITES WITHIN THE NEUROPIL The precise structural and functional identification of nerve cells within a variety of central ganglia in invertebrates has been extended to the neurites of these cells by the use of a dye-injection technique developed by Stretton and Kravitz (1968). The fluorescent dye Procion Yellow can be injected into the soma or large branches of a central neuron and will diffuse throughout the cell, including the fine neurites within the neuropil. The dye becomes fixed to the interior of the neuron and, when examined with fluorescence microscopy either in whole mount or in section, the detailed configuration of the soma and its branches can be determined. By stimulating and recording from various parts of the neuron with a dye-filled electrode, one can resolve the precise function of the cell or of its component parts. The dye is then injected into the experimental cell and the configuration of the fine branches within the neuropil determined. Figure 7 (page 804) is a 10-$\mu$ section from the metathoracic ganglion of the cockroach (*Periplaneta*), showing the soma and processes within the neuropil of a motor neuron to the extensor tibia muscle of the leg. We penetrated the neuron with a dye-filled electrode through which a current was passed to stimulate the neuron and observed which leg muscle was activated (Rowe et al., 1969). Figure 8 (page 805) shows a cockroach motor neuron reconstructed from 10-$\mu$ serial sections. Our work on insect central neurons indicates a consistent species-specific branching pattern of processes within the neuropil for the same identified neuron. There also appear to be differences between major functional groups, i.e., between neurons innervating the extensor of the tibia and those innervating the flexor.

This type of investigation has been carried out in considerable detail on the well-studied motor neurons and interneurons of the crayfish abdominal cord (Kennedy et al., 1969; Selverston and Kennedy, 1969). The particular

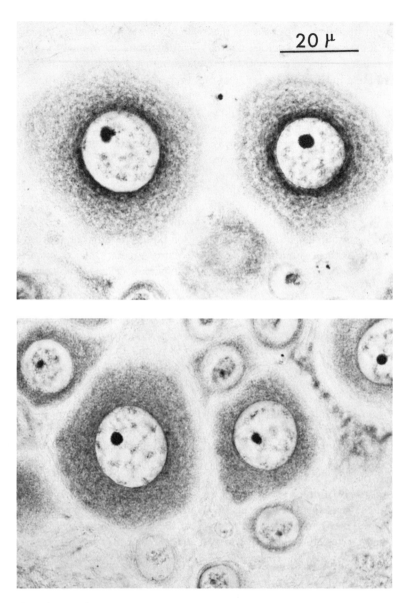

20 μ

FIGURE 5  Top: Cockroach central neurons the axons of which were cut two days before fixation and stained with pyronine-malachite green to show RNA. Note the dense perinuclear RNA ring in these cells. Bottom: Bilaterally matched control cells with intact axons from the opposite side of the same ganglion. Note the relative absence of the perinuclear RNA ring in these cells. (From Cohen, 1967b.)

power of this work has been the detailed correlation of electrophysiological properties of the neuron with the microstructure of its dendritic tree. These investigators again find a high degree of consistency from one animal to the next in the dendritic branching pattern of a particular identified neuron. In the reconstructed crayfish interneuron seen in Figure 9 (page 806), note again that the soma is far removed from the densely branched dendritic tree within the neuropil.

## The differentiation of neurites

The geographic isolation of the soma of the central neuron in arthropods from the dendritic branches within the neuropil appears to reflect a functional, as well as a structural, isolation of the soma from the electrogenic region of the neuron. The single process arising from the soma may run for several hundred microns into the neuropil before it expands in diameter and gives rise to dendritic branches

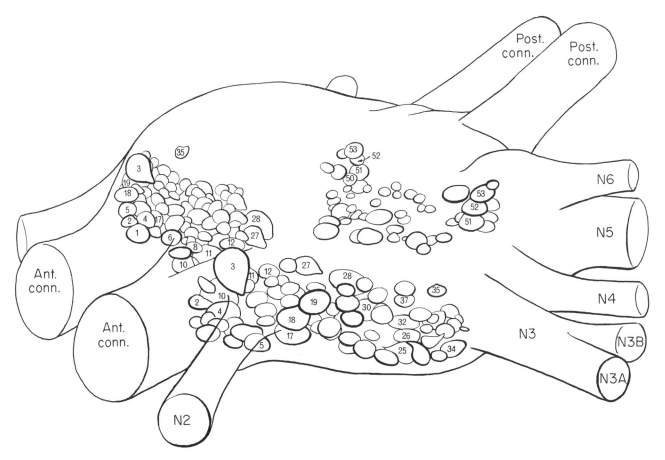

FIGURE 6　A three-dimensional rendering showing the distribution of some identified motor nerve cell bodies in the cockroach metathoracic ganglion. Members of a bilaterally matched pair of cells are given the same number on each side of the ganglion. Ant. conn., anterior connectives; Post. conn., posterior connectives; N2 to N6, peripheral nerve trunks. Length of ganglion is approximately 1 mm. (Modified from Cohen and Jacklet, 1967.)

(see Figure 9). I propose that the length of neurite between the soma and the expanded region yielding the dendritic tree be called the *link segment*. Sandeman (1969b) has termed the expanded process from which the fine dendrites emanate the *integrating segment*. The smooth, often tenuous, link segment therefore joins the neuronal soma to the highly branched integrating segment within the neuropil.

THE LINK SEGMENT　The relationship between the thin link segment arising immediately from the soma and the expanded integrating segment is shown in Figure 10 (page 807) for cell 3 in the metathoracic ganglion of *Periplaneta* (see Figure 6 cell map). In the cockroach, the length and diameter of the link segment vary from one cell type to the next. Kennedy et al. (1969) have pointed out that there is also variation in this part of the crayfish neuron; the inter-

neuron has a longer and thinner link segment separating the soma from the rest of the cell as compared with the motor neuron. The longer and thinner the link segment, the more attenuated are the passively conducted electrical potentials reaching the soma. In some crayfish cells, the soma is completely isolated electrically from the rest of the neuron (Selverston and Kennedy, 1969). For the most part, in cells of insect thoracic ganglia the isolation of the soma from electrical events occurring within the neuropil is virtually complete. This has held true, in our laboratory and elsewhere, for the cockroach (Rowe, 1969) and for the cricket (Bentley, personal communication).

No definitive intracellular work has been done on the electrical properties of the link segment. Judged from the degree to which the soma reflects electrical activity taking place in the synapses of the neuropil, the link segment at

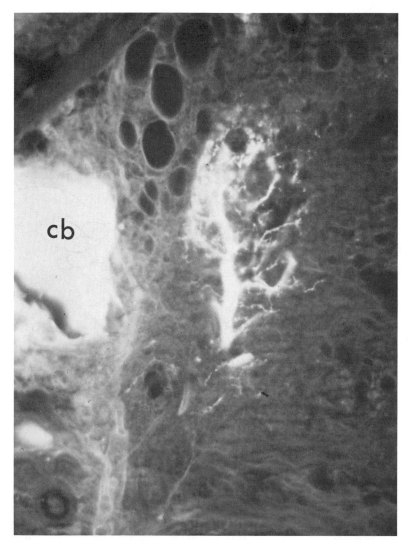

FIGURE 7    A 10 μ section of a cockroach metathoracic motor neuron innervating the extensor tibia muscle. The cell has been injected with the fluorescent dye Procion Yellow. The cell body (cb) is approximately 30 μ in its longest dimension. Note the expanded integrating segment and the fine branches of the dendritic tree emanating from it. (From Rowe and Cohen, in preparation.)

best permits some electrotonic spread of neuropil activity to reach the soma. In many neurons, however, the resistance of this region is apparently so high that no electrical sign of synaptic or propagated activity is detectable in the cell body.

THE INTEGRATING SEGMENT    The expansion of the 5–10-μ link segment into the 10–30-μ region giving rise to branches of the dendritic tree has been pictured in earlier work on arthropod central neurons (Zawarzin, 1924). It has been suggested that this area may serve as an integrating region (Maynard, 1966). However, its full significance as a struc-

tural and functional entity has only recently been realized and most elegantly defined in the crab by Sandeman (1969a, 1969b). He has termed this region the *integrating segment* and has studied it in detail both structurally and with intracellular recording. He finds the membrane of this segment to be electrically inexcitable, as judged by the decreasing amplitude of invading antidromic spikes. The side branches of the integrating segment conduct propagated spikes, which then evoke depolarizing excitatory potentials in the integrating segment. The presynaptic inhibitory input, which produces slightly depolarizing potentials in the integrating segment, appears to end directly on the

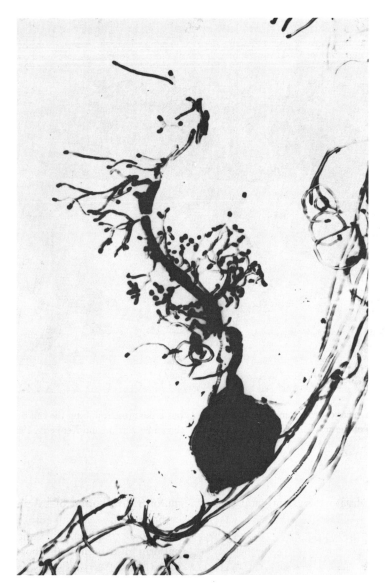

FIGURE 8  A reconstruction of a cockroach metathoracic motor neuron innervating the extensor tibia muscle. The cell was injected with Procion Yellow and reconstructed from 10 $\mu$ serial sections. Diameter of cell body, approximately 30 $\mu$. (From Rowe and Cohen, in preparation.)

membrane of this segment rather than on the fine branches emanating from it. The spike-initiating locus appears to be at the point where the integrating segment once again narrows distally to give rise to the axon proper. Figure 11 (page 808) summarizes the findings of Sandeman for an oculomotor neuron of the crab.

The integrating segment of the arthropod neuron, together with its fine, branching neurites, appears to be analogous to the soma-dendritic complex of the vertebrate neuron. Both anatomical entities have areas of electrically inexcitable membrane (Grundfest, 1969). They integrate excitatory and inhibitory presynaptic input that gives rise to a propagated signal at a spike-generating locus distal to the integrating membrane. The soma-dendritic complex of the vertebrate neuron places the genetic apparatus and the major site of macromolecular synthesis directly in the path of chemical transmitters and ionic changes associated with synaptic transmission and impulse initiation. In contrast, the arthropod central neuron isolates the soma with its genetic apparatus from the electrochemical events of synaptic and propagated activity. This isolation is such that in many neurons the soma reflects little or no electrical sign

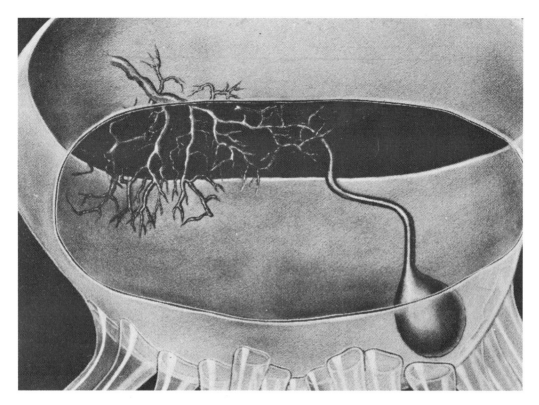

FIGURE 9   Model of a second-order sensory interneuron from the sixth ganglion of the crayfish. Note the long tenu-ous link segment joining the cell body to the dendritic tree. (From Selverston and Kennedy, 1969.)

of events occurring in the distant integrating and spike-generating regions.

## Cytological comparisons

The cytoplasmic basiphilic masses characteristic of many vertebrate central neurons (Figure 12, page 809) were first described by Nissl (1892). These Nissl bodies were later found to be composed of layers of rough endoplasmic reticulum (Palay and Palade, 1955). Their basiphilia is presumably due to the affinity of the ribosomal RNA in these structures for such classic basic dyes as cresyl violet and toluidine blue (Bodian, 1947). The response of vertebrate neurons to injury of their axons consists of a breakdown of these basiphilic masses (chromatolysis) within the first week after injury, as seen in Figure 12B (Nissl, 1892; Bodian and Mellors, 1945). This may also be accompanied by a shift of the nucleus to an eccentric location within the soma. If the neuron survives and regenerates an axon, there may be a gradual re-formation of the Nissl bodies; the nucleus eventually returns to a central location.

In contrast, when the central neuron of a normal adult cockroach is stained with basic dye, it shows a fine-grained uniform distribution of basiphilic material throughout its cytoplasm, as seen in Figure 12. This is correlated with a scarcity of rough endoplasmic reticulum and a relatively uniform distribution of ribosomes throughout the cytoplasm (Hess, 1960; Wigglesworth, 1960; Cohen, 1967a).

Examination, in the electron microscope, of the perinuclear RNA ring evoked by injury to the axon of the cockroach central neuron reveals that this is associated with an increase in rough endoplasmic reticulum (rough ER). This is shown in Figure 13 (page 810), which compares the distribution of rough ER in a bilaterally symmetrical pair of cockroach motor neurons from the same ganglion. There is a generalized increase of rough ER in the injured cell, with some concentration in the perinuclear area. The cytoplasm of the injured cell takes on the appearance of a crudely organized Nissl body in a vertebrate. Within one week, the augmented rough ER of the injured cell disappears, and the nucleus shifts to an eccentric position, as seen in Figure 12D. At this time the injured neuron starts to regenerate a new axon. The disappearance of the rough ER in the cockroach cell appears similar to vertebrate chromatolysis, when the ribosomes become dissociated from the cisternae of the rough ER (Porter and Bowers, 1963).

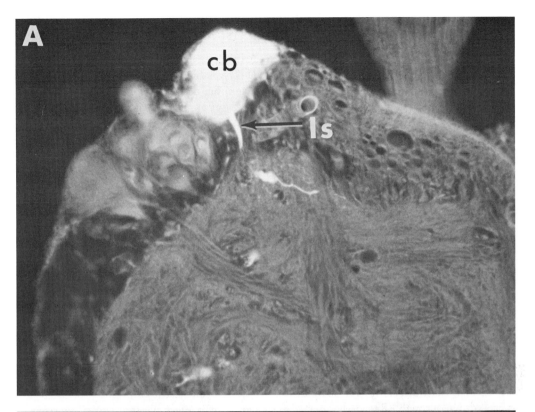

FIGURE 10 Procion Yellow injected cell from the cockroach metathoracic ganglion. This is cell number 3 as designated in Figure 6. A: 10 μ section showing the cell body (cb) with the link segment (ls) emanating from it. B: The next 10 μ section shows the thin link segment expanding into the broad integrating segment (is). The longest dimension of the cell body in A is approximately 80 μ. (From Rowe and Cohen, in preparation.)

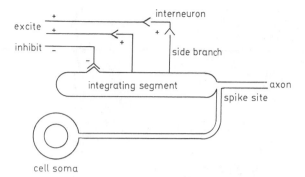

FIGURE 11   A diagram illustrating the various functional and morphological regions of an oculomotor neuron in the brain of a crab. Note the separation of the soma from the integrating region of the cell. The different modes by which excitatory and inhibitory input may terminate upon the cell are clearly shown. (From Sandeman, 1969b.)

The vertebrate neuron appears to be continually primed and involved in a relatively high level of protein synthesis (Hydén, 1960). This is associated with large amounts of rough ER characteristic of other actively synthesizing cells. The precipitous breakdown of rough ER, together with the dispersion of ribosomes in the vertebrate cell undergoing chromatolysis, is seen as throwing the cell into a super-active state of synthetic activity associated with axon repair (Brattgård et al., 1957). The intact central neuron of a cockroach differs from the central nerve cell of a vertebrate in lacking large aggregates of rough ER. This may be associated with a lower state of readiness for and participation in macromolecular synthesis. Only when extraordinary synthetic demands are placed on the cockroach neuron, as in axon regeneration, does this cell produce a crude approximation of the vertebrate neurocytological organization. It does so by first forming rough ER and then dispersing it in a manner similar to the chromatolysis response of the vertebrate nerve cell. Such arthropods as the crayfish (Hoy et al., 1967) and the locust (Usherwood, personal communication) do not show the type of axonal regeneration described for the cockroach (Bodenstein, 1957) and the vertebrate (Bodian, 1947). The neurons of these animals do not show the perinuclear RNA ring response to injury. Other invertebrates, such as the cricket (Edwards and Sahota, 1967) and the mollusk *Anodonta* (Salánki and Gubicza, 1969), show typical axon degeneration and regeneration following injury, and their central neurons have the perinuclear RNA ring response. This substantiates the concept that the formation and subsequent dispersion of the rough ER in the perinuclear ring are associated with an increase in the level of macromolecular synthesis associated with axon regeneration.

## Functional considerations

A major structural difference between the central neurons of invertebrates and those of vertebrates is the almost ubiquitous occurrence of monopolar neurons in the invertebrate groups above the Coelenterata (Bullock and Horridge, 1965). In contrast, the vertebrate central neuron is predominantly multipolar and heteropolar. It is divided into an integrating soma-dendritic complex and an axon that generates an all-or-none signal. At the unit level, a primary functional consequence of these differences is that the soma of the invertebrate central neuron, unlike that of the vertebrate cell, is removed from the path of synaptic transmission and impulse initiation.

Let us make the assumption that in neuronal systems specialized to store and reflect a sign of their functional history, such storage and reflection may be accomplished by coupling the bioelectrical signals to the genetic apparatus of the soma. This provides the possibility that some aspects of the transient electrochemical signals may be reflected more permanently in the neuron by an alteration in macromolecular synthesis (Hydén, this volume). Several lines of evidence from cells other than neurons suggest that the type of bioelectrical events associated with neuronal signals may affect the synthesis of macromolecules. In the salivary-gland chromosomes of *Chironomus*, Kroeger (1966) has demonstrated a correlation between "puffing" and the degree of depolarization of the gland-cell membrane. The insect growth hormone ecdysone also has an effect on the ion permeability of the nuclear membrane in the salivary gland cells of *Chironomus* (Ito and Loewenstein, 1965). Lubin (1967) reports a decrease in the rate of DNA and protein synthesis in cultured mammalian cells when the potassium concentration is lowered. Thus the tendency to isolate the invertebrate central nerve-cell body from the transient ionic and transmitter fluxes associated with information processing may be viewed as a means of stabilizing the function of neuronal circuits over the lifetime of the organism. This could account for a major evolutionary trend among the invertebrates that has taken the direction of relatively complex but highly stereotyped behavior.

The isolation of the neuronal soma in invertebrates may also allow the control of such complex but rigid behavioral patterns as flight, walking, and swimming to be reliably invested in very small numbers of neurons. Kennedy et al. (1969) described single command interneurons in the crayfish that control the activity of large segments of abdominal musculature involved in swimming. Willows (1967) has shown that stimulating a single central neuron can evoke the turning of the entire body of a sea snail.

The Mollusca provide a major exception to the concept that the central-neuron soma of invertebrates is relatively isolated from the bioelectric activity of the cell. In the sea

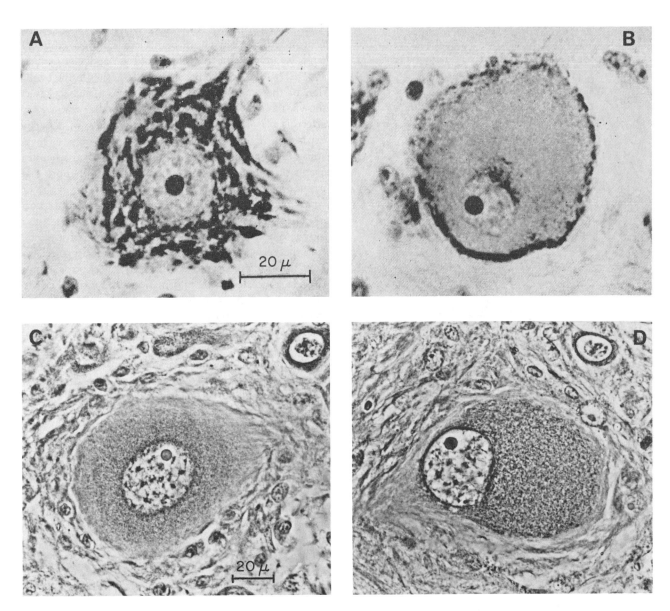

FIGURE 12  A comparison of injury responses in central neurons of a rhesus monkey (A and B) and a cockroach (C and D). A: Normal ventral horn cell, showing the basiphilic aggregates of Nissl bodies in the cytoplasm. B: Ventral horn cell seven days after lumbar spinal-root section. Chromatolysis is indicated by the absence of Nissl bodies and the eccentric location of the nucleus. C and D are bilaterally matched cells from the same metathoracic ganglion. C: Normal cell, showing a uniform cytoplasmic basiphilia and a centrally located nucleus. D: Cell axon was cut four weeks prior to fixation. The distribution of basiphilic material in the cytoplasm is unchanged, but the nucleus has assumed an eccentric position. (Plate from Cohen, 1967a. A and B modified from Bodian and Mellors, 1945; C and D from Cohen and Jacklet, 1967.)

snail *Aplysia,* the site of spike initiation may be as much as 1.5 mm from the soma (Frazier et al., 1967). Although the membrane excitability decreases from the site of spike initiation proximally toward the cell body, overshooting spikes can frequently be recorded from the soma (Tauc, 1962). More highly evolved members of the group, such as the cephalopods, are capable of complex rapid learning, particularly within a visual framework. Young (1964) summarized many of the behavioral studies in *Octopus* and correlated these with the anatomy of the brain and visual system. His drawings of silver preparations in the optic lobe of *Octopus* indicate cells that are multipolar and heteropolar.

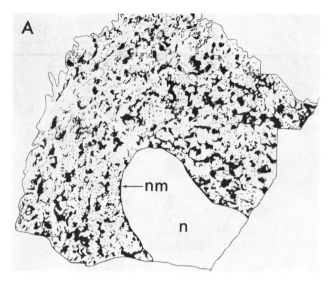

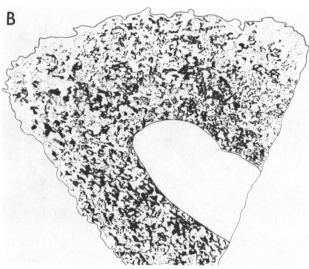

FIGURE 13 Tracings of montages made from electron micrographs taken at 17,000 magnification. The shaded area indicates endoplasmic reticulum and free ribosomes. A and B are tracings from a pair of bilaterally matched motor neurons situated on opposite sides of the same metathoracic ganglion in the cockroach. The cells were approximately 35 μ in diameter and roughly one half of each cell is pictured. n, nucleus; nm, nuclear membrane. A: normal cell with intact axon. B: matched cell whose axon was cut 48 hours previous to fixation. Note the increased amount of endoplasmic reticulum in the injured cell. (Modified from Young et al., in press.)

These neurons, such as the centrifugal cells, appear to have dendritelike processes emanating from the soma. This is one of the few instances of multipolar-heteropolar nerve cells reported in the central nervous system of an invertebrate.

The more primitive mollusks, such as *Aplysia,* have neurons whose somata are both electrically excitable and sensitive to synaptic transmitters (Tauc and Gershenfeld, 1961). These may have been crucial requisites in the evolution of the extraordinary labile and modifiable behavior shown by the Cephalopoda. Based on the prime characteristic of soma excitability, the appearance of dendritic expansions from the soma places the neuronal genetic apparatus of the Cephalopoda subject to the electrochemical events of transmission and conduction. It is of more than passing interest that this type of neuron appears in the invertebrate group with the most highly evolved type of modifiable behavior, and in that part of the nervous system shown to be intimately involved in the learning process.

In the vertebrates, the body of the central nerve cell has been thrust directly into the path of the electrochemical changes associated with information processing. The elements controlling macromolecular synthesis within the soma are thus exposed to fluxes of inorganic ions and chemical transmitters that may influence their function and provide some long-term sign of past activity in the form of a macromolecular residue. Another major advance in the neural organization is the relatively large number of neurons available in vertebrates as compared with invertebrates (Wiersma, 1952; Bullock and Horridge, 1965). This can provide the segments of the nervous system involved in low-threshold, highly plastic behavior with the degree of redundancy required to function in a reliable and adaptively significant manner. These two factors, (1) the exposure of the genetic apparatus to the electrochemical events of signal processing and (2) relatively large numbers of neurons, may have been critical in the evolution of the extraordinary plastic capabilities characteristic of the most highly developed portions of the vertebrate central nervous system.

## Acknowledgments

This work was supported in part by United States Public Health Service Grants NB 01624 and NB 08996-01 from the Institute of Neurological Diseases and Stroke.

### REFERENCES

BODENSTEIN, D., 1957. Studies on nerve regeneration in *Periplaneta americana. J. Exp. Zool.* 136: 89–115.

BODIAN, D., 1947. Nucleic acid in nerve cell regeneration. *Symp. Soc. Exp. Biol.* 1: 163–178.

BODIAN, D., 1967. Neurons, circuits, and neuroglia. *In* The Neurosciences: A Study Program (G. C. Quarton, T. Melnechuk, and F. O. Schmitt, editors). The Rockefeller University Press, New York, pp. 6–24.

BODIAN, D., and R. C. MELLORS, 1945. The regenerative cycle of motoneurons, with special reference to phosphatase activity. *J. Exp. Med.* 81: 469–488.

BRATTGÅRD, S. O., J. E. EDSTRÖM, and H. HYDÉN, 1957. The chemical changes in regenerating neurons. *J. Neurochem.* 1: 316–325.

BULLOCK, T. H., and G. A. HORRIDGE, 1965. Structure and Function in the Nervous System of Invertebrates. W. H. Freeman and Co., San Francisco and London.

COHEN, M. J., 1967a. Correlations between structure, function and RNA metabolism in central neurons of insects. *In* Invertebrate Nervous Systems (C. A. G. Wiersma, editor). University of Chicago Press, Chicago, pp. 65–78.

COHEN, M. J., 1967b. Some cellular correlates of behavior controlled by an insect ganglion. *In* Chemistry of Learning (W. C. Corning and S. C. Rotner, editors). Plenum Press, New York, pp. 407–424.

COHEN, M. J., 1969. Neuronal change in the regenerating and developing insect nervous system. *Symp. Int. Soc. Cell Biol.* 8: 263–275.

COHEN, M. J., and J. W. JACKLET, 1965. Neurons of insects: RNA changes during injury and regeneration. *Science (Washington)* 148: 1237–1239.

COHEN, M. J., and J. W. JACKLET, 1967. The functional organization of motor neurons in an insect ganglion. *Phil. Trans. Roy. Soc. (London), ser. B, biol. sci.* 252: 561–569.

DORSETT, D. A., 1967. Giant neurons and axon pathways in the brain of *Tritonia. J. Exp. Biol.* 46: 137–151.

EDWARDS, JOHN S., and T. S. SAHOTA, 1967. Regeneration of a sensory system: The formation of central connections by normal and transplanted cerci of the house cricket *Acheta domesticus. J. Exp. Zool.* 166: 387–396.

FRAZIER, W. T., E. R. KANDEL, R. W. KUFFERMANN, R. WAZIRI, and R. E. COGGESHALL, 1967. Morphological and functional properties of identified neurons in the abdominal ganglion of *Aplysia california. J. Neurophysiol.* 30: 1288–1351.

GRUNDFEST, H., 1969. Synaptic and ephaptic transmission. *In* Structure and Function of Nervous Tissue (G. H. Bourne, editor). Academic Press, New York, vol. 2, pp. 463–491.

GUTHRIE, D. M., 1962. Regenerative growth in insect nerve axons. *J. Ins. Physiol.* 8: 79–92.

HARTLINE, H. K., H. G. WAGNER, and E. F. MACNICHOL, JR., 1952. The peripheral origin of nervous activity in the visual system. *Cold Spring Harbor Symp. Quant. Biol.* 17: 125–141.

HESS, A., 1960. The fine structure of degenerating nerve fibers, their sheaths, and their terminations in the central nerve cord of the cockroach (*Periplaneta americana*). *J. Biophys. Biochem. Cytol.* 7: 339–344.

HOY, R. R., G. D. BITTNER, and D. KENNEDY, 1967. Regeneration in crustacean motoneurons: Evidence for axonal fusion. *Science (Washington)* 156: 251–252.

HUGHES, G. M., and L. TAUC, 1962. Aspects of the organization of central nervous pathways in *Aplysia depilans. J. Exp. Biol.* 39: 45–69.

HYDÉN, H., 1960. The neuron. *In* The Cell, Vol. 4 (J. Brachet and A. E. Mirsky, editors). Academic Press, London and New York, pp. 215–323.

ITO, S., and W. R. LOEWENSTEIN, 1965. Permeability of a nuclear membrane: Changes during normal development and changes induced by growth hormone. *Science (Washington)* 150: 909–910.

KANDEL, E. R., W. T. FRAZIER, R. WAZIRI, and R. E. COGGESHALL, 1967. Direct and common connection among identified neurons in *Aplysia. J. Neurophysiol.* 30: 1352–1376.

KENNEDY, D., A. I. SELVERSTON, and M. P. REMLER, 1969. Analysis of restricted neural networks. *Science (Washington)* 164: 1488–1496.

KROEGER, H., 1966. Potentialdifferenz und Puff-muster. *Exp. Cell Res.* 41: 64–80.

KUFFLER, S. W., 1954. Mechanisms of activation and motor control of stretch receptors of lobster and crayfish. *J. Neurophysiol.* 17: 558–574.

LUBIN, M., 1967. Intracellular potassium and macromolecular synthesis in mammalian cells. *Nature (London)* 213: 451–453.

MAYNARD, D. M., 1966. Integration in crustacean ganglia. *Symp. Soc. Exp. Biol.* 20: 111–149.

NICHOLLS, J. G., and D. A. BAYLOR, 1968. Specific modalities and receptive fields of sensory neurons in CNS of the leech. *J. Neurophysiol.* 31: 740–756.

NISSL, F., 1892. Ueber die Veranderungen der Ganglienzellen am Facialiskern des Kaninchens nach Ausreissung der Nerven. *Allg. Z. Psychiat.* 48: 197–198.

OTSUKA, M., E. A. KRAVITZ, and D. D. POTTER, 1967. Physiological and chemical architecture of a lobster ganglion with particular reference to gamma-aminobutyrate and glutamate. *J. Neurophysiol.* 30: 725–752.

PALAY, S. L., and G. E. PALADE, 1955. The fine structure of neurons. *J. Biophys. Biochem. Cytol.* 1: 69–88.

PEARSON, K. G., and S. J. BERGMAN, 1969. Common inhibitory motoneurones in insects. *J. Exp. Biol.* 50: 445–471.

PORTER, K. R., and M. B. BOWERS, 1963. A study of chromatolysis in motor neurons of the frog *Rana pipiens. J. Cell Biol.* 19: 56A–57A (abstract).

PRINGLE, J. W. S., 1939. The motor mechanism of the insect leg. *J. Exp. Biol.* 16: 220–231.

ROWE, E. C., 1969. Microelectrode records from a cockroach thoracic ganglion: Synaptic potentials and temporal patterns of spike activity. *Comp. Biochem. Physiol.* 30: 529–539.

ROWE, E. C., B. J. MOBERLY, H. M. HOWARD, and M. J. COHEN, 1969. Morphology of branches of functionally identified motoneurons in cockroach neuropile. *Amer. Zool.* 9: 1107.

SALÁNKI, J., and A. GUBICZA, 1969. Histochemical evidence of direct neuronal pathways in *Anodonta cygnea* L. *Acta Biol. Acad. Sci. Hung.* 20: 219–234.

SANDEMAN, D. C., 1969a. The synaptic link between the sensory and motoneurones in the eye-withdrawal reflex of the crab. *J. Exp. Biol.* 50: 87–98.

SANDEMAN, D. C., 1969b. Integrative properties of a reflex motoneuron in the brain of the crab *Carcinus maenas. Z. Vergl. Physiol.* 64: 450–464.

SELVERSTON, A. I., and D. KENNEDY, 1969. Structure and function of identified nerve cells in the crayfish. *Endeavour* 28: 107–113.

STRETTON, A. O. W., and E. A. KRAVITZ, 1968. Neuronal geometry: Determination with a technique of intracellular dye injection. *Science (Washington)* 162: 132–134.

TAUC, L., 1962. Site of origin and propagation of spike in the giant neuron of *Aplysia. J. Gen. Physiol.* 45: 1077–1097.

TAUC, L., and H. M. GERSCHENFELD, 1961. Cholinergic transmission mechanisms for both excitation and inhibition in molluscan central synapses. *Nature* (*London*) 192: 366–367.

WIERSMA, C. A. G., 1962. The neuron soma. Neurons of arthropods. *Cold Spring Harbor Symp. Quant. Biol.* 17: 155–163.

WIGGLESWORTH, V. B., 1960. Axon structure and the dictyosomes (Golgi bodies) in the neurones of the cockroach, *Periplaneta americana*. *Quart. J. Microscop. Sci.* 101: 381–388.

WILLOWS, A. O. D., 1967. Behavioral acts elicited by stimulation of single, identifiable brain cells. *Science* (*Washington*) 157: 570–574.

YOUNG, D., D. E. ASHHURST, and M. J. COHEN, 1970. Injury response of the neurons of *Periplaneta americana*. *Tissue and Cell* 2: 387–398.

YOUNG, J. Z., 1964. A Model of the Brain. Oxford University Press, London.

ZAWARZIN, A., 1924. Zur Morphologie der Nervenzentren. Das Bauchmark der Insekten. Ein Beitrag zur vergleichenden Histologie (Histologische Studien über Insekten VI). *Z. Wiss. Zool.* 122: 323-424.

# 71 Some Aspects of Gene Expression in the Nervous System

## E. M. SHOOTER

THE PRIMARY INSTRUCTIONS for the development of the nervous system in terms of cell specialization, cell migration, and the formation of intercellular connections are located in the genome and give rise to patterns of innate behavior. The programing of these events is fortunately not on a one-to-one basis. On the contrary, it is evident that the behavior of whole cell populations is dictated by single genes. In the neurological mutant mouse strain, reeler, which is homozygous for the allele rl, the granule cells fail to migrate normally inward through the Purkinje cells of the cerebellum (Sidman, 1968). Not only does this mutation affect the majority of the granule cells; it also affects the only other two neuroblast populations, those in the cerebral isocortex and the hippocampus, which migrate in the same way. The altered gene product of the allele rl results, therefore, either directly or indirectly, in the failure of the cell-to-cell signaling required for normal migration. It is also clear that throughout development internal and external environmental factors continue to modify gene expression in the nervous system and that, in turn, variable gene expression is likely to be one of the neural components that are the basis of learning.

This brief introduction illustrates the two general ap-

proaches to the study of gene expression in the nervous system I discuss here. First, one can examine the primary products of gene action, the proteins and enzymes of neurons and glia, follow their changing patterns during development, and delineate the consequences of mutations. Alternatively, one can see how internal stimuli such as hormones and growth factors modify or regulate gene expression in the target cells. These two categories are not exclusive and do overlap. For instance, the selective regulation of neuronal activity by hormones requires that these cells be able to distinguish between such inducers. One way to study this is through the analysis of nerve-cell-specific macromolecules, the primary product of the gene action of those cells.

## The primary products of gene action

SITES OF PROTEIN SYNTHESIS Because of the unusual morphology of nervous-tissue cells compared with other eukaryotic cells, it is important to consider the sites within these cells at which protein may be synthesized. In addition to the usual ribosomal and mitochondrial protein-synthesizing systems, two other possible systems, the synaptosomal and axonal, should be considered.

RIBOSOMAL PROTEIN SYNTHESIS In large neurons, the cytoplasmic ribosomes are concentrated near the endoplasmic reticulum of the Nissl substance in the perinuclear

E. M. SHOOTER Departments of Genetics and Biochemistry and the Lt. Joseph P. Kennedy, Jr. Laboratories for Molecular Medicine, Stanford University School of Medicine, Stanford, California

region, the initial segment, and the axonal hillock (Sotelo and Palay, 1968). A high proportion of these ribosomes are not attached to the membrane of the endoplasmic reticulum and could thus function directly in the synthesis of protein involved in axoplasmic transport to the nerve ending (see Weiss, Davison, this volume). The membrane-bound ribosomes, on the other hand, are probably involved in the synthesis of proteins for storage in the cell or of conjugated proteins. In many smaller neurons, Nissl bodies and ribosomes appear in the most peripheral cytoplasm, underlying either synaptic terminals (Bodian, 1965) or neuroglial processes adjacent to the surface of the neuron. In contrast, in glia, as in other eukaryotic cells, the ribosomes are more evenly distributed through the cytoplasm, and a greater proportion are bound to the endoplasmic reticulum (Mugnaini and Walberg, 1964). Glia contain less RNA than do neurons, which appears to be directly related to the ribosomal content of the two types of cells (Hydén and Egyházi, 1963; Gray, 1964). Recent reports suggest that glial ribosomes are as active as neuronal ribosomes in protein synthesis (Roberts et al., 1970).

A ribosomal preparation from immature rat brain has many of the characteristics of typical mammalian ribosomal systems in its requirement for pH 5 enzymes and adenosine triphosphate (ATP) and its inhibition by ribonuclease (Campbell et al., 1966). Its activity is higher than that of a similar liver preparation, although the activity decreases significantly with the age of the animal (Orrego and Lipmann, 1967). The level of activity will depend on, among other factors, the availability of the initiating, binding, translocation, and dissociation factors now defined for the bacterial ribosomal system (see Davis, this volume) and about which little is known in the nervous system. Maximum incorporation with the brain ribosomal system occurs at somewhat higher $Mg^{2+}$ concentrations than with liver or reticulocyte preparations (Table I). Although mammalian ribosomal systems, in general, are more resistant to such agents as high $Mg^{2+}$ concentrations, which increase ambiguity in translation (Weinstein et al., 1966), it is not known whether this is also true for brain ribosomes. The requirement of the latter for high $K^+$ concentrations links ribosomal protein synthesis with active $K^+$ transport and, in turn, with the bioelectric phenomena. Zomzely et al. (1968) have reported that polysome-mRNA complexes from brain are more sensitive to endogenous ribonuclease than is a liver preparation, release endogenous mRNA more readily on incubation, and correspondingly show a greater stimulation on subsequent addition of polyuridylic acid (poly-U). Nuclear RNA from brain is more active in stimulating an *E. coli* ribosomal system than is hepatic nuclear RNA, contains a high proportion of small (i.e., <12S) species, and has a higher turnover rate in young than in adult animals, giving rise to the suggestion that, during development, relatively short-lived mRNA species with high template activity are replaced by longer-lived species with a lower activity (Roberts et al., 1970; Bondy and Roberts, 1968).

MITOCHONDRIAL PROTEIN SYNTHESIS Mitochondria contain their own machinery for protein synthesis (see Lehninger, this volume), and, although mitochondrial protein synthesis is quantitatively less significant than ribosomal, it is especially important in the nervous system because of the presence of mitochondria in nerve endings and the question of local protein synthesis in these regions. It is, therefore, of even greater interest that brain mitochondria are among the most active of the known mammalian mitochondrial systems (Campbell et al., 1966). Protein synthesis by brain mitochondria does not require an external source of ATP and is not inhibited by ribonuclease, but, typically, is inhibited by chloramphenicol. In a medium that is optimal for oxidative phosphorylation, incorporation increases by a factor of ap-

TABLE I

*Synthetic Capacity of Various Subcellular Fractions of Brain\**
*(19 to 21-day-old Rat Cerebral Cortex)*

| Fraction | Incorporation $\mu\mu$mole leu/mg protein/hr | Ionic conc., mM |
|---|---|---|
| Ribosomal\*\* | 300 | $K^+$ 100, $Na^+$ 40, $Mg^{2+}$ 10 |
| Mitochondrial\*\* | 17 | $K^+$ 120, $Na^+$ 40 |
| | 31 | $K^+$ 20 |
| Synaptosomal† | 4 | $K^+$ 20 |
| | 30 | $K^+$ 25, $Na^+$ 125 |

\* The figures presented are average values to indicate only the relative levels of activity.
\*\* Campbell et al., 1966.
† Morgan and Austin, 1968, 1969; Austin et al., 1970; Autilio et al., 1968; Appel et al., 1969.

proximately two (Table I) to a level that is about one tenth that of ribosomal incorporation. Conversely, under these conditions, incorporation is inhibited by substances like rotenone and antimycin A, which are specific inhibitors of oxidative phosphorylation. The DNA in mitochondria is of limited size and is potentially capable of coding for only about 40 polypeptides, each of 150 amino acid residues. In keeping with this are the reports that mitochondria synthesize very few proteins, among which are certain of their own membrane proteins (Neupert et al., 1968). However, the fact that *Neurospora* mitochondria can synthesize protein in vivo (see Lehninger, this volume), suggesting the rather massive synthesis of a few specific proteins, prompts a re-examination of the role of both cell-body and nerve-ending mitochondria in the synthesis of extra-mitochondrial protein. Not all mitochondria are alike, those in the nerve ending being typically smaller than those in the cell body. Also Hajós and Kerpel-Fronius (1969a, 1969b) have now shown by electron histochemical techniques that mitochondria that are some distance from the cell body, particularly those in certain postsynaptic regions and unlike those in the perikarya, lack the enzyme succinic dehydrogenase and accumulate considerably less $Ca^+$ and $Sr^{2+}$. Similar differences can be detected in isolated mitochondrial pellets. This type of biochemical differentiation may well be of critical importance to the functions of mitochondria in various regions and sites of the nervous system.

SYNAPTOSOMAL PROTEIN SYNTHESIS The studies of Weiss (this volume) and Barondes (1966), among others, have shown that at least a portion of the mitochondrial and vesicular proteins (if not also whole mitochondria and vesicles) in the nerve endings derive from the cell body. It is also true that some protein in nerve-ending organelles and membranes is rapidly labeled, in vivo after intraventricular or intraperitoneal injection of radioactive amino acids (von Hungen et al., 1968), in vitro in brain slices (Austin and Morgan, 1967), or in vitro in isolated ciliary ganglia (Droz and Koenig, 1969), and the question arises as to where this particular protein is synthesized. Attempts to

answer this question have been made by examinations of the metabolic properties of isolated synaptosomes (Autilio et al., 1968; Morgan and Austin, 1968; Gordon and Deanin, 1968) and axons (Koenig, 1970). Isolated synaptosomes incorporate labeled amino acids into protein at very low levels, as compared with brain ribosomes (Table I.) Bacterial contamination has been excluded, as has also contamination by free neuronal or glial microsomes or ribosomes, because the system is not sensitive to ribonuclease and does not require ATP. Furthermore, labeled brain microsomes separate quantitatively from synaptosomes in the preparative gradients used to isolate the latter (Austin and Morgan, 1967). The data of Autilio et al. (1968) and Austin et al. (1970) have shown that only a portion of the protein-synthetic capacity of the synaptosome is due to synaptosomal mitochondria. In the presence of chloramphenicol, total incorporation is reduced to only about 75 per cent of control, whereas incorporation into mitochondrial protein, but not into soluble or other membrane protein, is almost totally inhibited (Table II). Conversely, cycloheximide markedly reduces total incorporation and incorporation into soluble and nonmitochondrial membrane protein but does not inhibit protein synthesis in synaptosomal mitochondria. In line with these results is the finding (Autilio et al., 1968) that some 20 per cent of the label is normally incorporated into the soluble protein of the synaptosome, 25 per cent into mitochondrial protein, and the remainder into nonmitochondrial membrane protein.

A characteristic feature of synaptosomal protein synthesis is its synergistic stimulation by appropriate concentrations of $Na^+$ and $K^+$ (Table I). The ionic concentrations required for maximal incorporation also result in maximal sodium-potassium adenosine triphosphatase activity, $K^+$ uptake, and oxygen uptake (Appel et al., 1969; Escueta and Appel, 1969; Morgan and Austin, 1969; Austin et al., 1970). It may be noted that these concentrations are quite different from those required for maximum incorporation with the brain ribosomal system (Table I). Inhibition of the sodium-potassium adenosine triphosphatase activity by ouabain also markedly inhibits synaptosomal protein synthesis. These

TABLE II

*Inhibition of Incorporation of $^{14}C$-leucine into Synaptosomal Fractions by Antibiotics\**

| Antibiotic | Incorporation as percentage of control into | | | |
| --- | --- | --- | --- | --- |
| | Whole synaptosome | Soluble protein | Membrane protein | Mitochondrial protein |
| Puromycin, 50 $\mu$M | 24 | 35 | 31 | 23 |
| Cycloheximide, 100 $\mu$M | 29 | 13 | 27 | 92 |
| Chloramphenicol, 300 $\mu$M | 74 | 98 | 91 | 5 |

* Adapted from Austin et al., 1970.

results not only suggest a unique character for synaptosomal protein synthesis, but also point to a close coupling of the synthetic activity with ionic flux and energy metabolism in the nerve ending. There remains the problem of the mechanism of protein synthesis in the nerve ending other than in the mitochondria. The presence of both RNA and enzymes that activate amino acids has been reported (Austin et al., 1970), and the mechanism is certainly sensitive to cycloheximide, a known inhibitor of mammalian ribosomal protein synthesis. In one way, these findings again raise the problem of contamination, possibly by elements of the ribosomal system which enter the synaptosome before membrane closure or which remain outside the synaptosome but are also membrane enclosed. At least a part answer to this has been given by the demonstration that the pattern of labeling of proteins in a microsomal system is grossly different from that in the synaptosome (Autilio et al., 1968), and a final answer may be available when it is known specifically which proteins are synthesized in the nerve ending.

*Axonal protein synthesis* The evidence for local protein synthesis by axons has been reviewed by Koenig (1970). Myelin-free axons will incorporate labeled RNA precursors into an acid-insoluble product, a major portion of which is digested by ribonuclease. This RNA does not appear either to be mitochondrial in origin or to be transferred from the Schwann cell to the axon (Koenig, 1969). The amount of RNA present in axons is, however, extremely small—1/500th to 1/1000th of that in the cell body. As with the synaptosomal system, a more rigorous demonstration of the synthetic capacity of axons will come from the identification of the products of this activity.

## Identification of the primary products and their function

If a protein possesses some catalytic or specific biological or physical property, it can be identified and the regulation of its synthesis can be followed with the aid of that particular property. McKhann and Ho (1967) have shown, for example, that the development pattern of the enzyme galactocerebroside sulfatransferase, responsible for the last step in the biosynthesis of the myelin lipid sulfatide, parallels that of myelination. Again $D(-) = \beta$-hydroxybutyrate dehydrogenase, the activity of which is high in developing brain but low in adult brain, can be induced in the latter by starvation, enabling the adult brain to use an alternate oxidizable substrate (Smith et al., 1969b). There are also many examples of altered gene function in the nervous system, as expressed in the behavior of enzymes, some of which give rise to disease states. Here the effect of a mutation in a catabolic enzyme is equally as dramatic as that in an anabolic enzyme, as is shown in metachromatic leuco-

dystrophy by the accumulation of large amounts of a normal component of the myelin sheath, sulfatide, as a result of the functional absence of an arylsulfatase catabolic enzyme (Mehl and Jatzkewitz, 1965).

Changes in the number and composition of brain-membrane proteins arising from the modification of gene expression during development can now also be followed (Grossfeld, 1968). With increasing age in the mouse a greater preponderance of species with higher molecular weight appears (Figure 1); these may result from modification of the original proteins with carbohydrate or other prosthetic groups or from the shift from a neuronal to a largely glial cell population. Mahler and Cotman (1970) have shown that these or similar methods can be used to characterize the proteins of the membranes of nerve endings.

NERVOUS SYSTEM SPECIFIC PROTEINS Nerve cells obviously have unique function compared with other cells. Gene activity in nerve cells subserves these functions; it is therefore possible to learn about gene activity by searching out and studying its unique products.

Brain extracts contain a number of relatively acidic proteins that do not appear in extracts of other organs, and two of these, S100 and 14–3–2, have now been shown by immunological criteria to be specific to the nervous system (Moore and Perez, 1968). Antiserum to beef brain S100 cross-reacted with brain extracts from all vertebrates and some invertebrates, and antiserum to 14–3–2 reacted with extracts from all vertebrates except fish and reptiles, so they appear to have a high degree of evolutionary stability and thus, presumably, an important function. Because of its presence in cultured gliomas (Benda et al., 1968) but not in cultured neuroblastomas, its high content in cerebral white and low content in cerebral gray matter, and a constant or slightly increased amount in degenerating optic nerve, S100 is thought to be localized primarily in glia (Moore and Perez, 1968). By similar criteria, 14–3–2 is a neuronal protein. A possible clue to the function of S100 comes from studies of its interactions with calcium, sodium, and potassium ions (Calissano et al., 1969). The S100 protein specifically binds two calcium ions, resulting in local unfolding of the structure. Other divalent ions, such as magnesium, do not produce this conformational change, although potassium and, to a lesser extent, sodium ions tend to inhibit the effect. In so doing they insure that the conformational change occurs in the range of calcium-ion concentration that probably exists in the nervous system. The protein 14–3–2 does not bind calcium ion, has a higher molecular weight than S100, and appears to exist also in polymeric forms. Antigen $\alpha$, the soluble protein specific to the nervous system that was isolated by Bennett and Edelman (1968), is unrelated by immunological or chemical

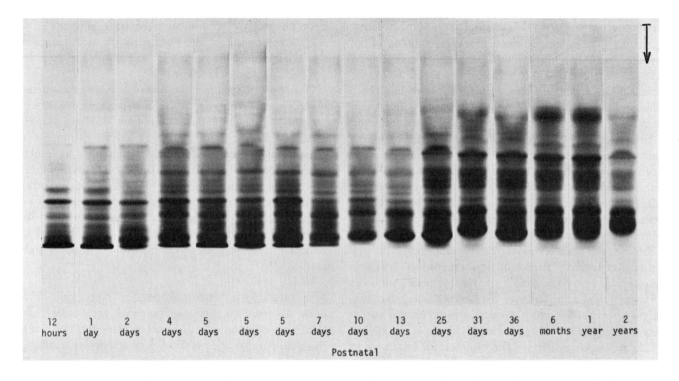

FIGURE 1 Developmental changes in membrane proteins from mouse brain. Mouse brains from animals of different ages extracted with 1 per cent Triton X-100 in 0.01 M tris-Cl buffer pH 8.1. Residue extracted with 0.1 per cent sodium dodecyl sulfate in the same buffer with sonication. These extracts dialyzed against 0.025 per cent sodium dodecyl sulfate in the same buffer and applied to polyacrylamide gels. Electrophoresis was carried out in the tris glycinate discontinuous buffer system, pH 9.6, with 0.025 per cent sodium dodecyl sulfate in all solutions. Stained after electrophoresis with Naphthol Blue Black (Grossfeld, 1968).

criteria to S100 protein, but it forms aggregates stabilized by disulfide bonds as does protein 14-3-2.

Myelin also contains a number of specific proteins, including a 165-residue basic protein responsible for the encephalitogenic activity of nervous-system tissue (Eyhlar and Hashim, 1969). Smaller fragments derived from it by the action of a brain acid protease or pepsin also have the activity, and the sequence of one of these has been reported (Eyhlar and Hashim, 1969; Kibler et al., 1969). Other myelin proteins are the proteolipid proteins described by Folch and Lees (1951) and Wolfgram (1966). These three groups have a functional role in the maintenance of the myelin membrane.

## Modification of gene expression

Both internal factors, such as hormones and growth factors, as well as external stimuli regulate gene expression during development of the nervous system. One of the best-known examples of such agents is the nerve growth factor (NGF), discovered and studied extensively by Levi-Montalcini and her colleagues (Levi-Montalcini and Angeletti, 1968). The nature of this growth factor and of its mechanism of action is obviously of great interest as a model system for the study of the modification of neuronal behavior. NGF stimulates the growth of sympathetic and sensory ganglia in chick embryos or young mice, producing marked increases in the size of the responsive neurons and in fiber outgrowth from such neurons. If chick sensory or sympathetic neurons of appropriate age are cultured in a hanging drop of semi-solid plasma containing optimum concentrations of NGF, a uniform and dense halo of nerve fibers grows out from the ganglia. The varying degree of fiber growth under these conditions in response to differing amounts of NGF forms the basis of a bioassay for this factor (Levi-Montalcini et al., 1954).

Antiserum to NGF inhibits the activity of NGF in the tissue culture assay, and when it is injected into newborn animals of several species it causes a rapid and almost complete destruction of the nerve cells in the sympathetic ganglia (Cohen, 1960; Levi-Montalcini and Booker, 1960). The first property of the ganglia to be affected in vivo is that of electrical transmission, the action potential being markedly depressed a few hours after injection and before significant changes take place in either metabolic function or ganglion size (Halstead *in* Larrabee, 1969, p. 108).

THE MACROMOLECULAR STRUCTURE OF NGF   What is the nature of the factor that produces these specific growth effects? It has been shown that in the two common sources of this material, snake venom and the adult mouse sub-maxillary gland, NGF activity is associated with protein (Cohen, 1959, 1960). Subsequent work confirmed these initial observations and showed that it was possible to isolate, from either source, small, rather basic proteins that had NGF activity (Cohen, 1959, 1960; Varon et al., 1967a; Angeletti, 1968; Banks et al., 1968; Zanini et al., 1968). It is now clear, however, that these basic proteins are derived, as a result of the isolation procedure, from larger protein species that exist in both venom and extracts of the mouse gland. The macromolecular composition of these larger species is therefore of considerable interest in relation to their role in stimulating neuronal growth.

In the submaxillary gland extracts, the biologically active basic protein is only one type of subunit, the $\beta$ subunit, in a multisubunit protein, 7S NGF (Varon et al., 1967b, 1968). 7S NGF, which elicits the same in vivo and in vitro effects as the smaller basic protein, can be isolated from the gland in a reasonable degree of homogeneity (Figure 2A), and its subunit composition can be demonstrated by its reversible dissociation at extremes of pH. This process is readily visualized by the isoelectric focusing procedure when 7S NGF is dissociated by applying it to the acidic end of the pH gradient in the gel (Figure 3b). The undissociated 7S NGF can also be analyzed by the same procedure if it is initially applied near the center of the gradient, and its isoionic point lies between the isoionic points of the $\alpha$ and $\gamma$ sub-units (Figure 3a, c). The three groups of subunits are separable by ion-exchange chromatography and show the same distribution of protein species within each group after isolation (Varon et al., 1968; Smith et al., 1968). The analysis of isolated $\beta$ subunits is shown in Figure 3d as an example. The separated subunits recombine rapidly when mixed at neutral pH and in appropriate amounts to re-generate a protein the physical properties of which are the same as those of 7S NGF (Figure 2).

The specific biological activity of the $\beta$ subunit isolated from 7S NGF at acid pH is similar to that of 7S NGF but is somewhat higher when isolated from the penultimate material in the 7S NGF preparation (Varon et al., 1968). Both Zanini et al. (1968) and Fenton and Edwards (quoted in Vernon et al., 1969) also reported that the specific

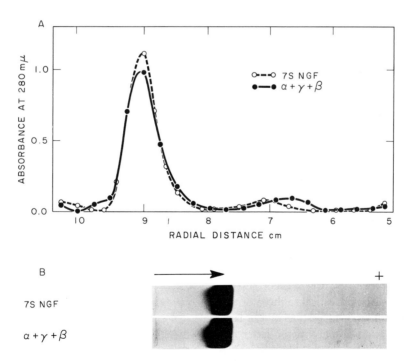

FIGURE 2   Comparison of the sedimentation and electro-phoretic properties of 7S NGF and the product of the recombination of $\alpha$, $\beta$, and $\gamma$ subunits. A. Sedimentation of 200 $\mu$g 7S NGF or a mixture of 72 $\mu$g $\alpha$, 118 $\mu$g $\gamma$, and 42 $\mu$g $\beta$ subunits mixed in 0.05 M tris-Cl buffer (pH 7.4). Sed-imentation was for 13 hours at 60,000 rpm in a 5–20 per cent sucrose gradient in 0.05 M tris-Cl buffer (pH 7.4) at 5° C. B. Electrophoresis in the bistris-tes buffer system (pH 7.55) of 150 $\mu$g 7S NGF and a mixture of 60 $\mu$g $\alpha$, 78 $\mu$g $\gamma$, and 35 $\mu$g $\beta$ subunits in 0.05 M tris-Cl buffer (pH 7.4).

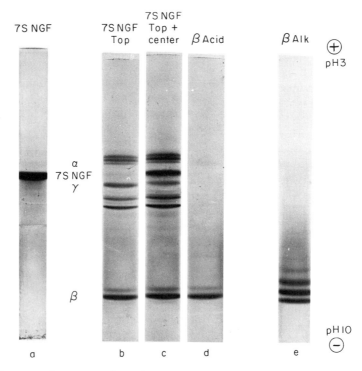

FIGURE 3 Isoelectric focusing of 7S NGF and its subunits. A. 7S NGF applied near center of gel. B. 7S NGF applied at top. C. 7S NGF applied at top and near center. D. β subunit isolated by chromatography at acid pH. E. β subunit isolated by chromatography at alkaline pH. The acrylamide gels were 7.5 per cent made in the ampholine buffer system pH 3–10. After electrophoresis they were stained with Coomassie Blue.

activity of the β subunit (NGF, in the terminology of the former authors), isolated by the same procedure but by using milder acid conditions, is higher than that of 7S NGF. The biological activity of the β subunit is therefore modified by its interaction with the α and γ subunits, and both biologically inactive subunits are required to produce this specific effect with the β subunits (Varon et al., 1968) as they are to produce 7S NGF. The 7S NGF protein also exhibits esterase or peptidase activity relatively specific for argininyl esters or amides, an activity that resides only in the γ subunits, and likewise the specific enzymatic activity of these subunits is modified (decreased) by its interaction with the α and β subunits (Greene et al., 1969). Again, both of the subunits are obligatory for this effect (Table III). The biological and enzymatic properties of 7S NGF reconstituted from the three subunits are identical to those of the original 7S NGF preparation, and, in particular, the lag-phase enzyme kinetics characteristic only of the latter reappear upon the aggregation of the three subunits (Greene et al., 1968). Other evidence that pertains to the specificity of the interaction of the three subunits is summarized by Shooter and Varon (1970).

Although 7S NGF migrates as a single band on electrophoresis, it is actually a family of 7S proteins, each com-

posed of the same number of α, γ, and β subunits but differing in the type of α or γ subunits they contain (Smith et al., 1969a). It is the dissociation of these multiple forms of 7S NGF that produces the variety of individual α and γ subunits (Figure 3). Furthermore, these multiple 7S NGF proteins are in equilibrium with one another through an exchange of free α subunits (Smith et al., 1969). This may be demonstrated in the observation that [125]I-labeled α sub-

TABLE III

*The Effect of the α and β Subunits of 7S NGF on the Enzymatic Activity of the γ Subunit*

| Subunits | Relative enzymatic activity* |
|---|---|
| γ | 1.00 |
| γ + β | 0.99 |
| γ + α | 1.10 |
| γ + α + β | 0.20 |
| | (lag-phase kinetics) |

* Enzymatic activities are given as ratios of the activity of each mixture to that of the γ subunit alone. Activity was measured with 1.0 mM benzoyl arginine ethyl ester in 0.05 M tris-Cl buffer (pH 7.0). (Adapted from Greene et al., 1969.)

units are incorporated into 7S NGF when mixed with the latter at neutral pH, or by the appearance of all the individual α subunits in the free subunit pool when an excess of any one α subunit, e.g., the α² subunit, is again added to 7S NGF. It follows that the time lag observed before 7S NGF hydrolyzes appropriate substrates at maximum rate is caused by the time required for the dissociation equilibrium of 7S NGF to re-establish itself at a new equilibrium position after dilution for the assay, the principal enzymatic species being the uncomplexed forms of the γ subunits (possibly γβ complex, although other species are possible) rather than 7S NGF itself. Conversely, the lag can be eliminated if the diluted 7S NGF solution stands for some time before the substrate is added, which allows 7S NGF to dissociate.

The basic NGF protein, separated by Zanini et al. (1968), which has the physical properties of the β subunit, on standing in dilute solution apparently disaggregates into half molecules that give identical immunodiffusion and electrophoretic patterns and complement fixation curves. That dissociation of the β subunit of 7S NGF can take place is demonstrated by its behavior on electrophoresis in the presence of sodium dodecyl sulfate. In line with previous sedimentation data, both α and γ subunits migrate as single bands in positions corresponding to a molecular weight of about 30,000 (Figure 4). 7S NGF dissociates completely in this solvent and also, as expected, gives a major band in the same position as the subunits; in addition, a faster minor band appears, which can arise only from the dissociation of the β subunit (Figure 4). Analysis of a β subunit isolated by chromatography at alkaline rather than at acid pH (Figure 3E) also demonstrates how derivatives of this subunit of differing isoionic points can arise under various conditions.

A recent method for isolating NGF from *Crotalus adamanteus* venom (Angeletti, 1968) closely parallels the one devised for the mouse gland NGF (Varon et al., 1967b) and suggests that NGF aggregates in the venom have molecular weights lower than that of 7S NGF. However, a major final product is still a basic protein similar in size and electrophoretic properties to the β subunit of 7S NGF. Another component is a basic protein about half of the size and identical to the major component in amino acid composition and immunochemical properties (Levi-Montalcini and Angeletti, 1968). Immunochemical analysis confirms that the NGF basic protein from venom is structurally related to the β subunit from the mouse gland (Zanini et al., 1968). Using an alternate isolation procedure, Banks et al. (1968) obtained from the venom of *Vipera russelli* a basic protein, homogeneous by sedimentation analysis, although not by other criteria, and of molecular

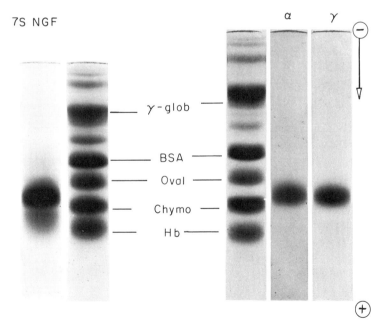

FIGURE 4 Molecular size and composition of the subunits of 7S NGF. 7S NGF (100 μg) or the α and γ subunits (30 μg) were subjected to electrophoresis in acrylamide gels in 0.1 M phosphate buffer (pH 7.0) containing 0.1 per cent sodium dodecyl sulfate. The mixtures of standard proteins were made from 30 μg each of γ globulin (γ-glob), bovine serum albumin (BSA), ovalbumin (oval), chymotrypsinogen (chymo), and human adult hemoglobin (Hb). All proteins were denatured first in 1 per cent sodium dodecyl sulfate in the same buffer. Stained after electrophoresis with Coomassie Blue.

weight 40,000, with high NGF activity. The increase in biological activity of their preparation resulted only from ion-exchange chromatography on a strongly basic resin and exceeded the actual protein purification obtained in this step, so these authors suggest that NGF normally is associated with some other protein, removed in this particular step, which masks its intrinsic activity. The relationships of the various basic NGF proteins from venom remain to be clarified, as does the molecular composition of the larger NGF species in the venom itself.

Why is NGF activity in the venom and the submaxillary gland extracts associated with these subunit-containing or aggregated proteins, and what part do these extra proteins, e.g., the $\alpha$ and $\gamma$ subunits, play in the molecular mechanisms that finally produce fiber outgrowth? It is well known that the $\beta$ subunit or the small basic proteins from venom are alone sufficient to elicit the NGF effect and also, as mentioned earlier, that antibodies made against these species inhibit fiber growth in the bioassay in vitro and result in the destruction of ganglia in the sympathetic chain in vivo (Levi-Montalcini and Angeletti, 1966). Moreover, it is an apparent paradox that, although the NGF effect is elicited in vitro, at concentrations of $10^{-8}$ grams per milliliter, in which at least some subunit-containing proteins, e.g., oxyhemoglobin, are fully dissociated, it is still possible to distinguish between the level of activity of 7S NGF and that of the $\beta$ subunit. The availability of $^{125}$I-labeled 7S NGF makes it possible to determine which molecular species are present at those concentrations. Sedimentation analyses show that even at $10^{-8}$ grams per milliliter, more than 50 per cent of the protein is present at 7S NGF, suggesting a dissociation constant of approximately $10^{-10} - 10^{-11}$M (Smith, unpublished data). Although these experiments were performed at low temperatures and in dilute buffer solutions and do not relate directly to in vivo or bioassay conditions, they do demonstrate the remarkable stability of 7S NGF, a property it shares with at least one other protein, aspartate transcarbamylase (Schachman and Edelstein, 1966). In keeping with this, measurements of the rate of $\alpha$ subunit exchange and of the rate of association of the three subunits suggest that the rate of dissociation of 7S NGF is relatively slow even at elevated temperatures (Smith et al., 1969a). Thus the time required for 7S NGF to dissociate and re-establish a new equilibrium on dilution for enzymatic assay is of the order of 15 minutes at room temperature (Greene et al., 1969). As far as they go, the present data do not discount the possibility that 7S NGF functions as an entity in eliciting biological activity, which would be in keeping with the ability to distinguish biologically between 7S NGF and $\beta$ subunits. Although it is clear that the biological and enzymatic activities of the $\beta$ and $\gamma$ subunits are modified by their interactions with the other subunits—that of the $\gamma$ subunit to such an extent that

it may be inactive in 7S NGF—there is no direct information as to what role the biologically inactive subunits may play in the growth-stimulating properties of NGF. It is of interest to note, however, that at least two other esteropeptidases have been shown to have effects on the growth of certain cells (Attardi et al., 1967; Grossman et al., 1969) and that thrombin, itself an esteropeptidase, exhibits an NGF-like activity (Hoffman and McDougall, 1968).

THE MECHANISM OF ACTION OF NGF   The question of whether NGF acts to stimulate fiber outgrowth by directly affecting transcription of new mRNA has been examined in some detail. The earlier evidence, which argued in favor of the hypothesis, was the finding that, using sensory ganglia, NGF stimulated the incorporation of $^3$H-uridine in vitro prior to its effect on the incorporation of $^{14}$C-amino acids (Angeletti et al., 1965). The finding that actinomycin-D blocks the NGF effect on amino acid incorporation, while puromycin does not affect the NGF-induced increase of uridine incorporation into RNA, also fits in with the concept of a direct effect on mRNA synthesis, as does the observation that actinomycin-D fails to prevent the NGF stimulation of amino acid incorporation if the ganglia are pre-incubated with NGF. At the morphological level, the structural components of the nucleoli and nuclei alter under the influence of NGF, undergoing a process of condensation and separation into masses of different density and structure (Levi-Montalcini et al., 1968, 1969). On the other hand, attempts to detect increased mRNA synthesis by changes in the distribution pattern of labeled RNA or in RNA base ratios in the presence or absence of NGF have not been successful (Toschi et al., 1966; Burdman, 1967).

More recently, Larrabee and his colleagues (Larrabee, 1969) have emphasized that it is difficult to demonstrate unequivocally that NGF stimulates certain metabolic pathways in ganglia in tissue culture because of the known deterioration of control ganglia in the absence of NGF (Angeletti et al., 1964; Levi-Montalcini et al., 1968). When the oxidation of glucose by sympathetic ganglia is monitored continuously, it is found that NGF does not increase the utilization of the Cl of labeled glucose in the first few hours. The difference in Cl utilization only becomes significant after about five hours, by which time the Cl utilization in the control ganglia is actually falling (Figure 5). Thus these experiments may only be measuring the other known property of NGF, which is the preservation of cell viability (Levi-Montalcini and Angeletti, 1963), or they may be measuring both the better maintenance and increased growth. In addition, these workers have been able to demonstrate that increased fiber outgrowth is not necessarily dependent on increased RNA synthesis. Using sympathetic ganglia from 14-day-old chick embryos, they confirmed that actinomycin-D not only inhibits RNA syn-

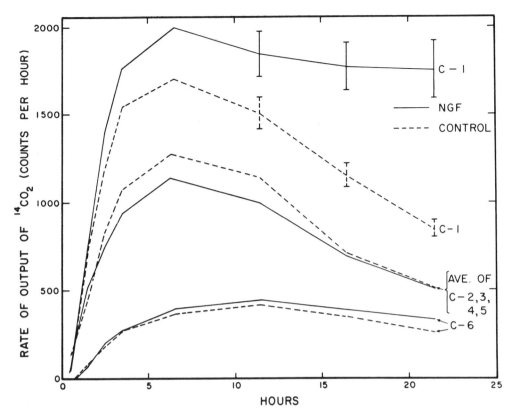

FIGURE 5 Effect of NGF on the glucose metabolism of sympathetic ganglia. Release of labeled $CO_2$ from sympathetic ganglia excised from chick embryos and supplied with glucose labeled in various positions with $^{14}C$ in the presence and absence of NGF. The average output for carbons 2, 3, 4, and 5 was calculated by subtracting the output of C1 and C6 from the output with uniformly labeled glucose (Brown in Larrabee, 1969, p. 110).

thesis in control ganglia but eliminates the usual NGF-induced stimulation of RNA synthesis (Figure 6). In spite of the almost complete cessation of uridine incorporation over the entire incubation period, fiber growth, measured by the width of the fiber halo, was reduced by only about 25 per cent. Experiments with the protein synthesis inhibitor cycloheximide showed that this reduction in growth was attributable to the minor effect of actinomycin-D on protein synthesis. The converse of this situation holds in sympathetic ganglia from younger, nine- to 10-day-old embryos. Although RNA synthesis is increased considerably in the presence of NGF, these ganglia show only a very limited fiber outgrowth (Partlow and Larrabee, 1969; Partlow, 1969). Because the two major effects of NGF, stimulating RNA synthesis and causing substantial fiber outgrowth, can now be separated, it appears unlikely that NGF acts only on the transcriptional level.

A COMPARISON OF THE EFFECTS OF INSULIN AND NGF
The separation of the effects of NGF on fiber outgrowth and on the maintenance or stimulation of metabolic activity is reminiscent of the effect of insulin on embryonic ganglia. Insulin, as does NGF, will preserve the normal histology and, therefore, presumably the metabolic activity, of eight-day sensory ganglia for at least one day in vitro (Levi-Montalcini, 1966). Insulin also parallels NGF in its effect on the conversion of glucose into $CO_2$, lactate, and glutamate (Liuzzi et al., 1968) and on the stimulation of the incorporation of labeled acetate or mevalonic acid into such ganglia (Liuzzi and Foppen, 1968). Partlow (1969) has also observed that the time course of the incorporation of labeled uridine or amino acid into the 14-day-old sympathetic ganglia is similar with both NGF and insulin, these ganglia also being histologically well preserved by either agent. However, although insulin and NGF have these identical anabolic effects, they differ in their action in one significant respect. Insulin does not cause fiber outgrowth from either sensory (Levi-Montalcini, 1966) or sympathetic ganglia (Partlow, 1969). Thus, NGF might act like insulin to maintain the normal metabolic activity of the responsive neurons and, in addition, have the extra capability of stimulating fiber growth. It is pertinent to note that an anabolic effect of insulin is to stimulate the production of proteins already synthesized by an organism without the

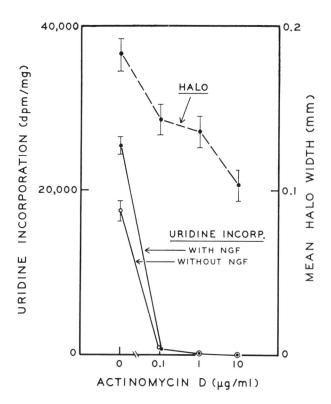

FIGURE 6 Effects of actinomycin-D on fiber outgrowth and incorporation of labeled uridine into RNA of embryonic sympathetic ganglia with and without NGF. Fiber outgrowth was measured by the width of the halo. Incubation was for 15 hours (Partlow and Larrabee, 1969; Larrabee 1969).

requirement for prior RNA synthesis. It is known, for example, that insulin acts at the translational level to correct defects in the large ribosomal subunit from diabetic animals (Wool et al., 1968; Martin and Wool, 1968). Other hormones, such as growth hormone, act in a similar way, in this instance, to overcome a defect in the small ribosomal subunit from hypophysectomized animals (Korner, 1968; Barden and Korner, 1969). The long-term effects of these two hormones result in the synthesis of more ribosomes and endoplasmic reticulum (Wyatt and Tata, 1968), which is exactly the situation found for NGF-treated sensory ganglia (Levi-Montalcini et al., 1968).

Another growth factor has been isolated from the submaxillary gland, and progress has been made in determining its mechanism of action.

THE EPIDERMAL GROWTH FACTOR The EGF stimulates epidermal growth in intact animals or in tissue culture and concurrently stimulates incorporation into protein and all forms of cytoplasmic RNA (Cohen and Stastny, 1968). The initial cellular event induced by EGF that is responsible for this sequence of events appears to be a conversion of pre-

existing monomer ribosomes into polysomes. Because the EGF-induced polysome formation also occurs in the presence of cycloheximide or even a mixture of cycloheximide and puromycin, which inhibits protein synthesis more than 99 per cent, it is not dependent on the synthesis of new proteins. Nor, as the effect is obtained with cells preincubated in the absence of amino acids or glucose, does EGF appear to stimulate transport of these compounds. In addition, earlier work (Hoober and Cohen, 1967) also demonstrated that, in the presence of EGF, ribosomes from epidermis were twice as active as controls in an incorporation system using either endogenous mRNA or poly-U. In this respect, EGF action closely resembles the actions of insulin and growth hormone. Recent work indicates that EGF from the mouse submaxillary gland is a complex of 70,000 molecular weight composed of two different types of subunits (Taylor et al., 1970). One subunit has a molecular weight of 6,000, an isoionic point of 4.6, and carries the EGF activity. The other subunit is an esterase whose size (molecular weight 27,000), isoionic point (pH 5.6), and enzymatic specificity make it very similar to the γ subunit of 7S NGF. The EGF complex contains two subunits of each type.

A COMPARISON OF NGF AND PHYTOHEMAGGLUTININ There also exists another growth factor, phytohemagglutinin (PPHA), which as a protein shows some remarkable similarities to NGF, but whose mechanism of action apparently differs from that of EGF. PPHA, as isolated from red kidney beans, is a homogeneous glycoprotein of molecular weight 128,000 that may be reversibly dissociated in 8 M urea into about eight subunits of equal size but differing charge properties (Rigas et al., 1966). The parent protein has, as has NGF, two biological activities. One causes the agglutination of erythrocytes (Li and Osgood, 1949); the other, the induction of mitosis in normal human blood lymphocytes (Nowell, 1960). Rigas and Head (1969) have now shown that the two activities reside in different subunits. Although it has been reported that [3]H-PPHA is taken up by the condensed heterochromatin of lymphocyte nuclei (Stanley et al., 1968), Kornfeld and Kornfeld (1970) have, in contrast, isolated from human erythrocyte membranes a glycopeptide which binds specifically to PPHA, abolishing both its agglutinating and mitosis-stimulating properties, and therefore having the properties of a membrane receptor site for PPHA. Correspondingly, a major part of the first RNA synthesized in response to PPHA is a heterogeneous form with high molecular weight, most of which remains in the nucleus and turns over fairly rapidly (Cooper, 1968; Darnell, 1968). Aronson and Wilt (1969) have suggested, on the basis of experiments with pregastrular sea-urchin embryos, that this RNA may be present and functioning as membrane-bound nuclear polysomes and also that the remaining nuclear RNA is a precursor to cytoplasmic RNA of the messenger type. In the PPHA-treated lymphocytes, the synthesis of the heterogeneous RNA species is rapidly overshadowed by the much greater synthesis of precursors for ribosomal RNA; the formation of ribosomes follows the pathways delineated in HeLa cells.

HORMONES AND GENE ACTIVITY    If PPHA interacts with the nucleus, then it functions like a number of the steroid hormones. Aldosterone, which regulates sodium transport across epithelial cells, is firmly bound to a specific nuclear protein, and the complex appears to act as an inducer at the transcriptional level (Herman et al., 1968). Binding to the nuclear protein is highly specific, even stereospecific, because the biologically inactive isomer 17-isoaldosterone is not retained. The identity of this protein is not known, but there are at least two candidates—the histones and the acidic nuclear proteins. The latter show high rates of labeling when new RNA synthesis is intense (Church and McCarthy, 1967). The histones appear to be tissue and animal specific (Bustin and Cole, 1968) and to show a sequence asymmetry of the basic amino acids that might provide a DNA binding site (Bustin et al., 1969; DeLange et al., 1969; Bonner et al., 1969). Alternatively, histones may couple with chromosomal RNA, which itself is capable of hybridizing with homologous native DNA (Bekhor et al., 1969). It is now clear that the histone composition of a tissue responds to an altered physiology; when casein synthesis begins in the mammary gland, for example, the synthesis of two of the tissue-specific histones changes dramatically (Stellwagen and Cole, 1968; Hohmann and Cole, 1969).

Although it is not yet known if NGF acts at the transcriptional or translational level or by some entirely different mechanism, the information provided by the other model hormone systems may lead to further experimental clarification of the situation.

MODIFICATION BY SENSORY STIMULATION OR DURING LEARNING    It has been reported that sustained afferent stimulation involving constant firing for many hours of an isolated stretch-receptor organ of the crayfish does not alter the amount of RNA present, although it does change the RNA base ratios during the first few hours (Grampp and Edström, 1963). Most other studies, however, indicate that stimulation of the cortex lowers the amount of both RNA and protein present (Geiger et al., 1956). In cortex slices, electrical stimulation decreases the rate of incorporation or precursors into RNA and protein, but kidney slices show no changes (Orrego, 1967; Orrego and Lipmann, 1967). Unilateral spreading cortical depression decreases $^3$H-leucine incorporation into soluble cortical protein of the depressed hemisphere, compared with that of the control hemisphere,

although soluble brainstem protein from both sides had the same specific activity (Bennett and Edelman, 1969). Because the levels of many other metabolites also vary under these conditions, it seems likely that this and other phenomena noted above arise from alterations in precursor or energy pools rather than from direct effects on the protein synthetic machinery. On the other hand, there is good evidence that electroshock, which impairs memory consolidation, can disaggregate polysomes in the rabbit cortex (Vesco and Giuditta, 1968).

Although visual deprivation results in decreases in RNA and protein and also in their rates of synthesis in most cells of the optic system, the effects of light after a period of dark are stimulatory. Appel et al. (1967), for example, have reported significant increases in cortical polysomes and in the incorporation of radioactive leucine in adult rats exposed to light after three days in the dark. No change was noted in the level of total RNA, leucyl transfer RNA, or in activating enzymes.

Changes in the rates of RNA and protein synthesis occur when an animal undergoes training, but the question as to whether they are directly related to learning rather than to the other stimuli is difficult to answer (Glassman, 1969). The subject is discussed in detail by Hydén and Lange (this volume).

## Acknowledgments

It is a pleasure to acknowledge that the work on the nerve growth factor described here was carried out in collaboration with Dr. Silvio Varon (Department of Biology, University of California at San Diego) and our colleagues Drs. Junichi Nomura, Andrew P. Smith, Lloyd A. Greene, and H. Ronald Fisk. The studies on the solubilization and examination of mouse brain membrane proteins are those of Drs. Robert M. Grossfeld and Thomas W. Waehneldt. The research was supported by a United States Public Health Service grant from the National Institute of Neurological Disease and Stroke, NB 04270, and by a grant from the National Science Foundation, GB 6878.

### REFERENCES

ANGELETTI, R. A. H., 1968. Studies on the nerve growth factor (NGF) from snake venom. Molecular heterogeneity. *J. Chromatogr.* 37: 62–69.

ANGELETTI, P. U., D. GANDINI-ATTARDI, G. TOSCHI, M. L. SALVI, and R. LEVI-MONTALCINI, 1965. Metabolic aspects of the effect of nerve growth factor on sympathetic and sensory ganglia: Protein and ribonucleic acid synthesis. *Biochim. Biophys. Acta* 95: 111–120.

ANGELETTI, P. U., A. LIUZZI, and R. LEVI-MONTALCINI, 1964. Stimulation of lipid biosynthesis in sympathetic and sensory ganglia by specific nerve growth factor. *Biochim. Biophys. Acta* 84: 778–781.

APPEL, S. H., L. AUTILIO, B. W. FESTOFF, and A. V. ESCUETA, 1969. Biochemical studies of synapses *in vitro*. III. Ionic activation of protein synthesis. *J. Biol. Chem.* 244: 3166–3172.

APPEL, S. H., W. DAVIS, and S. SCOTT, 1967. Brain polysomes: Response to environmental stimulation. *Science (Washington)* 157: 836–838.

ARONSON, A. I., and F. H. WILT, 1969. Properties of nuclear RNA in sea urchin embryos. *Proc. Nat. Acad. Sci. U. S. A.* 62: 186–193.

ATTARDI, D. G., M. J. SCHLESINGER, and S. SCHLESINGER, 1967. Submaxillary gland of mouse: Properties of a purified protein affecting muscle tissue *in vitro*. *Science (Washington)* 156: 1253–1255.

AUSTIN, L., and I. G. MORGAN, 1967. Incorporation of $^{14}$C-labelled leucine into synaptosomes from rat cerebral cortex *in vitro*. *J. Neurochem.* 14: 377–387.

AUSTIN, L., I. G. MORGAN, and J. J. BRAY, 1970. The biosynthesis of proteins within axons and synaptosomes. *In* Protein Metabolism of the Nervous System (A. Lajtha, editor). Plenum Press, New York, pp. 271–287.

AUTILIO, L. A., S. H. APPEL, P. PETTIS, and P. L. GAMBETTI, 1968. Biochemical studies of synapses *in vitro*. I. Protein synthesis. *Biochemistry* 7: 2615–2622.

BANKS, B. E. C., D. V. BANTHORPE, A. R. BERRY, H. ff. S. DAVIES, S. DOONAN, D. M. LAMONT, R. SHIPOLINI, and C. A. VERNON, 1968. The preparation of nerve growth factors from snake venoms. *Biochem. J.* 108: 157–158.

BARDEN, N., and A. KORNER, 1969. A defect in the 40S ribosomal subunit after hypophysectomy of the rat. *Biochem. J.* 114: 30P (abstract).

BARONDES, S. H., 1966. On the site of synthesis of the mitochondrial protein of nerve endings. *J. Neurochem.* 13: 721–727.

BEKHOR, I., J. BONNER, and G. K. DAHMUS, 1969. Hybridization of chromosomal RNA to native DNA. *Proc. Nat. Acad. Sci. U. S. A.* 62: 271–277.

BENDA, P., J. LIGHTBODY, G. SATO, L. LEVINE, and W. H. SWEET, 1968. Differentiated rat glial cell strain in tissue culture. *Science (Washington)* 161: 370–371.

BENNETT, G. S., and G. M. EDELMAN, 1968. Isolation of an acidic protein from rat brain. *J. Biol. Chem.* 243: 6234–6241.

BENNETT, G. S., and G. M. EDELMAN, 1969. Amino acid incorporation into rat brain proteins during spreading cortical depression. *Science (Washington)* 163: 393–395.

BODIAN, D., 1965. A suggestive relationship of nerve cell RNA with specific synaptic sites. *Proc. Nat. Acad. Sci. U. S. A.* 53: 418–425.

BONDY, S. C., and S. ROBERTS, 1968. Hybridizable ribonucleic acid of rat brain. *Biochem. J.* 109: 533–541.

BONNER, J., J. GRIFFITH, T. SHIH, and A. SADGOPAL, 1969. Properties of chromosomal proteins. *Biophys. J.* 9: A–87.

BURDMAN, J. A., 1967. Early effects of a nerve growth factor on the RNA content and base ratios of isolated chick embryo sensory ganglia neuroblasts in tissue culture. *J. Neurochem.* 14: 367–371.

BUSTIN, M., and R. D. COLE, 1968. Species and organ specificity in very lysine-rich histones. *J. Biol. Chem.* 243: 4500–4505.

BUSTIN, M., S. C. RALL, R. H. STELLWAGEN, and R. D. COLE, 1969. Histone structure: Asymmetric distribution of lysine residues in lysine-rich histone. *Science (Washington)* 163: 391–393.

CALISSANO, P., B. W. MOORE, and A. FRIESEN, 1969. Effect of calcium ion on S-100, a protein of the nervous system. *Biochemistry* 8: 4318–4326.

CAMPBELL, M. K., H. R. MAHLER, W. J. MOORE, and S. TEWARI, 1966. Protein synthesis systems from rat brain. *Biochemistry* 5: 1174–1184.

CHURCH, R. B., and B. J. McCARTHY, 1967. Ribonucleic acid synthesis in regenerating and embryonic liver. I. The synthesis of new species of RNA during regeneration of mouse liver after partial hepatectomy. *J. Mol. Biol.* 23: 459–475.

COHEN, S., 1959. Purification and metabolic effects of a nerve growth promoting protein from snake venom. *J. Biol. Chem.* 234: 1129–1137.

COHEN, S., 1960. Purification of a nerve-growth promoting protein from the mouse salivary gland and its neuro-cytotoxic antiserum. *Proc. Nat. Acad. Sci. U. S. A.* 46: 302–311.

COHEN, S., and M. STASTNY, 1968. Epidermal growth factor. III. The stimulation of polysome formation in chick embryo epidermis. *Biochim. Biophys. Acta* 166: 427–437.

COOPER, H. L., 1968. Ribonucleic acid metabolism in lymphocytes stimulated by phytohemagglutinin. II. Rapidly synthesized ribonucleic acid and the production of ribosomal ribonucleic acid. *J. Biol. Chem.* 243: 34–43.

DARNELL, J. E., JR., 1968. Ribonucleic acids from animal cells. *Bacteriol. Rev.* 32: 262–290.

DeLANGE, R. J., D. M. FARNGROUGH, E. L. SMITH, and J. BONNER, 1969. Calf and pea histone IV. II. The complete amino acid sequence of calf thymus histone IV; presence of ε-N-acetyllysine. *J. Biol. Chem.* 244: 319–334.

DROZ, B., and H. L. KOENIG, 1969. Turnover of protein in nerve endings as revealed by electron microscopic radioautography. *In* Second International Meeting of the International Society for Neurochemistry (R. Paoletti, R. Fumagalli, and C. Galli, editors). Tamburini Editore, Milan, Italy, pp. 158–159.

ESCUETA, A. V., and S. H. APPEL, 1969. Biochemical studies of synapses *in vitro*. II Potassium transport. *Biochemistry* 8: 725–733.

EYHLAR, E. H., and G. A. HASHIM, 1969. The structure of the encephalitogenic protein of myelin. *In* Second International Meeting of the International Society for Neurochemistry (R. Paoletti, R. Fumagalli, and C. Galli, editors). Tamburini Editore, Milan, Italy, pp. 53–54.

FOLCH, J., and M. LEES, 1951. Proteolipides, a new type of tissue lipoproteins. Their isolation from brain. *J. Biol. Chem.* 191: 807–817.

GEIGER, A., S. YAMASAKI, and R. LYONS, 1956. Changes in nitrogenous components of brain produced by stimulation of short duration. *Amer. J. Physiol.* 184: 239–243.

GLASSMAN, E., 1969. The biochemistry of learning: An evaluation of the role of RNA and protein. *Annu. Rev. Biochem.* 38: 605–646.

GORDON, M. W., and G. G. DEANIN, 1968. Protein synthesis by isolated rat brain mitochondria and synaptosomes. *J. Biol. Chem.* 243: 4222–4226.

GRAMPP, W., and J. E. EDSTRÖM, 1963. The effect of nervous activity on ribonucleic acid of the crustacean receptor neuron. *J. Neurochem.* 10: 725–731.

GRAY, E. G., 1964. Tissue of the central nervous system. *In* Electron Microscopic Anatomy (S. M. Kurtz, editor). Academic Press, New York, pp. 369–417.

GREENE, L. A., E. M. SHOOTER, and S. VARON, 1968. Enzymatic activities of mouse nerve growth factor and its subunits. *Proc. Nat. Acad. Sci. U. S. A.* 60: 1383–1388.

GREENE, L. A., E. M. SHOOTER, and S. VARON, 1969. Subunit interaction and enzymatic activity of mouse 7S nerve growth factor. *Biochemistry* 8: 3735–3741.

GROSSFELD, R. M., 1968. The extraction and fractionation of mouse brain proteins. Stanford University, Stanford, California, doctoral thesis.

GROSSMAN, A., K. P. LELE, J. SHELDON, I. SCHENKEIN, and M. LEVY, 1969. The effect of esteroproteases from mouse submaxillary gland on the growth of rat hepatoma cells in tissue culture. *Exp. Cell Res.* 54: 260–263.

HAJÓS, F., and S. KERPEL-FRONIUS, 1969a. Electron histochemical observation of succinic dehydrogenase activity in various parts of neurons. *Exp. Brain Res.* 8: 66–78.

HAJÓS, F., and S. KERPEL-FRONIUS, 1969b. Ultracytochemical evidences for functional heterogeneity of mitochondria in the nervous system. *In* Second International Meeting of the International Society for Neurochemistry (R. Paoletti, R. Fumagalli, and C. Galli, editors). Tamburini Editore, Milan, Italy, pp. 205–206.

HERMAN, T. S., G. M. FIMOGNARI, and I. S. EDELMAN, 1968. Studies on renal aldosterone-binding proteins. *J. Biol. Chem.* 243: 3849–3856.

HOFFMAN, H., and J. McDOUGALL, 1968. Some biological properties of proteins of the mouse submaxillary gland as revealed by growth of tissues on electrophoretic acrylamide gels. *Exp. Cell Res.* 51: 485–503.

HOHMANN, P., and R. D. COLE, 1969. Hormonal effects on amino-acid incorporation into lysine-rich histones. *Nature (London)* 223: 1064–1066.

HOOBER, J. K., and S. COHEN, 1967. Epidermal growth factor. II. Increased activity of ribosomes from chick embryo epidermis for cell-free protein synthesis. *Biochim. Biophys. Acta* 138: 357–368.

HUNGEN, K. VON, H. R. MAHLER, and W. J. MOORE, 1968. Turnover of protein and ribonucleic acid in synaptic subcellular fractions from rat brain. *J. Biol. Chem.* 243: 1415–1423.

HYDÉN, H., and E. EGYHÁZI, 1963. Glial RNA changes during a learning experiment in rats. *Proc. Nat. Acad. Sci. U. S. A.* 49: 618–624.

KIBLER, R. F., R. SHAPIRA, S. McKNEALLY, J. JENKINS, P. SELDEN, and F. CHOU, 1969. Encephalitogenic protein: Structure. *Science (Washington)* 164: 577–580.

KOENIG, E., 1969. Membrane protein synthesizing machinery of the axon. *In* Second International Meeting of the International Society for Neurochemistry (R. Paoletti, R. Fumagalli, and C. Galli, editors). Tamburini Editore, Milan, Italy, pp. 12–13.

KOENIG, E., 1970. The axon as a heuristic model for studying membrane protein synthesizing machinery. *In* Protein Metabolism of the Nervous System (A. Lajtha, editor). Plenum Press, New York, pp. 259–267.

KORNER, A., 1968. Anabolic action of growth hormone. *Ann. N. Y. Acad. Sci.* 148: 408–418.

KORNFELD, R., and S. KORNFELD, 1970. Structure of a phytohemagglutinin receptor site from human erythrocytes. *J. Biol. Chem.* 245: 2536–2545.

LARRABEE, M. G., 1969. Metabolic effects of nerve impulses and nerve-growth factor in sympathetic ganglia. *Progr. Brain Res.* 31: 95–110.

LEVI-MONTALCINI, R., 1966. The nerve growth factor: Its mode of action on sensory and sympathetic nerve cells. *Harvey Lect. (1964–1965)*, ser. 60: 217–259.

LEVI-MONTALCINI, R., and P. U. ANGELETTI, 1963. Essential role of the nerve growth factor in the survival and maintenance of dissociated sensory and sympathetic embryonic cells *in vitro*. *Develop. Biol.* 7: 653–659.

LEVI-MONTALCINI, R., and P. U. ANGELETTI, 1966. Immunosympathectomy. *Pharmacol. Rev.* 18: 619–628.

LEVI-MONTALCINI, R., and P. U. ANGELETTI, 1968. Nerve growth factor. *Physiol. Rev.* 48: 534–569.

LEVI-MONTALCINI, R., and B. BOOKER, 1960. Destruction of the sympathetic ganglia in mammals by an antiserum to a nerve-growth protein. *Proc. Nat. Acad. Sci. U. S. A.* 46: 384–391.

LEVI-MONTALCINI, R., F. CARAMIA, and P. U. ANGELETTI, 1969. Alterations in the fine structure of the nucleoli in sympathetic neurons following NGF-antiserum treatment. *Brain Res.* 12: 54–73.

LEVI-MONTALCINI, R., F. CARAMIA, S. A. LUSE, and P. U. ANGELETTI, 1968. *In vitro* effects of the nerve growth factor on the fine structure of the sensory nerve cells. *Brain Res.* 8: 347–362.

LEVI-MONTALCINI, R., H. MEYER, and V. HAMBURGER, 1954. *In vitro* experiments on the effects of mouse sarcomas 180 and 37 on the spinal and sympathetic ganglia of the chick embryo. *Cancer Res.* 14: 49–57.

LI, J. G., and E. E. OSGOOD, 1949. A method for the rapid separation of leukocytes and nucleated erythrocytes from blood or marrow with a phytohemagglutinin from red beans (*Phaseolus vulgaris*). *Blood J. Hematol.* 4: 670–675.

LIUZZI, A., and F. H. FOPPEN, 1968. Sterol-like compound from sensory ganglia: effect of a nerve growth factor and insulin on its biosynthesis. *Biochem. J.* 107: 191–196.

LIUZZI, A., F. POCCHIARI, and P. U. ANGELETTI, 1968. Glucose metabolism in embryonic ganglia: Effect of nerve growth factor (NGF) and insulin. *Brain Res.* 7: 452–454.

McKHANN, G. M., and W. HO, 1967. The *in vivo* and *in vitro* synthesis of sulfatides during development. *J. Neurochem.* 14: 717–724.

MAHLER, H. R., and C. COTMAN, 1970. Insoluble proteins of the synaptic plasma membrane. *In* Protein Metabolism of the Nervous System (A. Lajtha, editor). Plenum Press, New York, pp. 151–183.

MARTIN, T. E., and I. G. WOOL, 1968. Formation of active hybrids from subunits of muscle ribosomes from normal and diabetic rats. *Proc. Nat. Acad. Sci. U. S. A.* 60: 569–574.

MEHL, E., and H. JATZKEWITZ, 1965. Evidence for the genetic block in metachromatic leucodystrophy (ML). *Biochem. Biophys. Res. Commun.* 19: 407–411.

MOORE, B. W., and V. J. PEREZ, 1968. Specific acidic proteins of the nervous system. *In* Physiological and Biochemical Aspects of Nervous Integration (F. D. Carlson, editor). Prentice-Hall, Englewood Cliffs, New Jersey, pp. 343–359.

MORGAN, I. G., and L. AUSTIN, 1968. Synaptosomal protein synthesis in a cell-free system. *J. Neurochem.* 15: 41–51.

MORGAN, I. G., and L. AUSTIN, 1969. Ion effects and protein synthesis in synaptosomal fraction. *J. Neurobiol.* 2: 155–167.

MUGNAINI, E., and F. WALBERG, 1964. Ultrastructure of neuroglia. *Ergeb. Anat. Entwickl.* 37: 194–236.

NEUPERT, W., D. BRDICZKA, and W. SEBALD, 1968. Incorporation of amino acids into the outer and inner membrane of isolated rat-liver mitochondria. *In* Biochemical Aspects of the Biogenesis of Mitochondria (E. C. Slater, J. M. Tager, A. Papa, and E. Quagliariello, editors). Adriatica Editorice, Bari, Italy, pp. 395–408.

NOWELL, P. C., 1960. Phytohemagglutinin: An initiator of mitosis in cultures of normal human leukocytes. *Cancer Res.* 20: 462–466.

ORREGO, F., 1967. Synthesis of RNA in normal and electrically stimulated brain cortex slices in vitro. *J. Neurochem.* 14: 851–858.

ORREGO, F., and F. LIPMANN, 1967. Protein synthesis in brain slices: effects of electrical stimulation and acidic amino acids. *J. Biol. Chem.* 242: 665–671.

PARTLOW, L. M., 1969. Metabolic effects of nerve growth factor on cultured sympathetic ganglia from chick embryos. The Johns Hopkins University, Baltimore, Maryland, doctoral thesis.

PARTLOW, L. M., and M. G. LARRABEE, 1969. Metabolic effects of nerve growth factor on chick sympathetic ganglia *in vitro*. Inhibitory effects of actinomycin-D and cycloheximide. *Fed. Proc.* 28: 886 (abstract).

RIGAS, D. A., and C. HEAD, 1969. The dissociation of phytohemagglutinin of *Phaseolus vulgaris* by 8.0 M urea and the separation of the mitogenic from the erythroagglutinating activity. *Biochem. Biophys. Res. Commun.* 34: 633–639.

RIGAS, D. A., E. A. JOHNSON, R. T. JONES, J. D. MCDERMED, and V. V. TISDALE, 1966. The relationship of the molecular structure to the hemagglutinating and mitogenic activities of the phytohemagglutinin of *Phaseolus vulgaris*. *Journées Hellènes de séparation immédiate et de chromatographie (IIIèmes J.I.S.I.C.)*, pp. 151–223.

ROBERTS, S., C. E. ZOMZELY, and S. C. BONDY, 1970. Protein synthesis in the nervous system. *In* Protein Metabolism of the Nervous System (A. Lajtha, editor). Plenum Press, New York, pp. 3–35.

SCHACHMAN, H. K., and S. J. EDELSTEIN, 1966. Ultracentrifuge studies with absorption optics. IV. Molecular weight determinations at the microgram level. *Biochemistry* 5: 2681–2705.

SHOOTER, E. M., and S. VARON, 1970. Macromolecular aspects of the nerve growth factor proteins. *In* Protein Metabolism of the Nervous System (A. Lajtha, editor). Plenum Press, New York, pp. 419–437.

SIDMAN, R. L., 1968. Development of interneuronal connections in brains of mutant mice. *In* Physiological and Biochemical Aspects of Nervous Integration (F. D. Carlson, editor). Prentice-Hall, Englewood Cliffs, New Jersey, pp. 163–193.

SMITH, A. P., L. A. GREENE, H. R. FISK, S. VARON, and E. M. SHOOTER, 1969a. Subunit equilibria of the 7S nerve growth factor protein. *Biochemistry* 8: 4918–4926.

SMITH, A. P., H. S. SATTERTHWAITE, and L. SOKOLOFF, 1969b. Induction of brain D(—)-$\beta$-hydroxybutyrate dehydrogenase activity by fasting. *Science (Washington)* 163: 79–81.

SMITH, A. P., S. VARON, and E. M. SHOOTER, 1968. Multiple forms of the nerve growth factor protein and its subunits. *Biochemistry* 7: 3259–3268.

SOTELO, C., and S. L. PALAY, 1968. The fine structure of the lateral vestibular nucleus in the rat. I. Neurons and neuroglial cells. *J. Cell Biol.* 36: 151–179.

STANLEY, D. A., J. H. FRENSTER, and D. A. RIGAS, 1968. Subnuclear localization of tritiated phytohemagglutinin during gene de-repression within human lymphocytes. *J. Cell Biol.* 39 (no. 2, pt. 2): 129a (abstract).

STELLWAGEN, R. H., and R. D. COLE, 1968. Comparison of histones obtained from mammary gland at different stages of development and lactation. *J. Biol. Chem.* 243: 4456–4462.

TAYLOR, J., S. COHEN, and W. MITCHELL, 1970. Epidermal growth factor (EGF): properties of a high molecular weight species. *Fed. Proc.* 29 (no. 2): 670 (abstract).

TOSCHI, G., E. DORE, P. U. ANGELETTI, R. LEVI-MONTALCINI, and CH. DE HAËN, 1966. Characteristics of labelled RNA from spinal ganglia of chick embryo and the action of a specific growth factor (NGF). *J. Neurochem.* 13: 539–544.

VARON, S., J. NOMURA, and E. M. SHOOTER, 1967a. The isolation of the mouse nerve growth factor protein in a high molecular weight form. *Biochemistry* 6: 2202–2209.

VARON, S., J. NOMURA, and E. M. SHOOTER, 1967b. Subunit structure of a high-molecular-weight form of the nerve growth factor from mouse submaxillary gland. *Proc. Nat. Acad. Sci. U. S. A.* 57: 1782–1789.

VARON, S., J. NOMURA and E. M. SHOOTER, 1968. Reversible dissociation of the mouse nerve growth factor protein into different subunits. *Biochemistry* 7: 1296–1303.

VERNON, C. A., B. E. C. BANKS, D. V. BANTHORPE, A. R. BERRY, H. FF. S. DAVIES, D. M. LAMONT, F. L. PEARCE, and K. A. REDDING, 1969. Nerve growth and epithelial growth factors. *In* Homeostatic Regulators/A Ciba Foundation Symposium (G. E. W. Wolstenholme and J. Knight, editors). J. and A. Churchill Ltd., London, pp. 57–70.

VESCO, C., and A. GIUDITTA, 1968. Disaggregation of brain polysomes induced by electroconvulsive treatment. *J. Neurochem.* 15: 81–85.

WEINSTEIN, I. B., S. M. FRIEDMAN, and M. OCHOA, JR., 1966. Fidelity during translation of the genetic code. *Cold Spring Harbor Symp. Quant. Biol.* 31: 671–681.

WOLFGRAM, F., 1966. A new proteolipid fraction of the nervous system. I. Isolation and amino acid analyses. *J. Neurochem.* 13: 461–470.

WOOL, I. G., W. S. STIREWALT. K. KURIHARA, R. B. LOW, P. BAILEY, and D. OYER, 1968. Mode of action of insulin in the regulation of protein biosynthesis in muscle. *Recent Progr. Hormone Res.* 24: 139–208.

WYATT, G. R., and J. R. TATA, 1968. The hybridization capacity of ribonucleic acid produced during hormone action. *Biochem. J.* 109: 253–258.

ZANINI, A., P. ANGELETTI, and R. LEVI-MONTALCINI, 1968. Immunochemical properties of the nerve growth factor. *Proc. Nat. Acad. Sci. U. S. A.* 61: 835–842.

ZOMZELY, C. E., S. ROBERTS, C. P. GRUBER, and D. M. BROWN, 1968. Cerebral protein synthesis. II. Instability of cerebral messenger ribonucleic acid-ribosome complexes. *J. Biol. Chem.* 243: 5396–5409.

# 72 Mitochondria and Their Neurofunction

## ALBERT L. LEHNINGER

SINCE THE LAST NRP Intensive Study Program, in 1966, some illuminating advances have been made in the study of cell organelles, particularly of mitochondria. These developments not only are relevant to the role of mitochondria in neuronal and glial cells, but they also may yield new insight into the evolutionary origin of organelles and into the nature of energy-transforming mechanisms in membranes. In the present paper two general areas are selected for special emphasis. The first includes the recently accumulated evidence that mitochondria possess a distinctive form of DNA as well as a distinctive apparatus for synthesis of protein. These new findings bring us squarely to the problem of the origin and biogenesis of organelles and their membranes, as well as the mechanism of non-Mendelian cytoplasmic inheritance. The second area concerns the molecular mechanisms by which energy transformations occur in the mitochondrial membrane, a question that has very broad implications for all energy-dependent, membrane-linked phenomena.

## Mitochondrial DNA

That mitochondria undergo a division process after division of the cell is now strongly supported by isotopic experiments on *Neurospora* (Luck, 1963a, 1963b, 1965). In this organism, mitochondria do not arise *de novo*. They grow by accretion of precursors of small molecular weight, such as phospholipids and proteins, to pre-existing mitochondrial structure, and divide by formation of a septum and pinch-

ing-off, much as bacteria divide. This view has also been strongly supported by direct electron-microscope observations in liver and other tissues of higher species.

However, the most compelling and dramatic evidence for mitochondrial continuity has come from the discovery of mitochondrial DNA and the presence in mitochondria of the various molecular components required in the replication, transcription, and translation of DNA. Although evidence for the occurrence of DNA in mitochondria goes back more than 20 years, early observers considered it to be the result of nuclear contamination of mitochondrial fractions. In the last few years, however, with greatly improved methodology, mitochondria from more than 50 different types of cells have been found to possess a distinctive type of DNA; no mitochondria have yet been found that do not contain DNA.

Mitochondrial DNA (M-DNA) has been isolated and studied in detail (for review, see Nass, 1969) from a number of mammalian, avian, and amphibian tissues, from at least two higher plants, and from yeast and *Neurospora*. It is readily extracted and purified, although it represents only about 0.2 per cent of the total cell DNA in most somatic cells. In oocytes, however, it may represent up to 50 per cent of the total DNA. The buoyant density and, thus, the base composition of M-DNA are usually different from those of the nuclear DNA, and the two forms can therefore be separated readily and identified on density gradients. M-DNA has been examined in detail by both hydrodynamic and electron-microscope approaches. Its predominant form in nearly all types of mitochondria is open or twisted (supercoiled) circular duplexes with a contour length of about 4.5 to 5.5 $\mu$ and a mass of 9.0 to $10.0 \times 10^6$ daltons. An exception is the DNA of yeast mitochondria, which largely occurs in the form of linear molecules 4.0 to 4.5 $\mu$ in

ALBERT L. LEHNINGER  Department of Physiological Chemistry, The Johns Hopkins University School of Medicine, Baltimore, Maryland

length. Yeast M-DNA, however, has been found to possess complementary cohesive ends with the capacity to form hydrogen-bonded circles; presumably, it can be converted into covalent circles by action of DNA-ligase. Although the great bulk of mitochondrial DNA is found in the circular monomeric forms described above, small amounts of circular dimers, trimers, and tetramers have been found, particularly in mitochondria from HeLa cells, fibroblasts, and leucocytes from leukemia patients, in which they are often catenated or topologically linked (reviewed by Nass, 1969). Mitochondrial DNA does not hybridize with nuclear DNA of the same species, indicating that it has little similarity in base sequence. However, it does hybridize readily with mitochondrial RNA, indicating that the latter may be transcribed from mitochondrial DNA (Wood and Luck, 1969).

It has been estimated that each mitochondrion contains anywhere from two to six monomeric molecules of DNA. They are presumably identical and homogeneous, because the melting curve of M-DNA is very sharp and melted M-DNA reanneals very quickly, even at $0°$ C. Individual filaments of DNA are easily recognized in mitochondria of a number of higher cells (Nass et al., 1965; Nass and Nass, 1963a, 1963b).

## Protein Synthesis in Mitochondria

For some time it has been known that isolated mitochondria are capable of incorporating amino acids into mitochondrial protein in vitro, but there have been a number of conflicting observations as to the characteristics of this process and its similarity to cytoplasmic protein synthesis in isolated ribosomes (reviewed by Slater et al., 1968; Roodyn and Wilkie, 1968). For example, it has been claimed that high-energy intermediates generated in electron transport are required for mitochondrial protein synthesis. However, more recent work indicates that ATP and GTP suffice as energy sources, although inhibitors of respiration and phosphorylation can profoundly modify mitochondrial protein synthesis (Wheeldon and Lehninger, 1966). There are other complications. Isolated mitochondria contain a substantial pool of endogenous amino acids, their membranes present permeability barriers to the entry of external amino acids, and they very likely contain specific amino acid carriers or transport mechanisms. Study of protein synthesis in intact mitochondria at present is therefore a little like studying protein synthesis in intact bacterial cells.

Despite these difficulties, it is now clear that isolated mitochondria contain all the familiar enzymatic apparatus for carrying out DNA-directed protein synthesis, including a distinctive DNA polymerase, different from that in the nucleus, and a distinctive DNA-directed, actinomycin-sensitive RNA polymerase. Moreover, ribosomes, as well as ribosomal RNA, have been isolated from mitochondria. There has been some disagreement concerning the sedimentation coefficient of mitochondrial ribosomes, values between 73S and 81S having been reported (see Schmitt, 1969), but Küntzel (1969) has found what appears to be the basis of this variation. He has reported that the larger subunit of mitochondrial ribosomes is of the bacterial, or 50S, type, whereas the small subunit is of the type found in cytoplasm of eukaryotes, namely, 37S. Mitochondrial ribosomes contain the 16S and 23S species of rRNA characteristic of bacteria. Ribosomes have been visualized by the electron microscope in the matrix of Neurospora and in yeast mitochondria, in which they are rather closely associated with the inner membrane (Vignais et al., 1969).

Mitochondria have also been found to contain distinctive tRNAs and aminoacyl tRNA synthetases for various amino acids. In Neurospora, the mitochondria contain tRNAs that differ from homologous tRNAs in the cytoplasm, presumably in base composition and sequence, as they differ in buoyant density and can be separated by sedimentation methods (Barnett and Brown, 1967). Moreover, Neurospora mitochondria contain distinctive activating enzymes, which react readily with the corresponding mitochondrial tRNAs as aminoacyl acceptors, but do not react with the homologous cytoplasmic tRNAs (Barnett et al., 1967). Conversely, cytoplasmic activating enzymes do not react readily with mitochondrial tRNAs.

That mitochondrial protein synthesis is distinctively different from cytoplasmic protein synthesis is also shown by the observation that chloramphenicol, a characteristic inhibitor of bacterial protein synthesis, inhibits protein synthesis in mitochondria but not in the cytoplasm of eukaryotic cells (Kroon, 1965; Clark-Walker and Linnane, 1966). Conversely, cycloheximide, a characteristic inhibitor of protein synthesis in the 80S cytoplasmic ribosomes of eukaryotic cells, has no effect on mitochondrial protein synthesis. The similarity between mitochondrial and bacterial protein synthesis has been strikingly confirmed by the recent demonstration that the initiating amino acid in protein synthesis in mitochondria is N-formylmethionine (fMet), which is also the initiating amino acid in bacterial protein synthesis (Smith and Marcker, 1968). This observation contrasts with cytoplasmic protein synthesis, in which fMet has been excluded as the initiating amino acid. In many respects (Table I), the mitochondrial genetic apparatus closely resembles that of bacteria.

These recent advances have led to a number of interesting questions concerning the origin and biogenesis of mitochondria, the nature of the proteins synthesized by mitochondria, and the effect of mutation on mitochondrial function.

## Evolutionary origin of mitochondria

The similarity between the bacterial and the mitochondrial systems for translation of genetic information has given strong support to the old concept that mitochondria may have had their evolutionary origin in bacteria that parasitized the cytoplasm of larger host cells, the precursors of present-day eukaryotes. This concept was already inherent in the visionary speculations of Altmann (1894). Mitochondria and bacteria have common denominators not only in their apparatus for protein synthesis, but also in their ultrastructure and enzyme distribution. Both bacteria and mitochondria possess an inner membrane of limited permeability and a sievelike outer wall or coat. The mesosomes of bacteria have their counterpart in the cristae of the mitochondria; both are infoldings of the inner membrane. Moreover, the enzymes of electron transport and oxidative phosphorylation are situated on the inner membrane in both bacteria and mitochondria.

In bacteria and mitochondria, the DNA is "bare" and is localized in nuclear zones or nucleoids. Because mitochondrial DNA is much smaller than that of bacteria, however, it appears possible that the primordial parasitizing bacteria lost many genes—presumably those that were redundant—during the course of evolution and development of eukaryotic cells.

## Gene products of mitochondrial DNA

The nature of the proteins synthesized by mitochondria is a matter of some interest, as the relatively small DNA

### TABLE I
Comparison of Basic Elements of Genetic Apparatus

| | Prokaryote (E. coli) | Eukaryote (Neurospora) Cytoplasm | Mitochondria |
|---|---|---|---|
| DNA | $2 \times 10^9$ | $2 \times 10^9$ | $1 \times 10^7$ |
| Ribosomes s20, w | 70S | 80S | ~80S |
| Large subunit | 50S | 60S | 50S |
| Small subunit | 30S | 37S | 37S |
| Initiating amino acid | fMet | N-acetyl-X | fMet |
| tRNAs Activating enzymes | | distinctive specific for cytoplasmic tRNA | distinctive specific for mitochondrial tRNA |
| Chloramphenicol sensitivity | + | 0 | + |
| Cycloheximide sensitivity | 0 | + | 0 |

molecules in mitochondria are able to code for only about 40 polypeptide chains of 150 amino acid residues each. This figure must represent a maximum, because it has been shown that mitochondrial RNA hybridizes readily with mitochondrial DNA (Wood and Luck, 1969), which strongly suggests that a significant fraction of M-DNA codes for mitochondrial tRNAs and rRNAs.

Nearly all the amino acid incorporation by isolated mitochondria in vitro is into the proteins of the inner mitochondrial membrane, in a crude fraction called "structure protein." This fraction is now known to contain many polypeptide species, but only a very few of these become labeled in vitro in isolated mitochondria. To date, no evidence exists, from in vitro experiments, for incorporation of amino acids into any mitochondrial enzyme. Actually, it appears very likely that most mitochondrial proteins are coded by nuclear chromosomes. It is possible, however, that the mitochondrial genome may code for a few enzymes required by the "host" eukaryotic cell in some vital process. For example, two enzymes required in the biosynthesis of heme are localized in mitochondria, particularly ferrochelatase, which is found exclusively in the inner mitochondrial membrane (Schnaitman and Greenawalt, 1968). It is also possible that one or more of the polypeptide subunits of mitochondrial ribosomes are coded by the mitochondrial chromosome.

One of the most promising approaches to identification of the proteins specified by mitochondrial genes is the study of mutations in mitochondrial structure or function, which may be the result of changes in either the nuclear or mitochondrial chromosomes; a number of mutants defective in energy-coupling and ATP synthesis have been identified (Sherman and Slonimski, 1964; Mattoon and Sherman, 1966; Beck et al., 1968; Parker and Mattoon, 1969).

Recent work in our department (Hawley and Greenawalt, 1970) has shown that isolated *Neurospora* mitochondria tested in vitro show only a very small fraction of their in vivo activity, suggesting that protein synthesis in mitochondria is dependent upon as-yet-unidentified factor(s) contributed by the extramitochondrial cytoplasm or other intracellular structures. These are under examination.

## The outer and inner mitochondrial membranes

Since the 1966 Intensive Study Program, several investigators have succeeded in separating the outer and inner membranes, a development made possible by the discovery of marker enzymes for the outer membrane, monoamine oxidase in particular, but also kynurenine hydroxylase and rotenone-insensitive cytochrome c reductase (Schnaitman and Greenawalt, 1968; Schnaitman, et al., 1967; Chan et al.,

1970; Sottocasa et al., 1967a, 1967b). Such markers are essential for identification, because the outer membrane mass is usually only a small fraction of the inner membrane—about 10 per cent or less in liver mitochondria and as little as 2 per cent in heart. Direct analysis has shown that the outer membrane is richer in total lipids than the inner membrane and differs in lipid distribution (Stoffel and Schiefer, 1968). The inner membrane is remarkable in possessing only about 20 per cent lipid, of which a considerable fraction is cardiolipin, a lipid also present in large amounts in bacteria. The inner membrane is very thin (ca. 60 Å), although much of its mass is made up of the enzyme molecules involved in electron transport and oxidative phosphorylation. On the average, the inner membrane of mitochondria contains one assembly of respiratory carriers for each 40,000 Å², or a square 200 Å on each side, and from three to five of the elementary particles of Fernández-Morán, also called inner membrane spheres, which are now known to be adenosine triphosphate (ATP) synthetase molecules (Racker, 1967). The latter structures are visualized on the inner surface of the cristae after negative contrast procedures, but are not visible after osmium fixation. The spheres are about 80 to 90 Å in diameter in negative staining and the inner membrane is but 60 Å thick in osmium-fixed sections, so it has been suggested that the spheres do not normally protrude but may possess a flattened conformation in the intact membrane. Whether either of the mitochondrial membranes possesses a phospholipid bilayer core is not clear; however, the inner membrane does appear to have a continuous nonpolar phase.

## Carrier systems in the mitochondrial membranes

Figure 1 shows the nature of the rather considerable metabolic traffic across the mitochondrial membranes in the intact cell. During normal mitochondrial function, fuels must pass from the cytoplasm into the inner compartment, in which oxidation and phosphorylation take place; accompanying this flow, adenosine diphosphate (ADP) and phosphate must also enter. Among the end-products leaving are ATP and $HCO_3^-$. The traffic across the mitochondrial membrane is remarkable in that most of the molecules that must cross it do not readily penetrate other biological membranes. For example, it is well known that the plasma membrane is completely impermeable to ATP and ADP, and is only very slowly permeable to citrate and phosphate, whereas these components cross the mitochondrial membrane readily in most cells. Actually, however, the mitochondrial membrane is also intrinsically impermeable to these ionic solutes, which are able to cross it only through the intervention of ion-specific carriers or transporters. A number of such specific carriers have recently been iden-

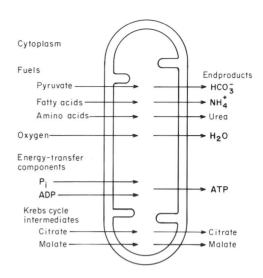

FIGURE 1  Metabolic exchanges across the mitochondrial membranes.

tified (Table II) (reviewed by Chappell, 1968). There are two distinctive types: fixed and mobile.

The ATP-ADP carrier is typical of the fixed carriers. It facilitates a stoichiometric exchange-diffusion process across the inner membrane, where one molecule of ADP going in is exchanged for one of ATP coming out; adenosine monophosphate (AMP) is not transported. No net transport of adenine nucleotide occurs—only exchange. The recognition of this carrier was made possible by the discovery of the toxic agent atractyloside, which blocks phosphorylation

TABLE II
*Mitochondrial Membrane Transport Systems*

| Fixed carriers | Inhibitor |
|---|---|
| *substrate(s)* | |
| ADP-ATP | Atractyloside |
| Phosphate | Mersalyl |
| Ca++ | Sr++ |
| Malate | n-Butylmalonate |
| HCO₃⁻ | Diamox |
| | |
| Mobile carriers (shuttles) | Carrier |
| *substrates* | |
| Electron equivalents from NADH* | Dihydroxyacetone phosphate |
| Electron equivalents from NADPH** | Oxaloacetate |
| Fatty acids | Carnitine |
| | |
| Ion-carrying antibiotics | |
| K+ | Valinomycin, nonactin |
| Na+ | Gramicidin |

\* Reduced form of nicotinamide adenine dinucleotide.
\*\* Reduced form of nicotinamide adenine dinucleotide phosphate.

of ADP by mitochondria, not through any intrinsic effect on the phosphorylation mechanism, but because it inhibits the transfer of ADP and ATP across the membrane. We have shown that atractyloside binds to the ADP-ATP carrier to "freeze" it—presumably by inducing a conformational change (Winkler and Lehninger, 1968)—in such a form that it can neither load nor unload ADP or ATP. Also among the fixed carriers are a phosphate carrier, amino acid carriers, and a bicarbonate carrier; the last appears to be carbonic anhydrase. Recently we have succeeded in extracting from the membrane in soluble form a specific $Ca^{++}$-binding protein of high affinity, which we had earlier concluded to be a specific $Ca^{++}$ carrier (Reynafarje and Lehninger, 1969; Lehninger and Carafoli, 1969). The $Ca^{++}$-binding protein is a relatively large molecule, over 100,000 in molecular weight. It also binds $Sr^{++}$ and $Mn^{++}$, but not $Mg^{++}$.

The mobile carriers of rat-liver mitochondria include the well-known $\alpha$-glycerophosphate carrier for electrons or reducing equivalents and the carnitine carrier for fatty acids (Tubbs and Garland, 1968). In the former, dihydroxyacetone phosphate accepts electrons and protons enzymatically; the resulting glycerol phosphate serves as a membrane-permeant carrier. Carnitine serves as a lipid-soluble carrier of fatty acids, after enzymatic transfer of the fatty acyl group from fatty acyl CoA. Another type of mobile carrier is now of special interest, namely, the ion-carrying or ionophorous antibiotics (Pressman, 1970). More than 50 different antibiotics are now known which are capable of inducing permeability of mitochondria to alkali metal cations. The best known among these are gramicidin, valinomycin, nonactin, and nigericin; some are polypeptides and others are macrolides. Some of these antibiotics, such as gramicidin, induce permeability to all alkali metal cations and do not distinguish between $Na^+$ and $K^+$, whereas others, such as valinomycin, induce permeability to $K^+$ but not $Na^+$. Most of these antibiotics are annular molecules, with a relatively hydrophobic exterior and a highly polar "doughnut-hole." It has been postulated that such annular antibiotics may coordinate alkali metal cations within the polar doughnut hole. This hypothesis has been amply supported by recent work, particularly with nonactin. The crystalline $K^+$ salt of this antibiotic has been subjected to X-ray analysis, which has shown the $K^+$ to be coordinated within the annular ring (Kilbourn et al., 1967); nonactin does not form a complex with $Na^+$. Valinomycin not only induces the capacity for $K^+$ transport in mitochondria, but also transports $K^+$ across artificial phospholipid bilayers and through nonbiological apolar phases, such as chloroform (reviewed by Mueller and Rudin, 1969).

Although the ionophorous antibiotics are toxic agents elaborated by microorganisms, it appears possible that they may serve as instructive molecular prototypes of $K^+$ or $Na^+$ gates in excitable membranes. Mueller and Rudin (1968) have shown that the antibiotic alamethicin will confer on synthetic phospholipid bilayers the property of electrical excitability, providing a $K^+$ gradient is imposed. Evidently alamethicin allows $K^+$ to pass only when the system is electrically perturbed. The uncoupling agent 2, 4-dinitrophenol also can act as a mobile transmembrane carrier, specifically of $H^+$ ions (Hopfer et al., 1968). The intrinsic impermeability of the inner mitochondrial membrane to $H^+$, $Na^+$, and $K^+$, and the increase in permeability induced by agents such as valinomycin and 2, 4-dinitrophenol, are important elements in the transformation of electron transport energy (see below).

## Modalities of energy coupling in mitochondria

We still have very little information as to the molecular mechanism by which ATP is regenerated from ADP and phosphate during electron transport in mitochondria. Although it had appeared for some years that the solution to this problem was just around the corner, more recently a theoretical and experimental trilemma has developed. There are currently three major hypotheses for the mechanism of oxidative phosphorylation: the chemical coupling hypothesis; the chemiosmotic coupling hypothesis; and the conformational coupling hypothesis. These hypotheses are by no means of parochial interest to "mitochondriacs" only, because they are relevant to the action of all types of membranes that form or utilize ATP. It must be recalled that phosphorylation of ADP coupled with electron transport is a fundamental process not only in respiration in mitochondria, but also in photosynthetic phosphorylation of chloroplasts and in respiration in prokaryotic bacteria, in which the properties of the membrane-linked, energy-conserving mechanisms appear to be very similar. Moreover, they are also relevant to the action of the $Na^+$-transporting adenosine triphosphatase systems of plasma membranes, which show many similarities to the $H^+$-transporting adenosine triphosphatase system of the mitochondrial membrane.

The free energy of electron transport in mitochondria can be utilized not only to cause coupled synthesis of ATP from ADP and phosphate, but also can be used to drive at least two other types of energy-requiring processes—ion transport and conformational and/or volume changes (reviewed by Lehninger, 1964). For some time it has been known that mitochondria can accumulate certain cations, such as $Ca^{++}$ and $K^+$ (in the presence of valinomycin), from the suspending medium at the expense of electron transport, in a process that is stoichiometric with electron transport but alternative to oxidative phosphorylation. Such ion movements are uncoupled by 2, 4-dinitrophenol, as is oxidative phosphorylation of ADP. During cation accumulation in mitochondria, there is a simultaneous energy-linked extrusion of $H^+$ into the medium. Proton ejection (or its equivalent hydroxyl ion accumulation) is currently regarded as the primary event to which is coupled the accumulation of external cations such as $Ca^{++}$ and $K^+$.

The other modality of energy coupling is less well known in molecular terms, but is conspicuous ultrastructurally. Mitochondria undergo two types of energy-dependent conformational change (Lehninger, 1964). The first is a relatively slow, "large-amplitude," swelling-contraction cycle, in which the mitochondria undergo large increases and decreases in volume; not all mitochondria show this property. The second is a very rapid, "small-amplitude" change, in which the total mitochondrial volume does not change but in which an internal rearrangement of mitochondrial structure takes place. This type of structural change, which takes place in all mitochondria, is characteristically influenced by respiratory inhibitors and uncoupling agents. These three modalities of energy coupling, i.e., phosphorylation of ADP, active ion transport, and conformational changes, are the basic processes on which the three current hypotheses of mitochondrial energy coupling are constructed (Figure 2).

## The chemical coupling hypothesis

This hypothesis proposes that electron transport generates a high-energy chemical intermediate, often symbolized $X \sim I$, which is the energetic precursor of the high-energy phosphate bond in ATP, in a sequence of linked reactions that have common high-energy intermediates. One variant of the chemical coupling hypothesis is shown in Figure 3. The hypothesis has been the working assumption of most

investigators since it was first postulated in 1953–1954 (see Lehninger, 1964). It is similar in principle to the common-intermediate reaction schemes of most multienzyme systems, in particular that of the ATP-yielding glycolytic cycle. This hypothesis thus has ample biochemical precedent. The chemical coupling hypothesis can also be made to account for mitochondrial $H^+$ transport, if it is assumed that the postulated high-energy intermediate $X \sim I$ can undergo asymmetric hydrolysis across the membrane by the action of a vectorial enzyme:

$$H^+_{inside} + OH^-_{outside} + X \sim I \rightarrow X\text{-}OH + H\text{-}I.$$

Such a reaction would leave the outer compartment more acid and the inner more alkaline. The chemical intermediate $X \sim I$ can also account for swelling-contraction cycles, if $X \sim I$ breakdown or discharge is accompanied by a conformational change transmitted to the membrane. In fact, the chemical coupling hypothesis is capable of explaining nearly all existing experimental observations on mitochondrial energy coupling (see Greville, 1969).

However, the hypothesis has two conspicuous flaws. First, there is at present no significant evidence for the formation or existence of the imputed high-energy chemical intermediates, after many years of effort devoted to finding them. Second, the chemical coupling hypothesis does not account for the widely confirmed observation that the mitochondrial membrane is essential for oxidative phosphorylation, and must be present as an enclosed vesicle. Actually, the chemical coupling hypothesis provides no role for the membrane other than as "floor space" for the respiratory assemblies. It is, of course, still possible that the imputed high-energy chemical intermediates may yet be found and that someone will succeed in reconstructing

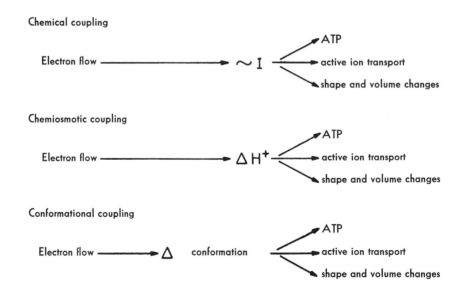

FIGURE 2 Energy-coupling hypotheses.

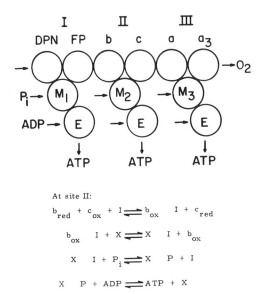

$$b_{red} + c_{ox} + I \rightleftharpoons b_{ox} \quad I + c_{red}$$

$$b_{ox} \quad I + X \rightleftharpoons X \quad I + b_{ox}$$

$$X \quad I + P_i \rightleftharpoons X \quad P + I$$

$$X \quad P + ADP \rightleftharpoons ATP + X$$

FIGURE 3   A chemical coupling hypothesis.

oxidative phosphorylation in a membrane-independent system, but the probability appears rather low.

## The chemiosmotic hypothesis

The chemiosmotic hypothesis, which has earlier forerunners, was explicitly postulated in 1961 by Mitchell in a rudimentary form and later developed in a much more detailed manner (Greville, 1969; Mitchell, 1961, 1966, 1968). It is perhaps more accurately termed the electrochemical or proton-translocation hypothesis. The chemiosmotic hypothesis (Figure 4) proposes that the membrane is an essential element in energy coupling, across which a proton gradient or "proton-motive" force is generated by electron transport via the membrane-linked carriers. Because the mitochondrial membrane is postulated to be impermeable to protons, the proton gradient generated by electron transport thus constitutes a high-energy state. The proton-motive force so generated is then used to drive the dehydrating synthesis of ATP from ADP and phosphate through the action of an asymmetric or vectorial adenosine triphosphatase, also situated in the membrane, which is capable of removing the elements of $H_2O$ from ADP and phosphate to form ATP, in such a way that $H^+$ enters the inside compartment and the $OH^-$ the outer compartment. The proton-motive force generated by electron transport thus "pulls" the synthesis of ATP from ADP and phosphate; the result is a cyclic proton current across the membrane, outwardly directed by electron transport and inwardly directed by ATP formation. In more recent refinements of the chemiosmotic hypothesis, Mitchell has proposed that there need

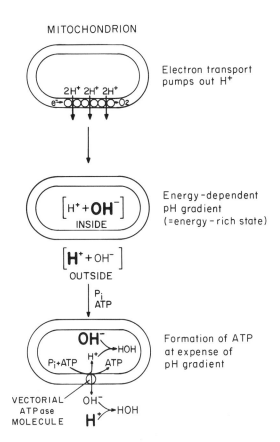

FIGURE 4   The chemiosmotic hypothesis.

not be a large proton gradient across the membrane if a transmembrane potential difference that is energetically equivalent is generated by electron transport (Mitchell and Moyle, 1969).

An interesting feature of the chemiosmotic hypothesis is the prediction that uncoupling agents act by virtue of their capacity to serve as lipid-soluble proton carriers that make the lipid-phase of the membrane permeable to protons, thus short-circuiting the proton current. It was further postulated that the electrochemical gradient generated by electron transport is the ultimate driving force for other mitochondrial ion-transport processes and that it also is responsible for conformational changes in membranes.

A number of features of the chemiosmotic hypothesis have been given direct experimental support. There is good evidence for the postulated impermeability of the mitochondrial membrane to $H^+$ ions, although there are those who choose to interpret existing data otherwise. The capacity of the respiratory chain to pump protons outward is particularly well established; moreover, studies on the stoichiometry of proton transport indicate that two $H^+$ are ejected per pair of electrons per site (see Greville, 1969). That ATP synthesis and hydrolysis are also accompanied by proton uptake and ejection, respectively, has also been established; there is a molar stoichiometry of about three H⁺/ATP at pH 7.0 (Bielawski and Lehninger, 1966). The mitochondrial membrane enzyme systems therefore have the predicted sidedness in relation to one another and also the minimum required stoichiometry. Moreover, there is considerable evidence that 2, 4-dinitrophenol increases the permeability of mitochondria to H⁺ (Greville, 1969). In fact, 2, 4-dinitrophenol added to synthetic phospholipid bilayers not only increases the permeability to H⁺ dramatically (Bielawski et al., 1966), but this effect is specific for H⁺; a synthetic phospholipid bilayer in the presence of 2, 4-dinitrophenol comes very close to giving a theoretical membrane potential of 59 millivolts for a gradient of 1.0 pH unit at 25° C (Hopfer et al., 1968). One of the most interesting pieces of evidence supporting the chemiosmotic hypothesis is Jagendorf's finding that an artificially created pH gradient across the chloroplast membrane in the dark causes ATP to be formed from ADP and phosphate (Jagendorf, 1967).

The chemiosmotic hypothesis has been vigorously opposed, particularly by Chance and by Slater. Chance and his colleagues failed to observe a significant change in the intramitochondrial or intracristal pH measured with an adsorbed pH indicator (bromothymol blue) in the transition from resting (State 4) to active (State 3) respiration induced by adding ADP. However, their interpretations of this evidence have been criticized. There is probably no necessity to expect a large change in transmembrane pH in a tightly coupled system under steady state conditions. In interesting recent experiments in which the transmembrane potential across the membrane of large mitochondira has been measured directly with an electrode inserted into the matrix, no evidence of a substantial potential difference was found (Tupper and Tedeschi, 1969). There are still many technical difficulties in experimental procedures designed to test the occurrence of chemical or electrical gradients across the mitochondrial membrane, and it is perhaps wisest at the moment not to accept all such data at face value.

At the moment, the chemiosmotic hypothesis is unfashionable in some quarters, but, on any objective grounds, it cannot be dismissed from the most serious consideration. Moreover, it appears to account particularly well for energy coupling in chloroplasts.

## The conformational-coupling hypothesis

That conformational changes of the electron carrier molecules may be involved in the transformation of respiration energy was first suggested in 1961, to account for the mechanism of respiration-linked swelling-contraction cycles of mitochondria (Lehninger, 1960, 1961, 1964). Conformational changes were later invoked more explicitly as a vehicle of energy coupling by Boyer (1968). Not until recently, however, did a firmer experimental basis for this concept arise, particularly from electron-microscope investigations of Hackenbrock (1966, 1968; Hackenbrock and Gamble, in press) on mitochondria sampled in various stages of the respiratory cycle. Mitochondria respiring in State 4 (the resting state; no ADP) show the classical or "orthodox" conformation of the cristae and inner matrix. On the addition of ADP, which stimulates respiration to yield State 3, the mitochondria quickly (within one second) undergo transformation to the "condensed" conformation, in which the inner compartment becomes contracted to about 50 per cent of its State-4 volume and the inner membrane becomes contorted and shrunken. When all the added ADP is phosphorylated, the mitochondria return to the orthodox conformation. There is no over-all volume change of the mitochondria; only the inner membrane undergoes such rapidly reversible changes in conformation. These changes are characteristically inhibited (or promoted) by inhibitors of respiration and phosphorylation. Similar observations have been made more recently on other types of mitochondria in Green's laboratory (Penniston et al., 1968).

One possible interpretation of the inner compartment changes is that they are merely passive and secondary to respiration-dependent ion movements and are thus the result of changes in internal osmolar concentration. It has recently been shown, however, that the total osmolarity of the inner compartment of the mitochondria does not change during State 4 to State 3 transitions. Hackenbrock has therefore postulated that the respiration-linked conformational changes in the inner membrane represent the vehicle by which the energy of electron transport is converted into the phosphate-bond energy of ATP. Conformational coupling has also been vigorously championed by Green and his colleagues (Penniston et al., 1968).

The conformational-coupling hypothesis has also been given strong support in recent work of Chance and his colleagues, who have used the fluorescent dye 1-anilino-8-naphthalenesulfonic acid (ANS) as a probe of changes in mitochondrial structure during respiratory transitions (Azzi et al., 1969). ANS bound to mitochondria undergoes extremely rapid changes in fluorescence during respiratory transitions, which Chance has interpreted as reflecting energy-dependent conformational changes in membrane proteins. Such fluorescence changes may also be caused by changes in the dielectric properties of the medium or the surface to which it is bound, or in the degree or manner of binding. It is especially noteworthy that the fluorescence changes are extremely rapid; in fact, they appear to be the earliest physical or chemical changes yet observed during respiratory transitions and thus may be closely associated with the primary act of energy transformation.

Despite the considerable recent attention given the conformational-coupling hypothesis, specific molecular formulations have not been described, probably because there are some thermodynamic, kinetic, and geometric restrictions on possible mechanisms (Williams, 1969). The model in Figure 5 may be proposed to yield at least a schematic idea of one possible mechanism. It assumes that the electron-carrier molecules have different conformations in their oxidized and reduced states; this has in fact been established for cytochrome c. Similarly, it is assumed that the ATP synthetase molecule may also exist in a taut or energized form, in which ATP formation readily takes place, and in a relaxed or energy-poor form, which is inactive. When ADP becomes available to any given respiratory assembly, the carriers will become oxidized and thus undergo transition into their energized forms. It is then suggested that this all-or-none conformational transition is to be transmitted sterically or mechanically to the ATP synthetase molecule, either directly or via membrane structure proteins, through cooperative interactions of the membrane protomers. The ATP synthetase is thus converted into its active or taut form, causing it to make ATP. In this way a conformational change in the carriers may be converted into the chemical change of ATP formation (Figure 6, page 838).

It must be pointed out that this hypothesis requires that any given respiratory assembly undergoes all-or-none transitions between two steady states (States 4 and 3), and that energy recovery is associated with the transition of the assembly rather than with the steady state itself. Because each mitochondrion has many thousands of such assemblies, the transitions may average out over time to account for the observed linearity of ATP formation with electron transport. In the light of present knowledge of protein conformational transitions, conformational-coupling hypotheses must necessarily have such an on-off character. It is otherwise very difficult, if not actually impossible, to construct a conformational-coupling mechanism that can function in one continuous steady state. It is interesting that such a mechanism puts respiratory energy conversion into the same class of mechanochemical or chemomechanical processes as those taking place in actomyosin, in ribosomes, in microtubules, and in flagella.

At present it is not possible to choose among these hypotheses; airtight experiments capable of destroying one or the other have proved to be very difficult to design. Whatever the molecular basis for the energy-coupling process, it now appears highly probable that the same basic mechanism will prove to operate in mitochondria, in chloroplasts, in bacterial membranes, and may be directly relevant to the ATP-dependent electrogenic pumps of the plasma membrane.

## Mitochondria in neurofunction

Brain mitochondria are extremely active in protein synthesis, and exceed mitochondria from liver and certain other organs in this activity. Work in our laboratory (Hamberger et al., 1970) has shown that mitochondria isolated from neuronal cells are considerably more active in protein synthesis than mitochondria from glial cells. Protein synthesis in mitochondria isolated from whole rabbit brain is dramatically stimulated if the rabbits are first forced to swim in a tank of water for 30 to 40 minutes (Hamberger et al., 1969). Whether this increase is due to a systemic effect of muscular activity or to psychic activity is not clear. The possible role of mitochondria in the synthesis of enzymes at nerve endings, particularly enzymes concerned in the secretion and metabolism of neurotransmitter substances, deserves further study.

The extraordinary affinity and capacity of most mitochondria, including those of the brain, to segregate $Ca^{++}$ during respiration (Lehninger et al., 1967) have suggested that mitochondria may play a role in excitation-contraction coupling and in recovery. Recent investigations have revealed that mitochondria rather than sarcoplasmic reticulum may be the major $Ca^{++}$-segregating organelle in some types of muscle, particularly "red" muscles, which are rich in mitochondria; similarly, mitochondria may play a role in sequestration and release of $Ca^{2+}$ at nerve endings.

It has been pointed out that the mitochondrial membrane may be a useful model of the excitable neuronal membrane (Lehninger, 1968), particularly with respect to the interplay between $Ca^{2+}$ and $Na^+$ or $K^+$, which is now considered to be a central process in the transmission of impulses along the axon. For one thing, the molecular basis for passage of monovalent cations across membranes can be approached through study of the characteristic action of valinomycin and other ionophorous antibiotics on mitochondrial membranes.

It has often been suggested that the neuronal membrane is normally poised in a metastable state that responds in an all-or-none manner to triggering impulses, leading to movements of $Ca^{2+}$ and $Na^+$. Like the neuronal membrane, the mitochondrial membrane also possesses specific binding sites for $Ca^{++}$ and also interacts with monovalent cations (Lehninger, 1968). Under certain conditions of pH and $K^+$ concentration, rat-liver mitochondria will respond to additions of $Ca^{++}$ (the stimulus) in a manner suggesting that they are poised in a metastable state, i.e., they rapidly take up $Ca^{++}$ and eject $H^+$ and then proceed to reverse and overshoot, to produce a highly damped oscillation resembling a slow-motion action potential (Carafoli et al., 1966). Chemical exchanges across the mitochondrial membrane can now be measured with great sensitivity and ease, and much of the basic enzymology of mitochondria is understood. The recent success of Tupper and Tedeschi (1969) in inserting electrodes into mitochondria should now make it possible to measure electrical activity of mitochondrial membranes, and thus to supplement the extensive biochemical information.

Although it may be stretching facts to compare the inner

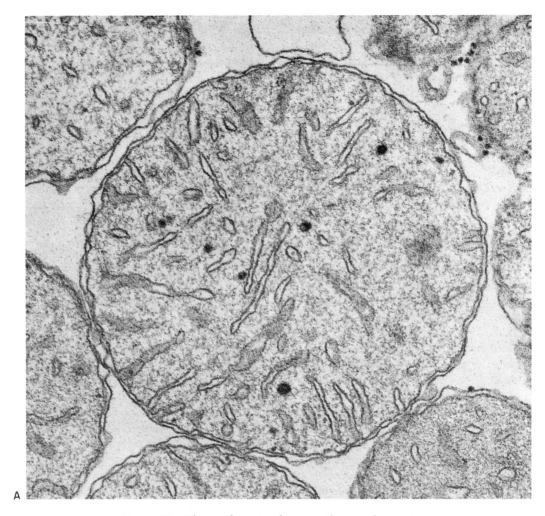

FIGURE 5A The conformational states of mouse-liver mitochondria. The orthodox conformation (in the absence of ADP).

mitochondrial membrane with the neuronal membrane,
such similarities should not be dismissed lightly, as they may
yield new insights. Some day we must come closer to a
unifying theory of membrane structure and function, and
it may be that the easily accessible mitochondrial membranes
will yield important clues to the function and behavior of
the neuronal membrane.

## REFERENCES

ALTMANN, R., 1894. Die Elementarorganismen und ihre Beziehungen zu den Zellen. Veit and Co., Leipzig.

AZZI, A., B. CHANCE, G. K. RADDA, and C. P. LEE., 1969. A
fluorescence probe of energy-dependent structure changes in
fragmented membranes. *Proc. Nat. Acad. Sci. U. S. A.* 62: 612–
619.

BARNETT, W. E., and D. H. BROWN., 1967. Mitochondrial transfer
ribonucleic acids. *Proc. Nat. Acad. Sci. U. S. A.* 57: 452–458.

BARNETT, W. E., D. H. BROWN, and J. L. EPLER., 1967. Mito-
chondrial-specific aminoacyl-tRNA synthetases. *Proc. Nat. Acad.
Sci. U. S. A.* 57: 1775–1781.

BECK, J. C., J. R. MATTOON, D. C. HAWTHORNE, and F. SHERMAN,
1968. Genetic modification of energy-conserving systems in
yeast mitochondria. *Proc. Nat. Acad. Sci. U. S. A.* 60: 186–193.

BIELAWSKI, J., and A. L. LEHNINGER, 1966. Stoichiometric relationships in mitochondrial accumulation of calcium and phosphate supported by hydrolysis of adenosine triphosphate. *J.
Biol. Chem.* 241: 4316–4322.

BIELAWSKI, J., T. E. THOMPSON, and A. L. LEHNINGER, 1966. The
effect of 2, 4-dinitrophenol on the electrical resistance of phospholipid bilayer membranes. *Biochem. Biophys. Res. Commun.*
24: 948–954.

BOYER, P. D., 1968. Oxidative phosphorylation. *In* Biological
Oxidations (T. P. Singer, editor). John Wiley and Sons, New
York, pp. 193–235.

CARAFOLI, E., R. L. GAMBLE, and A. L. LEHNINGER, 1966. Rebounds
and oscillations in respiration-linked movements of $Ca^{2+}$ and
$H^+$ in rat liver mitochondria. *J. Biol. Chem.* 241: 2644–2652.

CHAN, T. L., J. W. GREENAWALT, and P. L. PEDERSEN, 1970.
Biochemical and ultrastructural properties of a mitochondrial
inner membrane fraction deficient in outer membrane and
matrix activities. *J. Cell Biol.* 45: 291–305.

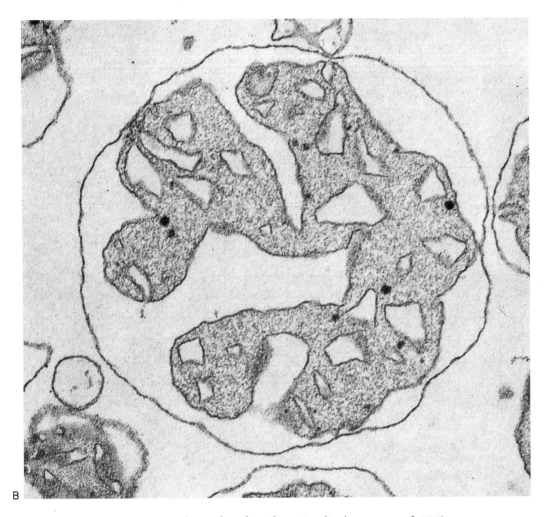

FIGURE 5B   The condensed conformation (in the presence of ADP).
(Electron micrographs courtesy of Dr. Charles Hackenbrock.)

CHAPPELL, J. B., 1968. Systems used for the transport of substances into mitochondria. *Brit. Med. Bull.* 24: 150–157.

CLARK-WALKER, G.-D., and A. W. LINNANE, 1966. In vivo differentiation of yeast cytoplasmic and mitochondrial protein synthesis with antibiotics. *Biochem. Biophys. Res. Commun.* 25: 8–13.

GREVILLE, G. D., 1969. A scrutiny of Mitchell's chemiosmotic hypothesis of respiratory chain and photosynthetic phosphorylation. *Curr. Top. Bioenerg.* 3: 1–78.

HACKENBROCK, C. R., 1966. Ultrastructural bases for metabolically linked mechanical activity in mitochondria. I. Reversible ultrastructural changes with changes in metabolic steady state in isolated liver mitochondria. *J. Cell Biol.* 30: 269–297.

HACKENBROCK, C. R., 1968. Chemical and physical fixation of isolated mitochondria in low-energy and high-energy states. *Proc. Nat. Acad. Sci. U. S. A.* 61: 598–605.

HACKENBROCK, C. R., and J. L. GAMBLE. Conformational activity versus ion induced osmotic perturbation in the transformation of mitochondrial ultrastructure. *In* Structural Probes for the Function of Proteins and Membranes (B. Chance, editor). Academic Press, New York (in press).

HAMBERGER, A., N. GREGSON, and A. L. LEHNINGER, 1969. The effect of acute exercise on amino acid incorporation into mitochondria of rabbit tissues. *Biochim. Biophys. Acta* 186: 373–383.

HAMBERGER, A., C. BLOOMSTRAND, and A. L. LEHNINGER, 1970. Comparative studies on mitochondria isolated from neurone-enriched and glia-enriched fractions of beef brain. *J. Cell Biol.* 45: 221–234.

HAWLEY, E., and J. W. GREENAWALT. An assessment of in vivo mitochondrial protein synthesis in *Neurospora crassa. J. Biol. Chem.* (in press).

HOPFER, U., A. L. LEHNINGER, and T. E. THOMPSON, 1968. Protonic conductance across phospholipid bilayer membranes induced by uncoupling agents for oxidative phosphorylation. *Proc. Nat. Acad. Sci. U. S. A.* 59: 484–490.

JAGENDORF, A. T., 1967. Acid-base transitions and phosphorylation by chloroplasts. *Fed. Proc.* 26: 1361–1369.

KILBOURN, B. T., J. D. DUNITZ, L. A. R. PIODA, and W. SIMON, 1967. Structure of the $K^+$ complex with nonactin, a macrotetrolide antibiotic possessing highly specific $K^+$ transport properties. *J. Mol. Biol.* 30: 559–563.

KROON, A. M., 1965. Protein synthesis in mitochondria. III. On the effect of inhibitors on the incorporation of amino acids into

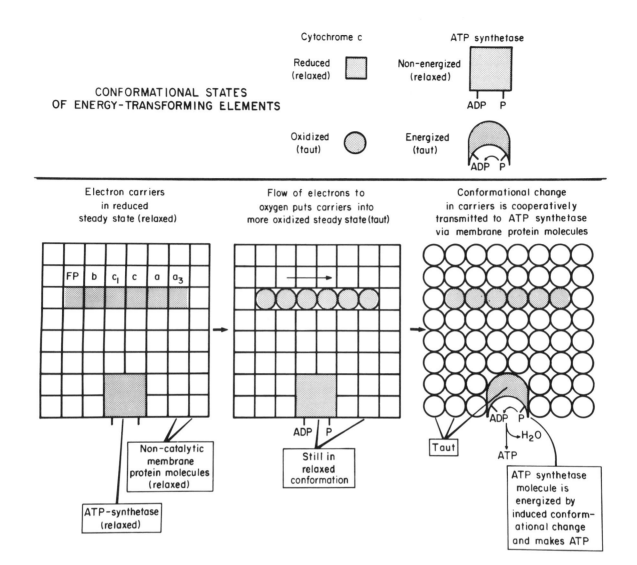

FIGURE 6 Schematic representation of conformational-coupling hypothesis (above). Conformational states of electron carriers and ATP synthetase (below). Conformational transition of carriers to more oxidized state causes ATP synthetase molecule to go into energized state.

protein by intact mitochondria and digitonin fractions. *Biochim. Biophys. Acta* 108: 275–284.

KÜNTZEL, H., 1969. Mitochondrial and cytoplasmic ribosomes from *Neurospora crassa:* Characterization of their subunits. *J. Mol. Biol.* 40: 315–320.

LEHNINGER, A. L., 1960. Oxidative phosphorylation in submitochondrial systems. *Fed. Proc.* 19: 952–962.

LEHNINGER, A. L., 1961. Components of the energy-coupling mechanism and mitochondrial structure. *In* Biological Structure and Function, Vol. II (T. W. Goodwin and O. Lindberg, editors). Academic Press, New York, pp. 31–51.

LEHNINGER, A. L., 1964. The Mitochondrion. W. A. Benjamin, New York.

LEHNINGER, A. L., 1968. The neuronal membrane. *Proc. Nat. Acad. Sci. U. S. A.* 60: 1069–1080.

LEHNINGER, A. L., and E. CARAFOLI, 1969. The reaction of $Ca^{++}$ with the mitochondrial membrane: A comparative study on mitochondria from different sources. *Fed. Proc.* 28: 664 (abstract).

LEHNINGER, A. L., E. CARAFOLI, and C. S. ROSSI, 1967. Energy-linked ion movements in mitochondrial systems. *Adv. Enzymol.* 29: 259–320.

LUCK, D. J. L., 1963a. Genesis of mitochondria in *Neurospora crassa. Proc. Nat. Acad. Sci. U. S. A.* 49: 233–240.

LUCK, D. J. L., 1963b. Formation of mitochondria in *Neurospora crassa.* A quantitative radioautographic study. *J. Cell Biol.* 16: 483–499.

LUCK, D. J. L., 1965. Formation of mitochonrdia in *Neurospora crassa.* A study based on mitochondrial density changes. *J. Cell Biol.* 24: 461–470.

Mattoon. J. R., and F. Sherman, 1966. Reconstitution of phosphorylating electron transport in mitochondria from a cytochrome c-deficient yeast mutant. *J. Biol. Chem.* 241: 4330–4338.

Mitchell, P., 1961. Coupling of phosphorylation to electron and hydrogen transfer by a chemi-osmotic type of mechanism. *Nature (London)* 191: 144–148.

Mitchell, P., 1966. Chemiosmotic coupling in oxidative and photosynthetic phosphorylation. *Biol. Rev. (Cambridge)* 41: 445–502.

Mitchell, P., 1968. Chemiosmotic Coupling and Energy Transduction. Glynn Research Ltd., Bodmin, England.

Mitchell, P., and J. Moyle, 1969. Estimation of membrane potential and pH difference across the cristae membrane of rat liver mitochondria. *Eur. J. Biochem.* 7: 471–484.

Mueller, P., and D. O. Rudin, 1968. Action potentials induced in biomolecular lipid membranes. *Nature (London)* 217: 713–719.

Mueller, P., and D. O. Rudin, 1969. Translocators in biomolecular lipid membranes: Their role in dissipative and conservative bioenergy transductions. *Curr. Top. Bioenerg.* 3: 157–249.

Nass, M. M. K., 1969. Mitochondrial DNA: advances, problems, and goals. *Science (Washington)* 165: 25–35.

Nass, M. M. K., and S. Nass, 1963a. Intramitochondrial fibers with DNA characteristics. I. Fixation and electron staining reactions. *J. Cell Biol.* 19: 593–611.

Nass, M. M. K., S. Nass, and B. A. Afzelius, 1965. The general occurrence of mitochondrial DNA. *Exp. Cell Res.* 37: 516–539.

Nass, S., and M. M. K. Nass, 1963b. Intramitochondrial fibers with DNA characteristics. II. Enzymatic and other hydrolytic treatments. *J. Cell Biol.* 19: 613–629.

Parker, J. H., and J. R. Mattoon, 1969. Mutants of yeast with altered oxidative energy metabolism: Selection and genetic characterization. *J. Bacteriol.* 100: 647–657.

Penniston, J. T., R. A. Harris, J. Asai, and D. E. Green, 1968. The conformational basis of energy transformations in membrane systems. I. Conformational changes in mitochondria. *Proc. Nat. Acad. Sci. U. S. A.* 59: 624–631.

Pressman, B. C., 1970. Energy-linked transport in mitochondria. *In* Membranes of Mitochondria and Chloroplasts (E. Racker, editor). Reinhold Publishing Corp., New York, pp. 213–250.

Racker, E., 1967. Resolution and reconstitution of the inner mitochondrial membrane. *Fred. Proc.* 26: 1335–1340.

Reynafarje, B., and A. L. Lehninger, 1969. High affinity and low affinity binding of $Ca^{++}$ by rat liver mitochondria. *J. Biol. Chem.* 244: 584–593.

Roodyn. D. B., and D. Wilkie, 1968. The Biogenesis of Mitochondria. Methuen, London.

Schmitt, H., 1969. Characterization of mitochondrial ribosomes from *Saccharomyces cerevisiae*. *Fed. Eur. Biochem. Soc. Lett.* 4: 234–238.

Schnaitman, C., V. G. Erwin, and J. W. Greenawalt, 1967. The submitochondrial localization of monoamine oxidase. An enzymatic marker for the outer membrane of rat liver mitochondria. *J. Cell Biol.* 32: 719–735.

Schnaitman, C., and J. W. Greenawalt, 1968. Enzymatic properties of the inner and outer membranes of rat liver mitochondria. *J. Cell Biol.* 38: 158–175.

Sherman, F., and P. P. Slonimski, 1964. Respiration-deficient mutants of yeast. II. Biochemistry. *Biochim. Biophys. Acta* 90: 1–15.

Slater, E. C., J. M. Tager, S. Papa, and E. Quagliariello (editors), 1968. Biochemical Aspects of the Biogenesis of Mitochondria. Adriatica Editrice, Bari, Italy.

Smith, A. E., and K. A. Marcker, 1968. N-formylmethionyl transfer RNA in mitochondria from yeast and rat liver. *J. Mol. Biol.* 38: 241–243.

Sottocasa, G. L., B. Kuylenstierna, L. Ernster, and A. Bergstrand, 1967a. Separation and some enzymatic properties of the inner and outer membranes of rat liver mitochondria. *In* Methods in Enzymology, Vol. X (R. W. Estabrook and M. E. Pullman, editors). Academic Press, New York pp. 448–463.

Sottocasa, G. L., B. Kuylenstierna, L. Ernster, and A. Bergstrand, 1967b. An electron-transport system associated with the outer membrane of liver mitochondria. *J. Cell Biol.* 32: 415–438.

Stoffel, W., and H.-G. Schiefer, 1968. Biosynthesis and composition of phosphatides in outer and inner mitochondrial membranes. *Hoppe-Seyler's Z. Physiol. Chem.* 349: 1017–1026.

Tubbs, P. K., and P. B. Garland, 1968. Membranes and fatty acid metabolism. *Brit. Med. Bull.* 24: 158–164.

Tupper, J., and H. Tedeschi, 1969. Electrical properties of mitochondrial membranes as determined by microelectrodes. *Fed. Proc.* 28: 462 (abstract).

Vignais, P. V., J. Huet, and J. André, 1969. Isolation and characterization of ribosomes from yeast mitochondria. *Fed. Eur. Biochem. Soc. Lett.* 3: 177–181.

Wheeldon, L. W., and A. L. Lehninger, 1966. Energy-linked synthesis and decay of membrane proteins in isolated rat liver mitochondria. *Biochemistry* 5: 3533–3545.

Williams, R. J. P., 1969. Electron transfer and energy conservation. *Curr. Top. Bioenerg.* 3: 79–156.

Winkler, H. H., and A. L. Lehninger, 1968. The atractylosidesensitive nucleotide binding site in a membrane preparation from rat liver mitochondria. *J. Biol. Chem.* 243: 3000–3008.

Wood, D. D., and D. J. L. Luck, 1969. Hybridization of mitochondrial ribosomal RNA. *J. Mol. Biol.* 41: 211–224.

# 73 Neuronal Dynamics and Neuroplasmic Flow

## PAUL A. WEISS

IN RECENT DECADES, interest in the dynamics of the neuron has been enlivened by the realization that nerve fibers are not static; that they are more than "live wires" spun out once and for all during development for service to the finished body as carriers for the propagation and transmission of impulses. In the older static-structural image, the neuron was tacitly denied some of the essential features of what in those days was referred to as "protoplasm." The neuron was conceded some limited power of respiratory metabolism as fuel for "conductivity," but was denied the powers of contractility and, above all, of postnatal growth; for growth was measured in terms of cell enlargement or cell multiplication, neither of which was observed in mature neurons of higher vertebrates, except in post-traumatic regeneration. The very fact that regeneration was considered to be a reawakening from dormancy, rather than simply the intensified expression of a continuously ongoing growth process, as which it now has been recognized, is telling testimony of the change of view that has occurred.

This change, which conceded to the mature neuron the properties of growth and contractility, thus readmitting it to the ranks of full-fledged living cells, has come about by the confluence of two streams of facts and inferences. One stream has been the dramatic progress of molecular biology in resolving "protoplasm" into rather well-defined populations of specific macromolecules and more ubiquitous smaller molecules and ions, and especially in the detailed identification of the processes of "protoplasmic contractility" and "protoplasmic reproduction" in molecular terms. The other contributing stream was the demonstration of the phenomenon of axonal flow or, more broadly, "neuroplasmic flow." While the nucleated soma of the neuron has been conclusively shown to be the major production plant for the specific macromolecular systems needed in the neuronal household, the studies on axonal flow have revealed the facts and modes of traffic from the localized production centers to sites of consumption and distant destinations.

As my presentation will indicate, there is distinct "hybrid vigor" to be expected from the continued blending of these two lines of investigation, for, as in an industrial operation, the managerial considerations of the production process cannot be separated from those of the distribution and marketing. Biochemical research on the neuron has focused

predominantly on the former, that is, on the synthesis of specific compounds from raw materials, which is the task of a chemical industry. Yet, a cell can more aptly be compared to an industry that shapes its diverse chemical materials into such ordered structures as machines or buildings or pipelines. This aspect of the cell as an industry of structural devices has only lately come into prominence.

Turning to what, in the nerve fibers, corresponds to the distribution and marketing problems of an industry, there are, first, the *economical* considerations of how much of a given product is manufactured in the factory, how much is stored, how much is shipped out, where it ends, how long it lasts, and if and how demand controls production and shipping schedules. Second, there are the *technological* aspects of distribution, i.e., the means by which the transport from source to destination is carried out and regulated. We know that in the neuron this is effected by axonal flow, but unfortunately, with the exception of transport of ions and relatively small molecules across membranes and the ingestion of matter in phagocytosis and pinocytosis, our knowledge of the mechanism of cellular transport is dismally poor. This lack of proper models accounts for the striking disproportion in the amount of attention given to the economics of the translocation of neuronal material, as against its technological problems, as is well illustrated by such technologically naive terms as "migratory protein," as if the molecule had legs with which to "migrate."

We evidently are just at the beginning of realizing, identifying, and disentangling the complex web of intimately interrelated processes that operate in the industrial system of the neuron. In this rudimentary state of knowledge, it seems far more important to gather further information with care and accuracy and to correlate the data from the diverse technical approaches involved, such as microanatomy and submicroanatomy, cytochemistry, cinemicrography, hydrodynamics, pharmacology, and so forth, than to indulge in sweeping generalizations, or even to give an air of finality to reasonably validated ones that are subject to further verification. It is my aim to let the following discussion reflect this cautionary attitude.

## Neuronal cell dynamics

Much detailed information on neuroplasmic synthesis will be given in subsequent chapters. Detailed accounts of axoplasmic flow have been given in several recent reviews

---

PAUL A. WEISS   The Rockefeller University, New York

(Barondes, 1969; Lubińska, 1964; Weiss, 1963, 1967a, 1969a). To avoid repetition, I shall confine myself to sketching the emergence of an integrated and unified concept of neuronal dynamics from the variegated collection of data.

PROTEIN SYNTHESIS   Using the rate of protein synthesis as indicator of the intensity of the renewal of the macromolecular population, the mature nerve cell must be rated as close to the most active of all cell types, which include intestinal mucosa, bone marrow, or certain tumor cells. The use of the term "protein" in the singular, disregarding the great diversity of protein species, is but a tactical convenience. The transfer of "genetic information" to a cellular feature is safely established through the standard formula of transcription of nucleic-acid pair sequences from DNA to RNA, followed by translations of messenger RNA series into conforming seriations of amino acids to yield polypeptides, which then combine to proteins in various conformations from linear to three-dimensional patterns. From this point on, the safe track of information is lost. The continuation of chemical differentiation as the result of enzyme activity and of compounding new products in bulk, that is, in purely scalar terms, can still be extrapolated from the same model, but the *physical order* restraining the chemical interactions in the cell in coordinated fashion calls for the introduction (often no more than an invocation) of organizing principles generally lumped under the term "control," the nature of which is not only quite obscure but often not even properly recognized, let alone critically investigated.

DIFFERENTIATION VS. PROLIFERATION   The whole problem of "differentiation" is in this indefinite state. I am not talking here about the various ingenious attempts to explain the differential changes in the bulk composition of a given cell by "gene repression" and "de-repression" which have been derived from the study of prokaryotes. The generative and the somatic cell are one and the same in these simple forms, so they can offer no model for cell differentiation in the higher metazoans, whose many disparately differentiated cell types, once they have acquired special properties, can keep on multiplying almost indefinitely, each true to its own differentially specialized cell type even in an indifferent environment, and even though all of them continue to carry the same common genome (Defendi, 1964; Ursprung, 1968; Weiss, 1953). Neither the manner in which this relative stability of self perpetuation is insured, nor the range within which it can be modified, are understood; the only well-attested facts are the undiluted propagation of specificity in the offspring of an established cell line and the limited degrees of transformability of the strain.

This state of uncertainty reflects directly on a prominent feature of the mature nerve cell, namely, the absence of neuronal cell multiplication. At definite times, varying for different parts of the nervous system, neuronal cell division ceases in vertebrates in postmetamorphic (or, in the higher forms, in perinatal) stages. Now, in most proliferating tissues, a given cell has the dual capacity of directing its synthetic machinery either toward the production of another cell of the same type, or toward manufacturing specialized cell products (Weiss, 1969b). The decision between these alternatives is made either by conditions in the microenvironment of the cell or by an internal time clock indigenous to the particular cell strain. The question is what switches the neuronal machinery from proliferation into production.

Proliferation stops when DNA replication in the nucleus ceases, which can be determined by the cessation of the incorporation of thymidine. Mature brain neurons no longer incorporate thymidine.* However, Gurdon (1967) has shown in a most striking experiment that the nucleus of such a nonproliferating neuron of an adult frog can again be made to start incorporating thymidine massively if it is excised and transplanted into the cytoplasm of an enucleated egg. Capacity for DNA replication, therefore, is ever present in the neuronal nucleus, but is actively suppressed by the nuclear environment (Jacobson, 1968). In my own experience with many thousands of amphibian central nervous systems, I remember having seen only two instances of mitotic figures in mature neurons with fully developed axons and neurofibrils; in both cases they were in ectopic position in the central canal. Perhaps the capacity of cerebrospinal fluid to unblock DNA stasis in mature neurons might be worth testing.

At any rate, the absence of DNA replication restricts the observed high rate of protein synthesis to the role of maintaining the existing cellular unit and to producing specific products assigned to that cell type, which, in the present case, include "neurohumors" and "neurosecretions" of the older literature as well as the enzyme systems instrumental in their manufacture. However, while in most other terminal cells of the mature organism the internal catabolic degradation of the macromolecules is spatially so closely intermingled with their replenishment by neosynthesis that bulk determinations reveal only the balance sheet of the account, the sharp localization of the production site in the neuron makes the latter into a natural separator between the two phases.

NUCLEAR PRIMACY   The cell dynamics of the neuron brings to mind that of another cell that also has an eccentrically located nucleus, namely, the alga *Acetabularia* (Brachet, 1965). The nucleus in this single-celled alga lies at

---

* Sporadic residual straggler neurons have been reported, however, by Altman (1967).

the end of a long stalk, the other end of which forms characteristic structures roughly comparable to nerve endings. When such an alga is enucleated by amputation of the lower nucleated end, it can survive and even regenerate ablated tops, although its viability declines as does also the rate of residual protein synthesis in the enucleated fragment. At any rate, the fact that protein is still synthesized for some time in the absence of the nucleus is noteworthy. A more detailed analysis of this synthetic process, however, has reduced its comparability to the neuron by showing that autonomous DNA-containing chloroplasts in the enucleate fragments permit the continuation of RNA synthesis. This, in turn, actuates continued protein synthesis, as evidenced by the active incorporation of isotopically labeled amino acids into isolated chloroplasts comparable to such other cell organelles as mitochondria (Reich and Luck, 1966) and basal bodies of cilia (Randall et al., 1967), which have their own DNA machinery.

By contrast, the axon offers a much clearer and cleaner demarcation from the nucleated soma. The only DNA-equipped organelles in axoplasm, the mitochondria, assume less than one per cent of the axonal volume. The main axonal mass from the axon hillock down is free of ribosomes, and although low amounts of RNA not attributable to contaminations have been reported to be present in a few types of nerve fibers (Edström, 1969), evidence of active protein synthesis along the course of the axon (Singer, 1968) has generally been questioned (Droz, 1969). The burden of protein synthesis in the neuron thus rests conclusively upon the soma. This primacy of the nucleated territory does not necessarily imply exclusivity, and the synthesis of some proteins at nerve endings has been asserted (Austin and Morgan, 1967). In this connection, one must also consider that, daily, each nerve ending receives hundreds of mitochondria through the axonal flow. These then break down, raising the question of whether, perhaps, the resulting release of mitochondrial DNA and RNA might not be instrumental in the reported protein synthesis at nerve terminals (see below).

AXONAL FLOW AS DRAINAGE   When axonal flow was first demonstrated, all that was certain was that substance leaves the cell body at a rate that advances the cylindrical column of the axon from one to a few millimeters daily (Weiss and Hiscoe, 1948). Fate, function, and composition of the flow were at first wholly conjectural. It was sheer assumption when I suggested, in contradiction to Ramón y Cajal, that the axon itself cannot synthesize the replacements for that population of its macromolecules that is subject to continual catabolic degradation, and that axonal flow therefore represents a system in stationary balance between its own self-consumption and its replenishment from the manufacturing center located in the cell soma.

Enzymes and other proteins suggested themselves as the most indicative markers for measuring the flowing contingent. The high ammonia liberation from nerve, recorded in the older literature, could then be regarded as an index of protein de-amination. A calculation based on this tenuous argument yielded an average nerve protein half-life of a few weeks, which corresponds to the more direct and accurate later determinations (Lajtha, 1964) of protein turnover in brain. Calculating further, from this short life expectancy, the rate at which a supply stream would have to compensate for the continual depletion to maintain a steady state, we calculated an advance of the feeder column at an average daily rate of one millimeter (Weiss and Hiscoe, 1948). The identity of this value with the observed rate of the axonal flow was suggestive, but not conclusive, evidence for our interpretation of axonal flow and of its premises. The clinching proof came only from the study of translocations of isotopically marked protein.

Before turning to them, let me mention a few points on method. Bulk measurements are particularly uninformative when one deals with inhomogeneous sources—for instance, mixtures of nerve cell bodies, nerve fibers, glia cells, and vascular tissue, such as are being used in metabolic assays of slices or homogenates of central or peripheral nerve tissue. The delicate micro-techniques of separating pure neuronal cell bodies from their associated glial cells (Hydén, 1967) and of extruding pure axonal core from the enveloping myelin and Schwann cell (Edström, 1964; Koenig, 1965) promise to circumvent this difficulty, provided the trauma involved and the microscopically indistinct residual contaminants prove to be negligible. Some of these difficulties can be bypassed by comparing intact peripheral nerve with nerves transformed into cords of Schwann cells as a result of earlier transection (see below).

A further cautionary note pertains to the frequent treatment of protein as if it were in free soluble form. While identifiable microscopic and submicroscopic aggregates of structural proteins are clearly recognizable, systems devoid of such conspicuous criteria are still rather elusive, and the empirical division, after extraction, into soluble and insoluble proteins can hardly represent the graded scale of complex formations. This consideration has a crucial bearing on our understanding of the cohesiveness of the substance flow from the cell body. Sufficient evidence has been accumulated by now to prove conclusively that the one-millimeter type of axonal flow consists of flow of the whole axon as a *structurally cohesive semisolid system,* rather than as substance flow within an axon that is consolidated in a fixed position. There is also considerable evidence for liquid transport channels within the cohesive axonal matrix, but this is a second system *sui generis,* which must not be confounded with the moving reference system within which it lies enclosed. Until we have more concrete information about the structural properties of the axoplasmic matrix beyond what is implied in the general term of "thixotropic gel," we cannot take it for granted that "nonstructural" proteins are necessarily freely mobile. Just as an isotope-labeled amino acid serves as a marker for the protein in which it is incorporated, so it may be necessary to consider the protein prima facie as a marker for a larger substance continuum with which it is physically and chemically linked.

As in the coacervation of gels in general, this aspect immediately raises the question of *water* as a structural component of the neuronal matrix. I know of no data on how much of the approximately 85 per cent water content of the neuron is freely exchangeable, how much is chemically bound to hydrated compounds, and how much is physically enmeshed in the tangled structure of the large macromolecules. Preliminary explorations in my laboratory (Kreutzberg, unpublished) have shown that tritiated water, locally applied to the soma of spinal-ganglion cells, gradually appears in significant proportions in the peripheral axons of the spinal nerves issuing from those cells; it is distinguishable from water exchanged peripherally across the axonal surface. Systematic explorations of this problem on a broader scale are urgently needed.

## Analysis by radioisotopes

We can now take up briefly the rather spectacular advances made in tracking protein down the axon by isotope labeling. The early experiments using $^{32}$P and $^{14}$C as markers did not go much beyond confirming the phenomenon of axonal flow and, in some cases, its average velocity of one mm per day (Weiss, 1961b). A major step forward came only with the introduction of tritiated molecules as tracers by Droz and Leblond (1963). Amino acids as protein markers have been applied (1) by single-pulse injections into an animal; (2) by regional injections (Ochs et al., 1962; Miani, 1964); (3) by strictly localized application to the cell bodies of origin of the test neurons (Taylor and Weiss, 1965; Weiss and Holland, 1967; McEwen and Grafstein, 1968; Rahmann, 1968); and (4) by explanting neurons into the labeling solution (Weiss, 1967b). The following paragraph summarizes results with (1) by Droz (1969).

A few minutes after intravenous injection of tritiated leucine in the rat, radioactivity is detectable in the nucleus of the neuronal soma. The evidence seems adequate that its source is protein-incorporated, rather than free, amino acid. Sampling at graded intervals after injection reveals the sequential spreading of the labeled protein from the nucleolus to the rest of the nucleus, and from there to the perikaryon. After two to five minutes, the concentration of labeled protein has become highest in the RNA-rich Nissl substance, reaching a peak in the Golgi body at about half an hour, and reaching the smooth reticulo-endothelium a few hours later, where some of it may remain for days. Some of the protein, however, bypasses the complex cytoplasmic machinery and moves directly from the nucleus into the axon, at the base of which the first traces of the label appear as early as 20 to 30 minutes after the injection, which, one notes, corresponds to the one-millimeter daily rate (40 micra per hour). These observations thus have succeeded in separating at least two different geographic pathway systems for the newly synthesized proteins. Correlated with these data, Droz has found different residence times of two classes of protein within the cell soma: a transitory population that leaves the soma within one day, and a second group that occupies the cell space for about two weeks. Tentatively, one might identify the short residence group with the protein components of the moving axonal matrix, perhaps including neurofilaments, whereas the population of longer residence would go through the serial changes in which, according to Droz, newly synthesized protein is being "accumulated, segregated and conjugated" in the Golgi complex. While the establishment of this clear-cut dichotomy among classes of protein marks a significant forward step, our technical inability to discern finer subspecifications within each class remains a major challenge for the future.

FUNCTIONAL ADAPTATION Thus far I have dealt with the chemical production machinery as if it were autonomous. Ever since the classical experiments of Hydén and his collaborators, however, it has been evident that the intensity of protein synthesis in the neuron varies sensitively with the level of functional activity (Hydén, 1967). Increased activity expresses itself not only in chemical terms in a rise of RNA and protein, but also in morphological changes, such as enlargement of nucleolus, nucleus, and perikaryon. This hypertrophy is evoked by prolonged excitation (Hamberger and Hydén, 1945), as well as by increased trophic demands upon the neuron, such as during the regenerative regrowth of a severed nerve fiber (Murray and Grafstein, 1969) or when neurons are made to take over a larger area of terminal innervation by profuse peripheral branching (Wohlfahrt and Swank, 1941; Edds, 1950). Reciprocally, reductions of the load or demands placed upon the neuronal soma manifest themselves in a lowering of the production rate, which in favorable cases (e.g., after permanent severance of a nerve fiber with blockage of the regeneration of the stump) is marked by size regression (Weiss et al., 1945) proceeding from nucleolus to nucleus to perikaryon and finally to the caliber of the axon (Cavanaugh, 1951). The so-called "axon reaction" after the transection of a nerve fiber is a composite of this regressive phase, superseded rapidly by the progressive changes subserving the regenerative process.

PROTEIN SYNTHESIS IN EXPLANTS In order to extricate the neuronal cell bodies from the network of complex relations in the whole animal, we have tried to extend the analysis to isolated nerve centers in vitro with a minimum of trauma. Postnatal spinal ganglia of mice or rats, explanted with their capsule into nutrient solutions, proved to be adequate for the task. In the presence of tritiated leucine, their radioactivity mounted according to a regular, repeatable time course, becoming stabilized two to three hours after explantation (Weiss, 1967b). At the time, more than half of the labeled amino acid was found to be actually incorporated in ganglionic protein. Unexpectedly, it took as long as a three-hour chase in ordinary nonradioactive leucine or even leucine-free balanced salt solution to rid the ganglion of the nonincorporated remainder. Extending the tests to other tissue samples of about one cubic millimeter or less, we found that spinal ganglia, spinal cord, peripheral nerve, liver, and muscle fragments had time courses and saturation levels of protein neosynthesis that were characteristically different for each type of tissue (Weiss et al., in preparation).

Fragments of peripheral nerve, the axons of which had to be assumed to be nonparticipating in protein synthesis, gave the highest values. This made us suspect that the Schwann cells surrounding the axons have an inordinately high rate of protein synthesis. To test this presumption, we compared fragments of normal nerve with those of distal stumps of nerves severed two weeks previously, so that their axons had degenerated and been replaced by hypertrophic Schwann cells. Counts of radioactive protein per microgram (after immersion for two hours in $^3$H leucine, followed by a one-hour chase) were proportionately higher in the degenerated, axon-free, samples than in the normal, axon-containing ones. The high rate of protein neosynthesis in peripheral nerve must therefore be allocated to the Schwann cell (and associated endoneurial cells), without any demonstrable participation of the axon.

The most puzzling results, however, were obtained when we compared protein synthesis in live tissue samples with that in tissues frozen solid on dry ice, kept frozen over night, and then thawed (Weiss et al., in preparation). It turned out that the specific differentials according to tissue types among the different parts of the nervous system, liver, and muscle were fully retained in the devitalized samples, whereas the differences between the frozen samples and the live controls for the same tissue were negligible; the only exception was peripheral nerve, in which the values for the frozen fragments were distinctly lower than those for their live counterparts.

These few examples may illustrate the puzzles and surprises we have encountered in our tests of explanted tissue fragments. I therefore shall not dwell on them further except for one set of experiments of such potential significance for pharmacology that it deserves to be added here. In the presence of sodium barbital, the radioactive count after two

hours was only about half that of the control; and neural tissue fragments that were first labeled and then supplemented with barbiturate, leaked appreciable amounts of protein into the medium, the correspondence between the exuded and the retrieved amounts having been verified by radioactivity counts. Comparable tests with liver fragments indicated that the leakage is distinctive of neural tissue, being about 15 times greater from spinal cord and sciatic nerve than from liver. This investigation on a hitherto unsuspected action of narcotics should be extended with more refined techniques.

Having now briefly exemplified both the potentialities and the limitations of studies of the production plant of the neuronal cell body, with all their capriciousness and variability according to the techniques used (e.g., pulse labeling vs. homogenate assay vs. explantation), ostensibly aggravated by our lumping measurements of many disparate component processes under a common blanket, let us now look at least at some of the neuronal components that can be singled out because of their structural characteristics—the organelles.

## Neuronal organelles

The study of neuronal organelles and other inclusions is, of course, subject to the uncertainties of the electron microscope technique, which, depending on the state of the tissue and the methods of fixation, exaggerates certain fine-structural features, while failing to record others. The extent to which physiological fluctuations of the internal milieu, for instance, might affect the electron-microscopic visibility of a given fine structure in nerve has rarely been taken into consideration. Due allowance for such variability is made in the following brief account.

MITOCHONDRIA  During unimpeded axonal flow, ultrathin sections of myelinated axons show an average of two mitochondria per section. However, when the flow is locally dammed up, many hundreds of mitochondria pile up within a few days in the sections immediately upstream to the obstruction (Weiss and Pillai, 1965), proving that the axonal flow carries the mitochondria from the perikaryon cellulifugally (Kapeller and Mayor, 1967). Besides passive carriage, mitochondria may exhibit some independent local motility within narrow range (Pomerat et al., 1967). Arrested in their convection, they rapidly degenerate (Weiss and Pillai, 1965). We find the same accumulations and disintegrations also at the nerve endings (Mayr and Weiss, in preparation). Density and cellulifugal convection rate of mitochondria at the axonal flow rate (1 mm = about 1,000 times mitochondrial length, per day, or one mitochondrial length per minute) indicate a daily production rate of mitochondria within the cell body in the neighborhood of 500. Mitochondrial enzymes, such as succinic acid dehydrogenase (Friede, 1959) and DPN-diaphorase (Kreutzberg, 1963), have been recovered in high local concentrations at the upstream ends of blockages of axonal flow, which now is readily explained by the local accumulation of their carriers.

After disruptions of nerve continuity more violent than the mild constrictions used in our method, e.g., after a tight ligature, crush, or transection (Lasek, 1967), fleeting accumulations of mitochondrial and other enzymes have been observed at the distal end of what in these cases is a traumatized tissue segment (Lubińska et al., 1963). To ascribe this, without further evidence, to a reversal of axonal flow in the distal stump, is gratuitous: the upstream navigation of a power-driven riverboat does not signify that the river is flowing backwards. Conceivably, the distal accumulation, which subsides within a few days, could be explained as follows. The breakdown of the damaged nerve segment establishes a local center of high acidity and electronegativity, which attracts mitochondria toward the lesion by galvanotaxis, and perhaps positively charged molecules, including some enzymes, by the familiar "demarcation," or injury current, of severed nerve fibers. Although as yet unsupported by direct experiments, this or some similar explanation seems superior to "flow reversal," which is contradicted by all direct observations on *mature* nerve fibers (Mayor and Kapeller, 1967).

The breakdown of mitochondria at the blind distal end of axons raises a crucial question, already alluded to above: as they disintegrate, to what extent does their content of DNA, RNA, and nucleotide molecules undergo decomposition? Considering the evidence for some autonomy of protein synthesis in nerve endings (Barondes 1969), is it not conceivable that some residual, undecomposed transcription and translation machinery for protein synthesis is released from the disintegrated mitochondria and activated in situ? If true, this notion would still leave the nucleated soma of the neuron with a monopoly for macromolecular reproduction, expanded by shipping a subsidiary branch factory to an outlying location.

NEUROTUBULES The "neurotubules" (neuronal microtubules) are relatively stiff structures of ± 220-Ångström diameter that run the length of the axon in bundles and are frequently associated with mitochondria. Their composition and structure are dealt with by other contributors to this book. As a tentative concept of their origin, I submit that they grow forth continuously from bases in the nucleated soma by a process of stacking onto their open central ends the near-globular subunits (Schmitt and Samson, 1969) of which they are known to be composed; these units might well be part of that protein population of prolonged cytoplasmic residence commutated on its passage through the Golgi body. *Pari passu* with this accretion at its root, the finished stem of the tubule would be drawn out continuously toward the periphery by the axonal flow. Such growth of elongate structures from their basal ends is common in organisms (for instance, the growth of cilia from their basal bodies). This concept implies that the number of neurotubules along the whole length of a given axon cannot increase. In other words, peripheral branching of neurotubules on an extensive scale could not be expected, although an occasional apical fission might be envisaged.

To test this premise, we have made counts in serial sections of axons with near-terminal branchings (Mayr et al., in preparation). As predicted by our concept, the sum of neurotubules in the branches was equal, within the limits of counting error, to the total number of tubules in the common stem. Thus, even though the axon branches, the individual neurotubules do not. Another requirement for the concept is that if the driving mechanism for the axonal advance is paralyzed, while the central addition to the base of the neurotubules continues, a tangle of twisted neurotubules should be observable in the cell body.*

Notions about the configuration of the distal ends of neurotubules are vague. Mechanical considerations intimate that the tubule is closed. If it serves as a canal for substances to be discharged at the end, that distal lid must be assumed either to undergo intermittent rupture or else to become pinched off, forming the wall of a vesicle. Evidence for the latter alternative, directly bearing on the formation of synaptic vesicles, has been advanced (Pellegrino de Iraldi and De Robertis, 1968) and the actual pinching off of such vesicles from the blind ends of neurotubules, closed by a constriction of the nerve, has been observed (van Breemen et al., 1958). However, the identification of such vesicles with true synaptic vesicles is still debatable.

NEUROFILAMENTS In contrast to the straight neurotubules, the neurofilaments, 70–100 Ångström units in diameter, show a more wavy course. They also seem to run continuously and unbranched from the cell body to their terminations. The evidence for this is found in our recent observation that, at least in certain types of peripheral nerve fibers, the neurofilaments are spaced at very regular lateral intervals from one another, the histogram of the intervals showing a sharp peak at 428 Ångström (Weiss et al., in preparation). The geometry of this grid suggests either the presence of longitudinal channels (unresolved in the electron microscope) in the matrix between them or a sort of linear macrocrystalline organization of the matrix itself, as seen in other ground substances (Weiss, 1968, pp. 24–122), but nothing further can as yet be said about this regularity.

* Isotope labeling of colchicine-binding protein of neurotubules has confined their advance at the one-mm rate (J.-O. Karlsson, unpubl.).

INTERTUBULAR CANALS   The suggestion of such 400-Å-wide spaces parallels another system, better resolvable by the electron microscope, of longitudinal canals of about 600–800-Ångström diameter ("inter-tubular canals"), which we have seen in the matrix of certain axons in overosmicated preparations (Weiss, 1969a); both in location and size order, these correspond to the highly regular "honeycomb" cross sections observed in axons, both central and peripheral, in a considerable variety of pathological conditions (Morales and Duncan, 1966; Hirano et al., 1968). In line with our earlier comments about electron-microscopic visibility, one might assume that this canalicular system is a true microstructural system, the electron-microscopic demonstration of which is marginal in normal nerve, but is improved by more favorable local conditions of a pathologically altered internal milieu.

## Fast traffic

The presence of such preformed channels in different types of neurons is intimately related to the demonstration in recent years of fast transport routes between cell body and nerve endings. Various radioactively labeled substances were found to span that distance at speed some 100 times as fast as the one-millimeter advance of the axonal column itself (Miani, 1964; Dahlström and Häggendal, 1967; Karlsson and Sjöstrand, 1968; Lasek, 1968; McEwen and Grafstein, 1968; Ochs et al., 1969). In such cases, the velocity curve of labeled proteins showed two separate peaks, one registering the standard one-millimeter daily rate of the axon, the other signaling the faster traffic. There are thus at least two distinct traffic systems in operation. Whether the traffic rate in the fast system of a given axon type is uniform is not yet known. The velocity curve of the axonal advance itself, however, spreads out with time (Taylor and Weiss, 1965), which indicates rate differences among the individual fibers of a given nerve. Moreover, according to cinemicrographic recordings of axonal flow, the velocity within a given axon seems to decline transversely from the central axis toward the surface.

In this connection, it should be noted that since the axonal advance represents *volume* flow (ca. 10 cubic micra per minute), that is, displacement of three-dimensional mass, measurements in linear (e.g., *millimeter*) units disregard variations in the cross-sectional area. In first approximation, this has been acceptable. However, with greater refinement of our knowledge, the caliber of nerve fibers will have to be taken into account. This point is illustrated, for instance, by the comparison between axonal flow rates in young and old rats, which, in measurements on the linear advance scale, suggested a marked decline with age (Droz, 1969). However, since the caliber of nerve fibers increases steadily with growth, an average increment of 60 per cent would readily account for the observed linear deceleration, even if the daily volume outflow from the cell body remained constant. In the same sense, linear velocities in submicroscopic narrow channels within the axonal matrix need not indicate particularly high production rates of the substances in question. This is one reason among many why the fast transport observed must be viewed as an intra-axonal phenomenon, not to be confounded with the flow of the axonal matrix. Neurotubules, intertubular spaces, and similar continuous channels from cell body to periphery are logical candidates for the role of pipelines for substances to be shipped in solution from the manufacturing center to distant work stations.

As Kreutzberg (1969) in my laboratory has shown, the neurotubules qualify in that sense. On the premise that neurotubules share their main properties with the microtubules of other cell types, and considering that the motility of the microtubules that constitute the spindle fibers of the mitotic process can be immobilized by colchicine, minute amounts of this drug were injected into peripheral nerves. This blocked the passage down the axons of acetylcholinesterase locally within the injected stretch, but did not interrupt the continued flow of the axon itself, as evidenced by the continued downward progress of the mitochondria, assayed by their enzymes. The piling up of catecholamines in their production center in the neuronal soma after colchicine treatment (Dahlström, 1968; Karlsson and Sjöstrand, 1969) leads to a similar conclusion.

## Transport mechanisms

These considerations of possible traffic routes are intimately related to the question of the physical conveyor mechanisms effecting the cellulifugal substance transfer. The bimodal rate distribution between the slow axonal convection and the fast intra-axonal conduction (see above), indicates that each is actuated by a separate and independent mechanism. The blockage, by moderate dosages of colchicine, of one but not the other supports this view. There is, moreover, evidence for a retrograde, cellulipetal transmission of "messages" along the axon, not explicable in terms of standard impulse transmission, which calls for a third axonal communication system, although not necessarily one involving molecular transfer. The three systems, therefore, are reviewed here separately.

NEUROPLASMIC FLOW   As stated in the beginning, the quantitative mechanostructural analysis of the deformations resulting from the constriction of axons has conclusively proved that the solid axonal matrix—and, as mentioned earlier, presumably also part of the associated water—move as a single cohesive, viscous, somewhat plastic, body within the Schwann (or glia) cell, with or without an interposed laminated sheath of myelin. On elementary mechanical principles, a unidirectional translatory movement of such a column can result only from propagated waves of pressure differentials (e.g., traveling contraction-relaxation oscillations) sweeping unidirectionally over its length. The prototype of such a drive is a *peristaltic wave*. The pertinence of this model for axonal flow is documented (a) by the fact that the movement continues in isolated fragments of nerve, and (b) by the observation that when two ligatures are placed around a nerve in tandem, exoplasm dams up at the "upstream" sides of both the upper and the lower block (Weiss and Hiscoe, 1948; Dahlström, 1967; Mayor and Kapeller, 1967). This evidence would not be fully conclusive if it were based only on nerves studied in toto (Dahlström, 1967), as part of the fibers might have remained unconstricted, hence unblocked, at the upper level; but our original demonstration (Weiss and Hiscoe, 1948) of tandem damming in *single* fibers of doubly constricted nerves removes that reservation. We know, therefore, that the contractile energy must be available at every level of the axon, but be activated "domino"-fashion in cellulifugal direction. As a model for such a self-propagating process, I have suggested a chemomechanical energy transduction in incompressible fluid bodies covered by lipid-protein membranes (Weiss, 1964). The nerve fiber could satisfy the conditions of the model.

Although actual proof for this or any other hypothetical mechanism of the drive is still lacking, the sheer fact of its peristaltic character has been rather well established, as it could be directly recorded visually by time-lapse cinemicrography of living, mature, myelinated nerve fibers (Weiss et al., 1962; Weiss, 1963). Representative scenes from the vast footage of film taken in this manner over the years by my collaborators (mostly A. C. Taylor, M.D. Rosenberg, and A. Bock) have been shown on many occasions and are being prepared for more general circulation. Only the most pertinent observations can be reported here. (1) The constrictive wave definitely travels over the *surface* of the axon, involving visibly both the myelin sheath and the axolemma. (2) The axonal *content* itself is propelled passively by the surface wave; for instance, if a local obstruction (e.g., a kink or a fold) impedes the smooth advance of the content, the axonal column on the upstream side of the block becomes longitudinally compressed and dilated until its pressure head at the block overcomes the obstruction and the whole column can again proceed. (3) In the early phases of Wallerian degeneration of fibers severed from their central cell bodies, each of the fragments (Ramón y Cajal's "ovoids"), into which the fiber breaks up, continues to show peristalsis. (4) In all axons thus far observed, whether intact or fragmented, the frequency of the surface waves is the same and remarkably constant, with a mean duration of 17 minutes per cycle (i.e., three to four

pulses per hour), at a mean velocity of propagation adequate to account for the common axonal flow rate of 50–100 micra per hour. In contrast to the continuity and regularity of the drive, the actual advance of the content varies with local conditions of the traffic channel, and hence appears to proceed in spurts. (5) No records have as yet been taken of unmyelinated fibers. (6) The paucity of mitochondria in the axon, in contrast to their great abundance in the Schwann cell, makes one favor the latter as the source of the *energy supply* for the peristaltic drive. On the other hand, many observations, calculations, and mechanical considerations tend to discount the possibility that the Schwann cell with its pulsation (Pomerat, 1961), which it has in common with other cells observed in vitro (Weiss, 1961a), is of itself the driving *mechanism;* it just provides the fuel, as it were, but not the machine.

With the cinemicrographic data on hand, we have gone on now to construct a hydrodynamic model to reproduce the peristaltic axonal drive to allow a quantitative study of the major variables. Save for a few theoretical treatments of very recent date (Fung and Yih, 1968; Shapiro et al., 1969), the assumption that a superficial pressure wave and deformation of small amplitude could set and keep the interior mass in flowing motion has not been substantiated. Our mechanical model (constructed with the collaboration of Professor Martin Levy and Mr. Robert Biondi of my staff) has now verified that assumption, opening a direct technological approach to the finer analysis of the parameters of axonal flow. In combination with our cinemicrographic studies (not reviewed here) of shifts of the axonal content produced artificially by localized pressure on explanted nerve pieces, we have reasons to expect an early substitution of solid information for the current vague conjectures about the mechanism of axonal flow.

FAST-TRAFFIC MECHANISM Unfortunately, the same optimism hardly seems warranted in respect to the mechanism of fast intra-axonal transport. I have given above the fine-structural indications for preformed transport channels, such as the neurotubules and intertubular spaces, but even if confirmed in their role, their operative mechanism would still remain conjectural. The physical preconditions for traffic in those submicrodimensions (Weiss, 1969a) are so exacting that no incontrovertible explanation is presently in sight. An ingenious speculation, implying the faculty of microtubules for propelling larger particulates on their outsides, has been suggested by Schmitt (1969). However, the general applicability of that hypothesis seems questionable in the light of the extensive observations made in our laboratory by Green (1968) on the role of the microtubules of pigment cells in the massive concentration and dispersion movements of the large (one micron) pigment granules in response to drugs; in these experiments, a constraining role of microtubules on the *orientation* of the granular displacements was unmistakable, but analyses by the electron microscope and by cinemicrography failed to substantiate a *motile* role of the tubules in the propulsion of the granules. My own, no less speculative, proposal for a motile mechanism of tubules for *intra*tubular convection is the following. If we accept the current thesis that globular protein subunits compose the spiral chains of which the walls of the tubules presumably consist (Shelanski and Taylor, 1968), we need only presume that (1) a chemical, electrical, or mechanical energy input at one end produces a change in molecular conformation, which entails a reduction of molecular volume, and, hence, a closer packing, i.e., a local constriction of the tubule; (2) some tubular content is thereby squeezed out into the neighboring sector of the tubule; (3) the conformational change, which started at the base of the wall, triggers a corresponding change at the adjacent level, resulting in a constriction of the latter; (4) the content is thereby propelled further; and so forth, in what amounts to a peristaltic advance.

Because this scheme, although plausible, is no less unsubstantiated than any other, the safest answer to the question of fast intra-axonal transport mechanisms remains "We just do not know."

RETROGRADE MESSAGE TRANSFER  In the matter of ascending information passed on to the neuronal cell body from its peripheral extensions in ways other than impulse conduction, we are just as much in the dark. The postulate of such a cellulipetal message service rests on (a) the "chromatolytic reaction" of the perikaryon of a neuron, which arises whenever a nerve fiber has been severed; and (b) the fact that the central cell body "knows" the specific type of peripheral organ on which its axon happens to have terminated (Weiss, 1965).

As I have indicated above, the notion of a reversed bulk flow as message carrier can be dismissed as unfounded. In fact, there is no reason for presuming that the communication relies at all on massive substance transfer. One possible mode of action might be the rapid passage of charged molecules along interfaces by an electrically actuated "molecular bucket brigade." We have found that in tendon, which consists of rhythmically banded collagen fibers embedded in a matrix of mucopolysaccharide, the spreading of molecules that carry a net electrical charge, but not of neutral ones, occurs preferentially and rapidly along the interfaces between fibers and matrix (Weiss, 1961a). Electron microscope analysis (Grover, 1966) suggested a saltatory transfer of the molecules from one heavy dark band of the collagen fiber to the next (a distance of 640 Ångström units), those bands marking the clustered positions of the polar amino-acid side chains. If a similar phenomenon could be demonstrated for nerve, the interfaces between neurofilaments and axonal matrix might qualify as the "message carriers," which, incidentally, would also allow us to assign a useful function to the neurofilaments.

Whether applicable to nerve or not, the unexpected expedition of molecular transfer along interfaces in tendon once more intimates how many more principles of basic relevance to the understanding of cellular dynamics may still be waiting to be discovered and how unwarranted it would be to rely solely on our present stock of knowledge for the interpretation of such novel problems as those which keep arising in the study of neuronal dynamics.

## The fate of the neuronal substance

One of the problem areas is, of course, the fate of the axonal material that is incessantly flowing down from the perikaryon. It makes up about 15 per cent of the mass of the axonal column, the rest being water and small molecules diffusing in and out. Of that 15 per cent, an unknown fraction must be allocated to constituents of known fate, namely (a) transmitters of central origin, which are discharged, and (b) mitochondria, which break down. Furthermore, if we accept the concept that the neurotubules and neurofilaments are perpetually renewed from their perikaryal bases and commensurately disintegrated at their distal ends, we can also subtract their mass from the account sheet. Accordingly, the fate of no more than perhaps 10 per cent of centrally produced axonal substance remains to be accounted for.

This would be essentially the protein component engaged in the sustenance of the living state of the axon itself, i.e., largely the enzyme systems needed in local metabolism along the line. In view of the short half-life of neuronal proteins, counting by weeks (Lajtha, 1964), on the one hand, and the absence of local sources for enzyme replenishment for the enormous metabolizing mass of the axon on the other, one might have had to postulate a supply stream from a far-off factory, such as has now been empirically discovered. We, therefore, feel entitled to assign the major portion of the missing balance in our account to the repletion of the catabolically degrading units of the axonal enzyme population. If the *slow* axonal flow were the supply vehicle, uniform delivery over the whole length would require not only a flow-velocity gradient declining from axonal core to surface, for which there is some evidence, but also insulation of the traveling molecules against proteolysis during their long trek (circa one year from spinal cord to toe in man). If, on the other hand, enzymes are transported by the *rapid* intra-axonal system (note, for instance, the blockage of acetylcholinesterase by colchicine, reported above), questions arise as to their ubiquitous availability along the whole axon. So, it would seem best, for the present, not to speak of "enzymes" in that generality, ignoring the different needs, routes, distributions, and destinations for each class and, above all, to restrain commitment to speculative interpretations as long as no firmer ground on which to base it is at hand.

## Acknowledgment

Original work cited in the text has been supported by grants from the National Cancer Institute and the National Institute for Neurological Diseases and Blindness of the U. S. Public Health Service and from the Faith Foundation in Houston.

## REFERENCES

ALTMAN, J., 1967. Postnatal growth and differentiation of the mammalian brain, with implications for a morphological theory of memory. *In* The Neurosciences: A Study Program (G. C. Quarton, T. Melnechuk, and F. O. Schmitt, editors). The Rockefeller University Press, New York, pp. 723–743.

AUSTIN, L., and I. G. MORGAN, 1967. Incorporation of $^{14}$C-labelled leucine into synaptosomes from rat cerebral cortex *in vitro*. *J. Neurochem.* 14: 377–387.

BARONDES, S. H., 1969. Axoplasmic transport. *In* Neurosciences Research Symposium Summaries, vol. 3 (F. O. Schmitt, T. Melnechuk, G. C. Quarton, and G. Adelman, editors). M.I.T. Press, Cambridge, Massachusetts, pp. 191–299.

BRACHET, J., 1965. Le controle de la synthèse des proteines en l'absence du noyau cellulaire. Faits et hypothèses. *Bull. Acad. Roy. Belg.* 51: 257–276.

BREEMEN, V. L. VAN, E. ANDERSON, and J. F. REGER, 1958. An attempt to determine the origin of synaptic vesicles. *Exp. Cell Res.* (*Suppl.*) 5: 153–167.

CAVANAUGH, M. W., 1951. Quantitative effects of the peripheral innervation area on nerves and spinal ganglion cells. *J. Comp. Neurol.* 94: 181–219.

DAHLSTRÖM, A., 1967. The transport of noradrenaline between two simultaneously performed ligations of the sciatic nerves of rat and cat. *Acta Physiol. Scand.* 69: 158–166.

DAHLSTRÖM, A., 1968. Effect of colchicine on transport of amine storage granules in sympathetic nerves of rat. *Europe J. Pharmacol.* 5: 111–113.

DAHLSTRÖM, A., and J. HÄGGENDAL, 1967. Studies on the transport and life-span of amine storage granules in the adrenergic neuron system of the rabbit sciatic nerve. *Acta Physiol. Scand.* 69: 153–157.

DEFENDI, V., 1964. Retention of functional differentiation in cultured cells. *Wistar Inst. Symp. Monogr.* 1.

DROZ, B., 1969. Protein metabolism in nerve cells. *Int. Rev. Cytol.* 25: 363–390.

DROZ, B., and C. P. LEBLOND, 1963. Axonal migration of proteins in the central nervous system and peripheral nerves as shown by radioautography. *J. Comp. Neurol.* 121: 325–346.

EDDS, M. V., JR., 1950. Hypertrophy of nerve fibers to functionally overloaded muscles. *J. Comp. Neurol.* 93: 259–275.

EDSTRÖM, A., 1964. The ribonucleic acid in the Mauthner neuron of the goldfish. *J. Neurochem.* 11: 309–314.

EDSTRÖM, A., 1969. RNA and protein synthesis in Mauthner fiber components of fish. *Symp. Int. Soc. Cell Biol.* 8: 51–72.

FRIEDE, R. L., 1959. Transport of oxidative enzymes in nerve fibers; a histochemical investigation of the regenerative cycle in neurons. *Exp. Neurol.* 1: 441–466.

FUNG, Y. C., and C. S. YIH, 1968. Peristaltic transport. *J. Appl. Mech.* 90: 669–675.

GREEN, L., 1968. Mechanism of movements of granules in melanocytes of fundulus heteroclitus. *Proc. Nat. Acad. Sci. U. S. A.* 59: 1179–1186.

GROVER, N. B., 1966. Anisometric transport of ions and particles in anisotropic tissue spaces. *Biophys. J.* 6: 71–85.

GURDON, J. B., 1967. On the origin and persistence of a cytoplasmic state inducing nuclear DNA synthesis in frogs' eggs. *Proc. Nat. Acad. Sci. U. S. A.* 58: 545–552.

HAMBERGER, C.-A., and H. HYDÉN, 1945. Cytochemical changes in the cochlear ganglion caused by acoustic stimulation and trauma. *Acta Oto-Laryngol.* (*Suppl.*) 61: 5–89.

HIRANO, A., R. RUBIN, C. H. SUTTON, and H. M. ZIMMERMAN, 1968. Honeycomb-like tubular structure in axoplasm. *Acta Neuropathol.* 10: 17–25.

HYDÉN, H., 1967. Dynamic aspects on the neuron-glia relationship—A study with micro-chemical methods. *In* The Neuron (H. Hydén, editor). Elsevier Publishing Co., Amsterdam, pp. 179–219.

JACOBSON, C.-O., 1968. Reactivation of DNA synthesis in mammalian neuron nuclei after fusion with cells of an undifferentiated fibroblast line. *Exp. Cell Res.* 53: 316–318.

KAPELLER, K., and D. MAYOR, 1967. The accumulation of mitochondria proximal to a constriction in sympathetic nerves. *J. Physiol.* (*London*) 191: 70P–71P.

KARLSSON, J.-O., and J. SJÖSTRAND, 1968. Transport of labelled proteins in the optic nerve and tract of the rabbit. *Brain Res.* 11: 431–439.

KARLSSON, J.-O., and J. SJÖSTRAND, 1969. The effect of colchicine on the axonal transport of protein in the optic nerve and tract of the rabbit. *Brain Res.* 13: 617–619.

KOENIG, E., 1965. Synthetic mechanisms in the axon—II. RNA in myelin-free axons of the cat. *J. Neurochem.* 12: 357–361.

KREUTZBERG, G. W., 1963. Enzymhistochemische Veranderungen in Axonen des Ruckenmarks nach Durchschneidung der langen Bahnen. *Deut. Z. Nervenheilk.* 185: 308–318.

KREUTZBERG, G. W., 1969. Neuronal dynamics and axonal flow, IV. Blockage of intra-axonal enzyme transport by colchicine. *Proc. Nat. Acad. Sci. U. S. A.* 62: 722–728.

LAJTHA, A., 1964. Protein metabolism of the nervous system. *Int. Rev. Neurobiol.* 6: 1–98.

LASEK, R. J., 1967. Bidirectional transport of radioactively labelled axoplasmic components. *Nature* (*London*) 216: 1212–1214.

LASEK, R. J., 1968. Axoplasmic transport in cat dorsal root ganglion cells: As studied with ($^3$H)-L-Leucine. *Brain Res.* 7: 360–377.

LUBIŃSKA, L., 1964. Axoplasmic streaming in regenerating and in normal nerve fibres. *Progr. Brain Res.* 13: 1–71.

LUBIŃSKA, L., S. NIEMIERKO, B. ODERFELD, L. SZWARC, and J. ZELENÁ, 1963. Bidirectional movements of axoplasm in peripheral nerve fibers. *Acta Biol. Exp.* (*Warsaw*) 23: 239–247.

McEWEN, B. S., and B. GRAFSTEIN, 1968. Fast and slow components in axonal transport of protein. *J. Cell Biol.* 38: 494–508.

MAYOR, D., and K. KAPELLER, 1967. Fluorescence microscopy and electron microscopy of adrenergic nerves after constriction at two points. *J. Roy. Microscop. Soc.* 87: 277–294.

MIANI, N., 1964. Proximo-distal movement of phospholipid in the axoplasm of the intact and regenerating neurons. *Progr. Brain Res.* 13: 115–126.

MORALES, R., and D. DUNCAN, 1966. Multilaminated bodies and other unusual configurations of endoplasmic reticulum in the cerebellum of the cat. An electron microscopic study. *J. Ultrastruct. Res.* 15: 480–489.

MURRAY, M., and B. GRAFSTEIN, 1969. Changes in the morphology and amino acid incorporation of regenerating goldfish optic neurons. *Exp. Neurol.* 23: 544–560.

OCHS, S., D. DALRYMPLE, and G. RICHARDS, 1962. Axoplasmic flow in ventral root nerve fibers of the cat. *Exp. Neurol.* 5: 349–363.

OCHS, S., M. I. SABRI, and J. JOHNSON, 1969. Fast transport system of materials in mammalian nerve fibers. *Science (Washington)* 163: 686–687.

PELLEGRINO DE IRALDI, A., and E. DE ROBERTIS, 1968. The neurotubular system of the axon and the origin of granulated and nongranulated vesicles in regenerating nerves. *Z. Zellforsch. Mikroskop. Anat.* 87: 330–344.

POMERAT, C. M., 1961. Cinematography, indispensable tool for cytology. *Int. Rev. Cytol.* 11: 307–334.

POMERAT, C. M., W. J. HENDELMAN, C. W. RAIBORN, JR., and J. F. MASSEY, 1967. Dynamic activities of nervous tissue *in vitro. In* The Neuron (H. Hydén, editor). Elsevier Publishing Co., Amsterdam, pp. 119–178.

RAHMANN, H., 1968. Syntheseort und Ferntransport von Proteinen im Fischhirn. *Z. Zellforsch. Mikroskop. Anat.* 86: 214–237.

RANDALL, J. R., T. CAVALIER-SMITH, A. McVITTIE, J. R. WARR, and J. M. HOPKINS, 1967. Developmental and control processes in the basal bodies and flagella of *Chlamydomonas reinhardii. Dev. Biol. (Suppl.)* 1: 43–83.

REICH, E., and D. J. L. LUCK, 1966. Replication and inheritance of mitochondrial DNA. *Proc. Nat. Acad. Sci. U. S. A.* 55: 1600–1608.

SCHMITT, F. O., 1969. Fibrous proteins–Neuronal organelles. *In* Neurosciences Research Symposium Summaries, Vol. 3 (F. O. Schmitt, T. Melnechuk, G. C. Quarton, and G. Adelman, editors). M. I. T. Press, Cambridge, Massachusetts, pp. 566–575.

SCHMITT, F. O., and F. E. SAMSON, 1969. Neuronal fibrous proteins. *In* Neurosciences Research Symposium Summaries, Vol. 3 (F. O. Schmitt, T. Melnechuk, G. C. Quarton, and G. Adelman, editors). M.I.T. Press, Cambridge, Massachusetts, pp. 301–403.

SHAPIRO, A. H., M. Y. JAFFRIN, and S. L. WEINBERG, 1969. Peristaltic pumping with long wavelengths at low Reynolds number. *J. Fluid Mech.* 37: 799–825.

SHELANSKI, M. L., and E. W. TAYLOR, 1968. Properties of the protein subunit of central-pair and outer-doublet microtubules of sea urchin flagella. *J. Cell Biol.* 38: 304–315.

SINGER, M., 1968. Penetration of labelled amino acids into the peripheral nerve fibre from surrounding body fluids. *In* Ciba Foundation Symposium on Growth of the Nervous System (G. E. W. Wolstenholme and M. O'Connor, editors). J. and A. Churchill, Ltd., London, and Little, Brown and Co., Boston, pp. 200–215.

TAYLOR, A. C., and P. WEISS, 1965. Demonstration of axonal flow by the movement of tritium-labelled protein in mature optic nerve fibers. *Proc. Nat. Acad. Sci. U. S. A.* 54: 1521–1527.

URSPRUNG, H., 1968. The stability of the differentiated state. *In* Results and Problems in Cell Differentiation, Vol. 1 (W. Beermann, J. Reinert, and H. Ursprung, editors). Springer-Verlag, New York.

WEISS, P., 1953. Some introductory remarks on the cellular basis of differentiation. *J. Embryol. Exp. Morphol.* 1: 181–211.

WEISS, P., 1961a. Structure as the coordinating principle in the life of a cell. *In* Proceedings of the Robert A. Welch Foundation Conferences on Chemical Research, Vol. 5. (W. O. Milligan, editor). Robert A. Welch Foundation, Houston, Texas, pp. 5–31.

WEISS, P., 1961b. The concept of perpetual neuronal growth and proximodistal substance convection. *In* Regional Neurochemistry (S. S. Kety and J. Elkes, editors). Pergamon Press, Oxford, pp. 220–242.

WEISS, P., 1963. Self-renewal and proximo-distal convection in nerve fibers. *In* The Effect of Use and Disuse on Neuromuscular Functions (E. Gutmann and P. Hník, editors). Elsevier Publishing Co., Amsterdam, pp. 171–183.

WEISS, P., 1964. The dynamics of the membrane-bound incompressible body: A mechanism of cellular and subcellular motility. *Proc. Nat. Acad. Sci. U. S. A.* 52: 1024–1029.

WEISS, P., 1965. Specificity in the neurosciences. *Neurosci. Res. Program Bull.* 3 (no. 5): 1–64.

WEISS, P., 1967a. Neuronal dynamics. *Neurosci. Res. Program Bull.* 5 (no. 4): 371–400.

WEISS, P., 1967b. Neuronal dynamics and axonal flow, III. Cellulifugal transport of labeled neuroplasm in isolated nerve preparations. *Proc. Nat. Acad. Sci. U. S. A.* 57: 1239–1245.

WEISS, P., 1968. Dynamics of Development: Experiments and Inferences. Academic Press, New York.

WEISS, P., 1969a. Neuronal dynamics and neuroplasmic ("axonal") flow. *Symp. Int. Soc. Cell Biol.* 8: pp. 3–34.

WEISS, P., 1969b. Principles of Development. A Text in Experimental Embryology. Hafner Publishing Co., New York.

WEISS, P., M. V. EDDS, JR., and M. CAVANAUGH, 1945. The effect of terminal connections on the caliber of nerve fibers. *Anat. Rec.* 92: 215–233.

WEISS, P., and H. B. HISCOE, 1948. Experiments on the mechanism of nerve growth. *J. Exp. Zool.* 107: 315–395.

WEISS, P., and Y. HOLLAND, 1967. Neuronal dynamics and axonal flow, II. The olfactory nerve as model test object. *Proc. Nat. Acad. Sci. U. S. A.* 57: 258–264.

WEISS, P., and A. PILLAI, 1965. Convection and fate of mitochondria in nerve fibers: Axonal flow as vehicle. *Proc. Nat. Acad. Sci. U. S. A.* 54: 48–56.

WEISS, P., A. C. TAYLOR, and P. A. PILLAI, 1962. The nerve fiber as a system in continuous flow: Microcinematographic and electron-microscopic demonstrations. *Science (Washington)* 136: 330 (abstract).

WOHLFART, G., and R. L. SWANK, 1941. Pathology of amyotrophic lateral sclerosis; fiber analysis of the ventral roots and pyramidal tracts of the spinal cord. *Arch. Neurol. Psychiat.* 46: 783–799.

# 74   Axoplasmic Transport:
## Physical and Chemical Aspects

PETER F. DAVISON

THE SUBJECT OF axoplasmic transport has become popular recently and has been reviewed by several authors (Barondes and Samson, 1967; Schmitt and Samson, 1969; Weiss, 1969; Grafstein, 1969). My purpose is not to add to these reviews but to discuss certain aspects of the transport phenomena that merit further investigation and interpretation.

Neurons synthesize and turn over protein rapidly; after the local administration of radioactive amino acids to nervous tissue, neurons become densely labeled in comparison with satellite and connective-tissue cells (Droz, 1969). In studies on roach and crayfish, maximal protein labeling is achieved within hours and then much of the protein is catabolized again, but 10 to 20 per cent is long lived, and of this much moves from the cell body and along the axon (Smith, 1967; Fernandez and Davison, 1969). Proteins labeled in this way have been used by many investigators to study the process of axoplasmic transport in various animals. If cycloleucine is substituted for leucine or if the incorporation of $^3$H-leucine into protein is blocked by puromycin, the movement of a labeled protein is not observed.

Many proteins move along axons at from 1 to 3 mm per day as Weiss and his coworkers found originally, but certain materials, often particulates, move more rapidly, and rates of as high as 2800 mm per day have been reported (Jasinski et al., 1966). The existence of "slow" and "fast" transport has led to the belief that there must exist at least two molecular mechanisms that move the contents of the axon distally. The characteristics of these processes are discussed below, but the justification for discriminating two mechanisms is considered in a later section with reference to the constraints that the experimental findings place on any mechanism that is postulated to explain the transport.

### Experimental parameters of axoplasmic transport

DISCRETE LOCALIZATION   Many experimenters have observed that labeled protein moves along the axon as a broad

peak. Fernandez and Davison (1969), however, found that a small injection of very high specific activity $^3$H-leucine into a ganglion of the crayfish nerve cord leads to the synthesis of labeled protein over a brief period, and some of this protein is subsequently detectable moving caudad along the cord at 1.1 mm per day. For periods of up to 14 days, at 20°C, this moving column of protein is confined to a 1-mm length of cord (Figure 1). One characteristic of laminar flow

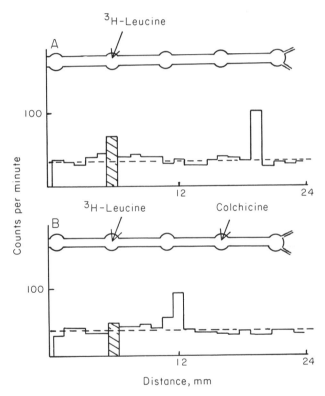

FIGURE 1   The distribution of $^3$H-leucine incorporated into protein in 1-mm segments of the crayfish nerve cord. A. 12 days after injection of label into the third abdominal ganglion (the injected segment is hatched in the histogram and marked with an arrow is the diagrammatic representation of the cord above). B. 12 days after the label injection, as above, and 11 days after the injection of colchicine into the fifth ganglion. The normal 1.1 mm per day caudad movement has been interrupted at a distance of 4 mm from the colchicine injection site. The dashed line shows background counts.

PETER F. DAVISON   Department of Biology, Massachusetts Institute of Technology, Cambridge, Massachusetts. *Present Address:* Boston Biomedical Research Institute, Boston, Massachusetts

in a tube is the parabolic velocity profile within the fluid: the liquid at the center of the tube travels at twice the average velocity of the contents of the tube, and the velocity approaches zero at the wall itself. This flow profile may be readily demonstrated by sucking a small volume of dye into a fluid-filled capillary (Figure 2). Such a dispersion of labeled protein certainly does not take place in the slowly moving protein of the axon, so we may dismiss the concept that this transport represents *flow* of dissolved protein driven by any mechanism. Furthermore, the sharp "front" of rapidly transported protein in the cat sciatic nerve, described by Ochs et al. (1969), precludes any flow process in this instance of rapid transport, as well. The coherence of the slow transport suggests that the axoplasm is gel-like, entirely consistent with the physical characteristics of extruded squid axoplasm (Huneeus and Davison, 1970). Alternatively, parallel mechanisms averaged over the whole cross section of the axon might effect a concerted transport that maintains a linear front.

CONSTANT VELOCITY    The displacement of labeled protein along the axon from the site of synthesis can be mea-

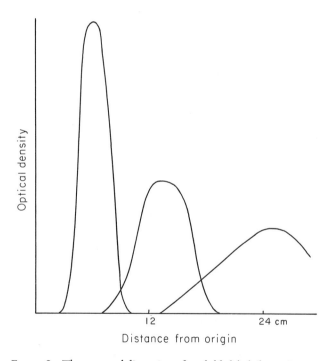

FIGURE 2  The expected dispersion of a soluble labeled protein perfusing an axon under laminar flow is shown by the distribution, assayed at three different times, of dye injected into and then pumped along a water-filled tube 0.7 mm in diameter, 300 mm long (equivalent to the dimensions of a 14-mm-long, 30-micron axon in a crayfish cord). Each tube was cut into 1-cm segments, and the dye was eluted from each and measured. The spread results from the parabolic velocity across the tube.

sured with some precision as a function of time. Experiments on the crayfish have shown that the slow movement is linear with time up to at least 16 days. Many axons narrow progressively with distance from the soma, which may imply that transport rate is "independent" of axon diameter.

CONCENTRATION GRADIENT    Peterson et al. (1967) have shown that no concentration gradient of protein can be detected in the cat optic nerve to account for protein transport. Furthermore, such a mechanism cannot explain a constant transport rate, and it could not be effective in maintaining flow over distances greater than a few centimeters.

PROTEIN GROWTH    The progressive displacement of protein along the axon could be explained by the growth of filamentous protein structures from the soma. This possibility is negated by the failure of an inhibitor of protein synthesis acting in the soma to block slow transport (Peterson et al., 1967).

TEMPERATURE DEPENDENCE    The rate of movement of protein in the crayfish cord as a function of temperature has been found to be proportional to the temperature (in degrees centigrade), except that movement ceased at 3° C, although it was readily observable (0.3 mm per day) at 5° C (Fernandez et al., 1970). The linearity with temperature could suggest that physical, as well as chemical, factors limit the rate, and the cessation of movement between 3° and 5° C could indicate the disruption of the transport mechanism. It may be noted that certain microtubules dissociate at low temperatures (Tilney and Porter, 1967).

INTRINSIC PROCESS    Several investigators (e.g., Dahlström, 1967) have shown that rapid transport persists in a segment of axon defined by two ligations. The source of movement must therefore be sought among the structures within the axon, without immediate reference or reaction to the activity of the cell soma or termini.

COLCHICINE EFFECTS    The interruption of the rapid transport process by local application of colchicine to a nerve has been reported by Dahlström (1968). Inhibition of the fast, and a lesser effect on the slow, processes have been reported by Kreutzberg (1969) and Karlsson and Sjöstrand (1969). On the other hand Fernandez et al. (1970) did not study the rapid transport, but observed that a local injection of colchicine inhibited slow transport for as much as 14 days and for 2 to 4 mm around the injection site (Figure 1). In view of the demonstrated binding specificity of colchicine for microtubule protein (Borisy and Taylor, 1967), the experiments strongly suggest the involvement of microtubules in the axon with some aspect of the fast transport process and the slow.

COMPLEXITY OF TRANSPORTED PROTEINS   Several authors have compared the rapidly and slowly transported axoplasmic constituents and have found a preponderance of particulate materials in the rapidly moving and soluble proteins in the slowly moving labeled fractions. In view of the tendency of certain of the axoplasmic proteins of squid to aggregate (Huneeus and Davison, 1970), this discrimination between particulate and soluble protein could be unreliable; therefore, Fernandez et al. (1970) dissolved the crayfish cord in 8 M urea and separated at least eight components in urea by gel electrophoresis. The specific activity of these proteins was roughly constant. In particular, no dip in specific activity was observed in the region of the gel where the microtubules subunit (a major protein component of the crayfish axon) was expected to migrate (Figure 3).

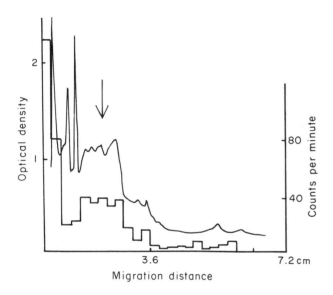

FIGURE 3   Gel electropherogram of protein labeled by injected ³H-leucine and isolated four days later from the slow-moving protein in the crayfish cord. The protein was extracted and analyzed in 8 M urea, 0.05 M mercaptoethanol (Fernandez et al., 1970). The histogram shows the concentration of tritium in 3-mm slices of gel; the graph shows the distribution of protein as revealed by dye-staining and densitometry. Clearly, a complex population of proteins of roughly uniform specific activity is present. The position where the microtubule protein was expected to move is shown by the arrow. The second and third sharp peaks on the left of the diagram correspond to the interfaces in the discontinuous gel.

It may be concluded that the microtubules form part of the population of slowly moving proteins that includes all the major protein constituents of the axoplasmic gel. It follows that the faster-moving proteins must be minor constituents, as has indeed been noted (McEwen and Grafstein, 1968).

BIDIRECTIONAL TRANSPORT   Several workers have reported the accumulation of specific axonal constituents on both sides of a ligature, and have concluded that transport may be directed both toward and from the soma, although it is generally agreed that net transport is consistently somatofugal (see Lubińska, 1964; Dahlström 1967). The significance of these observations has been questioned (Weiss, 1969), so we do not take issue here on the implications of these observations for the mechanisms of axoplasmic transport. The observations of Lasek (1968), however, on axoplasmic transport in the bipolar cells originating in the cat dorsal-root ganglion show that, although transport is somatofugal, it is not always orthodromic. Therefore, if the transport is in any way correlated with, perhaps to sustain, electrical or transsynaptic activity, the direction of the action potential does not define the direction of protein movement.

These observations on transport listed above clearly limit the hypotheses that are devised to explain the mechanisms of transport, if the constraints can be generalized to all species.

*Possible mechanisms of axoplasmic transport*

From the preceding discussion, we must conclude that the mechanism underlying the fast and, possibly, also the slow transport processes is present in any isolated segment of axon. It also appears probable that any energy-transducing mechanism capable of propelling a significant weight of axoplasm should be sufficiently structured to be resolvable by electron microscopy. The only consistent entities discernable by electron microscopy of axons are the axonal membrane, neurofilaments, microtubules, and occasional vesicles and mitochondria. The probability that any of the first three structures is involved is discussed below.

THE NERVE MEMBRANE   Weiss has suggested that the propulsion of the contents of the axon may be effected by a slow (17-minute period) somatofugal, peristaltic wave that traverses the nerve membrane (see Weiss, 1969). The origin of this contractile process has not been defined precisely. It could arise from the neuronal or Schwann-cell membranes, although thus far no contractile proteins have been isolated from these sources. The propulsive efficacy of a peristaltic wave depends on the shape and constancy of the waveform and the consistency and thixotropy of the tube contents. An investigation of the correlation between the temperature dependence of the wave velocity and the transport process, and between the transport velocity and the axonal diameters, should be able to determine if the wave is responsible for the slow-transport process, but these experiments have not yet been performed. It appears plausible that peristalsis could account for the process by which the bulk of the axoplasmic constituents are moved (in the form of a gel). If such is the case, it appears that we must look for a second mechanism that transports selected components, which may be primarily vesicular, at a more rapid rate.

The possibility remains that a transport mechanism other than peristalsis originates at the cell inner membrane; the geometry of the situation requires that if this mechanism exists it probably drives the bulk transport rather than the fast, for the vesicular contents of the axon generally are not peripherally located. The mitochondria are an exception to this statement; they are often distributed conspicuously close to the axon membrane, a fact that supports the idea that energy-demanding processes occur at the surface. We do not know whether these processes include those concerned with axoplasmic transport or only those concerned with the maintenance of transmembrane bioelectrical processes.

MICROTUBULES Microtubules have been found to be prominent in many cellular regions where transport processes are found in various types of cells. They delineate the path followed by chromosomes in mitosis, by melanin granules in melanocytes, and by nuclei in a virus-induced cell syncytium (Holmes and Choppin, 1968). They also line the tentacles of the protozoan *Tokophrya,* where a rapid flow of cytoplasm occurs. These and other examples have been discussed elsewhere (Schmitt, 1968; Schmitt and Samson, 1969); in many instances, no organized structure but the cell membrane and the microtubules are detectable where materials are transported, and the possibility has been frequently voiced that the microtubules are the propelling agency. However, Porter and his colleagues, who have investigated microtubules intensively in many kinds of tissue, have interpreted the role of microtubules as cytoskeletal in most instances (Porter, 1966). Although the flexing of the tails of sperm, of flagella, or of the protozoan axostyle appears to relate to the microtubular structures they contain, the structures in these instances are paired or aggregated and adenosine triphosphatase activity has been observed in certain attendant structures (Gibbons and Rowe, 1965). It is, therefore, not obvious that the chemomechanical functions of these more complex microtubular structures can be extrapolated to the dispersed populations found in neurons and most other cells (see also Tilney and Gibbins, 1968). The only definitive claim that tubular structures in the axoplasm mediate transport of protein is by Kasa (1968). In this instance, however, the electron micrographs do not show characteristic microtubules but rather show irregular structures resembling smooth endoplasmic reticulum.

Very recently Tilney and Gibbins (1969) concluded that microtubules may be not a static cytoskeleton but the precursor and directive agency responsible for eccentric cellular outgrowth. This opinion leaves open the question of their participation. Do they serve as a structure that merely constrains the direction of cytoplasmic movement, like a railway track, while the chemomechanical forces from some other structure power the movement? Or do they, as their specific binding of energy-rich GTP suggests, provide the motive power and direction for transport? Such a transport, moving stepwise along a filament, is familiar in the progression of a ribosome along ribonucleic acid during protein synthesis. Schmitt (1968) has recently postulated a translation mechanism involving an adenosine triphosphatase-generating interaction between the subunits of the microtubule surface and selected particulates in the axoplasm that have the appropriate surface structure. The fastest transport rate in axoplasm would require a displacement across each subunit of the microtubule surface in approximately one millisecond.

NEUROFILAMENTS The possibility of the participation in transport along axons of neurofilaments, 100 Å in diameter, is suggested by their prominence in most axons and their lack of other known function. Moreover, other acidic protein filaments (albeit of smaller diameter, 30–60 Å) in the cytoplasm of amoeba (Morgan et al., 1967), *nitella,* and slime molds are believed to be implicated in cytoplasmic flow or streaming (Nagai and Rebhun, 1966). Indeed, Adelman and Taylor (1969) have isolated from *Physarum* the actin-like filaments and a myosin-like component with which they interact in a contractile process. Neurofilaments, however, are notably sparse in the narrow, sympathetic nerve fibers through which norepinephrine and its synthesizing enzymes are rapidly transported. Furthermore, they are also infrequent in developing neurites that, in their growth, must place heavy demands on the axoplasmic transport systems (Peters and Vaughn, 1967). (Droz, 1969, reported that transport is three times as rapid in young as in mature rats). In mature axons, the microtubules are partly replaced by an increase in the number of neurofilaments. This circumstantial evidence militates against, but does not disprove, the involvement of neurofilaments in axoplasmic transport. It is also improbable that neurofilaments and microtubules interact in any consistent manner, because they often are disposed in discrete assemblies and are not intermixed over an axon cross section. Schmitt and Davison (1961) suggested that material might move down the hollow core of these filaments, as they appear to run uninterruptedly from the cell soma to axon termini, but such a movement in a narrow channel presents a novel and unexplored hydrodynamic situation. If means of communication between the cell periphery and soma must be sought, the filaments remain a possibility, but their participation in somatofugal transport remains dubious. It is possible that their function is purely structural, providing the framework of the axoplasmic gel, for the maintenance of an extended tube of cytoplasm without constrictions or collapse would appear to be a demanding process.

The replacement of microtubules by neurofilaments in the axons of a maturing animal, and the appearance of filamentous structures in the place of microtubules in colchicine- and vinblastine-treated animals (Wisniewski et al., 1968), have led to the speculation that the tubules and neurofilaments might represent two conformations of the same protein subunit. Such a speculation, at least, of all the speculations discussed in this paper, can be laid to rest. In experiments on the axoplasm of squid we have isolated the protein subunits of microtubules and neurofilaments after guanidine hydrochloride denaturation and shown them to be distinct by the following criteria: amino-acid composition (Table I), gel electrophoresis, and molecular weight (Davison and Huneeus, 1970). Antibody has been made to the denatured protein subunit (filarin) of the neurofilament; this antibody reacts with native and denatured filarin, but not detectably with the microtubule protein or with muscle actin. The difference between these filamentous structures is also emphasized by their markedly different ease of dissociation.

TABLE I

*Amino-acid compositions of fibrous proteins from squid axoplasm*

(moles per 100 moles amino acid)

| | Neurofilament * | Microtubules † |
|---|---|---|
| Aspartic acid | 13.5 | 9.4 |
| Threonine | 5.2 | 5.8 |
| Serine | 8.7 | 8.6 |
| Glutamic acid | 16.9 | 12.6 |
| Proline | 2.3 | 6.1 |
| Glycine | 5.4 | 8.8 |
| Alanine | 7.7 | 7.3 |
| Valine | 5.1 | 5.5 |
| Cysteine | 1.4 | 1.5 |
| Methionine | 5.9 | 2.6 |
| Isoleucine | 3.5 | 4.0 |
| Leucine | 6.5 | 6.5 |
| Tyrosine | 2.0 | 3.0 |
| Phenylalanine | 1.6 | 3.0 |
| Lysine | 9.0 | 4.1 |
| Histidine | 1.1 | 1.7 |
| Arginine | 5.5 | 4.6 |

* The averages of highly concordant analyses on three separate preparations of electrophoretically pure neurofilament preparations.
† The average of two preparations, one obtained by preparative gel electrophoresis. In certain cases, agreement was not satisfactory, so that these values may need to be amended.

OTHER MECHANISMS  The fact that the prevalence of microtubules in cells was not generally recognized until glutaraldehyde was used as a fixative for electron microscopy should give us warning that we are not necessarily yet aware of all the apparatus with which the neuron is endowed. Indeed, J. D. Robertson (this volume) reports that Peracchia has found previously unrecognized septae across crayfish axons, the visualization of which depends on improved fixation. Therefore, we may not yet be aware of neuronal structures that may be responsible for one or another aspect of axoplasmic transport.

## How many transport mechanisms?

Although the consistent observation that axoplasm moves 1 to 3 mm per day and the frequent detection of a faster transport process have led to the discrimination of "slow" and "fast" transport systems, it is noteworthy that the latter ranges from 10 to 3000 mm per day, and it is by no means obvious that these velocities can be mediated by one mechanism; and if they are, it must also be recognized that any mechanism that can span such a range of speeds might also drive the slow transport. To take an example, Burdwood (1965) has reported that the brief displacements of mitochondria in neurons by saltation have a velocity similar to the highest velocities reported in axonal transport. Thus, if there were some continuously functioning molecular machinery, such as a conveyor belt, with which the various axonal constituents associated intermittently and with characteristic frequencies, this same mechanism could drive material selectively with various speeds, and the normal "slow" movement of the mitochondrion would result from irregular and infrequent jumps at the "fast" rate. The two transport rates cannot be discriminated on the grounds that the slow movement is a bulk process, because the gel character of the axoplasm would insure bulk transport even if only one component of the matrix—the neurofilaments, for example—were directly propelled.

The existence of two independent mechanisms would be irrefutable, however, if they could be discriminated by function or drug sensitivity. Lasek (1968) has shown that the slow movement in a bipolar neuron occurs both orthodromically and antidromically at the bifurcation. If the rapid transport, which in some axons, at least, pertains to neurotransmitter, were directed only to the presynaptic region, there could be no doubt about the existence of at least two mechanisms, but this experiment has yet to be reported.

The pharmacological evidence is uncertain; studies show that both slow and fast transport processes are inhibited by colchicine (albeit, the fast at a lower concentration). Furthermore, microtubules were still observed by electron microscopy in the crayfish nerve cord treated with colchicine (although it remains a possibility that they were non-functional). Therefore, colchicine does not unambiguously distinguish two mechanisms or implicate microtubule function exclusively in either process. If, however, the slow transport is driven by peristalsis, it is probable that a second mechanism is available for the faster transport, and present evidence would suggest strongly that the microtubules are involved; in this case, the effect of higher concentrations of colchicine on slow transport could reflect the cytotoxicity of the drug (Angevine, 1957).

## Conclusions

In summary, although there is convincing evidence for axoplasmic transport, the mechanism, or mechanisms, are unknown. It is likely, but not certain, that more than one distinct transport process occurs; the *slow* may be mediated by peristalsis in the nerve membrane and the *fast* may involve microtubules—but whether the microtubules define the direction of transport or the direction *and* propulsive power, and how they may do so, is not yet clear. Experiments have been discussed that could amplify our present knowledge.

An even larger problem remains unsolved: What is the function of the rapid turnover of protein in the neuronal soma, and what are the purpose and destiny of the proteins that are transported along the axons? Of the former we can say little. Of the latter, the slowly transported proteins are clearly a complex mixture representing, as mentioned above, most of the proteins in the axoplasm. Weiss (1969) has suggested that the slowly transported proteins may be consumed, during their passage down the axon, in replacing catabolized enzyme in the membrane and elsewhere, but our experiments (Fernandez and Davison, 1969) on the rate of depletion of labeled protein in its passage along the crayfish cord show that at least a significant fraction of the transported protein reaches the terminal segments of the axons. This result is not surprising, in that, if most of the slow proteins are destined to replace enzymes and other proteins, many should reach the synaptic regions because the synapse is clearly the site of considerable metabolic activity, whereas the only known function fulfilled by the axoplasm in the length of the axon is the preservation of the flow mechanisms and the viability of the axoplasm and the membrane. However other nonenzymic proteins reach the axon termini. A significant fraction of the protein traversing the axon is microtubule protein; whether its function is completed on reaching the presynaptic extremities or the subunits provide the raw material for other structures in the synapse is unknown. The intriguing possibility remains that certain proteins have functions not yet understood, and with further study new dimensions of neuronal function may be discovered. Among the compounds transported must be those trophic factors influencing the postsynaptic cells, and any other agencies that may facilitate, restrict, or direct chemical transmission. Considerable evidence has been found relating drug-induced interruption of protein synthesis to a block in learning consolidation (e.g., Agranoff et al., 1967). The effect of such inhibitors does not demonstrate that protein synthesis itself is the mechanism by which memory is stored, but merely that it is a necessary concomitant to the processes by which storage occurs. Thus, further study of the protein metabolism of the nervous system gives promise of a better understanding of all aspects of neuronal function, and a characterization of transsynaptic activity may give insight into the coordinate activity of the brain that is the basis for higher nervous functions, including integration and learning.

## Acknowledgments

The experimental work by the author and his co-workers described in this paper was supported by Grant NB 00024 from the National Institute of Neurological Diseases and Blindness, United States Public Health Service. During the preparation of the manuscript the author was supported by Grant NB 08525 from the same Institute.

### REFERENCES

ADELMAN, M. R., and E. W. TAYLOR, 1969. Isolation of an acto-myosin-like protein complex from slime mould plasmodium and the separation of the complex into actin- and myosin-like fractions. *Biochemistry* 8: 4964–4988.

AGRANOFF, B. W., R. E. DAVIS, L. CASOLA, and R. LIM, 1967. Actinomycin D blocks formation of memory of shock-avoidance in goldfish. *Science* (*Washington*) 158: 1600–1601.

ANGEVINE, J. B., JR., 1957. Nerve destruction by colchicine in mice and golden hamsters. *J. Exp. Zool.* 136: 363–391.

BARONDES, S. H., and F. E. SAMSON, 1967. Axoplasmic transport. *Neurosci. Res. Program Bull.* 5 (no. 4): 307–419.

BORISY, G. G., and E. W. TAYLOR, 1967. The mechanism of action of colchicine. Binding of colchicine-³H to cellular protein. *J. Cell Biol.* 34: 525–533.

BURDWOOD, W. O., 1965. Rapid bidirectional particle movement in neurons. *J. Cell Biol.* 27: 115A (abstract).

DAHLSTRÖM, A., 1967. The transport of noradrenaline between two simultaneously performed ligations on the sciatic nerves of rat and cat. *Acta Physiol. Scand.* 69: 158–166.

DAHLSTRÖM, A., 1968. Effect of colchicine on transport of amine storage granules in sympathetic nerves of rat. *Europe J. Pharmacol.* 5: 111–113.

DAVISON, P. F., and F. C. HUNEEUS, 1970. Fibrillar proteins from squid axons: II. Microtubule protein. *J. Mol. Biol.* (in press).

DROZ, B., 1969. Protein metabolism in nerve cells. *Int. Rev. Cytol.* 25: 363–390.

FERNANDEZ, H. L., and P. F. DAVISON, 1969. Axoplasmic transport in the crayfish nerve cord. *Proc. Nat. Acad. Sci. U. S. A.* 64: 512–519.

FERNANDEZ, H. L., F. C. HUNEEUS, and P. F. DAVISON, 1970. Studies on the mechanism of the axoplasmic transport in the crayfish cord. *J. Neurobiol.* (in press).

GIBBONS, I. R., and A. J. ROWE, 1965. Dynein: A protein with adenosine triphosphatase activity from cilia. *Science (Washington)* 149: 424–426.

GRAFSTEIN, B., 1969. Axonal transport: Communication between soma and synapse. *In* Advances in Biochemical Psychopharmacology, Vol. 1 (E. Costa and P. Greengard, editors). Raven Press, New York, pp. 11–25.

HOLMES, K. V., and P. W. CHOPPIN, 1968. On the role of microtubules in movement and alignment of nuclei in virus-induced syncytia. *J. Cell Biol.* 39: 526–543.

HUNEEUS, F. C., and P. F. DAVISON, 1970. Fibrillar proteins from squid axons: I. Neurofilament protein. *J. Mol. Biol.* (in press).

JASINSKI, A., A. GORBMAN, and T. J. HARA, 1966. Rate of movement and redistribution of stainable neurosecretory granules in hypothalamic neurons. *Science (Washington)* 154: 776–778.

KARLSSON, J. O., and J. SJÖSTRAND, 1969. The effect of colchicine on the axonal transport of protein in the optic nerve and tract of the rabbit. *Brain Res.* 13: 617–619.

KASA, P., 1968. Acetylcholinesterase transport in the central and peripheral nervous tissue: The role of tubules in the enzyme transport. *Nature (London)* 218: 1265–1267.

KREUTZBERG, G. W., 1969. Neuronal dynamics and axonal flow, IV. Blockage of intra-axonal enzyme transport by colchicine. *Proc. Nat. Acad. Sci. U. S. A.* 62: 722–728.

LASEK, R., 1968. Axoplasmic transport in cat dorsal root ganglion cells: As studied with [³H]-L-leucine. *Brain Res.* 7: 360–377.

LUBIŃSKA, L., 1964. Axoplasmic streaming in regenerating and in normal nerve fibres. *Progr. Brain Res.* 13: 1–71.

McEWEN, B. S., and B. GRAFSTEIN, 1968. Fast and slow components in axonal transport of protein. *J. Cell Biol.* 38: 494–508.

MORGAN, J., D. FYFE, and L. WOLPERT, 1967. Isolation of microfilaments from *Amoeba proteus. Exp. Cell Res.* 48: 194–198.

NAGAI, R., and L. I. REBHUN, 1966. Cytoplasmic microfilaments in streaming *Nitella* cells. *J. Ultrastruct. Res.* 14: 571–589.

OCHS, S., M. I. SABRI, and J. JOHNSON, 1969. Fast transport system of materials in mammalian nerve fibers. *Science (Washington)* 163: 686–687.

PETERS, A., and J. E. VAUGHN, 1967. Microtubules and filaments in the axons and astrocytes of early postnatal rat optic nerves. *J. Cell Biol.* 32: 113–119.

PETERSON, R. P., R. M. HURWITZ, and R. LINDSAY, 1967. Migration of axonal protein: Absence of a protein concentration gradient and effect of inhibition of protein synthesis. *Exp. Brain Res.* 4: 138–145.

PORTER, K. R., 1966. Cytoplasmic microtubules and their functions. *In* Principles of Biomolecular Organization/A Ciba Foundation Symposium (G. E. W. Wolstenholme and M. O'Connor, editors). Little, Brown, and Co., Boston, Massachusetts, pp. 308–356.

SCHMITT, F. O., 1968. Fibrous proteins—Neuronal organelles. *Proc. Nat. Acad. Sci. U. S. A.* 60: 1092–1101.

SCHMITT, F. O., and P. F. DAVISON, 1961. Biologie moléculaire des neurofilaments. *Actual. Neurophysiol., Ser.* 3: 355–369.

SCHMITT, F. O., and F. E. SAMSON, 1969. Neuronal fibrous proteins. *Neurosci. Res. Program Bull.* 6 (no. 2): 113–219.

SMITH, B. H., 1967. Neuroplasmic transport and bioelectrical activity in the nervous system of the cockroach, *Periplaneta americana.* Massachusetts Institute of Technology, Cambridge, Massachusetts, doctoral thesis.

TILNEY, L. G., and J. R. GIBBINS, 1968. Differential effects of antimitotic agents on the stability and behavior of cytoplasmic and ciliary microtubules. *Protoplasma* 65: 167–179.

TILNEY, L. G., and J. R. GIBBINS, 1969. Microtubules in the formation and development of the primary mesenchyme in *Arbacia punctulata* II. An experimental analysis of their role in development and maintenance of cell shape. *J. Cell Biol.* 41: 227–250.

TILNEY, L. G., and K. R. PORTER, 1967. Studies on the microtubules in heliozoa II. The effect of low temperature on these structures in the formation and maintainance of the axopodia. *J. Cell Biol.* 34: 327–343.

WEISS, P., 1969. Neuronal dynamics and neuroplasmic ("axonal") flow. *Symp. Int. Soc. Cell Biol.* 8: 3–34.

WISNIEWSKI, H., M. L. Shelanski, and R. D. Terry, 1968. Effects of mitotic spindle inhibitors on neurotubules and neurofilaments in anterior horn cells. *J. Cell Biol.* 38: 224–229.

# 75 Molecular Biological Approaches to Neurophysiological Problems: A Correlation

PHILLIP G. NELSON

IT WAS MY TASK at Boulder to suggest to the participants of the symposium on Molecular Neurobiology subjects at the cellular or system level of nervous-system functions that would be particularly rewarding as judged from the viewpoint of the neurophysiologist. In this brief essay, some of those points are discussed, as indicated at the outset of the symposium, and a few additional ideas are gleaned from the proceedings.

On the broad front of the symposium (and of the Intensive Study Program as a whole), the methods of molecular biology are being applied fruitfully to neurobiology, although particular psychological and neurophysiological problems, such as sensory coding, the mechanism of memory, and learning, are less susceptible at this time to that application. A case in point is the role of protein, generally considered to be the effector of the cell. No neuroprotein has been found that plays a central role in neuroscience, as, for example, myosin plays in myoscience. In this connection, G. M. Edelman argued that it would be better to mount a long-range program of characterizing brain-specific proteins one by one as they are discovered, rather than to seek a protein that, for hypothetical reasons, may be thought to play that central role. S. H. Barondes expressed similar views regarding brain glycolipids and glycoproteins. If, in the course of such investigations, he said, a protein of demonstrable importance is discovered, it will then be appropriate to apply the full armamentarium of molecular biological methods to its characterization.

## The action potential mechanism

A problem central to neurophysiology and neurobiology, and one with a very long history, is that of the mechanism of the production of the action potential and its modulations that have been observed in neurophysiology. The importance of sodium, potassium, and calcium ions in the generation of the action wave itself has long been known. Hodgkin and Huxley (1952) have provided a quantitative treat-

ment of the ionic movements involved in the action potential, and the ionic hypothesis has been of enormous value to virtually every practicing neurophysiologist. Tasaki (1968) has proposed an alternative hypothesis—the two-stable-state model of membrane excitability—because some of his experimental observations, particularly with internally perfused squid axons, seemed inconsistent with the Hodgkin-Huxley model. Tasaki has also attempted to incorporate some structural or molecular features into the bistable-state model of the neuronal membrane. A large number of elegant electrical measurements and manipulations have been made of internal and external ionic constituents surrounding the nerve membrane, but it is probably fair to say that these have not discriminated really decisively between the ionic hypothesis and the two-stable-state hypothesis of how nerve membranes work. It is interesting that in the laboratory represented by Keynes, and in Tasaki's laboratory, the techniques now being used for studying excitable membranes are directed explicitly toward the question of the membrane structures, the molecular species, responsible for excitability (Tasaki et al., 1968; Keynes, this volume). These include the optical methods, described by Keynes and Tasaki, which include light-scattering and birefringence studies and changes in fluorescent materials that have been incorporated in the membrane. McConnell's work on spin labeling of excitable membranes was also directed at this type of question. In addition, the studies of artificial systems involving lipid bilayers, which have been shown to exhibit some degree of excitability when treated with oligopeptides such as alamethicin, seem to offer great promise in demonstrating the molecular basis of excitability (Mueller and Rudin, 1968; Eigen, this volume).

In the analysis of central nervous system function, the mechanism of action-potential generation is frequently taken for granted. It is assumed that the action potential is required and is sufficient to conduct information between points in the nervous system; the various coding methods for conveying information by this all-or-none action potential are much studied. In general, however, the interpretations of such information transmission in the nervous system have not been concerned with the mechanisms of the action potential as such. Transmission of information by

PHILLIP G. NELSON National Institute of Child Health and Human Development, Bethesda, Maryland

axons across relatively large distances undoubtedly relies on an all-or-none, reliable, transmission mechanism. It is probable, however, that in central structures involving highly branched systems of axons—in particular at the axonal terminations—there are many points of low safety factor at which blocking of the action potential does take place. In fact, there is good evidence that at least intermittent blockading of conduction in branched axonal systems is, perhaps, common. Krnjević and Miledi (1958) showed that at the neuromuscular junction there is blockade at axonal branch points.

Old observations by Barron and Matthews (1935), and recent work by Raymond (1969) at the Massachusetts Institute of Technology, have demonstrated that intermittent conduction or periodic failure of conduction in axons occurs in the spinal cord. In the dendritic trees of neurons the question of partial conduction of action potentials is certainly very real. Llinás et al. (1968), Nelson and Frank (1964), and Nelson and Burke (1967), among others, have studied dendritic spike generation and have discussed the possibility that there is a wide variation in the coding function performed by neurons and that the variation depends on the different degrees of spike generation in the arborization of the dendrites.

With respect to modulations of the excitatory mechanism required in physiological function, it may be mentioned that the processes of accommodation and adaptation in neurons can play an important role in determining the behavior of a given neural system. "Accommodation" refers to a property of excitable membrane that causes the effectiveness of an excitatory stimulus to be strongly dependent on the rate of application of that excitatory stimulus. This property of accommodation varies considerably between different types of neural tissue, and in the motor control system of the mammalian nervous system it is differentiated in a functionally meaningful way. The motor neurons in the lumbosacral spinal cord can be distinguished in terms of the kinds of muscle fibers they innervate. Some motor neurons innervate slow-twitch, red muscle fibers, such as are commonly found in the soleus muscle; at the other extreme, some motor neurons innervate fast-twitch muscle fibers, such as are found in the gastrocnemius muscle. The mechanical properties of the soleus muscle unit seem optimally designed to provide the steady tonic contractions that are required for maintaining posture, whereas the mechanical properties of gastrocnemius fast-twitch muscle units seem much better designed for initiating movement, providing rapid bursts of high tension development. Almost none of the motor neurons supplying the slow-twitch soleus muscle fibers (type S) exhibit any significant degree of accommodation to slowly increasing excitatory input.

With regard to the motor neurons innervating the large amplitude, fast-twitch muscle units (type F), on the other hand, about half of the cells do exhibit substantial degrees of accommodation; that is, they respond much better to transient or rapidly changing excitatory stimuli and tend to be much less responsive to very slowly varying excitatory stimuli (Burke and Nelson, in preparation). In addition, when steady excitatory currents are passed across the cell membrane in these two classes of motor neurons, the motor neurons innervating the tonically functioning soleus muscle cells tend to show much more sustained repetitive firing than do the class of motor neurons innervating the fast-twitch fibers of the gastrocnemius muscle (Mishelevich, 1969). Thus, at the level of the membrane properties of the individual motor neurons there has been a differentiation in the property of the excitable neuronal structures such that the motor system has divided into a group of tonically firing muscle cells and a group of muscle cells that tend to be activated only transiently. Similar and probably functionally significant differences in the membrane properties of cortical pyramidal cells have been described by Takahashi (1965).

## In vitro studies

Detailed study by molecular biological methods of neural systems such as that described above is extremely difficult, but there has recently been developed an in vitro preparation that shows some interesting properties and possibilities from the standpoint of studies of excitable membranes. Augusti-Tocco and Sato (1969) have worked with a tissue culture preparation of the mouse neuroblastoma cell line. I have been working with this preparation in collaboration with M. Nirenberg and B. W. Ruffner at the National Institutes of Health, and many of the cells of these cultures have been found to exhibit electrical excitability (Nelson et al., 1969). Examples of these properties are shown in Figures 1 and 2. Although the responses obtained did depend on the conditions of the recording, it appears that there is a true range of excitable properties in these cells. That is, under standard conditions of membrane potential and culture conditions, some cells show much more marked spike-generating activity than others, and some cells, indeed, showed essentially passive membrane properties. This preparation, therefore, may be a favorable one in which to study biochemical correlates of varying degrees of electrical excitability.

## The physiological role of neuronal structural geometry

An area that should be of great concern to neurobiologists relates to those processes that govern the genesis and maintenance of the complex neuronal geometry so characteristic of the central nervous system. Both the elaborate dendritic trees and the highly complex axonal ramifications are

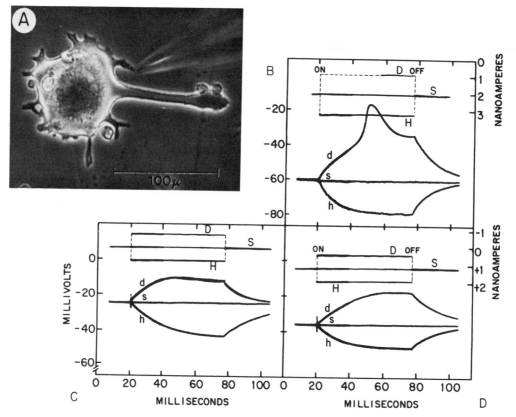

FIGURE 1 Mouse neuroblastoma cell. A. The cell is from a culture that had been treated with trypsin to dissociate the cells five days earlier. The cell was examined electrophysiologically soon after the culture had passed the logarithmic phase of growth. Total time in vitro was 108 days. B. An active response. C, D. Examples of rectification. The lines labeled D and H indicate pulses of current that evoke changes in voltage across the cell membrane labeled d and h, respectively. S designates the steady current used to adjust the voltage across the cell membrane, s.

among the most striking features of the nervous system. The Scheibels (1969) have provided an elegant picture of neuronal configurations throughout the nervous system, and Ramón-Molinar (1968) has attempted to generalize about the enormously varied dendritic structures that he has studied in various portions of the nervous system. Possible common functions have been suggested for neurons that exhibit relatively regularly branched dendritic trees and those with more specialized tufted, wavy, or "idiodendritic" trees, which are seen in other portions of the brain.

## Neuroplasmic flow

Both Weiss and Davison (this volume) have discussed the problem of neuroplasmic flow and the possible role of surface contractile processes or neurofilaments and microtubules in that flow. This neuroplasmic flow would certainly participate in the replenishment of materials at the synaptic junctions at the end of cellular processes, either dendrites or axons, and this would seem to be one of the important functions of the flow. Lumsden (1968) has raised the question of

whether neurofilaments and microtubules may play some role in determining the branching patterns of axons. It might be that some relatively simple molecular instruction mechanisms utilize these fibrous proteins of dendritic and axonal processes to determine the form of even very complicated branched dendritic or axonal systems. Weiss has emphasized that surface contact and guidance phenomena undoubtedly are important in the generation of both branched systems and processes. It is obvious that the shapes of neurons in different parts of the nervous system are instrumental in determining the functions of neuronal syssstems, and the molecular basis for this highly elaborate cellular morphology is a crucial problem for molecular neurobiology.

## Electrical properties of dendritic structures

An important problem in electrophysiology has been to determine some of the electrical properties of the neuronal dendritic structures. Wilfrid Rall (1967) has provided a quantitative mathematical model that is capable of dealing

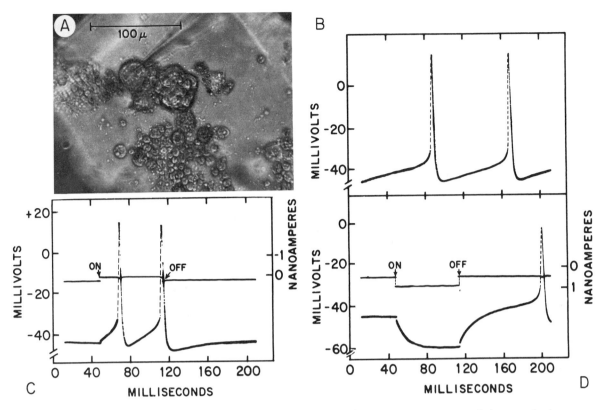

FIGURE 2   A. The large, round cell, indicated in the photograph by the microelectrode, was studied. The culture was in the stationary phase of growth; total time in vitro was 66 days. B. Repetitive action potentials recorded soon after the cell was penetrated by the microelectrode. The action potentials took place in the absence of stimulating current. The broken lines represent portions of the records that were filled in during reproduction. C. Two action potentials elicited by a pulse of current. D. An example of "off-excitation." An action potential was evoked by turning *off* the pulse of current. The top of the action potential is not shown.

with these structures. Lux and I (Nelson and Lux, 1970) set out to answer two questions about these neuronal dendrites by examining motor neurons in the spinal cord of the cat. We wished to know, first, what percentage of the physiologically effective surface area of the cell was represented by dendritic membranes; second, how far away, electrically speaking, the ends of the dendrites were from this trigger zone in the cell body, i.e., what percentage of the synapses that occurred on the surface of the motor neurons were on the dendrites and how effective the dendritic synapses would be in initiating spike activity in the cell body.

It has turned out that between 80 and 90 per cent of the effective receptive surface of the motor neuron is on the dendrites and that even the farthest portions of the dendritic tree are only between one and two equivalent electrotonic lengths away from the cell body. The synapses appearing on the farthest portions of the dendritic tree thus have a steady-state attenuation factor of between about 3 and 9. The effectiveness of these synapses is attenuated, but even the synapses situated most distally have a significant effect on the spike-initiating zone in the cell body. The entire dendritic tree, therefore, is significant in terms of spike initiation and must be taken into account in attempts to relate dendritic structure to the possible functional properties of a given cell.

## Specificity of interneuronal connections

Crucial for an understanding of nervous system function is the degree of specificity of connections between neurons. How carefully are the complicated axonal terminal ramifications and the dendritic trees of neurons interdigitated and how specifically are synaptic connections between the nerve cells regulated? Burke (1967a, 1968a) has looked carefully at the connections between sensory fibers from muscle and the motor neurons that innervate those same muscles. The experimental arrangement used in studying this problem is shown in Figure 3. This arrangement made it possible to record intracellularly with microelectrodes from individual motor neurons, to stimulate

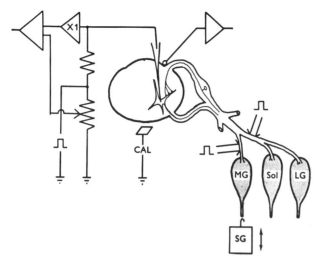

FIGURE 3 Experimental arrangement for studying properties of cat spinal motor neurons and the muscle fibers they innervate. Circuitry to the left allows stimulation and recording of single motor neurons through intracellular electrodes (center). Strain gauge (SG) records mechanical activity of muscles. Abbreviations: MG, medial gastrocnemius muscle; SOL, soleus muscle; LG, lateral gastrocnemius muscle; CAL, calibration pump. (From Burke, 1967a.)

these motor neurons and record the tension developed by the muscle fibers they activate, and to record synaptic activity produced when different muscle nerves were stimulated either by electrical stimulation or by stretching the muscles to which the motor neurons were connected. As has been pointed out above, motor neurons can be divided on the basis of the characteristics of the muscles they innervate. If one looks at these groups of motor neurons on the basis of the nature of the synaptic connections from afferent sensory fibers, there also appears to be some differentiation. The synaptic potentials produced in motor neurons innervating the slow-twitch type of muscle units (type S) were larger and of somewhat longer time course than were the synaptic potentials produced in the motor neurons innervating the fast-twitch muscle fibers (type F).

An extensive study of the shapes of unitary postsynaptic potential generated by activity in a single afferent fiber has indicated that the time course is related to the location of the synaptic endings on the dendritic tree of the motor neurons (Burke, 1967b; Rall et al., 1967). Synapses that occur in the peripheral parts of the dendritic tree tend to produce slower, longer-lasting synaptic potentials because of electrotonic effect. Synaptic potentials generated by endings that are close to the recording site in the cell body are much sharper and faster. The longer time course of the synaptic potentials that are produced by stimulating the population of afferent fibers in the peripheral nerve indicates that a higher percentage of the endings on slow-

twitch motor neurons are in the peripheral parts of the dendritic tree.

The differences in the time course of the postsynaptic potentials do not seem to be very dramatic, but if one looks at the summation of asynchronously occurring, unitary, postsynaptic potentials generated by muscle stretch, the effect of this differential localization of synaptic endings is considerable (Burke, 1968b). More substantial and sustained depolarizations take place in slow-twitch motor neurons, and repetitive activation occurs much more commonly and readily (Figures 4 and 5).

Thus, in this final output loop of the nervous system, in which the muscle fibers and the sensory fibers come from the muscle to end on the motor neurons, we have a functionally significant differentiation. The motor neuron membrane properties, the cell size, the axonal conduction velocity, the contractile properties of the muscle fiber, the specific connections of the afferent fibers in terms of the motor neuron with which they connect, and even the location of those endings on the dendritic tree have all been specified in such a way as to optimize the performance of the system. The slow-twitch motor system is optimized for the steady development and maintenance of contractions and fine gradations of this steady tension, whereas the fast-twitch system appears to be optimized for providing rapidly changing, high increments of tension that last for only a brief period; that is, their purpose is to initiate movement.

In addition, segregation of synapses on different parts of cortical cells, depending on the source of the synaptic endings, has been shown by Globus and Scheibel (1967a, 1967b).

## Plasticity in neuronal function: use – disuse

An area of great interest to neurobiologists is that of plasticity in nervous function. Perhaps, in its simplest form, this may be put as the question of alteration of synaptic function as a result of use or disuse of the synaptic region. Can electrical activation of a synapse produce a long-term change in the efficacy of that synapse? Functional and morphological effects of altered activity have been demonstrated in the visual system (Hubel and Wiesel, 1965; Globus and Scheibel, 1967c; Cragg, 1969).

Recent work by Norman Robbins and Gerald Fischbach (in preparation) indicates that such synaptic plasticity does take place in a relatively simple preparation that allows detailed study of the relevant synaptic mechanisms. To demonstrate this, they used the neuromuscular junction of the soleus muscle of the rat. They showed that when the hind limb of a rat is immobilized by pinning of the ankle and knee joint, the activity of the soleus muscle of the limb, as recorded electromyographically, goes down by a factor of 20 or so. Because the electromyographic activity is the

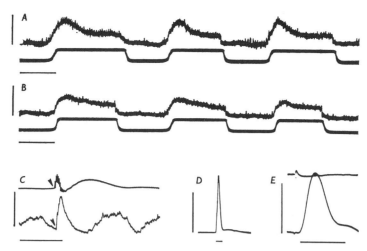

FIGURE 4 Records from a type F medial gastrocnemius muscle (MG) motor unit illustrating "phasic" response pattern to maximum MG muscle stretch (about 14 mm extension from resting length). Records A and B: intracellular potential records (upper trace in each record), showing depolarizations produced by summated asynchronous EPSP activity during MG muscle stretch (signaled by deflections of the myograph tension record, the lower trace in each record). The unit did not fire to the stimulus either before (record A) or after (record B) a 15-second tetanus to the MG nerve at 250 pulses/sec. Calibration for A and B: 10 mV and 1–0 sec. Record C: EPSP response, again without spike (lower trace, arrow), elicited by sharp tap to the string connecting muscle and myograph (arrow in upper trace, which is MG tension record; tension change of about 100 g produced by a tap). Note that a reflex twitch of the MG muscle follows the tap, indicating that other units were activated by the stimulus. Calibrations: 5 mV and 50 msec. Record D: antidromic spike; calibration 50 mV and 1–0 msec. Record E: muscle unit twitch (lower trace) produced by intracellular stimulation (at artifact in upper trace); muscle at about 100 g passive initial tension, with background active tension suppressed by deep peroneal tetanus (see Figure 1B). Calibrations: 20 g and 50 msec.

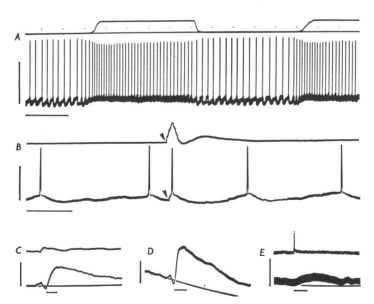

FIGURE 5 Records from a type S MG motor unit (same animal as the unit in Figure 4) illustrating "tonic" response to MG stretch. Record A: intracellular record (lower trace) shows spontaneous firing without muscle stretch and more rapid sustained discharge during MG stretch of about 12 mm (signaled by deflection of tension record in upper trace). Calibrations: 50 mV and 1–0 sec. Record B: regular unit firing with small (about 20 g) tension on MG muscle, with an interposed EPSP and spike response (lower arrow) elicited by tap to the myograph string (at arrow in the upper tension trace). Calibration: 50 mV and 50 msec. Record C: submaximal MG monosynaptic EPSP (lower trace) for EPSP wave form measurements. Upper trace is potential at the dorsal root entry. Record D: maximum MG EPSP recorded during blockade of antidromic invasion of the motor neuron with hyperpolarizing current passed through the microelectrode (hence the sloping baseline). Calibrations for C and D: 5 mV and 1–0 msec. Record E: muscle unit twitch response (lower trace) to intracellular stimulation. Calibrations: 2 g and 50 msec.

direct consequence of neural activity, the experiment shows that pinning of the limb reduces the activity of soleus motor neurons. They then looked at acetylcholine receptors in the postsynaptic regions in the muscles. The chemosensitive region or the number of acetylcholine receptors in the muscle increased. Thus, disuse leads to an increase in the number of receptors or a spread in sensitivity. They then studied the output of transmitter from the nerves coming from these disused muscles and found a drop-off in the ability of the disused nerves to maintain the output of synaptic transmitter in the face of repetitive stimulation. The number of quanta released per second is expressed as a function of the frequency of stimulation for the normal and disused junctions, as shown in Figure 6.

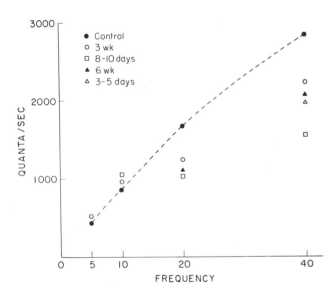

FIGURE 6 The relationship between frequency of nerve stimulation and transmitter output (in quanta per second) in the rat soleus nerve muscle preparation in vitro. Results are from normal muscles and muscles immobilized for the periods indicated.

This shows a fall-off of about 20 per cent in the ability of disused nerve to maintain transmitter output in the face of rapid repetitive stimulation. Quantal output may have been higher in the disused junctions at lower rates of stimulation.

## Metabolic regulation of synaptic function

One of the most exciting aspects of this symposium was the discussion of the various ways in which synaptic function is controlled. There are several possible mechanisms by which electrical activity might relate to the metabolic events at the synaptic junctions and thereby regulate the effectiveness of the synaptic regions. Action potentials in extremely small structures, such as the presynaptic axons or boutons, would

produce rather large changes in the ionic composition of both the intracellular and the extracellular fluids. With activity, sodium ingresses, which would be expected to stimulate the sodium- and potassium-activated adenosine triphosphatase. The consequent splitting of adenosine triphosphate could well affect a number of metabolic pools in the tissue. It has been shown (Birks and Cohen, 1968a; Dunant, 1969) that a blockade of the sodium pump by inhibitors such as ouabain promptly affects synaptic transmission, presumably because of intracellular accumulation of Na ions in the presynaptic terminal (Birks and Cohen, 1968b). Complex changes in spontaneous and evoked endplate potentials begin within a matter of minutes, and in an hour or so the spike-generating mechanism is blockaded.

The protein synthetic mechanisms in the synaptosome have been shown to be sensitive to the concentration of Na or K ions in the incubating medium (Shooter, this volume; Morgan and Austin, 1968; Appel et al., 1969). In contrast to the sensitivity of the ribosomal protein-synthesizing system, in synaptosomes the optimum concentration of sodium for incorporation of radioactive amino acid into protein is high (50mM or more), which is more than one would expect to find intracellularly under normal conditions. If repetitive spike activity occurred in a very small structure, such as a synaptic ending, intracellular concentration of sodium ion could increase appreciably. This increase could reasonably be expected to stimulate protein synthesis in the synaptic ending, which would provide a natural way to couple electrical activity to biochemical activity. Kakiuchi et al. (1969) have shown that electrical and chemical stimulation of brain slices increases the level of cyclic adenosine monophosphate in the slices by a factor of 10 or so. The mechanism of this stimulation is not worked out, but production of cyclic AMP clearly could be expected to affect many metabolic processes in brain tissue. This, therefore, represents another mechanism whereby electrical activity could be coupled to metabolic activity. Whittaker (this volume) has discussed the preparation of synaptosomes and of constituents of synaptosomes. These preparations are favorable for the study of synaptic function and plasticity. The relationship of electrical activity to metabolic activity in neural structures can clearly be approached by presently available techniques.

## Connectionistic versus distributed, parallel processing in brain function

The kinds of questions discussed here arise naturally and inevitably from one concept of the nature of brain function, namely, the cellular, connectionistic model of central nervous function. This view of the central nervous system holds that the processing of information by the brain is determined by the specific patterns of converging and diverging connections between different nerve cells at different levels of the brain. That framework has undoubtedly been successful when investigators are dealing with the handling of information in the specific sensory systems (Hubel and Wiesel, 1962) and motor control systems. It is reflected in the theory of the "command neuron" in invertebrates (Kennedy et al., 1969) and perhaps has been carried to its extreme by Wickelgren (1969) in supposing the existence of "concept neurons." Various patterns of synaptic convergence and divergence are considered to result finally in the firing of a single cell, the activity of which could relate to a single specific, even a very complex, concept or high-order mental construct.

Experimentally, however, the single-cell approach to understanding brain function encounters severe difficulty as one goes from the specific sensory motor system to the higher associative areas of the brain. Even if the firing of a single cell were uniquely related to some complex configuration of events in the external or internal world, the possible combinations of events become effectively infinite and experimentally unapproachable. As the possible kinds of stimuli become larger than the stimulus represented by particular sensory modalities, it becomes impossible to determine the unique set of events that would be most appropriate to trigger the firing of a given cell. This is an *ad hoc,* or nontheoretical, quality of the specific connectionistic model of the brain that makes it extremely difficult to generalize, except in a vague way, about how the "associational" brain functions. In addition, some observations seem to pose a serious challenge to the specific connectionistic view.

The experiments of Lashley (1929) have been taken to pose such a challenge. The deficits in psychological functions produced by circumscribed lesions appear to be not nearly so specific as one might expect. That is, similar deficits may result when similar amounts of tissue are taken from rather different areas of the brain, so that a degree of equal potentiality of cerebral tissue was postulated by Lashley. Several criticisms of these experiments and their interpretation have been made (see Kandel and Spencer, 1968). John (1967) and John and Morgades (1969) made electrical recordings from widely distributed sites in the brain, and found relatively similar electrical activity as a correlate of various discriminative performances of trained animals. This was interpreted as indicating a more global, distributed, homogeneous response of the brain in distinction to what might be expected in a highly specific connectionistic model of brain function. An analogy has been drawn between this sort of distributed storage of information and that represented by the photographic technique of holography. Longuet-Higgins (1968) has shown that principles similar to those underlying the hologram can provide a basis for a distributed storage of temporal information.

Adey et al. (1969) have suggested that the extracellular space in the brain has a highly specific macromolecular composition and that important aspects of brain functions are mediated by this extracellular matrix of macromolecules. It might be expected that mucopolysaccharides and other macromolecules in the extracellular space and large molecules on the surface of neurons would be altered by electrical activity of nerve cells, and thus represent a possible substrate for the storage of information. That is, memory could reside in this extracellular compartment. Experiments of Baylor and Nicholls (1969a, 1969b, 1969c) on the central nervous system of the leech indicate clearly that interactions between nerve cells and glia may be mediated by extracellular accumulation of potassium resulting from the activity of nerve cells, and they raise the question of whether there may be a relationship between glial cells and specific groups of neurons that may be coupled by this glia-potassium mechanism. Extracellular current flows certainly mediate some degree of interaction between nerve cells, as has been shown in spinal motor neurons (Grinnell, 1966; Nelson, 1966).

Specific connections between nerve cells undoubtedly play a dominant part in the processing of information by the nervous system. Nerve cells, however, exist in a complex milieu, and extracellular ionic and macromolecular constituents undoubtedly play an important, still largely unexplored, role in modulating and modifying the function of the neurons. To discover if extracellular constituents and glial cells play a *primary* role in neural transactions remains a challenging aspect of molecular neurobiology.

### REFERENCES

ADEY, W. R., B. G. BYSTROM, A. COSTIN, R. T. KADO, and T. J. TARBY, 1969. Divalent cations in cerebral impedance and cell membrane morphology. *Exp. Neurol.* 23: 29–50.

APPEL, S. H., L. AUTILIO, B. W. FESTOFF, and A. V. ESCUETA, 1969. Biochemical studies of synapses *in vitro.* III. Ionic activation of protein synthesis. *J. Biol. Chem.* 244: 3166–3172.

AUGUSTI-TOCCO, G., and G. SATO. 1969. Establishment of functional clonal lines of neurons from mouse neuroblastoma. *Proc. Nat. Acad. Sci. U. S. A.* 64: 311–315.

BARRON, D. H., and B. H. C. MATTHEWS, 1935. Intermittent conduction in the spinal cord. *J. Physiol.* (*London*) 85: 73–103.

BAYLOR, D. A., and J. G. NICHOLLS, 1969a. Changes in extracellular potassium concentration produced by neuronal activity in the central nervous system of the leech. *J. Physiol.* (*London*) 203: 555–569.

BAYLOR, D. A., and J. G. NICHOLLS, 1969b. After-effects of nerve impulses on signalling in the central nervous system of the leech. *J. Physiol.* (*London*) 203: 571–589.

BAYLOR, D. A., and J. G. NICHOLLS, 1969c. Chemical and electrical synaptic connexions between cutaneous mechanoreceptor neurones in the central nervous system of the leech. *J. Physiol.* (*London*) 203: 591–609.

BIRKS, R. I., and M. W COHEN, 1968a. The action of sodium

pump inhibitors on neuromuscular transmission. *Proc. Roy. Soc., ser. B, biol. sci.* 170: 381–399.

BIRKS, R. I., and M. W. COHEN, 1968b. The influence of internal sodium on the behaviour of motor nerve endings. *Proc. Roy. Soc., ser. B, biol. sci.* 170: 401–421.

BURKE, R. E., 1967a. Motor unit types of cat triceps surae muscle. *J. Physiol. (London)* 193: 141–160.

BURKE, R. E., 1967b. Composite nature of the monosynaptic excitatory postsynaptic potential. *J. Neurophysiol.* 30: 1114–1137.

BURKE, R. E., 1968a. Group $I_a$ synaptic input to fast and slow twitch motor units of cat triceps surae. *J. Physiol. (London)* 196: 605–630.

BURKE, R. E., 1968b. Firing patterns of gastrocnemius motor units in the decerebrate cat. *J. Physiol. (London)* 196: 631–654.

CRAGG, B. C., 1969. The effects of vision and dark-rearing on the size and density of synapses in the lateral geniculate nucleus measured by electron microscopy. *Brain Res.* 13: 53–67.

DUNANT, Y., 1969. Presynaptic spike and excitatory postsynaptic potential in sympathetic ganglion; their modifications by pharmacological agents. *Progr. Brain Res.* 31: 131–139.

GLOBUS, A., and A. B. SCHEIBEL, 1967a. Synaptic loci on parietal cortical neurons: Terminations of corpus callosum fibers. *Science (Washington)* 156: 1127–1129.

GLOBUS, A., and A. B. SCHEIBEL, 1967b. Synaptic loci on visual cortical neurons of the rabbit: The specific afferent radiation. *Exp. Neurol.* 18: 116–131.

GLOBUS, A., and A. B. SCHEIBEL, 1967c. The effect of visual deprivation on cortical neurons: A Golgi study. *Exp. Neurol.* 19: 331–345.

GRINNELL, A. D., 1966. A study of the interaction between motoneurones in the frog spinal cord. *J. Physiol. (London)* 182: 612–648.

HODGKIN, A. L., and A. F. HUXLEY, 1952. A quantitative description of membrane current and its application to conduction and excitation in nerve. *J. Physiol. (London)* 117: 500–544.

HUBEL, D. H., and T. N. WIESEL, 1962. Receptive fields, binocular interaction and functional architecture in the cat's visual cortex. *J. Physiol. (London)* 160: 106–154

HUBEL, D. H., and T. N. WIESEL, 1965. Binocular interaction in striate cortex of kittens reared with artificial squint. *J. Neurophysiol.* 28: 1041–1059.

JOHN, E. R., 1967. Mechanisms of Memory. Academic Press, New York.

JOHN, E. R., and P. P. MORGADES, 1969. Neural correlates of conditioned responses studied with multiple chronically implanted moving microelectrodes. *Exp. Neurol.* 23: 412–425.

KAKIUCHI, S., T. W. RALL, and H. McILWAIN, 1969. The effect of electrical stimulation upon the accumulation of adenosine 3′, 5′-phosphate in isolated cerebral tissue. *J. Neurochem.* 16: 485–491.

KANDEL, E. R., and W. A. SPENCER, 1968. Cellular neurophysiological approaches in the study of learning. *Physiol. Rev.* 48: 65–134.

KENNEDY, D., A. I. SELVERSTON, and M. P. REMLER, 1969. Analysis of restricted neural networks. *Science (Washington)* 164: 1488–1496.

KRNJEVIĆ, K., and R. MILEDI, 1958. Failure of neuromuscular propagation in rats. *J. Physiol. (London)* 140: 440–461.

LASHLEY, K. S., 1929. Brain Mechanisms and Intelligence. University of Chicago Press, Chicago, pp. xiv, 186.

LLINÁS, R., C. NICHOLSON, J. A. FREEMAN, and D. E. HILLMAN, 1968. Dendritic spikes and their inhibition in alligator Purkinje cells. *Science (Washington)* 160: 1132–1135.

LONGUET-HIGGINS, H. C., 1968. The non-local storage of temporal information. *Proc. Roy. Soc., ser. B, biol. sci.* 171: 327–334.

LUMSDEN, C. E., 1968. Nervous tissue in culture. *In* Structure and Function of Nervous Tissue (G. H. Bourne, editor). Academic Press, New York, vol. I, pp. 67–140.

MISHELEVICH, D. J., 1969. Repetitive firing to current in cat motoneurons as a function of muscle unit twitch type. *Exp. Neurol.* 25: 401–409.

MORGAN, I. G., and L. AUSTIN, 1968. Synaptosomal protein synthesis in a cell-free system. *J. Neurochem.* 15: 41–51.

MUELLER, P., and D. O. RUDIN, 1968. Action potentials induced in biomolecular lipid membranes. *Nature (London)* 217: 713–719.

NELSON, P. G., 1966. Interaction between spinal motoneurons of the cat. *J. Neurophysiol.* 29: 275–287.

NELSON, P. G., and R. E. BURKE, 1967. Delayed depolarization in cat spinal motoneurons. *Exp. Neurol.* 17: 16–26.

NELSON, P. G., and K. FRANK, 1964. Orthodromically produced changes in motoneuronal extracellular fields. *J. Neurophysiol.* 27: 928–941.

NELSON, P. G., and H. D. LUX, 1970. Some electrical measurements of motoneuron parameters. *Biophys. J.* 10: 55–73.

NELSON, P. G., B. W. RUFFNER, and M. NIRENBERG, 1969. Neuronal tumor cells with excitable membranes grown *in vitro*. *Proc. Nat. Acad. Sci. U. S. A.* 64: 1004–1010.

RALL, W., 1967. Distinguishing theoretical synaptic potentials computed for different soma-dendritic distributions of synaptic input. *J. Neurophysiol.* 30: 1138–1168.

RALL, W., R. E. BURKE, T. G. SMITH, P. G. NELSON, and K. FRANK, 1967. Dendritic location of synapses and possible mechanisms for the monosynaptic EPSP in motoneurons. *J. Neurophysiol.* 30: 1169–1193.

RAMÓN-MOLINAR, E., 1968. The morphology of dendrites. *In* Structure and Function of Nervous Tissue (G. H. Bourne, editor). Academic Press, New York, vol. I, pp. 205–267.

RAYMOND, S. A., 1969. Physiological influences on axonal conduction and distribution of nerve impulses. Massachusetts Institute of Technology, Cambridge, Massachusetts, doctoral thesis.

SCHEIBEL, M. E., and A. B. SCHEIBEL, 1969. Terminal patterns in cat spinal cord. III. Primary afferent collaterals. *Brain Res.* 13: 417–443.

TAKAHASHI, K., 1965. Slow and fast groups of pyramidal tract cells and their respective membrane properties. *J. Neurophysiol.* 28: 908–924.

TASAKI, I., 1968. Nerve Excitation; A Macromolecular Approach. Charles C Thomas, Springfield, Illinois.

TASAKI, I., A. WATANABE, R. SANDLIN, and L. CARNAY, 1968. Changes in fluorescence, turbidity, and birefringence associated with nerve excitation. *Proc. Nat. Acad. Sci. U. S. A.* 61: 883–888.

WICKELGREN, W. A., 1969. Learned specification of concept neurons. *Bull. Math. Biophys.* 31: 123–142.

# 76 Molecular Neurobiology: An Interpretive Survey

## FRANCIS O. SCHMITT

THE SYMPOSIUM ON MOLECULAR NEUROBIOLOGY, with its associated plenary lectures, was productive of many new facts and concepts brought out in the lectures and informal discussions. This summational essay—not a review—must perforce be selective in the subjects and material to be treated; many interesting aspects will not even be mentioned, although some ideas generated after the symposium are included.

The division of the material into four sections—membranes, neuronal junctions, neurobiosynthesis, and neuroplasmic kinesis—is arbitrary, but the manuscripts group agreeably in this way. The order of discussion of individual papers is not the same as that in which the papers were presented.

### Membrane systems

Until recently, ideas about the molecular organization of cell membranes followed a rather simple model. In 1925, Gorter and Grendel maintained that the cell membrane was only a few molecules thick. That it contained thin layers of "protein" as well as a bilayer of mixed lipid was suggested by Danielli and Davson (1935). Their theory was made highly probable by polarization optical and X-ray diffraction analyses of myelin membranes (Schmitt and Bear, 1939; Schmitt et al., 1941). The structure, seen in the electron microscope as two parallel dense lines, was popularized as the "unit" membrane by Robertson (1967).

The lipid-protein trilamina, which forms the continuous portion of the cell membrane, may be referred to as the *matrix*. From the outer surface of the matrix, stringy chains ("fuzz") of glycoprotein or glycolipid may extend for varying lengths (see Figure 1). This integral part of the membrane plays an important role in the cell's interaction with its microenvironment and with other cells (Schmitt and Samson, 1969). Upon or within the film matrix are situated such molecular devices as ion carriers and permeases, ion pumps, receptors, and other equipment to carry out special functions, e.g., the propagation of the nerve impulse.

FRANCIS O. SCHMITT Chairman, Neurosciences Research Program, Massachusetts Institute of Technology, Brookline, Massachusetts

Much remains to be discovered about cell membranes. For example, do they contain DNA (only recently discovered in mitochondria) or RNA? Where is adenyl cyclase situated relative to receptors and the sources of its substrate adenosine triphosphate (ATP)? The special devices (pumps, permeases, and so forth) are thought to be distributed sparsely (about several dozen per square micron) over the surface (see Figure 2). The trilaminar matrix, as such, may be primarily a convenient interfacial barrier to the cell's microenvironment and a bearer of the molecular machines. This model of the real neuronal membrane is doubtless oversimplified.

Delbrück (this volume), a leader in molecular genetics for a generation, considers that the present state of knowledge of membranes resembles that of chromosome organization 30 years ago, and he predicts that discoveries of membrane organization will rank with the greatest in molecular biology. It is from such a viewpoint that the membrane section of the symposium is here surveyed.

Should neuroscientists, like molecular geneticists of two decades ago, seek a macromolecular type that performs specialized functions by virtue of particular composition and structure? Do we seek enzymes and other substances which carry out topochemical reactions that integrate and "cyberneticize" membrane functions, and do we relate these to vital processes in the cell center and in the cellular microenvironment? Do allosteric interactions between membrane subunits and enzyme assemblies play an important role in this integration? From our knowledge of other membranes, especially those of mitochondria, can we pose productive technical and conceptual strategies for investigating the membranes of neurons and glial cells?

THE MEMBRANE MATRIX The molecular organization of the membrane matrix is fruitfully studied crystallographically in nerve myelin. Recently there has been a recrudescence of interest in the investigation of myelin structure with improved X-ray diffraction methods: about 15 orders of the large (approximately 180 Å) repeating period have been recorded (Kirschner and Caspar, 1969; Worthington and Blaurock, 1969), and detailed crystallographic data have been obtained. At the same time, there has been a saltation in our understanding of the protein of myelin, the subunits of which have been isolated in water-soluble form (Folch-Pi and Sherman, 1969; Sherman and

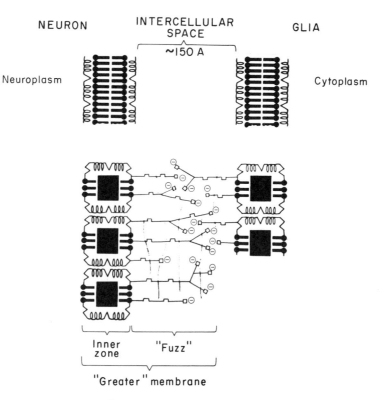

NEURON      INTERCELLULAR SPACE      GLIA

~150 A

Neuroplasm      Cytoplasm

Inner zone    "Fuzz"

"Greater" membrane

FIGURE 1 Diagrammatic representation of membrane structure. Upper figure: Matrix with lipid bilayer, unit membrane structure. Lower figure: "Greater" membrane consisting of lipid-protein matrix (after the Wallach-Singer model) and glycoprotein coat ("fuzz"), and intercellular microenvironment also containing macromolecules (probably mucopolysaccharides).

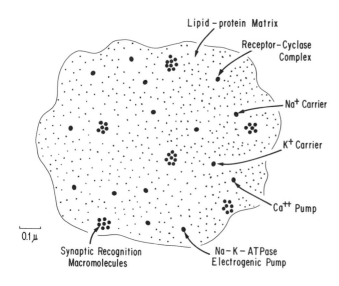

Lipid - protein Matrix

Receptor-Cyclase Complex

Na⁺ Carrier

K⁺ Carrier

Ca⁺⁺ Pump

0.1 μ

Synaptic Recognition Macromolecules

Na – K – ATPase Electrogenic Pump

FIGURE 2 Diagrammatic illustration of sparsity of pump, carriers, and other membrane effectors in the matrix of the membrane.

Folch-Pi, 1970), and the role of lipid in determining myelin stability has been assessed (O'Brien, 1965; O'Brien and Sampson, 1965).

A point of major interest is whether the current analyses, as did those of the early studies (Schmitt et al., 1941), require the assumption of continuous lipid bilayers. It seems improbable that the data could be adapted to models that depict the membrane as composed of protein subunits with lipid molecules inserted in molecular folds and oriented in a relatively unspecified manner, although such a structure may exist in the mitochondrial membrane for which the model was devised (Green and MacLennan, 1969). In some membranes, cooperative allosteric interaction may occur between protein molecules capable of transition between two stable states, as suggested by Changeux et al. (1967; Changeux, 1969).

Optical rotatory dispersion and circular dichroism data have been interpreted to indicate that protein in the membrane matrix structure interacts with lipid not through electrostatic bonds, as held in the conventional trilaminar theory, but rather by hydrophobic interaction of protein chains extending throughout the thickness of the membrane. In this view, the lipid does not form a continuous bilayer (see Schmitt and Samson, 1969). The consensus of the symposium participants was that the theoretical basis for any of these models is not rigorous and that lipid bilayers, at least in portions of the membrane, are not excluded by the evidence.

The preparation of a variety of specially designed electron spin labels offers much hope for our gaining important information about the fluidity of membrane matrix, presumably in the lipid-rich portions. McConnell (this volume) sees the protein and the carbohydrate components as conferring solidity to the matrix, whereas unsaturated lipid chains favor fluidity. Myelin lipids are rich in unsaturated chains, and these may promote chain fluidity (see Luzzati et al., 1969). What protects these unsaturated linkages from auto-oxidation? Can fundamental membrane properties (e.g., maintenance of fluid regions that permit transmembrane penetration of carriers) be controlled by alteration of the state of oxidation of lipid double bonds? Whether blisters seen in electron micrographs of freeze-etched preparations indicate regions of fluidity in membrane matrix is not yet certain; electron microscopists tend to associate such regions with processes of endocytosis or exocytosis.

Experiments on various types of membrane models, as described by Delbrück (this volume) and Eigen (this volume), will doubtless provide important new insights into membrane structure and dynamics, particularly when kinetic aspects are investigated by fast relaxation methods and applied to bilayer and vesicular models (Eigen, 1968; Eigen and De Maeyer, 1970).

MEMBRANE-MOUNTED MOLECULAR MACHINES AS SPECIFIC ION CARRIERS  The mechanism by which cells, especially neurons, distinguish $Na^+$ from $K^+$ has eluded cell physiologists for generations. The conventional notion that ions are transported through the membrane by ion-specific "channels" explains little at the molecular level, although Onsager (personal communication) expressed the opinion that channels formed by specific chemical groupings (e.g., protein side-chains) of membrane molecules may suffice for ion-specific transport. The concept of polypeptide cages, highly specific with respect to the ions that they enfold and carry across the cell membrane, bids fair to solve this age-old problem. It is as if the "channel" (peptide cage) itself moves across the membrane with its specific ion cargo.

The picture conveyed in the recent work on bilayer models (Tosteson, 1969; Eigen and De Maeyer, 1970) is illustrated in Figure 3. The cation is stripped of its solvate water molecules seriatim, one by one, as the ion enters the carrier cage and bonds coordinately with oxygen atoms of the cage. A conformational shift causes the cage to close. The side chains of the peptide cage are directed outward and, being hydrophobic, minimize interaction with lipid and facilitate transport of the peptide-enclosed ion through the membrane. In lipid bilayer model systems, the addition of certain peptides and antibiotics changes the electrical resistance of the bilayer by orders of magnitude and does so specifically with respect to particular ions, e.g., $Na^+$, $K^+$, $Ca^{2+}$, and $Mg^{2+}$.

Whether such small-molecule carriers actually exist in brain-cell membranes is unknown, and it will be difficult to isolate and chemically characterize carriers because they are probably present in the membrane in very low concentrations—perhaps one part per million. Eigen (this volume) suggests that the carrier function may be borne in cells by a small molecule linked with an allosteric pump protein, by conformational changes of a specific membrane protein, or by other membrane constituents. The energy source and the coupling for this transport are unknown. The possibility was posed that, by recycling, the carrier might make round trips across the membrane with sufficient rapidity to account for the data of Hodgkin and Huxley. Eigen suggests that, at least in bilayer model systems, recycling might occur in about one millisecond.

The question was raised at the symposium whether transmitter molecules (e.g., quaternary ammonium compounds) might cross the synaptic membrane by a rapid-transit carrier mechanism. It was not immediately apparent how such an unorthodox view might be tested, nor did participants press for an alternative to the canonical quantal-exocytosis hypothesis of transmitter release, discussed in more detail below.

The significance of the $Na^+$ pump in determining postsynaptic membrane potentials was emphasized, especially as related to the accumulation of ions in thin dendritic branches in which even moderate activity might substantially change the $Na^+$ concentration of the neuroplasm (Carpenter and Alving, 1968; Marmor and Gorman, 1970; Pinsker and Kandel, 1969). It is believed that the frequency of firing in thin C-fibers is limited by the capacity of the $Na^+$ pump.

EXCITABLE MEMBRANES  The excitable mechanism in neuronal membranes involves molecular devices that are not only sparsely distributed in the membrane (about 10 to 20 per $\mu^2$) and difficult to detect by physical means, but also are fast operating ($10^{-4}$ seconds and maybe much faster). On the assumption that the process involves change in orientation or conformation of macromolecules or their parts, Schmitt and Schmitt (1940) attempted to discover such changes by use of a highly sensitive method for detecting transient changes in birefringence associated with the action wave. These experiments yielded negative results. Recently, in the laboratories of Keynes (Cohen et al., 1968) and of Tasaki (Tasaki et al., 1968, 1969), positive correlations were found between bioelectrical parameters and changes in birefringence, light scattering, and emission by fluorescent probes; in these experiments the signal-to-noise ratio was

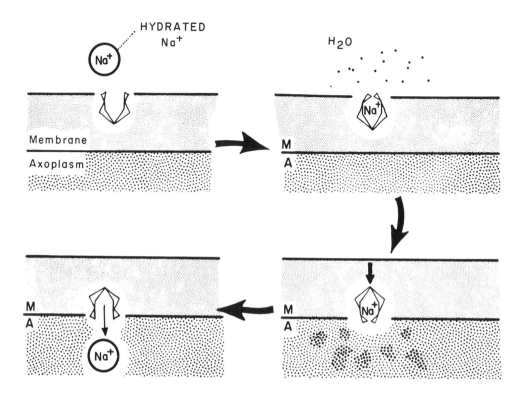

FIGURE 3 Diagrammatic representation of peptide ion carrier to indicate role of change in solvation of ion and in conformation of peptide carrier.

greatly augmented by the use of time-averaging computers. The correlations are well described and evaluated by Keynes (this volume). The small (0.01 to 0.1 per cent) changes of birefringence, light scattering, and fluorescence emission depend on the membrane potential rather than on $Na^+$ conductance, and hence tell little about the molecular mechanism of the action wave. Also, current-dependent changes in light scattering and in fluorescence emission may be referable not to membrane changes, but to changes in external compartments of the neuronal microenvironment. This discouraging conclusion of much brilliant experimental work over many years is tempered by the hope that, as knowledge is accumulated about ion carriers, optical methods may yet be devised for ion-carrier detection and characterization. For example, Frank et al. (1969) found birefringence changes in bilayer systems to which lipid-soluble carrier molecules had been added.

The fast-spike wave is important in conveying information rapidly from one neuron to the next in neural circuits. However, slow-wave processes, particularly in dendrites, may play a significant role in neural transactions and information processing in the brain. Here is an area of great promise for biophysical and biochemical investigation. The problem of the postsynaptic potential, usually explained in terms of $Na^+$, $K^+$, and $Cl^-$ conductances, i.e., the ionic

effectors rather than the molecular-reacting system, should be re-examined in the light of recent studies of specific ion carriers and of membrane ultrastructure.

### Neuronal junctions

Recent electron-microscope investigations suggest that certain synaptic structures, still poorly characterized, may be functionally important not only in guiding synapse formation in the developing nervous system, but also in controlling the moment-to-moment synaptic activity of the adult nervous system. Also worth noting is the powerful role of scanning techniques, clearly indicated in the work of Lewis et al. (1969), which show details of synaptic connectivity.

The end bulb, cleft, and postsynaptic tissue possess a rich ultrastructure. More is involved than vesicles and membranes whose variations of polarization underlie impulse transmission in neuronal circuits. The end bulb is full of equipment, and there is what may be called a "synaptic apparatus" that may provide a degree of continuity between presynaptic and postsynaptic junctions across the cleft. The term "synaptic apparatus" is here used to denote a proteinaceous structure, including presynaptic dense projections, cleft structure, and subsynaptic web.

THE SYNAPTIC APPARATUS   From the plenary lectures of Robertson (this volume) and of Bloom (this volume), as well as from descriptions by Akert (Pfenninger et al., 1969; Akert and Pfenninger, 1970; Bloom et al., 1970), we learn that a synaptic apparatus exists which, to judge from its staining properties, may extend from far inside the presynaptic terminal to deep within the postsynaptic neuroplasm and may occur in brain synapses in a wide spectrum of animal forms. The nature of the cleft-material, which is more difficult to preserve, remains conjectural.

Little is known about the chemical nature or function of the synaptic apparatus. In all probability, it is proteinaceous and contains basic groups that stain with ethanolic phosphotungstic acid and bismuth iodide ($BiI_4$). Zinc iodide (ZnI) apparently complexes with osmium tetroxide ($OsO_4$) inside vesicles.

Perhaps, as in the case of the mitotic apparatus, it will be possible, through the development of special methods, to isolate the synaptic apparatus, to characterize the protein, and to deduce the function of the apparatus. The amount and the organization of the synaptic apparatus increase with development until, in maturity, a complex structure is formed that consists of dense projections extending from the presynaptic membrane well inside the terminal; a hexagonal, weblike plaque facing the presynaptic membrane; and a fibrous web penetrating into the postysnaptic cytoplasm. Some of the properties suggest the presence of elongate (protein?) macromolecules or filaments parallel to the axis of the synaptic apparatus and normal to synaptic membranes.

The trilaminar membranes limiting the presynaptic and the postsynaptic structures are readily distinguishable from the synaptic apparatus, but it is possible that, at particular regions, the membranes are pierced by the fibrous synaptic apparatus. Preliminary experiments described by Akert (see Bloom et al., 1970) suggest that $Ca^{2+}$ may be important in binding presynaptic and postsynaptic membranes and that this binding is reduced by the high ionic strength of the extraction medium.

The intercellular cleft, considered by physiologists primarily as a medium for the diffusion of transmitter, is shown by special techniques to contain a substantial complement of macromolecular material, including glycoprotein, glycolipid, and perhaps mucopolysaccharide. According to Bogoch (1965, 1968, 1969) and Barondes (1969a, 1969b, this volume), this material commonly found in intercellular space and in the microenvironment of neurons may provide specificity and molecular recognition important in the ontogenetically correct linking of neurons and perhaps in the consolidation of experiential information. Examples of molecular surface recognition in various cell types are well known. An impressive example is the specification of pathways for the homing of lymphocytes in the vascular-lymphatic system; alteration of the terminal groups of the carbohydrate chains by treatment with glycosyl or fucosyl transferases changes the pattern of the lymphocyte homing (Gesner and Ginsburg, 1964).

Barondes (1968) found that a portion of the protein synthesized in the soma and transported down the axon is conjugated with carbohydrate moities in the synaptic terminal; the glycoprotein thus formed presumably passes through the synaptic membrane and becomes part of the surface of the presynaptic membrane or is contained in the intercellular cleft material. In the view of Bogoch (1968; personal communication), alteration of carbohydrate (glycoprotein) end groups is a fast topochemical enzymatic reaction that plays a central role in molecular specification of synaptic function, hence also in phenomena such as information processing and memory consolidation.

The synaptic cleft may be the locus of a kind of valving of synaptic function through intrusion of glial membranes and processes into the space (see Bunge, this volume).

Synaptic receptors for biodynamically active substances such as transmitters, hormones, and perhaps certain trophic substances represent a fascinating and important subject for molecular neurobiological investigation. Receptors are usually pictured as being macromolecular, probably protein in nature, and situated on the cleft surface of the postsynaptic membrane. Two strategies of research have been considered to characterize receptors:

1. Affinity labeling, in which the molecular topography of the receptor is deduced from molecular parameters of substances able, in varying degrees, to bind to the receptor and to elicit the appropriate physiological response. This type of investigation is being employed profitably by Karlin (1969).

2. Direct isolation of receptor molecules, in the manner effectively demonstrated by Gilbert (this volume) for genetic repressors. Such a possibility, still little explored, is mentioned by Shooter (this volume).

The membrane-bound enzyme adenyl cyclase is thought to be in close interaction with receptors and to be allosterically activated by them (Figure 4). This enzyme converts ATP to cyclic adenosine monophosphate (AMP), a nucleotide containing energy (of hydrolysis) greater than that of ATP (Greengard et al., 1969). Cyclic AMP appears to be a common effector in cells, a second messenger, bringing about results usually attributed to hormones or other highly biodynamic substances, including synaptic and neuromuscular transmitters. The investigation of the biological and biochemical role of cyclic AMP and adenyl cyclase has expanded rapidly in the last few years. Its role in brain-cell function was the subject of a recent Work Session of the Neurosciences Research Program (see Rall and Gilman, 1970). Bloom and his colleagues (Siggins et al., 1969; Bloom, this volume) have discovered important facts about

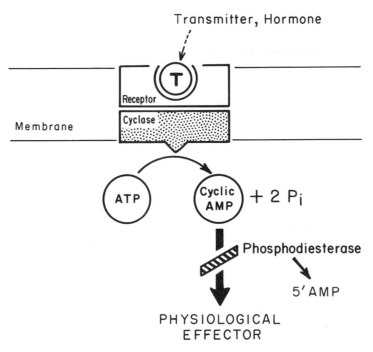

FIGURE 4 Diagrammatic representation of the role of membrane-bound adenyl cyclase in the postulated linkage between transmitter and intracellular effector (3′, 5′, cyclic AMP).

the regulatory role of cyclic AMP in cerebellar function and have also demonstrated an antagonistic effect of prostaglandin on norepinephrine action on Purkinje cells.

TRANSFER OF BIODYNAMIC MOLECULES AT THE SYNAPSE; TRANSMITTER DYNAMICS   The synapse—once it is formed developmentally—carries out a number of information-processing functions (see Bloom et al., 1970), one of which is transferring excitation across junctions through release by the action wave of a chemical transmitter. The transmitter diffuses rapidly across a 200-Å strait of "saline" and activates receptors, which, in turn, mediate Na⁺ or Cl⁻ permeation and cause excitation (depolarization) or inhibition (hyperpolarization). Another part of the dogma is that vesicles, acting as storage repositories and capable of moving, by yet unspecified means, to the synaptic membrane, discharge their quantal contents of transmitter into the cleft by a process of exocytosis. Certain aspects of this simplistic, canonical view of synaptic function were questioned during the symposium in the light of new facts discovered in recent pharmacological and biochemical investigations; new and interesting alternatives were suggested.

Only a small fraction of brain synapses has been clearly specified with respect to the transmitter involved. It is possible that, in addition to at least five transmitters known to satisfy rigorous criteria, many other putative transmitters may exist. In the usual view, specificity of the transmission is determined primarily by the receptor apparatus rather than by the transmitter. In some synapses (e.g., *Aplysia*) both excitation and inhibition may be mediated (Wachtel and Kandel, 1967).

The quantal hypothesis that Katz (Castillo and Katz, 1956; Katz, 1965) proposed for the nerve-muscle junction has not been definitively proved to apply to cortical synapses; the role of bioelectrical factors in effecting the release, presumably by exocytosis, of transmitter from presynaptic vesicles, is far from clear. Spike waves are not required for release, nor is Na⁺ ion conductance (Moore and Narahashi, 1967); reduction of membrane potential by as little as 6 millivolts, in the case of amacrine cells, may liberate transmitters (Rall, this volume). Other dissident facts have often been quoted: cases in which vesicles are on the "wrong" side of a junction physiologically, are on both sides of a junction, or are in electrotonic junctions at which function presumably requires no transmitter.

Interest centered particularly on release mechanisms. It seems clear that, even according to the exocytosis hypothesis, the vesicle, in emptying its contents into the synaptic cleft, does not fuse the lipid and the protein of its wall with those of the presynaptic membrane. Whittaker (this volume) found that the chemical compositions of the two membranes differ; they are immunochemically distinguishable; moreover, extensive stimulation does not exhaust, or even demonstrably reduce, the number of vesicles, as attested by electron microscopy (see Bloom, this volume; Kety, this volume). No evidence was found at the electron-microscope level for the view that vesicles move to, and touch, the presynaptic membrane to discharge transmitter (see Gray, 1966). The observation of "blisters" in freeze-etched preparations suggested that the transmitter may be liberated laterally from the end bulb rather than at the region of synaptic adhesion (see also Peters and Kaiserman-Abramof, 1969).

Valuable clues to an understanding of the transmitter-release mechanism, as well as other synaptic functions, come from recent investigations of the release of neurohormones from the neurosecretory cells, of epinephrine and norepinephrine from the chromaffin cells of the adrenal medulla, and of norepinephrine from adrenergic neurons, as in the splenic nerve. The material released resembles in kind and proportion the contents of the vesicle, i.e., transmitter (norepinephrine), protein (chromogranin proteins, dopamine $\beta$-hydroxylase), and ATP. These investigations support the exocytosis theory and the notion that, upon excitation, the vesicle at the membrane surface briefly punctures the presynaptic membrane, discharges its soluble content, recedes from the membrane, and re-enacts the cycle of filling with, or synthesizing, transmitter in preparation for another release (see Iversen, this volume). However, Whittaker, in fractionation studies, found transmitter (acetylcholine) not only in the vesicles, but also in solution in the terminal cytoplasm; the evidence indicated that the most recently synthesized transmitter is in the soluble cytoplasmic pool and that it is from this pool that the transmitter is released (see also Robertson, this volume).

As an alternative to the vesicle-exocytosis hypothesis, Kety (this volume) proposes that, although vesicles act as storage receptacles, the presynaptic membrane may also store and release transmitters. K. Rosenheck (personal communication) suggested that not only may the vesicles function as stores for transmitter, but also the soluble proteins may have a specific function in facilitating transmitter movement during stimulation.

Synaptic physiology offers much challenge to the molecular neurobiologist. Not the least of these challenges concerns the transmission from neurons to innervated tissue of substances not involved in transmission of excitation, but required for the specification, differentiation, or maintenance of the innervated cells. These "trophic"—the word is probably as inadequate as the word "transmitter"—substances, as well as the processes by which they are released or delivered from presynaptic endings, remain almost completely unknown (see Guth, 1969).

Discoveries of neuroplasmic flow and fast somatofugal transport of synthesized products to endings relate closely to the problem of the synapse and its possible role in processes other than those of impulse transmission. The fate and the function of nontransmitter, nonhormonal materials, including mitochondria, which are not consumed en route to the endings but are nevertheless discharged into extraneuronal space at the synapse, are regrettably still unknown, A substantial amount of organic material must be catabolized or otherwise removed. A fraction of this material is protein. Useful clues are furnished by a study of neurosecretion. The peptide hormones pass down the axon in granules that also contain specific proteins, the neurophysins, and ATP. The chromogranins and several other proteins (enzymes) in small quantities accompany catecholamine in dense core vesicles both in the adrenal medulla and in the sympathetic neurons (e.g., splenic and vas deferens), which are the only ones thus far studied. Why the neuronal soma produces protein to accompany catecholamine down the axon remains a mystery, as does the function of the protein liberated at endings in amounts greater than that of transmitter. Why should ATP (present in stoichiometric, 1:4, proportions to catecholamine in the chromaffin granule) presumably, after making the intravesicular journey down the adrenergic axon, be extruded along with transmitter into the synaptic extraneuronal space? Do still-undiscovered, energy-requiring reactions take place there? Are they related to the metabolic maintenance of brain tissue or to still-uncharacterized neurophysiological processes?

Could we be entering a period in which the theoretical basis of neurophysiology may be as thoroughly revised as it was when neurohumoral, rather than bioelectrical, transmission of the impulse was demonstrated to occur at most synapses (see Schmitt and Samson, 1969)? Will macromolecules, proteins (e.g., chromogranins), conjugated proteins, polysaccharides, and enzymes play an important role? Could they, through interaction with transmitters or with synaptic structures, give rise to chemical modulation of function, especially to physiological or psychological plasticity? These possibilities were discussed in some detail at a recent NRP Work Session on Macromolecules in Synaptic Function (Bloom et al., 1970).

## Biosynthesis and gene expression in brain cells

Little is known about mechanisms of gene expression specifically relevant to the functioning of neurons and glia (see Ebert, 1967). Processes may take place in eukaryotes, and especially in their nervous tissue, that are markedly different from the prototype that has emerged from a decade of brilliant investigations of prokaryotes. Many new discoveries along this line were brought out in the symposium organized and chaired by Edelman (this volume). Especially interesting to students of plasticity are transcriptional and translational variants that may be especially sensitive to ionic strength, particular ions, and certain biodynamic molecules.

Recent experiments indicate that physiological (i.e., electrical or transmitter) stimulation of individual, charted, *Aplysia* ganglion cells increases RNA production that is related to the postsynaptic potential, but not to spikes produced by antidromic invasion (Peterson, 1970; Peterson and Kernell, 1970; Kernell and Peterson, 1970; but see Berry, 1969).

Synthesis of specific proteins such as tyrosine hydroxylase is decreased by the administration of inhibitors of protein synthesis, according to Weiner (personal communication) and Axelrod (see Mueller et al., 1969; Thoenen et al., 1969). Although nothing is yet known about the linkage between membrane and biosynthetic processes, this line of investigation seems very promising (see Bloom et al., 1970). An interesting suggestion made by Weiner (1970) is that inhibitors of protein synthesis that block permanent memory storage may act by blocking the synthesis of enzymes necessary for the synthesis of transmitters, without which synaptic transmission in the relevant circuits would be abolished along with learning and consolidation.

Approaches to the problem of gene expression from the comparative and the evolutionary views may also prove profitable (M. J. Cohen, this volume). For example, electrical depolarization of *Chironomus* gland cells leads to chromosomal puffing, i.e., to greatly accelerated production of RNA; the linkage between bioelectrical and synthetic processes is unknown.

Textbook portrayals of the neuron—usually the vertebrate motor-effector neuron—are probably far from typical; patterns of dendrites, axons, and terminal branchings vary greatly in the central nervous system and in various animal forms. Cohen points out that the pattern of many invertebrate neurons differs importantly from that characteristic of vertebrate neurons in the relative isolation of the soma, which contains the biosynthetic apparatus, from the bioelectrical signal-conducting dendrite and from axonal mechanisms. Correlated with this isolation is the stereotyped behavior of many invertebrates.

Physiological and psychological plasticity apparently require active participation of biosynthetic and metabolic regulatory processes in neurophysiological function. The role of a few "command" neurons (Wiersma and Ikeda, 1964; Kennedy et al., 1969; Wilson, this volume) in invertebrates is substituted in vertebrates by a multiplicity of neurons, highly differentiated in structure and connectivity. Important in the present context, however, is the close linkage between gene expression and bioelectrical parameters. In the study of this linkage in fortunately chosen invertebrate forms (e.g., *Aplysia* and certain insects) lies promise for rewarding molecular neurobiological research. Such work may establish a base line for the investigation of more complex vertebrate neurons capable of plastic alteration subserving learning and memory.

In the broad spectrum of neurochemical research problems, the search for brain-specific macromolecules, particularly proteins, produced by gene expression in brain cells offers much promise. The catalogue of these proteins presently includes the S100, 14-3-2, antigen α, and certain myelin proteins. As more are added to the list, understanding of their localization and specific function will provide important growth points in neurochemistry. Nevertheless, the brain-cell problem is more subtle than that of other tissues (e.g., muscle, which depends for function on relatively high concentrations of effector proteins, actin and myosin). Much more may be involved in brain functioning than is implied in impulse propagation and transmission; processes different from those commonly met in other tissues may be responsible for the storage, retrieval, and psychological utilization of experiential information of the brain.

Equally interesting and profitable, as Shooter (this volume) has ably shown, is the study of the response of brain cells to stimulation or inhibition of gene expression by highly potent biodynamic agents such as hormones (of which insulin provides an excellent investigative model), transmitters (and enzymes necessary for their synthesis), still ill-defined trophic substances, and, most spectacularly, nerve growth factor, the study of which has been brilliantly developed by Levi-Montalcini (see Levi-Montalcini and Angeletti, 1968). From the work of Shooter (this volume) and his colleagues (see A. P. Smith et al., 1969) has come a better understanding of the biochemical characterization of this substance, its constituent subunits, and their role in neurobiological phenomena. Meanwhile, it remains to be definitively demonstrated that nerve-growth-factor action is not an epiphenomenon.

ORGANELLES  The study of the specifically neuronal role of mitochondria and other organelles is in an early phase. In brain cells, unlike other tissues, these organelles may travel relatively large distances from cell center to cell periphery at the endings of axons and dendrites. The most obvious purpose of this transport process is to provide ATP energy along the way, to bring metabolites, proteins, specific enzymes (such as transmitter-synthesizing enzymes carried in vesicles), and trophic and other regulatory substances to axonal endings. What becomes of the organelle residues at endings? Are endings graveyards in which organelles, the usefulness of which has ended, are dismembered and disposed of, or does organelle-borne cargo (e.g., enzymes, proteins, ATP) participate in dynamic processes at endings?

Lehninger (this volume) points out that mitochondria provide a valuable model for neuronal membranes with respect to ultrastructure (role of enzyme assemblies, structure proteins, and other machinery), to ion transfer (effects of valinomycin and other carriers), and even to bioelectrical properties. Mitochondria have been impaled by electrodes, and membrane potentials have been measured as a function of the composition of the microenvironment (Tupper and Tedeschi, 1969a, 1969b, 1969c).

## Neuroplasmic kinesis

One of the most active areas of molecular neurobiology today and one that brings into full play the wide range of metabolic, biogenetic, and bioelectrical dynamics of neurons is that of neuroplasmic flow and fast transport of neuroplasmic constituents. Although we are still largely ignorant about the nature, fate, and function of the transported material and the mechanism of transport, investigation of this newly opened territory is spurred by the hope of finding experimental answers not only to problems of neuronal dynamics already posed, but to the possibility that the problem of memory and learning may also be illuminated. The major issues are discussed by Weiss (this volume, Chapter 6), who first opened up the field experimentally, and Davison (this volume) provides a critical evaluation of theories of mechanism of flow and transport. It remains but to mention certain points raised at the symposium and to indicate my own views regarding some of the issues.

Cytoplasmic flow and fast specific transport are apparently ubiquitous in neurons and probably in many other types of cells, but the phenomenology varies over a wide range of cell types, depending on their specialized function. Similarly, the ultrastructural factors, e.g., ratio of microtubules to neurofilaments and the presence of vesicles and mitochondria, may vary widely. Therefore, Weiss's caution against premature attempts at explanations and generalizations is warranted.

It seems probable, as Weiss long ago proposed, that slow (about 1 mm per day) flow is a mass movement of viscid, semisolid neuroplasm generated by mechanisms that remain turned on genetically perhaps throughout the life of the neuron. A possible explanation of this flow is Weiss's suggestion of peristaltic contraction of membranes—neuronal, glial, or both in combination. No specific biophysical or biochemical mechanism for the peristalsis has, however, yet been provided. Membranous systems (e.g., mitochondria and certain kinds of vacuoles) may contract. The axon-Schwann cell interface is in some cases (as in some crustacean and molluscan axons) highly, although irregularly, serrated, mitochondria being concentrated at the interface and oriented with respect to the inpocketings of the serrations, as though energy were required for local processes (see Geren and Schmitt, 1954, 1955). This may be a reflection of peristalsis-like contraction.

Peristaltic contractions that move constituents by hydrostatic pressure, whether at the cellular or organelle level,

cannot explain fast transport or other properties of cytoplasmic flow that move particulates, such as dense-core vesicles, specifically with respect to other neuroplasmic components, such as microtubules and neurofilaments. Present evidence favors the view that mechanisms underlying the slow bulk flow differ from those producing fast transport of particular materials, such as transmitter-containing, dense-core vesicles in adrenergic neurons. Velocities of fast transport range from about 20 to 2000 mm per day; different axon types manifest different velocities, as, for example, various kinds of striated and smooth muscles show different contraction velocities and frequencies.

Materials most rapidly transported seem to be packaged in particulate, vesicular form (neurosecretory granules, dense-core vesicles, lysosomes). As in fast transport in other cell types, including movement of chromosomes, pigment granules, and various kinds of vesicles, so in neurons the transport occurs in close association with microtubules. The evidence is admittedly circumstantial, but it is parsimonious to assume that microtubules are essential in the transport process and perhaps act as directional guides and possibly as energizing substrates for the transport of vesicles and membrane-bound particulates.

In a speculative model, along the lines of modern muscle theory, I have postulated (Schmitt, 1968; Schmitt and Samson, 1968) that chemomechanical energy coupling takes place between subunits of the microtubules and those of the vesicle surface membrane: a sliding or saltating vesicle theory is suggested, analogous to the sliding filament theory of muscle contraction (Huxley, 1969). Some evidence has been adduced to support this hypothesis. Recently D. S. Smith et al. (1970) have obtained electron micrographs of unmyelinated axons and synapses in the central nervous system of lamprey (*Petromyzon marinus*) ammocoetes. They reveal almost diagrammatic clustering of vesicles around microtubules. It remains to be seen whether such instances represent isolated cases with no general significance or whether clustering is widespread, but, as do microtubules themselves, requires special preparative techniques for preservation and visualization in the electron microscope. The momentary contact between vesicles and microtubules during the postulated saltation might be highly metastable and difficult to preserve and demonstrate in that microscope.

The protein of brain microtubules, present in substantial amounts in developing neurons (Dutton and Barondes, 1969), has been obtained in purity (Marantz and Shelanski, 1970), and vesicles have been isolated from nerve tissue by various workers. Therefore, simple systems containing microtubules and vesicles can now be investigated and a model for saltatory transport can be developed. For example, supercontraction of purified actomyosin was used by Szent-Györgyi as a model for muscle contraction, and the addition of ATP to glycerinated preparations (muscle, dividing cells, ciliated cells, and so forth) permitted investigation of details of contractile mechanisms.

Because of the relation between microtubules and neurofibrils (see Schmitt and Samson, 1968) and because microtubules may reveal transport routes in the perikarya, dendrites, and axons of neurons, the descriptions of neurofibrils in the classical neurohistological and neurophysiological literature may suggest functional significance.

The directional transport of materials in cells by vesicles or particulates may, in evolution, possibly antedate muscle as a means of chemomechanical energy coupling to transport protoplasmic components. Transport was readily demonstrable and measurable in neurons because their elongated axons greatly facilitate testing with radioactive and fluorescent tracers. Clearly the investigation of transport in neuroplasm ranks high in the list of priorities for molecular neurobiologists and may have general biological significance.

## Summary

The symposium demonstrated that, among the neurosciences, molecular neurobiology is in a fruitful period, not only in re-examining time-honored tenets in the light of modern molecular biological concepts and techniques, but also in perceiving opportunities for new neurophysiological and electron-microscopic discoveries. Significant advances have been made since the 1966 NRP Intensive Study Program.

Axoplasmic movement and fast transport have been examined in many neuron types, the parameters governing the phenomena have been better defined, and progress has been made in understanding physicochemical mechanisms. The study of trophic effects on innervated tissue has benefited from this upsurge of activity, although little is yet known about the identity of trophic substances—other than acetylcholine—and the mechanism of transfer from presynaptic to postsynaptic structure.

Synapse study is in active ferment: new discoveries have been made of ultrastructure and of the neurochemistry of synaptosomes and other particulates isolated after fragmentation; new data have been obtained on the mechanism of release of transmitter and the possible information processing of macromolecules in synaptic function. The interposition of adenyl cyclase into the receptor-effector train via cylic AMP suggests that a substantial revision of views tacitly accepted by neurophysiologists during the last two decades may be imminent.

Membrane neurobiology has been greatly stimulated by widespread studies of membrane structure and properties in general, of lipid bilayers with and without activating additives, and by new concepts of specific transport of ions through membranes.

Little has been learned about the molecular or ultrastructural mechanisms involved in distributed, parallel-processing, holistic activity of the brain, which is thought by some investigators to occur with the neurophysiologically more conventional connectionistic processing of experiential information in learning, memory, and other brain processes (see Nelson, 1967; this volume). Attention has been called to the outer surface of brain-cell membranes and to macromolecular (polysaccharide, glycoprotein, and glycolipid) constituents of intercellular space—a continuous, interconnecting compartment throughout the brain—as a possible substratum for such phenomena. However, other than changes in impedance and slow potentials (Adey, 1969), little has yet been adduced to explain these central phenomena of brain function. This situation is understandable; the reductionist, physicochemical approach requires that the higher-level phenomena be reasonably well characterized before mechanisms can be deduced effectively.

Lessons learned from the evolution of genetic theory can be profitably applied to problems of brain function (Schmitt, 1965) whether or not molecular coding proves a primary process in the latter as in the former. Developmental and genetic biologists of the 1920s and 1930s—the elite of those days—were sure that key clues to the bewilderingly complex phenomenology of genetics and development must be sought at the cellular or higher-systems level. To be sure, the ground rules of this important segment of life science had to be established at the cellular and organismic level, but the conceptual breakthrough came from a combination of finding the simplest systems (microorganisms and viruses) that displayed the phenomena and of discovering the macromolecules (DNA and RNA), their structure, and chemodynamic properties that subserve the genetic process.

Is any particular macromolecular type likely to prove as important a key to unlock the mysteries of brain function as it was in the case of genetics? Interestingly, molecular biologists—the elite of our day—also seem certain that the key to brain function is to be sought at the cellular or higher-systems levels. The ground rules in the brain sciences, comparable with those of genetics of the mid-twentieth century, have not yet been deduced. Much, however, has been learned in recent years about the properties of macromolecules and their complexes and of fast reactions that may govern macromolecular interaction; these may well prove relevant to major brain problems.

Many workers from other fields the world over are joining the ranks of investigators of neuroscience. The intellectual atmosphere today contains the distinct intimation of great discoveries of utmost importance to man—the functioning of his own brain and the nature of his being. To realize the potential of neuroscience requires not only concentration on particular levels—molecular, cellular, neural, or behavioral—but, importantly, on a mutual sharing of the precepts and conceptual insights of scientists of demonstrated originality at each level, under conditions that foster intellectual and conceptual "allosterism." This is the *raison d'être* of the Neurosciences Research Program.

## Acknowledgments

The author gratefully acknowledges the invaluable aid of Drs. F. E. Samson, P. G. Nelson, B. H. Smith, J. H. Bigelow, H. H. Wang, and L. N. Irwin in interpreting, collating, and integrating information communicated during the symposium on Molecular Neurobiology.

This work was supported by grants from the United States Public Health Service, National Institutes of Health (GM10211), National Aeronautics and Space Administration (Nsg 462), the Rogosin Foundation, and the Neurosciences Research Foundation.

## REFERENCES

ADEY, W. R., 1969. Slow electrical phenomena in the central nervous system. *Neurosci. Res. Program Bull.* 7 (no. 2): 75–180.

AKERT, K., and K. PFENNINGER, 1969. Synaptic fine structure and neural dynamics. *Symp. Int. Soc. Cell Biol.* 8: 245–260.

BARONDES, S. H., 1968. Incorporation of radioactive glucosamine into macromolecules at nerve endings. *J. Neurochem.* 15: 699–706.

BARONDES, S. H., 1969a. *In* Brain Cell Microenvironment (F. O. Schmitt and F. E. Samson, editors). *Neurosci. Res. Program Bull.* 7 (no. 4): 349–351.

BARONDES, S. H., 1969b. Two sites of synthesis of macromolecules in neurons. *Symp. Int. Soc. Cell Biol.* 8: 351–364.

BERRY, R. W., 1969. Ribonucleic acid metabolism in a single neuron: Correlation with electrical activity. *Science (Washington)* 166: 1021–1023.

BLOOM, F. E., L. L. IVERSEN, and F. O. SCHMITT, 1970. Macromolecules in synaptic function. *Neurosci. Res. Program Bull.* (in press).

BOGOCH, S., 1965. Brain proteins in learning: Findings in the experiments in progress discussed at the NRP Work Session, June 27–29, 1965. *In* Brain and Nerve Proteins: Functional Correlates (F. O. Schmitt and P. F. Davison, editors). *Neurosci. Res. Program Bull.* 3 (no. 6): 38–41. Also *in* Neurosciences Research Symposium Summaries, Vol. 2 (F. O. Schmitt, T. Melnechuk, G. C. Quarton, and G. Adelman, editors). The M.I.T. Press, Cambridge, Massachusetts, 1967, pp. 374–376.

BOGOCH, S., 1968. The Biochemistry of Memory. Oxford University Press, New York.

BOGOCH, S., 1969. Brain mucoids (discussion). *In* Brain Cell Microenvironment (F. O. Schmitt and F. E. Samson, editors). *Neurosci. Res. Program Bull.* 7 (no. 4): 351–354.

CARPENTER, D. O., and B. O. ALVING, 1968. A contribution of an electrogenic $Na^+$ pump to membrane potential in *Aplysia* neurons. *J. Gen. Physiol.* 52: 1–21.

CASTILLO, J. DEL, and B. KATZ, 1956. Biophysical aspects of neuro-muscular transmission. *Progr. Biophys. Biophys. Chem.* 6: 121–170.

CHANGEUX, J. P., 1969. Remarks on the symmetry and cooperative properties of biological membranes. *In* Nobel Symposium 11, Symmetry and Function of Biological Systems at the Macromolecular Level (A. Engström and B. Strandberg, editors). Almqvist and Wiksell, Stockholm, and John Wiley and Sons, New York, pp. 235–256.

CHANGEUX, J. P., J. THIÉRY, Y. TUNG, and C. KITTEL, 1967. On the cooperativity of biological membranes. *Proc. Nat. Acad. Sci. U. S. A.* 57: 335–341.

COHEN, L. B., R. D. KEYNES, and B. HILLE, 1968. Light scattering and birefringence changes during nerve activity. *Nature (London)* 218: 438–441.

DANIELLI, J. F., and H. DAVSON, 1935. A contribution to the theory of permeability of thin films. *J. Cell .Comp. Physiol.* 5: 495–508.

DUTTON, G. R., and S. BARONDES, 1969. Microtubular protein: Synthesis and metabolism in developing brain. *Science (Washington)* 166: 1637–1638.

EBERT, J. D., 1967. Gene expression. *Neurosi. Res. Program Bull.* 5 (no. 3): 223–306. Also *in* Neurosciences Research Symposium Summaries, Vol. 3 (F. O. Schmitt, T. Melnechuk, G. C. Quarton, and G. Adelman, editors). The M. I. T. Press, Cambridge, Massachusetts ,1969, pp. 109–189.

EIGEN, M., 1968. Die "unmessbar" schnellen Reaktionen. *In* Les Prix Nobel en 1967. Imprimerie Royale P. A. Norstedt and Söner, Stockholm, pp. 151–180.

EIGEN, M., and L. C. M. DE MAEYER, 1970. Carriers and specificity in membranes. *Neurosci. Res. Program Bull.* (in press).

FOLCH-PI, J., and G. SHERMAN, 1969. Preparation and study of brain proteolipid apoprotein *In* Second International Meeting of the International Society of Neurochemistry, Milan, September 1–5, 1969 (R. Paoletti, R. Fumgalli, and C. Galli, editors). Tamburini Editore, Milan, p. 169 (abstract).

FRANK, G. M., G. N. BEPESTOVSKY, and V. Z. LUNEVSKY, 1969. Changes in birefringence of an axon membrane accompanying excitation. *In* Third International Biophysics Congress of the International Union for Pure and Applied Biophysics, Cambridge, Massachusetts, August 29 to September 3, 1969, p. 102 (abstract).

GEREN, B. B., and F. O. SCHMITT, 1954. The structure of the Schwann cell and its relation to the axon in certain invertebrate nerve fibers. *Proc. Nat. Acad. Sci. U. S. A.* 40: 863–870.

GEREN, B. B., and F. O. SCHMITT, 1955. Electron microscope studies of the Schwann cell and its constituents with particular reference to their relation to the axon. *In* Symposium of Fine Structure of Cells, 8th Congress of Cell Biology, Leiden, 1954. John Wiley and Sons, New York, pp. 251–260.

GESNER, B. M., and V. GINSBURG, 1964. Effect of glycosidases on the fate of transfused lymphocytes. *Proc. Nat. Acad. Sci. U. S. A.* 52: 750–755.

GORTER, E., and F. GRENDEL, 1925. On bimolecular layers of lipoids on the chromocytes of the blood. *J. Exp. Med.* 41: 439–443.

GRAY, E. G., 1966. Problems of interpreting the fine structure of vertebrate and invertebrate synapses. *Intl Rev. Gen. Exp. Zool.* 2: 139–170.

GREEN, D. E., and D. H. MACLENNAN, 1969. Structure and function of the mitochondrial cristael membrane. *BioScience* 19: 213–222.

GREENGARD, P., S. A. RUDOLPH, and J. M. STURTEVANT, 1969. Enthalpy of hydrolysis of the 3′ bond of adenosine 3′, 5′-monophosphate and guanosine 3′, 5′-monophosphate. *J. Biol. Chem.* 244: 4798–4800.

GUTH, L., 1969. "Trophic" effects of vertebrate neurons. *Neurosci. Res. Program Bull.* 7 (no. 1): 1–73.

HUXLEY, H. E., 1969. The mechanism of muscular contraction. *Science* (*Washington*) 164: 1356–1366.

KARLIN, A., 1969. Chemical modification of the active site of the acetylcholine receptor. *J. Gen. Physiol.* 54: 245s–264s.

KATZ, B., 1965. The physiology of motor nerve endings. *In* Proceedings of the 23rd International Union of Physiological Sciences, Vol. 4, Tokyo, 1965, (D. Noble, editor). Excerpta Medica Foundation, Amsterdam, pp. 110–121.

KENNEDY, D., A. I. SELVERSTON, and M. P. REMLER, 1969. Analysis of restricted neural networks. *Science* (*Washington*) 164: 1488–1496.

KERNELL, D., and R. P. PETERSON, 1970. The effect of spike activity versus synaptic activation on the metabolism of ribonucleic acid in a molluscan giant neurone. *J. Neurochem.* (in press).

KIRSCHNER, D. A., and D. L. D. CASPAR, 1969. Myelin structure. *In* Third International Biophysics Congress of the International Union for Pure and Applied Biophysics. Cambridge, Massachusetts, August 29-September 3, 1969, p. 237 (abstract).

LEVI-MONTALCINI, R., and P. U. ANGELETTI, 1968. Nerve growth factor. *Physiol. Rev.* 48: 534–569.

LEWIS, E. R., T. E. EVERHART, and Y. Y. ZEEVI, 1969. Studying neural organization in *Aplysia* with the scanning electron microscope. *Science* (*Washington*) 165: 1140–1143.

LUZZATI, V., T. GULIK-KRZYWICKI, A. TARDIEU, E. RIVAS, and F. REISS-HUSSON, 1969. Lipids and membranes. *In* The Molecular Basis of Membrane Function (D. C. Tosteson, editor). Symposium of the Society of General Physiologists, Durham, North Carolina, August 20–23, 1968. Prentice-Hall, Englewood Cliffs, New Jersey, pp. 79–93.

MARANTZ, R., and M. L. SHELANSKI, 1970. Structure of microtubular crystals induced by vinblastine in vitro. *J. Cell Biol.* 44: 234–238.

MARMOR, M. F., and A. F. GORMAN, 1970. Membrane potential as the sum of ionic and metabolic components. *Science* (*Washington*) 167: 65–67.

MOORE, J. W., and T. NARAHASHI, 1967. Tetrodotoxin's highly selective blockage of an ionic channel. *Fed. Proc.* 26 (no. 6): 1655–1663.

MUELLER, R. A., H. THOENEN, and J. AXELROD, 1969. Increase in tyrosine hydroxylase activity after reserpine administration. *J. Pharmacol. Exp. Ther.* 169: 74–79.

NELSON, P. G., 1967. Brain mechanisms and memory. *In* The Neurosciences: A Study Program (G. C. Quarton, T. Melnechuk, and F. O. Schmitt, editors). The Rockefeller University Press, New York, pp. 772–775.

O'BRIEN, J. S., 1965. Stability of the myelin membrane. *Science* (*Washington*) 147: 1099–1107.

O'BRIEN, J. S., and E. L. SAMPSON, 1965. Fatty acid and fatty aldehyde composition of the major brain lipids in normal human gray matter, white matter, and myelin. *J. Lipid Res.* 6: 545–551.

PETERS, A., and I. R. KAISERMAN-ABRAMOF, 1969. The small pyramidal neuron of the rat cerebral cortex. *Z. Zellforsch. Mikroskop. Anat.* 100: 487–506.

PETERSON, R. P., 1970. RNA in single identified neurons of *Aplysia. J. Neurochem.* 17: 325–338.

PETERSON, R. P., and D. KERNELL, 1970. Effects of nerve stimulation on the metabolism of ribonucleic acid in a molluscan giant neurone. *J. Neurochem.* (in press).

PFENNINGER, K., C. SANDRI, K. AKERT, and C. H. EUGSTER, 1969. Contribution to the problem of structural organization of the presynaptic area. *Brain Res.* 12: 10–18.

PINSKER, H., and E. R. KANDEL, 1969. Synaptic activation of an electrogenic sodium pump. *Science* (*Washington*) 163: 931–935.

RALL, T. W., and A. G. GILLMAN, 1970. The role of cyclic AMP in the nervous system. *Neurosci. Res. Program Bull.* (in press).

ROBERTSON, J. D., 1967. Origin of the unit membrane concept. *Protoplasma* 63: 219–245.

SCHMITT, F. O., 1965. The physical basis of life and learning. *Science* (*Washington*) 149: 931–936.

SCHMITT, F. O., 1968. Fibrous proteins—Neuronal organelles. *Proc. Nat. Acad. Sci. U. S. A.* 60: 1092–1101.

SCHMITT, F. O., and R. S. BEAR, 1939. The ultrastructure of the nerve axon sheath. *Biol. Rev.* (*Cambridge*) 14: 27–50.

SCHMITT, F. O., R. S. BEAR, and K. J. PALMER, 1941. X-ray diffraction studies on the structure of the nerve myelin sheath. *J. Cell. Comp. Physiol.* 18: 31–42.

SCHMITT, F. O., and F. E. SAMSON, JR. 1968. Neuronal fibrous proteins. *Neurosci. Res. Program Bull.* 6 (no. 2): 113–219.

SCHMITT, F. O., and F. E. SAMSON, JR., 1969. Brain cell microenvironment. *Neurosci. Res. Program Bull.* 7 (no. 4): 277–417.

SCHMITT, F. O., and O. H. SCHMITT, 1940. Partial excitation and variable conduction in the squid giant axon. *J. Physiol.* (*London*) 98: 26–46.

SHERMAN, G., and J. FOLCH-PI, 1970. Rotatory dispersion and circular dichroism of brain "proteolipid" protein. *J. Neurochem.* 17: 597–605.

SIGGINS, G. R., B. J. HOFFER, and F. E. BLOOM, 1969. Cyclic adenosine monophosphate: Possible mediator for norepinephrine effects on cerebellar Purkinje cells. *Science* (*Washington*) 165: 1018–1020.

SMITH, A. P., L. A. GREENE, H. R. FISK, S. VARON, and E. M. SHOOTER, 1969. Subunit equilibria of the 7S nerve growth factor protein. *Biochemistry* 8: 4918–4926.

SMITH, D. S., U. JÄRLFORS, and R. BERANEK, 1970. The organization of synaptic axoplasm in the lamprey (*Petromyzon marinus*) central nervous system. *J. Cell Biol.* 46 (no. 2): 199–219.

TASAKI, I., L. CARNAY, and A. WATANABE, 1969. Transient changes in extrinsic fluorescence of nerve produced by electric stimulation. *Proc. Nat. Acad. Sci. U. S. A.* 64: 1362–1368.

TASAKI, I., A. WATANABE, R. SANDLIN, and L. CARNAY, 1968. Changes in fluorescence, turbidity, and birefringence associated with nerve excitation. *Proc. Nat. Acad. Sci. U. S. A.* 61: 883–888.

THOENEN, H. R. A. MUELLER, and J. AXELROD, 1969. Transsynaptic induction of adrenal tyrosine hydroxylase. *J. Pharmacol. Exp. Ther.* 169: 249–254.

TOSTESON, D. C. (editor), 1969. The Molecular Basis of Membrane Function. Symposium of the Society of General Physiologists, Durham, North Carolina, August 20–23, 1968. Prentice-Hall, Englewood Cliffs, New Jersey.

TUPPER, J. T., and H. TEDESCHI, 1969a. Microelectrode studies on the membrane properties of isolated mitochondria. *Proc. Nat. Acad. Sci. U. S. A.* 63: 370–377.

TUPPER, J. T., and H. TEDESCHI, 1969b. Microelectrode studies on the membrane properties of isolated mitochondria. II. Absence of a metabolic dependence. *Proc. Nat. Acad. Sci. U. S. A.* 63: 713–717.

TUPPER, J. T., and H. Tedeschi, 1969c. Mitochondrial membrane potentials measured with microelectrodes: Probable ionic basis. *Science (Washington)* 166: 1539–1540.

WACHTEL, H., and E. R. KANDEL, 1967. A direct synaptic connection mediating both excitation and inhibition. *Science (Washington)* 158: 1206–1208.

WEINER, N., 1970. Regulation of norepinephrine biosynthesis. *Annu. Rev. Pharmacol.* 10: 273–290.

WIERSMA, C. A. G., and K. IKEDA, 1964. Interneurons commanding swimmeret movements in the crayfish, *Procambarus clarki* (Girard). *Comp. Biochem. Physiol.* 12: 509–525.

WORTHINGTON, C. R., and A. E. BLAUROCK, 1969. A structural analysis of nerve myelin. *Biophys. J.* 9: 970–990.

# RECOGNITION AND CONTROL AT THE MOLECULAR LEVEL

# 77 Prefatory Comments on Recognition and Control at the Molecular Level

G. M. EDELMAN

SOME JUSTIFICATION, BUT NO APOLOGY, must be made for the fact that this symposium does not include a single paper explicitly concerned with the neurosciences. One of the major accomplishments of the last ISP was to dispel naive expectations that the nervous system worked by encoding information about the world directly into macromolecules. Nonetheless, a detailed understanding of the workings of the nervous system will obviously depend on a knowledge of the structure, function, and interaction of specific proteins, nucleic acids, and small molecules. Only beginnings have been made in isolating and determining the function of specific neural proteins, and the role of protein synthesis in neuronal behavior is not yet understood. Moreover, much remains to be understood about simpler systems before an attempt can be made to apply our knowledge of molecular biology successfully to the more complicated problems of the nervous system.

This symposium on recognition and control at the molecular level was planned, therefore, with two major purposes in mind. The first was to provide neuroscientists with some views of the relation between the structure and function of proteins and nucleic acids, particularly in terms of problems of gene expression. The second was to emphasize recent explorations into the properties of DNA and special pro-

G. M. EDELMAN   The Rockefeller University, New York, New York

tein systems in eukaryotic organisms, because the nervous system has evolved in eukaryotes.

There is sufficient justification for considering the molecular genetics of eukaryotes as a subject in itself. Recent investigations of a number of systems in eukaryotes suggest that other modes of control may be operative in addition to those described for micro-organisms. There are a number of phenomena in eukaryotes that may provide clues to the evolution and workings of complex systems of differentiation. They include: (1) multiple copies of genes and the occurrence of differential gene amplification; (2) chromosome diminution and elimination; (3) systems of metastable and transposable control elements at the chromosomal level, such as those found in maize; and (4) allelic exclusion and evidence for translocation of DNA in the biosynthesis of immunoglobulins. The immune response represents a selective system that has evolved to contend with unexpected molecular interactions. Although none of the detailed mechanisms of the immune response may have correlates in nervous tissue, the concepts developed to analyze immunity at the molecular level will be useful in molecular neurobiology.

The list of phenomena I have mentioned could be extended but, even so, it might be considered a collection of evolutionary curiosities and byways without general significance. It may be more valuable, however, to consider these phenomena as suggestive of the idea that control over differentiation and specific protein synthesis in eukaryotes requires mechanisms *additional* to those described for prokaryotes.

With these phenomena in mind, we may ask a number of questions about levels of control in gene expression. It is clear from the work of Jacob and Monod that protein repressors function as negative control elements for specific operons. Searching efforts are being made to determine whether detailed systems of translational control exist. The intriguing new possibilities are that replication of portions of the eukaryotic genome, as well as somatic recombination and transposition of DNA, may also serve as mechanisms in the control of gene expression. The stability and reversibility of such processes and their direct roles in cellular differentiation are being examined. The classic idea of constancy of the DNA in somatic cells must be reconsidered in terms

of the precision of the measurements as well as the relatively small proportion of the genome that must be affected to exert a large change in gene expression. It is pertinent to ask whether the existence of multiple copies of genes in the eukaryotic genome is a *fundamental* development in evolution. Not only could it change the over-all amounts of synthesis of tRNA or messenger RNA, but it might provide preadapted pools of genetic information which could respond to unexpected selective pressures during evolution. In this sense, the existence of multiple copies of genes may change the rate of evolution and may have been one of the chief reasons for the evolution of the complex chromosomes of eukaryotes.

Characteristically, molecular biologists will go to any species for an answer, because of their belief that there is a general pattern in the fundamental processes of all organisms. For that reason, this symposium was also designed to review and extend our knowledge of the control of gene expression in prokaryotes. Here we are on more solid ground, although there are still many unsolved questions. Molecular interactions of DNA and repressors, as well as the mechanisms of recombination and replication in phage and bacteria, are currently under active study. Such processes also occur in eukaryotes and undoubtedly they will bear resemblances to their counterparts in prokaryotes.

Ultimately, these processes must be understood in terms of detailed chemical interaction of specific macromolecules. The papers in this symposium were arranged to proceed from the primary and tertiary structure of proteins and protein-small molecule interactions, through the intricacies of protein-nucleotide interactions required in protein synthesis on the ribosome, and, finally, to a consideration of immense structures, such as the chromosomes themselves.

Such a program sounds overambitious when dissected out of the particular fabric made up of the following series of papers. Clearly, many details have been given inadequate coverage, and no attempt to be inclusive could possibly succeed. The main hope, however, is to reveal the molecular biological approach to several current problems. At least some of these problems have a direct bearing on neurobiology, and all of them will eventually bear in one way or another on the analysis of the development and functioning of nervous tissues.

# 78 The Structure and Genetics of Antibodies

G. M. EDELMAN

THE VERY EXISTENCE of brains capable of learning indicates that certain biological systems can deal with enormous amounts of information without proportionately great increases in the amount of genetic material. Systems such as the primate brain can cope with unexpected constellations of events, all the information for which certainly could not have been incorporated in a prior genetic program. At the present stage of our knowledge, we do not know how the nervous system accomplishes such a feat, but examination of simpler systems capable of dealing with information that the organism has not seen either in its somatic lifetime or in the history of its evolution may give us a few clues.

Perhaps the most extensively analyzed example of such a system is the adaptive immune response, which is found only in vertebrate species. (For a brief review, the reader can consult Nossal, 1967; Edelman, 1967; and Jerne, 1967). Consideration of the special characteristics of the antibody response leads to some general conclusions about the nature of biological systems that can contend with unexpected events. In addition, it suggests that unusual genetic mechanisms have evolved to deal with such events. The production of antibodies by plasma cells is an example of a system of differentiation in eukaryotic organisms that requires molecular control mechanisms in addition to those that have been so well studied in bacteriophage and bacteria. We can hardly say that the immune system provides anything more than a provocative example, however, for none of the detailed genetic mechanisms have been worked out.

Nevertheless, there are other reasons for beginning this section on recognition and interaction at the molecular level with a discussion of antibodies. First, antibodies represent the only well-studied case of a molecular recognition system with selective properties, and there are reasons to expect that other examples will be discovered, particularly in the nervous system. Second, the analysis of their primary structures and the thermodynamics of antigen binding indicate that antibodies are as good an example of protein specificity as can be found. Third, there is evidence to indicate that antibodies, which are very large protein molecules, have evolved by mechanisms of gene duplication that may be more general than had been previously suspected from the results of analyses of smaller protein molecules, such as ribonuclease and lysozyme. A consideration of antibody structure thus leads naturally into discussions of the problem of multiple genes in eukaryotic organisms.

The specific aim of the present paper is to review certain features of the immune response in terms of knowledge of the chemistry, genetics, and evolution of antibodies, particularly the prevalent class of mammalian antibodies known as $\gamma$G immunoglobulins. An attempt is also made to relate recent findings on the chemistry of antibodies to other more general problems of protein structure and differentiation.

## The sequence of events in humoral antibody production

Several detailed studies have been made of the kinetics and heterogeneity of humoral antibody production after the administration of antigens. Classical studies of the immune response have shown that introduction of an antigenic molecule into vertebrate organisms elicits the synthesis of antibodies that can bind specifically to that antigen (Landsteiner, 1945). Different systems differ in detail, but the set of studies (Uhr, 1964) schematized in Figure 1 will serve as an example.

Following the injection of antigen and after a variable latent period, antibody activity rises exponentially in the serum for four to five days. The first antibody molecules to appear belong to the $\gamma$M globulins, which have a sedimentation coefficient of 19S and a molecular weight of about $10^6$. Synthesis continues at a diminishing rate for another four to six days, and then the level of $\gamma$M antibodies falls with a half-life of 24 hours.

About one week after the initial injection, there follows an exponential increase of 7S antibodies of the $\gamma$G immunoglobulin class (molecular weight 150,000), which lasts for about four days. Synthesis of $\gamma$G immunoglobulin may continue thereafter at lower rates for periods ranging from two years to the entire life span of the animal.

One month after the first injection, when the primary 7S response has reached a plateau, a second injection of antigen yields a *secondary response*, consisting of an accelerated increase in $\gamma$G immunoglobulin production. Although a secondary response is not seen in $\gamma$M antibody production, a second dose of antigen nine days after the first (when $\gamma$M

G. M. EDELMAN   The Rockefeller University, New York, New York

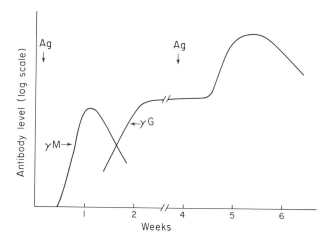

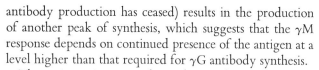

FIGURE 1 Sequence of antibody responses to two injections of antigen (Ag) spaced about three weeks apart. The antibody level is measured in the serum of the animal, and synthesis of two classes of immunoglobulin is stimulated: γM immunoglobulin and γG immunoglobulin. The second antigenic stimulus provokes an accelerated and increased production of specific γG immunoglobulins. This is known as the specific anamnestic response, or immunological memory.

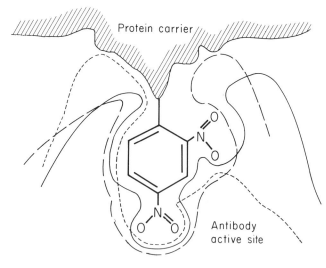

FIGURE 2 Diagram of the complementarity between antibody and hapten antigen (the dinitrophenyl group) coupled to a protein carrier. The various contours of the antibody active site indicate the greater or lesser degrees of fit between sites of *different* antibody molecules and the same antigen.

antibody production has ceased) results in the production of another peak of synthesis, which suggests that the γM response depends on continued presence of the antigen at a level higher than that required for γG antibody synthesis.

The major point to emphasize here is that, regardless of their class, the antibodies are capable of binding to the injected antigen but not to chemically unrelated antigens. Moreover, there appears to be a real synthesis by antibody-forming cells, not just a release from some store. Upon a second encounter with the same antigen, the system shows "immunological memory," i.e., the response is more rapid and larger amounts of antibodies specific for that antigen are synthesized. Thus, antibodies constitute a molecular recognition system that is remarkable for its specificity, range of response, and control.

## Specificity and range in antigen binding: complementarity and degeneracy

The variety of chemically different antigenic structures is enormous: antibodies may be produced by the injection of protein, carbohydrate, nucleic acid, or artificial antigens known as haptens. The capacity of the immune system to respond to so wide a range of stereochemically distinct molecules is as astonishing as the specificity with which each antigen is bound. In the past, many immunologists (see Karush, 1962) were preoccupied with analyzing the thermodynamics and mechanisms of antigen binding by antibody molecules. From their studies, it is clear that the

binding depends on the close complementarity of the shapes of the antigen and the antigen-binding site on the antibody molecule. This complementarity (Figure 2) has been deduced from studies of cross reactions (Landsteiner, 1945) and the nature of the binding: the free energy ranges from approximately −4 to −12 kcal/mole, and hydrophobic interactions, hydrogen bonds, electrostatic interactions, and steric repulsion contribute in varying degrees, depending on the particular antigen-antibody reaction. Because this type of binding does not differ from that between enzymes and substrates, it is tempting to think of antibodies in the same way as one thinks of enzymes.

There are, however, several features of the immune response that indicate that this analogy holds only in the most superficial sense. For example, unlike most substrates for enzymes, antigens that never existed during the evolutionary history of the organism may be synthesized (Landsteiner, 1945). Moreover, antibodies may be elicited by virtually any protein molecule from a variety of animal species. It is unlikely that all these proteins played a direct role in natural selection during the evolution of the immune response in each vertebrate species. Furthermore, unlike the induction of enzymes by substrates, a single antigen usually elicits the synthesis of a large population of different antibody molecules that, with greater or lesser degrees of complementarity, fit the antigen (Fahey, 1962; Haber, 1968). Thus, the antibody response is *degenerate*, i.e., there is a many-one relationship between antibodies and any single antigen, and no single "best" antibody has evolved for each particular three-dimensional antigenic structure. This heterogeneity of antibodies is one of the most important features of the immune response, and the analysis of its structural basis and biosynthetic origins leads directly to several of the most fundamental problems of immunology.

## The idea of selection

Early attempts to explain antibody synthesis resulted in the formulation of a number of so-called instructive theories (Breinl and Haurowitz, 1930; Mudd, 1932; Pauling, 1940). The fundamental notion underlying these theories was that information on the three-dimensional structure of the antigen was necessary at some stage of the synthesis and folding of the polypeptide chains of the antibody molecule. In their simplest forms (Pauling, 1940), instructive theories suggested that the antigen served as a template around which was folded that portion of the antibody polypeptide chain destined to become the antigen-binding site (see Figure 2).

Instructive theories have now been abandoned (Haurowitz, 1967), largely because the results of analyses of the structure of antibody molecules and the cellular dynamics of the immune response have made them untenable. They have been supplanted by selective theories (Jerne, 1955; Burnet, 1959; Lederberg, 1959), which state that the information required for specificity and range *already* resides in the organism before exposure to the antigen. The antigen serves to stimulate (or select) only those cells that synthesize complementary antibodies. As shown in Figure 3, selective theories require a minimum of three conditions: (1) the organism must contain information for the synthesis of an *enormous* repertoire of different antigen-binding sites, most of which may never be used; (2) there must be a mechanism

for antigen trapping to favor encounter with the appropriate lymphoid cells; and (3) stimulated lymphoid cells must constitute an amplifier of high gain, which is triggered by the antigen, so that, after selection of the appropriate cells, a significant number of antibody molecules of the correct specificity are produced. This is realized by cell division and an increased rate of antibody synthesis in the progeny cells.

Such a system has a number of interesting properties. It is clear, for example, that the specificity of the antigen-binding function (ABF) is a property of the entire system rather than of single antibody molecules. Moreover, the system is clonal, i.e., each cell synthesizes a single variety of antibody molecule and, after encounter with the appropriate complementary antigen, is stimulated to divide and form progeny cells, which synthesize the identical type of antibody at greatly increased rates (Mäkelä, 1967). The released antibodies can bind antigen in the fluid spaces of the organism or bind to cells, and they can also carry out a number of effector functions (EF) which do not depend on the specificity of antigen binding. These include complement fixation, skin fixation, opsonization, and placental transfer (Cohen and Porter, 1964). Effector functions are, in one sense, the important functions of antibodies, for they carry out the major tasks of the immune system. Nevertheless, it is the binding or recognition function that is of more fundamental concern to us here, for it appears to have evolved in an unusual way. One additional set of hypothetical functions could be classified as effector functions, and are of fundamental interest because they may be absolutely required to permit the selective system to operate: particular structures on antibody molecules may serve to trigger mitogenesis and cell maturation after interaction with the antigen. Similar or identical structures on antibodies may also serve to block these essential steps and lead to a specific failure to respond to a particular antigen, a state known as

FIGURE 3 A scheme of events in a selective antibody response. The letter n inside stem cells indicates that they are uncommitted and pluripotential. For clarity, diversification and commitment are presented as separate events, although a detailed model of differentiation from various stem cells is not intended. Different Arabic numbers outside each cell represent unique immunoglobulins produced after clonal commitment. These "receptor" antibodies can interact with various antigenic determinants. The degeneracy of the immune response is indicated by the recognition of a single antigenic determinant (e.g., A or B) by several different cells, each producing different antibodies which more or less fit (see Figure 2). Interaction between a committed cell and antigen stimulates amplification, which consists of maturation, mitosis, and increased antibody synthesis.

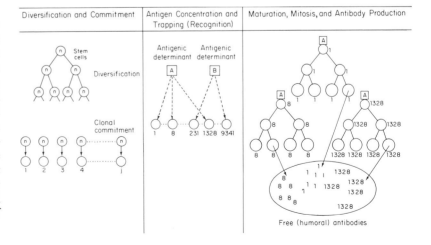

immune tolerance. So far, however, these structures have not been identified.

Although this picture of a selective molecular recognition system is qualitatively satisfactory, we lack detailed *quantitative* information on many of its features, and we are far from a complete quantitative description of the immune response. We do not know the number of different antibody specificities in an organism, the mechanism of their generation, the efficiency of the processes of diversification, trapping, and amplification, or the features of control required to allow the whole system to function (Möller, 1967). One of the least-understood features of the system is the rate at which the various lymphoid cells encounter the antigen and each other. Moreover, very little is known of the chemical state of the antigen and the functioning of various anatomical and cellular components which insure that the antigen will encounter the proper cells. Thus, the antigen-trapping step shown in Figure 3 is not simply an encounter between an "antigen gas" and a "cell gas" but, more likely, depends on special mechanisms that have evolved to insure efficiency of encounter.

Studies on the molecular structure and genetics of antibodies have yielded detailed information on the steps concerned with diversification and clonal commitment, and the remainder of this paper is devoted to these subjects.

## The structure of γG immunoglobulin

The major emphasis is placed on the structure of γG immunoglobulin, for there are good reasons to believe that this most-prevalent class is representative of the main features of antibodies. We now know the covalent structure (amino-acid sequence and disulfide bonds) of an entire γG immunoglobulin molecule (Edelman et al., 1969). Although its three-dimensional structure and the details of the antigen-binding sites remain to be determined, we can say a good deal about the nature of diversity and specificity by comparing the entire structure with previously analyzed portions of other immunoglobulin molecules (for reviews of previous work, see Cairns, 1967; Killander, 1967; Edelman and Gall, 1969). This is meaningful because the three-dimensional structures of the molecule and its combining sites depend on the amino-acid sequences of its polypeptide chains (Haber, 1964).

A schematic diagram showing the arrangement and features of the polypeptide chains in a typical human γG immunoglobulin molecule (Edelman et al., 1969) is given in Figure 4. The structure contains two identical light polypeptide chains (molecular weight 23,000) and two identical heavy chains (molecular weight 55,000). The disulfide bonds are linearly and periodically disposed in both chains, and it is striking that the light and heavy chains can be aligned so that their disulfide bonds are in register. Such

an arrangement has implications for the evolution and function of the molecule, which I discuss below. Each light chain is linked to its neighboring heavy chain by non-covalent interactions and by a disulfide bond consisting of half-cystines V of each chain (Figure 4). Half-cystines VI and VII of the heavy chain form the two bonds that link these chains and the half-molecules together. Limited hydrolysis of the molecule with proteolytic enzymes, such as papain or trypsin, cleaves the molecule into Fab fragments, which contain the antigen-combining sites, and Fc fragments, which mediate certain effector functions (Porter, 1959).

Together, the light and heavy chains make up 660 amino-acid residues, and the symmetry of the molecule gives a total of 1320 residues in the entire structure. In considering the whole structure, one is struck by the linear regional differentiation of the molecule: the amino-terminal portions of light and heavy chains contain so-called variable regions, which appear to be concerned with antigen binding, and the remainder of the molecule consists of constant regions, which are devoted to effector functions. This sharp distinction must be qualified in the absence of detailed knowledge of the three-dimensional structure, but the chemical evidence strongly favors the demarcation. The definition of variable and constant regions relies on extensive analyses of the amino-acid sequence of different immunoglobulins that have been made in a number of laboratories. This subject has been critically reviewed (Cairns, 1967; Killander, 1967; Edelman and Gall, 1969), and no attempt is made here to analyze the subject critically or historically. For convenience, the protein Eu, the complete sequence of which has been determined, serves as a reference; obviously other proteins could be used as well.

Before I discuss the sequence data in detail, I should emphasize again that all the information for the specificity of antibodies is contained in the amino-acid sequences of their constituent polypeptide chains. It follows from the selection dogma that there must be many different sequences. Moreover, the bulk of the evidence suggests that the information is in both light and heavy chains (Singer and Doolittle, 1966). These facts pose operational problems, for antibodies even to a single antigen are heterogeneous, and it is not possible to determine an unequivocal amino-acid sequence on a complex mixture of proteins. Fortunately for research purposes, certain plasma-cell tumors (myelomas) of mouse and man produce large amounts of homogeneous immunoglobulins. Each myeloma protein appears to have a unique amino-acid sequence, although its general structure resembles that of heterogeneous immunoglobulin. Unfortunately, however, there is no general way of determining the activity or specificity of myeloma proteins for antigens. Therefore, we are faced with a peculiar dilemma in structure-activity determination: either we must study a

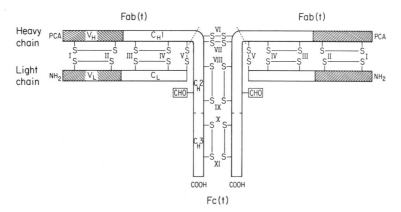

Heavy chain ... Light chain

Fab(t)　　　　　　Fab(t)

Fc(t)

Figure 4 Structure of a human γG immunoglobulin molecule Eu from myeloma patient (Edelman et al., 1969). Site of cleavage by trypsin to form Fab(t) and Fc(t) fragments is indicated by dotted line. Half-cystines contributing to disulfide bonds are designated by Roman numerals. Abbreviations: $V_H$ and $V_L$, variable regions of heavy and light chains; $C_L$, constant region of light chain; $C_H1$, $C_H2$, and $C_H3$, homology regions in constant region of heavy chain; PCA, pyrollidonecarboxylic acid; CHO, carbohydrate.

myeloma globulin, the activity of which is unknown, or we must attempt to determine the amino-acid sequences of an indeterminately large mixture of specific antibodies. Some efforts are being made to circumvent this dilemma by searching for homogeneous antibodies (Haber, 1968) or for myeloma proteins with the capacity to bind antigens (Eisen et al., 1967). Regardless of the structure-activity problem, the evidence suggests that myeloma proteins are typical

immunoglobulins and they are the main source of data on antibody diversity.

## Amino-acid sequences of light and heavy polypeptide chains: variable and constant regions

The amino-acid sequence of the light, or κ, chains of protein Eu is shown in Figure 5. In multiple myeloma, light chains

```
1                              10                                    20
ASP-ILE-GLN-MET-THR-GLN-SER-PRO-SER-THR-LEU-SER-ALA-SER-VAL-GLY-ASP-ARG-VAL-THR-
                              30                                    40
ILE-THR-CYS-ARG-ALA-SER-GLN-SER-ILE-ASN-THR-TRP-LEU-ALA-TRP-TYR-GLN-GLN-LYS-PRO-
                              50                                    60
GLY-LYS-ALA-PRO-LYS-LEU-LEU-MET-TYR-LYS-ALA-SER-SER-LEU-GLU-SER-GLY-VAL-PRO-SER-
                              70                                    80
ARG-PHE-ILE-GLY-SER-GLY-SER-GLY-THR-GLU-PHE-THR-LEU-THR-ILE-SER-SER-LEU-GLN-PRO-
                              90                                    100
ASP-ASP-PHE-ALA-THR-TYR-TYR-CYS-GLN-GLN-TYR-ASN-SER-ASP-SER-LYS-MET-PHE-GLY-GLN-
                              110                                   120
GLY-THR-LYS-VAL-GLU-VAL-LYS-GLY-THR-VAL-ALA-ALA-PRO-SER-VAL-PHE-ILE-PHE-PRO-PRO-
                              130                                   140
SER-ASP-GLU-GLN-LEU-LYS-SER-GLY-THR-ALA-SER-VAL-VAL-CYS-LEU-LEU-ASN-ASN-PHE-TYR-
                              150                                   160
PRO-ARG-GLU-ALA-LYS-VAL-GLN-TRP-LYS-VAL-ASP-ASN-ALA-LEU-GLN-SER-GLY-ASN-SER-GLN-
                              170                                   180
GLU-SER-VAL-THR-GLU-GLN-ASP-SER-LYS-ASP-SER-THR-TYR-SER-LEU-SER-SER-THR-LEU-THR-
                              190                                   200
LEU-SER-LYS-ALA-ASP-TYR-GLU-LYS-HIS-LYS-VAL-TYR-ALA-CYS-GLU-VAL-THR-HIS-GLN-GLY-
                              210          214
LEU-SER-SER-PRO-VAL-THR-LYS-SER-PHE-ASN-ARG-GLY-GLU-CYS
```

Figure 5 Amino-acid sequence of the light chain of Eu. Positions in which amino acids differ in κ chains of the same subgroup (see Figure 6) are indicated by dots under the appropriate residues. Half-cystinyl residues contributing to disulfide bonds (see Figure 4) are placed in boxes; methionyl residues are underlined. The $V_L$ region extends to residue 108. Variant position at 191 is known to represent genetic polymorphism of the constant region, $C_L$. The allelic form has a leucine at this position.

are excreted in the form of Bence-Jones proteins (Edelman and Gally, 1962), each with a unique primary structure. Because of their purity and relatively low molecular weight, the amino-acid sequences of a number of these urinary proteins have been determined (Hilschman and Craig, 1965; Titani et al., 1965; see also Cairns, 1967; and Killander, 1967). Comparison of the sequences of different Bence-Jones proteins reveals an unusual structure. The first 108 residues have 41 positions in which different amino acids may be replaced in different molecules, whereas the last 106 residues show a replacement only at position 191. Thus, the chain may be divided into a variable, or V, region and a constant, or C, region. The replacement in the C region at position 191 is a genetic polymorphism similar to that found in other proteins. The evidence indicates that a single autosomal gene specifies the C region; for example, individuals may be homozygous for valine or leucine at position 191 (Baglioni et al., 1966; Herzenberg et al., 1968).

The genetics of the V region are more complicated. No genetic markers have been identified, but examination of various proteins has indicated that V regions of $\kappa$ chains fall into three subgroups which are nonallelic, i.e., each individual has at least three genes specifying the subgroups (Milstein et al., 1969). Indeed, on the basis of this finding and the evidence for only one C gene per haploid set, it appears that each chain is specified by one V gene and one C gene (Hood and Ein, 1968). The implications of this unusual situation are discussed in detail below.

Individual proteins within each V-region subgroup differ from one another, on the average, by about 10 residues. The sites of amino-acid variation and some examples of the types of substitution found in one subgroup are shown in Figure 6. In a sample of 15 proteins compared, there are 41 positions in the $V_\kappa$ region in which replacements have been observed. On the average, there are somewhat more than two replacements per variable position, but certain positions (e.g., 96) have as many as five amino-acid substitutions. Other positions show no variation and, so far, no other amino acid has been found to replace a half-cystinyl residue. Most of the substitutions can be accounted for by single base changes in the codons corresponding to the amino acids replaced.

Although fewer examples of heavy chains have been studied, they also have V and C regions (Figure 7), as well as subgroups of V regions (Gottlieb et al., 1968; Hood, 1969, personal communication). The sequence of the Eu heavy chain (Edelman et al., 1969) has been compared with two other partial sequences, Daw (Press and Piggot, 1967) and He (Cunningham et al., 1969), and the data suggest that the variable region of the heavy chain ($V_H$) is about the same size as that of the light chain, so the constant region ($C_H$) is about three times as large. Trypsin cleaves the heavy chain at lysyl residue 222 to produce the Fc fragment.

Isolation of a single glycopeptide indicates that the polysaccharide portion of the molecule (see Figure 2) is attached at asparaginyl residue 297. By comparison with the data of Thorpe and Deutsch (1966), it has been suggested (Edelman et al., 1969) that glutamyl residue 356 and methionyl residue 358 are associated with the Gm specificity and the genetic polymorphism of the Fc region. Similarly, the arginyl residue at position 214 probably is also associated with variation at the Gm locus. Thus, like the $C_L$ region of the light chain, the heavy-chain C regions contain a few variants that are inherited in a classical Mendelian fashion. This fact poses a paradox: How can we account for the diversity of the V regions and the simultaneous constancy of the C regions?

## Origins of diversity

The data on the amino-acid sequences provide a molecular basis for the diversity requirement in the selective immune response: the different amino-acid sequences of V regions result in different three-dimensional structures for the antigen-combining sites. How such a result is actually achieved will depend on the results of X-ray crystallography of antibodies. In the meanwhile, we are left with the equally challenging problem of the genetic origins of the diversity. Three theories (see Figure 8) have been proposed at the gene level: (1) There is a single V gene for each different V region and thus each organism contains an enormous number of V genes (Dreyer and Bennett, 1965). Although there is probably sufficient DNA for this purpose, the major problem posed by the theory is how such a large number of duplicated genes remained so alike. It is obvious that the individual antigens cannot have been responsible for the selection. Because of the great similarity of V genes (see Figure 6), very strong selective forces must be imposed to maintain gene constancy. In view of the fact that we can make many immunoglobulins which never serve as antibodies, the selection pressure must be small. Moreover, it is difficult to explain the evolution of V-gene subclasses and species-specific amino acids (Kabat, 1967) from a large number of similar genes. (2). There is one V gene for each type of chain, and it undergoes somatic hypermutation. After cell division, the mutants are selected for, by an as-yet-unknown principle, according to their capacity to make antibody molecules (Brenner and Milstein, 1966). This theory requires rapid and efficient somatic selection to favor those mutants that make properly folded antibodies (Cohn, 1968). The somatic mutation rate must be very high in order to produce sufficient variation in a small number of cell generations. This requires a special mechanism which affects V genes but not C genes or other genes that specify different proteins in the cells. Because of the nonrandom nature of the interchanges found in V regions, a highly efficient selec-

```
          1                                    10
EU  ASP - ILE - GLN - MET - THR - GLN - SER - PRO - SER - THR - LEU - SER - ALA - SER - VAL - GLY - ASP - ARG - VAL - THR - ILE - THR - CYS - ARG - ALA -
         VAL   LEU   LEU                     THR   SER               VAL         LEU   ARG                     ILE               ALA               GLN
                                                   PHE


         30                                    40                                    50
    SER - GLN - SER - ILE - ASN - THR - TRP - LEU - ALA - TRP - TYR - GLN - GLN - LYS - PRO - GLY - LYS - ALA - PRO - LYS - LEU - LEU - MET - TYR - LYS -
         ASP         SER   SER   TYR               ASN               GLY         LYS                           ILE         ILE         ASP
                     LYS   ILE   PHE
                           LYS
                           ASN


         ALA - SER - SER - LEU - GLU - SER - GLY - VAL - PRO - SER - ARG - PHE - ILE - GLY - SER - GLY - SER - GLY - THR - GLU - PHE - THR - LEU - THR - ILE -
              ASN               THR                           SER         THR         PHE               ASP         PHE
              LYS               ALA


         80                                    90                                   100
    SER - SER - LEU - GLN - PRO - ASP - ASP - PHE - ALA - THR - TYR - TYR - CYS - GLN - GLN - TYR - ASN - SER - ASP - SER - LYS - MET - PHE - GLY - GLN -
         GLY               GLU         ILE                           PHE   ASP   THR   LEU   PRO   ARG   THR               GLY
                                                                           GLU   ASN               LEU               PRO
                                                                                 ASP               PRO
                                                                                                   TYR


         110                                   120
    GLY - THR - LYS - VAL - GLU - VAL - LYS - GLY - THR - VAL - ALA - ALA - PRO - SER - VAL - PHE - ILE - PHE - PRO - PRO - SER - ASP - GLU - GLN - LEU -
        LEU   ASP   ILE         ARG
              LYS   PHE
                    LEU


         130                                   140                                   150
    LYS - SER - GLY - THR - ALA - SER - VAL - VAL - CYS - LEU - LEU - ASN - ASN - PHE - TYR - PRO - ARG - GLU - ALA - LYS - VAL - GLN - TRP - LYS - VAL -


         160                                   170
    ASP - ASN - ALA - LEU - GLN - SER - GLY - ASN - SER - GLN - GLU - SER - VAL - THR - GLU - GLN - ASP - SER - LYS - ASP - SER - THR - TYR - SER - LEU -


         180                                   190                                   200
    SER - SER - THR - LEU - THR - LEU - SER - LYS - ALA - ASP - TYR - GLU - LYS - HIS - LYS - VAL - TYR - ALA - CYS - GLU - VAL - THR - HIS - GLN - GLY -
                                                                                                                    LEU


         210
    LEU - SER - SER - PRO - VAL - THR - LYS - SER - PHE - ASN - ARG - GLY - GLU - CYS
```

FIGURE 6   Types of amino-acid substitutions found on comparing 15 κ chains of the same subgroup. The reference sequence was chosen to be the light chain of protein Eu.

tive mechanism is required by their theory. So far, no convincing mechanism has been proposed. (3). There are a few V genes that have evolved to contain the mutations, and these genes recombine in the precursors of antibody-forming cells to provide the diverse variants required (Edelman and Gally, 1967, 1969). Such V genes would be similar but not identical. In certain positions, they are assumed to contain mutations that differ from gene to gene, and these mutations would have been selected for in the ensemble to yield diverse sequences after somatic recombination. The genes, which have been preselected in evolution, would yield enormous diversity if they recombined frequently in the soma. Recombination of the tandem array of genes would occur intrachromosomally and would not occur between tandem arrays corresponding to V-region subgroups. This theory must explain how a small set of multiple V genes can have evolved as a unit; an attempt at such an explanation is given below.

ANTIBODY STRUCTURE AND GENETICS     891

```
1                                                   10                                                  20
PCA - VAL - GLN - LEU - VAL - GLN - SER - GLY - ALA - GLU - VAL - LYS - LYS - PRO - GLY - SER - SER - VAL - LYS - VAL -
      •           •     •     •           •           •     •                 •           •     •     •     •     •
                                              30                                                  40
SER - CYS - LYS - ALA - SER - GLY - GLY - THR - PHE - SER - ARG - SER - ALA - ILE - ILE - TRP - VAL - ARG - GLN - ALA -
      •     •     •     •                 •     •                 •           •     •           •     •     •     •
                                        50                                                  60
PRO - GLY - GLN - GLY - LEU - GLU - TRP - MET - GLY - GLY - ILE - VAL - PRO - MET - PHE - GLY - PRO - PRO - ASN - TYR -
      •     •           •     •     •           •     •     •     •     •           •     •           •     •     •
ALA - GLN - LYS - PHE - GLN - GLY - ARG - VAL - THR - ILE - THR - ALA - ASP - GLU - SER - THR - ASN - THR - ALA - TYR -
      •     •     •     •     •     •     •     •     •     •     •     •     •     •     •     •     •     •     •
                                          90                                                 100
MET - GLU - LEU - SER - SER - LEU - ARG - SER - GLU - ASP - THR - ALA - PHE - TYR - PHE - CYS - ALA - GLY - GLY - TYR -
      •     •           •           •     •                       •           •           •     •     •     •
GLY - ILE - TYR - SER - PRO - GLU - GLU - TYR - ASN - GLY - GLY - LEU - VAL - THR - VAL - SER - SER - ALA - SER - THR -
      •     •           •     •     •     •     •     •           •           •           •
LYS - GLY - PRO - SER - VAL - PHE - PRO - LEU - ALA - PRO - SER - SER - LYS - SER - THR - SER - GLY - GLY - THR - ALA -
                                            130                                                 140
ALA - LEU - GLY - CYS - LEU - VAL - LYS - ASP - TYR - PHE - PRO - GLU - PRO - VAL - THR - VAL - SER - TRP - ASN - SER -
                                      150                                                 160
GLY - ALA - LEU - THR - SER - GLY - VAL - HIS - THR - PHE - PRO - ALA - VAL - LEU - GLN - SER - SER - GLY - LEU - TYR -
                                          170                                                 180
SER - LEU - SER - SER - VAL - VAL - THR - VAL - PRO - SER - SER - SER - LEU - GLY - THR - GLN - THR - TYR - ILE - CYS -
                                            190                                                 200
ASN - VAL - ASN - HIS - LYS - PRO - SER - ASN - THR - LYS - VAL - ASP - LYS - ARG - VAL - GLU - PRO - LYS - SER - CYS -
                                            210                                      •            220
ASP - LYS - THR - HIS - THR - CYS - PRO - PRO - CYS - PRO - ALA - PRO - GLU - LEU - LEU - GLY - GLY - PRO - SER - VAL -
                                    230                                                 240
PHE - LEU - PHE - PRO - PRO - LYS - PRO - LYS - ASP - THR - LEU - MET - ILE - SER - ARG - THR - PRO - GLU - VAL - THR -
                                          250                                                 260
CYS - VAL - VAL - VAL - ASP - VAL - SER - HIS - GLU - ASP - PRO - GLN - VAL - LYS - PHE - ASN - TRP - TYR - VAL - ASP -
                                    270                                                 280
GLY - VAL - GLN - VAL - HIS - ASN - ALA - LYS - THR - LYS - PRO - ARG - GLU - GLN - GLN - TYR - ASX - SER - THR - TYR -
                                          290                                                 300
ARG - VAL - VAL - SER - VAL - LEU - THR - VAL - LEU - HIS - GLN - ASN - TRP - LEU - ASP - GLY - LYS - GLU - TYR - LYS -
                                    310                                                 320
CYS - LYS - VAL - SER - ASN - LYS - ALA - LEU - PRO - ALA - PRO - ILE - GLU - LYS - THR - ILE - SER - LYS - ALA - LYS -
                                          330                                                 340
GLY - GLN - PRO - ARG - GLU - PRO - GLN - VAL - TYR - THR - LEU - PRO - PRO - SER - ARG - GLU - GLU - MET - THR - LYS -
                                    350                                    •                  360
ASN - GLN - VAL - SER - LEU - THR - CYS - LEU - VAL - LYS - GLY - PHE - TYR - PRO - SER - ASP - ILE - ALA - VAL - GLU -
                                          370                                                 380
TRP - GLU - SER - ASN - ASP - GLY - GLU - PRO - GLU - ASN - TYR - LYS - THR - THR - PRO - PRO - VAL - LEU - ASP - SER -
                                    390                                                 400
ASP - GLY - SER - PHE - PHE - LEU - TYR - SER - LYS - LEU - THR - VAL - ASP - LYS - SER - ARG - TRP - GLN - GLU - GLY -
                                          410                                                 420
ASN - VAL - PHE - SER - CYS - SER - VAL - MET - HIS - GLU - ALA - LEU - HIS - ASN - HIS - TYR - THR - GLN - LYS - SER -
                                    430                                                 440
LEU - SER - LEU - SER - PRO - GLY
446
```

FIGURE 7  Amino-acid sequence of the heavy chain of protein Eu. Positions in which amino acids differ in different heavy chains so far studied are indicated by dots. Because these proteins are probably in different subgroups and therefore coded by different genes, the number of variable residues in a subgroup is expected to be fewer than indicated.

Although it has been proposed (Smithies, 1967) that only two genes are required for somatic recombination, such a suggestion seems to be ruled out by the existence of three amino acids in one position in light chains of inbred mice and by the existence of nonallelic (Milstein et al., 1969) V-region subgroups.

So far, none of the theories has been tested rigorously. Some confusion has arisen from the assumption that certain theories are easier to test than others. The obvious critical experiments are to count immunoglobulin genes by DNA-RNA hybridization techniques and to search for a genetic marker in V regions. Unfortunately, these experiments are

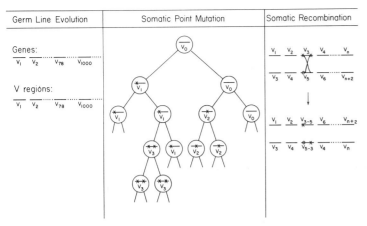

| Germ Line Evolution | Somatic Point Mutation | Somatic Recombination |
|---|---|---|

FIGURE 8  Theories of diversity at the gene level. Germ line evolution: large set of V genes, one for each V region. Somatic point mutation: hypermutation of one or a few genes during development of organism followed by selection of products in the soma; x indicates different mutations in the V gene. Somatic recombination: a small set of V genes with mutations already selected for during evolution recombines in the lymphoid cell to give new combinations of sequences.

difficult to carry out. If a homozygous marker were found, however, the multigene germ-line theory (Dreyer and Bennett, 1965) and pauci-gene recombination theory (Edelman and Gally, 1967, 1969) would be weakened. The existance of a homozygous marker would suggest that there are at most a few V genes, because recombination and mutation of a large set of genes during evolution would tend to destroy homozygosity.

## *The translocation hypothesis, allelic exclusion, and the possible bases of clonal commitment*

Besides providing a molecular basis for the diversification requirement in the selective immune response, studies of antibody structure have suggested ways in which the clonal nature of the system may have arisen (Edelman and Gally, 1969). As pointed out in the section on V and C regions, there are at least three V genes and only one C gene for κ chains in each haploid set. Because the heavy and light chains appear to be synthesized from a single starting point (Lennox et al., 1967; Fleischman, 1967), a mechanism must exist for transposing one of the V genes to the C gene so that a complete VC gene exists (Dreyer and Bennett, 1965; Gottlieb et al., 1968; Edelman and Gally, 1969). One mechanism by which such a transposition might occur is shown in Figure 9. By translocation and recombination, a

V-gene episome may be integrated into the genome, and a VC gene with the appropriate start signals may be formed. If this event were followed by transcription and translation, it would effectively represent an irreversible differentiation step. Of the vast set of sequences for which it has the genetic information, the cell could make only one or, at most, two of the heavy or light chains. In fact, as pointed out above, it has been observed that in a heterozygous individual with κ chains (see Figure 5) having valine and leucine at position 191, only valine *or* leucine is actually seen in immunoglobulin molecules from any given single lymphoid or plasma cell. This unusual phenomenon of *allelic exclusion* can be explained by the translocation hypothesis, for it is highly unlikely that both V genes from each parent would be simultaneously translocated.

Translocation is consistent with the observation that a committed cell has only one specificity (Mäkelä, 1967). The commitment to one type of antibody guarantees that there will be no loss of specificity in the system, and thus diversification is coupled to amplification (see Figure 3) in an unequivocal fashion. This, in turn, would assure that the cell is most efficient in amplifying the synthesis of its antibody molecules. It should be pointed out that the translocation hypothesis has not been proved nor has the mechanism of translocation been studied; nevertheless the hypothesis appears to be compatible with the evidence on antibody structure.

## *Evolution of antibodies: gene duplication and the domain hypothesis*

An examination of the structure of γG immunoglobulin provides evidence to support a scheme for the evolution of the antibody molecule to serve both antigen-binding functions and effector functions. Protein sequences may be compared for identical or similar residues by visual inspection. In addition, a scheme has been developed (Fitch, 1966a, 1966b) for comparing polynucleotide sequences that correspond to the protein sequences. If due caution is used

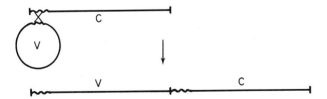

FIGURE 9  A hypothetical model for translocation of V genes to C genes. Episomal V gene crosses over with one of the two C genes, and the product is a single VC gene capable of being transcribed. Wavy lines indicate regions of recombination and integration.

EU V_L    (RESIDUES 1-108)
EU V_H    (RESIDUES 1-114)

```
                                                 1                                        10
                                ASP ILE GLN MET THR GLN SER PRO SER THR
                                PCA VAL GLN LEU VAL GLN SER GLY  -  ALA

                                            20
LEU SER ALA SER VAL GLY ASP ARG VAL THR ILE THR CYS ARG ALA SER GLN SER ILE ASN   30
GLU VAL LYS LYS PRO GLY SER SER VAL LYS VAL SER CYS LYS ALA SER GLY GLY THR PHE

                                 40
THR  -   -  TRP LEU ALA TRP TYR GLN GLN LYS PRO GLY LYS ALA PRO LYS LEU LEU MET
SER ARG SER ALA ILE ILE TRP VAL ARG GLN ALA PRO GLY GLN GLY LEU GLU TRP MET GLY

    50                                    60
TYR LYS ALA SER SER  -  LEU GLU SER GLY VAL PRO SER ARG PHE ILE GLY SER GLY SER
GLY ILE VAL PRO MET PHE GLY PRO PRO ASN TYR ALA GLN LYS PHE GLN GLY  -  ARG VAL

    70                                                                 80
GLY THR GLU PHE THR  -   -   -   -   -   -  LEU THR ILE SER SER LEU GLN PRO
THR ILE THR ALA ASP GLU SER THR ASN THR ALA TYR MET GLU LEU SER SER LEU ARG SER

                                    90
ASP ASP PHE ALA THR TYR TYR CYS GLN GLN  -  TYR ASN SER ASP SER LYS MET PHE GLY
GLU ASP THR ALA PHE TYR PHE CYS ALA GLY GLY TYR GLY ILE TYR SER PRO GLU GLU TYR

100
GLN GLY THR LYS VAL GLU VAL LYS GLY
ASN GLY GLY LEU VAL THR
```

FIGURE 10    Homology in V regions of protein Eu. Identical residues in the amino-acid sequences of V_L and V_H are shaded. Deletions indicated by dashes have been introduced to maximize the homology.

(Nolan and Margoliash, 1968), one may find evidence for evolutionary homology, which in general suggests evolution from a common gene precursor. Comparisons (Figures 10 and 11) of the V_L, V_H, C_L, and C_H regions of protein Eu show that there are a number of homologies in the sequences. These are schematized in Figure 12. V_L and V_H regions are homologous to each other and C_L is homologous to three adjoining regions of C_H known as C_H1, C_H2, and C_H3. V regions do not show any obvious homology to C regions.

In accord with these findings, we may modify the original suggestion of Hill et al. (1966) on the evolution of immunoglobulins in the fashion shown in Figure 13. In this hypothesis, immunoglobulins arose by divergence of V and C genes, each capable of specifying a polypeptide chain of about 110 residues. Gene duplication and subsequent variation then accounted for the development of the various kinds of V regions and C regions. Evolution of a translocation mechanism to join V and C genes would result in a complete gene specifying a molecule with linear differentiation of function: V regions for ABF and C regions for EF (Figure 13).

Such a proposal is in accord with the covalent chemistry and symmetry of the molecule. There is a rotational symmetry axis in the Fc region which passes through the interheavy-chain disulfide bonds and contains the homology regions C_H2 and C_H3; in addition, a pseudosymmetry axis may exist between V_H and V_L and C_L and C_H1 (see Figure 4). On these and other grounds, it has been proposed (Edelman and Gall, 1969; Edelman et al., 1969) that the immunoglobulin molecule is folded in compact domains, each consisting of a homology region containing at least one active site and serving a separate molecular function (Figure 14). Each domain is stabilized by one disulfide bond and is linked to neighboring domains by less compactly folded stretches of the polypeptide chain. Within each group of V or C homology regions, the domains would have similar but not identical tertiary structures. V_H and V_L serve antigen-binding functions and are the only domains for which the function is known. C_L and C_H1 may

```
                               110                                    120
EU C_L  (RESIDUES 109-214)   THR VAL ALA ALA PRO SER VAL PHE ILE PHE PRO PRO SER
EU C_H1 (RESIDUES 119-220)   SER THR LYS GLY PRO SER VAL PHE PRO LEU ALA PRO SER
EU C_H2 (RESIDUES 234-341)   LEU LEU GLY GLY PRO SER VAL PHE LEU PHE PRO PRO LYS
EU C_H3 (RESIDUES 342-446)   GLN PRO ARG GLU PRO GLN VAL TYR THR LEU PRO PRO SER

                                             130
ASP GLU GLN  -   -  LEU LYS SER GLY THR ALA SER VAL VAL CYS LEU LEU ASN ASN PHE
SER LYS SER  -   -  THR SER GLY GLY THR ALA ALA LEU GLY CYS LEU VAL LYS ASP TYR
PRO LYS ASP THR LEU MET ILE SER ARG THR PRO GLU VAL THR CYS VAL VAL VAL ASP VAL
ARG GLU GLU  -   -  MET THR LYS ASN GLN VAL SER LEU THR CYS LEU VAL LYS GLY PHE

140                              150
TYR PRO ARG GLU ALA LYS VAL  -   -  GLN TRP LYS VAL ASP ASN ALA LEU GLN SER GLY
PHE PRO GLU PRO VAL THR VAL  -   -  SER TRP ASN SER  -  GLY ALA LEU THR SER GLY
SER HIS GLU ASP PRO GLN VAL LYS PHE ASN TRP TYR VAL ASP GLY  -  VAL GLN VAL HIS
TYR PRO SER ASP ILE ALA VAL  -   -  GLU TRP GLU SER ASN ASP  -  GLY GLU PRO GLU

ASN SER GLN GLU SER VAL THR GLU GLN ASP SER LYS ASP SER THR TYR SER LEU SER SER
 -  VAL HIS THR PHE PRO ALA VAL LEU GLN SER  -  SER GLY LEU TYR SER LEU SER SER
ASN ALA LYS THR LYS PRO ARG GLU GLN GLN TYR  -  ASP SER THR TYR ARG VAL VAL SER
ASN TYR LYS THR THR PRO PRO VAL LEU ASP SER  -  ASP GLY SER PHE PHE LEU TYR SER

    180                              190
THR LEU THR LEU SER LYS ALA ASP TYR GLU LYS HIS LYS VAL TYR ALA CYS GLU VAL THR
VAL VAL THR VAL PRO SER SER SER LEU GLY THR GLN  -  THR TYR ILE CYS ASN VAL ASN
VAL LEU THR VAL LEU HIS GLN ASN TRP LEU ASP GLY LYS GLU TYR LYS CYS LYS VAL SER
LYS LEU THR VAL ASP LYS SER ARG TRP GLN GLN GLY ASN VAL PHE SER CYS SER VAL MET

    200                              210
HIS GLN GLY LEU SER SER PRO VAL THR  -  LYS SER PHE  -  ASN ARG GLY GLU CYS
HIS LYS PRO SER ASN THR LYS VAL  -  ASP LYS ARG VAL  -   -  GLU PRO LYS SER CYS
ASN LYS ALA LEU PRO ALA PRO ILE  -  GLU LYS THR ILE SER LYS ALA LYS GLY
HIS GLU ALA LEU HIS ASN HIS TYR THR GLN LYS SER LEU SER LEU SER PRO GLY
```

FIGURE 11   Homology in C regions of protein Eu. Identical residues are darkly shaded. Dark and light shadings are used to indicate identities that occur in pairs in the same positions.

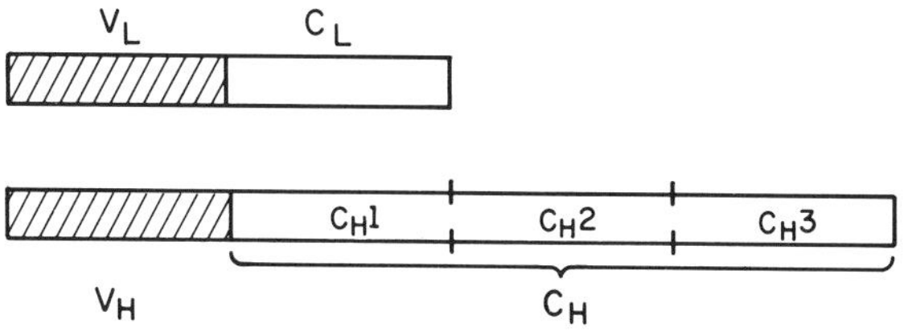

FIGURE 12   Diagram summarizing internal homologies in the structure of γG immunoglobulin. Variable regions V_H and V_L are homologous. The constant region of the heavy chain (C_H) is divided into three regions, C_H1, C_H2, and C_H3, which are homologous to one another and to the C region of the light chain (C_L).

I.  EVOLUTION OF V AND C GENES

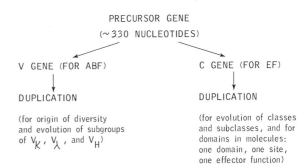

II.  EVOLUTION OF TRANSLOCATION MECHANISM TO JOIN V AND C GENES

FIGURE 13   Hypothetical scheme for evolution of antigen-binding functions (ABF) and effector functions (EF) of immunoglobulins.

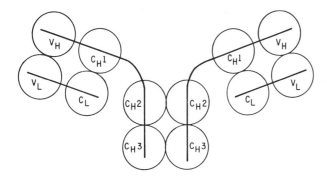

FIGURE 14 The domain hypothesis. Homology regions (see Figure 11) that constitute each domain are indicated. $V_L$, $V_H$: domains made up of variable homology regions; $C_L$, $C_H1$, $C_H2$, $C_H3$: domains made up of constant homology regions. Within each of these groups, domains are assumed to have similar three-dimensional structures, and each is assumed to contribute to an active site. The V domain sites contribute to antigen recognition functions and the C domain sites to effector functions.

serve to stabilize the interaction of the V domains but may also play a role in triggering the cell to mature after encounter with the antigen. $C_H2$ domains contain the carbohydrate that appears to be required for export of the molecule from the cell (Melchers and Knopf, 1967; Moroz and Uhr, 1967). $C_H3$ domains may contain sites for skin fixation, but so far no function has been determined for these regions.

Verification of the domain hypothesis will depend on electron microscopic and X-ray crystallographic analysis of the three-dimensional structure of antibodies. This hypothesis and the evolutionary scheme discussed above suggest an over-all sequence of events in the evolution of the selective immune response. The first evolutionary event may have been the development of a diversification mechanism to yield a sufficiently wide variety of antibodies. These antibodies may have been fixed to cells similar to those concerned with delayed hypersensitivity reactions. Subsequent development of C genes, translocation mechanisms, and carbohydrate fixation allowed for export of antibodies and for diversification of effector functions, each mediated by a specific domain. This would account for the evolution of classes of antibodies other than the γG immunoglobulins. These classes have similar light chains, but each has a distinctive kind of heavy chain and different effector functions. At present, however, too little is known of the amino-acid sequences of the immunoglobulins of lower forms to specify the detailed order of emergence of classes. Nevertheless, there is some suggestion that γM immunoglobulins (see Figure 1) were among the

first antibodies to evolve (Marchalonis and Edelman, 1965, 1966, 1968).

The sequence of events described above suggests that, once a mechanism of diversification was present, the major evolutionary developments to improve the selective immune response were focused on the amplification mechanism (see Figure 3) and on the refinement of effector functions (Marchalonis and Edelman, 1968).

The foregoing remarks have been concerned with the antibody system itself. In the three last sections of this paper, some possible implications of the data for more general problems of protein structure and evolution and differentiation are discussed briefly.

## Implications for protein structure in general

Gene duplication has been implicated in the evolution of proteins other than immunoglobulins, such as ferredoxin (Eck and Dayhoff, 1966) and haptoglobin (Black and Dixon, 1968). In addition, evidence has been obtained for multiple structural genes for polypeptide chains of hemoglobin (Schroeder et al., 1968; Hilse and Popp, 1968). These proteins do not have the peculiar periodic structure seen in immunoglobulin, however. Certain other proteins made up of polypeptide chains of great length, such as β-galactosidase, may have arisen by gene duplication and are arranged in domains. Tandem homology regions in multichain structures can be arranged in a variety of ways across symmetry and pseudosymmetry axes, and such arrangements may be evolutionarily advantageous—particularly if there is a need for linear differentiation of function. It would not be surprising, therefore, if tandem homology regions were found more generally in large proteins. It is already clear that the classical division of proteins into "globular" and "fibrous" may require additional categories, for γG immunoglobulin is not a classical globular protein (Valentine and Green, 1967).

The structural findings on Bence-Jones proteins and immunoglobulins may also have a direct bearing on the profound problem of determining tertiary structure by primary structure (Perutz et al., 1965). Bence-Jones proteins may be uniquely suited for the experimental analysis of this problem. They represent the only large family of related proteins of low molecular weight that have been selected to have *variable* three-dimensional structures. They represent a diverse set of amino-acid sequences disposed around several basic patterns, such as the subgroups of κ and λ chains. Thus, their V regions have presumably evolved for variation in shape, and the C regions for conservation of shape. Although not conclusive, there is evidence (Edelman and Gally, 1962) that each Bence-Jones protein has a different tertiary structure. Moreover, Bence-Jones proteins can be crystallized and, because of their low molecular weight, it should not take an excessive time to analyze them by X-ray crystallography, provided that suitable isomorphous derivatives are obtained.

From a comparison of a number of Bence-Jones protein sequences and three-dimensional structures, rules may emerge which will shed light on the folding problem. Moreover, if the domain hypothesis is correct, the three-dimensional structure of Bence-Jones proteins should resemble that of corresponding portions of antibody molecules.

## Antibodies and the problem of multiple genes: a hypothetical mechanism for democratic gene conversion

Multigene theories and the pauci-gene somatic recombination theory for the origin of antibody diversity require the existence of a number of very similar V genes in a tandem array. Such an array could have evolved by nonhomologous crossing over between misaligned genes. Such a process also may have led to the development of homology regions $C_H1$, $C_H2$, and $C_H3$ of C gene subclasses (Kunkel et al., 1964) and the evolution of classes of heavy chains. Obviously, the rate of duplication must have been higher than in other proteins. Once two genes are tandemly arranged by this means, however, the number of genes could increase rapidly by crossing over among genes paired in phase but out of register.

The existence of multiple genes poses the problem of explaining how the set appears to evolve as a single unit. This problem is not unique to the antibody system, as is pointed out elsewhere in this book (Edelman and Gally; Thomas). For example, structural genes for ribosomal RNA (Ritossa and Spiegelman, 1965; Wallace and Birnstiel, 1966) and for 5 S and transfer RNA (Ritossa et al., 1966) appear to be repeated many times in higher organisms, yet they give relatively homogeneous products (Forget and Weissman, 1967; Pinder et al., 1969).

A mechanism based on the recombination process itself has been proposed (Edelman and Gally, 1969, and this volume) to account for the evolution of multiple genes as units. It is assumed that, in this mechanism, recombining genes in a tandem, duplicated set are compared with one another by heteroduplex DNA formation (Figure 15). As shown in Figure 15, recombination is assumed to occur by breakage, reunion, and repair, and DNA synthesis during repair leads to gene conversion (Holliday, 1964; Fogel and Mortimer, 1969; Stahl, 1969) or nonreciprocal recombination. In this way, genetic information can be transferred among genes within a set as in the "master-slave" system proposed by Whitehouse (1967). The process described

here is more democratic, for, although there is a strong tendency for conformity, no one sequence is imposed on the V genes. In any of the V genes, a point mutation that is selectively advantageous could spread rapidly along the set rather than having to occur independently. Chromosomal rearrangements could separate sets of V genes so that each set could become differentiated, while genes within a set remained relatively similar to one another.

This gene-conversion mechanism could explain how relatively few tandem duplicated genes might remain similar. On the other hand, it is difficult to imagine how this process could account for rapid changes in a very large set of genes. This problem remains as one of the major obstacles to understanding the evolution of duplicated genes in eukaryotic genomes.

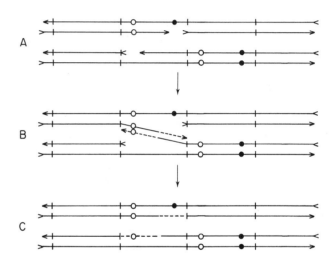

FIGURE 15 Scheme of democratic gene conversion. Solid horizontal lines represent strands of DNA that comprise two sister chromatids containing a set of immunoglobulin V genes. The V genes are aligned in register, but there is pairing between nonidentical genes. Solid and open circles represent point variation in one of the genes of the set; other genes contain other variations, which are omitted for the sake of simplicity. A. Single-strand nicks and excisions are made in complementary strands of opposite chromatids. B. Double helix is formed between two partially excised strands, allowing DNA repair synthesis to occur, as shown by dotted lines. C. Newly synthesized regions are incorporated into original two chromatids, resulting in formation of heteroduplex DNA. The net result is elimination of point variation from one strand of one chromatid and duplication of other point mutations of the set. By frequent repetition, point mutations could become represented multiply along both strands or be eliminated from the set, depending on selective pressures. Similar processes could occur intrachromosomally or between tandem genes of paired homologous chromosomes. The only essential feature of this model is repair synthesis in strands of hybrid DNA.

## The relevance to differentiation in eukaryotic organisms

In one sense, commitment of an antibody-forming cell can be considered a well-marked and irreversible differentiation event. Compared with other organic differentiation events at the cellular level, differentiation of antibody-forming cells appears to be distinguished by a kind of microscopic singularity. In other differentiated organ systems, such as the liver, one can certainly find more than two cells that are alike. In the kind of differentiation event seen in antibody production, however, each lymphoid cell appears to differentiate to produce one particular kind of antibody molecule. At the stage of commitment, therefore, one is likely to find a high degree of cellular singularity, with each cell producing a different antibody. Of course, after stimulation of committed cells by antigens, there are whole clones of similar or identical cells. In view of the changing patterns of antigenic exposure, no two individuals, even uniovular twins, would be expected to have similar populations of lymphoid cells. The analysis of the population dynamics of these cells is one of the challenging areas of modern immunology.

At present, it is not clear whether the mechanisms of cellular maturation and clonal commitment of antibody-forming cells will have a general bearing on differentiation processes in eukaryotic organisms. The indication that eukaryotic genomes contain multiple copies of genes (Britten and Kohne, 1967) and the long-standing example of transposition of genes in maize (McClintock, 1965) have certain resemblances to some of the unusual features of the antibody system. If these processes are related to differentiation in eukaryotes, it is clear that a study of somatic genetics and of events at the DNA level of somatic cells should be particularly rewarding. Inasmuch as the study of somatic genetics is still difficult, analyses (Lindahl and Edelman, 1968; Lindahl et al., 1969) of enzymatic mechanisms of repair and recombination in eukaryotic cells may also provide valuable information both on the immune response and on other specialized manifestations of cellular differentiation.

## Summary

Antibodies constitute a molecular recognition system that carries out two major functions in the immune response: (1) binding to a great variety of sterically different antigen molecules (antigen-binding function or ABF); and (2) mediating essential immune reactions within the organism, such as complement fixation (effector functions or EF). The ABF is *selective*, i.e., the information for synthesizing antibodies to a large range of chemically different antigens *already* exists in the organism before exposure to these antigens. Particular antigens serve only to stimulate (and thereby select) those antibody-forming cells that synthesize antibodies having appropriately complementary antigen-binding sites. This process requires that the number of sites must greatly exceed the number of different antigens. Furthermore, there must be an exceedingly efficient system for amplifying the production of specific antibodies after stimulation by injection of a particular antigen. This appears to be achieved by maturation and cell division of stimulated cells, each of which synthesizes only one kind of antibody. Division and maturation are followed by an increased synthesis of that particular antibody by the clonal progeny of the stimulated cell.

Studies of the molecular structure of antibodies or immunoglobulins have been useful in an analysis of some of the mechanisms of the selective immune response. The covalent structure of an entire $\gamma$G immunoglobulin (molecular weight 150,000) has been determined and compared with portions of other immunoglobulin molecules analyzed by other workers. The light and heavy polypeptide chains of the molecule each consists of a variable, or V, region for antigen binding and a constant, or C, region for effector functions. The amino-acid sequences of V regions differ from molecule to molecule, whereas the sequence of C regions is relatively invariant. The mechanism by which diversification of amino-acid sequence is achieved in antibody V regions has not been established, but several theories on the origin of diversity have been proposed and are reviewed.

The amino-acid sequence of $\gamma$G immunoglobulin shows a linear periodic arrangement which suggests the evolutionary origin of the molecule by gene duplication. It is proposed that the molecule is arranged in successive compact domains organized about axes of symmetry and pseudo-symmetry. According to this proposal, *domains* of V regions mediate antigen binding, and successive domains of the C regions mediate different effector functions. This is in accord with the hypothesis that the antibody molecule evolved from two genes, V and C, each of which underwent successive duplication.

The analysis of antibody structure suggests the possibility that a number of unusual genetic mechanisms may be required to account for the evolution of selective recognition systems. Evidence from several laboratories has led to the conclusion that each polypeptide chain is specified by a V gene and a C gene. It is suggested that the V gene may be translocated in the lymphoid cell to form a single VC gene. Translocation could account for the clonal nature of the selective response, for, although each cell may have information for many V sequences, genetic analysis indicates that it has at most only two C genes for each chain. Translocation would commit the cell to the expression of the smallest possible number of antibody chains and help to maintain the specificity and efficiency of the selective immune response.

## Acknowledgments

This work was supported by Grant GB 8371 from the National Science Foundation and by Grant AM 04256 from the National Institutes of Health.

## REFERENCES

BAGLIONI, C., L. A. ZONTA, D. CIOLI, and A. CARBONARA, 1966. Allelic antigenic factor Inv (a) of the light chains of human immunoglobulins: Chemical basis. *Science (Washington)* 152: 1517–1519.

BLACK, J. A., and G. H. DIXON, 1968. Amino acid sequence of alpha chains of human haptoglobin. *Nature (London)* 218: 736–741.

BREINL, F., and F. HAUROWITZ, 1930. Chemische Untersuchung des Präzipitates aus Hämoglobin und Anti-Hämoglobin-Serum und Bemerkungen über die Natur der Antikörper. *Hoppe-Seyler's Z. Physiol. Chem.* 192: 45–57.

BRENNER, S., and C. MILSTEIN, 1966. Origin of antibody variation. *Nature (London)* 211: 242–243.

BRITTEN, R. J., and D. KOHNE, 1967. Repeated nucleotide sequences. *Carnegie Inst. Wash. Year Book* 66: 73–88.

BURNET, F. M., 1959. The Clonal Selection Theory of Acquired Immunity. Cambridge University Press, Cambridge, England.

CAIRNS, J. (editor), 1967. *Cold Spring Harbor Symp. Quant. Biol.* 32.

COHEN, S., and R. R. PORTER, 1964. Structure and biological activity of immunoglobulins. *Advan. Immunol.* 4: 287–349.

COHN, M., 1968. The molecular biology of expectation. *In* Nucleic Acids in Immunology (O. J. Plescia and W. Braun, editors). Springer Verlag, New York, pp. 671–715.

DREYER, W. J., and J. C. BENNETT, 1965. The molecular basis of antibody formation: A paradox. *Proc. Nat. Acad. Sci. U.S.A.* 54: 864–869.

ECK, R. V., and M. O. DAYHOFF, 1966. Evolution of the structure of ferredoxin based on living relics of primitive amino acid sequences. *Science (Washington)* 152: 363–366.

EDELMAN, G. M., 1967. Antibody structure and diversity: Implications for theories of antibody synthesis. *In* The Neurosciences: A Study Program (G. C. Quarton, T. Melnechuk, and F. O. Schmitt, editors). The Rockefeller University Press, New York, pp. 188–200.

EDELMAN, G. M., B. A. CUNNINGHAM, W. E. GALL, P. D. GOTTLIEB, U. RUTISHAUSER, and M. J. WAXDAL, 1969. The covalent structure of an entire γG immunoglobulin molecule. *Proc. Nat. Acad. Sci. U.S.A.* 63: 78–85.

EDELMAN, G. M., and W. E. GALL, 1969. The antibody problem. *Annu. Rev. Biochem.* 38: 415–466.

EDELMAN, G. M., and J. A. GALLY, 1962. The nature of Bence-Jones proteins. Chemical similarities to polypeptide chains of myeloma globulins and normal γ globulins. *J. Exp. Med.* 116: 207–227.

EDELMAN, G. M., and J. A. GALLY, 1967. Somatic recombination of duplicated genes: An hypothesis on the origin of antibody diversity. *Proc. Nat. Acad. Sci. U. S. A.* 57: 353–358.

EDELMAN, G. M., and J. A. GALLY, 1969. Antibody structure, diversity, and specificity. *Brookhaven Symp. Biol.* 21: 328–344.

EISEN, H. N., J. R. LITTLE, C. K. OSTERLAND, and E. S. SIMMS, 1967. A myeloma protein with antibody activity. *Cold Spring Harbor Symp. Quant. Biol.* 32: 75–81.

FAHEY, J. L., 1962. Heterogeneity of γ-globulins. *Advan. Immunol.* 2: 41–109.

FITCH, W. M., 1966a. An improved method of testing for evolutionary homology. *J. Mol. Biol.* 16: 9–16.

FITCH, W. M., 1966b. Evidence suggesting a partial internal duplication in the ancestral gene for heme containing globulins. *J. Mol. Biol.* 16: 17–27.

FLEISCHMAN, J. B., 1967. Synthesis of the γG heavy chain in rabbit lymph node cells. *Biochemistry* 6: 1311–1320.

FOGEL, S., and R. K. MORTIMER, 1969. Informational transfer in meiotic gene conversion. *Proc. Nat. Acad. Sci. U. S. A.* 62: 96–103.

FORGET, B. G., and S. M. WEISSMAN, 1967. Nucleotide sequence of KB cell 5S RNA. *Science (Washington)* 158: 1695–1699.

GOTTLIEB, P. D., B. A. CUNNINGHAM, M. J. WAXDAL, W. H. KONIGSBERG, and G. M. EDELMAN, 1968. Variable regions of heavy and light polypeptide chains of the same γG immunoglobulin molecule. *Proc. Nat. Acad. Sci. U. S. A.* 61: 168–175.

HABER, E., 1964. Recovery of antigenic specificity after denaturation and complete reduction of disulfides in a papain fragment of antibody. *Proc. Nat. Acad. Sci. U. S. A.* 52: 1099–1106.

HABER, E., 1968. Immunochemistry. *Annu. Rev. Biochem.* 37: 497–520.

HAUROWITZ, F., 1967. The evolution of selective and instructive theories of antibody formation. *Cold Spring Harbor Symp. Quant. Biol.* 32: 559–567.

HERZENBERG, L. A., H. O. McDEVITT, and L. A. HERZENBERG, 1968. Genetics of antibodies. *Annu. Rev. Genet.* 2: 209–244.

HILL, R. L., R. DELANEY, R. E. FELLOWS, JR., and H. E. LEBOVITZ, 1966. The evolutionary origins of the immunoglobulins. *Proc. Nat. Acad. Sci. U. S. A.* 56: 1762–1769.

HILSCHMANN, N., and L. C. CRAIG, 1965. Amino acid sequence studies with Bence-Jones proteins. *Proc. Nat. Acad. Sci. U.S.A.* 53: 1403–1409.

HILSE, K., and R. A. POPP, 1968. Gene duplication as the basis for amino acid ambiguity in alpha-chain polypeptides of mouse hemoglobins. *Proc. Nat. Acad. Sci. U. S. A.* 61: 930–936.

HOLLIDAY, R., 1964. A mechanism for gene conversion in fungi. *Genet. Res.* 5: 282–304.

HOOD, L., and D. EIN, 1968. Immunoglobulin lambda chain structure: Two genes, one polypeptide chain. *Nature (London)* 220: 764–767.

JERNE, N. K., 1955. The natural-selection theory of antibody formation. *Proc. Nat. Acad. Sci. U. S. A.* 41: 849–857.

JERNE, N. K., 1967. Antibodies and learning: Selection versus instruction. *In* The Neurosciences: A Study Program (G. C. Quarton, T. Melnechuk, and F. O. Schmitt, editors). The Rockefeller University Press, New York, pp. 200–205.

KABAT, E. A., 1967. The paucity of species-specific amino acid residues in the variable regions of human and mouse Bence-Jones proteins and its evolutionary and genetic implications. *Proc. Nat. Acad. Sci. U. S. A.* 57: 1345–1349.

KARUSH, F., 1962. Immunologic specificity and molecular structure. *Advan. Immunol.* 2: 1–40.

KILLANDER, J. (editor), 1967. Nobel Symposium 3, Gamma Globulins. Structure and Control of Biosynthesis. Almqvist and Wiksell, Stockholm.

KUNKEL, H. G., J. C. ALLEN, and H. M. GREY, 1964. Genetic characters and the polypeptide chains of various types of gammaglobulin. *Cold Spring Harbor Symp. Quant. Biol.* 29: 443–447.

LANDSTEINER, K., 1945. The Specificity of Serological Reactions. Revised edition. Harvard University Press, Cambridge, Massachusetts.

LEDERBERG, J., 1959. Genes and antibodies. *Science (Washington)* 129: 1649–1653.

LENNOX, E. S., P. M. KNOPF, A. J. MUNRO, and R. M. E. PARKHOUSE, 1967. A search for biosynthetic subunits of light and heavy chains of immunoglobulins. *Cold Spring Harbor Symp. Quant. Biol.* 32: 249–254.

LINDAHL, T., and G. M. EDELMAN, 1968. Polynucleotide ligase from myeloid and lymphoid tissues. *Proc. Nat. Acad. Sci. U.S.A.* 61: 680–687.

LINDAHL, T., J. A. GALLY, and G. M. EDELMAN, 1969. Deoxyribonuclease IV: A new exonuclease from mammalian tissues. *Proc. Nat. Acad. Sci. U. S. A.* 62: 597–603.

McCLINTOCK, B., 1965. The control of gene action in maize. *Brookhaven Symp. Biol.* 18: 162–184.

MÄKELÄ, O., 1967. The specificity of antibodies produced by single cells. *Cold Spring Harbor Symp. Quant. Biol.* 32: 423–430.

MARCHALONIS, J., and G. M. EDELMAN, 1965. Phylogenetic origins of antibody structure. I. Multichain structure of immunoglobulins in the smooth dogfish (*Mustelus canis*). *J. Exp. Med.* 122: 601–618.

MARCHALONIS, J., and G. M. EDELMAN, 1966. Phylogenetic origins of antibody structure. II. Immunoglobulins in the primary immune response of the bullfrog, *Rana catesbiana*. *J. Exp. Med.* 124: 901–913.

MARCHALONIS, J. J., and G. M. EDELMAN, 1968. Phylogenetic origins of antibody structure. III. Antibodies in the primary immune response of the sea lamprey, *Petromyzon marinus*. *J. Exp. Med.* 127: 891–914.

MELCHERS, F., and P. M. KNOPF, 1967. Biosynthesis of the carbohydrate portion of immunoglobulin chains: Possible relation to secretion. *Cold Spring Harbor Symp. Quant. Biol.* 32: 255–262.

MILSTEIN, C., C. P. MILSTEIN, and A. FEINSTEIN, 1969. Non-allelic nature of the basic sequences of normal immunoglobulin κ chains. *Nature (London)* 221: 151–154.

MÖLLER, G., 1967. Control of cellular antibody synthesis by antibody and antigen. *In* Nobel Symposium 3, Gamma Globulins. Structure and Control of Biosynthesis (J. Killander, editor). Almqvist and Wiksell, Stockholm, pp. 473–503.

MOROZ, C., and J. W. UHR, 1967. Synthesis of the carbohydrate moiety of γ-globulin. *Cold Spring Harbor Symp. Quant. Biol.* 32: 263–264.

MUDD, S., 1932. A hypothetical mechanism of antibody formation. *J. Immunol.* 23: 423–427.

NOLAN, C., and E. MARGOLIASH, 1968. Comparative aspects of primary structures of proteins. *Annu. Rev. Biochem.* 37: 727–790.

NOSSAL, G. J. V., 1967. The biology of the immune response. *In* The Neurosciences: A Study Program (G. C. Quarton, T. Melnechuk, and F. O. Schmitt, editors). The Rockefeller University Press, New York, pp. 183–187.

PAULING, L., 1940. A theory of the structure and process of formation of antibodies. *J. Amer. Chem. Soc.* 62: 2643–2657.

PERUTZ, M. F., J. C. KENDREW, and H. C. WATSON, 1965. Structure and function of hæmoglobin. II. Some relations between polypeptide chain configuration and amino acid sequence. *J. Mol. Biol.* 13: 669–678.

PINDER, J. C., H. J. GOULD, and I. SMITH, 1969. Conservation of the structure of ribosomal RNA during evolution. *J. Mol. Biol.* 40: 289–298.

PORTER, R. R., 1959. The hydrolysis of rabbit γ-globulin and antibodies with crystalline papain. *Biochem. J.* 73: 119–126.

PRESS, E. M., and P. J. PIGGOT, 1967. The chemical structure of the heavy chains of human immunoglobulin G. *Cold Spring Harbor Symp. Quant. Biol.* 32: 45–51.

RITOSSA, F. M., K. C. ATWOOD, and S. SPIEGELMAN, 1966. On the redundancy of DNA complementary to amino acid transfer RNA and its absence from the nucleolar organizer region of *Drosophila melanogaster*. *Genetics* 54: 663–676.

RITOSSA, F. M., and S. SPIEGELMAN, 1965. Localization of DNA complementary to ribosomal RNA in the nucleolus organizer region of *Drosophila melanogaster*. *Proc. Nat. Acad. Sci. U.S.A.* 53: 737–745.

SCHROEDER, W. A., T. H. J. HUISMAN, J. R. SHELTON, J. B. SHELTON, E. F. KLEIHAUER, A. M. DOZY, and B. ROBBERSON, 1968. Evidence for multiple structural genes for the γ chain of human fetal hemoglobin. *Proc. Nat. Acad. Sci. U. S. A.* 60: 537–544.

SINGER, S. J., and R. F. DOOLITTLE, 1966. Antibody active sites and immunoglobulin molecules. *Science (Washington)* 153: 13–25.

SMITHIES, O., 1967. Antibody variability. *Science (Washington)* 157: 267–273.

STAHL, F. W., 1969. One way to think about gene conversion. *Genetics (Suppl.)* 61 (no. 1, pt. 2): 1–13.

THORPE, N. O., and H. F. DEUTSCH, 1966. Studies on papain produced subunits of human γG-globulin—II. Structures of peptides related to the genetic Gm activity of γG-globulin Fc fragments. *Immunochemistry* 3: 329–337.

TITANI, K., E. WHITLEY, JR., L. AVOGARDO, and F. W. PUTNAM, 1965. Immunoglobulin structure: Partial amino acid sequence of a Bence Jones protein. *Science (Washington)* 149: 1090–1092.

UHR, J. W., 1964. The heterogeneity of the immune response. *Science (Washington)* 145: 457–464.

VALENTINE, R. C., and N. M. GREEN, 1967. Electron microscopy of an antibody-hapten complex. *J. Mol. Biol.* 27: 615–617.

WALLACE, H., and M. L. BIRNSTIEL, 1966. Ribosomal cistrons and the nucleolar organizer. *Biochim. Biophys. Acta* 114: 296–310.

WHITEHOUSE, H. L. K., 1967. A cycloid model for the chromosome. *J. Cell Sci.* 2: 9–22.

# 79

## The Origin of Specificity in Binding: A Detailed Example in a Protein-Nucleic Acid Interaction

F. M. RICHARDS, H. W. WYCKOFF and N. ALLEWELL

ENZYMES INTERACT with a vast array of small molecules and macromolecules and cause changes in the covalent structure of these substances. Individual enzymes differ widely in both substrate specificity and turnover number. For the over-all catalytic process, the simple enzymes manage to carry out the series of binding and bond-altering steps with structures composed solely of the amino acids as building blocks. There is no compelling evidence for the existence of any fundamental forces or types of interaction that are not found in small molecules (Kauzmann, 1959; Jencks, 1966; Koshland and Neet, 1968). The mystery appears to lie in the three-dimensional assembly of the basic units into a cooperative system, the dynamic behavior of which provides simultaneously the required degree of specificity toward potential interactants and the enhanced reactivity, required for catalysis, of functional groups in the enzyme, the substrate, or both. Our understanding of these systems does not yet permit prediction of the structure required for a given reaction. We are beginning, however, to be able to look in detail at the actual structures of a few enzymes and of some of their complexes with small ligands, and we hope that a limited set of structural ground rules will eventually emerge.

General and detailed three-dimensional structural information on proteins is available only from diffraction studies on single crystals (Dickerson, 1964; Phillips, 1966). The best that can be done presently is the specification, within certain limits of confidence, of the average position of the atoms in the structure. Considerable prior chemical information is normally assumed in the interpretation of the data, i.e., covalent-bond distances and angles and amino-acid sequences. Although the diffraction patterns are affected by vibrations of atoms and librations of groups of atoms, such information cannot be extracted with confidence from data of the extent and accuracy now available. For such problems in molecular dynamics, the various spectroscopic techniques are much more promising, especially with the X-ray structures as starting points.

### Noncovalent forces affecting conformation

The tendency of water to try to squeeze hydrocarbons out of solution is responsible for a large fraction of the energy involved in stabilizing the compact shape of most globular proteins (Kauzmann, 1959; Davidson, 1967). On a more detailed level, this tendency may be thought of as the energy derived from decreasing the surface area of the solvent cavity surrounding the solute molecule; the more compact and spherical the molecule, the smaller the surface area and the lower the energy (Sinanoğlu, 1968). Although the emphasis is different, the solvent entropy effects (Kauzmann, 1959; Némethy and Scheraga, 1962; Scheraga, 1963) are contained in this more recent theory. Unless markedly different polarizabilities are involved, dispersion forces probably do not make a major contribution, because roughly the same number of contacts will be made between atoms in the folded protein and between solvent and atoms in the unfolded chain. For the same reason, the interactions between polar groups (for example, hydrogen bonds) also may not contribute to the stabilization energy, but they can have a dramatic influence on the final structure because of their directional nature and the requirements for orientation and group pairing when solvent interactions are removed. In this sense, such interactions make incorrect structures highly improbable instead of providing deep-energy minima for the correct structure. Coulombic forces between formal charges are of uncertain general importance but are clearly significant in certain special cases. Quantitative calculations present formidable problems, especially as net stabilization energy in each of these cases is invariably the small difference between large numbers. Superimposed on these attractive "forces" are the geometrical requirements for packing (van der Waals' repulsion) and the steric limitations imposed by the covalent-bond distances and angles implied by the amino-acid sequence. The combination of these many effects that are attributable to individual atoms or groups of atoms leads to the structure of the "resting" enzyme at the point of the lowest accessible free-energy minimum. The same considerations apply to the interaction of ligands with the "resting" enzyme. The specificity must be controlled by certain directional forces and the geometrical packing requirements.

F. M. RICHARDS, H. W. WYCKOFF, and N. ALLEWELL  Department of Molecular Biophysics and Biochemistry, Yale University, New Haven, Connecticut

## Specificity

The term specificity is used in many ways, even within the restricted field of protein chemistry. Enzyme-catalyzed rates of reaction may differ markedly among a series of related substrates. For a given enzyme, a fast relative rate for a particular substrate among a group of substrates is usually cited as evidence of a high degree of specificity. The term may refer, however, to the maximum turnover number or to a velocity at some fixed substrate concentration, a number that may contain both the turnover rate and binding parameters. With antigen-antibody reactions, specificity is more clearly tied to differences in the binding constant for a given antibody preparation and a series of related antigens. In this case, high specificity is usually thought to be synonymous with large association constants. The word specificity is ambiguous unless qualified.

STATIC SPECIFICITY   This term applies to associating systems at thermodynamic equilibrium.

1. From one point of view the *free energy of association,* as reflected in the *binding constant,* may be used to describe static specificity. The degree of specificity is then related to the *differences in free energy* for the associations of a given molecule with a series of different ligands. The absolute value of the free energy is *not* related to specificity. Very tight binding of a number of ligands does not constitute specificity. The latter term refers only to differences in binding energy. For analytical tests or chromatographic separations, for example, such considerations alone may be adequate. For other problems, however, the free energy of binding may be of secondary importance and, in fact, may not be a useful aspect of specificity at all.

2. From this second point of view, the *geometry* of the complex is the important point. Here the energy is relevant only in that the binding constant must be such that the partners are brought together. The specificity is now considered in relation to the detailed structures of the various associated pairs in a series of complexes. Discrimination can occur, even if all the binding constants are identical. The differences in geometry may show up in any of a number of physical or chemical properties of the complex, i.e., spectral perturbations, titration behavior, and so forth.

KINETIC SPECIFICITY   This adds to static specificity the consideration of the activation parameters of reacting systems.

1. In *transient-state* kinetics, specificity implies the existence of multiple paths for a given reaction, with the principal path governed by the properties of the reactants. There is an intimate correlation between geometry and reaction mechanism. Thus, the differences in the activation-free energies and the transition-state geometries are probably inextricably intermingled (Jencks, 1966). A pertinent example is the folding of an extended peptide chain to produce an apparently unique functional protein. The kinetic path determines the useful structure and a local energy minimum. On an infinite time scale, this structure may or may not represent the lowest energy conformer demanded by thermodynamics.

2. In *steady-state* kinetics, the possibility of rate-limiting steps and pools of intermediates in quasi-equilibrium requires a mixture of all the above considerations. In studies of enzyme catalysis, the geometrical factors become particularly clear when a slight shift in ligand structure changes a substrate to an inhibitor or when apparently related inhibitors with similar inhibition constants can be shown to be bound at spatially separate sites.

In the present paper, we focus on packing geometry and on the hydrogen bond as a specific type of dipole interaction. The complexes of lysozyme and ribonuclease with certain inhibitors are used as examples.

## Hydrogen-bonding groups in proteins

The principal hydrogen bonds in proteins involve only oxygen and nitrogen atoms as donors, D, acceptors, A, or both. A hydrogen bond, D-H---A, is assumed to exist when the hydrogen-acceptor atom distance is significantly ($>0.2$ Å) less than the sum of the van der Waals' radii. For N-H---O bonds, the N-O distance is usually less than 3.0 Å. The D-H---A bond angle is frequently not 180 degrees and may differ from that value by as much as 30 degrees. The acceptor atom is attached to another group, R, and the angle D---A-R is found to be highly variable and not significantly correlated with the electron orbital structures proposed for the acceptor atoms. The hydrogen bond is clearly electrostatic in origin, but its directional character can be specified only within rather broad ranges. Donor and acceptor atoms are easily recognized, and the required pairing of such groups produces rather severe restrictions on protein structures, even though the individual bond geometry is quite flexible. Bifurcated hydrogen bonds, in cases in which a single hydrogen atom is shared between two acceptor groups, are rare, but they do exist, and must be kept in mind when one is considering possible structures. Donohue discussed these points in a recent review (1968).

The amino-acid residues that can form hydrogen bonds are indicated in Figure 1. A survey of this list shows a variety of possible types of behavior, but also some clear restrictions. The nitrogen of the *tryptophan* ring, A, can serve only as a hydrogen donor. The partner must be an acceptor atom, its position is closely defined with respect to the ring system, and this, in turn, is always found firmly fixed in the protein structure. *Arginine,* B, is also an obligate hydrogen donor, but with many possible acceptor sites. A particularly good cooperative bond might be made between the guanidinium group and an ionized carboxyl group. A pair of hydrogen bonds would be formed with the coulomb energy of the ion pairs as a bonus. Other factors, however, seem to be more important in crystals of arginine salts (Mazumdar et al., 1966), and, although such bonds may be found in papain (Drenth et al., 1968), other proteins—for example, ribonuclease (Kartha et al., 1967; Wyckoff et al., 1967) and lysozyme (Blake et al., 1965)—appear to have none. The *main chain*, C, can serve as both donor and acceptor, but the chain geometry will normally be fixed, so the external acceptor and donor atoms will have fixed positions and cannot be interchanged. The *terminal amide* groups of *asparagine* and *glutamine*, D, usually on the surface, may or may not be firmly fixed by the rest of the protein structure. Rotation about the terminal C-C bond may be sterically possible, and either a donor or acceptor group at a given position could be accommodated. The *aliphatic hydroxyl* group of *serine* and *threonine*, E, can easily be either a donor or acceptor, with only the movement of a single proton. (The change is shown symbolically

as rotation, although it could equally involve exchange with water molecules—a process that takes place rapidly with OH groups in aqueous solution [Hvidt and Nielsen, 1966].) The ionizable functional groups, F through I, show the same range of characteristics, but have, in addition, the variations implied by the conjugate acid and base forms. Oxygen and nitrogen atoms behave in a reciprocal manner. Protonation of an oxygen atom converts an obligate acceptor to a donor-acceptor group (D/A group). Protonation of a nitrogen atom converts an acceptor or D/A group to an obligate donor.

The particular properties of imidazole, I, should be noted. If, in the free base form, the single proton in either position 1 or 3 of the ring is bonded to an obligate acceptor atom in some other fixed part of the protein structure, the remaining nitrogen atom can be protonated with no resulting structural change. If the free nitrogen is bonded to an obligate donor, however, the imidazole group is forced to stay in the conjugate base form, unless a substantial change in ring position is permitted in order to allow space for the added proton in the conjugate acid. As a minimum con-

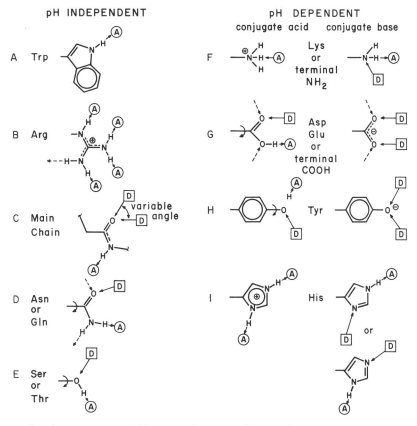

FIGURE 1  Hydrogen bonding groups available in simple proteins. The symbol A stands for an acceptor atom, the symbol D for a donor atom. The hydrogen atom provided by the donor is not shown. The diagram is approximately to scale, to indicate the geometry implied by these interactions. The variation in hydrogen-bond geometry is discussed in the text.

formational change, rotation around Cα-Cβ or Cβ-Cγ of the histidine residue would be required.

## Lysozyme-inhibitor complexes

The elegant work of Phillips and his colleagues on the structure of lysozyme (Blake et al., 1965) and some lysozyme-saccharide complexes (Blake et al., 1967) provides specific examples of some of these points. The hen egg-white enzyme catalyzes the hydrolysis of polymers of N-acetylglucosamine (NAG) and appropriate copolymers containing this group, and N-acetylmuramic acid (NAM) as found in certain bacterial cell walls. The structure of a non-productive complex of the enzyme and the trisaccharide tri-NAG has been determined by X-ray diffraction techniques. A schematic diagram showing a single NAG residue and some of the surrounding parts of the enzyme is given in Figure 2. This C site makes the largest contribution to the association constant of the trisaccharide (Chipman et al., 1967). The hexose is in the "chair" conformation with the 6-hydroxyl group bonded to tryptophan 62 and the 3-

hydroxyl group to tryptophan 63. The most interesting interaction is shown by the acetamido group of this NAG residue. The carbonyl function is bonded to the chain NH of residue 59, and the NH function is bonded to the chain carbonyl of residue 107. The obligate donor and acceptor character of the two main chain groups is fixed. The peptide unit is the only one that could be attached to the sugar and that would have the necessary geometry, dimensions, and H-bonding character. The methyl group of the acetyl function fits into a pocket in the protein. Bulky groups would be sterically excluded.

Such an example is one of *strict geometrical specificity*. The effect is provided by a combination of the three clearly identifiable interactions—two hydrogen bonds and the packing requirements of the methyl group. The single amide carbonyl bond to NH 57 localizes the oxygen atom fairly well, but by itself would put little restriction on the more remote atoms of the NAG group. The addition of the NH bond to CO 107 markedly restricts the possibilities, but there might still be free rotation about an axis connecting the two oxygen atoms without violating the geometry of the H bonds. The final addition of the methyl-group packing locks the group in three dimensions and closely defines the N-C2 bond direction. The combination of the three unit interactions of low specificity can thus produce a group interaction of high specificity. In general, many interactions will be involved in the ligand-enzyme complex as a whole. If a particular group is absent or is situated in a different position, the ligand might still interact with the enzyme in much the same way, but its binding constant, its orientation, or both might be affected. The difference in the binding of the stereoisomers α-NAG and β-NAG is a good example of this effect (Blake et al., 1967). The acetamido side chain is in the same location in each case, but the sugar ring is oriented quite differently.

From this binding study, Phillips deduced the binding of a hexasaccharide substrate and proposed a mechanism for the catalytic splitting of the chain, which has been validated by all subsequent chemical work on lysozyme up to the present. The substrate binding produces a strained conformation, the half-chair, in the appropriate sugar ring—the D residue in the hexasaccharide. The glycosidic oxygen is protonated by a neighboring carboxyl group. The transient carbonium ion is stabilized by a carboxylate function on the opposite side of the susceptible bond. After one product is replaced by a water molecule, the steps are reversed to complete the reaction. During all these binding and catalytic steps, no part of the resting enzyme has been observed to move, or need be assumed to move, more than an Ångstrom unit. The catalysis is assumed to be intimately related to the strained conformation of the substrate induced by its association with the enzyme. This example is an excellent one of the principle, suggested by Pauling (1946, 1948), that an enzyme

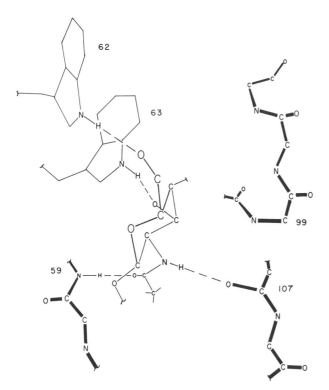

FIGURE 2  Schematic diagram of part of the hexose binding site C in egg-white lysozyme. Parts of the main peptide chain are shown as heavy lines, with the residue numbers indicated. Two tryptophan residues of the enzyme and an N-acetylglucosamine residue of the inhibitor in the chair conformation are shown. The dashed lines are hydrogen bonds. This figure is adapted from Blake et al. (1967).

should be designed to interact with the transition-state structure and not with either substrate or product.

## Structure of ribonuclease

Bovine pancreatic ribonuclease is a hydrolytic enzyme with a single chain of 124 amino-acid residues. The sequence has been reported by Smyth et al. (1963). The three-dimensional structure of the native enzyme, ribonuclease-A, has been worked out by Kartha et al. (1967); that of the modified but catalytically active form, ribonuclease-S, by Wyckoff et al. (1967). The ribonuclease-S system is shown schematically in Figure 3. Cleavage of a peptide bond near residue 20 per-

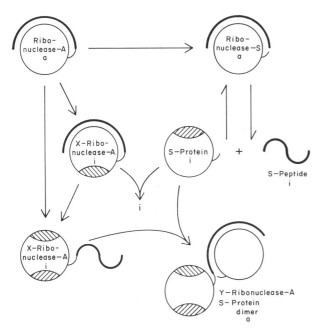

FIGURE 3 Schematic diagram of the ribonuclease-S system. The irreversible cleavage of a single peptide bond is shown at the top, followed by the reversible dissociation of the product into S-protein and S-peptide. Appropriate denatured forms of ribonuclease-A will interact with S-protein to regenerate activity as a hybrid dimer (an example of in vitro complementation). (This Figure taken from Richards, 1964)

mits separation of the enzyme into two parts, S-peptide (residues 1–20) and S-protein (residues 21–124). The two parts can be dissociated and recombined with concomitant loss and restoration of enzymic activity.

A number of derivatives of S-peptide have been synthesized by K. Hofmann and E. Scoffone and their respective colleagues (Hofmann et al., 1966; Scoffone et al., 1968). The ability of these derivatives to bind to S-protein and to

regenerate catalytic activity has been tested. The substitution of histidine 12 with the closely related analogues $\beta$(1-pyrazolyl)alanine or $\beta$(3-pyrazolyl)alanine completely eliminates enzymic activity (Hofmann and Bohn, 1966). The significance of histidine 12 for the activity of the enzyme is discussed below. The replacement of phenylalanine 8 by a tyrosine residue has little effect on the peptide-protein binding or on the activity of the complex (Marchiori et al., 1966). This observation is interesting in connection with general protein structure problems.

A stereoscopic view of the positions of the $\alpha$-carbon atoms of the peptide chain of ribonuclease-S is shown in Figure 4A. The numbered planes refer to sections in the electron-density map from which this model was derived (Wyckoff et al., 1970). The contents of part of sections 17 through 24 can be seen in Figure 4B, in which the viewing direction is normal to the planes of Figure 4A. The rings of all three phenylalanyl residues can be seen in this diagram: 46, 120, and 8 (8 is just below 120 but unnumbered). Histidine 12, threonine 45, and valine 43, which are important in the active site of this enzyme, are also seen. One can get some impression of the complexity of the packing geometry from this diagram. The simplicity of regular structures, such as the $\alpha$-helix or $\beta$-pleated sheets, which do occur in this protein, is lost in a welter of detailed interactions between side chains. Phenylalanine 46, for example, is seen to be completely surrounded. Residues at the top and bottom of the ring cavity (not shown) complete the enclosure and prevent any access from external solvent.

Phenylalanine 8 is in a similar cavity, but one that is less tightly packed than that surrounding phenylalanine 46. The implication of the finding of Marchiori et al. (1966) on the substitution of tyrosine for phenylalanine at position 8 is that the structure of this modified enzyme must be very similar to that of ribonuclease-S and that tyrosine 8 must fit into the same spot as phenylalanine 8 normally does. There is probably space to accommodate the extra OH group, but there are no accessible groups to satisfy the presumed H-bonding requirements of this phenolic OH. Possibilities of this kind increase the difficulty of developing reliable packing rules.

Wyckoff (1968) has considered the possible structure of pancreatic ribonuclease from the rat in the light of the known sequence (Beintema and Gruber, 1967) and the three-dimensional structure of the bovine enzyme. There are 41 differences in the sequences of the rat and bovine enzymes. If one assumes that the two enzymes have the same backbone conformation, one can examine each residue in turn to see if the changes are sterically permissible or reasonable. All the changes can be accommodated in the bovine structure. Beyond that there appears to be a significant pairing of the charges. Residues that are spatially close frequently change by the simultaneous loss (or gain) of both

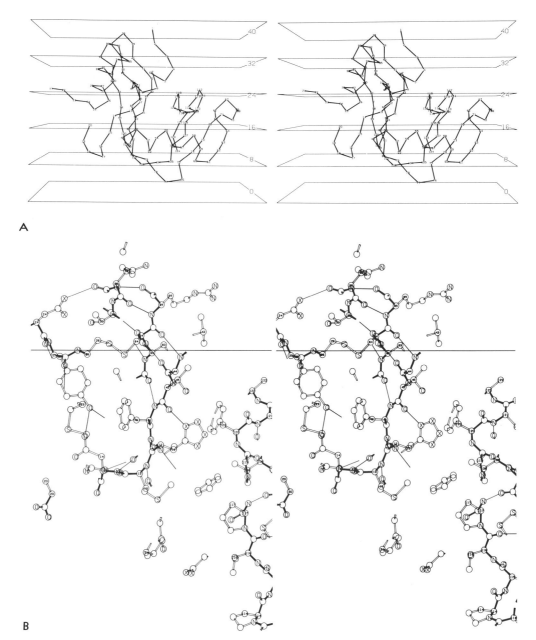

A

B

FIGURE 4  Stereodiagrams of ribonuclease-S. A. Outline of peptide chain. Residue numbers are shown at the positions of the carbon-$\alpha$ atoms. The horizontal planes are sections through the molecule corresponding to sections in the electron density map (reprinted from Figure 4A in Wyckoff et al., 1969). B. Part of the volume between sections 17 and 24. All nonhydrogen atoms are shown. The main chain bonds are black. N and O atoms are labeled. X refers to N or O or N or C where the assignment is uncertain. Residue numbers are given on $\alpha$ and $\beta$ carbon atoms. The view is perpendicular to the sections shown in A and down from the top. Stereoviewers may be obtained from various suppliers, for example: Abrams Instrument Company, Lansing, Michigan, model CF8; Wards Natural Science Establishment, Inc., Rochester, New York, model 25 W 2951.

a negative and a positive group. Thus, not only the net charge but the charge distribution remain remarkably constant. In the interior of the molecule, addition of a meth-ylene group to one residue is accompanied by the removal of such a group from a neighboring residue. If, in fact, the three-dimensional structures of the two enzymes are so similar, these observations lead to the concept of *paired isosteric replacements*. Whether this will be the exception or the rule remains to be seen. (See, for example, the comparison of lysozyme and α-lactalbumin discussed by Browne et al., 1969.)

## Interaction of ribonuclease with nucleotides

Pancreatic ribonuclease hydrolyzes single-strand RNA chains in a two-step process: (1) cleavage of the chain with the formation of a pyrimidine 2′, 3′ cyclic phosphodiester on one side of the bond cleaved and a free 5′ OH group on the other side; and (2) hydrolysis of the cyclic diester end group to give the free 3′ monophosphate ester. The enzyme is specific for pyrimidine bases on the 3′ side of the bond that is cleaved, and the 2′ OH of the ribose is an absolute requirement for activity. The crystallographic study of complexes of the enzyme with various nucleotides is now well under way (Allewell, 1969). At the moment, this system provides the only available detailed information close to the atomic level on the interaction of a protein and a nucleic-acid moiety.

The free 3′ pyrimidine nucleotides are strong competitive inhibitors of ribonuclease. Both uridine-3′-phosphate and cytidine-3′-phosphate appear to bind in almost precisely the same place on the enzyme. A schematic diagram is shown in Figure 5; a stereoscopic view, in Figure 6. Both uracil and

cytosine have a carbonyl group in position 2. The chain NH group, an obligate donor, is seen to form a bond to the oxygen atom on C2. Positions 3 and 4 are either donors or acceptors, depending on whether the base is uracil or cytosine. The protein partners are of the D/A variety, both being aliphatic OH groups. The binding site thus provides an example of *partial geometrical specificity*. The N-H---O bond to position 2 is nearly straight, but is not in the plane of the pyrimidine ring. In position 3, the bond to both uracyl and cytosine is not quite straight, although it is probably within acceptable limits. The angle can be improved by slight rotation of the side chain of threonine 45 around the carbon α-carbon β bond. The bond to the $NH_2$ group in position 4 of cytosine is markedly nonlinear and probably is not acceptable for a bond of any strength. Substantial motion of serine 123 would be required to straighten this bond. This residue is near the C-terminus of the molecule, in an area not well defined in the electron density map, so such motion is possible but not yet proved. This particular interaction may not be of great importance, as the enzyme lacking both valine 124 and serine 123 still has substantial catalytic activity. (No consideration has been given here to other possible tautomeric forms of the bases.)

Behind and somewhat below the plane of the pyrimidine ring is phenylalanine 120, with the plane of its ring at an angle to that of the base. This packing arrangement is reminiscent of the herringbone pattern commonly found in the crystal structures of aromatic ring compounds. (The analogy should not perhaps be pushed too far, for the pyrimidine can be reduced to the dihydro compound without marked change in its ability to interact with the enzyme.) In front of the ring is valine 43, which, with phenylalanine 120, forms a groove into which the pyrimidine ring appears to go (see also Figure 4B). Positions 5 and 6 on the base are accessible to solvent and not in contact with the surface of the enzyme, according well with the fact that pyrimidines substituted in positions 5 or 6 still interact strongly with the enzyme.

In Figures 5 and 6 the sugar residue is shown in the *anti*-conformation with respect to the base, as expected from the structures of the free nucleotides (Haschemeyer and Rich, 1967). The hydrogen bond to N3 of histidine 12 is easily formed from the 2′ OH of the sugar. N1 of histidine 12 is bonded to a backbone carbonyl oxygen atom. The geometry of histidine 12 is thus fixed, and N3 can either extract a proton from, or donate one to, the 2′ oxygen as required in the various steps of the catalytic mechanism. Note, however, that, in the actual binding of the product 3′-CMP, the NMR data of Meadows et al. (1969) show that both histidine 12 and histidine 119 are protonated at neutral pH. Thus the H bond to the 2′ OH would not be formed, as shown in Figures 5 and 6. Recent interpretation of the X-ray electron-density maps of this complex shows that both histidine

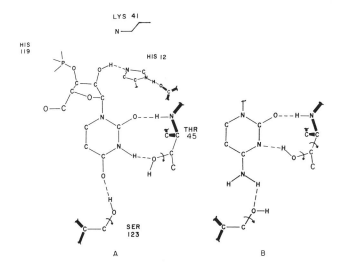

FIGURE 5    Schematic diagram of the interaction of A, 3′-uridylic acid and B, 3′-cytidylic acid with ribonuclease-S. Selected parts of the protein are shown. The hydrogen bonding flexibility provided by the aliphatic OH donor/acceptor groups is shown, together with the fixed requirements of the chain NH group (an example of partial specificity). The nature of the histidine 12 involvement is discussed in the text.

A

B

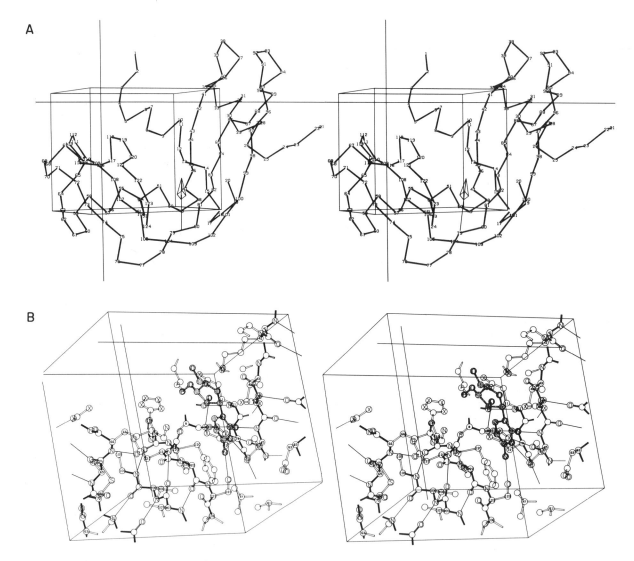

FIGURE 6 Stereodiagrams of the binding of 3' CMP to ribonuclease-S. A. α-carbon atom chain (different view from Figure 4A) with inscribed box and coordinate axes— Y, vertical; X, horizontal. The arrow shows the viewing direction for the pair below. B. Contents of the box shown in A. The same coding scheme is used as in Figure 4, except that the 3' CMP molecule is stippled to distinguish it clearly from the protein.

12 and histidine 119 appear to be hydrogen bonded to the phosphate group, and that the sugar hydroxyl group serves as the donor to the side-chain carbonyl oxygen of asparagine 44. The oxygen of glutamine 11 may be involved in holding the attacking water molecule in the hydrolytic step of catalysis.

At the present stage of the X-ray work, histidine 119 is not clearly defined, and is shown merely as being near the phosphate group in Figures 5 and 6. On the opposite side and a little above is lysine 41. Both of these nitrogen-containing groups are potentially near the phosphate group,

but they are sufficiently flexible to allow considerable latitude in the exact geometry of their interaction. For the same reason, their role in the catalytic mechanism is difficult to specify in detail from purely steric considerations. Rotation about the Cα-Cβ bond puts the imidazole ring in a position in which a hydrogen bond can be established between aspartic acid 121 and one N of the ring, and the other is close to the 5' oxygen of the leaving nucleotide in a dinucleotide substrate. The X-ray evidence with ribonuclease-S for such a convenient position is unfortunately weak.

Because of the shape and nature of the base binding site,

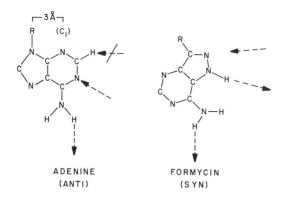

FIGURE 7 The purine nucleotides containing adenine and formycin, with the conformation indicated. They are shown in the same orientation as the pyrimidine nucleotides in Figure 5. The arrows represent possible hydrogen bonds to or from the protein groups shown in Figure 5.

one might expect adenine-containing nucleotides to bind in the same way as pyrimidines. In Figure 7, it can be seen that, if so, the sugar residue would be moved about 3 Å away from the position it has with pyrimidine nucleotides. No interaction with histidine 12 or lysine 41 would be possible, and the relation to histidine 119 would be markedly changed. Purines do bind to the enzyme and act as inhibitors, but none have so far been found to bind in the pyrimidine site. Recent studies by W. Carlson (unpublished) indicate that 3′ AMP is bound above and to the left (cf. Figure 6) of the pyrimidine site in ribonuclease-S, which would force histidine 119 out of the position in which it is shown in Figure 4B. The precise mode of binding has not yet been established, but it is assumed that this is also the site of binding of the second nucleotide in a dinucleotide substrate.

Polymers of the purine analogue formycin are exceptional, as reported by Ward et al. (1969). Surprisingly, these polymers are very good substrates for ribonuclease. Ward and Reich (1968) have also shown that the nucleotides have the *syn* conformation. If so, the base can go into the binding site "backward," as shown in Figure 7. The fit is close to that of a pyrimidine, with the C′₁ atom of the sugar in almost exactly the same position in the two cases. The glycosidic bond direction is, however, distinctly different, and a change in the ribose conformation would be required to put the 2′ OH close to histidine 12. There appears to be sufficient flexibility in the molecule for the accommodation of such a change.

The observed geometry of the principal nucleotide binding site in ribonuclease thus appears to fit the known requirement for a pyrimidine base rather nicely and to explain the lack of action on the normal purine nucleotides. If only

the three-dimensional structure of the protein were known, however, and no information on the location of the active site were available, it seems most unlikely that one could have properly located the pyrimidine binding site. This difficulty is emphasized by present attempts to define the binding site for the second nucleotide on the other side of the susceptible bond in an RNA chain. In spite of the evidence of Carlson and Tsernoglou (unpublished) on the approximate location of the purine site, the bonding details are not obvious on inspection of the model. The substrate specificity of this site is much less than that of the pyrimidine site, as almost any substituent in a pyrimidine-containing diester will be hydrolyzed. There are rate differences of more than three orders of magnitude, however, depending on whether the second ester position is represented by a methyl group or a purine nucleotide. Some interaction of this part of an oligonucleotide ligand must be postulated. Direct crystallographic study of substrate binding is not possible, as the enzyme is catalytically active in the crystal lattice and diffusion limitation over distances of hundreds of microns is too severe (Doscher and Richards, 1963).

The recent elegant studies of Jardetsky and his colleagues on the NMR spectra of ribonuclease-A, ribonuclease-S, and some nucleotide complexes complement and extend the X-ray structure information. The work has been specifically directed to the four histidine residues: 12, 48, 105, and 119. Histidine 105 is a normal, freely accessible residue unconnected with enzymic action (Roberts et al., 1969b). Histidine 48 is related to some conformational perturbations in the protein, but is not directly connected with activity. Histidine 12 and histidine 119, however, are markedly affected by the binding of nucleotides (Meadows et al., 1969). The structural conclusions made on the basis of these extensive NMR studies concerning the probable conformation and environment of the nucleotides are in gratifying agreement with the proposals put forward from the X-ray studies (Allewell, 1969). In addition, the direct involvement of histidine 119 has been much more definitely shown in the NMR work.

The detailed mechanism of action of ribonuclease has been the subject of much speculation. The proposals, roughly in order, have been by Findlay et al. (1962), Witzel (1963), Wang (1968), Hammes (1968), and Usher (1969). The most recent entry is by Roberts et al. (1969b). These authors discuss briefly the earlier ideas and the recent work on phosphate ester hydrolysis, in particular that of Westheimer and his associates. The phosphate ester group goes through a pentacoordinate intermediate during the course of hydrolysis. This transition-state complex is represented by trigonal bipyrimid, on which entering or leaving groups must occupy the apical positions. In the "linear mechanism," the entering and leaving groups are both apical, and no difficulty arises. If the geometry requires a potential leaving group to be equatorial, a process of "pseudorotation" must take place to put it in an apical position. Both the linear and pseudorotation mechanisms are possible and known in model systems. Roberts et al. (1969a) conclude that, on the basis of present structural evidence (both X-ray and NMR) and the clear involvement of two histidine residues, the linear mechanism is applicable in ribonuclease.

Their summary diagram is shown in Figure 8, which outlines the hydrolysis of a dinucleotide. In the first, or depolymerization, step, the diagram implies a linear mechanism when the attacking alkoxide formed from the ribose 2' OH is opposite the leaving 5' O of the next nucleotide. In the second, or hydrolytic, step, the diagram shows the incoming water molecule opposite the leaving 2' O in the

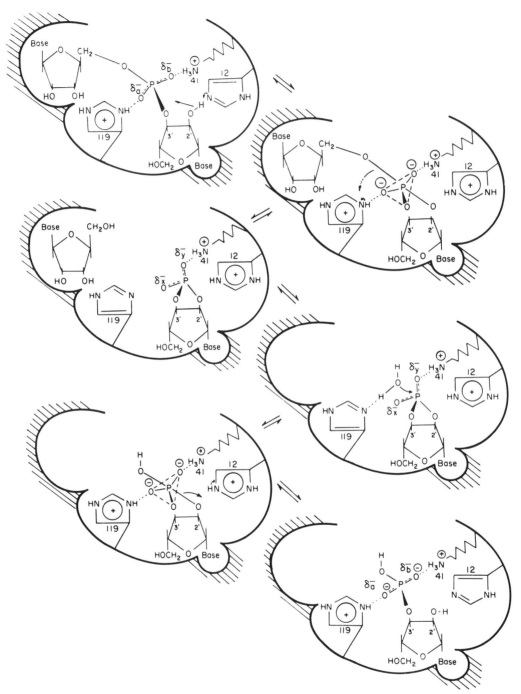

FIGURE 8  Possible mechanism for the catalytic action of ribonuclease. (Reprinted from Roberts et al., 1969a.)

cyclic phosphate ring of the intermediate compound. This may or may not be true. The X-ray data are ambiguous. The kinetic data of Usher (unpublished) on various nucleotide derivatives point to adjacent attack and the subsequent involvement of pseudorotation.

In this example, there is no evidence for the production of an obviously strained substrate conformation. Indeed, the conformational flexibility of the substrate and at least parts of the enzyme make it difficult to establish a geometry-requiring strain. In this sense, the mechanism is quite different in character from that proposed for lysozyme by Blake et al. (1967).

## Summary

At close to atomic resolution, the structures of a rapidly increasing number of proteins are becoming available from X-ray diffraction studies. Complexity in detail is still the order of the day. Generalizations, other than those assumed for many years, are so far elusive. Studies of stable complexes of these proteins with small ligands are generally possible. Detailed postulates of enzymic mechanisms can be developed. The origin of specificity, both strict and partial, begins to appear. Although the over-all binding energy for the interaction of a ligand and a protein may frequently be nonpolar association caused by the properties of water, the specificity will be provided by steric repulsion and by a combination of directional polar interactions, such as hydrogen bonds. These latter interactions, in effect, make incorrect associations less favorable energetically, rather than correct associations more favorable.

A binding site on ribonuclease has been identified, and the reasons for its specificity seem clear. Through adjustable interactions, the site can handle uracil or cytosine nucleotides equally well, but it discriminates against adenine- and guanine-containing nucleotides. These purine compounds bind at a different location on the enzyme. The appetite is whetted for details of the much more involved interaction sites that must exist in the enzymes concerned with the synthesis and functional control of proteins and nucleic acids, and the even more remote possibility of explaining specificity in cellular adhesion.

### REFERENCES

ALLEWELL, N., 1969. Crystallographic analysis of the binding of substrates and inhibitors to ribonuclease-S. Yale University, New Haven, Connecticut, Ph.D. thesis.

BEINTEMA, J. J., and GRUBER, M., 1967. Amino acid sequence in rat pancreatic ribonuclease. *Biochim. Biophys. Acta* 147: 612–614.

BLAKE, C. C. F., L. N. JOHNSON, G. A. MAIR, A. C. T. NORTH, D. C. PHILLIPS, and V. R. SARMA, 1967. Crystallographic studies of the activity of hen egg-white lysozyme. *Proc. Roy. Soc., ser. B, biol. sci.* 167: 378–388.

BLAKE, C. C. F., D. F. KOENIG, G. A. MAIR, A. C. T. NORTH, D. C. PHILLIPS, and V. R. SARMA, 1965. Structure of hen egg-white lysozyme. *Nature (London)* 206: 757–761.

BROWNE, W. J., A. C. T. NORTH, D. C. PHILLIPS, K. BREW, T. C. VANAMAN, and R. L. HILL, 1969. A possible three-dimensional structure of bovine α-lactalbumin based on that of hen's egg-white lysozyme. *J. Mol. Biol.* 42: 65–86.

CHIPMAN, D. M., V. GRISARO, and N. SHARON, 1967. The binding of oligosaccharides containing N-acetylglucosamine and N-acetylmuramic acid to lysozyme. *J. Biol. Chem.* 242: 4388–4394.

DAVIDSON, N., 1967. Weak interactions and the structure of biological macromolecules. *In* The Neurosciences: A Study Program (G. C. Quarton, T. Melnechuk, and F. O. Schmitt, editors). The Rockefeller University Press, New York, pp. 46–56.

DICKERSON, R. E., 1964. X-ray analysis and protein structure. *In* The Proteins (H. Neurath, editor). Academic Press, New York, vol. 2, pp. 603–778.

DONOHUE, J., 1968. Selected topics in hydrogen bonding. *In* Structural Chemistry and Molecular Biology (A. Rich and N. Davidson, editors). W. H. Freeman and Co., San Francisco, pp. 443–465.

DOSCHER, M. S., and F. M. RICHARDS, 1963. The activity of an enzyme in the crystalline state: Ribonuclease S. *J. Biol. Chem.* 238: 2399–2406.

DRENTH, J., J. N. JANSONIUS, R. KOEKOEK, H. M. SWEN, B. G. WOLTHERS, 1968. Structure of papain. *Nature (London)* 218: 929–932.

FINDLAY, D., D. G. HERRIES, A. P. MATHIAS, B. R. RABIN, and C. A. ROSS, 1962. The active site and mechanism of action of bovine pancreatic ribonuclease. 7. The catalytic mechanism. *Biochem. J.* 85: 152–153.

HAMMES, G. G., 1968. Relaxation spectrometry of enzymatic reactions. *Accounts Chem. Res.* 1: 321–329.

HASCHEMEYER, A. E. V., and A. RICH, 1967. Nucleoside conformation: An analysis of steric barriers to rotation about the glycosidic bond. *J. Mol. Biol.* 27: 369–384.

HOFMANN, K., and H. BOHN, 1966. Studies on polypeptides. XXXVI. The effect of pyrazoleimidazole replacements on the S-protein activating potency of an S-peptide fragment. *J. Amer. Chem. Soc.* 88: 5914–5919.

HOFMANN, K., F. M. FINN, M. LIMETTI, J. MONTIBELLER, and G. ZANETTI, 1966. Studies on polypeptides. XXXIV. Enzymic properties of partially synthetic de (16–20)- and de (15–20)-ribonucleases S'. *J. Amer. Chem. Soc.* 88: 3633–3639.

HVIDT, A., and S. O. NIELSEN, 1966. Hydrogen exchange in proteins. *Advan. Protein Chem.* 21: 287–386.

JENCKS, W. P., 1966. Strain and conformation change in enzymatic catalysis. *In* Current Aspects of Biochemical Energetics (N. O. Kaplan and E. P. Kennedy, editors). Academic Press, New York, pp. 273–298.

KARTHA, G., J. BELLO, and D. HARKER, 1967. Tertiary structure of ribonuclease. *Nature (London)* 213: 862–865.

KAUZMANN, W., 1959. Some factors in the interpretation of protein denaturation. *Advan. Protein Chem.* 14: 1–63.

KOSHLAND, D. E., JR., and K. E. NEET, 1968. The catalytic and regulatory properties of enzymes. *Annu. Rev. Biochem.* 37: 359–410.

MARCHIORI, F., R. ROCCHI, L. MORODER, and E. SCOFFONE, 1966. Synthesis of peptides analogous to the N-terminal eicosapeptide of ribonuclease-A. VI. Synthesis of 8-tyrosine-10-ornithine peptide S. *Gazz. Chim. Ital.* 96: 1549–1559.

MAZUMDAR, S. K., and R. SRINIVASON, 1966. The crystal structure of L-arginine monohydrobromide monohydrate. *Zeit. Kristallog.* 123: 186–205.

MEADOWS, D. H., G. C. K. ROBERTS, and O. JARDETZKY, 1969. Nuclear magnetic resonance studies of the structure and binding sites of enzymes. VIII. Inhibitors binding to ribonuclease. *J. Mol. Biol.* 45: 491–511.

NÉMETHY, G., and H. A. SCHERAGA, 1962. The structure of water and hydrophobic bonding in proteins. III. The thermodynamic properties of hydrophobic bonds in proteins. *J. Phys. Chem.* 66: 1773–1789.

PAULING, L., 1946. Molecular architecture and biological reactions. *Chem. Eng. News* 24: 1375–1377.

PAULING, L., 1948. Nature of forces between large molecules of biological interest. *Nature (London)* 161: 707–709.

PHILLIPS, D. C., 1966. Advances in protein crystallography. *In* Advances in Structure Research by Diffraction Methods (R. Brill and R. Mason, editors). Interscience Publishers, New York, vol. 2, pp. 75–140.

RICHARDS, F. M., 1964. Structure and activity of ribonuclease. *In* Structure and Activity of Enzymes (T. W. Goodwin, J. I. Harris, and B. S. Hartley, editors). Academic Press, London, pp. 5–12.

ROBERTS, G. C. K., E. A. DENNIS, D. H. MEADOWS, J. S. COHEN, and O. JARDETZKY, 1969a. The mechanism of action of ribonuclease. *Proc. Nat. Acad. Sci. U. S. A.* 62: 1151–1158.

ROBERTS, G. C. K., D. H. MEADOWS, and O. JARDETZKY, 1969b. Nuclear magnetic resonance studies of the structure and binding sites of enzymes. VII. Solvent and temperature effects on the ionization of histidine residues of ribonuclease. *Biochem.* 8: 2053–2056.

SCHERAGA, H. A., 1963. Intramolecular bonds in proteins. II. Non-covalent bonds. *In* The Proteins. Second edition (H. Neurath, editor). Academic Press, New York, vol. I, pp. 477–594.

SCOFFONE, E., F. MARCHIORI, L. MORODER, R. ROCCHI, and A. SCATTURIN, 1968. *In* Peptides, 1968. *Proc. IX European Peptide Symposium,* Orsay, France. North-Holland, Amsterdam, pp. 325–329.

SINANOĞLU, O., 1968. Solvent effects on molecular associations. *In* Molecular Associations in Biology (B. Pullman, editor). Academic Press, New York, pp. 427–445.

SMYTH, D. G., W. H. STEIN, and S. MOORE, 1963. The sequence of amino acid residues in bovine pancreatic ribonuclease: Revisions and confirmations. *J. Biol. Chem.* 238: 227–234.

USHER, D. A., 1969. On the mechanism of ribonuclease action. *Proc. Nat. Acad. Sci. U. S. A.* 62: 661–667.

WANG, J. H., 1968. Facilitated proton transfer in enzyme catalysis. *Science (Washington)* 161: 328–334.

WARD, D. C., W. FULLER, and E. REICH, 1969. Stereochemical analysis of the specificity of pancreatic RNase with polyformycin as substrate: Differentiation of the transphosphorylation and hydrolysis reaction. *Proc. Nat. Acad. Sci. U. S. A.* 62: 581–588.

WARD, D. C., and E. REICH, 1968. Conformational properties of polyformycin: A polyribonucleotide with individual residues in the *syn* conformation. *Proc. Nat. Acad. Sci. U. S. A.* 61: 1494–1501.

WITZEL, H., 1963. The function of the pyrimidine base in the ribonuclease reaction. *Progr. Nucl. Acid Res.* 2: 221–258.

WYCKOFF, H. W., K. D. HARDMAN, N. M. ALLEWELL, T. INAGAMI, L. N. JOHNSON, and F. M. RICHARDS, 1967. The structure of ribonuclease-S at 3.5 A resolution. *J. Biol. Chem.* 242: 3984–3988.

WYCKOFF, H. W., 1969. Discussion. *Brookhaven Symp. Biol.* 21: 252–257.

WYCKOFF, H. W., D. TSERNOGLOU, A. W. HANSON, J. R. KNOX, B. LEE, and F. M. RICHARDS, 1970. The three-dimensional structure of ribonuclease-S. Interpretation of an electron density map at a nominal resolution of 2 Å. *J. Biol. Chem.* 245: 305–328.

# 80  Assembly of Ribosomes

## MASAYASU NOMURA

EVERY INVESTIGATOR in biological science knows that the living cell is not merely a random mixture of many molecules dissolved in certain media and surrounded by a membrane, but that it has a highly developed organization. The cell has many different kinds of subcellular structures, called organelles, and many of the major cellular functions are carried out by such organelles in a highly efficient, orderly, and regulated way. For example, cellular energy production is carried out by the mitochondria; and the translation of genetic information, i.e., protein synthesis, by the ribosomes. In addition, the cytoplasmic membrane itself is an organized structure with which several important functions are associated. In order to study the mechanism of cellular functions, one must first understand the structure of its organelles. The complexity of the structures has provided a great challenge to biochemical investigators. In this article, our recent work on one such organelle, the ribosome, is summarized. Specifically, we describe how we are approaching the two major problems encountered in such studies: first, elucidation of the structure of the organelle and its relationship to function and, second, the mechanism of assembly of the completed structure from its component molecules.

### Gross structure and function of the ribosome

The gross structure of the *E. coli* ribosome is shown in Figure 1. There are two subunits of unequal size, the 50S and 30S subunits (S = Svedberg unit $\times$ 10$^{13}$). These two particles join under suitable conditions to make a 70S particle, which is the functioning unit in protein synthesis. As described by B. Davis in this volume, one of the major functions of the 30S ribosomal subunit is its role in polypeptide-chain initiation. By itself, the 30S subunit is able to bind a special tRNA (formylmethionyl tRNA$_F$ or fMet-tRNA$_F$) in response to a messenger RNA (mRNA) containing an initiator codon and, in the presence of several protein factors (initiation factors), the initiator codon (the trinucleotides AUG and GUG) signals the ribosome where to begin translation of the mRNA. A second function of the 30S subunit is to bind other aminoacyl tRNAs in response

to mRNA; this function requires the participation of the 50S subunit. The 50S subunit also catalyzes the formation of the actual peptide bond. It has been shown that the "enzyme" responsible for this activity, peptidyl transferase, is an integral part of the 50S subunit. There are several other reactions catalyzed by the ribosome, including translocation and chain termination; these are also discussed by B. Davis elsewhere in this volume. In our routine assays, we analyze the over-all activity of the 30S subunit by measuring the rate of polypeptide synthesis directed by either synthetic or natural mRNA in the presence of the 50S subunit. Conversely, the over-all activity of the 50S subunit can be determined by measuring the rate of polypeptide synthesis in the presence of the 30S subunit. There are several other functional assays available to us. For example, the binding of tRNA directed by various mRNAs can be used to assay a more specific functional capability of the 30S ribosome preparation.

From what we know about the function of the ribosome, we can draw the following picture: the two ribosomal subunits, either independently or together, bind many com-

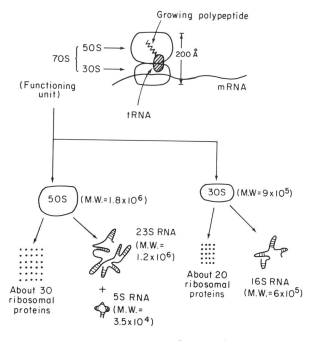

FIGURE 1   Gross structure of *E. coli* ribosomes.

MASAYASU NOMURA Laboratory of Genetics, University of Wisconsin, Madison, Wisconsin

ponents—mRNA, various aminoacyl tRNAs, several non-ribosomal protein factors—and somehow assure the proper interaction of all these components for the efficient and accurate synthesis of proteins. The fine details of this picture must still be filled in, and it is clear that this requires an equally detailed description of ribosomal structure. Our knowledge of this subject is still quite elementary (Figure 1). We know that each ribosomal subunit consists of about two-thirds RNA and one-third protein (Tissières and Watson, 1958). The 30S subunit contains one 16S RNA molecule and about 20 different proteins (Kurland, 1960; Traut et al., 1967; Hardy et al., 1969); the 50S subunit contains one 23S RNA molecule, one 5S RNA molecule, and about 30 or more different proteins (Kurland, 1960; Brownlee et al., 1967; Kurland, personal communication). The proteins contained in the 30S subunit have been separated and purified. Analysis of their amino-acid compositions, peptide fragments after trypsin digestion, and molecular weights has convincingly shown that these proteins are chemically different from one another. (Traut et al., 1967; Craven et al., 1969; Fogel and Sypherd, 1968; Kaltschmidt et al., 1967). One problem that has emerged in these studies is concerned with identification of all the separated proteins from a given ribosome preparation as genuine ribosomal components rather than tightly bound contaminants. The problem can perhaps be solved only by the demonstration of a functional requirement for each of these protein components.

Complete separation and chemical characterization of the 50S ribosomal proteins have not yet been achieved.

## Approaches to study the structure and function of the ribosome

The complex structure of the ribosome must reflect the highly organized and sophisticated functions it carries out in protein synthesis. Our problem is how to study this complicated structure and to find its relation to the function of the ribosome. In analogy to studies on the relationship between structure and function of enzyme molecules, effects of specific chemical modification on ribosome activity have

been studied in the past. However, identification of the modified components responsible for the inactivation was difficult. Similarly, studies on mutants with altered ribosomal functions gave only limited, although useful, information on the structure-function relationship; the major limitation to this approach was again the difficulty encountered in identifying the altered ribosomal components. Clearly, then, the primary limitation in all these studies has been the lack of an assay for the function of each of the ribosomal components.

One obvious way to assay the functional role of a component is to dissociate the ribosome into its molecular constituents, separate these, and then reconstitute the ribosomal particles from the fractionated components to see if a given component is required for the reconstitution of the active ribosome. We have attempted and succeeded in this approach, although to date we have been able to reconstitute only the 30S ribosome (Traub and Nomura, 1968a).

## Reconstitution of 30S ribosomal subunits from RNA and proteins

The standard method of reconstitution that we have worked out is described in Figure 2. The components of the system are purified 16S RNA and a mixture of ribosomal proteins obtained from purified 30S ribosomes; the necessary conditions include appropriate salt concentration, pH, and incubation temperature (Traub and Nomura, 1969). No other components are required. The reconstituted 30S ribosomes have been shown to be identical to the original intact 30S ribosomes with respect to sedimentation properties, protein composition (Figure 3), and the several functional abilities examined (Traub and Nomura, 1968a). Thus, at least in this in vitro system, the remarkably complex 30S ribosomal structure can be spontaneously self-assembled. One can conclude that all the information needed for correct assembly is contained in the structure of its molecular components and not in some other nonribosomal factors.

With this reconstitution system available, one can now perform unambiguous identification and functional analysis of all the essential components of the 30S subunit. We have

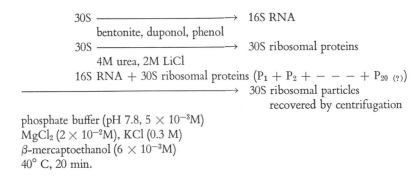

FIGURE 2 The standard procedure for reconstitution of 30S ribosomal subunits.

separated and purified each of the proteins contained in the 30S subunit. Altogether, we have isolated 19 major components as pure proteins (Nomura et al., 1969). After having obtained these components, we asked whether the 19 proteins comprise all the essential 30S ribosomal proteins. To answer this question, we performed the reconstitution, using the 19 purified proteins together with 16S RNA, and compared the degree of reconstitution with that obtained by using unfractionated proteins. We found that the efficiency of the reconstitution was only about half of that obtained with the unfractionated proteins (Nomura et al., 1969). It is possible that some essential components are lost during the fractionation. Nonetheless, the fact that reconstitution could be achieved with even this efficiency encouraged us to begin investigating the functional roles of these separated protein components.

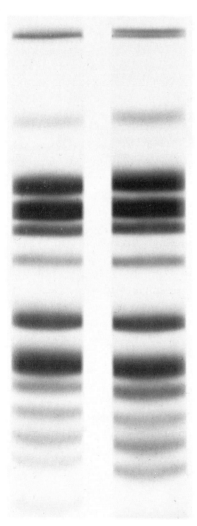

FIGURE 3  Polyacrylamide gel electrophoresis of the proteins of the original 30S particles, A, and the reconstituted 30S particles, B. Some bands represent more than one protein, and altogether about 20 proteins are found in the 30S particle. (From Traub and Nomura, 1968a.)

## Mutationally altered ribosomes

The first protein we studied in detail is the one responsible for sensitivity or resistance to the antibiotic streptomycin (Sm). It is known that Sm inhibits bacterial growth by inhibiting protein synthesis (for reviews on Sm actions, see Jacoby and Gorini, 1967; Weisblum and Davies, 1968). Two major effects of Sm have been observed in vitro in systems that synthesize protein: (1) Sm inhibits protein synthesis directed by natural mRNA, such as RNA from phage f2 and, to a lesser extent, that directed by certain synthetic mRNAs, such as poly-U (polyphenylalanine synthesis); (2) Sm causes misreading of synthetic mRNA. For example, poly-U normally stimulates the incorporation of phenylalanine into polypeptides, leading to formation of polyphenylalanine. In the presence of Sm, this "normal" incorporation is partially inhibited and, instead, poly-U stimulates the incorporation of leucine, isoleucine, tyrosine, and serine into polypeptide products. This latter observation is presumably related to the fact that Sm can suppress amino-acid requirements phenotypically in certain mutants of *E. coli*.

It is known that both in vitro effects of Sm are abolished or greatly reduced in cell-free systems when 30S ribosomes from Sm-resistant strains are used. Therefore, the site of action of Sm is the 30S ribosomal subunit. It has been postulated that Sm interacts with the ribosome to cause a distortion of its structure, which results in frequent violation of the codon-anticodon rule, leading to the observed misreading of mRNA in vitro and the suppression of mutational effects in vivo. It is clear that the component altered by the Sm-resistance mutation is important in the ribosome functions of protein synthesis and in controlling accuracy of translation of the genetic message. Using the reconstitution technique, we have identified this protein.

We first showed that the component altered by the mutation is in the protein fraction of the 30S subunit and is not the 16S RNA (Traub and Nomura, 1968b). The 30S particles reconstituted with the use of the protein from a resistant mutant and RNA from a sensitive strain were resistant to streptomycin in a cell-free, protein-synthesizing system, whereas the reverse combination produced 30S particles sensitive to streptomycin. We then purified 30S ribosomal proteins from both streptomycin-sensitive and streptomycin-resistant bacteria, and systematically substituted single proteins from a resistant strain into a mixture of proteins from a sensitive strain. We then assayed these various reconstituted ribosomes for their in vitro response to streptomycin (Table I). In this way, we have shown that the origin of a single protein (P10) determines the sensitivity of the entire 30S ribosomal particle to the inhibitory action of streptomycin, its sensitivity to streptomycin-induced misreading, and its ability to bind the antibiotic (Ozaki et al., 1969).

TABLE I

*Identification of P10 as the Protein Altered by the Mutation to Streptomycin (Sm) Resistance*

| Origin of Proteins Used for Reconstitution | | Inhibition by Sm of Protein Synthesis Directed by Phage f2 RNA (%) | Relative Degree of Misreading of Poly-U Message (%) | Binding of H³-DHSm (Relative Values) |
|---|---|---|---|---|
| Pi-P10 | P10 | | | |
| s | s | 74 | (100) | (100) |
| s | r | 10 | 8 | 5 |
| r | r | 3 | 3 | 5 |
| r | s | 78 | (100) | (100) |
| Control 30S | (s) | 69 | (100) | (100) |
| Control 30S | (r) | 10 | 7 | 4 |

s, derived from Sm-sensitive cells; r, derived from Sm-resistant cells.
ΣPi-P10, a mixture of 30S ribosomal proteins other than P10. H³-DHSm, H³-labeled dihydrostreptomycin.
This is a summary of experimental data published by Ozaki et al. (1969).

We then studied particles that do not contain this protein. In the absence of P10, 16S RNA and other 30S ribosomal proteins can still assemble into a particle sedimenting at 30S. The P10-deficient particles thus produced were found to have several interesting properties. Under the conditions used for the assays, the P10-deficient particles had a high activity when a synthetic messenger RNA, poly-U, was used as a template, but their activity was very weak when directed by RNA from a natural source. It is known that a special initiation mechanism is required for in vitro polypeptide synthesis directed by natural mRNA (see the article by Davis in this volume). In fact, it was shown that P10-deficient particles are very weakly active in this initiation function (Ozaki et al., 1969).

Another unique feature of the P10-deficient particles is related to the misreading of poly-U induced by Sm. A drastic decrease in Sm-induced misreading relative to the normal reading of poly-U was observed with these particles. As is described below, P10 is the only protein, out of 19 proteins examined, which showed such a decrease. Surprisingly, the frequency of error in translation with P10-deficient particles was very much reduced, not only in the presence of streptomycin, but also in the presence of other antibiotics, ethyl alcohol, or high concentrations of Mg⁺⁺, all of which are known to induce translational errors. In fact, these 30S ribosomal particles that are deficient in P10 are able to read synthetic mRNA more accurately than can 30S ribosomal particles containing that protein, even in the *absence* of error-inducing agents. These observations suggest that the P10 protein has a unique function that influences error frequency in the information-transfer process.

The success of this study indicates the usefulness of such an approach. The isolation of mutants with altered ribosomes should provide considerable insight into the relationship between structure and function.

## Functional analysis of 30S ribosomal proteins

We have performed functional analyses similar to those used with P10 on the remaining proteins (Nomura et al., 1969). The basic approach is to perform the reconstitution, omitting one protein, and then to ask whether physically intact 30S particles are formed and, if so, whether they are active in several available functional assays. We determined physical intactness of the particles by analyzing their sedimentation behavior by sucrose gradient sedimentation analysis. The protein-deficient particles were then examined for their activities in several known 30S functions. The following assays were routinely performed: (1) incorporation of phenylalanine into proteins directed by poly-U; (2) Sm-induced misreading of poly-U; (3) phenylalanyl tRNA binding directed by poly-U; and (4) binding of the initiator tRNA, formylmethionyl tRNA, directed by the triplet AUG. From such analyses we have obtained considerable information; Table II summarizes the effects of the omission of each of the 19 proteins on the gross physical assembly and the functional capability of the particles. In the Table, a given protein is indicated as being required for function when significant reduction of activity is observed in any of the assays used.

The results of sedimentation analysis of the protein-deficient reconstituted particles have allowed division of the proteins into two groups. One comprises proteins which apparently are not essential for the self-assembly reaction in vitro, and the other includes proteins (P4a, P4b, P5, P8, and P9) which are essential for the formation of particles with a sedimentation coefficient of 30S. Particles formed by the omission of one of these latter proteins either moved with a sedimentation coefficient of 20S to 25S, rather than 30S, or sedimented as a heterogeneous mixture of variously sized particles. In these cases, it appears that the particles formed are also deficient in several other proteins, even though these were present in the reconstitution mixture. Thus, the presence of some proteins is essential for the binding of other proteins, which suggests that the assembly process is cooperative.

Although detailed analyses of the protein-deficient reconstituted particles are still under way, the results obtained so far have allowed us to draw several conclusions.

1. Most of the proteins, except P1 and P3a, are clearly required for the intact functional activity of the ribosome. Omission of any of these proteins caused decreases in some or all of the functional activities assayed. Thus, these 17

TABLE II

*Nomenclatures of 30S Ribosomal Proteins and Tentative Conclusion on Functional Requirements*

| Our Code | Kurland's Code | Requirement for | | Note |
|---|---|---|---|---|
| | | Assembly | Function | |
| 1 (A$_1$) | 1 | − | − | |
| 2 (A$_2$) | 4a | − | + | |
| 3 (B$_1$) | 9 | ± | ++ | |
| 3a | 2 | − | − | |
| 4 (B$_2$) | 3 | ± | + | Spc sensitivity |
| 4a | 10 | + | (++) | |
| 4b | 2a | + | (++) | |
| 5 | 8 | + | (++) | K character |
| 6 (B$_3$) | 4 | ± | ++ | |
| 7 | 11 | ± | ++ | Fidelity of translation |
| 8 (B$_4$) | 12 | + | (++) | |
| 9 | 6(and 7?) | + | (++) | |
| 10 | 15 | − | ++ | Str sensitivity; ambiguity of translation |
| 10a | 14 | − | + | |
| 11 (B$_5$) | 12b | ± | ++ | |
| 12 | 12a | − | + | |
| 13 | 13 | ± | ++ | |
| 14 | 16 | − | + | |
| 15 | 15a | − | + | |
| 3b | | − | (±) | |
| 3c | | − | (±) | |
| | 5 | | | |
| | 7 | | | |

Symbols for assembly requirement: +, omission of the protein produces particles sedimenting at 20–25S; ±, omission of the protein produces a 27–28S particle: −, the protein is not required for formation of the 30S particle.

Symbols for functional requirement: (++), strong requirement presumably because of requirement in the assembly reaction; ++, strong requirement; +, partial requirement; −, no requirement demonstrated; (±), weak effect on the initiation reaction.

For details, see the text.

proteins are not irrelevant contaminating proteins and, in all likelihood, must represent genuine ribosomal components. On the other hand, we have not yet reached any definite conclusion with respect to P1 and P3a. They might be not genuine ribosomal proteins, but merely proteins tightly bound to the 30S ribosomal particles. Alternatively, it is possible that they are ribosomal proteins the functions of which cannot be detected by the activity assays we have used.

2. Proteins P4a, P4b, P5, P8, and P9, which are essential for physical assembly, were also shown to be essential in almost all the functional assays employed. Particles lacking anyone of these proteins are devoid of function. Therefore, "intact" physical assembly is apparently prerequisite to the appearance of 30S ribosomal functions.

3. Some other proteins, which apparently are not required for assembly, were also shown to be required for some or most of the ribosomal functions. Conversely, any of several known 30S ribosomal functions studied were shown to be affected by omission of any of a number of proteins. Many of the 30S ribosomal functions seem to require the presence of more than one protein component, so one can say that the 30S ribosomal proteins function cooperatively.

4. Two proteins, P10 and P7, were found to be of special interest with respect to translational fidelity. As mentioned above, P10 is the protein altered, as a result of mutation, to streptomycin resistance, and the P10-deficient particles showed a great decrease in the extent of translational errors induced by various agents. On the other hand, the particles reconstituted in the absence of P7 showed a dramatic *increase* in a misreading of the poly-U message. Again, out of 19 proteins examined, P7 is the only protein that showed such a marked increase in misreading. Thus, these two proteins apparently act in opposite ways; the presence of P10 decreases fidelity and that of P7 increases fidelity. Mutational alterations in these proteins may confer various additional specificities in the function of these proteins in translational fidelity. It is conceivable, therefore, that a sophisticated and subtle control of translational fidelity is effected genetically, as well as physiologically, by these two proteins.

## Functional analysis of 16S ribosomal RNA

The functional role of ribosomal RNA has long been a mystery. The availability of the 30S reconstitution system now permits us to investigate this problem.

First, we have found that RNA must be present for the assembly reaction to occur. In the absence of 16S RNA, no soluble particles resembling the ribosome are formed. Thus, the situation is quite different from the assembly of the TMV particle, in which case particles resembling mature virus are formed from protein subunits even in the absence of viral RNA. Moreover, 16S RNA from yeast or 18S ribosomal RNA from rat liver cannot replace *E. coli* 16S RNA to make particles sedimenting at 30S. Only very heterogeneous nonspecific aggregates or insoluble particles are formed under these circumstances (Traub and Nomura, 1968a). The presence of some specific RNA is, therefore, essential for assembling ribosomal proteins into a particle sedimenting at 30S. Specifically, then, the base sequence of the RNA should be an important factor. To test this, we initiated a series of experiments in which we performed reconstitutions, using 16S RNA from one species of bacteria and 30S ribosomal proteins from a distantly related species (Nomura et al., 1968). We found, in fact, that such "artificial" ribosomes are, in many cases, as active as the respective homologous RNA and protein combinations. Although 16S RNAs from these different bacterial species have some portions of their base sequence in common, a large part of this sequence is different. We conclude that, although there is definitely a specificity requirement of the RNA, such a requirement is not absolute. It appears that only certain small regions of 16S RNA are directly involved in the specific interaction with the ribosomal protein. Identification of such regions and the nature of the interactions are the principal goals in the study of ribosomal assembly.

As mentioned above, chemical modification is one approach to the question of the relationship between the structure and function. With the reconstitution technique available, it is now possible to identify the altered components in the inactive ribosomes treated with various specific reagents. It is also possible to modify separated components individually, and then to examine their functional alterations in the reconstituted ribosomes. Using such an approach, we have studied nitrous acid modification of RNA and of ribosomes. We have found that amino groups in the bases of 16S RNA are important but are probably primarily required for the assembly reaction (Nomura et al., 1968). In conclusion, the only positively identified function of 16S RNA to date is its role in the assembly of the ribosomal particle. The question of whether ribosomal RNA is directly involved in any of the known ribosomal functions must await further studies.

### Assembly of 30S ribosomes studied in vitro

The 30S ribosome reconstitution system has been used to study the mechanism of the self-assembly reaction (Traub and Nomura, 1969). After we had defined the optimum conditions for reconstitution with respect to ionic strength, $Mg^{++}$ concentration, pH, and temperature, we studied the kinetics of reconstitution at different temperatures. Two major conclusions were obtained: (1) the reaction is first

order with respect to formation of active 30S ribosomes, and (2) the rate of reaction is strongly dependent on the temperature of incubation and has an Arrhenius activation energy of 38 Kcal/mole. The first conclusion was obtained by our following the time course of the reaction and examining the effect of dilution of the incubation mixture on the kinetics of the reaction. Thus, although the over-all assembly reaction involves as many as 20 reactants, the rate-limiting reaction appears to be unimolecular.

The rate-limiting step in the assembly process could occur at any time before the formation of functionally active 30S ribosomes, i.e., only after none or some proteins are bound to the RNA, or after all are bound. In any case, our observations would have been the same. Although we have not yet rigorously identified the intermediate undergoing the unimolecular structural rearrangement reaction, we have obtained evidence that the intermediate particle is a ribonucleoprotein particle deficient in certain proteins. As mentioned above, the assembly reaction is extremely slow at low temperatures. We have found that, under such conditions, a particle that has some, but not all, of the proteins accumulates. We were able to isolate such a particle, activate it by warming the solution, and observe the relatively instantaneous binding of the rest of the proteins, with consequent formation of a completed ribosome. In this way, we could elucidate the general nature of the "pathway" of the self-assembly reaction: rapid binding of some of the proteins to the RNA, a slow structural rearrangement of this intermediate with the use of thermal energy, then rapid binding of the rest of the proteins.

We have already seen that the presence of one of a few proteins (P4a, P4b, P5, P8, and P9) is essential for some other proteins to bind. Systematic studies with separated proteins should soon reveal more details of the pathway of the self-assembly reaction, and may even give some insight into the nature of the geometrical relationship of the different molecular components.

### Assembly of ribosomes in vivo

After obtaining all this information on assembly in vitro, one must ask the obvious question: Are these same principles operating in the living cell? The problem of ribosome biosynthesis in vivo was being studied long before in vitro assembly was even seriously considered. Intricate experiments were performed in which short pulses of radioactive RNA precursor compounds were administered to growing bacterial cells, and their flow into mature ribosomes was followed (McCarthy et al., 1962; Mangiarotti et al., 1968; Osawa, 1968). In this way, several classes of precursor particles were postulated, but their characterization was difficult in most cases. It is clear that such an approach has obvious limitations in providing the detailed mechanism of in vivo ribosome assembly. Another approach to this problem is to isolate mutants defective in the assembly of ribosomes and to identify the nature of the defects and the genes responsible for them. In analogy to mutants defective in biosynthetic pathways of small molecular weight metabolities, some of these mutants may accumulate intermediate particles in ribosome assembly. In fact, several bacterial mutants have been isolated previously which show an accumulation of some 50S "precursor" particles (Lewandowski and Brownstein, 1966; MacDonald et al., 1967), but the block in the biosynthetic pathway was not complete, and conditions for "precursor" accumulation were not defined. Moreover, no systematic method to isolate these ribosome-assembly, defective mutants has been available.

In order to confirm our in vitro results, we decided to use a genetic approach. As mentioned before, in our detailed study of the in vitro assembly reaction, we observed the remarkable dependence of the rate of reaction on temperature. The conversion of the inactive intermediate particle to an active configuration, described above, is almost infinitely slow at 10° C or below, resulting in the accumulation of these particles. Thus, the assembly reaction is inherently easier at higher temperatures in this in vitro system. If the biosynthesis of ribosomes in vivo also involves essentially the same principle, then many mutational defects, either in ribosomal structural components or in extra-ribosomal components affecting the assembly process, should manifest themselves more clearly at lower temperatures. Such assembly-defective mutants should be viable at high temperatures, but unviable at lower temperatures; that is, they could be isolated as cold-sensitive mutants. This reasoning proved to be correct. We have isolated a large number of cold-sensitive mutants of *E. coli,* a significant fraction of which appear to be defective in ribosome assembly (Guthrie et al., 1969). We have already identified three distinct classes of particles, from three different mutants, which accumulate in cells grown at 20° C (Table III). Two of these particles appear to be 50S precursors, and the third is a 30S precursor. We have done some preliminary studies on the characterization of these particles, as well as on their formation and fate under various conditions both in vivo and in vitro. In addition, we are currently isolating many more mutants and are performing genetic analysis of the various mutants in the hope of obtaining much-needed information on the genetic organization and control of the ribosome and ribosomal assembly.

## Summary

Several years ago, our knowledge of ribosome structure and its relation to function was meager. The success of the reconstitution of 30S ribosomes from their molecular components has provided the means to attack these problems. Coupled with biochemical and genetic techniques, the reconstitution technique has now started to produce much information on the structure, function, and assembly of the 30S ribosome. This situation makes us optimistic that even more complex cellular structures, including nerve-membrane structures, will some day also be understood on a truly molecular level.

## Acknowledgment

Our work reviewed in this article was supported by United States Public Health Research Grant GM-15422 from the National Center for Urban and Industrial Health, and by National Science Foundation Grant GB-6594. This paper is No. 1322 from the laboratory of Genetics, University of Wisconsin.

REFERENCES

BROWNLEE, G. G., F. SANGER, and B. G. BARRELL, 1967. Nucleotide sequence of 5S-ribosomal RNA from *Escherichia coli. Nature (London)* 215: 735–736.

CRAVEN, G. R., P. VOYNOW, S. J. S. HARDY, and C. G. KURLAND, 1969. The ribosomal proteins of *Escherichia coli* II. Chemical and

TABLE III

*Ribosomal Subunit Assembly Defective Mutants (sad Mutants)*

| Strains | Growth | | Ribosome Biosynthesis at 20° C (3 hrs.) | | |
|---|---|---|---|---|---|
| | 42° C | 20° C | 50S | 30S | Other Particles Accumulated |
| parent (AB1133) | + | + | + | + | — |
| sad-19 | + | − | − | + | 32S |
| sad-68 | + | − | − | + | 43S |
| sad-38 | + | − | ± | ± (?) | 32S, 21S |

±, reduced synthesis.

physical characterization of the 30S ribosomal proteins. *Biochemistry* 8: 2906–2915.

FOGEL, S., and P. S. SYPHERD, 1968. Chemical basis for heterogeneity of ribosomal proteins. *Proc. Nat. Acad. Sci. U. S. A.* 59: 1329–1336.

GUTHRIE, C., H. NASHIMOTO, and M. NOMURA, 1969. Structure and function of *E. coli* ribosomes, VIII. Cold-sensitive mutants defective in ribosome assembly. *Proc. Nat. Acad. Sci. U. S. A.* 63: 384–391.

HARDY, S. J. S., C. G. KURLAND, P. VOYNOW, and G. MORA, 1969. The ribosomal proteins of *Escherichia coli* I. Purification of the 30S ribosomal proteins. *Biochemistry* 8: 2897–2905.

JACOBY, G. A., and L. GORINI, 1967. The effect of streptomycin and other aminoglycoside antibiotics on protein synthesis. *In* Antibiotics, Vol. I (D. Gottlieb and P. D. Shaw, editors). Springer-Verlag, New York, pp. 726–747.

KALTSCHMIDT, E., M. DZIONARA, D. DONNER, and H. G. WITTMANN, 1967. Ribosomal proteins. I. Isolation, amino acid composition, molecular weights and peptide mapping of proteins from *E. coli* ribosomes. *Molec. Gen. Genetics* 100: 364–373.

KURLAND, C. G., 1960. Molecular characterization of ribonucleic acid from *Escherichia coli* ribosomes. I. Isolation and molecular weights. *J. Mol. Biol.* 2: 83–91.

LEWANDOWSKI, L. J., and B. L. BROWNSTEIN, 1966. An altered pattern of ribosome synthesis in a mutant of *E. coli. Biochem. Biophys. Res. Commun.* 25: 554–561.

MCCARTHY, B. J., R. J. BRITTEN, and R. B. ROBERTS, 1962. The synthesis of ribosomes in *E. coli*. III. Synthesis of ribosomal RNA. *Biophys. J.* 2: 57–82.

MACDONALD, R. E., G. TURNOCK, and J. FORCHHAMMER, 1967. The synthesis and function of ribosomes in a new mutant of *Escherichia coli. Proc. Nat. Acad. Sci. U. S. A.* 57: 141–147.

MANGIAROTTI, G., D. APIRION, D. SCHLESSINGER, and L. SILENGO, 1968. Biosynthetic precursors of 30S and 50S ribosomal particles in *Escherichia coli. Biochemistry* 7: 456–472.

NOMURA, M., S. MIZUSHIMA, M. OZAKI, P. TRAUB, and C. V. LOWRY, 1969. Structure and function of ribosomes and their molecular components. *Cold Spring Harbor Symp. Quant. Biol.* 34: 49–62.

NOMURA, M., P. TRAUB, and H. BECHMANN, 1968. Hybrid 30S ribosomal particles reconstituted from components of different bacterial origins. *Nature (London)* 219: 793–799.

OSAWA, S., 1968. Ribosome formation and structure. *Annu. Rev. Biochem.* 37: 109–130.

OZAKI, M., S. MIZUSHIMA, and M. NOMURA, 1969. Identification and functional characterization of the protein controlled by the streptomycin-resistant locus in *E. coli. Nature (London)* 222: 333–339.

TISSIÈRES, A., and J. D. WATSON, 1958. Ribonucleoprotein particles from *Escherichia coli. Nature (London)* 182: 778–780.

TRAUB, P., and M. NOMURA, 1968a. Structure and function of *E. coli* ribosomes, V. Reconstitution of functionally active 30S ribosomal particles from RNA and proteins. *Proc. Nat. Acad. Sci. U. S. A.* 59: 777–784.

TRAUB, P., and M. NOMURA, 1968b. Streptomycin resistance mutation in *Escherichia coli:* Altered ribosomal protein. *Science (Washington)* 160: 198–199.

TRAUB, P., and M. NOMURA, 1969. Structure and function of *Escherichia coli* ribosomes VI. Mechanism of assembly of 30S ribosomes studied *in vitro. J. Mol. Biol.* 40: 391–413.

TRAUT, R. R., P. B. MOORE, H. DELIUS, H. NOLLER, and A. TISSIÈRES, 1967. Ribosomal proteins of *Escherichia coli*, I. Demonstration of different primary structures. *Proc. Nat. Acad. Sci. U. S. A.* 57: 1294–1301.

WEISBLUM, B., and J. DAVIES, 1968. Antibiotic inhibitors of the bacterial ribosome. *Bacteriol. Rev.* 32: 493–528.

# 81 Recent Advances in Understanding Ribosomal Action

## BERNARD D. DAVIS

THREE YEARS AGO in Boulder Rich (1967) reviewed the state of our knowledge of protein synthesis. It was certain by then that a linear messenger RNA (mRNA) molecule is simultaneously translated by a number of ribosomes, on each of which the growing polypeptide chain is covalently attached to a transfer RNA (tRNA) molecule, which in turn is noncovalently held on the ribosome. There were also good grounds for inferring that the elongation of the polypeptide involves its shuttling back and forth between two sites, as described below. Finally, the analysis of the genetic code, by the groups of Nirenberg, Ochoa, Brenner, and Khorana, revealed that the ribosome moving along the

BERNARD D. DAVIS Bacterial Physiology Unit, Harvard Medical School, Boston, Massachusetts mRNA reads its sequence of nucleotides in units of three: each such trinucleotide codon pairs with the complementary trinucleotide anticodon of a particular aminoacyl-tRNA (aa-tRNA) and thus specifies the addition of a particular amino acid.

Further work has provided strong support for the two-site shuttle mechanism. The studies have employed two of the classic approaches of biochemistry: fractionation of the system, and the use of compounds (in this case, mostly antibiotics) that inhibit the process at various stages. The coherence of the results obtained has generated considerable confidence in this model (but does not exclude the possibility, for which there is suggestive evidence, of one or more additional sites for prealigning aa-tRNAs). In the two-site model, the process of adding an amino acid to the nascent chain can be described in terms of a cycle of three steps (Figure 1). In the first, or *recognition*, step, the entering

moved from the A to the P site; in this major shift, the donor tRNA is displaced from the P site, the A site is freed, and the movement of the mRNA brings the next codon up to that site.

## Summary of recent advances

In addition to verifying the two-site model, work of the past three years has also revealed a number of significant new features of ribosome structure and function, which I now review briefly. Most of this work is documented in a recent Cold Spring Harbor Symposium (Cairns, 1969) and has been reviewed in more detail by Lipmann (1969) and by Lengyel and Söll (1969).

DIRECTION OF TRANSCRIPTION, TRANSLATION, AND mRNA DEGRADATION It has been known for some time

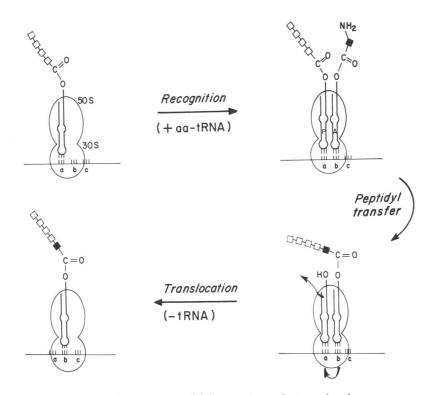

FIGURE 1 The two-site model for protein synthesis on the ribosome.

aa-tRNA is bound to the "A" (aminoacyl) site of the ribosome, while the polypeptidyl-tRNA is held in the "P" (polypeptidyl) site. In the next step, *peptidyl transfer,* the polypeptide is shifted from its tRNA to the amino group of the amino acid on the adjacent tRNA, thus becoming one residue longer. The cycle is completed by the *translocation* reaction, in which the peptidyl-tRNA, still complexed with its complementary codon in the mRNA, is

that the synthesis of mRNA (transcription) and its translation both proceed in the $5' \rightarrow 3'$ direction. In bacteria, in which mRNA turnover is rapid, Morikawa and Imamoto (1969) and Morse et al. (1969) have recently shown that mRNA degradation takes place in the same direction. Moreover, further evidence indicates that while the head end of a growing mRNA molecule is being transcribed on the DNA the tail end is simultaneously being destroyed, and in between as many as 100 or more ribosomes are closely packed on the mRNA and evidently protect it from destruction (Figure 2). In this respect, protein synthesis in bacteria differs from that in eukaryotic cells, in which most mRNA clearly cannot remain attached to the nuclear DNA because it is translated in the cytoplasm.

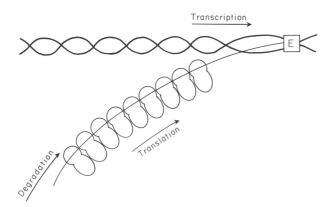

FIGURE 2   Schematic view of the simultaneous formation, translation, and degradation, in the same direction, of the mRNA chain. E: Enzyme (RNA polymerase) transcribing DNA.

PROTEIN COMPOSITION OF RIBOSOMES   It is now generally agreed that the 30S ribosomal subunit contains 19 or 20 different protein molecules; the 50S subunit, about 30. Moreover, no protein is represented more than once; hence the entire surface of the ribosome is highly differentiated. Surprisingly, some ribosomal proteins have been found in fractional amounts (Kurland et al., 1969; Traut and Delius, 1969), but the significance of this observation is not clear: it may mean either that the ribosomes of a cell are heterogeneous in their protein composition, or that the process of purification may have partly removed certain proteins.

RIBOSOME RECONSTITUTION   In an important development, Traub and Nomura (1968) have succeeded in reconstituting fully active 30S subunits from their separated RNA and proteins. The separation requires rather drastic conditions (2 M LiCl + 4 M urea), and even the proteins alone require a high salt concentration to prevent spontaneous aggregation and precipitation. The key to successful reconstitution appears to lie in allowing the proteins to reaggregate with the RNA and with one another at a high enough temperature (40° C) to permit the various molecules to shift position until they have reached the best fit.

THE INITIATING tRNA   In E. coli, the polypeptide chain is initiated by a codon, AUG, that causes the binding not of an aa-tRNA but of formylmethionyl-tRNA (f-met-tRNA) (Adams and Capecchi, 1966); the terminal formyl group is later deleted enzymatically from the growing chain. This feature of initiation is clearly part of a mechanism for preventing the generation of incomplete polypeptides by initiation at codons within a gene: the aa-tRNAs that these codons call for differ from both polypeptidyl-tRNA and f-met-tRNA in having a charged α-amino group; and loss of this charge, which changes the configuration of the tRNA, is evidently required before it can be accepted by the P site.

INITIATION BY SUBUNITS   The construction of the 70S bacterial ribosome in the form of a readily dissociated pair of subunits (50S and 30S) has long been a puzzle. The solution was provided by experiments of Kaempfer et al. (1968) with mixtures of heavy and light ribosomes: in the course of protein synthesis, the subunits were found to exchange partners, producing hybrids of intermediate density. Moreover, Nomura and Lowry (1967) showed at the same time that in a system initiating protein synthesis physiologically, in contrast to the familiar system with synthetic homopolynucleotides as messenger, f-met-tRNA complexes with the initiating codon on a 30S subunit rather than on a complete 70S ribosome. Thus, the macrocycle of ribosome release from the reattachment to messenger involves a sequence of dissociation, formation of an initiation complex by the 30S subunit, and reassociation of the 50S subunit with this complex.

PROTEIN FACTORS   In the course of ribosomal function, several protein factors are attached and detached at specific stages. 1. In the microcycle of chain extension, as Nishizuka and Lipmann (1966) have shown, three separable factors (Tu, Ts, and G) are required. 2. In the release of the completed polypeptide, Capecchi (1967) discovered that the termination codons are read not by a special tRNA but by a release protein, which directly or indirectly causes the peptidyl-tRNA bond to be cleaved. More recently, two proteins, of different codon specificity, have been shown each to have this release function (Scolnick et al., 1968). 3. The use of natural (viral) mRNA revealed that the initiation process involves three protein factors, called F1, F2, and F3 (Stanley et al., 1966; Iwasaki et al., 1968), or factors A, C, and B (Revel and Gros, 1966; Revel et al., 1968), respectively.

This multiplicity of removable proteins functioning on the ribosome suggested at first that the definitions of true ribosomal proteins and of separate protein factors might be arbitrary. This question has been resolved, however, by further work on the roles of the factors, and I wish now to consider this important development in somewhat more detail.

## Role of chain elongation factors

The function of the chain elongation factors (Ts, Tu, and G) has recently been reviewed (Lipmann, 1969). These proteins are present in abundance in the cytoplasm of *E. coli*, constituting in all about 2 per cent of its soluble protein. Ts catalyzes the formation of a Tu-GTP-aa-tRNA complex, which then enzymatically binds the aa-tRNA to the ribosome in the proper position, with the utilization of energy derived from hydrolysis of the GTP. The G factor is required for translocation, which involves the hydrolysis of another molecule of GTP. The Tu factor that is bound along with a GTP is released after the hydrolysis of the latter to GDP and phosphate, and the same is probably true of the bound G factor; these factors thus appear to have only a transient attachment to the ribosome.

The antibiotic puromycin has played an important role in the development of this picture. It serves as an analogue for the aminoacyl end of aa-tRNA and has provided a useful model for the peptidyl transfer reaction. Moreover, it has been helpful in our recognizing and characterizing the translocation step, for it releases the nascent polypeptide as polypeptidyl-puromycin when the peptidyl-tRNA is in the P site but not when it is in the A site. Additional antibiotics with a well-established action on the ribosome are also listed in Table I.

versible synthesis, but also to reconcile firm and highly specific attachment with freedom of movement on the ribosome surface. For example, because the specificity of binding is determined by the relatively small energy contribution of the codon-anticodon interaction, the total energy involved in the initial binding cannot be very large. After this binding, however, the energy contribution from GTP hydrolysis could be used to lock the aa-tRNA in place and thus to permit it to accept peptidyl transfer without danger of losing the incomplete chain.

## Role of initiation factors

In contrast to the large supply of elongation factors in the cytoplasm, the three protein initiation factors are found in a smaller supply. Moreover, this supply is largely attached to the 30S subunits (Eisenstadt and Brawerman, 1967).

In considering the functions of the initiation factors, we must know whether these functions begin before or after the dissociation step that takes place after runoff. For a time, it was widely believed that this dissociation is spontaneous and immediate, for certain kinds of gently prepared extracts were found to contain only polysomes and subunits (Mangiarotti and Schlessinger, 1966); the 70S ribosomes observed earlier would therefore be fragmented polysomes, in which the subunits are held together by

TABLE I

*Specific Actions of Various Antibiotics on the Ribosome*

| Antibiotic | Sensitive Ribosomes* | Action |
|---|---|---|
| Puromycin | A, B | Analogue of aminoacyl end of tRNA: releases polypeptide |
| Tetracycline | B | Blocks recognition |
| Chloramphenicol | B | Blocks peptidyl transfer |
| Sparsomycin | A, B | Blocks peptidyl transfer |
| Fusidic acid | A, B | Blocks translocation |
| Streptomycin | B | (1) Distorts recognition site (with poly-U messenger); (2) Impairs effective binding of aminoacyl-tRNA or pep-tRNA in A site (with natural mRNA). |

* A, ribosomes from cytoplasm of various animal cells; B, ribosomes from bacteria.

The amount of energy expended by the cell in protein synthesis is surprisingly large: two high-energy phosphate bonds are hydrolyzed in attaching an amino acid to tRNA, plus one in aa-tRNA binding and one in translocation. Thus, about 32 kcal are expended to create a peptide bond, the energy of which is about 3 kcal. The surplus of energy is presumably used not only to promote rapid and irre-

complexing with peptidyl-tRNA and mRNA. However, we and others have obtained several kinds of evidence that the 70S runoff ribosome is a significant and stable component of the cell. 1. A method of lysis that avoided certain artifacts yielded polysome-containing extracts of *E. coli*, in which about 10 per cent of the ribosomes were present as 70S particles (Kohler et al., 1968). 2. Under various conditions that caused accumulation of the products of runoff, by promoting runoff or preventing reinitiation, the net increment in these products appeared entirely as 70S particles and not as subunits (Kohler et al., 1968; Algranati et al., 1969). 3. Such runoff ribosomes, and also the "normal" 70S ribosomes in extracts of growing cells, could be distinguished from fragmented polysomes by their greater ease of dissociation at a low $Mg^{++}$ concentration (Ron et al., 1968).

The unexpected constancy of the subunit concentration, in the face of a large increment in the runoff ribosomes, had the further significance of suggesting that the dissociation of runoff ribosomes in the cell might require stoichiometric complexing with a dissociation factor, present in limiting supply. Moreover, because the mixture of known initiation factors can support initiation by 70S ribosomes, and because initiation requires dissociation, this mixture should include the dissociation factor. This prediction was readily verified: the usual preparation of initiation factors, extracted by 1 M $NH_4Cl$ from the ribosomal pellet (or from the 30S fraction), caused dissociation of runoff ribosomes (Subramanian et al., 1968) but not of polysomal ribosomes (Davis et al., 1969). This finding led to the model for the ribosomal macrocycle depicted in Figure 3.

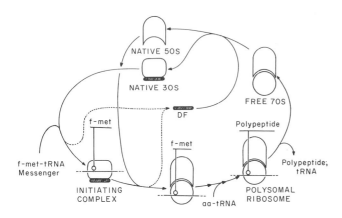

FIGURE 3    Model for the macrocycle of ribosome runoff, dissociation, formation of initiation complex, and chain extension (Subramanian et al., 1968). Not only the dissociation factor (DF), as depicted, but also the other initiation factors (Eisenstadt and Brawerman, 1967) are attached to the native 30S subunit and then are detached at some stage in its conversion to a complete polysomal ribosome.

Subsequent work with purified preparations of the three initiation factors, kindly provided by A. J. Wahba and S. Ochoa, has shown (Subramanian et al., 1969) that the dissociation factor is present in the $F_3$ (Factor B) fraction. Because this factor is known to promote the binding of natural mRNA (Revel et al., 1969; Wahba et al., 1969), it appears that complexing of $F_3$ with the 30S moiety of a 70S ribo-

some may cause both a decrease in its affinity for the 50S moiety and an increase in its affinity for the messenger. This linkage of functions seems reasonable, as dissociation and binding of messenger are evidently the first two steps in initiation. The other two initiation factors are known to promote the next step, the binding of f-met-tRNA by the messenger-30S complex (Revel et al., 1969; Wahba et al., 1969; Kolakofsky et al., 1969).

An unexpected further finding was that the effectiveness of the $F_3$ fraction in promoting dissociation can be increased severalfold by ATP or GTP, in physiological concentrations (Subramanian et al., 1969). A non-hydrolyzable analogue of ATP or GTP also proved to be effective, so it appears that the dissociation reaction does not require an input of energy. These findings raise the possibility of a regulatory response of the dissociation reaction to the "energy level" in the cell.

## General conclusions and speculations

A common feature of the initiation factors and the chain elongation factors is that each factor appears to go on and off the ribosome during the addition of a single amino acid. A major difference is that the elongation factors are found largely free, but the initiation factors are largely bound to the 30S subunits; moreover, the total supply of the latter factors (assuming one molecule of each per 30S subunit) is considerably smaller. This pattern is clearly economical: the initiation factors are available for instant use, as they are stored largely on the 30S subunits; the elongation reaction, which occurs several hundred times as frequently as initiation, requires a large supply of its free factors, comparable to that of aa-tRNAs, in order to provide a high rate of contact.

Although the reason for physiological initiation by one subunit followed by the other, rather than by the whole ribosome, is not certainly known, the striking applicability of simple mechanical principles to other aspects of molecular genetics tempts one to speculate that the subunits form a clamp around the mRNA. In comparison with simple adsorption, such a mechanism could obviously result in firmer attachment without loss of ability to move during translocation.

Because the free ribosomes do not appear to play an active role in the cycle, in contrast to polysomal units and subunits, their existence in the cell requires teleonomic justification. The explanation, I suggest, may lie in the finely adjusted affinities that permit the ribosomes to dissociate at one stage in the macrocycle and to reassociate at another. For the dissociation reaction is not an all-or-none reaction, driven to completion by a large drop in free energy. Rather, it appears to be a reversible equilibrium. The optimal conditions for the succession of dissociation and association may thus require a narrow range of subunit concentration; the pool of runoff ribosomes could then provide a buffer that permits the subunit level to remain constant, despite wide shifts in the rate of protein synthesis (and hence in the level of polysomes).

## Mammalian ribosomes

In eukaryotic cells, in organisms ranging from fungi to mammals, the cytoplasmic ribosomes are known to differ in a number of respects from the ribosomes of the prokaryotic bacteria: they are larger (ca. 80S); they dissociate less readily as the $Mg^{++}$ concentration is lowered; they are inhibited by a different set of antibiotics (with some overlap); and the mRNA that they translate is considerably more stable. Nevertheless, the process of chain extension similarly involves three protein factors and GTP, and the process of runoff and reinitiation similarly involves subunit exchange. However, the initiating tRNA in this system, and the initiation factors (if any), have not been identified.

We may note that evolution has handled the machinery of protein synthesis conservatively: not only is the genetic code universal, but prokaryotic and cytoplasmic eukaryotic ribosomes, despite the difference in their size, can function with some tRNAs and even with the G factor from either class of organisms.

UNSOLVED PROBLEMS   Major problems for the future include definition of the structure of the ribosome (e.g., mapping the relation of the various proteins to one another and to various parts of the RNA), correlation of this structure with the sites inferred from functional studies, and analysis of the conformational changes that are associated with the various steps in ribosomal function. Of these steps, translocation is the most challenging, because it is clearly much more complex than any known enzyme reaction.

At present, there is a wide gap between the chemical studies on ribosomal proteins and the functional studies that have suggested the several sites and steps depicted in Figure 1. I hope that electron micrographs of increased resolution will help to bridge this gap, but the pictures so far available (Figure 4) suggest only a rather complex shape for the 50S subunit, with no definitely recognizable binding sites.

Several possible implications for neurobiology may be seen in these studies on bacterial ribosomes. First, as already noted, much the same mechanism of protein synthesis is found in mammalian cells—and in neurons this process is apparently important for function as well as for growth. Second, the formation of the 30S ribosomal subunit by spontaneous aggregation of its component molecules, described by Nomura in greater detail elsewhere in this volume, encourages the hope that the morphogenesis of cell

membranes can also be reproduced in vitro. Finally, the extension of the study of biosynthesis to longer and longer sequences of enzymatic reactions failed to explain protein synthesis. The solution was found instead in a much more elaborate piece of molecular machinery—one in which a stable aggregate of some 50 different proteins and three RNA molecules cooperates with several transiently attached ligands, and expends considerable free energy, to provide specificity in the basically simple formation of a peptide bond. Perhaps an equally novel and elaborate machinery is involved in some of the mysteries in the development and the function of the nervous system.

As an epilogue, I wish to contrast the central role of genetics in modern biology, including the contributions of molecular genetics to our knowledge of protein synthesis, with the slight role of genetics thus far in the neurosciences. In neurophysiology, to be sure, the possibility of finding useful genetic approaches to most of the current problems seems limited, but in the developmental biology of the nervous system the prospect is exciting and has already attracted several investigators with brilliant records in molecular genetics. Moreover, genetics seems to me to have an even more definite future role in the understanding of behavior, including human behavior. We cannot pretend that the intellectual plasticity provided by the neocortex is infinite, and it would be sad if the application of the advancing science of genetics to the problems of human behavior were restricted by political considerations based on this assumption. If our increasingly complex society builds its social policies on the basis of either liberal or reactionary preconceptions about "human nature," and if this foundation does not correspond to reality, we surely will eventually pay a cruel price. If the education of students in the behavioral sciences regularly included a substantial exposure to genetics, it might help to avoid such an unfortunate development.

## Summary

Recent advances in the study of the ribosome have confirmed the conclusion that the polypeptide chain adds an amino acid by shuttling back and forth between two sites, remaining always attached to the tRNA of the last amino acid added. It has also been shown that this process involves the participation of a protein factor that goes on and off the ribosome in the recognition step in each addition of an amino acid, and another that participates in the translocation step.

The process of chain initiation similarly involves transient attachment of several protein factors, which participate in the several steps of initiation: the dissociation of the runoff ribosomes into subunits; the formation of an initiation complex of the 30S subunit, messenger RNA, and formyl-

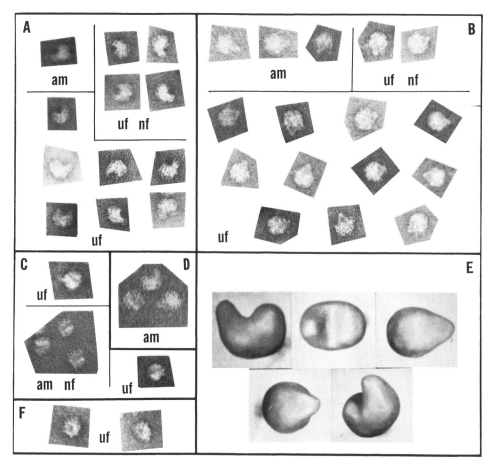

FIGURE 4   Electronmicrographs of 50S ribosomal subunits of *E. coli,* viewed from various angles (Lubin, 1968). Particles were fixed with buffered formaldehyde unless labeled nf; negative stain was ammonium molybdate (am) or uranyl formate (uf). Also included (Panel E) is a model that synthesizes these views.

methionyl-tRNA; and reassociation with the 50S subunit. The cell has thus developed an elaborate machinery, and expends a large amount of energy (four high-energy phosphate bonds per peptide bond formed), not only to promote speed but also to insure specificity, correct initiation, and correct termination in the translation of an RNA message into a polypeptide product.

## Acknowledgment

The research in our laboratory included in this presentation was supported by National Science Foundation Grant #7654.

### REFERENCES

ADAMS, J. M., and M. R. CAPECCHI, 1966. N-formylmethionyl-sRNA as the initiator of protein synthesis. *Proc. Nat. Acad. Sci. U. S. A.* 55: 147–155.

ALGRANATI, I. D., N. S. GONZÁLEZ, and E. G. BADE, 1969. Physiological role of 70S ribosomes in bacteria. *Proc. Nat. Acad. Sci. U. S. A.* 62: 574–580.

CAIRNS, J. (editor), 1969. The mechanism of protein synthesis. *Cold Spring Harbor Symp. Quant. Biol.* 34.

CAPECCHI, M. R., 1967. Polypeptide chain termination in vitro: Isolation of a release factor. *Proc. Nat. Acad. Sci. U. S. A.* 58: 1144–1151.

EISENSTADT, J. M., and G. BRAWERMAN, 1967. The role of the native subribosomal particles of *Escherichia coli* in polypeptide chain initiation. *Proc. Nat. Acad. Sci. U. S. A.* 58: 1560–1565.

IWASAKI, K., S. SABOL, A. J. WAHBA, and S. OCHOA, 1968. Translation of the genetic message. VII. Role of initiation factors in formation of the chain initiation complex with *Escherichia coli* ribosomes. *Arch. Biochem. Biophys.* 125: 542–547.

KAEMPFER, R. O. R., M. MESELSON, and H. J. RASKAS, 1968. Cyclic dissociation into stable subunits and re-formation of ribosomes during bacterial growth. *J. Mol. Biol.* 31: 277–289.

KOHLER, R. E., E. Z. RON, and B. D. DAVIS, 1968. Significance of the free 70S ribosomes in *E. coli* extracts. *J. Mol. Biol.* 36: 71–82.

KOLAKOFSKY, D., K. DEWEY, and R. E. THACH, 1969. Purification and properties of initiation factor f². *Nature (London)* 223: 694–697.

KURLAND, C. G., P. VOYNOW, and S. J. S. HARDY, 1969. Physical and functional heterogeneity of *E. coli* ribosomes. *Cold Spring*

*Harbor Symp. Quant. Biol.* 34: 17–24.

LENGYEL, P., and D. SÖLL, 1969. Mechanism of protein biosynthesis. *Bacteriol. Rev.* 33: 264–301.

LIPMANN, F., 1969. Polypeptide chain elongation in protein biosynthesis. *Science (Washington)* 164: 1024–1031.

LUBIN, M., 1968. Observations on the shape of the 50S ribosomal subunit. *Proc. Nat. Acad. Sci. U. S. A.* 61: 1454–1461.

MANGIAROTTI, G., and D. SCHLESSINGER, 1966. Polyribosome metabolism in *Escherichia coli*. I. Extraction of polyribosomes and ribosomal subunits from fragile, growing *Escherichia coli*. *J. Mol. Biol.* 20: 123–143.

MORIKAWA, N., and F. IMAMOTO, 1969 On the degradation of messenger RNA for the tryptophan operon in *Escherichia coli*. *Nature (London)* 223: 37–40.

MORSE, D. E., R. MOSTELLER, R. F. BAKER, and C. YANOFSKY, 1969. Direction of *in vivo* degradation of tryptophan messenger RNA—A correction. *Nature (London)* 223: 40–43.

NISHIZUKA, Y., and F. LIPMANN, 1966. Comparison of guanosine triphosphate split and polypeptide synthesis with a purified *E. coli* system. *Proc. Nat. Acad. Sci. U. S. A.* 55: 212–219.

NOMURA, M., and C. V. LOWRY, 1967. Phage f2 RNA-directed binding of formylmethionyl-tRNA to ribosomes and the role of 30S ribosomal subunits in initiation of protein synthesis. *Proc. Nat. Acad. Sci. U. S. A.* 58: 946–953.

REVEL, M., and F. GROS, 1966. A factor from *E. coli* required for the translation of natural messenger RNA. *Biochem. Biophys. Res. Commun.* 25: 124–132.

REVEL, M., M. HERZBERG, and H. GREENSHPAN, 1969. Initiator protein dependent binding of mRNA to ribosomes. *Cold Spring Harbor Symp. Quant. Biol.* 34: 261–275.

REVEL, M., J. C. LELONG, G. BRAWERMAN, and F. GROS, 1968. Function of three protein factors and ribosomal subunits in the initiation of protein synthesis in *E. coli*. *Nature (London)* 219: 1016–1021.

RICH, A., 1967. The ribosome—A biological information translator. *In* The Neurosciences: A Study Program (G. C. Quarton, T. Melnechuk, and F. O. Schmitt, editors). The Rockefeller University Press, New York, pp. 101–112.

RON, E. Z., R. E. KOHLER, and B. D. DAVIS, 1968. Magnesium ion dependence of free and polysomal ribosomes from *E. coli*. *J. Mol. Biol.* 36: 83–89.

SCOLNICK, E., R. TOMPKINS, T. CASKEY, and M. NIRENBERG, 1968. Release factors differing in specificity for terminator codons. *Proc. Nat. Acad. Sci. U. S. A.* 61: 768–774.

STANLEY, W. M., JR., M. SALAS, A. J. WAHBA, and S. OCHOA, 1966. Translation of the genetic message: Factors involved in the initiation of protein synthesis. *Proc. Nat. Acad. Sci. U. S. A.* 56: 290–295.

SUBRAMANIAN, A. R., B. D. DAVIS, and R. J. BELLER, 1969. The ribosome dissociation factor. *Cold Spring Harbor Symp. Quant. Biol.* 34: 223–230.

SUBRAMANIAN, A. R., E. Z. RON, and B. D. DAVIS, 1968. A factor required for ribosome dissociation in *Escherichia coli*. *Proc. Nat. Acad. Sci. U. S. A.* 61: 761–767.

TRAUB, P., and M. NOMURA, 1968. Structure and function of *E. coli* ribosomes. V. Reconstitution of functionally active 30S ribosomal particles from RNA and proteins. *Proc. Nat. Acad. Sci. U. S. A.* 59: 777–784.

TRAUT, R. R., H. DELIUS, C. AHMAD-ZADEH, T. A. BICKLE, P. PEARSON, and A. TISSIÈRES, 1969. Ribosomal proteins of *E. coli*: Stoichiometry and implications for ribosome structure. *Cold Spring Harbor Symp. Quant. Biol.* 34: 25–38.

WAHBA, A. J., Y.-B. CHAE, K. IWASAKI, R. MAZUMDER, M. J. MILLER, S. SABOL, and M. A. G. SILLERO, 1969. Initiation of protein synthesis in *Escherichia coli*. I. Purification and properties of the initiation factors. *Cold Spring Harbor Symp. Quant. Biol.* 34: 285–290.

# 82 Extracellular Evolution of Replicating Molecules

## S. SPIEGELMAN

### *The gene and the concept of "self-duplication"*

As a result of experiments to be described, we can, for the first time, explore certain aspects of precellular biological evolution. Before these results are detailed, it may be in-

structive to consider in somewhat general terms the concept of evolution and its significance for "living" and "nonliving" matter.

We begin with a discussion of the minimal conditions required for evolution and the central contributions made to this problem by geneticists at the beginning of the present century.

A necessary condition for evolution is the existence of an entity capable of "self-duplication" or "self-replication."

S. SPIEGELMAN Institute of Cancer Research, Columbia University, New York, New York

We use these terms because of their historical prevalence. They are semantically unfortunate, because they have generated a plethora of unnecessary difficulties. To some, these terms have implied that such entities must be able to function by themselves, unaided by other devices, in the production of copies. This line of reasoning led to holistic arguments that the cell was the smallest unit capable of self-duplication, in accord with pregenetic thinking, which held that the cell and its capacities to produce copies of itself were consequences of a concatenation of interlocking reactions which served to maintain the phenotypic characteristics from one cell generation to the next.

With the recognition of mutations and their implications, this view became increasingly difficult to maintain. To support it, one would have to argue that a single modification in the complex of interacting reactions so modified the whole as to maintain and transmit this unique event—a viewpoint tenable logically, but implausible practically. Further, if this view were correct, unraveling the details of the duplicating mechanism would constitute an inordinately difficult task. In contrast, the geneticists offered a more optimistic alternative. They proposed the existence of cellular entities called *genes,* which alone possessed the capacity of self-duplication. All cellular properties, including self-duplication, were the consequence of the autosynthetic and heterosynthetic instructive capabilities of the genetic material. The implications of the geneticists were therefore both more optimistic and more understandable; they were saying that, if we ever identify the chemical nature of the gene, and thereby gain an understanding of its mode of replication, we will have understood the essence of replication. This prediction was fully confirmed in 1953 with the announcement of the Watson-Crick model of DNA.

## *An operational definition of "self-duplication"*

To be useful to biologists, one must extend the concept of self-duplication beyond that of copy generation. Any copying mechanism that is *absolutely* faithful has no evolutionary future and therefore serves no purpose as a biological model. Selection cannot operate on identical copies, so the potential for evolution demands variation. The existence of variants that permits selection demands a source of variability, which ultimately must come from inaccuracies in the copies. These inaccuracies can, in principle, occur either in the copying process or by modification of some of the copies. Whatever their origin, the changed copies must *retain* the capacity of replication if they are to have evolutionary consequences.

A usable operational definition of a self-duplicating entity has a similar requirement. It is important to recognize at the outset that self-duplication always involves two components. One is the entity being copied and the other is the

copying machine. Essentially, we want to know which one of these two components is the *instructive* agent in the replicating process. It is conceivable that the copying machine is designed to turn out copies only of the particular entity being considered, the duplicated object serving merely as a stimulus for synthetic activity. Were such the case, the entity being copied would not be the instructive agent in the duplicating process, and any mistakes made in the copying process would not be transmitted to future generations.

By a self-duplicating entity, we mean a self-instructive agent that carries the information in its structure for its own synthesis and can transmit this information to a passive synthesizing machine, which fabricates the kinds of copies it is told to. A requirement for an operational definition of a self-instructive replicating agent is the availability of at least two distinguishable variants. Each is then given to the synthesizing machine. If the product produced is always the same and is independent of the particular entity used to initiate the synthesis, we do not have a self-duplicating object. On the other hand, if the copies generated by the synthesizing machine always correspond to the *initiating* variant, we have satisfied the operational definition of an object capable of self-instructive replication. We see, then, that evolutionary potential and a useful operational definition of the class of self-duplicating objects both require a source of variants.

To summarize, we define a self-duplicating—or, better, a "self-instructive replicating" (SIR)—object as one that, in the presence of a suitable synthesizing machine, possesses the following two properties: (1) at least one copy is required to get another, and (2) variants are occasionally produced and can function as instructive agents to initiate copies of themselves.

It should be noted that nothing is said about whether or not the synthesizing machine is also copied. Both situations satisfy our definition. In biological systems, the copying machines are also copied, but other devices are required for *their* fabrication, and all the information needed to synthesize these machines and other ancillary devices required must be contained in the *structure* of the self-instructive replicating object. These and other issues can be seen more clearly in terms of comparatively simple objects we consider below.

## *"Living" and "nonliving," "self-instructive" replicating objects*

It is important to emphasize that our operational definition of self-instructive replication is not meant to provide an operational device to distinguish "living" from "nonliving" matter. In practice, it is not certain that an operational distinction is possible; we return to this question later.

To make the central issue clear, it is instructive to consider some examples of objects that satisfy the two conditions specified and that are not generically derivable from what we would all accept as "living" material.

A somewhat trivial case is provided by Xerox copies in a universe of Xerox machines. One needs one copy to get another. Further, if a particular copy is modified and then fed back into the Xerox machine, the modifications will be reproduced in future copies. Of course, the permissible changes here are limited to those that can be made on paper and that can, in turn, be Xeroxed. All copying devices, however, have similar restrictions of greater or lesser complexity. This system clearly has an evolutionary potential, but of a relatively limited nature.

The Xerox example gains in interest to us, and greatly increases its evolutionary range, when we recognize that the material being copied can, in fact, contain information. In particular, this information could constitute the set of directions for making Xerox machines. We now have a potentially more interesting relation between the Xerox machine and the copies it is generating. If modifications are produced on individual copies and these modifications are, in fact, used to generate copies employed by engineers to build new Xerox machines, the consequence will be an evolution of the Xerox machines themselves. This sort of extension brings us close in principle to another class of devices, the "self-duplicating automata" conceived of by von Neumann (1966).

The von Neumann machines represent logical extensions of devices commonly found in the automatic tool-making industry. In principle, it is possible to devise a machine that can make copies of itself. Such a machine can be placed upon a pile of its component parts. The machine then selects the one required at the particular stage in the manufacture. So that the operations leading to the building of the new machine may contain no inconsistencies, the set of instructions can be put serially on a tape. In such a way, one avoids the possibility that the machine will attempt to do something physically impossible. Thus, if the machine must put a bolt in a hole, it is important that the instruction for drilling the hole precede that for the insertion of the bolt.

After the set of instructions leading to the fabrication of the new machine are finished, one will find a replica of the original machine sitting beside the old one. This is a copying device, but it does not have the essential property of transmitting mistakes, which we require of a self-instructive replicating object. Suppose, however, that, subsequent to the instructions for the fabrication of the new machine, we give another set of instructions that say, "Make a copy of me, the tape, and then transfer it to the new machine." Once this last set of instructions is completed, the newly completed machine contains the set of instructions required to fabricate the next one. We now have a self-instructive replicating device in which everything, including the copying mechanism, is copied.

Imagine that a biologist were to come upon a group of machines such as these busily engaged in making new ones. If he were a geneticist, he would look for variants in the machines (e.g., an abnormal positioning of a hole on a particular plate). Once it was identified, he would isolate it and see if the next generation made by the variant machine would, in fact, inherit the variation. If the new machine produced by the variant did not exhibit the variation, the geneticist would label it a "phenotypic" variant and would conclude that the mistake was not made in *copying* the set of instructions, but rather in *following* the instructions. If, on the other hand, the variation was, in fact, transmitted to the next generation, he would conclude that this variation was genetic, i.e., the mistake was made in copying the set of instructions transmitted to the new machine.

It should be noted that such machines could be instructed to undergo a process akin to that of recombination. Two machines could approach each other and align their tapes so that the two ends coincided. The tapes could then be compared and, if differences were found, the tapes could be cut at some point between the differences and rejoined so that each possessed a portion of the other's tape.

It seems clear that, if a geneticist carried through the kinds of operations on these machines that he does with living matter, he would identify the tapes as the genetic material. The instructional tapes would satisfy his definition of the self-duplicating entities, and copying errors would be identified as genetic mutations.

We now come to the question of the relation of von Neumann machines, and other similar devices, to living material. Perhaps it is important to note at the outset that any mechanism involving the transfer of information from one generation to the next can possess the basic feature of a self-instructive replicating entity. In fact, this feature has generated one of the most important difficulties facing man. When the first man scratched a piece of information on the wall of a cave, he invented a system that has its own genetics. Furthermore, the resulting evolution is *directed* and, as such, is far faster than the selection of random mutations, which has thus far characterized the biological evolution of man. As an example, one might cite the speed with which the memory capacity of computing machines has evolved in the past decade and a half. To achieve the same quantitative improvement by biological evolution would have required hundreds of millions of years. This dramatic difference in the rate of man's evolution compared to the evolution of the devices he invents poses a central issue of man's survival.

Although we are willing to place such devices as the von Neumann self-duplicating automata and living material in the class of self-instructive replicating agents, there is an important difference of origin. We entertain as plausible the hypothesis that the archetype of living objects originated from nonreplicating subunits by polymerization into chains capable of making copies of themselves by a mechanism that initially employed only their own chemical reactivities. We see today, at least dimly, how the same sort of reactions that generate new copies of DNA enzymatically could have taken place nonenzymatically by chemical reactions, which we understand reasonably well. Thus, we are willing to grant the *spontaneous* origin of *living,* self-instructive, replicating entities. It is difficult to see how the same sort of thing could occur with the tapes of the von Neumann automata. In other words, the basic difference is that, in the inanimate case, we are virtually forced to invoke an act of creation by some intelligent agent. The origin of living material demands no such postulate.

## Precellular evolution

We now come to the central issue of our discussion, namely, the evolutionary events that preceded the appearance of cells. Basically, we assume that the first self-duplicating objects were genes, i.e., DNA or RNA replicating by means of the hydrogen-bond recognition device. These are the only cellular macromolecules we know that contain in their chemical structures the inherent potential of carrying through a copying process. A given polynucleotide strand can, in principle, serve as a catalytic surface to attract complementary activated nucleotides that would be automatically aligned in a complementary array on the surface of the strand, ready for subsequent polymerization. That this can happen without the aid of protein catalysts is indicated by recent experiments of Orgel and his collaborators (Sulston et al., 1969).

Cells, as we know them, can be looked upon as inventions of the nucleic acid genetic elements, made to provide a local environment optimally suitable for the production of more genetic material. The progress of cells to multicellular organisms can be interpreted as devices that permit DNA to occupy more and more terrestial space, culminating with the invention of man, who provided DNA with the opportunity to explore extraterrestial possibilities.

One of the most difficult evolutionary phases to examine is the period of precellular evolution. What were the rules of the game? In that period, the environment was looking not at gene products but at the genetic material itself in making decisions about which genetic variant was more suitable for survival.

An approach to such problems would be available if the nucleic acid replicating mechanism could be isolated so that the replication of genetic material could be studied in the test tube. If this could be achieved, in terms of extensive and continuous synthesis, one could begin to vary the environmental parameters in an attempt to uncover the kinds of selective forces that can be imposed directly on replicating genetic material.

In the biological universe, there are two types of organisms: one that uses DNA and another that uses RNA as genetic material. The RNA system was the first to yield an experimental situation in which evolutionary problems could be explored. The probable reason is that, although the general mechanism of replication of RNA is similar to that of DNA, it differs in certain important simplifying details.

The event that stimulated our group to institute an effort to study the RNA replicating system was the discovery by Loeb and Zinder (1961) of the RNA bacteriophage, f2. This was rapidly followed by the isolation of related phages such as MS2 (Strauss and Sinsheimer, 1963), R17 (Paranchych and Graham, 1962), and others. The armamentarium accumulated during several decades of T-phage technology thus became available to persons dealing with RNA viruses. Some of the difficulties and disadvantages inherent in the use of plant and animal systems could now be obviated, and a number of laboratories, including our own, quickly took advantage of the opportunities provided.

In what follows I describe briefly the key experiments that led to the successful isolation of the RNA replicating system, and summarize our efforts to study the extracellular evolution that it made possible.

## The problems of communication between an RNA virus and its host cell

All organisms that use RNA as their genomes are mandatory intracellular parasites. Therefore, they must carry out a major portion of their life cycle in cells that use DNA as genetic material and RNA as genetic messages. On entry, the viral RNA is faced with a problem of inserting itself into the flow pattern of cellular information in order to communicate its own instructions to the synthesizing machinery. One method of so doing depends on whether an RNA virus employs the host DNA-to-RNA-to-protein pathway of information flow. This could take place either because the DNA of the host already contains a sequence homologous to the viral RNA (i.e., the escaped genetic message hypothesis) or because such DNA sequences are generated subsequent to infection by a reversal of transcription (i.e., of the RNA synthesizing reaction that is dependent on DNA). Both hypotheses predict that the DNA of infected cells should contain sequences complementary to viral RNA.

It is clear that a decision on the existence or nonexistence of homology between viral DNA and the infected host DNA is a necessary prelude to further experiments designed to delineate the molecular details of the life history of an RNA genome.

In an attempt to settle this issue, we (Doi and Spiegelman, 1962) employed the specific DNA-RNA hybridization test (Hall and Spiegelman, 1961) combined with the subsequently developed (Yankofsky and Spiegelman, 1962a) use of ribonuclease to eliminate nonspecific pairing. The sensitivity required had already been attained in the course of experiments that identified the DNA complements of ribosomal RNA (Yankofsky and Spiegelman, 1962a, 1962b) and of tRNA (Giacomoni and Spiegelman, 1962; Goodman and Rich, 1962). Under conditions in which the expected hybrid complexes were observed between 23S rRNA and DNA, none was detected between the RNA of the bacteriophage MS2 and the DNA derived from cells infected with this virus. The viral RNA used in these experiments was labeled with $^{32}$P at a specific activity, which would have permitted hybrid detection even if the DNA had contained only one-tenth of an RNA equivalent per genome. This insured a meaningful interpretation of the negative answer.

The negative outcome of the hybridization test implied that the DNA-RNA pathway is not employed in the life cycle of the RNA phages. Further support for this view came from the experiments of Cooper and Zinder (1962), who showed that infection of a mutant that lacked thymine under conditions in which DNA synthesis was suppressed to the extent of 97 per cent resulted in undiminished yields of virus.

It must be recognized that, in a logically rigorous sense, all the evidence provided is negative and, as such, cannot be used to eliminate a proposed mechanism. Nevertheless, the absence of any evidence of DNA involvement was generally accepted to imply that RNA bacteriophages had evolved a DNA-independent mechanism of generating RNA copies from RNA. The next obvious step was to identify and then to isolate the new type of RNA-dependent RNA polymerase (replicase) predicted by this line of reasoning.

## The search for RNA replicase and template specificity

The search for a replicase unique to cells infected with an RNA virus is complicated by the presence of a variety of enzymes which can incorporate the ribonucleotides either terminally or subterminally into pre-existent RNA chains. Further, there are others (e.g., transcriptase, RNA phosphorylase, polyadenylate synthetase) which can mediate extensive synthesis of polynucleotide chains. The complications introduced by these and other enzymes can be ameliorated by suitable adjustments of the assay conditions to minimize their activity. It was clear at the outset, however, that claims for a new type of RNA polymerase would ultimately

have to be supported by evidence for RNA dependence and a demonstration that the enzyme possesses some unique characteristic that differentiates it from the known RNA polymerases.

In addition to these enzymological difficulties, we recognized another potential complication inherent in the fact that an RNA virus must always operate in a heterogenetic environment replete with strange RNA molecules. The point at issue may perhaps best be described in rather naive and admittedly anthropomorphic terms. Consider an RNA virus approaching a cell some $10^6$ times its size and into which the virus is going to inject its only strand of genetic information. This strand codes for the new kind of enzyme required for the replication of the viral genome. Even if the protein-coated ribosomal RNA molecules are ignored, the cell cytoplasm still contains thousands of RNA molecules of various sorts. *If the new RNA replicase were indifferent and copied any RNA molecule it happened to meet, what chance would the original injected strand have of multiplying?*

Admittedly, there are several ways out of this dilemma. One could, for example, segregate the new enzyme and the viral RNA in some corner in which the replication could be carried out undisturbed by the mass of the cellular RNA components. Because it had experimental consequences, however, we entertained the then-unusual hypothesis that the virus is ingenious enough to design a replicase which would recognize the genome of its origin and ignore all other RNA molecules.

Initially, we could not know which solution had been adopted by the virus to solve the dilemma we posed, or even if the dilemma were real. The possibility that it did exist, however, and that template selectivity by the replicase might be the chosen solution meant that the implications could not be ignored; *if true, to disregard them would guarantee that the attempt to isolate the relevant enzyme would inevitably end in failure.* In particular, this meant that, in the search for replicase, we could not afford the luxury of employing any RNA conveniently available in the usual biochemical laboratory. It demanded the use of the specific viral RNA as the challenging template at all stages of enzyme purification. Finally, this line of reasoning could be pushed to its ultimate pessimistic conclusion. It might be that fragments do not possess the proper recognition structures; in that case, a further demand would have to be imposed—the viral RNA employed in the assays during enzyme purification must be monitored for its size to insure intactness. This meant that assays in the presence of contaminating nucleases might be completely meaningless.

THE MS2 REPLICASE  Despite all these potential obstacles, many of which were actually realized, our first success was achieved in 1963 with *E. coli* infected with the MS2 phage (Haruna et al., 1963). A procedure involving negative protamine fractionation combined with column chromatography yielded what resembled a relevant enzyme activity. Most important of all, the preparation exhibited a virtually complete dependence on added RNA for activity, permitting a test of the expectation of specific template requirements. The results of these tests are summarized in Table I. It will be seen that in the responses of the MS2 replicase to the various kinds of nucleic acids a striking preference for its own RNA was revealed. No significant response was observed with the tRNA or the rather similar RNA of the host cell. It would appear that our intuitive guess was confirmed. Producing a polymerase which ignores cellular RNA components guarantees that replication is focused on the single strand of the incoming viral RNA, the ultimate origin of progeny.

The announcement of the specific template requirement of the RNA replicase was greeted with what may best be

## TABLE I

*Template specificity of purified RNA-dependent polymerase*

The standard reaction of 0.25 ml contained the following in $\mu$M: Tris-HC1 pH 7.5, 21; MgCl$_2$, 1.4; MnCl$_2$, 1.0; KCl, 3.75; mercaptoethanol, 0.65; spermine, 2.5; phosphoenolpyruvate (PEA), 1.0; (NH$_4$)$_2$LO$_4$, 70; CTP, ATP, GTP, and UTP, 0.5 each. In addition, it contained pyruvate kinase (PEA kinase), 5 $\mu$g; DNAase, 2.5 $\mu$g and, where indicated, 10 $\mu$g of the polynucleotide being tested as template. Enzyme was assayed at levels between 50 and 300 $\mu$g of protein per sample. DNAase was always omitted in assays for DNA-dependent polymerase activity. Incubations were carried out at 35° C for 10 minutes and terminated by placing the reaction mixture in an ice bath and by the addition of 0.15 ml of neutralized saturated pyrophosphate, 0.15 ml of neutralized orthophosphate, and 0.1 ml of 80 per cent of trichloracetic acid (TCA). The precipitate was washed onto a millipore filter and washed five times with 10 ml of cold 10 per cent TCA containing 0.9 per cent of Na pyrophosphate. The millipore membrane was then dried and counted in a liquid scintillation counter. The pyrophosphate is included in the wash to depress the zero time backgrounds to acceptable levels (40–70 cpm per sample containing input counts of 1 × 10$^6$ cpm). (Haruna et al., 1963).

| Template<br>All at 10 $\gamma$/0.25 ml | NT incorporated in<br>m$\mu$M/10 min/mg protein |
| --- | --- |
| 0 | .08 |
| MS2RNA | 8.5 |
| sRNA | .09 |
| Ribosomal RNA | .06 |
| Ribosomal RNA + MS2 RNA | 8.0 |
| TMV RNA | 0.3 |
| TYMV RNA | 2.2 |
| CT-DNA * | 0.11 |

*DNAase omitted from assay mixture

described as "well-controlled enthusiasm." It should be noted that, thus far, template specificity had not been observed in any of the known nucleic acid polymerases, including the Kornberg enzyme, the RNA phosphorylase, and the RNA polymerase that is dependent on DNA. It should be noted, however, that all these cellular enzymes were evolved in an essentially closed genetic system. These enzymes were rarely faced (except in virus infection) with the problem of deciding whether a particular nucleic acid was genetically related to it or not. In contrast, the RNA-dependent RNA polymerases induced by RNA viruses *always* had to function in a heterogenetic environment and therefore evolved under conditions in which they were continually required to ask the question, "Do you belong to me?"

The fact that we could offer a plausible argument to explain why viral RNA replicases might possess the property of specific template requirements probably only served to increase skepticism. In addition, doubt was simultaneously introduced by August and his coworkers (August et al., 1963; August et al., 1965; Shapiro and August, 1965), who questioned the specificity of the replicase reaction. They prepared an enzyme from *E. coli* infected with the SU-11 mutant of the f2 bacteriophage. The preparation showed partial dependence on RNA, and appeared to respond to a variety of RNA molecules, including ribosomal RNA. As we pointed out (Spiegelman and Haruna, 1966a, 1966b), however, the very large amounts required for stimulation in their systems made it uncertain whether the RNA was functioning as a template or was serving as a scavenger to protect the product of an endogenous reaction from nuclease digestion.

It was not until 1967 (Fiers et al., 1967) that our isolation of the MS2 replicase was repeated and its specific template requirements were confirmed.

CONFIRMATION OF SPECIFIC TEMPLATE REQUIREMENTS WITH THE Q$\beta$ REPLICASE Template specificity was not accepted in 1963, and further evidence was clearly needed to support so novel a concept. Our line of reasoning predicted that replicases induced by other viruses would possess a similar preference for their homologous templates. It seemed desirable, therefore, to check this with another virus, preferably one unrelated to the MS2 group. The Q$\beta$ phage of Watanabe (1964) was chosen because of its immunological and other chemical differences (Overby et al., 1966a, 1966b). In addition, it had the advantage of an excellent difference in the molar ratios of adenine (A) and uracil (U). As a result, one could readily distinguish the viral (plus) strand from its complement (minus) strand, a possibility not available with any of the RNA phages discovered previously. Finally, an unexpected windfall was the remarkable stability of the Q$\beta$ replicase, which made it much easier to handle enzymologically than the replicases of the MS2 group.

The isolation and purification of the Qβ replicase followed, with slight modifications, the procedures worked out earlier for the MS2 replicase. The general properties of the Qβ replicase were similar to those observed in the MS2 replicase, including requirements for all four riboside triphosphates and magnesium. The ability of various RNA molecules to stimulate the Qβ replicase to synthetic activity at saturation concentrations of homologous RNA is recorded in Table II. The response of the Qβ replicase is in accord

TABLE II

*Response of Qβ replicase to different templates*

As in all cases, assay for DNA-dependent activity is carried out at 10 μg of DNA per 0.25 ml of reaction mixture. Input levels of template QβRNA were 1 μg per 0.25 ml. Control reactions containing no template yielded an average of 30 cpm. Numbers represent counts per minute (cpm). (Haruna and Spiegelman, 1965a.)

| Template | Incorporation |
|---|---|
| Qβ | 4929 |
| TYMV | 146 |
| MS2 | 35 |
| Ribosomal RNA | 45 |
| sRNA | 15 |
| Bulk RNA from infected cells | 146 |
| Satellite virus | 61 |
| DNA (10 μg) | 36 |

with what we have previously noted for the MS2 enzyme, the preference being clearly for its own template. The heterologous viral RNAs, MS2, and STNV are completely inactive, as are the ribosomal and transfer RNA species of the host cell. It is important to note that, as in the case of MS2 replicase, the absence of response in DNA shows that our purification procedure eliminates detectable evidence of active transcriptase from our enzyme preparations.

INTACTNESS REQUIREMENT AND THE "AMPHORA" MODEL OF TEMPLATE RECOGNITION An obvious device to explain the requirement for a specific template would invoke the recognition of a beginning sequence, a possibility open to the simple test of challenging the replicase with fragmented preparations of homologous RNA. If the initial sequence is the sole requirement, RNA fragmented to half and quarter pieces should serve as templates. One might perhaps expect that the initial synthetic rate with fragments would be the same as with intact templates, although the reaction might terminate sooner because release of the enzyme from the template might require the terminal se-

quence. A different response would, however, be predicted if a secondary structure requiring intact molecules were involved in the recognition mechanism. It was found (Haruna and Spiegelman, 1965a, 1965b) that randomly fragmented material is unable to stimulate the replicase to anywhere near its full activity. Half pieces achieve approximately 10 per cent of the rate of the intact strand, and the rate obtainable with quarter pieces is only 2 per cent of normal. Further, in both cases the reaction terminates quickly.

The replicase is unable to employ properly the fragments of its own genome as templates, which argues against a recognition mechanism in only the beginning sequence. The enzyme can apparently "sense" when it is confronted with an intact RNA molecule, which implies that some element of secondary structures is involved. A plausible, formal explanation can be proposed in terms of "functional circularity." A decision on the intactness of a polynucleotide chain can be made readily by an examination of both ends for the proper sequences. Such an examination would be physically aided by forming an "amphora"-like structure, using terminal sequences of overlapping complementarity. The enzyme could then recognize the resulting double-standard region. This amphora model is offered only as an example of how an enzyme could simultaneously decide on both sequence and intactness. Whatever the details turn out to be, it appears that the RNA replicases are designed to minimize the futility of replicating either foreign sequences or incomplete copies of their own genome.

## The nature of the product synthesized

PHYSICAL AND CHEMICAL PROPERTIES By 1965, the purification of Qβ replicase had been brought to a stage at which it was largely freed of contaminating nuclease. The enzyme was virtually dependent on added RNA and could mediate almost unlimited syntheses of RNA over long periods of time, as may be seen from Figure 1. It seemed desirable, therefore, to begin a study of the product produced in extensive synthesis. As can be seen in Figure 2, the synthetic material was indistinguishable in size from the RNA obtained from virus, and its base composition was also identical chemically. Naturally, we wanted more subtle information, and turned to our enzyme to provide it.

RECOGNITION OF PRODUCT BY THE REPLICASE The ability of replicase to distinguish one RNA molecule from another can be used to provide information pertinent to the question of the similarity between the original template and the synthesized product. Two sorts of readily performed experiments can show whether the product is recognized by the enzyme as a template. One approach is to examine the kinetics of RNA synthesis at template con-

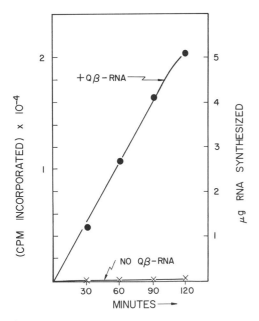

FIGURE 1 Kinetics of replicase activity. Each 0.25 ml contained 40 μg of protein and 1 μg of Qβ RNA. See Figure 3 for further details. The specific activity of the ³²UTP (uridinetriphosphate) was such that the incorporation of 4000 cpm corresponds to the synthesis of 1 μg of RNA. (From Haruna and Spiegelman, 1965a.)

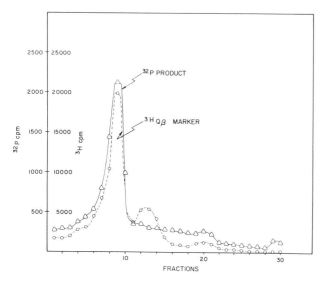

FIGURE 2 Sedimentation analysis of product of extensive syntheses by a purified replicase. A standard reaction mixture contained 50 μg of "purified" enzyme protein and 0.2 μg of Qβ RNA. The reaction was continued for 40 min. at 35° C, during which a 30-fold synthesis had occurred. To 0.1 ml of the reaction mixture were added 0.01 ml of 20 per cent sodiumdodecylsulphate, 0.005 ml ³H-Qβ RNA; 0.20 ml 0.01 м tris, pH 7.4; and 0.005 м MgCl₂. It was then layered onto a linear gradient of 2.5 per cent to 15 per cent sucrose in 0.01 м tris, pH 7.4; 0.005 м MgCl₂; 0.1 м NaCl, and centrifuged at 25,000 rpm for 12 hours at 10° C in the Spinco SW-25 rotor. (Pace and Spiegelman, 1966a.)

centrations that start below those required to saturate the enzyme. If the product can serve as a template, a period of autocatalytic increase of RNA should be observed. Exponential kinetics should continue as long as there are enzyme molecules unoccupied by template. However, when the product saturates the enzyme, the synthesis should become linear.

A second type of experiment is a direct test of the ability of the synthesized product to function as an initiating template. Here a synthesis of sufficient extent is carried out to insure that the initial input of RNA becomes a quantitatively minor component of the end product. The synthetic RNA can then be purified and examined for its template functioning capacities, a property readily examined by a saturation curve. If the response of the enzyme to variations in concentration of product is the same as that observed with authentic viral RNA, one would have to conclude that the product generated in the reaction is as effective a template for the replicase as is RNA obtained from the mature virus particle. We performed both types of experiments (Haruna and Spiegelman, 1965c). Figure 3 shows the

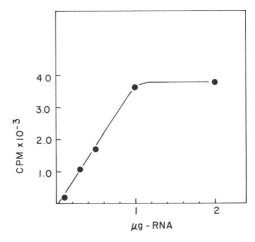

FIGURE 3 Template saturation of "replicase." In addition to 40 μg of enzyme protein, each standard reaction volume of 0.25 ml contained the following: 21 μM tris HCl, pH 7.4; 3.2 μM MgCl₂; 0.2 μM each of CTP, ATP, UTP, and GTP. The reaction is run for 20 min. at 35° C and terminated in an ice bath by the addition of 0.15 ml of neutralized saturated pyrophosphate, 0.15 ml of neutralized saturated orthophosphate, and 0.1 ml of 80 per cent trichloracetic acid. The precipitate is transferred to a membrane filter and washed 7 times with 5 ml of cold 10 per cent TCA. The membrane is then dried and counted in a liquid scintillation spectrometer. The washing procedure yields zero time values of 80 cpm with input counts of 1 x 10⁶ cpm. The radioactively labeled ³²UTP was synthesized as detailed earlier (Figure 1) and was used at a level of 1 x 10⁶ cpm/0.2 μM. The enzyme was isolated from E. coli (Q 13) infected with bacteriophage Qβ as detailed by Haruna and Spiegelman (1965c).

saturation curve we obtained with a fixed amount of enzyme and varying amounts of template input. It will be observed that saturation of the enzyme occurs at a level of 1 microgram of viral RNA per reaction mixture. Syntheses were then started at well below saturation levels; the results obtained are summarized in Figure 4. We note that auto-catalytic synthesis takes place until the amount of RNA produced reaches the 1-microgram level, whereupon it becomes linear. The results are plotted both arithmetically and semilogarithmically to permit easier comparison.

Figure 5 compares the enzyme-saturation ability of RNA synthesized in the test tube with that of RNA obtained

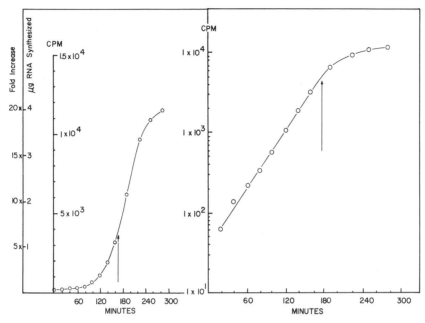

FIGURE 4 The kinetics of RNA synthesis. A 2.5-ml reaction mixture was set up containing the components at the concentrations specified in the legend for Figure 1. The mixture contained 400 $\mu$g of enzyme protein and 2 $\mu$g of input Q$\beta$RNA, so that the starting ratio of template to enzyme was one-fifth of the saturating level. At the indicated times, 0.19-ml aliquots were removed and assayed for radioactive RNA, as detailed in Figure 3. The ordinates for cpm and $\mu$g of RNA synthesized refer to those found in 0.09-ml samples. The data are plotted against time arithmetically on the right, and semilogarithmically on the left. The arrows indicate change from autocatalytic to linear kinetics. (From Haruna and Spiegelman, 1965c.)

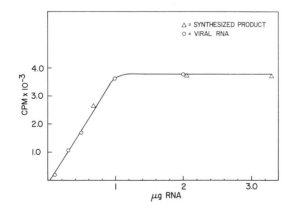

FIGURE 5 Saturation of enzyme by synthesized RNA compared to viral RNA. The experiment was carried out exactly as detailed in the legend for Figure 1. The circles refer to the values obtained with RNA isolated from virus particles; the triangles, to the rates obtained with the RNA synthesized in the experiment of Table II. Because, in the latter case, the template used was labeled with [32]P, [3]H-UTP at 1 x 10[6] cpm per 0.2 $\mu$M was used to follow the synthesis. All preparations and counting of samples were carried out as described in Figure 3. (From Haruna and Spiegelman, 1965c.)

from mature virus particles. The synthesis in this case was 60-fold, so the initial RNA constituted less than 2 per cent of the material being added. It is clear from the results shown in Figure 5 that the response of the enzyme to increasing levels of synthetic RNA is identical with that observed in authentic viral RNA isolated from the Qβ particles.

The data support the assertion that the reaction produces a polynucleotide of the same molecular weight ($1 \times 10^6$) as viral RNA, and one the replicase cannot distinguish from its homologous genome. Evidently the enzyme in the test tube is, at the very least, faithfully copying the recognition sequences employed by the replicase to distinguish one RNA molecule from another.

EVIDENCE FOR THE SYNTHESIS OF AN INFECTIOUS VIRAL RNA The next question concerns the extent of the similarity between product and template. Have identical duplicates been produced? The most decisive test would determine whether the product contains all the information required to program the synthesis of complete virus particles in a suitable test system. The success recorded above encouraged us to attempt the next phase of the investigation, i.e., to subject the synthesized RNA to this more rigorous challenge (Spiegelman et al., 1965).

The ability to perform these experiments depended on the fact that one could introduce viral nucleic acid into protoplasts of bacteria, i.e., bacterial cells from which the cell wall had been removed. Once inside the cells, the RNA is translated and replicated to form complete virus particles which can be assayed by the usual plaque assay on intact cells.

As a first approach, the appearance of newly synthesized RNA was compared with the number of infectious RNA strands during the course of an extensive synthesis. We took aliquots from a reaction mixture to determine the amount of radioactive RNA that was synthesized, and purified the phenol for the infectivity assay. As is illustrated in Figure 6, increase in RNA is paralleled by a rise in the number of infectious units.

This kind of experiment offers plausible evidence for the infectivity of newly synthesized and radioactively labeled RNA. It is not, however, conclusive, because there is still a possibility that the agreement is fortuitous. It must be recalled that the infective efficiency of RNA strands is low—of the order of 1 in $10^6$. One could therefore argue, however implausibly, that the enzyme is activating the input RNA to higher levels of infective efficiency while it is synthesizing new, noninfectious RNA. The concordance of the rather complex combination of exponential and linear kinetics of the two processes would then be just an unlucky coincidence.

To answer issues raised by such arguments, one must design experiments to eliminate the possibility that the input RNA is involved in the infectivity tests. There are several possible approaches to this problem, including the use of heavy isotopes and the separation of newly formed strands by density difference. One can, however, perform a much simpler experiment by taking advantage of the fact that we are dealing with a presumed self-propagating entity.

Consider a series of tubes, each containing 0.25 milliliter of the standard reaction mixture, but with no added template. The first tube is inoculated with 0.2 microgram of Qβ RNA and incubated long enough to synthesize 10 times as much radioactive RNA. A one-tenth aliquot is transferred to the second tube, which is permitted to synthesize about the same amount of RNA, a portion of which is transferred to a third tube, and so on. If each successive synthesis produces RNA that can serve to initiate the next one, the experiment can be continued indefinitely and, in

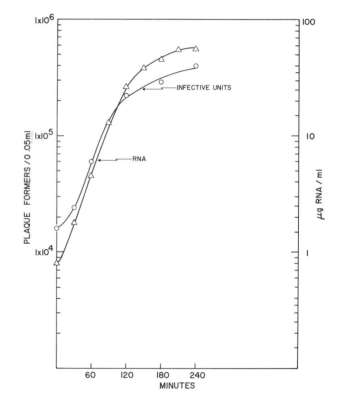

FIGURE 6 Kinetics of RNA synthesis and formation of infectious units. An 8-ml reaction mixture was set up, and samples were taken as follows: one ml at 0 time and 30 min.; 0.5 ml at 60 min.; 0.3 ml at 90 min.; and 0.2 ml at all subsequent times. Twenty λ were removed for assay of incorporated radioactivity. The RNA was purified from the remainder, radioactivity being determined on the final product to monitor recovery. (From Spiegelman et al., 1965.)

particular, until the point is reached at which the initial RNA of tube one has been diluted to an insignificant level. *If in all the tubes, including the last one, the number of infectious units corresponds to the amount of radioactive RNA found, convincing evidence is offered that the newly synthesized RNA is indeed infectious.*

A complete account of such a serial transfer experiment

unit ascribable to the initiating RNA, and the fifteenth tube contained less than one strand of the initial input. Nevertheless, every tube, including the last, showed an increment in infectious units corresponding to the amount of radioactive RNA found. In other words, the newly synthesized RNA was as infectious as the material obtained originally from the virus particle.

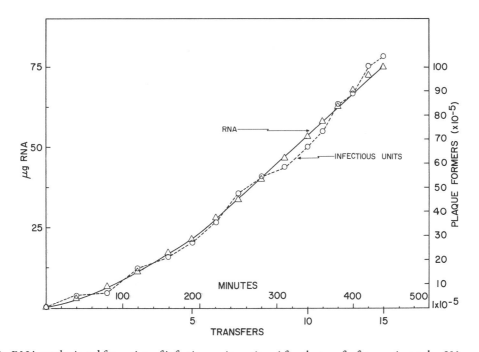

FIGURE 7   RNA synthesis and formation of infectious units in a serial transfer experiment. Sixteen reaction mixtures of 0.25 ml were set up, each containing $40\gamma$ of protein and the other components specified for the "standard" assay. $0.2\gamma$ of template RNA was added to tubes 0 and 1; RNA was extracted from the former immediately, and the latter was allowed to incubate for 40 minutes. Then $50\lambda$ of tube 1 was transferred to tube 3, and so on, each step after the first involving a 1 to 6 dilution of the input material. Every tube was transferred from an ice bath to the 35° C water bath a few minutes before use to permit temperature equilibra- tion. After the transfer from a given tube, $20\lambda$ was removed to determine the amount of $^{32}$P-RNA synthesized, and the product was purified from the remainder. Control tubes incubated for 60 minutes without the addition of the $0.2\gamma$ of RNA showed no detectable RNA synthesis, nor any increase in the number of infectious units. All recorded numbers are normalized to 0.25 ml. The ordinates represent cumulative increases of infectious units and radioactive RNA in each transfer; the abscissae record elapsed time and the transfer number. Further details are to be found in Spie- gelman et al. (1965).

can be found in Spiegelman et al. (1965), and Figure 7 describes the outcome. We knew the molecular weight and the amount of RNA put into the initial tube, so we could readily calculate how many tubes should be used to insure that the last tube contained less than one strand of the initial input. In this experiment, aside from the controls, 15 transfers were made, each resulting in a dilution of one to six. By the eighth tube there was less than one infectious

## A rigorous proof that the added RNA is the self-duplicating entity

The central issue I now consider is one I have already noted in my introductory remarks. It stems from the fact that two informed components are present in the reaction mixture— replicase and RNA template. None of the experiments thus far described *proved* that the RNA synthesized in this system is a self-duplicating entity. As we have seen, what is required in such situations is a rigorous demonstration that the RNA, not the replicase, is the instructive agent in the replicative process.

A definitive decision would be provided by an experimental answer to the following question: If the replicase is provided alternatively with two distinguishable RNA molecules, is the product always identical to the initiating template? A positive outcome would establish that the RNA is directing its own synthesis and simultaneously eliminate any remaining possibility that the RNA present as a contaminant of the enzyme preparation is activated. Experiments to settle these issues were undertaken by Pace and Spiegelman (1966a, 1966b).

The discriminating selectivity of the replicase for its own genome as a template makes it impossible to employ any kind of heterologous RNA in the test experiments, and we took recourse in mutants. For ease in isolation and simplicity in distinguishing between mutant and wild phenotype, temperature-sensitive (ts) mutants were chosen. The ts mutant grows poorly at 41° C as compared with 34° C. The wild type grows equally well at both temperatures.

We should be able to determine whether the product produced by a normal replicase primed with ts-Qβ RNA is mutant or wild type. As in previous investigations, this is done best by a serial transfer experiment, to avoid the ambiguity of examining reactions containing significant quantities of initiating RNA. The basic experiment is as follows: The mutant virus is grown at the permissible temperature and the RNA is extracted. This RNA is then presented to a wild-type enzyme, and the virus particles produced from the newly synthesized RNA in bacterial protoplasts are examined. If they retain the temperature-sensitive phenotype, it is clear that the RNA is the instructive agent. If, on the other hand, the RNA is wild type, it is the enzyme that is geared to synthesize RNA, and the initiating template simply serves as a stimulator of synthesis.

Accordingly, we prepared seven standard reaction mixtures, each containing 60 micrograms of Qβ replicase, isolated from cells infected with *normal* virus. To the first reaction mixture we added 0.2 microgram of tsRNA and allowed synthesis to proceed at 35° C. After a suitable interval, we used one-tenth of this reaction mixture to initiate a second reaction, which, in turn, was diluted into a third, and so on for seven transfers. A control series was carried in a manner identical to that just described, except that we added no RNA to the first tube. We examined aliquots from each reaction for radioactivity in the material that was precipitated by trichloracetic acid and also assayed for infectious RNA at 34° and 41° C. Figure 8 summarizes the outcome of the experiment in a cumulative plot of the RNA synthesized and the plaque formers detected at the two test temperatures. It is clear that the RNA synthesized

has the ts phenotype. Although the plaque formation tested at 34° C increases in parallel with the new RNA that is synthesized, no such increase takes place when the tests are carried out at 41° C. It should be noticed (on the upper panel of Figure 8) that no significant synthesis of either RNA or infectious units was observed in the control series of tubes that lacked initiating templates.

## Extracellular Darwinian experiments with the replicating RNA molecules

The experiments just described demonstrated a specific response of one and the same enzyme preparation to the particular template added. This proved that the RNA is the instructive agent in the replicative process and hence

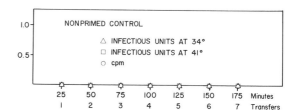

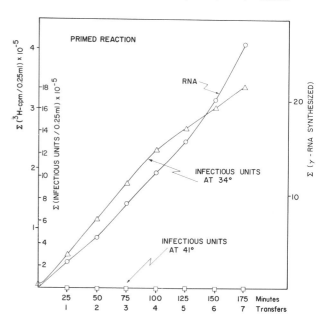

FIGURE 8 Synthesis of mutant RNA. Each 0.25-ml reaction contained 60 μg of Qβ replicase purified through CsCl and sucrose centrifugation. The first reaction was initiated by addition of 0.2 μg of tsRNA. Each reaction was carried out at 35° C for 25 min., whereupon 0.02 ml was withdrawn for counting and 0.025 ml was used to prime the next reaction. All samples were stored frozen at −70° C until infectivity assays were carried out at 41° C and 34° C. A control series was carried out in which no initiating RNA was added. (From Pace and Spiegelman, 1966b.)

satisfies the operational definition of a self-duplicating entity. An opportunity is thus provided for studying the evolution of a self-replicating nucleic acid molecule outside a living cell. It should be noted that this situation mimics an early, precellular evolutionary event, when the environmental selection presumably operated directly on the genetic material, rather than on the gene product. The comparative simplicity of the system and the accessibility of its known chemical components to manipulation permit the imposition of a variety of selection pressures during growth of replicating molecules. As we noted earlier, this experimental situation should permit us to explore at least some of the rules of the game that must have been played in the evolution of living material before the advent of cells.

In the universe provided to them in the test tube, the RNA molecules are liberated from many of the restrictions derived from the requirements of the complete viral life cycle. The only restraint imposed is that they retain whatever sequences are involved in the recognition mechanism employed by the replicase. Thus, sequences which code for the coat protein, replicase, and attachment protein components may now be dispensable. Under these circumstances, it is of no little interest to design an experiment which attempts an answer to the following question: What will happen to the RNA molecules if the only demand made on them is the biblical injunction, "Multiply," with the biological proviso that they do so as rapidly as possible? The conditions required are readily obtained by a serial transfer experiment in which the intervals of synthesis between transfers are adjusted to select the earliest molecules completed. If the sequences coding for the three viral proteins are unnecessary for the replicative act, these sequences become so much excess genetic baggage. Because the longer a chain is, the longer it takes to complete, molecules would gain an advantage by discarding any unneeded genetic information to achieve a smaller size and therefore a more rapid completion.

*Isolation of Fast-Growing Variant (V-1)* Mills et al. (1967) performed a series of 75 transfers (Figure 9), during which each reaction mixture was diluted 12.5 times in the next tube. The incubation intervals were reduced periodically as faster and faster growth was achieved. As can be seen from the insert in Figure 9, although RNA continued to be synthesized, the formation of infectious RNA ceased after the fourth transfer. A dramatic increase in the rate of RNA synthesis occurred after the eighth transfer.

It should be noted that this sort of evolutionary experiment has its own built-in paleology, for each tube can be frozen and its contents expanded whenever one decides to examine what happened at that particular period in the evolutionary process. We illustrate this by analyzing the products of several of the transfers in a sucrose gradient. The product of the first reaction (Figure 10) shows the 28S

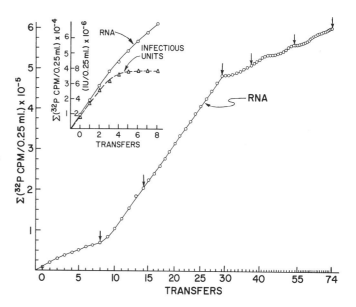

FIGURE 9 Serial transfer experiment. Each 0.25-ml standard reaction mixture contained 40 μg of Qβ replicase and ($^{32}$P) UTP. The first reaction (0 transfer) was initiated by the addition of 0.2 μg ts-1 (temperature-sensitive RNA) and incubated at 35° C for 20 min., whereupon 0.02 ml was drawn for counting and 0.02 ml was used to prime the second reaction (first transfer), and so on. After the first 13 reactions, the incubation periods were reduced to 15 min. (transfers 14–29). Transfers 30–38 were incubated for 10 min. Transfers 39–52 were incubated for 7 min., and transfers 53–74 were incubated for 5 min. The arrows above certain transfers (0, 8, 14, 29, 37, 53, and 73) indicate where 0.001–0.1 ml of product was removed and used to prime reactions for sedimentation analysis on sucrose. The inset examines both infectious and total RNA. The results show that biologically competent RNA ceases to appear after the fourth transfer. (From Mills et al., 1967.)

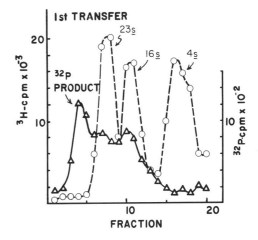

FIGURE 10 Sedimentation analysis of first transfer reaction. 0.002 ml of the 0 reaction was used to initiate a reaction for a first transfer reaction product. After completion, this reaction was adjusted to 0.2 per cent SDS, an aliquot was withdrawn, diluted to 0.2 ml in TE (tris-EDTA) buffer (0.01 M tris, pH 7.4, 0.003 M EDTA), and then layered onto a 5-ml linear sucrose (2–20 per cent in 0.1 M tris, pH 7.4, and 0.003 M EDTA). $^3$H-labeled bulk RNA of *E. coli* was included as internal size markers. (From Mills et al., 1967.)

peak characteristic of mature Qβ RNA, as well as the peaks corresponding to the usual replicative complexes observed during in vitro synthesis (Mills et al., 1966). Products of subsequent transfers showed a gradual shift in the RNA to smaller S values. By the thirtieth transfer no molecules larger than 16S were found. By the fifty-fourth transfer, a single peak was found at about 15S, and, by the seventy-fifth transfer, this peak had moved to a value of 12S (Figure 11). This single peak is due to the poor resolving

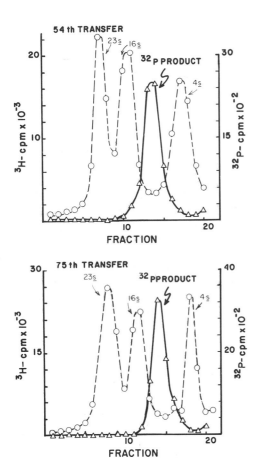

FIGURE 11 Sedimentation analysis of the 54th and 75th transfers. Details as in Figure 10. (From Mills et al., 1967.)

power of sucrose gradients and is composed of several components, as is easily shown by analysis on acrylamide gels, in which the replicative intermediates are readily resolved from the single-stranded forms. We determined the size of the variant by using acrylamide gel electrophoresis with internal markers. The position of the variant found in the seventy-fourth transfer (V-1) indicates that it has a molecular weight of $1.7 \times 10^5$ daltons, corresponding to

550 nucleotides. This means that, in response to the selection pressure for fast growth, the replicating molecule discarded 83 per cent of its genetic information.

The principal purpose of the experiment was to illustrate the potentialities of the replicase system for our examining the extracellular evolution of a self-replicating nucleic acid molecule. The seven samples we examined indicate that progress to a smaller size occurs in a series of steps.

Eighty-three per cent of the original molecule has been discarded, which means that neither the specific recognition nor the replicating mechanism requires the entire original sequence. In this connection, it should be noted that these variants, although abbreviated, are not equivalent to random fragments, because, as we have already seen, the latter are unable to complete the replicative act.

The availability of a molecule that has eliminated large and unnecessary segments provides an object with obvious experimental advantages for the analysis of many aspects of the replicative process. Finally, these abbreviated RNA molecules have a high affinity for the replicase but are no longer able to direct the synthesis of virus particles. This opens up a pathway toward a highly specific device for interfering with viral replication.

*Cloning of Self-replicating Molecules* In principle, the Qβ replicase system should be capable of generating clones descended from individual strands. The resulting clones would provide the sort of uniformity required for sequence and genetic studies. We (Levisohn and Spiegelman, 1968) succeeded in demonstrating that this can be attained.

As a first step, we isolated a new mutant, characterized by an ability to initiate synthesis at a low level of input. This was clearly necessary for our attempt at cloning. The selection of this type of mutant was accomplished by modifying the earlier, serial-selection procedures described in Figure 7. As shown in this Figure, the time intervals between transfers were decreased, but the dilution remained constant. In the experiments to be described, the time was kept constant and the dilution was increased as the transfers were continued. During a serial transfer experiment, the incubation interval was held constant at 15 minutes, and increasing selection pressure was achieved by recurrent sharp decreases in the dilution experienced by successive transfers from $1.25 \times 10^{-1}$ to $2.5 \times 10^{-10}$.

The variant RNA that evolved by the seventeenth transfer was selected for further studies and was called Variant 2 (V-2). Figure 12 compares the response of the two variants, V-1 and V-2, to low-level inputs of template RNA in a 15-minute incubation at 30 degrees. V-2 is clearly superior in growth at levels below 100 μμg of RNA. In fact, V-2 can initiate synthesis with 0.29 μμg of RNA which corresponds to the weight of one strand. A kinetic analysis reveals that, during the exponential growth phase, V-2 has a doubling time of 0.403 minute as compared with 0.456 minute for V-1. In a 15-minute period of logarithmic growth, V-2 can experience more doublings (i.e., a 16-fold increase) than can V-1. V-1 and V-2 have a similar electrophoretic mobility on polyacrylamide gels and the same base composition. They do, however, differ in sequence, as determined by oligonucleotide fingerprint patterns.

We obtained clones of RNA molecules in vitro by using an approach that depends on a straightforward comparison of the observed frequency distribution in a series of replicate syntheses with the frequency expected from the Poisson distribution. If one strand is sufficient to initiate synthesis, the proportion of tubes showing no synthesis should correspond to $e^{-m}$, m being the average number of strands per tube. Further, if the onset and syntheses are adequately synchronized, one should be able to identify tubes that received initially one, two, or three strands. These tubes should appear with frequencies corresponding to $me^{-m}$, $(m^2/2!)e^{-m}$ and $(m^3/3!)e^{-m}$, respectively.

A reaction mixture containing 0.29 $\mu\mu$g of V-2 RNA per 0.1 milliliter was distributed at 0° C in 0.1-milliliter portions into each of 82 tubes. The tubes were placed simultaneously in a bath of 38° C. and, after a 30-minute interval, the reactions were stopped simultaneously. If the assumption underlying the Poisson distribution has been satisfied by the conditions of the experiment, 36.8 per cent of the 82 tubes, that is to say 30.2, should show no synthesis. This is in excellent agreement with the 30 tubes found in Table III.

### Table III

*Distribution of template strands at an average input of one template strand per tube*

A reaction mixture in which 13.2 $\mu$g Variant-2 RNA were synthesized was diluted to a final concentration of 2.9 $\mu\mu$g/ml RNA into standard reaction mixture containing $7.8 \times 10^5$ cpm $^{32}$P-GTP per 0.25 ml. Quantities of 0.1 ml were distributed into each of 82 tubes. After incubation of 30 minutes at 38° C, the acid-insoluble $^{32}$P-GTP was determined. The average is for the actual number of counts after subtraction of the average of nine control tubes (22 cpm) to which no variant RNA was added. In this experiment 3240 cpm is equivalent to 1 $\mu$g of variant RNA. (Levisohn and Spiegelman, 1968.)

| Strands per Tube | P(r) | Among 82 tubes | |
| --- | --- | --- | --- |
| | | Expected | Found |
| 0 | 0.368 | 30.2 | 30 |
| 1 | 0.368 | 30.2 | 29 |
| 2 | 0.184 | 15.1 | 19 |
| 3 | 0.0613 | 5.4 | 3 |
| 4 | 0.0153 | 1.3 | 1 |

To identify tubes inoculated with one, two, three, or more strands, the sum of the counts observed in all tubes was divided by the number of template strands. The result, 309 counts per minute, is the average amount ascribable to a single template strand. The actual incorporation observed in each tube was divided by 309, and the result was rounded out to the nearest integer to yield an approximation of the number of strands that initiated synthesis in that tube. Table III shows a good agreement between the results expected from the Poisson distribution and those actually found. It is highly probable that a tube exhibiting an incorporation close to 309 cpm was, in fact, initiated by a single strand.

An interesting outcome of the last experiment is that 0.29 $\mu\mu$g indeed satisfied the Poisson expectation for an average of one strand of RNA. This permits an independent estimation of molecular weight as 174,000 daltons for V-2 RNA, in good agreement with the value deduced from gel electrophoresis.

The ability to clone variant RNA molecules in vitro has already proved useful for sequence analysis of V-2 (Bishop et al., 1968). The molecules also provide a necessary prerequisite for a genetic analysis.

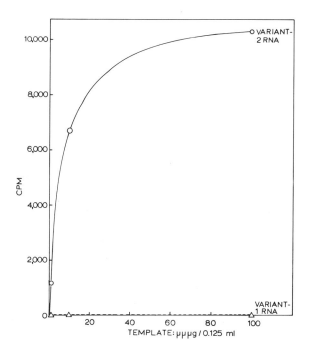

FIGURE 12 Comparison between the synthesis of Variant-1 (V-1) RNA and V-2 RNA when used as templates at very low concentrations. The indicated concentrations of V-1 RNA and V-2 RNA were added as templates to 0.125 ml of standard reaction mixture. After a 15-min. incubation at 38° C, the $^{32}$P-UTP incorporated into acid-insoluble material was determined. (From Levisohn and Spiegelman, 1968.)

## Diverse variants isolated under different selective conditions

The experiments thus far described were concerned with the isolation of mutants possessing increased growth rates under standard conditions. We now turn our attention to a question of no little theoretical and practical interest and inquire whether other mutant types can be isolated. In effect, we are asking the following question: can qualitatively distinguishable phenotypes be exhibited by a nucleic acid molecule under conditions in which its information is replicated but never translated? We (Levisohn and Spiegelman, 1969) undertook to answer this question and, as will be seen, our results show that numerous distinguishable variants can be isolated, the number depending on the ingenuity expended in designing the appropriate selective conditions.

SELECTION OF "NUTRITIONAL MUTANTS" One general approach for obtaining a variety of mutants is to run the syntheses under less-than-optimal conditions with respect to a component or parameter of the reaction. If a variant arises that can cope with the imposed suboptimal condition, continued transfers should lead to its selection over the wild type. This can be done with variations in the level of the ribonucleoside triphosphates. It can be seen in Figure 13 that the rate of synthesis of V-2 begins to decrease sharply as the level of cytidine triphosphate (CTP) drops below 20

millimicromoles per 0.125 milliliter. At a CTP concentration of 2 millimicromoles, the rate of synthesis of V-2 is only 25 per cent of normal. At 1 millimicromole of CTP, the rate decreases to 5 per cent of normal. With such information available, a search was made for variants that could replicate better than V-2 on limiting levels of CTP. A serial transfer experiment at 2 millimicromoles of CTP per reaction was initiated, with Qβ RNA culminating after 10 transfers in the appearance of V-4. A second series of transfers at 1 millimicromole of CTP per reaction was then started with V-4 and after 40 transfers led to the isolation of V-6.

Variants V-4 and V-6 replicate 28 per cent and 56 per cent better, respectively, than does V-2 at low levels of CTP. This increased capacity might be explained on the basis of smaller sizes or modification of base composition toward a lower cytosine content. On polyacrylamide gel V-2 and V-6 showed no significant difference in chain length. Furthermore, as can be seen in Table IV, there is no significant difference in base composition.

### TABLE IV

*Base ratios of variants*

The base composition of purified plus strands of Variant-2 and Variant-6. The numbers represent mole per cent. (Levisohn and Spiegelman, 1969.)

| Base | V-2 | V-6 |
|------|------|------|
| C | 24.8 | 24.8 |
| A | 23.2 | 23.5 |
| G | 26.6 | 26.7 |
| U | 25.4 | 25.1 |

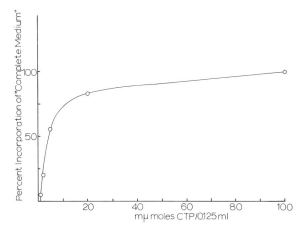

FIGURE 13   CTP concentration curve. Reaction mixtures of 0.125 ml containing the indicated concentrations of CTP and otherwise identical to "standard reaction mixture" were incubated for 30 min. at 38° C with 20 μg Qβ replicase and 0.01 μg V-2 RNA. Subsequently, the acid-insoluble radioactivity was determined. In the "complete medium" (containing 100 mμM CTP) 1.6 μg of V-2 RNA were synthesized. (From Levisohn and Spiegelman, 1969.)

Evidently the modifications leading to the properties possessed by variants 4 and 6 do not involve massive modifications in the composition of the molecule. The identification of the changes will require more subtle examination by nucleotide fingerprint patterns, and these are being carried out.

The most obvious pathways for solving the problem of low CTP were not employed, which leads one to consider more sophisticated devices for achieving the desired end result. It is useful here to recall that the mutant RNA molecules must form a complex with replicase, so changes in sequence that would leave such gross features as base composition and size unmodified could, nevertheless, lead to different secondary structures of the mutant molecules. These, in turn, could have allosteric effects on the replicase, permitting the complex to employ CTP more effectively at suboptimal concentrations. If such were the case, and there were a common site for the four riboside triphosphates as there is in DNA polymerase, as shown by Atkinson et al.

(1969), it might be expected that a mutant selected for better replication on low CTP would also exhibit increased capacities to accommodate to low levels of the other riboside triphosphates. Table V summarizes data comparing

ATP concentration. Variant 6 was chosen to start a series of transfers in the reaction mixture that contained 1.5 milli-micromoles of ATP, which led to the isolation of variant 8. The doubling time of variant 8 in the reaction mixture with

TABLE V

TABLE V

*Logarithmic synthesis of variant RNA on limiting media*

The data are based on experiments performed in standard reaction mixtures and in reaction mixtures with only 4 m$\mu$moles of ATP, 9 m$\mu$moles of GTP, or 2 m$\mu$moles of UTP. (Levisohn and Spiegelman, 1969.)

| Limiting Nucleotides | m$\mu$moles | Variant | Doubling Time Min. | Relative Slope | |
|---|---|---|---|---|---|
| | | | | Compared with V-2 | Compared with V-4 |
| None | 100 | 2 | 0.42 | 1.00 | |
| | | 4 | 0.42 | 1.00 | 1.00 |
| | | 6 | 0.35 | 1.21 | 1.21 |
| ATP | 4 | 2 | 2.41 | 1.00 | |
| | | 4 | 1.54 | 1.56 | 1.00 |
| | | 6 | 1.47 | 1.64 | 1.05 |
| CTP | 1 | 2 | 1.81 | 1.00 | |
| | | 4 | 1.41 | 1.28 | 1.00 |
| | | 6 | 1.16 | 1.56 | 1.21 |
| GTP | 9 | 2 | 3.05 | 1.00 | |
| | | 4 | 2.25 | 1.35 | 1.00 |
| | | 6 | 2.31 | 1.31 | 0.97 |
| UTP | 2 | 2 | 2.06 | 1.00 | |
| | | 4 | 1.69 | 1.22 | 1.00 |
| | | 6 | 1.54 | 1.33 | 1.09 |

the logarithmic synthesis of variants V-4 and V-6 with that of V-2 on limiting levels of each of the four riboside triphosphates. The data show that the two variants selected on low CTP also do much better on limiting concentrations of the other three substrates. More definitive delineation of the underlying mechanism will require studies of substrates that bind with enzyme-complex and mutant and wild-type templates.

SELECTION OF A VARIANT RESISTANT TO AN INHIBITORY ANALOGUE Tubercidin is an analogue of adenosine in which the nitrogen atom in position 7 is replaced by a carbon atom. Tubercidin triphosphate (TuTP) inhibits the synthesis of Q$\beta$ RNA in vitro. However, TuTP cannot completely replace adenosine triphosphate in the reaction, although it can be incorporated into the growing chains. It was of some interest to see whether one could derive a mutant that would show resistance to the presence of this agent. In such experiments, it was desirable to have the ratio of the analogue to ATP as high as possible. To attain this more readily, we isolated a new variant by limiting the

1.5 millimicromoles of ATP was 2.8 minutes as compared with 8.4 minutes for V-6, the starting variant.

The replication rate of V-8 in a reaction mixture that contained 5 millimicromoles of ATP was inhibited fourfold with the additon of 30 millimicromoles of TuTP. A serial transfer was initiated with V-8 in the inhibitory medium and led to the isolation of V-9. The doubling time of V-9 in the presence of TuTP was 2.0 minutes as compared with 4.1 minutes for V-8. In the absence of TuTP, both variants synthesized with a 1.0-minute doubling time. It is clear that V-9 exhibits a specifically increased resistance to the inhibitory effect of TuTP. The resistance mechanism does not involve a more effective exclusion of TuTP. At 30 millimicromoles of TuTP and 5 millimicromoles of ATP, the ratio of U to A in the product was 3.6 for V-8 and 3.5 for V-9, the resistant mutant.

Table VI lists the variants isolated in the experiments described and summarizes the relevant information on their origins and conditions of selection. It will be noted that V-4 is an independent derivative from the parental Q$\beta$RNA. Another variant, V-3 (not listed), was isolated

## TABLE VI

*Conditions used in isolation of variants*

Variants were selected on standard reaction mixture, or on a standard reaction mixture modified to contain one of the four nucleoside-triphosphates at the indicated concentration. Starting with the RNAs indicated in column 3, a series of transfers were made with reaction product, diluted $1.25 \times 10^4$. Subsequently, the dilution factor between transfers was gradually increased to about $1 \times 10^{11}$. (Levisohn and Spiegelman, 1969.)

| Variant | Selective Limitations | RNA Used to Start Selection | No. of Transfers at Dilution of $1.25 \times 10^4$ | Total No. of Transfers |
|---------|----------------------|------------------------------|------------------------------------------------------|------------------------|
| V-2[6] | None | Qβ | 0 | 17 |
| V-4 | 2 mμmoles CTP | Qβ | 5 | 10 |
| V-6 | 1 mμmoles CTP | V-4 | 30 | 40 |
| V-8 | 1.5 mμmoles ATP | V-6 | 11 | 16 |
| V-9 | 5 mμmoles ATP +30 mμmoles TuTP | V-8 | 15 | 19 |

with limiting CPT, starting with V-2 instead of Qβ RNA. V-3 possesses phenotypic properties indistinguishable from those of V-4; thus one can arrive at the V-4 phenotype either from Qβ RNA or from V-2. It seems probable that Qβ RNA passes through the V-2 stage before arriving at the V-4 phenotype.

It will be of no little interest to compare the actual sequence changes among the mutants that differ in their relatedness and phenotypic properties. With this information available, one can begin to construct the probable modifications of the secondary structure. Only then will we be in a position to begin the attempt to understand the molecular basis of these new phenotypes.

## Some theoretical and practical implications of extracellular Darwinian experiments

We pointed out previously that extracellular Darwinian selections may mimic that aspect of precellular evolution when environmental selection operated only on the replicating gene and not on the gene product. Such experiments provide some insight into the rules of these early stages of evolution. It was not obvious *a priori* what kinds of selective forces could be operative, as much depended on how many different ways a molecule could be selected by the environment as superior. The experiments reported here reveal an unexpected wealth of phenotypic differences that a replicating nucleic acid molecule can exhibit. It is true that many of these involve interactions between nucleic acid molecules and a highly evolved protein catalyst. It is possible, however, to imagine similar types of interactions with a primitive surface catalyst. Sequence changes which would increase the catalytic capacity slightly could have powerful selective effects in these precellular stages of evolving genetic material.

It should not escape the attention of the reader that this phenomenon provides a possible solution to the following puzzling question: What pressures could have forced replicating nucleic acids toward greater complexity before they invented cells or subcellular components to help them replicate?

It is apparent, from the limited number of examples described, that a host of new mutant types possessing predetermined phenotypes can be isolated by varying other parameters of the system. In addition, one can expand the possibilities by introducing neutral agents (for example, proteins) with which the replicating molecules may interact. Selection can then be exerted to favor variants that can induce these foreign agents to become participants in the replicative process. Indeed, one might be able to persuade the RNA molecules themselves to acquire a weak catalytic function that would aid their replication.

Finally, I wish to note a practical implication of the mutant that is resistant to the inhibitory analogue, TuTP. These abbreviated variants possess various features that make them potentially powerful chemotherapeutic agents. They combine a high affinity for the replicase and a rapid growth rate. They compete effectively with a normal, viral nucleic acid for the replicase and thus halt the progress of virus production. To these properties we can now add a third—resistance to a chemotherapeutic agent that is effective against the original virus particle. All these features can be built into one variant by the kinds of serial selection described here, adding another dimension to the potential use of these agents as chemotherapeutic devices and opening up a novel pathway for the specific control of viral diseases.

## Acknowledgments

The investigations reported here and originating in the author's laboratory were supported by Public Health Service Research Grant No. CA-01094 from the National

Cancer Institute and Grant No. GB-4876 from the National Science Foundation.

## REFERENCES

ATKINSON, M. R., J. A. HUBERMAN, R. B. KELLY, and A. KORNBERG, 1969. The active center of DNA polymerase. *Fed. Proc.* 28: 374 (abstract).

AUGUST, J. T., S. COOPER, L. SHAPIRO, and N. D. ZINDER, 1963. RNA phage induced RNA polymerase. *Cold Spring Harbor Symp. Quant. Biol.* 28: 95–97.

AUGUST, J. T., L. SHAPIRO, and L. EOYANG, 1965. Replication of RNA viruses. I. Characterization of a viral RNA-dependent RNA polymerase. *J. Mol. Biol.* 11: 257–271.

BISHOP, D. H. L., D. R. MILLS, and S. SPIEGELMAN, 1968. The sequence at the 5′ terminus of a self-replicating variant of viral Qβ ribonucleic acid. *Biochemistry* 7: 3744–3753.

COOPER, S., and N. D. ZINDER, 1962. The growth of an RNA bacteriophage: The role of DNA synthesis. *Virology* 18: 405–411.

DOI, R. H., and S. SPIEGELMAN, 1962. Homology test between the nucleic acid of an RNA virus and the DNA in the host cell. *Science* (*Washington*) 138: 1270–1272.

FIERS, W., H. VERPLANCKE, and B. VAN STYVENDAELE, 1967. The synthesis of bacteriophage MS₂ RNA in vitro. *In* Regulation of Nucleic Acid and Protein Biosynthesis (V. V. Koningsberger and L. Bosch, editors). Elsevier Publishing Co., Amsterdam, pp. 154–166.

GIACOMONI, D., and S. SPIEGELMAN, 1962. Origin and biologic individuality of the genetic dictionary. *Science* (*Washington*) 138: 1328–1331.

GOODMAN, H. M., and A. RICH, 1962. Formation of a DNA-soluble RNA hybrid and its relation to the origin, evolution, and degeneracy of soluble RNA. *Proc. Nat. Acad. Sci. U. S. A.* 48: 2101–2109.

HALL, B. D., and S. SPIEGELMAN, 1961. Sequence complementarity of T₂-DNA and T₂-specific RNA. *Proc. Nat. Acad. Sci. U. S. A.* 47: 137–146.

HARUNA, I., K. NOZU, Y. OHTAKA, and S. SPIEGELMAN, 1963. An RNA "replicase" induced by and selective for a viral RNA: Isolation and properties. *Proc. Nat. Acad. Sci. U. S. A.* 50: 905–911.

HARUNA, I., and S. SPIEGELMAN, 1965a. Specific template requirements of RNA replicases. *Proc. Nat. Acad. Sci. U. S. A.* 54: 579–587.

HARUNA, I., and S. SPIEGELMAN, 1965b. Recognition of size and sequence by an RNA replicase. *Proc. Nat. Acad. Sci. U. S. A.* 54: 1189–1193.

HARUNA, I., and S. SPIEGELMAN, 1965c. The autocatalytic synthesis of a viral RNA in vitro. *Science* (*Washington*) 150: 884–886.

LEVISOHN, R., and S. SPIEGELMAN, 1968. The cloning of a self-replicating RNA molecule. *Proc. Nat. Acad. Sci. U. S. A.* 60: 866–872.

LEVISOHN, R., and S. SPIEGELMAN, 1969. Further extracellular Darwinian experiments with replicating RNA molecules: Diverse variants isolated under different selective conditions. *Proc. Nat. Acad. Sci. U. S. A.* 63: 805–811.

LOEB, T., and N. D. ZINDER, 1961. A bacteriophage containing RNA. *Proc. Nat. Acad. Sci. U. S. A.* 47: 282–289.

MILLS, D. R., N. R. PACE, and S. SPIEGELMAN, 1966. The in vitro synthesis of a noninfectious complex containing biologically active viral RNA. *Proc. Nat. Acad. Sci. U. S. A.* 56: 1778–1785.

MILLS, D. R., R. L. PETERSON, and S. SPIEGELMAN, 1967. An extracellular Darwinian experiment with a self-duplicating nucleic acid molecule. *Proc. Nat. Acad. Sci. U. S. A.* 58: 217–224.

NEUMANN, J. VON, 1966. Theory of Self-reproducing Automata (A. W. Burks, editor). University of Illinois Press, Urbana, Illinois.

OVERBY, L. R., G. H. BARLOW, R. H. DOI, M. JACOB, and S. SPIEGELMAN, 1966a. Comparison of two serologically distinct ribonucleic acid bacteriophages. I. Properties of the viral particles. *J. Bacteriol.* 91: 442–448.

OVERBY, L. R., G. H. BARLOW, R. H. DOI, M. JACOB, and S. SPIEGELMAN, 1966b. Comparison of two serologically distinct ribonucleic acid bacteriophages. II. Properties of the nucleic acids and coat proteins. *J. Bacteriol.* 92: 739–745.

PACE, N. R., and S. SPIEGELMAN, 1966a. The synthesis of infectious RNA with a replicase purified according to its size and density. *Proc. Nat. Acad. Sci. U. S. A.* 55: 1608–1615.

PACE, N. R., and S. SPIEGELMAN, 1966b. In vitro synthesis of an infectious mutant RNA with a normal RNA replicase. *Science* (*Washington*) 153: 64–67.

PARANCHYCH, W., and A. F. GRAHAM, 1962. Isolation and properties of an RNA-containing bacteriophage. *J. Cell. Comp. Physiol.* 60: 199–208.

SHAPIRO, L., and J. T. AUGUST, 1965. Replication of RNA viruses. II. The RNA product of a reaction catalyzed by a viral RNA-dependent RNA polymerase. *J. Mol. Biol.* 11: 272–284.

SPIEGELMAN, S., and I. HARUNA, 1966a. Problems of an RNA genome operating in a DNA-dominated biological universe. *J. Gen. Physiol.* 49 (no. 6, pt. 2): 263–304.

SPIEGELMAN, S., and I. HARUNA, 1966b. A rationale for an analysis of RNA replication. *Proc. Nat. Acad. Sci. U. S. A.* 55: 1539–1554.

SPIEGELMAN, S., I. HARUNA, I. B. HOLLAND, G. BEAUDREAU, and D. R. MILLS, 1965. The synthesis of a self-propagating and infectious nucleic acid with a purified enzyme. *Proc. Nat. Acad. Sci. U. S. A.* 54: 919–927.

STRAUSS, J. H., and R. L. SINSHEIMER, 1963. Purification and properties of bacteriophage MS2 and of its ribonucleic acid. *J. Mol. Biol.* 7: 43–54.

SULSTON, J., R. LOHRMANN, L. E. ORGEL, H. SCHNEIDER-BERNLOEHR, B. J. WEIMANN, and T. H. MILES, 1969. Non-enzymic oligonucleotide synthesis on a polycytidylate template. *J. Mol. Biol.* 40: 227–234.

WATANABE, I., 1964. Persistent infection with an RNA bacteriophage. *Nihon Rinsho* 22: 1187–1195.

YANKOFSKY, S. A., and S. SPIEGELMAN, 1962a. The identification of ribosomal RNA cistron by sequence complementarity. I. Specificity of complex formation. *Proc. Nat. Acad. Sci. U. S. A.* 48: 1069–1078.

YANKOFSKY, S. A., and S. SPIEGELMAN, 1962b. The identification of the ribosomal RNA cistron by sequence complementarity. II. Saturation of and competitive interaction at the RNA cistron. *Proc. Nat. Acad. Sci. U. S. A.* 48: 1466–1472.

# 83 Repressors and Genetic Control

## WALTER GILBERT

How are genes controlled? The Jacob and Monod (1961) model of genetic control proposed that control genes would make repressors which would turn off other genes. At the last Boulder meeting, Gunther Stent (1967) described this picture of control that had emerged from genetic studies in bacteria and bacterial viruses. However, since then our knowledge about the control of gene function has been given an explicit biochemical base through the isolation of several repressors (Gilbert and Müller-Hill, 1966; Ptashne, 1967; Riggs and Bourgeois, 1968; Pirrotta and Ptashne, 1969).

The particular model for genetic control put forward by Jacob and Monod contains two germinal ideas. One of these is the concept of separate, specific genes involved in the control of other genes—regulatory genes as distinct from "structural" genes. According to the model, these regulatory genes made substances that would directly affect some rate-determining step in the synthesis of the products of other genes. This rate-determining step defines a target for the control. Such a target, named an "operator," could be, for example, a specific point on a DNA molecule. The binding of a substance to the region of the DNA molecule that serves as an initiation point for the synthesis of messenger RNA clearly could block (or affect or enhance) the synthesis of that messenger. Alternatively, the target for control could be a region on a messenger RNA molecule. The binding of a control substance to such a messenger could easily be imagined to block the attachment of ribosomes and hence the synthesis of proteins. (I have used the term "bind" to supply a specific physical picture. As far as genetic arguments are concerned, any biochemical process that leads to the same result would serve. The control substance could modify bases, achieving the same ends through a chemical alteration of structure.)

The other idea is that the controlling substance itself will act as an intermediate to make a connection (a physical connection) between a small molecule—a compound acting as a signal—and the target for the control. A biochemical realization would be that the molecule serving as a signal could bind to the control substance to alter the affinity of the con-

trol substance for the operator. This proposal is a complete shift away from instructive theories of control, which require that there be a way in which the signal is like the act controlled: the enzyme to be called forth by the substrate (even formed upon the substrate in the primitive form of such theories), the antibody shaped by the antigen, or the sperm to contain a homunculus. The interposition between signal and act of a controlling substance, which interacts separately with a small molecule and the operator, permits the relationship of the small molecule to the enzyme called forth to be dictated only by evolutionary accident; there is no necessary structural relationship.

The explicit mechanism that Jacob and Monod suggested, that the control should involve a gene product that represses gene function, is an accident of the systems that were analyzed originally. It is true for phage lambda, for the enzymes of lactose utilization, for galactose, for glycerol, for the tryptophan biosynthetic enzymes, and for others, that the controlling gene makes a gene product that behaves as a repressor. (This has an explicit genetic meaning: in the absence of the control gene—if the gene is deleted, for example—the structural genes run at full rate. In a genetic complementation experiment, a cytoplasmic product of the control gene turns off the structural genes. General reviews of the various systems are by Epstein and Beckwith, 1968, and Dove, 1968.)

There are a number of other cases, however—genes $N$ and $Q$ of phage lambda that turn on the early and late functions of that phage, and the enzymes for the metabolism of arabinose and maltose, in which the structural genes will not function without the product of the control gene. These are examples of positive control, the regulatory gene product being necessary for expression. How such genes act is unknown. If such control is exerted at the DNA level, a positive control gene might make a factor (like the sigma factor, Burgess et al., 1969) that binds to the RNA polymerase and permits that enzyme to initiate on a new region of DNA, or to make a factor that binds to DNA, to open it for reading, or to act negatively to prevent a stopping of the reading of DNA by interfering with an RNA stop signal. Alternatively, a variety of steps at the RNA-to-protein level could be affected. It is still possible, however, that these positive control genes may act in a more trivial way by generating or shaping a small molecule that acts to interfere with a repressor made by some still-unidentified gene.

WALTER GILBERT  Departments of Physics, Biology, and Biochemistry and Molecular Biology, Harvard University, Cambridge, Massachusetts

If the cell is to respond to a chemical signal coming from outside, a number of steps will intervene before that signal affects the synthesis of proteins. The chemical signal must penetrate, be concentrated, possibly altered, possibly trigger the release of other chemicals, which, in turn, serve as signals—each step requiring a specific enzyme or function, and so requiring the product of a specific gene. Genetically, each of these genes can be altered by mutation; each such mutation can appear as a "control" or "regulatory" effect. Exhaustive genetic analysis can only suggest which control genes affect the transmission of the signal and which specify the actual way in which the process is controlled. Only biochemical analysis can ultimately prove the mechanism. Now that several repressors have been isolated, we can understand their functioning in detail.

## The lactose operon

One of the genetic systems that has been exploited most successfully involves the genes for the enzymes of lactose metabolism in *E. coli*. Figure 1 shows a schematic genetic map of this region of the *E. coli* chromosome. There are three structural genes: *z*, the gene for $\beta$-galactosidase, the enzyme that splits lactose into glucose and galactose; *y*, the gene producing a protein that is involved in the permeation (and active concentration) of lactose; and *a*, the gene for the thiogalactoside transacetylase, an enzyme that has no known function in vivo, strains deleted for the *a* gene behaving normally in all tests. These three contiguous genes are con-

trolled together, the levels of the three products changing in a coordinate fashion. Such a set of genes is called an operon, and is thought of as being under the control of an operator.

Three controlling elements are shown on the map: the *i* gene; *p*, the promoter; and *o*, the operator. The *i* gene makes a controlling substance, a repressor. We know this through the variety of mutations that exist in this gene. The basic defining mutation is the change to $i^-$: when no product is made by the *i* gene—if it were deleted, for example—the structural genes function at full rate at all times; in the wild-type ($i^+$) cell (in the absence of sugar), the enzymes are made only one-thousandth of the full rate. The crucial experiment that formed the basis of the Jacob and Monod theory was a genetic complementation experiment. If both a wild-type *i* gene and a defective *i* gene were put into the same cell, the wild-type gene dominated and turned off not only the adjacent piece of DNA but also all other pieces of DNA carrying the lactose genes in the same cell.

One infers that the *i* gene makes some product that acts through the cytoplasm to prevent the expression of the lactose genes. Thus the $i^-$ mutations are recessive constitutives, constitutive meaning that the enzymes are made without being induced by the sugar, and the distribution of these $i^-$ mutations defines the *i* gene. The two other controlling regions do not appear to make any product; both are defined through mutations that have only *cis* effects, changing the behavior only of the piece of DNA bearing the mutation.

The operator, conceptually the target for the *i* gene

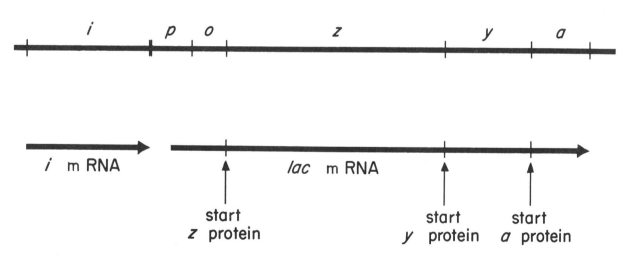

FIGURE 1  A schematic map of the lactose (*lac*) region of the *E. coli* chromosome. The Figure shows the order of the regions on the DNA making up the lactose operon. The products of the three structural genes *z*, *y*, and *a* ($\beta$-galactosidase, permease, and transacetylase) are synthesized coordinately under the control of a repressor gene, *i*, of which the product works at the operator region, *o*, to prevent the synthesis of messenger RNA (mRNA) for the structural genes. The messenger for the *lac* enzymes is synthesized, starting in the promoter region, *p*. The *i* product (the *lac* repressor) is made from the *i* messenger RNA, which is synthesized in the direction shown, starting at an *i*-gene promoter at the far left of the gene.

product, is a site for *cis*-dominant constitutive mutations. These $o^C$ (operator constitutive) mutations permit a partial escape from the control by the product of the *i* gene, but in a complementation experiment only the structural genes physically adjacent to the mutation function.

The promoter is defined genetically by mutations that prevent the expression of all three structural genes simultaneously. These mutations, $p^-$, occur outside the structural genes and affect only the physically adjacent genes in the *cis* position (Jacob et al., 1964; Ippen et al., 1968). The order of genes is *p-o-z-y-a*. RNA for this operon is read in the direction from *z* to *a* (Kumar and Szybalski, 1969). It is generally believed, although there is no explicit biochemical proof as yet, that the promoter region provides the starting site of the RNA for the operon, read out as one piece, and thus that the $p^-$ mutations change the rate of initiating RNA synthesis, for example, by being base changes that change the affinity of the RNA polymerase for this special region.

How does the sugar induce the enzymes? The hypothesis of Jacob and Monod was that the sugar, or an analogue of the sugar, would block the action of the repressor, prevent the attachment of the repressor to the operator, and thus leave the genes open to function. The inducer might bind to the repressor in order to make it physically unable to interact with the operator. Such a model is supported by the existence of another class of mutants of the *i* gene, mutants in which the *i*-gene product behaves as if it has lost its affinity for the inducer. Such a defective repressor should be able to bind only to the operator; a cell containing such repressors should never be induced. Such mutations, $i^s$ mutations, were found: mutations to a trans-dominant uninducible phenotype. Because the *i* cells could be induced at a thousandfold greater concentration of inducer than was needed for the wild type, and because one could supply an argument based on the interaction between $i^s$ and $o^C$ in complementation to show that no more repressor was synthesized in the $i^s$ than in the wild type, one would believe that these mutations were adequately explained as loss of affinity of the repressor for the inducer (Willson et al., 1964).

The earlier comment about the penetration and modification of a small molecule signal has an explicit realization in the lactose operon. The sugar lactose itself is not an inducer (Burstein et al., 1965). The sugar must be acted upon by $\beta$-galactosidase, which, as a transgalactosidase, transfers the galactose to some (unknown) receptor to make the inducer. Thus, the basal level of $\beta$-galactosidase is needed to provide the first molecules of inducer. A $z^-$ cell cannot be induced. Furthermore, a $y^-$ cell cannot be induced by lactose. The permease must be functional to keep a high enough level of lactose in the cell to maintain the induction by the sugar. Thus, if only the "natural" inducer were known, both $z^-$ and $y^-$ mutations woud appear to have some "regulatory" properties. These complexities are avoided by the use of

unmetabolized inducers such as IPTG (isopropyl-thiogalactoside), which cannot be split by $\beta$-galactosidase.

## The nature of the repressor

In 1961, a general view was held that repressors would turn out to be RNA molecules. That thought was based on an explicit experiment, a chloramphenicol inhibition experiment (Pardee and Prestidge, 1959), which was believed to show that the *lac* (lactose) repressor could be made in the presence of chloramphenicol and thus was an RNA molecule. The experiment is false, for several years later it could be shown that the repressor cannot be made in the presence of inhibitors of protein synthesis (Horiuchi and Ohshima, 1966). During that period, however, people talked themselves into the view that repressors were RNA molecules, overcoming the attitude that the easiest way of providing a connection between a small molecule and a nucleic-acid sequence was to use a well-shaped protein molecule. Ever since, one wondered whether the control molecule would turn out to be RNA, or protein, or both.

Further reasons for believing that the active structure required protein were that one found a variety of temperature-sensitive mutations of the *i* gene, both thermolabile ($i^{TL}$) and temperature-sensitive synthesis ($i^{TSS}$) mutations (described in detail in Sadler and Novick, 1965). More convincing, however, were the isolations of nonsense mutations in the *i* gene (changes to the nonsense codon UAG, which interrupts polypeptide synthesis). These mutations produce an inactive repressor (Müller-Hill, 1966; Bourgeois et al., 1965). Ultimately, the proof of the nature of this control substance waited on its isolation and purification as a protein.

We (Gilbert and Müller-Hill, 1966) isolated the product of the lactose *i* gene by using only that one property most central to the picture of repressor control—the repressor must bind to the inducer. The experiments were not biased by any requirement as to the composition of the repressor, nor did they depend on any model for its action. The inducer that was used is an analogue of the sugar IPTG, the best inducer known. The binding of radioactive IPTG to the repressor was followed by equilibrium dialysis: 0.1 ml samples of protein solutions were dialyzed in an ordinary dialysis sack against a solution of radioactive IPTG. Although the interaction is rather tight, a dissociation constant of $1.3 \times 10^{-6}$ M, there is so little repressor in the wild-type cell that the molecule must be purified blindly before its presence can be detected in the protein solution that is put into the dialysis sack (even directly in the bacterial cell, only a few per cent excess of IPTG can be bound). After an ammonium sulfate fractionation of the bacterial extract, however, we observed material that bound IPTG. This material could be purified, and it turned out to be protein. But was this material, which bound the inducer, relevant? It would be only if we could prove that it was, in fact, the product of the *i* gene. Even though the affinity of binding is comparable to that which one would estimate the repressor to have from an in vivo argument, this is not sufficient identification. Only the variety of mutant forms of the *i* gene yields the necessary proof. The first tests performed to show that the binding material was the *i*-gene product indicated that the material was not made in *i*⁻ cells (nonsense *i*⁻ cells in which no gene product would be made), although all the lactose enzymes were being made at a high level, and that the material was not observed in *i*ˢ cells. These cells differ from the uninduced wild type only by a point mutation in the *i* gene that abolishes the affinity of the repressor for IPTG. (The various mutant forms of the *i* gene were put into identical genetic backgrounds, so that unknown variations from strain to strain could not confuse the issue.) These tests, although successful, were negative in character; a better test would be to show that a modification in the *i* gene would yield a modified (but identifiable) product. We had a further mutation available, a mutant strain which induced more easily than does the wild type and which thus had a repressor with a different affinity for the inducer. In vitro, this mutant repressor bound IPTG with a different dissociation constant than that of the wild type.

More recently, a number of further proofs have appeared. Ohshima et al. (1968) have shown that temperature-sensitive mutations of the *i* gene produce an IPTG-binding protein that has a different temperature stability than does the wild type. Furthermore, there are mutations in the *i* gene that change the amount of gene product, both as estimated from in vivo tests and as determined physically. The amount of repressor made by the wild-type cell is very small, about 0.002 per cent of the cell's protein: about 10 copies per haploid genome. Benno Müller-Hill (Müller-Hill et al., 1968) sought and found mutant forms of the *i* gene which made more repressor (*i*^Q, Q for quantity, mutations). The original *i*^Q makes 10 times more repressor. The mutation lies at the end of the *i* gene farthest from the *z* gene (at the left in Figure 1); the *i* gene is read from left to right (Miller et al., 1968; Kumar and Szybalski, 1969); and so we believe that this *i*^Q mutation is most likely a promoter mutation, producing a tenfold faster synthesis of *i*-gene messenger.

The technology has now reached the point at which the *lac* repressor is an easily obtainable protein. The very low level in the wild-type cell, 0.002 per cent of the protein, has been raised more than 1000 times. The *i*^Q mutation raised the level by a factor of 10, to 0.02 per cent (or 0.05 per cent in a diploid). To get still more repressor, we put the *i*^Q on a defective phage, a derivative of phage λ that carries the *lac* genes as a replacement of the late phage functions. This phage can be triggered by heat to multiply within the cell.

Several hundred copies of the phage genome are made, but the phage is unable to lyse the cell. These multiple copies of the *i* gene make, in practice, about 25 times more repressor, so these cells are harvested with 0.5 per cent of their protein *lac* repressor. Very recently, Jeffrey Miller (unpublished) isolated a new mutant, an *i* ^super Q that makes five times more repressor than does the *i*^Q parent. This is 50 times the basal level. When this mutation is placed upon the phage, cells can be harvested with 2.5 per cent of their protein *lac* repressor. This is a yield of several grams per kilogram of cells. Large amounts of repressor have been made in Cologne, and work on the amino-acid sequence is being started there.

The material is purified easily by ammonium sulfate fractionation, elution from a phosphocellulose column at pH 7.5, and, if necessary, a DEAE-column step. The *lac* repressor is an acidic protein (but it has a basic region capable of binding to phosphocellulose at neutral pH). There is nothing unusual about its amino-acid composition. It is a tetrameric protein, 150,000 in molecular weight, with four identical subunits of 38,000. It binds four molecules of IPTG. How does it work?

## The repressor binds to operator DNA

Ultimately to understand how a repressor works, we wish to know the physical target and how the interaction of the repressor with that target blocks protein synthesis. The experiments that are presently feasible are to examine the interaction of the repressor with candidates for the operator. One can show that the *lac* repressor binds specifically to the *lac* operator DNA (Gilbert and Müller-Hill, 1967; Riggs et al., 1968). Such an experiment is done most easily by using DNA from the defective variant of phage λ that carries the *lac* genes. This phage will yield DNA molecules of a uniform size, each about $30 \times 10^6$ in molecular weight and each carrying one copy of the *lac* operator. (This is a gene dosage about 70 times higher than would be true for bacterial DNA. Nonetheless, a $3\,\gamma$/ml solution of such DNA is only $10^{-10}$ M in operators.) The DNA sediments at 40 S; the *lac* repressor itself sediments at 7 S. Thus, if we mix radioactive *lac* repressor, made by purifying the *lac* repressor protein from radioactively labeled *E. coli,* with the purified phage DNA, and sediment the two together on a glycerol gradient, if the protein binds to the DNA, it obviously sediments faster, moving in a band with the DNA.

Such experiments can show that the *lac* repressor binds only to DNA carrying the *lac* region. If one uses phage DNA that does not have the *lac* region, the protein will not bind, nor will it bind to denatured *lac* DNA. But far more specifically, if the phage DNA carries an *o*ᶜ mutation, a point mutation, or a small deletion mutation in the operator, that mutation will abolish or weaken the binding. Thus we infer that the protein is binding only to a single sequence that occurs only in the operator (as defined genetically).

The physical size of the objects are shown in Figure 2. The phage DNA molecule is 15 $\mu$ long; the entire *lac* region is 2 $\mu$ long; but we are implicating only one point along the molecule at which a protein, about 70 Å across, binds.

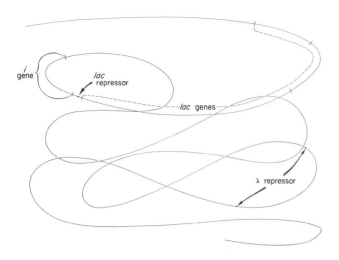

FIGURE 2 The relative sizes of the *lac* repressor, the *lac* genes, phage lambda DNA, and the lambda repressor. The phage DNA is represented as it might be free in solution, with a *lac* repressor bound to the *lac* operator and with two lambda repressors bound to the phage operators.

## The phage λ repressor

Phage λ makes a repressor that turns off the early genes of the phage. Mark Ptashne (1967a) isolated this repressor. It is a protein about 30,000 in molecular weight, the size of the subunit of the *lac* repressor. This material has not yet

been completely purified physically, but it has been purified radiochemically. This protein binds to two specific regions on phage λ to block two operons (Ptashne, 1967b; Ptashne and Hopkins, 1968). Figure 3 shows this region of the phage. The gene that makes its repressor is called the $C_I$ gene, and its product binds to two operators (again identified genetically by operator-constitutive mutations), which, in turn, can be shown to prevent the binding of the λ repressor controlling operons that run out in opposite directions along the λ DNA. These repressors are represented to scale in Figure 2.

## Induction in vitro

Although for λ the story stops with the specific interaction of the repressor with the operator, the *lac* repressor can be pushed further—what does the inducer, IPTG, do to the binding? It prevents the binding. Or, more specifically, if the repressor is first bound to the DNA, then the inducer, IPTG, will cause the complex to fall apart. These experiments directly support the notion that the repressor, on binding to the operator sequence on the DNA, prevents the functioning of the promoter. The inducer weakens the binding to the operator, the repressor comes off, and the operon can function. We do not know explicitly, however, that the repressor blocks (sterically) the attachment of the RNA polymerase to the promoter.

## The affinity of the repressor for the operator

We estimated how tightly the repressor interacts with the operator by asking how low a concentration of components could be used before the complex fell apart as the DNA moved down the centrifuge tube. That estimate for the dissociation constant is $2 \times 10^{-12}$ M, or tighter, in $10^{-2}$M Mg$^{++}$ and $10^{-2}$ M monovalent salt. At higher salt concentrations, the affinity weakens to $10^{-10}$ M in 0.15 M KCl. These salt

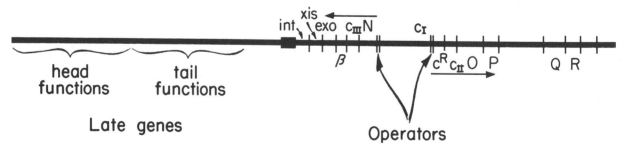

FIGURE 3 Schematic map of the genes of phage lambda. On the right half of the phage genetic map lies the gene for the phage repressor, $C_I$, and the early genes that function when the repressor is inactivated. One operon, read to the left, contains genes N (early positive control) and a set of genes—*beta, exo* (early positive control), *xis* (excision), *int* (integration)—involved in recombination and in the integration of the phage DNA into the host chromosome. To the right of $C_I$, another operon is read to the right; this contains genes O and P involved in phage DNA replication. Gene Q turns on late functions, such as R (lysozyme), and the other late genes.

concentrations span the physiological range and yield affinities in the range of in vivo estimates.

Such in vivo estimates can be made by arguing that the rate of enzyme synthesis depends only on the number of operators free from repressors. If the concentration of free operators is governed by a mass-action formula:

$$[O] [R] = K [OR],$$

relating the concentrations of free operators and repressors to the concentration of complexes of operators with repressors, one can estimate K from the statements that there are about 10 repressors per cell (and thus $[R] \sim 10^{-8}$M) and that the basal level of enzyme synthesis is one 1000th of the full level ($[O]/[OR]$ is 1/1000). The dissociation constant, K, would then be on the order of $10^{-11}$ M. (The argument depends on the basal level's reflecting of the maximal amount of repression. If the basal level is caused by some escape process, then the actual affinity could be much higher.)

A far more accurate approach has been developed and exploited by Arthur Riggs and Suzanne Bourgeois (Riggs et al., 1968, 1970). They discovered that the repressor can cause the trapping of DNA molecules on cellulose-nitrate filters. In the absence of repressor, the entire native DNA molecule passes through the filter. When one repressor molecule binds to a DNA molecule, that DNA molecule adheres to the base of the cellulose-nitrate filter. (Possibly, because the repressor protein itself binds to the filter, the DNA molecule is glued to the filter material by the repressor.) By labeling the DNA and varying the concentration of repressor, they can observe a saturation curve and estimate a binding constant. Their estimates are now that the dissociation constant is $10^{-13}$ M in low salt (0.01 $Mg^{++}$, 0.01 KCl, 0.01 tris pH 7.4, 5 per cent DMSO). One can measure such high affinities directly because of the amplification resulting from the size of the DNA molecule, the binding of one repressor being measured by the fixation of $30 \times 10^6$ molecular weight of DNA (or $10^5$ phosphates). Gene concentrations in the range of $10^{-14}$ M are easily measurable.

The in vitro and in vivo estimates for the affinity are so high that they raise an immediate question about the rates of formation and dissociation of the repressor-operator complex. The dissociation constant might be thought of as the ratio of two rates—the ratio of the rate of decay of the complex $k_d$, to the forward rate of formation, $k_f$: $K = k_d/k_f$. The repressor must diffuse up to the DNA, must find a particular point on the DNA molecule, and might have to adjust its shape or wait for the operator to assume the correct shape before the complex can form. This over-all process cannot go faster than the repressor can diffuse up to the neighborhood of the operator. If we are generous with our estimate, the diffusion limit for a molecule the size of the lac repressor to hit the surface of a 20 A sphere would be one or two times $10^9$ per molar second. Thus, one would expect the decay rate to be about $10^{-4}$ per molar second, if the dissociation constant is to be $10^{-13}$M. A characteristic decay of $10^4$ seconds (which is several hours) would be quite an appreciable time for the components to stay together before the complex falls apart, but this is an immediate consequence of the tight binding.

Is it possible to detect this slow decay experimentally? Yes, it has been seen by Riggs and Bourgeois (unpublished) using the filter-binding assay. They complexed the lac repressor to radioactive lac DNA. Then they mixed that solution with a great excess of unlabeled lac DNA. As the repressor came off a labeled molecule and went onto an unlabeled one, that labeled molecule no longer could be trapped on the filter. By filtering samples of such a mixture at different times, they could observe the loss over time of the label that could be trapped, and this loss reflects directly the decay of the original repressor-operator complex. Their actual estimate for the half-time is 30 minutes, a rate of about $5 \times 10^{-4}$ per second. The complex is indeed slow to decay in vitro.

## The nature of induction

This slow decay immediately raises a problem, because we know that in vivo one can induce promptly on adding the inducer, with no detectable lag before the inducer-triggered event. If the kinetic process of induction requires the repressor to fall off the operator and then to be caught by the inducer, one should not be able to induce any faster than the repressor can leave the operator: a 30-minute delay. There is no reason, however, to accept this picture unless we wish to believe that the inducer and the operator use the same site on the repressor, for only then must competition be the method of induction.

If the inducer and operator use different sites, the induction can be a two-step process in which the inducer first binds to the repressor stuck to the DNA and only afterward does that complex dissociate. These two models are shown in Figure 4. Although they can have indistinguishable equilibrium properties, the kinetics will be different. In one model, the natural decay of the repressor-operator complex is unaffected by the inducer, which blocks only the rate of reformation of the complex; in the other, the inducer interacts directly with the repressor-operator complex, and binds to and changes slightly the repressor moiety of the complex, producing a fast decay of that complex. Such a change in the decay can be seen experimentally in vitro. Riggs and Bourgeois (personal communication) have examined the decay of the repressor-operator complex in the presence of inducer and see a faster decay. There are also compounds that are competitive inhibitors of induction—molecules analogous to the sugar but that prevent induction. Such molecules, like ONPF (ortho-nitro-phenyl-fucoside), bind to the repressor

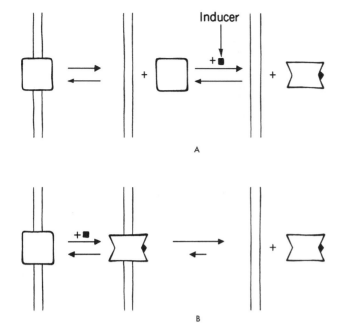

A

B

FIGURE 4 Two models for induction. Alternate possibilities for the induction are shown. The inducer ultimately removes the repressor from the DNA so that the genes can function. This process could be thought of as either (A) the equilibrium between the repressor and the operator being perturbed by the free repressor's binding to the inducer, so that it cannot return to the operator, or (B) the inducer's binding to the repressor while still on the operator and perturbing the structure of the repressor, so that the binding to the operator is weakened.

and, in fact, prevent the dissociation: experimentally, one sees a slower decay for the complex. We can understand these effects by thinking of the bound IPTG changing the repressor so that its affinity to the DNA is less, and the binding is weaker, leading to a faster decay. With ONPF the binding is actually stronger, the decay slower. Because the inducer changes the decay rates, the second model in Figure 4 must be correct: a ternary complex is actually formed through which the inducer pushes the repressor off the operator.

There is a catch in these numbers because they require that the forward rate of formation of the complex be $5 \times 10^9$ per molar second. This is, in fact, a very high rate constant. Riggs and Bourgeois (personal communication) have also measured this rate directly by diluting the components to the level at which the complex takes tens of minutes to form. They observe, over these long times, a rate consistent with this figure of $5 \times 10^9$ per molar second. This rate is so fast that one begins to expect that the critical object up to which the repressor diffuses will be trapped eventually in some region on the DNA molecule appreciably bigger than

the operator. The repressor may need to hit only within several thousand Ångstroms of the operator and then can move along the DNA to find the correct region. This is not unreasonable, because the repressor has a general affinity for DNA in that it can stick to the phosphate backbone. (This affinity in the case of the *lac* repressor can be enhanced by leaving out the magnesium ion to the extent that the nonspecific binding to the phosphates completely dominates the specific binding to the operator.)

The picture of induction as passing through a ternary complex would argue that there should be an interaction between the repressor-inducer complex and the operator. One should be able to detect the interaction between the "fully induced" form of the repressor and operator DNA. So far this has not been seen in vitro, but in vivo we can estimate an affinity. If there is a residual affinity of the repressor-inducer complex for DNA, then, if there is enough of that complex in the cell, it will bind to the operator and keep the genes from functioning. This actually happens in vivo in those mutants that make more and more repressor. The $i^Q$ is inducible only to 65 per cent of the full rate; the $i^{super\ Q}$, only to 25 per cent (Jeffrey Miller, personal communication). The argument that the rate of expression should be proportional to the concentration of free operators can be used to show that the affinity of the operator for the fully induced complex is a factor of $2 \times 10^4$ weaker than that for the uncharged repressor. In the wild-type cell, these numbers mean that the amount of repressor which is $10^3$ times the dissociation constant, and which keeps the operator closed 99.9 per cent of the time, is only one 20,000th effective after full induction and keeps the operator closed 5 per cent of the time.

## The nature of the operator

We do not yet know the structure of the operator or the details of the interaction between the operator and the repressor. The experiments so far show only that double-stranded DNA is required, and that the interaction sees either the outside of the native helix or requires the participation of both strands. There is enough information available, visible in the large groove, to distinguish all the base pairs from the outside when the DNA is double-stranded. It is not necessary to separate the strands to see the hydrogen bonds used in the Watson-Crick pairing in order to read a sequence. How big is the operator? To the extent that this region is a stretch along the DNA to which a protein molecule the size of one subunit of the *lac* repressor (38,000 mol. wt.) or of the size of the lambda repressor (30,000 mol. wt.) can bind, it can only be between one and two turns of the DNA molecule (35 to 70 Å, 10 to 20 bases).

There is a general argument as to how large such a region must be, based simply on the realization that there are very few *lac* repressors in an *E. coli* cell. These repressors bind to one site on the *coli* DNA with very high affinity. If they find other regions of the DNA with equally high affinity, they will be sopped up and made unable to function. This means that the region to which they bind must be unique in the chromosome ($3 \times 10^6$ base pairs) and thus must have *at least* the specificity of 12 bases. We might think of a region 12 to 15 bases long, 40 to 50 Å long, associated with a binding energy in the range of 16 to 18 kilocalories. These are not unreasonable numbers to be associated with one another. An interaction with this specificity, "seeing" 12 to 15 bases, must make about 12 to 15 contacts; to the extent that each contact yields between 1 and 2 kilocalories, the energetics (i.e., the tightness of binding) and the specificity are commensurate. The change in binding energy involved on induction, a change by $2 \times 10^4$ in affinity, or 6 kilocalories, is brought about by the binding of up to four IPTGs, each one of which can bind with about 8 kilocalories. Clearly, there is enough free energy available in the binding of the IPTGs to produce a distortion in the repressor capable of weakening its total binding to the operator by 6 kilocalories.

## Phenotypic complexities

Several phenomena are explained by our knowledge that the *lac* repressor is a tetramer (see Müller-Hill et al., 1968). One of these phenomena is the existence of dominant $i^-$ mutations ($i^{-d}$). Although the original defining characteristic of the *i* gene was that constitutive mutations were recessive, a further class of point mutations in the *i* gene was found a few years ago and were later shown to be trans-dominant constitutives. These mutations, however, are recessive to the $i^Q$: 10 times more repressor overcomes the effect of the mutations. The $i^{-d}$ can be interpreted as having bad subunits which join together to form an IPTG-binding, but inactive, repressor and which can mix with good subunits to form hybrid, inactive molecules. The bad subunits distort the good, so that the molecule will no longer bind to DNA. Another complementation effect is that the $i^s$ which makes a repressor that cannot be induced and which dominates the $i^+$ is dominated in turn by the $i^Q$. In an environment that contains a great excess of good subunits, the $i^s$ subunits will be tied up in hybrids with one $i^s$ and three $i^+$ subunits, and these hybrids will be inducible: their affinity for the operator can be weakened by IPTG.

## Possible implications for higher cells

What aspects of these repressor control systems might appear again in higher cells? The control in *lac* is exercised by a small number of molecules, made by a gene that appears to run at a fixed rate under the control of a promoter changeable only by mutation. These control molecules, acidic proteins, interact with high affinity with a specific site on the DNA. These properties may appear again, but the volume of cells and the amount of DNA increase about 1000 times; thus, the specificity and affinity should also increase 1000 times. The control is efficient. One *lac* repressor of 150,000 daltons controls about $4 \times 10^6$ daltons of DNA; the control material bound to the DNA is about 3 per cent by weight of the genes controlled. Phage lambda provides a still more vivid example of efficiency, because $30 \times 10^6$ daltons of DNA are controlled by a repressor constituting only a few tenths of one per cent by weight of the DNA that is controlled. This is a better analogue for higher cells—control through elements of high specificity but there, in the chromosome, in trace amounts. Not only does lambda provide an example of efficiency; the control here is more sophisticated than in the simple case of *lactose*.

Eisen et al. (1968) have shown that there is a specific separate control of the amount of lambda repressor. While the repressor is present, it shuts off the two operators controlling the two early operons of the phage. Once the repressor has been inactivated, however, and the operon on the right begins to run, no further $C_I$ repressor can be made. The most recent experiments by Harvey Eisen suggest that the first product made by this operon is a repressor of the repressor ($C^{repressor}$), which turns off any further synthesis of the $C_I$ repressor. The phage can have two stable intracellular states which can be displayed if the phage genes that kill the cell are removed. Either it makes the $C_I$ and never the $C^R$ product, or it makes the $C^R$ (and other enzymes from the right-hand operon) and not the $C_I$.

There are also two steps of positive control. The operon that runs to the left makes the gene $N$ product that is required for the high-level expression of all the early phage functions. The $N$ product is a positive control element; if the left-hand operator is open, the $N$ product turns on a further set of genes to the left. These are involved in recombination, integration, and excision functions. If the right-hand operator is open, the $N$ product enhances the level of gene products made to the right. These are the genes involved in phage DNA synthesis, which permits phage to multiply. There is a further separate control on the late phage functions. The $N$ product also turns on gene $Q$, which itself turns on the late functions such as coat proteins, tail structures, and a lysozyme. Here, indeed, is a model of sufficient complexity, put together out of simple control loops, to provide an example of what might be going on in a higher cell.

## Acknowledgment

This work was supported in part by the National Institute of General Medical Sciences, Grant GM 09541.

# REFERENCES

BOURGEOIS, S., M. COHN, and L. E. ORGEL, 1965. Suppression of and complementation among mutants of the regulatory gene of the lactose operon of *Escherichia coli*. *J. Mol. Biol.* 14: 300–302.

BURGESS, R. R., A. A. TRAVERS, J. J. DUNN, and E. K. F. BAUTZ, 1969. Factor stimulating transcription by RNA polymerase. *Nature (London)* 221: 43–46.

BURSTEIN, C., M. COHN, A. KEPES, and J. MONOD, 1965. Rôle du lactose et de ses produits métaboliques dans l'induction de l'opéron lactose chez *Escherichia coli*. *Biochim. Biophys. Acta* 95: 634–639.

DOVE, W., 1968. The genetics of the lambdoid phages. *Annu. Rev. Genet.* 2: 305–340.

EISEN, H., L. PEREIRA DA SILVA, and F. JACOB, 1968. The regulation and mechanism of DNA synthesis in bacteriophage λ. *Cold Spring Harbor Symp. Quant. Biol.* 33: 755–764.

EPSTEIN, W., and J. R. BECKWITH, 1968. Regulation of gene expression. *Annu. Rev. Biochem.* 37: 411–436.

GILBERT, W., and B. MÜLLER-HILL, 1966. Isolation of the lac repressor. *Proc. Nat. Acad. Sci. U. S. A.* 56: 1891–1898.

GILBERT, W., and B. MÜLLER-HILL, 1967. The *lac* operator in DNA. *Proc. Nat. Acad. Sci. U. S. A.* 58: 2415–2421.

HORIUCHI, T., and Y. OHSHIMA, 1966. Inhibition of repressor formation in the lactose system of *Escherichia coli* by inhibitors of protein synthesis. *J. Mol. Biol.* 20: 517–526.

IPPEN, K., J. H. MILLER, J. SCAIFE, and J. BECKWITH, 1968. New controlling element in the *lac* operon of *E. coli*. *Nature (London)* 217: 825–827.

JACOB, F., and J. MONOD, 1961. Genetic regulatory mechanisms in the synthesis of proteins. *J. Mol. Biol.* 3: 318–356.

JACOB, F., A. ULLMAN, and J. MONOD, 1964. Le promoteur, élément génétique nécessaire à l'expression d'un opéron. *Compt. Rend. Acad. Sci.* 258: 3125–3128.

KUMAR, S., and W. SZYBALSKI, 1969. Orientation of transcription of the *lac* operon and its repressor gene in *Escherichia coli*. *J. Mol. Biol.* 40: 145–151.

MILLER, J. H., J. BECKWITH, and B. MÜLLER-HILL, 1968. Direction of transcription of a regulatory gene in *E. coli*. *Nature (London)* 220: 1287–1290.

MÜLLER-HILL, B., 1966. Suppressible regulator constitutive mutants of the lactose system in *Escherichia coli*. *J. Mol. Biol.* 15: 374–376.

MÜLLER-HILL, B., L. CRAPO, and W. GILBERT, 1968. Mutants that make more lac repressor. *Proc. Nat. Acad. Sci. U. S. A.* 59: 1259–1264.

OHSHIMA, Y., J. TOMIZAWA, and T. HORIUCHI, 1968. Isolation of temperature-sensitive *lac* repressors. *J. Mol. Biol.* 34: 195–198.

PARDEE, A. B., and L. S. PRESTIDGE, 1959. On the nature of the repressor of β-galactosidase synthesis in *Escherichia coli*. *Biochim. Biophys. Acta* 36: 545–547.

PIRROTTA, V., and M. PTASHNE, 1969. Isolation of the 434 phage repressor. *Nature (London)* 222: 541–544.

PTASHNE, M., 1967a. Isolation of the λ phage repressor. *Proc. Nat. Acad. Sci. U. S. A.* 57: 306–313.

PTASHNE, M., 1967b. Specific binding of the λ phage repressor to λ DNA. *Nature (London)* 214: 232–234.

PTASHNE, M., and N. HOPKINS, 1968. The operators controlled by the λ phage repressor. *Proc. Nat. Acad. Sci. U. S. A.* 60: 1282–1287.

RIGGS, A. D., and S. BOURGEOIS, 1968. On the assay isolation and characterization of the *lac* repressor. *J. Mol. Biol.* 34: 361–364.

RIGGS, A. D., S. BOURGEOIS, R. F. NEWBY, and M. COHN, 1968. DNA binding of the *lac* repressor. *J. Mol. Biol.* 34: 365–368.

RIGGS, A. D., H. SUZUKI, and S. BOURGEOIS, 1970. *Lac* repressor-operator interaction. *J. Mol. Biol.* 48: 67–83.

SADLER, J. R., and A. NOVICK, 1965. The properties of repressor and the kinetics of its action. *J. Mol. Biol.* 12: 305–327.

STENT, G. S., 1967. Induction and repression of enzyme synthesis. *In* The Neurosciences: A Study Program (G. C. Quarton, T. Melnechuk, and F. O. Schmitt, editors). The Rockefeller University Press, New York, pp. 152–161.

WILLSON, C., D. PERRIN, M. COHN, F. JACOB, and J. MONOD, 1964. Non-inducible mutants of the regulator gene in the "lactose" system of *Escherichia coli*. *J. Mol. Biol.* 8: 582–592.

# 84 Two Sequence-Specific DNA-Protein Recognition Systems

A. D. KAISER

LYSOGENY FOR bacteriophage λ depends on the addition of viral DNA to the DNA of its host and removal of the viral DNA at a later time. Many temperate bacteriophages, a number of episomal bacterial genetic elements (Campbell, 1969), and possibly DNA tumor viruses (Dulbecco, 1968) attach to their host's chromosome. The addition and removal of λ DNA are made with such high precision that neither phage nor bacterium gains or loses base pairs in the process, except in the rare instances of transducing phage production, in which a fragment of the viral DNA is excised with an adjoining fragment of bacterial DNA. A complete cycle of addition and removal requires two different DNA joining and cutting steps, both of which are sequence-specific and require phage-induced proteins. These steps provide us with examples of DNA-protein recognition amenable to experimental exploration, and the purpose of the present paper is to give a molecular description of those steps.

Particles of phage λ contain a single linear DNA molecule, 50,000 base pairs long. How is this molecule added to the hundred-times-larger circular molecule of bacterial DNA? The correct solution was proposed by Campbell (1962, 1969); he suggested that a linear molecule of phage DNA first joins its ends to become a circle, then recombines with the circular bacterial DNA to form a single large circle, as illustrated in Figure 1.

When lysogenic bacteria produce active phage, these processes are reversed, in a formal sense. Phage DNA is excised from the chromosome, and, after replication, linear DNA molecules are encapsulated in the new phage particles. The two steps depicted in Figure 1 involve the joining and the cutting of DNA molecules, as follows. In the forward direction of Step 1, two ends of one infecting DNA molecule join and, in the reverse direction, a newly synthesized DNA molecule is cut to create ends. In the forward direction of Step 2, two circular molecules are cut and joined to each other. In the reverse direction of Step 2, one DNA molecule is cut in two places, and the four ends join in pairs

to recreate a molecule of phage DNA and a molecule of bacterial DNA. The base sequence-specificity inherent in the DNA-protein recognition in both steps is very high. Cutting in Step 1 occurs at just one place in the phage DNA molecule of $5 \times 10^4$ base pairs. The forward direction of Step 2 occurs at just one place in the phage DNA molecule and at just one place in a bacterial DNA molecule of $5 \times 10^6$ base pairs. We turn now to a discussion of the molecular events.

PHAGE DNA ENDS, THE FIRST RECOGNITION SYSTEM
DNA molecules extracted from λ phage particles are linear. If these linear molecules are heated to about 70° C and allowed to cool slowly, they spontaneously join their ends and become circles (Hershey et al., 1963; Ris and Chandler, 1963; MacHattie and Thomas, 1964). The joining of ends is specific, for λ DNA will not join to other types of DNA molecules. If circular molecules are heated to 70° C (λ DNA retains its native double-stranded structure up to 90°) and cooled rapidly, they open into linear structures. The structural basis of the reversible joining and separating is a pair of single strands which protrude from the two ends of a molecule of λ DNA (Wu and Kaiser, 1968; Wu, manuscript in preparation) as shown in Figure 2. The two protruding single strands have complementary base sequences which

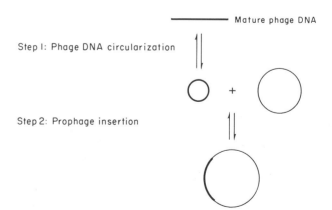

FIGURE 1 Changes in DNA structure during lysogenization by phage λ. A line represents a double-stranded DNA molecule. The thick line represents λ DNA; and the thin line, bacterial DNA.

A. D. KAISER Department of Biochemistry, Stanford University School of Medicine, Stanford, California

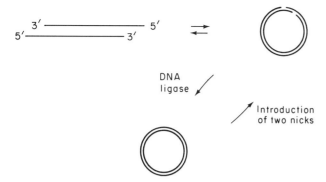

FIGURE 2 Joining and cutting of phage DNA ends. The ends of the strands of the linear phage DNA molecule are labeled 3′ and 5′ to indicate the hydroxyl group on the terminal nucleotide that is *not* engaged in phosphodiester linkage. Both 5′ termini carry phosphate groups, and both 3′ termini are free hydroxyl groups. Note that the circular DNA molecule at the top is held together by base pairing only and that the one below is held together by base pairing and covalent bonds.

join to each other to form a helix of base pairs. The cohesive ends "melt" at a lower temperature (65° C) than does the rest of the molecule (92° C), because only 12 base pairs hold the ends together, whereas 50,000 pairs hold the rest of the molecule. Cohesive ends of this type are not unique to λ DNA. They have been found in nine other bacteriophages and in the DNA isolated from the mitochondria of yeast.

Joining of cohesive ends can be viewed as a DNA-DNA recognition process and is interesting as such. Ends join rather slowly in vitro. For example, at physiological temperature, pH, and ionic strength the half-time for joining is about one hour (Wang and Davidson, 1968). It seems that the reaction that limits this rate is not the diffusion of ends toward each other, but the base pairing and stacking that follow (Wang and Davidson, 1966). In vivo, however, the ends join about 10 times faster, suggesting the existence of an intracellular catalyst.

When the ends of a molecule of λ DNA are joined by base pairing, the 5′ phosphate end of each strand is adjacent to its own 3′ hydroxyl end (Wu and Kaiser, 1968). In vitro, the enzyme DNA ligase will close both nicks with covalent bonds to form a double-stranded circular molecule with no interruptions in the polynucleotide chains (Gellert, 1967). In vivo, DNA of the infecting phage circularizes and is covalently closed within five minutes after injection. DNA ligase is present in the host cells before infection (Gellert, 1967; Olivera and Lehman, 1967a), so it is likely that the ends cohere and are sealed with covalent bonds by DNA ligase in vivo, but there is no direct evidence on the point. DNA ligase is not sequence-specific. It requires only a 3′OH adjacent to a 5′P held in the proper positions by a complementary strand.

After an infecting DNA molecule circularizes, it may add to the host DNA to create a lysogenic bacterium or it may replicate manyfold, produce new, mature, phage particles, and lyse the infected cell. Cohesive ends of mature phage DNA play no further role in the lysogenization process. On the other hand, when new phage particles are produced, new cohesive ends are generated. The structures of the intermediates in λ DNA replication are not yet firmly established, although there is evidence for replicating circular molecules (Young and Sinsheimer, 1967; Tomizawa and Ogawa, 1968) and for circular molecules with linear polymeric tails (Gilbert and Dressler, 1968; Eisen et al., 1968; Kiger and Sinsheimer, 1969). Neither of these structures would be expected to have cohesive ends, and the failure of extracted intracellular λ DNA to infect helper-infected bacteria suggests that none of the replicative intermediates have the mature phage cohesive ends. In fact, new cohesive ends appear only with the formation of new phage heads, complete with their protein membrane (Dove, 1966; Mackinlay and Kaiser, 1969). Thus, the act of recognition that picks out the proper base sequence and transforms it into a cohesive end is not performed until the end of a cycle of virus multiplication, coupled to the formation of phage heads.

A genetic and a chemical argument demonstrate the sequence-specificity involved in creating phage DNA ends. The joined cohesive ends are situated on the phage recombination map between genes *A* and *R* (Kaiser and Inman, 1965). Many different deletion mutants of λ have been isolated; these have included, in one mutant or another, all the known genes of λ. (Incomplete, and therefore defective, phage genomes can be preserved and replicated as prophage, which is one reason that λ and its relatives are useful genetic tools.) Some deletions remove gene *A* but not *R*; others remove gene *R* but not *A*; and some, both *A* and *R*. Only the first two classes of deletion DNAs can be packaged into phage particles (Kayajanian and Campbell, 1966; Kayajanian, 1968). Members of the third class, deleted for *A* and *R* and, therefore, for the intervening DNA, can be propagated as defective prophage, but cannot be matured into complete phage particles, presumably because they lack the proper base sequence.

The chemical argument for specificity is that the cohesive ends have a defined base sequence. Part of the sequence has been determined by the limited addition of nucleotides catalyzed by *E. coli* DNA polymerase and is summarized in Figure 3. Of the 12 base pairs in the cohesive ends, 10 are guanine-cytosine (G-C) and two are adenine-thymine (A-T). Part of the sequence is a pure G-C run of eight base pairs. The existence of a defined sequence implies that all molecules of λ DNA have the same sequence and suggests that the bases in the cohesive ends themselves are part of the recognition sequence. Were this

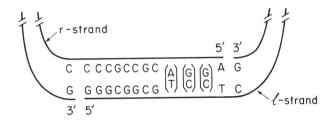

FIGURE 3 Base sequence at the phage DNA ends. The ends are shown cohered to each other. The r-strand has its 5′ terminus at the right end of the λ DNA molecule, drawn as the genetic map is conventionally drawn with the R gene at the right end, and the ℓ-strand has its 5′ terminus at the left end. Brackets around the A–T pair and two G–C pairs in the center of the Figure indicate that their orientation is not known, that is, whether A or T is in the ℓ-strand.

not so, one would have expected heterogeneity in the cohesive sequence between different λ molecules.

The basic chemical event in creating phage DNA ends may be the introduction of a pair of single-stranded breaks in the proper places in a circular or linear, polymeric DNA molecule. The a priori probability that a particular sequence of 12 bases will occur in a random sequence of $5 \times 10^4$ base pairs, the length of λ DNA, is $10^{-2}$. Therefore, it is improbable that the cohesive-end sequence repeats elsewhere in the molecule. It should be noted that the single-stranded break in phage ΦX174 replicative form II DNA, the product of another DNA-protein recognition, is also between two G–C pairs (Knippers et al., 1969). It has been suggested that G-C runs might have a three-dimensional shape that facilitates their recognition (Szybalski et al., 1966; Dove, 1968).

PROPHAGE DNA ENDS, A SECOND RECOGNITION SYSTEM
The addition of a circular molecule of λ DNA to its host DNA and the subsequent removal of the viral DNA utilize another recognition system with at least four specific elements. Two are specific base sequences, one in λ DNA, situated 28,700 base pairs from the left cohesive end (Parkinson, 1969) and indicated by P·P′ in Figure 4. Another is a base sequence in *E. coli* DNA, situated approximately 12,000 base pairs from the galactose operon (Kayajanian and Campbell, 1966) and indicated by B·B′ in Figure 4. The two remaining elements are gene products, probably proteins, specified by the *int* (for integrate) and *xis* (for excise) genes of λ.

An overview of the operation of this system is presented first, followed by a summary of the supporting evidence; finally, a molecular mechanism is suggested.

The *int* protein recognizes the P·P′ and B·B′ sequences in phage and *E. coli* DNA. As a consequence, both sequences are broken at the dots, and a reciprocal transfer occurs, P

joining to B′ and B joining to P′, inserting the phage DNA into the bacterial DNA to form one large circular molecule. An important feature of the system is that *int* protein cannot recognize the two new sequences at the ends of the prophage B·P′ and P·B′, so the adduct of viral and host DNA is stable. A combination of the *xis* protein with the *int* protein can, however, recognize the new sequences. Acting together, they catalyze a breaking of the B·P′ and P·B′ sequences at the dots, and again a reciprocal transfer occurs. Now P rejoins to P′ and B to B′, recreating a circular molecule of λ DNA and a nonlysogenic *E. coli* chromosome. According to this scheme, the reaction goes toward insertion or toward excision, depending on whether *int* protein alone or *int* and *xis* proteins are present.

The scheme just outlined might seem to violate the principle of microscopic reversibility, because different catalysts have been proposed for the forward and reverse directions. Note, however, that it is a genetic scheme and not a complete chemical reaction. It seems likely that chemi-

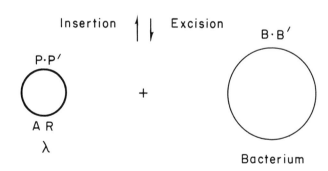

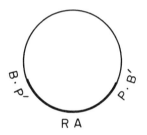

FIGURE 4 Insertion and excision of λ DNA into and from bacterial DNA. The thick line represents phage DNA, and the thin line, bacterial DNA. P and P′ represent the left and right recognition base sequences in λ DNA. B and B′ represent the left and right recognition sequences in bacterial DNA. The dots in P·P′ and B·B′ indicate the places where the sequences are cut and rejoined. R and A are the phage genes on either side of the cohesive end sequence, which is joined by covalent bonds in all the structures of this Figure.

cal energy would be required for the joining reactions. One kind of DNA ligase is known to employ nicotinamide-adenine dinucleotide (NAD) and another adenosine triphosphate (ATP) (Olivera and Lehman, 1967b; Weiss et al., 1968). Hydrolysis of the energy donor coupled to the transformations of DNA might drive the reaction to completion in both directions. There may be an analogy with glycogen metabolism in which different enzymes, utilizing different small molecules, catalyze either extensive glycogen synthesis or extensive breakdown. We turn now to examine the evidence on which this scheme is based.

## The int gene product

The first element in the recognition system, the int gene, is defined by missense and nonsense (amber) point mutants and by deletions. Int mutants cannot lysogenize although they can establish immunity. They cannot perform the insertion step (Zissler, 1967; Gingery and Echols, 1967; Gottesman and Yarmolinsky, 1968). In the presence of a coinfecting int⁺ phage, an int⁻ mutant can insert. This complementation implies that an int⁻ mutant has normal recognition sequences but is unable to synthesize a diffusible protein, probably an enzyme. Once installed by complementation as a prophage, an int⁻ mutant cannot excise, again unless an int⁺ phage is also present to provide active int protein. Thus, the product of the int gene is a cytoplasmic protein necessary for insertion and excision.

The int protein is sequence-specific. Phage lambda, phage 21, and phage Φ80 occupy different sites, as prophage, on the chromosome of E. coli K12. Neither Φ80 int⁺ nor 21 int⁺ can complement a λ int⁻ and cause it to lysogenize (Signer and Beckwith, 1966; Gottesman and Yarmolinsky, 1968). The λ, Φ80, and 21 int proteins, therefore, each recognize different sequences. Deletion mutations place the int gene within 750 base pairs of the sequence recognized by its protein product.

## The xis gene product

The xis gene is defined by missense, nonsense, and deletion mutations. Xis⁻ mutants lysogenize at normal frequency, but they cannot excise (Guarneros and Echols, 1970; Kaiser and Masuda, 1970a). For this reason, they are trapped in the bacterial chromosome. Xis⁻ mutants can, however, excise if an xis⁺ phage is present, showing that the xis gene product is diffusible. In particular, an xis⁻ mutant can excise if an int⁻ mutant is present, demonstrating that an int⁻ mutant makes an effective xis gene product. Xis⁻ mutants can also be complemented by a deletion mutant, called bio 16A, which terminates in the int gene, but not by another deletion, called bio 7-20, which extends 1750 base pairs farther on. Therefore the xis gene lies very close to the int

gene and to the base sequence on which it functions. This clustering of the specific base sequence, the int, and the xis gene is yet another example of the clustering of genes with related functions that characterizes bacteriophage λ (Dove, 1968). As is the int protein, the product of the xis gene is sequence-specific. For example, the xis protein of phage 21 will not work with the int protein of λ on the sequence at the ends of prophage λ, nor will the 21 int protein work with the xis protein of phage λ on the ends of prophage λ (Kaiser and Masuda, 1970b).

## The recognition sequences in E. coli and λ DNA

The third element in this recognition system is a sequence of bases in E. coli DNA, called attλ for attachment λ, and represented in Figure 4 by B·B′. The sequence is situated between galE and bioA (genes in the galactose and biotin operons) on the E. coli map, because λ can attach to F′ gal bio, which carries just this portion of the map. Deletion of the chromosomal region containing attλ reduces the frequency of λ lysogenization by at least 100 times (Gingery and Echols, 1968; Shapiro and Adhya, 1969; Adhya, unpublished); therefore the base sequence attλ is specific. A number of other specific prophage attachment sites have been mapped in E. coli; att80 and att21, for example, are near the tryptophan operon, and att424 is near the histidine operon (Jacob and Wollman, 1961).

The fourth element in the recognition system is a specific sequence of bases in λ DNA, represented by P·P′ in Figure 4. If a part of this sequence is deleted, as in the mutant λb2, a phage is produced which multiplies normally, but cannot lysogenize (Zichichi and Kellenberger, 1963). Unlike λ int⁻, the lysogenization defect in λb2 cannot be overcome by a diffusible product of a coinfecting λb2⁺ (Campbell, 1965). By appropriate maneuvers λb2 can be made a prophage. But, as a prophage, it cannot be excised, and, again, its excision defect cannot be complemented by coinfecting b2⁺. Thus λb2 possesses a structural defect for insertion and excision, which, because some of its DNA has been deleted, is believed to be the loss of part of the recognition sequence.

## Locating the recognition sequence in λ DNA

Rare mistakes in excision give rise to λ gal or λ bio transducing phage (Campbell, 1962, 1969). As diagrammed in Figure 5, the two transducers excise one of the prophage ends with adjacent bacterial genes. Lambda gal contains the left end of the prophage, indicated by B·P′ in Figures 4 and 5, and λ bio carries the right end, P·B′. A λ gal bio has been constructed in two steps and contains a pure bacterial attachment site, B·B′ (Kayajanian, 1968). This λ gal bio carries no phage genes apart from the recognition sequence for the cohesive ends and the recognition sequence for insertion-excision; the rest of its DNA is bacterial. It can be replicated as a prophage and excised and incorporated into phage particles in the presence of a coinfecting normal λ.

The DNA from λ *gal* and λ *bio* phages can be used to locate the junction between phage and bacterial DNA. (This junction is represented by a dot in the symbols B·P′ and P·B′.) A DNA-DNA hybrid between λ *gal* and λ is constructed, then examined in the electron microscope. Transition from a double-stranded, λ/λ, structure to a single-stranded, *gal*/λ, structure indicates the junction. Hradecna and Szybalski (1969) have found the position of the transition point to be the same in all λ *gal* and λ *bio*, namely, 0.57 of the length of a molecule of λ DNA, measured from the left cohesive end. The invariance of this distance means that λ DNA always breaks at the same point, within a measurement error of about 250 base pairs, when it inserts into the *E. coli* chromosome.

The bacterial and the phage recognition sequences are different, as is most clearly seen from the properties of λ *gal* and λ *bio*, which have recognition sequences that are part bacterial and part phage. Lambda *gal* lysogenizes with low efficiency and the defect is structural, because it cannot be complemented in trans, that is, the function cannot be supplied by a second phage (Weisberg and Gottesman 1969). The same holds for λ *bio* (Manly et al., 1969). If the phage and bacterial recognition sequences were identical, there would be no structural defect in either λ *gal* or λ *bio*. Because both are defective, there must be sequence recognition on both sides of the exchange point, and phage and bacterium must differ in both.

If the *int* and *xis* products recognize and recombine two specific base sequences, it should be possible to observe

*int-xis* catalyzed recombination between two phage which carry the proper recognition sequences. The experimental problem in observing this effect is that both *E. coli* and λ specify general recombination systems the activity of which would hide attachment-site specific recombination. Using a bacterial mutant that lacks the general recombination system, however, Weil and Signer (1968) and Echols, Gingery, and Moore (1968) demonstrated recombination between two phage mutants also deficient for the phage-specified general recombination system. The observed recombination has the properties expected of the integration-excision system. First, it is site specific: it occurs only in the region containing the insertion exchange point. Second, it is reciprocal: single, mixedly-infected bacteria tend to produce equal numbers of the two reciprocal recombinants (Weil, 1969). Using this type of recombination in an elegant way, Echols (1970) has delineated the specificity of the *int* and *xis* products. Some of his data are reproduced in Table I. It shows that recombination in the forward, or insertion,

TABLE I

*Recognition Specificity of* Int *and* Xis
*Determined by Phage Crosses*

1. "Insertion" recombination: P·P′ × B·B′ → P·B′ + B·P′

| Cross | % recombination |
|---|---|
| A⁻ gal bio × P⁻ red⁻ | 2.9 |
| A⁻ gal bio × P⁻ red⁻ int⁻ | 0.01 |
| A⁻ gal bio × P⁻ red⁻ xis⁻ | 1.9 |

2. "Excision" recombination: P·B′ × B·P′ → P·P′ + B·B′

| Cross | % recombination |
|---|---|
| A⁻ bio × P⁻ gal red⁻ | 10.0 |
| A⁻ bio × P⁻ gal red⁻ int⁻ | 0.05 |
| A⁻ bio × P⁻ gal red⁻ xis⁻ | 0.03 |

These data are abridged from Echols (1970). The host bacteria for the crosses were *rec⁻*. λ *bio* and λ *gal bio* are both deleted for *int*. *Red⁻* phage are deficient for the general phage recombination system.

direction of Figure 4 requires *int* function but is independent of *xis* to the first approximation. On the other hand, recombination in the reverse, or excision, direction requires both *int* and *xis* function.

By what mechanism does *int* protein or *int* plus *xis* cut and join DNA molecules with the proper sequences? Undoubtedly, the answer awaits isolation and investigation of these proteins in vitro. One possibility is suggested by analogy with the structure of phage DNA cohesive ends. In this scheme, formulated in Figure 6, *int* protein would be a specific DNA endonuclease, which, recognizing the same short sequence of base pairs present in both *att*λ of *E. coli* and

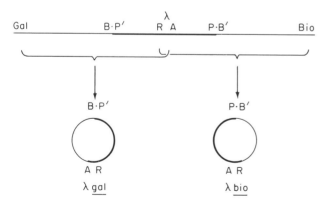

FIGURE 5 Origin of λ *gal* and λ *bio* transducing phage. The thick line represents λ DNA; and the thin line, bacterial DNA. B, P, R, and A are defined in the legend of Figure 4. The two braces indicate the DNA segments which are excised from the chromosome and joined at their ends to produce λ *gal* and λ *bio*. The mechanism that produces these rare, abnormal excisions is not known.

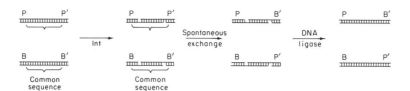

FIGURE 6 A hypothetical molecular mechanism for the insertion reaction. Between P and P′ in λ DNA, and B and B′ in bacterial DNA, the same base sequence, called the common sequence, is assumed to exist. The *int* protein is assumed to be an endonuclease specific for the common sequence and for the configuration P·P′/B·B′. An analogous series of reactions would be initiated by the *int* protein plus the *xis* protein, except that the starting configuration would have to be P·B′/B·P′.

in λ DNA, would cleave two phosphodiester bonds on opposite strands. (The common sequence would be represented by the dot in P·P′ and B·B′.) Cleavage would thus create an identical pair of cohesive ends in bacterial DNA and phage DNA. The ends could then spontaneously exchange pairing partners, and DNA ligase could re-form the missing phosphodiester bonds to complete the insertion of λ DNA into *E. coli* DNA. Further, the scheme would have it that the action of *int* enzyme is also controlled by base pairs flanking the common sequence. These base pairs, to the left and to the right of the common sequence, represented by P, P′, B, and B′, differentiate phage from bacterial DNA and prophage ends from both. The role of the *xis* protein would then be to change the specificity of *int* enzyme with respect to the flanking base pairs. Two testable deductions flow from this hypothesis. One is that there should be a common sequence of base pairs in λ and *E. coli* DNA at the site of insertion and excision exchange. Let us call the sequence C. The second deduction is that the recognition sequences P and P′ flanking C must be short, so that both can be recognized simultaneously by one enzyme or enzyme complex.

Parkinson (1969) has isolated many deletion mutants of λ that remove segments of the attachment region. By electron microscopy of DNA-DNA hybrids formed between these mutants, he can place upper limits on the sizes of C, P, and P′. His current estimates are that P′ must be less than 500 base pairs long and that P and C are each less than 50 base pairs. His experiments do not rule out the possibility that C is zero, that is, that the insertion-excision mechanism does not use cohesive ends. Further refinement will be necessary to settle this point.

## General implications of the insertion-excision system

General genetic recombination resembles prophage insertion-excision in two respects. Both processes result in an exchange of parts of DNA molecules in such a way that no base pairs are gained or lost. Prophage insertion-excision, the general recombination system of eukaryotes, and the general recombination system in bacteria are all reciprocal. But resemblance seems to end there. In general recombination, two homologous DNA sequences recognize each other, most likely through base pairing (see, for example, Thomas, 1967), and general recombination enzymes are not sequence-specific. In insertion-excision, on the other hand, two specific, nonhomologous DNA sequences are recognized by sequence-specific proteins. Why, we might ask, do not temperate phage use their host's general recombination system for insertion-excision rather than synthesizing special proteins and providing special sequences? The answer is, I think, that, if they did so, lysogenic bacteria would be unstable because recombination would also excise the prophage. Bacterial episomes, which depend on homology and host recombination enzymes for their integration, are, in fact, less stable than a prophage in a single lysogenic bacterium. A prophage owes its stability, I would suggest, to the use of specific excision proteins, synthesis of which is repressed in the lysogenic state. Lysogenic bacteria are induced to produce phage by lifting this repression and releasing the synthesis of excision proteins along with other proteins needed for virus multiplication.

The ultimate effect of the two recognition systems described in this paper is a specific and stable translocation of genetic material. Similar molecular mechanisms might be used by higher cells during development to create new gene arrangements in somatic cells. For example, in antibody synthesis the association of a particular variable region with a constant region may be the consequence of gene translocation (Edelman, this volume). Control of hemoglobin synthesis offers another possible application. How is the production of α chains coordinated with γ chains in the fetus and with β chains in the adult animal? It is conceivable that the genes for the α and β chains, which are on separate chromosomes in germ cells, might be brought together by specific translocation in somatic cells at the appropriate time in development for their joint expression. Differentiating nerve cells may also exploit this possibility for stable and directed change. New methods for genetic analysis of somatic cells (Nabholz et al., 1969) may permit an experimental test of these possibilities.

## Acknowledgments

The work from the author's laboratory was supported by a grant from the National Institute of Allergy and Infectious Disease and by a grant from the National Science Foundation.

## REFERENCES

CAMPBELL, A. M., 1962. Episomes. *Advances Genet.* 11: 101–145.

CAMPBELL, A. M., 1965. The steric effect in lysogenization by bacteriophage lambda. II. Chromosomal attachment of the $b_2$ mutant. *Virology* 27: 340–345.

CAMPBELL, A. M., 1969. Episomes. Harper and Row, New York.

DOVE, W. F., 1966. Action of the lambda chromosome. I. Control of functions late in bacteriophage development. *J. Mol. Biol.* 19: 187–201.

DOVE, W. F., 1968. The genetics of the lambdoid phages. *Annu. Rev. Genet.* 2: 305–340.

DULBECCO, R., 1968. The state of the DNA of polyoma virus and SV40 in transformed cells. *Cold Spring Harbor Symp. Quant. Biol.* 33: 777–783.

ECHOLS, H., 1970. Integrative and excisive recombination by bacteriophage λ: Evidence for an excision-specific recombination protein. *J. Mol. Biol.* 47: 575–583.

ECHOLS, H., R. GINGERY, and L. MOORE, 1968. Integrative recombination function of bacteriophage λ: Evidence for a site-specific recombination enzyme. *J. Mol. Biol.* 34: 251–260.

EISEN, H., L. PEREIRA DA SILVA, F. JACOB, 1968. The regulation and mechanism of DNA synthesis in bacteriophage λ. *Cold Spring Harbor Symp. Quant. Biol.* 33: 755–764.

GELLERT, M., 1967. Formation of covalent circles of lambda DNA by *E. coli* extracts. *Proc. Nat. Acad. Sci. U.S.A.* 57: 148–155.

GILBERT, W., and D. DRESSLER, 1968. DNA replication: The rolling circle model. *Cold Spring Harbor Symp. Quant. Biol.* 33: 473–484.

GINGERY, R., and H. ECHOLS, 1967. Mutants of bacteriophage λ unable to integrate into the host chromosome. *Proc. Nat. Acad. Sci. U.S.A.* 58: 1507–1514.

GINGERY, R., and H. ECHOLS, 1968. Integration, excision, and transducing particle genesis by bacteriophage λ. *Cold Spring Harbor Symp. Quant. Biol.* 33: 721–727.

GOTTESMAN, M. E., and M. B. YARMOLINSKY, 1968. Integration-negative mutants of bacteriophage lambda. *J. Mol. Biol.* 31: 487–505.

GUARNEROS, G., and H. ECHOLS, 1970. New mutants of bacteriophage λ with a specific defect in excision from the host chromosome. *J. Mol. Biol.* 47: 565–574.

HERSHEY, A. D., E. BURGI, and L. INGRAHAM, 1963. Cohesion of DNA molecules isolated from phage lambda. *Proc. Nat. Acad. Sci. U.S.A.* 49: 748–755.

HRADECNA, Z., and W. SZYBALSKI, 1969. Electron micrographic maps of deletions and substitutions in the genomes of transducing coliphages λ dg and λ bio. *Virology* 38: 473–477.

JACOB, F., and E. L. WOLLMAN, 1961. Sexuality and the Genetics of Bacteria. Academic Press, New York, pp. 129–154.

KAISER, A. D., and R. B. INMAN, 1965. Cohesion and the biological activity of bacteriophage lambda DNA. *J. Mol. Biol.* 13: 78–91.

KAISER, A. D., and T. MASUDA, 1970a. Evidence for a prophage excision gene in λ. *J. Mol. Biol.* 47: 557–564.

KAISER, A. D., and T. MASUDA, 1970b. Specificity in curing by heteroimmune superinfection. *Virology* 40: 522–529.

KAYAJANIAN, G., 1968. Studies on the genetics of biotin-transducing, defective variants of bacteriophage λ. *Virology* 36: 30–41.

KAYAJANIAN, G., and A. CAMPBELL, 1966. The relationship between heritable physical and genetic properties of selected $gal^-$ and $gal^+$ transducing λ dg. *Virology* 30: 482–492.

KIGER, J. A., and R. L. SINSHEIMER, 1969. Vegetative lambda DNA. V. Evidence concerning single-strand elongation. *J. Mol. Biol.* 43: 567–579.

KNIPPERS, R., A. RAZIN, R. DAVIS, and R. SINSHEIMER, 1969. The process of infection with bacteriophage ΦX174 XXIX. *In vivo* studies on the synthesis of the single stranded DNA of progeny ΦX174 bacteriophage. *J. Mol. Biol.* 45: 237–263.

MacHATTIE, L. A., and C. A. THOMAS, JR., 1964. DNA from bacteriophage lambda: molecular length and conformation. *Science (Washington)* 144: 1142–1144.

MACKINLAY, A. G., and A. D. KAISER, 1969. DNA replication in head mutants of bacteriophage λ. *J. Mol. Biol.* 39: 679–683.

MANLY, K. F., E. R. SIGNER, and C. M. RADDING, 1969. Nonessential functions of bacteriophage λ. *Virology* 37: 177–188.

NABHOLZ, M., V. MIGGIANO, and W. BODMER, 1969. Genetic analysis using human-mouse somatic cell hybrids. *Nature (London)* 223: 358–363.

OLIVERA, B. M., and I. R. LEHMAN, 1967a. Linkage of polynucleotides through phosphodiester bonds by an enzyme for *Escherichia coli. Proc. Nat. Acad. Sci. U.S.A.* 57: 1426–1433.

OLIVERA, B. M., and I. R. LEHMAN, 1967b. Diphosphopyridine nucleotide: A cofactor for the polynucleotide-joining enzyme from *Escherichia coli. Proc. Nat. Acad. Sci. U.S.A.* 57: 1700–1704.

PARKINSON, J. S., 1969. Organization and function of phage lambda chromosome. California Institute of Technology, Pasadena, doctoral thesis.

RIS, H., and B. L. CHANDLER, 1963. The ultrastructure of genetic systems in prokaryotes and eukaryotes. *Cold Spring Harbor Symp. Quant. Biol.* 28: 1–8.

SHAPIRO, J. A., and S. L. ADHYA, 1969. The galactose operon of *E. coli* K-12. II. A deletion analysis of operon structure and polarity. *Genetics* 62: 249–264.

SIGNER, E. R., and J. R. BECKWITH, 1966. Transposition of the *lac* region of *Escherichia coli*. III. The mechanism of attachment of bacteriophage Φ80 to the bacterial chromosome. *J. Mol. Biol.* 22: 33–51.

SZYBALSKI, W., H. KUBINSKI, and P. SHELDRICK, 1966. Pyrimidine clusters on the transcribing strand of DNA and their possible role in the initiation of RNA synthesis. *Cold Spring Harbor Symp. Quant. Biol.* 31: 123–127.

THOMAS, C. A., JR., 1967. The recombination of DNA molecules. *In* The Neurosciences: A Study Program (G. C. Quarton, T. Melnechuk, and F. O. Schmitt, editors). The Rockefeller University Press, New York, pp. 162–182.

TOMIZAWA, J.-I., and T. OGAWA, 1968. Replication of phage

lambda DNA. *Cold Spring Harbor Symp. Quant. Biol.* 33: 533–551.

WANG, J. C., and N. DAVIDSON, 1966. Thermodynamic and kinetic studies on the interconversion between the linear and circular forms of phage lambda DNA. *J. Mol. Biol.* 15: 111–123.

WANG, J. C., and N. DAVIDSON, 1968. Cyclization of phage DNAs. *Cold Spring Harbor Symp. Quant. Biol.* 33: 409–415.

WEIL, J., 1969. Reciprocal and non-reciprocal recombination in bacteriophage λ. *J. Mol. Biol.* 43: 351–355.

WEIL, J., and E. R. SIGNER, 1968. Recombination in bacteriophage λ. II. Site-specific recombination promoted by the integration system. *J. Mol. Biol.* 34: 273–279.

WEISBERG, R. A., and M. E. GOTTESMAN, 1969. The integration and excision defect of phage λ *dg. J. Mol. Biol.* 46: 565–580.

WEISS, B., A. JACQUEMIN-SABLON, T. R. LIVE, G. C. FAREED, and C. C. RICHARDSON, 1968. Enzymatic breakage and joining of deoxyribonucleic acid VI. Further purification and properties of polynucleotide ligase from *Escherichia coli* infected with bacteriophage T4. *J. Biol. Chem.* 243: 4543–4555.

WU, R., and A. D. KAISER, 1968. Structure and base sequence in the cohesive ends of bacteriophage lambda DNA. *J. Mol. Biol.* 35: 523–537.

YOUNG, E. T., II, and R. L. SINSHEIMER, 1967. Vegetative bacteriophage λ DNA. II. Physical characterization and replication. *J. Mol. Biol.* 30: 165–200.

ZICHICHI, M. L., and G. KELLENBERGER, 1963. Two distinct functions in the lysogenization process: The repression of phage multiplication and the incorporation of the prophage in the bacterial genome. *Virology* 19: 450–460.

ZISSLER, J., 1967. Integration-negative (*int*) mutants of phage λ. *Virology* 31: 189.

# 85 Arrangement and Evolution of Eukaryotic Genes

G. M. EDELMAN and J. A. GALLY

IT IS A CLASSIC OBSERVATION that the architecture of eukaryotic chromosomes is quite unlike that of prokaryotes (see papers by C. A. Thomas and J. H. Taylor in this volume). The genetic information in the nuclear genome is packaged in a morphologically distinct set of organelles, the exact structure of which is still unknown. It appears that the eukaryotic chromosome is longitudinally arranged into structural subunits (Beermann, 1967), which may also serve as units of biosynthesis (Plaut et al., 1966; Huberman and Riggs, 1968), function, and evolution (Keyl, 1965). The proper functioning and replication of eukaryotic cells require additional structures, such as the mitotic apparatus, synaptonemal complexes, and nucleoli, none of which is found in bacteria.

Moreover, unlike bacteria, metazoan cells usually can be classified into two lines—germ cells and somatic cells. Although some single somatic cells of higher plants can be induced to give rise to complete, differentiated plants (Stewart et al., 1964), and nuclei of differentiated cells from the frog can provide the nuclear genetic material for regenerated animals (Gurdon and Woodland, 1968), irreversible alterations of somatic genomes may yet be found to play a major role in metazoan cell differentiation. Outside the laboratory, the genome of the somatic cell would never be called upon to reproduce a whole organism, and it might be free to undergo irreversible changes during differentiation. The many examples in which such changes are visible, such as polytenization, gene amplification (Pavan and da Cunha, 1969) and loss (Serra, 1968), breakage (McClintock, 1951), diminution (Beermann, 1966), or irreversible inactivation (Lyon, 1968) of chromosomes, all suggest the possibility that small, but significant, irreversible changes occur in the DNA from cell to cell. Recent studies suggesting the existence of "metabolic DNA" are in accord with such a suggestion (Pelc, 1968).

Rather than discuss the question of how gross chromosomal rearrangements or modifications might affect eukaryotic genetics, we shall review a number of specific phenomena that suggest that the arrangement and behavior of some of the genes in eukaryotes are fundamentally different from those in prokaryotes. The models proposed to account for these phenomena have two features in common: they all invoke a more-or-less directed or programed alteration in DNA sequences in the nuclear genome, and they presume the existence in the DNA of multiply repeated base sequences.

1. *Antibodies.* The experimental investigations on structural, genetic, and biosynthetic properties of immunoglob-

G. M. EDELMAN   The Rockefeller University, New York, New York; J. A. GALLY   Meharry Medical College, Nashville, Tennessee

ulins are discussed elsewhere (Edelman, this volume) and are summarized in a recent review (Edelman and Gall, 1969). The data suggest that each immunoglobulin polypeptide chain is coded for in the genome by two separate nucleotide sequences. Furthermore, these two sequences must fuse prior to the synthesis of mRNA for the chain. It has been suggested that this gene translocation also plays a role in the differentiation of the cell, committing it to the production of a single immunoglobulin molecule. In addition, there are good reasons for believing that the genes for the immunoglobulin polypeptide chains must diversify in some manner during ontogeny, requiring other sorts of programed alterations in somatic genes.

2. *Transposable control elements.* We refer here to the complex genetic systems in maize, described by McClintock (1965) and others (e.g., Peterson, 1961), and those in *Drosophila,* described by Green (1969). Although elements in these systems resemble certain aspects of genetic control in bacteria, e.g., diffusible repressors (Jacob and Monod, 1961) or an episome that can be translocated (Smith-Keary and Dawson, 1964), these analogies do not account for many of the details of the systems. McClintock's studies demonstrate that there are genetic elements in higher plants that can regulate gene expression. These elements do not remain localized in any one locus of the genome, however, but can be transposed from place to place on the chromosomes. Sometimes they control only those loci near which they are located, and sometimes they regulate many different loci distributed throughout the genome. These changes in position and inheritable "changes of state" (McClintock, 1951) imply molecular rearrangements in the genetic material that are sometimes made visible by localized chromosome breakage.

3. *Paramutation.* Brink (1964) has reported evidence that the genes controlling the rate of expression at certain loci in maize are inherently unstable and undergo changes during somatic-cell growth. As a result, the phenotype of the daughter cells changes toward a greater or lesser degree of phenotypic expression of the structural gene at the affected locus. The rate and direction of the change can be regulated by controlling the genetic background of the inherently unstable allele; in particular, the *rate* of this so-called paramutation is affected by the other alleles present at the paramutable locus. Other examples of similar phenomena are known in plants, and a seemingly analogous phenomenon has been observed in the bobbed loci of *Drosophila* (Ritossa, 1968). In both cases it has been presumed that the heritable changes involve alteration in the actual number of similar DNA sequences.

4. *Gene amplification.* Certain sites within eukaryotic chromosomes undergo differential replication. Pavan and da Cunha (1969), for example, have described "DNA puffs" in the dipteran species *Rhynchosciara angeli*. In certain dif-

ferentiating cells of this species, the DNA at particular chromosomal loci replicates to such an extent that visibly enlarged amounts of DNA appear. The function of the DNA at these loci is not known. Other examples are provided by the evidence that the genes for rRNA replicate differentially during the oogenesis of a number of species that are distant from one another in the evolutionary tree (Gall, 1969). The production of the increased amounts of rDNA has been rationalized in terms of the large cytoplasm the egg nucleus must provide, as well as the lack of rRNA production during the early cleavage of the embryo (Brown and Gurdon, 1964).

## Repeated DNA sequences

It is noteworthy that models proposed for all the genetic systems mentioned above postulated the existence in the genome of repeated DNA sequences before any direct evidence for these sequences was available. Multiple copies of genes for immunoglobulins were postulated in order to account for the enormous variety and specificity of antibodies that vertebrates can synthesize (Dreyer and Bennett, 1965). Because of the possibility of crossing over between paired, nonhomologous genes, instability is inherent in all repeated DNA sequences. Semi-stable genetic loci of this type readily allow quantitative variations and, therefore, multiple genes were invoked in models of both transposable control elements (McClintock, 1950) and paramutation (Smith-Keary and Dawson, 1964).

This potential instability led Thomas (1966) to suggest that the genome of prokaryotes would tend to evolve in order to eliminate or minimize the repetition of any long base sequences. His work, and that of others, provided evidence to support this hypothesis, for it was found, in a study of the rate of renaturation of DNA from bacteria and viruses, that repeated nucleotide sequences are rarely found in these genomes. In viruses, repetitive DNA appears at the end of recurrently redundant genomes, but apparently nowhere else. In bacteria, also, almost all the DNA sequences appear to be "unique," i.e., they occur only once per genome. A single bacterial chromosome can contain a number of genes coding for rRNA (Kohne, 1969), but these genes appear to be maintained by the requirement for large amounts of rRNA in the bacterial cell (Mueller and Bremer, 1968).

It has been suspected for several years, however, that such constraints on repetitive DNA sequences may not apply strictly to eukaryotes. Direct studies of the rate of renaturation of nuclear DNA (both somatic and germ-line) have indicated that repeated DNA sequences occur with high frequency (Martin and Hoyer, 1966; Britten and Kohne, 1965). Repetitious DNA has been detected in the genomes of all eukaryotes whose DNA has been studied by this method, including a few organisms from most of the major phyla of plants and animals (Britten and Kohne, 1968). The amount of repetitious DNA present is hardly negligible, amounting to about 40 per cent of the DNA in calf thymus, for example. Obviously, this implies that the rest of the DNA represents unique, i.e., unrepeated, genes. It now appears that the existence of easily detectable amounts of repetitious DNA is as characteristic of eukaryotes as is the nuclear membrane or the mitotic apparatus. In view of the possibility that repetitious DNA may confer characteristic genetic properties upon eukaryotes, the acquisition of the ability to maintain the repeated nucleotide sequences in a stable form may have marked a very important step in the evolution of all nucleated cells. Indeed, chromosomes themselves may have evolved mainly to allow multiple genes to exist and function in a stable fashion.

The use of a single term, such as "repetitious DNA," to refer to all examples of repeated nucleotide sequences in the genome may be misleading. It is probable, for example, that repeated DNA sequences include genes for ribosomal and transfer RNA, nucleotide sequences having other functions, such as those concerned with replication and recombination of chromosomes, and multiple, tandem-duplicated, structural genes for proteins.

That rRNA genes fall into the class of repetitious DNA has been demonstrated by DNA-RNA hybridization experiments, in which labeled rRNA was used as one of the hybridizing species. The number of rRNA genes has been estimated to range from 100 to 400 in a haploid set of chromosomes in species as varied as yeasts (Retel and Planta, (1968), *Drosophila* (Ritossa and Spiegelman, 1965), frogs (Wallace and Birnstiel, 1966), pea seedlings (Chipchase and Birnstiel, 1963), and man (Attardi et al., 1965). As in bacteria, a single rRNA gene per cell could not be transcribed at a rate sufficiently rapid to provide enough of this essential structural component of the protein-synthetic apparatus. This alone might account for the multiple copies found in the genome (Kohne, 1969). Similarly, the genes for tRNA (Ritossa et al., 1966) and 5S RNA (Brown and Weber, 1968) molecules appear to be repeated in the chromosomes, in some cases to a very high degree (27,000 5S RNA genes in germ-line and somatic cells of certain frogs [Brown and Weber, 1968]).

In addition to tRNA and rRNA, other RNA species of unknown function are also transcribed from the repetitious DNA. Because of the large size of eukaryotic genomes, RNA transcribed from a *unique* DNA sequence would not be expected to hybridize with cellular DNA at an easily detectable rate. Because it is known that much of the non-ribosomal RNA in the cell does indeed hybridize rapidly with nuclear DNA, it has been concluded that some of the repetitious DNA is being transcribed (Britten and Kohne, 1968). It should be pointed out, however, that a portion of the repetitious DNA in eukaryotic genomes may never be transcribed into RNA. For example, Flamm et al. (1969) report that the extremely repetitious DNA of the mouse does not appear to hybridize with any RNA found in a number of different somatic tissues. This satellite DNA accounts for about 10 per cent of the mouse genome and appears to consist of roughly a million copies, each about 400 nucleotides long (Waring and Britten, 1966).

In contrast to ribosomal and tRNA genes, there is no evidence to suggest that the structural genes for the more commonly studied eukaryotic proteins, such as cytochrome c (Margoliash et al., 1969), are specified by many gene copies in the genome—although, in some cases, indirect arguments for multiple genes can be formulated, based on the observed rates of synthesis of these proteins (Jerne, 1967). The primary structures of a number of nonallelic proteins are quite similar, however, and it appears probable that the genes for their constituent polypeptide chains arose through gene duplication (presumably as a result of unequal crossing over), followed by evolutionary divergence. Examples of such proteins include the various hemoglobins (Ingram, 1963), the immunoglobulins (Edelman, this volume), and such functionally distinct proteins as lysozyme and lactalbumin (Hill et al., 1969). Obviously, such a process could not proceed if duplicated DNA sequences were *rapidly* eliminated from the gene pool; some mechanism must exist to prevent such rapid elimination. The homogeneity of the proteins isolated from homozygous animals, the kinds of polymorphism observed (for example, in human hemoglobins), and the nature of the evolutionary differences in the amino-acid sequences of homologous proteins from different animal species, all provide strong arguments for the existence of a single structural gene copy for most of the proteins so far studied. Experimental methods such as hybridization with purified messengers may, when they become available, permit a direct estimate of the number of genes involved.

At present, one can only speculate on the functions of much of the repeated DNA in eukaryotes. For example, repetitious DNA might contain binding sites for chromosomal proteins or chromosomal RNA, as suggested by the studies of Huang and Bonner (1965) on protein-linked chromosomal RNA. Chromosomal RNA has been reported to affect the specificity of histone-DNA interactions (Bekhor et al., 1969b; Huang and Huang, 1969) and appears to bind to repetitious DNA sequences (Bonner and Widholm, 1967). This possible function of repetitious DNA has been recently reviewed (Britten and Davidson, 1969). The peculiarities of this hybridization reaction, however, leave some doubt as to whether the usual linkage between complementary base-pairs of the nucleic acid strands is involved (Bekhor et al., 1969a).

A certain portion of the repetitious DNA may play a role

in determining how DNA fibers fold in the chromosome, either by Watson-Crick pairing or by serving as binding sites for proteins that stabilize metaphase chromosomes and the synaptonemal complex. The finding of Maio and Schildkraut (1969), that repetitious DNA is evenly distributed in the mouse chromosomes and is tightly bound to their structural components, is consistent with this hypothesis. As separated strands of repetitious DNA renature much more rapidly than strands of DNA of unique sequence, repetitious DNA may function to aid chromosomal recombination or to aid in the repair of chromosomal breakage. Either function would provide a ready explanation for the observed ability of broken ends of chromosomes to rejoin one another. Similarly, repeated DNA sequences might aid pairing during meiosis.

A fraction of repetitious DNA may not contain any genetic "information" at all but may have evolved to function merely as a space-filler in the genome. This possibility would at least provide an explanation for the great disparity between the large number of genes that can be encoded in the nuclear DNA and the much smaller number of genes that have been estimated to be present in the chromosomes. (This discrepancy is discussed at length by C. A. Thomas elsewhere in this volume.) There appears to be no great correlation between the structural complexity of an organism and the amount of DNA in its genome. A good correlation has been reported between the amount of DNA per nucleus and the nuclear volume, however (Commoner, 1964), and this correlation holds true both for a number of vertebrate species and for bacteria (DNA/cell volume). The large differences in the amount of DNA in closely related species (Keyl, 1965; Fox, 1969) are also consistent with the idea that the amount of genetic material present is controlled by some property other than information content. The controlling factor might be rather nonspecific and trivial, such as the maintenance of the nuclear viscosity. If so, the ease with which repetitious DNA can expand or contract in amount, by mechanisms that are discussed below, might make it an excellent "fill" material, inert and innocuous. This possible function for repetitious DNA would be most difficult to prove directly and, in any case, is esthetically displeasing.

Finally, some repetitious DNA may consist not of identical copies, but rather of a large number of *similar* genes capable of storing large amounts of genetic information. There may exist, in the antibody system, the nervous system, and other systems of like complexity, a requirement for a very large store of preadapted genetic information. Much of this information would function only at rare, critical times in the lifetime of the organism, and thus would remain inactive much of the time. As suggested by Zuckerkandl and Pauling (1962), repetitious DNA might also represent a storehouse of evolving genes. Sheltered in these

duplications, there might exist evolving variants of structural genes for important enzymes. This rather romantic view of the multiple-gene copies is difficult to reconcile, however, with certain fundamental theorems of population biology.

We can summarize these speculations about the genetic functions of repetitious DNA by concluding that, whatever they are, they are probably diverse. Regardless of the function of repetitious DNA, however, its very existence raises fundamental problems for evolutionary theory.

## The evolution of repetitious DNA

The existence of repetitious DNA poses several new and difficult problems for classical Mendelian genetics and evolutionary theory. The key observations to be accounted for include:

1. The genomes of all investigated eukaryotic cells contain large numbers of DNA sequences which resemble one another far more than would be expected if they had evolved independently and by chance (Martin and Hoyer, 1966; Britten and Kohne, 1965). Similar sequences are not found in prokaryotic genomes (Thomas, 1966).

2. In any one species, these repeated DNA sequences can be divided into "families" of genes. In some families, the duplicated genes resemble one another very closely; in other families the members have diverged. All the families also differ in both number of members and length of the reiterated sequence (Britten and Kohne, 1965, 1968).

3. Denatured repetitious DNA strands from the nucleus of one species will renature more rapidly with one another than they will with the analogous strands from another species. Even closely related species can contain distinctively different repetitious DNA sequences (Martin and Hoyer, 1966; Britten and Kohne, 1966; Walker, 1968).

It is perhaps this third point that is most perplexing and, in fact, as we attempt to show, it suggests a new idea: in certain instances, all the members of the gene family change together, so that certain sets of closely related genes evolve to *different* sets of genes which are *also closely related*. For convenience, we call this process *coevolution of DNA sequences*.

Given the existence of multiple genes, how does natural selection operate at all? One might suppose that duplicated gene sequences would be buffered against rapid changes during evolution. If a certain gene is represented a great many times (100 or more) in the genome, any mutation in one of the copies could not greatly affect the selection pressures acting on a given individual, unless it led to the production of an actively deleterious gene product. For example, if a mutation occurs in *just one* rRNA gene, making the RNA transcribed from that gene unable to fold into a functional ribosome, no detectable effect on the viability of the animal would be expected. Frogs that lack half of their rRNA genes are known to be quite viable (Brown and Gurdon, 1964). The problem is even greater if all the genes in the family do not function simultaneously: what selection pressure could operate to maintain the constancy of an inert gene? Moreover, any rare "favorable" mutation that increases the efficiency of its gene product would be diluted in the cell with products of hundreds of its nonmutant sister genes. Therefore, it is difficult to see how natural selection could increase the gene frequency of these favorable mutants.

In trying to explain the evolutionary behavior of highly reiterated identical genes, one might assume that most single-base changes in the repetitious DNA represent truly "neutral" mutations, i.e., mutations that could confer no detectable selective advantage or disadvantage on the organisms. The possible fate of such neutral mutations is at present a matter of spirited controversy (King and Jukes, 1969). Depending on the viewpoint taken, one might expect that: (a) members of a family of like genes would gradually diverge from one another by the action of genetic drift acting on random neutral mutations (King and Jukes, 1969; Kimura, 1968), or (b) a family of a large number of like genes would remain constant from species to species throughout evolutionary time, because neutral mutations can never attain an appreciable gene frequency in a large, randomly breeding population (Simpson, 1964).

In fact, the data are not in accord with either expectation. The DNA representing the repeated sequences apparently *can* evolve, for it differs considerably from species to species (Martin and Hoyer, 1966; Zuckerkandl and Pauling, 1962; Britten and Kohne, 1966). This evolution is not necessarily accompanied by a general divergence of the members of each duplicated gene family within a species, however, and the genes thus can coevolve. What is the evidence that such a process occurs? The most clear-cut example is provided by the rRNA genes. We have already reviewed the evidence that these genes are multiply represented in nuclear chromosomes. It has also been shown that these genes do differ from species to species (Pinder et al., 1969). Although some hybridization experiments suggest that rRNA genes in a single nucleus are not all identical (Moore and McCarthy, 1968), they appear to be more closely homologous to one another than to rRNA genes of other species. The exact degree of similarity of the base sequences of rRNA among different species has not yet been determined, but human and mouse 28S rRNA can be separated on the basis of physicochemical properties (Eliceiri and Green, 1969), suggesting that none of the rRNA genes in one species is identical to any in the other species.

So far, it has not been shown that the genes for the 5S RNA or transfer RNA do, in fact, differ from species to species within eukaryotes. If they do differ, they, too, must coevolve. The highly reiterated satellite DNA sequences isolated from any one species are also more closely related to one another than they are to similar fractions from other species (Walker, 1968). In the absence of knowledge about their function, however, the assumption that these fractions are evolutionarily homologous appears less compelling. The data on immunoglobulin sequences and genetics (Edelman, this volume) may also provide an example for multiple-gene coevolution.

## Possible mechanisms to account for coevolution of DNA sequences

What mechanisms account for coevolution of multiple genes? Perhaps the simplest answer is to rely on the "invisible hand" of natural selection, and to leave it at that. If, for example, there is a single optimal rRNA sequence for mouse ribosomes, all the gene copies might have undergone exactly parallel evolution to the same end-product by random-point mutations acted upon by selective forces. We believe this to be most unlikely. It appears that a very specific mechanism is necessary, and we briefly describe some of those that have been proposed.

*1. Reiteration or incorporation of multiple copies.* Britten and Kohne have suggested that coevolution can be accounted for by genetic "bursts" (Britten and Kohne, 1965, 1968), i.e., single-gene copies undergo a process of reiterative replication in the germ-line of certain individuals (Figure 1A). The gene frequencies of the duplicated genes increase, presumably because they confer some selective advantage and, therefore, eventually become fixed in the population. Following such "saltatory evolution," *which must occur fairly frequently* to account for the observed differences between closely related species, the initially identical genes gradually diverge from one another by random drift, becoming more and more heterogeneous. Presumably a compensatory gene loss also occurs frequently to make space for the genes introduced by the "bursts," because the total genome size does not increase proportionately. According to this picture, each gene family would arise from a different initial "burst." The age of a family would be correlated with the amount of heterogeneity of its gene sequences, and, therefore, the newest families would consist of almost identical genes.

The molecular mechanism underlying the "burst" event has not been specified, but presumably it might resemble that required for gene amplification of oocyte rRNA genes (Gall, 1969) or DNA "puffs" (Pavan and da Cunha, 1969). The even distribution of the family of highly repetitive DNA throughout the karyotype of mice argues against this proposal, unless the genes can somehow translocate swiftly through the genome soon after being generated (Bekhor et al., 1969a; 1969b).

Another mechanism suggested by Britten and Kohne (1966), viz., incorporation of multiple copies of lysogenic viral genomes, might account for the even distribution. The

# GENE REITERATION

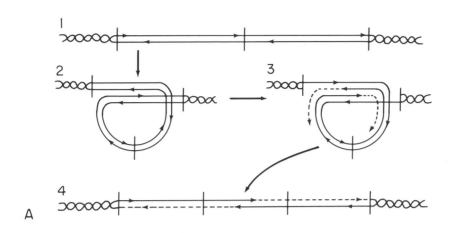

# NON-HOMOLOGOUS RECOMBINATION
## interchromosomal

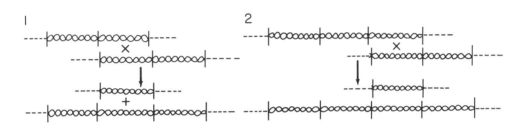

## intrachromosomal

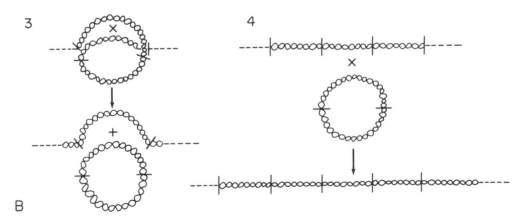

FIGURE 1   A. One possible molecular mechanism for gene reiteration. (1) DNA sequences within vertical bars are similar to one another. For clarity the double-helix is drawn unwound in these sequences only. The process involves pairing of like sequences (2), nucleolytic attack, hybrid-DNA formation, and repair synthesis of the DNA (3). These steps lead to a larger family of like genes (4). (Other mechanisms for reiteration are possible; see Thomas, this volume; and Pavan and da Cunha, 1969.) B. (1) Recombination between paired, tandem, similar but non-homologous DNA sequences in different chromatids increases the number of members of one gene family and decreases the number in the paired chromatid. (2) This process, if continued, can lead to progressively larger gene families. (3) Pairing and recombination between two members of a gene family on a single chromatid would lead to formation of an episome which might be able to integrate elsewhere in the genome (4).

actual mechanism would resemble that illustrated in Figure 1, B3. The recent demonstration that many copies of the genomes of human oncogenic viruses can become incorporated into the nucleus of a single transformed cell provides a model for this mechanism (Dulbecco, 1968). The small size of the repeated base sequences in the highly repetitious DNA of mouse (Waring and Britten, 1966), however, would not appear to be able to code for a complete viral genome.

The other mechanisms that have been proposed to account for gene coevolution all involve the ability of repeated DNA base sequences to mispair, i.e., to join in Watson-Crick base-pairing with nonhomologous members of the family located on the same or different chromosomes. Genetic recombination events that take place in these mispaired regions might allow base-sequence information to be transmitted from sequence to sequence along a tandemly duplicated gene family, thus aiding in their coevolution. A number of different mechanisms could account for this phenomenon, depending on the exact nature of the recombination and conversion steps postulated. Some specific examples are discussed below.

2. *Reciprocal, unequal crossing over*. The mispaired regions might undergo a breakage and reunion process in which both strands of the DNA exchange partners (Figure 1B). If the mispaired genes are on different chromosomes, or on different chromatids of replicated chromosomes, an increase in the number of tandem-repeated genes in one chromosome and an equivalent decrease in the other would result. If the mispaired genes were on the same chromosome, a loop of tandem genes might be deleted. This loop might be lost from the nucleus, or it might be re-incorporated elsewhere in the genome in a lysogenic fashion (Figure 1B).

The potentialities inherent in unequal crossing over have long been known. It is the most commonly accepted hypothesis to account for the incorporation of new structural genes into the genome, and the interesting instabilities to which it might give rise have been investigated by a number of workers (Lewis, 1967). Paramutation phenomena (Brink, 1964) have been explained in this way and a similar mechanism may operate in the bobbed mutants of *Drosophila* (Ritossa, 1968). Bobbed mutants apparently arise as a result of deletions of a large number of linked rRNA genes. Some of these deletions are in turn unusually unstable, and revert to normal, a change that is accompanied by an increase in the number of detectable rRNA genes in the genome. It has been suggested that these genes are held in a dynamic equilibrium in the population. The number of genes can either increase or decrease as a result of unequal crossing over, and natural selection operates to maintain the gene fraction found in pooled individuals (Ritossa et al., 1966). In a dynamic system of this sort, an allele arising from a point mutation in an rRNA gene which confers selective advantage to an organism can increase in gene frequency, both in the population *and* in the tandem set of genes in one organism. This might well account for the coevolution of genes in a species.

3. *Master and slave genes*. As a result of a careful investigation of the lampbrush chromosomes of amphibian oocytes, Callan (1967) concluded that each of the lateral loops represented a functional unit of the chromosome, and that each loop might produce only a single gene product. This is discussed at length in Thomas's paper in this volume. Because these loops are far longer than would be expected for a single structural cistron, Callan suggested that they consist of a series of tandem-repeated identical genes (Figure 2A). To account for the coevolution of the gene sequences implied by such a structural arrangement, he suggested that one member of each sequence has some distinctive property (perhaps its position in the series) which would cause it to act as a "master" gene. The other members of the duplicated set are referred to as "slave" genes. At some time during the growth of the germ cells of each individual, the nucleotide strands of the DNA sequences coding for the slave genes would unwind in order to allow these strands to form double helixes with the complementary strands of the master gene. If the sequences of the master and the slave were identical, the slave would dissociate and renature with its original strand. If the slave gene had incorporated a point mutation prior to the last comparison with the master gene, a distortion would be created in the master-slave hybrid that could be enzymatically corrected by changing the slave strand to complement the master strand. Mutations in the master gene would be incorporated in each of the slave copies. As a result of these hypothetical steps, all the repeated genes would evolve in exactly the same way and at the same rate as the single master gene.

Detailed mechanisms for the master-slave comparison process have been proposed by Whitehouse (1967a) and Thomas (this volume), and variant schemes have been advanced to account both for antibody variability (Whitehouse, 1967b) and for gene instability in plants (Fincham, 1967). A possible variant of the master-slave scheme is to propose that all slave genes are lost by intrachromosomal recombination at some stage in germ-cell growth and replaced by gene amplification of the single master gene remaining in the chromosome (Markert, 1968). No evidence for any *general* gene amplification in germ cells has been reported, however, despite a great deal of work on DNA content and biosynthesis in the nuclei of these cells.

In contrast to the other mechanisms discussed here, the master-slave mechanism predicts that all members of a gene family would remain identical as long as they remain under a single master. Accordingly, this model would be the only one consistent with the hypothesis (Thomas, this volume) that structural genes for the commonly studied proteins

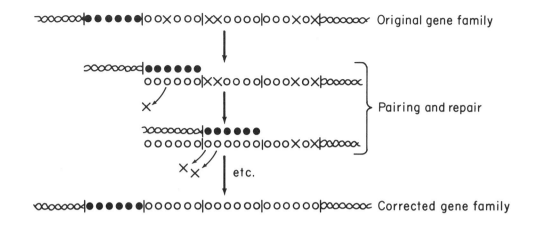

## A MASTER-SLAVE

Original gene family

} Pairing and repair

etc.

Corrected gene family

## B DEMOCRATIC GENE CONVERSION

1

2

3

FIGURE 2   A. The master gene (filled circles) pairs with each slave gene (open circles). Any differences in the nucleotide sequences of the slave genes (indicated by crosses) are excised and repaired, so that each slave gene in a corrected gene family is identical to the master gene. B. (1) A family of like genes can accumulate point mutations (closed circles and triangles), causing the tandem genes to be nonidentical.

(2) *Nonreciprocal* recombination among these genes may increase the frequency of some of these mutant sequences and decrease or eliminate others. (For details, see Edelman, page 885.) (3) A point mutation that confers selective advantage to the species might spread to all members of the gene family by means of conversion and natural selection.

have a high multiplicity. The gene duplications postulated to account for the evolution of various hemoglobin polypeptide chains, for example, would then require the duplication of whole sets of both master and slave genes, followed by divergence of the master gene. The master-slave mechanism would not easily account, however, for partial gene duplications or for the unequal crossing over which has been postulated to give rise to the genes for Lepore hemoglobin (Baglioni, 1963) and variant gamma chains of human immunoglobulins (Kunkel et al., 1969).

4. *"Democratic" gene conversion.* We have suggested that the observed sequences of variable regions of immunoglobulin chains might arise through unequal intrachromosomal crossing over among tandem duplicated genes (Edelman and Gally, 1967; Edelman, this volume). Such recombination might be either reciprocal or, more likely, nonreciprocal. Recent experiments on recombination suggest that the recombining DNA molecules join by forming hybrid helixes containing complementary strands from different origins (Meselson, 1967). Enzymatic steps after the

formation of such hybrid duplexes are assumed to excise and repair stretches of the molecular hybrid. The result would be nonreciprocal recombination or conversion. We have found several enzymes (Lindahl and Edelman, 1968; Lindahl et al., 1969a, 1969b) in the somatic cells of mammals that might play roles in such a mechanism. Gene conversion has been observed in many eukaryotic (Holliday, 1964; Whitehouse and Hastings, 1965) and some prokaryotic (Sermonti and Carere, 1968) genomes, and it has been suggested that conversion plays a major role in evolution (Fogel and Mortimer, 1969).

One of us (Gally) has proposed that democratic gene conversion (Figure 2B) might provide an additional mechanism to account for the coevolution of repeated genes. It is assumed that mispairing and nonreciprocal recombination take place frequently between the members of a family of like but nonidentical genes. Each member of a family would repeatedly compare itself with other members by hybrid DNA formation, and mismatched base-pairs in the hybrid DNA would tend to be eliminated by enzymatic excision and repair processes. In this mechanism, neither DNA strand would be treated preferentially, and a new point mutation would be as likely to be copied as to be eliminated. Thus, a selectively advantageous mutation in only one member of a large family has the potential of distributing itself throughout the family *much* more rapidly than it could by independent mutation and selection.

Because the rate of appearance of point mutants is proportional to the length of the nucleotide sequence in which they can occur, we might think of families of democratically converting, tandem genes as mutation "nets" capable of trapping favorable mutations, which can then spread laterally through the family. Disadvantageous mutants would be rapidly eliminated.

The rate of coevolution would depend on the phenotypic function of the gene product and on the rate of the conversion process. Depending on the nature of the selective forces acting on the final gene product, members of a family of like genes may be similar or they may diverge with a high degree of heterogeneity. Democratic gene conversion would not allow for the simultaneous alteration of the homologous base-pairs in *all* members of the family, as would occur in the master-slave gene conversion model. Therefore, we would not expect to find polymorphism among democratically converting genes.

If a gene family were split up by a translocation event so that members of one part of the family could no longer recombine with the rest, the two parts of the original family would no longer evolve together. If gene conversion accounts for the evolutionary behavior of rRNA genes, for example, it would predict that the rRNA genes in one autosome of a haploid set would resemble one another more closely than those on a nonhomologous chromosome.

A gene family on one chromosome which is prevented from crossing over with its homologue in a diploid set (e.g., by chromosomal inversion) would likewise not be able to coevolve with the corresponding genes on the homologous chromosome. This would give rise to *apparent* polymorphism of multiple genes in a population, but such "pseudo-polymorphism," *would not resemble* that described for structural genes. For example, the gene family would not differ from its "polymorphic" alternative by a single nucleotide at a certain position of each gene. Instead, the differences would resemble those found in the genes for the amino terminal position of the immunoglobulin heavy chains of rabbits of different allotypes (Koshland, 1967; Wilkinson, 1969).

If gene conversion occurred, many of the questions that have been asked concerning the frequencies and changes of frequencies of genes within a population could be asked about gene frequencies within an organism. Many of the concepts developed to account for the evolution of sexually recombining populations, e.g., the prevalence of heterosis, might be expected to apply to the populations of genes found within repeated gene families in a single chromosome.

The mechanisms to account for the coevolution of multiple genes discussed above are not exhaustive, and none of them may be correct. They do, however, show how unstable, dynamic systems, which would greatly increase evolutionary possibilities, might be generated by means of known biochemical rearrangements and transformations of DNA. These systems could behave as "genetic algorithms," i.e., by establishing a relatively short basic nucleotide sequence and a number of rules to govern its rearrangement, vast new arrays of information may be created upon which natural selection might be able to act. At present, the task is to provide experimental evidence for or against any of the hypothetical models that have been presented to account for the evolution of multiple genes.

## Acknowledgment

This work was supported by Grant GB 8371 from the National Science Foundation and by Grants AM 04256 and AI 09273 from the National Institutes of Health.

REFERENCES

ATTARDI, G., P-C. HUANG, and S. KABAT, 1965. Recognition of ribosomal RNA sites in DNA. II. The HeLa cell system. *Proc. Nat. Acad. Sci. U. S. A.* 54: 185–192.

BAGLIONI, C., 1963. Correlations between genetics and chemistry of human hemoglobins. *In* Molecular Genetics (J. H. Taylor, editor). Academic Press, New York, pt. 1, pp. 405–475.

BEERMANN, S., 1966. A quantitative study of chromatin diminution in embryonic mitoses of *Cyclops furcifer*. *Genetics* 54: 567–576.

BEERMANN, W., 1967. Gene action at the level of the chromosome. *In* Heritage from Mendel (R. A. Brink, editor). University of Wisconsin Press, Madison, pp. 179–201.

BEKHOR, I., J. BONNER, and G. K. DAHMUS, 1969a. Hybridization of chromosomal RNA to native DNA. *Proc. Nat. Acad. Sci. U. S. A.* 62: 271–277.

BEKHOR, I., G. M. KUNG, and J. BONNER, 1969b. Sequence-specific interaction of DNA and chromosomal protein. *J. Mol. Biol.* 39: 351–364.

BONNER, J., and J. WIDHOLM, 1967. Molecular complementarity between nuclear DNA and organ-specific chromosomal RNA. *Proc. Nat. Acad. Sci. U. S. A.* 57: 1379–1385.

BRINK, R. A., 1964. Genetic repression of R action in maize. *In* Role of Chromosomes in Development (M. Locke, editor). Academic Press, New York, pp. 183–230.

BRITTEN, R. J., and E. H. DAVIDSON, 1969. Gene regulation for higher cells: A theory. *Science (Washington)* 165: 349–357.

BRITTEN, R. J., and D. E. KOHNE, 1965. Nucleotide sequence repetition in DNA. *Carnegie Inst. Wash. Year Book* 65: 78–106.

BRITTEN, R. J., and D. E. KOHNE, 1966. Repeated nucleotide sequences. *Carnegie Inst. Wash. Year Book* 66: 73–88.

BRITTEN, R. J., and D. E. KOHNE, 1968. Repeated sequences in DNA. *Science (Washington)* 161: 529–540.

BROWN, D. D., and J. B. GURDON, 1964. Absence of ribosomal RNA synthesis in the anucleolate mutant of *Xenopus laevis. Proc. Nat. Acad. Sci. U. S. A.* 51: 139–146.

BROWN, D. D., and C. S. WEBER, 1968. Gene linkage by RNA-DNA hybridization. I. Unique DNA sequences homologous to 4S RNA, 5S RNA, and ribosomal RNA. *J. Mol. Biol.* 34: 661–680.

CALLAN, H. G., 1967. The organization of genetic units in chromosomes. *J. Cell. Sci.* 2: 1–7.

CHIPCHASE, M. I. H., and M. L. BIRNSTIEL, 1963. On the nature of nucleolar RNA. *Proc. Nat. Acad. Sci. U. S. A.* 50: 1101–1107.

COMMONER, B., 1964. Roles of deoxyribonucleic acid in inheritance. *Nature (London)* 202: 960–968.

DREYER, W. J., and J. C. BENNETT, 1965. The molecular basis of antibody formation: A paradox. *Proc. Nat. Acad. Sci. U. S. A.* 54: 864–869.

DULBECCO, R., 1968. The state of the DNA of polyoma virus and SV40 in transformed cells. *Cold Spring Harbor Symp. Quant. Biol.* 33: 777–783.

EDELMAN, G. M., and W. E. GALL, 1969. The antibody problem. *Annu. Rev. Biochem.* 38: 415–466.

EDELMAN, G. M., and J. A. GALLY, 1967. Somatic recombination of duplicated genes: An hypothesis on the origin of antibody diversity. *Proc. Nat. Acad. Sci. U. S. A.* 57: 353-358.

ELICEIRI, G. L., and H. GREEN, 1969. Ribosomal RNA synthesis in human-mouse hybrid cells. *J. Mol. Biol.* 41: 253–260.

FINCHAM, J. R. S., 1967. Mutable genes in the light of Callan's hypothesis of serially repeated gene copies. *Nature (London)* 215: 864–866.

FLAMM, W. G., P. M. B. WALKER, and M. McCALLUM, 1969. Some properties of the single strands isolated from the nuclear satellite of the mouse (*Mus musculus*). *J. Mol. Biol.* 40: 423–443.

FOGEL, S., and R. K. MORTIMER, 1969. Informational transfer in meiotic gene conversion. *Proc. Nat. Acad. Sci. U. S. A.* 62: 96–103.

FOX, D. P., 1969. The relationship between DNA value and chromosome volume in the coleopteran genus *Dermestis. Chromosoma* 27: 130–144.

GALL, J. G., 1969. The genes for ribosomal RNA genes during oögenesis. *Genetics (Suppl.)* 61 (no. 1, pt. 2): 121–132.

GREEN, M. M., 1969. Controlling element mediated transpositions of the *white* gene in *Drosophila melanogaster. Genetics* 61: 429–441.

GURDON, J. B., and H. R. WOODLAND, 1968. The cytoplasmic control of nuclear activity in animal development. *Biol. Rev. (Cambridge)* 43: 233–267.

HILL, R. L., K. BREW, T. C. VANAMAN, I. P. TRAYER, and P. MATTOCK, 1969. The structure, function, and evolution of α-lactalbumin. *Brookhaven Symp. Biol.* 21: 139–154.

HOLLIDAY, R., 1964. A mechanism for gene conversion in fungi. *Genet. Res.* 5: 282–304.

HUANG, R. C., and J. BONNER, 1965. Histone-bound RNA, a component of native nucleohistone. *Proc. Nat. Acad. Sci. U. S. A.* 54: 960–967.

HUANG, R. C., and P. C. HUANG, 1969. Effect of protein-bound RNA associated with chick embryo chromatin on template specificity of the chromatin. *J. Mol. Biol.* 39: 365–378.

HUBERMAN, J A., and A. D. RIGGS, 1968. On the mechanism of DNA replication in mammalian chromosomes. *J. Mol. Biol.* 32: 327–341.

INGRAM, V. M., 1963. The Hemoglobins in Genetics and Evolution. Columbia University Press, New York.

JACOB, F., and J. MONOD, 1961. Genetic regulatory mechanisms in the synthesis of proteins. *J. Mol. Biol.* 3: 318–356.

JERNE, N. K., 1967. Summary: Waiting for the end. *Cold Spring Harbor Symp. Quant. Biol.* 32: 591–603.

KEYL, H. G., 1965. A demonstrable local and geometric increase in the chromosomal DNA of *Chironomus. Experientia (Basel)* 21: 191–193.

KIMURA, M., 1968. Evolutionary rate at the molecular level. *Nature (London)* 217: 624–626.

KING, J. L., and T. H. JUKES, 1969. Non-Darwinian evolution. *Science (Washington)* 164: 788–798.

KOHNE, D. E., 1969. Isolation and characterization of bacterial ribosomal RNA cistrons. *Carnegie Inst. Wash. Year Book* 69: 310–320.

KOSHLAND, M. E., 1967. Location of specificity and allotypic amino acid residues in antibody Fd fragments. *Cold Spring Harbor Symp. Quant. Biol.* 32: 119–127.

KUNKEL, H. G., J. B. NATVIG, and F. G. JOSLIN, 1969. A "Lepore" type of hybrid γ globulin. *Proc. Nat. Acad. Sci. U. S. A.* 62: 144–149.

LEWIS, E. B., 1967. Genes and gene complexes. *In* Heritage from Mendel (R. A. Brink, editor). University of Wisconsin Press, Madison, pp. 17–47.

LINDAHL, T., and G. M. EDELMAN, 1968. Polynucleotide ligase from myeloid and lymphoid tissues. *Proc. Nat. Acad. Sci. U. S. A.* 61: 680–687.

LINDAHL, T., J. A. GALLY, and G. M. EDELMAN, 1969a. Properties of deoxyribonuclease III from mammalian tissues. *J. Biol. Chem.* 244: 5014–5019.

LINDAHL, T., J. A. GALLY, and G. M. EDELMAN, 1969b. Deoxyribonuclease IV: A new exonuclease from mammalian tissues. *Proc. Nat. Acad. Sci. U. S. A.* 62: 597–603.

Lyon, M. F., 1968. Chromosomal and sub-chromosomal inactivation. *Annu. Rev. Genet.* 2: 31–52.

McClintock, B., 1950. The origin and behavior of mutable loci in maize. *Proc. Nat. Acad. Sci. U. S. A.* 36: 344–355.

McClintock, B., 1951. Chromosome organization and genic expression. *Cold Spring Harbor Symp. Quant. Biol.* 16: 13–47.

McClintock, B., 1965. The control of gene action in maize. *Brookhaven Symp. Biol.* 18: 162–182.

Maio, J. J., and C. L. Schildkraut, 1969. Isolated mammalian metaphase chromosomes. II. Fractionated chromosomes of mouse and Chinese hamster cells. *J. Mol. Biol.* 40: 203–216.

Margoliash, E., W. M. Fitch, and R. E. Dickerson, 1969. Molecular expression of evolutionary phenomena in the primary and tertiary structure of cytochrome c. *Brookhaven Symp. Biol.* 21: 259–305.

Markert, C. L., 1968. Panel discussion: Present status and perspectives in the study of cytodifferentiation at the molecular level. I. Initial remarks. *J. Cell. Physiol.* 72 (suppl. 1): 213–230.

Martin, M. A., and B. H. Hoyer, 1966. Thermal stabilities and species specificities of reannealed animal deoxyribonucleic acids. *Biochemistry* 5: 2706–2713.

Meselson, M., 1967. The molecular basis of genetic recombination. *In* Heritage from Mendel (R. A. Brink, editor). University of Wisconsin Press, Madison, pp. 81–104.

Moore, R. L., and B. J. McCarthy, 1968. Related base sequences in the DNA of simple and complex organisms. III. Variability in the base sequence of the reduplicated genes for ribosomal RNA in the rabbit. *Biochem. Genet.* 2: 75–86.

Mueller, K., and H. Bremer, 1968. Rate of synthesis of messenger ribonucleic acid in *Escherichia coli*. *J. Mol. Biol.* 38: 329–353.

Pavan, C., and A. B. da Cunha, 1969. Gene amplification in ontogeny and phylogeny of animals. *Genetics (Suppl.)* 61 (no. 1, pt. 2): 289–304.

Pelc, S. R., 1968. Turnover of DNA and function. *Nature (London)* 219: 162–163.

Peterson, P. A., 1961. Mutable $a_1$ of the En system in maize *Genetics* 46: 759–771.

Pinder, J. C., H. J. Gould, and I. Smith, 1969. Conservation of the structure of ribosomal RNA during evolution. *J. Mol. Biol.* 40: 289–298.

Plaut, W., D. Nash, and T. Fanning, 1966. Ordered replication of DNA in polytene chromosomes of *Drosophila melanogaster*. *J. Mol. Biol.* 16: 85–93.

Retel, J., and R. J. Planta, 1968. The investigation of the ribosomal RNA sites in yeast DNA by the hybridization technique. *Biochim. Biophys. Acta* 169: 416–429.

Ritossa, F. M., 1968. Unstable redundancy of genes for ribosomal RNA. *Proc. Nat. Acad. Sci. U. S. A.* 60: 509–516.

Ritossa, F. M., K. C. Atwood, D. L. Lindsley, and S. Spiegelman, 1966. On the chromosomal distribution of DNA complementary to ribosomal and soluble RNA. *Nat. Cancer Inst. Monogr.* 23: 449–472.

Ritossa, F. M., and S. Spiegelman, 1965. Localization of DNA complementary to ribosomal RNA in the nucleolus organizer region of *Drosophila melanogaster*. *Proc. Nat. Acad. Sci. U. S. A.* 53: 737–745.

Sermonti, G., and A. Carere, 1968. Mechanism for polarized recombination in *Streptomyces*. *Mol. Gen. Genet.* 103: 141–149.

Serra, J. A., 1968. Modern Genetics, Vol. 3. Academic Press, New York, pp. 99–243.

Simpson, G. G., 1964. Organisms and molecules in evolution. *Science (Washington)* 146: 1535–1538.

Smith-Keary, P. F., and G. W. P. Dawson, 1964. Episomic suppression of phenotype in *Salmonella*. *Genet. Res.* 5: 269–281.

Stewart, F. C., M. O. Mapes, A. E. Kent, and R. D. Holsten, 1964. Growth and development of cultured plant cells. *Science (Washington)* 143: 20–27.

Thomas, C. A., Jr., 1966. Recombination of DNA molecules. *Progr. Nucl. Acid Res. Mol. Biol.* 5: 315–337.

Walker, P. M. B., 1968. How different are the DNAs from related animals? *Nature (London)* 219: 228–232.

Wallace, H., and M. L. Birnstiel, 1966. Ribosomal cistrons and the nucleolar organizer. *Biochim. Biophys. Acta* 114: 296–310.

Waring, M., and R. J. Britten, 1966. Nucleotide sequence repetition: A rapidly reassociating fraction of mouse DNA. *Science (Washington)* 154: 791–794.

Whitehouse, H. L. K., 1967a. A cycloid model for the chromosome. *J. Cell Sci.* 2: 9–22.

Whitehouse, H. L. K., 1967b. Crossover model of antibody variability. *Nature (London)* 215: 371–374.

Whitehouse, H. L. K., and P. J. Hastings, 1965. The analysis of genetic recombination on the polaron hybrid DNA model. *Genet. Res.* 6: 27–92.

Wilkinson, J. M., 1969. Variations in the N-terminal sequence of heavy chains of immunoglobulin G from rabbits of different allotype. *Biochem. J.* 112: 173–185.

Zuckerkandl, E., and L. Pauling, 1962. Molecular disease, evolution, and genic heterogeneity. *In* Horizons in Biochemistry (M. Kasha and B. Pullman, editors). Academic Press, New York, pp. 189–225.

# 86    The Theory of the Master Gene

C. A. THOMAS, JR.

THE THEORY of the master gene was first advanced in a remarkable paper by Callan and Lloyd that appeared in the Philosophical Transactions of the Royal Society of London in 1960. This theory was based on the morphological study, by optical phase-contrast microscopy, of the lampbrush chromosomes from the oocytes of newts. The studies of lampbrush chromosomes by Callan, Gall, Macgregor, and others that appeared during the 1950s combine to form an important series of observations. The theory itself must be considered a triumph of reason. It is outrageous in every respect. Its only direct support, even to this day, is based on morphological considerations. When first proposed, the nature of the chromatid was not known, although the Taylor, Woods, and Hughes paper (1957) had appeared, touching off the debate between the "uninemists" and the "polynemists" that persists to this day. Callan supposed a chromatid to be fundamentally a duplex DNA molecule, a view that must now be considered essentially correct. To account for his observations, Callan supposed that genes were often found in multiple representation—"serially repeated," perhaps 100 to 10,000 times along the chromatid or DNA duplex. This model challenged the conventional wisdom of two generations of Mendelian geneticists, who have supplied overwhelming evidence that alleles are inherited in an all-or-nothing manner. To answer this profound objection, Callan supposed that the nucleotide sequence of each of the repeated genes is brought into accord with that of a "master copy" by some unspecified mechanism. Thus, only mutations in the master copy would be recovered; those occurring in other copies would be brought into accord with the master copy—that is, would revert to wild type.

In view of the unfamiliar character of the material (the oocytes of newts) and the apparent incongruity of this theory with both Mendelian and molecular genetics, together with the *ad hoc* character of the master gene and its role in specifying the thousands of identical serial copies, it is not surprising that this theory was greeted with skepticism. Now, however, converging lines of evidence lead one seriously to consider the idea. If correct, the theory of the master gene will directly affect every aspect of our present understanding of biology. Therefore, it is worthwhile to

study the lines of evidence that support the theory and then to discuss some of its predictions.

## The theory

The exposition that follows is based on the ideas found in the 1960 paper and on their extensions (Callan, 1967). Substantial modifications have, however, been made and new vocabulary has been added, so Callan should not be held totally responsible for what follows. The scheme is presented in starkest form for clarity, even though the resulting outline is not adequate to represent the variety and variation seen in nature.

The eukaryotic chromatid (the chromosome after mitosis, but before a round of DNA synthesis) is fundamentally a single DNA duplex in which the entire genetic text is represented in nucleotide sequence only once. Most or all genes (or operons) specifying RNA or protein molecules are represented not once but 100 to 10,000 times. This is accomplished by the serial repetition of each gene type an exact number of times. On passing along the linear DNA molecule from gene to gene in a given series (or family), a miniature observer of nucleotide sequence would be unable to tell that he was not repeatedly moving around the same circular DNA molecule containing the single gene (Figure 1). The sequence of each serially repeated gene is identical to every other serially repeated gene in the same family.

A special member of each gene family is called the *master gene*; all other members are called *slaves*. (In Figure 1, the master gene is depicted as the first member of the series.) In the germ line of cells, and probably in somatic cells as well, the master gene specifies the sequence of every slave in the family. It does so by a process called *rectification*. Whether rectification is accomplished by repair synthesis, replacement, or replication need not be specified at this moment. Rectification must occur about as frequently as cell division, although it could be more frequent. Because of rectification, only mutational alterations in the master gene are ever detected. Both masters and slaves are transmitted through the germ line (to sperm and egg) without an increase or decrease in their number. The "one gene-one protein theory" is to be modified to read "one gene family-one-protein type." Because the sequence of all slaves is identical, those sequences controlling the rate of transcription of RNA (the promoter and operator regions) are identical. Thus, it is to be expected that entire families of genes are

C. A. THOMAS, JR.   Department of Biological Chemistry, Harvard Medical School, Boston, Massachusetts

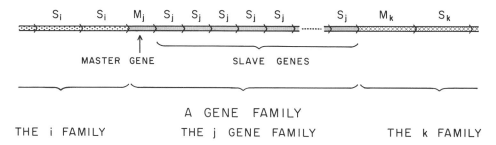

FIGURE 1  The eukaryotic chromosome: masters and slaves. The chromatid is pictured to be a single DNA molecule containing gene families, each made up of a defined (large) number of identical genes. One gene, called the *master,* is depicted as the first of a series of identical slaves. The genes of one family can differ in sequence from those in another, although they may be related sequences.

turned on or off together. They would be expected to function as a unit. In this case, the family size would determine the abundance of the messenger RNA it specifies. If repressor molecules or RNA polymerase were limiting, a graded response that is controlled by their concentration could be achieved.

Gene families are considered to be found only in the "higher forms." In the lower forms, such as bacteria and viruses, genes are usually represented only once. Every gene is a "master gene." A general comparison of the two basic biological forms is revealing (Table I). As this comparison suggests, there is a profound difference between the miniature organisms and the comparatively monstrous cells that make up the so-called higher forms. The major differences are the *size* of the cell and the way it manages its hereditary apparatus.

## The morphology and function of lampbrush chromosomes

Although this evidence is, in many ways, the most difficult to recount, the lampbrush chromosome played the significant role historically. In order to understand the arguments of Callan, one must review what is known about these wonderful structures.

Lampbrush chromosomes are found in the oocytes (and in some species, the spermatocytes) of almost every species studied—from arrow worms to humans (Baker and Franchi, 1967). They were first encountered by Flemming and studied by Rückert, who gave them their name before the turn of the century. With the advent of phase optics, unfixed preparations could be studied. Other innovations by Gall, Callan, and others have made it possible to study the morphology of these chromosomes in considerable detail. Miller and Beatty (1969a, 1969b) have recently made a significant advance in the preparation of lampbrush chromosomes for electron microscopy that allows them to be studied at the molecular level.

Generally speaking, oocytes are arrested in the process of meiosis. The chromosomes are held at the *diplotene* stage, which means that all chromatids are double, as they were replicated just prior to meiotic prophase. The homologous two-stranded chromosomes have paired with each other to form the four-stranded complex (*pachytene*), and have begun to separate and are visibly double, thereby revealing any chiasmata that may have taken place (*diplotene*). In newts, for example, they may wait in this condition for up to two years; in humans for up to 40 years or more. During this time, the oocyte accumulates yolk, ribosomes, and other ingredients for the egg. Just prior to ovulation, the oocyte engages in two divisions to produce the egg. This process reduces the chromosomal DNA from a value of 4C to 1C (C refers to the haploid complement of DNA).

The chromosomes in the arrested diplotene stage are not compacted; rather, they are expanded and elongated to lengths ranging from 200 to 800$\mu$ in *Triturus*. Each of the

TABLE I

*A Comparison of Lower and Higher Forms*

| Lower Forms (Prokaryotes) | Higher Forms (Eukaryotes) |
| --- | --- |
| Low DNA content (small cell size and weight) | Very high DNA content (large cell size and weight) |
| No organized chromosomes (chromonemal) | Organized chromosomes (chromosomal) |
| No basic proteins | DNA united with basic proteins |
| No nuclear membrane | Nuclear membrane |
| Little differentiation | Elaborate differentiation |
| All cells master cells (aging unknown) | Germ line and somatic cells (somatic cells age) |
| All genes master genes | Most genes in large "families" of identical members |

two homologous chromosomes that form the so-called meiotic *bivalent* appear to be composed of rows of dense bodies called *chromomeres*. Two lateral loops project from many, but not all, chromomeres (Figure 2). The two loops from a given chromomere are invariably identical in length and appearance. We now know, mainly from electron micrographs and nuclease sensitivity, that the axis of these

loop (and chromomere) itself—a local phenotypic expression of local DNA nucleotide sequence (Figure 4). This idea is given dramatic support by the electron micrographs of Oscar Miller that show the DNA molecule in the act of being transcribed into long RNA chains: An example is shown in Figure 5 (page 978). Autoradiographs reveal that radiolabeled ribonucleotides are incorporated in all regions of the

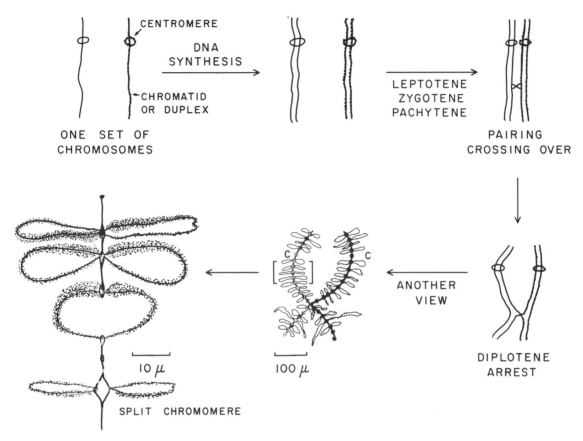

FIGURE 2 The morphology of lampbrush chromosomes. The chromatid (DNA duplex) replicates to become "double"; it then unites with its homologue, crosses over, begins to separate, and is arrested. Each chromosome is still double, and each loop contains a DNA duplex as its axis. Tension splits the chromomere laterally. In some species, longitudinal chromomere splitting is seen to take place naturally.

loops is a single DNA duplex: the pair of loop axes are the two chromatids produced by the pre-meiotic replication event (Figure 3). The bulk of the loop material is RNA and protein. All loops that can be studied carefully show a thin and a thick end. The thin ends always point either toward or away from the centromere, depending on the loop in question.

The appearance of many of the larger loops is specific to the chromosome and the subspecies or race. Callan's work demonstrates conclusively that the morphology of these loops is genetically specified, and leaves little doubt that the DNA that specifies the loop morphology resides in the

loop. Certain unusual loops, however, such as the giant loop of chromosome XII of *Triturus cristatus*, seem to synthesize RNA in a special location near the pointed end of the loop (Gall and Callan, 1962). Thus, there is now direct evidence that the loops are macroscopic manifestations of underlying genes.

Many loops appear to slough off particulate matter into the nuclear sap. Callan noticed that these particles often were shed from the thick end of the loop *and* from intermediate positions as well. Moreover, sometimes a given loop is found that is not fully extended from the chromomere, yet it, too, sheds the same kind of particulate matter.

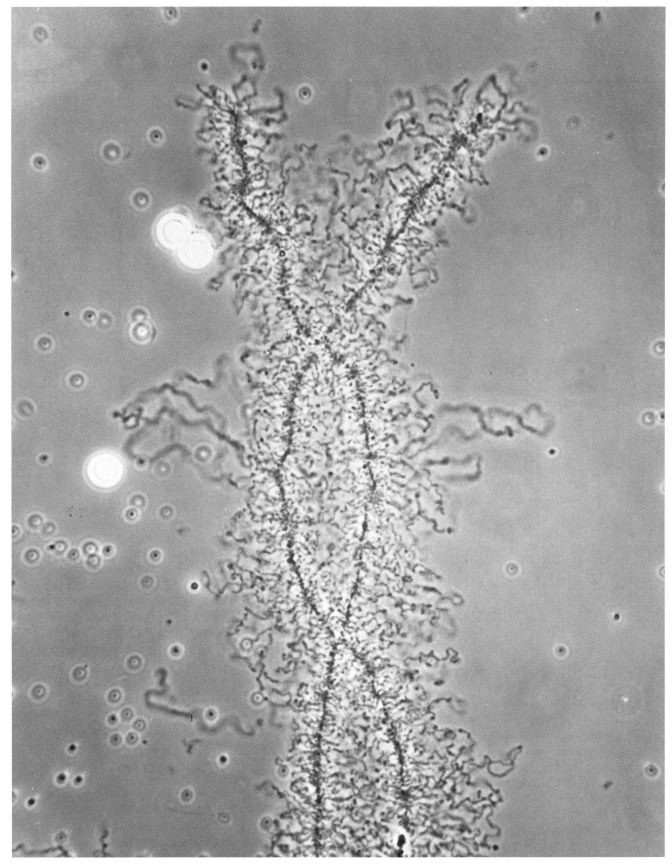

FIGURE 3   Phase contrast photograph of a lampbrush chromosome. (Courtesy of J. Gall.)

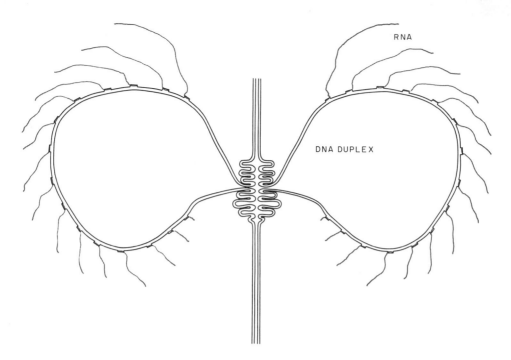

RNA

DNA DUPLEX

FIGURE 4  Schematic diagram of how the DNA in the loops of a lampbrush chromosome is thought to be "expressed"—transcribed. It is not yet clear whether the RNA is translated into protein as a general rule, although some protein synthesis is thought to occur in the oocyte nucleus.

If we assume that particles that detach from a loop are "fully fashioned" gene products, the question presents itself: How can different regions of the same loop produce what appears to be the same material? Weighing carefully some alternative explanations, Callan (Callan and Lloyd, 1960) advanced the most likely interpretation: that there is "no diversity of genetic information within the loop"—that "the information carried by a particular region of the chromosome may be serially repeated along the loop axis."

In order to account for the asymmetry of the loops, Callan supposed that the loop axis moves—being continuously spun out from one part of the chromomere and reeled into the other. If, indeed, the loop axis moves, as seems to be the case in the giant loops of chromosome XII, the idea of repeating genes provides an easy explanation for the fact that the loops have a more or less constant morphology, irrespective of the age of the oocyte, which, as mentioned above, can extend to years! The loop appearance remains the same, because each new gene that is spun into the loop from the chromomere is identical with the others that preceded it.

It is not certain if all the elements of Callan's argument will be supported by future work. In particular, the extent to which protein synthesis occurs on the loops is not known. It could be possible that proteins made in the cytoplasm associate with loop material (RNA) to give it a characteristic appearance. Callan's conclusion could still apply, irrespective of how this important question is resolved.

## Mutation and recombination

MUTATION RATE   Only mutational alterations affecting the master gene will be observed, because those taking place in the slave genes will be rectified by an unaltered master. Thus, we would expect the mutation rate per gene to be much smaller in eukaryotes than in prokaryotes, in which every gene acts independently. Such proves to be the case, as shown in Table II. Here we see that the spontaneous, forward mutation rate, calculated per base pair per replication, for two eukaryotes (*Neurospora* and *Drosophila*) is about 1000 times lower than the rate seen in bacteriophage, which accords with expectation. Unfortunately, this evidence is

TABLE II

*Spontaneous Forward-Mutation Rate in Prokaryotes and Eukaryotes Per Nucleotide Pair, Per Replication*

| | |
|---|---|
| phage ($\lambda$, T$_4$) | $1.7 - 2.4 \times 10^{-8}$ |
| *E. coli, S. typhimurium* | $2 \times 10^{-10}$ |
| *Neurospora crassa* | $0.7 \times 10^{-11}$ |
| *Drosophila melanogaster* | $7 \times 10^{-11}$ |

(From Drake, 1969.)

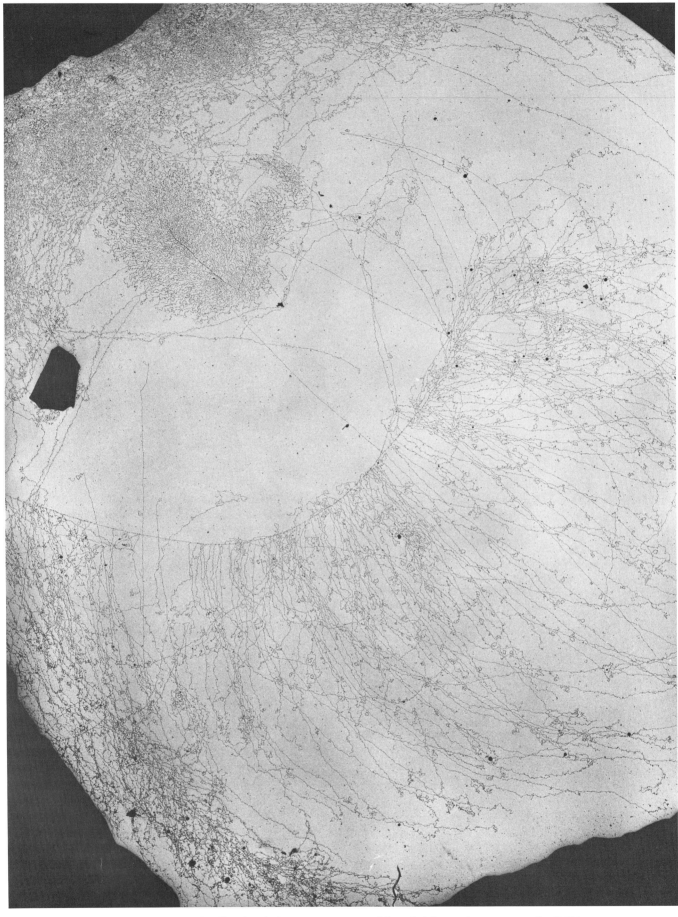

FIGURE 5 Electron micrograph of lampbrush chromosome loop. DNA duplex is the traverse thread; long RNA chains extend from it. The numbers correspond to DNA measurements in arbitrary units. (Courtesy of O. L. Miller, Jr.)

not conclusive. Not only is the estimation of forward mutation rate subject to considerable uncertainty (Drake, 1969), but the known effects of antimutagenic DNA polymerase, and the possibilities of excision and repair of mutational heterozygotes, may affect the frequency of observed mutants to unknown extents. Indeed, the observed mutation rate of lambda replicating as a prophage is 20 to 100 times lower than when replicating during a cycle of vegetative growth.

RECOMBINATION The concept that the chromosome is constructed of gene families leads to an unusual prediction regarding intergenic and intragenic recombination. As illustrated in Figure 6A, a recombination event among slave genes is effective in recombining gene families, but the process of rectification quickly eliminates recombinant slaves. On the other hand, a recombination event *within* a master gene leads to recombinant slaves as well as recombinant gene families (Figure 6B). Callan supposed that re-

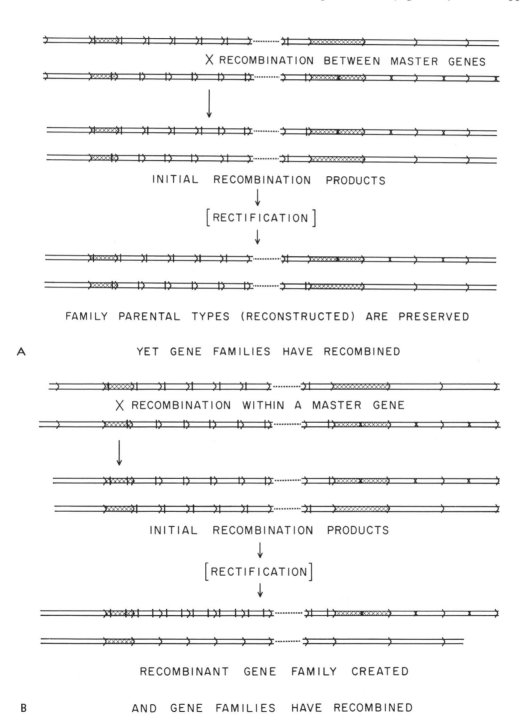

FIGURE 6  A. Recombination between two master genes. B. Recombination within a single master gene.

combination is totally restricted to the master genes, because chiasmata are never seen in the lateral loops of the lampbrush chromosomes.

In phage and bacteria (in which every gene is a master), it is generally agreed that there is no discontinuity in the genetic map on passing from one gene to another. Intragenic and intergenic recombination proceeds at roughly the same rate (Figure 7A). The minimum intergenic recombination frequency can be less than the maximum intragenic recombination frequency. On the other hand, the situation is not equivalent in eukaryotes, as is diagrammed in Figure 7B.

genic recombination values could mean simply that intervening genes have not yet been recognized by mutations. Although these objections cannot be ruled out completely, I think it is very unlikely that the objections are valid. Thus, we find a second prediction of The Theory fulfilled: recombination between genes occurs more frequently than recombination within genes. But there is more.

*Chromosomal rearrangements in* Drosophila Such rearrangements (deletions, inversions, translocations) generally do not damage the genes at the point of refusion, although in some cases they do. In phage and *E. coli,* deletions

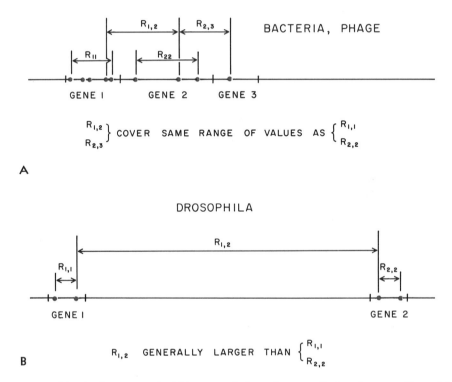

FIGURE 7 Recombination between and within genes. A. In prokaryotes. B. In eukaryotes (*Drosophila*).

Among the most closely spaced independent genes in *Drosophila,* the intergenic recombination values are about 10 times the *maximum* intragenic recombination values. This is illustrated in Table III, which contains information adapted from Pontecorvo's (1958) book. Evidence against this point is supplied by Chovnick's study of the *rosy* locus (1966). The *White–Notch* region, however, displays about 100 times more recombination than can be accounted for by the three genes it is known to contain (E. B. Lewis, personal communication, 1969).

Now, the molecular geneticist is quick to point out that the number of mutants available to the *Drosophila* geneticist may not be sufficient to mark truly the maximum intragenic distances that exist. Furthermore, large minimum inter-

nearly always damage the gene at the terminus of the deletion. In microorganisms, this effect is so predictable that it is the standard method for mapping the location of unknown markers. Why is it that chromosomal rearrangements can be made with relative impunity in higher organisms? One explanation could be that the inert material between genes recombines to form the rearranged chromosome. Another could be that unusual rejoining generally takes place among slave genes. On the next round of rectification, the full family of slaves is reconstituted, and the original slave family now finds itself neighbors to a new family of genes, with both families functional.

*The Number of Nucleotide Pairs Per Gene* Finally, we come to the amount of recombination in relation to DNA

TABLE III

Recombination Between and Within Genes

| | | | Minimum R % Between Genes | Maximum R % Within Genes | Ratio |
|---|---|---|---|---|---|
| Drosophila | w-rst | | 0.2, 0.5 | w(4) .056 | |
| 280 map units | sp-bl | | 0 3 | bx(5) .03 | |
| | ey-ci | | 0.2 | lz(3) .14 | |
| range | | | 0.2 — 0.5 | 0.03 — .14 | 1.4 — 16 |
| Aspergillus | y-ad16 | | 0.05 | bi(3) 0.1 | |
| | ad15-pabal | | 0.5 | ad8(6) 0.18 | |
| range | | | 0.05 — 0.5 | 0.1 — 0.18 | 0.28 — 5.0 |
| T4 | $r_{II}(A)$-$r_{II}(B)$ | | — .04(?) | $r_{II}(A)$ 4.3 | .01 |

Conclusion: It appears that the minimum recombination that occurs between (presumably) neighboring genes in Drosophila is on the order of 10 times the maximum observed within a gene. This is contrasted with situation in T4, in which this ratio can be as small as 0.01. (Adapted from Pontecorvo, 1958.)

content. The total length of the genetic map is equal (after dividing by 100) to the average number of recombination events that have occurred per genome. If it be assumed that recombination events take place with equal probability in all intervals (which is wrong), and that the frequency of recombinants depends only on the length of the interval between the markers (which is also wrong), it is possible to calculate the total number of genes. This calculation is made in Table IV for T4, Drosophila, and Aspergillus from data assembled in Pontecorvo's (1958) book, somewhat revised. Here we see that T4 has about 100 to 200 genes, which is a reasonable value. About 80 have so far been identified (Sober, 1968). When compared with the number of nucleotide pairs in the molecule, one calculates about 1000 nucleotide pairs per gene. A gene of this length could specify a protein of 330 amino acids, or about 36,000 molecular weight. When we come to Drosophila, however, the esti-

mated number of genes is only 10 times larger, yet the DNA content is 1000 times larger. The number of nucleotide pairs per gene calculates to 200,000, or 20 times larger than makes structural sense.

Now, there are many ways to explain away these results. The estimation of the number of genes is naive and ignores known effects that could make great differences. The increased DNA per gene could result from the possible polynemic nature of Drosophila chromatids or from the presence of nonfunctional "spacer DNA" between genes. All this may be true, but it is equally true that these facts are in exact accord with the idea that the genes in Drosophila (and all higher forms) are represented by gene families.

The Role of Histones   The mechanism of genetic recombination between DNA molecules is just now becoming susceptible to mutational and biochemical study in phage. Ultimately, the recombination event must involve the annealing of complementary polynucleotide chains to form a duplex DNA molecule. No other known process has the specificity equal to the feat of producing recombinant molecules that (in general) have neither gained nor lost a single nucleotide. If this is assumed, the repetitive nature of a series of slave genes would provide multiple opportunities for crossover events, which would be between identical sequences, yet not homologous from the point of view of the whole chromosome. Such aberrant crossovers are depicted in Figure 8. This process, if allowed to happen, would lead to recombinants of unequal family size.

A more severe event is shown in Figure 9. Internal recombinations made possible by the repeating sequences lead to circular molecules (circles of slaves) and a reduction in family size. These events would have disastrous consequences and must be prevented. I suppose that this role is played by histones—the ever-present escort of all eukaryotic DNAs. Histones seem to complex with DNA and never irreversibly dissociate from it, following the polynucleotide chains through many cycles of replication and mitosis (Hancock, 1969). This view, if provisionally accepted, provides an explanation for the mysterious, superhelical DNA mole-

TABLE IV

DNA Content Per Gene

Calculated from length of the genetic map, the map length of a gene, and the amount of DNA per haploid nucleus. This calculation must be taken in a qualitative sense only. An important issue, namely, high negative interference, has been ignored. However, when this is properly taken into consideration (Stahl et al., 1964), one comes to the same number of genes per T4 chromosome.

| Species | Total Map | 1 Gene | Calculated Total No. of Genes | Total NTP | NTP/Gene |
|---|---|---|---|---|---|
| T4 | 400 | 2–4% | 100–200 | $1.8 \times 10^5$ | 1000 |
| Drosophila | 280 | 0.2–0.3% | 1000 | $2 \times 10^8$ | 200,000 |
| Aspergillus | 660 | 0.5% | 1320 | $4 \times 10^7$ | 30,000 |

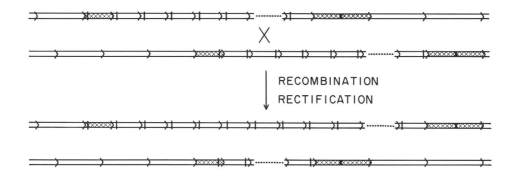

RECOMBINANTS OF UNEQUAL FAMILY SIZE

FIGURE 8  Possible unequal crossing over made possible by tandem duplications. This kind of crossing over produces recombinants of unequal family size.

INTERNAL RECOMBINATION DELETES THE SLAVES

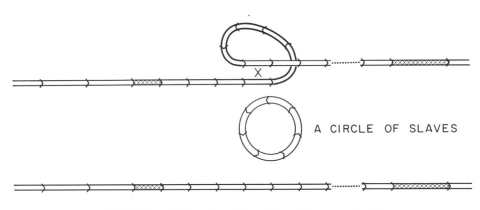

A CIRCLE OF SLAVES

RECOMBINANT OF REDUCED FAMILY SIZE

FIGURE 9  Internal recombination deletes the slaves. This process, if allowed to proceed, could delete all but a single gene.

cules of unusually short contour length, discovered by Radloff, Bauer, and Vinograd (1967) in *HeLa* cells. They could be "circles of slaves" that result from the recombination of tandemly-repeated genes that are imperfectly protected by histone.

These suggestions are supported by the known facts in *E. coli*. Here the DNA is *not protected* by histones. Known tandem duplications are rare. When they do occur, they suffer from a high degree of instability. They tend spontaneously to produce mutants that contain only a single copy of the gene. Such is the case with su$_{III}$ genes (suppressors) that produce tyrosine tRNAs. In the su$_{III}^+$ strains, the anti-

codon of one of two tandemly-located tyrosine tRNA genes is altered from GUA to CUA, which presumably allows tyrosine to be inserted for the amber codon UAG. The important feature of this situation for the present argument is: (1) two copies of the tRNA gene reside in tandem in the *coli* chromosome; and (2) in the original su$_{III}^+$ strain, only *one* of them is altered (C for G) in one place (the anticodon).

The fact that *both* forms can exist for many generations means that *rectification does not take place* in *E. coli* at this locus.

The su$_{III}^+$ strains are unstable and spontaneously revert at a high rate to the su$_{III}$⁻ condition, largely because of the deletion of one of the su$_{III}$ genes, through an internal crossover event that deletes the su$_{III}$⁺ anticodon (Russell et al., 1970). A similar situation exists with *E. coli* strains capable of suppressing missense mutations in the A protein of tryptophan-synthetase (Hill et al., 1969, 1970; Brody and Yanofsky, 1963). Again, when one of a series of genes can be recognized by the existence of a mutational alteration, a high spontaneous reversion frequency is observed. Further examples of this can be found in the *E. coli* strains bearing multiple *gal* genes (Campbell, 1965).

Thus, in *E. coli* and phage, in which the chromosome (DNA) is not complexed with protective histone, the DNA is "freely recombining" (Thomas, 1966). There is some selective pressure (of significant but uncertain magnitude) to eliminate repetitive sequences by internal recombination. This idea is supported by the general finding that these strains are unstable, showing a high reversion rate. I formerly supposed that repetitions were prohibited for this reason (Thomas, 1966), but, as seen above, there is good evidence that some low level of tandem duplication is tolerated. Presumably, the strain balances the disadvantages of instability against the benefits of gene duplication, and the balance point is at *one* or a small number of duplications in bacteria and phage. A similar situation may occur at the ribosomal RNA locus; annealing experiments indicate several genes for rRNA (Yankofsky and Spiegelman, 1962). However, 100 or 1000 tandem repeats present a graver instability problem. Genomes of this type might have been impossible, had not histones developed to prevent internal recombination.

The protection of slaves from recombination may account for the observation that recombination between gene families is only 10 times more frequent than recombination within master genes. If both masters and slaves were equally sensitive to recombination, intergenic recombination should be 100 to 1000 times more frequent than intragenic recombination.

*Summary* In this section, the recombination rate between genes and within genes in *Drosophila* is shown to be in accord with The Theory. Chromosomal rearrangements are in general nondamaging in eukaryotes, but they are generally damaging in phage and *E. coli*. When the estimated number of genes is compared with the DNA content per genome, eukaryotes display an inordinately high amount of DNA per gene, whereas in phage one arrives at a reasonable value. Finally, a plausible role is suggested for histones—namely, to prevent recombination events among the slaves. Were they not so protected, they would suffer deletions by recombination, as is seen to be the case in prokaryotes. All these features of eukaryotes can be explained in terms of The Theory of the Master Gene. Unfortunately, none of these arguments is a strong reason for accepting The Theory, because each is weakened by counter arguments that I have largely ignored.

## The chromomeric organization of chromosomes

CHROMOMERES AND MENDELIAN GENES Perhaps the most salient feature of the morphology of chromosomes is that they appear to be linear arrays of chromomeres. As the name implies, these are densely staining bodies, as seen in the light microscope. As mentioned above, the lampbrush chromosomes are organized into chromomeres. Belling (1931) observed a total of about 2500 pairs of chromomeres in the meiotic chromosomes of *Lilium*. He considered the chromomere as the structure in which individual genes were housed. The dipteran salivary chromosome is the classic exaggerated example of the chromomeric architecture of chromosomes (Figure 10). There can now be little doubt that these structures consist of 4000 to 8000 chromatids (DNA molecules) arranged in a parallel package, with all genetic regions in register. (This extreme example of polyteny is found in *Chironomus* and *Rhynchosciara;* in *Drosophila* there are probably 512 to 1024 chromatids per chromosome.) The banded pattern is thought to result from the fusion of the chromomeres of the individual chromatids. Bridges (1938) counted 1024 bands in the x chromosome from the salivary gland of *D. melanogaster*. Contemporary thinking held that each band represented a gene. Thus the x chromosome had about 1024 or more genes. There is now evidence that this may be a maximum estimation of the number of Mendelian genes, rather than a minimum (Beermann, 1967). Notice that this estimate of the number of genes agrees with the estimates based on uniform recombination (Table IV) and the other early estimates of gene number (Muller, 1929). An excellent account of this is to be found in Demerec's last article (1967). It is strange, indeed, as pointed out by Beermann (1967), that the chromomeric pattern of chromosomes has been ignored in speculations about their architecture.

The identification of chromomeres with Mendelian genes becomes more convincing in view of the fact that mutants can be correlated with the deletion of single chromomeres. Labeling with ³H-thymidine followed by autoradiography has revealed that the chromomeres themselves are units of replication (Plaut, 1969). Those chromomeres that contain more DNA seem to require a longer time to label, which suggests a limited number of growing points per chromomere. Finally, a study of the puffed regions shows that the chromomere seems to function in an all-or-nothing manner. Thus, as units of structure, mutation, replication, and expression, the identification of chromomeres with Mendelian genes would appear to be complete (Beermann, 1967).

So what's the problem? The chromomere contains too much DNA to be consistent with the 1000 nucleotide pairs

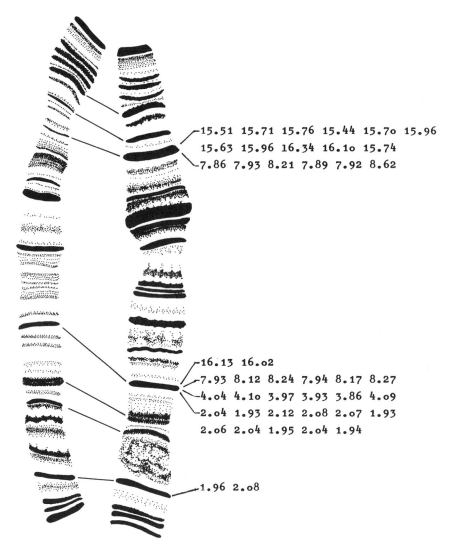

15.51  15.71  15.76  15.44  15.70  15.96
15.63  15.96  16.34  16.10  15.74
7.86  7.93  8.21  7.89  7.92  8.62

16.13  16.02
7.93  8.12  8.24  7.94  8.17  8.27
4.04  4.10  3.97  3.93  3.86  4.09
2.04  1.93  2.12  2.08  2.07  1.93
2.06  2.04  1.95  2.04  1.94

1.96  2.08

FIGURE 10   A comparison of the DNA contents in homologous bands of *Chironomus thummi thummi* and *C. thummi piger*. (Courtesy of H.-G. Keyl.)

required to specify an "average" protein. Rudkin (1961) estimated that there are 45,000 nucleotide pairs per chromatid in the smallest bands to 450,000 in the larger ones. This would correspond to 45 to 450 structural genes. From the presumed rate of movement of the giant-loop axis of chromosome XII in *T. cristatus,* Gall and Callan (1962) estimated that this chromomere contains a millimeter of DNA or more. This would correspond to 3000 genes; the number may be two or three times larger. The conclusion is clear: the observable chromomeres contain enough DNA to make hundreds and thousands of genes from the point of view of the molecular geneticist, yet they contain only *one* gene from the point of view of the Mendelian geneticist.

The resolution of this dilemma is easy in terms of The Theory. The chromomeres contain *gene families,* each composed of hundreds or thousands of slaves. This resolution makes structural sense, as well. The tandemly-repeating sequences seen in Figure 1 would be expected to interact with the same kinds of histones to form a chromatin of regular repeating structure. Regular polymeric structures form helixes according to the general rules that govern the generation of such secondary structures as the α-helix, or double helix. Thus, chromomeres could be thought to be the natural structural consequence of a linear array of repeating units. Whether the unorganized interchromomeric regions represent "spacers" between gene families is not known. It is significant that the interchromomeric space is rather uniform—0.1 to 0.2 $\mu$ (see Beermann, 1967).

MUTATIONAL ALTERATIONS IN FAMILY SIZE Before leaving the subject of polytene chromosomes, I must mention the work of H. G. Keyl (1965a, 1965b). *Chironomus thummi thummi* and *Chironomus thummi piger* are so closely related that most of the bands in the salivary chromosomes can be correlated with one another. This identification is facilitated by the somatically paired salivary chromosomes in the $F_1$ hybrids.

Most homologous bands have the same DNA content, as measured by microspectrophotometry of Feulgen-stained preparations, but in 30 pairs of homologous bands there are striking differences. The *C. thummi thummi* bands sometimes contain 2, 4, 8, or 16 times more DNA than their homologous *C. thummi piger* bands. An example can be seen in Figure 10. Sometimes there is a variability within the "pure" *C. thummi thummi* population, in which certain homologous bands differ by 1:2 or 1:4. It appears that some relatively frequent event results in the doubling and redoubling of the DNA content of bands, yet no multiple other than $2^N$ is seen. Keyl interpreted this as the result of an error in replication—rejection of unequal recombination on the grounds that such a mechanism would lead to triplications or other nondoubling values. The results, however, are precisely what is expected from a recombination event if the chromomere contains a family of slave genes that form a single replicon (Figure 11). This model requires that the *ends* of the replicon be near each other, and that these most recently replicated ends be not yet fully protected from recombination by histone. As seen in Figure 11D, a single, nonreciprocal, recombination event leads to the (near) doubling of the family size. Notice that a nonreciprocal recombination of the opposite type leads to the deletion of all but a single gene (Figure 11E). Deletion of bands is, of course, a frequently recorded event—one that has played an important role in the correlation of the genetic map with the banded pattern of the salivary chromosome (see Swanson, 1957).

As can be seen in Figure 11E, nonreciprocal crossing over at the extremities of a gene family can either lead to a doubling of the family size, or to the deletion of all but one family member. Figure 11F-H shows how a crossover involving the alternate sister strands can account for the production of circular DNA molecules that contain almost a full gene family. Such a mechanism could account for the production of multiple copies of the nucleolar DNA, which is seen to be in the form of circular necklaces in *Triturus* oocytes (Miller, 1966). The amplification of special regions of the chromosome is seen in the "micronucleoli" of *Hybosciara* (Da Cunha et al., 1969). Thus, the chromomere

doubling observed by Keyl, chromomere deletion, and the production of free circular DNA from certain regions of the chromosome may be alternative manifestations of a common mechanism—recombination between the terminal genes in a family.

There is another kind of increase in the DNA content of certain bands. Occasionally this is seen at the refusion point of certain inversions (see Keyl, 1965a). This observation is also in accord with The Theory. Consider that an aberrant breakage and rejoining event takes place between two different gene families. There is no assurance that this event will occur at a specific place, so subsequent rectification will lead to substantially *increased* or decreased family size, depending on the point of rejoining (see Figure 8). This process would not destroy any gene family that still retained its master. Thus, this interpretation is in accord with the finding mentioned above, that many (or most) chromosomal rearrangements are not damaging to genes situated near the refusion point.

## The amount of nuclear DNA, evolution, and morphological complexity

According to current understanding, DNA is the sole carrier of hereditary information from one generation to the next. This being so, it would seem likely that there should be some simple relation between DNA content and information content. Although it is quite possible to measure the amount of DNA in cells, there is no known method of measuring the hereditary "information" in a biological object. As an only recourse, one turns to phylogenetic position or some qualitative feeling of "complexity." For example, the neural nets in a man are more complex than those in a fly and therefore must require more genetic information to specify them.

In Figure 12, the haploid DNA contents for a number of species are plotted against complexity (as I judge this mysterious quality). I suppose there would be no doubt that sponges are more complex than *E. coli,* that *Arbacia* and *Drosophila* are more complex than sponges. By the same token, a man is more complex than the lungfish. But birds and trigger fish are not obviously more complex than man! Nonetheless, Figure 12 is plotted this way. Even so, it makes no sense; it is a joke.

Such evidence has caused some persons to propose a nongenetic role for DNA. In an attempt to determine with what feature of the cell DNA was correlated, an interesting and obvious finding was made: DNA content is generally correlated with cell size and weight (for a recent contribution to this idea, which dates back to the last century, see Holm-Hansen, 1969), meaning that the amount of DNA is proportional to the amount of RNA and protein it specifies.

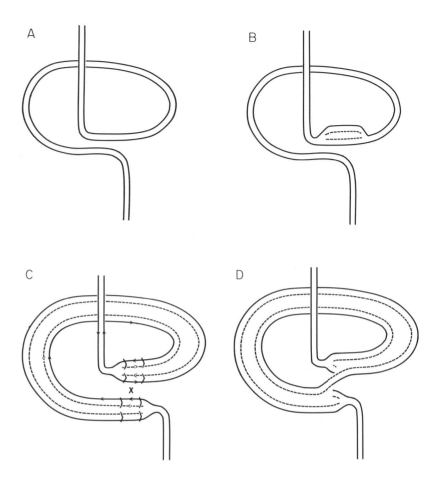

This finding does not support the notion that DNA is playing a nongenetic role: just the contrary.

Returning to DNA content and phylogeny, one can see certain trends if one stays within a single genus. Fishes provide an excellent example. There are more than 20,000 species of fish and room for more work; however, Hinegardner (1968) has reported assays performed on 200 species of teleost fish. DNA contents range from 4.4 to 0.4 picograms (pg) per 1C. Trusting that the lungfish has retained the DNA content (50 pg/1C) of its primitive ancestor, we can see that the DNA content of fish spans a 100-fold range. Hinegardner pointed out that the more ornamented and "specialized" the species is, the lower its DNA content. The same point, made by Mirsky and Ris (1951), is that advanced snakes contain less DNA per 1C than do more primitive snakes. On a larger scale, it is generally thought that reptiles have evolved from amphibians and birds from reptiles, yet as we pass from "primitive" amphibians to reptiles and then to birds, the DNA contents decrease by more than 60 times.

Barring the not unthinkable possibility that the present representatives do not resemble their archetypical forms in DNA content, we are led to the following generalization: as evolution proceeds and the forms become more specialized, DNA is discarded.

FIGURE 11 An interpretation of Keyl's results by recombination. A. A chromomere is shown schematically as a gene family, looped so that the termini are in apposition. B. Replication commences. C. Replication reaches the termini that are not yet fully protected by nucleoproteins. D. Recombination takes place, producing a doubling in gene-

DNA is discarded, but is information discarded as well? The loss of DNA could be attributed solely to a reduction in gene-family size, which would not lower the information content of the cell. Possibly, the amount of information could increase even with a decreasing DNA content. Finally, with our present level of understanding, there is no reason to deny the notion that, as specialization proceeds, genetic information is lost. Because of this latter possibility, it is not possible to build a strong case that the observed loss of DNA during evolution of species is due solely to discarding slaves.

On the other hand, some species, virtually identical in appearance and function, differ sharply in DNA content. For example, it is said that the spermatids of the amphipod crustacean *Gammarus pulex* contain three times as much DNA as those of *Gammarus chevreuxi*. The rhabdocoele

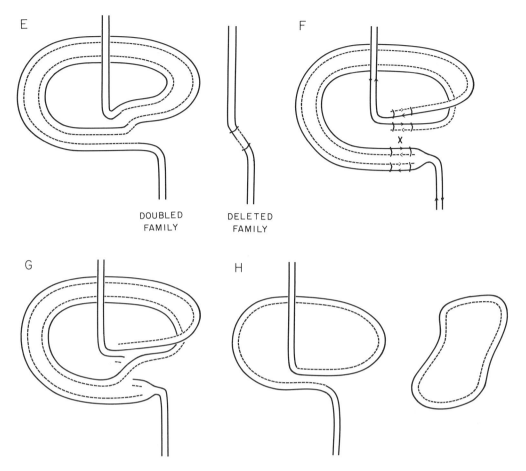

E

F

DOUBLED
FAMILY

DELETED
FAMILY

X

G

H

family size. E. A nonreciprocal recombination of opposite type deletes all but a single gene in the family. F. A recombination involving the alternate pair of sister-strands; compare with 11C. G. Produces a circular molecule containing nearly a full gene family. H. Products of such a crossover.

planarian *Mesostoma ehrenbergi* contains about 11 times as much DNA as does *M. lingua,* yet both species of planarian have the same number of chromosomes (Keyl, cited in Callan, 1967). The same kind of variation is observed among subspecies of *Allium* (Jones and Rees, 1968). In these cases, it seems very unlikely that the amount of genetic information varies substantially between the subspecies. This, again, is in accord with the idea that genes exist as gene families. The difference in DNA content can be attributed to a difference in the number of slaves per family, a proposal that is open to test.

A striking example that the differences in nuclear DNA content are due to differences in all the chromosomes is afforded by the work of Rees and Jones (1967). *Allium cepa* has 27 per cent more DNA per nucleus than does *Allium fistulosum,* yet they are so closely related that it is easy to

FIGURE 12 Haploid DNA contents per cell.

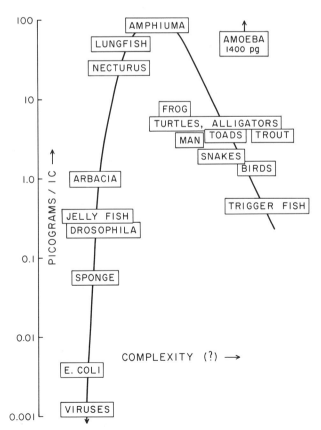

form fertile F$_1$ hybrids. In these hybrids, the meiotic chromosomes show aberrations in pairing at pachytene. Large, unpaired loops and overlaps are seen. At the first metaphase, all bivalents are asymmetrical, indicating that the chromosomes from one parent contain more DNA than those from the other parent (Figure 13), suggesting that rather insignificant mutational events can be responsible for large changes in DNA content. This observation resembles those of Keyl on polytene chromosomes (see above), and is susceptible to the same kind of interpretation.

*Summary* Variations in DNA content per nucleus appear to be unrelated to morphological or developmental complexity, possibly for three reasons: (1) it is not possible to relate the "complexity" of a biological object with genetic information; (2) specialization can be accompanied by a reduction in the number of different gene families; and (3) significant alterations in family size can occur without change in the number of families.

## The evidence from annealing of polynucleotide chains

As is generally known, polynucleotide chains that are "complementary" in sequence will rejoin and reform a duplex molecule. Under favorable conditions, this process is extremely rapid (Studier, 1969a, 1969b), which is to be expected if "nucleation," the formation of the first few complementary base pairs, is the rate-limiting step (Thomas, 1966; Wetmur and Davidson, 1968). The "stability" of the reformed duplex depends on the length of the duplex, the type of base pairs, and the number and location of noncomplementary base pairs. There are only two ways known of testing the stability of a reformed duplex: (1) by measuring its thermal stability and (2) by testing its sensitivity to ribonuclease. The latter is useful only for DNA-RNA hybrid duplexes, and its specificity in this situation is not understood.

Even if the mechanism of the reaction is not understood,

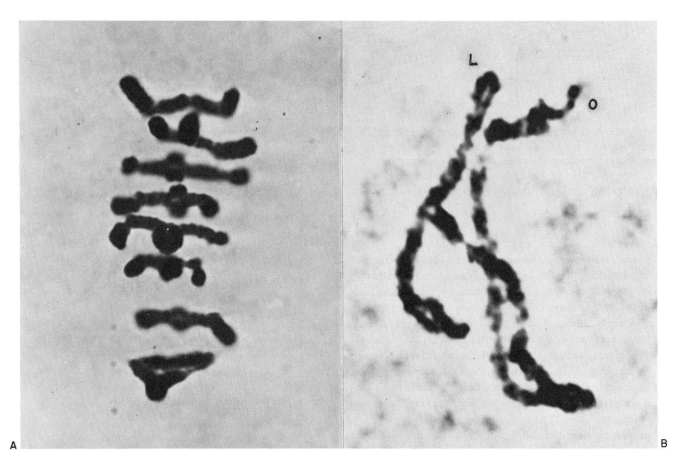

A                                                                 B

FIGURE 13  Asymmetric meiotic bivalents of *Allium cepa* and *A. fistulosum*. *Cepa* has 27 per cent more DNA than *fistulosum*. A. Meiotic bivalents at first metaphase. Note that each is asymmetrical. B. An isolated pair of chromosomes. Note the loop (L) and the terminal overlap (O). (Courtesy of H. Rees; see Rees and Jones, 1967.)

annealing experiments have been used widely for a variety of purposes, because there can be no doubt that the reaction depends primarily on complementary nucleotide sequences. Generally speaking, two different experimental situations have proved useful: (1) denatured DNA of high molecular weight is immobilized on nitrocellulose filters and labeled RNA or short, single-chain segments of DNA are annealed to the immobilized DNA; and (2) annealing of sheared, denatured DNA in bulk solution. The first type of experiment is generally concerned with the maximum amount of RNA that can be complexed with the fixed DNA in a state that survives ribonuclease treatment (Gillespie and Spiegelman, 1965). Presuming that the complexed RNA is united with the DNA chain to form a duplex hybrid, one calculates the fraction of the DNA that is complexed with RNA. Remembering the factor of two that enters because the known RNA species are complementary to only one of the two complementary DNA chains, the molecular weight of the RNA chains, and the molecular weight of the genome, one may calculate the maximum number of RNA molecules that can complex per genome. This is sometimes called "titrating the number of genes," which is permissible with a smile. Table V includes some representative findings

doses of the NO region contain proportional increases in the observed number of RNA genes. In the case of *Xenopus,* the normal animal has 450 rRNA genes per 1C, but the anucleolar mutant has none.

Table V is in accord with The Theory. It predicts that, when purified messenger RNA molecules (such as those for hemoglobin or even synthetic RNAs) become available, they will complex with hundreds of genes per genome, just as is true for the other RNA species in Table V.

*Visualizing Transcription* The DNA-RNA hybridization experiments do not reveal the location or distribution of the genes in question. Nor do these experiments say anything about the arrangement of genes along the chromosome, although other lines of evidence indicate that they are arranged in tandem. The most unusual and compelling evidence is supplied by the recent electron micrographs of O. L. Miller, Jr., which show ribosomal precursor RNA in the act of being transcribed from nucleolar DNA (Figure 14). Here we see regular "plumes" of RNA, each pointing in the same direction and, with certain reservations, spaced at regular intervals along the DNA strand. Thus, it seems pretty clear that the rRNA genes are clustered and arranged in tandem. The current debate centers on the amount and

TABLE V

*Number of Genes Per Genome for Known RNA Species As Measured
By Saturation-Hybridization Studies*

|  | *E. coli* | Yeast | *Drosophila* | *Xenopus* | *HeLa* | Chicken |
|---|---|---|---|---|---|---|
| rRNA | 4–5 | 140 | 130 | 450 | 400–600 | 100 |
| tRNA | 40 | 6 × 60 | 15 × 60 | 15 × 60 | | |
| 5S RNA | 2<br>(*B. subtilis*) | | | (27,000)<br>est. | | |
| | Yankofsky and Spiegelman (1962, 1963)<br>Goodman and Rich (1962)<br>Morell et al. (1967)<br>Brownlee et al. (1968) | Schweizer et al. (1969) | Ritossa et al. (1966) | Birnsteil et al. (1968)<br><br>Brown and Dawid (1968)<br><br>Brown and Weber (1968) | Attardi et al. (1965a, 1965b) | Ritossa et al. (1966) |

on ribosomal RNA, transfer RNAs, and the 5S RNA. As can be seen, in each case tested, there appear to be *hundreds* of genes for these RNA molecules in the higher organisms examined, yet only a few in *E. coli*. It might be thought that the large amount of rRNA that is complexed results from fortuitous complementary sequences at many places in the genome. Such is not the case, because mutants with specific deletions of the nucleolar organizer (NO) region have lost the ability to complex rRNA. Strains containing multiple

function of the DNA intervening between the rRNA genes.

What can be said about the rest of the eukaryotic genome? Does it consist of repeated segments as well? When DNA from eukaryotes is highly fragmented, denatured, and annealed, it is now fairly clear that some kind of secondary structure is reformed. That structure has a lower absorbance at 260 m$\mu$, and a thermal stability that is nearly (but not quite) as high as the undenatured DNA. This renatured material fractionates in a manner similar to that of

duplex DNA on hydroxyapatite, and a portion of it is resistant to nucleases that are specific for single chains.

Annealing experiments conducted in bulk solution are usually performed as follows: (1) the duplex DNA is sheared into segments containing several hundreds of nucleotide pairs (a rather substantial heterogeneity of size can be expected); (2) the segments are denatured by heat or alkali; (3) the solution is incubated at relatively high salt concentrations at a temperature 20° to 25° C below the melting temperature; and (4) the fraction of the DNA "renatured" is estimated by the absorbance (260 m$\mu$) decrease or by altered chromatographic properties on hydroxyapatite. The fraction of material remaining in the denatured form (F) decreases as the incubation proceeds.

To interpret the rate of annealing, it is necessary to assume the following: (1) that the shear breakages occur at precisely identical points in all molecules, so that there are no single-chain overlaps on the reformed duplexes; and (2) that the length (L) of the single-chain segments is small in comparison with the length of the *nonrepeating* nucleotide sequence from which they are broken. With these assumptions, Wetmur and Davidson (1968) made the following formulation:

$$1/F = 1 + 1/2\,k_2\,P_0\,t,$$

in which $P_0$ is the initial concentration of nucleotide; $k_2$ the second order rate constant for the reaction, and $t$ the time. A plot of $1/F$ against time generally gives a straight line for *E. coli* and phage DNAs, which are thought to be mostly nonrepeating. Assuming that each pair could be a potential nucleation point, they calculated

$$k_2 = k_N\,L/N,$$

in which $k_N$ is the average rate constant for nucleation at each nucleation point. This equation means that the observed rate of annealing of chains of identical length cut from nonrepeating genomes should be inversely proportional to the number of nucleotide pairs in their genome. This expectation is generally confirmed with T7, T4, and *E. coli* DNA. With eukaryotic DNAs, the plots of $1/F$ against time are curved, indicating a heterogeneity in annealing rates. Part of the DNA appears to anneal very rapidly, another part more slowly, another part even more slowly. Components and rates must be estimated. In spite of the conceptual and experimental difficulties, certain qualitative conclusions emerge: a fraction of the single-chained fragments from many eukaryotic sources anneals at a rate that indicates a high and variable degree of repetition, unlike the nonrepetitive DNAs just mentioned (Britten and Kohne, 1968). In the case of mouse and hamster satellite DNA, the rate of annealing is very rapid indeed (Flamm et al., 1969a, 1969b).

A number of problems are presented in annealing-rate experiments. There are defects in the theory by which these experiments are interpreted. These come from the assumption that the re-formed duplex segments have no single-chain termini that can react further, whereas, in fact, re-annealing short segments can produce structures of very high molecular weight. The experimental means of assessing the extent of renaturation are wanting: a lowered absorbance, or retention by hydroxyapatite, is not a faithful measure of the fraction of the nucleotides in duplex structure.

Finally, a conceptual problem, which applies equally to DNA-DNA and DNA-RNA annealing experiments, needs attention. It has been shown that ribonucleotide oligomers that are 12 or more units long will form a ribonuclease-resistant complex with homologous single-chained DNA (Niyogi and Thomas, 1967; Niyogi, 1969). It is likely that the same number applies for deoxyoligomers. There are only 8,388,608 different noncomplementary oligomers 12 units long. Yet mammals have at least 3,000,000,000 such oligomers (on one chain) per haploid nucleus, and amphibians have as many as 10 times more than do mammalian DNAs. Thus, each 12-unit oligomer must, on the average, be repeated 360 times. Will the presence of such oligomers, which are capable of forming stable duplexes, increase or decrease the rate of reformation of duplex segments? How do they affect the saturation levels in RNA-DNA annealing experiments? I don't think the answers are known.

## A test for serial repetition of nucleotide sequences in duplex DNA

The experiments described above tell nothing about the arrangements of repeating sequences, because the fragments are *smaller* than the nonrepeating segment (L $\ll$ N). In order to test the idea that sequences are tandemly-repetitious in eukaryotic DNAs, the length of the DNA fragments under study must be *longer* than the length of the repetitious unit (L $\gg$ N). This scheme is depicted in Figure 15. When segments of this type are denatured and annealed, circular DNA duplexes would be expected. Experiments of this kind have been performed to check this expectation (Thomas et al., 1970). Salmon and trout sperm DNA were sheared by being passed through a fine needle. Electron microscopy of the resulting preparation reveals linear molecules, 2 to 5$\mu$ long. Only rarely were circular structures seen with contour lengths in the 5-$\mu$ range. When this preparation is denatured, by either heat or alkali, and annealed, and then

FIGURE 14 Electron micrograph of ribosomal precursor RNA being transcribed from nucleolar DNA found in oocytes of *Triturus*. (Courtesy of O. L. Miller, Jr.)

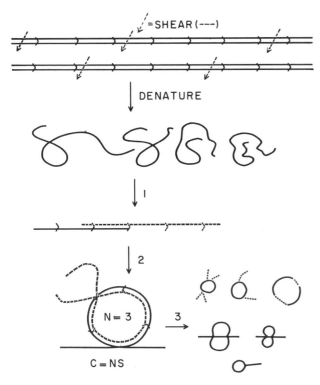

FIGURE 15 Denaturation and annealing of tandemly-repeating DNA.

is examined by electron microscopy, many circular molecules are seen. Those from trout and salmon sperm DNA are shown in Figures 16 and 17. Their contour lengths ranged from 0.2 to 2 μ. To date we have photographed and

measured about 200. If this experiment is repeated with prokaryotic DNAs, such as *E. coli* DNA or T7 DNA, essentially no circles are found. The few examples found have relatively long contour lengths (more than 2 μ) and are probably accidental arrangements.

We have attempted to measure the frequency and contour length of circles as a function of annealing time. Sheared trout sperm DNA was denatured and annealed for various periods of time, then passed through hydroxyapatite, and the second peak (containing partly-duplex DNA) was examined by electron microscopy. Circular molecules could be found after as little as 15 minutes of annealing. While the abundance of circular molecules increased with longer annealing, the distribution of contour lengths remained the same.

Some preliminary experiments indicate that segments of DNA of differing G+C composition will produce circles upon annealing. For this purpose, salmon sperm DNA was fractionated with respect to its melting temperature by "thermal chromatography" on hydroxyapatite (Miyazawa and Thomas, 1965). Each thermal fraction generated circular molecules, indicating that circle formation is a general property of all the chromosomal DNA, not just a certain compositional class of segments.

Special figures, more elaborate than circles, may be formed. What appears to be the next most complex structure is shown in Figure 18. Some of these, and their degraded derivatives, can be seen in Figures 16 and 17. These are the "mops" and tangled structures in annealed preparations of sperm DNA that are so hard to decipher. The prokaryotic DNA seems to produce cleaner linear structures.

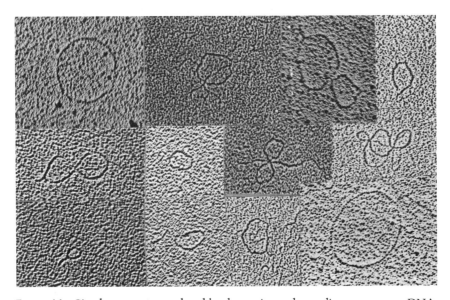

FIGURE 16 Circular structures produced by denaturing and annealing trout sperm DNA.

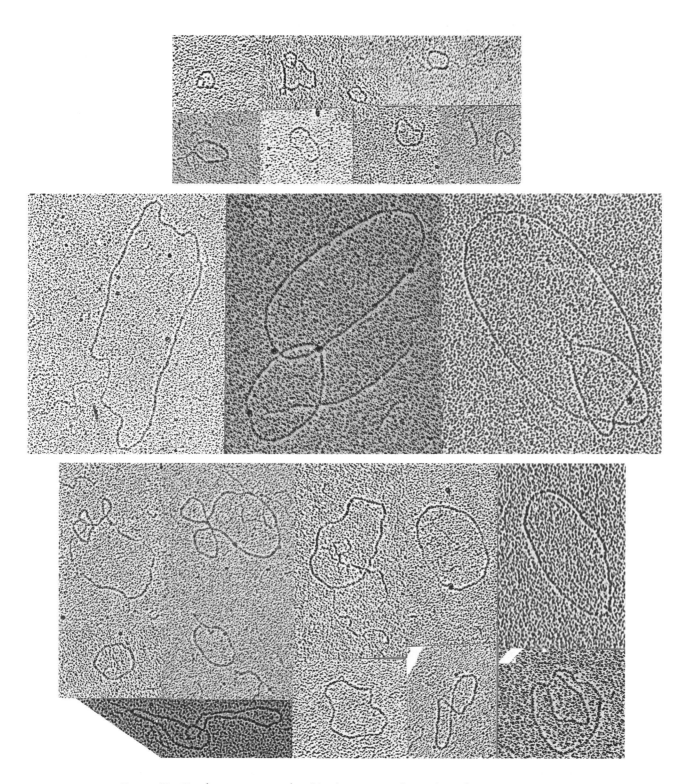

FIGURE 17   Circular structures produced by denaturing and annealing salmon sperm DNA.

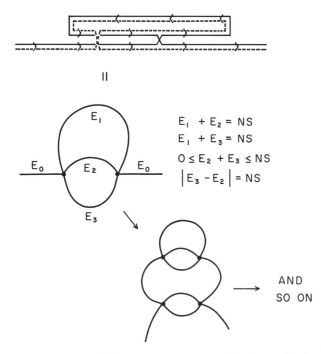

$$E_1 + E_2 = NS$$
$$E_1 + E_3 = NS$$
$$0 \leq E_2 + E_3 \leq NS$$
$$|E_3 - E_2| = NS$$

AND SO ON

FIGURE 18 More elaborate structures expected when tandemly-repeating DNA is denatured and annealed.

*The Analogy with Phage DNAs* A phage DNA molecule is one that is cut to a length slightly longer than the non-repeated segment. This is generally seen from the opposite point of view—that phage DNA molecules are terminally repetitious (see Thomas et al., 1968). When a nonpermuted collection of DNA molecules possessing a very long terminal repetition is denatured and annealed, many circles are found. An example of such a situation is afforded by phage T1 DNA (MacHattie et al., 1969). Indeed, phage DNA molecules that are permuted and repetitious provide a fragmentary analogy of the tandemly-repeating slave genes to be found in each gene family.

*Summary* In this section, some preliminary evidence is described that supports the notion that tandemly-repeating nucleotide sequences are a frequent occurrence in sperm DNA.

## Rectification

Unless we can find some other logical alternative, the presence of tandemly-repeating genes means that some way must be found to account for the fact that recessive point mutants can be found in eukaryotes, and that alleles are inherited in an all-or-nothing manner as discrete Mendelian genes. To accomplish this, we hypothesize the existence of a master gene that *impresses* its nucleotide sequence on every slave in the family, about once every cell division. This

process is called *rectification* in this paper. Rectification must take place in the germ line to account for Mendelian genetics. In view of the fact that crossing over can occur in somatic cells, and that the genes seem to be segregated as discrete units in this situation, as well (Stern, 1936, 1968; Pontecorvo, 1958), it is likely that rectification also must occur in somatic cells.

Such a concept is not pleasant—a master gene rectifying slaves. Some critics have suggested to me that there is no need for a master gene: it could all be done by democratic consensus! If so, it would be difficult to explain the appearance of recessive point mutants. For similar reasons, the idea of rectification seems to collide with preconceptions that are not completely scientific. Therefore, I wish to present a few possible models for this yet-to-be discovered process, and show that they are not at all unthinkable.

As a starting point, let us consider the "rolling circle" model for DNA replication that seems to account for some features of the replication of $\phi x$ DNA (Gilbert and Dressler, 1968) (Figure 19A). It has been shown that one intermediate in the replication of mature T7 DNA molecules is a long, linear duplex up to four or more genomes in length. These molecules are almost surely concatenated genomes of T7 (Kelly and Thomas, 1969). One likely possibility is that they are produced by a "rolling circle" mechanism (Figure 19B). As can be seen from these diagrams, the nucleotide sequence in the tandemly-connected genomes is completely determined by the sequence in the closed "rolling circle." This closed polynucleotide loop is therefore the master gene that determines the sequence of the slave series. In the examples just given, the rolling circle contains an entire genome, but there is no mechanical reason why it could not contain only

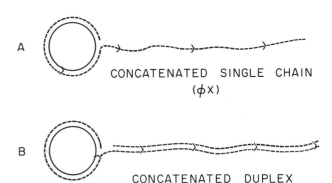

A. CONCATENATED SINGLE CHAIN ($\phi x$)

B. CONCATENATED DUPLEX (T7)

FIGURE 19 Rolling circle models of DNA replication. A. To produce a repeating single chain as has been suggested for $\phi x$ (Gilbert and Dressler, 1968). B. To produce a repeating duplex as has been suggested for T7 (Kelly and Thomas, 1969).

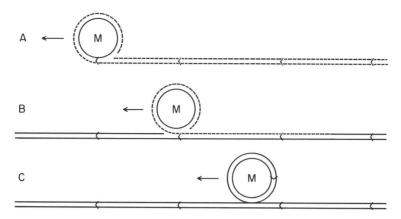

FIGURE 20  Models for rectification. A. Rectification by replication. B. Rectification by replacement and repair. C. Rectification by repair alone. The master gene is pictured as a closed polynucleotide loop that rolls down the slave series, affecting replication or strand-selective repair.

a thousand or so nucleotides. This gives rise to the first model for Rectification by Replication (Figure 20A).

Rectification by Replication presents some difficulties, one of which is that there is no convenient way to tell it to stop! The gene family would be of uncertain size, and we know that total DNA contents (and the DNA content in the bands of the polytene chromosomes) is precisely specified. This scheme in unmodified form is a conservative mode of replication and does not fit with density-label shift experiments (Taylor, 1969).

To avoid these objections, we might suppose that Rectification by Replacement could take place (Figure 20B). In this model, a rolling circle (a master gene) rolls down a series of slaves, templating new DNA as it goes, and hydrolyzing nucleotides from the 5' end of the existing chain. In this model, one chain is specified by the master gene; the other is pre-existent. At this point, we must picture strand-selective repairing to remove any existing nucleotide mismatches. This model would call for large amounts of DNA turnover, which does not appear to happen in somatic cells (Siminovitch and Graham, 1956; Healy et al., 1956) or during spermatogenesis (Lam et al., 1969).

To avoid this difficulty, we might suppose that Rectification by Repair occurred, which calls for a circular master chain to be looped into one of the two chains of the DNA duplex (Figure 20C). The circular master chain is pictured to roll down the slave family. When a base pair mismatch occurs, a strand-selective repairing system brings the sequence of the long chain into complementarity with the master chain. After the circular master chain is passed, a second step of strand-selective repair eliminates any mismatches with the first. Although strand-selective repair is not known at present, other strand-selective processes have been demonstrated, e.g., RNA transcription. Therefore, strand-selec-

tive repair is not outside the bounds of reasonable conjecture.

The purpose of these models is to show that it is possible to think of *rectification* in physical terms that are not unreasonable. I think that it will be possible to test some of these proposals. The most interesting prediction is that it should be possible to visualize the rectification apparatus in the electron microscope.

*Evidence*  At present, there is no firm evidence for the existence of rectification. It is a logical requirement if we suppose that genes exist in multiple and functional form. There is, however, an observation made by Forget and Weissman (1969) who have sequenced the 5S RNA from a human epidermoid tumor, which would be expected to have multiple copies of the 5S RNA gene. They find *no* evidence for any heterogeneity in sequence; all the molecules appear to be identical. In sharp contrast, Brownlee et al. (1968) found that *E. coli* contains two forms of 5S RNA, and even further heterogeneity in sequence can be detected, as would be expected, if rectification were occurring in the human cells but not in *E. coli*.

## Some implications of the theory

The implications of this theory are profound and touch almost every area of biology. The notion that there are about the same number of different genes in a fruit fly as there are in *E. coli* will not quickly be embraced by all biologists. The idea that a man has only about 10 times as many different genes as *E. coli* will receive an even less enthusiastic welcome. There are many reasons to be skeptical, because not a single argument that I have brought to support The Theory is completely convincing.

*Immunoglobulins*  Clearly, The Theory impinges on our

current understanding of the specification of immunoglobulin sequences. In a sense, it inverts the question from What renders the variable portions of the light and heavy chains variable? to What causes the constant portions to remain constant? The Theory supplies the answer to the last question: rectification renders all genes in a family identical to the master gene. The origin of the variable gene, or genes, must result from a suspension aberration of the rectification process. In this way, mutational alterations could accumulate in this gene. Indeed, as was suggested to me by Edelman (see Edelman, this volume), if the rectification process could be held in abeyance for certain periods for special gene families, new opportunities for experimentation and subsequent selection would be made available at the subcellular level. Whether all this has anything to do with the specification of immunoglobulins is not clear. If anything, The Theory will supply new fuel for the model builders who are attempting to account for antibody specificity.

*The Specificity of Self-assembling Neural Nets*  Those persons working on the organization of the nervous system are finding increasing evidence for the precise specification of very elaborate patterns of connectivity (Sperry and Hibbard, 1968; Jacobson, 1969, and this volume). If all these cellular connections were specified by different single genes, the number of genes required would be at least equal to the number of neurons (some $10^{11}$ to $10^{12}$ in man). On the basis of the DNA content of $3 \times 10^9$ nucleotide pairs, it might be imagined that there are $3 \times 10^6$ structural genes. Such a number is far too few, if each connection is to require a different gene product. According to The Theory, we must reduce this to perhaps 30,000 different gene families—or fewer! The problem is then rendered even more acute. A solution is obvious; nature must employ a code (see Griffith, 1967). Only 10 different gene products arranged in ordered sets of 12 will produce $10^{12}$ unique arrangements; or, 100 gene products arranged in ordered sets of six will give the same number. After all, thousands of different proteins are specified by only four nucleotides. Therefore, there is no reason to reject the notion that a man has only 30,000 different genes simply because his neural net is highly specified.

*Prospect*  If it be presumed that The Theory is generally correct, the most significant implication is excitement. If the genetic complexity of man is only 10 times greater than *E. coli,* there is a real chance that we shall soon understand as much about his biochemistry as we now do of *E. coli.* The task before us may be 1000 times easier than we feared.

## Acknowledgments

Writing this paper has been an exercise in learning. Many people have given me assistance and sometimes encouragement. Oscar Miller, Jr., and through him J. G. Gall, have been most helpful. Ed Lewis and Sol Spiegelman have brought many items to my attention. I thank Professors H. Rees, H. G. Keyl, J. G. Gall, and O. L. Miller, Jr., for their photographs. This work was supported by the National Science Foundation (GB-8611) and the National Institutes of Health (GM-08186). I am also indebted to my colleagues who debated these issues with me over many months: Garret Ihler, Barbara Hamkalo, Barbara Otto, Roy Barzilai, Lorne MacHattie, Larry Okun, and Gene Kennedy. A former laboratory technician, Mrs. Cindy Kelly, took some of the electron micrographs included here; Dr. Hamkalo, the others. Jackie Vaughan deserved a medal for seeing the various drafts through to this stage. Finally, I express my appreciation to Professor H. G. Callan, whose papers stimulated my interest in this subject.

## REFERENCES

ATTARDI, G., P. C. HUANG, and S. KABAT, 1965a. Recognition of ribosomal RNA sites in DNA. I. Analysis of the E. coli system. *Proc. Nat. Acad. Sci. U.S.A.* 53: 1490–1498.

ATTARDI, G., P. C. HUANG, and S. KABAT, 1965b. Recognition of ribosomal RNA sites in DNA. II. The hela cell system. *Proc. Nat. Acad. Sci. U.S.A.* 54: 185–192.

BAKER, T. G., and L. L. FRANCHI, 1967. The structure of the chromosomes in human primordial oocytes. *Chromosoma* 22: 358–377.

BEERMANN W., 1967. Gene action at the level of the chromosome. *In* Heritage from Mendel (R. A. Brink and E. D. Styles, editors). University of Wisconsin Press, Madison, pp. 179–201.

BELLING, J., 1931. Chromosomes of liliaceous plants. *Univ. Calif. Publ. Bot.* 16: 153–170.

BIRNSTIEL, M., J. SPEIRS, I. PURDOM, K. JONES, and U. E. LOENING, 1968. Properties and composition of the isolated ribosomal DNA satellite of *Xenopus laevis. Nature* (London) 219: 454–463.

BRIDGES, C. B., 1938. A revised map of the salivary gland X-chromosome of *Drosophila melanogaster. J. Hered.* 29: 11–13.

BRITTEN, R. J., and D. E. KOHNE, 1968. Repeated sequences in DNA. *Science* (*Washington*) 161: 529–540.

BRODY, S., and C. YANOFSKY, 1963. Suppressor gene alteration of protein primary structure. *Proc. Nat. Acad. Sci. U.S.A.* 50: 9–16.

BROWN, D. D., and I. B. DAWID, 1968. Specific gene amplification in oocytes. *Science* (*Washington*) 160: 272–280.

BROWN, D. D., and C. S. WEBER, 1968. Gene linkage by RNA-DNA hybridization. I. Unique DNA sequences homologous to 4S RNA, 5S RNA and ribosomal RNA. *J. Mol. Biol.* 34: 661–680.

BROWNLEE, G. G., F. SANGER, and B. G. BARRELL, 1968. The sequence of 5S ribosomal ribonucleic acid. *J. Mol. Biol.* 34: 379–412.

CALLAN, H. G., 1967. The organization of genetic units in chromosomes. *J. Cell Sci.* 2: 1–7.

CALLAN, H. G., and L. LLOYD, 1960. Lampbrush chromosomes of crested newts *Triturus cristatus* (Laurenti). *Phil. Trans. Roy. Soc.* (*London*), ser. B, biol. sci. 243: 135–219.

CAMPBELL, A., 1965. The steric effect in lysogenization by bacteriophage lambda. I. Lysogenization of a partially diploid strain of *Escherichia coli* K12. *Virology* 27: 329–339.

CHOVNICK, A., 1966. Genetic organization in higher organisms. *Proc. Roy. Soc., ser. B, biol. sci.* 164: 198–208.

DA CUNHA, A. B., C. PAVAN, J. S. MORGANTE, and M. C. GARRIDO, 1969. Studies on the cytology and differentiation in Sciarae. II. DNA redundancy in salivary gland cells of *Hybosciara fragiles* (Diptera, Sciaridae). *Genetics (Suppl.)* 61 (no. 1, pt. 2): 335–349.

DEMEREC, M., 1967 Properties of genes. *In* Heritage from Mendel. (R. A. Brink and E. D. Styles, editors). University of Wisconsin Press, Madison, pp. 49–61.

DRAKE, J. W., 1969. Comparative rates of spontaneous mutation. *Nature (London)* 221: 1132.

FLAMM, W. G., P. M. B. Walker, and M. McCallum, 1969a. Some properties of the single strands isolated from the DNA of the nuclear satellite of the mouse (*Mus. musculus*). *J. Mol. Biol.* 40: 423–443.

FLAMM, W. G., P. M. B. WALKER, and M. McCALLUM, 1969b. Renaturation and isolation of single strands from the nuclear DNA of the guinea pig. *J. Mol. Biol.* 42: 441–455.

FORGET, B. G., and S. M. WEISSMAN, 1969. The nucleotide sequence of ribosomal 5S ribonucleic acid from KB cells. *J. Biol. Chem.* 244: 3148–3165.

GALL, J. G., and H. G. CALLAN, 1962. [3]H uridine incorporation in lampbrush chromosomes. *Proc. Nat. Acad. Sci. U.S.A.* 48: 562–570.

GILBERT, W., and D. DRESSLER, 1968. DNA replication: The rolling circle model. *Cold Spring Harbor Symp. Quant. Biol.* 33: 473–484.

GILLESPIE, D., and S. SPIEGELMAN, 1965. A quantitative assay for DNA-RNA hybrids with DNA immobilized on a membrane. *J. Mol. Biol.* 12: 829–842.

GOODMAN, H. M., and A. RICH, 1962. Formation of a DNA-soluble RNA hybrid and its relation to the origin, evolution and degeneracy of soluble RNA. *Proc. Nat. Acad. Sci. U.S.A.* 48: 2101–2109.

GRIFFITH, J. S., 1967. A View of the Brain. Clarendon Press, Oxford.

HANCOCK, R., 1969. Conservation of histone in chromatin during growth and mitosis in vitro. *J. Mol. Biol.* 40: 457–466.

HEALY, G. M., L. SIMINOVITCH, R. C. PARKER, and A. F. GRAHAM, 1956. Conservation of desoxyribonucleic acid phosphorus in animal cells propagated in vitro. *Biochim. Biophys. Acta* 20: 425–426.

HILL, C. W., J. FOULDS, L. SOLL, and P. BERG, 1969. Instability of a missense suppressor resulting from a duplication of genetic material. *J. Mol. Biol.* 39: 563–581.

HILL, C. W., D. SCHIFFER, and P. BERG, 1970. Transduction of partial heterozygosity: Induced duplication of recipient genes. *J. Mol. Biol.* (in press).

HINEGARDNER, R., 1968. Evolution of cellular DNA content in teleost fishes. *Amer. Natur.* 102: 517–523.

HOLM-HANSEN, O., 1969. Algae: Amounts of DNA and organic carbon in single cells. *Science (Washington)* 163: 87–88.

JACOBSON, M., 1969. Development of specific neuronal connections. *Science (Washington)* 163: 543–547.

JONES, R. N., and H. REES, 1968. Nuclear DNA variation in *Allium. Heredity.* 23: 591–605.

KELLY, T. J., JR., and C. A. THOMAS, JR., 1969. An intermediate in the replication of bacteriophage T7 DNA molecules. *J. Mol. Biol.* 44: 459–475.

KEYL, H.-G., 1965a. Duplikationen von Untereinheiten der Chromosomalen DNS während der Evolution von *Chironomus thummi. Chromosoma* 17: 139–180.

KEYL, H.-G., 1965b. A demonstrable and geometric increase in the chromosomal DNA of *Chironomus. Experientia (Basel)* 21: 191–193.

LAM, D. M. K., R. FURRER, and W. R. BRUCE, 1970. The separation, physical characterization and differentiation kinetics of spermatogonial cells of the mouse. *Proc. Nat. Acad. Sci. U.S.A.* 65: 192–199.

MILLER, O. L., JR., 1966. Structure and composition of peripheral nucleoli of salamander oocytes. *Nat. Cancer Inst. Monogr.* 23: 53–66.

MILLER, O. L., JR., and B. R. BEATTY, 1969a. Visualization of nucleolar genes. *Science (Washington)* 164: 955–957.

MILLER, O. L., JR., and B. BEATTY, 1969b. Portrait of a gene. *J. Cell. Physiol.* 74 (suppl. 1): 225–232.

MIRSKY, A. E., and H. RIS, 1951. The desoxyribonucleic acid content of animal cells and its evolutionary significance. *J. Gen. Physiol.* 34: 451–462.

MIYAZAWA, Y., and C. A. THOMAS, JR., 1965. Nucleotide composition of short segments of DNA molecules. *J. Mol. Biol.* 11: 223–237.

MORELL, P., I. SMITH, D. DUBNAU, and J. MARMUR, 1967. Isolation and characterization of low molecular weight ribonucleic acid species from *Bacillus subtilis. Biochemistry* 6: 258–265.

MULLER, H. J., 1929. The gene as the basis of life. *Proc. Int. Cong. Plant Sci.* 1: 897–921.

NIYOGI, S. K., 1969. The influence of chain length and base composition on the specific association of oligoribonucleotides with denatured deoxyribonucleic acid. *J. Biol. Chem.* 244: 1576–1581.

NIYOGI, S. K., and C. A. THOMAS, JR., 1967. The specific association of ribooligonucleotides of known chain length with denatured DNA. *Biochem. Biophys. Res. Commun.* 26: 51–57.

PLAUT, W., 1969. On ordered DNA replication in polytene chromosomes. *Genetics (Suppl.)* 61 (no 1, pt. 2): 239–244.

PONTECORVO, G., 1958. Trends in Genetic Analysis. Columbia University Press, New York.

RADLOFF, R., W. BAUER, and J. VINOGRAD, 1967. A dye-buoyant-density method for the detection and isolation of closed circular duplex DNA: The closed circular DNA in hela cells. *Proc. Nat. Acad. Sci. U.S.A.* 57: 1514–1521.

REES, H., and R. N. JONES, 1967. Structural basis of quantitative variation in nuclear DNA. *Nature (London)* 216: 825–826.

RITOSSA, F. M., K. C. ATWOOD, D. L. LINDSLEY, and S. SPIEGELMAN, 1966. On the chromosomal distribution of DNA complementary to ribosomal and soluble RNA. *Nat. Cancer Inst. Monogr.* 23: 449–472.

RUDKIN, G. T., 1961. Cytochemistry in the ultraviolet. *Microchem. J. Symp.* 1: 261–276.

RUSSELL, R. L., J. ABELSON, A. LANDY, S. BRENNER, M. L. GEFTER, and J. D. SMITH, 1970. Duplicate genes for tyrosine transfer RNA in *E. coli. J. Mol. Biol.* 47: 1–13.

SCHWEIZER, E., C. MacKECHNIE, and H. O. HALVORSON, 1969. The redundancy of ribosomal and transfer RNA genes in *Saccharomyces cerevisiae*. *J. Mol. Biol.* 40: 261–277.

SIMINOVITCH, L., and A. F. GRAHAM, 1956. Significance of ribonucleic acid and deoxyribonucleic acid turnover studies. *J. Histochem. Cytochem.* 4: 508–515.

SOBER, H. A. (editor), 1968. Handbook of Biochemistry: Selected Data for Molecular Biology. The Chemical Rubber Company, Cleveland, Ohio.

SPERRY, R. W., and E. HIBBARD, 1968. Regulative factors in the orderly growth of retino-tectal connexions. *In* Growth of the Nervous System/A Ciba Foundation Symposium (G. E. W. Wolstenholme and M. O'Connor, editors). Little Brown and Company, Boston, Massachusetts, pp. 41–52.

STAHL, F. W., R. S. EDGAR, and J. STEINBERG, 1964. The linkage map of bacteriophage T4. *Genetics* 50: 539–552.

STERN, C., 1936. Somatic crossing over and segregation in *Drosophila melanogaster*. *Genetics* 21: 625–730.

STERN, C., 1968. Genetic Mosaics and Other Essays. Harvard University Press, Cambridge, Massachusetts.

STUDIER, F. W., 1969a. Conformational changes of single-stranded DNA. *J. Mol. Biol.* 41: 189–197.

STUDIER, F. W., 1969b. Effects of the conformation of single-stranded DNA on renaturation and aggregation. *J. Mol. Biol.* 41: 199–209.

SWANSON, C. P., 1957. Cytology and Cytogenetics. Prentice-Hall, Inc., Englewood Cliffs, New Jersey, Chapter 6.

TAYLOR, J. H., 1969. Replication and organization of chromosomes. *In* 12th International Congress of Genetics, Tokyo, 1968, vol. 3, pp. 177–189.

TAYLOR, J. H., P. S. WOODS, and W. L. HUGHES, 1957. The organization and duplication of chromosomes as revealed by autoradiographic studies using tritium-labeled thymidine. *Proc. Nat. Acad. Sci. U.S.A.* 43: 122.

THOMAS, C. A., JR., 1966. Recombination of DNA molecules. *Progr. Nucl. Acid Res. Mol. Biol.* 5: 315–337.

THOMAS, C. A., JR., B. A. HAMKALO, D. N. MISRA, and C. S. LEE, 1970. The cylization of eucaryotic DNA fragments. *J. Mol. Biol.* (in press).

THOMAS, C. A., JR., T. J. KELLY, JR., and M. RHOADES, 1968. The intracellular forms of T7 and P22 DNA molecules. *Cold Spring Harbor Symp. Quant. Biol.* 33: 417–424.

WETMUR, J. G., and N. DAVIDSON. 1968. Kinetics of renaturation of DNA. *J. Mol. Biol.* 31: 349–370.

YANKOFSKY, S. A., and S. SPIEGELMAN, 1962. The identification of the ribosomal RNA cistron by sequence complementarity, II. Saturation of and competitive interaction at the RNA cistron. *Proc. Nat. Acad. Sci. U.S.A.* 48: 1466–1472.

YANKOFSKY, S. A., and S. Spiegelman, 1963. Distinct cistrons for the two ribosomal RNA components. *Proc. Nat. Acad. Sci. U.S.A.* 49: 538–544.

# 87 Structure and Replication of Eukaryotic Chromosomes

## J. HERBERT TAYLOR, WILLIAM A. MEGO, and DONALD P. EVENSON

DURING THE LAST 15 years, enough new information has been obtained to allow reasonable predictions concerning the organizational and regulatory features of chromosomes. Although much of our understanding of the properties of DNA and genetic regulation has come about through the analyses of viruses and bacteria, attention in this presentation is given to higher cells, i.e., the cells of eukaryotes. The principal evidence is drawn from mammalian cells, with background information from prokaryotes in mind, but seldom referred to directly. Perhaps the most important generalization about these two cell types is that prokaryotes, which lack true nuclei and are represented by a number of viruses and a few bacterial species that have been studied extensively, maintain a single copy of most genes; in eukaryotes, which have true nuclei, mechanisms have evolved for handling multiple copies. In addition, eukaryotes possess nuclear membranes, nucleoli, histones associated with their DNA, and chromosomes that condense into rod-shaped bodies during mitosis.

Britten and Kohne (1968) have reported that 10 per cent or more of the DNA of mammalian cells reanneals after

J. HERBERT TAYLOR, WILLIAM A. MEGO, and DONALD P. EVENSON Institute of Molecular Biophysics, Florida State University, Tallahassee, Florida

shearing and denaturing at a rate which indicates that each genome contains about a million copies of very similar sequences. No such repeating sequences are found in viruses, bacteria, blue-green algae, or other prokaryotes. However, an examination of more than 50 species of eukaryotes distributed among many classes of both the plant and animal kingdoms revealed fast reannealing fractions of DNA. This observation indicates a striking difference between the prokaryotes and the eukaryotes. According to these authors, most other base sequences in the DNA of eukaryotes are represented by no more than 10 copies. Although these data give us only an order of magnitude for redundancy in the genetic apparatus, they lend a rather convincing argument to other circumstantial evidence, cited by Thomas in this volume, that higher cells contain numerous copies of each genetic locus. The number of copies is assumed to be relatively small in an organism such as *Neurospora,* which has only 10 times as much DNA per genome as does the prokaryote *Escherichia coli.* Mammalian cells have nearly 1000 times the amount of DNA that is in a cell of *E. coli,* but the record is held by the lungfishes and certain amphibians, which have 30 to 40 times as much DNA as does the mammalian cell (Taylor, 1969a).

The problem posed in the maintenance of multiple copies has been considered by Thomas. He has elaborated and modified the concept of the master gene, a hypothesis originally suggested by Callan and Lloyd (1960) and treated in detail by Callan in 1967. This proposal is that multiple identical copies of genes can exist in spite of independent mutation, because the master gene periodically corrects to its base sequence the whole cluster of genes associated with it. This is done by some repair or base substitution process to make the slaves (multiple working copies) match the master copy.

It is still difficult to explain the fast reannealing fraction by this concept of multiple copies, for there are too many similar or identical copies of one type—about one million per genome for the mammalian cell. Some animals have a satellite band of DNA, i.e., a density species that bands in CsCl or other cesium salts at a different position from the main band of DNA. A light satellite in the mouse has been studied extensively by Flamm et al. (1969). It reanneals more rapidly than does the main-band DNA, and may provide a clue to the fast reannealing fraction. Maio and Schildkraut (1969) have found that the satellite DNA is distributed about equally among the various chromosomes of the mouse. Flamm et al. (1969) also found that satellite DNA hybridizes to some extent with main-band DNA. However, one of the most revealing features may be their finding that cytoplasmic messenger RNA does not hybridize with the satellite DNA. In addition, they found some indication that the fast reannealing fraction in rodents drifts faster in evolution than do the less numerous sites. For example, satellite DNAs

in mice and guinea pigs hybridize proportionately less well than do their main-band DNAs.

We have examined the subunits of DNA released from chromosomes by various treatments and manipulations. We believe we have found the basic repeating unit, which is about 2 microns long. Circumstantial evidence indicates that it contains five segments united by more flexible regions of the DNA. We propose that the fast reannealing fraction of DNA is produced by one of the five segments that consists primarily of adenine- and thymine-containing nucleotides, which we refer to as the A–T segment. Some additional assumptions are necessary to explain the satellite DNAs, but we shall come back to this problem after presenting some new information on the size and characteristics of DNA subunits in chromosomes. Because of space limitation, the material must be presented with a minimum of documentation. After presenting the concepts on the organization of chromosomes, we shall make some preliminary suggestions concerning possible functions of the genetic apparatus peculiar to neurons.

## Subunits of DNA from chromosomes

80S SUBUNITS  The chromosomal DNA in Chinese hamster cells can be dissociated into large subunits, which sediment as a single, rather narrow band in a sucrose gradient (Figure 1). For example, nearly all the DNA from rapidly

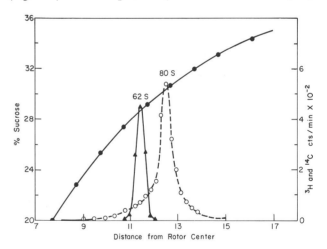

FIGURE 1  Profile of DNA in an isokinetic sucrose gradient, i.e., one designed so that the change in density and viscosity of the solution compensates for the increase in centrifugal force along the length of the tube. The distance of a sedimenting band from the center of rotation is linearly related to the sedimentation coefficient (S). All subsequent sucrose gradients are of this type. This particular gradient shows the sedimentation of dissociated hamster DNA (open circles) and a marker, phage T-4 DNA (closed triangles). The DNA was spun in the Spinco SW 27 rotor at 21,960 rpm for 10 hours at 20° C in 1 M NaCl buffered at pH 9.5 with tris-EDTA and sucrose.

growing, log-phase cells or from metaphase chromosomes and G₁-phase cells can be dissociated into these subunits at pH 9.0 to 10.0 in 0.15 M NaCl after deproteinization by exhaustive pronase digestion. The band of hamster DNA is not appreciably wider than that formed by the marker T-4 DNA used in the gradients when care has been taken to avoid mechanical shear in the lysis and subsequent handling. These results have been obtained many times after a variety of treatments to bring the DNA into solution, including lysis at 0° C with sodium dodecyl sarcosinate in viscous sucrose solutions. However, lysis at 37° C without sucrose or other viscous materials gives similar results, if the DNA is handled carefully and pipetted very slowly through a wide-mouth pipette. Preparations have also been made by lysing the cells in a centrifuge tube and floating the lysate on a sucrose solution delivered into the bottom of the tube to avoid pipetting the DNA. All methods give similar results, except that aggregation of the DNA often leads to the loss of much of the material by rapid sedimentation to the bottom of the tube. The pronase digestion is not necessary to obtain dissociation if the pH is raised to between 10 and 10.5. Without deproteinization, however, the yield of material that remains in the gradient is usually poor. One of the most important considerations is the lysis of cells at a dilution of not more than about 10⁶ cells per 10 milliliters, i.e., at about 1 microgram of DNA per milliliter. More concentrated lysates can be diluted, but the sedimentation properties are often variable. DNA should not be stored in the cold in the presence of a detergent that will precipitate, because the DNA also comes out of solution. Once out of solution, the DNA is very difficult, if not impossible, to dissolve without shear and consequent breakage. Entanglement of the long strands must be avoided. Even very slow rolling of the tube usually leads to entanglement and aggregation.

These large subunits of DNA sediment with a peak at 80S, which indicates that they are about 100 to 110 microns in length, i.e., nearly twice the length and molecular weight of intact molecules of T-4 DNA. Although the exact molecular weight and uniformity of these subunits cannot yet be proved, we refer to them for convenience as the 80S, or 110-micron, subunits. The mechanism by which they are dissociated cannot be stated with any assurance, but to us it appears unlikely that they are linked by protein. Perhaps a double-stranded DNA in a region with single-stranded interruptions on either side melts, or a region that is extremely fragile to shear exists at the sites of dissociation. Longer pieces of DNA have been indicated by autoradiography (Cairns, 1966; Huberman and Riggs, 1968), but the relation of these pieces (1 to 3 mm long) to the subunits obtained by dissociation is not evident. Larger aggregates probably occur in our lysates, but no quantitative information can be obtained until these dissociate as described. It is possible that 80S subunits represent structural units of the chromosome,

for they contain the amount of DNA that might be in a loop formed by the nucleoprotein fibrils (230 Å in diameter) that have been identified in chromosomes and interphase nuclei by DuPraw (1968). On the basis of mass per unit length and the estimated composition of nucleoprotein in the fibrils, he has estimated that the packing ratio is 56:1 for DNA. Because the nucleoprotein loops are about 2 microns long in metaphase chromosomes of mammalian cells, they contain perhaps one of the 110-micron subunits of DNA.

We have spent considerable time trying to separate the strands of these 80S subunits of double-stranded DNA, but the results so far have been confusing. If they are banded in an alkaline sucrose gradient (pH 11.6 to 12.0) without digestion of the lysate with pronase, they have nearly the same sedimentation rate as have the native particles (Taylor, 1969b). Under similar conditions, the chains of the marker T-4 DNA will separate and produce a band with a peak at about 45S to 46S. We conclude that strand separation does not take place, although the hypochromic shift, which indicates denaturation, occurs between pH 10.5 and 11.5 in these sucrose solutions. When lysates are digested exhaustively in pronase (1 milligram per milliliter) at 37° C at pH 9.0 for from six to 12 hours, the strands apparently separate and produce a rather wide band, with a peak at about 30S (Figure 2). These pieces have an average length of about 20 to 22 microns, and their lengths indicate that each chain of the 80S subunit can be dissociated into five subunits in alkaline gradients. However, care must be taken to obtain this separation. The pH must be above 11.6, and the best results are obtained by raising the pH before layering lysates on gradients, preferably by dialysis. Our results also indicate

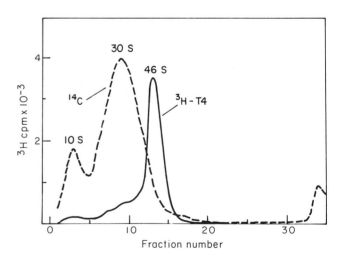

FIGURE 2 Profile of DNA in a sucrose gradient at pH 11.8. The hamster DNA (broken line) was from a stationary phase cell culture, and the marker (solid line) is DNA from phage T-4.

that the DNA of higher cells is broken by alkaline solution, in contrast to the marker T-4 DNA, which appears stable at pH 11.5 to 12.0 for many hours at 10° to 20°C. The hamster DNA is broken very slowly at 10°C and rather extensively at 37°C.

Attempts to separate the chains by nonalkaline solutions have so far met with only limited success. A gradient of sodium trichloracetate (TCAate) was prepared that was nearly isokinetic, 2.1 M to 3.18 M. Cells were lysed in a solution of 1.5 M TCAate, with EDTA (ethylene-diaminotetracetic acid) and SDS (sodium dodecyl sulfate). The DNA was exposed briefly to 3 M TCAate at 37° C and then centrifuged through the gradient at pH 9.5 at 20° C. The sedimentation coefficient of the marker T-4 DNA was calculated to be 43S and that of the hamster DNA to be 57S (Figure 3). However, the sedimentation coefficient of the hamster DNA is actually 58S to 59S (if one assumes that the T-4 strands separated and had the expected S value of 45 to 46); this value of 58 to 59S indicates a particle about one-half of the size of the 80S double-stranded unit. Therefore, our tentative conclusion is that most strands in the 110-micron (80S) subunit are uninterrupted. An attempt to sediment DNA from $G_1$ cells through a DMSO (dimethyl sulfoxide) sucrose gradient gave a rather broad distribution, with a peak about 62S to 63S. This should be the denaturing solvent of choice for deciding the question of the size of the single-stranded units. The large polymer is difficult, however, to keep in solution during transfer to these nonaqueous solvents, and more work is necessary to get reliable evidence.

TWO-MICRON SUBUNITS We have usually used synchronized cells for our studies so that we could characterize the DNA in relation to the stages of the cell cycle. In Chinese hamster cells, these stages are as follows: division stages (30 to 45 minutes), $G_1$ (the gap between division and the beginning of DNA replication, five to seven hours), S phase (the period of DNA replication, six hours), and $G_2$ (the gap between the end of DNA replication and the next division stages, two to three hours). We noted rather early in our studies that DNA in alkaline gradients frequently showed a small band of particles with a peak at 10S to 12S (Figure 2). These were seen in DNAs from interphase stages and telophase, but not in DNA from metaphase chromosomes. The small segments were also seen in cells synchronized in S phase, but the short chains were composed primarily of template DNA (labeled the previous cell cycle) rather than newly formed strands. Therefore, it was not originally considered a functional unit in replication, but was thought to be a unit of transcription. Later, one of us (Mego) found that the short pieces were much more abundant in static cultures, i.e., cells inhibited by crowding and perhaps by a deficiency of essential nutrients. The cells could recover when transferred to new medium, and the small pieces would decrease in frequency before the cells entered the S phase. Even more revealing was his discovery that double-stranded pieces about 2 microns long, sedimenting with a peak at 15S, could be dissociated from the DNA of these static cells by being heated to about 65° C in 0.1 to 0.15 M NaCl and cooled before centrifugation through a sucrose gradient (Figure 4). In most instances, formaldehyde was added dur-

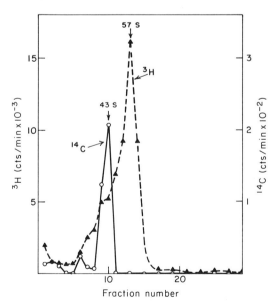

FIGURE 3 Profile of DNA in a TCAate gradient at pH 9.5. The hamster DNA (closed triangles) was isolated from cells at the $G_1$ phase, and the marker (open circles) is DNA from phage T-4.

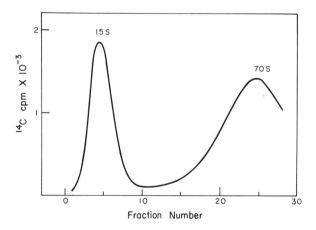

FIGURE 4 Sedimentation profile of hamster DNA in a gradient at pH 9.5. The DNA was isolated from Chinese hamster cells (strain B14FAF 28-G3) in a culture that had reached a static condition or nearly stationary growth phase because of crowding and partial depletion of nutrients in the medium. The DNA was heated to 65° C with 1 per cent formaldehyde before being cooled for centrifugation.

ing or after heating to increase the yield. This indicated to us that single-stranded nicks occurred at regular intervals in each strand, but alternated in position so that two breaks were not opposite each other (Mego and Taylor, 1969). The overlap regions melted out at a low temperature, which suggested that they contained a high proportion of adenine-thymine (A-T) base pairs. The rather uniform size of the subunits suggested that these high A-T regions occurred at regular intervals along the DNA strands.

More recent results (see section on replication below) have revealed small amounts of a 15S to 18S fraction, which contains newly replicated DNA after a pulse label of cells synchronized in the S phase. This leads to the concept that the 2-micron subunit is the basic unit in replication. There would be about 55 of these units (2 microns in length) per large 110-micron segment of DNA; these are probably organized into subgroups of 10 to 11 units each, which have a tendency to separate under alkaline conditions to give single-stranded fragments about 20 to 22 microns long (30S). There would be five of these subgroups of 11 per 110-micron subunit. As mentioned below, this unit of 11 2-micron subunits is thought to be a structural and functional grouping of some significance in regulation of replication, but, as stated above, the 2-micron segments are probably the basic replicating unit. Each is believed to be composed of a high A-T region and four other regions about 0.4 micron long, with a flexible joint between. We have seen these joints occasionally in our electron micrographs, and they were predicted on the basis of hydrodynamic evidence (Gray and Hearst, 1968). If one assumes that the region of high A-T content is the same size as the other four units, which are assumed to have the four bases in nearly equimolar ratios, the peculiar base composition of eukaryotic chromosomal DNA is predicted. One-third of the thymine is in the A-T segment, and about one-sixth is in each of the other four segments. The A-T segment almost certainly has a small number of G-C base pairs, because a few G-C base pairs have been found even in the A-T satellite obtained from the crab (Davidson et al., 1965). This distribution of segments predicts an average base composition of about 40 per cent G-C, which, with few exceptions, is typical of eukaryotes.

SMALLER SUBUNITS OF REPLICATION When Chinese hamster cells are given a two-minute pulse of $^3$H-thymidine, a large part of the label is incorporated into a fraction that is detached from the high molecular weight DNA in lysates. The small pieces of DNA can be isolated in sucrose gradients of lysates, but a better preparation is obtained by removing the high molecular weight DNA with hydroxyapatite. One fraction (I) does not attach to the hydroxyapatite in 0.05 M phosphate buffer at pH 7.0 (Figure 5). This fraction sediments with a peak near the top of the gradient

(about 2S), as shown in Figure 6. Another fraction elutes off the hydroxyapatite in phosphate buffer, pH 7.0, at concentrations between 0.10 and 0.16 M (Figure 5). This is the region over which single-stranded DNA is eluted. These pieces sediment with a peak at about 7S (Figure 6). The small pieces of DNA that are precipitated in 5 to 10 per cent cold trichloracetic acid (TCA) are held up in BioGel P-30 and P-10 when raised to pH 12.0 to denature before being filtered through the column. They are excluded from Bio-Gel P-2. On this basis, a size of about 10 nucleotides in length is suggested (Hohn and Schaller, 1967). The 7S material is excluded from BioGel P-60 and appears in the fraction equivalent to the void volume of the gel. It is retarded by an agarose gel designed to retard globular molecules with molecular weights between $10^4$ and $5 \times 10^6$ daltons. No conclusions concerning its maximum size can yet be drawn from the gel filtration. From its sedimentation, one can estimate the length of the 7S material as equivalent to a polynucleotide of 1000 or more nucleotides.

To summarize, then, we think there is a hierarchy of significant subunits. To begin with the two smallest, which have been detected only during replication, there is a presumptive initiator-primer unit of perhaps 20 to 30 nucleotides, in addition to the much larger 7S unit, which is released as a single strand in lysates. Probably the initiators consist of at least two partially complementary polynucleo-

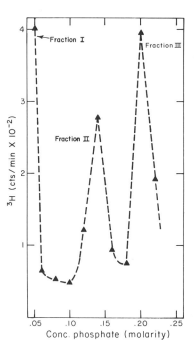

FIGURE 5 Elution of pulse-labeled DNA from a hydroxyapatite column. Fraction I did not adhere to the hydroxyapatite in 0.05 M phosphate. Fraction II is single-stranded DNA, and fraction III is double-stranded. (From Schandl and Taylor, 1969.)

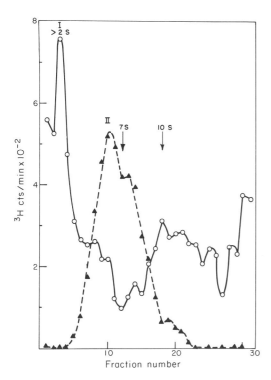

FIGURE 6  Sedimentation profiles of fractions I and II (similar to those shown in Figure 5) run separately in two sucrose gradients, pH 11.8. Only fractions from the top of the tube are shown, as most of the labeled DNA consisted of these rather small pieces. The material that failed to adsorb on hydroxyapatite was rather heterogeneous with respect to sedimentation properties, but a striking peak produced by small polynucleotides is evident near the top of the gradient, fraction I. The single-stranded DNA (fraction II) is more uniform in size (from Schandl and Taylor, 1970).

tides, as shown in Figure 7. We have assumed that a group of nonmatching base pairs is maintained to produce the flexible joints. The mechanism proposed to maintain this irregularity is the insertion of a primer-initiator subunit at the junctions between each 7S subunit.

A group of five 7S subunits (9S, if double stranded) is assumed to make up one 2-micron (15S) subunit (Figure 7). An electron micrograph of a 2-micron subunit is shown in Figure 7A. Many short segments of DNA were observed in a protein film (cytochrome-C spread on a water-air interface) prepared from a sucrose gradient similar to that shown in Figure 4. Other stiff, double-stranded DNA segments shorter and longer than this one were observed, but most of the particles were 1.6 to 2.4 microns in length. A spacing of readily denaturable regions 0.3 to 0.5 micron apart was observed in the DNA, which was heated to 55° C in 11 per cent formaldehyde dissolved in 0.01 M phosphate buffer at pH 7.0. The DNA was cooled quickly, spread on a water-air interface in a film of cytochrome-C for two minutes, picked up on a carbon-coated grid, and stained with uranyl

acetate. The most frequent center-to-center distance between melted regions was about 0.5 micron, but another modal frequency occurred at 0.9 to 1.0 micron. However, long segments of the DNA failed to open in this fashion. Only additional studies can indicate whether the failure was the result of fluctuations produced by the manipulations or actual variations in structural features.

In some cases a large denatured segment, 0.3 to 0.5 micron, was found adjacent to three or four smaller openings (Figure 7B). We have assumed that the structural features may have a regularity, and have designated the five segments assumed to compose a 2-micron subunit as 1, 2, 3, 4, and 0 (Figure 8). One, 2, 3, and 4 are assumed to be coding sequences, and the presumed high A-T regulatory subunit is designated 0. A cluster of 10 or 11 2-micron subunits forms a group that can be dissociated and is seen as single strands in alkaline gradients (30S). When double-stranded, the pieces sediment with a peak at 40S to 41S. The largest dissociated subunit is the 80S (110-micron) subunit, which consists of about 55 of the 2-micron units in five clusters. These 2-micron subunits would be our choice for the repeating unit in the chromosome, but the larger clusters may have some regulatory features in common, probably in both replication and transcription.

The compact folding of the 2-micron DNA subunits is visualized as consisting of a stack of five flat platelets with a considerable portion of the protein between them. Each platelet is formed from one of the 0.4-micron regions between the flexible joints. The DNA is assumed to be folded eight or 10 times in each platelet to form nine or 11 parallel cylindrical rods, a type of packing indicated for crystals that have been prepared from a solution of DNA (Giannoni et al., 1969). The high A-T segment forms the first layer, and a flexible joint allows transition to the next layer of identical size. Five layers finish the nucleoprotein segment, and the attachment of a replication guide (probably a nonhistone protein) marks the transition to the next 2-micron subunit. The platelets formed of the high A-T DNA would alternate along the nucleoprotein fibrils so that, in each stack of five, the A-T platelet is on the opposite side of the fibril from the platelets in its two neighboring 2-micron subunits. Although one must concede that the picture drawn is a bit fanciful and idealized, it fits the available evidence and gives us a pleasing model to test.

## Replication of chromosomal DNA

Replication takes place over an interval of the cell cycle that is remarkably uniform in many kinds of mammalian cells, both in vivo and in vitro. The interval is about six hours in cell cycles, which may vary from 12 hours to many days. Both the S phase and the length of the $G_2$ and mitotic stages are rather uniform. The variable stage is the $G_1$ phase, i.e.,

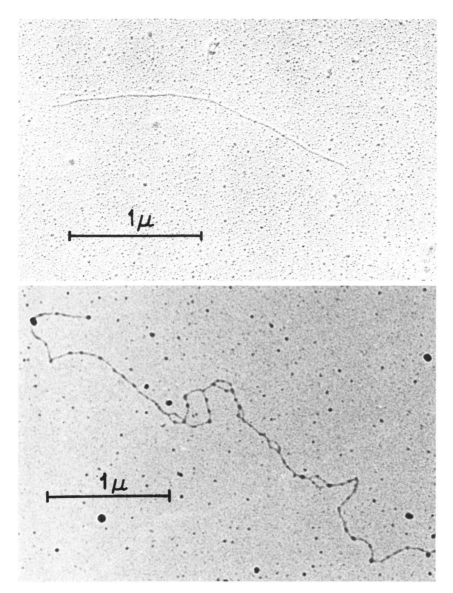

FIGURE 7 Top. Electron micrograph of a segment of native DNA from the 15S peak of a sucrose gradient similar to that shown in Figure 4. The DNA was prepared for microscopy by spreading in a film of cytochrome C on 0.15 M ammonium acetate by the method of Kleinschmidt (1968). Bottom. Electron micrograph of a segment of DNA which was heated at 55° C in 0.01 M phosphate buffer, containing 11 per cent formaldehyde pH 7.0, for 10 minutes, quickly cooled and spread in a film of cytochrome C on 0.15 M ammonium acetate.

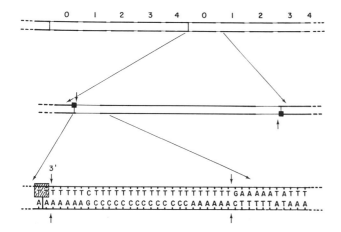

FIGURE 8 Diagram of the proposed structural arrangement in the 2-micron subunits of DNA. Each subunit is shown to be composed of five 7S segments, which are shown in greater detail in the second figure. These segments are coupled by the initiator segments shown in detail in the lower figure. The squares show the position of the hypothetical replication guides, presumably protein with a high affinity for the DNA, at the opposite ends of each 0 segment.

the interval between the shift from telophase to interphase and the beginning of DNA synthesis. Once synthesis is initiated, most cells appear to be committed to a programed sequence of events that leads to another mitosis, unless some unusual intervention takes place. Few, if any, measurable parameters reveal a sequence of events in the $G_1$ phase, even though a number of steps appear to be necessary before the cell is committed to a new cycle of division. Perhaps the cell remains in an uncommitted stage—which Lajtha (1963) would call the $G_0$ phase—until a stimulus initiates events leading to DNA replication and division. As a result of autoradiographic studies of chromosome reproduction, it became clear a number of years ago that not all chromosomes or sectors of the same chromosome have their DNA replicated over the relatively long time interval of six hours (see a review by Taylor, 1969). The long arm of the X-chromosome of the Chinese hamster, for example, completes replication in the last three hours of the S phase. Within this and other chromosomal segments are many units that replicate independently, some simultaneously and others out of phase. Precise programing of the various units may occur during the regular replication cycle, and a few sectors have been demonstrated to replicate repeatedly at a precise stage related to development in specialized cells. Examples are the ribosomal-RNA coding portions of the DNA in oocytes and a few regions of salivary-gland chromosomes of *Sciara, Rhynchosciara,* and perhaps other dipteran genera. In any case, there is enough information to indicate a highly evolved regulatory system for replication that is intimately involved in differentiation and various functional aspects of transcription.

Several studies have indicated that, during replication, new DNA chains in mammalian cells increase in length at the rate of 1 micron to 2 microns per minute. One of the most direct methods is the measurement of the increase in length of new polynucleotide chains in cells synchronized at the beginning of the S phase. The new chains can be detached from the template in alkaline sucrose gradients (Taylor, 1969b), and the lengths estimated from sedimentation rates. Elongation is more-or-less linear with time up to about 20 or 30 minutes. However, when cells are synchronized by blocking, for a few hours at the beginning of the S phase, with fluorodeoxyuridine (FUdR) and then released with thymidine, the initial rate of chain growth may be faster than usual. Our recent data indicate that rates can vary from 1 micron to 2 microns per minute. The faster rates may result from the simultaneous replication of several small subunits, which are later joined enzymatically. Huberman and Riggs (1968) reported autoradiographic studies on patterns of labeling which indicated that chain growth may proceed from a point and move in both directions. They exposed cells, partially synchronized by blocking with FUdR for 12 hours, to ³H-thymidine of very high

specific activity for 30 minutes. When cells were lysed and the DNA was extended on a membrane filter, an autoradiograph showed streaks of grains in tandem. Each group of grains presumably represented segments of DNA, which grew in the 30-minute interval. The lengths were consistent with growth rates of 0.5 micron to 2 microns per minute, but a more interesting observation was the labeling patterns of DNA from cells removed from the ³H-thymidine so that the specific activity of the DNA precursor dropped while the DNA chains continued to grow. The relatively dense rows of grains then showed a decreasing number of grains per unit length at both ends of many segments. From this observation they proposed that growth was taking place from both ends of the new chains. If they are correct, a measure of the rate of increase in length of new chains by sedimentation in a sucrose gradient would give a rate twice the true rate of chain growth.

The discovery that replication proceeds by the production of many short segments, the 7S subunits, could make our concepts of chain growth obsolete. A variety of evidence in both prokaryotes and eukaryotes, however, indicates sequential linear growth over many microns of DNA. These observations, in addition to our recent evidence that the 2-micron subunit may be a unit of replication, leads to the following proposed scheme of replication. Other circumstantial evidence, cited below, also fits into the proposed scheme.

Our present picture of replication is that initiation occurs by opening one strand of the template at the right or left terminus of an A-T segment. These regions are assumed to be self-priming, as shown in Figure 9, i.e., initiation does not require a small polynucleotide to fit on the template to initiate chain growth. Yet, we find small polynucleotides in pulse-labeled cells, as pointed out in the previous section. The evidence for this assumption of self-priming segments is not strong, but it allows us to explain partial replicas,

FIGURE 9 Diagram of a segment of DNA containing one 2-micron subunit and its two adjoining subunits. Replication has been initiated and is proceeding from left to right. Note that strand growth is always produced by addition of nucleotides at the 3′ OH end of chains as described in the text.

which are stable for hours in cells with perturbations in replication (see evidence presented in the next section). Furthermore, Bollum's studies (1967) of in vitro DNA replication indicate that denatured DNA is self-priming and forms new chains covalently linked to the template.

One plausible model is shown in Figures 8 and 9. A sequence of adenine-containing nucleotides to the right of an initiator segment could provide a region complementary to the terminal nucleotides at the end of each 2-micron subunit. A replication guide, pictured as a protein coupled to specific nucleotides beyond the terminus of the 2-micron subunit, might specify the locus at which a nuclease opens the chain and, at the same time, permanently blocks the movement of a growing point past its locus. The only alternative would be for the initiator end to bend back and allow its five terminal thymines and one cytosine to pair with the guanine and five adenines just to the right. This self-priming feature is assumed to be unique for the A-T segment. At the opposite end of each A-T segment is another initiator segment assumed to be capable of self-priming in a similar way. Operation of both sites simultaneously, however, would probably be excluded, because that would probably lead to the fragmentation of the DNA and chromosome breakage. Simultaneous attachment of a replicase and pairing as indicated above would initiate replication that would proceed to the end of the A-T segment. At that point, growth would stop unless an initiator-primer segment were available. In other words, segments 1, 2, 3, and 4 are not assumed to be self-primed. Presumably they are initiated in sequence, although such a requirement is perhaps not necessary. It is presumably these segments (7S subunits) that are displaced when pulse-labeled cells are lysed and yield the 5S to 7S single-stranded segments (Schandl and Taylor, 1969, 1970). The labeled initiator-primer segments are also detected in pulse-labeled cells, and separate as a distinct band near the top of sucrose gradients (~ 2S). The 2-micron subunits are assumed to be initiated in sequence, and growth probably takes place in both directions from a single 2-micron subunit.

Data, which support this model and indicate the importance of the 2-micron subunits in replication, are presented in Figure 10. It shows the profile in a preformed sucrose gradient at pH 9.5 of the particles of DNA labeled in a two-minute pulse with [3]H-thymidine in Chinese hamster cells synchronized at the beginning of the S phase. The DNA was heated to 65° C in the presence of 11 per cent formaldehyde at pH 9.0 to dissociate subunits held together by pairing of base sequences, which melted well below the average melting temperature of the chromosomal DNA. Near the top of the gradient are the small subunits typically found after a short pulse of [3]H-thymidine. The region of the gradient of interest here, however, is that with particles 18S to 20S, i.e., to the right of this band of smaller particles,

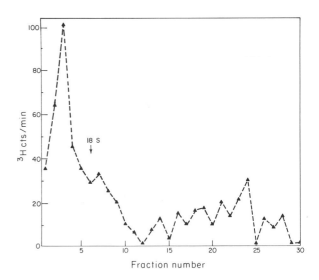

FIGURE 10 A sedimentation profile in a preformed sucrose gradient of [3]H-thymidine-labeled DNA isolated after a two-minute pulse of Chinese hamster cells in culture. The cells were lysed and the proteins digested as completely as possible with pronase. A dilute solution of DNA (1 µg/ml) was heated for 15 minutes to 65° C in 11 per cent formaldehyde dissolved in 0.10 M NaCl buffered at pH 9.0. The solution was rapidly cooled before layering on a preformed sucrose gradient.

7S to 9S. Examination of DNA from this region by means of electron microscopy reveals segments about 2 microns long with single strands attached at the ends of the double-stranded positions. We believe these are the subunits caught in the process of replication, but far enough advanced to have part of the new, growing chains attached to the template, as shown in the diagram in Figure 9.

*Partial replicas of the 2-micron subunits*

Although the 2-micron subunits are assumed to be initiated in sequence to produce the long chains of DNA detected in alkaline gradients, the evidence from incorporation of a density label into DNA indicates that interruptions in such sequential growth can occur. When cells are forced to substitute bromouracil for thymine during an interval by blocking the synthesis of thymidylate with FUdR and supplying bromodeoxyuridine, a density hybrid DNA appears, but various partially substituted satellite bands also can be demonstrated (Haut and Taylor, 1967; Taylor and Miner, 1968). Even when the lighter or heavier bands do not separate clearly from the main band of DNA in a CsCl gradient, their presence can be detected as a shoulder on the light or heavy side of the main band. In a typical experiment, a density hybrid DNA labeled with [14]C is produced by forcing the complete substitution of bromouracil for thymine. Figure 11 shows such a distribution of Chinese hamster DNA in a CsCl gradient produced by spinning a solution of DNA in the appropriate concentration of CsCl for 48 hours. Hybrid DNA and any other partially substituted DNA is labeled with ¹⁴C. In addition, a scan of the optical density shows the distribution of the total DNA in the gradient. A significant shoulder appears on each side of the main band of DNA. The component with the highest density contains a significant amount of ¹⁴C-bromouracil, and there is a detectable amount of bromouracil in the light fraction. Previous studies have revealed that when the high-density satellite is rebanded it again occupies the same density position in the gradient (Miner, 1967). The light component is more difficult to follow, and its rebanding properties and composition are still in doubt. Shearing the heavy satellite breaks out segments which appear to be fully substituted with bromouracil as shown in Figure 12 (Miner, 1967; Taylor and Miner, 1968). The original distribution of DNA (Figure 12A) shows clearly a labeled heavy satellite on the left of the main band as well as a very slightly labeled light satellite on the right. The hybrid density band is shown distinctly at the far left. Rebanding one-half of the unsheared DNA in fractions 40 to 42 (Figure 12A) indicates that the satellite is a distinct fraction that maintains its density upon

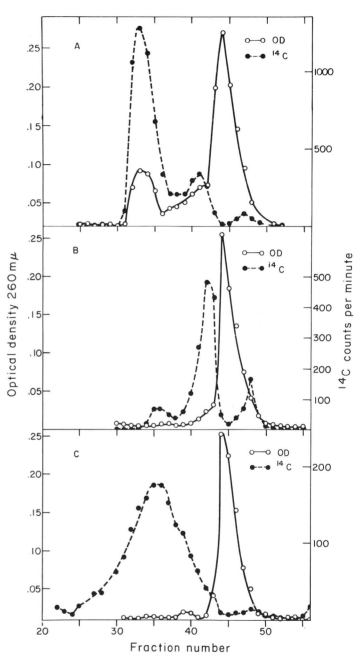

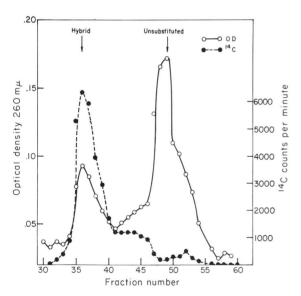

FIGURE 11  Profile of DNA in a CsCl gradient produced by spinning DNA in a Spinco SW 39 rotor for 48 hours. The DNA was extracted from hamster cells after they grew for four hours in a medium containing ¹⁴C-bromodeoxyuridine which substituted for thymidine. Thymidylate synthesis was blocked by fluoro-deoxyuridine ($10^{-6}$ M). The position of the bands of DNA is indicated by the variation in optical density (open circles), and the density hybrid is also indicated by the radioactivity (closed circles). Note the atypical shoulders on each side of the main band of DNA which in less steep gradients are resolved into satellite bands. (From Miner, 1967.)

FIGURE 12  Profiles of hamster DNA in CsCl gradients after four hours of replication in which substitution of bromodeoxyuridine for thymidine occurred after a period of one hour in which replication was blocked with FUdR. A. The original profile of the chromosomal DNA, showing the density hybrid as a heavy satellite to the left of the main band. B. Rebanding of fractions 40–42 with unlabeled marker DNA without shearing. C. Rebanding of a portion of the same fractions sheared to a molecular weight of about $10^6$ daltons along with unlabeled DNA which was not sheared. (From Miner, 1967.)

rebanding (Figure 12B). After shearing, however, this same DNA can be demonstrated to contain short segments with enough bromouracil to make the average density nearly as high as the regular density hybrid (Figure 12C).

We originally interpreted the satellite as an indication of interrupted growth because of these properties and because such a satellite does not occur during replication of DNA with thymidine instead of bromodeoxyuridine (Taylor and Miner, 1968). Now we can think of the interruption in terms of the new model of replication (Figures 8 and 9). If one strand of each A-T segment in the DNA were predominantly adenine and the other segment predominantly thymine, the peculiar satellites could be explained. Let us assume that the segment with the high adenine content alternates in a regular sequence with the high-thymine segment, as shown in Figure 13A. If initiation is restricted to one strand of each A-T segment and occurs at opposite ends of each 2-micron subunit, a sequence of 2-micron subunits will have either the A strands or the T strands replicated, but not both. Those regions with replicated T strands would be heavily substituted with bromouracil in short segments and would band as a heavy satellite. The regions with replicated A strands would be light and would contain a small amount of ¹⁴C-bromouracil either if random breaks isolated single T segments with a group of adenine-containing regions, or

if the high-A strands contained a small amount of bromouracil, which would substitute for a naturally occurring thymine in the predominantly A segment. The assumption is made that the partially replicated segments form a rather stable three-stranded region (Syzbalski, 1969), which behaves in a way similar to native DNA in a solution of CsCl.

If the hypothesis is correct, the light satellite should disappear after replication is complete, but the heavy satellite should persist because of its bromouracil content. Figure 14 presents an example that shows the survival of the heavy satellite through a division stage before which all DNA may be assumed to have completed replication. The DNA was labeled with ¹⁴C-thymidine over a long period to get essentially uniform label in chromosomal DNA. Then a portion of the replicating DNA was labeled with nonradioactive bromouracil by growing cells in the corresponding nucleoside for three hours along with FUdR to block thymidylate synthesis. By removal of FUdR, cells were then allowed to

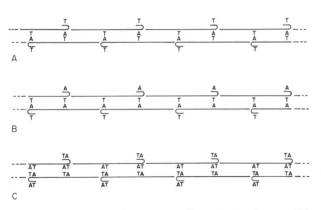

FIGURE 13 Diagram of segments of DNA showing possible initiation of replication at the 0 segments with various arrangements of these high A-T segments. A. Alternating adenine- and thymine-rich segments along a polynucleotide chain with all partial replicas composed of thymine-rich segments. B. All the thymine segments in one chain but with alternate initiation of thymine- or adenine-rich segments. C. Mixed A-T polymer in each segment along the polynucleotide chains. Note that if alternate sites along a chain are initiated, the structure will remain intact. A, B, and C show various extreme arrangements of nucleotides. Probably no DNA reaches any of these extreme compositions, but, if species variability in these segments is high, all three situations might be approached by some organisms.

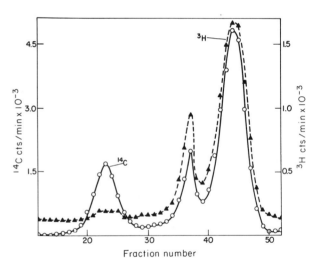

FIGURE 14 Profile of DNA in a CsCl gradient which was labeled rather uniformly with ¹⁴C-thymidine and then, during the late part of a subsequent S phase, was forced to incorporate the density label, bromodeoxyuridine. Cells were collected at the following metaphase, and further synchrony in entering the next S phase was induced by FUdR given during the last four hours of $G_1$. Release of the cells into the S phase with ³H-thymidine allowed the new DNA to be seen in the gradient. The point of interest here is the observation that during bromodeoxyuridine incorporation a density hybrid was formed, as well as a heavy satellite band, which shows a rather narrow distribution in the gradient in spite of the fact that random shearing of the DNA reduced the particle weight to 5–10 × 10⁶ daltons before banding. In spite of considerable variation in length of DNA particles, the ratio of bromouracil to thymine in the partial replicas is remarkably uniform.

recover the ability to make their own thymidylate. Cells labeled in this way were collected by shaking them off the glass surface at metaphase; they then were allowed to proceed synchronously to the S phase. Synchrony was increased by treatment with FUdR during the last four hours of the G₁ phase. Note that the heavy satellite persists and replicates without changing its density. The density of this heavy satellite is about one-third of that of the density hybrid, which is fully substituted in one strand. If the T segments alternated with A segments, as shown in Figure 13A, they could be fully substituted and, after replication, would represent one-third of the thymine in one polynucleotide chain of each DNA duplex. This would be consistent with the density position of the satellite band in Figure 14.

The density of partially substituted DNAs before replication is completed would be difficult to predict because not only is the pairing of the third strand uncertain; the proportion of the partially replicated sites might vary. The observation is that satellites of variable densities are found. Very short periods of substitution produce only a small change in density—about 15 per cent (Taylor and Miner, 1968)—but, after longer periods of substitution, the density of the satellite increases to one-third or one-half of that of hybrid DNA and finally merges with the hybrid band when bromodeoxyuridine remains available to the cell to complete replication. Figures 15 to 17 show three stages in the replication of DNA with bromouracil substituted for thymidine. Cells were prelabeled with ¹⁴C- and then ³H-thymidine for part of one S phase. They were then synchronized by shaking off metaphases and treating them with FUdR during the last four hours of G₁. Release of the cells into S phase with bromodeoxyuridine allowed replication. Some partial replicas, however, are indicated by the heavy satellite that appeared after one hour (Figure 15). Its density indicates substitution to the equivalent of about one-third or less of the thymine in one strand. After three hours (Figure 16), the average density indicates the equivalent of one-half substitution in one strand. By five hours, all the partially replicated DNA had become hybrid, with one strand fully substituted (Figure 17).

The partial replicas are consistent with our model of replication, and their presence is explained by the model, but of course such data can never prove the correctness of the model. Other more direct proof and evidence for or against the model are being assembled and, we hope, can be presented in a few months.

The partial replicas almost certainly exist, even if the model is incorrect. Their existence under these rather unusual conditions of replication might be less important if they did not occur under conditions that the cell probably meets during its normal replication and growth. Some evidence suggests that such partial replicas can be induced by per-

turbing DNA replication in other ways. If these partial replicas can be shown to be induced in cells at certain stages or to happen regularly during replication, they certainly provide the raw materials for various types of recombination between chromosomes. Some of the recombination in phages appears to be associated with partial replicas of the whole genome. As was pointed out previously, such partial replicas of DNA subunits in the chromosome provide a

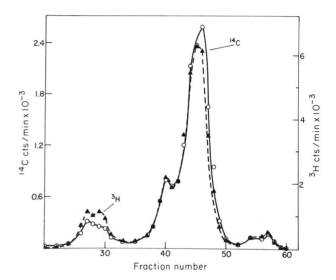

FIGURE 15  Profile of Chinese hamster DNA in a CsCl gradient with a uniform label of ¹⁴C-thymidine and a part of an S phase in ³H-thymidine followed by synchronization and incorporation of the density label, bromodeoxyuridine, for one hour. Notice that the heavy satellite constitutes a significant proportion of the total DNA.

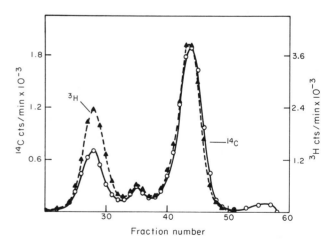

FIGURE 16  Profile of DNA in a CsCl gradient similar to that in Figure 15 except that the incorporation of bromodeoxyuridine continued for three hours. Note that the heavy satellite is now of a higher density, nearly one-half substituted compared with the hybrid band.

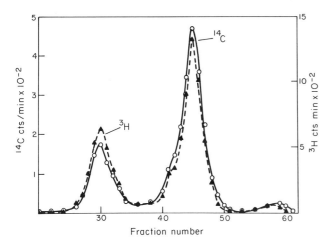

FIGURE 17 Profile of another sample of DNA similar to that in Figure 15, except that incorporation of bromodeoxyuridine had continued for five hours, and all the DNA has replicated that would do so under these conditions of growth. Note that the heavy satellite has disappeared, presumably because it had finally become fully substituted hybrid DNA.

means for nonreciprocal recombination (Taylor, 1967) as well as associated events that could lead to reciprocal recombination without loss of viability of either of the four chromatids of the two chromosomes involved.

Although the regulation of replication and transcription is not evident either from the proposed model of DNA or from its mechanism of replication, several possibilities are presented. One level might be the role of some nuclease that opens the 2-micron subunit by nicking one strand. Another is possibly the regulation of initiator-primer production. If the hypothesis presented is correct, replication of DNA follows a scheme similar to that indicated for RNA, *coli* phage $\phi$X DNA, and other systems in which a three-stranded intermediate is believed to exist during replication. A difference between the RNA and DNA systems would be the requirement for a primer (short-chain initiators) in DNA replication, but this could have evolved as a mechanism that preserves a necessary structural irregularity in DNA that could not be copied by the regular base-pairing mechanism. We have suggested mismatched base pairs as the structural basis for the flexible joints, but this is only one possibility. The basis for flexible joints can be decided only when the base composition and sequence in the initiators are known and their role in replication is confirmed.

## The role of proteins in chromosomal organization and regulation

The histones form complexes with DNA in vitro and presumably have some resemblance to the complexes that exist in the intact cell. Bonner et al. (1968) and Georgiev (1968) have shown that histone I is removed completely by 0.6 M NaCl, but the other known classes of histones are not removed so readily. However, higher salt concentrations completely remove histones from DNA and there are probably no covalent linkages between the two molecular species. Bonner's group (1968) has identified a small RNA molecule that is believed to be linked to a nonhistone protein. The histones are thought to form a complex with this protein. Their suggestion is that the RNA finds a complementary site on the DNA and thereby guides the positioning of the histones on the DNA. It is interesting that our presumed initiator DNA is about the same size as this RNA.

Histone I is easily removed by low salt concentrations, and also has some interesting properties with respect to maintenance of the structure of the 230-Å fibril and the regulation of RNA transcription. According to Bonner et al. (1968), the histones will reassociate with the DNA in what they believe to be the native nucleohistone state. This blocks the transcription of RNA. Furthermore, removal of the histone allows synthesis of RNAs that have a high affinity for the DNA in molecular hybridization tests. Georgiev (1968) reported, however, that the removal of only the histone I from DNA gives maximum synthesis of this type of RNA. Replacement of histone I prevents this synthesis by RNA polymerase, although Georgiev does not agree that it restores nucleoprotein to its original native state. He also pointed out that removal of histone I causes the nucleoprotein fibrils to unfold and be reduced in diameter to about 40 Å.

These findings lead us to suggest that histone I is complexed to the A-T segments in our model of the chromosomal DNA. It is primarily the copying of these segments by RNA polymerase that yields the RNA with a high affinity for DNA, as suggested by Georgiev. Such RNA should have a high uracil content similar to or higher than certain nuclear RNAs. That is not to say that other segments would not be copied to some extent, but the RNA from other segments should have a low affinity in the hybridization tests. Flamm et al. (1969) reported that messenger RNA (cytoplasmic RNAs, other than ribosomal, that have high molecular weight) do not bind to the light-satellite DNA of the mouse, which has a high content of A-T (66–67 per cent). We propose that the light-satellite DNA in most, if not all, cases contains two 0.4-micron segments of A-T DNA for each four heterogeneous segments in contrast to main-band DNA, which, we think, contains one A-T segment per four segments of heterogeneous base composition (nearly equimolar ratios of A + T and G + C). The base composition of mouse satellite reported by Flamm et al. (1969) is consistent with this hypothesis, for the percentage of G-C in the satellite is 33.3 per cent rather than the 40 per cent characteristic of main-band DNA.

The addition of these extra segments could be by intercalation or, possibly, by addition of segments as shown in partially replicated DNA (Figure 10). Such segments of DNA either would not be transcribed as a part of messenger RNA or, if transcribed, would be split off before delivery of this RNA to the cytoplasm. On the other hand, the heterogeneous nuclear RNA (Darnell, 1968) would contain copies of these repeating high A-T segments. This would explain its high affinity for DNA in molecular hybridization tests and its high content of uracil.

## Recombination and correction mechanisms

A mechanism of recombination based on the pairing of incompletely replicated DNA subunits in chromosomes has been proposed by Taylor (1967). At that time no unit of the size proposed (about 2 microns) was known. Now, with the discovery of a unit of replication of an appropriate size and the usefulness of the model to explain nonreciprocal recombination in ascomycetes (Herbert Gutes, personal communication), the model promises to become a reality instead of a speculative scheme. The block in synthesis could occur after the synthesis of the A-T segments or some other portion of 2-micron units. Less frequently, a three-stranded region would be formed by the growth of one strand of a 2-micron subunit, but the other strand would be blocked after the formation of the A-T segment only. The three-stranded segments in two homologous chromosomes would be the basis for nonreciprocal recombination in which segregation is in the ratio of 3:1 or 6:2. The A-T segments, which should be the most common type of interrupted chain growth, could also provide a mechanism for specific pairing, if the assumption is made that base sequence of these subunits varies along the chromosome in rather subtle ways. Their similarity could provide a mechanism of nonhomologous and homologous pairing when the exact partner was absent, as in haploids, some hybrids, and allopolyploids. Considerably more information on the similarity of the A-T segments within and between phylogenetically related genomes is necessary before valid predictions can be made.

The work of Waring and Britten (1966) and Flamm et al. (1969) indicates that many similar sequences occur in each complement, but the degree of similarity is not clear. Are they exact copies and, if so, what mechanism keeps them similar or identical within a species? The persistence of multiple copies of any locus implies some kind of correction mechanism. We have few clues as to how this correction operates. An internal recombination system that continually produces exchanges of A-T segments between different parts of the genome does not appear very attractive. A periodic replacement of the segments from some limited number of copies at certain stages of the life cycle, at meiosis, or in early embryonic development is an interesting possibility. Another plausible, but unorthodox, possibility is the production of RNA by transcription of a master DNA copy. By excision and replacement of mismatched base pairs similar to DNA repair replication, DNA chains, rather than RNA chains, might be altered selectively. Such a mechanism might provide a role for the heterogeneous nuclear RNA that appears to have such a high turnover and never to leave the nucleus, but any precise scheme is too speculative now to merit recording. Nevertheless, the role of RNA in transformation or modification of DNA in higher cells should not be dismissed too lightly. Admittedly, the evidence so far presented is only suggestive, but better-designed experiments are called for. A case in point is the report of Temin (1964) that the nucleotide sequences of viral RNA in Rous sarcoma are found in the DNA of transformed cells. If the transfer can be demonstrated in one clear case, the way would be open for an understanding of many puzzling phenomena relating to long-lasting immunity and, perhaps, to long-term memory mechanisms.

## Transcription and differentiation of neuronal nets

One of the most interesting and provocative reports on nucleic acids, and one that may have significance for memory mechanisms, is that which indicates that some cell membranes may contain RNA and, furthermore, that this RNA may extend out and make contact with the environment of the cell and its neighbors. Mayhew and Weiss (1968) have reported that cells have negative surface charges that affect their electrophoretic mobilities. The charge is changed when the cells are placed in ribonuclease; this suggests that RNA which extends out from the membrane is removed. Furthermore, cells in culture that are attached to glass or plastic surfaces leave radioactive "footprints" if the RNA is labeled with $^3$H-uridine (Weiss and Mayhew, 1966). Although these bits of evidence cannot be taken as conclusive, the possibility that RNA is found in plasma membranes raises some interesting questions. Could such RNA be the basis for cell-to-cell contact specificity in differentiation? We need only imagine that cells of a particular type produce copies of an RNA, which becomes attached to a structural component of the membrane so that chains extend out and form base pairs with complementary chains from another cell that is an acceptable associate. Cells without complementary sequences would not be accepted. One mechanism to insure complementarity would be a cell programed to transcribe the plus strand at a DNA locus while its sister cell would be programed to transcribe the minus strand at the same locus. If there were several independent loci, a high degree of specificity could be attained by the appropriate regulation of such a genetic system.

If specific RNAs at surfaces were a property of cells in which a mechanism of differentiation has evolved, it would be easy to see how the system could be specialized for neuronal recognition. The puzzling recognition system that optic nerve axons have for their specific neurons in the optic tectum of the frog (Sperry, 1951) could have such a basis. The specificity of such a recognition system could be subtle and precise, and the variability could be sufficient for an almost infinite number of pair recognitions. The proper regulation in such a system could give a large population of interacting cells or only pair mates. If the locus held in common were fixed at the division from which two daughters were produced, these would be the only two to interact.

Such mechanisms could be extended to explain the inheritance of neural nets that predetermine certain behavior patterns or instinctual behavior in animals. The mechanism would require precise regulation associated with segregation of the genome during critical cell divisions in which pairs or more complex associations of sensory and motor neurons were being produced. Speculation on precise regulatory mechanisms that could achieve such patterns, however, would not appear to be fruitful without some evidence that properties of RNA in cell surfaces and in junctions of the type involved actually indicate determination of specificity. The challenging problem is to devise experiments that could yield further clues concerning the existence of this type of RNA.

If long-term memory involves the change in any molecule, and one must assume it does, the record of the change would almost have to be recorded in a nucleic acid, and chromosomal DNA is probably the best candidate. The easiest change to imagine would be one that alters an RNA molecule, not in such a way as to code a message or bit of information directly, but in such a way as to alter the future behavior of the cell, perhaps in a way that would change the spike-train pattern in a subtle way or alter its responsiveness to an impulse of a particular type, frequency, or magnitude. We believe the most plausible feedback to the chromosomal DNA would be through an RNA that could transform a specific locus which originally produced it. This would have to be done in such a way that molecules of RNA identical to the one effecting transformation would be transcribed from the locus thereafter. These statements are made with the full knowledge that molecular biologists have not discovered such mechanisms and are not likely to until higher cells can be studied by rigorous genetic and chemical techniques that have been successful in demonstrating DNA transformation in bacteria with DNAs of viral and bacterial origin.

## Acknowledgment

We wish to acknowledge the contribution of unpublished data obtained in our laboratory by Dr. Philip Miner (Figures 10 and 11) and by Dr. Terry L. Myers (Figures 13–16), and to thank Mrs. Nancy Straubing for her technical assistance. The experimental work was supported by Contract AT(401–1)2690 with the Division of Biology and Medicine, U. S. Atomic Energy Commission, and by National Science Foundation Grant GB-6568.

### REFERENCES

BOLLUM, F. J., 1967. Enzymatic replication of polydeoxynucleotides. *In* Genetic Elements: Properties and Function (D. Shugar, editor). Academic Press, New York, pp. 3–15.

BONNER, J., M. E. DAHMUS, D. FAMBROUGH, R. C. HUANG, K. MARUSHIGE, and D. Y. H. TUAN, 1968. The biology of isolated chromatin. *Science* (*Washington*) 159: 47–56.

BRITTEN, R. J., and D. E. KOHNE, 1968. Repeated sequences in DNA. *Science* (*Washington*) 161: 529–540.

CAIRNS, J., 1966. Autoradiography of HeLa cell DNA. *J. Mol. Biol.* 15: 372–373.

CALLAN, H. G., 1967. The organization of genetic units in chromosomes. *J. Cell Sci.* 2: 1–7.

CALLAN, H. G., and L. LLOYD, 1960. Lampbrush chromosomes of crested newts *Trituris cristatus Laurenti. Phil. Trans. Roy. Soc.* (*London,*) *ser. B, biol. sci.,* 243: 135–219.

DARNELL, J. E., 1968. Ribonucleic acids from animal cells. *Bacteriol. Rev.* 32: 262–290.

DAVIDSON, N., J. WIDHOLM, U. S. NANDI, R. JENSEN, B. M. OLIVERA, and J. C. WANG, 1965. Preparation and properties of native crab dAT. *Proc. Nat. Acad. Sci. U. S. A.* 53: 111–118.

DUPRAW, E. J., 1968. Cell and Molecular Biology. Academic Press, New York, p. 533.

FLAMM, W. G., P. M. B. WALKER, and M. McCALLUM, 1969. Some properties of the single strands isolated from the DNA of the nuclear satellite of the mouse (*Mus musculus*). *J. Mol. Biol.* 40: 423–443.

GEORGIEV, G. P., 1968. The regulation of the biosynthesis and transport of messenger RNA in animal cells. *In* Regulatory Mechanisms for Protein Synthesis in Mammalian Cells (A. San Pietro, M. R. Lamborg, and F. T. Kenney, editors). Academic Press, New York, pp. 25–48.

GIANNONI, G., F. J. PADDEN, JR., and H. D. KEITH, 1969. Crystallization of DNA from dilute solution. *Proc. Nat. Acad. Sci. U. S. A.* 62: 964–971.

GRAY, H. B., JR., and J. E. HEARST, 1968. Flexibility of native DNA from the sedimentation behavior as a function of molecular weight and temperature. *J. Mol. Biol.* 35: 111–129.

HAUT, W. F., and J. H. TAYLOR, 1967. Studies of bromouracil deoxyriboside substitution in DNA of bean roots (*Vicia faba*). *J. Mol. Biol.* 26: 389–401.

HOHN, TH., and H. SCHALLER, 1967. Separation and chain-length determination of polynucleotides by gel filtration. *Biochim. Biophys. Acta* 138: 466–473.

HUBERMAN, J. A., and A. D. RIGGS, 1966. Autoradiography of chromosomal DNA fibers from Chinese hamster cells. *Proc. Nat. Acad. Sci. U. S. A.* 55: 599–606.

HUBERMAN, J. A., and A. D. RIGGS, 1968. On the mechanism of DNA replication in mammalian chromosomes. *J. Mol. Biol.* 32: 327–341.

KLEINSCHMIDT, A., 1967. Structural aspects of the genetic apparatus of viruses and cells. *In* Molecular Genetics, Part II (J. H. Taylor, editor). Academic Press, New York, pp. 47–93.

LAJTHA, L. G., 1963. On the concept of the cell cycle. *J. Cell. Comp. Physiol.* 62 (suppl. 1): 143–145.

MAIO, J. J., and C. L. SCHILDKRAUT, 1969. Isolated mammalian metaphase chromosomes. II. Fractionated chromosomes of mouse and Chinese hamster cells. *J. Mol. Biol.* 40: 203–216.

MAYHEW, E., and L. WEISS, 1968. Ribonucleic acid at the periphery of different cell types, and effect of growth rate on ionogenic groups in the periphery of cultured cells. *Exp. Cell Res.* 50: 441–453.

MEGO, W., and J. H. TAYLOR, 1969. Alterations in the pattern of single-strand scissions in the DNA of stationary cell cultures. *Biophys. J.* 9: A-125 (abstract).

MINER, P., 1967. Replication of DNA in cultured Chinese hamster cells. Columbia University, New York, doctoral thesis.

OKAZAKI, R., T. OKAZAKI, K. SAKABE, and K. SUGIMOTO, 1967. Mechanism of DNA replication: Possible discontinuity of DNA chain growth. *Jap. J. Med. Sci. Biol.* 20: 255–260.

SCHANDL, E. K., and J. H. TAYLOR, 1969. Early events in the repli-cation and integration of DNA into mammalian chromosomes. *Biochem. Biophys. Res. Commun.* 34: 291–300.

SCHANDL, E. K., and J. H. TAYLOR, 1970. A presumptive *in vivo* primer for DNA replication in pulse labeled mammalian cells. *J. Biol. Chem.* (in press).

SPERRY, R. W., 1951. Regulative factors in the orderly growth of neural circuits. *Growth (Suppl.)* 15: 63–87.

SZYBALSKI, W., 1969. Initiation and patterns of transcription during phage development. *Proc. Can. Cancer Res. Conf.* 8: 183–215.

TAYLOR, J. H., 1967. Patterns and mechanisms of genetic recombination. *In* Molecular Genetics, Part II (J. H. Taylor, editor). Academic Press, New York, pp. 95–135.

TAYLOR, J. H., 1969a. The structure and duplication of chromosomes. *In* Genetic Organization (E. W. Casperi and A. Raven, editors). Academic Press, New York, pp. 165–221.

TAYLOR, J. H., 1969b. Replication and organization of chromosomes. *In* 12th International Congress of Genetics, Tokyo, 1968, Vol. 3. Dai Nippon Printing Co., Tokyo, pp. 177–189.

TAYLOR, J. H., and P. MINER, 1968. Units of DNA replication in mammalian chromosomes. *Cancer Res.* 28: 1810–1814.

TEMIN, H. M., 1964. Homology between RNA from Rous sarcoma virus and DNA from Rous sarcoma virus-infected cells. *Proc. Nat. Acad. Sci. U. S. A.* 52: 323–329.

WARING, M., and R. J. BRITTEN, 1966. Nucleotide sequence repetition: a rapidly reassociating fraction of mouse DNA. *Science (Washington)* 154: 791–794.

WEISS, L., and E. MAYHEW, 1966. The presence of ribonucleic acid within the peripheral zones of two types of mammalian cell. *J. Cell. Physiol.* 68: 345–360.

# ASSOCIATES, NEUROSCIENCES RESEARCH PROGRAM

ADEY, W. ROSS  Professor of Anatomy and Physiology; Director, Space Biology Laboratory, Brain Research Institute, University of California at Los Angeles

BODIAN, DAVID  Professor and Director, Department of Anatomy, The Johns Hopkins University School of Medicine, Baltimore, Maryland

BULLOCK, THEODORE H.  Professor of Neurosciences, University of California at San Diego School of Medicine, La Jolla, California

CALVIN, MELVIN  Professor of Chemistry, University of California at Berkeley

DE MAEYER, LEO C. M.  Professor, Max Planck Institute for Physical Chemistry, Göttingen, Germany

EDDS, MAC V., JR.  Professor of Biology, Division of Biological and Medical Sciences, Brown University, Providence, Rhode Island

EDELMAN, GERALD M.  Professor, The Rockefeller University, New York, New York

EIGEN, MANFRED  Director, Max Planck Institute for Physical Chemistry, Göttingen, Germany

FERNÉNDEZ-MORÁN, HUMBERTO  A. N. Pritzker Professor of Biophysics, Department of Biophysics, University of Chicago School of Medicine, Chicago, Illinois

GALAMBOS, ROBERT  Professor of Neurosciences, University of California at San Diego School of Medicine, La Jolla, California

GOODENOUGH, JOHN B.  Research Physicist and Group Leader, Lincoln Laboratory, Massachusetts Institute of Technology, Cambridge, Massachusetts

HYDÉN, HOLGER V.  Director, Institute of Neurobiology, University of Göteborg, Faculty of Medicine, Göteborg, Sweden

KAC, MARK  Professor, The Rockefeller University, New York, New York

KATCHALSKY, AHARON  Professor and Head, Polymer Department, Weizmann Institute of Science, Rehovoth, Israel

KETY, SEYMOUR S.  Chief, Psychiatric Research Laboratories, Massachusetts General Hospital; Professor of Psychiatry, Harvard Medical School, Boston, Massachusetts

KLÜVER, HEINRICH  Sewell L. Avery Distinguished Service Professor Emeritus of Biological Psychology, The University of Chicago, Chicago, Illinois

LEHNINGER, ALBERT L.  Professor and Director, Department of Physiological Chemistry, The Johns Hopkins University School of Medicine, Baltimore, Maryland

LIVINGSTON, ROBERT B.  Professor and Chairman, Department of Neurosciences, University of California at San Diego School of Medicine, La Jolla, California

LONGUET-HIGGINS, H. CHRISTOPHER  Royal Society Research Professor, Department of Machine Intelligence and Perception, University of Edinburgh, Edinburgh, Scotland

MACKAY, DONALD M.  Professor of Communication, The University of Keele, Keele, Staffordshire, England

McCONNELL, HARDEN M.  Professor of Chemistry, Stanford University, Stanford, California

MILLER, NEAL E.  Professor. The Rockefeller University, New York, New York

MORRELL, FRANK  Visiting Professor of Psychiatry, New York Medical College, New York, New York

MOUNTCASTLE, VERNON B.  Professor and Director, Department of Physiology, The Johns Hopkins University School of Medicine, Baltimore, Maryland

NAUTA, WALLE J. H.  Professor of Neuroanatomy, Department of Psychology, Massachusetts Institute of Technology, Cambridge, Massachusetts

NIRENBERG, MARSHALL  Chief, Laboratory of Biochemical Genetics, National Heart Institute, National Institutes of Health, Bethesda, Maryland

ONSAGER, LARS J.  Willard Gibbs Professor of Theoretical Chemistry, Sterling Chemistry Laboratory, Yale University, New Haven, Connecticut

PLOOG, DETLEV,  Clinical Institute, Max Planck Institute for Psychiatry, Munich, Germany

QUARTON, GARDNER C.  Director, Mental Health Research Institute, Professor of Psychiatry, University of Michigan, Ann Arbor, Michigan

REICHARDT, WERNER E.  Director, Max Planck Institute for Biological Cybernetics, Tübingen, Germany

ROBERTS, RICHARD B.  Chairman, Biophysics Section, Department of Terrestrial Magnetism, Carnegie Institution of Washington, Washington, D.C.

SCHMITT, FRANCIS O.  Chairman, Neurosciences Research Program, Massachusetts Institute of Technology, Cambridge, Massachusetts

SWEET, WILLIAM H.  Chief, Neurosurgical Service, Massachusetts General Hospital, Professor of Surgery, Harvard Medical School, Boston, Massachusetts

WEISS, PAUL A.  Professor Emeritus, The Rockefeller University, New York, New York

---

Leroy G. Augenstein served as an NRP Associate until his death, November 8, 1969.

# PARTICIPANTS IN ISP 1969

ABELES, MOSHE Lecturer in Physiology, Department of Physiology, Hebrew University, Hadassah Medical School, Jerusalem, Israel

ADELMAN, GEORGE Librarian and Managing Editor, Neurosciences Research Program, Massachusetts Institute of Technology, Brookline, Massachusetts 02146

ADEY, W. ROSS Professor of Anatomy and Physiology, Department of Anatomy, University of California, Center for the Health Sciences, Los Angeles, California 90024

ANGEVINE, JAY B., JR. Associate Professor of Anatomy, Department of Anatomy, University of Arizona College of Medicine, Tucson, Arizona 85721

BAKER, MARY ANN Assistant Professor, University of Southern California Medical School, Los Angeles, California 90024

BALDESSARINI, ROSS J. Department of Psychology, Massachusetts General Hospital, Boston, Massachusetts 02114

BARONDES, SAMUEL H. Department of Psychiatry, University of California at San Diego School of Medicine, La Jolla, California 92037

BAUMGARTEN, RUDOLF J. VON Professor of Physiology, Mental Health Research Institute, University of Michigan, Ann Arbor, Michigan 48104

BERGMANN, FRED H. Health Scientist Administrator, National Institute of General Medical Sciences, National Institutes of Health, Bethesda, Maryland 20014

BIGELOW, JULIAN H. The Institute for Advance Study, Princeton, New Jersey 97549

BISHOP, PETER O. Professor of Physiology, John Curtin School of Medical Research, Australian National University, Canberra, A.C.T., Australia

BITTNER, GEORGE D. Department of Zoology, University of Texas, Austin, Texas 78712

BLACK, IRA B. Research Associate, Section on Pharmacology, Laboratory of Clinical Science, National Institute of Mental Health, National Institutes of Health, Bethesda, Maryland 20014

BLOOM, FLOYD E. Acting Chief, Laboratory of Neuropharmacology, St. Elizabeth's Hospital, National Institute of Mental Health, National Institutes of Health, Bethesda, Maryland 20014

BODIAN, DAVID Professor and Director, Department of Anatomy, The Johns Hopkins University School of Medicine, Baltimore, Maryland 21205

BULLOCK, THEODORE H. Professor of Neurosciences, University of California at San Diego School of Medicine, La Jolla, California 92037

BUNGE, RICHARD P. Department of Anatomy, Washington University School of Medicine, St. Louis, Missouri 63110

CALVIN, MELVIN Director, Laboratory of Chemical Biodynamics; Professor of Chemistry, Professor of Molecular Biology, Department of Chemistry, University of California, Berkeley, California 94720

COHEN, HIRSH Research Staff Member, IBM, Thomas J. Watson Research Center, Yorktown Heights, New York 10598

COHEN, MELVIN J. Department of Biology, Kline Tower, Yale University, New Haven, Connecticut 06520

COULOMBRE, ALFRED J. Head, Section on Experimental Embryology, Ophthalmology Branch, National Institute of Neurological Diseases and Stroke, National Institutes of Health, Bethesda, Maryland 20014

CRAMER, HINRICH Laboratory of Chemical Pharmacology, Building 10, Room 7N-108, National Heart and Lung Institute, National Institutes of Health, Bethesda, Maryland 20014

CREUTZFELDT, OTTO D. Head, Department of Neurophysiology, Max Planck Institute for Psychiatry, Munich, Germany

CROWELL, ROBERT M. Research Associate, National Institute of Neurological Diseases and Stroke, Building 10, Room 4N-262, National Institutes of Health, Bethesda, Maryland 20014

CUNNINGHAM, BRUCE A. Assistant Professor of Biochemistry, The Rockefeller University, New York, New York 10021

DAVIS, BERNARD D. Professor of Bacterial Physiology, Department of Bacteriology and Immunology, Harvard Medical School, Boston, Massachusetts 02115

DAVISON, PETER F. Director, Department of Fine Structure Research, Boston Biomedical Research Institute, Boston, Massachusetts 02114

DELBRÜCK, MAX Professor of Biology, Division of Biology, California Institute of Technology, Pasadena, California 91104

DELIUS, JUAN D. Department of Psychology, Durham University, Durham, England

DE LUCA, ALDO Research Associate, Italian National Council of Research, Laboratory of Cybernetics, Naples, Italy

DEMAEYER, LEO C. M. Scientific Fellow, Max Planck Institute for Physical Chemistry, Göttingen, Germany

DESCARRIES, LAURENT Department of Physiology, University of Montreal Faculty of Medicine, Montreal, Quebec, Canada

DICARA, LEO VINCENT Associate Professor, The Rockefeller University, New York, New York 10021

EDDS, MAC V., JR. Professor of Biology, Division of Biological and Medical Sciences, Brown University, Providence, Rhode Island 02912

EDELMAN, GERALD M. Professor, The Rockefeller University, New York, New York 10021

EGGER, M. DAVID Department of Anatomy, University College London, Gower Street, London W.C. 1, England

EIGEN, MANFRED Director, Max Planck Institute for Physical Chemistry, Göttingen, Germany

ERVIN, FRANK R. Director, Stanley Cobb Laboratories for Psychiatric Research, Massachusetts General Hospital, Boston, Massachusetts 02114

EVANS, EDWARD FRANK Senior Research Fellow, Medical Research Council Group, Department of Communication, University of Keele, Keele, Staffordshire, England

EVANS, JOHN WYNNE Assistant Professor of Mathematics, University of California at San Diego, La Jolla, California 92037

FAMIGLIETTI, EDWARD V. Department of Anatomy, Boston University Medical School, Boston, Massachusetts 02118

FERNÁNDEZ-MORÁN, HUMBERTO A.N. Pritzker Professor of Biophysics, Department of Biophysics, University of Chicago, Chicago, Illinois 60637

FIELDS, HOWARD Research Neurologist, Department of Neurophysiology, Walter Reed Army Institute of Research, Walter Reed Army Medical Center, Washington, D.C. 20012

FOX, STEPHEN S. Associate Professor, Department of Psychology, University of Iowa, Iowa City, Iowa 52240

GALAMBOS, ROBERT Professor of Neurosciences, University of California at San Diego School of Medicine, La Jolla, California 92037

GERSTEIN, GEORGE L. Associate Professor of Biophysics and Physiology, Department of Biophysics and Physical Chemistry, University of Pennsylvania School of Medicine, Philadelphia, Pennsylvania 19104

GILBERT, WALTER Professor of Molecular Biology, Biological Laboratories, Harvard University, Cambridge, Massachusetts 02138

GOODENOUGH, JOHN B. Research Physicist and Group Leader, Lincoln Laboratory, Massachusetts Institute of Technology, Lexington, Massachusetts 02173

GORSKI, ROGER A. Associate Professor and Vice-Chairman for Graduate Affairs, Department of Anatomy, University of California School of Medicine, Los Angeles, California 90024

GOTTWALD, PETER J. Max Planck Institute for Psychiatry, 8 Munich 23, Germany

GOY, ROBERT W. Chairman, Department of Reproduction, Physiology, and Behavior, Oregon Regional Primate Research Center; Professor of Medical Psychology, University of Oregon Medical School, Beaverton, Oregon 97005

GRASSO, ALFONSO Laboratorio di Neurobiologia, Istituto Superiore di Sanita, viale Regina Elena 299, Rome, Italy

HAMBERGER, ANDERS Institute of Neurobiology, University of Göteborg, Göteborg, Sweden

HAMBURGER, VIKTOR Professor of Biology; Chairman, Department of Biology, Washington University, St. Louis, Missouri 63130

HAMPRECHT, BERND Scientific Assistant, Biochemistry Institute, Max Planck Institute for Cell Chemistry, 8 Munich 2, Germany

HARMON, LEON D. Member of Technical Staff, Systems Theory Research Department, Bell Telephone Laboratories, Inc., Murray Hill, New Jersey 07971

HAYES, ARTHUR National Institute of Mental Health, Bethesda, Maryland 20014

HEISENBERG, MARTIN A. Max Planck Institute for Biological Cybernetics, Tübingen, Germany

HELD, RICHARD M. Professor of Experimental Psychology, Department of Psychology, Massachusetts Institute of Technology, Cambridge, Massachusetts 02139

HELLERSTEIN, DAVID Senior Fellow, Department of Physiology and Biophysics, University of Washington, Seattle, Washington 98105

HERSCHMAN, HARVEY Assistant Research Biologist, Laboratory of Nuclear Medicine; Assistant Professor, Department of Biological Chemistry, University of California, Los Angeles, California 90024

HILLMAN, PETER Department of Zoology, The Hebrew University, Jerusalem, Israel

HILLYARD, STEVEN A. Department of Neurosciences, University of California at San Diego School of Medicine, La Jolla, California 92037

HIRSH, RICHARD Division of Biology, California Institute of Technology, Pasadena, California 91109

HODOS, WILLIAM Chief, Neuropsychology Laboratory, Department of Experimental Psychology, Walter Reed Army Institute of Research, Walter Reed Army Medical Center, Washington, D.C. 20012

HOFFER, BARRY J. Division of Special Mental Health Research, IRP, National Institute of Mental Health, William A. White Building, Saint Elizabeth's Hospital, Washington, D.C. 20032

HÖKFELT, TOMAS G. M. Department of Histology, Karolinska Institute, Stockholm, Sweden

HONGO, TOSHINORI Associate Professor, Tokyo Medical and Dental University, Bunkyo-ku, Tokyo, Japan

HUBBELL, WAYNE L. Department of Chemistry, Stanford University, Stanford, California 94305

IRWIN, LOUIS N. Bureau of Child Research and Development of Physiology and Cell Biology, University of Kansas, Lawrence, Kansas 66045

IVERSEN, LESLIE L. Department of Pharmacology, University of Cambridge, Cambridge, England

JACKLET, JON W. Assistant Professor of Biological Sciences, Department of Biological Sciences, State University of New York, Albany, New York 12203

JACOBSON, MARCUS Thomas C. Jenkins Department of Biophysics, The Johns Hopkins University, Baltimore, Maryland 21218

JANSEN, JAN K. S. Associate Professor, Institute of Physiology, University of Oslo, Oslo, Norway

JOHNSON, HOWARD W. President, Massachusetts Institute of Technology, Cambridge, Massachusetts 02139

KAISER, ARMIN DALE Professor of Biochemistry, Department of Biochemistry, Stanford University Medical School, Palo Alto, California 94304

KARPLUS, MARTIN  Professor, Department of Chemistry, Harvard University, Cambridge, Massachusetts 02138

KARTEN, HARVEY J.  Research Associate, Department of Psychology, Massachusetts Institute of Technology, Cambridge, Massachusetts 02139

KETY, SEYMOUR S.  Professor of Psychiatry, Harvard Medical School; Director of Psychiatric Research Laboratories, Massachusetts General Hospital, Boston, Massachusetts 02114

KEYNES, RICHARD D.  Director, Agricultural Research Council, Institute of Animal Physiology, Babraham, Cambridge, England

KLIVINGTON, KENNETH  Alfred P. Sloan Foundation, 630 Fifth Avenue, New York, New York 10020

KOMISARUK, BARRY R.  Associate Professor, Institute of Animal Behavior, Rutgers University, Newark, New Jersey 07102

KŘIVÁNEK, JIŘÍ  Physiological Institute, Czechoslovak Academy of Sciences, Prague, Czechoslovakia

KUETHER, CARL A.  Program Administrator, National Institute of General Medical Sciences, National Institutes of Health, Bethesda, Maryland 20014

LANGE, G. DAVID  Assistant Professor of Neurosciences, University of California at San Diego School of Medicine, La Jolla, California 92037

LEHNINGER, ALBERT L.  Professor and Director of the Department of Physiological Chemistry, The Johns Hopkins University School of Medicine, Baltimore, Maryland 21205

LENNEBERG, ERIC H.  Professor of Psychology and Neurobiology, Department of Psychology, Cornell University, Ithaca, New York 14850

LEWIS, EDWIN R.  Associate Professor, Department of Electrical Engineering and Computer Sciences, University of California at Berkeley, Berkeley, California 94720

LIVINGSTON, ROBERT B.  Professor and Chairman, Neurosciences Department, University of California at San Diego School of Medicine, La Jolla, California 92037

LLINÁS, RODOLFO  Division of Neurobiology, Department of Physiology and Biophysics, University of Iowa, Oakdale Campus, Iowa City, Iowa 52240

LYSER, KATHERINE M.  Associate Professor, Department of Biological Sciences, Hunter College of the City University of New York, New York, New York 10021

McCONNELL, HARDEN M.  Professor of Chemistry, Stanford University, Stanford, California 94305

MacKAY, DONALD M.  Professor of Communications, University of Keele, Keele, England

MacLEAN, PAUL D.  Chief, Section on Limbic Integration and Behavior, Laboratory of Neurophysiology, National Institute of Mental Health, National Institutes of Health, Bethesda, Maryland 20014

MALETTA, GABE J.  Staff Physiologist and Health Scientist Administrator, Adult Development and Aging Branch, National Institute of Child Health and Human Development, National Institutes of Health, Bethesda, Maryland 20014

MARCHALONIS, JOHN J.  Assistant Professor of Medical Sciences, Brown University, Providence, Rhode Island 02912

MELNECHUK, THEODORE  Director of Communications, Neurosciences Research Program, Massachusetts Institute of Technology, Brookline, Massachusetts 02146

MORRELL, FRANK  Brain Research Laboratory, Department of Psychiatry, New York Medical College, Fifth Avenue and 105th Street, New York, New York 10029

MÜLLER, EUGENIO E.  Associate Professor of Pharmacology, Istituto di Farmacological di Terapia, Universita Degli Studi, Via Vanvitelli 32, 20129 Milan, Italy

MURPHY, JOHN T.  Assistant Research Professor of Physiology, State University of New York at Buffalo, Buffalo, New York 14226

NAUTA, WALLE J. H.  Professor of Neuroanatomy, Psychology Department, Massachusetts Institute of Technology, Cambridge, Massachusetts 02139

NELSON, PHILLIP GILLARD  Chief, Behavioral Biology Branch, National Institute of Child Health and Human Development, National Institutes of Health, Bethesda, Maryland 20014

NIRENBERG, MARSHALL W.  Chief, Laboratory of Biochemical Genetics, National Heart Institute, National Institutes of Health, Bethesda, Maryland 20014

NOMURA, MASAYASU  Professor of Genetics, University of Wisconsin, Madison, Wisconsin 53706

NORTON, ALLEN C.  Assistant Professor of Physiology, Brain Information Service, University of California, The Center for Health Sciences, Los Angeles, California 90024

NOTTEBOHM, FERNANDO  Assistant Professor, The Rockefeller University, New York, New York 10021

OLIVERIO, ALBERTO  Laboratory of Psychobiology and Psychopharmacology, Consiglio Naxionale delle Ricerche, Via Reno 1, 00198 Rome, Italy

ONSAGER, LARS  J. W. Gibbs Professor of Theoretical Chemistry, Yale University, New Haven, Connecticut 06520

OSHIMA, TOMOKAZU  Department of Physiology, Toho University School of Medicine, 5–21 Omori-Nishi, OTA-Ku, Tokyo, Japan

PARTLOW, LESTER MARTIN  Postdoctoral Fellow, Department of Biology, The Johns Hopkins University, Baltimore, Maryland 21218

PERKEL, DONALD H.  Department of Biological Sciences, Stanford University, Stanford, California 94305

PFAFF, DONALD  Assistant Professor, Biomedical Division of the Population Council, The Rockefeller University, New York, New York 10021

PFEIFFER, STEVEN E.  Postdoctoral Fellow, Graduate Department of Biochemistry, Brandeis University, Waltham, Massachusetts 02154

PFENNINGER, KARL H.  Research Associate, Brain Research Institute, University of Zurich, 8008 Zurich, Switzerland

PLOOG, DETLEV  Professor of Psychiatry, Max Planck Institute for Psychiatry, 8 Munich 23, Germany

PRESTIGE, MARTIN C. Lecturer in Physiology, University of Edinburgh, Edinburgh, 8, Scotland

PURPURA, DOMINICK P. Professor and Chairman, Department of Anatomy, Albert Einstein College of Medicine, Bronx, New York 10061

QUARTON, GARDNER C. Director, Mental Health Research Institute, University of Michigan, Ann Arbor, Michigan 48104

RALL, WILFRID Senior Research Physicist, Mathematical Research Branch, National Institute of Arthritis and Metabolic Diseases, National Institutes of Health, Bethesda, Maryland 20014

REDMAN, STEPHEN JOHN Department of Electrical Engineering, Monash University, Clayton, Victoria 3186, Australia

REICHARDT, WERNER E. Professor and Director, Max Planck Institute for Biological Cybernetics, 7400 Tübingen, Germany

RICHARDS, FREDERIC M. Ford Professor of Molecular Biophysics, Yale University, New Haven, Connecticut 06520

ROBERTSON, J. DAVID Professor and Chairman, Anatomy Department, Duke University, Durham, North Carolina 27706

ROSENHECK, KURT Polymer Department, Weizmann Institute of Science, Rehovoth, Israel

SAMSON, FREDERICK E., JR. Professor and Chairman, Bureau of Child Research and Department of Physiology and Cell Biology, University of Kansas, Lawrence, Kansas 66044

SCHARRER, BERTA Professor of Anatomy, Albert Einstein College of Medicine, Bronx, New York 10461

SCHEIBEL, ARNOLD B. Associate Professor of Anatomy and Psychiatry, University of California Medical Center, Los Angeles, California 90024

SCHEIBEL, MADGE E. Associate Professor of Anatomy and Psychiatry, University of California Medical Center, Los Angeles, California 90024

SCHMITT, FRANCIS O. Institute Professor, Professor of Biology, Chairman, Neurosciences Research Program, Massachusetts Institute of Technology, 280 Newton Street, Brookline, Massachusetts 02146

SCHMITT, OTTO H. Professor of Biophysics-Bioengineering, Department of Electrical Engineering, Institute of Technology, University of Minnesota, Minneapolis, Minnesota 55414

SCHNEIDER, DIETRICH Director, Max Planck Institute for Behavioral Physiology, Seewiesen über Starnberg, Germany

SCHNEIDER, GERALD E. Assistant Professor, Department of Psychology, Massachusetts Institute of Technology, Cambridge, Massachusetts 02139

SEGUNDO, JOSÉ, P. Professor of Anatomy, University of California, Los Angeles, California 90024

SHAW, LESLIE M. J. Postdoctoral Fellow, Mergenthaler Laboratory for Biology, The Johns Hopkins University, Baltimore, Maryland 21218

SHELANSKI, MICHAEL L. Assistant Professor of Neuropathology, Albert Einstein College of Medicine, Bronx, New York 10461

SHEPHERD, GORDON M. Assistant Professor, Department of Physiology, Yale University School of Medicine, New Haven, Connecticut 06520

SHOFER, ROBERT J. Assistant Professor, Department of Anatomy, Albert Einstein College of Medicine, Bronx, New York 10461

SHOOTER, ERIC M. Professor of Genetics and Biochemistry, Department of Genetics, Stanford University School of Medicine, Stanford, California 94304

SIDMAN, RICHARD LEON Professor of Neuropathology, Harvard Medical School, Boston, Massachusetts 02115

SINSHEIMER, ROBERT L. Chairman, Division of Biology, California Institute of Technology, Pasadena, California 91109

SIZER, IRWIN W. Dean of the Graduate School, Massachusetts Institute of Technology, Cambridge, Massachusetts 02139

SMITH, ANDREW P. Graduate Student, Department of Genetics, Stanford Medical Center, Stanford, California 94305

SMITH, BARRY HAMILTON Rogosin Laboratories, Cornell University Medical College, 525 East 68th Street, New York, New York 10021

SPIEGELMAN, SOL Director, Institute of Cancer Research, Columbia University, New York, New York 10032

SPOONER, CHARLES E. Assistant Professor of Neurosciences, University of California at San Diego School of Medicine, La Jolla, California 92037

STEIN, RICHARD B. Associate Professor, Department of Physiology, University of Alberta, Edmonton, Alberta, Canada

STENT, GUNTHER S. Laboratory of Neurobiology, Harvard Medical School, 25 Shattuck Street, Boston, Massachusetts 02115

STONE, FREDERICK L. President, New York Medical College, Flower and Fifth Avenue Hospital, New York, New York 10029

SWEET, WILLIAM H. Chief, Neurosurgical Services, Massachusetts General Hospital, Professor of Surgery, Harvard Medical School, Boston, Massachusetts 02114

SZENTÁGOTHAI, JOHN Professor and Director, Department of Anatomy, University Medical School, Budapest IX, Hungary

TAYLOR, J. HERBERT Professor, Institute of Molecular Biophysics, Florida State University, Tallahassee, Florida 32306

TAYLOR, WILLIAM M. Research Training Grants Branch, National Institute of General Medical Sciences, National Institutes of Health, Bethesda, Maryland 20014

TERZUOLO, CARLO A. Professor and Director, Laboratory of Neurophysiology, University of Minnesota Medical School, Minneapolis, Minnesota 55455

THOMAS, CHARLES A., JR. Professor of Biological Chemistry, Harvard Medical School, Boston, Massachusetts 02115

VALENSTEIN, ELLIOT S. Senior Research Associate, Professor of Psychology, Fels Research Institute, Antioch College, Yellow Springs, Ohio 45387

VARON, SILVIO Associate Professor, Department of Biology, University of California at San Diego, La Jolla, California 92037

VAUGHN, JAMES E. Assistant Professor, Department of Anatomy, Boston University School of Medicine, Boston, Massachusetts 02118

VIDAL, JACQUES J. Associate Professor, Department of Engineering, University of California, Los Angeles, California 90024

WALLØE, LARS Lecturer, Institute of Physiology, Oslo University, Oslo 1, Norway

WANG, HOWARD Assistant Professor, Division of Natural Sciences, University of California, Santa Cruz, California 95060

WASHBURN, SHERWOOD L. Professor of Anthropology, University of California, Berkeley, California 94720

WECHSLER, WOLFGANG Max Planck Institute for Brain Research, Ostmerheimer Strasse 200, 5 Koln-Merheim, Germany

WEISS, PAUL A. Professor Emeritus, The Rockefeller University, New York, New York 10021

WERNER, GERHARD Professor and Chairman, Department of Pharmacology, University of Pittsburgh School of Medicine, Pittsburgh, Pennsylvania 15213

WHITTAKER, VICTOR P. Professor, Department of Biochemistry, University of Cambridge, Cambridge, England

WILLIAMS, HAROLD L. Research Professor of Psychology, Department of Psychiatry and Behavioral Sciences, University of Oklahoma School of Medicine, Oklahoma City, Oklahoma 73104

WILSON, DONALD M. Professor of Biology, Stanford University, Stanford, California 94304

WORDEN, FREDERIC G. Executive Director, Neurosciences Research Program; Professor of Psychiatry, Massachusetts Institute of Technology, Brookline, Massachusetts 02146

WURTMAN, RICHARD J. Professor, Department of Nutrition and Food Sciences, Massachusetts Institute of Technology, Cambridge, Massachusetts 02139

# NAME INDEX

## A

Abdel-Latif, A. A., 761, 767
Abdulaev, N. D., 684, 696
Abeles, M., 1016
Abelson, J., 997
Abercrombie, M., 59, 60
Abood, L. G., 751, 760
Abplanalp, P., 21, 25
Abzug, C., 469
Adam, G., 514, 517, 678, 683, 703, 705
Adams, J. A., 167, 174
Adams, J. M., 922, 926
Adams, R. D., 176
Adams, R. T., 577, 586, 588, 596
Adamson, L., 25
Adelman, G., 849, 850, 877, 1016
Adelman, M. R., 854, 856
Ades, H. W., 259
Adey, W. R., 25, 162, 224, 225, 226, 227, 228,
    229, 230, 231, 232, 233, 234, 236, 239, 240,
    241, 242, 243, 244, 255, 257, 258, 282, 288,
    295, 596, 793, 795, 865, 876, 877, 1015, 1016
Adhya, S. L., 958, 961
Adler, J., 123, 128
Adorjani, C., 485
Adrian, E. D., 244, 257, 450, 456, 587, 588,
    596, 598, 604, 605, 615
Afzelius, B. A., 839
Aghajanian, G. K., 732, 733, 734, 735, 736, 737,
    738, 739, 740, 743, 744
Agranoff, B. W., 163, 174, 272, 274, 277, 856
Agtarap, A., 685, 696
Ahmad-Zadeh, C. C., 927
Ahrens, R., 357, 360
Aitken, J. T., 80, 81, 111, 113
Ajmone Marson, C., 244, 257, 259, 456
Ajuriaguerra, J. de, 335
Akasaki, K., 685, 696
Akert, K., 724, 726, 727, 729, 731, 735, 737,
    739, 743, 746, 871, 877, 878
Akiyama, T., 408
Aladzhalova, N. A., 578, 585
Albe-Fessard, D., 465, 467, 469
Albers, R. W., 538
Albert, A., 621, 622, 629
Albert, D. J., 274, 277
Alberts, W. W., 316
Aldanova, N. A., 696
Alexandroff, P., 612, 615
Alexandru, C., 107
Algranati, I. D., 924, 926
Allen, J. C., 900
Allen, J. M., 107
Allen, R. A., 324
Allewell, N. M., 901, 907, 909, 911, 912
Allison, A. C., 540, 551
Allison, J. E., 197, 205
Alnaes, E., 617, 628
Altman, J., 21, 25, 65, 67, 69, 71, 136, 140,
    163, 174, 289, 296, 333, 334, 841, 849
Altmann, R., 63, 350, 355, 360, 829, 836
Altmann, S. A., 47, 360

Alousi, A., 325, 327, 335
Alving, B. O., 869, 877
Ambache, N., 81
Ambler, M., 101, 106
Ambrose, J. A., 357, 360
Anand, B. K., 208, 216
Anastasi, A., 293, 296
Andén, N.-E., 327, 331, 334
Anders, T. R., 162, 163, 295
Andersen, P., 438, 440, 442, 446, 450, 451, 456,
    459, 460, 461, 469
Anderson, B., 208, 210, 216
Anderson, E., 722, 728, 849
Anderson, J. W., 425
Andersson, S. A., 450, 451, 456, 460, 461, 469
André, J., 839
Andres, K. H., 527, 546, 551, 560, 564
Andriesse, P., 738
Angel, A., 433, 442
Angeletti, P. U., 84, 86, 99, 104, 107, 784, 796,
    816, 819, 820, 823, 825, 826, 827, 874, 878
Angeletti, R. A. H., 817, 819, 820, 823
Angell, J. B., 389, 396
Angevine, J. B., Jr., 52, 62, 64, 65, 66, 67, 68,
    69, 70, 71, 72, 100, 102, 103, 104, 106, 136,
    856, 1016
Angulo y González, A. W., 147, 150
Ansell, G. B., 99, 791, 795
Anstis, S. M., 308, 315
Antonov, A. M., 696
Apirion, D., 920
Apostol, V., 470
Appel, J. B., 336
Appel, S. H., 107, 729, 743, 813, 814, 823, 824
Appleton, P. I., 195
Aprison, M. H., 770, 782
Apter, M. J., 615, 616
Arbib, M. A., 615, 616
Arbit, J., 169, 174
Arceneaux, C. J., 727
Arduini, A. A., 342, 348, 637, 647
Arfel, G., 469
Argyle, M., 357, 360
Ariëns Kappers, C. U., 28, 37, 411, 424
Ariëns Kappers, J., 537
Armett, C. J., 601, 604
Arnold, W. J., 335
Aronson, A. I., 822, 824
Aronson, N. N., 749, 758
Arora, H. L., 80, 82, 110, 113, 115, 155, 157
Arqueros, L., 781
Arvanitaki, A., 266, 267, 270
Asai, J., 839
Asanuma, H., 460, 469
Aschoff, J. C., 470
Ashhurst, D. E., 812
Aslanov, A. S., 226, 242
Asratyan, E. A., 243
Assem, A. van den, 194, 195
Astbury, W. T., 675
Åström, K.-E., 64, 71, 137, 140
Atkinson, M. R., 942, 945
Atkinson, R. C., 164, 174
Attardi, D. G., 123, 128, 154, 156, 756, 758,
    820, 823, 824

Attardi, G., 964, 970, 989, 996
Atwood, H. L., 402, 408
Atwood, K. C., 900, 972, 997
Atwood, R. P., 102, 106
August, J. T., 932, 945
Augusti-Tocco, G., 859, 865
Austin, C. M., 242
Austin, L., 761, 767, 813, 814, 815, 824, 826,
    842, 849, 864, 866
Autilio, L. A., 743, 813, 814, 815, 824, 865
Autrum, H., 495, 509
Averbach, E., 165, 169, 174
Avogardo, L., 900
Axelrod, J., 325, 327, 334, 336, 529, 531, 533,
    538, 747, 781, 874, 878
Axelsson, J., 111, 113
Azcurra, J. M., 761, 766
Azmitia, E. C., 333, 334
Azoulay, L., 128
Azzi, A., 834, 836

## B

Babakov, A. V., 684
Bach-y-Rita, P., 314, 315
Bacon, F., 382
Baddeley, A. D., 167, 174
Bade, E. G., 926
Baer, D. M., 176, 186
Bagg, H. J., 58, 61
Bagley, C., Jr., 18, 25
Baglioni, C., 890, 899, 969, 970
Bailey, P., 361, 371, 826
Baird, J. J., 109, 113
Bajusz, E., 525, 527
Bak, I. J., 734, 743
Baker, M. A., 1016
Baker, P. F., 711, 714
Baker, R. E., 117, 188, 119, 120, 121, 122, 128
Baker, R. F., 927
Baker, T. G., 974, 996
Balaban, M., 151
Baldessarini, R. J., 328, 334, 1016
Balla, D., 188
Ballantine, H. T., 175
Balogh, G., 114
Balthasar, K., 270
Baly, W., 596
Bamberg, E., 684
Banks, B. E. C., 817, 819, 824, 826
Banks, P., 776, 780
Banthorpe, D. V., 824, 826
Banuazizi, A., 218, 220, 221, 222, 223
Barajas, L., 520, 527
Barbizet, J., 170, 174
Barchas, J. D., 327, 334
Barcroft, J., 147, 150
Bard, P., 208, 617
Barden, N., 822, 824
Barer, R., 523, 527
Bargmann, W., 442, 520, 521, 522, 526, 527,
    529, 732, 743
Barker, D., 156
Barkley, D. S., 123, 128
Barlow, G. H., 945

Barlow, H. B., 308, 315, 472, 473, 476, 477, 485, 492, 493, 550, 551, 600, 604, 628
Barlow, J. S., 241, 242
Barnard, J. W., 415, 424
Barnett, S. A., 42, 46
Barnett, W. E., 828, 836
Barondes, S. H., 156, 162, 272, 273, 275, 276, 277, 278, 282, 294, 295, 296, 333, 334, 443, 676, 729, 744, 747, 751, 753, 754, 755, 758, 759, 760, 761, 766, 767, 768, 814, 824, 841, 845, 849, 851, 856, 858, 871, 875, 877, 1016
Barr, L., 722, 726
Barr, M. L., 622, 628
Barraclough, C. A., 199, 206
Barrell, B. G., 919, 996
Barrnett, R. J., 734, 736, 744, 746
Barron, D. H., 76, 79, 81, 147, 150, 859, 865
Barry, W. H., 710, 713, 714
Bartlett, F., 195
Bartlett, M. S., 595, 596
Bartter, F. C., 538
Barzilai, R., 996
Baskin, D. G., 528
Bass, L., 240, 241
Basser, L. S., 365, 371
Bastian, J., 182
Batenchuk, C., 222
Bateson, P. P. G., 191, 195
Battersby, W. S., 316, 371
Bauer, B., 279, 289
Bauer, H. J., 751, 760
Bauer, J. A., Jr., 321, 323
Bauer, W., 982, 997
Baumgarten, H. G., 520, 527
Baumgarten, R. J. von, 162, 260, 263, 264, 265, 270, 271, 276, 295, 296, 542, 551, 1016
Baumgartner, G., 473, 485
Bautz, E. K. F., 954
Bayless, W. A., viii
Baylor, D. A., 550, 801, 811, 865
Bayly, E. J., 527, 577, 581, 585, 586, 629, 661, 663, 664, 666, 667, 668, 670, 671
Bazanova, I. S., 741, 744
Beach, F. A., 197, 198, 199, 206
Beament, J. W. L., 529, 746
Bean, R. C., 683, 709, 714
Bear, R. S., 711, 714, 769, 780, 867, 878
Beatty, B. R., 974, 997
Beaudoin, A. R., 74, 79, 81, 109, 113
Beaudreau, G., 945
Bechmann, H., 920
Bechterew, W. V., 265, 270
Beck, J. C., 829, 836
Becker, M. C., 551
Beckmann, R., 517
Beckwith, J. R., 946, 954, 958, 961
Beeman, E. A., 201, 206
Beermann, W., 850, 962, 970, 971, 983, 985, 996
Beggs, J., 797
Behar, I., 35, 38
Beintema, J. J., 905, 911
Békésy, G. von, 394, 395
Bekhor, I., 823, 824, 964, 966, 971
Bell, C. C., 594, 596
Bellairs, A. d'A., 151
Beller, R. J., 927
Belling, J., 983, 996
Bellman, R. E., 616
Bello, J., 911
Benda, P., 278, 288, 815, 824

Bender, M. B., 264, 271, 316, 371
Benitez, H. R., 85, 99
Bennett, E. L., 296, 423, 745
Bennett, G. S., 279, 288, 815, 823, 824
Bennett, H. S., 715, 726
Bennett, J. C., 890, 893, 899, 963, 971
Bennett, M. V. L., 721, 727
Bensch, K. G., 746
Benson, D. F., 171, 172, 174
Bentley, D., 803
Bepestovsky, G. N., 877
Beranek, R., 878
Berde, B., 521, 527
Berg, P., 997
Bergland, R. M., 112, 113
Bergman, S. J., 801, 811
Bergmann, F. H., 1016
Bergstrand, A., 839
Berkeley, G., 337
Berkhout, J., 225, 232, 233, 236, 237, 238, 239, 241, 243
Berquist, H., 58, 60
Berlyne, D. E., 178, 181, 183, 184, 186
Bern, H. A., 520, 521, 523, 525, 527, 528, 777, 778, 780
Bernhard, C. G., 396
Bernstein, B., 605, 616
Bernstein, J. J., 112, 113
Berry, A. R., 824, 826
Berry, M., 104, 106
Berry, R. W., 333, 334, 874, 877
Bertler, A., 771, 780
Bertrand, G., 467, 469
Besson, M. J., 326, 334
Bessey, O. A., 58, 61
Best, J. B., 563, 564, 736, 744
Bickford, R., 233, 241
Bickle, T. A., 927
Biedebach, M. C., 510
Bielawski, J., 834, 836
Biemiller, R., viii
Bierce, A., 486, 493
Bigelow, J. H., 349, 377, 877, 1016
Bignami, G., 333, 334
Bijou, S. W., 176, 186
Billingham, R. E., 82
Biondi, R., 847
Birks, R. I., 772, 780, 864, 865, 866
Birnstiel, M. L., 900, 964, 971, 972, 989, 996
Bishop, D. H. L., 941, 945
Bishop, G. H., 244, 257
Bishop, L. G., 501, 509
Bishop, P. O., 307, 310, 379, 471, 473, 474, 475, 476, 477, 479, 480, 481, 482, 483, 485, 491, 492, 493, 494, 605, 616, 643, 1016
Bittar, E. E., 767
Bittar, N., 767
Bitterman, M. E., 34, 38
Bittner, G. D., 577, 582, 583, 585, 811, 1016
Björklund, A., 522, 527
Black, A. H., 221, 222
Black, I. B., 1016
Black, J. A., 896, 899
Blackburn, K. J., 777, 780
Blackett, D. W., 612, 616
Blackman, R. B., 225, 241
Blackstad, T. W., 331, 334, 345, 348
Blake, C. C. F., 903, 904, 911
Blakemore, C., 485, 493
Blakemore, W. F., 792, 795
Bland, J. O. W., 91, 99

Blaurock, A. E., 867, 879
Blest, A. D., 500, 510
Blinzinger, K., 156, 795
Bliss, E. L., 296, 297
Blitz, J., 360
Block, B. C., 518
Bloedel, J., 563, 564, 742, 744
Bloedel, J. R., 414, 425
Blomquist, A. J., 38
Bloom, F. E., 266, 290, 329, 335, 336, 520, 529, 676, 729, 732, 733, 734, 736, 737, 738, 739, 740, 742, 743, 744, 746, 871, 872, 873, 874, 877, 878, 1016
Bloomstrand, C., 837
Blundell, J. E., 208, 216
Boadle, M. C., 736
Bock, A., 847
Bock, P., 224, 241
Bock, R., 525, 527
Bodenstein, D., 801, 808, 810
Bodian, D., 52, 129, 130, 132, 133, 134, 135, 137, 140, 147, 150, 547, 548, 551, 723, 726, 739, 744, 798, 800, 801, 806, 809, 810, 813, 824, 1015, 1016
Bodmer, W., 961
Boeckh, J., 512, 513, 514, 515, 516, 517, 518
Bogdanski, D. F., 326, 334, 761, 766
Bogen, J. E., 315
Bogoch, S., 195, 278, 296, 747, 750, 753, 758, 871, 877
Bohdanecky, Z., 277
Bohn, H., 905, 911
Bollum, F. J., 1006, 1012
Bolt, J. R., 28, 38
Bondareff, W., 240, 241, 325, 334, 732, 738, 744, 751, 758
Bondy, S. C., 813, 824, 826
Bone, Q., 155, 156
Bonin, G. von, 361, 371
Bonner, J., 823, 824, 964, 971, 1010, 1012
Bonner, J. F., 289, 296, 615, 616
Bonner, J. T., 123, 128, 526, 527
Bonnevie, K., 58, 60
Booker, B., 816, 825
Boole, G., 368, 371
Boord, R. L., 23, 25
Borenstein, P., 246, 257
Boring, E. G., 317, 323, 337, 348
Borisy, G. G., 852, 856
Bornstein, M. B., 91, 92, 97, 99, 106, 137, 140, 744
Borys, H. K., 335
Bosch, L., 945
Boschek, C. B., 495
Bosmann, H. B., 748, 758, 759
Bossom, J., 320, 321, 322, 323, 324
Bostelmann, W., 528
Boulder Committee, 52, 63, 71, 100, 106
Bourgeois, S., 946, 948, 951, 952, 954
Bourn, S., 81
Bourne, G. H., 107, 528, 768, 786, 796, 797, 811, 866
Bovet, D., 334
Bovet, D. F., 189, 195
Bovet-Nitti, F., 195
Bowden, D., 351, 360
Bower, E., 241
Bower, G. H., 164, 174
Bowers, C. Y., 529
Bowers, M. B., 806, 811
Bowman, R. J., 666, 670

Dennis, E. A., 912
Denny-Brown, D., 111, 113
Dent, J. N., 520, 527
De Robertis, E. D. P., 715, 722, 726, 734, 740, 746, 761, 766, 767, 776, 781, 845, 850
Descarries, L., 327, 334, 731, 734, 744, 1016
Descartes, R., 317, 318, 323, 337
Desiraju, T., 463, 466, 467, 469
Desjardins, C., 201, 206
Desoer, C. A., 384, 396
Detwiler, S. R., 51, 79, 81, 108, 109, 110, 111, 112, 113, 114
Deutsch, H. F., 890, 900
Deutsch, J. A., 275, 277
Devine, C. E., 526, 527, 734, 744
De Vore, I., 46, 195
Dewey, K., 926
Dewey, M. M., 722, 726
Dhariwal, A. P. S., 778, 781
Diamond, I., 762, 767
Diamond, I. T., 21, 26
Diamond, J., 111, 113
Diamond, M., 200, 206, 529
Diamond, M. C., 290, 296
DiCara, L. V., 162, 218, 219, 220, 221, 222, 223, 1016
Dichburn, R. W., 224, 241
Dichgans, J., 308, 316
Dickens, F., 81
Dickerson, R. E., 901, 911, 972
Didio, J., 241, 288
Diebler, H., 688, 696
Dill, R. C., 243, 257, 259
Dilworth, M., 456
Dische, Z., 747, 758
Dixon, G. H., 896, 899
Dixon, W. P., 579, 585
Doane, B., 258
Dobbing, J., 290, 296
Dobzhansky, T. G., 39, 46
Döcke, F., 520, 529
Dodenheimer, T. S., 728
Dodge, F. A., 666, 670
Dodson, J. W., 80, 81
Doerr-Schott, J., 520, 528
Doggenweiler, C. F., 722, 725, 726, 727, 791, 796
Dohner, R. E., 696
Doi, R. H., 931, 945
Dolhinow, P. C., 46
Donahue, S., 796
Donaldson, A. A., 469
Donaldson, I. M. L., 469
Donfoffer, H., 242
Donner, D., 920
Donohue, J., 902, 911
Doolittle, R. F., 888, 900
Doonan, S., 824
Dore, E., 826
Dorsett, D. A., 801, 811
Doscher, M. S., 909, 911
Dostoyevsky, F., 342, 348
Doty, B. A., 35, 38
Doty, L. A., 38
Doty, R. W., 605, 616
Douglas, W. W., 525, 528, 742, 744, 774, 775, 779, 780
Doutt, R. L., 518
Dove, W. F., 946, 954, 956, 957, 961
Dowling, J. E., 395, 396, 550, 551, 552, 597, 604, 631, 647

Dozy, A. M., 900
Drachman, D. A., 169, 174
Drachman, D. B., 776, 780
Drake, J. W., 977, 979, 997
Drenth, J., 903, 911
Dressler, D., 956, 961, 994, 997
Dreyer, W. J., 890, 893, 899, 963, 971
Droz, B., 327, 334, 734, 814, 824, 842, 843, 846, 849, 851, 854, 856
Dubnau, D., 997
Dubois, M., 105, 107
Dudel, J., 774, 780
Dulbecco, R., 955, 961, 968, 971
Dunant, Y., 864, 866
Duncan, C. P., 167, 174
Duncan, D., 63, 71, 846, 850
Dunitz, J. D., 684, 696, 837
Dunlap, J. B., 199
Dunlop, C. W., 241, 244, 257
Dunn, J. J., 954
DuPraw, E. J., 1000, 1012
Durkovic, R. G., 249, 257
Dusser de Barenne, J. C., 443, 456
Dutta, C. R., 424, 442
Dutton, G. R., 751, 753, 754, 755, 758, 759, 875, 877
Duve, C. de, 749, 758
Duyff, J. W., 442
Dwyer, S. J., 241
Dzionara, M., 920

**E**

Eagle, H., 76, 81
Earle, K. M., 797
Easley, R. B., 207
Eastlick, H. L., 111, 113
Eaton, G. G., 199, 206
Ebbeson, S. O. E., 72, 107
Ebert, J. D., 62, 71, 873, 877
Ebling, F. J. G., 529
Ebner, F. F., 21, 25
Eccles, J. C., 45, 46, 108, 111, 112, 113, 116, 117, 128, 241, 244, 257, 261, 263, 270, 315, 316, 413, 414, 415, 419, 420, 424, 429, 440, 441, 442, 446, 450, 451, 456, 461, 469, 542, 544, 547, 548, 551, 564, 583, 585, 620, 625, 626, 628, 665, 666, 670, 723, 726, 731, 742, 744
Eccles, R. M., 113, 128
Echols, H., 958, 959, 961
Eck, R. V., 896, 899
Eckert, H., 499, 505, 510
Edds, M. V., Jr., 51, 62, 71, 82, 108, 111, 113, 116, 125, 128, 152, 154, 156, 189, 676, 843, 849, 850, 1015, 1016
Edelman, G. M., 156, 279, 288, 815, 823, 824, 858, 873, 883, 885, 888, 889, 890, 891, 893, 894, 896, 897, 898, 960, 962, 963, 964, 966, 969, 970, 971, 996, 1015, 1016
Edelman, I. S., 825
Edelsack, E. A., 395, 616
Edelstein, S. J., 820, 826
Edgar, R. S., 998
Edinger, T., 27, 38
Edström, A., 791, 796, 842, 849
Edström, J.-E., 81, 811, 823, 824
Edwards, D. A., 207
Edwards, D. C., 817
Edwards, J. S., 123, 128, 808, 811
Efremova, T. M., 225, 241, 242

Efron, D. H., 277
Egger, M. David, 1016
Eggers, F., 693, 696
Eguchi, E., 497, 510
Egyházi, E., 261, 271, 281, 287, 288, 813, 825
Ehrenberg, A., 706
Eibl-Eibesfeldt, I., 188, 195, 214, 216
Eichberg, J., Jr., 761, 767
Eichner, J., 683, 714
Eide, E., 618, 619, 621, 622, 623, 626, 628
Eigen, M., 676, 680, 682, 685, 687, 696, 858, 869, 877, 1015, 1017
Ein, D., 890, 899
Einstein, A., 336
Eisen, H. N., 889, 899, 953, 954, 956, 961
Eisenstadt, J. M., 923, 924, 926
Eisenstein, E. M., 163, 174, 264, 270, 272, 277
Ekdahl, M., 169, 175
Elazar, Z., 224, 227, 229, 230, 231, 232, 241, 242, 246, 258
Eldefrawi, M. E., 787, 796
Eldred, E., 669, 671
Eleftheriou, B. L., 190, 195
Elekessy, E., 485
Elfvin, L.-G., 787, 796
Eliceiri, G. L., 966, 971
Elkes, J., 336, 538, 768, 850
Ellis, N. R., 188
Ellory, J. C., 709, 714
Ellsworth (Ruth H. and Warren A.) Foundation, x
Elsasser, W. M., 117, 128
Elsberg, C. A., 111, 113
Elul, R., 224, 225, 226, 227, 228, 241, 247, 257, 721, 726
Elwyn, A., 12, 26
Encabo, H., 586
Enemar, A., 520, 527, 528
Eneström, S., 734, 744
Engel, B. T., 218, 223
Engel, W. K., 111, 114
Engelbert, V. E., 99
Enger, P. S., 618, 628
Engström, A., 877
Enroth-Cugell, C., 485, 632, 647
Enslein, K., 257, 661
Eoyang, L., 945
Epler, J. L., 836
Epstein, R., 263, 271
Epstein, W., 946, 954
Eränkö, O., 327, 334, 730, 744
Eriksen, C. W., 165, 174
Ernst, K.-D., 513, 514, 517
Ernster, L., 839
Erspamer, V., 771, 780
Ervin, F., 277
Ervin, F. R., 162, 163, 195, 295, 1017
Erwin, V. G., 839
Escueta, A. V., 743, 814, 824, 865
Espir, M. L. E., 362, 371
Esser, A. H., 360
Essick, C. R., 101, 102, 106
Essman, W. B., 333
Estabrook, R. W., 839
Esterhuizen, A. C., 520, 528
Estrada-O., S., 696
Eugster, C. H., 727, 746, 878
Euler, C. von, 226, 242, 244, 257, 492, 493
Euler, U. S. von, 325, 334, 335, 442, 493, 520, 528, 629, 647, 671, 767, 768, 770, 780, 782
Evans, D. H. L., 80, 81

Hagadorn, I. R., 521, 528
Haggar, R. A., 622, 628
Häggendal, J., 773, 780, 846, 849
Hagiwara, S., 572, 585, 707, 714
Hagiware, D. R., 591, 596
Hagopian, A., 748, 758, 759
Hahn, G. M., 74, 81
Haigler, H. J., 264, 270
Hajós, F., 443, 814, 825
Hake, T., 738, 745
Hakuhara, T., 264, 270
Halas, E. S., 257
Haldane, J. B. S., 194, 196
Hale, E. B., 41, 46
Halkerston, I. D., 330, 334
Hall, B. D., 931, 945
Hall, J. L., 660, 661
Hall, M. O., 324
Hall, V. E., 242, 348, 383, 469
Hall, W. C., 21, 25
Hall, Z. W., 781
Halle, M., 314, 315
Halstead, D. C., 817
Halvorson, H. O., 998
Hama, K., 560, 721, 728
Hamberger, A., 835, 837, 1017
Hamberger, B., 331, 334, 773, 781
Hamberger, C.-A., 843, 849
Hamburg, D. A., 44, 46
Hamburg, M. D., 277
Hamburger, V., 51, 52, 58, 59, 61, 62, 67, 69,
    71, 72, 76, 79, 80, 81, 84, 97, 108, 109, 111,
    113, 114, 116, 129, 141, 142, 144, 145, 148,
    149, 151, 157, 290, 825, 1017
Hamburgh, M., 104, 106
Hamilton, C. L., 697, 706
Hamilton, C. R., 322, 323
Hamilton, D. W., 787, 797
Hamilton, W. J. III, 188, 195
Hamkalo, B. A., 996, 998
Hammes, G. G., 909, 911
Hámori, J., 95, 411, 413, 425, 427, 429, 441,
    442, 443, 647
Hamprecht, B., 1017
Hampton, T., 792, 796
Hanawalt, J. T., 408
Hanbery, J., 451, 456
Hancock, R., 981, 997
Handelman, E., 671
Handy and Harman, x
Hanley, J., 225, 234, 235, 236, 242
Hansen, S. P., 218, 223
Hanson, A. W., 912
Hara, T. J., 857
Hård, E., 205, 206
Harding, R. S., 7, 39
Hardman, K. D., 912
Hardy, S. J. S., 914, 919, 920, 926
Harker, D., 911
Harkmark, W., 80, 81, 126, 128
Harlan, P. M., 107
Harlow, H. F., 34, 35, 36, 38, 191, 195, 201,
    205, 206
Harlow, M. K., 36, 38, 39, 191, 195, 324
Harmon, L. D., 225, 389, 394, 396, 415, 425,
    486, 490, 491, 493, 591, 1017
Harper, L. V., 339, 348
Harrington (Charles A.) Foundation, x
Harris, A. E., 79, 82, 109, 114
Harris, C. S., 308, 315
Harris, G. W., 199, 200, 206

Harris, H., 123, 126
Harris, J. I., 912
Harris, R. A., 839
Harrison, R. G., 51, 84, 91, 99, 108, 109, 111,
    112, 114
Harrison, R. J., 520, 529
Harrison, W. H., 734, 745
Hartley, B. S., 912
Hartley, C., 163
Hartline, D. K., 387, 396, 582, 585, 629
Hartline, H. K., 798, 811
Haruna, I., 931, 932, 933, 934, 935, 945
Harvey, R. J., 583, 585
Harwood, C. W., 218, 223
Haschemeyer, A. E. V., 907, 911
Hashim, G. A., 816, 824
Hashimoto, Y., 734, 745
Hasselbach, W., 701, 706, 765, 767
Hassenstein, B., 501, 510
Hassler, R., 71, 106, 234, 242, 348, 467, 469
Hastings, P. J., 970, 972
Haurowitz, F., 887, 899
Haut, W. F., 1006, 1012
Hawley, E., 829, 837
Hawthorne, D. C., 836
Hay, E. D., 80, 82
Hayes, A., 1017
Hayes, C., 38
Hayes, K. J., 35, 38
Hayes, T. L., 389, 396
Hayhow, W. R., 741, 744
Head, C., 822, 826
Head, H., 305, 315, 605, 616
Healy, G. M., 995, 997
Hearst, J. E., 1002, 1012
Hebb, C. O., 764, 767, 771, 780
Hebb, D. O., 164, 175, 295, 296
Hecht, M. K., 28, 38
Hechter, O., 330, 334
Hecker, E., 512, 517, 518, 703, 705
Hedquist, P., 781
Hegstad, H., 577, 578, 579, 582, 586, 629, 671
Heimer, L., 540, 551
Hein, A., 291, 296, 321, 322, 323, 324
Heinmets, F., 747, 759
Heisenberg, M. A., 1017
Held, R., 291, 296, 302, 305, 317, 319, 320, 321,
    323, 324, 377, 383, 1017
Helfenstein, M., 143, 151
Helle, K. B., 780
Hellerstein, D., 1017
Hellner, K. A., 585
Helmholtz, H. von, 319, 320, 321, 324, 336
Hendelman, W. J., 791, 796, 850
Hendrickson, A., 742, 745
Hendrix, C. E., 241
Henle, J., 419, 425
Henneman, E., 622, 629, 669, 670
Henry, G. H., 473, 474, 475, 476, 478, 480,
    481, 485
Herberg, L. J., 208, 216
Herlant, M., 525, 528
Herman, L., 63, 71, 100, 106
Herman, M. M., 441, 442, 446, 457
Herman, T. S., 823, 825
Hernandez, W., 759
Herrick, C. J., 12, 28, 29, 38, 444, 446, 456
Herries, D. G., 911
Herschman, H., 1017
Hershey, A. D., 955, 961
Hertz, L., 795, 796

Herz, A., 272, 476, 485, 639, 647
Herz, M. J., 278
Herzberg, M., 927
Herzenberg, L. A., 890, 899
Hess, A., 751, 753, 755, 759, 791, 796, 806, 811
Hess, E. L., 675, 677
Hess, W. R., 208
Heuser, G., 241
Heuser, J. E., 722, 725, 726, 727, 791, 796
Hibbard, E., 110, 114, 123, 128, 996, 998
Hicks, R. G., 241
Hicks, S. P., 100, 106
Higuchi, T., 760
Hilali, S., 619, 628
Hild, W., 91, 99
Hilfer, E. K., 82
Hilfer, S. R., 80, 82
Hilgard, E. R., 260, 271
Hill, A. V., 725, 727
Hill, B. J., 745
Hill, C. W., 983, 997
Hill, D. K., 712, 714
Hill, R. L., 894, 899, 911, 964, 971
Hill, R. M., 308, 315, 485
Hillarp, N.-Å., 112, 114, 327, 334, 526, 774,
    780
Hille, B., 395, 708, 709, 877
Hillman, D. E., 409, 410, 411, 412, 413, 414,
    420, 422, 424, 425, 442, 469, 866
Hillman, P., 1017
Hillyard, S. A., 195, 289, 1017
Hilschmann, N., 890, 899
Hilse, K., 896, 899
Hiltz, F. F., 649, 661
Himwick, W. A., 290, 296
Hind, J. E., 596, 661
Hinde, R. A., 188, 195, 348
Hinds, J. W., 64, 65, 67, 69, 71, 136, 137, 140
Hinegardner, R., 986, 997
Hirano, A., 722, 727, 787, 788, 790, 791, 795,
    796, 846, 849
Hirata, Y., 546, 551, 560, 564
Hirsch, J., 189, 195
Hirsh, R., 1017
His, W., 63, 70, 100, 106
Hiscoe, H. B., 79, 82, 86, 99, 842, 847, 850
Hník, P., 111, 113, 850
Ho, W., 99, 815, 825
Hoag, A., 80, 82, 112, 116
Hoag, W. G., 104, 107
Hoagland, H., 236, 242
Hoch, P. H., 360, 371
Hodgkin, A. L., 391, 396, 554, 564, 597, 604,
    630, 647, 708, 713, 714, 787, 792, 796, 858,
    866, 869
Hodos, W., 7, 20, 21, 25, 26, 27, 36, 38, 194,
    1017
Hoebel, B. G., 208, 211, 216
Hofer, H., 42, 46, 360, 371, 521, 528
Höfer, I., 696
Hoffer, B. J., 336, 742, 746, 878, 1017
Hoffman, H., 111, 114, 820, 825
Hofmann, K., 905, 911
Hofstätter, P., 356, 360
Hohmann, P., 823, 825
Hohn, Th., 1002, 1013
Hökfelt, T., 325, 331, 334, 335, 526, 528, 732,
    734, 744, 745, 773, 780, 1017
Holden, A. B., 104, 107, 782, 797
Holland, I. B., 945
Holland, Y., 843, 850

Witschi, E., 205, 206
Wittkower, E., 348
Wittmann, H. G., 920
Witzel, H., 909, 912
Wohlfart, G., 843, 850
Wohlwill, J. F., 180, 187, 188
Wolbach, S. B., 58, 61
Wolf, G., 222, 223
Wolf, K. M., 357, 360
Wolf, M. K., 104, 105, 107, 782, 797
Wolfe, D., 537
Wolfe, D. E., 733, 747, 782, 792, 797
Wolfe, L. S., 747, 759
Wolff, H. A., 616
Wolff, J., 793, 795, 797
Wolff, P. H., 361
Wolfgram, F., 816, 826
Wolken, J., 683
Wollman, E. L., 958, 961
Wolman, E., 493
Wolpert, L., 126, 128, 615, 616, 857
Wolstenholme, G. E. W., 82, 97, 99, 106, 113, 115, 151, 157, 443, 797, 826, 850, 857, 998
Wolthers, B. G., 911
Wong, C. L., 200, 206
Wood, 286
Wood, D. D., 828, 829, 839
Woodburne, R. T., 109, 116
Woodland, H. R., 962, 971
Woodruff, J. J., 750, 760
Woods, P. S., 973, 998
Woodward, D. J., 742
Wool, I. G., 822, 825, 826
Woolley, D. W., 750, 760
Woolsey, C. N., 324, 605, 606, 615, 616, 617
Worden, F. G., 1020
Wortham, R. A., 111, 113
Worthington, C. R., 867, 879
Wragg, L. E., 528

Wright, E. W., Jr., 316
Wright, J. C., 186, 187
Wright, W. E., ix
Wu, R., 596, 955, 962
Wu, S.-Y., 296
Wurtman, R. J., 520, 527, 529, 530, 531, 532, 533, 534, 535, 538, 1020
Wyatt, G. R., 822, 826
Wyckoff, H. W., 901, 903, 905, 906, 912
Wyers, E. J., 241
Wyman, H., 586, 629, 671
Wyman, R. J., 398, 399, 400, 405, 409
Wyrwicka, W., 257, 259

Y

Yahr, M. D., 72, 334, 442, 456, 457, 468, 469, 470
Yamamoto, C., 543, 544, 552
Yamamoto, T., 205, 207, 552
Yamasaki, S., 824
Yang, S.-J., 81
Yankelevich, G., 586
Yankofsky, S. A., 931, 945, 983, 989, 998
Yanofsky, C., 927, 983, 996
Yarmolinsky, M. B., 958, 961
Yih, C. S., 847, 849
Yokota, D. T., 345, 347, 348, 349
Yonehara, H., 696
Yoshida, M., 423, 425
Yoshii, N., 243, 258, 259
Yoshino, S., 585
Young, B. A., 520, 529
Young, E. T., II, 962
Young, D., 810, 812
Young, G. A., 222
Young, I. J., 751, 760
Young, J. Z., 27, 39, 40, 47, 81, 194, 196, 330, 336, 563, 564, 743, 747, 809, 812

Young, L. R., 494, 956
Young, W. C., 199, 205, 206, 207, 294, 297

Z

Zacchei, A. M., 99
Zadeh, L. A., 384, 396
Zadorozhny, A. G., 670
Zaltzman-Nirenberg, P., 336
Zanetti, G., 911
Zangwill, O. L., 169, 171, 172, 176, 605, 616
Zanini, A., 817, 819, 827
Zankov, L. V., 185, 188
Zapfe, H., 43, 47
Zaporozhets, A. V., 184, 185, 188
Zatman, L. J., 81
Zatz, M., 751, 753, 754, 760
Zawarzin, A., 798, 800, 804, 812
Zeaman, D., 177, 188
Zeevi, Y. Y., 271, 396, 878
Zelená, J., 111, 116, 849
Zhegalkina, N. G., 241
Zichichi, M. L., 958, 962
Ziegler, H. P., 35, 39
Zieher, L. M., 732, 735, 736, 745, 746
Zigler, E., 186, 187, 188
Zikmund, V., 265, 270
Zimmerman, H. M., 796, 849
Zimmerman, R. R., 34, 39
Zinchenko, V. P., 184
Zinder, N. D., 930, 931, 945
Zissler, J., 958, 962
Zomzely, C. E., 813, 826, 827
Zonta, L. A., 899
Zotterman, Y., 518, 598, 604
Zubin, J., 360, 371
Zucker, R. S., 555, 565
Zuckerkandle, E., 965, 966, 972

# SUBJECT INDEX

*Page references to figures and tables appear in italics*

## A

*Acetabularia*
protein synthesis in, 841, 842
acetoxycycloheximide
glycoprotein and, 751, 753, *753*
memory, effect on, 272-273, *272, 273*, 275
protein synthesis inhibition by, 272ff, *272*
acetylcholine, 531, 770ff, *770*
biosynthetic pathway, 772
insect nervous system, effects in, 792
localization, 764-765, 771
metabolism, drugs and, 340
miniature endplate potential and, 715, 774
muscle fiber sensitivity to, 776
as neurotransmitter, 329
in synaptic vesicles, 722-723, 725, 771
acetylcholinesterase
colchicine blocking of, 846, 848
localization, 765-766
actinomycin
memory impairment effects, 275, *275*
RNA synthesis, effect on, 275
ACTH (adrenocorticotropic hormone)
effect on PNMT activity, 533
localization of, 764ff, *765*
in mammalian adrenal medulla, 533, *533*
subcellular and released, 765
action potential
mechanism for, 858-859
of nerve tissue in vitro, 89, *94*
neuronal, extracellular recording, 649
neuronal, intracellular recording, 648-649
prolongation, 707, *707*
activating system
nonspecific, 377
adaptive behavior, 188
affective state, role in forming, 330
of birds, 41-42
genetic influences, 192ff
innate and learned responses, 193-194
phylogenetic selection pressures and, 194
of primates, 42ff
protracted parental care and, 194
of rats, 42
reinforcing events, 193
ritualization as, 350
stimulus significance and, 330
adenohypophysis, 525
neurosecretomotor junctions in, 523
adenyl cyclase, 535
function, 871, *872*
ADP (adenosine diphosphate)
transport across mitochondrial membrane, 830-831
*see also* oxidative phosphorylation
ATP (adenosine triphosphate), 830ff
as energy source, 828
in mitochondria, 828ff
adrenal medulla
ATP release during stimulation, 775, *775*
catecholamine secretion, 775, *775*
chromaffin cells in, 532, 533
chromogranin A release, 775, *775*

derivation from neural crest, 62
exocytosis in, 775-776
as a neuroendocrine transducer, 533
adrenalin, 770
adrenomedullary chromaffin cells, 530
affect, 324, 337-338, *338*, 341
adaptive behavior, role in, 330
amine roles, 324ff, *332*
behaviors guided by, 338
brain correlates, 302
brain mechanisms, 344-345
defined, 337
during epileptic seizure, 341ff
humoral and neural mediation, 331
individuality and, 346
limbic system role, 302
norephinephrine role, 327
schematic view of, 338
self-orienting nature of, 346
*see also* emotion; individuality, sense of
afterimages, 306, *307*
complementary images, *307*
waterfall effect, 308
*see also* optical illusions
afternystagmus, 264-265, *265, 266*
aggregation culture, 105-106, *105*
aldosterone
binding to specific nuclear protein, 823
*see also* hormones
alga, *see Acetabularia*
alkali ion
ligand binding and, 685ff, *685, 686*
alkali-ion carrier complexes, 693-696
equilibration, 689ff
macrotetrolide, structure, 691, *692*
stability constants, 685-686, *686*, 687
alkali-ion carriers, 685-696
allelic exclusion, 893
alligator
parallel fiber activation, 420
allocortex
neurogenesis, 65, *68*
*see also* isocortex
amacrine cells, 563
*see also* axonless cell; granule cell
amines
appetitive behavior, role in, 328, 329
biogenic, 324-336
emotion, role in, 324
learning, role in, 330ff, *332*
memory, role in, 329-330
metabolism, drugs and, 340
neurotransmitters in brain, 327
in non-neuronal tissues, 771
*see also* catecholamine; dopamine; epinephrine; norepinephrine
amino acid sequences
of light chains of protein Eu, 889, *889, 891, 892, 895*
C region, 890ff
V region, 890ff, *893*
amnesia, anterograde, 164, 168, 169, 173
classical conditioning in, 170
due to Papez circuit lesions, 172

motor skills in, 169, 170
operant conditioning in, 171
perceptual learning in, 171
primary memory in, 169
secondary memory in, 169
sensory memory in, 169
amnesia, retrograde, 164, 171, 172, 173, 273-274
electroconvulsive shock effects, 273-274
primary memory in, 171
puromycin and, 275
secondary memory in, 171
tertiary memory in, 167, 171
AMP (adenosine monophosphate), cyclic function, 330, 871-872, *872*
norepinephrine, effects on, 333
amphetamine
memory, effects on, 273, *273*
amphibians
motor system organization, 109
optic nerve regeneration, 154
phylogeny, 28-29, *29*
*see also* frog; salamander; toad; urodeles
amputation, larval limb
effect on peripheral nervous system, 76ff
effect on ventral-horn cells, 76, 78, *78*
amygdala, 464, 465
androgen
alteration of nerve tissue hormone sensitivity, 201, 205
central nervous system action, 294
male social behavior and, 201-205, *202, 203*
*see also* testosterone
androstenedione, 200
*see also* androgen; testosterone
animal behavior
social communication, 349-359
stereotypic, 291, 339
*see also* adaptive behavior; comparative psychology; ethology
annealing
of polynucleotide chains, 951, 988, 989, 991
ANS (1-anilino-8-naphthalenesulfonic acid), 834
ansa lenticularis, 465, 466, *466*, 467
anterograde amnesia, *see* amnesia, anterograde
antibiotics
complex formation constants, 685-686, *686*
effects on ribosome, 923, *923*
"hole punchers," 682
ion-selective, 685, *685*
ionophorous, 831
ligand role in complex formation, 686-687
mitochondrial membrane permeability, effect on, 831
antibodies, 882-900
complementarity with hapten antigen, 886
constant regions, 889, 890, 893
degenerate response of, 886-887
diversity of, 889ff
domain hypothesis, 896
eukaryotic differentiation and, 898
evolution, 893, 894, *895*, 896
gamma globulin, 962-963, 969
genetics of, 885-900
heterogeneity of, 887

cell adhesion, 125-126
    cell migration and, 126
    specificity of neuronal connections and, 126
cell alignment, 104ff, 106
    hippocampal formation, 105, *105*
    mechanism, 106
cell asynchrony, 76
cell count
    by autoradiography, 74
    by mitosis, 74
    of ventral-horn cells, 74ff, *75*
cell culture, 74, 83-84
    aggregation cultures, 105-106, *105*
    bioelectric maturation, 89-95
    cerebral tissue, embryonic, 94-95, *96*
    CNS tissue, 94-95
    enzymatic dissociation, 85, *87*, *92*
    Feulgen staining, 63
    growth cones in, 85
    histiotypic aggregation in, 86, 89
    mechanical dissociation, 95
    mutants, 104-105
    monolayer, 86, 95
    reaggregation, 105
    silver stain, 89, *91*
        *see also* culture media; explant culture; nerve
        tissue culture
cell death, 58, 71, 73
    function, 79
    histogenetic, 79, 109
    morphogenetic, 79, 109
    ventral-horn cells, 74, 76ff, *77*
cell degeneration
    significance of, 79
cell differentiation, 69-71
    ablation effects, 79
    independence of neurogenetic events, 69ff
    of neural nets, 1011, 1012
    spinal cord regions, 73
    vs proliferation of neurons, 841
cell division, *see* cell proliferation
cell evolution, 930
cell interaction, 100-107
cell membrane, *see* membranes
cell metabolism
    peripheral influences, 80-81
cell migration, 69, 100-107
    cell adhesiveness and, 126
    cerebellar granular layer, 103, *103*, 105
    in cerebral isocortex, 101, *101*, 104, *104*
    in CNS development, 100, 102, *102*, 103-104
    into ventral horn, 79
    neuroblast behavior, 69, 70
    process, 103-104
    trapezoid nucleus, 103-104
cell proliferation, 62-69, 100-107
    autoradiographic analysis, 64
    brain components, 65ff, *68*, *70*
    cell death and, 102
    control of, 102-103
    historical perspective, 62-63
    histology of, 63
    neural tube, 62-63
    neuroglia, 69
    patterned, 65ff, 101, 103
    rate, 102-103
        *see also* ventricular cell
cell shape, 104-105
cell turnover, 76
cell types
    germ, 962, 968

somatic, 962, 963
central canal
    ciliary beat in, 58
central nervous system (CNS)
    aberrations, 58
    amines in, 324-336
    androgen actions on, 294
    cell aggregation patterns in, 106
    cell types in, 100ff
    conduction pathways, *9*
    cortical system, defined, 539
    developing, in mammals, 100-107
    electrical activity in, 222
    functional operations, 379-382, *381*
    mammalian, 100-107
    morphogenesis, 58
    morphology, nonmammalian, 8-10, *9*, 13
    neurons, quantity in, 12
    nonmammalian vertebrate, *9*
    norepinephrine release in, 327-328
    systems analysis applied to, 380-382, *381*
    tissue culture, 91-95, *96*
    vertebrate and invertebrate compared, 798-
        801, *799*, *800*
central nervous system development, 62
    cell migration, 100ff, *101*, *102*, 103-104
    cell proliferation, 100, *101*, 102-103
    dynamics, 53, 54, 56
    macrodeterminacy, 54
    microdeterminacy, 54
    selectivity, 58, 59
    zones, 100
cephalopods
    dendritic field, 810
cerebellar cortex
    cell aggregation in vitro, 106
    functional organization, 413-419
    glomerulus in, *430*
    granular layer, cell development in, 103, *103*
    internuncial neurons, 419
    morphological organization, 409-413
    Purkinje cells, 103, *103*
    stimulation in alligator, *421*
cerebellar glomeruli, 427, *429*, *431*, 435, 440, 441
    as an excitatory synapse, 429
    as an inhibitory synapse, 429
    axon terminals, 427
    dendritic terminals, 427
    ultrastructure, *429*
cerebellum, 409-426, 466, 468, 469
    as a closed-loop system, 424
    as an open-loop system, 424
    basic circuit, 409-411, 413
    function, 379, 423-424
    and motor skills, 415
    neuronal fabric of, *55*
    neuronal operations in, 409-426
    neurons in, 55
    relationships of neural elements, *416*
cerebellum, fetal
    cell proliferation and migration, *101*
    neuron differentiation sequence, *139*
cerebral cortex, *445*, *449*, *450*
    brain waves, 244-246ff
    evoked potential, 244-246ff, 250-254, *255*
    evolution of, 7-26
    function in visual pathway, 377
    homology, 22, 361
    mammalian and nonmammalian, 10, *11*, 22ff,
        *24*
    neuron differentiation sequence, *139*

neurogenesis, 65, *139*
    ontogenetic derivation of neurons, 24
    photically responsive areas, 342, *343*
    phylogeny, 19ff, 339
    sensory representation, properties of, 613
        *see also* allocortex; corpus striatum isocortex;
        external striatum; visual cortex
cerebral dominance
    aphasia and, 362
    handedness and, 362, *362*
    language disorders and, 365
cerebral hemisphere, nonmammalian, *9*
    morphology, 8-10
    sensory conduction to, *9*, 16
cerebral lesions, *see* brain lesions
cerebral tissue
    conductive changes and information storage,
        225
    electrical impedance, regional modifications,
        240
    nerve membrane polyanions, 239-240
cerebrospinal fluid
    ionic environment, 787
    turgor, 58
cerebrum
    neurogenesis, 64
    tissue culture, 92, 94-95, *96*, *98*
chain elongation factors
    in protein synthesis, 923, 924
chelation
    free energy, alkali ion and, 687, *687*
chemoreception, 511, 516, 517
    silk moth (*Bombyx mori*), 514, *515*
        *see also* insect olfaction
chemotaxis
    in nervous system, 123, 126
chick embryo
    brachial nerve plexus, 112
    deafferentation, 143-144, *143*, *144*, 147
    dissociated sensory ganglia of, 84-99, *87*, *88*,
        *90-93*
    electrical activity, 144-147, *146*, *148*
    hatching, preparation for, 149
    innervation of limbs, 112
    integrated movement, 149
    motility, 142-144, *142*, *144*
    stimulated movements, 149-150
chicken
    hatching behavior, 149
    pecking behavior genetically determined, 193
child development, 176ff, 179ff
    verbal stimuli role, 183
        *see also* cognitive development
chimpanzee
    EEG patterns of decision-making, *233*, 234-
        236, *236*
    object use by, 43, 44
    social behavior, 43, 44
    speech, lack of, 54
        *see also* apes (*Pongidae*)
cholesterol
    hydrocarbon chain motions inhibited by, 703
choline acetyltransferase
    localization, 765
cholinergic terminals
    organization of, 764
chordamesoderm, 62
chromaffin cells of mammalian adrenal medulla,
    532, 533
    PNMT in, 532, 533
chromatolysis, 73

in chickens, 149
heart rate training, operant, 218-220, *219, 220,* 221
 behavioral consequences, 221-222
Hering's law, 472
hermaphroditism, 205
heterosaccharides, 756
 surface, of retinal and tectal cells, 756
heterosynaptic facilitation (HSF), 263-264
 episynaptic action and, 263
 as a memory mechanism, 263-264
 minimal latency, 264
 optimal interstimulus interval, 264
 specificity, 263-264
HIAA, *see* hydroxyindole acetic acid
hind-limb
 cortical representation, 606, 608, *608, 609,* 610, *611*
HIOMT, *see* hydroxyindole-O-methyl transferase
hippocampal formation
 cell alignment, 105
 cell pattern, reaggregated, 105-106, *105*
 cell proliferation, *68*
 differential action on neuron inputs, 345ff, *346*
 intracellular recordings, *347*
 neurogenesis, 64, 65
 olfactory stimuli elicitation of EPSPs, 346, *347*
 pyramidal cell firing, 226-227
  *see also* area dentata; hippocampus; retro-hippocampal formation; septum; subiculum
hippocampal gyrus, *345*
 functions, 342-343
 photic stimulation of, 342, *343, 344*
hippocampus
 amnesia from electrical stimulation, 172
 amnesia from lesions, 172, 173
 EEG data from, *231*
 epileptogenic focus, 341
 location of, *68*
 memory, role in, 231-232
 memory stimulation in, 173
 neural circuit interactions, role in, 229ff, *233, 346, 347*
 neuronal origins in, *68*
 neurons and learning, 282ff, *284, 287*
 protein in, neuronal, 282, *284, 285,* 288
  *see also* hippocampal gyrus
histiotypic aggregation, 86, *88, 89*
histograms, cross-interval, 651ff, *654-658*
 peristimulus-time, 649ff, *651, 659*
histones, 981, 982, 983, 985, 1010
homeostasis
 visceral conditioning and, 222
*Homo erectus,* 44
homologous response, 110, 153-155
 to homonymous muscles, 109-110
homology
 antibody regions, 894, *894, 895*
 brain, *11*
 cerebral cortex, *361*
 corpus striatum, 19-20, 22-25
 forebrain, 10, 18-20, 22
 lemniscus, 14
 mesencephalon, 16-17
 neocortex, 18
 neuronal systems, 8
 regions, 896, *896*
 responses, 153-155
 rhombencephalon, 10
 thalamus, 14

vocal system, 357
homosexuality
 hormonal influences on, 200
hormones
 adrenocorticotrophic (ACTH), 333
 behavioral maturation, role in, 293-294
 cortical synapses and, 333
 deficiencies, 197-199
 gene activity and, 823
 influences on sexual behavior, 196-207
 sexual development affected by, 196-207
 and social behavior, 201-205, *202-204*
 steroid, 333
 testicular, 199-205, *204*
  *see also names of specific hormones;* neuro-hormones; testosterone
horopter, 477ff, *478*
hydrogen bonds
 in lysozyme-inhibitor complexes, 904
 in proteins, 902, 903, *903*
 in ribonuclease-nucleotide interactions, 907, 908
hydrocarbon chains, 701-703
 double bonds, 702-703, *702*
 low-energy conformation, 702, *702*
 melting, 701-702
 molecular fluidity and, 703
 motion increased by cholesterol, 703
 unsaturation and perpendicular transport, 704-705
hydroxyindole acetic acid (HIAA), 535
hydroxyindole-O-methyl transferase (HIOMT), 534, 535
hyperplasia, 78, 79, 80
hypophysiotropic hormones, 531, 532
hypophyseal portal system, 525, 526
hypoplasia, 78
hypothalamus
 electrolytic lesions between brainstem and, 211-212
 neural circuitry for drives, 291
 neurohormones released by, 521, 527, 530ff
 neurosecretory system, 778-779, *778*
 stimulation effects, 208, *209, 211, 212, 215, 216*
hysteresis
 in insect muscle control, 406-407, *407*

## I

identical retinal points, *see* corresponding retinal points
immune response, 885ff
 adaptive, 885
 diversification in, 893
 translocation hypothesis, 893
immunity
 mechanisms, 884
immunoglobulin, 885ff, 995-996
 7S antibody increase in, 885
 antigen-binding function, 894, *895*
 effector functions, *895*
 gamma G, 885ff, *886, 895*
 gamma M, 885, *885*
 synthesis of, 885ff
  *see also* amino acids; antibodies
imprinting, 190-191
individuality, sense of, 342
 aura and, 342
 limbic system and, 346
 neural mechanisms, 342ff

self-orienting nature of affect and, 346
information
 from spike trains, 587
information processing, 486-487, 489
 hierarchical, 492ff
information registration
 defined, 164
information retention
 defined, 164
information retrieval, 173
 defined, 164
 in primary memory, 166, 168
 in retrograde amnesia, 164
 in secondary memory, 167-168
 in tertiary memory, 167-168
 puromycin effects in, 275
information storage, 163
 and amnesia, 164
 in CNS, 397
 in primary memory, 166
 in secondary memory, 166, 168
 in sensory memory, 165
 in teritary memory, 167
 nerve membrane structure and, 240
 neuronal sites of, 260-261
 nonsynaptic, 264-270
 by synapses, 261-264, *262, 263*
information theory, 355-356, 572
 dominance behavior and, 356-357, *356*
 neural transmission and, 624
 neurophysiological application, 572-574, *573, 574*
 social communication and, 355-357
 stochastic analysis of bidirectional communication, 355-356, *355, 356*
information transfer, 173
 brain electrical rhythms and, 224
 in cerebral systems, 232
 efficiency, firing pattern and, 627-628
 interneuronal, 665-666, *667*
 retrograde, 79, 80
information transmission (in nervous system), *see* neural transmission
information updating
 in ventrobasal cell matrix, 446
inhibition, 492
 excitation and, 491-492
 reciprocal, *393*
 recurrent, 490
 functional, 491
 lateral, of mitral cells, 562
 self-, of mitral cells, 562
  *see also* IPSP; lateral inhibition
inhibitory postsynaptic potential, *see* IPSP
initiation factors
 in protein synthesis, 923, 924, *924*
innervation
 limbs, 112
 muscle cells, 111
 rotated skin graft, 117-118, *119, 121*
 and spinal cord of chick embryo, 80
  *see also* nerve-fiber connections
insect olfaction, 511-518
 Poisson distribution and, 515-517, *515*
 receptors, 511, 512-513, *513*
 signal-noise ratio, 515
 transduction, 516-517
 receptor cell response threshold, 512, 514-516, *515*
insects
 acetylcholine effects in nervous system, 792

antennae, 512, 516
eye of, 494-511
ganglionic neurons, 798, *799*, *800*
motor-control systems, 490
muscle, *407*
olfactory receptor cells, 513ff, *513*
sensillum, 513, *513*
sensillum liquor, 513, 514
sexual attractant, 512-513, 517
*see also* pheromones; silk moth (*Bombyx mori*)
insertion-excision system
in prophage, 960
instinct, 189
instrumental learning, 218-223
*see also* conditioning, operant
insular cortex, 344-345, 346
auditory units, 344
somatic units, 344
insulin, 531, 532
and NGF, 821-822
intelligence
concrete, 179-180
formal operational, 180
human, 176-188
preoperational, 179-180
*see also* cognition
Intensive Study Programs, v-viii
interference theory, 166-167
proactive inhibition, 166-167
retroactive inhibition, 166-167
*see also* forgetting; learning
intermediate zone, *see* mantle zone
interneuron, 130, 411, 413, 420
brain development, role in, 67
crayfish, 399, 402-404, *402*
defined, 409
ontogeny, 411, 413
origin, 65
phylogeny, 411
interneuronal interface
thalamic, 465-469
interneuronal recognition
glycomacromolecule role, 750, 758
internuncial neurons
of cerebellar cortex, 419
intertubular canals, 845, 846, 847
intestinal contraction
operant conditioning of, 220, *221*
invertebrates
central neurons of, 798-811, *799*, *800*
nerve-cell maps, 801
nerve ensheathment, 791
nervous systems of, 260, 276
neural conduction in, 791
neural transmission in, 720-721
*see also* insects
ion carrier
peptide, *870*
solvation change, 869, *870*
ion injection, 681-682
ion transport
alkali-, 691ff
membrane fluidity role, 703-704
mitochondrial, 830-831, *830*
rate-limiting step of, 695
relaxation spectrometry for study of, 695
solvation and conformation changes, 869, *870*
*see also* membrane; permeation
IPSP (inhibitory postsynaptic potential), 452, 453, 458ff

alternation with EPSPs, 450, 461-462, *461*, *462*, 467, 468, 578, *578*
attenuation, 462
mean rate influence on, *579*
mitral cell membrane hyperpolarization and, 544
neuron discharge synchronization, role in, 460-462, *460*, *461*, *462*
thalamic stimulation and, 463ff
in thalamocortical afferents, 459
*see also* EPSP; PSP
isocortex, 464-466
homology, 17, 22, 24
limbic system, dichotomy of function with, 341
neurons, 65, *66*
phylogeny, 19, 22-25, *24*
stimulus discrimination in, 345, 347
*see also* external striatum

J

*jamais vu* phenomenon, 366
*see also* cognitive disorders
juxtaglomerular cells
of mammalian kidney, 532

K

Klein bottle
characteristics, 612
as cortical model, 612, *612*, *613*
Korsakoff's psychosis, 169, 170, 173

L

L-ascorbic acid synthesis
in birds, evolutionary trends, 31ff, *33*
labor, division of, 44
labyrinthodonts, 28-29, *29*
lactose (*lac*)
genes, relative size of, *950*
operon, 947-948
phenotypic complexities, 953
region of *E. coli*, 947ff, *947*
repressor, 948-950, *950*, 952, *952*, 953
transport, 704
lactose operon, 947-948, *947*
Land effect, 304
language, 181-182, 184, 302, 361-371
biological origins, 181-182
brain correlates of, 361-371
brain topography and, 362, 367
characteristics of, 181, 368-370
cognition and, 182-183, 264-365, *364*, *365*, 367-368
color recognition and, 182
concept formation facilitated by, 182-183
critical periods of learning, 292
development of, 181-182, 183, 291, 292
etiology, 365-366
evolution of, 45
"naming" adaptation, 45
neurological correlates, 363ff
origin of, 45
recognition and, 366-368
second signal system of, 183
simian, 45
skills, dissociability of, 364, *364*, 365, *365*
sound patterns, processing of, 369
speech and, 364, 365

speech praxia, 365
stimulus-response (S-R) models of behavior and, 177
symbolic processes, 183
vocal system and, 357, 359
word-presence, 365
*see also* cognition; linguistics; speech; vocalization
language capacity, 291
characteristics of, 368-370
ineradicability of, 365
vocal-tract structure and, 363
neuroanatomy and, 361-363
brain activity patterns and, 362, 365
brain size and, 361-362
*see also* speech
language disorders, 363, 367
brain lesions and, 361ff, 367
cerebral dominance and, 362, 365
cognitive disorder similarities, 366-367
etiology, 365-366
language-specific computation, test of capacity for, 370
logorrhea, 365
symptom dissociability, 364-365, *364*, *365*
*see also* aphasia; cognitive disorders; speech disorders
language knowledge, 370
defined, 364, 365
lateral geniculate body (LGB)
degeneration in, 345
evoked potential in, 256, *256*
glomerular synapses in, 431, 433, *433*, *435*, 441
presynaptic inhibition in, 433
recurrent inhibition in, 640, *642*
response to inhibitory input, 640, *642*
receptive-field inhibition, 640, *642*
receptive fields, 638, 639
synaptic organization of transmission in, 638-642, 646
*see also* medial geniculate body
lateral inhibition, 419
analysis by circuit theory, 389, *392*
in flies, 399
of mitral cells, 562, 563
in olfactory bulb, 549
in peripheral sensory systems, 394
in retinal ganglion cell, 636-638
learned glandular responses, 218-223
learned responses
electrical brain rhythms and, 224-243
learned visceral responses, 218-223
heart rate, 218-220, *219*, *220*, *221*
homeostasis and, 222
intestinal contractions, 220, *221*
reinforcement of, 218, 219-220
specificity of learning, 220-221
learning, 162, 193-194, 218-223
affective state and, 331, *332*
biogenic amines and, *332*, 324-336
memory and, 164
biochemical changes in neurons during, 278, 281-288
brain response patterns and, 295
brain rhythms and, 224-243
critical periods for, 292
disorders, 292
genetic differences and, 189
glandular, 222
innate propensities and, 194
instrumental, 218-223

memory consolidation and, 189
memory establishment following, 272-277
motor, 167
neural mechanisms, 260
norepinephrine release and, 333
observational, 191
perceptual, 171
phylogenetic comparisons, 194
protein changes in, 278-289, *283*
skeletal, 218-220
specificity, 220-221
plasticity and, 162
S100 protein and, 285-286, *285, 286,* 287ff
protein synthesis during, 333, 823
RNA synthesis during, 823
self-preservation, role in, 193
stability and, 162
trial-and-error, 192-193
verbal, 167
*see also* cognitive development; conditioning; memory
learning retention
actinomycin effects, *273*
protein synthesis effects, 272, *272*
*see also* memory
learning theories
American, 177-179
education and, 185-186
Soviet theories, 183-184
stimulus-response theories, 177
*see also* cognitive development theory; Piaget
learning set, 34
comparative development, *35*
evolutionary trends, 34-36, *36*
lemniscus, *11, 13*
glomeruli and, 433
homology, 14
nerve fiber terminal patterns, 446
neurons, functional characteristics, 446
lemniscus, medial
tactile-kinesthetic function, 605ff
Lewis blood group
fucosyl transferases and, 748
leucine
incorporation into acidic protein during learning, 281-285, *284, 285*
incorporation into neuronal protein during conditioning, 286-287, *287, 288*
LGB, *see* lateral geniculate body
ligand
substitution in metal complexes, 687-688
ligand binding
free energy of, 686, *686,* 687
limb
cortical representation of hind-limb, 606, 608, *608, 609,* 610, *611*
limb movements
reflexive, misdirected, 117-118, *120, 121*
visual deprivation effects, 321-322, *322*
*see also* skin grafts
limb transplant
innervation, 112
mytotypic respecification, 154
urodele spinal cord and, 154
limbic system, 10, *314,* 377-378
attitudes, influence on, 346-347
brain lesion effects, 377
bull-taming experiment, 377
electrical stimulation effects, 377
emotion, role in, 302, 339ff
insular cortex, 344-345, *346*

isocortical dichtotomy of function, 341
limbic cortex, 339-340, *341*
limbic lobe, mammalian, 339, *340*
morphology, 339-340, *341*
nomenclature, 10
nonmammalian morphology, 10
olfactory impulses, 345, 346, *347*
psychic functions, 347
serotonin in, 340
thalamic link to, 458
*Limulus*
eye model, *392, 393*
linguistics, 178
semantics, 369
syntax, 178, 369
transformational grammar, 178, 368-369, *369*
*see also* language
lipid
hydrocarbon chain packing in, 702-703
in membranes, role of, 703
*see also* glycolipid; hydrocarbon chains
lipid bilayers, 677-684
additives, effects on conductance, 682-683
antibiotics and, 682
bubbles, 679, *679*
construction of, *680*
field dissociation effect, 682
formation, 679-680, *679, 680*
gateable additives in, 682-683
ion carriers in, 682
ion transport in, 682, 869
microvesicles in, *680*
nonlinear conductance, 681ff
potassium ion permeability, 682
proton carriers in, 681
study of, 678ff
structure of membrane fluid regions, 701
lizard (*Iguana iguana*)
head-nodding display, 350, *350*
lobster (*Homarus americanus*)
walking-leg nerve fiber, paramagnetic resonance in, 698, *700*
walking-leg nerve fiber, membrane fluidity, 701
logorrhea, 365
long-term memory, *see* memory, long-term
lordosis reflex
castration effects on, 198
estrogen effects on, 199
suppression by hormones, 200
testosterone effects on, 199
*see also* sexual behavior
lysogenization
by bacteriophage lambda, 955, *955,* 956
lysozyme-inhibitor complexes, 904, *904*
binding in, 904
conformation in, 904
hydrogen bonds in, 904
specificity in, 904

## M

mDNA, *see* DNA, mitochondrial
macaque, *see* monkey
Mach band, 394
macrodeterminacy
in CNS development, 54
*see also* microdeterminacy
macroglia, 63
macromolecular nets, vi ff
macrotetrolides

chemical structure, *691*
*see also* monactin
mammals
fetal behavior in, 149ff
neuroendocrine transducer cells in, 530-538
phylogeny of, 29-30
*see also* individual species
mammals, placental
phylogeny, 30, *30*
man
evoked potential, conditioning of, 250, *250,* 252, *252*
evolution, 44-46
maturation rate, 42, 44
object use, 44
nonverbal communication, 357-358, *358*
phylogeny, 42ff
social behavior, 44-45
stress, EEG patterns during, 236-239
*see also Australopithecus: Homo erectus;* primates
mantle zone, 63, *64*
marginal zone, 63
masculine social behavior
androgen effects, 201-205, *202, 203*
gonadectomy effects, 201, *204*
neural structure influence on, 205
master genes, *see* gene, master
mathematical models
dendritic branching, 552, 553
neurons, 552-554, *553, 554*
maturation, 176, 184, 293
cognitive development and, 180, 184, 290
eye movements and, 184
genetic determinants, 293
genetic "switching," 294
hormonal influences, 294
language development and, 182, 291
*see also* behavioral ontogeny
Mauthner neurons
circuit system, 393, 394
of fish, 398, *398,* 403
maze-learning, 291
genetic factors, 189
medial geniculate body
glomerular synapses in, 433
*see also* lateral geniculate body
medulla oblongata, fetal
cell proliferation and migration, *101*
melanocytes, 62
melatonin, 531, 532, 535
biosynthesis in mammalian pineal organ, *534*
neural and photic regulation of synthesis of, 534, 535
norepinephrine effects on, 535
membrane
bilayer models of, 678-681, *679, 680,* 869
binding sites, *704*
circular dichroism in, 868
cyclic antibiotics and, 682
DNA in, 867
excitation in, 678, 685, 708-713, 869-870
fluidity of, 701-705
gateable additives and, 682
hydrophobic regions of, 697, *697,* 701
information transmission in, 681, 870
ion carriers in, 678, 681, 682, 685-696, 869, *870*
ionic theory of, 708
lipid bilayers in, 677-684
lipid-protein interaction in, 683

lipids in, 867, 869
matrix of, 867
mitochondrion, 677
molecular components of, 867, *868*, 869
molecular motion in, 697ff, 701
molecular organization of, 867
myelin, 677
and optical rotatory dispersion, 868
organization of, 867
permeation of, 678, 703
protein in, *816*, 867
RNA in, 867
spin labels in, 697ff, 869
structure of, 704, *704*, 867-869, *868*
systems, 867-870
transport in, 704-705
two-stable-states theory of, 707
types of, 677
"unit," model of, 867
variations in, 677
    *see also* biological membranes; lipid bilayers;
    mitochondria
membrane excitation
    methods of studying, 858
    two-stable-state hypothesis, 858
membrane fluidity
    biological functions, 703-705
    molecular origin, 701-703
    perpendicular transport and, 703-704
membrane matrix
    molecular organization, 867-869
    pump, carriers, and other membrane effectors,
    867, *868*
membrane models
    of neurons, 554, *554*
membrane polyanions
    role of in electrical rhythm, 239-240
membrane potential, 583
    EPSP effects on, 583, *584*
    of mitral cells, 543-544
    neuroglial, 792
membrane systems, 867
memory, 342
    amine role in, 329-333
    amphetamine effects on, 273, *273*
    biology of, 272-278
    criteria for study of, 164
    epileptic seizure and, 342
    evocation by brain stimulation, 173
    extracellular space in brain and, 865
    formation, 272
    heterosynaptic facilitation and, 263-264
    human, 164-168, *168*
    learning and, 164
    neural mechanisms, 172-173, 330
    neuronal mechanisms, 260, 270
    nomenclature, 164
    normal, 163-176
    pathological, 163-176
    processes, 272-277
    processes, human, 164-168, *168*
    storage mechanisms, 275-276, *276*
    study of, 164
    temporal parameters, 264
    time and, 164
    vestibular, 264-265, *265*
        *see also* forgetting; information retrieval;
        learning
memory, long-term, 272-274, 277
    acetoxycycloheximide effects, 272-273, *273*
    arousal role, 273

conditions for, 273
corticosteroids and, 273
defined, 164
protein synthesis, dependence on, 272-273, 277
    *see also* memory, remote
memory, pathological, 168-173
memory, primary, 166-168, 172
    in anterograde amnesia, 169, 173
    in motor skills, 170
    in retrograde amnesia, 171
memory, recent
    defined, 164
memory, recoded, 165-166
memory, remote
    defined, 164
memory, secondary, 166-168, 170, 172
    in anterograde amnesia, 169, 173
    in retrograde amnesia, 171, 173
memory, sensory, 164-166
    in anterograde amnesia, 169
    auditory mode, 165
    defined, 164
    visual mode, 165
memory, short-term, 189, 272-274, 277
    acetoxycycloheximide, effects, 272, *272*, 273
    defined, 164
        *see also* memory, recent
memory, tertiary, 167-168
    in retrograde amnesia, 171
memory, vestibular, 264-265, *265*
memory consolidation
    learning performance and, 189
Mendelian genes, 983-985
mentation, 346
mesencephalon, *see* midbrain (mesencephalon)
mesenchyme, 62
    epithelio-mesenchymal interactions, 80
messenger RNA, *see* mRNA (messenger RNA)
metaphase arrest, 63
micodeterminacy
    in CNS development, 54
metal complexes
    ligand substitution rates in, 687-688, *688*
    stability constants, 687, *688*
methanol
    field effect relaxation of $Na^+$-murexide com-
    plex in, *691*
    rate and stability constants of alkali complexes
    in, 689, *689*
    as solvent for carriers, 689
methodology
    autoradiography, 62, 64, 74, 103, 122, 1005
    behavior measurement, 36-37, 248-249
    circuit theory, 389
    coincident-current selection techniques, 454
    EEG pattern recognition and analysis by com-
    puter, 234ff, *236*
    electroantennogram (EAG), 512, 516
    field analysis, 444
    language-specific computation, test of capacity
    for, 370
    membrane excitation, study of, 858
    neural circuit mapping, 389
    neuronal electrical activity, recording methods,
    648-649
    neurophysiological research, basic approaches
    569-570, *570*, *571*
    radioactive labeling, 64ff, *66*, *68*, 69
    radioisotopic analysis, 843, 844, 846
    relaxation amplitude method for enthalpy,
    691-692

relaxation spectrometry, 695
sensory physiology, 494
spike-train analysis, 572-573, 595-596, 648-660
stimulus-response matrix, *571*, *573*, *573*
subcellular fractionation, 761-764
synaptic staining procedures, 735-736, *736*
    *see also* electrical brain stimulation; electron
    microscope; mathematical models; nerve
    tissue culture; statistical techniques; systems
    analysis
microtubules, 854
    amino-acid composition of protein in, 855, *855*
    in axoplasmic transport, 854
    neuronal, 845-848
    in oligodendrocytes, 788, *790*, 791
    synaptic vesicle contacts, 875-876
midbrain (mesencephalon)
    homology, 16-17
    tectum, visual pathway function, 377
    vocalization, role in, 363
mind, vii, 486, 493
miniature end-plate potential, 715
    acetylcholine release and, 715, 774
    neurotransmitter release and, 774, *775*
mitochondria
    and bacterial protein synthesis, 828-829
    biochemical differences among, 814
    chemical coupling in, 832, *833*
    chemiosmotic process and, 833, *833*
    conformational-coupling hypothesis, 834
    DNA in, 827, 828
    electron transport, 831-832
    energy-coupling modalities, 831-832, *832*
    evolutionary origin, 829
    functions, 874-875
    gene products of, 828
    genetic apparatus, 828, *829*
    growth, 827
    ion transport, 830-832, *830*
    membranes, 678, 718, *718*, 829-830, *830*, 835-
    836
    membranes, metabolic exchanges across, 830-
    831, *830*
    in myelinated axons, 844
    neurofunction of, 827-839, 874-875
    origin, 678
    protein, 829
    protein synthesis, 813-814, *813*, 828, 829, 835
    ribosomes, 828
    RNA, 828
    separation from myelin and glial fragments,
    763
    transport systems, 830-831, *830*
    uncouplers, effects of, 681
        *see also* DNA, mitochondrial; oxidative
        phosphorylation
mitochrondria, neuronal
    in axonal flow, 844, 845, 847
    breakdown, 842, 845, 848
    enzymes of, 845
mitochondrial DNA, *see* DNA, mitochondrial
mitral cells
    axon, 541, 544, 546
    axon, soma, and dendrites, 554, *554*
    dendrites in, 539-541, 544, 546-549, *546*
    excitation and inhibition of, 541-544, 542
    hyperpolarization in, 544
    impulse generation in, *542*, *543*
    inhibition, 562
    lateral inhibition in, 562, 563
    model of, *554*

of olfactory bulbs, 539, 540, 549, 550
population, 557
simulation of antidromic activation in, 554-555
synchrony and symmetry of, 555, *555*
voltage transients from, *558*
modularization
of cortical pyramid, 448
molar behavior, 178
brain-wave modifications and, 254, 255
molecular biology
approaches to neurophysiology and, 858-866
molecular genetics
contributions to neuroscience, vi
molecular motion in biological membranes, 697-706
molecular neurobiology, 675-879
molecules, replicating, 927ff
mollusks
central neuron functions in, 808-810
ganglionic neurons, 54
*see also Aplysia*
monactin, 691-693, *692*
field-effect amplitudes, *694*
murexide absorption and, 691, *692*
monkey
communication by, 352ff, *355, 356*
destriate, and vision, 322
fine-structure development, 130-139
genital display in, 351ff
motor behavior onset and fetal growth, *131*
New World, 41, 352
Old World, 41, 352
phylogeny, 40, 42-43
septal stimulation, 346, *347*
topological mapping of, 606ff, *608ff*
vocal repertoires, 352ff, *353, 354*
*see also* apes (*Pongidae*)
monkey embryo
motility, spontaneous, 130, *131*
motor-neuron neuropil, synaptic bulbs in, 134ff, *134, 135, 136, 137*
reflex movements, 130, *131*
spinal-cord neurogenesis, 130, 135
molecular vision, 471
deprivation effects on visual-motor coordination, 321, *321, 322,* 323
response to horizontal movement, 480, *480*
Moody-Parriss model, 497-498
morphogenesis, 615
brain, 58
cell adhesiveness role in, 126
CNS, 58
genetic control, 126
mossy fibers, 411, 414-415, 418-419, 427, 429
and cellular glomerulus, *412*
functional properties, 415
schematic of, *417*
mossy rosette, 427, 440
excitation of granule cell and Golgi cell dendrites, 429
motility, embryonic
analysis of, 130, *131*
cyclic nature of, 142-143, *142*
integrated, 149
nonreflexogenic, 141-144, 148, *148*
onset, 130, *131*
random, 148
sensory input, 143-144
spontaneous, 141-144, 147, 148
stimulated, 149-150
in vertebrates, 141-151

motivation
plasticity of, 207-208
prepotent responses and, 214
stability of, 207-208
*see also* drive state
motivational systems
stability and plasticity of, 207-217
history of, 208
motoneuron, *see* motor neuron
motor coordination, 177-178
motor end plate, 715, *716*
*see also* miniature end-plate potential
motor neuron
of crayfish, 399, *410,* 402-403
differentiation, 131, *132*
epithelio-mesenchymal interactions, 80
firing pattern and synaptic noise, 624
of flies, 399, *400*
function-specific groupings, 109-110
membrane properties, 859
of nonmammalian CNS, *9,* 13
subspecification, 59
synaptic potentials in, 862, *863*
*see also* nerve-muscle interaction; reflex
motor pathways, 17-18
establishment of, 111
motor periphery
differentiation, 110-111
connections with center, 111-112
motor skills, 169-170
motor systems, 17-18
nerve pathways, 17-18
nonmammalian, *9,* 18
vertebrate, development of, 108-116
movement
control of, 318, 319, 321
movement perception, 308
in insects, 501, 502, 509
multiplexing
of thalamic output, 452-453, 456
murexide, 689, 690-691
field-effect amplitudes, *694*
spectral properties, 689-690, *689, 690*
spectrophotometric titration of, 689, *689*
*Musca, see* fly
muscle
antagonistic, generation of drive for, 393, *393*
homonymous, homologous response to, 109-110
muscle fiber
contraction times, 111
differentiation, nerve-dependent, 110-111
"fast" and "slow," 111
innervation, 859, 862
skeletal, motor axons, 112
muscle spindles
firing patterns, 623, *623*
spike frequency and output, 667, *668*
muscle-nerve interaction, 81, 153ff
mutation, 977, 979
alteration in gene family size by, 985
ribosome alteration by, 915, 916, *916*
mychronogram, 155
myelin
fluidity, 701
formation, 788, *790*
metabolic properties, 787-791
protein, 816
structure, 867-868
myelin sheath, 62, 677-678, 787ff
axolemma contacts with, 787, *788*

axon nourished by, 791
extracellular space and, 787, *788*
formation, 112
function, 135-136
neuroglial delay in acquiring, 135-136
oligodendrocyte relationship, 788, *789, 790*
specializations, 787
*see also* neurilemma; Schwann cells
myotypic specification, 109, 110
in limb transplants, 154
in neurons, 155

## N

natural code
defined, 574
neural change and, 577
natural selection, 965, 966
neocortex, *see* isocortex
neopallium, *see* isocortex
nerve activity
structural changes during, 707-714
nerve cell, *see* neuron
nerve endings, 723
nerve fiber, 17-18, 59
adhesiveness variations, 126
command fibers, 402-404, *403,* 408, 490ff
differentiation of muscle fibers by, 110-111
fasciculation, 85, 89, 111-112
giant, of crayfish, 399, *401,* 402-404
growth, 123ff
pathfinder, 111
peripheral nervous system, 62
optic tract, 15, 639
regeneration, in optic nerve, 120
sensory, 16-17
size, neural transmission and, 597-598
space constant, 597
*see also* axon; dendrite; nerve pathways
nerve fiber connections
after eye rotation, 120-122
chemotropism role in, 123ff
CNS, sensory, 117-118
eye-brain, 118-123
functional validation, 126-127
genetic determination, 126-127
growth and, 123-126
limb-spinal cord, formation, 112
mechanisms, 123-126
nonspecific afferentation, 16-17
regenerated, 118, 120, 123
retinotectum, 118-122, 123
sensory, rearrangement of, 118
skin graft, 117-127
specification, 117-127
specificity of, 861-862, *862,* 865
nerve-fiber growth
actinomycin-D effects on, 821, *822*
contact guidance, 111
in vitro, 85, 86
mechanisms, 123-126
nerve growth factor (NGF) effects on, 820-821, *822*
nerve growth factor (NGF), 62, 80, 84-86, *87,* 89, 816
fiber outgrowth and, 85-86
inhibition, effects in ganglia, 816-817
insulin and, 821-822
macromolecular structure, 817-820
mechanisms of action, 820-821
neuron survival and, 104

chemoaffinity of neurons, 120
displacement, 104, *104*
ganglia, retinal, 120-122
microdeterminacy vs. macrodeterminacy, 54
pathways in developmental dynamics of, *57*, 60
retrohippocampal formation, 65
spinal cord, 130-140
of synaptic connections, 104-106
visual system, 120-123
*see also* neuron, differentiation
neuroglia, 782, 782fn, 842
antigen composition, 279ff
development, 782
differentiation, 64, 130
extracellular space and, 784, *785*, 787
function, 782-797
gantlet concept of function, 792-793
histogenesis, 63, 67, 69
ionic environment, 787
neuronal interactions, 792-795
neurons and, 782, *785*
proteins in, 278-279
resting potential, 792
ribosomes in, 813
sheathing in, 135, *783*, 784ff, *785*, 787
structure, 782-797
*see also* astrocyte; glial capsule; glioblast; macroglia; nerve ensheathment; oligodendrocyte; Schwann cells
neurohemal organ, 521, 777
neurohormones, 521, 768-782
intrahypophyseal functions, 521
neuroendrocrine communication and, 526
released by neurosecretory neurons, 779
*see also* oxytocin; vasopressin
neurommatidia *see* optic cartridges
neuron
activation, changes after, 330-331
afferent, development of, 131
analysis of, 649-652
antigen composition, 279ff
astrocytic processes and, 792, *793*
binocular gate, *482*
biochemical changes during learning, 278, 281-288
birefringence, 710-712, *710*, *711*
central, 801
chemoaffinity, 120
cholinergic, 765ff
chromatolysis, 801
coding, 569-586
command neurons, 489, 490-491, 865
common excitation effects, 592-594, *593*
communication properties, 530-531, 569-586
connections, 116-129, 861-862
depth distributions of potential in, 558-559, *559*
dendritic structures of, 860-861
differentiation sequence, 130-131
differentiation vs. proliferation, 841
dynamics, 840-850
efferent, development of, 131
electrical activity, recording methods, 648-649
endocrine action on, 78
information storage in, 260-261
in frog tectum, 378-379
functional adaptation, 843
functional association of, 648-661
functional plasticity, 862-864
ganglion, excitation of, *593*
glycoprotein transport in, 753, *754*

hippocampal, specific activities of, *284*, *285*
histograms of, *651*, *654-658*
identification by Procion Yellow injection, 801, *804*, *807*
information content, 378ff
information storage sites, 260-261, 264-270
information theory on, 572ff
information transfer, *531*, 661ff, *667-669*
information transfer function, 666, *667*
information transmission, 572-583, 597-598
inhibition, effects of, 625-627
inputs and outputs, *530*
interaction, stimulus effects on, 653-658, *656-658*
interaction, temporal structure of, 658-660, *659*
interconnections, 652-653, *653*, 861-862, *862*, *863*
intracellular recordings of, 648, *649*
invertebrate, 676, 798-811
ionic environment, 787
pyramidal, protein of, *284*, *285*
mathematical models, 552-554, *553*, *554*
maturation, 782
meaning of signals in, 378, 379
membrane model of, 554ff, *554*
methods of recording electrical activity, 648-649
modifiability, 127
mollusk and crustacean ganglia, 54
neuroglia and, 782, *785*
neuroglial interactions, 792-795
organelles, 827ff
organic anions in, 768
origins, 70
pathways between, *653*
presynaptic and postsynaptic, cross-correlation, 591-592, *592*
proteins in, 278-279
protein synthesis in, 841-845
radioactive labeling, 64
recordings of, 649ff, *650*
regeneration, 840, 843
"sameness," 378-379
second-order, 620ff, *620*, *621*, 625-627, *626*, *627*
self-re-excitation, 267ff
sensory, *94*, 117-118
sheathing onset, 135-136
signal transmission between, 617-629
simplified models of, 553, *553*
small molecules in, 768-782
soma and spike initation sites, 798, 808-810
soma-dendritic complex, 805
soma isolation from bioelectric activity, 805, 808-810, *808*
spatial averaging of, 601-603
as spike-train analyzer, 577-582
stimulus duration of, *600*
structure, physiological role and, 859-860
types, defined, 130
vertebrate and invertebrate, 798-811, *799*, *800*
*see also* axonless cells; circuits; Mauthner neuron; motor neuron; nerve fiber; nerve fiber connections; neural conduction; neurogenesis; neuron specification; neurosecretory neuron; Nissl bodies; pacemaker neuron; sensory neuron
neuronal circuitry, 487, 563
circuit theory, applicability, 389, 391, 393
decomposition, 387-389
development, 131-136

for drives, 291
feedback loops, 54
functional readiness, 131
hippocampal role in cerebral interactions, 231-232, *233*
mapping methods, 389
maturation and, 293
nonspecific afferentation, 16-17
synaptic closure, *132*
*see also* nerve fiber connections; reflex
neuronal conduction
birefringence, voltage-dependency of, 711-712, *711*
disulfide bonds and, 718-719, *719*
entropy changes during, 709-710, *709*
fluorescence emission changes during, 713
invertebrate, 791
ionic theory, 708, *708*
light intensity change during, 710-711, *710*
light-scattering changes during, 712-713, *712*, *713*
temperature changes during, 709, *709*
structural change during, 707ff, 713-714
neuronal discharge, *see* firing
neuronal discharge pattern, *see* firing patterns
neuronal hypertrophy, 843
neuronal junctions, *see* synapse
neuronal membrane, 260-261, 676
"accommodation," 859
axoplasmic transport in, 853-854
binding sites, 701
cerebral, structural limits, 239-240
divalent cations and excitation, 240
entropy changes, 709, 710
excitability, 710-712, *710*, *711*, 869-870
excitation, 685
information storage and, 240
input-output relations, experimental approaches, 662-663, *663*, 669
ionic theory, 708
linear input-out relationship and, 618
permeability, molecular mechanisms of, 708
potassium channels, 708, 709, 714
sodium channels, 708, 709, 714
spin-label motion, 701
stable-states theory, 707-708
structural changes during conduction, 707ff, *712*, 713-714
surface, polysaccharides on, 747, *748*
*see also* neuronal conduction; biological membrane
neuronal microtubules, *see* microtubules
neuronal mitochondria, *see* mitochondria, neuronal
neuronal modulation, 110
*see also* neuronal specificity
neuronal nets
transcription and differentiation, 1011, 1012
neuronal organelles, 842, 844, 845, 846
*see also* mitochondria; neurotubules; neurofilaments; intertubular canals
neuronal protein, *see* protein, neuronal
neuronal specificity, 59ff, 62, 110, 116, 127, 152-156
degrees of, 117, 127
genetic determination, 117
mechanisms, 156
peripheral, of cutaneous nerves, 117-118, *121*, *122*
subspecification, 59
*see also* nerve fiber connections; neuronal

modulation
neuronal transport mechanisms
fast traffic mechanism, 847, 848
neuroplasmic flow, 846, 847
retrograde message transfer, 848
see also axonal flow; neuroplasmic flow
neuronal waves
amplitude distribution studies, 225-226
neurobiosynthesis
epidermal growth factor (EGF) and, 822
gene activity, 823
gene expression and, 812ff, 873, 874
hormones and, 823
mitochondria and, 827-839
nerve growth factor (NGF) and, 817ff, 817-820
organelles and, 874
protein products in, 815-816
of specific proteins, 812-816, 816, 874
in vertebrates and invertebrates, 798ff, 799-800
see also gene; mitochondria
neurophysins, 779
neurophysiology, 486
experimental approaches in, 569-570, 570, 571
information theory application to, 572-574, 573, 574
molecular biological approaches to, 858-866
operations research and, 380
revolution in, 376
neuropil
cervical motor, macaque embryo, 136
crayfish, 399
intersection of fibers in, 391
neurite distribution within, 801-802
neuroglial processes, 136, 138
synaptic bulbs in, 136
neuroplasmic fast transport, 677
neuroplasmic flow, 677, 840ff, 846-847, 860, 875ff
axonal material in, 848
chemomechanical energy coupling in, 875
fast, 875
mechanisms, 847
microtubules and, 786
mitochondria in, 844-845, 847
peristaltic contractions and, 875
proteins and, 875
Schwann cell role, 847
slow, 875
synaptic functions and, 873
velocity, 846
see also axoplasmic transport
neuroscience, vi
genetics and, 925
Neurosciences Research Program (NRP), v, vii
Intensive Study Programs, v-viii
neurosecretion, 521ff, 777ff
classical concept of, 521-525
Dale's law, 779
defined, 777
exocytosis as mechanism for, 776
neurotransmission and, 774, 777, 778
into portal vascular system, 777, 778
into systemic circulation, 778
glandular secretion and, 779
in vertebrate hypothalamo-neurohypophyseal region, 777-778, 778
neurosecretory neuron
A-type fibers, 522
B-type fibers, 522, 526
control of endocrine cells by, 521
diagram of, 523
neurohormone release mechanism, 524-525

neurohormone release sites, 524-525, 524
see also neurohormones
neurotransmission
glandular secretion and, 774
neurosecretion and, 774, 777, 778
structures related to, 715
neurotransmitters, 326, 530-531, 533-535, 768-782, 770, 771
acetylcholine, 329, 770ff
adrenalin, 770
at adrenergic synapse, 325
aminobutyric acid, 770
biosynthesis, 771-773, 772
catabolism, 776-777
chemical structures, 771
dopamine, 770
dynamics, 872
glutamate, 770
glycine, 770
5-hydroxytryptamine, 770
inactivation, 777, 778
identification, 761
neuroendocrine integration and, 521, 524
neurohormones and, 526
neurosecretion and, 778
norepinephrine function as, 326-327, 329, 770
output from disused nerves, 864, 864
physiological functions, 769-770
postsynaptic receptors for, 770-771, 779-780
quantal release, 715, 720, 723, 725, 774
receptor site, interaction with, 770
storage in synaptic vesicles, 734, 774
transport system properties, 777, 777
see also Dale's law; neurohormones
neurotransmitter release, 773-776
exocytosis hypothesis, 726, 775-776, 776, 872-873
mechanisms, 774, 776, 872-873
motor endplate potential and, 774, 775
neurotubules, 845-848
neurulation, 62
NGF, see nerve growth factor (NGF)
Nissl bodies, 806, 809
differentiation of, 134
nodes of Ranvier
myelin and, 787, 788
saltatory conduction, 722
nomenclature
brain cells, 52fn
limbic system vs rhinencephalon, 10
memory, 164
neuroembryology, 52ff
neuroepithelial zone, 52ff
neurons, 130
peripheral nervous system, 73
spinal cord cells, 52fn
synaptic, 771
noradrenaline, see norepinephrine
norepinephrine, 531-534, 532, 769, 770, 771ff
amounts in brain, 327, 329, 331
AMP effects, 333
appetitive behavior, induced by, 329
arousal and, 329
changes in endogenous, 328
deficiency in brain and depression, 328
effects on melatonin synthesis, 535
electroconvulsive shock and, 328, 328
emotion and, 327
interneuronal distribution in sympathetic adrenergic neuron of cat and rat, 773
metabolism, drug effects on, 327, 328, 333

neurotransmitter function, 326-327, 329
rage and, 327
storage in adrenergic synaptic vesicles, 773, 776
storage in large granular synaptic vesicles, 734
storage in presynaptic vesicles, 325
transport, 854
turnover in brain, 327, 328, 328
norepinephrine release
appetitive stimulation and, 328
in central nervous system, 327-328
learning and, 333
nerve stimulation and, 325-326, 326
synaptic, 326, 327
norepinephrine synthesis
electroconvulsive shock simulation of, 327-328
nerve stimulation effects, 326
rate, 325, 325
in sympathetic nerve endings, 325, 325
sympathetic nerve-impulse activity and, 325, 325
tyrosine hydroxylase role, 325
notocord, 58, 62
nuclear double membrane, 677, 677
nucleic acid
protein interaction, 901-912
nucleoprotein, 1000, 1003
nucleotides
purine, 909, 909
pyramidine, 909
nucleotide sequences
serial repetition in duplex DNA, 991, 992
nucleotides
interaction with ribonuclease, 907, 907, 908, 908, 909, 909, 910, 911
nucleus reticularis thalami, 451
as a frequency sensitive gate, 452
nystagmus, 264ff
labyrinthectomy effects, 265
optomotoric, 264-265, 265
see also afternystagmus

O

object use, 44
odorants
acceptors, 514, 516
molecule transfer, 516
see also insect olfaction; pheromones
olfaction
chemoreception and, 511ff
see also insect olfaction
olfactory bulb
anatomy, 539-541
as cortical system, 539-552
functional implications of neuronal interactions in, 547-550
mitral cell excitation and, 541-544
neuron differentiation sequence, 139
neuronal interactions in, 546
neuronal pathways and connections in, 545, 546
neuronal systems in, 544, 546-547
neurons, origin of, 64-65
organization similarities to retina, 549, 549, 550
potential divider effect in, 555
punctured symmetry, 555
in rabbit, 540, 542
repetitive responses in, 545
rhythmic potentials in, 549
self and lateral inhibition in, 549

synaptic responses in, *545*
synchrony and symmetry, 555
theories on, 552ff
olfactory nerve fibers, 539-541
olfactory receptors
in *Bombyx mori*, 511-518
olfactory system
nonmammalian morphology, 8ff
olfactory tract, 539, 541
oligodendrocyte, 788, *789*, *790*
microtubules in myelin-related cytoplasm, 788,
*790*, 791
ommatidia
of fly, 494-496, *495*, 500, 501, 505, 509
*see also* rhabdomeres
ON/OFF mechanisms, 491-492
ontogeny
behavioral, 129-140, 188-196
neuronal specification, 152
synaptic, 129-140
operations research, *see* systems analysis
optic afferent, 431, 433
optic cartridges
of fly's compound eye, 500, 501, *502*, 509
optic chiasm
retinocerebral pathways and, 472
optic nerve
adhesiveness of fibers, 126
connections, selectivity of, 123
fiber growth, 122
regeneration, 120, 123, 154
specification of connections, 122, 156
optic radiations, 342ff, *343*, 345
Meyer's loop, 343, *343*
optic tectum, 70
cell specification, 756ff, *757*
function, 15
intracentral connection, 59
regenerated nerve fibers in, 154
retinal connections, 120-122, *123*, *124*, 756ff,
757
"sameness" neurons in frog, 378-379
optic tract
nerve fibers, 639
optic vesicle
cell differentiation, 122-123
DNA synthesis, 122-123
neuroepithelium, 122-123
optical data
uptake, transduction, and processing, 494-511
optical ganglion lamina
of fly's compound eye, 500, *503*, 509
optical illusions, 304, 312-313
ambiguous figures, 312, *314*
brightness contrast, 304, *305*
complementary images, *307*
contour direction, 306
Frazier "spiral," 310, *311*, 313
Helmholtz's checkerboard, 319, *319*
Land effect, 304
moiré figure, 311, *313*
of motion, 308, 311, *313*
Necker cube, *314*
self-luminous objects, 308
simultaneous stimulation and, 306
stroboscopic illumination and, 308
visual noise effects, 306, *307*, *309*
*see also* afterimages; perceptual anomalies
optical rearrangement, *see* visual rearrangement
optical representation
in fly's compound eye, 500-501

optomotor responses
of fly, 498-499, 501, 505, 509
organismic psychology, 176-177
behavior theory and, 181
cognitive growth, 179
*see also* Piaget, Jean
organotypic maturation, 84, 92
*see also* fasciculation
orientation
monocular deprivation and, 323
scotopic vision and, 321
visual rearrangement effects on, 320
ovariectomy
hormonal changes, 199
overflow-preventing system in cerebellar cortex,
419
owl
optic tract, 22
Wulst region of brain, *21*, 22
*see also* avian brain
oxidative phosphorylation, 831
chemical-coupling hypothesis, 832-833, *833*
chemiosmotic hypothesis, 833-834, *833*
conformational-coupling hypothesis, 834-835,
*836-837*, *838*
uncoupling in mitochondria, 681
oxytocin, 532, 779, *779*

## P

pacemaker neurons
plastic properties, 265, 266-267, *267*, *268*, 269
packing geometry
of proteins, 902-904
of ribonuclease, 905
paleontology, 40, 45
*see also* fossils
pallial mantle, *see* cerebral cortex
Panum's fusional area, 472, 479, 482
Papez circuit
lesions and anterograde amnesia, 172
parallel fibers
in cerebellar cortex, 411, *413*
inputs from, 415-416
Purkinje cells and, 418, 419, 422-423, *422*
synapses of, 416, *417*
tonic activation in, 419
parallel processing
of thalamic output, 464-465, *465*
in thalamus, 456
parallel sensory systems, 377
lemniscal-ventrobasal pathway, 377
spinothalamic-posterior thalamic group pathway, 377
visual pathway, 377
paramutation, 963
paraventricular hypothalamic nuclei, 532
parenchymal cells of pineal organ, *see* pinealocytes
pars intermedia
endocrine effector cell innervation, 520, *522*,
526
partial replicas
of 2-micron subunits, 1005, 1006, 1007, 1008,
1009, 1010
pathfinder fibers, 111
pathological memory, *see* memory, pathological
pattern generation
in CNS, 380, 381
pattern generators
arthropod, 399-408
central, 397-399

in fly, 399
in grasshopper flight control, 397, 398, *398*,
404-406, 408
pattern recognition
visual cortical neuron role, 643, 647
Pavlovian conditioning, *see* conditioning, classical
peptidyl transfer
in protein synthesis, 921, 923
perception, 302, 303-316
ambiguity of perceived modality, 314
analysis by synthesis, 314
brain correlates, 303-315
brain function and, 302, 303-315
brain lesions and, 305
as conscious experience, neural correlates for,
310, 315
development, 184
*see also* cognitive development
evaluation criteria for sensory input, 310
"feature detectors" of, 306-308
feedback in, 312
of form, 310
functional approach to, 305
hierarchic neural structure for, 313ff
just noticeable difference (JND), 303, 315
levels, 313-314
"matching response" mechanism, 306, *306*,
312-314
"motion detectors," 308
of movement, by insects, 501-502, 509
neural representation, 306
readiness state and, 305, 313
stability of perceived world, 306, 308, 310, *310*
structural inhomogeneity and, *309*
van Holst's "cancellation" model, 310, *310*
*see also* auditory perception; cognition;
depth perception; form perception; perceptual anomalies; speech perception; visual
perception
perception, visual, *see* depth perception; visual
perception
perceptual anomalies, 304-305
ambiguity of perceived modality, 314
auditory contrast enhancement, 304
brain lesions and, 305
hallucination, 366
lateral inhibition and, 304, *305*
neural information processing and, 304-305
motion without displacement, 308
phosphenes, 366
retinal rivalry, 308
simultaneous stimulation and, 304
superimposed sights, 366
visual pattern and motion, 311
*see also* optical illusions
perceptual disorders, *see* cognitive disorders
perceptual phenomena, 303
*see also* optical illusions
perikaryal synthesis
neurotransmitter storage and, 773
perineurium, *785*, 786
peripheral nervous system, 73-82
amputation effects on, 76, 78
cellular composition of tissue, 84
cellular metabolism, effect on, 80, 81
CNS cell populations, effect on, 73-79
degradation in, 73-82
differentiation in, 73-82
nomenclature, 73
tissue culture, 84-91
peristalsis

in axonal flow, 847
peristimulus-time (PST) histograms, 649ff, *651, 659*
permease protein, *704*
permeation (membrane), 678
phage, *see* bacteriophage
phasic control system, 415
phenotypic complexities, 953
phenylethanolamine-N-methyl transferase (PNMT), 532, *532*, 533
  ACTH effect on, *533*
  hypophysectomy effect on, *533*
phenylketonuria (PKU), 189-190
pheromone, 511
  degradation, 517
  sexual attractant, 512
  *see also* bombykol; odorants
phospholipids
  in membrane, *704*
  vesicles of, 680, *680*
photic stimulation, 342-343, *343, 344*
photoreceptors, *see* rhabdomeres
phylogenetic comparison, 30-37
  behavioral plasticity, 34-35, *35, 36*
  L-ascorbic acid synthesis, 31ff, *33*
  vertebrate brains, 30-31, *31, 32*
phylogenetic scale
  defined, 27
phylogenetic tree
  defined, 27
phylogeny
  primates, 30, 40ff, 42-46
  vertebrates, 27-30, *28, 29, 30*
  *see also* evolution
phytohemagglutinin (PPHA), 822-823
  nerve growth factor (NGF) similarities, 822
  RNA synthesis and, 822-823
Piaget, Jean
  cognitive development theory, 179-181, 290ff
pigeon
  auditory conduction pathways, *23*
  brain, *15*
  forebrain, *20*
  *see also* avian brain
pineal organ, 532ff, *536, 537*
  hypophysiotrophic factors in CNS control, 525
  melatonin synthesis in, 534, 535
  serotonin synthesis in, 534, 535
pinealocytes, 532
  as neuroendocrine transducers, 534, 535
pinocytosis, 725
pituitary gland, 530
plasmalemma, *see* cell membrane
plasticity
  *Aplysia* neurons, 266-267
  in autonomic nervous system, 218-222, 260-271
  behavioral, 167-297
  determinants of, 161-297
  environment and, 289-294
  evolutionary trends, 34ff, *36*
  heredity and, 289-294
  of motivational systems, 207-208ff
  of nervous system, 161-297, 660
  of networks, 489
  pacemaker neurons, 266-267, *268, 269*
  respiratory centers, 265-266
  synaptic, 740-742
  at unitary level, 260-271
  *see also* learning set; stability

PNMT, *see* phenylethanolamine-N-methyl transferase
polymerase, RNA-dependent, 932, *932*
polymorphism, 964, 970
polynucleotide chains
  annealing, 981ff
  complementary, 988
polypeptide chains, 887ff
  amino acid sequences, 888, 889, *889*, 890, *891, 892*
polysaccharides
  interneuronal synaptic relationships and, 747ff, 757-758
  in retina, 757
*Pongidae, see* apes (*Pongidae*)
positive feedback network, 398
  in decapod crustaceans, 398
  in *Tritonia*, 398
postsynaptic adhesions, 761, *762*
  of synaptosomes, 761, *762*
postsynaptic discharge, 577, *578*
postsynaptic inhibition, 440
postsynaptic potential (PSP), 459ff, 577-578, 582
  patterns, 459
  shape differences, 648-649, 652
  spike generation probabilities and, 574
postsynaptic potential (PSP) trains, *see* spike trains
potential divider effect
  in olfactory bulb, 555
PPHA, see phytohemagglutinin
prechordal plate, 62
predication, 368-369, *369*
*presque vu* phenomenon, 366
  *see also* cognitive disorders
presynaptic disinhibition, 433, 440
presynaptic facilitation, 264
presynaptic inhibition, 433, 440
presynaptic membrane
  catecholamines in, 326
presynaptic terminals, 578ff
  discharge rates, 578ff, *579*
presynaptic vesicles
  amine storage, 325, 326, *326*
  neurotransmitter synthesis in, 325
primary memory, *see* memory, primary
primates
  behavioral evolution, 39-47
  hind-limb map, 606-608, *607*
  language communication, 45
  neonate behavior (nonhuman), 351-352
  phylogeny, 30, 40ff, 42-46
  social communication, 357
  socialization, 352, 359
  spinal cord ontogeny, 130
  tactile-kinesthetic system, 605-606
  *see also* apes; evolution, primate; man; monkey
problem solving
  in CNS, 382
progesterone, 531
prokaryotes
  defined, 998
  DNA in, 1005
  eukaryotes and, 873, *974*, 999
  gene conversion in, 970
  gene expression in, 884, 980, 962-963
  mutation rate in, 977, *977*
prophage, 956, 959, 960
  DNA ends, 957, 958
  insertion-excision system in, 960

proportional error servomechanism
  in grasshopper flight control, 405-406
proprioceptive feedback, 404, 408
  in grasshopper flight control, 405
  in grasshopper leg control, 407
proprioceptive reflex, 54
protein, 883
  axoplasmic transport, 851
  binding, 904, 907
  chain elongation factors in, 923
  changes in learning, 278-289
  in eukaryotic organisms, 883-884
  factors, of ribosomes, 922
  gene expression and, 883, 884
  hydrogen-bonding groups in, 902, 903, *903*, 907
  initiation factors of, 923-924
  interaction, 901ff
  membrane binding sites, 704
  membrane, transport, 704, *704*
  in membranes, developmental changes, 815, *816*
  in mitochondria, 829
  in myelin, 816
  neural excitation, role in, 296
  in neuroglia, 278ff
  noncovalent forces affecting conformation of, 901
  nucleic acid interaction, 901-912
  packing geometry, 902-904, 907
  release from synaptic vesicles, 326
  repressors, 884
  role in chromosomal organization, 1010
  sensory conditioning, changes during, 286-287, *287*, 288
  structure, 896-897
  structure-function relationships, 883
  transport, function of, 856
  *see also* glycoprotein; immunoglobulin
protein, acidic, 279ff, 287
  in brain, 815-816
protein, antigen-alpha, 815
protein, Bence-Jones, 890, 896, 897
protein, neuronal, 278ff
  amino acid compositions, 855, *855*
  radioisotopic analysis, 843, 844, 846
protein, permease, *704*
protein, soluble
  brain-specific, 278-279, 755
S100 protein, 278, *279*, 815-816
  antiserum, learning impaired by, 285-286, *286*, 287, 295-296
  learning and, 295
  synthesis during learning, 285-286, *285*, 288
14-3-2 protein, 815-816
protein synthesis
  in *Acetabularia*, 842
  altered, during training, 281-285
  acetoxycycloheximide inhibition of, 272ff, *272, 273*
  axonal, 815, 843
  chain elongation factors in, 923, 924
  cycloheximide inhibition, 273, *273*
  cytoplasmic, 828
  dissociation factor in, 923, 924, *924*
  during learning, method for tracing, 281-283, *283*
  in explants, 843
  hormone stimulation of, 333
  inhibitors, 874
  initiation factors in, 923, 924, *924*

learning and, 295, 333, 823
memory and, 272ff, *272*, 275, 277, 294-295
mitochondrial, 813-814, *813*, 828, 829, 835
neuronal, 86, 841, 842, 843, 844, 845
neuronal, endoplasmic reticulum and, 808
neuronal behavior and, 883
peptidyl transfer in, 921, 923
rate of, 841
recognition in, 921
ribosomal, 812-813, *813*
in Schwann cells, 844
sites of, 812
synaptic, neuron activity and, 330-331, *332*
synaptosomal, *813*, 814-815
two-site model for, 921, *921*
psychological functions
complex, 301-371
psychological stress
EEG index of, 238, *239*
EEG patterns during, 236-239
PSP *see* postsynaptic potential (PSP)
PST (peristimulus-time) histogram, *see* peri-stimulus-time (PST) histogram
psychology, *see* behavior theory; child psychology; comparative psychology; organismic psychology; stimulus response (S-R) theories of behavior
psychophysiology
neurosciences and, 303-304, 314-315
perceptual, 303-304
*see also* Fechner's law; Weber's law
psychosexual differentiation, 196ff
pulvinar
glomerular synapses in, 435, *436*, 441
punctured symmetry
of olfactory bulb, 555
Purkinje cells
activation of, *418*
climbing fibers and, 409, *410*, 414-415
dendrites of, 69, *413*, 415-416, 419, 421, 422, *422*
disfacilitation in, 423
in elasmobranchs, *418*
extracellular recordings of, 414, 416
lateral inhibition in, 394
in mammals, 423
maturation, 105
mossy fibers and, 411, 429
neuronal connectivities, 415
parallel fibers and, 415-416, 440
schematic drawing of, *417*
sequential activation of, 416, *418*, 419
soma of, 411
synapses with, 413, *416*
in teleosts, 420, 422, *422*
transverse integration, 420-422, *422*, 423
vertical integration, 420, 422, *422*, 423
puromycin
information retrieval, effects on, 275
retrograde amnesic effects, 274, *275*
putamen, 465
pyramidal cells, 65, 448-449, 464-465, *464*
focusing, 450
output biasing, 449-450

## Q

Qβ replicase
kinetics of, *934*
properties of, 933ff
response to templates, *933*, *934*

sedimentation analysis of, *934*
template specificity and, 932ff, *933*
transfer experiments, 938ff, *938*, *939*
synthesis of, 933ff, *938*, *939*, 943ff

## R

rabbit, 539ff
diagram of head, *540*
olfactory bulb, *542*, 549
radiations, optic, *see* optic radiations
radioactive labeling
of neurons, 64ff, *66*, *68*, *69*
*see also* radioisotopic analysis; thymidine-$^{14}$C; thymidine-$^3$H
radioactive time scales, 41
radioisotopic analysis
of neuronal protein, 843, 844, 846
*see also* radioactive labeling; thymidine-$^{14}$C; thymidine-$^3$H
rage
norepinephrine and, 327
Ramón y Cajal, S.
neuron connection theory, 123ff, 136
*Rana pipiens, see* frog
rat
forebrain, *19*
neural and photic regulation of melatonin synthesis in, 534, 535
norepinephrine in, 773
social behavior, 42
readiness
perception and, 305, 310, 313
readout system, 414
recall, *see* information retrieval
recent memory, *see* memory, recent
reciprocal inhibition network, 398, *398*, 399
in Mauthner neurons of fish, 398, *398*
receptive field
on cortical map of somatosensory area I, 606, *608*, *609*
cutaneous nerves, 153
hind limb, 606, *608*, *609*
inhibition effects, 492
patterns, 609ff, *612*
skin graft, reinnervated, 153
visual cortex, 318, *318*
receptive field, visual
alignment, *478*, 479ff
binocular stimulus specificities, 476-477
complex, 483-484
complex units, binocular responses, 483-484, *483*
complex units, specificities of, 483ff
inhibitory regions, 474, *475*, 476
monocular stimulus specificities, 474-476
organization, 472-473
simple cells, binocular stimulus specificities, 476-477
simple cells, monocular stimulus specificities, 474-476
subliminal excitation zones, 474
receptive field disparity (vision), 477-478, *477*
depth perception, role in, 477-478
*see also* retinal disparity
receptors
antennal, 512
firing rate and activity level, 666
reciprocal inhibition, *393*
recognition
molecular, 883-1013

interneuronal, 747-760
in protein synthesis, 901
systems, DNA-protein, 955ff
recognition, perceptual, *see* cognition
recognition sequences
in *E. coli* and bacteriophage lambda DNA, 958, 959, 960
recognition systems
DNA-protein, 955-962
rectification, 973, 979, 994, 995, *995*
recurrent inhibition
in olfactory bulb, 548, *548*, 549
reeler mouse
brain development, 104
cell alignment, 106
reflex
arm flexion, 132
carotid, 54
conditioned, 261, *262*, 264
cutaneous, onset of local, 132, *134*
development, 133
knee-jerk, 293
local, onset of, 132-133, *133*
misdirected in frog, 117-118, *120*, *121*
in rotated skin graft in frog, 117-118, *120*, *121*, 153
synaptogenesis and, 132-133
wiping, 153
regeneration, neuronal, 840, 843
reinforcement
of motor patterns, *215*
of self-stimulated behavior, 213-214
stimulus-bound behavior and, 214-216, *215*, *216*
of visceral response, 218, 219-220
remote memory, *see* memory, remote
renin, 531, 532
Renshaw pathway, 544, 546, 548, 549
repetitious DNA, *see* DNA, repetitious
replication
of DNA, 1001, 1083, 1005-1010
of eukaryotic chromosomes, 998-1013
of molecules, 927-945
of RNA, 935, 939, *939*, 940, *940*, 941, *941*
replication, "self-instructive"
"living and nonliving," 928, 929
replicase, RNA, 931
MS2, 931ff, *931*
sedimentation analysis and, *934*
template specificity and, 931ff, *932*, *934*
repressors
affinity for operators, 950-951
binding and, 949ff
and genetic control, 950-951
*lac*, 949ff, *950*
lambda, *950*
nature of, 948ff
reptiles
behavior patterns, 339
phylogeny, 29
social signals, 350, *350*
resolution
in fly's compound eye, 504, 505, 509
respiratory centers
assimilation of rhythms, 266, *267*
plastic changes in, 265-266
respondent behavior, *see* stimulus-bound behavior
reticular formation
functions, 17
morphology, 16

sensory pathways, 16-17
reticulocortical activation, 462
retina
    cat, 630ff, *633*ff
    cell specification, 756ff, *757*
    dendrites of a cone, *390*
    functional organization, 631ff, *631*
    ganglion cells of, 632ff, *633*ff
    intracentral connection, 59
    lateral inhibition effects, 636-637
    neural circuit interactions, 394
    neuron differentiation sequence, *139*
    optic tectum, neuron connections to, 120-122,
        *123, 124*
    polysaccharides in, 757
    schema of, *630*
    sensitivity, Weber's law and, 638, *638*
    tectum connections, 756, *757*
    *see also* eye; ganglia, retinal; visual system
retinal disparity, 471, 472
    *see also* receptive field disparity (vision)
retinotectal system
    carbohydrates in, 757, *757*
    interneuronal recognition in, 756-758
    polysaccharides and, 757-758
retrograde amnesia, *see* amnesia, retrograde
retrohippocampal formation
    neurogenesis, 65
rhabdomeres
    angular sensitivity distributions, 496, 509
    arrangement of in upper eye, *496*
    dichroic absorption properties, 497, 498
    of fly, 494-501, *496*, 505, 508, 509
    spectral sensitivities, 494-497
rhinencephalon, *see* limbic system
rhombencephalon
    homology, 10
    neuromeres, 69
    sensory mechanisms, *9*, 13-14
rhythm assimilation, 264
    in respiratory centers, 266, *267*
    by *Aplysia*, 266-267, *268*
rhythmic potentials
    in olfactory bulb, 549
ribonuclease
    catalytic action of, *910*
    interaction with nucleotides, 907, *907*, 908,
        *908*, 909, *909*, 910, 911
    packing geometry of, 905
    pancreatic, 905
    structure, 905, *905*, 906, *906*, 907
robonuclease-A, 905
ribonuclease-S, *905*
    cleavage, 905
    interactions, *907, 908*
    structure, 905, *906*
ribosomal RNA, *see* RNA, ribosomal
ribosomes
    action of, 920-927
    altered by mutation, 915, 916, *916*
    antibiotics actions on, 923, *923*
    assembly of, 913-920
    dissociation factor, *924*
    of *E. coli*, 913ff, *913*
    functional analysis of 16S RNA, 917, 918
    functional analysis of 30S proteins, 916, 917,
        *917*
    gross structure and function, 913, *913*, 914
    initiation by subunits, 922, 924
    in vivo assembly of, 918, 919
    mammalian, 925

mitochondrial, 828
mutationally altered, 915-916
neuroglial, 813
protein composition, 916ff
protein factors, 922
protein synthesis in, 812-813, *813*, 921ff, *921*
reconstitution, 922
runoff, *924*
streptomycin resistance, 915, 916
structure and function, 913-914, 921, 922
subunit assembly defective mutants (*sad*
    mutants), *919*
30S assembly in vitro, 918
30S subunit, 913, *917*, 922, 923, 924, 925
30S subunit reconstitution from RNA and
    proteins, 914, *914*, 915, *915*
50S subunit, 913, 919, 922, 924, 925, *926*
70S, 922
Ricco's law, 632
ritualization, 349-350, 351, *351*
RNA
    eukaryotic genome and, 884
    evolution of, 930
    genes per genome in, *989*
    infectious viral, 936ff
    molecules, 938ff
    mutants, 937ff, 938, *938*
    polymerase, 947ff
    self-duplicating, 937-938
    self-instructive replication, 938, 939, *939*,
        940, *940*
    sensory stimulation, changes during, 823
    species, 989, *989*
    transfer from glia to neuron, 287
    variants, 939ff, *939, 940*, 942, *942*, 943,
        *943*, 944, *944*
    viral, 931ff
mRNA (messenger RNA), 920, 921
    degradation, 921, *922*
    formation, *922*
    transcription, 820, 921, *922*
    translation, 921, *922*
RNA, mitochondrial, 828, 829
    gene products of, 829
Qβ, RNA, 934, *934*, 942ff
    replicase activity in, *934*
    response to different, *933*
    transfers and, 942ff, *942*
RNA, neuronal
    synthesis, 81, 86
RNA, ribosomal, 828, 963, 964, 965, 966, 970,
    989, *989*
    16S, analysis of, 917
    precursor, *990*
    transcription, 989
5S RNA, 989, *989*, 995
30S RNA(subunit), 913ff, *914, 915, 916, 917*
50S RNA(subunit), 913ff
RNA synthesis, 946ff
    infectious viral, 936ff, *936, 937*
    kinetics, 934ff, *934-935*
    messenger, 947
tRNA (transfer)
    aminoacyl-, 921ff
    codon requirement for, 922
    and DNA repetition, 964
    genes in, 966, 989, *989*
    initiating, 922
    in mitochondria, 922
    polypeptide relationship, 920
RNA, viral, 936ff

communication of, 930-931
and enzyme responses, 934
formation of, *936, 937*
and host cell, 930-931
radioactivity and, 937
replicase for, 931-933
synthesis of, 936-937, *936, 937*, 943-944
transfer experiments, 939ff, *939, 940, 941*
RNA replicase
    activity kinetics 932ff, *934*
    MS 2 as, 931ff
    purified, *934*
    Qβ as, 932ff, *938, 939*
    template saturation of, *934*
    and template specificity, 931-933, *932-933*
    and viral RNA, 931
    of bacteriophage, 931, 932, 933
    template specificity, 931, 932, *932*, 933, *933*
RNA synthesis, 933ff
    actinomycin-D inhibition, 820-821, *822*
    kinetics, *935*
    nerve growth factor (NGF) and, 821-822, *822*
    phytohemagglutinin (PPHA) and, 822-823
    and viral RNA, 935-937, *936, 937*, 941, *941*

S

salamander (*Ambystoma*)
    motility, embryonic, 141, 148, 149, 150
    mytotypic respecification, 153-154, 155
    transplanted limb, 112, 153-154
saltations, 376-378
"sameness" neurons, 378-379
sarcoplasm
    membrane fluidity, 701
*Saimiri sciureus*, *see* squirrel monkey
satellite cells, 84
Schwann cells
    axons and, 112, *786*
    birefringence and, 711
    cytotypic maturation and, 84
    derivation, 62
    destruction in culture, 782, 782fn
    generation cycle, 63
    intercellular clefts, 784ff, *786*
    location in culture, 89, *92*
    myelin and, 842
    nerve fibers and, *783*
    nerve regeneration and, 112
    nodes of Ranvier and, 722
    oligodendrocyte pulsatility in, 91
    organotypic maturation and, 84
    in peripheral nervous system, *788*
    protein synthesis in, 844
    selectivity, 59
    surface membrane complexes of, 716
    in vitro, 85-86, 847
        *see also* myelin sheath; neurilemma; neu-
        roglia; spindle cells
secondary memory, *see* memory, secondary
second signal system, 183, 184
    *see also* cognitive development; language
selective antibody response, 887, *887*, 888
self-assembling neural nets, *see* neural networks,
    self-assembling
self-duplication
    defined, 928
    gene and, 927, 928
self-inhibition
    of mitral cells, 562
self-instructive replication, *see* replication, self-

instructive
self-preservation
  affects of, 338
  learning role, 193
  neural mechanisms, 340
  *see also* adaptive behavior
self-stimulation, 210
  duration, 214
  in embryos, 143, 150
  reinforcement of stimulus-bound behavior
    and, 213-214
sensation, 317
  *see also* perception
sensillum liquor, 513, *513*, 514
sensory filter, 397, 408
  in grasshopper wing stretch receptors, 404
  in *Limulus* eye, 399
sensory ganglia, *see* ganglia, spinal (sensory)
sensory input
  electrical activity of spinal cord, 146
  in embryonic motility, 143, 144
  role of, 215-216, *216*
sensory memory, *see* memory, sensory
sensory neuron
  differentiation, 131, *132*
  information transmission, 573
  specificity, 59
    *see also* nerve fiber connections; neural
    circuitry
sensory pathway, 13-16
  nonmammalian vertebrate, *9*
  nonspecific activating nervous system and, 377
  *see also* nerve pathway
sensory physiology
  methodology, 494
sensory relay nuclei
  glomerular synapses in, 440-441
    *see also* medial geniculate body; lateral geni-
    culate body; ventralis posterolateralis
sensory signals
  neuronal response to, 598, *598*
  spike-train transmission and distortion, 597-604
sensory stimulation
  RNA changes during, 823
sensory systems, *9*, 13-17
  parallel, 377
septum
  classical conditioning, role in, 346
  electrical stimulation, 346, *347*
  EPSPs in monkey and, 346, *347*
  phylogeny, 340
sequence-specificity
  in bacteriophage DNA ends, 956
serotonin
  behavior, role in, 329
  biosynthesis in mammalian pineal organ, *534*
  central nervous system, distribution in, 327
  distribution, 771
  fluorescence, 327
  in limbic system, 340
  pineal function and, 535
  sleep production and, 329
set (psychological)
  "attention" set, EEG patterns, 230
  duration, 228
sex-related behavior
  testosterone effects on, 201-205, *202, 203, 204*
sexual attractant
  insect, 511-518
  *see also* pheromone
sexual behavior, 196-207

gonadectomy effects, 198-199, *198*
hormonal influences on development, 196ff
mounting, 201ff, *204*
social experience and, 201, *204*
testosterone effects on, 200
  *see also* lordosis reflex
sexual development
  anhormonal, 198-199
  genetic disorders, 196, 197ff, 200
  hormonal influences, 196ff
  perinatal testis influence, 198-199
  social experience and, 201, *204*
  testosterone effects on, 200, 205
sham rage, 208
short-axoned cells, *see* stellate cells
short-term memory, *see* memory, short-term
signal-to-noise ratio
  in fly's compound eye, 499, *499*, 500
simple nervous systems, *see* nervous system
single quantum effects
  in fly's compound eye, 498, 499, *499*, 509
silk moth (*Bombyx mori*), 511-517, *513*
  antenna, 513
  antenna, axon fusion, 513
  chemoreception, 514, 515
  electroantennogram (EAG), 512
  olfactory receptors, 511-514, *515*
  sensillum trichodium, 513-514, *513*
  sex attractant system, 512, 517
  single-cell response in, 514-516
simultagnosia, 366
simultaneous contrast, 304, *305*
sinusoidal oscillator, one-Hertz, 387
skin grafts
  dorsoventrally reversed, 153ff
  in frog, *119-122*
  receptive fields, 118, *119, 122,* 153
  reflex development, 117-118, *120, 121,* 153
  reinnervation, 117-118, *119, 121, 122*
  rotated, 153ff
  size, 118
  wiping reflex and, 153
    *see also* cutaneous nerves; frog
slave genes, *see* gene, slave
sleep
  EEG amplitude distribution during, 226, *227*
snake venom
  nerve growth factor (NGF) in, 819, 820
social behavior
  dominance, 356-357, *358*
  gonadectomy effects on, 201, *204*
  in *Iguana, 350*
  in monkeys, 350-359
  phylogenetic aspects of, 349-351
  play, *202-203*
  ritualization, 349-350, 351, *351*
  septum and, 340
  streotyped, evolutionary aspects, 339
  sexual differences, 201, *204*
    *see also* social signals
  strategies of control, 356-357, *356*
    *see also* male social behavior; social com-
    munication; social signals; squirrel monkey
social communication, 302, 349
  animal, 349-359
  between partners, 355-357, *355, 356*
  determinants, 358-359
  information theory and, 355-357
  innate and learned components, 352, 357-358
  nonverbal in man, 357-358, *358*
  ontogenetic aspects, 351-352

perceptual analysis by synthesis and, 314
phylogenetic aspects, 349-351, 359
primate, 357
ritualization, 349-350, 351, *351*
  *see also* behavior, information theory and;
  language
social signals
  auditory, response to, 358-359
  back-rolling, 350, *351*
  electrical brain stimulation analysis, 355
  gaze, 357-358, *358, 359*
  genital display in monkeys, 351, *351*, 352
  head-nodding of lizards, 350, *350*
  nonverbal in man, 357-358, *358*
  ontogenetic aspects, 351-352
  visual signals, response to, 358-359
  vocalization, 350, 352-353, *353, 354,* 359
    *see also* ritualization; social behavior; social
    communication
social skills
  chimpanzees, 43
Society for Neuroscience, vi
solvation
  free energy, alkali ion and, 687, *687*
somatic afferent pathway, 605-617
somatosensory area I
  cortical map, peripheral stimulus configura-
    tion, 613, *614*
  hind-limb representation in, 606ff, *608,* 609
  microelectrode penetration through, 606, *607*
  topology, 608-611
somatosensory area II
  cortical mapping in, 615
somatosensory mapping, *see* cortical mapping
somatosensory projection, 606, 608
  dermatomal trajectory role, 609-611, *611,* 613,
    *614*
    *see also* cortical mapping
somatosensory spaces
  abstract mapping and, 613-615
somesthetic systems, 377
species-typical behavior
  hypothalamic stimulation and, 213, 214
specificity
  in binding, 901-912
  defined, 902
  kinetic, 902
  in lysozyme-inhibitor complexes, 904
  neuronal, 62
  in ribonuclease-nucleotide interaction, 907
  of RNA replicase, 931-933, *931-933*
  static, 902
spectral analysis
  of spike trains, 592, 594-596
spectral sensitivities
  of rhabdomeres, 496, 497, 509
speech
  brain mechanisms, 363
  cerebral representation, 357
  control of behavior by, 183
  language knowledge and, 364, 365
  motor coordination for, 363
  peri-aqueductal gray, 363
speech disorders
  anarthria, 363ff
  cortical topography and, 362
  symptoms, 364
    *see also* language disorders
speech perception, 314
  analysis by synthesis, 314
    *see also* auditory perception

speech praxia
  defined, 365
spike
  duration, calcium level of membrane and, 707-708
  latency, prediction of, 244, *245*
  probability-predicted, 244-248, *247*, *247*
  prolongation, 707
  shape-sorting methods, 648-649, *650*, 654
  in visual cortex, *246*
  waveforms, 649, *650*
  *see also* firing; neural conduction
spike potential, 598
  discharge, cross-correlation with, *580*
  generating probabilities, 574, *574*
  generation, 581-583
  information transfer by, 597ff
  presynaptic discharge and, 574
spike shapes
  sorting of, *650*
spike trains
  analysis of, 649ff, 661ff
  analysis techniques, 389
  autocorrelation of intervals, 590
  autocovariance, 594
  axon filtering, 377
  codes, variety of, 587-590
  communication channel probabilities and, 598-599, *598*
  comparison of simultaneously observed, 591-594
  cross-correlation histograms, 591-592, *592*
  data transmission by, 661-671
  density, information transmission and, 600-601
  demodulated, 664ff, *664*
  firing times, 665ff, *666*, *667*
  Fourier transform, analysis by, 592, 595-596
  functional interaction categories, 652-660
  impulse frequency and information transmission, 662-663
  information-carrying parameter, 662-663, *663*
  information transmission, role in, 587-596, 597-604
  input patterns and, 578, *579*
  input rates and, *576*, 577-578
  instantaneous frequency, 662
  interspike intervals, *570-571*
  interval codes, 588ff, *589*
  interval histogram, 590
  joint peri-stimulus-time histogram, 652, *659*
  low-pass filter in, *665*, *666*
  measurement, 572-573
  measurement, descriptive statistics, 649, 651-652
  neural codes in, 587-588
  neuron as analyzer of, 577-582
  neuronal analysis, 577
  neuronal response and, 598-599, *598*
  random, power spectrum of, 668, *669*
  rate codes, 588ff, *589*
  regular rates, effects on single-channel transmission, 666-668, *668*
  sensory signals and, 597-604
  serial correlogram of intervals, 590-591
  spectral analysis of, 594-596
  spike intervals, 662, 666
  statistical measures for, 590-596
  time intervals, measurement, 651-652
  *see also* firing pattern; neural codes
spin labels, 697ff, 869
  amphiphilic labels, 698, *700*, 701

anisotrophic motions, 700-701
  Hamiltonian, 698
  motions in membranes, 698, 700ff
  nitroxide free radicals, 697
  nitroxide group, paramagnetic resonance, 698, 700
  orientation, 698, 700, 701
  paramagnetic resonance spectra, 698, *699*, 700, *700*, 701
  Pauling-Corey-Koltun model, 698, *699*, 702
  *see also* TEMPO
spinal cord
  brachial enlargement, 109
  central canal, 58
  conduction pathways, 17-18
  deafferentation, in chick embryo, 143-144, *143*, *144*, *147*
  development of, 136-139, *139*
  electrical activity, in embryo, 144-147, *146*
  embryonic, *144*, *147*
  limb movement, role in, 154
  lumbar enlargement, 109
  mechanisms, intrinsic, 10
  mechanisms, sensory, *9*, 13-14
  morphogenesis, 58
  motor column differentiation, 109
  ontogeny, 130-136
  primate ontogeny, 130
  somatotopic representation, 154
  synaptic connections, 130
spindle cells, 64, 84, *88*, *90*
  migration, in vitro, 85
  in spinal ganglia, 86, *88*, *90*
  ventricular zone, orientation, 64
  *see also* fibrocytes; Schwann cells
spinocerebellar pathway, dorsal
  neuronal signal transmission in, 617-629
split-brain studies, 322
spongioblasts
  germinal cell origin, 63
  *see also* neural tube
squid
  axoplasm, 855, *855*
squirrel monkey (*Saimiri sciureus*)
  back-rolling, 350, *351*
  dominance behavior, 356-357
  genital display, 351, *351*, 352
  social display, 350ff, *351*
  vocal repertoire, 352, *353*, 359
stability, neurobehavioral, 162
  learning and, 162
  of motivation systems, 207-208ff
  *see also* plasticity
statistical techniques
  autocorrelation of spike intervals, 590
  autocorrelation of spike discharges of a neuron, 582, *582*
  autocovariance, 594
  correlogram, 652
  cross-correlation, 578, *580*, 581, 591-592
  cross-covariance, 595
  cross-interval histogram, 651-652, 653, *654*, *655*
  Fourier transform, 595-596
  joint-peri-stimulus-time scatter diagram, 652, *659*
  serial correlogram, 590-591
  spectral analysis, 592, 594-596
  of spike-train analysis, 590-596
  *see also* systems analysis
stellate cells

lateral inhibition and, 419
  morphology of, 411
  as part of module, 449
  origin of, 411, 413
stereopsis, *see* depth perception
stereoscopic vision, *see* depth perception
Steven's law, 303-304, 306
*Stimulationsorganen*
  defined, 404
stimulation, brain, *see* brain stimulation
stimulation, visual
  contour analysis, 319ff, *319*
  global features, 318
  locus-specific analysis, 319-321, *319*, 322
  spatially distributed processing, 317-323
stimulus-response (S-R) theories of behavior, 177ff, 185
  concept formation, 177
  habit-family hierarchy, 178
  hierarchical organization of performance, 177-178
  language and, 177, 178
  reversal shift, 177
  verbal mediation, 177
stimulus, sensory
  emotional state and novelty, 330
  genetically significant, 330
stimulus persistence, 165-166
stimulus-bound behavior, 208, 214
  drinking, 210ff
  drive state and, 210
  eating, 210ff
  electrode size and, 210
  electrolytic brain lesions and, 211-213
  environmental conditions and, 214
  hypothalamic site and, 209ff, 211, *212*, 213, 215
  modifiability, 208-211, *209*
  prepotent responses, 213-214, 215
  reinforcement of, 214, 215-216, *215*, *216*
  stimulus duration and, 214
  *see also* electrical brain stimulation
stimulus-response matrices, *571*, 573, *573*
streptomycin
  resistance of mutationally altered ribosomes, 915, 916
stress, *see* psychological stress
striate cortex
  binocular receptive fields of neurons, 476-477
  neurons, classification of, 473
  receptive fields of single neurons in, 472
  retinocerebral fibers in, 472
subcellular fractionation, 761-764, *763*
subjectivity, 336, 337
  *see also* affect; individuality
submaxillary gland
  7S nerve-growth factor in, 817ff
substantia gelatinosa
  synaptic complexes in, 438, *439*, 440-441
subsystem
  defined, 486ff
  thalamic cortical, 427-443
  *see also* neural subsystems
subventricular zone
  glial cell, formation of, 69
supraoptic hypothalamic nuclei, 532
symmetry
  of mitral cell population 555, *555*
synapse
  adrenocortical hormones and, 333
  apparatus, 871
  amine release and, 331

chemical transmission at, 715-720
cholinergic terminal of, 764ff, 786ff, 770, 773, 775
circuitry development in, 134, 135
classification of, 739-740
cleft material, 761-762
closure, sequence of, 132-136, 132
connections, formation of, 740-741
consolidation of, 261-262, 262, 263
differentiation rate, 134
electrical activity of, 715ff, 718, 719, 768ff
end-plate potentials and, 775
electron coupling, 721-722
"epoxy theory" of consolidation, 262-263
E-PTA staining, 735-738, 737, 740
excitatory and inhibitory, 723-724
glycoproteins and, 747ff, 752-756
information storage by, 261-264
information-transmitting capacity, 572
interactions, 263-264
investigation techniques, 761ff, 763
juxtamembranal specializations, 724-726
link development, 104ff, 132-136, 132
linkage sequence, 132, 134
membrane complexes at, 715ff
molecular biology of, 676, 715-782
molecular transfer at, 872
morphological features, 738
neuroplasmic flow and, 873
nomenclature, 771
number of, 739
plasticity of, 740-742, 862, 864
polysaccharides and, 747ff
postsynaptic structures, 739
presynaptic microstructure, 739
release mechanisms in, 872
retinal ganglion cells and, 757, 757
specializations in, 724, 724, 735-739
staining techniques for, 722
structures at, 870-872
subcellular structure and, 762
transmission, invertebrate, at, 720-721, 769ff
transmission structures at, 715
transmission, vertebrate, at, 721-722
transmitter release mechanisms at, 776ff, 777, 778, 872-873
transposition, 741
ultrastructure of, 715-728, 729-747, 870-873
"unspecific afferents" and, 331, 333
variations in, 740
vesicles at, 723, 724ff, 724, 725, 729ff, 730, 731, 733, 737, 738
see also motor end plate; polysaccharides;
synapse, adrenergic, 326
amine storage, 326, 326
endocrine-cell innervation and, 520, 522
neurotransmitter at, 325
neurotransmitter inactivation at, 778
synapse, axo-axonic, 440, 441
in LGB glomeruli, 433
presynaptic inhibition in, 433
in substantia gelatinosa, 438
in VPL glomeruli, 433
synapse, axodendritic
in substantia gelatinosa, 438
synapse, cholinergic
neurotransmitter inactivation at, 778
organization, 764-766
synapse, dendrodendritic, 552-565
graded activation, 563
in olfactory bulb, 541, 546, 546, 547

synapse, excitatory
cerebellar glomerulus as, 429
morphological features, 723-724
synapse, glomerular, 427-443
in lateral geniculate body, 431, 433, 433, 435, 441
in lateralis posterior nuclear group of thalamus, 435
in medial geniculate body, 433
operation of, 427-443
in pulvinar, 435, 441
in sensory relay nuclei, 440-441
in ventralis posterolateralis, 433, 440
see also synaptic complex
synapse, inhibitory
cerebellar glomerulus and, 429
morphological features, 723-724
synapse, retinal
specializations, 723, 723
synapse, retinotectal, 756
"synapse gap," 729, 742-743
synaptic activation, graded
of dendrodendritic synapses, 563
synaptic apparatus, 870, 871-872
synaptic boutons, 715, 721
synaptic vesicles in, 723
synaptic bulb
development rate, 137, 134ff
in neuropil, 136, 137
reflex onset and, 133, 134, 135
synaptic cleft
neurotransmitter in, 776
synaptosomes and, 761
synaptic complex
in substantia gelatinosa, 438, 438, 440-441
see also synapse, glomerular
synaptic coupling
specificities of DSCT, 617-618
synaptic disk, 721, 721
synaptic function, 761-768
metabolic regulation, 864
synaptic junction, 411
density, maturation and, 738, 739
morphology, 739
"tight" junctions, 721-722
synaptic membrane
morphological changes with electrical stimulation, 715-720, 717-719
osmiophilia, 718-719, 719, 720
synaptic membrane complex
in crayfish ganglia, 716, 717-719
synaptic organization
in visual system, 630-647
synaptic potential
of motor neurons, 862, 863
synaptic receptors, 871
synaptic transmission
in invertebrates, 720-721
in vertebrates, 721-723
see also neural transmission
synaptic vesicles, 715ff, 729-735
acetylcholine in, 722-723, 725, 764, 771
catecholamine content, 732
differentiation of excitatory from inhibitory endings and, 724
derivation from interconnected tubular system, 725ff, 726
electron-lucent (agranular), 730-732, 731
electron-opaque (granular), 731, 732-735
esterase activity, 765-766
exocytosis and, 725

fixation, variations with, 734-735, 736
isolated, 726, 766
large granular (LGV), 733-734, 733
microtubule contacts, 875-876
neurotransmitter storage in, 730, 774
origin, 729
presynaptic densities, 724, 724
quantal release and, 725
in rat cerebellar cortex, 737
size, in terminal axons, 715, 724
small granular (SGV), 732-733, 732, 734-735
types of, 731
synaptogenesis 130-140
onset, 134
synaptoid configurations, 524, 524
release sites for neurohormones, 525
synaptoid vesicles, 525
synaptology, 715, 722-723
synaptosomes, 761ff
leucine incorporation inhibited by antibiotics, 814, 814
as miniature cells, 761
protein synthesis in, 813, 814-815, 864
separation from myelin and glial fragments, 763
stability of, 763
synchrony
of mitral cell population, 555, 555
syntax, 369
hierarchical organization, 178
systems
components, 487ff
general-purpose, 489
motivational, 207-217
simple cortical, 539-552, 552-565
special-purpose, 489
see also neural subsystems
systems analysis, 375-383, 384, 395
abstract element, 384
functional context, 395
of neuronal functions, 378-382, 381
of neural transmission, 661-662, 669
signal context, 395
of social institutions, 380, 381
structural context, 395
see also information theory; neural subsystems

T

tactile-kinesthetic system, 605-606, 613
tectum, see optic tectum
teleology, 488
template
specificity of RNA replicase, 931, 932, 932, 933, 933
TEMPO (tetramethylpiperidine-1-oxyl)
paramagnetic resonance spectrum, 697-698, 697
Pauling-Corey-Koltun model, 698ff
see also biological membranes; spin label
terminology, see nomenclature
tertiary memory, 166, 167ff
Testicular Feminization, 197, 199
testosterone
lordosis and, 199
in plasma of newborn rats, 198, 198
see also androgen
tetrapods
phylogeny, 28, 29
tetrodotoxin (TTX)

# ERRATA

A bullet (●) indicates the corrections of text

*page 35*  FIGURE 7  A family of "ideal" curves representing the development of learning set in various animals. The curves were obtained by the fitting of smooth curves to empirical data points. The data of the human children and chimpanzees were from Hayes et al. (1953); those of the gorillas, from Fischer (1962); those of the rhesus monkeys, from Harlow (1959); those of the mangabeys, from Behar (1962); those of the squirrel monkeys, from Miles (1957); those of the marmosets, from Miles and Meyer (1956); those of the cebus monkeys and spider monkeys, from Shell and Riopelle (1958); those of the squirrels and rats, from Rollin, cited in Warren (1965); those of the ferrets, minks, skunks, and cats, from Doty et al. (1967); those of the tree shrews, from Leonard et al. (1966); and those of the pigeons, from Ziegler (1961). As the human, gorilla, and chimpanzee in studies each reported data on only two subjects, the curves ● of each subject are shown separately.

*page 268*  FIGURE 7  Effect of rhythmic intracellular stimulation of a burster cell (R15) of *Aplysia*. The average spontaneous burst interval is 67 seconds. Rhythmic intracellular depolarization ● is applied (black bars) six times, with an interstimulus interval of 56 seconds. After the end of the stimulation period, the first spontaneous burst appears after the entrained interstimulus interval. Maximal spontaneous variability of the mean interburst interval, 3.5 per cent. Induced change, 15 per cent.

*page 272*  ● FIGURE 1  Effect of acetoxycycloheximide on protein synthesis in the cerebrum and memory. Mice were injected subcutaneously with 240 $\mu$g of acetoxycycloheximide 30 minutes before training to escape shock by choosing the lighted limb of a T-maze to a criterion of five out of six consecutive correct responses. Different groups were tested for retention (per cent savings) at each of the indicated times. (For details, see Barondes and Cohen, 1968a.)

*page 421*  FIGURE 8  Current sources and sinks calculated from the field potentials evoked by local stimulation of the surface of the cerebellar cortex in the alligator (*Caiman sclerops*). The field potentials were recorded simultaneously by an array of five microelectrodes (see photograph and diagram to the right). A shows a computer average of the fields produced by a set of eight successive local stimuli, each set administered at a different depth and recorded from electrode #5. Upward arrows represent stimulus artifacts. In B, current densities calculated from potentials recorded with the five microelectrodes at depths indicated to the left of each record. First downgoing arrow indicates current generated by the parallel-fiber volley; second arrow the current generated by the activation of parallel fiber and Purkinje cell. Upgoing transients indicate current sinks; downgoing transients indicate current sources. Second downgoing arrows in B demonstrate that, after local stimulation, a sink of current is generated and moves downward to the level of the soma. Note increase in latency of the current sink with depth. Time calibration as in D. In C and D, a series of field potentials recorded by electrode #5 after a weak stimulus to the cerebellar cortex. C was recorded immediately in line, and D was recorded $200\mu$ out of beam. E and F show the current densities associated with the potentials shown in C and D. Note that the weak local stimulus is able to generate a dendritic sink which moves downward in time and which seems to move transversely so that in F a source-sink relation is generated which does not, however, invade antidromically but instead produces a dendritic current source at higher levels (see arrow at $200\mu$ depth). Abbreviations: PC, Purkinje cell; ML, molecular layer. (From Llinás and Nicholson, 1969.) ●

*page 495*  FIGURE 1  Cross section through an ommatidium of *Musca*. Seven sensory cells with their attached rhabdomeric structures are shown. The diameter of each of the peripheral rhabdomeres (1–6) is about twice that of the central one (7). Rhabdomere 8 cannot be seen in the Figure, as it is situated proximal to 7. Note the different orientations of the microvilli in the different rhabdomeres. Fixation: glutaraldehyde-osmium. (From C. B. Boschek, unpublished.) ●

*page 978*  FIGURE 5  Electron micrograph of lampbrush chromosome loop. DNA duplex is the traverse thread; long RNA chains extend from it. (Courtesy of O. L. Miller, Jr.) ●